Encyclopedia of Environmental Management

Volume III

Invasion—Rare Earth

Encyclopedias from Taylor & Francis Group

Agriculture Titles

Encyclopedia of Agricultural, Food, and Biological Engineering, Second Edition (Two-Volume Set)
Edited by Dennis R. Heldman and Carmen I. Moraru
ISBN: 978-1-4398-1111-5   Cat. No.: K10554

Encyclopedia of Animal Science, Second Edition (Two-Volume Set)
Edited by Duane E. Ullrey, Charlotte Kirk Baer, and Wilson G. Pond
ISBN: 978-1-4398-0932-7   Cat. No.: K10463

Encyclopedia of Biotechnology in Agriculture and Food
Edited by Dennis R. Heldman, Dallas G. Hoover, and Matthew B. Wheeler
ISBN: 978-0-8493-5027-6   Cat. No.: DK271X

Encyclopedia of Pest Management *and* Encyclopedia of Pest Management, Volume II
Edited by David Pimentel
Edition: ISBN: 978-0-8247-0632-6   Cat. No.: DK6323
Volume II: ISBN: 978-1-4200-5361-6   Cat. No.: 53612

Encyclopedia of Plant and Crop Science
Edited by Robert M. Goodman
ISBN: 978-0-8247-0944-0   Cat. No.: DK1190

Encyclopedia of Soil Science, Second Edition (Two-Volume Set)
Edited by Rattan Lal
ISBN: 978-0-8493-3830-4   Cat. No.: DK830X

Encyclopedia of Water Science, Second Edition (Two-Volume Set)
Edited by Stanley W. Trimble
ISBN: 978-0-8493-9627-4   Cat. No.: DK9627

Business Titles

Encyclopedia of Public Administration and Public Policy, Second Edition (Three-Volume Set)
Edited by Jack Rabin and David H. Rosenbloom
ISBN: 978-1-4200-5275-6   Cat. No.: AU5275

Encyclopedia of Supply Chain Management (Two-Volume Set)
Edited by James B. Ayers
ISBN: 978-1-4398-6148-6   Cat. No.: K12842

IT Titles

Encyclopedia of Information Assurance (Four-Volume Set)
Edited by Rebecca Herold and Marcus K. Rogers
ISBN: 978-1-4200-6620-3   Cat. No.: AU6620

Encyclopedia of Library and Information Sciences, Third Edition (Seven-Volume Set)
Edited by Marcia J. Bates and Mary Niles Maack
ISBN: 978-0-8493-9712-7   Cat. No.: DK9712

Encyclopedia of Software Engineering (Two-Volume Set)
Edited by Phillip A. Laplante
ISBN: 978-1-4200-5977-9   Cat. No.: AU5977

Encyclopedia of Wireless and Mobile Communications, Second Edition (Three-Volume Set)
Edited by Borko Furht
ISBN: 978-1-4665-0956-6   Cat. No.: K14731

Environmental Titles

Encyclopedia of Environmental Management (Four-Volume Set)
Edited by Sven Erik Jorgensen
ISBN: 978-1-4398-2927-1   Cat. No.: K11434

Encyclopedia of Environmental Science and Engineering, Sixth Edition (Two-Volume Set)
Edited by Edward N. Ziegler
ISBN: 978-1-4398-0442-1   Cat. No.: K10243

Engineering Titles

Dekker Encyclopedia of Nanoscience and Nanotechnology, Second Edition (Six-Volume Set)
Edited by Cristian I Contescu and Karol Putyera
ISBN: 978-0-8493-9639-7   Cat. No.: DK9639

Encyclopedia of Energy Engineering and Technology (Three-Volume Set)
Edited by Barney L. Capehart
ISBN: 978-0-8493-3653-9   Cat. No.: DK653X

Encyclopedia of Optical Engineering
Edited by Ronald G. Driggers
ISBN: 978-0-8247-0940-2   Cat. No.: DK9403

Chemistry Titles

Encyclopedia of Chemical Processing (Five-Volume Set)
Edited by Sunggyu Lee
ISBN: 978-0-8247-5563-8   Cat. No.: DK2243

Encyclopedia of Chromatography, Third Edition (Three-Volume Set)
Edited by Jack Cazes
ISBN: 978-1-4200-8459-7   Cat. No.: 84593

Encyclopedia Of Corrosion Technology
Edited by Philip A. Schweitzer, P.E.
ISBN: 978-0-8247-4878-4   Cat. No.: DK1295

Encyclopedia of Supramolecular Chemistry (Two-Volume Set)
Edited by Jerry L. Atwood and Jonathan W. Steed
ISBN: 978-0-8247-5056-5   Cat. No.: DK056X

Encyclopedia of Surface and Colloid Science, Second Edition (Eight-Volume Set)
Edited by P. Somasundaran
ISBN: 978-0-8493-9615-1   Cat. No.: DK9615

Medical Titles

Encyclopedia of Biomaterials and Biomedical Engineering, Second Edition (Four-Volume Set)
Edited by Gary E. Wnek and Gary L. Bowlin
ISBN: 978-1-4200-7802-2   Cat. No.: H7802

Encyclopedia of Biopharmaceutical Statistics, Second Edition
Edited by Shein-Chung Chow
ISBN: 978-0-8247-4261-4   Cat. No.: DK261X

Encyclopedia of Dietary Supplements, Second Edition
Edited by Paul M. Coates, Joseph M. Betz, Marc R. Blackman, Gordon M. Cragg, Mark Levine, Joel Moss, and Jeffrey D. White
ISBN: 978-1-4398-1928-9

Encyclopedia of Clinical Pharmacy
Edited by Joseph T. DiPiro
ISBN: 978-0-8247-0752-1   Cat. No.: DK7524

Encyclopedia of Medical Genomics and Proteomics (Two-Volume Set)
Edited by Jürgen Fuchs and Maurizio Podda
ISBN: 978-0-8247-5564-5   Cat. No.: DK2208

Encyclopedia of Pharmaceutical Technology, Fourth Edition (Six-Volume Set)
Edited by Gary E. Wnek and Gary L. Bowlin
ISBN: 978-1-8418-4819-8

These titles are available both in print and online. To order, visit:
www.crcpress.com - Telephone: 1-800-272-7737 - Fax: 1-800-374-3401 - E-Mail: orders@taylorandfrancis.com

Encyclopedia of Environmental Management

Volume III

Invasion—Rare Earth

Edited by
Sven Erik Jørgensen

CRC Press is an imprint of the
Taylor & Francis Group, an **informa** business

CRC Press
Taylor & Francis Group
6000 Broken Sound Parkway NW, Suite 300
Boca Raton, FL 33487-2742

© 2013 by Taylor & Francis Group, LLC
CRC Press is an imprint of Taylor & Francis Group, an Informa business

No claim to original U.S. Government works

Printed in the United States of America on acid-free paper
Version Date: 20121011

International Standard Book Number: 978-1-4398-2930-1 (Hardback)

This book contains information obtained from authentic and highly regarded sources. Reasonable efforts have been made to publish reliable data and information, but the author and publisher cannot assume responsibility for the validity of all materials or the consequences of their use. The authors and publishers have attempted to trace the copyright holders of all material reproduced in this publication and apologize to copyright holders if permission to publish in this form has not been obtained. If any copyright material has not been acknowledged please write and let us know so we may rectify in any future reprint.

Except as permitted under U.S. Copyright Law, no part of this book may be reprinted, reproduced, transmitted, or utilized in any form by any electronic, mechanical, or other means, now known or hereafter invented, including photocopying, microfilming, and recording, or in any information storage or retrieval system, without written permission from the publishers.

For permission to photocopy or use material electronically from this work, please access www.copyright.com (http://www.copyright.com/) or contact the Copyright Clearance Center, Inc. (CCC), 222 Rosewood Drive, Danvers, MA 01923, 978-750-8400. CCC is a not-for-profit organization that provides licenses and registration for a variety of users. For organizations that have been granted a photocopy license by the CCC, a separate system of payment has been arranged.

Trademark Notice: Product or corporate names may be trademarks or registered trademarks, and are used only for identification and explanation without intent to infringe.

Visit the Taylor & Francis Web site at
http://www.taylorandfrancis.com

and the CRC Press Web site at
http://www.crcpress.com

Brief Contents

Volume I

Acaricides	1
Acid Rain	5
Acid Rain: Nitrogen Deposition	20
Acid Sulfate Soils	26
Acid Sulfate Soils: Formation	31
Acid Sulfate Soils: Identification, Assessment, and Management	36
Acid Sulfate Soils: Management	55
Adsorption	61
Agricultural Runoff	81
Agricultural Soils: Ammonia Volatilization	85
Agricultural Soils: Carbon and Nitrogen Biological Cycling	88
Agricultural Soils: Nitrous Oxide Emissions	96
Agricultural Soils: Phosphorus	100
Agricultural Water Quantity Management	105
Agriculture: Energy Use and Conservation	118
Agriculture: Organic	125
Agroforestry: Water Use Efficiency	129
Air Pollution: Monitoring	132
Air Pollution: Technology	149
Alexandria Lake Maryut: Integrated Environmental Management	164
Allelochemics	175
Alternative Energy	179
Alternative Energy: Hydropower	201
Alternative Energy: Photovoltaic Modules and Systems	215
Alternative Energy: Photovoltaic Solar Cells	226
Alternative Energy: Solar Thermal Energy	241
Alternative Energy: Wind Power Technology and Economy	258
Aluminum	267
Animals: Sterility from Pesticides	277
Animals: Toxicological Evaluation	280
Antagonistic Plants	288
Aquatic Communities: Pesticide Impacts	291
Aral Sea Disaster	300
Arthropod Host-Plant Resistant Crops	304
Bacillus thuringiensis: Transgenic Crops	307
Bacterial Pest Control	321
Bioaccumulation	324
Biodegradation	328
Biodiversity and Sustainability	333
Bioenergy Crops: Carbon Balance Assessment	345
Biofertilizers	349
Bioindicators: Farming Practices and Sustainability	352
Biological Controls	366
Biological Controls: Conservation	370
Biomass	373
Biopesticides	381

Volume I (cont'd)

Bioremediation	387
Bioremediation: Contaminated Soil Restoration	401
Biotechnology: Pest Management	407
Birds: Chemical Control	413
Birds: Pesticide Use Impacts	416
Boron and Molybdenum: Deficiencies and Toxicities	424
Boron: Soil Contaminant	431
Buildings: Climate Change	435
Cabbage Diseases: Ecology and Control	442
Cadmium and Lead: Contamination	446
Cadmium: Toxicology	453
Carbon Sequestration	456
Carbon: Soil Inorganic	462
Chemigation	469
Chesapeake Bay	474
Chromium	477
Climate Policy: International	483
Coastal Water: Pollution	489
Cobalt and Iodine	502
Community-Based Monitoring: Ngarenanyuki, Tanzania	505
Composting	512
Composting: Organic Farming	528
Copper	535
Cyanobacteria: Eutrophic Freshwater Systems	538
Desertification	541
Desertification: Extent of	545
Desertification: Greenhouse Effect	549
Desertification: Impact	552
Desertification: Prevention and Restoration	557
Desertization	565
Developing Countries: Pesticide Health Impacts	573
Distributed Generation: Combined Heat and Power	578
Drainage: Hydrological Impacts Downstream	584
Drainage: Soil Salinity Management	588
Ecological Indicators: Eco-Exergy to Emergy Flow	592
Ecological Indicators: Ecosystem Health	599
Economic Growth: Slower by Design, Not Disaster	614
Ecosystems: Large-Scale Restoration Governance	624
Ecosystems: Planning	632
Endocrine Disruptors	643
Energy Commissioning: Existing Buildings	656
Energy Commissioning: New Buildings	665
Energy Conservation	677
Energy Conservation: Benefits	691
Energy Conservation: Industrial Processes	697
Energy Conservation: Lean Manufacturing	705
Energy Conversion: Coal, Animal Waste, and Biomass Fuel	714

Brief Contents (*cont'd*)

Volume II

Energy Efficiency: Low-Cost Improvements	735
Energy Efficiency: New and Emerging Technology	742
Energy Efficiency: Strategic Facility Guidelines	752
Energy Master Planning	763
Energy Sources: Natural versus Additional	770
Energy Use: Exergy and Eco-Exergy	778
Energy Use: U.S.	790
Energy: Environmental Security	798
Energy: Physics	808
Energy: Renewable	824
Energy: Solid Waste Advanced Thermal Technology	830
Energy: Storage	853
Energy: Walls and Windows	860
Energy: Waste Heat Recovery	869
Environmental Legislation: Asia	874
Environmental Legislation: EU Countries Solid Waste Management	892
Environmental Policy	914
Environmental Policy: Innovations	922
Erosion	930
Erosion and Carbon Dioxide	934
Erosion and Global Change	939
Erosion and Precipitation	943
Erosion and Sediment Control: Vegetative Techniques	947
Erosion by Water: Accelerated	951
Erosion by Water: Amendment Techniques	958
Erosion by Water: Assessment and Control	963
Erosion by Water: Empirical Methods	974
Erosion by Water: Erosivity and Erodibility	980
Erosion by Water: Process-Based Modeling	991
Erosion by Water: Vegetative Control	994
Erosion by Wind: Climate Change	1004
Erosion by Wind: Global Hot Spots	1010
Erosion by Wind: Principles	1013
Erosion by Wind: Source, Measurement, Prediction, and Control	1017
Erosion by Wind-Driven Rain	1031
Erosion Control: Soil Conservation	1034
Erosion Control: Tillage and Residue Methods	1040
Erosion: Accelerated	1044
Erosion: History	1050
Erosion: Irrigation-Induced	1053
Erosion: Snowmelt	1057
Erosion: Soil Quality	1060
Estuaries	1063
Eutrophication	1074
Everglades	1080
Exergy: Analysis	1083
Exergy: Environmental Impact Assessment	1092
Farming: Non-Chemical and Pesticide-Free (European Council Regulations [EED/209/91])	1103
Farming: Organic	1109
Farming: Organic Pest Management	1115
Food Quality Protection Act	1118
Food: Cosmetic Standards	1120
Food: Pesticide Contamination	1124
Fossil Fuel Combustion: Air Pollution and Global Warming	1127
Fuel Cells: Intermediate and High Temperature	1138

Volume II (*cont'd*)

Fuel Cells: Low Temperature	1145
Genotoxicity and Air Pollution	1154
Geographic Information System (GIS): Land Use Planning	1163
Geothermal Energy Resources	1167
Giant Reed (*Arundo donax*): Streams and Water Resources	1182
Global Climate Change: Carbon Sequestration	1189
Global Climate Change: Earth System Response	1192
Global Climate Change: Gas Fluxes	1202
Global Climate Change: Gasoline, Hybrid-Electric and Hydrogen Fueled Vehicles	1206
Global Climate Change: World Soils	1213
Globalization	1218
Green Energy	1226
Green Processes and Projects: Systems Analysis	1242
Green Products: Production	1253
Groundwater: Arsenic Contamination	1262
Groundwater: Contamination	1281
Groundwater: Mining	1284
Groundwater: Mining Pollution	1289
Groundwater: Modeling	1295
Groundwater: Nitrogen Fertilizer Contamination	1302
Groundwater: Numerical Method Modeling	1312
Groundwater: Pesticide Contamination	1317
Groundwater: Saltwater Intrusion	1321
Groundwater: Treatment with Biobarrier Systems	1333
Heat Pumps	1346
Heat Pumps: Absorption	1356
Heat Pumps: Geothermal	1363
Heavy Metals	1370
Heavy Metals: Organic Fertilization Uptake	1374
Herbicides	1378
Herbicides: Non-Target Species Effects	1382
Human Health: Cancer from Pesticides	1394
Human Health: Chronic Pesticide Poisonings	1397
Human Health: Consumer Concerns of Pesticides	1401
Human Health: Hormonal Disruption	1405
Human Health: Pesticide Poisonings	1408
Human Health: Pesticide Sensitivities	1411
Human Health: Pesticides	1414
Hydroelectricity: Pumped Storage	1418
Industrial Waste: Soil Pollution	1430
Industries: Network of	1434
Inland Seas and Lakes: Central Asia Case Study	1436
Inorganic Carbon: Composition and Formation	1444
Inorganic Carbon: Global Carbon Cycle	1448
Inorganic Carbon: Modeling	1451
Inorganic Compounds: Eco-Toxicity	1455
Insect Growth Regulators	1459
Insecticides: Aerial Ultra-Low-Volume Application	1471
Insects and Mites: Biological Control	1474
Insulation: Facilities	1478
Integrated Energy Systems	1493
Integrated Farming Systems	1507
Integrated Nutrient Management	1510
Integrated Pest Management	1521
Integrated Weed Management	1524

Brief Contents (cont'd)

Volume III

Invasion Biology	1531
Irrigation Systems: Sub-Surface Drip Design	1535
Irrigation Systems: Sub-Surface History	1539
Irrigation Systems: Water Conservation	1542
Irrigation: Efficiency	1545
Irrigation: Erosion	1551
Irrigation: Return Flow and Quality	1557
Irrigation: River Flow Impact	1563
Irrigation: Saline Water	1568
Irrigation: Sewage Effluent Use	1570
Irrigation: Soil Salinity	1572
Lakes and Reservoirs: Pollution	1576
Lakes: Restoration	1588
Land Restoration	1600
Laws and Regulations: Food	1609
Laws and Regulations: Pesticides	1612
Laws and Regulations: Rotterdam Convention	1615
Laws and Regulations: Soil	1618
Leaching	1621
Leaching and Illuviation	1624
Lead: Ecotoxicology	1630
Lead: Regulations	1636
LEED-EB: Leadership in Energy and Environmental Design for Existing Buildings	1647
LEED-NC: Leadership in Energy and Environmental Design for New Construction	1654
Manure Management: Compost and Biosolids	1660
Manure Management: Dairy	1663
Manure Management: Phosphorus	1667
Manure Management: Poultry	1670
Mercury	1676
Methane Emissions: Rice	1679
Minerals Processing Residue (Tailings): Rehabilitation	1683
Mines: Acidic Drainage Water	1688
Mines: Rehabilitation of Open Cut	1692
Mycotoxins	1696
Nanomaterials: Regulation and Risk Assessment	1700
Nanoparticles	1711
Nanoparticles: Uncertainty Risk Analysis	1720
Nanotechnology: Environmental Abatement	1730
Natural Enemies and Biocontrol: Artificial Diets	1746
Natural Enemies: Conservation	1749
Nematodes: Biological Control	1752
Neurotoxicants: Developmental Experimental Testing	1755
Nitrate Leaching Index	1761
Nitrogen	1768
Nitrogen Trading Tool	1772
Nitrogen: Biological Fixation	1785
Nuclear Energy: Economics	1789
Nutrients: Best Management Practices	1805
Nutrients: Bioavailability and Plant Uptake	1817
Nutrient–Water Interactions	1823
Oil Pollution: Baltic Sea	1826
Organic Compounds: Halogenated	1841
Organic Matter: Global Distribution in World Ecosystems	1851

Volume III (cont'd)

Organic Matter: Management	1857
Organic Matter: Modeling	1863
Organic Matter: Turnover	1872
Organic Soil Amendments	1878
Ozone Layer	1882
Permafrost	1900
Persistent Organic Compounds: Wet Oxidation Removal	1904
Persistent Organic Pesticides	1913
Pest Management	1919
Pest Management: Crop Diversity	1925
Pest Management: Ecological Agriculture	1930
Pest Management: Ecological Aspects	1934
Pest Management: Intercropping	1937
Pest Management: Legal Aspects	1940
Pest Management: Modeling	1949
Pesticide Translocation Control: Soil Erosion	1955
Pesticides	1963
Pesticides: Banding	1983
Pesticides: Chemical and Biological	1986
Pesticides: Damage Avoidance	1997
Pesticides: Effects	2005
Pesticides: History	2008
Pesticides: Measurement and Mitigation	2013
Pesticides: Natural	2028
Pesticides: Reducing Use	2033
Pesticides: Regulating	2035
Pests: Landscape Patterns	2038
Petroleum: Hydrocarbon Contamination	2040
Pharmaceuticals: Treatment	2060
Phenols	2071
Phosphorus: Agricultural Nutrient	2091
Phosphorus: Riverine System Transport	2100
Plant Pathogens (Fungi): Biological Control	2107
Plant Pathogens (Viruses): Biological Control	2111
Pollutants: Organic and Inorganic	2114
Pollution: Genotoxicity of Agrotoxic Compounds	2123
Pollution: Non-Point Source	2136
Pollution: Pesticides in Agro-Horticultural Ecosystems	2139
Pollution: Pesticides in Natural Ecosystems	2145
Pollution: Point and Non-Point Source Low Cost Treatment	2150
Pollution: Point Sources	2166
Polychlorinated Biphenyls (PCBs)	2172
Polychlorinated Biphenyls (PCBs) and Polycyclic Aromatic Hydrocarbons (PAHs): Sediments and Water Analysis	2186
Potassium	2208
Precision Agriculture: Engineering Aspects	2213
Precision Agriculture: Water and Nutrient Management	2217
Radio Frequency Towers: Public School Placement	2224
Radioactivity	2234
Radionuclides	2243
Rain Water: Atmospheric Deposition	2249
Rain Water: Harvesting	2262
Rare Earth Elements	2266

Brief Contents (*cont'd*)

Volume IV

Remote Sensing and GIS Integration	2271
Remote Sensing: Pollution	2275
Rivers and Lakes: Acidification	2291
Rivers: Pollution	2303
Rivers: Restoration	2307
Runoff Water	2320
Salt-Affected Soils: Physical Properties and Behavior	2334
Salt-Affected Soils: Plant Response	2345
Salt-Affected Soils: Sustainable Agriculture	2349
Sea: Pollution	2357
Sodic Soils: Irrigation Farming	2364
Sodic Soils: Properties	2367
Sodic Soils: Reclamation	2370
Soil Degradation: Global Assessment	2375
Soil Quality: Carbon and Nitrogen Gases	2388
Soil Quality: Indicators	2393
Soil Rehabilitation	2396
Solid Waste Management: Life Cycle Assessment	2399
Solid Waste: Municipal	2415
Stored-Product Pests: Biological Control	2423
Strontium	2426
Sulfur	2431
Sulfur Dioxide	2437
Sustainability and Planning	2446
Sustainable Agriculture: Soil Quality	2457
Sustainable Development	2461
Sustainable Development: Ecological Footprint in Accounting	2467
Sustainable Development: Pyrolysis and Gasification of Biomass and Wastes	2482
Thermal Energy: Solar Technologies	2498
Thermal Energy: Storage	2508
Thermodynamics	2525
Tillage Erosion: Terrace Formation	2536
Toxic Substances	2542
Toxic Substances: Photochemistry	2547
Toxicity Prediction of Chemical Mixtures	2572
Vanadium and Chromium Groups	2582

Volume IV (*cont'd*)

Vertebrates: Biological Control	2595
Waste Gas Treatment: Bioreactors	2611
Waste: Stabilization Ponds	2632
Wastewater and Water Utilities	2638
Wastewater Treatment: Biological	2645
Wastewater Treatment: Conventional Methods	2657
Wastewater Treatment: Wetlands Use in Arctic Regions	2662
Wastewater Use in Agriculture	2675
Wastewater Use in Agriculture: Policy Issues	2681
Wastewater Use in Agriculture: Public Health Considerations	2694
Wastewater: Municipal	2709
Water and Wastewater: Filters	2719
Water and Wastewater: Ion Exchange Application	2734
Water Harvesting	2738
Water Quality and Quantity: Globalization	2740
Water Quality: Modeling	2749
Water Quality: Range and Pasture Land	2752
Water Quality: Soil Erosion	2755
Water Quality: Timber Harvesting	2770
Water Supplies: Pharmaceuticals	2776
Water: Cost	2779
Water: Drinking	2790
Water: Surface	2804
Water: Total Maximum Daily Load	2808
Watershed Management: Remote Sensing and GIS	2816
Weeds (Insects and Mites): Biological Control	2821
Wetlands	2824
Wetlands: Biodiversity	2829
Wetlands: Carbon Sequestration	2833
Wetlands: Conservation Policy	2837
Wetlands: Constructed Subsurface	2841
Wetlands: Methane Emission	2850
Wetlands: Petroleum	2854
Wetlands: Sedimentation and Ecological Engineering	2859
Wetlands: Treatment System Use	2862
Wind Farms: Noise	2867
Yellow River	2884
Appendixes	2887

Encyclopedia of Environmental Engineering

Editor-in-Chief

Sven Erik Jørgensen
Institute A, Section of Environmental Chemistry, Copenhagen University, Copenhagen, Denmark

Editorial Advisory Board

Marek Biziuk
Gdansk University of Technology, Gdansk, Poland

Ni-Bin Chang
University of Central Florida, Orlando, Florida, U.S.A.

Krist V. Gernaey
Technical University of Denmark, Kongens Lyngby, Denmark

Velma I. Grover
York University, Hamilton, Ontario, Canada

Bhola Gurjar
Indian Institute of Technology Roorkee, Roorkee, India

William Hogland
Linnaeus University, Kalmar, Sweden

Sandeep Joshi
Shrishti Eco Research Unit (SERI), Pune, India

Puangrat Kajitvichyanukul
Naresuan University, Phitsanulok, Thailand

Tarzan Legovic
Rudjer Boskovic Institute, Zagreb, Croatia

Saburo Matsui
Kyoto University, Kyoto, Japan

Anil Namdeo
Newcastle University, Newcastle, U.K.

Roberta Sonnino
Cardiff University, Cardiff, U.K.

Abhishek Tiwary
Newcastle University, Newcastle, U.K.

Katherin von Stackelberg
Harvard School of Public Health, Boston, Massachusetts, U.S.A.

Reviewers

Christopher Amrhein
William Andreen
M. Angulo-Martinez
Kwame Badu Antwi-Boasiako
Yongan Ao
Andres Arnalds
James Aronson
Karen L. Bailey
Jim Barbour
Henryk Bem
Francesco Berti
G. Ronnie Best
Marek Biziuk
Enrique Roca Bordello
Alessandro Pietro Brivio
Zbigniew Brzózka
Gary D. Bubenzer

Cornelia Ada Bulucea
James Burger
R.T. Bush
Thomas J. Butler
David Butterfield
Gemma Calamandrei
Patrick Carr
Gerry Carrington
F.P. Carvalho
Yunus A. Çengel
Ying Chen
M.A. Chitale
Jock Churchman
Gary Clark
Brian Cooke
Charles H. Culp
Jean-Claude Dauvin

Larry Degelman
M. Rifat Derici
Ibrahim Dincer
Marisa Domingos
George Ekström
Daniel D. Euhus
Greg Evanylo
Wayne Fairchild
Delvin S. Fanning
Nerilde Favaretto
Gary Feng
Daniel Fiorino
Fred Fishel
Robert W. Fitzpatrick
Ignazio Floris
Ronal F. Follett
J.D. Fontana

Reviewers (cont'd)

Richard Frankel
Alan J. Franzluebbers
Michael H. Glantz
Sabine Goldberg
Neva Goodwin
Tadeusz Górecki
Shela Gorinstein
Scott Grosse
Velma I. Grover
Silvio J. Gumiere
Heli Haapasaari
Douglas Hale
Ed Hanna
Chunyan Hao
P.S. Harikumar
Paul Hatcher
Michaela Hegglin
George Helz
Ivan Holoubek
A.R. Horowitz
Lucas Hyman
Harish Jeswani
Buzz Johnson
Jodi L. Johnson-Maynard
Bill Jokela
Jan Åke Jönsson
Sandeep Joshi
Puangrat Kajitvichyanukul
Ioannis K. Kalvrouziotis
Mehmet Kanoglu
Douglas L. Karlen
Gilbert Kelling
Ben Keraita
Thomas Kinraide
Holger Kirchmann
Andreas Klumpp
Gordana Kranjac-Berisavljevic
John M. Laflen
Paul J. Lamothe
Matthew Langholtz
E.F. Legner

Tarzan Legovic
Ronnie Levin
Guy Levy
Witold Lewandowski
Jianbing Li
Alessandro Ludovisi
Tapas Malick
Rob Malone
Andreas Mamolos
Stanley E. Manahan
Duncan Mara
Dan Marion
E.J.P. Marshall
David McBride
David McKenzie
Bernd Markert
Rick Miller
Lesley Mills
Roger Minear
Virginia Moser
Amitava Mukherjee
Deborah A. Neher
Kristian Fog Nielsen
Darrell Norton
Krystyna Olanczuk-Neyman
David Olszyk
A.D. Patwardhan
Janusz Pempkowiak
Robert Percival
Sandra Perez
Tanapon Phenrat
David Pimentel
Jon K. Piper
I. Popescu
Federico M. Pulselli
X.S. Qin
Philip S. Rainbow
Barnett Rattner
E. Remoundaki
Sergio Revah
Marc Ribaudo

H. Rodhe
Art Rose
Don Ross
Diederik Rousseau
John R. Ruberson
Mario Russo
Yvonne Rydin
Barbara J.S. Sanderson
Marvin Schaffer
James S. Schepers
Steven Sehr
Supapan Seraphin
Balwinder Singh Sra
Bogdan Skwarzec
Scott Slocombe
Shaul Sorek
Irena Staneczko-Baranowska
B.A. Stewart
Kristian Syberg
Piotr Szefer
Moses M. Tenywa
Khi V. Thai
Daniel L. Thomas
Abhishek Tiwary
José Torrent
Jeff N. Tullberg
Marco Vighi
Earl Vories
Maurizio Vurro
Maria Waclawek
J.Y. Wang
Waldemar Wardencki
Apichon Watcharenwong
Bernard Weiss
Fern Wickson
L.T. Wilson
Bofu Yu
Frank G. Zalom
Rick Zartman

Contributors

Diana Aga / *Department of Chemistry, State University of New York at Buffalo, Buffalo, New York, U.S.A.*
Matthew Agarwala / *Geography and Environment Department and Grantham Research Institute on Climate Change and the Environment, London School of Economics, London, U.K.*
Shaikh Ziauddin Ahammad / *School of Civil Engineering and Geosciences, Newcastle University, Newcastle, U.K.*
Imad A.M. Ahmed / *Lancaster Environment Center, Lancaster University, Lancaster, U.K.*
Erhan Akça / *Technical Programs, Adiyaman University, Adiyaman, Turkey*
Claude Amiard-Triquet / *French National Center for Scientific Research (CNRS) and University of Nantes, Nantes, France*
Ronald G. Amundson / *College of Natural Resources, University of California—Berkeley, Berkeley, California, U.S.A.*
Jirapat Ananpattarachai / *Center of Excellence for Environmental Research and Innovation, Faculty of Engineering, Naresuan University, Phitsanulok, Thailand*
Samson D. Angima / *Oregon State University, Oregon, U.S.A.*
Kalyan Annamalai / *Paul Pepper Professor of Mechanical Engineering, Texas A&M University, College Station, Texas, U.S.A.*
Massimo Antoninetti / *Institute for Electromagnetic Sensing of the Environment (IREA), National Research Council of Italy (CNR), Milan, Italy*
George F. Antonious / *Department of Plant and Soil Science, Water Quality/Environmental Toxicology Research, Kentucky State University, Frankfort, Kentucky, U.S.A.*
Ramón Aragues / *Agronomic Research Service, Government of Aragon, Zaragoza, Spain*
Senthil Arumugam / *Enerquip, Inc., Medford, Wisconsin, U.S.A.*
G.J. Ash / *E.H. Graham Center for Agricultural Innovation, Industry and Investment, NSW and Charles Sturt University, Wagga Wagga, New South Wales, Australia*
Muhammad Asif / *School of the Built and Natural Environment, Glasgow Caledonian University, Glasgow, U.K.*
William Au / *Department of Preventitive Medicine and Community Health, University of Texas Medical Branch, Galveston, Texas, U.S.A.*
Lu Aye / *Renewable Energy and Energy Efficiency Group, Department of Infrastructure Engineering, Melbourne School of Engineering, University of Melbourne, Melbourne, Victoria, Australia*
Thomas Backhaus / *Department of Plant and Environmental Sciences, University of Gothenburg, Gothenburg, Sweden*
Seungyun Baik / *Department of Chemistry, State University of New York at Buffalo, Buffalo, New York, U.S.A.*
Kenneth A. Barbarick / *Colorado State University, Fort Collins, Colorado, U.S.A.*
Javier Barragan / *Department of Agroforestry Engineering, University of Lleida, Lleida, Spain*
Simone Bastianoni / *Ecodynamics Group, Department of Chemistry, University of Siena, Siena, Italy*

Anders Baun / *Department of Environmental Engineering, Technical University of Denmark, Kongens Lyngby, Denmark*

Susana Bautista / *Department of Ecology, University of Alicante, Alicante, Spain*

Lindsay Beevers / *Lecturer in Water Management, School of the Built Environment, Heriot Watt University, Edinburgh, U.K.*

Richard W. Bell / *School of Environmental Science, Murdoch University, Perth, Western Australia, Australia*

Suha Berberoğlu / *Departments of Soil Science, Landscape Architecture, and Agricultural Engineering, University of Cukurova, Adana, Turkey*

Sanford V. Berg / *Director of Water Studies, Public Utility Research Center, University of Florida, Gainesville, Florida, U.S.A.*

Lars Bergström / *Department of Soil Science, Swedish University of Agricultural Sciences (SLU), Uppsala, Sweden*

Angelika Beyer / *Department of Analytical Chemistry, Chemical Faculty, Gdansk University of Technology, Gdansk, Poland*

Jerry M. Bigham / *Ohio State University, Columbus, Ohio, U.S.A.*

Marek Biziuk / *Department of Analytical Chemistry, Chemical Faculty, Gdansk University of Technology, Gdansk, Poland*

David L. Bjorneberg / *Northwest Irrigation and Soils Research Lab, Agricultural Research Service (USDA-ARS), U.S. Department of Agriculture, Kimberly, Idaho, U.S.A.*

John O. Blackburn / *Professor Emeritus of Economics, Duke University, Maitland, Florida, U.S.A.*

Frederick Paxton Cardell Blamey / *School of Land, Crop and Food Sciences, University of Queensland, St. Lucia, Queensland, Australia*

Elke Bloem / *Institute for Crop and Soil Science, Julius Kuhn Institute (JKI), Braunschweig, Germany*

W.E.H. Blum / *Institute of Soil Research, University of Natural Resources and Life Sciences, Vienna, Austria*

Pascal Boeckx / *Faculty of Agricultural and Applied Biological Sciences, University of Ghent, Ghent, Belgium*

Ben Boer / *School of Law, University of Sydney, Sydney, New South Wales, Australia*

Julia Boike / *Water and Environmental Research Center, University of Alaska—Fairbanks, Fairbanks, Alaska, U.S.A.*

Nadia Bernardi Bonumá / *Federal University of Santa Maria, Santa Maria, Brazil*

J.D. Booker / *Plant and Soil Science, Texas Tech University, Lubbock, Texas, U.S.A.*

John Borden / *Department of Biological Sciences, Simon Fraser University, Burnaby, British Columbia, Canada*

Virginie Bouchard / *School of Natural Resources, Ohio State University, Columbus, Ohio, U.S.A.*

Céline Boutin / *Science and Technology Branch, Environment Canada, Carleton University, Ottawa, Ontario, Canada*

John T. Brake / *Department of Soil Science, North Carolina State University, Raleigh, North Carolina, U.S.A.*

Vince Bralts / *Agricultural and Biological Engineering, Purdue University, West Lafayette, Indiana, U.S.A.*

James R. Brandle / *School of Natural Resource Sciences, University of Nebraska—Lincoln, Lincoln, Nebraska, U.S.A.*

D.R. Bray / *Department of Animal Sciences, University of Florida, Gainesville, Florida, U.S.A.*

Zachary T. Broome / Bowen Radson Schroth, P. A., Eustis, Florida, U.S.A.
Dominic A. Brose / University of Maryland, College Park, Maryland, U.S.A.
Steven N. Burch / University of Maryland, College Park, Maryland, U.S.A.
Benjamin Burkhard / Institute for the Conservation of Natural Resources, University of Kiel, Kiel, Germany
David Burrow / Agriculture Victoria (Tatura), Tartura, Victoria, Australia
E.D. Burton / Southern Cross GeoScience, Southern Cross University, Lismore, New South Wales, Australia
Alan Busacca / Department of Crop and Soil Sciences, Geology, Washington State University, Pullman, Washington, U.S.A.
R.T. Bush / Southern Cross GeoScience, Southern Cross University, Lismore, New South Wales, Australia
Liana Buzdugan / Center for Sustainable Exploitation of Ecosystems (CESEE), Alexandru Ioan Cuza University of Iasi, Iasi, Romania
Nídia Sá Caetano / Chemical Engineering Department, School of Engineering (ISEP), Polytechnic Institute of Porto (IPP), and Laboratory for Process, Environmental and Energy Engineering, Porto, Portugal
James Call / President, James Call Engineering, PLLC, Larchmont, New York, U.S.A.
Carl R. Camp, Jr. / Agricultural Research Service (USDA-ARS), U.S. Department of Agriculture, Florence, South Carolina, U.S.A.
Kenneth L. Campbell / Agricultural and Biological Engineering Department, University of Florida, Gainesville, Florida, U.S.A.
Barney L. Capehart / Department of Industrial and Systems Engineering, University of Florida College of Engineering, Gainesville, Florida, U.S.A.
Kristi Denise Caravella / Florida Atlantic University, Boca Raton, Florida, U.S.A.
Jesus Carrera / Technical University of Catalonia (UPC), Barcelona, Spain
Vera Lucia S.S. de Castro / Ecotoxicology and Biosafety Laboratory, Environment, Brazilian Agricultural Research Corporation (Embrapa), São Paulo, Brazil
Nina Cedergreen / Faculty of Life Sciences, University of Copenhagen, Frederiksberg, Denmark
Carlos C. Cerri / University of São Paulo, São Paulo, Brazil
Dipankar Chakraborti / Director (Research), School of Environmental Studies, Jadavpur University, Calcutta, India
David Chandler / Plant, Soils, and Biometeorology Department, Utah State University, Logan, Utah, U.S.A.
Ni-Bin Chang / Department of Civil, Environmental, and Construction Engineering, University of Central Florida, Orlando, Florida, U.S.A.
Guangnan Chen / Faculty of Engineering and Surveying, University of Southern Queensland, Toowoomba, Queensland, Australia
Alexander H.-D. Cheng / Department of Civil Engineering, University of Mississippi, Oxford, Mississippi, U.S.A.
Angelique Chettiparamb / School of Real Estate and Planning, University of Reading, Reading, U.K.
Tarit Roy Chowdhury / School of Environmental Studies, Jadavpur University, Calcutta, India
Torben Røjle Christensen / Climate Impacts Group, Department of Ecology, Lund University, Lund, Sweden
Jock Churchman / Land and Water, Commonwealth Scientific and Industrial Research Organization (CSIRO), Adelaide, South Australia, Australia
Maria V. Cilveti / Department of Entomology, Cornell University, Ithaca, New York, U.S.A.

Ioan Manuel Ciumasu / *ECONOVING International Chair, University of Versailles Saint-Quentin-en-Yvelines, Guyancourt, France, and Center for Sustainable Exploitation of Ecosystems (CESEE), Alexandru Ioan Cuza University of Iasi, Iasi, Romania*

David E. Claridge / *Department of Mechanical Engineering, Energy Systems Laboratory, College Station, Texas, U.S.A.*

Sharon A. Clay / *Plant Science Department, South Dakota State University, Brookings, South Dakota, U.S.A.*

Matthew Cochran / *School of Natural Resources, Ohio State University, Columbus, Ohio, U.S.A.*

Gretchen Coffman / *Department of Environmental Health Sciences, University of California—Los Angeles, Ventura, California, U.S.A.*

Alexandra Robin Collins / *Department of Biology and Ecology, University of Fribourg/Perolles, Fribourg, Switzerland*

Ray Correll / *Commonwealth Scientific and Industrial Research Organization (CSIRO), Adelaide, South Australia, Australia*

Luca Coscieme / *Ecodynamics Group, Department of Chemistry, University of Siena, Siena, Italy*

Richard Cowell / *School of City and Regional Planning, Cardiff University, Cardiff, U.K.*

Robin Kundis Craig / *Attorneys' Title Professor of Law and Associate Dean for Environmental Programs, Florida State University College of Law, Tallahassee, Florida, U.S.A.*

Gemma Cranston / *Global Footprint Network, Geneva, Switzerland*

Eric T. Craswell / *Fenner School of Environment and Society, College of Medicine, Biology, and Environment, Australian National University, Canberra, Australian Capital Territory, Australia*

Richard Cruse / *Iowa State University, Ames, Iowa, U.S.A.*

Marc A. Cubeta / *Center for Integrated Fungal Research, Plant Pathology, North Carolina State University, Raleigh, North Carolina, U.S.A.*

Keith Culver / *Okanagan Sustainability Institute, University of British Columbia, Kelowna, British Columbia, Canada*

Marianna Czaplicka / *Institute of Non-Ferrous Metals, and Department of Analytical Chemistry, Silesian University of Technology, Gliwice, Poland*

Seth M. Dabney / *National Sedimentation Laboratory, U.S. Department of Agriculture, Agricultural Research Service (USDA-ARS), Oxford, Mississippi, U.S.A.*

Abhijit Das / *School of Environmental Studies, Jadavpur University, Calcutta, India*

Bhaskar Das / *School of Environmental Studies, Jadavpur University, Calcutta, India*

Franck Dayan / *National Center for Natural Products Research, Agricultural Research Service (USDA-ARS), U.S. Department of Agriculture, University, Missouri, U.S.A.*

Patrick De Clerq / *Department of Crop Protection, Ghent University, Ghent, Belgium*

Ana Maria Evangelista de Duffard / *Laboratorio de Toxicologia Expiremental, National University of Rosario, Rosario, Argentina*

J.R. de Freitas / *Department of Soil Science, University of Saskatchewan, Saskatoon, Saskatchewan, Canada*

Victor de Vlaming / *Aquatic Toxicology Laboratory, University of California—Davis, Davis, California, U.S.A.*

Bernd Delakowitz / *Faculty of Mathematics and Natural Sciences, University of Applied Sciences, Zittau, Germany*

Kathleen Delate / *Departments of Agronomy and Horticulture, Iowa State University, Ames, Iowa, U.S.A.*

Jorge A. Delgado / *Soil Plant Nutrient Research Unit, Agricultural Research Service (USDA-ARS), U.S. Department of Agriculture, Fort Collins, Colorado, U.S.A.*

Detlef Deumlich / *Institute of Soil Landscape Research, Leibniz Center for Agricultural Landscapre Research (ZALF), Muncheberg, Germany*

Malcolm Devine / *Aventis CropScience Canada Co., Saskatoon, Saskatchewan, Canada*

Harvey E. Diamond / *Energy Management International, Conroe, Texas, U.S.A.*

Jan Dich / *Unit of Cancer Epidemiology, Institution of Oncology-Pathology, Karolinska Institutet and Radiumhemmet, Karolinska University Hospital, Stockholm, Sweden*

Christina D. DiFonzo / *Department of Entomology, Michigan State University, East Lansing, Michigan, U.S.A.*

Peter Dillon / *Commonwealth Scientific and Industrial Research Organization (CSIRO), Adelaide, South Australia, Australia*

Ibrahim Dincer / *Faculty of Engineering and Applied Science, University of Ontario Institute of Technology (UOIT), Oshawa, Ontario, Canada*

Barbara Dinham / *Eurolink Center, Pesticide Action Network U.K., London, U.K.*

Craig Ditzler / *National Leader for Soil Classification and Standards, Lincoln, Nebraska, U.S.A.*

Jan Dolfing / *School of Civil Engineering and Geosciences, Newcastle University, Newcastle, U.K.*

Douglas J. Dollhopf / *Department of Land Resources and Environmental Sciences, Montana State University, Bozeman, Montana, U.S.A.*

Cenk Dönmez / *Departments of Soil Science, Landscape Architecture, and Agricultural Engineering, University of Cukurova, Adana, Turkey*

John W. Doran / *U.S. Department of Agriculture, Agricultural Research Service (USDA-ARS), Agronomy, University of Nebraska—Lincoln, Lincoln, Nebraska, U.S.A.*

Steve Doty / *Colorado Springs Utilities, Colorado Springs, Colorado, U.S.A.*

Pay Drechsel / *International Water Management Institute (IWMI), Colombo, Sri Lanka*

Svetlana Drozdova / *Institute of Chemical Technologies and Analytics, Vienna University of Technology, Vienna, Austria*

J.K. Dubey / *Regional Center, National Afforestation and Eco-Development Board, Dr. Y.S. Parmar University of Horticulture and Forestry, Solan, India*

Rathindra Nath Dutta / *Department of Dermatology, Institute of Post Graduate Medical Education and Research, SSKM Hospital, Calcutta, India*

Bahman Eghball / *Agricultural Research Service (USDA-ARS), U.S. Department of Agriculture, Lincoln, Nebraska, U.S.A.*

Reza Ehsani / *Ohio State University, Columbus, Ohio, U.S.A.*

Anna Ekberg / *Department of Ecology, Plant Ecology, Lund University, Climate Impacts Group, Lund, Sweden*

George Ekström / *Swedish National Chemicals Inspectorate (KEMI), Solna, Sweden*

Krisztina Eleki / *Department of Agronomy, Iowa State University, Ames, Iowa, U.S.A.*

Mehmet Akif Erdoğan / *Departments of Soil Science, Landscape Architecture, and Agricultural Engineering, University of Cukurova, Istanbul, Turkey*

Sarina J. Ergas / *Department of Civil and Environmental Engineering, University of South Florida, Tampa, Florida, U.S.A.*

Gunay Erpul / *Department of Soil Science, Ankara University, Ankara, Turkey*

Shannon Estenoz / *Office of Everglades Restoration Initiatives, U.S. Department of the Interior, Davie, Florida, U.S.A.*

Sara Evangelisti / *Interuniversity Research Center for Sustainable Develepment (CIRPS), Sapienza University of Rome, Rome, Italy*

Anna Eynard / *Ohio State University, Brookings, South Dakota, U.S.A.*
Delvin S. Fanning / *Department of Environmental Science and Technology, University of Maryland, College Park, Maryland, U.S.A.*
Norman R. Fausey / *Soil Drainage Research Unit, Agricultural Research Service (USDA-ARS), U.S. Department of Agriculture, Columbus, Ohio, U.S.A.*
David Favis-Mortlock / *Environmental Change Institute, University of Oxford, Oxford, U.K.*
Jolanta Fenik / *Department of Analytical Chemistry, Chemical Faculty, Gdansk University of Technology, Gdansk, Poland*
David N. Ferro / *Department of Entomology, University of Massachusetts, Amherst, Massachusetts, U.S.A.*
Charles W. Fetter / *C. W. Fetter, Jr. Associates, Oshkosh, Wisconsin, U.S.A.*
Maria Finckh / *Department of Ecological Plant Protection, University of Kassel, Witzenhausen, Germany*
Guy Fipps / *Agricultural Engineering Department, Texas A&M University, College Station, Texas, U.S.A.*
Dennis C. Flanagan / *National Soil Science Research Laboratory, Agricultural Research Service (USDA-ARS), U.S. Department of Agriculture, West Lafayette, Indiana, U.S.A.*
Stefan Fraenzle / *Department of Biological and Environmental Sciences; Research Group of Environmental Chemistry, International Graduate School, Zittau, Zittau, Germany*
Alan J. Franzluebbers / *Agricultural Research Service (USDA-ARS), U.S. Department of Agriculture, Watkinsville, Georgia, U.S.A.*
Gary W. Frasier / *U.S. Department of Agriculture (USDA), Fort Collins, Colorado, U.S.A.*
Dwight K. French / *Director, Energy Consumption Division, Energy Information Administration, U.S. Department of Energy, Washington, District of Columbia, U.S.A.*
John R. Freney / *Commonwealth Scientific and Industrial Research Organization (CSIRO), Campbell, Australian Capital Territory, Australia*
Martin V. Frey / *Department of Soil Science, University of Stellenbosch, Matieland, South Africa*
Brenda Frick / *Bluebur Fluent Organics, Saskatoon, Saskatchewan, Canada*
Monika Frielinghaus / *Institute of Soil Landscape Research, Leibniz Center for Agricultural Landscape Research (ZALF), Muncheberg, Germany*
W. Friesl-Hanl / *Health and Environment Department, Environmental Resources & Technologies, AIT Austrian Institute of Technology GmbH, Tulln, Austria*
Roger Funk / *Institute of Soil Landscape Research, Leibniz Center for Agricultural Landscape Research (ZALF), Muncheberg, Germany*
Donald Gabriels / *Department of Soil Management and Soil Care, Ghent University, Ghent, Belgium*
Wendy B. Gagliano / *Clark State Community College, Springfield, Ohio, U.S.A.*
Renata Gaj / *Institute of Soil Science, Agricultural University, Poznan, Poland*
Alessandro Galli / *Global Footprint Network, Geneva, Switzerland*
Ján Gallo / *Department of Plant Protection, Slovak University of Agriculture, Nitra, Slovak Republic*
Agniezka Gałuszka / *Division of Geochemistry and the Environment, Institute of Chemistry, Jan Kochanowski University, Kielce, Poland*
Anurag Garg / *Center for Environmental Science and Engineering, Indian Institute of Technology, Bombay, Mumbai, India*
Anja Gassner / *Institute of Science and Technology, University of Malaysia—Sabah, Kota Kinabalu, Malaysia*

David K. Gattie / *Biological and Agricultural Engineering Department, University of Georgia, Athens, Georgia, U.S.A.*

M.H. Gerzabek / *Institute of Soil Research, University of Natural Resources and Life Sciences, Vienna, Austria*

Stephen R. Gliessman / *Program in Community and Agroecology, Department of Environmental Studies, University of California—Santa Cruz, Santa Cruz, California, U.S.A.*

Fredric S. Goldner / *Energy Management & Research Associates, East Meadow, New York, U.S.A.*

Dan Golomb / *Department of Environmental, Earth and Atmospheric Sciences, University of Massachusetts—Lowell, Lowell, Massachusetts, U.S.A.*

Ragini Gothalwal / *Institute of Microbiology and Biotechnology, Barkatullah University, Bhopal, India*

Andrew S. Goudie / *St. Cross College, Oxford, U.K.*

Simon Gowen / *Department of Agriculture, University of Reading, Reading, U.K.*

David W. Graham / *School of Civil Engineering and Geosciences, Newcastle University, Newcastle, U.K.*

Timothy C. Granata / *Department of Civil and Environmental Engineering and Geodetic Science, Ohio State University, Columbus, Ohio, U.S.A.*

Alex E.S. Green / *Professor Emeritus, University of Florida, Gainesville, Florida, U.S.A.*

Ed G. Gregorich / *Easter Cereal and Oilseed Research Center, Agriculture and Agri-Food Canada, Ottawa, Ontario, Canada*

Simon Grenier / *Functional Biology, Insects and Interactions, National Institute for Agricultural Research (INRA), Villeurbanne, France*

Khara D. Grieger / *Department of Environmental Engineering, Technical University of Denmark, Kongens Lyngby, Denmark*

Lisa Guan / *School of Chemistry, Physics and Mechanical Engineering, Science and Engineering Faculty, Queensland University of Technology, Brisbane, Queensland, Australia*

Eliane Tigre Guimarães / *Experimental Air Pollution Laboratory, Department of Pathology, School of Medicine, University of São Paulo, São Paulo, Brazil*

Silvio J. Gumiere / *Department of Soil Science, Laval University, Quebec City, Quebec, Canada*

Umesh C. Gupta / *Crops and Livestock Research Center, Agriculture and Agri-Food Canada, Charlottetown, Prince Edward Island, Canada*

Andrew Paul Gutierrez / *Center for the Analysis of Sustainable Agricultural Systems, University of California—Berkeley, Berkeley, California, U.S.A.*

Ann E. Hajek / *Department of Entomology, Cornell University, Ithaca, New York, U.S.A.*

Ardell D. Halvorson / *U.S. Department of Agriculture (USDA), Fort Collins, Colorado, U.S.A.*

Denis Hamilton / *Department of Animal and Plant Health Service, Queensland Department of Primary Industries, Brisbane, Queensland, Australia*

Silvia Haneklaus / *Institute for Crop and Soil Science, Julius Kuhn Institute (JKI), Braunschweig, Germany*

Ian Hannam / *Center for Natural Resources, Department of Infrastructure, Planning and Natural Resources, Sydney, New South Wales, Australia*

Chris Hanning / *Sleep Medicine, University Hospitals of Leicester, Leicester, U.K.*

Lise Stengård Hansen / *Danish Pest Infestation Laboratory, Danish Institute of Agricultural Sciences, Konigs Lyngby, Denmark*

Steffen Foss Hansen / *Department of Environmental Engineering, Technical University of Denmark, Kongens Lyngby, Denmark*
Peter Harris / *Agriculture and Agri-Food Canada, Lethbridge, Alberta, Canada*
Kelsey Hart / *College of Veterinary Medicine, University of Georgia, Athens, Georgia, U.S.A.*
James D. Harwood / *Department of Entomology, University of Kentucky, Lexington, Kentucky, U.S.A.*
John V. Headley / *Water Science and Technology Directorate, Saskatoon, Saskatchewan, Canada*
Steven D. Heinz / *Good Steward Software, State College, Pennsylvania, U.S.A.*
Arif Hepbasli / *Department of Energy Systems Engineering, Faculty of Engineering, Yaşar University, Bornova, Izmir, Turkey*
Keith E. Herold / *Fischell Department of Bioengineering, University of Maryland, College Park, Maryland, U.S.A.*
James G. Hewlett / *Energy Information Administration, U.S. Department of Energy, Washington, District of Columbia, U.S.A.*
Robert W. Hill / *Biological and Irrigation Engineering Department, Utah State University, Logan, Utah, U.S.A.*
Philippe Hinsinger / *Sun and Environment Unit, National Institute for Agricultural Research (INRA), Montpellier, France*
Michael C. Hirschi / *University of Illinois, Urbana, Illinois, U.S.A.*
Rusty T. Hodapp / *Vice President and Sustainability Officer, Energy and Transportation Management, Dallas/Fort Worth International Airport Board, Dallas/Forth Worth Airport, Texas, U.S.A.*
Laurie Hodges / *Department of Agronomy and Horticulture, University of Nebraska—Lincoln, Lincoln, Nebraska, U.S.A.*
Glenn J. Hoffman / *Biological Systems Engineering, University of Nebraska—Lincoln, Lincoln, Nebraska, U.S.A.*
Heikki Hokkanen / *Department of Applied Biology, University of Helsinki, Helsinki, Finland*
John Holland / *Head of Entomology, Game Conservancy Trust, Hants, U.K.*
David J. Horn / *Department of Entomology, Ohio State University, Columbus, Ohio, U.S.A.*
Lloyd R. Hossner / *Soil and Crop Sciences Department, Texas A&M University, College Station, Texas, U.S.A.*
Terry A. Howell / *Conservation and Production Research Laboratory, Agricultural Research Service (USDA-ARS), U.S. Department of Agriculture, Bushland, Texas, U.S.A.*
Hei-Ti Hsu / *Floral and Nursery Plants Research, Agricultural Research Service (USDA-ARS), U.S. Department of Agriculture, Beltsville, Maryland, U.S.A.*
Zhengyi Hu / *Institute of Soil Science, Chinese Academy of Sciences, Nanjing, China*
Nathan E. Hultman / *University of Maryland, Maryland, U.S.A.*
Hayriye Ibrikci / *Soil Science and Plant Nutrition Department, Cukurova University, Adana, Turkey*
Craig Idso / *Center for the Study of Carbon Dioxide and Global Change, Tempe, Arizona, U.S.A.*
Keith E. Idso / *Center for the Study of Carbon Dioxide and Global Change, Tempe, Arizona, U.S.A.*
Sherwood Idso / *U.S. Department of Agriculture (USDA), Tempe, Arizona, U.S.A.*
R. Cesar Izaurralde / *Battelle Pacific Northwest National Laboratory, Washington, District of Columbia, U.S.A.*
Alexandra Izosimova / *St. Petersburg Agricultural Physical Research Institute, St. Petersburg, Russia*

Pierre A. Jacinthe / *Ohio State University, Columbus, Ohio, U.S.A.*
C. Rhett Jackson / *Daniel B. Warnell School of Forest Resources, University of Georgia, Athens, Georgia, U.S.A.*
Bruce R. James / *University of Maryland, College Park, Maryland, U.S.A.*
Philip M. Jardine / *Oak Ridge National Laboratory, Oak Ridge, Tennessee, U.S.A.*
David Jasper / *Center for Land Rehabilitation, University of Western Australia, Nedlands, Western Australia, Australia*
Julie D. Jastrow / *Environmental Research Division, Argonne National Laboratory, Argonne, Illinois, U.S.A.*
Ike Jeon / *Department of Animal Science and Industry, Kansas State University, Manhattan, Kansas, U.S.A.*
Kui Jiao / *Department of Mechanical Engineering, University of Waterloo, Waterloo, Ontario, Canada*
Blanca Jimenez / *Engineering Institute, National Autonomous University of Mexico (UNAM), Coyoacan, Mexico*
Sven Erik Jørgensen / *Institute A, Section of Environmental Chemistry, Copenhagen University, Copenhagen, Denmark*
Sandeep Joshi / *Shrishti Eco-Research Unit (SERI), Pune, India*
Sayali Joshi / *Shrishti Eco-Research Institute (SERI), Pune, India*
Puangrat Kajitvichyanukul / *Center of Excellence for Environmental Research and Innovation, Faculty of Engineering, Naresuan University, Phitsanulok, Thailand*
Gabriella Kakonyi / *Kroto Research Institute, Sheffield University, Sheffield, U.K.*
Inger Källander / *Swedish Ecological Farmers Association, Uppsala, Sweden*
Marion Kandziora / *Institute for the Conservation of Natural Resources, University of Kiel, Kiel, Germany*
Douglas Kane / *Water and Environmental Research Center, University of Alaska—Fairbanks, Fairbanks, Alaska, U.S.A.*
Chih-Ming Kao / *Institute of Environmental Engineering, National Sun-Yat Sen University, Kaohsiung, Taiwan*
Burçak Kapur / *Department of Biosystems Engineering, University of Yuzuncu Yil, Van, Turkey*
Selim Kapur / *Departments of Soil Science, Landscape Architecture, and Agricultural Engineeering, University of Çukurova, Adana, Turkey*
Gholamreza Karimi / *Department of Mechanical Engineering, University of Waterloo, Waterloo, Ontario, Canada*
Douglas L. Karlen / *U.S. Department of Agriculture (USDA), Ames, Iowa, U.S.A.*
Subhankar Karmakar / *Center for Environmental Science and Engineering (CESE), Indian Institute of Technology Bombay, Mumbai, India*
Marianne Karpenstein-Machan / *Institute of Crop Science, University of Kassel, Witzenhausen, Germany*
Janey Kaster / *Yamas Controls West, San Francisco, California, U.S.A.*
Anthony P. Keinath / *Coastal Research and Education Center, Clemson University, Charleston, South Carolina, U.S.A.*
Keith A. Kelling / *Professor Emeritus, Department of Soil Science, University of Wisconsin—Extension, Madison, Wisconsin, U.S.A.*
George Kennedy / *Department of Entomology, North Carolina State University, Raleigh, North Carolina, U.S.A.*
Rami Keren / *Agricultural Research Organization of Israel, Bet-Dagan, Israel*
Matthias Kern / *GTZ Pilot Project on Chemicals Management, Bonn, Germany*

Peter Kerr / Commonwealth Scientific and Industrial Research Organization (CSIRO) Ecosystem Sciences, Canberra, Australian Capital Territory, Australia
Gregory Kiker / University of Florida, Gainesville, Florida, U.S.A.
Peter I.A. Kinnell / School of Resource, Environmental and Heritage Sciences, University of Canberra, Holt, Australian Capital Territory, Australia
Ronald L. Klaus / VAST Power Systems, Elkhart, Indiana, U.S.A.
Peter Kleinman / Pasture Systems and Watershed Management Research Unit, U.S. Department of Agriculture (USDA), University Park, Pennsylvania, U.S.A.
Andreas Klik / Department of Water, Atmosphere and Environment, University of Natural Resources and Life Sciences, Vienna, Austria
Ewa Klugmann-Radziemska / Chemical Faculty, Gdansk University of Technology, Gdansk, Poland
Birgitta Kolmodin-Hedman / Department of Public Health Sciences, Karolinska Institute, Stockholm, Sweden
Rai Kookana / Commonwealth Scientific and Industrial Research Organization (CSIRO), Adelaide, South Australia, Australia
Monika Kosikowska / Department of Analytical Chemistry, Chemical Faculty, Gdansk University of Technology, Gdansk, Poland
John Kost / Department of Agronomy, Iowa State University, Ames, Iowa, U.S.A.
Andrey G. Kostianoy / P.P. Shirshov Institute of Oceanology, Russian Academy of Sciences, Moscow, Russia
Milivoje M. Kostic / Department of Mechanical Engineering, Northern Illinois University, DeKalb, Illinois, U.S.A.
William L. Kranz / Northeast Research and Extension Center, University of Nebraska—Lincoln, Norfolk, Nebraska, U.S.A.
David P. Kreutzweiser / Canadian Forest Service, Natural Resources Canada, Sault Sainte Marie, Ontario, Canada
Tore Krogstad / Department of Plant and Environmental Sciences, Norwegian University of Life Science, Aas, Norway
Atif Kubursi / Department of Economics, McMaster University, Hamilton, Ontario, Canada
Umesh Kulshrestha / School of Environmental Sciences, Jawaharlal Nehru University, New Delhi, India
Amit Kumar / Research Group Environmental Organic Chemistry Group (EnVOC), Faculty of Bioscience Engineering, Ghent University, Ghent, Belgium, and Department of Environmental Engineering and Water Technology, Institute for Water Education (UNESCO-IHE), Delft, the Netherlands
Klaus Kümmerer / Institute of Environmental Medicine and Hospital Epidemiology, University Hospital Freiburg, Freiburg, Germany
Yu-Chia Kuo / Institute of Environmental Engineering, National Sun Yat-Sen University, Kaohsiung, Taiwan
Witold Kurylak / Institute of Non-Ferrous Metals, Gliwice, Poland
John M. Laflen / Agricultural Research Service (USDA-ARS), U.S. Department of Agriculture, Buffalo Center, and Iowa State University, Ames, Iowa, U.S.A.
Rattan Lal / School of Environment and Natural Resources, Ohio State University, Columbus, Ohio, U.S.A.
Freddie L. Lamm / Research Extension Center, Kansas State University, Colby, Kansas, U.S.A.
Judith Lancaster / Desert Research Institute, Reno, Nevada, U.S.A.

Doug Landis / *Department of Entomology, Michigan State University, East Lansing, Michigan, U.S.A.*

Tomaz Langenbach / *Federal University of Rio de Janeiro, Rio de Janeiro, Brazil*

David B. Langston, Jr. / *Rural Development Center, University of Georgia, Tifton, Georgia, U.S.A.*

Robert J. Lascano / *Agricultural Research Service (USDA-ARS), U.S. Department of Agriculture, Lubbock, Texas, U.S.A.*

Jean-Claude Lefeuvre / *Laboratory of the Evolution of Natural and Modified Systems, University of Rennes, Rennes, France*

E.F. Legner / *Department of Entomology, University of California—Riverside, Riverside, California, U.S.A.*

Henry Noel LeHouerou / *Center for Functional and Evolutionary Ecology (CEFE), National Center for Scientific Research (CNRS), France*

Reynald Lemke / *Agriculture and Agri-Food Canada, Swift Current, Saskatchewan, Canada*

Rocky Lemus / *Mississippi State University, Starkville, Mississippi, U.S.A.*

Piet Lens / *Department of Environmental Engineering and Water Technology, Institute for Water Education (UNESCO-IHE), Delft, the Netherlands*

Xianguo Li / *Department of Mechanical Engineering, University of Waterloo, Waterloo, Ontario, Canada*

Shu-Hao Liang / *Institute of Environmental Engineering, National Sun Yat-Sen University, Kaohsiung, Taiwan*

Mingsheng Liu / *Architectural Engineering Program, Peter Kiewit Institute, University of Nebraska—Lincoln, Omaha, Nebraska, U.S.A.*

K.F. Andrew Lo / *Department of Natural Resources, Chinese Culture University, Taipei, Taiwan*

Hugo A. Loaiciga / *Department of Geography, University of California—Santa Barbara, Santa Barbara, California, U.S.A.*

Leslie London / *Occupational and Environmental Health Research Unit, University of Cape Town, Observatory, South Africa*

Richard Lowrance / *Agricultural Research Service (USDA-ARS), U.S. Department of Agriculture, Tifton, Georgia, U.S.A.*

John Ludwig / *Ecosystem Sciences, Commonwealth Scientific and Industrial Research Organization (CSIRO), Winnellie, Northern Territory, Australia*

Rory O. Maguire / *Virginia Tech, Blacksburg, Virginia, U.S.A.*

Barbara Manachini / *Departments of Environmental Biology and Biodiversity, University of Palermo, Palermo, Italy*

Kyle R. Mankin / *Department of Biological and Agricultural Engineering, Kansas State University, Manhattan, Kansas, U.S.A.*

Gamini Manuweera / *Secretariat of the Stockholm Convention, United Nations Environmental Program, Chatelaine, Switzerland*

Tek Narayan Maraseni / *Australian Center for Sustainable Catchments, School of Accounting, Economics and Finance, University of Southern Queensland, Toowoomba, Queensland, Australia*

Bernd Markert / *Environmental Institute of Scientific Networks (EISN), Haren-Erika, Germany*

Rudolf Marloth / *San Diego State University, San Diego, California, U.S.A.*

Jay F. Martin / *Department of Food, Agricultural, and Biological Engineering, Ohio State University, Columbus, Ohio, U.S.A.*

Jocilyn Danise Martinez / *University of South Florida, Tampa, Florida, U.S.A.*
Graça Martinho / *Department of Environmental Sciences and Engineering, Faculty of Sciences and Technology, New University of Lisbon, Caparica, Portugal*
Saburo Matsui / *Graduate School of Global Environmental Studies, Kyoto University, Kyoto, Japan*
Thomas J. Mbise / *Tanzania Association of Public Occupational and Environmental Health Experts, Dar-es-Salaam, Tanzania*
Ann McCampbell / *Chair, Multiple Chemical Sensitivities Task Force of New Mexico, Santa Fe, New Mexico, U.S.A.*
Donald K. McCool / *Agricultural Research Service (USDA-ARS), U.S. Department of Agriculture, Pullman, Washington, U.S.A.*
Richard McDowell / *AgResearch Ltd., Invermay Agricultural Center, Mosgiel, New Zealand*
Leslie D. McFadden / *Department of Earth and Planetary Sciences, University of New Mexico, Albuquerque, New Mexico, U.S.A.*
Sean McGinn / *Agriculture and Agri-Food Canada, Lethbridge, Alberta, Canada*
Ronald G. McLaren / *Soil, Plant, and Ecological Sciences Division, Lincoln University, Canterbury, New Zealand*
Mike J. McLaughlin / *Land and Water, Commonwealth Scientific and Industrial Research Organization (CSIRO), Glen Osmond, South Australia, Australia*
Agata Mechlińska / *Department of Analytical Chemistry, Chemical Faculty, Gdansk University of Technology, and Department of Environmental Toxicology, Interdepartmental Institute of Maritime and Tropical Medicine, Medical University of Gdansk, Gdansk, Poland*
Mallavarapu Megharaj / *Commonwealth Scientific and Industrial Research Organization (CSIRO), Adelaide, South Australia, Australia*
D. Paul Mehta / *Department of Mechanical Engineering, Bradley University, Peoria, Illinois, U.S.A.*
Michael D. Melville / *School of Biological, Earth, and Environmental Sciences, University of New South Wales, Sydney, New South Wales, Australia*
Neal William Menzies / *School of Land, Crop and Food Sciences, University of Queensland, St. Lucia, Queensland, Australia*
Gustavo Enrique Merten / *Federal University of Rio Grande do Sul, Porto Alegre, Brazil*
Andrea Micangeli / *Interuniversity Research Center for Sustainable Development (CIRPS), Sapienza University of Rome, Rome, Italy*
Małgorzata Michalska / *Institute of Maritime and Tropical Medicine, Gdynia, Poland*
Adnan Midili / *Department of Mechanical Engineering, Faculty of Engineering, Nigde University, Nigde, Turkey*
Zdzisław M. Migaszewski / *Division of Geochemistry and the Environment, Institute of Chemistry, Jan Kochanowski University, Kielce, Poland*
J. David Miller / *Department of Chemistry, Carleton University, Ottawa, Ontario, Canada*
Pierre Mineau / *Science and Technology Branch, Environment Canada, Ottawa, Ontario, Canada*
Jean Paulo Gomes Minella / *Federal University of Santa Maria, Santa Maria, Brazil*
I.M. Mishra / *Department of Chemical Engineering, Indian Institute of Technology, Roorkee, Mumbai, India*
J. Kent Mitchell / *Agricultural Engineering, University of Illinois, Urbana, Illinois, U.S.A.*
Luisa T. Molina / *Massachusetts Institute of Technology, Cambridge, Massachusetts, U.S.A.*

P. Mondal / *Department of Chemical Engineering, Indian Institute of Technology, Roorkee, Roorkee, India*

H. Curtis Monger / *Department of Agronomy and Horticulture, New Mexico State University, Las Cruces, New Mexico, U.S.A.*

Lynn E. Moody / *Earth and Soil Sciences Department, California Polytechnic State University, San Luis Obispo, California, U.S.A.*

David A. Mouat / *Division of Earth and Ecosystem Sciences, Desert Research Institute, Reno, Nevada, U.S.A.*

Martin A. Mozzo, Jr. / *M and A Associates Inc, Robbinsville, New Jersey, U.S.A.*

Subhas Chandra Mukherjee / *Department of Neurology, Medical College, Calcutta, India*

Felix Müller / *Ecology Center, University of Kiel, Kiel, Germany*

Heinz Müller-Schärer / *Department of Biology and Ecology, University of Fribourg/Perolles, Fribourg, Switzerland*

Tariq Muneer / *School of Engineering, Napier University, Edinburgh, U.K.*

Rafael Munoz-Carpena / *University of Florida, Florida, U.S.A.*

Joji Muramoto / *Program in Community and Agroecology, Department of Environmental Studies, University of California—Santa Cruz, Santa Cruz, California, U.S.A.*

Stephen D. Murphy / *Faculty of Environment, University of Waterloo, Waterloo, Ontario, Canada*

O.M. Musthafa / *Center for Pollution Control and Environmental Engineering, Pondicherry University, Pondicherry, India*

Ravendra Naidu / *Commonwealth Scientific and Industrial Research Organization (CSIRO), Adelaide, South Australia, Australia*

D.V. Naik / *Biofuels Division, Indian Institute of Petroleum, Dehradun, India*

Jacek Namieśnik / *Department of Analytical Chemistry, Chemical Faculty, Gdansk University of Technology, Gdansk, Poland*

Inés Navarro / *Engineering Institute, National Autonomous University of Mexico (UNAM), Coyoacan, Mexico*

Alexandra Navrotsky / *Department of Chemical Engineering and Materials Science, University of California—Davis, Davis, California, U.S.A.*

Mark A. Nearing / *Southwest Watershed Research Center, Agricultural Research Service (USDA-ARS), U.S. Department of Agriculture, Tucson, Arizona, U.S.A.*

Jerry Neppel / *Department of Agronomy, Iowa State University, Ames, Iowa, U.S.A.*

Elena Neri / *Ecodynamics Group, Department of Chemistry, University of Siena, Siena, Italy*

Aiwerasia V.F. Ngowi / *Tanzania Association of Public Occupational and Environmental Health Experts, and Department of Environmental and Occupational Health, Muhimbili University of Health and Allied Sciences (MUHAS), Dar-es-Salaam, Tanzania*

Valentina Niccolucci / *Ecodynamics Group, Department of Chemistry, University of Siena, Siena, Italy*

Niels Erik Nielsen / *Plant Nutrition and Soil Fertility Laboratory, Department of Agricultural Sciences, Royal Veterinary and Agricultural University, Frederiksberg, Denmark*

Egide Nizeyimana / *Department of Agronomy and Environmental Resources Research Institute, Pennsylvania State University, University Park, Pennsylvania, U.S.A.*

L. Darrell Norton / *National Soil Science Research Laboratory, Agricultural Research Service (USDA-ARS), U.S. Department of Agriculture, West Lafayette, Indiana, U.S.A.*

John J. Obrycki / *Department of Entomology, University of Kentucky, Lexington, Kentucky, U.S.A.*

Philip Oduor-Owino / *Department of Botany, University of Kenyatta, Nairobi, Kenya*
K.E. Ohrn / *Cypress Digital Ltd., Vancouver, British Columbia, Canada*
Michael T. Olexa / *Center for Agricultural and Natural Resource Law, Institute of Food and Agricultural Sciences, University of Florida, Gainesville, Florida, U.S.A.*
Jacob Opadeyi / *Department of Surveying and Land Information, Faculty of Engineering, University of the West Indies, St. Augustine, Trinidad and Tobago*
Bohdan W. Oppenheim / *U.S. Department of Energy Industrial Assessment Center, Loyola Marymount University, Los Angeles, California, U.S.A.*
Barron Orr / *Office of Arid Lands Studies, School of Natural Resources and the Environment, University of Arizona, Tucson, Arizona, U.S.A.*
Lloyd B. Owens / *U.S. Department of Agriculture (USDA), Coshocton, Ohio, U.S.A.*
Margareta Palmborg / *Swedish Poisons Information Center, Stockholm, Sweden*
Maurizio G. Paoletti / *Department of Biology, University of Padova, Padova, Italy*
J. Parikh / *Department of Chemical Engineering, S.V. National Institute of Technology, Surat, India*
David R. Parker / *Department of Soil and Environmental Sciences, University of California—Riverside, Riverside, California, U.S.A.*
Steven A. Parker / *Pacific Northwest National Laboratory, Richland, Washington, U.S.A.*
Shyamapada Pati / *Department of Obstetrics and Gynaecology, Calcutta National Medical College, Calcutta, India*
Timothy Paulitz / *Department of Plant Science, MacDonald Campus of McGill University, Ste-Anne-de-Bellevue, Quebec, Canada*
Judith F. Pedler / *Department of Environmental Sciences, University of California—Riverside, Riverside, California, U.S.A.*
Meir Paul Pener / *Department of Cell and Developmental Biology, Hebrew University of Jerusalem, Jerusalem, Israel*
Mark B. Peoples / *Plant Industry, Commonwealth Scientific and Industrial Research Organization (CSIRO), Canberra, Australian Capital Territory, Australia*
Kerry M. Peru / *Water and Technology Directorate, Saskatoon, Saskatchewan, Canada*
Julie A. Peterson / *Department of Entomology, University of Kentucky, Lexington, Kentucky, U.S.A.*
Mark A. Peterson / *Sustainable Success LLC, Clementon, New Jersey, U.S.A.*
Robert Pietrzak / *Department of Chemistry, Adam Mickiewicz University, Poznan, Poland*
David Pimentel / *Department of Entomology, Cornell University, Ithaca, New York, U.S.A.*
Ana Pires / *Department of Environmental Sciences and Engineering, Faculty of Sciences and Technology, New University of Lisbon, Caparica, Portugal*
Peter W. Plumstead / *North Carolina State University, Raleigh, North Carolina, U.S.A.*
Paola Poli / *Department of Genetics, Biology of Microorganisms, Anthropology, and Evolution, University of Parma, Parma, Italy*
Zaneta Polkowska / *Gdansk University of Technology, Gdansk, Poland*
Wendell A. Porter / *Department of Agricultural and Biological Engineering, University of Florida, Gainesville, Florida, U.S.A.*
Wilfred M. Post / *Oak Ridge National Laboratory, Oak Ridge, Tennessee, U.S.A.*
Franco Previtali / *Department of Environmental Science, University of Milan—Bicocca, Milan, Italy*
Odo Primavesi / *Brazilian Agricultural Research Corporation (Embrapa), São Paulo, Brazil*
Soyuz Priyadarsan / *Texas A&M University, College Station, Texas, U.S.A.*

Federico M. Pulselli / *Ecodynamics Group, Department of Chemistry, University of Siena, Siena, Italy*

Manzoor Qadir / *International Center for Agricultural Research in the Dry Areas (ICARDA), Aleppo, Syria, and International Water Management Institute (IWMI), Colombo, Sri Lanka*

X.S. Qin / *School of Civil and Environmental Engineering, Nanyang Technical University, Singapore*

Quazi Quamruzzaman / *Dhaka Community Hospital, Dhaka, Bangladesh*

Anna Rabajczyk / *Independent Department of Environment Protection and Modeling, Jan Kochanowski University of Humanities and Sciences in Kielce, Kielce, Poland*

Martin C. Rabenhorst / *University of Maryland, College Park, Maryland, U.S.A.*

Mohammad Mahmudur Rahman / *School of Environmental Studies, Jadavpur University, Calcutta, India*

Shafiqur Rahman / *Department of Agricultural and Biosystems Engineering, North Dakota State University, Fargo, North Dakota, U.S.A.*

Liqa Raschid-Sally / *International Water Management Institute (IWMI), Colombo, Sri Lanka*

Abdul Rashid / *Pakistan Atomic Energy Commission, Islamabad, Pakistan*

José Miguel Reichert / *Federal University of Santa Maria, Santa Maria, Brazil*

Pichu Rengasamy / *Department of Soil and Water Science, University of Adelaide, Waite, South Australia, Australia*

James D. Rhoades / *Agricultural Salinity Consulting, Riverside, California, U.S.A.*

Lisa A. Robinson / *Independent Consultant, Newton, Massachusetts, U.S.A.*

Mark Robinson / *Center for Ecology and Hydrology, Wallingford, U.K.*

Philippe Rochette / *Soils and Crops Research Center, Agriculture and Agri-Food Canada, Saint-Foy, Quebec, Canada*

Justyna Rogowska / *Department of Analytical Chemistry, Chemical Faculty, Gdansk University of Technology, and Department of Environmental Toxicology, Interdepartmental Institute of Maritime and Tropical Medicine, Medical University of Gdansk, Gdansk, Poland*

Alexandru V. Roman / *School of Public Administration, Florida Atlantic University, Boca Raton, Florida, U.S.A.*

Larama M.B. Rongo / *Muhimbili University of Health and Allied Sciences, Dar-es-Salaam, Tanzania*

Stephen A. Roosa / *Energy Systems Group, Inc., Louisville, Kentucky, U.S.A.*

Marc A. Rosen / *Faculty of Engineering and Applied Science, University of Ontario Institute of Technology (UOIT), Oshawa, Ontario, Canada*

Erwin Rosenberg / *Institute of Chemical Technologies and Analytics, Vienna University of Technology, Vienna, Austria*

Clayton Rubec / *Center for Environmental Stewardship and Conservation, Ottawa, Ontario, Canada*

Jennifer Ruesink / *Department of Zoology, University of Washington, Seattle, Washington, U.S.A.*

Gunnar Rundgren / *Grolink AB, Hoeje, Sweden*

John Ryan / *International Center for Agricultural Research in the Dry Areas (ICARDA), Aleppo, Syria*

D.W. Rycroft / *Formerly at Department of Civil and Environmental Engineering, Southampton University, Southampton, U.K.*

Paul G. Saffigna / *Tropical Forestry, University of Queensland, Gatton, Queensland, Australia*

Khitish Chandra Saha / *School of Environmental Studies, Jadavpur University, Calcutta, India*

Alka Sapat / *School of Public Administration, Florida Atlantic University, Boca Raton, Florida, U.S.A.*

Danilo Sbordone / *Interuniversity Research Center for Sustainable Develepment (CIRPS), Sapienza University of Rome, Rome, Italy*

Claus Schimming / *Institute for the Conservation of Natural Resources, University of Kiel, Kiel, Germany*

William H. Schlesinger / *Deptartment of Geology and Botany, Duke University, Durham, North Carolina, U.S.A.*

Ewald Schnug / *Institute for Crop and Soil Science, Julius Kuhn Institute (JKI), Braunschweig, Germany*

A. Paul Schwab / *Department of Agronomy, Purdue University, West Lafayette, Indiana, U.S.A.*

Cetin Sengonca / *Department of Entomology and Plant Protection, Institute of Plant Pathology, University of Bonn, Bonn, Germany*

Hamid Shahandeh / *Texas A&M University, College Station, Texas, U.S.A.*

A.V. Shanwal / *Department of Soil Science, Chaudhary Charan Singh Haryana Agricultural University, Hisar, India*

Andrew N. Sharpley / *University of Arkansas, Fayetteville, Arkansas, U.S.A.*

Brenton S. Sharratt / *Land Management and Water Conservation Research Unit, Agricultural Research Service (USDA-ARS), U.S. Department of Agriculture, Pullman, Washington, U.S.A.*

Daniel Shepherd / *Auckland University of Technology, Auckland, New Zealand*

Paul K. Sibley / *School of Environmental Science, University of Guelph, Guelph, Ontario, Canada*

Brian Silvetti / *CALMAC Manufacturing Corporation, Fair Lawn, New Jersey, U.S.A.*

Bal Ram Singh / *Department of Plant and Environmental Sciences, Norwegian University of Life Sciences, Aas, Norway*

S.P. Singh / *National Bureau of Soil Survey and Land Use Planning, Indian Agricultural Research Institute, New Delhi, India*

Johan Six / *Department of Agronomy and Range Science, University of California—Davis, Davis, California, U.S.A.*

Jeffrey G. Skousen / *Division of Plant and Soil Sciences, West Virginia University, Morgantown, West Virginia, U.S.A.*

Bogdan Skwarzec / *Faculty of Chemistry, University of Gdansk, Gdansk, Poland*

Edward H. Smith / *Cornell University, Ithaca, New York, U.S.A.*

Matt C. Smith / *Biological and Agricultural Engineering Department, University of Georgia, Athens, Georgia, U.S.A.*

Pete Smith / *Institute of Biological and Environmental Sciences, School of Biological Sciences, University of Aberdeen, Aberdeen, U.K.*

Sean M. Smith / *Ecosystem Restoration Center, Maryland Department of Natural Resources, Annapolis, Maryland, U.S.A.*

Hwat Bing So / *Center for Environmental Systems Research, Griffith University, Nathan, Queensland, Australia*

Robert E. Sojka / *Northwest Irrigation and Soils Research Lab, U.S. Department of Agriculture (USDA), Kimberly, Idaho, U.S.A.*

Leslie A. Solmes / *LAS and Associates, Mill Valley, California, U.S.A.*

Rolf Sommer / *International Center for Agricultural Research in the Dry Areas (ICARDA), Aleppo, Syria*
Roy F. Spalding / *Water Science Laboratory, University of Nebraska—Lincoln, Lincoln, Nebraska, U.S.A.*
Graham P. Sparling / *Landcare Research, Hamilton, New Zealand*
Gerd Sparovek / *College of Agriculture, Graduate School of Agriculture Luiz de Queiroz (ESALQ), University of São Paulo, São Paulo, Brazil*
Eric B. Spurr / *Department of Wildlife Ecology, Landcare Research New Zealand, Ltd., Lincoln, New Zealand*
Victor R. Squires / *Dryland Management Consultant, South Australia, Australia*
Amanda Staudt / *Climate Scientist, National Wildlife Federation, U.S.A.*
Nicolae Stefan / *Center for Sustainable Exploitation of Ecosystems (CESEE), Alexandru Ioan Cuza University of Iasi, Iasi, Romania*
Joshua Steinfeld / *Florida Atlantic University, Boca Raton, Florida, U.S.A.*
Larry D. Stetler / *Department of Geology and Geological Engineering, South Dakota School of Mines and Technology, Rapid City, South Dakota, U.S.A.*
F. Craig Stevenson / *University of Saskatchewan, Department of Soil Science, Saskatoon, Saskatchewan, Canada*
B.A. Stewart / *Dryland Agriculture Institute, West Texas A&M University, Canyon, Texas, U.S.A.*
Therese Stovall / *Oak Ridge National Laboratory, Oak Ridge, Tennessee, U.S.A.*
Evamarie Straube / *Institute of Occupational Medicine, University of Greifswald, Greifswald, Germany*
Sebastian Straube / *Department of Physiology, University of Oxford, Oxford, U.K.*
Wolfgang Straube / *Department of Gynecology and Obstetrics, University of Greifswald, Greifswald, Germany*
Tanja Strive / *Commonwealth Scientific and Industrial Research Organization (CSIRO) Ecosystem Sciences, Canberra, Australian Capital Territory, Australia*
Scott J. Sturgul / *Outreach Program Manager, Nutrient and Pest Management Program, University of Wisconsin, Madison, Wisconsin, U.S.A.*
Donald L. Suarez / *U.S. Salinity Laboratory, Agricultural Research Service (USDA-ARS), U.S. Department of Agriculture, Riverside, California, U.S.A.*
Praful Suchak / *Sneha Plastics Pvt. Ltd., Suchak's Consultancy Services, Mumbai, India*
L.A. Sullivan / *Southern Cross GeoScience, Southern Cross University, Lismore, New South Wales, Australia*
Matthew O. Sullivan / *Department of Food, Agricultural, and Biological Engineering, Ohio State University, Columbus, Ohio, U.S.A.*
Rao Y. Surampalli / *Department of Civil Engineering, University of Nebraska—Lincoln, Lincoln, Nebraska, U.S.A.*
Claus Svendsen / *Center for Ecology and Hydrology, Wallingford, U.K.*
John M. Sweeten / *Texas A&M University, Amarillo, Texas, U.S.A.*
Piotr Szefer / *Department of Food Sciences, Medical University of Gdansk, Gdansk, Poland*
Kenneth K. Tanji / *Department of Land, Air, and Water Resources, University of California—Davis, Davis, California, U.S.A.*
Maciej Tankiewicz / *Department of Analytical Chemistry, Chemical Faculty, Gdansk University of Technology, Gdansk, Poland*
Meena Thakur / *Department of Environment Sciences, Dr. Y.S. Parmar University of Horticulture and Forestry, Solan, India*

Daniel L. Thomas / *Louisiana State University, Baton Rouge, Louisiana, U.S.A.*
Bob Thorne / *Massey University, New Zealand*
Thomas L. Thurow / *Department of Renewable Resources, University of Wyoming, Laramie, Wyoming, U.S.A.*
Jill S. Tietjen / *Technically Speaking, Inc., Greenwood Village, Colorado, U.S.A.*
Ralph W. Tiner / *National Wetlands Inventory Program, U.S., Fish and Wildlife Service, Hadley, Massachusetts, U.S.A.*
Greg Tinkler / *RLB Consulting Engineers, Houston, Texas, U.S.A.*
Abhishek Tiwary / *Newcastle University, Newcastle, U.K.*
Marek Tobiszewski / *Department of Analytical Chemistry, Chemical Faculty, Gdansk University of Technology, Gdansk, Poland*
David Tongway / *Ecosystem Sciences, Commonwealth Scientific and Industrial Research Organization (CSIRO), Canberra, Australian Capital Territory, Australia*
Alberto Traverso / *Thermochemical Power Group, Department of Mechanical, Energy, Management and Transportation Engineering (DIME), University of Genoa, Genoa, Italy*
David Tucker / *National Energy Technology Laboratory, Department of Energy, Morgantown, West Virginia, U.S.A.*
W.D. Turner / *Department of Mechanical Engineering, Energy Systems Laboratory, College Station, Texas, U.S.A.*
Wayne C. Turner / *Industrial Engineering and Management, Oklahoma State University, Stillwater, Oklahoma, U.S.A.*
Robert E. Uhrig / *Department of Nuclear Engineering, University of Tennessee, Knoxville, Tennessee, U.S.A.*
Oswald Van Cleemput / *Faculty of Agricultural and Applied Biological Sciences, University of Ghent, Ghent, Belgium*
H.F. van Emden / *Department of Agriculture, University of Reading, Reading, U.K.*
H.H. Van Horn / *Department of Animal Science, University of Florida, Gainesville, Florida, U.S.A.*
Herman Van Langenhove / *Department of Environmental Engineering and Water Technology, Institute for Water Education (UNESCO-IHE), Delft, the Netherlands*
R. Scott Van Pelt / *Wind Erosion and Water Conservation Research Unit, Agricultural Research Service (USDA-ARS), U.S. Department of Agriculture, Big Spring, Texas, U.S.A.*
George F. Vance / *Department of Ecosystem Sciences and Management, University of Wyoming, Laramie, Wyoming, U.S.A.*
Andrea Nunes Vaz Pedroso / *Nucleus Research in Ecology, Institute of Botany, São Paulo, Brazil*
Peter Victor / *Faculty of Environmental Studies, York University, Toronto, Ontario, Canada*
Dalmo A.N. Vieira / *National Sedimentation Laboratory, Arkansas State University, U.S. Department of Agriculture, Agricultural Research Service (USDA-ARS), Jonesboro, Arkansas, U.S.A.*
Denise Vienne / *School of Public Administration, Florida Atlantic University, Boca Raton, Florida, U.S.A.*
Jan Vymazal / *Faculty of Environmental Sciences, Department of Landscape Ecology, Czech University of Life Sciences Prague, Prague, Czech Republic*
Leszek Wachowski / *Department of Chemistry, Adam Mickiewicz University, Posnan, Poland*
Mathis Wackernagel / *Global Footprint Network, Oakland, California, U.S.A.*
Joel T. Walker / *Department of Food, Agricultural, and Biological Engineering, Ohio State University, Columbus, Ohio, U.S.A.*

Doug Walsh / *Washington State University, Prosser, Washington, U.S.A.*
Ivan A. Walter / *Ivan's Engineering, Inc., Denver, Colorado, U.S.A.*
A. Wang / *E.H. Graham Center for Agricultural Innovation, Industry and Investment, NSW and Charles Sturt University, Wagga Wagga, New South Wales, Australia*
Xingxiang Wang / *Institute of Soil Science, Chinese Academy of Sciences, Nanjing, China*
Ynuzhang Wang / *Academy of Yellow River Conservancy Science, Zhengzhou, China*
Waldemar Wardencki / *Department of Chemistry, Gdansk University of Technology, Gdansk, Poland*
Andrew Warren / *Department of Geography, University College London, London, U.K.*
Apichon Watcharenwong / *School of Environmental Engineering, Suranaree University of Technology, Nakhon Ratchasima, Thailand*
Thomas R. Way / *National Soil Dynamics Laboratory, Agricultural Research Service (USDA-ARS), U.S. Department of Agriculture, Auburn, Alabama, U.S.A.*
Johannes Bernhard Wehr / *School of Land, Crop and Food Sciences, University of Queensland, St. Lucia, Queensland, Australia*
W.W. Wenzel / *Institute of Soil Research, University of Natural Resources and Life Sciences, Vienna, Austria*
Catharina Wesseling / *Central American Institute for Studies on Toxic Substances (IRET), National University, Heredia, Costa Rica*
Ian White / *College of Science, Fenner School of Environment and Society, College of Medicine, Biology and Environment, Australian National University, Canberra, Australian Capital Territory, Australia*
Dennis Wichelns / *Principal Economist, International Water Management Unit, Columbo, Sri Lanka*
Keith D. Wiebe / *Resource Economics Division, U.S. Department of Agriculture (USDA), Washington, District of Columbia, U.S.A.*
Gerald E. Wilde / *Department of Entomology, Kansas State University, Manhattan, Kansas, U.S.A.*
Larry P. Wilding / *Department of Soil and Crop Sciences, Texas A&M University, College Station, Texas, U.S.A.*
Wilhelm Windhorst / *Institute for the Conservation of Natural Resources, University of Kiel, Kiel, Germany*
Wanpen Wirojanagud / *Department of Environmental Engineering, Faculty of Engineering, Khon Kaen University, Khon Kaen, and Center of Excellence on Hazardous Substance Management, National Centers of Excellence (PERDO), Bangkok, Thailand*
Lidia Wolska / *Department of Analytical Chemistry, Chemical Faculty, Gdansk University of Technology, and Department of Environmental Toxicology, Interdepartmental Institute of Maritime and Tropical Medicine, Medical University of Gdansk, Gdansk, Poland*
Eric A. Woodroof / *Profitable Green Solutions, Plano, Texas, U.S.A.*
Brent Wootton / *Center for Alternative Wastewater Treatment, Fleming College, Lindsay, Ontario, Canada*
Steve D. Wratten / *Division of Soil, Plant and Ecological Sciences, Lincoln University, Canterbury, New Zealand*
I. Pai Wu / *College of Tropical Agriculture and Human Resources, University of Hawaii, Honolulu, Hawaii, U.S.A.*
Simone Wuenschmann / *Environmental Institute of Scientific Networks (EISN), Haren-Erika, Germany*

Y. Xu / *MOE Key Laboratory of Regional Energy Systems Optimization, S-C Energy and Environmental Research Academy, North China Electric Power University, Beijing, China*

Kazuyuki Yagi / *National Institute for Agro-Environmental Sciences, Japan*

Colin N. Yates / *Faculty of Environment, University of Waterloo, Waterloo, Ontario, Canada*

Bofu Yu / *Griffith School of Engineering, Griffith University, Nathan, Queensland, Australia*

Taolin Zhang / *Institute of Soil Science, Chinese Academy of Sciences, Nanjing, China*

X.-C. (John) Zhang / *Grazinglands Research Laboratory, Agricultural Research Service (USDA-ARS), U.S. Department of Agriculture, El Reno, Oklahoma, U.S.A.*

He Zhong / *Pesticide Environment Impact Section, Public Health Entomology Research and Education Center, Florida A&M University, Panama City, Florida, U.S.A.*

Xinhua Zhou / *School of Natural Resources, University of Nebraska—Lincoln, Lincoln, Nebraska, U.S.A.*

Yifei Zhu / *Gemune LLC, Fremont, California, U.S.A.*

Zixi Zhu / *Henan Institute of Meteorology, Zhengzhou, China*

Andrew R. Zimmerman / *Department of Geological Sciences, University of Florida, Gainesville, Florida, U.S.A.*

Claudio Zucca / *Department of Agriculture, University of Sassari, Sassari, Italy*

Contents

Editorial Advisory Board ix
Contributors xi
Topical Table of Contents xli
Foreword lxvii
Preface lxix
How to Use This Encyclopedia lxxi
About the Editor-in-Chief lxxvii

Volume I

Acaricides / *Doug Walsh* 1
Acid Rain / *Umesh Kulshrestha* 5
Acid Rain: Nitrogen Deposition / *George F. Vance* 20
Acid Sulfate Soils / *Delvin S. Fanning* 26
Acid Sulfate Soils: Formation / *Martin C. Rabenhorst, Delvin S. Fanning, and Steven N. Burch* 31
Acid Sulfate Soils: Identification, Assessment, and Management / *L.A. Sullivan, R.T. Bush, and E.D. Burton* 36
Acid Sulfate Soils: Management / *Michael D. Melville and Ian White* 55
Adsorption / *Puangrat Kajitvichyanukul and Jirapat Ananpattarachai* 61
Agricultural Runoff / *Matt C. Smith, David K. Gattie, and Daniel L. Thomas* 81
Agricultural Soils: Ammonia Volatilization / *Paul G. Saffigna and John R. Freney* 85
Agricultural Soils: Carbon and Nitrogen Biological Cycling / *Alan J. Franzluebbers* 88
Agricultural Soils: Nitrous Oxide Emissions / *John R. Freney* 96
Agricultural Soils: Phosphorus / *Anja Gassner and Ewald Schnug* 100
Agricultural Water Quantity Management / *X.S. Qin and Y. Xu* 105
Agriculture: Energy Use and Conservation / *Guangnan Chen and Tek Narayan Maraseni* 118
Agriculture: Organic / *Kathleen Delate* 125
Agroforestry: Water Use Efficiency / *James R. Brandle, Laurie Hodges, and Xinhua Zhou* 129
Air Pollution: Monitoring / *Waldemar Wardencki* 132
Air Pollution: Technology / *Sven Erik Jørgensen* 149
Alexandria Lake Maryut: Integrated Environmental Management / *Lindsay Beevers* .. 164
Allelochemics / *John Borden* 175
Alternative Energy / *Bernd Markert, Simone Wuenschmann, Stefan Fraenzle, and Bernd Delakowitz* 179
Alternative Energy: Hydropower / *Andrea Micangeli, Sara Evangelisti, and Danilo Sbordone* 201
Alternative Energy: Photovoltaic Modules and Systems / *Ewa Klugmann-Radziemska* 215
Alternative Energy: Photovoltaic Solar Cells / *Ewa Klugmann-Radziemska* 226
Alternative Energy: Solar Thermal Energy / *Andrea Micangeli, Sara Evangelisti, and Danilo Sbordone* 241
Alternative Energy: Wind Power Technology and Economy / *K.E. Ohrn* 258
Aluminum / *Johannes Bernhard Wehr, Frederick Paxton Cardell Blamey, and Neal William Menzies* 267
Animals: Sterility from Pesticides / *William Au* 277
Animals: Toxicological Evaluation / *Vera Lucia S.S. de Castro* 280
Antagonistic Plants / *Philip Oduor-Owino* 288

Volume I (cont'd)

Aquatic Communities: Pesticide Impacts / *David P. Kreutzweiser and Paul K. Sibley* 291
Aral Sea Disaster / *Guy Fipps* 300
Arthropod Host-Plant Resistant Crops / *Gerald E. Wilde* 304
Bacillus thuringiensis: Transgenic Crops / *Julie A. Peterson, John J. Obrycki, and James D. Harwood* 307
Bacterial Pest Control / *David N. Ferro* 321
Bioaccumulation / *Tomaz Langenbach* 324
Biodegradation / *Sven Erik Jørgensen* 328
Biodiversity and Sustainability / *Odo Primavesi* 333
Bioenergy Crops: Carbon Balance Assessment / *Rocky Lemus and Rattan Lal* 345
Biofertilizers / *J.R. de Freitas* 349
Bioindicators: Farming Practices and Sustainability / *Joji Muramoto and Stephen R. Gliessman* 352
Biological Controls / *Heikki Hokkanen* 366
Biological Controls: Conservation / *Doug Landis and Steve D. Wratten* 370
Biomass / *Alberto Traverso and David Tucker* 373
Biopesticides / *G.J. Ash and A. Wang* 381
Bioremediation / *Ragini Gothalwal* 387
Bioremediation: Contaminated Soil Restoration / *Sven Erik Jørgensen* 401
Biotechnology: Pest Management / *Maurizio G. Paoletti* 407
Birds: Chemical Control / *Eric B. Spurr* 413
Birds: Pesticide Use Impacts / *Pierre Mineau* 416
Boron and Molybdenum: Deficiencies and Toxicities / *Umesh C. Gupta* 424
Boron: Soil Contaminant / *Rami Keren* 431
Buildings: Climate Change / *Lisa Guan and Guangnan Chen* 435
Cabbage Diseases: Ecology and Control / *Anthony P. Keinath, Marc A. Cubeta, and David B. Langston, Jr.* 442
Cadmium and Lead: Contamination / *Gabriella Kakonyi and Imad A.M. Ahmed* 446
Cadmium: Toxicology / *Sven Erik Jørgensen* 453
Carbon Sequestration / *Nathan E. Hultman* 456
Carbon: Soil Inorganic / *Donald L. Suarez* 462
Chemigation / *William L. Kranz* 469
Chesapeake Bay / *Sean M. Smith* 474
Chromium / *Bruce R. James and Dominic A. Brose* 477
Climate Policy: International / *Nathan E. Hultman* 483
Coastal Water: Pollution / *Piotr Szefer* 489
Cobalt and Iodine / *Ronald G. McLaren* 502
Community-Based Monitoring: Ngarenanyuki, Tanzania / *Aiwerasia V.F. Ngowi, Larama M.B. Rongo, and Thomas J. Mbise* 505
Composting / *Nídia Sá Caetano* 512
Composting: Organic Farming / *Saburo Matsui* 528
Copper / *David R. Parker and Judith F. Pedler* 535
Cyanobacteria: Eutrophic Freshwater Systems / *Anja Gassner and Martin V. Frey* 538
Desertification / *David Tongway and John Ludwig* 541
Desertification: Extent of / *Victor R. Squires* 545
Desertification: Greenhouse Effect / *Sherwood Idso and Craig Idso* 549
Desertification: Impact / *David A. Mouat and Judith Lancaster* 552
Desertification: Prevention and Restoration / *Claudio Zucca, Susana Bautista, Barron Orr, and Franco Previtali* 557

| Contents | xxxiii |

Desertization / *Henry Noel LeHouerou* 565
Developing Countries: Pesticide Health Impacts / *Aiwerasia V.F. Ngowi, Catharina Wesseling, and Leslie London* 573
Distributed Generation: Combined Heat and Power / *Barney L. Capehart, D. Paul Mehta, and Wayne C. Turner* 578
Drainage: Hydrological Impacts Downstream / *Mark Robinson and D.W. Rycroft* 584
Drainage: Soil Salinity Management / *Glenn J. Hoffman* 588
Ecological Indicators: Eco-Exergy to Emergy Flow / *Simone Bastianoni, Luca Coscieme, and Federico M. Pulselli* 592
Ecological Indicators: Ecosystem Health / *Felix Müller, Benjamin Burkhard, Marion Kandziora, Claus Schimming, and Wilhelm Windhorst* 599
Economic Growth: Slower by Design, Not Disaster / *Peter Victor* 614
Ecosystems: Large-Scale Restoration Governance / *Shannon Estenoz, Denise Vienne, and Alka Sapat* 624
Ecosystems: Planning / *Ioan Manuel Ciumasu, Liana Buzdugan, Ioan Manuel Ciumasu, Nicolae Stefan, and Keith Culver* 632
Endocrine Disruptors / *Vera Lucia S.S. de Castro* 643
Energy Commissioning: Existing Buildings / *David E. Claridge, Mingsheng Liu, and W.D. Turner* 656
Energy Commissioning: New Buildings / *Janey Kaster* 665
Energy Conservation / *Ibrahim Dincer and Adnan Midili* 677
Energy Conservation: Benefits / *Eric A. Woodroof, Wayne C. Turner, and Steven D. Heinz* 691
Energy Conservation: Industrial Processes / *Harvey E. Diamond* 697
Energy Conservation: Lean Manufacturing / *Bohdan W. Oppenheim* 705
Energy Conversion: Coal, Animal Waste, and Biomass Fuel / *Kalyan Annamalai, Soyuz Priyadarsan, Senthil Arumugam, and John M. Sweeten* 714

Volume II

Energy Efficiency: Low-Cost Improvements / *James Call* 735
Energy Efficiency: New and Emerging Technology / *Steven A. Parker* 742
Energy Efficiency: Strategic Facility Guidelines / *Steve Doty* 752
Energy Master Planning / *Fredric S. Goldner* 763
Energy Sources: Natural versus Additional / *Marc A. Rosen* 770
Energy Use: Exergy and Eco-Exergy / *Sven Erik Jørgensen* 778
Energy Use: U.S. / *Dwight K. French* 790
Energy: Environmental Security / *Muhammad Asif* 798
Energy: Physics / *Milivoje M. Kostic* 808
Energy: Renewable / *John O. Blackburn* 824
Energy: Solid Waste Advanced Thermal Technology / *Alex E.S. Green and Andrew R. Zimmerman* 830
Energy: Storage / *Rudolf Marloth* 853
Energy: Walls and Windows / *Therese Stovall* 860
Energy: Waste Heat Recovery / *Martin A. Mozzo, Jr.* 869
Environmental Legislation: Asia / *Wanpen Wirojanagud* 874
Environmental Legislation: EU Countries Solid Waste Management / *Ni-Bin Chang, Ana Pires, and Graça Martinho* 892
Environmental Policy / *Sanford V. Berg* 914
Environmental Policy: Innovations / *Alka Sapat* 922
Erosion / *Dennis C. Flanagan* 930
Erosion and Carbon Dioxide / *Pierre A. Jacinthe and Rattan Lal* 934
Erosion and Global Change / *Taolin Zhang and Xingxiang Wang* 939

Volume II (cont'd)

- Erosion and Precipitation / *Bofu Yu* 943
- Erosion and Sediment Control: Vegetative Techniques / *Samson D. Angima* 947
- Erosion by Water: Accelerated / *David Favis-Mortlock* 951
- Erosion by Water: Amendment Techniques / *X.-C. (John) Zhang* 958
- Erosion by Water: Assessment and Control / *José Miguel Reichert, Nadia Bernardi Bonumá, Gustavo Enrique Merten, and Jean Paolo Gomes Minella* 963
- Erosion by Water: Empirical Methods / *John M. Laflen* 974
- Erosion by Water: Erosivity and Erodibility / *Peter I.A. Kinnell* 980
- Erosion by Water: Process-Based Modeling / *Mark A. Nearing* 991
- Erosion by Water: Vegetative Control / *Seth M. Dabney and Silvio J. Gumiere* 994
- Erosion by Wind: Climate Change / *Alan Busacca and David Chandler* 1004
- Erosion by Wind: Global Hot Spots / *Andrew Warren* 1010
- Erosion by Wind: Principles / *Larry D. Stetler* 1013
- Erosion by Wind: Source, Measurement, Prediction, and Control / *Brenton S. Sharratt and R. Scott Van Pelt* 1017
- Erosion by Wind-Driven Rain / *Gunay Erpul, L. Darrell Norton, and Donald Gabriels* 1031
- Erosion Control: Soil Conservation / *Eric T. Craswell* 1034
- Erosion Control: Tillage and Residue Methods / *Richard Cruse, Jerry Neppel, John Kost, and Krisztina Eleki* 1040
- Erosion: Accelerated / *J. Kent Mitchell and Michael C. Hirschi* 1044
- Erosion: History / *Andrew S. Goudie* 1050
- Erosion: Irrigation-Induced / *Robert E. Sojka and David L. Bjorneberg* 1053
- Erosion: Snowmelt / *Donald K. McCool* 1057
- Erosion: Soil Quality / *Craig Ditzler* 1060
- Estuaries / *Claude Amiard-Triquet* 1063
- Eutrophication / *Sven Erik Jørgensen and Claude Amiard-Triquet* 1074
- Everglades / *Kenneth L. Campbell, Rafael Munoz-Carpena, and Gregory Kiker* 1080
- Exergy: Analysis / *Marc A. Rosen* 1083
- Exergy: Environmental Impact Assessment / *Marc A. Rosen* 1092
- Farming: Non-Chemical and Pesticide-Free (European Council Regulations [EED/209/91]) / *Inger Källander and Gunnar Rundgren* 1103
- Farming: Organic / *Brenda Frick* 1109
- Farming: Organic Pest Management / *Ján Gallo* 1115
- Food Quality Protection Act / *Christina D. DiFonzo* 1118
- Food: Cosmetic Standards / *David Pimentel and Kelsey Hart* 1120
- Food: Pesticide Contamination / *Denis Hamilton* 1124
- Fossil Fuel Combustion: Air Pollution and Global Warming / *Dan Golomb* 1127
- Fuel Cells: Intermediate and High Temperature / *Xianguo Li, Gholamreza Karimi, and Kui Jiao* 1138
- Fuel Cells: Low Temperature / *Xianguo Li and Kui Jiao* 1145
- Genotoxicity and Air Pollution / *Eliane Tigre Guimarães and Andrea Nunes Vaz Pedroso* 1154
- Geographic Information System (GIS): Land Use Planning / *Egide Nizeyimana and Jacob Opadeyi* 1163
- Geothermal Energy Resources / *Ibrahim Dincer and Arif Hepbasli* 1167
- Giant Reed (*Arundo donax*): Streams and Water Resources / *Gretchen Coffman* 1182
- Global Climate Change: Carbon Sequestration / *Sherwood Idso and Keith E. Idso* 1189
- Global Climate Change: Earth System Response / *Amanda Staudt and Nathan E. Hultman* 1192
- Global Climate Change: Gas Fluxes / *Pascal Boeckx and Oswald Van Cleemput* 1202

Global Climate Change: Gasoline, Hybrid-Electric and Hydrogen Fueled Vehicles / *Robert E. Uhrig*	1206
Global Climate Change: World Soils / *Rattan Lal*	1213
Globalization / *Alexandru V. Roman*	1218
Green Energy / *Ibrahim Dincer and Adnan Midili*	1226
Green Processes and Projects: Systems Analysis / *Abhishek Tiwary*	1242
Green Products: Production / *Puangrat Kajitvichyanukul, Jirapat Ananpattarachai, and Apichon Watcharenwong*	1253
Groundwater: Arsenic Contamination / *Abhijit Das, Bhaskar Das, Subhas Chandra Mukherjee, Shyamapada Pati, Rathindra Nath Dutta, Khitish Chandra Saha, Quazi Quamruzzaman, Mohammad Mahmudur Rahman, Tarit Roy Chowdhury, and Dipankar Chakraborti*	1262
Groundwater: Contamination / *Charles W. Fetter*	1281
Groundwater: Mining / *Hugo A. Loaiciga*	1284
Groundwater: Mining Pollution / *Jeffrey G. Skousen and George F. Vance*	1289
Groundwater: Modeling / *Jesus Carrera*	1295
Groundwater: Nitrogen Fertilizer Contamination / *Lloyd B. Owens and Douglas L. Karlen*	1302
Groundwater: Numerical Method Modeling / *Jesus Carrera*	1312
Groundwater: Pesticide Contamination / *Roy F. Spalding*	1317
Groundwater: Saltwater Intrusion / *Alexander H.-D. Cheng*	1321
Groundwater: Treatment with Biobarrier Systems / *Chih-Ming Kao, Shu-Hao Liang, Yu-Chia Kuo, and Rao Y. Surampalli*	1333
Heat Pumps / *Lu Aye*	1346
Heat Pumps: Absorption / *Keith E. Herold*	1356
Heat Pumps: Geothermal / *Greg Tinkler*	1363
Heavy Metals / *Mike J. McLaughlin*	1370
Heavy Metals: Organic Fertilization Uptake / *Ewald Schnug, Alexandra Izosimova, and Renata Gaj*	1374
Herbicides / *Malcolm Devine*	1378
Herbicides: Non-Target Species Effects / *Céline Boutin*	1382
Human Health: Cancer from Pesticides / *Jan Dich*	1394
Human Health: Chronic Pesticide Poisonings / *Birgitta Kolmodin-Hedman and Margareta Palmborg*	1397
Human Health: Consumer Concerns of Pesticides / *George Ekström and Margareta Palmborg*	1401
Human Health: Hormonal Disruption / *Evamarie Straube, Sebastian Straube, and Wolfgang Straube*	1405
Human Health: Pesticide Poisonings / *Margareta Palmborg*	1408
Human Health: Pesticide Sensitivities / *Ann McCampbell*	1411
Human Health: Pesticides / *Kelsey Hart and David Pimentel*	1414
Hydroelectricity: Pumped Storage / *Jill S. Tietjen*	1418
Industrial Waste: Soil Pollution / *W. Friesl-Hanl, M.H. Gerzabek, W.W. Wenzel, and W.E.H. Blum*	1430
Industries: Network of / *Sven Erik Jørgensen*	1434
Inland Seas and Lakes: Central Asia Case Study / *Andrey G. Kostianoy*	1436
Inorganic Carbon: Composition and Formation / *Larry P. Wilding and H. Curtis Monger*	1444
Inorganic Carbon: Global Carbon Cycle / *William H. Schlesinger*	1448
Inorganic Carbon: Modeling / *Leslie D. McFadden and Ronald G. Amundson*	1451
Inorganic Compounds: Eco-Toxicity / *Sven Erik Jørgensen*	1455
Insect Growth Regulators / *Meir Paul Pener*	1459
Insecticides: Aerial Ultra-Low-Volume Application / *He Zhong*	1471
Insects and Mites: Biological Control / *Ann E. Hajek*	1474
Insulation: Facilities / *Wendell A. Porter*	1478
Integrated Energy Systems / *Leslie A. Solmes and Sven Erik Jørgensen*	1493

Volume II (cont'd)

Integrated Farming Systems / *John Holland* 1507
Integrated Nutrient Management / *Bal Ram Singh* 1510
Integrated Pest Management / *H.F. van Emden* 1521
Integrated Weed Management / *Heinz Müller-Schärer and Alexandra Robin Collins* 1524

Volume III

Invasion Biology / *Jennifer Ruesink* 1531
Irrigation Systems: Sub-Surface Drip Design / *Carl R. Camp, Jr. and Freddie L. Lamm* 1535
Irrigation Systems: Sub-Surface History / *Norman R. Fausey* 1539
Irrigation Systems: Water Conservation / *I. Pai Wu, Javier Barragan, and Vince Bralts* 1542
Irrigation: Efficiency / *Terry A. Howell* 1545
Irrigation: Erosion / *David L. Bjorneberg* 1551
Irrigation: Return Flow and Quality / *Ramón Aragues and Kenneth K. Tanji* 1557
Irrigation: River Flow Impact / *Robert W. Hill and Ivan A. Walter* 1563
Irrigation: Saline Water / *B.A. Stewart* 1568
Irrigation: Sewage Effluent Use / *B.A. Stewart* 1570
Irrigation: Soil Salinity / *James D. Rhoades* 1572
Lakes and Reservoirs: Pollution / *Subhankar Karmakar and O.M. Musthafa* 1576
Lakes: Restoration / *Anna Rabajczyk* 1588
Land Restoration / *Richard W. Bell* 1600
Laws and Regulations: Food / *Ike Jeon* 1609
Laws and Regulations: Pesticides / *Praful Suchak* 1612
Laws and Regulations: Rotterdam Convention / *Barbara Dinham* 1615
Laws and Regulations: Soil / *Ian Hannam and Ben Boer* 1618
Leaching / *Lars Bergström* 1621
Leaching and Illuviation / *Lynn E. Moody* 1624
Lead: Ecotoxicology / *Sven Erik Jørgensen* 1630
Lead: Regulations / *Lisa A. Robinson* 1636
LEED-EB: Leadership in Energy and Environmental Design for Existing Buildings / *Rusty T. Hodapp* 1647
LEED-NC: Leadership in Energy and Environmental Design for New Construction / *Stephen A. Roosa* 1654
Manure Management: Compost and Biosolids / *Bahman Eghball and Kenneth A. Barbarick* 1660
Manure Management: Dairy / *H.H. Van Horn and D.R. Bray* 1663
Manure Management: Phosphorus / *Rory O. Maguire, John T. Brake, and Peter W. Plumstead* 1667
Manure Management: Poultry / *Shafiqur Rahman and Thomas R. Way* 1670
Mercury / *Sven Erik Jørgensen* 1676
Methane Emissions: Rice / *Kazuyuki Yagi* 1679
Minerals Processing Residue (Tailings): Rehabilitation / *Lloyd R. Hossner and Hamid Shahandeh* 1683
Mines: Acidic Drainage Water / *Jerry M. Bigham and Wendy B. Gagliano* 1688
Mines: Rehabilitation of Open Cut / *Douglas J. Dollhopf* 1692
Mycotoxins / *J. David Miller* 1696
Nanomaterials: Regulation and Risk Assessment / *Steffen Foss Hansen, Khara D. Grieger, and Anders Baun* 1700
Nanoparticles / *Alexandra Navrotsky* 1711
Nanoparticles: Uncertainty Risk Analysis / *Khara D. Grieger, Steffen Foss Hansen, and Anders Baun* 1720
Nanotechnology: Environmental Abatement / *Puangrat Kajitvichyanukul and Jirapat Ananpattarachai* 1730

Natural Enemies and Biocontrol: Artificial Diets / *Simon Grenier and Patrick De Clerq*	1746
Natural Enemies: Conservation / *Cetin Sengonca*	1749
Nematodes: Biological Control / *Simon Gowen*	1752
Neurotoxicants: Developmental Experimental Testing / *Vera Lucia S.S. de Castro*	1755
Nitrate Leaching Index / *Jorge A. Delgado*	1761
Nitrogen / *Oswald Van Cleemput and Pascal Boeckx*	1768
Nitrogen Trading Tool / *Jorge A. Delgado*	1772
Nitrogen: Biological Fixation / *Mark B. Peoples*	1785
Nuclear Energy: Economics / *James G. Hewlett*	1789
Nutrients: Best Management Practices / *Scott J. Sturgul and Keith A. Kelling*	1805
Nutrients: Bioavailability and Plant Uptake / *Niels Erik Nielsen*	1817
Nutrient–Water Interactions / *Ardell D. Halvorson*	1823
Oil Pollution: Baltic Sea / *Andrey G. Kostianoy*	1826
Organic Compounds: Halogenated / *Marek Tobiszewski and Jacek Namieśnik*	1841
Organic Matter: Global Distribution in World Ecosystems / *Wilfred M. Post*	1851
Organic Matter: Management / *R. Cesar Izaurralde and Carlos C. Cerri*	1857
Organic Matter: Modeling / *Pete Smith*	1863
Organic Matter: Turnover / *Johan Six and Julie D. Jastrow*	1872
Organic Soil Amendments / *Philip Oduor-Owino*	1878
Ozone Layer / *Luisa T. Molina*	1882
Permafrost / *Douglas Kane and Julia Boike*	1900
Persistent Organic Compounds: Wet Oxidation Removal / *Anurag Garg and I.M. Mishra*	1904
Persistent Organic Pesticides / *Gamini Manuweera*	1913
Pest Management / *E.F. Legner*	1919
Pest Management: Crop Diversity / *Marianne Karpenstein-Machan and Maria Finckh*	1925
Pest Management: Ecological Agriculture / *Barbara Dinham*	1930
Pest Management: Ecological Aspects / *David J. Horn*	1934
Pest Management: Intercropping / *Maria Finckh and Marianne Karpenstein-Machan*	1937
Pest Management: Legal Aspects / *Michael T. Olexa and Zachary T. Broome*	1940
Pest Management: Modeling / *Andrew Paul Gutierrez*	1949
Pesticide Translocation Control: Soil Erosion / *Monika Frielinghaus, Detlef Deumlich, and Roger Funk*	1955
Pesticides / *Marek Biziuk, Jolanta Fenik, Monika Kosikowska, and Maciej Tankiewicz*	1963
Pesticides: Banding / *Sharon A. Clay*	1983
Pesticides: Chemical and Biological / *Barbara Manachini*	1986
Pesticides: Damage Avoidance / *Aiwerasia V.F. Ngowi and Larama M.B. Rongo*	1997
Pesticides: Effects / *Ana Maria Evangelista de Duffard*	2005
Pesticides: History / *Edward H. Smith and George Kennedy*	2008
Pesticides: Measurement and Mitigation / *George F. Antonious*	2013
Pesticides: Natural / *Franck Dayan*	2028
Pesticides: Reducing Use / *David Pimentel and Maria V. Cilveti*	2033
Pesticides: Regulating / *Matthias Kern*	2035
Pests: Landscape Patterns / *F. Craig Stevenson*	2038
Petroleum: Hydrocarbon Contamination / *Svetlana Drozdova and Erwin Rosenberg*	2040
Pharmaceuticals: Treatment / *Diana Aga and Seungyun Baik*	2060
Phenols / *Leszek Wachowski and Robert Pietrzak*	2071
Phosphorus: Agricultural Nutrient / *John Ryan, Hayriye Ibrikci, Rolf Sommer, and Abdul Rashid*	2091
Phosphorus: Riverine System Transport / *Andrew N. Sharpley, Peter Kleinman, Tore Krogstad, and Richard McDowell*	2100

Volume III (cont'd)

- Plant Pathogens (Fungi): Biological Control / *Timothy Paulitz* 2107
- Plant Pathogens (Viruses): Biological Control / *Hei-Ti Hsu* 2111
- Pollutants: Organic and Inorganic / *A. Paul Schwab* 2114
- Pollution: Genotoxicity of Agrotoxic Compounds / *Vera Lucia S.S. de Castro and Paola Poli* 2123
- Pollution: Non-Point Source / *Ravendra Naidu, Mallavarapu Megharaj, Peter Dillon, Rai Kookana, Ray Correll, and W.W. Wenzel* 2136
- Pollution: Pesticides in Agro-Horticultural Ecosystems / *J.K. Dubey and Meena Thakur* 2139
- Pollution: Pesticides in Natural Ecosystems / *J.K. Dubey and Meena Thakur* 2145
- Pollution: Point and Non-Point Source Low Cost Treatment / *Sandeep Joshi and Sayali Joshi* 2150
- Pollution: Point Sources / *Ravendra Naidu, Mallavarapu Megharaj, Peter Dillon, Rai Kookana, Ray Correll, and W.W. Wenzel* 2166
- Polychlorinated Biphenyls (PCBs) / *Marek Biziuk and Angelika Beyer* 2172
- Polychlorinated Biphenyls (PCBs) and Polycyclic Aromatic Hydrocarbons (PAHs): Sediments and Water Analysis / *Justyna Rogowska, Agata Mechlińska, Lidia Wolska, and Jacek Namieśnik* 2186
- Potassium / *Philippe Hinsinger* 2208
- Precision Agriculture: Engineering Aspects / *Joel T. Walker, Reza Ehsani, and Matthew O. Sullivan* 2213
- Precision Agriculture: Water and Nutrient Management / *Robert J. Lascano and J.D. Booker* 2217
- Radio Frequency Towers: Public School Placement / *Joshua Steinfeld* 2224
- Radioactivity / *Bogdan Skwarzec* 2234
- Radionuclides / *Philip M. Jardine* 2243
- Rain Water: Atmospheric Deposition / *Zaneta Polkowska* 2249
- Rain Water: Harvesting / *K.F. Andrew Lo* 2262
- Rare Earth Elements / *Zhengyi Hu, Gerd Sparovek, Silvia Haneklaus, and Ewald Schnug* 2266

Volume IV

- Remote Sensing and GIS Integration / *Egide Nizeyimana* 2271
- Remote Sensing: Pollution / *Massimo Antoninetti* 2275
- Rivers and Lakes: Acidification / *Agniezka Gałuszka and Zdzisław M. Migaszewski* 2291
- Rivers: Pollution / *Bogdan Skwarzec* 2303
- Rivers: Restoration / *Anna Rabajczyk* 2307
- Runoff Water / *Zaneta Polkowska* 2320
- Salt-Affected Soils: Physical Properties and Behavior / *Hwat Bing So* 2334
- Salt-Affected Soils: Plant Response / *Anna Eynard, Keith D. Wiebe, and Rattan Lal* 2345
- Salt-Affected Soils: Sustainable Agriculture / *Pichu Rengasamy* 2349
- Sea: Pollution / *Bogdan Skwarzec* 2357
- Sodic Soils: Irrigation Farming / *David Burrow* 2364
- Sodic Soils: Properties / *Pichu Rengasamy* 2367
- Sodic Soils: Reclamation / *Jock Churchman* 2370
- Soil Degradation: Global Assessment / *Selim Kapur, Suha Berberoğlu, Erhan Akça, Cenk Dönmez, Mehmet Akif Erdoğan, and Burçak Kapur* 2375
- Soil Quality: Carbon and Nitrogen Gases / *Philippe Rochette, Sean McGinn, and Reynald Lemke* 2388
- Soil Quality: Indicators / *Graham P. Sparling* 2393
- Soil Rehabilitation / *David Jasper* 2396
- Solid Waste Management: Life Cycle Assessment / *Ni-Bin Chang, Ana Pires, and Graça Martinho* 2399
- Solid Waste: Municipal / *Angelique Chettiparamb* 2415

Stored-Product Pests: Biological Control / *Lise Stengård Hansen*	2423
Strontium / *Silvia Haneklaus and Ewald Schnug*	2426
Sulfur / *Ewald Schnug, Silvia Haneklaus, and Elke Bloem*	2431
Sulfur Dioxide / *Marianna Czaplicka and Witold Kurylak*	2437
Sustainability and Planning / *Richard Cowell*	2446
Sustainable Agriculture: Soil Quality / *John W. Doran and Ed G. Gregorich*	2457
Sustainable Development / *Mark A. Peterson*	2461
Sustainable Development: Ecological Footprint in Accounting / *Simone Bastianoni, Valentina Niccolucci, Elena Neri, Gemma Cranston, Alessandro Galli, and Mathis Wackernagel*	2467
Sustainable Development: Pyrolysis and Gasification of Biomass and Wastes / *P. Mondal, J. Parikh, and D.V. Naik*	2482
Thermal Energy: Solar Technologies / *Muhammad Asif and Tariq Muneer*	2498
Thermal Energy: Storage / *Brian Silvetti*	2508
Thermodynamics / *Ronald L. Klaus*	2525
Tillage Erosion: Terrace Formation / *Seth M. Dabney and Dalmo A.N. Vieira*	2536
Toxic Substances / *Sven Erik Jørgensen*	2542
Toxic Substances: Photochemistry / *Puangrat Kajitvichyanukul*	2547
Toxicity Prediction of Chemical Mixtures / *Nina Cedergreen, Claus Svendsen, and Thomas Backhaus*	2572
Vanadium and Chromium Groups / *Imad A.M. Ahmed*	2582
Vertebrates: Biological Control / *Peter Kerr and Tanja Strive*	2595
Waste Gas Treatment: Bioreactors / *Amit Kumar, Piet Lens, Sarina J. Ergas, and Herman Van Langenhove*	2611
Waste: Stabilization Ponds / *Sven Erik Jørgensen*	2632
Wastewater and Water Utilities / *Rudolf Marloth*	2638
Wastewater Treatment: Biological / *Shaikh Ziauddin Ahammad, David W. Graham, and Jan Dolfing*	2645
Wastewater Treatment: Conventional Methods / *Sven Erik Jørgensen*	2657
Wastewater Treatment: Wetlands Use in Arctic Regions / *Colin N. Yates, Brent Wootton, Sven Erik Jørgensen, and Stephen D. Murphy*	2662
Wastewater Use in Agriculture / *Manzoor Qadir, Pay Drechsel, and Liqa Raschid-Sally*	2675
Wastewater Use in Agriculture: Policy Issues / *Dennis Wichelns*	2681
Wastewater Use in Agriculture: Public Health Considerations / *Blanca Jimenez and Inés Navarro*	2694
Wastewater: Municipal / *Sven Erik Jørgensen*	2709
Water and Wastewater: Filters / *Sandeep Joshi*	2719
Water and Wastewater: Ion Exchange Application / *Sven Erik Jørgensen*	2734
Water Harvesting / *Gary W. Frasier*	2738
Water Quality and Quantity: Globalization / *Kristi Denise Caravella and Jocilyn Danise Martinez*	2740
Water Quality: Modeling / *Richard Lowrance*	2749
Water Quality: Range and Pasture Land / *Thomas L. Thurow*	2752
Water Quality: Soil Erosion / *Andreas Klik*	2755
Water Quality: Timber Harvesting / *C. Rhett Jackson*	2770
Water Supplies: Pharmaceuticals / *Klaus Kümmerer*	2776
Water: Cost / *Atif Kubursi and Matthew Agarwala*	2779
Water: Drinking / *Marek Biziuk and Małgorzata Michalska*	2790
Water: Surface / *Victor de Vlaming*	2804
Water: Total Maximum Daily Load / *Robin Kundis Craig*	2808
Watershed Management: Remote Sensing and GIS / *A.V. Shanwal and S.P. Singh*	2816
Weeds (Insects and Mites): Biological Control / *Peter Harris*	2821
Wetlands / *Ralph W. Tiner*	2824

Volume IV (cont'd)

- Wetlands: **Biodiversity** / *Jean-Claude Lefeuvre and Virginie Bouchard* 2829
- Wetlands: **Carbon Sequestration** / *Virginie Bouchard* 2833
- Wetlands: **Conservation Policy** / *Clayton Rubec* 2837
- Wetlands: **Constructed Subsurface** / *Jan Vymazal* 2841
- Wetlands: **Methane Emission** / *Anna Ekberg and Torben Røjle Christensen* 2850
- Wetlands: **Petroleum** / *John V. Headley and Kerry M. Peru* 2854
- Wetlands: **Sedimentation and Ecological Engineering** / *Timothy C. Granata and Jay F. Martin* 2859
- Wetlands: **Treatment System Use** / *Kyle R. Mankin* 2862
- Wind Farms: **Noise** / *Daniel Shepherd, Chris Hanning, and Bob Thorne* 2867
- **Yellow River** / *Zixi Zhu, Ynuzhang Wang, and Yifei Zhu* 2884
- **Appendixes** 2887

Topical Table of Contents

The entries have been classified according to the presented procedure for integrated, holistic environmental and ecological management. The content uses the following topical classifications:

CLT: means that the solutions are based on cleaner technology
COV: indicates that the articles give comparative overviews of important topics for environmental management or background knowledge that is important for the evaluation of environmental problems
DIA: means that the articles are about diagnostic tools: monitoring, ecological modelling, ecological indicators and ecological services
ECT: covers solutions of the problems based on ecotechnology
ELE: focuses on the use of environmental legislation to solve environmental problems
ENT: refers to solutions of environmental problems by the use of environmental technology
IMS: are articles uncovering the possibilities to integrate the various tool boxes to make an integrated and holistic management
PSS: covers entries that focus on a pollution problem and its sources

The classical environmental classification indicates the sphere that is touched by the environmental problem. Is it a water problem? – The hydrosphere is involved. Is it a air pollution problem? – The atmosphere is involved. Or is a terrestrial problem? – The lithosphere is involved. This more traditional classification has been used as an additional topical classification, and the following abbreviations are used in this context:

AIR: air pollution problems
WAT: water pollution problems
TER: terrestrial pollution problems
GEN: general pollution problems that may involve more than one sphere
GLO: global pollution problems

All articles are marked with both topological classifications.

Agriculture (AGR)

Cleaner Technology (CLT)

Agriculture: Organic / *Kathleen Delate* 125
Arthropod Host-Plant Resistant Crops / *Gerald E. Wilde* 304
***Bacillus thuringiensis*: Transgenic Crops** / *Julie A. Peterson, John J. Obrycki, and James D. Harwood* 307
Cabbage Diseases: Ecology and Control / *Anthony P. Keinath, Marc A. Cubeta, and David B. Langston, Jr.* 442
Farming: Organic / *Brenda Frick* 1109
Land Restoration / *Richard W. Bell* 1600
Manure Management: Compost and Biosolids / *Bahman Eghball and Kenneth A. Barbarick* 1660
Manure Management: Dairy / *H.H. Van Horn and D.R. Bray* 1663
Manure Management: Poultry / *Shafiqur Rahman and Thomas R. Way* 1670
Nutrients: Best Management Practices / *Scott J. Sturgul and Keith A. Kelling* 1805
Nutrients: Bioavailability and Plant Uptake / *Niels Erik Nielsen* 1817

Agriculture (AGR) (cont'd)

- Pesticide Translocation Control: Soil Erosion / *Monika Frielinghaus, Detlef Deumlich, and Roger Funk* 1955
- Plant Pathogens (Fungi): Biological Control / *Timothy Paulitz* 2107
- Plant Pathogens (Viruses): Biological Control / *Hei-Ti Hsu* 2111
- Precision Agriculture: Engineering Aspects / *Joel T. Walker, Reza Ehsani, and Matthew O. Sullivan* 2213
- Salt-Affected Soils: Sustainable Agriculture / *Pichu Rengasamy* 2349
- Sustainable Agriculture: Soil Quality / *John W. Doran and Ed G. Gregorich* 2457

Comparative Overviews (COV)

- Agricultural Runoff / *Matt C. Smith, David K. Gattie, and Daniel L. Thomas* 81
- Agricultural Water Quantity Management / *X.S. Qin and Y. Xu* 105
- Agriculture: Energy Use and Conservation / *Guangnan Chen and Tek Narayan Maraseni* 118
- Erosion: History / *Andrew S. Goudie* 1050
- Erosion: Soil Quality / *Craig Ditzler* 1060
- Nitrate Leaching Index / *Jorge A. Delgado* 1761
- Nitrogen / *Oswald Van Cleemput and Pascal Boeckx* 1768
- Nitrogen: Biological Fixation / *Mark B. Peoples* 1785
- Nutrients: Best Management Practices / *Scott J. Sturgul and Keith A. Kelling* 1805
- Nutrients: Bioavailability and Plant Uptake / *Niels Erik Nielsen* 1817
- Salt-Affected Soils: Physical Properties and Behavior / *Hwat Bing So* 2334
- Salt-Affected Soils: Plant Response / *Anna Eynard, Keith D. Wiebe, and Rattan Lal* 2345
- Sodic Soils: Properties / *Pichu Rengasamy* 2367
- Soil Degradation: Global Assessment / *Selim Kapur, Suha Berberoğlu, Erhan Akça, Cenk Dönmez, Mehmet Akif Erdoğan, and Burçak Kapur* 2375

Diagnostic Tools (DIA)

- Acid Sulfate Soils: Identification, Assessment, and Management / *L.A. Sullivan, R.T. Bush, and E.D. Burton* 36
- Animals: Toxicological Evaluation / *Vera Lucia S.S. de Castro* 280
- Bioindicators: Farming Practices and Sustainability / *Joji Muramoto and Stephen R. Gliessman* 352
- Drainage: Soil Salinity Management / *Glenn J. Hoffman* 588
- Erosion by Water: Assessment and Control / *José Miguel Reichert, Nadia Bernardi Bonumá, Gustavo Enrique Merten, and Jean Paolo Gomes Minella* 963
- Erosion by Water: Empirical Methods / *John M. Laflen and John M. Laflen* 974
- Erosion by Water: Erosivity and Erodibility / *Peter I.A. Kinnell* 980
- Erosion by Water: Process-Based Modeling / *Mark A. Nearing* 991
- Erosion by Wind: Source, Measurement, Prediction, and Control / *Brenton S. Sharratt and R. Scott Van Pelt* 1017
- Nitrate Leaching Index / *Jorge A. Delgado* 1761
- Soil Degradation: Global Assessment / *Selim Kapur, Suha Berberoğlu, Erhan Akça, Cenk Dönmez, Mehmet Akif Erdoğan, and Burçak Kapur* 2375
- Soil Quality: Indicators / *Graham P. Sparling* 2393

Ecotechnology (ECT)

- Acid Sulfate Soils: Management / *Michael D. Melville and Ian White* 55
- Agricultural Water Quantity Management / *X.S. Qin and Y. Xu* 105
- Agriculture: Organic / *Kathleen Delate* 125

Agroforestry: Water Use Efficiency / *James R. Brandle, Laurie Hodges, and Xinhua Zhou*	129
Arthropod Host-Plant Resistant Crops / *Gerald E. Wilde*	304
Biofertilizers / *J.R. de Freitas*	349
Biological Controls / *Heikki Hokkanen*	366
Biological Controls: Conservation / *Doug Landis and Steve D. Wratten*	370
Bioremediation / *Ragini Gothalwal*	387
Cabbage Diseases: Ecology and Control / *Anthony P. Keinath, Marc A. Cubeta, and David B. Langston, Jr.*	442
Composting / *Nídia Sá Caetano*	512
Composting: Organic Farming / *Saburo Matsui*	528
Desertification: Prevention and Restoration / *Claudio Zucca, Susana Bautista, Barron Orr, and Franco Previtali*	557
Erosion and Sediment Control: Vegetative Techniques / *Samson D. Angima*	947
Erosion by Water: Amendment Techniques / *X.-C. (John) Zhang*	958
Erosion by Water: Vegetative Control / *Seth M. Dabney and Silvio J. Gumiere*	994
Erosion Control: Soil Conservation / *Eric T. Craswell*	1034
Erosion Control: Tillage and Residue Methods / *Richard Cruse, Jerry Neppel, John Kost, and Krisztina Eleki*	1040
Farming: Organic / *Brenda Frick*	1109
Insects and Mites: Biological Control / *Ann E. Hajek*	1474
Land Restoration / *Richard W. Bell*	1600
Manure Management: Compost and Biosolids / *Bahman Eghball and Kenneth A. Barbarick*	1660
Manure Management: Dairy / *H.H. Van Horn and D.R. Bray*	1663
Manure Management: Phosphorus / *Rory O. Maguire, John T. Brake, and Peter W. Plumstead*	1667
Manure Management: Poultry / *Shafiqur Rahman and Thomas R. Way*	1670
Natural Enemies and Biocontrol: Artificial Diets / *Simon Grenier and Patrick De Clerq*	1746
Nutrients: Best Management Practices / *Scott J. Sturgul and Keith A. Kelling*	1805
Nutrients: Bioavailability and Plant Uptake / *Niels Erik Nielsen*	1817
Organic Soil Amendments / *Philip Oduor-Owino*	1878
Pesticide Translocation Control: Soil Erosion / *Monika Frielinghaus, Detlef Deumlich, and Roger Funk*	1955
Plant Pathogens (Fungi): Biological Control / *Timothy Paulitz*	2107
Plant Pathogens (Viruses): Biological Control / *Hei-Ti Hsu*	2111
Sodic Soils: Reclamation / *Jock Churchman*	2370
Soil Rehabilitation / *David Jasper*	2396
Sustainable Agriculture: Soil Quality / *John W. Doran and Ed G. Gregorich*	2457
Tillage Erosion: Terrace Formation / *Seth M. Dabney and Dalmo A.N. Vieira*	2536
Weeds (Insects and Mites): Biological Control / *Peter Harris*	2821

Environmental Legislation (ELE)

Laws and Regulations: Food / *Ike Jeon*	1609
Laws and Regulations: Rotterdam Convention / *Barbara Dinham*	1615
Laws and Regulations: Soil / *Ian Hannam and Ben Boer*	1618
Nitrogen Trading Tool / *Jorge A. Delgado*	1772

Environmental Technology (ENT)

Agricultural Water Quantity Management / *X.S. Qin and Y. Xu*	105
Agroforestry: Water Use Efficiency / *James R. Brandle, Laurie Hodges, and Xinhua Zhou*	129

Agriculture (AGR) (cont'd)

Precision Agriculture: Engineering Aspects / *Joel T. Walker, Reza Ehsani, and Matthew O. Sullivan* 2213

Integrated and Holistic Management (IMS)

Integrated Farming Systems / *John Holland* 1507
Integrated Nutrient Management / *Bal Ram Singh* 1510
Integrated Weed Management / *Heinz Müller-Schärer and Alexandra Robin Collins* 1524
Nutrients: Best Management Practices / *Scott J. Sturgul and Keith A. Kelling* 1805
Nutrients: Bioavailability and Plant Uptake / *Niels Erik Nielsen* 1817

Pollution Problems and Their Sources (PSS)

Acid Sulfate Soils / *Delvin S. Fanning* 26
Acid Sulfate Soils: Formation / *Martin C. Rabenhorst, Delvin S. Fanning, and Steven N. Burch* 31
Agricultural Soils: Ammonia Volatilization / *Paul G. Saffigna and John R. Freney* 85
Agricultural Soils: Nitrous Oxide Emissions / *John R. Freney* 96
Agricultural Soils: Phosphorus / *Anja Gassner and Ewald Schnug* 100
Animals: Toxicological Evaluation / *Vera Lucia S.S. de Castro* 280
Cabbage Diseases: Ecology and Control / *Anthony P. Keinath, Marc A. Cubeta, and David B. Langston, Jr.* 442
Desertification / *David Tongway and John Ludwig* 541
Desertification: Extent of / *Victor R. Squires* 545
Desertification: Greenhouse Effect / *Sherwood Idso and Craig Idso* 549
Desertification: Impact / *David A. Mouat and Judith Lancaster* 552
Erosion / *Dennis C. Flanagan* 930
Erosion and Carbon Dioxide / *Pierre A. Jacinthe and Rattan Lal* 934
Erosion and Global Change / *Taolin Zhang and Xingxiang Wang* 939
Erosion and Precipitation / *Bofu Yu* 943
Erosion by Water: Accelerated / *David Favis-Mortlock* 951
Erosion by Wind-Driven Rain / *Gunay Erpul, L. Darrell Norton, and Donald Gabriels* 1031
Erosion: Accelerated / *J. Kent Mitchell and Michael C. Hirschi* 1044
Erosion: Irrigation-Induced / *Robert E. Sojka and David L. Bjorneberg* 1053
Erosion: Snowmelt / *Donald K. McCool* 1057
Irrigation: Soil Salinity / *James D. Rhoades* 1572
Leaching / *Lars Bergström* 1621
Leaching and Illuviation / *Lynn E. Moody* 1624
Manure Management: Phosphorus / *Rory O. Maguire, John T. Brake, and Peter W. Plumstead* 1667
Manure Management: Poultry / *Shafiqur Rahman and Thomas R. Way* 1670
Nitrogen / *Oswald Van Cleemput and Pascal Boeckx* 1768
Phosphorus: Agricultural Nutrient / *John Ryan, Hayriye Ibrikci, Rolf Sommer, and Abdul Rashid* 2091
Phosphorus: Riverine System Transport / *Andrew N. Sharpley, Peter Kleinman, Tore Krogstad, and Richard McDowell* 2100
Sodic Soils: Irrigation Farming / *David Burrow* 2364
Soil Degradation: Global Assessment / *Selim Kapur, Suha Berberoğlu, Erhan Akça, Cenk Dönmez, Mehmet Akif Erdoğan, and Burçak Kapur* 2375
Soil Quality: Carbon and Nitrogen Gases / *Philippe Rochette, Sean McGinn, and Reynald Lemke* 2388

Air Pollution Problems (AIR)

Comparative Overviews (COV)
Soil Quality: Carbon and Nitrogen Gases / *Philippe Rochette, Sean McGinn, and Reynald Lemke*............2388

Diagnostic Tools (DIA)
Air Pollution: Monitoring / *Waldemar Wardencki*..132

Environmental Technology (ENT)
Adsorption / *Puangrat Kajitvichyanukul and Jirapat Ananpattarachai*...61
Air Pollution: Technology / *Sven Erik Jørgensen*...149
Waste Gas Treatment: Bioreactors / *Amit Kumar, Piet Lens, Sarina J. Ergas, and Herman Van Langenhove*..2611

Pollution Problems and Their Sources (PSS)
Acid Rain / *Umesh Kulshrestha*...5
Acid Rain: Nitrogen Deposition / *George F. Vance*..20
Agricultural Soils: Ammonia Volatilization / *Paul G. Saffigna and John R. Freney*........................85
Agricultural Soils: Nitrous Oxide Emissions / *John R. Freney*..96
Fossil Fuel Combustion: Air Pollution and Global Warming / *Dan Golomb*...................................1127
Genotoxicity and Air Pollution / *Eliane Tigre Guimarães and Andrea Nunes Vaz Pedroso*............1154
Rain Water: Atmospheric Deposition / *Zaneta Polkowska*..2249
Rivers and Lakes: Acidification / *Agniezka Gałuszka and Zdzisław M. Migaszewski*....................2291
Soil Quality: Carbon and Nitrogen Gases / *Philippe Rochette, Sean McGinn, and Reynald Lemke*............2388
Sulfur Dioxide / *Marianna Czaplicka, Marianna Czaplicka, and Witold Kurylak*...........................2437

Energy Issues (ENE)

Cleaner Technology (CLT): Alternative Energy
Agriculture: Energy Use and Conservation / *Guangnan Chen and Tek Narayan Maraseni*..............118
Alternative Energy / *Bernd Markert, Simone Wuenschmann, Stefan Fraenzle, and Bernd Delakowitz*..........179
Alternative Energy: Hydropower / *Andrea Micangeli, Sara Evangelisti, and Danilo Sbordone*................201
Alternative Energy: Photovoltaic Modules and Systems / *Ewa Klugmann-Radziemska*.....................215
Alternative Energy: Photovoltaic Solar Cells / *Ewa Klugmann-Radziemska*......................................226
Alternative Energy: Solar Thermal Energy / *Andrea Micangeli, Sara Evangelisti, and Danilo Sbordone*........241
Alternative Energy: Wind Power Technology and Economy / *K.E. Ohrn*..258
Carbon Sequestration / *Nathan E. Hultman*..456
Distributed Generation: Combined Heat and Power / *Barney L. Capehart, D. Paul Mehta, and Wayne C. Turner*............578
Energy Efficiency: New and Emerging Technology / *Steven A. Parker*..742
Energy Sources: Natural versus Additional / *Marc A. Rosen*...770
Energy: Renewable / *John O. Blackburn*..824
Energy: Solid Waste Advanced Thermal Technology / *Alex E.S. Green and Andrew R. Zimmerman*.........830
Energy: Storage / *Rudolf Marloth*...853
Geothermal Energy Resources / *Ibrahim Dincer and Arif Hepbasli*..1167
Global Climate Change: Carbon Sequestration / *Sherwood Idso and Keith E. Idso*......................1189

Energy Issues (ENE) (*cont'd*)

 Global Climate Change: Gasoline, Hybrid-Electric and Hydrogen-Fueled Vehicles /
  Robert E. Uhrig 1206
 Green Energy / *Ibrahim Dincer and Adnan Midili* 1226
 Green Processes: Systems Analysis / *Abhishek Tiwary* 1242
 Green Products: Production / *Puangrat Kajitvichyanukul, Jirapat Ananpattarachai, and*
  Apichon Watcharenwong 1253
 Heat Pumps / *Lu Aye* 1346
 Heat Pumps: Absorption / *Keith E. Herold* 1356
 Heat Pumps: Geothermal / *Greg Tinkler* 1363
 Hydroelectricity: Pumped Storage / *Jill S. Tietjen* 1418
 Integrated Energy Systems / *Leslie A. Solmes and Sven Erik Jørgensen* 1493
 Thermal Energy: Solar Technologies / *Muhammad Asif and Tariq Muneer* 2498
 Thermal Energy: Storage / *Brian Silvetti* 2508

Cleaner Technology (CLT): Energy Conservation

 Buildings: Climate Change / *Lisa Guan and Guangnan Chen* 435
 Energy Commissioning: Existing Buildings / *David E. Claridge, Mingsheng Liu, and W.D. Turner* ... 656
 Energy Commissioning: New Buildings / *Janey Kaster* 665
 Energy Conservation / *Ibrahim Dincer and Adnan Midili* 677
 Energy Conservation: Benefits / *Eric A. Woodroof, Wayne C. Turner, and Steven D. Heinz* 691
 Energy Conservation: Industrial Processes / *Harvey E. Diamond* 697
 Energy Conservation: Lean Manufacturing / *Bohdan W. Oppenheim* 705
 Energy Conversion: Coal, Animal Waste, and Biomass Fuel / *Kalyan Annamalai,*
  Soyuz Priyadarsan, Senthil Arumugam, and John M. Sweeten 714
 Energy Efficiency: Low-Cost Improvements / *James Call* 735
 Energy Efficiency: Strategic Facility Guidelines / *Steve Doty* 752
 Energy Master Planning / *Fredric S. Goldner* 763
 Energy Use: U.S. / *Dwight K. French* 790
 Energy: Environmental Security / *Muhammad Asif* 798
 Energy: Walls and Windows / *Therese Stovall* 860
 Energy: Waste Heat Recovery / *Martin A. Mozzo, Jr.* 869
 Insulation: Facilities / *Wendell A. Porter* 1478
 LEED-EB: Leadership in Energy and Environmental Design for Existing Buildings / *Rusty T. Hodapp* ... 1647
 LEED-NC: Leadership in Energy and Environmental Design for New Construction /
  Stephen A. Roosa 1654
 Sustainable Development: Pyrolysis and Gasification of Biomass and Wastes / *P. Mondal,*
  J. Parikh, and D.V. Naik 2482
 Waste Gas Treatment: Bioreactors / *Amit Kumar, Piet Lens, Sarina J. Ergas, and*
  Herman Van Langenhove 2611
 Wind Farms: Noise / *Daniel Shepherd, Chris Hanning, and Bob Thorne* 2867

Comparative Overviews (COV)

 Agriculture: Energy Use and Conservation / *Guangnan Chen and Tek Narayan Maraseni* 118
 Alternative Energy / *Bernd Markert, Simone Wuenschmann, Stefan Fraenzle, and*
  Bernd Delakowitz 179
 Alternative Energy: Wind Power Technology and Economy / *K.E. Ohrn* 258
 Bioenergy Crops: Carbon Balance Assessment / *Rocky Lemus and Rattan Lal* 345

Buildings: Climate Change / *Lisa Guan and Guangnan Chen* 435
Carbon Sequestration / *Nathan E. Hultman* 456
Cost Control: Consequences / *Peter Victor* 535
Energy: Physics / *Milivoje M. Kostic* 808
Energy: Renewable / *John O. Blackburn* 824
Fuel Cells: Intermediate and High Temperature / *Xianguo Li, Gholamreza Karimi, and Kui Jiao* 1138
Fuel Cells: Low Temperature / *Xianguo Li and Kui Jiao* 1145
Global Climate Change: Carbon Sequestration / *Sherwood Idso and Keith E. Idso* 1189
Global Climate Change: Earth System Response / *Amanda Staudt and Nathan E. Hultman* 1192
Global Climate Change: Gas Fluxes / *Pascal Boeckx and Oswald Van Cleemput* 1202
Global Climate Change: World Soils / *Rattan Lal* 1213
Globalization / *Alexandru V. Roman* 1218
Green Processes: Systems Analysis / *Abhishek Tiwary* 1242
Green Products: Production / *Puangrat Kajitvichyanukul, Jirapat Ananpattarachai, and Apichon Watcharenwong* 1253
Heat Pumps / *Lu Aye* 1346
Hydroelectricity: Pumped Storage / *Jill S. Tietjen* 1418
Inorganic Carbon: Composition and Formation / *Larry P. Wilding and H. Curtis Monger* 1444
Inorganic Carbon: Global Carbon Cycle / *William H. Schlesinger* 1448
Methane Emissions: Rice / *Kazuyuki Yagi* 1679
Permafrost / *Douglas Kane and Julia Boike* 1900
Thermal Energy: Solar Technologies / *Muhammad Asif and Tariq Muneer* 2498
Thermodynamics / *Ronald L. Klaus* 2525

Diagnostic Tools (DIA)

Bioenergy Crops: Carbon Balance Assessment / *Rocky Lemus and Rattan Lal* 345
Energy Use: Exergy and Eco-Exergy / *Sven Erik Jørgensen* 778
Inorganic Carbon: Composition and Formation / *Larry P. Wilding and H. Curtis Monger* 1444
Inorganic Carbon: Global Carbon Cycle / *William H. Schlesinger* 1448
Inorganic Carbon: Modeling / *Leslie D. McFadden and Ronald G. Amundson* 1451
Permafrost / *Douglas Kane and Julia Boike* 1900

Ecotechnology (ECT)

Bioenergy Crops: Carbon Balance Assessment / *Rocky Lemus and Rattan Lal* 345
Carbon Sequestration / *Nathan E. Hultman* 456
Desertification: Prevention and Restoration / *Claudio Zucca, Susana Bautista, Barron Orr, and Franco Previtali* 557
Global Climate Change: Carbon Sequestration / *Sherwood Idso and Keith E. Idso* 1189
Global Climate Change: Earth System Response / *Amanda Staudt and Nathan E. Hultman* 1192
Green Processes: Systems Analysis / *Abhishek Tiwary* 1242

Environmental Legislation and Policy (ELE)

Buildings: Climate Change / *Lisa Guan and Guangnan Chen* 435
Climate Policy: International / *Nathan E. Hultman* 483
Cost Control: Consequences / *Peter Victor* 535
Energy Commissioning: Existing Buildings / *David E. Claridge, Mingsheng Liu, and W.D. Turner* 656
Energy Commissioning: New Buildings / *Janey Kaster* 665
Nuclear Energy: Economics / *James G. Hewlett* 1789

Energy Issues (ENE) (cont'd)

Environmental Technology (ENT)

- **Alternative Energy** / *Bernd Markert, Simone Wuenschmann, Stefan Fraenzle, and Bernd Delakowitz* 179
- **Alternative Energy: Hydropower** / *Andrea Micangeli, Sara Evangelisti, and Danilo Sbordone* 201
- **Alternative Energy: Photovoltaic Modules and Systems** / *Ewa Klugmann-Radziemska* 215
- **Alternative Energy: Photovoltaic Solar Cells** / *Ewa Klugmann-Radziemska* 226
- **Alternative Energy: Solar Thermal Energy** / *Andrea Micangeli, Sara Evangelisti, and Danilo Sbordone* 241
- **Alternative Energy: Wind Power Technology and Economy** / *K.E. Ohrn* 258
- **Buildings: Climate Change** / *Lisa Guan and Guangnan Chen* 435
- **Energy Conservation: Industrial Processes** / *Harvey E. Diamond* 697
- **Energy Conservation: Lean Manufacturing** / *Bohdan W. Oppenheim* 705
- **Energy Conversion: Coal, Animal Waste, and Biomass Fuel** / *Kalyan Annamalai, Soyuz Priyadarsan, Senthil Arumugam, and John M. Sweeten* 714
- **Energy Efficiency: New and Emerging Technology** / *Steven A. Parker* 742
- **Nuclear Energy: Economics** / *James G. Hewlett* 1789

Integrated and Holistic Management (IMS)

- **Energy Efficiency: Low-Cost Improvements** / *James Call* 735
- **Energy Efficiency: New and Emerging Technology** / *Steven A. Parker* 742
- **Energy Efficiency: Strategic Facility Guidelines** / *Steve Doty* 752
- **Energy Master Planning** / *Fredric S. Goldner* 763
- **Energy Sources: Natural versus Additional** / *Marc A. Rosen* 770
- **Energy: Solid Waste Advanced Thermal Technology** / *Alex E.S. Green and Andrew R. Zimmerman* 830
- **Energy: Storage** / *Rudolf Marloth* 853
- **Green Energy** / *Ibrahim Dincer and Adnan Midili* 1226
- **Integrated Energy Systems** / *Leslie A. Solmes and Sven Erik Jørgensen* 1493
- **LEED-EB: Leadership in Energy and Environmental Design for Existing Buildings** / *Rusty T. Hodapp* 1647
- **LEED-NC: Leadership in Energy and Environmental Design for New Construction** / *Stephen A. Roosa* 1654

Pollution Problems and Their Sources (PSS)

- **Desertification: Greenhouse Effect** / *Sherwood Idso and Craig Idso* 549
- **Desertization** / *Henry Noel LeHouerou* 565
- **Erosion and Global Change** / *Taolin Zhang and Xingxiang Wang* 939
- **Erosion and Precipitation** / *Bofu Yu* 943
- **Erosion by Wind: Climate Change** / *Alan Busacca and David Chandler* 1004
- **Erosion by Wind: Global Hot Spots** / *Andrew Warren* 1010
- **Methane Emissions: Rice** / *Kazuyuki Yagi* 1679
- **Wind Farms: Noise** / *Daniel Shepherd, Chris Hanning, and Bob Thorne* 2867

General Pollution Problems (GEN)

Cleaner Technology (CLT)

- **Industries: Network of** / *Sven Erik Jørgensen* 1434
- **Nanotechnology: Environmental Abatement** / *Puangrat Kajitvichyanukul and Jirapat Ananpattarachai* 1730

Comparative Overviews (COV): Pollution and Contamination

Agricultural Soils: Carbon and Nitrogen Biological Cycling / *Alan J. Franzluebbers* 88
Aluminum / *Johannes Bernhard Wehr, Frederick Paxton Cardell Blamey, and Neal William Menzies* 267
Coastal Water: Pollution / *Piotr Szefer* 489
Cyanobacteria: Eutrophic Freshwater Systems / *Anja Gassner and Martin V. Frey* 538
Estuaries / *Claude Amiard-Triquet* 1063
Fossil Fuel Combustion: Air Pollution and Global Warming / *Dan Golomb* 1127
Giant Reed (*Arundo donax*): Streams and Water Resources / *Gretchen Coffman* 1182
Inorganic Compounds: Eco-Toxicity / *Sven Erik Jørgensen* 1455
Mercury / *Sven Erik Jørgensen* 1676
Nanoparticles / *Alexandra Navrotsky* 1711
Persistent Organic Pesticides / *Gamini Manuweera* 1913
Pollution: Non-Point Source / *Ravendra Naidu, Mallavarapu Megharaj, Peter Dillon, Rai Kookana, Ray Correll, and W.W. Wenzel* 2136
Pollution: Point Sources / *Ravendra Naidu, Mallavarapu Megharaj, Peter Dillon, Rai Kookana, Ray Correll, and W.W. Wenzel* 2166
Rain Water: Atmospheric Deposition / *Zaneta Polkowska* 2249
Rain Water: Atmospheric Deposition / *Zaneta Polkowska* 2249
Rivers and Lakes: Acidification / *Agniezka Gałuszka and Zdzisław M. Migaszewski* 2291
Rivers and Lakes: Acidification / *Agniezka Gałuszka and Zdzisław M. Migaszewski* 2291
Rivers: Pollution / *Bogdan Skwarzec* 2303
Rivers: Pollution / *Bogdan Skwarzec* 2303
Sea: Pollution / *Bogdan Skwarzec* 2357
Sea: Pollution / *Bogdan Skwarzec* 2357

Comparative Overviews (COV): Processes, Environmental Quality and Conditions

Antagonistic Plants / *Philip Oduor-Owino* 288
Bioaccumulation / *Tomaz Langenbach* 324
Biodegradation / *Sven Erik Jørgensen* 328
Biodiversity and Sustainability / *Odo Primavesi* 333
Biomass / *Alberto Traverso and David Tucker* 373
Cost Control: Consequences / *Peter Victor* 535
Erosion / *Dennis C. Flanagan* 930
Erosion: History / *Andrew S. Goudie* 1050
Groundwater: Contamination / *Charles W. Fetter* 1281
Industries: Network of / *Sven Erik Jørgensen* 1434
Inorganic Carbon: Composition and Formation / *Larry P. Wilding and H. Curtis Monger* 1444
Inorganic Carbon: Global Carbon Cycle / *William H. Schlesinger* 1448
Irrigation: Efficiency / *Terry A. Howell* 1545
Irrigation: Erosion / *David L. Bjorneberg* 1551
Irrigation: Return Flow and Quality / *Ramón Aragues and Kenneth K. Tanji* 1557
Nanotechnology: Environmental Abatement / *Puangrat Kajitvichyanukul and Jirapat Ananpattarachai* 1730
Solid Waste: Municipal / *Angelique Chettiparamb* 2415
Toxic Substances / *Sven Erik Jørgensen* 2542
Toxic Substances: Photochemistry / *Puangrat Kajitvichyanukul* 2547
Toxicity Prediction of Chemical Mixtures / *Nina Cedergreen, Claus Svendsen, and Thomas Backhaus* 2572

General Pollution Problems (GEN) (cont'd)

Vanadium and Chromium Groups / *Imad A.M. Ahmed* 2582
Water Quality: Range and Pasture Land / *Thomas L. Thurow* 2752
Water Quality: Soil Erosion / *Andreas Klik* 2755
Water: Cost / *Atif Kubursi and Matthew Agarwala* 2779
Water: Drinking / *Marek Biziuk and Małgorzata Michalska* 2790
Water: Surface / *Victor de Vlaming* 2804
Water: Total Maximum Daily Load / *Robin Kundis Craig* 2808

Diagnostic Tools (DIA)

Birds: Chemical Control / *Eric B. Spurr* 413
Community-Based Monitoring: Ngarenanyuki, Tanzania / *Aiwerasia V.F. Ngowi, Aiwerasia V.F. Ngowi, Larama MB Rongo, and Thomas J. Mbise* 505
Ecological Indicators: Eco-Exergy to Emergy Flow / *Simone Bastianoni, Luca Coscieme, and Federico M. Pulselli* 592
Ecological Indicators: Ecosystem Health / *Felix Müller, Benjamin Burkhard, Marion Kandziora, Claus Schimming, and Wilhelm Windhorst* 599
Exergy: Analysis / *Marc A. Rosen* 1083
Exergy: Environmental Impact Assessment / *Marc A. Rosen* 1092
Geographic Information System (GIS): Land Use Planning / *Egide Nizeyimana and Jacob Opadeyi* 1163
Inorganic Carbon: Composition and Formation / *Larry P. Wilding and H. Curtis Monger* 1444
Inorganic Carbon: Global Carbon Cycle / *William H. Schlesinger* 1448
Inorganic Carbon: Modeling / *Leslie D. McFadden and Ronald G. Amundson* 1451
Remote Sensing and GIS Integration / *Egide Nizeyimana* 2271
Remote Sensing: Pollution / *Massimo Antoninetti* 2275
Solid Waste Management: Life Cycle Assessment / *Ni-Bin Chang, Ana Pires, and Graça Martinho* 2399
Sustainable Development: Ecological Footprint in Accounting / *Simone Bastianoni, Valentina Niccolucci, Elena Neri, Gemma Cranston, Alessandro Galli, and Mathis Wackernagel* 2467

Ecotechnology (ECT)

Adsorption / *Puangrat Kajitvichyanukul and Jirapat Ananpattarachai* 61
Biofertilizers / *J.R. de Freitas* 349
Biological Controls / *Heikki Hokkanen* 366
Biological Controls: Conservation / *Doug Landis and Steve D. Wratten* 370
Bioremediation / *Ragini Gothalwal* 387
Ecosystems: Large-Scale Restoration Governance / *Shannon Estenoz, Denise Vienne, and Alka Sapat* 624
Ecosystems: Planning / *Ioan Manuel Ciumasu, Ioan Manuel Ciumasu, Liana Buzdugan, Nicolae Stefan, and Keith Culver* 632
Industries: Network of / *Sven Erik Jørgensen* 1434
Land Restoration / *Richard W. Bell* 1600
Manure Management: Phosphorus / *Rory O. Maguire, John T. Brake, and Peter W. Plumstead* 1667
Nutrients: Best Management Practices / *Scott J. Sturgul and Keith A. Kelling* 1805
Nutrients: Bioavailability and Plant Uptake / *Niels Erik Nielsen* 1817
Pollution: Point and Non-Point Source Low Cost Treatment / *Sandeep Joshi and Sayali Joshi* 2150

Environmental Legislation and Policy (ELE)

Cost Control: Consequences / *Peter Victor* 535
Environmental Legislation: Asia / *Wanpen Wirojanagud and Wanpen Wirojanagud* 874

Environmental Legislation: EU Countries Solid Waste Management / *Ni-Bin Chang, Ana Pires, and Graça Martinho*	892
Environmental Policy / *Sanford V. Berg*	914
Environmental Policy: Innovations / *Alka Sapat*	922
Food Quality Protection Act / *Christina D. DiFonzo*	1118
Food: Cosmetic Standards / *David Pimentel and Kelsey Hart*	1120
Laws and Regulations: Food / *Ike Jeon*	1609
Laws and Regulations: Rotterdam Convention / *Barbara Dinham*	1615
Water: Cost / *Atif Kubursi and Matthew Agarwala*	2779

Environmental Technology (ENT)

Adsorption / *Puangrat Kajitvichyanukul and Jirapat Ananpattarachai*	61

Integrated and Holistic Management (IMS)

Cost Control: Consequences / *Peter Victor*	535
Ecological Indicators: Ecosystem Health / *Felix Müller, Benjamin Burkhard, Marion Kandziora, Claus Schimming, and Wilhelm Windhorst*	599
Ecosystems: Large-Scale Restoration Governance / *Shannon Estenoz, Denise Vienne, and Alka Sapat*	624
Ecosystems: Planning / *Ioan Manuel Ciumasu, Ioan Manuel Ciumasu, Liana Buzdugan, Nicolae Stefan, and Keith Culver*	632
Geographic Information System (GIS): Land Use Planning / *Egide Nizeyimana and Jacob Opadeyi*	1163
Integrated Pest Management / *H.F. van Emden*	1521
Pest Management / *E.F. Legner*	1919
Solid Waste Management: Life Cycle Assessment / *Ni-Bin Chang, Ana Pires, and Graça Martinho*	2399
Sustainability and Planning / *Richard Cowell*	2446
Sustainable Development / *Mark A. Peterson*	2461
Sustainable Development: Ecological Footprint in Accounting / *Simone Bastianoni, Valentina Niccolucci, Elena Neri, Gemma Cranston, Alessandro Galli, and Mathis Wackernagel*	2467

Pollution Problems and Their Sources (PSS): Generally Point Sources

Aquatic Communities: Pesticide Impacts / *David P. Kreutzweiser and Paul K. Sibley*	291
Boron and Molybdenum: Deficiencies and Toxicities / *Umesh C. Gupta*	424
Cadmium and Lead: Contamination / *Gabriella Kakonyi and Imad A.M. Ahmed*	446
Cadmium: Toxicology / *Sven Erik Jørgensen*	453
Chromium / *Bruce R. James and Dominic A. Brose*	477
Coastal Water: Pollution / *Piotr Szefer*	489
Cobalt and Iodine / *Ronald G. McLaren*	502
Copper / *David R. Parker and Judith F. Pedler*	535
Endocrine Disruptors / *Vera Lucia S.S. de Castro*	643
Groundwater: Mining / *Hugo A. Loaiciga*	1284
Groundwater: Mining Pollution / *Jeffrey G. Skousen and George F. Vance*	1289
Heavy Metals: Organic Fertilization Uptake / *Ewald Schnug, Alexandra Izosimova, and Renata Gaj*	1374
Herbicides / *Malcolm Devine*	1378
Herbicides: Non-Target Species Effects / *Céline Boutin*	1382
Human Health: Hormonal Disruption / *Evamarie Straube, Sebastian Straube, and Wolfgang Straube*	1405

General Pollution Problems (GEN) (cont'd)

- Irrigation: Erosion / *David L. Bjorneberg* 1551
- Irrigation: Return Flow and Quality / *Ramón Aragues and Kenneth K. Tanji* 1557
- Mercury / *Sven Erik Jørgensen* 1676
- Petroleum: Hydrocarbon Contamination / *Svetlana Drozdova and Erwin Rosenberg* 2040
- Phenols / *Leszek Wachowski and Robert Pietrzak* 2071
- Pollution: Point Sources / *Ravendra Naidu, Mallavarapu Megharaj, Peter Dillon, Rai Kookana, Ray Correll, and W.W. Wenzel* 2166
- Radio Frequency Towers: Public School Placement / *Joshua Steinfeld* 2224
- Radioactivity / *Bogdan Skwarzec* 2234
- Radionuclides / *Philip M. Jardine* 2243
- Rare Earth Elements / *Zhengyi Hu, Gerd Sparovek, Silvia Haneklaus, and Ewald Schnug* 2266
- Toxic Substances / *Sven Erik Jørgensen* 2542
- Vanadium and Chromium Groups / *Imad A.M. Ahmed* 2582

Pollution Problems and Their Sources (PSS): Generally Non-Point and Diffuse Sources

- Adsorption / *Puangrat Kajitvichyanukul and Jirapat Ananpattarachai* 61
- Agricultural Soils: Carbon and Nitrogen Biological Cycling / *Alan J. Franzluebbers* 88
- Allelochemics / *John Borden* 175
- Aluminum / *Johannes Bernhard Wehr, Frederick Paxton Cardell Blamey, and Neal William Menzies* 267
- Antagonistic Plants / *Philip Oduor-Owino* 288
- Bioaccumulation / *Tomaz Langenbach* 324
- Cyanobacteria: Eutrophic Freshwater Systems / *Anja Gassner and Martin V. Frey* 538
- Erosion / *Dennis C. Flanagan* 930
- Erosion: History / *Andrew S. Goudie* 1050
- Estuaries / *Claude Amiard-Triquet* 1063
- Fossil Fuel Combustion: Air Pollution and Global Warming / *Dan Golomb* 1127
- Giant Reed (*Arundo donax*): Streams and Water Resources / *Gretchen Coffman* 1182
- Groundwater: Contamination / *Charles W. Fetter* 1281
- Inland Seas and Lakes: Central Asia Case Study / *Andrey G. Kostianoy* 1436
- Methane Emissions: Rice / *Kazuyuki Yagi* 1679
- Mycotoxins / *J. David Miller* 1696
- Nanoparticles / *Alexandra Navrotsky* 1711
- Phosphorus: Agricultural Nutrient / *John Ryan, Hayriye Ibrikci, Rolf Sommer, and Abdul Rashid* 2091
- Pollutants: Organic and Inorganic / *A. Paul Schwab* 2114
- Pollution: Genotoxicity of Agrotoxic Compounds / *Vera Lucia S.S. de Castro and Paola Poli* 2123
- Pollution: Non-Point Source / *Ravendra Naidu, Mallavarapu Megharaj, Peter Dillon, Rai Kookana, Ray Correll, and W.W. Wenzel* 2136
- Potassium / *Philippe Hinsinger* 2208
- Rain Water: Atmospheric Deposition / *Zaneta Polkowska* 2249
- Rivers and Lakes: Acidification / *Agniezka Gałuszka and Zdzisław M. Migaszewski* 2291
- Rivers: Pollution / *Bogdan Skwarzec* 2303
- Sea: Pollution / *Bogdan Skwarzec* 2357
- Water Quality: Soil Erosion / *Andreas Klik* 2755

Global Pollution Problems (GLO)

Cleaner Technology (CLT)

Alternative Energy / *Bernd Markert, Simone Wuenschmann, Stefan Fraenzle, and Bernd Delakowitz* 179
Buildings: Climate Change / *Lisa Guan and Guangnan Chen* 435
Carbon Sequestration / *Nathan E. Hultman* 456
Energy Conservation / *Ibrahim Dincer and Adnan Midili* 677
Energy Conservation: Benefits / *Eric A. Woodroof, Wayne C. Turner, and Steven D. Heinz* 691
Energy Conservation: Industrial Processes / *Harvey E. Diamond* 697
Energy Efficiency: New and Emerging Technology / *Steven A. Parker* 742
Energy Master Planning / *Fredric S. Goldner* 763
Energy: Environmental Security / *Muhammad Asif* 798
Energy: Physics / *Milivoje M. Kostic* 808
Energy: Renewable / *John O. Blackburn* 824
Energy: Storage / *Rudolf Marloth* 853

Comparative Overviews (COV)

Agriculture: Energy Use and Conservation / *Guangnan Chen and Tek Narayan Maraseni* ... 118
Alternative Energy / *Bernd Markert, Simone Wuenschmann, Stefan Fraenzle, and Bernd Delakowitz* 179
Carbon Sequestration / *Nathan E. Hultman* 456
Energy: Environmental Security / *Muhammad Asif* 798
Energy: Physics / *Milivoje M. Kostic* 808
Energy: Renewable / *John O. Blackburn* 824
Energy: Storage / *Rudolf Marloth* 853
Energy: Waste Heat Recovery / *Martin A. Mozzo, Jr.* 869
Global Climate Change: Carbon Sequestration / *Sherwood Idso and Keith E. Idso* 1189
Global Climate Change: Earth System Response / *Amanda Staudt and Nathan E. Hultman* ... 1192
Global Climate Change: Gas Fluxes / *Pascal Boeckx and Oswald Van Cleemput* 1202
Global Climate Change: Gasoline, Hybrid-Electric and Hydrogen-Fueled Vehicles / *Robert E. Uhrig* 1206
Globalization / *Alexandru V. Roman* 1218
Green Processes: Systems Analysis / *Abhishek Tiwary* 1242
Oil Pollution: Baltic Sea / *Andrey G. Kostianoy* 1826
Organic Compounds: Halogenated / *Marek Tobiszewski and Jacek Namieśnik* 1841
Organic Matter: Global Distribution in World Ecosystems / *Wilfred M. Post* 1851
Organic Matter: Turnover / *Johan Six and Julie D. Jastrow* 1872
Ozone Layer / *Luisa T. Molina* 1882
Pesticides / *Marek Biziuk, Jolanta Fenik, Monika Kosikowska, and Maciej Tankiewicz* 1963
Pesticides: Effects / *Ana Maria Evangelista de Duffard* 2005
Rain Water: Atmospheric Deposition / *Zaneta Polkowska* 2249
Rivers and Lakes: Acidification / *Agniezka Gałuszka and Zdzisław M. Migaszewski* 2291
Rivers: Pollution / *Bogdan Skwarzec* 2303
Runoff Water / *Zaneta Polkowska* 2320
Sea: Pollution / *Bogdan Skwarzec* 2357

Global Pollution Problems (GLO) (cont'd)

Soil Degradation: Global Assessment / Selim Kapur, Suha Berberoğlu, Erhan Akça,
  Cenk Dönmez, Mehmet Akif Erdoğan, and Burçak Kapur..........2375
Water Quality and Quantity: Globalization / Kristi Denise Caravella and
  Jocilyn Danise Martinez..........2740
Water Quality: Range and Pasture Land / Thomas L. Thurow..........2752
Water Quality: Soil Erosion / Andreas Klik..........2755
Water: Drinking / Marek Biziuk and Małgorzata Michalska..........2790
Water: Surface / Victor de Vlaming..........2804
Water: Total Maximum Daily Load / Robin Kundis Craig..........2808
Wetlands / Ralph W. Tiner..........2824
Wetlands: Biodiversity / Jean-Claude Lefeuvre and Virginie Bouchard..........2829
Wetlands: Carbon Sequestration / Virginie Bouchard..........2833
Wetlands: Conservation Policy / Clayton Rubec..........2837

Diagnostic Tools (DIA)

Community-Based Monitoring: Ngarenanyuki, Tanzania / Aiwerasia V.F. Ngowi,
  Aiwerasia V.F. Ngowi, Larama MB Rongo, and Thomas J. Mbise..........505
Geographic Information System (GIS): Land Use Planning / Egide Nizeyimana and
  Jacob Opadeyi..........1163
Inorganic Carbon: Composition and Formation / Larry P. Wilding and H. Curtis Monger..........1444
Inorganic Carbon: Global Carbon Cycle / William H. Schlesinger..........1448
Inorganic Carbon: Modeling / Leslie D. McFadden and Ronald G. Amundson..........1451
Remote Sensing and GIS Integration / Egide Nizeyimana..........2271
Remote Sensing: Pollution / Massimo Antoninetti..........2275
Soil Degradation: Global Assessment / Selim Kapur, Suha Berberoğlu, Erhan Akça, Cenk
  Dönmez, Mehmet Akif Erdoğan, and Burçak Kapur..........2375
Sustainable Development: Ecological Footprint in Accounting / Simone Bastianoni,
  Valentina Niccolucci, Elena Neri, Gemma Cranston, Alessandro Galli, and Mathis Wackernagel..........2467
Water Quality and Quantity: Globalization / Kristi Denise Caravella and
  Jocilyn Danise Martinez..........2740
Water Quality: Modeling / Richard Lowrance..........2749

Ecotechnology (ECT)

Carbon Sequestration / Nathan E. Hultman..........456
Nutrients: Best Management Practices / Scott J. Sturgul and Keith A. Kelling..........1805
Nutrients: Bioavailability and Plant Uptake / Niels Erik Nielsen..........1817
Wetlands / Ralph W. Tiner..........2824
Wetlands: Biodiversity / Jean-Claude Lefeuvre and Virginie Bouchard..........2829
Wetlands: Carbon Sequestration / Virginie Bouchard..........2833
Wetlands: Conservation Policy / Clayton Rubec..........2837
Wetlands: Constructed Subsurface / Jan Vymazal..........2841
Wetlands: Sedimentation and Ecological Engineering / Timothy C. Granata and Jay F. Martin..........2859
Wetlands: Treatment System Use / Kyle R. Mankin..........2862

Environmental Legislation and Policy (ELE)

Climate Policy: International / Nathan E. Hultman..........483
Environmental Legislation: Asia / Wanpen Wirojanagud and Wanpen Wirojanagud..........874

Topical Table of Contents

Environmental Legislation: EU Countries Solid Waste Management / *Ni-Bin Chang, Ana Pires, and Graça Martinho* 892
Environmental Policy / *Sanford V. Berg* 914
Environmental Policy: Innovations / *Alka Sapat* 922

Integrated and Holistic Management (IMS)

Energy: Environmental Security / *Muhammad Asif* 798
Energy: Renewable / *John O. Blackburn* 824
Energy: Storage / *Rudolf Marloth* 853
Geographic Information System (GIS): Land Use Planning / *Egide Nizeyimana and Jacob Opadeyi* 1163
Global Climate Change: Carbon Sequestration / *Sherwood Idso and Keith E. Idso* 1189
Global Climate Change: Earth System Response / *Amanda Staudt and Nathan E. Hultman* 1192
Global Climate Change: Gas Fluxes / *Pascal Boeckx and Oswald Van Cleemput* 1202
Globalization / *Alexandru V. Roman* 1218
Green Processes: Systems Analysis / *Abhishek Tiwary* 1242
Oil Pollution: Baltic Sea / *Andrey G. Kostianoy* 1826
Organic Matter: Management / *R. Cesar Izaurralde and Carlos C. Cerri* 1857
Organic Matter: Turnover / *Johan Six and Julie D. Jastrow* 1872
Sustainability and Planning / *Richard Cowell* 2446
Sustainable Development / *Mark A. Peterson* 2461
Sustainable Development: Ecological Footprint in Accounting / *Simone Bastianoni, Valentina Niccolucci, Elena Neri, Gemma Cranston, Alessandro Galli, and Mathis Wackernagel* 2467

Pollution Problems and Their Sources (PSS)

Erosion by Wind: Climate Change / *Alan Busacca and David Chandler* 1004
Erosion by Wind: Global Hot Spots / *Andrew Warren* 1010
Fossil Fuel Combustion: Air Pollution and Global Warming / *Dan Golomb* 1127
Pesticides / *Marek Biziuk, Jolanta Fenik, Monika Kosikowska, and Maciej Tankiewicz* 1963
Pesticides: Effects / *Ana Maria Evangelista de Duffard* 2005
Rain Water: Atmospheric Deposition / *Zaneta Polkowska* 2249
Rivers and Lakes: Acidification / *Agniezka Gałuszka and Zdzisław M. Migaszewski* 2291
Rivers: Pollution / *Bogdan Skwarzec* 2303
Sea: Pollution / *Bogdan Skwarzec* 2357
Soil Degradation: Global Assessment / *Selim Kapur, Suha Berberoğlu, Erhan Akça, Cenk Dönmez, Mehmet Akif Erdoğan, and Burçak Kapur* 2375
Water Quality and Quantity: Globalization / *Kristi Denise Caravella and Jocilyn Danise Martinez* 2740
Water Quality: Range and Pasture Land / *Thomas L. Thurow* 2752
Water Quality: Soil Erosion / *Andreas Klik* 2755
Water: Drinking / *Marek Biziuk and Małgorzata Michalska* 2790
Water: Surface / *Victor de Vlaming* 2804

Terrestrial Pollution Problems (TER)

Cleaner Technology (CLT)

Acid Sulfate Soils: Management / *Michael D. Melville and Ian White* 55
Pesticide Translocation Control: Soil Erosion / *Monika Frielinghaus, Detlef Deumlich, and Roger Funk* 1955
Salt-Affected Soils: Sustainable Agriculture / *Pichu Rengasamy* 2349
Sustainable Agriculture: Soil Quality / *John W. Doran and Ed G. Gregorich* 2457

Terrestrial Pollution Problems (TER) (*cont'd*)

Comparative Overviews (COV)

Agricultural Soils: Carbon and Nitrogen Biological Cycling / *Alan J. Franzluebbers* 88
Erosion by Wind: Principles / *Larry D. Stetler* 1013
Erosion: History / *Andrew S. Goudie* 1050
Erosion: Soil Quality / *Craig Ditzler* 1060
Global Climate Change: World Soils / *Rattan Lal* 1213
Global Climate Change: World Soils / *Rattan Lal* 1213
Industrial Waste: Soil Pollution / *W. Friesl-Hanl, M.H. Gerzabek, W.W. Wenzel, and W.E.H. Blum* 1430
Salt-Affected Soils: Physical Properties and Behavior / *Hwat Bing So* 2334
Salt-Affected Soils: Plant Response / *Anna Eynard, Keith D. Wiebe, and Rattan Lal* 2345
Sodic Soils: Properties / *Pichu Rengasamy* 2367
Soil Degradation: Global Assessment / *Selim Kapur, Suha Berberoğlu, Erhan Akça, Cenk Dönmez, Mehmet Akif Erdoğan, and Burçak Kapur* 2375
Soil Quality: Carbon and Nitrogen Gases / *Philippe Rochette, Sean McGinn, and Reynald Lemke* 2388
Solid Waste: Municipal / *Angelique Chettiparamb* 2415

Diagnostic Tools (DIA)

Acid Sulfate Soils: Identification, Assessment, and Management / *L.A. Sullivan, R.T. Bush, and E.D. Burton* 36
Drainage: Soil Salinity Management / *Glenn J. Hoffman* 588
Erosion by Water: Assessment and Control / *José Miguel Reichert, Nadia Bernardi Bonumá, Gustavo Enrique Merten, and Jean Paolo Gomes Minella* 963
Erosion by Water: Empirical Methods / *John M. Laflen and John M. Laflen* 974
Erosion by Water: Erosivity and Erodibility / *Peter I.A. Kinnell* 980
Erosion by Water: Process-Based Modeling / *Mark A. Nearing* 991
Erosion by Wind: Source, Measurement, Prediction, and Control / *Brenton S. Sharratt and R. Scott Van Pelt* 1017
Soil Quality: Indicators / *Graham P. Sparling* 2393
Solid Waste Management: Life Cycle Assessment / *Ni-Bin Chang, Ana Pires, and Graça Martinho* 2399

Ecotechnology (ECT)

Acid Sulfate Soils: Management / *Michael D. Melville and Ian White* 55
Bioremediation: Contaminated Soil Restoration / *Sven Erik Jørgensen* 401
Composting / *Nídia Sá Caetano* 512
Composting: Organic Farming / *Saburo Matsui* 528
Desertification: Prevention and Restoration / *Claudio Zucca, Susana Bautista, Barron Orr, and Franco Previtali* 557
Erosion and Sediment Control: Vegetative Techniques / *Samson D. Angima* 947
Erosion by Water: Amendment Techniques / *X.-C. (John) Zhang* 958
Erosion by Water: Vegetative Control / *Seth M. Dabney and Silvio J. Gumiere* 994
Erosion Control: Soil Conservation / *Eric T. Craswell* 1034
Erosion Control: Tillage and Residue Methods / *Richard Cruse, Jerry Neppel, John Kost, and Krisztina Eleki* 1040
Organic Soil Amendments / *Philip Oduor-Owino* 1878

Topical Table of Contents

Pesticide Translocation Control: Soil Erosion / *Monika Frielinghaus, Detlef Deumlich, and Roger Funk* 1955
Sodic Soils: Reclamation / *Jock Churchman* 2370
Soil Rehabilitation / *David Jasper* 2396
Sustainable Agriculture: Soil Quality / *John W. Doran and Ed G. Gregorich* 2457
Tillage Erosion: Terrace Formation / *Seth M. Dabney and Dalmo A.N. Vieira* 2536

Environmental Legislation and Policy (ELE)

Laws and Regulations: Soil / *Ian Hannam and Ben Boer* 1618

Environmental Technology (ENT)

Tillage Erosion: Terrace Formation / *Seth M. Dabney and Dalmo A.N. Vieira* 2536

Integrated and Holistic Management (IMS)

Integrated Farming Systems / *John Holland* 1507
Integrated Nutrient Management / *Bal Ram Singh* 1510
Integrated Weed Management / *Heinz Müller-Schärer and Alexandra Robin Collins* 1524
Solid Waste Management: Life Cycle Assessment / *Ni-Bin Chang, Ana Pires, and Graça Martinho* 2399

Pollution Problems and Their Sources (PSS)

Acid Sulfate Soils / *Delvin S. Fanning* 26
Acid Sulfate Soils: Formation / *Martin C. Rabenhorst, Delvin S. Fanning, and Steven N. Burch* 31
Agricultural Soils: Ammonia Volatilization / *Paul G. Saffigna and John R. Freney* 85
Agricultural Soils: Nitrous Oxide Emissions / *John R. Freney* 96
Agricultural Soils: Phosphorus / *Anja Gassner and Ewald Schnug* 100
Boron: Soil Contaminant / *Rami Keren* 431
Carbon: Soil Inorganic / *Donald L. Suarez* 462
Desertification / *David Tongway and John Ludwig* 541
Desertification: Extent of / *Victor R. Squires* 545
Desertification: Greenhouse Effect / *Sherwood Idso and Craig Idso* 549
Desertification: Impact / *David A. Mouat and Judith Lancaster* 552
Desertization / *Henry Noel LeHouerou* 565
Erosion / *Dennis C. Flanagan* 930
Erosion and Carbon Dioxide / *Pierre A. Jacinthe and Rattan Lal* 934
Erosion and Global Change / *Taolin Zhang and Xingxiang Wang* 939
Erosion and Precipitation / *Bofu Yu* 943
Erosion by Water: Accelerated / *David Favis-Mortlock* 951
Erosion by Wind: Climate Change / *Alan Busacca and David Chandler* 1004
Erosion by Wind: Global Hot Spots / *Andrew Warren* 1010
Erosion by Wind-Driven Rain / *Gunay Erpul, L. Darrell Norton, and Donald Gabriels* 1031
Erosion: Accelerated / *J. Kent Mitchell and Michael C. Hirschi* 1044
Erosion: Irrigation-Induced / *Robert E. Sojka and David L. Bjorneberg* 1053
Erosion: Snowmelt / *Donald K. McCool* 1057
Irrigation: Soil Salinity / *James D. Rhoades* 1572
Leaching / *Lars Bergström* 1621
Leaching and Illuviation / *Lynn E. Moody* 1624
Sodic Soils: Irrigation Farming / *David Burrow* 2364
Solid Waste: Municipal / *Angelique Chettiparamb* 2415

Toxic Substances in the Environment (TOX)

Cleaner Technology (CLT)

Bacterial Pest Control / *David N. Ferro* 321
Chemigation / *William L. Kranz* 469
Invasion Biology / *Jennifer Ruesink* 1531
Nanotechnology: Environmental Abatement / *Puangrat Kajitvichyanukul and Jirapat Ananpattarachai* 1730
Pest Management: Crop Diversity / *Marianne Karpenstein-Machan and Maria Finckh* 1925
Pest Management: Ecological Agriculture / *Barbara Dinham* 1930
Pest Management: Ecological Aspects / *David J. Horn* 1934
Pest Management: Intercropping / *Maria Finckh and Marianne Karpenstein-Machan* 1937
Pesticide Translocation Control: Soil Erosion / *Monika Frielinghaus, Detlef Deumlich, and Roger Funk* 1955
Pesticide Translocation Control: Soil Erosion / *Monika Frielinghaus, Detlef Deumlich, and Roger Funk* 1955
Pesticides: Damage Avoidance / *Aiwerasia V.F. Ngowi, Aiwerasia V.F. Ngowi, and Larama MB Rongo* 1997
Pesticides: Reducing Use / *David Pimentel and Maria V. Cilveti* 2033
Stored-Product Pests: Biological Control / *Lise Stengård Hansen* 2423

Comparative Overviews (COV)

Acaricides / *Doug Walsh* 1
Aluminum / *Johannes Bernhard Wehr, Frederick Paxton Cardell Blamey, and Neal William Menzies* 267
Antagonistic Plants / *Philip Oduor-Owino* 288
Bioaccumulation / *Tomaz Langenbach* 324
Biodegradation / *Sven Erik Jørgensen* 328
Biodiversity and Sustainability / *Odo Primavesi* 333
Biofertilizers / *J. R. de Freitas* 349
Biological Controls / *Heikki Hokkanen* 366
Biological Controls: Conservation / *Doug Landis and Steve D. Wratten* 370
Biomass / *Alberto Traverso and David Tucker* 373
Bioremediation / *Ragini Gothalwal* 387
Chemigation / *William L. Kranz* 469
Industrial Waste: Soil Pollution / *W. Friesl-Hanl, M.H. Gerzabek, W.W. Wenzel, and W.E.H. Blum* 1430
Inorganic Carbon: Composition and Formation / *Larry P. Wilding and H. Curtis Monger* 1444
Inorganic Carbon: Global Carbon Cycle / *William H. Schlesinger* 1448
Inorganic Compounds: Eco-Toxicity / *Sven Erik Jørgensen* 1455
Invasion Biology / *Jennifer Ruesink* 1531
Mercury / *Sven Erik Jørgensen* 1676
Nanoparticles / *Alexandra Navrotsky* 1711
Nanotechnology: Environmental Abatement / *Puangrat Kajitvichyanukul and Jirapat Ananpattarachai* 1730
Persistent Organic Pesticides / *Gamini Manuweera* 1913
Pest Management / *E.F. Legner* 1919
Pesticides / *Marek Biziuk, Jolanta Fenik, Monika Kosikowska, and Maciej Tankiewicz* 1963
Pesticides: Chemical and Biological / *Barbara Manachini* 1986

Pesticides: Damage Avoidance / *Aiwerasia V.F. Ngowi, Aiwerasia V.F. Ngowi, and Larama MB Rongo*	1997
Pesticides: Effects / *Ana Maria Evangelista de Duffard*	2005
Pesticides: History / *Edward H. Smith and George Kennedy*	2008
Toxic Substances / *Sven Erik Jørgensen*	2542
Toxic Substances: Photochemistry / *Puangrat Kajitvichyanukul*	2547
Toxicity Prediction of Chemical Mixtures / *Nina Cedergreen, Claus Svendsen, and Thomas Backhaus*	2572
Water Supplies: Pharmaceuticals / *Klaus Kümmerer*	2776

Diagnostic Tools (DIA)

Animals: Toxicological Evaluation / *Vera Lucia S.S. de Castro*	280
Birds: Chemical Control / *Eric B. Spurr*	413
Inorganic Carbon: Composition and Formation / *Larry P. Wilding and H. Curtis Monger*	1444
Inorganic Carbon: Global Carbon Cycle / *William H. Schlesinger*	1448
Inorganic Carbon: Modeling / *Leslie D. McFadden and Ronald G. Amundson*	1451
Nanomaterials: Regulation and Risk Assessment / *Steffen Foss Hansen, Khara D. Grieger, and Anders Baun*	1700
Nanoparticles: Uncertainty Risk Analysis / *Khara D. Grieger, Steffen Foss Hansen, and Anders Baun*	1720
Neurotoxicants: Developmental Experimental Testing / *Vera Lucia S.S. de Castro*	1755
Organic Matter: Management / *R. Cesar Izaurralde and Carlos C. Cerri*	1857
Organic Matter: Modeling / *Pete Smith*	1863
Organic Matter: Turnover / *Johan Six and Julie D. Jastrow*	1872
Pest Management: Modeling / *Andrew Paul Gutierrez*	1949
Pesticides: Measurement and Mitigation / *George F. Antonious*	2013
Polychlorinated Biphenyls (PCBs) and Polycyclic Aromatic Hydrocarbons (PAHs): Sediments and Water Analysis / *Justyna Rogowska, Agata Mechlińska, Lidia Wolska, Lidia Wolska, and Jacek Namieśnik*	2186

Ecotechnology (ECT)

Biological Controls / *Heikki Hokkanen*	366
Biological Controls: Conservation / *Doug Landis and Steve D. Wratten*	370
Bioremediation / *Ragini Gothalwal*	387
Bioremediation: Contaminated Soil Restoration / *Sven Erik Jørgensen*	401
Invasion Biology / *Jennifer Ruesink*	1531
Natural Enemies: Conservation / *Cetin Sengonca*	1749
Nematodes: Biological Control / *Simon Gowen*	1752
Pesticide Translocation Control: Soil Erosion / *Monika Frielinghaus, Detlef Deumlich, and Roger Funk*	1955
Stored-Product Pests: Biological Control / *Lise Stengård Hansen*	2423
Toxic Substances / *Sven Erik Jørgensen*	2542
Vertebrates: Biological Control / *Peter Kerr and Tanja Strive*	2595
Wetlands / *Ralph W. Tiner*	2824
Wetlands: Petroleum / *John V. Headley and Kerry M. Peru*	2854

Environmental Legislation and Policy (ELE)

Developing Countries: Pesticide Health Impacts / *Aiwerasia V.F. Ngowi, Aiwerasia V.F. Ngowi, Catharina Wesseling, and Leslie London*	573
Farming: Non-Chemical and Pesticide-Free (European Council Regulations [EED/209/91]) / *Inger Källander and Gunnar Rundgren*	1103

Toxic Substances in the Environment (TOX) (cont'd)

- Food Quality Protection Act / Christina D. DiFonzo 1118
- Food: Cosmetic Standards / David Pimentel and Kelsey Hart 1120
- Food: Pesticide Contamination / Denis Hamilton 1124
- Human Health: Cancer from Pesticides / Jan Dich 1394
- Human Health: Chronic Pesticide Poisonings / Birgitta Kolmodin-Hedman and Margareta Palmborg 1397
- Human Health: Consumer Concerns of Pesticides / George Ekström and Margareta Palmborg 1401
- Laws and Regulations: Pesticides / Praful Suchak 1612
- Lead: Regulations / Lisa A. Robinson 1636
- Nanomaterials: Regulation and Risk Assessment / Steffen Foss Hansen, Khara D. Grieger, and Anders Baun 1700
- Pest Management: Legal Aspects / Michael T. Olexa and Zachary T. Broome 1940
- Pesticides: Banding / Sharon A. Clay 1983
- Pesticides: Damage Avoidance / Aiwerasia V.F. Ngowi, Aiwerasia V.F. Ngowi, and Larama MB Rongo 1997
- Pesticides: Regulating / Matthias Kern 2035

Environmental Technology (ENT)

- Biopesticides / G.J. Ash and A. Wang 381
- Biotechnology: Pest Management / Maurizio G. Paoletti 407
- Farming: Organic Pest Management / Ján Gallo 1115
- Lead: Ecotoxicology / Sven Erik Jørgensen 1630
- Lead: Regulations / Lisa A. Robinson 1636
- Minerals Processing Residue (Tailings): Rehabilitation / Lloyd R. Hossner and Hamid Shahandeh 1683
- Mines: Rehabilitation of Open Cut / Douglas J. Dollhopf 1692
- Nanotechnology: Environmental Abatement / Puangrat Kajitvichyanukul and Jirapat Ananpattarachai 1730
- Persistent Organic Compounds: Wet Oxidation Removal / Anurag Garg and I.M. Mishra 1904
- Pesticides: Natural / Franck Dayan 2028
- Pests: Landscape Patterns / F. Craig Stevenson 2038
- Pharmaceuticals: Treatment / Diana Aga and Seungyun Baik 2060

Integrated and Holistic Management (IMS)

- Integrated Pest Management / H. F. van Emden 1521
- Nanomaterials: Regulation and Risk Assessment / Steffen Foss Hansen, Khara D. Grieger, and Anders Baun 1700
- Nanoparticles: Uncertainty Risk Analysis / Khara D. Grieger, Steffen Foss Hansen, and Anders Baun 1720
- Organic Matter: Global Distribution in World Ecosystems / Wilfred M. Post 1851
- Organic Matter: Management / R. Cesar Izaurralde and Carlos C. Cerri 1857
- Organic Matter: Turnover / Johan Six and Julie D. Jastrow 1872
- Pest Management / E.F. Legner 1919

Pollution Problems and Their Sources (PSS): Generally Point Sources

- Aluminum / Johannes Bernhard Wehr, Frederick Paxton Cardell Blamey, and Neal William Menzies 267
- Cadmium and Lead: Contamination / Gabriella Kakonyi and Imad A.M. Ahmed 446
- Cadmium: Toxicology / Sven Erik Jørgensen 453

Chromium / *Bruce R. James and Dominic A. Brose* 477
Cobalt and Iodine / *Ronald G. McLaren* 502
Copper / *David R. Parker and Judith F. Pedler* 535
Endocrine Disruptors / *Vera Lucia S.S. de Castro* 643
Heavy Metals: Organic Fertilization Uptake / *Ewald Schnug, Alexandra Izosimova, and Renata Gaj* 1374
Human Health: Cancer from Pesticides / *Jan Dich* 1394
Human Health: Chronic Pesticide Poisonings / *Birgitta Kolmodin-Hedman and Margareta Palmborg* 1397
Human Health: Consumer Concerns of Pesticides / *George Ekström and Margareta Palmborg* 1401
Human Health: Hormonal Disruption / *Evamarie Straube, Sebastian Straube, and Wolfgang Straube* 1405
Human Health: Pesticide Poisonings / *Margareta Palmborg* 1408
Human Health: Pesticide Sensitivities / *Ann McCampbell* 1411
Human Health: Pesticides / *Kelsey Hart, David Pimentel, and David Pimentel* 1414
Lead: Regulations / *Lisa A. Robinson* 1636
Mercury / *Sven Erik Jørgensen* 1676
Methane Emissions: Rice / *Kazuyuki Yagi* 1679
Mycotoxins / *J. David Miller* 1696
Nanoparticles / *Alexandra Navrotsky* 1711
Neurotoxicants: Developmental Experimental Testing / *Vera Lucia S.S. de Castro* 1755
Organic Compounds: Halogenated / *Marek Tobiszewski and Jacek Namieśnik* 1841
Ozone Layer / *Luisa T. Molina* 1882
Persistent Organic Pesticides / *Gamini Manuweera* 1913
Petroleum: Hydrocarbon Contamination / *Svetlana Drozdova and Erwin Rosenberg* 2040
Phenols / *Leszek Wachowski and Robert Pietrzak* 2071
Polychlorinated Biphenyls (PCBs) / *Marek Biziuk and Angelika Beyer* 2172
Potassium / *Philippe Hinsinger* 2208
Radioactivity / *Bogdan Skwarzec* 2234
Radionuclides / *Philip M. Jardine* 2243
Rare Earth Elements / *Zhengyi Hu, Gerd Sparovek, Silvia Haneklaus, and Ewald Schnug* 2266
Strontium / *Silvia Haneklaus and Ewald Schnug* 2426
Sulfur / *Ewald Schnug, Silvia Haneklaus, and Elke Bloem* 2431
Sulfur Dioxide / *Marianna Czaplicka, Marianna Czaplicka, and Witold Kurylak* 2437
Toxic Substances / *Sven Erik Jørgensen* 2542
Toxic Substances: Photochemistry / *Puangrat Kajitvichyanukul* 2547
Vanadium and Chromium Groups / *Imad A.M. Ahmed* 2582
Water Supplies: Pharmaceuticals / *Klaus Kümmerer* 2776

Pollution Problems and Their Sources (PSS): Generally Non-Point and Diffuse Sources

Acaricides / *Doug Walsh* 1
Animals: Sterility from Pesticides / *William Au* 277
Animals: Toxicological Evaluation / *Vera Lucia S.S. de Castro* 280
Antagonistic Plants / *Philip Oduor-Owino* 288
Aquatic Communities: Pesticide Impacts / *David P. Kreutzweiser and Paul K. Sibley* 291
Bioaccumulation / *Tomaz Langenbach* 324
Birds: Pesticide Use Impacts / *Pierre Mineau* 416
Boron and Molybdenum: Deficiencies and Toxicities / *Umesh C. Gupta* 424

Toxic Substances in the Environment (TOX) (*cont'd*)

    Boron: Soil Contaminant / *Rami Keren* 431
    Carbon: Soil Inorganic / *Donald L. Suarez* 462
    Developing Countries: Pesticide Health Impacts / *Aiwerasia V.F. Ngowi, Aiwerasia V.F. Ngowi,*
        Catharina Wesseling, and Leslie London 573
    Food: Pesticide Contamination / *Denis Hamilton* 1124
    Groundwater: Pesticide Contamination / *Roy F. Spalding* 1317
    Herbicides / *Malcolm Devine* 1378
    Herbicides: Non-Target Species Effects / *Céline Boutin* 1382
    Inorganic Carbon: Composition and Formation / *Larry P. Wilding and H. Curtis Monger* 1444
    Inorganic Compounds: Eco-Toxicity / *Sven Erik Jørgensen* 1455
    Insect Growth Regulators / *Meir Paul Pener* 1459
    Insecticides: Aerial Ultra-Low-Volume Application / *He Zhong* 1471
    Lead: Ecotoxicology / *Sven Erik Jørgensen* 1630
    Oil Pollution: Baltic Sea / *Andrey G. Kostianoy* 1826
    Pesticides / *Marek Biziuk, Jolanta Fenik, Monika Kosikowska, and Maciej Tankiewicz* 1963
    Pollutants: Organic and Inorganic / *A. Paul Schwab* 2114
    Pollution: Genotoxicity of Agrotoxic Compounds / *Vera Lucia S.S. de Castro and Paola Poli* 2123
    Pollution: Pesticides in Agro-Horticultural Ecosystems / *J.K. Dubey and Meena Thakur* 2139
    Pollution: Pesticides in Natural Ecosystems / *J.K. Dubey and Meena Thakur* 2145
    Toxicity Prediction of Chemical Mixtures / *Nina Cedergreen, Claus Svendsen, and*
        Thomas Backhaus 2572

Water Pollution Problems (WAT)

Cleaner Technology (CLT)

    Agricultural Water Quantity Management / *X.S. Qin and Y. Xu* 105
    Irrigation Systems: Water Conservation / *I. Pai Wu, Javier Barragan, and Vince Bralts* 1542
    Irrigation: Sewage Effluent Use / *B.A. Stewart* 1570
    Precision Agriculture: Water and Nutrient Management / *Robert J. Lascano and J.D. Booker* 2217
    Wastewater and Water Utilities / *Rudolf Marloth* 2638
    Wastewater Use in Agriculture / *Manzoor Qadir, Pay Drechsel, and*
        Liqa Raschid-Sally 2675
    Wastewater Use in Agriculture: Policy Issues / *Dennis Wichelns* 2681
    Water Harvesting / *Gary W. Frasier* 2738

Comparative Overviews (COV)

    Adsorption / *Puangrat Kajitvichyanukul and Jirapat Ananpattarachai* 61
    Agricultural Runoff / *Matt C. Smith, David K. Gattie, and Daniel L. Thomas* 81
    Agricultural Water Quantity Management / *X.S. Qin and Y. Xu* 105
    Agroforestry: Water Use Efficiency / *James R. Brandle, Laurie Hodges, and Xinhua Zhou* 129
    Allelochemics / *John Borden* 175
    Aquatic Communities: Pesticide Impacts / *David P. Kreutzweiser and Paul K. Sibley* 291
    Coastal Water: Pollution / *Piotr Szefer* 489
    Cyanobacteria: Eutrophic Freshwater Systems / *Anja Gassner and Martin V. Frey* 538
    Estuaries / *Claude Amiard-Triquet* 1063
    Giant Reed (*Arundo donax*): Streams and Water Resources / *Gretchen Coffman* 1182

Groundwater: Contamination / *Charles W. Fetter*	1281
Groundwater: Mining / *Hugo A. Loaiciga*	1284
Groundwater: Mining Pollution / *Jeffrey G. Skousen and George F. Vance*	1289
Inland Seas and Lakes: Central Asia Case Study / *Andrey G. Kostianoy*	1436
Irrigation Systems: Sub-Surface History / *Norman R. Fausey*	1539
Irrigation Systems: Water Conservation / *I. Pai Wu, Javier Barragan, and Vince Bralts*	1542
Irrigation: Efficiency / *Terry A. Howell*	1545
Irrigation: Erosion / *David L. Bjorneberg*	1551
Irrigation: Return Flow and Quality / *Ramón Aragues and Kenneth K. Tanji*	1557
Lakes and Reservoirs: Pollution / *Subhankar Karmakar and O.M. Musthafa*	1576
Rain Water: Atmospheric Deposition / *Zaneta Polkowska*	2249
Rivers and Lakes: Acidification / *Agniezka Gałuszka and Zdzisław M. Migaszewski*	2291
Rivers: Pollution / *Bogdan Skwarzec*	2303
Rivers: Restoration / *Anna Rabajczyk*	2307
Runoff Water / *Zaneta Polkowska*	2320
Sea: Pollution / *Bogdan Skwarzec*	2357
Wastewater and Water Utilities / *Rudolf Marloth*	2638
Wastewater Treatment: Conventional Methods / *Sven Erik Jørgensen*	2657
Wastewater Use in Agriculture / *Manzoor Qadir, Manzoor Qadir, Pay Drechsel, and Liqa Raschid-Sally*	2675
Wastewater Use in Agriculture: Public Health Considerations / *Blanca Jimenez and Inés Navarro*	2694
Wastewater: Municipal / *Sven Erik Jørgensen*	2709
Water Quality and Quantity: Globalization / *Kristi Denise Caravella and Jocilyn Danise Martinez*	2740
Water Quality: Range and Pasture Land / *Thomas L. Thurow*	2752
Water Quality: Soil Erosion / *Andreas Klik*	2755
Water: Cost / *Atif Kubursi and Matthew Agarwala*	2779
Water: Drinking / *Marek Biziuk and Małgorzata Michalska*	2790
Water: Surface / *Victor de Vlaming*	2804
Water: Total Maximum Daily Load / *Robin Kundis Craig*	2808

Diagnostic Tools (DIA)

Groundwater: Modeling / *Jesus Carrera*	1295
Groundwater: Numerical Method Modeling / *Jesus Carrera*	1312
Polychlorinated Biphenyls (PCBs) and Polycyclic Aromatic Hydrocarbons (PAHs): Sediments and Water Analysis / *Justyna Rogowska, Agata Mechlińska, Lidia Wolska, Lidia Wolska, and Jacek Namieśnik*	2186
Water Quality: Modeling / *Richard Lowrance*	2749

Ecotechnology (ECT)

Agricultural Water Quantity Management / *X.S.Q. and Y. Xu*	105
Agroforestry: Water Use Efficiency / *James R. Brandle, Laurie Hodges, and Xinhua Zhou*	129
Alexandria Lake Maryut: Integrated Environmental Management / *Lindsay Beevers*	164
Groundwater: Treatment with Biobarrier Systems / *Chih-Ming Kao, Shu-Hao Liang, Yu-Chia Kuo, and R.Y. Surampalli*	1333
Irrigation Systems: Sub-Surface Drip Design / *Carl R. Camp, Jr. and Freddie L. Lamm*	1535
Irrigation: Sewage Effluent Use / *B.A. Stewart*	1570
Lakes: Restoration / *Anna Rabajczyk*	1588
Rain Water: Harvesting / *K.F. Andrew Lo*	2262

Water Pollution Problems (WAT) (cont'd)

Rivers: Restoration / *Anna Rabajczyk*..2307
Waste: Stabilization Ponds / *Sven Erik Jørgensen*..2632
Wastewater Treatment: Wetlands Use in Arctic Regions / *Colin N. Yates, Brent Wootton, Sven Erik Jørgensen, and Stephen D. Murphy*..2662
Wastewater Use in Agriculture / *Manzoor Qadir, Manzoor Qadir, Pay Drechsel, and Liqa Raschid-Sally*..2675
Water and Wastewater: Filters / *Sandeep Joshi*..2719
Water and Wastewater: Ion Exchange Application / *Sven Erik Jørgensen*..2734
Water Harvesting / *Gary W. Frasier*..2738
Wetlands / *Ralph W. Tiner*...2824
Wetlands: Biodiversity / *Jean-Claude Lefeuvre and Virginie Bouchard*...2829
Wetlands: Carbon Sequestration / *Virginie Bouchard*..2833
Wetlands: Conservation Policy / *Clayton Rubec*..2837
Wetlands: Constructed Subsurface / *Jan Vymazal*..2841
Wetlands: Methane Emission / *Anna Ekberg and Torben Røjle Christensen*.....................................2850
Wetlands: Petroleum / *John V. Headley and Kerry M. Peru*..2854
Wetlands: Sedimentation and Ecological Engineering / *Timothy C. Granata and Jay F. Martin*......2859
Wetlands: Treatment System Use / *Kyle R. Mankin*..2862

Environmental Legislation and Policy (ELE)

Wastewater Use in Agriculture: Policy Issues / *Dennis Wichelns*..2681
Wastewater Use in Agriculture: Public Health Considerations / *Blanca Jimenez and Inés Navarro*..................2694
Water: Cost / *Atif Kubursi and Matthew Agarwala*...2779

Environmental Technology (ENT)

Adsorption / *Puangrat Kajitvichyanukul and Jirapat Ananpattarachai*...61
Wastewater Treatment: Biological / *Shaikh Ziauddin Ahammad, David W. Graham and Jan Dolfing*........2645
Wastewater Treatment: Conventional Methods / *Sven Erik Jørgensen*...2657
Wastewater: Municipal / *Sven Erik Jørgensen*..2709
Water and Wastewater: Filters / *Sandeep Joshi*..2719
Water and Wastewater: Ion Exchange Application / *Sven Erik Jørgensen*..2734
Water Harvesting / *Gary W. Frasier*..2738

Integrated and Holistic Management (IMS)

Alexandria Lake Maryut: Integrated Environmental Management / *Lindsay Beevers*.........................164
Chesapeake Bay / *Sean M. Smith*..474
Eutrophication / *Claude Amiard-Triquet and Sven Erik Jørgensen*...1074
Everglades / *Kenneth L. Campbell, Rafael Munoz-Carpena, and Gregory Kiker*...........................1080
Wastewater and Water Utilities / *Rudolf Marloth*..2638
Wastewater Use in Agriculture / *Manzoor Qadir, Manzoor Qadir, Pay Drechsel, and Liqa Raschid-Sally*..2675
Wastewater Use in Agriculture: Policy Issues / *Dennis Wichelns*..2681
Wastewater Use in Agriculture: Public Health Considerations / *Blanca Jimenez and Inés Navarro*..................2694
Water: Cost / *Atif Kubursi and Matthew Agarwala*...2779
Watershed Management: Remote Sensing and GIS / *A.V. Shanwal and S.P. Singh*..........................2816
Wetlands: Biodiversity / *Jean-Claude Lefeuvre and Virginie Bouchard*...2829
Wetlands: Carbon Sequestration / *Virginie Bouchard*..2833
Yellow River / *Zixi Zhu, Ynuzhang Wang, and Yifei Zhu*..2884

Pollution Problems and Their Sources (PSS): Generally Point Sources

- Aral Sea Disaster / *Guy Fipps* 300
- Groundwater: Mining / *Hugo A. Loaiciga* 1284
- Groundwater: Mining Pollution / *Jeffrey G. Skousen and George F. Vance* 1289
- Irrigation: Erosion / *David L. Bjorneberg* 1551
- Irrigation: Return Flow and Quality / *Ramón Aragues and Kenneth K. Tanji* 1557
- Irrigation: River Flow Impact / *Robert W. Hill and Ivan A. Walter* 1563
- Irrigation: Saline Water / *B.A. Stewart* 1568
- Wastewater Use in Agriculture / *Manzoor Qadir, Manzoor Qadir, Pay Drechsel, and Liqa Raschid-Sally* 2675
- Wastewater Use in Agriculture: Public Health Considerations / *Blanca Jimenez and Inés Navarro* 2694
- Wastewater: Municipal / *Sven Erik Jørgensen* 2709
- Water Quality: Timber Harvesting / *C. Rhett Jackson* 2770
- Water Supplies: Pharmaceuticals / *Klaus Kümmerer* 2776

Pollution Problems and Their Sources (PSS): Generally Non-Point and Diffuse Sources

- Acid Rain / *Umesh Kulshrestha* 5
- Acid Rain: Nitrogen Deposition / *George F. Vance* 20
- Alexandria Lake Maryut: Integrated Environmental Management / *Lindsay Beevers* 164
- Aquatic Communities: Pesticide Impacts / *David P. Kreutzweiser and Paul K. Sibley* 291
- Chesapeake Bay / *Sean M. Smith* 474
- Coastal Water: Pollution / *Piotr Szefer* 489
- Cyanobacteria: Eutrophic Freshwater Systems / *Anja Gassner and Martin V. Frey* 538
- Drainage: Hydrological Impacts Downstream / *Mark Robinson and D.W. Rycroft* 584
- Estuaries / *Claude Amiard-Triquet* 1063
- Eutrophication / *Claude Amiard-Triquet and Sven Erik Jørgensen* 1074
- Everglades / *Kenneth L. Campbell, Rafael Munoz-Carpena, and Gregory Kiker* 1080
- Giant Reed (*Arundo donax*): Streams and Water Resources / *Gretchen Coffman* 1182
- Groundwater: Arsenic Contamination / *Abhijit Das, Bhaskar Das, Subhas Chandra Mukherjee, Shyamapada Pati, Rathindra Nath Dutta, Khitish Chandra Saha, Quazi Quamruzzaman, Mohammad Mahmudur Rahman, Tarit Roy Chowdhury, and Dipankar Chakraborti* 1262
- Groundwater: Contamination / *Charles W. Fetter* 1281
- Groundwater: Nitrogen Fertilizer Contamination / *Lloyd B. Owens and Douglas L. Karlen* ...1302
- Groundwater: Pesticide Contamination / *Roy F. Spalding* 1317
- Groundwater: Saltwater Intrusion / *Alexander H.-D. Cheng* 1321
- Inland Seas and Lakes: Central Asia Case Study / *Andrey G. Kostianoy* 1436
- Lakes and Reservoirs: Pollution / *Subhankar Karmakar and O.M. Musthafa* 1576
- Lakes: Restoration / *Anna Rabajczyk* 1588
- Mines: Acidic Drainage Water / *Jerry M. Bigham and Wendy B. Gagliano* 1688
- Nutrient—Water Interactions / *Ardell D. Halvorson* 1823
- Rain Water: Atmospheric Deposition / *Zaneta Polkowska* 2249
- Rivers and Lakes: Acidification / *Agniezka Gałuszka and Zdzisław M. Migaszewski* 2291
- Rivers: Pollution / *Bogdan Skwarzec* 2303
- Sea: Pollution / *Bogdan Skwarzec* 2357
- Water Quality and Quantity: Globalization / *Kristi Denise Caravella and Jocilyn Danise Martinez* 2740
- Water Quality: Soil Erosion / *Andreas Klik* 2755
- Yellow River / *Zixi Zhu, Ynuzhang Wang, and Yifei Zhu* 2884

Foreword

Environmental management started at the beginning of the Neolithic Era, when man began modifying his natural environment to deter predators and prepare land for agriculture. Nevertheless, it was not until the second half of the 20th century that environmental management was identified as a discipline in need of systematic study. During most of written history, management cases in agriculture have been researched and implemented, but these were isolated, local in character, and dependent upon the technology of the day. The international component was added at the 1972 Stockholm United Nations Conference on the Human Environment.

During the next four decades, university programs covering environmental management began to flourish, resulting in the full spectrum of university education available today—spanning undergraduate, graduate, and post-graduate programs. Parallel to increased academic activity, the International Organization for Standardization (ISO) formed Technical Committee No. 207 on Environmental Management, in response to objectives of sustainable development adopted at the United Nations Conference on Environment and Development (Rio de Janeiro, 1992). Gradually, over the next two decades, the committee issued the ISO 14000 family of international standards on environmental management. The work is not yet complete and more standards will be issued in the future.

Presently, environmental management is maturing into a truly comprehensive interdisciplinary activity including natural, medical, technical, economic, and social sciences. Although some cases have considered (and implemented) almost all of these aspects, the majority of previous studies propose a specific solution, which would have been different if the missing disciplines had been included.

This encyclopedia presents a unique collection of almost 400 issues, case studies, and practices. As such, it will be of critical importance for the sustainability of human development as examinations of current state-of-the-art techniques have at last been made available, thus making possible a much faster development towards more refined solutions in the future. Whether discussing the management of air, water, soil quality, or indeed any other issue, including those of global importance, this encyclopedia will be used frequently for years to come by all who hold a stake in the environment, including educational programs related to environmental management, industry, environmental regulatory bodies, and non-governmental organizations.

It is with greatest pleasure that I congratulate the editor S.E. Jorgensen, all contributors, and the editorial office for creating this monumental work.

Tarzan Legović
Secretary General, International Society for Ecological Modelling
Professor and Chairman, Division of Marine and Environmental Research, R. Bošković Institute
Zagreb, Croatia

Preface

The aims and scope of this encyclopedia are to present the basic knowledge for performance of an integrated environmental and ecologically sound management system. This encyclopedia cannot include all environmental problems, but it has been attempted to cover at least more than 90–95% of the most important problems and their sources. Environmental problems can be considered as health problems of ecosystems and solutions of the problems require therefore an as detailed as possible diagnosis, which we can develop by the use of the following toolboxes: ecological models, ecological indicators, and ecological services. The encyclopedia presents all three diagnostic tools, but a more comprehensive overview of the tool boxes can be found for ecological models in *Handbook of Ecological Models Used in Ecosystem and Environmental Management* (2011), edited by S.E. Jørgensen; if surface modelling would be beneficial to use, in *Surface Modeling, High Accuracy and High Speed Methods* (2011), edited by Tian-Xiang Yue; and for ecological indicators in *Handbook of Ecological Indicators for Assessment of Ecosystem Health* (2010), edited by S.E. Jørgensen, Fu-Liu Xu, and R. Costanza. It can therefore be recommended also to consult these handbooks to obtain more knowledge about the diagnostic tools. In addition to the problems and their sources, emphasis is also placed on solutions to the problems, where we distinguish four possibilities: environmental technology, ecological technology, cleaner technology, and environmental legislation. All four tools have been presented with reference to the environmental problems that they can solve, but the most comprehensive overview is given for the first three possibilities, while environmental legislation is covered with less detail, because it is more difficult due to significant variations from country to country.

Integrated and ecologically sound environmental management means that the definition of the problems, the sources and the ecosystems affected, the diagnosis, and the possible solutions are integrated and used as an entity to ensure that we do not solve one problem while creating two others and that we obtain a clear improvement of the health problems of the affected ecosystem. The encyclopedia also contains articles or entries that focus on the integration of the problem, the source, the ecosystem, the diagnosis, and the solution. Moreover, a few entries cover background knowledge needed to interpret the problem and the diagnosis. The details of the topical classification are given in the introduction to the Topical Table of Contents. The scope is that the user of the encyclopedia can find an overview of the above mentioned steps to integrated and ecologically sound environmental management. Each article has a comprehensive reference list, which the user of course should utilize to implement the proposed procedure.

To launch an encyclopedia requires a very wide knowledge of many details, which of course is impossible for one person to provide. This encyclopedia would therefore not have been possible without a very knowledgeable and skilled editorial board, covering different aspects of environmental management. Furthermore, without the contribution of all the authors, it would of course not have been possible to produce this encyclopedia. I would therefore underline my appreciation of the tremendous and important work by the editorial board and by the authors of the articles. In this context I also mention that the staff of Taylor & Francis has been very valuable in providing all the pieces and placing them correctly in the large mosaic of the encyclopedia.

Sven Erik Jørgensen
Editor
Copenhagen

How to Use This Encyclopedia

Integrated Ecological and Environmental Management

Integrated ecological and environmental management means that the environmental problems are viewed from a holistic angle considering the ecosystem as an entity and considering the entire spectrum of solutions, including all possible combinations of proposed solutions. Articles in the encyclopedia focusing on *integrated* ecological and environmental management using *holistic* approaches are indciated with IMS. The experience gained from environmental management over the last forty years has clearly shown that it is important not to consider solutions of single problems but to consider *all* the problems associated with a considered ecosystem simultanously and evaluate *all* the solution possibilities proposed by the relevant disciplines at the same time, or, expressed differently: to observe the forest and not the single trees. The experience has clearly underlined that there is no alternative to *integrated* management, at least not on a long-term basis. Fortunately, new ecological sub-disciplines have emerged that offer tool boxes to perform integrated ecological and environmental management.

Integrated ecological and environmental mangement of today consists of a seven-step procedure that is proposed in Jørgensen and Nielsen:[1]

- Define the problem
- Determine the ecosystems involved
- Find and quantify all the sources to the problem
- Set up a diagnosis to understand the relation between the problem and the sources
- Determine all the tools that could be used to solve the problem(s)
- Select a solution or a combination of solutions and implement the selected solutions
- Follow the recovery process to ensure that the problems have been solved

When an environmental problem has been detected, it is necessary to determine and quantify the problem and all the sources to the problem. The entries dealing with this step of the procedure are marked with PPS. It requires the use of analytical methods or a monitoring program. To solve the problem a clear diagnosis has to be developed: what is the problem that the ecosystems are facing and what are the relationships between the sources and their quantities and the determined problem? Or expressed differently: to what extent do we solve the problems by reducing or eliminating the different sources to the problems? Entries dealing with these two questions are denoted DIA. A holistic integrated approach is needed in most cases because the problems and the corresponding ecological changes in the ecosystems are most often very complex, particularly when several environmental problems are interacting. When the first green wave started in the mid-1960s, the tools to answer these questions, which we today consider as very obvious questions in an environmental management context, were not yet developed. We were able to carry out the first three points on the above shown list but had to stop at point 4 and could at that time only recommend eliminating the source completely or almost completely by using the methods that were available at that time—that is, environmental technology (covered by entries marked with ENT) which at that time was at a slightly lower level than today.

Due to the development of several new ecological sub-disciplines, it is today possible to accomplish the fourth to sixth points, presented here tool boxes that we can apply today to carry out the the fourth to sixth points. They are the result of the emergence of six new ecological sub-disciplines: for a better diagnosis, ecological modelling, ecological indicators, ecological services (all three sub-disciplines are covered by the entries denoted DIA) and for more tools to solve the problems, ecological engineering (also denoted ecotechnology, covered by entries named ECT), cleaner production (covered by entries marked CLT), and environmental legislation (articles focusing on environmental legislation are named ELE).

Tool Boxes Available Today to Develop an Ecological-Environmental Diagnosis

A massive use of ecological models as an environmental management tool was initiated in the early 1970s. The idea was to answer the question: what is the relationship between a reduction of the impacts on ecosystems and the observable, ecological improvements? The answer could be used to select the pollution reduction that the society would require and could effort economically. Ecological models were developed as early as the 1920s by Steeter–Phelps and Lotka Volterra (see for instance Jørgensen and Fath),[2] but in the 1970s started a much more consequential use of ecological models and many more models of different ecosystems and different pollution problems were developed. Today we have at least a few models available for all combinations of ecosystems and environmental problems. The journal *Ecological Modelling* was launched in 1975 with an annual publication of 320 pages and about 20 papers. Today, the journal publishes 20 times as many papers. This means that ecological modelling has been adopted as a very powerful tool in ecological-environmental management to cover particularly the fourth point in the integrated ecological and environmental mangement procedure proposed previously. The encyclopedia contains several articles about the use of ecological and environmental models, but a more comprehensive overview of available models can be found in Jørgensen.[3] If surface modelling would be more beneficial to use, see Tian-Xiang Yue.[4]

Ecological models are powerful management tools but they are not easily developed. They require in most cases good data, which are resource- and time-consuming to provide. About 20 years ago, it was therefore proposed to use another tool box, one that required less resources to provide a diagnosis, namely, ecological indicators (see, for instance, Costanza, Norton and Haskell[5]). Ecological indicators can be classified as shown in Table 1 according to the spectrum from a more detailed or reductionistic view to a system or holistic view (see Jorgensen[16]). The reductionistic indicators can, for instance, be a chemical compound that causes pollution or specific species. A holistic indicator could, for instance, be a thermodynamic variable or biodiversity. Indicators can either be measured or they can be determined by the use of a model. In the latter case, time consumption is of course not reduced by the use of indicators instead of models, but the models get a more clear focus on one or more specific state variables, namely, the selected indicator, which best describes the problems. In addition, indicators are usually associated with very clear and specific health problems of the ecosystems, which of course is beneficial in environmental management. Several articles in the encyclopedia focus on the use of indicators for developing a diagnosis, but a more comprehensive overview of the application of indicators can be found in Jørgensen, Xu and Costanza.[7]

Over the last 10–15 years, the services offered by ecosystems to society have been discussed and attempts have been made to calculate the economic values of these services.[8] A diagnosis that would focus on the services actually reduced or eliminated due to environmental problems could be developed. Another possibility of using ecological services to assess the environmental problems and their consequences could be to determine the economic values of the overall ecological services offered by the ecosystems and then compare them with what is normal for the type of ecosystems considered. Jørgensen[9] has determined the values of all the services offered by various ecosystems by the use of the ecological holistic indicator eco-exergy expressing the total work capacity. It is a good measure of the total amount of ecological services as all services require a certain amount of free energy, i.e. energy that can do work. The values published in Jørgensen[9] are shown in Table 2 and can be used for the above indicated comparison. The eco-exergy is found as presented in the articles about this indicators, but see also Jørgensen et al.[10] and Jørgensen.[11] The use of eco-exergy as an indicator to find the value of the ecosystem services in this context is beneficial, because the development of sustainability can be described as maintenance of the total work capacity that is at our disposal.[12]

Assessments of ecosystem services frequently use ecological indicators. The indicators can be determined and followed by the use of models, and models can determine the reduced or lost ecological services of ecosystems. The three diagnostic tool boxes are closely related with other words and obviously the use of all three tool boxes will give the most complete diagnosis. They are, however, all based on observations, which means that they are dependent on a solid monitoring program. Articles in the encyclopedia about monitoring are denoted DIA, as monitoring works hand in hand with the diagnostic tools. On the other hand, the resources available for environmental management are always limited, which means that it is

Table 1 Classification of ecological indicators.

Level	Example
Reductionistic (single) indicators	PCB, Species present/absent
Semiholistic indicators	Odum's attributes
Holistic indicators	Biodiversity/ecological network
"Super-holistic"	Thermodynamic indicators as eco-exergy and energy

Table 2 Work capacity used to express the ecosystem services for various types of ecosystems.

Ecosystem	Biomass (MJ/m² y)	Information factor (β-value)	Work capacity (GJ/ha y)
Desert	0.9	230	2,070
Open sea	3.5	68	2,380
Coastal zones	7.0	69	4,830
Coral reefs, estuaries	80	120	960,000
Lakes, rivers	11	85	93,500
Coniferous forests	15.4	350	539,000
Deciduous forests	26.4	380	1,000,000
Temperate rainforests	39.6	380	1,500,000
Tropical rainforests	80	370	3,000,000
Tundra	2.6	280	7,280
Croplands	20.0	210	420,000
Grassland	7.2	250	18,000
Wetlands	18	250	45,000

It is calculated as biomass * the information factor.

hardly possible to apply all three tool boxes in all cases, and also because in most cases it would require a comprehensive monitoring program. It is therefore necessary in many cases to make a choice. If an ecological model is developed, anyhow, to be able to give more reliable prognoses, it is of course natural to apply the developed model and it may be beneficial in addition to select one or a few indicators to focus more specifically on a well-defined problem. If a model is not available but a monitoring program has to be developed, one would have to direct the observations to encompass the state variables that can be applied to assess the indicators that are closely related to the defined health problems. If society is dependent on specific ecological services of the ecosystem, it would be natural to assess to what extent these services are reduced or lost, possibly supplemented with health indicators that are particularly important for the maintenance of these services. The choice of tool boxes is therefore a question about the available resources and the specific case and problem.

Tool Boxes Available Today to Solve Environmental Problems

The tool box environmental technology (articles are denoted ENT) was the only methodological discipline available to solve environmental problems 45 years ago, when the first green waves started in the 1960s. This tool box was able to solve only point source problems, and sometimes at a very high cost. Today, fortunately, we have additional tool boxes that can solve diffuse pollution problems or find alternative solutions at lower costs when environmental technology would be too expensive to apply. As for diagnostic tool boxes, they are developed on basis of new ecological sub-disciplines.

To solve environmental problems today we have four tool boxes:

- Environmental technology (denoted ENT)
- Ecological engineering, also denoted ecotechnology (articles are marked with ECT)
- Cleaner production; under this heading we would also in this context include industrial ecology (see the articles denoted CLT)
- Environmental legislation (articles denoted ELE)

Environmental technology came about with the emergence of the first green waves about 45 years ago. Since then, several new environmental-technological methods (covered by articles named ENT) have been developed, and all the methods have been streamlined and are generally less expensive to apply today. There is and has been, however, an urgent need for other alternative methods to solve the entire spectrum of environmental problems at an acceptable cost. Environmental management today is more complicated than it was 45 years ago because of the many more tool boxes that should be applied to find the optimal solution and because global and regional environmental problems have emerged. The use of tool boxes and the more complex situation today is illustrated in Fig. 1.

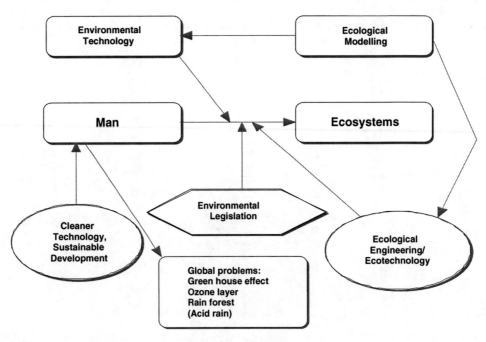

Fig. 1 Conceptual diagram of the complex ecological-environmental management of today, where there are various tool boxes available to solve the problems and where the problems are local, regional and global.

The tool box containing ecological engineering methods (the articles are indicated with ECT) was developed in the late 1970s. Ecological engineering is defined as the designing of sustainable ecosystems that integrate human society with its natural environment to the benefit of both.[13] It is an engineering discipline that operates in ecosystems, which implies that it is based on both design principles and ecology. The tool box contains four classes of tools:

- Tools that are based on the use of natural ecosystems to solve environmental problems (e.g., the use of wetlands to treat agricultural drainage water)
- Tool that are based on imitations of natural ecosystems (e.g., construction of wetlands to treat waste water)
- Tools that are applied to restore ecosystems (e.g., restoration of lakes by the use of biomanipulation)
- Ecological planning of the landscape (e.g., the use of agro-forestry)

The introduction of ecological engineering has made it possible to solve many problems that environmental technology could not solve, such as non-point pollution problems and a fast restoration of deteriorated ecosystems.

Some environmental problems, however, cannot be solved without more strict environmental legislation and for some problems a global agreement may be needed to achieve a proper solution, for instance by out-phasing the use of Freon to stop or reduce the destruction of the ozone layer. Notice also that environmental legislation (articles are de noted ELE) requires an ecological insight to assess the required reduction of the emission that is needed through the introduction of environmental legislation.

As environmental legislation has been tightened, it has been more and more expensive to treat industrial emissions, and the industry has of course considered whether it is possible to reduce the emissions by other methods at a lower cost. That has led to the development of what is called cleaner production (CLT is used to indicate the articles focusing on cleaner technology), which means the idea to produce the same product by a new method that would give a reduced emission and therefore less costs for the pollution treatment. New production methods have been developed by the use of innovative technology that has created a completely new method to produce the same product with less environmental problems. This has been the case particularly in energy technology, where a wide spectrum of methods are being developed as alternatives to the use of fossil fuel. Other emission reductions have been developed with the use of ecological principles on the industrial processes, for instance, recycling and reuse. In many cases it has also been possible to reduce environmental problems by identifying unnecessary waste. Industrial ecology could, in the author's opinion, be defined as the use of ecological principles in production, such as recycling, reusing and using holistic solutions to achieve a high efficiency in the general use of the resources. Industrial ecology today is, however, used to cover the use of waste from one production in another production.

Today, with the four tool boxes with environmental management solutions, it is possible to solve any environmental problem and often at a moderate cost and sometimes even at a cost which makes it beneficial to solve the problem properly. As is the case for diagnostic tool boxes, tool boxes with problem solution tools are, as indicated, rooted in recently developed ecological sub-disciplines that are named after the tools: ecological engineering, environmental legislation and cleaner technology.

Follow the Recovery Process

Environmental management is only complete if the environmental problem and the ecosystem are followed carefully after the tool boxes have been applied. It is usually not a problem because it is a question of providing the observations needed to follow the prognoses of the

- eventually developed ecological model
- the selected ecological indicators
- the recovery of the ecological services of the ecosystem (which can be done by focusing on a specific service or on the values of all the ecological services offered by the ecosystem)

Conclusion

From this review of the up-to-date integrated environmental management procedure, it is possible to conclude that the following steps are recommended:

- Define, preferably quantitatively, the problem(s), the ecosystem affected and the sources of the problem. Use articles here marked PSS. They correspond to the first three steps in the beginning of this introduction.
- Set up a diagnosis by using ecological models, ecological indicators and assessments of ecological services. Use the articles here marked by DIA.
- Go through all the possible tools that could be implemented to solve the problem. The encyclopedia presents here a particularly wide spectrum of articles, covering:
    - environmental technology, marked by ENT
    - ecotechnology, marked by ECT
    - cleaner technology, included industrial ecology, marked by CLT
    - environmental legislation, marked by ELE

Integrated, up-to-date environmental management requires the use of all seven of the presented tool boxes and would not be possible if these tool boxes were not developed as a result of recently emergent ecological sub-disciplines: ecological modelling, ecological engineering, application of ecological indicators, cleaner technology and industrial ecology. These ecological sub-disciplines are therefore crucial for environmental management today and they form an indispensable bridge between ecology and environmental management—between the basic science of ecology and its application in practical environmental management. IMS denotes articles that uncover the possibilities of integrating the various tool boxes to make an integrated and holistic management.

The encyclopedia gives significant background knowledge to be able to cover the seven presented steps to achieve an integrated, holistic environmental and ecological management of many of the actual environmental problems, from pollution problems rooted in our extensive use of fossil fuel and our water-, air- and solid-waste problems to the erosion and non-point agricultural pollution of pesticides and fertilizers. COV, which has not been mentioned yet, denotes articles that give comparative overviews of important topics for environmental management or background knowledge that are important for the evaluation of environmental problems.

References

1. Jørgensen, S.E.; Nielsen, S.N. Tool boxes for an integrated ecological and environmental management. Ecological Indicators, in press.
2. Jørgensen, S.E.; Fath, B. *Fundamentals of Ecological Modelling*, 4th ed., Elsevier: Amsterdam, 400 pp., 2011.
3. Jørgensen, S.E. *Handbook of Ecological Models Used in Ecosystem and Environmental Management*, 2011.
4. Yue, T.-X. Surface modeling. High Accuracy and High Speed Methods, 20.

5. Costanza, R.; Norton, B.G.; Haskell, B.D. *Ecosystem Health: New Goals for Environmental Management*, Island Press: Washington, 270 pp., 1992.
6. Jørgensen, S.E. *Integration of Ecosystem Theories: A Pattern*, Kluwer: Dordrecht, 386 pp.
7. Jørgensen, S.E.; Xu, F.-L.; Costanza, R. *Handbook of Ecological Indicators for assessment of Ecosystem Health*, 2nd ed., CRC: Boca Raton, FL, 484 pp., 2010.
8. Costanza, R.; d'Arge, R.; de Groot, R.; Farber, S.; Grasso, M.; Hannon, B.; Naeem, S.; Limburg, K.; Paruelo, J.; O'Neill, R.V.; Raskin, R.; Sutton, P.; van den Belt, M. The value of the world's ecosystem services and natural capital. Nature **1997**, *387*, 252–260.
9. Jørgensen, S.E. Ecosystem services, sustainability and thermodynamic indicators. Ecological Complexity **2010**, *7*, 311–313.
10. Jørgensen, S.E.; Ladegaard, N.; Debeljak, M.; Marques, J.C. Calculations of exergy for organisms. Ecological Modelling **2005**, *185*, 165–176.
11. Jørgensen, S.E. *Introduction of Systems Ecology*, CRC: Boca Raton, FL, 320 pp., 2012.
12. Jørgensen, S.E. *Eco-Exergy as Sustainability*, WIT: Southampton, 220 pp., 2006.
13. Mitsch, W.J.; Jørgensen, S.E. *Ecological Engineering and Ecosystem Restoration*, John Wiley: New York, 410 pp., 2004.

About the Editor-in-Chief

Sven Erik Jørgensen

Institute A, Section of Environmental Chemistry, Copenhagen University, Copenhagen, Denmark

Dr. Sven Erik Jørgensen is a professor of environmental chemistry at Copenhagen University. He received a doctorate of engineering in environmental technology and a doctorate of science in ecological modeling. He is an honorable doctor of science at Coimbra University, Portugal, and at Dar es Salaam University, Tanzania. He was editor-in-chief of *Ecological Modelling* from the journal's inception in 1975 to 2009. He has also been the editor-in-chief of the *Encyclopedia of Ecology*. In 2004 Dr. Jørgensen was awarded the prestigious Stockholm Water Prize and the Prigogine Prize. He was awarded the Einstein Professorship by the Chinese Academy of Science in 2005. In 2007 he received the Pascal medal and was elected a member of the European Academy of Science. He has written close to 350 papers, most of which have been published in international peer-reviewed journals. He has edited or written 64 books. Dr. Jørgensen has given lectures and courses in ecological modeling, ecosystem theory and ecological engineering worldwide.

Encyclopedia of Environmental Management

Volume III
Pages 1531–2270
Invasion–Rare Earth

Invasion Biology

Jennifer Ruesink
Department of Zoology, University of Washington, Seattle, Washington, U.S.A.

Abstract
Biological invasions occur when species or distinct populations breach biogeographical barriers and extend their ranges to areas where they were not historically present. Invasion biology concerns the causes and consequences of these new species, which are also referred to as invasive, introduced, alien, exotic, nonnative, or nonindigenous species. Biological invasions occur in five steps: arrival, establishment, population growth, population spread, and impact. Only a small proportion of species that arrive actually reach the first step; thus each step acts as a filter for the invasion process. No widely accepted method currently exists for identifying the characteristics that promote invasion a priori.

INTRODUCTION

Biological invasions occur when species or distinct populations breach biogeographical barriers and extend their ranges to areas where they were not historically present.[1–3] Invasion biology concerns the causes and consequences of these new species, which are also referred to as invasive, introduced, alien, exotic, nonnative, or nonindigenous species.

Biological invasions occur in five steps: arrival, establishment, population growth, population spread, and impact (Table 1). Only a small proportion of species that arrive actually establish, and so forth; thus each step acts as a filter for the invasion process. Successful invasion depends on characteristics of the invading species, of the recipient environment, and of the process by which the two are brought together.[4] However, no widely accepted method currently exists for identifying the characteristics that promote invasion a priori.[5]

WHY DO NEW SPECIES INVADE?

Species have always expanded their ranges, but the pace of invasion has accelerated recently due to increased human travel and trade.[5–7] Humans transport species in three ways: 1) on purpose, with the intention that they will grow in outdoor environments (fish and game, plantation trees, and biological control agents), 2) on purpose but with no intention that they will establish (pets, horticultural and agricultural plants, aquaculture, and sterile releases), and 3) accidentally (hitchhiking on packages, live imports, and people). The contribution of each of these main pathways varies among taxa. Of South Africa's weeds, 89% were intentionally introduced[8] but only 11% of insect invaders in North America were intentional.[9] Ducks, pheasants, pigeons, finches, and parrots have more introduced species worldwide than would be expected by chance because these bird families include many pet or game species.[10]

WHICH SPECIES INVADE?

Propagule Pressure

Species can invade a new area if abiotic (especially climatic) conditions are suitable and an exploitable resource exists. Those species that can invade, will invade if given sufficient opportunity.[11–13] High rates of arrival have been termed 'propagule pressure' and can occur either through numerous releases or releases of many individuals. Some of the best evidence that propagule pressure affects invasion comes from compilations of biocontrol introductions: the successful establishment of insect predators rose seven times when the number of introductions doubled, and releases >31,200 individuals were eight times more successful than those of <5000 individuals.[14]

Species Traits

For a given propagule pressure, some species may be more likely to invade than others. Traits promoting invasion could include the ability to increase rapidly from low density, a generalist diet, and broad climatic tolerance. Although statistical relationships between species traits and invasibility are often weak,[15] analyses of certain taxa introduced to particular environments have been successful.[16,17] For instance, invasive species of pines in South Africa tend to have small seeds, short intervals between reproductive bouts, and short times to maturity, whereas noninvasive species show the opposite traits.[18] For woody plants, those that have become invasive in North America often reproduce vegetatively, germinate easily, and have a long fruiting period.[19]

Table 1 Reasons for and responses to five steps of an invasion.

Step	Why?	Control
Arrival	Intentional releases, intentional imports, accidental hitchhiking (by product)	Risk assessment—choice of species, treatment or quarantine of vectors
Establishment	Suitable abiotic conditions, available resources, propagule pressure	Reduce pathways, containment
Population growth	Intrinsic rate of reproduction, apomixis/vegetative reproduction	Early eradication
Population spread	Mobility—dispersal, home range size, transport by humans	Eradicate new populations, reduce human transport
Impact	Density (abundant resources), few enemies, per capita effect (new role), alteration of resource base	Effective screening prior to introduction; mechanical, chemical, or biological control; make environment less suitable

WHERE DO SPECIES INVADE?

In addition to species traits, characteristics of the recipient community may also influence the ease with which new species invade. Indeed, particular conditions in native environments (e.g., disturbance and species interactions) may select for species that are effective invaders of other areas, thus resulting in asymmetric patterns of invasion. For instance, European insects have invaded forests in North America but not vice versa,[20] and there has been a unidirectional appearance of tillering grasses in bunchgrass habitats worldwide.[21] Assemblages may be more resistant to invasion if they are undisturbed[22,23] or contain many natural enemies.[24,25] For many years it was a rule of thumb that species poor islands were invasible and species rich tropics were not.[1] However, surveys of plant assemblages indicate that areas of naturally high species richness tend also to have numerous invaders, perhaps because soil and climate conditions are generally conducive to plant growth.[26] Although species richness per se may not influence invasion, loss of species could make systems easier to invade.

Habitat alteration (flooding, drought, fire, wind, eutrophication, and channelling) creates new conditions that are suitable for a new suite of species. These species are often introduced. For instance, western North America hosts many conspicuous invasive fishes in part because once fast-flowing rivers are now lakes separated by dams.[27] Roadsides in North America contain a high proportion of European plant species, probably because European plants have had millenia to evolve to take advantage of disturbance, whereas disturbance in North American habitats has risen recently.[28] Based on an Australian study of two invaders, both physical disturbance and nutrient addition improve the performance of invasive plants.[29]

WHICH SPECIES HAVE EFFECTS AND WHERE?

Impacts of invaders relevant to pest management include the ecological and/or economic damage from pests and the effectiveness of biocontrol agents.[30] Impacts are expected to be particularly pronounced when species reach high abundances or have high per capita effects.[31] High abundances can occur when species escape limits to population growth (abundant resources or few enemies). High per capita effects may arise when a species plays a new role, especially by altering the resource base. For instance, many of the worst plant invaders of natural areas are nitrogen fixers (which alter nutrients) or climbing vines (which alter light).[32]

A disproportionate number of invasive species have harmful effects, more so if introduced accidentally rather than intentionally. In Japan, for instance, 8% of native insects are considered pests, but 72% of introduced insects are pests.[33] Of agricultural weeds in North America, 50%–75% are nonindigenous.[6] On the other hand, only 1%–6% of nonindigenous plant species in Great Britain have become weedy or widespread.[8] Some introduced species may be problematic due to an absence of natural enemies, but this escape from control does not appear to be entirely general. Based on a compilation of life tables for 124 holometabolous insects, mortality rates due to parasitoids, predators, and diseases do not differ for native and introduced species.[34]

Species introduced accidentally often have harmful effects, and, conversely, species introduced intentionally have beneficial effects less often than desired. Of 463 grasses and legumes introduced to Australia to improve pasture, only 5% have raised productivity.[35] About a third of established biocontrol insects actually reduce the target organism.[36] Although impact in the native environment is not necessarily a useful indicator of what a species will do once introduced, one indicator of potential impact is the fate of prior introductions.[19]

Pest status tends to be based on economic considerations, but introduced species also cause ecological damage. Invasive species contribute to endangerment of nearly 50% of species listed under the United States Endangered Species Act,[37,38] and they have dramatically altered the structure and function of ecosystems.[39,40] Ecological and economic effects have the same root causes—abundance and high per capita effects of invaders—but the affected habitats can be quite distinct. Only 25% of plants that cause problems in natural areas are also agricultural weeds.[32] Thus, screening procedures that keep out economic pests would fail to restrict many species that cause ecological harm.

Time Lags and Surprises

Introduced species have occasionally surprised researchers by expanding to previously intolerable places or by irrupting after remaining localized and rare for many years. "Boom and bust" patterns have also been observed, in which an invader initially reaches high abundance and then declines, sometimes even going extinct. Tolerance of new conditions (e.g., temperature or host plants) may require genetic adaptation, which could result in time lags before invasion.[41] Of 184 woody species currently considered invasive near Brandenburg, Germany, 51% did not appear to be invasive until >200 years after their initial introduction.[42] However, only 7% of 627 cases involving biocontrol introductions showed time lags before population increase, whereas 28% increased and 27% went extinct immediately.[43] Regardless of frequency, cases in which invaders have unexpected impacts have become well known, especially for biocontrol agents with nontarget effects such as feeding on endangered species (e.g., the weevil *Rhinocyllus conicus* on thistles and the moth *Cactoblastis* on *Opuntia* cacti)[44–46] or competing with natives (e.g., the ladybird beetle *Coccinella septumpunctata*).[5]

CAN INVASIONS BE CONTROLLED?

The first steps of an invasion can be controlled by limiting entry or by vigilantly eliminating newly established populations of invaders. For instance, an assessment of species associated with raw logs indicated a potential loss of billions of dollars due to forest pests if Siberian larch was not treated prior to import into the United States.[47] For many years, medflies (*Ceratitus capitata*) have epitomized the notion that "an ounce of prevention is worth a pound of cure." California spent $100 million to eradicate an incipient invasion in 1981, thereby preventing nine times that amount of crop damage. By 1996, however, medflies were apparently established and spreading. The extent of the invasion makes eradication unlikely.[48] Humans simply have to learn to live with these naturalized species.[49]

Control during the last steps of an invasion usually involves chemical, biological, or mechanical reduction of unwanted species in areas where effects are most serious. Control efforts at this stage would benefit from considerations of demography and behavior of invaders. For instance, seedling competition among annual plants is often fierce, so efforts to reduce seed production will not reduce plant numbers. Instead, control efforts should be directed at reducing seedling growth and survival.[50] Knowing how insects move among microhabitats could aid in trap placement or in crafting habitats that promote desirable species and discourage undesirable ones.[51] Knowing encounter rates and feeding rates could aid in calculating the number of consumers necessary for effective biological control.

Species invasions are a form of ecological gambling in which the consequences of any particular introduction are uncertain, despite an emerging framework of factors contributing to high risk invasions. The influx of new species can be slowed by reducing pathways for introduction and by intentionally introducing species only when beneficial effects will be large and native alternatives do not exist.[52] Distinct biotas are valuable and intriguing but increasingly difficult to maintain under pressures of globalization.

REFERENCES

1. Elton, C.S. *The Ecology of Invasions by Animals and Plants*; Chapman and Hall: London, 1985.
2. Williamson, M. *Biological Invasions*; Chapman and Hall: London, 1996.
3. Mack, R.N.; Simberloff, D.; Lonsdale, W.M.; Evans, H.; Clout, M.; Bazzaz, F.A. Biotic invasions: causes, epidemiology, global consequences, and control. Ecol. Appl. **2000**, *10* (3), 689–710.
4. Lodge, D.M. Biological invasions: lessons for ecology. Tr. Ecol. Evol. **1993**, *8*, 133–137.
5. Ruesink, J.L.; Parker, I.M.; Groom, M.J.; Kareiva, P.M. Reducing the risks of nonindigenous species introductions: guilty until proven innocent. Bioscience **1995**, *45* (7), 465–477.
6. Office of Technology Assessment. *Harmful Non-Indigenous Species in the United States*; U.S. Government Printing Office: Washington, DC, 1993.
7. Cohen, A.N.; Carlton, J.T. Accelerating invasion rate in a highly invaded estuary. Science **1998**, *279*, 555–558.
8. Crawley, M.J.; Harvey, P.H.; Purvis, A. Comparative ecology of the native and alien floras of the British isles. Phil. Trans. R. Soc. Lond. B. **1996**, *351*, 1251–1259.
9. Sailer, R.I. History of Insect Introductions. In Exotic Plant Pests and North American Agriculture; Wilson, C.L., Graham, C.L., Eds.; Academic Press: New York, 1983; 15–38.
10. Lockwood, J.L. Using taxonomy to predict success among introduced avifauna: relative importance of transport and establishment. Conserv. Biol. **1999**, *13* (3), 560–567.
11. Veltman, C.J.; Nee, S.; Crawley, M.J. Correlates of introduction success in exotic New Zealand birds. Amer. Nat. **1996**, *147* (3), 542–557.
12. Green, R.E. The influence of numbers released on the outcome of attempts to introduce exotic bird species to New Zealand. J. Anim. Ecol. **1997**, *66*, 25–35.
13. Duncan, D.P. The role of competition and introduction effort in the success of passeriform birds introduced to New Zealand. Amer. Nat. **1997**, *149* (5), 903–915.
14. Beirne, B. Biological control attempts by introductions against pest insects in the field in Canada. Can. Entomol. **1975**, *107*, 225–236.
15. Goodwin, B.J.; McAllister, A.J.; Fahrig, L. Predicting invasiveness of plant species based on biological information. Conserv. Biol. **1999**, *13* (2), 422–426.

16. Panetta, F.D. A system of assessing proposed plant introductions for weed potential. Plant Prot. Quarterly **1993**, *8*, 10–14.
17. White, P.S.; Schwarz, A.E. Where do we go from here? The challenges of risk assessment for invasive plants. Weed Technol. **1998**, *12* (4), 744–751.
18. Rejmanek, M.; Richardson, D.M. What attributes make some plant species more invasive. Ecology **1996**, *77* (6), 1655–1661.
19. Reichard, S.H.; Hamilton, C.W. Predicting invasions of woody plants introduced to North America. Conserv. Biol. **1997**, *11* (1), 193–203.
20. Niemela, P.; Mattson, W.J. Invasions of North American forests by European phytophagous insects. Bioscience **1996**, *46* (10), 741–753.
21. Mack, R.N. Alien Plant Invasion into the Intermountain West. A Case History. In *Ecology of Biological Invasions of North America and Hawaii*; Mooney, H.A., Drake, J.A., Eds.; Springer-Verlag: New York, 1986; 191–213.
22. Orians, G.H. Site Characteristics Favoring Invasions. In *Ecology of Biological Invasions of North America and Hawaii*; Mooney, H.A., Drake, J.A., Eds.; Springer-Verlag: New York, 1986; 133–148.
23. Hobbs, R.J.; Huenneke, L.F. Disturbance, diversity, and invasion: implications for conservation. Conserv. Biol. **1992**, *6*, 324–337.
24. Goeden, R.D.; Louda, S.M. Biotic interference with insects imported for weed control. Annu. Rev. Entomol. **1976**, *21*, 325–342.
25. Mack, R.N. Predicting the identity and fate of plant invaders: emergent and emerging approaches. Biol. Conserv. **1996**, *78*, 107–121.
26. Stohlgren, T.J.; Binkley, D.; Chong, G.W.; Kalkhan, M.A.; Schell, L.D.; Bull, K.A.; Otsuki, Y.; Newman, G.; Bashkin, M.; Son, Y. Exotic plant species invade hot spots of native plant diversity. Ecol. Monogr. **1999**, *69* (1), 25–46.
27. Moyle, P.B.; Light, T. Fish invasions in California: do abiotic factors determine success? Ecology **1996**, *77* (6), 1666–1670.
28. Pysek, P. Is there a taxonomic pattern to plant invasions? Oikos **1998**, *82*, 282–294.
29. Hobbs, R.J. The Nature and Effects of Disturbance Relative to Invasions. In *Biological Invasions: A Global Perspective*; Drake, J.A., Mooney, H.A., di Castri, R., Kruger, F., Groves, R., Rejmanek, M., Williamson, M., Eds.; John Wiley and Sons: Chichester, U.K., 1986; 389–405.
30. Pimentel, D.; Lach, L.; Zuniga, R.; Morrison, D. Environmental and economic costs of nonindigenous species in the United States. Bioscience **2000**, *50*, 53–65.
31. Parker, I.M.; Simberloff, D.; Lonsdale, W.M.; Goodell, K.; Wonham, M.; Kareiva, P.M.; Williamson, M.H.; Von Holle, B.; Moyle, P.B.; Byers, J.E.; Goldwasser, L. Impact: toward a framework for understanding the ecological effects of invaders. Biological Invasions **1999**, *1* (1), 3–19.
32. Daehler, C.C. The taxonomic distribution of invasive angiosperm plants: ecological insights and comparisons to agricultural weeds. Biol. Conserv. **1998**, *84*, 167–180.
33. Morimoto, N.; Kiritani, K. Fauna of exotic insects in Japan. Bull. National Institute of AgroEnvironmental Sci. **1995**, (12), 87–120.
34. Hawkins, B.A.; Cornell, H.V.; Hochberg, M.E. Predators, parasitoids, and pathogens as mortality agents in phytophagous insect populations. Ecology **1997**, *78* (7), 2145–2152.
35. Lonsdale, W.M. Inviting trouble: introduced pasture species in Northern Australia. Australian J. Ecol. **1994**, *19* (3), 345–354.
36. Williamson, M.; Fitter, A. The varying success of invaders. Ecology **1996**, *77* (6), 1661–1666.
37. Foin, T.C.; Riley, S.P.D.; Pawley, A.L.; Ayres, D.R.; Carlsen, T.M.; Hodum, P.J.; Switzer, P.V. Improving recovery planning for threatened and endangered species. Bioscience **1998**, *48*, 177–184.
38. Wilcove, D.S.; Rothstein, D.; Dubow, J.; Phillips, A.; Losos, E. Quantifying threats to imperiled species in the United States. Bioscience **1998**, *48*, 607–615.
39. Vitousek, P.M.; D'Antonio, C.M.; Loope, L.L.; Westbrooks, R. Biological invasions as global environmental change. Am. Scientist **1996**, *84* (5), 468–478.
40. Mack, M.C.; D'Antonio, C.M. Impacts of biological invasions on disturbance regimes. Tr. Ecol. Evol. **1998**, *13* (5), 195–198.
41. Secord, D.; Kareiva, P. Perils and pitfalls in the host specificity paradigm. Bioscience **1996**, *46* (5), 448–453.
42. Kowarik, I. Time Lags in Biological Invasions with Regard to the Success and Failure of Alien Species. In *Plant Invasions—General Aspects and Special Problems*; Pysek, P., Prach, K., Rejmanek, M., Wade, M., Eds.; SPB Academic Publishing: Amsterdam, 1995; 15–38.
43. Crawley, M.J. The population biology of invaders. Phil. Trans. R. Soc. Lond. B. **1986**, *314*, 711–731.
44. Louda, S.M.; Kendall, D.; Connor, J.; Simberloff, D. Ecological effects of an insect introduced for the biological control of weeds. Science **1997**, *277* (5329), 1088–1090.
45. Johnson, D.M.; Stiling, P.D. Distribution and dispersal of Cactoblastis cactorum (Lepidoptera: Pyralidae), an exotic opuntia-feeding moth, in Florida. Florida Entomol. **1998**, *81*, 12–22.
46. Cory, J.S.; Myers, J.H. Direct and indirect ecological effects of biological control. Tr. Ecol. Evol. **2000**, *15* (4), 137–139.
47. U.S. Department of Agriculture. *Pest Risk Assessment of the Importation of Larch from Siberia and the Soviet Far East*; USDA Forest Service, Misc. Publ. No. 1495. U.S. Government Printing Office: Washington, DC, 1991.
48. Carey, J.R. The future of the Mediterranean fruit fly *Ceratitus capitata* invasion of California: a predictive framework. Biol. Conserv. **1996**, *78* (1–2), 35–50.
49. Myers, J.H.; Savoie, A.; Van Randen, E. The irradication and pest management. Annu. Rev. Entomol. **1998**, *43*, 471–491.
50. McEvoy, P.B.; Rudd, N.T. Effects of vegetation disturbance on insect biological control of tansy ragwort, *Senecio jacobea*. Ecol. Appl. **1993**, *3* (4), 682–698.
51. Holway, D.A.; Suarez, A.V. Animal behavior: an essential component of invasion biology. Tr. Ecol. Evol. **1999**, *14* (8), 328–330.
52. Ewel, J.J.; O'Dowd, D.J.; Bergelson, J.; Daehler, C.C.; D'Antonio, C.M.; Gomez, L.D.; Gordon, D.R.; Hobbs, R.J.; Holt, A.; Hopper, K.R.; Hughes, C.E.; LaHart, M.; Leakey, R.B.; Lee, W.G.; Loope, L.L.; Lorence, D.H.; Louda, S.M.; Lugo, A.E.; McEvoy, P.B.; Richardson, D.M.; Vitousek, P.M. Deliberate introductions of species: research needs. Bioscience **1999**, *49* (8), 619–630.

Irrigation Systems: Sub-Surface Drip Design

Carl R. Camp, Jr.
Agricultural Research Service (USDA-ARS), U.S. Department of Agriculture, Florence, South Carolina, U.S.A.

Freddie L. Lamm
Research Extension Center, Kansas State University, Colby, Kansas, U.S.A.

Abstract

Subsurface drip irrigation (SDI) is generally defined as the application of water below the soil surface through emitters, with discharge rates in the same range as drip irrigation. Most SDI laterals are installed at a depth sufficient to prevent interference with surface traffic or tillage implements, and to provide a useful life of several years as opposed to annual replacement of surface or near-surface drip laterals.

INTRODUCTION

Subsurface drip irrigation (SDI) is generally defined as the application of water below the soil surface through emitters, with discharge rates in the same range as drip irrigation.[1] While this definition is not specific regarding depth below the soil surface, most SDI laterals are installed at a depth sufficient to prevent interference with surface traffic or tillage implements, and to provide a useful life of several years as opposed to annual replacement of surface or near-surface drip laterals.

Development of drip irrigation accelerated with the availability of plastics following World War II, primarily in Great Britain, Israel, and United States. SDI was part of drip irrigation development in the United States beginning about 1959, especially in Hawaii and California. While early drip irrigation products were relatively crude by modern standards, SDI devices were being installed in both experimental and commercial farms by the 1970s. As drip irrigation products improved during the 1970s and early 1980s, surface drip irrigation grew at a faster rate than SDI, probably because of emitter plugging problems and root intrusion. However, interest in SDI increased during the early 1980s, increased rapidly during the last half of the 1980s, and continues today, especially in areas with declining water supplies, with environmental issues related to irrigation, and where wastewater is used for irrigation. Initially, SDI was used primarily for sugarcane, vegetables, tree crops, and pineapple in Hawaii and California. Later, SDI use was expanded to other geographic areas and to agronomic and vines crops, including cotton, corn, and grapes.

SDI has the advantage of multiple-year life, reduced interference with cultural practices, dry plant foliage, and a dry soil surface. Multiple-year life allows amortization of the entire system cost over several years, often more than ten. If all system components are installed below tillage depth, surface cultural practices can be accomplished with minimal concern for system damage. Dry soil surfaces can reduce weed growth in arid climates and may reduce evaporation losses of applied water. Because the plant canopy is not irrigated, the foliage remains dry, which may reduce incidence of disease. SDI is also very adaptable to irregularly shaped fields and low-capacity water supplies that may provide design limitation with other irrigation systems.

The major disadvantages of SDI include system cost, difficulty in locating and repairing system leaks and plugged emitters, and poor soil surface. Most system components are installed below the soil surface and are neither easy to locate nor directly observable. In a properly designed and managed SDI system, the soil surface should seldom be wet. Consequently, seed germination, especially for small seeds, can be very difficult.

SDI systems offer considerable flexibility, both in design and operation. For example, SDI systems can apply small, frequent water applications, often multiple times each day, to very specific sites within the soil profile and plant root zone. Fertilizers, pesticides, and other chemical amendments can be applied via the irrigation system directly into the active root zone, often at a modest increase in equipment cost. In many cases, the operational cost may be less than that for applying these chemicals via conventional surface equipment.

SYSTEM DESIGN

Site, Water Supply, and Crop

Design of subsurface drip systems is similar to that of surface drip systems, especially with regard to hydraulic characteristics.[2] Specific crop and soil characteristics are used in the design process to select emitter spacing and flow rate, lateral depth and spacing, and the required system capacity. Emitter properties and lateral location are influenced by soil properties such as texture, soil compaction, and soil layering because these affect the rate of water movement through the soil profile and the subsequent wetting pattern for each emitter.

The water supply capacity directly affects the design of a SDI system. The size of the irrigated field or zone is often determined by the water supply capacity. For example, in some humid areas, high-capacity wells are not available but multiple low-capacity wells can be distributed throughout a farm. Fortunately, the design of SDI systems can be economically adjusted to correspond to the field size and shape, to the available water supply capacity, and to other factors. Water supply quality should be tested by an approved laboratory before proceeding with system design. This information is needed for the proper design and management of the water filtration and treatment system. Some water supplies require frequent or intermittent injection of acids and/or chlorine. Other saline and/or sodic water supplies may require treatment or special management. As water supplies become more limited, treated wastewater is becoming an increasingly important alternative water supply that can be applied through SDI systems. Camp[3] listed several reports that emphasized water supplies (saline, deficit, and wastewater) for SDI systems.

The SDI system is usually designed to satisfy peak crop water requirements, which vary with specific site, soil, and crop conditions. When properly designed and managed, SDI is one of the most efficient irrigation methods, providing typical application efficiencies exceeding 90%. In comparison with other methods of irrigation, reported yields with SDI were equal to or greater than those with other irrigation methods. Generally, water requirements with SDI are similar or slightly lower than those with other irrigation methods. In some cases, water savings of up to 40% have been reported.[3] However, unless more specific information is available, it is usually best to use standard net water requirements for the location when designing SDI systems.

Lateral Type, Spacing, and Depth

SDI lateral depth for various cropping systems is normally optimized for prevailing site conditions and soil characteristics.[3] Where systems are used for multiple years and tillage is a consideration, lateral depths vary from 0.20 m to 0.70 m. Where tillage is not a consideration (e.g., turfgrass, alfalfa) depth is sometimes less (0.10–0.40 m). Lateral spacing also varies considerably (0.25–5.0 m), with narrow spacing used primarily for turfgrass and wide spacing used for vegetable, tree, or vine crops. In uniformly spaced row crops, the lateral is usually located under either alternate or every third midrow area (furrow). For crops with alternating row spacing patterns, the lateral is located about 0.8 m from each row, usually in the narrow spacing of the pattern.

The lateral should be installed deep enough to prevent damage by tillage or injection equipment but shallow enough to supply water to the crop root zone without wetting the soil surface. Generally, laterals in SDI systems are placed at depths of 0.1–0.5 m, at shallower depths in coarse-textured soils and at slightly deeper depths on finer-textured soils. The selection of emitter spacing and flow rate are influenced by crop rooting patterns, lateral depth, and soil characteristics. It is also desirable to select an emitter spacing that provides overlapping subsurface wetted zones along the lateral for most row crops. For wider spaced crops such as trees and vines, emitters are normally located near each plant and may have wider spacings that do not provide overlapping patterns. Lateral spacing is determined primarily by the soil, crop, and cultural practice, and should be narrow enough to provide a uniform supply of water to all plants.

Special Requirements

Site topography must be considered in system design and selection of components as with any irrigation system, but SDI is suitable for most sites, ranging from flat to hilly. For sites with considerable elevation change, especially along the lateral, pressure-compensating emitters should be used.

Two special design requirements for SDI systems, which are significantly different from those for surface drip systems, are the needs for flushing manifolds and air entry valves. Flushing manifolds are needed to allow frequent flushing of particulate matter that may accumulate in laterals. Air relief valves are needed to prevent aspiration of soil particles into emitter openings when the system is depressurized. These valves must be located in sufficient number and at the higher elevations for each lateral or zone to prevent negative pressures within the laterals.

Emitter plugging caused by root intrusion is a major problem with some SDI systems, but can be minimized by chemicals, emitter design, and irrigation management. Chemical controls include the use of herbicides, either slow-release compounds embedded into emitters and filters or periodic injection of other chemical solutions (concentrated and/or diluted) into the irrigation supply. Periodic injection of acid and chlorine for general system maintenance can also modify the soil solution immediately adjacent to emitters and reduce root intrusion. In some cases, emitters plugged by roots may be cleared via injection of higher concentrations of chemicals, such as acids and chlorine.

Emitter design may also affect root intrusion. Smaller orifices tend to have less root intrusion but are more susceptible to plugging by particulate matter. Some emitters are constructed with physical barriers to root intrusion. Root intrusion appears to be more severe when emitters are located along dripline seams, which can be an area of preferential root growth. However, root intrusion problems appear to be greater for emitters, driplines, and porous tubes that are not chemically treated.

Irrigation management can also be used to influence root intrusion by controlling the environment immediately adjacent to the emitter. High frequency pulsing that frequently saturates the soil immediately surrounding the emitter can discourage root growth in that area for some crops but not others. Conversely, deficit irrigation sometimes practiced to increase quality or maturity, or to control vegetative growth, can increase root intrusion in lower rainfall areas because of high root concentrations in the soil zone near emitters.

SYSTEM COMPONENTS

Pumps, Filtration, and Pressure Regulation

Pump requirements for SDI are similar to those for other drip irrigation systems, meaning water must be supplied at a relatively low pressure (170–275 kPa) and flow rate in comparison to other irrigation methods. Because of the flushing requirement for SDI systems, a flow velocity of about $0.3 \, \text{m} \, \text{sec}^{-1}$ must be achievable, either by reducing the zone size while using the same pumping rate or by increasing the pumping rate without changing the zone size.

Water filtration is more critical for SDI systems than for surface drip systems because the consequences of emitter plugging are more severe and more costly. Generally, the better the water quality, the less complex the filtration system required. Surface and recycled or wastewater supplies require the most elaborate filtration systems. However, good filtration is the key to good system performance and long life, and should be a major emphasis in system design. Filtration systems range from simple screen filters for relatively clean water to more elaborate and complex disc and sand media filters for poorer quality water.

The pressure regulation requirement in SDI systems is similar to that in surface drip systems. When non-pressure-compensating emitters are used on relatively flat areas, pressure is typically regulated within the system supply lines (main and/or submain) using pressure-regulating valves. When pressure-compensating emitters are used, typically on more hilly terrain, the pressure within the system supply lines is controlled at a higher, but more variable, pressure that is within the recommended input pressure range for the emitters used. Water pressure should be monitored on a regular basis at the pump or supply port and at various locations throughout the SDI system, especially at the both ends of laterals.

Laterals and Emitters

Many types of driplines have been used successfully for SDI and most have emitters installed as an integral part of the dripline. This is accomplished by one of three methods: 1) molded indentions created during the fusing of dripline seams; 2) prefabricated emitters welded inside the dripline; or 3) circular prefabricated in-line emitters installed during extrusion. Regardless of the emitter used, dripline wall thickness and expected longevity must be considered along with other design factors in selecting the lateral depth. Flexible, thin-walled driplines typically are installed at shallow depths and normally have a shorter expected life. Thicker-walled, flexible driplines have been used successfully for several years provided they are installed deep enough to avoid tillage, cultivating, and harvesting machinery, but shallow enough to prevent excessive deformation or permanent collapse of the dripline by machinery or soil weight. Rigid tubing with thicker walls can be installed at deeper depths without deformation, and is often used on perennial crops or on annual crops for longer time periods (>10 years). Some driplines are impregnated with bactericides or other chemicals to reduce the formation of sludge or other material that could plug emitters.

Chemical Injection

Subsurface drip systems offer the potential for precise management of water, nutrients, and pesticides if the system is properly designed and managed. The marginal cost to add chemical injection equipment is generally competitive with other, more conventional application methods. Water and fertilizers can be applied in a variety of modes, varying from multiple continuous or pulsed applications each day to one application in several days. Choice of application frequency depends upon several factors, including soil characteristics, crop requirements, water supply, system design, and management strategies. If labeled for the purpose, some systemic pesticides and soil fumigants can be safely injected via SDI systems. Use of the SDI system for chemical applications has the potential to minimize exposure to workers and the environment, to reduce the cost of pesticide rinse water disposal, and improve precision of application to the desired target (root pests). Injection of other chemicals, such as acids and chlorine, is often required to clean and maintain emitters in optimum condition. However, a high level of management with system automation and feedback control is required to minimize chemical movement to the ground water when chemicals are used.

Air Entry and Flushing

Air entry valves must be installed at higher elevations in SDI systems to prevent the emitter from ingesting soil particles that could plug emitters when the system is depressurized. Typically, air entry valves are located in water supply lines near the head works or control station, and in both the supply and flushing manifolds. In some cases, such as turf or pasture, air entry valves may be installed below the soil surface and enclosed within a protective box. Flushing valves installed on the flushing manifold are required to control periodic system flushing.

OPERATION AND MAINTENANCE

Operation

SDI systems can be operated in several modes, varying from manual to fully automated. Overall, SDI systems are probably more easily automated than many other types of irrigation. One reason is that most are controlled from a central point using electrical or pneumatic valves and controllers that vary from a simple clock system to microprocessor systems, which are capable of receiving external inputs to initiate and/or terminate irrigation events.

Irrigation scheduling is as important for SDI systems as for any other type of irrigation. Choosing to initiate an irrigation event and how much water to apply during each event depends on crop, soil, and irrigation system type and design. Factors that affect those decisions include soil water storage volume, sensitivity of the crop to water stress, irrigation application rate, weather conditions, and water supply capacity. Camp[3] discussed several irrigation scheduling methods that have been used successfully with SDI. However, the important point is that a science-based scheduling method can conserve the water supply and increase profit.

If seed germination and seedling establishment and growth are critical, especially in arid climates when initial soil water content is not adequate, either sprinkler or surface irrigation is often used for germination. However, the need for two systems increases cost and decreases economic return. If subsurface drip is used for germination, an excessive amount of irrigation is often required to wet the seed zone for germination, which could result in excessive leaching and off-site environmental effects as well as increased cost. Surface wetting can also occur when the emitter flow rate exceeds the hydraulic conductivity of the soil surrounding the emitter, but wetted areas are often not uniform.

Because salts tend to accumulate above the lateral, high salt concentrations may occur between the lateral and soil surface in arid areas where rainfall is not available to leach the salts downward. Salts may also be moved under the row when laterals are placed under the furrow.[4] Supplemental sprinkler irrigation may be required in some areas to control salinity if precipitation is inadequate for leaching during several consecutive years.

Maintenance

Often, SDI systems must have a long life (>10 years) to be economical for lower value crops. Thus, appropriate management strategies are required to prevent emitter plugging and protect other system components to ensure proper system operation. Locating and repairing/replacing failed components is much more difficult and more expensive with SDI systems than with surface systems because most system components are buried, difficult to locate, and cannot be directly observed by managers. Consequently, operational parameters such as flow rate and pressure must be measured frequently and used as indicators of system performance. Good system performance requires constant attention to maintain good water quality, proper filtration, and periodic system flushing to remove particulate matter that could plug emitters. Periodic evaluation of SDI system performance in relation to design performance can identify problems before they become serious and significantly affect crop yield and quality.

CONCLUSION

Although there is general consensus that use of SDI is increasing, this growth is difficult to document. A recent survey of irrigation in the United States reported 156,070 ha of SDI, which is about 0.6% of the total irrigated area of 25,501,831 ha.[5] Use of SDI should increase in the future, depending primarily upon the economic and water conservation benefits in comparison to other irrigation methods. As water supplies become more limited, the high application efficiency and water conserving features of SDI should increase its application. Also, SDI offers potential advantages such as reduced odors and exposure to pathogens when using recycled domestic and animal wastewater. The SDI technology offers the capability to precisely place water, nutrients, and other chemicals in the plant root zone at the time and frequency needed for optimum crop production. With proper design, installation, and management, SDI systems can provide excellent irrigation efficiency and reliable performance with a system life of 10–20 years.

REFERENCES

1. S526.2. Soil and water terminology. *ASAE Standards*, 49th Ed.; ASAE: St. Joseph, MI, 2001; 970–990.
2. EP405.1. Design and installation of microirrigation systems. *ASAE Standards*, 49th Ed.; ASAE: St. Joseph, MI, 2001; 903–907.
3. Camp, C.R. Subsurface drip irrigation: a review. Trans. ASAE **1998**, *41* (5), 1353–1367.
4. Ayars, J.E.; Phene, C.J.; Schoneman, R.A.; Meso, B.; Dale, F.; Penland, J. Impact of bed location on the operation of subsurface drip irrigation systems. In *Microirrigation for a Changing World*, Proc. Fifth International Microirrigation Congress, Orlando, FL, April 2–6, 1995; Lamm, F.R., Ed.; ASAE: St. Joseph, MI, 1995; 141–146.
5. Anonymous. 1999 Annual Irrigation Survey. Irrig. J. **2000**, *50* (1), 16–31.

Irrigation Systems: Sub-Surface History

Norman R. Fausey
Soil Drainage Research Unit, Agricultural Research Service (USDA-ARS), U.S. Department of Agriculture, Columbus, Ohio, U.S.A.

Abstract
Water management has quite a broad scope, including all practices that influence any component of the hydrologic cycle. Within this broad scope are many practices, including things such as cloud seeding to increase precipitation, reservoir management to minimize flood events and to store water for municipal use, use of plastic mulches to reduce evaporation, and the use of infiltration basins to enhance recharge of groundwater aquifers. Municipal water supply, public safety issues related to flood forecasting, minimization, and urban storm water management, recreational needs, and agricultural production are examples of why water may need to be managed. Agricultural water management practices generally fall into one of five primary categories: irrigation, drainage, soil erosion control, water supply for animal needs, and waste water disposal. Irrigation involves adding water to assure an adequate supply for crop needs. Irrigation water may be applied on the soil surface or below the soil surface by various methods. One of the methods used to apply water below the soil surface is sub-irrigation.

DEFINITION AND DESCRIPTION OF SUB-IRRIGATION

Sub-irrigation is the practice of adding water to the soil by means of subsurface drains that are also used to drain water from the soil during periods when the soil is too wet. The drains may be open drains (ditches) or closed drains (drainpipes). Water may be supplied from a surface or a subsurface source, and is delivered into the subsurface drains and allowed to redistribute within the soil from these subsurface drains. Control structures within the ditches or at the outlet of the closed drains are used to block the water from leaving through the outlet, and, thereby, to establish a pressure gradient to cause water to flow from the drains into the soil.

Sub-irrigation is only applicable to areas needing subsurface drainage and having an adequate water supply. These areas typically have high water tables during some times of the year that can be lowered by subsurface drainage. Sub-irrigation depends upon being able to re-establish an elevated water table; this requires a substantial amount of water not only to raise the water table, but also to meet the evapotranspiration demand of the crop to maintain the water table at a raised position within the soil.

THE PAST

Agricultural water management using sub-irrigation is not new. It seems reasonable to presume that once the idea to construct drains to remove excess water from the soil proved successful, the idea of putting water back into the soil through the drains could not have been too far behind. Providing an adequate source of water and a means to move the water against the gravitational gradient would certainly have limited the feasibility prior to the advent of efficient pumping systems. Although not well documented, efforts and progress toward applying sub-irrigation were certainly made, as suggested by an anonymous[1] quote found in an early publication on drainage, stating: "I want the drains to irrigate with as much as to drain." An extension bulletin providing guidance for sub-irrigation in Florida was published in 1938,[2] indicating a recurring early demand for this information in this region. Renfro[3] summarized and discussed the use of sub-irrigation in U.S.A. up to the mid-1950s. Most of the early applications were on very permeable organic or sandy soils, using open ditches, for high value crops (vegetables and citrus), and in areas with a readily available water supply, including the Sacramento–San Joaquin Delta in central California, the Everglades of southern Florida, the San Luis Valley in Colorado, the Flatwoods of the Florida coastal plain, the Cache Valley in northern Utah, the Egin Bench in southern Idaho, and the Great Lakes states.

THE PRESENT

During the past 30–40 years, agriculture has evolved rapidly. World population pressures, scientific advances, and changing economic and social values have accelerated a shift in agriculture; diversity has given way to specialization. There are fewer farms and fewer farmers; production per unit of land is greater and is increasing. Risk reduction is a strong driving force in management decisions. Water management has become an important tool for reducing the risk of too much and too little water. This environment has led to a greater awareness of water management options and impacts, and, as a result, sub-irrigation has become a topic of increased interest for farmers and researchers.

In 1991, an international conference on sub-irrigation and controlled drainage water management was held in East Lansing, Michigan, U.S.A. The book[4] that resulted from this conference gives an excellent overview of the present status of sub-irrigation around the world. Reports of studies and experience from Canada, China, England, Finland, Italy, the Netherlands, and various locations within U.S.A. are included and illustrate a high level of interest in sub-irrigation.

Sub-irrigation, when properly managed in concert with subsurface drainage, can produce consistently high yields every year regardless of the weather conditions during the growing season. The subsurface drainage function of the system is used to remove excess soil water to assure trafficability for early planting, lengthening the growing season, and to avoid flooding and lack of oxygen in the root zone, which causes root damage and stunting or death of the plants. The sub-irrigation function of the system is used to avoid deficit water conditions in the root zone, causing stunting and premature senescence of plants. An assured adequate supply of water also allows planning and management for high yields that would not occur when relying on natural rainfall. High yield management involves higher plant populations and more fertilizer application to take full advantage of the available water. With sub-irrigation, yield goals can be raised and still be reached consistently.

Sub-irrigation water management is also beneficial to the environment. An adequate supply of water to meet crop needs encourages maximum growth and, therefore, efficient use of applied nutrients. Nutrients are taken up rather than being left in the soil where they would be subject to being transported to surface water and ground water as non-point-source pollutants. Sub-irrigation promotes greater production and increases the amount of organic residue returned to the soil. Sustained soil quality and soil health depend upon maintaining or increasing the amount of organic matter in the soil. Present environmental goals also promote the capture and sequestration of carbon in the soil rather than the release of carbon dioxide into the atmosphere. Increasing the soil organic matter content stores more carbon in the soil.

THE FUTURE

Economic and social pressures will likely continue to have a significant impact on agriculture and, consequently, on agricultural water management practices. As society becomes more environmentally aware and sensitive overall, an ethic will emerge for the elimination, or certainly the reduction, of delivery of non-point-source contaminants from agriculture to surface and ground waters. Some of this has already begun and has resulted in the promotion of uncultivated vegetated corridors along streams designed to slow and filter runoff waters moving to streams. Cost share programs are available in some states to offset the annual loss of production for farmers willing to establish permanent vegetation in these corridors or to assist with installation of structures to control subsurface drainage discharge to reduce nitrate delivery to streams.

Realistically, it is difficult to control water quality during storm events. Subsurface drainage increases infiltration and decreases surface runoff. The use of sub-irrigation during the summer months would result in less available storage in the soil for rainwater and, therefore, increased runoff, unless the drains are opened quickly and the water table is allowed to fall rapidly ahead of the infiltrating water. Anything that increases runoff encourages the transport of sediments and other pollutants with the runoff water. Water that can be retained in the soil or on the land during storm events will not contribute to runoff and pollutant transport.

Historically, the landscape included more wetland areas where runoff waters were retained or slowed and filtered before reaching streams. Agricultural and cultural development resulted in the loss of many of these wetlands that were also barriers to transportation or breeding grounds for diseases. Removal of the wetlands caused rapid delivery of runoff water to streams resulting in increased flooding and increased transport of sediments and other pollutants to the streams. One way to improve water quality would be to re-establish more wetland areas back into the landscape. Surface runoff and subsurface drainage waters could be directed to these wetlands for treatment and volume reduction before being discharged to streams. Some or all of the water leaving a wetland could be captured/harvested and stored on site in lieu of continuing offsite and downstream. The stored water could meet irrigation or other water supply needs.

Water supply is a critical need for sub-irrigation systems to be feasible and economical. The concept of capture of water during periods when excess water is available and its reuse to meet crop needs during deficit water periods offers an opportunity to make sub-irrigation affordable and practical in many areas. A holistic approach involving the integration of constructed wetlands, water storage facilities, and sub-irrigation of crops offers a unique opportunity to realize consistent and high crop yields. Such a system will generate more wetland habitat and protect water quality—two of society's current high priority goals. Therefore, the future for sub-irrigation is bright, considering the value to many segments of society to develop and use such integrated systems.

Sub-irrigation as a part of an overall water management plan that protects water quality, increases wetland habitat, and stabilizes crop yields will become the best water management practice.

REFERENCES

1. Anonymous. Tile to irrigate. Drainage J. **1890**, *12* (10), 283.
2. Spencer, A.P. Sub-irrigation. Florida Agric. Ext. Bull. **1938**, *99*.
3. Renfro, G. Applying water under the surface of the ground. In *Water: The 1955 Yearbook of Agriculture*; U.S. Government Printing Office: Washington, DC, 1955; 273–278.
4. Belcher, H.W.; D'Itri, F.M., Eds. *Sub-irrigation and Controlled Drainage*; Lewis Publishers: Boca Raton, FL, 1995.

Irrigation Systems: Water Conservation

I. Pai Wu
College of Tropical Agriculture and Human Resources, University of Hawaii, Honolulu, Hawaii, U.S.A.

Javier Barragan
Department of Agroforestry Engineering, University of Lleida, Lleida, Spain

Vince Bralts
Agricultural and Biological Engineering, Purdue University, West Lafayette, Indiana, U.S.A.

Abstract
A drip irrigation system is a form of localized irrigation that delivers water directly into the root zone of a crop. When properly designed and managed, a drip irrigation system can eliminate surface runoff, minimize deep seepage, and achieve high uniformity of water distribution and irrigation application efficiency. The development of drip irrigation in the late 1960s marked a period of tremendous improvement in irrigation science and technology in which water use is done more beneficially for agricultural production.

INTRODUCTION

With the increasing consequence of limited water resources and the increasing need for environmental protection, drip irrigation will play an even more important role in the future. Drip irrigation systems can be used for many different types of agricultural crops, including fruit trees, vegetables, pastures, specialty crops such as sugarcane, ornamentals, golf course grasses, and high economic value crops grown in greenhouses. An understanding of drip irrigation systems, irrigation scheduling, crop response, and economic ramifications will encourage greater use of drip irrigation in future agricultural production.

UNIFORMITY OF WATER APPLICATION AND DESIGN CONSIDERATIONS

The desired uniformity of water application and the specific crops to be grown guides the creation of drip irrigation systems. There are two types of drip irrigation uniformity: system uniformity and spatial uniformity in the field. The consistency of system distribution of water into the field describes the system uniformity. The spatial uniformity is the regularity of water distribution considering overlapping emitter flow and translocation of water in the soil. For drip irrigation systems designed for trees with large spacing, the system uniformity is equal to the water application uniformity in the field. For high-density plantings, the emitter spacing should be designed considering overlapped wetting patterns and the spatial uniformity in the field. The uniformity of a drip irrigation system depends primarily on the hydraulic design, but must also consider the manufacturer's variation, temperature effects, and potential emitter plugging. The effect of water temperature is generally negligible when using turbulent flow emitters. A combination of proper filtration and turbulent emitters can control emitter plugging. When grouping a number of emitters together as a unit, such as those designed to irrigate an individual plant's root system, the uniformity of water application with respect to the plant will improve.

Many expressions have been used to describe uniformity. The system uniformity, or emitter flow uniformity, can be expressed as the range or variation of water distribution in the field. This term was initially used for hydraulic design of drip irrigation systems given that the minimum and maximum emitter flows could be calculated and determined.[1] When more emitter flows are used or more samples are required for determining variation or spatial uniformity in the field, the Christiansen uniformity coefficient (UCC)[2] and coefficient of variation (CV), which is the ratio of standard deviation and the mean, are used. Each of the uniformity expressions are highly correlated with one other.

HYDRAULIC DESIGN OF DRIP IRRIGATION SYSTEMS

Once selection of the type of drip irrigation emitter is complete, the hydraulic design can be made to achieve the expected uniformity of irrigation application.

The hydraulic design of a drip irrigation system involves designing both the submain and lateral lines. Early research in drip irrigation hydraulic design concentrated mainly on the single lateral line approach,[1,3,4] but in 1985 Bralts and Segerlind developed a method to design a submain unit. The hydraulic design is based on the energy relations in the drip tubing, the friction drop, and energy changes due to slopes in the field. Direct calculations of water pressures along a lateral line or in a submain unit are made by using an energy gradient line approach.[1] All emitter flows along a lateral line and in a submain can be determined based on their corresponding water pressures. Once the emitter flows are determined, the emitter flow variation, q_{var} is expressed by

$$q_{var} = \frac{q_{max} - q_{min}}{q_{max}} \quad (1)$$

where q_{max} is the maximum emitter flow and q_{min} is the minimum emitter flow. Based on these data, other uniformity parameters such as UCC and CV can also be determined. There is a strong correlation between any two of the three uniformity parameters in the hydraulic design of drip irrigation systems, thus any one of the uniformity parameters can be used as a design criterion. This correlation also justifies using the simple emitter flow variation q_{var} for hydraulic design. The emitter flow variation q_{var} is converted to the CV when it is combined with the manufacturer's variation of emitter flow.

The total emitter flow variation caused by both hydraulic and manufacturer's variation can be expressed by[5]

$$CV_{HM} = \sqrt{CV_H^2 + CV_M^2} \quad (2)$$

where CV_{HM} is the coefficient of variation of emitter flows caused by both hydraulic and manufacturer's variation; CV_H and CV_M are the coefficients of variation of emitter flows caused by hydraulic design and manufacturer's variation, respectively.

The design criterion for emitter flow variation q_{var} for drip irrigation design is arbitrarily set as 10.0%–20.0%, which is equivalent to a CV, from 0.033 to 0.076, or 3.0%–8.0%. Based on the research of last 30 years, the manufacturer's variation of turbulent emitters is maintained only in a range 3.0%–5.0%, expressed by CV. When this variation is combined with emitter flow variation caused by hydraulic design with a range 3.0%–8.0% in CV, the total

Table 1 Design criteria for uniformity of drip irrigation system design.

Design consideration	CV (%)	UCC (%)
Water is abundant and no environmental pollution problems	30–20 20–10	75–85 80–90
Water is abundant but with environmental protection considerations	25–15	80–90
Limited water resources but with no environmental pollution problems	15–5	85–95
Considerations for both water conservation and environmental protection		

variation determined by the equation above will be limited to a CV of less than 10.0%. This variation illustrates that the drip irrigation systems are designed to achieve high uniformity and irrigation application efficiency.

Economic return can also be the basis of design criteria for drip irrigation. A new set of design criteria for drip irrigation was developed,[6] based on achieving an expected economic return with various water resources and environmental considerations (Table 1).

DRIP IRRIGATION FOR OPTIMAL RETURN, WATER CONSERVATION, AND ENVIRONMENTAL PROTECTION

When the uniformity of a drip irrigation system is designed with a UCC of 70.0%, 30.0% or less in CV, the irrigation application is expressed as a straight-line distribution,[7,8] as shown in Fig. 1. This figure was plotted using percent of area (PA) against a relative irrigation depth, X, which is the ratio of required irrigation depth to mean irrigation application. The straight-line distribution in the dimensionless plot can be specified by a minimum value, a, a maximum value, $(a + b)$, in the X-scale and a slope b, where b specifies the uniformity of water application.[9]

When a drip irrigation system is designed with fixed uniformity, it is possible to determine the sloped straight line with known value of a and b. A value (X) can then be selected between value a and $(a + b)$ and plotted (Fig. 1). The triangle formed above the horizontal line (X) results in an irrigation deficit and yield reduction. The triangle below the horizontal line results in over-irrigation and deep seepage.

An important irrigation scheduling parameter, the relative irrigation depth, (X) indicates how much irrigation water is applied. The effectiveness of drip irrigation is shown not only by the high uniformity of the drip irrigation system, but also by the irrigation requirement and the strategy of irrigation scheduling. As illustrated in Fig. 1, the irrigation scheduling parameter (X) affects the areas of over-irrigation and water deficit conditions in the field and is directly related to the economic return. Practically speaking, the X parameter is selected in a range from a to $(a + b)$, as shown in Fig. 1. Three typical irrigation schedules can be expressed by X and are as follows:

$X = a$ This schedule is a conventional irrigation schedule, which is based on the minimum emitter or minimum water application. The field is fully irrigated and whole field is over-irrigated except the point of minimum irrigation application.

$X = X_0$ For an optimal return there is a value of X for the irrigation scheduling parameter between a and $(a + b)$.

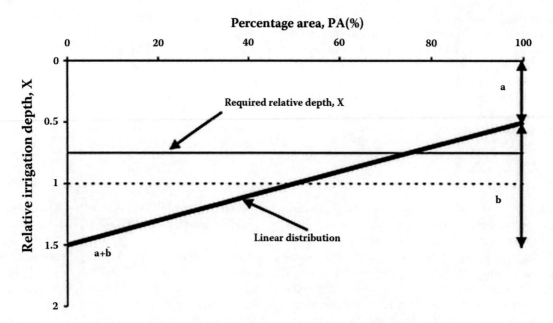

Fig. 1 A linear water application model for drip irrigation.

$X = (a + b)$ This irrigation schedule is based on the maximum emitter flow or maximum irrigation application.
The whole field is under deficit condition except the point of maximum water application.
There is no deep percolation.

An optimal irrigation schedule for maximum economic return was determined[9] based on cost of water, price of the yield, and damage such as environmental pollution and groundwater contamination caused by over irrigation. Different irrigation strategies require different amounts of water application. Water conservation and environmental protection are realized by comparing any two of the irrigation strategies.[10]

CONCLUSION

Drip irrigation is an irrigation method that can distribute irrigation water uniformly and directly into the root zone of crops. It is one of the most efficient irrigation methods and can be designed and scheduled to meet the water requirement of crop and produce maximum yield in the field.

When the drip irrigation system is designed with high uniformity, the slope b of the straight line of water application function (Fig. 1) can be controlled to achieve the desired variation. In this case the conventional irrigation schedule, $X = a$, optimal irrigation schedule, X_0, which is a location between a and $(a + b)$, and the irrigation schedule for environmental protection, $X = a + b$, are in close proximity. This closeness shows that the drip irrigation system can achieve optimal economic return, water conservation, and environmental protection.

REFERENCES

1. Wu, I.P.; Gitlin, H.M. *Design of Drip Irrigation Lines*; HAES Technical Bulletin 96, University of Hawaii: Honolulu, HI, 1974; 29 pp.
2. Christiansen, J.E. The uniformity of application of water by sprinkler systems. Agric. Eng. **1941**, *22*, 89–92.
3. Howell, T.A.; Hiler, E.A. Trickle irrigation lateral design. Trans. ASAE **1974**, *17* (5), 902–908.
4. Bralts, V.F.; Segerlind, L.J. Finite elements analysis of drip irrigation submain unit. Trans. ASAE **1985**, *28*, 809–814.
5. Bralts, V.F.; Wu, I.P.; Gitlin, H.M. Manufacturing variation and drip irrigation uniformity. Trans. ASAE **1981**, *24* (1), 113–119.
6. Wu, I.P.; Barragan, J. Design criteria for microirrigation systems. Trans. ASAE **2000**, *43* (5), 1145–1154.
7. Seginer, I. A note on the economic significance of uniform water application. Irrig. Sci. **1978**, *1*, 19–25.
8. Wu, I.P. Linearized water application function for drip irrigation schedules. Trans. ASAE **1988**, *31* (6), 1743–1749.
9. Wu, I.P. Optimal scheduling and minimizing deep seepage in microirrigation. Trans. ASAE **1995**, *38* (5), 1385–1392.
10. Barragan, J.; Wu, I.P. Optimal scheduling of a microirrigation system under deficit irrigation. J. Agric. Eng. Res. **2001**, *80* (2), 201–208.

Irrigation: Efficiency

Terry A. Howell
Conservation and Production Research Laboratory, Agricultural Research Service (USDA-ARS), U.S. Department of Agriculture, Bushland, Texas, U.S.A.

Abstract
Irrigation efficiency is a basic engineering term used in irrigation science to characterize irrigation performance, evaluate irrigation water use, and to promote better or improved use of water resources, particularly those used in agriculture and turf/landscape management. It affects the economics of irrigation, the amount of water needed to irrigate a specific land area, the spatial uniformity of the crop and its yield, the amount of water that might percolate beneath the crop root zone, the amount of water that can return to surface sources for downstream uses or to groundwater aquifers that might supply other water uses, and the amount of water lost to unrecoverable sources.

INTRODUCTION

Irrigation efficiency is a critical measure of irrigation performance in terms of the water required to irrigate a field, farm, basin, irrigation district, or an entire watershed. The value of irrigation efficiency and its definition are important to the societal views of irrigated agriculture and its benefit in supplying the high quality, abundant food supply required to meet our growing world's population. "Irrigation efficiency" is a basic engineering term used in irrigation science to characterize irrigation performance, evaluate irrigation water use, and to promote better or improved use of water resources, particularly those used in agriculture and turf/landscape management.[1–4] Irrigation efficiency is defined in terms of: 1) the irrigation system performance, 2) the uniformity of the water application, and 3) the response of the crop to irrigation. Each of these irrigation efficiency measures is interrelated and will vary with scale and time. Fig. 1 illustrates several of the water transport components involved in defining various irrigation performance measures. The spatial scale can vary from a single irrigation application device (a siphon tube, a gated pipe gate, a sprinkler, a microirrigation emitter) to an irrigation set (basin plot, a furrow set, a single sprinkler lateral, or a microirrigation lateral) to broader land scales (field, farm, an irrigation canal lateral, a whole irrigation district, a basin or watershed, a river system, or an aquifer). The timescale can vary from a single application (or irrigation set), a part of the crop season (preplanting, emergence to bloom or pollination, or reproduction to maturity), the irrigation season, to a crop season, or a year, partial year (premonsoon season, summer, etc.), or a water year (typically from the beginning of spring snow melt through the end of irrigation diversion, or a rainy or monsoon season), or a period of years (a drought or a "wet" cycle). Irrigation efficiency affects the economics of irrigation, the amount of water needed to irrigate a specific land area, the spatial uniformity of the crop and its yield, the amount of water that might percolate beneath the crop root zone, the amount of water that can return to surface sources for downstream uses or to groundwater aquifers that might supply other water uses, and the amount of water lost to unrecoverable sources (salt sink, saline aquifer, ocean, or unsaturated vadose zone).

The volumes of the water for the various irrigation components are typically given in units of depth (volume per unit area) or simply the volume for the area being evaluated. Irrigation water application volume is difficult to measure, so it is usually computed as the product of water flow rate and time. This places emphasis on accurately measuring the flow rate. It remains difficult to accurately measure water percolation volumes groundwater flow volumes, and water uptake from shallow groundwater.

IRRIGATION SYSTEM PERFORMANCE EFFICIENCY

Irrigation water can be diverted from a storage reservoir and transported to the field or farm through a system of canals or pipelines; it can be pumped from a reservoir on the farm and transported through a system of farm canals or pipelines; or it might be pumped from a single well or a series of wells through farm canals or pipelines. Irrigation districts often include small to moderate size reservoirs to regulate flow and to provide short-term storage to manage the diverted water with the on-farm demand. Some on-farm systems include reservoirs for storage or regulation of flows from multiple wells.

Water Conveyance Efficiency

The conveyance efficiency is typically defined as the ratio between the water that reaches a farm or field and

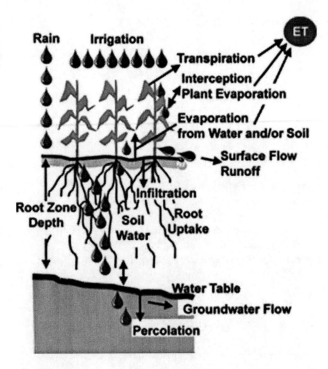

Fig. 1 Illustration of the various water transport components needed to characterize irrigation efficiency.

that diverted from the irrigation water source.[1,3,4] It is defined as

$$E_c = 100 \frac{V_F}{V_t} \quad (1)$$

where E_c is the conveyance efficiency (%), V_f is the volume of water that reaches the farm or field (m³), and V_t is the volume of water diverted (m³) from the source. E_c also applies to segments of canals or pipelines, where the water losses include canal seepage or leaks in pipelines. The global E_c can be computed as the product of the individual component efficiencies, E_{ci}, where i represents the segment number. Conveyance losses include any canal spills (operational or accidental) and reservoir seepage and evaporation that might result from management as well as losses resulting from the physical configuration or condition of the irrigation system. Typically, conveyance losses are much lower for closed conduits or pipelines[4] compared with unlined or lined canals. Even the conveyance efficiency of lined canals may decline over time due to material deterioration or poor maintenance.

Application Efficiency

Application efficiency relates to the actual storage of water in the root zone to meet the crop water needs in relation to the water applied to the field. It might be defined for individual irrigation or parts of irrigations (irrigation sets). Application efficiency includes any application losses to evaporation or seepage from surface water channels or furrows, any leaks from sprinkler or drip pipelines, percolation beneath the root zone, drift from sprinklers, evaporation of droplets in the air, or runoff from the field. Application efficiency is defined as

$$E_a = 100 \frac{V_s}{V_F} \quad (2)$$

where E_a is the application efficiency (%), V_s is the irrigation needed by the crop (m³), and V_f is the water delivered to the field or farm (m³). The root zone may not need to be fully refilled, particularly if some root zone water-holding capacity is needed to store possible or likely rainfall. Often, V_s is characterized as the volume of water stored in the root zone from the irrigation application. Some irrigations may be applied for reasons other than meeting the crop water requirement (germination, frost control, crop cooling, chemigation, fertigation, or weed germination). The crop need is often based on the "beneficial water needs."[5] In some surface irrigation systems, the runoff water that is necessary to achieve good uniformity across the field can be recovered in a "tailwater pit" and recirculated with the current irrigation or used for later irrigations, and V_f should be adjusted to account for the "net" recovered tailwater. Efficiency values are typically site specific. Table 1 provides a range of typical farm and field irrigation application efficiencies[6–8] and potential or attainable efficiencies for different irrigation methods that assumes irrigations are applied to meet the crop need.

Storage Efficiency

Since the crop root zone may not need to be refilled with each irrigation, the storage efficiency has been defined.[4] The storage efficiency is given as

$$E_s = 100 \frac{V_s}{V_{rz}} \quad (3)$$

where E_s is the storage efficiency (%) and V_{rz} is the root zone storage capacity (m³). The root zone depth and the water-holding capacity of the root zone determine V_{rz}. The storage efficiency has little utility for sprinkler or microirrigation because these irrigation methods seldom refill the root zone, while it is more often applied to surface irrigation methods.[4]

Seasonal Irrigation Efficiency

The seasonal irrigation efficiency is defined as

$$E_i = 100 \frac{V_b}{V_F} \quad (4)$$

where E_i is the seasonal irrigation efficiency (%) and V_b is the water volume beneficially used by the crop (m³). V_b is somewhat subjective,[4,5] but it basically includes the

Table 1 Example of farm and field irrigation application efficiency and attainable efficiencies.

Irrigation method	Field efficiency (%)			Farm efficiency (%)		
	Attainable	Range	Average	Attainable	Range	Average
Surface						
Graded furrow	75	50–80	65	70	40–70	65
w/tailwater reuse	85	60–90	75	85	–	–
Level furrow	85	65–95	80	85	–	–
Graded border	80	50–80	65	75	–	–
Level basins	90	80–95	85	80	–	–
Sprinkler						
Periodic move	80	60–85	75	80	60–90	80
Side roll	80	60–85	75	80	60–85	80
Moving big gun	75	55–75	65	80	60–80	70
Center pivot						
Impact heads w/end gun	85	75–90	80	85	75–90	80
Spray heads wo/end gun	95	75–95	90	85	75–95	90
LEPA[a] wo/end gun	98	80–98	95	95	80–98	92
Lateral move						
Spray heads w/hose feed	95	75–95	90	85	80–98	90
Spray heads w/canal feed	90	70–95	85	90	75–95	85
Microirrigation						
Trickle	95	70–95	85	95	75–95	85
Subsurface drip	95	75–95	90	95	75–95	90
Microspray	95	70–95	85	95	70–95	85
Water table control						
Surface ditch	80	50–80	65	80	50–80	60
Subsurface drain lines	85	60–80	75	85	65–85	70

[a]LEPA is low energy precision application.
Source: Howell,[6] Merriam and Keller,[7] and Hart.[11]

required crop evapotranspiration (ET_c) plus any required leaching water (V_l) for salinity management of the crop root zone.

Leaching Requirement (or the Leaching Fraction)

The leaching requirement,[9] also called the leaching fraction, is defined as

$$L_r = \frac{V_d}{V_F} = \frac{EC_i}{EC_d} \quad (5)$$

where L_r is the leaching requirement, V_d is the volume of drainage water (m³), V_f is the volume of irrigation (m³) applied to the farm or field, EC_i is the electrical conductivity of the irrigation water (dS m⁻¹), and EC_d is the electrical conductivity of the drainage water (dS m⁻¹). The L_r is related to the irrigation application efficiency, particularly when drainage is the primary irrigation loss component. The L_r would be required "beneficial" irrigation use ($V_l \equiv L_r V_i$), so only V_d greater than the minimum required leaching should reduce irrigation efficiency. Then, the irrigation efficiency can be determined by combining Eqs. 4 and 5.

$$E_i = 100\left(\frac{V_b}{V_F} + L_r\right) \quad (6)$$

Burt et al.[5] defined the "beneficial" water use to include possible off-site needs to benefit society (riparian needs or wildlife or fishery needs). They also indicated that V_f should not include the change in the field or farm storage of water, principally soil water but it could include field (tailwater pits) or farm water storage (a reservoir) that wasn't used within the time frame that was used to define E_i.

IRRIGATION UNIFORMITY

The fraction of water used efficiently and beneficially is important for improved irrigation practice. The uniformity of the applied water significantly affects irrigatic efficiency. The uniformity is a statistical property of the applied water's distribution. This distribution depends on many factors that are related to the method of irrigation, soil topography, soil hydraulic or infiltration characteristics, and hydraulic characteristics (pressure, flow rate, etc.), of the irrigation system. Irrigation application distributions are usually based on depths of water (volume per unit area); however, for microirrigation systems they are usually based on emitter flow volumes because the entire land area is not typically wetted.

Christiansen's Uniformity Coefficient

Christiansen[10] proposed a coefficient intended mainly for sprinkler system based on the catch volumes given as

$$C_U = 100 \left[\frac{1 - (\sum |X - \bar{x}|)}{\sum X} \right] \quad (7)$$

where C_U is the Christiansen's uniformity coefficient in percent, X is the depth (or volume) of water in each of the equally spaced catch containers in mm or mL, and $\bar{x}$ is the mean depth (volume) of the catch (mm or mL). For C_U values >70%, Hart[11] and Keller and Bliesner[8] presented

$$C_U = 100 \left[1 - \left(\frac{\sigma}{x}\right)\left(\frac{2}{\pi}\right)^{0.5} \right] \quad (8)$$

where σ is the standard deviation of the catch depth (mm) or volume (mL). Eq. 8 approximates the normal distribution for the catch amounts.

The C_U should be weighted by the area represented by the container[12] when the sprinkler catch containers intentionally represent unequal land areas, as is the case for catch containers beneath a center pivot. Heermann and Hein[12] revised the C_U formula Eq. 8 to reflect the weighted area, particularly intended for a center pivot sprinkler, as follows:

$$C_{U(H\&H)} = 100 \left\{ 1 - \left[\frac{\sum S_i \left| V_i - \left(\frac{\sum V_i S_i}{\sum S_i}\right) \right|}{\sum (V_i S_i)} \right] \right\} \quad (9)$$

where S_i is the distance (m) from the pivot to the ith equally spaced catch container and V_i is the volume of the catch in the ith container (mm or mL).

Low-Quarter Distribution Uniformity

The distribution uniformity represents the spatial evenness of the applied water across a field or a farm as well as within a field or farm. The general form of the distribution uniformity can be given as

$$D_{U_p} = 100 \left(\frac{\bar{V}_p}{\bar{V}_f} \right) \quad (10)$$

where D_{Up} is the distribution uniformity (%) for the lowest p fraction of the field or farm (lowest one-half $p = 1/2$, lowest one-quarter $p = 1/4$), $\bar{V}_p$ is the mean application volume (m³), and $\bar{V}_f$ is the mean application volume (m³) for the whole field or farm. When $p = 1/2$ and $C_U > 70\%$, then the D_U and C_U are essentially equal.[13] The USDA-NRCS (formerly, the Soil Conservation Service) has widely used D_{Ulq} ($p = 1/4$) for surface irrigation to access the uniformity applied to a field, i.e., by the irrigation volume (amount) received by the lowest one-quarter of the field from applications for the whole field. Typically, D_{Up} is based on the postirrigation measurement[5] of water volume that infiltrates the soil because it can more easily be measured and better represents the water available to the crop. However, the postirrigation infiltrated water ignores any water intercepted by the crop and evaporated and any soil water evaporation that occurs before the measurement. Any water that percolates beneath the root zone or the sampling depth will also be ignored.

The D_U and C_U coefficients are mathematically interrelated through the statistical variation (coefficient of variation, $\sigma/\bar{x}$, C_v) and the type of distribution. Warrick[13] presented relationships between D_U and C_U for normal, log-normal, uniform, specialized power, beta- and gamma-distributions of applied irrigations.

Emission Uniformity

For microirrigation systems, both the C_U and D_U concepts are impractical because the entire soil surface is not wetted. Keller and Karmeli[14] developed an equation for microirrigation design as follows

$$E_U = 100 \left[1 - 1.27(C_{vm})n^{-1/2} \right] \left(\frac{q_m}{\bar{q}} \right) \quad (11)$$

where E_U is the design emission uniformity (%), C_{vm} is the manufacturer's coefficient of variability in emission device flow rate (1/h), n is the number of emitters per plant, q_m is the minimum emission device flow rate (1/h) at the minimum system pressure, and $\bar{q}$ is the mean emission device flow rate (1/h). This equation is based on the D_{Ulq} concept,[4] and includes the influence of multiple emitters per plant that each may have a flow rate from a population of random flow rates based on the emission device manufacturing variation. Nakayama, Bucks, and Clemmens[15] developed a design coefficient based more closely on the C_U concept for emission device flow rates from a normal distribution given as

$$C_{Ud} = 100\left(1 - 0.798(C_{vm})n^{-1/2}\right) \quad (12)$$

where C_{Ud} is the coefficient of design uniformity in percent and the numerical value, 0.798, is

$$\left(\frac{2}{\pi}\right)^{0.5}$$

from Eq. 8.

Many additional factors affect microirrigation uniformity including hydraulic factors, topographic factors, and emitter plugging or clogging.

WATER USE EFFICIENCY

The previous sections discussed the engineering aspects of irrigation efficiency. Irrigation efficiency is clearly influenced by the amount of water used in relation to the irrigation water applied to the crop and the uniformity of the applied water. These efficiency factors impact irrigation costs, irrigation design, and more important, in some cases, the crop productivity. Water use efficiency (WUE) has been the most widely used parameter to describe irrigation effectiveness in terms of crop yield. Viets[16] defined WUE as

$$\text{WUE} = \frac{Y_g}{\text{ET}} \quad (13)$$

where WUE is water use efficiency ($kg\,m^{-3}$), Y_g is the economic yield ($g\,m^{-2}$), and ET is the crop water use (mm). Water use efficiency is usually expressed by the economic yield, but it has been historically expressed as well in terms of the crop dry matter yield (either total biomass or aboveground dry matter). These two WUE bases (economic yield or dry matter yield) have led to some inconsistencies in the use of the WUE concept. The transpiration ratio (transpiration per unit dry matter) is a more consistent value that depends primarily on crop species and the environmental evaporative demand,[17] and it is simply the inverse of WUE expressed on a dry matter basis.

Irrigation Water Use Efficiency

The previous discussion of WUE does not explicitly explain the crop yield response to irrigation. Water use efficiency is influenced by the crop water use (ET). Bos[3] defined a term for WUE to characterize the influence of irrigation on WUE as

$$\text{WUE} = \frac{(Y_{gi} - Y_{gd})}{(\text{ET}_i - \text{ET}_d)} \quad (14)$$

where WUE is irrigation water use efficiency ($kg\,m^{-3}$), Y_{gi} is the economic yield ($g\,m^{-2}$) for irrigation level i, Y_{gd} is the dryland yield ($g\,m^{-2}$; actually, the crop yield without irrigation), ET_i is the evapotranspiration (mm) for irrigation level i, and ET_d is the evapotranspiration of the dryland crops (or of the ET without irrigation). Although Eq. (14) seems easy to use, both Y_{gd} and ET_d are difficult to evaluate. If the purpose is to compare irrigation and dryland production systems, then dryland rather than non-irrigated conditions should be used. If the purpose is to compare irrigated regimes with an unirrigated regime, then appropriate values for Y_{gd} and ET_d should be used. Often, in most semiarid to arid locations, Y_{gd} may be zero. Bos[3] defined irrigation WUE as

$$\text{IWUE} = \frac{(Y_{gi} - Y_{gd})}{\text{IRR}_i} \quad (15)$$

where IWUE is the irrigation efficiency ($kg\,m^{-3}$) and IRR_i is the irrigation water applied (mm) for irrigation level i. In Eq. 15, Y_{gd} may be often zero in many arid situations.

CONCLUSION

Irrigation efficiency is an important engineering term that involves understanding soil and agronomic sciences to achieve the greatest benefit from irrigation. The enhanced understanding of irrigation efficiency can improve the beneficial use of limited and declining water resources needed to enhance crop and food production from irrigated lands.

REFERENCES

1. Israelsen, O.R.; Hansen, V.E. *Irrigation Principles and Practices*, 3rd Ed.; Wiley: New York, 1962; 447 pp.
2. ASCE. Describing irrigation efficiency and uniformity. J. Irrig. Drain. Div., ASCE **1978**, *104* (IR1), 35–41.
3. Bos, M.G. Standards for irrigation efficiencies of ICID. J. Irrig. Drain. Div., ASCE **1979**, *105* (IR1), 37–43.
4. Heermann, D.F.; Wallender, W.W.; Bos, M.G. Irrigation efficiency and uniformity. In *Management of Farm Irrigation Systems*; Hoffman, G.J., Howell, T.A., Solomon, K.H., Eds.; Am. Soc. Agric. Engrs.: St. Joseph, MI, 1990; 125–149.
5. Burt, C.M.; Clemmens, A.J.; Strelkoff, T.S.; Solomon, K.H.; Bliesner, R.D.; Hardy, L.A.; Howell, T.A.; Eisenhauer, D.E. Irrigation performance measures: efficiency and uniformity. J. Irrig. Drain. Eng. **1997**, *123* (3), 423–442.
6. Howell, T.A. Irrigation efficiencies. In *Handbook of Engineering in Agriculture*; Brown, R.H., Ed.; CRC Press: Boca Raton, FL, 1988; Vol. I, 173–184.

7. Merriam, J.L.; Keller, J. *Farm Irrigation System Evaluation: A Guide for Management*; Utah State University: Logan, UT, 1978; 271 pp.
8. Keller, J.; Bliesner, R.D. *Sprinkle and Trickle Irrigation*; The Blackburn Press: Caldwell, NJ, 2000; 652 pp.
9. U.S. Salinity Laboratory Staff. *Diagnosis and Improvement of Saline and Alkali Soils*; Handbook 60; U.S. Government Printing Office: Washington, DC, 1954; 160 pp.
10. Christiansen, J.E. *Irrigation by Sprinkling*; California Agric. Exp. Bull. No. 570; University of California: Berkeley, CA, 1942; 94 pp.
11. Hart, W.E. Overhead irrigation by sprinkling. Agric. Eng. **1961**, *42* (7), 354–355.
12. Heermann, D.F.; Hein, P.R. Performance characteristics of self-propelled center-pivot sprinkler machines. Trans. ASAE **1968**, *11* (1), 11–15.
13. Warrick, A.W. Interrelationships of irrigation uniformity terms. J. Irrig. Drain. Eng., ASCE **1983**, *109* (3), 317–332.
14. Keller, J.; Karmeli, D. *Trickle Irrigation Design*; Rainbird Sprinkler Manufacturing: Glendora, CA, 1975; 133 pp.
15. Nakayama, F.S.; Bucks, D.A.; Clemmens, A.J. Assessing trickle emitter application uniformity. Trans. ASAE **1979**, *22* (4), 816–821.
16. Viets, F.G. Fertilizers and the efficient use of water. Adv. Agron. **1962**, *14*, 223–264.
17. Tanner, C.B.; Sinclair, T.R. Efficient use of water in crop production: research or re-search? In *Limitations to Efficient Water Use in Crop Production*; Taylor, H.M., Jordan, W.R., Sinclair, T.R., Eds.; Am. Soc. Agron., Crop Sci. Soc. Am., Soil Sci. Soc. Am.: Madison, WI, 1983; 1–27.

Irrigation: Erosion

David L. Bjorneberg
Northwest Irrigation and Soil Research Lab., Agricultural Research Service (USDA-ARS), U.S. Department of Agriculture, Kimberly, Idaho, U.S.A.

Abstract
Irrigation is essential for global food production. However, irrigation erosion can limit the ability of irrigation systems to reliably produce food and fiber in the future. The factors affecting soil erosion from irrigation are the same as rainfall—water detaches and transports sediment. However, there are some unique differences in how the factors occur during irrigation and in our ability to manage the application of water that causes the erosion. All surface irrigation entails water flowing over soil. Soil type, field slope, and flow rate all affect surface irrigation erosion, with flow rate being the main factor that can be managed. Ideally, sprinkler irrigation will have no runoff, but application rates on moving irrigation systems can exceed the soil infiltration rate, resulting in runoff and erosion. Using tillage practices to increase soil surface storage and selecting sprinklers with lower application rates will reduce sprinkler-irrigation runoff. Irrigation can be managed to minimize erosion and maintain productivity.

INTRODUCTION

Irrigation is vital to food production in the world. However, irrigation-induced soil erosion reduces productivity of irrigated land and can cause off-site water quality problems. Surface irrigation utilizes the soil to distribute water through the field. Water flowing over soil inherently detaches and transports sediment. Sprinkler and drip irrigation distribute water through fields in pipes, eliminating erosion from water distribution, but erosion can still occur if water is applied faster than it can infiltrate into the soil. This entry will briefly discuss the importance of irrigation to global food production and then discuss the important factors affecting soil erosion for surface- and sprinkler-irrigated land. Much of the information will focus on the United States, with international information included when possible.

IMPORTANCE OF IRRIGATION

Irrigated agriculture contributes a disproportionate amount to global food production. The most cited statistics indicate that irrigated cropland produces about one-third of the world's crop production on only 16% of the cropland that is irrigated.[1] In the United States, farms with all cropland irrigated account for only 8% of the total cropland and about half of the total irrigated land.[2] These farms produce 33% of the market value of crops and 12% of the market value of livestock. Over half of the crop value (55%) is produced on farms with some irrigated land, and these farms account for only 26% of the total cropland in the United States.[3] In some areas, irrigation provides essentially all of the water necessary for crop growth. In other areas, irrigation provides only a small portion of the total crop water requirement but reduces the potential for water stress during critical periods.

While irrigation is critical to global food production, applying water to soil can cause erosion. This is especially true with surface irrigation, where the soil conveys and distributes water through a field by gravity. Sprinkler irrigation and microirrigation use pipes to distribute water through the field. Surface irrigation is generally thought to cause more erosion than sprinkler irrigation; however, erosion can occur any time water flows over soil. Water can be applied with sprinkler irrigation so no runoff occurs, and therefore, no erosion will occur. However, there are situations, especially with moving irrigation systems like center pivots, where water is applied faster than it can infiltrate into the soil, resulting in ponding and, possibly, runoff.

UNIQUE ASPECTS OF IRRIGATION EROSION

The factors affecting soil erosion from irrigation are the same as rainfall. Water detaches and transports sediment in both situations. However, there are some unique differences in how the factors occur with irrigation.[4] For example, rainfall occurs relatively uniformly over an entire field, whereas irrigation is seldom applied to an entire field at the same time. Irrigation is a controlled procedure where water is applied to a specific field, or portion of a field, at a specific time. This can affect the hydrology of the erosion processes on surface- and sprinkler-irrigated fields. A center pivot, for example, is essentially a moving storm

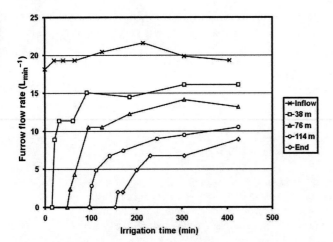

Fig. 1 Furrow flow rate with time at five points in a 150 m–long furrow.

that covers only 1%–2% of the field at any given time. This results in unique runoff conditions where water can do the following: 1) flow parallel to the lateral under similar conditions as rainfall; 2) flow from wet soil onto dry soil if the lateral is moving downhill; or 3) flow onto wet soil if the lateral is moving uphill.

In surface irrigation, water flow rate decreases with distance during surface irrigation as water infiltrates. Furrow flow rates also increase with time as infiltration rate decreases (Fig. 1). This creates a condition where sediment can be detached on the upper end of the field and deposited on the lower end. Trout[4] documented erosion rates on the upper end of a field that were 6 to 20 times greater than the field-average erosion rates. Fig. 2 shows eroded furrows on the upper end of a field after one furrow irrigation. During rainfall, raindrops wet the soil surface and detach soil particles. As runoff begins, rills form in wet soil. In contrast, irrigation furrows are formed prior to irrigation, and water flows onto initially dry soil. Furrows with initially dry soil

Fig. 2 Eroded furrows on the upper end of a furrow-irrigated field in Idaho with approximately 1% slope.

have greater soil erosion than furrows that were prewet immediately before furrow irrigation.[5] Irrigation water flowing in furrows is not exposed to falling raindrops that can increase sediment detachment and decrease deposition.

The quality of irrigation water can vary dramatically among water sources, or even within an irrigation tract if drainage water is reused. Conversely, electrolyte concentration of rainfall is quite consistent. Electrolyte concentration in irrigation water affects erosion for both surface and sprinkler irrigation. Furrow-irrigation erosion was greater on a silt loam when irrigation water had low electrical conductivity (EC = 0.7 dS m^{-1}) and high sodium adsorption ratio (SAR = 9.1) compared with low EC (0.5 dS m^{-1}) and low SAR (0.9), high EC (2.1 dS m^{-1}) and low SAR (0.5), and high EC (1.7 dS m^{-1}) and high SAR (9.3).[6] Soil erosion was also greater with low-EC water in laboratory and field rainfall simulation studies.[7,8] Lower electrolyte concentrations in water cause greater dispersion of soil particles, which tends to reduce infiltration and increase soil loss.[9]

SURFACE-IRRIGATION EROSION

Surface irrigation continues to be the most common method of irrigation in the world. The four countries with the most irrigated land are India (60.8 Mha), China (57.8 Mha), United States (22.4 Mha), and Pakistan (19.6 Mha).[10] These four countries account for 58% of the irrigated area in the world. All other countries have less than 10 Mha of irrigated land.[10] According to the country fact sheets on the Food and Agriculture Organization's Aquastat Web site,[11] surface irrigation is used on 97% of the irrigated land in India, 94% in China, 44% in the United States, and 100% in Pakistan.

Koluvek et al.[12] provided a good overview of soil erosion from irrigation in the United States. Unfortunately, this information has not been updated, and similar information is not readily available from other countries, so it is difficult to track erosion trends on irrigated lands. Some early studies documented erosion rates as great as 145 Mg ha^{-1} in 1 h[13] and 40 Mg ha^{-1} in 30 min.[14] While these rates represent extreme conditions that can occur, not typical season-long soil loss rates, these studies indicate the potential severity of the problem. One study measured annual soil losses of 1 to 141 Mg ha^{-1} from 33 fields with silt loam soils.[15] The greatest soil loss occurred on a sugar beet (*Beta vulgaris* L.) field with 4% slope. The authors noted that erosion increased sharply when field slope was greater than 1%. Close-growing crops like alfalfa (*Medicago sativa* L.) or wheat (*Triticum aestivum* L.) on fields with 1% slope had annual soil loss of less than 1 Mg ha^{-1}. A recent study in the same area documented that average soil loss from an 80,000 ha irrigated watershed decreased from 450 kg ha^{-1} in 1970 to less than 50 kg ha^{-1} in 2005.[16] This watershed was approximately 90% furrow irrigated in 1970 and 60% furrow irrigated in 2005. Another study measured daily sediment loads of 0.4 kg ha^{-1} in a watershed with no furrow irrigation compared to 19 kg ha^{-1} in a watershed with 58% of the cropland furrow irrigated.[17] Irrigation method explained 67% of the variation in soil loss measured in April and May in these nine watersheds.

The main factors affecting surface-irrigation erosion are soil type, field slope and flow rate. Soil erosion is typically not a concern where field slopes are less than 0.5% (Fig. 3). However, erosion tends to increase exponentially for increasing inflow rate and field slope, with an exponent between 1 and 3 for flow rate, and between 2 and 3 for slope.[12,18,19,20] Increasing inflow rate 20% increased erosion 30% and 70% on the upper quarter of two fields.[4] Increasing inflow rate another 20% increased erosion 50% and 100%, which indicates that the exponent between erosion and flow rate was between 2 and 3.[4] Fig. 4 shows soil loss from 10 furrows during a 4 h irrigation at Kimberly, Idaho, with inflow rates randomly set for each furrow.

Reducing field slope by grading the land is a costly practice that is not feasible in most situations compared with alternatives like installing a sprinkler-irrigation system. Reducing inflow rate is a good practice as long as the water advances down the field fast enough to uniformly irrigate the field. Slow water advance rates from low inflow rates cause overirrigation on the inflow end of the field and underirrigation on the lower end of the field due to differences in infiltration opportunity time. This results in poor distribution uniformity but little runoff. Soil loss decreases as distribution uniformity decreases.[20] An excellent practice for reducing irrigation erosion without affecting irrigation uniformity is applying small amounts of polyacrylamide (PAM) with irrigation water.[21,22] Dissolving 10 mg L^{-1} of high-molecular-weight, anionic PAM in furrow-irrigation inflow can reduce soil loss 60%–99% compared with untreated furrows. Other technologies like filter strips and sediment ponds on the lower end of the field remove sediment from the water rather than reducing erosion from occurring on the field.

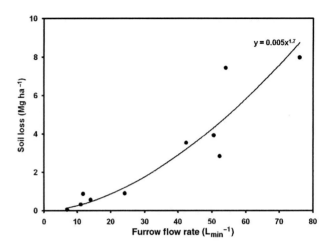

Fig. 4 Relationship between furrow flow rate and soil loss for 10 and 30 m-long furrows with randomly set inflow rates.

SPRINKLER-IRRIGATION EROSION

Ideally, sprinkler-irrigation systems are designed and managed to have all applied water infiltrate into the soil where it was applied. When all water infiltrates, there is no runoff or soil erosion. Solid-set (sprinklers located in the same position for the entire irrigation season) or set-move (sprinklers remaining in a location for 12 to 24 h, then moved to the next set) irrigation systems usually apply water at a low rate (e.g., 3 to 6 mm h^{-1}), so irrigation application rate does not exceed the soil infiltration rate and no soil erosion occurs. Moving irrigation systems, like center pivots, traveling guns, and lateral-move systems, often apply water faster than the infiltration rate. This occurs because the irrigation system must apply enough water as it moves across the field to meet crop water needs until the next time it irrigates that portion of the field. For example, a center pivot operating at 60 h per revolution needs to apply 20 mm per revolution to meet an 8 mm d^{-1} crop water requirement. The irrigation application rate increases with distance from the center pivot because the lateral irrigates more land as the radial distance from the pivot point increases.[23] Near the pivot point, the mean application rate could be about 4 mm h^{-1} (assuming 15 m wetting diameter). Near the end of the pivot, about 400 m, the mean application rate would be about 60 mm h^{-1}. An important fact about moving irrigation systems is that the application rate is a function of irrigation system capacity, or system flow rate. Operating the system faster decreases only the application depth, not the application rate. For example, the same center pivot operating at 48 h per revolution will apply 16 mm of water at the same application rates.

Fig. 3 Level furrow irrigation in Arizona. (Photo by Jeff Vanuga, USDA Natural Resources Conservation Service [NRCSAZ02037]).

Center-pivot irrigation is the most popular type of irrigation system in the United States. According to the United States Department of Agriculture (USDA) National Agricultural Statistics Service, center-pivot irrigation was used on 47% of the irrigated land and 83% of the sprinkler-irrigated land in 2008, an increase from 25% of the irrigated land in 1988.[2] More land was irrigated by center pivots in the United States in 2008 (10.4 Mha) than all types of gravity irrigation combined (8.9 Mha). As center pivots gained popularity, researchers began to consider runoff potential, mainly to efficiently apply irrigation water. Most sprinkler-irrigation studies were not concerned with soil erosion, probably because the effects of sprinkler-irrigation erosion tend to occur within the field rather than off site.

A 1969 study evaluated center-pivot runoff from a theoretical perspective and showed the importance of modifying infiltration parameters, determined from pond infiltration tests, for the low initial application rate that occurs with moving irrigation systems.[23] Their theoretical evaluation showed that 0%–40% of the applied water could run off with typical operating conditions. A 1971 field study documented 11%–41% runoff on four center pivots operated by farmers.[24] Runoff with center-pivot irrigation became a more important issue as low-pressure sprinklers began to be used to reduce energy costs. Early types of low-pressure sprinklers applied water to a smaller area, which increased application rates and potential runoff.[25] Low-pressure sprinklers (40 and 100 Pa) averaged 69 or 70 mm of runoff compared with 8 to 10 mm of runoff for high-pressure sprinklers (170 and 345 Pa) during a 4-year field study.[26] Reducing pressure from 380 to 140 Pa increased irrigation runoff 30% for a center pivot with impact sprinklers.[27] Peak application rate at the outer end of a center pivot would be about 30 mm h^{-1} for a high-pressure impact sprinkler with 20 m wetted radius and more than 100 mm h^{-1} for a low-pressure spray sprinkler with 5 m wetted radius.[25] Fig. 5 shows two sprinkler application rate curves with time and an infiltration rate curve. The volume of water applied when application rate exceeds infiltration is potential runoff. All of this water may not run off if some is ponded or stored on the soil surface.

Many types of sprinklers are now available for center pivots. Some apply water in defined streams with a wetted diameter over 20 m with nozzle pressure of 200 Pa or less. Others distribute water evenly over the wetted area with various combinations of droplet sizes. Various sprinkler designs are the result of manufacturers trying to reduce the kinetic energy applied to the soil during irrigation, so all applied water can infiltrate. Kincaid[28] developed a model in 1996 to estimate kinetic energy per unit drop volume for common sprinkler types. Calculating area-weighted kinetic energies per unit drop volume for individual sprinklers showed that sprinklers with the smallest drop size distributions had the lowest kinetic energy. Sprinklers with smaller sized drops tend to have smaller

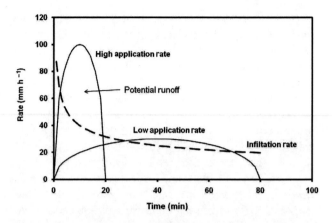

Fig. 5 Example of soil infiltration rate and sprinkler application rates for high- and low-application-rate sprinklers. Runoff potentially occurs when sprinkler application rate exceeds the soil infiltration rate.

wetted diameters because small drops cannot travel as far as large drops. Larger drops travel farther and therefore cover a greater portion of the circular wetted area for an individual sprinkler. A smaller wetted diameter also results in a higher application rate when sprinkler application patterns are overlapped like occurs on a center pivot. An alternative method for characterizing sprinkler kinetic energy is calculating the rate that energy is applied to the soil, or specific power, as a function of radial distance from the sprinkler.[29] The specific power distribution is energy per drop volume multiplied by the application rate and can be overlapped, like water application rate, to develop a composite specific power profile for a sprinkler system. A flat-plate sprinkler with small-sized drops had higher average composite specific power than two other sprinklers with larger drop sizes and larger wetted diameter.[29] Recent field research on small plots showed that soil erosion was significantly greater with the flat-plate spray sprinkler compared with the two other sprinklers with larger drop size distributions.[30] This directly contradicts previous conventional thinking that sprinklers with smaller drops caused less erosion.

The most effective way to control sprinkler-irrigation erosion is to eliminate runoff, which also increases water application efficiency. One way to control runoff is to increase water storage on the soil surface.[25] Reservoir tillage is a practice that forms small pits in the soil to store water. Each pit can hold 5 to 10 L of water.[31] This is especially important on sloping fields. Reservoir tillage reduced runoff 68% and soil erosion 92% during a 50 mm simulated irrigation on a field with 10% slope.[32] Runoff was not different when field slope was only 1%. Increasing surface residue also decreases sprinkler-irrigation runoff similar to rainfall runoff.[27] Disking corn stubble prior to planting, which left approximately 30% of the soil surface covered with crop residue, reduced runoff to 17% of applied irrigation compared with 25% runoff for moldboard plow plots in a 4 years study.[33] In addition to reducing runoff, disking also reduced soil loss about 50% compared with moldboard plowing.

Applying PAM with sprinkler irrigation can also improve infiltration, which reduces runoff and soil erosion. Field studies have shown that erosion decreased under moving sprinkler systems when 20 kg PAM ha^{-1} was applied to the soil before irrigation.[34,35] Lower PAM application rates can be effective when PAM is applied with irrigation water rather than sprayed directly on the soil surface. In laboratory studies with 2 m^2 soil boxes, applying 2 to 4 kg PAM ha^{-1} at 10 to 20 mg L^{-1} with sprinkler-irrigation water reduced soil erosion 75% compared with untreated soil, but these benefits decreased with subsequent irrigations without PAM.[36] In a similar laboratory study, applying 1 kg PAM ha^{-1} with three consecutive irrigations reduced cumulative runoff 50% compared with untreated soil, while applying 3 kg PAM ha^{-1} with one irrigation only reduced runoff by 35%.[37] Field tests in the United States showed that applying PAM with four irrigations (2 to 3 kg ha^{-1} total applied) significantly reduced soil erosion from 52 and 34 kg ha^{-1} for the control to 21 and 5 kg ha^{-1} for the PAM treatment during the 2 years of the study.[38] Soil erosion was not significantly different for a similar field study in Portugal with lower PAM application rates (0.3 kg ha^{-1}).[38]

CONCLUSIONS

Irrigation is vital to world food production, but soil erosion during irrigation threatens the long-term productivity of irrigation. Soil erosion is generally greater from surface irrigation because water flows over the soil during irrigation. Surface-irrigation management is often a tradeoff between irrigation uniformity and erosion. High flow rates can cause erosion; low flow rates can cause poor irrigation uniformity. Ideally, sprinkler irrigation should not have any runoff; however, moving irrigation systems, like center pivots, often apply water faster than it can infiltrate into the soil. Current research is attempting to quantify runoff and erosion potential for various types of center-pivot sprinklers so manufacturers can improve sprinkler designs. Irrigation can be managed to minimize erosion and maintain productivity.

REFERENCES

1. Kendall, H.W.; Pimentel, D. Constraints on the expansion of the global food supply. Ambio **1994**, *23* (3), 198–205.
2. USDA National Agricultural Statistics Service. *Farm and Ranch Irrigation Survey*, available at http://www.agcensus.usda.gov (accessed December 2010).
3. Bjorneberg, D.L.; Kincaid, D.C.; Lentz, R.D.; Sojka, R.E.; Trout, T.J. Unique aspects of modeling irrigation-induced soil erosion. Int. J. Sediment Res. **1999**, *15* (2), 245–252.
4. Trout, T.J. Furrow irrigation erosion and sedimentation: On-field distribution. Trans. ASAE **1996**, *39* (5), 1717–1723.
5. Bjorneberg, D.L.; Sojka, R.E.; Aase, J.K. Pre-wetting effect on furrow irrigation erosion: A field study. Trans. ASAE **2002**, *45* (3), 717–722.
6. Lentz, R.D.; Sojka, R.E.; Carter, D.L. Furrow irrigation water-quality effects on soil loss and infiltration. Soil Sci. Soc. Am. J. **1996**, *60* (1), 238–245.
7. Kim, K.H.; Miller, W.P. Effect of rainfall electrolyte concentration and slope on infiltration and erosion. Soil Technol. **1996**, *9* (3), 173–185.
8. Flanagan, D.C.; Norton, L.D.; Shainberg, I. Effect of water chemistry and soil amendments on a silt loam soil. Part II: Soil erosion. Trans. ASAE **1997**, *40* (6), 1555–1561.
9. Levy, G.J.; Levin, J.; Shainberg, I. Seal formation and interrill soil erosion. Soil Sci. Soc. Am. J. **1994**, *58* (1), 203–209.
10. International Commission on Irrigation and Drainage, available at http://www.icid.org/imp_data.pdf (accessed November 2010).
11. FAO Aquastat, http://www.fao.org/nr/water/aquastat/main/index.stm (accessed November 2010).
12. Koluvek, P.K.; Tanji, K.K.; Trout, T.J. Overview of soil erosion from irrigation. J. Irrig. Drain. Eng. **1993**, *119* (6), 929–946.
13. Israelson, O.W.; Clyde, G.D.; Lauritzen, C.W. *Soil erosion in small irrigation furrows*. Bull. 320. Utah Agricultural Experiment Station: Logan, UT, 1946.
14. Mech, S.J. Effect of slope and length of run on erosion under irrigation. Agric. Eng. **1949**, *30* (8), 379–383, 389.
15. Berg, R.D.; Carter, D.L. Furrow erosion and sediment losses on irrigated cropland. J. Soil Water Conserv. **1980**, *35* (6), 267–270.
16. Bjorneberg, D.L.; Westermann, D.T.; Nelson, N.O.; Kendrick, J.H. Conservation practice effectiveness in the irrigated Upper Snake River/Rock Creek watershed. J. Soil Water Conserv. **2008**, *63* (6), 487–495.
17. Ebbert, J.C.; Kim, M.H. Relation between irrigation method, sediment yields, and losses of pesticides and nitrogen. J. Environ. Qual. **1998**, *27* (2), 372–380.
18. Kemper, W.D.; Trout, T.J.; Brown, M.J.; Rosenau, R.C. Furrow erosion and water and soil management. Trans. ASAE **1985**, *28* (5), 1564–1572.
19. Mailapalli, D.R.; Raghuwanshi, N.S.; Singh, R. Sediment transport in furrow irrigation. Irrig. Sci. **2009**, *27* (6), 449–456.
20. Fernandez-Gomez, R.; Mateos, L; Giraldez, J.V. Furrow irrigation erosion and management. Irrig. Sci. **2004**, *23* (3), 123–131.
21. Lentz, R.D.; Sojka, R.E. Long-term polyacrylamide formulation effects on soil erosion, water infiltration, and yields of furrow-irrigated crops. Agron. J. **2009**, *101* (2), 305–314.
22. Lentz, R.D.; Sojka, R.E. Field results using polyacrylamide to manage furrow erosion and infiltration. Soil Sci. **1994**, *158* (4), 274–282.
23. Kincaid, D.C.; Heermann, D.F.; Kruse, E.G. Application rates and runoff in center-pivot sprinkler irrigation. Trans. ASAE **1969**, *12* (6), 790–794,797.
24. Aarstad, J.S.; Miller, D.E. Soil management to reduce runoff under center-pivot sprinkler systems. J. Soil Water Conserv. **1973**, *28* (4), 171–173.

25. Gilley, J.R. Suitability of reduced pressure center-pivots. J. Irrig. Drain. Eng. **1984**, *110* (1), 22–34.
26. DeBoer, D.W.; Beck, D.L.; Bender, A.R. A field evaluation of low, medium, and high pressure sprinklers. Trans. ASAE **1992**, *35* (4), 1185–1189.
27. Mickelson, R.H.; Schweizer, E.E. Till-plant systems for reducing runoff under low-pressure, center-pivot irrigation. J. Soil Water Conserv. **1987**, *42* (2), 107–111.
28. Kincaid, D.C. Spraydrop kinetic energy from irrigation sprinklers. Trans. ASAE **1996**, 39 (*3*), 847–853.
29. King, B.A.; Bjorneberg, D.L. Characterizing droplet kinetic energy applied by moving spray-plate center-pivot irrigation sprinklers. Trans. ASABE **2010**, 53 (*1*), 137–145.
30. King, B.A.; Bjorneberg, D.L. Evaluation of potential runoff and erosion of four center pivot irrigation sprinklers. Trans. ASABE **2011**, in press.
31. Oliveira, C.A.S.; Hanks, R.J.; Shani, U. Infiltration and runoff as affected by pitting, mulching and sprinkler irrigation. Irrig. Sci. **1987**, *8* (1), 49–64.
32. Kranz, W.L.; Eisenhauer, D.E. Sprinkler irrigation runoff and erosion control using interrow tillage techniques. Appl. Eng. Agric. **1990**, *6* (6), 739–744.
33. DeBoer, D.W.; Beck, D.L. Conservation tillage on a silt loam soil with reduced pressure sprinkler irrigation. Appl. Eng. Agric. **1991**, *7* (5), 557–562.
34. Levy, G.J.; Ben-Hur, M.; Agassi, M. The effect of polyacrylamide on runoff, erosion, and cotton yield from fields irrigated with moving sprinkler systems. Irrig. Sci. **1991**, *12* (2), 55–60.
35. Stern, R.; Van Der Merwe, A.J.; Laker, M.C.; Shainberg, I. Effect of soil surface treatments on runoff and wheat yields under irrigation. Agron. J. **1992**, *84* (1), 114–119.
36. Aase, J.K.; Bjorneberg, D.L.; Sojka, R.E. Sprinkler irrigation runoff and erosion control with polyacrylamide—Laboratory tests. Soil Sci. Soc. Am. J. **1998**, *62* (6), 1681–1687.
37. Bjorneberg, D.L.; Aase, J.K. Multiple polyacrylamide applications for controlling sprinkler irrigation runoff and erosion. Appl. Eng. Agric. **2000**, *16* (5), 501–504.
38. Bjorneberg, D.L.; Santos, F.L.; Castanheira, N.S.; Martins, O.C.; Reis, J.L.; Aase, J.K.; Sojka, R.E. Using polyacrylamide with sprinkler irrigation to improve infiltration. J. Soil Water Conserv. **2003**, *58* (5), 283–289.

Irrigation: Return Flow and Quality

Ramón Aragues
Agronomic Research Service, Government of Aragon, Zaragoza, Spain

Kenneth K. Tanji
Department of Land, Air, and Water Resources, University of California—Davis, Davis, California, U.S.A.

Abstract

Irrigation return flows (IRF) are considered the major diffuse or non-point contributor to the pollution of surface and groundwater bodies. Water pollution standards and emerging policies regulating the discharge of the IRF are being implemented in developed countries. The degree of the off-site irrigation-induced pollution depends on the hydrogeological characteristics of the irrigated land and substrata, the agricultural production technologies used, and the water supply and drainage conveyance systems. This entry reviews these issues in IRF, describes the main components and chemical constituents of IRF, and summarizes recommended management practices aimed at reducing the off-site water quality impact from irrigated agriculture.

INTRODUCTION

The return flows from irrigated agriculture (i.e., Irrigation Return Flows, IRF) are considered the major diffuse or "non-point" contributor to the pollution of surface and groundwater bodies.[1] This off-the-farm discharge ("off-site" contamination) is inevitable since irrigated agriculture cannot survive if salts and other constituents accumulate in excessive amounts in the crop's root zone ("on-site" contamination), and so they must be reached and exported with the drainage waters.[2] Thus, the major task concerning the viability and the long-term sustainability of irrigated agriculture is the attainment of a proper balance for optimizing crop production while minimizing both the "on-site" and the "off-site" environmental damages or impacts and, ultimately, finding an acceptable disposal of the IRF.[3,4]

As a consequence of this increasing "off-site" environmental problem, water pollution standards and emerging policies regulating the discharge of the IRF are being implemented in developed countries. The key policies for mitigating the negative environmental impacts of irrigation are incorporated in the Water Pollution Control Act in United States,[5] and in the Nitrates, Habitats and Environmental Impact Assessment, and Water Framework directives in European Union.[6]

The degree of the "off-site" irrigation-induced pollution depends on the hydrogeological characteristics of the irrigated land and substrata, the agricultural production technologies used, and the water supply and drainage conveyance systems.[1] This entry reviews these issues in IRF, describes the main components and chemical constituents of IRF, and summarizes recommended management practices aimed at reducing the off-site water quality impact from irrigated agriculture.

COMPONENTS OF IRF

Fig. 1 gives a schematic diagram of a typical irrigation-crop-soil-drainage system, composed of the water delivery, the farm, and the water removal subsystems.[2] The water removal subsystem (i.e., the IRF) may be divided into the surface drainage, consisting of the overflow or by-pass water and surface runoff or tailwater, and the collected subsurface drainage components. Since IRF are mixtures of these components, their proportions determine the final quality of IRF. Table 1 summarizes the expected water quality changes of the three IRF components (overflow, tailwater, and subsurface drainage) relative to the quality of the applied irrigation water.

Overflow is the result of operational spill waters from distribution conveyances that are directly discharged into the drainage system and its quality is generally similar to that of the irrigation water (Table 1).

Tailwater is the portion of the applied irrigation water that runs off over the soil and discharges from the lower end of the field directly into the drain system. Because of its limited contact and exposure to the soil surface, its quality degradation is generally minor. Even so, these waters may increase slightly in salinity and may pick up considerable amounts of sediments and associated nutrients (phosphorus in particular) as well as water-applied agricultural chemicals such as pesticides and nitrogen fertilizers (anhydrous ammonia in particular) (Table 1).

Subsurface drainage is the portion of the infiltrating water that flows through the soil and is collected by the under drainage system. Because of its more intimate contact with the soil and the dynamic soil–plant–water interactions, its quality degradation is generally substantial. These subsurface drain waters carry any anthropogenic chemicals

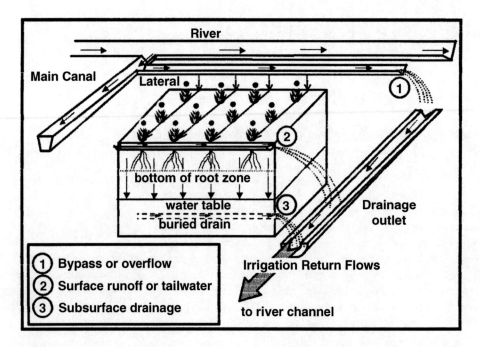

Fig. 1 Idealized sketch showing the diversion of irrigation water through a main canal, its distribution through a lateral, and its application to croplands. The three main components of the irrigation return flows to the river channel are shown. The deeper groundwater zone, a second receiving water system, is not shown.

present in a soluble form in the soil water as well as any salts and other soluble elements present in the soil and parent geologic material and intercepted shallow groundwaters. The salinity and agrochemicals in subsurface drainage are the primary source of pollution associated with irrigated agriculture (Table 1).

WATER QUALITY CONSTITUENTS IN IRF

Irrigation return flows provide the vehicle for conveying the pollutants to a receiving stream or groundwater reservoir. It is therefore necessary to characterize their most important water quality constituents (namely, inorganic

Table 1 Quality parameters of the three irrigation return flow (IRF) components and their expected quality changes as related to the quality of the irrigation water.

Quality parameters	Components of IRF		
	Overflow	Tailwater	Subsurface drainage
General quality degradation	0	+	++
Salinity	0	0, +	++
Nitrogen	0	0, +, ++	++, +
Phosphorus	0, +	++	0, −, +
Oxygen demanding organics	0	+, 0	0, −, −−
Sediments	0, +, −	++	−−
Pesticide residues	0	++	0, −, +
Trace elements	0	0, +	0, −, +
Pathogenic organisms	0	0, +	−, −−

0: Negligible quality changes expected.
+, −: Expected to be slightly higher (i.e., pick up), lower (deposition).
++: Expected to be significantly higher due to concentrating effects, application of agricultural chemicals, erosional losses, pick up of natural geochemical sources, etc.
−−: Expected to be significantly lower due to filtration, fixation, microbial degradation, etc.

salts, agrochemicals and trace elements) and to develop management strategies aimed at alleviating their detrimental effects on the receiving water bodies.

Salts

Salts are a major quality factor since they can restrict the municipal, industrial and agricultural uses of water and can dramatically decrease the productivity and sustainability of irrigated agriculture in arid zones.

The primary source of dissolved mineral salts (also referred to as salinity) is the chemical weathering of rocks, minerals, and soils. Salinity is reported in terms of total dissolved solids (TDS in $mg\,L^{-1}$) or Electrical Conductivity (EC in $dS\,m^{-1}$ at $25°C$). The main solutes contributing to salinity are the cations calcium (Ca), sodium (Na), and magnesium (Mg), and the anions chloride (Cl), sulfate (SO_4), and bicarbonate (HCO_3). These solutes are reactive in waters and soil solutions participating, among others, in cation exchange and mineral solubility. The excessive accumulation of Na (i.e., sodicity, generally expressed by the Sodium Adsorption Ratio or SAR) in the soil solution and exchange complex may impair poor soil physical properties and is a critical factor in the sustainability of irrigated soils.[7]

Growing plants extract water through evapotranspiration and leave behind most of the dissolved salts, increasing its concentration in the soil water ("evapoconcentration effect"). Irrigation also adds to the salt load in IRF by leaching natural salts arising from weathered minerals occurring in the soil profile, or deposited below ("weathering effect").[2] As a consequence of both effects, it follows that the salinity and chemical composition of IRF depend basically on the characteristics of the irrigation water, the soil and subsoil, and the hydrogeology, as well as on the management of the irrigation water or Leaching Fraction (LF) defined as the fraction of infiltrated water that percolates out of the root zone. Thus, high LFs promote the weathering effect and the salt load carried out with the IRF (i.e., increased "off-site" pollution) whereas low LFs promote the evapoconcentration effect and the concentration of salts in the crop's root zone (i.e., increased "on-site" pollution).

In conclusion, the mass of salts or salt loading in IRF depends mainly on the salinity of the irrigation water, the minerals present in the soil and subsoil, and the water management (LF). The salt loading values may vary widely, from values similar to those of the irrigation water to values one order of magnitude higher. Thus, typical salt loading values in IRF from arid-land irrigated agriculture vary between $2\,Mg\,ha^{-1}\,yr^{-1}$ and $20\,Mg\,ha^{-1}\,yr^{-1}$.[1,2,7] The quantification of salt loading is critical to ascertain the "off-site" contamination of irrigated agriculture, since the prediction of the resultant salt concentration in a body of water after mixing with the IRF requires knowledge of the mass of salts (i.e., concentration and flow) in each contributing body.

Nitrogen

Nitrogen can be in either the organic or the inorganic (ammonium, nitrate and nitrite) form. Organic N is predominant in surface drainage (although it is not usually an issue in arid areas), whereas inorganic N is predominant in subsurface drainage water. Although nitrite is considered more hazardous than nitrate, it is in general a transient form of N present in small quantities. Nitrate is thus the dominant form of N in IRF and should be the focus of the water quality evaluation.[3]

High nitrate (NO_3) concentrations in IRF are a major concern since they may cause eutrophication (excessive algal growth) and hypoxia (decline in dissolved oxygen from decay of algae) problems. When nitrate is ingested in substantial amounts by humans and animals, it may cause methemoglobinemia (blue-bay like symptoms from oxygen starvation exhibited by infants and elderly) and certain cancers.[7] Thus, USEPA has set the maximum allowable concentration of nitrate in public water supplies at $45\,mg\,L^{-1}$,[5] whereas the European Union has limited it to $50\,mg\,L^{-1}$.[6]

The three major sources of nitrate found in IRF are leaching from croplands, land disposal of urban sewage, and concentrated animal feeding (beef feedlots, dairies, swine, chicken houses) wastes. The potential for nitrate leaching is a function of soil type, weather conditions and crop management system. In general, the higher the N application rate, the greater the amount of N available to be lost, since fertilizer N recovery by harvested crops averages about 50% and tends to be even lower when high N application rates are used. In addition, mineralization of organic N, followed by nitrification of NH_4 may also increase the N losses.[4]

Drainage has a large influence upon losses of nitrogen. The N loss from poorly drained soils is generally much less than from soils with improved drainage systems. As previously indicated, much of the N transported in surface runoff is organic N associated with the sediment, although the amount lost is usually small and poses little threat to the environment except in pristine waters. On the other hand, nitrate concentrations in subsurface drainage water are much higher and variable, depending on the N fertilization rates and time of applications, and on water and soil management.[4]

Phosphorus

Phosphorus (P), present in both organic and inorganic forms, is a relevant water quality constituent in IRF because of its contribution to eutrophication of surface waters. Most of the P in surface drainage is in particulate (i.e., sediment and organic matter-bound) form whereas most of the P in subsurface drainage water is in soluble phosphate form.

The release of P depends on such biogeochemical processes as adsorption/desorption of phosphate, precipitation/dissolution of inorganic P forms, and mineralization of organic P forms.[7] Phosphorus in subsurface drainage waters is typically low in concentration because of its strong adsorption in arid zone soils. Thus, although P discharge from agricultural fields vary considerably, it is usually in the range of $0.2\,kg\,P\,ha^{-1}\,yr^{-1}$ to less than $3\,kg\,P\,ha^{-1}\,yr^{-1}$. Even though P loading in IRF is minor, the P concentrations measured in many agricultural IRF may be orders of magnitude above the soluble ($10\,\mu g\,P\,L^{-1}$) and total ($20\,\mu g\,P\,L^{-1}$) critical levels assumed to accelerate the eutrophication of freshwater aquatic ecosystems.[5]

Pesticide Residues

Pesticide contamination in IRF is of concern in some agricultural areas, although it is in general less significant than the salinity or nitrogen pollution problem.[3]

Pesticides used in irrigated agriculture include herbicides, insecticides, fungicides, and nematicides. These various types make it difficult to assess their potential impacts on water quality. Pesticide concentrations in surface drainage are usually much greater than those in subsurface drainage due to the filtering action of the soil. Thus, the total loss of pesticides via subsurface drainage is usually 0.15% or less of the amount applied, whereas losses via surface drainage can be up to 5% or more.[4]

The environmental fate of pesticides is quite complex. Chemical-specific properties influence the reactivity of pesticides. Pesticides can be degraded by microbes, chemical and photochemical reactions, adsorbed on to soil organic matter and clay minerals, lost to the atmosphere through volatilization, and lost through surface runoff and leaching.[4] Once a pesticide enters into the soil, its fate is largely dependent on sorption (evaluated by use of a sorption coefficient based on the organic carbon content of soils) and persistence (evaluated in terms of the half-life or the time it takes for 50% of the chemical to be degraded or transformed). Pesticides with low sorption coefficient (such as atrazine, DBCP, and aldicarb) are likely to leach readily, whereas pesticides with long half-lives (such as DDT, lindane, and endosulfan) are so persistent that many of them banned various decades ago are still found in stream sediments or are now being detected in the groundwaters.[7]

Trace Elements

High concentrations of trace elements in soils and waters pose a threat to agriculture, wildlife, drinking water, and human health. The trace elements of most importance, documented as pollutants associated with irrigated agriculture, are barium (Ba) and lithium (Li) (alkali and alkali earth metals), chromium (Cr), molybdenum (Mo), and vanadium (V) (transition metals), arsenic (As), boron (B), and selenium (Se) (non-metals), and cadmium (Cd), copper (Cu), lead (Pb), mercury (Hg), nickel (Ni), and zinc (Zn) (heavy metals).[3] Those trace elements such as As, Cd, Hg, Pb, B, Cr, and Se are especially harmful to aquatic species because of biological magnification.[5] Due to the generally narrow window between deficiency and toxicity of trace elements, it is essential to have an adequate information on their concentrations in soils and waters.

The sources of trace element contamination may be divided into natural (i.e., geologic materials) and agricultural-induced (i.e., fertilizers, irrigation waters, soil and water amendments, animal manures, sewage effluent and sludge, and pesticides). Increases in trace element concentrations in surface runoff are generally not expected, whereas the presence of trace elements in groundwaters is influenced by the nature of the sources, the speciation and reactivity of the trace elements, and the mobility and transport processes. Thus, high concentrations of trace elements in subsurface drainage water appear to be strongly associated with the geologic setting of the irrigated area and may be affected by the same processes that affect the soil and groundwater salinity.[3,7]

An illustrative example of trace element contamination is the selenium toxicosis of waterfowl at Kesterson reservoir (California, U.S.A.), a terminal evaporation pond for drainage waters high in Se (300 ppb average) originating from the Moreno shale, a geologic formation of the Coast Range Mountains in the west side of the San Joaquin Valley.[7]

MANAGEMENT OPTIONS TO REDUCE OFF-SITE WATER QUALITY IMPACTS FROM IRRIGATED AGRICULTURE

The basic idea behind the control of irrigation-induced environmental problems is the change in focus from a "water resource development" to a "water resource management" approach. This new "thinking" involves both policy changes, such as reducing the applied water through economic and regulatory policies (i.e., water metering, water pricing, licenses and time-limited abstraction permits), and developing farmer's incentives for promoting best management practices (i.e., compensation and agri-environment payments for irrigated crops), and a variety of technical measures.[5,6,8]

Since a detailed description of the technical measures is too lengthy for this entry, Table 2 summarizes some of the recommended strategies aimed at reducing the off-site water quality impacts from irrigated agriculture. However, it should be cautioned that these measures should be applied in a "case-by-case" basis, since some of them could aggravate the "on-site" pollution problems. Typical exam-

Table 2 Summary of recommended management practices at the water delivery, farm, and water removal subsystems to reduce off-site water quality impacts from irrigated agriculture.

Water delivery subsystem

Designed to meet the farm water requirements while reducing undesirable water losses

Canal lining and/or closed conduits and reservoir lining: prevent seepage losses, phreatophyte ET losses, soil waterlogging, and groundwater recharge; improve irrigation water quality (i.e., suspended solids).

Installation of flow measuring devices: water control; appropriate water charges and penalties; reduce bypass losses; attain high water-conveyance efficiencies.

Construction of regulation reservoirs at the irrigation district level to increase flexibility in water delivery.

Implement an efficient institutional framework, service-oriented besides its regulatory character; scheduled maintenance programs.

Farm subsystem

Designed to maintain or increase crop productivity while improving source control

Improve cultural practices: rate and timing of fertilizers; slow-release fertilizers; fertigation; pest control; seeding and tillage practices.

Adopt less environmentally damaging agricultural practices: integrated management systems; mixed cropping practices; organic farming.

Increase irrigation application efficiency and uniformity: proper design of the farm irrigation layout; choice of irrigation system; optimum irrigation scheduling; reduce evaporation through mulching and reduced tillage.

Minimize the Leaching Fraction according to the leaching requirement of crops: reduce drainage volume; maximize mineral precipitation; minimize pick up of salts.

Provide training and technical services to farmers; eliminate institutional constraints.

Water removal subsystem

Designed to improve sink control and minimize loading in IRF

Constraints in disposal of IRF to meet quality objectives in the receiving water body.

Reuse for irrigation drainage waters, municipal wastewaters and sewage effluents; integrated on-farm drainage management (i.e., on-farm cycling of drainage waters through biological materials-agroforestry systems).

Ocean and inland (i.e., evaporation ponds; solar evaporators; deep well injection) disposal of drainage waters.

Design and management of drainage systems: include water quality as a design parameter; depth and distance of placement of drains; integrated drain flow and irrigation management; crop water use from shallow watertables (i.e., subirrigation); controlled drainage (i.e., management of the water level in the drainage outlet); reduce nitrate effluxes by maintaining a high water table to increase denitrification losses.

Pumping and disposal of groundwater to reduce intercepted groundwater by the drainage network.

Flowing of surface drainage water through vegetated filters and riparian vegetation (removal of sediments and sediments-associated contaminants), flowing of subsurface drainage water through riparian zones (removal of nitrate due to plant uptake and denitrification); flowing of drainage water through constructed wetlands (sink for sediment, nutrients, trace elements, and pesticides).

Physical, chemical, and biological treatment of drainage waters: particle removal; adsorption, air stripping; desalination (membrane processes and distillation); coagulation and flocculation; chemical precipitation; ion exchange; advanced oxidation processes; biofiltration (irrigation of specific crops that accumulate large quantities of undesirable constituents such as Se, Mo, B, NO_3, etc.); algal–bacterial treatment facilities (removal of NO_3 and Se).

---

ples will be (i) the "minimum leaching fraction concept," that could promote soil sodification and structural stability problems due to the precipitation of calcium minerals such as calcite and gypsum; (ii) the reuse of drainage water for irrigation, which is only sustainable if it is of sufficient good quality; and (iii) the disposal of drainage water in evaporation ponds, which may eventually lead to other environmental problems.

The reader is referred to the references given at the end of the entry for further information on the myriad of technical management options developed in the last decades and on details of their advantages and limitations.

REFERENCES

1. Law, J.P.; Skogerboe, G.W.; Eds. *Irrigation Return Flow Quality Management*, Proceedings of National Conference, Fort Collins, Co, May 16–19, 1977, 451 pp.
2. Yaron, D.; Ed.; *Salinity in Irrigation and Water Resources*, Civil Engineering No. 4; Marcel Dekker: New York, 1981; 432 pp.
3. Food and Agriculture Organization of the United Nations, *Management of Agricultural Drainage Water Quality*; Madramootoo, CA., Johnston, W.R., Willardson, L.S., Eds.; FAO Water Report No. 13; FAO: Rome, 1997; 94 pp.

4. Skaggs, R.W.; van Schilfgaarde, J.; Eds. *Agricultural Drainage*, Agronomy No. 38, ASA, CSSA and SSSA, Inc.: Madison, WI, 1999; 1328 pp.
5. National Research Council, *Soil and Water Quality, an Agenda for Agriculture*, Batie, S.A. Chair; Committee on Long-Range Soil and Water Conservation, National Academy Press: Washington, DC, 1993; 516 pp.
6. Institute for European Environmental Policy. *The Environmental Impacts of Irrigation in the European Union*; Report to the Environment Directorate of the European Commission: London, 2000; 138 pp.
7. Tanji, K.K.; Ed. *Agricultural Salinity Assessment and Management*, ASCE Manuals and Reports on Engineering Practices No. 71; American Society of Civil Engineers New York, 1990; 619 pp.
8. Food and Agriculture Organization of the United Nations, *Control of Water Pollution from Agriculture*, Ongley, E.D., Ed.; FAO Irrigation and Drainage Paper No. 55; FAO: Rome, 1996; 85 pp.

Irrigation: River Flow Impact

Robert W. Hill
Biological and Irrigation Engineering Department, Utah State University, Logan, Utah, U.S.A.

Ivan A. Walter
Ivan's Engineering, Inc., Denver, Colorado, U.S.A.

Abstract
The practice of irrigation necessitates developing a water source, conveying the water to the field, application of the water to the soil and collection and reuse or disposal of tailwater and subsurface drainage. These processes alter river basin hydrology and water quality in space and time. To sum up the effect of irrigation on a watershed in a word, it would be: *depletion*.

INTRODUCTION

In hydrologic studies it is common engineering practice to quantify the impact upon the stream(s) from which the irrigation water is diverted. The impact upon the stream is actually of two kinds: 1) diversions that decrease the streamflow and 2) return flows that increase the streamflow. The engineering term used to describe the overall impact is "streamflow depletion" which means the net reduction in streamflow resulting from diversion to irrigation uses. Actual stream depletions are a function of many factors including the amount and timing of diversions, the type of diversion structure (well vs. ditch), crops grown, soil type, depth to groundwater, irrigation method, irrigation efficiency, properties of the alluvial aquifer, area irrigated, and evapotranspiration of precipitation, groundwater, and irrigation water.

Depletion

Depletion, in this context, is the consumptive abstraction of water from the hydrologic system as a result of irrigation. It is in addition to consumptive water use that would have occurred in the unmodified natural situation. As an example, waters of the Bear River Basin of Southern Idaho, Northern Utah, and Western Wyoming, because it is an interstate system, are administered by a federally established commission under the authority of the Bear River Compact.[1] Depletion is the basis, in the compact, for allocating Bear River water use among the three states. It is defined by a "Commission Approved Procedure" which includes consideration of land use and incorporates an equation for estimating depletion based on evapotranspiration. In a study for the commission, Hill[5] defined crop depletion as

$$Dpl = Et - Smco - Pef \quad (1)$$

where Dpl is estimated depletion for a given site or sub-basin; Et is calculated crop water use; SMco is moisture which is "carried over" from the previous non-growing season (October 1–April 30) as stored soil water in the root zone available for crop water use subsequent to May 1; and Pef is an estimate of that portion of precipitation measured at an NWS station during May–September, which could be used by crops.

The carry-over soil moisture (SMco) was estimated by assuming that 67% of adjusted precipitation from October through April could be stored in the root zone. If this exceeded 75% of the available soil water-holding capacity of the average root zone in the sub-basin, the excess was considered as lost to drainage or runoff and not available for crop use. Growing season precipitation was considered to be 80% effective in contributing to crop water use. The effectiveness factor of 80% allowed for precipitation depths throughout a sub-basin that might differ from NWS rain–gage amounts. It also included a reduction for mismatches in timing between rainfall events and irrigation scheduling.

HYDROGRAPH MODIFICATION

Diversion of significant amounts of water from rivers and streams for irrigation and subsequent return flows alters the shape and timing of downstream hydrographs. In watersheds where mountain snowmelt provides the irrigation supply, such as in the Western United States, diversion during the spring runoff attenuates the peak flow rate while later return flows extend the flow duration into late summer and early fall.

Reservoir Storage

Storage of water in reservoirs can significantly modify the natural stream hydrograph depending on the timing and quantity of the storage right. Irrigators with junior rights may only be able to store during time periods with low irrigation demand, such as during the winter, or during peak flow periods. Reductions of stream flow during the winter time may have considerable impact on downstream in-stream flows. Whereas, storage during periods of peak runoff may not affect minimum in-stream flow needs, but could deposit considerable amounts of sediment in the reservoir.

Irrigation Return Flows

Irrigation return flows are comprised of surface runoff and/or subsurface drainage that becomes available for subsequent rediversion from either a surface stream or a groundwater aquifer downstream (hydrologically) of the initial use. Reusable return flow can be estimated as irrigation diversion minus crop related depletions minus additional abstractions. Additional abstractions include incidental consumptive use from water surfaces as in open drains, along with non-crop vegetation. The timing of return flow varies from nearly instantaneous (recaptured tailwater) to delays of weeks and months or perhaps longer with deep percolation subsurface drainage. In a hydrologic model study of the Bear River Basin[4] delay times between diversion and subsequent appearance of the return flow at the next downstream river gage varied from 1.5 months to as long as 6 months. The delay appeared to be related to sub-basin shape and size.

Irrigation Methods

Four general irrigation methods are used: surface, subsurface, sprinkler, and trickle (also known as low flow or drip). Surface methods include wild or controlled flooding, furrow, border-strip, and ponded water (basin, paddy, or low-head bubbler). Hand move, wheel move, and center pivot are examples of sprinkler irrigation. Trickle irrigation includes point source emitters, microspray, bubbler, and linesource drip tape (above or below ground). Whereas the efficiency of surface irrigation is dependent upon the skills and experience of the irrigator, the performance of trickle and sprinkler systems is more dependent on the design. Generally, the more control that the system design (hardware) has on the irrigation system performance, the higher the application efficiency (E_a) can be. Thus, typical wheel move sprinklers have higher E_a values than surface irrigation, but lower values than for center pivots or trickle, assuming better than average management practices for each method.

The impact on river flows can be quite different among the various irrigation methods. The nature of furrow and border surface irrigation generally produces tail water runoff, which can be immediately recaptured and reused, as well as deep percolation, which may not be available for reuse until after a period of time. Tailwater is essentially eliminated and deep percolation reduced with sprinklers (Fig. 1) compared to conventional surface irrigation. Whereas, with drip methods, deep percolation can be further reduced. The reduction of deep percolation implies increased salt concentration in the root zone leachate, but, perhaps significant reduction in salt pick-up potential from geologic conditions.

Irrigation Efficiencies

Although a full discussion of the several variations of irrigation efficiency is beyond the scope herein, two terms will be defined and discussed. More complete discussions relating to irrigation efficiencies and water requirements are given elsewhere.[6–9,13] Keller and Bliesner[9] give a particularly thorough presentation of distribution uniformity and efficiencies.

Application efficiency (E_a):

$$E_a = 100 \times \frac{\text{Volume of water stored in the root zone}(V_s)}{\text{Volume of water delivered in the farm or field}(V_f)}$$

Distribution uniformity:

The distribution uniformity is a measure of how evenly the on-farm irrigation system distributes the water across the field. The definition of DU is:

$$DU = 100 \times \frac{\text{Average of the lowest 25\% of infiltrated water depth}}{\text{Average of all infiltrated water depths across the field}}$$

On-farm or field application efficiencies can be affected by the distribution uniformity and vary widely for both surface and sprinkle irrigation methods. This is largely due to difference in management practices, appropriateness of design in matching the site conditions (slope, soils, and wind), and the degree of maintenance. In addition, for a given system uniformity, the higher the proportion of the field that is adequately irrigated (i.e., infiltrated water refills the soil water deficit) the lower will be the application efficiency. This is due to greater deep percolation losses in the overirrigated portions of the distribution pattern. Some values determined in recent Utah field evaluations are:

Method	Observed		
	High (%)	Low (%)	Typical (%)
Surface irrigation			
E_a	72	24	50
Tailwater	55	5	20
Deep percolation	65	20	30
Sprinkler irrigation			
E_a	84	52	70
Evaporation	45	8	12
Deep percolation	37	8	18

The E_a for a particular field may vary greatly during the season. Cultivation practices, microconsolidation of

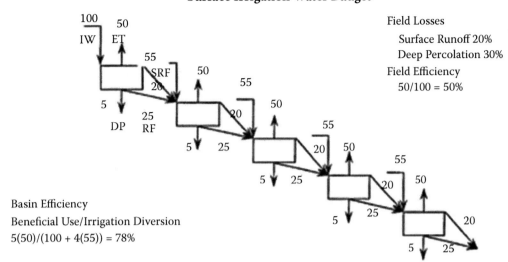

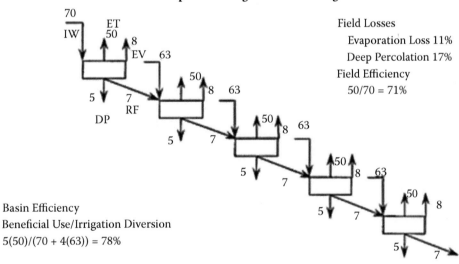

Fig. 1 Comparison of basin efficiencies between surface and sprinkler irrigation methods with four return flow reuse cycles.

the soil surface and vegetation will alter surface irrigation efficiency both up and down from the seasonal average. Seasonal and diurnal variations in wind, humidity, and temperature will also affect sprinkle application efficiencies.

BASIN IRRIGATION EFFICIENCY

The actual irrigation efficiency realized for several successive downstream fields where capture and reuse of return flows is experienced is higher than the E_a of an individual field. This notion of "Basin Irrigation Efficiency"[12,13] is illustrated in Fig. 1. This simple example comparison of surface and sprinkle methods assumes four reuse cycles. In each of the five "fields" Et is assumed to be 50 units. The surface runoff is captured for reuse on the next field. All of the irrigation-related evaporation is assumed "lost" as well as 5 units of deep percolation. After the fifth field, all surface and subsurface flows are lost. The basin efficiency for surface is 78%, which is the same as for sprinkle. The surface irrigation basin efficiency increase is dependent upon the surface return flow reuse, which is 20 units in this example. However, the depletion is greater for sprinkler due to the extra evaporation. In a Colorado field study, Walter and Altenhofen[11] found a progressive increase in irrigation efficiencies from field (average E_a of 45%), to farm, to efficiency of ditch or sectors (average of 83%). This was due to the reuse of tailwater (10%–20% of delivery) and deep percolation (46% of delivery).

ENVIRONMENTAL CONCERNS

The process of evapotranspiration, or crop water use, extracts pure water from the soil water reservoir, which leaves behind the dissolved solids (salts) contained in the applied irrigation water. The "evapoconcentration" of salts is an inevitable result of irrigation for crop production. As stated by Bishop and Peterson:[2]

> . . . Other uses add something to the water, but irrigation basically takes some of the water away, concentrating the residual salts. Irrigation may also add substances by leaching natural salts or other materials from the soil or washing them from the surface. Irrigation return flow is a process by which the concentrated salts and other substances are conveyed from agricultural lands to the common stream or the underground water supply . . .

Water Quality Implications for Agriculture

Irrigated agriculture is dependent upon adequate, reasonably good quality water supplies. As the level of salt increases in an irrigation source, the quality of water for plant growth decreases. Since all irrigation waters contain a mixture of natural salt, irrigated soils will contain a similar mix to that in the applied water, but generally at a higher concentration. This necessitates applying extra irrigation water, or taking advantage of non-growing season precipitation, to leach the salts below the root zone.

Salt Loading Pick-Up

Water percolating below the root zone or leaking from canals and ditches may "pick-up" additional salts from mineral weathering or from salt–bearing geologic formations (such as the Mancos shale of Western Colorado and Eastern Utah). This salt pick-up will increase the salt load of return flows and consequently increase the salinity of receiving waters.

In the Colorado River Basin in the United States and Mexico salinity is a concern because of its adverse effects on agricultural, municipal, and industrial users.[10] The Salinity Control Act of 1974 (Public Law 93-320) created the Colorado River Basin Salinity Control Program to develop projects to reduce salt loading to the Colorado River. Salinity control projects include lining open canals and laterals (or replacing with pipe) and installing sprinklers in place of surface irrigation for the purpose of decreasing salt loading caused by canal leakage and irrigated crop deep percolation. Recently selenium in irrigation return flow has become a concern[3] and may also be reduced by salinity reduction projects.

In-Stream Flow Requirements

Diversions in some reaches in some Western United States streams are "dried up" immediately downstream of diversion structures during times of peak irrigation demand. This condition eliminates any use of the reach for fisheries and other uses which depend on in-stream flow. In some instances, negotiated agreements with senior water rights users have allowed for bypass of minimal amounts of water to sustain the fishery or habitat, and for control of tailwater runoff to reduce agricultural related chemicals in the receiving water.

REFERENCES

1. Bear River Commission. *Amended Bear River Compact*, S.B. 255, 1979 Utah Legislature Session. 1979.
2. Bishop, A.A.; Peterson, H.B. (team leaders). Characteristics and Pollution Problems of Irrigation Return Flow. Final Report Project 14-12-408, Fed. Water Pollution Control Adm. U.S. Dept. of Interior (Ada, Oklahoma). Utah State University Foundation, Logan, Utah. May 1969.
3. Butler, D.L. *Effects of Piping Irrigation Laterals on Selenium and Salt Loads, Montrose Arroyo Basin, Western Colorado*; U.S. Geological Survey Water-Resources Investigations Report 01-4204 (in cooperation with the U.S. Bureau of Reclamation): Denver, Colorado, 2001.
4. Hill, R.W.; Israelsen, E.K.; Huber, A.L.; Riley, J.P. *A Hydrologic Model of the Bear River Basin*; PRWG72-1, Utah Water Resource Laboratory, Utah State University: Logan, Utah, 1970.
5. Hill, R.W.; Brockway, C.E.; Burman, R.D.; Allen, L.N.; Robison, C.W. *Duty of Water Under the Bear River Compact: Field Verification of Empirical Methods for Estimating Depletion. Final Report*; Utah Agriculture Experiment Station Research Report No. 125, Utah State University: Logan, Utah, January 1989.
6. Hoffman, G.J.; Howell, T.; Solomon, K. *Management of Farm Irrigation Systems*; The American Society of Agricultural Engineers: St. Joseph, MI, 1990.
7. Jensen, M.E. *Design and Operation of Farm Irrigation Systems*; ASAE Monograph, American Society of Agricultural Engineers: St. Joseph, MI, 1983.
8. Jensen, M.E.; Burman, R.D.; Allen, R.G.; Eds. *Evapotranspiration and Irrigation Water Requirements*; ASCE Manual No. 70, American Society of Civil Engineers: New York, 1990.

9. Keller, J.; Bliesner, R.D. *Sprinkle and Trickle Irrigation*; Van Nostrand Reinhold: New York, 1990.
10. U.S. Department of the Interior. *Quality of Water—Colorado River Basin: Bureau of Reclamation, Upper Colorado Region*; Progress Report no 19; Salt Lake City, Utah, 1999.
11. Walter, I.A.; Altenhofen, J. Irrigation efficiency studies—Northern Colorado. Proceedings USCID Water Management Seminar. Sacramento, CA Oct 5–7; USCID: Denver, CO, 1995.
12. Willardson, L.S.; Wagenet, R.J. *Basin-Wide Impacts of Irrigation Efficiency*; Proceedings of an ASCE Specialty Conference on Advances in Irrigation and Drainage, Jackson, WY, 1983.
13. Willardson, L.S.; Allen, R.G.; Frederiksen, H.D. *Elimination of Irrigation Efficiencies*; Proceedings 13th Technical Conference, USCID: Denver, CO, 1994.

Irrigation: Saline Water

B.A. Stewart
Dryland Agriculture Institute, West Texas A&M University, Canyon, Texas, U.S.A.

Abstract
As water becomes more limited, there is increasing use of saline waters for irrigation that were previously considered unsuitable. As salinity in the root zone increases, the osmotic potential of the soil solution decreases and therefore reduces the availability of water to plants. The extent that plant growth is affected by saline water is dependent on the crop species. Soils are also negatively impacted by salt, particularly sodium salts. Sodium ions tend to disperse clay particles and this has deleterious effects on infiltration rate, structure, and other soil physical properties.

INTRODUCTION

As water becomes more limited, there is increasing use of saline waters for irrigation that were previously considered unsuitable. Rhoades, Kandiah, and Marshali[1] classified saline waters as shown in Table 1. Electrical conductivity is a convenient and practical method for classifying saline waters because there is a direct relationship between the salt content of the water and the conductance of an electrical current through water containing salts. Electrical conductivity values are expressed in siemens (S) at a standard temperature of 25°C.

Most waters used for irrigation have electrical conductivities less than $2\,dS\,m^{-1}$.[1] When water higher than this level is used, there can be serious negative effects on both plants and soils. As salinity in the root zone increases, the osmotic potential of the soil solution decreases and therefore reduces the availability of water to plants. At some point, the concentration of salts in the root zone can become so great that water will actually move from the plant cells to the root zone because of the osmotic effect. Salts containing ions such as boron, chloride, and sodium can also be toxic to plants when accumulated in large quantities in the leaves. The extent that plant growth is affected by saline water is dependent on the crop species. Some plants, such as barley and cotton, are much more resistant to salt than crops like beans. Rhoades, Kandiah, and Marshali[1] list the tolerance levels of a wide range of fiber, grain, and special crops; grasses and forage crops; vegetable and fruit crops; woody crops; and ornamental shrubs, trees, and ground cover. Soils are also negatively impacted by salt, particularly sodium salts. Sodium ions tend to disperse clay particles and this has deleterious effects on infiltration rate, structure, and other soil physical properties.

IRRIGATING WITH SALINE WATERS

Water limitations and the need to increase food and fiber production in many parts of the world have resulted in the use of water for irrigation containing increasing levels of salts. The United States, Israel, Tunisia, India, and Egypt have been particularly active in irrigating with saline waters.[1] Rhoades, Kandiah, and Marshali[1] published an extensive paper on the use of saline waters for crop production and it is a valuable guide for anyone interested in the subject. They reported that many drainage waters, including shallow ground waters underlying irrigated lands, fall in the range of $2\,dS\,m^{-1}$ to $10\,dS\,m^{-1}$ in electrical conductivity. Such waters are in ample supply in many developed irrigated lands and have good potential even though they are often discharged to better quality surface waters or to waste outlets. These waters can be successfully used in many cases with proper management. Reuse of second-generation drainage waters with electrical conductivity values of $10\,dS\,m^{-1}$ to $25\,dS\,m^{-1}$ is also sometimes possible but to a much lesser degree because the crops that can be grown with these waters are atypical and much less experience exists upon which to base management recommendations.

Miller and Gardiner[2] suggest that successful irrigation with saline water requires three principles. First, the soil should be maintained near field capacity to keep the salt concentration as low as possible. Second, application techniques should avoid any wetting of the foliage. Third, salts accumulating in the soil should be periodically leached. To accomplish these objectives, Miller and Gardiner[2] recommend the following general rules:

- Apply water at or below soil surface. Sprinklers should be used only if they avoid wilting the foliage (such as sprinkling before plant emergence or below-canopy to avoid salt-burn damage).
- Keep water additions almost continuous, but at or below field capacity so that most flow is unsaturated. This maintains adequate aeration.
- Enough water should be added to keep salts moving downward, thus avoiding salt buildup in the root zone.

Table 1 Classification of saline waters.

Water class	Electrical conductivity (dS m^{-1})	Salt concentration (mg L^{-1})	Type of water
Non-saline	<0.7	<500	Drinking and irrigation
Slightly saline	0.7–2	500–1500	Irrigation
Moderately saline	2–10	1500–7000	Primary drainage and groundwater
Highly saline	10–25	7000–15,000	Secondary drainage and groundwater
Very highly saline	25–45	15,000–35,000	Very saline groundwater
Brine	>45	>45,000	Seawater

Source: Rhoades et al.[1]

Miller and Gardiner[2] stress that these rules are difficult to meet and are best satisfied by some form of drip irrigation. They also state that due to the need for high water levels and because of high sodium ratios that sandy soils are more adaptable to the use of saline waters than soils containing high percentages of silt and clay particles.

Rhoades, Kandiah, and Marshali[1] also list specific management practices for producing crops with salty waters. Their list includes the following guidelines:

- Selection of crops or crop varieties that will produce satisfactory yields under the existing or predicted conditions of salinity or sodicity.
- Special planting procedures that minimize or compensate for salt accumulation in the vicinity of the seed.
- Irrigation to maintain a relatively high level of soil moisture and to achieve periodic leaching of the soil.
- Use of land preparation to increase the uniformity of water distribution and infiltration, leaching and removal of salinity.
- Special treatments (such as tillage and additions of chemical amendments, organic matter and growing green manure crops) to maintain soil permeability and tilth. The crop grown, the quality of water used for irrigation, the rainfall pattern and climate, and the soil properties determine to a large degree the kind and extent of management practices needed.

BLENDING LOW-SALT AND SALTY WATERS

Miller and Gardiner[2] reported that countries such as Israel have developed extensive canal and reservoir systems where both low-salt and salty waters are mixed to obtain usable water. Rhoades, Kandiah, and Marshali,[1] however, state that blending or diluting excessively saline waters with good quality water supplies should only be undertaken after consideration is given to how this affects the volumes of consumable water in the combined and separate supplies. They suggest that blending or diluting drainage waters with good quality waters in order to increase water supplies or to meet discharge standards may be inappropriate under certain situations. More crop production can usually be achieved from the total water supply by keeping the water components separated. Serious consideration should be given for keeping saline drainage waters separate from the good quality water, especially when the good quality waters are used for irrigation of salt-sensitive crops. The saline waters can be used more effectively by substituting them for good quality water to irrigate certain crops grown in the rotation after seeding establishment.

CONCLUSION

There is ample evidence that saline waters once considered unacceptable for irrigation can be used successfully provided that they are properly managed. There is also ample evidence, however, to show that these waters can be highly damaging to the environment and to the soil resource base when improperly managed. Therefore, saline waters should be only used for irrigation after careful study and considering as many factors as possible. Then, when the waters are used for irrigation, a careful monitoring program should be implemented of both the crops produced and of the resulting soil and environmental changes.

REFERENCES

1. Rhoades, J.D.; Kandiah, A.; Marshali, A.M. The use of saline water for crop production. *FAO Irrigation and Drainage Paper 48*; Food and Agriculture Organization of the United Nations: Rome, 1992.
2. Miller, R.W.; Gardiner, D.T. *Soils in Our Environment*, 8th Ed.; Prentice Hall: Upper Saddle River, NJ, 1998.

Irrigation: Sewage Effluent Use

B.A. Stewart
Dryland Agriculture Institute, West Texas A&M University, Canyon, Texas, U.S.A.

Abstract
Soils store, decompose, or immobilize nitrates, phosphorus, pesticides, and other substances that can become pollutants in air or water. Consequently, soil has, for centuries, been used for the application of sewage effluents. It should not, however, be assumed that irrigation is always the best solution for wastewater disposal. While disposal is the primary objective in many cases, the need of water for irrigation is becoming more often the driver for using sewage effluent on land. In general, however, sustainable and environmentally sound systems can be developed in most situations provided proper management practices are followed.

INTRODUCTION

One of the primary functions of soil is to buffer environmental change. This is the result of the biological, chemical, and physical processes that occur in soils. The soil matrix serves as an incubation chamber for decomposing organic wastes including pesticides, sewage, solid wastes, and many other wastes. Soils store, decompose, or immobilize nitrates, phosphorus, pesticides, and other substances that can become pollutants in air or water. Consequently, soil has, for centuries, been used for the application of sewage effluents. Sewage effluent provides farmers with a nutrient-enriched water supply and society with a reliable and inexpensive means of wastewater treatment and disposal. It should not, however, be assumed that irrigation is always the best solution for wastewater disposal. Disposal by irrigation should always be compared with alternative options based on environmental, social, and economic costs and benefits.

While disposal is the primary objective in many cases, the need of water for irrigation is becoming more often the driver for using sewage effluent on land. This is particularly true in areas like the Middle East where population growth is resulting in severe water shortages. The guidelines for using effluent for irrigation vary considerably among countries and other governing bodies. Cameron[1] conducted a literature review and found wide differences of guidelines for effluent irrigation projects being used throughout the world. In general, however, sustainable and environmentally sound systems can be developed in most situations provided proper management practices are followed.

CONCERNS OF IRRIGATING WITH SEWAGE EFFLUENT

In spite of the documented benefits associated with the use of sewage effluent for irrigation, there are numerous concerns. Many industrial wastewaters have been routinely dumped into municipal sewage lines. While this issue has been addressed in some jurisdictions, it has not in many others. In the United States, the Environmental Protection Agency requires that wastewaters be treated prior to disposal into municipal treatment plants or back into groundwater. Irrigating with wastewaters partially cleans water by percolation through the soil, but soluble salts and some inorganic and organic chemicals may continue to flow with the water to groundwater or surface supplies. In general, the Environmental Protection Agency allows sewage effluents to be used for irrigation only if it does not cause: 1) extensive groundwater pollution; 2) a direct public health hazard; 3) an accumulation in the soil or water of hazardous substances that can get into the food chain; 4) an accumulation of pollutants such as odors into the atmosphere; and 5) other aesthetic losses, within the limits.[2]

Bouwer[3] has also expressed concerns about the use of sewage effluent for irrigation. He is particularly concerned with pathogens and warns that complete removal of viruses, bacteria, and protozoa and other parasites should be required before the effluent can be used to irrigate fruits/vegetables consumed raw or brought into the kitchen, or parks, playgrounds and other areas with free public access. Bouwer also stresses that long-term effects of sewage effluent irrigation on underlying groundwater should be considered in addition to the changes in nitrate and salinity. Ground water in low rainfall regions can be highly affected by percolating sewage effluent because much of the water is used by the growing crops and this greatly concentrates the chemicals in the small amounts of water that actually percolate to the groundwater. These chemicals can include disinfection byproducts, pharmaceutically active chemicals, and compounds derived from humic and fulvic acids formed by the decomposition of plant material. Bouwer claims that many of these chemicals are suspected carcinogens or toxic. Therefore, Bouwer concludes that while sewage irrigation looks good on the surface, a more extensive look reveals a potential for serious contamination of groundwater. He states that municipalities and other entities responsible for irrigation with sewage effluent should do a groundwater impact analysis to develop management protocols and be prepared for liability actions. Those who benefit are local and state institutions in water resources, environmental quality protection, public health, consultants, and operators of effluent irrigation projects.

REUSE STANDARDS

The standards for using sewage effluent for irrigation of agricultural crops vary widely among different countries of the world. Mexico and many South American countries, e.g., use untreated wastewater for irrigation.[4] Most of these countries do not have the resources or capital to treat sewage effluents. Wastewater is utilized after little or no treatment, and health risks are minimized by crop selection. Mexico does not allow wastewater to be used to irrigate lettuce, cabbage, beets, coriander, radishes, carrots, spinach, and parsley. Acceptable crops include alfalfa, cereals, beans, chili, and green tomatoes. In contrast, Israel has very stringent water reuse requirements. Effluent water requires a high level of treatment (large soil-aquifer recharge systems with dewatering) before the water can be reused for irrigation of vegetables to be consumed raw.[5] Health guidelines for irrigation with treated wastewater developed in California indicate that effluent waters used on food crops must be disinfected, oxidized, coagulated, clarified, and filtered.[6] Total coliform counts cannot exceed a median value of 2.2/100 mL or a single sample value of 25/100 mL. Total coliforms must be monitored daily and turbidity cannot exceed 2 nephelometric turbidity units and must be monitored continuously. Less restrictive guidelines developed by Shuval et al.,[7] and adopted by most of the international agencies, suggested that effluent water reuse was relatively safe to use if it contained less than 1 helminth egg L^{-1}, and less than 1000 fecal coliforms/100 mL.

MONITORING GUIDELINES

Site selection is a critical and necessary step in initiating a sewage effluent irrigation system. The U.S. Environmental Protection Agency[8] published detailed information on site characterization and evaluation. Information was provided on the design of systems, site characteristics, expected quality of the effluent water after land treatment, and typical permeabilities and textural classes suitable for each land treatment process. Information was provided for designing and monitoring site characteristics for slow rate processes (sprinkler and other typical farm irrigation systems), rapid infiltration basins, and overland flow systems. Monitoring requirements will vary considerably among projects depending on the cropping patterns, soil characteristics, and specific environmental concerns. In most cases, monitoring procedures and criteria will be site specific. In all cases, however, the objectives should be to use the resources effectively, protect the land, protect the groundwater, protect the surface water, and protect the community amenity.

REFERENCES

1. Cameron, D.R. *Sustainable Effluent Irrigation Phase 1: Literature Review International Perspective and Standards*, Technical Report Prepared for Irrigation Sustainability committee; Canada–Saskatchewan Agriculture Green Plan, 1996.
2. Miller, R.W.; Gardiner, D.T. *Soils in Our Environment*, 8th Ed.; Prentice-Hall Inc: Upper Saddle River, NJ, 1998.
3. Bouwer, H. *Groundwater Problems Caused by Irrigation with Sewage Effluent*; Irrigation and Water Quality Laboratory, USDA-ARS: Phoenix, AZ, 2000.
4. Strauss, M.; Blumenthal, U.J. *Human Waste Use in Agriculture and Aquaculture: Utilization Practices and Health Perspectives*; IRWCD Report No. 09/90; International Reference Centre for Waste Disposal: Deubendorf, 1990.
5. Shelef, G. The role of wastewater reuse in water resources management in Israel. Water Sci. Tech. **1990**, *23*, 2081–2089, Switzerland.
6. Ongerth, H.J.; Jopling, W.F. Water reuse in California. In *Water Renovation and Reuse*; Shuval, H.I., Ed.; Academic Press: New York, 1977.
7. Shuval, H.I.; Adin, A.; Fattal, B.; Rawitz, E.; Yekutiel, P. *Wastewater Irrigation in Developing Countries. Health Effects and Technical Solutions*; World Bank Tech. Pap.; 1986; Vol. 51, 325 pp.
8. U.S. EPA. *Process Design Manual: Land Treatment of Municipal Wastewater*; EPA 625/1-81-013; U.S. EPA Center for Environmental Research Information: Cincinnati, OH, 1981.

Irrigation: Soil Salinity

James D. Rhoades
Agricultural Salinity Consulting, Riverside, California, U.S.A.

Abstract

Irrigation is an ancient practice that predates recorded history. Irrigation has resulted in considerable salination of associated land and water. Surviving the salinity threat requires that the seriousness of the problem be recognized more widely, the processes contributing to salination of irrigated lands be understood, effective control measures be developed and implemented that will sustain the viability of irrigated agriculture, and that practical reclamation measures be implemented to rejuvenate the presently degraded lands.

INTRODUCTION

Irrigation is an ancient practice that predates recorded history. While irrigated farmland comprises only about 15% of the worlds' total farmland, it contributes about 36% of the total supply of food and fiber, and it stabilizes production against the vagaries of weather.[1] In 30 years time, irrigated agriculture is expected to have to supply 50% of the worlds' food production requirements.[1] However, over the last 20 years, irrigation growth has actually slowed to a rate that is now inadequate to keep up with the projected expanding food requirements.[1] Furthermore, irrigation has resulted in considerable salination of associated land and water. It has been estimated variably that the salinized area is as low as 20 and as high as 50% of the worlds' irrigated land.[2–4] Worldwide, about 76.6 Mha of land have become degraded by human-induced salination over the last 45–50 years.[3] It has been estimated that the world is losing at least three hectares of arable land every minute to soil salination (about 1.6 Mha per year), second only to erosion as the leading worldwide cause of soil degradation.[5–7] These data imply that the rate of salinization in developed irrigation projects now exceeds the rate of irrigation expansion.[8]

Surviving the salinity threat requires that the seriousness of the problem be recognized more widely, the processes contributing to salination of irrigated lands be understood, effective control measures be developed and implemented that will sustain the viability of irrigated agriculture, and that practical reclamation measures be implemented to rejuvenate the presently degraded lands.[9,10]

DELETERIOUS EFFECTS OF SALTS ON PLANTS, SOILS, AND WATERS

Salt-affected soils have reduced value for agriculture because of their content and proportions of salts, consisting mainly of sodium, magnesium, calcium, chloride, and sulfate and secondarily of potassium, bicarbonate, carbonate, nitrate, and boron. Saline soils contain excessive amounts of soluble salts for the practical and normal production of most agricultural crops. Sodic soils are those that contain excessive amounts of adsorbed sodium in proportion to calcium and magnesium, given the salinity level of the soil water. An example of a salt-affected irrigated soil is shown in Fig. 1.

Soluble salts exert both general and specific effects on plants, both of which reduce crop yield.[11] Excess salinity in the seedbed hinders seedling establishment and in the crop root zone causes a general reduction in growth rate. In addition, certain salt constituents are specifically toxic to some plants. For example, boron is highly toxic to susceptible crops when present in the soil water at concentrations of only a few parts per million. In some woody crops sodium and chloride may accumulate in the tissue over time to toxic levels. These toxicity problems are, however, much less prevalent than is the general salinity problem.

Salts may also change soil properties that affect the suitability of the soil as a medium for plant growth.[12] The suitability of soils for cropping depends appreciably on the readiness with which they conduct water and air (permeability) and on their aggregate properties (structure), which control the friability (ease with which crumbled) of the seedbed (tilth). In contrast to saline soils, which are well aggregated and whose tillage properties and permeability to water and air are equal to or higher than those of similar nonsaline soils, sodic soils have reduced permeabilities and poor tilth. These problems are caused by the swelling and dispersion of clay minerals and by the breakdown of soil structure (slaking and crusting), which results in loss of permeability and tilth. Sodic soils are generally less extensive but more difficult to reclaim than saline soils.

Beneficial use of water in irrigation consists of transpiration and leaching for salinity control (the leaching requirement). Plant growth is directly proportional to water consumption through transpiration.[13] From the point of view of irrigated agriculture, the ultimate objective of irrigation is to increase the amount of water available to

Fig. 1 Photograph of salt-affected irrigated field.

support transpiration. Salts reduce the fraction of water in a supply (or in the soil profile) that can be consumed beneficially in plant transpiration.[14] In considering the use of a saline water for irrigation and in selecting appropriate policies and practices of irrigation and drainage management, it is important to recognize that the total volume of a saline water supply cannot be consumed beneficially in crop production (i.e., transpired by the plant). A plant will not grow properly when the salt concentration in the soil water exceeds some limit specific to it under the given conditions of climate and management.[11] This is even true for halophytes.[15] Thus, the practice of blending or diluting excessively saline waters with good quality water supplies should be undertaken only after consideration is given to how it affects the volumes of consumable (usable) water in the combined and separated supplies.[14]

CAUSES OF SALINATION INDUCED BY IRRIGATION AND DRAINAGE

While salt-affected soils occur extensively under natural conditions, the salt problems of greatest importance to agriculture arise when previously productive soils become salinized as a result of agricultural activities (the so-called secondary salination). The extent and salt balance of salt-affected areas has been modified considerably by the redistribution of water (hence salt) through irrigation and drainage. The development of large-scale irrigation and drainage projects, which involves diversion of rivers, construction of large reservoirs, and irrigation of large landscapes, causes large changes in the natural water and salt balances of entire geohydrologic systems. The impact of such developments can extend well beyond that of the immediate irrigated area. Excessive water diversions and applications are major causes of soil and water salination in irrigated lands. It is not unusual to find that less than 60% of the water diverted for irrigation is used in crop transpiration.[9] This implies that about 40% of the irrigation water eventually ends up as deep percolation. This drainage water contains more salt than that added with the irrigation water because of salt dissolution and mineral weathering[14] within the root zone. It often gains additional salt-load as it dissolves salts of geologic origin from the underlying substrata through which it flows in its down-gradient path. This drainage water often flows laterally to lower lying areas, eventually resulting in shallow saline groundwaters of large areas of land (waterlogging). Salination occurs in soils underlain by saline shallow groundwater through the process of "capillary rise" as groundwater (hence, salt) is driven upwards by the force of evaporation of water from the soil surface. Correspondingly, saline soils and waterlogging are closely associated problems.

Seepage from unlined or inadequately lined delivery canals occurs in many irrigation projects and is often substantial. Law, Skogerboe, and Denit[16] estimated that 20% of the total water diverted for irrigation in the United States is lost by seepage from conveyance and irrigation canals. Biswas[17] estimated that 57% of the total water diverted for irrigation in the world is lost from conveyance and distribution canals. Analogous to on-farm deep percolation resulting from irrigation, these seepage waters typically percolate through the underlying strata (often dissolving additional salts in the process), flow to lower elevation lands or waters, and add to the problems of waterlogging and salt-loading associated with on-farm irrigation there. A classic example of the rise in the water table following the development of irrigation has been documented in Pakistan and is described by Jensen, Rangeley, and Dieleman[9] and Ghassemi, Jakeman, and Nix.[2] The depth to the water table in the irrigated landscape located between three major river-tributaries rose from 20 to 30 m over a period of 80–100 years, i.e., from preirrigated time (about 1860) to the early 1960s, until it was nearly at the soil-surface. In one region, the water table rose nearly linearly from 1929 to 1950, demonstrating that deep percolation and seepage resulting from irrigation were the primary causes. Ahmad[18] concluded that about 50% of the water diverted into irrigation canals in Pakistan eventually goes to the groundwater by seepage and deep percolation.

The role of irrigated agriculture in salinizing soil systems has been well recognized for hundreds of years. It is of relatively more recent recognition that salination of water resources from agricultural activities is a major and widespread phenomenon of likely equal concern to that of soil salination. The causes of water salination are essentially the same as those of soils, only the final reservoir of the discharged salt-load is a water supply in the former case.[14] The volume of the water supply is reduced through irrigation diversions and irrigation; thus, its capacity to assimilate such received salts before reaching use-limiting levels is reduced proportionately. Only in the past 15 years has it become apparent that trace toxic constituents, such as selenium, in agricultural drainage waters can also cause serious pollution problems.[19]

IRRIGATION AND DRAINAGE MANAGEMENT TO CONTROL SOIL SALINITY

The key to overall salinity control is strict control that maintains a net downward movement of soil water in the root zone of irrigated fields over time while minimizing excess irrigation diversions, applications, and deep percolation.[20] The direct effect of salinity on plant growth is minimized by maintaining the soil-water content in the root zone within a narrow range at a relatively high level, while at the same time avoiding surface-ponding and oxygen deletion and minimizing deep percolation. Combined methods of pressurized, high-frequency irrigation and irrigation scheduling have been developed that permit substantially the desired control to be achieved.[21,22] These systems transfer control of water distribution and infiltration from the soil to the irrigation equipment. This results in less excess water (and hence, less salt) being applied overall to the field to meet the needs of a part of the field area having lowest intake rate, as done in the more traditional gravity irrigated systems. However, gravity irrigation systems can be designed to achieve good irrigation efficiency and salinity control, even though surface ponding still does occur. The so-called level-basin, multi-set, cablegation, surge, and tailwater-return systems are among them.[21,22] The need for irrigation and the amount required to meet evapotranspiration and leaching requirement is determined from plant stress measurements, calculations of evapotranspiration amounts, measurements of soil-water depletion, measurements of soil (or soil-water) salinity, or a combination of them.[21,22]

In addition to effective methods of irrigation scheduling and application, appropriate irrigation and salinity management also require an effective delivery system. Delivery systems have generally been designed to provide water on a regular schedule. Efficient irrigation systems require more flexible deliveries that can provide water on demand as each crop and particular field have need of it. Delivery systems can be improved by lining the canals, by containing the water within closed conduits, and by implementing techniques that increase the flexibility of delivery.

As briefly discussed earlier, irrigated agriculture is a major contributor to the salinity of many rivers and groundwaters, as well as soils. Reducing deep percolation generally lessens the salt load that is returned to rivers or groundwater and their pollution.[14] Additionally, saline drainage waters should be intercepted before being allowed to mix with water of better quality. The intercepted saline drainage water should be desalted and reused, disposed of by pond evaporation or by injection into some suitably isolated deep aquifer, or better yet it should be used for irrigation in a situation where brackish water is appropriate. Various irrigation and drainage strategies have been developed for minimizing the pollution of waters from irrigation and for using brackish waters for irrigation.[14,23] Desalination of agricultural drainage waters is not now economically feasible, but improved techniques for doing this exist and some are being implemented. However, more needs to be done in this regard.

Traditionally, the concepts of leaching requirement and salt-balance index have been used to plan and judge the appropriateness of irrigation and drainage systems, operations and practices with respect to salinity control, water use efficiency, and irrigation sustainability. However, these approaches are inadequate. The recommended method is to monitor directly the root-zone salinity levels and distributions across fields as a means to evaluate the effectiveness of salinity, irrigation, and drainage management practices, to detect problems (current and developing), to help determine the underlying causes of problems, and to determine source areas of major water and salt-load contributions to the underlying groundwater. Theory, equipment, and practical technology have been developed for these purposes.[24] More information about irrigation and drainage management to control soil and water salinity is found elsewhere.[25–27]

REFERENCES

1. FAO. *World Agriculture Toward 2000*: An FAO Study; Alexandratos, N., Ed.; Bellhaven Press: London, 1988; 338 pp.
2. Ghassemi, F.; Jakeman, A.J.; Nix, H.A. *Salination of Land and Water Resources. Human Causes, Extent, Management and Case Studies*; CAB International: Wallingford, U.K., 1995; 526 pp.
3. Oldeman, L.R.; van Engelen, V.N.P.; Pulles, J.H.M. The Extent of human-induced soil degradation. In *World Map of the Status of Human-Induced Soil Degradation: An Explanatory Note*; Oldeman, L.R., Hakkeling, R.T.A., Sombroek, W.G., Eds.; International Soil Reference and Information Center (ISRIC): Wageningen, the Netherlands, 1991; 27–33.
4. Adams, W.M.; Hughes, F.M.R. Irrigation development in desert environments. In *Techniques for Desert Reclamation*; Goudie, A.S., Ed.; Wiley: New York, 1990; 135–160.
5. Buringh, P. Food production potential of the world. In *The World Food Problem: Consensus and Conflict*; Sinha, R., Ed.; Pergamon Press: Oxford, 1977; 477–485.
6. Dregne, H.; Kassas, M.; Razanov, B. A new assessment of the world status of desertification. Desertification Control Bull. **1991**, *20*, 6–18.
7. Umali, D.L. Irrigation-induced salinity. In *A Growing Problem for Development and Environment*; Technical Paper; World Bank: Washington, DC, 1993.
8. Seckler, D. *The New Era of Water Resources Management: From "Dry" to "Wet" Water Savings*; Consultative Group on International Agricultural Research: Washington, DC, 1996.
9. Jensen, M.E.; Rangeley, W.R.; Dieleman, P.J. Irrigation trends in world agriculture. In *Irrigation of Agricultural Crops*; American Society of Agronomy Monograph No. 30; ASA: Madison, WI, 1990; 31–67.
10. UNEP. *Saving Our Planet: Challenges and Hopes*; Nairobi, United Nations Environment Program: Nairobi, Kenya, 1992; 20 pp.

11. Maas, E.V. Crop salt tolerance. ASCE Manuals and Reports on Engineering No. 71. In *Agricultural Salinity Assessment and Management Manual*; Tanji, K.K., Ed.; ASCE: New York, 1990; 262–304.
12. Rhoades, J.D. Principal effects of salts on soils and plants. In *Water, Soil and Crop Management Relating to the Use of Saline Water*; Kandiah, A., Ed.; FAO (AGL) Misc. Series Publication 16/90; Food and Agriculture Organization of the United Nations: Rome, 1990; 1933.
13. Sinclair, T.R. Limits to crop yield? In *Physiology and Determination of Crop Yield*; Boone, K.J., Ed.; American Society of Agronomy: Madison, WI, 1994; 509–532.
14. Rhoades, J.D.; Kandiah, A.; Mashali, A.M. *The Use of Saline Waters for Crop Production*; FAO Irrigation and Drainage Paper 48; FAO: Rome, Italy, 1992; 133 pp.
15. Miyamoto, S.; Glenn, E.P.; Oslen, M.W. Growth, water use and salt uptake of four halophytes irrigated with highly saline water. J. Arid Environ. **1996**, *32*, 141–159.
16. Law, J.P.; Skogerboe, G.V.; Denit, J.D. The need for implementing irrigation return flow control. p. 1–17. In *Managing Irrigated Agriculture to Improve Water Quality*; Proc. Math. Conf. Manag. Irrig. Agric. Improve Water Avail., Denver, CO, May 1972; Graphics Manage. Corp.: Washington, DC, 16–18.
17. Biswas, A.K. Conservation and management of water resources. In *Techniques for Desert Reclamation*; Goudie, A.S., Ed.; Wiley: New York, 1990; 251–265.
18. Ahmad, N. Planning for Future Water Resources of Pakistan. Proceedings of Darves Bornoz Spec. Conference, National Committee of Pakistan; ICID: New Delhi, India, 1986; 279–294.
19. Letey, J.; Roberts, C.; Penberth, M.; Vasek, C. *An Agriculturl Dilemma: Drainage Water and Toxics Disposal in the San Joaquin Valley*; Special Publication 3319; University of California: Oakland, 1986.
20. Rhoades, J.D. Soil salinity—causes and controls. In *Techniques for Desert Reclamation*; Goude, A.S., Ed.; Wiley: New York, 1990; 109–134.
21. Hoffman, G.J.; Rhoades, J.D.; Letey, J.; Sheng, F. Salinity management. In *Management of Farm Irrigation Systems*; Hoffman, G.J., Howell, T.A., Solomon, K.H., Eds.; ASCE: St. Joseph, MI, 1990; 667–715.
22. Kruse, E.G.; Willardson, L.; Ayars, J. On-farm irrigation and drainage practices. In *Agricultural Salinity Assessment and Management Manual*; Tanji, K.K., Ed.; ASCE Manuals and Reports on Engineering No. 71; ASCE: New York, 1990; 349–371.
23. Rhoades, J.D. Use of saline drainage water for irrigation. In *Agricultural Drainage*; ASA Drainage Monograph 38; Skaggs, R.W., van Schilfgaarde, J., Eds.; ASA Drainage Monograph 38; American Society of Agronomy: Madison, WI, 1999; 619–657.
24. Rhoades, J.D.; Chanduvi, F.; Lesch, S. *Soil Salinity Assessment: Methods and Interpretation of Electrical Conductivity*; FAO Irrigation and Drainage Paper 57; FAO, United Nations: Rome, Italy, 1999; 152 pp.
25. Rhoades, J.D. Use of saline and brackish waters for irrigation: implications and role in increasing food production, conserving water, sustaining irrigation and controlling soil and water degradation. Proceedings of the International Workshop on "The Use of Saline and Brackish Waters for Irrigation: Implications for the Management of Irrigation, Drainage and Crops" at the 10th Afro-Asian Conference of the International Committee on Irrigation and Drainage, Bali, Indonesia, July 23–24; Ragab, R., Pearce, G., Eds.; International Committee on Irrigation and Drainage: Bali, Indonesia, 1998; 261–304.
26. Rhoades, J.D.; Loveday, J. Salinity in irrigated agriculture. In *Irrigation of Agricultural Crops*; Stewart, B.A., Nielsen, D.R., Eds.; Agron. Monograph. No. 30; American Society of Agronomy: Madison, Wisconsin, 1990; 1089–1142.
27. Tanji, K.K. Nature and extent of agricultural salinity. In *Agricultural Salinity Assessment and Management*; Tanji, K.K., Ed.; ASCE Manuals and Reports on Engineering No. 71, ASCE: New York, 1990; 1–17.

Lakes and Reservoirs: Pollution

Subhankar Karmakar
Center for Environmental Science and Engineering (CESE), Indian Institute of Technology Bombay, Mumbai, India

O.M. Musthafa
Center for Pollution Control and Environmental Engineering, Pondicherry University, Pondicherry, India

Abstract
Lakes and reservoirs are major resources as these hold about 90% of the world's surface freshwater and are the key freshwater resources for agriculture, fisheries, domestic, industrial, recreational, landscape entertainment, and energy production. However, these utilizations depend on the desirable water quality that should be based on a well-balanced environment in terms of its physical, chemical, and biological characteristics. Fresh surface water systems, such as rivers, streams, lakes, and ponds, have been severely affected by a multitude of anthropogenic as well as natural disturbances. This damage has caused serious negative impacts on the structures and functions of the entire ecosystem. The lentic surface water quality in reservoirs, lakes, or ponds is severely affected by anthropogenic pollution, and many efforts have already been made to assess and manage their water quality. Information and case studies reviewed in this entry indicate that the lakes and reservoirs are at risk from overexploitation, overenrichment, toxic contamination, and sedimentation. The entry may inspire future environmental and water resource professionals to take necessary actions to mitigate lake and reservoir pollution.

INTRODUCTION

Surface water is one of the most important natural resources in the world. It has been explicitly established that water of good quality is a fundamental element to sustainable socioeconomic development. It is the habitat for a large number of species and is a crucial component for metabolic activities of plants and animals. Aquatic ecosystems are endangered on a worldwide scale by a multitude of pollutants as well as damaging land-use or water-management practices. Some problems have been present for a long time but have only recently reached a critical level, while the rest are recently emerging. Oxygen balance in the aquatic systems is severely affected by organic pollution, which often results in severe pathogenic contamination. Enrichment of aquatic systems with nutrients from various origins, predominantly domestic sewage, agricultural runoff, and agro-industrial effluents, results in enhanced eutrophication, of which lakes and reservoirs are affected the most.

Lakes and reservoirs are the major resources of fresh surface water. They are larger and deeper than ponds and are not part of the ocean. Lakes and reservoirs are major resources as these hold about 90% of the world's fresh surface water and are the key freshwater resources for agriculture, fisheries, domestic, industrial, recreational, landscape entertainment, and energy production. Natural lakes are bodies of water, created by volcanic, tectonic, or glacial activity, whereas reservoirs are artificial impoundments. A lake is a relatively large lentic freshwater or saltwater body, which is localized in a basin surrounded by land. Natural lakes are generally found in mountainous areas, rift zones, and areas with ongoing glaciations. In some parts of the world, there are many lakes formed due to chaotic drainage patterns left over from the last ice age. They are the habitats of a variety of flora and fauna, making them a source of fish and a destination for migratory birds to reproduce or rest. A reservoir, which is known as an artificial lake, is constructed for the benefit of man's water needs, sometimes for one particular purpose, but more recently for multiple purposes. Reservoirs are different from lakes in many ways. They have usually larger drainage basins than lakes, and many are located in watersheds with extensive agricultural activities.

Direct contamination of surface waters with metals in discharges from mining, smelting, and industrial manufacturing is a long-lasting phenomenon. The emission of airborne pollutants has now reached such proportions that long-range atmospheric transport causes contamination, not only in the vicinity of industrial regions but also in more remote areas. Similarly, precipitation of acid rain occurs, when moisture in the atmosphere combines with gases such as sulfur dioxide, which are produced when fossil fuels are burnt. This may cause significant acidification of surface waters, especially lakes and reservoirs. Contamination of water by synthetic organic micropollutants and emerging contaminants results either from direct discharge into the surface waters through runoff or after transport through the atmosphere.

This entry briefly presents classification of lakes and reservoirs based on the flow of water in and out of the system and their utility, respectively, followed by the problems associated with lakes and reservoirs causing deterioration of water quality. A discussion on monitoring of water quality and various protective and restorative measures is also made through a literature survey, which may be useful to future research aspirants and water resource professionals in identifying economically and environmentally sustainable lake and reservoir management strategy.

CLASSIFICATION OF LAKES AND RESERVOIRS

All lakes are temporary over geologic time scales, as they will slowly fill in with sediments or spill out of the basin.[1] Water enters into lakes from a variety of sources such as seepage through groundwater storage, runoff from watershed, direct precipitation into the lake, and other surface waters bodies (like streams or rivers). Water may drain out from lakes through deep percolation to join groundwater table or through surface water flow and evaporation. Natural lakes can be classified into four major types based on how water enters and exits the lake. Water may enter into the lake through one source or multiple sources. The water quality of a lake and its biodiversity are significantly influenced by the type of lake. Depending upon the way of entrance and exit of water, lakes can be classified into four categories as shown in Fig. 1.

Seepage lakes do not have a distinct inlet or an outlet, and occasionally overflow. The major sources of water are direct precipitation, surface runoff from the immediate drainage area, and seepage through groundwater storage, as seepage lakes are landlocked water bodies. Since seepage lakes are sensitive to groundwater levels and local rainfall patterns, water levels may fluctuate seasonally. These lakes may have a less diverse fishery as the direct source of water is not a flowing water body or stream. Seepage lakes also have a smaller drainage area, which may help to account for lower nutrient levels. *Spring lakes* have no distinct inlet, but do have an outlet. The major source of water for spring lakes is groundwater flowing into the bottom of the lake from inside and outside the immediate surface drainage area. *Groundwater drained lakes* have no inlet, but similar to spring lakes, these may have an uninterruptedly flowing outlet. Drained lakes are not groundwater fed and their principal sources of water are precipitation and direct drainage from the surrounding land. The water levels in drained lakes fluctuate frequently depending on the supply of water. Under severe conditions, the outlets from drained lakes may become intermittent. *Drainage lakes* have both an inlet and an outlet where the main water source is stream drainage. These lakes support fish populations that are not necessarily identical to the streams connected to them. Drainage lakes mostly have higher nutrient levels than natural seepage or spring lakes.[2] Depending on the utility, reservoirs can be classified into four classes as shown in Fig. 2.

Storage/conservation reservoirs retain excess water supplies during peak flows and release water gradually during low flows as and when needed. A *flood control reservoir* stores a portion of the flood flows so that it can minimize the flood peaks in the areas to be protected downstream. A *multipurpose reservoir* is meant for serving multiple purposes such as water supply, flood and soil erosion control, hydroelectric power generation, recreation, irrigation, etc. A *distribution reservoir* is connected with a network of primary water supply and is used to supply water to the end users according to fluctuations in demand over a short time period and serves as local storage in the case of emergency. Such reservoirs, thus, support the water treatment plants to work at a uniform rate and can store water when there is less demand and, thus, supply water during high-demand periods. Water quality in both lakes and reservoirs is influenced by many factors such as water body type, ecosystem characteristics, land use and land cover, and human activities.[3]

Freshwater/Saline Lakes

Most lakes hold freshwater, but some, particularly those where water cannot discharge via a river, can be salty. As a matter of fact, some lakes such as the Great Salt Lakes are saltier than the oceans. Lakes whose salinity content is more than 3 g/L are considered as saline lakes. They are prevalent and present on all regions, including Antarctica

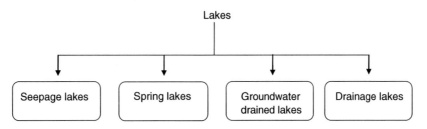

Fig. 1 Classification of lakes.

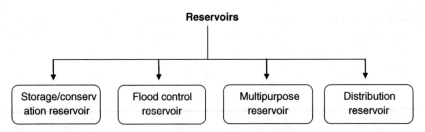

Fig. 2 Classification of reservoirs.

(e.g., Caspian Sea, Dead Sea, etc.). Though the inland saline water constitutes around 45% of the total inland water, a few deep lakes (mainly the Caspian Sea) occupy a significant volume of these saline waters. Salinity has a great influence on the freezing point of water, amount of dissolved oxygen, etc.[4]

Trophic Status

Lakes can be classified based on their trophic state as "eutrophic," "mesotrophic," and "oligotrophic." The word "trophic" means nutrition or growth. A eutrophic lake is characterized by the presence of a high concentration of plant nutrients and associated excess plant growth. On the other hand, an oligotrophic lake is characterized by low nutrient concentrations and low plant growth. Mesotrophic lakes fall in between these two. The major factors that regulate the trophic status of a lake are as follows: 1) rate of nutrient supply; 2) climatic condition (sunlight, temperature, precipitation, lake basin turnover time, etc.), and 3) morphometry/shape of lake basin (mean and maximum depth, volume and surface area, watershed-to-lake surface area ratio, etc.).[5]

PROBLEMS ASSOCIATED WITH LAKES AND RESERVOIRS

Although more than three-fourths of the earth is occupied with water, only less than 0.3% (including surface water and groundwater) is available for human consumption.[6] The total area of the lakes on earth amounts to approximately 2.5×10^6 km^2 or 1.8% of the continental area, containing 1.2×10^5 km^3 of water. The 253 largest lakes (larger than 500 km^2) of the planet contain an estimated 78% of the world's unfrozen fresh surface water and thus represent an essential global life support system.[7] Though the amount of water in lakes and reservoirs is very small, because of its rapid renewal, these habitats are the primary sustainable supplier of freshwater for most regions. Since freshwater on the earth is scarce, sincere efforts are required for the conservation of existing water resources to ensure the availability of sufficiently good quality water. The surface water resources are more susceptible to pollution as compared to the groundwater resources due to the ease of access of pollutants and contaminants in the former.[8,9]

Most of the lakes and reservoirs around the world face environmental stress, and the appropriate functioning of various vital ecosystems is in danger. Due to the explosion in human population, supplementary demand-related damage is forced on lakes and reservoirs. Water levels become lower as a result of higher consumption by households and industries; a growing number of human population results in shrinking and altering their water resources; inappropriate use of the land, particularly hills and mountains, results in increased sedimentation on their basins. Ultimately, pollution from agricultural lands and from domestic and industrial sources may produce eutrophication, resulting in undesirable effects such as the presence of toxic algae, reduction of oxygen, and generation of foul odor. The proliferation of contaminants within lake and reservoir systems can deteriorate water quality significantly. In many regions, lake ecosystems have already degraded and restoration to desirable water quality needs enormous effort and high cost. Following the present status, none can predict for how long these resources can serve as renewable sources of pure surface water for domestic, industrial, and agricultural uses, or as sources of protein-rich food.[10] Lakes and reservoirs have a more vulnerable and complex ecosystem than rivers as they do not have a self-purifying ability and, hence, readily accumulate pollutants. Because of their importance, their beauty, their religious and cultural significance, and their relative susceptibility to degradation, lakes and reservoirs require more concerted attention than is paid generally to river systems.[11] The next few paragraphs elaborate the major pollution problems of lakes and reservoirs.

Eutrophication

It can be defined as the process of enrichment of waters with plant nutrients that causes raw water quality loads, such as high primary production, low oxygen concentrations, and increased concentrations of hydrogen sulfide, carbon dioxide, dissolved iron, and manganese in the hypolimnion.[12] It is considered as one of the serious negative effects faced by the lentic water bodies such as lakes, ponds, and reservoirs and is one of the major water quality problems worldwide[13] and the most serious challenge for water management professionals in densely populated areas.[14] Water from eutrophic reservoirs may have meager taste, odor, and color, and some have high concentrations of naturally occurring organic compounds that may form trihalomethanes, which are carcinogenic and mutagenic and other by-products of disinfection.[15,16]

Seasonal pattern is prominently influenced by the availability of solar insolation and nutrients. Usually, the shallower the lake, the less the check of internal nutrient recycling by thermal stratification and light availability by critical depth as compared to mixing depth. During summer wind events, the lakes that are having an Osgood Index (OI)[17] less than 7 will show a strong tendency for mixing, as a result of which, nutrients from sediments/hypolimnion undergo recycling and enter into the photic zone. Such events entrain nutrients and cause summer blooms.[18] Thus, in highly dynamic lakes, the pattern of phytoplankton abundance/species composition differs with wind events.

Toxic Materials

It includes mainly heavy metals, toxin-producing microphytes, and pesticides from agricultural land. Toxic substances may enter water bodies directly as land runoff from urban streets and mining areas or as agricultural runoff including forestry drainage, discharge of inadequately treated sewage, and industrial effluents, and through deposition of airborne pollutants. Toxic substances can also be created in drinking water when chemicals from treatment plants interact with organic molecules in the raw water to form carcinogenic compounds such as trihalomethanes. A range of heavy metals such as mercury, arsenic, lead, and cadmium; chlorinated substances such as dichlorodiphenyltrichloroethane (DDT), dichlorodiphenyldichloroethylene (DDE), and polychlorinated biphenyls (PCBs); organic substances such as polyaromatic hydrocarbons (PAHs); Dieldrin; etc., create toxic conditions.

A large diversity of organic pollutants poses a serious threat to aquatic ecosystems. PCBs and 3,4-benzpyrene were confirmed as jeopardizing Lake Constance.[19] However, pesticide is the major group of contaminants in the top layers of sediments. The recognition of DDT as being harmful to human and animals served as the eye opener to serious consequences of various pesticides. PCBs are also considered as persistent organic pollutants and, thus, have huge potential hazards. The main sources of pesticides in lakes and reservoirs are 1) agriculture and forestry; 2) actions against aquatic weeds (e.g., water fern *Salvinia molesta*, *Eichhornia crassipes*); 3) actions against parasites and waterborne diseases (e.g., malaria and schistosomiasis); and 4) regulation of fish populations with rotenone (an active ingredient of derris, which is used as a fish poison for centuries). Atrazine, which is used for the protection of corn from weeds, is demonstrated as detrimental to human, and hence, it has recently been proscribed in many countries. Likewise, lindane, a persistent organochlorine insecticide, is toxic to fish in concentrations as low as 1 ppm. Pesticides are applied for curbing undesirable weeds such as water fern *Salvinia* sp. and *Azolla* sp., water hyacinth *E. crassipes*, water lettuce *Pistia stratiotes*, etc. Application of the pesticides should only be considered in cases where mechanical and biological approaches fail.[19]

Sedimentation

Sedimentation of lakes is a common phenomenon; in general, it takes place very slowly. Any process in a lake watershed that disturbs the soil can significantly hasten this process. Most of these activities are anthropogenic and comprised farming on fragile soils and on steep slopes, surface mining, and construction activities. Suspended sediment and sedimentation have many degrading effects on lakes and shoreline ecosystems. Peripheral wetlands can be completely covered by silt, eradicating their value as nutrient sinks, wave absorbers, nursery areas for fishers, and habitat. As silt settles in a lake, spawning areas are covered and lake volume is reduced, and this causes degradation in fishery production. As a result of reduced storage capacity, both the volume and extent of flooding can increase. In the case of hydroelectric amenities, generating capacity may reduce significantly. The increase in shoal areas as a result of sedimentation can enhance increased macrophyte growth and can interfere with recreational activities such as boating and fishing. Soil loss and suspended silt also contribute to eutrophication and lake contamination since the silt generally includes attached nutrients, herbicides, pesticides, and other chemicals. This increases water treatment costs and maintenance problems of water treatment plants.

Acidification

Atmospheric pollution by sulfur dioxide and nitrous oxides is the major cause of acidification in aquatic systems, through precipitation (dry or wet deposition) and, to a larger extent, through leaching from affected land. Other sources of acid deposits include industrial effluents and mining wastes.

Fish Depletion

Fish provides a substantial portion of animal protein consumed by humans. In tropical developing countries, 60% of the people depend on fish and 40% or more of total protein intake comes from fish.[20] The majority of the world's landed fish catch (87%) comes from marine areas. New developments in watersheds, such as dams and reservoirs to control the annual distribution of water, frequently cause large losses in floodplain fertility and species diversity. Floodplains are needed by many species of fish for reproduction and for refuge. The natural seasonal flooding and drying are signals used by some river-dwelling animals to begin to reproduce or migrate. Proper timing of water level drawdowns in upstream reservoirs is therefore important, especially to enhance spawning of certain species.

Stratification

Stratification is a significant feature influencing water quality in reasonably stagnant, deep waters, such as lakes and reservoirs, which occurs mainly because of the difference in temperature, leading to a variation in density. Occasionally, it can be due to the difference in solute concentrations. Water quality in various layers of stratified water body is subjected to different influences. Solar insolation will be more in the upper layer while the lower layer is physically detached from the atmosphere and may be in touch with decaying sediments that exert an oxygen demand. Because of these varying influences, the lower layer will usually have a reduced oxygen concentration relative to the upper layer. The anoxic condition thus produced will enhance the diffusion of constituents from sediments and form various compounds such as ammonia, nitrate, phosphate, sulfide, silicate, iron, and manganese.

During summer and spring, the surface layer of the water body in temperate regions becomes warmer and hence less dense. A resistance to vertical mixing is formed, because of the existence of warm water over cold water. The top warmer surface layer is known as the epilimnion and the colder water stuck underneath is the hypolimnion. There is a shallow zone called metalimnion or the thermocline, in between epilimnion and hypolimnion, where the temperature changes from warmer epilimnion to colder hypolimnion. Normally, wind- and surface-current-induced mixing is limited to epilimnion, while hypolimnion remains stagnant. The density difference between two layers (viz., epilimnion and hypolimnion) is diminished gradually when the weather becomes cooler. This will enhance the wind-induced vertical mixing between these two layers, which will result in the phenomenon known as "overturn," which can occur quite rapidly. The frequency of overturn and mixing governs predominantly on climate (temperature, solar insolation, and wind) and the characteristics of the lake and its surroundings (depth and exposure to wind). Lakes may be classified according to the frequency of overturn as follows:

- Monomictic: once a year—temperate lakes that do not freeze
- Dimictic: twice a year—temperate lakes that do freeze.
- Polymictic: several times a year—shallow, temperate, or tropical lakes
- Amictic: no mixing—arctic or high-altitude lakes with permanent ice cover, and underground lakes
- Oligomictic: Poor mixing—deep tropical lakes
- Meromictic: incomplete mixing—mainly oligomictic lakes but sometimes deep monomictic and dimictic lakes

Because of the action of crosswind and the flow of water, thermal stratification is not seen in lakes, which have depths less than 10 m. In shallow tropical lakes, complete mixing occurs several times a year, whereas in very deep lakes, stratification may continue all over the year, even in tropical and equatorial regions. This stable stratification results in "meromixis."

In the case of tropical lakes, as a result of moderately constant solar insolation, seasonal changes in water temperature are small. The annual water temperature range is only 2–3°C at the surface and even less at depths greater than 30 m.[21] Winds and precipitation, both play a vital role in mixing. Because of the large difference in rainfall between wet and dry seasons, large variation in water level is seen in some tropical lakes. Such variations have a prominent influence on dilution and nutrient supply, which, in turn, affect algal blooms, zooplankton reproduction, and fish spawning. Wind speeds are usually greater during the dry season and evaporation rates are at their highest. The subsequent heat losses, combined with the turbulence caused by wind action, stimulate the process of mixing.

As far as recreation is concerned, reservoirs are as important as natural lakes but have surplus scopes for flood control, hydropower generation, and water supply. Although both lakes and reservoirs are subjected to silt, organic, and nutrient loadings, reservoirs usually having hefty watersheds and peculiar morphometric conformations are subjected to more water quality problems. Even though lakes and reservoirs have biotic and abiotic processes in common and similar habitats, they have some significant differences. They differ in their geologic history and setting, basin morphology, and hydrologic factors.[22,23]

Non-point source of water pollution generated by expanding agricultural production is considered as a major environmental threat to some lakes. Many chemicals in common agricultural use have a strong affinity for fine soil particles. When the latter erode, these chemicals are carried with them into surface waters. The soil itself is a problem when it accumulates in great quantity in lakes. Lake Pittsfield lost nearly a quarter of its volume to sedimentation in only 24 years. Transport and deposition of eroded materials as well as substances dissolved in runoff and attached to soil particles lead to negative impacts on agricultural land and the Three Gorges Reservoir including water quality decline,[24,25] which are generally thought to be caused by land-use changes of converting forest resources to agriculture in watersheds. Conversion of cropland with a slope greater than 10° into forestland meets the reduction goal.[26]

Over recent decades, the water quality of lakes and reservoirs has been deteriorating rapidly due to external and internal pollution including that from the sediment, and the eutrophication phenomenon has become a more serious global threat. Some researchers suggest applying a plan of sediment dredging to this water body. After dredging, however, a vast amount of sediment would become solid pollutants containing high concentrations of heavy metals

Table 1 Major pollution issues of lakes and reservoirs.

Sl. no.	Pollution issue	Effects	Representative case studies
1	Eutrophication	High primary production, low oxygen concentrations and increased concentrations of hydrogen sulfide, carbon dioxide, dissolved iron, and manganese in the hypolimnion	West Twin Lake, Ohio[28]
2	Sedimentation	Volume will be decreased, macrophytes growth may be enhanced	Lake Superior at Superior Harbor, Wisconsin[29]
3	Acidification	Adverse effects on the most sensitive aquatic species	Many Scandanavian lakes[11]
4	Toxic substances and heavy metal contamination	Bioaccumulation of these toxicants poses health hazards to all members of the food chain including humans	Lake Nainital, India[30]
5	Fish depletion	As the fishery plays a vital role in the supply of animal protein, its depletion affects the food security and economy of a significant portion of humans	Lake Victoria, Uganda[11]

(mainly Hg and Cd), which would certainly cause secondary pollution. This pollution should not be ignored, and a further research is therefore needed on the elaborate restoration scheme for these precious drinking-water sources.[27] Major pollution issues of lakes and reservoirs with few representative past studies are tabulated in Table 1.

SOURCES OF LAKES AND RESERVOIR POLLUTION

Lakes and reservoirs tend to collect not only sediments but also most of the pollutants that are washed into them, and thus they function, in part, as environmental sinks. Eroded soil dissolves in the water and fills in lake bottoms—this activity has significantly degraded lake ecosystems across the world. The sources of pollution can be mainly classified into two: 1) point sources and 2) non-point sources. A point source of pollution is a single identifiable localized source of air, water, thermal, noise, or light pollution.[31] In earlier days, control of "point source" nutrients and toxic contaminants was the principal focus of exertions for the protection and restoration of lakes and reservoirs, but nowadays, the substantial contaminant and nutrient sources to lakes and reservoirs are "non-point" type such as agricultural runoff, erosion from urban or deforested areas, surface mining, or atmospheric depositions. Most lake water is rich in nutrients that support growth of many aquatic macrophytes and algal blooms. Besides, water is contaminated with metals like chromium (Cr), copper (Cu), iron (Fe), manganese (Mn), nickel (Ni), lead (Pb), and zinc (Zn). High concentrations of these metals are also found in sediments, but it is found that the level of metal concentrations of lake varies considerably in different seasons.

In earlier days, we believed that lakes and reservoirs were enriched with nutrients and organic matter from "point" sources such as industrial discharges and wastewater treatment plant outfalls. However, for many lakes, non-point or diffuse nutrient loading, both internal and external, is found to be momentous. These non-point sources are challenging to assess and regulate,[32] and water quality in many lakes has remained deteriorated following diversion or treatment of point sources.

The main pollutants that enter through non-point sources are various forms of phosphorus and nitrogen such as total phosphorus, phosphate phosphorus, nitrate nitrogen, nitrite nitrogen, ammoniacal nitrogen, organic nitrogen, chlorine, sodium, calcium, and suspended solids. The exports of all constituents occur mostly during rainfall- or snowmelt-generated runoff events during the spring runoff period. It has been shown that one of the major reasons for the enrichment of lakes is the conversion of previously cropped land into agricultural production while conversion to forests has very little impact on enrichment.

The runoff of nutrients from old fields depends on the nutrient status of the soils and soil water. This nutrient status reflects the soil type, fertilizer and cropping practices prior to abandonment, number of years since abandonment, and the succeeding vegetation present at any particular point. Soil data identify the reservoir of nutrients available for runoff, provide a means to relate runoff to nutrient content of that particular soil, and provide the data necessary for design of management schemes to prevent release of nutrients. Farm lands that have been abandoned for 15 to 20 years are not major non-point sources of pollution.

WATER QUALITY MONITORING

Water quality monitoring refers to the acquisition of quantitative and representative information on the physical, chemical, and biological characteristics of a water body over time and space.[33] It is a complex task, comprising all the activities to extract information with respect to the aquatic system. A variety of contaminants, in addition to a multitude of imprudent water quality management practices and destructive land uses, are currently threatening aquatic systems on a worldwide scale. In addition, it has been shown that water of good quality is a critical component for sustainable socioeconomic development.[34] The impact and behavior of contaminants in an aquatic ecosystem are complex and may involve adsorption–desorption, precipitation–solubilization, filtration, biological uptake, excretion, and sedimentation–suspension. Besides natural processes affecting water quality, there are also anthropogenic impacts, such as man-induced point and non-point sources, xenobiotic, and alteration of water quality due to unwise water use and river engineering projects (e.g., irrigation, damming, etc.).[35]

The degradation of water resources has increased the need for determining the ambient status of water quality, in order to provide an indication of changes induced by anthropogenic activities. To understand the process dynamics of a watershed, a well-designed water quality monitoring network identifies water quality problems while establishing baseline values for short- and long-term trend analysis. The need to evaluate observed water quality conditions and their suitability for the intended uses reflects a need for cost-effective and logistically practical water quality monitoring network design methods.

Types of Monitoring

Lake Sampling

It characterizes the water quality of the lake to identify status and trends. Two main types of lake monitoring are water-column sampling and near-shore (shallow water) sampling. Water-column sampling tries to quantify the overall response of the lake to contamination. The lake's trophic status, as designated by phosphorus, turbidity, chlorophyll-a, and dissolved oxygen, is of particular concern. Some invasive species, such as spiny water flea, can also be identified by water-column sampling. During seasons when the lake is stratified, the water-column sample should be sampled in both the epilimnion (the warm upper layer) and hypolimnion (lower layer of cold water).

Near-shore refers to the depth at which rooted plants can grow. Sampling can be adequately done at the end of a dock. Sampling for pathogens and pathogen indicators is important because of contact recreation such as swimming. Near-shore monitoring allows study of the lake bottom including sediment sampling for heavy metals, macroinvertebrates, and attached or rooted invasives such as zebra mussels and Eurasian watermilfoil.

Tributary Mass Load Sampling

It determines the tributary mass loads of water contaminants entering the lake. A significant portion of the lake's water pollution is brought by the tributaries flowing into it. Determination of tributary mass loads is particularly important for management of the lake's phosphorus and sediment problems.

Tributary Water Quality Sampling

It characterizes the water quality of tributaries to identify status and trends. Tributaries may be threatened by contaminants or stresses that affect the stream health but are not significantly detrimental to the lake. The tributaries are valued for recreation and aesthetics, drinking water, irrigation, and wildlife habitat and deserve protection.

Biological Integrity Sampling

It characterizes the long-term ecological health of the lake and tributaries. Ecological sampling is useful for detecting the effects of impairments that are not present at the time of sampling, for evaluating habitat health and for determining the biological integrity of surface waters. Ecological sampling may include bioassessments of fish and benthic macroinvertebrate communities, periphyton, and single-species monitoring (trout, salmon, and freshwater mussels are often used). Biological indices, a composite of different indicators, can be developed.

Citizen Monitoring

It encourages citizen participation in the measurement of watershed quality. To the extent that people care about the watershed's lands and waters, the watershed will be protected and enhanced for generations to come. One way to encourage such stewardship is through involvement of students and other citizens in water quality monitoring. Monitoring conducted by citizen volunteers increases public awareness and knowledge about water quality and its protection.

The design of efficient water quality monitoring network is essential for effective water management. To date, many water quality monitoring networks for surface freshwaters have been rather arbitrarily designed without a consistent or logical design strategy. Moreover, design practices in recent years indicate a need for cost-effective and logistically adaptable network design approaches.[36] Furthermore, the monitoring in a water quality management program is recognized as a statistical approach, so that both the assessment and the design problems can be addressed via a statistical method. In this view, the statistical methods have been found very efficient for redesigning and assessment of water quality monitoring networks (WQM).

The International Organization for Standardization (ISO) defines water quality monitoring as: "the programmed process of sampling, measurement and subsequent recording or signalling, or both, of various water characteristics, often with the aim of assessing conformity to specified objectives." This general definition can be differentiated into three types of monitoring activities that distinguish between long-term, short-term, and continuous monitoring programs:

- Long-term observation and standardized measurement of the aquatic environment for defining the water quality status and prediction of the trend are known as monitoring.
- Intensive programs for the measurement and observation of status of the aquatic environment for a specific purpose are known as surveys. These are of limited duration.
- Continuous measurement and observation of the aquatic environment for the management of the quality of the water and other operational activities are known as surveillance.

The constituents that decide the quality of the water are transported by water from the watershed. Therefore, we have to construct a perfect water budget, as it plays a key role in identifying a lake's (reservoir's) problem. By conducting a reconnaissance survey of water from the watershed, main tributaries can be selected. Since high flows are the chief segment of the water budget and huge volume influxes are followed by high concentration, continuous gauge recording is recommended for the determination of flow in major tributaries. From a successive record of inflow and outflow in the main tributaries, an annual water budget is constructed so that estimated inflows equal outflows with a correction for lake storage.

The water budget is formulated as:[37]

$$SF_i + GW + DP + WW = SF_o + EVP + EXF + WS + \Delta STOR$$

where SF_i and SF_o are stream flow in and out, respectively; GW is groundwater in (includes deep and subsurface seepage); DP is direct precipitation on the lake surface; WW is wastewater, if any; EVP is evaporation; EXF is exfiltration; WS is removal for water supply, if any; and $\Delta STOR$ is the change in lake volume.

PROTECTIVE AND RESTORATIVE MEASURES

Removal or treatment of direct input of wastewater, stormwater, or both constitutes the primary step in the restoration of water quality of eutrophic lakes and reservoirs, as these sources frequently contain comparatively high concentrations of phosphorus and nitrogen. For the realization of any long-term benefits from in-lake treatments, such external loadings should be reduced. In some cases, reduction of external loading is adequate to restore the water body (e.g., Lake Washington[38,39]), but in others, where internal loading of nutrients is significant, in-lake treatments may be indispensable to accomplish lake quality improvement (e.g., Lake Trummen[40]).

Advanced wastewater treatment (AWT) and diversions are two most commonly used techniques for the reduction of external inputs. In diversion, the treated sewage or industrial wastewater is carried away from the degraded water body to waters that are having high assimilative capacity, by the installation of interceptor lines. AWT involves the reduction of phosphorus concentration in wastewater effluent by using chemicals such as alum (aluminum sulfate), lime (calcium hydroxide), or ferric chloride. Stormwater runoff is the next dominant source of external enrichment. Even though the P concentration in stormwater is very low (2%–10%), and solubility is less than that of sewage effluent, such non-point sources can denote momentous contributions. P retention in wet detention basins and wetlands, rapid filtration through soil, and P removal in predetention basins are the principal P removal methods applied for runoff water. If internal loading of P is anticipated to hinder the recovery following primary treatments such as diversion or AWT, then supplementary in-lake treatments may be justified to accelerate reclamation. Lakes and reservoirs lose volume due to siltation. Sediment removal, together with land management and construction of device to trap silt, is an example of their restoration and protection. Some management approaches are institutional arrangement, formation and operation of lake association, sports fishery management, etc.

Eutrophication can be controlled if the phosphorus (P) concentrations in the water body are lowered to a level that will limit the growth of the algae. This can be achieved by diversion of external input, dilution, flushing, or a combination of these approaches. Where there is substantial loading reduction, comparatively augmented rate of algal flushing, and negligible recycling from sediment, in-lake phosphorus concentration can be reduced significantly and trophic state can be improved rapidly. Lake Washington is a good example for this approach.[40,41] However, for many lakes, internal phosphorus release sustained the lake's enriched trophic state and reinforced the state of continued eutrophication, in spite of the removal of a noteworthy fraction of external loading by diversion.[42,43] In those lakes, following nutrient diversion, supplementary in-lake treatment may be necessary, to avoid an extended eutrophic state.

Other methods are dilution and flushing, which are used interchangeably. Dilution involves the reduction in the concentration of the nutrients and a washout of algal cells, whereas flushing comprises only the latter. Where there is a high nutrient load, water should be diverted if possible for the low dilution rate to be most effective. This plan provides for a reduction in biomass primarily through nutrient limitation. If only moderate to high nutrient water is available, flushing may work well if the loss rate of cells is sufficiently great relative to the growth rate. Flushing rate on the order of 10%–15% per day will afford some control through washout. We can opt for the technique such as "phosphorus inactivation," which involves, usually, the application of salts of aluminum, such as aluminum sulfate (alum), sodium aluminate, etc., to precipitate phosphate as

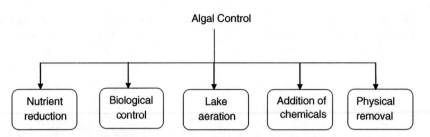

Fig. 3 Major algal control measures.

aluminum phosphate and thus to bind a significant fraction of P. Aluminum hydroxide floc, which is formed during this process, enhances the settling of the precipitate and it will continue to sorb and retain P in their molecular lattices.

Nutrient reduction mainly consists of processes such as source reduction in the watershed, issuance of fertilizer guidelines, setting up of shoreland buffer strips, and restriction of motorboats. *Biological control* involves processes such as use of bacteria for algae control, use of algae-eating fishes, biomanipulation, aquascaping, and bioscaping. *Lake aeration* comprises conventional aeration, solar powered aeration, wind-powered aeration, fountain aeration, and hypolimnetic aeration. *Addition of chemicals* includes use of barley straw, alum, buffered alum for sediment treatments, calcium compounds, liquid dyes, chlorine, and algicide[44] (Fig. 3). Copper is an efficient algicide. Copper sulfate application, the typical treatment for algal problems for many decades, is often effective for short-term solution to a current algae problem, predominantly in water supply reservoirs. However, there is substantial confirmation against the prolonged usage of this chemical. Its effects are found to be temporary, cost can be higher, there are major negative impacts to non-target organisms, and significant copper contamination of sediments can be possible. We can go for long-term and permanent options such as control of external and internal nutrient loading for the effective management of algal bloom. For a better water quality, and if the P is not estimated to reach an algal limiting level, an in-lake treatment for the control sediment should be established soon after external controls are in place.

Macrophytes includes all macroscopic aquatic flora, comprising macroalgae such as the stoneworts *Chara* and *Nitella*, aquatic liverworts, mosses, ferns, and flowering vascular plants. Aquatic plant management aims to curb annoyance species, to exploit the favorable features of plants in water bodies, and to reorganize plant communities. Its principal objective is the establishment of stable, diverse, aquatic plant communities comprising high percentages of desirable species. An exhaustive understanding of macrophytes biology is the foundation for evolving innovative management tactics. Continued research and development will advance our understanding of the relationship of aquatic plants to overall lake and reservoir quality and our capability to manage aquatic plant communities to preserve or improve that quality.[36,37] Table 2 shows major restoration measures for the lakes and reservoirs.

Table 2 Major restoration measures for the lakes and reservoirs.

Sl. no.	Restoration measure	Advantage(s)	Drawback(s)
1	Diversion	Reduce the overall nutrient loading to the system	Presence of another water body (having high assimilative capacity) is required in the near vicinity
2	Advanced wastewater treatment	Control eutrophication by limiting nutrient concentration with suitable chemicals	Accumulation of chemicals in the system and high cost
3	Dilution and flushing	Control eutrophication by limiting the P concentration	Cannot accommodate high nutrient loading
4	Application of appropriate algicide (such as copper sulfate)	Control eutrophication by curbing algal growth	Its effects are found to be temporary, higher cost, there are major negative impacts to non-target organisms, and significant copper contamination of sediments can be possible
5	Sediment removal (dredging)	Sediment removal, together with land management can restore the lake, affected by siltation	High cost and, in some cases, may cause environmental damage

Eutrophication Model Framework

Phosphorus-loading models are often employed for the evaluation of eutrophication problems in lakes and reservoirs. In these models, phosphorus loading is linked to the average total phosphorus concentration in the lake water and to other indicators of water quality that are associated to algal bloom, such as chlorophyll and transparency. Physical and hydrologic features influence the response of the lake to phosphorus loading and thus these models take in to account various characteristics such as lake volume, average depth, flushing rate, etc. The underlying principles behind the eutrophication model are as follows: 1) phosphorus acts as the limiting nutrient for algal growth; 2) any change in the amount of phosphorus discharged into a lake over an annual or seasonal period will alter the average concentration of P in lakes and hence the extent of algal bloom; and 3) the capacity of the lake to adjust with the P loading, without causing algal blooms increases with the volume, depth, and the flushing rate of the lake. Models recapitulate these relationships in mathematical forms, based upon observed water quality responses of large numbers of lakes and reservoirs. Eutrophication models are pitched for the prediction of average status of water quality over a season or a year. Averaging is mainly done over three dimensions: 1) depth; 2) sampling stations; and 3) season.[45]

CONVENTIONS FOR THE PROTECTION OF LAKES AND RESERVOIRS

Lakes and Wetlands—Ramsar Convention, Iran, February 2, 1971

The Ramsar Convention (The Convention on Wetlands of International Importance) is an international treaty for the conservation and sustainable utilization of wetlands. The convention was developed and implemented by participating countries at a meeting in Ramsar, hosted by the Iranian Department of Environment, and came into force on December 21, 1975. The Ramsar List of Wetlands of International Importance presently comprises 1950 sites (known as Ramsar Sites) covering around 1,900,000 km^2 up from 1021 sites in 2000. The nation with the highest number of sites is the United Kingdom at 168; the nation with the greatest area of listed wetlands is Canada, with more than 130,000 km^2. Presently, there are 161 contracting parties, up from 119 in 1999 and from 21 initial signatory nations in 1971. Signatories meet every 3 years as the Conference of the Contracting Parties (COP); the first was held in Cagliari, Italy, in 1981. There is a standing committee, a scientific review panel, and a secretariat. The headquarters is located in Gland, Switzerland, shared with the International Union for Conservation of Nature (IUCN).[46]

UNECE Water Convention, Helsinki, March 17, 1992

The Convention on the Protection and Use of Transboundary Watercourses and International Lakes (Water Convention) is intended to strengthen national measures for the protection and ecologically sound management of transboundary surface waters and groundwater. The Parties to this convention are obliged to prevent, control, and reduce transboundary impact; use transboundary waters in a reasonable and equitable way; and ensure their sustainable management. Parties bordering the same transboundary waters shall cooperate by entering into specific agreements and establishing joint bodies. The Convention includes provisions on monitoring, research and development, consultations, warning and alarm systems, mutual assistance, and exchange of information, as well as access to information by the public.

Protocol on Water and Health, London, June 17, 1999

The Protocol on Water and Health aims to protect human health and well-being by better water management, including the protection of water ecosystems, and by preventing, controlling, and reducing water-related diseases. It is the first international agreement of its kind adopted specifically to attain an adequate supply of safe drinking water and adequate sanitation for everyone and effectively protect water used as a source of drinking water. Parties to the Protocol commit to set targets in relation to the entire water cycle.[47]

CONCLUSION

Many of the lakes and reservoirs are situated in developing countries or industrializing countries that only now identify the devastating economic and social impact that pollution, overfishing, and habitat degradation are having on their water resources. Once these natural systems are severely polluted, they often cannot be restored completely; they can at best only be improved to a level where they can meet basic functions, and society must bear the increased costs and risks to human health.

The reviewed case studies in this entry and elsewhere illustrate that contaminated lakes and reservoirs can be restored, at least partially. The entry presented a number of examples where this process has been undertaken effectively. From an economic and environmental point of view, it appears wise to prevent the occurrence or exacerbation of problems, rather than going for radical and expensive restorative actions. Ample evaluations of water resources should be conducted to understand the state of lakes and reservoirs, so that we can provide the essential means for the establishment of imperative water resource management goals and objectives.

All-inclusive watershed management programs that comprise significant water quality objectives precise to the hydraulic characteristics of lakes and reservoirs should be prepared. The differences between temperate and tropical lake situations need to be highlighted. During the formulation of management strategies, all exertion should be made to avert the discharge of toxic substances to lakes and reservoirs. Introduction of exotic and invasive species should be banned, unless there has been ample environmental assessment.

We should incorporate the "precautionary" approach and the "polluter pays" principle in the management of lakes and reservoirs, which can benefit from various economic tools and ample financial policies. Primary action areas for protection and restoration of lakes and reservoirs involve watershed assessment, watershed control measures, and best management practices, which comprise management of pollution from agriculture, silviculture, mining, industrial pollution, urban runoff, and other pollutant sources, predominantly for nutrients and persistent toxic pollutants. Management programs for lakes and reservoirs should be comprehensive in scope and watershed-wide in nature. If necessary, joint governance institutions, such as joint commissions, composed of high-level government officials, should be established to improve management of transboundary lake and reservoir basins. The formulation of legal instruments, treaties, and a hierarchy for making the decision are essential for the advancement of the water resources management. It is clear that the world's lake reservoirs are threatened. In many cases, lake degradation is so advanced that populations depending on them are in great danger. The scope, magnitude, and dimension of the problem demand an international cooperation as well as national exertions.

A logical and consistent design methodology that allows more efficient and effective data collection and, hence, more useful information extraction should be developed. Such an approach not only permits better water pollution control recommendations and better allocation of financial resources but also, ultimately, a better understanding of the ecosystems. To appreciate the challenges of designing water quality networks, it is crucial to clearly define objectives and identify statistically acceptable assumptions. Assumptions are an inherent part of the monitoring network design process, mainly due to the stochastic influences on water quality variables in the aquatic environment. The number and type of simplifying assumptions made or allowed are dependent upon network objectives. Furthermore, assumptions in monitoring network design should be made relative to water quality hydrologic principles, applicable statistics, information utilization, and budget constraints.

REFERENCES

1. Bindler, R.; Renberg, I.; Brännvall, M.-L.; Emteryd, O.; El-Daoushy F. A whole-basin study of sediment accumulation using stable lead isotopes and flyash particles in an acidified lake, Sweden. Limnol. Oceanogr. **2001**, *46*, 178–188.
2. Available at http://www.wisconsinlakes.org/index.php/the-science-of-lakes/21-lake-types (accessed May 2012).
3. Available at http://www.unep.or.jp/ietc/publications/short_series/lakereservoirs-1/fwd.asp (accessed February 2011).
4. Available at http://ga.water.usgs.gov/edu/earthlakes.html (accessed February 2012).
5. Available at http://www.waterontheweb.org/under/lakeecology/16_trophicstatus.html (accessed February 2012).
6. Dodds, W.K. *Freshwater Ecology—Concepts and Environmental Applications*, 1st Ed.; Elsevier: New Delhi, India, 2006.
7. Tilzer, M.M.; Bossard, P. Large lakes and their sustainable development. J. Great Lakes Res. **1992**, *18* (3), 508–517.
8. Papatheodorou, G.; Demopoulou, G.; Lambrakis, N. A long-term study of temporal hydrochemical data in a shallow lake using multivariate statistical techniques. Ecological Modell. **2005**, *193*, 759–776.
9. Zhao, Y.; Yang, Z.; Li, Y. Investigation of water pollution in Baiyangdian Lake, China. Procedia Environ. Sci. **2010**, *2*, 737–748.
10. Yu, F.; Fang, G.; Ru, X. Eutrophication, health risk assessment and spatial analysis of water quality in Gucheng Lake, China. Environ. Earth Sci. **2010**, *59* (8), 1741–1748.
11. Dinar, A.; Seidl, P.; Olem, H.; Jordan, V.; Duda, A.; Johnson, R. *Restoring and Protecting the World's Lakes and Reservoirs*; World Bank Technical Paper Number 289, The World Bank: Washington, DC, 1995; 1–113.
12. Richards, R.P. The Lake Erie agricultural systems for environmental quality project: An introduction. J. Environ. Qual., **2002**, *31*, 6–16.
13. Zemenchik, R.A. Bio-available phosphorus in runoff from alfalfa, smooth bromegrass, and alfalfa-smooth bromegrass. J. Environ. Qual. **2002**, *31*, 280–286.
14. Ouyang, W.; Hao, F.H.; Wang, X.L.; Cheng H G. Nonpoint source pollution responses simulation for conversion cropland to forest in mountains by SWAT in China. Environ. Manage. **2008**, *41*, 79–89.
15. Cook, G.D.; Carlson, R.E. *Reservoir Management for Water Quality and THM Precursor Control*; American Water Works Association Research Foundation: Denver, CO, 1989.
16. Cook, G.D.; Kennedy, R.H. Managing drinking water supplies. Lakes Reservoirs Manage. **2001**, *17*, 157–174.
17. Osgood, R.A. Lake mixes and internal phosphorous dynamics. Arch. Hydrobiol. **1988**, *113*, 629–638.
18. Larsen, D.P.; Schultz, D.W.; Malueg, K.W. Summer internal phosphorous supplies in Shagava Lake, Minnesota. Limnol. Oceanogr. **1981**, *26*, 740–753.
19. Jørgensen, S.E. *Lake and Reservoir Management*, Revised Ed.; Developments in Water Science; Elsevier, 2005; Vol. 54.
20. WRI (World Resources Institute). World Resources 1992–93. Washington, DC, 1994.
21. Bartram, J; Balance, R. *Water Quality Monitoring—A Practical Guide to the Design and Implementation of Freshwater Quality Studies and Monitoring Programmes*; Published on behalf of United Nations Environment Programme and the World Health Organization UNEP/WHO ISBN 0 419 22320 7 (Hbk) 0 419 21730 4 (Pbk), 1996.
22. Kennedy, R.H.; Thornton, K.W.; Ford, D.E. Characterization of the reservoir ecosystem. In *Microbial Processes in*

Reservoirs; Gunnison, D., Ed.; Junk Publishers: The Hague, Netherlands, 1985; 27–38.
23. Kennedy, R.H. Consideration for establishing nutrient criteria for reservoirs. Lake Reservoir Manage. **2001**, *17*, 175–187.
24. Kurtz, D.A. *Long Range Transport of Pesticides*; Kurtz, Ed.; Lewis Publisher: Chelsea, MI, 1990; 440 pp.
25. Cotham, W.E; Bidleman, T.F. Estimating the atmospheric deposition of organochlorine contaminants to the Arctic. Chemosphere **1991**, *22* (1–2), 165–188.
26. Wolf, M.S; Toniolo, P.G. Environmental organochlorine exposure as potential etiologic factor in breast cancer. Environ. Health Perspect. **1995**, *103*, 141–145.
27. Wania, F.; Mackay, D. Global fractionation and cold condensation of low volatility of organochlorine compounds in Polar Regions. Ambio **1994**, *22*, 10–18.
28. Cook, G.D.; Heath, R.T.; Kennedy, R.H.; MComas, M.R. The Effect of Sewage Diversion and Aluminium Sulfate Application on Two Eutrophic Lakes. USEPA-600/3-78-033, 1978.
29. Kostic, S.; Parker, G. Progradational sand-mud deltas in lakes and reservoirs. Part 1. Theory and numerical modeling. J. Hydraulic Res. **2003**, *41* (2), 127–140.
30. Ali, M.B.; Tripathi, R.D.; Rai, U.N.; Pal, A.; Singh, S.P. Physico-chemical characteristics and pollution level of lake Nainital (U.P., India): Role of macrophytes and phytoplankton in biomonitoring and peytoremediation of toxic metal ions. Chemosphere, **1999**, *39* (12), 2171–2182.
31. Liu, Y.; Islam, M.A.; Gao, J. Quantification of shallow water quality parameters by means of remote sensing. Prog. Phys. Geogr. **2003**, *27*(1), 24–43.
32. Line, D.E.; Jennings, G.D.; McLaughlin, R.A.; Osmond, D.L.; Harman, W.A.; Lombardo, L.A.; Tweedy, K.L.; Spooner, J. Nonpoint sources. Water Environ. Res. **1999**, *71*, 1054–1069.
33. Sanders, T.G.; Ward, R.C.; Loftis, J.C.; Steele, T.D.; Adrian, D.D.; Yevjevich, V. *Design of Networks for Monitoring Water Quality*; Water Resources Publications LLC: Highlands Ranch, CO, 1983.
34. Bartram, J.; Balance, R. *Water Quality Monitoring—A Practical Guide to the Design and Implementation of Freshwater Quality Studies and Monitoring Programmes*; Published on behalf of United Nations Environment Programme and the World Health Organization 1996, UNEP/WHO ISBN 0 419 22320 7 (Hbk) 0 419 21730 4 (Pbk).
35. Chapman, D., Ed. *Water Quality Assessments. A Guide to the Use of Biota, Sediments and Water in Environmental Monitoring*; Chapman and Hall: London, 1996.
36. Strobl, R.O.; Robillard, P.D. Network design for water quality monitoring of surface freshwaters: A review. J. Environ. Manage. **2008**, *87*, 639–648.
37. Cooke, G.D.; Welch, E.B.; Peterson, S.A.; Nichols. S.A. *Restoration and Management of Lakes and Reservoirs*, 3rd Ed.; CRC Taylor and Francis Group, New York, 2005.
38. Edmondson, W.T. *Trophic Equilibrium of Lake Washington*. USEPA-600/3-77-087, 1978.
39. Edmondson, W.T. Sixty years of Lake Washington: A curriculum vitae. Lake Reservoir Manage. **1994**, *10*, 75–84.
40. Bjork, S. *European Lake Rehabilitation Activities*. Inst. Limnol. Rept. University of Lund: Sweden, 1974.
41. Edmondson, W.T. Phosphorus, nitrogen, and algae in Lake Washington after diversion of sewage. *Science* **1970**, *169*, 690–691.
42. Cullen, P.; Fosberg, C. Experiences with reducing point sources of phosphorous to lakes. Hydrobiologia **1988**, *170*, 321–336.
43. Scheffer, M. *Ecology of Shallow Lakes*; Chapman and Hall: New York, 1998.
44. McComas, S. *Lake and Pond Management Guidebook*; Lewis Publishers, A CRC Press Company: U.S., 2003.
45. Moore, L.; Thornton, K. *The Lake and Reservoir Restoration Guidance Manual*. EPA 440/5-88-002 First Edition. Washington, DC, 1988.
46. Available at http://www.ramsar.org/cda/ramsar/display/main/main.jsp?zn=ramsar&cp=1_4000_0_ (accessed February 2012).
47. Available at http://www.unece.org/env/water/ (accessed February 2012).

Lakes: Restoration

Anna Rabajczyk
Independent Department of Environment Protection and Modeling, Jan Kochanowski University of Humanities and Sciences in Kielce, Kielce, Poland

Abstract
Water reservoirs play a significant role in human environment and economy. However, several factors and substances of anthropogenic origin that condition their functioning affect the pace of change occurring therein. Water protection and adequate water resource management are among the fundamental responsibilities of each state. Consequently, plans for water and intake quality monitoring are developed, and a register of protected sources of pollution emissions is kept. One of the objectives of the state's environmental policy is to undertake actions aimed at restoration of utility value to particular water ecosystems, i.e., recultivation. The said actions must be based upon very thorough catchment inventories. Basing on the detailed information collected, it is possible to develop programs that enable the effective restoration of lake functionality.

INTRODUCTION

The 21st century is anticipated as the century of the environment. In this regard, it is crucial to remedy pollution and conserve resources, particular with respect to aqueous ecology, for fostering a healthy environment. Water reservoirs, whether natural or artificial, differ in methods of lake basin formation. The former exist as a result of physical and physicochemical processes occurring in nature; the latter appear due to human interference. Natural water reservoirs of relatively slow exchange of the liquid found therein are defined as lakes.[1,2]

Artificial reservoirs, in turn, also referred to as barrier lakes, are formed by closing river valleys with water dams or drops as damming structures. Several of these reservoirs may also be natural barrier lakes with regulated outflows.[3]

According to their functions, three main types of water reservoirs may be distinguished:[2]

- Dry reservoirs: periodical storage of water during the passage of flood peaks
- Flow-through reservoirs: maintenance of steady retention levels
- Retention reservoirs: storage of water at times of surplus in order to utilize it at other times (subtypes include flood control, navigation, power engineering, water equalization, municipal, industrial, agricultural, or rock waste control reservoirs)

Another classification, which takes into account land configuration, includes mountain, submountain, and lowland reservoirs. The first of these are usually the deepest, with short and high dams. The lowland ones occupy the largest areas and are less deep.[2]

Such considerable diversification of both formation methods and localities causes lakes to be characterized by varying parameters of lake basin morphometry, water transfer speed and routes, water quality therein, etc. Lake shape and dimensions condition lake susceptibility to degradation. Wind speed is higher over large lakes than over small ones, which results in stronger waving and better water oxygenation. Waters in lakes whose elongation follows the most frequent wind directions are also better mixed than in those with transverse orientation.[4]

Low values of the mean-to-maximum depth ratio suggest that large depths are found in a small part of the lake; hence, they may not be of considerable importance for lake functioning. The occurrence of thermal stratification and the size of the littoral zone that uses the biogenes found in the water are both dependent on lake depth.

Each lake, together with the surrounding land and watercourses, forms a catchment characterized by parameters such as topographic features, plant cover, basal complex, and climate conditions, which determine the volume of water resources in a given area. A catchment shaped by low atmospheric precipitation and lowland topographic features provides less water than one with high precipitation and mountainous topographic features, facilitating surface wash. The direction of precipitation waters, accompanied by a variety of compounds, is determined by the basal complex that conditions the river network density and, thus, inflow and outflow volumes. The vegetation found in a particular area stores the water and then gives it off to the atmosphere via the process of transpiration.[2]

At moderate latitudes, water exchange in reservoirs occurs twice yearly: in spring and in autumn. During spring circulation, winter-specific temperature distribution changes when bottom temperatures reach 4°C and upper layers of the reservoir become cooler. As a result of more intensive solar radiation and air warming, surface water temperatures rise. As they reach 4°C, the temperatures of the vertical section equalize because water temperature at the reservoir bottom is 4°C.[5]

As a consequence of wind-induced circulation currents, the entire water mass is mixed (spring homothermy). Further warming of surface waters causes near-surface waters to become lighter than the near-bottom waters whose temperature reaches 4°C (summer stagnation), and an upper warm layer called the epilimnion is formed. The lower layer of cold water is called the hypolimnion. In the metalimnion, i.e., the lake middle or transition water layer, a sharp temperature drop occurs at that time.[6]

In autumn, surface waters cool until they reach a state in which, upon reaching the temperature of 4°C, they attain density identical to that of near-bottom waters. In a way analogous to spring, waters are mixed by wind-induced motion (autumn homothermy). Further cooling of surface waters makes them obtain temperature that approximates 0°C, as well as lesser density than that of near-bottom waters. Subsequently, on the surface, the ice cover is formed (winter stagnation).[5]

Thermal stratification is related to oxygen content stratification in reservoir waters, thus affecting organic life growth. Because the epilimnion is a layer of intensive wind-induced water mixing as well as easy solar radiation access, it displays better oxygenation than deeper layers. As a result of intensive photosynthesis, in summer, water may become supersaturated with oxygen. Spring and autumn circulations cause oxygen to be distributed over the entire water mass.[5,7]

Temperature also conditions other processes that occur in the lake, including the rates of all chemical, biological, and physical reactions.[5] Higher temperature, for example, intensifies the processes of phosphorus release from bottom sediments (Fig. 1).

Water temperature increase in the reservoir decreases water viscosity and density. This is particularly significant for plankton organisms as it accelerates the sedimentation speed and changes locomotive conditions. Consequently, the rates of chemical and biochemical reactions increase, and solubility of most substances is enhanced as well, while gas solubility is reduced. The intensification of the processes occurring in the reservoir results in increased use of oxygen. Photosynthetic processes are accelerated, which may lead to cyclical supersaturation of upper water layers. A sudden water temperature rise by 10°C may lead to thermal shock or even death of organisms.[9]

Temperature increase may cause spontaneous discharge of bottom sediments and intensify gas liberation to the extent that the material collected at the bottom is loosened and discharged into water.

Access to solar radiation conditions the growth of autotrophs in the water, as well as that of the organisms situated behind them in the food chain, because the fluctuations in producer numbers affect level 1 (and further) consumer numbers. The propagation of solar radiation in the lake is affected by lake size and shape, i.e., morphometry. For instance, it determines thermal stratification of lake waters as well as chemical, physical, and biological processes that occur in reservoirs.[10]

Natural water reservoirs are efficiently operating ecosystems, capable of maintaining internal homeostasis despite the operation of adverse factors. Yet, human proximity and in particular the effects of human existence have far-reaching consequences (frequently negative) for the environment. Human activity takes forms against which a lake cannot defend naturally. Pollutants rich in biogenic elements from a variety of sources (including municipal waste, agricultural fertilizers, degraded woodland and communication routes) are discharged into lakes and rivers, causing considerable increase in the fertility of lake waters and rivers that serve as lake outflows. In raw municipal waste, for instance, nitrogen concentrations are

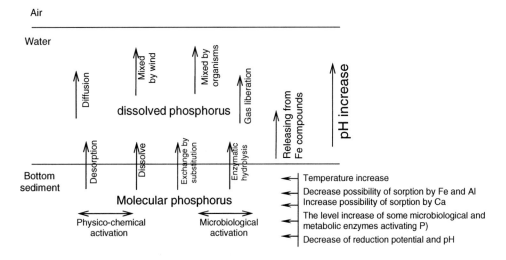

Fig. 1 Phosphorus release from bottom sediments affected by temperature.
Source: Bajtkiewicz-Grabowska and Zdanowski.[8]

contained in the 25–50 mg·dm^{-3} range, while total phosphorus concentrations are in the range 4.0–12 mg·dm^{-3}.[3] According to Vollenweider,[11] limiting concentrations of phosphorus and nitrogen compounds in waters, in excess of which mass algal growth may occur, amount to 0.01 mg P·dm^{-3} and 0.3 mg N·dm^{-3}, respectively.

For the assessment of external factors affecting a particular lake, an assessment of the catchment in terms of yearly delivery of biogenic (N and P) compounds into lake waters must be performed. The calculations are based on the formulas:[3,11]

$$L = I/P_j \ (\text{g·yr}^{-1}\text{·m}^{-2}) \quad (1)$$

$$I = I_{pr} + I_{pl} + I_{pz} + I_o + I_k + I_w \, I_l \ (\text{kg·yr}^{-1}) \quad (2)$$

where P_j is lake area (m^2), I is total load, I_{pr} is grassland load, I_{pl} is woodland load, I_{pz} is built-up area load, I_o is atmospheric precipitation load, I_k is bathers load, I_w is anglers load, and I_l is linear source load.

The values of external nitrogen and phosphorus loading in a particular lake are then compared with Vollenweider's criteria,[11] which define permissible and hazardous nitrogen and phosphorus loadings to a lake:

$$L_{d(N)} = 15 \ (25 \ z^{0.6})$$

$$L_{n(N)} = 15 \ (50 \ z^{0.6})$$

$$L_{d(P)} = 25 \ z^{0.6}$$

$$L_{n(P)} = 50 \ z^{0.6}$$

where z is mean lake depth, $L_{d(N)}$ is permissible reservoir N loading (mg·yr^{-1}·m^{-2}), $L_{d(P)}$ is permissible reservoir P loading (mg·yr^{-1}·m^{-2}), $L_{n(N)}$ is hazardous reservoir N loading (mg·yr^{-1}·m^{-2}), $L_{n(P)}$ is hazardous reservoir P loading (mg·yr^{-1}·m^{-2}), 25 is the limiting loading rate for oligo- and mesotrophic reservoirs, and 50 is the limiting loading rate for meso- and eutrophic reservoirs.

External loading of a reservoir with biogenic compounds per area unit is a major indicator specifying the reservoir's trophy. Trophy increase is based primarily on increased concentrations of biogenic compounds such as nitrogen and phosphorus, and this phenomenon is referred to as eutrophication. It is worth noting that the process of eutrophication is a most natural phenomenon. Organic substances are delivered to the lake starting from the moment of its formation and deposited at the bottom in the form of bottom sediments. Yet, due to intensified pollutant inflow, the phenomenon of eutrophication begins to intensify until it becomes a major threat to lakes.

RECULTIVATION METHODS

The growing eutrophication rate of water reservoirs and the increasing number of degraded reservoirs create the demand for effective prevention methods. In the 1960s, attempts at recultivation of degraded lakes were made. However, adequate know-how resulting from experience in this field was lacking, particularly because each water reservoir operates under different internal and external conditions. The current know-how and further developments in other fields of science, not only in natural sciences, have made it possible to explore theoretical issues related to water reservoir recultivation on the basis of the practical experience obtained.

As water ecosystems, lakes are located in land depressions; hence, they provide natural receiving water for the pollutants that come from the catchment area. Due to the functions that they perform in human environment and economy, it is crucial to take appropriate steps in order to improve the quality of degraded waters or protect those in better condition. To this end, scientific, organizational, and technological actions are undertaken jointly, known as recultivation. This process consists in elimination of possible chemical contamination of waters and sediments, improvement of oxygen and nutrient balance, maintenance of flora and fauna at a level appropriate for a given water ecosystem, and provision of engineering elements of flow regulation in the form of desired depths, erosion control, and speeds.[12–15]

The idea of lake recultivation is to restore the previous functions of lakes (e.g., water storage, recreation, household, and agriculture), as well as physical, chemical, and biological features approximating as closely as possible the natural ones. The selection of an appropriate method is determined by the diversity of individual lakes, differences in the ways and scopes of pollution, together with their location in a catchment. The type of the recultivation method used depends on the reservoir size, the nature of the fauna and flora resident therein, connection with watercourses, and proximity to clean water reservoirs in the vicinity of the reservoir under recultivation.[12,14]

Recultivation of water reservoirs is a four-phase process, including the phases of preparation, planning, implementation, and verification monitoring (Fig. 2). The duration of actions aimed at restoration of utility values to a particular area, from the initial phase to the final stage of implementation, is typically measured in years.

The first stage is the collection of information as well as the development of a detailed inventory of the immediate catchment. The following may also be helpful: soil maps; maps of localities with erosion-susceptible land and soils, utility or arable lands adjacent to rivers and reservoirs, and critical terrain (with its specific problems); maps of areas where surface waters supply groundwaters; and urbanization plans for the building in catchment areas and industrial development trends.

In view of the above, an individualized approach to each reservoir and planning a complete ecosystem restoration process are necessary. While taking any recultivation actions, it should be remembered that physical, chemical, external, and internal environmental factors may cause profound changes, destabilize the water ecosystem, and

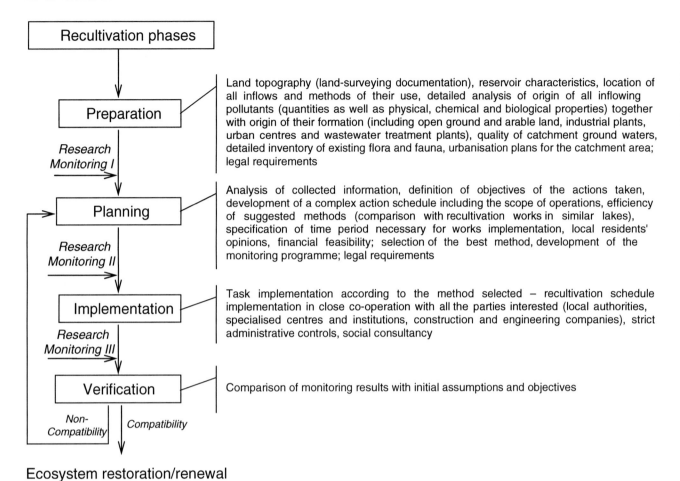

Fig. 2 Recultivation phases.

push it into one of the two alternative stable states. It must also be borne in mind that biological elements and processes conditioned thereby, which occur in a water ecosystem, stabilize and perpetuate its status quo.

A significant element of recultivation is the specification of a monitoring program, i.e., catchment control at critical points as well as in terms of registration of environmental quality indicators important for the recultivation process. This program assumes the establishment of observation sites, equipped with control and measurement apparatus. After a set monitoring time, all data from field observation sites are covered in a technical study that enables developing a representation of a catchment demonstrating the origin of possible pollutants and the problems to be tackled. Following from an extensive data base, including input from various sources, and complete with monitoring program results, it becomes possible to enter the definition phase for particular elements and actions necessary to achieve recultivation effect for a particular reservoir.

Nevertheless, it must be stressed that Research Monitoring I usually differs from the subsequent phases of monitoring, which results from task implementation at individual stages. During the inventorying process, maximum quantities of data concerning pollutant sources and migration routes, as well as the quantity and quality of substances delivered, ought to be collected. In contrast, the objective of Monitoring II is to obtain information on whether the selected recultivation method is well adjusted to the lake's individual characteristics.

To improve the situation in a degraded lake, it is not enough to restore its condition to that from before the disturbance: much feedback in the lake perpetuates the post-change status, and lakes tend to demonstrate resilience, i.e., immunity to recultivation actions. Very often, the condition for the application of any recultivation method is prior reduction of phosphorus concentration in the water to the 0.050–0.100 mg $PO_4 \cdot dm^{-3}$ level and phosphorus removal/inactivation in sediments (internal import inhibition).[12–14]

As a rule, each catchment and its environment are distinct, but from a practical viewpoint, it is possible to distinguish several actions, techniques, and methods enabling renewal or protection of the status quo of the lake at issue (Table 1).

The most popular methods that provide significant effects in biogene removal are those based on sparging and dredging. Reservoir water sparging enables reduction of

Table 1 Most commonly used lake recultivation methods and techniques with their characteristics.

Method/technique	Characteristics	Model application	Literature
Dredging	Bottom sediment removal from the entire lake, deepest waters and inflow (in flow-through lakes); complete sediment removal to reach mother rock guarantees radical improvement in water quality; necessity of thorough physico-chemical tests to determine the chemical composition of the sediments, their thickness and distribution in the reservoir (specification which sediment layer and in which lake parts ought to be removed); a sensitive issue is to find sites for storage and management of extracted sediments treated as hazardous	Mission Lake, Kansas; Lake Elkhorn, Columbia; Saluda Lake, South Carolina; Chain Lake, British Columbia	[16–19]
Macrophyte re-introduction	Bentonite is a good macrophyte substratum, which enables macrophyte layer reconstruction as well as restoration of destroyed trophy web structures	Alderfen, Barton, Belaugh and Cockshoot Broads, Norfolk;	[20]
Flushing, rewashing	Provision of clean, nutrient-deficient water, removal of strongly eutrophicated hypolimnion waters and replacing them with well-oxygenated waters from outside the lake; requires a source of clean water in the vicinity of the lake; lake water exchange ought to be performed a couple of times a year	Lake Veluwe, the Netherlands; Willow Lake, Arizona	[21,22]
Macrophyte removal	Comprises artificial lowering of lake water level, lake icing in winter, in-freezing of plant stalks into ice, raising the river level and plant uprooting, collection of floating plant remains and removal from the lake; the quantities of phosphorus thus removed are much lower than the loads introduced to the reservoir; effective in the case of lakes which are strongly polluted with biogenes, accumulated mostly in littoral vegetation	Lakes: Mary, Ida and Bass, Minnesota; Boreal Lake, Ontario	[23,24]
Hypolimnion water removal	Procedure possible in flow-through reservoirs; consists in liquid removal from the profundal layer via a hose; a disadvantage of this method is the pollution of the watercourse discharged from the lake; during near-bottom water removal, the sediments which remain in chemical balance with them become depleted; the process may cause reservoir pollution with H_2S, accumulated in the sediments and temperature increase near the bottom, which leads to hypolimnion oxygen depletion	Bled, Yugoslavia; Lake Kortowo, Olsztyn	[14,25]
Sparging, oxygenation (aeration)	Introduction of a duct into the water through which air is pumped, without disturbance (near-bottom liquid temperature is low: slow rate of organic matter decomposition) or with disturbance of natural lake stratification (entire water mass mixing and water oxygenation: near-bottom layer temperature increase, intensification of chemical and biological processes, including enhanced mineralisation and internal supply); enhances hypolimnion oxygen conditions as well as enforcing lake water circulation	Canyon Lake, Teksas; Lake Commabbio, Lombardy; Kieleckie Bay, Kielce	[26–28]
Artificial destratification	Destruction or prevention of the lake's thermal stratification, stimulating mass algal growth in the top (warmed and lighted) water layer; very often linked to sparging	Canyon Lake, California; Lake Catherine, Arkansas; Lake Starodworskie, Olsztyn	[29–31]
Nutrient deposition and deactivation in water and sediments	Possible in small reservoirs; does not guarantee permanent phosphorus removal from water; enduring results depend on coagulant properties ($FeCl_3$, $Al_2(SO_4)_3$, $FeSO_4$); optimum water pH 6–8 and high Redox potential level; phosphorus inactivation in sediments is possible due to application of a mixture of bentonite clay with lanthanum, which becomes a phosphorus bonding element upon adsorption on bentonite	Kielecki Bay, Kielce; Jessie Lake, Alberta; Lake Głęboczek, Tuchola; Lake Sønderby, Sønderby	[28,32–34]

(*Continued*)

Table 1 Most commonly used lake recultivation methods and techniques with their characteristics. (*Continued*)

Method/technique	Characteristics	Model application	Literature
Sediment isolation (capping)	Possible in small reservoirs where water transfer is low and does not cause mobility of the introduced layer; used in order to counter chemical exchange in the sediments-water system; physical isolation: sand and foil; chemical isolation: $+Al^{+3}$ diatomite, bentonite or other clayey minerals	Great Lakes, North America; Onondaga Lake, Central New York; Venice Lagoon, Porto Marghera	[35–37]
Precipitation (chemical methods)	Use of herbicides, including algaecides (e.g. barley straw, tree leaves) to fight algae; short-lasting procedure results due to appearance of other algae in place of the removed blooms; causes introduction of large amount of hazardous substances into water	Shoecraft Lake, Washington; Eight lakes located in the eastern and western portions of southern Michigan; Joliet Junior College Lake (JJC Lake), Illinois	[38–40]
Biomanipulation	Change in living conditions of organisms or quantitative ratios in a given ecosystem by means of several food chain dependencies: limiting the populations or complete elimination of individual groups of organisms (e.g. increase in the amount of zooplankton and introduction of selected fish species in order to reduce algal population, introduction of silver carp to the lake to limit the growth of phytoplankton; introduction of white amur to eliminate excess of macrophysical vegetation; introduction of predators such as pike, pike perch or perch in order to limit plankton-eating fish populations); use of biomanipulation requires thorough analysis of dependencies occurring in a given ecosystem	Lake Terra Nova, Loosdrecht Lakes ; Lake Eymir, Ankara	[41–43]
Fish catch	Regular catch of fast-growing fish (important is good co-operation with the lake management) enables reducing lake fertility (so-called trophy) and systemic enhancement of water quality; effective in the case of lakes which are strongly polluted with biogenes, accumulated mostly in the ichtiofauna	Lake Fure, Copenhagen; Lake Ringsjön, Skåne County	[44,45]
Seston removal/catch	Pumping out water at sites richest in seston, its subsequent filtration in appropriate apparatus; in barrier lakes, zooplankton removal may be accomplished by water inflowing through bottom culverts when zooplankton density is highest at the bottom	Lake Jussi, Pikkjärv	[46]

Source: Cooke,[12] Klapper,[13] Sengupta and Dalwani,[14] and Gupta et al.[15]

excessive algal and waterweed growth. Aeration is performed with the use of either mechanical apparatus or compressed air. Water sparging by means of mechanical apparatus may be performed by water spraying or direct penetration of air into the reservoir.[26,27]

Spraying occurs when water passes along a perforated hose through which it is catapulted into the air in the form of fine droplets. This effect can also be achieved by directing water to purpose-made diffusers. The most important element of this technique is the generation of adequately sized water droplets. The finer the water droplets that penetrate into the atmosphere, the larger the area of gas exchange between the droplet and the air, thanks to which better aeration effects are obtained—more oxygen is transferred into the water.[12–14]

The apparatus used for direct transfer of oxygen into a reservoir includes, for example, aerators with horizontal and vertical rotation axes, including surface, turbine-driven subsurface, and combination aerators. Aerators may also be classified according to paddle-wheel shapes: some may be straight, whereas others have single- or multisided curvatures.

The simplest is the surface aerator. Its operation relies on a turbine or rotor pumping the water upwards and then downwards, thus causing turbulence and splashing on water surface. Oxygen is delivered through water droplets in contact with atmospheric air. This type of aerator is used in ponds, water reservoirs, and rivers.

Combination aerators, which display features of both surface and turbine-driven aerators, are characterized by high capacity for oxygen transfer and provide good sparging results in deep reservoirs. In this type of devices, the surface rotor and the turbine are usually powered by the same aggregator, which causes high power demand, high noise levels, and possible icing during operation.[13,14]

The other types of aeration devices have vertical axes, which enables operation in both open and closed setups. In open aerators, the liquid is centrifugally catapulted, which causes the occurrence of the suction force that sucks the air into the so-called hopper and inter-paddle-wheel space. At the outlet of this space, a mixture of air and liquid appears and is then splashed on the surface of the aerating chamber. This construction provides for the mixing of the liquid and the air in the entire volume. In closed aerators, the liquid is propelled onto paddle-wheels by means of a pump. The air is thus sucked from the atmosphere by means of suction resulting from the flowing liquid at set sites in the inter-paddle-wheel channel.[47]

The compressed air aerator consists of a blower or a fan, sources of compressed air, and distribution ducts and sparging devices. In order to obtain the largest possible contact area between the two phases, a variety of porous elements are used for air dispersion. They are made of ceramic or artificial materials, and are pipe or plate shaped.[47]

In the process of selection of lake aeration systems and methods, the following factors ought to be taken into consideration:

- Oxygen concentration in the water and at the bottom of water reservoirs, which depends on water reservoir eutrophication degree, oxygen deficit volume in bottom sediments, with allowances being made for sediment resuspension, temperature, spring and autumn circulation intensity, and possible waving effects.
- Nominal result of aerator operation, which depends on total nominal hydraulic efficiency, reservoir depth at installation site, reservoir area, air volume fed into the aerator, and aerator construction type, such as the following:

  – Pump and surface dispersion
  – Pump and reversal-to-bottom with surface dispersion (with a thin water jacket)
  – Pump and reversal-to-bottom (with a thick water jacket)
  – Pump with deep suction—in all the above types[48–50]

Overall change to oxygen and dynamic conditions at aerator installation site, i.e., genuine efficiency, depends on lake type and operation mode selected, i.e., aerator construction, as exemplified by the air-stream bottom aerator (ASB) (Fig. 3).

During the method selection process, it is also necessary to answer the question of what immediate effect is expected from aeration since the procedure of aeration alone can bring positive results only in few cases.

An equally popular method of lake recultivation is dredging. During sediment removal, it is necessary to construct single or multiple thick-structure islands. Scarps, or island shores, should be equipped with bottom baffle piers that will constitute initial base for the benthos. At the early stage of island construction, the circumference wall should not be closed completely. Through the gap, waters overlying the sediment will be discharged. The gap ought to be secured with sedimentation barriers as well as a set of directional ASBs operating in the countercurrent to the outlet from the closed area. The applied cofferdam will only be temporary. At the end of island filling, the outlet gap ought to be closed with a cofferdam. Upon completion of sediment dredging, the whole ought to be covered with stable material.[48–50]

Upon completion of engineering work, red osier (for instance) needs to be planted. This type of islands can be used for recreation as well as natural purposes. Appropriate, separated parts of the islands can become nesting areas for many bird species. Sediments remaining at the bottom,

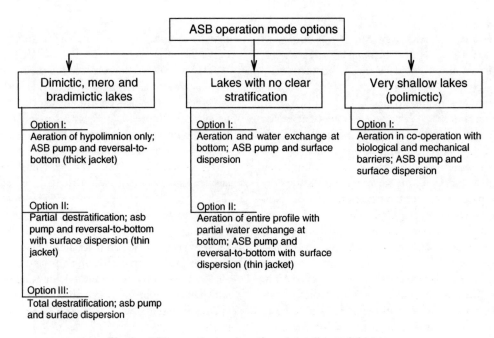

Fig. 3 ASB operation mode options depending on lake type.

whose thickness should not exceed 20 cm, may be immobilized by means of a baffle pier grid.

In the course of recultivation procedures, supply points ought to be perpetually monitored. If biogenes are found to be introduced via these routes, it is necessary to place biological barriers, possibly active (with aeration) at inflow inlets.[51,52]

The negative influence of surface wash can also be limited by barriers of appropriate plantings parallel to the shoreline. When forming the islands, it would be advisable to dredge a couple of local deepest waters (ca. 5–6 m). At those sites, an aerator is placed to disable the occurrence of oxygen deficits at windless periods. Artificial sparging enhancement ought to be sustained until balance is obtained between food and consumers, as well as between producers and reducers. Hence, in order to reduce power costs, in near-shore island zones and non-developed shore segments, installation of artificial reef as initial base for benthos is recommended. The ichtiopopulation status and composition also need to be considered.[48–50]

Another method aimed at restoration of good water quality is chemical bonding and inactivation of nitrogen and phosphorus. This system, used in Scandinavia, has been adapted to Polish conditions and tested at several sites, e.g., Lake Długie in Olsztyn. The phosphorus inactivation method is recommended for those lakes in which, despite cutoff of external sources of biogenic compounds, high fertility is sustained through deposition of these compounds from bottom sediments. Used primarily in shallow lakes, this method relies on removal of excess phosphorus from the pelagic zone and trapping it in bottom sediments. It also increases sorption capacity of degraded lake bottom sediments. Strongly eutrophicated lakes are characterized by high biogenic compound content, which causes blooms (increased phytoplankton population). At this degree of degradation, the lake no longer fulfills its recreational or economic functions. Chemical deposition of phosphorus by means of aluminum (PAX) and ferric (PIX) coagulants reduces the quantity of biogenic (nutrient) compounds, thus limiting algal growth intensity, which results in improvement of water quality and transparency.[53,54]

In the case of shallow reservoir recultivation, it is necessary to mitigate the deoxygenated layer of benthic water, the so-called oxicline. The oxicline generally occurs in the summer, appearing as a result of oxygen depletion in the course of chemical reactions. These reactions occur at the contact zone between bottom sediment and water, depending, among others, on ambient temperature: the warmer the water, the faster the reactions proceed. Under the circumstances, more and more frequently, actions are taken to place aerators equipped with a phosphorus inactivation system in reservoirs. This method has also been used in the Kielce Lake, where a pulverizing aerator was installed in May 2008. It is powered with wind energy and has a built-in installation dosage system for iron sulfate, which bonds phosphorus.[28]

Most efforts to alleviate the detrimental and undesirable effects of eutrophication on aquatic systems address the problem of P reduction in the inflows. Despite many cases of success in the reduction of nutrient loads of lakes in recent years, the expected corresponding reduction of phytoplankton abundances has often been delayed by many years. Sometimes, lakes have been reported to be resistant in their response to loading reductions. Recycling of P from the sediments becomes more important if the P in inflows is reduced.[55]

Phosphate inactivation in bottom sediments enables the elimination of internal transport in the reservoir (release of biogenes accumulated in bottom sediments into the pelagic zone), which is the cause of self-maintenance of high trophy levels—a feature of overfertile lakes, conditioning the occurrence of blue-green algae blooms. This means that as a result of the application of this method, the cause is removed, rather than the effects of excess of biogenes in the water. This testifies to greater efficiency of the above technique.[56]

The only alternative for application of the phosphorus inactivation method in bottom sediments is lake dredging. This is a technique that is equally effective, but very expensive, long-lasting, and accompanied by the problem of storage and further treatment of the removed sediments. It must be stressed that the storage site must not be located in the lake catchment area so that effluent waters together with their biogene load cannot find their way back to the reservoir. Similarly, treatment of the sediments is connected with the need for a recipient, transportation, etc. The phosphorus inactivation method in bottom sediments is far less costly, is incomparably faster at application, and provides enduring results of reduced phosphorus concentration in the water.[56,57]

The oldest method of phosphate inactivation in sediments, with simultaneous mineralization of organic matter, is the Riplox method, which yields good results. The required labor intensity and technical complexity make it extremely unpopular. Other techniques are its modifications, the Prote method being a case in point. The similarity lies in the fact that in both cases, air and the flocculant are simultaneously supplied to bottom sediments. A major difference is prior induction of intense sediment resuspension and flocculant application to a very carefully delineated sediment layer. The responsibility for the success of lake recultivation with this method lies with an effective device, providing precision dosage of flocculants to lake bottom sediments.[56,58]

To that end, several purpose-made mixes, containing a variety of substances as well as enzyme complexes, selected bacteria, biological activators, complete with extended-surface mineral carriers and stabilizers, have been developed. A preparation introduced into a water reservoir strongly activates the detritus food chain and increases the participation of autotrophic bacteria in matter circulation.[56,58]

The effects are so enduring that the following results are observed over 2 years:[56,58]

- Acceleration of organic matter decomposition in the sediments: quantitative reduction of accumulated bottom sediments and deceleration of natural reservoir shallowing
- Elimination of anaerobic zones in the hypolimnion and acceleration of sediment surface oxidation
- Enhanced elimination of biogenic compounds from matter circulation
- Blue-green algae bloom intensity control
- Boosting the growth of submerged water vegetation.
- Increased fish biomass[56,58]

Achieving enduring improvement of reservoir condition takes time and efficient implementation of the stages aimed at reduction of external supply, internal supply (elimination of anaerobic zones), and the quantity of biogenes in circulation (trophy reduction).

The experience in terms of external supply reduction has shown that the main condition that determines the efficiency of this technique is possibly the fastest reduction of external biogene supply. If this requires improvement of wastewater management in the reservoir catchment, the issue may seem problematic and distant in time due to the need for a wastewater treatment plant to be modernized or constructed. In contrast, the application of the Trigger-1 and Trigger-2 biopreparations in already-existing household, municipal, or industrial treatment plants enables limitation of pollutant inflow into the lake. Because the elimination of 1 kg of phosphorus and nitrogen in the process of wastewater treatment, even at plants with inadequate technical equipment, is far easier and cheaper than in the water environment, it is not advisable to wait for the construction to be completed, instead using all the methods to enable the reduction of external supply.[59]

In reservoirs where storage of large internal loads at stagnation periods has already occurred, the occurrence of anaerobic zones above the bottom is observed in the water. This triggers the mechanism of internal supply, i.e., releasing the previously accumulated biogenes from bottom sediments. This leads to accelerated degradation of the reservoir, although initially, only increased reservoir productivity is noted (primary and secondary productivity). Natural processes that occur in the reservoir are accelerated several times and result in rapid expansion of anaerobic zones. The methods applied to check this phenomenon, such as aeration, are expensive and carry a high risk of failure. A less costly method that does not require complex technical procedures is the application of a biopreparation, such as Trigger-3max, which, without changing water stratification, eliminates anaerobic zones in the hypolimnion, reducing the internal supply and eliminating anaerobic zones.

Recently, many biopreparations containing saprophytic microorganisms are commercially available and commonly offered for use in ponds, lakes, and reservoirs. These preparations usually consist of selected bacterial strains immobilized on a mineral carrier. Sometimes, the preparations are also enriched by bacterial enzymes such as biocatalyzators. Some preparations are even enriched by nutrients as growth stimulants at the beginning of growth after addition into the water. However, in most cases, parallel aeration or sediment oxidation by adding another electron acceptor such as nitrate will be necessary to support the growth and activity of added microbes. Otherwise, the method hardly brings any effects. With regard to many bacterial extracellular exudates that have been reported to inhibit the growth of cyanobacteria, a direct effect on cyanobacterial development may be possible as well. However, no scientific proof of such an effect in the case of application of these commercially available biopreparation is available.[56,60]

All biogenes that occur in a water reservoir ecosystem are in circulation, the major stages of which are transformations of the dissolved phase into the solid phase and back into the dissolved phase. The main mechanisms of this change are the initial placement of elements in the biomass (solid and molecular phase), and their subsequent release from mortified organisms and fecal matter (dissolved phase). Phosphorus is the quickest to be released, and consequently, it circulates in the epilimnion, passing many times from the dissolved phase to the solid before falling onto the sediments. The application of a biopreparation will thus enable a reduction of the quantity of biogenes in circulation, thus limiting lake trophy.[48]

More and more frequently, attempts are made to develop and apply sustainable methods of water reservoir qualitative enhancement, based on "bio" structures. A case in point is the bio-hydro structure (biofilter filling), as well as active biological and mechanical filters, which may be used in open water reservoirs, thanks to which the processes of mineralization and biomass growth are induced.[48,61]

Components of biofilter filling are manufactured as standard panels. They are five-layer grids, made of wide strips, with single net-mesh dimensions. Each subsequent layer is shifted in relation to the previous layer in two directions, by half a net-mesh module.[48]

When this packet is placed vertically, the vertical strip plane is perpendicular to the panel plane. The horizontal strip plane is differently inclined to the panel plane. This geometric arrangement enables directing the stream upward or downward, depending on flow direction. Consequently, this structure may be used as, for instance, a sedimentation element, a lift component, or a resuspension barrier and initial base for the complex of plant (phytobenthos) and animal (zoobenthos) organisms, which occupy the water reservoir bottom.[61]

Strip surface is rough (knurled) and made of chemically neutral plastic, which enables occupancy by a variety of sedentary species. Despite the fact that the extended surface of such filters amounts to $122\ m^2 \cdot m^{-3}$, the filter displays a low degree of flow suppression and relatively large vacant internal space. Due to this, upon placement in the pelagic zone, it very soon becomes the habitat of a very wide range of periphyton. By being an ecological niche in its own right, it also acts as an attractant to many sedentary species as well as fish fry. Above the pelagic zone, autonomously, on the top elements of the panels, a wide variety of hydrophilous plants settle.[48]

In the 1980s, at Biotechnika Company, the occupancy extent of these filters was studied in Lakes Karczemne and Klasztorne. On average, one standard panel came to be occupied by biomass in the amount of ca. 70 kg. If such an occupied filter is removed from the reservoir at the right moment, then, together with the biomass, the built-in biogenic compound load will also be removed. The certainty of the effect achieved, in contrast to chemical methods for example, consists in the fact that the application of chemical deposition does not guarantee that, under certain conditions (pH, redox potential), the sorption capacity of metals in the coagulants applied will not be destroyed. As data demonstrate, this is a very frequent occurrence and then, apart from the still unsolved problem of phosphorus removal, the additional "foreign" load in the form of metal compounds is found in the water. Then, the phenomenon of system adjustment to new conditions, which can moreover take a completely unknown direction, arises.[48]

Sparging, in connection with bio-hydro barriers, also offers several options, including passive biological barriers as well as active biological and mechanical filters. If a lake receives a strongly polluted watercourse, the inlet of the watercourse may be separated from the remaining area with a barrier (consisting of several layers to make it most advantageous) made of bio-hydro structures placed jointly, streamwise—first downwards and then upwards. Mechanically, this setup will function as a sedimentation barrier, which will, due to autonomous occupancy, after a short period of time, also and perhaps primarily become a biological barrier.

On account of its structure, despite initial passivity, as occupancy progresses, it becomes an attractant for an increasing variety of species, in most cases beneficial to further development of reservoir biocenosis and further recultivation procedures. During the observation of this type of barrier placed in Lake Rybnickie, it was noted that it had become a "refuge," initiating changes in both animal and plant populations. A double-positive effect was thus achieved: the inflowing pollutant load that degraded the lake has been set to work towards beneficial renewal of the reservoir.

In Lake Rybnickie, the positive result of barrier operation was visible as soon as after 1 year of working. Positive results of water quality tests restored the recreational value of the reservoir, and after many years, the State Sanitary and Epidemiological Board ultimately permitted bathing.[48]

CONCLUSIONS

Fertility or abundance of a water reservoir with biogenic mineral substances gradually increases in time, and may be either a natural process or one resulting from human activity. Natural eutrophication proceeds slowly, whereas anthropogenic eutrophication is characterized by a rapid change rate. At the initial stage, an increase of primary and secondary productivity is observed, but as eutrophication progresses, the efficiency of secondary productivity decreases and algal biomass gradually accumulates. The reservoir basin gradually fills with sediments and the water volume decreases, which, even at no biogenic inflow, simultaneously accelerates eutrophication. Further succession leads to the reservoir being turned into land.

In order to preserve a lake, human intervention becomes necessary. All recultivation strategies require sensible planning, time, and, frequently, considerable financial outlay. Their results are not always certain, but the current know-how concerning the subject enables appropriate selection and implementation of an effective recultivation process in a particular lake. The type of the method and technique used depends on the size of the reservoir, the nature of fauna and flora resident therein, the character of the catchment, and the type of adjacent land development. Effectiveness and efficiency of the work performed depend on accurate assessment of the reservoir status and adoption of an appropriate action plan.

While analyzing the efficiency of various procedures carried out in the world's lakes, it must be stated that before embarking on recultivation, it is necessary to

- Define the current trophic status of the lake as well as causes and sources of its degradation basing on physicochemical and biological tests of lake waters, inflows and outflows, complete with an analysis of sediment chemical composition, considering exchange processes between sediments and water.
- Specify the feasibility of the water reservoir recultivation in view of its natural, recreational, and economic value.
- Design and perform protective procedures consisting of the elimination, or at least limitation, of individual pollution sources.
- Develop the most advantageous concept of a recultivation method that ought to make allowances for the trophic status as well as morphometric and hydrological conditions of the reservoir, together with possible uses of margin land, its development, ownership status, access to the shore and a power supply, duration of the process, and financial outlay required.
- Ensure that recultivation will be performed by an expert team under the supervision of an experienced limnologist at all times.

- Document the course of recultivation and its effects through the results of monitoring carried out at particular stages.
- Deem recultivation unfeasible in view of its failing to offer an opportunity for water quality enhancement, if it is found that excess loading of the lake is impossible to eliminate.
- Have a close connection with appropriate regulations, enabling the protection of a lake subject to recultivation procedures from further degradation processes.

ACKNOWLEDGMENTS

This study has been financed in part by the State Committee for Scientific Research (KBN) (Grant No. N N305 306635).

REFERENCES

1. USEPA. *The Lake and Reservoir Restoration and Guidance Manual*, 2nd Ed. (OWRS, EPA 440/4-90-006); United States Environmental Protection Agency, Office of Water: Washington, DC, 1990.
2. Bajkiewicz-Grabowska, E.; Mikulski Z. *Hydrology*; PWN: Warszawa, 2006.
3. Leonard, J.; Crouzet, P. *Lakes and Reservoirs in the EEA Area*; European Environment Agency: Copenhagen, 1998.
4. Viekman, B.E.; Flood, R.D.; Wimbush, M.; Faghri, M.; Asako, Y.; Van Leer, J.C. Sedimentary furrows and organized flow structure: A study in Lake Superior. Limnol. Oceanogr. **1992**, *37* (4), 797–812.
5. French, R.H.; McCutcheon, S.C.; Martin, J.L. Environmental hydraulics (Ch. 5). In *Hydraulic Design Handbook*; Mays, L., Ed.; McGraw-Hill Professional: New York, 1999; 5.1–5.33.
6. Wetzel, R.G. *Limnology*, 2nd Ed.; Saunders College Publishing: Fort Worth, 1983.
7. Martin, K.L.; McCuthceon, S.C. *Hydrodynamics and Transport for Water Quality Modeling*; CRC Press: Boca Raton, FL, 1999.
8. Grabowska, E.; Zdanowski, B. Phosphorus retention in lake section of Struga Siedmiu Jezior. Limnol. Rev. **2006**, *6*, 5–12.
9. Tipton, M.J.; Brooks, C.J. The dangers of sudden immersion in cold water (Ch. 3). In *Survival at Sea for Mariners, Aviators and Search and Rescue Personnel (RTO-AG-HFM-152)*; RTO/NATO, NATO Research and Technology Organization: Neuilly-sur-Seine Cedex, France, 2008, 3.1–3.10.
10. Lenntech, Water Treatment Solutions. Water ecology FAQ frequently asked questions, http://www.lenntech.com/water-ecology-faq.htm (accessed May 2011).
11. Vollenweider, R. Advances in defining critical levels for phosphorus in lake eutrophication. Mem. Ist. Ital. Idrobiol. **1976**, *33*, 53–83.
12. Cooke, G.D. *Restoration and Management of Lakes and Reservoirs*, 3rd Ed.; CRC Press, Taylor and Francis Group: Boca Raton, 2005.
13. Klapper, H. Technologies for lake restoration. Papers from Bolsena Conference (2002). Residence time in lakes: Science, management, education. J. Limnol. **2003**, *62* (suppl. 1), 73–90.
14. Sengupta, M.; Dalwani, R., Eds. Hypolimnic Withdrawal for Lake Conservation, Proceedings of Taal2007: The 12th World Lake Conference, Jaipur, India, 2008; 812–818.
15. Gupta, A.K.; Shrivastva, N.G.; Shrama, A. Pollution Technologies for Conservation of Lakes; Proceedings of Taal2007: The 12th World Lake Conference, Jaipur, India; Sengupta, M.; Dalwani, R., Eds., 2008; 894–905.
16. Lake Dredging. Illinois EPA, Northeastern Illinois Planning Commission, Chicago, Illinois, 1998.
17. Helmke, D. Dredging: Kansas Takes First Steps to Reclaim Reservoirs, The Kansas Lifeline, Seneca, Kansas, 2010, 96–99.
18. Maxwell, A.O. Upstate's Saluda Lake Revived through Grassroots Conservation Effort. USDA-NRCS, http://www.sc.nrcs.usda.gov/news/saluda_lake.html (accessed May 2011).
19. Murphy, T.P.; Macdonald, R.H.; Lawrence, G.A.; Mawhinney, M. Chain lake restoration by dredging and hypolimnetic withdrawal. In *Aquatic Restoration in Canada*; Backhuys Publishers: Leiden, the Netherlands, 1999; 195–211.
20. Kelly, A. Appendix 5. History of lake restoration. In Lake Restoration Strategy for The Broads, Broads Authority, The Broads—A Member of the National Part Family, Norwich Norfolk, 2008.
21. Hosper, H.; Meyer, A.-L. Control of phosphorus loading and flushing as restoration methods for Lake Veluwe, the Netherlands. Aquat. Ecol. **1986**, *20* (1–2), 183–194.
22. Mankiewicz, P.S.; Mankiewicz, J.A. *Ecological Engineering and Restoration Study. Flushing Meadows Lakes and Watershed*. The Gaia Institute: City Island Avenue, Bronx, NY, 2002.
23. Cross, T.K.; McInerny, M.C.; Davis, R.A. Macrophyte Removal to Enhance Bluegill, Largemouth Bass and Northern Pike Populations. Minnesota Department of Natural Resources, St. Paul, Investigation Report 415, 1992.
24. Rabasco, R. Trophic effects of macrophyte removal on fish populations in a boreal lake. A thesis submitted to the Faculty of Graduate Studies of the degree of Master of Natural Resources Management, Natural Resources Institute, The University of Manitoba, Canada, 2000.
25. Dunalska, J. Influence of limited water flow in a pipeline on the nutrients budget in a lake restored by hypolimnetic withdrawal method. Pol. J. Environ. Stud. **2002**, *11* (6), 631–637.
26. Anderson, M.A.; Paez, C.; Men, S. Sediment nutrient flux and oxygen demand study for Canyon Lake with assessment of in-lake alternatives. Final report: Canyon Lake Nutrient Flux and In-Lake Alternatives, Dept. of Environmental Sciences, Univ. of California, 2007.
27. Riccardi, N.; Mangoni, M. Chemical consequences of oxygenation in a shallow eutrophic lake studied with mesocosms. J. Aquat. Ecosyst. Health **1996**, *5* (1), 63–71.

28. Rabajczyk, A.; Jóźwiak, M. Littoral zone flora versus quality of Kielecki Bay waters. Ecol. Chem. Eng. A, **2008**, *15* (12), 1359–1368.
29. Fast, A.W. Lake Aeration System for Canyon Lake, California, Lake Elsinore and San Jacinto Watersheds Authority (LESJWA), Riverside, CA, 2002.
30. Kothandaraman, V.; Roseboom, D.; Evans, R.L. Pilot Lake Restoration Investigations—Aeration and Destratification in Lake Catherine. Illinois State Water Survey, Urbana, Illinois, 1979.
31. Lossow, K.; Gawrońska, H.; Jaszczułt, R. Attempts to use wind energy for artificial destratification of Lake Starodworskie. Pol. J. Environ. Stud. **1998**, *7* (4), 221–227.
32. Itasca Soil and Water Conservation District, Jessie Lake Nutrient TMDL for Jessie Lake. Wenck Associates, Inc., Minnesota, 2011.
33. Łopata, M.; Gawrońska, H. Effectiveness of the polymictic Lake Głęboczek in Tuchola restoration by the phosphorus inactivation method. Pol. J. Nat. Sci. **2006**, *21* (2), 859–870.
34. Reitzel, K.; Hansen, J.; Andersen, F.Ø.; Hansen, K.S.; Jensen, H.S. Lake restoration by dosing aluminum relative to mobile phosphorus in the sediment. Environ. Sci. Technol. **2005**, *39* (22), 4234–4140.
35. Palermo, M.R.; Maynord, S.; Miller, J.; Reible, D.D. Guidance for In-Situ Subaqueous Capping of Contaminated Sediments, Assessment and Remediation of Contaminated Sediments (ARCS) Program, Great Lakes National Program Office, US EPA 905-B96-004, 1998.
36. Draft Onondaga Lake Capping and Dredge Area and Depth Initial Design Submittal. Parsons, Anchor QEA, 2009; P:\Honeywell-SYR\444576 2008 Capping\09 Reports\9.3 December 2009_Capping and Dredge Area and Depth IDS\Appendices\Appendix A -RA Delineation\Attachment A-1.doc (accessed May 2011).
37. Bona, F.; Cecconi, G.; Affiotti, A. An integrated approach to assess the benthic quality after sediment capping in Venice Lagoon, Aquat. Ecosyst. Health Manage **2000**, *3*, 379–386.
38. Wagner, K.J. The Practical Guide to Lake Management in Massachusetts. Executive Office of Environmental Affairs, Commonwealth of Massachusetts, Boston, 160 pp, 2004.
39. Madsen, J.D.; Getsinger, K.D.; Owens, Ch.S. Whole Lake Fluridone Treatments For Selective Control of Eurasian Watermilfoil: II. Impacts on Submersed Plant Communities, Lake and Reservoir Management, **2002**, *18* (3), 191–200.
40. Mitchell, J. STS CONSULTANTS, LTD. Joliet Junior College Lake. Diagnostic/Feasibility Study and Restoration Plan. Phase 1 B Report, 2007.
41. Van de Haterd, R.J.W.; Ter Heerdt, G.N.J. Potential for the development of submerged macrophytes in eutrophicated shallow peaty after restoration measures. Hydrobiologia **2007**, *584*, 277–290.
42. Bontes, B.M.; Per, R.; Ibelings, B.W.; Boschker, H.T.S.; Middelburg, J.J.; Van Donk, E. The effects of biomanipulation on the biogeochemistry, carbon isotopic composition and pelagic food web relations of a shallow lake. Biogeosciences **2006**, *3*, 69–83.
43. Beklioglu, M.; Ince, O.; Tuzun, I. Restoration of the eutrophic Lake Eymir, Turkey, by biomanipulation after a major external nutrient control I. Hydrobiologia **2003**, *489*, 93–105.
44. Frederiksborg County, Final report: Restoration of Lake Fure—A nutrient-rich lake near Copenhagen. LIFE02NAT/DK/8589, Hillerod, 2006.
45. Hamrin, S.F. Planning and execution of the fish reduction in Lake Ringsjön. Hydrobiologia, **1999**, *404*, 59–63.
46. Terasmaa, J. Seston Fluxes and Sedimentation Dynamics in Small Estonian Lakes. Tallinn University, Dissertations on Natural Sciences 11, Tallinn, 2005.
47. Wilson, D.E.; Beutel, M. Final Report: Review of the Feasibility of Oxygen Addition or Accelerated Upwelling in Hood Canal. Washington, Brown and Caldwell, 2005.
48. Sadecka, Z.; Waś, J. Non-invasive methods of reservoir restoration—Perspective. In *Wastewater and Sludge Treatment Utilization*; Sadecka, Z.; Myszograj, S., Eds.; University of Zielona Góra Press: Poland, 2008; 247–260.
49. Company WBWW-BIOPAX Sp. z o.o, http://wbww-biopax.pl/ (accessed May 2011).
50. Company Biopax.pl, http://www.biopax.pl/cm.php?id=16 (accessed May 2011).
51. Allied Biological Inc. New York, available at http://www.alliedbiological.com/lake_managementandrestoration_specialtyservices.htm (accessed May 2011).
52. Environmental Consulting and Technology, Inc., Carpenter Lake Restoration Project, Final Report, Corporate Offices—Gainesville, FL, 2008.
53. Brzozowska, R.; Gawrońska, H. Effect of the applied restoration techniques on the content of organic matter in the sediment of Lake Długie. Limnol. Rev. **2006**, *6*, 39–46.
54. Brzozowska, R.; Gawrońska, H. The influence of a long-term artificial aeration on the nitrogen compounds exchange between bottom sediments and water in Lake Długie. Oceanol. Hydrobiol. Stud., Int. J. Oceanogr. Hydrobiol. **2009**, *37* (1), 113–119.
55. Van Donk, E.; Hessen, D.O.; Verschoor, A.M.; Gulati, R.D. Re-oligotrophication by phosphorus reduction and effects on seston quality in lakes. Limnology **2008**, *38*, 189–202.
56. Drábková, M. Methods for control of the cyanobacterial blooms development in lakes. Dissertation thesis, RECETOX—Research Centre for Environmental Chemistry and Ecotoxicology, Brno, Czech Republic, 2007.
57. Salgot, M. Eutrophication in Lake Maryut: Diagnosis and actions proposal. ANNEX 1, Alexandria Lake Maryiut Integrated Management (ALAMIM), Alexandria, Egypt, 2009.
58. Quaak, M.; Does, J.; Boers, P.; Vlugt, J. A new technique to reduce internal phosphorus loading by in-lake phosphate fixation in shallow lakes. Hydrobiologia **1993**, *253* (1–3), 337–344.
59. VWP Individual Permit No. 08-0572, Part I—Special Conditions, DEQ, 2008, http://stauntonriverwatch.com/VWPP%20Final%20Part%20I%20special%20conditions.pdf (accessed May 2011).
60. Duval, R.J.; Anderson, L.J.W. Laboratory and greenhouse studies of microbial products used to biologically control algae. J. Aquat. Plant Manage. **2001**, *39*, 95–98.
61. Gulati, R.D.; Pires, L.M.D.; Van Donk, E. Lake restoration studies: Failures, bottlenecks and prospects of new ecotechnological measures. Limnology **2008**, *38*, 233–247.

Land Restoration

Richard W. Bell
School of Environmental Science, Murdoch University, Perth, Western Australia, Australia

Abstract

Restoration of degraded land is a largely untapped opportunity, worldwide, that can enhance the natural capital of land and the provision of soil ecosystem services. The most extreme cases of land disturbance causing degradation are commonly associated with mining. Conversely, best practice in land restoration has emerged from the land disturbed by mining. Determination of end land use is critical for developing a land restoration strategy, since it is a prerequisite for the clear identification of limiting factors and possible amelioration approaches, as well as for the development of success or completion criteria. Where land restoration managers seek to return the original vegetation, as well as the critical ecosystem functions, several cycles of research and adaptive management are needed to achieve a high level of success. In this case, a flexible approach to setting end land use is advisable. Sound principles for land restoration are emerging from site-specific research and adaptive management, but the future challenge is to scale up these practices to larger land areas degraded by a variety of processes.

INTRODUCTION

Land degradation is widespread globally[1–3] and is devaluing the natural capital of land as well as compromising the provision of soil ecosystem services. Arresting the degradation of land should be a high-priority goal. However, restoration of land that has previously been degraded represents an untapped opportunity to increase the natural capital of land and the provision of key soil and plant ecosystem services, particularly those related to plant productivity and water supply. Many forms of disturbance such as overgrazing, excessive tillage, overirrigation lead to land degradation including water erosion, wind erosion, decline of soil structure, acidification, salinization, waterlogging, and decline in soil fertility. Land disturbance after mining is comparatively extreme, but research and adaptive management on these sites have fostered the development of sound principles and practices of land restoration. In the present entry, most examples of land restoration are drawn from mine land restoration to illustrate the principles of the emerging discipline of land restoration. While land restoration can be quite site specific in its practice, there are general principles that can be applied and these have a degree of relevance to most forms of land degradation.

LAND AS A FINITE RESOURCE

For sustainable environmental management, land should be regarded as a finite global resource. The critical surface layer of the finite land resource is the soil profile that varies from a few millimeters' thickness to several meters. While soils are continually forming, the rate is slow, equivalent to 1 mm every 100 years[4] or even less.[5] Land may be created by reclamation of submerged areas along shorelines or in wetlands, and river deltas create new land surfaces as they deposit sediments when they reach the sea or lakes. Locally, these processes that create new land surfaces may contribute significantly to available land resources, but at a global scale, the amounts created are small relative to the total area and to the areas being degraded. Moreover, there is already evidence that in low-lying coastal zones, land is being inundated by rising sea levels or eroded by storm surges associated with rising sea levels. These processes are offsetting gains from natural or anthropogenic land creation. There seems little alternative but to regard land as a finite resource, to slow and prevent degrading processes, and to implement restoration programs on land that has suffered degradation.

Natural Capital and Soil Services

While most emphasis in the past was on food and fiber production from land, the concept of its value is broadening. Land is a form of natural capital that underpins several critical ecosystem services.[6] The soil ecosystem services outlined in Table 1 comprise supporting, regulating, provisioning, and cultural services.[6] The emergence of the concept of natural capital is beginning to redefine the notion of land and its value. Production of food is still a critical function of land, indeed one whose importance is predicted to rise over the next several decades as the challenge of providing sufficient quantity and nutritional quality of food for a rising population becomes more evident.[7,8] In

Table 1 Societal soil ecosystem services.

Supporting	Physical stability and support for plants
	Renewal, retention, and delivery of nutrients for plants
	Habitat and gene pool
Regulating	Regulation of major elemental cycles
	Buffering, filtering, and moderation of the hydrological cycle
	Disposal of wastes and dead organic matter
Provisioning	Building material
Cultural	Heritage sites, archaeological preserver of artifacts
	Spiritual value, religious sites, and burial grounds

Source: Adapted from Robinson et al.[6]

addition, the role of land and soil in regulating major elemental cycles, the hydrological cycle, and the disposal of wastes and dead organic matter is gaining increasing importance.

Land Degradation

Land as a finite resource is susceptible to a range of degrading processes that limit its productivity, its land use potential, and its value in providing ecosystem services. Conacher and Conacher[9] define land degradation broadly as "alteration of all aspects of the natural (biophysical) environment by human actions, to the detriment of vegetation, soils, landforms and water (surface and subsurface, terrestrial and marine) and ecosystems." More broadly, land degradation can be defined as those processes that lower the natural capital of land and compromise the provision of soil ecosystem services.

The processes of land degradation, their measurement, impact, and management have been dealt with elsewhere.[10–12] Most commonly, the impact of land degradation is described in terms of areas of land degraded.[13] However, this approach has several limitations. Apart from the difficulty of acquiring good quality data that are consistent across large areas to make such estimates, the different impacts of degrading processes makes it impossible to compare processes based only on area affected. For example, the degradation of soil structure by tillage may have an incremental effect on crop yields but cause a quantum increase in water runoff. The degradation of land by salinity, which raises soil salinity levels above a critical threshold for growth of most plants, prevents continuation of current land use options, while opening up others (such as salt-tolerant vegetation). By contrast, soil acidification may have no effect on plants until a critical threshold is crossed, leading to the release of toxic Al^{3+} concentrations that impair plant growth. Hence, the key to predicting the effects of land disturbance is to understand the threshold values of soil properties for plant growth and for ecosystem services. The differences among degrading processes are also critically important in planning the restoration of degraded land.

Alternatives to assessing and comparing only the land areas suffering degradation are to estimate the value of lost production from the degradation or to estimate the cost of restoration. The latter is generally considered to be critical since restoration costs will be a major impediment to reversing the effects of degradation. The cost per hectare for land restoration on mine sites is much larger than that on production lands not only because of the more extreme degradation but also because of the greater capacity to pay for restoration. Costing the restoration works for different forms of degradation and different areas of degraded land provides a basis for prioritization of restoration works. However, costs of restoration commonly only consider land productivity, rather than biodiversity and the full range of ecosystem services. If the natural capital of land in total was considered, there may be less variation in costs of land restoration among different forms of disturbance.

Land Restoration Definition

The term *land restoration* has a generic meaning, first proposed by Hobbs and Norton,[14] to indicate "that restoration occurs along a continuum and that different activities are simply different forms of restoration." The term *land rehabilitation*, as defined by Aronson et al.,[15] is also appropriate when applied to land degraded by mining. In the United States, the term *reclamation* is commonly used to describe activities that replace ecological functions by planting different vegetation to what previously grew.

Land restoration will usually focus on re-establishing ecological functions such as nutrient cycling, hydrological balance, and ecosystem resilience.[16] In some situations, restoring the original flora is in addition a realistic and appropriate goal. A case study based on the application of this goal is outlined below for restoration of eucalyptus forest after bauxite mining. Further expansion of the concept of land restoration should consider its role in the recovery of natural capital and soil ecosystem services.

LAND RESTORATION PRINCIPLES

The process of land restoration comprises the following: determination of the end land use, definition of the main limiting factors for restoration and means of alleviating them, and, finally, planning and implementation of a restoration program including monitoring and evaluation against success or completion criteria. This entry focuses on the first two components, while the latter component is described in more detail by Hobbs.[16]

End Land Use

Development of a land restoration strategy is generally considered to need a well-defined end land use. The key constraints that need to be alleviated depend on the land use envisaged. Hence, definition of the end land use is often considered to be a prerequisite to the restoration of degraded land. From the defined end land use, it is possible to identify the stakeholders whose interests need to be considered, the scope of the restoration challenge, and the prime constraints that will have to be alleviated.[16,17] It is also a prerequisite for the establishment of the measurable goals and targets for restoration that are used in setting success or completion criteria. End land use will largely define the complexity and difficulty of the restoration task and the costs associated with achieving a successful outcome. At one end of the spectrum, the complete restoration of the pre-existing ecosystem is a challenging goal. It may require decades of systematic research and continuous improvement of the restoration procedures before that goal is achieved and before the success criteria can be defined and validated. On the other hand, simply achieving a stable land surface or creating a pasture suited to low-intensity grazing would be more easily achieved.

Existing land use on surrounding areas, or prior land use at the site, often determine the end land use. However, degradation such as salinization may alter the substrate for plant growth so radically as to make the pre-existing land use impossible or undesirable. Furthermore, on mine sites, the alteration in landforms may require a change in land use in order to achieve land stability. In the case where land use change is necessary, stakeholders will need to be engaged to arrive at an acceptable alternative end land use.

A great diversity of end land uses has been applied during land restoration. Apart from natural ecosystems, agriculture, forestry, nature conservation, grazing, housing, wetlands, amenity and recreational facilities, and waste disposal are all possible options. Sociopolitical and economic factors will generally determine the selection of end land use particularly if the land is in a protected area, on production land, or in a densely settled zone.

While a well-defined end land use is necessary to set realistic goals and targets for restoration, there are risks associated with setting a highly prescriptive end land use if the restoration technology has not been well developed and based on solid research. The development of best practice for restoration of a particular type of degraded land may require decades of research and adaptive management. Hence, the premature setting of the end land use and targets for achievement may lock in an inferior set of outcomes. Using an adaptive management approach will allow continuous improvements in restoration practice to be made and tested. As improved practices and outcomes become possible, new benchmarks for completion can be set and new possibilities for end land use emerge. The bauxite-mining case study below illustrates this process.

Diagnosis of Limiting Factors

Degradation of land takes many forms and is triggered by many agents.[10] The types of constraints and their severity and the consequences for the restoration plan will clearly vary from site to site. Correct diagnosis of the key constraints and identification of likely feasible solutions are a prerequisite for successful restoration of degraded land. Identification of the key agents causing degradation is also essential. At a mine site, the active cause of degradation is mining and its associated disturbance. In other cases, the degrading agent may be tillage, overgrazing, wind erosion, saline groundwater discharge, etc. Apart from the biophysical constraints related to substrate properties, landform, climate, and hydrology, limiting factors may be associated with socioeconomic factors that prevent the degrading processes from being arrested. In the present entry, the focus is on the biophysical constraints to restoration (Table 2).

Biophysical Factors Limiting Land Restoration

Climate

The growing conditions for plants when restoring degraded land are determined primarily by climate. Climate is the main limitation on potential land uses for each site. Species that are indigenous to the site will usually be well adapted to the rainfall, temperature, and extreme conditions that occasionally occur in a given climatic regime such as drought, heat stress, and frost. However, if the substrate is unsuitable for the indigenous species, or the end land use requires a different selection of species such as agricultural species, then the chosen species need to be well adapted to the site climate.

When considering climate-related constraints, it is not just the average conditions that should be examined, but also the frequency and severity of extreme events, such as drought, heavy rain storms (e.g., cyclones), frost, snow, hail, etc. The coincidence of extreme events with commencement of a land restoration project may prevent seed germination or cause the loss of seed and topsoil due to erosion.

Landform

Mining creates voids below ground level and waste material that is stacked above ground level. Hence, the overall slope angle of land on a degraded mine site will be increased relative to the pre-existing landform. Hence, the creation of stable, non-eroding surfaces is generally the first major goal of restoration after mining. The key landform factors that limit the achievement of stable surfaces are slope angle, elevation, aspect, and surface drainage.

Table 2 Biophysical factors limiting land restoration, their consequences for site stability and plant growth, and common treatment methods.

Factor	Constraint	Consequence	Treatment
Climate	Drought	Failure of germination, poor emergence or establishment, plant death	Irrigation, drought-tolerant or drought-avoiding species, adjust time of sowing
	High temperature	Poor germination and emergence, plant death	Mulching, retention of crop/plant residue
	Low temperature/frost	Delayed emergence, plant death, poor seed set	Frost- or cold-tolerant species, adjust time of sowing
	Extreme rainfall/wind events	Water or wind erosion episodes, loss of seeds	Contour banks, soil cover by foliage, mulch or stubble, windbreaks
Landform	Slope	Land slippage, soil creep, unsafe conditions for machinery operation	Deep-rooted plants, reshape to lower slope angle, contour banks, engineering design, prevent water run-on from upslope
	Runoff	Water loss, sediment loss, downslope deposition	Contour banks, reshape to lower slope angle, improve infiltration and soil water storage
	Exposure	Drought, high winds, extreme temperatures	Tolerant plant species especially as windbreaks
	Aspect	High winds, extreme temperatures	Tolerant plant species especially as shade plants
Hydrology	Runoff	Reduced soil water storage, waterlogging, flooding downslope	Contour banks, increased drainage intensity
	Limited profile water storage	Drought, runoff, increased groundwater recharge	Tolerant plant species, deep ripping, treatment of subsoil chemical constraints
	Groundwater discharge	Water-filled voids, acid mine drainage, waterlogging, salinity	Containment of water, wetland treatment ponds, drainage
Substrate properties	Acidity	Poor plant growth especially roots, nutrient deficiencies, plant death	Lime, acid-tolerant species, P fertilizer
	Alkalinity	Poor plant growth, nutrient deficiencies, plant death	Gypsum, leaching, alkaline tolerant species, acidifying materials
	Salinity	Poor plant growth, plant death	Leaching, salt-tolerant species, drainage
	Sodicity	Soil dispersion, crusting, poor seedling emergence, runoff, water erosion	Gypsum, leaching, organic matter addition
	Nutrient deficiency	Poor plant growth	Fertilizer (mineral and organic)
	Metal toxicity	Poor plant growth, plant death	Lime, tolerant species, burial or capping of substrate, removal of substrate, phytoremediation
	Low water availability	Poor plant growth, plant death	Irrigation, mulching, organic matter, deep ripping, adjust time of sowing, drought tolerant species
	Waterlogging	Poor plant growth, plant death	Drainage, tolerant plant species
	Poorly structured soils	Crusting, poor water holding capacity, poor root growth	Mulching, organic matter, gypsum
	Mycorrhiza	Poor plant growth, nutrient deficiency	Topsoil management, inoculation of nursery plants
	Rhizobium	Nitrogen deficiency	Topsoil management, inoculation, liming acid soils
	Soil microbes	Slow mineralization of soil organic matter	Topsoil management

Source: Modified from Bell.[18]

Maximum slope angles are generally set by regulation and vary among jurisdictions. In West Australia, a maximum of 20° is the guideline for restored mine slopes such as the outer surface of waste rock dumps.[19] The change in landform is less of a consideration with other forms of disturbance apart from mining.

Hydrology

Any significant removal or disturbance of vegetation, change in surface soil properties, increase in average slope angle, or change in drainage density on a degraded site will alter the water balance. Reduced vegetation cover will increase runoff and there may also be increased deep drainage. The increase in slope angle on disturbed sites will generally increase the proportion of rainfall that becomes runoff. Erosion and downstream flooding and/or sedimentation are the likely consequences of altered hydrology unless precautions are taken to avoid these effects. Where sulfidic substrates are excavated by mining and stored in contact with moisture and oxygen, acid mine drainage may be discharged into groundwater or surface water. The discharge of acidic water alters downstream ecology and may damage the infrastructure it contacts, such as concrete structures in bridges.

Substrate Properties

Physical properties: Degrading processes such as erosion, overgrazing, or excessive tillage may alter the physical properties of the surface substrate in ways that decrease its suitability for plant establishment and growth, particularly by changing water storage and availability. Mining substrates and mineral processing residues are commonly poorly sorted, which alters physical properties such as available water capacity, porosity, soil strength, crusting, and susceptibility to wind or water erosion.

Water erosion strips away topsoil, decreases soil depth, and exposes subsoil material with different texture, lower organic matter levels, and degraded soil structure. The eroded soil has less favorable physical properties for water infiltration, water holding capacity, seed germination, and root growth. Wind erosion selectively removes clay and humus from the soil surface, increasing the prevalence of coarse materials. Fine sand deposits from wind erosion may bury topsoil and vegetation. Dust from bare surfaces can be a health hazard especially if it contains alkali salts or other toxic elements.

The passage of heavy machines on agricultural, forest, and mine sites causes compaction of the substrate, and this may be a major constraint to plant growth by restricting root depth. In drought-prone environments, the failure of roots to penetrate to depth may cause stunting or death of plants. Deep ripping to break the compacted layer is generally necessary in order to achieve deep root growth in mine pits.

Chemical properties: Plant growth may be hampered by low nutrient levels, acidity, alkalinity, salinity, sodicity, low organic matter levels, and excessive levels of toxic elements or compounds in the substrate. Therefore, effective land restoration depends on thorough chemical characterization of the substrate for these likely chemical constraints. Where feasible, any substrate that is likely to hinder plant growth should be isolated or buried under more benign materials so that root contact with it is avoided or minimized. This approach, which is practiced on mine sites, is clearly not possible on agricultural land. Soil and plant analysis are the most common methods for predicting likely nutrient deficiency or toxicity in the substrate.[20] Fertilizer applications can usually be effective in correcting nutrient deficiencies although determination of appropriate rates of application requires expert judgement or decision support systems and depends on the end land use. For acidity, alkalinity, salinity, or ion toxicities in the substrate, plant tolerance is the most cost-effective strategy for achieving successful plant growth. This may mean selecting a different suite of species or ecotypes to those that existed before mining unless the local species are already adapted to those constraints.

Biological properties: The disturbance of topsoil during mining, through stripping, transport, replacement, and/or storage, all have negative effects on soil biological activity.[21] Maximum soil microbial activity is retained when topsoil is used immediately after stripping without a period of storage. Storage for extended periods should be avoided, but in those mining operations where long-term storage is unavoidable, the stockpile should be uncompacted, should be <2 m deep, and should support a vegetative cover that maintains soil biological activity. Restoration of microbial biomass in replaced topsoil on revegetated mine sites may take 7–10 years.[21,22] In southwest Australia, the presence or absence of the pathogen *Phytopthora cinnamomi* in topsoil determines how it should be stored and reused.

Plant Establishment

Except in rare circumstances, vegetation cover is a requirement for land restoration. The alleviation of substrate constraints that limit plant establishment or growth is a prerequisite for successful vegetation establishment. Where a choice of substrate materials is available, the most benign of these should be placed at the surface to create a favorable root zone. This will minimize the need for expensive amelioration procedures for any adverse soil conditions. Topsoil, if available, is generally the most favorable substrate.

In land restoration, plant establishment is pursued through direct application of seed, the application of topsoil containing viable seed banks, transplanting of seedlings, or combinations of two or more approaches. Topsoil typically contains a substantial seed bank. Use of topsoil for revegetation can achieve large cost savings by avoiding seed collection and spreading. Topsoil is most commonly used for revegetation when the indigenous species of the area are to be replaced. However, where the topsoil contains seed of exotic species or has a large proportion of weeds in the seed bank, it may be preferable to avoid its use as the seedbed. In addition to the seed bank, topsoil contains soil microorganisms that are required for the restoration of soil functions such as organic matter decomposition and nutrient cycling. The nutrient stores in topsoil may be critically important for the production of adequate biomass in the restored ecosystem. The physical conditions in the topsoil are generally more favorable for seed germination and emergence than from other substrates. However, well-characterized subsoils or regolith materials may be suitable for revegetation when topsoil is absent or where topsoil is unsuitable due to chemical constraints or heavy weed infestation.

In some plant communities, seed is stored in the canopy of key species rather than in the topsoil. Placement of cut branches on the soil surface may be used in these cases to maximize recruitment of indigenous species.

Transplanting nursery-raised seedlings is a reliable approach for ensuring rapid plant establishment and canopy cover on a degraded site. Transplanted seedlings compete effectively against weeds and have greater survival under grazing by herbivores than plants recruited from direct seeding. For species that are difficult to propagate from seed, nursery-raised seedlings using tissue culture or other vegetative means of propagation may be the only reliable method for introducing those species during land restoration. However, raising nursery seedlings is relatively expensive, especially if only few plants of a large number of species need to be treated in this way.

Ecosystem Restoration

According to Hobbs and Norton,[14] the following are the ecosystem characteristics that should be considered when setting goals for land restoration: vegetation composition, structure, pattern, and heterogeneity; species interactions; and ecosystem function, dynamics, and resilience. A set of measures is needed to determine the success of restoration. They need to be not only low-cost and reliable indicators of present condition and function but also predictors of future trajectories for the restored ecosystem. Composition and structure of vegetation are the most commonly used indicators.[23] For pattern, heterogeneity, dynamics, and resilience, the indicators are less advanced in their development, in part because these indicators can only be identified and validated from long-term data. As a nutrient cycling indicator, microbial biomass has been proposed,[22] while Koch and Hobbs[23] concluded that soil organic matter was the best indicator of the restoration of nutrient cycling processes for bauxite mine restoration.

Increasingly, the land restoration activities in the mining industry are being assessed against completion criteria or success criteria.[17] These are legal instruments established following negotiation between regulators, mining companies, government advisers, and the community. Their aim is to provide certainty for all stakeholders about the restoration process to be followed and the expected outcome. They are designed to avoid future liability to government agencies or private landowners once mining has ceased and mine ownership has passed to new owners.

CASE STUDY: BAUXITE MINE RESTORATION

The restoration of land after bauxite mining in southwest Australia is recognized globally for its excellent practice.[24] Current practice is based on more than 40 years of research and development and adaptive management. There are important lessons to be learned about land restoration from this case study. Bauxite mining in the dry sclerophyll eucalyptus forest of southwest Australia is a shallow surface mining operation. The current restoration goal is to restore the forest values.[24] One hundred percent of the species are now routinely returned. Nutrient cycling appears to be on a trajectory towards restoring the nutrient stores and the fluxes of nutrients present in the premining forest.[22,25] Hydrological balance is disturbed for up to 12 years after land clearing for bauxite mining and subsequent revegetation.[26] Thereafter, water levels return to premining levels, but may drop below premining levels due to the increase in leaf area index relative to premining levels.[27] The restored forest is resilient to fire.[25] Completion criteria have been developed for bauxite mine restoration and several areas have been determined to meet the designated targets.[24]

The present restoration practice as described by Koch[28] is the result of four major revisions in the goals over 40 years and several other significant improvements in practice. The first rehabilitation simply planted *Pinus radiata* as a single species plantation. This was followed by a *Eucalyptus saligna* plantation and then by a goal to restore a diverse forest rather than a plantation. At this stage, it was thought that planting of local eucalyptus would not succeed because of the existence of *P. cinnamomi* in the soils and its threat to the survival of a wide range of native species. Further research demonstrated that the reconstructed profile would produce low risk of *P. cinnamomi* infection in susceptible species, and hence, it was decided to change the end land use goal to that of a forest compatible with the jarrah forest, using jarrah and marri as the overstory species.[24] Finally, it was decided to revise the end land use goal to achieve the restoration of the jarrah forest.[17,25] The selection of this goal was based on research breakthroughs that demonstrated that it was possible to stimulate the germination and emergence of recalcitrant species and hence reach close to 100% species return.

The main learning from this case study is that reaching the point when practices enable full ecosystem restoration takes several cycles of research and adaptive management. It will be based on a systematic program of research into biotic and abiotic constraints. A flexible approach from regulators enabled end land use goals to be revised over time as new research demonstrated the potential to achieve more challenging goals and targets. The end land use goals set in 1963 would have only resulted in exotic pine plantation on the mined bauxite pits. The present end land use goal is a fully functioning jarrah forest that can be integrated into existing forest management programs and achieve the multiple land use goals for the forest estate.[24]

A comprehensive report of the background research on bauxite mine restoration is found in *Restoration Ecology Special Issue* of 2007. It is a worthy model for study and emulation in land restoration on mine sites.

CASE STUDY: RESTORATION OF LAND AFFECTED BY DRYLAND SALINITY

Unlike the first case study of restoration on mined land, where disturbance is localized, dryland salinity has a more widespread impact and requires a landscape-scale response for land restoration. This case study is illustrated using examples of dryland salinity in southwest Australia and Northeast Thailand. By 2003, about 1 million ha in the wheatbelt of southwest Australia was affected by dryland salinity.[29] In Northeast Thailand, up to 30% of land could potentially become salt affected.[30]

Dryland salinity demonstrates how perturbation in water balance can have devastating consequences for the natural capital of landscapes. Whereas the native plants were predominantly deep-rooted perennial species, those agricultural species that replaced them were predominantly shallow rooted and annual. Annual plants use water only during their growing season, and their usage is limited by the fact that roots are generally confined to the surface 50–100 cm. Thus, the additional water under agricultural species is distributed to increased runoff, causing erosion and waterlogging, and to increased recharge to groundwater (Table 3). Williamson et al.[31] similarly concluded that dryland salinity in Northeast Thailand was triggered by deforestation of the uplands to produce crops like kenaf and cassava.

Two attributes of the landscape that gave rise to dryland salinity in southwest Australia were the deeply weathered regolith and the accumulation of salts from rainfall accretion.[33] Salt contents in extreme cases of up to 20,000 tons/ha have been reported,[34] most of it stored below 5 m depth.[32] Prior to clearing the native vegetation, plant roots in the upper 5 m of the regolith were largely separated from the salt bulge below, and the semipermeable aquifer at the base of the regolith was often dry.[32] However, with increased recharge, the aquifers have filled, causing water levels to rise at the rate of 0.2–1 m/yr. After a 20–30 years period of groundwater rise, saline groundwater discharge is observed commencing generally in valley floor landforms. In Northeast Thailand, the origin of salt discharge is not rainfall accretion but salt mobilized from halite sequences in the Mesozoic sediments that underlie the Korat Plateau.[31]

Reversing Dryland Salinity

Ultimately, restoring the preclearing water balance is the only complete solution to the dryland salinity problem. This requires treatments in recharge zones of landscapes to decrease recharge rates. The species that can mimic recharge rates that existed before clearing will therefore probably need to be deep rooted and perennial. They will also have to be adapted to a variety of soil conditions and climatic regimes across the affected environment. Finally, it is imperative that the species chosen to fulfill the above functions are economically viable within the farm enterprise to accelerate their adoption by land managers.

In order to manage dryland salinity, it is necessary to understand the groundwater systems as well as water balance components. With intense winter rainfall, in landscapes extensively covered by shallow-rooted annual species, recharge can occur virtually anywhere in the landscape that is not actively discharging.[33] Computer modeling adds weight to the conclusion that the only fully effective revegetation solutions for salinity control in southwest Australia are with deep-rooted perennial vegetation over most of the whole catchment.[33] Even systems like agro-forestry that place a high density of woody shrubs and small trees in rows 30 m apart were insufficient in the modeling scenarios to restore water balance and achieve complete control

Table 3 Changes in water balances for cleared catchments before and after clearing.

Catchment	Year	Rainfall (mm)	Interception (mm)	Evapotranspiration (mm)	Change in water storage (mm)	Change in groundwater storage (mm)	Stream flow (mm)
Wights forested	1975	1027	130	855	−28	−11	81
Wights cleared	1985	1147	0	565	−	−21	115
Lemon forested	1975	739	74	656	4	−1	5
Lemon cleared	1983	821	38	708	−	−19	56

Source: Williamson.[32]

of salinity. Continued reliance in the farming system on annual shallow-rooted crops such as cereals is problematic because these crops will allow continued recharge. Loss of these species is also problematic because they are the main source of income for farmers.

Until recharge control treatments start to decrease saline groundwater discharge, treatments are also needed in the discharge areas. These may include both engineering treatments to alleviate waterlogging[35] as well as vegetation options that cope with saline waterlogged conditions.[36] In Northeast Thailand, there has been considerable investigation of salt-tolerant species that can be grown on various classes of salt-affected soil.[30]

The case study on dryland salinity may therefore serve as a useful model for landscape-scale restoration. As with the case of land restoration after bauxite mining, the present approach has involved background research to understand the underlying physical processes (landscape water balance, water fluxes, hydrogeology) and to develop effective solutions (the effect of land use and vegetation type on water balance). The present set of strategies to control dryland salinity has evolved out of several phases of research and adaptive management leading to current understanding and solutions.

CONCLUSIONS

Land is a form of natural capital that is essentially finite and non-renewable. Every effort must be exerted to avoid degradation of land because degradation diminishes its natural capital value and compromises the ecosystem services provided by soil. The restoration of degraded land has the potential to increase its natural capital value and enhance ecosystem services. Since large areas of land globally are degraded, there is substantial scope for increasing ecosystem services by restoring degraded land. Land restoration is a relatively new discipline, which, along with restoration ecology, consists of successful practices for land restoration at a site-specific scale. Examples of best practice can be found in restored mine sites. The challenge remaining is to scale up land restoration from site-specific cases on mine sites to regional or landscape scales for a variety of degrading land disturbances. Most success at both scales has, to date, been concerned with restoring key ecosystems functions such as organic matter accumulation, nutrient cycling, and water balance.

REFERENCES

1. Lal, R.; Stewart, B.A., Eds. Soil Restoration. In *Advances in Soil Science*; CRC Press: Boca Raton, Florida, 1992; Vol. 17.
2. Kaiser, J. Wounding Earth's fragile skin. Science **2004**, *304*, 1616–1618.
3. Reich, P.; Eswaran, H. Soil and trouble. Science **2004**, *304*; 1614–1615.
4. McKenzie, N.; Jacquier, D; Isbell, R.F.; Brown, K. *Australian Soils and Landscapes. An Illustrated Compendium*; CSIRO Publishing: Collingwood, 2004; 416.
5. Pillans, B. Soil development at a snail's pace: Evidence from a 6 Ma soil chronosequence on basalt in north Queensland. Geoderma **1997**, *80*, 117–128.
6. Robinson, D.A.; Lebron, I.; Vereecken, H. On the definition of the natural capital of soils: A framework for description, evaluation, and monitoring. Soil Sci. Soc. Am. J. **2009**, *73*, 1904–1911.
7. Lal, R. Soil degradation as a reason for inadequate human nutrition. Food Secur. **2009**, *1*, 45–57.
8. Schaffnit-Chatterjee, C. *The Global Food Equation. Food Security in an Environment of Increasing Scarcity*; Deutsche Bank Research: Frankfurt, Germany 2009.
9. Conacher, A.J.; Conacher, J. *Rural Land Degradation in Australia*; Oxford Univ. Press: Melbourne, Australia, 1995.
10. Lal, R.; Blum, W.H.; Valentine, C.; Stewart, B.A., Eds. *Methods for Assessment of Soil Degradation*; CRC Press: Boca Raton, Florida, 1998.
11. Stocking, M. A. Land degradation. In *International Encyclopedia of the Social & Behavioral Sciences*; Smelser, N.J., Baltes, P.B., Eds.; 2001; 8242–8247.
12. Bossio, D.; Geheb, K.; Critchley, W. Managing water by managing land: Addressing land degradation to improve water productivity and rural livelihoods. Agric. Water Manage. **2010**, *97*, 536–542.
13. FAO; UNDP; UNEP. Land degradation in south Asia: Its severity, causes and effects upon the people; World Soil Resources Reports. Food and Agriculture Organization of the United Nations: Rome, 1994.
14. Hobbs, R.J.; Norton, D.A. Towards a conceptual framework for restoration ecology. Restor. Ecol. **1996**, *4*, 93–110.
15. Aronson, J.; LeFoc'h, E.; Floret, C.; Ovalle, C.; Pontanier, R. Restoration and rehabilitation of degraded ecosystems in arid and semiarid regions. II. Case studies in Chile, Tunisia and Cameroon. Restor. Ecol. **1993**, *1*, 168–187.
16. Hobbs, R.J. Restoration ecology. In *Encyclopedia of Soil Science*; Lal, R., Ed.; Marcel Dekker: New York, 2002; 1153–1155.
17. Ward, S. Success criteria. In *Encyclopedia of Soil Science*; Lal, R., Ed.; Marcel Dekker: New York, 2002; 1156–1160.
18. Bell, R.W. Principles of land restoration. In: *Encyclopedia of Soil Science*; Lal, R., Ed.; Marcel Dekker: New York, 2002; 766–769.
19. Department of Minerals and Petroleum. *Draft Guidelines for Preparing Mine Closure Plans*; Department of Minerals and Petroleum: Perth, 2010.
20. Bell, R.W. Diagnosis and prognosis of soil fertility constraints for land restoration (Ch. 16). In *Remediation and Management of Degraded Lands*; Wong, M.H., Wong, J.W.C., Baker, A.J.M., Eds.; Lewis Publishers: Boca Raton, Florida, 1999, 163–173.
21. Jasper, D.A.; Sawada, Y.; Gaunt, E.; Ward, S.C. Indicators of reclamation success—Recovery patterns of soil biological activity compared to remote sending of vegetation. In *Land Reclamation: Achieving Sustainable Benefits*; Fox, H.R., Moore, H. M., McIntosh, A.D., Eds.; A.A. Balkema: Rotterdam, 1998; 21–24.

22. Jasper, D.A. Beneficial soil microorganisms of the Jarrah forest and their recovery in bauxite mine restoration in southwestern Australia. Restor. Ecol. **2007**, *15*, S74–S84.
23. Koch, J.M.; Hobbs, R.J. Synthesis: Is Alcoa successfully restoring a Jarrah forest ecosystem after bauxite mining in Western Australia? Restor. Ecol. **2007**, *15*, S137–S144.
24. Gardner, J.H.; Bell, D.T. Bauxite mining restoration by Alcoa World Alumina Australia in Western Australia: Social, political, historical, and environmental contexts. Restor. Ecol. **2007**, *15*, S3–S10.
25. Grant, C.D.; Ward, S.C.; Morley, S.C. Return of ecosystem function to restored bauxite mines in Western Australia. Restor. Ecol. **2007**, *15*, S94–S103.
26. Croton, J.T.; Reed, A.J. Hydrology and bauxite mining on the Darling Plateau. Restor. Ecol. **2007**, *15*, S40–S47.
27. Bari, M.A.; Ruprecht, J.K. *Water yield response to land use change in south-west Western Australia*, Salinity and Land Use Impacts Series Report No. SLUI 31; Department of Environment: Perth, Australia, 2003.
28. Koch, J.M. Alcoa's mining and restoration process in south western Australia. Restor. Ecol. **2007**, *15*, S11–S16.
29. McFarlane, D.; George, R.J.; Cacetta, P.A. The extent and potential area of salt-affected land in Western Australia estimated using remote sensing and digital terrain models. In *1st National Salinity Engineering Conference, 9–12 November 2004, Perth Australia*, Conference Proceedings; Dogramaci, S., Waterhouse, A., Eds.; Institution of Engineers: Australia, 2004; 55–60.
30. Yuvaniyama, A. Managing problem soils in northeast Thailand. In *Natural Resource Management Issues in the Korat Basin of Northeast Thailand: An Overview*; Kam, S.P., Hoanh, C.T., Trebuil G., Hardy, B., Eds.; IRRI Limited Proceedings No. 7; 2001; 147–156.
31. Williamson, D.R.; Peck, A.J.; Turner, J.V.; Arunin, S. Groundwater hydrology and salinity in a valley in Northeast Thailand. Groundwater Contam. Int. Assoc. Hydrogeologists Publ. **1989**, *185*, 147–154.
32. Williamson, D.R. Land degradation processes and water quality effects: Waterlogging and salinisation (Ch. 17). In *Farming Action Catchment Reaction*; Williams, J., Hook, R.A., Gascoigne, H., Eds.; CSIRO: Melbourne, 1998; 162–190.
33. Clarke, C.J.; George, R.J.; Bell, R.W.; Hatton, T.J. Dryland salinity in southwestern Australia: Its origins, remedies, and future research directions. Aust. J. Soil Res. **2002**, *40*, 93–113.
34. Moore, G.W. Salinity. In *Soil Guide. A Handbook for Understanding and Managing Agricultural Soils*; Moore, G.W., Ed.; Bulletin 4343; Department of Agriculture: Western Australia, 2004; 146–158.
35. Bell, R.W.; Mann, S. Amelioration of salt and waterlogging-affected soils: Implications for deep drainage. In *1st National Salinity Engineering Conference, 9–12 November 2004, Perth Australia*, Conference Proceedings; Dogramaci, S., Waterhouse, A., Eds.; Institution of Engineers: Australia, 2004; 95–100.
36. Barrett-Lennard, E.G. *Saltland Pastures in Austalia. A Practical Guide*, 2nd Ed.; Land, Water and Wool Sustainable Grazing on Saline Lands Sub-program: Canberra, 2003.

Laws and Regulations: Food

Ike Jeon
Department of Animal Science and Industry, Kansas State University, Manhattan, Kansas, U.S.A.

Abstract
The food supply in the United States is considered the safest in the world, although chemical residues such as pesticide residues on food have been of great concern to the consumer. Food labels on consumer packages do not contain any statements relative to the pesticide residues or other matters such as insect fragments. This is because foods the consumer buys at supermarkets or grocery stores should be free from these contaminants. In a practical sense, however, producing foods absolutely free from chemical residues or insect fragments is not possible with the practices of modern agricultural production. The U.S. basic food law (Food, Drug, and Cosmetic Act) allows the regulatory agencies to establish tolerance limits for various food products. A tolerance limit is defined as the maximum quantity of a substance allowable on food. There are two major categories for tolerance limits. One is for poisonous or deleterious substances (e.g., pesticide residues) in human food and animal feed, and the other is for natural or unavoidable defects (e.g., insect fragments) in foods that present no health hazards for humans. These tolerance limits are enforced by the federal government during food processing, packaging, and distribution.

TOLERANCE LIMITS FOR PESTICIDE RESIDUES

The responsibility for ensuring that pesticide residues in foods are not present above the limits is shared by three major government agencies.[1] The Environment Protection Agency (EPA) determines the safety of pesticide products and sets tolerance levels for pesticides. The Food and Drug Administration (FDA) enforces the tolerances in all foods except meat and poultry products. The U.S. Department of Agriculture's Food Safety and Inspection Service (FSIS) regulates commercially processed egg, meat, and poultry products including combination products (e.g., stew, pizza). In addition, any products containing 2% or more poultry or poultry products, or 3% or more red meat or red meat products are also under jurisdiction of the FSIS. The pesticides of concern usually include insecticides, fungicides, herbicides, and other agricultural chemicals. Table 1 illustrates examples of tolerance levels for pesticide residues in several food categories.[2,3] These tolerance levels are extremely low, usually below parts per million, but do not represent permissible levels of contamination where it is avoidable. In addition, blending of a food (or feed) containing a substance in excess of an action level or tolerance with another food (or feed) is not permitted, and the final product from blending is unlawful, regardless of the level of the contaminant.

Regulatory Inspection and Enforcement

The FDA monitors the levels of pesticide residues in processed foods. For imported products, the FDA checks a sample of the food at entry into the United States and can stop shipments at the entry. If illegal residues are found in domestic samples, FDA can take regulatory actions, such as seizure or injunction.

The U.S. Department of Agriculture also monitors pesticide residues in food.[4] The Department was charged in 1991 with implementing a program to collect data on pesticide residues on various food commodities. The program has become a critical component of the Food Quality Protection Act of 1996 and currently is known as the Pesticide Data Program. The data on pesticides in selected commodities are used by the EPA to support its dietary risk assessment process and pesticide registration and by the FDA to refine sampling for enforcement of tolerances.

If a product is in violation of the tolerance limits, it is *adulterated* under the food law. The product may be destroyed or recalled from the market by the manufacturer or shipper. The recall may be initiated voluntarily by the manufacturer (or shipper) or at the request of the regulatory agency. The responsible agency also may seize the product on orders obtained from the Federal courts and may prosecute persons or firms responsible for the violation.

TOLERANCE LIMITS FOR INSECT FRAGMENTS

Many food materials may contain natural but unwanted debris that cause no health hazards for humans. These debris may include insects, insect fragments, and rodent hairs and are considered unavoidable defects in foods with the current agricultural practices. In fact, the use of chemical substances

Table 1 Examples of tolerance limits for pesticide residues in human food.

Substance	Commodity	Action level (Parts per million) 0.03	Remark
Aldrin and dieldrin	Asparagus		
	Fish	0.3	Edible portion
	Peanuts	0.05	
Chlordane	Carrots	0.1	
	Fish	0.3	Edible portion
	Lettuce	0.1	
	Poultry	0.3	Fat basis
DDT[a]	Carrots	3.0	
	Citrus fruits	0.1	
	Tomatoes	0.05	
Lindane	Beans	0.5	
	Corn	0.1	
	Milk	0.3	Fat basis
	Beef	7.0	Fat basis

[a]Dichlorodiphenyltrichloroethane.
Source: FDA[2] and USDA.[3]

to control insects, rodent, and other contaminants has little, if any, impact on natural and unavoidable defects in foods. The FDA contends that the use of pesticides does not effectively reduce the presence of these food defects. This has led the regulatory agencies to establish maximum levels of natural or unavoidable defects allowable in foods for human use. The FDA currently lists over 100 products from fruits to fish,[5] and Table 2 shows only several examples. If no defect action level exists for a product, the FDA evaluates and decides on a case-by-case basis using criteria of reported findings such as length of hairs and size of insect fragments.

The FDA sets these action levels under the premise that it is economically impractical to grow, harvest, or process raw products that are totally free of nonhazardous, naturally occurring, unavoidable defects. It is incorrect, however, to assume that because the FDA has an established defect action level for a food, the manufacturer needs only keep defects just below that level. The defect levels do not represent averages of the defects that occur in any of the products. The levels represent limits at which FDA will regard the food product as *adulterated* and, therefore, subject to enforcement action. Like pesticide residues, blending of food with a defect at or above the current defect action level with another lot of the same or another food is not permitted. That practice renders the final food unlawful regardless of the defect level of the finished food.

RESPONSIBILITY OF FOOD MANUFACTURERS

Food manufacturers are required to follow the standard manufacturing procedures under a federal regulation, known as good manufacturing practice (GMP), during food production.[6] The GMP guidelines imply that all food materials used must not exceed the tolerance limits set for pesticide residues or any other poisonous or deleterious substances. The GMP also calls for the same regulatory requirement for natural or unavoidable defects in all food materials. The food materials susceptible to contamination may be tested for compliance or relied on a supplier's guarantee or certification that they are in compliance. In addition, the GMP regulation stipulates that food manufacturers and distributors must utilize at all times quality control operations that reduce natural or

Table 2 Examples of tolerance limits for natural or unavoidable defects in foods.

Product	Defect	Action level
Sweet corn, canned	Insect larvae	2 or more 3 mm or longer larvae
Macaroni	Insect filth	225 insect fragments or more per 225 g
	Rodent filth	4.5 rodent hairs or more per 225 g
Peaches, canned and frozen	Mold/insect damage	Wormy or moldy on 3% or more fruits
	Insects	1 or more larvae and/or larval fragments whose aggregate length exceeds 5 mm in 12 one-pound cans
Peanut butter	Insect filth	30 or more insect fragments per 100 g
	Rodent filth	1 or more rodent hairs per 100 g
Popcorn	Rodent filth	1 or more rodent excreta pellets or rodent hairs in 1 or more subsamples
Tomato juice	Drosophila fly	10 or more fly eggs per 100 g
	Mold	24% of mold counts in 6 subsamples
Wheat flour	Insect filth	75 or more insect fragments per 50 g
	Rodent filth	1 or more rodent hairs per 50 g

Source: FDA.[5]

unavoidable defects to the lowest level feasible with the current technology.

POTENTIAL CONSUMER BENEFITS

Through conducting a monitoring program, the federal government agencies work together to improve consumer protection. The EPA will continue to review scientific data on all pesticide products, while the FDA and U.S. Department of Agriculture will closely monitor levels of pesticide residues in all foods including both domestic and imported products. The U.S. Department of Agriculture's data for 1998 suggest that violation of the pesticide tolerance limits was very low in all raw products including fruit and vege, wheat, and milk samples. In 1993, the FDA reported that no pesticide residues were found in infant formulas, and no residues over EPA tolerances or FDA action levels were found in any of the foods that were prepared as consumers normally would prepare them at home.[7]

ACKNOWLEDGMENTS

Contribution No. 00-231-B, Kansas Agricultural Experiment Station, Manhattan, Kansas 66506, U.S.A.

REFERENCES

1. FDA. *FDA's Food and Cosmetic Regulatory Responsibilities*; U.S. Food and Drug Administration: Washington, DC, 1998; 1–5, http://vm.cfsan.fda.gov/~dms/regresp.html (accessed June 2000).
2. FDA. *Action Levels for Poisonous or Deleterious Substances in Human Food and Animal Feed*; U.S. Food and Drug Administration: Washington, DC, 1998; 1–17, http://vm.cfsan.fda.gov/~lrd/fdaact.html (accessed June 2000).
3. USDA. *Domestic Residue Book (Appendix I)*; U.S. Department of Agriculture, Food Safety and Inspection Service: Washington, DC, 1998; 1–30, http://www.fsis.usda.gov:80/OPHS/redbook1/appndx1.htm (accessed June 2000).
4. USDA. *Pesticide Data Program Annual Summary—Calendar Year of 1998*; U.S. Department of Agriculture, Agricultural Marketing Service: Washington, DC, 2000; 1–19.
5. FDA. The Food Defect Action Levels—Levels of Natural or Unavoidable Defects in Foods that Present No Health Hazards for Humans. In *FDA/CFSAN Food Defect Action Level Handbook*; U.S. Food and Drug Administration: Washington, DC, 1998; 1–36, http://vm.cfsan.fda.gov/~dms/dalbook.html (accessed June 2000).
6. CFR. Current good manufacturing practice in Manufacturing, Packing, or Holding Human Food. In *Code of Federal Regulations, Title 21, Part 110*; U.S. Government Printing Office: Washington, DC, 1999; 206–215.
7. FDA. *FDA Reports on Pesticides in Foods*; U.S. Food and Drug Administration: Washington, DC, 1993; 1–5, http://vm.cfsan.fda.gov/~lrd/pesticid.html (accessed June 2000).

Laws and Regulations: Pesticides

Praful Suchak
Sneha Plastics Pvt. Ltd., Suchak's Consultancy Services, Mumbai, India

Abstract
Chemical or biological pesticides have target specific toxicity that controls or eradicates pests falling under different groups. These products, though developed for specific usage, could have adverse effects on living beings and the environment and unchecked use can cause havoc. Regulating pesticides, therefore, would assure reasonable safety in use of these toxic substances and ensure that risks from pesticides to humans and their environment are minimized and are consistent with the benefits achieved by their use in terms of reduced losses.

INTRODUCTION

Why Regulate Pesticides?

Chemical or biological pesticides have target specific toxicity that controls or eradicates pests falling under different groups. These products, though developed for specific usage, could have adverse effects on living beings and the environment and unchecked use can cause havoc. Regulating pesticides, therefore, would assure reasonable safety in use of these toxic substances and ensure that risks from pesticides to humans and their environment are minimized and are consistent with the benefits achieved by their use in terms of reduced losses.

Regulating pesticides at the international and national level should consider social costs in line with social benefits. Pesticides impose costs on society, such as health risks and environmental degradation, which are not borne by the user. The available policy remedies include bans on individual or classes of chemicals that prohibit the introduction of hazardous compounds into the environment, and economic instruments such as taxes, registration fees, and import duties that work to redistribute and adjust the social costs occurring for pesticide use and also provide the government with revenues that can be used to cover health costs and environmental clean-up activities.

HISTORY

The United States in 1910 introduced the Federal Pesticide Act that underwent complete metamorphosis to become the Federal Insecticide, Fungicide and Rodenticide Act (FIFRA) in 1947, which since 1970 is under the auspices of the Environmental Protection Agency.

Australia initiated pesticide legislation with one state in 1925 and by 1945 all states had their individual laws. The Industry Association brought law common to all states in 1995. By the end of 1999 about 95% of the countries in the world had adopted full/partial regulatory systems.

Early in-depth studies were not carried out on the long-term effects of: 1) repeated exposures, 2) residual toxicity, 3) accumulated toxicity, and 4) the impact on environment. With additional knowledge on the cumulative toxicity of chlorinated hydrocarbons such as DDT having come to light, the regulating authorities have started demanding the generation of additional critical toxicological data to assess short-term, long-term, and environmental toxicity of earlier registered pesticides. The European Union has already undertaken reviews of 90 molecules in the first phase by a Commission regulation dated December 11, 1992, to be completed in 12 years, and a further 148 molecules in the second phase effective March 1, 2000. The remaining substances in the European Union would be included in third phase.

Regulatory requirements for pesticides have undergone a change over the past half a century. With the advent of highly sophisticated testing equipment, more knowledge about harmful effects of the toxic chemicals has come to light. Consistent watch by environmentalists and organizations like the Pesticides Action Network (PAN), Greenpeace, Save the Planet groups, and other nongovernmental organizations has resulted in added awareness resulting in hosts of data requirements for registration/reregistration of pesticides.

Although all developed countries and most of the developing countries have their own legislation to regulate pesticides, there have been vast variations in data requirements for registrations between these countries. With globalization it has became imperative to have harmonized data requirements so that the registrant can hope for faster registration in different (pesticide consuming) countries.

AVAILABLE INTERNATIONAL GUIDELINES

1. Agenda 21 of the United Nations Conference on Environment and Development (UNCED)
2. The *Codex Alimentarius*
3. The FAO International Code of Conduct and Prior Informed Consent (PIC)

4. WTO and International Trade with respect to pesticides
5. Agreement on Persistent Organic Pollution (POP)
6. Guidelines of Minor Donor Institutions on the purchase of pesticides

IMPLEMENTATION PROBLEMS

Although FAO took the lead to harmonize data requirements in participating nations for registrations of pesticides, certain problems and practical difficulties have occurred such as

1. The original registrant, having invested huge amounts in data generation, is unable to protect the data
2. Absence of confidentiality assurance by the registering country, creating difficulties in multiple country registrations
3. Recommended uses differ from country to country, resulting in difficulties
4. Unchecked dumping of unsafe or banned pesticides in less-developed countries
5. New registrations by a company other than the original registrant by providing data generated by such a company could not be checked

STEPS UNDERTAKEN

Although though PIC entry of banned pesticides could be prevented, this instrument has not been fully effective. Once it becomes fully operational legally things should improve.

With the United States implementing the Food Quality Protection Act and fixing maximum residue limits for 3000 toxic compounds, countries worldwide would need to harmonize their registrations on toxic chemicals so as to meet the residue levels in food.

The formation of the European Union with 15 member countries, OECD with 29 members, and the Technical Working Group having EPA, Canada, and Mexico, has accelerated the pace towards harmonization. However, since a vast disparity exists between developed countries on one side and developing countries on the other side, it is rather difficult to have a unified data requirement, particularly in case of risk assessment.

Acceptance of electronic data submission and dossier/monograph submissions and joint reviews by EU would also pave the way toward harmonization and would address questions in the nondietary exposure area.

Apart from studies related to bioefficiency of the product, the toxicological studies of the toxicant, its analogues, impurities and breakdown products, residual toxicity, etc., as listed in Appendix 1 would help understanding and regulating pesticides.

PRESENT SCENARIO AND PROBABLE REMEDIES

Substantial evidence exists that pesticides are being applied in a technically and economically inefficient manner. Many developing countries subsidize pesticides and equipment, resulting in excessive use of pesticides.

Also in developing countries, the current legal environment and enforcement capabilities have been inadequate and dysfunctional, thus exerting a significant impact on current levels of pesticide use. This is partly due to lack of resources and partly due to manipulation by vested interests.

The inadequacies of the existing regulatory framework, institutional rigidities, and a bias in favor of pesticide-dependent paths also contribute to improper use of pesticides.

A major problem confronting many countries is the absence of well-established procedural mechanisms for public involvement in the decision making process including crop protection policy. Competing interests with a stake in the process, including farmers, the pesticide industry, and policy makers responsible for food security, argue for a more liberal regulatory stance. On the other hand, environmentalists, public health workers, and consumers demand strict regulation and reduced pesticide volumes.

To be more effective, pesticide regulation and implementation should be handled by a neutral agency like the Ministry of Environment or similar organization and not the Ministry of Agriculture or other interested ministry.

Pesticide policy needs to be integrated into the broader public policy debate concerning the nations' agricultural, environmental, and health strategies.

Nevertheless, two general principles should apply. First, dispassionate analysis of the costs and benefits of pesticide use would provide a useful tool for the formulation of normal policies; and second, the broader and more inclusive the debate, the more likely it is that the outcome will serve the public rather than specific private interests.

FUTURE GLOBAL POLICY

A uniform global regulatory system needs to ensure

1. Agricultural chemical use increases agricultural output
2. Food supplies are safe from harmful toxicants/residues
3. Reduced-risk chemical pesticides, biopesticides, and nonchemical alternatives are encouraged
4. Uniform MRLs to eliminate trade barriers
5. Uniform health-based safety standards for pesticide residues
6. Special provisions for certain groups of the population including infants and children

APPENDIX 1

Toxicological and Other Data Requirements for Pesticide Registration

1. Identity of active substance
    - Chemical name
    - Empirical and structural formula
    - Molecular mass
    - Method of manufacture (synthesis pathways)
    - Purity
    - Identity and content of isomers
    - Impurity and additives

2. Physical and chemical properties
    - Melting point
    - Boiling point and relative density
    - Vapor pressure
    - Volatility
    - Appearance
    - Absorption spectra-molecular extinction at relevant wavelength
    - Solubility in water/organic solvents
    - Partitioning coefficient N-octanol/water
    - Stability and hydrolysis rate in water
    - Photochemical degradation on surface, in water, and in air
    - Thermal stability and stability in air

3. Analytical method
    - Analytical method for the determination of the pure active substance in the technical grade.
    - For breakdown products and additives in plant products, soil, water, animal body fluids, and tissues.

4. Toxicological and metabolism studies
    - Studies on acute toxicity—oral, percutaneous, inhalation, intraperitoneal, skin and, where appropriate, eye irritation, and skin sensitization.
    - Short-term toxicity—oral, cumulative toxicity, and other routes inhalation or dermal.
    - Chronic toxicity—oral, long-term toxicity, and carcinogenicity.
    - Mutagenicity—reproductive toxicity-teratogenicity and multigeneration studies in mammals.
    - Metabolism studies in mammals—absorption, distribution, and excretion studies, elucidation of metabolic pathways.
    - Supplementary studies—neurotoxicity studies—toxic effects of metabolites from treated plants and toxic effects on livestock and pests.
    - Medical data—medical surveillance on manufacturing plant personnel, clinical cases, poisoning incidents from industry and agriculture sensitization/allergenicity observations, observations on exposure of the general population, and epidemiological studies if appropriate. Diagnosis and specific signs of poisoning, clinical tests, and prognosis of expected effects of poisoning. Proposed treatment: first aid measures, antidotes, and medical treatment.
    - Summary of toxicological studies and conclusions, critical scientific evaluation with regard to all toxicological data, and other information concerning the active substance.

5. Residues in or on treated products, food and feed metabolism in plants and livestock
    - In treated plants (distribution, metabolism, binding constituents, etc.).
    - In livestock (uptake, distribution, metabolism, binding constituents, etc.).

6. Fate and behavior in the environment
    - Studies on aerobic and anaerobic degradation under laboratory conditions in different soil types.
    - Adsorption and desorption in different soil types including metabolites.
    - Mobility of the active ingredients in different soil types.
    - Behavior in water and air, rate and route of degradation.

7. Ecotoxicological studies
    - Effects on birds, fish, aquatic organisms such as Daphnia magna, algae, honeybees, earthworms, other nontarget macroorganisms and microorganisms.

8. Information concerning the labeling including indication of danger and safety measures.

BIBLIOGRAPHY

1. Pesticides Policies in Developing Countries—Do They Encourage Excessive Use? In *World Bank Discussions Paper No. 238;* 1994.
2. Asian Development Bank. In *Handbook on the Use of Pesticides in Asia Pacific Region*; ADB: Manila, Philippines, 1987.
3. *Pesticide Policy Project Hannover*; Publication Serial No. 1, January 1995; No. 2, November 1995; No. 3, December 1995; No. 4, December 1996; No. 5, December 1996; No. 6, 1998; No. 7, April 1999; No. 8, April 1999.
4. EC Directives 91/414/EEC and Subsequent Directives Including 1999/80/EC.
5. Proceedings of Asia Pacific Crop Protection Conference 1997 and 1999, PMFAI: Mumbai, India.
6. *Global Pesticides Directory*, 2nd Ed.; Suchak's Consultancy Services: Mumbai, India, 1997. suchakgr@vsnl.net
7. *Pesticides News*; No. 20–47, Pesticides Action Network (PAN): London, 1993 to 1999.
8. Guidelines on the Operation of Prior Informed Consent (PIC) Rome FAO 1990, Guidance to Government in PIC Rome 1991, and Other FAO Publications.
9. U.S. EPA Pesticides Information Network. http://www.cdpr.ca.govt/docs/epa/epachim.htm.

Laws and Regulations: Rotterdam Convention

Barbara Dinham
Eurolink Center, Pesticide Action Network U.K., London, U.K.

Abstract
The Rotterdam Convention takes an important step toward protecting humans and the environment from highly toxic chemicals. For the first time, it will help monitor and control trade in dangerous substances, circulate better information about health and environmental problems of chemicals, and prevent unwanted imports of certain hazardous chemicals.

INTRODUCTION

When chemical pesticides were introduced 50 years ago, little attention was paid to the environmental and health impacts. With the rapid expansion of use in the 1950s, understanding gradually increased of the consequences of exposure to certain chemicals. Wide-ranging impacts began to be identified, including: environmental persistence and effects on birds and wildlife; residues in soil, water, and air; residues in food; human poisonings from acutely toxic pesticides or long-term health impacts such as cancer; and pest resistance, often leading to dramatic crop losses.

With almost 1000 different pesticides and thousands of formulations on the market to control insects, diseases, weeds, and other pests, action was clearly needed to protect human health and the environment. International standards recommended that governments establish a registration system to authorize each formulation of a pesticide for each specific crop or other use. Concern with some pesticides led governments to ban or restrict them to a limited number of uses. Few developing countries can fully implement a registration scheme, and they are often unaware of bans imposed elsewhere. Recognizing these problems, in the early 1980s, governments, international organizations, and public interest groups began to demand action to provide a warning system to help developing countries regulate or ban the use of hazardous pesticides.

The Rotterdam Convention on Prior Informed Consent Procedure for Certain Hazardous Chemicals and Pesticides in International Trade[1] is the outcome of 15 years of activity on trade in hazardous chemicals. Adopted on 10 September 1998 in Rotterdam, the Netherlands, the Convention was signed by 73 countries[2] and by June 2001 had been ratified by 14 parties. It will become legally binding after 50 countries have ratified.

The Convention takes an important step toward protecting humans and the environment from highly toxic chemicals. For the first time, it will help monitor and control trade in dangerous substances, circulate better information about health and environmental problems of chemicals, and prevent unwanted imports of certain hazardous chemicals.

Central to the Rotterdam Convention is the system of Prior Informed Consent (PIC), a means of obtaining and disseminating decisions of importing countries about their willingness to receive shipments of certain chemicals, and ensuring compliance to these decisions by the exporter. To be included in PIC, a pesticide must be banned or severely restricted for health or environmental reasons by two countries in two different regions of the world—indicating that its adverse effects are a "global concern."

But focusing on banned or severely restricted pesticides may only touch the tip of the iceberg. Industrialized countries rely on trained and informed users able to apply good practice as safeguards: in developing countries where pesticides are often used under conditions of poverty, these measures cannot be applied. Furthermore, older—and often more hazardous—pesticides are often cheaper, making them attractive to poorer farmers. The Convention recognizes that "severely hazardous pesticide formulations" should be included in PIC if they cause health or environmental problems in developing countries or in Eastern Europe—termed "countries with economies in transition"—in the Convention.

HISTORY OF PIC

A PIC system was first proposed in the early 1980s as part of the International Code of Conduct on the Distribution and Use of Pesticides, negotiated by governments in the Food and Agriculture Organization (FAO) of the UN. Some governments resisted the concept, and the Code was adopted in 1985 without any reference to PIC. But intense pressure from nongovernmental organizations (NGOs) and others won support, and the principle was accepted in 1987. It took until 1989 to establish the wording and issue a revised version of the Code.[3] That same year, the UN Environment Programme (UNEP) included an identical provision in the London Guidelines on the Exchange of Information on Chemicals in International Trade, and a voluntary system was put in place with the FAO acting as the Secretariat for pesticides and UNEP for industrial chemicals.

The first pesticides were added in 1991, and by 1995, 22 pesticides and five industrial chemicals were included.

From Voluntary to Legally Binding

The issue of transforming the voluntary scheme into a legally binding international Convention was first mooted in 1992 at the United Nations Conference on Environment and Development (UNCED).[4] In November 1994, the FAO Council meeting agreed to proceed, and this was followed in May 1995 by a decision of the UNEP Governing Council. The two organizations convened an Intergovernmental Negotiating Committee (INC) to draft and agree international legally binding instrument.

Banning Exports of Banned Pesticides

An alternative to PIC strongly advocated at the time was to stop all exports of banned pesticides. However, unless action to limit the market for a banned pesticide could be taken, banning exports could encourage companies to relocate production, possibly in a country with less stringent controls. Preventing the export of banned pesticides would have no effect on severely restricted chemicals. Without a PIC system, a developing country could unwittingly allow the import of banned or severely restricted pesticides, ignorant of action taken by some governments. Many developing countries maintained that an export ban could limit their development, as alternatives were more expensive, and that import decisions should rest with them. PIC does not prevent individual countries from deciding that their banned pesticides should not be exported, but does ensure that regulatory actions are widely shared.

HOW THE CONVENTION IS OPERATED

In negotiating the text of the Rotterdam Convention, governments built on the experience gained in the voluntary PIC. As a mark of its importance, the Convention began immediately on a voluntary basis, with FAO and UNEP continuing as an interim Joint Secretariat.

Designated National Authorities

To participate in PIC, governments must appoint a Designated National Authority (DNA). By December 2000, 170 governments had appointed a DNA or a focal point. When ratifying the Convention, DNAs must be authorized to carry out administrative functions such as receiving, transmitting, and circulating information.

Notifying Regulatory Actions

When a government bans or severely restricts a pesticide, it must notify the Joint Secretariat within 90 days. Governments need to demonstrate that their action is final and that it was based on a risk evaluation, including a review of scientific data, and the Secretariat will validate the notification. Once two valid notifications from different PIC regions have been received for the same pesticide, it becomes a candidate for PIC.

Chemical Review Committee

The Convention set up a Chemical Review Committee to consider notifications, and advise the Conference of the Parties (CoP—this will replace the INC after ratification). A parallel structure operates in the voluntary phase, with an Interim Chemical Review Committee (ICRC). The Committee will review PIC notifications, and—when they meet the agreed criteria—draft a Decision Guidance Document (DGD).

Table 1 Pesticides covered by the interim PIC procedure, November 2000.

Banned or severely restricted pesticides[a]
2,4,5-T (dioxin contamination)
Aldrin
Binapacryl (INC6)[a]
Captafol
Chlordane
Chlordimeform
Chlorobenzilate
DDT
Dieldrin
Dinoseb and dinoseb salts
1,2-Dibromoethane (EDB, or ethylene dibromide)
Ethylene dichloride (INC7)[a]
Ethylene oxide (INC7)[a]
Fluoroacetamide
HCH, mixed isomers
Heptachlor
Hexachlorobenzene
Lindane
Mercury compounds
    mercuric oxide
    mercurous chloride, Calomel
    other inorganic mercury compounds
    alkyl mercury compounds
    alkoxyalkyl/aryl mercury compounds
Pentachlorophenol
Toxaphene (INC6)[a]

Severely hazardous pesticide formulations[b]
Monocrotophos
Methamidophos
Phosphamidon
Methyl parathion
Parathion

[a]Indicates that these four pesticides were added to the PIC list at the 6th and 7th International Negotiating Committee meetings.
[b]Only certain formulations of these severely hazardous pesticides are included.
Source: http://www.pic.int/[5]

Two Routes to be "PIC-ed"

Pesticides in the voluntary PIC were carried forward, and new pesticides continue to be added. By June 2001, the process included 26 pesticides and five industrial chemicals (Table 1).

There are two routes for adding pesticides to the Convention. Under Article 5, a ban or severe restriction in any two regions triggers PIC if the action is taken for health or environmental reasons. Governments have decided that the PIC regions would be: Africa (48 countries), Latin America, and the Caribbean (33 countries), Asia (23 countries), Near East (22 countries), Europe (49 countries), North America (2 countries: Canada and US), Southwest Pacific (16 countries).

The second route is covered in Article 6, and addresses "severely hazardous pesticide formulations." This category applies only to pesticide formulations found to be causing health or environmental problems under conditions of use in developing countries, or countries with economies in transition. These pesticides may not have been banned, but—generally because of high toxicity—cause poisonings and deaths when used without extreme caution. Governments must submit evidence based on a "clear description of incidents related to the problem, including the adverse effects and the way in which the formulation was used." Nevertheless, this kind of evidence is rare, and collecting information is difficult: incidents take place far from medical facilities; many farmers are unaware of the active ingredients of pesticides they use; and it is common to use mixtures of several pesticides. The ICRC is investigating how to deal with these problems.

Import Decisions, Information, and Website

Once a pesticide is included in PIC, the DGD is circulated to all governments who must decide whether to consent to or prohibit its import. Import decisions are posted on the PIC website, and circulated biannually. Governments in exporting countries must ensure that their exporters comply. Of course, many countries are both importers and exporters and under the rules of international trade, a country cannot ban the import of a pesticide that is manufactured and used nationally.

An important tool is the PIC Circular, updated every six months by the Secretariat. Circulated in hard copy and on the website,[5] it includes new bans and severe restrictions, importing country responses, and general progress reports. For the first time, it is easy to access sound information on government regulatory actions, even if these do not meet all the full PIC criteria.

The Convention—More Than PIC

Information exchange is an important principle promoted under Article 14 of the Convention. Developing countries lack resources to undertake extensive evaluations of pesticides and governments are encouraged to share scientific, technical, economic, and legal information on chemicals within the scope of the Convention, as well as other information on their regulatory actions.

BUILDING CAPACITY/IMPROVING REGULATIONS

The process of identifying problem pesticides through PIC will be slow, and there are limitations. In some cases, for example, governments will have no easy substitute, although this may increase the incentive to seek safer and more appropriate alternatives, including Integrated Pest Management strategies.

Financial resources are needed, not only to allow the Secretariat to meet its obligations, but also to ensure that regulators in developing countries can participate in workshops and training sessions. In poorer countries, with competing demands on scarce resources, chemical regulation is not always a priority. The status of an international Convention gives PIC the attention it requires to be effective, and should help attract the necessary funds.

PIC is just one tool, although an important one, in the regulation of pesticides. With good training and additional resources, PIC can play a central role as part of capacity-building initiatives to help governments improve their ability to regulate pesticides, and to look for products and strategies that reduce the dependence on hazardous chemicals.

REFERENCES

1. Rotterdam Convention on the Prior Informed Consent Procedure for Certain Hazardous Chemicals and Pesticides in International Trade, UNEP and FAO, Text and Annexes, January 1999.
2. The signatory countries can be found on the PIC website: http://www.pic.int/. The Convention closed for signatures in September 1999: countries which have not signed accede to, rather than ratify, the Convention, to the same effect.
3. *International Code of Conduct on the Distribution and Use of Pesticides (Amended Version)*; FAO, 1989. The Code is currently being revised and updated.
4. United Nations Conference on Environment and Development, Agenda 21, Chapter 19, Environmentally Sound Management of the Toxic Chemicals, Including Prevention of International Illegal Traffic in Toxic and Dangerous Products, UNEP, Nairobi, 1992.
5. Convention text and PIC website (http://www.pic.int/).

Laws and Regulations: Soil

Ian Hannam
Center for Natural Resources, Department of Infrastructure, Planning and Natural Resources, Sydney, New South Wales, Australia

Ben Boer
School of Law, University of Sydney, Sydney, New South Wales, Australia

Abstract
This entry provides a short background to the current international and national soil law and outlines recent moves to promote reform of soil law including progress made on developing appropriate frameworks for successful management of soil as an ecological element.

INTRODUCTION

At a national level, soil law means a body of law to promote soil conservation enacted by a legislature, e.g., an act, decree, regulation, or other formal legal instrument that is legally enforceable. Soil law, or "soil legislation" as it may also be referred, includes those laws that have primary responsibility for soil conservation, soil and water conservation, and land rehabilitation. They are generally characterized by provisions to mitigate and manage soil erosion and soil degradation and methods to conserve soil resources. Internationally, the legal framework for the conservation of soil can include conventions, protocols, agreements, and covenants, which are expressed to be legally binding. Worldwide, soil law is managed by a variety of legal and institutional systems, which are the individual organizational and operational regimes that have the administrative authority over soil.

WHY LAW FOR SOIL?

Soil bodies are effectively large ecosystems and comprise fundamental components of the earth's biodiversity. Soil is thus seen as the basis for the conservation of terrestrial biological diversity and the sustenance of all terrestrial organisms, including people. The ongoing and widespread soil degradation as a result of human use of soil provides the imperative for enactment of soil law. The ever-increasing demand for food by rapidly growing populations in many countries in the past few decades has exerted increasing environmental stress on the soil leading to widespread soil degradation.[1] The following definitions provide the context for soil law.

Soil

Soil forms an integral part of the earth's ecosystems and is situated between the earth's surface and bedrock. It is subdivided into successive horizontal layers with specific physical, chemical, and biological characteristics. From the standpoint of history of soil use, and from an ecological and environmental point of view, the concept of soil also embraces porous sedimentary rocks and other permeable materials together with the water that these contain and the reserves of underground water.[2]

Soil Degradation

Soil degradation is a loss or reduction of soil functions or soil uses. It includes aspects of physical, chemical, and biological deterioration, including loss of organic matter, decline in soil fertility, decline in structural condition, erosion, adverse changes in salinity, acidity, or alkalinity, and the effects of toxic chemicals, pollutants, or excessive flooding.[1]

Sustainable Use of Soil

The sustainable use of soils preserves the balance between the processes of soil formation and soil degradation while maintaining the ecological functions and needs of soil. In this context, the use of soil means the role of soil in the conservation of biological diversity and the maintenance of human life.[3]

INTERNATIONAL LAW AND SOIL

International environmental law is an essential component for setting and implementing global, regional, and national policy on environment and development. There is an increasing recognition of the role of international environmental law to overcome the global problems of soil degradation, including its ability to provide a juridical basis for action by nations and the international community.[4] A number of international and regional instruments introduced in the past 10 years contain elements that can contribute to achieving sustainable use of soil. None are sufficient on their own. Some of the instruments could assist by promoting the management of some of the activities that can control soil degradation. However, this role is not readily apparent except for those that include provisions specifically directed to soil (e.g., see Article IV "Soil"—1968 *African Convention on the Conservation of Nature and Natural Resources*, final revision text adopted by the African Union Assembly on July 11, 2003).

Declarations

A number of nonbinding declarations and charters draw attention to the fact that soil degradation and desertification are reaching alarming proportions and seriously endangering human survival. They call on states to cooperate and develop the tools to conserve soils. Key declarations relevant to soil include the 1972 *Stockholm Declaration on the Human Environment*, the 1981 FAO World Soil Charter, the 1982 *World Charter for Nature*, the 1982 *Nairobi Declaration*, the 1992 *Rio Declaration on Environment and Development*, and the 2002 *Johannesburg Declaration on Sustainable Development*. Also of relevance is the Programme for the Development and Periodic Review of Environmental Law for the First Decade of the 21st Century, known as the Montevideo Programme; this program includes provisions to improve the conservation, rehabilitation, and sustainable use of soils.[5]

International Conventions, Covenants, Treaties, and Agreements

Many multilateral agreements include provisions that could be used to promote sustainable use of soil, but the provisions are generally tangential to the needs of soil as such. Key global instruments relevant to soil include the 1992 *Convention on Biological Diversity*, the 1992 *United Nations Framework Convention on Climate Change* and the 1997 *Kyoto Protocol*, and the 1994 *United Nations Convention to Combat Desertification*. Relevant regional instruments include the 1968 *African Convention on the Conservation of Nature and Natural Resources* (Revised July 2003), the 1985 *ASEAN Agreement on the Conservation of Nature and Natural Resources*, the 1986 *Convention for the Protection of the Natural Resources and Environment of the South Pacific Region*, the 1986 *European Community Council Directive*, the 1995 *Convention Concerning the Protection of the European Alps*, and the 1998 *Protocol for the Implementation of the Alpine Convention of 1991 in the Area of Soil Protection*.[6]

NATIONAL SOIL LAW

Legislation has been used for some 60 years in many countries to control soil degradation problems and to manage soil. A worldwide examination of national legal and institutional frameworks indicates that most countries approach the management of soil in a fragmented manner. The term "soil law" also covers those situations where comprehensive provisions for soil protection and management have been integrated in legislation that protects other aspects of the environment, such as forests, water, biodiversity, and desertification. In general, soil law thus provides for farm planning, implementation of soil erosion control measures, establishing community groups, planning catchment schemes, and compliance and enforcement. Some jurisdictions, such as the United Kingdom, have multiple soil legislation mechanisms that cover a broad range of functions including soil planning, access to sensitive land types, organic farming practices, nitrate sensitive areas, and soil restoration. On the other hand, federally organized countries often have a system where each state or province has its own soil legislation and supportive legal mechanisms. Hybrid situations also exist, such as in the People's Republic of China, which has enacted the *Water and Soil Conservation Law 1991* and the *Desertification Law 2002* at a national level, but causes them to be implemented through a comprehensive provincial system of law and regulations.

There is a wide variety of types of legal mechanisms used to protect and manage soil, including acts, decrees, resolutions, ordinances, codes, regulations, circulars, decisions, orders, and bylaws. Whereas these are generally appropriate, many need to be applied in more inventive ways to effectively manage the soil in an ecosystem context.[3]

EFFECTIVENESS OF SOIL LAW

The effectiveness of international and national soil law is generally dependent on two matters: first, the capacity of a legal and institutional framework to manage soil—which is measured by the ability of a legislative and institutional system to achieve sustainable use of soil—and second, by the number and type of essential legal and institutional elements present in a soil statute in a format that enables soil degradation issues to be identified. These need to be backed by the legal, administrative, and technical capability in the particular instrument as a basis of some form of effective action. Capacity is also represented in the form of legal rights, the type of legal mechanisms, and importantly, the number and comprehensiveness of the essential elements and their functional capabilities. Legal and institutional "elements" for soil are the basic, essential components of a legal and institutional system. An individual law can include a number of legal mechanisms in a well thought-out structure that gives an organization the power it needs, through its executive and administrative structure, to address soil degradation. It is also possible that the necessary elements may be distributed among a number of individual laws within a comprehensive national legal and institutional system.[7]

Most key soil management issues are multifactorial (i.e., many include a sociological, a legal, and a scientific component), so it is obvious that generally more than one piece of environmental legislation (along with detailed regulations) and many types of legal and institutional elements will be needed to effectively manage soil degradation issues.[7] Legal and institutional elements can be used to assist in the evaluation of an existing law or legal instrument to determine its capacity to meet certain prescribed standards of performance for the sustainable use of soil. They can also be used to guide the reform of an existing soil law or to develop new legislation for the sustainable use of soil. The manner and degree in which an "essential element" is applied will vary according to the particular type of legal mechanism concerned and its expected role in a particular jurisdiction. For example, an international legal instrument may include a provision for dispute resolution, but the actual implementation of this provision between states might not rely on, or be influenced by, the existence of similar provisions within a law of either of the disputing states.[7]

IUCN COMMISSION ON ENVIRONMENTAL LAW

The Commission on Environmental Law of IUCN (The World Conservation Union) has carried out extensive investigation into the options for a new international instrument focusing on soil. The commission has also identified a variety of ways available for states to approach the task of a detailed legal and institutional analysis and the design of appropriate legal and institutional systems that provides for the effective management of soil. Arising from this work, in which the authors have been centrally involved, two principal strategies can be considered for the development of legal and institutional arrangements for soil. These are:

- A nonregulatory strategy which is characterized by elements for education, participatory approaches, soil management, and incentive schemes
- A regulatory approach that is characterized by statutory soil use plans that prescribe legal limits and targets of soil and land use, issue of licenses or permits to control soil use, and the use of restraining orders and prosecution for failure to follow prescribed standards of sustainable soil use.

These strategies can be approached on a short-term time frame for implementation or a longer-term time frame, which involves substantial reform of existing laws, policies, and institutional and sectoral change.[7]

CONCLUSIONS

Soil law in the past has been neglected at the international level and, in many of the world's regions, at the domestic level. However, the growing recognition of soil degradation as a major international environmental issue in the context of the conservation of biological diversity is gradually being addressed, and this is starting to change attitudes toward the benefits of improved international and national legal and institutions for soil.[8] Soil bodies represent complex terrestrial ecosystems. They require careful management of their ecological characteristics through the medium of soil law at a national and international level. This approach is essential for the long-term sustainable use of soil and to meet the food production requirements of the expanding human population of the world, as well as to meet the needs of all flora and fauna that depend on the soil for sustenance.

REFERENCES

1. Bridges, E.M., Hannam, I.D., Oldeman, L.R., Penning deVries, F., Scherr, S.J., Sombatpanit, S., Eds.; *Response to Land Degradation*; Science Publishers Inc.: Enfield, NH, 2001.
2. Council of Europe European conservation strategy. Recommendations for the 6th European Ministerial Conference on the Environment; Council of Europe: Strasbourg, 1990.
3. Hannam, I.D.; Boer, B.W. *Legal and Institutional Frameworks for Sustainable Soils*; The World Conservation Union: Gland, U.K., 2002.
4. Khan, R. International law of land degradation. In *International Studies. 30:3*; Sage Publications: New Delhi, India, 1993.
5. UNEP. The Montevideo Programme III—The Programme for the Development and Periodic Review of Environmental Law for the First Decade of the 21st Century; 2001 Decision 21/23 of the Governing Council of UNEP, UNEP: Nairobi, February 2001.
6. Sands, P. *Principles of International Environmental Law*; Cambridge University Press: Cambridge, 2003.
7. Boer, B.W.; Hannam, I.D. Legal aspects of sustainable soils: International and National. Rev. Eur. Commun. Int. Environ. Law **2003**, *12* (2), 149–163.
8. WSSD (World Summit on Environment and Development). *A Framework for Action on Agriculture*; WEHAB Working Group: New York, 2002.

Leaching

Lars Bergström
Department of Soil Science, Swedish University of Agricultural Sciences (SLU), Uppsala, Sweden

Abstract
The extent of groundwater contamination mainly depends on the degree to which pesticides leach through the unsaturated zone of soils on which they have been applied. Pesticide movement through the unsaturated zone in tile-drained fields may also be a source of pesticides in surface waters. Knowledge of pesticide movement in soil above the groundwater table is very important. The most important factors influencing pesticide leaching are soil properties, inherent properties of the pesticide molecules, climatic conditions, and management practices.

INTRODUCTION

The contamination of groundwater by pesticides is of concern mainly because it may limit its use as drinking water. The extent of groundwater contamination to a large extent depends on the degree to which pesticides leach through the unsaturated zone of soils on which they have been applied.[1] Pesticide movement through the unsaturated zone in tile-drained fields may also be a source of pesticides in surface waters, which support aquatic ecosystems and are used for drinking water in many areas of the world. Therefore, knowledge of pesticide movement in soil above the groundwater table is very important.[2] This has also been the focus of a large number of studies performed during the past couple of decades.[3,4] The most important factors influencing pesticide leaching are soil properties, inherent properties of the pesticide molecules, climatic conditions, and management practices.[3]

FACTORS INFLUENCING LEACHING

Soil and Hydrological Conditions

The rate and direction of water flow in the unsaturated zone are determined by the hydraulic gradient and the hydraulic conductivity. The presence of air-filled pores restricts the pathways through which water percolates downward, which means that the hydraulic conductivity in the unsaturated zone also depends on the level of water saturation. As the soil dries out, water becomes more strongly bound within the matrix of the soil, and the volume of water and the rate with which water percolates through soil decrease. This relationship between water retention and hydraulic conductivity varies considerably with soil type (soil texture and structure, and organic matter content).[5]

A complication, which has a major impact on pesticide leaching, is the fact that water, and pesticides dissolved in the water phase, often move through large pores in soil (e.g., earthworm and root channels, cracks etc.), a process commonly referred to as preferential flow.[6,7] Under such conditions, an equilibrium pesticide concentration throughout the soil profile cannot be obtained. This phenomenon primarily occurs in fine-textured soils with high clay contents, especially those that have the potential to swell and shrink. Through preferential flow, pesticides can be transported rapidly through large portions of the unsaturated zone and bypass biologically active layers in which they otherwise would be degraded or sorbed. Exposure to preferential flow is most pronounced soon after application of the pesticide, when high concentrations occur in the soil solution in upper soil layers, in combination with intensive rainfall.[8] Once the pesticide is mixed in with the soil matrix, water moving through preferential flow paths does not interact with the soil, and leaching is therefore reduced. In other words, preferential flow can both increase and decrease pesticide leaching depending on the time when it occurs in relation to pesticide application. The final result is that (although transient flow peaks shortly after application causing elevated concentrations in water leaching through soils in preferential flow paths) the leaching loads over extended periods are typically quite small in such soils. Indeed, leaching loads are often larger in sandy soils, in which water and pesticide movement mainly occurs between individual soil particles within the main soil matrix.[8] Pesticide concentrations are typically lower in sandy soils than in clay soils, but the water volumes displacing the pesticide are often much larger in sandy soils. The principal difference in pesticide leaching patterns in sand and clay soils is illustrated in Fig. 1. Irrespective of which leaching mechanism prevails, the total amount of the majority of pesticides that reach groundwater after normal agricultural use rarely exceeds 1% of the applied amount and is commonly well below 0.1%.

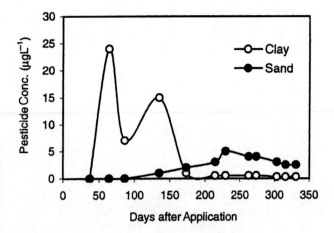

Fig. 1 Concentrations of a pesticide in water leaching from 1-m undisturbed soil columns of a clay and a sandy soil.
Source: Modified from Bergström and Stenström.[8]

Pesticide Properties

The physicochemical properties of pesticides have a major impact on their leachability. In this context, the rate with which they are degraded and how strongly they are sorbed to soil are the most important factors. As a general rule of thumb, leaching decreases with increasing sorption affinity and faster degradation, and increases when the opposite conditions prevail. However, it is important to note that degradation rates often become slower with residence time in the soil as a result of decreased availability due to sorption in the soil. This means that strongly sorbed pesticides, which are less mobile in soil than weakly sorbed compounds, are typically quite persistent. An example of a leaching classification scheme, based on the sorption strength of some pesticides, is shown in Table 1.

A factor that complicates the picture of pesticide movement in soil in relation to sorption affinity is the possibility that pesticide mobility is enhanced by adsorption to various mobile colloids, a process often referred to as "facilitated transport." It is known that organic solutes, such as nonpolar pesticides with very low water solubilities, can form complexes with dissolved organic carbon and clay colloids that move through soil. Even though the role of colloids in facilitating pesticide transport is still relatively poorly understood, there is little doubt that failure to account for this mode of transport can lead to underestimates of both amounts and distances that strongly sorbed pesticides may migrate through the unsaturated zone.

Climatic Conditions

The amount and intensity of precipitation are the most important climatic factors influencing pesticide leaching. Water, in excess of evapotranspiration and what is required to maintain field capacity (i.e., the water content in soil when it is freely drained), leaches through the unsaturated zone and can thereby potentially move pesticides to groundwater. As mentioned above, in clay soils, it is primarily high intensity rainfall soon after application that may displace pesticides to depth in soil. In other words, the timing of precipitation is critical.

Soil temperature also has impact on leaching of pesticides, mainly by influencing the persistence of pesticides in soil and by affecting flow processes. Up to a certain level, degradation rates increase with increasing temperatures, which means that less of the compound will be available for leaching. Increasing temperatures will also increase evapotranspiration rates, which will reduce the amounts of water that can potentially move pesticides downward in soil. In climates with subzero temperatures in the winter season, soil will be frozen during extended periods. Under such conditions, pesticide movement in soil is very restricted, if it occurs at all. Leaching then will occur mainly during autumn and spring, when the soil is unfrozen and the evapotranspiration demand is low. Temperatures will also indirectly affect pesticide leaching by affecting the sorption/desorption process, although this influence is not yet thoroughly investigated and, therefore, less well recognized.

Management Practices and Strategies to Prevent Pollution

Management practices that have a major impact on the amount of pesticides that can move through soils can be grouped into the following categories: cropping/tillage, irrigation, and pesticide application practices.

Due to increased concern over soil erosion and input of pesticides to rivers and lakes, agricultural practices with reduced tillage or no-till management have been introduced. Such practices will also affect water infiltration rates, and therefore pesticide leaching through the unsaturated zone. In the short term, reduced tillage may decrease soil permeability compared with a conventionally tilled soil. However, over a whole growing season, infiltration rates tend to be higher under reduced tillage, especially in clay soils. This is largely due to the fact that reductions in tillage lead

Table 1 Classification of pesticide mobility in soil based on their sorption strength, which in this case is expressed by their K_{oc} values (soil sorption coefficient, normalized to the soil organic carbon content).

K_{oc} value	Expected mobility	Type of pesticide
0–50	Very high	Bentazone, Dicamba
50–150	High	Atrazine, 2,4-D
150–500	Medium	Simazine, Metolachlor
500–2000	Low	Lindane, Linuron
2000–5000	Very low	Phenmedipham, Fenpropimorph
>5000	Immobile	DDT, Paraquat

Source: Modified from Torstensson.[9]

to less disruption of macropores in which pesticides can be rapidly transported through soil (see above). Reduced tillage also leaves more crop residues on the soil surface and contributes to reducing compaction of the subsoil caused by heavy equipment; both tend to increase permeability, and thus pesticide leaching. From the standpoint of reducing pesticide leaching, reduced tillage is therefore, in most cases, not a good management option.

As expected, irrigation increases leaching of pesticides by increasing the amount of water that potentially can move through soil. The amount of water, the rate at which it is applied, and the timing of irrigation are important for the same reasons as discussed previously for precipitation. Different irrigation methods (e.g., sprinkler, and drip and furrow irrigation) have also been shown to affect leaching.

Pesticide application strategies that influence pesticide residue levels in soil and thereby potential leaching include pre- vs. postapplication, split applications, placement methods, and use of different pesticide formulations. However, their influence on pesticide leaching is quite unclear and often overshadowed by other factors. Nevertheless, available data indicate that dividing the dose into two applications instead of one tends to reduce pesticide concentrations and the depth of migration in the subsoil. For similar reasons, pesticide leaching can be restricted by use of "slow-release" formulations in which the active ingredient is mixed with a solid matrix from which it gradually diffuses into the soil over an extended period. Placement of the pesticide instead of broadcasting, which reduces the soil surface area to which the pesticide is applied, also tends to reduce pesticide leaching.

FUTURE CONCERNS

Leaching of pesticides will undoubtedly continue to be of concern in the foreseeable future, and something that will be considered in various regulatory assessment schemes. In this context, it is important not only to look at the leachability, but also to evaluate the risks associated with leaching and the occurrence of pesticides in groundwater both from a human health and an ecotoxicological point of view.

In the future, there is reason to believe that fewer toxic compounds will be allowed, especially those that show high leachability in soil. There is also reason to believe that, in line with the increasing awareness of problems associated with leaching of pesticides, improved management strategies will be developed that reduce pesticide leaching further.

REFERENCES

1. Enfield, C.G.; Yates, S.R. Organic Chemical Transport to Groundwater. In *Pesticides in the Soil Environment: Processes, Impacts, and Modeling*; Cheng, H.H. Ed.; Soil Science Society of America Book Series: Madison, WI, 1990; 2, 271–302.
2. *The Lysimeter Concept—Environmental Fate of Pesticides*; Führ, F.; Hance, R.J.; Plimmer, J.R.; Nelson, J.O., Eds. ACS Symposium Series: Washington, DC, 1998; 699, 284.
3. Barbash, J.E.; Resek, E.A. *Pesticides in Ground Water—Distribution, Trends, and Governing Factors*; Ann Arbor Press: Chelsea, MI, 1996; 588.
4. Flury, M. Experimental evidence of transport of pesticides through field soils—a review. J. Environ. Qual. **1996**, *25* (1), 25–45.
5. Carter, A.D. Leaching Mechanisms. In *Pesticide Chemistry and Bioscience—The Food–Environment Challenge*; Brooks, G.T., Roberts, T.R., Eds.; Royal Society of Chemistry: Milton Road, U.K., 1999; 8, 291–301.
6. Bergström, L.F.; Jarvis, N.J. Leaching of dichlorprop, bentazon, and 36Cl in undisturbed field lysimeters of different agricultural soils. Weed Sci. **1993**, *41* (2), 251–261.
7. Brown, C.D.; Carter, A.D.; Hollis, J.M. Soils and Pesticide Mobility. In *Environmental Behaviour of Agrochemicals* Roberts, T.R., Kearney, P.C., Eds.; John Wiley and Sons: Chichester, England, 1995; 9, 131–184.
8. Bergström, L.; Stenström, J. Environmental fate of pesticides in soil. Ambio. **1998**, *27* (1), 16–23.
9. Torstensson, L. Kemiska Bekämpningsmedel—Transport, Bindning och Nedbrytning i Marken. In *Aktuellt från Sveriges Lantbruksuniversitet*; Swedish University of Agricultural Sciences: Uppsala, Sweden, 1987; 357, 36, (in Swedish).

Leaching and Illuviation

Lynn E. Moody
Earth and Soil Sciences Department, California Polytechnic State University, San Luis Obispo, California, U.S.A.

Abstract

Soils provide essential ecological services, including the absorption, filtration, storage, and transmission of water. Maintaining soil quality is essential for these services to continue to be provided, and understanding how soils form and function in the ecosystem is essential for maintenance of soil quality. Among the many soil-forming processes are translocations of materials through soils. Translocation processes include leaching and the paired processes of eluviation and illuviation. As well as natural materials—solutes and particles produced by weathering—there are many contaminants that enter soil by natural or anthropogenic means. Contaminants either translocate through soils as solutes or adsorbed onto soil mineral and organic particles by various physicochemical reactions. These physicochemical reactions include outer-sphere and inner-sphere adsorption and chelation of metals by dissolved organic matter. Contaminants fastened to soil particles are effectively removed from the soil solution and are, thus, not reactive. Desorption, however, enables contaminants to become biologically active. Recent and current research involves quantification of sorption behavior of various contaminants, determining the mechanisms and kinetics of adsorption and desorption reactions, as well as the soil and environmental conditions conducive to adsorption and desorption of contaminants. Thus, environmental quality relates to leaching and illuviation, as the processes by which contaminants move through soil.

INTRODUCTION

Soils affect ecosystem composition and function and are, therefore, fundamental to every terrestrial ecosystem and many aquatic ecosystems as well. Soils absorb, filter, store, and transmit water; provide a substrate and nutrients for plants; serve as habitat for animals and microorganisms; store carbon; and cycle wastes and organic matter. These ecological services of soils create an imperative to maintain and enhance the quality or "health" of soils. This imperative is an essential consideration in any action plan regarding environmental quality and management.

The dynamic system of soils includes various types of translocation processes, involving movements of materials in aqueous solution or suspension, from one part of the solum into another part of the solum. Organic and inorganic materials may be produced in or liberated from the upper part of a soil profile by physical and biogeochemical weathering or added to the soil surface either accidentally or deliberately. Many of these materials subsequently are translocated deeper in the profile. The dominant translocation process in a soil profile depends partially on the nature of the contaminant (inorganic or organic, for example). At a given stage of soil development, one process may be dominant or several may operate simultaneously or sequentially. Percolating water is the mechanism of translocation, and rates and intensities of translocation processes are controlled by the factors that regulate water percolation: climate, topography, and tortuosity and size of soil pores. Pore size and tortuosity usually are quantified and expressed as hydraulic conductivity.

In addition to natural materials such as weathering products, contaminants and nutrients supplied by fertilization may be translocated; any material that can be dissolved or suspended can travel through soils in the percolating water. The issue becomes the likelihood that contaminants either stay in the upper part of the soil profile or reach the groundwater, and if they enter groundwater, the rate at which they do so.

The processes of leaching and eluviation/illuviation affect environmental quality and management because, in concert with soil chemical properties, they control the mobility of contaminants. This entry divides translocations into leaching (translocations of solutes), including a discussion of podzolization (a specific case of leaching of organometallic complexes), and the paired processes of eluviation and illuviation (translocations of suspended particles).[1] Each major process discussion is further subdivided into, first, the natural processes related to soil formation and, second, the environmental implications of each process. By understanding the natural processes of leaching and illuviation, we can also better understand how contaminants move through or are retained in soils.

LEACHING

Leaching is the translocation of solutes. Some authorities specify that leaching is the removal of solutes entirely out of the solum, representing a loss of materials from the soil profile,[2] but to many experts, on the other hand, leaching includes the translocation of solutes within the solum.[3] In a natural system, solutes arise mainly by biogeochemical weathering of minerals and decomposition of organic matter, and solutes include ions, complex ions, and ion pairs.

Mineral weatherability depends on several factors of mineral and soil chemistry. It also depends on particle size (smaller → more soluble). If chemical bonds in a mineral are ionic in character, then the ions readily detach from the mineral surface and enter solution,[4] as in the congruent dissolution of halite into $Na^+(aq)$ and $Cl^-(aq)$. Chemical bonds with some covalent character are less easily disrupted. Because they generally require a strong polarizer such as H^+,[5] dissolution proceeds less readily, as in the incongruent weathering of feldspar:

$$2KAlSi_3O_8 + 11H_2O + 2CO_2 \rightarrow 2K^+(aq) + Al_2Si_2O_5(OH)_2 + 2HCO_3^-(aq) + 4H_4SiO_4(aq).$$

In this reaction, particulate kaolinite is produced, as well as several solutes.

The relative weatherability of the common silicate minerals was deduced by Goldich.[6] The ultramafic and mafic silicates olivine, pyroxenes, and Ca-plagioclases are most readily weathered, and quartz is most resistant. Of the common nonsilicates, sodium salts (including sodium carbonates) and chlorides of Ca, Mg, and K are readily soluble in water, sulfates less so, and carbonates are slowly soluble. The solubility of Fe and Mn oxides depends on the oxidation state of the metals; reduction of Fe and Mn in saturated soils causes their oxides to dissolve. As minerals weather, organic matter decomposes, and elements are leached away, the soil unit decreases in mass and volume.[7] Silicate and nonsilicate minerals may contain potential contaminants in the crystal structures, as a major component (barium in barite, $BaSO_4$, for example) or present by isomorphous substitution (nickel for magnesium in serpentine minerals, for example), but these are unlikely to be released in toxic quantities by mineral weathering under natural conditions,[8] with some exceptions discussed below.

Solutes produced by decomposition of organic matter include any substances that are mineralized by microbial activity in the processes of decomposition. Organic matter decomposition likewise depends on several factors, including climate (temperature and moisture), biota, the type of organic matter residue, and soil factors such as texture, structure, drainage class, nutrient status, and salinity. Briefly, soil animals fragment organic matter, and soil microorganisms decompose it, breaking it down to simpler compounds, mineralizing and releasing plant-available nutrients and evolving CO_2 (in most soil systems).

When chemical conditions in the soil are suitable, solutes precipitate, usually deeper than they originate. One or more of several conditions, including the following, may initiate precipitation of solutes. Desiccation, the removal of water by evaporation or sorption by roots, may cause precipitation of soluble salts, gypsum, carbonates, or silica. Reduction in CO_2 partial pressure below the zone of maximum biological activity is an additional cause of pedogenic carbonate precipitation. Since the solubility of many minerals is partially dependent on pH, a change in soil pH may also cause precipitation. For example, pH determines solubility of the various polymorphs of $Al(OH)_3$ and $AlOOH$.[9] Finally, the oxidation state of Mn and Fe determines whether the oxides and oxyhydroxides of these metals dissolve or precipitate. As materials precipitate in the soil, mass is added and volume increases.[7]

Podzolization is a specific term for eluviation and illuviation of Fe and Al chelated by organic matter within the solum. Such leaching may occur in humid climates, acidic soil conditions, and characteristically under coniferous, mixed coniferous–deciduous forests, or ericaceous plant communities. The types of soils formed by podzolization are podzols, generally equivalent to Spodosols in the U.S. taxonomy. The morphological expression of podzolization is a Bs, Bh, or Bhs (spodic) horizon, where the h symbolizes humus and s denotes Al and Fe sesquioxides, often overlain by a highly leached E (eluvial) horizon. A spodic horizon may be cemented by enrichment in humus, Al, Fe, and Si, in which case it is known as ortstein. A placic horizon is a thin, wavy, cemented layer enriched in humus and Fe or Fe and Mn that is often associated with podzols.

Two models have been developed to describe the podzolization process. The first, favored by most authorities, involves the translocation of Al and Fe as organometallic complexes.[10] Fulvic acids, components of humus, are formed during the decomposition of acidic leaf litter, especially litter from conifers and ericaceous (heath) plants. Humification is complex and probably a combination of processes involving the breakdown of lignin and the synthesis of phenols and aromatics from carbohydrates and amino acids.[11] Fulvic acids are water-soluble anions under acidic soil conditions and, thus, mobile with percolating water. They are rich in phenols and carboxyls; the juxtaposition of these functional groups enables the organic matter to chelate trivalent cations Al and Fe. The organometallic complexes are carried downward in the percolating water. In an aerated, acidic environment, the metal cation: organic anion ratio mainly controls the solubility of the organometallic complex. If the cation:anion ratio is low, the complexes are soluble and, thus, mobile; they become less soluble by increases in the cation:anion ratio. Thus, precipitation of the organometallic complexes may be initiated at depth in the solum by shift in the cation:anion ratio brought about by microbial decomposition of the fulvic acids or by saturation of the complexes by greater concentrations of Al and Fe released by weathering. As available chelation sites are filled, the molecules lose polarity, become hydrophobic, and precipitate. Precipitation also may be initiated by desiccation, sorption of the complexes onto mineral surfaces, or flocculation in the presence of divalent cations. As the organic matter eventually decomposes, Al and Fe are released from complexation and are free to form oxyhydroxides or sesquioxides. With the addition of silica, Al can form imogolite and allophane.

A second, contrasting model of podzol formation holds that the dominant translocation process is inorganic and that organic acids are involved mainly in weathering rather than in transport.[12,13] Weathering of primary minerals by carbonic and organic acids at the soil surface releases Al, Fe, and Si. At pH < 5, hydroxyaluminum cations react with silica to form stable and mobile, inorganic "proto-imogolite" complexes. The complexes precipitate in deeper horizons as imogolite and allophane, by desiccation or reaction to higher pH. Evidence supporting this model consists of podzols containing Al predominantly as imogolite, allophane, and proto-imogolite, and Fe as oxides, with Al-fulvic acid chelates sorbed and precipitated on allophane surfaces.[14]

The two models may not be exclusive. Early podzolization in some Greenland soils involved translocation of Al and Fe by inorganic processes, whereas later stages involved their transport in complexes with organic acids.[15] However, most studies of podzols favor the organic model,[16,17] even in early stages of podzolization.[18]

Environmental Implications

Environmental contaminants generally are characterized based on type, such as organic compounds including volatile and semivolatile organic compounds, liquid petroleum, and some pesticides; inorganic contaminants such as heavy metals, radioactive isotopes, and salts; nutrients (considered contaminants where, for example, they are present in greater-than-natural quantities and cause eutrophication); and pathogenic microorganisms (bacteria and viruses). Contaminants may be added to soil accidentally, as in spills, or intentionally as in pesticides or fertilizers. Contaminants may escape from landfills or toxic waste repositories. Pathogenic viruses and bacteria may originate from landfills, feedlots, farmyards, or land application of biosolids and manures.

Isomorphous substitution can determine that some potentially toxic elements be found in common minerals. For example, the metals Ni, Co, and Cr each can substitute for Mg in serpentine, but no toxic effects of these elements in serpentinitic soils have been noted.[9] However, some toxic elements have accumulated in bedrock, and when these rocks weather, the element can be released into the environment with deleterious result. Selenium, found in soils and highly concentrated in agricultural drainage water in the Central Valley of California, originated from weathering of petroleum-rich sedimentary rocks of the Kreyenhagen and Moreno Formations in the adjacent Coast Ranges on the western edge of the Central Valley.[19,20] Selenium released from these rocks by weathering accumulated in the Central Valley soils, was further concentrated in agricultural drainage by repeated irrigation events, and eventually accumulated in constructed wetlands—constructed specifically to receive the drainage—to toxic quantities, causing embryonic defects in waterfowl. Another example is the toxic level of arsenic in shallow groundwater in South and Southeast Asia. The primary source of As was weathering of sulfide minerals in the Himalayas and subsequent transport of the following As-bearing weathering products: Fe oxides, hydroxides, and oxyhydroxides.[21] Subsequent dissolution of the secondary Fe minerals by microbial activity and in saturated conditions released the As in bioactive form into the groundwater.

After a given contaminant has been introduced into the soil, its fate depends on its nature and the nature of the soil. Most contaminants will be sorbed onto particles through one of several types of physicochemical interactions. These interactions include outer-sphere and inner-sphere adsorption, polymerization, surface precipitation, and lattice diffusion; lattice diffusion may progress to isomorphous substitution.[9] These sorption mechanisms (discussed later) effectively remove the contaminants from the soil solution and inhibit leaching of the contaminant into groundwater. This filtering of water is one of the ecological services of soils.

Desorption of a contaminant depends on soil chemical conditions, whereupon it diffuses into the soil solution and is then subject to leaching. After desorption, leaching of the contaminant depends on the homogeneity, porosity, texture, structure, and mineralogy of the soil, all of which affect the hydraulic conductivity of the soil. For example, copper adsorbed onto particulate organic matter entered the soil solution as that organic matter dissolved and then the Cu formed mobile complexes with dissolved organic carbon in podzols in Australia.[22] Arsenic was released from solid soil particles under anaerobic, reducing environments in Southeast Asia; the As then freely leached into aquifers and eventually into surface waters.[23]

Coliphage (*Escherichia coli*) attached to or embedded in solid particles in biosolids from wastewater treatment showed limited tendency to leach through sandy soils in column studies.[24] In contrast, lysimeter studies have demonstrated *E. coli* from land application of cattle manure leached through poorly drained soils under natural rainfall by preferential flow through macropores.[25] Brennan et al.[25] suggest that the higher velocity of water through macropores may shear off bacteria attached to soil particles lining the macropores.

Eluviation and Illuviation

As the finest particles in soils, clays (<0.002 mm = 2 μm diameter) are most susceptible to eluviation (removal) and illuviation (deposition). Fine clays (0.2 μm), in particular, are small enough to be mobile in soil profiles. Larger particles, even silts and very fine sands, may be mobile in some soils, depending on size and geometry of pores and velocity of percolating water.[26] Clays originate in soils by physical reduction of particles to clay size, by chemical transformation or precipitation, or by introduction as wind-transported dust. Once clays are present, their translocation in a soil profile involves mobilization, removal and transport, and deposition in a lower part of the profile.

Clay particles must be detached from the greater soil mass to enter into aqueous suspension. Slaking is physical detachment, by swelling during wetting, dislodgement by the shear of water flow, raindrop impact, or wind shear. Dispersion is chemical detachment that can result from the introduction of waters with low electrolyte concentration, dominance of Na compared to divalent cations on the exchange complex, or high pH. Clay minerals differ in their mobility: of the phyllosilicates, smectites are most easily dispersed, and kaolinite and illite, less so.[27] Once detached, particles are then free to move in percolating water.

Pore space in most soils consists of micropores and macropores. Macropores may be animal burrows, channels left by decayed roots, large interstitial pores between coarse fragments, or spaces between peds. Under saturated conditions or during infiltration of free water from the surface, percolating water carrying detached clays flows preferentially, and with greater velocity, through macropores than through micropores of the soil matrix. Water in the large pores is drawn into drier and microporous fabric by matric suction. The micropores serve a filtering function, and clay particles coat the walls of the large pores and the faces of peds as clay coatings (known as argillans, clay films, clay skins, or clay linings). Phyllosilicates assume a face-to-face configuration, with the c crystallographic axis approximately perpendicular to the ped face or pore wall. The orientation of phyllosilicate particles gives the argillan a strong optical orientation. To the unaided eye, and at low magnification, the argillan appears shiny or waxy. At high magnification of petrographic microscopy, the argillan displays undulatory extinction as the microscope stage is rotated under crossed polarizers.

Pore walls that do not allow appreciable water absorption, such as between rock fragments and sand grains, also develop clay coatings by illuviation. Drying of clay suspensions following downward flow of suspensions draws in the water interface progressively closer to the grain surfaces and to contact points between grains. Clay particle deposition follows the pattern of drying, argillans are thicker at the contact points than away from them, and phyllosilicate particles orient with faces parallel and crystallographic axis perpendicular to the surface.[28] Thus, clay coatings on grains, and as bridges between grains, have a strong optical orientation.

The depth to which the suspension percolates determines the depth at which deposition of clay particles occurs. The wetting front, driven by a potential gradient, stops where and when gravitational, pressure, and matric forces acting on the water reach equilibrium. The water evaporates or is absorbed by roots, and suspended particles are deposited at the depth of water percolation.[29,30]

Soil scientists accept the strongly oriented argillan as evidence of clay illuviation. However, phyllosilicate clays, which are deposited by flocculation (upon encountering high electrolyte concentrations or in the presence of di- and trivalent cations), tend to assume edge-to-edge or edge-to-face configuration and may not develop strong optical orientation. Thus, their illuvial origin may be difficult to identify. Finally, after deposition, argillans may be modified or destroyed by shrink–swell or mixing of soil material by animal burrowing. In addition to argillans, the ratio of fine clay to coarse clay or to total clay provides evidence of clay illuviation, particularly if eluvial horizons have substantially or significantly lower fine clay:coarse clay ratios compared to Bt horizon in the same profile.[31]

Iron and aluminum oxides, hydroxides, and oxyhydroxides readily enter complexes with metals and various ligands under a range of soil conditions. These minerals tend to be of very small size, with correspondingly high amounts of surface area. In a well-drained and aerated soil, they coat phyllosilicate clay particles and are transported with the phyllosilicates. Evidence for this includes Bt horizons with redder hues than eluvial horizons and parent material, and Bt lamellae enriched in oxalate- and dithionite-extractable Fe and Al along with the higher clay contents.[32]

Environmental Implications

Outer-sphere complexation involves an ion with a hydration shell being attracted to a charged site on a particle surface. An example of this is the attraction of hydrated magnesium cation to a negatively charged site on a phyllosilicate particle surface. Many phyllosilicates tend to develop permanent, negatively charged sites via isomorphous substitution (for example, Al^{3+} for Si^{4+} in tetrahedral sheets). Other colloids, such as iron and aluminum oxides, kaolinite, and humus, develop negative charges dependent on soil pH (higher pH corresponds with greater magnitude of negative charge, hence greater cation exchange capacity). These outer-sphere complexation reactions are reversible, and cations held in this manner tend to be relatively easily exchanged. Since many cations are plant nutrients, our agricultural systems depend on cation exchange. In addition, some positively charged contaminants may be exchangeable. Some organic contaminants, such as the herbicide paraquat, are adsorbed onto negatively charged phyllosilicates. The strength of adsorption varies from weak to very strong and depends on the compound and on soil pH.[1] Low pH encourages protonation of some organic compounds and thus increases the positive charge. A small percentage of heavy metals from sewage sludge applications are exchangeable in soils, although most tend to form complexes with organic matter or with carbonates and iron oxides.[33] Radioactive strontium (^{90}Sr) behaves in soil as does calcium and undergoes cation exchange. As such, ^{90}Sr can be absorbed by forage plants and enter into dairy products.[1] Radioactive cesium (^{137}Cs) also is a cation and is exchangeable in some soils, but in smectitic and vermiculitic soils, it tends to become "fixed" into the interlayer position as is potassium. When fixed in this way, the ^{137}Cs is immobile and unavailable to plants or to leaching into groundwater.[1]

Inner-sphere adsorption involves loss of the hydration shell, and covalent and/or hydrogen bonding of the contaminant to a functional group on a particle surface. For example, As(V), as arsenate (AsO_4^{3-}), or Cr(VI), as chromate (CrO_4^{2-}), forms an inner-sphere complex on goethite surfaces.[21] Many contaminants behave similarly, including the metals Cd, Co, Cu, and Pb, especially at low pH values,[8] Se as selenite (SeO_3^{2-}),[34] and U as uranyl on ferrihydrite surfaces.[35] Copper also adsorbs onto organic matter.[22] Boron, as $B(OH)_3$ or $B(OH)_4$, adsorbs readily onto poorly crystalline Fe and Al hydroxides;[8] adsorption is maximized at pH 8.5.[36] While adsorbed onto soil particles as inner-sphere complexes, these contaminants are effectively isolated from plant uptake or from leaching into groundwater as solutes. Desorption of contaminants, as a result of decomposition of the organic particles or dissolution of the iron or aluminum oxide or hydroxide, would then release the contaminant into the soil solution and mobilize the contaminant.

CONCLUSIONS

The ecological services provided by soils include water absorption, filtration, and transmission. Because the processes of leaching and eluviation/illuviation are highly dependent on water movement into and through the soil, any activity or land use that interrupts water infiltration and percolation will affect those processes. Contaminants adsorbed onto soil particles are removed from the soil solution and effectively rendered inactive. Sorption can take place by several possible mechanisms, and resulting complexes can involve phyllosilicate clays; iron and aluminum oxides, hydroxides, and oxyhydroxides; and particulate organic matter. Dissolved organic matter also can form complexes with metal ions. Desorption by various processes allows the contaminants to enter the soil solution whereupon they are mobile and potentially bioactive. Desorption can take place under saturated (reducing) soil conditions, as secondary iron minerals dissolve. The movement of contaminants, either in solution or attached to illuviating particles, takes place mostly through macropores. Much recent and current research on this topic involves quantification of sorption behavior of various contaminants, determining the kinetics of the various adsorption and desorption reactions, the mechanisms, and the soil and environmental conditions conducive to adsorption and desorption of contaminants.

REFERENCES

1. Brady, N.C.; Weil, R.R. *The Nature and Properties of Soils*, 12th Ed.; Prentice Hall: Upper Saddle River, NJ, 1999.
2. Buol, S.W.; Hole, F.D.; McCracken, R.J.; Southard, R.J. *Soil Genesis and Classification*, 4th Ed.; Iowa State University Press: Ames, 1997.
3. Tan, K.H. *Environmental Soil Science*; Marcel Dekker: New York, 1994.
4. Sposito, G. *The Chemistry of Soils*; Oxford University Press: New York, 1989.
5. Malmström, M.E.; Destouni, G.; Banwart, S.A.; Strömberg, B.H.E. Resolving the scale-dependence of mineral weathering rates. Environ. Sci. Technol. **2000**, *34*, 1375–1378.
6. Goldich, S.S. A study in rock weathering. J. Geol. **1938**, *46*, 17–58.
7. Chadwick, O.A.; Nettleton, W.D. Quantitative relationships between net volume change and fabric properties during soil evolution. In *Soil Micromorphology: Studies in Management and Genesis*; Ringrose-Voase, A.J., Humphreys, G.S., Eds.; Developments in Soil Science; Elsevier: Amsterdam, 1994; Vol. 22, 353–359.
8. Sparks, D.L. *Environmental Soil Chemistry*; Academic Press, Elsevier: Amsterdam, 2003.
9. White, G.N.; Dixon, J.B. Kaolin-serpentine minerals. In *Soil Mineralogy with Environmental Applications*; Dixon, J.B., Schulze, D.G., Eds.; Soil Science Society of America Book Series Number 7; Soil Science Society of America, Inc.: Madison, WI, 2002; 389–414.
10. DeConinck, F. Major mechanisms in formation of spodic horizons. Geoderma **1980**, *24*, 101–123.
11. Oades, J.M. An introduction to organic matter in mineral soils. In *Minerals in Soil Environments*; Dixon, J.B., Weed, S.B., Eds.; Soil Science Society of America: Madison, Wisconsin, 1989; 89–159.
12. Farmer, V.C.; Russell, J.D.; Berrow, M.L. Imogolite and proto-imogolite in spodic horizons: Evidence for a mobile aluminum silicate complex in podzol formation. J. Soil Sci. **1980**, *31*, 673–684.
13. Farmer, V.C. Significance of the presence of allophane and imogolite in podzol Bs horizons for podzolization mechanisms: A review. Soil Sci. Plant Nutr. **1982**, *28*, 571–578.
14. Anderson, H.A.; Berrow, M.L.; Farmer, V.C.; Hepburn, A.; Russell, J.D.; Walker, A.D. A reassessment of podzol formation processes. J. Soil Sci. **1982**, *33*, 125–136.
15. Jakobsen, B.H. Multiple processes in the formation of subarctic podzols in Greenland. Soil Sci. **1991**, *152*, 414–426.
16. Barrett, L.R.; Schaetzel, R.J. An examination of podzolization near Lake Michigan using chronofunctions. Can. J. Soil Sci. **1992**, *72*, 526–541.

17. Gustafsson, J.P.; Bhattacharya, P.; Bain, D.C.; Fraser, A.R.; McHardy, W.J. Podzolisation mechanisms and the synthesis of imogolite in Northern Scandinavia. Geoderma **1995**, *66*, 167–194.
18. Certini, G.; Ugolini, F.C.; Corti, G.; Agnelli, A. Early stages of podzolization under Corsican pine (*Pinus nigra* Arn. ssp. *laricio*). Geoderma **1998**, *83*, 103.
19. Johnson, C.L.; Graham, S.A. Middle tertiary stratigraphic sequences of the San Joaquin Basin, California (Ch. 6). In *Petroleum Systems and Geologic Assessment of Oil and Gas in the San Joaquin Basin Province, California*; Scheirer, A.H., Ed.; U.S. Geological Survey Professional Paper 1713, 2007.
20. Scheirer, A.H.; Magoon, L.B. Age, distribution, and stratigraphic relationship of rock units in the San Joaquin Basin Province, California (Ch. 5). In *Petroleum Systems and Geologic Assessment of Oil and Gas in the San Joaquin Basin Province, California*; Scheirer, A.H., Ed.; U.S. Geological Survey Professional Paper 1713, 2007.
21. Fendorf, S.; Michael, H.A.; van Geen, A. Spatial and temporal variations of groundwater arsenic in South and Southeast Asia. Science **2010**, *328*, 1123–1127.
22. Burton, E.D.; Phillips, I.R.; Hawker, D.W.; Lamb, D.T. Copper behaviour in a podosol. 2. Sorption reversibility, geochemical partitioning, and column leaching. Aust. J. Soil Res. **2005**, *43*, 503–513.
23. Polizzotto, M.L.; Kocar, B.D.; Benner, S.G.; Sampson, M.; Fendorf, S. Near-surface wetland sediments as a source of arsenic release to ground water in Asia. Nature **2008**, *454*, 505–509.
24. Chetochine, A.S.; Brusseau, M.L.; Gerba, C.P.; Pepper, I.L. Leaching of phage from Class B biosolids and potential transport through soil. Appl. Environ. Microbiol. **2006**, *72*, 667–671.
25. Brennan, F.P.; O'Flaherty, V.; Kramers, G.; Grant, J.; Richards, K.G. Long-term persistence and leaching of *Escherichia coli* in temperate maritime soils. Appl. Environ. Microbiol. **2010**, *76*, 1449–1455.
26. Nettleton, W.D.; Brasher, B.R.; Baumer, O.W.; Darmody, R.G. Silt flow in soils. In *Soil Micromorphology: Studies in Management and Genesis*; Ringrose-Voase, A.J., Humphreys, G.S., Eds.; Developments in Soil Science; Elsevier: Amsterdam, 1994; Vol. 22, 361–371.
27. Stern, R.; Ben-Hur, M.; Shainberg, I. Clay mineralogy effect on rain infiltration, seal formation, and soil losses. Soil Sci. **1991**, *152*, 455–462.
28. Sullivan, L.A. Clay coating formation on impermeable materials: Deposition by suspension retention. In *Soil Micromorphology: Studies in Management and Genesis*; Ringrose-Voase, A.J., Humphreys, G.S., Eds.; Developments in Soil Science; Elsevier: Amsterdam, 1994; Vol. 22, 373–380.
29. Dijkerman, J.M.; Cline, M.G.; Olson, G.W. Properties and genesis of textural subsoil lamellae. Soil Sci. **1967**, *104*, 7–16.
30. Moody, L.E.; Graham, R.C. Pedogenic processes in thick sand deposits on a marine terrace, central California. In *Whole Regolith Pedology*; Cremeens, D.L., Brown, R.B., Huddleston, J.H., Eds.; Special Pub.; Soil Science Society of America: Madison, Wisconsin, 1994; Vol. 34, 41–55.
31. Calero, N.V.; Barron, V.; Torrent, J. Water dispersible clay in calcareous soils of Southwestern Spain. Catena **2008**, 22–30.
32. Moody, L.E.; Graham, R.C. Geomorphic and pedogenic evolution in coastal sediments, central California. Geoderma, **1995**, *67*, 181–201.
33. Chang, A.C.; Page, A.L.; Warneke, J.E.; Grgurevic, E. Sequential extraction of soil heavy metals following a sludge application. J. Environ. Qual. **1984**, *13*, 33–38.
34. Coppin, F.; Chabroullet, C.; Martin-Garin, A. Selenite interactions with some particulate organic and mineral fractions isolated from a natural grassland soil. Eur. J. Soil Sci. **2009**, *60*, 369–376.
35. Stewart, B.D.; Nico, P.S.; Fendorf, S. Stability of uranium incorporated into Fe (hydr)oxides under fluctuating redox conditions. Environ. Sci. Technol. **2009**, *43*, 4922–4927.
36. Communar, G.; Keren, R.; Li, F.H. Deriving boron adsorption isotherms from soil column displacement experiments. Soil Sci. Soc. Am. J. **2004**, *68*, 481–488.

Lead: Ecotoxicology

Sven Erik Jørgensen
Institute A, Section of Environmental Chemistry, Copenhagen University, Copenhagen, Denmark

Abstract

Lead is dispersed globally mainly due to the combustion of leaded gasoline, which has now been phased out in industrialized countries. The use of leaded ammunition has also been banned in most industrialized countries. The global dispersion of lead can be seen by the concentration of lead in the glacial ice on Greenland, having increased 10-fold during the last 250 years. Lead can be found everywhere in the ecosphere, including in food items. The regional lead concentrations in river water may often be several hundred times higher than the concentrations found in the open sea. One of the consequences of the application of leaded gasoline has been that the lead concentration in the blood of people living in big towns is significantly higher than that in the blood of a rural population. The elevated lead concentration in the blood of humans has two sources: atmospheric pollution and the contamination of food items. The two sources have a different uptake efficiency of respectively approximately 50% and 10%. The lead uptake from food and the atmosphere by humans has declined during the last 25 years due to the phasing out of leaded gasoline. The spectrum of toxicological and ecotoxicological effects of lead is very wide. The pollution abatement methods and the integrated environmental management of lead are shortly discussed with references to other entries.

INTRODUCTION: DISPERSION AND APPLICATION

Lead is found in all environmental components—both living and non-living. The use of lead has dispersed the metal worldwide due to its long-term use in gasoline, batteries, solders, pigments, ammunition, paint, ceramic, and even piping. It is found for instance on the glacial ice and snow of Greenland, one of the most uncontaminated places on the Earth. Many toxic substances are generally widely dispersed and a global increase in the concentration of heavy metals and pesticides has been recorded, as exemplified for lead in Fig. 1. The concentration of lead is shown in micrograms per ton (µg/t) as a function of time, which can be found by the use of analyses of ice cores. As it can be seen, the lead concentration has, since the mid-eighth century, increased 10 times—from about 10 µg/t to about 200 µg/t of snow. The dispersion of lead is caused by the many uses of this metal: in mining and smelting, in batteries, in lead-based paints, in electronic devices, in leaded gasoline, and in shots applied in hunting and target shooting, The last two applications were phased out in most industrialized countries more than 25 years ago, but lead gasoline is still in use in many developing countries. In the United States, the manufacture of batteries is the dominant use of lead today.[1] The global dispersion of lead is particularly caused by the combustion of leaded gasoline, which today is inconceivable because the organic lead chemicals were applied in the gasoline to obtain a sufficiently high octane number at the lowest cost, and the less harmful alternatives could only have increased the gasoline price by about a quarter of an American cent per liter.

The concentrations of lead in food items are shown in Table 1 to illustrate the presence of lead in our food—another illustration of the consequences of the global dispersion of lead. The concentrations in the table are taken from the mid-1980s, i.e., before the introduction of lead-free gasoline had shown any significant effect. The differences between the three countries are explained by the differences in the traffic (and population) densities.

The concentration of lead in completely uncontaminated water is about 1 ng/L, while concentrations of 20 ng/L are often found when only minor discharge of lead has taken place. In contaminated and very contaminated water, a lead concentration of 100–200 ng/L is often found.

Heavy metals are dispersed globally, but the regional concentrations of most heavy metals may of course be much higher regionally than globally. The relationship between a global and a regional pollution problem and the role of dilution for this relationship are illustrated in Table 2, where the ratios of heavy metal concentrations in the River Rhine and the North Sea are shown. Notice that the amount of nickel and lead used in the region of the River Rhine is the same but the ratio is 70 times higher for lead than for nickel due to the application of leaded gasoline, which disperses the lead uncontrolled, while nickel has more closed applications, which allow recycling of the metal. Notice also that lead is transported in the atmosphere primarily in the particulate phase.[4]

Table 1 Lead in food (see Jørgensen[2]).

Food items	Typical lead concentration (mg/kg fresh weight)		
	England	Holland	Denmark
Milk	0.03	0.02	0.005
Cheese	0.10	0.12	0.05
Meat	0.05	<0.10	<0.10
Fish	0.27	0.18	0.10
Eggs	0.11	0.12	0.06
Butter	0.06	0.02	0.02
Oil	0.10	–	–
Corn	0.16	0.045	0.05
Potatoes	0.03	0.1	0.05
Vegetables	0.24	0.065	0.15
Fruits	0.12	0.085	0.05
Sugar	–	0.01	0.01
Soft drinks	0.12	0.13	–

Table 2 Heavy metal pollution in the River Rhine (from 1985) (see Jørgensen[2]).

	River Rhine (t/yr)	Ratio: conc. in the Rhine/ conc. in the North Sea
Cr	1,000	20
Ni	2,000	10
Zn	20,000	40
Cu	200	40
Hg	100	20
Pb	2,000	700

ECOTOXICITY AND ENVIRONMENTAL PROBLEMS OF LEAD

The toxicity of Pb is mainly associated with the free ions Pb^{2+} (the +2 oxidation state), but lead can form complexes with hydroxide ions, carbonate, chloride ions, and many organic compounds, for instance, humic acid and amino acids. The toxicity of the complexes is generally lower than the toxicity of the free ions, because the uptake of the complexes is slower than the uptake of the free ions. It implies that it is necessary in every case study to determine by analytical methods or chemical calculation the concentrations of lead as free ions and in the form of the various complexes to determine the toxicity. The toxicity of lead declines with higher concentrations of the hardness ions, calcium, and magnesium. Lead shows, as other heavy metals, bioaccumulation and biomagnification, which is more pronounced at lower pH, because the solubility and the relative concentration of the free ions are increasing with decreasing pH.

Contaminated aquatic ecosystems have a significantly elevated concentration of lead in the sediment. Generally, the sediment has higher concentrations of heavy metals than the water, and as it is possible to analyze sediment core,[5,6] it is possible to find the contamination of heavy metals as functions of the time, provided the settling rate is known or can be estimated. This is particularly informative in the case of lead, because the use of leaded gasoline started shortly after the Second World War and was banned in industrialized countries before or around the mid-1980s. It entails that the sediment, from the approximately 40 years when leaded gasoline was used, will show a particularly high lead concentration, which of course facilitates the dating of the sediment. The lead concentration in sediment is usually 10–200 mg/kg dry weight, but as much as 3000–10,000 mg/kg dry weight can be found in contaminated areas.

A filter feeding bivalve mollusc shows a contamination of lead (and other heavy metals) that is proportional to the concentration in the sediment. The proportional constant is dependent on the composition of the sediment, but it is frequently between 0.01 and 0.05—the highest values for sediment with a high concentration of organic matter.[7]

A major source of lead exposure and toxicity for wild birds is the ingestion of lead-based ammunition. For birds, concentration of lead is in the order of 0.2 mg/kg dry weight, while a toxic effect would correspond to 100 times as much and death to 250 times as much.[5] LD_{50} for rats is 130 mg/kg dry weight (see *Inorganic Compounds: Eco-Toxicity*, p. 1479). There is a primary lead shot poisoning from direct ingestion of lead-based ammunition and a secondary lead shot poisoning when birds (and of course also other animals) ingest lead shotgun pellets and bullet fragment embedded in the flesh of dead or wounded animals shot with lead-based ammunition.

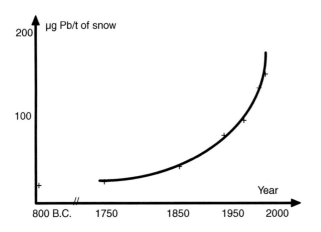

Fig. 1 The lead concentration in the glacial ice, Greenland, as a function of time, from 800 B.C. to year 2000. The lead concentration has increased tenfold during the last 250 years, mainly due to combustion of leaded gasoline.
Source: After Jørgensen[2] and Chemistry.[3]

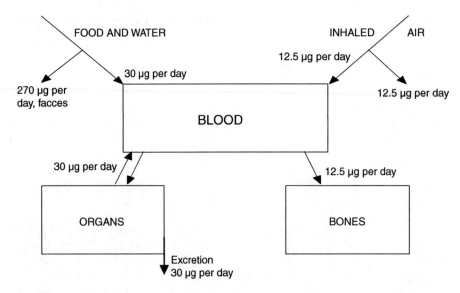

Fig. 2 A steady-state model of an average European in 1985. The uptake from food was 30 μg/day out of 300 μg/day (10% efficiency) and 12.5 μg lead per day by respiration (uptake efficiency, 50%). The excretion balances the uptake from the food, and the lead taken up by respiration is accumulated in the bones.

Lead is bound to SH- groups in the proteins and can therefore generally be taken up by plants with high concentrations of proteins more effectively than by plants with low protein concentrations; see also the entry about "Bioremediation." Removal of lead from areas that have been applied for target shooting with ammunition containing lead is possible by bioremediation.[8]

With other heavy metals, the uptake of lead from food is relatively low—about 7%–10%,[9,10] while lead is taken up from the atmosphere by the lungs with a higher efficiency. Fig. 2 shows a steady-state model of the uptake for an average European in the mid-1980s, just before the use of leaded gasoline was banned for all new cars. The uptake from food is as seen 10%, namely, 30 μg/day out of the 300 μg/day that is in the food (compare with Table 1). The amount of lead in the 20 m³ of air that is daily used for respiration is about 25 μg, but as it is taken up with an efficiency of 50%, as much as 12.5 μg lead per day was accumulated in the body of an average European 25 years ago due to direct atmospheric pollution. The 30 μg/day is excreted mainly through the urine and therefore the 12.5 μg lead per day is accumulated in the body—mainly in the bones, where their effect fortunately is very minor. Due to the reduced use of leaded gasoline, the average European will today have less lead in the body. Food contains roughly half as much lead today than 25 years ago and the atmospheric pollution is also one half of the level today compared with 1985. It means that the amount of lead from food today is 15 μg lead per day and that from respiration is 6.25 μg lead per day. Excretion is also reduced to a level about 15–20 μg lead per day, which implies that less lead is accumulated in the bones today. The indicated amounts of lead for an average European today are found by use of a model that was calibrated and validated by the use of the amounts from 1985—it means the values that are shown in Fig. 2.

A conceptual diagram for a lead model of a food chain in an aquatic ecosystem is shown in Fig. 3. The boxes represent state variables, which in this case are lead in water, lead in phytoplankton, lead in zooplankton, lead in planktivorous fish (fish I), lead in carnivorous fish (fish II), and lead in the sediment. The two latter concentrations are the highest in most cases. At each level in the food chain, lead is taken up from the water and from the food. The process rates are dependent on the organisms, the temperature, the pH, and the concentrations of the free lead ions and lead complexes. Quantification of the all the processes will reveal that the lead concentration is increasing at a factor of 10–100 through the food chain. This is in contrast to a factor of about 10,000 for DDT. The difference is due to the less effective uptake of lead from the food—as mentioned, the uptake efficiency for heavy metals in food items is only about 7%–10%. DDT is taken up from food with an efficiency of 90% due to the low solubility in water and high solubility in fat tissue.

The spectrum of toxicological and ecotoxicological effects of lead is very wide. Acute toxicity of lead can cause headaches, irritability, and loss of appetite. Chronic toxicity can cause brain damage, reduced memory, anemia, liver and kidney damage, and possibilities of cancerous tumors of the kidney. It has been shown that if children are exposed to high lead concentrations, it will have a pronounced effect on their learning ability.[11] Elevated lead concentrations have also teratogenic effects,[12] and prenatal

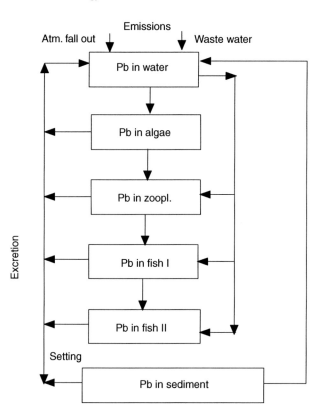

Fig. 3 Conceptual diagram of biomagnifications of lead through the food chain in an aquatic ecosystem. Notice that each level in the food chain takes lead up from the water and from the food, but each level also excretes lead to the water.

lead exposure has been demonstrated to be associated with an increased risk of malformations.[13] Genetic effects on animals have also been observed.[12] Epidemiological studies have shown that lead is related to the risk of elevated blood pressure.[14]

Significantly different lead concentrations have been found in the blood of people living far from towns and those in urban areas. Indians in the Amazons have about 8 ng lead per gram of blood. Farmers living in the European countryside have about 50 ng/g of blood and people living in very big cities (New York for instance) have about 150 ng/g of blood.[15] The highest concentrations of lead have been found in policemen who regulated traffic in industrialized countries before lead was phased out: 150–700 ng/g blood. Lead in clean air is as low as 0.0005 ng/L, while in urban areas, it is mostly between 2 ng/L and 25 ng/L.[16] In very polluted urban areas, the concentration may reach 50 ng/L, which could give a daily uptake of 0.345 mg per person and a lead concentration in blood about 720 ng/g.[16] Fig. 2 shows a model that could be used to find these values in the blood, when the air pollution level is known.

Lead in soil shows a similar wide range of values. Generally, a lead concentration of 0.2–5 mg/kg dry matter in agricultural areas is found,[17] but as a concentration as high as 100 mg/kg dry matter is found in particularly contaminated soil,[18] and in some extreme cases, concentrations as high as 10,000 mg/kg dry matter have been found.[19] A typical average soil in an industrialized country contains about 20 mg/kg dry matter of lead, and 95% of soil samples randomly sampled will show concentrations between 10 and 60 mg/kg dry matter. Lead concentrations above 1000 mg/kg dry weight in soil can kill earthworms and springtails.

The half-life of lead in human is about 6 years (whole body) and about 3 times as much for the skeletal system. It has been shown that skeletal burdens of lead increase linearly with age, while the non-skeletal burden is eliminated faster (as indicated, whole-body lead has a half-life of 6 years, which means that the non-skeletal lead is exchanged faster than once every 6 years). It is possible to reduce the body burden for patients clinically affected by lead by ethylene-diamine-tetra-acetate (EDTA).[20]

ABATEMENT METHODS FOR REDUCTION OF LEAD POLLUTION

Two major sources of lead pollution, namely, the combustion of leaded gasoline and the use of lead ammunition, have been eliminated in most industrialized countries by environmental legislation. The use of lead in ceramic and paints has similarly been reduced significantly by legislation or by agreement between environmental agencies and the industry.

Lead pollution has been reduced considerably during the last 25–30 years due to environmental legislation and the treatment of wastewater and contaminated smoke and air. Environmental technological solutions have been increasingly applied for the treatment of water and air. A description of the environmental technological methods can be found in the following entries of this encyclopedia:

- *Ion Exchange Application for Treatment of Water and Wastewater*
- *Wastewater Treatment: Overview of Conventional Methods*
- *Municipal Wastewater and Its Treatment*
- *Air Pollution and Environmental Technology: An Overview*

Contaminated land has also been treated, both by environmental technological methods and by bioremediation (see *Bioremediation*, p. 408).

A few cases of successful application of cleaner technology (replacing lead by less harmful components) have also been reported in the journal *Cleaner Technology*. The results of these efforts are encouraging for the use of a consequent environmental management. When environmental legislation, environmental technology, cleaner technology, and ecotechnology are working hand in hand, it is possible

USE OF INTEGRATED ENVIRONMENTAL MANAGEMENT IN THE CASE OF LEAD CONTAMINATION

As with all pollution problems, lead contamination is complex. This section briefly discusses how to go around this complexity to propose an integrated and holistic environmental management and thereby solve the problems properly. The discussion in this section is in principle valid for all pollution problems (compare to the Topical Table of Contents and the How to Use This Encyclopedia in the frontmatter) although particularly for all heavy metal contaminations. The following crucial questions require answers in the case of lead contamination:

1. Which forms have the lead-free ions, or which complexes in which concentrations?
2. What are the sources to the problem? Quantitatively?
3. What possibilities do we have to eliminate which sources? Will that be sufficient to solve the problem?
4. How can we best combine the methods to solve the problem?

It is recommended to consider the following points to be able to arrive at an answer for the crucial questions:

1. The forms of lead can be found either analytically or by chemical calculations;[21] where straightforward chemical calculations are clearly shown for heavy metals with illustrative examples.
2. It is advantageous to set up a mass balance. Fig. 4 shows a mass balance for the lead contamination of 1 ha Danish agricultural land. The mass balance clearly reveals the important sources. Atmospheric fallout is the dominant source, although it may also be beneficial to reduce the lead contamination coming from sludge and fertilizers. Particularly the first one of these two could be eliminated.
3. It is possible to eliminate the air pollution of lead as it has been discussed by phasing out the use of lead in gasoline. The second most important air pollution source, in many countries, is coal-fired power plants, which can be eliminated by changing to other forms of fossil fuel or to alternative energy sources. This change of the energy policy will, however, often be prohibitively expensive, and it is therefore as indicated a political question. The phasing out of lead in gasoline may be sufficient. The core question is this: how much will the lead contamination in food be reduced if we reduce the atmospheric fallout so and so much? The answer requires calculations and sometimes the use of models. Lead models have been developed by Jørgensen.[7,9] Lead models considering hydrodynamics, bioconcentration, bioaccumulation, excretion, and sedimentation can furthermore be found in Lam and Simons[22] and Aoyama et al.[23] Another core question is: if we are able to reduce the concentration in food, sediment, or soil to desired levels, would this reduction be sufficient to reduce or even eliminate the effect? This question will require a comprehensive overview of the toxicological and

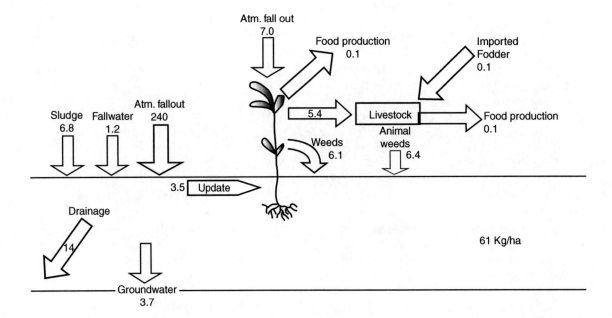

Fig. 4 Lead balance of average Danish agriculture land. All rates are expressed as grams of Pb per hectare per year.

ecotoxicological literature about lead and its effect. If information about the effect is not available, it is necessary to perform bioassay.[24,25] These references describe in detail how to perform bioassay for heavy metals including lead.

4. A discussion of the answer to the fourth question is covered in "The Contents of the Encyclopedia and How to Use the Encyclopedia for Integrated Ecological and Environmental Management." Other examples are given in Jørgensen et al.[21]

REFERENCES

1. Laws, E.A. *Aquatic Pollution: An Introductory Text*, 2nd Ed.; John Wiley and Sons: New York, 1993; 380 pp.
2. Jørgensen, S.E. *Principles of Pollution Abatement*; Elsevier: Amsterdam, 2000; 520 pp.
3. Chemistry. The lead we breathe. Chemistry **1968**, *41*, 7–12.
4. Haygarth, P.M; Jones, K.C. Atmospheric deposition of metals to agricultural surfaces. In *Biochemistry of Trace Metals*; Adriano, D.C., Ed.;. Lewis Publ.: Boca Raton, 1992; 423 pp.
5. Jørgensen, S.E.; Fath, B. *Ecotoxicology*; Elsevier: Amsterdam, Oxford, 2010; 390 pp.
6. Forstner, U.; Wittmann, G.T.W. *Metal Pollution in the Aquatic Environment*; Springer Verlag: Heidelberg, Berlin, New York, 1979; 490 pp.
7. Jørgensen, S.E.; Fath, B. *Fundamentals of Ecological Modelling*, 4th Ed.; Elsevier: Amsterdam, Oxford, 2011; 396 pp.
8. Mitsch, W.J.; Jørgensen, S.E. *Ecological Engineering and Ecosystem Restoration*; John Wiley: New York, 2004; 410 pp.
9. Jørgensen, S.E.; Bendoricchio, G. *Fundamentals of Ecological Modelling*, 3rd Ed.; Elsevier: Amsterdam, Oxford, 2001; 530 pp.
10. Newman, M.C.; Unger, M.A. *Fundamentals of Ecotoxicology*; 2nd Ed.; CRC Press: Boca Raton, London and New York, 2003.
11. Bellinger, D.; Leviton, A.; Needleman, H.L.; Waternaux, C.; Rabinowitz, H.B. Low-level lead exposure and infant development in the first year. Neurobehav. Toxicol. Teratol. **1986**, *8*, 151–161.
12. Swedish EPA. *Om Metaller*, 2nd Ed.; Statens Naturvårdsverk: Sweden,1980.
13. Needleman, H.L.; Rabinowitz, M.; Leviton, A.; Linn, S.; Schoenbaum, S. The relationship between prenatal exposure to lead and congenital anomalies. J. Am. Med. Assoc. **1984**, *251*, 2956.
14. Schwartz, J.; Angle, C.; Pitcher, H. Relationship between childhood blood lead level and stature. Pediatrics **1986**, *77*, 281–288.
15. Jørgensen, S.E.; Jørgensen, L.A.; Nors Nielsen, S. *Handbook of Ecological and Ecotoxicological Parameters*; Elsevier: Amsterdam, 1991; 1380 pp.
16. Francis, B.M. *Toxic Substances in the Environment*; John Wiley and Sons: New York, 1994; 362 pp.
17. Prost, R. *Contaminated Soil*; Institut National de la Recherche Agronomique: Ann Arbor, MI, 1995; 525 pp. +CD.
18. Hutschinson, T.C.; Gordon, C.A.; Meema, K.M. *Global Perspectives on Lead Mercury and Cadmium Cycling in the Environment*; Wiley Eastern Limited: Ann Arbor, MI, 1994; 412 pp.
19. Kabata-Pendias, A. *Effects of Trace Metals Excess in Soils and Plants*; CRC: Boca Raton, 1986; 330 pp.
20. Nigel, B. *Environmental Chemistry*; 2nd and 3rd Eds.; Wuerz Publ. Ltd.: Winnepeg, 1994 and 2000; 378 pp. and 420 pp.
21. Jørgensen, S.E.; Tundisi, J.G.; Tundisi, T. *Handbook of Inland Aquatic Ecosystem Management*; CRC: Boca Raton, 2012; 480 pp.
22. Lam, D.C.L.; Simons, T.J. Computer model for toxicant spills in Lake Ontario. In *Metals Transfer and Ecological Mass Balances, Environmental Biochemistry*; Nriago, J.O., Ed.; Ann Arbor Science; Ann Arbor, MI, 1976; Vol. 2, 537–549.
23. Aoyama, I.; Inoue, Yos; Inoue, Yor. Simulation analysis of the concentration process of trace heavy metals by aquatic organisms from the viewpoint of nutrition ecology. Water Res. **1978**, *12*, 837–842.
24. Newman, M.C.; Jagoe, C.H. *Ecotoxicology. A Hierarchical Treatment*; CRC: Boca Raton, 1996; 412 pp.
25. Landis, W.G.; Yu, M.H. *Introduction to Environmental Toxicology*; Lewis Publ.: Boca Raton, 1995; 330 pp.

Lead: Regulations

Lisa A. Robinson
Independent Consultant, Newton, Massachusetts, U.S.A.

Abstract

Evidence that lead poses hazards even at relatively low exposure levels is mounting. As a result, the methods used to estimate the monetary value of the benefits of reducing these exposures need to be improved and updated, so that these benefits can be compared to the costs of potential new control strategies. In previous regulatory analyses, the most significant quantifiable benefits included the effects of lead-related neurological damage on future earnings and lead's effect on the risk of premature mortality from cardiovascular disease. However, many of the monetary values are now outdated. In addition, recent assessments include a large number of additional benefit categories that could be included in future regulatory analyses.

INTRODUCTION

Phasing out lead in gasoline is often identified as one of the U.S. Environmental Protection Agency's (USEPA's) greatest achievements.[1] As part of the regulatory process, USEPA economists developed innovative approaches for expressing the benefits of the resulting risk reductions in monetary terms, so that they could be directly compared to the costs of different control options.[2] As evidence mounts that lead may pose hazards even at relatively low exposure levels,[3,4] the methods used to estimate these monetary benefit values are in need of continual improvement and updating for application in future regulatory assessments. This entry reviews how these benefit values have been developed and identifies challenges for subsequent work. While it focuses on the approaches used in U.S. regulatory analyses, these approaches can be adapted for use in other countries and have global relevance.

The entry first reviews previous assessments, summarizing the framework for regulatory benefit–cost analysis and providing examples of how this framework has been implemented to assess the benefits of lead regulations. Because these analyses indicate that the most significant quantifiable benefits include the relationship between lead-related neurological damage and future earnings (for children) and between lead exposure and cardiovascular disease (for adults), it next discusses these benefit categories in more detail. It then describes benefits that have not been fully incorporated into previous regulatory analyses, but that may be desirable to explore in the future. The concluding section summarizes the discussion and highlights related challenges.

REVIEW OF PREVIOUS ANALYSES

The USEPA's pathbreaking work on valuing lead benefits was initiated in the late 1970s and early 1980s, as the United States moved to phase out lead in gasoline.[2,5] Lead was completely banned from gasoline in 1996. The valuation approaches evolved as the USEPA subsequently considered regulating lead in drinking water, sewage sludge, and paint, as well as ambient air. Typically, the quantified benefits are dominated by the effect of lead on earnings (due to its effect on IQ) and, in some cases, by its effects on cardiovascular disease, particularly the risks of premature mortality. However, analysts have been unable to quantify many other potential impacts, due largely to gaps or inconsistencies in the research on the association between lead exposure and additional health outcomes.

General Framework

When the USEPA began assessing the benefits of reducing lead exposure, requirements for conducting regulatory benefit–cost analysis were in their infancy.[6] The first comprehensive Executive Order mandating such analysis was issued by President Reagan in 1991.[7] The requirements for these analyses have since been refined, updated, and expanded, although the basic principles remain the same. Thus, the early work on lead was both innovative and pioneering, evolving into approaches used in numerous subsequent analyses.

Currently, benefit–cost analysis is required for major rules under President Clinton's Executive Order 12866,[8] as supplemented by President Obama's Executive Order

13563.[9] These executive orders mandate that Federal agencies assess alternative policies for actions that may be economically significant, i.e., that may lead to a rulemaking that has an annual economic effect of $100 million or more or has important adverse effects. To support implementation of these requirements, the U.S. Office of Management and Budget (OMB) issued Circular A-4, *Regulatory Analysis*,[10] in 2003 and summarized and clarified related requirements in a 2010 checklist for agencies.[11]

Typically, the resulting analyses contain five major components, in addition to sections discussing the rationale for the rulemaking and the regulatory options considered, as summarized in Table 1. These analyses also include both quantitative and qualitative assessment of related uncertainties, as well as information on effects that could not be quantified or valued.

This entry is primarily concerned with the economic valuation of benefits, under step 4b in Table 1. As discussed later, the ability to carry out such valuation is dependent on the extent to which the changes in risks can be assessed under step 4a. Building a strong research base that allows analysts to estimate the changes in risks of different types associated with various changes in lead exposures has been a major challenge.

Ideally, economic values reflect individuals' willingness to pay (WTP) for the risk reductions or other benefits they would receive from a regulation or policy. This approach is consistent with the theoretical framework underlying benefit–cost analysis, which is based on respect for individual preferences (often referred to as "consumer sovereignty"), assuming that each individual is the best judge of his or her own welfare. The consideration of WTP also reflects the types of trade-offs being considered in policy decisions. Given constrained resources, pollution abatement policies inevitably require trading off increased expenditures on risk reductions against decreased expenditures on other desired goods and services on a society-wide level. WTP, the maximum amount of income (or wealth) an individual is willing to exchange for a beneficial outcome, represents this type of exchange. Willingness to accept compensation (WTA), or the smallest amount an individual would accept to forego the improvement, is also consistent with this framework. WTA is used infrequently in regulatory analysis, however, both because it can be difficult to measure and because regulations often involve paying for improvements rather than compensating for harms.

In some cases, individual WTP can be estimated based on consumer demand for market goods. For outcomes not directly bought and sold in the marketplace (such as health risk reductions), WTP is instead estimated from revealed or stated preference studies. Revealed preference studies use data from market transactions or observed behaviors to estimate the value of related nonmarketed goods or outcomes. For example, the value of mortality risk reductions is often estimated from the relationship between earnings and job-related risks, controlling for other influencing factors. Alternatively, stated preference studies rely on responses to survey questions or similar approaches. For example, a survey respondent may be asked whether he or she would be willing to pay "$X" for a 1-in-10,000 reduction in the risk of death associated with decreased exposure to an air pollutant. Each approach has advantages and limitations. In particular, revealed preference studies rely on data from actual markets but often address scenarios that differ from those of concern in policy analysis. Stated preference studies enable researchers to better tailor the scenario to the risks of concern, but the responses are hypothetical and must be carefully elicited.

For many outcomes, estimates of WTP are lacking, and analysts often rely instead on averted costs as rough proxies. In particular, health risk reductions may be valued using cost of illness estimates, including expenditures on medical treatment and often lost productivity.[12,13] The latter is typically valued using the human capital approach, which assumes that workers are paid the value of their marginal product.[12–16] Thus, compensation data are used to value illness-related lost work time. The human capital approach also can be used to assess the value of changes

Table 1 Overview of analytic components.

1. Estimate current and potential future *baseline conditions* in the absence of government intervention.
2. Predict *responses* to each policy option under consideration.
3. Estimate the *national costs* associated with each option, summing the costs of the predicted compliance actions across those subject to the provisions and accounting for resulting market impacts (e.g., changes in consumption due to price increases).
4. Estimate the *national benefits* associated with each option, including the effects on human health and the natural and built environment. For environmental hazards, this generally consists of two steps:
    a. A *risk assessment*, which considers the link between changes in pollution levels and each health or environmental outcome of concern.
    b. An *economic analysis*, which includes monetary valuation of each outcome to the extent possible.
5. Assess the *distribution* of the impacts across subpopulations of concern. Such subpopulations typically include small businesses as well as sensitive and/or vulnerable subgroups (such as children or low-income individuals) whom policymakers wish to protect against disproportionate adverse effects.

in unpaid work time (i.e., household production and volunteer work), based on estimates of either the wages foregone when an individual chooses to engage in unpaid rather than paid work or the cost of replacing the unpaid worker with one who is compensated.

Another, less commonly used approach is the friction cost method, which assumes that productivity will decrease temporarily while the employer implements measures to replace the absent individual rather than over the full course of the illness.[14,16,17] The "friction period" is defined as the time it takes to find and train a new employee or reallocate duties among existing employees. However, this approach is likely to understate the loss particularly during periods of full employment, because it does not take into account the additional loss that accrues if the new employee was previously working in a different job or involved in nonmarket production.

WTP to reduce the risk of illness may differ from these averted costs. Costs reflect incurred cases, not expected risk reductions, and being ill and treated is typically worse than not being ill at all. Medical costs can be difficult to estimate accurately due to the distorting effects of insurance and other third-party payments, and analysts often rely on average per case costs, which may differ from the marginal costs of small changes in the risk of illness.

Approaches for valuing time losses also often rely on a number of simplifying assumptions regarding the functioning of the labor market and individual choices between paid and unpaid work and leisure. In reality, productivity losses at a societal level will depend on unemployment rates and on the extent to which coworkers compensate for the loss of the ill worker's time. At an individual level, sick leave and disability insurance may reduce the impact of illness on earnings. In addition, inflexible work hours and other factors limit workers' ability to choose jobs that reflect their willingness to trade off paid and unpaid work and leisure time; hence, wages may over- or understate their WTP for changes in time use.

For these and other reasons, estimates of averted costs may be less than, equal to, or greater than WTP to avoid the illness. For example, medical costs may frequently understate WTP because they do not reflect the value of avoiding pain and suffering, but may overstate WTP in cases where the availability of insurance leads individuals to seek treatment that they would not be willing to pay for themselves. In addition, although average costs per case are often reported on an yearly basis, few studies track individuals longitudinally and provide data on the lifetime costs associated with particular illnesses. Thus, while the concept of averted costs seems straightforward, the direct and indirect costs of illness can be difficult to estimate. However, they provide a reasonable and widely-used proxy when WTP estimates are not available..

As discussed below, the valuation approaches used in lead regulatory analyses rely largely on averted costs, due to the lack of suitable WTP estimates. Very few WTP studies directly address lead exposures[18] or IQ-related decrements;[19] more generally, WTP for morbidity risks has not been well-studied.[20] In particular, the approach typically used to value lead-related IQ decreases is an averted cost approach, focusing on lost earnings; some analyses also consider related educational costs. The logic is that individual WTP to avoid the IQ loss would be at least equal to the income foregone. Medical costs are typically not included, because no fully effective treatment now exists. While chelation therapy has been used to address high blood lead levels, its effectiveness has been questioned in recent years.[21]

Averted costs are often also used to estimate the value of morbidity risk reductions, including reductions in nonfatal cardiovascular disease associated with decreased lead exposure. WTP estimates are available for mortality risk reductions, however.

Types of Benefits Assessed

The approaches used to value lead-related lost earnings and cardiovascular effects in regulatory analysis have evolved over the past 30 years, reflecting new data as well as increasingly sophisticated methods. One of the first comprehensive benefit analyses of reduced lead exposures was published in 1985 for USEPA regulations developed under the Clean Air Act, addressing the phase down of the amounts of lead allowed in gasoline.[5] The outcomes assessed include reductions in the costs of lead screening and treatment; compensatory education; the risks of hypertension, myocardial infarction, stroke, and premature mortality; and damages to pollution control equipment, vehicle maintenance, and fuel economy; as well as health and welfare effects associated with other pollutants. While the authors indicate that the majority of the benefits (75%) stem from the reduced risks of cardiovascular disease (with mortality risk reductions dominating the results), these effects were not considered in the USEPA's decision making because the underlying epidemiological studies had not yet undergone widespread review. However, even if the benefits of cardiovascular risk reductions are not counted, the costs of the regulations are less than the benefits.

Soon after, the USEPA's 1986 analysis of a drinking water rule added an important new benefit category, the effects of changes in IQ on earnings, while also assessing many of the other benefit categories included in the 1985 analysis of lead in gasoline.[22] Rather than summing the estimates of lost earnings and educational costs, this analysis compares them as alternative measures of cognitive damage. For this rulemaking, the majority of the benefits result from reduced water system corrosion. Of the health-related benefits, lost earnings and total cardiovascular risk reductions are similar in magnitude.

These approaches were refined in subsequent analyses. A particularly significant example is the USEPA's 1997 *Retrospective Analysis of the Clean Air Act*.[23] That analysis covers the effects of the full set of regulations implemented under the Clean Air Act over the years 1970 to 1990. For lead, the benefits assessed include compensatory education and other educational costs, future earnings, neonatal mortality, and cardiovascular disease (hypertension, coronary heart disease, stroke, and premature mortality). The results in this case are dominated by the value of reducing the mortality risks associated with cardiovascular disease.

More recent USEPA analyses focus on the effects of lead on IQ and earnings, including its 2008 assessment of the Lead Renovation, Repair, and Painting Program rule developed under the Toxic Substances Control Act (TSCA) and its 2008 analysis of the lead National Ambient Air Quality Standards (NAAQS) under the Clean Air Act.[24,25] The benefits associated with reducing cardiovascular risks were not quantified in these analyses, because the USEPA found that data on adult blood lead levels were outdated and that more information on the association between these levels and related risks was needed. For IQ and earnings, the two analyses use somewhat different assumptions. For the renovation rule, USEPA relies solely on 1995 estimates of the relationship of IQ to earnings developed by Salkever, while for the NAAQS rule, it also considers the effects of a lower 1994 estimate developed by Schwartz, because recent research suggests that the Salkever estimates may be overstated. These estimates are discussed in more detail in the next section.

In general, these regulatory analyses indicate that the monetized benefits of reductions in lead exposure often far exceed the costs. The most significant benefit categories are the impact of lead on future earnings (for children) and its impact on mortality risks from cardiovascular disease (for adults). Regulations that focus on lead in ambient air or drinking water will affect both adults and children, while those that primarily address the ingestion of lead paint or contaminated soil will largely affect children.

The benefit categories addressed are similar across analyses in part because of the epidemiological research available. The benefit assessments focus on those areas where the risk assessments have found the most consistently strong associations, and where the outcomes are amenable to monetary valuation. The number of lead epidemiological studies is large, and they consider a much wider range of endpoints than reflected in these benefit analyses.[4] Some outcomes are excluded from the benefit–cost analysis because the risk data are weak or inconsistent; for others, the risk assessment shows strong associations but addresses endpoints that are difficult to value in monetary terms. For example, small changes at the cellular level cannot be easily monetized unless information is also available on the likelihood that such changes will lead to noticeable health impairments. As a result, these analyses are likely to understate benefits, but the degree of understatement is uncertain.

VALUING IQ-RELATED BENEFITS

The benefit category most frequently assessed in lead-related regulatory analyses is the effect of IQ on earnings. This assessment generally includes two components: determining the percent change in lifetime earnings associated with each one-point change in IQ and estimating the dollar value of lifetime earnings for each cohort affected. Because the first component is far more difficult to address, the discussion that follows focuses on how the percent change in earnings is estimated. The second component, lifetime earnings, can be estimated based on readily accessible data on earnings and survival probabilities, as demonstrated by Grosse et al.,[15] who provide estimates by year of age as of 2007.

Describing these benefits as the effect of lead-associated IQ decrements on earnings is somewhat misleading, however. The underlying epidemiological studies report associations between lead and varying measures of cognitive abilities, as well as measures of behavioral and other problems that may affect both school- and work-related achievement. Thus, this category may include a number of interrelated neurological effects that impact educational attainment, the likelihood of employment, the amount of earned income, and household production.

This entry does not discuss the effects of lead or IQ on school-related costs in detail, because they tend to be a much smaller proportion of total benefits. Reductions in lead exposure may have counterbalancing impacts on these costs. To the extent that reductions improve IQ and other aspects of neurological functioning, the years of education may increase and the need for compensatory education may decrease. These changes may not be completely offsetting, however, due to differences in per-student costs and in the number of students affected. The change in the need for compensatory education is likely to affect a relatively small number of very low IQ children, while the increase in regular education may affect a larger number of children throughout the IQ spectrum. In addition, to the extent that IQ gains increase the number of years of schooling, they will also defer full employment while increasing compensation once employed.

Previous Regulatory Analyses

In simple terms, expected earnings are the product of the likelihood of employment (i.e., labor force participation) and the wages earned if employed. Cognitive ability can affect both wage levels and the probability of employment directly. In addition, cognitive ability can affect the number of years of schooling, which in turn also affects wages and participation rates. Again, the effects of lead may be greater than the effects of cognitive ability alone, because of associated behavioral problems, attentional difficulties, and other neurological effects that can influence functioning in school and at work. These relationships are illustrated conceptually in Fig. 1.

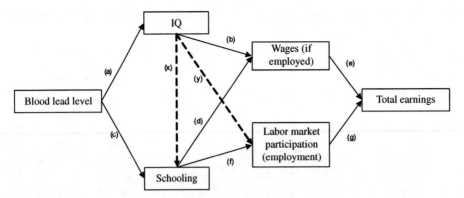

Fig. 1 Conceptual model of relationship between lead and earnings.

Assessing the pathways that lead from blood lead levels to changes in earnings is a complicated task. The foundation for the approach used in USEPA regulatory analyses is described in a 1994 article by Schwartz.[26] These relationships were then re-estimated in 1995 by Salkever,[27] whose estimates have been used in many, if not most, subsequent analyses. However, more recent work by Grosse[28,29] suggests that Salkever may overstate these effects. Disentangling the direct and indirect effects on earnings, to avoid double counting while capturing lead's impacts on earnings to the fullest extent possible, has been a major challenge.

Schwartz focuses on the relationships marked as solid lines (a) through (g) in Fig. 1: the direct effect of lead on IQ and earnings (a → b → e), the indirect effect of lead on earnings through schooling and wages (c → d → e), and the indirect effect of lead on earnings through schooling and labor force participation (c → f → g). While these effects are interrelated, they can be summed as long as the underlying studies control for each effect separately. In total, Schwartz estimates that these factors lead to a 1.76% change in earnings for each one-point change in IQ. However, because his analysis includes estimates of the direct effects of lead on schooling, it likely captures some of the effects of lead on behavior (e.g., on attention span) rather than purely measuring the effects of IQ. While the percent change may appear small on an individual basis, the large number of individuals affected by many regulations leads to high total benefits.

Salkever then revisits the approach developed by Schwartz, using data from the National Longitudinal Study of Youth (NLSY) to extend and re-estimate many of the components of the analysis. Salkever adds the two dashed arrows marked (x) and (y) in Fig. 1, estimating the effect of IQ on schooling and on labor force participation. He develops estimates separately for women and men.

When weighted to reflect the relative contribution of women and men to total earnings, the Salkever estimates average about a 2.4% change in earnings per IQ point, significantly higher than the Schwartz estimate (1.76%). For example, Schwartz calculated that a 1 μg/dl reduction in 1984 childhood blood lead concentrations would yield increased earnings of $5.1 billion annually (1989 dollars). If we instead apply the factors from Salkever, this amount increases by more than 30%, to close to $7 billion. While lead exposure has dropped in recent years, newer epidemiological studies suggest that the effects of lead on IQ may be larger than suggested by this older research, which would further increase these 1984 benefits estimates.

Recent Research Findings

Linking lead-related IQ changes to changes in earnings is difficult for two reasons. First, the longitudinal databases used in the econometric analyses often rely on tests of cognitive ability that differ from those used in epidemiological studies of the associations between lead and cognitive ability. The epidemiological studies generally rely on the Wechsler Intelligence Scale for Children (WISC), which is a standard and widely used test of IQ administered in childhood. Many of the econometric studies (including the Salkever analysis) rely on data from the NLSY, which instead includes results from the Armed Forces Qualification Test (AFQT) administered in 1980 when NLSY participants were between ages 15 and 23. The AFQT measures achievement as well as ability. Thus, while IQ tends to be relatively stable after early childhood, the inclusion of achievement means that AFQT scores can change as a result of subsequent education and work experience.

Second, the causal relationships are complex and not necessarily well understood, posing challenges in terms of both measurement and statistical analysis. A wide range of personal, family, and community factors (including education system and job market characteristics) may affect earnings, and also may be affected by IQ. The causal relationships can be somewhat circular. Green and Riddell[30] provide a simple example: if we are trying to estimate the independent effects of IQ and years of schooling on earnings, but higher-IQ individuals stay in school longer because their psychic costs of schooling (e.g., the difficulty of studying) are lower, then IQ is in part causing the increased schooling and the contributions of the two factors to earnings are difficult to separate.

More recent evidence on the relationship between IQ and earnings is reviewed by Grosse.[28,29] Some of these studies use the same NLSY data set as the Salkever study but implement various types of adjustments. The NLSY began tracking individuals in 1979 when they were between ages 14 and 22; Salkever and the more recent studies discussed below use 1990 NLSY data, which include information on educational attainment, hourly wages, and hours worked as of ages 25 to 33. As noted earlier, the measure of cognitive ability in the NLSY is the AFQT, administered in 1980, when many respondents were already working or pursuing postsecondary education.

Recognizing that postsecondary job experience and education are likely to increase AFQT scores, in 1996, Neal and Johnson[31] re-estimated the relationship of IQ to earnings as part of their study of earnings differentials between blacks and whites. They restrict their NLSY sample to those who were age 18 or younger when they took the AFQT, so that their estimates of the relationship between ability and earnings are less inflated by the effects of further education and experience than the estimates used by Salkever.

In 2002, Grosse, Schwartz, and colleagues[28] used the results from the Neal and Johnson study to assess the economic gains associated with the U.S. reduction in blood lead levels that occurred from 1976 to 1999. According to Grosse et al., the Neal and Johnson study indicates that a one-point IQ loss results in a 1.15% difference in earnings for men and a 1.52% difference for women. These results are from models that do not control for schooling; hence, Grosse et al. do not add separate estimates of the effects of schooling on earnings to avoid double counting. However, because Neal and Johnson do not estimate the impact of ability on labor force participation, Grosse et al. add the estimate of this effect from Salkever. Once weighted to reflect the relative share of earnings by men and women in the labor force, Grosse et al. find that a one-point IQ loss is associated with a 1.66% change in earnings, slightly lower than the Schwartz estimate of 1.76% and substantially less than the Salkever estimate of 2.4%.

In his 2007 review, Grosse also discusses a newer study by Heckman et al.[32] These authors rely on the same NLSY data as discussed previously, considering the factors that influence wages at age 30. The modeling is substantially more sophisticated than prior studies, considering a wide array of personal, family, and community characteristics, including standard demographic variables as well as behavioral choices (such as smoking and participation in crime). It takes into account latent (i.e., unobserved) factors representing cognitive and non-cognitive ability and considers how these factors influence schooling choices as well as labor market choices and earnings.

Comparing the Heckman et al. findings to the findings from other studies requires some conversion. Heckman et al. multiply their coefficients by the standard deviation of the associated variable in reporting their results. Because IQ is standardized so that one standard deviation equals 15 IQ points, this entry follows Grosse[29] and divides the Heckman et al. coefficients on cognitive ability by 15 to determine the effect per IQ point. However, it uses results from Heckman et al.'s model 2 (with corrected AFQT scores) in Table 5 of Heckman[32] rather than their model 3 (as in Grosse, which relies on latent measures of ability), because the latter are more difficult to interpret. Depending on the controls included in the statistical modeling, the results suggest that a one-unit change in cognitive ability raises wages for males by 0.44% to 0.95% per IQ point.

This result is lower than the results from Neal and Johnson (1.15% for men and 1.52% for women), which do not control for schooling. Adding the pathways not included in the Heckman et al. analysis, i.e., the effect of ability on labor force participation, would increase the Heckman et al. estimates somewhat, as would including women (for whom returns to ability appear higher than for men). Perhaps more importantly, Heckman et al. find that the effects of non-cognitive factors are significant. Lead has many behavioral effects not captured in the epidemiological studies of the association between lead exposure and IQ; considering the effect of these factors could noticeably increase the impact of lead exposure on earnings.

Grosse also identifies another study, by Zax and Rees,[33] that uses an alternative data source: the Wisconsin Longitudinal Study (WLS) of Social and Psychological Factors in Aspiration and Attainment. This study includes IQ scores based on the Henmon–Nelson Test of Mental Ability administered in the 11th grade. The authors explore the effects of factors measured at age 17 (in 1957) on male earnings at ages 35 and 53. Zax and Rees find that household, community, and peer characteristics all affect earnings at both ages. Depending on the extent to which the model controls for these factors, the effect of a one-IQ-point change on earnings ranges from 0.36% to 0.74% at age 35, rising to 0.90%–1.39% at age 53, for the subsample who had earnings at both ages (Tables 5 and 6 of Zax and Rees[33]). The effect of IQ on earnings for the full sample at age 35 was almost identical to the effect for this subsample at that age.

These findings have four important implications for considering the effects of lead on earnings. First, the socioeconomic status of those affected by reduced lead exposures may affect the relationship between IQ and earnings. The Zax and Rees results for age 35 are somewhat lower than the factors found by Heckman et al. at age 30. However, the Zax and Rees sample is from a relatively affluent state, which may not be representative of the national population, and they rely on data for high school graduates. In contrast, the NLSY data used by Heckman et al. represent a wider population. Different relationships between IQ and earnings may hold if lead exposures disproportionately affect lower-income individuals, who may be less likely to complete high school and for whom the returns to IQ may be influenced dissimilarly by household and community characteristics.

Second, these findings suggest that using a constant factor, based on the IQ-to-earnings relationship for ages around 30 to 35, will understate the effect of IQ on earnings at older ages. The effect of this changing relationship on the present value of lifetime earnings is unclear, however, because it depends on three factors: the extent to which the IQ-and-earnings relationship varies across all years of age, the extent to which earnings change with age, and the discount rate used to estimate the present value of future earnings.

Third, the Zax and Rees analysis further emphasizes some of the difficulties inherent in sorting out causal links. Their model with the most controls (which leads to the lowest proportional changes; i.e., the 0.36% and 0.90% above) includes variables such as whether parents encouraged the individual to attend college. Parents may be more encouraging if their child has a higher IQ, or this encouragement may be related to other factors (such as whether the parents attended college or believe that college is affordable). To the extent that IQ leads to such encouragement, then controlling for this encouragement will understate the effects of IQ.

Finally, applying a longitudinal study to the present time increases uncertainty. To track individuals from childhood to their peak earning years requires starting at least 30 years in the past. Whether these past relationships apply to children exposed to lead in the current year is uncertain, given changes in the educational system, labor market, and other factors over time.

In sum, newer research suggests that both the Salkever estimate of a 2.4% change in earnings per one-point change in IQ and the older Schwartz estimate of 1.76% are overstated. The degree of overstatement is uncertain, however, due to the difficulties inherent in disentangling the effects represented in Fig. 1. In the near term, analysts may want to test the sensitivity of their results to a range of factors, as does the USEPA in its 2008 NAAQS analysis.[25]

In the longer term, more detailed review of recent research on the relationship between IQ and earnings would be desirable, given both the complexity of the relationships and their importance in assessing the benefits of lead abatement. This review could include determining whether it is possible to use the data that underlie the reported results to estimate factors that are more directly applicable to the populations most affected by changes in lead exposures. New research may be needed that directly addresses these populations and focuses on the data needed to estimate earnings-related benefits. The research cited above suggests that the impacts of lead on non-cognitive functioning may also have important effects on education and earnings that should be further investigated. In addition, more work is needed to estimate individual WTP to avoid these risks, given that WTP is a more conceptually correct indicator of the value of related benefits.

VALUING CARDIOVASCULAR RISK REDUCTIONS

For cardiovascular risks, suitable WTP estimates that incorporate both morbidity and mortality are not currently available; thus, these two components are typically valued separately in regulatory analysis and then summed. Well-established approaches exist for valuing mortality risk reductions, based on estimates of WTP, but few studies address WTP for morbidity risks.[20] As a result, cost-of-illness estimates are generally used to value reductions in nonfatal cardiovascular risks.

The suite of lead-related cardiovascular conditions assessed varies across USEPA's benefit analyses but has at times included hypertension, stroke, and myocardial infarction or coronary heart disease more generally. The approaches for valuing these conditions have evolved over time, reflecting changes in the available data.

For mortality risk reductions, the USEPA currently follows a valuation approach based on a review conducted in the early 1990s and described in both its 2000 and 2010 *Guidelines for Preparing Economic Analyses*.[34,35] The estimates are typically expressed as the "value per statistical life" or VSL, to parallel how risk reductions are typically expressed in policy analyses.[36,37] Most regulations lead to small risk changes at the individual level, which are often presented as "statistical" cases aggregated across the affected population. For example, a 1-in-10,000 risk reduction affecting 10,000 individuals can be expressed as a statistical case (1/10,000 risk reduction × 10,000 individuals = 1 statistical case), as can a 1-in-100,000 risk reduction affecting 100,000 individuals (1/100,000 risk reduction × 100,000 individuals = 1 statistical case). For most regulations, the specific individuals who would avoid illness or whose lives would be extended by the policy cannot be identified in advance. A regulation that is expected to "save" a statistical life is one that is predicted to result in one less death in the affected population during a particular time period. "Saving" a statistical life is not the same as saving an identifiable individual from certain death.

The VSL is typically calculated by dividing individual WTP for a small risk change (in a defined time period) by the size of the risk change. For example, if an individual is willing to pay $700 for a 1-in-10,000 reduction in his or her risk of dying in the current year, the VSL is $7 million ($700 ÷ 1/10,000 = $7 million). Alternatively, individual WTP for small risk reductions could be aggregated across a population. A $7 million VSL also results if each member of a population of 10,000 is willing to pay an average of $700 for a 1-in-10,000 annual risk reduction ($700 × 10,000 = $7 million).

The USEPA has devoted considerable attention to research on the value of mortality risk reductions. At present, it recommends applying a mean VSL of 7.4 million dollars with a standard deviation of $4.7 million (in 2006 dollars), updated to the appropriate dollar year.[35] Analysts typically adjust this value to reflect changes in real income over time as well as any significant delays between changes in exposure and changes in mortality incidence (i.e., latency or cessation lag).

This approach is likely to evolve in the near future, as a result of new research and expert review. The USEPA asked its Science Advisory Board to review a 2010 White Paper it drafted on changes in its approach.[38] In 2011, the Science Advisory Board recommended that EPA 1) change the VSL terminology to avoid confusion between the value of small risk reductions and the value of saving an individual's life; 2) recognize that values vary across contexts and, to the extent possible, use values tailored to the particular risk to be regulated; 3) apply enhanced criteria to determine which valuation studies are of sufficient quality for application in policy analysis; and 4) conduct additional research on topics such as whether individuals are willing to pay differing amounts for risk reductions with differing characteristics.[39] In the near term, the extent to which these recommendations will significantly change the values used in USEPA regulatory analyses is unclear; over the longer term, an ambitious research agenda will be needed to fully meet the challenges raised by the report.

For reductions in morbidity risks, the USEPA has long relied on cost-of-illness estimates due to the lack of suitable estimates of WTP. These estimates have changed over time, but typically include the direct costs of a wide range of inpatient and outpatient services, supplies, and medications. For indirect costs, the estimates include lost market work time and, at times, lost household production.

The most recent (2008) USEPA analyses do not include cardiovascular risks in the monetized benefits estimates, because of concerns about the data available on adult blood lead levels and on the effects of lead on these risks. Previous lead analyses, such as a 2006 assessment of the USEPA's proposed renovations rule,[40] relied on relatively old cost data to value nonfatal risks, collected in the 1970s and 1980s. Given that medical treatment and labor market characteristics have changed substantially since that time, more review of research is needed to determine the appropriate values for these impacts, regardless of whether WTP or cost-of-illness methods are used.

OTHER CHALLENGES AND OPPORTUNITIES

While there are many opportunities to improve the approaches used previously to assess the benefits of increased regulation of lead exposure, perhaps the largest challenge is expanding the types of outcomes considered. When comparing benefits to costs, policy analysts and decision makers understand that monetized benefits are likely to understate total benefits, because they exclude the value of many potential outcomes. However, without more data and analysis, it is difficult to determine whether the excluded outcomes would have a relatively small or relatively large impact on the benefit estimates, and hence whether they suggest that a substantial increase in spending on these regulations might be worthwhile.

Expanding the types of benefits that are quantified poses challenges in terms of both the risk assessment and the valuation approach. As illustrated in Box 1, estimating the benefits of lead regulations first requires data on the associated risk reductions (step 4a). The discussion of previous regulatory analyses indicates how concerns about the quality of the risk data can limit the inclusion of some benefit categories (e.g., cardiovascular risk reductions), despite the availability of a valuation method. In addition, double counting may be a concern when various epidemiological studies cover overlapping outcomes, as illustrated by the discussion of the relationship between IQ, schooling, labor force participation, and earnings, as well as the relationship to other types of neurological damage associated with lead.

The USEPA periodically reviews the scientific evidence linking lead exposures to different outcomes. The most recent draft of its *Integrated Science Assessment for Lead*,[4] which is now being reviewed, covers seven health-related categories: neurological effects, cardiovascular effects, renal effects, immune system effects, effects on heme synthesis and red blood cell function, reproductive effects and birth outcomes, and cancer. It concludes that there is a causal relationship for all of these outcomes except cancer, for which the causal relationship is described as "likely." In addition, the report covers a number of ecological effects and determines that lead affects growth, mortality, and other factors related to the health of terrestrial and aquatic systems. Careful review of these data is needed to develop estimates of the relationships between exposure reductions and risk reductions that are suitable for use in regulatory analysis.

Another challenge is determining which outcomes will be noticeably affected by a particular regulation. For example, while complete elimination or major decreases in lead exposure might avert the need for periodic testing of young children, an individual rulemaking may not have widespread or large enough effects to change the testing requirements, in which case related savings would not be expected. Regulations that address childhood lead exposures may affect their cardiovascular risks when they become adults; cardiovascular risks will be more immediately affected by a rule that addresses adult exposures. Careful comparison of baseline and postregulatory effects (steps 1 and 2 in Box 1) is needed to identify those outcomes that may be most important within the context of a particular analysis.

In addition, developing values for some outcomes included in the risk assessment may be difficult. Valuation requires information on discernable impacts, the effects of which can be weighed by individuals in determining how much they would be willing to pay to reduce related risks.

For example, to value changes in red blood cell function, information on how these changes might ultimately affect an individual's activities and quality of life is needed. Similarly, valuing changes in blood pressure requires linking these changes to the need for treatment of hypertension and to the likelihood of stroke or other serious illnesses. Generally, in these cases, analysts must develop a model that estimates how these types of changes affect the likelihood of more significant health impairments as well as premature death.

Once data on the risk reductions attributable to a change in lead exposure are available, the effects can be valued using a number of different approaches. Scholars have assessed many lead-related effects that are not typically addressed in the USEPA's regulatory analyses, due in part to differences in the nature of the risk change and in part to concerns about the quality of the data.

Many of these analyses focus on elimination of lead-related hazards, rather than the smaller marginal changes likely to be associated with a federal regulation. For example, Landrigan et al.[41] assess the effects of eliminating all childhood asthma, cancer, and developmental disabilities (as well as lead poisoning) associated with environmental pollutants, using an averted cost approach that could be updated and adapted for use in other contexts. Another example is an analysis by Gould,[42] based in part on earlier work by Korfmacher, which explores the costs and benefits of eliminating the hazards that household lead paint poses to young children. She explores lead's effects on crime and attention deficit–hyperactivity disorder (ADHD) as well as other outcomes, using an averted cost approach. For example, for ADHD, Gould notes that the effects are lifelong and can include a range of types of misconduct and antisocial behavior, including drug use and crime. Considering only medical treatment and parental lost work time, she estimates that lead-associated ADHD costs $267 million annually (1996 dollars). She also estimates the effects of reducing average preschool blood lead levels on the incidence of burglaries, robberies, aggravated assaults, rapes, and murders. She concludes that the total direct cost of these crimes is approximately $1.8 billion, including victim costs, costs of legal proceedings and incarceration, and lost earnings for both the criminal and victim. Finally, she estimates that an additional $11.6 billion is lost in indirect costs, including psychological and physical damage that requires treatment and measures taken to prevent criminal actions. In addition to suggesting that the benefits associated with these large changes in exposure are significant, these and other studies indicate that data are available that can be used to estimate the value of additional benefit categories.

factors, such as the characteristics and size of the exposed population, the magnitude of the averted exposures, the types and severities of the health effects, and the monetary values of the impacts. However, this entry indicates that the effects of lead on future earnings, and on premature mortality from cardiovascular disease, are the most significant benefit categories monetized in previous regulatory analyses. It further indicates that these analyses could be expanded to include numerous other potentially significant benefit categories.

For childhood exposures, the impact of lead on earnings (due to its effects on IQ and schooling) dominates the results. On a per-person basis, the changes in earnings are small. However, because USEPA regulations often affect a substantial number of individuals, the earnings-related benefits become large when aggregated nationally. More work is needed, however, to update and refine the estimates of the effect of a change in IQ on earnings, as well as to better understand the relationship between lost earnings and WTP for IQ changes.

For adult cardiovascular effects, premature mortality dominates the results, due in part to the fact that the value of mortality risk reductions is generally measured in millions of dollars while morbidity values are often at least an order of magnitude less. While the value of mortality risk reductions is well studied and current practices are evolving as a result of recent research and expert review, WTP estimates for many types of morbidity risks are not currently available.[20] Work currently underway (e.g., by Hammitt and Haninger[43] and Cameron and DeShazo.[44]) may eventually provide useful estimates of WTP for cardiovascular risks. More research on the lifetime costs of cardiovascular disease would also be informative.

Regulatory analyses currently do not include quantified and monetized estimates for several other benefit categories. The expansion of the epidemiological research base, represented by USEPA's recent *Integrated Science Assessment*, provides opportunities for better characterizing the relationships between lead and those outcomes traditionally included in the monetary valuation of benefits, as well as opportunities for expanding the analysis to include important additional outcomes. Some areas worthy of additional exploration include lead's non-cognitive effects on educational attainment and earnings and its effects on delinquent behavior and crime. Analysts will need to pay careful attention to the interrelationships between outcomes as the number of endpoints considered expands. Tracing these relationships is important both to avoid double counting and to ensure that the most significant benefit categories are included in the analysis to the greatest extent possible.

CONCLUSIONS

The types and magnitude of benefits from lead regulations are determined by the complex interactions of a number of

ACKNOWLEDGMENTS

I would like to thank Scott D. Grosse (Centers for Disease Control and Prevention), James E. Neumann (Industrial

Economics, Incorporated), and two anonymous reviewers for their helpful comments on earlier drafts of this entry.

REFERENCES

1. Aspen Institute. *40 Years: EPA 40th Anniversary*; Aspen Institute: Washington, DC, 2010.
2. Robinson, L.A. *Benefits of Reduced Lead Exposure: A Review of Previous Studies.* Prepared for the U.S. Environmental Protection Agency under subcontract to Industrial Economics, Incorporated, 2007.
3. Lanphear, B.P.; Hornung, R.; Khoury, J.; Yolton, K.; Baghurst, P.; Bellinger, D.C.; Canfield, R.L.; Dietrich, K.N.; Bornschein, R.; Greene, T.; Rothenberg, S.J.; Needleman, H.L.; Schnaas, L.; Wasserman, G.; Graziano, J.; Roberts, R. Low-level environmental lead exposure and children's intellectual function: An international pooled analysis. Environ. Health Perspect. **2005**, *113* (7), 894–899.
4. U.S. Environmental Protection Agency. Integrated Science Assessment for Lead. U.S. Environmental Protection Agency: Washington, DC, forthcoming. Available at http://www.epa.gov/ncea/isa/index.htm.
5. U.S. Environmental Protection Agency. *Costs and Benefits of Reducing Lead in Gasoline*; EPA 230-05-85-006. U.S. Environmental Protection Agency: Washington, DC, 1985.
6. U.S. Office of Management and Budget. *Report to Congress on the Costs and Benefits of Federal Regulations,* 1997.
7. Reagan, R. Executive Order 12291: Federal regulation. Fed. Regist. **1981**, *46* (190), 13193–13198.
8. Clinton, W.J. Executive Order 12866: Regulatory planning and review. Fed. Regist. **1993**, *58* (190), 51735–51744.
9. Obama, B. Executive Order 13563: Improving regulation and regulatory review. Fed. Regist. **2011**, *76* (14), 3821–3823.
10. U.S. Office of Management and Budget. *Circular A-4: Regulatory Analysis*; U.S. Office of Management and Budget: Washington, DC, 2003.
11. U.S. Office of Management and Budget. *Agency Checklist: Regulatory Impact Analysis*; U.S. Office of Management and Budget: Washington, DC, 2010.
12. Rice, D.P. Estimating the cost of illness. Am. J. Public Health **1967**, *57*, 424–440.
13. Yabroff, K.R. et al., Eds. Health care costing: Data, methods, future directions. Med. Care **2009**, *47* (7), Sup 1.
14. Grosse, S.D.; Krueger, K.V. The income-based human capital valuation methods in public health economics used by forensic economics. J. Forensic Econ. **2011**, *22* (1), 43–57.
15. Grosse, S.D.; Krueger, K.V.; Mvundura, M. Economic productivity by age and sex: 2007 estimates for the United States. Med. Care **2009**, *47* (7), Sup. 1.
16. Zhang, W.; Bansback, N.; Anis, A.H. Measuring and valuing productivity loss due to poor health: A critical review. Soc. Sci. Med. **2011**, *72*, 185–192.
17. Koopmanschap, M.A.; Rutten, F.F.H.; van Ineveld, B.M.; van Roijen, L. The friction cost method for measuring indirect costs of disease. J. Health Econ. **1995**, *4*, 171–189.
18. Agee, M.D.; Crocker, T.D. Parental altruism and child lead exposure: Inferences from demand for chelation therapy. J. Hum. Resour. **1996**, *31* (3), 677–691.
19. von Stackelberg, K.; Hammitt, J.K. Use of contingent valuation to elicit willingness-to-pay for the benefits of developmental health risk reductions. Environ. Resour. Econ. **2009**, *43*, 45–61.
20. Robinson, L.A.; Hammitt, J.K. Valuing health and longevity in regulatory analysis: Current issues and challenges (Ch. 30). In *Handbook on the Politics of Regulation*; Levi-Faur, D., Ed.; Edward Elgar: Cheltenham, U.K. and Northampton, 2011.
21. Rischitelli, G.; Nygren, P.; Bougatsos, C.; Freeman, M.; Helfand, M. Screening for elevated lead levels in childhood and pregnancy: An updated summary of evidence for the U.S. Preventive Services Task Force. Pediatrics **2006**, *118*, 1867–1895.
22. U.S. Environmental Protection Agency. *Reducing Lead in Drinking Water: A Benefit Analysis;* EPA 230-09-86-019. U.S. Environmental Protection Agency: Washington, DC, 1986 (revised 1987).
23. U.S. Environmental Protection Agency. *The Benefits and Costs of the Clean Air Act: 1970–1990*; EPA 410-R-97-002. U.S. Environmental Protection Agency: Washington, DC, 1997.
24. U.S. Environmental Protection Agency. *Economic Analysis for the TSCA Lead Renovation, Repair, and Painting Program Final Rule for Target Housing and Child-Occupied Facilities*; U.S. Environmental Protection Agency: Washington, DC, 2008.
25. U.S. Environmental Protection Agency. *Regulatory Impact Analysis of the Proposed Revisions to the National Ambient Air Quality Standards for Lead*; U.S. Environmental Protection Agency: Washington, DC, 2008.
26. Schwartz, J. Societal benefits of reducing lead exposure. Environ. Res. **1994**, *66*, 105–124.
27. Salkever, D.S. Updated estimates of earnings benefits from reduced exposure of children to environmental lead. Environ. Res. **1995**, *70*, 1–6.
28. Grosse, S.D.; Matte, T.D.; Schwartz, J.; Jackson, R.J. Economic gains resulting from reduction in children's exposure to lead in the United States. Environ. Health Perspect. 2002, *110* (6), 563–569.
29. Grosse, S.D. How much does IQ raise earnings? Implications for regulatory impact analyses. AERE Newsl. **2007**, *27* (2), 17–21.
30. Green, D.A.; Riddell, W.C. Literacy skills, non-cognitive skills, and earnings: An economist's perspective. In *Towards Evidence-Based Policy for Canadian Education*; de Broucker, P., Sweetman, A., Eds.; McGill-Queens University Press: Montreal and Kingston, 2002.
31. Neal, D.A.; Johnson, W.R. The role of premarket factors in black–white wage differences. J. Political Econ. **1996**, *104*, 869–895.
32. Heckman, J.J.; Stixrud, J.; Urzua, S. The effects of cognitive and noncognitive abilities on labor market outcomes and social behavior. J. Labor Econ. **2006**, *24*, 411–482.
33. Zax, J.S.; Rees, D.I. IQ, academic performance, environment, and earnings. Rev. Econ. Stat. **2002**, *84*, 600–616.
34. U.S. Environmental Protection Agency. *Guidelines for Preparing Economic Analysis*; EPA 240-R-00-003. U.S. Environmental Protection Agency: Washington, DC, 2000.

35. U.S. Environmental Protection Agency (EPA). *Guidelines for Preparing Economic Analysis*; EPA 240-R-10-001.U.S. Environmental Protection Agency: Washington, DC, 2010.
36. Hammitt, J.K. Valuing mortality risk: Theory and practice. Environ. Sci. Technol. **2000**, *34*, 1396–1400.
37. Robinson, L.A. How US government agencies value mortality risk reductions. Rev. Environ. Econ. Policy. **2007**, *1* (2), 283–299.
38. U.S. Environmental Protection Agency. *Valuing Mortality Risk Reductions for Environmental Policy: A White Paper (Review Draft)*. Prepared by the National Center for Environmental Economics for consultation with the Science Advisory Board–Environmental Economics Advisory Committee, U.S. Environmental Protection Agency: Washington, DC, 2010.
39. Kling, C. et al. *Review of Valuing Mortality Risk Reductions for Environmental Policy: A White Paper (December 10, 2010)*; EPA-SAB-11-011. U.S. Environmental Protection Agency: Washington, DC, 2011.
40. U.S. Environmental Protection Agency. *Economic Analysis for the Renovation, Repair, and Painting Program Proposed Rule*; U.S. Environmental Protection Agency: Washington, DC, 2006.
41. Landrigan, P.J.; Schechter, C.B.; Lipton, J.M.; Fahs, M.C.; Schwartz, J. Environmental pollutants and disease in American children: Estimates of morbidity, mortality, and costs for lead poisoning, asthma, cancer, and developmental disabilities. Environ. Health Perspect. **2002**, *110* (7), 721–728.
42. Gould, E. Childhood lead poisoning: Conservative estimates of the social and economic benefits of lead hazard control. Environ. Health Perspect. **2009**, *117* (17), 1162–1167.
43. Hammitt, J.K.; Haninger, K. Valuing morbidity risk: Willingness to pay per quality-adjusted life year. Toulouse School of Economics, LERNA Working Paper Series, 2011.
44. Cameron, T.A.; DeShazo, J.R. Demand for health risk reductions. Unpublished manuscript, 2010.

LEED-EB: Leadership in Energy and Environmental Design for Existing Buildings

Rusty T. Hodapp
Energy and Transportation Management, Dallas/Fort Worth International Airport Board, Dallas/Fort Worth Airport, Texas, U.S.A.

Abstract
The United States Green Building Council's LEED for Existing Buildings: Operations and Maintenance (LEED-EB O&M) Green Building Rating System™ is presented in overview. LEED-EB O&M is a means of guiding, measuring, and verifying green building operation, maintenance, and management. Distinct from the other products in the LEED family of rating systems, EB focuses on building operations and maintenance from a whole-building perspective. As a road map to improving building performance, LEED-EB O&M (and green O&M practices in general) should be of interest to facility managers, energy managers, owners, and other stakeholders. With new construction typically amounting to only 3%–5% of the total stock of buildings in any year, existing buildings represent the greatest opportunity available to reduce the industry's significant consumption of energy and other resources as well as to improve the indoor environment's impact on human health.

INTRODUCTION

Existing buildings comprise a significant proportion of the total building stock in the United States and building operations consume large amounts of resources (energy, water, building materials, land, etc.), while generating great amounts of waste. For example, in the United States, commercial and industrial buildings alone are estimated to be responsible for the following[1]:

- 38.9% of primary energy use
- 38% of all CO_2 emissions
- 72% of all electricity use
- 13.6% of all potable water use
- 170 million tons of construction and demolition debris
- Using 40% of raw materials globally

Furthermore, because the average person spends 90% of their time indoors, the quality of a building's interior environment impacts virtually everyone. This suggests a very personal interest in better buildings in addition to the national implications of large-scale resource consumption. Issues such as these have driven government, corporate, and personal interest in sustainability and "green" topics. Applied to the building industry, this interest is forcefully seen in the rise of green building certification programs and, in particular, the Leadership in Energy and Environmental Design (LEED) Green Building Rating System™ of the United States Green Building Council (USGBC). Intended to guide the development and verify the performance of green buildings, LEED rating systems have become well accepted as a national standard. Consisting of a number of products, the LEED Green Building Rating System largely focuses on design and construction of new buildings. One rating system, however, known as LEED for Existing Buildings: Operations and Maintenance (LEED-EB O&M) is oriented at the operation, maintenance, and management of existing buildings. LEED-EB O&M is of particular interest to facility managers, energy managers, owners, or others interested in reducing operating costs, improving indoor environmental quality, and minimizing the environmental impact of buildings as a growing body of case study and research evidence suggests that these outcomes are linked to green building practices.

This entry presents an overview of the LEED-EB O&M Green Building Rating System, its benefits and distinctions from other LEED rating systems, how it is organized and implemented, and the value of high-performance green buildings.

INTRODUCTION TO LEED

LEED stands for the USGBC's family of standards for rating "green buildings." The LEED Green Building Rating System is USGBC's effort to provide a national standard to define a green building. Used as guideline for design, construction, operation, and maintenance and with third-party certification, LEED provides a consistent, credible means of developing and operating high-performance, environmentally sustainable buildings.

United States Green Building Council

Formed in 1993, the USGBC has become perhaps the most prominent "green building" organization in America.

With more than 18,000 member organizations, a network of 78 local affiliates, and more than 140,000 LEED Professional Credential holders, the non-profit organization works through leaders in all sectors of the building industry to advance buildings that are environmentally responsible, profitable, and healthy places to live and work. Driving its mission to transform the building marketplace to sustainability is the Council's LEED Green Building Rating System and related training and professional accreditation programs. USGBC also supports education, research, and advocacy programs as well as strategic alliances with key industry, research organizations, and federal, state, and local government agencies to transform the building market.[2]

LEED Green Building Rating System

The LEED Green Building Rating System is a voluntary, consensus-based, market-driven building rating system. LEED assesses sustainability from a whole-building perspective by evaluating five key areas of a building's performance in terms of economic, environmental, and human health impact: sustainable site development, water savings, energy efficiency, materials selection, and indoor environmental quality. Buildings are awarded different levels of certification (Certified, Silver, Gold, Platinum) based upon the amounts of credits satisfied and points earned. These credits are performance oriented and intended to address specific impacts inherent in the design, construction, operation, and maintenance of buildings.

The initial LEED rating system (referred to as LEED Version 1.0) was released in August 1998. LEED 1.0 was extensively modified and released as Version 2.0 in March 2000 with LEED Version 2.1 following in 2002 and Version 2.2 in 2005. LEED has continued to evolve, undertaking new initiatives and expanding into a family of products.

As of April 2009, the portfolio of LEED rating systems consists of several products targeting specific sectors of the buildings market:

- LEED 2009 for New Construction and Major Renovation
- LEED 2009 for Commercial Interiors
- LEED 2009 for Core and Shell Development
- LEED 2009 for Existing Buildings: Operations and Maintenance
- LEED 2009 for Schools
- LEE 2009 for Retail
- LEED for Healthcare
- LEED for Homes
- LEED 2009 for Neighborhood Development

With new products, technologies, and design innovations coming into the green building marketplace daily, the Rating Systems and Reference Guides will continue to evolve as necessary to stay current and relevant.[3]

Benefits of Green Building Certification

LEED provides a guidebook for the design, construction, operation, and maintenance of green buildings, the general benefits of which will be described in more detail later in this entry. The LEED rating system is flexible in order to provide owners and design teams the ability to accommodate circumstances or goals specific to their project. The rigorous and independent certification process provides firm and compelling proof that the building has achieved the sustainable goals established for it and is performing as intended. The credible assurance that a building is in fact green can be valuable to owners, occupants, investors, and other key stakeholders in the industry as well as the public at large.

LEED and Existing Buildings

With the exception of LEED-EB O&M, the LEED family of products is intended to address the design and construction phases of buildings. The primary users of these products are architects, engineers, construction contractors, and building owners. A building designed and constructed to LEED standards has verifiably incorporated green or sustainable features and, therefore, should perform better in the key impact areas (economic returns, environmental impact, and occupant health and comfort) than a typical building built to basic code standards. Addressing the design and construction phase of a building's life is extremely important because it is in these phases that many irreversible decisions with long-term impacts on the building's performance are made.

Fig. 1 depicts graphically what has become well established regarding the life-cycle cost of buildings: 1) the majority (75% or more) of total life-cycle costs occur after construction (i.e., during the O&M phase); and 2) many of the decisions that drive long-term cost occur during programming and design.[4]

However, the fact is that new buildings represent a very small percentage of the total commercial building stock in the United States. According to some sources, new construction amounts to 2% of the total stock of build-

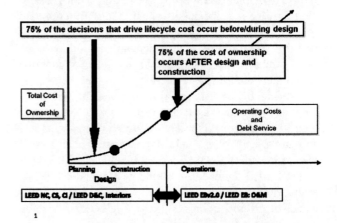

Fig. 1 Building life-cycle cost curve.

ings in any one year.[5] No doubt it is safe to assume that the majority of these existing buildings were not designed or are being operated and maintained to green standards. In order to improve the building sector's performance in terms of sustainability, clearly the existing stock must be addressed in addition to new construction. Similarly, since the post-construction phase of a building's life cycle contributes disproportionately to its total cost, resource consumption, and impact on users, standards for operations and maintenance are necessary to maximize the benefits of green practices in the building sector. This is exactly the focus of the LEED-EB O&M Rating System.

LEED for Existing Buildings: Operations and Maintenance

The LEED-EB O&M Rating System is a set of performance standards for sustainable operations and maintenance of existing buildings of various types and all sizes. It is intended to advance high-performance, healthy, durable, affordable, and environmentally sound practices in existing buildings. LEED-EB O&M provides an entry point into the LEED certification process for the existing building stock. It can be used for buildings new to LEED certification as well as those previously certified under LEED-NC. The USGBC began developing LEED-EB in 2000 and it was tested in a pilot phase involving more than 100 buildings in 2002. The final version (Version 2.0) was released in October 2004. The current version, LEED-EB O&M, was released in April 2009 under the suite of 2009 LEED rating systems and has been further updated in April 2010. The introduction to the LEED 2009 for Existing Buildings: Operations and Maintenance Green Building Rating System states "LEED for Existing Buildings Operations and Maintenance encourages owners and operators of existing buildings to implement sustainable practices and reduce the environmental impacts of their buildings over their functional life cycles." To achieve this, LEED-EB O&M addresses exterior building site maintenance, water and energy use, environmentally preferred products and practices for cleaning and alternations, waste stream management, and ongoing indoor environmental quality.[6]

Issues Addressed by LEED-EB O&M

LEED-EB O&M addresses all the key facets of building operations and maintenance that impact total cost of ownership, the environment, and building occupants. Some examples include the following:

- Energy use
- Water use
- Building operations and maintenance
- Building systems (e.g., mechanical, electrical, plumbing) performance
- Maintenance of building exterior and site
- Ventilation and indoor air quality
- Lighting quality
- Thermal comfort of spaces
- Green cleaning
- Recycling programs
- Green product purchasing programs
- Management of indoor pollutants and toxic substances
- Systems upgrades[7]

LEED-EB O&M seeks to address sustainability on an ongoing basis. This largely falls under the scope of those involved in managing and operating buildings, and clearly, their involvement and expertise are necessary to successfully certify a building under LEED-EB O&M. To the extent the benefits of green buildings and popularity of standards like LEED continue growing, market forces will create new opportunities for facility managers, energy managers, etc., demand for their services, and highlight the overall value of their contributions.

Key Distinctions between LEED-EB O&M and Other LEED Products

Although sharing many common features in terms of structure and process with other LEED products, LEED-EB O&M is fundamentally distinct in three key ways. LEED-NC (and the other new building-oriented products) is essentially a one-time event whereas LEED-EB O&M represents an ongoing process. Second, with LEED-NC, the green building process ends after the design and construction phase. For LEED-EB O&M, the green building phase is a continuous process that deals with ongoing operations, maintenance, and upgrades of a building over its life cycle. Buildings certified under LEED-EB O&M require recertification at least once every 5 years. Finally, given their focus on different phases of a building's life cycle, LEED-NC is primarily a capital budget event while LEED-EB O&M deals with operating budgets.[8]

OVERVIEW OF LEED-EB O&M

In the same manner as all LEED products, the LEED-EB O&M Rating System is based on evaluations of a building in seven categories:

1. Sustainable sites
2. Water efficiency
3. Energy and atmosphere
4. Materials and resources
5. Indoor environmental quality
6. Innovation (in this case, in operations)
7. Regional priority

Minimum Program Requirements

All projects must meet certain minimum program requirements (MPRs) to be eligible for certification under the LEED 2009 rating systems. MPRs define the minimum characteristics that a project must possess in order to be certified and are intended to: 1) provide clear guidance to users; 2) protect the integrity of LEED program; and 3) reduce challenges during the certification process. The LEED 2009 MPRs for EB O&M are as follows:

1. Must comply with environmental laws
2. Must be a complete, permanent building or space
3. Must use a reasonable site boundary
4. Must comply with minimum floor area requirements
5. Must comply with minimum occupancy rates
6. Must commit to sharing whole-building energy and water usage data
7. Must comply with a minimum building area-to-site area ratio

The ongoing performance data from buildings required as part of the certification will be compiled and used to establish benchmarks for building performance and provide operators an idea of how their building compares on water and energy use. To further its commitment to improving building performance, the USGBC launched the Building Performance Initiative (BPI) in August 2009 to complement the MPR for ongoing performance data. The BPI will make the data collected available to building owners for analysis and feedback.

Prerequisites

Also consistent with other LEED rating systems, LEED-EB O&M requires every project to meet certain prerequisites in order to be considered for certification (see the list of prerequisites by category in Table 1). All prerequisites must be satisfied for a project to be eligible for certification.

The prerequisites include such items as minimum levels of water and energy efficiency, building commissioning, no CFC refrigerants, no-smoking policy, and other basic elements of a high-performance, green building operation. A key prerequisite involves a minimum performance period for the building. LEED-EB O&M requires buildings to be in operation for a minimum of 12 continuous months before certifying (3 months for all prerequisites and credits except Energy and Atmosphere Prerequisite 2 and Credit 1, which require a minimum of 12 months).

Credits and Points

Buildings achieve certification under all LEED products by accumulating a certain number of credit points. Points can be obtained in any combination within and among the credits and categories (see Table 2 for the credit and point breakdown for LEED-EB O&M). All LEED rating systems have 100 base points, and up to 10 bonus points can be earned through Innovation and Regional Priority credits.

Each LEED rating system uses the same format for prerequisites and credits. The sections include the following:

- *Intent*—describes the main goal of the prerequisite or credit.
- *Requirements*—specifies the criteria needed to satisfy the prerequisite or credit.
- *Submittals*—specifies the documentation required to demonstrate compliance with the prerequisite or credit.
- *Potential Technologies and Strategies*—identifies means and methods that project teams may consider to achieve the prerequisite or credit.

Certification Levels

LEED rating systems allow buildings to achieve various levels of certification based on points achieved (see Table 3 for the certification levels for LEED-EB O&M).

Registration Process

With the launch of LEED Version 3 in 2009, USGBC implemented a new certification model. LEED v3 consists of three components:

- LEED 2009 rating systems.
- An upgrade to LEED Online to make it faster and easier to use.
- New building certification model—an expanded infrastructure based on ISO standards, administered by the

Table 1 LEED 2009 for EB O&M prerequisites.

Category	Prerequisites
Sustainable sites	0
Water efficiency	1
Energy and atmosphere	3
Materials and resources	2
Indoor environmental quality	3
Total	9

Table 2 LEED-EB O&M credits and points.

Category	Credits	Points
Sustainable sites	9	26
Water efficiency	4	14
Energy and atmosphere	9	35
Materials and resources	9	10
Indoor environmental quality	15	15
Innovation in operations	3	6
Regional priority	1	4
Total	50	110

Table 3 LEED-EB O&M certification levels.

Certification level	Points required
Certified	40–49
Silver	50–59
Gold	60–79
Platinum	80+

Green Building Certification Institute (GBCI) for improved capacity, speed, and performance.

LEED Online is the primary resource for managing the LEED documentation process. It allows project teams to manage project details, complete documentation requirements for LEED credits and prerequisites, upload supporting files, submit applications for review, receive reviewer feedback, and ultimately earn LEED certification. The GBCI is an independent, third-party organization that has assumed administration of LEED certification for all commercial and institutional projects registered under any of the LEED rating systems.

The process of certifying a building under any of the LEED rating systems is essentially the same—the project is first registered with GBCI using LEED Online. Once a project is registered, access to software tools, supplemental resources, sample documentation, credit interpretation rulings, and other essential information is provided. For LEED-EB O&M, the initial application (application for standard review) must be submitted within 60 calendar days of the performance periods used. The application has to include complete documentation for all prerequisites and enough points for certification. GBCI reviews the application and designates each credit and prerequisite as anticipated pending or denied. This preliminary standard review is targeted (but not guaranteed) for completion within 25 business days of receipt of the application. Within 25 days of receiving GBCI's preliminary standard review, the owner may submit a response including any revised documentation. GBCI will then review and return comments for all credits and prerequisites in response to the preliminary Standard Review and designate each as awarded or denied. This final Standard Review is targeted for completion within 15 business days of receipt of the completed application. The owner then accepts or appeals the final review. Following acceptance of the final certification review, LEED projects:

1. Will receive a formal certificate of recognition
2. Will receive information on how to order plaque and certificates, photo submissions, and marketing
3. May be included (at the owner's discretion) in the online LEED Project Directory of registered and certified projects
4. May be included (along with photos and other documentation) in the U.S. Department of Energy High Performance Buildings Database[9]

IMPLEMENTATION PROCESS

From a practical standpoint, the process of implementing LEED-EB O&M should generally involve the following steps:

1. Become familiar with the LEED-EB O&M Rating System
2. Gain the support of key decision makers and stakeholders
3. Form a project team
4. Conduct a preliminary building audit and identify corrective actions required to meet prerequisites and/or opportunities
5. Establish project goals related to target certification level, credits to be pursued, and budget
6. Register the project
7. Create and adopt policies and procedures, implement upgrades, make operational changes, etc., in accordance with the project goals
8. Track performance
9. Assemble and submit required documentation (preliminary and any required corrections or resubmittals)
10. Achieve certification

The minimum performance period for initial certifications under LEED-EB O&M is 12 months. During this period, actual operational performance must be tracked and reported. The performance tracking period can be as long as 2 years depending upon the project goals and/or implementation strategy.

The USGBC provides project teams with numerous resources including the LEED Online system for managing and preparing the certification application, credit templates that define supporting documentation needed and compliance calculations, credit interpretation rulings that can help answer questions on credits and implementation strategies, and the LEED Reference Guides.[10,11]

BENEFITS OF GREEN BUILDINGS

The premise inherent in the LEED rating systems is that "green" buildings provide superior value to owners, occupants, and other stakeholders. Typically, the value proposition construct for sustainability, green buildings, etc., is the well-known triple bottom line of economic returns, environmental impact, and social benefit. Green buildings, also known (perhaps more accurately) as high-performance buildings, are premised as providing superior economic returns, reduced environmental impact, and enhanced social benefits. Such buildings in theory:

- Were properly built and/or are well operated and maintained

- Use resources (e.g., energy, water, building materials, O&M supplies) more efficiently
- Provide a safer, more comfortable, and productive working environment for occupants

In fact, there is a robust and growing body of evidence in research and case study that supports these claims. The following examples are illustrative.

A report commissioned by California's Sustainable Building Task Force found that energy savings alone exceeded the average cost increase associated with 33 different LEED buildings studied. When adding the life-cycle cost benefits of water savings, reduced emissions, operations and maintenance efficiencies, and improved occupant productivity and health, the 20 years net present value of the financial benefits of green buildings exceeded the implementation costs by as much as 10–15 times.[12] Case studies on commissioning alone show that construction and operating costs can be reduced from 1 to 70 times the initial cost of commissioning.[13] Improved thermal comfort, reduced indoor pollutants, enhanced ventilation rates, and other characteristics of green buildings have been found to have positive impacts on occupant productivity, student test scores, absenteeism, and incidences of various sicknesses.[14]

Other benefits continue to be demonstrated in case studies and research, including the following:

- Increased building value
- Risk mitigation
- Employee loyalty and recruitment
- Brand image and public relations
- Environmental stewardship[15]

ENERGY EFFICIENCY POTENTIAL OF GREEN BUILDINGS

In LEED 2009, points are allocated among credits based on the potential environmental impacts and human benefits of each credit. As a result, the allocation of points significantly changed in comparison to previous versions of the LEED rating systems. These changes increased the relative importance of reducing energy consumption and building-related greenhouse gas emissions. Reflecting this, one credit in LEED 2009 EB O&M comprises the largest potential amount of points—Energy and Atmosphere Credit 1 Optimize Energy Efficiency Performance provides the opportunity for 18 possible points. Furthermore, with its emphasis (and associated requirements) on demonstrated performance, LEED-EB O&M presents tremendous potential to reduce energy consumption throughout the commercial building sector. When considering the fact that existing buildings comprise some 95% of the commercial building stock in the United States, the magnitude of potential reduction is immense. To put this potential in context, consider U.S. Energy Information Administration (EIA) projections of the impact of energy efficiency on per capita commercial energy consumption. In their *Annual Energy Outlook 2010*, EIA estimates per capita commercial energy consumption in 2035 could be decreased by as much as 17.5% depending upon the degree to which technology-based efficiency improvements are deployed throughout the sector.[16] McKinsey and Company estimates that, by 2020, the United States could reduce annual energy consumption by 9.1 quadrillion British thermal units (BTUs) (23%) of end-use energy (18.4 quadrillion BTUs in primary energy) from a business-as-usual baseline by deploying an array of energy efficiency measures with the commercial sector accounting for 25% of this potential. At full potential, the projected efficiency improvements could reduce greenhouse gas emissions by as much as 1.1 gigatons of CO_2 by 2020 and serve as a bridge to low-carbon energy sources.[17] Similarly, the Electric Power Research Institute (EPRI) estimates that the combination of energy efficiency and demand response programs has the potential to reduce summer peak electric demand in the United States by 157 GW to 218 GW (14% to 20%) by 2030.[18] Finally, consider the potential national impact of LEED energy savings presented in the Green Building Market and Impact Report 2009. This report projects that given the acceleration of LEED adoption, energy savings in the United States could reach 1.75 quadrillion BTUs by 2020 and 3.9 quadrillion BTUs by 2035 (8.3% and 17.3%, respectively, of national annual commercial building energy use). A best-case scenario could see those savings rise to 22.3% by 2030.[19]

CONCLUSION

The USGBC's LEED Green Building Rating System has become a well-recognized standard for guiding the development of green, high-performance buildings and for verifying that established green building goals have been accomplished. The LEED family of rating systems is focused on the design and construction process for new buildings. However, one rating system—LEED-EB O&M—focuses on operations and maintenance of existing buildings. LEED-EB O&M is of particular interest to facility managers, energy managers, and other professionals involved in building operation and management. LEED-EB O&M provides a guidebook for those interested in "greening" their existing building stock. Implementing these green processes and practices can be an effective means of reducing a building's life-cycle costs, reducing its environmental impact, and improving occupant health and productivity.

REFERENCES

1. United States Green Building Council. Green Building Facts. Available at http://www.usgbc.org/ (accessed August 2007).

2. United States Green Building Council. About USGBC. Available at http://www.usgbc.org/ (accessed August 2007).
3. United States Green Building Council. *LEED Reference Guide for Green Building Design and Construction 2009 Edition*; USGBC: Washington, 2009; xii.
4. National Research Council. *Investments in Federal Facilities, Asset Management Strategies for the 21st Century*; The National Academies Press: Washington, 2004; 27.
5. Architecture2030. Available at http://architecture2030.org/the_solution/buildings_solution_how (accessed November 2010).
6. United States Green Building Council. *LEED 2009 for Existing Buildings: Operations and Maintenance Rating System (Updated April 2010)*; USGBC: Washington, 2010; xvi.
7. Opitz, M. What LEED-EB Is and Why to Use It, available at http://www.fmlink.com/ProfResources/Sustainability/Articles/article.cgi?USBGC:200604-01.html (accessed August 2007).
8. United States Green Building Council. LEED-EB Presentation. Available at http://www.usgbc.org/ (accessed August 2007).
9. Green Building Certification. GBCI LEED Certification Manual. Available at http://www.gbci.org/main-nav/building-certification/leed-certification.aspx (accessed July 2010).
10. Opitz, M. Starting and Managing Your LEED-EB Project, available at http://www.fmlink.com/ProfResources/Sustainability/Articles/article.cgi?USBGC:200609-20.html (accessed September 2007).
11. United States Green Building Council. *LEED 2009 for Existing Buildings: Operations and Maintenance Rating System (Updated April 2010)*. USGBC: Washington, 2010; xvi–xxiv.
12. Kats, G.; Alevantis L.; Berman, A.; Mills, E.; Perlman, J. The Costs and Financial Benefits of Green Buildings: A Report to California's Sustainable Building Task Force, available at http://www.cap-e.com/ewebeditpro/items/O59F3259.pdf (accessed September 2006).
13. ASHRAE. *ASHRAE GreenGuide*; Elsevier: Burlington, 2006; 14.
14. Callan, D. Green Building Report: Studies Relate IAQ and Productivity. Build. Operating Manage. **2006**, *52* (11), 66–68.
15. Yudelson, J. *Marketing Green Buildings, Guide for Engineering, Construction and Architecture*; The Fairmont Press: Lilburn, 2006; 51–67.
16. U.S. Energy Information Administration, U.S. Department of Energy. *Annual Energy Outlook 2010*, 2010, 59.
17. Choi Granade, H.; Creyts, J.; Derkach, A.; Farese, P.; Nyquist, S.; Ostrowski, K. Unlocking Energy Efficiency in the U.S. Economy, available at http://www.mckinsey.com/USenergyefficiency (accessed July 2010).
18. Faruqui, A.; Hledik, S.; Rohmund, I.; Sergici, S.; Siddiqui, O.; Smith, K.; Wikler, G.; Yoshida, S. *Assessment of Achievable Potential from Energy Efficiency and Demand Response Programs in the U.S. (2010–2030)*. EPRI: Palo Alto, CA, 2009; xi.
19. Watson, R. Green Building Market and Impact Report, 2009, available at http://www.greenbiz.com/business/research/report/2009/11/05/green-building-market-and-impact-report-2009 (accessed July 2010).

LEED-NC: Leadership in Energy and Environmental Design for New Construction

Stephen A. Roosa
Energy Systems Group, Inc., Louisville, Kentucky, U.S.A.

Abstract
The Leadership in Energy and Environmental Design (LEED™) Green Building Rating System is a set of rating systems for various types of construction projects. Developed by the U.S. Green Building Council (USGBC), the rating systems evolved with the intent of helping to "fulfill the building industry's vision for its own transformation to green building." The rating systems developed by the USGBC are for new construction, existing buildings, core and shell construction, commercial interiors, homes, and residences. This entry considers one of these systems, LEED-NC. It discuses the influences that shaped development of LEED-NC for new construction and major renovations, details how the process works, and considers how new projects are scored. The entry concludes by providing a brief assessment of the program's strengths and weaknesses.

INTRODUCTION

Land development practices have yielded adverse environmental consequences, urban dislocation, and changes in urban infrastructure. Urban development in particular has long been associated with reduced environmental quality and environmental degradation.[2] The rate at which undeveloped land is being consumed for new structures—and the growing appetite of those structures for energy and environmental resources—has contributed to ecosystem disruption and has fostered impetus to rethink how buildings are sited and constructed. While urban developmental patterns have been associated with environmental disruptions at the local and regional scales, the scientific assessments of global impacts have yielded mixed results. In part as a reaction to U.S. development patterns that have traditionally fostered suburbanization and subsidized automobile-biased transportation infrastructure, design alternatives for structures with environmentally friendly and energy efficient attributes have become available.

According to the United Nations Commission on Sustainable Development, "air and water pollution in urban areas are associated with excess morbidity and mortality ... Environmental pollution as a result of energy production, transportation, industry or lifestyle choices adversely affects health. This would include such factors as ambient and indoor air pollution, water pollution, inadequate waste management, noise, pesticides and radiation."[3] It has been demonstrated that a relationship exists between the rates at which certain types of energy policies are adopted at the local level and select indicators of local sustainability.[4] As more urban policies focus on the built environment, buildings continue to be the primary building blocks of urban infrastructure. If buildings can be constructed in a manner that is less environmentally damaging and more energy efficient, then there is greater justification to label them as "green" buildings.

The concept of sustainability has evolved from considerations of land development, population growth, fossil fuel usage, pollution, global warming, availability of water supplies, and the rates of resource use.[5] Thankfully, a vocabulary of technologies and methodologies began to develop in the 1970s and 1980s that responded to such concerns. Driven by ever increasing energy costs, energy engineers began to apply innovative solutions, such as use of alternative energy, more efficient lighting systems and improved electrical motors. Controls engineers developed highly sophisticated digital control systems for heating, ventilating and air conditioning systems. With growing concerns about product safety and liability issues regarding the chemical composition of materials, manufacturers began to mitigate the potential adverse impacts of these materials upon their consumers. Resource availability and waste reduction became issues that began to influence product design. In the span of only 25 years, local governments made curbside recycling programs in larger U.S. cities nearly ubiquitous. Terms and phrases such as "mixed use planning," "brownfield redevelopment," "alternative energy," "micro-climate," "systems approach," "urban heat island effect," "energy assessments," "measurement and verification," and "carrying capacity" created the basis for a new vocabulary which identifies potential solutions. All of these concerns evolved prior to the 1992 U.N. Conference on the Environment and Development, which resulted in the Rio Agenda 21 and clarified the concept sustainability.

In regard to the built environment, architectural designers renewed their emphasis on fundamental design issues, including site orientation, day lighting, shading, landscaping, and more thermally cohesive building shells. Notions of "sick building syndrome" and illnesses like Legionnaires' disease, asthma and asbestosis, jolted architects and engineers into re-establishing the importance of the indoor environmental conditions in general and indoor air quality (IAQ) in particular when designing their buildings.

The decisions as to what sort of buildings to construct and what construction standards to apply are typically made locally. Those in the position to influence decisions in regard to the physical form of a proposed structure include the builder, developer, contractors, architects, engineers, planners, and local zoning agencies. In addition, all involved must abide by regulations that apply to the site and structure being planned. The rule structure may vary from one locale to another. What is alarming is that past professional practice within the U.S. building industry has only rarely gauged the environmental or energy impact of a structure prior to its construction. Prior to the efforts of organizations like the U.S. Green Building Council (USGBC) (established in 1995), the concept of what constituted a "green building" in the United States lacked a credible set of standards.

CONCEPT OF GREEN BUILDINGS

Accepting the notion that sustainable, environmentally appropriate, and energy efficient buildings can be labeled "green," the degree of "greenness" is subject to multiple interpretations. The process of determining which attributes of a structure can be considered "green" or "not green" is inconclusive and subjective. Complicating the process, there are no clearly labeled "red" edifices with diametrically opposing attributes. While it is implied that a green building may be an improvement over current construction practice, the basis of attribute comparison is often unclear, subjective, and confusing. It is often unclear as to what sort of changes in construction practice, if imposed, would lead the way to greener, more sustainable buildings. If determinable, the marketplace must adjust and provide the technologies and means by which materials, components, and products can be provided to construction sites where greener buildings can arise. Since standards are often formative and evolving, gauging the degree of greenness risks the need to quantify subjective concepts.

There are qualities of structures, such as reduced environmental impact and comparatively lower energy usage, which are widely accepted as qualities of green construction practices. For example, use of recycled materials with post-consumer content that originates from a previous use in the consumer market and post-industrial content that would otherwise be diverted to landfills is widely considered an issue addressable by green construction practices. However, evaluation of green building attributes or standards by organizations implies the requirement that decisions be based on stakeholder consensus. This process involves input to the decision-making processes by an array of representative stakeholders in often widely diverse geographic locations. For these and other reasons, developing a rating system for green buildings is both difficult and challenging.

RATING SYSTEMS FOR BUILDINGS

Rating systems for buildings with sustainable features began to emerge in embryonic form in the 1990s. The most publicized appeared in the United Kingdom, Canada, and the United States. In the United Kingdom, the Building Research Establishment Environmental Assessment Method (BREEAM) was initiated in 1990. BREEAM™ certificates are awarded to developers based on an assessment of performance in regard to climate change, use of resources, impacts on human beings, ecological impact, and management of construction. Credits are assigned based on these and other factors. Overall ratings are assessed according to grades that range from pass to excellent.[6]

The International Initiative for a Sustainable Built Environment, based in Ottawa, Canada, has its Green Building Challenge program with more than 15 countries participating. The collaborative venture is geared toward the creation of an information exchange for sustainable building initiatives and the development of "environmental performance assessment systems for buildings."[7] In the United States, agencies of the central government co-sponsored the development of the Energy Star™ program, which provides "technical information and tools that organizations and consumers need to choose energy-efficient solutions and best management practices."[8] Expanding on their success, Energy Star™ developed a building energy performance rating system which has been used for over 10,000 buildings.

Entering the field at the turn of the new century, the USGBC grew from an organization with just over 200 members in 1999 to 3500 members by 2003.[9] The LEED™ rating system is a consensus-developed and reviewed standard, allowing voluntary participation by diverse groups of stakeholders with interest in the application and use of the standard. According to Boucher, "the value of a sustainable rating system is to condition the marketplace to balance environmental guiding principles and issues, provide a common basis to communicate performance, and to ask the right questions at the start of a project."[10] The first dozen pilot projects using the rating system were certified in 2000.

LEED-NC RATING SYSTEM

The USGBC's Green Building Rating System is a voluntary, consensus-developed set of criteria and standards.

This rating system evolved with a goal of applying standards and definition to the idea of high-performance buildings. The use of sustainable technologies is firmly established within the LEED project development process. LEED loosely defines green structures as those that are "healthier, more environmentally responsible and more profitable."[1]

LEED-NC 2.1 is the USGBC's current standard for new construction and major renovations. It is used primarily for commercial projects such as office buildings, hotels, schools, and institutions. The rating system is based on an assessment of attributes and an evaluation of the use of applied standards. Projects earn points as attributes are achieved and the requirements of the standards are proven. Depending on the total number of points a building achieves upon review, the building is rated as Certified (26–32 points), Silver (33–38 points), Gold (39–51 points) or Platinum (52 or more points).[11] Theoretically, there are a maximum of 69 achievable points. However, in real world applications, gaining certain credits often hinders the potential of successfully meeting the criteria of others. While achieving the rating of Certified is relatively easily accomplished, obtaining a Gold or Platinum rating is rare and requires both creativity and adherence to a broad range of prescriptive and conformance-based criteria.

The LEED process involves project registration, provision of documentation, interpretations of credits, application for certification, technical review, rating designation, award, and appeal. Depending on variables such as project square footage and USGBC membership status, registration fees can range up to $7500 for the process.[12]

LEED PREREQUISITIES CATEGORIES AND CRITERIA

To apply for the LEED labeling process, there are prerequisite project requirements which earn no points. For example, in the Sustainable Sites category, certain procedures must be followed to reduce erosion and sedimentation. In the category of Energy and Atmosphere, minimal procedures are required for building systems commissioning. Minimal energy performance standards must be achieved (e.g., adherence to ANSI/ASHRAE/IESNA Standard 90.1-1999, Energy Standard for Buildings Except Low-Rise Residential Buildings, or the local energy code if more stringent), and there must be verification that CFC refrigerants will not be used or will be phased out. In addition, there are prerequisite requirements outlining mandates for storage and collection of recyclable material, minimum IAQ performance (the requirements of ASHRAE Standard 62-1999, Ventilation for Acceptable Indoor Air Quality must be adhered to), and the requirement that non-tobacco smokers not be exposed to smoke.

In addition to the prerequisite requirements, the LEED process assigns points upon achieving certain project criteria or complying with certain standards. The total points are summed to achieve the determined rating. Projects can achieve points from initiatives within the following sets of categories: Sustainable Sites (14 points), Water Efficiency (5 points), Energy and Atmosphere (17 points), Materials and Resources (13 points), and Indoor Environmental Quality (15 points). Use of a LEED Accredited Professional (1 point) to assist with the project[13] earns a single point. Additional points are available for Innovation and Design Process (maximum of 4 points).

Within each category, the specific standards and criteria are designed to meet identified goals. In the category of Sustainable Sites, 20.2% of the total possible points are available. This category focuses on various aspects of site selection, site management, transportation and site planning. The goals of this category involve reducing the environmental impacts of construction, protecting certain types of undeveloped lands and habitats, reducing pollution from development, conserving natural areas and resources, reducing the heat island impacts, and minimizing light pollution. Site selection criteria are designed to direct development away from prime farmland, flood plains, habitat for endangered species and public parkland. A development density point is awarded for projects that are essentially multi-story. If the site has documented environmental contamination or is designated by a governmental body as a brownfield, another point is available. In regard to transportation, four points are available for locating sites near publicly available transportation (e.g., bus lines or light rail), providing bicycle storage and changing rooms, provisions for alternatively fueled vehicles and carefully managing on-site parking. Two points in this category are obtained by limiting site disturbances and by exceeding "the local open space zoning requirement for the site by 25%."[14] In addition, points are available by following certain storm water management procedures, increasing soil permeability, and attempting to eliminate storm water contamination. Potential urban heat island effects are addressed by crediting design attributes such as shading, underground parking, reduced impervious surfaces, high albedo materials, reflective roofing materials, or vegetated roofing. Finally, a point is available for eliminating light trespass.

Water efficiency credits comprise 7.2% of the total possible points. With the goal of maximizing the efficiency of water use and reducing the burden on water municipal systems, points are credited for reducing or eliminating potable water use for site irrigation, capturing and using rainwater for irrigation, and using drought tolerant or indigenous landscaping. This section of the LEED standard also addresses a building's internal water consumption. Points are available for lowering aggregate water consumption and reducing potable water use. Reducing the wastewater quantities or providing on-site tertiary wastewater treatment also earns points.

Energy and Atmosphere is the category that offers the greatest number of points, 24.6% of the total possible. The intents of this category include improving the calibration of equipment, reducing energy costs, supporting alternative energy, reducing the use of substances that cause atmospheric damage, and offering measurement and verification criteria. Optimizing the design energy cost of the regulated energy systems can achieve a maximum of ten points. To assess the result, project designs are modeled against a base case solution which lacks certain energy-saving technologies. Interestingly, the unit of measure for evaluating energy performance to achieve credits is not kilocalories or million Btus, but dollars. Points are awarded in whole units as the percentage of calculated dollar savings increases incrementally. In addition to the ten points for energy cost optimization, a maximum of three additional points is available for buildings that use energy from on-site renewable energy generation. Purchased green power is allocated a single point if 50% of the electrical energy (in kWh) comes from a two year green power purchasing arrangement. This category provides points for additional commissioning and elimination of the use of HCFCs and halon gases. Measurement and Verification (M&V) is allowed a point, but only if M&V options B, C, and D, as outlined in the 2001 edition of the International Measurement and Verification Protocol (IPMVP), are used.

The Materials and Resources category represents 18.8% of the total possible points. This category provides credit for material management; adaptive reuse of structures; construction waste management; resource reuse; use of material with recycled content; plus the use of regionally manufactured materials, certain renewable materials and certified wood products. A point is earned for providing a space in the building for storage and collection of recyclable materials such as paper, cardboard, glass, plastics and metals. A maximum of three points is available for the adaptive reuse of existing on-site structures and building stock. The tally increases with the extent to which the existing walls, floor, roof structure, and external shell components are incorporated into the reconstruction. LEED-NC 2.1 addresses concerns about construction waste by offering a point if 50% of construction wastes (by weight or volume) are diverted from landfills and another point if the total diversion of wastes is increased to 75%. A project that is composed of 10% recycled or refurbished building products, materials, and furnishings gains an additional two points. Another two points are available in increments (one point for 5%, two points for 10%) if post-consumer or post-industrial recycled content (by dollar value) is used in the new construction. To reduce environmental impacts from transportation systems, a point is available if 20% of the materials are manufactured regionally (defined as being within 500 miles or roughly 800 km of the site), and an added point is scored if 50% of the materials are extracted regionally. A point is available if rapidly renewable materials (e.g., plants with a ten year harvest cycle) are incorporated into the project, and yet another point is earned if 50% of the wood products are certified by the Forest Stewardship Council.

The category of Indoor Environmental Quality allows 21.7% of the possible total points available. The goals include improving IAQ, improving occupant comfort, and providing views to the outside. With ASHRAE Standard 62-1999 as a prerequisite, an additional point is available for installing CO_2 monitoring devices in accordance with occupancies referenced in ASHRAE Standard 62-2001, Appendix C. A point is also available for implementing technologies that improve upon industry standards for air change effectiveness or that meet certain requirements for natural ventilation. Systems that provide airflow using both underfloor and ceiling plenums are suggested by LEED documentation as a potential ventilation solution. Points are available for developing and implementing IAQ management plans during construction and prior to occupancy. The requirements include using a Minimum Efficiency Reporting Value (MERV) 13 filter media with 100% outside air flush-out prior to occupancy. There are points available for use of materials that reduce the quantity of indoor air pollutants in construction caused by hazardous chemicals and by volatile organic compounds in adhesives, sealants, paints, coatings, composite wood products, and carpeting. A point is offered for provision of perimeter windows and another for individual control of airflow, temperature, and lighting for half of the non-perimeter spaces. Points are available for complying with ASHRAE Standard 55-1992 (Thermal Environmental Conditions for Human Occupancy), Addenda 1995, and installing permanent temperature and humidity control systems. Finally, points are gained for providing 75% of the spaces in the building with some form of daylighting and for providing direct line of-sight vision for 90% of the regularly occupied spaces.

In the category of Innovation and Design Process, 7.2% of the total possible points are available. The innovation credits offer the opportunity for projects to score points as a result of unusually creative design innovations, such as substantially exceeding goals of a given criteria or standard.

ASSESSING LEED-NC

The LEED-NC process has numerous strengths. Perhaps the greatest is its ability to focus the owner and design team on addressing select energy and environmental considerations early in the design process. The LEED design process brings architects, planners, energy engineers, environmental engineers, and IAQ professionals into the program at the early stages of design development. The team adopts a targeted LEED rating as a goal for the project. A strategy evolves based on selected criteria. The team members become focused on fundamental green design practices that have often been overlooked when traditional design development processes were employed.

Furthermore, the LEED program identifies the intents of the environmental initiatives. Program requirements are stated and acceptable strategies are suggested. Scoring categories attempt to directly address certain critical environmental concerns. When appropriate, the LEED-NC program defers to engineering and environmental standards developed outside of the USGBC. The components of the program provide accommodation for local regulations. Case study examples, when available and pertinent, are provided and described in the LEED literature. To expedite the process of documenting requirements, letter templates and calculation procedures are available to program users. The educational aspects of the program, which succinctly describe select environmental concerns, cannot be understated. A Web site provides updated information on the program with clarifications of LEED procedures and practice. The training workshops sponsored by the USGBC are instrumental in engaging professionals with a wide range of capabilities.

These considerations bring a high degree of credibility to the LEED process. Advocates of the LEED rating system have hopes of it becoming the pre-eminent U.S. standard for rating new construction that aspires to achieve a "green" label. To its credit, it is becoming a highly regarded standard and continues to gain prestige. Nick Stecky, a LEED accredited professional, firmly believes that the system offers a "measurable, quantifiable way of determining how green a building is."[15]

Despite its strengths, the LEED-NC has observable weaknesses. The LEED-NC registration process can appear to be burdensome, and has been perceived as slowing down the design process and creating added construction cost. Isolated cases support these concerns. Kentucky's first LEED-NC school, seeking a Silver rating, was initially estimated to cost over $200/ft^2 ($2152/m^2) compared to the local standard costs of roughly $120/ft^2 ($1290/m^2) for non-LEED construction. However, there are few comparative studies available to substantially validate claims of statistically significant cost impact. Alternatively, many case studies suggest that there is no cost impact as a result of the LEED certification process. It is also possible that the savings resulting from the use of certain LEED standards (e.g. reduced energy use) can be validated using life-cycle costing procedures. Regardless, LEED-NC fails as a one-size-fits-all rating system. For new construction, Kindergarten to 12th-grade (K-12), school systems in New Jersey, California, and elsewhere have adopted their own sustainable building standards.

There are other valid concerns in regard to the use of LEED-NC. In an era when many standards are under constant review, standards referenced by LEED are at times out of date. The ASHRAE Standard 90.1-1999 (without amendments) is referenced throughout the March 2003 revision of LEED-NC. However, ASHRAE 90.1 was revised, republished in 2001, and the newer version is not used as the referenced standard. Since design energy costs are used to score Energy and Atmosphere points, and energy use comparisons are baselined against similar fuels, cost savings from fuel switching is marginalized. In such cases, the environmental impact of the differential energy use remains unmeasured, since energy units are not the baseline criteria. There is no energy modeling software commercially available that has been specifically designed for assessing LEED buildings. LEED allows most any energy modeling software to be used, and each has its own set of strengths and weaknesses when used for LEED energy modeling purposes. It is possible for projects to comply with only one energy usage prerequisite, applying a standard already widely adopted, and still become LEED certified. In fact, it is not required that engineers have specialized training or certification to perform the energy models. Finally, LEED documentation lacks System International (SI) unit conversions, reducing its applicability and exportability.

A number of the points offered by the rating system are questionable. While indoor environmental quality is touted as a major LEED concern, indoor mold and fungal mitigation practices, among the most pervasive indoor environmental issues, are not addressed and are not necessarily resolvable using the methodologies prescribed. It would seem that having a LEED-accredited professional on the team would be a prerequisite rather than a optional credit. Projects in locations with abundant rainfall or where site irrigation is unnecessary can earn a point by simply documenting a decision not to install irrigation systems. The ability of the point system to apply equally to projects across varied climate classifications and zones is also questionable and unproven.

While an M&V credit is available, there is no requirement that a credentialed measurement and verification professional be part of the M&V plan development or the review process. Without the rigor of M&V, it is not possible to determine whether or not the predictive preconstruction energy modeling was accurate. The lack of mandates to determine whether or not the building actually behaves and performs as intended from an energy cost standpoint is a fundamental weakness. This risks illusionary energy cost savings. Finally, the M&V procedures in the 2001 IPMVP have undergone revision and were not state-of-the-art at the time that LEED-NC was updated in May 2003. For example, there is no longer a need to exclude Option A as an acceptable M&V alternative.

The LEED process is not warranted and does not necessarily guarantee that in the end, the owner will have a "sustainable" building. While LEED standards are more regionalized in locations where local zoning and building laws apply, local regulations can also preempt certain types of green construction criteria. Of greater concern is that it is possible for a LEED certified building to devolve into a building that would lack the qualities of a certifiable building. For example, the owners of a building may choose to remove bicycle racks, refrain from the purchase of green energy after a couple of years, disengage control systems, abandon their M&V program, and remove recycling centers—yet retain the claim of owning a LEED certified building.

CONCLUSION

The ideal of developing sustainable buildings is a response to the environmental impacts of buildings and structures. Developing rating systems for structures is problematic due to the often subjective nature of the concepts involved, the ambiguity or lack of certain standards, and the local aspects of construction. While there are a number of assessment systems for sustainable buildings used throughout the developed world, LEED-NC is becoming a widely adopted program for labeling and rating newly constructed "green" buildings in the United States. Using a point-based rating system, whereby projects are credited for their design attributes, use of energy, environmental criteria, and the application of select standards, projects are rated as Certified, Silver, Gold, or Platinum.

The LEED-NC program has broad applicability in the United States and has been proven successful in rating roughly 150 buildings to date. Its popularity is gaining momentum. Perhaps its greatest strength is its ability to focus the owner and design team on energy and environmental considerations early in the design process. Today, there are over 1700 projects that have applied for LEED certification. Due to the program's success in highlighting the importance of energy and environmental concerns in the design of new structures, it is likely that the program will be further refined and updated in the future to more fully adopt regional design solutions, provide means of incorporating updated standards, and offer programs for maintaining certification criteria. It is likely that the LEED program will further expand, perhaps offering a separate rating program for K-12 educational facilities. Future research will hopefully respond to concerns about potential increased construction costs and actual energy and environmental impacts.

REFERENCES

1. U.S. Green Building Council. *LEED green building rating system*. 2004.
2. Spirn, A.W. *The Granite Garden: Urban Nature and Human Design*; Basic Books: New York, 1984.
3. United Nations. *Indicators of Sustainable Development: Guidelines and Methodologies*; United Nations: New York, 2001; 38.
4. Roosa, S.A. *Energy and Sustainable Development in North American Sunbelt Cities*; RPM Publishing: Louisville, KY, 2004.
5. Koeha, T. What is Sustainability and Why Should I Care? *Proceedings of the 2004 world energy engineering congress*, Austin, TX, Sept 22–24, 2004; AEE: Atlanta 2004.
6. URS. http://www.urseurope.com/services/engineering/engineering-breeam.htm (accessed Feb 2005).
7. iiSBE. International Initiative for a Sustainable Built Environment http://iisbe.org/iisbe/start/iisbe.htm (accessed Feb 2005).
8. United States Environmental Protection Agency. *Join energy star—Improve your energy efficiency*. http://www.energystar.gov. (accessed June 2003).
9. Gonchar, J. Green building industry grows by leaps and bounds; Engineering News-Record; 2003.
10. Boucher, M. Resource efficient buildings—Balancing the bottom line. *Proceedings of the 2004 world energy engineering congress*, Austin, TX, Sept 22–24, 2004; AEE: Atlanta 2004.
11. U.S. Green Building Council. *Green Building Rating System for New Construction and Major Renovations Version 2.1 Reference Guide*; May 2003; 6.
12. U.S. Green Building Council. *LEED—Certification Process*. http://www.usgbc.org/LEED/Project/certprocess.asp. (accessed Oct 2004).
13. U.S. Green Building Council. *Green Building Rating System for New Construction and Major Renovations Version 2.1*: Nov 2002.
14. U.S. Green Building Council. *Green Building Rating System for New Construction and Major Renovations Version 2.1*: Nov 2002; 10.
15. Stecky, N. Introduction to the ASHRAE greenguide for LEED, *Proceedings of the 2004 world energy engineering congress*, Austin, TX, Sept 22–24 2004; AEE: Atlanta 2004.

Manure Management: Compost and Biosolids

Bahman Eghball
Agricultural Research Service (USDA-ARS), U.S. Department of Agriculture, Lincoln, Nebraska, U.S.A.

Kenneth A. Barbarick
Colorado State University, Fort Collins, Colorado, U.S.A.

Abstract

Manure, compost, and biosolids (municipal sludge) are organic residuals that contain nutrients and organic matter. They are excellent substitutes for chemical fertilizers. The organic matter in these renewable organic residuals can significantly improve the chemical and physical properties of soil and enhance biological activities. Because manure, compost, and biosolids contain nutrients and organic matter, they can be used to improve degraded, eroded, or less productive soils as soil amendments. If not used properly, manure, compost, and biosolids can be sources of environmental pollution.

MANURE

Manure (animal waste) is generated in beef cattle feedlots, swine operations, dairy barns, poultry houses, and other livestock operations. The number of animals and the number of large production facilities in U.S.A. have significantly increased in the past 10 years.[1] Manure, as well as composts and biosolids, is a renewable resource and an excellent source of macro- (N, P, K, Ca, Mg, and S) and micronutrients (Zn, Cu, Fe, Mn, etc.), that are essential for growing plants. For centuries manure was used throughout the world for improving soil fertility and enhancing crop productivity. However, with the advent of synthetic fertilizers after World War II, manure was considered more a liability than a nutrient resource for crop production.

When animals are grazing on pastures and rangelands, manure is dispersed across a large area and little management is needed because the material is not concentrated and decomposes rapidly. However, when animals are concentrated in small feeding areas, the quantity of manure requiring proper management increases greatly. Significant amounts of manure are generated each year in U.S.A. from the confined feedlots of major livestock species (Table 1). The amount of N, P, and K present in the manure from these species would replace 25%, 25%, and 45% of the purchased N, P, and K fertilizers, respectively, in U.S.A., if utilized at agronomic application rates (Table 1). However, because of the high hauling cost, replacement of fertilizer is presently limited to specific areas in the country where the animal feeding operations are located. Crop producers are also reluctant to use manure because of factors such as hauling and spreading costs, potential introduction of weed seeds, nonavailability of manure where needed, uncertainty about availability of manure nutrients to plants, and problems of odor and application uniformity.

The global numbers of major livestock species are given in Table 2.

Even though manure is an excellent source of multiple nutrients and organic matter, it can also contribute to water, air, and land pollution because of the potential for environmental loading with excess phosphorus, nitrate, salts, undesirable microorganisms, pathogens (disease-causing organisms), and greenhouse gases. Manure application in excess of crop needs can cause a significant build-up of P, N, trace elements (As, Cd, Pb, Hg, Mo, Ni, Cu, Fe, Mn, Se, and Zn), and salts in soils. Trace-element limits in soil are given in Table 3. The elevated P and N levels in soil are of environmental concern when these nutrients are carried by runoff to streams and lakes and cause "eutrophication," which is the nutrient enrichment of water that can promote algal growth and depletion of dissolved oxygen in water. This oxygen is essential for aquatic animals. Pathogens (such as bacteria, viruses, and parasites) in runoff from fields treated with manure can be another source of water pollution. Pathogens and odorous materials can also be carried by wind from the feeding operations to neighboring areas. Excess manure application can contaminate the groundwater with nitrate-N. Nitrate is a water-soluble ion that moves with water into the soil and can reach the groundwater within a few days after application. The U.S. Environmental Protection Agency (USEPA) has set a 10 mg NO_3–N/L standard for drinking water.

COMPOST

Composting is the aerobic decomposition of organic materials in the thermophilic temperature range of 40–65°C. The composted material should be an odorless, fine textured, low-moisture content material that can be bagged

Table 1 Annual manure, N, P, and K generated by animals confined in beef cattle feedlots, dairy barns, and poultry and swine operations, and fertilizer use in U.S.A.

Animal species	Animals on feed[a] (million)	Manure (dry weight) (million Mg)	N[b] (Mg × 1000)	P[b] (Mg × 1000)	K[b] (Mg × 1000)
Beef cattle	13.22	31.67	602	206	633
Dairy cows	13.14	22.01	782	140	522
Chickens (broilers and layers)[c]	8263.00	13.26	544	186	278
Turkeys[c]	284.00	3.09	142	65	65
Swine	59.41	15.15	709	451	709
Total		85.18	2779	1048	2207
1999 Fertilizer use in U.S.A.[d]			12436	4345	5016
1996 Global fertilizer use[d]			78353	13543	17516
(Manure nutrient/U.S. fertilizer use) × 100 (%)			25	25	45

[a]From United States department of agriculture.[1]
[b]Manure weight and N, P, and K contents taken from United States department of agriculture.[2]
[c]Yearly production numbers.
[d]From Fertilizer Statistics.[3]

and sold for use in gardens, potting, and nurseries or used as a source of nutrients and organic matter on cropland with little fly-breeding potential. Other advantages of composting include improving the handling characteristics of any organic residue by reducing its volume and weight. Composting also has the potential to kill pathogens and weed seeds. Disadvantages of composting organic residues include loss of N and other nutrients during composting, the time taken for processing, cost of handling equipment, need for available land for composting, odors during composting, marketing, diversion of manure or residue from cropland, and slow release of available nutrients. Similar to manure or biosolids, composts can cause water, air, and land pollution if not used properly.

Temperature, water content, C:N ratio, pH level, aeration rate, and the physical structure of organic materials are important factors influencing the rate and efficiency of the composting process. Ideal values for these factors include a temperature of 54–60°C, C:N ratio of 25:1–30:1, 50%–60% moisture content, oxygen concentration > 5%, pH of 6.5–8.0, and particle size of 3–13 mm. The requirement set by the U.S. Environmental Protection Agency regulations for composting municipal waste is that the temperature should be maintained at 55°C or above for at least three days so as to destroy the pathogens. A temperature of 63°C within the compost pile is needed to destroy the weed seeds.

Homogeneous manure solids can be composted alone without mixing with bulk materials. Bulking agents are required to provide structural support when manure solids, or other organic residues, are too wet to maintain air space within the composting pile, and to reduce water content and/or to change the C:N ratio. Dry and fibrous materials, such as saw dust, leaves, and finely chopped straw or peat moss, are good bulking agents for composting wet manure or organic residues. Depending on the ambient temperature, a complete composting process may take two to six months.

There are a number of methods for composting organic materials. These include active windrow (with turning), passive composting piles, passively aerated windrow (supplying air through perforated pipes embedded in the windrow), active aerated windrow (forced air), bins, rectangular agitated beds, silos, rotating drums, containers, anaerobic digestion, and vermicompost (using earthworms). Carcass composting can be done by using all types of animals. Mortality composting can be accomplished in backyard-type bins, indicator composter bins, and in temporary open bins using layers of saw dust or chopped straw and dead animals. Water content is an important factor to be considered when composting dead animals, and should be maintained at about 40–50%.

Table 2 Global numbers of major livestock species in 1997.

Animal species	Animals[a] (million)
Cattle	1333
Chickens	14156
Sheep and goats	1754
Swine	837

[a]From Food and Agricultural Organization.[4]

BIOSOLIDS

Treatment of municipal wastewater results in a mostly organic by-product known as "biosolids." Land application of biosolids for beneficial use has been practiced since the early 20th century in U.S.A. The USEPA announced requirements regarding beneficial use of biosolids with

Table 3 United States Environmental Protection Agency (40 CFR 503.13, revised July 1, 1999) trace element limits, and concentrations in Littleton/Englewood, CO biosolids, July 27, 1999.

Trace element	Agronomic rate concentration limit (mg/kg)	Ceiling concentration limit (mg/kg)	Annual soil loading limit (kg/ha)	Cumulative soil loading limit (kg/ha)	Littleton/Englewood biosolids (mg/kg)
Arsenic (As)	41	75	2.0	41	2.7
Cadmium (Cd)	39	85	1.9	39	5.6
Copper (Cu)	1500	4300	75	1500	256
Lead (Pb)	300	840	15	300	46
Mercury (Hg)	17	57	0.85	17	1.2
Molybdenum (Mo)	—	75	—	—	8.0
Nickel (Ni)	420	420	21	420	15
Selenium (Se)	100	100	5.0	100	4.6
Zinc (Zn)	2800	7500	140	2800	198

Note: Concentration and quantities are on dry weight basis.

promulgation of the 40 CFR503 regulations in February 1993. The USEPA and the state agencies that control land application of biosolids encourage the judicious recycling of biosolids on crop- or rangeland, as they contain essential plant nutrients and organic matter.

A key aspect of USEPA and Colorado Department of Health (CDH) regulations requires the application of biosolids at an agronomic rate. The CDH[5] defines agronomic rate as "the rate at which biosolids are applied to land such that the amount of nitrogen required by the food crop, feed crop, fiber crop, cover crop or vegetation grown on the land is supplied over a defined growth period, and such that the amount of nitrogen in the biosolids which passes below the root zone of the crop or vegetation grown to groundwater is minimized." The USEPA trace-element limits for land application of biosolids are shown in Table 3. State agencies that control biosolids recycling on land are required to adopt these limits as minimum requirements to protect the environment and public health. Risk assessment of different biological pathways served as the foundation for establishing the trace-element restrictions. For example, the concentrations for Littleton/Englewood biosolids shown in Table 3 indicate that it meets the agronomic-rate limits and can, therefore, be applied at an agronomic rate with minimal restriction. New, aggressive pretreatment programs have significantly reduced trace-element concentration in biosolids since about 1970, and therefore, environmental and public health risks are even more minimal.

The USEPA requires municipal wastewater treatment facilities to reduce pathogens and to reduce the attraction of insects and animals before applying biosolids to land. Most municipal wastewater treatment plants use heat and attack by beneficent micro-organisms through anaerobic (without air) or aerobic (with air) digestion to kill potential pathogens and reduce odors that may reside in wastewater. Municipalities accomplish further reduction of pathogens and stabilization by composting, drying, or other techniques.

The major reason that the USEPA promotes land application of biosolids is that the plant nutrients and organic matter can benefit the soil–plant agroecosystem. For example, Littleton/Englewood biosolids used at two research locations contained up to 5.0% organic–N, 1.3% ammonium–N, 140 mg/kg nitrate–N, 3.7% P, and 0.30% K. Biosolids can also provide plant micronutrients such as Fe and Zn. The organic carbon in biosolids can help to develop and stabilize soil structure with a concomitant increase in precipitation capture and decrease in soil erosion. Efficacious land application of biosolids changes the perspective from disposal of a waste (i.e., a nuisance) to recycling a valuable resource (i.e., a beneficial process).

CONCLUSIONS

Organic residuals (wastes) can serve as excellent sources of plant nutrients such as N, P, and micronutrients such as Fe and Zn. Proper management is required, however, to match nutrient amounts supplied by the organic materials with crop needs so as to avoid potential pollution problems.

REFERENCES

1. United States department of agriculture. In *Agricultural Statistics*; Government Printing Office: Washington, DC, 2000.
2. United States department of agriculture. In *Agricultural Uses of Municipal, Animal, and Industrial Byproducts*; Conserv. Research Report No.44; Government Printing Office: Washington, DC, 1998.
3. *Fertilizer Statistics*; The Fertilizer Institute: Washington, DC, 2001; Available at http://www.tfi.org.
4. Food and Agricultural Organization. In *Statistical database*; United Nations: New York, 2001; Available at http://apps.fao.org.
5. Colorado Department of Health. In *Biosolids Regulation 4.9.0*; 1993, 46 pp.

Manure Management: Dairy

H.H. Van Horn
D.R. Bray
Department of Animal Sciences, University of Florida, Gainesville, Florida, U.S.A.

Abstract
Water use is essential for all dairies. Drinking water is indispensable for cattlelife; some amount of water is necessary for cleaning and sanitation procedures; moderate amounts are important during periods of heat stress for evaporative cooling of cows to improve animal production and health; additional amounts can be used in labor-saving methods to move manure and clean barns by flushing in properly designed facilities; and the recovered wastewater can be recycled to supplement water requirements of forage crops grown to meet roughage requirements of the dairy herd. Extensive water use, however, increases the potential of surface runoff and its penetration into the ground with possible environmental impacts offsite. Heightened environmental concerns and the need for resource conservation, in many cases, have caused implementation of water-use permits. Thus, it is important to determine various essential uses of water, other uses that are important to management, and also consider whether reuse of some water is possible and if it is necessary to do so.

INTRODUCTION

Water use is essential for all dairies. Drinking water is indispensable for cattlelife; some amount of water is necessary for cleaning and sanitation procedures; moderate amounts are important during periods of heat stress for evaporative cooling of cows to improve animal production and health; additional amounts can be used in labor-saving methods to move manure and clean barns by flushing in properly designed facilities; and the recovered wastewater can be recycled to supplement water requirements of forage crops grown to meet roughage requirements of the dairy herd. Extensive water use, however, increases the potential of surface runoff and its penetration into the ground with possible environmental impacts offsite. Heightened environmental concerns and the need for resource conservation, in many cases, have caused implementation of water-use permits. Thus, it is important to determine various essential uses of water, other uses that are important to management, and also consider whether reuse of some water is possible and if it is necessary to do so.

Some of the useful unit conversions are listed as follows:

1 gal of water = 8.346 lb.
1 ft^3 of water = 7.48 gal.
1 acre = 43,560 ft^2.
1 acre in. of water = 27,152 gal.

Calibration methods to estimate use: Water flow meters should be installed on major water supply lines. If water meters are not in place to measure gallons pumped, it becomes necessary to estimate the usage. This can be achieved by capturing flow through various water lines for specified times and multiplying by the time the water flows through these lines every day.

DRINKING

Table 1 provides estimates of drinking water requirements in gallons per cow per day. Consumption of 25–30 gal of water per day by lactating cows is common, which varies depending on milk yield, dry matter intake (DMI), temperature, and other environmental conditions.[1]

COW WASHING

Presently most dairies, in warm climates, bring cows to be milked into a holding area equipped with floor-level sprinklers, which spray water upward to wash cows. Each cow usually has a holding area of about 15 ft^2 and are typically washed for 3 min. Amount of water used per cow should be calculated for each dairy. An estimate for conservative use is that a holding area for 300 cows is 30 × 150 ft^2 (15 ft^2 per cow) and is equipped with sprinklers with 5-ft spacing (say 7 across and 30 rows) having 210 sprinklers. If each sprinkler applies 5 gal min^{-1}, total usage is 1050 gal min^{-1} or 3150 gal for 3 min, the average consumption per cow would be 3150/300 = 10.5 gal per cow per wash cycle. If cows are milked three times this would require 31.5 gal per cow per day.

The washing system previously described also helps in cooling of cows while they are crowded together waiting to be milked. However, the cooling effect could be achieved

Table 1 Predicted daily water intake of dairy cattle as influenced by milk yield, DMI, and season.[a,b]

Milk yield (lb)	Cool season (e.g., February)		Warm season (e.g., August)	
	DMI (lb)	Water intake (gal)	DMI (lb)	Water intake (gal)
0	25	11.5	25	16.3
60	45	22.2	44	26.8
100	55	28.6	48	31.9

[a] Drinking water intake predicted from equation of Murphy et al., J. Dairy Sci., **1983**, 66, 35: Water intake (lb day^{-1}) = 35.2 × DMI (lb day^{-1}) + 0.90 × milk produced (lb day^{-1}) + 0.11 × *sodium* intake (g day^{-1}) + 2.64 × weekly mean minimum temperature [°C = (°F − 32) × 5/9]. For examples above, diet dry matter was assumed to contain 0.35% Na. Predicted water intakes (lb) from formula calculations were divided by 8.346 lb water per gal to convert to gallons.

[b] Average minimum monthly temperatures for February (43.5°F) and August (71°F) used with prediction equation were 70 years averages for specified months at Gainesville, FL (Whitty et al., Agronomy Dept, Univ. FL, 1991).

by sprinkling a little amount of water from above, alternatively with fans to give evaporative cooling, if cows were clean enough so that extensive washing was not required and water conservation was necessary.

WASHING MILKING EQUIPMENT AND MILKING PARLOR

Use of water for these purposes is not as directly related to the number of cows as for other uses. For washing milking equipment, a common wash vat volume is 75 gal. If this is filled for rinse, wash, acid rinse, and sanitizing at each of three milkings, this amounts to 900 gal for the herd, e.g., with 300 cows, only 3 gal per cow per day. This is an extremely small component of the total water budget. The amount used to wash out the milking parlor varies largely. If only hoses are used, the amount may be as little as 2 gal per cow per milking or 6 gal per cow per day if cows are milked three times daily. If flush tanks are used, the amount may be more, i.e., nearly 3000 gal per milking or 9000 gal day^{-1} for three times, equivalent to 30 gal per cow per day for a 300-cow system.

SPRINKLING AND COOLING

Sprinklers along with fans are used for evaporative cooling to relieve heat stress in dairy cows during hot periods of the year. Their use has shown increased cow comfort (lowered body temperature and respiration rates) and economic increases in milk production and reproductive performance.[2,3] Application rates used by dairymen vary. Florida experiments compared application rates of 51 gal per cow per day, 88 gal per cow per day, and 108 gal per cow per day at 10 psi in one experiment and 13 gal per cow per day, 25 gal per cow per day, and 40 gal per cow per day in another experiment. The application rate, 13 gal per cow per day, is close to the estimated evaporation rate from the cow and surrounding floors. This component should be considered in water use but not in runoff water that must be managed in the manure management system. We estimate 25 gal per cow per day as the minimum practical application rate in order to get adequate coverage of cows to cool them because often they are not in the sprinkled area. Total application days per year vary from 120 days to 240 days. A separate water well, or reserve tank and booster pump, may be needed to supply short-term high demand required by the sprinkler system.

FLUSHING MANURE

Flushing manure can be made a clean and labor-saving process, if facilities include concrete floors with enough slope so that water flow propelled by gravity could be used to move manure. Amounts of water used per cow vary widely depending on size and design of facilities and frequency of flushing. However, usually a flush of about 3000 gal is required to clean an alley width of 10–16 ft. If 4 alleys are common for every 400 cows and alleys are flushed twice daily, this would amount to an average use of 60 gal per cow per day. Many dairies use more flushings per day.

RECYCLING DAIRY WASTEWATER THROUGH IRRIGATION OF FORAGE CROPS

Most often nitrogen is the nutrient on which manure application rates are budgeted. To maximize nutrient uptake, crop growth should be as vigorous as possible. This requires irrigation during most of the year in many dairy regions for the disposal of flushed wastewater. In southern regions, multiple cropping systems are possible, which will recycle effectively nitrogen excretions from 100 cows on a sprayfield or manure application field of about 30 acres.[4]

Tentative estimates of total water needs of the growing crops in warm climates average about 1.75 in. of water per week (0.25 in. per day) from irrigation plus rainfall with a minimum of 0.5 in. per week tolerated even in rainy season on sandy soils.[5,6] Table 2 provides estimates of water requirements for two triple cropping forage systems that are common in southern climates. In sandy soils that hold only about 1.0 in. of water per foot of soil depth, some amount of rainfall cannot be stored. Therefore, even in heavy rainfall seasons, judicious irrigation is often needed during lower rainfall weeks. Limited data are available on the maximum amount of water that could be applied and not reduce yield or quality of forage and not result in pollution of groundwater with nitrates and other minerals. However, the maximum probably is at least 35–45 in. per year above the acre totals in Table 2.

Table 2 Crop yield and water requirement estimates for two triple cropping forage systems.[a]

		Silage yield			Water required			
Crop No.	Name	Ton/A 35% DM	Ton/A DM	lb/A DM	lb/lb DM	lb/A Total	gal/A Total	A-in. Total
1	Wheat	10	3.5	7000	500	3,500,000	419,362	15.4
2	Corn	24	8.4	16,800	368	6,182,400	740,762	27.3
3	Corn	14	4.9	9800	368	3,606,400	432,111	15.9
	Total	48	16.8	33,600		13,288,800	1,592,235	58.6
1	Rye	10	3.5	7000	500	3,500,000	419,362	15.4
2	Corn	24	8.4	16,800	368	6,182,400	740,762	27.3
3	F. Sorghum	18	6.3	12,600	271	3,414,600	409,130	15.1
	Total	52	18.2	36,400		13,097,000	1,569,254	57.8

[a]A = acre; No. = number; DM = dry matter.

RAINWATER FROM ROOFS AND CONCRETE AREAS

Rainwater entering wastewater holding areas can be significant. For example in the dairy representing typical minimum water usage with a flush system in southeast United States (Table 3), the net accumulation during the hot season was calculated as follows: assumed wastewater holding area is 1 acre surface area per 100 cows, net rainfall accumulation in holding area is 3 in. more than evaporation per month, concrete areas and/or undiverted roof areas that capture rainfall are 15,000 ft^2 per 100 cows that divert 15,000/43,560 ft^2 per acre of the 3 in. to the wastewater holding facility. Thus, 3 in. + 0.344 × 3 = 4.03 acre in. mo^{-1} or essentially 1.0 acre in. per week per 100 cows (approximately 27,000 gal per 100 cows).

Table 3 Estimated water budgets for three example dairies.

	Flush systems		Non-flush Theoretical minimum	Worksheet for your dairy
Water use in the dairy	Typical need during hot season	Common usage on some dairies		
Drinking (cows)	25	25	25	
Cleaning cows	32	150	0	
Cleaning milking equipment	3	5	3	
Cleaning milking parlor	30	30	6	
Sprinklers for cooling	25	130	12	
Flushing manure	60	80	0	
Total use per cow per day	175	400	46	
Total use per 100 cows per day	17,500	40,000	4600	
Use per 100 cows per week	122,500	280,000	32,200	
Water in milk per 100 cows per week	4500	4500	4500	
Estimated evaporation (at 20% of use)	24,500	56,000	6440	
Average rainfall and watershed drainage into storage facility per 100 cows per week	27,000	27,000	13,000	
Wastewater produced from 100 cows/week	120,500	246,500	38,760	
Acre in. per 100 cows per week	4.44	9.08	1.43	
in. per week if 30 acre in sprayfield	0.15	0.30	0.05	

All values are in gal unless otherwise noted.
Example calculations (column 1): Total use per cow per day = 175 gal; total use per 100 cows per week = 122,500 gal less 4500 in milk and 24,500 gal evaporation = 93,500 gal week^{-1}; net rainfall and watershed drainage to storage per 100 per cows per week = 27,000; acre in. per 100 cows per week = (93,500 + 27,000)/27,152 gal per acre in. = 4.44.; if 30 acre were in sprayfield, 4.44/30 = 0.15 in. week^{-1}.; if crop needed 1.75 acre in. week^{-1} (a common average), a total of 1.75 in. × 30 acre × 27,152 gal per acre in. = 1,425,480 gal is needed of which only 120,500 gal (8.5%) would come from dairy wastewater. The remaining (91.5% of total) would have to come from rainfall or fresh irrigation water.

DEVELOPING A WATER BUDGET

A wide range exists in water usage on dairy farms. For most dairy waste management systems designed to utilize flushed manure nutrients through cropping systems grown under irrigation, water amounts are small in relation to irrigation needs for crop production. Costs for construction of storage structures for holding wastewater until used for irrigation warrant consideration. For example, water-use budgets given in Table 3 show that water usage is small in comparison to irrigation needs when there are 30 acre of sprayfield crop production per 100 cows. Conversely, the amounts used in most dairy systems would be large and unmanageable if application through irrigation is not an option or if less acreage for irrigation is available than needed for application of all manure nutrients.

If a dairy does not have acreage available close by to utilize manure nutrients and water through an environmentally accountable sprayfield application system, it would be necessary to export nutrients off the farm, preferably as solid wastes to avoid excessive hauling or pumping costs. If the water and manure nutrients cannot be used through irrigation, a non-flush system should be utilized. However, usually some irrigation is possible, permitting dairymen to use cow washers and limited flushing if they scrape and haul manure from some areas.

Strategies to minimize water usage: Table 3 presents one column indicating a theoretical minimum amount of water use in a dairy. This system implies that cows are clean and cool enough so that sprinkler washers are not required to clean and cool cows while being held for milking. In addition, it is assumed that all of the manure is scraped and hauled to manure disposal fields or transported off the dairy in some other fashion. Intermediate steps that might be taken include the following:

1. Scraping and hauling manure from high use areas such as the feeding barn so that this manure can be managed off the dairy.
2. Using wastewater rather than fresh water to flush manure from feeding areas and freestall barns.
3. Using a housing system that will keep cows clean enough so that cow washers are not required to clean cows before milking. This system, however, may require use of alternating sprinklers and fans to keep crowded cows cool during hot weather conditions.

If flushing is desired in conjunction with scraping and hauling from heavy use areas, perhaps the feeding area could be flushed with recycled water after scraping to clean the area. These procedures would reduce total nutrient loads retained in wastewater and would significantly reduce the size of the sprayfield needed for water and manure nutrient recycling.

REFERENCES

1. Beede, D.K. Water for dairy cattle. *Large Dairy Herd Management*; American Dairy Science Assoc.: Champaign, IL, 1992; 260–271.
2. Bray, D.R.; Beede, D.K.; Bucklin, R.A.; Hahn, G.L. Cooling, shade, and sprinkling. *Large Dairy Herd Management*; American Dairy Science Assoc.: Champaign, IL, 1992; 655–663.
3. Van Horn, H.H.; Bray, D.R.; Nordstedt, R.A.; Bucklin, R.A.; Bottcher, A.B.; Gallaher, R.N.; Chambliss, C.G.; Kidder, G. Water Budgets for Florida Dairy Farms; Circular 1091; Florida Coop. Ext., Univ. Florida: Gainesville, 1993.
4. Van Horn, H.H.; Nordstedt, R.A.; Bottcher, A.V.; Hanlon, E.A.; Graetz, D.A.; Chambliss, C.F. Dairy Manure Management: Strategies for Recycling Nutrients to Recover Fertilizer Value and Avoid Environmental Pollution; Circular 1016; Florida Coop. Ext., Univ. Florida: Gainesville, 1998; 1–24.
5. North Florida Research and Education Center. AREC Research Report 77–2; IFAS, University of Florida, Gainesville, 1977.
6. Wesley, W.K. Irrigated Corn Production and Moisture Management; Bul. 820; Coop. Ext. Serv., Univ. Georgia College of Agric. and USDA, 1979.

Manure Management: Phosphorus

Rory O. Maguire
Virginia Tech, Blacksburg, Virginia, U.S.A.

John T. Brake
Department of Soil Science, North Carolina State University, Raleigh, North Carolina, U.S.A.

Peter W. Plumstead
North Carolina State University, Raleigh, North Carolina, U.S.A.

Abstract

Since the earliest development of agriculture, animal manures have been applied to cropland to supply essential nutrients and increase crop yield. Modernization of agriculture has been typified by greater yields of both crop and animal products from smaller land areas, and a greater emphasis on external inputs such as chemical fertilizers and concentrated animal feeds. Intensification and confinement of animal production have now led to a situation in some areas where there are more nutrients available in animal manure than what the local crops require. These excesses of manure have raised concerns about the fate of the nutrients they contain, as losses of nutrients such as nitrogen (N) and phosphorus (P) from agricultural land can cause problems such as eutrophication in rivers and lakes. To address these regional excesses of P, research efforts have recently been focused on reducing the P concentrations in diets fed to animals, as well as on improving the availability thereof, both of which result in a reduction in the P concentration in manure generated. These dietary alterations affect not only the concentrations of P in manures produced, but also the forms of P, their fertilizer value, and environmental impact.

RECENT PROGRESS IN NUTRIENT MANAGEMENT PRACTICES

With increasing human population and awareness of environmental problems associated with pollution of rivers and streams with P, the end of the 20th century saw a refocusing of nutrient management strategies from increasing production toward environmental protection, in developed countries.[1] Initially, nutrient management strategies were aimed at avoiding overapplication of nutrients and voluntary best management practices were developed. However, manure still tended to be applied at a rate to supply crops with sufficient N, which typically leads to overapplication of P, as manures have N:P ratios of around 2:1 to 4:1 (depending on species) while crop requirements have an N:P ratio of 8:1. When P is overapplied to soils, it builds up the soil test P (STP) concentration and over a period of many years, this leads to STP rising to levels in excess of crop requirement.[2] Soil test P in excess of crop requirement is undesirable, as studies have shown that when STP rises above agronomically optimum values P losses to rivers and streams increase to the detriment of water quality,[2] and it can also lead to a negative effect on some micronutrients, e.g., Zn.

Nutritional Strategies to Reduce the Manure Phosphorus Concentration

In the past, dietary P levels frequently included a safety margin that allows for any potential variation in both the requirement of the animals and the degree of dietary P utilization, with little attention paid to the resultant manure P concentration. However, recent research in broilers and broiler breeders demonstrates that levels of inorganic P, supplemented to both broiler and broiler breeder diets may be considerably reduced relative to minimum requirements published by the National Research Council without an adverse effect on performance.[3,4]

When considering strategies to increase P digestion and retention, species can be split into two categories according to their ability to digest phytate-P: ruminants (cattle, sheep) can digest phytate-P, while monogastric species (poultry, swine) can digest little. As approximately 66% of the P in corn and soybean is in the form of phytic acid, and these ingredients frequently constitute about 60%–90% of the diet, up to 80% of the total fecal P output may originate directly from corn and soybean meal in poultry diets.[5] The primary strategy to address the low digestibility of phytate-P in monogastric species has been the application of exogenous phytase enzymes to these diets and is done on the basis of replacing 0.1% of available P from supplemental calcium phosphate with a more or less standard amount of enzyme. Alternate strategies other than the use of phytase include the development of low phytate variants of corn and soybean cultivars, because these possess similar, if not greater, potential in reducing manure P excretion. However, the commercial application of these new cultivars has to date been limited as a result of the practical complications arising from the identity preservation of these ingredients from harvest to point of feeding.

IMPACT OF DIET MODIFICATION ON NUTRIENT MANAGEMENT

The total P and soluble P concentrations in manures have been shown to be affected by dietary P concentration, P contributing ingredients, and dietary phytase supplementation. Total P is important because applications of manure total P contribute to long-term changes in STP. Once optimum STP for crop production is attained, it is beneficial to balance manure total P applications with crop removal of P. This will maintain crop production without raising STP to levels of environmental concern; although in some soils, P can revert to less available forms so P availability needs to be monitored with soil testing. Applications of soluble P in manure have been correlated to soluble P losses in runoff from manure-amended soils, so minimizing the soluble P concentration in manure is important when P losses to rivers and streams are of concern. Reducing the P concentration in animal feeds (high vs. low nonphytate P diets) and using phytase (phytase vs. no phytase diets) can reduce the total and soluble P concentrations in poultry wastes (Fig. 1).[6] It is estimated that using currently available technology, dietary amendment can potentially decrease the total P concentration in poultry by 40%, swine by 50% and dairy by 30%, while future advances could raise these figures to 60% for both poultry and swine.[7] Research on the impact of dietary amendments on soluble P in manure and P losses in runoff from manure-amended soils is at an early stage; however, trends are starting to appear. Supplying P at an adequate level to satisfy animal requirements is generally achieved through a reduction in the added inorganic dietary calcium phosphate component, and results in a reduction in both total and soluble P in the manures generated. While most studies show that feed additives such as phytase reduce manure total P while having little effect on soluble P, this effect is very dependent on the level of dietary P relative to the requirements of the animals.[6,7] Phytase supplementation to diets containing P in excess of animal requirements, or without an adequate reduction in the P from supplemental calcium phosphate, will result in a potential increase in the water soluble P when expressed on a percentage basis.[5]

Traditionally, manure application rates to cropland have been decided by individual farmers. However, increasing concerns over the impact of the land application of nutrients in excess of crop requirement on water quality

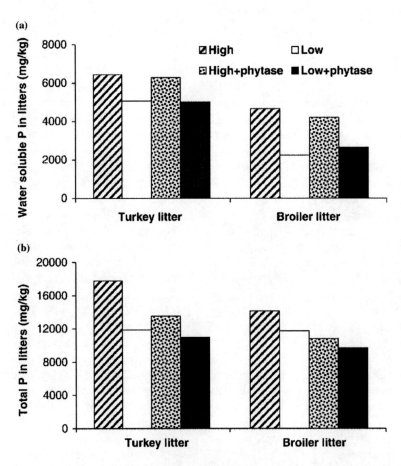

Fig. 1 Effects of diets high and low in dietary nonphytate P, and the impact of dietary phytase on (a) water soluble P in litters, and (b) total P in litters produced.
Source: Maguire et al.[6]

in rivers and streams, have led to increasing regulation of land application of manures in many developed countries. These regulations are primarily based on N and P application rates that are often linked to problems identified in localized areas and are based on the current levels of soil N and P, and the annual rate of removal of these nutrients through their incorporation into harvested crops. As forage crops typically have an N:P ratio of 8:1 and manure contains N:P ratios of approximately 2:1 to 4:1, nutrient management regulations based on P application may require up to four times more cropland than previous N-based application guidelines. The limitation of available land in areas proximate to intensive livestock production can cause problems associated with what to do with the manure if it cannot be land applied, as other uses of manure are currently fairly limited and transport costs limit the potential transportation of the wastes over any appreciable distance. If nutrient management regulations are based on P, then reducing the P concentration in manures could greatly alleviate these problems. For example, if one could reduce the total P concentration in a manure by 40%, then application of 40% more manure by weight to meet the same total P application limit would be possible. This may eliminate problems with excess manure P in some areas of intensive animal production, but may not be sufficient for other areas that already have high STP.

CONCLUSIONS

Dietary amendment to reduce the P concentration in feeds and hence manures generated will help address concerns over the fate of manure nutrients in areas of intensive animal production. If manures have the same N concentrations but reduced P concentrations, then these manures will come closer to meeting crop N needs without oversupply of P. However, dietary amendment alone will not solve the problem of excessive manure P in all areas of animal production. Dietary amendment will, however, certainly help alleviate the current situation where applying animal manures to meet crop N requirements leads to overapplication of P, buildup of STP, and eventually increased losses of P to rivers and streams.

REFERENCES

1. Beegle, D.B.; Carton, O.T.; Bailey, J.S. Nutrient management planning: justification, theory, practice. J. Environ. Qual. **2000**, *29*, 72–79.
2. Sims, J.T.; Edwards, A.C.; Schoumans, O.F.; Simard, R.R. Integrating soil phosphorus testing into environmentally based agricultural practices. J. Environ. Qual. **2000**, *29*, 60–71.
3. National research council. In *Nutrient Requirements of Poultry*, 9th revised ed.; National Academy Press: Washington, DC, 1994.
4. Brake, J.; Williams, C.V.; Lenfestey, B.A. Optimization of dietary phosphorus for broiler breeders and their progeny. In *Nutritional Biotechnology in the Feed and Food Industries*, Proceedings of Alltech 19th International Symposium; Lyons, T.P., Jacques, K.A., Eds.; Nottingham University Press: Nottingham, U.K., 2003, 77–84.
5. Brake, J.; Plumstead, P.; Gernat, A.; Maguire, R.; Kwanyuen, P.; Burton, J. Reducing phytate versus adding phytase to reduce fecal phosphorus. In *Session PCP 3: Phytate in Soybeans and Related Environmental Concerns*, Proceedings of the 95th AOCS Meeting, Cincinnati, OH, 2004, 118.
6. Maguire, R.O.; Sims, J.T.; Saylor, W.W.; Turner, B.L.; Angel, R.; Applegate, T.J. Influence of phytase addition to poultry diets on phosphorus forms and solubility in litters and amended soils. J. Environ. Quality **2004**, *33*, 2306–2316.
7. CAST (council for agricultural science and technology). In *Animal Diet Modification to Decrease the Potential for Nitrogen and Phosphorus Pollution*. Issue Paper No. 21; CAST: Washington, DC, 2002.

Manure Management: Poultry

Shafiqur Rahman
Department of Agricultural and Biosystems Engineering, North Dakota State University, Fargo, North Dakota, U.S.A.

Thomas R. Way
National Soil Dynamics Laboratory, Agricultural Research Service (USDA-ARS), U.S. Department of Agriculture, Auburn, Alabama, U.S.A.

Abstract

Poultry manure management poses challenges at different steps of manure management, including manure collection, storage and handling, and utilization or disposal. Utilization or disposal of manure in an environmentally safe way is one of the biggest challenges. This entry discusses various management options for poultry manure.

INTRODUCTION

The U.S. Environment Protection Agency (EPA) includes a number of species, including chicken, turkey, and duck, in their definition of poultry. In the United States, approximately 8.55 billion broilers, 337 million laying hens, and 247 million turkeys were raised in 2009.[1] The total revenue from the poultry industry in 2008 was $35.9 billion, which was up 11% as compared with 2007 ($32.2 billion). Of the combined total revenue, the broiler, layer, and turkey industries contributed 64%, 23%, and 12%, respectively. Modern poultry production occurs primarily in confined facilities, which allows larger intensive production in smaller areas. Owing to intensive poultry production systems, large amount of manure (i.e., feces, urine, undigested feed, spilled water, bedding used in poultry houses, etc.), is produced in a smaller area, which leads to air pollution (including odors and pollutant gas emissions) and water pollution (i.e., eutrophication) problems. It is the producer's responsibility to manage manure in an environmentally safe way. Therefore, poultry manure management is the focus of this entry.

MANURE PRODUCTION AND MANAGEMENT SYSTEM

Table 1 lists the total number of broilers, layers, and turkeys produced and the estimated manure production by bird type. A large quantity of manure or litter is produced daily, which needs to be managed properly to maximize agronomic benefits (i.e., nutrient values) and minimize environmental concerns (i.e., water and air pollution). Poultry manure management is composed of manure collection and handling, storing, and utilization. Manure collection and handling depend on manure properties, and they are influenced by bird type, bird age, diet, bird productivity, and management system.[2] Manure can be handled as a solid (>20 total solids), semi-solid (10%–20% total solids), slurry (4%–10% solids), or liquid (<4% solids).[3] Solid manure can be stacked, while liquid manure can be stored in earthen anaerobic lagoons or storage structures (e.g., concrete or steel). For economical reasons, liquid manure is typically stored in anaerobic lagoons.

Table 1 Poultry manure production, as excreted.

Bird type	Live market weight (kg)[a]	Manure per 1000 birds/day (kg)[a]	Total number of birds (1000s)[b]	Total manure production/day (tons)
Broiler	2.0	80	8,550,500	684,040
Layer	1.8	118	337,376	39,810
Turkey (tom, light)	10.0	267	249,914	66,727
Total manure production				790,577

[a]From Poultry Waste Management Handbook, NRAES-132.[4]
[b]From National Agricultural Statistics Service.[1]

In broiler and turkey facilities, bedding or litter materials (e.g., sawdust, wood shavings, peanut hulls, rice hulls, sunflower hulls, etc.), are used, and litter is handled as a solid (Fig. 1). Litter is sometimes completely removed after each flock or partially cleaned between flocks when litter is wet and packed around feeders and waterers. After partial cleaning, a thin layer of fresh bedding material is added.[2] It is also common that producers decake litter between flocks using a decaking implement to remove the larger clumps of litter while leaving the smaller clumps as bedding for the next flock. Depending on the management practices, litter cleanout varies significantly from farm to farm.

High-rise layer facilities (elevated cages) (Fig. 2a) produce solid manure, and that manure is stored in the portion of the poultry house beneath the cages, as shown in Fig. 2b. Manure is removed once or twice a year depending on individual farm management practices and regulatory requirements.

Fig. 2 (a) High-rise layer house and (b) manure storage pit (stockpiled beneath cages).

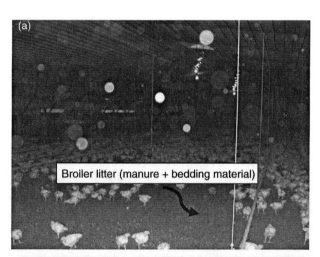

Fig. 1 Poultry production facility and litter for (a) broiler and (b) turkey.

Some layer facilities with a shallow pit remove manure two or three times per week using a scraper (Fig. 3a) and may produce semisolid or liquid manure and store it in an earthen anaerobic lagoon or outdoor storage (Fig. 3b). However, manure cleaning or removing frequency may vary depending on the storage capacity, manure storage conditions (dry or wet), and environmental (indoor air quality) conditions.

MANURE STORAGE

Manure or litter can be stored both indoors and outdoors depending on manure type and management practices. Both options have some pros and cons as discussed below.

Indoor Storage

Most of the high-rise layer houses (Fig. 2a) stockpile manure directly beneath the birdcages, i.e., deep-pit storage, as shown in Fig. 2b. This option minimizes runoff from

Fig. 3 (a) Layer house with mechanical scraper and (b) anaerobic earthen lagoon system.

precipitation, and manure can be stored for a longer time to allow scheduling of land application around crop nutrient requirements.[5] Although this option minimizes water pollution from runoff, it may cause air quality issues with the house, if not managed properly. Typically, high-rise layer houses are cleaned of manure once or twice a year; however, a manure storage pit should be inspected daily for any sign of water leakage. Ammonia (NH_3) and hydrogen sulfide (H_2S) emissions from indoor stockpiled manure may cause poor air quality and may affect bird health and productivity as well as worker health and safety. It is important to maintain recommended ventilation rates and to keep manure as dry as possible to minimize anaerobic conditions in manure stockpiles. Fans are commonly used in the pit to keep the manure pile relatively dry. If any portion of the manure stockpile is wet from water leakage, it is important to promptly remove manure from indoors and correct any water leakage problem. Typically, wet manure will require more frequent removal than dry manure.

Outdoor Storage

Poultry litter should be removed from the house between flocks. The litter should be stored outside if no land is available for immediate manure application or if there is insufficient indoor storage room for a period of 6 to 9 mo. If solid manure is stored outside, proper management is necessary to minimize runoff and groundwater pollution. Outdoor stockpiles should be at least 30 m from the nearest surface water and should be separated from groundwater by an impervious soil layer having a minimum thickness of 1.2 m. To minimize runoff resulting from rainfall and to reduce nutrient losses, a cover on top of the litter pile is required to limit manure exposure to air. If regulatory guidelines and time permit, and if cropping conditions are appropriate, outside stockpiled manure should be applied to cropland as soon as possible.

Liquid manure or slurry is removed from under caged birds by mechanical scrapers (Fig. 3) or mechanical belts or flushed with water and stored in an outdoor anaerobic lagoon or containment pond (Fig. 3b). Manure is typically stored in an anaerobic earthen lagoon for a period of 6 to 9 mo and applied to cropland near or during the growing season to optimize use of manure nutrients. However, during manure pumping, it is very important to maintain the minimum design volume or treatment volume of an anaerobic lagoon to provide a dual function of storing and treating the manure.[6] Otherwise, anaerobic lagoons may be a significant source of nuisance odor.

MANURE UTILIZATION

Poultry manure contains essential nutrients (e.g., nitrogen, phosphorus, potassium, etc.), and can be applied to land as fertilizer to meet crop nutrient requirements. However, when land application of poultry manure as fertilizer is not an environmentally and economically sound option, manure (especially solid manure) can be composted (Fig. 4), and the compost can be used as a fertilizer source for crops, gardens, and nurseries. In that case, nuisance odor can be minimized, and nutrients become bound in organic forms. However, for proper composting, several parameters (carbon–nitrogen ratio, pile moisture content, pile temperature, oxygen concentration, etc.), need to be maintained appropriately. Otherwise, anaerobic conditions may prevail, and the pile may become a source of nuisance odor.

During storage, nutrient losses are a normal phenomenon. A portion of nutrient losses occurs during land application depending on the application method, such as surface vs. subsurface injection. Typically, most solid manure is applied on the surface, often followed by incorporation. Solid manure is commonly applied using truck-mounted spreaders (Fig. 5). There are many options for solid manure spreaders, including side delivery and rear delivery (spinner spreader; single or double horizontal beater; vertical beaters). With a single beater, it is difficult to have uniform manure distribution to take full advantages of nutrients in manure, and it is difficult to break litter cake or large clumps. Researchers[6] concluded that distribution of manure across the application swath of a spreader needs

Fig. 4 Turkey litter composting pile and temperature sensors.

to be relatively uniform to take full advantage of the fertilizer value of solid manure. To overcome this issue, double vertical beaters might be a better option (Fig. 5b and c).

Traditional surface application of manure may cause nutrient losses, nuisance odor, and nutrient runoff, and reduces nutrient availability. Although incorporation will minimize some of these concerns, delayed incorporation (more than 24 hr) can result in increased nuisance odor, nitrogen nutrient losses through volatilization, and surface runoff.

To overcome problems associated with surface application of solid manure, manure can be subsurface banded. Until recently, subsurface application has been available for only liquid manure or slurry. Research has been conducted in both Canada and the United States to develop implements for subsurface band application of solid manure from poultry and feedlots.[7–9] Applying the manure in a controlled way decreases odor and greenhouse gas emissions while improving air quality and the social acceptance of the practice.[7] The U.S. Department of Agriculture–Agricultural Research Service (USDA-ARS) National Soil Dynamics Laboratory (Auburn, Alabama) has developed a four-row prototype implement for subsurface band application of poultry litter (Fig. 6) and has used the implement to apply poultry litter in row crops and pastures.

The implement is equipped with conveyors, one for each trencher. Each conveyor drops the litter into a trencher, where the litter falls by gravity down into the trencher. The band spacing is adjustable from 0.25 to 1.0 m in increments of 25 mm. The implement is capable of applying poultry litter in a side-dressing manner to row crops, and the band spacing is then typically equal to the crop row spacing. Also, the implement is capable of applying litter to pastures, and typical band spacings that have been used for pastures are 250 to 380 mm. One disadvantage of the implement is that litter clumps measuring larger than 25 mm across have to be run through a hammer mill before being loaded into the hopper. Also, the litter moisture content should be <40% (wet basis). The implement is being modified to include on-the-go grinding as part of the machine. Subsurface banding of broiler litter to cotton with this implement has been shown to increase lint yield relative to surface broadcast application of litter.[10]

Fig. 5 Rear delivery manure spreaders: (a) spinner spreader, (b) double horizontal beater, and (c) double vertical beater.

Fig. 6 Side view of prototype subsurface band applicator implement for poultry litter developed by USDA-ARS at Auburn, Alabama.
Source: Dr. Thomas Way.

Subsurface band application of broiler litter to tall fescue and bermudagrass pastures using band spacings of 250 and 380 mm was found to produce forage yields equivalent to those for surface broadcast application of litter.[11] When broiler litter has been subsurface banded in a pasture with the implement, and rainfall has been applied using rainfall simulation, concentrations of phosphorus and nitrogen in runoff water have typically been reduced by 80%–95% relative to those for surface broadcast application of litter.

In contrast, liquid manure can be applied on both the surface and subsurface, and technologies are improving very fast. Injection of liquid manure limits the exposure of manure to the surface water and air, resulting in reduced nutrient losses and nuisance odor. For a liquid manure application system, a liquid storage tank with a flush spreader or injector assembly is used. However, a fully loaded tank may cause soil compaction, especially in wet clay soil. To overcome compaction issues, manure can be injected using a drag-hose system (Fig. 7a). One new technology, namely, AerWay SSD® (Holland Equipment Inc., Norwich, Ontario, Canada), can place liquid manure in the active root zone and can substantially reduce any risk of groundwater contamination (Fig. 7b).

Energy Uses

As of April 2010, there are 151 anaerobic digester systems operating at commercial livestock farms in the United States. Of these, only three anaerobic digester systems were using poultry manure. Poultry manure has higher biodegradable organic matter content than other livestock wastes,[12] and a substantial amount of water would be required to dilute poultry litter. They also found that efficiency of organic matter conversion to methane decreased with increasing organic loads to the digesters. As a result, limited anaerobic digester systems are using poultry manure.

Fig. 7 Liquid manure injection system: (a) drag-hose system and (b) AerWay SSD system.

CONCLUSIONS

Poultry manure management poses challenges at different steps of manure management, including manure collection, storage and handling, and utilization or disposal. Utilization or disposal of manure in an environmentally safe way is one of the biggest challenges. Poultry manure can be handled as either a solid or a liquid. Owing to enriched nutrient content in manure, it may be applied to crops to meet crop nutrient requirements. However, selection of manure application tools and methods is important in minimizing environmental concerns and maximizing agronomic benefits from manure.

REFERENCES

1. National Agricultural Statistics Service (NASS). 2010, available at http://usda.mannlib.cornell.edu/MannUsda/viewDocumentInfo.do?documentID=1509 (accessed August 16, 2010).

2. Collins, E.R.; Barker, J.C.; Carr, L.E.; Brodie, H.L.; Martin, J. H., Jr. *Natural Resource*; Agriculture and Engineering Service (NRAES), NRAES-132, Ithaca, NY, 1999.
3. MWPS-18. *Livestock Waste Facilities Handbook*, 2nd Ed.; Midwest Plan Service, Iowa State University: Ames, Iowa, 1985.
4. *Poultry Waste Management Handbook, NRAES-132*; Natural Resource, Agriculture, and Engineering Service (NRAES), Cooperative Extension, Ithaca, NY, 1999.
5. Fulhage, C.; Hoehne, J.; Jones, D.; Koelsch, R. *Manure Storages*, MWPS-18, Section 2; Midwest Plan Service, Iowa State University: Ames, IO, 2001.
6. Norman-Ham, H.A.; Hanna, H.M.; Richard, T.L. Solid manure distribution by rear- and side-delivery spreaders. Trans. ASABE **2008**, *51* (3), 831–843.
7. Laguë, C.; Agnew, J.M.; Landry, H.; Roberge, M.; Iskra, C. Development of a precision applicator for solid and semi-solid manure. Appl. Eng. Agric. **2006**, *22* (3), 345–350.
8. Farm Show Publishing, Inc. Prototype applicator buries poultry litter. Farm Show **2009**, *33* (2), 25; Farm Show Publishing, Inc.: Lakeville, MN.
9. Way, T.R.; Siegford, M.R.; Rowe, D.E. Applicator system and method for the agricultural distribution of biodegradable and non-biodegradable materials. U.S. Patent Number 7,721,662, May 25, 2010.
10. Tewolde, H.; Armstrong, S.; Way, T.R.; Rowe, D.E.; Sistani, K.R. Cotton response to poultry litter applied by subsurface banding relative to surface broadcasting. Soil Sci. Soc. Am. J. **2009**, *73*, 384–389.
11. Warren, J.G.; Sistani, K.R.; Way, T.R.; Mays, D.A.; Pote, D.H. A new method of poultry litter application to perennial pasture: Subsurface banding. Soil Sci. Soc. Am. J. **2008**, *72*, 1831–1837.
12. Bujoczek, G.; Oleszkiewicz, J.; Sparling, R.; Cenkowski, S. High solid anaerobic digestion of chicken manure. J. Agric. Eng. Res. **2000**, *76*, 51–60.

Mercury

Sven Erik Jørgensen
Institute A, Section of Environmental Chemistry, Copenhagen University, Copenhagen, Denmark

Abstract

Mercury caused one of the most significant environmental catastrophes in history, Minamata disease. In the 1950s, when the disease was discovered, it was not possible to explain but today, we know that it was a combination of mercury discharge from a chemical factory, accumulation in the sediment, formation of organic mercury compounds, and biomagnifications. Due to this discovery, several applications—particularly open applications—have been phased out during the last 20–40 years. By far the most significant mercury pollution today comes from coal-fired power plants, although natural processes, mainly of volcanic origin, also distribute as much as 1500 tons of mercury globally per year. Mercury can cause damage to the central nervous system and has both teratogenic and genetic effects. In environmental management, it is necessary to consider all the relevant processes of mercury in the environment.

INTRODUCTION

Mercury is an extremely toxic element. It has no role as biological element beyond its toxicity and is not essential for any organism. It is element number 80 in the periodic table and has an atomic weight of 200.59. Mercury caused one of the most significant environmental catastrophes in history. Mercury was discharged with the wastewater from a chemical factory in Minamata Bay, Japan, in the 1950s.[1] Mercury accumulated in the sediment, where it could react microbiologically with the organic matter and form methyl mercury ions and dimethyl mercury, which has a low boiling point and can be transferred from the sediment to the water phase or even to the atmosphere. Organic mercury compounds can be taken up by fish, where concentrations 3000 times higher than in water can be recorded. These processes explain why the fish caught in the Minamata Bay had a very high mercury concentration. As a result, hundreds of people died, and over a thousand became invalids due to mercury contamination. In the 1950s, the victims were considered to have a new, previously unknown disease, the Minamata disease. Later in the 1960s, it was detected that the disease was caused by high mercury concentration in fishermen's families. The accumulation in the sediment, the biomagnifications through the food chain (water–phytoplankton–zooplankton–fish–fisherman), and the ability of mercury to form organic compounds by microbiological reactions explain the emergence of Minamata disease and mercury poisoning.

SOURCES OF MERCURY POLLUTION

The use of mercury has declined during the last decades due to its extreme toxicity. Mercury compounds were previously applied as fungicides, as dyestuff (cinnabar red), and for the production of chlorine, but these applications are now banned in all industrialized countries. Chlorine and sodium hydroxide are unfortunately still produced in some developing countries by a method based on a dripping mercury electrode that causes mercury to be discharged with wastewater to the environment. Mercury still is applied by dentists and is used for some electrical instruments. Today's major mercury contamination is, however, caused by coal-fired power plants due to a small concentration of mercury in all coal. Each year, 3300 tons of mercury is discharged as air pollution to the atmosphere from coal-fired power plants and the incineration of solid waste.[2] The natural emission of mercury is about 1500 tons/yr, mainly from volcanic activity.

POLLUTION EFFECTS

The most important acute effects of mercury are on the lungs and the central nervous system. The chronic effect is complete damage of the central nervous system, which was the main symptom of the Minamata fishermen. In addition to central nervous system symptoms, the victims also lose weight and appetite. The LD50 value for mice is 5 mg/kg, and the World Health Organization recommends a maximum concentration of 1 ppm in food items. (Mercury has further teratogenic and genetic effects. For details about all effects, see Jørgensen et al.[3–5])

IMPORTANT MERCURY PROCESSES

Evaluation of mercury pollution requires that environmental management consider the environmental processes in

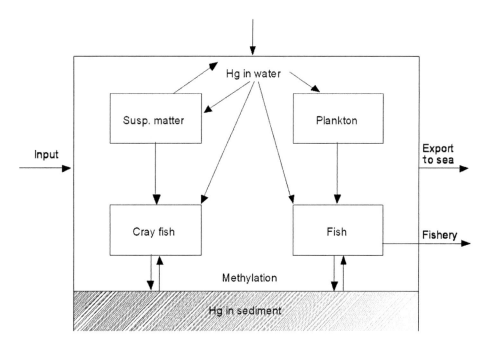

Fig. 1 Conceptual diagram of the mercury model for Mex Bay, Egypt. Notice that methylation is considered for the release of mercury from the sediment. Notice also that the fish mercury is determined by uptake directly from the water (including the concentration of organic mercury) and through the food chain—here indicated only as water–phytoplankton–fish, but one to two additional steps could be considered.

which mercury can participate. It is clear from Section 1 that the formation of organic mercury compounds mainly in the form of methyl and dimethyl mercury is extremely important, and these processes should therefore be included in the environmental considerations, included when an ecotoxicological model for the distribution and effect of mercury is developed.

Fig. 1 shows a conceptual diagram of mercury distribution and effects in Mex Bay, close to Alexandria. The details of the model can be found in Jørgensen and Bendoricchio.[6] From the figures, it is clear that the methylation processes are included and that the uptake from water and biomagnifications through the food chain are included to account for the most important state variable of the model: the mercury concentration in the top carnivorous fish (tuna fish), which is an important consumer fish in the Mediterranean region. Fig. 2 shows the input/output processes that are included in the model.

The uptake of mercury by fish is highly dependent on the pH, due to the pH dependence of mercury solubility. This dependence is of course of interest mainly when the mercury pollution is inorganic. Fig. 3 shows the pH dependence for the mercury in fish when exposed to 1.5 ppm inorganic mercury.[7] The uptake of mercury from food is very different for organic mercury and inorganic mercury, namely, 90% and 20%, respectively.[7] This difference entails that a mercury model should follow and include the processes for both inorganic and organic mercury to be able to account for the final concentration, for instance, in fish as a result of a mercury contamination, whether the discharge is inorganic or organic mercury. The transfer between the two forms should of course be included, too.

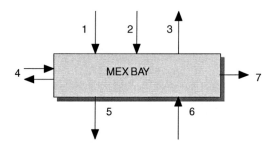

Fig. 2 A model of the distribution of mercury in an aquatic ecosystem is a bio-geo-chemical model that must include all inputs and outputs. The numbers indicate the following: 1) discharge from waste and tributary to the bay; 2) deposition of mercury from the atmosphere (for instance, from coal-fired power plants); 3) evaporation of mercury; 4) input and output from the open sea; 5) sedimentation; 6) release from the sediment; and 7) fishery.

ABATEMENT OF MERCURY POLLUTION

The mercury pollution caused by coal-fired power plants is reduced for the part of the mercury adsorbed to particulate matter by the methods used for treatment of air pollutants to the extent that they are removing particulate matter. However, oil and natural gases are depleted within 60–80

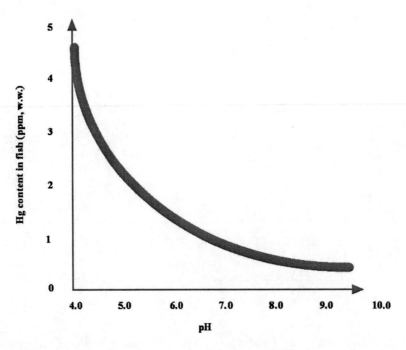

Fig. 3 The concentration of mercury in fish expressed as ppm wet weight versus pH. The concentrations are measured at steady state and for a concentration of mercury chloride in the water of 1.5 ppm (mg/L). The values are based on many measurements for a wide spectrum of different fish species, and the concentrations in the fish are therefore indicated as a range.

years, which implies that a coal-fired power plant will still emit mercury if the power plant is using coal as fossil fuel. A significant part of mercury emission is not adsorbed to the particulate matter. The question is if we, due to emission of carbon dioxide, can continue to use coal-fired power plants after years 2050–2080. Today's emission of mercury is, as indicated above, 3300 tons/yr, and it is therefore a question of whether we can use coal-fired power plants after oil and natural gas have been depleted, not only due to the continuous emission of carbon dioxide but also due to the continuous emission of mercury, which is a consequence of the use of coal. This problem has not yet been examined in sufficient detail to give a clear answer—so this question is still open and requires further attention. It will probably be necessary, due to the greenhouse effects of carbon dioxide, to shift partially or completely to alternative energy sources before year 2080.

The mercury pollution causing wastewater problems can be solved by ion exchange—and ion exchangers with a particularly high efficiency of mercury removal are available. Furthermore, adsorption on activated carbon (see *Wastewater Treatment: Conventional Methods*, p. 2677) removes mercury, even organic mercury compounds, rather effectively. In this context, it should be mentioned that bioremediation methods are also able to reduce mercury concentration in contaminated soil (see *Bioremediation*, p. 408).

In conclusion, mercury pollution problems can be solved by today's technology, except for the emission of mercury from coal-fired power plants, which requires particular attention if the use of coal as an energy source continues to grow, unless the problem is solved by shifts to alternative energy. It is furthermore expected that general restrictions in the use of mercury compounds will be tightened in the years to come.

REFERENCES

1. Newman, M.C.; Unger, M.A. *Fundamentals of Ecotoxicology*, 2nd Ed.; CRC Press: Boca Raton, London and New York, 2003.
2. NESCAUM. *Mercury Emissions from Coal-Fired Power Plants, The Case for Regulatory Action*; 2003.
3. Jørgensen, S.E.; Halling-Sørensen, B.; Mahler, H. *Handbook of Estimation Methods in Ecotoxicology and Environmental Chemistry*; Taylor and Francis Publ: Boca Raton, Boston, London, New York, Washington, D.C.; **1998**. 230 pp.
4. Jørgensen, S.E.; Jørgensen L.A.; Nors Nielsen, S. *Handbook of Ecological and Ecotoxicological Parameters*. Elsevier: Amsterdam, 1991; 1380 pp.
5. Jørgensen, L.A.; Jørgensen, S.E.; Nors Nielsen, S. *Ecotox*, CD; Elsevier: Amsterdam, 2000; 4000 pp.
6. Jørgensen, S.E.; Bendoricchio, G. *Fundamentals of Ecological Modelling*, 3rd Ed.; Elsevier: Amsterdam; 530 pp.
7. Jørgensen, S.E. *Principles of Pollution Abatement*; Elsevier: Amsterdam, 2000; 520 pp.

Methane Emissions: Rice

Kazuyuki Yagi
National Institute for Agro-Environmental Sciences, Japan

Abstract
The atmospheric concentration of methane has increased rapidly in recent years. Because it is a radiative trace gas and takes part in atmospheric chemistry, the rapid increase could be of significant environmental consequence. Of the wide variety of sources, rice fields are considered an important source of atmospheric methane because the harvest area of rice has increased greatly in the last 50 years. Because of the possibility of controlling the emission by agronomic practices, rice cultivation must be one of the most hopeful sources for mitigating methane emission.

INTRODUCTION

The atmospheric concentration of methane (CH_4) has increased rapidly in recent years. Because it is a radiative trace gas and takes part in atmospheric chemistry, the rapid increase could be of significant environmental consequence. Of the wide variety of sources, rice fields are considered an important source of atmospheric CH_4, because the harvest area of rice has increased by about 70% during last 50 years and it is likely that CH_4 emission has increased proportionally. Recent estimates suggest that global emission rates of CH_4 from rice fields account for about 4%–19% of the emission from all sources.[1] Due to the large amount of the global emission from rice cultivation, reduction of CH_4 emission from this source is very important in order to stabilize atmospheric concentration. In addition, because of the possibility of controlling the emission by agronomic practices, rice cultivation must be one of the most hopeful sources for mitigating CH_4 emission.

PROCESSES CONTROLLING CH_4 EMISSIONS FROM RICE FIELDS

Table 1 provides a summary of measured methane emissions at a number of specific research sites around the world.[2] It should be noted that methane fluxes from rice fields show pronounced diel and seasonal variations and vary substantially with different climate, soil properties, agronomic practices, and rice cultivars.

Processes involved in CH_4 emission from rice fields are illustrated in Fig. 1. Like other biogenic sources, CH_4 is produced by the activity of CH_4 producing bacteria, or methanogens, as one of the terminal products in the anaerobic food web in paddy soils. Methanogens are known as strict anaerobes that require highly reducing conditions. After soil is flooded, the redox potential of soil decreases rapidly by sequential biochemical reactions.

Flooded paddy soils have a high potential to produce CH_4, but part of CH_4 produced is consumed by CH_4 oxidizing bacteria, or methanotorophs. In rice fields, it is possible that a proportion of CH_4 produced in the anaerobic soil layer is oxidized in the aerobic layers, such as the surface soil–water interface and the rhizosphere of rice plants.

The emission pathways of CH_4 that is accumulated in flooded paddy soils is: diffusion into the flood water, loss through ebullition, and transport through the aerenchyma system of rice plants. In the temperate rice fields, more than 90% of CH_4 is emitted through plants,[5] while significant amounts of CH_4 may evolve by ebullition, in particular during the early part of the season in the tropical rice fields.[6] Therefore, it is concluded that possible strategies for mitigating CH_4 emission from rice cultivation can be made by controlling either production, oxidation, or transpor processes.

OPTIONS FOR MITIGATING CH_4 EMISSION

Water Management

Mid-season drainage (aeration) in flooded rice fields supplies oxygen into soil, resulting in a reduction of CH_4 production and a possible enhancement of CH_4 oxidation in soil.[7,8] A study using an automated sampling and analyzing system clearly showed that short-term drainage had a strong effect on CH_4 emission, as shown in Fig. 2. Total emission rates of CH_4 during the cultivation period were reduced by 42%–45% by short-term drainage practices compared with continuously flooded treatment.[9] These results indicate that improvement in water management can be one of the most promising mitigation strategies for CH_4 emission from rice fields. Increasing the rate of water percolation in rice fields by installing underground pipe drainage may also have an influence on CH_4 production and emission.

Table 1 Methane emission from rice fields in various world locations.[a]

Country	Daily average (g/m² day)	Flooding period (days)	Season total Average (g/m²)	Season total Range (g/m²)
China	0.19–1.39	75–150	13	10–22
India	0.04–0.46	60	10	5–15
Italy	0.10–0.68	130	36	17–54
Japan	0.01–0.39	110–130	11	3–19
Spain	0.10	120	12	
Thailand	0.04–0.77	80–110	16	4–40
U.S.A.	0.05–0.48	80–100	25	15–35

[a]The data are for the fields without organic fertilizer.

Soil Amendments and Mineral Fertilizers

The progress of soil reduction can be retarded by adding one of several electron acceptors in the sequential soil redox reactions. Sulfate is one of the most promising candidates for this strategy because it is commonly used as a component of mineral fertilizer and soil amendment. Field measurements have shown that CH_4 emission rate decreased by at most 55%–70% by application of ammonium sulfate or gypsum.[10,11]

Additions of other oxidants, such as nitrate and iron-containing materials, may influence CH_4 emission from rice fields. As well as adding oxidants, dressing paddy fields with other soils that contain a large amount of free iron and manganese may decrease CH_4 emission. Other chemical candidates are nitrification inhibitors and acetylene releasing materials.

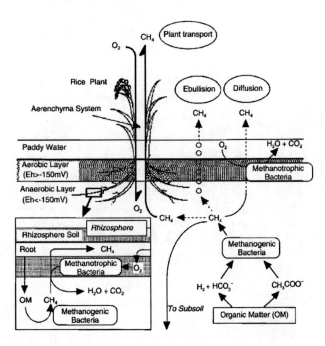

Fig. 1 Production, oxidation and emission of CH_4 in rice paddy fields.
Source: Conrad[3] and Knowles.[4]

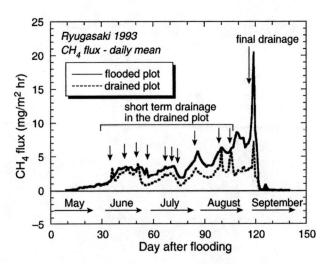

Fig. 2 Effect of water management on CH_4 emission from a rice paddy field. The arrows indicate period of midseason drainage in the intermittent irrigation plot and the timing of final drainage in both of the plots.

Organic Matter Management

In rice cultivation, fresh organic matter and animal wastes are often applied as fertilizers. In the fields, a proportion of the biomass of previous crops and weeds remains in soils at the start of rice cultivation. Such organic matter is decomposed in soils and acts as a substrate for fermentation reactions. Many researchers have demonstrated that incorporation of rice straw and green manure into rice paddy soils dramatically increases CH_4 emission.[7,10,12] The impact of organic amendments on CH_4 emissions can be described by a dose–response curve which adopts correction factors for composted and fermented organic matter.[6] Mitigation of CH_4 emission requires that the quantities of organic amendments be minimized.

Field experiments also indicated that composted or fermented organic matter increased CH_4 emission much less than fresh organic matter, due to a lower content of easily decomposable carbon.[6,7] Therefore, stimulation of composting organic amendments appears to be a promising mitigation option. Plowing the fields during the fallow period and promoting aerobic degradation of organic matter is also likely to reduce CH_4 emission.

Others

Different tillage and cropping practices change the physical, chemical, and microbiological properties of the plow layer soil and may reduce CH_4 emission. These include deep tillage, no tillage, and flooded rice-upland crop rotation.

Selecting and breeding rice cultivars that emit lower CH_4 is a desirable approach because it is easy to adopt. There are four points to consider for selecting cultivars: 1) they should exude low levels carbon from their roots; 2) they should have a low level of CH_4 transport and a high level of CH_4 oxidation in the rhizosphere; 3) they should have a higher harvest index, in order to reduce organic matter input into soil after harvest; and 4) they should be suitable and have a high productivity when other mitigation options are performed.

PROBLEMS AND FEASIBILITY OF THE OPTIONS

If the above mitigation options could be applied to world's rice cultivation, global CH_4 emission from rice fields could decrease significantly. However, there are several formidable obstacles to adopting the mitigation options into local rice farming. Table 2 summarizes the problems and feasibility of the individual mitigation options along with the efficiency of the options.

Application of some options is limited to specific types of rice fields. In particular, altering water management practices may be limited to rice paddy fields where the irrigation system is well equipped. Long midseason drainage and short flooding may cause possible negative effects on grain yield and soil fertility. Improving percolation by underground pipe drainage requires laborious engineering work. The increased water requirement is another problem

Table 2 Evaluation of the mitigation options for methane emission from fields.

	CH_4 mitigation efficiency	Problem for application							Other trade-off effects
		Applicability		Economy		Effects on		Time span	
		Irrigated	Rainfed	Cost	Labor	Yield	Fertility		
Water management									
Midseason drainage	□	o	•	~	↑	+	~	o	May promote N_2O emission
Short flooding	□	o	•	~	~	−	−	o	May promote N_2O emission
High percolation	□	o	•	↑	↑	+	~	o	May promote nitrate leaching
Soil amendments									
Sulfate fertilizer	□	o	o	↑	~	Δ	−	o	May cause H_2S injury
Oxidants	□	o	o	↑	↑	Δ	−	o	
Soil dressing	o	o	o	↑	↑	−	−	o	
Organic matter									
Composting	□	o	o	↑	↑	+	+	o	
Aerobic decomposition	□	o	o	~	↑	~	~	o	
Burning	o	o	o	~	↑	~	~	o	Causing atmospheric pollution
Others									
Deep tillage	o	o	o	↑	↑	−	−	o	
No tillage	?	o	o	~	↓	−	~	o	
Rotation	o	o	Δ	~	↑	−	−	o	
Cultivar	o	o	o	~	~	~	~	•	

Key:
□ Very effective
o Effective/applicable
Δ Case by case
• Not applicable/require long time
? No information
↑ Increase
↓ Decrease
~ About equal to previous situation
+ Positive
− Negative
Source: Ranganathan et al.,[13] Neue et al.,[14] and Yagi et al.[15]

in the water management options because water is a scarce commodity in many regions.

Cost and labor are serious obstacles for applying each option to local farmers. Most of the mitigation options will decrease profitability and the farmer net returns in the short run. To overcome these obstacles, an effort to maximize net returns by joining CH_4 mitigation and increased rice production will be needed, as well as political support.

It is recognized that the mitigation options should not have any significant trade-off effects, such as decreased rice yield, a decline in soil fertility, or increased environmental impact by nitrogen compounds. The development of anaerobic conditions in soil by flooding decreases decomposition rates of soil organic matter compared with aerobic soils, resulting in soil fertility being sustained for a long time. Flooded rice cultivation shows very little growth retardation by continuous cropping. Some mitigation options may reduce these advantages of rice fields. Application of sulfate-containing fertilizer may cause a reduction in rice yield due to the toxicity of hydrogen sulfide. Mid-season aeration and soil amendments may induce nitrogen transformation resulting in enhanced N_2O emissions.[16,17]

REFERENCES

1. Prather, M.; Derwent, R.; Ehhalt, D.; Fraser, P.; Sanhueza, E.; Zhou, X. Other trace gases and atmospheric chemistry. In *Climate Change 1994, Radiative Forcing of Climate Change and an Evaluation of the IPCC IS92 Emission Scenarios*; Houghton, J.T., Meira Filho, L.G., Bruce, J., Lee, H., Callander, B.A., Haites, E., Harris, N., Maskell, K., Eds.; Cambridge University Press: Cambridge, England, 1995; 73–126.
2. International panel on climate change. *Greenhouse Gas Inventory Reference Manual*; IPCC Guidelines for National Greenhouse Gas Inventories, OECD: Paris, France, 1997; Vol. 3, 46–60.
3. Conrad, R. Control of methane production in terrestrial ecosystems. In *Exchange of Trace Gases Between Terrestrial Ecosystems and the Atmosphere*; Andreae, M.O., Schimel, D.S., Eds.; John Wiley and Sons Ltd.: New York, 1989; 39–58.
4. Knowles, R. Processes of production and consumption. In *Agricultural Ecosystem Effects on Trace Gases and Global Climate Change*; Harper, L.A., Mosier, A.R., Duxbury, J.M., Rolston, D.E., Eds.; American Society of Agronomy: Madison, WI, 1993; 145–156.
5. Cicerone, R.J.; Shetter, J.D. Sources of atmospheric methane: measurements in rice paddies and a discussion. J. Geophys. Res. **1981**, *86*, 7203–7209.
6. Denier van der Gon, H.A.C.; Neue, H.-U. Influence of organic matter incorporation on the methane emission from a wetland rice field. Global Biogeochem. Cycles **1995**, *9*, 11–22.
7. Yagi, K.; Minami, K. Effect of organic matter application on methane emission from some Japanese paddy fields. Soil Sci. Plant Nutr. **1990**, *36*, 599–610.
8. Sass, R.L.; Fisher, F.M.; Wang, Y.B.; Turner, F.T.; Jund, M.F. Methane emission from rice fields: the effect of floodwater management. Global Biogeochem. Cycles **1992**, *6*, 249–262.
9. Yagi, K.; Tsuruta, H.; Kanda, K.; Minami, K. Effect of water management on methane emission from a Japanese rice paddy field: automated methane monitoring. Global Biogeochem. Cycles **1996**, *10*, 255–267.
10. Schütz, H.; Holzapfel-Pschorn, A.; Conrad, R.; Rennenberg, H.; Seiler, W. A 3 years continuous record on the influence of daytime, season, and fertilizer treatment on methane emission rates from an Italian rice paddy. J. Geophys. Res. **1989**, *94*, 16405–16416.
11. Denier van der Gon, H.A.C.; Neue, H.-U. Impact of gypsum application on methane emission from a Wetland rice field. Global Biogeochem. Cycles **1994**, *8*, 127–134.
12. Sass, R.L.; Fisher, F.M.; Harcombe, P.A.; Turner, F.T. Mitigation of methane emission from rice fields: possible adverse effects of incorporated rice straw. Global Biogeochem. Cycles **1991**, *5*, 275–287.
13. Ranganathan, R.; Neue, H.-U.; Pingali, P.L. Global climate change: role of rice in methane emission and prospects for mitigation. In *Climate Change and Rice*; Peng, S., Ingram, K.T., Neue, H.-U., Ziska, L.H., Eds.; Springer-Verlag: Berlin, Germany, 1995; 122–135.
14. Neue, H.-U.; Wassmann, R.; Lantin, R.S. Mitigation options for methane emissions from rice fields. In *Climate Change and Rice*; Peng, S., Ingram, K.T., Neue, H.-U., Ziska, L.H., Eds.; Springer-Verlag: Berlin, Germany, 1995; 137–144.
15. Yagi, K.; Tsuruta, H.; Minami, K. Possible options for mitigating methane emission from rice cultivation. Nutr. Cycling Agro-Ecosys. **1997**, *49*, 213–220.
16. Cai, Z.; Xing, G.; Yan, X.; Xu, H.; Tsuruta, H.; Yagi, K.; Minami, K. Methane and nitrous oxide emissions from rice paddy fields as affected by nitrogen fertilizers and water management. Plant Soil **1997**, *196*, 7–14.
17. Bronson, K.F.; Neue, H.-U.; Singh, U.; Abao, E.B., Jr. Automated chamber measurements of methane and nitrous oxide flux in a flooded rice soil: I. Residue, nitrogen, and water management. Soil Sci. Soc. Am. J. **1997**, *61*, 981–987.

Minerals Processing Residue (Tailings): Rehabilitation

Lloyd R. Hossner
Soil and Crop Sciences Department, Texas A&M University, College Station, Texas, U.S.A.

Hamid Shahandeh
Texas A&M University, College Station, Texas, U.S.A.

Abstract
Metals have been mined and exploited with the growth of world industry. In particular, iron, lead, zinc and copper have been mined extensively in regions all over the world. In the process of metal benefication, massive heaps of spoil (tailings and waste rock) have been created at mine sites or areas distant from mines. Tailings and waste rock create esthetic problems in the landscape and affect water, soil, plant, and public health.

INTRODUCTION

Metals have been mined and exploited with the growth of world industry. In particular, iron, lead, zinc and copper have been mined extensively in regions all over the world. In the process of metal benefication, massive heaps of spoil (tailings and waste rock) have been created at mine sites or areas distant from mines (Fig. 1). Tailings and waste rock create esthetic problems in the landscape and affect water, soil, plant, and public health. Tailings is defined as the solid waste product of the milling and mineral concentration process.[1] Mill tailings are the finely ground host rock materials from which the desired mineral values have been extracted during the concentration process. Generally, tailings are transported from the mill to their place of disposal as a water slurry containing 15%–50% solids by weight and discharged by impoundment on land in settling ponds adjacent to the mills, used as backfill in the open pit or underground mine, disposed in deep lakes or offshore, or processed for secondary metal recovery followed by disposal. Tailings impoundments range in size from <10 to >2000 ha, and may be stacked as high as 50 m.[2]

There are a large number of abandoned mine waste and tailings deposits from a wide variety of industries around the world. The total area of land disturbed by mining in China is estimated to be about 2 million ha.[3] In the U.S. between 1930 and 1980 more than 2 million ha of land were affected by mining operations.[4] The mine tailings produced in Malaysia, England, Thailand, and Canada are estimated in billions of tons occupying hundreds of thousands of hectares.[2]

POTENTIAL ENVIRONMENTAL PROBLEMS

Abandoned spoil and tailings contain the waste products of both mining and ore processing operations. Chemical extractants such as sulfuric acid and sodium bicarbonate are used for uranium ores, cyanide for gold ores, sodium hydroxide for aluminum ores, and sulfuric acid and hydrochloric acids for copper, nickel, and cobalt ores. In addition to these toxic solvents and the dissolved heavy metals, tailings from uranium and phosphate operations can contain radionuclides such as thorium and radium. These materials are often a major source of pollution in the local environment due to dust blow and the potential leaching of the products of mineral weathering into water sources.

Tailings are also subject to the process of weathering and, over time, changes may occur in their properties which could be hazardous to the environment. For example, sulfides are associated minerals that are readily oxidized in the tailings when exposed to air, water, and iron oxidizing bacteria. One of the oxidation products is sulfuric acid. Tailings containing sulfide minerals may eventually have a pH of 1.5–3.5.[2] Toxic ions may also contaminate soils and waters adjacent to smelters through seepage, runoff waters, and eroded sediments. Effluents arising from tailings seepage could be toxic in varying degrees to man, animals, and plant life. Also, there is a concern for the amount of heavy metal uptake by plants growing on tailings and its effect on the food chain.[5]

MINE WASTE REHABILITATION PRACTICES

Engineering Approach

Mine waste residue and soils contaminated with toxic metals and radionuclides can be remediated and stabilized using engineering approaches including chemical, physical, and thermal techniques (e.g., in situ mobilization, immobilization, degradation, and burial; or removal and reburial, vitrification, vacuum extraction, steam flooding, pumping and leaching, electroosmosis, and electroacoustic extraction).[6] The use of chemical and physical techniques to stabilize mineral wastes against wind and water erosion are limited because of the cost and maintenance.[7]

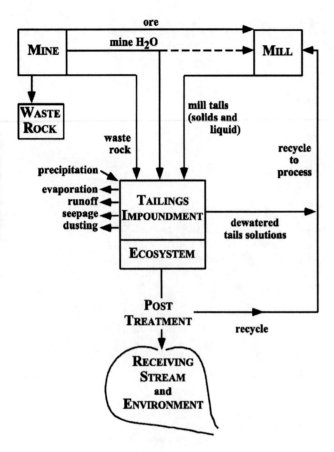

Fig. 1 Mine/mill environment.
Source: Fom Ritcey.[2]

Physical stabilization of materials such as waste rock from strip mining can be used to reduce wind and water erosion.

Chemical stabilization requires a chemical agent, such as lignin sulfate or resinous adhesive, to react with mine waste to provide a crust resistant to wind and water erosion. Physical stabilization can be performed by protecting or isolating the waste from the environment with physical barriers such as liners and clay caps. Usually, the design and installation of a cover system consists of three zones: an erosion resistant layer, a moisture retention layer, and an underlying clay barrier to prevent entrance of water into the waste material. The design of the capping sequence and incorporation of a coarse layer to break capillarity are shown to be very important. When waste material poses a threat to groundwater quality or to plant growth, a layer of impermeable plastic or compacted clay may be placed between the waste and the soil cap to prevent movement of water into the toxic wastes. In other materials it may be necessary to place a diffusion barrier of very coarse material between the toxic waste and the cover soil to break a water column that might permit diffusion of the toxic materials from the waste into the cover soil. Soil caps placed over nontoxic materials may be thinner than soil caps placed over toxic materials.[8]

Ecological Approach

According to the Center for Minesite Rehabilitation in Australia the strategies for rehabilitation of mine waste should be based on stabilization and sustainable revegetation of tailings and waste rock in a reconstructed ecosystem.[9] Tailings must be stabilized before or concurrent with full-scale revegetation. The keys to successful revegetation of metalliferous mill tailings are: 1) obtain a complete understanding of the physical and chemical properties of the tailing material; 2) select an appropriate final land use of the area under consideration; 3) select the most cost effective combinations of amendments, nutrient supplements, cover materials, and plant species to be employed; and 4) monitor the performance of the vegetation and make adjustments as needed. It is now widely accepted that the establishment of vegetation is a desirable method for stabilization of mine wastes.[10,11] Vegetation reduces wind velocity at the surface, captures dust particles, reduces raindrop impact, reduces runoff by increasing infiltration, and reduces the overland flow of water and sediment. Vegetative stabilization also improves the chemical, biological, and physical properties of the mine wastes by increasing the organic matter content, nutrient level, cation exchange capacity, and biological activity.

Vegetation established on mine tailings has several advantages over physical and chemical methods of stabilization. Vegetation is esthetically pleasing and accepted by the public, relatively inexpensive, less disruptive of site remediation, creates a beneficial habitat for wildlife, plant roots and shoots can take up heavy metals, and plants may stimulate microbial immobilization of heavy metals in the rhizosphere. Thus, establishment of vegetation in mining areas has the potential of reducing contamination of adjacent soil, surface water, and ground water.[8]

The main approaches to revegetation have been summarized in Table 1.[7] The waste characteristics of a site determine which approach is most suitable. Generally there are two approaches to revegetation which have been used in combination or separately. They are adaptive and ameliorative approaches. The adaptive approach to revegetation is to combat the toxicity of the waste by direct seeding with metal tolerant cultivars. The metal tolerant plants may be of great benefit to developing countries for low-cost revegetation. The other approach is ameliorative and the toxicity is avoided or diluted, rather than tolerated by using some form of covering system. This approach has been widely used and subjected to considerable research. The two types of covering material are ameliorants and inert amendments. Ameliorants are materials such as sewage sludge, compost, domestic refuse, peat, and topsoil. Inert ameliorants include materials such as clay, shale, or gypsum. Inert ameliorants are commonly wastes from other industry or mining activities. This approach has the added benefit of using one category of waste to overcome the problems of other forms of waste.[12]

Table 1 Approaches to revegetation of minerals processing residue (tailings).

Waste characteristics	Reclamation technique	Problems encountered
Low toxicity: total metal content <0.1%. No major acidity or alkalinity problems	Amelioration and direct seeding with agricultural or amenity grasses and legumes. Using traditional or specialized techniques	Probable commitment to a medium/long-term maintenance program. Grazing management must be monitored
Low toxicity and climatic limitations: toxic metal content <0.1%. No major acidity or alkalinity problems, extreme of temperature, rainfall, etc.	Amelioration and direct seeding with native species. Seed or transplant ecologically adapted native species using amelioration treatments where appropriate	Irrigation often necessary at establishment. Expertise required on the characteristics of native flora
High toxicity: toxic metal content >0.1%. High salinity in some cases	1) Amelioration and direct seeding with tolerant ecotypes. Apply lime, fertilizer and organic matter, as necessary, before seeding	Regular fertilizer application. Few species have evolved tolerance and few are available commercially. Grazing management not possible
	2) Surface treatment and seeding with agricultural or amenity grasses and legumes. Amelioration with 10–50 cm of innocuous mineral waste and/or organic material. Apply lime and fertilizer as necessary	Regression will occur if depths of amendments are shallow or if upward movement of metals occurs. Availability and transport costs may be limiting
Extreme toxicity: very high toxic metal content. Intense salinity or acidity	Isolation: surface treatment with 30–100 cm of innocuous barrier and surface binding with 10–30 cm of a suitable rooting medium. Apply lime and fertilizer as necessary	Susceptibility to drought according to the nature and depth of amendments. High cost and potential limitations of materials availability

Source: Tordoff.[7]

Although revegetation is desirable, metal wastes can present a very unfavorable environment for plants because of the presence of many growth limiting factors such as high salinity, metal toxicity, or nutrient deficiency in the mine tailings and soils.[13]

CHARACTERISTICS OF MINE WASTE THAT LIMIT REHABILITATION

Characteristics of tailings from 43 selected mine waste sites are shown in Table 2.[14] Alleviation or modification of each chemical and physical limitation of the tailing is required prior to the establishment of vegetation.

Chemical Properties

The chemical composition of tailings depends on the original ore mineralogy, extraction techniques, and associated minerals. Among the minerals present in tailings, sulfides are often an important constituent that must be considered in tailings management.

Mining often exposes the sulfide-bearing minerals (pyrite, marcasite, pyrohotite, chalcopyrite, arsenopyrite, cobalite) to the atmosphere. In the presence of water, iron sulfide oxidation by weathering [Eq. (1)] and by iron oxidizing bacteria catalysis [Eq. (2)] will convert the sulfides to sulfuric acid:[15]

$$2FeS_2 + \frac{15}{2}O_2 + 7H_2O = 2Fe(OH)_3 + 4H_2SO_4 \quad (1)$$

$$FeS_2 + 14Fe^{3+} + 8H_2O = 15Fe^{2+} + 2SO_4^{2-} + 16H^+ \quad (2)$$

Large amounts of lime, between 10–150 t ha^{-1}, may be required to neutralize the acidity produced in sulfidic tailings.[5]

Various salts may appear on the surface of tailings depending on the nature of the original mill process. The factors which are involved in the appearance of salts on the tailings surface include: excess concentrations of soluble salts in tailings materials, availability of shallow subsurface water, salt concentrations within the tailing water, and temperature and chemical potential gradients between the surface and the interior of the tailing. The upward migration of salts and their inhibiting effect on root growth and revegetation have been major concerns in the reclamation of many mine tailings. For example, a common problem associated with the establishment of vegetation in asbestos tailings, bentonite tailings, and red mud from bauxite ore processing for alumina is salinity and/or sodicity. These wastes can be highly alkaline (pH > 10), saline (EC > 30 dS m^{-1}), and sodic. Vegetation establishment on these tailings is difficult because of salinity, alkalinity, clay dispersion and low hydraulic conductivity.[5]

Mine wastes are usually deficient in major plant nutrients and almost universally deficient in nitrogen. Nutrient deficiencies, especially nitrogen, phosphorus and potassium, are often the principal limiting constraints to revegetation of kaolinitic china clay, iron, copper, gold, silver, and other heavy metal mine wastes.[5] The success of establishing plants on tailings depends on providing an

Table 2 Mean and range of values for selected physical and chemical characteristics of tailings.

Property	Unit	Mean	Range
Particle size distribution			
<2 mm	%	95	20–100
Sand	%	51	1–97
Silt	%	43	0–96
Clay	%	7	0–40
Moisture retention (bar)			
0.1	%	22	0–55
0.3	%	18	0–55
15	%	4	0–20
Available water holding capacity	%	16	0–35
Bulk density	g cm^{-3}	1.5	0.2–3.1
Particle density	g cm^{-3}	2.91	0.01–4.29
pH		6.2	1.8–9.4
Cation exchange capacity	cmolc kg	2.63	0.19–46.5
Organic matter	%	2	0.02–25
Electrical conductivity	dS/m	2	0.1–22.4
Available nutrients			
Phosphorus (P)	mg/kg	10	1–400
Potassium (K)	mg/kg	63	1–564
Calcium (Ca)	mg/kg	11,930	40–52,480
Magnesium (Mg)	mg/kg	230	15–1328
Total analysis			
Nitrogen (N)	%	0.013	0.001–0.166
Sulfur (S)	%	4.02	0.001–38.87
Iron (Fe)	%	15.5	0.4–56.81
Aluminum (Al)	%	2.8	0.1–8
Calcium (Ca)	%	1.7	0.01–10.95
Magnesium (Mg)	%	1.2	0.04–5.0
Sodium (Na)	%	0.5	0.01–2.9
Potassium (K)	%	0.7	0.04–3.32
Manganese (Mn)	%	0.2	0.01–4.0
Silicon (Si)	%	22	4–37
Cadmium (Cd)	mg/kg	38	2–280
Chromium (Cr)	mg/kg	1000	70–7000
Cobalt (Co)	mg/kg	1140	100–9999
Molybdenum (Mo)	mg/kg	70	10–800
Nickel (Ni)	mg/kg	96	10–546
Lead (Pb)	mg/kg	340	0.3–2810
Titanium (Ti)	mg/kg	2500	200–10,000
Zinc (Zn)	mg/kg	510	1–5000
Copper (Cu)	mg/kg	130	1–750

adequate supply of plant nutrients. High levels of the major essential nutrients may reduce the harmful effects of metal ions.

Toxic ions are frequently present in tailings of heavy metals in sufficient concentrations to prevent plant growth unless considerable amelioration is undertaken. Toxic ions decrease root respiration, limit water and nutrient uptake, reduce enzymatic activity and microbial populations, and inhibit cell mitosis in root meristematic regions.[15] Toxic ions may also contaminate soils and waters adjacent to tailings through seepage, runoff waters, and eroded sediments. Another problem with heavy metal contaminated tailings is the possible uptake of metals by plants in quantities that could be toxic in the general food chain. Radioactivity associated with uranium and phosphate tailings and high concentrations of arsenic, mercury and cyanide in silver and gold mine tailings are other environmental concerns.

Physical Properties

The physical properties of tailings vary with the mineral being processed, the origin of the ore body, and the process used for mineral concentration. Physical properties are the most important in determining productivity of vegetated mine wastes because of the cost involved in trying to ameliorate particle size distribution and water-holding capacity. Nonuniform texture is the main physical problem of mine wastes which limits the availability of water to plants. Mining generates a wide range of particle size materials. This includes coarse mine wastes, fine clays, flotation tailings, chemical precipitates, and slimes. In the mineral industry the slime size fraction is <5 μm.

Fine texture is a major problem in gold tailings. Gold tailing is finely crushed (0.01–0.1 μm) because of the large surface area needed for fast reaction with processing chemicals. It is relatively easy to establish vegetation on these tailings but it is difficult to obtain full ground cover because young plants are sand blasted by finely crushed, wind blown materials. Laying large rock fragments on the slope of sand deposits and installing reed wind breaks on the entire sand deposit has helped the establishment of vegetation.[2] A critical factor in reclaiming and vegetating a tailing is the availability of moisture for plant growth. Moisture curve characteristics of tailings derived from iron, nickel, copper, lead, zinc, and gold mines showed properties similar to a sandy loam soil, an indication of potential water deficiency for plant growth.[2]

Vegetation establishment and maintenance on coarse tailings, where there will be little water retention and rapid drainage, is also related to water availability. Establishment of vegetation on very coarse tailings is difficult but it can be achieved by adding a soil cover followed by hydromulching and hydroseeding.[15]

Crusting, cracking, and a general lack of structure are common characteristics of mine tailing brought about by differences in texture, lack of organic matter, and variable mineralogy. Structure determines the bulk density of tailings. Water infiltration is often limited in fine textured tailings due to poor structural characteristics. Root penetration and moisture stress of plants due to limited rooting generally becomes a problem with dry bulk density values above 1.5 Mg m^{-3} in tailings.[15]

Tailings have a high heat capacity. Tailings exposed to direct solar radiation can have temperatures of 55–65°C at 1–2 cm depth. Internal temperature in a silver mine in Canada reached 45°C during the oxidation process. Hay or straw mulch has been an effective insulator for stabilizing the tailings temperature.[2]

CONCLUSIONS

There are large areas of abandoned mine tailings from a variety of industrial operations around the world. Tailings contain a large proportion of the original ore that was mined. Tailings are deposited in impoundments with dimensions ranging from several square meters to some that are many square kilometers in area. In the past, these tailings have been largely abandoned and allowed to revegetate under natural conditions. In recent years there have been both public and scientific awareness of the possible consequences of long term disposal of tailings. Now many countries have enacted legislation to ensure reclamation of tailings of disposal sites once the mining and milling operations have ceased. A complete understanding of the physical and chemical properties of tailings is essential for planning successful rehabilitation programs.

REFERENCES

1. Richmond, T.C. The revegetation of metalliferous tailings. In *Reclamation of Drastically Disturbed Lands*; Barnhisel, R.I.; Darmody, R.G.; Daniels, W.L., Eds.; American Society of Agronomy: Madison, WI, 2000; 801–818.
2. Ritcey, G.M. *Tailings Management: Problems and Solutions in the Mining Industry*; Elsevier: New York, USA, 1989.
3. Ye, Z.H.; Wong, J.W.C.; Wong, M.H. Vegetation response to lime and manure compost amendments on acid lead = zinc mine tailings. Restor. Ecol. **2000**, *8*, 289–295.
4. Johnson, W.; Paone, J. *Land Utilization and Reclamation in the Mining Industry*, 1930–1980; U.S. Bureau of Mines Information Circular 8862; United States Printing Office: Washington, DC, 1982.
5. Hossner, L.R.; Shahandeh, H. Chemical and physical limitations to mine tailings reclamation. Trends Soil Sci. **1991**, *1*, 291–305.
6. Francis, A.J.; Dodge, C.J. Remediation of soils and wastes contaminated with uranium and toxic metals. Environ. Sci. Technol. **1998**, *32*, 3993–3997.
7. Tordoff, G.M.; Baker, A.J.M.; Willis, A.J. Current approaches to the revegetation and reclamation of metalliferous mine wastes. Chemosphere **2000**, *41*, 219–228.
8. Menzies, N.W.; Mulligan, D.R. Vegetation dieback on clay-capped mine waste. J. Environ. Qual. **2000**, *29*, 437–442.
9. Bell, L.C. The Australian centre for minesite rehabilitation research: an initiative to meet the strategic research needs for sustainable mining rehabilitations. Water, Air, Soil Pollut **1996**, *91*, 125–133.
10. Harris, J.A.; Birch, P.; Palmer, J. *Land Restoration and Reclamation: Principles and Practices*; Addison-Wesley/Longman: Harlow/Essex, England, 1996.
11. Munshower, F.F. *Practical Handbook of Disturbed Land Revegetation*; Lewis/CRC Press: London/Boca Raton, FL, 1994.
12. Bradshaw, A.D.; Johnson, M.S. Revegetation of metalliferous mine waste: the range of practical techniques used in western Europe. In *Minerals, Metals and the Environment*; Institute of Mining and Metallurgy: London, 1992.
13. Johnson, M.S.; Cooke, J.K.W.; Stevennson, J.K.W. Revegetation of metalliferous wastes and land after metal mining. In *Mining and Environmental Impact; Issues in Environmental Science and Technology*; Hester, R.E.; Harrison, R.M., Eds.; Royal Society of Chemistry: London, 1994; 31–48.
14. Murray, D.R. Pit Slope Manual. Supplement 10–1; Reclamation by Revegetation; Report No. 77-31, Vol. 1. Mine Waste Description and Case Histories; CANMET (Can. Center Min. Energy Technol.). Toronto, Canada, 1977.
15. Hossner, L.R.; Hons, F.M. Reclamation of mine tailings. Adv. Soil Sci. **1992**, *17*, 311–349.

Mines: Acidic Drainage Water

Wendy B. Gagliano
Clark State Community College, Springfield, Ohio, U.S.A.

Jerry M. Bigham
Ohio State University, Columbus, Ohio, U.S.A.

Abstract
Acid mine drainage refers to metal-rich sulfuric acid solutions released from mine tunnels, open pits, and waste rock piles. These solutions can seriously degrade water quality and harm aquatic life when discharged into streams and lakes. Acid production arises from the oxidative dissolution of sulfide minerals, such as pyrite [FeS$_2$], through exposure to air, water, and populations of acid tolerant bacteria. Successful prevention and treatment of mine drainage requires an understanding of many hydrological, geochemical, and biological principles.

INTRODUCTION

What Is Acid Mine Drainage?

Acid mine drainage refers to metal-rich sulfuric acid solutions released from mine tunnels, open pits, and waste rock piles (Table 1). Similar solutions are produced by the drainage of some coastal wetlands, resulting in the formation of acid sulfate soils. Acid mine drainage typically yields pH values ranging from 2 to 4; however, extreme sites such as Iron Mountain, California, have produced pH values as low as –3.6.[1] Neutral to alkaline mine drainage is also common in areas where the surrounding geologic units contain carbonate rocks to buffer acidity (Table 1).

Why Is Acid Mine Drainage a Problem?

Landscapes exposed to acid mine drainage do not support vegetation and are susceptible to erosion. When acid mine drainage enters natural waterways, changes in pH and the formation of voluminous precipitates of metal hydroxides can devastate fish populations and other aquatic life (Fig. 1). The corrosion of engineered structures such as bridges is also greatly accelerated. There may be as many as 500,000 inactive or abandoned mines in the United States, with mine drainage severely impacting approximately 19,300 km of streams and more than 72,000 ha of lakes and reservoirs.[2,3] Once initiated, mine drainage may persist for decades, making it a challenging problem to solve.

What Causes Acid Mine Drainage?

Mine drainage results from the oxidation of sulfide minerals such as pyrite (cubic FeS$_2$), marcasite (orthorhombic FeS$_2$), pyrrhotite (Fe$_{1-x}$S), chalcopyrite (CuFeS$_2$), and arsenopyrite (FeAsS). These minerals are commonly found in coal and ore deposits and are stable until exposed to oxygen and water. Their oxidation causes the release of metals and the production of sulfuric acid. This process can occur as a form of natural mineral weathering but is exacerbated by mining because of the sudden, large-scale exposure of unweathered rock to atmospheric conditions.

MINE DRAINAGE CHEMISTRY

Mine drainage is a complex biogeochemical process involving oxidation-reduction, hydrolysis, precipitation, and dissolution reactions as well as microbial catalysis.[1] The entire sequence is commonly represented by Reaction 1, which describes the overall oxidation of pyrite by oxygen in the presence of water to form iron hydroxide [Fe(OH)$_3$] and sulfuric acid.

$$\text{FeS}_{2(s)} + 3\frac{3}{4}\text{O}_{2(g)} + 3\frac{1}{2}\text{H}_2\text{O}_{(l)} \rightarrow \text{Fe(OH)}_{3(s)} + 2\text{H}_2\text{SO}_{4(aq)} \quad (1)$$

The actual oxidation process is considerably more complicated.

Pyrite and related sulfide minerals contain both Fe and S in reduced oxidation states. When exposed to oxygen and water, the sulfur moiety is oxidized first, releasing Fe^{2+} and sulfuric acid to solution (Reaction 2). The rate of oxidation is dependent on environmental factors like temperature, pH, Eh, and relative humidity, as well as mineral surface area and microbial catalysis.

$$\text{FeS}_{2(s)} + 3\frac{1}{2}\text{O}_{2(g)} + \text{H}_2\text{O}_{(l)} \rightarrow \text{Fe}^{2+}_{(aq)} + 2\text{SO}^{2-}_{4(aq)} + 2\text{H}^+_{(aq)} \quad (2)$$

Reaction 2 is most important in the initial stages of mine drainage generation and can be either strictly abiotic or

Table 1 Summary of mine drainage chemistry from 101 bituminous coal mine sites in Pennsylvania.

	Range	Median	Mean
pH	2.7–7.3	5.2	3.6
Fe (mg/L)	0.16–512.0	43.0	58.9
Al (mg/L)	0.01–108.0	1.3	9.8
Mn (mg/L)	0.12–74.0	2.2	6.2
SO_4 (mg/L)	120–2000	580.0	711.2

Unpublished data from C. Cravolta, III, 2001. USGS, Lemoyne, PA.

mediated by contact with sulfur-oxidizing bacteria.[4] The Fe^{2+} released by pyrite decomposition is rapidly oxidized by oxygen at pH > 3 as per Reaction 3.

$$Fe^{2+}_{(aq)} + \frac{1}{4}O_{2(aq)} + H^{+}_{(aq)} \rightarrow Fe^{3+}_{(aq)} + \frac{1}{2}H_2O_{(l)} \qquad (3)$$

If acidity generated by Reaction 2 exceeds the buffering capacity of the system, the pH eventually decreases. Below pH 3, Fe^{3+} solubility increases and a second mechanism of pyrite oxidation becomes important[5] (Reaction 4).

$$FeS_{1(2)} + 14Fe^{3+}_{(aq)} + 8H_2O_{(l)} \rightarrow 15Fe^{2+}_{(aq)} + 2SO^{2-}_{4(aq)} + 16H^{+}_{(aq)} \qquad (4)$$

In this case, pyrite is oxidized by Fe^{3+} resulting in the generation of even greater acidity than when oxygen is the primary oxidant. Pyrite decomposition is thus controlled by the rate at which Fe^{2+} is converted to Fe^{3+} at low pH.[6] At pH < 3, Fe^{2+} oxidation is very slow unless it is catalyzed by populations of iron-oxidizing bacteria like *Acidithiobacillus ferrooxidans* or *Leptospirillum ferrooxidans*. These acidophilic bacteria oxidize Fe^{2+} as a means of generating energy to fix carbon. In doing so, they supply soluble Fe^{3+} at a rate equal to or slightly greater than the rate of pyrite oxidation by Fe^{3+}.[5] Pyrite oxidation then regenerates Fe^{2+} (Reaction 4), creating a cyclic situation that leads to vigorous acidification of mine drainage water.

MINE DRAINAGE MINERALOGY

The hydrolysis of Fe^{3+} causes the precipitation of various iron minerals—generally represented as ($Fe[OH]_3$)—that are often the most obvious indicators of mine drainage contamination (Reaction 5).

$$Fe^{3+}_{(aq)} + 3H_2O_{(l)} \rightarrow Fe(OH)_{3(s)} + 3H^{+}_{(aq)} \qquad (5)$$

These precipitates are yellow-to-red-to-brown in color and have long been referred to by North American miners as "yellow boy." The actual mineralogy of the precipitates is determined by solution parameters like pH, sulfate, and metal concentration and can vary both spatially and temporally. Some of the most common mine drainage minerals are goethite (α-FeOOH), ferrihydrite ($Fe_5HO_8 \cdot 4H_2O$), schwertmannite ($Fe_8O_8[OH]_6SO_4$), and jarosite ($[H,K,Na]Fe_3[OH]_6[SO_4]_2$).[7]

Goethite is a crystalline oxyhydroxide that occurs over a wide pH range, is relatively stable, and may represent a final transformation product of other mine drainage minerals.[8] Ferrihydrite is a poorly crystalline ferric oxide that forms in higher pH (>6.5) environments. Schwertmannite is commonly found in drainage waters with pH ranging from 2.8 to 4.5, and with moderate to high sulfate contents. It may be the dominant phase controlling major and minor element activities in most acid mine drainage. Jarosite group minerals form in more extreme environments with pH < 3, very high sulfate concentrations, and in the presence of appropriate cations like Na and K.

Fig. 1 Mixing of acid mine drainage (right) with a natural stream resulting in the formation of voluminous precipitates of iron minerals.

MINE DRAINAGE MICROBIOLOGY

The most studied bacterial species in mine drainage systems belong to the genus *Acidithiobacillus* (formerly

Thiobacillus).[9] Species like *Acidithiobacillus thiooxidans* and *A. ferrooxidans* are important to sulfur and iron oxidation in acid drainage; however, many other microorganisms may also be involved.[10] Bacteria have been found in close association with pyrite grains and may play a direct role in mineral oxidation, but they most likely function indirectly through oxidation of dissolved Fe^{2+} as described previously. In low pH systems (<3), *A. ferrooxidans* can increase the rate of iron oxidation as much as five orders of magnitude relative to strictly abiotic rates.[6]

Iron-oxidizing bacteria are chemolithotrophic, meaning they oxidize inorganic compounds, like Fe^{2+}, to generate energy and use CO_2 as a source of carbon. Iron oxidation, however, is a very low energy yielding process. It has been estimated that the oxidation of 90.1 mol of Fe^{2+} is required to assimilate 1 mol of C into biomass.[11] Thus, large amounts of Fe^{2+} must be oxidized to achieve even modest growth.

In addition to mediating iron oxidation, bacteria may play an additional role in mineral formation. Bacteria in mine drainage systems have been shown to be partially encrusted with mineral precipitates.[12] Bacterial cell walls provide reactive sites for the sorption of metal cations, which can accumulate and subsequently develop into precipitates, using the bacterial surface (living or dead) as a template.[13,14]

ENVIRONMENTAL IMPACTS OF MINE DRAINAGE

Mine drainage is primarily released from open mine shafts or from mine spoil left exposed to the atmosphere. The drainage produced can have devastating effects on the surrounding ecosystem. Chemical precipitates can obstruct water flow, dramatically increase turbidity, and ruin stream aesthetics. Dissolved metals and acidity can also affect plant and aquatic animal populations.

Besides iron, Al is the most common dissolved metal in acid mine drainage. The primary source of Al is the acid dissolution of aluminosilicates found in soil, spoils, tailings deposits, and gangue material.[7] At high concentrations, Al can be toxic to plants, and colloidal aluminum precipitates can irritate the gills of fish, causing suffocation. Aluminum occurs as a dissolved species at low pH but rapidly hydrolyzes at about pH 5 to form felsöbanyáite $[Al_4(SO_4)(OH)_{10} \cdot 4H_2O]$ or gibbsite $[Al(OH)_3]$.[7,15] Aluminum precipitates are white in color but are readily masked by associated iron compounds.

Elevated levels of trace elements like As, Cu, Ni, Pb, and Zn may be released during the oxidation of sulfide minerals. These elements can play a role in mineralization processes by forming coprecipitates[16,17] but occur primarily as sorbed species.[18] Mine drainage precipitates can retain both anions and cations, depending on pH. While coprecipitation and sorption function to immobilize trace elements by removing them from solution, this effect may not be permanent. Dissolution of precipitates and shifts in pH can result in the release of sorbed species, providing a latent source of pollution.[19]

DEALING WITH MINE DRAINAGE

Successful control of mine drainage usually involves elements of both prevention and treatment.

Prevention

Prevention techniques include sealing mine shafts, burying or submerging spoil piles, and adding bactericides to limit the function of iron-oxidizing bacteria. These techniques often have limited success. Sealing of mines is extremely difficult due to fractures and the permeability of surrounding rocks. Covering spoil with soil material can decrease the degree of sulfide oxidation by limiting exposure to oxygen, but establishment of a vegetative cover is necessary to prevent erosion from re-exposing the spoil. Inhibition of iron-oxidizing bacteria with bactericides can decrease sulfide oxidation and reduce metal mobility; however, reapplication is necessary and adequate distribution to all affected areas is difficult. In addition, target bacteria may develop resistance, and beneficial bacteria may be harmed.[20]

Treatment

Solution pH usually underestimates the total acidity of mine drainage. Total acidity is the sum of "proton acidity" and "mineral acidity" generated upon oxidation and hydrolysis of metals like Fe^{2+}, Fe^{3+}, Mn, and Al^{3+}.[21] The traditional approach to treatment of acid mine drainage involves neutralization of total acidity by the addition of alkaline agents like caustic soda (NaOH) or hydrated lime $[Ca(OH)_2]$. This method is effective in neutralizing acidity and precipitating dissolved metals; however, it requires continuous oversight and produces large amounts of waste sludge that require disposal. Newer remediation strategies focus on low-cost, sustainable methods for treatment of drainage waters. For example, limestone drains coupled with settling ponds or compost wetlands have shown some promise as passive remediation technologies.[22] In these systems, drainage is channeled through either oxic or anoxic limestone substrates to neutralize active acidity. Dissolved metals are then allowed to hydrolyze and precipitate in ponds or wetland cells. A major difficulty is the loss of reactive surface by armoring of limestone particles with precipitates of Fe and Al that eventually obstruct flow.

Compost wetlands are designed to stimulate the development of anaerobic microbial populations, particularly sulfate-reducing bacteria. The bacteria use the compost as an organic substrate and remove sulfate from solution, either by converting it to H₂S, which is lost to the atmosphere, or by forming insoluble iron sulfides (Reactions 6 and 7).

$$2CH_3CHOHCOO^-_{(aq)} + SO_4^{2-}_{(aq)} \rightarrow 2CH_3COO^-_{(aq)} + 2HCO_3^-_{(aq)} + H_2S_{(g)} \quad (6)$$

$$H_2S_{(s)} + Fe^{2+}_{(aq)} \rightarrow FeS_{(s)} + 2H^+_{(aq)} \quad (7)$$

Bicarbonate is formed as a by-product of sulfate reduction and functions to buffer acidity. These systems have also shown limited success in the field. The sulfate removal rates are usually low (<10%), and pH often remains unchanged or decreases within the wetland.[23]

REFERENCES

1. Nordstrom, D.K.; Alpers, C.N. Geochemistry of acid mine waters. In *The Environmental Geochemistry of Mineral Deposits*; Plumlee, G.S., Logsdon, M.J., Eds.; Society of Economic Geologists Inc: Littleton, CO, 1999; Chapt. 6, 133–160.
2. Kleinmann, R.L.P. Acid mine drainage in the United States: Controlling the impact on streams and rivers. *4th World Congress on the Conservation of Built and Natural Environments*; Toronto, Ontario, 1989; 1–10.
3. Lyon, J.S.; Hilliard, T.J.; Bethel, T.N. *Burden of Guilt*; Mineral Policy Center: Washington, DC, 1993; 68.
4. Rojas, J.; Giersig, M.; Tributsch, H. Sulfur colloids as temporary energy reservoirs for *Thiobacillus ferrooxidans* during pyrite oxidation. Arch. Microbiol. **1995**, *163*, 352–356.
5. Nordstrom, D.K. Aqueous pyrite oxidation and the consequent formation of secondary iron minerals. In *Acid Sulfate Weathering*; Kittrick, J.A., Fanning, D.S., Hossner, L.S., Eds.; Soil Science Society of America: Madison, WI, 1982; 37–56.
6. Singer, P.C.; Stumm, W. Acidic mine drainage: The rate determining step. Science **1970**, *167*, 1121–1123.
7. Bigham, J.M.; Nordstrom, D.K. Iron and aluminum hydroxysulfates from acid mine waters. In *Sulfate Minerals: Crystallography, Geochemistry and Environmental Significance*; Alpers, C.N., Jambor, J.L., Nordstrom, D.K., Eds.; Reviews in Mineralogy and Geochemistry; The Mineralogical Society of America: Washington, DC, 2000; Vol. 40, 351–403.
8. Gagliano, W.B.; Brill, M.R.; Bigham, J.M.; Jones, F.S.; Traina, S.J. Chemistry and mineralogy of ochreous sediments in a constructed mine drainage wetland. Geochim. Cosmochim. Acta **2004**, *68*, 2119–2128.
9. Kelly, D.P.; Wood, A.P. Reclassification of some species of *Thiobacillus* to the newly designated genera *Acidithiobacilus* gen. nov., *Halothiobacillus* gen. nov. and *Thermithiobacillus* gen. nov. Int. J. Syst. Evol. Microbiol. **2000**, *50*, 511–516.
10. Gould, W.D.; Bechard, G.; Lortie, L. The nature and role of microorganisms in the tailings environment. In *The Environmental Geochemistry of Sulfide Mine-Wastes*; Blowes, D.W., Jambor, J.L., Eds.; Mineralogical Association of Canada Short Course, 1994; Vol. 22, 185–200.
11. Ehrlich, H.L. *Geomicrobiology*, 3rd Ed.; Marcel Dekker, Inc.: New York, 1996.
12. Clarke, W.A.; Konhouser, K.O.; Thomas, J.C.; Bottrell, S.H. Ferric hydroxide and ferric hydroxysulfate precipitation by bacteria in an acid mine drainage lagoon. FEMS Microbiol. Rev. **1997**, *20*, 351–361.
13. Schultze-Lam, S.; Fortin, D.; Davis, B.S.; Beveridge, T.J. Mineralization of bacterial surfaces. Chem. Geol. **1996**, *132*, 171–181.
14. Konhauser, K.O. Diversity of bacterial iron mineralization. Earth-Sci. Rev. **1998**, *43*, 91–121.
15. Nordstrom, D.K. Mine waters: Acid to circumneutral. Elements **2011**, *7*, 393–398.
16. Carlson, L.; Bigham, J.M.; Schwertmann, U.; Kyek, A.; Wagner, F. Scavenging of As from acid mine drainage by schwertmannite and ferrihydrite: A comparison with synthetic analogues. Environ. Sci. Technol. **2002**, *36*, 1712–1719.
17. Hita, R.; Torrent, J.; Bigham, J.M. Experimental oxidative dissolution of sphalerite in the Aznalcóllar sludge and other pyritic matrices. J. Environ. Qual. **2006**, *35*, 1032–1039.
18. Winland, R.L.; Traina, S.J.; Bigham, J.M. Chemical composition of ochreous precipitates from Ohio coal mine drainage. J. Environ. Qual. **1991**, *20*, 452–460.
19. Lee, G.; Faure, G.; Bigham, J.M.; Williams, D.J. Metal release from bottom sediments of Ocoee Lake No. 3, a primary catchment area for the Ducktown Mining District. J. Environ. Qual. **2008**, *37*, 344–352.
20. Ledin, M.; Pedersen, K. The environmental impact of mine wastes—Roles of microorganisms and their significance in treatment of mine wastes. Earth-Sci. Rev. **1996**, *41*, 67–108.
21. Kirby, C.S.; Cravotta, C.A., III. Net alkalinity and net acidity I: Theoretical considerations. Appl. Geochem. **2005**, *20*, 1920–1940.
22. Cravotta, C.A., III; Trahan, M.K. Limestone drains to increase pH and remove dissolved metals from acidic mine drainage. Appl. Geochem. **1999**, *14*, 581–606.
23. Mitsch, W.J.; Wise, K.M. Water quality, fate of metals, and predictive model validation of a constructed wetland treating acid mine drainage. Water Resour. **1998**, *32*, 1888–1900.

Mines: Rehabilitation of Open Cut

Douglas J. Dollhopf
Department of Land Resources and Environmental Sciences, Montana State University, Bozeman, Montana, U.S.A.

Abstract

Open pit mining includes quarries used to produce limestone, sandstone, marble, and granite; pits used to produce sand, gravel, and bentonite; and large excavations used to produce talc, copper, gold, iron, silver, and other metals. This mining method is distinguished by having one large pit or numerous small pits across the landscape. State or federal governments establish regulations that require the operator to restore the land to an approved land use that is equal to or better than the pre-mine land use. The approved post-mine land use is instrumental in determining the final graded topography and vegetation community.

INTRODUCTION

Mining directly disturbs approximately 240,000 km^2 of the Earth's surface.[1] Surface mining methods may be classified as: 1) open pit mining; 2) strip mining; 3) dredging; and 4) hydraulic mining.[2] Open pit mining includes quarries used to produce limestone, sandstone, marble, and granite; pits used to produce sand, gravel, and bentonite; and large excavations used to produce talc, copper, gold, iron, silver, and other metals. This mining method is distinguished by having one large pit or numerous small pits across the landscape. State or federal governments establish regulations that require the operator to restore the land to an approved land use(s) that is equal to or better than the premine land use. The approved postmine land use(s) is instrumental in determining the final graded topography and vegetation community.

LANDSCAPE REGRADING

Reclamation of an open pit mine must address the open pit itself, waste rock removed from the pit to gain access to the ore, tailing impoundments, and access roads (Fig. 1). For open pit mines, complete backfilling of the pit is not usually economically feasible. Because the pit remaining after mineral extraction may be hundreds of feet deep, the cost of moving waste rock back into the depression is prohibitive. Surface and groundwater may flow into the depression after mining is terminated and an impoundment develops. Pit slopes should be reduced to the limits required for safe access for humans, livestock, and wildlife. Pit slopes should not be permitted at gradients that jeopardize the success of postmine reclamation.

COVERSOIL RESOURCES

Successful plant community establishment on an open pit mine site is a direct function of the coversoil quality applied. Coversoil suitability criteria vary in the US from state to state, but are similar to those presented in Table 1. The coversoil resource emanates from two procedures: 1) salvaging the natural soil resource from the area to be disturbed and 2) mining unconsolidated geologic stratum when the soil resource is absent. Under present US federal and state mine regulatory programs, all portions of the soil resource shall be salvaged in a project area if it meets physicochemical suitability criteria. Soil materials that do not meet these suitability criteria are generally not used since they may impair plant establishment and growth. Once the soil resource is salvaged, it should be directly hauled to an area that has been backfilled and is ready to receive coversoil. Direct hauled soil contains a viable seed bank, mycorrhizal associations, organic matter, and nutrients that aid in plant establishment. Conversely, if the soil resource is stockpiled, these resources will deteriorate with time.

Prior to implementation of mine land reclamation regulations during the 1970s and 1980s, open pit US mine operations often did not institute land reclamation. The soil resource was not salvaged, the disturbed landscape was not graded to the approximate original contour, and the site was not seeded. Consequently, the soil resource was lost. Today, these lands are being coversoiled using unconsolidated geologic stratum. Valley fill areas adjacent to the disturbed landscape often contain alluvial materials beneath the soil resource that meet all soil suitability criteria (Table 1), except organic matter is absent. This alluvial stratigraphy may be 10–30 m thick or more. Earth moving equipment is used to stockpile the soil resource, separating true topsoil (A-horizon) from subsoil (B- and C-horizons), to enable excavation of deeper alluvial material. Following placement of this alluvial coversoil on the disturbed landscape, the pit created is contoured and the stockpiled soil resource is replaced. Lands receiving coversoil emanating from a geologic stratum should have organic matter applied to expedite establishment of nutrient cycles in the plant root zone. Cattle manure and municipal compost are two sources of organic matter used in land reclamation projects.

Fig. 1 Large open pit mine in the U.S. with water impounded in bottom.

COVERSOIL THICKNESS REQUIREMENTS

The thickness of coversoil required for maximum vegetation performance is dependent on several factors including the 1) vegetation species present; 2) local climate; 3) coversoil quality; and 4) physicochemical quality of the substrate beneath the coversoil. The maximum rooting depth of many rangeland ecosystem plant species is less than 45 cm and generally does not exceed 100 cm.[3] Investigators found that a minimum of 40 cm to a maximum of 150 cm of coversoil is needed for optimum plant growth.[4–8] This wide variation of findings is a function of site specific conditions identified before. The Barth[4] investigation is representative of most research. This investigator found that generic spoil, defined as nonalkaline and nonsaline loams with sodium adsorption ratio (SAR) below seven, showed an increase in plant production up to a coversoil depth of approximately 50 cm. Sodic spoil, or material with SAR of 26 or higher, required a minimum of 70 cm of coversoil to reach maximum cool season grass production. Coversoiled acid spoils, defined as spoils exhibiting pH values between 3.6 and 4.3, revealed increased grass production up to the maximum coversoil depth used in this study (152 cm).

IN SITU SOIL RECLAMATION

In situ soil reclamation at open cut mine sites means no coversoil resource is used. Chemical and/or organic amendments are applied to the graded spoil, tailings, or waste rock landscape to enable plant establishment and growth.

In Situ Sodic Minesoil Remediation

Thousands of hectares of land in Wyoming and Montana were open pit mined for the mineral bentonite prior to passage of state regulations requiring land reclamation. The absence of a soil resource to reclaim these lands means the plant ecosystem must be established on graded spoils. Spoils are overburden materials cast aside to access the underlying mineral, bentonite. These overburden materials have a sodic condition (ESP 20–40), clay texture (60% clay) dominated by swelling clay minerals such as smectite, and preclude plant establishment and growth. In situ soil remediation requires a two-fold approach: 1) permanently reduce exchangeable sodium percentage (ESP) to less than 10 within the 0–30 cm soil profile and 2) provide an organic amendment to immediately prevent soil crust development and increase water infiltration. Crust development in these clayey-sodic soil systems is precluded with applications of wood chips from saw mill waste or manure.[9] Applications of gypsum, calcium chloride, magnesium chloride, and sulfuric acid to these sodic soils have been shown to be effective at reducing the ESP to acceptable levels in minesoils and provide a root zone suitable for plant growth.[10]

In Situ Acid Minesoil Remediation

Open pit mines associated with gold, silver, and metal extraction from sulfide ore bodies frequently produce tailings impoundments and waste rock areas that are acidic (pH 2.5–5.5) and fail to support plant growth. Soil acidity is formed when sulfide minerals, such as pyrite, are exposed to oxidizing conditions in the presence of water. In the absence of a coversoil resource, tailings and waste rock material are treated with alkaline amendments to permanently neutralize the soil acidity. Soil analysis of the acid base account and active acidity enable determination of the calcium carbonate requirement.[11] Amendments used include calcium carbonate, calcium hydroxide, and calcium oxide. In situ treatment with amendments should be at least to the 45 cm soil depth and special engineered plowing equipment is required to attain this depth of incorporation.[12]

Table 1 Coversoil suitability criteria.

Soil parameter	Suitability criteria
pH	>6.5 and <8.0 standard units
Electrical conductivity	<4.0 dS/cm
Exchangeable sodium percentage	<12%
USDA textural classes (12 types)	All suitable except clay, loamy sand, sand
Rock content (particles >2 mm diameter)	
Slope gradients <25%	<35%, weight basis
Slope gradients >25%	>35 and <60%, weight basis
As, Cd, Cu, Pb, Zn, other metals	Near back ground levels, i.e., no enrichment
Organic matter	>0.5%, weight basis

ACID MINE DRAINAGE

Upon exposure of open pit mine wastes to water and oxygen, sulfide minerals oxidize to form acidic, sulfate rich drainage.[13] Iron-oxidizing bacteria, e.g., *Thiobacillus ferrooxidans*, expedite the acid producing reaction rate up to a million times. Metal composition (Al, Cu, Fe, Mn, Pb, Zn, and others) and concentrations in acid mine drainage (AMD) depend on the type and quantity of sulfide minerals present. Approaches to prevent or treat AMD include: 1) removal of air and water from the sulfide body that is disturbed; 2) wetland construction to precipitate contaminants with oxidation ponds and anaerobic-microbial reactions in plant root zones; 3) underground anoxic- and above ground open-limestone drains to raise water pH; and 4) treatment of drainage with neutralizing chemicals such as hydrated lime, quicklime, soda ash, caustic soda, and ammonia.[14]

STEEP SLOPE RECLAMATION

Open pit mines are frequently located in mountainous terrain and reconstructed slopes may have gradients as steep as 50%. Sediment yields tend to increase linearly with slope gradient. There is an inverse relationship between soil loss on slopes and rock cover.[15] High rock content in soils increases infiltration rate and surface roughness decreasing runoff and soil loss. As erosion occurs on rocky soil, rock cover increases as the coarse fragments below the surface are exposed. This armoring of the soil surface can help reduce soil erosion. It has been reported that plant growth may be impaired when soil rock content exceeds 35%.[16] High rock contents of 35%–60% applied on steep slopes (Table 1) may impair plant growth, but are considered a best management practice to facilitate plant establishment and slope stability.

Construction of shallow pits across steep slopes is commonly practiced to increase water storage on the slope and minimize runoff. Pitting techniques using equipment referred to as Dammer–Diker, gouger, and dozer basin blade have been shown to be effective in controlling sediment loss.[17]

CONCLUSIONS

During these past 30 years, land rehabilitation sciences at open pit mines developed at an exponential rate. Grading the mined landscape to the approximate original contour, use of adequate amounts of—and quality of—coversoil, and development of in situ soil treatments at sites where coversoil resources were absent have all provided means to establish a diverse plant community. Techniques have been developed to revegetate and stabilize slopes approaching a 50% gradient, which is an asset for mines located in mountainous terrain. Methods to control and treat AMD from open pit mines continue to improve, but this impact to surface water resources remains unresolved. Future research is required to develop better AMD treatment methods.

REFERENCES

1. Solomons, W. *Mining impacts worldwide*. Proceedings of the International Hydrology Program; UNESCO, Asian Institute of Technology: Bangkok, 1988; 7–9.
2. Paone, J.; Struthers, P.; Johnson, W. Extent of disturbed lands and major reclamation problems in the United States. In *Reclamation of Drastically Disturbed Lands*; Schaller, F.W., Sutton, P., Eds.; American Society of Agronomy: Madison, WI, 1978; 11–22.
3. Wyatt, J.W.; Dollhopf, D.J.; Schafer, W.M. Root distribution in 1- to 48-year old strip-mine spoils in southeastern Montana. J. Range Mgmt **1980**, *33*, 101–104.
4. Barth, R.C. *Soil-depth Requirements to Reestablish Perennial Grasses on Surface-Mined Areas in the Northern Great Plains*; No. 1, Publication 0192-6179/84/2701-0001; Mineral and Energy Resources Research Institute, Colorado School of Mines: Golden, 1984; Vol. 27.
5. Doll, E.C.; Merrill, S.D.; Halvorson, G.A. *Soil Replacement for Reclamation of Strip Mined Lands in North Dakota*; Agricultural Experiment Statement Bulletin 514; North Dakota State University: Fargo, 1984; 23 pp.
6. Halvorson, G.A.; Melsted, S.W.; Schroeder, S.A.; Smith, C.M.; Pole, M.W. Topsoil and subsoil thickness requirements for reclamation of nonsodic mined-land. Soil Sci. Soc. Am. J. **1986**, *50*, 419–422.
7. Power, J.F.; Sandoval, F.M.; Ries, R.E. Topsoil–subsoil requirements to restore North Dakota mined lands to original productivity. Mining Engng 1979, December, 23–27.
8. Schuman, G.E.; Taylor, E.M.; Pinchak, B.A. Revegetation of mined land: influence of topsoil depth and mulching method. J. Soil Water Conserv. **1985**, *40*, 249–252.
9. Schuman, G.E.; King, L.A.; Smith, J.A. Reclamation of bentonite mined lands. In *Reclamation of Drastically Disturbed Lands*; Barnhisel, R.I., Darmody, R.G., Daniels, W.L., Eds.; American Society of Agronomy: Madison, WI, 2000; Vol. 41, 687–707.
10. Dollhopf, D.J.; Rennick, R.B.; Smith, S.C. *Long-Term Effects of Physicochemical Amendments on Plant Performance at a Bentonite Mine Site in the Northern Great Plains*; Reclamation Research Unit Publication 88-02, Montana State University: Bozeman, 1988; 126 pp.
11. Sobek, A.A.; Schuller, J.R.; Freeman, J.R.; Smith, R.M. *Field and Laboratory Methods Applicable to Overburdens and Minesoils*; Publication 600/2-78-054; U.S. Environmental Protection Agency, Office of Research and Development: Cincinnati, Ohio, 1978; 47–67.
12. Dollhopf, D.J. *Deep Lime Incorporation Methods for Neutralization of Acidic Minesoils*; Reclamation Research Unit Publication 9201; Montana State University: Bozeman, 1992; 94.
13. Stumm, W.; Morgan, J.J. *Aquatic Chemistry*; Wiley: New York, 1970; 178 pp.
14. Skousen, J.G.; Sexstone, A.; Ziemkiewicz, P.F. Acid mine drainage control and treatment. In *Reclamation of Drastically Disturbed Lands*; Barnhisel, R.I., Darmody, R.G.,

Daniels, W.L., Eds.; American Society of Agronomy: Madison, WI, 2000; Vol. 41, 131–168.
15. Kapolka, N.M.; Dollhopf, D.J. Effect of slope gradient and plant growth on soil loss on reconstructed steep slopes. Int. J. Surf. Mining, Reclam. Environ. **2001**, *15* (2), 86–89.
16. Munn, L.; Harrington, N.; McGirr, D.R. Rock fragments. In *Reclaiming Mine Soils and Overburden in the Western United States; Analytical Parameters and Procedures*; Williams, R.D., Schuman, G.E., Eds.; Soil Conservation Society of America: Ankeny Iowa, 1987; 259–282.
17. Larson, J.E. *Revegetation Equipment Catalog*; Stock No. 001-00518-5 Forest Service; U.S. Department of Agriculture, U.S. Government Printing Office: Washington, DC, 1980.

Mycotoxins

J. David Miller
Department of Chemistry, Carleton University, Ottawa, Ontario, Canada

Abstract

Mycotoxins are chemicals that are produced by filamentous fungi that affect human or animal health. By convention, this excludes mushroom poisons. All of these species are Deuteromycetes, some of which have a known ascomycetous stage. The occurrence of mycotoxins is entirely governed by the existence of conditions that favor the growth of the fungi concerned. Under environmental conditions, different fungal species are favored as diseases of crop plants or as saprophytes on stored crops. When the conditions favor the growth of toxigenic species, it is an invariable rule that one or more of the compounds for which the fungus has the genetic potential are produced.

INTRODUCTION

The agriculturally important mycotoxins represent a major challenge for food and feed safety. They are unavoidable and no technology has been developed to make them disappear from our major food crops and none on the horizon. This briefly reviews the major toxins and provides an insight into the toxins, their major effects on human and animal health in fully developed market economies, the impact of fungicides and, in one case, pesticides on their reduction, and the impact on occupational health. This is one of the most intensively studied of any area of science because of the effect on the fully developed market economies and the great toll on human suffering in the developing world.[1] At the time of writing, a PubMed search of mycotoxins resulted in 35,000 entries, which serves to underscore both the importance of the issue and the selectivity of this entry.

FIVE IMPORTANT MYCOTOXINS

Although there are hundreds of fungal metabolites that are toxic in experimental systems, there are only five that are of major agricultural importance: deoxynivalenol, aflatoxin, fumonisin, zearalenone, and ochratoxin.[2] Deoxynivalenol occurs when wheat, barley, corn, and sometimes oats and rye are infected by *Fusarium graminearum* and *Fusarium culmorum*. These species cause Fusarium head blight in small grains, a major agricultural problem worldwide, as well as a similar disease in corn called Gibberella ear rot. Disease incidence is most affected by moisture at flowering, and most cultivars and hybrids used today lack genetic resistance to the disease. *F. graminearum* is common in wheat from North America and China. *F. culmorum* was the dominant species in cooler wheat-growing areas, such as Finland, France, Poland, and the Netherlands, but this trend has apparently changed in recent years as European summers have reached record warm temperatures such that *F. graminearum* largely dominates.[3] A third species, *Fusarium crookwellense*, can also cause head blight or corn ear rot and produces nivalenol and zearalenone.[4-6]

Humans are not less and are probably more sensitive to deoxynivalenol than swine.[3] Swine are the most sensitive domestic animal species to the effects of deoxynivalenol. Two mechanisms explain most of the variance associated with this response. The first is the potent neurotoxicity of this toxin[3] and the second relates to the fact that deoxynivalenol results in the suppression of insulin-like growth factor acid-labile subunit expression.[7] In addition, deoxynivalenol causes changes in immune system function in male mice, occurring at dietary concentrations often encountered by humans. As with other trichothecenes, high exposures increase susceptibility to facultative pathogens such as *Listeria*.[8] Cattle, cows, and poultry species are tolerant to deoxynivalenol and milk production and are not affected at typical field concentrations.[9] Deoxynivalenol is not a carcinogen. As noted by Miller,[3] the Provisional Maximum Tolerable Daily Intake (PMTDI) of the Joint Expert Committee on Food Additives and Contaminants (JECFA) of the World Health Organization (WHO) and Food and Agriculture Organization (FAO) for this toxin is reasonably secure. A minor change was made by the JECFA in early 2010 by making a group PMTDI for deoxynivalenol and the two acetylated precursors that co-occur with the parent compound in grains.[10]

Crops that are contaminated by deoxynivalenol can often contain zearalenone albeit at a lower frequency. Zearalenone is more common in maize than in small grains. Zearalenone is an estrogen analogue and causes hyperestrocism in female pigs at low levels; the dietary no-effect level is less than 1 mg/kg. Cows and sheep are also sensitive to the estrogenic effects of this toxin with depressed ovulation and lower lambing percentages.[9] Non-human primates and humans are also very sensitive to the estrogenic effects of zearalenone.[11,12]

Aflatoxin is mainly produced by *Aspergillus flavus*, which is a problem in many commodities, but most human exposure comes from contaminated corn, groundnuts, and rice. *Aspergillus parasiticus* is uncommon outside North and South America, and they are more associated with peanuts. There are also some rare species that produce aflatoxin. *A. flavus* contamination of corn or peanuts occurs in two basic ways. Either airborne or insect-transmitted conidia contaminate the silks and grow into the ear when the maize is under high-temperature stress or (more commonly) insect- or bird-damaged kernels become colonized with the fungus and accumulate aflatoxin. In corn and peanut plants, drought-, nutrient-, or temperature-stressed plants are more susceptible to colonization by *A. flavus* or *A. parasiticus*.[2,13,14]

In poultry, aflatoxin exposure results in liver damage, impaired productivity and reproductive efficiency, decreased egg production in hens, inferior egg-shell quality, inferior carcass quality, and increased susceptibility to disease. The effects of acute and chronic exposure in swine are largely attributable to liver damage. In cattle, the primary symptom is reduced weight gain, liver and kidney damage, and reduced milk production.[2] Aflatoxin is also immunotoxic in domestic and laboratory animals with oral exposures in the microgram-per-gram range. Cell-mediated immunity (lymphocytes, phagocytes, mast cells, and basophils) is more affected than humoral immunity (antibodies and complement).[15] Naturally occurring mixtures of aflatoxins were classified as class 1 human carcinogens, and aflatoxin B1 is also a class 1 human carcinogen.[16] As noted by Miller,[3] the JECFA PMTDI for aflatoxin is secure.

Fumonisins are produced by *Fusarium verticillioides* (formerly *moniliforme*), *Fusarium proliferatum*, and several uncommon fusaria. Fumonisins have been found as a very common contaminant of corn-based food and feed in the United States, China, Europe, southern Africa, South America, and Southeast Asia.[17]

F. verticillioides and *F. proliferatum* can be recovered from virtually all corn kernels including those that are healthy, which suggests that it may be an endophyte, i.e., a mutualistic relationship.[2,18,19] *F. verticillioides* and *F. proliferatum* cause a "disease" called Fusarium kernel rot. In parts of the United States and lowland tropics, this is one of the most important ear diseases and is associated with warm, dry years and insect damage. Corn plant disease stress also promotes the growth of *F. verticillioides* and fumonisin formation.[18,20]

Fumonisin causes equine leukoencephalomalacia (ELEM), which involves a liquefactive necrosis of the cerebral hemispheres. Clinical manifestations include abnormal movements, aimless circling, lameness, etc., followed by death.[9] In swine, high exposures of fumonisin results in porcine pulmonary edema (PPE) caused by fumonisin-induced heart failure. At lower exposures, both liver damage and kidney damage have been reported in swine. Fumonisin causes feed refusal and changes in carcass quality at dietary concentrations in the low milligram-per-kilogram range.[21]

Exposure to *F. verticillioides*-contaminated maize has been linked to the elevated rates of esophageal cancer in the Transkei for 25 years, and this has since been linked to fumonisin exposure (ICPS).[22] Fumonisin caused tumors in mice and rats in the U.S. National Toxicology Program study[17] and is currently an IARC class 2B carcinogen (possible human carcinogen).[16]

Fumonisin exposure results in neural tube birth (NTD) defects in rodents, which is associated with inhibiting folate transfer to neural crest cells.[23] There is evidence that this toxin results in NTDs in humans.[24,25] As noted by Miller,[3] the JECFA PMTDI for fumonisin is secure.

Human exposure to ochratoxin A (OTA) in the United States and Canada and in Europe mainly comes from eating foods made from barley and wheat in which *Penicillium verrucosum* has grown. A few percentage of surface-disinfected wheat and barley kernels collected at harvest in the United Kingdom and Denmark were contaminated by *Penicillium aurantiogriseum* and *P. verrucosum*, and this was similar in studies from Canada.[2,26,27]

Depending on consumption, ochratoxin exposure can arise from *Aspergillus carbonarius* and some strains of *Aspergillus niger*, and several related species also produce ochratoxin on grapes and coffee.[14,28] In Europe, minor exposures occur from meat, especially pork, from animals fed contaminated grain and from the consumption of moldy sausage, etc. The latter arises from *Penicillium nordicum*.[29,30]

Ochratoxin is a potent nephrotoxin in swine and causes kidney cancer in male Fisher 344 rats. Pigs appear to be the most sensitive species with respect to the nephrotoxic effects.[28,31] Low exposures result in kidney damage in swine but typically there are no overt signs. At higher concentrations (>2 µg/g), decreased weight gains occur. Poultry are affected, showing reduced growth rate and egg production at low ochratoxin concentrations >2 µg/g. Cattle are resistant to ochratoxin concentrations found in naturally contaminated grain.[9]

Ochratoxin is suspected as the cause of urinary tract cancers and kidney damage in areas of chronic exposure in parts of Eastern Europe. Despite considerable effort, no satisfactory conclusion has been reached regarding the linkage of ochratoxin with urinary tract cancers in humans. OTA has been the subject of recent JECFA evaluations.[31] The JECFA noted that a conclusive association between OTA intake and human cancer has not been demonstrated.

As noted by JECFA,[31] evidence for the carcinogenicity of OTA is confined to laboratory strains of rodents. OTA causes liver cancer in mice and kidney carcinomas in mice and rats. Mechanisms that explain the animal carcinogenicity of OTA not related to the production of OTA DNA adducts have been proposed.[31,32] The formation of DNA adducts by OTA and their potential role in cancer induction remain controversial.[32,33]

MYCOTOXINS AND PESTICIDES

Insect damage is known to promote accumulation of deoxynivalenol, fumonisin, aflatoxin, and probably zearalenone. Steps taken to reduce insect herbivory or drought stress on crops have the general effect of reducing mycotoxin contamination. Bt corn has lowered fumonisin content compared to similar non-Bt hybrids.[20,34]

It is also now known that some genotypes of corn and peanuts produce compounds that interfere with mycotoxin production as opposed to fungal growth per se. The first pure compound to be isolated with such activity was reported from corn.[35,36]

Modeling of mycotoxin accumulations is part of the risk management strategy for foliar pesticides in Fusarium head blight in Ontario[37] and by agreement with the European Commission, which has the effect of restricting the prophylactic use of fungicides when not warranted. This is enforced by pesticide companies to all the older members of the European Union. This is an increasingly important area of research for food protection and food security.[38]

OCCUPATIONAL EXPOSURES

Grain dusts represent an occupational hazard for a number of reasons including the possibility of the allergic disease hypersensitivity pneumonitis and organic dust toxic syndrome, a poorly understood disease. Inhalation of mycotoxins contained in airborne dusts is also a potential health risk. Inhalation of mycotoxins in spores and dust affects macrophage function and other aspects of lung biology. Inhalation of mycotoxins is a more potent route of exposure for systemic toxicity for aflatoxin, trichothecenes, and ochratoxin. Workplace exposure to airborne aflatoxin results in increased relative risk of liver cancer.[39] A risk assessment model suggested that exposure to airborne aflatoxin B1 may pose little significance during maize harvest and elevator loading/unloading, but a relatively high risk for swine feeding and storage bin cleaning.[40]

Urinary biomarkers for deoxynivalenol (DON, DOM-1) were higher in active farmers in France, particularly from larger farms, than from retired farmers where exposure was from the diet.[41] Workplace exposure to ochratoxin has been reported to result in severe kidney damage in a farmer.[39] However, only limited ochratoxin exposure estimates from occupational exposure are available.[42]

CONCLUSION

As noted, mycotoxin contamination of field crops is a problem that cannot be "solved" in the normal use of that term, results in important losses of human life, and has major economic and social consequences. Recognition that fungi produced toxins in crops began with ergot in the middle ages and has been recognized as low-molecular-weight chemicals since the 1880s and solving their structures and understanding appropriate management strategies took agonizingly long periods of study.[2,3] The current strategies for managing mycotoxins in the diet require modeling, crop breeding, better detection systems, and awareness.

REFERENCES

1. Wild, C.P.; Gong, Y.Y. Mycotoxins and human disease: A largely ignored global health issue. Carcinogenesis **2010**, *31*, 71–82.
2. Miller, J.D. Fungi and mycotoxins in grain: Implications for stored product research. J. Stored Prod. Res. **1995**, *31*, 1–6.
3. Miller, J.D. Mycotoxins in small grains and maize: Old problems, new challenges. Food Addit. Contam. **2008**, *25*, 219–230.
4. Boutigny, A.L.; Richard-Forget, F.; Barreau, C. Natural mechanisms for cereal resistance to the accumulation of Fusarium trichothecenes. Eur. J. Plant Pathol. **2008**, *121*, 411–423.
5. Miller, J.D. Epidemiology of Fusarium ear diseases. In *Mycotoxins in Grain: Compounds Other Than Aflatoxin*; Miller, J.D., Trenholm, H.L., Eds.; Eagan Press: St. Paul, MN, 1994; 19–36.
6. Xu, X.M.; Nicholson, P. Ecology of fungal pathogens causing wheat head blight. Annu. Rev. Phytopathol. **2009**, *47*, 83–103.
7. Amuzie, C.J.; Pestka, J.J. Suppression of insulin-like growth factor acid-labile subunit expression—A novel mechanism for deoxynivalenol-induced growth retardation. Toxicol. Sci. **2010**, *113*, 412–421.
8. Bondy, G.S.; Pestka, J.J. Immunomodulation by fungal toxins. J. Toxicol. Environ. Health, Part B **2000**, *3*, 109–143.
9. Prelusky, D.B.; Rotter, B.A.; Rotter, R.G. Toxicology of mycotoxins. In *Mycotoxins in Grain: Compounds Other Than Aflatoxin*; Miller, J.D., Trenholm, H.L., Eds.; Eagan Press: St. Paul, MN, 1994; 359–404.
10. World Health Organization (WHO). Safety evaluation of certain food additives and contaminants. Prepared by the Seventy-Seventh Meeting of the Joint FAO/WHO Expert Committee on Food Additives. World Health Organization: Geneva, 2010.
11. Massart, F.; Meucci, V.; Saggese, G.; Soldani, G. High growth rate of girls with precocious puberty exposed to estrogenic mycotoxins. J. Pediatr. **2008**, *152*, 690–695.
12. Massart, F.; Saggese, G. Oestrogenic mycotoxin exposures and precocious pubertal development. Int. J. Androl. **2010**, *33*, 369–376.
13. Guo, B.; Chen, Z.Y.; Lee, R.D.; Scully, B.T. Drought stress and preharvest aflatoxin contamination in agricultural commodity: Genetics, genomics and proteomics. J. Integr. Plant Biol. **2008**, *50*, 1281–1291.
14. Perrone, G.; Susca, A.; Cozzi, G.; Ehrlich, K.; Varga, J.; Frisvad, J.C.; Meijer, M.; Noonim, P.; Mahakarnchanakul, W.; Samson, R.A. Biodiversity of Aspergillus species in some important agricultural products. Stud. Mycol. **2007**, *59*, 53–66.

15. Pestka, J.J.; Bondy, G.S. Mycotoxin-induced immune modulation. In *Immunotoxicology and Immunopgarmacology*; Dean, J.H., Luster, M.I., Munson, A.E., Kimber, I., Eds. Raven Press: NY, 1994; 163–182.
16. IARC Monograph 82. *Some Traditional Herbal Medicines, Some Mycotoxins, Naphthalene and Styrene*; IARC, WHO: France, 2002.
17. Bolger M.; Coker, R.D.; DiNovi, M.; Gaylor, D.; Gelderblom, W.; Olsen, M.; Paster, N.; Riley, R.T.; Shephard, G.; Speijers, G.J.A. Fumonisins. In *Safety Evaluation Of Certain Mycotoxins in Food*; WHO Food Additives Series 47; 2001; Vol. 74, 281–415.
18. Miller, J.D. Factors that affect the occurrence of fumonisin. Environ. Health Perspect. **2000**, *109* (s.2), 321–324.
19. Yates, I.E.; Sparks, D. *Fusarium verticillioides* dissemination among maize ears of field-grown plants. Crop Prot. **2008**, *27*, 606–613.
20. De La Campa, R.; Hooker, D.C.; Miller, J.D.; Schaafsma, W.A.; Hammond, B.G. Modelling effects of environment, insect damage and BT genotypes on fumonisin accumulation in maize in Argentina and the Philippines. Mycopathologia **2005**, *159*, 539–552.
21. Marasas, W.F.O.; Miller, J.D.; Riley, R.T.; Visconti, A. *Fumonisins B1. Environmental Health Criteria 219*; International Program for Chemical Safety, World Health Organization: Geneva, 2000.
22. Fumonisins B1. Environmental Health Criteria 219. International Program for Chemical Safety, World Health Organization: Geneva, Switzerland, 2000.
23. Gelineau-van Waes, J.; Voss, K.A.; Stevens, V.L.; Speer, M.C.; Riley, R.T. Maternal fumonisin exposure as a risk factor for neural tube defects. Adv. Food Nutr. Res. **2009**, *56*, 145–181.
24. Marasas, W.F.O.; Riley, R.T.; Hendricks, K.A.; Stevens, V.L.; Sadler, T.W.; Gelineau-van Waes, J.; Missmer, S.A.; Cabrera, J.; Torres, O.; Gelderblom, W.C.A.; Allegood, J.; Martinez, C.; Maddox, J.; Miller, J.D.; Starr, L.; Sullards, M.C.; Roman, A.V.; Voss, K.A.; Wang, E.; Merrill, A.H., Jr. Fumonisins disrupt sphingolipid metabolism, folate transport, and neural tube development in embryo culture and in vivo: A potential risk factor for human neural tube defects among populations consuming fumonisin-contaminated maize. J. Nutr. **2004**, *134*, 711–716.
25. Missmer, S.A.; Suarez, L.; Felkner, M.; Wang, E.; Merrill, A.H., Jr.; Rothman, R.A.; Hendricks, K.A. Exposure to fumonisins and the occurrence of neural tube defects along the Texas–Mexico border. Environ. Health Perspect. **2006**, *114*, 237–241.
26. Elmholt, S.; Hestbjerg, H. Field ecology of the ochratoxin A-producing *Penicillium verrucosum*: Survival and resource colonization in soil. Mycopathologia **1999**, *147*, 67–81.
27. Mills, J.T.; Seifert, K.A.; Frisvad, J.C., Abramson, D. Nephrotoxigenic *Penicillium* species occurring on farm-stored cereal grains in western Canada. Mycopathologia **1995**, *30*, 23–28.
28. Kuiper-Goodman, T.; Hilts, C.; Billiard, S.M.; Kiparissis, Y.; Richard, I.D.; Hayward, S. Health risk assessment of ochratoxin A for all age-sex strata in a market economy. Food Addit. Contam, Part A **2010**, *27*, 212–240.
29. Sorensen, L.M.; Mogensen, J.; Nielsen, K.F. Simultaneous determination of ochratoxin A, mycophenolic acid and fumonisin B-2 in meat products. Anal. Bioanal. Chem. **2010**, *398*, 1535–1542.
30. Dall'Asta, C.; Galaverna, G.; Bertuzzi, T.; Moseriti, A.; Pietri, A.; Dossena, A.; Marchelli, R. Occurrence of ochratoxin A in raw ham muscle, salami and dry-cured ham from pigs fed with contaminated diet. Food Chem. **2008**, *120*, 978–983.
31. World Health Organization (WHO). Safety evaluation of certain food additives and contaminants. Prepared by the Sixty-Eighth Meeting of the Joint FAO/WHO Expert Committee on Food Additives. World Health Organization: Geneva, 2008.
32. Mally, A.; Dekant, W. Mycotoxins and the kidney: Modes of action for renal tumor formation by ochratoxin A in rodents. Mol. Nutr. Food Res. **2009**, *53*, 467–478.
33. Mantle, P.G.; Faucet-Marquis, V.; Manderville, R.A.; Squillaci, B.; Pfohl-Leszkowicz, A. Structures of covalent adducts between DNA and ochratoxin a: A new factor in debate about genotoxicity and human risk assessment. Chem. Res. Toxicol. **2010**, *23*, 89–98.
34. Ostry, V.; Ovesna, J.; Skarkova, J.; Pouchova, V.; Ruprich, J. A review on comparative data concerning *Fusarium* mycotoxins in Bt maize and non-Bt isogenic maize. Mycotoxin Res. **2010**, *26*, 141–145.
35. Miller, J.D.; Miles, M.; Fielder, D.A. Kernel concentrations of 4-acetylbenzoxazolin-2-one and diferuloylputrescine in maize genotypes and Gibberella ear rot. J. Agric. Food Chem. **1997**, *45*, 4456–4459.
36. Holmes, R.A.; Boston, R.S.; Payne, G.A. Diverse inhibitors of aflatoxin biosynthesis. Appl. Microbiol. Biotechnol. **2008**, *78*, 559–572.
37. Schaafsma, A.W; Hooker, D.C. Climatic models to predict occurrence of *Fusarium* toxins in wheat and maize. Int. J. Food Microbiol. **2007**, *119*, 116–125.
38. van der Fels-Klerx, H.J.; Booij, C.J.H. Perspectives for geographically oriented management of *Fusarium* mycotoxins in the cereal supply chain. J. Food Prot. **2010**, *73*, 1153–1159.
39. Miller, J.D. Mycotoxins. In: *Handbook of Organic Dusts*; Rylander, R., Pettersen, Y., Eds.; CRC Press: Boca Raton, FL, 1994; 87–92.
40. Liao, C-M.; Chen, S-C. A probabilistic modeling approach to assess human inhalation exposure risks to airborne aflatoxin B1 (AFB1). Atmos. Environ. **2005**, *39*, 6481–6490.
41. Turner, P.C.; Hopton, R.P.; Lecluse, Y.; White, K.L.; Fisher, J.; Lebailly, P. Determinants of urinary deoxynivalenol and de-epoxy deoxynivalenol in male farmers from Normandy, France. J. Agric. Food Chem. **2010**, *58*, 5206–5212.
42. Mayer, S.; Curtui, V.; Usleber, E.; Gareis, M. Airborne mycotoxins in dust from grain elevators. Mycotoxin Res. **2007**, *23*, 94–100.

Nanomaterials: Regulation and Risk Assessment

Steffen Foss Hansen
Khara D. Grieger
Anders Baun
Department of Environmental Engineering, Technical University of Denmark, Kongens Lyngby, Denmark

Abstract

The topics of regulation and risk assessment of nanomaterials have never been more relevant and controversial in Europe than they are at this point in time. In this entry, we present and discuss a number of major pieces of legislation relevant for the regulation of nanomaterials, including REACH, the Water Framework Directive, pharmaceuticals regulation, and the Novel Foods Regulation. Current regulation of nanomaterials entail three overall challenges: 1) limitations in regard to terminology and definitions of key terms such as a "substance," "novel food," etc.; 2) safety assessment requirements triggered by thresholds values not tailored to the nanoscale but based on bulk material; and 3) limitations related to lack of metrological tools, (eco)toxicological data, and environmental exposure limits as required by, e.g., REACH, the pharmaceuticals regulation, and the recast of the Novel Foods Regulation. Chemical risk assessment provides a fundamental element in support of existing legislation. Risk assessment is normally said to consist of four elements, i.e., hazard identification, dose–response assessment, exposure assessment, and risk characterization. Each of these four elements hold a number of limitations specific to nanomaterials, i.e., the fact that mass might not be the proper metric to describe the dose in dose–response assessment. These limitations are not easily overcome despite the fact that a lot of effort is being put into investigating the applicability of each of these four elements.

INTRODUCTION

The topics of regulation and risk assessment of nanomaterials have never been more relevant and controversial in Europe than they are at this point in time. As the first major piece of legislation to be amended in Europe, the cosmetics legislation was adopted in 2009 requiring all nanomaterial-containing cosmetics to be labeled after 2013 and producers to provide a safety assessment of the nanomaterial used.[1,2]

Concurrently with the recasting of various pieces of legislation, such as the Novel Foods Regulation, the European Commission has commissioned an expert-/multistakeholder investigation of whether nanospecific amendments are needed to the current technical guidelines on substance identification and chemical safety assessment, which lie at the core foundations of the European Chemical legislation known as REACH—Registration, Evaluation, Authorization, and Restriction of Chemicals.[3,4] It is the major piece of legislation concerning regulating the manufacturing and applications of nanomaterials, although the text in REACH itself has only been subject to minor changes thus far. A number of other pieces of legislation relevant to the manufacturing, use, and disposal of nanomaterial and products have furthermore not been subject to any nanospecific changes, although they might be revised in the future.

In the following, some of the major pieces of legislation relevant for the regulation of nanomaterials in Europe will be presented. Examples of both horizontal regulation as well as subject-specific legislation will be given. Some of these have yet to take nanospecific issues into consideration, and the focus will therefore be at explaining the limitations of these in handling nanomaterials. For others, nanospecific aspects have recently been taken into consideration and for these the focus will be at explaining how this has been done and what kinds of challenges still need to be addressed.

Chemical risk assessment plays a crucial role in many of these pieces of legislation, and hence a short introduction and discussion of the applicability of chemical risk assessment to nanomaterials will be included.

REGISTRATION, EVALUATION, AND AUTHORIZATION OF CHEMICALS (REACH)

One of the key pieces of European legislation affecting nanomaterials is the European chemical regulation known as REACH, which went into force in mid-2007.[5] REACH prescribes

1. The registration of chemicals commercialized by manufacturers and importers in Europe as well as the collection of data on their use and toxicity.
2. The evaluation and examination by governments of the need for additional testing and regulation of chemicals.

3. That authorization has to be sought and given to manufacturers in order for them to use chemicals of high concern.
4. European Union (EU)-wide restrictions or complete ban of certain chemicals that cannot be used safely.

The REACH regulation replaced more than 40 other directives and subsequently shifted the responsibility in the registration and authorization process of REACH onto manufacturers and importers (including downstream users of chemicals) to provide data of uses and hazard information. Industry, furthermore, has to show that chemicals of high concern can be used safely. The evaluation and restriction process is still the responsibility of the national authorities, the newly established European Chemical Agency, and the European Commission.

Registration of all the commercialized chemicals in the European market is a tremendous task that is expected to occur gradually. Substances produced or imported in the highest volumes or of the greatest (known) concern are to be registered first. Substances produced or imported in more than 1000 tonnes per year per manufacturer or importer had to be registered by November 30, 2010, by the latest date. This was also the case for substances marketed in 100 tons/yr that have been classified as very toxic to aquatic organisms and for substances produced/imported in more than 1 ton/yr and which have been classified as Category 1 or Category 2 carcinogens, mutagens, or reproductive toxicants. Furthermore, substances entering the European market in yearly quantities above 100 tons, and 10 tons per producers or importers have to be registered by June 1, 2013, and June 1, 2018, respectively.[5]

REACH does not specifically mention nanomaterials, but does cover chemicals in all their physical–chemical states, using the following definition of a substance: "a chemical element and its compounds in the natural state or obtained by any manufacturing process, including any additive necessary to preserve its stability and any impurity deriving from the process used, but excluding any solvent which may be separated without affecting the stability of the substance or changing its composition."[5] Therefore, REACH is formally the relevant legislative frame for industrially used nanomaterials, and the exemption registration of carbon and graphite was redrawn in 2008 to address concerns raised about carbonaceous nanomaterials.[6,7] Companies will now have to register these materials if produced in quantities above 1 ton per producer or manufacturer per year. However, for a number of nanomaterials, it is not evident whether a nanoequivalent of a substance with different physico-chemical and (eco)toxicological properties from the bulk substance would be considered as the same or as another substance under REACH.[8]

If a nanomaterial is considered to be a different substance under REACH, hazard information specifically related to the nanoform of the substance would have to be generated for the registration, if produced in more than 1 ton/yr. On the other hand, if a nanomaterial is considered to be the same as a registered bulk material, hazard information data generated for the registration might not be directly relevant for the nanoform of the substance and hence open to discussion.[8,9] In response to these concerns, the European Commission has launched a multistakeholder project on nanomaterials to look into substance identification under REACH in order to get recommendations on whether the nanoform of a substance should be considered different from the bulk form of the substance.[3,4]

If manufacturers and importers produce or import nanomaterials in volumes of more than 10 tons/yr and if it meets the criteria for classification as dangerous or a PBT (persistent, bioaccumulative, and toxic) or vPvB (very persistent and very bioaccumulative), a chemical safety assessment is required that includes information about uses, (eco)toxicological information, exposure assessment, and risk characterization(s).

Thus far, no nanomaterial has been classified as PBT or as vPvB, but if it was to be it is highly unclear how companies should do a chemical safety assessment. Both the Commission of the European Communities[10] as well as the its Scientific Committee on Emerging and Newly Identified Health Risks (SCENIHR)[11] have pointed out that current test guidelines in REACH are based on conventional methodologies for assessing chemical risks and may not be appropriate for assessing risks associated with nanomaterials.

It should be noted that a chemical safety assessment can also be required if a nanomaterial is selected for further evaluation by a member state or by the European Chemicals Agency due to specific concerns; or if a substance is a CMR (carcinogenic, mutagenic, or toxic for reproduction), PBT, vPvB, ED (endocrine disrupting), or substance of equivalent concern.

EU WATER FRAMEWORK DIRECTIVE

Whereas REACH deals with the manufacturing and import of chemicals, the EU Water Framework Directive (WFD) deals with improving water quality and reducing dangerous chemicals in European river basins. The key aim of the WFD, which was adopted in 2000, is to promote long-term sustainable water use, preventing further deterioration of surface waters, transitional waters, coastal waters, and groundwater, and to protect and enhance the status of aquatic ecosystems with regard to their water needs, terrestrial ecosystems, and wetlands directly depending on the aquatic ecosystems,[12] Article 1.

The WFD establishes water management by a river basin approach with cooperation and joint objective setting across member state borders and even in some cases beyond the EU territory. Geographical and hydrological formation of each river basin determines which member states need to establish and implement a so-called river basin management plan. The river basin management plan, which needs to be updated every 6 years, specifies the measures to be taken to meet the environmental objectives for surface waters, for groundwater, and for protected areas. The WFD prescribes the setting of the environmental quality standards ensuring the general protection of the aquatic ecology, specific protection of unique and valuable habitats, and protection of drinking water resources and bathing water. For instance, for surface waters, member states shall implement necessary measures to prevent deterioration, and promote restoration of artificial and heavily modified water bodies with the aim of achieving "good ecological potential" and "good surface water chemical status" in 2015 by the latest. This has to be done along with a progressive reduction of pollution from a set of "priority substances" and discontinue emissions of priority hazardous substances,[12] Article 4.

For all surface waters, the WFD set a number of "general requirements for ecological protection" as well as a "general minimum chemical standard" and defines "good ecological status" and "good chemical status" in terms of the quality of the biological community, the hydrological characteristics, and the chemical characteristics,[12] Article 4.

The definition of "good chemical status" is especially relevant in regard to nanomaterials as it is defined in terms of compliance with all the quality standards established for chemical substances at the European level. For "priority substances," member states are required to set environmental quality standards (EQSs) to monitor the chemical status of a water body (European Parliament and the Council of the European Union (EP & CEU),[12] Article 16). Thus, the EQS is taken as concentration below which the chemical status is referred to as "good" in the WFD terminology (European Parliament and the Council of the European Union (EP & CEU),[12] Article 2). Even for the so-called priority hazardous substances, only a few EQSs have been set, but more substances will follow with a specific focus at substances that are toxic toward humans and/or aquatic organisms, compounds with a widespread environmental distribution, and those that are discharged in significant quantities.

A key question in regard to WFD is whether nanomaterials are possible candidates as priority substances.[13] In favor of this speaks the widespread and diffuse use of nanoparticles in a range of consumer products along with the hazard characteristics of some nanomaterials such as functionalized carbon nanotubes (CNTs), nanoscale silver, and zinc oxide. Some applications of nanomaterials furthermore involve direct contact with the water cycle, e.g., in relation to their use for water disinfection[14] and wastewater treatment,[15] as well as in regard to the direct use for treating soil and groundwater contamination.[16]

If a given type of nanoparticles is included in the list of priority substances in the future based on environmental occurrence or hazard information, an EQS will have to be defined.[12] To derive an EQS for a priority substance, the WFD outlines that test results from both acute and chronic ecotoxicological standard tests should be used for the "base set" organisms, i.e., algae and/or macrophytes, crustacean, and fish. Estimating EQS for nanoparticles is currently hampered by lack of ecotoxicological data even for the most tested nanoparticles such as C60, CNTs, TiO_2, ZnO, and Ag. For instance, the degradability of C60 and CNTs and their ability to bioaccumulate in the aquatic environment remains to be studied, making it virtually impossible to set an EQS for these two kinds of nanoparticles.[13]

Not only are the number of studies very limited, but the number of tested taxa is also too few to be used in the context of setting an EQS. The reliability and interpretation of the available ecotoxicity data is furthermore impeded as a result of factors such as particle impurities, suspension preparation methods, release of free metal ions, and particle aggregation.[13,17,18]

Besides these issues, mainly related to the lack of relevant data, it is also questionable whether the principles for deriving EQSs for chemicals can be directly transferred to nanoparticles. The setting of EQS is based on a chemical safety assessment similar to the one required under REACH and, as noted above, the European Commission's SCENIHR have pointed out that amendments have to be made to the guidelines for chemical safety assessment.[11,19]

Another manner in which nanomaterials could meet the criteria to be included in a WFD list of priority substances is if there is "evidence from monitoring of widespread environmental contamination".[12] However, when it comes to nanomaterials, monitoring in natural waters represents some profound challenges.[20,21] While applicable methods for in situ monitoring remain to be developed and refined,[22] it is also challenging to set up a reliable monitoring program for nanoparticles since a number of issues still remain to be resolved, e.g., choice of suitable sampling materials, preconcentration/fractionation methods, and analytical methods to characterize and quantify collected particles.[21] Despite significant progress in recent years, reliable methods are not yet available to determine nanoparticle identity, concentrations, and characteristics in complex environmental matrices, such as water, soil, sediment, sewage sludge, and biological specimens.

PHARMACEUTICAL REGULATION

Liposomes, polymer–protein conjugates, polymeric substances, or suspensions are examples of well-described and understood medicinal products containing nanoparticles and have been given marketing authorizations within the EU under the existing regulatory framework.[23–26]

As in the case of REACH, nanomedicine and nanomaterials are not specifically mentioned in the EU legislation on medicinal products and devices, tissue engineering, and other advanced therapies. Although the scope of the various EU regulations and directives that constitutes this framework covers nanomedicine, they have been accused of being too general, non-specific, and fraught with difficulties in case of complex drugs.[27,28] Given this, it does seem that it is generally believed that the regulatory framework for medicine covers medical products based on nanotechnology, and that the extensive premarket safety assessment of medicine in general is sufficient to ensure that the benefits outweigh any identified risks or the adverse side effects.[29,30]

Concerns have, however, been raised that the risk assessment, safety, and quality requirements for medicine may not be designed to address nanomedicine and medical devices based on nanotechnology, as these have to be fulfilled by conformity to established quality systems and published product standards. This might be especially true for novel applications such as nanostructure scaffolds for tissue replacement, nanostructures enabling transport across biological barriers, remote control of nanoprobes, integrated implantable sensory nanoelectronic systems, and chemical structures for drug delivery and targeting of disease.[31]

Currently, the mechanism of action is key to decide whether a product should be regulated as a medicinal product or as a medical device. This could be problematic when it comes to many novel applications of nanomedicinal products as they are likely to span regulatory boundaries between medicinal products and medical devices.[29,31] This is due to the notion that they may exhibit a complex mechanism of action combining mechanical, chemical, pharmacological, and immunological properties, and combining diagnostic and therapeutic functions.

For new marketing authorization applications of pharmaceuticals, an environmental risk assessment has to be provided, which involves a rough calculation of the predicted environmental concentration (PEC) for surface water. Actions have to be initiated if the PEC is predicted to surpass 0.01 ppb.[32] However, this threshold cannot be interpreted as a safe concentration, and it is not based on a scientific evaluation.[32] It could furthermore be problematic when it comes to nanomedicine, as concentration in terms of mass per volume might not be the relevant metric to characterize the environmental hazard of nanomaterials.[33–35]

NANOFOOD REGULATION

Food and food packaging are regulated by a number of directives and regulations in the EU, such as the EU Food Law Regulation and the EU Novel Foods Regulation.[36] As an overarching principle, all food are required to be safe and this overarching principle of safety applies to all foods and food packaging that contain nanomaterials. This has, however, been criticized for being too loose.[37]

During the recent discussion related to the update of regulation regarding food additives, the European Parliament's Committee on Environment, Public Health, and Food Safety stated that it wanted separate limit values for nanotechnologies and that the permitted limits for an additive in nanoparticle form should not be the same as when it is in traditional form.[38] This demand, however, never made into the actual regulation and the final adopted regulation on food additives is limited to requiring that food additives that have been produced via nanotechnology or consist of/or include materials fulfill a number of criteria before it can be include on the list of approved food, food additives, food enzymes, and food flavorings. Nanotechnology and nanomaterials are not defined in the regulation, but these criteria include what use does not pose a safety concern to the health of the consumer at the level of use proposed on the basis of available scientific evidence. Furthermore, there has to be a reasonable technological need that cannot be achieved by other economically and technologically practicable means, and using the food additives should entail consumer advantages and benefits.[39]

Another important piece of legislation in regard to food regulation in the EU is the Novel Foods Regulation. This regulation requires mandatory premarket approval of all new ingredients and products. In 2008, the European Commission adopted a proposal to revise the Novel Foods Regulation with the purpose of improving the access of new and innovative foods to the EU market.[40] The definition of novel foods was broadened to include those modified by new production processes, such as nanotechnology and nanoscience, which might have an impact on the food itself. This proposal is currently being discussed in the European Parliament and is going through what is known as a "third reading," and has to be adopted after a co-decision wherein both the Council of the European Union and the Parliament has to agree on the final text of the regulation. If agreement cannot be reached, it goes to conciliation.

There are a number of areas on which the European Parliament and the Council of the European Union disagree in regard to nanomaterials. The requirement of having mandatory labeling is also controversial and so is the issue of whether to have premarket safety testing of nanotechnology and nanomaterials in food and packaging.[41]

In the first line of revisions suggested to the Novel Foods Regulation, both the Council and the Parliament mention the lack of adequate information and lack of test methods for assessing the risks of nanomaterials.[41] Once the European Commission receives an application for authorization of a novel food, the European Food Safety Authority (EFSA) is responsible for the evaluation of whether a novel food and its use as an ingredient presents a danger to or misleads consumers. By regulation, the EFSA is required to provide assessment on the composition, nutritional value, metabolism, intended use, and the level of microbiological and chemical contaminants. Studies on the toxicology, allergenicity, and details of the manufacturing process may also be considered. No distinction is, however, made in regulation in regard to particle size, and hence nanoparticles will not require new safety assessments if the substance has already been approved in bulk form.

RISK ASSESSMENT OF NANOMATERIALS

Three different kinds of limitations have been identified in various independent analysis of the applicability of existing regulatory frameworks when it comes to nanomaterials.

The first category of limitations are related to the limitations of definitions of what qualifies as a "substance," "novel food," etc., when it comes to nanomaterials. For instance, does the definition of a chemical substance cover both the bulk from as well as the nanoform of the substance, and does any given application of nanotechnology to manufacture a given food fall under the definition of a novel food? This issue is currently being discussed in a multistakeholder expert working group; however, this has failed to reach a consensus.[42]

In the second category fall requirements triggered by thresholds values not tailored to the nanoscale, but based on bulk material. For instance, for pharmaceuticals, the environmental concentration of medical products has to be estimated before marketing, and if it is below 0.01 ppb and "no other environmental concerns are apparent," no further actions are to be taken for the medical product in terms of environmental risk assessment.[32] Such a predefined action limit could potentially be problematic since the new properties of nanobased products are expected to also affect their environmental profiles, and this problem has yet to be addressed.[35]

The third category of limitations are related to lack of metrological tools, (eco)toxicological data, and environmental exposure limits as required by, e.g., REACH, the pharmaceuticals regulation, and the recast of the Novel Foods Regulation. The availability of (eco)toxicological data and chemical risk assessments is necessary to support existing legislation.

In regard to REACH, companies are urged to use already existing guidelines when performing chemical risk assessments, despite the fact that both the European Commission[10] and its SCENIHR,[11] as well as others,[9,10] have pointed out that current test guidelines supporting REACH are based on conventional methodologies for assessing chemical risks and may not be appropriate for the assessment of risks associated with nanomaterials.

Chemical risk assessment consists of four elements i.e., hazard identification, dose–response assessment, exposure assessment, and risk characterization. In Europe, legislation for controlling the production, use, and release of chemical substances is based on chemical safety assessment or risk assessment, as described in detail in the "Guidance on Information Requirements and Chemical Safety Assessment".[43] The guidance totals a staggering number of pages and is issued by the ECHA to help companies carry out chemical safety assessments. It includes extensive technical details for conducting hazard identification, dose (concentration)–response (effect) assessment, exposure assessment, and risk characterization in relation to human health and the environment.[43] Each of these four elements holds a number of limitations that are not easily overcome despite the fact that a lot of effort is being put into investigating the applicability of each of these four elements.

Hazard Identification of Nanomaterials

Toxicity and ecotoxicity have been reported on for multiple nanoparticles (metal and metal oxide nanoparticles, carbonaceous nanomaterials, and quantum dots) in scientific studies; however, many of these need further confirmation. Univocal hazard identification is currently impossible as it is hard to systematically link reported nanoparticle properties to the observed effects.

For instance, in regard to multiwalled CNTs (MWCNTs), Poland et al.[44] compared the toxicity of four kinds of MWCNTs of various diameters, lengths, shape, and chemical composition by exposing the mesothelial lining of the body cavity of three mice with 50 mg MWCNT for 24 hr or 7 days. This method was used as a surrogate for the mesothelial lining of the chest cavity. It was found that long MWCNTs "produced length-dependent inflammation, FBGCs, and granulomas that were qualitatively and quantitatively similar to the foreign body inflammatory response caused by long asbestos." Only the long MWCNTs caused significant increase in polymorphonuclear leukocytes or protein exudation. The short MWCNTs failed to cause any significant inflammation at 1 day or giant cell formation at 7 days. Poland et al.[44] also found that the water-soluble components of MWCNT did not produce significant inflammatory effects 24 hr after injection, which rules out that residue metals were the cause of the observed effects, as others previously had speculated on the basis on in vitro studies.[45,46] The findings by Poland et al.[44] have since then been supported by Ma-Hock et al.[47] and Pauluhn et al.[48] in 90-day inhalation toxicity studies.

Less work has been done in regard to exploring the ecotoxicological aspects of nanomaterials, but a number of significant studies have been published.

In 2004, Oberdorster[49] published the first ecotoxicological study and reported observed significant increase in lipid peroxidation of the brain of juvenile largemouth bass after exposure to uncoated fullerenes (99.5%) in concentrations of 0.5 and 1 ppm after exposure for 48 hr. C60 was dissolved in tetrahydrofuran (THF), which have since then led to some discussion about whether C60 or the THF was responsible for the effects observed.[50,51] The use of THF is no longer recommended.[18]

In regard to CNTs, Templeton et al.[52] compared "as prepared" single-walled CNTs (SWCNTs) with electrophoretically purified SWCNTs and the fluorescent fraction of nanocarbon by-products. They observed an average cumulative life cycle mortality of 13 ± 4%, while mean life cycle mortalities of 12 ± 3%, 19 ± 2%, 21 ± 3%, and 36 ± 11% were observed for 0.58, 0.97, 1.6, and 10 mg/L. Exposure to 10 mg/L showed: 1) significantly increased mortalities for the naupliar stage and cumulative life cycle; 2) a dramatically reduced development success to 51% for the nauplius to copepodite window, 89% for the copepodite to adult window, and 34% overall for the nauplius to adult period; and 3) a significantly depressed fertilization rate averaging only 64 ± 13%.

A number of studies have furthermore highlighted the need to investigate the potential interactions with existing environmental contaminants or what has become known as the "Trojan horse effect." For instance, Baun et al.[53] found that the toxicity of phenanthrene was increased toward algae and crustaceans following sorption to C60 aggregates. In contrast, Baun et al.[53] found that the toxic effect of pentachlorophenol decreased when C60 was added. After studying the ecotoxicity of cadmium to algae in the presence of 2 mg/L TiO_2 nanoparticles of three different sizes, Hartmann et al.[17] found that the presence of TiO_2 in algal tests reduced the toxicity of cadmium. This is thought to be due to decreased bioavailability of cadmium resulting from sorption/complexation of Cd^{2+} ions to the TiO_2 surface. However, the observed growth inhibition was, however, greater for the 30 nm TiO_2 nanoparticles than could be explained by the concentration of dissolved Cd(II) species alone, which indicates a possible carrier effect, or combined toxic effect of TiO_2 nanoparticles and cadmium.

Dose–Response Relationship in Regard to Nanomaterials

In regard to the second element of chemical risk assessment, it is fundamental that a dose–response relationship can be established so that no-effect concentrations or no-effect levels need to be predicted or derived. It is unclear whether a no-effect threshold can be actually be established and what the best hazard descriptor(s) of nanoparticles is, and what the most relevant dose metrics and the what the most sensitive endpoints are. Several studies have reported observing a dose–response relationship. This goes for, especially, in vitro studies on, among others, C60, SWCNTs and MWCNTs, and various forms of nanometals. Normally, dose refers to "dose by mass"; however, based on the experiences gained in biological tests of nanoparticles, it has been suggested that biological activity of nanoparticles might not be mass dependent, but dependent on physical and chemical properties not routinely considered in toxicity studies.[54] For instance, Oberdorster and colleagues[55,56] and Stoeger and colleagues[57,58] found that the surface area of the nanoparticles is a better descriptor of the toxicity of low-soluble, low-toxicity particles, whereas Wittmaack[59,60] found that the particle number worked best as dose metrics. Warheit et al.[61,62] found that toxicity was related to the number of functional groups in the surface of nanoparticles.

Exposure Assessment

Completing a full exposure assessment requires extensive knowledge about, among others, manufacturing conditions, level of production, industrial applications and uses, consumer products and behavior, and environmental fate and distribution. Such detailed information is not available, and thus far no full exposure assessment has been published for any one or more nanomaterials. This may partly be due to difficulties in monitoring nanomaterial exposure in the workplace and the environment, and partly due to the fact that the biological and environmental pathways of nanomaterials are still largely unexplored.[63] Some efforts have been made to assess occupational, consumer, and environmental exposure, however, both to assess the level of exposure and to assess the applicability of current exposure assessment methods and guidelines.

These are, however, hampered by the paucity of knowledge, lack of access to information, difficulties in monitoring nanomaterial exposure in the workplace and the environment, and by the fact that the biological and environmental pathways of nanomaterials are still largely unexplored. Hence, they should be seen as "proof of principle" rather than actual assessment of the exposure.[64–67]

Risk Characterization

All the information from the first three elements of the risk assessment come together in the fourth and final element of chemical risk assessment, namely risk characterization.[63] In the risk characterization process, exposure levels are compared with quantitative or qualitative hazard information, then suitable predicted no-effect concentrations or derived no-effect levels are determined in order to decide if risks are adequately controlled.[43]

Often, risk characterization boils down to the estimation of a risk quotient. For the environment, this is, for instance, defined as the PEC/predicted no effect concentration (PNEC). If the risk quotient is <1, no further testing or risk reduction measures are needed according to the European Chemical Agency.[43] If it is >1, further testing can be initiated to lower the PEC/PNEC ratio. If that is not possible, risk reduction could be implemented.

A number of studies reported having completed—or attempted to complete—risk assessments of various nanomaterials such as CeO_2, TiO_2, C60, and CNTs.[18,67–69] For instance, in regard to the use of CeO_2-based diesel fuel additive in the United Kingdom, Park et al.[67] assessed the risk of CeO_2 causing pulmonary inflammation. First, they estimated an internal dose of 3.8×10^{-7} cm^2/cm^2 by converting the retained dose into surface area units and then dividing by the area of the proximal alveolar region of the lung. Then, they compared this value to the highest no-observed-effect level found in a number of in vitro toxicity studies. This value was 26.75 cm^2/cm^2. Assuming that in vitro exposure data can be accurately projected to the in vivo situation, Park et al.[67] concluded that "it is highly unlikely that exposure to cerium oxide at the environmental levels (from both monitored and modeled experimental data) would elicit pulmonary inflammation."

Mueller and Nowack[69] reported having completed the first quantitative risk assessment of nanoparticles in the environment. In a first attempt to derive PEC values, Mueller and Nowack used the threshold concentrations of 20 and 40 mg/L reported in the literature for nano-Ag on *Bacillus subtilis* and *Escherichia coli*, and considered it to be equivalent to a no-observed-effect concentration. For nano-TiO_2 and CNT, the lowest value found in the literature was <1 mg/L for algae, daphnia, and fish.[69] Applying the assessment factor of 1000, the PNEC in water was found to be 0.04, <0.001, and <0.0001 mg/L for nano-Ag, nano-TiO_2, and CNT, respectively. Combining these PNEC values with the predicted exposure, Mueller and Nowack[69] calculated the environmental concentrations in Switzerland for nano-Ag, nano-TiO_2, and CNTs stemming from textiles, cosmetics, coatings, plastics, sports gear, electronics, etc. Assuming worse-case exposure levels, Mueller and Nowack[69] found that the risk quotient for nano-Ag and CNT is less than one-thousandth, and they state that their modeling suggests that currently little or no risk is to be expected from nano-Ag and CNT to organisms in water and air. Nano-TiO_2, on the other hand, might pose a risk to organism in water—according to Mueller and Nowack[69]—with risk quotients ranging from >0.7 to >16. The PNEC for soil could not be determined due to lack of information.

Despite the preliminary risk characterizations by Park et al.,[67] Mueller and Nowack,[69] Shinohara et al.,[68] and Stone et al.,[18] it is important to realize that risk characterization critically involves reflection of the data behind each step and determining what the overall risk will be.[63] As elaborated on previously, each of the three first steps of risk assessment holds a number of challenges, and since risk characterization is the fourth and final step where all the information is to come together, the sum or maybe even the power all of these limitations are conveyed to calculating risk quotients for nanomaterials under REACH.[9]

Revisions of the Technical Guidance of Risk Assessment

The European Commission has commissioned an expert/multistakeholder investigation of whether nanospecific amendments are needed to the current technical guidelines on chemical safety assessment. This has to develop, among others, specific advice on

1. How REACH information requirements on intrinsic properties of nanomaterials can be fulfilled
2. The appropriateness of the relevant test methods for nanomaterials
3. The possible specific testing strategies, if relevant
4. Information needed for safety evaluation and risk management of nanomaterials (especially, information beyond the current information requirements under REACH)
5. How to do exposure assessment for nanomaterials, hazard, and risk characterization for nanomaterials[3,4]

The latter will involve threshold/non-threshold considerations, analysis of existing evidence related to setting limit values for nanomaterials, identification of critical items for dose description (mass, number concentration, surface area, particle size(s) etc.), whether and how no-effect-levels for health and the environment could be established, and finally development of recommendations on the feasibility of whether categorization of nanomaterials (e.g., different types of CNTs) in the hazard assessment is compatible with the exposure assessment parameters/metrics in order to prepare a meaningful risk characterization.[3,4]

In regard to novel food, EFSA published a scientific opinion in 2008 on the potential risks arising from the use of nanotechnology in food, concluding that nanotechnology aspects shall be considered when risk assessment guidance documents in the food and feed area are reviewed, and, among others, recommend that risk assessment of nanomaterials in the food and feed areas should consider the specific properties of nanomaterials in addition to those common to the equivalent non-nanoforms.[70] Recently, EFSA closed for public comments on a draft guidance on risk assessment concerning potential risks arising from applications of nanoscience and nanotechnologies to food and feed. This guidance holds practical advice on how to complete risk assessments of nanomaterials used in food and food products.[71]

In the light of the limitations of chemical risk assessment, a number of alternative or complementary tools and methods, such the precautionary matrix[72] and multicriteria decision analysis,[73] have been proposed recently. Many of them hold great promises, but they need further evaluation and validation.[74]

CONCLUSION

In this entry, we presented a number of major pieces of legislation such as REACH, the WFD, and the Novel Foods Regulation, and discussed their relevance and limitations in regard to nanomaterials. Only a limited number of EU regulations, directives, etc., actually mention nanotechnology and/nanomaterials. In general, there seem to be three overall challenges when it comes to current regulation of nanomaterials: 1) limitations in regard to terminology and definitions of key terms such as a "substance," "novel food," etc.; 2) safety assessment requirements triggered by thresholds values not tailored to the nanoscale, but based on bulk material and; 3) limitations related to lack of metrological tools, (eco)toxicological data, and environmental exposure limits as required by, e.g., REACH, WFD, the pharmaceuticals regulation, and the recast of the Novel Foods Regulation. Chemical risk assessment provides a fundamental element in support of existing legislation. Risk assessment is normally said to consist of four elements, i.e., hazard identification, dose–response assessment, exposure assessment, and risk characterization. Each of these four elements holds a number of limitations that are not easily overcome, although a lot of effort is being put into investigating the applicability of each of them. However, political decisions to revise substance definition and the current thresholds that trigger safety evaluation are still needed as these are not tailored to the nanoscale, but based on bulk material.

REFERENCES

1. European Parliament and Council of the European Union (EP & CEU). Regulation (EC) No 1223/2009 of the European Parliament and of the Council of 30 November 2009 on cosmetic products (1). Off. J. Eur. Union L342, vol. 52, 22 December 2009. ISSN 1725-2555. L 342/59 L 342/209.
2. Bowman, D.M.; Calster, G.v.; Friedrichs, S. Nanomaterials and regulation of cosmetics. Nat. Nanotechnol. 2009, 5, 92.
3. Safenano. REACH-NanoInfo: Rip-oN 2, 2010a. Available at http://www.safenano.org/REACHnanoInfo.aspx (accessed October 2010).
4. Safenano. REACH-NanoHazEx: Rip-oN 3, 2010b. Available at http://www.safenano.org/REACHnanoHazEx.aspx (accessed October 2010).
5. European Parliament and the Council of the European Union (EP & CEU). Regulation (EC) No 1907/2006 of the European Parliament and of the Council of 18 December 2006 concerning the Registration, Evaluation, Authorisation and Restriction of Chemicals (REACH), establishing a European Chemicals Agency, amending Directive 1999/45/EC and repealing Council Regulation (EEC) No 793/93 and Commission Regulation (EC) No 1488/94 as well as Council Directive 76/769/EEC and Commission Directives 91/155/EEC, 93/67/EEC, 93/105/EC and 2000/21/EC. 30.12.2006. Off. J. Eur. Union L 2006, 396/1-L 396/849.
6. Führ, M.; Hermann, A.; Merenyi, S.; Moch, K.; Möller, M. Legal Appraisal of Nanotechnologies, 2007. Available at http://www.umweltdaten.de/publikationen/fpdf-l/3198.pdf (accessed December 2010). Umwelt Bundes Amt UBA-FB 000996. ISSN 1862-4804.
7. C&EN. Carbon losses its exemption statues under REACH. Chem. Eng. News 2008, June 23, p.9.
8. Chaundry, Q.; Blackburn, J.; Floyd, P.; George, C.; Nwaogu, T.; Boxall, A.; Aitken, R. *A Scoping Study to Identify Gaps in Environmental Regulation for the Products and Applications of Nanotechnologies*; Department for Environment, Food and Rural Affairs: London, 2006.
9. Hansen SF. *Regulation and Risk Assessment of Nanomaterials—Too Little, Too Late?*, PhD Thesis; Technical University of Denmark: Kgs. Lyngby, Denmark, 2009. Available at http://www2.er.dtu.dk/publications/fulltext/2009/ENV2009-069.pdf (accessed April 2009).
10. Commission of the European Communities (CEC). *Communication from the Commission to the European Parliament, the Council and the European Economic and Social Committee Regulatory Aspects of Nanomaterials*, [Sec(2008) 2036] Com(2008), 366 final; Commission of the European Communities: Brussels, Belgium, 2008a.
11. Scientific Committee for Emerging and Newly Identified Health Risks (SCENIHR). *The Appropriateness of the Risk Assessment Methodology in Accordance with the Technical Guidance Documents for New and Existing Substances for Assessing the Risks of Nanomaterials*; European Commission of Health and Consumer Protection Directorate-General: Brussels, Belgium, 2007.
12. European Parliament and the Council of the European Union (EP & CEU). Directive 2000/60/EC of the European Parliament and of the Council of 23 October 2000 establishing a framework for Community action in the field of water policy. Off. J. Eur. Commun. 2000, L 327, 1–72, Brussels, Belgium.
13. Baun, A.; Hartmann, N.B.; Greiger, K.D.; Hansen, S.F. Setting the limits for engineered nanoparticles in European surface waters. J. Environ. Monit. 2009, 11, 1774–1781.
14. Li, Q.; Mahendra, S.; Lyon, D.Y.; Liga, M.V.; Li, D.; Alvarez, P. Antimicrobial nanomaterials for water disinfection and microbial control: Potential applications and implications. Water Res. 2008, 42, 4591–4602.
15. Nano Iron. *Technical Data Sheet*. NANOFER 25S. Rajhrad, Czech Republic, 2009. http://www.nanoiron.cz/en/?f1/4technical_data (accessed February 2011).
16. Li, X.; Elliott, D.W.; Zhang, W.X. Zero-valent iron nanoparticles for abatement of environmental pollutants: Materials and engineering aspects. Crit. Rev. Solid State Mater. Sci. 2006, 31, 111–122.
17. Hartmann, N.B.; Von der Kammer, F.; Hofmann, T.; Baalousha, M.; Ottofuelling, S.; Baun, A. Algal testing of titanium dioxide nanoparticles—Testing considerations, inhibitory effects and modification of cadmium bioavailability. Toxicology 2010, 269 (2010), 190–197.
18. Stone, V.; Hankin, S.; Aitken, R.; Aschberger, K.; Baun, A.; Christensen, F.; Fernandes, T.; Hansen, S.F.; Hartmann, N.B.; Hutchinson, G.; Johnston, H.; Micheletti, G.; Peters, S.; Ross, B.; Sokull-Kluettgen, B.; Stark, D.; Tran, L. Engineered Nanoparticles: Review of Health and Environmental Safety (ENRHES), 2010. Available at http://nmi.jrc.ec.europa.eu/project/ENRHES.htm (accessed February 2010).
19. Scientific Committee for Emerging and Newly Identified Health Risks (SCENIHR). *Risk Assessment of Products of Nanotechnologies*; Scientific Committee on Emerging and Newly Identified Health Risks, European Commission of Health and Consumer Protection Directorate-General: Brussels, Belgium, 2009.
20. Tiede, K. *Detection and Fate of Engineered Nanoparticles in Aquatic Systems*, PhD Thesis, University of York,

Environment Department and Central Science Laboratory: York, U.K., 2008.
21. Hassellöv, M.; Readman, J.W.; Ranville, J.F.; Tiede, K. Nanoparticle analysis and characterization methodologies in environmental risk assessment of engineered nanoparticles. Ecotoxicology **2008**, *17*, 344–361.
22. Lead, J.R.; Wilkinson, K.J. Natural aquatic colloids: Current knowledge and future trends. Environ. Chem. **2006**, *3*, 159–171.
23. European Parliament and the Council of the European Union (EP & CEU).Regulation (EC) No 726/2004 of the European Parliament and of the Council of 31 March 2004 Laying Down Community Procedures for the Authorisation and Supervision of Medicinal Products for Human and Veterinary Use and Establishing a European Medicines Agency, 2004.
24. Council of the European Communities (CEC). Council Directive 90/385/EEC of 20 June 1990 on the approximation of the laws of the Member States relating to active implantable medical devices. Off. J. Eur. Commun. **1990**, *L 189*, 17–36.
25. Council of the European Communities (CEC). Council Directive 93/42/EEC of 14 June 1993 concerning medical devices. Off. J. Eur. Commun. **1993**, *L 169*, 1–43.
26. Council of the European Communities (CEC). Directive 2001/83/EC of the European Parliament and of the Council of 6 November 2001 on the Community code relating to medicinal products for human use. Off. J. Eur. Commun. **2001**, *L 311*, 67–128
27. Editorial. Regulating nanomedicine. Nat. Mater. **2007**, *6*, 249.
28. D'Silva, J.; Van Calster, G. Regulating Nanomedicine: A European Perspective, 2008. Available at http://ssrn.com/abstract=1286215 (accessed November 2008).
29. European Group on Ethics in Science and New Technologies (EGE). Opinion 21—On the Ethical Aspects of Nanomedicine, The European Group on Ethics in Science and New Technologies (EGE), 2007. Retrieved March 15, 2008, available at http://ec.europa.eu/european_group_ethics/publications/docs/final_publication_%20op21_en.pdf.
30. N&ET Working Group. Report on Nanotechnology to the Medical Devices Expert Group: Findings and Recommendations, 2007. Available at http://ec.europa.eu/enterprise/medical_devices/net/entr-2007-net-wg-report-nanofinal.pdf (accessed December 2008).
31. European Medicines Agency (EMEA). *European Medicines Agency: Reflection Paper on Nanotechnology-Based Medicinal Products for Human Use*, EMEA/CHMP/70769/2006; European Medicines Agency: London, 2006a.
32. European Medicines Agency (EMEA). *Guideline on the Environmental Risk Assessment of Medicinal Products for Human Use*, EMEA/CHMP/SWP/4447/00; European Medicines Agency: London, 2006b.
33. Zhang, X.; Sun, H.; Zhang, Z.; Niu, Q.; Chen, Y.; Crittenden, J.C. Enhanced bioaccumulation of cadmium in carp in the presence of titanium dioxide nanoparticles. Chemosphere **2007**, *67*, 160–166.
34. Baun, A.; Hartmann, N.B.; Grieger, K.; Kusk, K.O. Ecotoxicity of engineered nanoparticles to aquatic invertebrates—A brief review and recommendations for future toxicity testing. Ecotoxicology **2008a**, *17* (5), 387–395.
35. Baun, A.; Hansen, S.F. Environmental challenges for nanomedicine. Nanomedicine **2008**, *2* (5), 605–608.
36. European Parliament and the Council of the European Union (EP & CEU). Regulation (EC) No 178/2002 of the European Parliament and of the Council of 28 January 2002 laying down the general principles and requirements of food law, establishing the European Food Safety Authority and laying down procedures in matters of food safety. Off. J. Eur. Commun. **2002**, L 31/1-L 31/24.
37. Friends of the Earth (FOE). *Out of the Laboratory and on to Our Plates Nanotechnology in Food and Agriculture*; Friends of the Earth: Australia, Europe, and U.S.A, 2008.
38. Halliday J. EU Parliament Votes for Tougher Additives Regulation, Foodnavigator.com, July 12, 2008. Available at http://www.foodnavigator.com/Legislation/EU-Parliamentvotes-for-tougher-additives-regulation (accessed November 2008).
39. European Parliament and the Council of the European Union (EP & CEU). Directive 2008/98/EC of the European Parliament and of the Council of 19 November 2008 on waste and repealing certain directives. Off. J. Eur. **2008**, Union L 312/3-L 312/30.
40. Commission of the European Communities (CEC). *Proposal for a Regulation of the European Parliament and of the Council on Novel Foods and Amending Regulation (EC) No. XXX/XXXX#* [common procedure] (presented by the Commission) [SEC(2008) 12] [SEC(2008) 13]. COM(2007), 872 final 2008/0002 (COD). 14-1 2008; Commission of the European Communities: Brussels, Belgium, 2008b.
41. European Parliament. Draft Recommendation for Second Reading on the Council Position at First Reading for Adopting a Regulation of the European Parliament and of the Council on Novel Foods, Amending Regulation (EC) No 1331/2008 and Repealing Regulation (EC) No 258/97 and Commission Regulation (EC) No 1852/2001 (11261/2/2009–C7-0000/2010–2008/0002(COD)), 2010.
42. C&EN. Wrangling Over Substance ID Hits REACH Nano Project. Lack of Consensus will Lead to Decision Driven by Policy Not Science, 2011. Available at http://chemicalwatch.com/6324/wrangling-over-substance-id-hits-reach-nano-project?q=Wrangling%20over%20substance%20ID%20hits%20REACH%20nano%20project (accessed January 2011).
43. ECHA. Guidance Documents. Guidance on Information Requirements and Chemical Safety Assessment, 2010. Available at http://guidance.echa.europa.eu/docs/guidance_document/information_requirements_en.htm?time=1288375681 (accessed October 2010).
44. Poland, C.A.; Duffin, R.; Kinloch, I.; Maynard, A.; Wallace, W.A.H.; Seaton, A.; Stone, V.; Brown, S.; Macnee, W.; Donaldson, K. Carbon nanotubes introduced into the abdominal cavity of mice show asbestos-like pathogenicity in a pilot study. Nat. Nanotechnol. **2008**, *3*, 423–428.
45. Shvedova, A.A.; Kisin, E.R.; Mercer, R.; Murray, A.R.; Johnson, V.J.; Potapovich, A.I.; Tyurina, Y.Y.; Gorelik, O.; Arepalli, S.; Schwegler-Berry, D.; Hubbs, A.F.; Antonini, J.; Evans, D.E.; Ku, B.K.; Ramsey, D.; Maynard, A.; Kagan, V.E.; Castranova, V.; Baron, P. Unusual inflammatory and fibrogenic pulmonary responses to single-walled

carbon nanotubes in mice. Am. J. Physiol. Lung Cell. Mol. Physiol. **2005**, *289*, 698–708.
46. Kagan, V.E.; Tyurina, Y.Y.; Tyurin, V.A.; Konduru, N.V.; Potapovich, A.I.; Osipov, A.N.; Kisin, E.R.; Schwegler-Berry, D.; Mercer, R.; Castranova, V.; Shvedova, A.A. Direct and indirect effects of single walled carbon nanotubes on RAW 264.7 macrophages: Role of iron. Toxicol. Lett. **2006**, *165* (1): 88–100.
47. Ma-Hock, L.; Treumann, S.; Strauss, V.; Brill, S.; Luizi, F.; Mertler, M.; Wiench, K.; Gamer, A.O.; van Ravenzwaay, B.; Landsiedel, R. Inhalation toxicity of multi-walled carbon nanotubes in rats exposed for 3 months. Toxicol. Sci. **2009**, *112*, 273–275.
48. Pauluhn, J. Subchronic 13-week inhalation exposure of rats to multiwalled carbon nanotubes: Toxic effects are determined by density of agglomerate structures, not fibrillar structures. Toxicol. Sci. **2010**, *113* (1), 226–242.
49. Oberdorster E. Manufactured nanomaterials (fullerenes, C60) induce oxidative stress in juvenile largemouth bass. Environ. Health Perspect. **2004**, *112*, 1058–1062.
50. Zhu, S.Q.; Oberdorster, E.; Haasch, M.L. Toxicity of an engineered nanoparticle (fullerene, C-60) in two aquatic species, *Daphnia* and fathead minnow. Mar. Environ. Res. **2006**, *62*, S5–S9.
51. Henry, T.B.; Menn, F.M.; Fleming, J.T.; Wilgus, J.; Compton, R.N.; Sayler, G.S. Attributing effects of aqueous C_{60} nano-aggregates to tetrahydrofuran decomposition products in larval zebrafish by assessment of gene expression. Environ. Health Perspect. **2007**, *115* (7), 1059–1065.
52. Templeton, R.C.; Ferguson, P.L.; Washburn, K.M.; Scrivens, W.A.; Chandler, G.T. Life-cycle effects of single-walled carbon nanotubes (SWNTS) on an estuarine meiobenthic copepod. Environ. Sci. Technol. **2006**, *40* (23), 7387–7393.
53. Baun, A.; Sorensen, S.N.; Rasmussen, R.F.; Hartmann, N.B.; Koch, C.B. Toxicity and bioaccumulation of xenobiotic organic compounds in the presence of aqueous suspensions of aggregates of nano-C60. Aquat. Toxicol. **2008b**, *86*, 379–387.
54. Oberdorster, G.; Maynard, A.; Donaldson, K.; Castranova, V.; Fitzpatrick, J.; Ausman, K.; Carter, J.; Karn, B.; Kreyling, W.; Lai, D.; Olin, S.; Monteiro-Riviere, N.; Warheit, D.; Yang, H. Principles for characterizing the potential human health effects from exposure to nanomaterials: Elements of a screening strategy. Part. Fibre Toxicol. **2005**, *2*, 8.
55. Oberdorster, G. Significance of particle parameters in the evaluation of exposure dose–response relationships of inhaled particles. Part. Sci. Technol. **1996**, *14*, 135–151.
56. Oberdorster, G.; Stone, V.; Donaldson, K. Toxicology of nanoparticles: A historical perspective. Nanotoxicology **2007**, *1* (1), 2–25.
57. Stoeger, T.; Reinhard, C.; Takenaka, S.; Schroeppel, A.; Karg, E.; Ritter, B.; Heyder, J.; Schultz, H. Instillation of six different ultrafine carbon particles indicates surface area threshold dose for acute lung inflammation in mice. Environ. Health Perspect. **2006**, *114* (3), 328–333.
58. Stoeger, T.; Schmid, O.; Takenaka, S.; Schulz, H. Inflammatory response to TiO_2 and carbonaceous particles scales best with BET surface area. Environ. Health Perspect. **2007**, *115* (6), A290–A291.
59. Wittmaack, K. In search of the most relevant parameter for quantifying lung inflammatory response to nanoparticle exposure: Particle number, surface area, or what? Environ. Health Perspect. **2007a**, *115*, 187–194.
60. Wittmaack, K. Dose and response metrics in nanotoxicology: Wittmaack responds to Oberdoerster et al. and Stoeger et al. Environ. Health Perspect. **2007b**, *115* (6), A290–291.
61. Warheit, D.B.; Webb, T.R.; Colvin, V.L.; Reed, K.L.; Sayes, C.R. Pulmonary bioassay studies with nanoscale and fine-quartz particles in rats: Toxicity is not dependent upon particle size but on surface characteristics. Toxicol. Sci. **2007a**, *95* (1), 270–280.
62. Warheit, D.B.; Webb, T.R.; Reed, K.L.; Frerichs, S.; Sayes, C.M. Pulmonary toxicity study in rats with three forms of ultrafine-TiO_2 particles: Differential responses related to surface properties. Toxicology **2007b**, *230*, 90–104.
63. CCA. *Small Is Different: A Science Perspective on the Regulatory Challenges of the Nanoscale*; The Council of Canadian Academies: Ottawa, Canada, 2008.
64. Boxall, A.B.A.; Chaudhry, Q.; Sinclair, C.; Jones, A.; Aitken, R.; Jefferson, B.; Watts, C. *Current and Future Predicted Environmental Exposure to Engineered Nanoparticles*; Central Science Laboratory: York, U.K., 2008.
65. Hansen, S.F.; Michelson, E.; Kamper, A.; Borling, P.; Stuer-Lauridsen, F.; Baun, A. Categorization framework to aid exposure assessment of nanomaterials in consumer products. Ecotoxicology **2008**, *17* (5), 438–447.
66. Luoma, S.N. *Silver Nanotechnologies and the Environment Old Problems or New Challenges?* PEN 15. Project on Emerging Nanotechnologies; Woodrow Wilson International Center for Scholars: Washington, D.C., 2008.
67. Park, B.; Donaldson, K.; Duffin, R.; Tran, L.; Kelly, F.; Mudway, I.; Morin, J-P.; Guest, R.; Jenkinson, P.; Samaras, Z.; Giannouli, M.; Kouridis, H.; Martin, P. Hazard and risk assessment of a nanoparticulate cerium oxide-based diesel fuel additive—A case study. Inhal. Toxicol. **2008**, *20* (6), 547–566.
68. Shinohara, N. *Risk Assessment of Manufactured Nanomaterials—Fullerene (C60)*, Interim Report issued on October 16, 2009, Executive Summary. The Research Institute of Science for Safety and Sustainability, AIST available at http://www.aist-riss.jp/main/modules/product/nano_rad.html.
69. Mueller, N.; Nowack, B. Exposure modeling of engineered nanoparticles in the environment. Environ. Sci. Technol. **2008**, *42*, 4447–4453.
70. European Food Safety Authority (EFSA). Draft Opinion of the Scientific Committee on the Potential Risks Arising from Nanoscience and Nanotechnologies on Food and Feed Safety (Question No EFSAQ-2007-124). European Food Safety Authority: Parma, Italy, 2008.
71. European Food Safety Authority (EFSA). Endorsed for Public Consultation Draft Scientific Opinion Guidance on Risk Assessment Concerning Potential Risks Arising From Applications of Nanoscience and Nanotechnologies to Food and Feed. EFSA Scientific Committee, European Food Safety Authority (EFSA): Parma, Italy, 2011.
72. Höck, J.; Epprecht, T.; Hofmann, H.; Höhner, K.; Krug, H.; Lorenz, C.; Limbach, L.; Gehr, P.; Nowack, B.; Riediker, M.; Schirmer, K.; Schmid, B.; Som, C.; Stark, W.; Studer, C.; Ulrich, A.; von Götz, N.; Wengert, S.; Wick,

P.; Guidelines on the Precautionary Matrix for Synthetic Nanomaterials. Federal Office of Public Health and Federal Office for the Environment, Berne 2010, Version 2. Available at http://www.bag.admin.ch/themen/chemikalien/00228/00510/05626/index.html?lang=en (accessed 17 May 2010).

73. Linkov, I.; Satterstrom, F.K.; Steevens, J.; Ferguson, E.; Pleus, R.C. Multi-criteria decision analysis and environmental risk assessment for nanomaterials. J. Nanopart. Res. **2007**, *9* (4), 543–554.

74. Grieger, K.D.; Linkov, I.; Hansen, S.F.; Baun, A. A review of alternative frameworks and approaches for assessing environmental risks of nanomaterials. Nanotoxicology, **2011**, *in press*.

75. Harrington, R. Nano Risk Assessment: A Work in Progress, 2010. Available at http://www.foodproductiondaily.com/Quality-Safety/Nano-risk-assessment-a-work-in-progress/?c=lIxHi8W7vkWyj1EO6E28rw%3D%3D&utm_source=newsletter_daily&utm_medium=email&utm_campaign=Newsletter%2BDaily (accessed June 2010).

76. Höck, J.; Epprecht, T.; Hofmann, H.; Höhner, K.; Krug, H.; Lorenz, C.; Limbach, L.; Gehr, P.; Nowack, B.; Riediker M. et al. *Guidelines on the Precautionary Matrix for Synthetic Nanomaterials*, Version 2; Swiss Federal Office for Public Health and Federal Office for the Environment: Berne, Switzerland, 2010.

77. Lam, C.W.; James, J.T.; McCluskey, R.; Hunter R.L. Pulmonary toxicity of single-wall carbon nanotubes in mice 7 and 90 days after intratracheal instillation. Toxicol. Sci. **2004**, *77*, 126–134.

78. Paik, S.Y.; Zalk, D.M.; Swuste, P. Application of a pilot control banding tool for risk level assessment and control of nanoparticle exposures. Ann. Occup. Hyg. **2008**, *52* (6), 419–428.

79. Royal Society and the Royal Academy of Engineering. *Nanoscience and Nanotechnologies: Opportunities and Uncertainties*; Royal Society: London, 2004.

Nanoparticles

Alexandra Navrotsky
Department of Chemical Engineering and Materials Science, University of California—Davis, Davis, California, U.S.A.

Abstract
Understanding nanoparticle formation and properties requires sophisticated physics, chemistry, and materials science. Tailoring nanomaterials to specific applications requires both science and Edisonian inventiveness. Applying them to technology is state-of-the-art engineering. The purpose of this entry is to describe some of the unique features of nanoparticles and to discuss their occurrence and importance in the natural environment.

INTRODUCTION

Nanoparticles have no exact definition, but they are aggregates of atoms bridging the continuum between small molecular clusters of a few atoms and dimensions of 0.2–1 nm and chunks of solid containing millions of atoms and having the properties of macroscopic bulk material. In water, nanoparticles include colloids; in air, they include aerosols. Nanoparticles are ubiquitous. We pay to have them. We pay more to not have them. They occur as dust in the air, as suspended particles that make river water slightly murky, in soil, in volcanic ash, in our bodies, and in technological applications ranging from ultratough ceramics to microelectronics. They both pollute our environment and help keep it clean. Microbes feast on, manufacture, and excrete nanoparticles.

Understanding nanoparticle formation and properties requires sophisticated physics, chemistry, and materials science. Tailoring nanomaterials to specific applications requires both science and Edisonian inventiveness. Applying them to technology is state-of-the-art engineering. Tracing their transport and fate in the environment invokes geology, hydrology, and atmospheric science. Applying them to improving soil fertility and water retention links soil science and agriculture to surface chemistry. Understanding their biological interactions brings in fields ranging from microbiology to medicine. Probing the impact of nanoparticles on humans and of human behavior on the production and control of nanoparticles requires the behavioral and social sciences, e.g., in dealing with issues of automotive pollution. The purpose of this review is to describe some of the unique features of nanoparticles and to discuss their occurrence and importance in the natural environment.

Although we often think of the natural environment as that part of the planet which we can see, a somewhat broader definition includes the "critical zone": the atmosphere, hydrosphere, and shallow portion of the solid earth that exchange matter on a geologically short time scale, on the order of tens to thousands of years. This critical zone affects us directly, and our activities influence it. Because of the active chemical reactions continuously taking place in the critical zone, and because its temperatures and pressures are relatively low and it is dominated by water, solids are constantly being formed and decomposed. Many of these solids start out as nanoparticles; many remain so. In a yet broader sense, our entire planet from crust to core, the solar system, and the galaxy are part of our environment.

PHYSICAL CHEMISTRY OF NANOPARTICLES

A major feature of nanoparticles is their high surface-to-volume ratio. Fig. 1 shows the volume fraction within 0.5 nm of the surface for a spherical particle of radius r. One can think of this fraction either as the fraction of atoms likely to be influenced by processes at the surface, or as the

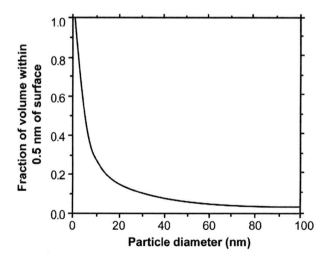

Fig. 1 Volume fraction of a nanoparticle within 0.5 nm of the surface as a function of particle radius.
Source: Navrotsky,[41] Kluwer Academic Publishers.

fraction of the volume of a material that could be taken up by a 0.5-nm coating of another material. In the first case, because the surface dominates chemical reactivity, the increased surface to volume ratio means that nanoparticles dominate chemical reactions. In the second case, the ability to carry a substantial coating offers a mechanism for the transport of nutrients or pollutants.

Many oxides are *polymorphic*, exhibiting several crystal structures as a function of pressure and temperature. Often, nanosized oxide particles crystallize in structures different from that of large crystals of the same composition.[1] Examples are γ-Al_2O_3, a defect spinel rather than α-Al_2O_3, corundum, γ-Fe_2O_3, the defect spinel maghemite rather than α-Fe_2O_3, hematite, and the anatase and brookite forms of TiO_2 rather than rutile. From arguments based on transformation sequences and the occur-

Table 1 Energetic parameters for oxide and oxyhydroxide polymorphs.

Formula	Polymorph	Metastability (kJ/mol)	Surface energy (J/m^2)
Al_2O_3[a]	Corundum (α)	0	2.6
	Spinel (γ)	13.4	1.7
Fe_2O_3[b]	Hematite	0	0.8
	Maghemite	20	0.8
TiO_2[c]	Rutile	0	2.2
	Brookite	0.7	1.0
	Anatase	2.6	0.4
AlOOH[d]	Diaspore	0	?
	Boehmite	4.9	0.5
FeOOH[b]	Goethite	0	0.3
	Lepidocrocite		0.3

[a]From McHale et al.[3]
[b]From Majzlan.[6]
[c]From Ranade.[4]
[d]From Majzlan.[5]

rence of phases, it was long argued that there may be a crossover in phase stability at the nanoscale if the structure which is metastable for large particles has a significantly lower surface energy.[2] This has been proven for alumina and titania in recent calorimetric studies (Fig. 2).[3,4] The resulting transformation enthalpies and surface energies, and those of other related systems are shown in Table 1. Another interesting feature is that the hydrous phases AlOOH boehmite and FeOOH goethite have significantly lower surface energies than their anhydrous counterparts, Al_2O_3 and Fe_2O_3.[5,6] Whether this is a general feature of hydrous minerals with hydroxylated surfaces is not yet known.

As particles become less than about 10 nm in size, their x-ray diffraction patterns are broadened sufficiently that they begin to appear "x-ray amorphous" (Fig. 3). This term lacks exact definition. High-resolution electron microscopy may still detect periodicity, and short-range order is certainly present.[7] The identification of structure in 1–10 nm particles is very difficult, and phases are empirically described as, for example, "two line ferrihydrite," based on x-ray diffraction patterns.[8]

NANOPARTICLES IN SOIL AND WATER

Soil is a complex aggregate of inorganic, organic, and biological material.[9] Its constituents of largest size are rocks and gravel, small animals, plant roots, and other debris. Smaller mineral grains, clumps of organic matter, and microorganisms make up an intermediate size fraction. The smallest particles, ranging into the nanoscale, are clays, iron oxides, and other minerals. These are often heterogeneous and coated by other minerals and organic matter. The

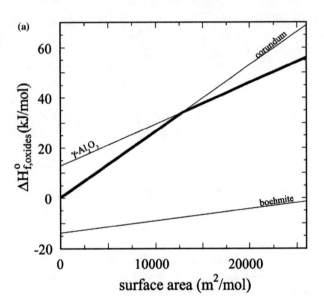

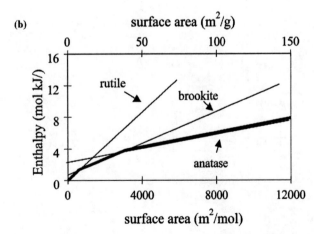

Fig. 2 (a) Enthalpies of alumina polymorphs as a function of surface area. Source: From McHale et al.[3] (b) Enthalpies of titania polymorphs as a function of surface area. The heavy lines show the stable polymorphs in each size range.
Source: Ranade,[4] PNAS.

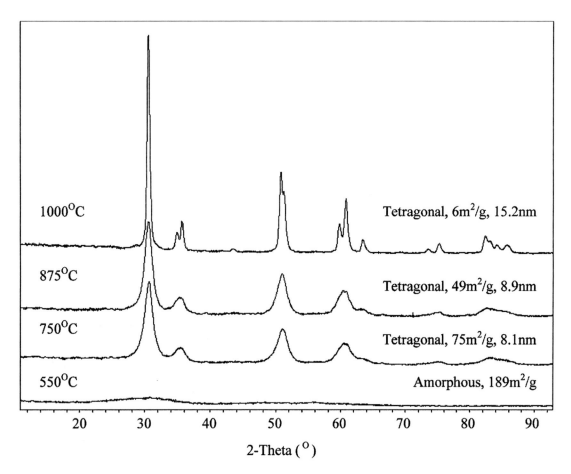

Fig. 3 Powder x-ray diffraction patterns of sol–gel zirconia heated at various temperatures. The structure and average particle diameters are indicated.
Source: Pitcher and Navrotsky[42]

entire composite is porous and hydrated. The percolation of water in soil transports both nanoparticles and dissolved organic and inorganic species. The texture and porosity, as well as the chemical composition and pH, are crucial to biological productivity. The surfaces of nanoparticles provide much of the chemical reactivity for both biological and abiotic processes.

Major aluminosilicate minerals in soils include clays, zeolites, and poorly crystalline phases (Table 2). These can change their water content in response to ambient

Table 2 Major soil minerals and constituents.

Type	Composition	Structures
Clay	Hydrated aluminosilicate	Layered
Zeolite	Hydrated aluminosilicate	Three-dimensional porous
Salts	NaCl, Na_2SO_4, $CaSO_4$	Ionic crystals
Carbonates	$CaCO_3$–$MgCO_3$–$FeCO_3$	Calcite, dolomite, others
Allophane	Hydrous aluminosilicate gel	Amorphous
Iron oxides	Fe_2O_3, FeOOH	Various polymorphs
Aluminum oxides	AlOOH, $Al(OH)_3$	Various polymorphs
Quartz	SiO_2	Quartz
Manganese oxides	Mn_2O_3, MnOOH, MnO_2	Various polymorphs
H_2O	H_2O	Water, ice, vapor
Organics	C–H–N–O	Large surface area amorphous colloids
Jarosite-alunite	Alkali (Fe, Al) sulfates	Ionic double salts

conditions, often swelling in wet seasons, and shrinking in dry seasons. These nanophase materials are major controllers of soil moisture and permeability. Iron and manganese oxides are another class of major sol minerals. Their extensive polymorphism at the nanoscale makes them highly variable. They sequester and/or transport and make available the essential plant nutrient iron, as well as other essential transition metals (cobalt, copper, zinc, etc.). They frequently carry coatings of other metal oxides and oxyhydroxides, including toxic metals such as lead and chromium. They also frequently have organic coatings. Sulfates, including the jarosite–alunite family of hydrated [(K, Na), (Al, Fe)] sulfates, are another important constituent. In alkaline and arid environments, other sulfates and halides form, and their formation, dissolution, and transport is a major issue in heavily irrigated regions. How much these processes are controlled by nanoscale phenomena is not known.

Groundwater is constantly in touch with soil and rock, and minerals are dissolving and precipitating as it flows. The load of fine sediments in streams and groundwater can be substantial, especially during spring floods. The Missouri River is called "the Big Muddy" because of its load of particulate matter, a large fraction of which is of nanoscale dimensions. The yearly flooding of the Nile, depositing fertile soil with its large nanoparticle content, made ancient Egyptian civilization flourish. Today, one of the major concerns of our system of dams, especially in the arid western United States, is interference with the normal cycle of sediment transport and "silting up" of the lakes behind the dams. Silt is partly nanoparticles.

Contaminants and pollutants in water can be transported as aqueous ions (dimensions <0.5 nm), as molecular clusters (0.5–2 nm), as nanoparticles (2–100 nm), as larger colloids (100–1000 nm), and as macroscopic particles (>1 μm). These size range distinctions are rather arbitrary and serve to illustrate the continuity between the dissolved and the solid state. Several examples illustrate this complexity. Aluminum oxyhydroxide particles can transport transition metals such as nickel, cobalt, and zinc, seemingly as adsorbed coatings. Initially thought to be loosely bound metal complexes at the surface of the aluminum oxyhydroxide mineral grain, these are now realized to be precipitates, only a few atomic layers thick, of mixed double hydroxides of the hydrotalcite family, in which anions such as carbonate play an essential role.[10] The transport of plutonium through groundwater is a concern in old plutonium processing facilities such as the Hanford, WA atomic energy reservation, in the Nevada nuclear test site, and in the planned nuclear waste repository at Yucca Mountain, Nevada. There remain questions of permeability and the adhesion of particles to the rock and engineered barrier walls, of colloid transport, of biological transport, and of mineral precipitation which can change the rate of progress of a contamination plume. Linking laboratory scale, field scale, and simulation studies of nanoparticle transport is an essential area of research for understanding radioactive and chemical contamination and geologic processes involving uranium and other actinides.[11]

When particles are below 5 nm in size, several other effects must be considered. Whereas for larger particles, most of the atoms are in specific planes or faces, for smaller ones, an increasing number of surface atoms must sit at the intersection of facets, in presumably even higher energy sites. An alternate, more macroscopic way of describing this is to consider the surface as curved, rather than as a series of planes. Then the surface energy per unit area is no longer a constant, but potentially increases quite rapidly with decreasing particle size. This unfavorable energy may be relaxed by the adsorption of various molecules on the surface, and there is evidence that the adsorption coefficient of organics rises steeply at very small particle size.[12]

The flocculation of colloids depends on the surface charge; the pH of which the surface is neutral is the "point of zero charge."[13] Does this depend on particle size? This is an area of active research.

How do nanocrystals form from solution? The classical picture of nucleation and growth by addition of single atoms or ions is probably inadequate.[14] There is increasing evidence for clusters of atoms or ions in solution which contain 5–50 atoms and clearly show some of the structural features of the solid. An example is the Keggin-molecular cluster containing 13 aluminum atoms shown in Fig. 4.[16] It appears stable over a wide range of neutral to basic pH, and is probably a major precursor to and a dissolution product of aluminum oxyhydroxides.[17] The growth of TiO_2 anatase may occur by the oriented attachment of ~3-nm particles.[18] The growth of zeolites templated by organics may involve 3-nm cuboctahedral clusters.[19] Nanoclusters have been invoked in the growth of sulfides in ore-forming solutions.[20] Characterization of such nanoscale precursors in aqueous solutions remains a major challenge.

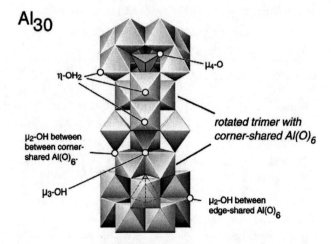

Fig. 4 $Al_2O_8Al_{28}(OH)_{56}(H_2O)_{24}^{18+}$(aq) (often called Al_{30}) cluster of 2 nm dimensions, which is intermediate in structure and properties between isolated ions and solid aluminum.
Source: W.H. Casey, personal communications. From Rowsell and Nazar.[15]

The coarsening and phase transformation of nanoscale precipitates upon heating is equally important to the sol–gel synthesis of ceramics and the geologic compaction and diagenesis of buried sediments. Loss of water, loss of surface area, and phase transformations to the stable bulk polymorph are closely interlinked.[14] A nanoparticle with dimensions below 5 nm probably cannot maintain defects or dislocations; they can migrate to the surface and be annihilated.[21] An aggregate of such single domain nanocrystals, with disorder and impurities at their surfaces, may be a common morphology in nature. Such aggregates give smaller average particle size from x-ray peak broadening than from gas adsorption [Brunauer–Emmett–Teller (BET)] measurements.

Magnetic nanoparticles cannot hold a direction of magnetization for a long time because of thermal fluctuations.[22] The magnetic iron oxides found in magnetotactic bacteria, which are single domain particles, neither too large nor too small, provide orientation in the Earth's magnetic field (see below). On a geologic time scale (millions of years), magnetization of largely nanophase iron oxides provides a record of the variation of the Earth's magnetic field through time, including periodic reversals of north and south poles. Thus the ability or inability of nanoparticle oxides to retain magnetization is of critical importance.

NANOPARTICLES IN THE ATMOSPHERE

Atmospheric particles include dust (rock and soil), sea salt, acids including sulfuric, organics (especially carbon), inorganics, and, of course, water and ice (Table 3). The atmosphere can carry particles of spherical equivalent diameters from 1 to 10^5 nm. Often, a trimodal distribution of particle sizes, with peaks in number density near 5, 50, and 300 nm, is seen.[23] The smaller particles account for most of the reactive surface area but little of the total mass.

Atmospheric particles affect the environment in many ways.[23,24] They reduce visibility (smog, haze) primarily through light scattering. They act as nuclei for water vapor condensation and cloud formation. They are involved in radiative forcing, changing the ratio of absorbed sunlight to reflected sunlight. Thus they are implicated in global climate change. Their effect on radiative forcing can be positive (more energy adsorbed) or negative (more energy reflected), leading to warming or cooling, respectively.[23,24] Their net effect is a subject of vigorous research and controversy.

Anthropogenic particles contribute disproportionately to the fine particle fractions.[23] These may have very significant effects on climate and (see below) health. Soot (carbon) from burning coal and oil and from automobile (especially diesel) emissions contributes greatly to the atmospheric load of nanoparticles.

Particles in the atmosphere travel a long way. Dust from Africa is seen in Florida; industrial emissions from China are detected in North America. Particles are removed from the atmosphere by diffusion and gravitational settling (aided by small particles coalescing into larger ones) and by rain. The residence time of nanoparticles in the atmosphere ranges from minutes to days.[23]

Atmospheric nanoparticles are more involved in gas phase reactions than particles in soil and water.[24] Their formation may involve combustion synthesis, as in industrial or automobile emission. Mineral nanoparticle surfaces may catalyze the oxidation of SO_2 and NO_2, leading to sulfuric and nitric acid. These acids can exist as gaseous species, liquids, or solid hydrates at low temperature. Nanoparticles are invoked in the depletion of atmospheric ozone by catalytic production of reactive chlorine compounds. Changes of phase (liquid to solid) are critical to the chemistry of sodium chloride and sodium nitrate particles, with their water content being controlled by available humidity.

Mineral dust particles may provide critical nutrients (e.g., iron) to the surface of the ocean far from land. The ocean's biological productivity is often limited by the availability of these nutrients; thus such inorganic nanoparticles may significantly influence the global cycling of carbon through ocean biomass.[25]

NANOPARTICLES IN SEDIMENTS, ROCKS, AND THE DEEP EARTH

The debris of rock weathering is brought down river to the ocean in sediments consisting of nanoscale particles of

Table 3 Atmospheric nanoparticles.

Liquid droplets
    Water
    Sulfuric acid
    Nitric acid
    Sea water and other salt solutions
    Organics
Solid particles
    Ice (H_2O)
    NaCl
    Na_2SO_4
    $CaSO_4 \cdot 2H_2O$
    $NaNO_3$
    $H_2SO_4 \cdot 4H_2O$
    $HNO_3 \cdot 3H_2O$
    C (graphite, amorphous, fullerenes, nanotubes)
    SiO_2
    Iron oxides
    Clays
    Organics
Many particles have core–shell structures and coatings

clay, small quartz grains, and other minerals. Indeed, the terms "clay" and "silt" have a classic connotation of size fraction, although the former also implies a structural group of minerals, the layered aluminosilicates. In the ocean, carbonates precipitate, dissolve, and reprecipitate as a complex function of depth.[26] Both silica and various polymorphs of calcium carbonate (calcite, aragonite, vaterite) are produced by organisms such as diatoms, foraminifera, and corals. Their debris rains down on the ocean bottom, forming sediments which often show annual cycles in composition and texture and which bear records of climate change, shifts in ecosystems, and catastrophic events such as meteor impacts.[27] These sediments start off largely nanoscale. They coarsen and dehydrate with time and depth of burial. The evolution of their organic matter leads to petroleum. The evolution of their minerals, involving coarsening and compaction, called diagenesis, leads to rocks such as limestones and shales. The nanoscale processes that take place (dehydration and organic loss, phase transformation, coarsening and densification) are natural analogs of ceramic processing which starts with nanoscale precipitates or gels.

Natural processes involving changes in temperature, pressure, acidity, and oxygen fugacity cause the concentration of trace metals into ore deposits. These often occur in hydrothermal systems, spatially contained circulations of hot, pressurized, metal-rich aqueous solutions. Our ability to mine low-grade deposits by chemical leaching techniques brings us into the world of nanoparticles and reactions at mineral surfaces. There is increasing evidence that microorganisms play an active role in ore deposition.[28,29] Hot springs at the surface produce deposits of nanoscale amorphous silica and other minerals, which may also be closely linked to microbial activity.[30]

At temperatures above a few hundred degrees Centigrade and pressures above a few kilobars, coarse-grained metamorphic and igneous rocks predominate. The interior of the Earth is layered, with seismic discontinuities delineating the crust, upper mantle, transition zone, lower mantle, and core. These discontinuities represent regions of rapidly changing density, mineralogy, and chemistry.[31] Ongoing phase transitions and chemical reactions can decrease the grain size of a material and render it easier to deform.[32] Thus nanoscale phenomena, occurring at specific locations, may play a disproportionate role in processes such as subduction, plate tectonics, earthquake generation, and volcanism. Shock processes, (e.g., meteor impact, nuclear detonation) also produce nanoparticles.

When a volcano erupts explosively, a plume of dust particle is sent into the atmosphere, sometimes reaching the stratosphere. These particles make beautiful sunsets but they also exert a significant cooling effect on climate for several years and pose a significant aviation hazard. Combining sedimentation, coarsening, subduction, volcanism, and weathering, there is an ongoing global geochemical cycle of nanoparticles, analogous in some ways to global geochemical cycles of elements such as carbon. However, the mass balances, or imbalances, in global nanoparticle production and consumption through time have not been characterized.

NANOPARTICLES BEYOND THE EARTH

In the early stages of planet formation, dilute and more or less uniform gas condensed to form a series of mineral particles, with the order of condensation described by thermodynamic calculations based on the volatility and stability of these phases.[37] The more refractory oxides condensed earlier than those with higher volatility. These particles accreted, under the influence of gravity, to form our solar system. What was the nature of these initial particles? What was their size distribution? Were they crystalline or amorphous? Were metastable polymorphs formed? While the initial high temperatures might argue against such metastability, the low pressures and condensation from a vapor argue for it. In technological processes, chemical vapor deposition produces nanoscale amorphous silica "snow," and combustion produces soot and inorganic nanoparticles. The role of nanoparticles in planetary accretion has not yet been explored. The change in stability at the nanoscale, which will be different for various compositions and polymorphs, may alter the sequence of condensation of phases. Are the particles now present in space as interplanetary dust partly or mostly nanoparticles?

The surfaces of the Moon and Mars, subject to "space weathering" by bombardment with meteorites of all sizes, contain an extensive fine grained dust or soil layer.[32] Samples of lunar soil, brought back by the Apollo missions, contain a distribution of particle sizes of spherules and irregular shards. Their particle size distribution appears not to have been a subject of active interest, but clearly a significant number are in the nanoregime. The red surface of Mars appears to be dominated by various fine-grained or nanophase iron oxides. Until Martian sample return missions, planned to occur in the next decade or two, bring some of this material to Earth, we must rely on remote sensing technology (spectroscopic techniques) and instrumentation on Martian landers (possibly Mossbauer spectroscopy, x-ray fluorescence, and x-ray diffraction) to obtain information on the composition and structure of Martian soil. Considering the difficulty of characterizing iron oxide nanoparticles in the best laboratories on Earth, definitive conclusions about the nature of Martian soil are unlikely until we have some samples in hand. Meteorites believed to be from Mars contain micron-sized spherules, which were proposed to be biological in origin. This sparked much recent controversy and it is by no means settled whether these structures are fossil microorganisms or the product of inorganic nanoscale crystal growth processes.[33,34]

NANOPARTICLES AND LIFE

Microbial communities are rich in the production and utilization of nanoparticles.[35] Table 4 lists some examples.

Table 4 Example of interactions of microorganisms and nanoparticles.

Class of organisms	Example	Nanoparticle interaction
Iron and manganese oxidizing bacteria	*Thiobacillus*	Oxidize soluble Mn^{2+} and Fe^{2+} to insoluble higher oxides
Iron and manganese reducing bacteria	*Shewenella*	Reduce insoluble Mn and Fe oxides to soluble forms oxidize sulfide to sulfate to sulfide precipitate
Sulfur reducing bacteria	*Thiobacillus*	Reduce sulfate to sulfur in sulfide
Magnetotactic bacteria	*Aquaspirillum magnetotacticum*	Nanoparticles of Fe_2O_3, Fe_3O_4, and/or iron sulfides
Uranium reducing bacteria	*Geobacter, shewenella*	Soluble $U^{6+} \rightarrow$ insoluble U^{4+}
Fungi	Specific strains unknown	Oxidize Mn^{2+}, precipitate MnO_2
Diatoms	Various	Precipitate silica
Foraminifera	Various	Precipitate $CaCO_3$ calcite and aragonite

In addition to aerobic respiration (the enzymatic oxidation of carbohydrates and other organics with molecular oxygen to produce water, carbon dioxide, and energy stored as high-energy phosphate linkages) organisms use many other strategies to extract energy from the environment. The following biological reactions produce or consume nanoparticles. Dissolved Mn(II) or Fe(II) can be oxidized by oxygen, producing Mn(III), Mn(IV), or Fe(III) oxide nanoparticles, while organics can be oxidized by manganese or iron oxides, producing soluble Mn(II) and Fe(II) species. Some bacteria can also utilize the U(IV)–U(VI) couple as an energy source. Because hexavalent uranium is much more soluble than tetravalent, biological processes that accelerate its production are of concern in modeling nuclear waste leaching. The sum of these two groups of redox processes is the oxidation of organics by oxygen, akin to respiration. The important difference is that the organic food source and the oxygen source can be spatially separated in the sharp gradients in oxygen and organic contents that frequently occur in sediments, and different communities of organisms participate in the two processes. In marine sediments, sulfate is the dominant biological electron acceptor and is more important than oxygen. Bacterial sulfate reduction produces sulfide which often precipitates as nanophase metal sulfide minerals. Sulfide and sulfur oxidizing bacteria typically live in specialized environments where there is enough oxygen to oxidize sulfur but not so much that chemical oxidation swamps biological oxidation. This oxidation consumes solid sulfur and sulfides, and produces soluble sulfate.

Organisms utilize nanoparticles in processes other than respiration. Bacterial precipitation of sulfide minerals, e.g., ZnS and UO_2, may also be a mechanism of detoxification.[36] Similar detox processes may occur in plants. Magnetotactic bacteria synthesize and align single domain magnetic iron oxide and iron sulfide particles in structures called magnetosomes.[37] Such bacteria align themselves both north–south and vertically in the Earth's magnetic field. The navigational (homing) capabilities of bees, pigeons, and probably other higher organisms utilize magnetic field orientation sensed by magnetic iron oxide particles in their brains. Similar particles, although at lower abundance, occur in many mammals, including *Homo sapiens*.[38] There has been a debate in the public sector whether the magnetic fields produced by high-voltage power lines are potentially dangerous to human health. In contrast, the use of magnets in alternative medicine, and the market for magnetic pillows, back supports, etc., suggests, or at least hopes for, a beneficial effect of the interaction of magnetic fields with animals. Key to either harmful or helpful biological effects is a mechanism for the magnetic field to interact with living cells. Interaction with biological magnetic nanoparticles may provide such a mechanism, but very little is known at present.

Nanoparticles have other documented health effects.[39] When inhaled into the lungs, particles cause an inflammatory response, which contributes to allergies, asthma, and cancer.[48] The detailed mechanism of this response, and how it depends on surface area, particle size, or specific particle chemistry, is not clear. The harmful effects of inhaled particles may be enhanced by other pollutants, particularly ozone, typically present in smog. Nanoparticles penetrate deep into the lungs. Many are returned with exhaled air, some stick to the surfaces of the alveoli, and some may even penetrate into general blood circulation and be transported to other organs. Studies linking detailed nanoparticle characterization, biochemical and physiological processes, and health effects are just beginning to be carried out. It is likely that not all particles have comparable effects, and understanding which are the most dangerous could lead to rational, rather than arbitrary, emission standards for automotive and industrial particulates.

In the early Earth, prebiotic processes culminated in the origin of life.[40] Because the synthesis of complex organic molecules competes with their destruction by hydrolysis and other degradation, it is possible that the most successful synthesis could have occurred in sheltered and catalytic environments, such as those provided by mineral surfaces, nanoparticle surfaces, and pores within mineral grains. Present-day organisms utilize a wide variety of elements (e.g., Fe, Co, Ni, Cr, Zn, Se) in specific enzymes. Although large amounts of such elements are toxic, trace amounts are essential. The active centers in enzymes utilizing these trace elements often consist of clusters of metal atoms, sometimes associated with sulfide. Are these fine-tuned by evolution from earlier simpler metal clusters and nanoparticles existing in the environment? Thus nanoparticles may play a role not just in the sustenance of life but in its origin.

CONCLUSIONS

Nanoparticles play diverse roles in the environment and are involved in both abiotic and biologically mediated chemical and physical processes. Their high surface area, chemical reactivity, polymorphism, and unique properties involve nanoparticles in a disproportionately large fraction of the chemical reactions occurring on and in the Earth and other planets. Understanding this involvement is itself evolving into a new field of study in the environmental and Earth sciences, which is beginning to be called "nanogeoscience." Nanogeoscience will take its place alongside other new areas such as astrobiology and biogeochemitry, fields that link physical, chemical, and biological processes viewed in the context of the long time and distance scales natural to the Geosciences.

REFERENCES

1. Navrotsky, A. Thermochemistry of Nanomaterials. In *Nanoparticles and the Environment*; Reviews in Mineralogy and Geochemistry; Banfield, J.F., Navrotsky, A., Eds.; Mineralogical Society of America: Washington, DC, 2001; Vol. 44, 73–103.
2. Garvie, R.C. The occurrence of metastable tetragonal zirconia as a crystallite size effect. J. Phys. Chem. **1965**, *69*, 1238–1243.
3. McHale, J.M.; Auroux, A.; Perrotta, A.J.; Navrotsky, A. Surface energies and thermodynamic phase stability in nanocrystalline alumina. Science **1997**, *277*, 788–791.
4. Ranade, M.R.; Navrotsky, A.; Zhang, H.Z.; Banfield, J.F.; Elder, S.H.; Zaban, A.; Borse, P.H.; Kulkarni, S.K.; Doran, G.S.; Whitfield, H.J. Energetics of nanocrystalline TiO_2. Proc. Natl. Acad. Sci. **2002**, *99* (Suppl. 2), 6476–6481.
5. Majzlan, J.; Navrotsky, A.; Casey, W.H. Surface enthalpy of boehmite. Clays Clay Miner. **2000**, *48*, 699–707.
6. Majzlan, J. Ph.D. Thesis; University of California at Davis, 2002.
7. Janney, D.E.; Cowley, J.M.; Buseck, P.R. Structure of synthetic 2-line ferrihydrite by electron nanodiffraction. Am. Mineral. **2002**, *85* (9), 1180–1187.
8. Schwertmann, U.; Cornell, R.M. *Iron Oxides in the Laboratory*, 2nd Ed.; Wiley-VCH, 2000; 188 pp.
9. Singer, M.J.; Munns, D.N. *Soils: An Introduction*, 3rd Ed.; Simon and Schuster Company: Upper Saddle River, NJ, 1991.
10. Thompson, H.A.; Parks, G.A.; Brown, G.E., Jr. Ambient-temperature synthesis, evolution, and characterization of cobalt–aluminum hydrotalcite-like solids. Clays Clay Mater. **1999**, *47*, 425–438.
11. Ragnarsdottir, K.V.; Charlet, L. Uranium behavior in natural environments. Environ. Mineral. **2000**, *9*, 245–289.
12. Zhang, H.; Penn, R.L.; Hamers, R.J.; Banfield, J.F. Enhanced adsorption of molecules on surfaces of nanocrystalline particles. J. Phys. Chem. B **1999**, *103*, 4656–4662.
13. Hunter, R.J. *Foundations of Colloid Science*; Oxford University Press: New York, 1993; Vol. 1.
14. Banfield, J.F.; Zhang, H. Nanoparticles in the Environment. In *Nanoparticles and the Environment*; Reviews in Mineralogy and Geochemistry; Banfield, J.F., Navrotsky, A., Eds.; Mineralogical Society of America: Washington, DC, 2001; Vol. 44, 2–58.
15. Rowsell, J.; Nazar, L.F. Speciation and thermal transformation in alumina sols: Structures of the polyhydroxyoxoaluminum cluster $[Al_{30}O_8(OH)_{56}(H_2O)_{26}]^{18+}$ and its Keggin moeté. J. Am. Chem. Soc. **2000**, *122*, 3777–3778.
16. Casey, W.H.; Phillips, B.L.; Furrer, F. Aqueous Aluminum Polynuclear Complexes and Nanoclusters: A Review. In *Nanoparticles and the Environment*; Reviews in Mineralogy and Geochemistry; Banfield, J.F., Navrotsky, A., Eds.; Mineralogical Society of America: Washington, DC, 2001; Vol. 44, 167–190.
17. Furrer, G.F.; Phillips, B.L.; Ulrich, K.U.; Poethig, R.; Casey, W.H. The origin of aluminum flocs in polluted streams. Science **2002**, in press.
18. Penn, R.L.; Banfield, J.F. Oriented attachment and growth, twinning, polytypism, and formation of metastable phases: Insights from nanocrystalline TiO_2. Am. Mineral. **1998**, *83*, 1077–1082.
19. de Moor, P.P.E.A.; Beelen, T.P.M.; Komanschek, B.U.; Beck, L.W.; Wagner, P.; Davis, M.E.; Van Santen, R.A. Imaging the assembly process of the organic-mediated synthesis of a zeolite. Chem. Eur. J. **1995**, *5*, 2083–2088.
20. Luther, G.W.; Theberge, S.M.; Richard, D.T. Evidence for aqueous clusters as intermediates during zinc sulfide formation. Geochim. Cosmochim. Acta **1999**, *64*, 579.
21. Jacobs, K.; Alivisatos, A.P. Nanocrystals as model systems for pressure-induced structural phase transitions. In *Nanoparticles and the Environment*; Reviews in Mineralogy and Geochemistry; Banfield, J.F., Navrotsky, A., Eds.; Mineralogical Society of America: Washington, DC, 2001; Vol. 44, 59–104.
22. Rancourt, D.G. Magnetism of Earth, Planetary, and Environmental Nanomaterials. In *Nanoparticles and the Environment*; Reviews in Mineralogy and Geochemistry; Banfield, J.F., Navrotsky, A., Eds.; Mineralogical Society of America: Washington, DC, 2001; Vol. 44, 217–292.
23. Anastasio, C.; Martin, S.T. Atmospheric Nanoparticles. In *Nanoparticles and the Environment*; Reviews in Mineralogy and Geochemistry; Banfield, J.F., Navrotsky, A., Eds.; Mineralogical Society of America: Washington, DC, 2001; Vol. 44, 293–349.
24. Ramanathan, V.; Crutzen, P.J.; Kiehl, J.T.; Rosenfeld, D. Aerosols, climate, and the hydrological cycle. Science **2001**, *294*, 2119–2124.
25. Martin, J.H.; Coale, K.H.; Johnson, K.S.; Fitzwater, S.E.; Gordon, R.M.; Tanner, S.J.; Hunter, C.N.; Elrod, V.A.; Nowicki, J.L.; Coley, T.L.; Barber, R.T.; Lindley, S.; Watson, A.J.; Vanscoy, K.; Law, C.S.; Liddicoat, M.I.; Ling, R.; Stanton, T.; Stockel, J.; Collins, C.; Anderson, A.; Bidigare, R.; Ondrusek, M.; Latasa, M.; Millero, F.J.; Lee, K.

Testing the iron hypothesis in ecosystems of the equatorial Pacific Ocean. Nature **1994**, *371*, 123–129.
26. Millero, F.J. *Chemical Oceanography*, 2nd Ed.; CRC: Boca Raton, FL, 1996.
27. Broecker, W.S. *The Great Ocean Conveyor*; AIP Conference Proceedings; Columbia University Palisades: New York, 1992; Vol. 347, 129–161.
28. Labrenz, M.; Druschel, G.K.; Thomsen-Ebert, T.; Gilbert, B.; Welch, S.A.; Kemner, K.M.; Logan, G.A.; Summons, R.E.; de Stasio, G.; Bond, P.L.; Lai, B.; Kelly, S.D.; Banfield, J.F. Formation of sphalerite (ZnS) deposits in natural biofilms of sulfate-reducing bacteria. Science **2000**, *290*, 1744–1745.
29. Ehrlich, H.L. Microbes as geologic agents: Their role in mineral formation. Geomicrobiol. J. **1999**, *16*, 135–153.
30. Konhauser, K.O.; Phoenix, V.R.; Bottrell, S.H.; Adams, D.G.; Head, I.M. Microbial–silica interactions in Icelandic hot spring sinter: Possible analogues for some Precambrian siliceous stromatolites. Sedimentology **2001**, *48*, 415–433.
31. Anderson, D.L. *Theory of the Earth*; Blackwell Scientific Publications: Brookline Village, MA, 1989.
32. Karato, S.; Li, P. Diffusion creep in perovskite: Implications for the rheology of the lower mantle. Science **1992**, *255*, 1238–1240.
33. Sasaki, S.; Nakamura, K.; Hamabe, Y.; Kurahashi, E.; Hiroi, T. Production of iron nanoparticles by laser irradiation in a simulation of lunar-like space weathering. Nature **2001**, *410*, 555–557.
34. Buseck, P.R.; Dunin-Borkowski, R.E.; Devouard, B.; Frankel, R.B.; McCartney, M.R.; Midgley, P.A.; Posfai, M.; Weyland, M. Magnetite morphology and life on Mars. Proc. Natl. Acad. Sci. U. S. A. **2001**, *98* (24), 13490–13495.
35. Gibson, E.K.; McKay, D.S.; Thomas-Keprta, K.L.; Wentworth, S.J.; Westall, F.; Steele, A.; Romanek, C.S.; Bell, M.S.; Toporski, J. Life on Mars: Evaluation of the evidence within Martian meteorites ALH84001, Nakhla, and Shergotty. Precambrian Res. **2001**, *106* (1–2), 15–34.
36. Nealson, K.H.; Stahl, D.A. Microorgansims and Biogeochemical Cycles: What Can We Learn from Layered Microbial Communities? In *Geomicrobiology: Interaction Between Microbes and Minerals*; Review in Mineralogy; Banfield, J.F., Nealson, K.H., Eds.; Mineralogical Society of America: Washington, DC, 1997; Vol. 35, 5–34.
37. Suzuki, Y.; Banfield, J.F. Geomicrobiology of Uranium. In. *Uranium: Mineralogy, Geochemistry and the Environment*; Review in Mineralogy; Burns, P.C., Finch, R., Eds.; Mineralogical Society of America: Washington, DC, 1999; Vol. 38, 388–432.
38. Stolz, J.F. Magnetotactic Bacteria: Biomineralization, Ecology, Sediment Magnetism, Environmental Indicator. In *Biomineralization Process of Iron and Manganese: Modern and Ancient Environments*; Catena Supplement 21; Skinner, H.C.W., Fitzpatrick, R.W., Eds.; Destedt. Germany, 1992; 133–146.
39. Kirschvink, J.L.; Walker, M.M.; Diebel, C.E. Magnetite-based magnetoreception. Curr. Opin. Neurobiol. **2001**, *11* (4), 462–467.
40. Guthrie, G.D.; Mossman, B.T. *Health Effects of Mineral Dusts*; Reviews in Mineralogy; Mineralogical Society of America: Chelsca, MI, 1993; Vol. 28.
41. Navrotsky, A. Nanomaterials in the environment, agriculture, and technology (NEAT). J. Nanopart. Res. **2000**, *2*, 321–323.
42. Pitcher, M.; Navrotsky, A. unpublished data.

BIBLIOGRAPHY

1. Nakashima, S.; Ikoma, M.; Shiota, D.; Nakazawa, K.; Maruyama, S. Geochemistry and the origin and evolution of life: A tentative summary and future perspectives. Precursors Chall. Investig. Ser. **2001**, *2*, 329–344.

Nanoparticles: Uncertainty Risk Analysis

Khara D. Grieger
Steffen Foss Hansen
Anders Baun
Department of Environmental Engineering, Technical University of Denmark, Kongens Lyngby, Denmark

Abstract
Scientific uncertainty plays a major role in assessing the potential environmental risks of nanoparticles. Moreover, there is uncertainty within fundamental data and information regarding the potential environmental and health risks of nanoparticles, hampering risk assessments based on standard approaches. To date, there have been a number of different approaches to assess uncertainty of environmental risks in general, and some have also been proposed in the case of nanoparticles and nanomaterials. In recent years, others have also proposed that broader assessments of uncertainty are also needed in order to handle the complex potential risks of nanoparticles, including more descriptive characterizations of uncertainty. Some of these approaches are presented and discussed herein, in which the potential strengths and limitations of these approaches are identified along with further challenges for assessing uncertainty pertaining to the potential environmental risks of nanoparticles. Currently, international research efforts are underway not only to assess these uncertainties but also to handle the embedded uncertainties within assessing the potential environmental risks of nanoparticles. However, it is clear that further research efforts are needed to sufficiently handle the extensive uncertainties associated with nanoparticle risks, given the diversity of materials, pace of innovation, and various environmental parameters to consider.

INTRODUCTION

As new technologies and new materials are developed and enter the marketplace, uncertainty in terms of their potential environmental risks is inherent in the process.[1] Only through time can scientists, engineers, and decision makers more comprehensively understand the environmental risks of,e.g., new chemicals or materials to a variety of organisms. The use and development of engineered nanoparticles in a variety of consumer products and other applications is no exception to this, as significant uncertainty currently exists within understanding their environmental (and health) risks in addition to uncertainty in even defining exactly what is "nanotechnology" and "nanoparticles"—a subject of ongoing debate.[2-4] For example, a report by the Scientific Committee on Emerging and Newly Identified Health Risks concluded in 2009 that "…nanotechnology has introduced new nanoparticulate forms of chemicals, of which properties, behavior and effects are largely unknown, and, hence, of concern".[5] As understanding and assessing the environmental risks of nanomaterials, defined herein as a "material having one or more externaldimensions in the nanoscale or which is nanostructured,"[6] including nanoparticles (defined as having three external dimensions on the nanoscale[7]), remain difficult yet extremely important tasks for scientists, governments, organizations, and decisionmakers, it is of utmost importance to more fully understand uncertainty in this field as well as approaches to handle it. This entry therefore aims to explore these aspects, specifically within the field of environmental risks of nanoparticles.

Defining Uncertainty

To begin, uncertainty is generally defined as "the state of being uncertain; doubt; hesitancy" (http://www.dictionary.reference.com) or "when something is not known, or something that is not known or certain" (http://dictionary.cambridge.org). There are other synonyms of uncertainty particularly within science and research that are also frequently used, such as knowledge gaps, ambiguities, "unknowns," and research needs. Similarly, uncertainty in the present entry refers to "any departure from the unachievable ideal of complete determinism",[8] similar to other proposed definitions.[1,9] In fact, there have been a wide range of models proposed that aim to describe and characterize uncertainty within science and environmental risk challenges, through for instance descriptive information regarding uncertainty and its various forms. These models have been based on either quantitative[10-12] or qualitative methods to characterize uncertainty.[8,9,13-16] While there have been a number of different models proposed to describe or characterize uncertainty, many of which have their particular advantages and limitations),[12] we have chosen to focus on the framework proposed by Walker et al.,[8] which relies on characterizing uncertainty according to its "location," "nature," and "level," as described in more detail in subsequent sections below.

In addition to various ways to describe uncertainty within science, there have also been a number of methods proposed to analyze uncertainty as a means of further describing uncertainty either quantitatively or qualitatively, and that may or may not be used in subsequent analyses, such as, e.g., Monte Carlo uncertainty analyses. In fact, quantitative uncertainty analyses such as Monte Carlo[10] and sensitivity analysis[11] are two of the most commonly used methods for analyzing uncertainty within environmental risk challenges. In addition to these, there are also a number of methods and strategies to handle uncertainty in order to cope or manage uncertainty in different contexts, such as various risk management strategies or stakeholder involvement.[1] For example, some have proposed that the ability for a technology to be reversed after introduction may be one way to handle uncertainty related to an environmental risk challenge.[17] These approaches to characterize, analyze, and handle uncertainty are further described in subsequent sections. In this entry, we aim to briefly explore both traditional, quantitative approaches and the broader, qualitative methods used to assess uncertainty within complex environmental challenges, and their applicability to environmental risks of nanoparticles.

UNCERTAINTY WITHIN ENVIRONMENTAL RISKS OF NANOPARTICLES

Importance of Understanding Uncertainty

The presence of uncertainty within environmental risks, particularly those that are complex in nature, is often a major challenge for a variety of stakeholders, including scientists, industry, regulators, and decision makers. For example, it is generally desired to avoid both false positives (i.e., falsely considering something as harmful when in fact it is harmless) as well as false negatives (i.e., falsely considering something harmless when it is in fact harmful).[18] In cases of false positives, there may be unrealized profits or societal benefits from not allowing the use of a particular product or substance, while in the case of false negatives, unanticipated adverse effects to human health or the environment may result from using a harmful product/substance when it was considered to be benign.[19] Therefore, there are often funding calls for research in order to better understand the consequences of using, e.g., a chemical or substance and hence reduce the associated uncertainty. This is also seen in the case of understanding the environmental, health, and safety (EHS) risks of nanomaterials, whereby, e.g., the U.S. National Nanotechnology Initiative has increased its total EHS research funding since 2007[20] and similar patterns are seen in Europe.[21]

Uncertainty is also important to consider in the case of engineered nanoparticles since it may be one of the main parameters that could limit nanotechnology (NT) development.[22] For example, Maynard[23] concluded that "…anyone who wants to see NT achieve its full potential has a vested interest in the establishment of a comprehensive NT risk research program. A thorough and open exploration of NT's potential threats to humans and the environment is the best way to keep concerns about NT rooted in objective, scientific review." In essence, the tasks of further understanding uncertainty within the potential environmental risks of nanomaterials as well as establishing approaches to handle this are extremely important and challenging, not only for scientists but also for industry and governments.

These challenges are furthermore heightened by the fact that environmental exposures to nanoparticles have most likely been increasing particularly within the last decade given the growth of consumer products and other applications that use nanoparticles,[24] including those aimed for environmental remediation.[25] For example, there are more than 1000 "manufacturer-identified nanotechnology-based consumer products" currently on the market in applications that range from electronic and automotive applications to cosmetics and food packing.[26,27] Engineered nanoparticles are also used in a number of environmental remediation treatments, including, for instance, zero-valent iron nanoparticles to clean up contaminated soil and groundwater.[28] Other nanoparticles that are also under development for environmental remediation include nanoscale zeolites, carbon nanotubes and fibers, enzymes, noble metals, and titanium dioxide.[25] For the most part, the environmental risks of these nanoparticles used in environmental remediation are generally unknown, although research efforts to explore this area have received growing attention in recent years.[28–31]

Approaches to Assessing Uncertainty

Standard approaches to understanding and assessing the potential environmental risks of an activity or exposure to, e.g., chemicals or substances, usually includes performing environmental risk assessments,[32] environmental impact assessments,[33] or similar approaches (i.e., Comprehensive Environment Assessment[34]; Integrated Environmental Health Impact Assessment[35]). For nanomaterials, the established (environmental) risk assessment framework that has been developed for chemicals has been proposed and has been the starting point thus far.[5,36] Briefly, the chemical risk assessment framework is designed to identify the potential adverse consequences of exposure to a, e.g., substance or chemical (i.e., effect); estimate the probability of these consequences (i.e., exposure); and subsequently combine these probabilities into a single numerical expression.[37] These probabilities are estimated by using a number of different methods, including the collection of statistical data and/or experimental studies.

Uncertainties in chemical or probabilistic risk assessments are for the most part dealt with through the identification of known knowledge gaps and/or quantitative attempts to represent uncertainty through, e.g., application of uncertainty factors. These approaches may be particularly useful in well-understood or controlled situations (e.g., rule-based systems) or highly repetitive events.[38] Furthermore, there are a number of different methods to separately analyze uncertainty outside of, or in addition to, risk assessment frameworks. These include quantitative methods such as Monte Carlo analysis, sensitivity analysis, data uncertainty engine, inverse modeling (parameter estimation or predictive uncertainty), as well as qualitative methods such as scenario analysis, expert elicitation, and stakeholder involvement (Refsgaard[12] and references within). In general, most of these uncertainty analyses are performed separately or subsequent to modeling or analysis rather than integrated through the modeling or analysis work.[12]

In addition, the European Chemicals Agency proposes three different approaches to performing an uncertainty analysis in chemical safety assessments.[39] This includes a qualitative uncertainty analysis (Level 1), deterministic uncertainty analysis (Level 2), and a probabilistic uncertainty analysis (Level 3).

For engineered nanoparticles, however, risk assessments have largely been absent thus far owing to extensive uncertainties and the lack of necessary data. To the authors' knowledge, only a handful of preliminary risk assessments have been performed thus far on a select number of nanoparticles, including fullerene C_{60}, carbon nanotubes, metal nanoparticles, and metal oxide nanoparticles.[40-42] In these preliminary assessments, uncertainty was treated either with the application of uncertainty factors (in Shinohara[40]) or with scenario analysis (i.e., realistic and worst-case scenarios in Stone et al.[42]).

Challenges within Assessing Uncertainty

Standard approaches to understanding and assessing scientific uncertainty, including those mentioned in the previous section, have also been criticized in recent years by a number of scientists and other authors. These criticisms have largely been centered around the fact that (environmental) risk assessments are mainly applicable to simple systems and do not handle the extensive, complex uncertainties present in many environmental and health challenges, especially those that are relevant for policy.[8,9,43] One of these criticisms is the oversimplification of the system(s) involved and reduction of the "true" state of uncertainty to (oversimplified) parameters that attempt to represent uncertainty. For instance, according to Martuzzi and Tickner,[44] "risk assessors often fail to distinguish among various kinds of uncertainty and tend to misclassify some model and fundamental uncertainty as statistical, to which they apply 'uncertainty' factors. When model and fundamental uncertainty predominate in a system, this approach becomes more likely to result in large errors in estimates of risk, failure to predict adverse effects removed in time and space, and complete failure to predict surprises or novel effects... Risk assessment can be a powerful tool if uncertainties are familiar and well understood, but ... many uncertainties are much more challenging than this. It is far from scientific to pretend otherwise." Other scientists and authors have expressed similar concerns about the traditional treatment of uncertainty in complex policy-related science,[45,46] including The National Academies of Science[47] in which they concluded that "uncertainty, an inherent property of scientific information, continues to lead to multiple interpretations and contribute to decision-making gridlock."

In relation to nanotechnology, Wallace[48] has claimed, "... the 20th-century emphasis on a classically linear, expert-driven, 'black box' approach to technological risk assessment and management is ill-suited to the unique nature and scale of 21st-century technologies—owing chiefly to the fact the relevant calculus for megasciences like nanotechnologies is complicated by a good deal of ambiguity and uncertainty that renders comprehensive risk quantification largely impossible." Similarly, others have also cited the challenges of applying standard (quantitative) uncertainty analysis tools to nanomaterials, including probabilistic modeling, predictive structure–activity analysis, etc.[49] These challenges may be further amplified given the fact that the definition of an engineered nanoparticle is still subject to intense debate,[4] and there are extreme difficulties in detecting and measuring them especially in environmental matrices.[50] Hence, there are great challenges not only in understanding and assessing the uncertainties in the potential environmental risks of nanoparticles, including which methods may be best for these assessments, but also in regards to problem definitions or framing.[51]

Alternative Approaches to Assessing or Handling Uncertainty

As a response to these aforementioned challenges in dealing with uncertainty in complex environmental issues in general, there have been many attempts to more comprehensively describe and characterize the different types of scientific uncertainties within a given environmental risk. Some more traditional tools, such as Monte Carlo uncertainty analysis (e.g., Poulter[10]) and sensitivity analysis,[11] are among the well-developed quantitative methods frequently proposed to analyze uncertainty in complex environmental risk challenges. In fact, these methods have been recently used in a modeling study that estimated predicted environmental concentrations of three nanoparticles in Switzerland.[52] In addition to performing uncertainty and sensitivity analyses of their modeling approaches, the authors also handled the extensive uncertainty in their estimates of nanoparticles' environmental concentrations by treating all parameters within a probabilistic material flow analysis model as probability distributions. However, it is also recognized that there may also be more fundamental types of uncertainties in this study thatare beyond statistical bounds, given the extensive data gaps used in this modeling analysis.

At the same time, others have argued that perhaps some qualitative methods may be more robust for complex environmental risks that have particularly high degrees of uncertainty, such as unknown interrelated and/or interdependent factors or the presence of ignorance (e.g., Funtowicz and Ravetz[9]) including the use of, e.g., expert elicitation,[53] stakeholder involvement,[54] and Numeral, Unit, Spread, Assessment and Pedigree (NUSAP).[9,14] These methods are designed to not necessarily analyze the uncertainty itself but rather to serve as approaches to handle uncertainty within health or environmental risk issues. In fact, some of these methods have also been used in the case of investigating the potential health and environmental risks of nanoparticles. For example, Morgan[53] used expert elicitation to develop a variety of influence diagrams, and there have been various stakeholder involvement projects initiated that focus on different aspects of nanomaterial risks.[54,55] In addition to these approaches, the International Risk Governance Council (IRGC) has also proposed that broader risk governance frameworks may be needed in the case of nanotechnology and the use of nanoparticles in food and cosmetics in order to handle extensive uncertainties.[56,57] The risk governance framework proposed by the IRGC relies heavily on communication among a range of stakeholders as well as reliance on a number of societal factors such as risk acceptance. In fact, the inclusion of stakeholder and citizen/public involvement have been frequently cited as important for inclusion in successful strategies to handle potential health and environmental risks of emerging technologies, such as nanotechnology, in which uncertainty is pervasive.[46,47,58–60]

Another approach that has been proposed in recent years to assess or handle uncertainty within complex environmental risks is to obtain a better understanding of the uncertainty itself in its many forms. Among other benefits, a qualitative characterization of uncertainty has been proposed to not only help increase understanding and transparency within the risk characterization process[61] but may also be important in democratic societies or deliberative processes and help quality control aspects.[62] Furthermore, a better understanding of the different types of uncertainties and the areas in which they are located may also help prioritize research efforts. These approaches been used to assess and understand the uncertainty within complex environmental risks of new technologies or materials, including genetically modified crops or effects from anthropogenic climate change.[16,63]

One qualitative, conceptual framework that handles uncertainty in model-based decision support activities in

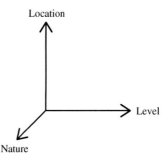

Fig. 1 Three dimensions of uncertainty (location, level, and nature) according to the Walker and Harremöes framework (2003).
Source: Krayer von Krauss.[62]

particular, including environmental risk assessment, was developed by Walker et al.,[8] termed the Walker and Harremöes framework. This framework was also applied by Grieger et al.[51] to characterize the uncertainty within health, environmental, and safety risks of nanoparticles as further detailed in subsequent sections. According to the Walker and Harremöes framework, there are three main dimensions that describe uncertainty: location, nature, and level (Fig. 1). "Location" (of uncertainty) describes where in the model uncertainties exist, such as identified knowledge gaps, while "level" describes the state of knowledge of degree of uncertainty ranging from "known unknowns" to "unknown unknowns" or ignorance. "Nature" describes the state of uncertainty according to if it may be reduced through further research (termed epistemic uncertainty) or if stochasticity is present (stochastic uncertainty), and further research will not likely reduce this. Finally, these three dimensions are combined in a matrix for better handling.

Characterizing Uncertainty within Environmental Risks of Nanoparticles

To the authors' knowledge, there has only been one study to date that has attempted to analyze or characterize the uncertainty within the field of environmental risks of nanoparticles (i.e., Grieger et al.[51]), which is discussed in the following paragraph, although a number of other studies and reports have listed some of the main areas of uncertainty or data gaps within the field. For example, the European Commission's Scientific Committee on Emerging and Newly Identified Health Risks (SCENIHR) highlighted the following as main areas of uncertainty within understanding the environmental risks of nanoparticles: environmental fate, behavior, and mobility; degradation, persistency, and bioaccumulation; and adverse effects to a variety of organisms.[5,64] Other important areas of uncertainty in this field include the following: 1) the persistence of nanoparticles in the atmosphere, which will depend on rates of agglomeration and deagglomeration, and on degradation; 2) the relevance of routes of exposure; 3) the metrics used for exposure measurements; 4) the mechanisms of translocation to different parts of the body and the possibility of degradation after nanoparticles enter the body; 5) the mechanisms of toxicity of nanoparticles; and 6) the phenomenon of transfer between various environmental media.[64] In addition to these data gaps, there are also a number of key research questions pertaining to the testing methods and equipment as well as the best metrics to use when conducting tests.[65] These findings have also been supported in a recent comprehensive review of the EHS data of nanoparticles,[42] while similar reviews of known knowledge gaps of the environmental risks of nanoparticles have also been conducted by a variety of scientists and organizations.[23,58,65–67]

In addition to these aforementioned areas of uncertainty, Grieger et al.[51] characterized the uncertainty within EHS risks of nanomaterials through the application of the Walker and Harremöes framework. This analysis attempted to describe and characterize the main areas of uncertainty in the field, as described in the subsequent paragraphs, given that the presence of uncertainty was a serious obstacle in not only assessing the potential risks of nanoparticles but also within other aspects, such as regulation. The analysis was conducted through an extensive review of 31 published and peer-reviewed reports and articles from scientists, government agencies, regulatory bodies, and international and national organizations that specifically focused on EHS risks of nanomaterials. The majority of the reports were published between 2006 and 2008 with the exception of four reports published in 2004–2005.

Among other results, it was found that extensive knowledge gaps were present in nearly all aspects of basic knowledge regarding nanomaterial EHS risks, including critical areas within human and environmental exposure and effect assessments, better characterization of nanomaterials, better testing procedures and equipment, and some uncertainties even within defining nanomaterials themselves (Fig. 2). Each of these main locations was further divided into sublocations of uncertainty, such as uncertainty within fate and behavior of nanomaterials within an organism, bioaccumulation and biomagnification, environmental exposures to nanomaterials, and environmental risk assessment testing strategies (Fig. 3). In fact, the most frequently cited "sublocations" of uncertainty included the lack of reference materials and standardization with 194 different citations, characterizing the environmental fate and behavior of nanomaterials (181), determining environmental effects and/or ecotoxicity (154), and a general lack of knowledge within characterizing nanomaterials (143) (Fig. 3).

It was also found that the "level" of uncertainty for these "locations" fell between what is termed as "scenario uncertainty" (i.e., known outcomes, unknown probabilities) and "recognized ignorance" (i.e., unknown outcomes and probabilities) (Table 1). This implies that the level of current knowledge in 2008–2009 was estimated to be at a relatively early state of development. Furthermore, the "nature" of uncertainty was also estimated to be mainly epistemic, indicating that further research is expected to reduce most of these uncertainties within the field of EHS risks of nanomaterials.

On the basis of the results of this study, it was recommended that research should be prioritized toward developing test procedures and equipment as well as full characterization of nanomaterials. This was primarily due to the minimal presence of stochastic uncertainty in these locations, whereby further empirical research is expected to reduce most uncertainty in these areas. It was also recommended that research should be prioritized toward better understanding the environmental fate and behavior of

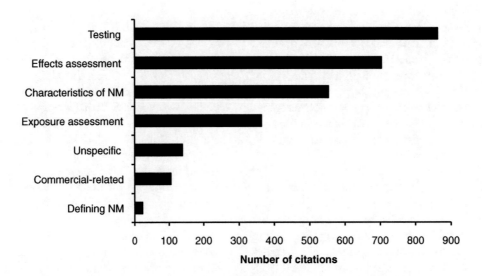

Fig. 2 Main locations of uncertainty related to the environmental, health, and safety risks of nanomaterials.
Source: Modified from Grieger, K.; Hansen, S.F.; Baun, A. The known unknowns of nanomaterials: Describing and characterizing uncertainty within environmental, health and safety risks. Nanotoxicology **2009**, *3* (3), 1–12.[51]

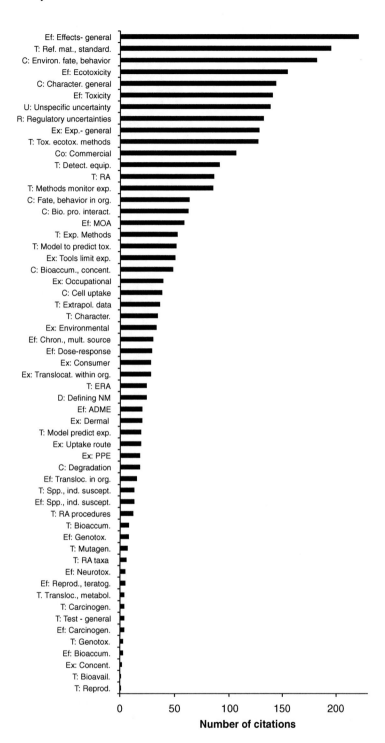

Fig. 3 Detailed view of sublocations of uncertainty related to the EHS risks of nanomaterials. C refers to the location "Characteristics of NM"; Co, "Commercial-related uncertainties"; Ex, "Exposure assessment"; Ef, "Effects assessment"; T, "Testing parameters"; and U, "Unspecific uncertainties."
Source: Modified from Grieger et al.[51]

nanomaterials since this was one of the most commonly cited knowledge gaps. In addition, it was also noted in this analysis that there was a general lack of attention given to the potential for bioaccumulation or persistency of nanomaterials in the screened literature. This was surprising, as these areas are extremely important aspects to consider given previous knowledge of persistent organic pollutants. In conclusion, this analysis described and characterized scientific uncertainty within the EHS risks of nanomaterials according to the current literature, and subsequent recommendations were made in regard to research prioritization.

Table 1 Overview of location, level, and nature of uncertainty related to the EHS risks of nanomaterials.

Location	%Total	Score	Level	Nature Epistemic	Stochastic
Characteristics of nanomaterials	21	2		✓	
Effects assessment	25	2		×	×
Exposure assessment	13	3		×	×
Testing considerations	31	1		✓	
Other areas[a]	10	3		✓	

▨ Statistical uncertainty ▨ Scenario uncertainty ▨ Recognized ignorance

[a]Combination of defining NM, commercial/industry-related, and unspecified uncertainties.
Source: Modified from Grieger et al.[51]

Further Perspectives

While the preceding sections discussed the concept of uncertainty and approaches to analyze or describe it, there are still many challenges in terms of how we can deal with the scientific uncertainty of nanoparticles in terms of their potential environmental risks. For instance, how can we try and reap the benefits of nanomaterials and nanotechnology while minimizing risks to health and the environment? These are some of the many questions that are currently asked by a variety of stakeholders, including scientists, engineers, governments, and regulatory agencies. In fact, international efforts to overcome the many uncertainties and data gaps within the field of environmental and health risks of nanomaterials are currently underway.[7,21,68–70] This is partially reflected in increased funding dedicated to understanding the health and environmental implications of nanotechnology and nanomaterials, which has increased from less than 1% of total research and development funds in 2004 to approximately 5% in 2009 within the United States,[20] and similar figures are seen in Europe.[71]

While these international efforts are essential for closing data gaps and better understanding the associated environmental risks of nanomaterials, it is expected that this process may take years and possibly decades in many cases. For example, Choi et al.[72] estimated that testing existing nanoparticles in the United States will most likely take 34–53 years to complete, and other estimates have also been on the order of decades.[23] In response, others have advocated for more research within nanomaterial risk governance mechanisms, particularly within timely decisions under conditions of uncertainty, as well as more research within alternative or complimentary tools to lengthy risk assessment processes.[73–76] Moreover, there is currently a variety of other tools that may be suited to handle uncertainty or for decision making within a specific context in terms of nanoparticle risk, such as multicriteria decision analysis,[77] the IRGC's Risk Governance Framework,[56] and Environmental Defense and Dupont's Nano Risk Framework.[78] However, a more rigorous assessment to their suitability is still needed in most cases.[60]

CONCLUSION

It is clear that assessing uncertainty pertaining to the potential environmental risks of nanoparticles remains a challenging yet extremely important task for scientists, researchers, governments, and policymakers. The presence of uncertainty may inhibit the full potential of nanotechnology, including the use of nanoparticles in a range of applications, while at the same time hinder risk or safety assessments of these products. Given the complexity involved in understanding both the behavior of nanoparticles and other nanomaterials, as well as understanding environmental risks associated with nanoparticles, robust approaches to assessing and handling uncertainty are clearly needed. This is especially the case given the fast pace of nano-innovation and increased number of products in which nanoparticles are being used.

REFERENCES

1. Renn, O. *Risk Governance: Coping with Uncertainty in a Complex World*; Earthscan: London, 2008; 455 pp.
2. International Organization for Standardization (ISO). *Nanotechnologies—Methodology for the Classification and Categorization of Nanomaterials*, Report No. ISO/TR 11360:2010; ISO: Geneva, Switzerland, 2010.
3. Lövestam, G.; Rauscher, H.; Roebben, G.; Klüttgen, B.; Gibson, N.; Putaud, J.P.; Stamm, H. *Considerations on a definition of nanomaterial for regulatory purposes*; European Commission Joint Research Centre: Luxembourg, 2010.
4. Scientific Committee on Emerging and Newly Identified Health Risks (SCENIHR). *Scientific Basis for the Definition of the Term "Nanomaterial"*; European Commission Health and Consumer Protection Directorate-General, Directorate C—Public Health and Risk Assessment, C7—Risk Assessment: Brussels, Belgium, 2010.
5. Scientific Committee on Emerging and Newly Identified Health Risks (SCENIHR). *Risk Assessment of Products of Nanotechnologies*; European Commission Health and Consumer Protection Directorate-General, Directorate C—Public Health and Risk Assessment, C7—Risk Assessment: Brussels, Belgium, 2009.

6. British Standards Institution. *Terminology for Nanomaterials*, Report No. PAS 136; British Standards Institution: London, 2007.
7. International Organization for Standardization (ISO). *Spécification Technique Internationale Nanotechnologies—Terminology and Definitions for Nano-Objects—Nanoparticle, Nanofibre and Nanoplate*, Report No. ISO/TS 27687; ISO: Geneva, Switzerland, 2008.
8. Walker, W.; Harremöes, P.; Rotmans, J.; Van der Sluijs, J.; Van Asselt, M.; Janssen, P.; Krayer von Krauss, M. Defining uncertainty: A conceptual basis for uncertainty management in model-based decision support. J. Integr. Assess. **2003**, *4* (1), 5–17.
9. Funtowicz, S.; Ravetz, J. *Uncertainty and Quality in Science for Policy*, 1st Ed.; Kluwer Academic Publishers: Dordrecht, the Netherlands, 1990; 235.
10. Poulter, S.R. Monte Carlo simulation in environmental risk assessment —Science, policy and legalissues. Risk **1998**, *7* (Winter), 7–26.
11. Pilkey, O.H.; Pilkey, J.L. *Useless Arithmetic: Why Environmental Scientists Can't Predict the Future*; Columbia University Press: New York, 2007; 230 pp.
12. Refsgaard, J.C.; van der Sluijs, J.P.; Hojberg, A.L.; Vanrolleghem, P.A. Uncertainty in the environmental modelling process—A framework and guidance. Environ. Model. Softw. **2007**, *22* (11), 1543–1556.
13. Janssen, P.H.M.; Petersen, A.C.; van der Sluijs, J.P.; Risbey, J.S.; Ravetz, J.R. *RIVM/MNP Guidance for Uncertainty Assessment and Communication: Quickscan Hints and Actions List*; Netherlands Environmental Assessment Agency, National Institute for Public Health and the Environment (RIVM): Bilthoven, the Netherlands, 2003.
14. van der Sluijs, J.P.; Craye, M.; Funtowicz, S.; Kloprogge, P.; Ravetz, J.; Risbey, J. Combining quantitative and qualitative measures of uncertainty in model-based environmental assessment: The NUSAP system. Risk Anal. **2005**, *25* (2), 481–492.
15. INTARESE. *Work Package 1.5: Cross-Cutting Issues in Risk Assessment: Integrating Uncertainty to Integrated Assessment*; Integrated Assessment of Health Risks of Environmental Stressors in Europe: Copenhagen, Denmark, 2006.
16. van der Sluijs, J.; Petersen, A.C.; Janssen, P.H.M.; Risbey, J.; Ravetz, J. Exploring the quality of evidence for complex and contested policy decisions. Environ. Res. Lett. **2008**, *3*, 1–9.
17. Collingridge, D. *The Social Control of Technology*, 1st Ed.; Frances Pinter Limited: London, 1980; 210.
18. Hansen, S.F.; von Krauss, M.P.K.; Tickner, J.A. Categorizing mistaken false positives in regulation of human and environmental health. Risk Anal. **2007**, *27* (1), 255–69.
19. European Environment Agency (EEA). *Late Lessons from Early Warnings: The Precautionary Principle 1896–2000*, Report No. 22; EEA: Copenhagen, Denmark, 2001.
20. National Nanotechnology Initiative (NNI). *FY Budget and Highlights*; NNI: Washington D.C., 2008a.
21. European Commission (EC). *EU Nanotechnology R&D in the Field of Health and Environmental Impact of Nanoparticles*; European Commission: Brussels, Belgium, 2008a.
22. Royal Society and Royal Academy of Engineering. *Nanoscience and Nanotechnologies: Opportunities and Uncertainties*; Clyvedon Press: Cardiff, U.K., 2004; ISBN 0 85403 604 0.
23. Maynard, A. *Nanotechnology: A Research Strategy for Addressing Risk*, Project on Emerging Nanotechnologies, Report No. PEN 3; Woodrow Wilson International Center for Scholars: Washington, D.C., 2006.
24. Boxall, A.; Chaudhry, Q.; Sinclair, C.; Jones, A.; Aitken, R.; Jefferson, B.; Watts, C.D. *Current and Future Predicted Environmental Exposure to Engineered Nanoparticles*; Central Science Laboratory for Department of Environment Food and Rural Affairs: York, U.K., 2007.
25. Karn, B.; Kuiken, T.; Otto, M. Nanotechnology and in situ remediation: A review of the benefits and potential risks. Environ. Health Perspect. **2009**, *117* (12), 1823–1831.
26. Woodrow Wilson Institute. *An inventory of nanotechnology-based consumer products currently on the market, Project on Emerging Nanotechnologies*; Washington, D.C., 2010. Available at http://www.nanotechproject.org/inventories/consumer/ (accessed May 23, 2012).
27. Hansen, S.F.; Michelson, E.S.; Kamper, A.; Borling, P.; Stuer Lauridsen, F.; Baun, A. Categorization framework to aid exposure assessment of nanomaterials in consumer products. Ecotoxicology **2008**, *17* (5), 438–447.
28. Grieger, K.D.; Fjordbøge, A.; Hartmann, N.B.; Eriksson, E.; Bjerg, P.L.; Baun, A. Environmental benefits and risks of zero-valent iron nanoparticles (nZVI) for in situ remediation: Risk mitigation or trade-off? J. Contam. Hydrol. **2010a**. *118* (3–4), 165–183. doi:10.1016/j.jconhyd.2010.07.011.
29. Li, H.; Zhou, Q.; Wu, Y.; Fu, J.; Wang, T.; Jiang, G. Effects of waterborne nano-iron on medaka (*Oryzias latipes*): Antioxidant enzymatic activity, lipid peroxidation and histopathology. Ecotoxicol. Environ. Saf. **2009**, *72*, 684–692.
30. Phenrat, T.; Saleh, N.; Sirk, K.; Tilton, R.D.; Lowry, G.V. Aggregation and sedimentation of aqueous nanoscale zerovalent iron dispersions. Environ. Sci. Technol. **2007**, *41* (1), 284–290.
31. Phenrat, T.; Liu, Y.; Tilton, R.D.; Lowry, G.V. Adsorbed polyelectrolyte coatings decrease FeO nanoparticle reactivity with TCE in water: Conceptual model and mechanisms. Environ. Sci. Technol. **2009**, *43* (5), 1507–1514.
32. European Chemicals Agency (ECHA). *Guidance on information requirements and chemical safety assessment*, 2010. Available at http://guidance.echa.europa.eu/docs/guidance_document/information_requirements_en.htm?time=1289468158 (accessed May 23, 2012).
33. European Commission (EC). *Environmental Assessment*, 2008b. Available at http://ec.europa.eu/environment/eia/home.htm (accessed May 23, 2012).
34. Davis, J.M. How to assess the risks of nanotechnology: Learning from past experience. J. Nanosci. Nanotechnol. **2007**, *7* (2), 402–409.
35. Briggs, D. A framework for integrated environmental health impact assessment of systemic risks. Environ. Health **2008**, *7* (1), 61, doi:10.1186/1476-069X-7-61.
36. Rocks, S.; Pollard, S.; Dorey, R.; Levy, L.; Harrison, P.; Handy, R. *Comparison of Risk Assessment Approaches for Manufactured Nanomaterials*, Report No. CB403; Cranfield University: Defra, 2008.
37. European Union (EU). *Technical Guidance Document in Support of Commission Directive 93/67/EEC on Risk Assessment for New Notified Substances and Commission Regulation (EC) 1488/94 on Risk Assessment for Existing Substances*; European Commission: Brussels, Belgium, 2003.

38. Krayer von Krauss, M.; Martuzzi, M. *Work Package 1: Evaluation and Exploitation of Research and Best Practices*; Health and Environment Network (HENVINET); Brussels, Belgium, 2007.
39. European Chemicals Agency (ECHA). Uncertainty analysis (Ch. 19). In *Guidance on Info Requirements and Chemical Safety Assessment*, European Chemical Agency: Helsinki, Finland. 2008.
40. Shinohara, N.; Gamo, M.; Nakanishi, J. *Risk Assessment of Manufactured Nanomaterials—Fullerene (C60)*, NEDO project Research and Development of Nanoparticle Characterization Methods, Report No. P06041, 2009.
41. Kobayashi, N.; Ogura, I.; Gamo, M.; Kishimoto, A.; Nakanishi, J. *Risk Assessment of Manufactured Nanomaterials—Carbon Nanotubes (CNTs)*, NEDO Project Research and Development of Nanoparticle Characterization Methods Report No. P06041, 2009.
42. Stone, V.; Aitken, R.; Aschberger, K.; Baun, A.; Christensen, F.M.; Fernandes, T.F.; Hansen, S.F.; Hartmann, N.B.; Hutchison, G.; Johnston, H., et al. *Engineered Nanoparticles: Review of Health and Environmental Safety (ENRHES)*, ENRHES EU FP 7 project, 2010. Final report.
43. Krayer von Krauss, M.; van Asselt, M.B.A.; Henze, M.; Ravetz, J.; Beck, M.B. Uncertainty and precaution in environmental management. Water Sci. Technol. **2005**, *52* (6), 1–9.
44. Martuzzi, M.; Tickner, J. *The Precautionary Principle: Protecting Public Health, the Environment and the Future of Our Children*; World Health Organization: Copenhagen, Denmark, 2004.
45. Ravetz, J. *The No-Nonsense Guide to Science*; New Internationalist Publications: Oxford, U.K., 2005; 42 pp.
46. Dale, V.H.; Biddinger, G.R.; Newman, M.C.; Oris, J.T.; Suter, G.W.; Thompson, T.; Armitage, T.M.; Meyer, J.L.; Allen-King, R.M.; Burton, G.A., et al. Enhancing the ecological risk assessment process. Integr. Environ. Assess. Manag. **2008**, *4* (3), 306–313.
47. The National Academies of Science. *Science and Decisions Advancing Risk Assessment*; The National Academies: Washington, D.C., 2008.
48. Wallace, D. Mediating the uncertainty and abstraction of nanotechnology promotion and control: "Late" lessons from other "Early warnings" in history. Nanotechnol. Law Bus. **2008**, *5* (3), 309–312.
49. Linkov, I.; Satterstrom, F.K.; Steevens, J.; Ferguson, E.; Pleus, R.C. Multi-criteria decision analysis and environmental risk assessment for nanomaterials. J. Nanopart. Res. **2007**, *9* (4), 543–554.
50. Hassellov, M.; Readman, J.W.; Ranville, J.F.; Tiede, K. Nanoparticle analysis and characterization methodologies in environmental risk assessment of engineered nanoparticles. Ecotoxicology **2008**, *17* (5), 344–361.
51. Grieger, K.; Hansen, S.F.; Baun, A. The known unknowns of nanomaterials: Describing and characterizing uncertainty within environmental, health and safety risks. Nanotoxicology **2009**, *3* (3), 1–12.
52. Gottschalk, F.; Sonderer, T.; Scholz, R.W.; Nowack, B. Possibilities and limitations of modeling environmental exposure to engineered nanomaterials by probabilistic material flow analysis. Environ. Toxicol. Chem. **2010**, *29* (5), 1036.
53. Morgan, K. Development of a preliminary framework for informing the risk analysis and risk management of nanoparticles. Risk Anal. **2005**, *25* (6), 1621–1635.
54. Gavelin, K.; Wilson, R.; Doubleday, R. *Democratic Technologies? The Final Report of the Nanotechnology Engagement Group (NEG)*; Involve: London, 2007; 172.
55. Mantovani, E.; Porcari, A.; Meili, C.; Widmer, M. *Mapping Study on Regulation and Governance of Nanotechnologies*, FramingNano project, Report nr D1.1 for Work Package 1, 2009.
56. International risk governance council (IRGC). *Nanotechnology Risk Governance: Recommendations for a Global, Coordinated Approach to the Governance of Potential Risks*; IRGC: Geneva, Switzerland, 2007.
57. International Risk Governance Council (IRGC). *Appropriate Risk Governance Strategies for Nanotechnology Applications in Food and Cosmetics*; IRGC: Geneva, Switzerland, 2009.
58. Environmental Protection Agency (US EPA). *Nanotechnology White Paper*, Report No. EPA 100/B-07/001; Science Policy Council, United States Environmental Protection Agency: Washington, D.C., 2007.
59. Wickson, F.; Gillund, F.; Myhr, A.I. Treating nanoparticles with precaution: Recognising qualitative uncertainty in scientific risk assessment. In *Nano Meets Macro: Social Perspectives on Nanoscale Sciences and Technologies*; Kjølberg, K., Wickson, F., Eds.; Pan Stanford Publishing: Singapore, 2010.
60. Grieger, K.; Linkov, I.; Hansen, S.F.; Baun, A. Environmental risk analysis for nanomaterials: Review and evaluation of frameworks. Nanotoxicology, **2011**, *6*(2), 196–212.
61. Stern, P.; Fineberg, H. *Understanding Risk: Informing Decisions in a Democratic Society*; National Academy Press: Washington, D.C., 1996.
62. Krayer von Krauss, M. *Uncertainty in Policy Relevant Sciences*, PhD thesis; Institute of Environment and Resources, Technical University of Denmark: Kongens Lyngby, Denmark, 2006.
63. Krayer von Krauss, M.; Kaiser, M.; Almaas, V.; van der Sluijs, J.; Kloprogge, P. Diagnosing and prioritizing uncertainties according to their relevance for policy: The case of transgene silencing. Sci. Total Environ. **2008**, *390* (1), 23–34.
64. Scientific Committee on Emerging and Newly Identified Health Risks (SCENIHR). *The Appropriateness of the Risk Assessment Methodology in Accordance with the Technical Guidance Documents for New and Existing Substances for Assessing the Risks of Nanomaterials*; European Commission Health and Consumer Protection Directorate-General, Directorate C—Public Health and Risk Assessment, C7—Risk Assessment: Brussels, Belgium, 2007.
65. Department for Environment, Food and Rural Affairs (DEFRA). *Characterising the Potential Risks Posed by Engineered Nanoparticles*; Second U.K. Government research report; Department for Environment, Food and Rural Affairs: London, 2007.
66. Organisation for Economic Co-operation and Development (OECD). *Current Developments/Activities on the Safety of Manufactured Nanomaterials/Nanotechnologies*, Tour de table at the 2nd meeting of the working party on manufactured nanomaterials, Berlin, Germany, April 25–27, 2007; Paris, Report No. ENV/JM/MONO(2007)16, 2007.
67. Baun, A.; Hartmann, N.B.; Grieger, K.; Kusk, K.O. Ecotoxicity of engineered nanoparticles to aquatic invertebrates: A brief review and recommendations for future toxicity testing. Ecotoxicology **2008**, *17* (5), 387–395.

68. International Council on Nanotechnology (ICON). *Towards Predicting Nano-Biointeractions: An International Assessment of Nanotechnology Environment, Health and Safety Research Needs*, Report No. 4; ICON: Houston, TX, 2008.
69. National Nanotechnology Initiative (NNI). *Strategy for Nanotechnology-Related Environmental, Health, and Safety Research*; NNI: Washington, D.C., 2008b.
70. Organisation for Economic Co-operation and Development (OECD). *Preliminary Review of OECD Test Guidelines for their Applicability to Manufactured Nanomaterials*; Paris: Environment Directorate, Joint Meeting of the Chemicals Committee and the Working Party on Chemicals, Pesticides and Biotechnology, 2009.
71. Aguar, P.; Murcia Nicolàs, J.J. *EU nanotechnology R&D in the field of health and environmental impact of nanoparticles*; European Commission, Research Directorate-General, 2008. Available at http://cordis.europa.eu/nanotechnology (accessed May 23, 2012).
72. Choi, J.; Ramachandran, G.; Kandlikar, M. The impact of toxicity testing costs on nanomaterial regulation. Environ. Sci. Technol. **2009**, *43* (9), 3030–3034.
73. Royal Commission on Environmental Pollution (RCEP). *Novel Materials in the Environment: The Case of Nanotechnology*, Report No. 27; TSO: Norwich, U.K., 2008.
74. Brown, S. The new deficit model. Nat. Nanotechnol. **2009**, *4* (10), 609–611.
75. Owen, R.; Baxter, D.; Maynard, T.; Depledge, M. Beyond regulation: Risk pricing and responsible innovation. Environ. Sci. Technol. **2009**. 43, 6902–6906. doi:10.1021/es803332u.
76. Grieger, K.; Baun, A.; Owen, R. Redefining risk research priorities for nanomaterials. J. Nanopart. Res. **2010b**, *2* (2), 383–392.
77. Environmental Defense (ED) and Dupont. *Nano Risk Framework*; Environmental Defense–Dupont Nano Partnership: Washington D.C., 2007. Available at http://apps.edf.org/documents/6496_nano%20risk%20framework.pdf (accessed May 23, 2012).

Nanotechnology: Environmental Abatement

Puangrat Kajitvichyanukul
Jirapat Ananpattarachai
Center of Excellence for Environmental Research and Innovation, Faculty of Engineering, Naresuan University, Phitsanulok, Thailand

Abstract

Currently, environmental nanotechnology is an important method for pollutant removal from the environment because of its ability to degrade toxic organic pollutants, dechlorinate chlorinated organic compounds, deactivate bacteria, and transform toxic heavy metals to nontoxic forms. Consequently, many environmental cleanup technologies that utilize several types of nanomaterials in removing several types of toxic chemicals for applications in wastewater treatment, drinking water treatment, and contaminated land and groundwater remediation have been proposed or developed. This entry reviews the significance of nanotechnology in environmental management. In addition, the latest nanomaterials, including nano-semiconductor catalysts (TiO_2, ZnO, Fe_2O_3, CdS, and ZnS), zero-valent iron nanoparticles, and noble nanoparticles, such as silver or gold nanoparticles, are reviewed, and their mechanisms in toxic pollutant removal and their potential applications for environmental abatement are demonstrated.

INTRODUCTION: SIGNIFICANCE OF NANOTECHNOLOGY IN ENVIRONMENTAL MANAGEMENT

Rapid developments in industrialization and population increases, which lead to increased demand for clean water, have become critical environmental issues worldwide. In conjunction with these issues, there is increased water contamination due to the overwhelming discharge of pollutants and contaminants into natural water bodies.[1–3] Many pollutants such as heavy metals and toxic organic substances are difficult to remove or degrade by natural processes. Also, these toxic pollutants can cause adverse effects on human health. The challenges of solving these problems are to seek and develop low-cost and highly efficient advanced technologies to treat and recycle the available water. The shortage of clean water and the pollution in natural waters are not the only problems, as increasingly, air pollution is becoming a significant problem. Currently, indoor air pollution is of considerable concern for many researchers. Indoor air pollutants mainly include nitrogen oxides (NO_x), carbon oxides (CO and CO_2), volatile organic compounds (VOCs), and particulates. VOCs are well-known indoor air pollutants. Many VOCs are known to be toxic and are considered to be carcinogenic, mutagenic, or teratogenic.[4]

Recently, many advanced environmental technologies have been developed to prevent and remediate pollution and to ensure sustainable growth. Because of their potential to contribute to the protection of the environment and to facilitate sustainable development, environmental nanotechnologies have been accepted by many scientists and researchers as a viable technique to abate pollution. The projected world market for applied environmental technologies was reported to be approximately $6 billion by 2010.[5] The global market for nanotechnology in environmental applications generated 1.1 billion dollars in 2008 and an estimated $2.0 billion in 2009. This figure is expected to increase at a compound annual growth rate of 61.8% to reach $21.8 billion in 2014. Environmental nanotechnologies contribute to remediation, protection, maintenance, and enhancement. From a BCC Research report, the environmental protection segment has the largest market share and was worth $661.4 million in 2008. This share is expected to increase to more than $1.0 billion in 2009 and $10.3 billion in 2014.[6]

Environmental nanotechnologies have gained significance as the standards for drinking water or clean air have been revised several times. For example, in the case of revisions to the drinking water standards by the U.S. Environmental Protection Agency from 1976 to 2001, the allowed lindane concentration was decreased from 4 ppb to 0.2 ppb, and the allowed arsenic concentration was decreased from 50 ppb to 10 ppb. Thus, the contaminant concentrations are now regulated to extremely low levels, even reaching molecular limits. Consequently, any promising technology to remove these contaminants should also be able to reach the same limits, even to the molecular level. This is an advantage that environmental nanotechnologies have over conventional techniques such as adsorption and chemical redox reactions. For example, when using nanomaterial for adsorption processes, the surface-to-volume ratio increases drastically with the reduction in the size of the adsorbent particle from bulk to nano dimensions. Higher numbers of atoms or molecules are available on the nanomaterial surface, which leads to a higher efficiency in contaminant adsorption. Moreover, nanomaterials, such as metals or semiconductors, also provide higher reactivity than other materials. The change in reactivity on the nanoscale was explained by the pioneering work of Henglein as the size-quantization effect.[7] The band structure in metals and semiconductors was explained as not an anatomic or molecular property but attributable to the arrangement of atoms in a specific order in the crystal lattice. In addition, as particles decrease in size, they acquire novel physical, chemical, and electronic properties. These unique catalytic properties can accelerate oxidation or reduction reactions with various pollutants and can gain high efficiencies in the removal of contaminants from air or water.

The nanomaterials for environmental abatement can be divided into many groups as shown in Table 1. These nanomaterials are mostly applied to remove the toxic contaminants, which include heavy metals and toxic organic contaminants such as polycyclic aromatic hydrocarbons (PAHs), persistent organic pollutants, and VOCs. These toxic pollutants are difficult to remove. Organic contaminants, in particular, are resistant to biodegradation and are persistent in water or air. For heavy metal removal, nanomaterials such as TiO_2 and ZnO can be used in redox reactions to change the oxidation state of the heavy metal to a nontoxic form. Halogenated hydrocarbons can be dechlorinated by using zero-valent iron (ZVI). Also, silver or gold nanoparticles can be applied to degrade pesticides such as endosulfan, malathion, and chlorpyrifos.[8] The mechanisms of the pollutant removal for each type of nanomaterial are discussed in the next section.

Table 1 Types of environmental nanomaterials and their mechanisms for pollutant removal.

Type	Pollutant removal mechanism	Shape of material	Reference
TiO_2	Oxidation reaction Reduction reaction	Nanopowders Nanoparticles Nanocrystalline Thin film Nanotube Nanopellet	[9–20]
ZnO	Oxidation reaction Reduction reaction	Nanopowders Nanoparticles Thin film Nanotube Nanorod Nanofiber	[21,22]
WO_3	Oxidation reaction Reduction reaction	Nanoparticles	[23]
Zero-valent iron (ZVI)	Adsorption Reduction reaction	Nanoparticles	[24–27]
Ag	Adsorption	Nanoparticles	[28–31]
Au	Adsorption	Nanoparticles	[32]

NANO-SEMICONDUCTOR CATALYSTS

Mechanism in Pollutant Removal

The application of nano-semiconductor catalysts and ultraviolet (UV) light to degrade a wide range of refractory organic compounds into biodegradable compounds and eventually to mineralize them into innocuous water (H_2O) and carbon dioxide (CO_2) molecules has received attention from many researchers across the world. In addition, their ability to transform toxic heavy metals, such as Cr(VI) or As(VI), into nontoxic species has excited the researchers in this field. Several types of catalysts (e.g., TiO_2, ZnO, Fe_2O_3, CdS, and ZnS) have been used in purifying air and water. Among these semiconductors, titanium dioxide (TiO_2) is thus far the most commonly used photocatalyst because of its exceptional optical and electronic properties, chemical stability, nontoxicity, and low cost. It is strongly resistant to chemical breakdown and photocorrosion. TiO_2 is the most active photocatalyst and remains stable after repeated catalytic cycles, whereas CdS and GaP are degraded and produce toxic products.[33] ZnO is generally unstable in illuminated aqueous solutions, especially at low pH values, and WO_3, although useful in the visible range, is generally less photocatalytically active than TiO_2.[34]

The fundamentals of heterogeneous photocatalysis that employs a nano-semiconductor catalyst, especially TiO_2, have been extensively reported in the literature.[35–37] The basic principles of this method are briefly described here. Semiconductors (e.g., TiO_2, ZnO, Fe_2O_3, CdS, and ZnS) can act as sensitizers for light-induced redox processes due to their electronic structure, which is characterized by a filled valence band and an empty conduction band. When a photon energy ($h\nu$) that is greater than or equal to the band gap energy (E_g) of that semiconductor is illuminated onto its surface, for example, 3.2 eV for anatase TiO_2 or 3.0 eV for rutile TiO_2, a lone electron is photoexcited to the empty conduction band, which leads to the formation of an electron–hole pair in the semiconductor particle. The reaction of the valence band "holes" (h_{vb}^+) with either the adsorbed H_2O or with the surface OH^- groups on the semiconductor can generate OH radicals (HO•), which have an important role in the degradation of organic pollutants. The reactions (1a–1c) that occur on the semiconductor surface are shown below using TiO_2 as a model.

$$TiO_2 \xrightarrow{h\nu} e_{cb}^-(TiO_2) + h_{vb}^+(TiO_2) \quad (1a)$$
$$TiO_2(h_{vb}^+) + H_2O_{ads} \to TiO_2 + HO^•_{ads} + H^+ \quad (1b)$$
$$TiO_2(h_{vb}^+) + HO^-_{ads} \to TiO_2 + HO^•_{ads} \quad (1c)$$

In general, donor (D) molecules, such as H_2O, will adsorb and react with a hole in the valence band, and an acceptor (A), such as dioxygen, will adsorb and react with the electron in the conduction band (e_{cb}^-), as shown in reactions 1d and 1e.

$$TiO_2 (h_{vb}^+) + D_{ads} \rightarrow TiO_2 + D^+_{ads} \quad (1d)$$

$$TiO_2 (e_{cb}^-) + A_{ads} \rightarrow TiO_2 + A^-_{ads} \quad (1e)$$

Oxygen can trap conduction-band electrons to form the superoxide ion ($O_2^{\bullet-}$). These superoxide ions can react with hydrogen ions (formed by splitting water) and form $HO_2^\bullet$, as shown in reactions 1f and 1g. The cleavage of H_2O_2 from reaction in 1g by another reaction (1h, 1i, and 1j) may yield an OH radical

$$TiO_2 (e_{cb}^-) + O_{2ads} + H^+ \rightarrow TiO_2 + HO_2^\bullet \rightarrow O_2^{\bullet-} + H^+ \quad (1f)$$

$$TiO_2 (e_{cb}^-) + HO_2^\bullet + H^+ \rightarrow H_2O_2 \quad (1g)$$

$$H_2O_2 + h\nu \rightarrow HO^\bullet \quad (1h)$$

$$H_2O_2 + O_2^{\bullet-} \rightarrow HO^\bullet + O_2 + HO^- \quad (1i)$$

$$H_2O_2 + TiO_2 (e_{cb}^-) \rightarrow HO^\bullet + HO^- + TiO_2 \quad (1j)$$

The oxidative species formed (in particular the hydroxyl radicals, $OH^\bullet$) from the above reactions can react with the majority of organic pollutants and cause degradation and further mineralization of the contaminants.

However, in the absence of suitable donor or acceptor molecules, the photoexcited electron recombines with the valence band hole in nanoseconds with the simultaneous dissipation of heat energy. If a suitable scavenger or surface-defect state is available to trap the electron or hole, the recombination is prevented and subsequent redox reactions may occur. A drawing depicts the mechanism of the electron–hole pair formation when the semiconductor is irradiated with adequate photon energy.

If the semiconductor is used in the presence of water or some other fluid, a spontaneous adsorption of water and pollutant in the water occurs. An electron transfer proceeds, according to the redox potential of each adsorbent, toward the acceptor molecules, and a positive hole is transferred to the donor molecule. Most organic photodegradation reactions utilize the oxidizing power of the holes either directly, by acting as an electron donor, or indirectly. It is well known that O_2 and water are essential for photooxidation. The degradation cannot occur in the absence of either, except for some simple organic molecules, such as oxalate and formic acid, which can be oxidized to CO_2 by direct electrochemical oxidation where the electrons are passed on to an alternative electron acceptor, such as metal ions in the solution.[38] Oxidative species such as $HO^\bullet$, $HO_2^\bullet$, and $O_2^{\bullet-}$ react with the majority of organic pollutants (as shown in Fig. 1). For example, with aromatic compounds, the aromatic ring is hydroxylated, and successive steps in oxidation/addition lead to the ring opening. The resulting aldehydes and carboxylic acids are decarboxylated and finally form CO_2.[33]

Having a large surface area-to-volume ratio, a semiconductor in nano-dimensions promotes the efficient charge separation and trapping at the physical surface. The light opaqueness of nanoscale TiO_2 catalysts has been reported to enhance the oxidation capability compared to bulk TiO_2 catalysts.[39] Nano-semiconductor catalysts can be synthesized in several forms such as nanopowder, nanotube, and thin film, as shown in Fig. 2.

Generally, two forms of catalysts in photocatalysis have been widely used, including highly dispersed fine particles or suspended particles in a liquid medium, and thin films on supportive materials.[14,40–42] When a suspension photocatalyst is used in a photocatalysis system, the suspended catalyst has to be separated upon completion of each reaction cycle. However, this problem can be avoided by using a thin film catalyst applied onto different types of supports. For a large-scale water treatment process, an immobilized photocatalyst is a necessity. Glass, woven cloths, ceramic, and alumina have been studied as support materials for the photooxidation of various organic contaminants and photoreduction of heavy metals in water purification.[14,43] Recently, many attempts have been focused on synthesizing some type of semiconductor, such as TiO_2, to have high reactivity under visible light excitation to allow for the utilization of the solar spectrum.[44] An anion-doped system is a well-known technique to obtain the visible light-activated

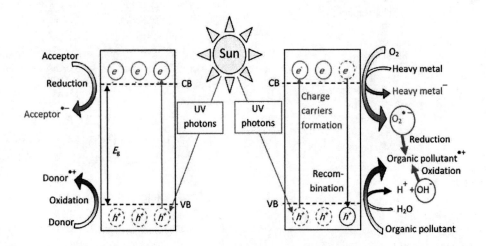

Fig. 1 The mechanism of the electron–hole pair formation when the semiconductor is irradiated with adequate photon energy.

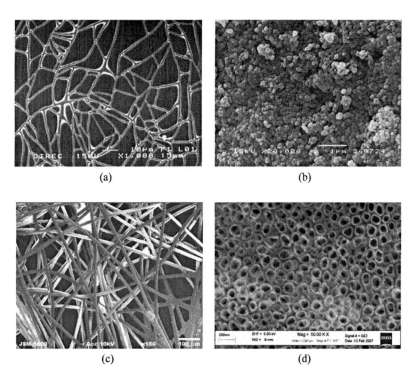

Fig. 2 Forms of nanoscale TiO$_2$ catalysts: (a) thin film, (b) nanopowder, (c) nanofiber, and (d) nanotube.

TiO$_2$.[45] Some anionic species, such as nitrogen, carbon, and sulfur, which potentially form new impurity levels closest to the valence band while maintaining the largest band gap for maximum efficiency, have been identified.[46–48] This new technique provides TiO$_2$ the ability to destroy toxic organic pollutants by sunlight.

Decontamination of Toxic Pollutants by Nano-Semiconductor Catalysts

Many nano-semiconductors catalysts, such as TiO$_2$, ZnO, ZrO$_2$, CdS, MoS$_2$, Fe$_2$O$_3$, and WO$_3$, have been examined and used as photocatalysts for the degradation of organic contaminants.[49] Normally, organic molecules can be degraded by photodegradation either by a direct reaction utilizing the oxidizing power of the holes or by an indirect reaction employing the oxidative species [hydroxyl radicals (OH$^{\bullet}$)]. Organic molecules that can adhere effectively to the surface of the photocatalyst will be more susceptible to direct oxidation.[50] Thus, the substituent groups on the organic molecule greatly affect the photocatalytic degradation of the organic molecule. For example, nitrophenol is a much stronger adsorbing substrate than phenol and therefore degrades faster.[51] A monochlorinated phenol degrades faster than a di- or trichlorinated one.[51] Organic substrates with an electron-withdrawing characteristic such as benzoic acid and nitrobenzene were found to strongly adhere to and be more susceptible to direct oxidation than those with electron donating groups.[51,52] The number of aromatic rings on the organic molecule is another factor for photocatalytic degradation. A comprehensive study of the photocatalytic degradation of phenanthrene, pyrene, and benzo[a]pyrene on soil surfaces using titanium dioxide (TiO$_2$) under UV light was conducted by Zhang et al.[53] They found that among these three types of organic pollutants, benzo[a]pyrene degraded the fastest. Under distinct UV wavelengths, the photocatalytic degradation rates of the PAHs were different. However, if the soil pH was changed, the pyrene and benzo[a]pyrene degradation rates were the fastest in acidic conditions, while phenanthrene was most significantly degraded in alkaline conditions.

Titanium dioxide (TiO$_2$) has been accepted to be the most widely used nano-semiconductor catalyst for environmental protection and remediation. Table 2 provides a list of the large number of classes of toxic organic compounds, and examples of each, that have been shown to be degraded by photocatalysis using TiO$_2$. Examples of the VOCs that can be degraded by TiO$_2$ are listed in Table 3.

In particular, TiO$_2$ is widely used in the degradation of toxic pollutants such as herbicides, pesticides, VOCs, and PAHs. Most of them are carcinogens and are often found in small but detectable amounts in water. Conventional wastewater treatment methods are generally not able to remove them. A very attractive feature of nano-semiconductor catalysts is their effectiveness against toxic pollutants, which are not easily removed by other water treatment processes. Moreover, these photocatalysts have been used to sensitize the photoconversion of toxic molecules to harmless or less toxic species. For example, TiO$_2$

Table 2 Some examples of toxic organic pollutants that can be degraded by TiO_2.

Chemicals	Examples	Initial concentration	Treatment conditions	Reaction products	References
Aldehydes	Acetaldehyde	240–300 ppm	Irradiation time = 50 min	CO_2	[54]
Carboxylic acids	Phenoxyacetic acid	1.0 mM	TiO_2 1 g/L, continuous O_2 purging and stirring, irradiation time = 120 min	Phenol and 1,2-diphenoxyethane	[55]
	2,4,5-Phenoxyacetic acid	0.5 mM	TiO_2 1 g/L, continuous O_2 purging and stirring, irradiation time = 120 min	2,4,5-Trichlorophenol, 2,4-dichlorophenol, and 1,2,4-trichloro-5-methoxy benzene	[55]
	Oxalic acid	0.8 M	TiO_2 1 g/L, irradiation time = 120 min	CO_2	[56]
Chloroanilines	2-Chloroaniline	94.85 mM	TiO_2 1.33 g/L and H_2O_2 0.01 mmol/L, pH 3.5, irradiation time = 25 min	Cl^- and CO_2	[57]
Chlorophenols	2-Chlorophenol	10–25 ppm	TiO_2/Ti anode, pH 3, irradiation time = 300 min, N-TiO_2 1 g/L, pH 7, irradiation time = 300 min	CO_2	[9,58]
	2,4-Dichlorophenol	125 ppm	TiO_2 2 g/L, pH 5, irradiation time = 360 min	Maleic acid, acetic acid	[59]
	Mixture of 4-chlorophenol, 2,4-dichlorophenol, 2,4,6-trichlorophenol, and pentachlorophenol	50 ppm (each)	1 TiO_2 g/L and/or 50 mM H_2O_2, irradiation time = 3 days	CO_2, Cl^-	[60]
Dyes	Acid orange 8 and acid red 1	2 to 7 × 10^{-5} M	TiO_2 0.1 g/L, pH 6, irradiation time = 166 min	n.a.	[61]
	Anazo dye and chrysoidine Y	0.25 mM	TiO_2 1 g/L, pH 3.5, irradiation time = 120 min	n.a.	[62]
	Acridine orange	1 mM	TiO_2 1 g/L, pH 3.5, irradiation time = 75 min	n.a.	[63]
	Ethidium bromide	2 mM	TiO_2 1 g/L, pH 3.5, irradiation time = 195 min	n.a.	[63]
	Methylene blue, methyl orange, indigo carmine, Chicago sky blue, mixed dye (mixture of the four dyes)	10 ppm (each)	TiO_2 glass plate, irradiation time = 270 min	n.a.	[64]
Ethers	Methyl *tert*-butyl ether (MBTE)	1.0 × 10^{-3} M	TiO_2 1 g/L, and H_2O_2 1.8 × 10^{-2} M, irradiation time = 166 min	*Tert*-butyl formate (TBF), *tert*-butyl alcohol (TBA), and acetone	[65]
Flourophenols	4-Fluorophenol	100 ppm	TiO_2 2 g/L, pH 7, irradiation time = 90 min	F^-, CO_2	[66]
Fungicides	Fenamidone	5 ppm	TiO_2-coated fiber, irradiation time = 300 min	Oxalic acid, fumaric acid, malonic acid, pyruvic acid, lactic acid, formic acid, acetic acid, and oxalacetic acid	[67]

(*continued*)

Table 2 Some examples of toxic organic pollutants that can be degraded by TiO_2. (*continued*)

Chemicals	Examples	Initial concentration	Treatment conditions	Reaction products	References
Herbicides	Isoproturon	0.5 mM	TiO_2 1 g/L, O_2 purging and stirring, irradiation time = 80 min	Chlorinated hydrocarbon	[68]
Ketones	Acetone	500 ppm	In air, feed rate 20 cm^3/min, Temp. 27°C, irradiation time = 250 min	CO_2	[69]
Perflouroaliphatics	Triflouroacetic acid, sulfonic acid of nonafluorobutane and heptadecaflourooctane	4.4 mM	TiO_2 2 g/L, purged with O_2 or N_2, irradiation time = 90 min	F^-, CO_2	[70]
Pharmaceuticals	Tetracycline	40 ppm	TiO_2 0.5 g/L,, irradiation time = 120 min	CO_2	[71]
	Lincomycin	10 and 75 µM	TiO_2 0.2 g/L, pH 6.3, irradiation time = 60 min	n.a.	[72]
Polymers	Polyvinylpyrrolidone (PVP)	0.2 ppm	TiO_2 2 g/L, irradiation time = 14 hr	Propanoic acid, acetic acid, formic acid, and CO_2	[73]

n.a., not available.

can be used to oxidize cyanide anions to from relatively harmless products, such as CO_2 and N_2, or to photosensitize the decomposition of bromate to bromide (Br^-) and oxygen. Because of their high activity in degrading and removing toxic organic pollutants from water and air, nano-semiconductor catalysts have become very popular in the new emerging nanotechnology field for environmental protection.

Table 3 Some examples of VOCs that can undergo photoconversion by TiO_2.

Class	VOC	Initial concentration	Treatment condition	Final product	Reference
Aldehydes	Acetaldehyde	930 ppmv	TiO_2 0.05 g, irradiation time = 300 min	CO_2	[74]
BTEX	Benzene, ethyl benzene, toluene, xylene	20 ppb	TiO_2 coated on glass 1.64 g/plate, humidity level = 21,000 ppmv, irradiation time = 3.7 min	CO_2	[75]
	Benzene, ethyl benzene, xylene	93, 21, 78 ppb	Thin film TiO_2 0.5 mg/cm^2, irradiation time = 2 sec	CO, CO_2	[76]
	Benzene, ethyl benzene, toluene, xylene	23 ± 2 ppb	Ln^{3+}–TiO_2 coated on glass 1.64 g/plate, humidity level = 15,700 ppmv, irradiation time = 72 sec	n.a.	[77]
	Benzene, ethyl benzene, toluene, xylene	3.72, 2.75, 3.41, 2.53 µmol m^{-3}	TiO_2 3%, flow rate = 100 mL min^{-1}, irradiation time = 44 hr	n.a.	[78]
Ethene	Perchloroethene (PCE)	77 ppm	Relative humidity = 40–60%, reactor residence time = 5 sec	CO_2	[79]
	Trichloroethene (TCE)	23 ppm	Relative humidity = 40–60%, reactor residence time = 5 sec	CO_2	[79]
Ethylene	Ethylene	227 ppm	ZrO_2/TiO_2 0.28 g in 30 × 15 × 2 mm reactor, flow rate = 20 mL/min, irradiation time = 250 min	n.a.	[80]
Ethylene	Tricholoroethylene	943 ppm	TiO_2 0.05 g, irradiation time = 300 min	CO_2	[74]
Ketones	Acetone	540 ppm	TiO_2 0.2 g coated on 60 × 60 × 2 mm plate, irradiation time = 50 hr	CO_2	[81]

n.a., not available.

In applying this technology in water treatment, there are two different types of photocatalytic applications: solar photocatalysis and photocatalytic systems equipped with artificial UV light. The Report from the Workshop on Nanotechnologies for Environmental Remediation identified solar photocatalysis as the main technological breakthrough for water treatment and purification. Photocatalysis with nanocatalysts is also a promising method for disinfection. In addition, photocatalysts combined with filtration membranes can reduce membrane fouling and thus increase efficiency and significantly enhance water cleaning. Photocatalysis has become an alternative method for water treatment applications and has almost become a mature market. The applications of photocatalytic systems for the disinfection of swimming pools are already commercially available. Small-scale photocatalytic systems with UV artificial light have been on the market for several years (http://www.ube.es/index.html). Solar photocatalytic water treatment plants are at the demonstration phase (http://www.raywox.com), and some pilot projects for drinking water purification have started (http://www.rcsi.ie/sodis/).

While metal oxide nanoparticles for photocatalysis are commercially available, there are still many challenges for researchers to explore, and many applications are still in the laboratory testing phase. One example is the development of a catalyst such as a pure nano-TiO_2 that can absorb visible wavelengths to be used in indoor applications. In addition, the most efficient system setup and substrate materials have to be determined for different applications to ensure maximal longevity, efficiency, and functionality of the photocatalyst. One of the major challenges for photocatalysis research is that nano-TiO_2 can destroy all organic materials and thus any organic matrix in which the nanoparticles are embedded. Nano-TiO_2 can thus only be applied in an inorganic environment. Currently, many researchers in this field are searching for ways to improve the efficiency and capabilities of this technology to improve our environment.

ZERO-VALENT IRON NANOPARTICLES

Mechanism in Pollutant Removal

Iron is the fourth most abundant element in the Earth's crust, and it exists in the environment predominantly in two valence states—the relatively water-soluble Fe(II) (ferrous iron) and the highly water-insoluble Fe(III) (ferric iron). Zero-valent (or elemental/native) iron [Fe(0)] is rarely formed on the Earth's surface due to the high reactivity of elemental iron.[82] Nanoscale zero-valent iron (nZVI) was extensively studied during the last decade. It is considered to be among the first generation of nanoscale environmental technologies.[24] Researchers have increasingly gained interest in it as an efficient sorbent for various types of aqueous pollutants, and nZVI has received considerable attention for its potential applications in groundwater treatment and site remediation, especially the transformation of halogenated organic contaminants and heavy metals in soils.[83,84] Two mechanisms, adsorption and reduction reactions, are important in the removal of these contaminants by nZVI.

The composition of nZVI has been reported[90] to likely be a core of mainly ZVI (Fe^0) with a shell that is largely made of iron oxides (i.e., FeO). In an aqueous solution, the surface is likely in the form of FeOOH. This hydrated surface may also cause the surface charge in water. The isoelectric point of nZVI was reported to be near a pH of 8.3, which is higher than that of magnetite (Fe_3O_4, 6.8) or maghemite (γ-Fe_2O_3, 6.6).[83,85] With this surface property, the surface of nZVI is positively charged at a pH lower than 8.3, which can induce several negatively charged ions on its surface. Thus, the nZVI surface charge may have important implications on its suspension stability and its mobility in soil and groundwater environments. Normally, nZVI tends to aggregate in water and has shown relatively low mobility in porous media or in an aquifer.

For the reduction reaction, ZVI (Fe^0) has long been used as a source of electrons in the reduction of various kinds of halogenated organic compounds. Normally, iron metal is a natural electron donor that readily loses electrons through corrosion as shown below:

$$Fe^0 \leftrightarrow Fe^{2+} + 2e^- \qquad (1k)$$

As the standard reduction potential (E^0) of ZVI is −440 mV, it can act as an electron donor or a reducing agent for redox reaction. When a redox reaction occurs in Fe^0–H_2O systems, there are three different pathways for the transformation of the oxidized compounds.[25] The first pathway was proposed by Metheson and Tratnyek and is the direct reduction reaction of adsorbed contaminants with electrons at the surface of iron, which is directly coupled with the oxidative dissolution of heavy metals.[86] In this pathway, the metallic iron is the reductant. In the second pathway, the reduction reaction occurs with dissolved ferrous ion (Fe^{2+}) produced from iron corrosion. However, it has been reported that reduction with Fe^{2+} is quite slow, and its contribution is considered relatively small.[87] The third pathway of reduction in Fe^0–H_2O systems involves a hydrogen-induced reductive reaction of the oxidized compounds. This reaction is rapid if some metals such as iron are present in the water system.[88] The surface of the iron, its defects, or other solid phases present in the system could provide the catalytic function. In Fe^0–H_2O systems, the ZVI may potentially function as both a reactant and a catalyst during the reduction of contaminants.[25]

In the degradation of organic contaminants such as chlorinated hydrocarbons, the following reaction takes place:

$$R\text{-}Cl + Fe^0 + H_2O \rightarrow R\text{-}H + Fe^{2+} + Cl^- + OH^- \qquad (1l)$$

The two pathways for the degradation of chlorinated hydrocarbons by nZVI are hydrogenolysis and beta elimination. Hydrogenolysis is possible through sequential dehalogenation, for example, from perchloroethylene to trichloroethylene, to dichloroethylene, to vinyl chloride, to ethene. An example of beta elimination is the degradation of trichloroethylene to chloroacetylene, to acetylene, to ethene.

If nZVI reacts with ionic heavy metals such as Pb^{2+}, the following reaction takes place:

$$Pb^{2+} + Fe^0 \rightarrow Pb^0 + Fe^{2+} \quad (1m)$$

In this case, the ionic heavy metal reacts into its elementary form or into an insoluble salt.

With its absorption characteristic and redox potential, nZVI has become an effective material in the treatment of groundwater and soil and has been used with a wide range of contaminants, including chlorinated organic compounds, inorganic anions, and a wide range of heavy metals. It can reduce nitrate to ammonia and perchlorate to chloride. It can also remove dissolved metals from solution (e.g., Pb, Ni); nZVI has also been used to immobilize Cr(VI) in chromium ore processing residue.[83]

Soil and Groundwater Remediation by nZVI

Recently, the rapid development of nanotechnology has driven considerable research into using nZVI to remove toxic pollutants.[82] Tremendous success in using nZVI has been found in the application of soil and groundwater remediation. The use of nZVI provides more rapid or cost-effective cleanup of wastes compared with conventional iron-based and other technologies.[89] Currently, technologies for land and groundwater remediation using nZVI have been rapidly transferred from the laboratory to field-scale application to full-scale commercial applications.[90] Because of their small particle size, nZVI is a promising, environment-friendly material used in removing contaminants from wastewater. It can effectively remediate a wide range of toxic pollutants. It was used for in situ dechlorination reduction of several chlorinated aliphatic compounds and polychlorinated biphenyls.[24,25] In addition, the reduction of aromatic nitro compounds, in which the nitro groups are transformed into amines groups, has been reported.[91] A summary of halogenated organic compounds, aromatic nitro compounds, phenol, benzene, and other pollutants in soil and groundwater that can be treated by nZVI is shown in Table 4.

nZVI is an effective material in both in situ and ex situ soil and groundwater remediation. It can treat dissolved chlorinated solvents and also remediate dense nonaqueous phase liquid (DNAPL) contaminant source zones.[105] The nZVI is typically injected as a slurry directly into the subsurface environment to remediate groundwater contamination plumes or contaminant source zones to prevent particle agglomeration and enhance reactivity and mobility. It is also used as a reactive material in permeable reactive barriers, which have been developed as alternatives for the conventional *pump-and-treat* technology. When ZVI is synthesized on a nanoscale, the uptake capacity is increased largely due to the increase in the surface area and the density of reactive sites.[106] The increase in the specific surface area results in high reaction rates with chlorinated organic compounds. In addition, with its small size and ability to resist aggregation and remain dispersed in water under the right conditions, nZVI can be easily delivered to contaminated aquifers.[105,107] nZVI has been used in several projects of in situ remediation of DNAPL source zones.[105,107] However, the potential influence of nZVI on the community of indigenous microorganisms that participate in the remediation process remains unknown.[108]

Several researchers have used nZVI in groundwater and soil remediation of DNAPL.[26,27,108,109] If the reductive treatment of DNAPL with nZVI and microbial dechlorination activity were synergistic, nZVI was able to do the following:[108]

1. Abiotically degrade a large fraction of the DNAPL mass in the source zone directly, and subsequent biological dechlorination might serve as a "polishing" step to remediate residual chloroethenes.
2. Provide electron donors for dechlorinating bacteria to eliminate the need for additional electron donor addition (e.g., lactate).
3. Reduce the time for site closure by quickly dechlorinating the DNAPL mass.
4. Mitigate the toxicity of DNAPL on dechlorinating bacteria by lowering the aqueous DNAPL concentration in the source zone.

While the use of nZVI for the removal of DNAPL and other organic contaminants has been intensively studied, the application of nZVI in removing or stabilizing several types of heavy metals and radioisotope has also been demonstrated in a variety of soil and water media. It was recently reported that nZVI was an ideal candidate for the in situ remediation of arsenic-contaminated groundwater.[110,111] It may also be a promising material for arsenic removal from drinking water due to its high arsenic adsorption capacity. The use of nanoscale magnetite crystals to remove Cr(VI) from wastewater has also been examined.[112] The potential to use stabilized Fe (ZVI and iron phosphate based) nanoparticles as soil amendments to reductively immobilize Cr(VI) and also to reduce the bioaccessibility of Cu in soil was demonstrated.[113,114] A list of heavy metals and radioisotopes that have been transformed to nontoxic species by nZVI is shown in Table 5.

According to several authors,[90,102,105] nZVI has already successfully been used in pilot-scale demonstrations, and field-scale commercial applications of nZVI are currently becoming common.[118] Granular ZVI has been used in reactive barriers at numerous sites all over the world for the removal of organic and inorganic contaminants. There are

Table 4 Some examples of contaminants that can be treated by nanoscale ZVI.

Class	Examples	Initial concentration	Treatment condition	Products	References
Halogenated organic compounds	Trichloroethylene (TCE)	20 ppm	nZVI 20 g/L, residence time = 1.7 hr	Dichloromethane, 1,2-dichloroethylene, methane, ethane	[24,27]
	Trichloroethylene (TCE)	0.04 mM	nZVI 2 g/L, experimental time = 8 mo	n.a.	[26]
	Polychlorinated biphenyls (PCBs) and a mixture of PCBs	5 ppm	nZVI 50 g/L, residence time = 17 hr	n.a.	[24,25,92]
	1,1,1-Trichloroethane (1,1,1-TCA)	0.1 mM	nZVI 0.4 g/L and magnetite 20 g/L, residence time = 5 hr	n.a.	[93]
	Polybrominated diphenyl ethers (PBDEs)	70 ppm	nZVI 20 g/L, residence time = 17 hr	Brominated diphenyl ethers (BDEs)	[94]
	Decabromodiphenyl ether (BDE 209)	10.5 ppm	nZVI 0.5 g/L, residence time = 244 hr	Octabromodiphenyl ether, nonabromodiphenyl ether	[95]
	Chlorinated ethenes	20 ppm	nZVI 5 g/L, residence time = 90 min	n.a.	[96]
Aromatic nitro compounds	Nitroaromatic compounds (nitrobenzene, nitrotoluene, dinitrobenzene and dinitrotoluene)	10 ppm (each)	nZVI 5 g/L, residence time = 30 min	CO_2	[25]
	2,4,6-Trinitrotoluene (TNT)	80 ppm	nZVI 5 g/L, residence time = 30 min	Dinitrotoluene	[97]
	2,4,6-Trinitrotoluene (TNT)	70 ppm	nZVI 1%, residence time = 8 hr	n.a.	[98]
Aromatic nitro compounds	RDX	32 ppm	nZVI 1% in soil, residence = time 96 hr	CO_2	[98]
Phenols	p-Chlorophenol	50 mg/L	nZVI 0.3 g/L, residence time = 6 hr	Phenol, CO_2	[99]
	Pentachlorophenol	1000 mg/kg soil	nZVI 1% in soil, residence = time 2 days	n.a.	[100]
	Carbothioate herbicide	100 ppb	nZVI 21.4 mM, pH 4, residence time = 3.5 hr	Oxygenated hexahydroazipine isocyanate ions	[101]
Pesticide, herbicide	Organochlorine pesticide	2 ppm	nZVI 1.7 kg on 4.5 m × 3.0 m area, experimental time = 2 days	n.a.	[102]
	1,2,4-Trichlorobenzene	170 μM	nZVI 1.65 g/L with 1.0% Pd, residence time = 90 min	Dichlorobenzene, benzene	[103]
Benzene and its derivatives	Hexachlorobenzene	90 ppm	nZVI 100 g/L with 1.0% Pd, residence time = 24 hr	Pentrachlorobenzene, tetrachlorobenzene, trichlorobenzene, dichlorobenzene	[104]

n.a., not available.

two possibilities for groundwater remediation on a field scale with nZVI:[90] 1) the injection of immobilized nZVI to form a zone of iron particles that adsorb onto aquifer solids and 2) the injection of mobile nZVI to form a plume of reactive iron to destroy any organic contaminants that dissolve from a DNAPL in an aquifer. Currently, numerous companies in the United States are working with nZVI for environmental remediation, such as Pars Environmental (http://www.parsenviro.com), GeoSyntec Consultants (http://www.geosyntec.com), RNAS, Inc. (http://www.rnasinc.com), and Toxicological & Environmental Associates (TEA), Inc. (http://www.teainconline.com). In South America, Nanoteksa (http://www.nanoteksa.com/) produces nanoscale iron and distributes these nanoparticles to many places. Some companies such as Weston Solutions, Inc. (http://www.westonsolutions.com) have office

Table 5 Some examples of heavy metals and radioisotope that can be transformed by nanoscale ZVI.

Pollutant	Type	Initial concentration	Treatment condition	Final product	Reference
Heavy metals	Cr(VI)	1.5 mM CrO_3	nZVI 1 g/L, initial pH: 2.67, experimental time = 60 days	$(Cr_{.67}Fe_{.33})(OH)_{3(s)}$	[112,115,116]
	Pb(II)	1.500 mM $Pb(NO_3)_2$	nZVI 1 g/L, initial pH: 2.92, experimental time = 60 days	$Pb(OH)_2$	[115]
	As(III)	1–10 ppm	nZVI 1 g/L, initial pH: 6.5, experimental time = 18 hr	n.a.	[110]
	As(V)	5 mg/L $NaAsO_2$,	nZVI 2 g/kg (0.2 wt.%), experimental time = 112 days at 12°C	$(Fe^{2+}(UO_2)(AsO_4) \cdot 8H_2O)$	[111,117]
	Cr	34 ppm	nZVI 0.08 g/L, pH 9, experimental time = 80 min	$Cr_{0.75}Fe_{0.25}(OH)_3$	[113,118]
	U	1000 ppm	nZVI 0.05 g/L, experimental time = 28 days	n.a.	[119,120]
	U	15 ppm $UO_2(NO_3)_2 \cdot 6H_2O$	nZVI 2 g/kg (0.2 wt.%), experimental time = 112 days at 12°C	$(Fe^{2+}(UO_2)(AsO_4) \cdot 8H_2O)$	[117]
	Ni	100 ppm	nZVI 5 g/L, experimental time = 3 hr	n.a.	[118]
	Cd	100 ppm	nZVI 5 g/L, experimental time = 3 hr	n.a.	[118]
	Ag	100 ppm	nZVI 5 g/L, experimental time = 3 hr	n.a.	[118]
	Zn	100 ppm	nZVI 5 g/L, experimental time = 3 hr	n.a.	[118]
Heavy metals	Ba	10^{-3} to 10^{-6} M	nZVI 10 g/L, experimental time = 24 hr	n.a.	[121]
Radioisotope	TcO_4	0.51 mM	nZVI 2.33 g/L, experimental time = 24 hr	n.a.	[116,122]
	Co	1 to 1000 mg/L	nZVI 0.5–2 g/L, experimental time = 24 hr	$CoOH^+$ and $Co(OH)_2$	[123]

n.a., not available.

locations in many countries such as Korea, Japan, the UAE, Afghanistan, Canada, and India. These companies are marketing nZVI for the purpose of remediation of soil and groundwater because it has become the most promising technique among the current advanced technologies.

Even though several companies are working with nZVI for the remediation of contaminated soil and groundwater, there are still issues that need to be solved by researchers, including the following:

- The technical difficulties of disseminating the iron belowground
- The nZVI technology suitable for large-scale applications in terms of a cost–benefit analysis
- The proper cost-effective methods of handling, mixing, and injecting the nZVI suspension for soil and water remediation
- Information from long-term experience with the technology (e.g., How ready is the technology? What kinds of risks are there? How effective is it?)

NOBLE METAL NANOPARTICLES (SILVER AND GOLD)

Mechanism in Pollutant Removal

Noble metals are metals that are resistant to corrosion and oxidation in moist air. They are ruthenium, rhodium, palladium, silver, osmium, iridium, platinum, and gold. The most well-known noble metal nanoparticles are silver nanoparticles and gold nanoparticles. These nanoparticles are widely used to remove three major types of contaminants, including halogenated organics such as pesticides, heavy metals, and microorganisms. The mechanism in pollutant removal is based on the chemical reactivity of these chemicals. The reason for the chemical reactivity of these noble metals is related to their standard reduction potentials that are different from other metals. Metals are usually electropositive and have a tendency to lose electrons depending on the corresponding ionization energy. The reduction potential is thus correlated with the electropositive nature of the metal. Thus, a metal with a high electropositive nature is likely to exist as an ion in the solution phase and thus is a strong reducing agent. Normally, metals belong to two groups. The d-block metals belong to the moderately reducing group ($Cd^{2+}|Cd = -0.40$ V, $Fe^{2+}|Fe = -0.44$ V), whereas s-block metals belong to the strongly reducing group ($Li^+|Li = -3.05$ V, $Na^+|Na = -2.71$ V). However, there are exceptions to this rule with gold ($Au^{3+}|Au = 1.5$ V), silver ($Ag^+|Ag = 0.80$ V), mercury ($Hg^{2+}|Hg = 0.87$ V), platinum ($Pt^{2+}|Pt = 1.2$ V), and palladium ($Pd^{2+}|Pd = 0.83$ V). The category of metals exhibiting this exception, thus, can exist in the metallic state, without any oxidative effects of oxygen or water ($O_2|OH^- = 0.40$ V). This property of existing in the metallic state renders the noble metals highly inactive for any chemical reactions.[32]

Some examples of the pollutant removal mechanisms by noble nanoparticles are described by Nair and Pradeep[8] who used silver nanoparticles in the dechlorination of carbon tetrachloride (CCl_4). The reductive dehalogenation of carbon tetrachloride can be represented by:

$$Ag + CCl_4 \rightarrow AgCl + C \quad \Delta H = 1.2 \text{ kJ mol}^{-1}$$

The reaction is nearly thermoneutral with the bulk noble metal at room temperature. The dechlorination reaction of carbon tetrachloride can be described by

$$Ag \rightarrow Ag^+ + e^-$$
$$CCl_4 + H^+ + 2e^- \rightarrow CHCl_3 + Cl^-$$
$$CHCl_3 + H^+ + 2e^- \rightarrow CH_2Cl_2 + Cl^-$$

One interesting application area of noble metal nanoparticles in drinking water purification is the sequestration of heavy metals. For heavy metal removal, there are two major processes: adsorption and redox reaction. In adsorption, the zero-valent form of heavy metals can be adsorbed by noble metal nanoparticles that have a high adsorption capacity. This behavior can be described by the reaction of silver particles and mercury.[32] Silver has shown its ability to participate in a reaction as a reducing agent for mercury and to exist with mercury in different phases.

As the noble metal decreases in size and approaches the nano-level, its properties change significantly from bulk chemical properties to nanoparticle properties. The properties of metals do not stay the same as the size of the crystal reaches the nanometer scale. For example, in a redox reaction, the redox potential of the microelectrode of silver nanoparticles has been reported to become more positive as the particles grow.[124]

$$Ag_n \leftrightarrow Ag_n^+ + e^-$$

For $n = 1$ (free silver atom), the potential is -1.8 V, and as n approaches infinity, the potential is 0.799 V. The change in the reduction potential of silver nanoparticles can also be illustrated with the Ag–HCl system.[125] In bulk form, silver is not attacked by HCl. However, silver nanoparticles prepared by the reduction of silver ions with sodium borohydride exhibit an unusually high reactivity with HCl. The dissolution of HCl occurs, and a white residue is obtained after the reaction of the silver nanoparticles with HCl. The reaction may be as follows:

$$Ag(S) + H^+ + Cl^- \rightarrow AgCl(s) + H_2 \quad E^0 = -0.22 \text{ V}$$

There is no driving force for the reaction with bulk silver, but at the nanoscale, the feasibility of the reaction (i.e., $E > 0$) confirms the increased reducing nature of the silver nanoparticle surfaces. With this redox property, noble nanoparticles, especially silver and gold nanoparticles, have a superb ability to reduce a large number of heavy metals.

Bactericidal Effect of Noble Nanoparticles

Among the several types of noble nanoparticles, silver nanoparticles have been shown for over a decade to have a superb antibacterial property. Silver in its ionic form (Ag^+) is commonly used against many species of bacteria, including *Escherichia coli*.[126–129] In comparison, silver, in the nanoparticle form, has demonstrated a similar effect at lower concentrations than with the ionic form (Ag^+).[28,29] This ability may come from the unique nanoparticle characteristics (such as large specific surface area, modified structure, and controlled surface composition and reactivity) that endow them with remarkable physical, chemical, and biological properties.[28–31] The size of the nanoparticles is important for the antibacterial property. As the particle size decreases to the nanoscale, a large surface area that improves the bactericidal effect, compared with the larger particles, is available for interaction. Silver nanoparticles have become well-known antibacterial materials owing to this superb ability.

The mechanism of the bactericidal effect of silver nanoparticles is discussed in many studies.[30,130] The bactericidal property of silver nanoparticles is partially similar to that of silver ions.[29] They may attach to phosphate and sulfur groups that are part of the phospholipid cell membrane or to membrane proteins and severely damage the cell and its major functions, such as cellular permeability, regulation of enzymatic signaling activity, and cellular oxidation and respiratory processes.[29,30,131–134] Silver nanoparticles may attach to the surface of the cell membrane and disturb the permeability and respiration functions of the cell.[130] It is also possible that silver nanoparticles not only interact with the surface of the membrane but also can penetrate inside the bacteria cell[30] and accumulate to toxic levels that may cause death of the organism.[133] In addition, silver nanoparticles can bind to the DNA inside the bacterial cells, preventing its replication or interact with the bacterial ribosomes.[30,126,135] Both Ag^+ and free radicals derived from the silver nanoparticles may be the reason for the antimicrobial activity.[30,136,137]

CONCLUSION

Nanotechnology for environmental abatement is an innovative technology, which is emerging from the rapid progress in nanoscience and nanoengineering. Currently, environmental nanomaterials include nano-semiconductor catalysts (TiO_2, ZnO, Fe_2O_3, CdS, and ZnS), ZVI nanoparticles, and noble nanoparticles, such as silver or gold nanoparticles, and are widely used to remove several toxic chemicals in an increasing number of applications, such as wastewater treatment, drinking water supply, and contaminated land and groundwater remediation. The increase in the surface-to-volume ratio with the reduction of the particle size from bulk to nanoscale leads to changes in several properties that differ from the original bulk material. The availability of higher numbers of atoms or molecules on the surface of the nanomaterial leads to the higher efficiency in contaminant adsorption and removal. Moreover, nanomaterial also provides higher reactivity than other materials. These advantages of nanomaterial improve the degradation of toxic organic pollutants, dechlorination of chlorinated organic compounds, deactivation of bacteria, and transformation of toxic heavy metals to nontoxic forms. For widespread application and adoption, these nanomaterials need to be shown to be low-cost and capable of being applied on the field scale with often relatively low contaminant concentrations.

The next research area in the application of nanomaterials for environmental abatement is the improvement of technology for practicality in fieldwork. The major challenges in implementing these environmental nanotechnologies are the same as those that apply to all new technology and their commercializations. While laboratory studies are unable to precisely simulate the environmental conditions, fieldwork can rarely yield reproducible results. The limitations of environmental nanotechnologies under different environmental conditions need to be clarified and quantified. In addition, the competitiveness of new technologies in comparison to existing technologies needs to be analyzed in several areas, such as chemical handling and storage, the operation and maintenance of the involved equipment, and cost–benefit analyses. Before applying these technologies on the industrial scale, it is important that the uncertainties over the health impacts and environmental fate of these nanoparticles be addressed and extensively researched.

REFERENCES

1. Wintgens, T.; Salehi, F.; Hochstrat, R.; Melin, T. Emerging contaminants and treatment options in water recycling for indirect potable use. *Water Sci.* Technol. **2008**, *57* (1), 99–107.
2. Richardson, S.D. Environmental mass spectrometry: Emerging contaminants and current issues. Anal. Chem. **2008**, *80* (12), 4373–4402.
3. Suárez, S.; Carballa, M.; Omil, F.; Lema, J.M. How are pharmaceutical and personal care products (PPCPs) removed urban wastewaters? Rev. Environ. Sci. Biotechnol. **2008**, *7* (2), 125–138.
4. Alberici, R.M.; Jardim, W.E. Photocatalytic destruction of VOCs in the gas-phase using titanium dioxide. Appl. Catal., B **1997**, *14* (1–2), 55–68.
5. Boehm, F. *Nanotechnology in Environmental Applications*, Report NAN039A; BCC Research: Norwalk, CT, 2006.
6. Boehm, F. *Nanotechnology in Environmental Applications: The Global Market*, Report NAN039B; BCC Research: Norwalk, CT, 2009.
7. Henglein, A. Small-particle research: Physicochemical properties of extremely small colloidal metal and semiconductor particles. Chem. Rev. **1989**, *89* (8), 1861–1873.
8. Nair, A.S.; Pradeep, T. Halocarbon mineralization and catalytic destruction by metal nanoparticles. Curr. Sci. **2003**, *84* (12), 1560–1563.
9. Ananpattarachai, J.; Kajitvichyanukul, P.; Seraphin, S. Visible light absorption ability and photocatalytic oxidation activity of various interstitial N-doped TiO_2 prepared from different nitrogen dopants. J. Hazard. Mater. **2009**, *168* (1), 253–261.
10. Chanmanee, W.; Watcharenwong, A.; Chenthamarakshan, C.R.; Kajitvichyanukul, P.; de Tacconi, N.R.; Rajeshwar, K. Titania nanotubes from pulse anodization of titanium foils. Electrochem. Commun. **2007**, *9* (8), 2145–2149.
11. Chanmanee, W.; Watcharenwong, A.; Chenthamarakshan, C.R.; Kajitvichyanukul, P.; de Tacconi, N.R.; Rajeshwar, K. Formation and characterization of self-organized TiO_2 nanotube arrays by pulse anodization. J. Am. Chem. Soc. **2008**, *130* (3), 965–974.
12. Jahromi, H.S.; Taghdisian, H.; Tasharrofi, S. Synthesis of TiO_2 nanopellet by $TiCl_4$ as a precursor for degradation of RhodamineB. Mater. Chem. Phys. **2010**, *122* (1), 205–210.
13. Kajitvichyanukul, P.; Amornchat, P. Effects of diethylene glycol on TiO_2 thin film properties prepared by sol-gel process. Sci. Technol. Adv. Mater. J. **2005**, *6* (3–4), 344–347.
14. Kajitvichyanukul, P.; Ananpattarachai, J.; Pongpom, S. Sol–gel preparation and properties study of TiO_2 thin film for photoreduction of chromium(VI) in photocatalysis process. Sci. Technol. Adv. Mater. J. **2005**, *6* (3–4), 352–358.
15. Macak, J.M.; Tsuchiya, H.; Ghicov, A.; Yasuda, K.; Hahn, R.; Bauer, S.; Schmuki, P. TiO_2 nanotubes: Self-organized electrochemical formation, properties and applications. Curr. Opin. Solid State Mater. Sci. **2007**, *11* (1–2), 3–18.
16. Pavasupree, S.; Jitputti, J.; Ngamsinlapasathian, S.; Yoshikawa, S. Hydrothermal synthesis, characterization, photocatalytic activity and dye-sensitized solar cell performance of mesoporous anatase TiO_2 nanopowders. Mater. Res. Bull. **2008**, *43* (1), 149–157.
17. Venkatachalam, N.; Palanichamy, M.; Arabindoo, B.; Murugesan, V. Enhanced photocatalytic degradation of 4-chlorophenol by Zr^{4+} doped nano TiO_2. J. Mol. Catal. A: Chem. **2007**, *266* (1–2), 158–165.
18. Vijay, M.; Selvarajan, V.; Sreekumar, K.P.; Yu, J.; Liu, S.; Ananthapadmanabhan, P.V. Characterization and visible light photocatalytic properties of nanocrystalline TiO_2 synthesized by reactive plasma processing. Sol. Energy Mater. Sol. Cells **2009**, *93* (9), 1540–1549.
19. Xie, M.; Jing, L.; Zhou, J.; Lin, J.; Fu, H. Synthesis of nanocrystalline anatase TiO_2 by one-pot two-phase separated hydrolysis-solvothermal processes and its high activity for photocatalytic degradation of rhodamine B. J. Hazard. Mater. **2010**, *176* (1–3), 139–145.
20. Zlamal, M.; Macak, J.M.; Schmuki, P.; Krýsa, J. Electrochemically assisted photocatalysis on self-organized TiO_2 nanotubes. Electrochem. Commun. **2007**, *9* (12), 2822–2826.
21. Daneshvar, N.; Aber, S.; Seyed Dorraji, M.S.; Khataee, A.R.; Rasoulifard, M.H. Photocatalytic degradation of the insecticide diazinon in the presence of prepared nanocrystalline ZnO powders under irradiation of UV-C light. Sep. Purif. Technol. **2007**, *58* (1), 91–98.
22. Kim, S.-J.; Park, D.-W. Preparation of ZnO nanopowders by thermal plasma and characterization of photo-catalytic property. Appl. Surf. Sci. **2009**, *255* (10), 5363–5367.

23. Watcharenwong, A.; Chanmanee, W.; Chenthamarakshan, C.R.; de Tacconi, N.R.; Kajitvichyanukul, P.; Rajeshwar, K. Anodic growth of nanoporous WO_3 films: Morphology, photoelectrochemical response and photocatalytic activity for methylene blue and hexavalent chrome conversion. J. Electroanal. Chem. **2008**, *612* (1), 112–120.
24. Wang, C.B.; Zhang, W.X. Synthesizing nanoscale iron particles for rapid and complete dechlorination of TCE and PCBs. Environ. Sci. Technol. **1997**, *31* (7), 2154–2156.
25. Choe, S.H.; Lee, S.H.; Chang, Y.Y. Rapid reductive destruction of hazardous organic compounds by nanoscale Fe^0. Chemosphere **2001**, *42* (4), 367–372.
26. Liu, Y.Q.; Lowry, G.V. Effect of particle age (Fe^0 content) and solution pH on NZVI reactivity: H_2 evolution and TCE dechlorination. Environ. Sci. Technol. **2006**, *40* (19), 6085–6090.
27. Kim, H.; Hong, H.-J.; Jung, J.; Kim, S.-H.; Yang, J.-W. Degradation of trichloroethylene (TCE) by nanoscale zerovalent iron (nZVI) immobilized in alginate bead. J. Hazard. Mater. **2010**, *176* (1–3), 1038–1043.
28. Lok, C.N.; Ho, C.M.; Chen, R.; He, Q.Y.; Yu, W.Y.; Sun, H.; Tam, P.K.H.; Chiu, J.F.; Che, C.M. Proteomic analysis of the mode of antibacterial action of silver nanoparticles. J. Proteome Res. **2006**, *5* (4), 916–924.
29. Pal, S.; Tak, Y.K.; Song, J.M. Does the antibacterial activity of silver nanoparticles depend on the shape of the nanoparticle? A study of the gram-negative bacterium *Escherichia coli*. Appl. Environ. Microbiol. **2007**, *73* (6), 1712–1720.
30. Morones, J.R.; Elechiguerra, J.L.; Camacho, A.; Holt, K.; Kouri, J.; Ramirez, J.T.; Yacaman, M.J. The bactericidal effect of silver nanoparticles. Nanotechnology **2005**, *16* (10), 2346–2353.
31. Parashar, U.K.; Saxena, P.S.; Srivastava, A. Role of nanomaterials in biotechnology. Dig. J. Nanomater. Biostruct. **2008**, *3* (2), 81–87.
32. Pradeep, T.; Anshup. Noble metal nanoparticles for water purification: A critical review. Thin Solid Films **2009**, *517* (24), 6441–6478.
33. Malato, S.; Fernández-Ibáñez, P.; Maldonado, M.I.; Blanco, J.; Gernjak, W. Decontamination and disinfection of water by solar photocatalysis: Recent overview and trends. Catal. Today **2009**, *147* (1), 1–59.
34. Litter, M.I. Heterogeneous photocatalysis: Transition metal ions in photocatalytic systems. Appl. Catal., B **1999**, *23* (2–3), 89–114.
35. Gaya, U.I.; Abdullah, A.H. Heterogeneous photocatalytic degradation of organic contaminants over titanium dioxide: A review of fundamentals, progress and problems. Journal of Photochemistry and Photobiology C: Photochemistry Reviews **2008**, *9* (1), 1–12.
36. Fujishima, A.; Rao, T.N.; Tryk, D.A. Titanium dioxide photocatalysis. J. Photochem. Photobiol., C **2000**, *1* (1), 1–21.
37. Chong, M.N.; Jin, B.; Chow, C.W.K.; Saint, C. Recent developments in photocatalytic water treatment technology: A review. Water Res. **2010**, *44* (10), 2997–3027.
38. Byrne J.A.; Eggins, B.R. Photoelectrochemistry of oxalate on particulate TiO_2 electrodes. J. Electroanal. Chem. **1998**, *457* (1–2), 61–72.
39. Siddiquey, I.A.; Furusawa, T.; Sato, M.; Honda, K.; Suzuki, N. Control of the photocatalytic activity of TiO_2 nanoparticles by silica coating with polydiethoxysiloxane. Dyes Pigm. **2008**, *76* (3), 754–759.
40. Lindner, M.; Bahnemann, D.W.; Hirthe, B.; Griebler, W.-D. Solar water detoxification: Novel TiO_2 powders as highly active photocatalysts. J. Sol. Energy Eng. **1997**, *119* (2), 120–125.
41. Kaitvichyanukul, P.; Chenthamarakshan, C.R.; Rajeshwar, K.; Qasim, S.R. Photo-catalytic reactivity of thallium(I) species in aqueous suspensions of titania. J. Electroanal. Chem. **2002**, *519* (1–2), 25–32.
42. Bhatkhande, D.S.; Pangarkar, V.G.; Beenackers, A. Photocatalytic degradation for environmental applications—A review. J. Chem. Technol. Biotechnol. **2001**, *77* (1), 102–116.
43. Pozzo, R.L.; Baltanás, M.A.; Cassano, A.E. Supported titanium dioxide as photocatalyst in water decontamination: state of the art. Catal. Today **1997**, *39* (3), 219–231.
44. Asahi, R.; Morikawa, T.; Ohwaki, T.; Aoki, A.; Yaga, Y. Visible-light photocatalysis in nitrogen-doped titanium oxides. Science **2001**, *293* (5528), 269–271.
45. Morikawa, T.; Asahi, R.; Ohwaki, T.; Aoki, K.; Suzuki, K.; Taga, Y. Visible-light photocatalyst-nitrogen-doped titanium dioxide. R&D Rev. Toyota CRDL **2005**, *40* (1) 45–50.
46. Ananpattarachai, J.; Kajitvichyanukul, P.; Seraphin, S. Visible light absorption ability and photocatalytic oxidation activity of various interstitial N-doped TiO_2 prepared from different nitrogen dopants. J. Hazard. Mater. **2009**, *168* (1), 253–261.
47. Valentin, D.D.; Finazzi, E.; Pacchioni, G.; Selloni, A.; Livraghi, S.; Paganini, M.C.; Giamiello, E. N-doped TiO_2: Theory and experiment. Chem. Phys. **2007**, *339* (1–3), 44–56.
48. Shanmugasundaram, S.; Horst, K. Daylight photocatalysis by carbon-modified titanium dioxide. Angew. Chem., Int. Ed. **2003**, *42* (40), 4908–4911.
49. Konstantinou, I.K.; Sakkas, V.A.; Albanis, T.A. Photocatalytic degradation of propachlor in aqueous TiO_2 suspensions. Determination of the reaction pathway and identification of intermediate products by various analytical methods. Water Res. **2002**, *36* (11), 2733–2742.
50. Tariq, M.A.; Faisal, M.; Muneera, M.; Bahnemann, D. Photochemical reactions of a few selected pesticide derivatives and other priority organic pollutants in aqueous suspensions of titanium dioxide. J. Mol. Catal. A: Chem. **2007**, *265* (1–2), 231–236.
51. Bhatkhande, D.S.; Kamble, S.P.; Sawant, S.B.; Pangarkar, V.G. Photocatalytic and photochemical degradation of nitrobenzene using artificial ultraviolet light. Chem. Eng. J. **2004**, *102* (3), 283–290.
52. Palmisano, G.; Addamo, M.; Augugliaro, V.; Caronna, T.; Di Paola, A.; García López, E.; Loddo, V.; Marcì, G.; Palmisano, L.; Schiavello, M. Selectivity of hydroxyl radical in the partial oxidation of aromatic compounds in heterogeneous photocatalysis. Catal. Today **2007**, *122* (1–2), 118–127.
53. Zhang, L.; Li, P.; Gong, Z.; Li, X. Photocatalytic degradation of polycyclic aromatic hydrocarbons on soil surfaces using TiO_2 under UV light. J. Hazard. Mater. **2008**, *158* (2–3), 478–484.
54. Kim, H.; Choi, W. Effects of surface fluorination of TiO_2 on photocatalytic oxidation of gaseous acetaldehyde. Appl. Catal., B **2007**, *69* (3–4), 127–132.

55. Singh, H.K.; Saquib, M.; Haque, M.M.; Muneer, M.; Bahnemann, D.W. Titanium dioxide mediated photocatalysed degradation of phenoxyacetic acid and 2,4,5-trichlorophenoxyacetic acid, in aqueous suspensions. J. Mol. Catal. A: Chem. **2007**, *264* (1–2), 66–72.
56. Harada, H.; Tanaka, H. Sonophotocatalysis of oxalic acid solution. Ultrasonics **2006**, *44* (1), e385–e388.
57. Chu, W.; Choy, W.K.; So, T.Y. The effect of solution pH and peroxide in the TiO_2-induced photocatalysis of chlorinated aniline. J. Hazard. Mater. **2007**, *141* (1), 86–91.
58. Ku, Y.; Lee, Y.-C.; Wang, W.-U. Photocatalytic decomposition of 2-chlorophenol in aqueous solution by UV/TiO_2 process with applied external bias voltage. J. Hazard. Mater. **2006**, *138* (2), 350–356.
59. Bayarri, B.; Giménez, J.; Curcó, D.; Esplugas, S. Photocatalytic degradation of 2,4-dichlorophenol by TiO_2/UV: Kinetics, actinometries and models. Catal. Today **2005**, *101* (3–4), 227–236.
60. Essam, T.; Amin, M.A.; Tayeb, O.E.; Mattiasson, B.; Guieysse, B. Sequential photochemical–biological degradation of chlorophenols. Chemosphere **2007**, *66* (11), 2201–2209.
61. Mrowetz, M.; Pirola, C.; Selli, E. Degradation of organic water pollutants through sonophotocatalysis in the presence of TiO_2. Ultrason. Sonochem. **2003**, *10* (4–5), 247–254.
62. Qamar, M.; Saquib, M.; Muneer, M. Semiconductor-mediated photocatalytic degradation of anazo dye, chrysoidine Y in aqueous suspensions. Desalination **2004**, *171* (2), 185–193.
63. Faisal, M.; Tariq, M.A.; Muneer, M. Photocatalysed degradation of two selected dyes in UV-irradiated aqueous suspensions of titania. Dyes Pigm. **2007**, *72* (2), 233–239.
64. Zainal, Z.; Hui, L.K.; Hussein, M.Z.; Taufiq-Yap, Y.H.; Abdullah, A.H.; Ramli, I. Removal of dyes using immobilized titanium dioxide illuminated by fluorescent lamps. J. Hazard. Mater. **2005**, *125* (1–3), 113–120.
65. Bertelli, M.; Selli, E. Kinetic analysis on the combined use of photocatalysis, H_2O_2 photolysis, and sonolysis in the degradation of methyl *tert*-butyl ether. Appl. Catal., B **2004**, *52* (3), 205–212.
66. Selvam, K.; Muruganandham, M.; Muthuvel, I.; Swaminathan, M. The influence of inorganic oxidants and metal ions on semiconductor sensitized photodegradation of 4-fluorophenol. Chem. Eng. J. **2007**, *128* (1), 51–57.
67. Danion, A.; Disdier, J.; Guillard, C.; Païssé, O.; Jaffrezic-Renault, N. Photocatalytic degradation of imidazolinone fungicide in TiO_2-coated optical fiber reactor. Appl. Catal., B **2006**, *62* (3–4), 274–281.
68. Haque, M.M.; Muneer, M. Heterogeneous photocatalysed degradation of a herbicide derivative, isoproturon in aqueous suspension of titanium dioxide. J. Environ. Manage. **2003**, *69* (2), 169–176.
69. Vorontsov, A.V.; Savinov, E.N.; Smirniotis, P.G. Vibrofluidized- and fixed-bed photocatalytic reactors: Case of gaseous acetone photooxidation. Chem. Eng. Sci. **2000**, *55* (21), 5089–5098.
70. Dillert, R.; Bahnemann, D.; Hidaka, H. Light-induced degradation of perfluorocarboxylic acids in the presence of titanium dioxide. Chemosphere **2007**, *67* (4), 785–792.
71. Reyes, C.; Fernández, J.; Freer, J.; Mondaca, M.A.; Zaror, C.; Malato, S.; Mansilla, H.D. Degradation and inactivation of tetracycline by TiO_2 photocatalysis. J. Photochem. Photobiol., A **2006**, *184* (1–2), 141–146.
72. Augugliaro, V.; García-López, E.; Loddo, V.; Malato-Rodríguez, S.; Maldonado, I.; Marcì, G.; Molinari, R.; Palmisano, L. Degradation of lincomycin in aqueous medium: Coupling of solar photocatalysis and membrane separation. Sol. Energy **2005**, *79* (4), 402–408.
73. Horikoshi, S.; Hidaka, H.; Serpone, N. Photocatalyzed degradation of polymers in aqueous semiconductor suspensions: V. Photomineralization of lactam ring-pendant polyvinylpyrrolidone at titania/water interfaces. J. Photochem. Photobiol., A **2001**, *138* (1), 69–77.
74. Li, D.; Haneda, H.; Hishita, S.; Ohashi, N. Visible-light-driven nitrogen-doped TiO_2 photocatalysts: Effect of nitrogen precursors on their photocatalysis for decomposition of gas-phase organic pollutants. Mater. Sci. Eng. B **2005**, *117* (1), 67–75.
75. Ao, C.H.; Lee, S.C. Enhancement effect of TiO_2 immobilized on activated carbon filter for the photodegradation of pollutants at typical indoor air level. Appl. Catal., B **2003**, *44* (3), 191–205.
76. Jo, W.K.; Park, J.H.; Chun, H.D. Photocatalytic destruction of VOCs for in-vehicle air cleaning. J. Photochem. Photobiol., A **2002**, *148* (1–3), 109–119.
77. Li, F.B.; Li, X.Z.; Ao, C.H.; Lee, S.C.; Hou, M.F. Enhanced photocatalytic degradation of VOCs using Ln^{3+}–TiO_2 catalysts for indoor air purification. Chemosphere **2005**, *59* (6), 787–800.
78. Strini, A.; Cassese, S.; Schiavi, L. Measurement of benzene, toluene, ethylbenzene and *o*-xylene gas phase photodegradation by titanium dioxide dispersed in cementitious materials using a mixed flow reactor. Appl. Catal., B **2005**, *61* (1–2), 90–97.
79. Jo, W.-K.; Park, K.-H. Heterogeneous photocatalysis of aromatic and chlorinated volatile organic compounds (VOCs) for non-occupational indoor air application. Chemosphere **2004**, *57* (7), 555–565.
80. Wang, X.C.; Yu, J.C.; Chen, Y.L.; Wu, L.; Fu, X.Z. ZrO_2-modified mesoporous manocrystalline TiO_2-xN_x as efficient visible light photocatalysts. Environ. Sci. Technol. **2006**, *40* (7), 2369–2374.
81. Ihara, T.; Miyoshi, M.; Iriyama, Y.; Matsumoto, O.; Sugihara, S. Visible-light-active titanium oxide photocatalyst realized by an oxygen-deficient structure and by nitrogen doping. Appl. Catal., B **2003**, *42* (4), 403–409.
82. Cundy, A.B.; Hopkinson, L.; Whitby, R.L.D. Use of iron-based technologies in contaminated land and groundwater remediation: A review. Sci. Total Environ. **2008**, *400* (1–3), 42–51.
83. Sun, Y.-P.; Li, X.-Q.; Cao, J.; Zhang, W.-X.; Wang, H.P. Characterization of zero-valent iron nanoparticles. Adv. Colloid Interface Sci. **2006**, *120* (1–3), 47–56.
84. Dickinson, M.; Scott, T.B. Kinetic and mechanistic examinations of reductive transformation pathways of brominated methanes with nano-scale Fe and Ni/Fe particles. J. Hazard. Mater. **2010**, *178* (1–3), 171–179.
85. Cornell, R.M.; Schwertmann, U. *The Iron Oxides: Structure, Properties, Reactions, Occurrences, and Uses*; 2nd Ed.; Wiley-VCH: Weinheim, 2003.
86. Metheson, L.J.; Tratnyek, P.G. Reductive dehalogenation of chlorinated methanes by iron metal. Environ. Sci. Technol. **1994**, *28* (12), 2045–2053.

87. Klecka, G.M.; Gonsior, S.J. Reductive dechlorination of chlorinated methanes and ethanes by reduced iron (II) porphyrins. Chemosphere **1984**, *13* (3), 391–402.
88. House, H.O. *Modern Synthetic Reactions*, 2nd Ed.; Benjamin: Menlo Park, CA, 1972, 145–227.
89. U.S. Environmental Protection Agency. *Nanotechnology white paper*, EPA 100/B-07/001. 20460: Science Policy Council, U.S. Environmental Protection Agency: Washington, DC, 2007.
90. Tratnyek, P.G.; Johnson, R.L. Nanotechnologies for environmental clean-up. Nanotoday **2006**, *1* (1), 44–48.
91. Zheng, J.L.; Li, J.W.; Chao, F.H. Study progress on aniline, nitrobenzene and TNT biodegradation. Microbiology **2001**, *28* (5), 85–88.
92. Nurmi, J.T. Characterization and properties of metallic iron nanoparticles: Spectroscopy, electrochemistry, and kinetics. Environ. Sci. Technol. **2005**, *39* (5), 1221–1230.
93. Bae, S.; Lee, W. Inhibition of nZVI reactivity by magnetite during the reductive degradation of 1,1,1-TCA in nZVI/magnetite suspension. Appl. Catal., B **2010**, *96* (1–2), 10–17.
94. Shih, Y.-H.; Tai, Y.-T. Reaction of decabrominated diphenyl ether by zerovalent iron nanoparticles. Chemosphere **2010**, *78* (10), 1200–1206.
95. Söderström, G.; Sellström, U.; de Wit, C.A.; Tysklind, M., Photolytic debromination of decabromodiphenyl ether (BDE 209). Environ. Sci. Technol. **2004**, *38* (1), 127–132.
96. Lien, H.L.; Zhang, W.X. Nanoscale iron particles for complete reduction of chlorinated ethenes. Colloids Surf., A **2001**, *191* (1–2), 97–105.
97. Zhang, X.; Lin, Y.-M.; Chen, Z.-L. 2,4,6-Trinitrotoluene reduction kinetics in aqueous solution using nanoscale zero-valent iron. J. Hazard. Mater. **2009**, *165* (1–3), 923–927.
98. Hundal, L.S.; Singh, J.; Bier, E.L.; Shea, P.J.; Comfort, S.D.; Powers, W.L. Removal of TNT and RDX from water and soil using iron metal. *Environ. Pollut.* **1997**, *97* (1–2), 55–64.
99. Cheng, R.; Wang, J.L.; Zhang, W.X. Comparison of reductive dechlorination of *p*-chlorophenol using Fe^0 and nanosized Fe^0. J. Hazard. Mater. **2007**, *144* (1–2), 334–339.
100. Liao, C.-J.; Chung, T.-L.; Chen, W.-L.; Kuo, S.-L. Treatment of pentachlorophenol-contaminated soil using nanoscale zero-valent iron with hydrogen peroxide. J. Mol. Catal. A: Chem. **2007**, *265* (1–2), 189–194.
101. Joo, S.H.; Feitz, A.J.; Waite, T.D. Oxidative degradation of the carbothioate herbicide, molinate, using nanoscale zero-valent iron. Environ. Sci. Technol. **2004**, *38* (7), 2242–2247.
102. Elliott, D.W.; Zhang, W.-X. Field assessment of nanoscale bimetallic particles for groundwater treatment. *Environ. Sci. Technol.* **2001**, *35* (24), 4922–4926.
103. Zhu, B.W.; Lim, T.T.; Feng, J. Reductive dechlorination of 1,2,4-trichlorobenzene with palladized nanoscale Fe^0 particles supported on chitosan and silica. Chemosphere **2006**, *65* (7), 1137–1145.
104. Shih, Y.-H.; Chen, Y.-C.; Chen, M.-Y.; Tai, Y.-T.; Tso, C.-P. Dechlorination of exachlorobenzene by using nanoscale Fe and nanoscale Pd/Fe bimetallic particles. Colloids Surf., A **2009**, *332* (2–3), 84–89.
105. Quinn, J.; Geiger, C.; Clausen, C.; Brooks, K.; Coon, C.; O'Hara, S.; Krug, T.; Major, D.; Yoon, W.S.; Gavaskar, A.; Holdsworth, T. Field demonstration of DNAPL dehalogenation using emulsified zero-valent iron. Environ. Sci. Technol. **2005**, *39* (5), 1309–1318.
106. Blowes, D.W.; Ptacek, C.J.; Benner, S.G.; McRae Che, W.T.; Bennett, T.A.; Puls, R.W. Treatment of inorganic contaminants using permeable reactive barriers. J. Contam. Hydrol. **2000**, *45* (1–2), 123–137.
107. Saleh, N.; Phenrat, T.; Sirk, K.; Dufour, B.; Ok, J.; Sarbu, T.; Matyjaszewski, K.; Tilton, R.D.; Lowry, G.V. Adsorbed triblock copolymers deliver reactive iron nanoparticles to the oil/water interface. Nano Lett. **2005**, *5* (12), 2489–2494.
108. Xiu, Z.-M.; Jin, Z.H.; Li, T.-L.; Mahendra, S.; Lowry, G.V.; Alvarez, P.J.J. Effects of nano-scale zero-valent iron particles on a mixed culture dechlorinating trichloroethylene. Bioresour. Technol. **2010**, *101* (4), 1141–1146.
109. Wang, C.B.; Zhang, W.X. Synthesizing nanoscale iron particles for rapid and complete dechlorination of TCE and PCBs. Environ. Sci. Technol. **1997**, *31* (7), 2154–2156.
110. Kanel, S.R.; Manning, B.; Charlet, L.; Choi, H. Removal of arsenic(III) from groundwater by nanoscale zero-valent iron. Environ. Sci. Technol. **2005**, *39* (5), 1291–1298.
111. Kanel, S.R.; Greneche, J.M.; Choi, H. Arsenic(V) removal from groundwater using nano scale zero-valent iron as a colloidal reactive barrier material. Environ. Sci. Technol. **2006**, *40* (6), 2045–2050.
112. Hu, J.; Lo, I.M.; Chen, G. Removal of Cr(VI) by magnetite nanoparticle. *Water Sci.* Technol. **2004**, *50* (12), 139–146.
113. Liu, R.Q.; Zhao, D.Y. In situ immobilization of Cu(II) in soils using a new class of iron phosphate nanoparticles. Chemosphere **2007**, *68* (10), 1867–1876.
114. Xu, Y.H.; Zhao, D.Y. Reductive immobilization of chromate in water and soil using stabilized iron nanoparticles. Water Res. **2007**, *41* (10), 2101–2108.
115. Ponder, S.M.; Darab, J.G.; Mallouk, T.E. Remediation of Cr(VI) and Pb(II) aqueous solutions using supported, nanoscale zero-valent iron. Environ. Sci. Technol. **2000**, *34* (12), 2564–2569.
116. Ponder, S.M.; Darab, J.G.; Bucher, J.; Caulder, D.; Craig, I.; Davis, L.; Edelstein, N.; Lukens, W.; Nitsche, H.; Rao, L.F.; Shuh, D.K.; Mallouk, T.E. Surface chemistry and electrochemistry of supported zerovalent iron nanoparticles in the remediation of aqueous metal contaminants. Chem. Mater. **2001**, *13* (2), 479–486.
117. Burghardt, D.; Simon, E.; Knöller, K.; Kassahun, A. Immobilization of uranium and arsenic by injectible iron and hydrogen stimulated autotrophic sulphate reduction. J. Contam. Hydrol. **2007**, *94* (3–4), 305–314.
118. Li, X.-Q.; Zhang, W.-X. Sequestration of metal cations with zerovalent iron nanoparticles: a study with high resolution x-ray photoelectron spectroscopy (HR-XPS). J. Phys. Chem. C **2007**, *111* (19), 6939–6946.
119. Dickinson, M.; Scott, T.B. The application of zero-valent iron nanoparticles for the remediation of a uranium-contaminated waste effluent. J. Hazard. Mater. **2010**, *178* (1–3), 171–179.
120. Riba, O.; Scott, T.B.; Ragnarsdottir, K.V.; Allen, G.C. Reaction mechanism of uranyl in the presence of zero-valent iron nanoparticles. Geochim. Cosmochim. Acta **2008**, *72* (16), 4047–4057.
121. Çelebi, O.; Üzüm, Ç.; Shahwan, T.; Erten, H.N. A radiotracer study of the adsorption behavior of aqueous Ba^{2+} ions on nanoparticles of zero-valent iron. J. Hazard. Mater. **2007**, *148* (3), 761–767.

122. Darab, J.G.; Amonette, A.B.; Burke, D.S.D.; Orr, R.D. Removal of pertechnetate from simulated nuclear waste streams using supported zerovalent iron. Chem. Mater. **2007**, *19* (23), 5703–5713.
123. Uzum, C.; Shahwan, T.; Eroglu, A.E.; Lieberwirth, I.; Scott, T.B.; Hallam, K.R. Application of zero-valent iron nanoparticles for the removal of aqueous Co^{2+} ions under various experimental conditions. Chem. Eng. J. **2008**, *144* (2), 213–220.
124. Henglein, A. Physicochemical properties of small metal particles in solution: "Microelectrode" reactions, chemisorption, composite metal particles, and the atom-to-metal transition. J. Phys. Chem. **1993**, *97* (21), 5457–5471.
125. Li, L.; Zhu, Y.-J. High chemical reactivity of silver nanoparticles toward hydrochloric acid. J. Colloid Interface Sci. **2006**, *303* (2), 415–418.
126. Feng, Q.L.; Wu, J.; Chen, G.O.; Cui, F.Z.; Kim, T.N.; Kim, J.O. A mechanistic study of the antibacterial effect of silver ions on *Escherichia coli* and *Staphylococcus aureus*. J. Biomed. Mater. Res. **2000**, *52* (4), 662–668.
127. Silvestry-Rodriguez, N.; Sicairos-Ruelas, E.E.; Gerba, C.P.; Bright, K.R. Silver as a Disinfectant. Rev. Environ. Contam. Toxicol. **2007**, *191*, 23–45 (Book Series: Reviews of Environmental Contamination and Toxicology, Springer: New York, 2007; Vol. 191).
128. Gentry, H.; Cope, S. Using silver to reduce catheter-associated urinary tract infections. Nurs. Stand. **2005**, *19* (50), 51–54.
129. Zhao, G.; Stevens, S.E. Multiple parameters for the comprehensive evaluation of the susceptibility of *Escherichia coli* to the silver ion. BioMetals **1998**, *11* (1), 27–32.
130. Kvitek, L.; Panacek, A.; Soukupova, J.; Kolar, M.; Vecerova, R.; Prucek, R.; Holecová, M.; Zbořil, R. Effect of surfactants and polymers on stability and antibacterial activity of silver nanoparticles (NPs). J. Phys. Chem. C **2008**, *112* (15), 5825–5834.
131. Shrivastava, S.; Bera, T.; Roy, A.; Singh, G.; Ramachandrarao, P.; Dash, D. Characterization of enhanced antibacterial effects of novel silver nanoparticles. Nanotechnology **2007**, *18* (22), 225103/1–225103/9.
132. Sondi, I.; Salopek-Sondi, B. Silver nanoparticles as antimicrobial agent: A case study on *E. coli* as a model for Gram-negative bacteria. J. Colloid Interface Sci. **2004**, *275* (1), 177–182.
133. Hatchett, D.W.; White, H.S. Electrochemistry of sulfur adlayers on the low-index faces of silver. J. Phys. Chem. **1996**, *100* (23), 9854–9859.
134. Holt, K.B.; Bard, A.L. Interaction of silver(I) ions with the respiratory chain of *Escherichia coli*: An electrochemical and scanning electrochemical microscopy study of the antimicrobial mechanism of micromolar Ag^+. Biochemistry **2005**, *44* (39), 13214–13223.
135. Yamanaka, M.; Hara, K.; Kudo, J. Bactericidal actions of a silver ion solution on *Escherichia coli*, studied by energy-filtering transmission electron microscopy and proteomic analysis. Appl. Environ. Microbiol. **2005**, *71* (11), 7589–7593.
136. Fernandez, E.J.; Garcia-Barrasa, J.; Laguna, A.; Lopez-de-Luzuriaga, J.M.; Monge, M.; Torres, C. The preparation of highly active antimicrobial silver nanoparticles by an organometallic approach. Nanotechnology **2008**, *19* (18), 185602/1–185602/6.
137. Kim, J.S.; Kuk, E.; Yu, K.N.; Kim, J.H.; Park, S.H.; Lee, H.J.; Kim, S.H. Antimicrobial effects of silver nanoparticles. Nanomed.: Nanotechnol., Biol. Med. **2007**, *3* (1), 95–101.

Natural Enemies and Biocontrol: Artificial Diets

Simon Grenier
Functional Biology, Insects and Interactions, National Institute for Agricultural Research (INRA), Villeurbanne, France

Patrick De Clerq
Department of Crop Protection, Ghent University, Ghent, Belgium

Abstract
The large-scale production of arthropod parasitoids and predators may be more convenient and cost-effective when using artificial diets or media. The promising results achieved in laboratory settings open up new prospects for natural enemy producers.

INTRODUCTION

Arthropod parasitoids and predators used in biological control strategies are at present mainly produced on natural or alternative hosts or prey. However, their large-scale production may be more convenient and cost-effective when using artificial diets/media. Studies aiming at the successful development of arthropod parasitoids and predators under artificial conditions have started a long time ago, but the practical use of insects and mites grown on artificial diets is still in its infancy. Besides their use for the production of natural enemies, artificial media may be valuable tools for physiological and behavioral studies of entomophagous arthropods due to a simplification of their environment. Different types of artificial diets with or without insect additives can support the development and/or reproduction of natural enemies. Successes have been achieved for several species of parasitoids and predators but these have mainly been restricted to an experimental level. Comparisons of the performances of artificially vs. naturally reared natural enemies (as quality control) have primarily been conducted in the laboratory, and only very rarely in the field. The promising results achieved in recent years open up new prospects for natural enemy producers.

ARTIFICIAL DIETS FOR PREDATORS AND PARASITOIDS

The culture of entomophagous insects and mites involves rearing not only of the host/prey, but often also of the host's/prey's plant food, and thus requires a tritrophic level system. Different steps were taken to try to reduce the production line for entomophagous arthropods. The complete line comprises plant growing, host/prey rearing, and parasitoid/predator rearing. The simplified line includes the use of artificial diets instead of plants for the phytophagous host/prey, or of factitious hosts/prey that are easier to rear in the laboratory than the natural food (e.g., eggs of *Ephestia kuehniella* or *Sitotroga cerealella*, larvae of *Galleria mellonella* or *Tenebrio molitor*). The ultimate reduction of the production line consists only of an artificial diet for direct parasitoid/predator rearing. Mass rearing entomophagous insects on artificial media, first suggested 60 years ago, holds the promise to increase the ease and flexibility of insect production, including automation of procedures, and to reduce cost. The early and subsequent efforts at developing artificial diets have extensively been reviewed.[1–3] The basic qualitative nutritional requirements of parasitoids and predators are similar to those of free-living insects. But the very fast growth of some parasitoids such as tachinid larvae requires a perfectly well-balanced diet[4] to minimize intermediate metabolism and toxic waste product accumulation.

Essentially, two types of artificial diets can be distinguished: Those including and those excluding insect components. The availability of media without insect components offers a greater independence from insect hosts/prey, even if in some countries insect components are cheap and easily available by-products, e.g., from silk production in Asia or South America.[5] In diets containing insect additives, such varied components as hemolymph, body tissue extract, bee brood extract or powder, egg juice, or homogenate of the natural host have been used. Products of insect cell culture have also been incorporated into diets as host factors. The composition of most media for in vitro rearing of *Trichogramma* egg parasitoids is based on lepidopterous hemolymph.[6] Media for the tachinid fly *Exorista larvarum*, the chalcid wasp *Brachymeria intermedia*, and the ichneumonid wasp *Diapetimorpha introita* contain various insect components. Bee extracts or bee brood have been commonly added in diets for predatory coccinellids.[1,5] Only few diets devoid of insect additives are composed of ingredients that are fully chemically defined in their composition and structure. Besides proteins or protein hydrolysates, most of such diets contain crude or complex components, e.g., hen's egg yolk, chicken embryo extract, calf serum, cow's milk, yeast extract or hydrolysate, meat or liver extract, or plant oils. Beef or pork meat and liver have extensively been used as basic components of diets for feeding coccinellids and several predatory heteropterans.[1,5]

SUCCESSES AND FAILURES WITH ARTIFICIAL DIETS

Both biochemical and physical aspects determine the success of an artificial diet. Artificial diets should be nutritionally adequate to support development and reproduction of an insect and should be formulated in such a manner that the medium is easily recognized and accepted for feeding or oviposition; the food should be readily ingested, digested, and absorbed.[7] For parasitoids, the diet must also allow the growing larvae to satisfy other physiological needs like respiration and excretion without diet spoiling. The best results on artificial media were obtained with idiobiontic parasitoids such as egg or pupal parasitoids and with polyphagous predators. Different tachinid species were also successfully grown in vitro, but the koinobiontic Hymenoptera appear the most difficult group to be reared in vitro, probably because of a close relationship with their living host that supplies them with crucial growth factors. Ectoparasitoids are generally easier to culture in vitro than endoparasitoids for which the diet is also the living environment of the immature stages.[2] Several predatory insects have been reared for successive generations on artificial diets, including heteropterans (e.g., *Geocoris punctipes*, *Orius laevigatus*, *Podisus maculiventris*), coccinellids (e.g., *Coleomegilla maculata*, *Harmonia axyridis*), and chrysopids (e.g., *Chrysoperla carnea*, *Chrysoperla rufilabris*).[3]

Artificial rearing of natural enemies has mostly remained at an experimental level, and the practical experience with natural enemies produced in artificial conditions has remained quite limited. Wasps of the genus *Trichogramma* reared on factitious host eggs are the most common agents used worldwide in biological control in many field crops and forests. In China, *Trichogramma* spp. and *Anastatus* spp. produced on a large scale in artificial host eggs have been released on thousands of hectares of different crops with a parasitization rate above 80%, leading to an effective pest control level equal to that of naturally reared parasitoids.[5] In the U.S.A., field tests with encouraging first results were conducted using the pteromalid parasitoid *Catolaccus grandis* reared for successive generations on artificial diet for the control of the cotton boll weevil *Anthonomus grandis*.[5] Since the late 1990s, biocontrol companies in the U.S.A. and Europe have started producing a number of natural enemies (partially) on artificial diets.

QUALITY CONTROL OF NATURAL ENEMIES PRODUCED ON ARTIFICIAL DIETS

Long-term rearing on artificial diets could lead to genetic bottleneck effects inducing high selection pressure on the entomophages and possible reduction of their effectiveness. Periodic population renewals from nature may circumvent this drawback. The use of natural enemies in augmentative biological control requires a reliable mass production of good quality insects. Therefore, quality control is a key element for the efficiency and the long-term viability of biological control. The quality control procedures developed for in vivo production of entomophages could be recommended as a first approach for in vitro production.[8] Many parameters can be used as quality criteria. Size, weight, life cycle duration, survival rate, and especially fecundity, longevity, and predation/parasitization efficiency are the most relevant characters.[5] Besides its value as a quality criterion, the biochemical composition (based upon carcass analyses) of the insects produced on artificial diets may be a powerful tool for improving the composition and performance of the diets through the detection of excess or deficiency of some nutrients. Often, different criteria are closely linked; hence, the quality control process may be simplified if one easily measured parameter can be used to predict another one that is more complex or time consuming to determine (e.g., fecundity). Arguably, excellent field performance of the artificially produced natural enemy against the target pest remains the ultimate quality criterion. However, quality assessments of artificially reared natural enemies have mostly been performed at a laboratory scale or in semifield conditions, and only rarely so in practical field conditions.

CONCLUSIONS

At present, rearing systems using natural or factitious foods remain the only effective way for industrial production of most entomophagous insects and mites. However, success achieved for a restricted number of species of parasitoids (e.g., *Trichogramma* spp., *Exorista larvarum*, *Catolaccus grandis*) and predators (e.g., *Orius* spp., *Geocoris punctipes*, *Chrysoperla* spp., *Harmonia axyridis*) has prompted producers to increasingly incorporate artificial diets into their mass rearing systems. Further behavioral and physiological investigations may lead to significant improvements in artificial rearing through a better knowledge of the host–parasitoid and predator–prey relationships. Besides an easier mechanization of the production line, the use of artificial diets opens new possibilities for preimaginal conditioning of parasitoids/predators to targeted hosts/prey by adding specific chemicals in their food. Artificial diets also seem the only way of mass rearing for some middle-sized egg parasitoids (Encyrtidae, Eulophidae, Eupelmidae,

Scelionidae to name a few) that are promising pest control agents but are unable to develop normally in the small lepidopteran substitution host eggs commonly used nowadays (*Ephestia kuehniella*, *Sitotroga cerealella*).

REFERENCES

1. Thompson, S.N. Nutrition and culture of entomophagous insects. Annu. Rev. Entomol. **1999**, *44*, 561–592.
2. Grenier, S.; Greany, P.D.; Cohen, A.C. Potential for mass release of insect parasitoids and predators through development of artificial culture techniques. In *Pest Management in the Subtropics: Biological Control—A Florida Perspective*; Rosen, D., Bennett, F.D., Capinera, J.L., Eds.; Intercept: Andover, U.K., 1994; 181–205.
3. Thompson, S.N.; Hagen, K.S. Nutrition of entomophagous insects and other arthropods. In *Handbook of Biological Control: Principles and Applications*; Bellows, T.S., Fisher, T.W., Eds.; Academic Press: San Diego, CA, 1999; 594–652.
4. Grenier, S.; Delobel, B.; Bonnot, G. Physiological interactions between endoparasitic insects and their hosts—Physiological considerations of importance to the success of in vitro culture: an overview. J. Insect Physiol. **1986**, *32* (4), 403–408.
5. Grenier, S.; De Clercq, P. Comparison of artificially vs. naturally reared natural enemies and their potential for use in biological control. In *Quality Control and Production of Biological Control Agents: Theory and Testing Procedures*; van Lenteren, J.C., Ed.; CABI Publishing: Wallingford, U.K., 2003; 115–131.
6. Grenier, S. Rearing of *Trichogramma* and other egg parasitoids on artificial diets. In *Biological Control with Egg Parasitoids*; Wajnberg, E., Hassan, S.A., Eds.; CAB International: Wallingford, U.K., 1994; 73–92.
7. Cohen, A.C. *Insect Diets—Science and Technology*; CRC Press: Boca Raton, U.S.A., 2003.
8. van Lenteren, J.C.; Hale, A.; Klapwijk, J.N.; van Schelt, J.; Steinberg, S. Guidelines for quality control of commercially produced natural enemies. In *Quality Control and Production of Biological Control Agents: Theory and Testing Procedures*; van Lenteren, J.C., Ed.; CABI Publishing: Wallingford, U.K., 2003; 265–303.

Natural Enemies: Conservation

Cetin Sengonca
Department of Entomology and Plant Protection, Institute of Plant Pathology, University of Bonn, Bonn, Germany

Abstract
Natural enemies suffer severely from the use of broad-spectrum pesticides in agroecosystems and absence or destruction of food resources, shelter sites, egg-laying places, etc., during the vegetation period. Furthermore, unfavorable environmental conditions and lack of overwintering shelters during hibernation are causing high mortality among beneficial insects. The careful use of pesticides and the proper modification of environmental conditions may conserve natural enemies and increase their efficacy.

INTRODUCTION

After the successful utilization of two methods of biological control, classical biological control (importation and establishment of exotic natural enemies against either exotic or native pests) and augmentation of natural enemies (either inundative or inoculative releases of mass reared natural enemies), the third method—conservation and enhancement—has become more and more important during recent years.[1,2] Conservation of natural enemies is probably the most important concept in the practice of biological control and, fortunately, is one of the easiest to understand and readily available to growers. Most authors consider conservation as an environmental modification to protect and enhance natural enemies.[3] This definition will be the main subject of this entry.

Natural enemies of arthropod pests, also known as biological control agents, include predators and parasitoids that occur in all production systems from commercial fields to backyards where they have adapted to the local environment and target pests. Their conservation is generally non-complicated and cost-effective.[4] With relatively little effort the activities of these natural enemies can be observed. Natural control agents are a major factor in controlling agricultural pests and need to be considered when making pest management decisions. Today, therefore, the conservation of natural enemies is considered inseparable from enhancement and together they represent a successful biological control method.

AVOID HARMFUL PRACTICES OF PESTICIDES

Pesticides used in agriculture are not only killing target pests but they also can have direct effects on natural enemies by killing them, or indirectly by eliminating their hosts or preys and causing them to starve. In contrast, the conservation concept of natural enemies attempts to avoid the application of particularly broad-spectrum, highly disruptive pesticides. Applying selective or specific and beneficially safe pesticides may contribute much toward preserving natural enemies. Pesticide selectivity to beneficial arthropods has been broadly classified into two forms. The first of these is physiological selectivity, that is, pesticides are less toxic to natural enemies than to their target pest when applied at the recommended rate. The second form is ecological selectivity that pertains to the means and domains in which pesticides are used. Systemic pesticides killing leaf-feeding herbivores, for example, may have little or no effect on the many natural enemies that have contact only with the leaf surface. In some cases, pesticides can be successfully integrated into pest management systems with little or no detrimental effect on natural enemies, and this trend is likely to increase substantially in the future. Nowadays, the pesticide industry places increasing emphasis on the development of beneficially safe and environmentally friendly pesticides that exhibit greater selectivity for natural enemies and have minimal environmental impacts. In the same way, governmental regulatory agencies increasingly consider the adverse effects of pesticides on natural enemies in their registration process for pesticides, reflecting the growing concern over negative effects on beneficial insects. Despite these important steps and the great progress that has been made, the latent effects of pesticides and the impact of "cocktail applications" on natural enemy populations are still not fully understood. Further research and implementation of research results are urgently needed.

Alternatively, when selective pesticides are unavailable, recommended conservation tactics usually involve exact timing of pesticide applications. Careful forecasting and observation of the occurrence and growth of pest populations can substantially reduce the number of pesticide applications. Forecasting systems should be based on a defined economic threshold for each pest, considering also the presence of natural enemies.[5] Another approach is the selective placement of pesticides in agricultural fields. Limiting pesticide application only to infested parts of the field will reduce costs and also conserve natural enemies. An ideal alternative approach is the use of microbial insecticides, such as commercially available *Bacillus thuringi-*

ensis and fungal and viral products that have little adverse effects on natural enemies and the environment.

HABITAT AND ENVIRONMENTAL MANIPULATION

Another form of natural enemy conservation is habitat and environmental manipulation. The agricultural landscape is currently so intensively managed that the species diversity of many natural habitats has disappeared or become endangered. A similar reduction can be observed among natural enemies too. It has been strongly suggested by many experts that natural enemies also can be conserved by simply encouraging vegetational diversity of the agroecosystem. In this context, hedgerows, cover crops, strips inside and bordering fields, and even in-field balks provide important refuges for parasitoids and predators of many pest species. There they find and benefit from safe shelters, sources of pollen and nectar, and also alternative prey or hosts in case of food scarcity in cultivated fields. And thus, at such sites, a long-lasting and self-regulating biocoenosis will develop. A higher acceptance of weeds in agricultural crops may also increase the efficacy of natural enemies. Similarly, mixed plantings, for example, Umbelliferae, mustard, and *Phacelia tanacetifolia*,[6] growing weed strips even within fields, and providing flowering field borders significantly increase habitat diversity. *P. tanacetifolia* has been cultivated widely between crops in the production system in Germany for more than a decade. At the same time this will provide shelters and alternative food sources for natural enemies. Another important concept in conservation by habitat manipulation is that of connectivity. Natural or less disturbed habitats are often scattered and isolated within the agricultural landscape. Connecting these habitats, for example, by hedgerows and woods, will establish a continuous network of corridors allowing movement of natural enemies between fields.

Experiments have shown that a constant population of natural enemies can be established and conserved by releasing their prey or hosts during periods of scarcity, for example, releasing the red mite, *Tetranychus urticae* Koch, to support the establishment of its predatory mite, *Phytoseiulus persimilis* Athias-Henriot, in cucumber and bean cultures widely in greenhouses in middle Europe.[7] Similarly, distributing sterilized *Eupoecelia ambiguella* Hb. eggs between the two generations of this lepidopteran pest preserved its egg parasitoid *Trichogramma semblidis* (Auriv.) in the Ahr valley in Germany. A classical example is the black scale, *Saissetia oleae* (Oliv.), which interrupts its development for a short period during summer in the hot arid areas of central California. Planting irrigated oleander plants adjacent to citrus orchards allowed the black scale a continuous development. As a result, this enabled its specific parasitoid *Metaphycus helvolus* (Comp.) to maintain its population, particularly during the hot summer months.[7] In another example, it has been found that preservation of nettles, *Urtica* spp., an important host plant of *Aglais urticae* L., can enhance the efficacy of *T. semblidis* on the second generation of *E. ambiguella*. The reason is that the parasitoid maintains its population on this alternative host during the two nonoverlapping generations of *E. ambiguella*.[8]

The manipulation of some simple cultural measures can also conserve natural enemies. A famous example is strip harvesting hay alfalfa, allowing mobile natural enemies to disperse from cut strips to half-grown strips. Similarly, *Trissolcus vasilievi* (Mayr) and *T. semistriatus* Nees, two important egg-parasitoids of the sunn pest, *Eurygaster integriceps* Put., were successfully conserved in Turkey by growing shade trees in hot arid areas, providing shade and thus suitable climatic conditions for these parasitoids.

OVERWINTERING AND SHELTER SITES

Natural enemies build up considerably high population densities during summer periods, but then suffer from lack of overwintering or shelter sites and unfavorable climatic conditions during winter. As a result of these detrimental environmental factors, extreme low entomophagous arthropod densities are often present the following year. This permits pest populations to explode and in consequence requires more pesticide applications. In contrast, by preserving existing or providing artificial hibernation sites or shelters, natural enemies can be conserved during overwintering periods. For example, planting of trees and perennial bunch grasses near agricultural sites, the use of burlap or cloth trees, and wrap and stones provided as hiding places at overwintering time allow coccinellid lady beetles higher survival rates during hibernation. In the same way "trunk traps" and "trap bands" can be used as artificial overwintering sites for predatory bugs and lacewings,[9] and felt belts can be wrapped around the trunks of fruit trees and vines for the predatory mite *Typhlodromus pyri* Scheuten. The green lacewing, *Chrysoperla carnea* (Stephens), overwinters as an adult in barns, roof trusses, houses, and under the bark of trees, where mortality rates during hibernation in middle European climatic conditions may still reach 60%–90%. By using specially designed, simple wooden shelters (hibernation boxes) the overwintering mortality was reduced to only 4%–8%.[10] On this ground, these hibernation boxes are now being accepted and commonly used by farmers, gardeners, and also environmental protectionists in Germany and Switzerland.

FUTURE CONCERNS

The majority of pest problems in agriculture are due to the elimination of natural enemies by the indiscriminate and intensive use of pesticides. Improper habitat manipulation and mismanagement of ecosystems have further intensified this problem by reducing the available flora and fauna. Conservation and enhancement of natural enemies is the easiest and least costly method of biological control offering solutions to most pest problems without harming and disturbing the natural ecosystem. Unfortunately, however, this field has received little attention and very little investment has been made in research. There is a serious need for research into the areas of direct conservation and enhancement of natural enemies during the vegetation period and hibernation. In a self-regulating mechanism focusing on conservation and enhancement, natural enemies can keep agricultural pests below their economic threshold and help to reduce the number and frequency of pesticide applications. Furthermore, integration of conservation and enhancement of natural enemies into existing IPM programs[11] will lead to a more sustainable and cost-effective agriculture system.

See also *Cosmetic Standards*, pages 152–154; *Conservation of Biological Controls*, pages 138–140; *Augmentative Controls*, pages 36–38; *Biological Controls*, pages 57–60; pages 61–63; pages 64–67; pages 68–70; pages 71–73; pages 74–76; pages 77–80; pages 81–84.

REFERENCES

1. Ehler, L.E. Conservation Biological Control: Past, Present and Future. In *Conservation Biological Control*; Barbosa, P., Ed.; Academic Press: New York, 1998; 1–8.
2. Bugg, R.L.; Pickett, C.H. Introduction: Enhancing Biological Control–Habitat Management to Promote Natural Enemies of Agricultural Pests. In *Enhancing Biological Control*; Pickett, C.H., Bugg, R.L., Eds.; University of California Press: Berkeley, 1998; 1–23.
3. DeBach, P. *Biological Control of Insect Pests and Weeds*; Chapman and Hall: New York, 1964; 844.
4. Barbosa, P. Agroecosystems and Conservation Biological Control. In *Conservation Biological Control*; Barbosa, P., Ed.; Academic Press: New York, 1998; 39–59.
5. Sengonca, C. Conservation and enhancement of natural enemies in biological control. Phytoparasitica **1998**, *26* (3), 187–190.
6. Sengonca, C.; Frings, B. Einfluss von phacelia tanacetifolia auf schaedlings- und nuetzlingspopulation in Zuckerruebe. Pedobiologia **1988**, *32* (5/6), 311–316.
7. Krieg, A.; Franz, J.M. *Lehrbuch der biologischen Schaedlingsbekaempfung*; Verlag Paul Parey: Berlin, 1989; 302.
8. Schade, M.; Sengonca, C. Foerderung des traubenwicklereiparasitoiden Trichogramma semblidis (Auriv.) (Hym., Trichogrammatidae) durch bereitstellung von ersatzwirten an brennesseln im weingebiet ahrtal. Vitic. Enol. Sci. **1998**, *53* (4), 157–161.
9. Beane, K.A.; Bugg, R.L. Natural and Artificial Shelter to Enhance Arthropod Biological Control Agents. In *Enhancing Biological Control*; Pickett, C.H., Bugg, R.L., Eds.; University of California Press: Berkeley, 1998; 240–253.
10. Sengonca, C.; Frings, B. Enhancement of green lacewing Chrysoperla carnea (Stephens) by providing artificial facilities for hibernation. Turk. Entomol. Derg. **1989**, *13* (4), 245–250.
11. *CRC Handbook of Pest Management in Agriculture*; Pimentel, D., Ed.; CRC Press: Boca Raton, FL, 1991; 2, 757.

Nematodes: Biological Control

Simon Gowen
Department of Agriculture, University of Reading, Reading, U.K.

Abstract
Nematodes are a difficult group of pests to manage because generally they are inhabitants of soil and roots and are not easily influenced by soil treatments or cultural practices. The interest in the exploitation of natural enemies of nematodes has increased in recent years because of the demise of soil fumigants and nematicides through restrictions and withdrawals of registration of some products.

INTRODUCTION

Many pathogens and predators of nematodes are known but few have the necessary characteristics of specificity, mobility, or speed of colonization to have a significant influence on a pest population. Attempts at their commercial exploitation as field treatments have not been successful largely because of their inconsistency. Understanding the subtleties associated with the deployment of biocontrol agents will require considerable research effort. Additionally, the recommended rates of application and the formulation on suitable carriers and nutrient sources poses a problem in practicability and in the interpretation of the biological processes involved.

Contemporary research has shown that natural control does exist and that in certain crop/nematode/pathogen situations nematode populations will decline as they are attacked by components of the soil microflora. Soils where this occurs are known as suppressive, but well-documented examples of naturally occurring suppressiveness to particular nematode pests are uncommon.

During the life of a crop the population densities of many of the serious nematode pests can increase by 1000-fold. Economic damage may result from initial population densities of one nematode per gram of soil. To be effective therefore a biocontrol agent must have an impact on the numbers of nematodes that would invade a host and not simply eliminate the surplus individuals that may never locate or invade a root. This being so, those pathogens and predators that are relatively unspecific (trapping, ingesting, or parasitizing all types of free-living nematodes in soil) may be considered less promising than those that parasitize specific pests.

Significant progress has been made in the recognition and deployment of such microorganisms parasitic on some of the species of sedentary nematodes such as the root-knot nematodes, *Meloidogyne* spp., and some of the cyst nematodes, *Heterodera* spp., and *Globodera* spp.

Root-knot and cyst nematodes produce eggs either in clusters on roots or contained within or attached to the cuticle of the female nematode. Biocontrol agents that prevent these nematodes from reproducing may have more impact from an epidemiological point of view than those that kill the free-living individuals in the soil.

BIOCONTROL AGENTS SPECIFIC TO CERTAIN NEMATODE PESTS

Verticillium chlamydosporium is a facultative, soil-dwelling fungus that parasitizes eggs in egg masses exposed on the root surface. Under the right conditions, such fungi will have a significant effect on nematode populations. The efficacy of *V. chlamydosporium* is partly dependent on its root colonizing ability; this can vary according to the plant host. Skill is required in selecting crops that support and/or increase the root colonization by the fungus but are also less favored hosts of root-knot nematodes. *V. chlamydosporium* may be less effective when it is deployed with plants that are highly susceptible and large galls are produced in response to the nematode infection. In such cases, many egg masses may not be exposed on the root surface and so escape infection.

Paecilomyces lilacinus is another fungus commonly found infecting the eggs of sedentary nematodes such as the root-knot and the cyst nematodes, and, like *V. chlamydosporium* being relatively easy to produce on defined growth media, has good potential for commercial development.

Pasteuria penetrans, an obligate bacterial parasite of root-knot nematodes begins its life cycle on free-living juveniles in the soil. Spores attach to the juveniles as they move in search of host roots. Parasitic development begins after the nematode enters a root and continues in synchrony with that of its host. The nematode eventually is overcome by its parasite; it fails to produce eggs; and its body, filled with the spores of the bacterium, eventually ruptures releasing spores into the soil.

The efficiency of *P. penetrans* as a biocontrol agent of root-knot nematodes depends on the concentrations of spores in the soil, the chances of contact with the juvenile stage, and the specificity of the particular *P. penetrans* population. Commercial success will depend therefore on finding techniques for mass-producing the bacterium and on developing populations with a broad spectrum of pathogenicity.

Other *Pasteuria* species parasitic on some sedentary *(Heterodera)* and migratory *(Pratylenchus)* species have been described.

Biological control agents such as *V. chlamydosporium*, *P. lilacinus*, and *P. penetrans* could provide an adequate replacement for nematicides in some cropping systems but the lack of immediate effects, such as are provided by nematicide or fumigant treatments, is a disadvantage. Protection is normally needed in the early stages of plant growth such as in nursery beds. In this situation, integration with other practices such as nematicides, rotation, solarization, and mulches is necessary.

There are several reports of the successful deployment of these biocontrol agents. Small field plots treated once with *P. penetrans* spores (produced by an in vivo system) caused a decline in numbers of root-knot nematodes and increases in yield over a series of crop cycles using root-knot nematode susceptible crops. In other locations, where treatments with *P. penetrans* were combined with *V. chlamydosporium*, organic manures and grass mulches showed similar declines in nematode populations. These two organisms acted against root-knot nematodes in a complementary fashion. As part of this strategy, root systems containing spore-filled cadavers were deliberately left to disintegrate in the soil after each crop. No field treatments were effective after only one crop indicating that some crop loss must be expected during the development of suppressiveness. *P. penetrans* was also effective when used in combination with a nematicide in permanent beds within a plastic polytunnel. Better control of root-knot nematodes was achieved if the biocontrol agent was combined with other control strategies. With such treatments, beneficial effects may develop over one crop cycle.

The chlamydospores of *V. chlamydosporium* do not have the persistence of the spores of *P. penetrans*, which can remain viable for many years.

NONSPECIFIC BIOCONTROL AGENTS

There is a long history of interest in the fungi that trap nematodes in soil such as species of *Arthrobotrys*. These are commonly found in all soils but despite much research effort the problems of the unreliability of soil applications have not been solved and none have become established as successful commercial products.

There are several rhizosphere colonists that have potential for alleviating nematode damage. The precise mechanisms are not clear. Some produce toxins but others may affect root exudation and thus indirectly the attractiveness of roots to nematodes. Experiments have shown that strains of *Pseudomonas fluorescens* can reduce root invasion by different plant parasites but as with the trapping fungi, poor consistency hinders successful development of these microorganisms as commercial products.

FUTURE PROSPECTS

Recently, the nematicidal (and insecticidal) effects of the toxins produced by the bacteria associated with entomopathogenic nematodes (*Photorhabdus* spp., *Xenorhabdus* spp., and *Pseudomonas oryzihabitans*) have been demonstrated.

Success in the commercial development of biocontrol agents does appear promising with those microorganisms that can be formulated as a standard product with proven reliability; others may have a future as single treatment introductions in the more intensively managed protected cropping systems but commercialization may be difficult.

Research is still needed to develop reliable methods of production, formulation, and application. The challenge is to provide a sufficient duration of protection. Such treatments will need to be part of a package of control measures.

BIBLIOGRAPHY

1. Aalten, P.M.; Vitour, D.; Blanvillain, D.; Gowen, S.R.; Sutra, L. Effect of rhizosphere fluorescent *Pseudomonas* strains on plant parasitic nematodes *Radopholus similis* and *meloidogyne* spp. Letters Appl. Microbiol. **1998**, *27*, 357–361.
2. Bourne, M.; Kerry, B.R.; De Leij, F.A.A.M. The importance of the host plant on the interaction between root-knot nematodes (*Meloidogyne* spp.) and the nematophagous fungus, *Verticillium chlamydosporium* Goddard. Biocon. SciTech. **1996**, *6*, 539–548.
3. Crump, D.J. A method for assessing the natural control of cyst nematode populations. Nematologica **1987**, *33*, 232–243.
4. Gowen, S.R.; Bala, G.; Madulu, J.; Mwageni, W.; Trivino, C.T. In *Field Evaluation of Pasteuria penetrans for the Management of Root-Knot Nematodes*, The 1998 Brighton Conference on Pests and Diseases, 1998; 3, 755–760.
5. Kerry, B.R.; Jaffee, B.A. Fungi as Biocontrol Agents for Plant Parasitic Nematodes. In *The Mycota IV Environmental and Microbial Relationships*; Wicklow, D.T., Soderstrom, B.E., Eds.; Springer-Verlag: Berlin, Heidelberg, 1997; 204–218.
6. Samaliev, H.Y.; Andreoglou, F.I.; Elawad, S.A.; Hague, N.G.M.; Gowen, S.R. The Nematicidal Effects of the Bacteria *Pseudomonas oryzihabitans* and *Xenorhabdus Nematophilus* on the Root-Knot Nematode, *Meloidogyne javanica*. Nematology, *in press*.
7. Stirling, G.R. Biological Control of Plant Parasitic Nematodes. CAB International: Wallingford, U.K., 1991; 282.

8. Tzortzakakis, E.A.; Channer, A.G. de R.; Gowen, S.R.; Ahmed, R. Studies on the potential use of *Pasteuria penetrans* as a biocontrol agent of root-knot nematodes (*Meloidogyne spp.*). Plant Pathol. **1997**, *46*, 44–55.
9. Tzortzakakis, E.A.; Gowen, S.R. Evaluation of *Pasteuria penetrans* alone and in combination with oxamyl, plant resistance and solarization for control of *Meloidogyne* spp. on vegetables grown in greenhouses in Crete. Crop Prot. **1994**, *13*, 455–462.
10. Weibelzahl-Fulton, E.; Dickson, D.W.; Whitty, E.B. Suppression of *Meloidogyne incognita* and *M. javanica* by *Pasteuria penetrans* in field soil. J. Nematol. **1996**, *28*, 43–49.

Neurotoxicants: Developmental Experimental Testing

Vera Lucia S.S. de Castro
Ecotoxicology and Biosafety Laboratory, Environment, Brazilian Agricultural Research
Corporation (Embrapa), São Paulo, Brazil

Abstract
The normal structure and function of the nervous system may be altered as a result of exposure to some pollutants before or after birth. The analysis of chemical pollutant effects is particularly relevant in assessing interference at the nervous system development. These pollutants, as pesticides, can be more toxic to the developing central nervous system than those affecting the adults depending on the exposure characteristics and period of vulnerability. Animal behavior might provide useful indicators or biomarkers for detecting harmful chemical contaminants. These biomarkers can evaluate alterations in sensory, motor, and cognitive functions in laboratory animals exposed to toxicants during nervous system development. However, the detection, measurement, and interpretation of developmental neurotoxicity effects depend on appropriate study design and execution.

INTRODUCTION

Neurotoxic substances may play a role in a number of neurodevelopmental disorders. They can be released by industrial facilities and by agricultural practices, much of which ends up in the air or groundwater. Since most neuroteratogens affect multiple regions and processes, they can result in various behavioral defects.

Given the importance placed on fostering optimal cognitive development and the fact that chemical exposures can perturb the exquisite spatial and temporal choreography of brain development, it is not surprising that neurodevelopmental deficit frequently serves as the critical adverse health effect in risk assessments.[1]

Detection and characterization of chemical-induced toxic effects in the central and peripheral nervous system represent a big challenge. Prediction of neurotoxic effects is a key feature in the toxicological profile of compounds and is therefore required by many regulatory testing schemes.[2]

Despite the increasing recognition of the need to evaluate developmental neurotoxicity (DNT) in safety assessment, only very few of the commercial chemicals in current use have been examined with respect to neurodevelopmental effects. Validated rodent models exist, but they are considered expensive and are only infrequently used. The neurodevelopmental disorders include learning disabilities, attention deficit hyperactivity disorder, autism spectrum disorders, developmental delays, and emotional and behavioral problems. The causes of these disorders are unclear, and interacting genetic, environmental, and social factors are likely determinants of abnormal brain development.[3,4]

In calculations of environmental burdens of disease in children, lead neurotoxicity to the developing brain is a major contributor. Pesticide effects could well be of the same magnitude, or larger, depending on the exposure levels.[4] For example, an emerging literature provides evidence of neurobehavioral consequences resulting from exposure to relatively low levels of organochlorine and organophosphate pesticides in infants and children.

Organophosphate pesticides continue to be applied widely in agriculture and in residences throughout the world, representing about half the total annual amount of insecticides used. One of the major concerns with these agents is their propensity to elicit DNT at exposures below the threshold for any systemic symptoms, so that potentially damaging fetal or childhood exposures may go undetected until persistent functional impairments become expressed.[5]

Organophosphate-poisoned populations have shown a consistent pattern of deficits when compared to a non-exposed or non-poisoned population on measures of motor speed and coordination, sustained attention, and information processing speed.[6] In experimental models, developing animals have been shown to be more susceptible than adult animals to the acute toxicity of the organophosphate pesticide chlorpyrifos, which can cause neurobehavioral abnormalities.[7] Neonatal diazinon exposure below the threshold for appreciable cholinesterase inhibition in a non-monotonic dose–effect caused persisting neurocognitive deficits in adulthood. The organophosphorous insecticide can affect transmitter systems supporting memory function, differently, implying participation of mechanisms other than their common inhibition of cholinesterase.[5]

Future studies should examine the neurodevelopment effects in human beings associated with pesticide mixtures and other classes of pesticides (e.g., carbamates, pyrethroids), and with pesticide mixtures, because there is increasing use of these pesticides in certain communities that are replacing the organophosphate and organochlorine pesticides.[8] In this direction, the U.S. Agency for Toxic Substances and Disease Registry (ATSDR) developed a program for chemical mixtures of which an integral part is a mixtures health risk assessment. ATSDR has completed evaluations for several simple mixtures of child-specific exposure concern.[9]

BEHAVIORAL ASPECTS OF DNT

Behavior represents an integrated response of the nervous system that can reveal functional changes important to the overall fitness and survival of the organism exposed to single pesticides or mixtures. Although some developmental neurotoxicants are structural teratogens as well, behavioral dysfunctions may be more serious than structural defects under certain circumstances.[10] The major developmental sensory systems of concern in toxicology include visual, auditory, olfactory, nociceptive (pain and other noxious stimuli), somatosensory, and vestibular. However, neurobehavioral functions are influenced by subject variables such as age, sex, education, and social and (especially in humans) cultural background.

BRAIN DEVELOPMENT AND MATURATION

The development and maturation of the mammalian brain is an extremely complex process. Brain development involves cell division, migration and differentiation, programmed cell death (apoptosis), cell-to-cell interactions (e.g., for migration and synaptic communication), and multiple other processes under different timetables for the various brain regions. Genetic, epigenetic, and environmental factors (e.g., exposure to toxic chemicals, including certain heavy metals, industrial chemicals, and pesticides), particularly during the susceptible periods of development and aging, can result in many possible adverse central nervous system (CNS) consequences, ranging from mild to severe and involving various functions (e.g., cognition, motor, or sensory dysfunction). DNT refers to any adverse effect of perinatal exposure to a toxic substance on the normal development of nervous system structure and/or function.[11]

The mammalian brain undergoes a period of rapid brain growth, which in humans occurs perinatally, spanning from the third trimester of pregnancy throughout the first 2 years of life. In rats and mice, the brain growth spurt occurs in the neonate, spanning the first 3–4 weeks of life and reaching its peak around postnatal day 10. This period is characterized by axonal and dendritic outgrowth and the establishment of neuronal connections, and during this period, animals acquire many new motor and sensory abilities. Neurotypic and gliotypic proteins can serve as sensitive indicators of time- and region-specific effects of chemicals on the developing nervous system. The presence of xenobiotics in the brain during this defined period of maturational processes is a critical factor for induction of persistent changes in behavior and transmitter systems.[12] For example, it has been reported that multiple neurotransmitter systems are altered following exposure to organophosphorous insecticides. Developmental chlorpyrifos exposure produces persistent deficiencies in cholinergic synaptic neurochemistry.[13] Also, there is increasing evidence that polychlorinated biphenyls (PCBs) and methyl mercury also have neurotoxic effects. An enhanced effect of these toxicants, due to either synergistic or additive effects, would be considered as a risk for fetal development. It is postulated that these neurotoxicants might interact.[14]

Epidemiological studies have demonstrated a relationship between perinatal exposure to persistent organic pollutants among others and neurological and behavioral disturbances in infants and children. Studies in animals have confirmed that contaminants like PCBs, metals, and pesticides can disrupt behavioral functioning.[15]

Evidence indicates that exposure to environmental chemicals could have an impact on children's health and development. The developing CNS of fetus and children is particularly susceptible to chemically induced damage compared with the brain of adults due to the different pharmacokinetic factors, diminished defense mechanisms, or the fact that the developing nervous system undergoes a highly complex series of ontogenetic processes that are vulnerable to chemical perturbation.[16]

PERIODS OF VULNERABILITY

The developing nervous system is particularly sensitive to environmental insults during critical periods that are dependent on the temporal and regional emergence of specific and sequential developmental processes (i.e., proliferation, migration, differentiation, synaptogenesis, myelination, and apoptosis). Evidence from numerous sources demonstrates that neural development extends from the embryonic period through adolescence. In general, the sequence of events is comparable among species, although the time scales are considerably different. Developmental exposure of animals or humans to numerous agents (e.g., x-ray irradiation, ethanol, lead, methyl mercury, or chlorpyrifos) demonstrates that interference with one or more of these developmental processes can lead to DNT.[17]

For many behaviors, a critical period exists during which the animal is sensitive to these organizational effects. Functional and structural life-lasting modifications can be induced by alterations of natural conditions during these adaptive developmental stages of maturation. The critical periods have been described for some cortical circuits involved in many different sensory systems such as the auditory, somatosensory, and olfactory systems. These critical periods occur also during postnatal life. Indeed, experience-dependent plasticity during critical periods of postnatal development shapes the adult brain anatomy and function.[18,19]

PROTOCOLS FOR EXPERIMENTAL STUDIES

In order to reduce the risk regarding the exposure to pollutants, addressing the behavioral aspects by appropriate investigation to sustain the safe use of the compounds is suggested. Evaluation of pre- and postnatal developmental parameters can be improved by including different tests on and safety assessment of chemicals to indicate the proper functioning of the sensory, motor, emotional, and cognitive domains.

Laboratory experimental studies suggest that many currently used pesticides such as organophosphates, carbamates, pyrethroids, ethylenebisdithiocarbamates, and chlorophenoxy herbicides can cause neurodevelopmental toxicity. Adverse effects on brain development can be severe and irreversible.[4]

Emotional processes can be viewed as adaptive events or states that are likely to occur across the animal kingdom, but that may or may not have subjective components, comprising physiological, behavioral, and subjective components, depending on the species and circumstances involved. Although these different components usually act in concert, they are potentially dissociable, not always operating as a functional whole.[20]

The utilization of an experimental protocol containing indices related to reproduction and animal development can identify initial damages due to exposure to environmental pollutants. Behavioral experimental methods are used to detect and characterize developmental neurotoxic effects on sensory, cognitive, and motor system functions. Neurobehavioral evaluations are widely used to examine the potential neurotoxicity of pesticides and other chemicals,[21] since neurobehavioral performance can be a sensitive biomarker of the neurodevelopmental consequences of exposure to environmental agents.

Prevention of possible damages due to pesticides during the development of young organisms, like newborns, requires an integrated strategy capable of monitoring the standard use of these products as well as the integration of the potential effects to improve evaluation. If available, biomarkers of exposure are useful for assessing the bioavailability of toxicants to the dam and offspring in utero and after birth. The evaluation of these biomarkers needs to differentiate normal variability from changes that are adverse in response.[22]

Animal models are used to understand neurophysiological processes on the basis of human exposure to xenobiotics. They represent a basis for understanding their pathophysiological traits. There is a variety of methodologies that can be utilized to assess these processes. Cross-species comparability between human and experimental animals supports the assumption that DNT effects in animals indicate a potential to affect development in humans.[10]

The first guideline specifically designed to evaluate DNT was developed and implemented by the U.S. Environmental Protection Agency (EPA) in 1991 and has later on been updated. The Organization for Economic Cooperation and Development (OECD) initiated the development of a DNT guideline (TG 426) following the recommendations of the OECD Working Group on Reproduction and Developmental Toxicity in Copenhagen in 1995. The first draft based on the U.S. EPA DNT guideline was prepared following a 1996 Expert Consultation Meeting that addressed a number of significant issues and incorporated improvements. The draft TG 426 was distributed to National Coordinators for comments in 1998, and significant technical issues in the comments were further discussed and revised.[10,23]

Developmental toxicity may result from either prenatal or postnatal exposure, may manifest at any life stage, and may be expressed as functional deficits. The DNT study is a specialized type of developmental toxicity study designed to screen for adverse effects of pre- and postnatal exposure on the development and function of the nervous system and to provide dose–response characterizations of those outcomes. The U.S. EPA and OECD DNT guidelines recommend administration of the test substance during gestation and lactation. Cohorts of offspring (typically rat) are randomly selected from control and treated litters for evaluations of gross neurologic and behavioral abnormalities during postnatal development and adulthood. These include assessments of physical development, behavioral ontogeny, motor activity, motor and sensory function, learning and memory, and postmortem evaluation of brain weights and neuropathology.[23]

There are a number of stimulus properties shared by all sensory systems, including intensity, frequency, duration, and location in space. In this way, behavioral tests of motor dysfunction in animals include those used to detect spontaneous movement disorders such as changes in gait, tremors, and myoclonus, and those used to detect changes in induced movement such as reflexes, reactions, and movements under operant control. Tests of motor function include observation of locomotion, measurement of locomotor activity, and tests of reflexes and reactions. Also, assessment of cognitive function is a critical component of a DNT assessment to address concerns over potential long-term consequences of exposures to toxicants during brain development. Cognitive function is thought to encompass learning, memory, and attention processes.[24–27]

In this way, more effort is needed to adequately evaluate the neurotoxic effects. More elaborated experimental protocols are continuously proposed. They will focus on the interpretation of data obtained in studies that link xenobiotics exposure and functional (behavioral) deficits due to specific neurotransmitter and synaptic mechanisms,[28] identifying possible chemical class-specific targets and biomarkers of effect. Above all, gene expression could be also used as a sensitive tool for the initial identification of DNT effects induced by different mechanisms of toxicity in both cell types (neuronal and glial) and at various stages of cell development and maturation.[16]

Recent literature have examined specific end points across multiple guideline DNT studies to demonstrate the value of current methods in hazard characterization and explore further opportunities for methodologic refinement, examining the interpretation of neurodevelopmental end points for human health risk assessment, data interpretation and variability, positive control data, and statistical analysis.[11,29,30]

Test method reliability, reproducibility, and relevance are attributable in part to the high level of standardization of the test methods;[23] in some cases, the variability of some end points (e.g., motor activity) is very large. Methods have been suggested to decrease such variability.[29-31] Sources of variability include factors related to environmental conditions, personnel, experimental procedures, and equipment.[21,31] The detection, measurement, and interpretation of DNT effects depend on appropriate study design and execution, using established methods with appropriate controls.

Furthermore, the nature and extent of developmental neurotoxic effects often are dependent on the timing of exposure to a toxic agent or combinations of agents and environmental conditions; i.e., organisms exhibit distinct temporal windows of susceptibility. Variations in neurotoxic outcomes across species are expected because stages of nervous system development can vary significantly between species in relation to the time of birth. Thus, the time and duration of exposure in animal models must also be selected carefully to match the window of exposure in the human situation and allow cross-species extrapolation.[32]

Detection and characterization of chemical-induced toxic effects in the central and peripheral nervous system represent a major challenge for employing newly developed technologies in the field of neurotoxicology. For example, those using specific brain cell types can produce results of general mechanism of action but not specific to the chemical tested. In addition, toxicokinetic models are to be developed in order to properly evaluate absorption, distribution, metabolism, and excretion, as well as the blood–brain barrier. Behavioral toxicologists will be needed to contribute for the experimental tests and computational models to anchor molecular initiating events to adverse outcomes. Therefore, an intensive search for the development of alternative methods using in silico models for neurotoxic hazard assessment is appropriate.[33] The following are some of the challenges that need to be overcome: predicting behavior using models of complex neurobiological pathways, standardizing study designs and dependent variables to facilitate creation of databases, and managing the cost and efficiency of behavioral assessments.[34]

CONCLUSION

There is growing evidence of the adverse impact of exposures to ambient and indoor air pollutants on fetal growth and both early childhood and animal neurodevelopment. The normal structure and function of the nervous system may be altered as a result of exposure to some pollutants before or after birth. Its analysis is particularly relevant in assessing the interference of a chemical pollutant with neuroendocrine maturation by behavioral methods, as it is a sensitive and broad marker of perturbation of both nervous and endocrine functions.

A number of methods can evaluate alterations in sensory, motor, and cognitive functions in laboratory animals exposed to toxicants during nervous system development. Assessment methods are being developed to examine other nervous system functions, including social behavior, autonomic processes, and biologic rhythms.[35] Fundamental issues underlying proper use and interpretation of these methods include 1) consideration of the scientific goal in experimental design; 2) selection of an appropriate animal model; 3) expertise of the investigator; 4) adequate statistical analysis; and 5) proper data interpretation.

ACKNOWLEDGMENTS

I wish to thank Dr. Stachetti for his precious contribution to this entry, consisting of the initial text, over which the present one has been constructed. Also, I would like to thank him for the opportunity offered to contribute to this prestigious publication, by referring my name as author. Without his help and his confidence on my work, this achievement would not have been possible. I also would like to thank Dr. Stachetti for the final text comments and suggestions.

REFERENCES

1. Bellinger, D.C. Interpreting epidemiologic studies of developmental neurotoxicity: Conceptual and analytic issues. Neurotoxicol. Teratol. **2009**, *31*, 267–274.
2. Crofton, K.M.; Makris, S.L.; Sette, W.F.; Mendez, E.; Raffaele, K.C. A qualitative retrospective analysis of positive control data in developmental neurotoxicity studies. Neurotoxicol. Teratol. **2004**, *26*, 345–352.
3. Dietrich, K.N.; Eskenazi, B.; Schantz, S.; Yolton, K.; Rauh, V.A.; Johnson, C.B.; Alkon, A.; Canfield, R.L.; Pessah, I.N.; Berman, R.F. Principles and practices of neurodevelopmental assessment in children: Lessons learned from the centers for children's environmental health and disease prevention. Res. Environ. Health Perspect. **2005**, *113* (10), 1437–1446.
4. Bjørling-Poulsen, M.; Andersen, H.R.; Grandjean, P. Potential developmental neurotoxicity of pesticides used in Europe. Environ. Health **2008**, *7*, 50.

5. Timofeeva, O.A.; Roegge, C.S.; Seidler, F.J.; Slotkin, T.A.; Levin, E.D. Persistent cognitive alterations in rats after early postnatal exposure to low doses of the organophosphate pesticide, diazinon. Neurotoxicol. Teratol. **2008**, *30*, 38–45.
6. Rohlman, D.S.; Lasarev, M.; Anger, W.K.; Scherer, J.; Stupfel, J.; McCauley, L. Neurobehavioral performance of adult and adolescent agricultural workers. NeuroToxicology, **2007**, *28*, 374–380.
7. Richardson, J.; Chambers, J. Effects of repeated oral postnatal exposure to chlorpyrifos on cholinergic neurochemistry in developing rats. Toxicol. Sci. **2005**, *84*, 352–359.
8. Eskenazi, B.; Rosas, L.G.; Marks, A.R.; Bradman, A.; Harley, K.; Holland, N.; Johnson, C.; Fenster, L.; Barr, D.B. Pesticide toxicity and the developing brain. Basic Clin. Pharmacol. Toxicol. **2008**, *102*, 228–236.
9. Pohl, H.R.; Abadin, H.G. Chemical mixtures: Evaluation of risk for child-specific exposures in a multi-stressor environment. Toxicol. Appl. Pharmacol. **2008**, *233*, 116–125.
10. Hass, U. The need for developmental neurotoxicity studies in risk assessment for developmental toxicity. Reprod. Toxicol. **2006**, *22*, 148–156.
11. Tyl, R.W.; Crofton, K.; Moretto, A.; Moser, V.; Sheets, L.P.; Sobotka, T.J. Identification and interpretation of developmental neurotoxicity effects: A report from the ILSI Research Foundation/Risk Science Institute expert panel on neurodevelopmental endpoints. Neurotoxicol. Teratol. **2008**, *30* (4), 349–381.
12. Viberg, H. Exposure to polybrominated diphenyl ethers 203 and 206 during the neonatal brain growth spurt affects proteins important for normal neurodevelopment in mice. Toxicol. Sci. 2009, *109* (2), 306–311.
13. Johnson, F.O.; Chambers, J.E.; Nail, C.A.; Givaruangsawat, S.; Carr R.L. Developmental chlorpyrifos and methyl parathion exposure alters radial-arm maze performance in juvenile and adult rats. Toxicol. Sci. **2009**, *109* (1), 132–142.
14. Andersen, I.S.; Voie, O.A.; Fonnum, F.; Mariussen, E. Effects of methyl mercury in combination with polychlorinated biphenyls and brominated flame retardants on the uptake of glutamate in rat brain synaptosomes: A mathematical approach for the study of mixtures. Toxicol. Sci. **2009**, *112* (1), 175–184.
15. Bowers, W.J.; Nakai, J.S.; Chu, I.; Wade, M.G.; Moir, D.; Yagminas, A.; Gill, S.; Pulido, O.; Meuller, R. Early developmental neurotoxicity of a PCB/organochlorine mixture in rodents after gestational and lactational exposure. Toxicol. Sci. **2004**, *77*, 51–62.
16. Hogberg, H.T.; Kinsner-Ovaskainen, A.; Coecke, S.; Hartung, T.; Bal-Price, A.K. mRNA expression is a relevant tool to identify developmental neurotoxicants using an in vitro approach. Toxicol. Sci. **2010**, *113* (1), 95–115.
17. Rice, D.; Barone, S., Jr. Critical periods of vulnerability for the developing nervous system: Evidence from humans and animal models. Environ. Health Perspect. **2000**, *108* (3), 511–533.
18. Soiza-Reilly, M.; Azcurra, J.M. Developmental striatal critical period of activity-dependent plasticity is also a window of susceptibility for haloperidol induced adult motor alterations. Neurotoxicol. Teratol. **2009**, *31*, 191–197.
19. Crews, F.; He, J.; Hodge, C. Adolescent cortical development: A critical period of vulnerability for addiction. Pharmacol. Biochem. Behav. **2007**, *86*, 189–199.
20. Paul, E.S.; Harding, E.J.; Mendl, M. Measuring emotional processes in animals: The utility of a cognitive approach. Neurosci. Biobehav. Rev. **2005**, *29*, 469–491.
21. Slikker, W., Jr.; Acuff, K.; Boyes, W.; Chelonis, J.; Crofton, K.; Dearlove, G.; Li, A.; Moser, V.; Newland, C.; Rossi, J.; Schantz, S.; Sette, W.; Sheets, L.; Stanton, M.; Tyl, S.; Sobotka, T. Behavioral test methods workshop. Neurotoxicol. Teratol. **2005**, *27*, 417–427.
22. Raffaele, K.C.; Rowland, J.; May, B.; Makris, S.L.; Schumacher, K.; Scarano, L.J. The use of developmental neurotoxicity data in pesticide risk assessments, Neurotoxicol. Teratol., **2010**, *32(5)*:563–572. doi:10.1016/j.ntt.2010.04.053.
23. Makris, S.L.; Raffaele, K.; Allen, S.; Bowers, W.J.; Hass, U.; Alleva, E.; Calamandrei, G.; Sheets, L.; Amcoff, P.; Delrue, N.; Crofton, K.M. A retrospective performance assessment of the developmental neurotoxicity study in support of OECD test guideline 426. Environ. Health Perspect. **2009**, *117* (1), 17–25.
24. Després, C.; Richer, F.; Roberge, M.C.; Lamoureux, D.; Beuter, A. Standardization of quantitative tests for preclinical detection of neuromotor dysfunctions in pediatric neurotoxicology. NeuroToxicology **2005**, *26*, 385–395.
25. Luft, J.; Bode, G. Integration of safety pharmacology endpoints into toxicology studies. Fundam. Clin. Pharmacol. **2002**, *16*, 91–103.
26. Moser, V.; Phillips, P.; Levine, A.; McDaniel, K.; Sills, R.; Jortner, B.; Butt, M. Neurotoxicity produced by dibromoacetic acid in drinking water of rats. Toxicol. Sci. **2004**, *79*, 112–122.
27. Sarter, M. Animal cognition, defining the issues. Neurosci. Biobehav. Rev. **2004**, *28*, 645–650.
28. Yanai, J.; Brick-Turin, Y.; Dotan, S.; Langford, R.; Pinkas, A.; Slotkin, T.A. A mechanism-based complementary screening approach for the amelioration and reversal of neurobehavioral teratogenicity. Neurotoxicol. Teratol., **2010**, *32* (1), 109–113.
29. Crofton, K.M.; Foss, J.A.; Hass, U.; Jensen, K.; Levin, E.D.; Parker, S.L. Undertaking positive control studies as part of developmental neurotoxicity testing. Neurotoxicol Teratol. **2008**, *30* (4), 266–287.
30. Raffaele, K.C.; Fisher, J.E.; Hancock, S.; Hazelden, K.P.; Sobrian, S.K. Determining normal variability in a developmental neurotoxicity test. Neurotoxicol. Teratol. **2008**, *30* (4), 288–325.
31. Castro, V.L.; Silva, P.A. Validation of neurobehavioral studies for evaluating the perinatal effects of single and mixture exposure to pesticides. In *Progress in Pesticides Research*; Kanzantzakis, C.M., Ed.; Nova Science Publishers: New York, 2009; 371–395.
32. Hines, R.N.; Sargent, D.; Autrup, H.; Birnbaum, L.S.; Brent, R.L.; Doerrer, N.G.; Cohen Hubal, E.A.; Juberg, D.R.; Laurent, C.; Luebke, R.; Olejniczak, K.; Portier, C.J.; Slikker ,W. Approaches for assessing risks to sensitive populations: Lessons learned from evaluating risks in the pediatric population. Toxicol. Sci. **2010**, *113* (1), 4–26.
33. Coecke, S.; Eskes, C.; Gartlon, J.; Kinsner, A.; Price, A.; van Vliet, E.; Prieto, P.; Boveri, M.; Bremer, S.; Adler, S.; Pellizzer, C.; Wendel, A.; Hartung, T. The value of

alternative testing for neurotoxicity in the context of regulatory needs. Environ. Toxicol. Pharmacol. **2006**, *21*, 153–167.

34. Bushnell, P.J.; Kavlock, R.J.; Crofton, K.M.; Weiss, B.; Rice, D.C. Behavioral toxicology in the 21st century: Challenges and opportunities for behavioral scientists—Summary of a symposium presented at the annual meeting of the Neurobehavioral Teratology Society, 2009. Neurotoxicol. Teratol. **2010**, *32*, 313–328.

35. Cory-Slechta, D.A.; Crofton, K.M.; Foran, J.A.; Ross, J.F.; Sheets, L.P.; Weiss, B.; Mileson, B. Methods to identify and characterize developmental neurotoxicity for human health risk assessment. I: Behavioral effects. Environ. Health Perspect. **2001**, *109* (1), 79–91.

Nitrate Leaching Index

Jorge A. Delgado
Soil Plant Nutrient Research Unit, Agricultural Research Service (USDA-ARS), U.S. Department of Agriculture, Fort Collins, Colorado, U.S.A.

Abstract
Nitrogen inputs contribute to many agricultural benefits, such as maximizing yields, improving product quality, and increasing economic returns for farmers. However, nitrogen inputs also increase the potential for losses of nitrogen via pathways such as nitrate leaching, a pathway that contributes to the off-site transport of nitrogen to surface water and groundwater bodies. Nitrate leaching is a significant pathway for losses, especially in coarser soils that are irrigated, or where large precipitation events occur when there are significant amounts of nitrate in the soil profile. In addition, tile systems have been identified as systems with higher potential for nitrate leaching. A nitrate leaching index is a simple, relatively quick approach (five minutes or less) to conducting a risk assessment of the potential for nitrate leaching. There are several different nitrate leaching indexes that are available. The Nitrogen Index is a new approach to assessing nitrate leaching losses, which not only includes a risk index for nitrate leaching, but also for other nitrogen loss pathways. The Nitrogen Index has been shown to be a new eco-technological approach users can take to quickly assess the risk of nitrate leaching, and the Nitrogen Index has the potential to be used as an effective tool for environmental conservation.

INTRODUCTION: IMPORTANCE OF NITROGEN

World agricultural production will have to increase significantly during the next few decades to keep up with the increasing demands for food brought about by a growing human population. Not only will increases in agricultural production be required to supply grain for a larger world population, but also a significant increase in grain and forage production will be needed to meet the higher demand for meat and dairy products that will be created as the world population grows in size and as large populations such as those in India and China become more affluent.

World trade in agricultural products could potentially increase as grain-import countries seek to meet increased demand by buying grain products from grain-export countries. World agricultural production will also have to increase as the bioenergy industry expands, increasing competition for land and water resources. Delgado et al.[1] reviewed some of the world's challenges with respect to population growth and increasing agricultural production and concluded that the best science and policies available will need to be implemented to help ensure that agricultural production is increased over the next decades.

Nitrogen will be at the center of the world's emerging sustainability challenges. Independent of what policies are implemented, best nitrogen inputs and management will have to be at the core of efforts to meet agricultural production needs. As the world moves closer to the middle of the 21st century, increased production per unit of land will be needed yet will have to be accomplished with limited soil and water resources. Since in general, across worldwide agroecosystems, agricultural production is increased with nitrogen inputs and best nitrogen management practices, nitrogen will be an essential component of efforts to increase productivity and sustainability.

NITROGEN LEACHING CONTRIBUTES TO ENVIRONMENTAL PROBLEMS

Use of nitrogen fertilizer has contributed to increased agricultural production. However, at the same time, inputs of nitrogen also contribute to increased losses of reactive nitrogen from agricultural systems, losses that can have environmental impacts.[2–4] The U.S. Environmental Protection Agency (EPA)[5] has reported that drinking water with nitrate concentrations above 10 mg NO_3-N L^{-1} can negatively impact humans and is thus unfit for human consumption. Follett et al.[6] recently reviewed the impacts of reactive nitrogen on the environment and human health and summarized that although improvement in nitrogen management will contribute to better production, management is also important to reduce the transport of reactive nitrogen from agricultural systems and thus reduce the negative impacts of reactive nitrogen on the environment and human health.

It has been documented across the literature that nitrogen losses via surface runoff and nitrate leaching can impact water bodies.[7,8] Nitrate leaching has been reported to impact aquifers across the world, from the Northern China Plain[9] to the European region and South America[10]) to across the United States.[11] Nitrate leaching from tile systems has also been reported to contribute to impacts on water bodies, and in some cases, it has even been reported to contribute to the transport of nitrates from agricultural fields to coastal systems.[12–16] Impact on domestic wells that are sources of drinkable water has been reported for significant areas across the United States by Dubrovsky et al.,[17] who conducted an extensive study on concentrations of nitrates and found that for a significant percentage of the wells, the nitrate levels were higher than the 10 mg NO_3-N L^{-1} safe limit. The literature shows that the nitrate leaching contributes to the transfer of reactive nitrogen from farm fields to surface water[12,18,19] and to underground waters. Since the U.S. EPA reports that concentrations above 10 mg NO_3-N L^{-1} are not safe for human consumption, new tools should be applied, and management practices should continue to be improved to minimize this transfer. There are many management practices that can be applied across worldwide agroecosystems to increase nitrogen use efficiencies and reduce nitrate leaching from agricultural fields, and the use of new, advanced tools such as the nitrogen index are among the new ecotechnologies that can be applied.[20,21,32]

NEW ECOTECHNOLOGY: QUICK TOOLS/ INDICATORS

A nitrate leaching index is a simple approach to estimating the risk of nitrate leaching to the environment. Several simple indexes have been developed and used in the past and have been discussed by Shaffer and Delgado[22] and, recently, Buczko et al.[23] Several of them are also discussed here and have been reported to be good indexes for assessing the risk of nitrate leaching.

One such index is the nitrogen index proposed by Shaffer and Delgado.[22] Shaffer and Delgado[22] proposed that there was a need to have easy-to-use risk assessment tools to quickly assess the risk of nitrate leaching for a given set of management practices. They proposed the development of a new nitrogen index that would be easy to use and that would consider the effects of management (e.g., manure, crops, fertilizer, irrigation), soils, climate, and off-site factors. This new nitrogen index, in other words, would serve as an indicator that assists users in making quick assessments of the risk of nitrate leaching and/or other nitrogen losses. This proposed nitrogen index was later published and integrates principles of ecological engineering such as hydrological factors.[9,20,32] The nitrogen index can be used as an indicator and/or new ecotechnology to assess the risk of nitrate leaching for a given set of management practices under given weather, soils, and off-site factors.

Shaffer and Delgado[22] reported that the nitrogen index can potentially help users assess the risk of environmental impacts. They reported that this type of index can be considered a tier-one approach to management, similar to the Phosphorus Index tool proposed by Lemunyon and Gilbert[24] and widely applied by Sharpley et al.[25] across over 40 states in the United States. Shaffer and Delgado defined each tier by considering the complexity of the tools and the need for detail. A quick and simple-to-use index or indicator was defined as a tier-one tool. A tier-two tool was defined as a tool of intermediate complexity, such as an application model. A tier-three tool would be a more complex research model, with very detailed simulations and supported by site-specific field data. Gross et al.[26] reviewed the tier approach and how it can contribute to conservation planning and to the development of nutrient management plans. They reported that the new nitrogen index has the potential to be used as a field assessment tool to provide guidance and to help assess the risk of nitrogen losses, including nitrate leaching losses.

A NITROGEN INDEX APPROACH TO ASSESS NITRATE LEACHING

There is potential to use the new nitrogen index as part of an ecological engineering approach that considers the risk of reactive nitrogen losses to the environment.[20,32] This approach has the potential to be integrated into nutrient management plans, contributing to conservation planning.[26] The nitrogen index is part of new advances in conservation science, soil science, and agronomy, and of the new ecotechnology approach developed in the last decade. As we move forward, the integration of this indicator with a Geographic Information System (GIS) framework and/or an Internet/Web approach using GIS will increase the capability and use of a tool that can assist in quick assessments of the potential for nitrogen losses via pathways such as nitrate leaching.[4,20,22,26,27,32] These types of tools can help with conservation efforts to reduce impacts on the environment.[26]

Shaffer and Delgado[22] recommended that a series of parameters be considered in developing a strong nitrate leaching index, including the need to simultaneously assess other nitrogen pathways such as atmospheric losses and surface runoff potential. In addition, they reported that a strong nitrogen index also needs to consider management practices, crop rotations, and off-site factors. The new nitrogen index developed by Delgado et al.[20,28,32] integrated all of these factors and even incorporated several indexes such as the water leaching index, a nitrogen mass balance approach similar to that of Pierce et al.,[29] and the Aquifer Risk Index, among others, to quickly evaluate the potential risk of nitrogen losses, including nitrate leaching.

Delgado et al.[20,28,32] reported that the new nitrogen index can be used to quickly estimate the amount of nitrogen available to leach after the growing season. Similarly, de Paz[4] tested the new nitrogen index approach joined with GIS capabilities across a Mediterranean region of Spain where vegetables, citrus, and grain crops are grown. This region has been identified as one of the regions of Europe that has been highly impacted by nitrate leaching.[4,10] de Paz et al.[4] found that the water leaching and nitrogen uptake by crops predicted by the nitrogen index were significantly correlated with measured values. Additionally, they also found that the nitrate leaching predicted by the nitrogen index was significantly correlated with the measured nitrate leaching across this Mediterranean region.

Using GIS, de Paz et al.[4] conducted a spatial analysis across this Mediterranean region and found that the areas with higher groundwater nitrate concentrations were correlated with the areas that the nitrogen index estimated as having higher nitrate leaching potential. Although the nitrogen index is a simple indicator of the nitrate leaching potential, assesses the nitrate leaching only below the root zone, and does not estimate the transfer of the mass of nitrate leaching to groundwaters, the fact that the areas that were identified as having higher nitrate leaching potential were correlated with the areas that had higher measured concentrations of groundwater nitrates suggests that this simple tool could be useful in assessing the risk of transfer of nitrates to water bodies. These results also suggest that the nitrogen index was able to integrate the spatially variable soils and management practices and accurately estimate groundwater nitrate risk impacts.

The new nitrogen index has also been tested in California, Mexico, the Caribbean, Bolivia, and Ecuador and transferred to each of these locations. There are several entries in review and in development about these efforts. This tool can be downloaded from the U.S. Department of Agriculture–Agricultural Research Service (ARS)–Soil Plant Nutrient Research Web page: http://www.ars.usda.gov/npa/spnr/nitrogentools.

Figueroa et al.[30] found that the effects of management practices in intensive manure systems in Mexico predicted by the nitrogen index correlated with the observed residual soil nitrate available to leach. Instituto Nacional de Investigaciones Forestales Agricolas y Pecuaris (INIFAP)[31] reported that the development of the Mexico Nitrogen Index, a tool that that can be used to assess the nitrate leaching risk in forage systems in this country, was one of the top five forage management achievements in Mexico. Results from Delgado et al.,[20,32] de Paz et al.,[4] and Figueroa et al.,[30] among other studies, suggests that quick tools such as the nitrogen index could be used as ecological indicators to assess the risk of potential nitrate leaching to the environment.

In addition to the nitrogen index, there are other approaches/nitrate leaching indexes that have been reported to be very successful in assessing the potential risk of nitrate leaching. Van Es et al.[33] reported that the New York Nitrate Leaching Index can be used to assess nitrate leaching losses in this state. In Canada, the Ontario Ministry of Agriculture, Food and Affairs (OMAFRA)[34] reported that their nitrogen index can also be successful in assessing the risk of nitrate leaching losses. In California, the Nitrate Leaching Hazard Index developed for irrigated agriculture by Wu et al.[35] has also been reported as a potential tool to assess the risk of nitrate leaching. Wu et al.[35] considered the rooting depth of crops in their index's assessments of nitrate leaching risk. Results from Delgado[36,37] found that the rooting depths were correlated with nitrate leaching potential.

There have been efforts to join the nitrogen index to a phosphorus index, as was initially proposed by Sharpley et al.[38,39] A joint index was developed by Heathwaite et al.[40] to assess nitrogen and phosphorous. Similarly, Delgado et al.[20,28,32] proposed to join the new nitrogen index to currently available P indexes. Their initial version of the nitrogen index was able to do both a nitrogen and a phosphorus index analysis. Their new version of the nitrogen index, which is currently in development, is written in JAVA and includes a phosphorus index. Another joint index was also proposed and developed by OMAFRA.[34]

The water leaching index by Williams and Kissel[41] has also been used to assess the potential for nitrate leaching. This index has been reported to be accurate in estimating the potential for nitrate leaching based on water leaching potential.[33,42] The disadvantage of this index is that it does not consider N inputs and N dynamics. Evaluations of advantages and disadvantages of other indexes were conducted by Shaffer and Delgado[22] and Buczko et al.[23,43]

COMPONENTS OF AN ECOTECHNOLOGICAL TOOL FOR ADDRESSING NITRATE LEACHING (NITROGEN INDEX)

The nitrogen index can be used to quickly assess the potential risk of nitrate leaching and other losses across different agroecosystems (Fig. 1). The nitrogen index was written in JAVA and has a series of screens organized by the type of data being entered as well as drop-down menus to facilitate quick entry of information. The newest version of the nitrogen index is available in metric and English units, and the user has the option to use the index in either the Spanish or the English language. The user could quickly navigate to different screens, where basic information as well as information on soil, manure, fertilizer, irrigation, crop, off-site factors, water management and hydrology, and qualitative factors can be manually entered and/or selected from drop-down menus. For additional details, the reader could download the nitrogen index user manual at http://www.ars.usda.gov/SP2UserFiles/ad_hoc/54020700NitrogenTools/Nitrogen%20Index%204.4%20User%20Manual%20final.pdf. The reader could also review Delgado et al.[20,28,32]

To use the nitrogen index, the user will need to enter basic information about the site/farm and soil properties such as initial soil inorganic nitrogen, organic matter content, and bulk density. The type of manure applied can be selected from a drop-down menu, and the amount applied for the current year or previous year can be entered. If the user has site-specific information about the manure nitrogen content, it can also be entered. The type of nitrogen fertilizer can be selected from a drop-down menu, and the

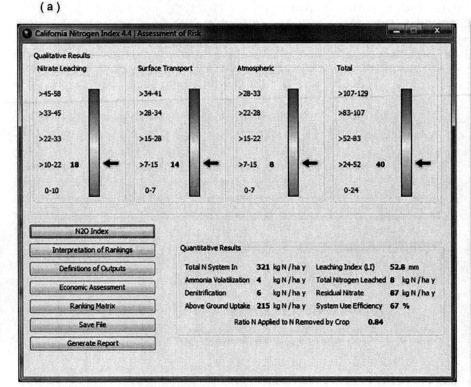
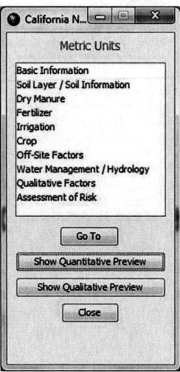

Fig. 1 Assessment of Risk screen of the nitrogen index, (a) where the risk of nitrate leaching is assessed for a given set of practices at a given set of soils, considering off-site factors, cropping system, nitrogen cycling, and other important properties. (See the Delgado et al.[44] user manual for the nitrogen index.) In the example shown in the figure, risk of nitrate leaching losses (only 8 kg NO_3-N ha^{-1}) is estimated to be low, as indicated by the high nitrogen use efficiency of the system (67%). Estimated risks of surface transport and atmospheric losses are also estimated to be low. Users can use the navigation window (b) to move between screens of the nitrogen index. The new version 4.4 of the nitrogen index is shown.

amount applied can be typed in. Total irrigation applied can be entered, including any background nitrogen that may be in the water. The crop planted can be selected from a drop-down menu, and the expected yield can be typed in for up to three crops. The amount of crop residue can also be entered.

The user can continue to the off-site factors, water management and hydrology, and qualitative factors screens, where a large number of drop-down menus minimize the amount of typing required, though it is needed in a few instances, such as when entering the annual precipitation during the growing season and non–growing season periods. For additional details about the nitrogen index, it is recommended that interested readers/users download the user manual from the above webpage and/or review Delgado et al.[20,28,32] and de Paz et al.[4]

The nitrogen index considers management practices (e.g., surface application and/or incorporation of nitrogen fertilizer), hydrology factors (e.g., hydrology soil type), off-site factors (depth to aquifer, distance to water bodies), as well as important crop properties or cropping systems (e.g., rooting depth, use of cover crops) to quickly generate a prediction/estimate of the quantitative nitrogen balance. Not only does the nitrogen index quickly generate a quantitative nitrogen mass balance to rank the risk of nitrogen losses (e.g., nitrate leaching, ammonia volatilization, denitrification), it also qualitatively ranks the risk of these losses.

This nitrogen index assists users in conducting risk assessments while considering field and off-site factors. For detailed analysis and/or site-specific quantification of nitrogen losses across different pathways, users could conduct ^{15}N studies and/or use more advanced models,[45] or for a quicker but less complex risk analysis, they could use the new nitrogen index.[20,22,28,32]

The user can quickly conduct an assessment of the nitrogen balance. The nitrogen index allows the user to consider general values of mineralization,[46] manure rate of turnover,[47] denitrification, and ammonia volatilization[48]; however, if the user has developed a set of site-specific coefficients for the site and/or region, the user's coefficients of rate of denitrification, ammonia volatilization, and mineralization can be entered into the nitrogen index evaluation and saved.

The nitrogen index ranks the risk of the nitrogen losses across different pathways as very low, low, medium, high, or very high. Although the nitrogen index also provides an overall risk ranking that can also range from very low to very high, it is recommended that users also look at the risk of nitrogen loss for each particular pathway. Additionally, even if the nitrogen index estimates that there is a medium risk of nitrate leaching losses, it is recommended that the user consider other estimates provided by the Index, such as the nitrogen available to leach. For example, even if the nitrogen index ranks the risk of nitrate leaching as medium, the amount of nitrate available to leach may be very high. In such a case, the user would want to assess whether too much nitrogen is still being applied.

CONCLUSION

The nitrogen index, which is available for download at the USDA-ARS Web site, has been tested and calibrated using data from a Mediterranean region of Spain, the North China Plain, Mexico, South America, and different sites across the United States. Results of these tests have shown that the nitrogen index is able to evaluate the effect of management on residual soil nitrate, water leaching, nitrate leaching, and crop nitrogen uptake. Additionally, nitrate leaching predicted by the nitrogen index has also been found to be correlated with measured groundwater nitrate concentrations. Although it is recommend that users get familiarized with the capabilities and limitations of the nitrogen index, the results from testing this tool across so many different agroecosystems, soils, management, and weather suggest that it can be used as a tool to assess the risk of management practices on potential losses of reactive nitrogen, including nitrate leaching. Results also suggest that the nitrogen index can be a part of an ecological engineering approach to soil and water conservation planning.[26,32]

It is recommended that users get familiarized with the Index, and that if necessary, they calibrate and/or validate the Index for their region. In 2010, INIFAP, a scientific research organization in Mexico, reported the Mexico Nitrogen Index as one of Mexico's top five forage management achievements because of the tool's usefulness in contributing to an assessment of the effects of nitrogen management on the potential for nitrogen losses such as nitrate leaching. There is potential to use this tool to assess the effects of management practices and to rank the risk as very low, low, medium, high, or very high. The nitrogen index can potentially be used as an indicator to assess the risk of nitrate leaching to the environment, so users and nutrient managers can then implement appropriate practices to reduce the reactive losses to the environment. This new tool and new approach shows promise for being used as an effective environmental conservation technology.

REFERENCES

1. Delgado, J.A.; Groffman, P.M.; Nearing, M.A.; Goddard, T.; Reicosky, D.; Lal, R.; Kitchen, N.; Rice, C.; Towery, D.; Salon, P. Conservation practices to mitigate and adapt to climate change. J. Soil Water Conserv. **2011**, *66*, 118A–129A.
2. Cowling, E.; Galloway, J.; Furiness, C.; Erisman, J.W., et al. Optimizing nitrogen management and energy production and environmental protection: Report from the Second International Nitrogen Conference, Bolger Center, Potomac, MD, Oct 14–18, 2001; 2002, available at http://www.initrogen.org/fileadmin/user_upload/Second_N_Conf_Report.pdf. (accessed May 2010).
3. Galloway, J.N.; Aber, J.D.; Erisman, J.W.; Seitzinger, S.P.; Howarth, R.W.; Cowling, E.B.; Cosby, B.J. The nitrogen cascade. BioScience **2003**, *53* (4), 341–356.
4. De Paz, J.M.; Delgado, J.A.; Ramos, C.; Shaffer, M.; Barbarick, K. Use of a new nitrogen index-GIS assessment for evaluation of nitrate leaching across a Mediterranean region. J. Hydrol. **2009**, *365*, 183–194.
5. USEPA (U.S. Environmental Protection Agency). *Federal Register*, 54 FR 22062, 22 May; USEPA: Washington, DC, 1989.
6. Follett, J.R.; Follett, R.F.; Herz, W.C. Environmental and human impacts of reactive nitrogen. In *Advances in Nitrogen Management for Water Quality*; Delgado, J.A., Follett, R.F., Eds.; Soil and Water Conservation Society: Ankeny, IA, 2010; 1–37.
7. Follett, R.F.; Delgado, J.A. Nitrogen fate and transport in agricultural systems. J. Soil Water Conserv. **2002**, *57* (6), 402–408.
8. Hatfield, J.L.; Follett, R.F. *Nitrogen Management in the Environment. Sources Problems, and Management*, 2nd Ed.; Elsevier Science Publ.: Amsterdam, Netherlands, 2008.
9. Li, X.; Hu, C.; Delgado, J.A.; Zhang, Y.; Ouyang, Z. Increased nitrogen use efficiencies as a key mitigation alternative to reduce nitrate leaching in North China Plain. Agric. Water Manage. **2007**, *89*, 137–147.
10. Lavado, R.S.; de Paz, J.M.; Delgado J.A.; Rimski-Korsakov, H. 2010 Evaluation of best nitrogen management practices across regions of Argentina and Spain. In *Advances in Nitrogen Management for Water Quality*; Delgado, J.A., Follett, R.F., Eds.; Soil and Water Conservation Society: Ankeny, IA, 2010; 313–342.
11. Rupert, M.G. Decadal-scale changes of nitrate in ground water of the United States, 1988–2004. J. Environ. Qual. **2008**, *37*, S240–S248.
12. Turner, R.E.; Rabalais, N.N. Linking landscape and water quality in the Mississippi River Basin for 200 years. BioScience **2003**, *53* (6), 563–572.
13. Mitsch, W.J.; Day, J.W. Restoration of wetlands in the Mississippi–Ohio–Missouri (MOM) River Basin: Experience and needed research. Ecol. Eng. **2006**, *26*, 55–69.
14. Rabalais, N.N.; Turner, R.E.; Wiseman, W.J., Jr. Gulf of Mexico hypoxia, a.k.a. "The Dead Zone." Annu. Rev. Ecol. Syst. **2002**, *33*, 235–263.
15. Rabalais, N.N.; Turner, R.E.; Scavia, D. Beyond science into policy: Gulf of Mexico hypoxia and the Mississippi River. BioScience **2002**, *52* (2), 129–142.

16. Goolsby, D.A.; Battaglin, W.A.; Aulenbach, B.T.; Hooper, R.P. Nitrogen input to the Gulf of Mexico. J. Environ. Qual. **2001**, *30*, 329–336.
17. Dubrovsky, N.M.; Burow, K.R.; Clark, G.M.; Gronberg, J.A.M.; Hamilton, P.A.; Hitt, K.J.; Mueller, D.K.; Munn, M.D.; Puckett, L.J.; Nolan, B.T.; Rupert, M.G.; Short, T.M.; Spahr, N.E.; Sprague, L.A.; Wilbur, W.G. *Nutrients in the Nation's Streams and Groundwater, 1992–2004*, Circular 1350; U.S. Geological Survey: Reston, VA, 2010.
18. Randall, G.W.; Delgado, J.A.; Schepers, J.S. Nitrogen management to protect water resources. In *Nitrogen in Agricultural Systems*; Schepers, J.S., Raun, W.R., Follett, R.F., Fox, R.H., Randall, G.W., Eds.; Agronomy Monograph 49; American Society of Agronomy: Madison, WI, 2008; 911–946.
19. Randall, G.W.; Goss, M.J.; Fausey, N.R. Nitrogen and drainage management to reduce nitrate losses to subsurface drainage. In *Advances in Nitrogen Management for Water Quality*; Delgado, J.A., Follett R.F., Eds.; Soil and Water Conservation Society: Ankeny, IA, 2010; 61–93.
20. Delgado, J.A.; Shaffer, M.; Hu, C.; Lavado, R.; Cueto Wong, J.A.; Joosse, P.; Sotomayor, D.; Colon, W.; Follett, R; Del Grosso, S.; Li, X.; Rimski-Korsakov, H. An index approach to assess nitrogen losses to the environment. Ecol. Eng. **2008**, *32*, 108–120.
21. Delgado, J.A.; Follett, R.F., Eds. *Advances in Nitrogen Management for Water Quality*; SWCS: Ankeny, IA, 2010.
22. Shaffer, M.J.; Delgado, J.A. Essentials of a national nitrate leaching index assessment tool. J. Soil Water Conserv. **2002**, *57*, 327–335.
23. Buczko, U.; Kuchenbuch, R.O.; Lennartz, B. Assessment of the predictive quality of simple indicator approaches for nitrate leaching from agricultural fields. J. Environ. Manage. **2010**, *91*, 1305–1315.
24. Lemunyon, J.L.; Gilbert, R.G. The concept and need for a phosphorus assessment tool. J. Prod. Agric. **1993**, *6*, 483–486.
25. Sharpley, A.N.; Weld, J.L.; Beegle, D.B.; Kleinman, P.J.A.; Gburuk, W.J.; Moore, P.A., Jr.; Mullins, G. Development of phosphorus indices for nutrient management planning strategies in the United States. J. Soil Water Conserv. **2003**, *53*, 137–151.
26. Gross, C.M.; Delgado, J.A.; Shaffer, M.J.; Gasseling, D.; Bunch, T.; Fry, R. A tiered approach to nitrogen management: A USDA perspective. In *Advances in Nitrogen Management for Water Quality*; Delgado, J.A., Follett, R.F., Eds.; Soil and Water Conservation Society: Ankeny, IA, 2010; 410–424.
27. Delgado, J.A.; Gagliardi, P.; Shaffer, M.J; Cover, H.; Hesketh, E.; Ascough, J.C.; Daniel, B.M. New tools to assess nitrogen management for conservation of our biosphere. In *Advances in Nitrogen Management for Water Quality*; Delgado, J.A., Follett, R.F., Eds.; SWCS: Ankeny, IA, 2010; 373–409.
28. Delgado, J.A.; Shaffer, M.; Hu, C.; Lavado, R.S.; Cueto Wong, J.; Joosse, P.; Li, X.; Rimski-Korsakov, H.; Follett, R.; Colon; W.; Sotomayor, D. 2006. A decade of change in nutrient management requires a new tool: A new nitrogen index. J. Soil Water Conserv. **2006**, *61*, 62A–71A.
29. Pierce, F.J.; Shaffer, M.J.; Halvorson, A.D. Screening procedure for estimating potentially leachable nitrate-nitrogen below the root zone. In *Managing Nitrogen for Groundwater Quality and Farm Profitability*; Follett, R.F., Keeny, D.R., Cruse, R.M., Eds.; Soil Science Society of America: Madison, WI, 1991; 259–283.
30. Figueroa, V.U.; Núñez, H.G.; Delgado, J.A.; Cueto, J.A.; Flores, M.J.P. Estimación de la producción de estiércol y de la excreción de nitrógeno, fósforo y potasio por bovino lechero en la Comarca Lagunera. In *Agricultura Orgánica*, 2nd Ed.; Orona et al., Eds.; Sociedad Mexicana de la Ciencia del Suelo, Consejo Nacional de Ciencia y Tecnología, and Universidad Autónoma del Estado de Durango-Facultad de Agricultura y Zootecnia: Gomez Palacio, Mexico, 2009; 128–151, 425.
31. Instituto Nacional de Investigaciones Forestales Agricolas y Pecuaris (INIFAP). Reporte Anual 2009 Ciencia y Tecnología para el Campo Mexicano Instituto Nacional de Investigaciones Forestales, Agrícolas y Pecuarias. Reporte Anual 2009 México, D. F. Publicación Especial Núm. 5; ISBN 978-607-425-316-0: Mexico 2010; 31 pp.
32. Delgado, J.A.; Shaffer, M.J.; Lal, H.; McKinney, S.; Gross, C.M.; Cover, H. Assessment of nitrogen losses to the environment with a Nitrogen Trading Tool (NTT). Comput. Electron. Agric. **2008**, *63*, 193–206.
33. Van Es, H.M.; Czymmek, K.J.; Ketterings, Q.M. Management effects on nitrogen leaching and guidelines for a nitrogen leaching index in New York. J. Soil Water Conserv. **2002**, *57*, 499–504.
34. Ontario Ministry of Agriculture, Food and Affairs (OMAFRA). *Fundamentals of Nutrient Management Training Course*; Queen's Printer of Ontario: Ontario, Canada, 2005.
35. Wu, L.; Letey, J.; French, C.; Wood, Y.; Bikie, D. Nitrate leaching hazard index developed for irrigated agriculture. J. Soil Water Conserv. **2005**, *60*, 90A–95A.
36. Delgado, J.A. Sequential NLEAP simulations to examine effect of early and late planted winter cover crops on nitrogen dynamics. J. Soil Water Conserv. **1998**, *53*, 241–244.
37. Delgado, J.A. Use of simulations for evaluation of best management practices on irrigated cropping systems. In *Modeling Carbon and Nitrogen Dynamics for Soil Management*; Shaffer, M.J., Ma, L., Hansen, S., Eds.; Lewis Publishers: Boca Raton, FL, 2001; 355–381.
38. Sharpley, A.N.; Daniel, T.; Sims, T; Lemunyon, J.; Stevens, R.; Parry, R. *Agricultural Phosphorus and Eutrophication*, U.S. Department of Agriculture Agricultural Research Service No. ARS-149; 1999.
39. Sharpley, A.N.; Kleinman, P.; McDowell, R. Innovative management of agricultural phosphorous to protect soil and water resources. Commun. Soil Sci. Plant Anal. **2001**, *32* (7 & 8), 1071–1100.
40. Heathwaite, L.; Sharpley, A.; Gburek, W. A conceptual approach for integrating phosphorous and nitrogen management at watershed scales. J. Environ. Qual. **2000**, *29*, 158–166.
41. Williams, J.R.; Kissel, D.E. Water percolation: An indicator of nitrogen-leaching potential. In *Managing Nitrogen for Groundwater Quality and Farm Profitability*; Follett, R.F., Keeney, D.R., Cruse, R.M., Eds.; Soil Science Society of America: Madison, WI, 1991; 59–83.
42. Van Es, H.M.; Delgado, J.A. Nitrate leaching index. In *Encyclopedia of Soil Science*; Lal, R., Ed.; Markel and Decker: New York, **2006**; 1119–1121.

43. Buczko, U.; Kuchenbuch, R.O. Environmental indicators to assess the risk of diffuse nitrogen losses from agriculture. Environ. Manage. **2010b**, *45*, 1201–1222.
44. Delgado, J.A., P.M. Gagliardi, E.J. Rau, R. Fry, U. Figueroa, C. Gross, J. Cueto-Wong, M. Shaffer, K. Kowalski, D. Neer, D. Sotomayor-Ramirez, J. Alwang, C. Monar, L. Escudero, and A.K. Saavedra-Rivera. 2011. Nitrogen Index 4.4 User Manual. USDA-ARS-SPNR, Fort Collins, CO. (USDA-ARS-SPNR & NRCS User Manual; available at official USDA ARS SPNR site at http://www.ars.usda.gov/npa/spnr/nitrogentools).
45. Delgado, J.A. Quantifying the loss mechanisms of nitrogen. J. Soil Water Conserv. **2002**, *57*, 389–398.
46. Vigil, M.F.; Eghball, B.; Cabrera, M.L.; Jakubowski, B.R.; Davis, J.G. Accounting for seasonal nitrogen mineralization: An overview. J. Soil Water Conserv. **2002**, *57*, 464–469.
47. Eghball, B.; Weinhold, B.J.; Gilley, J.E.; Eigenberg, R.A. Mineralization of manure nutrients. J. Soil Water Conserv. **2002**, *57*, 470–473.
48. Meisinger, J.J.; Randall, G.W. Estimating nitrogen budgets for soil-crop systems. In *Managing Nitrogen for Groundwater Quality*; Follett, R.F., Keeney, D.R., Cruse, R.M., Eds.; Soil Science Society of America: Madison, WI, 1991; 85–124.
49. Follett, R.F.; Walker, D.J. Groundwater quality concerns about nitrogen. In *Nitrogen Management and Groundwater Protection*; Follett, R.F., Ed.; Elsevier Sci. Pub.: Amsterdam, Netherlands, 1989; 1–22.
50. Shaffer, M.J.; Delgado, J.A. Field techniques for modeling nitrogen management. In *Nitrogen in the Environment: Sources, Problems, and Management*; Follett, R.F., Hatfield, J.L., Eds.; Elsevier: New York, 2001; 391–411.
51. Available at http://www.omafra.gov.on.ca/english/nm/nman/default.htm (accessed May 2011).

Nitrogen

Oswald Van Cleemput
Pascal Boeckx
Faculty of Agricultural and Applied Biological Sciences, University of Ghent, Ghent, Belgium

Abstract
Nitrogen is essential to all life. It is the nutrient that most often limits biological activity. In agricultural and natural ecosystems, nitrogen occurs in many forms. The use of nitrogen fertilizers has become essential to increase the productivity of agriculture, and has resulted in an almost doubling of the global food production in the past 50 years. However, this also implies that the natural nitrogen cycle has substantially been disturbed. This entry provides an overview of the different nitrogen transformation processes in the soil.

INTRODUCTION

Nitrogen (N) is essential to all life. It is the nutrient that most often limits biological activity. In agricultural and natural ecosystems, N occurs in many forms covering a range of valence states from −3 to +5. The change from one valence state to another depends primarily on environmental conditions. The transformations and flow from one form to another constitute the basics of the soil N cycle (Fig. 1). The use of N fertilizers has become essential to increase the productivity of agriculture, and has resulted in an almost doubling of the global food production in the past 50 years. However, this also implies that the natural N cycle has substantially been disturbed. In the following paragraphs an overview of the different N transformation processes in the soil is given.

NITROGEN CYCLE: GENERAL

Atmospheric N_2 gas (valence 0) can be converted by lightening to various oxides and finally to nitrate (NO_3^-) (valence +5), which can be deposited and taken up by growing plants. Also N_2 gas can be converted to ammonia (NH_3, valence −3) by biological N_2 fixation, with the NH_3 participating in a number of biochemical reactions in the plant. When plant residues decompose the N-compounds undergo a series of microbial conversions (mineralisation) leading first to the formation of ammonium (NH_4^+) (valence −3) and possibly ending up in NO_3^- (nitrification). Under anaerobic conditions NO_3^- can be converted to various N-oxides and finally to N_2 gas (denitrification). When mineral or organic N fertilizers are used they also undergo the same transformation processes and influence the rate of other N-transformations. In considering the soil compartment, there can be N gains (such as biological N_2 fixation) as well as N losses (such as leaching and denitrification). Furthermore N can be exported from the soil via harvest products, or immobilized in soil organic matter.

NITROGEN TRANSFORMATIONS IN THE SOIL

The principal forms of N in the soil are NH_4^+, NO_3^- or organic N-substances. At any moment, inorganic N in the soil is only a small fraction of the total soil N. Most of the N in a surface soil is present as organic N. It consists of proteins (20%–40%), amino sugars, such as the hexosamines (5%–10%), purine and pyrimidime derivates (1% or less), and complex unidentified compounds formed by reaction of NH_4^+ with lignin, polymerization of quinones with N compounds and condensation of sugars and amines. In the subsoil, an important fraction of the present N can be trapped in clay lattices (especially *illitic* clays) as nonexchangeable NH_4^+ and is consequently largely unavailable. Organic substances slowly mineralize by microorganisms to NH_4^+, which could be converted by other microorganisms to NO_3^- (see further).

The NH_4^+ can be adsorbed to negatively charged sites of clay minerals and organic compounds. This reduces its mobility in the soil compared to the more mobile NO_3^- ion. Microorganisms can use both NH_4^+ and NO_3^- to satisfy their need for N. This type of N transformation is called microbial immobilization.

The ratio between carbon (C) and N (C : N ratio) in organic matter determines whether immobilization or mineralization is likely to occur. When utilizing organic matter with a low N content, the microorganisms need additional N, decreasing the mineral N pool of the soil. Thus, incorporation of organic matter with a high C : N ratio (e.g., cereal straw) results in immobilization. Incorporation of organic matter with a low C : N ratio (e.g., vegetable or legume residues) results in N-mineralization. A value of the C : N ratio of 25 to 30 is often taken as the critical point toward either immobilization or mineralization.

Nitrification is a two-step process. In the first step NH_4^+ is converted to nitrite (NO_2^-) (valence +3) by a group of obligate autotrophic bacteria known as *Nitrosomonas* species. The second step is carried out by another group of obligate autotrophic bacteria known as *Nitrobacter* species.

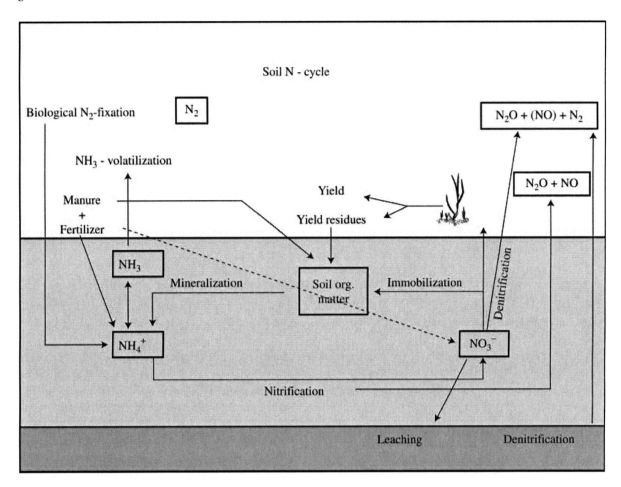

Fig. 1 The soil N cycle. The white compartment represents the atmosphere; the light gray compartment represents the biosphere and the dark gray compartment the subsoil.

Also a few heterotrophs can carry out nitrification, usually at much lower rates.

Soil water and aeration are crucial factors for nitrification. At a water potential of 0 kPa (saturation), there is little air in the soil and nitrification stops, due to oxygen limitation; nitrification is greatest near field capacity (−33 kPa in medium- to heavy-textured soils, to 0 to −10 kPa in light sandy soils). Also in dry soils NH_4^+ and sometimes NO_2^- accumulate presumably because *Nitrobacter* species are more sensitive to water stress than the other microorganisms.

Nitrification is slow in acid conditions with an increasing rate at increasing pH. Mainly under alkaline conditions, nitrite is also accumulating, because *Nitrobacter* is known to be inhibited by ammonia, which is formed under alkaline conditions. Nitrification is a process that acidifies the soil as protons (H^+) are liberated:

$$NH_4^+ + 2O_2 \rightarrow NO_3^- + 2H^+ + H_2O$$

During nitrification minor amounts of nitrous oxide (N_2O) (valence +1) and nitric oxide (NO) (valence +2) are formed. Both compounds have environmental consequences, discussed below.

The effect of temperature on nitrification is climate dependent. There is a climatic selection of species of nitrifiers, with those from cooler regions having lower temperature optima and less heat tolerance than species from warmer regions. All above-mentioned factors influencing nitrification also influence the nitrifying population. The population and activity of nitrifiers can be reduced by the use of nitrification inhibitors, such as dicyanodiamide, nitrapirin and neem (*Azadirachta indica*) seed cake. They are used mostly to retard the nitrification of manure; otherwise their practicality is controversial and they are not extensively used. More details about nitrification and nitrification inhibitors can be found in McCarty[1] and Prosser.[2]

NITROGEN INPUT PROCESSES

Atmospheric Nitrogen Deposition

The total atmospheric N (NH_4^+ and NO_3^-) deposition is in the order of 10–40 kg N ha^{-1} yr^{-1} in much of north-western and central Europe and some regions in North America. It ranges from 3–5 kg N ha^{-1} yr^{-1} in pristine areas.[3] It is originating from previously emitted NH_3 and NO_x from agricultural and industrial activities or traffic.

Biological Nitrogen Fixation

Rhizobium species living in symbiotic relationship in root nodules of legumes, e.g. clover (*Trifolium*), lucerne (*Medicago*), peas (*Pisum*) and beans (*Faba*)—can convert atmospheric N_2 gas to NH_3, which is further converted to amino acids and proteins. Parallel to this process, the *rhizobium* species receive from the legume the energy they need to grow and to fix N_2. Photosynthetic cyanobacteria are also N-fixing organisms and are especially important in paddy rice (*Oryza*). The amount of N fixed varies greatly from crop to crop, ranging from a few kg to a few hundred kg N ha^{-1} yr^{-1}. The process is depressed by ample N supply from other sources, and it is sensitive to lack of phosphorus. The amount of globally fixed N is almost the double the amount of applied fertilizer N. Next to symbiotic N fixing bacteria also non-symbiotic species (e.g. *Azotobacter*) occur in soils. In general, free-living diazotrophs make a small but significant contribution to the soil N status. Some nonleguminous trees and plants (e.g. alder (*Alnus*), sugarcane (*Saccharum*) host N-fixing bacteria as well. Much uncertainty exists about the association of N fixing bacteria with non-legumes (so called associative N fixing bacteria).

Mineral and Organic Nitrogen Fertilization

Theoretically plants should prefer NH_4^+ above NO_3^-, because NH_4^+ does not need to be reduced before incorporation into the plant. In most well-drained soils oxidation of NH_4^+ is fairly rapid and therefore most plants have developed to grow better with NO_3^-. However, a number of studies have shown that plants better develop when both sources are available. Rice, growing under submerged conditions must grow in the presence of NH_4^+ as NO_3^- is not stable under flooded conditions. When urea is applied it rapidly hydrolyzes under well-drained conditions, unless a urease inhibitor is being added; under submerged conditions rice plants may also absorb N directly as molecular urea. Organic manure can be of plant or animal origin or a mixture of both. However, most comes from dung and urine from farm animals. It exists as farmyard or stable manure, urine, slurry or as compost. Because its composition is not constant and because plant material (catch or cover crops, legumes) is often added freshly (green manure) to the soil, less than 30% of its nutrients becomes available for the next crop.

NITROGEN UPTAKE BY PLANTS

Growing plants get their N from fertilizer N as well as from organic soil N upon mineralization. Plants take up N compounds both as NO_3^- and as NH_4^+. In general, NO_3^- is the major source of plant N. There is some evidence that small amounts of organic N (urea or amino acids) can be taken up by plants from the soils solution. Plant uptake of N can be studied through the use of mineral fertilizers or organic matter labeled with the stable N isotope ^{15}N. The proportion of applied N taken up by the crop is affected by many factors, including crop species, climate and soil conditions. Above ground parts of the crop can recover 40%–60% of the fertilizer N applied.

NITROGEN LOSS PROCESSES

Ammonia Volatilization

Losses of N from the soil by NH_3 volatilization amount globally to 54 Mt (or 10^{12} g) NH_3-N yr^{-1} and 75% is of anthropogenic origin.[4] According to the ECETOC,[5] the dominant source is animal manure and about 30% of N in urine and dung is lost as NH_3. The other major source is surface application of urea or ammonium bicarbonate and to a lesser degree other ammonium-containing fertilizers. As urea is the most important N fertilizer in the world, it may lead to important NH_3 loss upon hydrolysis and subsequent pH rise in the vicinity of the urea till. The transformation of NH_4^+ to the volatile form NH_3 increases with increasing pH, temperature, soil porosity, and wind speed at the soil surface. It decreases with increasing water content and rainfall events following application. Ammonia losses from soils can be effectively reduced by fertilizer incorporation or injection instead of surface application.

Emission of Nitrogen Oxides (N_2O, NO) and Molecular Nitrogen (Nitrification and Denitrification)

Microbial nitrification and denitrification are responsible for the emission of NO and N_2O.[6] They are by-products in nitrification and intermediates during denitrification. Probably about 0.5% of fertilizer N applied is emitted as NO[7] and 1.25% as N_2O.[8] However, wide ranges have been reported. Intensification of arable agriculture and of animal husbandry has made more N available in the soil N cycle increasing the emission of N oxides. The relative percentage of NO and N_2O formation very much depends on the moisture content of the soil. At a water-filled pore space (WFPS, or the fraction of total soil pore space filled with water) below 40% NO is produced mainly from nitrification. Between a WFPS of 40% and 60% formation of NO and N_2O from nitrification occurs. Between a WFPS of 60% and 80% N_2O is predominantly produced from denitrification and the formation of NO is decreasing sharply. At a WFPS above 80% the formation of N_2 by denitrification is dominant. In practice these WFPS ranges will overlap anddepend on the soil type.[9] Next to water content, also temperature, land use and availability of N and decomposable organic matter are important determining factors for N_2O formation. Nitrous oxide is a greenhouse gas contributing 5%–6% to the enhanced greenhouse effect. Increased concentrations are also detrimental for the stratospheric ozone layer.[10] In the presence of sunlight, NO_x (NO and NO_2) react with volatile organic compounds from evaporated petrol and solvents and from vegetation and forms tropospheric ozone which is, even at low concentration, harmful to plants and human beings. The major gaseous end-product of denitrification is N_2. The ratio of N_2O to N_2 produced by denitrification depends on many environmental conditions. Generally the more anaerobic the environment the greater the N_2 production. Denitrification is controlled by three primary factors (oxygen, nitrate and carbon), which in turn are controlled by several physical and biological factors. Denitrification N loss can reach 10% of the fertilizer N input—more on grassland and when manure is also applied.[11] Chemical denitrification is normally insignificant and is mainly related to the stability of NO_2^- and acid conditions.[12] It is more difficult to reduce N_2O and NO from soils then NH_3 losses. A general principle is to minimize N surpluses in the soil profile via careful fertilizer adjustment, corresponding to the actual crop demands.

Leaching

Applied NO_3^- or NO_3^-, formed through nitrification from mineralized NH_4^+ or from NH_4^+ from animal manure, can leach out of the rooting zone. It is well possible that this leached NO_3^- can be denitrified at other places and returned into the atmosphere. The amount and intensity of rainfall, quantity and frequency of irrigation, evaporation rate, temperature, soil texture and structure, type of land use, cropping and tillage practices and the amount and form of fertilizer N are all parameters influencing the amount of NO_3^- leaching to the underground water. Nitrate leaching should be kept under control as it may influence the nitrate content in drinking water influencing human health and in surface water, causing eutrophication. Nitrate losses can be minimized by reducing the mineral N content in the soil profile during the winter period by careful fertilizer adjustment, growing of cover crops or riparian buffer areas.

REFERENCES

1. McCarty, G.W. Modes of action of nitrification inhibitors. Biol. Fert. Soils **1999**, *29*, 1–9.
2. Prosser, J.I., Ed. *Nitrification, Special Publications of the Society of General Microbiology*; IRL Press: Oxford, 1986; 20 pp.
3. Lagreid, M.; Bockman, O.C.; Kaarstad, O. *Agriculture, Fertilizers and the Environment*; CIBA Publishing: Oxon, U.K., 1999; 294 pp.
4. Sutton, M.A.; Lee, D.S.; Dollard, G.J.; Fowler, D. International conference on atmospheric ammonia: emission, deposition and environmental impacts. Atmospheric Environment **1998**, *32*, 1–593.
5. ECETOC *Ammonia Emissions to Air in Western Europe, (No. 62)*; European Centre for Ecotoxicology and Toxicology of Chemicals: Brussels, 1994; 196 pp.
6. Bremner, J.M. Sources of nitrous oxide in soils. Nutr. Cycl. Agroecosys. **1997**, *49*, 7–16.
7. Veldkamp, E.; Keller, M. Fertilizer-induced nitric oxide emissions from agricultural soils. Nutr. Cycl. Agroecosys. **1997**, *48*, 69–77.
8. Mosier, A.; Kroeze, C.; Nevison, C.; Oenema, O.; Seitsinger, S.; Van Cleemput, O. Closing the global N_2O budget: nitrous oxide emissions through the agricultural N cycle. Nutr. Cycl. Agroecosys. **1998**, *52*, 225–248.
9. Davidson, E.A. Fluxes of Nitrous Oxide and Nitric Oxide from terrestrial ecosystems. In *Microbial Production and Consumption of Greenhouse Gases: Methane, Nitrogen Oxides, and Halomethanes*; Rogers, J.E., Whitman, W.B., Eds.; American Society for Microbiology: Washington, DC, 1991; 219–235.
10. Crutzen, P.J. The influence of nitrogen oxides on the atmospheric ozone content. Quat. J. Royal Meteor. Soc. **1976**, *96*, 320–325.
11. von Rheinbaben, W. Nitrogen losses from agricultural soils through denitrification—A critical evaluation. Z. Pflanzenern. Bodenk. **1990**, *153*, 157–166.
12. Van Cleemput, O. Subsoils: chemo- and biological denitrification, N2O and N2 emissions. Nutr. Cycl. Agroecosys. **1998**, *52*, 187–194.

Nitrogen Trading Tool

Jorge A. Delgado
*Soil Plant Nutrient Research Unit, Agricultural Research Service (USDA-ARS),
U.S. Department of Agriculture, Fort Collins, Colorado, U.S.A.*

Abstract
Nitrogen is an important nutrient that contributes to increased crop yields, which are important for food security. However, when more nitrogen than necessary is applied, there is increased impact to the environment. There is potential to use environmental payments and/or credits in environmental trading and/or water or air quality programs to encourage farmers, nutrient managers, and organizations to apply best management practices that reduce the transport of reactive nitrogen to the environment. Numerous water and/or air quality programs have been established around the world. A nitrogen trading tool is a tool that is supported by algorithms to quickly assess the potential reduction in nitrogen losses that could result from using better management practices. The Nitrogen Trading Tool is defined as the NTT, which was released in 2010. Other tools that can be used to assess the benefits of management practices in terms of potential reductions of nitrogen losses to the environment are also nitrogen trading tools, though they may be referred to by another name (e.g., the Nutrient Tracking Tool). Nitrogen trading tools are ecotechnologies that will contribute to the development of marketplaces for trading ecosystem services that result from best management practices that contribute to air and water quality by reducing nitrogen losses to the environment.

IMPORTANCE OF NITROGEN

Nitrogen is essential for global sustainability. In thinking about nitrogen, it is important to realize that this element is essential for all living organisms: it is part of DNA and proteins and has important functions in plants and animals. Crops are no exception, and agricultural systems respond to application of nitrogen with higher yields. Agricultural systems around the world are under increased pressure to heighten productivity as food demand grows along with the human population. By increasing productivity, nitrogen inputs to agricultural systems make an important contribution to global sustainability; however, nitrogen inputs have also been reported to increase losses of reactive nitrogen via different pathways of the nitrogen cycle. Reactive nitrogen can impact the environment as it cycles through it, and the various effects that occur as reactive nitrogen cycles through the environment have been collectively referred to as the nitrogen cascade.[1,2]

NEED FOR NITROGEN LOSS MEASUREMENTS

The nitrogen cycle is impacted by human activities, including those that increase the use of nitrogen in agricultural systems, and this impact can be seen in effects such as increased nitrate (NO_3) levels in groundwater or surface water resources, increased concentration of nitrous oxide (N_2O) in the atmosphere, and increased ammonia (NH_3) deposition in natural areas. It has been reported that reactive nitrogen losses can potentially have negative impacts on human and ecosystem health.[3] In addition, more than two decades ago, the U.S. Environmental Protection Agency (USEPA) established 10 mg NO_3-N L^{-1} as the safe limit for nitrate in drinking water (USEPA[4]).

The food production benefits, on the one hand, and the environmental impacts from nitrogen applications, on the other, present a quandary for humans as to how to continue to maximize productivity across agricultural systems, which requires nitrogen inputs, while minimizing nitrogen losses to the environment. Recent and ongoing developments in the management of nitrogen to preserve water quality and reduce atmospheric emissions, such as the use of nitrogen trading tools (NTTs), can help provide some answers.[5] Improving nitrogen management requires that management be looked at within the context of the nitrogen cycle and with an understanding of nitrogen dynamics, pathways for nitrogen losses, and how management can serve as a tool to respond to the combination of site-specific factors (such as weather, crops, hydrology, and landscape) of a given field to reduce nitrogen losses.

WATER QUALITY TRADING PROGRAMS

The concept of using some type of environmental payments and/or credits in an environmental trading and/or water quality program to reduce the transport of reactive nitrogen to the environment is not new and was proposed earlier by several researchers.[6–10] The use of ecological engineering approaches to establish or develop wetlands to use as filters to reduce the transport of nitrate to the Gulf of Mexico was proposed by Hey[6] and Hey et al.[8] They suggested that strategically placed wetlands could filter the water carrying the nitrate as it enters the wetlands; farmers could then receive environmental credits (e.g., payments) for harvesting/removing nitrate from the water.

Berry et al.[11,12] and Delgado and Berry[13] suggested that precision conservation (target conservation) could be used to select the best locations to place these wetlands for nitrate removal. Delgado and Berry[13] reported that a precision conservation approach to nitrogen trading using robust tools (models, Geographic Information Systems [GIS], Global Positioning System ([GPS]) could contribute to the development of strategies that will maximize economic returns for trading (via maximizing reduction of reactive nitrogen losses, leading to greater potential for trade).

Lal et al.[14] reported that in the United States, there are some potential opportunities and markets for nitrogen trading such as in the Connecticut Department of Environmental Protection (DEP), Pennsylvania, Ohio, and Oregon. For example, in Ohio where it has been reported that about 40% of the rivers and streams are impacted, the Miami Conservation District Water Quality Trading Program receives proposals from farmers and funds projects to reduce the nutrient (e.g. nitrogen) loading of rivers, reducing losses (http://www.miamiconservancy.org/WQTP/index.asp?data=dataXML.asp). Lal et al.[14] reported that these markets might benefit from the use of a NTT to evaluate the potential to use conservation practices to reduce nitrogen losses. In Pennsylvania, the state's DEP has a nutrient trading program. Additional information can be found at http://www.portal.state.pa.us/portal/server.pt/community/chesapeake_bay_program/10513. The credits can be generated within the same watershed, and by applying these credits, the farmers can be paid, but the credits need to be certified, verified, and registered. For example, cover crops and other practices can be used to generate nitrogen credits.

Other authors have reviewed existing markets for noncommodity ecosystem services, which can be provided via practices that generate nitrogen savings that can potentially be traded. Ribaudo and Gottlie[15] reviewed several quality trading markets that included trading of reductions in pollutants from agriculture, including several projects at Massachussets Estuaries Project, Tar-Pamlico, North Carolina; Clermount County, Ohio; Conestoga River in Pennsylvania, and Chesapeake Bay Watershed. Ribaudo et al.[16,17] reported that the NTT is an example of the type of tool that can be used to reduce the information costs and uncertainty of such complex agricultural systems. Ribaudo et al.[16,17] reported that there is potential to use NTTs that can quantify non-commodity ecosystem services provided by the farmers if these tools can reduce the uncertainty and provide good information accepted by peers and the market system.

PAYMENTS FOR ENVIRONMENTAL SERVICES PROGRAMS

Lal et al.[14] reported that environmental services programs like water quality trading could provide payments to farmers and ranchers to help them implement conservation practices and that the type of payments received could depend on what type of conservation practice is implemented and on how effective the conservation practice is determined to be in providing a given service (e.g., minimizing transport of reactive nitrogen losses). Stanton et al.[18] reported that there are close to 300 marketplaces around the world that have been established covering some of the payments related to watershed services and water quality programs. They reported that, in 2008, there were programs spread across the globe and established in 24 different countries including the United States, Brazil, Canada, France, China, and New Zealand. They estimated that, in the same year, the total amount of payments from these programs was more than $9 billion ($1.35 billion in the United States) in transactions, impacting more than 3 billion ha worldwide (16.4 million ha in the United States) in conservation related to watershed services and water quality programs.

There are rules that can vary from state to state that govern payments for services or how the contracts are established. Stanton et al.[18] reported that the majority of the payments are from government programs but that there are private sector payment programs that are expected to increase in the future. To learn more about the specific programs and rules for trading among farmers, aggregators, buyers, and regulators, one could visit the homepages of several of the different U.S. nutrient programs that are available, such as the one from the Pennsylvania DEP: http://www.dep.state.pa.us/river/Nutrient%20Trading.htm. Other examples can be found for Ohio (http://www.miamiconservancy.org/water/surface.asp) and for Maryland (http://www.mda.state.md.us/nutrad/). Detailed information about the regulation of nutrient trading market-based programs and/or the trade of credits and payments can be found at such sites. Stanton et al.[18] described different payment mechanisms that could range from private transfer of funds in direct payments to cost share arrangements and/or arrangements where polluters try to meet standards by buying and selling pollution credits.

GREENHOUSE GAS OFFSET PROGRAMS

Delgado et al.[19,20] reported that the NTT, a new ecotechnology, can be used to assess the effects of management practices not only on water quality but also on air quality. Since N_2O can be expressed as CO_2 equivalents, reductions in emissions of N_2O can be expressed as carbon sequestration equivalents, increasing the potential to trade carbon savings in carbon markets. Delgado et al.[19,20] assessed the potential of using best management practices (BMPs) to reduce direct and even indirect N_2O emissions across no-till systems of the northeast, manure systems of the Midwest, and irrigated systems of the dry, irrigated western United States, and thus the potential to trade these savings in nitrogen losses in future air markets as non-commodity services. The assessment from Delgado et al.[19,20] showed that there is potential to increase carbon sequestration equivalents by applying practices that significantly increase nitrogen use efficiency such as using summer cover crops with limited irrigation in the irrigated west, or better manure management applications in the Midwest, and/or using leguminous crops and crediting nitrogen cycling in no-till systems of the northwestern United States. Lal et al.[14] and Ribaudo et al.[16,17] also reported on the potential of using BMPs to generate reductions in nitrogen losses for trade in current water quality markets and in future air markets.

Stanton et al.[18] reported that there are ongoing efforts where water quality trading programs (e.g., Ohio River Basin Trading Project) are assessing the benefits of better practices in reducing the off-site transport of nitrogen to water bodies, which can also result in benefits such as reductions in N_2O emissions. This is in agreement with results from Delgado et al.[19,20] that showed lower nitrate leaching and lower N_2O emissions with better manure management.

The American Carbon Registry is a leading nonprofit carbon market in the United States (http://www.americancarbonregistry.org/). This voluntary registry tracks carbon sequestration in soils as well as potential reductions in N_2O emissions from agricultural fields. For example, among their accepted methodologies for keeping track of greenhouse gas emissions, they include methodology for assessing the effects of better fertilizer management practices on reductions in N_2O emissions. This is another example of the potential for future connection of water quality trading, carbon markets, and air quality markets.

CONSERVATION PROGRAM PERFORMANCE

Having a means of assessing the efficiency of the conservation practices being implemented is important for the transparency of water and/or air quality trading programs.[18] Stanton et al.[18] reported that it is important to develop and/or have a performance metric that measures and/or assesses real improvements in the watershed, showing that true improvements in ecosystem health have been made. They reported that the development of procedures for payments for ecosystem services should also include performance metrics that connect the payments to standardized practices used in the marketplace. In a water quality trading market, the specific programs and rules for trading among farmers, aggregators, buyers, and regulators should also include a clear role for the administrators to contribute to a transparent process for monitoring performance and evaluating the system.[18]

Administrators could help develop realistic and viable performance measurements to help ensure that the services being paid for are contributing to ecosystem health. It is also important to measure performance over the long term: for example, how the concentrations of nitrate and total nitrogen are changing with time. The lack of solid performance measurements is seen as a weakness in water quality programs. Information about what conservation practices are registered, the efficiency of the conservation practices, and what general performance metrics are being used across the watershed to assess ecosystem health can be found at webpages such as http://www.dep.state.pa.us/river/Nutrient%20Trading.htm.

ROLE OF A MODELING TOOL SUCH AS NTT IN POLICY

Because there are so many pathways for reactive nitrogen loss (e.g., leaching, denitrification) and these pathways are affected by so many factors such as soil hydrology, water management, weather patterns that may not be possible to be manage or control, and other factors that can be controlled, such as time of application and method of application, Delgado et al.[19] suggested the use of a robust NTT as a method to assess the potential for trading nitrogen.

The concept of a robust NTT supported by algorithms based on research in biogeochemistry, chemistry, physics, hydrology, and other sciences that describes the processes of nitrogen transformation and pathways for losses and that is calibrated and validated for the region of interest to assess how management practices can reduce these losses was proposed by Delgado et al.[19] Delgado et al.[19] developed a set of equations to assess the effects of management and conservation practices on losses of reactive nitrogen via several different pathways. The complexity of the nitrogen cycle makes the task of quantifying nitrogen losses a difficult one,[21] but models and other new approaches to nitrogen loss assessments, such as a NTT, can help individuals quantitatively estimate nitrogen losses quickly. The USEPA[22] reported that modeling tools (e.g., the NTT) can help to ensure that nitrogen calculations are based on sound science. This approach of using new ecotechnologies such as the NTT appears to be a policy approach that could be supported by agencies such as the Natural Resources Conservation Service (NRCS) to provide an assessment of conservation practices for air and water quality trading (USEPA[22]).

Delgado et al.[20] also discussed the need to account for spatial and temporal variability. The concept of a NTT that can integrate many factors and that can be used with GIS standing alone and/or in an Internet setting was proposed by Delgado et al.[19,20,23] One example of a NTT is the one initiated by a United States Department of Agriculture Agricultural Research Service (USDA-ARS) effort, which can consider spatial variability by using GIS[19,20] and which uses the new Nitrogen Losses and Environmental Assessment Package (NLEAP).

Delgado et al.[19,20] showed how the technology is available to link models such as the new NLEAP-GIS[23,24] with the NRCS Web Soil Survey using the Internet. Delgado et al.[19,20] presented a prototype that was developed in a cooperative effort by USDA-NRCS and USDA-ARS where users could go to the Web Soil Survey ID for a given farm, bring the soils to the NTT web prototype, and run a comparison between a given baseline scenario and the new scenario to determine if there is potential to trade savings in nutrient loss (Figs. 3 and 4).

The NTT is an approach to assessing nitrogen losses where once the model has been calibrated and validated for a given region, it can produce an independent, quantitative assessment of the management practices at a given site. Delgado et al.[19,20] reported that with new advances in ecotechnology such as the NTT, conservation efforts can start to move from the older approach to cutting nitrogen losses, where some sources of nitrogen pollution can only be identified in a general way as non-point sources, to a new approach where every field that receives nitrogen inputs can be identified as a source of nitrogen losses by a model, and where nitrogen losses can potentially be cut using a targeted, site-specific approach (i.e., precision conservation).[19,20] The advances in the last few decades, from the development of desktop computers in the 1980s, to the rise of the Internet in the 1990s, to the extensive GIS applications in the 2000s, may lead to the elimination of non-point sources of nitrogen losses and towards the use of computer models to quantify nitrogen losses at a given farm and at a given time.[13] Ecotechnology also joins the fields of computer software engineering and space technologies (GIS, GPS) with the fields of agronomy, biology, biogeochemistry, hydrology, physics, and chemistry to develop advances such as models that are capable of assessing N losses and tools such as the NTT where there is even potential for farmers to trade reductions in N loss obtained from good management practices.

EXAMPLES OF OTHER MODELING TOOLS

Another nutrient trading tool is the NutrientNet spreadsheets, which were developed by the World Resource Institute, which has nutrient trading projects in the Chesapeake Bay (Pennsylvania) and in the Kalamazoo River Watershed;

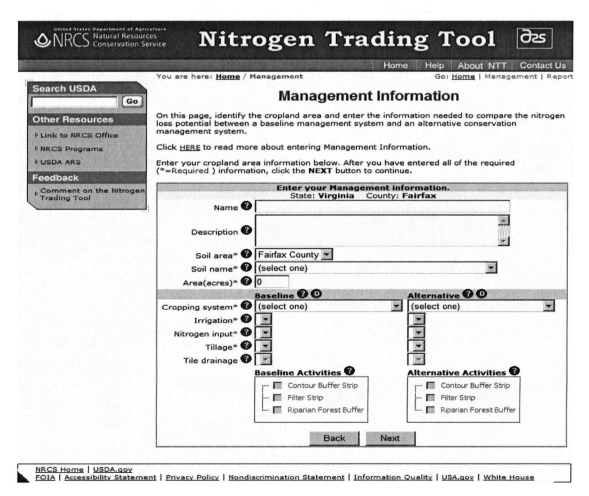

Fig. 1 Web-based NTT user interface.
Source: Gross et al.[25]

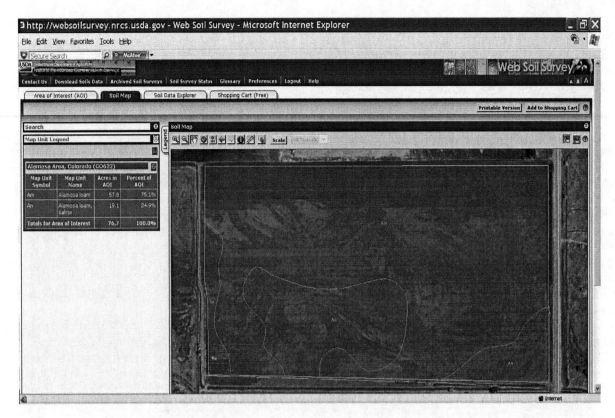

Fig. 2 A web prototype of the NTT can use the NRCS Web Soil Survey to identify a given farm or field from an available database. The highlighted soil information is copied and pasted into the NTT web prototype, and the NLEAP-GIS run is conducted for the two soils identified at the field site and the management practices to be evaluated.
Source: Delgado et al.[23]

both of these are voluntary programs. There is also a nutrient trading tool approach for Maryland; information about this trading tool can be found at http://www.mdnutrienttrading.org/.

A similar technology was applied to a Nutrient Tracking Tool (NTrT), which can evaluate not only nitrogen but also phosphorous, erosion, and other variables simultaneously.[26–28] Information about the NTrT website tool is available at http://tiaer.tarleton.edu/nttfactsheet.pdf. McKinney[26] reported that the goal is to have an ecosystem tool that combines the capabilities of COMET-VR, which uses Century DAYCENT, and the Nutrient Tracking Tool, which uses SWAT-APEX, to conduct a joint ecosystem analysis for trading nitrogen savings and other factors in a similar fashion to the initial NTT prototype. New advances are in development and will be forthcoming. It is expected that these tools will be capable of connecting to the Web Soil Survey to facilitate the assessment of management practices considering site-specific information at the farm level (see prototype shown in Figs. 1–3).[20,23]

Horan and Shortle[29] described one example of how markets can accept these tools. They reported that one of the goals of Pennsylvania's nutrient credit trading program is to reduce the nitrogen loads to the Chesapeake Bay from agricultural non-point sources. They reported that in Pennsylvania, farmers can use a spreadsheet validated by the Pennsylvania DEP to calculate the reduction in nitrogen loads they have achieved from implementing BMPs. The farmers indicate the BMPs they have implemented by selecting the ones that apply from an approved list, and the spreadsheet use this information, as well as a transport factor from USEPA's Chesapeake Bay model, to assess reductions in nitrogen losses to the Chesapeake Bay. Thus, there is the potential to use BMPs to estimate the nitrogen credits from non-point sources and the potential to trade.[15–17,19,20,29,30]

Independent of whether these new NTTs use a similar approach to the one developed by Delgado et al.,[19] or take a different approach to assess nitrogen savings for potential trade, it seems likely that they will compare one or multiple scenarios to a baseline scenario to determine whether there is potential to trade nitrogen similar to Delgado et al.[19,20] New ecotechnological tools are being developed and/or modified to expand their capabilities; these advances will continue to help us improve our ability to assess the spatial and temporal variability of the nitrogen cycle and other processes and help us to provide farmers with tools that can quantify these non-commodity ecosystem services and the potential to trade them.

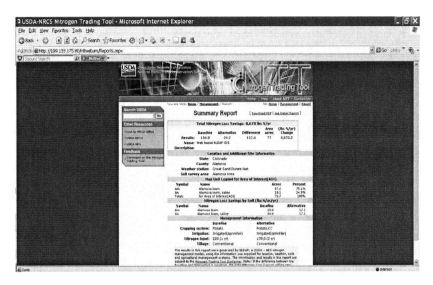

Fig. 3 A web prototype of the NTT can use the soil sites identified using the NRCS Web Soil Survey to run the NTT and generate an NTT GIS evaluation for a given farm.
Source: Delgado et al.[23]

NITROGEN CYCLE

Delgado et al.[19] reported that the NTT could help us to better understand the nitrogen cycle, how management can be used to reduce the losses of reactive nitrogen, and the potential to trade these savings (i.e., reductions in nitrogen losses from agricultural fields) resulting from improved management practices, in water and air quality markets.[19,20] Delgado and Follett[5] discussed recent advances in nitrogen management for conservation of water quality as well as some basic principles of management that can be implemented to reduce nitrogen losses across different world agroecosystems. Meisinger and Delgado[31] discussed principles of nitrogen management to reduce nitrate leaching, while Mosier et al.[32] discussed principles of nitrogen management specifically for reducing N_2O emissions.

IMPORTANT PROCESSES THAT NEED TO BE CAPTURED IN A MODELING TOOL

All pathways for nitrogen loss are affected by management practices. Not only do management practices affect these pathways as far as rate and quantity of nitrogen losses, but factors such as soil type, hydrology, weather, water balances, soil organic matter, crop type, and even crop rotations can affect the rate of losses via each of these pathways. Additionally, it is difficult to go out to fields and to measure nitrogen dynamics under each situation and quantify the effects of these practices on nitrogen losses.

Nitrogen is found in soil organic matter and as inorganic forms, especially as nitrate and ammonia. Nitrogen is found in close relation to carbon, and the mineralization of carbon and that of nitrogen are closely related. Although there are several other parameters that contribute to the rate of nitrogen mineralization, it is accepted that plant materials with high carbon-to-nitrogen ratios have a slower decomposition rate than plant materials with lower carbon-to-nitrogen ratios.

Plant materials with carbon-to-nitrogen ratios of 30 or lower are considered to have a faster decomposition rate, releasing ammonium (NH_4) and NO_3, while higher carbon-to-nitrogen ratios of 100 can even immobilize nitrogen while the plant material is decomposed and broken down to lower carbon-to-nitrogen ratios, when the nitrogen starts to be released again.[33–35] Inputs of inorganic nitrogen such as fertilizers, and/or organic fertilizers such as manure and compost, increase the amount of nitrogen entering the nitrogen cycle, and while nitrogen inputs help to maximize production, if they are not managed adequately, they can also contribute to greater losses of nitrogen from the agricultural system.

When nitrogen is in the form of nitrate, it can be moved easily in the environment, and as water moves and leaches out of the system, it can carry significant amounts of nitrate.[36–38] The surface transport of nitrogen can be due to water erosion and movement that can carry soil organic matter (including organic forms of nitrogen), inorganic nitrogen dissolved in water, or even soil particles that could have nitrogen bound to them.[39,40] Wind erosion is another pathway for surface transport. In the mineralization of nitrogen from soil organic matter to NH_4 to NO_3, as well as the denitrification process, there is also the production of N_2O,[41–44] a trace gas that has been reported to be one of the greenhouse gases that contribute to climate change.[45] It is reported that the Global Warming Potential of N_2O is about 310 times that of carbon dioxide (CO_2) over a 100 years time frame (http://www.epa.gov/otaq/climate/index.htm).[22]

One additional pathway of reactive nitrogen losses is NH_3 volatilization. The relationship of management with NH_3 volatilization has been described in detail by Meisinger and Randall,[46] Fox et al.,[47] Freney et al.,[48] and Sharpe and Harper,[49] among others. Management, method of application, soil pH, type of fertilizer, and other factors can affect the rate and magnitude of NH_3 volatilization. Although it is difficult to quantify the many soil, plant, and weather factors that affect the rate and magnitude of nitrogen losses across different pathways, nitrogen models can serve as tools to help users quickly integrate these factors with landscape management combinations across time and space to assess potential nitrogen losses under different scenarios.[21] Shaffer et al.[50] reported that there were more than 20 nitrogen models available that can potentially be used to assess potential nitrogen losses and study nitrogen dynamics.

NITROGEN TRADING TOOL

The concept of the NTT was reported and conceived by Delgado et al.,[19,20] and the tool uses a series of equations to assess the losses of reactive nitrogen losses associated with a given set of management practices, basically comparing the effect of a new management practice against a baseline scenario and estimating the reduction in reactive nitrogen losses.

In the NTT, the effect of management practices on reducing nitrogen losses is shown as a positive value. This value can be thought of as being similar to a bank account balance. For example, if the new management practice reduces nitrogen losses, and there is a positive nitrogen balance showing lower nitrate leaching, then this positive amount can be traded. However, if the new practice increases the nitrogen losses to the environment, then there will not be a positive balance, so there is no nitrogen to trade. Because nitrogen management is so complex, and activities during 1 year can impact the balance of the next year's available nitrogen (e.g., a large manure application of the previous year could still be mineralizing during the second year; such cycling and/or increased availability should be accounted for), only accounting for N fertilizer inputs for one given year may not be the best method of assessment. The new AgES-NLEAP-NTT prototype is another stand-alone NTT.[23] Ascough et al.[51] incorporated the NLEAP code into the AgES model prototype, which enables the user to conduct NLEAP analysis as well as use the NTT. The AgES-NLEAP model can be used for geospatial assessment and it is embedded in the open source GIS platform linked to the WorldWind web service from NASA. It is expected that new NTT will be further developed to support the new field of nitrogen trading for conservation of the biosphere.[23,51] When the prototype is released in its final version, it will be named NLEAP-GIS NT and will be a stand-alone version that is written in the programming language Java.

One of the limitations of the NTT is that it uses the NLEAP model, which does not simulate soil erosion, so it does not account for nitrogen that is lost as it is attached to particles of soil.[19] The NTT is effective in assessing N losses via all pathways under systems where soil erosion is minimal (e.g., minimum tillage).[19] If soil erosion is an important pathway for the system(s) under analysis, then the user could use a model such as the Revised Universal Soil Loss Equation (RUSLE 2)[52–55] to calculate the erosion potential, and then enter the erosion values for each system to be evaluated into the NTT; the NTT will then use these entered erosion values to simulate the N lost via erosion for each system (see Delgado et al.[20,23] for more

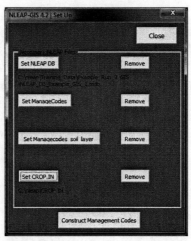

(a) (b)

Fig. 4 From the *Driver* window (a), the user can select the *Set Up NLEAP Files* option to access the *Set Up* screen (b) where databases can be selected for management comparisons.

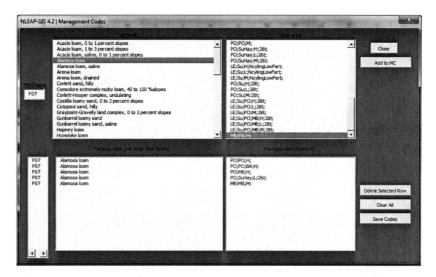

Fig. 5 From the *Management Codes* screen, users can select different combinations of soil types and management scenarios.

information on entering erosion values into the NTT). Users could also use tools such as the NTrT, which is capable of calculating N losses due to erosion. One of the goals of Ascough et al.[51] is to add the capability of simulating N losses due to erosion from AgES into the new NLEAP-GIS NT.

DESCRIPTION

The NTT can help nutrient managers and other decision-makers quickly compare a proposed new management practice, or even several scenarios, to a given baseline scenario and determine whether there is potential to improve nitrogen use efficiency and reduce losses of reactive nitrogen. If the new management practice increases the efficiency over the baseline and there is a reduction in nitrogen losses, these savings can potentially be traded in an ecosystem services market. For additional details on the capabilities and limitations of the NTT, see Delgado et al.[19,20,23]

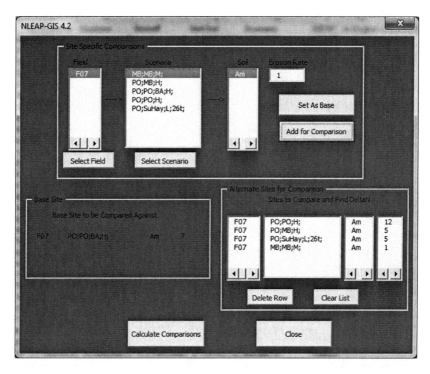

Fig. 6 In the screen for the NTT, the user can select a management scenario to be the baseline scenario as well as the scenarios that are to be compared to the baseline scenario. If available, the erosion rate can also be entered for each scenario. Once the scenarios for comparison have been selected, the user can click on *Calculate Comparisons* to assess the potential for nitrogen trading.

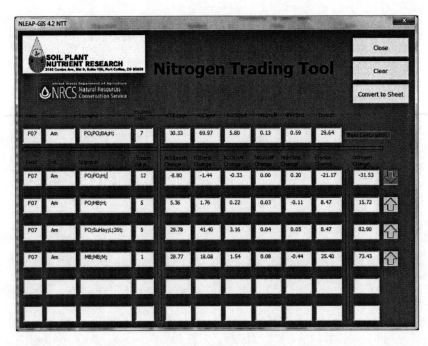

Fig. 7 Results for the example comparison calculated by the NTT. A red arrow indicates a scenario where there is no increase in efficiency and no potential for trading. A green arrow indicates higher nitrogen use efficiency, lower losses of reactive nitrogen, and potential to trade these savings (reductions in nitrogen losses) in water or air quality markets.

The material in the following section is an edited excerpt from the user guide for NLEAP-GIS (available at http://www.ars.usda.gov/Services/docs.htm?docid=20339), which includes detailed instructions on how to use the NTT.[20,23] The user will need to review how to use the NLEAP-GIS 4.2 in order to enter the management information. A stand-alone NTT is available at http://www.ars.usda.gov/npa/spnr/nitrogentools. Alternatively, interested peers could request that a (DVD) copy be mailed to them.

A WALKTHROUGH OF THE NTT

The user will need to know how to use the NLEAP-GIS 4.2. A training manual is available to download at http://www.ars.usda.gov/Services/docs.htm?docid=20334. The user will need to load the databases using the NLEAP setup window (Fig. 4) and proceed to the management windows to select the management scenarios to be compared. In the management windows, the user could select as many scenarios as desired to compare to the baseline. As an example, the user could compare a rotation of 1) potato–potato with high N input, (PO-PO-H); 2) potato–potato–malting barley with high N input, (PO-PO-BA-H); 3) potato–malting barley with high N input, (PO-BA-H); 4) potato–summer cover crop with limited irrigation with low N input (PO-SuHay-L); and 5) malting barley–malting barley rotation with medium N input (BA-BA-M) (Fig. 5). For this example, the baseline selected in the NTT screen was PO-PO-BA-H (Fig. 6). When the user clicks the button to calculate the comparison, all the scenarios are quickly compared against the baseline scenario. The user can also elect to have the scenarios printed in a sheet (Fig. 7).

OUTPUTS

In the example in Fig. 7, if the rotation is changed from a 3 years potato–potato–malting barley (PO-PO-BA-H) rotation to a continuous potato rotation (PO-PO-H), the losses of reactive N will increase. The nitrate leaching losses increase in the PO-PO-H rotation by about 9 kg N per hectare per year and the total N losses are close to 32 kg N per hectare per year (negative numbers indicate greater N losses, so there are no savings to trade). If the user changes to a rotation of 1 year of a potato crop with 1 year of small grain malting barley (PO-MB-H), then there are savings (reductions in N loss) to trade, and the average nitrate leaching per year is reduced by about 5 kg N per hectare per year and the total nitrogen losses are reduced by about 16 kg N per hectare per year. The savings (shown as positive numbers) are much higher if the user changes to a rotation of potato and a hay summer cover crop (PO-SuHay-L) or changes to a full rotation of small grain (MB-MB-M).

The user could also do a quick GIS analysis of the potential to apply management practices across a large number of fields. For example, Fig. 8 shows the N losses across several irrigated center pivots of the San Luis Valley with continuous rotations of potato and high N inputs (PO-PO-H). Fig. 9 shows the savings (nitrogen available to trade) across the same center pivot with use of summer cover crops with limited irrigation (PO-SuHay-L) and better N management practices.

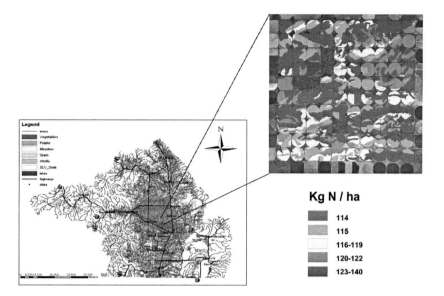

Fig. 8 A stand-alone NTT can be used to quickly evaluate the effects of management practices on total N losses and the resultant potential to trade across regions (hypothetical example). Nitrogen losses across several irrigated center pivots with continuous rotations of potato and high N inputs.

EXAMPLES ON HOW TO USE

Delgado et al.[20] discussed the importance of an NTT that accounts for spatial variability and differences among soils in reducing nitrogen losses across all pathways. Distances to water bodies and other important parameters related to precision conservation (also known as target conservation) are also important and need to be considered in order to implement BMPs that maximize trading potential (e.g., higher nitrogen use efficiencies).[13,20] BMPs such as buffers, wetlands, and other practices can be implemented using precision conservation to harvest nitrogen to trade in water and air markets.[13,20,28]

As new advances are developed, it will become clearer how systems to trade nitrogen that account for spatial variability will be used and/or developed. Currently, there are certified systems such as the one used in Pennsylvania, which uses coefficients for practices and transport factors.[29] There is potential to use models such as the NTT developed by Delgado et al.[19,20] and the AgES NLEAP

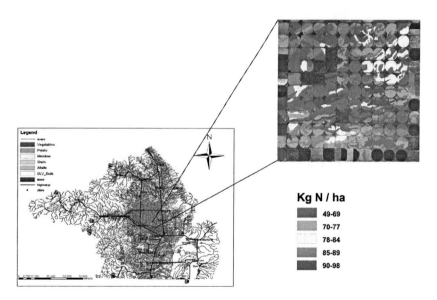

Fig. 9 A stand-alone NTT GIS prototype can be used to quickly evaluate the effects of management practices on total reactive N losses and the resultant potential to trade across regions (hypothetical example). Savings (reactive nitrogen available to trade) across irrigated center pivot with use of summer cover crops with limited irrigation.

prototype[23] to assess spatial variability and the potential to trade nitrogen. Other new models and approaches are also being developed such as the NTrT developed by Saleh et al.,[28] which builds on the NTT approach. These types of approaches have the potential to be certified for use in assessing the potential to trade nitrogen while considering spatial and temporal variability.[20,23,28] There is interest in these types of tools, which are powerful and are based on sound science and models, and as new advances in this type of ecotechnology are developed, these tools will be modified to increase precision conservation and the potential to trade nitrogen.[13,20,23,28]

There are several tools currently being used across different states to calculate nitrogen losses. The USDA-NRCS and USEPA reported that science-based tools (e.g., the NTT, as well as the enhanced version of the NTT, the Saleh et al.[28] NTrT, which assesses erosion and phosphorous losses) could potentially be used to assess the effects of management practices on losses of reactive N. Recent advances in computer modeling, GIS, and GPS allow the development of more precise tools that can assess site-specific practices. The coming decade looks very promising for improvements in ecotechnologies that will allow for detailed analyses of the effects of conservation practices and the potential to trade ecosystem services. These emerging ecotechnologies, based on scientific approaches, have the potential to reduce uncertainty and to be used to help establish performance metrics for trading programs.

CONCLUSION

The NTT[19,20] is a recent advance in the development of nutrient tools that shows how new ecotechnologies can help advance environmental conservation by moving the assessment approach from one that is limited to identifying some sources of pollution as non-point sources to an assessment approach that is site specific and can identify points of pollution.[20] Not only can new ecotechnological approaches such as the NTT be used to assess each field and quantify the nitrogen losses, they can also be used to help increase farms' nitrogen use efficiencies and economic returns for farmers. In addition, they can be used to help identify practices and strategies that farmers can use to provide these additional ecosystems services (e.g., reduced nitrogen losses) that could potentially be traded as non-commodity services in future air and water quality markets.

There is potential to assess, track, identify, and quantify the positive effects of conservation practices with these new ecotechnologies. The NTT is one recent example,[19,20] but other examples of new ecotechnologies include the World Resources Institute's NutrientNet (WRI) (http://www.nutrientnet.org/) and the new tracking tool NTrT, which can be used to assess nutrient trading.[28]

As we move forward and new advances in tools that can be used to conduct spatial assessment are developed, precision/target conservation approaches will be implemented. These new approaches will help increase conservation effectiveness by considering spatial and temporal variability. Nitrogen trading could potentially incorporate a precision conservation or targeted approach to assessing N losses, to maximize the trading of these ecosystem services. It is clear that with future climate change challenges and the need to produce food for a continuously growing world population, conservation will need to be maximized. The NTT and other new ecotechnologies being developed will help users to identify conservation practices that maximize nitrogen use efficiency while reducing nitrogen losses, providing an opportunity to trade the saved nitrogen in water and air markets and contributing to conservation of the biosphere.[19,20,23,56]

REFERENCES

1. Cowling, E.; Galloway, J.; Furiness, C.; Erisman, J.W. Optimizing nitrogen management in food and energy production and environmental protection: Report from the Second International Nitrogen Conference, Oct. 14–18, 2001; Bolger Center: Potomac, MD, 2002, available at http://www.initrogen.org/fileadmin/user_upload/Second_N_Conf_Report.pdf (accessed May 10, 2010).
2. Galloway, J.N.; Aber, J.D.; Erisman, J.W.; Seitzinger, S.P.; Howarth, R.W.; Cowling, E.B.; Cosby, B.J. The nitrogen cascade. BioScience **2003**, *53* (4), 341–356.
3. Follett, J.R.; Follett, R.F.; Herz, W.C. Environmental and human impacts of reactive nitrogen. In *Advances in Nitrogen Management for Water Quality*; Delgado, J.A., Follett, R.F., Eds.; Soil and Water Conservation Society: Ankeny, IA, 2010; 1–37.
4. U.S. Environmental Protection Agency. Federal Register, 54 FR 22062, 22 May; USEPA: Washington, DC, 1989.
5. Delgado, J.A., Follett, R.F., Eds. *Advances in Nitrogen Management for Water Quality*; Soil and Water Conservation Society: Ankeny, IA, 2010.
6. Hey, D.L. Nitrogen farming: Harvesting a different crop. Restor. Ecol. **2002**, *10*, 1–10.
7. Greenhalch, S.; Sauer, A. *Awakening the Dead Zone: An Investment for Agriculture, Water Quality, and Climate Change*; World Resources Institute: Washington, DC, 2003.
8. Hey, D.L.; Urban, L.S.; Kostel, J.A. Nutrient farming: The business of environmental management. Ecol. Eng. **2005**, *24*, 279–287.
9. Ribaudo, M.O.; Heimlich, R.; Peters, M. Nitrogen sources and Gulf hypoxia: Potential for environmental credit trading. Ecol. Econ. **2005**, *52*, 159–168.
10. Glebe, T.W. The environmental impact of European farming: How legitimate are agri-environmental payments? Rev. Agric. Econ. **2006**, *29*, 87–102.
11. Berry, J.R.; Delgado, J.A.; Khosla, R.; Pierce, F.J. Precision conservation for environmental sustainability. J. Soil Water Conserv. **2003**, *58*, 332–339.

12. Berry, J.K.; Delgado, J.A.; Pierce, F.J.; Khosla, R. Applying spatial analysis for precision conservation across the landscape. J. Soil Water Conserv. **2005**, *60*, 363–370.
13. Delgado, J.A.; Berry, J.K. Advances in precision conservation. Adv. Agron. **2008**, *98*, 1–44.
14. Lal, H.; Delgado, J.A.; Gross, C.M.; Hesketh, E.; McKinney, S.P.; Cover, H.; Shaffer, M. Market-based approaches and tools for improving water and air quality. Environ. Sci. Policy **2009**, *12*, 1028–1039.
15. Ribaudo, M.; Gottlie, J. Point–nonpoint trading—Can it work? J. Am. Water Resour. Assoc. **2011**, *47*, 5–14.
16. Ribaudo, M.; Delgado, J.; Hansen, L.; Livingston, M.; Mosheim, R.; Williamson, J. *Nitrogen in Agricultural Systems: Implications for Conservation Policy*; ERS: Economy Research Report, Washington DC, 2010.
17. Ribaudo, M.; Greene, C.; Hansen, L.; Hellerstein, D. Ecosystem services from agriculture: Steps for expanding markets. Ecol. Econ. **2010**, *69*, 2085–2092.
18. Stanton, T.; Echavarria, M.; Hamilton, K.; Ott, C. State of Watershed Payments: An Emerging Marketplace. Ecosystem Marketplace 2010, available at http://www.foresttrends.org/docu ments/files/doc_2438.pdf (accessed October 2011).
19. Delgado, J.A.; Shaffer, M.J.; Lal, H.; McKinney, S.; Gross, C.M.; Cover, H. Assessment of nitrogen losses to the environment with a Nitrogen Trading Tool. Comput. Electron. Agric. **2008**, *63*, 193–206.
20. Delgado, J.A.; Gross, C.M.; Lal, H.; Cover, H.; Gagliardi, P.; McKinney, S.P.; Hesketh, E.; Shaffer, M.J. A new GIS nitrogen trading tool concept for conservation and reduction of reactive nitrogen losses to the environment. Adv. Agron. **2010**, *105*, 117–171.
21. Delgado, J.A. Quantifying the loss mechanisms of nitrogen. J. Soil Water Conserv. **2002**, *57*, 389–398.
22. U.S. Environmental Protection Agency. *Water Quality Trading Toolkit for Permit Writers*; EPA 883-R-07-004; Office of Wastewater Management: Washington, DC, August 2007.
23. Delgado, J.A.; Gagliardi, P.; Shaffer, M.J.; Cover, H.; Hesketh, E.; Ascough, J.C.; Daniel, B.M. New tools to assess nitrogen management for conservation of our biosphere. In *Advances in Nitrogen Management for Water Quality*; Delgado, J.A., Follett, R.F., Eds.; Soil and Water Conservation Society: Ankeny, IA, 2010; 373–409.
24. Shaffer, M.J.; Delgado, J.A.; Gross, C.; Follett, R.F.; Gagliardi, P. Simulation processes for the Nitrogen Loss and Environmental Assessment Package (NLEAP). In *Advances in Nitrogen Management for Water Quality*; Delgado, J.A., Follett, R.F., Eds.; Soil and Water Conservation Society: Ankeny, IA, 2010.
25. Gross, C.; Delgado, J.A.; McKinney, S.; Lal, H.; Cover, H.; Shaffer, M. Nitrogen Trading Tool (NTT) to facilitate water quality credit trading. J. Soil Water Conserv., **2008**, *63*, 44A–45A.
26. McKinney, S. Calculating carbon, nitrogen and phosphorous for markets on agricultural lands. Carbon Markets: Expanding opportunities/Valuing Cobenefits. In Soil and Water Conservation Society 65th International Conference Final Program; Soil and Water Conservation Society: Ankeny, IA; 23, available at http://www.swcs.org/index.cfm?nodeID=24669&audienceID=1 (McKinney—NTT COMET-VR Integration.ppt; accessed May 2011).
27. Saleh, A. The Nutrient Trading Tool. In Soil and Water Conservation Society 65th International Conference Final Program, St. Louis, Missouri, July 17–20, 2010; Soil and Water Conservation Society: Ankeny, Iowa, 2010; 49.
28. Saleh, A.; Gallego, O.; Osei, E.; Lal, H.; Gross, C.; McKinney, S.; Cover, H. Nutrient Tracking Tool—A user-friendly tool for calculating nutrient reductions for water quality trading. J. Soil Water Conserv. **2011**, *66*, 400–410.
29. Horan, R.D.; Shortle, J.S. Economic and ecological rules for water quality trading. J. Am. Water Resour. Assoc. **2011**, *47*, 59–69.
30. Stephenson, K.; Shabman, L. Rhetoric and reality of water quality trading and the potential for market-like reform. J. Am. Water Resour. Assoc. **2011**, *47*, 15–28.
31. Meisinger, J.J.; Delgado, J.A. Principles for managing nitrogen leaching. J. Soil Water Conserv. **2002**, *57*, 485–498.
32. Mosier, A.R.; Doran, J.W.; Freney, J.R. Managing soil denitrification. J. Soil Water Conserv. **2002**, *57*, 505–513.
33. Doran, J.W.; Smith, M.S. Role of cover crops in nitrogen cycling. In *Cover Crops for Clean Water*; Hargrove, W.L., Ed.; SWCS: Ankeny, IA, 1991; 85–90.
34. Pink, L.A.; Allison, F.E.; Gaddy, U.L. Greenhouse experiments on the effect of green manures upon N recovery and soil carbon content. Soil Sci. Soc. Am. Proc. **1945**, *10*, 230–239.
35. Pink, L.A.; Allison, F.E.; Gaddy, U.L. Greenhouse experiments on the effect of green manures crops of varying carbon–nitrogen ratios upon nitrogen availability and soil organic matter content. Agron. J. **1948**, *40*, 237–248.
36. Gast, R.G.; Nelson, W.W.; Randall, G.W. Nitrate accumulation in soils and loss in tile drainage following nitrogen applications to continuous corn. J. Environ. Qual. **1978**, *18*, 258–261.
37. Goss, M.J.; Howse, K.R.; Lane, P.W.; Christian, D.G.; Harris, G.L. Losses of nitrate-nitrogen in water draining from under autumn-sown crops established by direct drilling or mouldboard ploughing. J. Soil Sci. **1993**, *44*, 35–48.
38. Follett, R.F., Ed. *Nitrogen Management and Ground Water Protection*; Elsevier Sci. Publishers: Amsterdam, the Netherlands, 1989.
39. Legg, J.O.; Meisinger, J.J. Soil nitrogen budgets. In *Nitrogen in Agricultural Soils*; Stevenson, F.J., Ed.; Agronomy Monograph 22; American Society of Agronomy: Madison, WI, 1982; 503–557.
40. Follett, R.F.; Delgado, J.A. Nitrogen fate and transport in agricultural systems. J. Soil Water Conserv. **2002**, *57* (6), 402–408.
41. Mosier, A.R.; Klemedtsson, L. Measuring denitrification in the field. In *Methods of Soil Analysis: Part 2. Microbiological and Biochemical Properties*; Weaver, R.W., Angle, S., Bottomley, P., Bezdicek, D., Smith, S., Tabatabai, A., Wollum, A., Eds.; Soil Science Society of America: Madison, WI, 1994; 1047–1065.
42. Davidson, E.A.; Kingerlee, W. A global inventory of nitric oxide emissions from soils. Nutr. Cycling Agroecosyst. **1997**, *48*, 37–50.
43. Mosier, A.; Kroeze, C.; Nevison, C.; Oenema, O.; Seitzinger, S.; Van Cleemput, O. Closing the global N_2O budget:

43. ...Nitrous oxide emissions through the agricultural nitrogen cycle. Nutr. Cycling Agroecosyst. **1998**, *52*, 225–248.
44. Hutchison, G.L. Biosphere–atmosphere exchange of gaseous N oxides. In *Soils and Global Change*; Lal, R., Kimble, J., Levine, E., Stewart B.A., Eds.; CRC Press, Inc.: Boca Raton, FL, 1995; 219–236.
45. Intergovernmental Panel on Climate Change (IPCC). Radiative forcing of climate change. The 1994 report to scientific assessment working group of IPCC; Bolin B., Houghton, J.; Filho, L.G.M., Eds.; Summary for Policymakers; WMO/UNEP: Geneva, 1994.
46. Meisinger, J.J.; Randall, G.W. Estimating nitrogen budgets for soil-crop systems. In *Managing Nitrogen for Groundwater Quality*; Follett, R.F., Keeney, D.R., Cruse, R.M., Eds.; Soil Science Society of America: Madison, WI, 1991; 85–124.
47. Fox, R.H.; Piekielek, W.P.; Macneal, K.E. Estimating ammonia volatilization losses from urea fertilizers using a simplified micrometeorological sampler. Soil Sci. Soc. Am. J. **1996**, *60*, 596–601.
48. Freney, J.R.; Simpson, J.R.; Denmead, O.T. Ammonia volatilization. Ecol. Bull. **1981**, *33*, 291–302.
49. Sharpe, R.R.; Harper, L.A. Soil, plant and atmospheric conditions as they relate it ammonia volatilization. Fert. Res. **1995**, *42*, 149–153.
50. Shaffer, M.J.; Ma, L.; Hansen, S., Eds. *Modeling Carbon and Nitrogen Dynamics for Soil Management*; CRC Press LLC: Boca Raton, FL, 2001.
51. Ascough, J.C., II; Delgado, J.A.; Shaffer, M.J.; Daniel, B.M.; Gagliardi, P.M. The AgES-NLEAP Tool for Precision Nitrogen Conservation Management. SWCS Abstract; 2011 Annual Meeting, Washington DC, 2011.
52. Renard, K.G.; Ferreira, V.A. RUSLE model description and database sensitivity. J. Environ. Qual. **1993**, *22* (3), 458–466.
53. Renard, K.G.; Foster, G.R.; Weesies, G.A.; Porter, J.P. RUSLE: Revised universal soil loss equation. J. Soil Water Conserv. **1991**, *46* (1), 30–33.
54. Wischmeier, W.H.; Smith, D.D. *Predicting rainfall-erosion losses from cropland east of the rocky mountains*; Agriculture Handbook No. 282; U.S. Department of Agriculture: Washington, DC, 1965; 1–47.
55. Wischmeier, W.H.; Smith, D.D. *Predicting rainfall erosion losses: a guide to conservation planning*; Agriculture Handbook No. 537.; U.S. Department of Agriculture: Washington, DC, 1978; 1–58.
56. Ribaudo, M.; Delgado, J.; Livingston, M. Preliminary assessment of the potential for nitrous oxide offsets in a cap and trade program. Agric. Resour. Econ. Rev. **2011**, *40*, 1–16.

Nitrogen: Biological Fixation

Mark B. Peoples
Plant Industry, Commonwealth Scientific and Industrial Research Organization (CSIRO), Canberra, Australian Capital Territory, Australia

Abstract

Although dinitrogen (N2) gas represents almost 80% of the Earth's atmosphere, it is not a source of nitrogen (N) that is readily available to plants. However, a number of procaryotic microorganisms have evolved that utilize the enzyme nitrogenase to reduce atmospheric N2 to ammonia, which can subsequently be used to support their growth. Biological N2 fixation (BNF) by some diazotrophs can occur in a free-living state, or via associative relationships with plants, while others can fix N2 only in symbiosis with specific plant hosts.

INTRODUCTION

Although dinitrogen (N_2) gas represents almost 80% of the Earth's atmosphere, it is not a source of nitrogen (N) that is readily available to plants. However, a number of procaryotic micro-organisms have evolved that utilize the enzyme nitrogenase to reduce atmospheric N_2 to ammonia, which can subsequently be used to support their growth. Biological N_2 fixation (BNF) by some diazotrophs can occur in a free-living state, or via associative relationships with plants, while others can fix N_2 only in symbiosis with specific plant hosts.[1] Although calculations of global contributions of BNF are subject to enormous approximations, annual inputs of fixed N into arable agroecosystems have been conservatively estimated to be 35–55 million metric tonnes of N, with a further 40–45 million tonnes of N per year occurring in permanent pastures.[2] This compares to 75–80 million tonnes of N applied to crops and grasslands as fertilizer each year.[2]

SOURCES OF FIXED NITROGEN IN AGRICULTURAL SYSTEMS

The N_2 fixation process can directly contribute to agricultural production where the fixed N is harvested in grain or other food for human or animal consumption. However, BNF can also represent an important renewable source of N that can help maintain or enhance the fertility of many agricultural soils. Examples of experimental estimates of amounts of N fixed by various N_2-fixing organisms are presented in Table 1. Values for associative and symbiotic systems have always been determined from measures of plant shoot biomass. Below-ground contributions of fixed N have generally been ignored. However, research now suggests that N associated with roots might represent between 25% and 60% of the total N accumulated by crops and pastures.[3–5] Therefore, total inputs of fixed N could be 50%–100% greater than has been traditionally determined from shoot-based measurements, such as those depicted in Table 1.

Contributions by Different N_2-Fixing Organisms

Free-living N_2-fixers probably contribute only small amounts of N to agricultural systems (Table 1). The data tend to be inconclusive concerning the role of diazotrophs associated with nonlegumes in temperate agriculture, but studies have demonstrated significant inputs of fixed N by tropical grasses and crops such as sugarcane (*Saccharum officinarum*).[1,6] Symbiotic associations between legumes and specific soil bacteria (*Rhizobium, Bradyrhizobium, Allorhizobium, Azorhizobium, Mesorhizobium*, or *Sinorhizobium* spp.) in specialized root structures (nodules) are generally responsible for the largest amounts of fixed N in farming systems (Table 1).

Inputs of Fixed Nitrogen by Legumes

BNF by legume systems play a key role in world crop production is irrefutable. The ability of legumes to progressively improve the N status of soils has been utilized for thousands of years in crop rotations and traditional farming systems.[1,2,7] The 163 million hectares of legume oilseeds (soybean—*Glycine max*; and groundnut—*Arachis hypogea*) and pulses sown globally each year, legume components of the 200 million hectares under temporary pastures or fodder crops, and the 10–12 million hectares of perennial legume cover crops in rubber (*Hevea brasiliensis*) and oil-palm (*Elaeis guineensis*) plantations contribute fixed N to farming systems. Most modern methods used to quantify inputs of fixed N by legumes separate the plant N into fractions originating from soil N or N_2 fixation.[1,5,8] Once the legume N can be partitioned into that proportion derived from atmospheric N_2 (%Ndfa, sometimes also described as %Pfix) and that coming from the soil, the amounts of N_2 fixed can be calculated from measures of shoot dry matter and N content. The formation of the symbiosis between legume and rhizobia is dependent upon many factors and cannot be assumed to occur as a matter of course. This is reflected in the range of values presented in Table 1. Such large variations in reported estimates of N_2 fixation make

Table 1 Experimental estimates of the amounts of N_2 fixed in different agricultural systems.

N_2-fixing organism	Range measured (kg N/ha per crop or per year)	Range commonly observed (kg N/ha per crop or per year)
Free-living		
Crops	0–80	0–15
Associative		
Tropical grasses	10–45	10–20
Crops	0–240	25–65
Symbiotic		
Azolla	10–150	10–50
Green manure legumes	5–325	50–150
Pasture/forage legumes	1–680	50–250
Crop legumes	0–450	30–150
Trees/shrubs	5–470	100–200

it difficult to generalize about how much N may be fixed by different legume species. Collectively, the data suggest maximum rates of N_2 fixation of 3–4 kg shoot N/ha/day[5] and potential inputs of fixed N by many legumes of several hundred kg of shoot N/ha each year (Table 1). However, much of the information in Table 1 was derived from research trials in which specific treatments were imposed to generate differences in %Ndfa values and legume growth as an experimental means of studying factors which regulate BNF. Therefore, these data may be of little relevance to what might actually be occurring in farmers' crops and pastures. Fortunately, measurement procedures have now been developed which allow on-farm measures of legume N_2 fixation to be conducted with confidence.[5,7,8]

Levels of Nitrogen Fixation Achieved in Farmers' Fields

Examples of the types of information which can be generated about BNF in farmers' fields are presented in Tables for farming systems in different regions of the world. These on-farm data and observations can be used to develop a picture of N_2 fixation within an individual country or region and provide insights into contributions of BNF to agriculture on a global scale. Collectively, the results in Table 2 indicate that the potential for BNF inputs can differ between legumes and countries, but they also suggest many commonalities. Although wide ranges in %Ndfa values have been observed, it seems that, on average, most winter pulses (e.g., chickpea—*Cicer arietinum*; lentil—*Lens culinaris*; field pea—*Pisum sativum*; fababean—*Vicia faba*; lupin—*Lupinus albus*) relatively satisfy higher proportions of their growth requirements from N_2 fixation (>65%) than do the summer legumes (e.g., mungbean—*Vigna radiata*; mashbean—*V. mungo*; soybean; ground-

Table 2 Summary of the proportion of plant N derived from N_2 fixation (%Ndfa) and the amounts of N_2 fixed by farmers' legume crops and pastures in different geographical regions.

Country and legume	Number of fields	Mean Ndfa (%)	Total N fixed (kg N/ha)[a]
Winter pulse crops			
Pakistan	126	78	79
Nepal	27	79	78
Syria	46	67	na[b]
Australia	90	65	170
Summer legume crops			
Pakistan	63	47	42
Nepal	50	55	77
Thailand	13	75	78
Vietnam	45	48	125
South Africa	14	58	na[b]
Australia	33	53	267
Annual pastures			
Australia	300	75	na[b]
Perennial pastures			
Australia	110	64	na[b]

[a]Includes an estimate of fixed N from the roots and nodules which assumes that below-ground N represents 33% of total plant N.
[b]Data not available for all fields.
Note: Ndfa = Nitrogen derived from atmospheric N_2.

nut) where %Ndfa values were commonly less than 60% (Table 2). Poor or variable nodulation observed in some summer legumes and the resulting increased reliance upon soil N may reflect greater N mineralization during summer, water stress, and/or low vegetative biomass accumulated by short duration legume crops.[8,9]

In grazed pastures, the competition for mineral N between legumes and companion grasses or vigorous broadleaf weeds growing within the pasture sward results in low levels of plant-available soil N throughout the growing season.[4,7] As a consequence, %Ndfa by the legume components of pastures tend to be high (Table 2). The slightly lower %Ndfa values detected in perennial legume species, such as alfalfa (*Medicago sativa*) and white clover (*Trifolium repens*), presumably result from a greater ability to scavenge soil mineral N from a larger rooting zone compared with annual pasture species.[4]

Although the levels of %Ndfa are important, the amounts of N_2 fixed are usually regulated by legume growth rather than %Ndfa in most farming systems,[7] and many legumes appear to fix approximately 20 kg of shoot N for every metric ton of shoot DM accumulated.[4,7,8]

Impact of Management

Factors that either enhance or depress N_2 fixation (Table 3) can generally be summarized in terms of environmental or

Table 3 Key factors influencing inputs of fixed N by legumes in farmers' fields.

Country	System	BNF regulated by		Primary factors
		DM	Soil nitrate	
Pakistan	Winter crop	+++		Rainfall nutrition, weed control
	Summer crop	+++	+++	Fertilizer N, no inoculation, insects, disease
Nepal	Winter crop	+++		Rainfall, nutrition
	Summer crop	+++	+	Total soil N, mineralized N, available P, legume species
Syria	Winter crop	+++	++	Soil nutrients, insects, disease
Thailand	Summer crop	+++		Available P
Vietnam	Summer crop	+++		Plant density, soil pH, available P, legume species
South Africa	Summer crop	++	++	Effective inoculation, nutrition, rotation, water availability
Australia	Winter crop	+++	+++	Rainfall, fallowing, legume species
	Summer crop	++	+++	Crop rotation, tillage, rainfall
	Pasture	+++	+	Soil pH, available P, legume density, grazing management

Note: BNF = biological N_2 fixation; DM = dry matter; P = phosphorus.

management constraints to crop growth (e.g., basic agronomy, nutrition, water supply, diseases, and pests). A number of strategies can be employed that specifically enhance BNF through increased legume biomass. These include the use of legume genotypes adapted to the prevailing edaphic and environmental conditions, procedures to improve legume plant density, irrigation (if available), the amelioration of soil nutrient toxicities or deficiencies, and the control of weeds and pests.[1,4,10] However, as the formation of an active symbiosis is dependent upon the compatibility of both the diazotrophic micro-organism and the legume host, local practices that limit the presence of effective rhizobia (no inoculation, poor inoculant quality) will also be crucial in determining the legume's capacity to fix N (Table 3), as will any management decisions that directly affect soil N fertility (excessive tillage, extended fallows, fertilizer N, and rotations), since mineral N is a potent inhibitor of the N_2 fixation process.[1,7,9,10]

CONCLUSIONS

Symbiotic associations between legumes and rhizobia are responsible for the greatest contributions of BNF in agricultural systems. Research trials suggest potential annual inputs of fixed N by most legumes equivalent to several hundreds of kg N/ha. However, data collected from pulses, legume oilseeds and pastures growing in farmers' fields generally indicate levels of BNF much lower than the potential values observed under experimental conditions. So while legumes should routinely be fixing >100 kg/ha each year, in reality they usually do not. Strategies are available to improve BNF beyond what is currently being achieved. For example, provided that a legume crop is abundantly nodulated and effectively fixing N_2, enormous benefits in terms of crop production and N_2 fixed can be derived from the application of good agronomic principles. But the ability to overcome constraints at the farm level may be limited because the relevant technologies are either not in the hands of the farmers, or they cannot readily adopt them because of lack of knowledge and information, economic constraints or operational imperatives. While the global inputs of fixed N by legumes may be considerably less than their genetic potential, and is lower than the amounts of N applied as fertilizer each year, some 15 million tonnes of N are harvested annually from legume crops and many million tonnes more will be consumed by animals in legume-based forage, and there are a number of environmental advantages in relying upon BNF over fertilizer N to produce such large quantities of high-quality protein.[11,12]

REFERENCES

1. Giller, K.E. *Nitrogen Fixation in Tropical Cropping Systems*, 2nd Ed.; CABI Publishing: Wallingford, England, 2001.
2. Peoples, M.B.; Herridge, D.F.; Ladha, J.K. Biological nitrogen fixation: an efficient source of nitrogen for sustainable agricultural production? Plant Soil **1995**, *174*, 3–28.
3. Khan, D.F.; Peoples, M.B.; Chalk, P.M.; Herridge, D.F. Quantifying below-ground nitrogen of legumes. 2. A comparison of 15N and non isotopic methods. Plant Soil **2002**, *239*, 273–289.
4. Peoples, M.B.; Baldock, J.A. Nitrogen dynamics of pastures: nitrogen fixation inputs, the impact of legumes on soil nitrogen fertility, and the contributions of fixed nitrogen to Australian farming systems. Aust. J. Expl. Agric. **2001**, *41*, 327–346.
5. Unkovich, M.J.; Pate, J.S. An appraisal of recent field measurements of symbiotic N_2 fixation by annual legumes. Field Crops Res. **2001**, *65*, 211–228.
6. Boddey, R.M.; de Oliveira, O.C.; Urquiaga, S.; Reis, V.M.; de Olivares, F.L.; Baldani, V.L.D.; Döbereiner, J. Biological nitrogen fixation associated with sugar cane and rice:

contributions and prospects for improvement. Plant Soil **1995**, *174*, 195–209.

7. Peoples, M.B.; Bowman, A.M.; Gault, R.R.; Herridge, D.F.; McCallum, M.H.; McCormick, K.M.; Norton, R.M.; Rochester, I.J.; Scammell, G.J.; Schwenke, G.D. Factors regulating the contributions of fixed nitrogen by pasture and crop legumes to different farming systems of eastern Australia. Plant Soil **2001**, *228*, 29–41.

8. Peoples, M.B.; Herridge, D.F. Quantification of biological nitrogen fixation in agricultural systems. In *Nitrogen Fixation: From Molecules to Crop Productivity*; Pedrosa, F.O., Hungria, M., Yates, M.G., Newton, W.E., Eds.; Kluwer Academic Publ: Dordrecht, the Netherlands, 2000; 519–524.

9. Herridge, D.F.; Robertson, M.J.; Cocks, B.; Peoples, M.B.; Holland, J.F.; Heuke, L. Low nodulation and nitrogen fixation of mungbean reduce biomass and grain yields. Aust. J. Expl. Agric. **2005**, *45*, 269–277.

10. Peoples, M.B.; Ladha, J.K.; Herridge, D.F. Enhancing legume N_2 fixation through plant and soil management. Plant Soil **1995**, *174*, 83–101.

11. Crews, T.E.; Peoples, M.B. Legume versus fertilizer sources of nitrogen: ecological tradeoffs and human needs. Agric. Ecosys. Environ. **2004**, *102*, 279–297.

12. Jensen, E.S.; Hauggaard-Nielsen, H. How can increased use of biological N_2 fixation in agriculture benefit the environment? Plant Soil **2003**, *252*, 177–186.

Nuclear Energy: Economics

James G. Hewlett
Energy Information Administration, U.S. Department of Energy, Washington, District of Columbia, U.S.A.

Abstract
This entry attempts to answer the following question: Is nuclear power economic? There are two major factors influencing the economics of nuclear power, and since major uncertainties with both of them exist, it is impossible give an unqualified answer. The first factor is the cost of building the nuclear power plant. On average, nuclear overnight capital costs derived from a small number of Public Utility Commission filings were about $4000 per kW of capacity. If these estimates are indicative of realized construction costs, nuclear power would not be economic. The analysis found that costs must fall to about $2500 to $3000 before nuclear power would be economic. The second factor is the environmental cost of the alternative. The economics of nuclear power would be greatly improved if all of the external costs related to global warming were included in the cost of generating electricity from fossil fuel–fired power plants. Nuclear power plants have their own set of environmental costs, but since they are incurred over hundreds of years, their present value "today" is very small. Some have objected to discounting expenses incurred over very long time periods because this procedure represents a strong incentive to impose large costs on future generations. However, such equity considerations are outside the realm of economic analysis.

INTRODUCTION

In an entry published in the *Encyclopedia of Energy Engineering and Technology*, I attempted to answer the following question: Is nuclear power economic? Unfortunately, I was unable to give a definitive answer to this question. Most of the work on that entry was done in the 2004–2005 time frame, and at that time, there was very little information on the realized costs of building nuclear power plants in the UnitedStates. There was some information on the realized cost of building nuclear power plants abroad (mainly in the Far East). However, that information was at best "sketchy" and difficult to use to estimate costs in the United States. Additionally, in the United States, a number of cost estimates of building hypothetical ("paper") plants at hypothetical ("Middletown USA") sites were published. History has shown that such estimates are always too low.[1] In short, at that time, there was a large amount of uncertainty about nuclear power plant construction costs. Thus, the best that could be done was to conclude that nuclear capital costs would have to be less than $1500 per kW before nuclear power would be economic. (This was substantially less than the cost of building nuclear power plants in the Far East.)

Over the last few years, the economic and political environment has changed, and thus, the question of whether nuclear power is economic needs to be reevaluated. First, a number of estimates of building actual nuclear power plants at actual sites have recently become available. This, by itself, would reduce some of the uncertainty about nuclear power plant construction costs. However, these estimates are two to three times higher than the ones made in the mid-2000s. Additionally, power plant construction costs in general have increased substantially. For example, the realized costs of building coal-fired power plants and wind farms have increased by about 80%–100%.[2] Such increases in construction costs would have major effects on the economics of nuclear power.

Second, unlike when the original entry was written, for a variety of reasons, utilities are now far more interested in nuclear power. In 2006, licensing activity at the U.S. Nuclear Regulatory Commission (NRC) was limited to four early site permit approvals. By issuing an early site permit (ESP), the U.S. NRC approves one or more sites for a nuclear power plant, independent of an application for a combined license to build and operate a power plant. This ESP can be valid for up to 40 years. Since then, 16 utilities have filed applications with the NRC to build and operate a total of 28 nuclear units. Given this increased licensing activity, a reexamination of the underlying economics of nuclear power is certainly in order.

Third, in the 2004–2005 time frame, most forecasts of coal and natural gas prices in 2015–2020 time frame were about $1.45 and $6 ($2009) per mmBtu, respectively.[3] Since then, fossil fuel prices increased substantially and then fell. As of 2010, most forecasts of fossil fuel prices are now greater than the ones made 5–6 years ago. Increased fossil fuel prices will clearly affect the economics of nuclear power.

The purpose of this entry is to reexamine the economies of nuclear power in light of the changed economic environment. Before proceeding with the analysis, two comments about the scope of the analysis will be made. First, in this entry, it is assumed that the investment decision—e.g., the decision to build a nuclear unit or coal-fired power plant—will be made by the owner of a traditional utility. This utility owns and operates other power plants, and the operation of all of them is interrelated. That is, if the utility chooses to operate power plant X less, it would have to operate power plant Y more to meet demand. Otherwise, "the lights would go out." In such cases, the decision maker's objective is to build the unit that minimizes the total cost of building the plant in question and operating all power plants. Suppose, for example, that the decision maker has the choice of building a nuclear or coal-fired power plant. The decision maker will calculate the total system costs if the nuclear plant is built and compare that estimate with the total system costs if the coal plant is built. The decision maker would choose to build the plant that yields the lower total system costs.

The available software to estimate total system costs is complex and expensive and requires many assumptions, and thus, using this approach is beyond the scope of this entry. However, if the alternative to the nuclear power plant is another baseload plant type operating in the same portion of the merit order (baseload demand), total system costs would be minimized by choosing the plant type that has the lower "stand-alone" or levelized cost. (Levelized costs will be defined below. Additionally, because electricity is costly to store, demand will vary over the day, month, and/or year. The portion of total demand that does not vary is called baseload, and the units that are used to meet this demand will run at close to full capacity over the entire year.) This is because the operation of both units under consideration will have the same effect on the operation of the other units. For example, suppose that the levelized cost of building and operating a nuclear power plant for 40 years is 6 cents per kWh, and the levelized cost of building and operating a combined-cycle natural gas–fired power plant is 8 cents per kWh. If both units are assumed to operate in the baseload mode, then the operation of both units will have the same effect on the operation of the other units. In such a case, total system costs would be minimized by building the nuclear plant.

Thus, in this entry, the alternatives to the nuclear power plant are two other baseload plant types—namely, coal-fired and combined-cycle natural gas–fired units. By limiting the comparisons to other baseload plant types, the analysis becomes much more tractable and transparent. Unfortunately, by just computing levelized stand-alone costs, many of the renewable technologies must be excluded from the analysis. The stand-alone cost of building and operating a wind farm, for example, can be computed. However, total system costs may not be minimized by building the wind farm even though that plant type has lower stand-alone costs. This is because the effects of the operation of the wind farm and the nuclear unit on the operation of the other units will not be the same.

Second, since fossil fuel prices will probably increase over time, there is a time dimension to the question of whether nuclear power is economic. Because of the recession and utility conservation programs, additional baseload capacity will probably not be needed until around 2020. Thus, in this analysis, the first year of a unit's operation is assumed to be 2020. By focusing on the mid-term, the carbon capture and storage (CCS) technologies will not be considered. Recently, a CCS task force was formed with a goal of bringing 5 to 10 commercial-size CCS units online by 2016.[4] Even if this goal, which is very ambitious, is met, to demonstrate that the technology works, the units would have to operate for probably 5–10 years. In all probably, it would take 15–20 years before the CCS technologies would be commercially available on a widespread basis.

CAPITAL COSTS FOR NUCLEAR AND COAL-FIRED POWERPLANTS

One of the major uncertainties in any analysis of the economics of nuclear power deals with the construction cost estimates. This section begins with a discussion about nuclear and coal-fired powerplants' overnight capital cost estimates. Overnight cost is defined as the cost of building a power plant instantaneously at some point in time. It is also a direct measure of the value of the land, labor, and materials needed to build a nuclear power plant. Thus, differences in overnight costs reflect differences in the values of the land, labor, and material needed to build the same unit.

It is obviously impossible to build a plant overnight, so the second part of this section describes how the total project costs are derived from the overnight costs. To do this, a number of important assumptions are needed, and in many studies, they are not articulated. This section will also show why comparisons of total project costs must be made with great care. The fuel costs for coal and natural gas–fired power plants are discussed in Section 4. The other assumptions will be briefly discussed in Section 6. These include nonfuel operating costs and nuclear fuel costs.

Nuclear and Coal-Fired Power Plant Overnight Costs

Prior to about 2007, most analyses/discussions about nuclear power plant overnight capital costs tended to focus on either realized costs of units built in the Far East (mainly Japan) or on the estimated costs of building generic units at generic sites in the United States.[5] Each of these sources had their own set of problems. There are always problems with transferring the experience of reactors built in foreign countries to the United States. Also, publicly available foreign overnight capital cost data are not well documented, so all the costs may not be included in the reported figures. Additionally, research has indicated that cost estimates of generic units built on generic sites were always too low, so there were problems with using the resulting estimates. Nonetheless, the analyses/discussions that based their cost estimates on foreign reactors tended to use overnight nuclear capital costs of about $2700 per kW (2009 dollars).[6] The ones that used the cost estimates of generic units tended to use overnight nuclear construction costs of about $1500–$2000 per kW (2009 dollars).

Over the last few years, as part of the process of getting approval from the state public utility commissions (PUCs) to proceed with their NRC licensing activities, utilities filed cost estimates of building actual powerplants at actual sites. In some cases, it was possible to determine what cost items were and were not included in the estimates. More important, if the overnight construction costs were not directly reported on the filings, they could be directly estimated. On average, these overnight nuclear construction cost estimates, shown in Table 1, were about $4000 per kW.

These estimates are clearly much better than the ones based on generic units built at generic sites. Because problems with the transfer of foreign cost information to the United States are avoided, they are much better than the estimates based on realized overnight costs of reactors built in Japan. However, they are also much higher than the ones based on generic designs at generic sites and were also much greater than the realized overnight costs of units built in Japan.

There are probably at least four reasons why the recent U.S. overnight cost estimates are greater than the realized costs of reactors built the Far East (mainly Japan). First, there are clearly cultural factors at play. Second, over the last 15 years, there has been a slow but relatively constant expansion of nuclear power in Japan, so Japanese builders are further down their learning curves than their counterparts in the United States. (The South Texas project is being built by Toshiba, a large Japanese firm that has built some nuclear power plants in Japan to another. The cost estimates for that project are not that much different from the others. Thus, the learning may not be transferred from one country to another). Third, it is not clear whether all the costs are being reported in the Japanese figures. This is especially true for the so-called owners' costs. Costs which are the ones incurred by the utility over and above the ones paid to the firms building the unit. In fact, it was never clear if the reported Japanese costs even included any owners' costs. Lastly, many of the major components for the proposed U.S. reactors, such as the reactor vessel, are being manufactured in Asia, so some exchange rate issues could exist.

There are also a number of reasons why the overnight nuclear construction cost estimates obtained from the PUC filings are higher than the ones for generic units. As was just noted, research has shown that generic cost estimates at generic sites are always too low. Additionally, a number of analysts have argued that the growth in overnight costs was due to increases in commodity prices—the prices of iron, steel, cement, and so on—that occurred from around 2005 to 2008.[7] However, about 40%–50% of the overnight cost of building a nuclear power plant is labor related, and therefore, the effects of increases in commodity prices on total construction costs are probably modest. Moreover, the cost estimates based on generic units might have assumed that all of the components were manufactured in the United States. Over the last 10 years, the value of the dollar relative to the yen has fallen. Since many of the major nuclear components for proposed U.S. plants will be imported from Japan, some of the increases in overnight costs could be due to the fall of the dollar. This, however, would depend upon how the contracts with the Japanese firms are structured, and there is very little public information about this. Lastly, with the rapid expansion of nuclear power in China, some bottlenecks in the production of the major components that would increase costs appeared.

The increases in nuclear power plant overnight construction cost estimates raise the question of whether the current cost levels will be permanent. The effects of commodity price increases and/or decreases in the value of the dollar on the cost estimates are very unclear. Thus,

Table 1 Estimated overnight costs and lead times of selected proposed nuclear plants.

Owner	Plant type	Plant	Capacity (mWe)	Costs (2009 dollars per kW of capacity)	Lead times (years)
Tennessee Valley Authority	ABWR	Bellefonte	1371	$3164	NA
Florida Power and Light	ESBWR	Turkey Point	3040	$3811	5–6
Progress Energy (Florida)	AP1000	Levy County	2212	$4541	5
South Carolina Electric and Gas	AP1000	Summer	2234	$4089	5 (unit 2) and 8 (unit 3)
Southern	AP1000	Vogtle	2200	$4535	6
NRG	ABWR	South Texas	2700	$3758	NA
Average				$3983	

Note: A correction was made to the Vogtle Estimate reported in Du and Parsons.[8]
ABWR, advanced boiling water reactor; ESBWR, economically simplified boiling water reactor; AP1000, advanced pressurized water reactor.
Source: Adapted from Du and Parsons.[8]

the effects of decreases in commodity prices (or increases in the value of the dollar relative to the yen) on costs would also be unclear. Additionally, even with the rapid expansion of nuclear power in the Far East, in the long run, markets would adjust, and the bottlenecks would no longer exist. Because of the complexity of the technology, it would take a number of years for existing firms to build new production facilities and for new firms to enter the market. Thus, the bottlenecks may not be removed for a number of years.

Lastly, currently, only a few firms produce the large nuclear components and build the power plants. In such cases, the behavior of one firm (e.g., Westinghouse) could possibly affect the behavior of its competitor (e.g., General Electric). In economist's jargon, such a market structure is called an oligopoly. One major characteristic of an oligopolistic market is price "stickiness"—i.e., when the firm's costs fall, prices will fall, but with a lag. Thus, even if the costs of the firms that produce the components and build the plants fall, it would take some time before reductions in costs would be reflected in reductions in the prices charged to the utilities.[9,10]

The availability of some actual overnight cost estimates from the PUC filing reduces some of the uncertainty in nuclear power plant construction costs. Unfortunately, the large increases in the estimates introduce another source of uncertainty—namely, whether the current cost levels will be permanent. Given current publicly available information, one can only speculate why overnight capital costs increased and whether they will fall in the future. Moreover, no nuclear power plants in the United States have been built on time and on budget, and cost overruns in the two units under construction in Europe have occurred. Thus, even with a number of detailed cost estimates of building actual units at actual sites, there is still a considerable amount of uncertainty about nuclear overnight construction costs. Indeed, until a few units are actually built, nuclear construction costs are essentially unknown.

Table 1 also shows estimated date of commercial operation leadtimes for a number of proposed nuclear units. As can be seen from Table 1, with the exception of one unit, the estimated leadtimes ranged from 6 to 7 years. It should be noted that these lead time estimates were made just before or at the beginning of the recession. Recently, the date of commercial operation for a number of these units was moved back simply because the capacity was not needed. In all probability, the utilities will also move back the construction start date and keep leadtimes the same. If, however, they begin construction and then revise their estimated date of commercial operation, lead times could actually increase. In the 1970s and 1980s, many utilities building nuclear power plants also increased leadtimes because of the lack of need for capacity. I have shown elsewhere that these actions also affected overnight costs.[11] Thus, this issue is important.

Table 2 Estimated overnight construction costs of selected proposed coal-fired power plants.

Owner	Plant	Capacity (mWe)	Costs (2009 dollars per kW of capacity)
Florida Power and Light	Glades	1960	$2130
Duke Power	Cliffside	800	$2124
AMP Ohio	Megis Co.	960	$3277
AEP Swepco	John W Turk Jr	600	$2508
Average			$2510

Source: Adapted from Du ans Parsons.[8]

Lastly, Table 2 shows that the average estimated overnight cost of building a number of recent coal-fired power plants was about $2500 per kW. A 2007 Massachusetts Institute of Technology (MIT) study estimated that the costs of building a wide range of coal-fired powerplants were about $1280 to $1360 per kW (2005 dollars).[12] Thus, the cost of building coal-fired power plants also increased substantially, and therefore, the cost growth was not limited to nuclear power.

Derivation of Total Project Costs

The focus of the preceding section was on overnight costs, because they are a direct measure of the values of the land, labor, and materials needed to build a power plant. However, the analysis will use total project (capital) costs. The best way to explain how total capital costs are computed is to outline the steps taken to derive an estimate of the total cost of building any power plant. The first step is to prepare a detailed "bottom-up" estimate using current commodity prices, labor wage rates, and so on. This bottom-up estimate is the product of the estimated quantities of the land, labor, and materials needed to build the unit times the prices of these inputs. Suppose, for example, that the firm is making an estimate in 2010 of the cost of building a nuclear unit that will become operational in 2020. Given a 6-yr lead time, construction must begin in 2015. Thus, the firm would first make a bottom-up estimate using 2010 prices and wage rates. This is the estimate of the overnight costs using 2010 prices. In the example shown in Table 3, the 2010 estimated overnight cost is $4000 per kW of capacity.

Next, the cost of building the unit overnight in 2015—the year that construction of the unit begins—must be estimated. To do this, the firm would make assumptions about how commodity prices, labor wage rates, and so on would change from 2010 to 2015. In the example shown in Table 3, commodity prices, labor wage rates, etc., are assumed to increase at a rate equal to the general inflation rate of 3%

Table 3 Derivation of total capital costs.

	(1)	(2)	(3)	Total capital costs			
				(4)	(5)	(6)	(7)
Year	Percent of total overnight costs spent in each year of construction Period	Expenditures in 2015 dollars (per kW)[a]	Expenditures in dollars of year funds expended (per kW)[a]	Financial costs not included (per kW)	Column 4 plus financing charges— lower financing rate (per kW)[b]	Column 4 plus financing charges— higher financing rate (per kW)[c]	Column 4 plus just debt component (per kW)[d]
2015	10.00%	463.7	$463.7	$463.7	$695.9	$925.1	$560.2
2016	15.00%	$695.6	$716.4	$716.4	$1004.8	$1273.9	$838.6
2017	20.00%	$927.4	$983.9	$983.9	$1289.7	$1559.3	$1116.0
2018	30.00%	$1391.1	$1520.1	$1520.1	$1862.2	$2147.1	$1670.8
2019	15.00%	$695.6	$782.9	$782.9	$896.3	$985.5	$833.8
2020	10.00%	$463.7	$537.6	$537.6	$575.2	$603.1	$554.8
Total Capital Costs per kW		$4637.1	$5004.6	$5004.6	$6324.1	$7494.1	$5574.1
Total Capital Costs Billions of Dollars		$ 10.2	$ 11.0	$ 11.0	$ 13.9	$ 16.5	$ 12.3

[a] Overnight capital costs in $2010 of $4000 per kW of capacity and a 3% annual escalation rate in costs were used to derive the data in columns 2 and 3.
[b] A 7% financing rate was used.
[c] A 12 % financing rate was used.
[d] An 8 % debt rate was used.

per year. Given this assumption, the 2015 overnight cost would be about $4600 per kW ($4000 × 1.03^5). Then, assumptions about how the funds are expended over each year of the construction period and how prices change over the construction period would be used to compute total costs in the dollars of the year the funds are expended. In the example shown in column 3 in Table 3, total construction costs excluding financing charges would be about $5000 per kW of capacity. Note that simply because of increases in prices and wages over the 2010–2020 period, costs increased by more than $1000 per kW.

Lastly, there is the issue of how to include the financing costs in total capital costs and analyses of the economics of nuclear power. Because of the complexity of this issue, there are wide variations in both the financing rates and the costs that are included in the total capital cost estimates. In some analyses, including the present one, for a variety of reasons, the total capital cost estimates that are computed do not include any financing costs. In such cases, financing issues are accounted for elsewhere.[8] In other analyses, the total capital cost estimates do include financing costs. Sometimes, the total capital cost estimates include just interest charges on the monies that are actually borrowed.[6] If the analysis is used by a utility that is subject to state-level rate-of-return regulation, the financing charges included in the total capital cost estimate often consist of interest on the funds that are borrowed (debt financing) and an implied charge for the funds that are internally generated (equity financing).

It is always tempting to compare total capital cost estimates that are reported in the media. Columns 4–7 in Table 3 illustrate that this must be done with great care. In particular, these columns show total capital cost estimates using four different assumptions about the financing costs that are included in them. As can be seen from this table, the total capital costs range from $5000 to $7500 per kW, depending upon how the financing costs are reported. Again, comparisons of cost estimates without detailed knowledge about what is included in them should not be done.

ENVIRONMENTAL COSTS AND REGULATIONS

In the long run, another important factor is the environmental costs of generating electricity from fossil fuel–fired and nuclear power plants. The fossil fuel–fired power plant's capital cost estimates shown in Table 2 include the expenses needed to meet sulfur dioxide and nitrogen oxide limits imposed by the Environmental Protection Agency. The estimates also include the costs of meeting current federal and state water discharge regulations. Additionally, in the analysis reported in Section 6, an explicit fee on carbon dioxide emissions will be included. The costs of all existing laws and regulations affecting fossil fuel–fired power plants, and an important proposed one are, therefore, included in the analysis.

Nuclear power has its own set of environmental costs—namely, the possibility of exposing the public to radiation, decommissioning, and radioactive waste disposal. The cost estimates of the units shown in Table 1 reflect designs that met NRC requirements as of about 2007 and 2008. These designs may, however, have to be changed because of additional NRC requirements, which could have cost implications. The design of the AP1000, a 1100-MW pressurized water reactor, was approved by the NRC in 2006. However, a number of changes to that design have been approved, and one additional change is currently under review by the NRC. Similarly, the design of the ABWR, a 1600-MW boiling water reactor developed by GE, was approved by the NRC in 1997. Currently, GE is in the process of renewing the NRC approval of that design. Additionally, in 2009, a number of design changes were submitted by GE to the NRC. The NRC is currently reviewing these changes.

Nuclear power decommissioning deals with the dismantlement of the plant and the decontamination of the site so it can be used for other purposes. The NRC must approve the utilities' plans, set residual radiation standards, and oversee the actual dismantlement of the plant. To insure that funds will be available when units are decommissioned, the state PUCs require that monies be placed in trusts. In the analysis presented in Section 6, it was assumed that decommissioning would cost $600 million and would occur 40 years after the plant's date of commercial operation.

The spent fuel from a light water reactor will be radioactive for millions of years, and the disposal of that waste is a major economic and political issue. There are two basic methods of disposing of the spent fuel from nuclear reactors: geological disposal and reprocessing/recycling. Reprocessing/recycling consists of extracting usable fuel from the waste and using it in other reactors. Currently, this is done in a few countries, most notably France and Japan. Needless to say, reprocessing/recycling is very controversial. It is very expensive and, at least in its current form, the reprocessed spent fuel could be used for military purposes. Thus, there are major proliferation concerns with reprocessing. Lastly, reprocessing has its own set of waste disposal problems.

With the passage of the Nuclear Waste Policy Act (NWPA) of 1982, the United States formally chose direct geological disposal of the waste. The initial act directed that the Department of Energy (DOE) study the feasibility of burying the waste at a number of sites. In 1988, however, Congress directed DOE to focus on just one site—Yucca Mountain, Nevada—and in 2003, the President formally chose Yucca Mountain as the country's high-level waste repository site. Under the NWPA, the state of Nevada could veto the President's decision, which they did. This veto was then overturned by Congress, and in 2008, the DOE submitted an application to the NRC for their approval to build the repository.

In 2010, the Obama Administration decided to stop work on Yucca Mountain, and DOE formally requested the NRC to permit them to withdraw the application. As of early 2011, the NRC has yet to publicly announce their decision. Additionally, the Administration abolished the Office of Civilian Radioactive Waste Management—the office within DOE that was managing the Yucca Mountain project. In response to the Obama Administration's actions, a number of lawsuits were filed by various states and localities in Federal Court. These lawsuits claim that DOE does not have the authority to unilaterally stop work on Yucca Mountain. As of early 2011, the ultimate outcomes of these lawsuits are unknown, and it is unclear what would happen if the courts ruled against DOE.

The NWPA requires that the government collect a fee of one mill (one-tenth of a cent) per kilowatt hour of electricity generated from nuclear power, and this charge is included in the analysis. Given the state of the U.S. spent fuel disposal policy, it is impossible to say anything about ultimate waste disposal costs. Regardless of what the ultimate cost is, at least for geological disposal, many of the costs will be incurred by future generations, and because of the "magic of compound interest," the total costs "today" will be small. To illustrate the effects of discounting, the yearly estimated costs of building and operating Yucca Mountain will be used. These costs were derived from a 2008 estimate prepared by DOE and the total undiscounted costs were about $100 billion. (Since Yucca Mountain has been abandoned, these estimates should be viewed as the cost of some hypothetical repository.) The cumulative costs are shown in Fig. 1. As can be seen from this figure, roughly 50% of the expenditures will be made from year 50 to year 150, and the bulk of the expenditures will be made from year 30 to year 80. The point here is that the bulk of the costs of a geological repository will probably be incurred many years in the future.

Because of the complexities involved in geological disposal of the waste, it is quite possible that the ultimate cost of such a repository will be much greater than $100 billion. For argument's sake, suppose that the yearly expenditures are five times those shown in Fig. 1. Fig. 2 shows the present value "today" of the yearly expenditures using various discount rates. Using a 2% discount rate, which is very low, the present value "today" of the yearly expenditures is about $200 billion. Many private sector discount rates range from 7.5% to 10%, and if these rates were used, the present value "today" of $500 billion s incurred over a 150 years period would be less than $50 billion—roughly one tenth of the undiscounted costs.

Because of the discounting process, very large costs imposed on future generations will appear to be very small "today." Consequently, some intergenerational equity issues dealing with evaluating the back-end costs exist. Some economists have attempted to include intergenerational fairness considerations in discounting, but unfortunately,

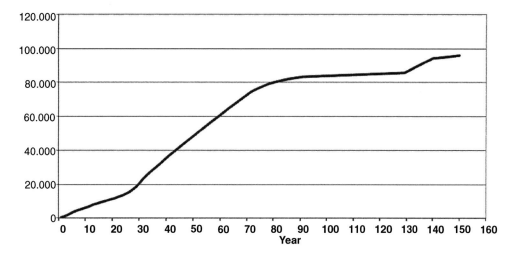

Fig. 1 Cumulative cost of building and operating a hypothetical geological nuclear waste repository (millions of 2007 dollars).
Source: Adapted from *Analysis of the Total System Life Cycle Cost of the Civilian Radioactive Waste Management Program, Fiscal Year 2007*.[13]

there is no consensus about how to incorporate equity issues in discounting.[14,15] Thus, there are some equity issues dealing with nuclear power that cannot be resolved with economic analysis.

DISCOUNT RATES

The choice of the appropriate discount rate is another very important assumption in any analysis of the economics of nuclear power. To compare the economics of different plant types, costs that are incurred in the future must be discounted back of the present ("today"). This discounting process will account for the fact that "a dollar today is worth more than a dollar tomorrow." Nuclear capital costs are much greater and the operating costs are much less when compared with fossil fuel–fired power plants, especially natural gas–fired units. Thus, a larger percentage of the total cost of building and operating a nuclear plant is incurred "upfront" when compared with a fossil-fired power plant. The higher the discount rate, the greater would be the weight placed on the "up-front" costs, and thus, higher discount rates would tend to favor natural gas and, to a lesser extent, coal-fired power plants relative to a nuclear unit.

Since most spreadsheets have routines that calculate present values, the mechanics of discounting are trivial. However, the same cannot be said for the choice of the discount rate. The discount rate should reflect the risk of the

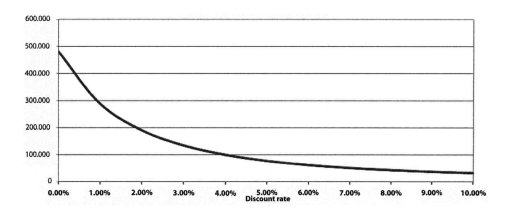

Fig. 2 The effects of discounting on the cost of building and operating a hypothetical nuclear waste repository (present value of costs in millions).
Note: The costs are assumed to be 5 times the ones shown in Fig. 1.
Source: Adapted from *Analysis of the Total System Life Cycle Cost of the Civilian Radioactive Waste Management Program, Fiscal Year 2007*.[13]

project—i.e., building and operating a power plant. There is a long literature on estimating discount rates, but unfortunately, in most cases, the necessary financial data (stock prices and bond yields) are at the utility and not at the project level.[16] This presents some problems when utilities own assets that have different risks. In particular, most but not all utilities own transmission and distribution facilities in addition to generating plants. The risks associated with building and operating the transmission/distribution (TD) system are very different from the risks of building and operating powerplants. Thus, the risks reflected in observed utility level data are some type of weighed average of the risks of the TD and generating plant assets. Additionally, the risks reflected in recent utility level data largely deal with operating factors because very few baseload power plants are currently under construction. Thus, project-specific risks (i.e., discount rates) for large-baseload power plants cannot be estimated with actual data. Indeed, this is one of the classic problems in finance—the choice of a discount rate when a low-risk firm undertakes a high-risk investment.

Since the choice of a discount rate is largely judgmental, the analysis described in Section 6 will use two of them. (See Table 4.) Financial theory states that investors will require higher returns for bearing risks that are nonrandom.[16] (Financial economists often refer to nonrandom risk as nondiversifiable or systematic risk. Even when observable data can be used, the estimation of such risks is very difficult). This is because random risks can be eliminated by constructing portfolios of diversified assets. Thus, the discount rate will be a direct function of the level of nonrandom risk. The higher discount rate assumes that the nonrandom risks of building and operating any power plant are 50% greater than the risk for the average investment. The lower discount rate assumes that the nonrandom risks are about 20% less than the risk for the average investment. The higher and lower discount rates are consistent with the nonrandom risks observed in the airlines/telecommunications and manufacturing industries, respectively. The higher discount rate is also slightly greater than the one used for nuclear power in the MIT study, and the lower rate is slightly less than the one used for fossil fuel–fired power plants in the MIT study.[6]

To properly interpret the differences in the two rates, the distinction between risk reduction and risk shifting becomes important. One example of risk shifting, as opposed to risk reduction, is the cost-based rate-of-return regulation of utilities. Under this form of regulation, utilities can recover all the costs for projects that are ex ante prudently expended but ex post uneconomic. In such cases, some of the risks are being shifted from utility shareholders to electricity consumers. State-level rate-of-return regulation, therefore, does not reduce risks but rather shifts some of the risks to consumers. Another example is long-term fixed-price purchase power contracts between deregulated generating companies and the regulated transmission and distribution firms. Again, such contracts do not reduce risks but instead shift the risks associated with volatile wholesale electricity prices from utility shareholders to consumers. In short, the difference in the two discount rates shown in Table 4 reflects differences in the underlying risks of building and operating power plants and not who bears that risk.

As Table 3 shows, the total cost of building a 2200-MW-nuclear power plant could be in the range of $10 to $15 billion. Given the size of many U.S. utilities, the failure of a $10-billion project may result in the firm's bankruptcy or insolvency. Thus, many utilities who are planning to build nuclear power plants are attempting to dilute bankruptcy possibilities by forming joint ventures or, in one case, a merger. The discussion above abstracts from bankruptcy risk per se because it a function of the size of the project relative to the size of the firm along with the underlying risk of the project. In other words, the risk of a project is not a function of the firm's size but instead is determined by the variability in the underlying costs and revenues.

COST OF THE ALTERNATIVE

A fourth factor affecting the decision to build a nuclear power plant is the cost of the alternative. As was noted in Section 2, coal-fired power plant construction costs have escalated over the last few years. This cost growth has introduced some uncertainties related to the permanency of the increases. However, environmental considerations aside, most of the uncertainty related to the cost of generating electricity from fossil fuel–fired power plants deals with fuel prices. Historical and projected coal prices are shown in Fig. 3. As this figure shows, after adjusting for inflation, coal prices fell until about the year 2000 and then increased by about 30% from 2005 to 2010. Two projections suggest that coal prices will remain relatively constant at about $2.00 per mmBtu over the 2010–2020 period, and the other one shows coal prices increasing to their 2009 levels. (In this section, all the prices are in 2009 dollar. Additionally, the first year of the unit's commercial operation is assumed to be 2020.)

Historically, natural gas prices have been much more volatile than coal prices. (See Fig. 4.) After natural gas prices were deregulated in the early 1980s, they fell from more than $6 to about $3–4 mmBtu. In about 1998, natural gas prices began to increase, and by 2008, they increased to more than $9.00 mmBtu. Then, partly because of the recession, they fell by more than 100% to about $5 mmBtu. All three projections shown in Fig. 4 have natural gas prices increasing to more than $6 mmBtu by about 2014. Over the 2015–2020 period, two of the three projections have natural gas prices increasing by relatively small amounts, whereas the third one has prices continuing to increase to more than $8.00 mmBtu by 2020.

Table 4 Assumptions used in the analysis.

Assumption	Plant type		
	Nuclear	Coal	Combined-cycle natural gas
Unit size (mWe)	1,000	1,000	1,000
Capacity factor	85.00%	85.00%	85.00%
Heat rate	10,400	8,870	6,800
Overnight costs (dollars per kW of capacity)	NA	2300	850
Lead times (years)	6	4	3
Fixed O&M costs (dollars per kW of capacity)	96	51	23
Variable O&M (dollars per kwh)	0.0004	0.00357	
Fuel costs (mills per kwh)	7	NA	NA
Waste fee (mills per kwh)	1	NA	NA
Decommissioning (million $s)	600	NA	NA
CO_2 fee for each dollar per metric ton CO_2 (2009 $/mmBtu)[a]	0	0.095	0.053
Escalation/inflation rates:			
Inflation rate	2.00%	2.00%	2.00%
Annual O&M real escalation rates	1.00%	1.00%	1.00%
Annual fuel cost real escalation rate	0.50%	0.20%	1.70%
Annual real capital cost escalation rate	0.00%	0.00%	0.00%
Financial:			
Tax rate	37.00%	37.00%	37.00%
Debt fraction—higher rate	40.00%	40.00%	40.00%
Debt fraction—lower rate	60.00%	60.00%	60.00%
Cost of debt capital—higher rate	8.00%	8.00%	8.00%
Cost of debt capital—lower rate	6.50%	6.50%	6.50%
Cost of equity capital—higher rate	15.00%	15.00%	15.00%
Cost of equity capital—lower rate	11.00%	11.00%	11.00%
Weighted after-tax cost of capital—higher rate[b]	11.02%	11.02%	11.02%
Weighted after-tax cost of capital—lower rate[b]	6.04%	6.04%	6.04%

[a] These values were used to compute the increase in fuel costs per kilowatt hour because of a carbon fee. For example, the increase in fuel costs for a coal plant caused by a fee of $20 per metric ton carbon was computed as follows: 20*.095*(8870/1000) = 16.86 mills per kwh.
[b] These values were used as the discount rates. These rates cannot be compared with the ones used in any study that excluded corporate income taxes.
Note: All costs are in 2009 dollars. Except for the financial assumptions, the others were generally obtained from Du and Parsons.[8]
Source: Adapted from Du and Parsons.[8]

As Table 5 shows, there are also wide regional variations in prices of coal and, to a lesser extent, natural gas. These variations are largely due to the cost of getting the fuel from the source of supply to the end users. Regions of the country that are far away from the source of supply tend to have higher costs, and the ones located close to the coal mines or gas fields have lower costs. Thus, the economics of nuclear power have a regional dimension.

RESULTS

As was noted in Section 2, the cost of building nuclear power plants is highly uncertain. There are issues dealing with the permanency of current construction cost levels and also their accuracy. Wide regional variations in coal prices and uncertainty about future natural gas prices also exist. Because of these factors, point estimates of the levelized

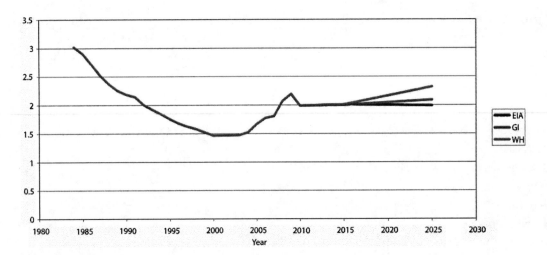

Fig. 3 Delivered price of coal to electric utilities, 1984–2025 (2009 dollars per mmBtu).
Note: EIA, Energy Information Administration; GI, IHS Global Insights; WH, Wood MacKenzie Company. For the GI and WH projections, only the years 2015 and 2025 were reported.
Source: *Annual Energy Outlook, 2010.*[17]

cost of generating electricity from coal, natural gas, and/or nuclear power plants have little value. The general approach used here is, therefore, to derive various combinations of nuclear overnight capital costs and "current" coal (natural gas) prices that would result in nuclear power and coal (natural gas) being equally economic—i.e., the levelized costs of coal (natural gas) and nuclear power are the same. The levelized cost is defined as the constant real price of electricity that would result in the net present value of the project (discounted revenues less discounted costs) equal to zero. The method used to compute the levelized costs can be found in the work of Du and Parsons.[8] Then, the combinations of fossil fuel prices and overnight nuclear capital costs that would result in nuclear power being economic (or uneconomic) could be determined. While this is not ideal, it is the best that can be done, given all the uncertainty.

To implement this approach, it was necessary to "fix" all the other variables affecting the economics of nuclear power, and unfortunately, there is some uncertainty in all of them. Some unreported sensitivity analyses suggested that variations in leadtimes, heatrates, and nonfuel

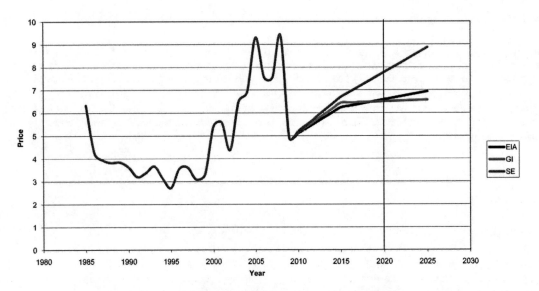

Fig. 4 Delivered price of natural gas to electric utilities, 1985–2025 (2009 dollars per mmBtu).
Note: EIA, Energy Information Administration; GI, IHS Global Insights; SE, Strategic Energy and Economic Research. For GI and SE projections, only the years 2015 and 2025 were used.
Source: *Annual Energy Outlook, 2010.*[17]

Table 5 2010 delivered price of coal and natural gas to electric utilities (2009 dollars per mmBtu).

Census region	Coal	Natural gas
New England	3.14	5.70
Mid Atlantic	2.23	5.03
South Atlantic	2.86	5.87
East North Central	1.88	4.30
East South Central	2.24	4.66
West North Central	1.25	4.62
West South Central	1.55	4.40
Mountain	1.55	5.00
Pacific	2.24	4.89
US average	1.99	4.85

Source: *Annual Energy Outlook, 2010.*[17]

Operations and Maintenance (O&M) costs had relatively minor effects on the basic conclusions of this analysis. Additionally, as was discussed in Section 2, there is uncertainty about the permanency of the increases in the coal-fired power plant overnight construction costs. As will be seen shortly, fixing the coal capital costs will not have any major impact on the basic conclusions of this analysis.

The results of a comparison of the economics of nuclear power relative to coal-fired powerplants using the higher discount rate are shown in Fig. 5a. (In this section, except where noted, all costs and prices are in 2009 dollars.) The solid line shows the combinations of overnight nuclear capital expenses and current coal prices that would result in both plant types having the same levelized cost. Thus, any combination of overnight nuclear capital costs and coal prices that fall in the region denoted as E would result in nuclear power being economic. Similarly, any combination of these two factors that fall in the region U would result in nuclear power being uneconomic relative to coal-fired power plants.

As was noted above, overnight nuclear capital costs derived from a small number of PUC filings averaged about $4000 per kW. Given overnight capital costs of $4000 per kW, projected coal prices would have to be about $6 per mmBtu in 2020 before nuclear power would be competitive with coal-fired power plants (point A in Fig. 5a). These coal prices are about two and a half to three times their 2010 levels and are also much greater than the projected values. Coal prices of $6 per mmBtu are also much greater than their 2010 levels in regions of the United States that are not close to the coal reserves.

Similarly, given projected coal prices in the mid-term of about $2 mmBtu, nuclear overnight capital costs would have to fall to levels roughly comparable with the ones for the coal-fired power plants, before nuclear power would be economic. Given $2 per mmBtu coal prices and the nuclear operating cost assumptions shown in Table 4, the operating costs of both plant types are about the same. Thus, for nuclear power to be competitive with coal-fired powerplants, the overnight capital costs of the two technologies would have to be similar (point B in Fig. 5a).

The results of the comparison of nuclear power to coal-fired power plants using the lower discount rate are shown in Fig. 5b. These results are qualitatively similar to the ones using the higher discount rate. That is, if the overnight nuclear capital costs would be about $4000 per kW, coal prices in 2020 would have to be much greater than the ones reported in a number of recent studies. Additionally, using projected coal prices in 2020 of about $2 per mmBtu, overnight nuclear capital costs would have to fall to about $2500 per kW before nuclear power would be competitive with coal-fired power plants.

To summarize, if nuclear overnight capital cost estimates found in a number of recent PUC filing are at all indicative of what it would actually cost to build a nuclear power plant, and if the external costs of carbon dioxide (CO_2) emissions are ignored, nuclear power is not competitive with efficient coal-fired power plants. Given nuclear capital costs of about $4000 per kW, coal prices would have to increase substantially before nuclear power would be competitive with coal-fired power plants. Also, given the assumptions shown in Table 4, these conclusions do not depend upon the assumed discount rate. As will be seen shortly, the same is not true for comparisons of nuclear power with natural gas–fired power plants.

The economics of nuclear power relative to an efficient combined-cycle natural gas–fired power plant using the higher discount rate is shown in Fig. 6a. Given overnight nuclear capital costs of $4000 per kW, natural gas prices would have to increase to about $9.50 per mmBtu before nuclear power would be economic relative to gas-fired power plants (point A in Fig. 6a). Natural gas prices of over $9 per mmBtu in 2020 are much higher than their 2010 levels and are also higher than their projected levels shown in Fig. 4. Similarly, using the higher discount rate and 2020 natural gas prices of about $7 per mmBtu, overnight nuclear capital costs would have to be about $3000 per kW before nuclear power would be competitive with natural gas–fired power plants.

Given overnight nuclear capital costs of $4000 per kW, roughly 75% of the total cost of generating electricity from the nuclear unit is incurred upfront in terms of construction costs. However, about 75% of the total cost of generating electricity from a combined-cycle natural gas–fired plant is incurred in future years. Again, the lower the discount rate, the lesser will be the weight placed on the up-front costs, and thus, lower discount rates would favor nuclear power. The comparison of the economics of nuclear power plants relative to gas-fired power plants using the lower discount rate, shown in Fig. 6b, suggests that this is the case. Using the lower discount rate and nuclear capital costs of about $4000 per kW, natural gas

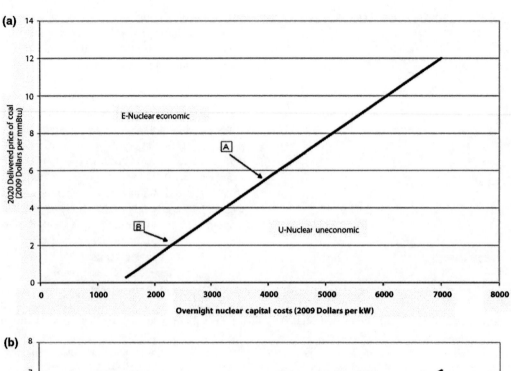

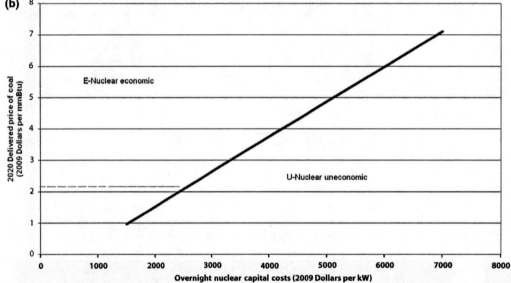

Fig. 5 The economics of a nuclear power plant relative to a coal-fired power plant: (a) higher discount rate (b) lower discount rate.
Note: The solid line shows the combinations of 2020 coal prices and nuclear capital costs that would produce the same levelized cost for both plants. The combinations of coal prices and nuclear capital costs that would fall in the region E (U) would produce levelized costs that are lower (higher) for the nuclear plant than for the coal plant.

prices would have to increase from their 2010 levels of about $5.00 to about $6.00 per mmBtu by 2020 before the nuclear power plant would be economic. Natural gas prices of about $6 per mmBtu in 2020 are slightly lower than two of the three projections shown in Fig. 4. Thus, if building and operating any power plant is perceived to be a relatively low-risk endeavor, even if the external costs related to CO_2 emissions are ignored, nuclear power is marginally economic relative to combined-cycle natural gas–fired power plants.

This result was based on the assumption that real natural gas prices would increase at an annual rate of .75% from 2021 to 2061. At least through 2035, this escalation rate is relatively high. If a lower escalation rate in annual real natural gas prices of 0.3% were assumed, and given nuclear overnight capital costs of $4000 per kW, 2020 real natural gas prices would have to be about $7 per mmBtu before a nuclear power plant would be competitive with a combined-cycle natural gas–fired power plant. Natural gas prices of $7 per mmBtu in 2020 are greater than two of the

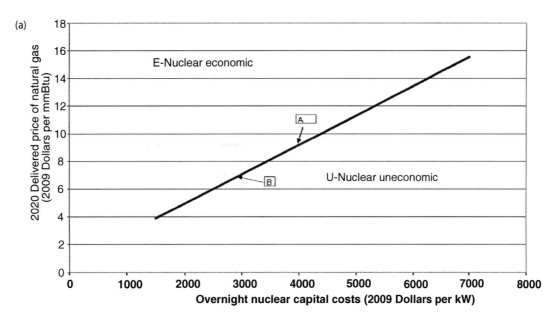

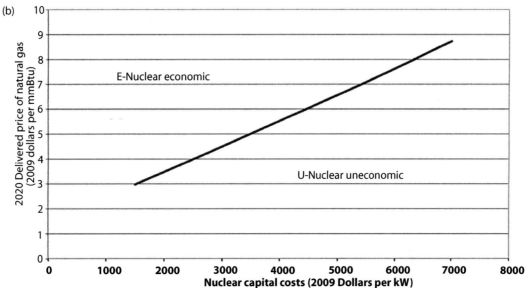

Fig. 6 The economics of a nuclear power plant relative to a combined cycle natural gas-fired power plant: (a) higher discount rate (b) lower discount rate.
Note: The solid line shows the combinations of 2020 natural–gas prices and nuclear capital costs that would produce the same levelized cost for both plants. The combinations of gas prices and nuclear capital costs that would fall in the region E (U) would produce levelized costs that are lower (higher) for the nuclear plant than for the gas–fired plant.

three projection shown in Fig. 4. Thus, using lower discount rate, nuclear power would be extremely marginally competitive with combined-cycle natural gas–fired power plants.

As was noted above, the fossil fuel capital costs include the expenses needed to meet current sulfur dioxide and nitrogen oxide emission levels. However, the external costs related to CO_2 emissions have not been included. The size of these costs is, however, highly uncertain. Indeed, the extent to which CO_2 emissions have resulted in global warming and the associated costs of the warming of the atmosphere are still being debated. However, one recent study recommended using external costs in 2020 of about $7 to $42 or perhaps about $80 per metric ton of CO_2 emissions if the Earth warms faster than expected.[18] Additionally, some recent analyses of the effects of recently proposed U.S. "cap-and-trade" programs estimated that in 2020, the "price" of CO_2 emissions would range from about $20 to $100 per metric ton in 2020.[19] Lastly, in Europe's CO_2 cap-and-trade program, over the last few years, CO_2 emission permits have been trading in the range of 15€–20€ per metric ton.

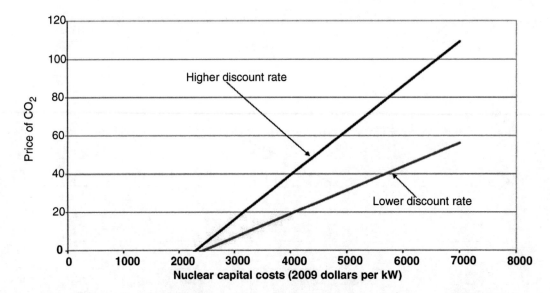

Fig. 7 The economics of a nuclear power plant relative to a coal-fired power plant: costs of CO_2 emissions included.
Note: The solid line shows the combinations of CO_2 prices and nuclear capital costs that would produce the same levelized cost for both plants. The combinations of coal prices and nuclear capital costs that would fall in the region above (below) or to the left (right) would produce levelized costs that are lower (higher) for the nuclear plant than for the coal plant. Real coal prices in 2020 were assumed to be about $2.10 per mmBtu. The carbon prices is in 2009 dollars per metric ton.

Because of these uncertainties, the best that could be done is to compute the combinations of nuclear capital costs and CO_2 prices that would result in nuclear power being economic relative to coal and combined-cycle natural gas–fired power plants. This, of course, requires that coal and natural gas prices in 2020 be fixed at $2.00 and $7.00 per mmBtu, respectively. The results of this exercise for coal and natural gas–fired power plants are shown in Figs. 7 and 8, respectively. As before, both the higher and lower discount rates were used.

These results suggest that if nuclear power construction costs were about $4000 per kW, CO_2 prices in 2020 would have to exceed $20–$40 per metric ton before nuclear power would be economic relative to coal-fired

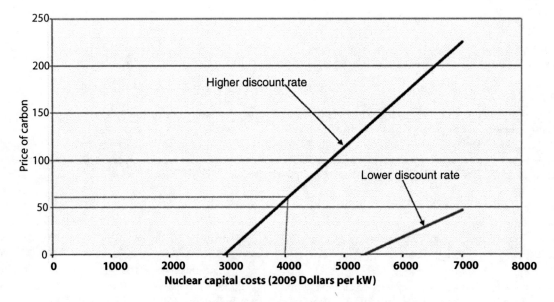

Fig. 8 The economics of a nuclear power plant relative to a combined-cycle gas-fired power plant: cost of CO_2 emissions included.
Note: The solid line shows the combinations of CO_2 prices and nuclear capital costs that would produce the same levelized cost for both plants. The combinations of natural gas prices and nuclear capital costs that would fall in the region above (below) or to the left (right) would produce levelized costs that are lower (higher) for the nuclear plant than for the gas plant. Real natural prices in 2020 were assumed to be about $7.00 per mmBtu. The carbon price is in 2009 dollars per metric ton.

power plants. (See Fig. 7.) These CO_2 prices are in the same range as ones used in a number of discussions. Thus, once the external costs related to CO_2 emissions from very efficient coal-fired power plants are considered, nuclear power could very well be economic relative to coal-fired power plants. However, using the higher discount rate, the price of CO_2 emissions would have to be somewhat higher (about $50–$60 per ton) before nuclear power would be competitive with combined-cycle natural gas–fired power plants. As was just noted, using the lower discount rate, nuclear power would be very marginally economic relative to gas-fired power plants even if the price of carbon were zero.

CONCLUSIONS

This entry attempted to answer the following question: Is nuclear power economic? Two major factors will influence the economics of nuclear power, and since there are major uncertainties with both of them, it is impossible to give an unqualified answer. The most obvious one is the cost of building a nuclear power plant. In a number of recent PUC rate cases, overnight nuclear power plant capital costs of around $4000 per kW were used. If these estimates are indicative of realized overnight costs, environmental considerations aside, then nuclear power would generally not be economic. The analysis in this entry suggests that overnight nuclear power plant capital costs would have to fall to between $2500 and $3000 per kW before nuclear power would be economic. The one exception to this conclusion is if a very low discount rate is used.

The second factor is the cost of the alternative, which would be either coal-fired power plants or combined-cycle natural gas–fired units. The major uncertainty here is the enactment of policies limiting CO_2 emissions. The economics of nuclear power would be greatly improved if all of the external costs related to global warming were included in the cost of generating electricity from fossil fuel–fired power plants. It must be noted that nuclear power has its own set of environmental costs in the form of nuclear waste disposal and decommissioning. Since these costs will be incurred over hundreds of years, because of discounting, the back-end costs are very small "today." Some have objected to discounting expenses incurred over very long time periods because this procedure represents a strong incentive to impose large costs on future generations. However, such equity considerations are outside the realm of economic analysis.

Lastly, there are probably at least two reasons why some utilities are interested in nuclear power even though the economics appears to be unfavorable. First, recent research using data from the United Kingdom found that nuclear power can be used as a hedge against volatile natural gas prices if the utility was subject to price regulation.[20] It is possible that utilities are using nuclear power as a hedge. This might partially explain why most of the utilities interested in nuclear power are located in states that have not been deregulated. Second, a number of utilities that are interested in nuclear power and their state-level regulators are assuming that eventually some type of explicit carbon price will be enacted. The analysis in the present entry does show that the economics of nuclear power is improved considerably when the cost of CO_2 emissions from fossil fuel–fired power plants is included.

ADDENDUM—THE IMPACT OF THE FUKUSHIMA NUCLEAR DISASTER ON THE COST OF NUCLEAR POWER

On March 11, 2011, as a result of a major earthquake and tsunami in Japan, the Fukushima nuclear power plant lost all of its on-site and off-site power, which is needed to operate the plant's safety systems. At least as of May 2011, it appears that the earthquake damaged the transmission facilities needed to supply off-site power, and the tsunami damaged all of the on-site power sources, including the emergency on-site backup diesel generators. As a result, without any source of power, the safety systems could not cool the reactor, and the fuel literally melted. The spent fuel is submerged in water in large storage pools, which cool the high-level waste. When the plant lost all power, and water could not cool the spent fuel, this resulted in a number of explosions. Because of these explosions and attempts to cool the reactor and spent fuel from off-site water sources, significant amounts of radiation were discharged into the air and water surrounding the plant.

As of May 2011, the owner of Fukushima and the Japanese government estimates that the plant will not be stabilized (cooled) until January 2012. Thus, it could take months or even years before authorities can determine what exactly happened to the plant. Indeed, it took 5 years before workers could enter the Three Mile Island plant to determine the exact nature of that accident. Consequently, at this point, it is impossible to determine with any degree of confidence what the impact of the Fukushima disaster will be on the cost of nuclear power. Nevertheless, it is possible that the disaster could affect the economics of nuclear power in a number of ways. First, there could be design changes to existing and new power plants that will increase capital costs, especially in the areas of backup power and spent fuel storage. Second, the disaster will clearly affect public acceptance of nuclear power, and the licensing hearings at the NRC could become more controversial, which would increase leadtimes. Third, in other research, I found evidence of a very small (1 to 2 percentage points) risk premium on the common stock of nuclear U.S. utilities resulting from the accident at Three Mile Island.[21] Over time, capital markets could have similar reactions to the Fukushima disaster. This would increase the cost of financing the construction and operation of a nuclear plant.

ACKNOWLEDGMENTS

I would like to thank my past and present colleagues at the Energy Information Administration for their many useful discussions with me about the economics of nuclear power and their assistance in my research in studying this subject. However, the views and opinions stated in this entry are the author's alone and do not represent the official position of the Energy Information Administration or the United States DOE. All errors are the sole responsibility of the author.

REFERENCES

1. Merrow, E.; Phillips, K.; Myers, C. *Understanding Cost Growth and Performance Shortfalls in Pioneer Process Plants*; The Rand Corporation: Santa Monica, CA, 1981.
2. Wiser, R.; Bolinger, M. *2008 Wind Technologies Market Report*; Lawrence Berkley National Laboratory: Berkley CA, 2009, available at http://www.nrel.gov/analysis/pdfs/46026.pdf (accessed April 2011).
3. *Annual Energy Outlook: 2004,* DOE/EIA-0383 (2004); U.S. Department of Energy, Energy Information Administration: Washington, DC, 2004.
4. *Report of the Interagency Task Force on Carbon Capture and Storage*, DOE/FE-0001 (2010); U.S. Department of Energy, Assistant Secretary for Fossil Energy: Washington, DC, 2010, available at http://fossil.energy.gov/programs/sequestration/ccstf/CCSTaskForceReport2010.pdf (accessed April 2011).
5. *Projected Costs of Generating Electricity: 2010 Update*; Nuclear Energy Agency-International Energy Agency, Organization of Economic Co-operation and Development: Paris, France, 2010.
6. Deutch, J.; Moniz, E.; Joskow, P. *The Future of Nuclear Power*; Massachusetts Institute of Technology: Cambridge, MA, 2003, available at http://web.mit.edu/nuclearpower/pdf/nuclearpower-full.pdf (accessed April 2011).
7. *World Energy Outlook*: *2008*; International Energy Agency: Paris, France, 2008.
8. Du, Y.; Parsons, J. *Update of the Cost of Nuclear Power*, Working Paper 09-004; Center for Energy and Environmental Research, Massachusetts Institute of Technology: Cambridge, MA, 2009.
9. McCabe, M. Principals, agents, and the learning curve: The case of steam electric power plant construction. *J.Ind. Econ.* **1996**, *20* (4), 240–270.
10. Cantor, R.; Hewlett, J. The economics of nuclear power: Further evidence of learning, economies of scale and regulatory effects. *Resour. Energy* **1988**, *10* (4), 315–335.
11. Hewlett, J. Why were the nuclear power plant cost and leadtime estimates so wrong?. In *Nuclear Power at the Crossroads*; Lowinger, T., Hinman, G., Eds.; University of Colorado Press: Bolder CO, 1994; 121–148.
12. Deutch, J.; Moniz, E. *The Future of Coal*; Massachusetts Institute of Technology: Cambridge, MA, 2007, available at http://web.mit.edu/coal/ (accessed April 2011).
13. *Analysis of the Total System Life Cycle Cost of the Civilian Radioactive Waste Management Program, Fiscal Year 2007*, DOE/RW-0591; U.S. Department of Energy, Office of Civilian Radioactive Waste Management: Washington, DC, 2008.
14. Portney. P.; Weyant, J. *Discounting and Intergenerational Equity*; Resources for the Future: Washington, DC, 1999.
15. Sumaila, U.; Walters, C. Intergenerational discounting: A new intuitive approach. *Ecol. Econ.* **2005**, *10* (2), 135–142.
16. Copeland, T.; Weston, F. *Financial Theory and Corporation Policy*, 3rd Ed.; Addison-Wesley: New York, 1988.
17. *Annual Energy Outlook, 2010*, DOE/EIA-0383 (2010); Energy Information Administration: Washington, DC, 2010.
18. Parry, I.; Williams, R. Is a carbon tax the only good climate policy. In *Resources*; Resources for the Future: Washington DC, Fall 2010, 176; 38–41, available at http://www.rff.org/resourcesno176/parry (accessed April 2011).
19. *Energy Market and Economic Impacts of the American Power Act of 2010*, SR/OIAF/2010-01; U.S. Department of Energy, Energy Information Administration: Washington, DC, 2010.
20. Roques, F.; Newbery, D. Nuclear power: A hedge against uncertain gas prices and carbon prices. *Energy J.* **2006**, *27* (4), 72–95.
21. Hewlett, J. *Investor Perceptions of Nuclear Power*, DOE/EIA-0446; Energy Information Administration, U.S. Department of Energy: Washington, DC, 1984.

Nutrients: Best Management Practices

Scott J. Sturgul
Outreach Program Manager, Nutrient and Pest Management Program, University of Wisconsin, Madison, Wisconsin, U.S.A.

Keith A. Kelling
Professor Emeritus, Department of Soil Science, University of Wisconsin—Extension, Madison, Wisconsin, U.S.A.

Abstract

Soil nutrients, like all agricultural inputs, need to be managed properly to meet the fertility requirements of crops without adversely affecting the quality of water resources. The nutrients of greatest concern relative to water quality are nitrogen (N) and phosphorus (P). Nitrogen not recovered by crops can add nitrate-N to groundwater through leaching. Surface water quality is the primary environmental concern with P, as runoff and erosion from fertile cropland add nutrients to water bodies that stimulate the excessive growth of aquatic weeds and algae. Best management practices for agricultural nutrients can vary widely from one region to another due to differences in cropping, topographic, environmental, and economic conditions. However, central to any nutrient management strategy would be an accurate assessment of nutrient need along with an accounting of on-farm nutrient resources such as manure, legumes, etc. The application of supplemental nutrients should be timed for maximum crop uptake and minimal chance of off-site movement. Nutrient management practices for optimizing crop production while protecting water quality are best summarized in a nutrient management plan that is regularly updated and tailored to the unique landscape characteristics and prevalent agricultural practices of individual farming operations.

INTRODUCTION

Soil nutrients, like all agricultural inputs, need to be managed properly to meet the fertility requirements of crops without adversely affecting the quality of water resources. The nutrients of greatest concern relative to water quality are nitrogen (N) and phosphorus (P).[1] Nitrogen not recovered by crops can add nitrate-N to groundwater through leaching. Nitrate is the most common groundwater contaminant found in the United States.[2–4] Nitrate levels that exceed the established U.S. drinking water standard of 10 ppm nitrate–N have the potential to adversely affect the health of infants and young livestock.[5] Furthermore, the movement of excessive nitrate to coastal estuaries has been linked to the development of hypoxic conditions affecting fisheries in the Gulf of Mexico.[6,7] Surface water quality is the primary environmental concern with P, as runoff and erosion from fertile cropland add nutrients to water bodies that stimulate the excessive growth of aquatic weeds and algae.[8] Of all crop nutrients, it is critical to prevent P from reaching lakes and streams since the biological productivity of aquatic plants and algae in fresh water environments is usually limited by this nutrient.[9] Consequences of increased aquatic plant and algae growth include reduced aesthetic and recreational value of lakes and streams as well as the seasonal depletion of water dissolved oxygen content, which may result in fish kills as well as other ecosystem disruptions.

Best management practices for agricultural nutrients can vary widely from one region to another due to differences in cropping, topographic, environmental, and economic conditions. With the variety of factors to consider, no single set of nutrient management practices can be recommended for all farms. Nutrient management practices for optimizing crop production while protecting water quality must be tailored to the unique conditions of individual farms. The following practices should be considered for any nutrient management strategy for optimizing both agricultural production and environmental protection.

NUTRIENT APPLICATION RATES

The most important management practice for environmentally—and economically—sound nutrient management is the application rate. Optimum nutrient application rates are identified through fertilizer response and calibration research for specific soils and crops. Economically optimum nutrient application rates provide maximum financial return, but as application rates near the economic optimum, the efficiency of nutrient use by the crop decreases, and the potential for loss to the environment increases.[10] Fig. 1 illustrates this concept for N applications to corn. Nutrient application above economically optimum rates (in this case, 160 lb N/acre) reduces profits and increases the likelihood of detrimental impact to the environment.

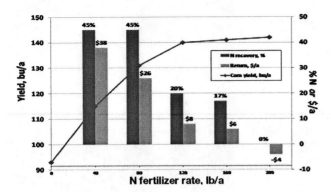

Fig. 1 Relationships between corn grain yield, economic return, and recovery of applied N.
Source: Bundy et al.[10]

As shown by many researchers, applications of N in excess of crop need can result in significantly higher N leaching losses.[11–15] A Wisconsin study[16] of the effects of various N fertilizer rates on nitrate-N leaching from several crop rotations found a direct relationship between nitrate-N loss by leaching and the amount of N applied in excess of crop needs (Fig. 2). Soil water nitrate-N concentration increased steadily as the amount of excess N increased, which strongly indicates a direct link between excessive N applications and the potential for nitrate-N loss to groundwater.

Because of the overall importance of nutrient application rates, accurate assessments of crop nutrient needs are essential for minimizing threats to water quality while maintaining economically sound production. Soil testing is the most widely used method to accurately estimate the existing fertility of soil as well as to determine the need for supplemental nutrients to meet the needs of crops. It has the further advantage of being performed prior to raising the crop, whereas other methods such as plant tissue analysis or optical scanners must have the target crop in place.

Optimum Nitrogen Applications

Most non-legume crops need supplemental N to improve crop yield and quality and to optimize economic return. It is imperative that N application rate recommendations accurately predict the amount of N needed to obtain profitable crop yields and minimize N losses to the environment. Application rate guidelines for N vary according to the crop to be grown, soil characteristics (including soil yield potential, organic matter, texture, etc.), and local climatic conditions.

Expected yield or yield goal estimates have been a primary input for determining grain N fertilizer recommendations for most of the United States since the 1970s.[17] However, research from several Midwest states over the past four decades shows that economic optimum N rates for corn are not best predicted using a yield parameter.[18–21]

More accurate N recommendations for corn and some other grain crops have been developed using the results of N rate response experiments conducted on the major soils of a given region. This soil-specific approach is based on corn yield response and associated economic return to incremental rates of N.[18,19] Two additional parameters influencing the economic optimum N application rate are the anticipated price of corn and the cost of fertilizer N. The data gathered from these experiments are the foundation for corn N recommendations throughout the upper Midwest United States.[22,23]

Additional Tests for Fine-Tuning Nitrogen Applications

The development of tests for assessing soil N levels provides additional tools for improving the efficiency of N fertilizer applications. These tests allow fertilizer recommendations to be adjusted to site-specific conditions that can influence N availability. Tests include the preplant soil profile nitrate test,[24] the pre–side-dress soil nitrate test,[25] plant analysis,[26,27] chlorophyll meters,[28] the end-of-season stalk nitrate test,[29] or the end-of-season soil nitrate test.[30]

Calibrated Soil Tests for Phosphorus and Potassium

In recent years, soil test recommendation programs for P, potassium (K), and other relatively immobile nutrients have tended to de-emphasize a soil buildup and maintenance

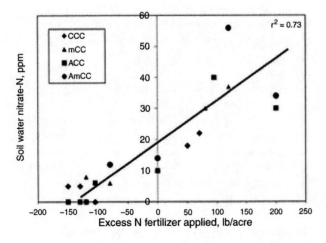

Fig. 2 Relationship between amount of excess N applied and soil water nitrate content for several cropping rotations (C = corn, A = alfalfa, m = manure).
Source: Andraski et al.[16]

philosophy in favor of a better balance between environmental and economic considerations by using a crop sufficiency approach.[23,31] These tests are calibrated by field experiments to obtain predictable crop yield responses. Such an approach adds extra emphasis to regular soil testing. It is recommended that soil tests be taken at least every 3 to 4 yr and more frequently on sandy and other soils of low buffering capacity.

Nutrient application recommendations based on soil test results can be accurate only if soil samples representative of the field of interest are collected. Samples that are unrepresentative of fields often result in recommendations that are misleading. Before collecting soil samples, relevant local guidance should be sought on the appropriate number samples to collect, as well as the methodology for collection. In addition, field history information should be provided with the soil samples in order to accurately adjust the fertility recommendations to account for nutrient credits from field-specific activities such as manure applications and legumes in the rotation.

Realistic Yield Goals

For many soil fertility programs, the recommendation of appropriate nutrient application rates is dependent on the establishment of realistic yield goals. Yield goal estimates that are too low will underestimate nutrient needs and can limit crop yield. Yield goal estimates that are too high will overestimate crop needs and result in soil nutrient levels beyond that needed by the crop, which in turn has the potential to increase nutrient contributions to water resources. Estimates should be based on field records and some cautious optimism—perhaps 10% above the recent 3 to 5 years average corn yield from a particular field. Yield goals reasonably higher than a multiyear average are suggested because annual yield variations due to factors other than nutrient application rates (primarily climatic factors) are often large.

Nutrient Credits

The integration of economic return and environmental quality protection requires that nutrients from all sources be considered. In the determination of supplemental fertilizer application rates, it is critical that nutrient contributions from manure, previous legume crops grown in the cropping rotation, and land-applied organic by-products are credited. Both economic and environmental benefits can result if the nutrient-supplying capacity of these nutrient sources is correctly estimated. Economically, commercial fertilizer application rates can often be reduced or eliminated entirely when nutrient credits are accounted. Environmentally, the prevention of overfertilization reduces potential threats to water quality.

Manure

Manure can supply crop nutrients as effectively as commercial fertilizers in amounts that can meet the total N, P, K, and sulfur need of many crops.[32] To utilize manure as a fertilizer resource, its application rate and nutrient-supplying capacity (i.e., plant-available nutrient content) need to be estimated.

Calibration of manure application equipment is key to estimating application rates. Calibration is a relatively easy task that can be done with platform scales or portable axle scales.[33] As a result of manure spreader calibration, an applicator will have a reasonable estimate of the manure application rate—provided the manure is uniformly applied across fields.

The most effective method for gauging the nutrient content of manure is to have samples analyzed by a commercial or university laboratory.[34] Large farm-to-farm variation can occur in manure nutrient content due to manure storage and handling techniques, livestock feed variations, or other farm management differences.[35] In instances when laboratory analysis is not convenient or available, estimates of crop nutrients supplied by animal manures can be made using published values for the average nutrient values of livestock manures common to a given state or region. These values are often provided by area universities. Note that not all the nutrients in manure are available to crops in the first year following application. When estimating the nutrient-supplying capabilities of manure applications, be certain that first-year crop-available nutrient content values are used to calculate the fertilizer value of the manure—not total nutrient content.

Legumes

Legume crops, such as alfalfa, clover, soybeans, and leguminous vegetables, have the ability to fix atmospheric N and convert it to a plant-available form. When grown in a rotation, some legumes can supply substantial amounts of N to a subsequent non-legume crop. For example, a dense stand of alfalfa can often provide most, if not all, of the N needed for a corn crop following it in a rotation.[36] An efficient nutrient management strategy needs to consider the N contributions of legumes to subsequent crops. The amount of legume N to credit varies regionally. Consult the local university extension service for appropriate recommendations.

Biosolids

The application of organic biosolids such as sewage sludge, whey, compost, or other organic wastes to cropland fields can be another source of potential crop nutrient credits. While the overall percentage of cropland acres receiving biosolids is relatively small when compared with manure or legumes, the nutrient contributions can be significant and should be accounted for prior to fertilizer applications. Special management and regulatory considerations pertain to the land application of these materials. Consult local regulations for further information.

TIMING OF NUTRIENT APPLICATIONS

The timing of application is a major consideration for the management of mobile nutrients such as N. For less mobile nutrients, application timing is not a major factor affecting water quality protection. However, nutrient applications on frozen sloping soils or surface applications prior to periods likely to produce runoff events should be avoided to prevent nutrient contributions to surface waters.

Nitrogen Applications

The period between application and crop uptake of N is an important factor affecting the efficient utilization of N by the crop and the potential for loss of N via leaching, denitrification, and other processes.[37,38] Loss of N can be minimized by supplying it just prior to the period of greatest crop uptake. However, several considerations, such as soil, equipment, labor, and fertilizer price and availability are involved in determining the most convenient, economical, and environmentally safe N fertilizer application period.

Fall Nitrogen Applications

The advantages and disadvantages of fall N fertilizer applications are commonly debated. An increased risk for N loss with fall applications needs to be weighed against the fertilizer price and time management advantages (greater window for fertilizer application and spring planting) that can be associated with fall-applied N. The agronomic concern with fall N applications is that losses between application and crop uptake the following growing season will lower recovery of N and reduce crop yield. The environmental concern with fall application is that the N lost prior to crop uptake will leach into groundwater. Fall to spring precipitation, soil texture, and soil moisture conditions influence the potential for fall-applied N losses. If a soil is wet in the fall, rainfall may cause either leaching of nitrate in coarse soils or denitrification of nitrate in heavy, poorly drained soils. Long-term studies indicate that fall applications on medium-textured soils are 10%–15% less effective than the same amount of N applied spring preplant.[37] For both agronomic and environmental reasons, fall applications of N fertilizers are not recommended on coarse-textured soils or on shallow soils over fractured bedrock. If fall applications are to be made on other soils, it is recommended that ammonium-N sources be used and that the applications be delayed until soil temperatures are below thresholds of biological activity (i.e., 50°F) in order to slow the conversion of ammonium to nitrate by soil organisms. If fall applications must be made when soil temperatures are higher than 50°F, a nitrification inhibitor should be used in conjunction with the N fertilizer.

Preplant Nitrogen Applications

Spring preplant applications of N are usually agronomically and environmentally efficient on medium-textured, well-drained soils. The potential for N loss prior to crop uptake on these soils is relatively low with spring applications. If spring preplant applications of N are to be made on sandy soils, ammonium forms of N treated with a nitrification inhibitor should be used. Likewise, nitrification inhibitors should be used if spring preplant N is applied to poorly drained soils. Use of nitrification inhibitors reduces the potential for N loss compared with preplant applications without them; however, side-dress or split applications can be more effective and cost efficient than preplant applications with nitrification inhibitors.

Side-Dress Nitrogen Applications

Side-dress applications of N to row crops during the growing season are effective on all soils with the greatest benefit on sandy or heavy-textured, poorly drained soils.[38,39] The greatest efficiency of side-dress N applications is achieved when the application of N occurs just prior to the period of rapid N uptake by crops. This results in a shorter period of exposure to potential losses of N from leaching or denitrification. Table 1 illustrates the higher yield and crop recovery of N on sandy soils with side-dress applications. In these trials, use of side-dress N applications improved average N recovery over preplant applications by 17%. The use of side-dress or delayed N applications on sandy soils is essential for minimizing N loss to groundwater since unrecovered N on these soils will be lost through leaching prior to the next growing season. Side-dress N applications may also be of benefit on shallow soils over fractured bedrock.

Table 1 Effect of rate and time of N application on corn yield and recovery of applied N on irrigated Plainfield sand.

N Rate (lb/a)	Relative yield increase[a]		N recovery	
	Preplant	Side-dress	Preplant	Side-dress
	--- (% over control) ---		--- (%) ---	
0	-	-	-	-
70	132	176	50	73
140	216	258	44	64
210	247	276	40	49
Average	197	237	45	62

[a] Side-dress treatments applied 6- weeks after planting.
Source: Bundy et al.[40]

Side-dressing N requires more management than pre-plant applications. To maximize efficiency, side-dress N applications must be properly timed to provide available N during the maximum N-uptake period for crops such as corn. An additional concern is that applications too late may result in lower yield and plant injury from root pruning and other physical damage.

Split Nitrogen Applications

Application of N fertilizer in several increments during the growing season can be an effective method for reducing N losses on sandy soils. However, a single well-timed side-dress application is often as effective as multiple applications.[41] Ideally, split applications supply N when needed by the crop and allow for N application rate adjustments based on early growing season weather or plant and soil tests. To be successful, the timing of application and placement of fertilizer materials are critical. Climatic factors, such as untimely rainfalls, may interfere with application schedules.

A common method for split N applications is via irrigation systems (fertigation). Multiple applications of fertilizer N can be injected into the irrigation water and applied to correspond with periods of maximum plant uptake. However, fertigation should not be relied upon as a sole method of applying N in a cropping season for the following reasons: 1) adequate rainfall during the early growing season could delay or eliminate the need for irrigation and subsequently delay fertilizer applications; and 2) leaching can result if N is applied through an irrigation system at a time when the crop does not need additional water.

Nitrification Inhibitors

Nitrification inhibitors are used with ammonium or ammonium-forming N fertilizers to improve N efficiency by slowing the conversion of ammonium to nitrate, thereby reducing the potential for losses of N that occur in the nitrate form (i.e., leaching and denitrification). The effectiveness of a nitrification inhibitor depends greatly on soil type, time of the year applied, N application rate, and soil moisture conditions that exist between the time of application and the time of N uptake by plants. Research has shown that the use of nitrification inhibitors on medium- and fine-textured soils with fall N applications, on poorly drained soils with fall or spring N applications, or on coarse-textured, irrigated soils with spring preplant N applications has the potential to increase corn yield and total crop recovery of N.[42] However, as noted earlier, side-dress applications alone are likely to be more effective on many of these soils. Fall applications of N with an inhibitor on sandy soils are not recommended. The cost of using nitrification inhibitors versus other strategies for minimizing N losses needs to be considered in an overall economic analysis of a grower's crop production system.

Controlled-Release Nitrogen Fertilizers

Controlled- (or slow-) release N fertilizers release their nutrients at gradual rates that, in theory, allow for increased plant uptake of N while minimizing losses due to leaching and volatilization. Although commonly used in high-value applications such as horticultural crops and turf, these fertilizer products have not been economical for widespread use in major agricultural crops due to relatively high cost and low crop prices. This may be changing due to cheaper controlled-release fertilizer products, higher N prices, and the demand for greater environmental protection.

Controlled-release fertilizers are broadly divided into uncoated and coated products. Uncoated products rely on inherent physical characteristics, such as low solubility, for their slow release. Coated products consist mostly of quick-release N sources surrounded by a barrier that prevents the N from releasing rapidly into the environment. Similar to earlier-developed materials, such as sulfur-coated urea, urea formaldehyde, and isobutylidene diurea, the critical concept is timing the release of N to correspond with crop need. Greenhouse and field studies have shown that polymer-coated urea can increase crop yield and N use efficiency in soils prone to leaching losses.[43–45]

Other Nutrient Applications

The timing of P and K applications is less critical as these nutrients are generally strongly held in the soil. Some water quality concerns may exist where P is surface broadcast on fields that have not been tilled as this results in less fertilizer–soil contact. Fall applications of K are not recommended on organic soils (peat or muck) since these soils do not effectively hold K against leaching losses. Considerations for P and K applications on other soils include the amount of material to be applied, the size of the application window, and the resources and available equipment.

NUTRIENT PLACEMENT

The placement of nutrients on cropland can influence their effectiveness as well as their potential ability to affect water quality. The concern with N placement focuses mainly on preventing N loss through ammonia volatilization. Applications of N in the form of urea or N solutions need to be incorporated into the soil by rainfall, irrigation, injection, or tillage. The amount of volatilization loss that occurs with surface N applications depends on factors such as soil pH, temperature, moisture, and crop residue. Minimal volatilization losses of N can be expected if surface applications are incorporated within 3 to 4 days—provided temperatures are low (<50°F) and the soil is moist.[46] A late spring or summer application should be incorporated within a day or two because higher temperatures and the chance of longer periods without rainfall could lead to significant N volatilization losses.

The placement of P nutrient sources can directly influence the amount of P transported to lakes and streams by surface runoff. If P fertilizer is broadcast on the soil surface and not incorporated, the amount of P in runoff water can rise sharply and have a greater potential impact on surface water quality than soil surfaces where P was incorporated.[47,48] Phosphorus is strongly bound to soil particles; however, adequate soil-to-P contact must occur to allow for adsorption. Incorporation by tillage or subsurface band placement of fertilizers is a very effective means of achieving this contact. To avoid enriching surface waters with soil nutrients, it is recommended that annual fertilizer applications for row crops, such as corn, be band-applied near the row as starter fertilizer at planting. Annual starter applications of P (and K) can usually supply all of the P required for corn. This practice reduces the chance for P enrichment of the soil surface and reduces potential P in cropland runoff. Band fertilizer placement ideally enriches about 20% of the plow layer volume.[49] If large broadcast P fertilizer applications are needed to increase low soil P levels, these applications should be followed by incorporation as soon as possible.

VARIABLE-RATE FERTILIZER TECHNOLOGIES

Nutrient availability in any field varies both spatially and temporally.[50,51] To address this, site-specific (precision agriculture) management techniques and tools have developed over the past 20 years. Global positioning systems and geographic information systems, along with crop nutrient-sensing systems, have allowed agricultural management decisions to be made with greater detail and precision. As a result, producers are able to manage nutrient variability within fields at increasingly finer resolution than in the past with inherent improvements in nutrient use efficiency, crop yields, and environmental stewardship.[52,53] While progress has been made, widespread acceptance of these technologies has been limited by costs, sampling requirements, required technical inputs, and the limited ability of current equipment to physically deliver nutrients at adjustable rates corresponding to field variability.[54]

MANURE MANAGEMENT

Manure applications to cropland provide nutrients essential for crop growth, add organic matter to soil, and improve soil physical and biological conditions. The major environmental concerns associated with manure application are related to its potential for overloading soils with nutrients if manure applications exceed crop needs and to direct runoff from manured fields to surface waters.

Manure Application Rates

Manure is often applied to cropland at rates that attempt to meet the N need of the intended crop. This strategy maximizes potential manure application rates and is preferred if the amount of land available for application is limited. In addition, a N-based strategy is usually time and labor efficient. A consequence of this approach can be the buildup of P in soils to excessive levels, which in turn increases the potential for P losses via runoff and soil erosion.[55–57] For example, the plant-available N and P contents of dairy manure are about equal. The N need of corn, however, is greater than the crop's need for P. A consequence of applying manure at rates to meet the N need of corn is that P applications will exceed crop removal (Fig. 3). The result is a buildup of P in cropland soils.[58] Long-term manure applications have elevated the soil P level of many soils above the range necessary for optimum crop growth.[59]

If maximum manure nutrient efficiency is the goal, rates of application need to be based on the nutrient present at the highest level relative to crop needs. For corn, this nutrient would be P. Manure application rates that meet the P requirement of corn are typically much lower than N-based rates. Subsequently, additional N will need to be supplied from other nutrient sources (Fig. 4). A P-based manure application strategy results in lower manure application rates, but it is less likely to elevate soil test P values. It has the disadvantages of being less efficient with respect to labor, energy, time, and economics.[60] A P-based strategy for manure applications requires spreading manure on a much larger acreage than is required for a N-based manure application.

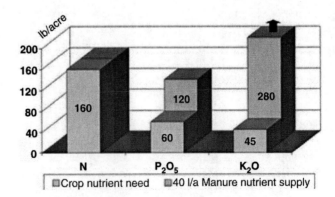

Fig. 3 Nitrogen-based manure application strategy for corn. Note: Standard convention in the United States is to express soil P and K levels in elemental form (i.e., ppm of P and K) while expressing P and K fertilizer application rates and analysis in oxide form (i.e., lb/acre of P_2O_5 and K_2O).
Source: *Understanding soil phosphorus.*[58]

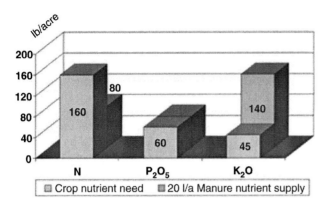

Fig. 4 Phosphorus-based manure application strategy for corn. **Source:** Sturgul and Bundy.[58]

Manure Application Timing

Manure application timing is an important management practice for minimizing nutrient contributions to surface waters. Manure should not be spread on sloping lands any time a runoff-producing event is likely. Unfortunately, runoff-producing events are difficult to predict, and the elimination of manure applications to sloping lands is seldom a practical option. For farmers in the upper Midwest, the period of major concern with manure spreading is late fall, winter, and early spring months. Manure applied on frozen ground has an increased likelihood for running off to surface waters due to snow melt and/or early spring rains.

If winter applications of manure must be made, the risk for nutrient loss should be minimized to the greatest extent possible. Manure applications to frozen fields should be limited to those of slight slope (generally less than 6%) that are preferably covered in previous crop residue, roughly tilled on the land contour, or protected from upslope runoff.[61,62] More steeply sloping fields that are intended for manure applications need to have soil and water conservation practices in place. Manure should not be applied to frozen soils on steeply sloping fields (generally 12% or greater).[61,62]

Site Considerations for Manure Applications

In addition to the slope criteria discussed previously, other site considerations for manure applications should include existing soil fertility levels, soil depth, soil texture, soil erodibility factors, and field proximity to water bodies.

In many areas, general recommendations, or even specific regulations, exist for reducing or eliminating manure applications to fields that have elevated levels of soil P. Numerous studies have found a correlation between elevated levels of soil P and the amount of P carried in runoff from agricultural fields.[55–57] As soil P levels become elevated, crop rotations should be diversified to include crops with a high demand for P (such as alfalfa), which can draw down soil P. When soil P levels become excessively high, manure applications should be discontinued until soil test levels decrease.[61,62] Soil runoff and erosion control practices such as residue management, conservation tillage, contour farming, and others are strongly recommended on soils with P levels in excess of crop needs. When planning manure applications, prioritize those fields low in soil fertility (particularly P) and strive to distribute manure across the available fields to avoid the excessive buildup of soil nutrients that results from repeated applications to the same sites.

Most soils have a high capacity for assimilating nutrients from manure. However, in locations of highly permeable or shallow soils over fractured bedrock, groundwater issues associated with the application of manure can result. Manure should not be applied to shallow soils (generally less than 10 in.) over fractured bedrock. Incorporation of manure shortly after application on moderately shallow soils will allow for increased soil adsorption of nutrients. Manure should not be applied to frozen, shallow soils.[61,62]

Movement of nitrate-N to groundwater is more likely on excessively drained (sandy) soils. Manure applications in early fall on these soils where no actively growing crop is present to utilize the N may allow for the conversion of organic N to nitrate, which is then subject to leaching losses. Manure should not be applied to sands or loamy sands in the fall when soil temperatures are greater than 50°F, unless there is an overwintering cover crop present to utilize the N. In the absence of a cover crop, manure applications to sandy soils should take place when soil temperatures are below 50°F.[61,62] The conversion of ammonium-N to nitrate-N is significantly reduced at soil temperatures below 50°F.

The main site characteristics affecting nutrient contributions to surface waters are those that affect soil runoff and erosion. These include slope, soil erodibility and infiltration, rainfall, cropping system, and the presence of soil conservation practices. Site-related management practices dealing specifically with manure placement to protect surface water include the following: 1) not applying manure (or other nutrients) to grassed waterways, terrace channels, open surface drains, or other areas where surface flow may concentrate; 2) restricting manure applications within designated floodplain or stream and lake setback distances; and 3) prohibiting manure applications in these areas when soils are frozen or saturated.

Manure Storage

During periods when suitable sites for land application of manure are not available (i.e., soils are frozen or seasonally saturated), the use of manure storage facilities is recommended. Storage facilities allow manure to be stored until conditions permit land application and incorporation.

In addition, storage facilities can minimize nutrient losses resulting from volatilization of ammonia and be more convenient for calibrated land applications. With the exception of those systems designed to filter leachate, storage systems should retain liquid manure and prevent runoff from precipitation on stored waste. It is imperative that manure storage facilities be located and constructed such that the risk of seepage to groundwater is minimized. With regards to maximum nutrient efficiency and water quality protection, it is critical that appropriate application techniques and accurate nutrient crediting of the manure resource are utilized when the storage facility is emptied.

Livestock Feed Management

On individual farms and in many areas of agricultural livestock production, inputs of P in feed and fertilizer exceed outputs of P contained in crop and animal produce leaving the farm or region (Fig. 5). This is especially true in areas where concentrated livestock production is prevalent.[63] The National Research Council[64] estimated that only 30% of the fertilizer and feed P imported onto farms is exported in crops and animal produce. The surplus 70% of the P is remaining on farm and leading to the excessive enrichment of soil P.

Livestock feed inputs have been found to be a major contributing factor to on-farm P surpluses.[65–67] Soil buildup of P is accelerated when livestock are overfed P in dietary rations. Phosphorus excretion in manure is directly related to the level of P intake.[68,69] High P in livestock dietary intake directly correlates with higher bypass P as reflected in elevated P content of livestock manure (Table 2). Overuse of dietary P supplements accelerates the buildup of soil test P to excessive levels and increases the potential for P losses from manured fields (Fig. 6).[70] Another consequence is an increase in land required for application of manure if P-based rate limitations are to be met.

Additional dietary P management options involve plant and livestock genetic manipulation for more effective

Table 2 Annual phosphorus fed to and excreted by a lactating cow.

Dietary P Level (%)	Supplemental P	Fecal P
	---------lb/cow/year---------	
0.35	0	42
0.38	5.5	47
0.48	23	65
0.55	36	78

Source: Powell et al.[65]

manure-P management from monogastric animals (nonruminants such as swine and poultry). All these techniques attempt to reduce the P content in manure of monogastric animals by improving the efficiency with which the animal extracts P from feed. An increase in P uptake by the animal from feed grains will reduce the amount of P that bypasses the animal via the manure. Increasing animal uptake of P can allow manure application rates to continue due to a slower buildup of soil P because of the reduced P content of the manure.

Reducing the phytate level of feed grains by use of low-phytate, high-available-phosphate (HAP) varieties is one feed management strategy for lowering manure P. In corn and most feed grain plants, P is stored in the phytate form, which is largely unavailable to nonruminant livestock. As a consequence, swine and poultry feed is routinely supplemented with P. The unutilized phytate-P from the plant is excreted by the animals, resulting in manure that is enriched in P content.[71] Low-phytate grain hybrids that will store P in the available phosphate form rather than as phytate are available.[72] Corn has been the crop most extensively developed. Phosphorus availability to monogastrics from low-phytate corn is about two to three times higher than from normal corn.[73] Subsequently, the P content of manure is reduced. Plant breeders are working to

Fig. 5 On-farm phosphorus cycle.
Source: Stargul and Bundy.[58]

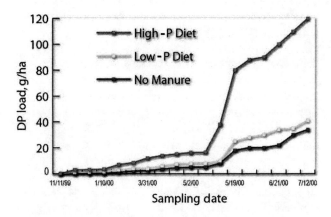

Fig. 6 Cumulative dissolved phosphorus (DP) load in natural runoff from manures differing in P content.
Source: Ebeling et al.[70]

incorporate the low-phytate trait into commercially competitive hybrids.

Another option for reducing the P content of manure from monogastric livestock is the use of commercially produced enzymes as a feed supplement. Phytase enzymes are capable of releasing phytate-P from plants into animal-available forms. Phytase enzymes occur naturally in some microorganisms, plants, and animals, such as ruminants (cattle). Monogastric animals lack phytase and can only poorly utilize the P reserves in many grains.[71] By adding phytase enzymes to nonruminant animal feed, the efficiency of P uptake during digestion can be increased with an associated reduction in the P content of monogastric manure.[74] In a study by Baxter et al.,[75] where phytase additives were combined with low-phytate corn, a 60% reduction in P excretion was recorded. While the phytase enzyme has been shown to decrease the need for mineral P additions, the economics of its use as a routine feed additive need to be considered.[63]

IRRIGATION WATER MANAGEMENT

When to apply irrigation water and how much to apply depend on crop, growth stage, and soil properties. Over-irrigation, or rainfall on recently irrigated soils, can leach nitrate and other contaminants below the root zone and into groundwater. Accurate irrigation scheduling that considers soil water holding capacity, crop growth stage, evapotranspiration, rainfall, and previous irrigation to determine the timing and amount of irrigation water to be applied can reduce the risk of leaching losses.

SOIL CONSERVATION PRACTICES

Land-use activities associated with agriculture can increase the potential for runoff and soil erosion. Consequences of cropland erosion include loss of fertile topsoil, accelerated eutrophication and sedimentation of surface waters, destruction of fish and wildlife habitat, and decreased recreational and aesthetic value of surface waters. The key to minimizing nutrient contributions to surface waters is to reduce the amount of runoff and eroded sediment reaching them. Numerous management practices for the control of runoff and soil erosion have been researched, developed, and implemented. Runoff and erosion control practices range from changes in agricultural land management (cover crops, diverse crop rotations, conservation tillage, contour farming, contour strip cropping, etc.), to the installation of structural devices (buffer strips, diversions, grade stabilization structures, grassed waterways, terraces, etc.), Substantial emphasis is currently being placed on the benefits and installation of vegetative buffer strips along riparian corridors, which can reduce the sediment and nutrient content of runoff waters reaching them.[76] The width of an effective buffer strip is often debated but varies according to land slope, type of vegetative cover, watershed characteristics, etc.[77]

Recently, the U.S. Department of Agriculture (USDA) reported that cropland conservation practices installed and applied by agricultural producers are indeed reducing sediment, nutrient, and pesticide losses from farm fields.[78] Key findings from this report on the effects of conservation practices on cropland in the Upper Mississippi River Basin illustrate the following: 1) suites of conservation practices work better than single practices; 2) targeting critical acres improves practice effectiveness significantly; and 3) the beneficial effect of conservation practices is greatest on the most vulnerable acres (highly erodible land and soils prone to leaching). The study also found that the use of soil conservation practices within the Upper Mississippi River Basin has reduced sediment loss by 69% compared with estimates of soil loss from the same land area without conservation practices in place. The study's authors concluded that the improved use of cropland conservation practices along with the consistent use of nutrient management practices would further reduce the risk of nutrient movement from fields to rivers and streams within the basin.

FARM NUTRIENT MANAGEMENT PLANS

A farm nutrient management plan is a dynamic, regularly updated strategy for obtaining the maximum economic return from both on- and off-farm nutrient resources in a manner that protects the quality of nearby water resources. While a plan is specific for an individual farming operation, there are common components to all nutrient management plans. These include the following.

Soil Test Reports

Complete and accurate soil tests are the starting point of any farm nutrient management plan. All cropland fields must be tested or have been tested recently to ensure the best possible fertilizer recommendations.

Assessment of On-Farm Nutrient Resources

The amount of crop nutrients supplied to cropland fields from on-farm nutrient resources such as manure, legumes, and organic by-products needs to be determined.

Nutrient Crediting

Once on-farm nutrient resources are determined, commercial fertilizer applications need to be reduced to reflect these nutrient credits. This action can lower fertilizer expenditures and protect water quality by eliminating nutrient applications in excess of crop need. Management skills come into play when determining nutrient credits. For example, to properly credit manure-supplied nutrients, both the manure application rate and the crop-available nutrient content of the manure must be known. To credit the N available to crops following alfalfa, the condition of the alfalfa stand as well as last cutting date need to be known.

Manure Inventory

Perhaps the most challenging aspect of developing and implementing a farm nutrient management plan is the advanced planning of manure applications to cropland fields. This involves estimating the amount of manure produced on a farm and then planning specific manure application rates for individual cropland fields.

Manure Spreading Plan

A major component of any nutrient management plan for livestock operations will deal with a manure spreading plan. The amount of manure the farm produces has to be applied to fields in a manner that considers both environmental and agronomic consequences. Manure applications rates should not exceed crop nutrient need as identified by the soil test report. The nutrient management plan should prioritize those fields that would benefit the most from the manure-supplied nutrients while posing little threat to water quality. The nutrient management plan must also identify fields with manure spreading restrictions.

The seasonal timing of manure applications to cropland should be identified in the nutrient management plan. The timing of planned manure applications will depend upon each farm's manure handling system. Manure application periods for farms with manure storage will be significantly different than those for farms that haul manure on a daily basis.

Consistency with Farm Conservation Plans

A nutrient management plan should be consistent with a farm's soil conservation plan. Operations participating in federal farm programs usually are required to have a soil conservation plan. Conservation plans contain needed information on planned crop rotations, slopes of fields (which are important when planning manure applications), and the conservation measures required to maintain soil erosion rates at tolerable levels.

Compliance with Nutrient Management Standards, Rules, and Regulations

Nutrient management plan criteria and requirements are often defined in standards developed by federal, state, or local government (i.e., the USDA–Natural Resources Conservation Service's Nutrient Management Standard 590). A nutrient management plan complying with a standard(s) is often a requirement for participation in federal and state government farm programs.

CONCLUSION

The previous text provides a brief summary of general nutrient management practices for crop production. This is not a complete inventory but, rather, an overview of soil fertility management options available to growers for protecting water quality and improving farm profitability. Best management practices for agricultural nutrients can vary widely from one region to another due to differences in cropping, topographic, environmental, and economic conditions. However, central to any nutrient management strategy would be an accurate assessment of nutrient need along with an accounting of on-farm nutrient resources such as manure, legumes, etc. The application of supplemental nutrients should be timed for maximum crop uptake and minimal chance of off-site movement. Nutrient management practices for optimizing crop production while protecting water quality are best summarized in a nutrient management plan that is regularly updated and tailored to the unique landscape characteristics and prevalent agricultural practices of individual farming operations.

REFERENCES

1. United States Environmental Protection Agency (US EPA). Nutrient Pollution: The Problem. http://www.epa.gov/nutrientpollution/problem/index.html. April 23, 2012.
2. Spalding, R.F.; Exner, M.E. Occurrence of nitrate in ground water—A review. J. Environ. Qual. **1993**, *22*, 392–402.
3. Blodgett, J.E.; Clark, E.H. Fertilizers, nitrates, and groundwater: An overview. In Proceedings of the Colloquium of Agrichemical Management to Protect Water Quality; Board of Agriculture, National Research Council: Washington, DC, 1986.
4. Madison, R.J.; Brunett, J.O. 1984. Overview of the occurrence of nitrate in groundwater of the United States. In *U.S. Geological Service National Water Summary, Water-Supply Paper 2275*; U.S. Government Printing Office: Washington, DC, 1984; 93–105.
5. United States Environmental Protection Agency (US EPA). *National Primary Drinking Water Regulations*, EPA pub. no. 816-F-09-004, http://water.epa.gov/drink/contaminants/upload/mcl-2.pdf (accessed March 6, 2012).
6. Rabalais, N.N.; Wiseman, W.I., Jr.; Turner, R.E. Comparison of continuous records of near-bottom dissolved oxygen from the hypoxic zone along the Louisiana coast. Estuaries **1994**, *14*, 850–861.
7. Goolsby, D.A.; Battaglin, W.A. *Nitrogen in the Mississippi Basin—Estimating sources and predicting flux to the Gulf of Mexico*, USGS Fact Sheet 135-00; USGS: Washington, DC, 2000, http://ks.water.usgs.gov/pubs/fact-sheets/fs.135-00.pdf (accessed May 14, 2012).

8. Parry, R. Agricultural phosphorus and water quality: A U.S. Environmental Protection Agency perspective. J. Environ. Qual. **1998**, *27*, 258–261.
9. Correll, D.L. The role of phosphorus in the eutrophication of receiving waters: A review. J. Environ. Qual. **1998**, *27*, 261–266.
10. Bundy, L.G.; Andraski, T.W.; Wolkowski, R.P. Nitrogen credits in soybean–corn crop sequences on three soils. Agron. J. **1993**, *85*, 1061–1067.
11. Delgado, J.A.; Mosier, A.R.; Valentine, A.W.; Schimel, D.S.; Parton, W.J. Long-term N studies in a catena of the shortgrass steppe. Biogeochemistry **1996**, *32*, 41–50.
12. Follett, R.F.; Keeney, D.R.; Cruise, R.M., Eds. *Managing Nitrogen for Groundwater Quality and Farm Profitability.* Soil Science Society of America: Madison, WI, 1991.
13. Newhould, P. The use of nitrogen fertilizer in agriculture. Where do we go practically and ecologically? Plant Soil **1989**, *115*, 297–311.
14. Randall, G.W.; Huggins, D.R.; Russelle, M.P.; Fuchs, A.J.; Nelson, W.W.; Anderson, J.L. Nitrate losses through subsurface tile drainage in conservation reserve program, alfalfa, and row crop systems. J. Environ. Qual. **1997**, *26*, 1240–1247.
15. Randall, G.W.; Iraqavarapu, T.K.; Schmitt, M.A. Nutrient losses in subsurface drainage water from dairy manure and urea applied to corn. J. Environ. Qual. **2000**, *29*, 1244–1252.
16. Andraski, T.W.; Bundy, L.G.; Brye, K.R. Crop management and corn nitrogen rate effects on nitrate leaching. J. Environ. Qual. **2000**, *29*, 1095–1103.
17. Meisinger, J.J. Evaluating plant available nitrogen in soil-crop systems. In *Nitrogen in Crop Production*; Hauck, R.D. et al., Eds.; ASA, CSSA, and SSSA: Madison, WI, 1984; 391–416.
18. Vanotti, M.B.; Bundy, L.G. An alternative rationale for corn nitrogen fertilizer recommendations. J. Prod. Agric. **1994**, *7*, 243–249.
19. Vanotti, M.B.; Bundy, L.G. Corn nitrogen recommendations based on yield response data. J. Prod. Agric. **1994**, *7*, 249–256.
20. Blackmer, A.M.; Binford, G.D.; Morris, T.; Meese, B. Effects of rates of nitrogen fertilization on corn yields, nitrogen losses from soils, and energy consumption. In *1991 Progress Report of the Integrated Farm Management Demonstration Program*; Iowa State University Pm-1467, 1992; 2.1–2.6.
21. Nafziger, E.D.; Sawyer, J.E.; Hoeft, R.G. Formulating N recommendations for corn in the corn belt using recent data. In Proceedings of the North Central Extension–Industry Soil Fertility Conference, Des Moines, IA, Nov. 17–18, 2004.
22. Sawyer, J.E.; Nafziger, E.D.; Randall, G.; Bundy, L.G.; Rehm, G.; Joern, B. *Concepts and Rationale for Regional Nitrogen Rate Guidelines for Corn*; Iowa State University Extension pub. PM 2015, 2006.
23. Laboski, C.A.M.; Peters, J.B.; Bundy, L.G. *Nutrient Application Guidelines for Field, Vegetable, and Fruit Crops in Wisconsin*; University of Wisconsin Extension pub. A2809, 2006.
24. Bundy, L.G.; Meisinger, J.J. Nitrogen availability indices. In *Methods of Soil Analysis: Biochemical and Microbial Properties*, Monogram 5; Weaver, R.W., Ed.; Soil Science Society of America: Madison, WI, 1994; 951–984
25. Magdoff, F.R.; Ross, D.; Amadon, J. A soil test for nitrogen availability to corn. Soil Sci. Soc. Am. J. **1984**, *48*, 1301–1304.
26. Kalra, Y.P., Ed. *Handbook of Reference Methods for Plant Analysis*. CRC Press, Inc; Boca Raton, Florida, 1998; 320 pp.
27. Jones, J.B. *Laboratory Guide for Conducting Soils Tests and Plant Analysis*. CRC Press, Inc; New York, Washington, D.C., 2001; 384 pp.
28. Varvel, G.E.; Schepers, J.S.; Francis, D.D. Chlorophyll meter and stalk nitrate techniques as complementary indices for residual nitrogen. J. Prod. Agric. **1997**, *10*, 147–151.
29. Blackmer, A.M.; Mallarino, A.P. *Cornstalk Testing to Evaluate Nitrogen Management*; Iowa State University Extension Pub. PM1584, 1996.
30. Sullivan, D.M.; Cogger, C.G. *Post-Harvest Soil Nitrate Testing for Manured Cropping Systems West of the Cascades*; Oregon State University Extension pub EM 8832-E, 2003, http://extension.oregonstate.edu/catalog/pdf/em/em8832-e.pdf (accessed May 14, 2012).
31. Sparks, D.L., Ed. *Advances in Agronomy*. Academic Press; San Diego, California, 1999; Vol. 67, 320 pp.
32. Midwest Plan Service. *Manure Characteristics–Manure Management Systems Series*; MWPS-18-S1D, Iowa State University; Ames, Iowa, 2004.
33. Nutrient and Pest Management (NPM) Program. *Know How Much You Haul*; University of Wisconsin Extension; 2 pp, http://ipcm.wisc.edu/Publications/tabid/54/Default.aspx (accessed May 14, 2012).
34. Peters, J.B., Ed. *Recommended Methods Of Manure Analysis*; University of Wisconsin Extension Pub. A3769, 2003.
35. Peters, J.B.; Combs, S.M. Variability in manure analysis as influenced by sampling and management. In Proceedings of the 1998 Wisconsin Fertilizer Dealers Update Meetings, University of Wisconsin, Madison, WI, Dec. 1998.
36. Bundy, L.G.; Kelling, K.A.; Good, L.W. *Using Legumes as a Nitrogen Source*; University of Wisconsin Extension; 8 pp, http://ipcm.wisc.edu/Publications/tabid/54/Default.aspx (accessed May 14, 2012).
37. Bundy, L.G. Review—Timing nitrogen applications to maximize fertilizer efficiency and crop response in conventional corn production. J. Fert. Issues **1986**, *3*, 99–106.
38. Vitosh, M.L. *Nitrogen Management Strategies for Corn Producers*; Michigan State University Cooperative Extension Service Bull. WQ06; 1985; 6 pp.
39. Randall, G.W. Improved N management can alleviate groundwater pollution. Solutions **1986**, *30* (5), 44–49.
40. Bundy, L.G.; Kelling, K.A.; Schulte, E.E.; Combs, S.M.; Wolkowski, R.P.; Sturgul, S.J. Nutrient management: practices for Wisconsin corn production and water quality protection; University of Wisconsin Extension Pub. A3557, 1994; 27 pp.
41. Bundy, L.G. Timing nitrogen applications to maximize fertilizer efficiency and crop response in conventional corn production. J. Fert. Issues **1986**, *3*, 99–106.
42. Nelson, D.W.; Huber, D. Nitrification inhibitors for corn production. In *National Corn Handbook 55*; Iowa State University Extension, 2001; 6 pp.

43. Wang, F.L.; Alva, A.K. Leaching of nitrogen from slow release urea sources in sandy soils. Soil Sci. Soc. Am. J. **1996**, *60*, 1454–1458.
44. Delgado, J.A.; Mosier, A.R. Mitigation alternatives to decrease nitrous oxide emissions and urea-nitrogen loss and their effects on methane flux. J. Environ. Qual. **1996**, *25*, 1105–1111.
45. Shoji, S.; Delgado, J.A.; Mosier, A.R.; Miura, Y. Use of controlled release fertilizers and nitrification inhibitors to increase nitrogen use efficiency and to conserve air and water quality. Comm. Soil Sci. Plant Anal. **2001**, *31*, 1051–1070.
46. Bundy, L.G. *Understanding Plant Nutrients: Urea—Its Use and Problems*; University of Wisconsin Cooperative Extension Service Bull. A2989, 1985; 4 pp.
47. Baker, J.L.; Laflen, J.M. Effects of corn residue and fertilizer management on soluble nutrient runoff losses. Trans. ASAE **1982**, *21*, 893–898.
48. Mueller, D.H.; Wendt, R.C.; Daniel, T.C. Phosphorus losses as affected by tillage and manure application. Soil Sci. Soc. Am. J. **1984**, *48*, 901–905.
49. Bundy, L.G. *Corn Fertilization*; University of Wisconsin Cooperative Extension Service Bull. A3340, 1998.
50. Legg, J.O.; Meisinger, J.J. Soil nitrogen budgets. In *Nitrogen in Agricultural Soils*, Agronomy Monogram 22; Stevenson, F.J., Brewner, J.M., Hanck, R.D., Keeney, D.R., Eds; American Society of Agronomy/Soil Science Society of America: Madison, WI, 1982; 503–566.
51. Jolkela, W.E.; Randall, G.W. Corn yield and residual nitrate as affected by time and rate of nitrogen application. Agron. J. **1989**, *81*, 720–726.
52. Fergusson, R.B.; Hergert, G.W.; Schepers, J.S.; Crawford, C.A.; Cahoon, J.E.; Peterson, T.A. Site-specific nitrogen management of irrigated maize: Yield and soil residual nitrate effect. Soil Sci. Soc. Amer. J. **2002**, *64*, 544–553.
53. Berry, J.R.; Delgado, J.A.; Khosla, R.; Pierce, F.J. Precision conservation for environmental sustainability. J. Soil Water Conserv. **2003**, *58*, 332–339.
54. Masek, T.J.; Schepers, J.S.; Mason, S.C.; Francis, D.D. Use of precision farming to improve application of feedlot waste to increase use efficiency and protect water quality. Commun. Soil Sci. Plant Anal. **2001**, *33*, 1355–1369.
55. Bundy, L.G.; Andraski, T.W.; Powell, J.M. Management practice effects on phosphorus losses in runoff in corn production systems. J. Environ. Qual. **2001**, *30*, 1822–1828.
56. Pote, D.H.; Daniel, T.C.; Sharpley, A.N.; Moore, P.A.; Edwards, D.R.; Nichols, D.J. Relating extractable soil phosphorus to phosphorus losses in runoff. Soil Sci. Soc. Am. J. **1996**, *60*, 855–859.
57. Sharpley, A.N. Identifying sites vulnerable to phosphorus loss in agricultural runoff. J. Environ. Qual. **1995**, *24*, 947–951.
58. Sturgul, S.J.; Bundy, L.G. *Understanding Soil Phosphorus*; University of Wisconsin Cooperative Extension Service Bull. A3771, 2004.
59. Sims, J.T. Environmental soil testing for phosphorus. J. Prod. Agric. **1993**, *6*, 501–506.
60. Bosch, D.J.; Zhu, M.; Kornegay, E.T. Net returns from microbial phytase when crop applications of swine manure are limited by phosphorus. J. Prod. Agric. **1998**, *11*, 205–213.
61. USDA–National Resources Conservation Service (NRCS). *Nutrient Management Code 590 Conservation Practice Standard*; National Resources Conservation Service: Washington, DC, 2005.
62. Madison, F.W.; Kelling, K.A.; Massie, L.; Ward-Good, L. *Guidelines for Applying Manure to Cropland and Pastures in Wisconsin*. University of Wisconsin Cooperative Extension Service Bull. A3392, 1998.
63. Daniel, T.C.; Sharpley, A.N.; Lemunyon, J.L. Agricultural phosphorus and eutrophication: A symposium overview. J. Environ. Qual. **1998**, *27*, 251–257.
64. National Research Council (NRC). *Nutrient Requirements of Dairy Cattle*, 7th Revised Ed; National Academy Press: Washington, DC, 2001.
65. Powell, J.M.; Wu, Z. Satter, L.D. Dairy diet effects on phosphorus cycles of cropland. J. Soil Water Conserv. **2001**, *56* (1), 22–26.
66. Satter, L.D.; Wu, Z. Reducing manure phosphorus by dairy diet manipulation. In Proceedings of the 1999 Fertilizer, Aglime and Pest Management Conference 38, Madison, WI; 1999; 183–192.
67. Sharpley, A.N.; Daniel, T.C.; Sims, J.T.; Lemunyon, J.; Parry, R. *Agricultural Phosphorus and Eutrophication*, USDA-ARS pub. no. ARS-149, 1999; 37 pp.
68. Khorasani, G.R.; Janzen, R.A.; McGill, W.B.; Kenelly, J.J. Site and extent of mineral absorption in lactating cows fed whole crop cereal grain silage or alfalfa silage. J. Animal Sci. **1997**, *75*, 239–248.
69. Metcalf, J.A.; Mansbridge, R.J.; Blake, J.S. Potential for increasing the efficiency of nitrogen and phosphorus use in lactating dairy cows. Anim. Sci. **1996**, *62*, 636.
70. Ebeling, A.M.; Bundy, L.G.; Andraski, T.W.; Powell, J.M. Dairy diet phosphorus effects on phosphorus losses in runoff from land applied manure. Soil Sci. Soc. Am. J. **2002**, *66*, 284–291.
71. Doerge, T.A. Low-phytate corn: A crop genetic approach to manure-P management. In Proceedings of the 1999 Wisconsin Fertilizer, Aglime, and Pest Management Conference, Madison, WI; 1999; 175–183.
72. Raboy, V.; Young, K.; Gerbasi, P. Maize low phytic acid (lpa) mutants. In the 4th International Congress of Plant Molecular Biology: Abs. No. 1827; 1994.
73. Ertl, D.S.; Young, K.A.; Raboy, V. Plant genetic approaches to phosphorus management in agricultural production. J. Environ. Qual. **1998**, *27*, 299–304.
74. Kornegay, E.T. Nutritional, environmental, and economic considerations for using phytase in pig and poultry diets. In *Nutrient Management of Food Animals to Enhance and Protect the Environment*; CRC Press, Inc; New York, 1996; 277–302.
75. Baxter, C.A.; Joern, B.C.; Adeola, L.; Brokish, J.E. Dietary P management to reduce soil P loading from pig manure. *Annual Progress Report to Pioneer Hi-Bred International, Inc*; 1998; 1–6.
76. Daniels, R.B.; Gilliam, J.W. Sediment and chemical load reduction by grass and riparian filters. Soil Sci. Soc. Am. J. **1996**, *60*, 246–251.
77. Schmitt, T.J.; Bosskey, M.G.; Hoagland, K.D. Filter strip performance and processes for different vegetation, widths, and contaminants. J. Environ. Qual. **1999**, *28*, 1479–1489.
78. USDA-NRCS. *Conservation Effects Assessment Project (CEAP): Assessment of the Effects of Conservation Practices on Cultivated Cropland in the Upper Mississippi River Basin*; 146 pp, http://www.nrcs.usda.gov/wps/portal/nrcs/detail/national/technical/nra/?&cid=nrcs143_014161 (accessed May 14, 2012).

Nutrients: Bioavailability and Plant Uptake

Niels Erik Nielsen
Plant Nutrition and Soil Fertility Laboratory, Department of Agricultural Sciences, Royal Veterinary and Agricultural University, Frederiksberg, Denmark

Abstract
Entry of nutrient elements into plants and, therefore, into the food web of human beings, depends on the capability of the soil to release and maintain a concentration bigger than the minimum concentration (c_{min}) of elements in the soil solution at the root surface, and on the uptake capacity of the plant roots. Root-induced rhizosphere processes influence nutrient availability, maintenance of the nutrient concentration in the soil solution and plant uptake. The entry into plants and the associated flux of a nutrient element toward root surface may be controlled by a number of rate-limiting and/or rate-determining processes. One of the rate-limiting processes may be the flux by diffusion from the solid constituents in the soil via the soil solution to the nutrient-absorbing cell membrane in root tissue near the root surface. The goal of good agronomy is always to obtain a useful and sustainable modification of rate-limiting and/or rate-determining processes in the soil plant atmosphere system, aiming at high yields and qualities of the crops. This goal requires identification of the rate-limiting and/or rate-determining steps, an understanding of their dynamics, and knowledge of how to obtain appropriate modifications of these steps.

SOIL PLANT SYSTEM

The movement of any nutrient element, M, from solid soil constituents to the root surface, and its entry into plants, can be divided into a sequence of processes (steps), as illustrated in Fig. 1, which also indicates major agronomical actions used to improve nutrition and growth of crop plants. The ↔ denotes solid-phase processes slowly approaching equilibrium, or microbial-mediated net mineralization of N, S and P, for example. Also denoted are the source/sink processes by which diffusible nutrients are being produced or removed by chemical, physical, and biological transformation processes. In Fig. 1, L denotes ligands that are any dissolved solute reacting with M to form ML, which are organic complexes and ion pairs dissolved in the soil solution. The occurrence of L and, therefore, of ML, increases the total concentration (M + ML) and mobility of the nutrient element. The ⇌ denotes reversible processes which are spontaneously approaching equilibrium. Depending on ion species, ion concentration at the root surface, and plant age, the symbol ⇌ denotes processes that may be irreversible. The irreversible processes are always rate limiting and/or rate determining, whereas reversible processes may be rate limiting, only. Processes 2 and 3 are in the vicinity of the soil particle, whereas process 7 is in the vicinity of the root. Processes 4, 5, and 6 are transport processes by mass-flow and diffusion due to water uptake and nutrient uptake (process 8) by cell membranes of root cells near the root surface or root hairs. Process 9 is the nutrient translocation in the plant. Process 10 is the plant growth that also integrates the absorbed nutrient into the plant tissue. Processes 8, 9, and/or 10 create the concentration (electro-chemical) gradients for irreversible net flux of nutrients from the soil-soil solution system into the plants. Hence, at any time, the rate-determining processes (Fig. 1) are then either the release of the nutrient into the pool of plant available nutrient (source/sink processes in the soil), or the nutrient uptake into the roots, its translocation/circulation in the plant, and/or the rate by which the nutrient is built into new tissue. Mass-flow and diffusion may be rate limiting only, and not rate determining. Usually only a small fraction of the plant-available nutrients is dissolved in the soil solution. This implies that the bioavailability of nutrients to plant roots is governed by several soil properties including, for example, the characteristics of process 2 in Fig. 1 and the possibilities for movement via soil solution to the root surface by mass-flow and diffusion. The concept of a *bioavailable nutrient* can then be defined as a nutrient element that is present in a pool of diffusible (available) nutrients which are close enough to arrive at water- and/or nutrient-absorbing root surfaces during a period of 10 days, for example. This seems to fit with the observation that most of the depletion zones of slowly moving nutrients, such as phosphorus, are created during the first 10 days after root growth into a new soil volume unit.[1] The bioavailable quantity of a nutrient in the soil is affected by at least five different groups of processes as indicated in Table 1.

DIFFUSION

Diffusion is the net movement of a solute or a gas from a region that has a higher concentration, to an adjacent

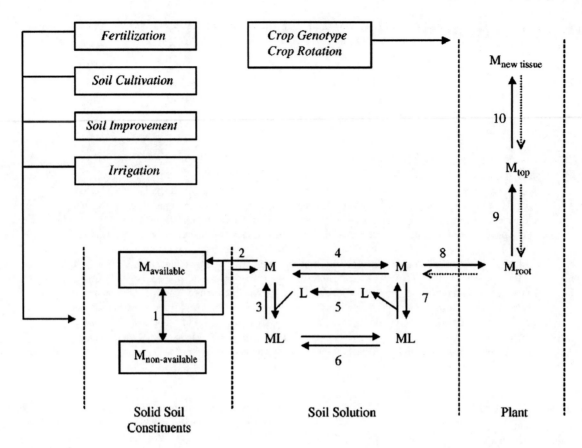

Fig. 1 Flux of a nutrient element in the soil plant atmosphere system and agronomy actions; nutrient element (M), ligands (L), reversible and irreversible processes.
Source: Nielsen.[22]

region that has a lower concentration. Diffusion is a result of the random thermal motions of molecules in the considered solids, solution, or air. The net movement caused by diffusion is a statistical phenomenon because the probability of the molecules' movement from the concentrated to the diluted region is greater than vice versa. Fick (1855) was one of the first to examine diffusion on a quantitative basis. The basic equation Eq. (1) to express diffusion, e.g., to a root is today known as Fick's first law of diffusion:

$$F = -D\frac{\partial C}{\partial r} \qquad (1)$$

in which F is the diffusive flux (mol cm^{-2}s^{-1} of a nutrient in the r-direction normal to the root cylinder. The driving

Table 1 Processes and factors involved in nutrient transfer from soil to plant roots.

Process	Factors
Release, mineralization, and dissolvement of the nutrient in the soil solution	Chemical and physical properties of the solid phases, temperature, soil water content, and activity of the microbial biomass
Root development	Root length, distribution of roots in the root zone, root morphology, root hairs, rate of root growth, and root surface contact area with soil solution
Solute movement by mass flow and diffusion to roots	Transpiration rate (w_o), concentration of the nutrient in the soil solution (c_b), effective diffusion coefficient (D_e), nutrient buffer power of the soil (b)
Rhizosphere processes, increasing the rate of nutrient release in the soil	Depletion of the soil solution for nutrients by the roots; root exudates as protons, reducing agents, chelates, organic anions, enzymes; modification of microbial activity; mycorrhizal and Rhizobium symbioses
Nutrient uptake	Concentration of the nutrient at the root surface (c_o); transport kinetic parameters of nutrient uptake by the roots ($\tilde{I}_{max}$, Km and c_{min})

Source: Nielsen.[22]

force (the gradient in the electro-chemical potential) is, in most cases, approximated by the concentration gradient

$$\frac{\partial C}{\partial r}(\text{mol cm}^{-4})$$

and D is the diffusion coefficient ($cm^2 s^{-1}$). They way to describe diffusion processes mathematically under various conditions has been presented by Jost[2] and Crank.[3] Great contributions to our understanding of solute movement in the soil root system by mass-flow and diffusion have been given or reviewed by Nye and Tinker[4,6] and Barber.[5] Recently, Willigen and colleagues[7] reviewed some aspects of the modeling of nutrient and water uptake by plant roots.

Nutrient Movement by Mass Flow and Diffusion from Soil to Plant Roots

Nutrients bound to the solid-soil phase are virtually immobile in the sense of its movements to roots. The nutrient has to be released into the soil solution as indicated in Fig. 1. Furthermore, contact between the root and nutrient-absorbing membranes in the root tissue near the root surface and the soil solution is a prerequisite for nutrient uptake. Contact to nutrient pools can be brought about by two means: 1) by growth of roots to the sites where nutrient pool are located (root interception); and 2) by movement of the nutrient from the bulk (the pool) of the soil to the root surface. Even so, nutrients may at any time move over a certain distance in the soil solution and cell wall before they reach the outside cell membrane of a root hair or a root cortical cell for uptake. The mechanisms for these transports are mass flow and diffusion.[4,8] The driving force for the net movement of nutrients (Fig. 1) is the water and the selective nutrient uptake by the plant root, creating a concentration gradient (dc/dr). The general equation of continuity (mass-balance) used to describe movements in a direction normal to a root cylinder at radial distance r and time t, may partly be developed from Eq. (1), extended and expressed as

$$\left[b\frac{\partial c}{\partial t}\right]_r = -\left[\frac{1}{r}\frac{\partial r F_T}{\partial r}\right]_t + U_{r,t} \quad (2)$$

in which

- U is the production/consumption term (mole $cm^{-3} s^{-1}$) at r (radial distance from the center of the root) and t (time).
- b is the buffer power (dC/dc) in which C is the total concentration of diffusible solute in the soil and c is the concentration of solute in the soil solution.
- F_T is total net flux of solute by mass flow and diffusion (mole $cm^{-2} s^{-1}$).

$$F_T = F_m + F_d$$

where $F_m = wc$ in which w is the flux of soil solution in the direction of the root ($cm^3 cm^{-2} s^{-1}$) and c is the nutrient concentration of soil solution (mole cm^{-3}). The expected rates of water flux at the root surface are 0.2–1 10^{-6} cm s^{-1}.[5] The flux, F_d, by diffusion can be expressed by Flick's first law

$$F_d = -D_c b \frac{dc}{dr}$$

The $b = dC/dc$ is the soil buffer power defined previously. The C is the sum of the amount of nutrient in the soil solution and the amount of adsorbed nutrient that is able to replenish the nutrient in the soil solution spontaneously. Hence, b is the parameter mediating the effects of the soil chemical conditions on nutrient uptake by plants. D_e denotes the effective diffusion coefficient in the soil. D_e differs between media, but it can be related to the diffusion coefficient D_o for the nutrient in free soil solution. The influences of soil on diffusion, and thereby the relation between D_e and D_o, can be expressed by Eq. (3).[9]

$$D_e = D_o \theta f / b \quad (3)$$

where θ is the volumetric water content expressed as a fraction, and f is the impedance factor that essentially allows for the increase in the actual diffusion distance because of the tortuous pathway of water filled soil pores and water films. The volumetric water content that allows a reasonable root activity is between 0.1 and 0.4. The value of f increases with increase in water content,[10] whereas the buffer power remains constant with changes in soil moisture at the same bulk density.[11] It has been observed[12] that the relation between f and θ can be expressed empirically by $f = 1.58\theta - 0.17$ for $\theta >$ about 0.11. From this it may be estimated that D_e decreases about 18 times if θ decreases from 0.40 to 0.15. Hence D_e is the parameter mediating the effects of soil moisture, soil chemical, and soil physical conditions on diffusion in soil.

Almost all studies on solute movement in the soil plant system neglect U in Eq. (2) because our understanding of the biology caused by root-induced processes and its effects on production or consumption of available nutrients is incomplete as yet.

To solve Eq. (2) for a given soil plant system is a complicated process, and in most cases, difficult or even impossible because of the lack of information on the soil root interactions and root behavior. The method for obtaining analytical and numerical solutions of Eq. (2) under a number of often simplified soil plant conditions has been summarized.[5–7] However, illustration of the importance of diffusion for the bioavailability of nutrients in soils may be based on Eq. (4)

$$\Delta r = \sqrt{2 D_e t} \quad (4)$$

Table 2 Expected effective diffusion coefficients of some nutrients in soil at field capacity of water content (e.g., $\theta = 0.40$), and estimated bioavailability as a fraction of diffusible (available) nutrient [Eqs. (4) and (5)] at a root density ($L_v = 5$ cm^{-2}) of roots without root hairs (Mean root radius, $r = 0.01$ cm; Time, $t = 10$ days).

Element	D_e^a (cm^2 s^{-1})	Bioavailable nutrient as a fraction of available nutrient (V, cm^3)	
		$\theta = 0.40$	$\theta = 0.15$
Nitrate	1*10^{-6}	27.558	1.574
Potassium	1*10^{-7}	2.847	0.180
Boron	1*10^{-7}	2.847	0.180
Magnesium	1*10^{-8}	0.314	0.026
Calcium	1*10^{-8}	0.314	0.026
Phosphorus	1*10^{-9}	0.042	0.006
Manganese	1*10^{-9}	0.042	0.006
Molybdenum	1*10^{-9}	0.042	0.006
Zinc	1*10^{-9}	0.042	0.006
Iron	1*10^{-10}	0.008	0.003

^{a}Values obtained from Barber[5] and Nielsen.[22]

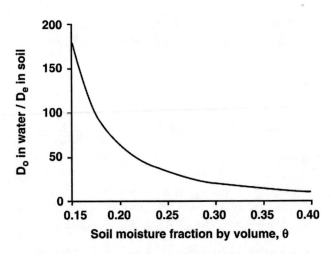

Fig. 2 Effects of soil and soil moisture on nutrient element mobility.

in which Dr is the average distance of diffusion; e.g., in a direction normal to a root. The mathematics behind Eq. (4) has been presented by Jost.[2] Based on Eq. (4), the equivalent soil volume (V in cm^3) of soil depleted for diffusible (available) nutrients—*the quantity of bioavailable nutrients*—can then be estimated as follows [Eq. (5)] for roots without root hairs:

$$V = \pi(\Delta r + r_o)^2 L_v \quad (5)$$

in which Δr is estimated from Eq. (4), r_o is the root radius and L_v is the root density in cm cm^{-3} of soil. The data in Table 2 show the expected effective diffusion coefficient of a number of plant nutrients in soil and corresponding influences on the nutrient bioavailability; in addition, the data show how a decrease of the soil moisture from $\theta = 0.40$ to $\theta = 0.15$ affects the bioavailability at a root density of 5 cm^{-2} of roots without root hairs. It may be calculated from $f = 1.58\theta - 0.17$ and Eq. (3) that the diffusive flux decreases by a factor of $18 = D_e^{\theta=0.40}/D_e^{\theta=0.15}$. At field capacity of water content, the expected, effective diffusion coefficient of nitrate in soil is 10^{-6} cm^2 s^{-1}. This is almost 10 times slower than in pure water. Hence, a pored media, such as soil, physically decreases the possibility for solute movement with a factor of nearly 10. As the soil dries out, this factor increases as illustrated in Fig. 2. Apart from nitrate and chlorine, nutrient elements are adsorbed more or less to the solid soil constituents. This is the main cause of the decrease of the diffusion coefficients below 10^{-6} cm^2 s^{-1}. The diffusion coefficient of phosphorus is as low as 10^{-9} cm^2 s^{-1} mainly because approximately 0.1% of the diffusible (available) phosphorus is dissolved in the soil solution only. This has a large effect on the bioavailability of the plant-available quantities of the various nutrient elements in soil as illustrated in columns 3 and 4 of Table 2. The V-values ≥ 1 indicate that the root at a density of 5 cm^{-2} is able to deplete all the available nutrient as seen for nitrate, even under dry conditions, whereas only 4% of the available phosphorus is bioavailable inside a period of 10 days. If the soil dries out to 1 indicate that the root at a density of 5 cm^{-2} is able to deplete all the available nutrient as seen $\theta = 0.15$, the bioavailability decreases to only 0.6%. This illustrates that the decrease of soil moisture may create nutrient deficiency even in soil with high phosphorus fertility. However, phosphorus uptake is increased by the activity of root hairs (discussed in the following).

IMPORTANCE OF ROOT HAIRS

Root hairs are outgrowths from specialized root epidermal cells (trichoblasts). Root hair length, diameter, and number per unit length of root, vary among plant species and among genotypes within the same species.[5,6,13] Frequency and size of root hairs are affected by many environmental factors, as well. In nature, the length of root hairs vary from 0.01–0.15 cm, the radius varies from 0.0005–0.002 cm, and the number per unit of length varies from 100–1000 per cm root. The importance of root hairs for phosphorus uptake has been demonstrated directly in the laboratory[14,15] and under field conditions.[13] It is reasonable to assume that the clusters of root hairs' outer tips form a fairly well-defined cylinder to which phosphorus diffuses, on average, a distance Δr in 10 days, and that root hair density and its period of function are long enough to withdraw the entire available nutrient in the soil penetrated by root hairs. The bioavailability of phosphorus, for example, as affected by root hair length, can then be estimated from the following extension of Eq. (10):

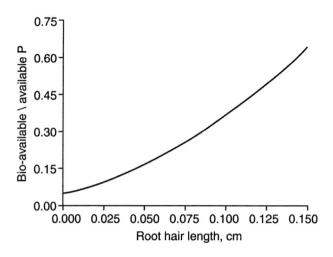

Fig. 3 The effect of root hair length on phosphorus bioavailability. Soil moisture ($\theta = 0.4$); root density 5 cm^{-2}; root radius 0.01 cm.

$$V = \pi(\Delta r + \sigma + r_o)^2 L_v \qquad (6)$$

in which σ is the root hair length in cm.

Fig. 3 illustrates that the bioavailability of phosphorus increases exponentially with the root hair length. Hence, root hairs play a very important role for the bioavailability of nutrients having a low effective diffusion coefficient (D_e) in soil.

BOUNDARY CONDITIONS AND NUTRIENT ENTRY

If depletion zones around the roots do not overlap, the solute concentration converges to the solute concentration c_b in the bulk solution, at which $F_T = F_m + F_d = 0$. The boundary conditions at the surface of the root are

$$F_T = F_m + F_d = \alpha c_o \qquad (7)$$

in which α (cm s^{-1}) is the root-absorbing power defined by Tinker and Nye[6] and c_o is the concentration of solute at the root surface. It can be learned from Eq. (7) that the actual concentration (c_o) at the root surface and, therefore, the rate of nutrient flow per unit length of root, is determined by the ratio F_T/α. The kinetics of net uptake of nutrients[5,16–18] may be expressed by:

$$\overline{In}L^* = \frac{L^*\overline{I}_{max}(c_o - c_{min})}{K_m + c_o - c_{min}} \qquad (8)$$

in which L^* is the root length per unit of plant biomass, $\overline{I}_{max}$ (mole cm^{-1} s^{-1}) is the mean maximal net influx, Km (mole cm^{-3}) is the Michaelis–Menten factor, c_o is the concentration of the nutrient at the root surface, and c_{min} is the nutrient concentration at which $\overline{In} = 0$. The values of the parameters $\overline{I}_{max}$, Km and c_{min} vary according to the plant nutrient, temperature, and plant species/genotype and plant age. Furthermore, kinetics of nutrient uptake by roots may be influenced by ion interactions. Determined values of L^*, $\overline{I}_{max}$, Km and c_{min} for uptake of several nutrients by several plant species or genotypes, obtained under conditions in which the rate-determining step of nutrient uptake was located in the roots, has been noted.[19] The data show that the values of L^*, $\overline{I}_{max}$, Km and c_{min} vary considerably among nutrients and among plant species and genotypes. This illustrates the efficiency by which these plants utilize soil as a source of nutrients. It is possible from $F_T = F_m + F_d = \alpha c_o$ and Eq. (8) to develop how α varies at varying solute concentration at the root surface by:

$$\alpha = \frac{\overline{I}_{max}(c_o - c_{min})}{2\pi r_o c_o (Km + c_o - c_{min})} \qquad (9)$$

Figs. 4a and 4b illustrate the variation of α for phosphorus uptake at low concentration (c_o) at the root surface of

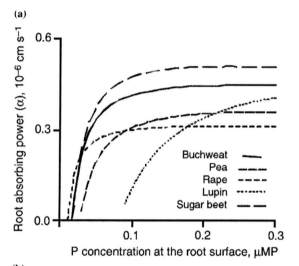

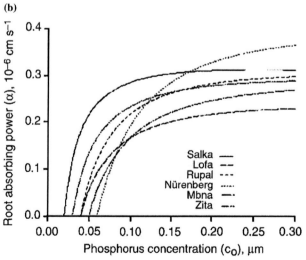

Fig. 4 Absorption power (α) at varying phosphorus concentration at the root surface of some plant species (a) and barley genotypes (b).
Source: Nielsen[19] and Nielsen.[22]

some plant species and barley genotypes. Hence, the root-absorbing power (α) varies also at low concentration (c_o) of solute at the root surface. This implies that phosphorus uptake at low P concentration is more under the control of the plant parameters determining the size of α than under the control of P diffusion in the soil, whereas at the range of c_o at which α has achieved its maximum, uptake is controlled by diffusion.

CONCLUSIONS

Even though the movement—and the main factors affecting the movement—of nutrient elements to root by mass-flow and diffusion is well known, the effect of soil conditions on crop growth is still not properly understood. It is obvious that the big variation (Table 2) of the effective diffusion coefficient (D_e), caused mainly by the variation of the soil chemistry of the various nutrient elements, has a large impact on the bioavailability of nutrient elements. The mobility of phosphorus and micronutrients is so low in most soils that the soil exploited by root hairs is the main source of these elements. The root-induced modifications to the soil in the rhizosphere would then have a considerable impact on the efficiency by which plants use the rhizosphere soil as a source of nutrients. The understanding of how root-induced processes accelerate solute movement and the transformation of non-available nutrients to bioavailable nutrients is increasing.[6,20,21] Root hair length and root-induced processes appear to vary between genotypes of our crop plants.[13] Hence, improvement of the efficiency by which plants use soil as a source of nutrients seems to be a possibility by targeted plant breeding.

REFERENCES

1. Gahoonia, T.S.; Raza, S.; Nielsen, N.E. Phosphorus depletion in the rhizosphere as influenced by soil moisture. Plant and Soil **1994**, *159*, 213–218.
2. Jost, W. *Diffusion in Solids, Liquids and Gasses*; Academic Press: New York, 1960.
3. Crank, J. *The Mathematics of Diffusion*, 2nd Ed.; Clarendon Press: Oxford, 1975.
4. Nye, P.H.; Tinker, P.B. *Solute Movement in the Soil-Root System*; Blackwell: Oxford, 1977.
5. Barber, S.A. *Soil Nutrient Bioavailability*, 2nd Ed.; Wiley: New York, 1995.
6. Tinker, P.B.; Nye, P.H. *Solute Movement in the Rhizosphere*; Oxford University Press: Oxford, 2000.
7. Willigen, P.; Nielsen, N.E.; Claassen, N.; Castrignanò, A.M. Modelling water and nutrient uptake. In *Root Methods a Handbook*; Smit, A.L., Bengough, A.G., Engels, C., Noordwijk, M., Pellerin, S., Geijn, S.C., Eds.; Springer: Berlin, 2000.
8. Barber, S.A. A diffusion and mass-flow concept of soil nutrient availability. Soil Science **1962**, *93*, 39–49.
9. Nye, P.H. The measurement and mechanism of ion diffusion in soil. I. the relation between self-diffusion and bulk diffusion. J. Soil Sci. **1966**, *17*, 16–23.
10. Rowell, D.L.; Martin, M.W.; Nye, P.H. The measurement and mechanisms of ion diffusion in soils. III. The effect of moisture content and soil solution concentration on the self-diffusion of ions in soils. J. Soil Sci. **1967**, *18*, 204–222.
11. Bhadoria, P.B.S.; Classen, J.; Jungk, A. Phosphate diffusion coefficient in soil as affected by bulk density and water content. Z. Pflanzenernaehr. Bodenk. 1991, 154, 53–57.
12. Barraclough, P.B.; Tinker, P.B. The determination of ionic diffusion coefficients in field soils. 1. Diffusion coefficient in sieved soils in relation to water content and bulk density. J. Soil Sci. **1981**, *32*, 225–236.
13. Gahoonia, T.S.; Nielsen, N.E.; Lyshede, O.B. Phosphorus acquisition of cereal cultivars in the field at three levels of P fertilization. Plant and Soil **1999**, *211*, 269–281.
14. Barley, K.P.; Rovira, A.D. The influence of root hairs on the uptake of phosphorus. Commun. Soil Sci. Plant Anal. **1970**, *1*, 287–292.
15. Gahoonia, T.S.; Nielsen, N.E. Direct evidence on the participation of phosphorus (P32) uptake from soil. Plant and Soil **1998**, *198*, 147–152.
16. Classen, N.; Barber, S.A. A method for characterizing the relation between nutrient concentration and flux into roots of intact plants. Plant Physiology **1974**, *54*, 564–568.
17. Nielsen, N.E. A transport kinetic concept for ion uptake by plants. III. Test of the concept by results from water culture and pot experiments. Plant and Soil **1976**, *45*, 659–677.
18. Nielsen, N.E.; Barber, S.A. Differences among genotypes of corn in the kinetics of P uptake. Agronomy Journal **1978**, *70* (5), 695–698.
19. Nielsen, N.E. Bioavailability, cycling and balances of nutrient in the soil plant system. In *Integrated Plant Nutrition Systems*; Fertilizer and Plant Nutrition Bulletin 12; Dudal, R., Roy, R.N., Eds.; FAO: Rome, 1995; 333–348.
20. Jungk, A.; Classen, N. Ion diffusion in the soil-root system. Adv. Agron. **1997**, *61*, 53–110.
21. Hensinger, P. How do plant roots acquire mineral nutrients? chemical processes involved in the rhizosphere. Adv. Agron. **1998**, *64*, 225–265.
22. Nielsen, N.E. Bioavailability of nutrients in soil. In *Roots and Nitrogen in Cropping Systems of the Semi-Arid Tropics*; Ito, O., Johansen, C., Adu-Gyamfi, J.J., Katayama, K., Kumar Rao, J.V.D.K., Rego, T.J., Eds.; International Crop Research Institute for the Semi-arid Tropics: India, 1996; 411–427.

Nutrient–Water Interactions

Ardell D. Halvorson
U.S. Department of Agriculture (USDA), Fort Collins, Colorado, U.S.A.

Abstract
Water is a major factor in nutrient availability to plants. It is the vehicle through which nutrients move through soil to access plant roots for uptake. Nutrients move via mass flow and diffusion in soil water to the root surface. Root interception is the third way in which plants obtain soil nutrients, as root hairs develop and contact the soil particles or solution.

INTRODUCTION

Water is a major factor in nutrient availability to plants.[1–4] It is the vehicle through which nutrients move through soil to access plant roots for uptake. Nutrients move via mass flow and diffusion in soil water to the root surface. Root interception is a third way in which plants obtain soil nutrients as root hairs develop and contact the soil particles and/or solution.

WATER AND NUTRIENT AVAILABILITY

In mass flow, nutrient ions are transported with water flow to the root as the plant absorbs water for transpiration. Many mobile nutrients, such as calcium (Ca), magnesium (Mg), nitrate-N (NO_3-N), and sulfate (SO_4), are transported to the root by mass flow. Diffusion of nutrients to the plant root occurs as ions move from high-concentration areas to low-concentration areas in the soil solution. Phosphorus (P) and potassium (K) are two nutrients that move by diffusion.

If soil water becomes limiting, as it frequently does under dryland or rainfed conditions, nutrient availability to plants can be affected.[5] Water is held as a film around soil particles. As the water content of the soil decreases, the thickness of the film decreases. Most plant nutrients are readily available when the soil is near field capacity, which is about the water content of the wet soil after two days of rain has saturated it and free drainage has ceased. Nutrient availability is at a minimum as the soil water content approaches the permanent wilting point, which is the water content at which plant roots cannot extract water from the soil. As soil water content diminishes, some less-soluble nutrients may precipitate out of the soil solution and become unavailable to plants. However, these minerals will dissolve and become available once again as the soil is rewetted. Thus, soil water content influences nutrient availability and plant growth.

Micronutrients are generally supplied to plant roots by diffusion in soil. Therefore, low soil moisture conditions will reduce micronutrient uptake. Plants require smaller quantities of micronutrients to optimize productivity than macronutrients such as P; thus, drought stress effects on micronutrient deficiency are not as serious as for P. However, iron (Fe) and zinc (Zn) deficiencies are frequently associated with high soil moisture conditions.[2]

Soil water content is an important factor in microbial activity in soils. Soil microbial activity is important in the breakdown of organic plant and animal residues, which release nutrients such as N and P for plant uptake. Microbial activity tends to be greatest when soil water is near field capacity with soil temperatures ranging from 25°C to 35°C. As soils dry, microbial activity decreases and lowers the rate of nutrient release from soil organic matter.[6,7]

NUTRIENT AND WATER USE EFFICIENCY

Adequate levels of plant nutrients are needed to optimize rooting depth and water extraction from the soil.[2,3,5] Healthy plants tend to root deeper into the soil profile, using more of the soil water in the root zone. Thus, plants not only need adequate water to optimize yield potential, but also require an adequate level of nutrients to allow the crop to take advantage of the available water supplies. Under dryland conditions, the crop will often use all of the available water (precipitation plus soil water in the root zone) during the growing season. Application of N and P fertilizers will frequently increase crop yields, thus increasing crop water use efficiency (WUE). Water use efficiency is the amount of crop produced per unit of available water from precipitation, soil, and irrigation. The influence of N fertilization on WUE of winter wheat, corn, and sorghum in a dryland wheat-corn or sorghum-fallow rotation is shown in Fig. 1.

When plant-available water is limited, overapplication of N can also result in reduced grain yields owing to increased vegetative growth and water use in the early growth

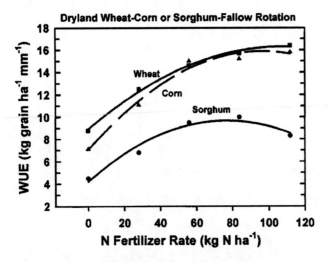

Fig. 1 Water use efficiency of wheat, corn, and sorghum as a function of N fertilizer rate in a dryland wheat-corn or sorghum-fallow rotation near Akron, Colorado, U.S.A.

stage, with insufficient water remaining to maximize grain development and yield. Application of N will not increase yields without adequate plant-available water, and increasing plant-available water will not increase crop yield without adequate N supply. The percentage increase in response of crops such as wheat to P fertilization tends to be greater in dry years than in wet years on P-deficient soils, while both N and P are needed to optimize yields in wetter years.

Water is important for activation and movement of fertilizer nutrients applied to soils.[1–3,7,8] Dry fertilizer granules must dissolve in the soil water before they become available to plants. When applied to dry soil, liquid fertilizers may become unavailable to plants until precipitation or irrigation water rewets the soil and they become part of the soil solution again. Rainfall affects the volatilization loss of N from ammonia-based fertilizers such as urea and urea ammonium nitrate (UAN). Rainfall received within 36 hr after surface applications of urea or UAN fertilizers will greatly reduce N volatilization losses and improve the N fertilizer use efficiency by crops. Rainfall moves the surface-applied N fertilizer into the soil where it can react and reduce NH_3 losses to the atmosphere. Excessive soil water, however, can result in anaerobic conditions and the loss of NO_3-N by denitrification. Nitrate-N is converted to various N gases, which are lost to the atmosphere under anaerobic conditions.

Water is essential for optimizing crop yields. Under irrigation, water is generally not a yield-limiting factor. Under dryland or rainfed conditions, crop yields are dependent on available soil water supplies and growing season precipitation. Adequate levels of essential plant nutrients are needed to optimize crop yields and WUE (i.e., kg grain produced/mm crop water use). Under rainfed conditions, crop water supplies during the growing season can vary weekly and annually. During periods of drought (i.e., low supply of plant-available water), less plant nutrients are needed to optimize crop yields than during years of average or above-average precipitation. In wetter years, both the crop yield potential and the nutrients needed to optimize crop yield increase.

Soil management practices, such as reduced- and no-till systems, that increase soil organic matter and improve soil physical quality also improve soil aggregation and porosity. This, in turn, improves water infiltration into the soil and water availability for increased crop productivity and improved nutrient use efficiency.

IRRIGATION WATER QUALITY AND FERTILIZER APPLICATION

Irrigation water quality can affect the application of fertilizer nutrients through irrigation systems.[3,8] For example, the addition of anhydrous NH_3 or liquid ammonium polyphosphate fertilizers to irrigation waters high in Ca can result in the formation of lime and calcium phosphate precipitates. The precipitates can plug sprinkler and drip irrigation systems. In some instances, precipitation of the Ca can result in a higher sodium (Na) hazard of the irrigation water, which may subsequently reduce the water intake capacity of the soil.

Applying fertilizers with both flood and furrow irrigation systems requires that a uniform distribution of water be achieved throughout the field to obtain a uniform distribution of fertilizer nutrients to the crop. With flood and furrow irrigation systems, fertilizer should not be applied with the initial flush of irrigation water because of the generally nonuniform distribution of water during the initial wetting of the soil surface by the irrigation water. The reactions of fertilizers with the irrigation water and the fertilizer distribution to the crop are affected by (irrigation) water quality. If fertigation (i.e., application of fertilizer nutrients through an irrigation system) is to be used, the compatibility of fertilizers to be applied with the quality of irrigation water available must be examined to avoid poor distribution of fertilizer nutrients.

ENVIRONMENTAL QUALITY

Nitrogen is generally transported from soils into surface and groundwater by runoff, erosion, and leaching.[7,9] Runoff water from watersheds with high levels of soluble N and P sources on the soil surface can contribute to eutrophication of streams, lakes, ponds, bays, and estuaries. Placing or positioning applied N and P sources below the soil surface and using soil management practices to minimize runoff will help reduce agriculture's impact on eutrophication of water bodies. Water erosion of soil not only carries soluble plant nutrients from a watershed, but also carries soil particles with sorbed nutrients, such as P, into water bodies that can then contribute to degradation of water quality.

Soil management practices such as no-till and other conservation tillage practices can reduce soil erosion by water.

Water moving through soil in excess of field capacity water content can move soluble nutrients, such as NO_3-N, below the root zone of crops and into groundwater. In summary, using cropping systems and an adequate fertility program to optimize crop WUE will help reduce loss of plant-available water and nutrients below the crop root zone.

CONCLUSIONS

Water plays a critical role in the availability of nutrients to plants. Adequate levels of both water and nutrients are needed to optimize plant growth and productivity. Fertilizer and water management practices can influence the efficient use of water and nutrients by plants and their subsequent impact on environmental quality.

REFERENCES

1. Engelstad, O.P., Eds. *Fertilizer Technology and Use*, 3rd Ed.; Soil Sci. Soc. Am.: Madison, WI, 1985.
2. Havlin, J.L.; Beaton, J.D.; Tisdale, S.L.; Nelson, W.L. *Soil Fertility and Fertilizers: An Introduction to Nutrient Management*, 6th Ed.; Prentice Hall: Upper Saddle River, NJ, 1999.
3. Mortvedt, J.J.; Murphy, L.S.; Follett, R.H. *Fertilizer Technology and Application*; Meister Publishing Co.: Willoughby, OH, 1999.
4. Troeh, F.R.; Thompson, L.M. *Soils and Soil Fertility*, 5th Ed.; Oxford University Press: New York, 1993.
5. Taylor, H.M.; Jordan, W.R.; Sinclair, T.R., Eds. *Limitations to Efficient Water Use in Crop Production*; Am. Soc. Agron., Crop Sci. Soc. Am., Soil Sci. Soc. Am.: Madison, WI, 1983.
6. Follett, R.F.; Stewart, J.W.B.; Cole, C.V., Eds. *Soil Fertility and Organic Matter as Critical Components of Production Systems*; Soil Sci. Soc. Am., Inc.: Madison, WI, 1987.
7. Pierzynski, G.M.; Sims, J.T.; Vance, G.F. *Soils and Environmental Quality*, 2nd Ed.; CRC Press: Boca Raton, FL, 2000.
8. Ludwick, A.E., Bonczkowski, L.C., Bruice, C.A., Compbell, K.B., Millaway, R.M., Petrie, S.E., Phillips, I.L., Smith, J.J., Eds.; *Western Fertilizer Handbook*; 8th Ed.; California Fertilizer Association, Interstate Publishers: Danville, IL, 1995.
9. Follett, R.F., Ed. *Nitrogen Management and Ground Water Protection*; Elsevier: New York, 1989.

Oil Pollution: Baltic Sea

Andrey G. Kostianoy
P.P. Shirshov Institute of Oceanology, Russian Academy of Sciences, Moscow, Russia

Abstract
Shipping activities in the Baltic Sea, including increasing oil transport and oil handled in harbors, have a number of negative impacts on the marine environment and coastal zone. Oil discharges from ships represent a significant threat to marine ecosystems. Oil spills cause the contamination of seawater, sediments, shores, and beaches, which may persist for several months and represent a threat to marine resources. Among the main tasks in the ecological monitoring of the Baltic Sea are an operational satellite and aerial detection of oil spillages, determination of their characteristics, establishment of the pollution sources, and forecast of probable trajectories of the oil spill transport. In this entry, we show the results of recent investigations of oil pollution in the Baltic Sea, including the application of satellite remote sensing technology.

INTRODUCTION

Detection of oil pollution is among the most important goals of monitoring of a coastal zone. Public interest in the problem of oil pollution arises mainly during dramatic tanker and oil platform catastrophes such as those that involved the following: *Amoco Cadiz* (France, 1978), *Ixtoc I* (Gulf of Mexico, 1979–1980), *Exxon Valdez* (Alaska, 1989), *The Sea Empress* (Wales, 1996), *Erica* (France, 1999), *Prestige* (Spain, 2002), and *Deepwater Horizon* (Gulf of Mexico, 2010). However, tanker and oil platform catastrophes are only one among the many causes of oil pollution. Oil and oil product spillages at sea take place all the time, and it would be a delusion to consider tanker accidents as the main environmental danger. According to the International Tanker Owners Pollution Federation (ITOPF), over the period of 1970–2009, spillages resulting from collisions, groundings, tanker holes, and fires amounted to 52% of total leakages during tanker loading/unloading and bunkering operations.[1] In the category 7–700 tons, some 38% of spills occurred during routine operations, most especially loading or discharging (31%). Accidents were the main cause of large spills (>700 tons), with groundings and collisions accounting for 65% of the total during the period 1974–2009.[1] Other significant causes included hull failures and fire/explosion. Discharge of wastewater containing oil products is another important source, by pollutant volume comparable to offshore oil extraction and damaged underwater pipelines. The greatest but hardest-to-estimate oil inputs come from domestic and industrial discharges, direct or via rivers, and from natural hydrocarbon seeps. The long-term effects of this chronic pollution are arguably more harmful to the coastal environment than a single, large-scale accident.

Each year, ships and industries damage delicate coastal ecosystems in many parts of the world by releasing oil or pollutants into ocean, coastal waters, and rivers. Offshore environments are polluted by mineral oil mainly due to tanker accidents, illegal oil discharges by ships, and natural oil seepage. Shipping activities in European coastal seas, including oil transport and oil handled in harbors, have a number of negative impacts on the marine environment and coastal zone. Oil discharges from ships represent a significant threat to marine ecosystems. Oil spills cause the contamination of seawater, sediments, shores, and beaches, which may persist for several months and even years, and represent a threat to marine resources.[2]

As highlighted by Oceana in its report *The Other Side of Oil Slicks*, chronic hydrocarbon contamination from washing out tanks and dumping bilge water and other oily waste represents a danger at least 3 times higher than that posed by the oil slicks resulting from oil tanker accidents.[3,4] For example, in the North Sea, the volume of illegal hydrocarbon dumping is estimated at 15,000–60,000 tons/yr, added to which are another 10,000–20,000 tons of authorized dumping. Oil and gas platforms account for 75% of the oil pollution in the North Sea via seepage and the intentional release of oil-based drilling muds.[5] In the Mediterranean Sea, it is estimated at 400,000–1,000,000 tons/yr. Of this, about 50% comes from routine ship operations and the remaining 50% comes from land-based sources via surface runoff.[5] In the Baltic Sea, this volume is estimated at 1,750–5,000 tons/yr.[3,4] In 2004, the Finnish Environment Institute[6] estimated the total annual number of oil spills in the Baltic Sea to be 10,000 and the total amount of oil running into the sea to be as much as 10,000 tons, which is considerably more than the amount of oil pouring into the sea in accidents. It is impressive to note that the total amount of oil spilled annually worldwide has been estimated at a level greater than 4.5 millions tons, equivalent to one full tanker disaster every week.[7]

One of the most important goals of the ecological monitoring of European seas is monitoring and detection of oil pollution.[8–10] After a tanker accident or illegal oil discharge, the biggest problem is to obtain an overall view of the phenomenon, getting a clear idea of the extent of the slick and predicting the way it will move. For natural and man-made oil spills, it is necessary to operate regular and operational monitoring. Oil pollution monitoring in the Mediterranean Sea, North Sea, and Baltic Sea is normally carried out by aircrafts or ships. This is expensive and is constrained by the limited availability of resources. Aerial surveys over large areas of the seas to check for the presence of oil are limited to the daylight hours, good weather conditions, and maritime boundaries between countries.

Satellite imagery can help greatly in identifying probable spills over very large areas and in guiding aerial surveys for precise observation of specific locations. The synthetic aperture radar (SAR) instrument, which can collect data independently of weather and light conditions, is an excellent tool to monitor and detect oil on water surfaces.[7–10] This instrument offers the most effective means of monitoring oil pollution: oil slicks appear as dark patches on SAR images because of the damping effect of the oil on the backscattered signals from the radar instrument. This type of instrument is currently on board the European Space Agency's (ESA's) *ENVISAT* and *ERS-2* satellites, the Canadian Space Agency's *RADARSAT-1* and *RADARSAT-2* satellites, and the German Earth observation satellite *TerraSAR-X*. The *ENVISAT* satellite was launched in March 2002 by the ESA. Operational systems, which include 10 instruments, have been developed to monitor oceans, ice, land, and the atmosphere. *ENVISAT* has a 35-day repeat cycle, but due to wide swaths by some of the instruments, the Earth is covered within a few days. ASAR (advanced synthetic aperture radar) instrument is used for mapping sea ice and oil slick monitoring, measurements of ocean surface features (currents, fronts, eddies, internal waves), ship detection, oil and gas exploration, etc. Users of remotely sensed data for oil spill applications include the Coast Guard, national environmental protection agencies and departments, oil companies, shipping, insurance and fishing industries, national departments of fisheries and oceans, and other organizations.

In this entry, we will focus on oil pollution in the Baltic Sea, briefly describing main sources of oil pollution, the number of oil spills observed yearly, the results of operational satellite monitoring and numerical modeling of oil spills, and the remaining problems.

MAIN SOURCES OF OIL POLLUTION IN THE BALTIC SEA

Crude oil and petroleum products account for about 40% of the total exports of Russia. Russian Federation stands as one of the leading operators in the international oil business, being the largest oil exporter after Saudi Arabia. In 2000, Russia exported approximately 145 million tons of crude oil and 50 million tons of petroleum products. Since 2000, exports of petroleum and petroleum products began to grow, and virtually doubled for the period from 1996 to 2005. According to the Federal Customs Service of Russia, in 2008, Russia exported 221.6 million tons, 7% less than in 2007. The Russian Ministry of Economic Development forecasts growth of oil exports from Russia in 2010 and 2011 to the level of 245.8 million tons.

The ports on the Baltic Sea play a huge role in the export of oil from Russia. The main oil terminals here used to be the Latvian port of Ventspils and the Port of Tallinn, Estonia. In the last 10 years, a number of new oil terminals have been built in the Baltic Sea area, resulting in increased transport of oil by ships and, consequently, an increased risk of accidents and increased risk of pollution of the marine environment. Today, in the Gulf of Finland, there are more than 18 oil terminals in Russia, Finland, and Estonia.[11] The following are the major existing and projected oil terminals in the Gulf of Finland from Russia: Primorsk, Vysotsk, Big Port of St. Petersburg, Ust-Luga, Batareinaya, Vistino, Gorki, and Lomonosov.[12,13]

Primorsk is the largest Baltic oil terminal located on Russian territory. In 2008, 75.6 million tons of oil products were exported from Primorsk, 13.6 million tons from Vysotsk, and 14.4 million tons from the St. Petersburg oil terminal. By 2015, the maximum export possibility of the Primorsk terminal is estimated at 120 million tons, while that of Vysotsk is at 20.5 million tons. In November 2000, Lukoil has opened an oil terminal in Kaliningrad. In 2001, the company built another terminal in Kaliningrad with a declared capacity of 2.5 million tons. These terminals can overload up to 3–5 million tons of oil annually.

According to estimates of the Centre for Maritime Studies at the University of Turku (Finland), in 2007, 263 million tons of cargo were transported through the Gulf of Finland, among which the share of oil is 56%.[13] Russian ports handled 60% of goods, Finnish ports handled 23%, and Estonian ports handled 17%. The share of imports was 22%, that of export was 76%, and that of local transportation was 2%. Russian ports held 68.6% of the total turnover of petroleum products, Estonian ports held 17.2%, and Finnish ports held 14.2%.[13] The major ports are the following: Primorsk (74.2 million tons), Saint Petersburg (59.5 million tons), Tallinn (35.9 million tons), Skoldvik (19.8 million tons), Vysotsk (16.5 million tons), and Helsinki (13.4 million tons). In 2007, the ports of the Gulf of Finland carried out about 53,600 ship calls, most of which were in St. Petersburg (14,651), Helsinki (11,727), and Tallinn (10,614). In 2009, vessels entered or left the Baltic Sea via Skaw 62,743 times; this is 20% more than in 2006. Approximately 21% of those ships were tankers, 46% were other cargo ships, and 4.5% were passenger ships.[14]

Forecasts of the Finnish Centre for Maritime Studies for the year 2015 according to three basic scenarios of economic development in Russia, Finland, and Estonia give a value of 322.4–507.2 million tons of cargo to be transported in the Gulf of Finland, which is 23%–93% more than in 2007, and under any scenario, growth in turnover will occur mainly due to Russia.[13] In addition, the share of oil and petroleum products among other goods will be an even greater increase in absolute terms—it can reach 158–262 million tons. For the transportation of petroleum products, 6655 to 7779 tankers will be used.

The growth of oil and other cargo through the terminals and the Baltic ports inevitably leads to an increase in the number of tankers and other types of vessels, which then leads to an increase in chronic sea pollution and a higher probability of ship accidents. According to statistics, shipping accounts for 45% of oil pollution in the ocean, while oil production at the shelf accounts for only 2%. In the Baltic Sea, about 2000 large ships and tankers are at sea every day; thus, shipping, including oil transport, has a major negative impact on the marine environment and coastal zone. Illegal discharges of oil and petroleum products from ships, ship accidents, collisions, and groundings represent a significant threat to the Baltic Sea.

According to Global Marine Oil Pollution Information Gateway,[15] major oil spills in the region in 1977–2002 resulted from the following ship accidents: *Tsesis* (1977, off Nynäshamn, Sweden, spill of 1000 tons), *Antonio Gramsci* (1979, off Ventspils, Latvia, spill of 5500 tons; another incident in 1985, off Porvoo, Finland, spill of 580 tons), *Jose Marti* (1981, off Dalarö, Sweden, spill of 1000 tons), *Globe Asimi* (1982, off Klaipeda, Lithuania, spill of 16,000 tons), *Sivona* (1984, in The Sound, Sweden, spill of 800 tons), *Volgoneft* (1990, off Karlskrona, Sweden, spill of 1000 tons), *Baltic Carrier* (2001, international waters between Denmark and Germany, spill of 2700 tons).

According to HELCOM data (http://www.helcom.fi), the total number of accidents on ships in 2000–2004 is 374, of which 29 resulted in the pollution of marine waters. The number of accidents has risen since 2006, which can be linked to the 20% increase in ship traffic. Now, there are 120–140 shipping accidents yearly in the Baltic Sea area.[14] The majority of accidents are groundings and collisions. The share of groundings in the total number of accidents is higher for the Baltic Sea than for other European waters. On average, 7% of the shipping accidents in the Baltic Sea result in some kind of pollution, usually containing not more than 0.1–1 tons of oil. For the last 6 years, no major accidental oil spill has happened in the Baltic Sea.[14]

As far as oil exploitation at sea and on the coast is concerned, offshore operations have been taking place for some years in Polish waters (two jack-up rigs); Germany operated two platforms very close to the coast; in March 2004, Russia started oil production at Lukoil D-6 platform in the waters between the Kaliningrad area (Russian Federation) and Lithuania, and Latvia plans to drill for oil in the waters between them and Lithuania.[15]

OIL SPILL OBSERVATIONS, STATISTICS, AND TENDENCIES

Every ship entering the Baltic Sea must comply with the antipollution regulations of the Helsinki Convention and MARPOL (marine pollution) Convention. Even though strict controls over ships' discharges have been established by the Baltic Sea countries, illegal spills and discharges continue to happen.[14] Fortunately, the number of illegal oil spills detected by aerial surveillance has been reduced significantly over the last 20 years, from 763 spills in 1989 to 178 spills in 2009 (Table 1 and Fig. 1). This is 60 less than in 2007 and 32 less than in 2008. Also, the volume of the spills has been decreasing—most are between 1 and 0.1 m^3 today.[14,16]

A decreasing trend in the number of observed illegal oil discharges despite rapidly growing density of shipping, increased frequency of surveillance flights (Fig. 1 and Table 2), and usage of satellite imagery, provided by the CleanSeaNet satellite service of the European Maritime Safety Agency (EMSA) (http://cleanseanet.emsa.europa.eu/), illustrate the positive results of the complex set of measures known as the Baltic Strategy, implemented by the Contracting Parties to the Helsinki Convention.[16]

In order to obtain the geographical distribution of oil spills, we put the spills observed in 1988–2002 on the same map (Fig.2). Analysis of the Tables 1 and 2 and Fig. 2 shows the following:

1. HELCOM data seem to be underestimated in figures in comparison with other estimates.[3,4,6]
2. Since 1993, Russia does not carry out aerial surveillance in the Southeastern Baltic Sea and in the Gulf of Finland.
3. Since 1994, Lithuania seems to have no regular and effective aerial surveillance, because no oil spills have been detected.
4. Since 2005, Latvia seems to have no effective aerial surveillance, because 0–5 oil spills yearly are not realistic figures.
5. Poland and Sweden respectively demonstrate an 80% and 50% increase in the number of oil spills observed during the last 2–3 years.
6. Traces of oil spills in Fig. 2 show the main ship routes in the Baltic Sea, as well as the approaches to major sea ports and oil terminals.
7. Fig. 2 proves that ships are primarily responsible for the oil pollution in the Baltic Sea.

Table 1 Country-wise data on the number of illegal oil discharges observed in national waters in the Baltic Sea in 1988–2009.

	Denmark	Estonia	Finland	Germany	Lithuania	Latvia	Poland	Russia	Sweden	Total
1988	129			90			40	82	168	509
1989	159			139			69	184	212	763
1990	34			45		73	88		184	424
1991	46			85	8	20	14	3	197	373
1992	18	18		76	34	15	92	13	278	544
1993	17	7		43	28	6	110		250	461
1994	30	4		75			104		375	588
1995	48	3	26	55			72		445	649
1996	36		42	44			50		241	413
1997	38	3	104	34			25		234	438
1998	53	10	53	23		33	33		249	454
1999	87	33	63	72		18	18		197	488
2000	68	38	89	51		17	51		158	472
2001	93	11	107	51	0	6	24		98	390
2002	54	8	75	44		21	25		117	344
2003	37	4	40	60		14	39		84	278
2004	30	19	36	42	0	13	10		143	293
2005	28	24	32	34	0	5	5	2	94	224
2006	41	31	29	22	0	0	3		110	236
2007	43	58	29	30		2	15		61	238
2008	41	46	28	24		5	22		44	210
2009	34	20	16	15		1	27		65	178

Source: *Illegal discharges of oil in the Baltic Sea during 2009.*[16]

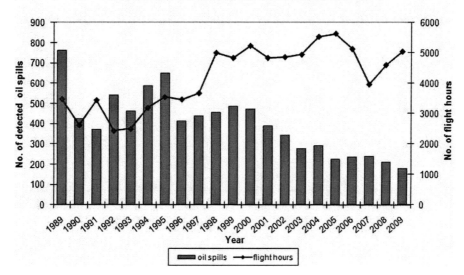

Fig. 1 Total number of flight hours and observed oil spills in the HELCOM area during aerial surveillance in 1989–2009.
Source: *Illegal discharges of oil in the Baltic Sea during 2009.*[16]

Table 2 Number of aerial surveillance flight hours performed by the HELCOM countries in 1989–2009.

	Denmark	Estonia	Finland	Germany	Lithuania	Latvia	Poland	Russia	Sweden	Total
1989				142			131	1618	1600	3491
1990	292			168		400	164		1600	2624
1991	199			129	348	408	140	629	1600	3453
1992	172			267	78	127	62	32	1700	2438
1993	153	40		201	133	24	49		1900	2500
1994	253	420		290		18	179		2038	3198
1995	225	420	355	291		8	301		1953	3553
1996	275	305	400	313	65	8	345		1763	3474
1997	209	284	355	288		64	291		2189	3680
1998	325	236	649	206		577	465		2544	5002
1999	416	268	603	286		320	375		2565	4833
2000	497	212	660	439	250	436	362		2374	5230
2001	463	161	567	466	300	412	187		2281	4837
2002	412	153	605	469		387	320		2518	4864
2003	510	201	615	446		414	228		2532	4946
2004	265	198	644	491	100	365	239		3231	5534
2005	251	178	625	549	54	384	141		3455	5638
2006	290	471	517	504	64	311	131		2842	5128
2007	271	410	529	598	41	343	380		1397	3969
2008	246	503	438	650		298	406		2063	4603
2009	240	371	351	638	66	61	561		2758	5046

Source: Illegal discharges of oil in the Baltic Sea during 2009.[16]

The Baltic Sea States aerial surveillance fleet today consists of more than 20 airplanes and helicopters, most of which are equipped with up-to-date remote sensing equipment—side-looking airborne radar, SAR, infrared (IR) and ultraviolet scanner, microwave radiometer, laser fluorosensor (lidar), and Forward Looking InfraRed (FLIR) high-resolution camera. The Baltic Sea States have to conduct aerial surveillance for detecting oil pollution and suspected ships at a minimum of twice per week over regular traffic zones including approaches to major sea ports as well as in regions with regular offshore activities. Other regions with sporadic traffic and fishing activities should be covered once per week.[16] Also, the Coordinated Extended Pollution Control Flights, which constitutes continuous surveillance of specific areas in the Baltic Sea for 24 hr or more, should be carried out twice a year.[16] Although the number of observations of illegal oil discharges shows a decreasing trend over 20 years, it should be noted that for some areas and countries, aerial surveillance is not evenly and regularly carried out and therefore there are no reliable figures for these areas.[16]

Since 2007, aerial surveillance in the Baltic Sea was supported by the satellite remote sensing technique for oil spill detection. CleanSeaNet is a near-real-time satellite-based oil spill and vessel monitoring service. It entered into operation on April 16, 2007.[16] The service is continually being expanded and improved and provides a range of different products to the Commission and to European Union member states. The legal basis for the CleanSeaNet service is Directive 2005/35/EC on ship-source pollution and on the introduction of penalties, including criminal penalties, for pollution offenses (as amended by Directive 2009/123/EC). EMSA has been tasked to "work with the Member States in developing technical solutions and providing technical assistance in relation to the implementation of this Directive, in actions such as tracing discharges by satellite monitoring and surveillance."[16]

OPERATIONAL SATELLITE MONITORING OF OIL POLLUTION

A satellite-based remote sensing system is capable of ensuring a relatively low-cost, high-standard observational system for oil pollution monitoring. SAR is the best instrument for the detection of oil slicks on the sea surface from space because slicks modify seawater viscosity and damp short waves measured by SAR. SAR images can be acquired regardless of cloud cover and light conditions. Along with 400 km wide swath, this is the main advantage

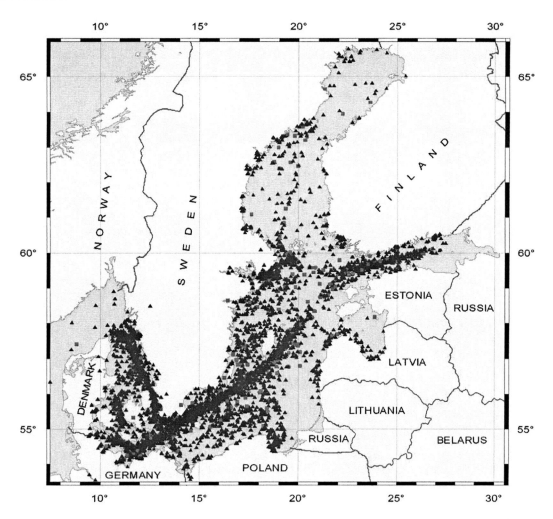

Fig. 2 Oil spills detected in the Baltic Sea by aerial survey in 1989–2002 based on HELCOM data.
Source: Kostianoy et al.[19]

of SAR in comparison with aerial surveillance. However, oil spill detection by SAR has a problem of distinguishing oil slicks from look-alikes, such as sea areas covered by organic films, algal bloom, sea ice, wind shadows, rain cells, and upwelling zones. Therefore, reliable automatic detection of oil spills on the basis of SAR data is not yet achieved and there is a risk of false alarms. This problem can be significantly reduced by a new approach, which consists in the combined use of all available quasi-concurrent satellite, oceanographic, and meteorological information, along with numerical modeling of oil spill transport. This operational system was specially elaborated in the beginning of 2004 for monitoring oil pollution in the vicinity of the Lukoil D-6 oil platform in the Southeastern Baltic, Russian Federation.[9,10,17–21]

Since 1993, regular aerial surveillance of the oil spills in the Russian sector of the Southeastern Baltic Sea and in the Gulf of Finland has stopped (Fig. 2). In June 2004, we organized daily service for monitoring of oil spills in the Southeastern Baltic Sea based on the operational receiving and analysis of ASAR *ENVISAT* and SAR *RADARSAT-1* data as well as of other satellite IR and optical (VIS) data, meteorological information, and numerical modeling of currents required for the identification of the slick nature in the sea and forecast of the oil spill drift.[9,10,17–21] This work was initiated and financed by Lukoil-Kaliningradmorneft (Kaliningrad, Russia) in connection with the start of oil production from the continental shelf of Russia in March 2004. The principal differences from the existing projects and satellite services were 1) an operational monitoring regime of 24 hours/day, 7 days/week for 18 mo and 2) a complex approach to the oil spill detection and forecast of their drift.

The general goals of the satellite oil pollution monitoring in the Baltic Sea were as follows:

1. Correct detection of oil spills in the vicinity of the D-6 oil platform as well as in the large area of the Southeastern Baltic Sea between 54°20′–58°N and 18°–22°E

2. Identification of sources of oil pollution
3. Forecast of the oil spill drift by different methods
4. Data systematization and archiving
5. Cooperation with authorities

Operational monitoring of oil pollution in the sea was based on the processing and analysis of ASAR *ENVISAT* (every pass over the Southeastern Baltic Sea, frame of 400 × 400 km, 75 m/pixel spatial resolution) and SAR *RADARSAT-1* (300 × 300 km, 25 m/pixel resolution) images received from KSAT Station (Kongsberg Satellite Services, Tromsø, Norway) in operational regime (1–2 hr after the satellite's overpass). For interpretation of ASAR *ENVISAT* imagery and forecast of the oil spill drift, IR and VIS AVHRR (the Advanced Very High Resolution Radiometer aboard the National Oceanic and Atmospheric Administration (NOAA) satellites) and Moderate Resolution Imaging Spectroradiometer (MODIS) (*Terra* and *Aqua*) images were received, processed, and analyzed, as well as the *QuikSCAT* scatterometer and the *Jason-1* altimeter data.[18–21] The total area covered by the monitoring was equal to about 60,000 km^2, which is almost one-sixth of the Baltic Sea total surface.

The satellite receiving station at the Marine Hydrophysical Institute (MHI) in Sevastopol (Ukraine) was used for operational (24 hours/day, 7 days/week) receiving of the AVHRR NOAA data for the construction of the sea surface temperature, optical characteristics of seawater, and currents maps. Sea surface temperature (SST) variability and intensive algae bloom (high concentration of blue-green algae on the sea surface in the summertime) allow one to highlight meso- and small-scale water dynamics in the Baltic Sea and to follow movements of currents, eddies, dipoles, jets, filaments, river plumes, and outflows from the Vistula and Curonian bays. Sequence of daily MODIS IR and VIS imagery allows reconstruction of a real field of surface currents (direction and velocity) with 0.25–1 km resolution, which is very important for a forecast of a direction and velocity of a potential pollution drift including oil spills. The combination of ASAR *ENVISAT* images with high-resolution VIS and IR MODIS images allows understanding of the observed form of the detected oil spills and prediction of their transport by currents.[18–21]

Sea wind speed fields were derived from scatterometer data from every path of the *QuikSCAT* satellite over the Baltic Sea (twice a day). These data were combined with data from coastal meteorological stations in Russia, Lithuania, Latvia, Estonia, Finland, Sweden, Denmark, Germany, and Poland, and numerical weather models. Altimetry data from every track of the *Jason-1* satellite over the Baltic Sea were used for compilation of sea wave height charts, which include the results of the FNMOC (Fleet Numerical Meteorology and Oceanography Center, United States) WW3 Model. Both data were used for the analysis of the ASAR *ENVISAT* imagery and estimates of the oil spill drift direction and velocity.[18–21]

In total, 274 oil spills were detected in 230 ASAR *ENVISAT* images and 17 SAR *RADARSAT-1* images received during 18 mo (June 2004 to November 2005).[9,10,20,21] One example from the oil spill gallery is shown in Fig. 3 where an illegal release of oil from three ships was detected on August 25, 2005. A map of all oil spills detected by the analysis of the ASAR *ENVISAT* imagery in the given area of the southeastern Baltic Sea from June 12, 2004, until November 30, 2005, is shown in Fig. 4. A real form and dimension of oil spills are shown. A square southwestward of Klaipeda shows the location of the D-6 oil platform. Oil spills clearly revealed the main ship routes in the Baltic Sea directed to ports of Ventspils, Liepaja, Klaipeda (routes from different directions), Kaliningrad, and along Gotland Island (Fig. 4). No spills originated from the D-6 oil platform were observed.

The interactive numerical model Seatrack Web of the Swedish Meteorological and Hydrological Institute (SMHI) was used for a forecast of the drift of satellite-detected oil spills and ecological risk assessment of the Lukoil D-6 oil platform.[18–21]

Since 2006, the satellite monitoring of the D-6 oil platform was transformed, reduced in size to about 24,000 km^2, and unfortunately, it lost its main peculiarity—a complex approach to the oil spill detection and forecast of their drift. In 2006–2009, 638 oil spills have been

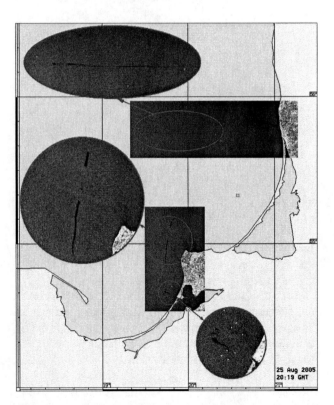

Fig. 3 A release of oil from three ships on August 25, 2005 (ASAR *ENVISAT*). The length of the spill in front of Klaipeda is 33.6 km, surface—8.6 km^2. The length of another long spill is 22 km.

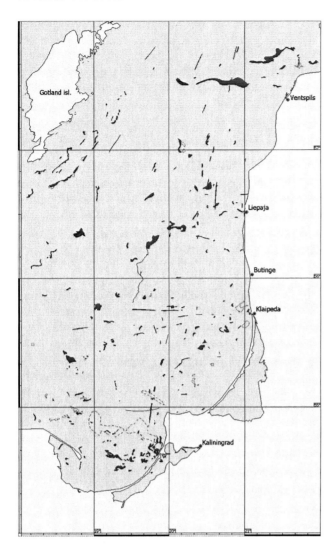

Fig. 4 Map of all oil spills detected by the analysis of the ASAR *ENVISAT* and SAR *RADARSAT* imagery in June 2004 to November 2005.

identified on 804 ASAR images, from which 319 spills were detected in this reduced area.[22] Combined analysis of the location and shape of the detected spills with location of the ships thanks to AIS (Automatic Identification System for ships) clearly indicates that the major source of sea pollution is shipping. In the Southeastern Baltic Sea, an area with no HELCOM statistical data, we also observe a decreasing trend in oil spill number and their total surface. In 2006, the total number of oil spills (and area of oil pollution) amounted to 114 (371.7 km^2); in 2007, 94 spills (213.7 km^2); in 2008, 67 spills (198.7 km^2); in 2009, 44 spills (81.7 km^2).[22]

A significant seasonal variability in oil spill detection is observed. During autumn and winter, we detect oil spills 4 times less than in spring and summer. This huge difference is explained by limitations of the SAR method to detect oil spills when the wind is stronger than 10 m/sec, which is very often during the cold season in the Baltic Sea. In addition, strong wind–wave mixing contributes to more rapid formation of emulsions ("water in oil" and "oil in water"), thus preventing formation of the oil slicks on the sea surface.

Image comparison of the number of detected oil spills in the morning (about 11:00 local time) and in the evening (about 22:00 local time) showed that the probability of finding oil pollution in the morning is about 40% higher than that in the afternoon and evening.[22] This fact indicates that the illegal discharge of oil from vessels occurs more often at night, when it is impossible for patrol aircrafts or ships to record this fact by photo and video camera. This once again confirms the advantages of satellite radar imagery for monitoring of oil pollution.

Based on the number of oil spills detected in 2004–2005 (about 180 spills yearly on 60,000 km^2) and in 2006–2009 (about 80 spills yearly on 24,000 km^2), we can estimate the total number of oil spills for the Baltic Sea (377,000 km^2) as 1100–1300 yearly and the total surface of oil pollution as 14,000 km^2 yearly, which is a half of the Gulf of Finland (we suppose that the density of oil spill is more or less uniform on the Baltic Sea area). These values may double and even triple if we take into account that the following: 1) SAR satellites pass over a specific area in the Baltic Sea every 2 days in average; 2) significant reduction of oil spill observation in autumn and winter; and 3) spatial resolution of SAR/ASAR imagery of 25–75 m/pixel.

NUMERICAL MODELING OF OIL SPILL DRIFT

The above-mentioned satellite monitoring of the Southeastern Baltic Sea was coupled with numerical modeling of oil spill transport. The interactive numerical model Seatrack Web SMHI was used for a forecast of the drift of 1) all large oil spills detected by ASAR *ENVISAT* in the Southeastern Baltic Sea and 2) virtual (simulated) oil spills from the Lukoil D-6 platform. The latter was done daily for the operational correction of the action plan for accident elimination at the D-6 and ecological risk assessment (oil pollution of the sea and the Curonian Spit).

This version of a numerical model on the Internet platform has been developed at the SMHI in close cooperation with the Danish Maritime Safety Administration, Bundesamt fur Seeschifffahrt und Hafen, and the Finnish Environment Institute.[23] The first version of Seatrack Web was introduced in 1995, and since then, Seatrack Web has been used successfully for oil spill drift forecast. It was developed to be a friendly tool for authorities responsible for oil spill response in the Baltic Sea region. Seatrack Web's main purpose is to calculate the drift and transformation of oil spills in the Gulf of Bothnia, the Gulf of Finland, the Baltic Sea Proper, the Sounds, the Kattegat, the Skagerack, and part of the North Sea (to 3°E).[23] The program can also be used for substances other than oil, such as chemicals, algae, and floating objects. In addition to an oil drift forecast, it is possible to make a backward calculation. Then, calculation starts at the position where a substance was found and the program calculates the drift backwards in time and traces the origin of the substance or an object.

The system uses two different operational weather models, ECMWF (European Centre for Medium-Range Weather Forecasts) and HIRLAM (High-Resolution Limited Area Model, 22 km grid), and the circulation model HIROMB (High-Resolution Operational Model for the Baltic Sea, 50 layers), which is a three-dimensional circulation model covering the whole Baltic Sea and part of the North Sea, driven by the two weather models, respectively, which calculates the current field at 1-nautical-mile grid. The wind forecasts used in Seatrack Web originate from the weather model at the ECMWF 5 days ahead and HIRLAM 2 days ahead.[23]

The forecasts are made daily and the model allows the forecasting of the oil drift 5 days ahead or making a hind cast (backward calculation) for 30 days in the whole Baltic Sea. When calculating the oil drift, wind and current forecasts are taken from the operational models. Every third hour, new current fields are used in the Seatrack Web. The oil spreading calculation is added to the currents, as well as oil evaporation, emulsification, sinking, stranding, and dispersion.[23]

The AIS functionality in Seatrack Web is a tool to identify the ship that causes an oil spill. The AIS function gives the ship tracks in the area where the spill was detected and backwards to its probable origin. By using the ship tracks simultaneously with the oil tracks, the probability of identifying a suspected ship increases. This powerful system was recommended by HELCOM for operational use in the Baltic countries.[23]

The Seatrack Web model was very useful for ecological risk assessment related to exploitation of the Lukoil D-6 oil platform, which is installed 22.5 km from the Curonian Spit, a UNESCO World Heritage Site. Virtual (simulated) oil spills of 10 m^3 were released daily from the platform during 6 months in order to calculate the shape, direction, distance, and velocity of their drift. Then, all 180 oil tracks were accumulated on one map showing the potential impact of oil pollution in case of an accident at the platform. Statistics, based on daily forecast of the oil spill drift in July–December 2004, shows potential probability (%) of the appearance of an oil spill in any point of the area during 48 h after an accidental release of 10 m^3 of oil (Fig. 5). The probability of the oil spill drift directed to the Curonian Spit (150° sector from D-6) is equal to 67%, but only in half of these cases did oil spills reach the coast due to a coastal current.[18,20] This new technology allowed a quantitative assessment of ecological risks.

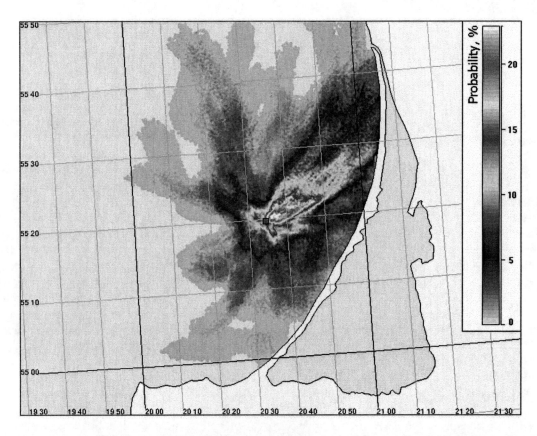

Fig. 5 Probability of observation of potential oil pollution from the D-6 platform during the first 48 hr after an accidental release of 10 m^3 of oil (based on 6 mo daily release of oil spill from the oil platform).

Later, the same methodology was applied to the risk assessment of the Nord Stream gas pipeline construction[24] and assessment of the impact of oil pollution along the ship routes on the Baltic Sea Marine Protected Areas (BSPAs).[25] Figs. 6 and 7 show two examples of oil spill drift modeling and a probability of the drift for specific points along main ship routes in the Gulf of Finland and southward of Gotland Island for July and August 2007. Fig. 6a shows a drift of a virtual oil spill of 10 m^3 during 48 hr, which was released on July 23, 2007, at a specific point (red square) of the ship route passing through the Gulf of Finland. The same numerical experiment was performed daily from July 1 to August 31, 2007. Thus, based on the compilation of 62 maps of oil spill drifts, we could construct Fig. 6b and calculate a probability of oil spill drift. Fig. 6b shows that for this time period, there is no impact of possible oil spill drift on the surrounding BSPAs along the coasts of Finland and Estonia, which are marked by blue and rose colors. This is explained by low wind speed and weak currents observed during July and August 2007. These weather conditions differ from those observed on July 26 to August 15, 2006, when a virtual oil spill could drift 33.5 nautical miles during 2 days with a velocity up to 50 cm/sec (Fig. 6c). Thus, potential releases of oil spills from the ships may represent a threat to seven protected areas located along the coasts of Finland and Estonia, as well as to the islands and coasts of these Baltic countries. The same dot lies on the trajectory of the Nord Stream gas pipeline (yellow line in Fig. 6c), which is under construction in the Gulf of Finland since May 2010. Risk assessment for the construction of the gas pipeline was performed using the same methodology in 2006 for seven key points of the pipeline.[24]

Fig. 7a shows a simultaneous release of oil from a long part of the ship route located southward of Gotland Island. Fig. 7b shows a significant impact of oil pollution produced by this virtual oil release. Both BSPAs were subjected to oil pollution, but at different degrees, so that a quantitative estimation was possible with the help of the Seatrack Web model (Fig. 7c and d). For instance, about 60% of the first BSPA and 95% of the second one will be polluted with the indicated probability. In this case, even the coastal zone of Gotland Island was potentially threatened by oil pollution with a clearly calculated probability.[25]

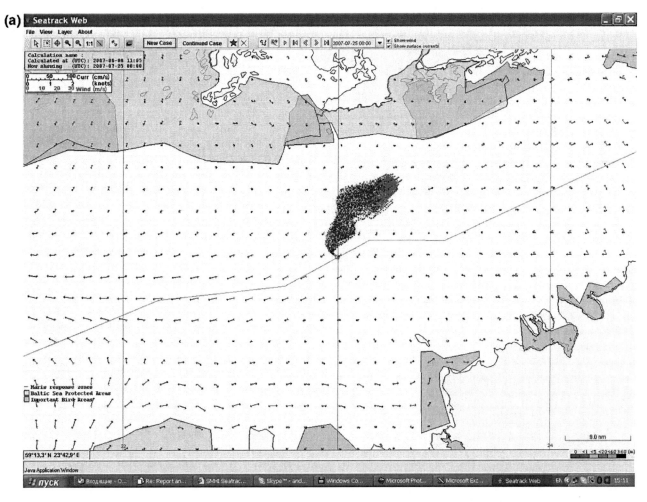

Fig. 6 Modeling of oil spill drift in the Gulf of Finland. Panel (a) shows oil spill drift on July 23, 2007.

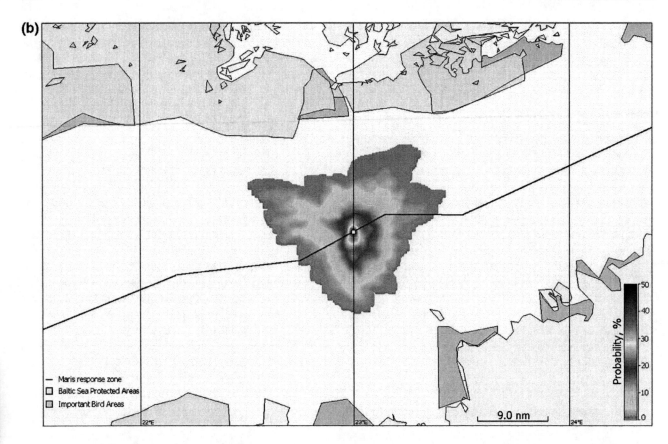

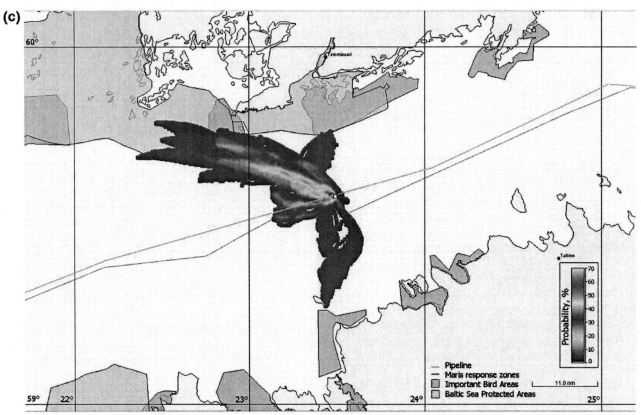

Fig. 6 (continued) Panel (b) shows probability (%) of oil spill drift calculated on the basis of daily modeling at this point for real wind and current conditions in July–August 2007. Panel (c) shows probability (%) of oil spill drift calculated for July 16 to August 15, 2006. BSPAs are shown in blue; important bird areas are in rose colors. The yellow line in panel (c) shows approximate position of the Nord Stream gas pipeline.

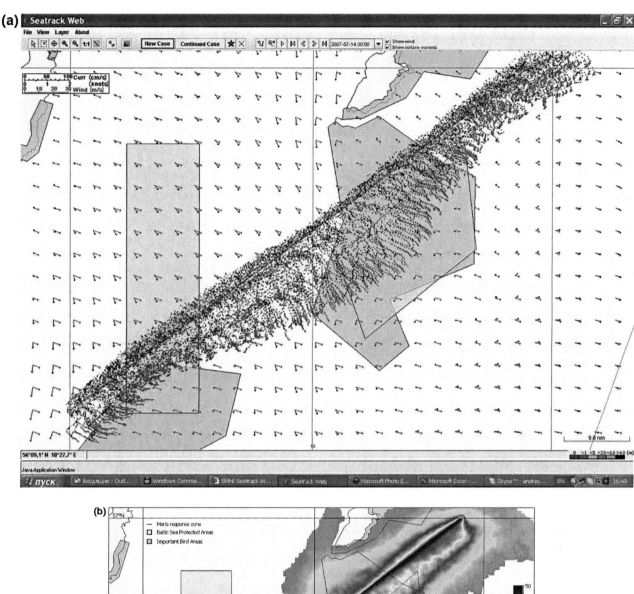
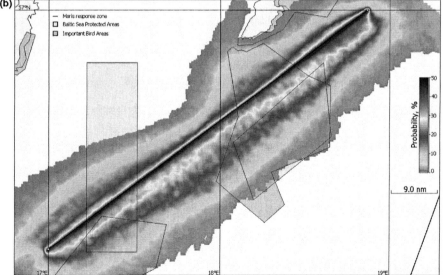

Fig. 7 Modeling of oil spill drift released from a long part of the ship route located southward of Gotland. Panel (a) shows oil spill drift on July 12, 2007. Panel (b) shows probability (%) of oil spill drift calculated on the basis of daily modeling at this line for real wind and current conditions in July–August 2007.

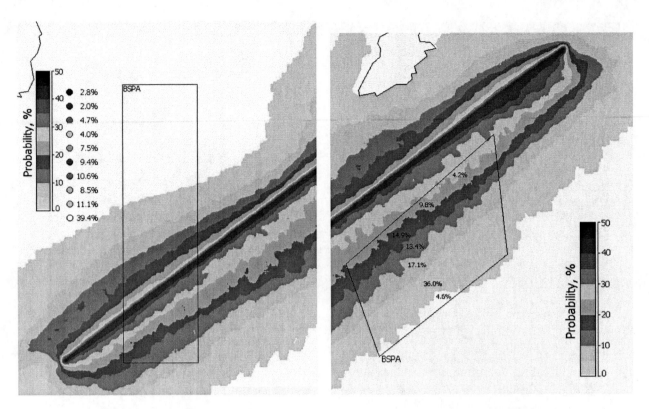

Fig. 7 (continued) Panels (c) and (d) show the impact of this part of the ship route on both BSPAs.

PROBLEMS AND SOLUTIONS

The main challenge is that the real degree of oil pollution in the Baltic Sea is unknown, because the number of observed oil spills and estimates differ significantly. Partially, this is explained by known differences in aerial surveillance and satellite monitoring. Both methods have their own advantages and limitations, and should complement each other. It is clear that statistics on oil spills is not complete and not comparable in different parts of the sea due to different efforts and methods applied for oil pollution monitoring (different number of oil patrol ships, aircrafts, and helicopters per country and per unit of the sea area; the number of flight hours per country and unit of the sea area; use of satellites; a number of ASAR/SAR images acquired and analyzed yearly; application of complex satellite monitoring based on the multisensor and multiplatform approach along with the analysis of metocean data; local peculiarities of the water area; and numerical modeling).

For example, since 1993, Russia does not carry out aerial surveillance in the Gulf of Finland and in the Southeastern Baltic Sea. The existing satellite monitoring is performed on a regular basis only in the Southeastern Baltic Sea and by a private company, Lukoil-Kaliningradmorneft. According to HELCOM data, since 1994 and 2005, respectively, Lithuania and Latvia seem to have had no regular aerial surveillance of oil pollution. Well-equipped regular aerial surveillance is very expensive, and it is clear that countries in economic recession reduce their aerial and in situ monitoring. Satellite monitoring may partially solve this problem, because satellites cover simultaneously very large areas of the Baltic Sea.

Organization of the Baltic International Satellite Monitoring Center in HELCOM could solve many problems in the operational monitoring of oil pollution in the Baltic Sea. It will:

1. Ensure full and uniform coverage of the Baltic Sea area by remote sensing control
2. Reinforce aerial surveillance and improve the oil pollution monitoring
3. Establish satellite monitoring for the countries where it is not yet applied
4. Remove duplication of satellite monitoring for the same area performed by neighboring countries
5. Significantly reduce the total cost of operational satellite monitoring for all countries
6. Provide data to all the Baltic Sea States in the same format
7. Solve the problem regarding different technologies, methods, and algorithms used for the analysis of satellite data in different countries
8. Solve the problem of the "night" oil spill pollution that is getting more and more acute

9. Stimulate exchange of data and cooperation between countries
10. Contribute to early warning in case of transboundary oil spill drift
11. Improve the ecological state of the Baltic Sea, coastal zones, and shores of the Baltic Sea States
12. Stimulate organization of analogous operational monitoring centers for the seas with a high density of shipping and/or oil/gas exploration/production industry, i.e., the North Sea, the Mediterranean Sea, the Black Sea, the Caspian Sea, the Gulf of Mexico, etc

Wide usage of the SMHI Seatrack Web model for oil spill drift forecast is required. Originally, the model was not devoted to the ecological risk assessment, but we found it very useful for this purpose as well.[24,25] The ecological risk assessment for all ports, oil terminals and platforms, subsurface oil pipelines, ship routes, the Baltic Sea Protected Areas, and any part of the 8000 km long coastline of the Baltic Sea can be performed based on the methodology we elaborated in 2004 and successfully used for Lukoil D-6 oil platform and Nord Stream gas pipeline construction.[18,20,24,25] This will quantitatively and precisely reveal the hot spots in the marine area, islands, and coastline of the Baltic Sea that are vulnerable to the impact of the shipping oil pollution. Such a general map of the Baltic Sea with calculated probability for any point of the sea and the coastline to be polluted may serve as a guideline for the Baltic Sea States to improve their monitoring systems.

CONCLUSIONS

Oil is a major threat to the Baltic Sea ecosystems. In the last decade, maritime transportation in the Baltic Sea region has been growing steadily, reflecting the intensified trade and oil export. Increase in the number of ships also means that we can expect a larger number of illegal oil discharges. Both oil tankers and all types of ships are responsible for oil pollution of the Baltic Sea. Any discharge into the Baltic Sea of oil, or diluted mixtures containing oil in any form including crude oil, fuel oil, oil sludge, or refined products, is prohibited. This applies to oily water from the machinery spaces of any ship, as well as from ballast or cargo tanks from oil tankers.[16] Every ship entering the Baltic Sea must comply with the antipollution regulations of the Helsinki Convention and MARPOL Convention. Even though strict controls over ships' discharges were established by the Baltic Sea countries, illegal spills and discharges still happen. The number of illegal oil spills has been reduced significantly over the last 20 years, from 763 spills in 1989 to 178 spills in 2009, and this is an evident and positive tendency, resulted from the long-term efforts of HELCOM.[16]

However, the actual total number of oil spills and their volume seem to be unknown because these values contradict significantly (10–20 times) with estimates of different organizations[3,4,6] and with the results of complex operational satellite monitoring performed in 2004–2009 in the Southeastern Baltic Sea.[20–22] Although the number of observations of illegal oil discharges shows a decreasing trend over the years, it should be kept in mind that for some areas and countries, aerial and satellite surveillance is not evenly and regularly carried out and therefore there are no reliable figures. We have to add to the uncertainties in the oil pollution statistics considerable seasonal variability in observations of oil spills on the sea surface and predominance of the "night" discharge of oil spills from the ships used to avoid any direct visible evidence of pollution, and responsibility for this fact.

So far, as the Baltic Sea ecosystem undergoes increasing human-induced impacts, especially associated with intensifying oil transport and production, further research on the link between physical, chemical, and biological parameters of the ecosystem, the complex monitoring of the Baltic Sea state, and, especially, oil spill monitoring are of great importance.[26] Oil spill behavior, modeling, prevention, effects, control, and cleanup techniques require supplementary information about a large number of complex physical, chemical, and biological processes and phenomena.

ASAR *ENVISAT* and SAR *RADARSAT* provide effective opportunities to monitor oil spills, particularly in the Baltic Sea, as well as in other European seas. Combined with satellite remote sensing (AVHRR NOAA, MODIS-*Terra* and -*Aqua*, QuikSCAT, *Jason-1*, etc.), of the SST, sea level, chlorophyll and suspended matter concentration, meso- and small-scale dynamics, and wind and waves, this observational system represents a powerful method for long-term monitoring of the ecological state of semi-enclosed seas especially vulnerable to oil pollution. Our experience in the operational oil pollution monitoring in the Baltic Sea could be easily applied to the Caspian, Black, Mediterranean, and other European seas.

Since 2004, we have elaborated several operational satellite monitoring systems for oil and gas companies in Russia and performed complex satellite monitoring of the ecological state of coastal waters in the Baltic, Black, Caspian, and Mediterranean seas.[27] The accident on the BP oil platform "Deepwater Horizon" on April 20, 2010, in the Gulf of Mexico showed that the absence of such a permanent complex satellite monitoring system makes all efforts related to cleaning operations at sea and on the shore during the first weeks after the accident less effective.[28]

A large number of discharges of hydrocarbons that annually take place in European waters, the vast quantity of waste generated by the sea traffic in Europe, the lack of adequate port installations for waste management, and the toxicity of compounds thrown into the sea make solving the chronic hydrocarbon pollution problem a priority for improving the environmental quality of European seas.[3,4] The growing availability of satellite and sea observation data should encourage interest, involvement, and investment into the complex operational monitoring systems from the side of the state authorities responsible for the environment, pollution control, meteorology, coastal protection, transport, fisheries, and hazard management, as well as from the side of private companies operating in the sea and coastal zone.

REFERENCES

1. ITOPF (International Tanker Owners Pollution Federation Limited), Handbook 2010/2011, available on line at http://www.itopf.com/information-services/publications/documents/itopfhandbook2010.pdf (accessed May 22, 2012).
2. European Environment Agency. *The European Environment—State and Outlook 2005*; EEA: Copenhagen, 2005.
3. Oceana. *The Other Side of Oil Slicks*; The Dumping of Hydrocarbons from Ships into the Seas and Oceans of Europe; 2003, http://www.oceana.org/north-america/publications/reports/the-other-side-of-oil-slicks (accessed May 22, 2012).
4. Oceana. *The EU Fleet and Chronic Hydrocarbon Contamination of the Oceans*, 2004, http://www.oceana.org/north-america/publications/reports/the-eu-fleet-and-chronic-hydrocarbon-contamination (accessed May 22, 2012).
5. UNESCO. *The integrated, strategic design plan for the coastal ocean observations module of the Global Ocean Observing System*, GOOS Report N 125, IOC Information Documents Series N 1183; UNESCO: Paris, 2003.
6. Finnish Environment Institute, 2004, available at http://www.ymparisto.fi (accessed March 15, 2005).
7. Migliaccio, M.; Gambardella, A.; Tranfaglia M. Oil spill observation by means of polarimetric SAR data. Proceedings of SEASAR 2006, Frascati, Italy, January 23–26, 2006; ESA SP-613; April 2006.
8. Lavrova, O.; Bocharova, T.; Kostianoy, A. Satellite radar imagery of the coastal zone: Slicks and oil spills. In *Global Developments in Environmental Earth Observation from Space*; Marçal, A., Ed.; Millpress Science Publishers: Rotterdam, Netherlands, 2006; 763–771.
9. Kostianoy, A.G. Satellite monitoring of oil pollution in the European Coastal Seas. Oceanis, **2008**, *34* (1/2), 111–126.
10. Kostianoy, A.G.; Lavrova, O.Y.; Mityagina, M.I. Integrated satellite monitoring of oil pollution in the Russian seas. In *Problems of Ecological Monitoring and Modelling of Ecosystems*; Izrael Yu, A., Ed.; 2009; Vol. 22, 235–266 (in Russian).
11. MARIS. *Maritime Accident Response Information System—MARIS. Baltic Marine Environment Protection Commission (HELCOM)*, 2004, http://www.helcom.fi/maris.html (accessed March 15, 2005).
12. Hanninen, S.; Rytkonen, J. Oil transportation and terminal development in the Gulf of Finland; VIT publication N 547; VIT Technical Research Center of Finland, 2004, 141 pp.
13. Kuronen, J.; Helminen, R.; Lehikoinen, A.; Tapaninen, U. *Maritime transportation in the Gulf of Finland in 2007 and in 2015*; Publications from the Centre for Maritime Studies, University of Turku, 2008, N A-45, 114 pp.
14. HELCOM. Maritime activities in the Baltic Sea—An integrated thematic assessment on maritime activities and response to pollution at sea in the Baltic Sea Region. Balt. Sea Environ. Proc. **2010**, *123*, 68.
15. Global Marine Oil Pollution Information Gateway, 2004, http://oils.gpa.unep.org/framework/region-2next.htm (accessed March 15, 2005).
16. HELCOM Response. *Illegal discharges of oil in the Baltic Sea during 2009*. HELCOM Indicator Fact Sheets 2009, http://www.helcom.fi/BSAP_assessment/ifs/ifs2010/en_GB/illegaldischarges/ (accessed November 20, 2010).
17. Kostianoy, A.G.; Lebedev, S.A.; Litovchenko, K.Ts.; Stanichny, S.V.; Pichuzhkina, O.E. Satellite remote sensing of oil spill pollution in the southeastern Baltic Sea. Gayana **2004**, *68* (2), 327–332.
18. Kostianoy, A.G.; Lebedev, S.A.; Soloviev, D.M.; Pichuzhkina, O.E. *Satellite monitoring of the Southeastern Baltic Sea. Annual Report 2004*; Lukoil-Kaliningradmorneft: Kaliningrad, 2005.
19. Kostianoy, A.G.; Lebedev, S.A.; Litovchenko, K.Ts.; Stanichny, S.V.; Pichuzhkina O.E. Oil spill monitoring in the Southeastern Baltic Sea. Environ. Res., Eng. Manage. **2005**, *3* (33), 73–79.
20. Kostianoy, A.G.; Litovchenko, K.Ts.; Lavrova, O.Y.; Mityagina, M.I.; Bocharova, T.Y.; Lebedev, S.A.; Stanichny, S.V.; Soloviev, D.M.; Sirota, A.M.; Pichuzhkina, O.E. Operational satellite monitoring of oil spill pollution in the southeastern Baltic Sea: 18 months experience. Environ. Res., Eng. Manage. **2006** (38), 70–77.
21. Kostianoy, A.G.; Lavrova, O.Y., Eds. *Oil Pollution in the Baltic Sea. The Handbook of Environmental Chemistry*; Springer-Verlag: Berlin, Heidelberg, New York, 2012.
22. Bulycheva, E.V.; Kostianoy, A.G. Results of the satellite monitoring of oil pollution in the Southeastern Baltic Sea in 2006–2009. Modern problems of remote sensing of the Earth from space, **2011**, *8* (2), 74–83 (in Russian).
23. Ambjorn, C. Seatrack Web, forecast of oil spills, a new version. Environ. Res., Eng. Manage. **2007**, *3* (41), 60–66.
24. Kostianoy, A.; Ermakov, P.; Soloviev, D. Complex satellite monitoring of the Nord Stream gas pipeline construction. Proc., US/EU Baltic 2008 International Symposium "Ocean Observations, Ecosystem-Based Management and Forecasting", May 27–29, 2008, Tallinn, Estonia, 2008.
25. Kostianoy, A.; Ambjorn, C.; Soloviev, D. Seatrack Web: A numerical tool to protect the Baltic Sea marine protected areas. Proc., US/EU Baltic 2008 International Symposium "Ocean Observations, Ecosystem-Based Management and Forecasting", May 27–29, 2008, Tallinn, Estonia, 2008.
26. Hopkins, C.C.E. *Overview of monitoring in the Baltic Sea*. Report to the Global Environment Facility/Baltic Sea Regional Project. AquaMarine Advisers: Denmark, 2000.
27. Kostianoy, A.G.; Solovyov, D.M. Operational satellite monitoring systems for marine oil and gas industry. Proc. 2010 Taiwan Water Industry Conference, October 28–29, 2010, Tainan, Taiwan, 2010, P. B173–B185.
28. Lavrova, O.Y.; Kostianoy, A.G. A catastrophic oil spill in the Gulf of Mexico in April–May 2010. Russian J. Remote Sensing (Issledovanie Zemli iz Kosmosa) **2010**, *6*, 67–72 (in Russian).

Organic Compounds: Halogenated

Marek Tobiszewski
Jacek Namieśnik
Department of Analytical Chemistry, Chemical Faculty, Gdansk University of Technology, Gdansk, Poland

Abstract

The entry focuses on halogenated organic compounds, mainly short-chain chlorinated aliphatic organic compounds. It describes their occurrence in the different compartments of the environment and their environmental fate, presents the basic toxicological mechanisms, and discusses the different approaches to remediation processes as applied to soil and water pollution. Finally, basic analytical solutions for the determination of halogenated organic compounds in different media are briefly characterized.

INTRODUCTION

Halogenated organic compounds are a group of contaminants of great environmental impact. They include polychlorinated and polybrominated biphenyls, chlorinated pesticides, short-chain aliphatic halogenated compounds, and halobenzenes; all are well-known pollutants. This entry will focus on halogenated solvents (short-chain halogenated aliphatic hydrocarbons); polyhalogenated biphenyls, pesticides, and flame retardants are dealt with in other entries.

Halogenated solvents have a high vapor pressure and hydrophobicity; thus, they are poorly soluble in water. If present in the environment, they tend to accumulate in living organisms, where their toxic, carcinogenic, mutagenic, and teratogenic properties are manifested.

TOXICITY

Organohalogen compounds can enter living organisms by inhalation, skin (dermal) contact, or oral ingestion. The last mentioned is connected with drinking of contaminated water or consumption of contaminated food.[1] Inhalation and skin contact arise mainly from the occupational exposure of employees to these substances in the chemical and polymer,[2] electronic parts,[3] paints, and metal industries.[1] Employees can also be exposed to halogenated organic compounds during gluing, spraying, degreasing, and dry-cleaning.[4]

The acute toxicity arising from high concentrations of halogenated solvents may manifest as narcosis and central nervous system depression.[5] Halogenated solvents are lipophilic: they readily diffuse across cell membranes and can be accumulated in the brain and other nerve cells. This results in headache, dizziness, reduced dexterity, and increased reaction time. Higher concentrations (e.g., >6 g m^{-3} in the case of trichloroethene [TCE]) can lead to coma and death.[6]

Chronic exposure to halogenated solvents reveals their toxic, carcinogenic, and mutagenic properties. Exposure to organic solvents during pregnancy may be a cause of cancer in the baby, especially leukemia.[7] Workers exposed to chlorinated solvents for up to 4 years showed increased incidence of leukemia, lymphoma, and urinary tract cancer.[8] The specificity of the target organ depends on the species and sex of the organism, the strain (in case of microorganisms), and the kind of halogenated solvent.[9] The solvent TCE causes primarily cancer of the liver and biliary tract, whereas tetrachloroethene (PCE) may be the cause of esophageal and cervical cancer.[10]

Halogenated solvents can be excreted from an organism in unchanged form or they may be metabolized. Two classes of enzymes are capable of biotransforming chloroethenes—cytochrome P-450 and glutathione S-transferase (GST) (see Fig. 1).[11] The major products of cytochrome P-450 oxidation are chlorinated epoxides and chloral, which are further oxidized to chlorinated ethanol and chloroacetic acid. The other metabolic pathway involves the conjugation of chlorinated ethene to glutathione, which is further metabolized to a cysteine conjugate, which is then degraded to a thiol.[12] Metabolism through cytochrome P-450 and the two initial steps of the GST pathway take place in various organs, but predominantly in the liver. The cysteine conjugate is metabolized mainly in the kidneys;[13] several of the metabolic products can form adducts with biomolecules.[14] At low concentrations, halogenated solvents are metabolized primarily through cytochrome P-450, and at higher concentrations by GST. Some halogenated compounds such as dichloromethane, especially at low concentrations, can be converted to carbon monoxide, which binds to hemoglobin.[15]

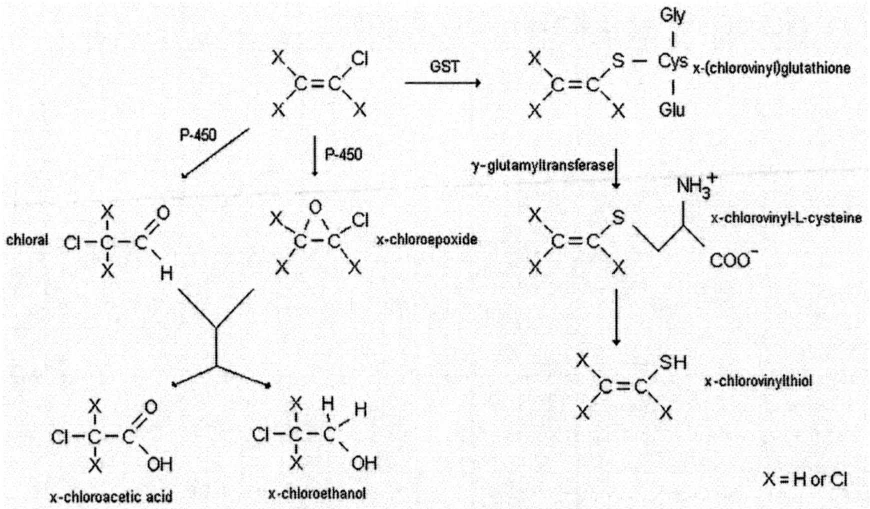

Fig. 1 Metabolic pathways of chlorinated solvents.

APPLICATIONS AND SOURCES

There are many natural and anthropogenic sources of organohalogen compounds in the environment. They are used as industrial solvents, as intermediates in organic and polymer synthesis, and as degreasing, lubricating, and cleaning agents, as well as in pesticide production and the pharmaceutical and medical industries.[16]

During water chlorination disinfection, by-products are formed that pose risk to human health. In the presence of chlorides or bromides and humic matter, formation of trihalomethanes, haloacetic acids, chloropicrin, haloacetonitriles, and halophenols occurs.

There are 3800 known organohalogen compounds naturally produced in the environment,[17] and these include fluorinated, chlorinated, brominated, iodinated aliphatic, and aromatic compounds. They range from short-chain compounds to highly complex molecular structures.[17] Organohalogen compounds can be produced by biota, mainly marine (algae, phytoplankton), but also by some evergreen trees, potato tubers, and wood-rotting fungi.[18] Marine micro- and macroalgae are the main emitters; the halogenated substances they emit include halomethanes and haloethanes.[19] In the case of bromoform, emission rates may be as high as 4 μg g^{-1} wet tissue per day.[20] The marine algae present in tropical regions emit more organohalogen compounds than those occurring in temperate and arctic regions.[21] Organohalogen compounds are reported to be emitted from the terrestrial environment—they have been found in the air and topsoil air of spruce forests.[22] Animals also contribute to the emission of such compounds. The activity of termites causes chloroform to be emitted: 15% of natural chloroform emissions are said to come from this source. Different tick species emit dichlorophenols, which are sex pheromones.[18] Organohalogen compounds (mainly short-chain aliphatic compounds) are emitted from pig[23] and dairy[24] farms; they are also formed in abiotic processes such as biomass fires, volcanic activity, and geothermal and rock weathering.

The utilization of chlorinated solvents involves their accidental or intentional release into the environment. The global anthropogenic emission of chloroform in 1990 was 10^5 tons. Annual emissions of chloromethane are 1.6×10^{11} g from combined coal combustion, incineration, and industrial sources;[25] however, this is only a fraction of the 3.5×10^{12} to 5×10^{12} g of chloromethane naturally produced in the environment each year, mainly by marine algae.[26] Non-natural sources of chloroform are estimated at 6.6×10^{11} g/yr.[27] The global anthropogenic emission of 1,1,1-TCE was estimated to have peaked in 1990 at a value of 7.17×10^{11} g; by 2000, this level had fallen to 1.97×10^{10} g.[28] The global production and consumption of chlorinated solvents is now decreasing.[29]

ENVIRONMENTAL FATE

As organohalogen compounds can be discharged into different compartments of the environment (air, water, or soil), their environmental fate depends on the pollution source and on the physical, chemical, and/or biological changes they undergo (Fig. 2).

Atmospheric organohalogen compounds contribute to stratospheric ozone depletion. Since they absorb light in the infrared spectrum, they contribute to global warming. Pollutants are removed from the atmosphere mainly by abiotic degradation. The predominant removal pathway of organohalogen compounds from the troposphere is the reaction with an OH radical, atomic chlorine,[30] ozone, or nitrate radicals.[31] The atmospheric lifetimes of TCE as regards reaction with a hydroxyl radical is 7 days, whereas those of PCE and dichloromethane are less than half a

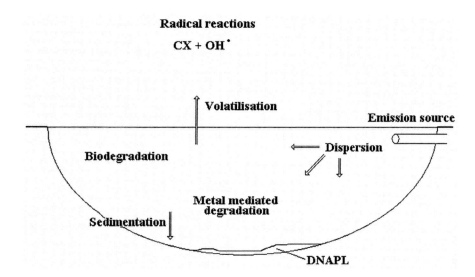

Fig. 2 Natural attenuation processes.

year.[32] The atmospheric transport of these pollutants with air masses depends on their lifetime in the air. Thus, PCE and TCE have been detected even in arctic regions. Their concentrations depend strongly on air fluxes and are affected mainly by pollution from European sources.[33] Atmospheric chlorinated solvents can be adsorbed on the surfaces of leaves and can be taken up by plants.[34]

Organohalogen compounds discharged to groundwater can form a non-aqueous phase liquid (NAPL). NAPLs are formed as a consequence of the poor solubility of halogenated solvents in water. If the solvent is less dense than water, a light NAPL (LNAPL) is formed, whereas denser solvents form dense NAPLs (DNAPLs); it is usually the latter that are formed in the environment. DNAPLs sink and spread until they reach a kind of niche—they do not necessarily move in the direction of the water flow. Because they dissolve slowly in the water, they are a serious threat to aquifers. DNAPLs can remain in groundwater from decades to centuries.[35] Halogenated solvents present in groundwater can also evaporate to the unsaturated zone of the soil.

Hydrolysis, hydrogenolysis, dehydrohalogenation, and dihaloelimination are the main degradation pathways of halogenated solvents (see Fig. 3).[36] Hydrodehalogenation and hydrolysis are abiotic reactions, whereas hydrogenolysis and dihaloelimination are reduction reactions, occurring in the presence of biologically derived compounds or metallic reducing agents. Hydrolysis involves the substitution of a halogen atom with a hydroxyl group, resulting in the formation of an alcohol as a product, a process that is favorable when the water pH is basic. Since environmental water is usually slightly acidic, hydrolysis of halogenated solvents is extremely slow, with a half-life of hundreds of years. Hydrodehalogenation implies the removal of halogen hydride (HX) from an organohalogen molecule, resulting in the formation of a double bond.

Similarly, dihaloelimination is the removal of two halogen atoms and the simultaneous addition of two electrons to form an unsaturated hydrocarbon. Hydrogenolysis is the addition of a proton and two electrons to the molecule with the release of chloride. Halogenated organic compounds can react in redox reactions, with other molecules serving as electron donors or acceptors. Such compounds can be biologically derived enzymes or transition group metals (e.g., iron, nickel, chromium, cobalt). The presence of halogen atoms makes the hydrocarbon molecule more oxidized. Thus, highly halogenated short-chain aliphatic

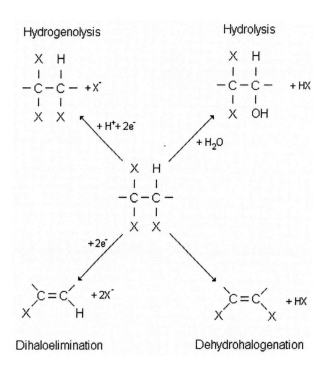

Fig. 3 Degradation pathways of chlorinated solvents.

compounds are more easily reduced than oxidized. Low-halogenated and non-halogenated degradation daughter compounds may accumulate at the contaminated site if reduction (usually biological) pathways are dominant. Thus, the emission of PCE leads to the presence of other halogenated compounds at the contaminated site—TCE, *cis*-dichloroethene (*cis*-DCE), *trans*-DCE, 1,1-DCE, and vinyl chloride (VC)[37]—daughter compounds that are more toxic than PCE itself. Abiotic degradation reactions occur much more slowly (their half-lives are usually measured in years) than biologically mediated reactions (with half-lives from days to months).

Halogenated solvents can be degraded by certain bacteria. In contrast to aliphatic hydrocarbons, they rarely serve as sole growth substrates; they are usually cometabolized.

Organohalogen compounds discharged to surface waters can degrade along much the same pathways as in groundwater. They can also evaporate into the atmosphere and undergo certain physical and photochemical processes there. Table 1 lists concentrations of selected chlorinated solvents in the atmosphere, and in marine water, surface water, and groundwater.

OCCURRENCE IN THE ENVIRONMENT

Halogenated compounds are present in saline water at nanogram per cubic decimeter (ng dm^{-3}) levels. Their concentrations in contaminated surface water can reach single-figure microgram per cubic decimeter (μg dm^{-3}) levels, and milligram per cubic decimeter (mg dm^{-3}) levels in contaminated groundwater. Higher concentrations of volatile halogenated contaminants are noted in urban air than in remote areas.

REMEDIATION STRATEGIES

The many possible environmental pathways of halogenated solvents make remediation of different environmental compartments difficult. The proper choice of remediation technology depends on the initial concentration of halogenated organic compounds, the spatial distribution of pollutants, soil porosity, and the time required for the process to take place.[51] Remediation of halogenated volatile organic compound (HVOC)-polluted sites is complex owing to the simultaneous occurrence of dissolved contaminants, vapors in unsaturated soil, and NAPLs. NAPLs are difficult to remove because they can be widely dispersed in the aquifer or soil.

Remediation is often performed by the pump-and-treat method, in which a series of extraction wells pump water for aboveground treatment.[52] Air sparging is based on pumping air into a saturated zone of soil or underground water. The air creates unsaturated zones in the soil, and volatile organic pollutants are vaporized and removed. Simultaneously, oxygen is pumped into the remediated site, stimulating the oxidative degradation of pollutants. The main advantage of air sparging is that water pumping can be dispensed with (unlike the pump-and-treat approach); the disadvantage is the poor performance of the process caused by the rebound of pollutant concentrations once sparging has been completed. Halogenated solvents are merely transferred from the aqueous phase to the atmosphere; no degradation occurs.[53] To prevent migration of contaminants, soil vapor extraction systems are applied in combination with air sparging.[54] Soil vapor extraction is a remediation technology that can be applied in isolation: a vacuum is applied at the extraction well, causing vapors to migrate toward it; pollutants are thus removed.[55]

Different classes of surfactants can remove chlorinated solvents from soil[56] and NAPLs from water.[57] Surfactant-enhanced aquifer remediation is capable of removing organic solvents as it improves the solubility and mobility of NAPLs by lowering interfacial tension and micellar solubilization. Surfactants are applied in soil remediation as they increase the desorption rates of pollutants. This technology is based on pumping a surfactant solution to the contaminated site to enhance extraction. When the solution comes to the surface, the surfactant has to be separated so that it can be reused. The basic requirements for surfactants are low cost, biodegradability, low toxicity, and sorption on soil.[58] The other remediation option is oxidation of chlorinated solvents with strong oxidants such as potassium permanganate (KMnO$_4$). KMnO$_4$ is released into groundwater through injection wells or can be placed on controlled-release solids that form permeable reactive barriers.[59] The reaction of permanganate with chlorinated solvents is

$$C_2H_{(4-n)}Cl_n + 2MnO_4^- \rightarrow 2MnO_2(s) + 2CO_2 + nCl^- + (4-n)H^+.$$

The reaction rates are faster for compounds containing fewer rather than more chlorine atoms, with half-lives ranging from 0.4 min (*trans*-DCE) to 18 min (PCE).[60] Permanganates are also applied in soil remediation, the main advantage of this method being the low cost of solid KMnO$_4$.[61]

Zero-valent iron and iron (II) are capable of reducing chlorinated short-chain aliphatic compounds. Iron (II) acts as a reductant in the Fenton reaction, which can be used in water and soil remediation. It reacts with hydrogen peroxide to yield the hydroxyl radical and the hydroxyl ion as products, both of which react with target pollutants. The reaction usually requires a low pH; however, in the presence of chelated iron catalysts, the Fenton reaction can be carried out without the local pH conditions having to be changed. The next generation of iron-based remediation technologies is based on iron nanoparticles. The advantages of these technologies include increased reactivity, a high surface area/volume ratio, and enhanced mobility due

Table 1 Concentrations of chlorinated solvents in different environmental compartments.

Compartment of the environment	Location	Compound	Concentration	Units	Reference
Air	Background for northern hemisphere	TCE	0.58–58	ng m^{-3}	[38]
	Urban background		41		
		PCE	34		
	Bristol (U.K.) urban mean	TCE	428		
		PCE	254		
	Fuji, Japan	1,1,1-TCA	0.051–0.195	µg m^{-3}	[39]
		PCE	0.083–5.23		
		TCE	0.126–1.27		
	North Atlantic	PCE	13–25	ng m^{-3}	[33]
		TCE	0.5–8.2		
	Yokohama, Japan	PCE	<68–18,000	ng m^{-3}	[40]
		TCE	<176–8,740		
		1,1,1-TCA	<60–774		
		1,2-DCA	<44–380		
Rainwater	Yokohama, Japan	PCE	<22–467	ng L^{-1}	[40]
		TCE	<22–767		
		1,1,1-TCA	<21–151		
		1,2-DCA	<47–982		
Surface water	Rivers in Osaka, Japan	VC	<0.64–56	µg L^{-1}	[41]
	Scheldt Estuary, Belgium/ Netherlands	PCE	1.4–3,800	ng L^{-1}	[42]
		TCE	<12–830		
		1,1-DCA	<3.1–21		
		1,1,1-TCA	<0.8–73		
		1,2-DCA	<16–140		
		1,1-DCE	<1.7–11		
		trans-DCE	<1–12		
	Rivers of northern Greece	PCE	<0.02–0.19	µg L^{-1}	[43]
		TCE	<0.02–40		
Groundwater	Area 6, Dover Air Force Base, United States	PCE	Up to 2,500	µg L^{-1}	[44]
		TCE	Up to 3,100		
		cis-DCE	Up to 6,500		
		VC	Up to 430		
	Bitterfeld/Wolfen, Germany	PCE	0.05–16,000		[45]
		TCE	0.6–410,000		
		cis-DCE	0.5–85,000		
		trans-DCE	0.5–18,000		
		VC	5–5,400		
	Copenhagen, Denmark	PCE	<0.05–340		[46]
		TCE	<0.01–0.44		
		cis-DCE	<0.1–140		
		trans-DCE	<0.1–1.4		
		1,1-DCE	<0.1–87		
		VC	<0.1–270		

(*Continued*)

Table 1 Concentrations of chlorinated solvents in different environmental compartments. (*continued*)

Compartment of the environment	Location	Compound	Concentration	Units	Reference
		PCE	<0.62–21		
		TCE	<0.89–231		
	Taiwan	1,1-DCE	<0.76–24		[47]
		cis-DCE	<0.53–82.4		
		VC	<0.97–651		
Saline water	Southern North Sea	1,1-DCE	1.2		[48]
		trans-DCE	1.6		
		1,1,1-TCA	6		
		1,1,2-TCA	2		
		1,2-DCP	0.9		
		1,2-DCA	7.8	ng L^{-1}	
		TCE	49		
		PCE	23		
	Ross Sea, Antarctica (from depth profiles)	1,1,1-TCA	<0.5–300		[49]
		TCE	<0.5–96		
		PCE	1.4–22		
Sediments	Scheldt Estuary, Belgium/ Netherlands	PCE	<50–350	pg g^{-1} wet weight	[50]
		TCE	<30–90		

to the small size of the particles. Iron-based nanoparticles are capable of removing not only dissolved halogenated solvents but also NAPLs. Another approach is the application of reactive barrier technologies. A barrier of reactive material is immersed in the water to be treated, and contaminants are intercepted and/or degraded. Iron can also be an important contributor to monitored natural attenuation processes.[62] Other metals capable of degrading chlorinated solvents are nickel and zinc.[63] Microorganisms can be used for remediation purposes. The mechanisms of degradation (reductive dechlorination, cometabolism, oxidation processes) are discussed elsewhere. Microbes can be used in reactors or in situ to stimulate natural processes.[64] Although much bacterial growth is inhibited at high concentrations of chlorinated solvents, some bacterial species are reported to dechlorinate these compounds near NAPL-saturated zones.[65]

An interesting solution is offered by biobarriers (a solid material support for microorganisms), which can be installed in situ in the path of an organohalogen contaminant plume. Support for a biobarrier could be provided by a synthetic polymeric or a natural material, e.g., peat, which offers good bioavailability of carbon; this is inexpensive, easily available, and easily installed in situ.

Monitored natural attenuation is a cost-effective (usually about one-third of the cost of traditional remediation) approach to the remediation of groundwater contaminated by chlorinated solvents. The method utilizes the natural self-purification potential of the contaminated site. Natural attenuation is defined as the sum of processes leading to a decrease in pollutant concentration at the contaminated site: physical—dilution, dispersion, volatilization, adsorption; biological—microbial activity; and chemical—hydrolysis, metal-mediated reactions. To assess the natural attenuation potential of a site, several conditions have to co-occur, e.g., reductive (preferably methanogenic) redox conditions, the presence of certain communities of bacteria and of daughter degradation products (less halogenated compounds or non-halogenated ones). Finally, the actual decrease in concentrations has to be measured and monitored.[44]

An important tool for distinguishing between natural chemical and physical attenuation processes is stable isotope fractionation analysis.[66] This distinction is important as natural physical attenuation processes involve only the transfer of a contaminant from the aqueous phase to another phase or dilution in the aqueous phase, whereas natural chemical attenuation involves chemical reactions leading to an actual decrease of contaminants. The method is based on investigating the ^{13}C/^{12}C ratio, which remains unchanged during physical processes. During abiotic and biotic degradation processes, this ratio is disturbed owing to the different reaction rates of ^{13}C- and ^{12}C-containing compounds. The chemical bonds of molecules containing the lighter isotope are broken more easily, resulting in the enrichment of molecules containing the heavier isotope.[67]

Phytoremediation can be applied to remove halogenated solvents from the soil and to some extent from water.

Rhizodegradation (the rhizosphere is the top layer of soil strongly influenced by plant roots) involves the degradation of pollutants by microorganisms living in symbiosis with plants or in a niche created by the roots. Halogenated solvents can be degraded by plant enzymes (phytodegradation); for instance, dehalogenase is responsible for the dechlorination of chlorinated derivatives of ethene. Plants can also transport halogenated compounds to the atmosphere through phytovolatilization.[68]

ANALYTICAL METHODS

These losses of volatile analytes can occur at the sample collection, transport, or storage stage, or during sample preparation before final analysis; thus, proper sample handling is required.[69] Liquid samples must be collected and stored without a headspace (HS) volume above the sample (the whole volume of the sampler must be filled with sample). Samples should be stored at low temperatures and preferably acidified for better preservation.

Air can be sampled in the form of whole air: a given volume of air is collected in a Tedlar of Teflon container, after which the samples are gas chromatographically analyzed by direct injection or after preconcentration. Samples can be collected by generating an initial subatmospheric pressure in the sampler; alternatively, they can be compressed to collect volumes larger than the container volume. Air samples can be taken by active methods—a given volume of air is pumped through a sorbent bed (e.g., charcoal) in which analytes are trapped.[70] The mode of operation of passive samplers is based on the diffusion of analytes from the air to the surface or bulk of the sampler. Passive samplers are simple to use, do not require pumps or flowmeters, and can be left unattended for a long time. Their main disadvantages are that they are unsuitable for measuring short-term variations and that the calibration process is relatively difficult. The analytes trapped in the passive samplers or in sorbent tubes are released by solvent extraction or thermal desorption.[71]

Direct methods are preferable because the number of operations between sample storage and final analysis are kept to a minimum, thus reducing possible sample contamination or analyte losses. The direct analytical methods for HVOC determination include direct aqueous injection—gas chromatography–electron capture detection (DAI–GC–ECD) and membrane inlet mass spectrometry. Water analysis by DAI–GC–ECD involves the injection of microliter samples into a chromatographic column. The injector is cooled to prevent analyte loss during injection. Detection is normally by electron capture, but mass spectrometry is also a possibility. Since chromatographic columns are sensitive to impurities present in water (e.g., suspended matter), the method is applicable only to samples with simple matrix compositions. Membrane inlet mass spectrometry (MIMS) incorporates the direct injection of analytes into the mass spectrometer: analytes are transferred through a selective polymeric membrane from the aqueous or gaseous sample present on one side of the barrier to the high vacuum of the mass spectrometer. The main advantages of MIMS are simplicity of the sample preparation step, high speed, sensitivity, and precision.[69]

The volatility of halogenated organic contaminants can be put to good use during the analysis. HS analysis is based on the extraction of analytes from liquid or solid samples to the gaseous phase over the sample and injection of the gaseous sample into the chromatographic column. Many approaches to HS analysis are available, but there are two basic ones, static and dynamic, the latter involving movement of a gaseous phase over the sample. The advantage of HS techniques is that a clean sample can be injected into the chromatographic column, thus minimizing the risk of column contamination. The purge-and-trap technique is based on purging the sample with a stream of inert gas and collecting the analytes in a sorbent trap. Once extraction has been completed, the trap is heated and the analytes are introduced into the chromatographic column. This method allows HVOCs to be determined at very low concentrations. A possible drawback is sample foaming, which renders analysis impossible. The standard extraction procedure involves the liquid–liquid extraction of HVOCs with methyl *tert*-butyl ether; however, this is laborious, is difficult to automate, and requires a large amount of solvent. Single-drop microextraction is the upshot of the miniaturization of liquid–liquid extraction. In this technique, analytes are transferred to one drop of organic solvent and placed at the tip of a microsyringe. After extraction, the solvent drop is transferred to the microsyringe from where it can be injected directly into the chromatographic column. The solvent drop can be placed directly in the sample or in the HS above the sample. Solid-phase microextraction (SPME) is based on the transfer of a tiny fraction of analyte onto a fiber placed in the sample solution or HS volume above the sample. After thermostatting, the SPME fiber is placed in the chromatographic injector, which introduces the analytes into the column. The advantage of this technique is simplicity of operation, while the main disadvantage is the sensitivity of the fiber to contamination, mainly by suspended matter or large-molecule compounds, which settle on it, causing a deterioration in the parameters of extraction.[72]

There are numerous extraction techniques using membranes, porous as well as nonporous ones. Analyses are usually performed using membrane extraction with a sorbent interphase, membrane-assisted solvent extraction, and microporous membrane liquid–liquid extraction. The main advantage of this group of methods is their high selectivity: samples with very complex matrix compositions can be analyzed. Their shortcomings, however, include poor extraction efficiency and the sensitivity of membranes to contamination.

SUMMARY

Halogenated volatile organic compounds are a serious threat to the environment, especially to freshwater resources. It is crucial to be in possession of knowledge about the environmental fate of these compounds so that proper measures can be applied to deal with pollution. Analytical methodologies are being developed for reliable assessments of the problem, and feasible remediation strategies are being improved for cleaning contaminated sites.

ACKNOWLEDGMENTS

M. Tobiszewski is grateful for grants awarded by the Ministry of Science and Higher Education N523 562838 and Iuventus Plus Grant IP2011 056271.

REFERENCES

1. Brautbar, N.; Williams, II, J. Industrial solvents and liver toxicity: Risk assessment, risk factors and mechanisms. Int. J. Hyg. Environ. Health **2002**, *205*, 479–491.
2. Giri, A.K. Genetic toxicology of vinyl chloride—A review. Mutat. Res. **1995**, *339*, 1–14.
3. Chang, Y.M.; Tai, C.F.; Yang, S.C.; Lin, R.S.; Sung, F.C.; Shih, T.S.; Liou, S.H. Cancer incidence among workers potentially exposed to chlorinated solvents in an electronics factory. J. Occup. Health **2005**, *47*, 171–180.
4. Blair, A.; Petralia, S.A.; Stewart, P.A. Extended mortality follow-up of a cohort of dry cleaners. Ann. Epidemiol. **2003**, *13*, 50–56.
5. Schenker, M.B.; Jacobs, J.A. Respiratory effects of organic solvent exposure. Tuber. Lung Dis. **1996**, *77*, 4–18.
6. Winek, C.L.; Wahba, W.W.; Huston, R.; Rozin, L. Fatal inhalation of 1,1,1-trichloroethane. Forensic Sci. Int. **1997**, *87*, 161–165.
7. Sung, T.I.; Wang, J.D.; Chen, P.-C. Increased risk of cancer in the offspring of female electronics workers. Reprod. Toxicol. **2008**, *25*, 115–119.
8. Zarchy, T.M. Chlorinated hydrocarbon solvents and biliary-pancreatic cancer: Report of three cases. Am. J. Ind. Med. **1996**, *30*, 341–342.
9. Cummings, B.S.; Parker, J.C.; Lash, L.H. Role of cytochrome P450 and glutathione S-transferase A in the metabolism and cytotoxicity of trichloroethylene in rat kidney. Biochem. Pharmacol. **2000**, *59*, 531–543.
10. Lynge, E.; Anttila, A.; Hemminki, K. Organic solvents and cancer. Cancer Causes Control **1997**, *8*, 406–419.
11. Tafazoli, M.; Kirsch-Volders, M. In vitro mutagenicity and genotoxicity study of 1,2-dichloroethylene, 1,1,2-trichloroethane, 1,3-dichloropropane, 1,2,3-trichloropropane and 1,1,3-trichloropropene, using the micronucleus test and the alkaline single cell gel electrophoresis technique (comet assay) in human lymphocytes. Mutat. Res. **1996**, *371*, 185–202.
12. Lash, L.H.; Putt, D.A.; Huang, P.; Hueni, S.E.; Parker, J.C. Modulation of hepatic and renal metabolism and toxicity of trichloroethylene and perchloroethylene by alterations in status of cytochrome P450 and glutathione. Toxicology **2007**, *235*, 11–26.
13. Lash, L.H.; Putt, D.A.; Hueni, S.E., Horwitz, B.P. Molecular markers of trichloroethylene-induced toxicity in human kidney cells. Toxicol. Appl. Pharmacol. **2005**, *206*, 157–168.
14. van Hylckama Vlieg, J.E.T.; Janssen, D.B. Formation and detoxification of reactive intermediates in the metabolism of chlorinated ethenes. J. Biotechnol. **2001**, *85*, 81–102.
15. Slikker, Jr., W.; Andersen, M.E.; Bogdanffy, M.S.; Bus, J.S.; Cohen, S.D.; Conolly, R.B.; David, R.M.; Doerrer, N.G.; Dorman, D.C.; Gaylor, D.W.; Hattis, D.; Rogers, J.M.; Setzer, R.W.; Swenberg, J.A.; Wallace, K. Dose-dependent transitions in mechanisms of toxicity: Case studies. Toxicol. Appl. Pharmacol. **2004**, *201*, 226–294.
16. Olaniran, A.O.; Pillay, D.; Pillay, B. Chloroethenes contaminants in the environment: Still a cause for concern. Afr. J. Biotechnol. **2004**, *3*, 675–682.
17. Gribble, G.W. The diversity of naturally produced organohalogens. Chemosphere **2003**, *52*, 289–297.
18. Gribble, G.W. The diversity of natural organochlorines in living organisms. Pure Appl. Chem. **1996**, *68*, 1699–1712.
19. Laturnus, F.; Wiencke, C.; Kloser, H. Antarctic macroalgae—Sources of volatile halogenated organic compounds. Mar. Environ. Res. **1996**, *41*, 169–181.
20. Laturnus, F. Release of volatile halogenated organic compounds by unialgal cultures of polar macroalgae. Chemosphere **1995**, *31*, 3387–3395.
21. Abrahamsson, K.; Choo, K.S.; Pedersen, M.; Johansson, G.; Snoeijs, P. Effects of temperature on the production of hydrogen peroxide and volatile halocarbons by brackish-water algae. Phytochemistry **2003**, *64*, 725–734.
22. Haselmann, K.F.; Ketola, R.A.; Laturnus, F.; Lauritsen, F.R.; Gron, C. Occurrence and formation of chloroform at Danish forest sites. Atmos. Environ. **2000**, *34*, 187–193.
23. Blunden, J.; Aneja, V.P.; Lonneman, W.A. Characterization of non-methane volatile organic compounds at swine facilities in eastern North Carolina. Atmos. Environ. **2005**, *39*, 6707–6718.
24. Filipy, J.; Rumburg, B.; Mount, G.; Westberg, H.; Lamb, B. Identification and quantification of volatile organic compounds from a dairy. Atmos. Environ. **2006**, *40*, 1480–1494.
25. Harper, D.B. The global chloromethane cycle: Biosynthesis, biodegradation and metabolic role. Nat. Prod. Rep. **2000**, *17*, 337–348.
26. Ballschmiter, K. Pattern and sources of naturally produced organohalogens in the marine environment: Biogenic formation of organohalogens. Chemosphere **2003**, *52*, 313–324.
27. McCulloch, A. Chloroform in the environment: Occurrence, sources, sinks and effects. Chemosphere **2003**, *50*, 1291–1308.
28. McCulloch, A.; Midgley, P.M. The history of methyl chloroform emissions: 1951–2000. Atmos. Environ. **2001**, *35*, 5311–5319.
29. Eurochlor, http://www.eurochlor.org/facts (accessed May 10, 2012).
30. Rudolph, J.; Koppmann, R.; Plass-Dülmer, C. The budgets of ethane and tetrachloroethene: Is there evidence for an impact of reactions with chlorine atoms in the troposphere? Atmos. Environ. **1996**, *30*, 1887–1894.

31. Klopffer, W. Environmental hazard assessment of chemicals and products. Part VI. Abiotic degradation in the troposphere. Chemosphere **1996**, *33*, 1083–1099.
32. McCulloch, A.; Midgley, P.M. The production and global distribution of emissions of trichloroethene, tetrachloroethene and dichloromethane over the period 1988–1992. Atmos. Environ. **1996**, *30*, 601–608.
33. Dimmer, C.H.; McCulloch, A.; Simmonds, P.G.; Nickless, G.; Bassford, M.R.; Smythe-Wright, D. Tropospheric concentrations of the chlorinated solvents, tetrachloroethene and trichloroethene, measured in the remote northern hemisphere. Atmos. Environ. **2001**, *35*, 1171–1182.
34. Cape, J.N. Effects of airborne volatile organic compounds on plants. Environ. Pollut. **2003**, *122*, 145–157.
35. Lucas, L.; Jauzein, M. Use of principal component analysis to profile temporal and spatial variations of chlorinated solvent concentration in groundwater. Environ. Pollut. **2008**, *151*, 205–212.
36. Vogel, T.M.; Criddle, C.S.; McCarty, P.L. Transformations of halogenated aliphatic compounds. Environ. Sci. Technol. **1987**, *21*, 722–736.
37. McNab, Jr., W.W. Forensic analysis of chlorinated hydrocarbon plumes in groundwater: A multisite perspective. Environ. Forensics **2001**, *2*, 313–320.
38. Rivett, A.C.; Martin, D.; Nickless, G.; Simmonds, P.G.; O'Doherty, S.J.; Gray, D.J.; Shallcross, D.E. In situ gas chromatographic measurements of halocarbons in an urban environment. Atmos. Environ. **2003**, *37*, 2221–2235.
39. Kume, K.; Ohura, T.; Amagai, T.; Fusaya, M. Field monitoring of volatile organic compounds using passive air samplers in an industrial city in Japan. Environ. Pollut. **2008**, *153*, 649–657.
40. Okochi, H.; Sugimoto, D.; Igawa, M. The enhanced dissolution of some chlorinated hydrocarbons and monocyclic aromatic hydrocarbons in rainwater collected in Yokohama, Japan. Atmos. Environ. **2004**, *38*, 4403–4114.
41. Yamamoto, K.; Fukushima, M.; Kakautani, N.; Tsuruho, K. Contamination of vinyl chloride in shallow urban rivers in Osaka. Water Res. **2001**, *35*, 561–566.
42. Huybrechts, T.; Dewulf, J.; Van Langenhove, H. State-of-the-art of gas chromatography-based methods for analysis of anthropogenic volatile organic compounds in estuarine waters, illustrated with the river Scheldt as an example. J. Chromatogr. A **2003**, *1000*, 283–297.
43. Kostopoulou, M.N.; Golfinopoulos, S.K.; Nikolaou, A.D.; Xilourgidis, N.K.; Lekkas, T.D. Volatile organic compounds in the surface waters of Northern Greece. Chemosphere **2000**, *40*, 527–532.
44. Witt, M.E.; Klecka, G.M.; Lutz, E.J.; Ei, T.A.; Grosso, N.R.; Chapelle, F.H. Natural attenuation of chlorinated solvents at Area 6, Dover Air Force Base: Groundwater biogeochemistry. J. Contam. Hydrol. **2002**, *57*, 61–80.
45. Nijenhuis, I.; Nikolausz, M.; Koth, A.; Felfoldi, T.; Weiss, H.; Drangmeister, J.; Großmann, J.; Kastner, M.; Richnow, H.H. Assessment of the natural attenuation of chlorinated ethenes in an anaerobic contaminated aquifer in the Bitterfeld/Wolfen area using stable isotope techniques, microcosm studies and molecular biomarkers. Chemosphere **2007**, *67*, 300–311.
46. Broholm, K.; Ludvigsen, L.; Feldthusen, T.; Řstergaard, J.H. Aerobic biodegradation of vinyl chloride and *cis*-1,2-dichloroethylene in aquifer sediments. Chemosphere **2005**, *60*, 1555–1564.
47. Fan, C.; Wang, G.S.; Chen, Y.C.; Ko, C.H. Risk assessment of exposure to volatile organic compounds in groundwater in Taiwan. Sci. Total Environ. **2009**, *407*, 2165–2174.
48. Huybrechts, T.; Dewulf, J.; van Langenhove H. Priority volatile organic compounds in surface waters of the southern North Sea. Environ. Pollut. **2005**, *133*, 255–264.
49. Zoccolillo, L.; Abete, C.; Cafaro, C.; Insogna, S. Evaluation of volatile chlorinated hydrocarbons distribution along depth profiles in the Ross Sea, Antarctica. Microchem. J. **2009**, *92*, 32–36.
50. Roose, P.; Dewulf, J.; Brinkman, U.A.; van Langenhove, H. Measurement of volatile organic compounds in sediments of the Scheldt estuary and the southern North Sea. Water Res. **2001**, *35*, 1478–1488.
51. Janda, V.; Vasek, P.; Bizova, J.; Belohlav, Z. Kinetic models for volatile chlorinated hydrocarbons removal by zerovalent iron. Chemosphere **2004**, *54*, 917–925.
52. Mackay, D.M.; Wilson, R.D.; Brown, M.J.; Ball, W.P.; Xia, G.; Durfee, D.P. A controlled field evaluation of continuous vs. pulsed pump-and-treat remediation of a VOC-contaminated aquifer: Site characterization, experimental setup, and overview of results. J. Contam. Hydrol. **2000**, *81*, 81–131.
53. Bass, D.H.; Hastings, N.A.; Brown, R.A. Performance of air sparging systems: A review of case studies. J. Hazard. Mater. **2000**, *72*, 101–119.
54. Waduge, W.A.P.; Soga, K.; Kawabata, J. Effect of NAPL entrapment conditions on air sparging remediation efficiency. J. Hazard. Mater. **2004**, *110*, 173–183.
55. Nobre, M.M.M.; Nobre, R.C.M. Soil vapor extraction of chlorinated solvents at an industrial site in Brazil. J. Hazard. Mater. **2004**, *110*, 119–127.
56. Mulligan, C.N.; Yong, R.N.; Gibbs B.F. Surfactant-enhanced remediation of contaminated soil: A review. Eng. Geol. **2001**, *60*, 371–380.
57. Qin, X.S.; Huang, G.H.; Chakma, A.; Chen, B.; Zeng, G.M. Simulation-based process optimization for surfactant-enhanced aquifer remediation at heterogeneous DNAPL-contaminated sites. Sci. Total Environ. **2007**, *381*, 17–37.
58. Paria, S. Surfactant-enhanced remediation of organic contaminated soil and water. Adv. Colloid Interface Sci. **2008**, *138*, 24–58.
59. Lee, E.S.; Schwartz, F.W. Characteristics and applications of controlled-release $KMnO_4$ for groundwater remediation. Chemosphere **2007**, *66*, 2058–2066.
60. Yan, Y.E.; Schwartz, F.W. Oxidative degradation and kinetics of chlorinated ethylenes by potassium permanganate. J. Contam. Hydrol. **1999**, *37*, 343–365.
61. Schnarr, M.; Truax, C.; Farquhar, G.; Hood, E.; Gonullu, T.; Stickney, B. Laboratory and controlled field experiments using potassium permanganate to remediate trichloroethylene and perchloroethylene DNAPLs in porous media. J. Contam. Hydrol. **1998**, *29*, 205–224.
62. Cundya, A.B.; Hopkinson, L.; Whitby, R.L.D. Use of iron-based technologies in contaminated land and groundwater remediation: A review. Sci. Total Environ. **2008**, *400*, 42–51.
63. Chen, S.; Wu, S. Feasibility of using metals to remediate water containing TCE. Chemosphere **2001**, *42*, 1023–1028.

64. Pant, P.; Pant, S. A review: Advances in microbial remediation of trichloroethylene (TCE). J. Environ. Sci. **2010**, *22*, 116–126.
65. Amos, B.K.; Suchomel, E.J.; Pennell, K.D.; Loffler, F.E. Microbial activity and distribution during enhanced contaminant dissolution from a NAPL source zone. Water Res. **2008**, *42*, 2963–2974.
66. Meckenstock, R.U.; Morasch, B.; Griebler, C.; Richnow, H.H. Stable isotope fractionation analysis as a tool to monitor biodegradation in contaminated aquifers J. Contam. Hydrol. **2004**, *75*, 215–255.
67. Hirrschorn, S.K.; Dinglasan, M.J.; Elsner, M.; Mancini, S.; Lacrampe-Couloume, G.; Edwards, E.A.; Sherwoodlollar, B. Pathway dependent isotopic fractionation during aerobic biodegradation of 1,2-dichloroethane. Environ. Sci. Technol. **2004**, *38*, 4775–4781.
68. Susarla, S.; Medina, V.F.; McCutcheon, S.C. Phytoremediation: An ecological solution to organic chemical contamination. Ecol. Eng. **2002**, *18*, 647–658.
69. Jakubowska, N.; Zygmunt, B.; Polkowska, Z.; Zabiegała, B.; Namieśnik, J. Sample preparation for gas chromatographic determination of halogenated volatile organic compounds in environmental and biological samples. J. Chromatogr. A **2009**, *1216*, 422–441.
70. Demeestere, K.; Dewulf, J.; De Witte, B.; Van Langenhove, H. Sample preparation for the analysis of volatile organic compounds in air and water matrices. J. Chromatogr. A **2007**, *1153*, 130–144.
71. Ras, M.R.; Borrull, F.; Marce, R.M. Sampling and preconcentration techniques for determination of volatile organic compounds in air samples. Trends Anal. Chem. **2009**, *28*, 347–361.
72. Tobiszewski, M.; Mechlińska, A.; Zygmunt, B.; Namieśnik, J. Green analytical chemistry in sample preparation for determination of trace organic pollutants. Trends Anal. Chem. **2009**, *28*, 943–951.

Organic Matter: Global Distribution in World Ecosystems

Wilfred M. Post
Oak Ridge National Laboratory, Oak Ridge, Tennessee, U.S.A.

Abstract
The exact ratio between living and dead organic matter in terrestrial ecosystems varies, depending on the ecosystem. The amount of carbon stored in soil is determined by the balance of two biotic processes—the productivity of terrestrial vegetation and the decomposition of organic matter. Each of these processes has strong physical and biological controlling factors. Interactions among these controlling factors are of particular importance. These biological and physical factors are the same as the ones that influence the above ground structure and composition of terrestrial ecosystems, so there are strong correspondences between soil organic matter content and ecosystem type.

INTRODUCTION

Globally, the amount of organic matter in soils, commonly represented by the mass of carbon, is estimated to be 1200–1500 Pg C (1 Pg C = 10^{15} g carbon) in the top 1 m of soil.[1,2] This is 2–3 times larger than the amount of organic matter in living organisms in all terrestrial ecosystems.[1] The exact ratio between living and dead organic matter in terrestrial ecosystems varies, depending on the ecosystem. The amount of carbon stored in soil is determined by the balance of two biotic processes—the productivity of terrestrial vegetation and the decomposition of organic matter. Each of these processes has strong physical and biological controlling factors. These include climate; soil chemical, physical, and biological properties; and vegetation composition. Interactions among these controlling factors are of particular importance. These biological and physical factors are the same as the ones that influence the above ground structure and composition of terrestrial ecosystems, so there are strong correspondences between soil organic matter content and ecosystem type.

ORGANIC MATTER INPUTS

Quantity

The amount of carbon stored in soils is to a great extent determined by the rate of organic matter input through litterfall, root exudates, and root turnover. The main factors that influence vegetation production are suitable temperatures for photosynthesis, available soil moisture for evapotranspiration, and rates of CO_2 and H_2O exchange. Dry and/or cold climates support low vegetation production rates and soils under such climates have low organic matter contents. Where climates are warm and moist, vegetation production is high and soil organic matter contents are correspondingly high. Fig. 1 shows the striking correspondence between soil organic matter content and general climate measurements that results from the relationship between vegetation production and suitable moisture and temperature conditions.

Vegetation production depends not only on climate but also on nutrient supply from decomposition and geochemical weathering. Walker and Adams[6] hypothesized that the level of available phosphorus during the course of soil development is the primary determinant of terrestrial net primary production. Numerous workers have examined this hypothesis. Tiessen, Stewart, and Cole[7] and Roberts, Stewart, and Bettany[8] found that available phosphorus explained about one-fourth of the variance in soil organic matter in many different soil orders. The relationship between phosphorus and carbon is strongest during the aggrading stage of vegetation–soil system development.[9] Initially, the production of acidic products by pioneer vegetation promotes the release of phosphorus by weathering of parent material. Organic matter builds up in the soil, increasing the storage of phosphorus in decomposing organic compounds. Nitrogen fixing bacteria populations, which depend on a supply of organic carbon and available phosphorus, can grow to meet ecosystem demands for nitrogen. Plant growth is enhanced by this increasing nitrogen and phosphorus cycling, resulting in increased rates of weathering. This process continues until the vegetation is constrained by other factors affecting phosphorus availability: Leaching losses become larger than the weathering inputs;[10] or an increasing fraction of the phosphorus becomes unavailable by adsorption or precipitation with secondary minerals;[11] or nitrogen availability (denitrification or leaching is affected) reaching or exceeding nitrogen inputs and fixation.[12] In mature soils, net primary production is more likely to be limited by nitrogen. Availability of other nutrients that are largely derived from parent materials, such as most base cations, may also influence soil organic matter accumulation during early soil development.[13] Soils derived from base cation rich volcanic parent materials (Andisols) have much higher carbon contents on average than soils from other parent materials.[4]

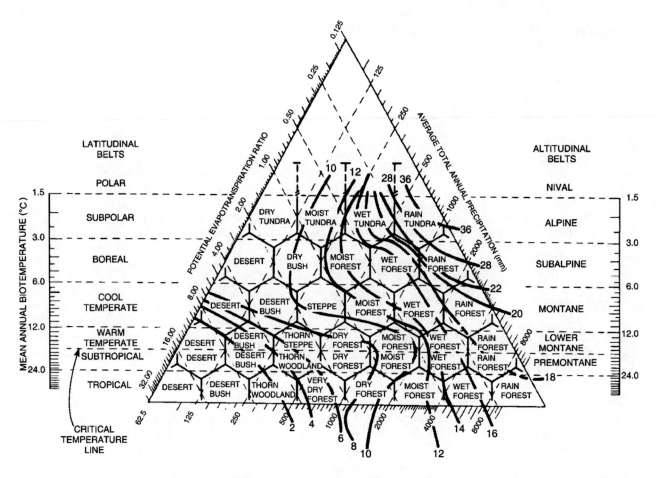

Fig. 1 Contours of soil carbon density (kg m^{-2}) plotted on Holdridge diagram[3] for world life-zone classification. Values of biotemperature and precipitation uniquely determine a life zone and associated vegetation. Contour lines for mean soil carbon content in the surface meter of soil are determined from data derived from over 3000 soil profiles.[4,5]

Species Composition

Biotic factors, in particular plant species composition, also affect soil organic matter dynamics. Production and decomposition rates are to some degree controlled by species composition. Each terrestrial plant species produces different amounts and chemical compositions of leaves, roots, branches, and wood of varying decomposability. This range of decomposability may be summarized by the lignin and nitrogen content of the organic material.[14,15] Litter decay rate is inversely related to C:N and lignin:N ratios and positively related to N content. Species with tissues that have low nutrient or high lignin content produce litter that is slow to decay. Nitrogen is made available to plants during the decomposition process. Nitrogen is a limiting element for productivity in most terrestrial ecosystems so the rate at which it is released during decomposition is an important factor in ecosystem production. Thus, the interactions between processes regulating plant populations and their productivity and microbial processes regulating nitrogen availability result in some of the observed variation in soil carbon and nitrogen storage.[16–19]

Placement

The deeper that fresh detritus is placed in the soil, the slower it decomposes. This is a result of declining decomposer activity and increased protection from oxidation with depth in the soil. Prairies have a somewhat lower productivity than forests and produce no slowly decomposing woody material. Nevertheless, prairies have a very high soil organic matter content because prairie grasses allocate twice as much production to belowground roots and tillers than to aboveground leaves.[20] The result is high soil organic matter contents with a uniform distribution in the upper 1 m of soil (Fig. 2). In contrast, a spruce–fir forest contains 50 percent of its soil organic matter in the top 10 cm. There are interesting exceptions to the rule that above-/belowground plant allocation determines soil organic matter distribution patterns in soil. Tropical moist forest soils show a uniform depth distribution similar to the depth distribution of temperate grasslands, however, in tropical forests this is largely due to a long-term accumulation of recalcitrant organic materials at lower depths in the soil rather than increased allocation to roots. Alpine tundra soils support a largely herbaceous flora but show a similar depth distribution

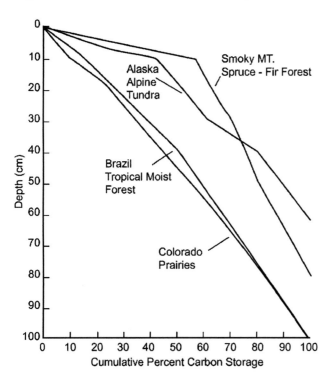

Fig. 2 Cumulative carbon storage as a function of depth for four ecosystems. Refer text for explanation of these patterns. **Source:** Zinke et al.[4]

as forest soils because of inhibition of surface litter decomposition by low temperatures and high water saturation.

DECOMPOSITION

Climate

Organic matter decay rates can be related to environmental parameters such as temperature and soil moisture. Climatic indices that correlate well with decay rates include plant moisture and temperature indices,[21,22] linear combinations of temperature and rainfall,[23] and actual evapotranspiration.[15] Warm temperatures and available soil moisture enhance microbial, and micro- and macro-invertebrate activity. These environmental conditions are also correlated with plant production. As a result, the amount of organic matter present in soil is highest in vegetation types with the highest rates of organic matter production. These are ones found in the warm, moist climate regions. The contours of soil carbon density displayed in Fig. 1 reflect the balance of input by vegetation production and loss from decomposition imposed by climate. Soil carbon content increases from lower left to the upper right in Fig. 1 as the temperature decreases in the cool temperate, boreal, and sub-polar life zones and as precipitation increases in the warm temperate, subtropical, and tropical life zones.

The combined influence of temperature and precipitation is presented by the third axis of the Holdridge diagram (Fig. 1) as the ratio of potential evapotranspiration (PET) to annual precipitation. When this ratio is less than 1.0, rainfall exceeds PET and vice versa. Life zones bordering the line with the PET is equal to precipitation (PET ratio = 1.0) have soil carbon contents around $10\,kg\,m^{-2}$ except in warm temperate and subtropical zones where strong seasonality limits production, but decomposition conditions are favorable for most of the year. Soil carbon content increases as the PET ratio decreases indicating that productivity increases faster than the rate of decomposition with increasing moisture availability.

Organic Matter Quality

On global scale, climate may be the most important factor controlling decay rates, but within a given region, substrate chemistry is the more important factor.[15,24,25] Decay rate is often negatively related to substrate C:N ratio. Litter C:N is initially much greater than microbial C:N but approaches microbial C:N as the microbes release the carbon as CO_2 while taking up nitrogen (nitrogen immobilization). The further the initial litter C:N is from microbial C:N, the slower the decay rate. Lignin content or lignin:N ratios may be better predictors of decay rates because lignin itself is difficult to decompose, and it shields nitrogen and other more easily degraded chemical fractions from microbes. Concise and simple models of decay rate are based on a combination of chemical and climatic indices.

The effect of litter quality on soil organic matter content is most dramatically expressed in Podzols (Spodosol in the United States Department of Agriculture classification). These occur over large areas in boreal zones dominated by evergreen conifers, but often occur in other regions on shallow or sandy soils. Low nitrogen content of organic matter inputs and cool temperatures reduce decomposition and soil animal activity. As a result, large surface organic matter accumulations occur over a thin A horizon. Low temperatures combined with leaching of organic acids result in podsolization as the predominant soil-forming process. Leaching of iron, aluminum oxides, and organic matter result in a distinct E horizon near the surface where these materials are removed and deposited in the B horizon. If the surface organic layers are included, these soils can have substantial organic matter contents, exceeding the expected amount for the climate conditions. Batjes[2] gives an average value for Podzols of $24.2\,kg\,m^{-2}$ for the surface meter which is considerably above the mean for most other soil types (see Table 1).

SIGNIFICANT PHYSICAL AND CHEMICAL INFLUENCES

There are several notable exceptions to the climate-based explanation of variation in soil carbon content. There are two in particular that have lower rates of decomposition

Table 1 Mean organic carbon contents (kg m^{-2}) by FAO–UNESCO soil units to 1 m depth.

Soil unit	Mean C (kg m^{-2})
Acrisols	9.4
Cambisols	9.6
Chernozems	12.5
Podzoluvisols	7.3
Ferrasols	10.7
Gleysols	13.1
Phaeozems	14.6
Fluvisols	9.3
Kastanozems	9.6
Luvisols	6.5
Greyzems	19.7
Nitosols	8.4
Histosols	77.6
Podzols	24.2
Arenosols	3.1
Regosols	5.0
Solonetz	6.2
Andisols	25.4
Vertisols	11.1
Planosols	7.7
Xerosols	4.8
Yermosols	3.0
Solochaks	4.2

These soil units generally span a wide range of climate conditions and therefore present a different view of soil organic matter content based on additional soil factors. In particular, the high C content of Podzols, Histosols, and Andisols is apparent. Refer text for additional explanation of biological, chemical and physical factors responsible.
Source: Batjes.[2]

and therefore higher accumulations of organic matter than expected (Table 1). These include Histosols due to hydrological conditions and Andisols due to parent material chemical effects.

Histosols

In landscape positions where water accumulates at or above the surface of the soil for an appreciable part of the growing season, decomposition can be reduced to such an extent that large amounts of undecomposed organic matter can accumulate. This soil type is called a Histosol and can be found in any region in wetlands where decomposition is restricted. The soil-surface of mature or old-growth boreal forests over shallow water tables are often covered with *Sphagnum* moss which may also lead to development of Histosols. Histosols with the largest areas and thickest accumulations occur in lowland tundra where a mixture of sedges, lichens, and mosses grow at the northern limit of vegetation in the northern hemisphere. Production, decomposition, and evaporation are limited by low temperatures and water-saturated soils. In these cold regions, deeper layers may freeze and not become thawed during the short growing season (permafrost). As a result, Histosols have carbon contents over 70 kg m^{-2} in the surface meter (Table 1). Some regions have been accumulating organic matter since the last glacial period without any substantial decomposition. Histosols in such regions may be several meters thick and contain over 250 kg C m^{-2}.[2] Globally it is estimated that boreal and sub-arctic Histosols contain 455 PgC that has accumulated during the postglacial period.[26]

Andisols

Andisols form on young volcanic stone (basalt lava) rich in nutrients and alkaline. Andisols are weakly weathered soils associated with pyroclastic parent materials that are rich in allophane, ferrihydrite, and other minerals that readily form complexes with humus molecules. These chemical constituents provide conditions promoting high vegetation production and also the retention of organic matter in soil. As a result, Andisols typically have higher soil carbon contents (25.4 kg m^{-2}, Table 1) than soils with the same environmental conditions but different parent materials.

CONCLUSIONS

Over long periods of time, organic matter in soils is the result of climatic, biological, and geological factors. These factors are not independent. In particular there exists a strong relationship between climate and vegetation type. In Fig. 1, the Holdridge climate based life zones have names that depict the dominant vegetation of climates. Jobbágy and Jackson[27] provide a summary of soil data based on biomes that demonstrates similar soil carbon distribution as that based on climate (Table 2).

Table 2 Mean organic carbon content (kg m^{-2}) by biome to 1 m depth.

Biome	Mean C (kg m^{-2})
Boreal forest	9.3
Crops	11.2
Deserts	6.2
Sclerophyllous shrubs	8.9
Temperate deciduous forest	17.4
Temperate evergreen forest	14.5
Temperate grassland	11.7
Tropical deciduous forest	15.8
Tropical evergreen forest	18.6
Tropical grassland/savanna	13.2
Tundra	14.2

Note: Biome classification is based on Whittaker.[28]
Source: Jobbágy and Jackson.[27]

Over shorter periods of time soil carbon varies with vegetation disturbances and changes in land use patterns that affect rates of organic matter input and its decomposition. Various land uses result in very rapid declines in soil organic matter from the native condition.[29–32] Losses of 50% in the top 20 cm and 30% for the surface 100 cm are average. Much of this loss in soil organic carbon can be attributed to erosion, reduced inputs of organic matter, increased decomposability of crop residues, and tillage effects that decrease the amount of physical protection to decomposition. Evidence from long-term experiments suggest that C losses due to oxidation and erosion can be reversed with soil management practices that minimize soil disturbance and optimize plant yield through fertilization. These experimental results are believed to apply to large regions and that organic matter is being restored as a result of establishment of perennial vegetation, increased adoption of conservation tillage methods, efficient use of fertilizers, and increased use of high yielding crop varieties.[33,34] Additionally, when agricultural land is no longer used for cultivation and allowed to revert to natural vegetation or replanted to perennial vegetation, soil organic carbon can accumulate by processes that essentially reversing some of the effects responsible for soil organic carbon losses initially—from when the land was converted from perennial vegetation—and return them to typical amounts for the climate, vegetation, landscape position, and parent material conditions.[35,36]

ACKNOWLEDGMENTS

Work sponsored by U.S. Department of Energy, Carbon Dioxide Research Program, Environmental Sciences Division, Office of Biological and Environmental Research and performed at Oak Ridge National Laboratory (ORNL). ORNL is managed by UT-Battelle, LLC, for the U.S. Department of Energy under contract DE-AC05-00OR22725.

REFERENCES

1. Post, W.M.; Peng, T.-H.; Emanuel, W.R.; King, A.W.; Dale, V.H.; DeAngelis, D.L. The global carbon cycle. American Scientist 1990, 78, 310–326.
2. Batjes, N.H. Total carbon and nitrogen in the soils of the world. European Journal of Soil Science 1996, 47, 151–163.
3. Holdridge, L.R. Determination of world plant formations from simple climatic data. Science 1947, 105, 367–368.
4. Zinke, P.J.; Stangenberger, A.G.; Post, W.M.; Emanuel, W.R.; Olson, J.S. Worldwide Organic Soil Carbon and Nitrogen Data; ORNL/TM-8857; Oak Ridge National Laboratory: Oak Ridge, TN, 1984.
5. Post, W.M.; Pastor, J.; Zinke, P.J.; Stangenberger, A.G. Global patterns of soil nitrogen storage. Nature 1985, 317, 613–616.
6. Walker, T.W.; Adams, A.F.R. Studies on soil organic matter: I. Influence of phosphorus content of parent materials on accumulations of carbon, nitrogen, sulfur, and organic phosphorus in grassland soils. Soil Science 1958, 85, 307–318.
7. Tiessen, H.J.; Stewart, W.B.; Cole, C.V. Pathways of phosphorus transformations in soils of differing pedogenesis. Soil Science Society of America Journal 1984, 48, 853–858.
8. Roberts, T.L.; Stewart, J.W.B.; Bettany, J.R. The influence of topography on the distribution of organic and inorganic soil phosphorus across a narrow environmental gradient. Canadian Journal of Soil Science 1985, 65, 651–665.
9. Anderson, D.W. The effect of parent material and soil development on nutrient cycling in temperate ecosystems. Biogeochemistry 1988, 5, 71–97.
10. Jenny, H. The Soil Resource; Springer: Berlin, 1980.
11. Walker, T.W.; Syers, J.K. The fate of phosphorus during pedogenesis. Geoderma 1976, 15, 1–19.
12. Schlesinger, W.H. Biogeochemistry: An Analysis of Global Change; Academic: New York, 1991.
13. Torn, M.S.; Trumbore, S.E.; Chadwick, O.A.; Vitousek, P.M.; Hendricks, D.M. Mineral control of soil organic carbon storage and turnover. Nature 1997, 389, 170–173.
14. Aber, J.D.; Melillo, J.M. Nitrogen immobilization in decaying hardwood leaf litter as a function of initial nitrogen and lignin content. Canadian Journal of Botany 1982, 58, 416–421.
15. Meentemeyer, V. Macroclimate and lignin control of litter decomposition rates. Ecology 1978, 59, 465–472.
16. Zinke, P.J. The pattern of influence of individual trees on soil properties. Ecology 1962, 42, 130–133.
17. Wedin, D.A.; Tilman, D. Species effects on nitrogen cycling: a test with perennial grasses. Oecologia 1990, 84, 433–441.
18. Hobbie, S.E. Effects of plant species on nutrient cycling. Trends in Ecology and Evolution 1992, 7, 336–339.
19. Hobbie, S.E. Temperature and plant species control over litter decomposition in Alaskan tundra. Ecological Monographs 1996, 66, 503–522.
20. Sims, P.L.; Coupland, R.T. Grassland Ecosystems of the World: Analysis of Grasslands and Their Uses; Coupland, R.T., Ed.; Cambridge University Press: Cambridge, 1979.
21. Olson, J.S. Energy storage and the balance of producers and decomposers in ecological systems. Ecology 1963, 44, 322–331.
22. Fogel, R.; Cromack, K. Effect of habitat and substrate quality on douglas fir litter decomposition in western Oregon. Canadian Journal of Botany 1977, 55, 1632–1640.
23. Pandey, V.; Singh, J.S. Leaf-litter decomposition in an oak–conifer forest in himalaya: the effects of climate and chemical composition. Forestry 1982, 55, 47–59.
24. Flanagan, P.W.; VanCleve, K. Nutrient cycling in relation to decomposition and organic matter quality in tiaga ecosystems. Canadian Journal of Forest Research 1983, 13, 795–817.
25. McClaugherty, C.A.; Pastor, J.; Aber, J.D.; Melillo, J.M. Forest litter decomposition in relation to soil nitrogen dynamics and litter quality. Ecology 1984, 66, 266–275.
26. Gorham, E. Northern peatlands: role in the carbon cycle and probable responses to climatic warming. Ecological Applications 1991, 1, 182–195.

27. Jobbágy, E.G.; Jackson, R.B. The vertical distribution of organic carbon and its relation to climate and vegetation. Ecological Applications 2000, *10* (2), 423–436.
28. Whittaker, R.H. *Communities and Ecosystems*; MacMillan: London, 1975.
29. Jenny, H. *Factors of Soil Formation*; McGraw-Hill: New York, 1941.
30. Davidson, E.A.; Ackerman, I.L. Changes in soil carbon inventories following cultivation of previously untilled soils. Biogeochemistry 1993, *20*, 161–193.
31. Mann, L.K. Changes in soil carbon after cultivation. Soil Science 1986, *142*, 279–288.
32. Schlesinger, W.H. Changes in soil carbon storage and associated properties with disturbance and recovery. In *The Changing Carbon Cycle: A Global Analysis*; Trabalka, J.R., Reichle, D.E., Eds.; Springer: New York, 1985.
33. Buyanovsky, G.A.; Wagner, G.H. Carbon cycling in cultivated land and its global significance. Global Change Biology 1998, *4*, 131–142.
34. Lal, R.; Kimble, J.M.; Follett, R.F.; Cole, C.V. *The Potential of U.S. Cropland to Sequester Carbon and Mitigate the Greenhouse Effect*; Ann Arbor Press: Ann Arbor, MI, 1998.
35. Post, W.M.; Kwon, K.C. Soil carbon sequestration and land-use change: processes and potential. Global Change Biology 2000, *6*, 317–328.
36. Silver, W.L.; Ostertag, R.; Lugo, A.E. The potential for carbon sequestration through reforestation of abandoned tropical agricultural and pasture lands. Restoration Ecology 2000, *8* (4), 394–407.

Organic Matter: Management

R. Cesar Izaurralde
Battelle Pacific Northwest National Laboratory, Washington, District of Columbia, U.S.A.

Carlos C. Cerri
University of São Paulo, São Paulo, Brazil

Abstract
Soil organic matter (SOM) consists of a complex array of living organisms such as bacteria and fungi, plant and animal debris in different stages of decomposition, and humus—a rather stable brown to black material showing no resemblance to the organisms from which it originates. The level of SOC in virgin soils reflects the action and interaction of the major factors of soil formation: climate, vegetation, topography, parent material, and age. These factors control SOC content by regulating the balance between carbon gains via photosynthesis and losses via autotrophic and heterotrophic respiration, as well as carbon losses in soluble and solid form.

INTRODUCTION

Soil organic matter (SOM) consists of a complex array of living organisms such as bacteria and fungi, plant and animal debris in different stages of decomposition, and *humus*—a rather stable brown to black material showing no resemblance to the organisms from which it originates. Because SOM is or has been part of living tissues, its composition is dominated by carbon (C), hydrogen, oxygen and—in lesser abundance—by nitrogen, phosphorus, sulfur among other elements. Levels of SOM are expressed in terms of soil organic carbon (SOC) concentration ($g\,kg^{-1}$) or mass per unit area ($g\,m^{-2}$) to a given depth. The level of SOC in virgin soils reflects the action and interaction of the major factors of soil formation: climate, vegetation, topography, parent material, and age. These factors control SOC content by regulating the balance between C gains via photosynthesis and losses via autotrophic and heterotrophic respiration, as well as C losses in soluble and solid form. The SOC content usually ranges between 5 and $100\,g\,kg^{-1}$ in mineral soils. These concentrations appear modest but at 1500 Pg, the amount of organic C stored globally in soils is second only to that contained in oceans and at least twice that found in either terrestrial vegetation or the atmosphere.

Cultivated soils usually contain less SOC than virgin soils[1] due to the magnification of two biophysical processes: 1) net nutrient mineralization accompanied by release of CO_2 due to microbial respiration and 2) soil erosion. SOC losses of up to 50% have been reported within 30–70 years of land use conversions under temperate conditions.[2–5] SOC losses reported in subtropical and tropical environments often match or even surpass those observed under temperate conditions.[6–8] In subtropical and tropical environments, shifting cultivation systems appear to conserve more SOC than forestlands permanently cleared for cultivation.[9]

MAJOR PROCESSES LEADING TO CARBON LOSSES FROM SOIL

Mineralization Processes

Depending on its frequency and kind, tillage changes the soil biophysical environment in ways that affect the net mineralization of nutrients and the release of carbon. These changes can be described in terms of increases or decreases in soil porosity, disruption of soil aggregates, and redistribution in the proportion of soil aggregate size, as well as alteration of energy and water fluxes. All these changes enhance, at least temporarily, the conversion of organic C into CO_2[10] and the net release of nutrients from SOM. Much of the success of past agricultural practices relied heavily on the control of decomposition processes through tillage operations to satisfy plant nutrient demands. All this came at a price, however, for a heavy reliance on soil nutrients to feed crops without proper replenishment led to the worldwide declines of SOM.[11]

Soil Erosion Processes

Agricultural ecosystems normally experience soil losses at rates considerably greater than natural ecosystems because of an incomplete plant or residue cover of the soil during rainy or windy conditions. When surface and environmental conditions are right (i.e., bare soil, sloping land, intense rain, windy weather), the kinetic energy embedded in wind and water is transferred to soil aggregates causing them to be detached and transported away from their original position across fields or downhill. Besides the physical loss of soil particles and on-site impact on soil productivity, the detachment and transport processes also cause aggregate breakdown, thereby exposing labile C to microbial activity. This aggregate breakdown also facilitates the preferential removal of soil materials comprised mainly of humus and clay or silt fractions. Consequently, water- and windborne sediments become enriched in C with respect to the contributing soil. Carbon enrichment ratios ranging from 3 to 360 have been reported.[12,13] The fate of these C-enriched sediments is not well known, for while transport and burial of C in eroded sediments may lead to "sequestration,"[14] it may also result in part of it being emitted back to the atmosphere as CO_2.[15]

RESTORING SOIL ORGANIC MATTER: THE EMERGING SCIENCE OF SOIL CARBON SEQUESTRATION

Role of Long-Term Field Experiments

SOM is an essential attribute of soil quality[16] and has an essential role in soil conservation and sustainable agriculture. Many practices—some involving land use changes—have been shown to increase SOM and thus received considerable attention for their possible role in climate change mitigation.[17–19] Carbon sequestration in managed soils occurs when there is a net removal of atmospheric CO_2 because C inputs (nonharvestable net primary productivity) are greater than C outputs (soil respiration, C costs related to fossil fuels and fertilizers). Soil C sequestration has the additional appeal that all its practices conform to principles of sustainable agriculture (e.g., reduced tillage, erosion control, diversified cropping systems, improved soil fertility). Long-term field experiments have been instrumental to increase our understanding of SOM dynamics.[20,21] The first and longest standing experiment was started at Rothamsted, England, by J. B. Lawes and J. H. Gilbert who in 1843 began documenting the impact of nutrient manipulation on crop yields and soil properties.[22] Other experiments were initiated thereafter in America, Europe, and Oceania with the goal of discovering interactions among climate, soil, and management practices. The knowledge that emerged from these experiments has been instrumental for the development and testing of agroecosystem and SOM models.[23]

Global Importance of Soil C Sequestration

There appears to be a significant opportunity for managed ecosystems to act as C sinks. For example, results from inverse modeling experiments suggest that during 1988–1992, terrestrial ecosystems may have been sequestering atmospheric C at rates of $1-2.2 \text{ Pg y}^{-1}$.[24] Some of the likely causes include the growth of new forest in previously cultivated land[25] and the "CO_2-fertilization effect."[26] Globally, agricultural soils have been estimated to have the capacity to sequester C at rates of 0.6 Pg y^{-1}[11] during several decades. The realization of this potential C sequestration would not be trivial since it would offset roughly about one-tenth of the current emissions from fossil fuels. In the U.S., annual gains in soil C from improved agricultural practices have been estimated at 0.14 Pg yr^{-1}.[25] Whether or not soil C sequestration practices are widely adopted will depend on their value relative to other C capture and sequestration technologies.

Mechanisms of Soil C Sequestration

Recent reviews of experimental results have contributed to organize our understanding of the environmental and management controls of soil C sequestration in grassland[27,28] and agricultural[29,30] ecosystems. The use of C balance, soil fractionation, and isotope techniques have been instrumental to reveal how new C (from crop residues, roots, and organic amendments) enters soil, resides shortly (for a few years) in labile soil fractions, and finally becomes a long-time constituent (for hundreds of years) of recalcitrant organo-mineral complexes.[31] Fig. 1 contrasts young (labile) organic matter fractions extracted from two cultivated soils with and without N fertilizer.[32] The amounts of labile organic matter—fine roots and other organic debris—present in each soil reflect differences in crop productivity induced by addition of N at annual rates of 50 kg ha^{-1} for 13 years. "Terra Preta" soil—in tropical regions of South

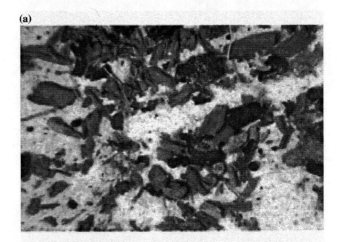
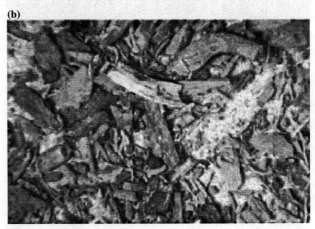

Fig. 1 Young organic matters (fine roots and other organic debris) extracted from two Cryoboralfs under cereal cropping for 13 years receiving N at annual rates of 0 (a) and 50 kg ha^{-1} (b). The black material is charcoal.
Source: Solberg et al.[32]

Organic Matter: Management

Table 1 Examples of worldwide land use and management impacts on SOC.

Region/country	Climate	Soil	Duration	Crop/land use	Treatment	Soil organic carbon (kg m^{-2}) Initial	Soil organic carbon (kg m^{-2}) Final	Depth (cm)	Reference
Argentina	Temperate humid	Argiudoll	17	Corn–wheat–soybean	Moldboard plow		4.95	20	[37]
					No tillage		5.46		
Chaco, Argentina	Subtropical semiarid	Alfisol	20	Highly restored			7.05	20	[6]
			10	Moderately restored			3.10		
			60	Highly degraded			1.50		
Rondonia, Brazil	Tropical humid			Forest			4.33	50	[38]
			5	Pasture			5.85		
			9	Pasture			5.26		
			20	Pasture			5.28		
			41	Pasture			6.56		
			81	Pasture			6.12		
Georgia, U.S.	Temperate humid	Hapludult	5	Bermudagrass	Unharvested	1.39	1.74	6	[39]
					Lightly grazed		2.01		
					Heavily grazed		2.00		
					Hayed		1.59		
Kentucky	Temperate humid	Paleudalf	20	Conventional tillage corn	0 kg N/ha^{-1}		4.89	30	[40]
					84 kg N/ha^{-1}		5.63		
					168 kg N/ha^{-1}		5.64		
					336 kg N/ha^{-1}		6.14		
				No till corn	0 kg N/ha^{-1}		5.54		
					84 kg N/ha^{-1}		5.84		
					168 kg N/ha^{-1}		5.89		
					336 kg N/ha^{-1}		6.63		
Kuztown, Pennsylvania	Temperate humid	Fragiudalf	15	Corn	Conventional	4.20	4.30	15	[35]
					Organic	4.40	5.00		
					Manure	4.10	5.30		
Michigan	Cool temperate humid	Hapludalf	7	No crops—natural succession	Tillage		2.06	15	[41]
					Control		2.22		
Swift Current, Canada	Cold semiarid	Haploboroll	10	Cont. wheat	Minimum tillage	3.05	3.52	15	[42]
				Fallow–wheat–wheat		2.99	3.34		
				Green manure–wheat–wheat		2.89	3.22		

(continued)

Table 1 Examples of worldwide land use and management impacts on SOC. (*Continued*)

Region/country	Climate	Soil	Duration	Crop/land use	Treatment	Soil organic carbon (kg m^{-2}) Initial	Soil organic carbon (kg m^{-2}) Final	Depth (cm)	Reference
Breton, Canada	Cold subhumid	Cryoboralf	51	Wheat–fallow	Nil	2.64	1.81	15	[34]
					Fertilizer		2.13		
					Manure		3.11		
				Wheat–oat–barley–hay–hay	Nil		2.91		
					Fertilizer		3.37		
					Manure		4.32		
Russia	Cool temperate humid	Mollisol	300	Native grassland			2.07	50	[5]
			50	Hay			2.13		
			100	Continuous cropping			1.59		
			50	Continuous fallow			1.51		
Punjab, India	Subtropical subhumid	Alluvial	6	Corn–wheat	Minimum tillage, residue retained		0.48	15	[43]
					Minimum tillage, residue removed		0.48		
					Conventional tillage		0.50		
Morocco	Warm temperate semiarid	Calcixeroll	11	Continuous wheat and other rotations	Conventional tillage	3.20	3.73	20	[44]
					No tillage		3.39		
Western Nigeria			20	Bush fallow			2.77	15	[45]
			25	Bush fallow			2.96		
			10	Bermudagrass			3.90		
			10	Cultivation			1.29		

America and West Africa—represents a prime example of ancient wisdom applied to develop sustainable agriculture through the improvement of soil fertility and SOM.[33]

The quantity and quality of C entering soil as well as the interaction of this C with the soil biophysical environment are major factors determining the rate and duration of soil C sequestration. The quantity of C added to soil in the form of roots, crop residues, and organic amendments has been shown to play a dominant role in defining the trajectory of SOC over time.[34] Management practices geared toward optimizing nutrient supply and building nutrient reserves (e.g., fertilization, use of legumes in crop rotations) are almost guaranteed to increase soil C stocks. The quality of crop residues and the timing of their incorporation to soil also have an influential role on C decomposition and, thus, on soil C storage.[35] The degree of soil disturbance—through its impact on soil aggregation—constitutes another major factor regulating C decomposition and retention in soil.[36] In this context, no tillage agriculture has come to represent one of the most significant technological innovations of the last 30 years because it allows farmers the possibility of growing crops economically while reducing erosion and improving both quantity and quality of SOM. A few examples of the management impacts on soil C sequestration from around the world are presented in Table 1.

Soil Organic Matter, Energy, and Full C Accounting

Land is the natural habitat of humans. Humans dwell on it and use it as a resource for the production of food, fiber, and other goods. Simply put, land is managed when there is a manipulation of energy and matter flows in order to meet certain economic and social objectives. Farm mechanization and fertilizers are two of the many technical innovations that—though they rely on the utilization of fossil energy—have brought dramatic increases in food production during the last century. Changes in management practices that include soil C sequestration as an objective require careful evaluation of their impact not only on soil C gains but also on C costs from the use of fossil energy (e.g., manufacture of fertilizers)[46,47] and on the net greenhouse gas emissions.[48]

ROLE OF SOIL ORGANIC MATTER IN THE 21ST CENTURY

SOM has played and will continue to play a central role in sustainable land management. The restoration of SOM at global scales offers a unique opportunity to mitigate global warming. As population levels and affluence increase, demands on land to produce food, fiber, biomass, and other products will remain high. Because land is finite, important decisions will have to be made in order to balance such demands with functional objectives such as the preservation of natural ecosystems. As part of any climate policy, the impact of land use changes and management on SOM storage should be included as a criterion for making these decisions. Depending on their degree of expansion, several evolving agricultural technologies—such as genetically modified crops, conservation tillage, organic farming, and precision farming—may have important implications for soil C sequestration.[19] Their ultimate impact on C sequestration will depend not only on the economic benefit realized by individual producers but also on whether society recognizes the value of soil C storage to mitigate global warming.

REFERENCES

1. Davidson, E.A.; Ackerman, I.L. Changes in soil carbon following cultivation of previously untilled soils. Biogeochemistry **1993**, *20*, 161–164.
2. Dalal, R.C.; Mayer, R.J. Long-term trends in fertility of soils under continuous cultivation and cereal cropping in southern Queensland. II. Total organic carbon and its rate of loss from the soil profile. Aust. J. Soil Res. **1986**, *24*, 281–292.
3. Mann, L.K. Changes in soil carbon after cultivation. Soil Sci. **1986**, *142*, 279–288.
4. Ellert, B.H.; Gregorich, E.G. Storage of carbon, nitrogen and phosphorus in cultivated and adjacent forested soils of Ontario. Soil Sci. **1996**, *161*, 587–603.
5. Mikhailova, E.A.; Bryant, R.B.; Vassenev, I.I.; Schwager, S.J.; Post, C.J. Cultivation effects on soil carbon and nitrogen contents at depth in the Russian chernozem. Soil Sci. Soc. Am. J. **2000**, *64*, 738–745.
6. Abril, A.; Bucher, E.H. Overgrazing and soil carbon dynamics in the western chaco of Argentina. Appl. Soil Ecol. **2001**, *16*, 243–249.
7. Lal, R. Deforestation and land use effects on soil degradation and rehabilitation in western Nigeria. II. Soil chemical properties. Land Degrad. Dev. **1996**, *7*, 87–98.
8. Lobe, I.; Amelung, W.; Du Preez, C.C. Losses of carbon and nitrogen with prolonged arable cropping from sandy soils of the South African highveld. Eur. J. Soil Sci. **2001**, *52*, 93–101.
9. Houghton, R.A. Changes in the storage of terrestrial carbon since 1850. In *Soils and Global Change*; Lal, R., Kimble, J., Levine, E., Stewart, B.A., Eds.; CRC/Lewis Publishers: Boca Raton, FL, 1995; 45–65.
10. Reicoski, D.C.; Lindstrom, M.J. Fall tillage method: effect from short-term carbon dioxide flux from soil. Agron. J. **1993**, *85*, 1237–1243.
11. Cole, V.; Cerri, C.; Minami, K.; Mosier, A.; Rosenberg, N.J.; Sauerbeck, D. Agricultural options for mitigation of greenhouse gas emissions. In *Climate Change 1995: Impacts, Adaptations and Mitigation of Climate Change*; Watson, R.T., Zinowera, M.C., Moss, R.H., Eds.; Report of IPCC Working Group II; Cambridge University Press: London, 1996; 745–771.
12. Sterk, G.; Herrmann, L.; Bationo, A. Wind-blown nutrient transport and soil productivity changes in southwest Niger. Land Degrad. Dev. **1996**, *7*, 325–336.
13. Zobeck, T.M.; Fryrear, D.W. Chemical and physical characteristics of wind-blown sediment. Trans. Am. Soc. Agric. Eng. **1986**, *29*, 1037–1041.
14. Stallard, R.F. Terrestrial sedimentation and the carbon cycle: coupling weathering and erosion to carbon burial. Global Biogeochem. Cycles **1998**, *12*, 231–257.
15. Lal, R. Global soil erosion by water and C dynamics. In *Soils and Global Change*; Lal, R., Kimble, J., Levine, E., Stewart, B.A., Eds.; CRC/Lewis Publishers: Boca Raton, FL, 1995; 131–141.
16. Doran, J.W.; Coleman, D.C.; Bezdicek, D.F.; Stewart, B.A., Eds.; *Defining Soil Quality for a Sustainable Environment*; Soil Science Society America Special Publication No. 35; SSSA: Madison, WI, 1994; 244 pp.
17. Batjes, N.H. Mitigation of atmospheric CO_2 concentrations by increased carbon sequestration in the soil. Biol. Fert. Soils **1998**, *27*, 230–235.
18. Post, W.M.; Kwon, K.C. Soil carbon sequestration and land-use change: processes and potential. Global Change Biol. **2000**, *6*, 317–327.
19. Izaurralde, R.C.; Rosenberg, N.J.; Lal, R. Mitigation of climatic change by soil carbon sequestration: issues of science, monitoring and degraded lands. Adv. Agron. **2001**, *70*, 1–75.
20. Powlson, D.S.; Smith, P.; Smith, J.U., Eds.; *Evaluation of Soil Organic Matter Models Using Existing Long-Term Datasets*; NATO ASI Series I; Springer: Heidelberg, 1996; Vol. 38, 429 pp.

21. Paul, E.A.; Paustian, K.; Elliott, E.T.; Cole, C.V., Eds.; *Soil Organic Matter in Temperate Agroecosystems: Long-Term Experiments in North America*; NATO ASI Series I; CRC/Lewis Publishers: Boca Raton, FL, 1997; 414 pp.
22. Jenkinson, D.S. The rothamsted long-term experiments: are they still of use? Agron. J. **1991**, *83*, 2–10.
23. Smith, P.; Smith, J.U.; Powlson, D.S.; McGill, W.B.; Arah, J.R.M.; Chertov, O.; Coleman, K.W.; Franko, U.; Frolking, S.; Jenkinson, D.S.; Jensen, L.S.; Kelly, R.H.; Klein-Gunnewiek, H.; Komarov, A.S.; Li, C.; Molina, J.A.E.; Mueller, T.; Parton, W.J.; Thornley, J.H.M.; Whitmore, A.P. A comparison of the performance of nine soil organic matter models using datasets from seven long-term experiments. Geoderma **1997**, *81*, 153–225.
24. Fan, S.; Gloor, M.; Mahlman, J.; Pacala, S.; Sarmiento, J.; Takahashi, T.; Tans, P. A large terrestrial carbon sink in North America implied by atmospheric and oceanic carbon dioxide data and models. Science **1998**, *282*, 442–446.
25. Houghton, R.A.; Hackler, J.L.; Lawrence, K.T. The U.S. carbon budget: contributions from land use change. Science **1999**, *285*, 574–578.
26. Kimball, B.A. Carbon dioxide and agricultural yield: an assemblage and analysis of 430 prior observations. Agron. J. **1983**, *75*, 779–782.
27. Conant, R.T.; Paustian, K.; Elliott, E.T. Grassland management and conversion into grassland: effects on soil carbon. Ecol. Appl. **2001**, *11*, 343–355.
28. Scott, N.A.; Tate, K.R.; Ford-Robertson, J.; Giltrap, D.J.; Smith, C.T. Soil carbon storage in plantations and pastures: land-use implications. Tellus **1999**, *51B*, 326–335.
29. Janzen, H.H.; Campbell, C.A.; Izaurralde, R.C.; Ellert, B.H.; Juma, N.; McGill, W.B.; Zentner, R.P. Management effects on soil c storage on the Canadian prairies. Soil Till. Res. **1998**, *47*, 181–195.
30. Paustian, K.; Collins, H.P.; Paul, E.A. Management controls on soil carbon. In *Soil Organic Matter in Temperate Ecosystems: Long-Term Experiments in North America*; Paul, E.A., Paustian, K., Elliott, E.T., Cole, C.V., Eds.; CRC/Lewis Publishers: Boca Raton, FL, 1997; 15–49.
31. Jastrow, J.D. Soil aggregate formation and the accrual of particulate and mineral-associated organic matter. Soil Biol. Biochem. **1996**, *28*, 665–676.
32. Solberg, E.D.; Nyborg, M.; Izaurralde, R.C.; Malhi, S.S.; Janzen, H.H.; Molina-Ayala, M. Carbon storage in soils under continuous cereal grain cropping: N fertilizer and straw. In *Management of Carbon Sequestration in Soil*; Lal, R., Kimble, J., Follett, R., Stewart, B.A., Eds.; CRC/Lewis Publishers: Boca Raton, FL, 1998; 235–254.
33. Glaser, B.; Haumaier, L.; Guggenberger, G.; Zech, W. The ''Terra Preta'' phenomenon: a model for sustainable agriculture in the humid tropics. Naturwissenschaften **2001**, *88*, 37–41.
34. Izaurralde, R.C.; McGill, W.B.; Robertson, J.A.; Juma, N.G.; Thurston, J.T. Carbon balance of the Breton classical plots after half a century. Soil Sci. Soc. Am. J. **2001**, *65*, 431–441.
35. Drinkwater, L.E.; Wagoner, P.; Sarrantonio, M. Legume-based cropping systems have reduced carbon and nitrogen losses. Nature **1998**, *396*, 262–265.
36. Six, J.; Elliott, E.T.; Paustian, K. Soil macroaggregate turnover and microaggregate formation: a mechanism for C sequestration under no-tillage agriculture. Soil Biol. Biochem. **2000**, *32*, 2099–2103.
37. Alvarez, R.; Russo, M.E.; Prystupa, P.; Scheiner, J.D.; Blotta, L. Soil carbon pools under conventional and no-tillage systems in the argentine rolling pampa. Agron. J. **1998**, *90*, 138–143.
38. Neill, C.; Cerri, C.C.; Melillo, J.M.; Feigl, B.J.; Steudler, P.A.; Moraes, J.F.L.; Piccolo, M.C. Stocks and dynamics of soil carbon following deforestation for pasture in Rondoônia. In *Soil Processes and the Carbon Cycle*; Lal, R., Kimble, J., Follett, R., Stewart, B.A., Eds.; CRC/Lewis Publishers: Boca Raton, FL, 1998; 9–28.
39. Franzluebbers, A.J.; Stuedemann, J.A.; Wilkinson, S.R. Bermudagrass management in the southern piedmont USA: I. Soil and surface residue carbon and sulfur. Soil Sci. Soc. Am. J. **2001**, *65*, 834–841.
40. Ismail, I.; Blevins, R.L.; Frye, W.W. Long-term notillage effects on soil properties and continuous corn yields. Soil Sci. Soc. Am. J. **1994**, *58*, 193–198.
41. Richter, D.D.; Babbar, L.I.; Huston, M.A.; Jaeger, M. Effects of annual tillage on organic carbon in a finetextured udalf: the importance of root dynamics to soil carbon storage. Soil Sci. **1999**, *149*, 78–83.
42. Curtin, D.; Wang, H.; Selles, F.; McConkey, B.G.; Campbell, C.A. Tillage effects on carbon fluxes in continuous wheat and fallow–wheat rotations. Soil Sci. Soc. Am. J. **2000**, *64*, 2080–2086.
43. Ghuman, B.S.; Sur, H.S. Tillage and residue management effects on soil properties and yields of rainfed maize and wheat in a subhumid subtropical climate. Soil Till. Res. **2001**, *58*, 1–10.
44. Mrabet, R.; Saber, N.; El-Brahli, A.; Lahlou, S.; Bessam, F. Total, particular organic matter and structural stability of a calcixeroll soil under different wheat rotations and tillage systems in a semiarid area of Morocco. Soil Till. Res. **2001**, *57*, 225–235.
45. Lal, R. Land use and soil management effects on soil organic matter dynamics on alfisols in western Nigeria. In *Soil Processes and the Carbon Cycle*; Lal, R., Kimble, J., Follett, R., Stewart, B.A., Eds.; CRC/Lewis Publishers: Boca Raton, FL, 1998; 109–126.
46. Schlesinger, W.H. Carbon and agriculture: carbon sequestration in soils. Science **1999**, *284*, 2095.
47. Izaurralde, R.C.; McGill, W.B.; Rosenberg, N.J. Carbon cost of applying nitrogen fertilizer. Science **2000**, *288*, 811–812.
48. Robertson, G.P.; Paul, E.A.; Harwood, R.R. Greenhouse gases in intensive agriculture: contributions of individual gases to the radiative forcing of the atmosphere. Science **2000**, *289*, 1922–1925.

Organic Matter: Modeling

Pete Smith
Institute of Biological and Environmental Sciences, School of Biological Sciences, University of Aberdeen, Aberdeen, U.K.

Abstract
Organic matter models have existed since the 1940s but most of the models commonly used today have their roots in models developed in the 1970s and 1980s. Organic matter models were originally developed to examine carbon (C) and nitrogen (N) dynamics at the site level, mostly in cropland and grassland systems, but have since become critical components of larger ecosystem models and, more recently, coupled carbon cycle climate models. Organic matter models play a critical role in science, by allowing us to integrate and test our best understanding of soil biogeochemistry, allowing us to simulate trends in organic matter dynamics, allowing us to extrapolate and project both temporally and spatially, and facilitating our exploration of the most critical questions in science today, such as biospheric carbon cycle feedbacks and the extent of future projected climate change.

INTRODUCTION

In this entry, I update earlier reviews of organic matter models and provide some of the key developments and remaining challenges in organic matter modeling. I begin by summarizing approaches to organic matter modeling, describing briefly process-based models, cohort models, and food-web models. I then describe factors affecting organic matter turnover in models and the performance of organic matter models, and then present a few examples of applications of organic matter models. In the last two sections before the conclusion, I describe recent advances in organic matter modeling and the challenges remaining in organic matter modeling.

APPROACHES TO MODELING ORGANIC MATTER DYNAMICS

There are a number of approaches to modeling organic matter turnover including process-based multicompartment models, models that consider each fresh addition of plant debris as a separate cohort that decays in a continuous way, and models that account for C and N transfers through various trophic levels in a soil food web. These approaches are described in more detail below.

Process-Based, Multicompartment Organic Matter Models

Most organic matter models are process based; i.e., they focus on the processes mediating the movement and transformations of matter or energy and usually assume first-order rate kinetics.[1] Early models simulated the organic matter as one homogeneous compartment.[2] Some years later, two-compartment models were proposed,[3,4] and as computers became more accessible, multicompartment models were developed.[5,6] Of the 33 organic matter models represented within the Global Change and Terrestrial Ecosystems (GCTE) Soil Organic Matter Network (SOMNET) database,[7–9] 30 are multicompartment, process-based models. Each compartment or organic matter pool within a model is characterized by its position in the model's structure and its decay rate. Decay rates are usually expressed by first-order kinetics with respect to the concentration (c) of the pool

$$dc/dt = -kc$$

where t is the time. The rate constant k of first-order kinetics is related to the time required to reduce by half the concentration of the pool *when there is no input*. The pool's half-life [$h = (\ln 2)/k$] or its turnover time ($\tau = 1/k$) are sometimes used instead of k to characterize a pool's dynamics: the lower the decay rate constant, the higher the half-life, the turnover time, and the stability of the organic pool.

The flows of C within most models represent a sequence of C going from plant and animal debris to the microbial biomass, then to soil organic pools of increasing stability. Some models also use feedback loops to account for catabolic and anabolic processes and microbial successions. The output flow from an organic pool is usually split. It is directed to a microbial biomass pool, another organic pool, and, under aerobic conditions, to CO_2. This split simulates the simultaneous anabolic and catabolic activities and growth of a microbial population feeding on one substrate. Two parameters are required to quantify the split flow. They are often defined by a microbial (utilization) efficiency and stabilization (humification) factor that controls the flow of decayed C to the biomass and humus pools, respectively. The sum of the efficiency and humification factors must be inferior to one to account for the release of CO_2. A thorough review of the structure and underlying assumptions of different process-based organic matter models is available.[6]

Cohort Models Describing Decomposition as a Continuum

Another approach to modeling organic matter turnover is to treat each fresh addition of plant debris into the soil as a cohort.[5] Such models consider one organic matter pool that decays with a feedback loop into itself. Q-SOIL,[10] for example, is represented by a single rate equation. The organic matter pool is divided into an infinite number of components, each characterized by its "quality" with respect to degradability as well as impact on the physiology of the decomposers. The rate equation for the model Q-SOIL represents the dynamics of each organic matter component of quality q and is quality dependent. Exact solutions to the rate equations are obtained analytically.[11]

Food-Web Models

Another type of model simulates C and N transfers through a food web of soil organisms;[1,12] such models explicitly account for different trophic levels or functional groups of biota in the soil.[13–18] Some models that combine an explicit description of the soil biota with a process-based approach have been developed.[19] Food-web models require a detailed knowledge of the biology of the system to be simulated and are usually parameterized for application at specific sites.

FACTORS AFFECTING ORGANIC MATTER TURNOVER IN MODELS

Rate "constants" (k) are constant for a given set of biotic and abiotic conditions. For non-optimum environmental circumstances, the simplest way to modify the maximum value of k is by multiplication by a reduction factor μ—ranging from 0 to 1. Environmental factors considered by organic matter models include temperature, water, pH, nitrogen, oxygen, clay content, cation exchange capacity, type of crop/plant cover, and tillage.

Many studies show the effect of temperature on microbially mediated transformations in soil, expressed as either a reduction factor or the Arrhenius equation, but the assumption that organic matter decomposition is temperature dependent was challenged by showing that old organic matter in forest soils does not decompose more rapidly in soils from warmer climates than in soils from colder regions.[20] Water and oxygen have a major impact on the microbial physiology. While some models simulate O_2 concentrations in soil explicitly,[21,22] many define the extent of anaerobiosis based on soil pore space filled with water (WFPS[23,24]). Soil clay content and total organic matter are correlated. Various schemes simulate the effect of clay on rate equations to obtain organic matter accumulation. Nitrogen is an essential element for microbial growth, which will be maximal when enough N is assimilated to maintain the microbial C:N ratio.[25] Table 1 presents an overview of the 33 models represented in the GCTE-SOMNET[8] including the factors affecting organic matter turnover.

ORGANIC MATTER MODEL PERFORMANCE

There are many reasons for evaluating the performance of an organic matter model. Model evaluation shows how well a model can be expected to perform in a given situation, how well it can help to improve the understanding of the system (especially where the model fails), how well it can provide confidence in the model's ability to predict changes in organic matter in the future or where there is no data, and how well it can be used to assess the uncertainties associated with the model's predictions. Models can be evaluated at a number of different levels. They can be evaluated at the individual process level or at the level of a subset of processes (e.g., net mineralization), or the models' overall outputs (e.g., changes in total organic matter over time) can be tested against measured laboratory and field data. Models can also be evaluated for their applicability in different situations, e.g., for scaling up simulated net C storage from a site-specific to a regional level.[59] Many examples of different forms of organic matter model evaluation are presented elsewhere.[6]

In the most comprehensive evaluation of organic models to date,[60] nine models were tested against 12 data sets from seven long-term experiments representing arable rotations, managed and unmanaged grassland, forest plantations, and natural woodland regeneration. The results showed that six models had significantly lower overall errors (root-mean-square error [RMSE]) than another group of three models (see Fig. 1).

The poorer performance of three of the models was related to failures in other parts of the ecosystem models, thus providing erroneous inputs into the organic matter module.[61]

EXAMPLES OF ORGANIC MATTER MODEL APPLICATION

Organic matter models are often used as research tools in that they are hypotheses of the dynamics of C and N in soil and can be used to distinguish between competing hypotheses.[47] Another increasing application of organic matter models is in agronomy; many organic matter models are now being used to improve agronomic efficiency and environmental quality through incorporation into decision

Table 1 Overview of organic matter models represented within GCTE-SOMNET.

Model	Time step	Inputs			Factors affecting decay rate constants	Outputs	Reference
		Meteorology	Soil and plant	Management		Soil outputs	
ANIMO	Day, week, month	P, AT, Ir, EvW	Des, Lay, Imp, Cl, OM, N, pH	Rot, Ti, Fert, Man, Res, Irr, AtN	T, W, pH, N, O	C, N, W, ST, gas	[26]
APSIM	Day	P, AT, Ir	Lay, W, C, N, BD, Wi, PG, PS	Rot, Ti, Fert, Irr	T, W, pH, N	C, N, W, ST, gas	[27]
Candy	Day	P, AT, Ir	D, Imp, W, N, C, Wi, PD, Nup	Rot, Ti, Fert, Man, Res, Irr, AtN	T, W, N, Cl	C, N, W, ST, gas	[28]
CENTURY	Month	P, AT	W, Cl, OM, pH, C, N	Rot, Ti, Fert, Man, Res, Irr, AtN	T, W, N, Cl, pH, Ti	C, BioC, ^{13}C, ^{14}C, N, W, ST, gas	[29]
Chenfang Lin model	Day	ST	OM, BD, W	Man, Res	T, W, F	C, BioC, gas	[30]
DAISY	Hour, day	P, AT, Ir, EvG	Lay, Cl, C, N, PG, PS	Rot, Ti, Fert, Man, Res, Irr, AtN	T, W, N, Cl	C, BioC, N, W, ST, gas	[31]
DNDC	Hour, day, month	P, AT	Lay, Cl, OM, pH, BD	Rot, Ti, Fert, Man, Res, Irr, AtN	T, W, N, Cl, Ti	C, BioC, N, W, ST, gas	[32]
DSSAT	Hour, day, month, year	P, AT, Ir	Des, Lay, Imp, W, Cl, PS, OM, pH, C, N	Rot, Ti, Fert, Man, Res, Irr	T, W, N, Cl, Ti	C, BioC, N, W, ST	[33]
D3R	Day	P, AT	Y, PS	Rot, Ti, Res	T, W, N, Cv, Ti	Decomp. of surface and buried residue	[34]
Ecosys	Minute, hour	P, AT, Ir, WS, RH	Lay, W, Cl, CEC, PS, OM, pH, N, BD, PG, PS	Rot, Ti, Fert, Man, Res, Irr, AtN	T, W, N, O, Cl, Cv	C, BioC, N, W, ST, pH, Ph, EC, gas, ExCat	[35]
EPIC	Day	P, AT	Lay, Imp, W, Cl, OM, pH, C, BD, Wi	Rot, Ti, Fert, Man, Res, Irr, AtN	T, W, N, pH, Cl, Ce, Cv	C, BioC, N, W, ST	[36]
FERT	Day	P, AT, WS	Des, Lay, Imp, W, Cl, OM, pH, C, N, BD, W, Ph, K, Nup, Y, PS	Rot, Ti, Fert, Man, Res, Irr	T, W, N, pH, Cv	C, N, Ph, K	[37]
ForClim-D	Year	P, AT	W, AG	None	T, W	C	[38]
GENDEC	Day, month	ST, W	W, InertC, LQ	Can be used—not essential	T, W, N	C, BioC, N, gas, LQ	[39]
HPM/EFM	Day	P, AT, Ir, WS	W, Cl, PS	Rot, Fert, Irr, AtN	T, W, N	C, BioC, N, W, gas	[40]
ICBM	Day, year	Combination of weather and climate	Many desirable: none essential	C inputs to soil	T, W, Cl	C	[41]
KLIMAT-SOIL-YIELD	Day, year	P, AT, ST, Ir, EvG, EvS, VPD, SH	Des, Lay, Imp, W, Cl, PS, OM, pH, C, N	Fert, Man, Res, Irr	T, W, N, Cl	C, BioC, N, W, ST	[42]
CNSP pasture model	Day	P, AT, Ir	Lay, Imp, W, Cl, CEC, OM, pH, C, N, PS, AS	Fert	T, W, N, pH	C, N, W, ST	[43]

(continued)

Model	Time step	Inputs			Factors affecting decay rate constants	Outputs	Reference
		Meteorology	Soil and plant	Management		Soil outputs	
Humus balance	Year	Climate based on P and AT	Des, Lay, PS, OM, pH, C, N	Rot, Fert, Man	N, H, Cl, Cv	C, N	[44]
MOTOR	User specified	P, AT, EvG	Des, OM	Rot, Ti, Fert, Man	T, W, N, Cl, Ti	C, BioC, ^{13}C, ^{14}C, gas	[45]
NAM SOM	Year	P, AT	Des, PS, OM, Ero	Man, Res	T, W, Cl, Cv	C, BioC	[46]
NCSOIL	Day	ST, (P, AT)	W, OM, C, N	Fert, Man, Res	T, W, N, pH, Cl, Ti	C, BioC, ^{14}C, N, ^{15}N, gas	[47]
NICCE	Hour, day	P, AT, Ir, WS	Imp, OM, C, N, W, TC, PG	Fert, Man, Res, Irr, AtN	T, W, Cl, N	C, BioC, ^{13}C, ^{14}C, N, ^{15}N, W, ST, gas	[48]
O'Brien model	Year	None	Lay, C, ^{14}C	None	None	C, ^{14}C	[49]
O'Leary model	Day	P, AT	Lay, W, Cl, pH, N	Ti, Fert, Res	T, W, N, Cl, Ti	C, BioC, N, W, ST, gas, ResC, ResN	[50]
Q-Soil	Year	Optional	C, N	Rot, Fert, Man, Res, AtN	T, W, N	C, BioC, ^{13}C, N	[10]
RothC	Month	P, AT, EvW	Cl, C, InertC (can be estimated)	Man, Res, Irr	T, W, Cl, Cv	C, BioC, gas, ^{14}C	[51]
SOCRATES	Week	P, AT	CEC, Y	Rot, Fert, Res	T, W, N, Cv, Ce	C, BioC, gas	[52]
SOMM	Day	P, ST	OM, N, AshL, NL	Man	T, W, N	C, N, gas	[53]
Sundial	Week	P, AT, EvG	Imp, Cl, W, Y	Rot, Fert, Man, Res, Irr, AtN	T, W, N, Cl	C, BioC, N, ^{15}N, W, gas	[54]
Verberne	Day	P, AT, Ir, WS, EvS	Des, W, Cl, PS, OM, C, N	Man, AtN	T, W, N, Cl	C, BioC, N, W	[55]
VOYONS	Day, week, month	P, ST	Cl, OM, C, N	Fert, Man, Res, Irr, AtN	T, W, Cl	C, BioC, ^{13}C, ^{14}C, N, gas	[56]
Wave	Day	P, AT, Ir, EvG	Lay, OM, C, N, W, PG	Rot, Ti, Fert, Man, Res, Irr, AtN	T, W, N	C, N, W, ST, gas	[57]

Key: Meteorology: P = precipitation, AT = air temperature, ST = soil temperature, Ir = irradiation, EvW = evaporation over water, EvG = evaporation over grass, EvS = evaporation over bare soil, WS = wind speed, RH = relative humidity, VPD = vapor pressure deficit, SH = sun hours. Soil and plant inputs: Des = soil description, Lay = soil layers, Imp = depth of impermeable layer, Cl = clay content, OM = organic matter content, N = soil nitrogen content/dynamics, C = soil carbon content/dynamics, InertC = soil inert carbon content, pH = pH, W = soil water characteristics, Wi = wilting point, PD = soil particle size distribution, CEC = cation exchange capacity, Ero = annual erosion losses, BD = soil bulk density, TC = thermal conductivity, PG = plant growth characteristics, PS = plant species composition, AS = animal species present, AG = animal growth characteristics, Y = yield, Nup = plant nitrogen uptake, LQ = litter quality, AshL = ash content of litter, NL = N content of litter. Management input details: Rot = rotation, Ti = tillage practice, Fert = Inorganic fertilizer applications, Man = Organic manure applications, Res = Residue management, Irr = Irrigation, AtN = Atmospheric nitrogen inputs. Factors affecting decay rate constants: T = temperature, W = water, pH = pH, N = nitrogen, O = oxygen, Cl = clay, Ce = cation exchange capacity, Cv = cover crop, Ti = tillage, F = fauna. Soil outputs: C, N, W, LQ, and ST as above. BioC = Biomass carbon, ^{13}C = ^{13}C dynamics, ^{14}C = ^{14}C dynamics, ^{15}N = ^{15}N dynamics, gas = gaseous losses (e.g., CO_2, N_2O, N_2), ResC = surface residue carbon, ResN = surface residue nitrogen, Ph = phosphorus dynamics, K = potassium dynamics. EC = electrical conductivity, ExCat = exchangeable cations. NB: N in the soil inputs and outputs section is used to denote all aspects of the N cycle. Further details regarding optimum decay conditions, soil organic matter components, rate constants, methods of pool fitting, and refractory soil organic matter are given elsewhere.[6,58] A metadatabase of all models is available.[7]

Source: Solberg et al.[8]

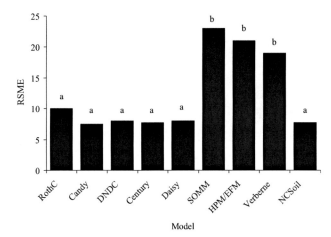

Fig. 1 Overall RMSE value for nine organic matter models when simulating changes in total soil organic carbon in up to 12 data sets from seven long-term experiments. The RMSE values of the models with the same letter (a or b) do not differ significantly (two-sample, two-tailed t test; $p > 0.05$), but the RMSE values of the two groups (a and b) do differ significantly (two-sample, two-tailed t test; $p < 0.05$).
Source: Smith et al.[60]

support systems, e.g., SUNDIAL-FRS,[54] DSSAT,[33] and APSIM.[43]

Organic matter models are now used, more than ever, to extrapolate our understanding of organic matter dynamics both temporally (in to the future) and spatially (to assess C fluxes from whole regions or continents). An early example of a regional-scale application was the use of the CENTURY model to predict the effects of alternative management practices and policies in agroecosystems of the central United States.[61] Since then, many studies have adopted similar methodologies to assess organic matter dynamics at the regional,[62] national,[63,64] and global scales.[29,65–73] Organic matter models are increasingly being used by policy makers at the national, regional, or global scales, for example, in the post-Kyoto debate on the ability of the terrestrial biosphere to store carbon.[74] With such an important role in society, it is important that organic matter models are transparent, well evaluated, and well documented. There is still a variety of understanding and different hypotheses incorporated in our current organic matter models. Future developments in organic matter models will further improve our understanding and allow models to be used in a predictive way, without the need for site-specific calibration. These developments will improve estimates of, and reduce, the uncertainty associated with organic matter model predictions.

RECENT ADVANCES IN ORGANIC MATTER MODELING

As described above, most existing organic matter models are based on conceptual pools. The need to match model conceptual pools with measurable fractions has long been recognized.[74] Recently, fractionation schemes have been developed that have allowed model pools to be directly measured.[75] Further advances in non-destructive soil analysis have allowed these fractions to be estimated without fractionating the soil,[76] raising the possibility that model organic matter pools could soon be measured non-destructively.

Many ecosystem models have been used to examine the potential impact of climate change on ecosystem components.[77] Many of these models include some description of the soil, though the detail with which soil processes are described tends to decrease with the scale of application. Site-specific applications of ecosystem models allow detailed mechanistic descriptions of soil processes to be described, whereas models applied at continental, biome, or global scales tend to have much more simplistic descriptions of the soil.[77] Models applied at the global scale often have a very simple description of the soil.[77] Many examples of the use of models to examine the impacts of climate change exist, but in most of these studies, climate change is used to drive the models, with no explicit feedback between the biosphere and the climate system.

Until recently, climate (general circulation) models (GCMs) had no organic matter component at all, relying instead upon a purely physical description of the earth's climate system. Recently, however, climate models with a fully coupled biospheric carbon module (including soil carbon) have been developed.[77–79] Though the coupled carbon cycle climate models differ widely, all suggest (to different degrees) that the feedback between climate and biospheric carbon leads to accelerated loss of carbon from the biosphere, which in turn leads to accelerated climate change.[78] One of the key processes leading to this acceleration is the response of soil respiration to increased temperature. The findings from studies using coupled biospheric carbon and climate models are subject to many uncertainties[79] but do highlight the need for including soils in assessments of future climate. Future developments require that the process-level understanding developed at the site level, using mechanistic process-based models, be incorporated in these coupled global climate models so that the full impact of the feedback of soil carbon change on climate can be quantified.[77]

REMAINING CHALLENGES IN ORGANIC MATTER MODELING

There is a need to include more biological and ecological understanding in organic matter models, since there are considerable gaps between the detailed process studies that biologists and ecologists undertake in the field and the ways that this understanding is represented within global-scale organic matter models. Ecological and biological research has crucial roles to play in the development of these models: first, it critically evaluates the current representation of plant–soil processes to ensure that key feedbacks are simulated; second, it supplies the theory and data to structure and parameterize global models; and third, it validates global model simulations against large-scale multifactor experiments specifically designed to study the interactive effects of simultaneous global change drivers and through the provision of data from across global gradients.[77]

Major uncertainties and areas for research and development of representations of organic matter in models relate to their inability to represent potentially important ecological phenomena including priming of soil C decomposition at depth,[80] the "gadkil" effect, non-equilibrium dynamics,[81–84] and assumptions about the temperature sensitivity of different soil carbon pools.[85,86] Other challenges include understanding the nature of impacts of land management change on the stability of different pools of soil carbon (e.g., the physical protection of organic matter[87,88]) and the inclusion of microbial feedbacks of decomposition that are not captured by first-order assumptions (e.g., soil methanogenesis and methanotrophy).

Perhaps the greatest challenge that remains is simulating organic matter turnover in highly organic soils, such as peatlands. Peatlands hold vast stocks of C and are vulnerable to climate change.[89] The vast majority of models described thus far have been developed for mineral soils, and while a few models have been developed to simulate organic matter dynamics in peatlands,[90–93] such models are in their infancy in terms of testing and predictive use.

CONCLUSION

Organic matter models have a long history and have been shown to perform well for mineral soils in most ecosystems. Many models exist. Organic matter models are increasingly being used to develop policy through many different applications and have recently found prominence as a key uncertainty in describing biospheric feedbacks in response to future climate change. Recent advances have allowed organic matter models to be coupled with climate models via land surface models to predict the size and nature of these biospheric feedbacks. In mineral soils, we can now begin to measure model pools, which will improve our confidence in organic matter models. There remain a number of key challenges for organic matter modeling, not least of which is the need for robust models to simulate and predict the behavior and response of peatlands to future climate change. The understanding we have brought to bear in our development of models for organic matter in mineral soils stands us in good stead for the challenges associated with modeling organic matter turnover in peatlands.

ACKNOWLEDGMENTS

Pete Smith is a Royal Society-Wolfson Research Merit Award holder.

REFERENCES

1. Paustian, K. Modelling soil biology and biochemical processes for sustainable agricultural research. In *Soil Biota. Management in Sustainable Farming Systems*; Pankhurst, C.E., Doube, B.M., Gupta, V.V.S.R., Grace, P.R., Eds.; CSIRO Information Services: Melbourne, 1994; 182–193.2.
2. Jenny, H. *Factors of Soil Formation. A System of Quantitative Pedology*; McGraw-Hill: New York, 1941.
3. Beek, J.; Frissel, M.J. *Simulation of Nitrogen Behaviour in Soils*; Pudoc: Wageningen, the Netherlands, 1973.
4. Jenkinson, D.S. Studies on the decomposition of plant material in soil. V. J. Soil Sci. **1977**, *28*, 424–434.
5. McGill, W.B. Review and classification of ten soil organic matter (SOM) models. In *Evaluation of Soil Organic Matter Models Using Existing, Long-Term Datasets*; Powlson, D.S., Smith, P., Smith, J.U., Eds.; NATO ASI I38, Springer-Verlag: Berlin, 1996; 111–133.
6. Molina, J.A.E.; Smith, P. Modeling carbon and nitrogen processes in soils. Adv. Agron. **1998**, *62*, 253–298.
7. Global Change and Terrestrial Ecosystems (GCTE) Soil Organic Matter Network (SOMNET) database.
8. Smith, P.; Falloon, P.; Smith, J.U.; Powlson, D.S., Eds. *Soil Organic Matter Network (SOMNET): 2001 Model and Experimental Metadata*, 2nd Ed.; GCTE Report 7; GCTE Focus 3 Office: Wallingford, Oxon, 2001; 224 pp.
9. Smith, P.; Powlson, D.S.; Smith, J.U.; Glendining, M.J. The GCTE SOMNET: A global network and database of soil organic matter models and long-term datasets. Soil Use Manage. **1996**, *108*, 57.
10. Bosatta, E.; Ågren, G.I. Theoretical analyses of the interactions between inorganic nitrogen and soil organic matter. Eur. J. Soil Sci. **1995**, *76*, 109–114.
11. Bosatta, E.; Ågren, G.I. Theoretical analysis of microbial biomass dynamics in soils. Soil Biol. Biochem. **1994**, *26*, 143–148.
12. Smith, P.; Andrén, O.; Brussaard, L.; Dangerfield, M.; Ekschmitt, K.; Lavelle, P.; Tate, K. Soil biota and global change at the ecosystem level: Describing soil biota in mathematical models. Global Change Biol. **1998**, *4*, 773–784.
13. Hunt, H.W.; Coleman, D.C.; Cole, C.V.; Ingham, R.E.; Elliott, E.T.; Woods, L.E. Simulation model of a food web with bacteria, amoebae, and nematodes in soil. In *Current Perspectives in Microbial Ecology*; Klug, M.J., Reddy, C.A., Eds.; American Society for Microbiology: Washington DC, 1984; 346–352.
14. Hunt, H.W.; Coleman, D.C.; Ingham, E.R.; Ingham, R.E.; Elliott, E.T.; Moore, J.C.; Rose, S.L.; Reid, C.P.P.; Morley, C.R. The detrital food web in a shortgrass prairie. Biol. Fertil. Soils **1987**, *3*, 57–68.
15. Hunt, H.W.; Trlica, M.J.; Redente, E.F.; Moore, J.C.; Detling, J.K.; Kittel, T.G.F.; Walter, D.E.; Fowler M.C.; Klein, D.A.; Elliott, E.T. Simulation model for the effects

16. de Ruiter, P.C.; Van Faassen, H.G. A comparison between an organic matter dynamics model and a food web model simulating nitrogen mineralization in agro-ecosystems. Eur. J. Agron. **1994**, *3*, 347–354.
17. de Ruiter, P.C.; Van Veen, J.A.; Moore, J.C.; Brussaard, L.; Hunt, H.W. Calculation of nitrogen mineralization in soil food webs. Plant Soil **1993**, *157*, 263–273.
18. de Ruiter, P.C.; Neutel, A.-M.; Moore, J.C. Energetics and stability in belowground food webs. In *Food Webs, Integration of Patterns and Dynamics*; Polis, G.A., Winemiller K.O., Eds.; Chapman & Hall: New York, 1995; 201–210.
19. McGill, W.B.; Hunt, H.W.; Woodmansee, R.G.; Reuss, J.O. PHOENIX, a model of the dynamics of carbon and nitrogen in grassland soil. In *Terrestrial Nitrogen Cycles. Processes, Ecosystem Strategies and Management Impacts*; Clark, F.E., Roswall, T., Eds.; Ecological Bulletins, NFR Publications: Stockholm, Sweden, 1981; Vol. 33, 49–115.
20. Giardina, C.P.; Ryan, M.G. Evidence that decomposition rates of organic carbon in mineral soil do not vary with temperature. Nature **2000**, *393*, 249–252.
21. Grant, R.F. A technique for estimating denitrification rates at different soil temperatures, water contents and nitrate concentrations. Soil Sci. **1991**, *152*, 41–52.
22. Sierra, J.; Renault, P. Respiratory activity and oxygen distribution in natural aggregates in relation to anaerobiosis. Soil Sci. Soc. Am. J. **1996**, *60*, 1428–1438.
23. Skopp, J.; Jawson, M.D.; Doran, J.W. Steady-state aerobic microbial activity as a function of soil water content. Soil Sci. Soc. Am. J. **1990**, *54*, 1619–1625.
24. Doran, J.W.; Mielke, L.N.; Stamatiadis, S. Microbial activity and N cycling as regulated by soil water-filled pore space. Proc. 11th Conf. Inter. Soil Tillage Res. Org. ISTRO Publications: Edinburgh, U.K., 1988; Vol. 1, 49–54.
25. Molina, J.A.E.; Clapp, C.E.; Shaffer, M.J.; Chichester, F.W.; Larson, W.E. NCSOIL, a model of nitrogen and carbon transformations in soil: Description, calibration, and behavior. Soil Sci. Soc. Am. J. **1983**, *47*, 85–91.
26. Rijtema, P.E.; Kroes, J.G. Some results of nitrogen simulations with the model ANIMO. Fert. Res. **1991**, *27*, 189–198.
27. McCown, R.L.; Hammer, G.L.; Hargreaves, J.N.G.; Holzworth, D.P.; Freebairn, D.M. APSIM: A novel software system for model development, model testing and simulation in agricultural systems research. Agric. Syst. **1996**, *50*, 255–271.
28. Franko, U. Modelling approaches of soil organic matter turnover within the CANDY system. In *Evaluation of Soil Organic Matter Models Using Existing, Long-Term Datasets*; Powlson, D.S., Smith, P., Smith, J.U., Eds.; NATO ASI I38, Springer-Verlag: Berlin, 1996; 247–254.
29. Parton, W.J.; Stewart, J.W.B.; Cole, C.V. Dynamics of C, N, P, and S in grassland soils: A model. Biogeochemistry **1987**, *5*, 109–131.
30. Lin, C.; Liu, T.S.; Hu, T.L. Assembling a model for organic residue transformation in soils. Proc. Natl. Council (Taiwan) Part B **1987**, *11*, 175–186.
31. Mueller, T.; Jensen, L.S.; Hansen, S.; Nielsen, N.E. Simulating soil carbon and nitrogen dynamics with the soil–plant–atmosphere system model DAISY. In *Evaluation of Soil Organic Matter Models Using Existing, Long-Term Datasets*; Powlson, D.S., Smith, P., Smith, J.U., Eds.; NATO ASI I38; Springer-Verlag: Berlin, 1996; 275–281.
32. Li, C.; Frolking, S.; Harriss, R. Modelling carbon biogeochemistry in agricultural soils. Global Biogeochem. Cycles **1994**, *8*, 237–254.
33. Hoogenboom, G.; Jones, J.W.; Hunt, L.A.; Thornton, P.K.; Tsuji, G.Y. An integrated decision support system for crop model applications. Paper 94-3025 presented at ASAE Meeting, Missouri, June, 1994; 23 pp.
34. Douglas, C.L., Jr.; Rickman, R.W. Estimating crop residue decomposition from air temperature, initial nitrogen content, and residue placement. Soil Sci. Soc. Am. J. **1992**, *56*, 272–278.
35. Grant, R.F. Dynamics of energy, water, carbon and nitrogen in agricultural ecosystems: Simulation and experimental validation. Ecol. Modell. **1995**, *81*, 169–181.
36. Williams, J.R. The erosion-productivity impact calculator (EPIC) model: A case history. Philos. Trans. R. Soc. London B **1990**, *329*, 421–428.
37. Kan, N.A.; Kan, E.E. Simulation model of soil fertility. Physiol. Biochem. Cultiv. Plants **1991**, *23*, 3–16 (in Russian).
38. Perruchoud, D.O. *Modeling the dynamic of non-living organic carbon in a changing climate: A case study for temperate forests*. PhD Thesis. ETH Diss. No. 11900, 1996; 196 pp.
39. Moorhead, D.L.; Reynolds, J.F. A general model of litter decomposition in the northern Chihuahuan Desert. Ecol. Modell. **1991**, *56*, 197–219.
40. Thornley, J.H.M.; Verberne, E.L.J. A model of nitrogen flows in grassland. Plant Cell Environ. **1989**, *12*, 863–886.
41. Andrén, O.; Kätterer, T. ICBM—The Introductory Carbon Balance Model for exploration of soil carbon balances. Ecol. Appl. **1997**, *7*, 1226–1236.
42. Sirotenko, O.D. The USSR climate–soil–yield simulation system. Meteorol. Gidrologia **1991**, *4*, 67–73 (in Russian).
43. McCaskill, M.; Blair, G.J. A model of S, P and N uptake by a perennial pasture. I. Model construction. Fert. Res. **1990**, *22*, 161–172.
44. Schevtsova, L.K.; Mikhailov, B.G. *Control of Soil Humus Balance Based on Statistical Analysis of Long-Term Field Experiments Database*; VIUA: Moscow, 1992 (in Russian).
45. Whitmore, A.P.; Klein-Gunnewiek, H.; Crocker, G.J.; Klír, J.; Körschens, M.; Poulton, P.R. Simulating trends in soil organic carbon in long-term experiments using the Verberne/MOTOR model. Geoderma **1997**, *81*, 137–151.
46. Ryzhova, I.M. Analysis of sensitivity of soil-vegetation systems to variations in carbon turnover parameters based on a mathematical model. Eurasian Soil Sci. **1993**, *25*, 43–50.
47. Molina, J.A.E.; Hadas, A.; Clapp, C.E. Computer simulation of nitrogen turnover in soil and priming effect. Soil Biol. Biochem. **1990**, *22*, 349–353.
48. Van Dam, D.; Van Breemen, N. NICCE—A model for cycling of nitrogen and carbon isotopes in coniferous forest ecosystems. Ecol. Modell. **1995**, *79*, 255–275.
49. O'Brien, B.J. Soil organic carbon fluxes and turnover rates estimated from radiocarbon measurements. Soil Biol. Biochem. **1984**, *16*, 115–120.
50. O'Leary, G.J. *Soil water and nitrogen dynamics of dryland wheat in the Victorian Wimmera and Mallee*. PhD Thesis, University of Melbourne, 1994; 332 pp.

51. Coleman, K.; Jenkinson, D.S.; Crocker, G.J.; Grace, P.R.; Klír, J.; Körschens, M.; Poulton, P.R.; Richter, D.D. Simulating trends in soil organic carbon in long-term experiments using RothC-23.6. Geoderma **1997**, *81*, 29–44.
52. Grace, P.R.; Ladd, J.N. *SOCRATES v2.00 User Manual*; PMB 2, Glen Osmond 5064; Co-operative Research Centre for Soil and Land Management: South Australia, 1995.
53. Chertov, O.G.; Komarov, A.S. SOMM—A model of soil organic matter and nitrogen dynamics in terrestrial ecosystems. In *Evaluation of Soil Organic Matter Models Using Existing, Long-Term Datasets*; Powlson, D.S., Smith, P., Smith, J.U., Eds.; NATO ASI I38, Springer-Verlag: Berlin, 1996; 231–236.
54. Smith, J.U.; Bradbury, M.J.; Addiscott, T.M. SUNDIAL: Simulation of nitrogen dynamics in arable land. A user-friendly, PC-based version of the Rothamsted nitrogen turnover model. Agron. J. **1996**, *88*, 38–43.
55. Verberne, E.L.J.; Hassink, J.; de Willigen, P.; Groot, J.R.R.; van Veen, J.A. Modelling soil organic matter dynamics in different soils. Neth. J. Agric. Sci. **1990**, *38*, 221–238.
56. André, M.; Thiery, J.M.; Courmac, L. ECOSIMP model: Prediction of CO_2 concentration changes and carbon status in closed ecosystems. Adv. Space Res. **1992**, *14*, 323–326.
57. Vanclooster, M.; Viaene, P.; Diels, J.; Feyen, J. A deterministic evaluation analysis applied to an integrated soil-crop model. Ecol. Modell. **1995**, *81*, 183–195.
58. Falloon, P.; Smith, P. Modelling refractory organic matter—A review. Biol. Fertil. Soils **2000**, *30*, 388–398.
59. Izarraulde, R.C.; Haugen-Kozyra, K.H.; Jans, D.C.; McGill, W.B.; Grant, R.F.; Hiley, J.C.; Soil organic carbon dynamics: Measurement, simulation and site to region scale-up. In *Assessment Methods for Soil Carbon. Advances in Soil Science*; Lal, R., Kimble, J.M., Follett, R.F., Stewart, B.A., Eds.; Lewis Publishers: Boca Raton, FL, 2000; 553–575.
60. Smith, P.; Smith, J.U.; Powlson, D.S.; McGill, W.B.; Arah, J.R.M.; Chertov, O.G.; Coleman, K.; Franko, U.; Frolking, S.; Jenkinson, D.S.; Jensen, L.S.; Kelly, R.H.; Klein-Gunnewiek, H.; Komarov, A.; Li, C.; Molina, J.A.E.; Mueller, T.; Parton, W.J.; Thornley, J.H.M.; Whitmore, A.P. A comparison of the performance of nine soil organic matter models using datasets from seven long-term experiments. Geoderma **1997**, *81*, 153–225.
61. Donigian, A.S., Jr.; Barnwell, T.O., Jr.; Jackson, R.B., IV; Patwardhan, A.S.; Weinrich, K.B.; Rowell, A.L.; Chinnaswamy, R.V.; Cole, C.V. *Assessment of Alternative Management Practices and Policies Affecting Soil Carbon in Agroecosystems of the Central United States*, U.S. EPA Report EPA/600/R-94/067; Athens, 1994; 194 pp.
62. Falloon, P.; Smith, P.; Smith, J.U.; Szabó, J.; Coleman, K.; Marshall, S. Regional estimates of carbon sequestration potential: linking the Rothamsted carbon model to GIS databases. Biol. Fert. Soil **1998**, *27*, 236–241.
63. Lee, J.J.; Phillips, D.L.; Liu, R. The effect of trends in tillage practices on erosion and carbon content of soils in the US Corn Belt. Water, Air, Soil Pollut. **1993**, *70*, 389–401.
64. Parshotam, A.; Tate, K.R.; Giltrap, D.J. Potential effects of climate and land-use change on soil carbon and CO_2 emissions from New Zealand's indigenous forests and unimproved grasslands. Weather Clim. **1996**, *15*, 3–12.
65. Post, W.M.; Emanuel, W.R.; Zinke, P.J.; Stangenberger, A.G. Soil carbon pools and world life zones. Nature **1982**, *298*, 156–159.
66. Post, W.M.; Pastor, J.; Zinke, P.J.; Staggenberger, A.G. Global patterns of soil nitrogen storage. Nature **1985**, *317*, 613–616.
67. Post, W.M.; King, A.W.; Wullschleger, S.D. Soil organic matter models and global estimates of soil organic carbon. In *Evaluation of Soil Organic Matter Models Using Existing, Long-Term Datasets*; Powlson, D.S., Smith, P., Smith, J.U., Eds.; NATO ASI I38, Springer-Verlag: Berlin, 1996; 201–222.
68. Potter, C.S.; Randerson, J.T.; Field, C.B.; Matson, P.A.; Vitousek, P.M.; Mooney, H.A.; Klooster, S.A. Terrestrial ecosystem production: A process model based on satellite and surface data. Global Biogeochem. Cycles **1993**, *7*, 811–841.
69. Schimel, D.S.; Braswell, B.H., Jr.; Holland, E.A.; McKeown, R.; Ojima, D.S.; Painter, T.H.; Parton, W.J.; Townsend, J.R. Climatic, edaphic, and biotic controls over storage and turnover of carbon in soils. Global Biogeochem. Cycles **1994**, *8*, 279–293.
70. Goto, N.; Sakoda, A.; Suzuki, M. Modelling soil carbon dynamics as a part of the carbon cycle in terrestrial ecosystems. Ecol. Modell. **1993**, *74*, 183–204.
71. Esser, G. Modelling global terrestrial sources and sinks of CO_2 with special reference to soil organic matter. In *Soils and the Greenhouse Effect*; Bouwman, A.F., Ed.; John Wiley & Sons: New York, 1990; 247–261.
72. Goldewijk, K.K.; van Minnen, J.G.; Kreileman, G.J.J.; Vloedbeld, M.; Leemans, R. Simulating the carbon flux between the terrestrial environment and the atmosphere. Water, Air, Soil Pollut. **1994**, *76*, 199–230.
73. Melillo, J.M.; Kicklighter, D.W.; McGuire, A.D.; Peterjon, W.T.; Newkirk, K.M. Global change and its effect on soil organic carbon stocks. In *Role of Nonliving Organic Matter in the Earth's Carbon Cycle*; Zepp, R.G., Sonntag, C.H., Eds.; John Wiley and Sons: New York, 1995; 175–189.
74. Elliott, E.T.; Paustian, K. Modelling the measurable or measuring the modelable: A hierarchical approach to isolating meaningful soil organic matter fractionations. In *Evaluation of Soil Organic Matter Models Using Long-Term Datasets*; Powlson, D.S., Smith, P., Smith, J.U., Eds.; NATO ASI Series 1: Global Environmental Change, 38; Springer-Verlag: Heidelberg, 1996; 161–179.
75. Zimmermann, M.; Leifeld, J.; Schmidt, M.W.I.; Smith, P.; Fuhrer, J. Measured soil organic matter fractions can be related to pools in the RothC model. Eur. J. Soil Sci. **2007**, *58*, 658–667.
76. Zimmermann, M.; Leifeld, J.; Fuhrer, J. Quantifying soil organic carbon fractions by infrared-spectroscopy. Soil Biol. Biochem. **2007**, *39*, 224–231.
77. Ostle, N.J.; Smith, P.; Fisher, R.; Woodward, F.I.; Fisher, J.B.; Smith, J.U.; Galbraith, D.; Levy, P.; Meir, P.; McNamara, N.P.; Bardgett, R.D. Integrating plant–soil interactions into global carbon cycle models. J. Ecol. **2009**, *97*, 851–863.
78. Cox, P.M.; Betts, R.A.; Jones, C.D.; Spall, S.A.; Totterdell, I.J. Acceleration of global warming due to carbon-cycle

79. Friedlingstein, P.; Cox, P.; Betts, R.; Bopp, L.; Von Bloh, W.; Brovkin, V. Climate-carbon cycle feedback analysis: Results from the (CMIP)-M-4 model intercomparison. J. Clim. **2006**, *19*, 3337–3353.
78. feedbacks in a coupled climate model. Nature **2000**, *408*, 184–187.
80. Fontaine, S.; Barot, S.; Barre, P.; Bdioui, N.; Mary, B.; Rumpel, C. Stability of organic carbon in deep soil layers controlled by fresh carbon supply. Nature **2007**, *450*, 277–280.
81. Schimel, J.P.; Weintraub, M.N. The implications of exoenzyme activity on microbial carbon and nitrogen limitation in soil: a theoretical model. Soil Biol. Biochem. **2003**, *35*, 549–563.
82. Fontaine, S.; Barot, S. Size and functional diversity of microbe populations control plant persistence and long-term soil carbon accumulation. Ecol. Lett. **2005**, *8*, 1075–1087.
83. Neill, C.; Gignoux, J. Soil organic matter decomposition driven by microbial growth: A simple model for a complex network of interactions. Soil Biol. Biochem. **2006**, *38*, 803–811.
84. Wutzler, T.; Reichstein, M. Soils apart from equilibrium—Consequences for soil carbon balance modelling. Biogeosciences **2007**, *4*, 125–136.
85. Davidson, E.A.; Janssens, I.A. Temperature sensitivity of soil carbon decomposition and feedbacks to climate change. Nature **2006**, *440*, 165–173.
86. Fang, C.; Smith, P.; Moncrieff, J.B.; Smith, J.U. Similar response of labile and resistant soil organic matter pools to changes in temperature. Nature **2005**, *433*, 57–59.
87. Denef, K.; Six, J.; Merckx, R.; Paustian, K. Carbon sequestration in microaggregates of no-tillage soils with different clay mineralogy. Soil Sci. Soc. Am. J. **2004**, *68*, 1935–1944.
88. Denef, K.; Zotarellia, L.; Boddey, R.M.; Six, J. Microaggregate-associated carbon as a diagnostic fraction for management-induced changes in soil organic carbon in two Oxisols. Soil Biol. Biochem. **2007**, *39*, 1165–1172.
89. Smith, P.; Fang, C. A warm response by soils. Nature **2010**, *464*, 499–500.
90. Clymo, R.S. Models of peat growth. Suo **1992**, *43*, 127–136.
91. Roulet, N.; Lafleur, P.M.; Richard, P.J.H.; Moore, T.R.; Humphreys, E.R.; Bubier, J. Contemporary carbon balance and late Holocene carbon accumulation in a northern peatland. Global Change Biol. **2007**, *13*, 397–411.
92. Smith, J.U.; Gottschalk, P.; Bellarby, J.; Chapman, S.; Lilly, A.; Towers, W.; Bell, J.; Coleman, K.; Nayak, D.R.; Richards, M.I.; Hillier, J.; Flynn, H.C.; Wattenbach, M.; Aitkenhead, M.; Yeluripurti, J.B.; Farmer, J.; Milne, R.; Thomson, A.; Evans, C.; Whitmore, A.P.; Falloon, P.; Smith, P. Estimating changes in national soil carbon stocks using ECOSSE—A new model that includes upland organic soils. Part I. Model description and uncertainty in national scale simulations of Scotland. Clim. Res. **2010**, *45*, 179–192.
93. Smith, J.U.; Gottschalk, P.; Bellarby, J.; Chapman, S.; Lilly, A.; Towers, W.; Bell, J.; Coleman, K.; Nayak, D.R.; Richards, M.I.; Hillier, J.; Flynn, H.C.; Wattenbach, M.; Aitkenhead, M.; Yeluripurti, J.B.; Farmer, J.; Milne, R.; Thomson, A.; Evans, C.; Whitmore, A.P.; Falloon, P.; Smith, P. Estimating changes in national soil carbon stocks using ECOSSE—A new model that includes upland organic soils. Part II. Application in Scotland. Clim. Res. **2010**, *45*, 193–205.

Organic Matter: Turnover

Johan Six
Department of Agronomy and Range Science, University of California—Davis, Davis, California, U.S.A.

Julie D. Jastrow
Environmental Research Division, Argonne National Laboratory, Argonne, Illinois, U.S.A.

Abstract

Soil organic matter (SOM) is a dynamic entity. The amount (stock) of organic matter in a given soil can increase or decrease depending on numerous factors including climate, vegetation type, nutrient availability, disturbance, land use, and management practices. But even when stocks are at equilibrium, SOM is in a continual state of flux; new inputs cycle—via the process of decomposition—into and through organic matter pools of various qualities and replace materials that are either transferred to other pools or mineralized. An understanding of SOM turnover is crucial for quantifying carbon and nutrient cycles and for determining the quantitative and temporal responses of local, regional, or global carbon and nutrient budgets to perturbations caused by human activities or climate change.

INTRODUCTION

Soil organic matter (SOM) is a dynamic entity. The amount (stock) of organic matter in a given soil can increase or decrease depending on numerous factors including climate, vegetation type, nutrient availability, disturbance, land use, and management practices. But even when stocks are at equilibrium, SOM is in a continual state of flux; new inputs cycle—via the process of decomposition—into and through organic matter pools of various qualities and replace materials that are either transferred to other pools or mineralized. For the functioning of a soil ecosystem, this "turnover" of SOM is probably more significant than the sizes of SOM stocks.[1] An understanding of SOM turnover is crucial for quantifying C and nutrient cycles and for determining the quantitative and temporal responses of local, regional, or global C and nutrient budgets to perturbations caused by human activities or climate change.[2]

DEFINITION OF SOIL ORGANIC MATTER TURNOVER

The turnover of an element (e.g., C, N, P) in a pool is generally determined by the balance between inputs (I) and outputs (O) of the element to and from the pool Fig. 1. Turnover is most often quantified as the element's mean residence time (MRT) or its half-life ($T_{1/2}$). The MRT of an element in a pool is defined as 1) the average time the element resides in the pool at steady state or 2) the average time required to completely renew the content of the pool at steady state. The term half-life is adopted from radioisotope work, where it is defined as the time required for half of a population of elements to disintegrate. Thus, the half-life of SOM is the time required for half of the currently existing stock to decompose.

The most common model used to describe the dynamic behavior or turnover of SOM is the first-order model, which assumes constant zero-order input with constant proportional mass loss per unit time[3,4]

$$\frac{\partial S}{\partial t} = I - kS \quad (1)$$

where S is the SOM stock, t is the time, k is the decomposition rate, and kS is equivalent to output O. Assuming equilibrium ($I = O$), the MRT can then be calculated as

$$\text{MRT} = \frac{1}{k} \quad (2)$$

and MRT and $T_{1/2}$ can be calculated interchangeably with the formula

$$\text{MRT} = T_{1/2} / \ln 2 \quad (3)$$

MEASURING SOIL ORGANIC MATTER TURNOVER

Most often the turnover of SOM, more specifically the turnover of SOM-C, is estimated by one of four techniques:

1. Simple first-order modeling
2. ^{13}C natural abundance technique
3. ^{14}C dating technique
4. "Bomb" ^{14}C technique.

This list does not include tracer studies where a substrate (e.g., plant material) enriched in ^{13}C, ^{14}C, and/or ^{15}N is added to soil, and its fate is followed over time. Most

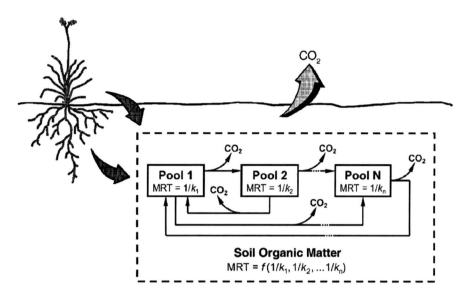

Fig. 1 The turnover of soil organic matter (SOM) is determined by the balance of inputs and outputs. Total SOM consists of many different pools that are turning over at different rates. The mean residence time (MRT) of total SOM is a function of the turnover rates of its constituent pools.

studies of this type (see Schimel[5] for a review) use the tracers to quantify the short-term (1–5 years) decomposition rate of freshly added material rather than the long-term turnover of whole-soil C.

Eqs. (1) and (2) form the basis for estimates of SOM turnover derived from first-order modeling; the unknown k is calculated as

$$k = \frac{I}{S}$$

by assuming a steady state

$$\frac{\partial S}{\partial t} = 0.$$

This approach requires estimates of annual C input rates, which can be assumed to be continuous or discrete.[3] The input can also be written as

$$I = hA$$

where A is the annual addition of C as fresh residue and h (the isohumification coefficient) represents the fraction that, after a rapid initial decomposition of A, remains as the actual annual input to S. An estimate of h is then necessary. A value of 0.3 is commonly used for agricultural crops, but the value can be higher for other materials such as grasses or peat.[6,7]

Another approach to estimate k by first-order modeling is "chronosequence modeling."[8] An increase (or decrease) in C across a chronosequence of change in vegetation, land use, or management practice can be fitted to a first-order model

$$S = S_e \left[1 - \left(\frac{S_e - S_0}{S_e}\right) e^{-kt}\right]$$

which is equivalent to

$$S = S_0 + (S_e - S_0)(1 - e^{-kt}) \qquad (4)$$

where t is the time since the change, S_e is the C content at equilibrium, and S_0 is the initial C content before the change ($t = 0$). An average value of I can then be calculated

$$I = kS_e,$$

but in this case I represents annual inputs of new SOM (hA) rather than inputs of fresh litter or detritus. This approach is also used for chronosequences of primary succession (e.g., on glacial moraines, volcanic deposits, river terraces, dune systems), in this case $S_0 = 0$.[4]

The ^{13}C natural abundance technique relies on 1) the difference in ^{13}C natural abundance between plants with different photosynthetic pathways (Calvin cycle [C_3 plants] vs. Hatch–Slack cycle [C_4 plants]); and 2) the assumption that the ^{13}C natural abundance signature of SOM is identical to the ^{13}C natural abundance signature of the plants from which it is derived.[9] Thus, where a change in vegetation type has occurred at some known point in time, the rate of loss of the C derived from the original vegetation and the incorporation of C derived from the new vegetation can be inferred from the resulting change in the ^{13}C natural abundance signature of the soil. The turnover of C derived from the original vegetation is then calculated by using the first-order decay model

$$\mathrm{MRT} = \frac{1}{k} = \frac{t}{\ln(S_t / S_0)} \qquad (5)$$

where t is the time since conversion, S_t is the C content derived from original vegetation at time t, and S_0 is the C content at $t=0$.[9,10]

The presence of ^{14}C with a half-life of 5570 years in plants and the transformation of this ^{14}C into SOM with little isotopic discrimination allows the SOM to be dated, providing an estimate of the age of the SOM. The ^{14}C dating technique is applicable within a time frame of 200–40,000 years; samples with an age less than 200 years are designated as modern (See Goh[11] for further details of the methodology.)

Thermonuclear bomb tests in the 1950s and 1960s caused the atmospheric ^{14}C content to increase sharply and then to fall drastically after the tests were halted. This sequence of events created an in situ tracer experiment; the incorporation of bomb-produced radiocarbon into SOM after the tests stopped allows estimates of the turnover of SOM.[2,12,13]

RANGE AND VARIATION IN ESTIMATES OF TOTAL SOIL ORGANIC MATTER TURNOVER

Comparisons of MRT values estimated by the four methods previously described (see, also, Table 1) reveal a wide range of MRTs. Although variations within each method are attributable to differences in vegetation, climate, soil type, and other factors, the largest variations in observed MRTs are method dependent. For example, MRTs estimated by simple first-order modeling and ^{13}C natural abundance are generally smaller by an order of magnitude than MRTs estimated by radiocarbon dating, because of the different time scales that the two methods measure. The ^{13}C method is generally used in medium-term observations or experiments (5–50 years); hence, this method gives an estimate of turnover dominated by relatively recent inputs and C pools that cycle within the time frame of the experiment. In contrast, the oldest and most recalcitrant C pools dominate estimates by radiocarbon dating because of the long-term time frame (200–40,000 years) that this method measures.[11]

FACTORS CONTROLLING SOIL ORGANIC MATTER TURNOVER

Primary production (specifically, the rate of organic matter transfer below-ground) and soil microbial activity (specifically, the rates of SOM transformation and decay) are recognized as the overall biological processes governing inputs and outputs and, hence, SOM turnover. These two processes (and the balance between them) are controlled by complex underlying biotic and abiotic interactions and feedbacks, most of which can be tied in some way to the state factor model of soil for-

Table 1 Range and average mean residence times (MRTs) of total soil organic C in various ecosystem types as estimated by four different methods.

Method and ecosystem	Sites and sources[a]	MRT (yr) Low[b]	MRT (yr) High[b]	Average ± SE[c]
First-order modeling				
Cultivated systems and recovering grassland or woodland systems	7/7	15 [14]	102 [15]	67 ± 12
^{13}C natural abundance				
Cultivated systems	20/10	18 [16]	165 [17]	61 ± 9
Pasture systems	12/10	17 [18]	102 [19]	38 ± 7
Forest systems	2/2	18 [20]	25 [21]	22 ± 4
Radiocarbon aging[d]				
Cultivated systems	21/8[e]	327 [22]	1770 [23]	880 ± 105
Grassland systems	4/3[f]	Modern [23]	1040 [24]	–[g]
Forest systems	4/3	422 [22]	1550 [25]	1005 ± 184
"Bomb" ^{14}C analysis				
Cultivated systems	1/1	1863 [13]	1863 [13]	1863[g]
Forest and grassland systems	14/12	36 [26]	1542 [27]	535 ± 134

[a]First value indicates the number of sites used to calculate average MRT values; second value indicates the number of literature sources surveyed (i.e., some sources provided data for multiple sites).
[b]Number in parentheses indicates reference to literature.
[c]SE, standard error.
[d]Values presented in MRT columns for this technique are radiocarbon ages in years B.P.
[e]Includes two sites dating as "modern."
[f]Includes three sites dating as "modern."
[g]Only one value available.

Table 2 Effect of tillage practices on mean residence time (MRT) of total soil organic C estimated by the ^{13}C natural abundance technique.

Site (Ref.)	Cropping system[a]	Depth (cm)	t^b (yr)	MRT (yr)
Sidney, NE[28]	Wheat–fallow (NT)	0–20	26	73
	Wheat–fallow (CT)			44
Delhi, Ont.[29]	Corn (NT)	0–20	5	26
	Corn (CT)			14
Boigneville, France[16]	Corn (NT)	0–30	17	127
	Corn (CT)			55
Rosemount, MN[30]	Corn (NT, 200 kg N ha^{-1} yr^{-1})	0–30	11	118
	Corn (CT, 200 kg N ha^{-1} yr^{-1})			73
	Corn (NT, 0 kg N ha^{-1} yr^{-1})			54
	Corn (CT, 0 kg N ha^{-1} yr^{-1})			72
Average ± SE[c]		NT		80 ± 19
		CT		52 ± 11

[a]NT, no tillage; CT, conventional (moldboard plow) tillage.
[b]Time period of experiment.
[c]SE, standard error.

mation.[4] Climate (especially temperature and precipitation) constrains both production and decomposition of SOM. Vegetation type affects production rates and the types and quality of organic inputs (e.g., below- vs. above-ground, amounts of structural tissue, C/N and lignin/N ratios), as well as the rates of water and nutrient uptake—all of which, in turn, influence decomposition rates. The types, populations, and activities of soil biota control decomposition and nutrient cycling/availability and hence influence vegetative productivity. Parent material affects SOM turnover as soil type, mineralogy, texture, and structure influence pH, water and nutrient supply, aeration, and the habitat for soil biota, among other factors. Topography modifies climate, vegetation type, and soil type on the landscape scale and exerts finer-scale effects on temperature, soil moisture, and texture. Lastly, time affects whether inputs and outputs are at equilibrium, and temporal scale influences the relative importance of various state factor effects on production and decomposition.

Disturbance or management practices also exert considerable influence on SOM turnover via direct effects on inputs and outputs and through indirect effects on the factors controlling these fluxes. An example of management effects on MRT is illustrated in Table 2; in most cases, the MRT of whole-soil C is significantly longer under no tillage agriculture than under conventional tillage practices.

TURNOVER OF DIFFERENT SOIL ORGANIC MATTER POOLS

The previous discussion is focused on the turnover and MRT of whole-soil C; hence, it treats SOM as a single, homogeneous reservoir. But, in fact, SOM is a heterogeneous mixture consisting of plant, animal, and microbial materials in all stages of decay combined with a variety of decomposition products of different ages and levels of complexity. Thus, the turnover of these components varies continuously, and any estimate of MRT for SOM as a whole merely represents an overall average value (Fig. 1).

Although average MRTs are useful for general comparisons of sites or the effects of different management practices, they can be misleading because soils with similar average MRTs can have very different distributions of organic matter among pools with fast, slow, and intermediate turnover rates.[2,31] Simulation models that account for variations in turnover rates for different SOM pools are now used to generate more realistic descriptions of SOM dynamics. A few models represent decomposition as a continuum, with each input cohort following a pattern of increasing resistance to decay,[32] but most models are multicompartmental, with several organic matter pools (often 3–5) that are kinetically defined with differing turnover rates. For example, the CENTURY SOM model[33] divides soil C into active, slow, and passive pools, with MRTs of 1.5, 25, and about 1000 years, respectively, and separates plant inputs into metabolic (readily decomposable; MRT of 0.1–1 years) and structural (difficult to decompose; MRT of 1–5 years) pools as a function of lignin : N ratio. Even though compartmental models are reasonably good at simulating changes in SOM, the compartments are conceptual in nature, and thus it has been difficult to relate them to functionally meaningful pools or experimentally verifiable fractions.[34,35]

The use of isotopic techniques to analytically determine the MRTs of physically and chemically separated SOM fractions has demonstrated the existence of various turnover rates for different pools. For example, low-density SOM (except for charcoal) invariably turns over faster than high-density, mineral-associated SOM, and hydrolyzable SOM turns over faster than nonhydrolyzable residues.[36,37] The MRTs of primary organomineral associations generally increase with decreasing particle size, although there are exceptions (particularly among fine gradations of silt- and clay-sized particles) that have been variously related to climate, clay mineralogy, and fractionation methodology.[34,38,39]

For a given set of biotic and abiotic conditions, the turnover of different SOM pools depends mechanistically on the quality and biochemical recalcitrance of the organic matter and its accessibility to decomposers. With other factors equal, clay soils retain more SOM with longer MRTs than do sandy soils.[40] Readily decomposable materials can become chemically protected from decomposition by association with clay minerals and by sorption to humic colloids.[38,41] Clay mineralogy also plays an important role. For example, montmorillonitic clays and allophanes generally afford more protection than illites and kaolinites.[42] In addition, the spatial location of SOM within the soil matrix determines its physical accessibility to decomposers. Relatively labile material may become physically protected by incorporation into soil aggregates[43] or by deposition in micropores inaccessible even to bacteria. Studies of the average MRTs of organic matter in macroaggregates vs. microaggregates show consistently slower turnovers in microaggregates (Table 3). Thus, a much higher proportion of the SOM occluded in microaggregates consists of stabilized materials with relatively long MRTs.

REFERENCES

1. Paul, E.A. Dynamics of organic matter in soils. Plant Soil **1984**, *76*, 275–285.
2. Trumbore, S.E. Comparison of carbon dynamics in tropical and temperate soils using radiocarbon measurements. Global Biogeochem. Cycles **1993**, *7*, 275–290.
3. Olson, J.S. Energy storage and the balance of producers and decomposers in ecological systems. Ecology **1963**, *44*, 322–331.
4. Jenny, H. *The Soil Resource—Origin and Behavior*; Springer: New York, 1980; 377 pp.
5. Schimel, D.S. *Theory and Application of Tracers*; Academic Press: San Diego, CA, 1993; 119 pp.
6. Buyanovsky, G.A.; Kucera, C.L.; Wagner, G.H. Comparative analyses of carbon dynamics in native and cultivated ecosystems. Ecology **1987**, *68*, 2023–2031.
7. Jenkinson, D.S. The turnover of organic carbon and nitrogen in soil. Phil. Trans. R. Soc. Lond. Ser. B **1990**, *329*, 361–368.
8. Jastrow, J.D. Soil aggregate formation and the accrual of particulate and mineral-associated organic matter. Soil Biol. Biochem. **1996**, *28*, 665–676.
9. Cerri, C.; Feller, C.; Balesdent, J.; Victoria, R.; Plenecassagne, A. Application du tracage isotopique natural en ^{13}C a l'etude de la dynamique de la matiere oganique dans les sols. C.R. Acad. Sci. Paris Ser. II **1985**, *300*, 423–428.
10. Balesdent, J.; Mariotti, A. Measurement of soil organic matter turnover using ^{13}C natural abundance. In *Mass Spectrometry of Soils*; Boutton, T.W., Yamasaki, S., Eds.; Marcel Dekker: New York, 1996; 83–111.
11. Goh, K.M. Carbon dating. In *Carbon Isotope Techniques*; Coleman, D.C., Fry, B., Eds.; Academic Press: San Diego, CA, 1991; 125–145.
12. Goh, K.M. Bomb carbon. In *Carbon Isotope Techniques*; Coleman, D.C., Fry, B., Eds.; Academic Press: San Diego, CA, 1991; 147–151.
13. Harrison, K.G.; Broecker, W.S.; Bonani, G. The effect of changing land use on soil radiocarbon. Science **1993**, *262*, 725–726.
14. Hendrix, P.F. Long-term patterns of plant production and soil carbon dynamics in a georgia piedmont agroecosystem. In *Soil Organic Matter in Temperate Agroecosystems: Long-Term Experiments in North America*; Paul, E.A., Paustian, K., Elliott, E.T., Cole, C.V., Eds.; CRC Press: Boca Raton, FL, 1997; 235–245.
15. Buyanovsky, G.A.; Brown, J.R.; Wagner, G.H. Sanborn field: effect of 100 years of cropping on soil parameters influencing productivity. In *Soil Organic Matter in Temperate Agroecosystems: Long-Term Experiments in North America*; Paul, E.A., Paustian, K., Elliott, E.T., Cole, C.V., Eds.; CRC Press: Boca Raton, FL, 1997; 205–225.
16. Balesdent, J.; Mariotti, A.; Boisgontier, D. Effect of tillage on soil organic carbon mineralization estimated from ^{13}C abundance in maize fields. J. Soil Sci. **1990**, *41*, 587–596.

Table 3 Mean residence time (MRT) of macro- and microaggregate-associated C estimated by the ^{13}C natural abundance technique.

Ecosystem (Ref.)	Aggregate size class[a]	μm	MRT (yr)
Tropical pasture[44]	M	>200	60
	m	<200	75
Temperate pasture grasses[19]	M	212–9500	140
	m	53–212	412
Soybean[45]	M	250–2000	1.3
	m	100–250	7
Corn[46]	M	>250	14
	m	50–250	61
Corn[47]	M	>250	42
	m	50–250	691
Wheat–fallow, no tillage[48]	M	250–2000	27
	m	53–250	137
Wheat–fallow, conventional tillage[48]	M	250–2000	8
	m	53–250	79
Average ± SE	M		42 ± 18
	m		209 ± 95

[a]M, macroaggregate; m, microaggregate.
[b]SE, standard error.

17. Vitorello, V.A.; Cerri, C.C.; Andreux, F.; Feller, C.; Victoria, R.L. Organic matter and natural carbon-13 distributions in forested and cultivated oxisols. Soil Sci. Soc. Am. J. **1989**, *53*, 773–778.
18. Desjardins, T.; Andreux, F.; Volkoff, B.; Cerri, C.C. Organic carbon and ^{13}C contents in soils and soil sizefractions, and their changes due to deforestation and pasture installation in Eastern Amazonia. Geoderma **1994**, *61*, 103–118.
19. Jastrow, J.D.; Boutton, T.W.; Miller, R.M. Carbon dynamics of aggregate-associated organic matter estimated by carbon-13 natural abundance. Soil Sci. Soc. Am. J. **1996**, *60*, 801–807.
20. Martin, A.; Mariotti, A.; Balesdent, J.; Lavelle, P.; Vuattoux, R. Estimate of organic matter turnover rate in a savanna soil by ^{13}C natural abundance measurements. Soil Biol. Biochem. **1990**, *22*, 517–523.
21. Trouve, C.; Mariotti, A.; Schwartz, D.; Guillet, B. Soil organic carbon dynamics under *eucalyptus* and *pinus* planted on savannas in the congo. Soil Biol. Biochem. **1994**, *26*, 287–295.
22. Paul, E.A.; Collins, H.P.; Leavitt, S.W. Dynamics of resistant soil carbon of midwestern agricultural soils measured by naturally-occurring ^{14}C abundance. Geoderma **2001**, *104*, 239–256.
23. Paul, E.A.; Follett, R.F.; Leavitt, S.W.; Halvorson, A.; Peterson, G.A.; Lyon, D.J. Radiocarbon dating for determination of soil organic matter pool sizes and dynamics. Soil Sci. Soc. Am. J. **1997**, *61*, 1058–1067.
24. Jenkinson, D.S.; Harkness, D.D.; Vance, E.D.; Adams, D.E.; Harrison, A.F. Calculating net primary production and annual input of organic matter to soil from the amount and radiocarbon content of soil organic matter. Soil Biol. Biochem. **1992**, *24*, 295–308.
25. Trumbore, S.E.; Bonani, G.; Wolfli, W. The rates of carbon cycling in several soils from AMS ^{14}C measurement of fractionated soil organic matter. In *Soils and the Greenhouse Effect*; Bouwman, A.F., Ed.; Wiley: London, 1990; 407–414.
26. O'Brien, B.J. Soil organic carbon fluxes and turnover rates estimated from radiocarbon enrichments. Soil Biol. Biochem. **1984**, *16*, 115–120.
27. Bol, R.A.; Harkness, D.D.; Huang, Y.; Howard, D.M. The influence of soil processes on carbon isotope distribution and turnover in the british uplands. Eur. J. Soil Sci. **1999**, *50*, 41–51.
28. Six, J.; Elliott, E.T.; Paustian, K.; Doran, J.W. Aggregation and soil organic matter accumulation in cultivated and native grassland soils. Soil Sci. Soc. Am. J. **1998**, *62*, 1367–1377.
29. Ryan, M.C.; Aravena, R.; Gillham, R.W. The use of ^{13}C natural abundance to investigate the turnover of the microbial biomass and active fractions of soil organic matter under two tillage treatments. In *Soils and Global Change*; Lal, R., Kimble, J., Levine, E., Stewart, B.A., Eds.; CRC Press: Boca Raton, FL, 1995; 351–360.
30. Clapp, C.E.; Allmaras, R.R.; Layese, M.F.; Linden, D.R.; Dowdy, R.H. Soil organic carbon and ^{13}C abundance as related to tillage, crop residue, and nitrogen fertilization under continuous corn management in Minnesota. Soil Till. Res. **2000**, *55*, 127–142.
31. Davidson, E.A.; Trumbore, S.E.; Amundson, R. Soil warming and organic carbon content. Nature **2000**, *408*, 789–790.
32. Ågren, G.I.; Bosatta, E. Theoretical analysis of the longterm dynamics of carbon and nitrogen in soils. Ecology **1987**, *68*, 1181–1189.
33. Parton, W.J.; Schimel, D.S.; Cole, C.V.; Ojima, D.S. Analysis of factors controlling soil organic matter levels in great plains grasslands. Soil Sci. Soc. Am. J. **1987**, *51*, 1173–1179.
34. Balesdent, J. The significance of organic separates to carbon dynamics and its modeling in some cultivated soils. Eur. J. Soil Sci. **1996**, *47*, 485–493.
35. Christensen, B.T. Matching measurable soil organic matter fractions with conceptual pools in simulation models of carbon turnover: revision of model structure. In *Evaluation of Soil Organic Matter Models*; Powlson, D.S., Smith, P., Smith, J.U., Eds.; Springer: Berlin, 1996; 143–159.
36. Martel, Y.A.; Paul, E.A. The use of radiocarbon dating of organic matter in the study of soil genesis. Soil Sci. Soc. Am. Proc. **1974**, *38*, 501–506.
37. Trumbore, S.E.; Chadwick, O.A.; Amundson, R. Rapid exchange between soil carbon and atmospheric carbon dioxide driven by temperature change. Science **1996**, *272*, 393–396.
38. Christensen, B.T. Physical fractionation of soil and organic matter in primary particle size and density separates. Adv. Soil Sci. **1992**, *20*, 1–90.
39. Feller, C.; Beare, M.H. Physical control of soil organic matter dynamics in the tropics. Geoderma **1997**, *79*, 69–116.
40. Sorensen, L.H. The influence of clay on the rate of decay of amino acid metabolites synthesized in soils during decomposition of cellulose. Soil. Biol. Biochem. **1974**, *7*, 171–177.
41. Jenkinson, D.S. Soil organic matter and its dynamics. In *Russell's Soil Conditions and Plant Growth*; Wild, A., Ed.; Wiley: New York, 1988; 564–607.
42. Dalal, R.C.; Bridge, B.J. Aggregation and organic matter storage in sub-humid and semi-arid soils. In *Structure and Organic Matter Storage in Agricultural Soils*; Carter, M.R., Stewart, B.A., Eds.; CRC Press: Boca Raton, FL, 1996; 263–307.
43. Tisdall, J.M.; Oades, J.M. Organic matter and waterstable aggregates in soils. J. Soil Sci. **1982**, *33*, 141–163.
44. Skjemstad, J.O.; Le Feuvre, R.P.; Prebble, R.E. Turnover of soil organic matter under pasture as determined by ^{13}C natural abundance. Aust. J. Soil Res. **1990**, *28*, 267–276.
45. Buyanovsky, G.A.; Aslam, M.; Wagner, G.H. Carbon turnover in soil physical fractions. Soil Sci. Soc. Am. J. **1994**, *58*, 1167–1173.
46. Monreal, C.M.; Schulten, H.R.; Kodama, H. Age, turnover and molecular diversity of soil organic matter in aggregates of a gleysol. Can. J. Soil Sci. **1997**, *77*, 379–388.
47. Angers, D.A.; Giroux, M. Recently deposited organic matter in soil water-stable aggregates. Soil Sci. Soc. Am. J. **1996**, *60*, 1547–1551.
48. Six, J.; Elliott, E.T.; Paustian, K. Aggregate and soil organic matter dynamics under conventional and no-tillage systems. Soil Sci. Soc. Am. J. **1999**, *63*, 1350–1358.

Organic Soil Amendments

Philip Oduor-Owino
Department of Botany, University of Kenyatta, Nairobi, Kenya

Abstract

Modern agriculture is faced with the challenge of becoming more productive and yet more sustainable. One important goal toward this end is to boost crop production through proper management of weeds, insect pests, and plant pathogens. These management tactics must be implemented without adversely affecting the ecosystem. Therefore, there is a need to change from the use of pesticides to safer pest management practices, which can be adopted in integrated pest management (IPM) programs. The use of organic soil amendments for the control of plant pathogens or pests may provide a viable alternative.

INTRODUCTION

Modern agriculture is faced with the challenge of becoming more productive and yet more sustainable. One important goal toward this end is to boost crop production through proper management of weeds, insect pests, and plant pathogens. These management tactics must be implemented without adversely affecting the ecosystem. Therefore, there is a need to change from the use of pesticides to safer pest management practices, which can be adopted in integrated pest management (IPM) programs. The use of organic soil amendments for the control of plant pathogens and/or pests may provide a viable alternative.[1–3]

ORGANIC SOIL AMENDMENTS AND THEIR MECHANISMS OF ACTION

Amending soil with organic matter such as chitin, oil cakes, compost, animal manures, and other industrial by-products in pest management studies is well recognized.[4–5] However, effects of these materials on disease development are not clear, and have been attributed, in part, to the factors discussed below.

Impacts of Organic Soil Amendments on Plant Health and Weeds

Soil amendments improve plant growth by enhancing plant nutrition.[1] The levels of nitrogen, phosphorus, potassium, and other essential elements are increased when organic matter is added to soil and is associated with better crop performance.[1] Changes in physical characteristics of soil may also enhance plant growth and the associated weeds, an attribute that should be utilized in disease management. Healthy plants produce higher yields, compete with weeds, and tolerate fungal, nematode, and insect damage better than unthrifty plants.[1,6]

Organic Soil Amendments and Plant Resistance

Materials such as oil cakes and sawdust have high phenolic content and alter the attractiveness of host plants to nematodes.[7] For example, seed treatment with ground oil cakes boosts plant resistance to *Tylenchulus semipenetrans* Cobb and root-knot nematodes due to increased levels of phenols in treated citrus and tomato roots, respectively.[4,7] In contrast, organic materials from *Tithonia diversifolia* (Hems) and chicken manure increase the severity of dry root-rot of French bean (*Phaseolus vulgaris* L.cv. Monel) caused by *Fusarium solani* f.sp. *phaseoli* (Mart) Sacc. (Table 1). This has been attributed to the formation of stimulatory ammonium compounds during decomposition.[8]

Release of Compounds Toxic to Insects and Plant Pathogens

Some organic materials release insecticidal, nematoxic, and/or fungitoxic chemicals during decomposition, for instance, neem oil and neem cake powder from the neem tree, *Azadirachtin indica*. A. Juss contains the limonoid azadirachtin, which is nematoxic and insecticidal in nature.[6,9] The black bean aphid, *Aphis fabae* (Scop), has been successfully controlled by this product (Table 2). The nematicidal activity of marigolds (*Tagetes* spp.) and castor (*Ricinus* spp.) has also been recognized, but in this case the toxic principles are Polythienyls and ricin, respectively.[4,7] Antimicrobial chemicals such as nitrites and hydrogen sulfide are also produced during decomposition and play an important role in disease control. Unfortunately, various changes in quality and quantity of these chemicals occur over time, making it difficult to obtain more than circumstantial evidence that any one compound is responsible for disease suppression.[2]

Table 1 Effect of cowdung (Cd) and organic soil amendments from *T. diversifolia* (Td) and their combination with metalaxyl (Mt) on plant growth and dry root rot of French beans 72 days after planting in soils inoculated with *F. solani* f.sp. *phaseoli* (Fs).

Soil treatment	Mean shoot dry weight (g)	Mean root dry weight (g)	Mean[a] L.D.E.T (mm)	Mean root rot index[b] (1–9)	Mean number pods per plant	Mean dry weight of 100 seeds
Cd + Fs	4.84b[c]	1.398a	30.0bc	3.13c	22.0a	25.05a
Cd + Mt + Fs	4.26b	1.166b	50.0abc	5.41b	21.0a	19.02b
Td + Fs	3.36b	0.232b	62.1ab	6.23b	5.0b	23.57b
Td + Mt + Fs	1.03c	0.014e	90.0ab	8.00a	0.0c	0.00c
Fs alone	3.12b	0.167e	97.5a	7.25a	4.0bc	21.37a
+ Cd; No Fs	7.84a	0.733c	6.11c	1.00d	22.0a	32.17a
Td alone	2.31c	0.796c	5.43c	1.00d	4.0bc	32.85a

[a]Length of discoured tissue (mm) (L.D.E.T).
[b]Mean root-rot index was based on a 0–10 rating scale, where, 0 = no symptoms and 10 = whole root system decayed.
[c]Numbers are means of five replicates. Means followed by the same letter within the same column are not significantly different at $P = 0.05$ level by Duncan's Multiple Range Test (DMRT).
Source: Wagichunge.[14]

Stimulation of Antagonistic Microorganisms

The hypothesis that organic soil amendments stimulate the activity of antagonistic microorganisms was proposed over 50 years ago.[5] When organic matter is added to soil, a sequence of microbial changes is initiated, none of which should be viewed in isolation. It is possible that the ability of nematophagous fungi such as *Paecilomyces lilacinus* Thom. (Samson) and *Verticillium chlamydosporium* (Goddard) to destroy/parasitize eggs of root-knot nematodes is stimulated by soil amendments.[2,10] Egg parasitism of up to 37% has been achieved with organic matter from castor plant or chicken manure (Table 3). Besides egg parasitism, the diverse range of microorganisms in amended soils competes with nematodes and other invertebrate pests for space and oxygen, thereby creating unfavorable anaerobic microsites in the soil. Bacteria such as *Streptomyces anulatus* (Beijerinck) Waksman, the collembolan, and *Entomobyroides dissimilis* (Moniez) are good examples.[5] Armillaria root rot of fruit and forest trees, caused by *Armillaria mellea* Vahl ex.fr, is minimized using coffee pulp that stimulates the antagonistic effects of *Trichoderma viride* link ex. Fries against a wide range of *Armillaria* spp.[11]

In conclusion, it is evident that various activities of soil microorganisms contribute significantly to the detrimental effects of organic matter on plant pathogens. However, it is difficult to determine whether any one activity or group of organisms is directly responsible for the suppression of specific diseases. The evidence available remains largely circumstantial.

USE OF ORGANIC SOIL AMENDMENTS IN THE 21ST CENTURY

Studies on the efficacy of organic soil amendments against plant pathogens should be intensified worldwide. Organic plant materials such as chitin, compost, and oil cakes have great nematode control potential but have remained unutilized in biological control systems due to inadequate and inconsistent information on their efficacy and compatibility with antagonistic microorganisms.[10,12] It is not known

Table 2 Weekly mean aphid scores on French beans following treatments with the insecticide Gaucho Neem Kernel (NKCP) and different neem products.

Treatment	Week 1	Week 2	Week 3	Week 4	Mean no. of pods/plant (Week 4)
Karate (2 mL/L)	0.1c[a]	0.1c	0.6c	1.2c	1.03c
Neem oil EC (3%)	0.1c	0.5c	0.5c	1.3c	9.3c
NKCP/WE (50 g/L)	1.1b	1.0c	0.7c	1.6c	6.2c
Gaucho (8 mL/kg)	0.9b	1.5b	2.1b	3.6b	5.0b
Control	3.4a	4.9a	6.7a	7.0a	3.3a

[a]Numbers are means of 10 replicates. Means followed by the same letter within columns do not differ significantly at $P = 0.05$ by Duncan's Multiple Range Test (DMRT).
Source: Maundu.[6]

Table 3 Effect of organic soil amendments, and soil treatments with captafol or aldicarb on the parasitism (%) of *Meloidogyne javanica* eggs with *P. lilacinus* and growth of tomato cv money maker plants.

Soil treatment[a]	Egg parasitism (%)	Juveniles/300 mL soil	Gall index[b] (0–4)	Shoot height (cm)	Shoot dry weight (g)
Tag + Mj	0.5f	188c	2.3c	36.0f[c]	2.8g
Dat + Mj	0.7f	189c	2.5c	35.0g	2.9g
Ric + Mj	1.2f	207c	2.4c	32.1h	3.5e
Ch.M + Mj	1.0f	217c	1.8d	46.8b	5.0b
Ald + Mj	0.8f	18e	1.3e	49.8a	5.4a
Cap + Mj	0.0f	521a	3.8a	32.5h	2.9g
F + Mj	21.2e	438b	3.5b	28.4T	1.7I
F + cap + Mj	1.3f	501a	3.6ab	30.2hi	2.6h
F + Ald + Mj	26.2d	10e	0.5f	46.6b	5.1b
F + Tag + Mj	30.9b	206c	2.0d	39.2e	3.0g
F + Dat + Mj	28.4c	201c	2.0d	42.8c	3.2f
F + Ric + Mj	37.2a	187c	2.5c	39.7e	3.6e
F + Ch.M + Mj	37.3a	147d	1.8d	42.0d	4.2d
Mj. "Only"	0.5f	425b	3.4b	23.5j	1.4j
Soil "Only"	0.0f	0.0e	0.0f	44.9bc	4.5c

[a]F = fungus; Mj = *M. javanica*; Cap = Captafol; Ald = aldicarb; Tag = *Tagetus minuta*; Dat = *Datura stramonium*; Ric = *Ricinus communis*; and Ch.M = chicken manure.
[b]Gall index was based on a 0–4 rating scale, where 0 = no galls and 4 = 76%–100% of the root system galled.
[c]Numbers are means of 10 replicates. Means followed by different letters within a column are significantly different ($P = 0.05$) according to Duncan's Multiple Range Test.
Source: Oduor-Owino.[1]

if these organic materials and fungal antagonists/predators can successfully be integrated into the same pest control systems. The future challenge in this case is to determine ways of boosting the antagonistic potential of specific beneficial organisms by using locally available amendments in quantities realistic for broad-scale agricultural use. The complexity of the soil environment may thwart efforts to achieve this, but previous studies[4,5,7] and recent work on the interaction between nematodes and organic soil amendments[1–3,13] suggest that this is a promising area for further research.

FUTURE CONCERNS

Organic soil amendments have a positive future in pest and disease control.[10,14] However, recent techniques used in the fields of biotechnology and molecular genetics[15] may dominate biological control research with a view of alleviating problems that are presently confronting researchers in an attempt to look for safe pest control alternatives. It is important that scientists, in their eagerness to embrace these new technologies, do not lose sight of the fact that the ultimate objective is the development of environmentally friendly pest control systems that can be applied in the field. We must strike the right balance between theoretical investigations and the more applied biological control studies aimed at developing viable pest management options.

REFERENCES

1. Oduor-Owino, P. Fungal Parasitism of Root-Knot Nematode Eggs and Effects of Organic Matter, Selected Agrochemicals and Intercropping on the Biological Control of *Meloidogyne javanica* on Tomato. Ph.D. thesis, Kenyatta University: Nairobi, 1996.
2. Oduor-Owino, P.; Waudo, S.W. Effects of delay in planting after application of chicken manure on *Meloidogyne javanica* and *Paecilomyces lilacinus*. Nematol. Mediter. **1996**, *24* (3), 7–11.
3. Oduor-Owino, P.; Sikora, R.A.; Waudo, S.W.; Schuster, R.P. Effects of aldicarb and mixed cropping with *Datura stramonium*, *Ricinus communis* and *Tagetes minuta* on the biological control and integrated management of *Meloidogyne javanica*. Nematologica **1996**, *42* (2), 127–130.
4. Bhattacharya, D.; Goswani, B.K. Comparative efficacy of neem and groundnut oil-cakes with aldicarb against *Meloidogyne incognita* in tomato. Rev. Nematol. **1987**, *10* (1), 467–470.
5. Linford, M.B. Stimulated activity of natural enemies of nematodes. Science **1937**, *85* (1), 123–124.
6. Maundu, M.E. Control of the Black aphid, *Aphis fabae* Scop of Beans Using Neem-Based Pesticides in Kenya. M.Sc. thesis, Kenyatta University, Nairobi, 1999.

7. Bandra, T.; Elgindi, D.M. The relationship between Phenolic content and *Tylenchulus semipenetrans* populations in nitrogen-amended citrus plants. Rev. Nematol. **1979**, *2* (3), 161–164.
8. Waudo, S.W.; Oduor-Owino, P.; Kuria, M. Control of Fusarium wilt of tomatoes using soil amendments. East Afr. Agric. For. J. **1995**, *60* (4), 207–217.
9. Oduor-Owino, P.; Waudo, S.W. Comparative efficacy of nematicides and nematicidal plants on root-knot nematodes. Trop. Agric. **1994**, *71* (4), 272–274.
10. Oduor-Owino, P.; Waudo, S.W.; Sikora, R.A. Biological control of *Meloidogyne javanica* in Kenya. Effect of plant residues, benomyl and decomposition products of mustard (*Brassica campestris*). Nematologica **1993**, *39* (3), 127–134.
11. Onsando, J.M.; Waudo, S.W.; Magambo, M.J.S. A biological control approach to root rot of tea *Armilleria mellea* in Kenya. Tea **1989**, *10* (2), 165–173.
12. Oduor-Owino, P.; Waudo, S.W.; Makhatsa, W.L. Effect of organic amendments on fungal parasitism of *Meloidogyne incognita eggs* and growth of tomato (*Lycopersicon esculentum* Mill) cv money maker. Int. J. Pest Manage. **1993**, *39* (4), 459–461.
13. Oduor-Owino, P.; Waudo, S.W. Medicinal plants of Kenya. Effects of *Meloidogyne incognita* and the growth of Okra. Afro-Asian J. Nematol. **1992**, *2* (1), 64–66.
14. Wagichunge, A.G. Efficacy of Seed-Dressing and Organic Amendments Against Fusarium Root-Rot of French Beans (*Phaseolus vulgaris* L.cv. Monel) in Kenya. M.Sc. thesis, Kenyatta University, Nairobi, Kenya, 2000.
15. Kerr, A. Commercial release of a genetically engineered bacterium for the control of crown gall. Agric. Sci. **1989**, *2* (1), 41–44.

Ozone Layer

Luisa T. Molina
Massachusetts Institute of Technology, Cambridge, Massachusetts, U.S.A.

Abstract

The Earth's ozone layer protects all life from the sun's harmful ultraviolet radiation. However, this fragile shield is being depleted since the late 1970s as a consequence of the emission of human-made chemicals, chlorofluorocarbons (CFCs), to the atmosphere. Observations of the ozone layer itself showed that the most dramatic loss was over Antarctica—the ozone hole—far from the emitted sources. Ozone was also being depleted in the Northern Hemisphere, particularly at high latitudes and in the winter and spring months, as well as in the lower stratosphere at mid-latitudes. A landmark international agreement, the Montreal Protocol on Substances that Deplete the Ozone Layer, has successfully reduced the global production, consumption, and emissions of CFCs and, more recently, the replacements, hydrochlorofluorocarbons (HCFCs). Furthermore, because these substances are also potent greenhouse gases, the Montreal Protocol has provided substantial climate benefits, in addition to environmental and health benefits. Assuming full compliance, the ozone layer outside the polar regions is projected to recover to its pre-1980 levels before the middle of this century, while the springtime ozone layer over the Antarctica is projected to recover much later. However, new challenges are emerging. Changes in climate are expected to have an increasing influence on stratospheric ozone in the coming decades. International efforts to protect the ozone layer would require improved understanding of the complex linkages between stratospheric ozone and climate change. Moreover, some new ozone-depleting substances (ODSs) replacements are extremely powerful global warming gases and represent a potential focal area within the overall climate change challenge. Effective control mechanisms for new ODSs and continued monitoring of the ozone layer are crucial to maintain momentum on recovering the ozone layer while simultaneously minimizing influence on climate.

INTRODUCTION

The Earth's ozone layer shields all life from the sun's harmful ultraviolet (UV) radiation. It is mainly located in the lower stratosphere, between 12 and 30 km above Earth's surface. Ozone is a gas that is present naturally in the Earth's atmosphere; it is continuously being made by the action of solar radiation on molecular oxygen, predominantly in the upper stratosphere and at low latitudes; it is also continuously being destroyed throughout the atmosphere by a variety of chemical processes. The ozone abundances in the atmosphere are therefore determined by the balance between chemical production and destruction processes.

The average concentration of ozone in the atmosphere is about 300 parts per billion by volume (ppbv); most of it (~90%) is contained in the stratosphere. Even though it occurs in such small quantities, ozone plays a vital role in sustaining life on Earth by absorbing most of the biologically damaging UV sunlight. However, the fragile ozone layer is being depleted since the late 1970s as a consequence of the emission of human-made chemicals, chlorofluorocarbons (CFCs), to the atmosphere. The CFCs are industrial chemicals that have been used in the past as coolants for refrigerators and air conditioners, propellants for aerosol spray cans, foaming agents for plastics, and cleaning solvents for electronic components, among other uses. They are thought of as "miracle" compounds because they are non-flammable, noncorrosive, and unreactive with most other substances. Ironically, it is their chemical inertness that creates a global scale problem by enabling them to reach the stratosphere, where they decompose, releasing chlorine atoms that deplete the ozone layer.

Observations of the ozone layer itself showed that depletion was indeed occurring; the most dramatic loss was discovered over Antarctica—far from the emitted sources. In response to the likelihood of increasing depletion of ozone in the stratosphere, an international agreement, the Montreal Protocol on Substances that Protect the Ozone Layer, was signed by most national governments of the world calling for an orderly phase out of all ozone-depleting substances (ODSs). Thus, ozone, a trace constituent in the atmosphere, has become an issue of global prominence and the model for international cooperation to protect the environment from unintended consequences of human activities.

While the Montreal Protocol has made great strides in phasing out ODSs, new challenges are emerging. Changes in climate are expected to have an increasing influence on stratospheric ozone in the coming decades. International efforts to protect the ozone layer would require improved understanding of the complex linkages between stratospheric ozone and climate change. Moreover, some new

ODSs replacements are extremely powerful global warming gases and represent a potential focal area within the overall climate change challenge. This entry describes the discovery of the ozone depletion phenomenon, the chemicals that cause ozone depletion in the stratosphere, polar ozone destruction processes, impacts of ozone depletion on human health and on ecosystems, international treaties that regulate the ODSs, and the linkage between stratospheric ozone and climate change.

ATMOSPHERIC OZONE

Ozone was discovered by the German chemist Christian Schöenbein in 1840 while observing an electrical discharge; he noted its distinctively pungent odor and named it "ozone," which means "smell" in Greek. An ozone (O_3) molecule is made of three oxygen atoms, instead of two of the normal oxygen molecule (O_2), which makes up 21% of the air we breathe. Ozone is found mainly in two regions of the Earth's atmosphere (Fig. 1):

1. Stratosphere, from about 10–16 km above Earth's surface up to 50 km. The ozone in this region is commonly known as the ozone layer.
2. Troposphere, the lowest region of the atmosphere, between Earth's surface and the stratosphere. Some of the tropospheric ozone is generated by atmospheric photochemical reactions (smog), and some is transported from the stratosphere.

The ozone molecules in the stratosphere and the troposphere are chemically identical; however, they have very

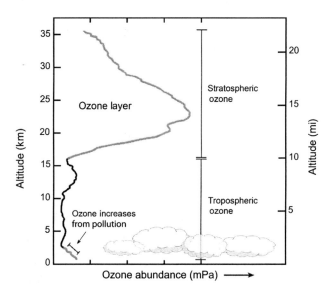

Fig. 1 Typical atmospheric ozone profile. Ozone abundances are shown here as the pressure of ozone at each altitude using the unit "milli-Pascals" (mPa) (100 million mPa = atmospheric sea-level pressure).

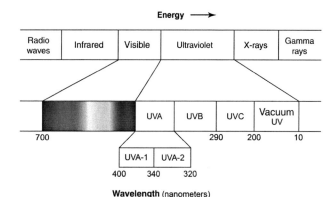

Fig. 2 Spectrum of electromagnetic radiation.

different roles in the atmosphere and very different effects on humans and the biosphere. Stratospheric ozone—sometimes referred to as "good ozone"—plays a beneficial role by absorbing most of the biologically damaging UV sunlight (UV-B), allowing only a small amount to reach the Earth's surface (Fig. 2). At the Earth's surface, ozone is a key component of urban smog and is harmful to human health, agriculture, and ecosystems; hence, it is often referred as "bad ozone." Furthermore, because ozone is a powerful greenhouse gas, increase in tropospheric ozone contributes to global warming.[1] Thus, depending on where ozone resides, it can have positive or negative impacts on human well-being and the environment.[2]

Origin of Ozone

The fundamental ozone formation–destruction mechanism consists of the following reactions, suggested initially by the British physicist Sidney Chapman[3] in the 1930s:

$$O_2 + UV\ light \rightarrow O + O \qquad (1)$$

$$O + O_2 + M \rightarrow O_3 + M \qquad (2)$$

$$O_3 + UV\ light \rightarrow O + O_2 \qquad (3)$$

$$O + O_3 \rightarrow O_2 + O_2 \qquad (4)$$

Molecular oxygen absorbs solar radiation at ~200 nm and releases oxygen atoms (Reaction 1), which rapidly combine with oxygen molecules to form ozone (Reaction 2). Ozone absorbs solar radiation very efficiently at wavelengths ~200 to 300 nm.[4] This absorption process leads to the decomposition of ozone, producing oxygen atoms (Reaction 3), which in turn regenerate the ozone molecule by Reaction 2. Thus, the net effect of Reactions 2 and 3 is the conversion of solar energy into heat, without the net loss of ozone. This process leads to an increase of temperature with altitude, which is the feature that gives rise to the stratosphere; the inverted temperature profile is responsible for its large stability toward vertical movements (Fig. 3). In contrast, the troposphere is characterized by decreasing temperature with altitude because of the heating effects from the absorption of the sun's energy at the Earth's surface. Because hot air rises, this causes rapid vertical mixing so that chemical substances emitted on the ground can rise to the tropopause (transition zone between troposphere and stratosphere) in a matter of days; they are also dispersed horizontally throughout the troposphere on the time scale of weeks to months by winds and convection. However, once they reach the stratosphere, the transport time scales in the stratosphere are much slower and mixing in the stratosphere can take months to years. Movement of air between the troposphere and stratosphere is very slow compared to the movement of air within the troposphere itself. However, this small air exchange is an important source of ozone from the stratosphere to the troposphere.

Most of the time, oxygen atoms react with molecular oxygen to make ozone (Reaction 2), but occasionally they destroy ozone (Reaction 4). The overall amount of ozone is determined by a balance between the production and the removal processes. Models based only on the Chapman's mechanism were found to overpredict stratospheric ozone levels; thus, there are other reactions that contribute to the destruction of ozone.

In the early 1970s, Crutzen[5] suggested that trace amounts of nitrogen oxides (NO_x = NO + NO_2) formed in the stratosphere through the decomposition of nitrous oxide (N_2O), which originates from soil-borne microorganisms, control the ozone abundance through the following catalytic cycle:

$$NO + O_3 \rightarrow NO_2 + O_2 \quad (5)$$

$$NO_2 + O \rightarrow NO + O_2 \quad (6)$$

$$O_3 + \text{UV light} \rightarrow O + O_2 \quad \text{(for Reaction 3)}$$

$$\text{Net: } 2\,O_3 \rightarrow 3\,O_2$$

The term "catalyst" refers to a compound that reacts with one or more reactants to form intermediates that subsequently give the final reaction product, in the process regenerating the catalyst. In the above cycle, the species NO (nitric oxide) and NO_2 (nitrogen dioxide) are still present after these three reactions have occurred, but two molecules of ozone have been destroyed. These species have an odd number of electrons; they are free radicals and are chemically very reactive. Although the concentration of NO and NO_2 is small (several ppbv), each radical pair can destroy thousands of ozone molecules before being temporarily removed, mainly by reaction with hydroxyl (OH) radical to form nitric acid:

$$OH + NO_2 \rightarrow HNO_3 \quad (7)$$

Independently, Johnston[6] suggested that the NO_x chain reaction can be initiated by the direct release of NO_x in the exhaust of supersonic transport (SST) aircraft and could disturb the delicate natural balance between ozone formation and destruction.

In the troposphere, NO_x, together with volatile organic compounds in the presence of sunlight, are the ingredients for the photochemical formation of ground-level ozone. Thus, NO_x plays a dual role, destroying or generating O_3 depending on the altitude.

Other free radicals that destroy stratospheric ozone are OH and HO_2, derived from the water molecule:

$$OH + O_3 \rightarrow HO_2 + O_2 \quad (8)$$

$$HO_2 + O_3 \rightarrow OH + 2\,O_2 \quad (9)$$

$$\text{Net: } 2\,O_3 \rightarrow 3\,O_2$$

Chlorine atoms are also very efficient catalysts for ozone destruction and may proceed in a similar cycle[7–9] as will be discussed below. Small amounts of chlorine compounds of natural origin exist in the stratosphere; the most important source is methyl chloride (CH_3Cl), which is emitted mainly from oceanic and terrestrial ecosystems. Most of the CH_3Cl is destroyed in the troposphere, but a few percent reaches the stratosphere. There are also large natural sources of inorganic chlorine compounds at the Earth's surface, e.g., NaCl and HCl from the oceans; however,

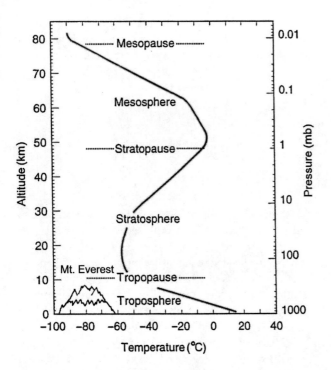

Fig. 3 Typical variation of atmospheric temperatures and pressures with altitude.

they are water soluble and are removed efficiently from the atmosphere by clouds and rainfall.

Measurements and Distribution of Stratospheric Ozone

In the 1920s, the British meteorologist G.M.B. Dobson developed the Dobson spectrophotometer that measures sunlight at two UV wavelengths: one that is strongly absorbed by ozone and one that is weakly absorbed. The difference in light intensity at the two wavelengths provides the total ozone above the location of the instrument.[10] In 1957, a global network of ground-based, total ozone observing stations was established as part of the International Geophysical Year; currently, there are about 100 sites distributed throughout the world. Ozone concentrations are also being routinely measured by a variety of instruments on balloons, aircraft, and satellites. One of the most commonly used units for measuring ozone concentration is called "Dobson unit" (DU), which is a measure of how much ozone is contained in a vertical column of air. The average amount of ozone in the atmosphere is about 300 DU, equivalent to a layer 3 mm thick. By comparison, if all of the air in a vertical column that extends from the ground up to space were collected and squeezed together at 0°C and 1 atm pressure, that column would be 8 km thick.

As shown in Fig. 4, the distribution of total ozone over the globe varies with latitude, longitude, and season.[11] The variations are caused by large-scale movements of stratospheric air and the chemical production and destruction of ozone. In general, the total ozone is highest in the polar regions and lowest at the equator. In the Northern Hemisphere, the ozone layer is thicker during spring and thinner during autumn.

CFCS AND THE OZONE LAYER

CFCs: The Miracle Compounds

In the 1930s, American mechanical engineer and chemist Thomas Midgley[12,13] invented the CFCs during a search for nontoxic and non-flammable substances that could be used as coolants in home refrigerators and air conditioners. The CFCs are compounds that contain only chlorine, fluorine, and carbon; they are also known under trademarks such as Freon (DuPont) and Genetron (Allied Signal).

The two important properties that make the CFCs commercially valuable are their volatility (they can be

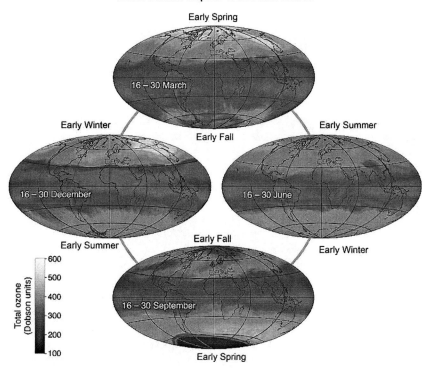

Fig. 4 Global satellite maps of total ozone in 2009 showing the variation with latitude, longitude, and season. The variations are demonstrated here with 2-week averages of total ozone in 2009 as measured with a satellite instrument. Total ozone shows little variation in the tropics (20°N–20°S latitudes) over all seasons. Total ozone outside the tropics varies more strongly with time on a daily to seasonal basis as ozone-rich air is moved from the tropics and accumulates at higher latitudes. The low total ozone values over Antarctica in September constitute the "ozone hole" in 2009. Since the 1980s, the ozone hole in late winter and early spring represents the lowest values of total ozone that occur over all seasons and latitudes.
Source: Fahey and Hegglin.[11]

readily converted from a liquid to a vapor and vice versa) and their chemical inertness (they are stable, nontoxic, and nonflammable). The CFCs were thought of as miracle compounds and soon replaced the toxic ammonia and sulfur dioxide as the standard cooling fluids. Subsequently, the CFCs found uses as propellants for aerosol sprays, blowing agents for plastic foam, and cleaning agents for electronic components. All of these activities doubled the worldwide use of CFCs every 6 to 7 years and eventually reached over 700,000 metric tons annually by the early 1970s.

CFCs and the Destruction of Stratospheric Ozone

In the early 1970s, James Lovelock showed that trichlorofluromethane (CCl_3F or CFC-11) was present in the air over Ireland using a newly developed electron capture detector.[14] Subsequently, Lovelock and coworkers[15] detected measurable levels of CCl_3F in the atmosphere over the South and North Atlantic and concluded that the CFCs were carried over by large-scale wind motions. Their interest in CFCs lies in the potential use of these compounds as inert tracers; they did not expect the CFCs to present any conceivable harm to the environment.

In 1973, Molina and Rowland[7,8,16–18] decided to investigate the ultimate fate of these new miracle compounds upon their release to the atmosphere. After carrying out a systematic search of chemical and physical processes that might destroy the CFCs in the lower atmosphere, they concluded that these compounds would break up in the middle stratosphere (~25–30 km) by solar UV radiation.

Because the CFCs are chemically inert and practically insoluble in water, they are not removed by the common cleansing mechanisms that operate in the lower atmosphere. Furthermore, the CFC molecules are transparent from 230 nm through the visible wavelengths; they are effectively protected below 25 km by the stratospheric ozone layer that shields the Earth's surface from UV light. Instead, they rise into the stratosphere, where they are eventually destroyed by the short-wavelength (~200 nm) solar UV radiation to yield radicals that can destroy stratospheric ozone through a catalytic process.[7] Because transport into the stratosphere is very slow, the atmospheric lifetime for the CFCs is about 50 to 100 years.

The destruction of CFCs by high-energy solar radiation leads to the release of chlorine atoms, as shown in Reaction 12 for CCl_3F:

$$CCl_3F + UV\ light \rightarrow Cl + CCl_2F \quad (10)$$

The chlorine atoms attack ozone within a few seconds and are regenerated on a time scale of minutes; the net result is the conversion of one ozone molecule and one oxygen atom to two oxygen molecules.

$$Cl + O_3 \rightarrow ClO + O_2 \quad (11)$$

$$ClO + O \rightarrow Cl + O_2 \quad (12)$$

$$\text{Net: } O + O_3 \rightarrow O_2 + O_2$$

In the above cycle, chlorine acts as a catalyst for ozone destruction because Cl and ClO react and are reformed, and ozone is simply removed. Oxygen atoms are formed when solar UV radiation reacts with ozone and oxygen molecule (Reactions 1 and 3). It is estimated that one Cl atom can convert 100,000 molecules of ozone into oxygen molecules before that chlorine becomes part of a less reactive compound, e.g., by reaction of ClO with HO_2 or NO_2 to produce hypochlorous acid (HOCl) or chlorine nitrate ($ClONO_2$), respectively, or by reaction of the Cl atom with methane (CH_4) to produce the relatively stable hydrogen chloride (HCl):

$$ClO + HO_2 \rightarrow HOCl + O_2 \quad (13)$$

$$ClO + NO_2 \rightarrow ClONO_2 \quad (14)$$

$$Cl + CH_4 \rightarrow HCl + CH_3 \quad (15)$$

The chlorine-containing product species HCl, $ClONO_2$, and HOCl function as temporary inert reservoirs: they are not directly involved in ozone depletion, but they are eventually broken down by reaction with other free radicals or by absorption of solar radiation, thus returning chlorine to its catalytically active free radical form. At low latitudes and in the upper stratosphere, where the formation of ozone is fastest, a few percent of the chlorine is in this active form; most of the chlorine is in the inert reservoir form, with HCl being the most abundant species. The temporary chlorine reservoirs remain in the stratosphere for several years before returning to the troposphere, where they are rapidly removed by rain or clouds. A schematic representation of these processes is presented in Fig. 5.

Other ODSs

Besides CFCs, there are other chlorine substances from human activities that destroy ozone in the stratosphere. Carbon tetrachloride and methyl chloroform are important ODSs that are used as fire extinguisher, cleaning agents, and solvents.

Another category of ODSs contains bromine. There are industrial sources of brominated ODSs as well as natural ones; the most important are the halons and methyl bromide (CH_3Br). Halons are halocarbon gases containing carbon, bromine, fluorine, and (in some cases) chlorine; they are produced industrially as fire extinguishers. Methyl bromide is both natural and human-made; it is used as an agricultural fumigant. These sources release bromine to the stratosphere at pptv levels, compared with ppbv for chlo-

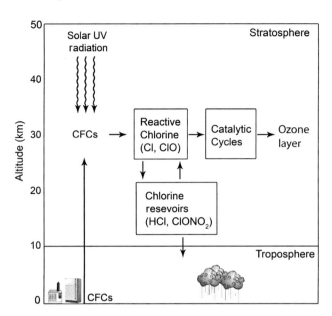

Fig. 5 Schematic representation of the CFC-ozone depletion hypothesis.

rine. On the other hand, bromine atoms are about 60 times more efficient than chlorine atoms for ozone destruction on an atom-per-atom basis;[19–21] a large fraction of the bromine compounds is present as free radicals because the temporary reservoirs are less stable and are formed at considerably slower rates than the corresponding chlorine reservoirs.

Many of the ODSs contain fluorine atoms in addition to chlorine and bromine. However, in contrast to chlorine and bromine, fluorine atoms abstract hydrogen atoms very rapidly from methane and from water vapor, forming the stable hydrogen fluoride (HF), which serves as a permanent inert fluorine reservoir. Hence, fluorine free radicals are extremely scarce and the effect of fluorine on stratospheric ozone is negligible. Halogen source gases that contain fluorine and no other halogens are not classified as ODSs. An important category is the hydrofluorocarbons (HFCs) discussed in the "International Response: Montreal Protocol" section.

The publication of the 1974 Molina–Rowland article[7] stimulated numerous scientific studies, including laboratory studies, computer modeling, and field measurements, to understand the impacts of chlorine and bromine on stratospheric ozone.[20,21] The U.S. National Academy of Sciences issued two reports in 1976, verifying the Molina–Rowland findings.[22,23]

STRATOSPHERIC OZONE DEPLETION

Discovery of the Antarctic Ozone Hole

In 1985, Joseph Farman and coworkers[24] reported that the ozone concentrations recorded at the Halley Bay Observatory in Antarctica has dropped dramatically in the spring months starting in the early 1980s, compared to the data obtained since 1957, when the British Antarctic Survey began ozone measurements using a Dobson Spectrometer. The 1984 October monthly ozone averaged less than 200 DU, about 35% lower than the 300 DU levels recorded in 1957–1958 and on through the 1960s. Farman et al.'s findings were subsequently confirmed by satellite observation from the total ozone mapping spectrometer (TOMS).[25] Furthermore, satellite measurements confirmed that the bulk of the chlorine in the stratosphere is of human origin.[26] Additional measurements from ground-based Dobson instruments[27] and from satellites indicate that the extent of ozone depletion over Antarctica in the spring months continued to increase after 1985, with concentrations as low as 85 to 95 DU reported from some of the polar stations.

Measurements show that ozone was also being depleted in the Northern Hemisphere, particularly at high latitudes and in the winter and spring months. Examination of the ozone records shows that significant changes have also occurred in the lower stratosphere at mid-latitudes.[20,21] Fig. 6 shows an acute drop in total atmospheric ozone during October in the early and mid-1980s measured by instruments from the ground and from a satellite.[28]

The discovery of the depletion of ozone over Antarctica—the ozone hole—was not predicted by the atmospheric scientists (see Fig. 7). The large magnitude of the depletion suggests that the stratospheric ozone is influenced by processes that had not been considered previously. Researchers all over the world raced to develop plausible explanation; the cause of this depletion soon became very clear. Laboratory experiments, field measurements over Antarctica, and model calculations showed unambiguously that the ozone hole over Antarctica can indeed be traced to the industrial CFCs.[20,21]

Polar Ozone Chemistry

Characteristics of the Polar Regions

ODSs emitted at Earth's surface are transported over great distances to the stratosphere by atmospheric air motions and are present throughout the stratospheric ozone layer. Yet, the most dramatic ozone depletion was over Antarctica—the ozone hole—far from the emitted sources. A major reason why an Antarctic ozone hole of the observed extent could happen is because of the unique atmospheric and chemical conditions that exist there. The very low winter temperatures in the Antarctic stratosphere cause polar stratospheric clouds (PSCs) to form. Special reactions that occur on PSCs, combined with the relative isolation of polar stratospheric air, allow chlorine and bromine reactions to produce the ozone hole in the Antarctica when the sunlight returns in the springtime.

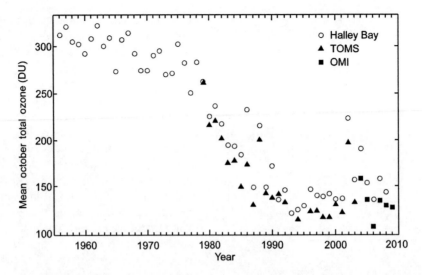

Fig. 6 Average total amount of ozone measured in October over Antarctica. Instruments on the ground (at Halley Bay) and high above Antarctica (the TOMS and ozone monitoring instrument [OMI] measured an acute drop in total atmospheric ozone during October in the early and middle 1980s).
Source: Adapted from NASA Ozone Hole Watch.[28]

Both polar regions of the earth are cold, primarily because they receive far less solar radiation than the tropics and mid-latitudes do. Moreover, most of the sunlight that does shine on the polar regions is reflected by the bright white surface. Winter temperatures at the North Pole can range from about −45°C to −25°C and summer temperatures can average around the freezing point (0°C). In comparison, the annual mean temperature at the South Pole is about −60°C in the winter and −28°C in the summer.

What makes the South Pole so much colder than the North Pole is that it sits on top of a very thick ice sheet, which itself sits on the continent of Antarctica. The surface of the ice sheet at the South Pole is more than 2700 m (9000 ft) above sea level; Antarctica is by far the highest continent on the earth. In contrast, the North Pole is at sea level in the middle of the Arctic Ocean, which also acts as an effective heat reservoir.

Stratospheric air in the polar regions is relatively isolated from other stratospheric regions for long periods in the winter months. The isolation comes about because of strong winds that encircle the poles, forming a polar vortex (or polar cyclone), which prevents substantial air masses

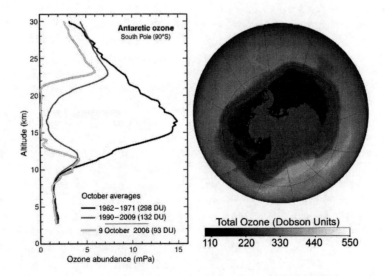

Fig. 7 Antarctic ozone hole. Left panel: Vertical distributions of Antarctic ozone. A normal ozone layer was observed to be present between 1962 and 1971. In more recent years, as shown here for October 9, 2006, ozone is almost completely destroyed between 14 and 21 km in the Antarctic in spring (Source: Adapted from Fahey and Hegglin[11]). Right panel: The darker shaded regions over the Antarctic continent show the severe ozone depletion or ozone hole on October 9, 2006, measured by satellite instrument. The hole reached 26.2 million km^2, the greatest extent recorded in the Antarctic.
Source: NASA Ozone Hole Watch.[28]

into or out of the polar stratosphere. This cyclonic circulation strengthens in winter as stratospheric temperatures decrease.[11] All through the long, dark winter, chemical changes occur in polar regions from reactions on PSCs inside the vortex; the isolation preserves those changes until the spring sunlight strikes the stratosphere above the frozen continent in late August. The result is massive ozone destruction inside the vortex forming an ozone hole, as described below. The polar vortex diminishes when the continent and the air above it begin to warm up and ozone-rich air from outside the vortex flows in, replacing much of the ozone that was destroyed.

The Antarctic polar vortex is more pronounced and persistent than its Arctic counterpart with the result that the isolation of air inside the vortex is much more effective in the Antarctic than in the Arctic. In addition to being significantly warmer, the Northern Hemisphere also has numerous mountain ranges and a more active tropospheric meteorology, giving rise to enhanced planetary wave and a less stable Arctic vortex.

Polar Stratospheric Clouds

The conditions in the polar stratosphere are unique in several ways. Firstly, ozone is not generated there because the high-energy solar radiation that is absorbed by molecular oxygen is scarce over the poles. Secondly, the total ozone column abundance at high latitudes is large because ozone is transported toward the poles from higher altitudes and lower latitudes. Thirdly, the prevailing temperatures over the stratosphere above the poles in the winter and spring months are the lowest throughout the atmosphere, particularly over Antarctica. Typically, average daily minimum values are as low as –90°C in July and August over Antarctica and near –80°C in late December and January over the Arctic (Fig. 8). Ozone is expected to be rather stable over the poles if one considers only gas-phase chemical and photochemical processes, because regeneration of ozone-destroying free radicals from the reservoir species would occur very slowly at those temperatures. However, another unique feature of the polar stratosphere is the seasonal presence of PSCs (Fig. 9). Different types of liquid and solid PSC particles form when stratospheric temperatures fall below –78°C. As a result, PSCs are often found over large areas of the winter polar regions and over significant altitude ranges. With a temperature threshold of –78°C, PSCs exist in larger areas and longer periods in the Antarctica than in the Arctic.

The stratosphere is normally very dry; water is present only at a level of a few ppmv, comparable to that of ozone itself. Over the poles, a somewhat larger amount of water is present, resulting from the oxidation of meth-

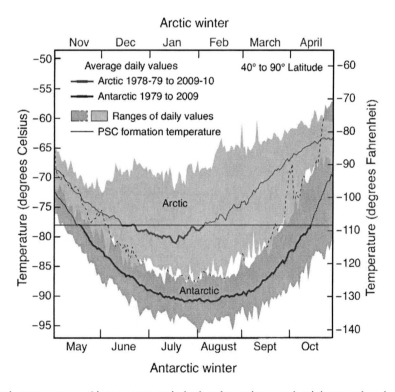

Fig. 8 Arctic and Antarctic temperatures. Air temperatures in both polar regions reach minimum values in the lower stratosphere in the winter season. Typically, average daily minimum values are as low as –90°C in July and August in the Antarctic and near –80°C in late December and January in the Arctic. PSCs are formed in the polar ozone layer when winter minimum temperatures fall below the formation temperature of about –78°C. Note that the dashed black lines denote the upper limits of the Antarctic temperature range where they overlap with the Arctic temperature range.
Source: Fahey and Hegglin.[11]

Fig. 9 Polar stratospheric clouds. The frozen crystals that make up PSCs provide a surface for the reactions that free chlorine atoms in the Antarctic stratosphere (left panel[28]) and in the Arctic stratosphere (right panel[111]).

ane. Furthermore, the temperature can drop below $-85°C$ over Antarctica in the winter and spring months, leading to the formation of ice clouds. The presence of trace amounts of nitric and sulfuric acids enables the formation of PSCs a few degrees above the frost point (the temperature at which ice can condense from the gas phase); these acids can form cloud particles consisting of crystalline hydrates.

Solomon et al.[29] first suggested that PSCs could play a major role in ozone depletion over Antarctica by promoting the release of photolytically active chlorine from its reservoir species (HCl, $ClONO_2$, and $HOCl$). This occurs mainly by the following reaction:

$$HCl + ClONO_2 \rightarrow Cl_2 + HNO_3 \qquad (16)$$

Indeed, laboratory studies have shown that this reaction occurs very slowly in the gas phase;[30] however, it proceeds with remarkable efficiency in the presence of ice surfaces.[31] The product, Cl_2, is immediately released to the gas phase while the other product, HNO_3, remains in the condensed phase. Cl_2 decomposes readily to Cl atoms even with the faint amount of sunlight present over Antarctica in the early spring:

$$Cl_2 + \text{sunlight} \rightarrow Cl + Cl \qquad (17)$$

When average temperatures begin increasing by late winter, PSCs form less frequently and their surface conversion reactions produce less ClO. Without continued ClO production, ClO amounts decrease and other chemical reactions re-form the reactive reservoirs, $ClONO_2$ and HCl. The most intense period of ozone depletion will end when PSC temperatures no longer occur.

An important feature for Reaction 16 in the presence of PSCs is the removal of NO_x from the gas phase; the source for these free radicals in the polar stratosphere is nitric acid, which condenses in the cloud particles. NO_x normally interfere with the catalytic ozone loss reactions by scavenging ClO to form chlorine nitrate (Reaction 14). In the absence of NO_x, the fraction of chlorine compounds in the form of free radical is larger and makes it possible for the ozone depletion reaction to occur. These experimental results have been corroborated by other studies.[32] Further laboratory studies[33,34] and theoretical calculations[35] indicate that HCl solvates readily on the ice surface, forming hydrochloric acid. Therefore, chlorine activation reactions on the surfaces of ice crystals proceed through ionic mechanisms analogous to those in aqueous solutions.

Another natural source of ozone depletion is the sulfate aerosols that come from volcanic eruptions. The most recent large eruption was that of Mt. Pinatubo in 1991, which ejected large amounts of sulfur dioxide into the stratosphere, resulting in up to a 10-fold increase in the number of particles available for surface reactions. These particles increased the global ozone depletion by 1%–2% for several years following the eruption.[11]

Polar Ozone Destruction

While the presence of the PSCs explains how chlorine can be released from the inactive reservoir chemicals, it remains unanswered how a catalytic cycle might be maintained to account for the large ozone destruction observed. The ClO_x cycles such as Reactions 11 and 12 are not efficient in the polar stratosphere because they require the presence of free oxygen atoms, which are scarce at high latitudes.

Several catalytic cycles have been suggested as occurring over Antarctica during winter and spring. One of the dominant cycles involving ClO dimer or chlorine peroxide (ClOOCl) was proposed by Molina and Molina.[36] The cycle is initiated by the combination of two ClO radicals forming ClOOCl, which then photolyzes to release molecular oxygen and free chlorine atoms:

$$ClO + ClO + M \rightarrow ClOOCl + M \quad (18)$$

$$ClOOCl + sunlight \rightarrow Cl + ClOO \quad (19)$$

$$ClOO + M \rightarrow Cl + O_2 + M \quad (20)$$

$$\underline{2\,[Cl + O_3 \rightarrow ClO + O_2] \quad \text{(for Reaction 11)}}$$

$$\text{Net: } 2\,O_3 \rightarrow 3\,O_2$$

No free oxygen atoms are involved in this cycle. Visible sunlight is required to complete and maintain the cycle; thus, this cycle can occur only in late winter/early spring when sunlight returns to the polar region. Laboratory studies have shown that the photolysis products of ClOOCl are indeed Cl atoms.[37,38]

Another important cycle operating in the polar stratosphere involves the reaction of bromine monoxide (BrO) with ClO suggested by McElroy et al.:[39]

$$BrO + ClO + sunlight \rightarrow Cl + Br + O_2 \quad (21)$$

$$Br + O_3 \rightarrow BrO + O_2 \quad (22)$$

$$\underline{Cl + O_3 \rightarrow ClO + O_2 \quad \text{(for Reaction 11)}}$$

$$\text{Net: } 2\,O_3 \rightarrow 3\,O_2$$

Field Measurements of Atmospheric Trace Species

Ground-based and aircraft expeditions were launched in the years following the ozone hole discovery to measure trace species in the stratosphere over Antarctica.[40] The results provided strong evidence for the crucial role played by industrial chlorine in the ozone depletion. One of the most convincing evidence was provided by the NASA ER-2 aircraft measurements in 1987, which flew into the Antarctic vortex. The flight data (Fig. 10) showed an anticorrelation between ClO measured by Anderson et al.[41] with in situ ozone measurements monitored by Proffitt et al.[42] The results show that the ClO + ClO cycle accounts for about three-quarters of the observed ozone loss, with the BrO + ClO cycle accounting for the rest. Furthermore, NO_x levels were found to be very low and nitric acid was shown to be present in the cloud particles, as expected from the laboratory studies.

Recent laboratory measurements of the dissociation cross section of ClO dimer[43] and analyses of observation from aircraft and satellites have reaffirmed that polar springtime ozone depletion is caused primarily by the ClO + ClO catalytic ozone destruction cycle, with substantial contributions from the BrO + ClO cycle.[44]

Field measurements have also been conducted in the Arctic stratosphere,[45] indicating that a large fraction of the chlorine is also activated there. Nevertheless, ozone depletion is less severe over the Arctic and is not as localized because the atmosphere above the Arctic is warmer than above the Antarctic; the active chlorine does not remain in contact with ozone long enough and at low enough temperatures to destroy it before the stratospheric air over the Arctic mixes with warmer air from lower latitudes. This warmer air also contains NO_2, which deactivates the chlorine. On the other hand, cold winters can lead to significant ozone depletion—30% or more—over large areas, as described below.[46]

Depletion of the Global Ozone Layer

Global total ozone levels are influenced not only by the concentrations of ODSs but also by atmospheric transport (winds), incoming solar radiation, aerosols (fine particles suspended in the air), and other natural compounds. Global total ozone has decreased beginning in the 1980s, reaching a maximum of about 5% in the early 1990s (Fig. 11a). The lowest global total ozone values occurred in the years following the eruption of Mt. Pinatubo in 1991, which resulted in up to a 10-fold increase in the number of sulfuric-acid-containing particles available for surface reaction in the stratosphere, thereby increased global ozone depletion for several years before they were removed from the stratosphere by natural processes. The depletion has lessened since then; the average global ozone for 2005–2009 is about 3.5% below the 1964–1980 average.

Observed total ozone loss varies significantly with latitude on the globe. Fig. 11b shows how the 2005–2009 ozone depletion varies with latitude. The ozone loss is very small near the equator and increases with latitude towards the poles. The largest decreases have occurred at

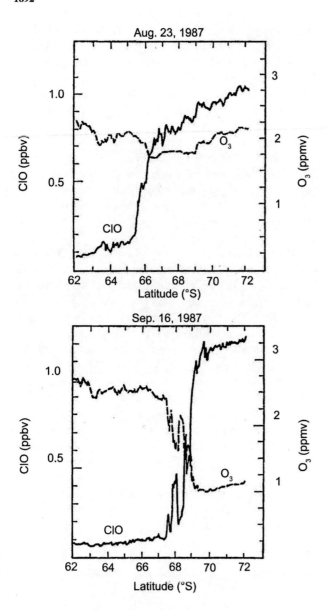

Fig. 10 Aircraft measurements conducted on August 23 and September 16, 1987, of chlorine monoxide (ClO) by Anderson et al.[41] and of ozone (O_3) by Proffitt et al.[42]
Source: Adapted from Anderson.[41]

high latitudes in both hemispheres because of the large winter/spring depletion in the polar regions; the losses are greater in the Southern Hemisphere because of the Antarctic ozone hole. Since the 1980s, Antarctic ozone loss in the springtime has been quite large, covering nearly the entire continent with virtually all of the ozone destroyed between 15 and 20 km. Ozone loss over the Arctic is smaller than its Antarctic counterpart; it is modulated strongly by variability in atmospheric dynamics, transport, and temperature. The degree of spring Arctic depletion is highly variable from year to year, but large Arctic depletion has also been observed recently with the most dramatic occurring in the spring of 2011. Observations over the Arctic region as well as from satellites show an unprecedented ozone column loss comparable to some Antarctic ozone holes. The formation of the "Arctic ozone hole" in 2011 was driven by an anomalously strong stratospheric polar vortex and an unusually long cold period, leading to persistent enhancement of active chlorine and severe ozone loss that exceeded 80% over 18–20 km altitude. This result raises the possibility of yet more severe depletion as lower stratospheric temperatures decrease. More acute Arctic ozone loss could exacerbate biological risks from exposure to increased UV radiation, especially if the vortex shifted over densely populated mid-latitudes, as it did in April 2011.[46]

Ozone depletion is also observed at the mid-latitudes between equatorial and polar latitudes. Total ozone averaged for 2005–2009 is about 3.5% lower in northern mid-latitudes

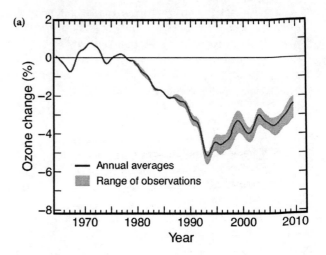

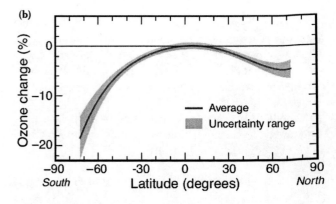

Fig. 11 Global total ozone changes. Satellite observations show depletion of global total ozone beginning in the 1980s. Panel (a) compares annual averages of global ozone with the average from the period 1964 to 1980 before the ozone hole appeared. Seasonal and solar effects have been removed from the observational data set. On average, global ozone decreased each year between 1980 and 1990. The depletion worsened for a few years after 1991 following Mt. Pinatubo eruption. Panel (b) shows variation of the 2005–2009 depletion with latitude over the globe. The largest decreases have occurred at high latitudes in both hemispheres because of the large winter/spring depletion in polar regions; the losses are greater in the Southern Hemisphere because of the Antarctic ozone hole.
Source: Fahey and Hegglin.[11]

fully extractive system, 144, 144f
point-type in situ system, 145, 145f
single-path-type in situ system, 145–146, 145f
Stafilinid, 409
Stand-alone cost. *See* Levelized cost
Stand-alone photovoltaic systems, 218–219, 218f, 220
Standardization, 374–375
Standard oil recovery, 458
Standard test conditions (STC), of photovoltaic systems, 217
State policy innovations, 923
State-space analysis, for crop water use, 2218, 2219
Static pile composting, 517, 517f. *See also* Composting
Stationary/bubbling fluidized bed (SFB) boilers, 375
Statistical models, in pest management strategies, 1949
Statistics
 for landscape crop water use, 2219
Steam gasification, 2490
Steel-framed walls, 863
Steel shots, 1432
Steep slope reclamation, 1694
Steinerema feltiae, 409
Stellaria media, 1928
Stem borers, effect on photosynthetic rate, 1951
Sterility
 caused by pesticides, 277–278
Sterling motor, 157
Sticky traps, 1112
Stir bar sorptive extraction (SBSE), 1970t, 2195
Stoats (*Mustela erminea*), 2596
Stochastic chance-constrained programming (SCCP), 111
Stochastic mathematical programming (SMP), 106–107
Stochastic (or random) models
 water quality, 2749
Stockholm Convention, 575
 on Persistent Organic Pollutants, 1914
Stoichiometric combustion, 774
Stoker firing, 726, 726f
Stokes' law, 151
Stone backfill, 2313
Storage, 2314
 and conservation reservoirs, 1577
 efficiency, 1546
 energy. *See* Energy storage
 of groundwater, 1284
 matrix, 1312
 tanks for rainwater, 2264
 variation (groundwater), 1312
 of water, 1563
Storage plants, 202–203
Stored-product pests, biological control of

classical biological control, 2423
description of, 2423
examples of pests and natural enemies, 2424t
externally feeding pests, biocontrol of, 2424–2425
factors for, 2423
internally feeding pests, biocontrol of, 2423–2424
Stories from a Heat Earth—Our Geothermal Heritage, 1167
Storm energy, 943
 and rain amount, 943–944, 944t
Stormwater
 collection in urbanized catchment, 2263
 treatment, 1081
Storm windows, 866
Stranglervine (*Morrenia odorata*), 1526t
Strategic environmental assessment (SEA), 2450
Strategus, 321
Stratification, 1580–1581
Stratification ratio of soil organic C, 92, 92f
Stratified chilled water storage tank, 2510, 2510f
Stratospheric ozone depletion, 681
 antarctic ozone hole, discovery of, 1887
 atmospheric trace species, field measurements of, 1891
 global ozone layer, depletion of, 1891–1893, 1892f
 measurements and distribution of, 1885
 polar ozone chemistry, 1887–1891
Straw bale walls, 865
Stream erosion, 958, 964
Streamflow depletion, 1563
Streamlining European Biodiversity Indicators (SEBI), 603
Streptomyces, 2029
Streptomyces anulatus, 1879
Streptomyces hygroscopis, 1378, 2030
Streptomyces kasugaensis, 2030
Streptomyces violaceoruber, 384
Strict liability, 1945
String cables, 222
String diodes
 in photovoltaic systems, 220
String fuses
 in photovoltaic systems, 219–220
String inverters, 222, 222f
Strip mining, 1692
Stripper column, 457
Strip tillage
 erosion process and, 1042–1043
Strobilurins, 2030, 2030f
Strontianite (SrCO3), 2426
Strontium
 adsorption of, 2427
 chemical speciation, 2427
 geogenic origin, 2427

isotopes of, 2426
soil parameters, correlation between, 2428f
in soils, 2426–2427, 2428t
 interaction with soil matrix, 2426–2427, 2428f
 sorption, 2427
 uptake by soils, 2428–2429
 impact of fertilizer use on, 2428f, 2429
Strontium chromate, 479t
Structural insulated panels (SIPs), 865, 1484–1486
Strychnine, 413
Students Against Cell Towers (SACT), 2226–2228
Sturnus vulgaris, 413
Submerged aquatic vegetation (SAV), 475
Submerged-macrophyte systems, 2863
Submerged orifice scrubbers, 154–155
Subsoiling, 570
Subsurface band applicator implement, for poultry litter, 1673–1674, 1674f
Subsurface drainage, 1557, 1558, 1558f, 1558t
Subsurface drip irrigation (SDI), 1535–1538
 air entry and flushing, 1537
 chemical injection, 1537
 defined, 1535
 development of, 1535
 laterals and emitters, 1537
 lateral type, spacing, and depth, 1536
 maintenance, 1538
 operation, 1538
 pumps, filtration, and pressure regulation, 1537
 site, water supply, and crop, 1535–1536
 special requirements, 1536–1537
 system components, 1537
 system design, 1535–1537
Subsurface-flow constructed wetlands, 2841–2846
 hybrid systems, 2845
 horizontal flow systems, 2841–2843
 vertical flow systems, 2843–2845
 use for wastewater, 2845–2846, 2845t
Subsurface flow wetlands. *See* Vegetated submerged bed (VSB) wetlands
SubWet, 2671
 2.0, 2672
Sugar beet (*Beta vulgaris* L.), 1307, 1552
Sugarberry. *See Celtis laevigate*
Sugarcane industry, New South Wales, 58, 59
Sulfate, 31
 in atmosphere, 14–15
Sulfate-reducing bacteria, 32
Sulfic Endoaquepts, 28
Sulfide mineral formation (sulfidization), 31–32

and accumulation, 32–33
oxidizable organic carbon, 31
reactive iron, 32
reducing/saturated conditions, 31–32
sulfate, 31
sulfate-reducing bacteria, 32
Sulfidic materials, 27
identification and assessment, 39
Sulfonamide antibiotics, 2061
Sulfonylureas, 409
Sulfur, 2431–2435
agro-ecological aspects, 2434–2435
global change, 2434–2435
interaction with other ecosystems, 2434
plant health and, 2435
sustainability, 2434
emissions, 2434
fertilizers, 2434
overview, 2431
in soils, 2431, 2432f
biological aspects, 2431–2433
capillary rise/leaching, 2433
crops and crop rotation, 2432–2433
mineralization, 2432
physico-chemical aspects, 2433
soil compaction, 2433–2434
soil water regime, 2433
Sulfur dioxide (SO_2), 1154, 2304, 2437
air pollution of, 157–158
ambient air quality, 2437–2438
anthropogenic sources, removal from, 2440
ammonia wet scrubber, 2441
characteristics of, 137
concentration of, from natural sources, 2439
anthropogenic sources, 2439–2440
dry techniques, 2442
circulating fluid bed dry scrubber, 2442
magnesium oxide process, 2443
sodium sulfite bisulfite process, 2442–2443
EN 14212, ultraviolet fluorescence for, 142
flue gases, concentration in, 2440
semidry techniques, 2441
duct sorbent injection, 2442
furnace sorbent injection, 2442
hybrid sorbent injection, 2442
spray dry scrubbers, 2441–2442
sources, 2438
biogenic, 2439
Sulfur hexafluoride, 486
Sulfuric horizon, 27
Sulfuricization, 33–34, 34f
Sulfuric material identification and assessment, 39
Sulphur hexafluoride (SF_6), 1204
Sumilarv. See Pyriproxyfen
Sunday soil, 2341
Sunnhemp (*Crotalaria juncea*), 1511

"Super bugs," 391
Supercritical fluid (SCF), 1257–1258
Supercritical fluid extraction, 2198
Superfund. See Comprehensive Environmental Response, Compensation, and Liability Act (CERCLA)
Supply-side pests, 1951
Supply side thermal storage, 2520–2521, 2521–2522f
Surface air temperature, global annual-mean, 1198f
Surface albedo feedback theory, 566
Surface area (SA), of biochar, 844
Surface coalmines, 916
Surface drainage, 1560
Surface energy budget effect, 1197t
Surface-flow wetlands. See Free-water surface (FWS) wetlands
Surface irrigation, 120, 1547t, 1551–1553, 1564, 1565f
Surface Mining Control and Reclamation Act (SMCRA), 1289
Surface mining methods classification, 1692
Surface moisture (SM), 717
Surface roughness
desertization and, 569
Surface runoff, 2262
limitation, 2314
Surface seals, soil, 960
Surface waters, 1576, 2804–2806
direct contamination of, 1576
effects of pesticides, 2805–2806
knowledge gaps, 2806
pesticide transport and paterns of occurence, 2804–2805
pesticide use and, 2804
runoff, 2358
Suspended growth processes, 2647
Suspended particle display (SPD), 867
Suspended particulate matter, 149
Suspended sediment, 2752
Suspended sediment concentration (SSC), 965, 965f
SUSpended Sediment TRAp (SUSTRA), 1022
Suspension firing, 725, 726f
Sustainability, 333, 2446
Brundtland Report on, 338
components of, 1950f
concept of, 1654
definition of, 338
filters, 639
importance of, 338–339
indicators for, 339
meaning of, 338
potential analysis, 639
reports, 2461, 2463
benefits of, 2463
steps for, 339

sulfur and, 2434
tools for, 340
Sustainability Assessment of Farming and the Environment (SAFE), 356
Sustainable agriculture, 2457–2459. See also Ecological agriculture
agroecosystem biodiversity, 2458
air and atmosphere, 2458
assessment of soil quality, 2458–2459, 2458f
management strategies, 2459t
soil, 2457
water, 2457–2458
Sustainable aquifer use, 1285–1287, 1287f
Sustainable communities, aspects of, 2464–2465
Sustainable development, 920, 1169–1170, 2446
aspects of, 2461
for community, 2464
creative, cooperative, design and planning teamwork, 2462–2463
defined, 2461
in developing countries, 2465
education for, 2463
geothermal energy for, 1170
renting vs. buying, 2463–2464
social interactions, 2462
Sustainable energy future, 2462
Sustainable forest management, 339
Sustainable land management (SLM), 558
SUTRA
for saltwater intrusion, 1329
SVE. See soil vapor extraction (SVE)
Swamp, 2825t, 2826f
Swedish Forest Agency, 2298
Swedish Poisons Information Centre, 1401, 1403
Swelling, clay
effect of clay–cation interaction on swelling and dispersion, 2338–2339
Swine, effects of deoxynivalenol on, 1696
Swirl burners, 725, 726
Swishing noise, 2871–2872
Switchgrass (SWG), 345
environmental benefits of, 346
SWRRB. See Simulator for Water Resources in Rural Basins (SWRRB)
Sycamore. see *Platanus occidentalis*
Synanthedon exitiosa, 1919
Synaptogenesis, 281
Synergism, in agroecosystems, 1990
Synergists, 1992
Synergy, 1988
Syngas, 376–377, 730
Synomone, 175
Synthesis gas, 457, 458
Synthetic aperture radar (SAR), 1827, 1830–1831, 1833, 2817t
advanced, 1827

Synthetic pesticide use, in United States, 1120, 1121f
Synthetic polymers, 960–961
Synthetic pyrethroids, 277, 1399
Syria, 2686
Syrphidae, 1116
System Analyzer software, 1365
System coefficient of performance (SCOP), heat pump, 1351
System energy, 811
System management approach, of biological weed control, 1525, 1525f
System networks, importance of, 1434
Syzygium aromaticum, 384

T

Tachinaephagus zealandicus, 1922
Tachinid species, 1747
Tagetes erecta L., 288, 289
Tagetes patula L., 288, 289
Tagetes spp., 288, 289, 1878
Tailings, 1683–1687
   characteristics, 1685–1687
   chemical properties, 1685–1686
   defined, 1683
   physical properties, 1686–1687
   potential environmental problems, 1683
   rehabilitation practices
      ecological approach, 1684–1685, 1685t
      engineering approach, 1683–1684
Tailwater, 1557, 1558t, 1564
Taiwan, pesticide poisonings incidence, 1408
Taiwan National Poison Center, 1408
Tall fescue (*Festuca arundinacea* cv. Kentucky 31), 425
Tandonia budapestensis, 383
Tandonia sowerbyi, 383
Tank storage, energy
   underground, 2518
Tannins
   condensed, 2082f
   hydrolyzable, 2082f
Tantalum, 2585, 2587–2588
Tapered element oscillating microbalance (TEOM) dust sampler, 1024, 1025f
Tapered element oscillating microbalance (TEOM), of particulate matter, 142
Tarnished plant bug, 368
Taum Sauk upper reservoir, 1422f
Tax
   on emissions, 919
   on polluting good, 919
Taxe d'Enlèvement des Ordures Ménagères (TEOM), 895
TCD. *See* Thermal conductivity detector (TCD)

TCDD. *See* 2,3,7,8-Tetrachlorodibenzo-paradioxin (TCDD)
TCE. *See* Trichloroethylene (TCE)
TDL. *See* Tuneable diode laser (TDL) technique
Tebufenozide, 1466
Technological exergy, 778–780, 779f
Technological impact ratio, 1235
Technologically enhanced naturally occurring radioactive materials (TENORM), 2240
Technological sequestration. *See* Carbon capture and storage (CCS)
Technology-forcing strategy, 919
Teflubenzuron, 1462
TEM. *See* Terrestrial Ecology Model (TEM)
Temkin isotherm, 65. *See also* Isotherm(s)
Temperature
   Antarctic, 1889, 1889f
   Arctic, 1889, 1889f
   control, of composting materials, 521, 522f
   desertization and, 566
Temporal stability
   of landscape patterns, 2038–2039
TEPP (tetraethyl pyrophosphate), 3
Terbufos, 419, 1318
Teretrius nigrescens, 2423, 2424t
Terminal insecticide concentration (TIC), 1472
Terminal restriction fragment length polymorphism (TRFLP), 2616, 2617
Terminator gene, 314
Termite-assisted hand-pitting, 570
Terra preta, 844
Terraces, 968–969, 968–969f, 2764, 2764f
   benefits and problems, 2539–2540
   engineered, 2537–2539
   formation, by tillage erosion, 2536–2540
      ancient lynchets, 2536
      contemporary lynchets, 2536–2538
Terrestrial carbon sinks, 460
Terrestrial ecology model (TEM), 2272
Terrestrial Ecosystem Function Index (TEFI), 604
Terrestrial environment, pesticide effect on, 2147
Tersilochus heterocerus, 1116
Tetanops yopaeformis, 1983
2,3,7,8-Tetrachlorodibenzo-paradioxin (TCDD), 1394, 1395
Tetrahedral oxyanion, 100
Tetrahedral structure, 431
Tetranychus urticae, 310, 1750, 2021
Tetronic acids, 3–4
Theba pisana, 383
The International Journal of Exergy, 1084
Thematic Mapper (TM), 2818
Therioaphis maculata, 1521

Thermal conductivity detector (TCD), 1203
Thermal design
   bottoming and topping cycles, 1357
   fundamentals, 1356–1357
Thermal energy, 772–773
   storage system, 1086
      energy efficiency of, 1086
      evaluation of, 1086
Thermal energy storage (TES), 2508–2522
   cool storage, 2511–2517
   technologies, 2509–2510
   heat storage, 2510–2511
   and renewable energy, 2517–2521
   types of, 2509
Thermal environmental conditions for human occupancy, 1657
Thermal exergy, 1084
Thermal mass, 861
Thermal power cycle, 1357
Thermal technologies
   advanced, 835–837
Thermally impaired waters, 2813
Thermionic specific detector (TSD), 1975
Thermochemical decomposition, 2483–2484. *See also* Pyrolysis
Thermo-chemical process, 730
Thermocline, 2510
Thermodynamic concepts as superholistic indicators, 778
Thermodynamic effect, 1197t
Thermodynamic information, 782
Thermodynamic process, 2526
Thermodynamic properties, calculation of
   ideal gas, 2532–2533
   phase equilibrium, 2533–2534
   real fluids, 2532–2533
   rigorous equations for, 2531–2532
   solutions, 2533
Thermodynamics, 2525–2534
Thermoelectric heat pump, 1348–1349, 1348f. *See also* Heat pumps
Thermogravimetric analysis (TGA), 723
Thermostatic expansion (T-X) valves, 1350
Thermotolerant coliforms, 2699
The World is Flat, 1503
Thin-film solar cells, 232
Thiobacillus ferrooxidans, 33, 43, 1694
Thiobencarb, 2806
Thiocarbamate herbicides, 1318
Thiocarbamate pesticides, 1317
"Third–party certification," 125
Thorium-232 decay series, 2237f
Three-dimensional ecological footprint geography, 2475–2476
Threshold agents, 1456
Ticks, 1, 2
   control of. *See* Acaricides
Tidal energy, 772
Tidal marsh, 2825t, 2826f

Tidal power, 824, 825
TIE. *See* Toxicity identification evaluation
Tiger. *See* Pyriproxyfen
Tilapia zillii, 1922
Tile drainage, nitrate-nitrogen concentrations in, 1303–1304
Tillage, 1111
   erosion process and, 1040, 1041f
      patterns of erosion and deposition, 2537f
         terrace formation by, 2536–2540
   impact on soil IC, 464
Tillage berms, 2539–2540, 2540f
Timber harvesting, 2770
   annual yield and storm flow response to planting and, 2773–2774
   hydrological behavior of forests and, 2770–2771
   modern forestry hydrology history, 2772
   variation in silvicultural practices and water quality effects, 2772–2773
   water quality effects mitigation through BMPs, 2773
Time-domain electromagnetic methods (TDEMs)
   for saltwater detection, 1327
Titania. *See* Titanium dioxide
Titanium dioxide (TiO_2), 235–236
   environmental pollution by, 2129
   semi conductor, 236
   solar cells, 238
      advantages, 238–239
Titanium dioxide (TiO_2) catalyst, 1731–1732
   decontamination of toxic pollutants by, 1733, 1735
   forms of, 1733f
Titanium white, 235
Tithonia diversifolia, 1878
T-max. *See* Noviflumuron
TMDL. *See* Total maximum daily load
Tolerance(s), 1124, 1125–1126, 1125t. *See also* Maximum residue limit (MRL)
   FQPA and, 1118
Tolerance limit
   defined, 1609
   for insect fragments, 1609–1610
      in food (examples), 1610t
   for pesticide residues, 1609
      in food (examples), 1610t
         regulatory inspection and enforcement, 1609
Tomato (*Lycopersicon esculentum*), 425
Ton, 1363–1364
Topping cycles, 1357
Topsoil, 1019
   loss of, 1019, 1019f
Total dissolved solids (TDS), 918, 1559
   for saltwater intrusion, 1325, 1329
Total maximum daily load (TMDL), 2750, 2752, 2808–2809
   in achieving water quality standards, 2812
      section 303 impaired waters list, 2812
      state water planning and, 2812–2814
   application of, 2814
   nonpoint source state management plans, 2811
      water quality standards before TMDL process, 2811–2812
   point source regulation and
      NPDES permit program, 2809–2810
      pollutant discharge, 2809
      section 404 dredge and fill permit program, 2810
   water quality standards in point source permitting before TMDL process, 2810–2811
Total petroleum hydrocarbons (TPHs), 2051–2056, 2054t
Total Petroleum Hydrocarbons Criteria Working Group (TPHCWG), 2049, 2052
Total primary energy supply (TPES), 834, 835f
Total project costs derivation, 1792–1793
Toxic Substances Control Act of 1976 (TSCA), 2125, 2181
Toxic substances removal, photochemistry of, 2547–2565
   heterogeneous TiO_2 photocatalysis, 2556–2565
      applications of, 2561, 2564
      mechanism of, 2557–2558
      operational parameters of, 2558, 2561, 2562–2563t
      optimum conditions and rate of, 2559–2560t
      for water treatment, research trends in, 2564–2565
   homogeneous photo-Fenton reaction, 2548–2556
      applications in water treatment, 2553–2556, 2554–2556t
      classification of, 2548f
      contaminant concentrations and characteristics, 2552
      iron concentration, impact of, 2549, 2552
      mechanism of, 2548–2549
      optimum conditions and rate of, 2550–2551t
      oxidant concentration, impact of, 2552
   technologies of, 2547–2548
Total suspended solids (TSS), 2865t
Toxaphene, 1318t
Toxicants
   for bird control, 413–414
Toxic chemicals, 280–281, 1570
   data quality control of experimental studies, 283
   developmental animal experimental studies, 282–283
   developmental impairment in children, 281–282
   international protocols, 283–284
      European Union's REACH legislation, 284–285
      Food Quality Protection Act, 284
   nervous system development, effects on, 281
   on non-target organisms in environment, 285–286
   oxidative stress, 281
   perspectives, 285
   reproductive and developmental protocols, 283
Toxicity, 1704, 2653
   of boron, 432, 433
   expressions for, 1457t, 1458
Toxicity identification evaluation (TIE), 2806
Toxic materials, 1579
Toxicological assays, 2170
Toxicological requirements
   for pesticides registration, 1614
Toxicological synergy, 1991
Toxins, 2357
Trace element contamination
   in groundwater, 1560
   sources of, 1560
   in subsurface drainage water, 1560
   in surface runoff, 1560
Trace elements, 1455. *See also* Heavy metals
   defined, 1370
Tradable emissions permit, 919
Trade issues
   pesticide residue, 1125–1126, 1125t
Trail formation, 2752
Trampling, 2752
Transformation, of polychlorinated biphenyls, 2178
Transgenic crops, 307. *See also Bacillus thuringiensis* (BT) Crops
Transgenic virus-resistant potatoes (Mexico), 409–412
Transition metals, 1560
Transition zone
   between saltwater and freshwater regions, 1324–1325, 1325f, 1328
Translocation
   herbicides, 1379
Transmission/distribution (TD) system, 1796
Transmutation, 187
Transparent conducting oxide (TCO), 236
Transparent coverage, 242
Transpiration ratio, 1549
Transpiration stream concentration factor (TSCF), 2117, 2117f
Transport, biofuels for, 377–378
Transport, soil erosion, 958
Transportation, energy use in, 716, 716f

Transport capacity, 1021
Trans stilbenes, general formula of, 2081f
Transuranic elements, 2237–2239
Trap crops
 effects on pests, 1925, 1928
Treated seeds, 419
Treatment systems use of wetlands. *See* Wetlands as treatment systems
Tree cover, permanent, 335
Trespass, 1945
Triazines, 1319
Tributary mass load sampling, 1582
Tributary water quality sampling, 1582
Tributyltin (TBT), 491
1281,1,1-trichloroethane (TCA), 1282
Trichloroethylene (TCE), 388, 1282, 1334, 1340, 2114, 2117
 federal drinking water standard for, 1282
Trichoderma, 2107, 2109
Trichoderma harzianum, 383
Trichoderma viride, 383, 1879
Trichogramma, 1930, 2424
Trichogramma semblidis, 1750
Trichogrammatidae, 2424
Trichoplusia ni (Hubner), 2021
Trickle bed reactor system, 1906
Trickle irrigation, 1564
Trickling filters, 2724–2725. *See also* Filter(s)
2-Tridecanone, 2021
Triffid weed. *See Chromolaena odorata*
Triflumuron, 1462
Trifluralin, 1957, 2024f, 2806
Trigard. *See* Cyromazine
Trihalomethanes (THMs), formation of, 2793, 2794f
Triketones, 2031, 2031f
Tripene, 1464
Triple bottom line, 2461
Triple cropping forage systems, 1664, 1665t
Trissolcus vasilievi, 1750
Triticum aestivum (Wheat), 1307, 1552
Triticum aestivum L. (Durum wheat), 425
Tritrophic interactions, 312
Trombe wall, 2499, 2519, 2520, 2520f
Trophic infaunal index, 603
Trophic status, lakes and reservoirs, 1578
Trueno. *See* Hexaflumuron
Tube wells, 1276
Tuneable diode laser (TDL) technique, 1204
Tungsten, 2586–2587, 2589–2590
Tunisia, 2682, 2689
Tunnel erosion, 964
Turf, NO_3-N contamination from, 1305–1306
Turning Off the Heat: Why America Must Double Energy Efficiency to Save Money and Reduce Global Warming, 1505
Tweed River fish kill, 56

Twomey effect, 1197t
Two-phase bioreactors (TPBRs), 2612f, 2625
Two-phase reactors, 1908
Tylenchulus semipenetrans, 288, 1878
Typhlodromus occidentalis, 2141
Typhlodromus pyri, 1750

U

$^{235/238}$U, 2243, 2244
U(VI), 2244, 2245
UASB reactor, 2651–2652, 2652f
UCC. *See* Christiansen uniformity coefficient
UGas, 2491
Ulmus americana, 2830
Ultimate analysis (ASTM D3176), 717, 719t, 720t–721t
Ultrafiltration. *See also* Filtration
 performance of, 2727
 system design, 2727
 in water treatment, 2727–2728
Ultramafic rocks, 2426
Ultrapyrolysis process, 2487
Ultrasound-assisted extraction (UAE), 2196
Ultraviolet–diode array detector (UV–DAD), 1975
Ultraviolet fluorescence, for SO_2 EN 14212, 142
Ultraviolet photometry, for O_3 EN 14625, 142
Ulva, 788
Umbrella, flagship, and keystone species, 603
UNCCD (United Nations Convention to Combat Desertification), 557, 558
UNCED. *See* United Nations Conference on Environment and Development
Uncertainty, 1720–1721
 in systems analysis, 1245
 three dimensions of, 1723f
Uncertainty analysis, 1720–1721
 alternative approaches, 1722–1723
 nanoparticles, within environmental characterizing, 1723–1725
2-Undecanone, 2021
Underground thermal energy storage (UTES), 2518–2519
 aquifer storage, 2518–2519
 borehole storage, 2518
 pit storage, 2518
 tank storage, 2518
Understoker furnaces, 375
UN Economic Commission for Europe (UNECE)
 Convention on Long-Range Transboundary Air Pollution (LRTAP), 1914

UN Environment Programme (UNEP), 1615
UNEP. *See* United Nations Environment Programme
Unglazed collectors, 242, 243, 243f
Uniformity in irrigation. *See* Irrigation efficiency
UNITAR. *See* United Nations Institute for Training and Research
United Arab Emirates, the, 2682
United Nations (UN)
 Human Poverty Index, 617
United Nations Children's Fund (UNICEF), 1265
United Nations Commission on Sustainable Development, 1654
United Nations Conference on Environment and Development (UNCED), 182, 1524, 1612, 1616
United Nations Convention on Biological Diversity, 333
United Nations Convention on Climate Change (UNCC), 1223
United Nations Convention to Combat Desertification (UNCCD), 1035
United Nations Environmental Programme (UNEP), 484, 1893, 2036, 2181, 2416–2417, 2671
United Nations Environment Programme (UNEP) to Combat Desertification, 565
United Nations Food and Agriculture Organization (FAO), 1108, 1397, 2035, 2036
United Nations framework convention on climate change (UNFCCC), 485, 486f, 845–846
United Nations Institute for Training and Research (UNITAR), 1409
United States, 2682, 2684
 acid rain in, 8
 control policy, 15
 saltwater intrusion in, 1321–1322
 trends in acidity, 9, 11–13
United States Army Corps of Engineers (USACE), 627
United States Department of Agriculture (USDA), 1940
United States Endangered Species Act, 1532
United States Environmental Protection Agency (USEPA), 924, 963, 1264, 1636–1643, 2013
United States Geological Survey (USGS) study, on quality of groundwater, 1283
United States Green Building Council (USGBC), 1647–1648
 LEED green building certification system, 2514
Unit energy costs, 1495

Unit energy equation, 943
Universal Index of Onchev, 943
Universal soil loss equation (USLE), 934, 943, 964, 966, 967, 975–977, 980–981, 991, 1048, 1059
  variants of, 981–983, 981–983f
University of Arizona
  district cooling, 2515, 2515f
Up-coning, 1323–1324, 1324f
Upland erosion, defined, 958
Uranium
  groundwater contamination by, 1292
Uranium-238 decay series, 2236f
Urban agriculture, 2695
Urban ecosystem health, indication of, 607–608, 608t
Urban planning, 638–639
Urban sewage management, 2150–2151
Urban wastes treatment, 1908
Urinary biomarkers, for deoxynivalenol, 1696
Urtica spp., 1750
Uruguay, 2774
U.S. Congress, 2010
USDA-NASS (U.S. Department of Agriculture–National Agricultural Statistics Service) report, 2142
U.S. Department of Agriculture, 1609, 1611
  Food Safety and Inspection Service (FSIS), 1609
  on pesticide residues in processed foods, 1609
U.S. Department of Agriculture (USDA), 1120
U.S. Department of Agriculture–Agricultural Research Service (USDA-ARS), 1673
U.S. Department of Energy (DOE), 215, 2243
U.S. Department of Housing and Urban Development (HUD), 1413
U.S. Endangered Species Act of 1973, 2823
U.S. Energy Information Administration (EIA), 1652
U.S. EPA National Acid Precipitation Assessment Program (NAPAP), 20
U.S. Geological Survey (USGS), 1983, 2272, 2804
U.S. National Cancer Institute, 1394
U.S. National Toxicology Program study, 1697
U.S. primary energy supplies, 834
U.S. Social Security Administration, 1413
USDA National Organic Standards Board, 125
USDA Natural Resources Conservation Services, 127
USEPA. *See* United States Environmental Protection Agency (USEPA)

U.S. Green Building Council (USGBC), 1654, 1655, 1658
USGS. *See* U.S. Geological Survey (USGS)
USLE. *See* Universal soil loss equation (USLE)
Utility-operated rainwater harvesting, 2262
Uttar Pradesh, groundwater arsenic contamination in, 1267–1268

V

Vaccinia virus, 2604
Vacuum pyrolysis, 2487
Validamycin, 2030, 2030f
Vanadium, 2585, 2587
Vanadium and chromium groups, 2582
  environmental levels, 2584
    chromium, 2585–2586
    molybdenum, 2586
    niobium, 2585
    tantalum, 2585
    tungsten, 2586–2587
    vanadium, 2585
  geochemical occurrences, 2582–2583
  metabolism and health effects
    chromium, 2588
    molybdenum, 2588–2589
    niobium, 2587–2588
    tantalum, 2587–2588
    tungsten, 2589–2590
    vanadium, 2587
  uses, 2583–2584
Vapor compression heat pump, 1350–1351, 1350f. *See also* Heat pumps
  classification of, 1351
  components in, 1350, 1350f
  compressors in, 1350
  condensing temperature effect on ideal COP, 1351f
  evaporating temperature effect on ideal COP, 1351f
  phase-changing processes in, 1350
Vapor compression refrigeration, 1358–1359, 1359f, 1360
Vapor pressure, 2533–2534
Variable-rate fertilizer technologies, 1810
Variogram maps, 103f
Varroa mites, 2
VC. *See* Vinyl chloride (VC)
Vector control technique, combining CP and BP for, 1990
Vegetable oil, 377
Vegetables
  pesticides in, 1972–1973, 1972t, 1973f, 1973t
  oils, for liquid fuels, 846

Vegetated submerged bed (VSB) wetlands, 2862, 2864f, 2865, 2865f
Vegetated wetlands, 2824
Vegetation, 2754
  for controlling soil water erosion, 994–1001
    buffers, 995–1000, 995f, 996–997t, 998–999f
    increased infiltration of water into soil, 995
    reduced soil erodibility, 995
    sediment trapping efficiency, 1000–1001, 999f
    slower runoff, 994–995
  cover, 560
  livestock impacts on, 2753
Vegetative deficiency, 433
Vegetative insecticidal (VIP) proteins, 308
Velvetleaf (*Abutilon theophrasti*), 1526t
Venturia canescens, 2424
Venturi scrubber, 155–156
Verdale-Simi Fire, 1185f
Vermicomposting, 519–520, 519f. *See also* Composting
Vermiculitic soils, 1628
Vertebrate pests, 2595
  biological control of, 2595–2605. *See also* Biological control, of vertebrate pests
  conventional control measures for, 2595
Vertical flow systems, for constructed wetlands, 2843–2845
Vertical mixing, 49
Verticillium chlamydosporium, 1752, 1753, 1879
Vesicular–arbuscular mycorrhizae (VAM), 1516
Veterinary pharmaceuticals, 2061
Vibrio cholerae, 2696, 2698
Vibroacoustic disease, 2877
Vicia faba, 411
Vienna Convention for the Protection of the Ozone Layer, 1893
Vinclozolin, 1405
Vinyl chloride (VC), 1334
Viola tricolor arvensis, 1928
Virally vectored immunocontraception (VVIC), 2604
  of foxes, 2604
  of mice, 2604
  prospects for, 2604–2605
  of rabbits, 2604
Vireo bellii pusillus, 1186
Viron/H, 382
Virus and viroid infections, 2111–2112
  biological control of, 2111–2112
  control measures, 2111–2112
  economic loss, 2111
  global impact, 2111
  prospects, 2112

Viruses, in waste and polluted water, 2698
Virus-resistant crops, 407, 408t
Visible infrared scanner (VIRS), 2817t
Visual deficiency, 433
Visual indicators, of soil water erosion, 966–967
Volatile matter (VM), 717, 723, 725
Volatile organic compounds (VOCs), 136, 2616
   in biofilters, 2616t
Volatile oxidation, 723–724
Volatilization, 1560, 2753
   of inorganic pollutants, 2120
   of polychlorinated biphenyls, 2175–2176
Volatilization, ammonia, 1770
   from agricultural soils, 85–87
   emissions
      animals and their wastes, 86
      biomass burning, 87
      cropping systems, 86–87, 86t
      global significance, 87
      mitigation, 87
   measurement, 86
   mechanism, 85–86
Volcanic forcing, 1195, 1199f
Volcanic sulfur-bearing gases, 2439
Volcanic variability, 1194–1195, 1196t
Volcanoes on weather and climate, 1196t
Volicitin, 177
Voluntary/regional programs for climate change, 487
V-O-R model, 601
VSB wetlands. *See* Vegetated submerged bed (VSB) wetlands

W

Wagon Wheel Gap Study, 2774
Wake zone, 129
Wall insulation, 1482–1486
   2 × 4, 1483–1484
      materials required, 1483–1484
      problems and solutions, 1484
   2 × 6, 1486
   blown foam insulation, 1484, 1485
   blown loose-fill insulation, 1484, 1485
   concrete, 1482
   concrete block cores, 1482
   exterior foam insulation, 1482, 1483
   insulated concrete forms, 1483
   interior foam, 1482, 1483
   interior framed, 1482, 1483
   lightweight concrete products, 1483, 1484
   side stapling, avoiding, 1484, 1485
Walls, 860–865, 861f. *See also* Windows
   building types, 860
   exterior insulation finish systems (EIFS), 864–865

masonry, 863–864, 864f
steel-framed, 863
straw bale, 865
structural insulated panels (SIPS), 865
utility of, 860
wood-framed, 861–862, 862f, 863f
Wsalsh, B. D., 2009
Washing
   of cows, 1663–1664
   of milking equipment, 1664
Waste, 2415
   batteries and accumulators, 902
   composition, 2652
   hierarchy, 2418f
   life cycle assessment (LCA), 2418
   management, institutionalization of, 2416
   oils, 908
Waste from electrical and electronic equipment (WEEE), 900, 906–907
   recovery rates of, 908t
   recycling rates of, 909t
Waste emissions, exergy contents of, 1095
Waste gas stream, characteristics of, 2620–2621
Waste gas treatment, bioreactors for. *See* Bioreactors, for waste gas treatment
Waste heat, 1356
   high-grade, 869
   low-grade, 869
   medium-grade, 869
   quality of, 869
   recovery, applications, 1356–1362
Waste heat recovery
   engineering concerns in, 870–871
   equipment
      selection of, 871
      types of, 872–873
   quality vs quantity, 869–870
   sample calculations, 871–872
Waste heat stream
   cleanliness and quality of, 871
   determining value of, 870
   dilution of, 870
   mass flow rate for, 870, 872
   quality of, 869
   quantifying, 870
   recovery, 873
Waste load allocation, 2813
Waste management, impacts on
   future perspectives, 911–912
   hazardous waste, 902
   solid waste, 900–902
   waste streams, 902
      biodegradable municipal waste, 908–909
      end-of-life tires, 909–910
      end-of-life vehicles, 902–903
      packaging waste, 903–906
      waste batteries and accumulators, 902

      waste electrical and electronic equipment, 906–907
      waste oils, 908
Waste stabilization pond system (WSPS)
   design of, 2635
      anaerobic ponds, 2636
      design parameters, 2635
      facultative ponds, 2636
      Helminth egg removal, 2636
      maturation ponds for fecal coliform removal, 2636
      water flows and BOD concentrations, 2636
   nutrient removal in, 2635
   oxygen tension in, 2634
   processes in, 2633
      anaerobic ponds, 2633
      facultative ponds, 2633–2634
      facultative ponds, kinetics, 2634
      maturation ponds, 2634–2635
   types of, 2632
   water quality and, 2636–2637
Waste streams, 902
   biodegradable municipal waste, 908–909
   end-of-life tires, 909–910
   end-of-life vehicles, 902–903
   packaging waste, 903–906
   waste batteries and accumulators, 902
   waste electrical and electronic equipment, 906–907
   waste oils, 908
Wastewater, 2681–2682
   exposure, motivating safe practices along, 2688–2689
   industrial, 1430–1431
   municipal. *See* Municipal wastewater
   policy interventions and risk reduction, 2686
      agricultural communities, 2686–2687
      farmers and families, 2686
      farm product consumers, 2687–2688
   policy issues
      in developed countries, 2684
      in developing countries, 2684–2685
   policy requirement, 2684
   public policy examples, 2689–2690
   purification of, 2729
   as resource in water-scarce settings, 2682
      in developed countries, 2682
      in developing countries, 2683
   treatment of, 2720–2721
      and non-treatment alternatives, 2685–2686
Wastewater irrigation, 2675–2679
   crop selection and diversification, 2675–2676
   irrigation management, 2676–2677
   soil-health-based interventions, 2678–2679

Wastewater recycling in dairy, 1664, 1665t
Wastewater treatment, 2645–2646, 2862
   activated sludge process, 2648, 2648f
      operating parameters in, 2649
   aeration tanks, 2648
   aerobic biological waste treatment processes, 2648–2650
   agricultural practices, 2701, 2704
   attached growth process, 2647–2648
   biological phosphorus removal, 2654
   biological removal of nitrogen, 2653
      anaerobic ammonium oxidation, 2653
      Canon process, 2654
      combined nitrogen removal, 2653–2654
      Sharon process, 2653
   biological treatment options, 2646
      aerobic processes, 2647
      anaerobic processes, 2647
      anoxic processes, 2647
   characteristics of, 2702t–2703t
   chemotherapy and immunization, 2705
   cleaner production and pretreatment discharge programs, 2700–2701
   crop selection restrictions, 2701
   educational and awareness campaigns, 2705
   food preparation, 2704–2705
   granule deterioration, 2653
   irrigation methods, 2701
   loading rate, 2652
   local technologies, 2704
   marketing, 2704
   pretreatment, 2646
   primary treatment, 2646
   retention time, 2652–2653
   secondary clarification, 2646
   secondary clarifiers, 2648–2649
   secondary treatment, 2646
   selection criteria for intervention measures, 2705
   solid retention time, 2648
   subsurface-flow constructed wetlands for, use of, 2845–2846, 2845t
   suspended growth processes, 2647
   temperature, 2652
   tertiary treatment, 2646
   toxicity, 2653
   transportation, 2704
   washing, packing, and on-site storage, 2704
   waste composition, 2652
Wastewater treatment/disposal. *See* Sewage effluent for irrigation
Wastewater treatment in Arctic regions, wetlands usage, 2662–2663
   Arctic Canada and its regions, map of, 2664f
   knowledge and practice, state of, 2663
   performance, 2663
   constructed wetlands, potential for, 2670–2671
   modeling treatment wetlands, 2671–2672
   Paulatuk treatment wetland, 2664–2670
Wastewater treatment plants (WWTP), 2060
Wastewater use, in agriculture, 2694
   assessment
      indicators, 2699
      monitoring, 2699
      risk assessment, 2699–2700
   future perspectives, 2707
   negative health impacts, 2695–2697
      diseases related to chemical exposure, 2696–2697
      exposed populations, 2695–2696
      exposure routes, 2695, 2696t
      infectious diseases, 2696
      secondary health problems, 2697
   pollutant sources, 2698–2699
   positive impacts, 2697–2698
   present situation, 2694–2695
   solutions, 2700–2707
      multiple-barrier concept. *See* Wastewater treatment
   policy framework, 2706–2707
   standards, setting, 2705–2706
Water, 916
   agricultural uses of, 1281
   aluminum in, 269–270
   budget in dairy farms, 1665t, 1666
   delivery subsystem, 1561t
   desalination, 2502
   efficiency, 1656
   environment, bioaccumulation and, 324–325
   erosion. *See* Erosion by water
      of soils, 1034
   erosion models, empirical
      ABAG (German USLE), 977
      erosion predictions, accuracy of, 977–978, 978f, 978t
      evolution of, 974–975, 976t
      Modified universal soil loss equation, 975–976
      Revised universal soil loss equation, 977
      Soil Loss Estimation Model for Southern Africa (SLEMSA), 976–977
      SOILOSS, 977
      Universal Soil Loss Equation, 975
   flow meters, 1663
   loss, 1184
   management, 1560–1561
   management in soil, 2217–2222
   nanoparticles in, 1712–1715, 1713t, 1714f
   nutrient deposition/deactivation in, 1592t
   pesticides in, 1968–1969, 1970t
   purification of, 2729
   as refrigerant, 1358
   soil erosion by. *See* Soil water erosion
   surface, soil erosion effect on. *See* Soil erosion, effect on surface water quality
   sustainable agriculture, 2457–2458
   treatment of, 2720–2721
   treatment, alternative systems for, 2062–2063
   usage minimization, 1665t, 1666
   use, 1184
   use in dairy farms. *See* Dairy, water use in
Water balance and groundwater mining, 1284–1285, 1285f, 1286f
Water cress (*Enhydra fluctuans*), 426
Water erosion, 1955–1956
Water Erosion Prediction Project (WEPP), 984–985, 966, 992, 1048 1959, 2761
Water erosion vulnerability map, 2376
Water-filled pore space (WFPS), 1770
Water filtering, 2704
Water flows and BOD concentrations, 2635–2636
Water Framework Directive (WFD), 1701–1702
Water harvesting
   classifications, 2738
   definition, 2738
   for domestic use, 2739
   failure, 2739
   for growing of crops, 2739
   for livestock drinking water, 2738–2739
Water hickory. *See* Carya aquatica
Water hyacinth (*Eichhornia crassipes*), 1527t
Water intake, 209–210
Water leaching index, 1763
Waterlogging, of sodic soils, 2341
Water management
   CH_4 emissions from rice fields and, 1679, 1680f
Water milfoil. *See* Myriophyllum spp.
Water pollution, 333, 2013, 2020. *See also* Environmental pollution, by pesticides; *See also* Pesticides
   prevention, 2138
Water Pollution Control Act in US, 1557
Water prices, 2779
   competitive markets and, 2780–2781
   multiple-criteria framework, 2785f
   optimizing model, 2781–2783
      scarcity rents, 2782
      shadow prices, 2782–2783
   structuring, 2783
      economic efficiency, 2785
      equity and fairness, 2785–2786
      full cost, 2784
      full economic cost, 2784
      full supply cost, 2784
      opportunity cost, 2784

simplicity, 2786
stability and quality, 2786
substitutability, 2784–2785
sustainability, 2786
water tariffs and pricing strategies, 2786–2788
Water Productor Program, 971
Water quality. *See also* Range and pasture lands
for agriculture, 1566
constituents in IRF, 1558–1560
nitrogen, 1559
pesticide contamination in, 1560
phosphorus, 1559–1560
salts, 1559
trace elements in, 1560
criteria, 2811
effects mitigation, through BMPs, 2773
monitoring, 1581–1582
biological integrity sampling, 1582
citizen monitoring, 1582–1583
lake sampling, 1582
trading programs, 1772–1773
tributary mass load sampling, 1582
tributary water quality sampling, 1582
wind erosion, effect of, 1020
and WSPS, 2636–2637
Water Quality Act (1965), 2808
Water quality modeling, 2749–2750
BMPS, 2750
classification, 2749–2750
large-scale systems behavior, 2750
overview, 2749
risk assessment of pesticides, 2750
roles, 2749
sources/impacts of pollutants, evaluation of, 2750
uses, 2750
Water removal subsystem, 1557, 1561t
Water retention
hydraulic conductivity and, relationship between, 1621
Watersheds
management
Geographic information system (GIS) for, 2816–2819, 2818t
implications to, 2102–2103
remote sensing (RS), 2816–2819, 2817t
water quality response, implications for, 2103, 2105
sediment transport in, 965, 965f
Water solubility, 1957
Water-stable aggregates, 91
Water supply buffer, 2262
Water table control, 1547t
Water table management, 49–50
Water transport components, 1545, 1546f
Water turbines, 212
Water use efficiency (WUE), 1549
agroforestry and, 129–130. *See also* Agroforestry, WUE and

Water vapor, greenhouse properties of, 1194, 1195f
Waterways, to control soil erosion, 969–970, 969f
Watery soft rot, 443, 443t
Wave energy, 772, 826
Weasels (*Mustela nivalis*), 2596
Weathering effect, 1559
Weather-resistive barrier on wall, 864
Web-based tools, for integrated energy investments. *See* Integrated energy systems, case study from ISU
Wedge Dust Flux Gauge (WDFG) samplers, 1022
Weed
diversity
effects on pests, 1928
herbicides resistance in, 1379
intercropping and, 1938
management and organic farming, 1107
organic soil amendments and, 1878
science, 1524
Weed(s), biological control of, 2821–2823
costs and agent success, 2822–2823, 2822t
history and impact of classical approach, 2821–2822
host-specificity tests and, 2822
legislation, 2823
Weed abundance, crop rotation and, 1928
Weed control, 119, 1524
biological, methods of, 1525f
inoculative/classical approach, 1525, 1525f
inundative/bioherbicide method, 1525, 1525f
system management approach, 1525, 1525f
integrating biological control with other methods, 1525, 1526t–1527t
ecological integration, 1525, 1528
horizontal integration, 1525
physiological integration, 1528
purpose-specific approaches, 1525
vertical integration, 1525
Well-posed initial and boundary value problem
for saltwater intrusion, 1329
WEPP. *See* Water Erosion Prediction Project (WEPP) equation
Werneckiella equi, 1462
West Bengal, groundwater arsenic contamination in, 1265–1267, 1266f, 1267f
Wet deposition, 149
Wetland microcosms, 426
Wetland(s), 607, 2854
and carbon sequestration, 2833–2835
case studies
effectiveness, evaluation of, 2854–2855

engineered/constructed wetland, treatment of petroleum hydrocarbons in, 2856–2857
petroleum hydrocarbons, abatement of, 2855f, 2855–2856
definitions, 2824, 2854
extent of, 2827, 2827t
fauna, 2831
flora, 2829–2831, 2830f, 2831f
hydrological patterns, 2829, 2830f
landscape perspective, 2829
as natural resources, 2824
types of, 2824–2825, 2825t, 2826f, 2827
Wetlands as treatment systems
defined, 2862
design considerations, 2864–2865, 2866t
FWS wetlands, 2865
VSB wetlands, 2865
influent concentrations, 2865t
in North America (1994), 2863t
operation and maintenance, 2865–2866
treatment processes, 2862–2863, 2865f
treatment wetland types
constructed *vs.* natural wetlands, 2862, 2863t
free-water *vs.* submerged-bed wetlands, 2863, 2863f
use of wetlands, 2862
wastewater, types of, 2865t, 2866t
Wet meadow, 2825t, 2826f
Wet oxidation (WO), 1905–1906, 1906t
advantages, 1907
applications, 1908
catalytic, 1909
using heterogeneous catalysts, 1910
using homogeneous catalysts, 1909–1910
challenges of, 1910–1911
commercial, 1908
limitations, 1907–1908
mechanism, 1906–1907, 1907f
non-catalytic, 1908–1909
POP removal using, 1904
Wet oxidation reactors, 1908
Wet scrubbers, 154–156
Wet techniques, 2440–2441
WFD. *See* Water Framework Directive
WFPS. *See* Water-filled pore space
Wheal Jane metal mine, Cornwall, 1291
Wheat (*T. aestivum* L.), 427, 1307, 1552
WHEELS, 1959
White mold, 443, 443t
WHO. *See* World Health Organization (WHO)
Whole systems thinking, 2461
Wildfire, 1184–1186, 1185f
Wildlife, biodiversity and, 1186
Wildlife management and husbandry, 570
Wild oat. *See Avena fatua* L.

Willingness to accept compensation (WTA), 1637
Willingness to pay (WTP), 1637–1638
Wind energy, 772, 824
  resources, 824
Wind erosion, 1013–1015, 1956
  affects of, 1017
  causes of, 1017–1019
  control strategies, 1027–1028
  definition of, 1017
  global hot spots, 1010–1012, 1011f
  implications of, 1019–1020
  induced, 1010–1012
  monitoring of, tools for, 1020–1021
    direct measurement, 1021, 1021f, 1022f
    horizontal mass flux, 1021–1024
    radioisotopic techniques, 1024–1025
    wind erosion models, 1025–1027
  natural, 1010
  overview, 1013
  particle entrainment, 1014, 1014f
  particles movement by wind and, 1017–1018
  processes, 1013
  self-balancing concept, 1015, 1015f
  source regions of, 1020, 1021f
  wind dynamics, 1013–1014, 1014f
Wind Erosion Assessment Model (WEAM), 1026t
Wind Erosion Equation (WEQ), 1025–1026, 1026t
Wind Erosion on European Light Soils model, 1025
Wind Erosion Prediction System (WEPS), 1025, 1026–1027, 1026t, 1959
Wind erosion vulnerability map, 2376
Windfarm noise, 2867, 2868
Wind generators, 916
Windows. *See also* Walls
  building types, 860
  energy transport, 865–866, 865f, 866f
  future improvements, 867
  solar heat gain through, 866
  utility of, 860
  window rating system, 866–867, 866f, 867t
Wind park, 2867
Wind power, 179, 192
  capacity distribution, 259
  capacity factor, 259–260
  capacity growth, 259
  climate protection, 182–183
  costs, 258–259, 260
  depletion of fossil fuels, 180–182
    data and predictions, 180f
    oil, delivery and detection of, 181f
    reserves/resources, regional distribution of, 180, 181f
  ecological economics
    Ems-axis, lower saxony, growth and

booming region, 196–199
renewable energy, 188–195
renewable energy in Germany and planned nuclear exit, 195–196
electrical production, 258
future
  hydrogen economy, 261
  maximum production, reaching, 261
  other issues, 261
  projected cost, 261
  projected growth, 261
  projected production, 261
geographical distribution, 258
governments and regulation, role of, 265
  environmental regulation, 265
  grid interconnection issues, 265
  improving wind information, 265
  subsidies, tax incentives, 265
history, 258
location
  favored geography, 260
  maximum production limits, 260–261
  sizing, 260
nuclear power, role of, 183–187
site, 260
strengths
  costs, 262
  environment, 261–262
  local and diverse, 262
  quick to build, easy to expand, 262
  renewable, 262
technology, 263–264
weaknesses
  bird impact, 263
  connection to grid, 262
  local resource shortage, 263
  natural variability, 262
  noise, 263
  visual impact, 263
Wind protection
  plant water status and, 130
Wind resources, of electricity, 1419
Windrows composting, 516–517, 516f. *See also* Composting
Wind strips, 1027–1028
WINTOX, 332
Wind turbine
  blades, 264
  components
    blade diameter, 263–264
    controls and generating equipment, 264
    tower height, 263
  control mechanisms, 264
  generators, 264
  nacelles, 264
  wind sensors, 264
Wind turbine noise
  acoustic profile of, 2868–2870
  horizontal-axis wind turbine, 2868f

human impacts of, 2870
  and annoyance, 2871–2873
  and low-frequency/infrasound components, 2876–2877
  psychological description, 2870
  quantifying the health impacts, 2870–2871
  and sleep, 2873–2876
  wind turbine syndrome, 2876
Winter cover crops, for reducing NO_3-N leaching, 1306–1307
Winter rye (*Secale cereale* L.), 1306–1307
Wirestem, 443, 443t
WO. *See* Wet oxidation
WOCAT (World Overview of Conservation Approaches and Technologies), 558
Women
  hormonal disruption in, 1406
Wood
  fuels, 375t
  preservatives
    used in homes and gardens, 1401
Wood-framed walls, 861–862, 862f, 863f
Wood pellets, 375, 375t
Work, 809, 811
Work and Health in Southern Africa (WAHSA), 506
  Action on Health Impacts of Pesticides, 507t
Work capacity, 778
Work–energy principle, 813
Work–heat–energy principle, 813–816
Working fluid, 243
World Atlas of Desertification, 552
World Bank, 2034
World Business Council for Sustainable Development project (WBCSD), 2477
World Commission on Environment and Development, 338
World energy production and consumption, 678–680
World Federation of Associations of Clinical Toxicology Centers and Poison Control Centers, 1409
World fossil fuel consumption (WFFC), 1229
World Geothermal Congress 2010, 1167
World green energy consumption (WGEC), 1229
World Health Organization (WHO), 132, 424, 507, 1263, 1264, 1397, 1400, 1408, 1409, 1917, 2001, 2033, 2686, 2687
  air quality guidelines, 134
  Guidelines for Drinking Water Quality, 2793
  pesticides chronic intoxication (statistics), 1397, 1398f

World Meteorological Organization (WMO), 484
World primary energy consumption (WPEC), 1229
World Reference Base (WRB), 28
World total fossil-fuel consumption and CO_2 production, 680, 681
World War II
 use of pesticides after, 1397
Worldwatch Institute, 437
World Wide Web (WWW), 1164
WUE. *See* Water use efficiency (WUE)
WWF International, 2477
WWW. *See* World Wide Web
Wyoming coal, 718t

X

Xanthobacter autotrophicus, 384
Xanthomonas campestris pathovar *campestris*, 442
Xenobiotics, 281
 effect of, 2005
Xenohormones, 1405
Xenopsylla cunicularis, 2597, 2600
Xenorhabdus, 321, 382
XPS. *See* Extruded polystyrene
"X-ray amorphous," 1712, 1713f
Xylocoris flavipes, 2424

Y

Year Average Common Air Quality Index (YACAQI), 136
Yellow boy, 1689
Yellow River, 2884–2886
 distributions of runoff, 2885
 harnessing of, 2885
 irrigation area, 2885
 overview, 2884
 terrain of, 2884
Yield goals, realistic, 1807
Yield sensors, 2215

Z

Zabrus gibus, 1116
Zea mays, 426, 1303, 1983
Zearalenone, 1696–1697
Zeolites, 70–72
"Zero tolerance," 1126
Zero-valent iron nanoparticles (nZVI)
 pollutant removal, mechanism in, 1736–1737
 soil and groundwater remediation by, 1737–1739, 1738t, 1739t
"Zero waste" initiatives, 2418
Zetzellia mali, 2141
Zinc (Zn), 1370
 contamination, in coastal waters, 490
 phytotoxicity, 2138
 use, and B toxicity, 426
Zinc yellow pigment, 479t
Zineb, 1318, 2806
Ziram, 415, 2806
Zn. *See* Zinc (Zn)
Zooplankton, 539

(35°N–60°N) and about 6% lower at southern mid-latitudes (35°S–60°S) compared with the 1964–1980 average (Fig. 11b). Chemical destruction processes occurring at mid-latitudes contributes to observed depletion in these regions, although it is much smaller than in the polar regions because the amounts of reactive halogen gases are lower and a seasonal increase of the most reactive halogen gases in Antarctic late winter does not occur in mid-latitude regions. Changes in mid-latitude ozone are also affected by changes occurring in the polar regions when the ozone-depleted air over both polar regions is dispersed away from the poles following the vortex breakdown, thus reducing average ozone concentrations at nonpolar latitudes.

OZONE DEPLETION AND BIOLOGINAL EFFECTS

Because ozone absorbs UV radiation from the sun, depletion of the stratospheric ozone layer is expected to lead to increases in the amount of solar UV radiation reaching the Earth's surface, predominantly in the wavelength range of 290 to 320 nanometers (UV-B radiation). UV-B radiation is also partially shielded by clouds, dust, and air pollutants. Large ozone losses in the Antarctica have produced a clear increase in surface UV radiation. Ground-based measurements show that the average spring erythemal (sunburning) irradiance for 1990–2006 is up to 85% greater than the modeled irradiance for 1963–1980, depending on site. The Antarctic spring erythemal irradiance is approximately twice that measured in the Arctic for the same season.

Analyses based on surface and satellite measurements show that erythemal UV irradiance over mid-latitudes has increased since the late 1970.[44] This is in qualitative agreement with the observed decrease in column ozone, although other factors (mainly clouds and aerosols) have influenced long-term changes in erythemal irradiance. Clear-sky UV observations from unpolluted sites in mid-latitudes show that since the late 1990s, UV irradiance levels have been approximately constant, consistent with ozone column observations over this period.

The environmental and health effects of stratospheric ozone depletion have been summarized in several assessment reports.[47–49] UV-B radiation can induce acute skin damage in humans, such as sunburn, as well as eye diseases and infectious diseases.[50,51] Human epidemiological studies and animal experiments have established that UV-B radiation is a key risk factor for development of skin cancer, both melanoma and non-melanoma, especially in light-skinned population. Non-melanoma (squamous cell carcinoma and basal cell carcinoma) is the more common form of skin cancer and can be readily treated and is rarely fatal,[51] whereas melanoma is the most dangerous and the leading cause of death from skin cancer. Absorption of strong UV radiation damages the DNA molecule, eventually leading to faulty replication and mutation.[52]

Numerous experiments have shown that the cornea and lens of the eye can also be damaged by UV-B radiation, and that chronic exposure to this radiation increased the likelihood of certain cataracts.[49] Studies in human subjects show that exposure to UV-B radiation can suppress proper functioning of the body's immune system. Animal experiments indicate that overexposure to UV-B radiation decreases the immune response to skin cancers and some infectious agents.[53,54]

Terrestrial plants can also be affected by UV-B radiation, although the response varies to a large extent among different species. In addition to plant growth, the changes induced by UV radiation can be indirect, for example, by affecting the timing of developmental phases or the allocation of biomass to the different parts of the plant.[55] Aquatic ecosystems can also be damaged by UV-B radiation; for example, there is evidence for impaired larval development and decreased reproductive capacity in some amphibians, shrimp, and fish.[56] There is also direct evidence of UV-B effects under the ozone hole on the productivity of natural phytoplankton communities in Antarctic waters.[57]

INTERNATIONAL RESPONSE: MONTREAL PROTOCOL

Following the publication of the Molina–Rowland article,[7] the U.S. National Academy of Sciences issued two reports in 1976 stating that the atmospheric sequence outlined by Molina and Rowland was essentially correct.[22,23] The United States, Canada, Norway, and Sweden responded in late 1970s by banning the sale of aerosol spray cans containing CFCs; this caused a temporary halt in the growing demands for CFCs. However, worldwide use of the chemicals continued and the production rate began to rise again.

The discovery of the massive ozone losses in Antarctica in 1985 spurred a rush of scientific research activities, leading to improved understanding of stratospheric chemistry and the evolution of the ozone layer. These new scientific developments have provided the foundation for the critical policy decisions that followed.

In 1985, under the auspices of the United Nations Environment Programme (UNEP), 20 nations signed the Vienna Convention for the Protection of the Ozone Layer.[58] In September 1987, the recognition that CFC use was increasing and the mounting scientific evidence that this increase would cause large ozone depletions led to an international agreement limiting the production of CFCs.[58] This agreement, the Montreal Protocol on Substances that Deplete the Ozone Layer, initially called for a reduction of only 50% in the manufacture of CFCs by the end of the century. In view of the scientific evidence that emerged in the following years, the initial provisions were strengthened through the London (1990), Copenhagen (1992), Montreal (1997), and Beijing (1999) amendments as well as several adjustments by both controlling additional ODSs and by moving up the date by which already controlled substances must be phased out. At the 19th Meeting of the Parties to the Montreal Protocol in 2007, the Parties agreed to adjust their commitments related to the phase out of HCFCs.[58] Fig. 12 shows the projected abundance of ODSs in the stratosphere according to the provisions of the Montreal Protocol and subsequent amendments.[44]

The Montreal Protocol is now more than 20 years old and has been ratified by 196 countries, although not all the parties have ratified the subsequent amendments. It is widely regarded as one of the most effective multilateral environmental agreements in existence. The production of CFCs in industrialized countries was phased out at the end of 1995, and other compounds such as the halons, methyl bromide, carbon tetrachloride, and methyl chloroform (CH_3CCl_3) were also regulated. Developing countries were allowed to continue CFC production until 2010, to facilitate their transition to the newer CFC-free technologies. An important feature of the Montreal Protocol was the establishment of a funding mechanism to help these countries meet the costs of complying with the protocol and with its subsequent amendments.[58]

A significant fraction of the former CFC usage is being dealt with by conservation and recycling. Some of the former use of CFCs is being temporarily replaced by hydrochlorofluorocarbons (HCFCs)—compounds that have similar physical properties to the CFCs, but their molecules contain hydrogen atoms and are less stable in the atmosphere. A large fraction of the HCFCs released industrially reacts in the lower atmosphere with the OH radical before reaching the stratosphere, forming water and an organic free radical that rapidly photo-oxidizes to yield water-soluble products, which are then removed from the atmosphere mainly by rainfall. Although HCFCs are more ozone friendly than CFCs, they still destroy some ozone. They are now also regulated under the Montreal Protocol; the concentrations of HCFCs are projected to grow for another two decades before decreasing.

Some HFCs, which do not contain chlorine atoms, are now being used as CFC replacements; they are ozone friendly because fluorine forms stable compounds in the stratosphere. However, they have the potential to contribute to global warming. HFCs are now regulated under the Kyoto Protocol, an agreement under the United Nations Framework Convention on Climate Change.

About half of the CFC usage has been replaced by not-in-kind alternatives; for example, CFC-113—used extensively as a solvent to clean electronic components—has been phased out by CFC-free cleaning technologies such as soap and water or terpene-based solvents; there are also new technologies to manufacture clean electronic boards. Other examples include the use of stick or spray pump deodorants to replace CFC-12 aerosol deodorants and the use of mineral wool to replace CFC, HFC, or HCFC insulating foam.

Overall, the provisions of the Montreal Protocol have been successfully enforced. Atmospheric measurements indicate that the abundance of chlorine contained in the CFCs and other halocarbons has declined in response to the Montreal Protocol regulations.[44] On the other hand, because of the long lifetime of the CFCs in the atmosphere, relatively high chlorine levels in the stratosphere—with the consequent ozone depletion—are expected to continue well into the 21st century.

The total tropospheric abundance of chlorine from ODSs and methyl chloride had declined in 2008 to 3.4 parts per billion (ppb) from its peak of 3.7 ppb between 1992 and 1994. However, the rate of decline in total tropospheric chlorine was only two-thirds as fast as was expected because of the increase in the HCFC abundances. The rapid HCFC increases are coincident with increased production in the developing countries, particularly in East Asia. The rate of decline of total tropospheric bromine from controlled ODSs was close to that expected and was driven by changes in methyl bromide.[44]

By the middle of the 21st century, the amounts of halogens in the stratosphere are expected to be similar to those present in 1980 prior to the onset of the ozone hole. However, the influence of climate change could accelerate or delay ozone recovery.[44]

STRATOSPHERIC OZONE AND CLIMATE CHANGE LINKAGE

Greenhouse Gases and Climate Change

The Earth absorbs energy from the Sun, and also radiates energy back into space. However, much of this energy go-

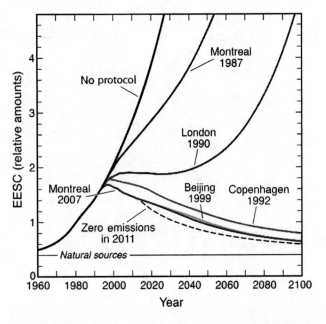

Fig. 12 Measured and projected chlorine concentrations in the stratosphere according to the provisions of the Montreal Protocol and subsequent amendments. EESC is the equivalent effective stratospheric chlorine.
Source: Fahey and Hegglin.[11]

ing back to space is absorbed by greenhouse gases occurring naturally in the atmosphere, such as CO_2, CH_4, water vapor, N_2O, and ozone. Because the atmosphere then radiates most of this energy back to the Earth's surface, our planet is warmer than it would be if the atmosphere did not contain these gases. This is the natural "greenhouse effect," which maintains the surface of the Earth at a temperature that is suitable for life as we know it today. However, this natural greenhouse effect has intensified since the start of the industrial era as human activities emit more greenhouse gases to the atmosphere, including industrial compounds such as CFCs, HFCs, perfluorocarbons (PFCs), and sulfur hexafluoride (SF_6), resulting in a shift in the radiative balance of the Earth's atmosphere.

According to the Intergovernmental Panel on Climate Change, there is now visible and unequivocal evidence of climate change impacts, and there is a consensus that greenhouse gas emissions from human activities are the main drivers of change. The Earth's average temperature has been recorded to have increased by approximately 0.74°C over the past century.[1] Climate change is now recognized as a major global challenge that will have significant and long-lasting impacts on human well-being and development.[59]

One way of quantifying the contribution of greenhouse gases to climate change is through the standard metric known as radiative forcing, which is defined as positive if it results in a gain of energy for the Earth system (warming) and negative if it results in a loss (cooling).[1] The largest radiative forcing comes from CO_2, followed by CH_4, tropospheric ozone, halocarbons, and N_2O. All of these gases absorb infrared radiation emitted from the Earth's surface and re-emit it at a lower temperature, thus decreasing the outgoing radiation flux and producing a positive forcing, leading to warming. In contrast, stratospheric ozone depletion represents a small negative forcing, which leads to cooling of Earth's surface (Fig. 13); however, this contribution is expected to decrease as ODSs are gradually removed from the atmosphere.

Halocarbons and Climate Change

The understanding of the interaction between ozone depletion and climate change has been strengthened in recent years.[44,49] Stratospheric ozone and tropospheric ozone both absorb infrared radiation emitted by Earth's surface. Stratospheric ozone also significantly absorbs solar radiation. Therefore, changes in the stratospheric ozone and tropospheric ozone are directly linked to climate change. Stratospheric ozone and climate change are also indirectly linked because both ODSs and their replacements are greenhouse gases and represent an important contribution to the radiative forcing (see Fig. 13).

One approach of comparing the influence of individual halocarbons on ozone depletion and climate change is to use ozone depletion potentials (ODPs) and global warming potentials (GWPs). The ODP and GWP of a gas quantify its effectiveness in causing ozone depletion and climate forcing, respectively. GWP is defined as the total forcing attributed to a mass of emitted pollutant during a specified time after emissions (typically 100 years) as compared to the same mass of CO_2. ODP is the relative value that indicates the potential of a substance to destroy ozone as compared with the potential of CFC-11, which is assigned a reference value of 1.

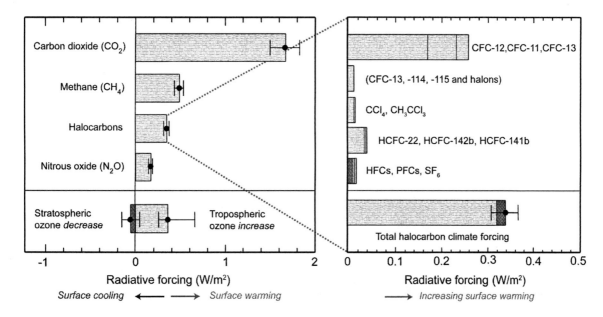

Fig. 13 Radiative forcing of climate change. Left panel: radiative forcing of major greenhouse gases. Right panel: radiative forcing of halocarbons (darker shading designates Kyoto Protocol gases).
Source: Adapted from Fahey and Hegglin.[11]

Table 1 Atmospheric lifetimes, global emissions, ODPs, and GWPs of some halogen source gases and HFC-substituted gases.

Gas	Atmospheric lifetime	Global emissions (2008) (kt/yr)[a]	ODP	GWP[b]
Halogen source gases				
Chlorine gases				
CFC-11	45	52–91	1	4,750
CFC-12	100	41–99	0.82	10,900
CFC-113	85	3–8	0.85	6130
Carbon tetrachloride (CCl_4)	26	40–80	0.82	1400
HCFCs	1–17	385–481	0.01–0.12	77–2220
Methyl chloroform (CH_3CCl_3)	5	<10	0.16	146
Methyl chloride (CH_3Cl)	1	3600–4600	0.02	13
Bromine gases				
Halon-1301	65	1–3	15.9	7140
Halon-1211	16	4–7	7.9	1890
Methyl bromide (CH_3Br)	0.8	110–150	0.66	5
Very short-lived gases (e.g., $CHBr_3$)	<0.5	c	Very low[c]	Very low[c]
HFCs				
HFC-134	13.4	149 ± 27	0	1370
HFC-23	222	12	0	14,200
HFC-143[a]	47.1	17	0	4180
HFC-125	28.2	22	0	3420
HFC-152[a]	1.5	50	0	133
HFC-32	5.2	8.9	0	716

[a] Includes both human activities (production and banks) and natural sources. Units are in kilotons per year.
[b] One-hundred-year GWPs. Values are calculated for emissions of an equal mass of each gas.
[c] Estimates are very uncertain for most species.
Source: Fahey and Hegglin[11] and WMO.[44]

Table 1 lists the atmospheric lifetime, global emissions, ODP, and GWP of some halogen source gases and the HFC replacement gases.[11,44] All ODSs and their substitutes shown here have a non-zero GWP, with values spanning the wide range of 4 to 14,000; they are far more effective than equivalent amount of CO_2 in causing climate forcing. Therefore, the future selection of specific HFCs as replacement for ODS will have important consequences for climate change.

The CFCs, halons, and HCFCs are ODSs; they are controlled under the Montreal Protocol. Thus, the Montreal Protocol has provided collateral benefit of reducing the contributions of ODSs to climate change. In 2010, the decrease of annual ODS emissions under the Montreal Protocol is estimated to be about 10 gigatons of avoided CO_2-equivalent emissions per year, which is about 5 times larger than the annual emissions reduction target for the first commitment period (2008–2012) of the Kyoto Protocol.[44]

The HFCs, used as ODSs substitutes, do not destroy ozone (ODPs equal zero) and are considered ozone safe and have become the major replacement for CFCs and HCFCs. However, like the ODSs they replace, many HFCs have high GWP. HFCs, together with CO_2, CH_4, N_2O, PFCs, and SF_6, are controlled under the Kyoto Protocol. According to a new UNEP report, emissions of HFCs are growing at a rate of 8% per year due to growing demand in emerging economies and increasing populations. Without intervention, the increase in HFCs emissions is projected to offset much of the climate benefit achieved by the earlier reduction in ODS emissions. It is therefore important to select HFCs with low GWP potential and short lifetimes to minimize the climate impact while protecting the ozone layer.[60] This is one of the focal areas for the newly launched Climate and Clean Air Coalition on Short-lived Climate Pollutants.[61]

Impact of Climate Change on Ozone

The ODSs have declined as a result of the Montreal Protocol and its subsequent amendments; it is expected that this will lead to the recovery in the stratospheric ozone abundances. However, it is difficult to attribute ozone increases to the decreases in ODSs alone during the next few years because of natural variability, observational uncertainty, and confounding factors, such as changes in stratospheric temperature or water vapor. In contrast to the diminishing role of ODSs, changes in climate are expected to have an increasing influence on the recovery of ozone layer from the effects of ODSs.

Stratospheric ozone is influenced by changes in temperatures and winds in the stratosphere. For example, lower temperatures and stronger polar winds could both increase the extent and severity of winter polar ozone depletion. Observations show that the global-mean lower stratosphere has cooled by 1–2°C and the upper stratosphere cooled by 4–6°C between 1980 and 1995. There have been no significant long-term trends in global-mean lower stratospheric temperatures since about 1995. The two main reasons for the cooling of the stratosphere are depletion of stratospheric ozone and increase in atmospheric greenhouse gases. Ozone absorbs solar UV radiation, which heats the surrounding air in the stratosphere. Loss of ozone means that less UV light gets absorbed, resulting in cooling of the stratosphere. A significant portion of the observed stratospheric cooling is also due to human-emitted greenhouse gases. While the Earth's surface is expected to continue warming in response to the net positive radiative forcing from greenhouse gas increases, the stratosphere is expected to continue cooling.

A cooler stratosphere would extend the time period over which PSCs are present in winter and early spring, leading to increased polar ozone depletion. In the upper stratosphere at altitudes above PSC formation regions, a cooler stratosphere is expected to increase ozone amounts because lower temperatures decrease the effectiveness of ozone loss reactions.

Climate change may also affect the stratospheric circulation, which will significantly alter the distribution of ozone in the stratosphere. These changes tend to decrease total ozone in the tropics and increase total ozone at mid- and high latitudes. Changes in circulation induced by changes in ozone can also affect patterns of surface wind and rainfall. The projected changes in ozone and clouds may lead to large decreases in UV at high latitudes, where UV is already low, and to small increases at low latitudes, where it is already high. This could have important implications for human and ecosystem health. However, these projections depend strongly on changes in cloud cover, air pollutants, and aerosols, all of which are influenced by climate change. It is therefore important to improve our understanding of the processes involved and to continue monitoring ozone and surface UV spectral irradiances both from the surface and from satellites.[62]

CONCLUSIONS

The depletion of the stratospheric ozone layer exemplifies the global environmental challenges human face: it is an unintended consequence of human activity. Strong involvement and cooperation of stakeholders at all levels (scientists, technologists, economic and legal experts, environmentalists, and policy makers); strengthening of human and institutional capacities, coupled with suitable mechanisms for facilitating technological and financial flows; and changes in human behavior have been critical to the success in phasing out the ODSs.

The 1987 Montreal Protocol on Substances that Deplete the Ozone Layer is a landmark agreement that has successfully reduced the global production, consumption, and emissions of ODSs. By protecting the ozone layer from much higher levels of depletion, it has provided direct benefits to human health and agriculture, which in turn provide economic benefits by decreasing health costs and increasing crop production. Furthermore, because many ODSs are also potent greenhouse gases, the Montreal Protocol has provided substantial co-benefits to climate change.

On the other hand, demand for replacement substances such as HFCs has increased. Many of these substances are Kyoto gases. Additional climate benefits could be achieved by managing the emissions of replacement fluorocarbon gases and by implementing alternative gases with lower GWPs, as well as designing buildings that avoid the need for air conditioning.

Assuming full Montreal Protocol compliance, mid-latitude ozone is expected to return to 1980 levels before mid-century. The recovery rate will be slower at high latitudes. Springtime ozone depletion is expected to continue to occur at polar latitudes, especially in Antarctica, in the next few decades. It is estimated that the ozone layer over the Antarctica will recover to pre-1980 levels between 2060 and 2075, and probably one or two decades earlier in the Arctic. However, effective control mechanisms for new chemicals threatening the ozone layer are essential; continued monitoring of the ozone layer is crucial to maintain momentum on recovering the ozone layer while simultaneously minimizing the influence on climate.

Changes in climate are expected to have an increasing influence on stratospheric ozone in the coming decades. An important scientific challenge is to project future ozone abundance based on an improved understanding of the complex linkages between stratospheric ozone and climate change.

Human activities will continue to change the composition of the atmosphere and new challenges that require international cooperation and collaboration will emerge. The ozone-hole phenomenon demonstrates the importance of long-term atmospheric monitoring and research, without which depletion of the ozone layer might not have been detected until more serious damage was evident. It is important for national and international agencies to continue their coordinated efforts on atmospheric monitoring, research, and assessment activities to provide sound scientific data needed to understand environmental changes on both regional and global scales.

ACKNOWLEDGMENTS

The author gratefully acknowledges the use of the material and figures from "WMO Scientific Assessment of Ozone Depletion: 2010" and NASA Ozone Hole Watch presented in this entry.

REFERENCES

1. IPCC (Intergovernmental Panel on Climate Change). Climate Change 2007: The Physical Science Basis, Contribution of Working Group I to the Fourth Assessment Report of the Intergovernmental Panel on Climate Change; Solomon, S., Qin, D., Manning, M., Chen, Z., Marquis, M., Avery, K.B., Tignor, M., Miller, H.L., Eds.; Cambridge University Press: Cambridge United Kingdom and New York, USA, 2007, 996 pp.
2. Molina M.J.; Molina L.T. Chlorofluorocarbons and destruction of the ozone layer. In *Environmental and Occupational Medicine*, 4th Ed.; Rom, W.N., Ed.; Lippincott, Williams and Wilkins: Philadelphia, 2007; 1605–1615.
3. Chapman, S. A theory of upper atmospheric ozone. Mem. R. Meteorol. Soc. **1930**, *3*, 103.
4. Molina, L.T.; Molina, M.J. Absolute absorption cross sections of ozone in the 185–350 nm wavelength range. J. Geophys. Res. **1986**, *91*, 14501–14508.
5. Crutzen, P.J. The influence of nitrogen oxides on atmosphere ozone content. Q. J. R. Meteorol. Soc. **1970**, *96*, 320–325.
6. Johnston H.S. Reduction of stratospheric ozone by nitrogen oxide catalysts from supersonic transport exhaust. Science **1971**, *173*, 517–522.
7. Molina, M.J.; Rowland, F.S. Stratospheric sink for chlorofluoromethanes: Chlorine-atom catalyzed destruction of ozone. Nature **1974**, *249*, 810–812.
8. Rowland, F.S.; Molina, M.J. Chlorofluoromethanes in the environment. Rev. Geophys. Space Phys. **1975**, *13*, 1–35.
9. Stolarski, R.S.; Cicerone, R. Stratospheric chlorine: A possible sink for ozone. Can. J. Chem. **1974**, *52*, 1610–1650.
10. Dobson, G.M.B.; Harrison, D.N. Measurement of the amount of ozone in the Earth's atmosphere and its relation to other geophysical conditions. Proc. R. Soc. London **1926**, *110*, 660–693.
11. Fahey, D.W.; Hegglin, M.I. Twenty Questions and Answers About the Ozone Layer: 2010 Update, Scientific Assessment of Ozone Depletion: 2010; World Meteorological Organization: Geneva, Switzerland, 2011; 72 pp.
12. Midgley, T. From the periodic table to production. Ind. Eng. Chem. **1937**, *29*, 241–244.
13. C&EN Special Issue **2008**, *86* (14), available at http://pubs.acs.org/cen/priestley/recipients/1941midgely.html (accessed May 29, 2011).
14. Lovelock, J.E. Atmospheric fluorine compounds as indicators of air movements. Nature **1971**, *230*, 379.
15. Lovelock, J.E.; Maggs, R.J.; Wade, R.J. Halogenated hydrocarbons in and over the Atlantic. Nature **1973**, *241*, 194–196.
16. Molina, M.J. Polar ozone depletion (Nobel lecture). Angew. Chem. Int. Ed. Engl. **1996**, *35*, 1778–1785.
17. Rowland, F.S. Stratospheric ozone depletion by chlorofluorocarbons (Nobel lecture). Angew. Chem. Int. Ed. Engl. **1996**, *35*, 1786–1798.
18. Rowland, F.S.; Molina, M.J. The CFC-Ozone Puzzle: Environmental Science in the Global Area. John H. Chafee Memorial Lecture on Science and the Environment. National Academy of Sciences, December 7, 2000.
19. DeMore, W.B.; Sander, S.D.; Golden, D.M. et al. Chemical kinetics and photochemical data for use in the stratospheric modeling. Evaluation no. 11, JPL publication no. 94-26. NASA Jet Propulsion Laboratory: Pasadena, CA, 1994.
20. World Meteorological Organization. Scientific assessment of ozone depletion: 1994. WMO Global Ozone Research and Monitoring Project, report no. *37*, Geneva: WMO, 1995.
21. World Meteorological Organization. Scientific assessment of ozone depletion: 1998. WMO Global Ozone Research and Monitoring Project, report no. *44*, Geneva: WMO, 1999.
22. National Research Council. *Halocarbon: Environmental Effects of Chlorofluoromethane Release*; National Academy of Sciences: Washington, DC, 1976.
23. National Research Council. *Halocarbons: Effects on Stratospheric Ozone*; National Academy of Sciences: Washington, DC, 1976.
24. Farman, J.C.; Gardiner, B.G.; Shanklin, J.D. Large losses of total ozone in Antarctica reveal seasonal ClO_x/NO_x interactions. Nature **1985**, *315*, 207–210.
25. Stolarski, R.S.; Bloomfield, P.; McPeters, R.D.; Herman, J.R. Total ozone trends deduced from Nimbus 7 TOMS data. Geophys. Res. Lett. **1991**, *18*, 1015–1018.
26. Russell, J.M. III; Luo, M.; Cicerone, R.J.; Deaver, L.E. Satellite confirmation of the dominance of chlorofluorocarbons in the global stratospheric chlorine budget. Nature **1996**, *379*, 526–529.
27. Jones, A.E.; Shanklin, J.D. Continued decline of total ozone over Halley, Antarctica, since 1985. Nature **1995**, *376*, 409–411.
28. NASA Ozone Hole Watch, available at http://ozonewatch.gsfc.nasa.gov/facts/hole.html (accessed May 28, 2011).
29. Solomon S.; Garcia, R.R.; Rowland, F.S.; Wuebbles, D.J. On the depletion of Antarctic ozone. Nature **1986**, *321*, 755–758.
30. Molina, L.T.; Molina, M.J.; Stachnick, R.A.; Tom, R.D. An upper limit to the rate of the $HCl + ClONO_2$ reaction. J. Phys. Chem. **1985**, *89*, 3779–3781.
31. Molina, M.J.; Tso, T.-L.; Molina, L.T.; Wang, F.C.-Y. Antarctic stratospheric chemistry of chlorine nitrate, hydrogen chloride and ice. Release of active chlorine. Science **1987**, *238*, 1253–1260.
32. Tolbert, M.A.; Rossi, M.J.; Malhotra, R.; Golden, D.M. Reaction of chlorine nitrate with hydrogen chloride and water at Antarctic stratospheric temperatures. Science **1987**, *238*, 1258–1260.
33. Abbatt, J.P.D.; Beyer, K.D.; Fucaloro, A.F.; McMahon, J.R.; Wooldridge, P.J.; Zhang, R. Interaction of HCl vapor

with water–ice: Implications for the stratosphere. J. Geophys. Res. **1992**, *97*, 15819–15826.
34. Molina, M.J. The probable role of stratospheric 'ice' clouds: Heterogeneous chemistry of the ozone hole. In *Chemistry of the Atmosphere: The Impact of Global Change*; Calvert, J.G., Ed.; Blackwell Scientific: Oxford, U.K., 1994; 27–38.
35. Gertner, B.J.; Hynes, J.T. Molecular dynamics simulation of hydrochloric acid ionization at the surface of stratospheric ice. Science **1996**, *271*, 1563–1566.
36. Molina L.T.; Molina M.J. Production of Cl_2O_2 from the self-reaction of the ClO radical. J. Phys. Chem. **1987**, *91*, 433–436.
37. McElroy, M.B.; Salawitch, R.J.; Wofsy, S.C.; Logan, JA. Reduction of Antarctic ozone due to synergistic interactions of chlorine and bromine. Nature **1986**, *321*, 759–762.
38. Cox, R.A.; Hayman, G.D. The stability and photochemistry of dimers of the ClO radical and implications for Antarctic ozone depletion. Nature **1988**, *332*, 796–800.
39. Molina, M.J.; Colussi, A.J.; Molina, L.T.; Schindler, R.N.; Tso. T-L. Quantum yield of chlorine-atom formation in the photodissociation of chlorine peroxide (ClOOCl) at 308 nm. Chem. Phys. Lett. **1990**, *173*, 310–315.
40. Tuck, A.F.; Watson R.; Condon, E.P.; Margitan, J.J.; Toon, O.B. The planning and execution of ER-2 and DC-8 aircraft flights over Antarctica, August and September 1987. J. Geophys. Res. **1989**, *94*, 181–222.
41. Anderson, J.G., Toohey, D.W., Brune, W.H. Free radicals within the Antarctic Vortex: the role of CFCs in Antarctic ozone loss. Science **1991**, *251*, 39–46.
42. Proffitt, M.H.; Steinkamp, M.J.; Powell, J.A. et al. In situ ozone measurements within the 1987 Antarctic ozone hole from a high-altitude ER-2 aircraft. J. Geophys. Res. **1989**, *94*, 547–555.
43. Chen, H.-Y.; Lien, C.Y.; Lin, W.-Y.; Lee, Y.T.; Lin, J.J. UV absorption cross sections of ClOOCl are consistent with ozone degradation models. Science **2009**, *324*, 781–784.
44. WMO (World Meteorological Organization). Scientific Assessment of Ozone Depletion: 2010, Global Ozone Research and Monitoring Project—Report No. 52, Geneva, Switzerland, 2011; 516 pp.
45. Turco, R.; Plumb, A.; Condon, E. The Airborne Arctic Stratospheric Expedition: Prologue. Geophys. Res. Lett. **1990**, *17*, 313–316.
46. Manney, G.L.; Santee, M.L.; Rex, M. et al. Unprecedented Arctic ozone loss in 2011. Nature **2011**, *478*, 469–475.
47. van der Leun, J.; Tang, X.; Tevini, M. Environmental effects of ozone depletion: 1994 assessment. Ambio **1995**, *24*, 138.
48. Biggs R.H.; Joyner, M.E.B., Eds. *Stratospheric Ozone Depletion/UV-B Radiation in the Biosphere*; Springer-Verlag: New York, 1994.
49. UNEP, Environmental Effects of Ozone Depletion and its Interaction with Climate Change: 2010 Assessment. United Nations Environment Programme, December 2010, Nairobi, Kenya, available at http://ozone.unep.org/Assessment_Panels/EEAP/eeap-report2010.pdf (accessed May 20, 2011).
50. Longstreth, J.D.; de Grujil, F.R.; Kripke, M.L.; Takizawa, Y.; van der Leun, J.C. Effects of solar radiation on human health. Ambio **1995**, *24*, 153–165.
51. Molina, M.J.; Molina L.T.; Fitzpatrick, T.B.; Nghiem, P.T. Ozone depletion and human health effects. In *Environmental Medicine*; Moller L., Ed.; Joint Industrial Safety Council Product 33: Sweden, 2000; 28–51.
52. Brash, D.E.; Rudolph, J.A.; Simon, J.A. et al. A role for sunlight in skin cancer: UV-induced p53 mutations in squamous cell carcinoma. Proc. Natl. Acad. Sci. U. S. A. **1991**, *88*, 124–128.
53. De Fabo, E.C.; Noonan, F.P. Mechanism of immune suppression by ultraviolet irradiation in vivo. I. Evidence for the existence of a unique photoreceptor in skin and its role in photo-immunology. J. Exp. Med. **1983**, *157*, 84–98.
54. Cooper, K.D.; Oberhelman, L.; Hamilton, T.A. et al. UV exposure reduces immunization rates and promotes tolerance to epicutaneous antigens in humans—Relationship to dose, CD1a-DR$^+$ epidermal macrophage induction and Langerhans cell depletion. Proc. Natl. Acad. Sci. U. S. A. **1992**, *89*, 8497–8501.
55. Caldwell, M.M.; Teramura, A.H.; Tevini, M.; Bomman, J.F.; Björn, L.O.; Kulandaivelu, G. Effects of increased solar ultraviolet radiation on terrestrial plants. Ambio **1995**, *24*, 166–173.
56. Häder, D.P.; Worrest, R.C.; Kumar, H.D.; Smith, R.C. Effects of increased solar ultraviolet radiation on aquatic ecosystems. Ambio **1995**, *24*, 174–180.
57. Smith, R.C.; Prézelin, B.B.; Baker, K.S. et al. Ozone depletion: Ultraviolet radiation and phytoplankton biology in Antarctic waters. Science **1992**, *255*, 952–959.
58. UNEP (United Nations Environment Program). *Handbook for the Montreal Protocol on Substances that Deplete the Ozone Layer*, 8th Ed.; UNEP Ozone Secretariat, 2009.
59. UNEP (United Nations Environment Program). *Global Environmental Outlook (GEO-4): Environment for Development*; Progress Press Ltd.: Malta, **2007**.
60. UNEP (United Nations Environment Program). HFCs: A Critical Link in Protecting Climate and the Ozone Layer, November 2011; 36 pp.
61. The Climate and Clean Air Coalition to Reduce Short-Lived Climate Pollutants, Fact Sheet. http://www.state.gov/r/pa/prs/ps/2012/02/184055.htm (accessed February 16, 2012)
62. McKenzie, R.L.; Aucamp, P.J.; Bais, A.F.; Björn, L.O.; Ilyas, M.; Madronich, S. Ozone depletion and climate change: impacts on UV Radiation. Photochem. Photobiol. Sci. **2011**, *10*, 182–198.

Permafrost

Douglas Kane
Julia Boike
Water and Environmental Research Center, University of Alaska, Fairbanks, Fairbanks, Alaska, U.S.A.

Abstract
This entry discusses those surface soils and deeper geologic layers that remain at or below freezing for a duration of two years or more and how they impact people living in this environment. Such frozen ground, both unconsolidated and bedrock, are commonly referred to as permafrost.

INTRODUCTION

On an extended south-to-north transect at northern latitudes, a transition from ground that never experiences seasonal freezing, to those that occasionally freeze during the winter, to those that freeze every year, to those that may remain frozen for an extended time could be encountered. The topic of discussion here is those surface soils and deeper geologic layers that remain at or below freezing for a duration of two years or more and how they impact people living in this environment. Such frozen ground, both unconsolidated and bedrock, are commonly referred to as permafrost.[1] Although at or below the freezing point of bulk water (0°C), the term "permafrost" neither implies that water is present or that water, if present, is frozen. In fact, it is possible for significant amounts of water to remain unfrozen in permafrost; this is also true for water in seasonal frost.

PERMAFROST CHARACTERISTICS

Spatially extensive permafrost can be found in Russia and Canada as far south as 45°N, and even farther south on the elevated Tibetan Plateau and Himalayan Mountains.[2] Approximately 25 million km^2 of permafrost exist in the northern hemisphere. In the higher latitudes, permafrost is continuous under the land surfaces. At intermediate latitudes permafrost is discontinuous or sporadic. Legget[3] reported that 20% of the land surface of the world is underlain by permafrost. More than 50% of Russia and Canada are underlain by permafrost. Alaska has continuous permafrost in the northern 1/3 of the state and discontinuous permafrost in the rest of the state, excluding the coastal areas from the Aleutian Islands to southeastern Alaska (Fig. 1). In the southern hemisphere, permafrost distribution is confined to Antarctica and high alpine or mountainous regions. Isolated permafrost is common at higher elevations, and evidence of past permafrost is common in areas that no longer have permafrost. The thickness of permafrost can vary from a thin lens of less than 1 m to greater than 1000 m (Fig. 1). Permafrost can also be found in coastal areas at the bottom of seas.

Permafrost ground is interesting to the engineer and scientist because the medium is usually composed of two solids (porous medium and ice) and two fluids (air and liquid water). The more components that are present in a mixture, the more difficult it is to predict the medium's response to an input of energy or mass. The amount of unfrozen water in saturated ground is strictly a function of the grain size (more specifically surface area) and the freezing temperature. A fine-grained soil such as clay has a very high surface area relative to coarse-grained soils such as sand; in frozen ground, this translates into much higher unfrozen water contents at the same temperature. The unfrozen water found in permafrost exists as a film of water around each soil particle. It is via these unfrozen films that water moves in permafrost. Frozen clay can have as much as 5%–7% unfrozen water by volume at −15°C. As the temperature decreases the amount of unfrozen water also decreases. Thermal and hydraulic properties of permafrost are quite variable and depend upon the percentages of the various ground components. Most heat transfer in permafrost is by conduction and can be modeled by Fourier's law, although simpler methods have been developed.

A typical temperature profile of permafrost appears in Fig. 2. Since there is a geothermal flux outward from the center of the earth, this heat has to be successfully transferred to the ground surface or the permafrost will warm and melt. In order to maintain the thermal integrity of the permafrost, the soils above the top of the permafrost table must completely freeze during the winter so there is a continuously decreasing thermal gradient along which the geothermal heat can be transferred to the surface by conduction (Fig. 2).

The seasonally thawed soil layer at the ground surface that goes through freezing/thawing annually and mantles the permafrost is called the active layer (Fig. 2). This layer acts as a buffer to heat and mass transfer to the permafrost. A typical active layer in the continuous permafrost zone would typically thaw to a maximum depth of 60 cm. The top 15–25 cm of these soils are generally composed of

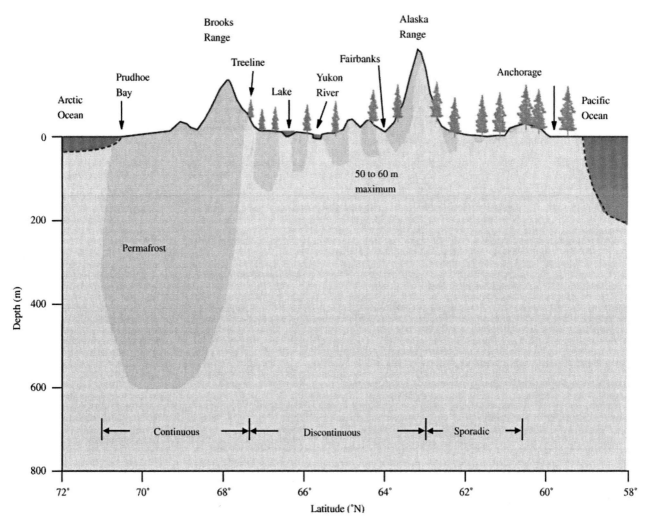

Fig. 1 Schematic of permafrost distribution in Alaska on a south-to-north transect.

organic material, with the deeper soils being mineral. Organic soils are good thermal insulators and, when coupled with snow cover during the winter months, they minimize heat loss from the ground. The thermal balance of the permanently frozen ground is maintained by heat loss to the atmosphere that occurs at high latitudes over the extended winter months.

Permafrost is usually considered to be impermeable to water movement. This is usually a good assumption for short durations of days or even a few months. For longer periods of time, the redistribution of water within permafrost can be appreciable. During snowmelt and major rainfall events, considerable water enters the active layer. Most of this water resides in the organic soils as the mineral soils are usually already near saturation. Since permafrost has relatively low hydraulic conductivity, there is no hydraulic connection between the perched water above the permafrost (suprapermafrost groundwater) and the subpermafrost groundwater below the permafrost (Fig. 2). In continuous permafrost, the subsurface hydrology is confined to the active layer. For areas of discontinuous permafrost, the subsurface hydrology is a combination of shallow flow over the frozen ground and deeper flow around and under it.

SURFACE ENERGY BALANCE

Any time the ground surface is disturbed, the surface energy balance that sustains permafrost is upset. This generally results in warming of the permafrost and thickening of the active layer. Much sporadic and discontinuous permafrost is maintained at temperatures just below freezing; climatic warming of just a few degrees would result in the melting of this frozen ground and warming of colder permafrost. Also where permafrost exists, surface disturbances such as removing vegetation or surface soils and ponding of water are sufficient to alter the surface energy balance and cause permafrost degradation. Permafrost generally appears where the mean annual surface temperature is at least a few degrees Celsius below freezing; for instance, Hay River, NWT, Canada has sporadic, shallow permafrost with a mean annual temperature of −3.4°C.

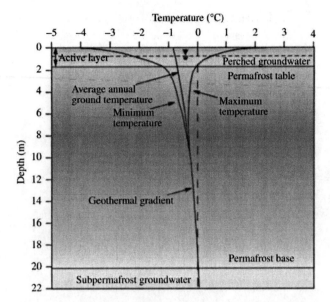

Fig. 2 Typical temperature profile for permafrost, with annual variation indicated near the ground surface.

In areas of discontinuous permafrost, the south-facing slopes are generally permafrost-free (Fig. 1) while north-facing slopes have permafrost; the east- and west-facing slopes and valley bottoms may or may not have permafrost, depending upon subtle site-specific conditions that impact the energy balance (vegetation, slope, moisture content, etc.). Because it took thousands of years for permafrost to develop, deep permafrost is a particularly good recorder of past climates, and changes in the temperature profile will reflect these impacts. Many general atmospheric circulation models (GCM) predict variable global warming in the high latitudes, and in some areas there is already field evidence of atmospheric warming[4] and permafrost warming.[5] To sustain permafrost, the geothermal heat radiating out from the core of the earth must be removed; this only occurs if the active layer is completely frozen during winter.

LIVING WITH PERMAFROST

Permafrost influences many facets of everyday living that are often taken for granted in warmer climates. Permafrost negatively impacts both the quality and quantity of the available groundwater and limits the potential for infiltration of wastewater. As a result, housing units with individual water and wastewater systems must be designed with innovative options to replace the traditional groundwater well and septic system for wastewater treatment. Utilities that are typically buried are often placed in aboveground utilidors. Also as roadways age, they become very bumpy as differential movement of the road surface occurs.

The significant property of permafrost from an engineering viewpoint is the ice content. The amount of ice can range from essentially none in well-drained ground to values that exceed the porosity of poorly drained ground. Within the permafrost, segregated ice with minimal entrained soil exists as ice lenses and wedges. As long as the permafrost stays frozen, the ground is relatively stable. However, thawing of ice-rich permafrost causes ground to settle or subside. This thaw settlement compromises the structural integrity of buildings, roadways, pipelines, and other structures built on it. For areas where construction must take place on frozen ground, special efforts—for example the use of insulation and thermosyphons—are necessary to maintain the structural integrity of the permafrost. Thermosyphons are vertical devices for removing heat from the ground, and they are used extensively under buildings and pipelines to maintain below freezing conditions in the ground.

Soil freezing causes additional engineering problems near the ground surface when strong thermal gradients induce water movement from warm ground to cold ground through unfrozen films of water on the surfaces of soil particles, regardless of whether it is frozen or not. This process causes "frost heave," as it results in an accumulation of ice in the form of lenses that can cause the ground to expand and deform. Surface heaving becomes significant over time periods of several years, and it is important for those engineering structures with design lives of tens of years to compensate for this additional stress.

CONCLUSIONS

Permafrost and seasonally frozen ground are very extensive in the high latitudes of Earth. Also, resource development and population in these areas is increasing. Initial efforts directed at building an infrastructure (roads, airports, water and sewer distribution, etc.), on permafrost assumed that the frozen ground was in thermal equilibrium. It is now obvious that because of climatic change that the permafrost is not in thermal equilibrium. Instead, it is warming in many areas and this needs to be considered in the present design criteria for all engineering structures built on permafrost. Warming, with subsequent thawing of permafrost, will affect the surface energy budget, water and gas fluxes (potential for release of greenhouse gases), and vegetation; hence, there is a direct feedback with the climate system. Just as it took a long time for permafrost to develop, it will take a long time for it to completely melt; but the consequences of thawing permafrost can impact our daily lives in the first or second season of melting.

REFERENCES

1. Muller, S.W. *Permafrost or Permanently Frozen Ground and Related Engineering Problems*. U.S. Engineers Office, Strategic Engineering Study, Special Report No. 62, 1943; 136 pp.

2. Brown, J.; Ferrians, O.J., Jr.; Heginbottom, J.A.; Melnikov, E.S. *Circum-Arctic Map of Permafrost and Ground-Ice Conditions*; U.S. Dept. of Interior, Geological Survey, Map CP-45, 1997.
3. Legget, R.F. Permafrost research. Arctic **1954**, *7* (3/4), 153–158.
4. Serreze, M.C.; Walsh, J.E.; Chapin, F.S.; Osterkamp, T.; Dyurgerov, M.; Romanovsky, V.; Oechel, W.C.; Morison, J.; Zhang, T.; Berry, R.G. Observational evidence of recent change in the northern high-latitude environment. Climate Change **2000**, *46*, 159–207.
5. Lachenbruch, A.H.; Marshall, B.V. Changing climate: geothermal evidence from permafrost in the Alaskan Arctic. Science **1986**, *234*, 689–696.

Persistent Organic Compounds: Wet Oxidation Removal

Anurag Garg
Center for Environmental Science and Engineering, Indian Institute of Technology, Mumbai, India

I.M. Mishra
Department of Chemical Engineering, Indian Institute of Technology, Roorkee, Mumbai, India

Abstract

Several industries are generating an enormous volume of wastewater contaminated with a number of toxic and persistent organic compounds (POPs) such as phenols, cresols, naphthenic compounds, lignin, mercaptans, and inorganic disulfides. These contaminants enter waste streams from chemical processing and handling operations and from refineries and petrochemical units. A number of such contaminants are biorefractory and toxic to microbial species, and therefore require some kind of pretreatment before being sent to conventional biological treatment methods. This entry focuses on the wet oxidation (WO) process that can be used for the treatment of waste streams containing POPs and other inorganics that are biorefractories. WO is an oxidation process mainly used for the degradation of organic and inorganic compounds at elevated temperature and pressure conditions (up to 320°C and 20 MPa, respectively), converting them into less harmful gaseous products and water-soluble innocuous products. The process is suitable for moderate- to high-strength wastewaters having a chemical oxygen demand ranging from 10,000 to 100,000 mg/L. The process is applicable for the treatment of low-volume wastewaters emanating from various chemical and process industries. The WO process is also an alternative solution for the regeneration of spent activated carbon and the desulfurization of coal. Several commercial WO processes are available in the market; however, their use is limited to developed countries as the processes are capital intensive and are carried out under severe operating conditions. This entry deals with the WO process, its mechanistic aspects and commercial processes, and the challenges for the process to become acceptable in wastewater treatment.

INTRODUCTION

With industrial development, we are encountering a number of environmental problems affecting the human population. Industrial waste streams contain a number of persistent organic pollutants (POPs) that are xenobiotic and persist in biotreatment systems without any chemical transformation. They are toxic and hazardous in nature, and affect the human population through biomagnification.[1] The toxicity of some POP compounds is given in Table 1.[2,3] The toxicity of such compounds adversely affects biological treatment systems, even potentially resulting in complete failure. Therefore, it is necessary to have a pretreatment process that can eliminate these compounds or transform them to such compounds that are amenable to biological treatment or to employ some alternative treatment strategy for converting POP-bearing wastewater streams into environmentally benign forms. There are several non-biological treatment methods, such as adsorption, coagulation, and chemical oxidation, for the treatment of such waste streams. Coagulation and adsorption processes are generally not used as stand-alone methods to treat highly polluted waste streams; these methods are used either as pretreatment or polishing methods. However, these methods are beset with other problems such as that of regeneration of the adsorbents, handling and disposal of spent adsorbents, and the handling and disposal of flocculated sludge. Chemical oxidation can be used as a complete or partial treatment step for treating such waste streams. The extent of treatment (pretreatment or full treatment) is highly dependent on the concentration and nature of pollutants, and the availability of facilities for posttreatment processes. Chemical oxidation processes are generally classified on the basis of the oxidants used. The oxidants may include air, molecular oxygen, hydrogen peroxide, ozone, etc. All of these processes have their own advantages and limitations.

In this entry, the main focus is on the removal of POP (toxic and recalcitrant organic pollutants) by chemical oxidation, using air or molecular oxygen as an oxidant. This process is also known as the wet oxidation (WO) process. WO fundamentals, its mechanism, and the pros and cons for its adoption as a treatment process are presented in the following sections. The major applications of the process are also reported. A synoptic view of the commercial WO processes is also presented. The challenges associated with

Table 1 Threshold limit values of phenolic compounds.

S. No.	Name of compound	Toxicological information	Threshold limit value (TLV)
1.	Phenol	Routes of entry—dermal, inhalation, ingestion, carcinogenic	19 mg/m^3
		Classified as A4 (not classifiable for humans) by ACGIH	
		LDL (human) oral route: 140 mg/L	
2.	Resorcinol	Routes of entry—absorbed through skin, dermal contact, inhalation, ingestion	10 ppm
		Classified as A4 (not classifiable for human or animal) by ACGIH	
		LDL (human): not available	
3.	Nitrophenol	Routes of entry—dermal, inhalation, ingestion	5 mg/m^3
		LDL (human): not available	
4.	Pentachlorophenol	Routes of entry—dermal contact, eye contact, inhalation, ingestion	0.5 mg/m^3
		LDL (human): not available	
5.	O-cresol	Routes of entry—absorbed through skin, dermal contact	19 mg/m^3
		Classified possible by IRIS	
6.	Phenolphthalein	Routes of entry—absorbed through skin, eye contact, inhalation	1000 ppm
		Classified as A4 (not classifiable for human/animal) by the ACGIH; may affect genetic material and cause adverse reproductive effect	

Source: Data from Material safety data sheet.[2,3]

the WO process are also reported in a separate section. At the end of the entry, a concise summary is provided.

WET OXIDATION PROCESS

Fundamentals

WO is a hydrothermal process generally used for the destruction of organic and inorganic compounds from wastewaters at high-temperature and -pressure conditions by using an oxidant. The temperature and total pressure may vary from 125°C to 320°C and 0.5 to 20 MPa, respectively.[4] The treatment time is comparatively much shorter than that for biological systems; however, it depends on the type and concentration of compounds, the type and amount of the oxidant, and the reaction conditions employed. Organic compounds are mineralized into CO_2 and H_2O on complete oxidation. Other organic compounds containing sulfur and nitrogen form SO_2 and NH_3 in addition to the abovementioned mineralized products. Inorganic compounds are converted into their stable forms with a little or no oxygen demand. The process is suitable for medium- to high-strength wastewaters having a chemical oxygen demand (COD) greater than 10,000 mg/L[5,6] and low biochemical oxygen demand (BOD) (or low biodegradability, i.e., BOD_5/COD <0.5). The severity of the process makes it an energy- and capital-intensive process. Therefore, low-volume wastewaters having a COD greater than 10,000 mg/L should be used so that that the process becomes self-sustaining without any energy supplement. For the initiation of the WO process, however, energy is to be provided from an external source. The energy produced during the process depends on the wastewater characteristics and the reaction conditions.

The WO process is carried out in a high-pressure reactor made of a corrosion-resistant material that could withstand high temperatures and pressures. Therefore, the selection of reactor material requires extensive care to avoid any operational problems. In a typical continuous WO system, other major appurtenances include a high-pressure pump, a heat exchanger, an oxidant source such as an air compressor or a pure oxygen cylinder, and a liquid–gas separator. All these units are interconnected by proper tubes, valves, and other fittings. The schematic flow diagram of a typical WO system is depicted in Fig. 1. The wastewater is introduced in the reactor vessel via a preheater (or a heat exchanger) through a high-pressure pump. The oxidant (air or oxygen) may be mixed with wastewater either before or after the preheater. The wastewater is allowed to flow in the reactor for a definite period to achieve the destruction of POP to a desired level. Due to exothermic reactions taking place in the WO reactor, treated wastewater and the reactor by-products contain significant amounts of heat. A part of the generated heat is used up in heating the reactor and sustaining it to the desired temperature. The reactor effluent is used in the preheater to heat the oxidant and/or the raw wastewater to the desired temperature before their

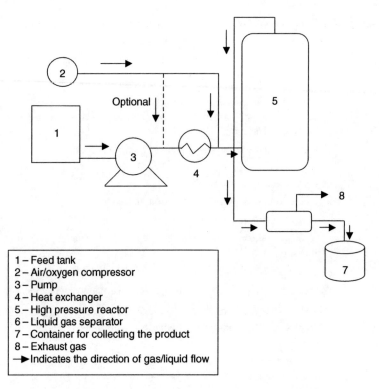

Fig. 1 Schematic diagram of a typical WO process.

entry into the WO reactor. The treated effluent is passed, thereafter, through the liquid–gas separator to separate the gas phase from the liquid phase. The gas-free treated waste stream is then sent either to a final polishing unit or to downstream biological treatment unit for further treatment, as the case may be. The separated gaseous stream is passed through appropriate treatment and emission devices to mitigate the adverse effects of the gaseous constituents in the atmosphere.

Various reactor configurations are used depending on the characteristics of the wastewater, the catalysts, and the oxidant. For instance, bubble column, jet agitator, and mechanically stirred reactors can be used for non-catalytic or homogenous catalytic WO involving gas–liquid phases. Trickle bed, bubble slurry column, bubble fixed bed, and fluidized bed reactors can be used for catalytic WO involving three phases—gas, liquid, and solid catalyst.[7,8] All reactor systems have their own advantages and disadvantages. For example, a trickle bed reactor system may encounter poor distribution of wastewater, incomplete liquid film on the catalyst particles, and difficulty in maintaining the temperature of the reactor. Therefore, the reactor system has to be designed with great care to overcome any problem associated with its operation. A major constraint in implementation of the WO process on a large scale is the confidence level in the design and operation of the system.[9] A micro–meso–macro method is generally used for three-phase WO reactions involving gas, liquid, and solid catalyst to integrate material (catalyst) synthesis and reactor selection.[9] Several issues related to solvent (liquid) evaporation during the reaction, incomplete catalyst wetting, and back-mixing effects are to be considered. Moving bed reactors may have an advantage in such situations where catalyst deactivation poses a major constraint. However, more investigations are needed for its assured use.

Mechanism of WO Reaction

The WO process is considered to consist of chain reactions that take place in several steps. In general, organic compounds transform into aldehydes, ketones, alcohols, and acids as intermediates that ultimately end up as CO_2 and H_2O as the major end products. The formation of various intermediates during the reaction depends on a number of factors such as pH of the reaction mixture, compounds in the wastewater and their chemical structures, catalyst (if any) properties and its mode of contact, and reaction conditions, viz. temperature and pressure, etc. Three major steps are considered for the degradation of organic pollutants by the WO process. These steps include chain initiation, propagation, and termination. The formation of hydroxyl radical (HO·) is considered to be the key event for effective WO reaction. The presence of hydroxyl radicals triggers the reaction in forward direction by generating more free radicals that react with the parent and intermediate compounds. The reaction is essentially carried out in wet conditions since H_2O itself acts as a catalyst and forms HO· free radicals. These radicals are formed by the reactions of different reactants (air, H_2O, O_2, and organic compounds) at high-temperature and -pressure conditions. The activation energy for these reactions is about 200 kJ/mol.[10] The propagation step is comparatively much easier to take place once hydroxyl radicals are formed. The activation energies for this step are generally around 11–12 kJ/mol. During this step, some organic intermediates may themselves act as catalysts and enhance the rate of reaction.[11] This is called auto-oxidation. Similarly, some intermediates may act as inhibitory compounds and reduce the rate of reaction.[12] In the termination step, the free radicals are destroyed owing to collision either with themselves or with the walls of the reactor. A simple scheme of WO reactions is shown in Fig. 2.

The WO process is a complex process because of the many parallel and series reactions occurring simultaneously. The overall process can be modeled in a number of pathways for the degradation of a single compound such as phenol and propionic acid.[13,14] The reaction pathways can be manipulated by changing the reaction conditions that may create favorable conditions for other reactions.

Use of a catalyst in the form of metals, their compounds/mixtures, or oxides during the WO process can facilitate the generation of hydroxyl and other free radicals at much lower energy inputs. The catalytic WO also occurs in the same way as non-catalytic WO. However, the participation of the catalyst in the reaction may occur in different ways. For instance, homogeneous catalysts become dissolved in the wastewater and can react with an oxidant and/or organic compounds to form complex compounds. These complexes may further react with other compounds/reactants and release active catalyst species along with the end products. On the other hand, if heterogeneous catalysts (solid catalysts) are used, the organic compounds and oxidant are adsorbed on the exterior and interior surfaces of the catalyst. In this way, the reaction takes place at the surface of the catalyst and the presence of metal species enhances the POP degradation rate. The products thus formed become desorbed from the surface of the catalyst, and the active sites thus become vacant for further adsorption of other reactant molecules and reaction. The metal species also regains its original oxidation state. The extent of adsorption depends on the active sites available on the catalyst surface. Hence, the active heterogeneous metal oxides are supported on a stable material to increase the effective surface area of the catalyst. Some of the supporting materials that have been used in earlier studies include CeO_2, Al_2O_3, SiO_2, TiO_2, and zeolites (such as ZSM-5, MCM-41, and 13X).[15–30] Some researchers have also explored activated carbon as a support to enhance the surface area and activity of the catalyst.[31–35]

The selected catalyst should be thermally stable and should provide large number of active sites. It should be resistant to sintering and leaching in high-temperature and corrosive environmental conditions. The pore size distribution of the catalyst is also very important since a pore size lower than that of organic or oxidant molecules will inhibit the ingress of these species in the pore space. In this case, most of the area will be wasted and the reaction between these species in the presence of a metal catalyst will not occur. At the same time, sufficient amount of active metal species should also be present in the catalyst. Metals having a high redox potential normally show good activity in WO reactions.[11] Transition metals (Cu, Fe, Co, Mn, etc.), and noble metals (Pt, Ru, Pd, etc.), have been found to be very active for the degradation of various organic and inorganic compounds.[4,8,11,12,17,35–52]

Advantages and Limitations

As stated earlier, the WO process is advantageous for the low-volume waste streams that contain recalcitrant and non-biodegradable toxic compounds in high concentrations. These advantages are as follows:

- Effectiveness in the destruction of toxic and recalcitrant compounds.
- Absence of dioxins and furans, which is a major concern during incineration.
- Amenability of the treated waste for further treatment in biological systems. Thus, the WO process may be an effective pretreatment option for toxic and biorefractory wastes.
- NO_x and SO_2 are not formed.
- Toxicity of the wastewater is eliminated.
- WO is a chemical destruction technique rather than a removal process.

However, the process is not simple and easy to implement owing to several constraints. The major limitations of the WO process are as follows:

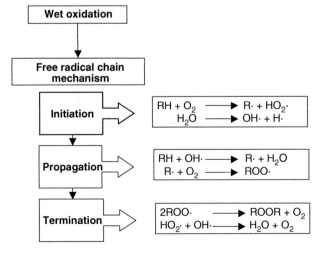

Fig. 2 General mechanism of a WO reaction.

- High capital and operating costs due to severe oxidation conditions
- Complex reaction mechanism
- Difficult design of the reactor vessel
- Not suitable for waste streams having a low to moderate COD (<10,000 mg/L)
- Unable to degrade low-molecular-weight organic compounds at less severe oxidation conditions
- Catalyst recovery, regeneration, and reuse may be problematic

Applications

The process can be used for the treatment of low-volume, high-COD-containing toxic wastewater streams or for the treatment of sludges. Several industries such as pulp and paper mills, textile mills, oil refineries, petrochemical units, coke oven plants, alumina refining industries, alcohol distilleries, food industries, leather industries, dye manufacturing and processing units, and pharmaceutical industries generate wastewaters that may contain large amounts of phenolic compounds, lignin, sulfides, dyes, and other benzene ring compounds that are xenobiotic, biorefractory, and toxic. These compounds will have to be either removed or destroyed and transformed into such compounds as are amenable to biological treatment. Sewage sludge and urban wastes can also be treated by this process to reduce their volumes and to make them nontoxic for further treatment and disposal. The intended degree of treatment decides the operating conditions of the process. The WO waste treatment spectrum is presented by Carlos and Maugans.[53] The WO process can also be used for the regeneration of adsorbents such as activated carbon and activated clays, as also for spent catalysts. This is also effective in desulfurization of coal.[54] The process has also been found useful for energy generation and chemical recovery from various biomass materials.

COMMERCIAL WO PROCESSES

There are a number of commercial WO processes (both catalytic and non-catalytic) available in the market and offered by various vendors. Some of these processes can be found in the literature.[8,9,55,56] Commercial WO processes can be categorized either on the basis of the number of phases involved and/or the type of catalysts used. Both of these classifications are depicted in Fig. 3. Wet oxidation reactors may be two- or three-phase reactors depending on the states of the reactants involved. Two-phase reactors are essentially gas–liquid reactors involving gas (oxidant)–liquid (wastewater constituents) reactions. Non-catalytic gas–liquid reactors fall into this category. The other example of a two-phase reaction is catalytic WO reaction using a homogenous catalyst. In this reaction, a solid catalyst is dissolved in the liquid completely and a homogenous aqueous mixture is obtained. To perform the WO process, air or oxygen (gaseous phase) is passed into the reactor. In three-phase reactions, the catalyst retains its solid state and does not dissolve in the liquid. The WO process can also be categorized under non-catalytic and catalytic reactions. Catalytic reactions can further be subcategorized into homocatalytic or heterocatalytic reactions depending on the type of catalyst used. Again, the catalyst can be homogeneous or heterogeneous in nature.

Non-Catalytic WO Processes

Non-catalytic WAO facilities generally employ Zimpro and VerTech processes. Moreover, there are some other processes, such as Wetox, Kenox, and Oxyjet, that do not use any catalyst. A brief summary about these non-catalytic processes is presented in Table 2. The processes are capable of treating the wastewater and sludges from various industries and urban areas; however, they require very severe operating conditions (such as temperature and

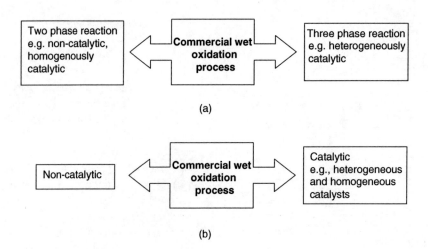

Fig. 3 Classification of commercial WO process (a) on the basis of phases and (b) on the basis of catalytic or non-catalytic reaction.

Table 2 Summary of commercial non-catalytic WO processes.

S. No.	Name of process	Substrate and oxidant	Operating conditions
1.	Zimpro	Sulfide-bearing wastewater, chemical industry effluents; compressed air	Temperature = 150–325°C, pressure = 12.5 MPa, and reaction time = 20–240 min
2.	VerTech	Urban sludge and industrial wastewater; cryogenic oxygen	Temperature = 180–280°C, pressure = 8.5–11 MPa, and reaction time = 60 min
3.	Wetox	Chemical industrial effluents; pure O_2	Temperature = 210–250°C, pressure = 5 MPa, and reaction time = 45 min
4.	Kenox	Industrial effluents; compressed air	Temperature = 200–240°C, pressure = 4.1–4.7 MPa, and reaction time = 40 min
5.	Oxyjet	Pharmaceutical waste, phenols, wood waste; ultrasound ejector	Temperature = 140–300°C, pressure = 2.5 MPa, and reaction time ≤5 min
6.	Linde	Sulfide-bearing waste	Temperature = 120–180°C, pressure = 4–8 MPa, and reaction time ≤45–120 min

Source: Data from Bhargava et al.,[8] Larachi,[9] Patria,[55] Luck,[56] and Mishra and Joshi.[57]

pressure). These processes are in operation in several parts of the world such as the United States, Brazil, Taiwan, and Europe (Italy, United Kingdom, the Netherlands, etc.).

Catalytic WO Processes

A list of catalyst (homogeneous and heterogeneous)-based WO processes is compiled in Table 3. A brief overview about both kinds of processes is given in the following two subsections.

CATALYTIC WO REACTIONS USING HOMOGENEOUS CATALYSTS

Among homocatalytic processes, LOPROX, Ciba-Geigy, and ATHOS are the major ones that use Fe^{2+} and Cu^{2+} as homogeneous catalysts. The processes are capable of degrading high-strength wastewaters and semisolid wastes up to the desired level (more than non-catalytic reactions). However, the Ciba-Geigy and ATHOS processes need high temperature and pressure. The LOPROX process is

Table 3 Summary of commercial catalytic WO processes.

S. No.	Name of process	Substrate, oxidant, and catalyst	Operating conditions
(a) With homogeneous catalysts			
1.	LOPROX	Low reactive industrial effluent and urban sludge; compressed air; Fe^{2+} catalyst (quinone-producing catalyst)	Temperature ~200°C, pressure = 0.5–2.0 MPa, and reaction time = 60–120 min
2.	Ciba-Geigy	Pharmaceuticals, sewage sludge; compressed air; Cu^{2+} catalyst	Temperature ~300°C, pressure = 15 MPa, and reaction time = 180 min
3.	ATHOS	Residual sludges; cryogenic O_2; Cu^{2+} catalyst	Temperature = 235–250°C, pressure = 4.4–5.5 MPa, and reaction time = 60 min
(b) With heterogeneous catalysts			
1.	NS–LC (Nippon Shokubai process)	Phenols, acetic acid, glucose, ammonia, formaldehyde; Pt-Pd/TiO_2-ZrO_2 catalyst and pure O_2	Temperature = 160–270°C, pressure = 0.9–8.0 MPa, and reaction time ~60 min
2.	Osaka Gas process (similar to the non-catalytic Zimpro process)	Coal gasifier and coke oven effluents, cyanide, sewage sludge, and residential wastes; mixture of precious and base metals (e.g., Fe, Co, Ni, Ru, Pd, Pt, Cu, Au, and W) supported on TiO_2 or TiO_2–ZrO_2; pure O_2	Temperature = 250–260°C, pressure ~6.9 MPa, and reaction time ~24 min
3.	CATOx2 process of the Engineers India Limited, New Delhi, India, and the Indian Institute of Technology, Roorkee, India	Sulfide, mercaptans-bearing wastewater, chemical industry effluents; COD up to 100,000 mg/L; copper-based catalysts; compressed air	Temperature = 150–170°C, pressure ~5–10 MPa, and reaction time = 120–180 min

Source: Data from Bhargava et al.,[8] Larachi,[9] Patria,[55] Luck,[56] and Mishra and Joshi.[57]

generally used to pretreat the wastewater so that it could become acceptable to biological treatment facility and, therefore, are conducted at lower temperatures (<200°C). The LOPROX and Ciba–Geigy processes are in operation at industrial scale in Switzerland, whereas full facilities of the ATHOS process are available in other European countries, namely, France, Belgium, and Italy.

There are other WO processes (e.g., WPO and ORCAN) that use homogeneous catalysts and hydrogen peroxide (in place of oxygen or air) as an oxidant. These processes can even be performed at lower-temperature and -pressure conditions (90–130°C temperature and 0.1–0.5 MPa pressure) in the presence of the Fe^{2+} catalyst. Nevertheless, there are some major issues associated with H_2O_2 that inhibit its use as an oxidant in the WO process.

CATALYTIC WO REACTIONS USING HETEROGENEOUS CATALYSTS

There are two heterogeneous catalytic WO processes, namely, NS-LC and Osaka Gas, which are being used in Japan. The Osaka Gas process is also used in the United Kingdom. These processes use pure oxygen as an oxidant and supported noble metals as heterogeneous catalysts. The Osaka process uses transition metals along with precious metals as catalysts. The catalysts, for both processes, are supported on titania or zirconia or their mixture. It can be seen from Table 3 that the catalytic WO process is capable of destroying a number of organic carbon and nitrogen compounds from the wastewater at elevated conditions within 20 to 60 min of reaction time. More than 99% of organics and sulfur- and nitrogen-containing compounds in the wastewater can be destroyed in the catalytic WO process in contrast to 5%–50% in non-catalytic WO.

Recently, a new process, named the CATOx2 process, has been developed jointly by IIT Roorkee and Engineers India Ltd.[57] for the treatment of spent caustic stream from petroleum refineries with a COD load of up to 100,000 mg/L. This process shows very high degradation efficiency for sulfide-bearing compounds present in the spent caustic from a sour-water-stripping unit in a petroleum refinery. The process uses air as an oxidant and a copper-based catalyst, and operates under mild to moderate operating conditions (pressure 5–10 barg; temperature 150–170°C). The process is able to remove nearly 90% COD and nearly 60% mercaptans. The effluent of the CATOx2 process is to be treated downstream chemically or biologically for the removal of residual compounds such as naphthenes, cresols, mercaptans, and phenols. The gaseous stream is condensed and the condensate is further treated in an adsorber, and the liquid stream is sent to the biological treatment unit. The process technology is being adopted by a refinery in India.

MAJOR CHALLENGES

Even though the WO process is in use for quite some time, it is still not being used on a large scale in wastewater treatment. Although several commercial processes are available either on industrial scale or have been tested for their efficacy on pilot plant scale, the process is still seen with some degree of concern basically because of severe operating conditions. To make the process widely acceptable, a number of issues related to the process have to be resolved. The major challenges may include the following:

1. A less expensive robust heterogeneous catalyst system must be developed that have a long life without showing significant decrease in the catalytic activity. Noble metals are generally very expensive in comparison with transition metals but have been found to have better performance. It necessitates the review of the catalyst preparation methods to produce a transition-metal-based catalyst with desired attributes.
2. In the case of homogeneous catalysts, a simple and cost-effective method for the recovery of dissolved metal species should be developed for process users so that the catalyst can be recycled for reuse in several cycles, and the toxic effect of the catalyst is avoided on the downstream treatment system or the discharge sink.
3. The catalyst should be effective in degrading the contaminants at mild to moderate pressure (<10 bar) and temperature (<170°C) conditions. Moderation in the operating conditions will significantly reduce the capital cost of the reactor. Even if the catalyst is used at milder temperature conditions (say up to 150°C), it may be possible to treat wastewaters having COD less than 10,000 mg/L. Since the process is exothermic, the heat released may be enough to meet the energy requirement of the system. It has been pointed out earlier that the heat release is a function of concentration and nature of the compounds to be degraded.
4. The process requires deeper understanding of the reaction mechanism under different operating conditions. Since the intermediates and the by-products can either enhance or retard the reaction rates, more reliable and robust techniques for sampling and analysis are to be explored. This will help in understanding the reaction pathways.
5. Alternatively, the process can be used as a pretreatment step before the conventional biological processes. It is also practiced for some commercial processes (e.g., LOPROX and CATOx2). The major intermediates (e.g., acetic acid) can either be treated in downstream adsorbent bed or biological system, or recovered, if economically viable. Another alternative may be to develop a low-cost catalyst that is capable of degrading acetic acid at low to moderate operating conditions.

6. The selection and design of the reactor is critical for successful WO operation. Several issues related to an individual reactor system need to be addressed. Corrosion is a major problem in the reactor operation. Hence, the reactor material should be chosen carefully so that it is able to withstand a highly acidic environment under moderate to severe operating conditions.

CONCLUSION

The WO process is attracting a number of researchers to test its efficacy for a number of POPs, although it has not been widely accepted in the field for the treatment of wastewaters containing POPs. The WO process is seen as an alternative solution to meet the strict requirements of the treated wastewater quality for its discharge into natural water sources. In the past, low-cost and eco-friendly biological processes were considered to be the most reliable processes as the treated wastewaters could meet less-stringent discharge standards. It is also recognized that the conventional biological processes alone are not capable of treating the highly polluted wastewaters containing toxic and non-biodegradable compounds such as POPs. The WO process offers an effective solution for such wastewaters as it destroys POPs and other recalcitrant pollutants to harmless products unlike other conventional processes in which a solid residue (e.g., sludge after biological or chemical [coagulation]/electrochemical processes) is generated. The reluctance in its adoption stems from its high capital and operating costs, as the process is carried out under severe operating conditions. The economic viability of the process depends on its effectiveness under milder operating conditions, use of efficient and low-cost catalysts having a long life and that are easily regenerated and/or recovered, and better design of the reactor along with its internals. Despite its limitations at present, the process has a promising potential to become one of the most effective treatment methods, especially for low-volume wastewaters contaminated by biorefractory and toxic compounds.

REFERENCES

1. U.S. Environmental Protection Agency. Persistent organic pollutants: A global issue, a global response, available at http://www.epa.gov/international/toxics/pop.html#table (accessed December 2011).
2. Material safety data sheet listing, available at http://www.sciencelab.com/msdsList.php (accessed December 2011).
3. Material safety data sheet, available at http://www.carolina.com/text/teacherresources/MSDS/phenolphth05.pdf (accessed December 2011).
4. Mishra, V.S.; Mahajani, V.V.; Joshi, J.B. Wet air oxidation. Ind. Eng. Chem. Res. **1995**, *34*, 2–48.
5. Debellefontaine, H.; Foussard, J.N. Wet air oxidation for the treatment of industrial wastes. Chemical aspects, reactor design and industrial applications in Europe. Waste Manage. **2000**, *20*, 15–25.
6. Levec, J.; Pintar, A. Catalytic wet-air oxidation processes: A review. Catal. Today **2007**, *124*, 172–184.
7. Iliuta, I.; Larachi, F. Wet air oxidation solid catalysis analysis of fixed and sparged three-phase reactors. Chem. Eng. Process. **2001**, *40*, 175–185.
8. Bhargava, S.K.; Tardio, J.; Prasad, J.; Föger, K.; Akolekar, D.B.; Grocott, S.C. Wet oxidation and catalytic wet oxidation. Ind. Eng. Chem. Res. **2006**, *45*, 1221–1258.
9. Larachi, F. Catalytic wet oxidation: Micro–meso–macro methodology from catalyst synthesis to reactor design. Top. Catal. **2005**, *33*, 109–134.
10. Li, L.; Chen, P.; Gloyna, E.F. Generalized kinetic model for wet oxidation of organic compounds. AIChE J. **1991**, *37*, 1687–1697.
11. Imamura, S. Catalytic and noncatalytic wet oxidation. Ind. Eng. Chem. Res. **1999**, *38*, 1743–1753.
12. Vaidya, P.D.; Mahajani, V.V. Insight into sub-critical wet oxidation of phenol. Adv. Environ. Res. **2002**, *6*, 429–439.
13. Devlin, H.R.; Harris, I.J. Mechanism of oxidation of aqueous phenol with dissolved oxygen. Ind. Eng. Chem. Fundam. **1984**, *23*, 387–392.
14. Rivas, F.J.; Kolaczkowaski, S.T.; Beltran, F.J.; McLurgh, D.B. Development of a model for the wet air oxidation of phenol based on a free radical mechanism. Chem. Eng. Sci. **1998**, *53*, 2575–2586.
15. Sadana, A.; Katzer, J.R. Catalytic oxidation of phenol in aqueous solution over copper oxide. Ind. Eng. Chem. Fundam. **1974**, *13*, 127–134.
16. Zhang, Q.; Chuang, K.T. Alumina-supported noble metal catalysts for destructive oxidation of organic pollutants in effluent from a softwood kraft pulp mill. Ind. Eng. Chem. Res. **1998**, *37*, 3343–3349.
17. Hamoudi, S.; Larachi, F.; Cerrella, G.; Cassanello, M. Wet oxidation of phenol catalyzed by unpromoted and platinum-promoted manganese/cerium oxide. Ind. Eng. Chem. Res. **1998**, *37*, 3561–3566.
18. Miró, C.; Alejandre, A.; Fortuny, A.; Bengoa, C.; Font, J.; Fabregat, A. Aqueous phase catalytic oxidation of phenol in a trickle bed reactor: Effect of the pH. Water Res. **1999**, *33*, 1005–1013.
19. Eftaxias, A.; Font, J.; Fortuny, A.; Giralt, J.; Fabregat, A.; Stüber, F. Kinetic modelling of catalytic wet air oxidation of phenol by simulated annealing. Appl. Catal. B **2001**, *33*, 175–190.
20. Wu, Q.; Hu, X.; Yue, P.-L.; Zhao, X.S.; Lu, G.Q. Copper/MCM-41 as catalyst for the wet oxidation of phenol. Appl. Catal. B **2001**, *32*, 151–156.
21. Neri, G.; Pistone, A.; Milone, C.; Galvagno, S. Wet air oxidation of p-coumaric acid over promoted ceria catalysts. Appl. Catal. B **2002**, *38*, 321–329.
22. Lin, S.S.; Chen, C.L.; Chang, D.J.; Chen, C.C. Catalytic wet air oxidation of phenol by various CeO_2 catalysts. Water Res. **2002**, *36*, 3009–3014.
23. Chang, D.J.; Lin, S.S.; Chen, C.L.; Wang, S.P.; Ho, W.L. Catalytic wet air oxidation of phenol using CeO_2 as the catalyst. Kinetic study and mechanism development. J. Environ. Sci. Health A **2002**, *37*, 1241–1252.

24. Massa, P.; Ayude, M.A.; Ivorra, F.; Fenoglio, R.; Haure, P. Phenol oxidation in a periodically operated trickle bed reactor. Catal. Today **2005**, *107–108*, 630–636.
25. Posada, D.; Betancourt, P.; Liendo, F.; Brito, J.L. Catalytic wet air oxidation of aqueous solutions of substituted phenols. Catal. Lett. **2005**, *106*, 81–88.
26. Kim, S.-K.; Kim, K.-H.; Ihm, S.-K. The characteristics of wet air oxidation of phenol over CuO_x/Al_2O_3 catalysts: Effect of copper loading. Chemosphere **2007**, *68*, 287–292.
27. Lopes, R.J.G.; Silva, A.M.T.; Quinta-Ferreira, R.M. Screening of catalysts and effect of temperature for kinetic degradation studies of aromatic compounds during wet oxidation. Appl. Catal. B **2007**, *73*, 193–202.
28. Yang, S.; Zhu, W.; Wang, J.; Chen, Z. Catalytic wet air oxidation of phenol over CeO_2–TiO_2 catalyst in the batch reactor and the packed-bed reactor. J. Hazard. Mater. **2008**, *153*, 1248–1253.
29. Liotta, L.F.; Gruttadauria, M.; Di Carlo, G.; Perrini, G.; Librando, V. Heterogeneous catalytic degradation of phenolic substrates: Catalysts activity. J. Hazard. Mater. **2009**, *162*, 588–606.
30. Garg, A.; Mishra, I.M.; Chand, S. Oxidative phenol degradation using non-noble metal based catalysts. Clean **2010**, *38*, 27–34.
31. Fortuny, A.; Font, J.; Fabregat, A. Wet air oxidation of phenol using activated carbon as catalyst. Appl. Catal. B **1998**, *19*, 165–173.
32. Stüber, F.; Font, J.; Fortuny, A.; Bengoa, C.; Eftaxias, A.; Fabregat, A. Carbon materials and catalytic wet air oxidation of organic pollutants in wastewater. Top. Catal. **2005**, *33*, 3–50.
33. Wu, Q.; Hu, X.; Yue, P.L. Kinetics study on heterogeneous catalytic wet air oxidation phenol using copper/activated carbon catalyst. Int. J. Chem. React. Eng. **2005**, *3*, A29.
34. Suárez-Ojeda, M.E.; Fabregat, A.; Stüber, F.; Fortuny, A.; Carrera, J.; Font, J. Catalytic wet air oxidation of substituted phenols: Temperature and pressure effect on the pollutant removal, the catalyst preservation and the biodegradability enhancement. Chem. Eng. J. **2007**, *132*, 105–115.
35. Garg, A.; Mishra, I.M.; Chand, S. Catalytic wet oxidation of the pretreated synthetic pulp and paper mill effluent under moderate conditions. Chemosphere **2007**, *66*, 1799–1805.
36. Katzer, J.R.; Ficke, H.H.; Sadana, A. An evaluation of aqueous phase catalytic oxidation. J. Water Pollut. Control Fed. **1976**, *48*, 920–933.
37. Pintar, A.; Levec, J. Catalytic liquid-phase oxidation of phenol aqueous solutions. A kinetic investigation. Ind. Eng. Chem. Res. **1994**, *33*, 3070–3077.
38. Matatov-Meytel, Y.I.; Sheintuch, M. Catalytic abatement of water pollutants. Ind. Eng. Chem. Res. **1998**, *37*, 309–326.
39. Zhang, Q.; Chuang, K.T. Lumped kinetic model for catalytic wet oxidation of organic compounds in industrial wastewater. AIChE J. **1999**, *45*, 145–150.
40. Hamoudi, S.; Belkacemi, K.; Larachi, F. Catalytic oxidation of aqueous phenolic solutions: Catalyst deactivation and kinetics. Chem. Eng. Sci. **1999**, *54*, 3569–3576.
41. Akolekar, D.B.; Bhargava, S.K.; Shirgoankar, I.; Prasad, J. Catalytic wet oxidation: An environmental solution for organic pollutant removal from paper and pulp industrial waste liquor. Appl. Catal. A-Gen. **2002**, *236*, 255–262.
42. Santos, A.; Yustos, P.; Quintanilla, A.; Rodríguez, S.; García-Ochoa, F. Route of the catalytic oxidation of phenol in aqueous phase. Appl. Catal. B **2002**, *39*, 97–113.
43. Brégeault, J.-M. Transition-metal complexes for liquid-phase catalytic oxidation: Some aspects of industrial reactions and of emerging technologies. Dalton Trans. **2003**, *17*, 3289–3302.
44. Arena, F.; Giovenco, R.; Torre, T.; Venuto, A.; Parmaliana, A. Activity and resistance to leaching of Cu-based catalysts in the wet oxidation of phenol. Appl. Catal. B **2003**, *45*, 51–62.
45. Vicente, J.; Rosal, R.; Díaz, M. Catalytic wet oxidation of phenol with homogeneous iron salts. J. Chem. Technol. Biotechnol. **2005**, *80*, 1031–1035.
46. Barbier, J., Jr.; Oliviero, L.; Renard, B.; Duprez, D. Role of ceria supported noble metal catalysts (Ru, Pd, Pt) in wet air oxidation of nitrogen and oxygen containing compounds. Top. Catal. **2005**, *33*, 77–86.
47. Cybulski, A. Catalytic wet air oxidation: Are monolithic catalysts and reactors feasible? Ind. Eng. Chem. Res. **2007**, *46*, 407–433.
48. Nousir, S.; Keav, S.; Barbier, J., Jr.; Bensitel, M.; Brahmi, R.; Duperz, D. Deactivation phenomena during catalytic wet air oxidation (CWAO) of phenol over platinum catalysts supported on ceria and ceria–zirconia mixed oxides. Appl. Catal. B **2008**, *84*, 723–731.
49. Lopes, R.; Quinta-Ferreira, R.M. Manganese and copper catalysts for the phenolic wastewaters remediation by catalytic wet air oxidation. Int. J. Chem. Reactor Eng. **2009**, *6*, A116 (p21).
50. Garg, A.; Mishra, I.M.; Chand, S. Catalytic oxidative treatment of diluted black liquor at mild conditions using CuO/CeO_2 catalyst. Water Environ. Res. **2008**, *80*, 136–141.
51. Garg, A.; Mishra, A. Wet oxidation—An option for enhancing biodegradability of leachate derived from municipal solid waste (MSW) landfill. Ind. Eng. Chem. Res. **2010**, *49*, 5575–5582.
52. Kim, K.-H.; Ihm, S.-K. Heterogeneous catalytic wet air oxidation of refractory organic pollutants in industrial wastewaters: A review. J. Hazard. Mater. **2011**, *186*, 16–34.
53. Carlos, T.M.S.; Maugans, C.B. Wet air oxidation of refinery spent caustic: A refinery case study. NPRA Conference, San Antonio, TX, Sept 12, 2000.
54. Pradt, L.A. Developments in wet oxidation. Chem. Eng. Prog. **1972**, *68*, 72–77.
55. Patria, L.; Maugans, C.; Ellis, C.; Belkhodja, M.; Cretenot, D.; Luck, F.; Copa, B. Wet air oxidation processes. In *Advanced Oxidation Processes for Water and Wastewater Treatment*, 1st Ed.; Parsons, S., Ed.; IWA Publishing: London, 2004; 247–274.
56. Luck, F. Wet air oxidation: Past, present and future. Catal. Today **1999**, *53*, 81–91.
57. Mishra, I.M.; Joshi, J.K. Personal communication, **2011**.

Persistent Organic Pesticides

Gamini Manuweera
Secretariat of the Stockholm Convention, United Nations Environmental Program, Chatelaine, Switzerland

Abstract

Persistent organic pesticides are part of a larger group of chemicals known as persistent organic pollutants or POPs. In addition to pesticides, POPs include industrial chemicals and unintentionally produced chemical substances or by-products of anthropogenic origin. POPs do not easily undergo common environmental degradation processes. Therefore, once released, these substances remain unchanged in the environment for a very long period of time. POPs are also highly toxic to living organisms. The toxic effects are mainly linked to long-term low-level exposure scenarios mostly resulting in chronic health problems. High persistence in the environment increases the availability of POPs for long-term exposure to human populations and ecosystems. Furthermore, POPs can undergo long-range transport in the environment. Combination of these effects drew the attention on this group of substances at the international level after it became apparent that they travel long distances across borders. Measures by individual countries were not sufficient to ensure satisfactory protection of human health and the environment from adverse effects of POPs. As a result, a global treaty, now known as the Stockholm Convention on Persistent Organic Pollutants, was created to address the issues related to POPs. The Stockholm Convention requires parties to take measures to reduce or eliminate releases from intentional production and use of POPs by taking necessary legal and administrative measures as recommended by the Convention. The latest concerns of POPs include interference of these substances with hormonal activities, acting as "endocrine disruptors," and possible interlinkages between climate change and POPs. Factors such as indiscriminate use of pesticides and lack of capacity for sound management of pests and disease vectors may lead to continued need for POP pesticides. The efforts on elimination of POP pesticides should seek alternative approaches that are sustainable. It will be important to ensure that POP pesticides are not simply replaced by other pesticides but that the principles of integrated pest and vector management are adopted with due consideration on related concerns such as resistance management.

INTRODUCTION

Persistent organic pesticides refer to a set of man-made organic chemical substances meant to control pests in the human environment including those concerning agriculture, veterinary health, and public health. Persistent organic pesticides are part of a larger group of chemicals known as persistent organic pollutants or POPs.[1] In addition to pesticides, POPs include industrial chemicals and unintentionally produced chemical substances or by-products of anthropogenic origin. Dioxins, for example, are unintentionally produced POPs formed during incomplete combustion processes involving organic matter and chlorine. Chemically, POPs include linear and cyclic halogenated hydrocarbons. In some POPs, functional moieties may also exist in the hydrocarbon molecule as in the case of perfluorooctane sulfonic acid and its salts.

POPs are highly toxic to living organisms. This group of chemicals does not easily undergo common environmental degradation processes including chemical, microbial, or photolytic reactions. Once released, POPs stay for a long period of time in the environment, posing higher risk of long-term exposure to human populations and ecosystems. POPs are lipophilic (has affinity to fat, lipids, etc.), due to the non-polar organic nature of the substances. The lipophilicity allows chemical substances to readily accumulate in fatty tissues of living organisms (the process referred to as bioaccumulation). Once accumulated, the concentration of POPs in the living organisms builds up through the food chain via biomagnification processes, increasing the risk of adverse effects at the higher tropic levels.

Some of the physicochemical properties of the POPs facilitate long-range transport in the environment. POPs are found in the alpine and mountainous regions, the Arctic, Antarctica, and remote Pacific islands far away from where activities associated with POPs are taking place. In the environment, POPs undergo sorption into organic matter, intramedia and intermedia dispersion (diffusion), and advection (transport mechanism of substances due to bulk motion of the medium). Most of the POPs are semivolatile in nature. With long residence time in transport media, POPs can travel very long distances across regions through environmental transportation processes, including atmospheric transport, making them available for human and environmental exposure at a global scale.

The toxic effects of POPs are mainly linked to long-term low-level exposure scenarios mostly resulting in chronic health problems, while some POPs could also exert acute effects. The toxic endpoints of POPs include cancer, birth defects, reproductive problems, damages to specific organs such as the liver and kidneys, among others.

Much attention was drawn to this group of substances at the international level after it became apparent that they travel long distances across borders. As a consequence, several countries started banning POPs in the 1970s. However, actions by a limited number of countries alone were unable to control continued environmental pollution and adverse health effects from such border-crossing substances. A regional legal agreement that specifically addresses POPs was adopted in 1998 with the Aarhus Protocol on Persistent Organic Pollutants under the regional Convention on Long-Range Transboundary Air Pollution (LRTAP) of the UN Economic Commission for Europe (UNECE).[3] As an agreement at the regional level was not sufficient to ensure satisfactory protection of human health and the environment from adverse effects of POPs, negotiations of an international legally binding instrument to reduce or eliminate, where possible, releases of POPs were initiated under the auspices of the United Nations Environmental Programme in 1998. In May 2001, more than 100 countries agreed and adopted a global treaty, now known as the Stockholm Convention on Persistent Organic Pollutants.[2,4] Some aspects of the life cycle of POPs are considered in other international legally binding instruments. POP wastes are included in the Basel Convention on the Control of Transboundary Movements of Hazardous Wastes and Their Disposal,[5] while the international trade of most POP pesticides is addressed by the Rotterdam Convention on the Prior Informed Consent Procedure for Certain Hazardous Chemicals and Pesticides in International Trade.[6] Several other international initiatives also address POPs, notably the Global Programme of Action for the Protection of the Marine Environment from Land-Based Activities (GPA)[7] and a number of regional seas agreements.

PERSISTENT ORGANIC PESTICIDES UNDER THE STOCKHOLM CONVENTION

As of 2011, there were 22 POPs included in the Stockholm Convention, of which 14 were pesticides. At the time of its entry into force in 2001, the Stockholm Convention had 9 pesticides among the total of 12 POPs listed therein.

The Convention provides provisions for Parties to make proposals to add new POPs. These proposals must contain information on the chemical relating to the screening criteria established under the Convention that include persistence, bioaccumulation, potential for long-range environmental transport, and adverse effects. The POPs Review Committee established by the Convention reviews submissions on candidate POPs, including related information from other sources, prepares a risk profile, and undertakes a risk management evaluation. If the candidate chemical satisfactorily meets the POP screening criteria of the Convention, the POPs Review Committee makes recommendations to the Conference of the Parties for considering the new chemical under the Convention. The Conference of the Parties then evaluates the recommendations made by the POPs Review Committee for the inclusion of the candidate chemical.

At its fourth meeting held in 2009, the Conference of the Parties to the Convention considered and included nine new chemical substances consisting five pesticides as POPs in the Convention. During the fifth Conference of the Parties held in 2011, the pesticide endosulfan was also included as the 14th POP pesticide of the Convention.

The Stockholm Convention requires Parties to take measures to reduce or eliminate releases from intentional production and use of POPs through two different approaches:

- Prohibit and/or take the legal and administrative measures necessary to eliminate the chemical substances listed in Annex A of the Convention.
- Restrict production and use of the chemicals listed in Annex B of the Convention in accordance with specific measures provided in that Annex.

The chemical substances for elimination are listed in Annex A of the Convention. When chemical substances with existing commercial uses are included in the Convention for elimination, some chemicals may still have certain specific use or uses for which parties to the Convention may require a transition period to completely eliminate the reliance on the chemical substance. For such chemicals, exemptions are provided for those specific uses and related production for a limited period of time. These specific exemptions are initially available for a 5 years period. The parties opt for the use of specific exemptions require conformation to a set of precautionary measures relevant its available uses established by the Convention to ensure reduced releases of the substance and effective elimination. Aldrin, for example, was listed in Annex A of the Convention for elimination with a specific exemption of use only as local ectoparasiticide and insecticide with no further production. At the end of the initial period, the exemptions for production and use are not available, if Parties have effectively eliminated the reliance on those POPs.

During the fourth meeting of the Conference of the Parties in 2009, it was noted that the specific exemptions provided for six POP pesticides initially listed in Annex

Table 1 Pesticides listed as POPs under the Stockholm Convention.

Aldrin
Alpha hexachlorocyclohexane
Beta hexachlorocyclohexane
Chlordane
Chlordecone
DDT
Dieldrin
Endosulfan
Endrin
Heptachlor
Hexachlorobenzene
Lindane
Mirex
Toxaphene

A of the Convention were no longer needed. Accordingly, the POP pesticides aldrin, chlordane, dieldrin, heptachlor, hexachlorobenzene (also found as a by-product and industrial chemical), and mirex listed in Annex A have no specific exemptions available for Parties anymore. In addition to those six POP pesticides, Annex A also contains alpha hexachlorocyclohexane (also found as a by-product), beta hexachlorocyclohexane (also found as a by-product), chlordecone, endrin, and toxaphene, for which no specific exemptions were provided at the time of listing them in the Convention. The two remaining POP pesticides listed in Annex A with specific exemptions are lindane and technical endosulfan (see Table 1) and its related isomers. The specific exemptions on lindane for use as a human health pharmaceutical for control of head lice and scabies as second-line treatment are available until 2015.

The chemical substances listed in Annex B are those identified with specific uses for which there are no alternatives available at present or that the alternatives are not accessible or effectively available under certain settings. Such uses are recognized under the Convention as acceptable purposes with no set timeline for elimination. Parties are allowed to produce and use the chemicals for those purposes according to the recommended practices provided with respective uses. The Convention has listed DDT, for example, under Annex B with an acceptable purpose for disease vector control. It requires activities associated with DDT to be in accordance with the World Health Organization's recommendations and guidelines on the use of DDT. Further, the use of DDT is allowed when locally safe, effective, and affordable alternatives are not available to the Party in question.

Following Article 4 of the Convention, the parties that use POP pesticides according to the specific exemptions and acceptable purposes provided in the Convention require notifying the Secretariat to register for that purpose.

The register of specific exemptions is made publicly available on the Convention's Web site (http://www.pops.int).

MAJOR ISSUES CONCERNING PERSISTENT ORGANIC PESTICIDES

Over the years, there has been an increase in both the general understanding and concern about adverse effects of POP pesticides. The latest concerns include interference of POPs with hormonal activities, acting as "endocrine disruptors," and possible interlinkages between climate change and POPs. Release, distribution, and degradation of POPs are highly dependent on environmental conditions. Climate change and increasing climate variability have the potential to affect POPs contamination via higher releases from primary sources and environmental reservoirs, changes in transport processes and pathways, and routes of degradation. Exposure to POPs and related impacts on environmental and human health can be further exacerbated by higher atmospheric temperatures.

Most of the POP pesticides included in the Stockholm Convention are first-generation pesticides discovered in the Second World War era. The subsequent advent of pesticides from the chemical families of organophosphates, carbamates, and synthetic pyrethroids challenged the continuity of favorable market for the first-generation pesticides. However, the remarkable successes achieved in the malaria eradication programs in the 1950s and 1960s using DDT for indoor residual spraying (IRS)[8] and the continued need to rely on DDT for disease vector control compelled the global community to place DDT under Annex B of the Convention, thereby allowing its continued production and use for disease vector control. The World Health Organization recommends DDT only for IRS. Countries can use DDT as long as necessary, in the quantity needed, provided that the guidelines and recommendations of the World Health Organization and the Stockholm Convention are all met, and until locally appropriate and cost-effective alternatives are available for a sustainable transition from DDT. The continued need for DDT for disease vector control is evaluated at regular conferences of the parties, held every 2 years. The evaluation is undertaken in consultation with the World Health Organization on the basis of available scientific, technical, environmental, and economic information. As a separate process, the World Health Organization also reviews new information on adverse health effects of DDT periodically to facilitate the evaluation of DDT by the Convention. In 2010, the World Health Organization expert consultation report on "DDT in Indoor Residual Spraying: Human Health Aspects" identified several potential hazards of DDT and its toxic metabolites. These include acute poisoning hazard for children with accidental ingestion, carcinogenicity, developmental toxicity, male reproductive effects, and concerns for women of childbearing age who live in DDT IRS-treated dwellings. However, in terms of relevant exposure scenarios for the general population in countries using IRS, the expert panel concluded that available evidence does not point to concern about levels of exposure for any of the endpoints that were assessed. The report demands further research to better evaluate risks that were suggested in the studies reviewed.[9]

The chemical substances currently added to the Convention as new POPs often have many active uses. For some of those uses, either alternatives are not currently found or cost-effective alternative products and options are not readily accessible under a certain setting, often under the conditions prevailing in developing countries. Therefore, it is not uncommon that new POPs added to the Convention consist a relatively longer list of use exemptions. The latest addition to POP pesticides, endosulfan, which is listed in Annex A, has a number of uses provided under its specific exemptions (see Table 2). Eliminating exposure to POPs is more challenging when the chemicals are included in Annex B of the Convention with many acceptable purposes, where there is no time-bound phase out requirement.

Indiscriminate use of pesticides could result in development of pest resistance and resurgence of new pests requiring increased dependence on pest control actions. Limitations in effective deployment of non-chemical pest control interventions, especially in developing countries, lead to further reliance on chemical control options. Even if the alternative pesticides are not highly toxic or do not possess POPs' characteristics, the increase of chemical load on the environment is inevitable, leading to undesirable consequences. Under certain settings where locally appropriate, cost-effective, and safer alternatives are not accessible, managing pest resistance has become a serious challenge demanding possible reintroduction or continued use of POP pesticides.

ALTERNATIVE APPROACHES

Alternatives to POP pesticides include chemical and nonchemical products as well as control interventions that focus on avoiding pest interference or creating conditions not favorable for the prevalence of the targeted pest. Often, any one of those potential alternatives would not fulfill all desirable features of the POP pesticides as a successful replacement. It requires formulating approaches with combination of viable options appropriate for the given situation. Availability of a wider range of choices from different pest control options is vital for the development of such alternative approaches. New developments in organic chemistry such as increased flexibility to modify known functional insecticide chemical backbones to produce new toxophores have expanded the prospects for new chemical alternatives. Current research on non-chemical options ranges from conventional biopesticide products to gene technology and interventions on physical ecosystem management. These initiatives should present promising opportunities for a wider selection of vector control options for efficient integration.

Major issues related to elimination of pesticides in the present list of POPs particularly concern the use of DDT for disease vector control. There are only three classes of pesticides currently available for public health vector con-

Table 2 Specific exemptions of crop–pest complexes of endosulfan available for parties under the Stockholm Convention.

Crop	Pest
Apple	Aphids
Arhar, gram	Aphids, caterpillars, pea semilooper, pod borer
Bean, cowpea	Aphids, leaf miner, whiteflies
Chilli, onion, potato	Aphids, jassids
Coffee	Berry borer, stem borers
Cotton	Aphids, cotton bollworm, jassids, leaf rollers, pink bollworm, thrips, whiteflies
Eggplant, okra	Aphids, diamondback moth, jassids, shoot, and fruit borer
Groundnut	Aphids
Jute	Bihar hairy caterpillar, yellow mite
Maize	Aphids, pink borer, stem borers
Mango	Fruit flies, hoppers
Mustard	Aphids, gall midges
Rice	Gall midges, rice hispa, stem borers, white jassid
Tea	Aphids, caterpillars, flushworm, mealybugs, scale insects, smaller green leafhopper, tea geometrid, tea mosquito bug, thrips
Tobacco	Aphids, oriental tobacco budworm
Tomato	Aphids, diamondback moth, jassids, leaf miner, shoot and fruit borer, whiteflies
Wheat	Aphids, pink borer, termites

trol as alternatives to DDT: synthetic pyrethroids, organophosphates, and carbamates. They represent two different modes of actions limiting the choice of pesticides available for the management of vector resistance. The situation demands urgent need for bringing new cost-effective public health pesticides to disease-endemic countries. All current public health pesticide active ingredients were initially developed for the agrochemical market. In the 1980s, the shift of the agrochemical target product profile, from broad-spectrum contact insecticides to stomach poisons, delivered a number of new agrochemicals that could be repurposed for public health. In efforts to eliminate DDT, governments should seek alternative approaches that are sustainable in situations prevailing in the country. It will be important to ensure that these POP pesticides are not simply replaced by other pesticides, but that the principles of integrated pest and vector management are adopted with due consideration on resistance management. Intergovernmental agencies such as the United Nations Environment Programme are promoting initiatives to demonstrate sustainable replacement of DDT in disease vector control using integrated multidisciplinary approach. It is also important to ensure that the strategies used will not be compromised by measures in other sectors. Efforts for effective management of resistance in disease vectors are hindered where the same insecticides are used in agriculture. Similarly, unplanned environmental modifications for developmental purposes could create more breeding grounds for malaria mosquitoes.

Assurance of close collaboration between sectors and key stakeholders is vital in endeavors by countries to find more sustainable solutions to POPs. These collaborations also help strengthen the base of the civil society in communities to increase other social benefits. Community participation, multisectoral initiatives on public awareness campaigns, and local social surveillance have been successfully integrated in a program implemented in Mexico and Central America on demonstrating the effectiveness of alternative methods to DDT for malaria control.[10] Such solutions must be based on the local conditions and can be best sustained through active community participation. Structures established under one sector such as Farmer Field Schools in agriculture may, for example, serve the purposes of public health and the environment. The interrelationship between the environment, agriculture, and health is, hence, a key for identifying sustainable strategies that will effectively and efficiently protect agriculture from pests, communities from diseases like malaria, and ecosystems from persistent pesticides.

GLOBAL ALLIANCE FOR ALTERNATIVES TO DDT FOR DISEASE VECTOR CONTROL

The fourth Conference of the Parties of the Stockholm Convention concluded that countries currently using DDT for disease vector control may need to continue such use until locally appropriate and cost-effective alternatives are available for a sustainable transition away from DDT. To support the countries still using DDT to reduce their reliance, it also endorsed the establishment of the Global Alliance for the development and deployment of products, methods, and strategies as alternatives to DDT for disease vector control.

The Global Alliance for alternatives to DDT provides an instrument for partnerships and collaboration among all stakeholders at the global and national level to increase momentum on achieving the common goals and to catalyze new initiatives for the development and deployment of alternatives towards the elimination of reliance on DDT. The work of the Global Alliance is organized in a manner that respects its non-involvement in funding and executing programs on the ground, yet addresses expectations that it will trigger significant actions in support of the development and deployment of alternatives to DDT.

The instrument is expected to stimulate the research community and chemical industry to accelerate the release of safer chemical alternatives to DDT. It draws global development initiatives to strengthen in-country capacity and knowledge base for efficient integration of vector control options into cost-effective and sustainable programs in disease-endemic countries. It involves global authorities and experts to review and develop support tools for countries to implement related activities and to support efficient and effective networks and communication to promote indigenous knowledge and innovative concepts on non-chemical approaches.

The platform has been established with partners from parties to the Convention with due consideration on malaria-disease-endemic countries and World Health Organization, including research and academic institutions, the donor community, civil society organizations, and the pesticide industry.[11]

CONCLUSION

The purpose of pesticides as a tool to control and, including in many cases, to kill living organisms presents the greatest drawback for its own existence and use. While the adverse effects of pesticides are many and diverse, some pesticides pose unique risks. Irrespective of the point of release, POP pesticides exert serious adverse health and environmental effects at the global scale, leaving little or no options except avoiding the reliance on them towards total elimination.

In spite of constant efforts, challenges for the development and deployment of sustainable solutions to avert continued reliance on some POP pesticides remain. The socioeconomic and ecological dimensions in specific settings may impede bringing in straightforward global solutions without compromising the benefits associated with certain uses of POP pesticides. The conventional approach of finding a chemical replacement with same pesticidal properties of the POP pesticide is not always viable. When long residual effects on the control of targeted pest, a property link to POPs' characteristics, is among the reasons for continued need for a POP pesticide, finding a chemical replacement with similar properties becomes even more challenging. The solution should therefore respect a multidisciplinary approach targeting interventions on a broader scope encompassing the life cycle of the pest consisting a series of control options. Any individual element of such multidisciplinary control approach shall not produce a complete control over the targeted pest. Implementation of properly formulated complementary control interventions offers a successful and sustainable outcome. Such approach requires collaboration of different sectors of the society in implementing respective control actions with proper coordination in a strategic framework. It also includes action by stakeholders at global, regional, and local levels to ensure enhanced sustainability of the initiatives. Benefits to the global community on such integrated approach are not limited to the protection of human health and the environment from adverse effects of POP pesticides. It also helps to achieve enhanced coordination and collaboration within the civil society for sustainable development especially under resource-limited settings.

REFERENCES

1. United Nations Environment Programme. Stockholm Convention on Persistent Organic Pollutants (POPs), 2009.
2. Persistent Organic Pollutant; United Nations Environmental Programme, available at http://www.chem.unep.ch/pops (accessed August 2011).
3. United Nations Economic Commission for Europe: Environment, available at http://www.unece.org/leginstr/env.html (accessed November 2011).
4. About the Convention. Stockholm Convention—Protecting Human Health and the Environment from Persistent Organic Pollutants, available at http://chm.pops.int/Convention/tabid/54/Default.aspx (accessed October 2011).
5. Basel Convention on the Control of Transboundary Movements of Hazardous Wastes and their Disposal, available at http://archive.basel.int/ (accessed March 2011).
6. Rotterdam Convention: Shared Responsibility, available at http://www.pic.int/ (accessed November 2011).
7. Global Programme of Action for the Protection of the Marine Environment for Land-based Activities (GPA); United Nations Environment Programme, available at http://www.gpa.unep.org/ (accessed November 2011).
8. Mörner, J.; Bos, R.; Fredrix, M. *Reducing and/or Eliminating Persistent Organic Pesticides—Guidance on Strategies for Sustainable Pest and Vector Management*; United Nations Environment Programme: Geneva, 2002.
9. World Health Organization, *DDT in Indoor Residual Spraying: Human Health Aspects, Environmental Health Criteria*; WHO Press, 2011; 241.
10. Regional Program of Action and Demonstration of Sustainable Alternative to DDT for Malaria Vector Control in Mexico and Central America (project DDT/UNEP/GEF/PAHP), available at http://www.paho.org/english/ad/sde/ddt-regionals.htm (accessed October 2011).
11. Global Alliance for Alternatives to DDT. Stockholm Convention: Protecting Human Health and the Environment from Persistent Organic Pollutants, available at http://www.pops.int/ga (accessed March 2011).

Pest Management

E.F. Legner
Department of Entomology, University of California—Riverside, Riverside, California, U.S.A.

Abstract
Improvements in the successful pest management of the agricultural ecosystem and public health sectors call for an overhaul of current procedures. The availability of specialist personnel to encourage effective management measures backed by technical research is indispensable. Without surveillance, management tends to descend to environmentally ineffective or harmful practices, and scheduled routines that do not respond to periodic environmental changes are counterproductive to sound management. Inadequacies of current practices in several examples illustrate the need for research institutions to augment their participation in management and a return to research funding by unbiased sources.

Pest management is a broad concept that involves considerations of genetics, climate, ecology, natural enemies, and cultural or chemical applications. Therefore, it is difficult to define this category exactly. A high level of sophistication is required to manage events in the environment for the efficient production of food and fiber and the abatement of public health and nuisance pests. A principal objective to the addition of sound environmental management is the reduction of pesticide usage albeit at the irritation of large commercial interests.[1,2]

Although scientific investigations in colleges and universities have led to a high level of production and pest abatement, deployment continues to face obstacles that are largely related to the absence of competent supervisory personnel. As expertise resides largely in the research community, this group is encumbered by an academic system that continues to stress research and teaching and to minimize the deployment aspect. The most successful programs in environmental management regularly require 5 or more years to develop. Investigator survival in the system demands frequent publication but not in the kind of journals that stress implementation. This distracts from the ultimate goal of deployment, which diminishes the amount of time an investigator has to be directly involved in an advisory capacity. Several examples of successful projects that have receded in the absence of this supervision but that could be reactivated with the proper advisory personnel present will explain some of the problems and difficulties involved.

NAVEL ORANGEWORM MANAGEMENT IN ALMOND ORCHARDS

The almond industry in California has suffered from the invasion of the navel orangeworm, *Amyelois transitella* (Walker), from Mexico and South America. Two external insect larval parasites, *Goniozus legneri* Gordh and *Goniozus emigratus* (Rohwer), and one internal egg-larval parasite, *Copidosomopsis plethorica* Caltagirone, which are dominant in south Texas, Mexico, Uruguay, and Argentina, were successfully established in irrigated and non-irrigated almond orchards in California.[3–4] Separate k-value analyses indicated significant regulation of their navel orangeworm host during the warm summer season. There is a diapause (hibernation) in the host triggered by several seasonally varying factors and a diapause in the parasites triggered by hormonal changes in the host. Possible latitudinal effects on diapause also are present. The ability of the imported parasites to diapause with their host enables their permanent establishment and ability to reduce host population densities to below economic levels.[5]

Although navel orangeworm infestations have decreased with the establishment of the three parasites,[6] the almond reject levels are not always below the economic threshold of 4%. Such rejects are sometimes due to other causes, such as ant damage and fungus infections. In certain years, the peach tree borer, *Synanthedon exitiosa* (Say), has been involved as its attacks stimulate oviposition by navel orangeworm moths and subsequent damage attributed to the latter.

In some orchards, the growers have sustained a reject level of 2.5% or less through 2008. Storing rejected almond mummies in ventilated sheds through winter allows for a buildup of natural enemies and their subsequent early entry into the fields to reduce orangeworm populations before the latter have an opportunity to increase. Commercial insectaries have harvested *G. legneri* from orchards for introductions elsewhere. *C. plethorica* and *G. legneri*, and to a lesser extent, *G. emigratus*, successfully overwinter in orchards year after year. However, only *Copidosomopsis* can consistently be recovered at all times of the year. The *Goniozus* species are not recovered in significant numbers until early summer. Therefore, pest management in almond orchards may require periodic releases of *G. legneri* to reestablish balances that were disrupted by insecticidal drift or by the absence of overwintering rejected almond refuges through aggressive sanitation practices. Although sanitation in this case may appeal to the grower, it is a costly procedure that also disrupts natural balances at low pest densities.

G. legneri has been reared from codling moth and oriental fruit moth in peaches in addition to navel orangeworm from almonds. A reservoir of residual almonds that remain in the trees after harvest is desirable to maintain a synchrony of these parasites with navel orangeworms in order to achieve the lowest pest densities. In fact, such reservoirs often exceed 1000 residual almonds per tree through the winter months and produce navel orangeworm densities at harvest that are below 1% on soft-shelled varieties. Superimposed upon the system is the diapausing mechanism in both the navel orangeworm and the parasites.[5] All of these forces must be considered for a sound, reliable integrated management. Almond producers have to make reasonable decisions on whether or not to remove residual almonds, a very costly procedure, or to use within-season insecticidal sprays. However, orchard managers rarely understand population stability through the interaction of natural enemies and their prey. Because the management of this pest with parasitic insects depends heavily on the perpetuation of parasites in orchards, it can be accomplished only by an understanding of the dynamics involved. Storing rejected almonds in protective shelters during winter months increases parasite abundance. This allows the parasites to reproduce in large numbers for subsequent spread throughout an orchard in the spring when outdoor temperatures rise. Complete sanitation of an orchard by removal of all rejected almonds is counterproductive to successful management as this also eliminates natural enemies.

AUSTRALIAN BUSH FLY MANAGEMENT IN MICRONESIA

Pestiferous flies in the Marshall Islands provide a classic example of the adaptation of invading noxious insects to an area with a salubrious climate. With nearly perfect temperature–humidity conditions for their development, an abundance of carbohydrate and protein-rich food in the form of organic wastes and excreta provided by humans and their animals, and a general absence of effective natural enemies, several species were able to reach maximum numbers.

There are principally four types of pestiferous flies in Kwajalein Atoll of the Marshall Islands, with the African–Australian bush fly, *Musca sorbens* Wiedemann, being by far the most pestiferous species. The common housefly, *Musca domestica* L., of lesser importance, frequents houses and is attracted to food in recreation areas. The remaining two types are the Calliphoridae (*Chrysomya megacephala* [Fab.] and [Wiedemann]) and the Sarcophagidae (*Parasarcophaga misera* [Walker] and *Phytosarcophaga gressitti* Hall and Bohart). These latter species are abundant around refuse disposal sites and wherever rotting meat and decaying fish are available. Most of the fly species differ from the common housefly and the bush fly in being more sluggish and noisy and by their general avoidance of humans. Because residents do not distinguish the different kinds of flies, non-pestiferous types are often blamed as nuisances when in fact they may be considered to fulfill a useful role in the biodegradation of refuse and rotting meat.

An initial assessment of the problem led to the expedient implementation of breeding source reduction to reduce the population of the housefly, *M. domestica* L., and both the Calliphoridae and Sarcophagidae to inconspicuous levels. These involved slight modifications of refuse disposal sites to disfavor fly breeding. These simple measures resulted in an estimated one-third reduction of total population of flies concentrating around beaches and residential areas. Because the housefly especially enters dwellings, the reduction in its numbers was desirable for the general health of the community, and fly annoyances indoors diminished. Thorough surveys of breeding sites and natural enemy complexes revealed that *M. sorbens* population reduction would not be quickly forthcoming, however. A schedule of importation of natural enemies was begun, and other integrated management approaches were investigated, e.g., baiting and breeding habitat reduction.

Bush Fly Origin and Habits

This species is known as the bazaar fly in North Africa, a housefly in India, and the bush fly in Australia.[7] It was first described from Sierra Leone in West Africa in 1830, where it is a notorious nuisance to humans and animals. The flies are attracted to wounds, sores, and skin lesions, searching for any possible food sources such as blood and other exudations. Although not a biting species, its habits of transmitting eye diseases, enteric infections, pathogenic bacteria, and helminth eggs make it a most important and dangerous public health insect.[8–11]

The bush fly has spread through a major portion of the Old World, Africa, and parts of Asia.[12] In Oceania, its distribution is in Australia,[13] New Guinea,[13] Samoa and Guam,[14] and the Marshall Islands.[15] In Hawaii, Joyce[16] first reported it in 1950. Later, Hardy[17] listed it in the *Catalog of Hawaiian Diptera*, and Wilton[18] reported its predilection for dog excrement. The importance of the bush fly increased in the 1960s, when it was incriminated as a potential vector of beta-hemolytic streptococci in an epidemic of acute glomerulonephritis.[8]

On the islands of Kwajalein Atoll, a substantial portion of the main density of *M. sorbens* emanated from dog, pig, and human feces. Inspections of pig droppings in the bush of 10 widely separated islets revealed high numbers of larvae (over 100 per dropping), making this dung, as in Guam,[15] a primary breeding source in the atoll. Pigs that are corralled on soil or concrete slabs concentrate and trample their droppings, making them less suitable breeding sites. In such situations, flies were able to complete their development only along the periphery of corrals. Coconut husks placed under pigs in corrals results in the production of greater numbers of flies by reducing the effectiveness of trampling. Kitchen and other organic wastes were not found to breed *M. sorbens*, although a very low percentage of the adult population could originate there judging from reports elsewhere. Nevertheless, this medium is certainly not responsible for producing a significant percentage of the adult densities observed in the Atoll.

Management Efforts Worldwide

Successful partial reduction of the bush fly population had been achieved only in Hawaii, through a combination of the elimination of breeding sites, principally dog droppings, and the activities of parasitic and predatory insects introduced earlier to combat other fly species, e.g., *M. domestica*.[19] The density of the bush fly population varies in different climatic zones in Hawaii, but the importance of this fly is minimal compared with Kwajalein. At times, hymenopterous parasites have been found to parasitize over 95% of flies sampled in the Waikiki area (H.S. Yu, unpublished data). Other parts of Oceania (e.g., Australia) either were not suitable for maximum effectiveness of known parasitic species or had principle breeding habitats that were not attractive to the natural enemies. Therefore, in Australia, a concerted effort has been made to secure scavenger and predatory insects from southern Africa that are effective in managing fly populations on range cattle and sheep dung, the principal fly-producing source.[20]

Kwajalein Atoll

Integrated fly management had reached a level of partial success by 1974. Initial surveys for natural enemies of *M. sorbens* revealed the presence of four scavenger and predatory insects: the histerid *Carcinops troglodytes* Erichson, the nitidulid *Carpophilus pilosellus* Motschulsky, the tenebrionid *Alphitobius diaperinus* (Panzer), and the dermapteran *Labidura riparia* (Pallas). Dog numbers were significantly reduced, and all privies were reconstructed or improved on one island, Ebeye. Dogs were reduced or tethered on Kwajalein Island and refuse fish, etc., disposed of thoroughly on Illeginni and other islands with American residents. Importations of natural enemies were made throughout the atoll, and the average population density of *M. sorbens* on Ebeye was subsequently reduced from an estimated 8.5 flies attracted to the face per minute to less than 0.5 fly per minute, which was readily appreciated by the inhabitants. The single most important cause appeared to be the partial elimination of breeding sources, with natural enemies playing a secondary role.

For the further reduction of bush fly numbers, the integration of a nondestructive insecticidal reduction measure was desirable. Sugar bait mixtures that have been used for houseflies in years prior to 1972 were wholly ineffective for killing adult *M. sorbens* due to their almost complete lack of attractiveness. However, a variety of decomposing foodstuffs including rotting eggs and rotting fish sauces were very highly attractive. Experiments using a 6-day-old mixture of one part fresh whole eggs to one part water[21] attracted over 50,000 bush flies that were then killed by a 0.5 ppm Dichlorvos additive. The poisoned mixture was poured in quantities of 100 mL each in flat plastic trays with damp sand at 20 sites in the shade and spaced every 10 m along a public beach on Kwajalein. Baits placed above the height of 1 m or against walls in open pavilions were only weakly attractive. After 48 hours, flies were reduced to inconspicuous levels all over Kwajalein Island. This condition endured for at least 3 days, after which newly emerging and immigrating flies managed to slowly increase to annoying levels as the baits ceased to be attractive. However, the former density of flies was never reached even one week after the baiting; these populations were subsequently reduced to even lower levels by applying additional fresh poisoned baits.

Baiting was extended to other islands in the atoll with the result of sustained reductions of bush fly numbers to below general annoyance levels (less than 0.01 attracted per minute on Kwajalein, Roi-Namur, Illeginni, and Meck Islands). A new attractant that augmented the rotting egg mixture consisted of beach sand soaked for one week in the decomposing body fluids of buried sharks. This new attractant was far superior to rotting eggs in both rate and time of attraction, the latter sometimes exceeding 5 days. The baiting method could be used effectively if applied initially twice a week, and only biweekly applications were necessary in the following months.

After January 2000, in the absence of specialist supervision, the baiting procedure in the atoll has not continued with the sophistication initially determined necessary. In the absence of supervision, the number of flies was not adequately reduced. Periodic personnel changes precluded the passing on of accurate information critical to managing the fly densities. Of vital importance is habitat reduction, the proper preparation of baits, and the latter's placement in shaded wind-calm areas of the islands. Because such sites are generally out of sight of the public, baiting has instead shifted to populated areas, where only very conspicuous but non-pestiferous species of flies are attracted to the baits in large numbers. Sometimes, even ammonia baits were substituted that attract harmless blow fly species but not the targeted bush fly.

AQUATIC WEED MANAGEMENT BY FISH IN IRRIGATION SYSTEMS

Imported fish species have been used for clearing aquatic vegetation from waterways, which has also reduced mosquito and chironomid midge abundance. In the irrigation systems, storm drainage channels, and recreational lakes of southern California, the California Department of Fish and Game authorized the introduction of three species of African cichlids: *Tilapia zillii* (Gervais), *Oreochromis (Sarotherodon) mossambica* (Peters), and *Oreochromis (Sarotherodon) hornorum* (Trewazas). These became established over some 2000 ha of waterways.[22] Their establishment reduced the biomass of emergent aquatic vegetation that was slowing down the distribution of irrigation water but that also provided a habitat for such encephalitis vectors as the mosquito *Culex tarsalis* Coquillet. Previous aquatic weed reduction practices had required an expensive physical removal of vegetation and/or the frequent application of herbicides.

One species, *T. zillii* can reduce mosquito populations by a combination of direct predation and the consumption of aquatic plants by these omnivorous fishes.[23–25] As Legner and Sjogren[22] indicated, this is a unique example of persistent biological suppression and probably applicable only for relatively stable irrigation systems where a permanent water supply is assured and where water temperatures are warm enough in winter to sustain the fish.[26] A threefold advantage in the use of these fish is as follows: 1) clearing of vegetation to keep waterways open; 2) mosquito abatement; and 3) a fish large enough to be used for human consumption. However, optimum management of these cichlids for aquatic weed reduction often is not understood by irrigation district personnel,[27–28,19] with the result that competitive displacement by inferior cichlids minimizes or eliminates *T. zillii*, the most efficient weed-eating species.[29]

The three imported fish species varied in their influence in different parts of the irrigation system. Each fish species possessed certain attributes for combating the respective target pests.[30,31] *T. zillii* was best able to perform as both a habitat reducer and an insect predator. It also had a slightly greater tolerance to low water temperatures, which guaranteed the survival of large populations through the winter months; at the same time, it did not pose a threat to salmon and other game fisheries in the colder waters of central California. It was the superior game species and most desirable as human food. Nevertheless, the agencies supporting the research (mosquito abatement and county irrigation districts) acquired and distributed all three species simultaneously throughout hundreds of kilometers of the irrigation system, storm drainage channels, and recreational lakes. The outcome was the permanent and semipermanent establishment of the two less desirable species, *S. mossambica* and *S. hornorum*, over a broader portion of the distribution range. This was achieved by the competitively advantaged *Sarotherodon* species that mouth-brood their fry, while *T. zillii* did not have this attribute strongly developed. It serves as an example of competitive exclusion such as conjectured by Ehler.[32] In the clear waters of some lakes in coastal and southwestern California, the intense predatory behavior of *S. mossambica* males on the fry of *T. zillii* could be easily observed, even though adults of the latter species gave a strong effort to fend off these attacks.

This outcome was not too serious for chironomid reduction in storm drainage channels because the *Sarotherodon* species are quite capable of permanently suppressing chironomid densities to below annoyance levels.[23,26] However, for the management of aquatic weeds, namely, *Potamogeton pectinatus* L., *Myriophyllum spicatum* var. *exalbescens* (Fernald) Jepson, *Hydrilla verticillata* Royle, and *Typha* species, they showed little capability.[31] Thus, competition excluded *T. zillii* from expressing its maximum potential in the irrigation channels of the lower Sonoran Desert and in the recreational lakes of southwestern California. Furthermore, as the *Sarotherodon* species were of a more tropical nature, their populations were reduced in the colder waters of the irrigation canals and recreational lakes. Although *T. zillii* populations could have been restocked, attention was later focused on a potentially more environmentally destructive species, the white amur, *Ctenopharyngodon idella* (Valenciennes), and other carps. The competitively advantaged *Sarotherodon* species are permanently established over a broad geographic area, which encumbers the reestablishment of *T. zillii* in storm drainage channels of southwestern California.

MANAGEMENT OF FILTH FLY ABUNDANCE IN DAIRIES AND POULTRY HOUSES

The most important of muscoid fly species are broadly defined as those most closely associated with human activities. Breeding habitats vary from the organic wastes of urban and rural settlements to those provided by various agricultural practices, particularly ones related to the management and care of domestic animals. Their degree of relationship to humans varies considerably with the ecology and behavior of the fly species involved. Some are more often found inside dwellings.

Research to reduce fly abundance has centered on the highly destructive parasitic and predatory species, such as the encyrtid *Tachinaephagus zealandicus* Ashmead, five species of the pteromalid genus *Muscidifurax*, and *Spalangia* species that destroy dipterous larvae and pupae in various breeding sources. The natural enemies are capable of successful fly suppression if the correct species and strains are applied in the right locality.[24,33–36] Other approaches have included the use of pathogens and predatory mites and inundative releases of parasites and predators.[37,38] Although partially successful, none of these strategies has become the sole method for fly abatement, and the choice of a ineffective parasite strain may have detrimental results.[19] Instead, the focus is on integrated management including habitat reduction, adult baiting, and aerosol treatments with short residual insecticides. Also, it is generally agreed that existing predatory complexes exert great influences on fly densities[39] and that many natural enemies of these flies have a potential to significantly reduce their abundance if managed properly.[29,40] Because climatic and locality differences dictate which abatement strategies are effective, simple instructions to the public are impossible, and the involvement of skilled personnel is required. Of primary importance for successful management is the provision of relatively stable breeding habitats and their natural enemy complexes. Periodic cleaning operations should stress the partial removal of breeding sites and the deposition of such waste into large stacks, which favors the generation of destructive heat while minimizing the area and attractiveness for fly oviposition. Nevertheless, this management procedure is difficult for abatement personnel to grasp in the absence of competent supervision.

REFERENCES

1. Garcia, R.; Legner, E.F. Biological control of medical and veterinary pests. In *Handbook of Biological Control: Principles and Applications*; Fisher, T.W., Bellows, T.S., Jr. Eds.; Academic Press: San Diego, CA, 1999; 935–953.
2. Pimentel, D.; McLaughlin, L.; Zepp, A.; Lakitan, B.; Kraus, T.; Kleinman, P.; Vancini, F.; Roach, W.J.; Graap, E.; Keeton, W.S.; Selig, G. Environmental and economic impacts of reducing U.S. agricultural pesticide use. In *Handbook of Pest Management in Agriculture*, 2nd Ed.; Pimentel, D., Ed.; CRC Press: Boca Raton, Florida, 1991; Vol. I., 679–718.
3. Caltagirone, L.E. A new *Pentalitomastix* from Mexico. Pan-Pac. Entomol. **1966**, *42*, 145–151.
4. Legner, E.F.; Silveira-Guido, A. Establishment of *Goniozus emigratus* and *Goniozus legneri* [Hym: Bethylidae] on navel orangeworm, *Amyelois transitella* [Lep: Phycitidae], in California and biological control potential. Entomophaga **1983**, *28*, 97–106.
5. Legner, E.F. Patterns of field diapause in the navel orangeworm (Lepidoptera: Phycitidae) and three imported parasites. Ann. Entomol. Soc. Am. **1983**, *76*, 503–506.
6. Legner, E.F.; Gordh, G. Lower navel orangeworm (Lepidoptera: Phycitidae) population densities following establishment of *Goniozus legneri* (Hymenoptera: Bethylidae) in California. J. Econ. Entomol. **1992**, *85* (6), 2153–2160.
7. Yu, H. The biology and public health significance of *Musca sorbens* Wied. in Hawaii, M.S. Thesis, University of Hawaii, 1971; 72 pp.
8. Bell, T.D. Epidemic glomerulonephritis in Hawaii [mimeograph]. Rep. Pediat. Serv., Dep. Med., Tripler Army Hospital: Honolulu, Hawaii, 1969; 25 pp.
9. Greenberg, B. *Flies and Disease*; Ecology; Classification and Biotic Associations; Princeton University Press: Princeton, NJ, 1971; Vol. I, 856 pp.
10. Hafez, M.; Attia, M.A. Studies on the ecology of *Musca sorbens* Wied. in Egypt. Bull. Soc. Entomol. Egypt **1958**, *42*, 83–121.
11. McGuire, C.D.; Durant, R.C. The role of flies in the transmission of eye disease in Egypt. Am. J. Trop. Med. Hyg. **1957**, *6*, 569–75.
12. Van Emden, F.I. The fauna of India and the adjacent countries. *Diptera*; Muscidae, Pt. I; Gov. Publ. India: Delhi, India, 1965; Vol. 7.
13. Patterson, H.E.; Norris, K.R. The *Musca sorbens* complex: The relative status of the Australian and two African populations. Aust. J. Zool. **1970**, *18*, 231–45.
14. Harris, A.H.; Down, H.A. Studies of the dissemination of cysts and ova of human intestinal parasites by flies in various localities on Guam. Am. J. Trop. Med. **1946**, *26*, 789–800.
15. Bohart, G.E.; Gressitt, J.L. *Filth inhabiting flies of Guam*; Bull. B.P. Bishop Museum, Honolulu 1951, *204*; 152 pp.
16. Joyce, C.R. Notes and exhibitions. Proc. Hawaii. Entomol. Soc. **1950**, *16* (3), 338.
17. Hardy, D.E. Additions and corrections to Bryan's check list of the Hawaiian Diptera. Proc. Hawaii. Entomol. Soc. **1952**, *14* (3), 443–84.
18. Wilton, D.P. Dog excrement as a factor in community fly problems. Proc. Hawaii. Entomol. Soc. **1963**, *28* (2), 311–17.
19. Legner, E.F. Diptera. Medical and veterinary pests. In *Introduced Parasites and Predators of Arthropod Pests and Weeds: A Review*; Clausen, C.P., Ed.; U.S. Department of Agriculture Technology Report, 1978; 1012–1019, 1043–1069.
20. Bornemissza, G.F. Insectary studies on the control of dung breeding flies by the activity of the dung beetle, *Onthophagus gazella* F. (Coleoptera: Scarbaeinae). J. Aust. Entomol. Soc. **1970**, *9*, 31–41.
21. Legner, E.F.; Sugerman, B.B.; Hyo-sok, Y.; Lum, H. Biological and integrated control of the bush fly, *Musca sorbens* Wiedemann, and other filth-breeding Diptera in Kwajalein Atoll, Marshall Islands. Bull. Soc. Vector Ecol. **1974**, (1), 1–14.
22. Legner, E.F.; Sjogren, R.D. Biological mosquito control furthered by advances in technology and research. J. Am. Mosq. Control Assoc. **1984**, *44* (4), 449–456.
23. Legner, E.F.; Fisher, T.W. Impact of *Tilapia zillii* (Gervais) on *Potamogeton pectinatus* L., *Myriophyllum spicatum* var. *exalbescens* Jepson, and mosquito reproduction in lower Colorado Desert irrigation canals. Acta Oecol., Oecol. Appl. **1980**, *1* (1), 3–14.
24. Legner, E.F.; Murray, C.A. Feeding rates and growth of the fish *Tilapia zillii* [Cichlidae] on *Hydrilla verticillata*, *Potamogeton pectinatus* and *Myriophyllum spicatum* var. *exalbescens* and interactions in irrigation canals in southeastern California. J. Am. Mosq. Control Assoc. **1981**, *41* (2), 241–250.
25. Legner, E.F.; Pelsue, F.W., Jr. Contemporary appraisal of the population dynamics of introduced cichlid fish in south California. Proc. Calif. Mosq. Vector Control Assoc., Inc. **1983**, *51*, 38–39.
26. Legner, E.F.; Medved, R.A.; Pelsue, F. Changes in chironomid breeding patterns in a paved river channel following adaptation of cichlids of the *Tilapia mossambica*–

hornorum complex. Ann. Entomol. Soc. Am. **1980**, *73* (1), 293–299.

27. Hauser, W.J.; Legner, E.F.; Medved, R.A.; Platt, S. *Tilapia*—A management tool for biological control of aquatic weeds and insects. Bull. Am. Fish. Soc. **1976**, *1*, 15–16.

28. Hauser, W.J.; Legner, E.F.; Robinson, F.E. Biological control of aquatic weeds by fish in irrigation channels. In Proceedings of Water Management for Irrigation and Drainage, ASC/Reno, Nevada, Jul 20–22, 1977; 139–145.

29. Legner, E.F. Biological control of aquatic Diptera; *Contributions to a Manual of Palaearctic Diptera*; Science Herald: Budapest, 2000; Vol. 1, 847–870.

30. Legner, E.F.; Medved, R.A. Influence of *Tilapia mossambica* (Peters), *T. zillii* (Gervais) (Cichlidae) and *Mollienesia latipinna* LeSueur (Poeciliidae) on pond populations of *Culex* mosquitoes and chironomid midges. J. Am. Mosq. Control Assoc. **1973**, *33*, 354–64.

31. Legner, E.F.; Medved, R.A. Predation of mosquitoes and chironomid midges in ponds by *Tilapia zillii* (Gervais) and *T. mossambica* (Peters) (Teleosteii: Cichlidae). Proc. Calif. Mosq. Control Assoc., Inc. **1973**, *41*, 119–121.

32. Ehler, L.E. Foreign exploration in California. Environ. Entomol. **1982**, *11*, 525–30.

33. Axtell, R.C.; Rutz, D.A. Role of parasites and predators as biological control agents in poultry production facilities. Misc. Publ. Entomol. Soc. Am. **1986**, *61*, 88–100.

34. Legner, E.F.; Greathead, D.J.; Moore, I. Equatorial East African predatory and scavenger arthropods in bovine excrement. Environ. Entomol. **1981**, *10*, 620–25.

35. Mandeville, J.D.; Mullens, B.A.; Meyer, J.A. Rearing and host age suitability of *Fannia canicularis* (L.) for parasitization by *Muscidifurax zaraptor* Kogan and Legner. Can. Entomol. **1988**, *120*, 153–59.

36. Pawson, B.M.; Petersen, J.J. Dispersal of *Muscidifurax zaraptor* (Hymenoptera: Pteromalidae), a filth fly parasitoid, at dairies in eastern Nebraska. Environ. Entomol. **1988**, *17*, 398–402.

37. Ripa, R. Survey and use of biological control agents on Easter Island and in Chile. In *Biological Control of Muscoid Flies*; Patterson, R.S., Rutz, D.A., Eds.; Misc. Publ. Entomol. Soc. Am. 61; 1986; 39–44.

38. Ripa, R. Biological control of muscoid flies in Easter Island. In *Biocontrol of Arthropods Affecting Livestock and Poultry*; Rutz, D.A., Patterson, R.S., Eds.; Westview Press: Boulder, CO, 1990; 111–119.

39. Geden, C.J.; Axtell, R.C. Predation by *Carcinops pumilio* (Coleoptera: Histeridae) and *Macrocheles muscaedomesticae* (Acarina: Macrochelidae) on the housefly (Diptera: Muscidae): Functional response, effects of temperature and availability of alternative prey. Environ. Entomol. **1988**, *17*, 739–44.

40. Mullens, B.A.; Meyer, J.A.; Mandeville, J.D. Seasonal and diel activity of filth fly parasites (Hymenoptera: Pteromalidae) in caged-layer poultry manure in southern California. Environ. Entomol. **1986**, *15*, 56–60.

Pest Management: Crop Diversity

Marianne Karpenstein-Machan
Institute of Crop Science, University of Kassel, Witzenhausen, Germany

Maria Finckh
Department of Ecological Plant Protection, University of Kassel, Witzenhausen, Germany

Abstract
In this entry, diversity by planting sequence (crop rotation), crop-border diversity, and crop-weed diversity are discussed.

INTRODUCTION

Improvement in crop management by the use of modern machinery equipment, fertilizers, and pesticides and progress in plant breeding within a few select crops has led in recent decades to highly specialized agricultural practices on farms. It is estimated that about 7000 crops are known worldwide; however, only seven crops are cultivated on more than % of the arable land. A general impoverishment of plant diversity and a high degree of genetic erosion in several important crops for human nutrition is documented worldwide.

Simplification of the ecosystem by using one-sided crop rotations, monoculture, and crop plants of uniform genotype and the elimination of weeds with herbicides result in a strong selection for adapted pests, pathogens, and weeds leading to frequent resistance breakdown and severe weed infestations.[1–3]

The chances that pests and weeds will develop resistance to any one of the control agents are reduced by increasing the diversity of selection pressures acting on the species. The establishment of diversity in the crop ecosystem with adapted crops in sequential cropping, multiline cultivars, and variety mixtures results in less damage by pests, pathogens, and weeds than when grown in monocultures. In a diverse crop rotation, weeds are less abundant but with a greater diversity in weed species. Outbreaks of weed, pest, and disease epidemics and the probability of losses are reduced in diverse rotations.[2–4]

In this entry, diversity by planting sequence (crop rotation), crop-border diversity, and crop-weed diversity are discussed.

INFLUENCE OF CROP ROTATION ON DISEASES AND PESTS

The positive effects of a diverse crop rotation on yield and agricultural stability have been well known since ancient times (Table 1). The risk of infestation by weeds, diseases, and pests can be decreased by a well-organized crop rotation. Crop rotation is most effective in the case of specialized pathogens and pests that are dependent on a host crop or have a narrow host range. For example, many soil-borne pathogens or insects survive on root residues and require the cultivation of a susceptible crop for continuous survival. The length of host crop interval for disease control will depend on how quickly the pest or pathogen can be destroyed by antagonistic effects.[1] When developing a crop rotation, certain criteria should be considered: maintenance of high levels of organic matter in the soil, tillage to expose the residues to weathering, and inclusion of crops which do not stimulate subsequent growth of the pathogen or crops that have direct negative effects on pests and pathogens (e.g., producing toxins).

As an example, the pigeon pea wilt fungus, *F. oxysporum* f. sp. *udum*, is specific to the pigeon pea. Many other legumes, such as cowpea and soybean, which are nonhosts, are able to stimulate the germination of this fungus. If one of these legumes is cultivated as a subsequent crop, the chlamydospores will germinate. In soils with high microbiological activity due to rich organic matter, the fungus may be destroyed by antagonists, but in the case of low biological activity the fungus may survive by forming secondary chlamydospores.[1,2] Cereals stimulate chlamydospores less compared with legumes. In addition, root exudates contain toxic components that may kill germinated chlamydospores before they can form spores again.[2,3]

INFLUENCE OF DECOY AND TRAP CROPS ON PESTS

Decoy crops are nonhost crops that are sown to stimulate the activation of dormant propagules of the pathogen in the absence of the host. In this way the soil-borne pathogens waste their inoculum potential. For example, *Lolium* spec., *Papaver rhoeas*, and *Reseda odorata* can act as decoy crops for the pathogen *Plasmodiophora brassicae* in *Brassica*.[3]

In the case of trap crops, the crop is host to the pathogen (nematode). The trap crop attracts nematodes to infect, but

Table 1 Crop diseases and pests in which crop rotation has a major effect in disease and pest control.

	Wheat	Barley	Sorghum	Maize	Rye	Oats	Crucifers	Beta-Beet	Soybean	Other legumes	Potatoes	Tobacco
Virus												
Beet necrotic yellow vein virus (BNYVV)								X				
Barley yellow mosaic virus (BYMY)		X										
Bacteria												
Scab (*Streptomyces scabies*)											X	
Fungi												
Take all and eyespot diseases of cereals (*Gaeumannomyces graminis, Pseudocercosporella herpotrichoides, Rhizoctonia cerealis*)	X	X			X							
Leaf and ear diseases of cereals (*Typhula incarnata, Rhyncho-sporium secalis, Septoria nodorum, Septoria tritici*)	X	X			X							
Fusarium diseases of cereals (*Fusarium avenaceum, F. culmorum, F. nivale*)	X	X		X	X							
Brown spot (*Septoria glycinea*)									X			
Fusarium root rot complex of legumes										X		
Stark rot (*Fusarium moniliforme*)			X									
Stem canker (*Leptosphaeria maculans*)							X			X	X	
Corn smut (*Ustilago maydis*)				X								
Cercospora leaf spot (*Cercospora beticola*)								X				
Black root rot (*Thielaviopsis basicola*)												X
Root rot (*Meloidogyne hapla*)			X									

Pest Management: Crop Diversity

Pest								
Nematodes								
Cereal cyst nematode (*Heterodera avenae*)	X	X	X^a	X				
Beet cyst nematode (*Heterodera schachtii*)					X	X		
White and golden cyst nematode (*Globodera pallida, G. rostochiensis*)								X
Stem and bulb nematode (*Ditylenchus dipsaci*)			X^b	X^b	X^b	X	X^b	
Cyst nematode (*Heterodera glycinea*)							X	
Insects								
Springtails (*Collembola* spp.)							X	
Pigmy mangold beetle (*Atomaria linearis*)							X	
Haplodiplosis marginata	X	X		X				
European corn borer (*Ostrinia nubilalis*)								X
Cabbage stem flea beetle (*Psylliodes chrysocephalus*)						X		
Swede seed midge (*Dasineura brassicae*)						X		

^{a}Infection without strong effect on crop yield.
^{b}Host specific races.

before the nematode can complete its life cycle the crop is harvested or destroyed. Therefore in Germany it is recommended to sow crucifers and plow before the beet cyst nematode can fully develop its life cycle.[5]

INFLUENCE OF CROP ROTATION ON WEED ABUNDANCE

The kind of crop rotation influences weed density and abundance of weed species. Main effects are caused by the sowing time of crops (winter or spring crops), sequence and placement of crops in the crop rotation, competitive ability of different crops against weeds, and weed management methods (herbicides and/or mechanical control).[6]

In monoculture and pure grain rotations weed infestation reaches a high level and weeds are more difficult to control due to herbicide resistance. Long-term trials in Germany have shown that in pure grain rotations and cereal monocultures the weed density was two and three times higher compared to a rotation in alternation with dicotyledonous plants, for example field beans. Especially the degree of infestation of problem weeds, e.g. *Apera spica venti*, *Viola tricolor arvensis*, and *Matricaria* spp., increases considerably.[7]

Allelopathic effects of certain crops (e.g., barley, rye, maize, sorghum, sudangrass, buckwheat, sunflowers, rape, soybeans, alfalfa, and hemp) may reduce germination of different weeds. The root exudates of barley, wheat, and rye have especially strong negative effects on the germination and development of *Stellaria media*, *Capsella bursa pastoris*. Wheat and rye hinder germination of *Anthemis arvensis* and *Matricaria perforate*.[8]

INFLUENCE OF BORDER DIVERSITY ON PESTS

The abundance and diversity of entomophagous insects within a field are closely related to the character of the surrounding vegetation. There are many examples that indicate that crops cultivated near hedgerows or uncultivated fields with flowering weeds sustain less damage by pest and disease organisms than crops cultivated in the absence of these flowering weeds. Brassicas were less damaged by the diamondback moth, *Plutella xylostella* (L.) and the cabbage aphid, *Brevicoryne brassicae* (L.) due to increased predation and parasitism caused by flowering weeds in bordering uncultivated fields. Less intensive management of hedgerows tends to increase the proportion of predacious insects.[9]

Potatoes grown near woods were less infected by the Colorado potato beetle *Leptinotarsa decemlineata* (Say), due to increased predation on the larvae.

The presence of nectar source plants in sugarcane fields in Hawaii allows for the development of higher levels of a sugarcane weevil parasitoid, *Lixophaga sphenophori*, thus increasing parasitoid efficiency. A complex vegetation in the crop borders was also the reason that the mean number of species of both herbivore and predator parasitoids per habit space in soybean fields was higher at the edge than in the center of the fields.[9]

Sometimes the neighboring vegetation can also contain host plants for diseases and crop pests.[4] An astute management of the surrounding crop fields is an option to reduce the amount of damage to crop plants.

INFLUENCE OF WEED DIVERSITY ON PESTS

Weeds not only compete with crops; they are also hosts and intermediate hosts for diseases and parasites and offer with their flowers and leaves the basic nutrition for many predators in an agricultural ecosystem. Within monocropped fields, weeds increase diversity and may be useful to improve the stability of the agricultural ecosystem. Certain weeds (mostly *Umbelliferae*, *Leguminosae*, and *Compositae*) play an important ecological role by supporting a complex of beneficial arthropods that aid in suppressing pest populations and improve the chances of the crop to escape the pest damage. Strip management with weeds or flowering crops influenced the rates of colonization of natural enemies within the fields, as an example from Switzerland shows. Weed strips sown between winter cereals increased ground beetle densities and the number of species considerably by providing these beneficial arthropods with better food supplies and more suitable overwintering sites, from which they can colonize cereals in spring. In spring their potential as pest control agents is greatest. A reduction of insect pests (aphids) due to the enhancing effect of beneficial arthropods was shown.[10,11]

In numerous insect studies with crops such as cotton, sugarcane, alfalfa, soybeans, and corn in addition to others, the reduction of population density of insect pests in weedy crops compared to weed-free fields has been verified.[2] In most cases natural enemies regulated pest populations, acting as predators and parasitoids.

FUTURE CONCERNS

Diversity provides an essential key to reduce the risk of losing crop yield to pest damage. Many field trials worldwide have shown that with an astute management of diversity improvement of agricultural stability is achievable. The realization of greater crop diversity by crop rotation, trap crops, and surrounding vegetation and high production of corn, wheat, soybean, cotton, and other crops on a large scale is feasible. The implementation of concepts based on crop diversity will preserve long-term stability and productivity of agricultural land and minimize environmental problems caused by intensive agriculture, e.g. soil erosion, groundwater and air pollution with nutrients and pesticides, and genetic erosion of both flora and fauna.[12]

See also *Intercropping for Pest Management*, pages 423–425, 100000391.

REFERENCES

1. Altieri, M.A. Biodiversity and Biocontrol: Lessons from Insect Pest Management. In *Advances in Plant Pathology*; Andrews, J.H., Tommerup, I., Eds.; Academic Press: London, 1995; 11, 191–209.
2. Baliddawa, C.W. Plant species diversity and crop pest control—an analytical review. Insect Sci. Appl. **1985**, *6* (4), 479–487.
3. Chaube, H.S.; Singh, U.S. Adjustment of Crop Culture to Minimize Disease. In *Plant Disease Management: Principles and Practices*; Chaube, H.S., Singh, U.S., Eds.; CRC Press: Boca Raton, FL, 1991; 199–214.
4. Finckh, M.R.; Wolfe, M.S. Diversification Strategies. In *The Epidemiology of Plant Diseases*; Jones, D.G., Ed.; Kluwer: Dordrecht, 1998; 231–259.
5. Mueller, J.; Steudel, W. The influence of cultivation period of different trap crops on the abundance of heterodera schachtii schmidt. Nachrichtenbl Deutscher Pflanzenschutzd **1983**, *35*, 103–108, (in German).
6. Liebmann, M.; Dyck, E. Crop rotation and intercropping strategies for weed management. Ecological Appl. **1993**, *3*, 92–122.
7. Kreuz, E. Late weed infestation in winter wheat stands in relation with intensification of the cultivation and with crop rotation. Archiv. Phytopath. Pflanz. **1993**, *28*, 379–388, (in German).
8. Narwal, S.S. Allelopathy: Future Role in Weed Control. In *Allelopathy in Agriculture and Forestry*; Narwal, S.S., Tauro, P., Eds.; Scientific Publishers: Jodhpur, 1994; 245–272.
9. Pollard, E.; Hedges, V.I. Habitat diversity and crop pests. A study of *brevicoryne brassicae* and its syrphid predators. J. Appl. Ecol. **1971**, *8*, 751–780.
10. Wyss, E.; Niggli, U.; Nentwig, W. The impact of spiders on aphid populations in a strip management apple orchard. J. Appl. Entomol. **1995**, *119*, 473–478.
11. Hausammann, A. The effects of weed strip-management on pests and beneficial arthropods on winter wheat fields. J. Plant Dis. Prot. **1996**, *103*, 70–81.
12. Karpenstein-Machan, M. *Chances of Pesticide Free Cultivation of Energy Crops for Thermal Uses*; DLG-Verlags-GmbH: Frankfurt, 1997; 183, (in German).

Pest Management: Ecological Agriculture

Barbara Dinham
Eurolink Center, Pesticide Action Network U.K., London, U.K.

Abstract
Ecological agriculture implies approaches to prevent or minimize application of chemical pesticides, which promote local inputs, and where increased farmer knowledge becomes the basis of managing pests and improving yields and sustainability. Organic agriculture is distinctive in requiring an approved body to certify that no chemical fertilizers or pesticides have been used. This provides a guarantee for consumers, and can enable farmers to receive a premium for their crops.

INTRODUCTION

Strategies to develop ecological farming methods stem from economic, health, environmental, and practical concerns with chemical pesticides. Ecological agriculture, also called sustainable agriculture,[1] agroecology, low-external input, regenerative agriculture,[2] and farmer-participatory integrated pest management (IPM), has no single definition.[3] The Food and Agriculture Organisation (FAO) of the United Nations suggests that such sustainable development (in the agriculture, forestry, and fisheries sectors) conserves land, water, plant, and animal genetic resources, is environmentally nondegrading, technically appropriate, economically viable, and socially acceptable.[4]

Ecological agriculture implies approaches to prevent or minimize application of chemical pesticides, which promote local inputs, and where increased farmer knowledge becomes the basis of managing pests and improving yields and sustainability.[5] Organic agriculture is distinctive in requiring an approved body to certify that no chemical fertilizers or pesticides have been used. This provides a guarantee for consumers, and can enable farmers to receive a premium for their crops.

PEST MANAGEMENT STRATEGIES

Ecological pest management practices have been adopted in both industrialized and developing countries. Most farming systems still use certain aspects: crop rotation, field clearance to destroy pest refuges, resistant varieties, early or late planting regimes. More specific ecological strategies employed by farmers vary widely, according to the cropping system, whether the farmer is in an industrialized or a developing country, and the locally available inputs.

As interest in ecological pest management grows, it is increasing the demand for biological technologies. The biological control agent in most widespread use, *Bacillus thuringiensis* (Bt) can be easily produced in large quantities, and can be used like a chemical. As a result it is applied in both conventional and organic agriculture in many countries. Bt has proved extremely effective in controlling persistent pests such as the diamondback moth, which plagues cabbages and related crops, and has led to a pesticide treadmill in many areas.[6]

Biological pest controls require the development of breeding centers for insect predators and parasitoids. Cuba has the most advanced program globally with a country-wide network of over 300 centers for the reproduction of entomopathogens and entomophages supplying bacteria, fungi, viruses, and insect parasitoids. These include *Lixophaga diatraege* for cane stem borer, the parasitic wasp *Trichogramma*, and the fungal disease *Beauvaria bassiana* against a total of seven pests.[7]

Particularly in developing countries, many indigenous plants and locally adapted technologies are used against a variety of pests. Most widely known is the neem tree *(Azardirachta indica)*, native to India and also found in parts of Africa, whose leaves are effective against many pests. Other common solutions are pyrethrum, chili peppers, wood ash, and castor oil seeds.[8,9]

SPREADING ECOLOGICAL PRACTICES

Ecological strategies for pest management are characterized by a holistic approach and not only the substitution of biological controls for chemical pesticides. Management strategies include soil conservation, seed selection, and maintenance of agricultural biodiversity. Participation of the women and men farmers to ensure cultural and local appropriateness is of central importance in the development of new strategies. Farmer field schools (FFS) have proved a successful training approach, where training takes place at field level, focusing on recognition of pests and predators and their life cycles. Designing in-field experiments, examining economic losses from pest damage, and encouraging observation of local plants that may act as a trap crop for pests or have repellent properties, become part of the armory.[10]

ECOLOGICAL DIVERSITY FOR DIVERSE CROPPING SYSTEMS

Fragile tropical soils may benefit most from ecological strategies, but these farming approaches are not restricted by crop, climate, or continent. Table 1 indicates a range of cropping systems and countries that have benefited from IPM programs. One of the most successful examples is in European glasshouses where 70% of commercial glasshouses have been managed through IPM based on biological controls for more than 15 years. In these intensive production systems, IPM enables growers to control all major pests and avoid most pesticide use.

Cotton is the crop that uses the most chemical insecticides worldwide. Nevertheless, cotton IPM and organic strategies have demonstrated that ecological alternatives are successful. An FFS trainer- and farmer-training program in Pakistan prevented insecticide applications in the first 8–10 weeks after planting, allowing natural enemy populations to build up, and giving higher yields in seven of the 10 demonstration plots.[11] Similar successes have been achieved in India, Zimbabwe, and elsewhere.

Adverse environmental impacts of pesticide use in conventional cotton systems and the problems of insect resistance to pesticides have encouraged farmers to invest in organic systems. While still a small proportion of overall cotton production, the growing consumer support for ecological fiber is likely to further encourage producers. The United States is the largest single producer with 32% of the total certified organic cotton fiber production in 1997, followed by Turkey with 22%. Organic cotton production is well suited to small-scale cropping systems and in the same year 15% of certified organic cotton fiber came from India, 19% from Africa, and 11% from Latin America (mainly Peru).[12]

In Indonesia IPM strategies based on improving farmers' knowledge of ecological pest management were highly successful against infestations of brown rice-hopper that decimated the rice crop in the mid-1980s. An IPM rice program supported by the FAO in South and South East Asia that began in 1980 has targeted farmers using high chemical inputs. More than 500,000 farmers have been trained and now save on average $10 per hectare per season, while maintaining or increasing yields.[13]

Farmers develop their own strategic improvements. In an area of low rainfall and high soil erosion in Burkina Faso, local groups and villagers worked with the government and local organizations to develop an ecological approach ranging from tree planting to increased use of manure. Covering more than 200 villages, farmers increased sorghum yields from 870 kg/ha to 1650–2000 kg/ha. In the semi-arid Machakos region of Kenya farmers developed appropriate agricultural techniques, building terracing, selective animal grazing, and manure collection, significantly increasing soil fertility and yields.[14]

The major physical constraints to adoption of ecological pest management relate to agricultural production systems rather than crops, regions, and climate. Large-scale monoculture, for example, does not easily lend itself to ecological approaches because of dependence on single varieties, loss of natural soil fertility, and emergence of specialized pests that kill natural enemies. The most sustained improvements can be found when government policies support ecological practices.

Table 1 Impact of selected IPM programs on pesticide use, crop yields, and annual savings.

Country and crop	Average change in pesticide use (as % of conventional treatments)	Changes in yields (as % of conventional treatments)	Annual savings of program $1000
Togo, cotton	50	90–108	11–13
Burkina Faso, rice	50	103	No data
Thailand, rice	50	No data	5–10,000
Philippines, rice	62	110	5–10,000
Indonesia, rice	34–42	105	50–100,000
Nicaragua, maize	25	93[a]	No data
United States, nine commodities	No. of applications up, volumes applied down	110–130	578,000
Bangladesh, rice	0–25	113–124	No data
India, groundnuts	0	100	34
China, rice	46–80	110	400
Vietnam, rice	57	107	54
India, rice	33	108	790
Sri Lanka, rice	26	135	1,000

[a]Lower yields, but higher net returns.
Source: Pretty.[2]

ENVIRONMENTAL AND ECONOMIC BENEFITS

Because of the many starting points, farmers have different motivations for adopting ecological approaches. In industrialized countries, where farmers generally rely heavily on external inputs, ecological practices may lower yields, but improve the environment.

In developing countries, some 2.3–2.6 billion people are supported by agriculture using the higher inputs of green revolution technologies and in these areas farmers would generally stabilize or achieve slightly higher yields with ecological pest management, also gaining environmental benefits. The remaining 1.9–2.2 billion people are largely fed by traditional agriculture and farmers. Here, ecological management strategies can substantially improve yields and income.[15]

The environmental benefits of ecological agriculture stem from reductions in chemical inputs, which threaten biodiversity, pollute water sources, and kill fish and other nontarget beneficials, often including cattle or domestic animals. Some entomologists believe that a significant proportion of the most serious insect pest problems have been introduced or worsened as pesticide use eliminates local natural enemies.

Economically, dependence on pesticides and poor management strategies can have devastating impacts, and the health and environmental costs of pesticide use are rarely calculated. In India cotton production accounts for more than 50% of pesticides used and poor application practices have resulted in insect pest resistance to chemicals. Farmers lacked the know-how to develop alternatives, and became deeply indebted to money lenders and pesticide dealers. The FAO has estimated that IPM could reduce pesticide use in Asian rice crops by 50% or more, without compromising yields, and maintain or improve net returns to the farmer—savings could amount to $1 billion.[16]

In areas of fragile and problem soil, intensive agriculture can lead to desertification, while resource-conserving systems in arid and semi-arid regions can deliver sustainable and often increasing yields. Farmers in these regions can rarely afford the external inputs and small improvements in yield would have profound impacts on the lives of the largely poorer populations that these agricultural systems support.

RISKS IN ADOPTING ECOLOGICAL STRATEGIES

The risks involved in adopting ecological pest management strategies are lower for small-scale farmers who are not yet using chemical inputs, and for farmers on a pesticide treadmill caused by high dependence and pest resistance to chemical inputs. In these instances the main risks are from short-term projects imposed without involvement of women and men in the farming communities who may swap one kind of dependency for another—the goodwill of donors. Farmers need knowledge to transfer to more ecological pest management strategies, and even successful IPM training can be limited by lack of funds for follow up.

Farming systems in industrialized countries that are effectively managing pesticide usage are likely to lose yields when adopting ecological strategies, particularly during the transitional stage. These farmers are also likely to face higher labor costs, though offset to some extent by lower input costs. The benefits are longer term and less tangible: improvements to the environment and to health, and in maintaining yields over time. For farmers who adopt organic practices and seek certification, the initial lower yields can be offset by a premium for their crops.

REFERENCES

1. In *Promoting Sustainable Agriculture and Rural Development, Agenda 21* Ch. 14, UNCED United Nations Conference on Environment and Development, Rio de Janiero, 1992. http://www.igc.apc.org/habitat/agenda21/ (accessed January 2001).
2. Pretty, J. Local Groups and Institutions for Sustainable Agriculture. In *Regenerating Agriculture: Policies and Practice for Sustainability and Self-Reliance*; Earthscan Publications: London, 1995; 131–162.
3. UNDP. In *Benefits of Diversity: An Incentive Toward Sustainable Agriculture*; UNDP: New York; 1992.
4. In *The den Bosch Declaration and Agenda for Action on Sustainable Agriculture and Rural Development*, Report of the Conference, FAO/Netherlands Conference on Agriculture and the Environment, FAO and Ministry of Agriculture, Nature Management and Fisheries of the Netherlands, the Netherlands, April 15–19, 1991.
5. Reijntjes, C.; Haverkort, B.; Waters-Bayer, A. Low-External-Input and Sustainable Agriculture (LEISA): An Emerging Option. In *Farming for the Future: An Introduction to Low-External-Input and Sustainable Agriculture*; Macmillan: London, 1992; 2–21.
6. CAB International. In *Global Crop Protection Compendium*; CAB International: Wallingford, U.K., 1999; (CD ROM).
7. Management of Insect Pests, Plant Diseases and Weeds and Soil Management: A Key to the New Model. In *The Greening of the Revolution: Cuba's Experiment with Organic Agriculture*; Benjamin, M., Rosset, P., Eds.; Ocean Press: Melbourne, 1994; 35–65.
8. Stoll, G. Methods of Crop and Storage Protection. In *Natural Crop Protection Based on Local Farm Resources in the Tropics and Subtropics*; Agrecol, Verlag Josef Margraf: Langen, Germany, 1986; 80–167.
9. Elwell, H.; Maas, A. Natural Pest and *Disease Control*; Natural Farming Network: Harare, Zimbabwe, 1995; 3–128.
10. http://www.communityipm.org/ (accessed January 2001).
11. Poswal, A.; Williamson, S. Off the 'Treadmill': Cotton IPM in Pakistan. In *Pesticides News*; 1998 June, 40, 12–13.

12. Myers, D.; Stolton, S. *Organic Cotton: From Field to Final Product*; Intermediate Technology and the Pesticides Trust: London, 1999; 1–120.
13. Farmer First–Field Schools are a Key to IPM Success. In *Growing Food Security: Challenging the Link between Pesticides and Access to Food*; Dinham, B., Ed.; Pesticide Action Network U.K.: London, 1996; 87–88.
14. Tiffen, M.; Mortimore, M.; Gichuki, F. Management and Managers. In *More People, Less Erosion—Environmental Recovery in Kenya Part III*; John Wiley and Sons: Chichester, U.K., 1994; 131–226.
15. Pretty, J. *Regenerating Agriculture: Policies and Practice for Sustainability and Self-Reliance*; (op.cit 2), Earthscan Publications: London, 1995; 1–25.
16. In FAO. *Rice and the Environment: Production Impact, Economic Costs and Policy Implications*, Committee on Commodity Problems, Intergovernmental Group on Rice, Seville, Spain, May 14–17, 1996. FAO: Rome, 1996; 1–13. CCP:RI 96/CRS1.

Pest Management: Ecological Aspects

David J. Horn
Department of Entomology, Ohio State University, Columbus, Ohio, U.S.A.

Abstract
Every animal and plant population is surrounded by an interactive biotic and physical environment, and these enormously complex interactions often stabilize population densities to reduce the probability of a pest outbreak. However, ecological interactions are often disrupted, simplified, or overridden in managed ecosystems, reducing the impact of naturally-occurring pest population regulation and leading to outbreaks. Understanding how ecological interactions may either cause or prevent pest outbreaks can lead to crop and landscape management activities that achieve relative stability of pest populations below damaging levels without resorting to widespread and environmentally disruptive intervention.

INTRODUCTION

Every animal and plant population is surrounded by an interactive biotic and physical environment, and these enormously complex interactions often stabilize population densities to reduce the probability of a pest outbreak. However, ecological interactions are often disrupted, simplified, or overridden in managed ecosystems, reducing the impact of naturally-occurring pest population regulation and leading to outbreaks. Understanding how ecological interactions may either cause or prevent pest outbreaks can lead to crop and landscape management activities that achieve relative stability of pest populations below damaging levels without resorting to widespread and environmentally disruptive intervention. Increasing environmental complexity can enhance pest management. Intelligent environmental management with due regard for the place of a pest within a complex and interconnected ecosystem reduces pest outbreaks when attention is given to 1) increasing species diversity within cropping or landscaped ecosystems, 2) crop rotations and use of short-term, rapidly-maturing cultivars, 3) reducing field size and encouraging intervening areas of uncultivated and undisrupted vegetation, 4) reduced tillage and tolerance of weedy backgrounds, and 5) increased genetic diversity within the crop.

PEST POPULATION DYNAMICS AND SPECIES DIVERSITY

Apparently simple ecosystems such as annual agricultural crop monocultures are complex, and the relative impact of alternate crops, weeds, natural enemies, competitors, and associated organisms on pests may be highly variable. Pest populations and their effective environments are constantly changing in space and time, and events impacting one population at one location and time interval usually do not duplicate events in the same population at another time and place. Despite ecosystem complexity, seasonal agricultural crops that are periodically disrupted due to harvesting and tilling may never achieve a steady, sustained state typical of later stages in ecological succession. Annual or more frequent disruption selects for pests that can locate and exploit resources quickly and efficiently. Colonizing species of plants and insects display rapid dispersal and an ability to increase numbers quickly when suitable habitat is located. The ancestors of many crop plants are typical of early stages of ecological succession, as are their associated insect pests. Conventional agriculture invites early-successional species that are likely to undergo uncontrolled population outbreaks. Populations of such pests are rarely in equilibrium at any given place and time, but are maintained by a loose balance of colonization and extinction over a large geographical area. Equilibrium is more likely to be reached in populations occupying longer-lasting ecosystems such as orchards and forests. Population fluctuations in these more complex ecosystems are partly buffered by the complex interactions within food webs, so there is less likelihood of outbreak of any particular pest species.

Species diversity is formally measured as an index combining numbers and proportion of each species present in an ecosystem. A species diversity index reflects the number of links in a food web, and the overall stability of any ecosystem is partly a function of the number of interactions among plant, pests, natural enemies, and pathogens. Ecologists and pest managers debate over whether there exists a direct relationship between species diversity and stability of individual populations within an ecosystem. It is often presumed that pest outbreaks are suppressed in more complex (and therefore more diverse) ecosystems. The so-called "diversity–stability hypothesis" holds that ecological communities with a higher species diversity are more stable because outbreaks of pest species are ameliorated by the checks and balances and alternative pathways that exist within a large and integrated food web. Evidence supporting this view comes from experiments that mix several plant species with the primary host of a specialist herbivore, resulting in reduced populations of the specialist herbivore. This observation is termed the *resource concentration hypothesis*, which holds that insect herbivores are more likely to locate and to remain on hosts growing in dense or pure stands, and the most specialized species frequently attain highest densities in ecosystems with low plant species diversity. Biomass becomes concentrated in a few species, with a concomitant decrease in species diversity of herbivores in monocultures. Increases in herbivore populations in crop monocultures generally result from higher rates of colonization and reproduction along with reductions in dispersal, predation, and parasitism. As species diversity increases in agroecosystems, more internal links result within food webs and these links promote greater stability, resulting in fewer pest outbreaks. Structural diversity is an important physical component of overall diversity; for instance, cropping systems with taller plants (such as corn among beans and squash) present more physical space to arthropods and this increases the variety of prey and provides greater shelter for predators.

MONOCULTURE AND POLYCULTURE

Monoculture is the planting of a single species of crop plant, which often results in increased populations of specialist herbivores, a result consistent with the resource concentration hypothesis. Polyculture, the planting of more than one crop species in the same local area (often the same field), may reduce impact of herbivorous pests because the presence of a variety of plants disrupts orientation of specialist herbivores to their hosts. For example, cabbage flea beetles and cabbage aphids that locate their hosts via specific chemical cues (such as the alkaloid sinigrin) are less effective in locating their host plants when these are intermingled with a variety of other plant species. The result is lower populations of the flea beetles and aphids. Local movement of cucumber beetles and lady beetles is enhanced when cucumbers are interplanted with corn and beans as compared with these insects' movement in monocultures, where they tend to remain on individual plants. Numbers of specialist herbivores (cabbage aphids, diamondback moth, and imported cabbageworm) on collards planted in weedy backgrounds are lower than populations of these same herbivores on collards planted against bare soil or plastic mulch. This influence of weeds intensifies once the weeds become as tall as the collards, effectively allowing the collards to escape herbivory by concealment among the weeds. The frequent replacement of one crop by another in crop rotation maintains populations of specialist herbivores below damaging levels. Field crop producers in the midwestern U.S. can prevent the increase of corn rootworm populations by rotating from corn to soybeans every 2 or 3 years.

OPEN AND CLOSED ECOSYSTEMS

Ecosystems may be considered to be *open* (subsidized) or *closed* depending on the amount of nutrient and energy exchange with ecosystems outside themselves. Open ecosystems are characterized by regular input of energy and nutrients, followed by removal of a large proportion of nutrients. In maize fields in the midwestern U.S. heavy importation of mineral fertilizer occurs at planting and subsequent energy is input when the crop is tilled and pesticides are applied. Most of the nutrients in a maize field are removed at harvest, as yield or crop residue. In landscaping towns and suburbs we fertilize (providing input) and rake leaves and remove mowed grass (exporting nutrients) to maintain a pleasing appearance.

The assemblage of species within a maize field is artificial, with novel interspecific associations. Many of the major insect pests are of exotic origin, and the association with maize is relatively recent. For instance, the European corn borer arrived in North America around 1910, before which maize (native to Mesoamerica) had no ecological association with the corn borer. It may take many years for native natural enemies to expand their host or prey range to include exotic organisms. Most anthropogenic agroecosystems and landscaped ecosystems are artificial assemblages and open ecosystems. The species assemblage in planned landscapes often includes a preponderance of exotic species.

In a closed ecosystem, such as a deciduous forest, most nutrient movement remains localized with little import and export. For example, the nutrients and energy in the forest canopy fall to the ground as leaves and frass, or leaves are converted into caterpillars, which in turn are eaten by insectivorous birds, predatory insects, and parasitic wasps. There is little "leakage" of nutrients from closed ecosystems into surrounding environments. The species assemblages of many closed ecosystems have been associated for millennia, resulting in multiple trophic links and close ecological associations among mostly native species. Such ecosystems are relatively "immune" to invasion by exotic species (although there are exceptions, such as the successful invasion of the gypsy moth into the forests of eastern North America).

EXAMPLES

The Mexican bean beetle overwinters in hedgerows and along field edges, so that soybean fields nearest overwintering sites are likely to become infested earlier and bean beetle populations subsequently will be higher. Soybeans located near bush and pole beans (which are more suitable hosts for the bean beetle) are also likely to develop economically damaging infestations earlier. Natural enemies often move from unmanaged field edges and nearby hedgerows and forests into adjacent farm fields and the nature of this movement may be very important to local suppression of pests. Many studies have shown that there are increased numbers and activity of natural enemies near field borders when there is sufficient natural habitat to provide cover and alternate prey and hosts, as well as food in the form of nectar and pollen. This function of wild border areas significantly enhances biological control.

Weedy vegetation in or near crop fields support a diverse fauna, including natural enemies of pests on the crop plants. This depends on the species of weeds present; for instance, if the weeds were particularly attractive to aphids and their natural enemies, this enhances aphid control on a commercial crop. Weeds such as pigweed (*Amaranthus*), lambs quarters (*Chenopodium*), and shepherd's purse (*Capsella*) when heavily infested with aphids serve as "nurseries" for production of aphid predators and parasitoids, which move onto neighboring crop plants when these are located near these weeds.

Growing several crops in the same space reduces pest problems relative to monocultures of the same species. When blackberries are planted among grapevines in central California, the parasitoid *Anagrus epos* attacks the eggs of both grape leafhopper and the leafhopper *Dikrella cruentata* on blackberry. By encouraging blackberries between alternate grape arbors, a constant supply of eggs of both leafhoppers are available to the parasitoid, which persists in populations high enough to bring the grape leafhopper under biological control.

Floral undergrowth in orchards provides resources to adult parasitic wasps and flies and increases parasitism of phytophagous insects (particularly Lepidoptera) on the trees. The presence of nectar and pollen along with alternate prey is particularly favorable to populations of generalist predators such as the lady beetle *Coleomegilla maculata* resulting in lower aphid populations. In relay cropping, two (or more) different crops are grown on the same area in successive seasons. The seasonal change from one crop to another prevents the increase of specialist pests especially if the crops are evolutionarily distantly related (such as legumes and grasses). When soybeans are relay cropped after winter wheat, pests of soybeans are less abundant than when soybeans are cropped alone.

Planting and harvesting corn and beans in alternating plots rather than solid monocultures reduces pest numbers on both crops, and this management approach is widely practiced in traditional agriculture. Where alfalfa can be grown throughout the year it is possible to harvest on a 3–4 week rotation when half the field is cut in strips. Natural enemies of the alfalfa weevil, alfalfa caterpillar, and aphids are conserved in the regrowth, and there are alternative food sources and hiding places for these predators and parasitoids all year, so they are always present to suppress pest populations below damaging levels. By planting alfalfa adjacent to cotton, control of *Lygus* bugs is achieved by allowing increase of natural enemies of *Lygus* in the alfalfa. These natural enemies move into the cotton and control *Lygus* bugs there.

BIBLIOGRAPHY

1. Altieri, M.A. *Biodiversity and Pest Management in Agroecosystems*; Food Products Press: New York, 1994; 185.
2. Collins, W.W.; Qualset, C.O. *Biodiversity in Agroecosystems*; CRC Press: Boca Raton, FL, 1999; 334.
3. Horn, D.J. *Ecological Approach to Pest Management*; Guilford Press: New York, 1988; 285.
4. National Research Council. *Ecologically Based Pest Management: New Solutions for a New Century*; National Research Council: Washington, DC, 1996; 160.

Pest Management: Intercropping

Maria Finckh
Department of Ecological Plant Protection, University of Kassel, Witzenhausen, Germany

Marianne Karpenstein-Machan
Institute of Crop Science, University of Kassel, Witzenhausen, Germany

Abstract

Intercropping can be practiced at the species, variety, and gene level with effects on pathogens, insect pests, and weeds. One of the most important considerations for the successful design of intercropping systems for pest control is the achievement of functional diversity, i.e., diversity that limits pathogen and pest expansion and that is designed to make use of knowledge about host–pest/pathogen interactions to direct pathogen evolution.

INTRODUCTION

Until the past few hundred years, agricultural systems were based on large numbers of different crops, crop varieties, and landraces that were heterogeneous in genetic make-up. In addition, farming systems included both animals and plants further increasing diversity. As a result of increasing specialization, mechanization, and modern plant breeding, diversity on the farming system level and crop level has been drastically reduced worldwide at an ever accelerating speed especially over the past 100 years. Fewer and fewer varieties that are genetically homogeneous are being grown in ever-larger fields.[1]

Monoculture refers usually to the continuous use of a single crop species over a large area. However, with respect to plant pathogens and pests it is important to differentiate between monoculture at the level of *species, variety,* or *resistance genes*.[2] For example, within a species there may be many different genotypes with different resistances to a specific pest or pathogen and great variation with respect to competitiveness with weeds and other crops. Within a variety, there is usually no diversity (but see Table 1) for resistance or morphological traits. Resistance gene monocultures are more difficult to conceptualize. Many different varieties may exist; however, they may all possess the same resistance (or susceptibility) gene(s). For example, in the late 1960s, virtually all hybrid maize cultivars in the southeastern United States possessed the cytoplasmatically inherited Texas male sterility (*Tms*). Unfortunately, *Tms* is closely linked to susceptibility to certain strains of the pathogen *Cochliobolus carbonum* (syn. *Helminthosporium maydis*). The monoculture for susceptibility (while different varieties had been planted) led to selection for these strains and in 1970 the pathogen caused more than $1 billion (= 10^9) losses.[3]

Intercropping[4] can be practiced at the species, variety, and gene level (Table 1) with effects on pathogens,[1,2,5] insect pests[6–8] and weeds[9,10] (Table 2). One of the most important considerations for the successful design of intercropping systems for pest control is the achievement of *functional diversity*, i.e., diversity that limits pathogen and pest expansion and that is designed to make use of knowledge about host–pest/pathogen interactions to direct pathogen evolution.[1,2]

PROTECTION MECHANISMS ACTING IN INTERCROPPED SYSTEMS

Pathogens, insect pests, and weeds differ fundamentally in their biology and their effects on crops, and different protection mechanisms act with respect to these organisms (Table 2).

Pathogens are mostly dispersed through wind, water splash, soil, and animals (vectors). In intercropped systems, the most important mechanisms for disease control are mechanical distance and barrier effects. In addition, resistance reactions induced by a virulent pathogen strains may prevent or delay infection by virulent strains. A large percentage of the reduction of airborne diseases such as the powdery mildews and rusts in cereal cultivar mixtures has been shown to be due to induced resistance. The protection mechanisms are universal with respect to airborne, splashborne, and some soilborne, diseases. Mixtures of plants varying in reaction to a range of diseases will lead to a multitude of additional interactions and the overall response in such populations will tend to correlate with the disease levels of the components that are most resistant to these diseases. In addition, less affected plants may compensate for yield losses due to reduced competition from diseased neighbors.[5]

In contrast to pathogens for which passive or vectored dispersal is the norm, insects often search actively for their hosts and behavioral, visual, and olfactory cues play an important role. While environmental factors and landing on

Table 1 Possibilities for intercropping at three levels of uniformity on which monocultures are commonly practiced.[a]

Level of uniformity	Intercropping possibilities
Species: Different individuals may differ in genetic make-up (resistance, morphology, etc.)	Arrangements among and within species, varieties, and resistances using intercropping
Variety: Usually genetically uniform, the same gene(s) in the same genetic background	Arrangements among and within varieties and resistances—includes variety mixtures, multilines, and populations
Resistance gene: The same gene may exist in different genetic backgrounds	Arrangements among resistances—multilines and populations

[a]From Finckh and Wolfe.[2]

a nonhost is likely the most important mortality factor for pathogens, natural enemies are at least as important for insect population dynamics.[7,8] Host dilution may affect an insect's ability to see and/or smell its hosts. Predators and parasitoids are dependent on the constant presence of prey and alternative food sources such as pollen and nectar in the absence of the hosts and it is critical that natural enemy populations are present in sufficient numbers to enable them to effectively control insect pests. The importance of natural enemies was often only recognized after insecticide applications induced pest resurgence due to the destruction of natural enemy populations. Intercrops and weeds therefore can play an important role in regulating insect pests.

Weeds usually are early successional plants adapted to colonize open, nutrient-rich spaces. Intercrops, especially cover and mulch crops directly compete with weeds for these spaces and also for light. As many weeds are adapted to certain crops and cropping patterns, changing these patterns (e.g., rotations) and management operations connected with different kinds of crops within the same field may make it difficult for weeds to cope.[9,10] An important consideration is that plants may be weeds only during certain phases of crop development. At other stages, the presence of the same "weeds" may be beneficial because they may provide food and habitat for beneficial insects and erosion control.

Besides the many positive effects of intercrops it is important to keep in mind that weeds may serve as alternative hosts for insect pests and pathogens and that insects often are disease vectors, especially for viruses that may reside symptomless in certain weeds.

Table 2 Mechanisms affecting pathogens, insect pests, and weeds in intercropped systems and selected additional interactions of importance.

Mechanisms reducing disease
Increased distance between susceptible plants
Barrier effects of intercrop
Induced resistance
Selection for most resistant and/or competitive genotypes
Interactions among pathogen strains on host plants
Mechanisms reducing insect pests
Enhancement of natural enemies
Reduction of host density (reduced resource concentration)
Reduction of plant apparency (visual or olfactory cues reduced)
Alteration of host quality (with respect to the insect pest) through plant–plant interactions
Mechanisms reducing weeds
Reduction of bare soil and layering of crops (increased competition for light, water, and nutrients)
Variation in tillage needs and operations of intercrops may disturb weeds
Other beneficial interactions
Yield enhancement through niche differentiation of hosts
Compensation for yield losses by less affected hosts
Better soil cover with intercrop (soil and water conservation, microclimatic effects)
Possible unwanted interactions
Weeds may serve as alternate hosts for pathogens and insects
Interactions among virus vectors and weeds
Greater difficulty to specifically reduce weeds with herbicides or mechanically
Microclimatic effects may enhance certain problems

Source: Wolfe and Finckh,[5] Andow,[7] and Liebman and Dyck.[9]

INTERCROPPING IN PRACTICE

Variety mixtures and multilines are used mainly to control diseases. For example, they are used in cereals on a commercial scale in the United States, Denmark, Finland, Poland, and Switzerland to control rusts, mildews, and certain soilborne diseases (e.g., *Cephalosporium* stripe). When barley cultivar mixtures were used on more than 300,000 ha in the former German Democratic Republic, powdery mildew of barley and consequently fungicide input was reduced by 80% within five years. Wheat cultivar mixtures and multilines are grown on several hundred thousand hectares in the US in the Pacific Northwest and in Kansas to protect against diseases and abiotic stresses. In Colombia, coffee multilines are grown on more than 400,000 ha to control coffee rust.[2]

Attention has also been called to possible beneficial effects of greater intravarietal diversity in the oat-frit fly (*Oscinella frit* L.) system. The flies can attack the host plants only at a particular growth stage and a higher degree of variability within an oat crop could allow for escape from attack and subsequent compensation.

Cereal species mixtures for feed production are currently grown on more than 1.4 million ha in Poland and have been shown consistently to restrict diseases. In Switzerland, the "maize-ley" system (i.e., maize planted without tillage into established leys), which is being promoted to reduce soil losses and nutrient leaching, has been shown to reduce smut disease and attacks by European stem borer and aphids.

The deliberate planting or maintenance of flowering weeds and grass in established vineyards in Switzerland and Germany greatly increases natural enemies while reducing soil erosion. This practice is becoming increasingly popular in Californian wine growing areas and in apple production. In the United Kingdom, a newly developed low-input system for growing wheat with a permanent understorey of white clover (*Trifolium repens*) greatly reduces the major pest aphid species and slugs and there should also be reductions in splash-dispersed diseases, such as those caused by *Septoria* spp.

The required reduction in insect populations for effective reduction of insect-transmitted diseases may be beyond that which can be achieved by diversity alone. However, simultaneous reduction of insect vectors and disease inoculum can be effective.[2]

FUTURE CONCERNS

There are many reasons why intercropping is not practiced more widely. First, modern crops are bred to be grown in monoculture and may not necessarily be well adapted to intercropping. Efforts of breeders to produce breeding lines adapted to intercropping need to be strengthened. Second, while intercropping clearly provides a means for reducing pesticide needs there is a lack of adapted machinery allowing for efficent management of intercropped cultures. Third, successful intercropping strategies have to be carefully designed as preventive measures while application of pesticides often can be done once a problem occurs; it is simple and usually very cheap. Fourth, a concern often raised is the quality of products raised as intercrops such as varietal mixtures of cereals. In some countries there is resistance in the food processing industry to such products. However, such problems could be overcome if breeders, producers, and processors work together. For example, in the 1980s, in the German Democratic Republic, first-quality malting barley was produced in large-scale mixtures in collaboration with breeders, growers, and processors.

REFERENCES

1. Finckh, M.R.; Wolfe, M.S. Diversification Strategies. In *The Epidemiology of Plant Diseases*; Jones, D.G., Ed.; Chapman and Hall: London, 1998; 231–259.
2. Finckh, M.R.; Wolfe, M.S. The Use of Biodiversity to Restrict Plant Diseases and Some Consequences for Farmers and Society. In *Ecology in Agriculture*; Jackson, L.E., Ed.; Academic Press: San Diego, 1997; 199–233.
3. Ullstrup, A.J. The impacts of the southern corn leaf blight epidemics of 1970–1971. Annu. Rev. Phytopathol. **1972**, *10*, 37–50.
4. Vandermeer, J.H. *The Ecology of Intercropping*; Cambridge University Press: Cambridge, New York, Melbourne, 1989; 235.
5. Wolfe, M.S.; Finckh, M.R. Diversity of Host Resistance within the Crop: Effects on Host, Pathogen and Disease. In *Plant Resistance to Fungal Diseases*; Hartleb, H., Heitefuss, R., Hoppe, H.H., Eds.; G. Fischer Verlag: Jena, 1997; 378–400.
6. Altieri, M.A.; Liebman, M. Insect, Weed and Plant Disease Management in Multiple Cropping Systems. In *Multiple Cropping Systems*; Francis, C.A., Ed.; Macmillan: New York, 1986; 183–218.
7. Andow, D.A. Vegetational diversity and arthropod population responses. Annu. Rev. Entomol. **1991**, *36*, 561–586.
8. Letourneau, D.K. Plant-Arthropod Interactions in Agroecosystems. In *Ecology in Agriculture*; Jackson, L.E., Ed.; Academic Press: London, 1997; 239–290.
9. Liebman, M.L.; Dyck, E. Crop rotation and intercropping strategies for weed management. Ecol. Appl. **1993**, *3*, 92–122.
10. Liebman, M.L.; Gallandt, E.R., Many Little Hammers: Ecological Management of Crop-weed Interactions. In *Ecology in Agriculture*; Jackson, L.E., Academic Press: San Diego, 1997; 291–343.

Pest Management: Legal Aspects

Michael T. Olexa
Center for Agricultural and Natural Resource Law, Institute of Food and Agricultural Sciences, University of Florida, Gainesville, Florida, U.S.A.

Zachary T. Broome
Bowen Radson Schroth, P. A., Eustis, Florida, U.S.A.

Abstract

To promote public health, personal safety, and environmental protection, pesticides and their use are extensively regulated. Pesticide regulation is primarily decided at the federal level and enforced by the Environmental Protection Agency (EPA). States comply with federal laws, but some states have enacted additional laws and regulations that establish state policies and positions on pesticide use within the state's jurisdiction. Both federal and state laws provide for criminal prosecution and can impose penalties such as fines or imprisonment. In addition, common law actions are used to regulate pesticide use. Common law actions are civil claims brought by private citizens or state agencies based on an allegation of improper pesticide production or use. When a pesticide manufacturer or user causes personal injury or property damage, the victim initiates a civil action (i.e., lawsuit) to recover financial compensation for the harm.

INTRODUCTION

To promote public health, personal safety, and environmental protection, pesticides and their use are extensively regulated. Pesticide regulation is primarily decided at the federal level and enforced by the Environmental Protection Agency (EPA). Although the states generally copy the federal laws, some state laws are more restrictive than federal laws. Both federal and state laws provide for criminal prosecution and can impose penalties such as fines or imprisonment. In addition, common law actions are used to regulate pesticide use. Common law actions are civil claims brought by private citizens based on an allegation of improper pesticide production or use. When a pesticide manufacturer or user causes personal injury or property damage, the victim initiates a civil action to recover financial compensation for the harm.

This entry will detail the federal regulatory framework of pesticides, including product registration and methods of enforcement. This entry will also address the conflicts between pesticide regulation and other federal environmental law, and the current preemption guideline for conflicts between federal and state regulations. Finally, this entry will address various common law remedies available as private forms of pesticide regulation, to ensure pesticide users and manufacturers act reasonably and limit potential harm to humans and the environment.

STATUTORY LAW AND REGULATION

Statutory laws are the formal acts of federal and state legislatures. Although statutory laws often provide general directions, technical details are assigned to a regulatory agency. The agency rules specify how the statute passed by Congress will be implemented and enforced by the agency. These rules have the force of law. Regulatory agencies implement their rules by requiring permits or licenses, and enforce those rules through civil and criminal penalties.[1]

FEDERAL LAW

Overview

Pesticides were first subject to federal laws by the Insecticide Act of 1910.[2] This law protected farmers from sellers of adulterated or misbranded pesticide products. Following a surge in the development of new pesticides during the Second World War, Congress repealed the Insecticide Act of 1910 and enacted the Federal Insecticide, Fungicide, and Rodenticide Act (FIFRA) of 1947.[3] FIFRA broadened federal control of pesticides by requiring the United States Department of Agriculture (USDA) to register any pesticide before its introduction into interstate commerce. In 1970, Congress transferred the administration of FIFRA to the newly created EPA. Thereafter, Congress enacted the 1972 Federal Environmental Pesticide Control Act (FEPCA) to address public concerns about pesticides' environmental impact and to ensure greater pesticide use regulation.[4] As a result, federal policy shifted from regulation of pesticides for reasonably safe use in agriculture to regulation of pesticides to prevent unreasonable risks to people and the environment. Subsequent amendments to FIFRA (1988, etc.), have clarified the EPA's duties and responsibilities.[5]

The Federal Food, Drug, and Cosmetic Act (FDCA), and 1954 Miller Amendments, required the government to establish tolerances (maximum allowable pesticide residue limits) for all pesticides used on food and feed crops.[6]

The 1996 Food Quality Protection Act (FQPA) amended the FDCA and FIFRA, especially the process of establishing tolerances for pesticide residues in food and feed.[7] For example, the EPA now uses a single standard to evaluate pesticide residues on raw and processed foods. FQPA represents the single largest shift in federal pesticide policy and process ever undertaken. The FQPA established a new safety standard to be met when establishing tolerances for a reasonable certainty of no harm from aggregate exposure. It also required all existing tolerances to be reassessed under the new standard and required the EPA to make an explicit determination that tolerances for residues in food are safe for infants and children.[7]

The heart of the regulatory scheme for pesticide application, storage, and disposal is the FIFRA provision making it unlawful to ". . . use any registered pesticide in a manner inconsistent with its labeling."[8] Thus, courts consider pesticide label instructions legislative regulations having the force of law. For pesticide users, labeling is the primary basis for enforcement of FIFRA. Hence, every person using a pesticide has a legal obligation to read and follow all label instructions attached to the product and all product usage directions contained in any printed materials mentioned on the label.

Pesticide Product Registration

The EPA retains primary oversight for providing guidance to major pesticide producers for what data are needed before the product is federally registered. Some states also conduct a review of scientific data before registering a pesticide in their state. State agencies generally identify product registration violations during their inspections and refer them to EPA for enforcement.[5] However, states also have the ability to regulate pesticide use with state-level registration, although the registration only applies within the regulating state.[9]

The federal government controls which pesticides are put on the market by requiring pesticide registration under FIFRA.[10] Unless exempt, all pesticides must be registered with the EPA administrator before being distributed or sold.[11] A pesticide will not be registered by EPA unless the agency's review of the registration data shows the pesticide will perform its intended function without "unreasonable adverse effects on the environment."[12] Pesticides are registered for general use or restricted use, depending upon the potential for adverse side effects.[13] If EPA determines that a pesticide generally will not have adverse effects on the environment or injure the applicator, then the pesticide is safe for general use.[14] This means that the product can be purchased by an unlicensed applicator. If EPA determines that a pesticide requires additional care or specific training in use, then the pesticide is defined as a restricted use product (RUP).[15] RUPs are subject to additional regulations and limited to use by a certified and licensed applicator.[16] Although many pesticides predated 1947, and were therefore exempt from registration, FIFRA was amended to require the reregistration of all pesticides "containing any active ingredient contained in any pesticide first registered before November 1, 1984."[17]

Most states require the federally regulated products to be registered in each state as well. A state also may regulate the registration of EPA-registered pesticides for additional distribution or use within that state, so long as the registered state use does not violate FIFRA.[18] For instance, a state can make it illegal to use a product registered by EPA, or limit the use of a registered pesticide. Unless the EPA administrator has denied the state's registered use, state registration of pesticides is equivalent to federal registration, but only within that state. State registrations are subject to disapproval by the administrator, who may also suspend a state's registration authority if the state issues registrations that violate either FIFRA or FDCA.[19]

Not all pesticides are regulated by the EPA. If the EPA administrator determines that a pesticide does not require regulation, then the administrator, by regulation, may exempt that pesticide from the registration requirements.[20] The pesticides that do not require regulation are natural substances and minimum-risk pesticides, such as cedar or geraniol, as well as substances treated with pesticides, pheromones, and biological preservatives.[21]

Pesticide Use Enforcement

Although EPA does have enforcement power, in many cases EPA delegates enforcement of pesticide use violations to the states, under cooperative agreements. Each state's agencies are responsible for administering pesticide regulations under state and federal statutes. States occasionally refer pesticide use violation cases to the EPA for enforcement.[22]

Enforcement of pesticide regulations is based on a hierarchy of state and federal agencies, beginning with the EPA Office of Pesticide Programs (OPP). OPP is the EPA entity responsible for registering all pesticides and formulating the label requirements that restrict pesticide use.[23] EPA in Washington works through 10 regional field offices (Regions) of EPA, which serve as liaisons between the states and OPP. The Regions work with the states to develop cooperative agreements, examine cases, answer questions, and conduct joint inspections.

Although EPA is the ultimate arbiter of pesticide regulation, states have the primary responsibility to enforce those regulations.[24] Each state has a designated lead agency responsible for meeting the OPP goals. The lead agencies investigate all major pesticide use sites, such as agricultural and urban operations, to ensure that registered pesticides are used according to the labels approved by OPP. The lead agencies are also responsible for investigating sites of pesticide misuse and taking appropriate enforcement information.[24]

Some Native American tribes have pesticide lead agencies that function the same as state lead agencies, and those tribes work with their EPA regions to develop pesticide field programs. Although the tribal enforcement programs are created in the same manner and with the same goals as state programs, OPP and the regions must treat the process with a high degree of deference to a tribe's sovereign treaty rights.[25]

If a pesticide user or manufacturer violates any FIFRA regulation, EPA and the state can stop the sale or use of the pesticide, seize the pesticide, and seek civil and criminal penalties.[26] If a pesticide manufacturer or distributor violates FIFRA, EPA can seek a $5000 civil penalty for each offense, or a criminal penalty of a $50,000 fine or 1-year jail term.[27] If a pesticide user violates FIFRA, EPA can seek a $1000 civil penalty for each offense, or a criminal penalty of a $1000 fine or 30-day jail term.[27] The most drastic enforcement procedures, in regard to pesticide manufacturers, are the cancellation or modification of a pesticide registration. Once a pesticide is registered under FIFRA, the applicant is not guaranteed permanent registration or classification. If EPA determines that a registered pesticide, or its labeling, does not comply with FIFRA or causes unreasonable adverse effects on the environment, the administrator may cancel the pesticide's registration, change its classification, or alter the label.[28]

Enforcement of Tolerances for Pesticides in Food

One important consideration in regulating pesticide use is the level of pesticide residue left on agricultural products that can be consumed by humans. The Food and Drug Administration (FDA) monitors raw and processed commodities for compliance with residue tolerances. The USDA monitors meat, milk, and eggs for residue tolerance compliance. The FDA and USDA programs cover both domestically produced and imported commodities. A few states, such as California and Florida, have additional residue monitoring and enforcement programs, but all states have the authority to enforce federal regulations against food located within each state.[29] States cannot impose regulatory limits on pesticide residue, however, unless the residue limits are identical to the federal levels or the state successfully petitioned the federal government for authorization to change the regulation.[30]

Under the FDCA, introducing adulterated food into interstate commerce is prohibited.[31] A claim for violating the FDCA can be brought in an appropriate federal district court, where the court can punish the guilty party with injunctive relief, criminal prosecution, or seizure.[32] The key for pesticide regulation is whether food containing pesticide residue constitutes adulterated food. Normally, a food is adulterated if it contains any poisonous substance or chemical.[33] Pesticide chemical residues, however, are specifically exempted from blanket regulation.[34] If the EPA has established a tolerance level for a pesticide, or exempted a pesticide from regulation, then the food product is not considered adulterated.[35] The EPA administrator can only establish a pesticide tolerance if consumption of the residue poses little risk to the consumer.[36] The consumption is safe so long as "there is a reasonable certainty that no harm will result from aggregate exposure to the pesticide chemical residue."[36] Once the EPA has established the tolerance level for a pesticide residue, the FDA enforces compliance because noncompliance residue that exceeds established tolerances constitutes food adulteration.

The FDA not only monitors U.S. pesticide residue but also develops cooperative agreements with foreign countries that export agricultural products into the United States.[37] These agreements require imported food to meet the FDCA standards for pesticide tolerance.[37]

While the EPA determines if the amount of pesticide residue found in the food is of "negligible risk" or below tolerable levels, there is a danger from consuming certain pesticide residues that the EPA has deemed carcinogenic.[38] Consumption of pesticide residue is also dangerous because of the pesticides that act as endocrine-disrupting chemicals (EDCs).[39] Evidence has established a link "between EDCs and decreased sperm counts; breast, testicular, and prostate cancer; and neurological disorders."[39] In fact, many of the chemicals found in pesticide residue have been banned in the European Union because of the danger to humans.[40]

Pesticide Regulation Interaction with Other Federal Regulations

One problem for the EPA is registering pesticides that conflict with other federal environmental law the EPA is charged with enforcing, especially the Endangered Species Act (ESA) and the Clean Water Act (CWA).[41]

Under section nine of the ESA, all persons, including federal agencies, are prohibited from causing the "take" of any endangered species.[42] While pesticides may threaten endangered species, EPA did not have a formal process to evaluate the potential impact of pesticides on threatened and endangered species, but EPA had never been liable for registering the pesticides.[43] In 1989, however, the Eighth Circuit held the EPA liable for approving the registration of a pesticide that later harmed a protected species.[44] According to the Eighth Circuit, an action solely based on pesticide regulation ordinarily should be brought under FIFRA, but a claim based on pesticide harm to an endangered species could be brought under the ESA.[45] Because EPA did not dispute the distributed pesticide caused the death of an endangered species, killing an endangered species is a taking, and pesticide distribution cannot legally occur without an EPA registration, the Eighth Circuit held EPA liable for a taking.[46]

Similarly, section seven of the ESA requires all federal agencies to conserve endangered species, which usually requires a consultation with the Fish and Wildlife Service or National Marine Fishery Service to ensure agency actions will not "jeopardize the continued existence of any endangered species."[47] In 2004, the Ninth Circuit ruled EPA violated section seven of the ESA by failing to obtain a "jeopardy" consultation before registering pesticide uses around listed salmon habitat.[48] EPA argued that any cancellation or modification of a pesticide's use must conform to FIFRA, and FIFRA's standard for registration and cancellation already accounted for an effect on listed species.[49] According to the Ninth Circuit, however, FIFRA does not exempt EPA from the ESA regulations, since the two statutes have different purposes, and EPA must comply with the ESA if its registration of pesticides will affect listed species.[49] In addition, EPA has an ongoing duty to ensure all registered pesticides do not violate the ESA, because EPA always has the discretion to cancel a pesticide's registration.[49]

EPA also has to consider the impact that registered pesticides may have on waterways, to ensure the pesticides will not violate the CWA.[50] Under the CWA, any discharge of a pollutant into the nation's waters requires the discharging entity to obtain a National Pollutant Discharge Elimination System (NPDES) permit.[51] The pesticide manufacturer claimed that EPA did not include a requirement for a NPDES permit in the FIFRA registration process, and since water-related usage restrictions were placed on the pesticide as part of the FIFRA registration, the manufacturers did not need additional permitting.[52] According to the Ninth Circuit, however, the CWA and FIFRA have different and complementary purposes, so the statutes should be treated as distinct entities.[52] FIFRA's objective is to protect human health and prevent environmental harm from pesticides through cost–benefit analysis, while the CWA's objective is to restore and maintain waterways by limiting pollution discharge.[52] Therefore, the court ruled that pesticide users are not exempt from CWA permitting requirements because they obtain FIFRA registration.[52]

This dual responsibility creates a tension for EPA, which is responsible for establishing water quality standards that dictate the permit requirements for a CWA permit.[53] Until recently, EPA had a Final Rule that exempted all FIFRA-compliant pesticides from NPDES permitting when those pesticides were put into water, or ran off property into water, as part of the pesticide's intended use.[53] The Sixth Circuit rejected this analysis, citing the CWA requirement that no pollutant could be discharged into any navigable water unless EPA determined the discharge would not cause undue harm to water quality and issued the discharger a NPDES permit.[54] Although the court agreed FIFRA-compliant chemical pesticides did not require a NPDES permit if put into water to work as a pesticide, because the pesticide was not being discharged as a waste and EPA only approves pesticides that "will not generally cause unreasonably adverse effects on the environment," the court cautioned that any introduced pollutant discharged with excess chemicals or residue required a NPDES permit.[55] In addition, all biological pesticides, regardless of whether or not it is excess pesticide or residue, always require NPDES permits because the CWA requires a permit for all biological material, not just waste.[56] While the court rejected EPA's Final Rule exempting pesticides from NPDES permitting, the court noted EPA and state authorities may grant general permits that allow for the discharge of a specific pollutant or type of pollutant across an entire region.[57] Once EPA or a state agency issues such a general permit, pesticide dischargers do not have to meet any additional requirements to comply with the CWA.[57] This issue is currently being worked on between EPA and the states, as they try to implement a national and state plan.

PREEMPTION

Because of the interaction between state and federal regulations, preemption is a fundamental concern for pesticide regulation. Under FIFRA, the states are allowed to regulate the sale and use of federal pesticides, but precluded from passing additional regulations on pesticide labels.[58] This preemption applied to both state regulations and common law tort claims that "directly or indirectly attacked the adequacy of the warnings on the EPA approved pesticide label."[59]

Background

When a manufacturer submits an application to register a new pesticide, the manufacturer must include a proposed label, and the EPA will only register the pesticide if the data support the registration and its label complies with the statute's misbranding prohibition. A pesticide is "misbranded" if its label contains a statement that is "false or misleading" or lacks adequate instructions or warnings.[60] Although preemption was originally restricted to state regulations, in 1992, the federal courts began applying preemption to common law tort claims.[61] Although the Supreme Court never ruled on FIFRA-specific preemption, after *Cipollone*, most state and federal courts would reject any common law claim that was premised on "failure to warn."[62] A common test for preemption, which led to *Bates v. Don Agrosciences, LLC*, was whether it was reasonably foreseeable that a tort claim against a pesticide manufacturer would "induce it [the manufacturer] to alter its product label."[63] If the litigation would give the manufacturer a "strong incentive" to change the pesticide label, then the claim was preempted.[63]

CURRENT LAW

In *Bates V. Don Agrosciences, LLC*, the U.S. Supreme Court held that FIFRA's express preemption clause did not preempt state common law claims merely because an adverse judgment on the claim might impel a manufacturer to change a product label.[64] According to the Court, "private remedies that enforce federal misbranding requirements would seem to aid, rather than hinder, the functioning of FIFRA" because the claims serve as catalysts for evaluating harm and pesticide performance.[65] Although the Court acknowledged the lack of uniformity in national FIFRA regulation, the Court maintained that Section 136v still prevents states from imposing labeling requirements different from FIFRA.[66]

Applying *Bates* to future pesticide litigation and regulation will be necessarily fact specific. In *Bates*, the Court parsed the plaintiffs' tort claims against the pesticide manufacturer into two categories, design defect and failure to warn.[67] The Court held that design defect claims, formerly preempted under the "inducement" test, were not preempted because a successful claim would incentivize a manufacturer to change its label or ingredients, but the claim placed no requirement on the labeling.[67] The plaintiffs' fraud and negligent failure-to-warn claims, however, were premised on common law rules that qualify as "requirements for labeling and packaging," but the Court established that Section 136v(b) only prohibits state law labeling and packaging requirements that are "in addition to or different from" the labeling and packaging requirements under FIFRA.[66] Therefore, any preemption analysis begins with the plaintiff's claim and "calls for an examination of the elements of the common law duty at issue."[68] After *Bates*, only tort claims that impose liability for a labeling requirement that is in addition to or different from EPA labeling requirements would be preempted.[59]

COMMON LAW

Common law actions do not depend upon statutes for their authority. Instead, common law arises from the generalized legal duty individuals in a law-abiding society owe to one another. Every adult person is obligated to a certain duty of care for the personal and property rights of others. A violation of this obligation can become a basis for a common law action. Common law theories generally encountered in actions resulting from pesticide use include negligence, trespass, nuisance, and strict liability.[68]

Negligence

Negligence is a legal standard applied to an individual who fails to act in a reasonably prudent manner.[69] To establish a negligence claim, the plaintiff must prove that the defendant owed the plaintiff a duty, the defendant breached that duty, the breach was the direct and proximate cause of harm to the plaintiff, and plaintiff's harm actually resulted in compensable damages.[69] Negligence actions can be used as regulatory tools against both pesticide users and the pesticide producers, to ensure pesticides are used reasonably to prevent damage to human health and the environment.[70]

A pesticide user can be liable for negligence if the user fails to exercise a reasonable duty of care in applying a toxic pesticide, and others are injured as a direct result of the negligent behavior.[70] A pesticide manufacturer can be liable for negligence if the manufacturer's label is defective, or if the manufacturer's label constitutes a failure to warn, and a person is injured because of the improper label.[67] The common law duty for pesticide use or production generally stems from the industry-established standard of care. However, because pesticides are heavily regulated at the state and federal level, statutes or administrative regulations generally define the applicable standard of care in a jurisdiction. These regulations also create a cause of action based on negligence per se, so any violation of a law regulating the manufacture, label, or use of a pesticide that results in harm is negligence.[71] In making a negligence claim for pesticide use, res ipsa loquitur can be a successful argument in cases where a pesticide, used as intended, was the direct cause of an injury.[72] Res ipsa can be difficult to prove, however, since many factors other than negligence of a manufacturer or user could cause or contribute to such injuries.

In negligence cases based on pesticide use, breach of duty and causation often relate to the same action—the pesticide dispersal. Pesticide users have a duty to act carefully when applying pesticides, so as to avoid applying pesticide on unintended targets. If a pesticide user does not breach that duty, then it is unlikely the pesticide caused a plaintiff's injury. To prove causation, the plaintiff must prove exposure to the pesticide and have an expert testify that exposure to the chemicals in the pesticide caused plaintiff's injury.[73] Generally, an expert can establish causation by using clinical studies where a certain level of pesticide caused certain chemical reactions, such that plaintiff's alleged exposure to that pesticide would likely result in plaintiff's injury. Experts can also establish causation by diagnosing the plaintiff's symptoms, and then eliminating all alternatives except the pesticide exposure.[70]

In actions based on pesticide use, damages are usually related to personal injury or property damage. Personal injury damages result from exposure to the toxic chemicals in pesticides and usually relate to health problems with the lungs and respiratory systems, skin rashes, swelling, discoloration, and lesions, as well as nervous system disorders, such as numbness or headaches, confusion, and memory loss.[70] Agricultural pesticide users often sue when a product malfunctions, so property damages are usually for harm to currently growing crops, although in certain cases, damages can be sought for the loss of property utility, if the pesticide causes property to lose agricultural value.[70] The same injury and damage rules apply to livestock and other "farmed" animals.[70]

Trespass

Trespass is an unauthorized entry onto the property of another by a person or thing that causes damage.[74] Pesticide application can result in liability for trespass if the pesticide, its residue, or container becomes deposited on another's land (through dumping, drift, runoff, incineration, or other means) and causes substantial damage to the property. Although trespass normally requires intent for liability, if the pesticide reached a plaintiff's property because the pesticide user behaved negligently or used the pesticide in an abnormally dangerous manner, then there is no requirement to prove intent.[75] The invasion element of pesticide occurs by either airborne drift or migration through soil or water.[76]

Nuisance

A nuisance is substantial interference with another's use and enjoyment of land.[77] A nuisance lawsuit requires no physical invasion, only a substantial interference with the possessor's enjoyment of land. Pesticide use resulting in offensive odors can be grounds for a nuisance suit. In almost all states, state "right-to-farm" statutes provide limited protection from nuisance actions. The farmer's "right-to-farm" defense is limited to nuisance actions.[78] Nuisance claims can be brought as either private or public actions, although most pesticide-based claims are for private nuisance, because a plaintiff cannot collect damages for a public nuisance claim.[79]

A private nuisance is a nontrespassory invasion of another's interest in the private use and enjoyment of land.[80] Generally, property owners bring private nuisance claims because some property use is prevented by a neighboring property owner's use of pesticides.[81] A private nuisance claim is often used to enjoin a defendant from continuing a certain action, although damages are often sought as compensation for harm already done by the defendant. Because nuisance is a common tort, and has been regularly applied to instances of pesticide drift, most courts will not bar a provable claim for nuisance. In addition, when the plaintiff is unable to prove which pesticide user out of several caused his or her particular harm, courts often allow the plaintiff to proceed on joint liability, so the defendants share the damages burden. However, a nuisance claim requires an ongoing pattern of pesticide application, so a single pesticide use is not actionable under this particular theory.[81]

A nuisance suit against a farmer for the use of pesticides often violates state right-to-farm laws.[78] Right-to-farm laws protect farmland in a variety of ways, including reducing farmer exposure to liability for common law claims based on farming operations.[78] One protection is the coming to the nuisance defense, which prevents a plaintiff from moving near a farm then claiming harm from an ongoing farm activity.[81] The coming to the nuisance defense only protects the farmer from reasonable actions, however, so using pesticides in a negligent manner could result in liability. In addition, coming to the nuisance only protects farm activities that were ongoing before the plaintiff's arrival. If the farmer did not apply pesticides until after the plaintiff's arrival, then the farmer could be subject to nuisance liability.[81]

Strict Liability

Some states hold pesticide applicators absolutely responsible for their pesticide application activities, regardless of fault, without a showing of negligence. This is known as strict liability.[82] Strict liability is normally associated with inherently dangerous or ultrahazardous activities. Whether an activity is inherently dangerous, and therefore subject to strict liability, usually depends on the degree of risk, likelihood of serious harm, inability to eliminate risk, commonness of usage, appropriateness of activity to area, and its value to the community.[83] Most misuse of pesticide claims are brought under the theory of negligence, but a few states allow strict liability claims against the aerial spraying of pesticides.[81]

A common strict liability claim is for a manufacturing defect in the pesticide, where some contaminant in the pesticide caused damage during the pesticide use.[84] Because FIFRA registration requires a manufacturer to register the exact chemical makeup of a pesticide product, the presence of any other chemical in the pesticide is a statutory violation that exposes the manufacturer to strict liability.[85] To bring a strict liability claim against a pesticide manufacturer, a plaintiff must be able to prove that he or she suffered injuries as a result of the use of a pesticide, the defendant was engaged in the manufacture and sale of the harmful pesticide products, the plaintiff purchased the particular pesticide that caused the alleged injuries, the pesticide had not been contaminated after leaving the manufacturer, and the pesticide was unreasonably dangerous to any user.[86]

REGULATORY TRENDS

For the foreseeable future, current and emerging societal issues will continue to stimulate regulatory action. EPA's regulatory focus is found in the National Pesticide Field Program, which establishes EPA's guideline for action to meet the statutory requirements to protect human health and the environment. Currently, EPA regulations focus on worker safety, water quality, and endangered species protection, and promoting pesticide stewardship.[87]

EPA regulation of worker safety is done to minimize occupational risks from pesticides, as outlined in the Worker Protection Standard. The potential danger of pesticide exposure is especially prevalent in agricultural operations, because workers have high occupational exposure to pesticides. In large-scale commercial agriculture operations, workers are often in such close contact with pesticides that they develop serious health problems.[88] Regulating worker safety also gives regulatory bodies an additional "watchdog" group of on-site workers trained in appropriate pesticide manufacture and application, as EPA hopes that the educated workers will report pesticide misuse to mitigate health risks.[89]

Another regulatory trend is reducing environmental risks from registered pesticides, in order to reduce EPA liability under federal environmental protection statutes. Courts have repeatedly held that EPA has a duty to comply with complementary environmental statutes when issuing FIFRA registrations, so two of the focal points for EPA regulation are compliance with the ESA and CWA water quality controls.[89]

Finally, EPA promotes better performance regulation as a part of an incentive plan to encourage agricultural pesticide users to move away from high-risk pesticides. EPA is currently collecting funds for grants and collaborating with the USDA to establish the best environmentally and socially sound pesticide use policies, and EPA plans to implement those policies by giving incentives to pesticide users. To establish which users are successful, and deserve grants, EPA is continually altering alternative regulations to establish baseline performance standards.[89]

CONCLUSION

To promote public health, personal safety, and environmental protection, pesticides and their use are extensively regulated. Pesticide regulation is primarily decided at the federal level, and enforced by the EPA. Although the states generally copy the federal laws, some state laws are more restrictive than federal laws. Both federal and state laws provide for criminal prosecution and can impose penalties such as fines or imprisonment. In addition, common law actions are used to regulate pesticide use. Common law actions are civil claims brought by private citizens based on an allegation of improper pesticide production or use. When a pesticide manufacturer or user causes personal injury or property damage, the victim can initiate a civil action to recover financial compensation for the harm.

Regulating the manufacture and use of pesticides is necessary to protect human health and the environment, but those regulations require enforcement. Although EPA penalties are a deterrent, the specters of cancelled registration or toxic tort litigation offer successful alternatives, especially in light of the Supreme Court's relaxation of the limitations on state tort claims. In addition, the recent bout of environmental litigation compelling increased EPA enforcement of pesticide regulations has clearly impacted EPA's model of enforcement, which will hopefully lead to overall social benefit.

ACKNOWLEDGMENTS

Special thanks to Fred Whitford, coordinator of Purdue Pesticide Programs, and Dr. Fred Fishel (University of Florida/Institute of Food and Agricultural Sciences) for reviewing this entry.

REFERENCES

1. Olexa, M.T.; Leviten, A., Eds. *Circular 1139: Farm and Ranch Handbook of Florida Solid and Hazardous Waste Regulation*; University of Florida Cooperative Extension Service, 1999; 10.
2. *The Insecticide Act of 1910*, Pub. L. No. 6–152, 36 Stat. 331, 1910.
3. *Fungicide and Rodenticide Act of 1947*, Pub. L. No. 80–104, 61 Stat. 163, 1947.
4. *The Federal Environmental Pesticide Control Act of 1972*, Pub. L. No. 92–516, 86 Stat. 973(codified at 7 U.S.C. § 136, 1988; see also 7 U.S.C. §§ 135-135y, as amended 7 U.S.C. §§ 136 et seq., 2009.
5. Olexa, M.T.; Kubar, S.; Cunningham, T.; Meriwether, P., Eds. *Bulletin 311: Laws Governing Use and Impact of Agricultural Chemicals: An Overview*; University of Florida Cooperative Extension Service, 1995; 7.
6. 21 U.S.C. §§ 301 et seq. seq., 2009.
7. *Food Quality Protection Act of 1996*, Pub. L. No. 104–170, 110 Stat. 1489, 1996.
8. 7 U.S.C. § 136 j(a)(2)(G), 2009.
9. 7 U.S.C. §§ 136v (a), (b), (c)(1), 2009.
10. 7 U.S.C. §§ 136 et seq., 2009.
11. 7 U.S.C. §§ 136a (a)–(b), 2009.
12. 7 U.S.C. § 136a (c)(5)(C), 2009.
13. 7 U.S.C. § 136a (d)(1) (A–C), 2009.
14. 7 U.S.C. § 136a (d)(1) (B), 2009.
15. 7 U.S.C. § 136a (d)(1) (C), 2009.
16. 7 U.S.C. § 136a-1 (a), 2009.
17. 7 U.S.C. § 136v (a), 2009.
18. 7 U.S.C. § 136v (c)(1), 2009.
19. 7 U.S.C. § 136v (c)(3–4), 2009; see also 40 CFR §162 (D) (regulations governing state pesticide registration authority and suspension, state registration procedures, and the disapproval of state registrations).
20. 7 U.S.C. § 136w (b), 2009; 40 CFR § 152.20, 2010 (pesticides regulated by another federal agency); 40 CFR § 152.25, 2010 (pesticides that don't require regulation).
21. 40 CFR § 152.25, 2010.
22. Olexa, M.T.; Kubar, S.; Cunningham, T.; Meriwether, P., Eds. *Bulletin 311: Laws Governing Use and Impact of Agricultural Chemicals: An Overview*; University of Florida Cooperative Extension Service, 1995; 11.
23. EPA. *Pesticide Field Programs: Ensuring Protection through Partnerships*, 2006; 5–6.
24. EPA. *Pesticide Field Programs: Ensuring Protection through Partnerships*, 2006; 16.
25. 7 U.S.C. §§ 136a (a)–(b), 2009.
26. 7 U.S.C. §§ 136k–1, 2009.
27. 7 U.S.C. § 136l, 2009.
28. 7 U.S.C. § 136d, 2009.

29. Olexa, M.T.; Kubar, S.; Cunningham, T.; Meriwether, P., Eds. *Bulletin 311: Laws Governing Use and Impact of Agricultural Chemicals: An Overview*; University of Florida Cooperative Extension Service, 1995; 14–16. [pubmed]; 21 U.S.C. § 337, 2009.
30. 21 U.S.C § 346a (n)(3–4), 2009.
31. 21 U.S.C. § 331 (a), 2009.
32. 21 U.S.C. §§ 332–334, 2009.
33. 21 U.S.C. § 342 (a)(1–2), 2009.
34. 21 U.S.C. § 342 (a)(2)(A–B), 2009; 21 U.S.C. § 346a (a) (2009).
35. 21 U.S.C. § 346a (a)(1) and (4), 2009.
36. 21 U.S.C. § 346a (b)(2)(A), 2009.
37. 21 U.S.C. § 1402 (a), 2009.
38. Eubanks, W.S., II. A rotten system: Subsidizing environmental degradation and poor public health with our nation's tax dollars. Stan. Environ. Law J. **2009**, 213, 277.
39. Watnick, V.J. Our toxics regulatory system and why risk assessment does not work: Endocrine disrupting chemicals as a case in point. Utah Law Rev. **2004**, 1305, 1307–1309, 1310–1311.
40. Pollan, M. *The Omnivore's Dilemma: A Natural History of Four Meals*; Penguin Press: New York, 2006; 178.
41. Angelo, M.J. The killing fields: Reducing the casualties in the battle between U.S. Species Protection Law and U.S. Pesticide Law, Harvard Environ. Law Rev. **2008**, *32*, 95, 97–98.
42. 16 U.S.C. § 1538 (a)(1), 2009; 16 U.S.C. § 1532 (19), 2009.
43. Angelo, M.J. The killing fields: Reducing the casualties in the battle between U.S. Species Protection Law and U.S. Pesticide Law, Harvard Environ. Law Rev. **2008**, *32*, 95, 98–101, 111.
44. *Defenders of Wildlife v. Administrator, EPA*, 882 F.2d 1294 (8th Cir. 1989).
45. *Defenders of Wildlife v. Administrator, EPA*, 882 F.2d 1294, 1296–1299 (8th Cir. 1989).
46. *Defenders of Wildlife v. Administrator, EPA*, 882 F.2d 1294, 1301 (8th Cir. 1989).
47. 16 U.S.C. § 1536 (a)(1–2), 2009.
48. *Wash. Toxics Coal. v. EPA*, 413 F.3d 1024, 1031–1032 (9th Cir. 2005) *cert. denied*, 546 U.S. 1090, 2006.
49. *Wash. Toxics Coal. v. EPA*, 413 F.3d 1024, 1031–1033 (9th Cir. 2005).
50. *Headwaters, Inc. v. Talent Irrigation Dist.*, 243 F.3d 526, 532 (9th Cir. 2001).
51. 33 U.S.C. § 1311 (a), 2009.
52. *Headwaters, Inc. v. Talent Irrigation Dist.*, 243 F.3d 526, 531–532 (9th Cir. 2001).
53. *National Cotton Council of America v. EPA*, 553 F.3d 927 (6th Cir. 2009).
54. *National Cotton Council of America v. EPA*, 553 F.3d 927, 929–932 (6th Cir. 2009); 33 U.S.C. §§ 1311 (a) and 1342 (2006).
55. *National Cotton Council of America v. EPA*, 553 F.3d 927, 936 (6th Cir. 2009).
56. *National Cotton Council of America v. EPA*, 553 F.3d 927, 937 (6th Cir. 2009).
57. *National Cotton Council of America v. EPA*, 553 F.3d 927, 931 (6th Cir. 2009).
58. 7 U.S.C. § 136v (a) and (b), 2009.
59. Mandler, J.P. FIFRA preemption in the wake of *Bates v. Dow*. Nat. Resour. Environ. **2005**, *20*, 64.
60. 7 U.S.C. § 136 (q)(1) and (2), 2009.
61. *Cipollone v. Liggett Group, Inc.*, 505 U.S. 504 (1992) cited by Bates v. Dow Agrosciences LLC, 544 U.S. 431, 441, 2005.
62. *Bates v. Dow Agrosciences LLC*, 544 U.S. 431, 441–442, 2005.
63. *Bates v. Dow Agrosciences LLC*, 544 U.S. 431, 436, 2005.66.
64. *Bates v. Dow Agrosciences LLC*, 544 U.S. 431, 443–445, (2005) (jury verdict is not a requirement to change, merely motivation).
65. *Bates v. Dow Agrosciences LLC*, 544 U.S. 431, 451, 2005.
66. *Bates v. Dow Agrosciences LLC*, 544 U.S. 431, 452, 2005 (pre-emption of any statutory or common-law rule that would impose a labeling requirement divergent from FIFRA, but not of state rules fully consistent with federal requirements).
67. *Bates v. Dow Agrosciences LLC*, 544 U.S. 431, 445–446, 2005.
68. Whitford, F.; Olexa, M.T.; Thornburg, M.; Gunter, D.; Ward, J.; Lejurne, L.; Harrison, G.; Becovitz, J., Eds. *Pesticides and the Law: A Guide to the Legal System*; Purdue University Cooperative Extension System: West Lafayette, Indiana, 1996; 12.
69. Restatement (Third) of Torts: Liab. Physical Harm § 6, 2010.
70. Payne, A. Causes of action for damages from pesticides. Causes of Action 2d **2009**, *39*, 579.
71. *Obendorf v. Terra Hug Spray Co., Inc.*, 188 P.3d 834 (Idaho, 2008).
72. *Farm Services, Inc. v. Gonzales*, 756 S.W.2d 747 (Tex. App. Corpus Christi 1988) (doctrine requires defect); Eaton Fruit Co. v. California Spray-Chemical Corp., 445 P.2d 437, 440 (Ariz. 1968) (plaintiff must establish defendant's exclusive control of the pesticide, as well as proper preparation and application of the pesticide); Stone's Farm Supply, Inc. v. Deacon, 805 P.2d 1109, 1114 (Colo. 1991) (supplier sold farmer chemical containing substances unsuitable for farmers' crops).
73. *Rhoten v. Dickson*, 223 P.3d 786 (Kan. 2010); Terry v. Caputo, 875 N.E.2d 72 (Ohio 2007).
74. Keeton, et al., Eds. *Prosser and Keeton on the Law of Torts*, 5th Ed.; Section 16; West Publishing Co.: St. Paul, Minnesota, 1984; 67.
75. Restatement (Second) of Torts § 165, 2010.
76. *Fisher v. Ciba Specialty Chemicals Corp.*, 245 F.R.D. 539, 69 Fed. R. Serv. 3d 119 (S.D. Ala. 2007).
77. Keeton, et al. Eds. *Prosser and Keeton on the Law of Torts*, 5th Ed. Section 87 1984; 622.
78. Fischer, J.; Olexa, M.T.; Leviten, A.; Saju, A., Eds. *Circular 1224: Handbook of Florida Agricultural Laws*; University of Florida Cooperative Extension Service: Gainesville, Florida, 1999; 4.
79. Restatement (Second) of Torts § 821B (2010) (public nuisance requires unreasonable interference with right common to the public); *see also* 58 Am. Jur. 2d, Nuisances § 39 (2010); 58 Am. Jur. 2d, Nuisances §§ 42–47 (private nuisance is an invasion of interest in private use and enjoyment of land).

80. Restatement (Second) of Torts § 821D, 2010.
81. Grossman, M.R. Biotechnology, property rights, and environment. Am. J. Comp. Law **2002**, *50*, 215.
82. Keeton, et al., Eds. *Prosser and Keeton on the Law of Torts*, 5th Ed. Section 75 1984; 536–537.
83. Restatement (Second) of Torts § 520, 2010.
84. Restatement (Third) of Torts: Products Liability § 2.
85. In re DuPont-Benlate Litigation, 859 F. Supp. 619 (D.P.R. 1994).
86. Am. L. Prod. Liab. 3d § 110:1, 2010.
87. EPA. *Pesticide Field Programs: Ensuring Protection through Partnerships*, 2006; 8.
88. Eubanks, W.S. II. A rotten system: Subsidizing environmental degradation and poor public health with our nation's tax dollars. Stan. Environ. Law J. **2009**, *28*, 213, 276.
89. EPA. *Pesticide Field Programs: Ensuring Protection through Partnerships*, 2006; 9–14.

Pest Management: Modeling

Andrew Paul Gutierrez
Center for the Analysis of Sustainable Agricultural Systems, University of California—Berkeley, Berkeley, California, U.S.A.

Abstract
Pest management is a key component of sustainable agriculture and may be defined as applied population ecology focusing on human managed populations of plants and/or domesticated animals and their pests and natural enemies in environments modified by weather and agro-technical inputs. The complexity of managing pests in an agro-ecosystem requires the development of models that enable the separation of losses due to pests from yield variation due to weather and agronomic practices. Such models provide the bases for evaluating the dual objectives in modern agriculture of *minimizing* inputs that cause adverse environmental, human, and animal health effects and *maximizing* net profits. In modern societies, these goals may result in conflicts between public and private interests, but this is usually not the case in subsistence agriculture where the goal is often yield stability. The components and methods commonly used in crop pest management research may also apply to research on medical and veterinary pests. The methods to analyze these problems fall under the ambit of agro-ecosystems analysis. The analysis must be tri-trophic in scope in most systems because natural processes such as biological and natural control, when correctly managed, can be used to replace disruptive pesticide inputs that may induce resurgence of target pests, outbreaks of secondary pests, and pesticide resistance. Agronomic inputs (e.g., pesticides, fertilizers, water) may also impact pest levels and must be considered as part of the pest management system. Several modeling approaches have been used in pest management, and they may be broadly classed as empirical, statistical, operations research, analytical, and simulation.

INTRODUCTION

Pest management is a key component of sustainable agriculture and may be defined as applied population ecology focusing on human managed populations of plants and/or domesticated animals and their pests and natural enemies in environments modified by weather and agro-technical inputs. The complexity of managing pests in an agro-ecosystem requires the development of models that enable separation of losses due to pests from yield variation due to weather and agronomic practice.[1] Such models provide the bases for evaluating the dual objectives in modern agriculture of *minimizing* inputs that cause adverse environmental, human, and animal health effects and *maximizing* net profits. In modern societies, these goals may result in conflicts between public and private interests, but this is usually not the case in subsistence ones[2] where the goal is often yield stability.

Some components of crop pest management research are illustrated in Fig. 1a, but the same approach applies to medical and veterinary pest management. The methods fall under the ambit of agro-ecosystem analysis.[3] The analysis must be tri-trophic in scope because natural processes such as biological control, when correctly managed, may replace disruptive pesticide inputs that may induce resurgence of target pests, outbreaks of secondary pests, pesticide resistance, and pollution. Misuse of other agronomic inputs (e.g., fertilizers, water) may also exacerbate pest levels and must be considered in the pest management system. Several modeling approaches have been used in pest management, and they may be broadly classed as empirical, statistical, operations research, analytical, and simulation.[1–6] To be truly sustainable, the system must be ecologically, economically, and socially sustainable (Fig. 1b).

MODELING APPROACHES

Empirical models based on trial and error have a long history and accumulated common wisdom. Traditional societies worldwide developed pest management strategies based on common wisdom that reduced pest damage and led to sustainable crop production. These systems may fail when modern agro-technical inputs are introduced and may require the accumulation of new experience to resolve management problem.

Population sampling and agronomic trials have been used to develop statistical models to assess the costs and benefits of pest levels and timing agronomic inputs (e.g., economic threshold studies). These models tend to be static and hence are time and place specific. They may not be able to be used to evaluate factors beyond the range of the data used to develop the model. Such models have, however, yielded useful pest management decision rules.

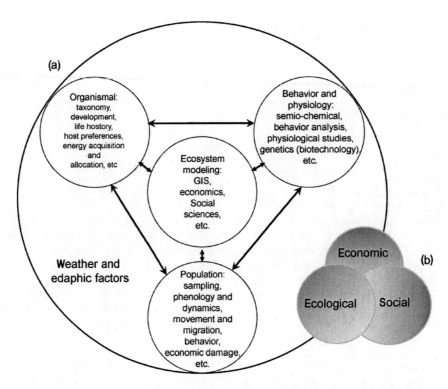

Fig. 1 Components of pest management research using (a) systems analysis as a unifying tool and (b) components of sustainability.
Source: Courtesy of Center for the Analysis of Sustainable Agricultural Systems (CASAS).

Attempts have been made to use operations research methods from engineering and economics to determine optimal pest management strategies. Such methods require sufficient knowledge of the system to formulate robust optimization models. Often this information is lacking, and/or the biology may be too complicated, requiring large reductions in the dimensions of the model. These simplifications often obviate their general utility except for general strategic analysis.

Pest management is *applied population ecology* and hence lends itself to demographic modeling methods. Mathematical models of the population dynamics may be simple descriptors of the system having analytical solutions or they may include considerable biological detail requiring numerical simulation. Analytical models may give strategic insights about the system, while richer simulation models may yield tactical recommendations for specific abiotic and biotic conditions. Modeling for tactical decision making requires a detailed understanding of the biology of the interacting species as modified by weather, edaphic, and agronomic factors. Biologically rich mathematical models are usually implemented as computer simulations.

There are many approaches to simulation modeling, but increasingly, mechanistic, physiologically based models of the energy dynamics of cropping systems are being developed because the important role of the plant as the integrator of all factors is being increasingly recognized. Energy is the currency of biology that when multiplied by price leads directly to economic models.

SIMULATION MODELS OF CROP SYSTEMS

Crops are age-structured populations of plants each with *time-varying mass-age-structured* subunit populations of fruits, leaves, stems, and roots. The number in each age-mass class may vary overtime in response to factors that affect birth–death rates of whole plants and/or their subunits as well as those of higher trophic level populations. The crop may be viewed as a canopy of average plants with populations of pests and natural enemies within, or as individual plants each with its own set of pest and natural enemy subpopulations with migration between plants or fields (i.e., a meta-population). Modeling of individuals in pest and natural enemy populations (individual-based models) is not recommended in pest management because the decision rules for individual behavior are rarely known and the computations may be extensive. Simpler approaches that capture the *meta* details of physiologically based approaches may also be used.[7]

A basic premise of the physiological approach is that all organisms, including the economic one, face the same processes of resource (energy) acquisition and allocation.[1,8] Physiological models often assume energy allocation priority first to respiration (maintenance costs in economics), then reproduction (profit) and if assimilate (revenues) remains, then allocation to growth and reserves (infrastructure and savings). The shapes of the acquisition functions are concave (photosynthesis, predations, etc.), and those for the maintenance costs are positive exponential (the Q_{10}

rule) with the concave net of the two functions being the amount of resources available for allocation. These analogies[3] allow the use of the same model to describe the dynamics of all species (including the economic one) at the per-capita, population and regional levels (Fig. 2). Each organism is assumed to try to satisfy a genetic (economic) demand for resources, but the process involves imperfect search, and the supply of a resource obtained is always less than (or equal to) the demand, and shortfall cause reductions in growth, reproduction, and survival rates from the maximum. In the model, all vital rates are controlled by one or more limiting supply/demand ratio with only the units and interpretation of the flow rates differing among species.

How biotic and abiotic factors affect plant growth and development is central to developing plant system models. These factors may affect either the supply (production) or the demand (sinks, e.g., fruits) side of the supply/demand ratio. Occasionally, both sides may be affected, and in other cases, the pests may attack the standing crop. The supply/demand paradigm simplifies model development, allowing assessment of yield loss and in some cases yield compensation in the face of pest damage. Such models facilitate evaluation of the costs and benefits of pest management options basic to the development of dynamic economic thresholds. Incorporating the biology of these factors in the model enables the development of highly realistic models.

SUPPLY-SIDE PESTS

Supply-side pests reduce the photosynthetic rate in various ways. Important supply-side pests are defoliators, sapsuckers, spider mites, nematodes, and diseases. Defoliation depletes leaves and may cause wound healing losses, but the effects on yield depend on the age of leaves attacked, the loss rate, and compensation due to increased light penetration to still intact leaves. In contrast, spider mites kill leaf cells, reducing photosynthesis in damaged leaves that are not shed, and reducing light penetration to lower leaves. Stem borers and vascular plant diseases may slow the photosynthetic rate by reducing the translocation of water and nutrients, and some may kill whole plants. Pests such as thrips and armyworms may damage the apical meristem, inducing developmental delays and reducing yield.

PLANT DEMAND-SIDE PESTS

Pests may attack fruit (e.g., net sinks), causing premature abscission, thereby altering present and future demands for photosynthate. High abscission rates may cause rank growth as the photosynthate is allocated to vegetative growth. Many plant species have reproductive capacity that allow varying degrees of compensation. Some plants may be determinant as all fruit are produced at roughly the same time (e.g., apple) and others are indeterminate

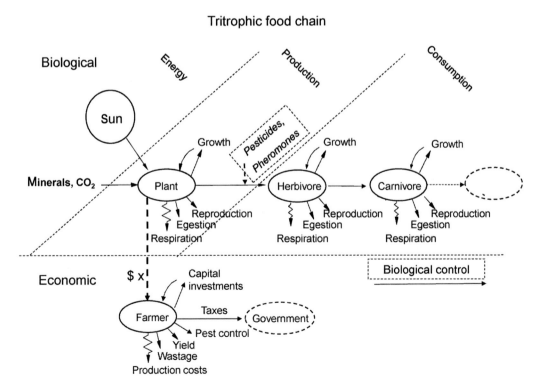

Fig. 2 Analogies between trophic levels including the economic one.
Source: Courtesy of John Wiley and Sons, New York.

(e.g., cotton), producing fruit until the maximum load is reached or the plant is limited by other factors. Of crucial importance in compensation is the time and energy lost in abscised fruits. Little time and energy may be lost when new buds and small fruit are abscised, and in indeterminate species such as cotton, replacement buds may be produced at rates sufficient for complete compensation. In determinate species such as apple, the time for compensation may be very short and compensation may be impossible if the attack rate is too high. In both growth types, attacks on older fruits may involve considerable losses in time and energy that often precludes compensation.

The ratio of the cumulative buds initiated to the cumulative numbers abscised may provide the basis for determining whether and how much compensation is possible. Such data yield a concave function that estimates the *compensation point* and may provide a rule of thumb for estimating the economic threshold. For example, losses of 30% of fruit bud in many cotton varieties do not affect yield.

OTHER KINDS OF PLANT PESTS

Some pests affect both sides of the supply/demand ratio. For example, pests attacking the apical meristem of a plant or branch kill primordial tissues that introduce time delays and alter future fruit and vegetative dynamics. Other pests may attack the standing crop of fruit without causing fruit to shed and without appreciable reductions in sink demands, making plant compensation in such cases unlikely. Furthermore, damage in such cases accrues over time.

MODELING THE IMPACT OF BIOTECHNOLOGY

Increasingly, crop plants are being genetically modified to protect them from insect pests and to make them tolerant to herbicides designed to kill weed competitors, and these developments have added an important genetic level to modeling pest management. One of the most common

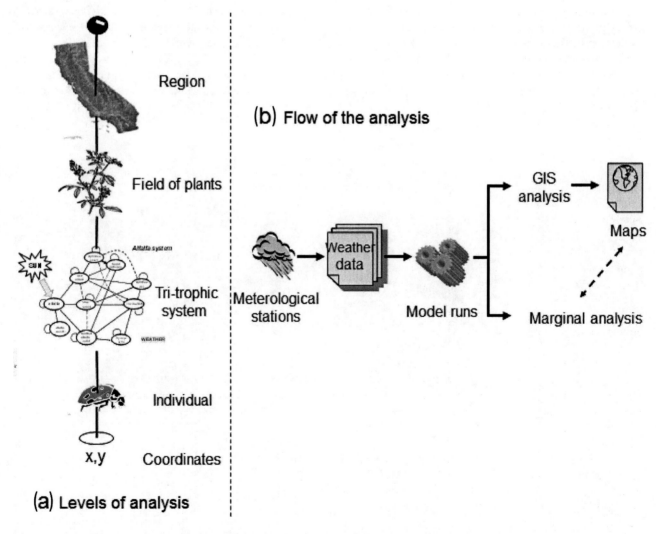

Fig. 3 (a) Application of the same dynamics model (see Fig. 2) at the individual, population, field, or larger geographic area, and (b) the use of the simulation model in regional GIS and marginal analysis.

genetic modifications is the insertions of genes from the bacterium *Bacillus thuringiensis* into transgenic crops for the production of Bt toxin. This innovation is thought to be environmentally friendly, and numerous recent economic studies concluded high benefits and good prospects for the technology (mainly in cotton and maize) in the United States, Australia, Argentina, China, South Africa, and other parts of Africa and Asia.[9] In many of these benefit studies, econometric methods were applied to cross-sectional data from farm surveys or experimental data, and while these methods can provide a good assessment of the static productivity of pest control agents (and other inputs), they are less suitable for capturing the interaction between control decisions and dynamic ecosystem reactions. The complexity introduced by biotechnology is exemplified by the cotton system: There are more than a dozen herbivore pests in cotton [e.g., pink bollworm, tobacco budworm, cotton bollworm, rough bollworm, fall armyworm, beet armyworm, cabbage looper, soybean looper, plant bugs (*Lygus* spp.), whitefly, cotton bollweevil, spider mites, and others], and each of them is affected differently by Bt toxin(s). In addition, their natural enemies may also be affected but at levels considerably less than by pesticides.[10] This complexity was examined using modeling that suggested the potential for secondary outbreaks of pests (i.e., plant bugs) induced by the use of Bt cotton.[10] Increases in secondary pests in Bt cotton have occurred in the Southeastern United States, India, and China,[11] with concomitant increased use of insecticides.[12] The pyramiding of new technologies (more Bt and other toxins) adds more complexity to the analysis. Modeling studies in transgenic crops are also made difficult by intellectual property rights that limit field scientific investigation.

The use of herbicide-resistant crop has increased herbicide use, led to the development of herbicide-resistant weeds, and increased environmental pollution and non-target effects (e.g., amphibious species).[13] Rather than simplifying the pest management system, biotechnology has greatly increased the complexity in the long run.

REGIONAL PEST MANAGEMENT

Modeling, geographic information system (GIS), and marginal analysis may be used to assess the regional distribution and abundance of pest and their biological control agents due to weather and over the long run of climate. Fig. 3a illustrates how the same model at the individual, tri-trophic, field, area, or regional level applies. Weather data [daily maximum and minimum temperatures (°C), solar radiation (kcal cm^{-2} day^{-1}), rainfall (mm), daily runs of wind (km day^{-1}), and relative humidity] for each location on a grid (or numerous locations) are used to run the model over given time periods (days, months, years). Appropriate summary variables from the model may be mapped using GIS technology (Fig. 3b) and used to show the aggregate regional effects of weather on biological system dynamics.[14] The GIS simulation data may also be analyzed using various statistical methods (e.g., linear multivariate regression), keeping only independent variables and interactions with slopes significantly different from zero (t values, $p <$ 0.05) in the statistical model. As in econometrics, marginal analysis may be used to estimate the average magnitude and direction of large effects ($\partial y/\partial x_i$).

CONCLUSIONS

Interactions among pests and higher trophic levels (top–down effects) are complicated, time varying, and best integrated via the plant's physiology and dynamics as modified by abiotic factors (bottom–up effects). Tri-trophic models based on the same supply/demand paradigm enable assembling and understanding the top–down biological relationships.[15,16] These models provide the bases for developing ecologically sound pest management strategies that may be implemented as part of a total farm expert system or regionally using GIS technology to develop region-wide pest management strategies. All of these methods fall under the ambit of agro-ecosystems analysis.

REFERENCES

1. Gutierrez, A.P., Ed. *Applied Population Ecology: A Supply–Demand Approach*; John Wiley and Sons, Inc.: New York, 1996.
2. Altieri, M.A. Sustainable agriculture. Encycl. Agric. Sci. **1994**, *4*, 239–247.
3. Gutierrez, A.P.; Curry, G.L. Conceptual framework for studying crop–pest systems. In *Integrated Pest Management Systems and Cotton Production*; Frisbie, R.E., El-Zik, K.M., Wilson, L.T., Eds.; John Wiley and Sons: New York, 1989; 37–64.
4. Curry, G.L.; Feldman, R.M., Eds. *Mathematical Foundations of Population Dynamics*, Series No. 3; TEES Monograph, 1987, Texas A&M Press, College Station, Tx.
5. deWit, C.T.; Goudriaan, J., Eds. *Simulation of Ecological Processes*, 2nd Ed.; PUDOC Publishers: the Netherlands, 1978.
6. DiCola, G.; Gilioli, G.; Baumgärtner, J. Mathematical models for age-structured population dynamics. In *Ecological Entomology*, 2nd Ed.; Huffaker, C.B., Gutierrez, A.P., Eds.; John Wiley and Sons: New York, 1998.
7. Gutierrez, A.P.; Daane, K.M.; Ponti, L.; Walton, V.L.; Ellis, C.K. Prospective evaluation of the biological control of the vine mealybug: Refuge effects. J. Appl. Ecol. **2007**, *44*, 1–13.
8. Regev, U.; Gutierrez, A.P.; Schreiber, A.P.; Zilberman, D. Biological and economic foundations of renewable resource exploitation. Ecol. Econ. **1998**, *26*, 227–242.
9. Qaim, M.; Zilberman, D. Yield effects of genetically modified crops in developing countries. Science **2003**, *299*, 900–902.

10. Gutierrez, A.P.; Adamcyzk, J.J. Jr.; Ponsard, S. A physiologically based model of *Bt* cotton–pest interactions: II. Bollworm–defoliator–natural enemy interactions. Ecol. Modelling **2006**, *191*, 360–382.
11. Lu, Y.; Wu, K.; Jiang, Y.; Xia, B.; Li, P.; Feng, H.; Wyckhuys, K.A.G.; Guo, Y. Mirid bug outbreaks in multiple crops correlated with wide-scale adoption of Bt cotton in China. Science **2010**, *328*, 1151–1154.
12. Pemsl, D.; Gutierrez, A.P.; Waibel, H. The economics of biotechnology under ecosystems disruption. Ecol. Econ. **2007**, *66*, 177–183.
13. Relyea, R.A. The lethal impact of roundup on aquatic and terrestrial amphibians. Ecol. Appl. **2005**, *15* (4), 1118–1124.
14. Gutierrez, A.P.; Ponti, L.; Ellis, C.K.; d'Oultremont, T. Analysis of climate effects on agricultural systems: A report to the Governor of California sponsored by the California Climate Change Center, **2006**, available at http://www.climatechange.ca.gov/climate_action_team/reports/index.html.
15. Gutierrez, A.P.; Neuenschwander, P.; van Alphen, J.J.M. Factors affecting the establishment of natural enemies: Biological control of the cassava mealybug in West Africa by introduced parasitoids: A ratio-dependent supply–demand driven model. J. Appl. Ecol. **1993**, *30*, 706–721.
16. Graf, B.; Gutierrez, A.P.; Rakotobe, O.; Zahner, P.; Delucchi, V. A simulation model for the dynamics of rice growth and development. II. Competition with weeds for nitrogen and light. Agric. Syst. **1990**, *32*, 367–392.

Pesticide Translocation Control: Soil Erosion

Monika Frielinghaus
Detlef Deumlich
Roger Funk
Institute of Soil Landscape Research, Leibniz Center for Agricultural Landscape Research (ZALF), Muncheberg, Germany

Abstract
The pesticide translocation due to wind and water erosion causes heavy environmental damages. Sediment transport by surface runoff or wind is the most important non-point pollutant generated by agriculture in many regions. The monitoring performed in agrarian landscapes with a high erosion risk is predicted upon the estimation of rain and wind erosivity, soil erodibility, and morphological pattern. Pesticide transport models are rare. Therefore, soil erosion models serve as a basis for risk assessment, completed by the characterization of pesticide management and erosion control.

SOIL EROSION: A GLOBAL PROBLEM

One-third of the world's agricultural soils are reported to be in a degraded state. Water and wind erosion contribute to approximately 84% of the observed deterioration. The worldwide soil loss from soil erosion is estimated at about 75 billion tons yr^{-1}.[1] The changes caused by soils by human-induced erosion over many years are significant and have often resulted in valuable land being abandoned.[2] Given a very slow rate of soil formation, any soil loss of more than 1 ton ha^{-1} yr^{-1} within a time span of 50–100 yr can be considered as irreversible. Soil from the Sahel zone in Africa is picked up in Florida and the Brazils Amazonas region and blown across the Atlantic Ocean each year. Regularly, when the Chinese till their land in spring, whirled-up soil is detected in Hawaii. Most soil loss can be traced to inappropriate land management practices. Viewed on a large scale, these scenarios create an agricultural dilemma of global proportions.

Pesticides are used in nearly all agricultural systems and provide essential help against diseases, pests, and weeds, thereby contributing substantially to securing global food supply. However, the pesticides' translocation due to wind and water erosion causes heavy environmental damages.[3] In many regions, sediment transport by surface runoff or wind is the most important non-point pollutant generated by agriculture. While pollutants may be too insoluble or soil bound to be transported in runoff water, erosion can mobilize them into runoff. Erosion sediment is also an important vehicle for the translocation of pesticides that are water insoluble or strongly bound to eroding soil. Although wind erosion has been overlooked in the past as a land degradation process, it has received more attention now as an important source of atmospheric pollution.[4] Mainly, this is attributed to the removal of fine soil particles and organic material, carrying the adsorbed pesticides.

WATER AND WIND EROSION AND SEDIMENT TRANSPORT

The processes of soil erosion involve detachment of material by two processes: raindrop impact and drag force traction. Material is then transported either by overland water flow or by saltation through the air.

Water Erosion

Runoff caused by heavy rainstorms (high intensity or long duration) is the most important direct driver of severe soil erosion by water (Fig. 1). Mean losses of 20 to 40 tons ha^{-1} in individual storms may happen once every 2 or 3 years. More than 100 tons ha^{-1} soil loss is measured after extreme storms.[5]

The water erosion rate is a complex result of high rainfall amount or intensity, soil erodibility (instability), slope steepness and length, and the type of land use. As to the latter, the main causes are inappropriate agricultural practices, deforestation, overgrazing, forest fires, and construction activities.

The formation of a water erosion system on agricultural areas without surface protection by crops or plant residues causes rainwater and sediment transport in rills or gullies. The first phase after ponding is a concentration of water at linear paths. The rill formation may be influenced by a variety of different factors like soil crusting, wheel tracks, and reduced infiltration by soil compaction. Rills and gullies often develop in close proximity to one another. The second phase is a heightened concentration of water at

Fig. 1 Water erosion with sediment translocation at a corn field.
Source: Photo courtesy of Deumlich.

Fig. 3 Wind erosion at a field after seedbed preparation.
Source: Photo courtesy of Schäfer.

morphological deep lines (thalways). The hydrological power of transport water as well as the off-site risk for sediment and pesticide transport into the lakes or rivers is increased with the thalways' catchment area (Fig. 2).

Often, the interaction between runoff and sediment transport complicates the process' description.

Wind Erosion

Wind erosion occurs when three conditions coincide: high wind velocity, a susceptible surface with loose particles, which can be picked up, and insufficient surface protection by plants or plant residues (Fig. 3).

Wind erosion results from wind moving across a dry soil surface and dislodging soil particles by pressure and lifting forces. The process is self-perpetuating; blowing sediment disturbs additional particles that are then lifted into the airstream.

The potential wind erosion rate is influenced by high wind velocity, soil erodibility (size/weight of particles), the degree to which the landscape is wind exposed, and the uses of land. The real wind erosion risk depends on the actual soil moisture and the real soil cover rate.

The modes of soil particle motion are closely related to particle size, density, and shape. It is important on agriculturally used soils of each textural class with organic material and absorbed pesticides (Fig. 4). Particles with a smaller terminal velocity than the turbulent motions become suspended. Particles smaller than 20 μm are subjected to long-term suspension whereby they can be carried across several hundred kilometres for several days. Particles with diameters between 20 and 70 μm remain suspended for only a few hours and cannot be transported very large distances. These kinds of small particles are an important factor regarding the adsorption of pesticides.[6]

Fig. 2 Sediment and pesticide transport into a channel caused by water erosion.
Source: Photo courtesy of Frielinghaus.

Fig. 4 Sediment and pesticide transport into a channel caused by wind erosion.
Source: Photo courtesy of Frielinghaus.

ASSESSMENT OF PESTICIDE TRANSLOCATION

Pesticide findings even during the non-spraying seasons indicate that the occurrence is not only due to direct contamination from accidental spills and incorrect handling but also to diffuse contamination originating from normal pesticide use.

Pesticide mobility may result in redistribution within the application site or movement of some amount of pesticide off-site.

Pesticide characteristics are relevant for determining the fate of transport by wind or water erosion. These properties include solubility in water, tendency to adsorb to the soil, and pesticide persistence in soil.[7]

Water solubility is measured in milligrams per liter. Pesticides with more than 30 mg L^{-1} solubility tend to move with surface runoff during an erosion event. Pesticides with solubility about less than 1 mg L^{-1} tend to remain on the soil surface and may move with soil sediment in surface runoff if soil erosion occurs.

Soil adsorption is measured by K_{oc}, which is the tendency of pesticides to be attached to soil particles. Higher values (>1000) indicate a pesticide that is strongly attached to soil and less likely to move unless sediment transport occurs. Lower values (<300–500) indicate pesticides that tend to move with surface runoff. Among other things, sorption is influenced by soil moisture, organic matter content, and texture. These parameters are also important to quantify soil erodibility. Soils high in clay or organic matter content, or both, have a higher potential to adsorb pesticides, because small particles have plenty of surface area and are chemically more active compared with sandy soil particles.

Pesticide persistence is measured in terms of the half-life, or the time in days required for a pesticide to degrade in soil to one-half of its original amount. A pesticide with a half-life >21 days may persist long enough to move with surface runoff or erosion before it degrades. For example, experiments with dieldrin and aldrin show half-life values of about 3 or 4 years. These two pesticides lost 90% of their original concentration in no less than 10 years.[8]

Field investigations to estimate the pesticide transport during an erosive rain were known by Gouy et al.[9] and Klöppel et al.[10]

In the first experiment, pesticides were applied onto the bar soil surface. The simulated rainfall with an intensity of 33 and 44 mm hr^{-1}, respectively, began 20 hr after pesticide application (Table 1).

Atrazine, simazine, and alachlor, for example, were mainly transported in surface runoff. In contrast, 90% of transported trifluralin was adsorbed on eroded particles. Lindane, for example, showed an intermediate status and the high dynamic of pesticide translocation.

The second rainfall simulation study on small field plots provided information about worst cases for pesticide translocation (rainfall intensity of about 70 mm hr^{-1} and 100 mm accumulated rainfall).[10] The pesticides of the first group had water solubility concentrations of 65 and 700 mg L^{-1}, e.g., isoproturon and dichlorprop-p, respectively. The second group had solubility concentrations <1 mg L^{-1}, e.g., bifenox. Two different soil cover conditions were tested.

The results demonstrate that the influence of soil cover characteristics on concentrations and total runoff losses of pesticides for the highly soluble group is restricted to the start of the rain event after pesticide application and the runoff rate. The sediment concentration of pesticides in the fairly soluble group was comparable to those of the first runoff event independent of the time lapse between the application and rainfall event. The fairly soluble pesticides are thus exiting the field sites only by being adsorbed to eroded sediment. The cumulative soil loss is the most reliable indicator that explains the decrease in sediment and pesticide concentrations during each rainstorm (Table 2).

It is not possible to compare these results with other data from literature, due to the great variety of parameters concerning the performance of experiments. However, the findings are consistent in that a rough estimation of pesticide concentration in runoff and total pesticide losses is possible when rainfall (duration, intensity) and soil erosion potential are known.

Pesticide transport by wind may occur through isolated pesticide displacement or in a sediment-bounded form. At present, only initial findings of the amount of translocated pesticides are available. It is assumed that the horizontal transport of particulate-bounded pesticides is more than 50% of the total loss resulting from extreme wind erosion events. The herbicide loss of about 1.5% of the total amount, which was applied and integrated into the upper humus horizon on steppe soils in Canada, was measured after 13 wind erosion events. Herbicide loss from a non-integrated application on soil surfaces was approximately 4.5%.[11]

Experiments in a wind tunnel with the pre-emergence herbicide flurochloridone showed <1% to <58% displacement dependent on soil erodibility and measuring height.[6]

Table 1 Total amounts of pesticide in runoff samples and their distribution between sediment and liquid phases.

Pesticide	Water solubility (mg L^{-1})	Adsorption, K_{oc} (mL g^{-1})	Total amount (% of applied)	Distribution On sediment	Distribution In solution
Atrazine	33	100	13	0.8	12.2
Simazine	6	130	9	0.6	8.4
Alachlor	240	170	14	0.8	13.2
Lindane	7	1100	17	6.0	11.0
Trifluralin	1	8000	14	12.6	1.4

Table 2 Average pesticide losses in relation to applied pesticide dependent on time (rainfall intensity: 70 mm hr^{-1}, accumulated rainfall: 100 mm).

Name of pesticide	Pesticide transporting medium	Time and concentration lapse between pesticide application and rainfall event				
		2 hr	1 day	3 days	5–7 days	14 days
Application on bare soil surface sites (pre-emergence application) % of applied amount						
Isoproturon	Total	4.5–12.4	4.0–17.2	5.4–13.3	2.1–11.2	n.e.
	In runoff water	3.8–9.8	2.9–13.9	4.0–9.8	1.4–9.0	n.e.
	In eroded sediment	0.7–2.6	1.7–3.3	1.4–3.5	0.7–2.2	n.e.
Dichlorprop-P	Total	2.1–10.4	2.2–16.7	5.9–10.4	1.2–8.3	n.e.
	In runoff water	1.9–9.8	1.8–15.3	5.3–9.3	0.9–7.0	n.e.
	In eroded sediment	0.2–0.6	0.4–1.4	0.6–1.1	0.3–1.3	n.e.
Bifenox	Total	15.6–19.0	19.3–21.6	14.3–17.2	9.3–13.8	n.e
	In runoff water	<0.1–0.9	<0.1–0.3	<0.1–0.3	<0.1–0.2	n.e.
	In eroded sediment	14.7–19.0	19.0–21.6	14.0–17.2	9.3–13.6	n.e.
Application on small covered soil surface (barley with 3–5 leaves) % of applied amount						
Isoproturon	Total	2.8–13.3	1.8–16.4	4.1–11.2	1.5–6.0	1.6
	In runoff water	2.2–11.4	1.5–13.5	3.3–8.6	1.0–4.6	1.1
	In eroded sediment	0.4–1.9	0.3–2.9	0.8–2.6	0.5–1.4	0.5
Dichlorprop-P	Total	0.8–10.0	0.9–9.2	1.0–8.3	0.7–4.1	0.8
	In runoff water	0.7–9.4	0.8–8.4	0.8–7.5	0.6–3.7	0.7
	In eroded sediment	0.1–0.6	0.1–0.8	0.2–0.8	0.1–0.4	0.1
Bifenox	Total	11.2–15.0	7.8–15.5	7.8–15.9	3.4–9.8	5.2
	In runoff water	<0.1–0.9	<0.1–0.2	<0.1–0.4	<0.1–0.3	<0.1
	In eroded sediment	11.2–14.1	7.8–15.5	7.4–15.9	3.4–9.5	5.2

Determination of pesticide content in the atmosphere is extremely important but very difficult. Additionally, pesticides enter the atmosphere via many different processes, particularly by volatilization (simulated with PEARL, PELMO), unconsidered in our context.[12]

The investigation of atrazine, alachlor, and acetochlor concentrations on soil surface and dissipation rates of wind-erodible sediment and larger fractions from two soil types was important.[13] Undisturbed and incorporated (5 cm deep) soil surface were analyzed. The surface (1 cm) of soil was removed by vacuum 1, 7, and 21 days after herbicide treatment. About 50% of the recovered material was classified as wind-erodible sediment. This erodible sediment contained about 65% (undisturbed soil surface) and 8% (incorporated soil) of the applied herbicides, respectively, after 1 day. The concentrations were similar after 7 and after 21 days. However, a 50% dissipation rate for each herbicide was found after 15 days for wind-erodible sediments compared with 30–55 days for greater fractions. These data indicate that wind-erodible size aggregates and particles could be a source of herbicide contamination, but there is currently no information about quantities.

It is important to recognize that only one fraction of pesticides in use is very strongly soil bound as to be transported in the sediment phase of runoff principally today (insecticides paraquat and pyrethroid, and other non-ionic hydrophobic species). Such pollutants may be too insoluble or soil bound to be transported in runoff or wind stream, but erosion can mobilize them.

MONITORING AND MODELING

Monitoring in landscapes with a high erosion risk (water and wind) is predicted upon the estimation of rain and wind erosivity, soil erodibility, and the morphological factors of the areas (slope steepness, field length, wind openness, thalways). In sites with an erosion risk, well-designed field studies are the best way to assess off-site transport paths. When these data are not available, estimation about modeling will be necessary.

Water Erosion Models

Pesticide transport models are rare. Therefore, currently, water and wind erosion models serve as a basis for risk assessment, as a means to an end (Table 3).[14]

Most of the models were developed to assess impacts of different agricultural management practices; they are

Table 3 Selected erosion models and hydrological models with integrated pesticide transport.

Erosion models	Models with pesticide transport
ANSWERS	CREAMS
KINEROS	GLEAMS
EUROSEM	AnnAGNPS
LISEM	SWAT
EROSION 3D	
WEPP	

Note: Wind Erosion Models: Wind Erosion Prediction System (WEPS); Revised Wind Erosion Equation (RWEQ).
Source: http://www.soilerosion.net/doc/models_menu.html.

not adapted to predict exact pesticide, nutrient, or sediment loading in an area. The curve number method is an event or field scale orientated model and limits the results of some other models (ANSWERS, GLEAMS, CREAMS, and SWAT). In most erosion models, runoff and sediment load are only computed for the catchment outlet. Most of the hydrological models predict total runoff better than sediment load. The models over- or underestimate empirical results for small erosion events especially. Also, these models do not consider any attenuation or partitioning during transport and therefore fail to predict loads of soluble pesticides to surface waters.

WEPP and EROSION 3D models simulate the erosion and sediment transport continuously. The WEPP model is based on fundamentals of erosion theory, soil and plant science, channel flow hydraulics, and rainfall–runoff relationship.

It is possible to extend erosion and sediment transport models to a transport model for pesticides as well as sediment particles. The pesticide transport behavior that includes interaction processes in solutions is unknown for watershed scale.

Wind Erosion Models

A physical-based process model is the wind erosion prediction aystem (WEPS). This is a continuous, daily time-step model for simulation of weather, field conditions, and wind erosion. It has the capability of simulating spatial and temporal variability or soil surface parameters and soil loss or deposition within a field. To aid in the evaluation of off-site impacts, the soil loss is subdivided into components and reported as saltation creep, total suspension, and fine particulate matter components (PM 10). The transport capacity for insoluble pesticides bounded in kind of suspension or PM 10 is appreciable.[5]

Further models are the revised wind erosion equation (RWEQ) and WHEELS, which can be used for single-event simulation, long-term risk assessment, and assessment of changing management strategies.

SOIL EROSION AND PESTICIDE TRANSLOCATION CONTROL

The best management strategy requires inventory and risk analysis.

A step-by-step analysis is the basis for an effective pesticide translocation control:

Step 1: Monitoring the pesticide content in lakes or rivers.
Step 2: Quantification and elimination of pesticide point sources.
Step 3: Definition of non-point pesticide sources.
Step 4: Evaluating the pesticide management in the catchment (water solubility, adsorption, persistence, time of application).
Step 5: Estimating the potential water erosion or wind erosion risk (erosivity of rainstorms and wind, erodibility of soil, landscape and field openness, composition, thalways).
Step 6: Estimating the actual water and wind erosion risk (soil cover by plants or plant residues, crop rotation, humus status, soil tillage practices).
Step 7: Development of an efficient concept to realize a sustainable control of soil erosion, runoff, and pesticide transport.

The most important soil-protection strategy is based on the principles of precaution that took into consideration the economic and ecological consequences of the high costs and the endangerment to life and biodiversity.

As a result, the risk of soil erosion and pesticide transport should be addressed by an appropriate selection and a carefully timed application of pesticides. Products with low water-soluble active substances and a very short persistence should be preferred for slopes with a high runoff potential, caused by a network of waterways nearby the lakes or rivers. For example, it is advisable to replace products with isoproturon or isopropylamine with products containing pendimethalin or bifenac. For purposes of crop protection, water-soluble products that infiltrate into the upper soil layer should be used on areas with a high risk of wind erosion. Pre-emergent application on bare soil surface should be avoided on areas with a high water or wind erosion risk.

The most effective system to prevent sediment transport and reduce runoff is a temporal and spatially closed vegetation or residue covering. This cover protects the soil surface from the initial soil detachment action caused by raindrop splash forces or wind power. The result is a lower volume of runoff and sediment transport. Plant roots retain soil particles and reduce the sediment load (Table 4).

The extent of soil cover material, green plant mass, or crop residues can be influenced by the farmer's management and soil tillage practices.

The farmers can estimate the risk of their tillage system independently (Table 5).

Table 4 Correlation between cover rate, runoff, and soil loss estimated from long-term experiments with conventional and conservation tillage treatments.

Soil cover with green plants (%)	Soil cover with plant residues (t ha^{-1} dry matter)	Runoff (% of rain)	Relative soil loss by sediment transport in runoff (%)	Relative soil loss caused by wind erosion (%)
0	0	45	100	100
Approx. 20–30	0.5	40	25	15
Approx. 30–50	1–2	35	8	3
Approx. 50–70	2–3	ca. 30	3	<1
>70	>3–4	ca. 30	<2	<1

Source: Deumlich et al.[5]

Table 5 Appraisal matrix to estimate soil cover of crops and crop rotations and reduced tillage systems.

	Evaluation criteria					
	Period between sowing and efficient soil cover	Cover rate dependent on distribution	Soil cover during summer time	Soil cover during winter time	High risk dependent on tillage system	Total evaluation
	Soil cover					
Crops						
Grass	1	1	1	1		1
Winter barley	1	1	1.5	1		1
Winter wheat						
Sowing before October 1	2	1	1.5	1.5		1.5
Sowing after October 1	3	1	3	2.5		2.5
Summer barley	1.5	1	2	3		2
Potatoes	2.5	3	3	3		3
Sugar beets	3	2.5	2.5	3		3
Maize	3	3	2.5	3		3
Sunflower	3	3	2.5	3		3
Crop rotations						
Grass rotation	1	1	1	1		1
Maize–winter wheat–winter barley	2.5	2	2	2.5	2.5	2.5
Sugar beet–winter wheat–winter barley	2.5	2	2	2.5	2.5	2.5
Conservation tillage systems						
Winter barley–cash crops–mulch (frozen cash crops)–maize–winter wheat	1.5	1	1.5	1.5	1.5	1.5
Winter rye with undersown crops–winter barley–cash crops–mulch (frozen cash crops)–maize–winter wheat	1.5	1	1.5	1.5	1.5	1.5
Winter barley–cash crops–mulch (frozen cash crops)–sugar beet–winter wheat–winter barley	2	1	1.5	2	1.5	2

Appraisal:

1) Efficient soil protection

2) Moderate soil protection

3) Increased erosion and pesticide translocation risk

The rate of soil cover is a highly effective indicator for assessing the risk of water and wind erosion. This indicator addresses the following questions: 1) how much soil cover is necessary to reduce the threat of erosion, runoff, and pesticide translocation for a high-risk area; and 2) how much cover can be realized dependent on crop type, crop rotation, tillage, and management practices in different regions. Based on this analysis, appropriate preventive management practices can be required.

The greatest wind and water erosion risk occurs after seedbed preparation, which is characterized by the lowest degree of soil surface roughness and bare soil. Crop selection, improved crop rotations, and a change in soil tillage practices are established methods for increasing the soil cover for a reduced pesticide transport by wind and water erosion and runoff.

More basic management methods such as contour tillage or strip cropping practiced in some parts of the world are not as effective as conservation tillage in minimizing soil erosion.

Other technical solutions to reduce pesticide transport in areas with high erosion risk are engineering measures such as small ponds or farm-track construction. Alteration of field design, hedge planting to reduce wind openness, and waterways with permanent grass-covered banks or buffer strips nearby the lakes or rivers are effective arrangements to interrupt pesticide transport.

Therefore, precautionary measures against water and wind erosion as well as pesticide translocation have to be targeted (for example, prevention of pesticide pollution), site specific (for example, adequate for water and wind erosion risk), and applied for a longer period.

EROSION AND RUNOFF CONTROL: PART OF THE ENVIRONMENTAL LEGISLATION

Precautionary measures play an important role in promoting sustainability and realizing the European Water Framework Directive (EU-Com, 2000, Water Framework Directive. Official Journal L 327). The Directive is a contribution to the progressive reduction of hazardous substances emission to water. Strategies against pollution of water are formulated in Articles 2 and 16: "Pollution means the direct or indirect introduction of substances or heats into the air, water or land as a result of human activity, which may be harmful to human health or the quality of aquatic ecosystems or terrestrial ecosystems directly depending on aquatic ecosystems . . ." The Commission identified priority hazardous substances like pesticides (Annex X). Quality standards as to the concentration of these substances in surface water, sediment, or biota exist.

There are diverse conditions and needs in the community members that require different specific solutions. The main principles of soil erosion control are formulated in the strategy entry of the EU-Com: "Strategy for soil protection 1995–2005."

A regional legislation is the German Soil Protection Act (1999).

The meaning of hazard prevention is defined in § 4: "(1) official directive for analysis of damage and erosion pattern after serious events; (2) legislative ordinance for conservation and land-use change to prevent future hazards."

The definition of precaution is paraphrased with the term "best management practice in agriculture" (§17 of the Act). The essentials of best practice in agricultural soil use are the monitoring of soil fertility and functional capacity as a natural resource. Best management practice is founded on the principle that "soil erosion shall be avoided wherever possible, by means of site-adapted use, especially use that takes slope, water and wind conditions and the soil cover into account; . . . the predominantly natural structural elements of field parcels that are needed for soil conservation, especially hedges, field shrubbery and trees, field boundaries and terracing, shall be preserved. . . ."

The German Federal Immission Control Act (2002) must also be considered in implementing the precautions necessary to protect the environment by wind-initialized pesticide transport in the form of PM 10 emission.

Similar acts and regional precautionary basis information exist in other European countries like Austria and Switzerland.

The Australian Pesticide Act (1999) aims to reduce the risks associated with the use of pesticides to human health, the environment, property, industry, and trade.

The U.S. Food, Conservation, and Energy Act of 2008 contains Title II, "Conservation" with the Conservation Reserve Program (Subtitle B) and the Environmental Quality Incentives Program (Subtitle F). The Environmental Protection Agency (EPA) regulates the sale and use of pesticides in the United States through registration and labeling of pesticide products. The EPA is directed to restrict the use of pesticides as necessary to prevent unreasonable adverse effects on people and the environment (Pesticide Registration Improvement Act of 2004, P.L.108–199).

CONCLUSION

Most soil loss caused by wind and water erosion can be traced to inappropriate land management practices. Pesticide findings even during the non-spraying season indicate that the occurrence is due not only to direct contamination from accidental spills and incorrect handling (point source). Non-point sources of pesticide water and air pollution are water and wind erosion. Pesticide solubility, soil adsorption of pesticides, and pesticide persistence are important parameters to estimate the translocation by runoff water and sediment movement. Rainfall simulation studies demonstrate the different kinds of translocation, dependent on the solubility: highly soluble pesticides were analyzed in runoff water and the concentration decreased after application (for example, atrazine, simazine, isoproturon, and dichlorprop); fairly soluble pesticides were analyzed, and they only adsorbed onto eroded sediment (for example, bifenox and trifluralin). The second group could be transported via water or wind erosion.

Only one fraction of pesticides in use is very strongly soil bounded and therefore too insoluble to be transported in runoff or wind stream, but erosion can mobilize them.

That means that soil erosion control includes non-point pollution by pesticide translocation. A step-by-step analysis is the basis for an effective, sustainable, and site-adapted land use and agriculture management. The risk for soil erosion and pesticide translocation should be addressed by an appropriate selection on a carefully timed application of pesticides. Products with low water-soluble active substances and a very short persistence should be preferred for slopes with water erosion risk. Water-soluble products that infiltrate into the upper soil layer should be used on areas with a high wind erosion risk. The most effective system to prevent sediment transport and runoff reduction is a temporal and spatial closed vegetation or residue covering. The farmers can evaluate their management self-contained to change the system.

REFERENCES

1. Hurni, H. Soil conservation policies and sustainable land management: A global overview. In *Soil and Water Conservation Policies and Programs*; Napier, T.L., Napier, S.M., Tvrdon, J., Eds.; CRC Press: New York, 2000; 19–30.
2. Pimentel, D. Soil erosion: A food and environmental threat. Environ. Dev. Sustainability **2006**, *8*, 119–137.
3. Reichenberger, S.; Bach, M.; Skitschak, A.; Frede, H.G. Mitigation strategies to reduce pesticide inputs into ground- and surface water and their effectiveness. A review. Sci. Tot. Environ. **2007**, *384*, 1–35.
4. Gobin, A.; Govers, G.; Jones, R.; Kirkby, M.; Kosmas, C. *Assessment and Reporting on Soil Erosion*, Technical Report 94; European Environment Agency: Copenhagen, 2003.
5. Deumlich, D.; Funk, R.; Frielinghaus, M.; Schmidt, W.A.; Nitzsche, O. Basics of effective erosion control in German agriculture. J. Plant Nutr. Soil Sci. **2006**, *169*, 370–381.
6. Funk, R.; Reuter, H. Wind erosion. In *Soil Erosion in Europe*; Boardman, J.; Poesen, J., Eds.; John Wiley & Sons Ltd, 2006; 563–582.
7. Stevenson, D.E.; Baumann, P.; Jackmann, J.A. Pesticide properties that affect water quality. Texas Agriculture Extension Service B—6050.
8. Ghadiri, H.; Rose, C.W. Water erosion processes and the enrichment of sorbed pesticides. J. Environ. Manage. **1993**, *37*, 23–35.
9. Gouy, V.; Dur, J.-C.; Calvet, R.; Belamie, R.; Chaplain, V. Influence of adsorption–desorption phenomena on pesticide run-off from soil using simulated rainfall. Pestic. Sci. **1999**, *55*, 175–182.
10. Klöppel, H.; Haider, J.; Kördel, W. Herbicides in surface runoff: A rainfall simulation study on small plots in the field. Chemosphere **1994**, *28* (4), 649–662.
11. Larney, F.J.; Cessna, A.J.; Bullock, M.S. Herbicide transport on wind-eroded sediment. J. Environ. Qual. **1999**, *28*, 1412–1421.
12. Wolters, A.; Linnemann, V.; Smith, K.E.C.; Klingelmann, E.; Park, B.J.; Vereecken, H. Novel chamber to measure equilibrium soil-air partitioning coefficients of low-volatility organic chemicals under conditions of varying temperature and soil moisture. Environ. Sci. Technol. **2008**, *42* (13), 4870–4876.
13. Clay, S.A.; DeSutter, T.M.; Clay, D.E. Herbicide concentration and dissipation from surface wind-erodible soil. Weed Sci. **2001**, *49* (3), 431–436.
14. Schulz, M.; Matthies, M. Runoff of pesticides: Achievements and limitations of modelling agrochemical dislocation from non-point sources at various landscape related scales. Living Rev. Landscape Res. **2007**, 1.

Pesticides

Marek Biziuk
Jolanta Fenik
Monika Kosikowska
Maciej Tankiewicz
Department of Analytical Chemistry, Chemical Faculty, Gdansk University of Technology, Gdansk, Poland

Abstract
Applied all over the world, pesticides are some of the most dangerous pollutants of the environment because of their mobility and ability to accumulate in the environment and their consequent long-term adverse effects on living organisms in general and human health in particular. For these reasons, it is essential to monitor and analyze pesticide residues in the environment. This entry discusses the classification of pesticides as well as their effects on and their fate in the environment. Since legal regulations have come into force stipulating highest permissible levels of pesticides and their residues in the environment, the analytical techniques applied are sensitive, selective, and appropriate to the low concentrations of the target analytes. The analysis of environmental samples for the presence of pesticides is very difficult: the processes involved in sample preparation are labor-intensive and time-consuming and may be a source of additional contamination and error; the low concentration levels of analytes and matrix complexity cause yet more problems. The extraction and determination of pesticide residues in environmental samples are discussed, as are the techniques most commonly used in these processes. The difficulties occurring at each stage in the analytical procedure are outlined.

INTRODUCTION

A numerous and diverse group of chemical compounds with an extensive range of action,[1] pesticides are very widely used, though primarily to eliminate insect pests and weeds and to limit their negative effects in agriculture and in the household.[2–4] They enable the quantities and quality of crops and food to be controlled and help to limit the many human diseases transmitted by insect or rodent vectors. Despite their many merits, however, pesticides are some of the most toxic, environmentally stable and mobile substances in the environment, which are able to bioaccumulate. They can also participate in various physical, chemical, and biological reactions, as a result of which even more toxic substances may be produced; by accumulating in living organisms, these can lead to irreversible, deleterious changes. The non-rational application of pesticides also adversely affects the environment and humans, increasing susceptibility to diseases and poisoning.[1]

CHARACTERISTICS OF PESTICIDES

The following are the current uses of pesticides:

- To remove, control the number of, or attract various kinds of pests destroying plants and plant products
- To kill weeds
- To destroy foliage and excess numbers of flowers
- To increase the production of animal and plant biomass
- To combat human, animal, and plant pathogens
- To control the growth of plants or their parts
- To combat microorganisms causing farm produce to rot and decay
- To combat insects and other animals occurring in homes, cellars, stores, etc.
- To protect fabrics in textile mills and dry-cleaning establishments
- To prevent the growth of algae in swimming pools
- To combat fungi in paints and paper products
- To protect museum exhibits against the action of pests
- To counteract growths on boats and ships[1]

Ideally, pesticides should be applied only in accordance with their purpose, so that they do not have a negative effect on humans, flora, or fauna. In practice, however, such complete selectivity is unattainable.[1,4] The factor determining whether a particular agent can be used is its rapid biodegradability and its minimal toxicity vis-à-vis the environment.

Chemical pesticides are of enormous importance in increasing the efficiency and quality of agricultural produce. They enter the environment in various forms: powders, moistened powders, powders for preparing aqueous solutions, and concentrates for making up

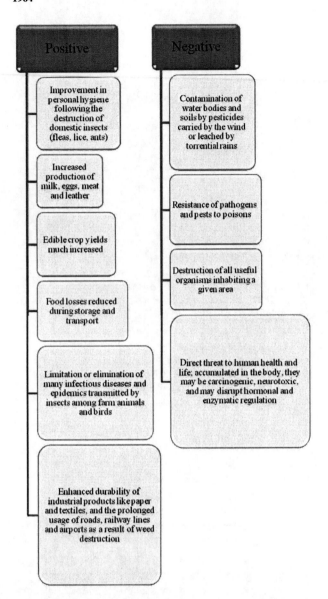

Fig. 1 The effects of using pesticides.
Source: Biziuk et al.,[1] Moreno et al.,[5] and Hajslova and Zrostlikova.[6]

emulsions or sprays. The ubiquitous application of pesticides not only contaminates water, soil, and air but also causes them to accumulate in farm produce like fruit and vegetables. Fig. 1 summarizes the effects of pesticide application.[1,5,6]

According to their Chemical Structure

Table 1 Classification of pesticides according to their chemical structure.

Class	Examples
Inorganic pesticides	Arsenic and fluoride insecticides, inorganic herbicides
Organic pesticides	Organochlorine and organophosphorus insecticides, carbamates

Source: Biziuk et al.[1]

According to their Chemical Class

Table 2 Classification of pesticides according to their chemical class.

Class	Examples
Organochlorine	DDT, endosulfan, methoxychlor
Organophosphorus	Malathion, dichlorvos, fonophos
Carbamates and their derivatives (urethanes)	Aldicarb, aminocarb, furathiocarb
Derivatives of phenoxycarboxylic acids	2,4-D, MCPA, dicamba
Triazines and their derivatives	Atrazine, simazine, anilazine

Source: Biziuk et al.[1]

The diversity of their chemical structure, action, and application makes any classification of pesticides difficult.[1] There are a number of criteria according to which they can be categorized: toxicity, purpose of application, chemical structure, environmental stability, and the pathways by which they penetrate target organisms. Tables 1–4 outline some of these classifications:[1]

According to their Application

Table 3 Classification of pesticides according to their application.

Class	Subclass	Application
Zoocides—agents for combating animal pests	Insecticides	Destruction of insects
	Aphicides	Destruction of aphids
	Acaricides	Destruction of plant mites
	Attractants	For attracting pests
	Bactericides	Destruction of bacteria
	Larvicides	Destruction of larvae
	Limacides	For killing slugs
	Molluscicides	For killing snails
	Nematocides	For killing nematodes
	Ovicides	Destruction of the eggs of insects and mites
	Repellents	For repelling insects
	Rodenticides	For combating rodents
Fungicides		Fungicidal and fungistatic agents
Herbicides		Weed killers
Plant growth regulators—stimulants or inhibitors of the life processes of plants	Deflorants	For removing excess flowers
	Defoliants	For removing excess leaves
	Desiccants	For drying plants
Synergetics		For potentiating the action of other substances

Source: Biziuk et al.[1]

According to their Toxicity

Table 4 Classification of pesticides according to their toxicity.

Toxicity class	Median lethal dose LD_{50} (mg/kg body mass) when administered via the digestive tract	Stability in the soil (time for degradation to harmless products)	Stability in the aquatic environment (time for degradation to harmless products), number of days
I—Highly toxic	≤ 25	More than 2 years	>30
II—Toxic	$25 < LD_{50} \leq 200$	0.5–2 years	11–30
III—Harmful	$200 < LD_{50} \leq 2000$	1–6 mo	6–10
IV—Not very harmful	>2000	Within 1 mo	<5

Source: Biziuk et al.[1]

CIRCULATION OF PESTICIDES IN THE ENVIRONMENT

Nowadays, pesticides are very widely applied, not just in agriculture. Their ubiquity and ever-increasing consumption pose a greater hazard to the environment to which they are transported. Pesticides can enter the environment in their primary form or as decomposition products. When in the environment, they are subject to various transformations. They may be borne by wind or rain from their points of application to neighboring areas and crops, where they are not required or may be harmful. The quantity of pesticides circulating in a given region depends to a large extent on their intensity of application or the type of crop being grown there. Both modern pesticides and older ones are present in the environment. Fig. 2 illustrates the circulation of pesticides in nature.

Pesticides in the Aquatic Environment

Pesticides are the most common contaminants in surface water and groundwater. There are also many reports of pesticides being found in drinking water, well water in farming areas, rainwater, subterranean water, and ice from the polar regions. Once in the aquatic environment, pesticides have a deleterious effect on the quality of waters used as sources of drinking water for the large majority of the Earth's human population. Concentrations are highest during the spring snow melt period and when pesticides are being applied to crops.[1,2] There are many sources from which pesticides get into the aquatic environment. Usually, these are area sources (e.g., precipitation, farming areas), but they may also come from point sources such as effluents of various kinds or leaking waste disposal sites. They can also be transported in the air for

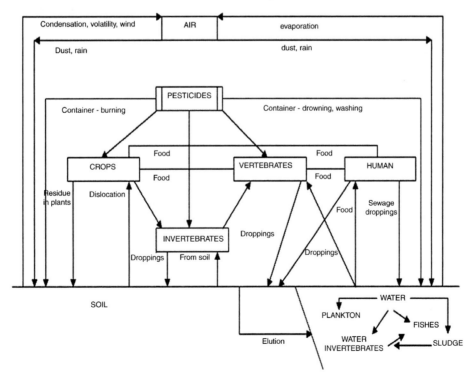

Fig. 2 The circulation of pesticides in nature.
Source: Biziuk et al.[1]

great distances.[1,7] When in the aquatic environment, pesticides are subject to a variety of transformations and processes:

- Physical (accumulation, deposition, dilution, diffusion)
- Chemical (hydrolysis, oxidation)
- Photochemical (photolysis, photodegradation)
- Biochemical (biodegradation, biotransformation, bioaccumulation)[3,8]

Substances of greater toxicity may be formed as a result of these processes, and when they accumulate in aquatic organisms, they can cause much irreversible damage to them.

Pesticides in the Air

The widespread application of pesticides contaminates not only waters and soils; the air is also affected. This is because pesticides rise into the air as they are being sprayed and thereafter (post-application emission).[1] An estimated 30%–50% of pesticides sprayed onto crops get into the atmosphere. By "post-application emission," we understand the evaporation of pesticides from soil or plant surfaces and due to the wind erosion of the soil. Depending on their stability, pesticides in the air may be degraded, or transported over long distances before being deposited. After their application, pesticides are present in the atmosphere in gaseous form, adsorbed onto solid particles, or dissolved in water vapor.

Pesticides in Crops

The presence of pesticides in the environment causes them to accumulate in crops like fruit and vegetables. The contamination of plant foods by pesticides is particularly dangerous because these compounds can reach every part of the plant, regardless of how they are applied. The degree of contamination of crops depends, among other things, on the dose and number of applications, the form of the pesticide preparation, the weather conditions, and the time elapsing between pesticide application and crop harvesting. Contamination is reduced by such factors as rainfall, wind, and chemical changes caused by oxygen, moisture, light, and plant enzymes. Plants with a large surface area relative to their mass retain larger amounts of pesticides.[1] Fruit and vegetables are capable of retaining larger quantities of pesticides. Pesticides can accumulate in fruit skins. The crops most exposed to the presence of pesticides are grapes, citrus fruits, and potatoes. Adsorbed pesticides can reduce the nutritious value of crops or alter their organoleptic properties. The contamination of crops with pesticides is due to their application in contradiction to good agricultural practice and the insufficient monitoring of their application. Their total elimination is, of course, often not possible, but the amounts applied can be limited to those that are harmless to human health.

Pesticides in Soil

Pesticides are transported to soil because of deliberate human activity. It is estimated that about 50% of pesticides come from washing processes of soils' surface. Transport of pesticides to plant material depends on their stability, and the main ways of absorption are via roofs and leaves. They can be also adsorbed on the molecules of clay. In this case, they do not penetrate the soil. Stability and fate of pesticides in soil depend on chemical structure, type of plant material, type of soil and their pH/temperature, weather conditions, etc.[1]

MODERN PESTICIDES

Organochlorine pesticides were widely used in agriculture and pest control. These pesticides are insecticides composed primarily of carbon, hydrogen, and chlorine (see Fig. 3). Most of them break down slowly and can remain in the environment long after application and in organisms long after exposure.

These chemicals were introduced for the first time (DDT [dichlorodiphenyltrichloroethane]) in the 1940s, and many of their uses have been nowadays cancelled or restricted because of their environmental persistence and potential adverse effects on wildlife and human health. Many organochlorines are no longer used in most countries, but other countries continue to use them (in Africa, South America, and Asia). Organochlorine pesticides can enter the environment after pesticide applications, disposal of contaminated wastes into landfills, and releases from manufacturing plants that produce these chemicals. Some organochlorines are volatile, and some can adhere to soil or particles in the air. In aquatic systems, sediments adsorb organochlorines, which can then bioaccumulate in fish and other aquatic mammals. These chemicals are fat soluble, so they are found at higher concentrations in fatty foods. Organochlorine pesticides are hydrophobic, lipophilic, and extremely stable. Organochlorines have a wide range of both acute and chronic health effects, including cancer, neurological damage, and birth defects. Many organochlorines are also suspected endocrine disruptors.[1,9,10]

The present-day trend is to move away from persistent pesticides and to apply agents with a short decomposition time and no tendency to bioaccumulate.[1] That is why organochlorine pesticides have been withdrawn.[11–14] Be-

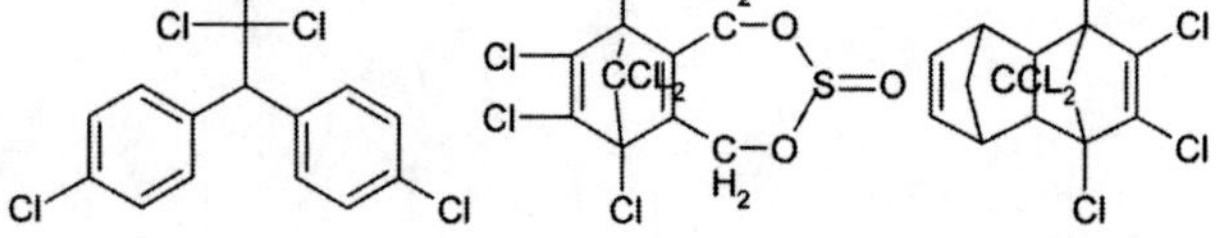

Fig. 3 The chemical structure of some organochlorine pesticides.

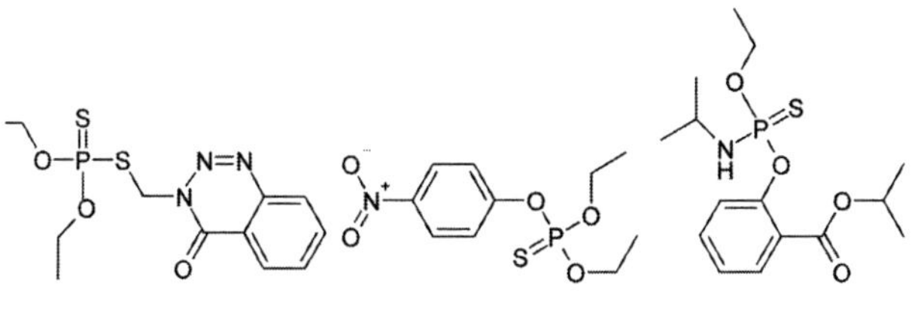

Fig. 4 The chemical structure of some organophosphorus pesticides.

cause of their great stability (as long as 30 years), however, they may still be present in the environment and be transported by air or water over great distances. Organochlorine pesticides have been replaced by organonitrogen and organophosphorus pesticides. They have become very popular because they are cheap and readily available, have a wide range of efficacy, are able to combat a large number of pest species, and have a shorter environmental half-life than their organochlorine predecessors.

Currently used pesticides in comparison with formerly used ones are less persistent in the environment and generally are more polar. Polarity of pesticides determines their presence, transport, and stability in different compartments of the environment. Polar pesticides are more soluble and penetrate the surface water and groundwater faster. Because they are less stable in the environment, their presence can be established in less time (since the application). By contrast, non-polar pesticides are typically sparingly soluble in water, which is associated with their greater persistence in the environment, and transport mechanisms of these compounds are different. They are mostly adsorbed on solid particle and only partially dissolve in water. In addition, pesticides accumulate in sediments or suspensions and are present there even for decades.[1]

One of the principal classes of compounds used for plant protection, organophosphorus pesticides (insecticides) embrace all organic compounds containing phosphorus. Usually taking the form of esters (Fig. 4) and degrading fairly easily, they are very poorly soluble in water, though better so in organic solvents and fats.

The umbrella term "organonitrogen pesticides" is a convenient way of referring to the large number of nitrogen-containing organic pesticides. In practice, however, these pesticides are known by the names of the various chemical classes. In the literature, the term "organonitrogen pesti-

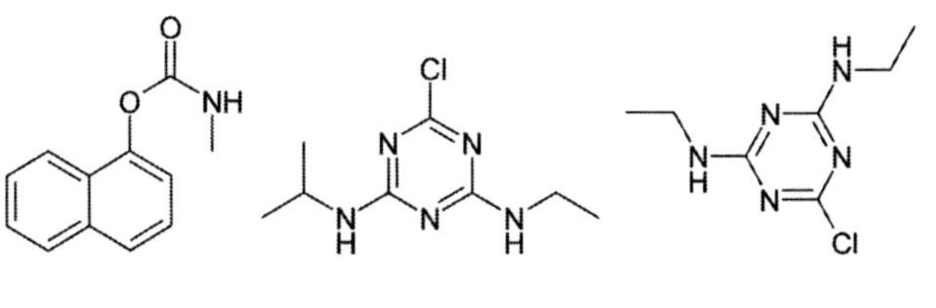

Fig. 5 The chemical structure of some organonitrogen pesticides.

Table 5 The characteristics of organonitrogen pesticides.

Chemical class	Characteristics
Carbamates	Mostly carbamic acid esters; decompose rapidly in the soil; poorly soluble in water, better soluble in organic solvents and fats; used worldwide to combat insects, fungi, and weeds, and also as plant growth regulators
Triazines	Used to combat weeds and to control the growth of maize, soybean, grains, and other crops; highly toxic towards mono- and dicotyledons, and extremely stable in the soil; one of the triazines is atrazine, 90% of which is applied to maize crops

Source: Biziuk et al.,[1] Zhang and Lee,[18] and Sabik et al.[19]

cides" usually refers to carbamates and triazines and their derivatives (see Fig. 5).[15–17]

Table 5 presents the characteristics of organonitrogen pesticides.[1,18,19]

In recent years, new insecticides—neonicotinoids—have been introduced to replace pyrethroids, organophosphorus compounds, and carbamates. This was necessary as insect pests were becoming extremely resistant to the latter agents. At present, neonicotinoids are used to combat the Colorado potato beetle (Mospilan 20 SP, active substance: acetamiprid).[20]

DANGERS RESULTING FROM THE APPLICATION OF PESTICIDES

Human beings come into contact with pesticides in a variety of situations.[7,21] Ubiquitous and constituting a risk to human health, these substances can enter the body through

- The skin
- The respiratory system (through the nose)
- The digestive system (accidental swallowing of the product)
- Consumption of food contaminated with pesticides

Penetration through the skin is easier because pesticides can remain on the skin for a long time, the area of contact is large, and absorption of these compounds is rapid, especially through damaged skin. The inhalation of pesticides in contaminated air is very dangerous, as harmful substances very quickly enter the lungs and blood, and thence reaching all the organs of the body. The symptoms of pesticide poisoning include headache, a feeling of cold, giddiness/dizziness, and skin rash. Organochlorine pesticides are capable of accumulating in the fatty tissues of living organisms, although their quantities in particular organs depend on the degree of fatness of the latter. When poisoning reaches a critical level, the most vulnerable organ is the brain. It was stated not long ago that some pesticides stimulate the production of microsomal hepatic enzymes responsible for the metabolism of some drugs (hypnotics, antiepileptics, analgesics), which reduces their therapeutic efficacy. Further symptoms of pesticide poisoning are as follows:

- Impairment of ion transport, leading to impaired neural conductivity
- Disturbances to neurotransmitter metabolism
- Abnormal immunological reactions
- Destruction of hepatocytes, leading to cirrhosis of the liver
- Cardiac arrhythmia

Organophosphorus and organonitrogen pesticides act as receptor inhibitors. They bind to the receptors of the enzyme acetylcholinesterase, which is essential for the correct functioning of the nervous system, preventing the decomposition of acetylcholine and acting through contact or systemically.[1,22] Blockage of cholinesterase activity causes the amount of acetylcholine at the synapses to increase, leading to a state of hyperarousal and paralysis of the muscles and central nervous system. If a xenobiotic binds to a receptor, one of the more important stages in metabolism may be blocked, elicit a harmful metabolic process, or alter the rate of transmission of nervous signals.[5,7,23,24] In addition, some carbamate insecticides, (e.g., carbaryl) can, in large doses, be teratogenic and can be nitrosated to form strongly carcinogenic nitroso compounds.

METHODOLOGIES FOR DETERMINING PESTICIDES

It is essential to monitor the levels of pesticide residues as they can easily make humans more vulnerable to different diseases. An assessment of the state of contamination by pesticides requires a knowledge of the maximum permitted levels of individual active substances. The full determination procedure consists of several stages (see Fig. 6). Because of the complex composition of samples and the different concentrations of the target analytes, the sample must be adequately prepared for analysis by the use of techniques for efficiently extracting the target compounds and for their cleanup prior to the quantitative determination stage.

The choice of methodology for determining pesticides depends in large measure on the sample matrix and the structure and properties of the target analytes. In view of the numerous legal regulations laying down highest permissible levels of pesticides in various matrices, sensitive and selective analytical techniques are used, appropriate to the low concentrations at which the target analytes occur in them.[27] In addition, each stage in the analytical procedure, as well as this process in its entirety, should be validated.[4,26]

The analysis of environmental samples for the presence of pesticides is fraught with difficulties, because arduous and time-consuming operations and processes have to be carried out in order to prepare the samples for analysis, and these could be a source of further contamination and error.

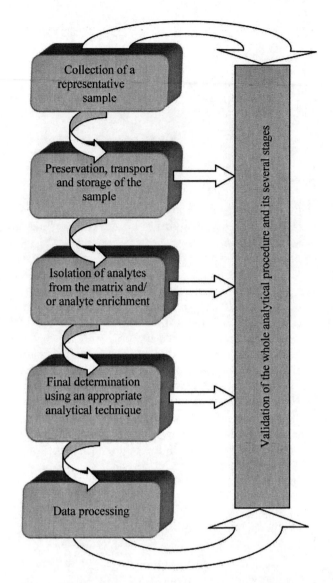

Fig. 6 The main stages in the analytical procedure for determining modern pesticides.
Source: Biziuk et al.,[1] Beyer and Biziuk,[4] Namieśnik,[25] and Namieśnik and Górecki.[26]

The normally low concentrations of target analytes and the often complex matrix composition are further problems the analyst has to face. Highly efficient extraction and cleanup techniques therefore need to be applied to the target compounds prior to their quantitative determination.

DETERMINATION OF MODERN PESTICIDES IN WATER SAMPLES

Collection and Preparation of Samples for Analysis

Environmental samples are diverse and complex materials to analyze because of the sampling site, type of matrix,

Table 6 The usual sample preparation processes and examples of their implementation.

Basic processes during the preparation of samples for analysis	Examples of implementation
Separation of suspended matter	Filtration
Chemical preservation	Reduction of the sample pH, addition of bactericides (e.g., saturated aq $HgCl_2$, formaldehyde, sodium azide)
Physical preservation	Storage of the sample at a low temperature (2–4°C), storage in the dark, sterilization by UV irradiation

Source: Biziuk et al.,[1] Biziuk et al.,[21] Namieśnik et al.,[25] Namieśnik and Górecki,[26] Namieśnik and Szefer,[28] Demkowska et al.,[29] and Łopuchin and Namieśnik.[30]

the presence of interferents, and the concentration of target analytes. To a large extent, it depends on the appropriate preparation of the sample whether the analysis will actually provide the desired information about the sample. That is why this stage is usually a complex task, and the operations and processes involved in this stage may be a cause of analyte loss as well as a source of additional contamination. Table 6 lists the usual sample preparation processes and examples of their implementation.[1,21,25,26,28–30]

A representative sample is taken for analysis using an appropriate sampler. Which type of sampler is used depends on the type of sample, the objective of the analysis, the site and frequency of sampling, and the sample size. Mains water is usually collected in dark glass bottles; surface waters or sewage (effluent) is placed in different containers.

Techniques for Isolating and Enriching Analytes from Water Samples

Because of the low concentrations of pesticides in the various compartments of the environment, it is essential not only to isolate the organic compounds from the complex matrix but also to enrich them prior to final determination. The use of passive dosimetry at the sample collection stage, with which analytes can be isolated and preconcentrated at the same time, is worth mentioning. The free transport of mass takes place across a membrane to the sorbent as a consequence of the difference in chemical potentials of the compounds in the sorbent and the water in which the sampler has been immersed.[30,31] For sampling pesticides from coastal waters and rivers, one uses polyethylene dosimeters packed with iso-octane or stainless steel dosimeters filled with cyclohexane as sorbent. Three or four samplers of one type are deployed, usually for 30 days. Most passive samplers are fairly small in size, which substantially reduces the amounts of organic solvents required (the volume of the dosimeter chamber is ca. 1 mL) and makes them easy to transport and to assemble at the deployment site. With the use of passive dosimeters, there are fewer steps in the analytical procedure, which means results are more reliable and reproducible. The shortcomings of such dosimeters, however, include an insufficient sensitivity to brief fluctuations in target analyte levels and a susceptibility to environmental factors like temperature and water movement, which means that samplers have to be calibrated in order to set the rate of analyte collection.[30,31]

During extraction and/or enrichment steps, the concentration of analytes is raised in order to make their determination at all possible. Moreover, analytes are transferred from the primary matrix to a secondary one with concomitant removal of interferents. The choice of technique depends on the properties of the target analytes, their volatility, polarity, and solubility in water and organic solvents. Table 7 lists the standard techniques for extracting modern pesticide analytes.

An equally important technique is membrane extraction, which may replace classical liquid–liquid extraction (LLE) as it uses smaller quantities of solvents or even none at all.[65] As a result, a number of drawbacks are eliminated, among them, the problem of emulsion formation. Table 8 lists the membrane extraction techniques used during the determination of pesticides in water samples.[65]

DETERMINATION OF MODERN PESTICIDES IN AMBIENT AIR

Sampling Techniques

Pesticides in the air can be determined in the following types of samples:

- Atmospheric dusts
- Analytes sampled from the gaseous phase

Dynamic[66] and passive (diffusional)[67,68] sampling methods are the usual ways of collecting air for determining its pesticide content.

Dynamic sampling enables the collection of pesticides present in the gaseous phase and in suspended form by pumping air through a filter coated with a solid adsorbent. Pesticides present in suspended dust particles are retained on the filter, while those in the gaseous phase are adsorbed by the sorbent.

The standard filters are of glass with different diameters (Ø = 30 cm, 10 cm, 90 mm, and 25 mm) and of quartz filters (Ø = 102 and 150 mm).

For collecting gas samples, the sorbents (XAD-2, XAD-4, Carbopack, Carbotrap, Carboxen, Tenax TA, Chromosorb, silica gel) are coated onto the filters, as mentioned above, or placed in stainless steel or glass vessels containing polyurethane foam (PUF).

The conventional methods of collecting gas samples for the determination of their pesticide content involve passing a

Table 7 The standard techniques for extracting modern pesticide analytes.

Extraction technique	Characteristics
LLE (liquid–liquid extraction)	Based on the partition of analytes between two immiscible liquids, usually an aqueous solution and an organic solvent. The most common extraction solvents are dichloromethane[17,32–36] and mixtures of petroleum ether and dichloromethane[37] and methylene chloride and hexane.[33] Though relatively simple and cheap, this method has many shortcomings: it requires relatively large quantities of often toxic solvents, and there is a risk of an emulsion forming during agitation. To achieve the desired enrichment coefficient, the excess solvent usually has to be evaporated, and the extracts often have to be cleaned up.
DLLME (dispersive liquid–liquid microextraction)	This is one example of the miniaturization of LLE, in which a dispersing solvent, e.g., acetone or methanol (0.5–2 mL) with added extraction solvent, e.g., C_2Cl_4, CS_2 (10–50 µL), is added to the water sample (5–10 mL). This procedure yields a turbid solution, which is then centrifuged to obtain ca. 5 mL of extract phase. By reducing the amounts of solvents, some of the inconveniences of LLE have been eliminated.
SPE (solid phase extraction)	This technique is based on the sorption of analytes on a sorbent—this is usually silica gel, aluminum oxide, Florisil, a porous polymer XAD-2, XAD-4, XAD-7, XAD-16, usually modified with octadecyl groups.[15,33,38–51] Desorption of adsorbed analytes is usually carried out using methanol,[15,37,40,43,46,51] ethyl acetate,[33,35,44] dichloromethane,[38,39,41,50] and mixtures of methanol and water,[45,49,51] methanol and acetonitrile,[42,47] and acetonitrile and water.[48] Research is in progress to develop selective sorbents (MISPE: *Molecularly Imprinted Solid Phase Extraction*; immunosorbents), which would be applied in the case of samples with a complex matrix composition such as sewage.
SPME (solid phase microextraction)	Simple to use and does not require solvents. Based on extraction from the liquid phase to an adsorption or absorption layer coated on an extraction fiber. It is also possible to introduce the fiber into the headspace (HS-SPME). The materials used for coating fibres include polydimethylsiloxane (PDMS),[53–58] polyacrylate (PA),[48,54–57,59,60] and also mixtures of polydimethylsiloxane and polydivinylbenzene (PDMS-DVB),[53–55,60–64] carbowax and polydivinylbenzene (CW-DVB),[53,57,64] and carbowax and molecularly imprinted resin (CW-TPR).[60] The analytes most frequently adsorbed are transferred to the GC injector where they are thermally desorbed and then determined.
SBSE (stir bar sorptive extraction)	The extractant, usually polydimethylsiloxane (PDMS), is used to coat a magnetic stir bar.[29] The adsorbed analytes are then thermally desorbed. The parameters determining the effectiveness of this type of extraction are the sample volume, the extraction time, and the stirring speed.

Table 8 Membrane extraction techniques used during the determination of pesticides in water samples.

Technique	Type of membrane	Combination of phases used: donor/membrane/acceptor
MASE: membrane-assisted solvent extraction; PME: polymer membrane extraction	Non-porous	Aqueous/polymer/aqueous Organic/polymer/aqueous Aqueous/polymer/organic
MMLLE: microporous membrane liquid–liquid extraction	Non-porous (microporous)	Aqueous/organic/organic Organic/organic/aqueous

Source: Jönsson and Mathiasson.[65]

fixed volume of air through a solid sorbent, for which pumps and flowmeters are needed. The expense of the former and the need to frequently calibrate the latter mean that it is difficult to collect gas samples in a fully professional manner. Analytes retained on sorbents require not only thermal desorption but also chemical desorption with expensive and potentially toxic solvents. The sampling time depends on the method's sensitivity and the breakthrough volume of the sorbents by the target analytes (Table 9). In order to collect samples without the need to use toxic solvents and expensive pumping equipment, we need an unpowered air sampler.[69]

Universal and inexpensive passive air samplers (PASs) have been developed as an alternative to the conventional active sampling methods. Passive sampling is based on the free flow of analytes from the environment sampled to a medium on which the analytes are sorbed (Table 10).

There are several types of PASs, using the following:

- SPMD semipermeable membranes.
- PUF discs; the passive samples consist of PUF discs placed inside stainless steel containers covered with a dome-shaped lid to reduce the effect of wind speed on

Table 9 Comparison of techniques for collecting air samples depending on the sampling time, type of filter, and adsorbent.

Sampling time (hr)	Type of filter	Adsorbent	References
5.5–9	GFF	XAD-2, s-PUF	[70]
24	GFF	XAD-2	[66–68,71,72]
24	GFF	PUF	[73–76]
48	GFF	XAD-2	[77]
18–67	GFF	PUF	[78]
84	GFF	XAD-2, s-PUF	[79–81]
168	GFF	XAD-2, s-PUF	[82–85]
12	QFF	XAD-2, s-PUF	[86]
23	QFF	PUF	[87]
24	QFF	PUF	[88–92]
24	QFF	XAD-2	[93–95]

the rate of sampling. This arrangement is also used to protect the PUF discs from precipitation, the direct deposition of dust, and UV light.
- Resins.
- A thin layer of ethylene/vinyl acetate as the sampling medium.

Solid-phase microextraction (SPME) is another way of sampling analytes from air. A solventless method, it is very convenient for use in the field, simple to operate, and, under optimal conditions, does not require the use of pumps.

When only suspended dust is to be sampled, the samples are often fractionated according to particle diameter, a process that is possible because the different fractions have different physical and chemical properties: the fine-grained fraction (<2.5 μm) is often acidic, whereas the coarse-grained fraction (>2.5 μm) is basic. Fractionation minimizes interactions between particles of different pH. Cascade impactors (absorbers) are also used. These operate on the principle that the dust-containing air is made to flow through nozzles of ever-decreasing diameter. At each nozzle, the linear velocity of the airflow increases and the relevant dust fractions are separated at the steps of the cascade impactor. If dust concentrations in gas samples are very high, cascade cyclones are used. These direct the dust-containing gas at a tangent to a cylinder, and the particles move under centrifugal force from the stream of gas to the inside wall of the cylinder.[111]

Methods of Extracting Analytes from Sorbents and Filters

The next step, after sampling but before the final determination, involves extracting the preconcentrated analytes from the solid sorbents or filters (Table 11). This is the most crucial step in the analytical process, since the aim of extraction is to liberate the greatest possible quantities of pesticides from the sorbent. To this end, both solvent and solventless techniques are employed.

The most commonly used solvents for extracting analytes from air samples are shown in Table 12.

The next step in sample preparation is evaporation and/or change of solvent. It is standard practice to evaporate the solvent in a vacuum evaporator, a quick and simple method. Another possibility is to evaporate the solvent in a stream of gas, usually nitrogen, again simple to carry out. These methods may be applied separately or in combination.

Table 10 List of techniques for the passive sampling of air.

Type	Sampling time	References
PUF discs	4 months	[96–98]
PUF discs	6 weeks	[99]
POG	7 days	[100,101]
PUF discs	2 months	[102]
XAD-2	1 year	[103]
XAD-2	5–8 months	[104]
PUF discs	28 days	[105–108]
PDMS	14 days	[109]
SPMD	7 days	[98,110]

Table 11 Techniques most frequently used for extracting analytes from air samples.

Sample	Extraction technique	References
$PM_{2.5}$ dust	Accelerated solvent extraction (ASE, PFE, PLE)	[76,112]
Dust and gaseous phase	Soxhlet apparatus	[66,67,71,72, 113–119,121]
Gaseous phase	Agitation assisted (LE)	[122]
Gaseous phase	Ultrasound assisted (UE)	[123]
$PM_{2.5}$ dust	Microwave assisted (MAE)	[124]
Gaseous phase	Thermal desorption	[125]

Table 12 Solvents most commonly used for extracting analytes from air samples.

Sample	Solvent	References
Suspended dust	Petroleum ether	[74,88,96]
Suspended dust	Acetone	[76,79,80,83]
Suspended dust/air	Hexane/acetone	[113]
Suspended dust	Hexane/dichloromethane	[66,67,71,72, 75,114–116]
Suspended dust/air	Dichloromethane/ petrochemical ether (diethyl ether, MTBE)	[117–119,123]
Suspended dust/air	Hexane/benzene	[121]
Suspended dust	Dichloromethane	[112]
Air	Ethyl acetate	[122]

Extract Cleanup

Extract cleanup involves fractionating the extract, which can be done in a number of ways:

- With column LC chromatography with normal phase (NP) or reversed phase (RP), including high-performance liquid chromatography (HPLC)
- With adsorption chromatography
- Passing the extract through a filter
- With gel chromatography (GPC)

In some cases, mainly when highly selective detectors like gas chromatography–tandem mass spectrometry (GC–MS/MS) and liquid chromatography–tandem mass spectrometry (LC–MS/MS) are used, the cleanup of extracts is not necessary. Before deciding whether the cleanup step can be omitted or not, however, matrix effects must be investigated for possible interferents and their influence on the apparatus.

DETERMINATION OF MODERN PESTICIDES IN FRUIT AND VEGETABLES

Sample Collection and Preparation of Samples for Analysis

It is extremely important that samples of material intended for analysis are homogeneous and representative; they should be stored frozen and in the dark.[28] The sample preparation process consists of several steps. In the case of fruit and vegetables, one of these steps involves the removal of surface contaminants by, for example, washing the sample in distilled water, after which the sample is dried (at an elevated temperature, at ambient temperature, or with the aid of a desiccant). Then, the sample has to be broken up, crumbled, minced, or ground with a pestle and mortar. It is then homogenized in special equipment. The sampling process depends on the type of biological material under investigation, and one has always to be mindful of possible losses of analytes and contamination of the sample as a result of human agencies.

Techniques for Isolating Pesticides from Fruit and Vegetable Samples

Isolation and/or enrichment involves the transfer of analytes from the primary matrix to a secondary one with the concomitant removal of interferents and the increase of analyte concentrations to levels above the limit of detection (LOD) of the analytical technique used. Since pesticide concentrations in fruit and vegetables are low, samples often require enrichment.[126,127]

The following are the usual techniques for isolating pesticides from fruit and vegetable samples:

- Liquid–liquid extraction (LLE)
- Accelerated solvent extraction (ASE)
- Supercritical fluid extraction (SFE)
- Soxhlet or Soxtec extraction
- Microwave-assisted extraction (MAE)

The merits of solvent extraction by agitation are ease of implementation and low cost; in addition, expensive and complicated apparatus can be dispensed with. Unfortunately, however, this technique requires a large quantity of toxic solvents and its selectivity leaves much to be desired. Soxhlet extraction are easy to carry out and enables the extract to be separated from the residue; on the downside, this type of extraction takes a long time with large amounts of solvent, and only single samples can be extracted. For this reason, the more efficient Soxtec apparatus was introduced; with this albeit very expensive apparatus, the extraction time is now much shorter, and several samples can be extracted simultaneously. Solvent extraction techniques are frequently assisted with microwaves or ultrasound in order to improve extraction yields. Acetone, acetonitrile, and ethyl acetate are the solvents (or their mixtures) (Table 13) most commonly used for the extraction of fruit and vegetable samples.

Table 13 Solvents for the extraction of samples of fruit and vegetables.

Sample	Solvent
Cabbage, grapes	Ethyl acetate, methanol, acetone[128]
Oranges, apples, grapes, pears	Ethyl acetate[126]
Apples, tomatoes	Acetone[127]
Tomatoes, pears, oranges	Acetonitrile[129]
Peppers, tomatoes, Brussels sprouts, melons, apples, lemons	Ethyl acetate[130]

Extract Cleanup

The cleanup of the extract is an essential step that should always precede its analysis. It is important, because during isolation, we obtain not only the target analytes but also interferents, which can distort the final result of the analysis. The usual cleanup techniques used with fruit and vegetable samples are as follows (Table 14):

- Solid-phase extraction (SPE)[131]
- Solid-phase microextraction (SPME)[132]
- Matrix solid-phase dispersion extraction (MSPDE)[133]
- Stir bar solvent extraction (SBSE)[134]
- Gel chromatography (GPC)[135–137]

SPE is currently the most popular extract cleanup technique.[138] It involves passing the sample through a bed of sorbent on which the target analytes are adsorbed. The retained compounds are then liberated by means of a solvent and analyzed. This method is easy to carry out and readily lends itself to automation. Its shortcomings, however, are that the sorbent bed has to be conditioned before each use and that analyte yields are low. A modification of SPE, i.e., SPME,[139] involves the adsorption of analytes on a fiber coated with a suitable solid phase that can be pushed out of a microsyringe. The analyte is then thermally desorbed and transferred to the GC injector. The benefits of this method are that solvents can be eliminated and that it is impossible to overload the column because of the limited volume of adsorbent. Optimization of the method is a problem, however. Depending on where the fiber is placed in relation to the sample, SPME can be divided into direct (Direct Immersion [DI]-SPME) or headspace (HS-SPME).

GPC is just as frequently used as a cleanup technique. It enables micromolecular pesticides to be separated from macromolecular substances present in the matrix. While the considerable longevity of the columns is an advantage, the poorer resolution in comparison with adsorption techniques, especially when gradient elution techniques are used, is a disadvantage.

Approaches are being sought to develop pesticide determination techniques that are quick, simple, cheap, effective, and safe. QuEChERS (quick, easy, cheap, effective, rugged, and safe)[148–157] is one such method. It is based on a number of stages (see Fig. 7).

The consumption of sample and toxic solvents with the QuEChERS method is minimal. By applying QuEChERS to the determination of pesticides in fruit and vegetables, matrix effects are eliminated and high recoveries of target analytes are possible. The method can be modified depending on the type of sample and the target analytes. To improve the extraction of polar organophosphorus pesticides, the method is modified by the addition of acetic acid. When samples of citrus fruit are under investigation, protective wax coatings can be removed by freezing the samples for at least 1 hr. For the analysis of citrus fruits, blackcurrants, and raspberries, it is recommended to add aq NaOH to reach pH = 5 and to improve the analysis.

Table 14 Extract cleanup techniques in the determination of pesticides in fruit and vegetables.

Sample	Extract cleanup technique
Oranges, apples, grapes, pears	MSPD[126]
Apples, tomatoes	HS-SPME[127]
Oranges, apples, peaches, pears, tomatoes, lettuce, potatoes, leeks	GPC[140]
Tomatoes, pears, oranges	SPE[129]
Vegetables	SPME[141–144]
Oranges	MSPD, SBSE[145]
Strawberries, cherries	HS-SPME[146]
Oranges, tangerines, grapefruits, lemons	MSPD[147]

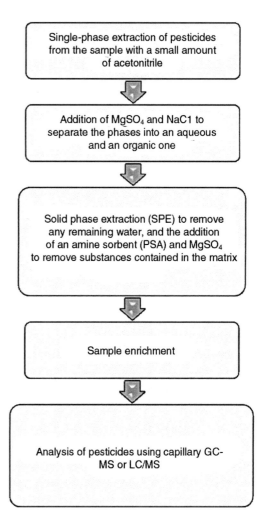

Fig. 7 Stages in the determination of pesticides using the QuEChERS method.

DETERMINATION OF RECENTLY APPLICABLE MODERN PESTICIDES IN SOIL SAMPLES[158–161]

Sample Collection and Preparation of Samples for Analysis

In order to determine the chemical composition of soil samples, it is necessary to preserve their natural structure. They are cut by manual scoop or a similar tool from the central, typical part of the material. In the next step, the material is introduced to a special cylinder of known volume. After this operation, the drying of the sample is desirable.

The drying step consists of two stages:

- Preliminary drying (also drying with circulation air masses) in temperature from 40°C to 100°C/105°C and drying in reduced temperature (lyophilization)
- Final drying, drying by chemical binding water, and drying using water adsorption

Techniques of Isolation and/or Enrichment Analytes from Soil Samples

For extracting the analytes from soil samples, the ultrasound-assisted liquid extraction (shaking)[162] or conventional Soxhlet extraction[163,164] can be used. In these cases, a mixture of non-polar and polar solvents is used. In addition, the extraction of analytes from soil samples can be investigated:

- Using supercritical fluid in-process—critical conditions are obtained by using highly compressed gases (the most compressed being CO_2) in the critical temperature range.[165,166]
- Using microwave process—microwave energy is absorbed by molecule compounds.
- As an ASE—the process runs at elevated temperature and pressure, providing appropriate solvent condition (liquid state of solvent).

Evaporation of the solvent is the next step, which provides the enrichment of the extract. Among the known methods for enriching the extract, the following are the most popular:

- Evaporation of the solvent in vacuum evaporator (vacuum distillation)
- Evaporation of the solvent in Kudern-Danish apparatus
- Evaporation of the solvent in a gas stream (nitrogen or air)

An important step in the preparation of the soil samples is removal of interfering substances. For elimination of sulfur, it is possible to use various chemical reactions, for example, reactions with sodium sulfate (IV), and aluminum oxides, or reactions with heavy metals. Aside from the use of chemical reactions, the methods of electrophoresis, chromatography, and saponification are also notable in the removal of interfering substances. To remove fats from soil samples, LLE,[167] adsorption chromatography, saponification, decomposition reaction using sulfuric acid at low temperature, precipitation, GPC,[168] and semipreparative HPLC can be applied.

Purification of the Extract

The purification process of extracts, which contain pesticides, includes fractionation of the extract. It can be carried out by different methods. Among the available methods are adsorption chromatography, GPC, and column chromatography.[169]

DERIVATIZATION OF PESTICIDES

By converting pesticide compounds into their derivatives, selectivity, analyte enrichment, target analyte resolution in the column, and selectivity of detection are all improved. Derivatization also enables a larger number of compounds to be determined in the sample. The inference is that the isolation of target analytes from the matrix and their enrichment are not always effective without changing the chemical structure of the analytes or the matrix.[1] Pesticides to be determined by GC are usually derivatized with pentafluorobenzyl bromide (PFBB).[67,72,75]

FINAL DETERMINATION

The last stage in the analytical procedure is the identification of compounds and their quantitative determination using an appropriate analytical technique. The usual techniques involved include the following:

- Capillary GC. Pesticides to be determined by GC should be volatile and thermally stable. During capillary GC, the injector can be operated in split mode (when the stream of carrier gas is divided) or splitless mode (when the carrier gas stream is not divided); either mode can be used for determining pesticides in environmental samples. The on-column injector, with which samples can be injected directly into the chromatographic column, is also used for the GC determination of pesticides.
- HPLC, usually in reversed-phase mode, is used for determining pesticides that cannot be determined by GC, for example, thermally unstable polar compounds like herbicides, carbamates, triazines, and also compounds that require derivatization.

Gas chromatography can be used to determine the residues of all pesticide classes. The choice of chromatographic column is extremely important as regards the separation of analytes and their qualitative and quantitative determination. The column should be highly efficient and be resistant to changes in separation parameters. The solid (stationary) phase should be thermally stable and be highly selective with respect to the constituents of the mixture being analyzed. The following detectors are used (Table 15):[157,170–172]

- ECD (electron capture detector)—highly sensitive in relation to compounds containing electronegative atoms
- FPD (flame photometric detector)—applied in the determination of organophosphorus compounds
- NPD (nitrogen phosphorus detector)—used for the simultaneous determination of organonitrogen and organophosphorus pesticides
- TSD (thermionic specific detector)—used to determine compounds containing nitrogen or phosphorus
- MS (mass spectrometry) or MS/MS (tandem mass spectrometry)
- UV–DAD (ultraviolet–diode array detector)—used in LC

Once a target analyte has been detected quantitatively, for example, by GC-NPD, the result must be confirmed by another independent method. The conditions of the process can be altered, say, by changing the temperature program or by using a different chromatographic column. It is crucial to obtain confirmation by another method as identification based solely on retention times is insufficient. MS/MS improves sensitivity and selectivity of analytical methods. In this technique, ions that were separated in the first analyzer are again fragmented and the derivative ions were analyzed in the second one. The chromatogram background is reduced, as a result of which the signal value is enhanced with respect to noise and the LOD of the target analytes is lowered.[174–176] Better chromatographic peak resolution and a smaller influence of the matrix on the final result can also be achieved using two-dimensional (2D) gas chromatography (GC × GC). This uses two columns: the partially separated constituents from the first column are further separated in the second one by a different mechanism.

Fast GC is equally frequently used to shorten the time of analysis and to obtain better peak resolution. Compared to classical GC, it requires shorter capillary columns with a smaller diameter and solid-phase films ca. 0.1 µm in thickness, as well as a faster flow rate and higher pressure of the carrier gas. These parameters yield determination results of a better precision.[174,177,178]

The trend at present is to develop analytical methods enabling a broad spectrum of analytes to be determined in a single analytical run (MRMs—multiresidue methods). However, the problem here is that the compounds to be determined simultaneously, often present at low concentrations, have different physicochemical properties depending on their chemical structure. Such a methodology, apart from being able to determine a large number of compounds in one run, should:

- Ensure maximum removal of interferents from extracts
- Give large recoveries of target compounds, high sensitivity, and good precision
- Be environmentally friendly, i.e., require the smallest possible quantities of samples and chemical reagents, especially organic solvents
- Be cheap, quick, and easy to carry out

Table 15 Characteristics of detectors used for the determination of pesticide analytes.

Type of detector	Merits	Drawbacks
Mass spectrometer (MS and tandem MS)	• Both quantitative and qualitative information obtainable • Highly sensitive • Used with both GC and LC • Applicable to trace determinations, in complex matrices • Can be universal or selective, as the need arises • Has a wide range of application used in multi-methods for determining pesticides of different classes • Better sensitivity and greater specificity sometimes obtainable with GC coupled to tandem MS (MS/MS) • Can operate in single ion monitoring (SIM) mode, which provides for greater sensitivity than the SCAN mode	• Expensive
Electron capture detector (ECD)	• A highly specific detector; in the case of pesticides, used to determine organochlorine compounds	• Applicable only to compounds containing electronegative atoms
Thermoionic detector (NPD)	• A detector suitable for selected classes of compounds	• Useful only for compounds containing nitrogen and phosphorus
Flame ionization detector (FID)	• Has a wide range of application • Suitable for almost all organic compounds • Highly sensitive and with a wide range of linearity	• The sample is destroyed

Source: http://www.pg.gda.pl.[173]

Research continues for the improvement of existing analytical methods and the development of new ones capable of supplying reliable results for a wide range of analytes.

SUMMARY

The increasingly widespread application of pesticides—substances with different physical and chemical properties—means that ever-larger amounts of these compounds are getting into the environment. As a result of the various processes they are subjected to, they may be converted to even more toxic compounds. They are currently regarded as some of the most dangerous environmental contaminants because of their stability, mobility, and long-term effects on living organisms. Moreover, they are usually present in the environment at very low concentrations in complex matrices, which makes their analysis difficult. Their determination is therefore prolonged by the need to perform time-consuming and often arduous processes in order to prepare samples for analysis, which themselves may be sources of further contamination and error. That is why for the analysis of pesticides in the environment, procedures that enable a large number of compounds to be detected simultaneously are required. It is often the case at present that target pesticide analytes have to be isolated from the matrix and enrichment before their final determination can be undertaken. To counteract the adverse effects of pesticides on human health, the continuous monitoring of their presence in the environment is fundamental. European Union recommendations are in place for analyzing the quality of water, air, and food in this respect. Existing techniques are being improved and new ones are being developed so that different classes of pesticides can be reliably determined in a quick, simple, cheap, and environmentally friendly manner.

REFERENCES

1. Biziuk, M.; Hupka, J.; Wardencki, W.; Zygmunt, B.; Siłowiecki, A.; Żelechowska, A.; Dąbrowski, Ł.; Wiergowski, M.; Zaleska, A.; Tyszkiewicz, H. *Pestycydy, występowanie, oznaczanie i unieszkodliwianie*; Wydawnictwo Naukowo-Techniczne WNT: Warszawa, 2001.
2. Sudo, M.; Kawachi, T.; Hida, Y.; Kunimatsu, T. Spatial distribution and seasonal changes of pesticides in lake Biwa, Japan. Limnology **2004**, *5*, 77–86.
3. Sosnowska, K.; Styszko-Grochowiak, K.; Gołaś, J. Nowe zanieczyszczenia w środowisku wodnym-źródła, zagrożenia, problemy analityczne. Analityka **2009**, *4*, 44–48.
4. Beyer, A.; Biziuk, M. Przegląd metod oznaczania pozostałości pestycydów i polichlorowanych bifenyli w próbkach żywności. Ecol. Chem. Eng. **2007**, *14*, 35–58.
5. Moreno, J.L.; Liebanas, F.J.; Frenich, A.G.; Vidal, J.L. Evaluation of different sample treatments for determining pesticide residues in fat vegetable matrices like avocado by low-pressure gas chromatography–tandem mass spectrometry. J. Chromatogr. A. **2006**, *1111*, 97–105.
6. Hajslova, J.; Zrostlikova, J. Matrix effects in (ultra)trace analysis of pesticide residues in food and biotic matrices. J. Chromatogr. A. **2003**, *1000*, 181–197.
7. Bartoszek-Pączkowska, A.; Bontemps-Gracz, M.; Mazerska, Z.; Biziuk, M.; Czerwiński, J.; Namieśnik, J.; Zasławska, L.; Żelechowska, A.; Czarnowski, W.; Jaśkowski, J., *Zarys Ekotoksykologii*; Eko-Pharma: Gdańsk, 1995.
8. Biziuk, M.; Buszewski, B.; Chrzanowski, W.; Czerwiński, J.; Gazda, K.; G,órecki, T.; Jamrógiewicz, Z.; Janicki, W.; Konieczka, P.; Kot-Wasik, A.; Łukasiak, J.; Makuch, B.; Namieśnik, J.; Pawliszyn, J.; Polkowska, Ż.; Przyjazny, A.; Sieńkowska-Zyskowska, E.; Wardencki, W.; Wolska, L.; Zasławska, L.; Zygmunt, B. *Fizykochemiczne metody kontroli zanieczyszczeń środowiska*; Wydawnictwo Naukowo-Techniczne WNT: Warszawa, 1998.
9. Ding, X.; Wang, X.; Wang, Q.; Xie, Z.; Xiang, C.; Mai, B.; Sun, L. Atmospheric DDTs over the North Pacific Ocean and the adjacent Arctic region: Spatial distribution, congener patterns and Skurce implication. Atmos. Environ. **2009**, *43*, 4319–4326.
10. Available at http://www.minddisrupted.org/documents/MD%20 Pesticide%20Fact.pdf (accessed May 16, 2012).
11. Papadakis, E.N.; Vryzas, Z.; Papadopoulou-Mourkidou, E. Rapid method for the determination of 16 organochlorine pesticides in sesame seeds by microwave-assisted extraction and analysis of extracts by gas chromatography–mass spectrometry. J. Chromatogr. A. **2006**, *1127*, 6–11.
12. Chen, S.; Shi, L.; Shan, Z.; Hu, Q. Determination of organochlorine pesticide residues in rice and human and fish fat by simplified two-dimensional gas chromatography. Food Chem. **2007**, *104*, 1315–1319.
13. Kodba, Z.C.; Koncina, D.B. A rapid method for the determination of organochlorine, pyrethroid pesticides and polychlorobiphenyls in fatty foods using GC with electron capture detection. Chromatographia **2007**, *66*, 619–624.
14. Escuderos-Morenas, M.L.; Santos-Delgado, M.J.; Rubio-Barroso, S.; Polo-Diez, L.M. Direct determination of monolinuron, linuron and chlorbromuron residues in potato samples by gas chromatography with q nitrogen–phosphorus detection. J. Chromatogr. A. **2003**, *1011*, 143–153.
15. Guardia Rubio, M.; Ruiz Medina, A.; Pascual Reguera, M.I.; Fernández de Córdova, M.L. Multiresidue analysis of three groups of pesticides in washing waters from olive processing by solid-phase extraction–gas chromatography with electron capture and thermionic specific detection. Microchem. J. **2007**, *85*, 257–264.
16. Santos, F.J.; Galceran, M.T. The application of gas chromatography to environmental analysis. Trends Anal. Chem. **2002**, *21*, 672–685.
17. Lartiges, S.B.; Garrignes, P. Gas chromatographic analysis of organophosphorus and organonitrogen pesticides with different detectors. Analysis **1995**, *23*, 418–421.
18. Zhang, J.; Lee, H.K. Application of liquid-phase microextraction and on-column derivatization combined with gas chromatography–mass spectrometry to the determination of carbamate pesticides. J. Chromatogr. A **2006**, *1117*, 31–37.
19. Sabik, H.; Jeannot, R.; Rondeau, B. Multiresidue methods using solid-phase extraction techniques for monitoring pri-

ority pesticides, including triazines and degradation products, in ground and surface waters. J. Chromatogr. A **2000**, *885*, 217–236.
20. Available at http://www.farmer.pl/srodki-produkcji/ochrona-roslin/jak_zwalczac_stonke_ziemniaczana_,d1f3f3d8d5452695e416.html (accessed February 2010).
21. Biziuk, M.; Przyjazny, A.; Czerwiński, J.; Wiergowski, M. Occurrence and determination of pesticides in natural and treated waters. J. Chromatogr. A. **1996**, *754*, 103–123.
22. Jokanovic, M. Medical treatment of acute poisoning with organophosphorus and carbamate pesticides. Toxicol. Lett. **2009**, *190*, 107–115.
23. Jokanović, M. Medical treatment of acute poisoning with organophosphorus and carbamate pesticides. Toxicol. Lett. **2009**, *190*, 107–115.
24. Jokanović, M. Biotransformation of organophosphorus compounds. Toxicology **2001**, *166*, 139–160.
25. Namieśnik, J.; Jamrógiewicz, Z.; Pilarczyk, M.; Torres, L. *Przygotowanie próbek środowiskowych do analizy*; Wydawnictwo Naukowo-Techniczne WNT: Warszawa, 2000.
26. Namieśnik, J.; Górecki, T. Preparation of environmental samples for the determination of trace constituents. Pol. J. Environ. Stud. **2001**, *10*, 77–84.
27. Directive of the European Union 98/83/EC of 3.XII.1998 r. on the quality of water intended for human consumption, Available at http://www.scribd.com/doc/8264629/Directive-9883EC-Drinking-water-standards (accessed February 2010).
28. Namieśnik, J.; Szefer, P. Preparing samples for analysis— The key to analytical success. Ecol. Chem. Eng. **2008**, *15*, 167–249.
29. Demkowska, I.; Polkowska, Ż.; Namieśnik, J. Ekstrakcja z wykorzystaniem ruchomego elementu sorpcyjnego typu Twister™ (SBSE). Analityka **2009**, *4*, 8–11.
30. Łopuchin, E.; Namieśnik, J. Dozymetria pasywna, dogodne narzędzie do badania zanieczyszczeń środowiska wodnego. Analityka **2009**, *1*, 29–36.
31. Namieśnik, J.; Chrzanowski, W.; Szpinek, P.; Bode, P.; Brzózka, Z.; Bulska, E.; Buszewski, B.; Gazda, K.; Górecki, T.; Jönsson, J.Å.; Kościelniak, P.; Levsen, K.; Lopez de Alda, M.J.; Matusiewicz, H.; Nawrocki, J.; Paschke, A.; Sandra, P.; Szefer, P.; Szpunar, J.; Trojanowicz, M.; Walkowiak, A.; Wardencki, W. *New Horizons and Challenges in Environmental Analysis and Monitoring*; CEEAM: Gdańsk, 2003.
32. Sankararamakrishnan, N.; Sharma, A.K.; Sanghi, R. Organochlorine and organophosphorous pesticide residues in ground water and surface waters of Kanpur, Uttar Pradesh, India. Environ. Int. **2005**, *31*, 113–120.
33. *Standardized Analytical Methods for Environmental Restoration Following Homeland Security Events-Revision 5.0*; United States Environmental Protection Agency (EPA): Cincinnati, USA, 2009.
34. Tse, H.; Comba, M.; Alaee, M. Method for the determination of organophosphate insecticides in water, sediment and biota. Chemosphere **2004**, *54*, 41–47.
35. Sabik, H.; Jeannot, R. Determination of organonitrogen pesticides in large volumes of surface water by liquid–liquid and solid-phase extraction using gas chromatography with nitrogen–phosphorus detection and liquid chromatography with atmospheric pressure chemical ionization mass spectrometry. J. Chromatogr. A **1998**, *818*, 197–207.
36. Sabik, H.; Fouquet, A.; Proulx, S. Ultratrace determination of organophosphorus and organonitrogen pesticides in surface water. Analysis **1997**, *25*, 267–273.
37. Tahboub, Y.R.; Zaater, M.F.; Al-Talla, Z.A. Determination of the limits of identification and quantitation of selected organochlorine and organophosphorous pesticide residues in surface water by full-scan gas chromatography/mass spectrometry. J. Chromatogr. A **2005**, *1098*, 150–155.
38. Polkowska, Ż.; Tobiszewski, M.; Górecki, T.; Namieśnik, J. Pesticides in rain and roof runoff waters from an urban region. Urban Water J. **2009**, *6*, 441–448.
39. Polkowska, Ż.; Górecki, T.; Namieśnik, J. Quality of roof runoff waters from an urban region (Gdańsk, Poland). Chemosphere **2002**, *49*, 1275–1283.
40. Grynkiewicz, M.; Polkowska, Ż.; Górecki, T.; Namieśnik, J. Pesticides in precipitation from an urban region in Poland (Gdańsk–Sopot–Gdynia Tricity) between 1998 and 2000. Water Air Soil Pollut. **2003**, *149*, 3–16.
41. Polkowska, Ż.; Kot, A.; Wiergowski, M.; Wolska, L.; Wołowska, K.; Namieśnik, J. Organic pollutants in precipitation: Determination of pesticides and polycyclic aromatic hydrocarbons in Gdańsk, Poland. Atmos. Environ. **2000**, *34*, 1233–1245.
42. Grynkiewicz, M.; Polkowska, Ż.; Górecki, T.; Namieśnik, J. Pesticides in precipitation in the Gdańsk region (Poland). Chemosphere **2001**, *43*, 303–312.
43. El-Kabbany, S.; Rashed, M.M.; Zayed, M.A. Monitoring of the pesticide levels in some water supplies and agricultural land, in El-Haram, Giza (A.R.E.). J. Hazard. Mater. A **2000**, *72*, 11–21.
44. Ballesteros, E.; Parrado, M.J. Continuous solid-phase extraction and gas chromatographic determination of organophosphorus pesticides in natural and drinking waters. J. Chromatogr. A **2004**, *1029*, 267–273.
45. Baugros, J.B.; Giroud, B.; Dessalces, G.; Grenier-Loustalot, M.F.; Cren-Olivé, C. Multiresidue analytical methods for the ultra-trace quantification of 33 priority substances present in the list of REACH in real water samples. Anal. Chim. Acta **2008**, *607*, 191–203.
46. Lyytikäinen, M.; Kukkonen, J.V.K.; Lydy, M.J. Analysis of pesticides in water and sediment under different storage conditions using gas chromatography. Arch. Environ. Contam. Toxicol. **2003**, *44*, 437–444.
47. Nogueira, J.M.F.; Sandra, T.; Sandra, P. Multiresidue screening of neutral pesticides in water samples by high performance liquid chromatography–electrospray mass spectrometry. Anal. Chim. Acta **2004**, *505*, 209–215.
48. Martínez, R.C.; Hermida, C.G.; Gonzalo, E.R.; Miguel, L.R. Behaviour of carbamate pesticides in gas chromatography and their determination with solid-phase extraction and solid-phase microextraction as preconcentration steps. J. Sep. Sci. **2005**, *28*, 2130–2138.
49. Liu, W.; Lee, H.K. Quantitative analysis of pesticides by capillary column high performance liquid chromatography combined with solid-phase extraction. Talanta **1998**, *45*, 631–639.
50. Potter, T.L.; Mohamed, M.A.; Ali, H. Solid-phase extraction combined with high-performance liquid chromatography-atmospheric pressure chemical ionization-mass spectrometry analysis of pesticides in water: Method performance and application in a reconnaissance survey of

residues in drinking water in Greater Cairo, Egypt. J. Agric. Food Chem. **2007**, *55*, 204–210.

51. López, F.J.; Beltran, J.; Forcada, M.; Hernández, F. Comparison of simplified methods for pesticide residue analysis use of large-volume injection in capillary gas chromatography. J. Chromatogr. A **1998**, *823*, 25–33.

52. Hu, J.Y.; Aizawa, T.; Magara, Y. Analysis of pesticides in water with liquid chromatography/atmospheric pressure chemical ionization mass spectrometry. Water Res. **1999**, *33*, 417–425.

53. Gonçalves, C.; Alpendurada, M.F. Multiresidue method for the simultaneous determination of four groups of pesticides in ground and drinking waters, using solid-phase microextraction-gas chromatography with electron-capture and thermionic specific detection. J. Chromatogr. A **2002**, *968*, 177–190.

54. Yao, Z.; Jiang, G.; Liu, J.; Cheng, W. Application of solid-phase microextraction for the determination of organophosphorous pesticides in aqueous samples by gas chromatography with flame photometric detector. Talanta **2001**, *55*, 807–814.

55. Frías, S.; Rodríguez, M.A.; Conde, J.E.; Pérez-Trujillo, J.P. Optimisation of a solid-phase microextraction procedure for the determination of triazines in water with gas chromatography–mass spectrometry detection. J. Chromatogr. A **2003**, *1007*, 127–135.

56. Su, P.; Huang, S. Determination of organophosphorus pesticides in water by solid-phase microextraction. Talanta **1999**, *49*, 393–402.

57. Natangelo, M.; Tavazzi, S.; Fanelli, R.; Benfenati, E. Analysis of some pesticides in water samples using solid-phase microextraction–gas chromatography with different mass spectrometric techniques. J. Chromatogr. A **1999**, *859*, 193–201.

58. Rocha, C.; Pappas, E.A.; Huang, Ch. Determination of trace triazine and chloroacetamide herbicides in tile-fed drainage ditch water using solid-phase microextraction coupled with GC–MS. Environ. Pollut. **2008**, *152*, 239–244.

59. Magdic, S.; Boyd-Boland, A.; Jinno, K.; Pawliszyn, J.B. Analysis of organophosphorus insecticides from environmental samples using solid-phase microextraction. J. Chromatogr. A **1996**, *736*, 219–228.

60. Volmer, D.A.; Hui, J.P.M. Rapid SPME/LC/MS/MS analysis of N-methylcarbamate pesticides in water. Arch. Environ. Contam. Toxicol. **1998**, *35*, 1–7.

61. Tomkins, B.A.; Ilgner, R.H. Determination of atrazine and four organophosphorus pesticides in ground water using solid phase microextraction (SPME) followed by gas chromatography with selected-ion monitoring. J. Chromatogr. A **2002**, *972*, 183–194.

62. Gonçalves, C.; Alpendurada, M.F. Solid-phase microextraction–gas chromatography–(tandem) mass spectrometry as a tool for pesticide residue analysis in water samples at high sensitivity and selectivity with confirmation capabilities. J. Chromatogr. A **2004**, *1026*, 239–250.

63. Gonçalves, C.; Alpendurada, M.F. Comparison of three different poly(dimethylsiloxane)–divinylbenzene fibres for the analysis of pesticide multiresidues in water samples: Structure and efficiency. J. Chromatogr. A **2002**, *963*, 19–26.

64. Sng, M.T.; Lee, F.K.; Lakso, H.Å. Solid-phase microextraction of organophosphorus pesticides from water. J. Chromatogr. A **1997**, *759*, 225–230.

65. Jönsson, J.Å.; Mathiasson, L. Liquid membrane extraction in analytical sample preparation. Trends Anal. Chem. **1999**, *18*, 318–325.

66. Sanusi, A.; Millet, M.; Mirabela, P.; Wortham, H. Comparison of atmospheric pesticide concentrations measured at three sampling sites: Local, regional and long-range transport. Sci. Total Environ. **2000**, *263*, 263–277.

67. Scheyer, A.; Morville, S.; Mirabel, P.; Millet, M. A multiresidue method using ion-trap gas chromatography–tandem mass spectrometry with or without derivatisation with pentafluorobenzylbromide for the analysis of pesticides in the atmosphere. Anal. Bioanal. Chem. **2005**, *381*, 1226–1233.

68. Peck, A.M.; Hornbuckle, K.C. Gas-phase concentrations of current-use pesticides in Iowa. Environ. Sci. Technol. **2005**, *39*, 2952–2959.

69. Wilson, W.E.; Chow, J.C.; Claiborn, C.; Fusheng, W.; Engelbrecht, J.; Watson, J.G. Monitoring of particulate matter outdoors. Chemosphere **2002**, *49*, 1009–1043.

70. White, L.M.; Ernest, W.R.; Julen, G.; Garron, C.; Leger, M. Ambient air concentrations of pesticides used in potato cultivation in Prince Edward Island, Canada. Pest Manage. Sci. **2006**, *62*, 126–136.

71. Sanusi, A.; Millet, M.; Mirabel, P.; Wortham, H. Gas-particle partitioning of pesticides in atmospheric samples. Atmos. Environ. **1999**, *33*, 4941–4951.

72. Scheyer, A.; Morville, S.; Mirabel, P.; Millet, M. Variability of atmospheric pesticide concentrations between urban and rural areas during intensive pesticide application. Atmos. Environ. **2007**, *41*, 3604–3618.

73. Harrad, S.; Mao, H. Atmospheric PCBs and organochlorine pesticides in Birmingham, U.K.: Concentrations, sources, temporal and seasonal trends. Atmos. Environ. **2004**, *38*, 1437–1445.

74. Alegria, H.; Bidleman, T.F.; Figueroa, M.S. Organochlorine pesticides in the ambient air of Chiapas, Mexico. Environ. Pollut. **2006**, *140*, 483–491.

75. Scheyer, A.; Graeff, C.; Morville, S.; Mirabel, P.; Millet, M. Analysis of some organochlorine pesticides in an urban atmosphere (Strasbourg, east of France). Chemosphere **2005**, *58*, 1517–1524.

76. Coscolla, C.; Yusa, V.; Marti, P.; Pastor, A. Analysis of currently used pesticides in fine airborne particulate matter (PM 2,5) by pressurized liquid extraction and liquid chromatography–tandem mass spectrometry. J. Chromatogr. A **2008**, *1200*, 100–107.

77. Schummer, C.; Mothiron, E.; Appenzeller, B.M.R.; Rizet, A.; Wennig, R.; Millet, M. Temporal variations of concentrations of currently used pesticides in the atmosphere of Strasbourg, France. Environ. Pollut. **2010**, *158*, 576–584.

78. Alegria, H.A.; Bidleman, F.F.; Shaw, T.J. Organochlorine pesticides in the ambient air of Belize, Central America. Environ. Sci. Technol. **2000**, *34*, 1953–1958.

79. Cessna, A.J.; Waite, D.T.; Kerr, L.A.; Grover, R. Duplicate sampling reproducibility of atmospheric residues of herbicides for paired pan and high-volume air samplers. Chemosphere **2000**, *40*, 795–802.

80. Waite, D.T.; Cessna, A.J.; Grover, R.; Kerr, L.A.; Snihura, A.D. Environmental concentrations of agricultural herbicides: 2,4-D and triallate. J. Environ. Qual. **2002**, *31*, 129–144.

81. Waite, D.T.; Cessna, A.J.; Grover, R.; Kerr, L.A.; Snihura, A.D. Environmental concentrations of agricultural herbicides: Bromoxynil, dicamba, diclofop, MCPA and trifluralin. J. Environ. Qual. **2004**, *33*, 1616–1628.
82. Yao, Y.; Tuduri, L.; Harner, T.; Blanchard, P.; Waite, D.; Poissantd, L.; Murphy, C.; Belzerf, W.; Aulagnierd, F.; Li, Y.; Sverko, E. Spatial and temporal distribution of pesticide air concentrations in Canadian agricultural regions. Atmos. Environ. **2006**, *40*, 4339–4351.
83. Waite, D.T.; Bailey, P.; Sproull, J.F.; Quiring, D.V.; Chau, D.F.; Bailey, J.; Cessna, A.J. Atmospheric concentrations and dry and wet deposits of some herbicides currently used on the Canadian Prairies. Chemosphere **2005**, *58*, 693–703.
84. Raina, R.; Sun, L. Trace level determination of selected organophosphorus pesticides and their degradation products in environmental air samples by liquid chromatography–positive ion electrospray tandem mass spectrometry. J. Environ. Sci. Health B **2008**, *43*, 323–332.
85. Bailey, R.; Belzer, W. Large volume cold on-column injection for gas chromatography–Negative chemical ionization–mass spectrometry analysis of selected pesticides in air samples. J. Agric. Food Chem. **2007**, *55*, 1150–1155.
86. Sofuoglu, A.; Odabasi, M.; Tasdemirc, Y.; Khalilid, N.R.; Holsen, T.M. Temperature dependence of gas-phase polycyclic aromatic hydrocarbon and organochlorine pesticide concentrations in Chicago air. Atmos. Environ. **2001**, *35*, 6503–6510.
87. Lammel, G.; Ghim, Y.; Grados, A.; Gao, H.; Huhnerfuss, H.; Lohmann, R. Levels of persistent organic pollutants in air in China and over the Yellow Sea. Atmos. Environ. **2007**, *41*, 452–464.
88. Gioia, R.; Offenberg, J.H.; Gigliottib, C.L.; Tottena, L.A.; Dua, S.; Eisenreich, S.J. Atmospheric concentrations and deposition of organochlorine pesticides in the US Mid-Atlantic region. Atmos. Environ. **2005**, *39*, 2309–2322.
89. Xu, D.; Dan, M.; Song, Y.; Chai, Z.; Zhuang, G. Concentration characteristics of extractable organohalogens in PM2.5 and PM10 in Beijing, China. Atmos. Environ. **2005**, *39*, 4119–4128.
90. Yang, Y.; Li, D.; Mu, D. Levels, seasonal variations and sources of organochlorine pesticides in ambient air of Guangzhou, China. Atmos. Environ. **2008**, *42*, 677–687.
91. Dvorska, A.; Lammel, G.; Klanova, J.; Holoubek, I. Kosetice, Czech Republic—Ten years of air pollution monitoring and four years of evaluating the origin of persistent organic pollutants. Environ. Pollut. **2008**, *156*, 403–408.
92. Contamination de l'air par les produits phytosanitaires en region Centre, Rapport final, 2006, available at http://www.ligair.fr (accessed May 16, 2012).
93. Burhler, S.; Basu, I.; Hites, R.A. A Comparison of PAH, PCB, and pesticide concentrations in air at two rural sites on Lake Superior. Environ. Sci. Technol. **2001**, *35*, 2417–2422.
94. Burhler, S.; Basu, I.; Hites, R.A. Causes of variability in pesticide and PCB concentrations in air near the Great Lakes. Environ. Sci. Technol. **2004**, *38*, 414–422.
95. Baraud, L.; Tessier, D.; Aaron, J.J.; Quisefit, J.P.; Pinart, J. A multi-residue method for characterization and determination of atmospheric pesticides measured at two French urban and rural sampling sites. Anal. Bioanal. Chem. **2003**, *377*, 1148–1152.
96. Motelay-Massei, A.; Harner, T.; Shoeib, M.; Diamond, M.; Stern, G.; Reosenberg, B. Using passive air samplers to assess urban–rural trends for persistent organic pollutants and polycyclic aromatic hydrocarbons. 2. Seasonal trends for PAHs, PCBs, and organochlorine pesticides. Environ. Sci. Technol. **2005**, *39*, 5763–5773.
97. Gouin, T.; Harner, T.; Blanchard, P.; Mackay, D. Passive and active air samplers as complementary methods for investigating persistent organic pollutants in the Great Lakes Basin. Environ. Sci. Technol. **2005**, *39*, 9115–9122.
98. Harner, T. Using passive air samplers to assess urban–rural trends for persistent organic pollutants. 1. Polychlorinated biphenyls and organochlorine pesticides. Environ. Sci. Technol. **2004**, *38*, 4474–4483.
99. Jaward, F.; Farrar, N.; Harner, T.; Sweetman, A.J.; Jones, A. Passive air sampling of PCBs, PBDEs, and organochlorine pesticides across Europe. Environ. Sci. Technol. **2004**, *38*, 34–41.
100. Farrar, J.; Harner, T.; Shoeib, M.; Sweetman, A.; Jones, C. Field deployment of thin film passive air samplers for persistent organic pollutants: A study in the urban atmospheric boundary layer. Environ. Sci. Technol. **2005**, *39*, 42–48.
101. Farrar, N.J.; Prevedouros, K.; Harner, T.; Sweetman, A.J.; Jones, K.C. Continental scale passive air sampling of persistent organic pollutants using rapidly equilibrating thin films (POGs). Environ. Pollut. **2006**, *144*, 423–433.
102. Pozo, K.; Harner, T.; Shoeib, M.; Urrutia, R.; Barra, R.; Parra, O.; Focardi, S. Passive-sampler derived air concentrations of persistent organic pollutants on a north–south transect in Chile. Environ. Sci. Technol. **2004**, *38*, 6529–6537.
103. Shen, L.; Wania, F.; Lei, Y.D.; Teixeira, C.; Bidleman, T.F. Atmospheric distribution and long-range transport behavior of organochlorine pesticides in North America. Environ. Sci. Technol. **2005**, *39*, 409–420.
104. Wania, F.; Shen, L.; Lei, Y.D.; Texeira, C.; Muir, D.C.G. Development and calibration of a resin-based passive sampling system for monitoring persistent organic pollutants in the atmosphere. Environ. Sci. Technol. **2003**, *37*, 1352–1359.
105. Klanova, J.; Kohoutek, J.; Hamplova, L.; Urbanova, P.; Holoubek, I. Passive air sampler as a tool for long-term air pollution monitoring: Part 1. Performance assessment for seasonal and spatial variations. Environ. Pollut. **2006**, *144*, 393–405.
106. Gouin, T.; Shoeib, M.; Harner, T. Atmospheric concentrations of current-use pesticides across south-central Ontario using monthly-resolved passive air samplers. Atmos. Environ. **2008**, *42*, 8096–8104.
107. Cupr, P.; Klanova, J.; Bartos, T.; Flegrova, Z.; Kohoutek, J.; Holoubek, I. Passive air sampler as a tool for long-term air pollution monitoring: Part 2. Air genotoxic potency screening assessment. Environ. Pollut. **2006**, *144*, 406–413.
108. Kot-Wasik, A.; Zabiegała, B.; Urbanowicz, M.; Dominiak, E.; Wasik, A.; Namieśnik, J. Advances in passive sampling in environmental studies. Anal. Chim. Acta **2007**, *602*, 141–163.
109. Wennrich, L.; Poppb, P.; Hafner, C. Novel integrative passive samplers for the long-term monitoring of semivolatile organic air pollutants. J. Environ. Monit. **2002**, *4*, 371–376.
110. Esteve-Turillas, F.A.; Pastor, A.; de la Guardia, M. Evaluation of working air quality by using semipermeable

membrane devices: Analysis of organophosphorus pesticides. Anal. Chim. Acta **2008**, *626*, 21–27.
111. Namieśnik, J.; Łukasiak, J.; Jamrógiewicz, Z. *Pobieranie próbek środowiskowych do analizy*; PWN: Warszawa, 1995.
112. Wu, S.; Tao, S.; Zhang, Z.; Lan, T.; Zuo, Q. Distribution of particle-phase hydrocarbons, PAHs and OCPs in Tianjin, China. Atmos. Environ. **2005**, *39*, 7420–7432.
113. Sadiki, M.; Poissant, L. Atmospheric concentrations and gas-particle partitions of pesticides: Comparisons between measured and gas-particle partitioning models from source and receptor sites. Atmos. Environ. **2008**, *42*, 8288–8299.
114. Sauret, N.; Wortham, H.; Putaud, J.; Mirabel, P. Study of the effects of environmental parameters on the gas/particle partitioning of current-use pesticides in urban air. Atmos. Environ. **2008**, *42*, 544–553.
115. Scheyer, A.; Morville, S.; Mirabel, P.; Millet, M. Gas/particle partitioning of lindane and current-used pesticides and their relationship with temperature in urban and rural air in Alsace region (east of France). Atmos. Environ. **2008**, *42*, 7695–7705.
116. Sauret, N.; Millet, M.; Herckes, P.; Mirabel, P.; Wortham, H. Analytical method using gas chromatography and ion trap tandem mass spectrometry for the determination of S-triazines and their metabolites in the atmosphere. Environ. Pollut. **2000**, *110*, 243–252.
117. Millet, M.; Wortham, H.; Sanusi, A.; Mirabel, P. A multiresidue method for determination of trace levels of pesticides in air and water. Arch. Environ. Contam. Toxicol. **1996**, *31*, 543–556.
118. Sofuoglu, A.; Cetin, E.; Bozacioglu, S.S.; Sener, G.D.; Odabasi, M. Short-term variation in ambient concentrations and gas/particle partitioning of organochlorine pesticides in Izmir, Turkey. Atmos. Environ. **2004**, *38*, 4483–4493.
119. Qiu, X.; Zhu, T.; Li, J.; Pan, H.; Li, Q.; Miao, G.; Gong, J. Organochlorine pesticides in the air around the Taihu Lake, China. Environ. Sci. Technol. **2004**, *38*, 1368–1374.
120. Schummer, C.; Mothiron, E.; Appenzeller, B.M.R.; Rizet, A.; Wennig, R.; Millet, M. Temporal variations of concentrations of currently used pesticides in the atmosphere of Strasbourg, France. Environ. Pollut. **2010**, *158*, 576–584.
121. Batterman, S.A.; Chernyak, S.M.; Gounden, Y.; Matooane, M.; Naidoo, R.N. Organochlorine pesticides in ambient air in Durban, South Africa. Sci. Total Environ. **2008**, *397*, 119–130.
122. Seiber, J.N.; Glotfelty, D.E.; Lucas, A.D.; McChesney, M.M.; Sagebiel, J.C.; Wehner, T.A. A multiresidue method by high performance liquid chromatography-based fractionation and gas chromatographic determination of trace levels of pesticides in air and water. Arch. Environ. Contam. Toxicol. **1990**, *19*, 583.
123. Briand, O.; Bertrand, F.; Seux, R.; Millet, M. Comparison of different sampling techniques for the evaluation of pesticide spray drift in apple orchards. Sci. Total Environ. **2002**, *288*, 199–213.
124. Coscolla, C.; Yusa, V.; Beser, M.I.; Pastor, A. Multi-residue analysis of 30 currently used pesticides in fine airborne particulate matter (PM 2.5) by microwave-assisted extraction and liquid chromatography–tandem mass spectrometry. J. Chromatogr. A **2009**, *1216*, 8817–8827.
125. Clement, M.; Arzel, B.; Le Bot, B.; Seux, R.; Millet, M. Adsorption/thermal desorption-GC/MS for the analysis of pesticides in the atmosphere. Chemosphere **2000**, *40*, 49–56.
126. Ramos, J.J.; González, M.J.; Ramos, L. Comparison of gas chromatography-based approaches after fast miniaturised sample preparation for the monitoring of selected pesticide classes in fruits. J. Chromatogr. A. **2009**, *1216*, 7307–7313.
127. Cai, L.; Gong, S.; Chen, M.; Wu, C. Vinyl crown ether as a novel radical crosslinked sol–gel SPME fiber for determination of organophosphorus pesticides in food samples. Anal. Chim. Acta **2006**, *559*, 89–96.
128. Mol, H.G.J.; van Dam, R.C.J.; Streijger, O.M. Determination of polar organophosphorus pesticides in vegetables and fruits using liquid chromatography with tandem mass spectrometry: Selection of extraction solvent. J. Chromatogr. A. **2003**, *1015*, 119–127.
129. Kmellár, B.; Fodor, P.; Pareja, L.; Ferrer, C.; Martinnez-Uroz, M.A.; Valwerde, A.; Fernăndez-Alba, A.R. Validation and uncertainty study of a comprehensive list of 160 pesticide residues in multi-class vegetables by liquid chromatography–tandem mass spectrometry J. Chromatogr. A. **2008**, *1215*, 37–50.
130. Ferrer, I.; Garcĭa-Reyes, J.F.; Mezcua, M.; Thurman, E.M.; Fernăndez-Alba, A.R. Multi-residue pesticide analysis in fruits and vegetables by liquid chromatography–time-of-flight mass spectrometry. J. Chromatogr. A. **2005**, *1082*, 81–90.
131. Melo, L.F.C.; Collins, C.H.; Jardim, I.C.S.F. New materials for solid-phase extraction and multiclass high-performance liquid chromatographic analysis of pesticides in grapes. J. Chromatogr. A. **2004**, *1032*, 51–58.
132. Sanusi, A.; Guillet, V.; Montury, M. Advanced method using microwaves and solid-phase microextraction coupled with gas chromatography–mass spectrometry for the determination of pyrethroid residues in strawberries. J. Chromatogr. A. **2004**, *1046*, 35–40.
133. Blasco, C.; Font, G.; Pico, Y. Comparison of microextraction procedures to determine pesticides in oranges by liquid chromatography–mass spectrometry. J. Chromatogr. A. **2002**, *970*, 201–212.
134. Kende, A.; Csizmazia, Z.; Rikker, T.; Angyal, V.; Torkos, K. Combination of stir bar sorptive extraction–retention time locked gas chromatography–mass spectrometry and automated mass spectral deconvolution for pesticide identification in fruits and vegetables. Microchem. J. **2006**, *84*, 63–69.
135. Sanchez, A.G.; Martos, N.R.; Ballesteros, E. Multiresidue analysis of pesticides in olive oil by gel permeation chromatography followed by gas chromatography–tandem mass-spectrometric determination. Anal. Chim. Acta **2006**, *558*, 53–61.
136. Gelsomino, A.; Petroviowi, B.; Tiburtini, S.; Magnani, E.; Felici, M. Multiresidue analysis of pesticides in fruits and vegetables by gel permeation chromatography followed by gas chromatography with electron-capture and mass spectrometric detection. J. Chromatogr. A. **1997**, *782*, 105–122.
137. Vreuls, J.J.; Swen, R.J.J.; Goudriaan, V.P.; Kerkhoff, M.A.T.; Jongenotter, G.A.; Brinkman, U.A.Th. Automated on-line gel permeation chromatography-gas chromatography

for the determination of organophosphorus pesticides in olive oil. J. Chromatogr. A. **1996**, *750*, 275–286.
138. Stajnbaher, D.; Zupancic-Kralj, L. Multiresidue method for determination of 90 pesticides in fresh fruits and vegetables using solid-phase extraction and gas chromatography-mass spectrometry. J. Chromatogr. A. **2003**, *1015*, 185–198.
139. Reinhard, H.; Sager, F.; Zoller, O. Citrus juice classification by SPME-GC–MS and electronic nose measurements. LWT—Food Sci. Technol. **2008**, *41*, 1906–1912.
140. Knezewvic, Z.; Serdar, M. Screening of fresh fruit and vegetables for pesticide residues on Croatian market. Food Control **2009**, *20*, 419–422.
141. Wu, J.; Luan, T.; Lan, Ch.; Wai Hung Lo, T.; Yuk Sing Chan, G. Removal of residual pesticides on vegetable using ozonated water. Food Control **2007**, *18*, 466–472.
142. Lambropoulou, D.A.; Albanis, T. A headspace solid-phase microextraction in combination with gas chromatography–mass spectrometry for the rapid screening of organophosphorus insecticide residues in strawberries and cherries. J. Chromatogr. A. **2003**, *993*, 197–203.
143. Juan-Garcĭa, A.; Picó, Y.; Font, G. Capillary electrophoresis for analyzing pesticides in fruits and vegetables using solid-phase extraction and stir-bar sorptive extraction. J. Chromatogr. A. **2005**, *1073*, 229–236.
144. Liu, W.; Hu, Y.; Zhao, J.; Xu, Y.; Guan, Y. Determination of organophosphorus pesticides in cucumber and potato by stir bar sorptive extraction. J. Chromatogr. A. **2005**, *1095*, 1–7.
145. Soler, C.; Mañes, J.; Picó, Y. Routine application using single quadrupole liquid chromatography–mass spectrometry to pesticides analysis in citrus fruits. J. Chromatogr. A. **2005**, *1088*, 224–233.
146. Riu-Aumatell, M.; Castellari, M.; Lopez-Tamames, E.; Galassi, S.; Buxaderas, S. Characterisation of volatile compounds of fruit juices and nectars by HS/SPME and GC/MS. Food Chem. **2004**, *87*, 627–637.
147. Soler, C.; Mañes, J.; Picó, Y. Routine application using single quadrupole liquid chromatography–mass spectrometry to pesticides analysis in citrus fruits. J. Chromatogr. A. **2005**, *1088*, 224–233.
148. Mezcua, M.; Ferrer, C.; García-Reyes, J.F.; Martínez-Bueno, M.J.; Sigrist, M.; Fernández-Alba, A.R. Analyses of selected non-authorized insecticides in peppers by gas chromatography/mass spectrometry and gas chromatography/tandem mass spectrometry. Food Chem. **2009**, *112*, 221–225.
149. Lehotay, S.T.; Ae Son, K.; Kwon, H.; Koesukwiwat, U.; Fu, W.; Mastovska, K.; Hoh, E.; Leepipatpiboo, N. Comparison of QuEChERS sample preparation methods for the analysis of pesticide residues in fruits and vegetables. J. Chromatogr. A. **2010**, *1217*, 2548–2560.
150. Thanh Dong, N.; Ji Eun, Y.; Dae Myung, L.; Gae-Ho, L. A multiresidue method for the determination of 107 pesticides in cabbage and radish using QuEChERS sample preparation method and gas chromatography mass spectrometry. Food Chem. **2008**, *110*, 207–213.
151. Lesueur, C.; Knittl, P.; Gartner, M.; Mentler, A.; Fuerhacker, M. Analysis of 140 pesticides from conventional farming foodstuff samples after extraction with the modified QuECheRS method. Food Control **2008**, *19*, 906–914.

152. Mayer-Helm, B. Method development for the determination of 52 pesticides in tobacco by liquid chromatography–tandem mass spectrometry. J. Chromatogr. A. **2009**, *1216*, 8953–8959.
153. Lacina, O.; Urbanova, J.; Poustka, J.; Hajslova, J. Identification/quantification of multiple pesticide residues in food plants by ultra-high-performance liquid chromatography-time-of-flight mass spectrometry. J. Chromatogr. A. **2010**, *1217*, 648–659.
154. Gilbert-López, B.; García-Reyes, J.F.; Molina-Díaz, A. Sample treatment and determination of pesticide residues in fatty vegetable matrices: A review. Talanta **2009**, *79*, 109–128.
155. Cunha, S.C.; Fernandes, J.O.; Oliveira, M.B.P.P. Fast analysis of multiple pesticide residues in apple juice using dispersive liquid–liquid microextraction and multidimensional gas chromatography–mass spectrometry. J. Chromatogr. A. **2009**, *1216*, 8835–8844.
156. Economou, A.; Botitsi, H.; Antoniou, S.; Tsipi, D. Determination of multi-class pesticides in wines by solid-phase extraction and liquid chromatography–tandem mass spectrometry. J. Chromatogr. A. **2009**, *1216*, 5856–5867.
157. Moral, A.; Sicilia, M.D.; Rubio, S. Anal. Chim. Acta **2009**, *650*, 207–213.
158. Gevao, B.; Semple, K.T.; Jones, K.C. Bound pesticide residues in soils: A review. Environ. Pollut. **2000**, *108*, 3–14.
159. Arias-Estevez, M.; Lopez-Periago, E.; Martınez-Carballo, E.; Simal-Gandara, J.; Mejuto, J.C.; Garcıa-Rıo, C. The mobility and degradation of pesticides in soils and the pollution of groundwater resources. Agric. Ecosyst. Environ. **2008**, *123*, 247–260.
160. Köhne, J.M.; Köhne, S.; Šimõnek, J. A review of model applications for structured soils: b) Pesticide transport. J. Contam. Hydrol. **2009**, *104*, 36–60.
161. Muller, K.; Magesan, G.N.; Bolan, N.S. A critical review of the influence of effluent irrigation on the fate of pesticides in soil. Agric. Ecosyst. Environ. **2007**, *120*, 93–116.
162. Hussen, A.; Westbom, R.; Megersa, N.; Mathiasson, L.; Bjorklund, E. Selective pressurized liquid extraction for multi-residue analysis of organochlorine pesticides in soil. J. Chromatogr. A **2007**, *1152*, 247–253.
163. Harner, T.; Wideman, J.L.; Jantunen, L.M.M.; Bidleman, T.F.; Parkhurst, W.J. Residues of organochlorine pesticides in Alabama soils. Environ. Pollut. **1999**, *106*, 323–332.
164. Oldal, B.; Maloschik, E.; Uzinger, N.; Anton, A.; Székács, A. Pesticide residues in Hungarian soils. Geoderma **2006**, *135*, 163–178.
165. Goncalves, C.; Carvalho, J.J.; Azenha, M.A.; Alpendurada, M.F. Optimization of supercritical fluid extraction of pesticide residues in soil by means of central composite design and analysis by gas chromatography–tandem mass spectrometry. J. Chromatogr. A **2006**, *1110*, 6–14.
166. Koinecke, A.; Kreuzig, R.; Bahadir, M. Effects of modifiers, adsorbents and eluents in supercritical fluid extraction of selected pesticides in soil. J. Chromatogr. A **1997**, *786*, 155–161.
167. Wang, X.; Zhao, X.; Liu, X.; Li, Y.; Fu, L.; Hu, J.; Huang, Ch. Homogeneous liquid–liquid extraction combined with gas chromatography–electron capture detector for the determination of three pesticide residues in soils. Anal. Chim. Acta **2008**, *620*, 162–169.

168. Wanner, U.; Burauel, P.; Fuehr, F. Characterisation of soil-bound residue fractions of the fungicide dithianon by gel permeation chromatography and polyacrylamide gel electrophoresis. Environ. Pollut. **2000**, *108*, 53–59.
169. Bi, E.; Schmidt, T.C.; Haderlein, S.B. Practical issues relating to soil column chromatography for sorption parameter determination. Chemosphere **2010**, *80*, 787–793.
170. Dugo, G.; Di Bella, G.; La Torre, M.; Saitta, M. Rapid GC-FPD determination of organophosphorus pesticide residues in Sicilian and Apulian olive oil. Food Control **2005**, *16*, 435–438.
171. Sobel, T.; Gul, O.; Buket, A. Simultaneous determination of various pesticides in fruit juices by HPLC-DAD. Food Control **2005**, *16*, 87–72.
172. Cunha, S.C.; Fernandes, J.O.; Alves, A.; Oliveira, M.B.P.P. Fast low-pressure gas chromatography–mass spectrometry method for the determination of multiple pesticides in grapes, musts and wines. J. Chromatogr. A. **2009**, *1216*, 119–126.
173. Available at http://www.pg.gda.pl/chem/Katedry/Analityczna/index.php?option=com_content&task=view&id=130&Itemid=86 (accessed February 2010).
174. Sadowska-Rociek, A.; Cieślik; E. Stosowane techniki i najnowsze trendy w oznaczaniu pozostałości pestycydów w żywności metodą chromatografii gazowej. Metrologia **2008**, *13*, 33–38.
175. Walorczyk, S. Różne możliwości wykorzystania chromatografii gazowej połączonej ze spektrometrią mas w analizie pozostałości środków ochrony roślin. Prog. Plant Protect. **2007**, *47*, 111–114.
176. Alder, L.; Greulich, K.; Kempe, G.; Vieth, B. Residue analysis of 500 high priority pesticides: Better by GC–MS or LC–MS/MS? Mass Spectrom. Rev. **2006**, *25*, 838–865.
177. Namieśnik, J. Modern trends in monitoring and analysis of environmental pollutants. Pol. J. Environ. Stud. **2001**, *10*, 127–140.
178. Beyer, A.; Biziuk, M. Methods for determining pesticides and polychlorinated biphenyls in food samples—Problems and challenges. Crit. Rev. Food Sci. **2008**, *48*, 888–904.

Pesticides: Banding

Sharon A. Clay
Plant Science Department, South Dakota State University, Brookings, South Dakota, U.S.A.

Abstract
Pesticides are applied to field crops to reduce insect, disease, and weed losses. The extensive use of pesticides is costly and of environmental concern. Band placement of pesticides is an alternative application practice that places chemicals only over crop rows, leaving interrows untreated.

INTRODUCTION

Pesticides are applied to field crops to reduce insect, disease, and weed losses. In 1996, more than 97% of United States row crops, such as corn, *Zea mays* and soybean *Glycine max* were treated with one or more pesticides[1] that were often applied in a uniform, broadcast pattern. The extensive use of pesticides is costly and of environmental concern. Band placement of pesticides is an alternative application practice that places chemicals only over crop rows, leaving interrows untreated. Banding reduces pesticide usage in proportion to the band width; treated areas typically range from 25% to 75% of the total field acreage. Banding also reduces pesticide input costs and environmental impacts. Insecticides are band applied more commonly than herbicides. Control of pests outside the application band may still need to be accomplished. For example, cultivation and residue management are nonchemical techniques used in interrows for weed control.

BAND APPLICATION USAGE

Chemicals applied in a band are restricted to a linear strip on or along a crop row as opposed to broadcast applications, where chemicals are applied uniformly over the entire area. About 20% of the total 1996 U.S. crop acreage was treated with a band application of insecticide over the row or soil injected next to the row to control insects that attack the crop seed, roots, or foliage.[2] The importance of banding varied by crop with 10% of soybean acreage, 60% of corn acreage, and 86% of tomato, *Lycopersicon esculentum* acreage treated with a band application. Insecticides can be band applied by several methods.[3,4] Seed treatments protect crop seeds from foraging insects such as the fire ant, *Solenopsis invicta*. Pre-emergent band applications may be applied in the open furrow (T-band) during planting or prior to crop emergence as an in-furrow spray and are used to reduce losses from soil dwelling insects such as corn rootworm *Dibrotica* spp. and sugar beet (*Beta vulgaris*) root maggot, *Tetanops yopaeformis*. Soil applied insecticides also may be systemic to control foliage feeders. Postemergent foliar band applications control stem and leaf pests such as aphids and mites in potato, *Solanum tuberosum* beetles and loopers in soybean, and bores and earworms in corn.

In contrast to insecticides that are banded to protect only the crop, herbicides generally are applied uniformly as a broadcast treatment to control all weeds. Herbicide band application has been proposed for row crops but is not a standard practice in United States crop production (Table 1). An average of 10% of all herbicide treated acres had a band application in 1996,[2] with 5% of soybean acres, 9% of the corn acres, and 38% of cotton *Gossypium hirsutum*, acres treated with a band. Herbicide banding was much greater in cotton grown in southern U.S. regions (40%) compared to cotton in the western region (9%). The overall low usage of herbicide band application may be due to the fact that weeds, even if limited to interrow areas, have the potential to reduce yield by competition or cause problems at harvest by slowing combine speed or plugging combines.

Advantages of Banding Pesticide Applications

Banding reduces chemical amounts applied to fields. For example, if a band covers 38 cm of a 76-cm row, 50% of the pesticide will be applied compared to a broadcast application reducing chemical costs. Pest control in the band is comparable to control in broadcast areas. Crop yields from band and broadcast herbicide treatments are similar if interrow weeds are controlled in a timely manner.[5,6]

Another major advantage of band applications is the reduction of environmental impact through leaching and runoff losses. The U.S. Geological Survey[7] estimated that more than 100,000 metric tons of soluble pesticides were applied to the U.S. Mississippi River Basin in 1996. Of this amount, most were herbicides, and more than 60% were soil applied. A problem with soil-applied pesticides is that only about 10% of the applied amount reaches the target. The rest of the pesticide is needed to counteract

Table 1 Percent of herbicide treated acres using herbicide broadcast vs. band application methods and percent of all planted acres using mechanical cultivation for weed control in the United States.

Crop	Region	Broadcast (% of herbicide-treated acres)	Band (% of herbicide-treated acres)	(% of planted acres)
Corn	Northeast	83	2	
	North Central	85	9	
	South	82	9	51
	All corn states	85	9	
Soybean	North Central	89	3	
	South	83	12	29
	All soybean states	88	5	
Cotton	South	43	40	
	West	71	9	89
	All cotton states	45	38	

Source: Fernandez-Cornejo and Jans.[1]

sorption to soil, microbial degradation, volatilization, photodecomposition, and soil dilution. However, a portion may cause environmental problems from off-site movement. The frequency of detection, maximum concentration, and mass loss of pesticide have been reported to be three to four times lower from band applications where 50% of the area was treated compared to losses from broadcast applications.[5]

Disadvantages of Band Applications

Band applications may allow pests to take shelter in interrow areas causing crop damage or pest populations to increase. For example, corn rootworm larvae may not attack roots directly under insecticide bands. However, as the roots expand into untreated areas, larvae may feed on these roots resulting in stressed plants and an adult population, although yield loss may not occur. Untreated interrow areas also may become refuges for novel pests permitting new outbreaks. The threat of outbreaks must be balanced against control and potential crop loss costs.

Supplemental methods, such as mechanical cultivation done once or twice prior to crop canopy closure, generally are needed to control weeds outside the herbicide band application area. Interrow cultivation of row crops is more common than banded herbicide applications. In 1996, 51%, 29%, and 89% of all corn, soybean, and cotton acres, respectively, were cultivated at least once for weed control (Table 1). Cultivation and band applications are compatible and may help explain the high frequency of band applications in southern U.S. cotton. However, a large discrepancy between cultivated and banded acres exists and points to other disadvantages of band applications.

If herbicide bands are too narrow, or cultivation not wide enough, strips of weeds may remain[8] However, cultivation too close to rows may prune crop roots possibly resulting in yield loss. The timing of cultivation is important. Adverse weather may cause unsuitable conditions for cultivation when weeds are vulnerable to mechanical control. Interrow weeds growing unrestrained for whatever reason often result in reduced yields, harvest problems, and increased weed seeds in the soil that perpetuate weed problems.

High fuel prices increase cultivation costs so that band plus cultivation cost may be similar if not more expensive than a broadcast application. This may discourage the use of banding techniques. However, interrow seeding of cover crops, such as short season annual medic (*Medicago* spp.) varieties that suppress weeds,[9] or residue management may be alternatives to cultivation.

Equipment Requirements

Broadcast applications rely on spray patterns that overlap for uniform application rates. Broadcast nozzles are flat fan types with feathered spray patterns resulting in more output directly below the nozzle than on either side. Even nozzles are used for band applications and provide uniform output across the spray pattern. Spray from adjacent nozzles should not overlap, as output in overlap areas would be double the amount desired.

Spray boom height must be appropriate and consistent, as boom height and nozzle output angle regulate band width. The band becomes wider as boom height or nozzle output angle increases. A 5-cm increase in boom height increases band width of either a 80° or 95° series nozzle by about 10 cm. The boom must be 2.5 cm higher for an 80° than a 95° nozzle to get the same band width.

Band applications of granular or dry material can be applied in several ways. Some pesticides can be applied as a seed treatment. Other banding methods include shanking the pesticide into the soil as a liquid, placing the pesticide in the seed furrow (T-band) that is subsequently covered with soil, or applying dry granules to the soil surface.

FUTURE APPLICABILITY OF BANDING PESTICIDES

Reliance on soil-applied pesticides has been reduced, but not eliminated, by the introduction of genetically modified crops for insect resistance and postemergence herbicide tolerance. However, soil-applied pesticides, which are often targeted for band application, will continue to be used for early control of crop pests or in situations where weather limits postemergent options. Band application use will increase despite concerns of uncontrolled pests and need for supplemental control methods. Band applications reduce offsite leaching into tile lines and groundwater, and runoff into surface water.[6] Nonpoint source reductions in pesticide residues are needed to tackle major environmental problems such as aquifer and surface water quality, and pesticide residue and exposure regulations, based on U.S. Food Quality Protection Act.

REFERENCES

1. Fernandez-Cornejo, J.; Jans, S. *Pest Management in U.S. Agriculture*; ERS/USDA: Washington, DC, 1999. http://www.ers.usda.gov/epubs/pdf/ah717 (accessed Dec 2000).
2. Padgitt, M.; Newton, D.; Penn, R.; Sandretto, C. *Production Practices for Major Crops in U.S. Agriculture, 1990–97*; ERS/USDA, SB-969, 2000. http://www.ers.usda.gov/epubs/pdf/sb969/index.htm (accessed Dec 2000).
3. Drees, B.M.; Cavazos, R.; Berger, L.A.; Vinson, S.B. Impact of seed-protecting insecticides on sorghum and corn seed feeding by red imported fire ants (Hymenoptera: Formicidae). J. Econ. Entomol. **1992**, *85* (3), 993–997.
4. Woodford, J.A.; Gordon, S.C.; Foster, G.N. Side-band application of systemic granular pesticides for the control of aphids and potato leafroll virus. Crop Prot. **1988**, *7* (2), 96–105.
5. Clay, S.A.; Clay, D.E.; Koskinen, W.C.; Berg, R.K., Jr. Application method: impacts on atrazine and alachlor movement, weed control, and corn yield in three tillage systems. Soil and Tillage Res. **1998**, *48* (3), 215–224.
6. Anderson, J.L.; Allmaras, R.R. Tillage systems and agricultural management: water quality effects. Soil and Tillage Res. **1998**, *48* (3), 141–257.
7. Thurman, E.M.; Meyer, M.T. *Herbicide Metabolites in Surface Water and Groundwater*; ACS Symposium Series 630, American Chemical Society: Washington, DC, 1996; 320.
8. Paarlberg, K.R.; Hanna, H.M.; Erbach, D.C.; Hartzler, R.G. Cultivator design for interrow weed control in notill corn. Appl. Eng. Agric. **1998**, *14* (4), 353–361.
9. Vos, R.J. *Effect of Spring-seeded Annual Medics on Weed Management in Zea mays Production*; Ph.D. Dissertation, South Dakota State University: Brookings, SD, 1999; 173.

Pesticides: Chemical and Biological

Barbara Manachini
Departments of Environmental Biology and Biodiversity, University of Palermo, Palermo, Italy

Abstract

The World Health Organization (WHO), Food and Agriculture Organization (FAO), and other international organizations are now calling for the development of environmentally sustainable systems that are less reliant on chemical pesticides (CPs) as the primary management tool for pest control. However, at the moment, biological control alone cannot solve all pest problems and must be considered an instrument to be used in combination with other methods. The compatibility of CPs and biological pesticides (BPs) is a key factor for the implementation of Integrated Pest Management (IPM) programs. Testing the compatibility of CP and BP is essential if these two agents are to be applied together in integrated management as synergistic, neutral, and antagonistic responses have been reported for some commercial BPs used with chemical ones. Although the compatibility of such pest control methods has been outlined as being a contentious and complicated issue, they are being increasingly used in combination, or even in tandem, to control pests in agricultural and urban settings, as well as for the control of invasive species and insect vectors of human and animal diseases. It is not always possible to achieve compatibility, but combinations can sometimes result in an additive or even synergistic effect, improving pest control and reducing CP use. There has been some discussion on the need for a wider international standard to achieve a much needed formal compatibility with respect to environmental and human health. Although an approach combining CP and BP can provide a safer and more comprehensive management program, which results in reducing CP, an exhaustive and standardized method is still needed to evaluate the environmental risk assessment (ERA) of such combinations. Indeed, carrying out an ERA is important as synergism on a targeted pest can result in toxicological synergism for non-target organisms. Most of the information documenting adverse environmental effects comes from studies focused on exposure to single pesticides, but in normal agricultural practice, the use of only a single pesticide is rare. The combination of BP and CP is an example of when an ERA is needed. Such an ERA for CP–BP combinations could be useful to predict an outcome, also in cases where combination was not fully intended. Thus, future goals should be directed towards evaluating and standardizing the ERA and cost analysis of CP and BP compatibility.

INTRODUCTION

In June 2010, the Food and Agriculture Organization (FAO) issued the publication *Guidance on Pest and Pesticide Policy Development*, a topic linked to the International Code of Conduct on the Distribution and Use of Pesticides.[1] Special attention was paid to Integrated Pest Management (IPM) as an approach to sustainable pest management and a viable avenue towards reducing reliance on chemical pesticides (CPs). The guidelines emphasize the importance and the scope of policies impacting pest management and encourage both governments and stakeholders to consider whether the present level and extent of CP use are actually justified.[2]

Reducing CP use is also an issue of the 6th Environment Action Program adopted by the European Parliament and the Council.[3] Several European countries have introduced specific policies to decrease pesticide use[4] and propose to label farms using an environmental compatibility value (ECV), an indicator of the farm's pesticide intensity.[5] However, complete biological control and subsequent CP elimination has not always been achieved, neither in agroecosystems,[6] nor in postharvest[7] and public health programs,[8] nor in controlling veterinary pests.[9,10] Moreover, there is also an increasing number of new threats from non-indigenous (i.e., invasive) pest species where biological control is advisable but not always available.[11,12]

An important approach in IPM programs could be the use of biological pesticides (BPs) together with the rational use of CPs that could lead to a reduction in the non-indigenous (i.e., invasive) pest species. BPs are mass-produced, biologically based agents used for pest control.[12] According to the EPA,[13] BPs include both mass-produced biocontrol agents (BCAs) (i.e., viruses, bacteria, fungi, protozoans and nematodes, parasitoids, predators) and natural materials derived from animals, plants, bacteria, and certain minerals.[12,13] However, to date, a consistent regulatory definition has still not been agreed upon, nor is there a standard definition for biopesticides or BPs.[14]

The definition of BPs is still under revision in Europe. Over the years, as Europe has moved towards a harmonized European Union (EU) system, it has been necessary to take into account the wide range of scientific and policy perspectives of the different countries, and the registration of BPs could still take years. The more used term, "biological" plant protection products, can be divided into the following categories: 1) biochemicals (plant extracts, naturally occurring chemicals, plant strengthener); 2) semiochemicals (chemicals that affect insect behavior: pheromones, allomones, kairomones); and 3) microorganisms and viruses. BPs are regulated under the same European legislation as chemical crop protection substances, which means that the active ingredient is registered at the EU level and that products containing it are registered in member states.[14,15] BPs are considered under the fourth review of Directive 91/414/EEC.[15,16] Products based on insect parasitic nematodes, or other macroorganisms, can be referred to as BPs but are exempt from registration.[14,16]

BPs are usually inherently less toxic than conventional pesticides and target a narrow range of pests, in contrast to the broad spectrum of CPs that can affect other non-target organisms. BPs are often effective in very small quantities and tend to decompose quickly, thereby resulting in lower pesticide exposures and largely avoiding the pollution problems caused by conventional pesticides.[12–17] New BPs are approved every year, (e.g., 15 in the United States in 2009[18]).

However, as mentioned before, CPs are not ruled out[7,12] when, for example, a range of pests is present, or when only one method is not efficient; combining two or more control methods often results in economic and environmental advantages.[7–10,19,20] CPs and BPs need to be compatible with each other as incompatibility can lead to loss of effectiveness, increased toxicity to humans and other non-target organisms, the development of pesticide resistance, major product loss, and crop injury. Some information on the selectivity of most pesticides towards natural enemies of pests is already available, but data on the compatibility of CP and specific BP are often limited and sometimes conflicting.[19] In the majority of crop systems today, emphasis is still placed on single technologies such as the use of pesticides, host plant resistance, and biocontrol, consideration rarely being given to their interaction.[17,19–21] We already have data on the compatibility of certain pesticides and natural enemies (mainly predators and parasitoids),[19,22–24] but there are no detailed reviews on the compatibility of CP and BP, even though the number of publications has recently been increasing.[8,9,11,12,21] Moreover, despite this and the experience gained in these last decades of CP and BP use, there still remain unanswered inevitable questions with regard to the wide-scale environmental impact of their combined use, including non-target effects on other organisms. These questions on their wide-scale impact can be addressed using recognized frameworks for environmental risk assessment (ERA),[1,3,21,22] which need to be adapted for CP and BP combinations or for use in a more extensive IPM.

POSSIBILITY OF INTEGRATION

Increasing problems with CP have stimulated the search for alternative control measures such as the use of BP. As BPs have quite a narrow target, there are still many situations for which they are not suitable. In fact, chemical control agents are probably still more frequently used and are the most widespread means of achieving effective and reliable pest reduction. However, nonselective CP can eliminate the natural enemies of pests and induce unintended side effects such as secondary pest outbreaks and pest resurgence. An intensive use of most pesticides will often also lead to pesticide resistance in target pests. Nevertheless, insecticides are used worldwide for insect pest control and often play a major role in IPM in many cropping systems. As such, pesticide compatibility with biological control agents is a major concern to practitioners of IPM. After considering the costs and benefits, as well as the ecological and environmental impact (including the impact on biodiversity), the effects of economically damaging pest populations may be reduced to acceptable levels through biological and chemical control measures applied individually or in combination.[18,21] Investigating the compatibility of CP and BP as derived natural materials (e.g., the most common botanical pesticides neem and pyrethrum) seems to be easier than using CP combined with living organisms. In fact, in the first case, compatibility studies would be similar to the tests used for chemical combinations. Thus, BP and CP are compatible if no adverse effects occur as a result of mixing them together; however, they are often applied at different times, in accordance with the target pest they need to reach (e.g., type of pest such as weeds or insects and their life cycles). In the case of a mixture, even if applied in tandem, the deactivation of an active ingredient can often occur because of chemical incompatibility, or generally because it is affected by temperature, pH, and duration of the mixture in the tank. Physical and chemical incompatibility results in an unstable state, with the formation of crystals, flakes, or a sludge mixture.

In the case of a combination of CP and BP based on living organisms, the effect will depend on the interaction between the products and the type of the species of interest. A variety of factors influence BP–CP compatibility, including whether the BP is a living organism, and if so which one (virus, bacteria, fungus, or protozoan); its species and strain sensitivity; its life-stage sensitivity, rate, timing, and mode of application; and the CP mode of action. CP can impact on living organisms by affecting several biological parameters: survival, infectivity, sporulation, germination, reproduction, development time, and host acceptance. When CP and BP are applied together, the resultant effect on pests or pathogens can be antagonistic (negative) or additive synergistic (positive). Antagonistic effects result in the efficacy of the integrated measures being lower than the sum of those of the individual components, while additive and synergistic effects are equal to or larger than the sum of the single effects of each component. The literature shows a number of CPs to be compatible with BPs, but case-by-case studies are needed. Synergistic, neutral, or antagonistic responses were recorded for some BPs with CPs, and examples are given in the next paragraphs and in Table 1.

However, when evaluating the potential of a CP for compatibility with BPs, it is important to consider whether they are to be applied together (e.g., in mixture) or subsequently. In the latter case, the persistence of the CP needs to be taken into account. Consideration must be given not only to the direct effects (including short- and long-term impact) of CP and their residues but also to their indirect effects (trophic-web perturbation, change in biodiversity in the natural community of antagonists), but few data are available on this aspect.

Incompatibility: Negative Integration

BP and CP can be incompatible as chemical compounds can severely reduce the activity of the live organisms used in the biological control, inducing mortality, low reproduction rates, and reduced infectivity capacity and changing the host searching behavior (e.g., for protozoan). Some chemical insecticides (chlorpyrifos, propetamphos, and cyfluthrin) have been found to have no effect on the conidial germination of the fungus *Metarhizium anisopliae* used against the German cockroach (*Blattella germanica*), but they do have an adverse effect on the growth and sporulation of the fungus. Because of the direct relationship between *M. anisopliae* growth and insecticide concentration, it has been necessary to greatly limit the in-field use of these insecticides.[25]

It has been found that CPs like fenitrothion, carbofuran, glyphosate, and azoxystrobin and insecticides like hexaflumuron have a negative effect on the sporulation vegetative growth and conidial germination of *M. anisopliae* while pyrazosulfuron-ethyl, bentazon, fipronil, pyriproxyfen, and clomazone have less impact.[26,27]

Fungicides did not show compatibility with several entomopathogenic fungi, particularly *Beauveria bassiana*;[28,29] there was inhibition of the growth of mycelia and germination of fungal conidia at all concentrations and reduced efficacy against the pest. With regard to other fungi species and CP, it would not be advisable to mix alive fungus with inorganic fungicides as the latter could nullify the effect of the microbial agent; however, *B. bassiana* is more compatible with the azoxystrobin fungicides and with fertilizers and some insecticides.[25–31] Insecticides, especially second-generation synthetic pyrethroids and neonicotinoids, have been found fairly compatible with *B. bassiana*, compared to others like the insecticide/acaricide *flufenoxuron* (benzoylurea) that completely, or strongly, inhibited its development.[25–31]

These results demonstrate that by carefully selecting the pesticides and fungicides, an integrated pest and disease control can be achieved with a microbial pesticide.

Another problem is that often there is incompatibility only for some of the species and strains used in BP. Recently proposed, however, was a quick method for evaluating the in vitro compatibility of *B. bassiana* and organophosphorous pesticides commonly used.[32] Based on these in vitro results (chemical concentrations permitting fungal survival) and biochemical analysis by SDS-PAGE, isolates of *B. bassiana* have now been categorized as more tolerant and less tolerant.[32] Thus, further biochemical analyses would be welcome for the detection of markers indicating resistance to pesticides and the susceptibility of BP containing living microorganisms; this would enable a quick screening for compatible potential strains.

Whereas mortality levels from direct chemical exposure have been measured and tested for several species using laboratory bioassays, possible sublethal effects resulting from indirect effects or secondary compounds are relatively unknown. For example, in the presence of sunlight, xanthene-based insecticides have been found to have a negative effect on useful species of fungi such as *B. bassiana*, *M. anisopliae*, and *Paecilomyces fumosoroseus*.[33]

The above examples show the need for detailed research into the possible interactions of CP and BP before their effective integration.

Compatibility: Positive Integration and Synergism

Although the combined use of CP and BP is a contentious and complicated issue,[34,35] the two methods are being increasingly used in tandem to control pests in agricultural settings, as well as for the control of invasive species.[36] Such use in tandem could reduce CP doses, improve application times, reach different life stages of the pathogen/pest, and control resistant pest strains. Often, the compatibility of the use of CP and BP results in a synergic effect.

Synergy can occur when one control measure directly improves the efficacy of the other, or when one control measure induces host resistance or predisposes the pathogen to increased susceptibility. Thus, the combination or alternation of CP with BP can enhance and stabilize the efficacy of biological control. In addition, this strategy can give even better control over resistant pathogen strains and can enable commercial growers and packinghouses to reduce the amount of CP used, thus lowering the amount of chemical residue on marketed products. The biological pathways and modes of action of synergistic treatment are still under investigation.[35] However, the synergistic

Table 1 Some other examples of type of interactions between chemical and biological pesticides.

Type	Chemical pesticides	Biological pesticides	Type of interaction	References
Acaricide	Fenitrotion	*Bacillus thuringiensis*	Negative: inhibition of replication but probably due more to the presence of emulsifiers	[58]
Fungicide	Benomyl	Entomopathogenic fungus *Nomuraea rileyi*	Negative	[59]
Fungicides	Benfungin	Entomopathogenic bacteria *Bacillus thuringiensis* var. *kurstaki*	Positive/Negative: harmful effect on the viability of Bt spores but not the adverse effect on crystal toxin activity	[60]
Fungicides	Aromex, Captafol, Captan, Chlorothalonil, Dinocap, Metalaxyl, sulphur, Triadimefon	Entomopathogenic bacteria *Bacillus* sp.	Neutral: activity of chitinase was not affected	[61]
Fungicides	Imazalil, Thiabenazole	*Pichia guillerimondii*	Positive: increase control of fungus in post harvest	[7]
Fungicide	Iprodione	Yeast *Metschnikowia fructicola*,	Positive: synergistic effects of sublethal treatments applied sequentially to control postharvest disease	[62]
Fungicide	Azoxystrobin	Bacteria *Pseudomonas fluorescens*	Positive: Synergic effect in control downy mildew.	[63]
Herbicide	Bentazone	Entomopathogenic fungus *Nomuraea rileyi*	Negative	[59]
Herbicide	Glyphosate	Bioherbicidal fungus, *Myrothecium verrucaria*	Positive: they effectively control the weeds redvine and trumpetcreeper	[64]
Insecticide	Allosamidin (inhibitor of chitinases)	*Bacillus* sp.	Negative inhibition of the activity of the chitinase	[61]
Insecticide	Carbaryl	Entomopathogenic fungus *Nomuraea rileyi*	Neutral/negative	[59]
Insecticide	Diflubenzuron	Entomopathogenic fungus *Nomuraea rileyi*	Neutral/negative low	[59]
Insecticide/ Acaricide	Cyhalothrin (pyrethroid insecticide)	Entomopathogenic fungus *Nomuraea rileyi*	Neutral/negative: lower inhibitory effect on the fungus.	[59]
Insecticides	Acephate, Chloropyriphos, Monocrotophos (Organophosphate)	*Bacillus* sp.	Neutral	[61]
Insecticide	Acephate (Organophosphate)	Entomopathogenic bacteria *Bacillus thuringiensis*	Neutral: no inhibition effect on replication, spore germination and crystal size of the bacterium	[65]
Insecticides	Methylparathion (Organophosphate)	Entomopathogenic bacteria *Bacillus thuringiensis* var. *kurstaki*	Neutral: compatible but without synergetic effects	[66]
Insecticides	Permethrin	Entomopathogenic bacteria *Bacillus thuringiensis* var. *kurstaki*	Neutral: compatible but without synergetic effects	[66]
Insecticide	Methomyl (Carbamate)	*Bacillus* sp.	Neutral	[61]
Insecticide	Carbofuran (Carbamate)	Entomopathogenic fungus *Beauveria bassiana*	Positive: increase mortality of *Ostrinia nubilalis*	[67]
Insecticide	Actellic. (Organophosphate)	*Bacillus thuringiensis* var. *israelensis*	Positive: larvae and adults *Aedes* mosquitoes	[68]
Insecticide	Aqua Resigen, Resigen (Permethrin/ S+bioallethrin)	*Bacillus thuringiensis* var. *israelensis*	Positive: larvae and adults *Aedes* mosquitoes	[68]
Insecticide	Fendona SC (Pyrethroid)	*Bacillus thuringiensis* var. *israelensis*	Positive	[68]

effects of using sublethal doses in tandem, or sequentially with a BP, have the potential of reducing the use of CP.

Examples of Compatibility: Vector of Diseases and Control of Resistant Strains

In the area of IPM, the best example of what can be accomplished by combining BP and CP comes from studies to combat the vector of malaria and dengue fever.[8,37,38] Simultaneous applications of CP and BP were made to control *Aedes aegypti* and *Aedes albopictus*, the vectors of dengue fever. Adulticide (pirimiphos-methyl) and larvicide (*Bacillus thuringiensis* var. *israelensis*) dispersed in thermal fog were effective against the mosquito adults and larvae, immediately causing almost complete mortality as well as long-term efficacy. However, as adulticide and larvicide treatments are both labor-intensive and time-consuming, it was considered that a chemical adulticide plus a biological larvicidal mixture could be a better option as such treatment is quick operationally and more economic.[8] Indeed, malaria infects hundreds of thousands of people each year and is the cause of over a million deaths worldwide. A major strategy currently used to control the transmission of malaria is the use of insecticides, but evolving resistance to such treatment continues to have a marked impact on such efforts.

The use of a combination of fungal spores and chemical insecticides is an effective way of combating insecticide-resistant malaria mosquitoes. Researchers have shown that fungi and insecticides reinforce each other's efficacy, and the effect of using their combination is greater than the sum of using the two methods separately. A fungal infection with *B. bassiana* and *M. anisopliae* made the wild mosquitoes more vulnerable to permethrin, and exposure to permethrin reinforced the efficacy of the fungi. Simultaneous exposure to both fungi and insecticides had the greatest impact on the insecticide-resistant mosquitoes. Indeed, the effect was even higher than expected, indicating that fungi and permethrin enhance each other's efficacy.[37]

Moreover, BP based on fungi entomopathogenic to mosquitoes could be an effective means of reducing malaria transmission, particularly if used in combination with insecticide-treated bednets.[35] These findings show that combining CP and BP for a vector control technique, and integrating this into existing management programs, could substantially reduce malaria transmission rates and help manage insecticide resistance.

Several studies confirm the hypothesis that CP and BP compatibility gives better efficiency in the treatment of some chemical-resistant, or less susceptible, arthropods.[10,36–39] Indeed, CP combined with entomopathogenic agents can result in sublethal doses that facilitate infectious processes, leading to the arthropod suffering debility and stress, and causing it to become more susceptible to the action of entomopathogens. The association of deltamethrin and the entomopathogenic fungus *M. anisopliae*, used against the pyrethroid-resistant tick *Boophilus microplus*, resulted in higher larvae mortality rates than those obtained with the respective non-associated concentrations.[10]

Examples of Synergism in Agroecosystems

There is some evidence that sublethal insecticide doses promote, when combined with BPs, synergetic effects, probably because the pest is already debilitated by the chemical, thus allowing the biological agent to work better. Extensive experiments with mixtures of *B. thuringiensis* var. *kurstaki* (Bt) and pyrethroids (cypermethrin, deltamethrin, fenvalerate, and permethrin) have shown the efficacy of these mixtures against the winter moth, *Operophtera brumata*, and their compatibility with integrated mite control in apple orchards, especially for cypermethrin. In fact, the level of winter moth damage to harvested fruit was found to be just as low with Bt and pyrethroid mixtures as with full-rate pyrethroid treatments.[39] Furthermore, several authors highly recommend the judicious use of Bt products combined with insect growth regulators (IGRs) and other biotechnology techniques to achieve the most effective and sustainable use of Bt.[40] Synergistic interaction was recorded in the simultaneous use of low doses of Bt and thiodicarb in controlling the chickpea pod borer *Helicoverpa armigera*.[41] Also, the joint action of nucleopolyhedrovirus (HearNPV) and synthetic pyrethroids at sublethal concentrations resulted in a synergic effect in controlling of *H. armigera*.[42]

For many years, it was generally thought that biological control and CP are incompatible. While this is likely to be true for broad-spectrum CPs such as the organophosphates and carbamates, it is not necessarily the case for newer CPs that have higher selectivity and a more desirable ecotoxicological profile.[35,36,39–41] Thus, there is the possibility that novel insecticides with a highly targeted delivery and mechanism of action could be compatible, or biochemically synergistic, with different BPs.

Some synergism mechanisms have been described, but many remain unknown.[35] The entomopathogenic virus *Oedaleus asiaticus* entomopoxvirus (OaEPV) enhances the insecticidal effects of the chemical insecticides malathion, chlorpyrifos, beta-cypermethrin, cyfluthrin, and deltamethrin against the grasshopper *Oedaleus asiaticus* by inhibiting the specific activity of the carboxylesterase enzyme in the midgut.[43]

The above examples demonstrate that the integration of appropriate CP and BP can sometimes provide better control than using one method alone.

Practical Approaches to Increase Compatibility

Although the replacement of CP by BP would be auspicious,[1–3] it is not always possible; factors it depends on are the type of pests, the crop, the costs involved, and the social and cultural possibilities of the countries. Furthermore, what Goettel claimed is still valid: "if we are to succeed in long sustainable agriculture, we must move swiftly to put the 'I' back into IPM."[44] In fact, it is essential that an IPM strategy integrate all the techniques into a single coordinated program. IPM should not be "chemically dependent" but neither should it neglect the judicious use of CP. Consideration must be given to many parameters in an endeavor to increase compatibility, to achieve the most efficient control, and to do so by adopting the least disruptive interventions.

Prerequisites for adopting a CP–BP strategy are as follows: 1) the identification of all pathogens, weeds, and pests (the targets) and any associated natural enemies and pathogens (the antagonists) that may be affected by the combined use of CP and BP; 2) knowledge of the behavior and life cycle of the target/s and antagonists; 3) dosage response of pest/s and antagonists, at least for CP but also for BP if possible; and 4) dosage response of pests and natural antagonists to the resulting outcome of CP and BP compatibility. All this information provides the basic foundation for maximizing the benefits of CP–BP compatibility and minimizes any undesirable effects on non-target pests and the resurgence of secondary pests. Selectivity is a key factor, and must be considered. When incompatibility occurs in the tandem, both physiological and ecological selectivity can be considered. Physiological selectivity is the result of the physiological differences that occur from the living BP strain's susceptibility to CP,[7,21,32] or by a correct choice of the BP. For example, yeasts are generally tolerant to many of the fungicides used in postharvest: *Metschnikowia pulcherrima* is tolerant to relatively high concentrations of benzimidazoles (benomyl and thiabendazole) and dicarboximides (vinclozolin and procymidone), fungicides used to control postharvest pathogens such as *Botrytis cinerea* and *Penicillium expansum*.[7] Ecological selectivity results from the differential exposure of BP to CP,[6,20] that is to say from, for example, habitat, area, mode, and time of application, and considering also the persistence of CP and the BP life cycle. However, selectivity need to be evaluated not only as a case of CP versus BP but also as a result of their combination. Table 2 resumes some practical indications for enhanced CP–BP compatibility.

The pesticide labels (or Web site) of several companies report the impact of their product on natural enemies commonly found in the crop system where the materials will be applied, but in the case of BP–CP compatibility, a similar approach is lacking.

To date, little has been done to identify the effects of such combinations on non-target organisms, a situation that has arisen from the lack of a methodological standardization, despite several techniques having been developed to determine compatibility; significant progress is needed in this field. Some studies report that CPs do not affect BP; however, other studies using different techniques, and different combinations, have reported that they do.[20,23,28] Methods need to be more standardized. Indeed, when evaluating the potential of the use of chemicals integrated with biological control agents (especially macroorganisms such as predator and parasitoid arthropods or entomopathogenic nematodes), the International Organization of Biological Control (IOBC) developed a sophisticated approach based on a tiered hierarchy made up of threshold values for lethal and sublethal effects on non-target antagonists.[15,19,22] The IOBC recommends a tiered approach whereby initial pesticide screening is done in the laboratory and, depending on the results obtained, the experiment can be followed by semi-field tests and field trials. The IOBC classification of pesticides is based on the degree of damage they cause to beneficial species in laboratory studies and falls into the following four categories: 1) harmless; 2) slightly harmful; 3) moderately harmful; and 4) harmful. IOBC also proposes a list of preventive and highly selective direct control measures to be used in IPM ("green list") and a list of pesticides to be used with restrictions ("yellow list");[22] actually, this involves more than 300 records, concerning 120 different products and 20 different beneficial species.[45] Despite this situation, CP-developing industries frequently do not test product toxicity to entomopathogens or to other living BP. For example, in most cases, as for commercial bacteria like *Bacillus sphaericus*, *B. thuringiensis*, and *Serratia entomophila*, there are limited data on the interaction between these bacterial control agents and conventional chemical insecticides.[35]

An IOBC similar, but modified, approach would also be advisable to evaluate BP–CP compatibility.

What Are the Consequences for Ecology?

Even though CP and BP combinations can result in synergistic pest management effects, we cannot exclude that they also represent a new "stress factor" for the environment. This means that more research is needed to clarify the real impact of such pesticide combinations on ecosystems, including their unintended lethal and sublethal effects on non-target organisms. While the issues of the synergism, efficacy, and selectivity of compatible CP–BP use in resistant target organisms are also of economical relevance, and they are becoming more and more investigated by several authors, the potential change in toxicity for non-target organisms has, until now, been little reflected in the literature. Toxicological synergy is a matter of concern for both the public and regulatory agencies because BPs (including biological ones) and CPs, considered safe individually, might pose an unacceptable health or ecological risk when the exposure is to their combination.[19,44,46,47] Definitions of toxicological synergy are sometimes vague, or may even contrast each other. In U.S. Environmental Protection Agency (EPA) guidance documents,[48] the no-interaction default assumption is dose addition, so synergy means a mixture response that exceeds that predicted from dose addition. The EPA emphasizes that synergy does not always make a mixture dangerous, but nor does an-

Table 2 Practical indications to promote compatibility and integration of chemical and biological pesticides.

1. If CP and BP are not compatible consider the CP persistence when BP is applied successively
2. Use physiologically selective CP to support BP
3. Evaluate the probable effect of CP secondary compounds
4. Reduce dosages of CP
5. Reduce the area of applying CP (e.g. treatments in alternate rows)
6. If CP and BP are not compatible or little compatible educe the contact between CP and BP (e.g. applications in strips or inside traps)
7. Time of application CP considering the minor interference possible if they are not totally compatible
8. Avoid the periods of greatest susceptibility of organisms composing the biological pesticides
9. Use and eventually create through biotechnology pesticide-resistant organisms for BP
10. When possible select BP resistant strains to the applied CP.

tagonism always make the mixture safe. All the pesticides have determined a regulatory level at which they would be considered individually safe; however, combinations are rarely tested. The EC reports "substances or preparations which, while showing no or only weak activity against plant pathogens, can give enhanced activity to the active substance(s) in a plant protection product, referred to as 'synergists.'"[16] In European regulations, the problem of the ecological effects of synergism is mentioned. In fact, it is indicated that plant protection products should not have harmful effects on human health, or on animal health, and it is necessary taking into account known and unknown cumulative and synergistic effects. In addition, "synergist" will be approved where it complies with environmental safety guidelines.[16]

The potential for synergy is thus unknown, or is estimated from data on similar combinations. This lack of information also applies to many other combinations to which the environment is exposed (temperature, humidity, pH, other xenobiotics, and other pathogens).[46–48] Moreover, short-term tests at high concentrations (acute scenarios) are more common than long-term tests at lower concentrations (chronic scenarios).[19,35,47] Also, the IOBC database comes in for some criticism as more than 90% of the entries are acute tests in the laboratory, the remainder being chronic tests; only some of the data are based on field test data,[49] and there is some inability in fully accounting for the sublethal effects of insecticides.[35] Moreover, despite the continued focus of the most widely used ecotoxicological tests on the survival or sublethal performance of individuals, it is generally recognized that such an approach is a considerable oversimplification of real ecological conditions. New developments in ecotoxicology are changing the way pesticides and other toxicants are evaluated. Emphasis is being placed on life histories and population fitness through the use of demography, while other measures of population growth rate, field studies, and modeling are being exploited to derive better estimates of the pesticide impact on both target and non-target species, rather than on traditional lethal dose estimates.[50] Communities are often very dynamic in composition, and complex interrelationships between this composition and ecosystem processes occur.[35,47,49,50] Communities are therefore not so easily characterized as organisms, which is why the concept of "health" is not always appropriate, though it is welcome.[47,49,50]

Pesticides and other toxicants have myriads of effects on insects and other arthropods, and such effects have been the subject of an enormous number of studies over the past several decades,[50] but few data are available on the effect of the combined use of CP and BP. Another crucial issue is that the scientific literature not only reports open questions concerning the general mode of action of BP and derived products[49] but also indicates that BP–CP combinations could be influenced by several factors. On the one hand, it is known that, in general, the toxicity of BP, such as *B. thuringiensis* in target organisms, depends on factors like pH, proteases, and receptors.[51] On the other hand, and more specifically, extrinsic factors and specific cofactors can also influence the efficacy of BP, and its integration with CP in resistant target organisms might also impact on selectivity and toxicity in non-target organisms.[52–55] Many factors can influence the toxicity and selectivity of pesticides, both chemical and biological, such as a combination of biotic and abiotic stress factors, enzymatic processing infectious diseases, normal gut bacteria, and interactivity with other toxins.[53] Additional factors such as cadmium and nematodes can enable Bt toxins to also impact on organisms such as snails (*Helix aspersa*), which show no effects on being exposed to the Bt toxin alone.[54] Moreover, indirect effects of the application of Bti spraying result in detectable effects at higher trophic levels, ultimately affecting vertebrate populations, particularly birds that feed on Diptera.[55] These findings highlight that the observed effects are relevant not only for target organisms but also for non-target organisms and should therefore be reflected in any ecological risk assessment (ERA); this shows the relevance of considering the effects of BP–CP combinations in ERA.

Any successful application of an ERA paradigm for new stressors derived by the CP–BP combination entails several key attributes including 1) the case-specific use of a proper problem formulation for the design of risk assessment; 2) the recognized need to establish a causal relationship of stressor exposure to receptor, resulting in a measurable consequence of exposure (e.g., exposure of an insect species of concern to a CP–BP combination results in an adverse effect, such as lethality or growth reduction); 3) conducting the risk assessment on comparative terms through the use of an appropriate comparator, environment, and management. Fig. 1 gives an example of the scheme of ERA (considering non-target species) in the case of releasing combined CP and BP.

Accumulated experience by researchers and practitioners indicates that CP and BP can result in effective pest control,[35–43] but in the scientific literature, there is a lack of *meta-analyses* of detectable negative impact on the environment and the long-term impact of these combinations. Providing such an analysis could be of considerable benefit to the environment as well as to the registration of new products.

Although detectable risks of BP–CP combinations once associated with BP are considered as likely to be low,[12,19,21] the lack of evidence of negative effects does not mean that combinations should be exempt from safety testing, since the resulting effects can vary significantly in host range, pathogenicity, and other biological characteristics, which can potentially affect environmental safety.[56]

In fact, the conservation of biological control agents within agroecosystems is a strategy adopted to exploit entomopathogens. Equally important are the techniques of inoculative, inundative, and incremental introduction. In all cases, either to preserve the entomopathogen or to use it in combination with CPs, it is necessary to know how these products act on the microorganism, and then determine their compatibility accordingly.[57] The compatibility of several CPs with the major entomopathogenic fungi naturally occurring in agroecosystems was evaluated. The action of the CP on the vegetative growth and sporulation of the microorganisms varied as a function of the chemical nature of the products, its concentration, and the microbial species. However, the laboratory results led to the conclusion that thiamethoxam is compatible with naturally occurring major fungi controlling insects and mites, even considering the maximum exposure of the microorganism to the action of the insecticide when submitted to in vitro studies. On the other hand, other CPs such as diazinon and monocrotophos inhibited conidial production by microorganisms. Moreover, the action of pesticides on the vegetative growth and sporulation of entomopathogenic fungi varies with product concentration and microorganism species and possibly affects fungal reproduction without reducing vegetative growth.[57]

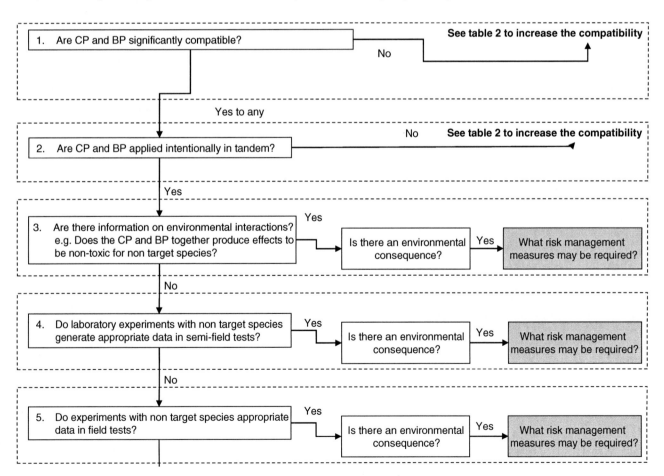

Fig. 1 Example of scheme of Environmental Risk Assessment (ERA) for non-target species, in the case of releasing combined CP and BP.

The ecological interactions determining the persistence and effects of BP and CP together are highly complex and include processes such as effect on immune systems, sublethal effect, competition, and predation/parasitism. Moreover, control of the strength and nature of the impact will depend on the structure and dynamics of the community to which they will be applied. Thus, it could, in practice, be difficult to make an ERA for all CP and BP combinations, but ERA would be most achievable for those that have a synergetic effect in pest control.

CONCLUSION

Over the last 10 years, significant progress has been made in integrating CP and BP. However, the progress made to date is insufficient considering the wide range of crop/health systems that still rely only on CP for pest suppression. Combining different CPs with BPs can improve control efficacy, increase the spectrum of controlled pathogens, and reduce the possibility of resistance development. However, to be successful, the different methods need to be compatible, the first treatment should not have any deleterious effect on the succeeding one; preferably, it should contribute to its efficacy.

Consideration must always be given to the effectiveness of CP–BP combinations in controlling target pests and the toxicity of this "combined product" to non-target pests.

Therefore, before using chemical compounds, it is essential to verify their selectivity and their compatibility with BPs. Moreover, by exposing pests to multiple control tactics, pest adaptability should be reduced. Thus, CP and BP compatibility is useful for insecticide resistance management programs, and their combined use can also result in reduced chemical doses, saving the environment and biodiversity. However, progress to date is insufficient considering the wide range of crop systems and associated pests and pathogens. Data are especially needed for CP–BP integration used against different targets (e.g., bioherbicides and chemical fungicides). Greater effort is necessary to develop ecologically selective techniques in the areas of physiology and ecology selectivity. Thus, future research should be aimed at improving knowledge of both the BP–pest–CP interaction and the mechanisms involved. Moreover, for highly compatible pesticides and those with high synergism, their evaluation on non-target organisms and, more in general, an ERA of them not only could be profitable for regulatory issues but also could lead to their adoption, helping to raise their profile among public policy makers and thus enabling them to realize their contribution to sustainability.

ACKNOWLEDGMENTS

I would like to thank Prof. Vincenzo Arizza (University of Palermo) and Dr. Filippo Castiglia (Regional Forest Agency) for their useful comments, Dr. Barbara Carrey for the English revision, the University of Palermo Progetti di Ateneo (ORPA07S7TR) and PRIN 2008 (200847CA28_002), and the referees for the useful remarks.

REFERENCES

1. Available at http://www.fao.org/fileadmin/templates/agphome/documents/Pests_Pesticides/Code/Policy_2010.pdf (accessed August 2010).
2. Available at http://www.ipmnet.org/IPMNews/2010/news181.html (accessed August 2010).
3. European Commission (EC). Towards a thematic strategy on the sustainable use of pesticides, 2002, Electronic citation available at http:// europa.eu.int/eur-lex/en/com/pdf/2002/com2002_0349en01.pdf (accessed June 2010).
4. Sattler, C.; Kaächele, H.; Verch, G. Assessing the intensity of pesticide use in agriculture. Agric., Ecosyst. Environ. **2007**, *119*, 299–304.
5. Burth, U.; Gutsche, V.; Freier, B.; Roßberg, D. Defining the 'necessary minimum' of pesticide use, 2003, available at http://europa.eu.int/comm/environment/ppps/pdf/bbareaction.pdf (accessed June 2010).
6. Johnson, M.W.; Tabashnik, B.E. Enhanced biological control through pesticide selectivity. In *Handbook of Biological Control*; Fisher, T.W., Bellows, T.S., Caltagirone, L.E., Dahlsten, D. L., Huffaker, C.B., Gordh, G., Eds.; Academic Press: San Diego, California (USA), 1999; 297–317.
7. Spadaro, D.; Gullino, M.L. State of the art and future prospects of the biological control of postharvest fruit diseases. Int. J. Food Microbiol. **2004**, *91*, 185–194.
8. Chung, Y.K.; Lam-Phua, S.G.; Chua, Y.T.; Yatiman, R. Evaluation of biological and chemical insecticide mixture against *Aedes aegypti* larvae and adults by thermal fogging in Singapore. Med. Vet. Entomol. **2001**, *15*, 321–327.
9. Wenzel, I.M.; Barci, L.A.G.; Almeida, J.E.M.; Gassen, M.H.; Prado, A.P. Compatibility of the entomopathogenic fungus *Beauveria bassiana* with chemical carrapaticides used to control *Boophilus microplus* (Acari: Ixodidae). Arq. Inst. Biol. (São Paulo) **2004**, *71*, 643–645.
10. Bahiense, T.C.; Fernandes, E.K.K.; Bittencourt, V.R.E.P. Source compatibility of the fungus *Metarhizium anisopliae* and deltamethrin to control a resistant strain of *Boophilus microplus* tick. Vet. Parasitol. **2006**, *141* (3/4), 319–324.
11. Pimentel, D.; Zuniga, R.; Morrison, D. Update on the environmental and economic costs associated with alien-invasive species in the United States. Ecol. Econ. **2005**, *52*, 273–288.
12. Chandler, D.; Davidson, G.; Grant, W.P.; Greaves, J.; Tatchell, G.M. Microbial biopesticides for integrated crop management: An assessment of environmental and regulatory sustainability. Trends Food Sci. Technol. **2008**, *19*, 275–283.
13. Available at http://www.epa.gov/opp00001/biopesticides/whatarebiopesticides.htm (accessed May 2010).
14. Available at http://www.fao.org/teca/system/files/biopesticides_registration.pdf (accessed December 2010).

15. Council Directive 91/414/EEC (July 15, 1991) concerning the placing of plant protection products on the market, Available at http://eur-lex.europa.eu/smartapi/cgi/sga_doc (accessed December 2010).
16. Regulation (EC) No 1107/2009 of the European Parliament and of the Council of October 21, 2009 concerning the placing of plant protection products on the market and repealing Council Directives 79/117/EEC and 91/414/EEC. Official Journal of the European Union. 24.11.2009, L309, 1–50.
17. Van Driesche, R.G.; Bellows, T.S. *Biological Control*; Chapman and Hall: New York, 1996.
18. Available at http://www.epa.gov/oppbppd1/biopesticides/product_lists/new_ai_2009.html (accessed June 2010).
19. Thomas, M.B. Ecological approaches and the development of "truly integrated" pest management. Proc. Natl. Acad. Sci. U. S. A. **1999**, *96*, 5944–5951.
20. Croft, B.A. *Arthropod Biological Control Agents and Pesticides*; John Wiley and Sons: New York, 1990.
21. Saucke, H. Selective bioinsecticides, selective chemical insecticides: Important options for integrated pest management (IPM) in cabbage. Harvest **1994**, *16* (1/2), 16–19.
22. Available at http://ec.europa.eu/environment/ppps/meeting 040609.htm (accessed June 2010).
23. Boller, E.F.; Vogt, H.; Ternes, P.; Malavolta, C. Working document on selectivity of pesticides, 2005, available at http://www.IOBC.Ch/2005/Working/Document/Pesticides_Explanations.pdf (accessed July 2010).
24. Boller, E.F.; Avilla, J.; Joerg, E.; Malavolta, C.; Wijnands, F.G.; Esbjerg, P. Integrated production principles and technical guidelines. IOBC/WPRS Bull. **2004**, *27*, 1–54.
25. Pachamuthu, P.; Kamble, S.T.; Yuen, G.Y. Virulence of *Metarhizium anisopliae* (Deuteromycotina: Hyphomycetes) strain ESC-1 to the German cockroach (Dictyoptera: Blattellidae) and its compatibility with insecticides. J. Econ. Entomol. **1999**, *92* (2), 340–346.
26. Rashid, M.; Baghdadi, A.; Sheikhi, A.; Pourian, H.R.; Gazavi, M. Compatibility of *Metarhizium anisopliae* (Ascomycota: Hypocreales) with several insecticides. J. Plant Prot. Res. **2010**, *50* (1), 22–27.
27. Rampelotti-Ferreira, F.T.; Ferreira, A.; Prando, H.F.; Tcacenco, F.A.; Grutzmacher, A.D.; Martins, J.F. Selectivity of chemical pesticides used in rice irrigated crop at fungus *Metarhizium anisopliae*, microbial control agent of *Tibraca limbativentris*. Cienc. Rural **2010**, *40* (4), 745–751.
28. Pinnamaneni, R.; Kalidas, P. Compatibility of *Beauveria bassiana* with certain agro chemicals. J. Appl. Biosci. **2009**, *35*(2), 155–158.
29. Gatarayiha, M.C.; Laing, M.D.; Miller, R.M. In vitro effects of flutriafol and azoxystrobin on *Beauvaria bassiana* and its efficacy against *Tetranychus urticae*. Pest Manage. Sci. **2010**, *66* (7), 773–778.
30. JeongJun, K.; KyuChin, K. Compatibility of entomopathogenic fungus *Lecanicillium attenuatum* and pesticides to control cotton aphid, *Aphis gossypii*. Int. J. Ind. Entomol. **2007**, *14* (2), 143–146.
31. Alizadeh, A.; Samih, M.A.; Khezri, M.; Riseh, R.S. Compatibility of *Beauveria bassiana* (Bals.) Vuill. with several pesticides. Int. J. Agric. Biol. **2007**, *9* (1), 31–34.
32. Palem, P.C.P.; Padmaja, V.; Vadlapudi, V. Evaluation of *Beauveria bassiana* isolates for tolerance to organophosphorous pesticides by in vitro and SDS-PAGE analysis. Arch. Appl. Sci. Res. **2010**, *2* (1), 202–210.
33. Krasnoff, S.B.; Faloon, D.; Williams, J.E.; Gibson, D.M. Toxicity of xanthene dyes to entomopathogenic fungi. Biocontrol Sci. Technol. **1999**, *9* (2), 215–225.
34. Devine, G.J.; Furlong, M.J. Insecticide use: Contexts and ecological consequences. Agric. Hum. Values **2007**, *24*, 281–306.
35. Gentz, M.C.; Murdoch, G.; King G.F. Tandem use of selective insecticides and natural enemies for effective, reduced-risk pest management. Biol. Control **2010**, *52*, 208–215.
36. Oi, D.H.; Williams, D.F.; Pereira, R.M.; Horton, P.; Davis, T.S.; Hyder, A.H.; Bolton, H.R.; Zeichner, B.C.; Porter, S.D.; Hoch, A.L.; Boswell, M.L.; Williams, G. Combining biological and chemical controls for the management of red imported fire ants (Hymenoptera: Formicidae). Am. Entomol. **2008**, *54*, 46–55.
37. Farenhorst, M.; Knols, B.G.; Thomas, M.B.; Howard, A.F.V.; Takken, W.; Rowland, M.; Guessan, R.N.; Gregson, A. Synergy in efficacy of fungal entomopathogens and permethrin against West African insecticide-resistant *Anopheles gambiae* mosquitoes. PLoS ONE **2010**, *5* (8), e12081 DOI: 10.1371/journal.pone.0012081.
38. Hancock, P.A. Combining fungal biopesticides and insecticide-treated bednets to enhance malaria control. PLoS Comput. Biol. **2010**, *5* (10), doi: 10.1371/journal.pcbi.1000525.
39. Hardman, J.M.; Gaul, S.O. Mixtures of *Bacillus thuringiensis* and pyrethroids control winter moth (Lepidoptera: Geometridae) in orchards without causing outbreaks of mites. J. Econ. Entomol. **1990**, *83* (3), 920–936.
40. Valentine, B.J.; Gurr, G.M.; Thwaite, W.G. Efficacy of the insect growth regulators tebufenozide and fenoxycarb for lepidopteran pest control in apples, and their compatibility with biological control for integrated pest management. Aust. J. Exp. Agric. **1996**, *36* (4), 501–506.
41. Khalique, F.; Khalique, A. Compatibility of bioinsecticide with chemical insecticide for management of *Helicoverpa armigera* Huebner. Pak. J. Biol. Sci. **2005**, *8* (3), 475–478.
42. Duraimurugan, P.; Regupathy, A. Synergistic effect of nucleopolyhedrovirus with synthetic pyrethroids against *Helicoverpa armigera* (Hübner). Pestic. Res. J. **2008**, *20* (1), 83–86.
43. Xin-Hua, Y.; Yong-Dan, L.; Zhao-Feng, T.; Xue-Yan, S.; Xi-Wu, G. Insecticidal effects of *Oedaleus asiaticus* entomopoxvirus in combination with chemical insecticides and effects on the main activities of detoxification enzymes of its host. Acta Entomol. Sin. **2008**, *51* (5), 498–503.
44. Goettel, M. Whatever happened to the "I" in "IPM"? Soc. Invertebr. Pathol. Newslett. **1992**, *24*, 5–6.
45. Jansen, J.P. Beneficial arthropods and pesticides: Building selectivity lists for IPM. IOBC/WPRS Bull. **2010**, *55*, 23–47.
46. Sinclair, C.J.; Boxall, A.B.A. Assessing the ecotoxicity of pesticide transformation products. Environ. Sci. Technol. **2003**, *37*, 4617–4625.
47. Calow, P.; Forbes, V.E. Does ecotoxicology inform ecological risk assessment? Environ. Sci. Technol. 2003, *4* (1), 146–151.
48. Available at http://www.epa.gov/oswer/riskassessment/pdf/vdmanual.pdf (accessed September 2010).

49. Then, C. Risk assessment of toxins derived from *Bacillus thuringiensis*—Synergism, efficacy, and selectivity. Environ Sci. Pollut. Res. **2010**, *17*, 791–797.
50. Stark, J.D.; Banks, J.E. Population-level effects of pesticides and other toxicants on arthropods. Annu. Rev. Entomol. **2003**, *48*, 505–519.
51. deMaagd, R.A.; Bravo, A.; Crickmore, N. How *Bacillus thuringiensis* has evolved specific toxins to colonize the insect world. Trends Genet. **2001**, *17*, 193–199.
52. Ito, A.; Sasaguri, Y.; Kitada, S.; Kusaka, Y.; Kuwano, K.; Masutomi, K.; Mizuki, E.; Akao, T.; Ohba, M. *Bacillus thuringiensis* crystal protein with selective cytocidal action on human cells. J. Biol. Chem. **2004**, *279*, 21282–21286.
53. Koppenhöfer, A.M.; Kaya, H.K. Additive and synergistic interaction between entomopathogenic nematodes and *Bacillus thuringiensis* for scarab grub control. Biol. Control **1997**, *8*, 131–137.
54. Kramarz, P.E.; Vaufleury, A.; Zygmunt, P.M.S.; Verdun, C. Increased response to cadmium and *Bacillus thuringiensis* maize toxicity in the snail *Helix aspersa* infected by the nematode *Phasmarhabditis hermaphrodita*. Environ. Toxicol. Chem. **2007**, *26* (1), 73–79.
55. Poulin, B.; Lefebvre, G.; Pax, L. Red flag for green spray: Adverse trophic effects of Bti on breeding birds. J. Appl. Ecol. **2010**, *47*, 884–889.
56. Stark, J.D.; Vargas, R.; Banks, J.E. Incorporating ecologically relevant measures of pesticide effect for estimating the compatibility of pesticides and biocontrol agents. J. Econ. Entomol. **2007**, *100* (4), 1027–1032.
57. Filho, A.B.; Almeida, J.E.M.; Lamas, C. Effect of thiamethoxam on entomopathogenic microorganisms. Neotrop. Entomol. **2001**, *30* (3), 437–447.
58. Kuzmanova, I. Study on the compatibility of *Bacillus thuringiensis* Berliner with three organophosphorus insecticides. Gradinar. Lozar. Nauka **1981**, *18*, 23–27.
59. Terribile, S.; Monteiro de Barros, N. Compatibility between pesticides and the fungus *Nomuraea rileyi* (Farlow) Samson. Rev. Microbiol. **1991**, *23* (1), 48–50.
60. Tabakovic-Toshic, M.; Rajkovic, S.; Golubovic-Curguz, V. Compatibility of fungicide benfungin and biological insecticide D-Stop in synchronised suppression of oak mildew and gypsy moth. Nauka za Gorata **2008**, *45* (3), 51–58.
61. Bhushan, B.; Hoondal, G.S. Effect of fungicides, insecticides and allosamidin on a thermostable chitinase from *Bacillus* sp. BG-11 world. J. Microbiol. Biotechnol. **1999**, *15* (3), 403–404.
62. Eshel, D.; Regev, R.; Orenstein, J.; Droby, S.; Gan-Mor, S. Combining physical, chemical and biological methods for synergistic control of postharvest diseases: A case study of Black Root Rot of carrot. Postharvest Biol. Technol. **2009**, *54*, 48–52.
63. Anand, T.; Chandrasekaran, A.; Kuttalam, S.; Raguchander, T.; Samiyappan, R. Management of cucumber (*Cucumis sativus* L.) mildews through azoxystrobin-tolerant *Pseudomonas fluorescens*. J. Agric. Sci. Technol. **2009**, *11* (2), 211–226.
64. Boyette, C.D.; Hoagland, R.E.; Weaver, M.A.; Reddy, K.N. Redvine (*Brunnichia ovata*) and trumpetcreeper (*Campsis radicans*) controlled under field conditions by a synergistic interaction of the bioherbicide, *Myrothecium verrucaria*, with glyphosate. Weed Biol. Manage. **2008**, *8* (1), 39–45.
65. Morris, O.N. Effect of some chemical insecticides on the germination and replication of commercial *Bacillus thuringiensis*. J. Invertebr. Pathol. **1975**, *26* (2), 199–204.
66. Habib, M.E.M.; Garcia, M.A. Compatibility and synergism between *Bacillus thuringiensis* (Kurstaki) and two chemical insecticides. Z. Angew. Entomol. **1981**, *91* (1), 7–14.
67. Lewis, L.C.; Berry, E.D.; Obrycki, J.J.; Bing, L.A. Aptness of insecticides (*Bacillus thuringiensis* and carbofuran) with endophytic *Beauveria bassiana*, in suppressing larval populations of the European corn borer. Agric. Ecosyst. Environ. **1995**, *57* (1), 27–34.
68. Seleena, P.; Lee, H.L.; Chiang, Y.F. Compatibility of *Bacillus thuringiensis* serovar *israelensis* and chemical insecticides for the control of *Aedes* mosquitoes. J. Vector Ecol. **1999**, *24* (2), 216–223.

Pesticides: Damage Avoidance

Aiwerasia V.F. Ngowi
Tanzania Association of Public Occupational and Environmental Health Experts, and Department of Environmental and Occupational Health, Muhimbili University of Health and Allied Sciences (MUHAS), Dar-es-Salaam, Tanzania

Larama M.B. Rongo
Muhimbili University of Health and Allied Sciences, Dar-es-Salaam, Tanzania

Abstract

To enhance the quality of life for farmers and society, getting rid of pesticides could remove the accompanying health and environmental problem. Pesticides are a big and growing business benefiting multinational companies in developed countries; however, the developing countries act as pesticide markets left to deal with the adverse effects. Pesticide control legislation was started to establish rules and principles for the management of pesticides so as to protect people and the environment against the harmful effects of pesticides. Exploring the weakness in the implementation of effective regulations in developing countries shows that it is mostly compounded by weak capability, infrastructure, and the international trade that hinder implementation of regulatory controls. To avoid damage, knowledge of hazards posed by the available pesticides could be developed and appropriate control measures established. Although it is known that use of personal protective equipment during pesticide handling is very important, farmers do not use it because it is too expensive, not available, or uncomfortable. Organic farming is practiced by farmers, especially in horticulture, to grow vegetables and fruits consumed by family members; however, they believe that by using pesticides, they increase yield and improve the appearance of the vegetables grown for sale. Surveillance is important for providing evidence of pesticide poisoning and its extent of pesticide poisoning to inform policy for intervention. Participatory approaches work very well in educating and also learning from the farmers' indigenous knowledge. In promoting sustainable agriculture, pesticide damage can be avoided by eliminating the most hazardous substances and reducing dependence on pesticides. This entry is organized into the following sections: general introduction, avoiding pesticide hazards, and pesticide poisoning surveillance. Illustrations are included in the sections on pesticide regulation, avoiding pesticide hazards, and encouraging judicious use of pesticides. The main goal of the entry is to explore how damage due to pesticides could be avoided, particularly in Africa, by using less harmful pesticides, more careful application, or not using at all.

INTRODUCTION

Human beings create hazards such as pesticides that threaten the health of the environment. Pesticides are spread across fields in either agriculture or vector control and enter the environment over large areas. An extremely small percentage (less than 0.3%) of the amount applied goes into direct contact with or is consumed by target pests; therefore, 99.7% goes where it is not wanted in the environment. The literature does not support the concept that some pesticides are safer than others; it simply points to different health effects with different latency periods for the different classes.

It is very difficult to give precise estimates of pesticide poisoning worldwide due to underreporting; however, it is generally recognized that a considerable number of people continue to be exposed to and affected by pesticides. Pesticides pose a real danger to the environment and human health, more so in Africa, which accounts for only about 3% of the worlds' pesticide consumption.[1] Underestimations of acute and long-term effects of pesticides in Africa occur due to underdiagnosis and/or underreporting. The impact of pesticide poisoning is also unknown because of weak surveillance for hazards and impact; import/export of banned or restricted compounds; lack of technical and laboratory capacity; weak regulations and enforcement; low level of worker and community awareness; as well as inappropriate pest control policies.[2,3] There is no such thing as a "safe" level of contamination because people die from direct and indirect poisoning with tiny amounts of pesticide exposure during pregnancy, with the possibility of long-term impact on offspring.[2] Pesticide damage can be avoided by eliminating the most hazardous substances and reducing dependence on pesticides in promoting sustainable agriculture. At a signal of harm, either through bioassay or epidemiological study, product(s) or process(es) that could lead to exposure should be identified. Existing alternative(s) that best reduce overall risk in a cost-effective way should be used to replace the harmful product(s). This entry explores the situations and ideas that could help in efforts to avoid harmful effects of pesticides in developing countries, particularly Africa.

Pesticide Development

Pesticides are a big growing business benefiting multinational companies in developed countries; however, the developing countries, as pesticide markets, are left to deal with the adverse effects, which must be avoided by all means. History tells us that the same pests that led to the use of pesticides centuries ago are still the same today (insects, fungi, weeds) and that synthetic organic pesticides with wide-spectrum activity and environmental stability desired centuries ago are no longer needed. Dichloro diphenyl trichloroethane (DDT) earned a Nobel prize for the discoverer of its insecticidal property, and later on, the high degree of chlorination was considered to be advantageous, hence the synthesis in succession of chlorinated compounds such as aldrin, dieldrin, heptachlor, endrin, toxaphene, and lindane. Organochlorines were followed by compounds such as parathion and malathion that are similar to nerve gases that have phosphorus as their basic unit, called organophosphates. These compounds are highly toxic to man, characterized by a narrower spectrum of activity, and more readily degraded. Adverse effects of organochlorines started to come to light with the publication of Rachel Carson's *Silent Spring*, and organophosphates were considered a good alternative to organochlorines.[4] Carbamates such as carbaryl were later discovered and were considered to be safer.

With the advancement in technology in insecticide toxicology and chemistry, scientists made modifications in structure to obtain better compounds. Pyrethrins appeared to be promising compounds for modification and resulted in the introduction of pyrethroids. Pesticide control legislation was started to establish rules and principles for the management of pesticides so as to protect people and the environment against the harmful effects of pesticides.

Pesticide Regulation

The weakness in the implementation of effective regulations in developing countries is compounded mostly by weak capability, infrastructure, and the international trade, which hinder implementation of regulatory controls.[3] In the history of the development of chemical pesticides, nearly all attention was focused on the ability of a substance to kill pests; concern for safety was secondary and followed somewhat later. Things have not changed in Africa. To make matters worse, the pesticide industry puts a lot of pressure on governments to allow for the use of pesticides that are not even accepted where they are manufactured. The same industry in collaboration with local governments has spearheaded the ineffective "safe use" campaigns[5,6] and pesticide stewardship programs in developing countries to promote more pesticide use in the pretext of safety, focusing on preventing acute health effects among agricultural workers and farmers. Safe use campaigns targeting the vulnerable population give them a false sense of security and make them use more pesticides, causing them to end up highly exposed and in danger of acute or chronic effects.

The industry provides data on toxicological and environmental properties of pesticides in the process of registration, as is demanded by the regulatory authorities, which, though useful, may not be appropriate under local conditions since the pesticides were tested under different agroclimatic and sociocultural conditions. The regulators also might not have the skills or expertise to interpret toxicological and environmental data provided by registrants; hence, approvals are, in most cases, based on efficacy rather than environmental health impact. Regulators, who in most cases are government officials, also tend to regulate pesticides from their offices, assuming that people follow the instructions and advice on use when they approve substances and that this mitigates the harm. In reality, many users do not follow best practice, which thus places people's health and the environment at risk.

Small-scale farmers, farm workers and their families, and pest control operators in Africa are at higher risk compared with the general population. They have a very rudimentary understanding of the danger associated with pesticides and put a lot of trust in regulatory authorities and scientists/experts in matters related to pesticides and their health and safety. Their attitude toward protecting the environment from pesticide pollution is masked by their interest in efficacy. Farmers are not aware of how they contribute to pesticide resistance and the resulting increased use of pesticides to combat resistant pests.[7]

In countries with laws to regulate pesticides, the law usually stipulates that no person, business, or organization shall be allowed to distribute or apply any pesticide that has not been registered with, or at least provisionally approved by, a responsible authority. The registration requirements for pesticides define their quality, range of application, labeling, packaging, and applicable safety measures.[4] In this context, protection of the users and the environment, as well as the proprietary rights of the manufacturers, is taken into account. However, pesticide registration does not consider the small holder specifically.[8]

Despite the existence of laws and regulations in some countries, marketing and advertising of pesticides are often uncontrolled. Considering that the label is the foremost source of information for the user with regard to indication, dosing, and safety practices, proper labeling is of central importance for the proper use of the pesticides.[9] The label should address local specificity in end-user comprehension of hazard communication for it to be effective. Nevertheless, incorrectly labeled or unlabeled formulations, including ready-made solutions in soft drink bottles and other containers, are commonly sold at open markets and in the streets.[10] Even when the local legislation specifies that the labels should be in local languages, it appears that there are some laxity and general consensus between the industry and regulators to overlook this requirement. The regulators should be held responsible on the issue of labels by the public and other stakeholders. In a country like Tanzania, where farmers know pesticides only by their trade names, they are confused when names are similar, as those shown in Fig. 1, and end up mixing the same active ingredient in single mixtures.

Global as opposed to local pesticide control initiatives should be more appropriate in dealing with the problems related to pesticide trading due to porous borders. Each regulator should establish a reporting system for pesticides sold and used in communities and their harmful effects. Pesticide regulations in many developing countries do not have a reporting system that informs regulators on types, quantity, and sources of pesticides used in communities. They receive data on imports that are declared by manufacturers/suppliers that facilitate fee payment. Collecting data on use will help in checking the reliability of declared types and quantities for imported or manufactured products and their destination. With data on types and quantities of pesticides used, feedback on distribution and harmful effects encountered in communities should inform regulators on which products are implicated in damage, so that they can be phased out. Regulators should also develop and facilitate actions to deal with the harmful effects of pesticides used in communities. Harmful effects of pesticides are established through research and surveys, but a clear procedure on what action to take or who is responsible is not usually there. In order to facilitate action, the relevant authorities should cooperate with communities to develop mechanisms and ensure that action is taken.

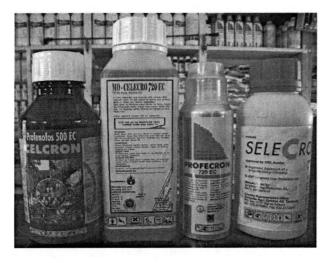

Fig. 1 Similar trade names confuse illiterate users, Lake Eyasi Basin, 2010.

Developing mechanisms to monitor implementation of legislation pertaining to import, sale, distribution, and use should contribute to risk reduction. It is known that pesticide laws and regulations in developing countries originate from without. The Food and Agriculture Organization of the United Nations (FAO) is responsible for advising member countries on regulation frameworks.[11] The code of conduct on the distribution of pesticides is used by most countries. The Globally Harmonized System of Classification and Labelling of Chemicals (GHS) is now advocated to prevent exposure to pesticides;[12] however, the proposed systems are not understood by the targeted audience. Interpretation of signs and symbols is problematic even to experts who are supposed to advise users. Through monitoring, it will be possible to know what works and what does not work. There are some signs and symbols that are understood easily, but others are not understood, and some are misinterpreted. Redesigning those would improve protection for users and the environment. Monitoring will also help in discovering the bad practices of storage in homes, where farmers store pesticides inside living room, kitchen, and toilets.

AVOIDING PESTICIDE HAZARDS

Knowledge of hazards posed by the available pesticides could be developed in order to minimize or control their impact. Regulators could reach out to educate and train stakeholders on policies and regulations in place to safeguard the environment and health. Experience has shown that regulators regulate pesticides from their offices; they never visit shops they authorize to sell pesticides, and neither do they visit farmers or farms to see how pesticides are applied. As a result, shops are managed by incompetent individuals who are not able to advise buyers on what is appropriate to use and how to use it safely. The main interest of pesticide dealers is selling and making money.[7] They go to the extent of advising farmers to mix more than one type of insecticide, for example, combining profenofos and chlorpyrifos (Fig. 2) in one mix in the Lake Eyasi Basin to kill pests, without realizing that in doing so, they are polluting the environment, creating resistance. Visiting farmers would enable regulators to see how the pesticides are endangering the environment and people's lives. Through monitoring, the reasons for the practice of mixing pesticides would be discovered and dealt with appropriately. The regulators would also learn what works for the farmers and use it to improve their regulation guidelines.

Some of the farmers are innovative and use storage facilities that are not what is recommended but are relatively safe. These could be used as models and be introduced introduced to other farmers. Some farmers also find their own protective gear that could be improved and used by others.

Reinforcement of adherence to guidelines and procedures for pesticide use is important at all stages of pesticide handling. There is a tendency for farmers to carry on with

Fig. 2 Two different insecticides mixed in one spray, Lake Eyasi Basin, 2010.

business as usual, not realizing that they are endangering their lives and the environment. Refreshing their knowledge through appropriate communication media will serve the purpose of protecting them. In Tanzania, August 8, called "nane nane," is farmers' day, in which exhibitions and demonstrations for farmers are held throughout the country. Regulators in the country could set up pavilions where farmers could learn the hazards of pesticides and best approaches to avoid damage. They could also use this day to highlight the guidelines and procedures that farmers could use to protect themselves. At times, pesticides are deregistered, but end users are not aware because there is no mechanism to inform them. For example, DDT was banned from use in agriculture in Tanzania, but very few people are aware of this.

Farmers could be made to understand the health hazards posed by pesticides they use if they are made to change their perception. Farmers are not aware of the hazards due to the way they perceive pesticides. In Tanzania, they call pesticides "dawa," which is synonymous to "medicine." It becomes difficult to believe that medicine that is meant to cure does kill. Reaching out to farmers to convey the notion that pesticides are medicine could be dealt with. Farmers also tend to think that pesticides get into their body only through ingestion. They try to be careful not to drink or eat from containers that have pesticides. Inhalation is also understood as a means of entry; pesticides can be smelled. Entry through skin is the least understood, although it is a very important route of pesticide entry into the body. Farmers and farm workers use the least protection when spraying because they do not realize that pesticides can enter their bodies through the skin. In hot climates, the pores open up, and skin becomes wet from perspiration, which becomes an important route of exposure. The farmers need to be told that this happens, and dangers should be explained to them.

Farmers could improve pesticide use practices to avoid unwanted exposure. They should be taught how to transport, store, use, and dispose of pesticides and empty containers. Studies in developing countries have shown that farmers buy pesticides from retailers far from their farms and have to transport products from one area to another using public transport.[13] They carry pesticides in public transport and mix these with cargo that contains food and other household items, which could harm people. By studying and advising farmers and transport owners on how to pack and place pesticides on public transport, the danger should be abated.

Farmers are known to store pesticides in areas that are not appropriate. Pesticides could easily contaminate food and feed and be accessed by children and mentally derailed or challenged persons. Volatile pesticides stored in living houses are likely to contaminate the air and expose people living in these houses, some of whom might be in poor health caused by other ailments, including those whose immunity is compromised, babies, or the elderly. Exposure to pesticides from poor storage exacerbates harm already experienced from field application. Farmers might be aware of the hazards, but due to low income, they would be unable to construct a proper storage facility. Some innovate and come up with storage facilities that, although not perfect, reduce exposure. It would be helpful if investigations were made into the way farmers store pesticides, and those storage facilities that could be improved and used for safe storage should be adopted.

Pesticide and empty container disposal contributes to unacceptable exposures to pesticides. Pesticides are disposed of because they are either ineffective or leftovers from a previous spray. Many countries do not have facilities to ensure that quality products are supplied on the market. Fake and substandard pesticides are common and contribute to losses encountered by farmers in terms of loss of productivity, increase in costs for inputs, loss of beneficial organisms such as bees for pollination, and ill health from exposures to pesticides. Improving the quality control system to ensure quality products on the market will reduce the need for disposal of pesticides and, hence, the dangers associated with haphazard disposal. Empty containers are left where mixing takes place near water sources/waterways; thrown in the field; or sold for reuse as containers for household fuel (kerosene), cooking oil, milk, drinking water, and local brew in the case of bottles, and containers for storing food and animal feed in the case of paper, plastic, and drums. Farmers need education and guidance on proper ways of handling these containers. FAO has issued a number of guidelines[11] to that effect, but they are left in files in ministries instead of being translated and transmitted in a language that will reach those who really need it.

Encouraging Judicious Pesticide Use

According to instructions given under the industry stewardship (Table 1), farmers in developing countries should not be handling pesticides. The authorities are aware that farmers access pesticides that are banned, restricted, or not registered for general use and handle pesticides with minimum protection and precautions, but they do not take steps to find out why and develop appropriate actions that would enhance adherence or prevent exposure.

The use of appropriate and well-maintained spraying equipment along with taking all precautions that are required in all stages of pesticide handling could minimize human exposure to pesticides and their potential adverse effects on the environment.[4] However, in Africa, there is little or no behavior change even where there is high awareness of harmful effects of pesticides. Farmers have a high acceptance of risk because there are often other priorities for immediate survival. The governments' preoccupation with food security has been shown to conflict with interests in pesticide safety, and hence, their efforts to prevent adverse effects fail. The industry takes advantage of the situation to market pesticides heavily[14] and forms alliances with the government ministries of agriculture to subsidize costs of pesticides to increase affordability by end users. Also, it form alliances with the Ministry of Health to use obsolete pesticides such as DDT for vector control.

The notion that pesticides can be used safely if label instructions are followed places the full responsibility and consequences of inadequate use on pesticide users. To avoid the damage caused by the pesticides, end users have an option of not using them at all or using nonchemical or other less harmful pesticides.

Farmers could develop pest control strategies that reduce the dependence on pesticides or use less toxic products in pest control. Pesticides used by farmers have been classified by the World Health Organization (WHO) according to hazard. Efforts should be made to translate this classification to a language that can be understood and used by farmers. They could use this classification when choosing pesticides, so as to buy the least hazardous. Farmers should be informed on how to choose products based on their toxicity and hazardous nature. Farmers choose a product because it has worked for their neighbors or because the pesticide dealer has told them that the product is new and will solve their problem. If they knew how to find out the toxicity or hazard class, it would be an added advantage, for they would know what the dangers are and what precautions to take.

Farmers rarely look at the label to find out how to avoid exposure or harm. They use the label to identify the product name, as they were told by their neighbors and peers, and the dosage if they are using it for the first time. Surveys carried out in a number of countries show that the standardized signs and pictograms are not understood or interpreted correctly.[15] Perceptions and cultures have great influence on how the pictograms can be interpreted. For example, a survey conducted by environmental health science students (not published) in Dar-es-Salaam to assess knowledge and practices regarding signs and symbols revealed that a sign indicating "Dangerous to wild animals and birds" was interpreted as "Can make wild animals run away" and "Can be used to control pests disturbing wild animals and birds." In Côte d'Ivoire, a study carried out to assess farmers' understanding of pesticide safety labels showed that 17% partially understood and 33% misunderstood the labels.[16]

Integrated pest management (IPM) programs in developing countries are not new, and many were donor driven and introduced by scientists from developed countries. The outcomes for such programs were set in proposals that did not include adoption of IPM by farmers. Once the funding was exhausted, the scientists left, and farmers resorted to their old farming practices. If the principles of IPM were internalized by the farmers who were introduced to IPM, it is unlikely that they would have gone back to the dangerous practices. The government policies should be geared toward encouraging judicious use of pesticides rather than allowing the pesticide industry to take an upper hand. Agricultural extension services do not promote IPM for lack of capacity and because of the incentives received from pesticide industry. With political will and extension service providers encouraged to observe ethical standards in their practice, implementation and follow-up of IPM practices in communities should reduce to a great extent the damage caused by pesticides.

Integrated pest management farmer field schools in some developing countries made farmers appreciate pesticide impact on natural enemies, markedly reducing pesticide applications (frequency) yet still achieving higher crop yields.[8]

Table 1 What pesticide stewardship entails.

Ensuring production, formulation, importation, exportation, and use of only appropriate types of pesticides
Encouraging distribution or sale of appropriate quantities of products
Promoting and ensuring market access to registered products only
Blocking market access by old, cancelled, or banned products via legal and backdoor channels
Encouraging and promoting spray operations only by certified applicators
Promoting the use of appropriate PPEs

Promoting Nonchemical Strategies

Organic farming is practiced by farmers, especially in horticulture, to grow vegetables and fruits consumed by the family; however, they believe that by using pesticides, they increase yield and improve the appearance of the vegetables grown for sale. Consumers also have a great influence on the use of pesticides in horticulture because they look for a blemish-free tomato, cabbage, or rose, thus encouraging pesticide application. Since farmers learn more from seeing what works and what does not, demonstration plots would be a good strategy to show that with good agronomy, crops could be grown without using pesticides.

In many developing countries, farmers are encouraged to use chemical fertilizers without consideration of the soil fertility. In some areas, the chemical fertilizers stress crops that grow on the soil because they are applied without investigating the soil deficiency. When crops are stressed, they sometimes exhibit symptoms that could be mistaken for disease, and farmers normally rush to spray with pesticides without knowing that it would cause further damage and that it would only contribute to endangerment of the environment and human health. Manure or compost, when used properly, could alleviate the problem of soil fertility.

Provision of Personal Protection

Farmers do not use personal protective equipment (PPE) because they are too expensive, not available, or uncomfortable to use; however, use of PPE during pesticide handling is very important. A mechanism for provision of PPE when needed should be set up by governments. The incorporation of PPE costs into registration fees and making it mandatory for suppliers of pesticide to supply appropriate PPE with their products would address the problem of cost and availability. Although these costs will be passed down to the farmers, it will mean that if they can afford to buy pesticides, then they will be assured of protection; otherwise, they will not be using pesticides. Farmers feel uncomfortable using PPE because they use them for too long a duration and they do not use them properly. Farmers should be encouraged to spray for short durations, particularly early in the morning and late in the evening in the tropics, to allow them to use PPE. They should also be taught how to put on protective clothing. Some tuck trousers in boots and sleeves in gloves and endanger their health when pesticides get trapped in the warm boots and gloves.

Farmers buy PPE such as respirators, without knowing that there are specific types based on chemicals they are exposed to. They use dust masks instead of respirators with chemical filters. Appropriate PPE should be supplied by the pesticide dealers, and farmers should be trained on how to use them effectively.

PESTICIDE POISONING SURVEILLANCE

Surveillance is important for providing evidence of pesticide poisoning and its extent, to inform policy for intervention. The WHO pesticide poisoning estimates are normally based on acute poisoning reported in health care facilities. These are underestimates due to the fact that only the serious cases of poisoning are reported to health care facilities. Furthermore, these few cases are misdiagnosed and under-reported.[3,14] Out of ignorance, history of ill health, and inefficient health care services, farmers and families do not consult in cases of poisoning but treat themselves or seek services of traditional healers. If communities are guided on how to self-monitor, record the signs and symptoms encountered during pesticide handling, and develop actions to deal with the problems, some of these long-term impacts will be reduced.[17]

Introducing community self-surveillance for pesticide poisoning requires the introduction of a data collection and reporting system that will give feedback on the effects of pesticides on health and the environment to inform policy for action.[8] The reporting system existing in many African countries, which relies on medical records from health care facilities, does not take into consideration the moderate and minor ailments suffered by people working with pesticides. Establishment of a system that will receive data collected by self-reporting community members and feeding into the health information systems should go an extra mile in assessing the extent to which pesticides contribute to ill health and thus inform actions to be taken to reduce the burden. Actions taken to avoid or reduce the impact on health should also lead to savings in the health care budget.

Medical practitioners could learn proper diagnosis and treatment of pesticide poisoning, and communities have to learn how to recognize signs and symptoms of pesticide poisoning. Surveys carried out in farming communities have shown that health care providers are not well trained to deal with pesticide poisoning incidences.[18] In Tanzania, health care providers do not learn much about pesticides in learning institutions to enable them make appropriate diagnosis and treatment. The health practitioners who are already out in the field should, together with farmers in their respective areas, learn to understand the exposures in order to enhance the capability to recognize signs and symptoms of poisoning. Also, health care workers should be provided with an inventory of pesticides available in their service area and how they can diagnose and treat poisoning incidences. They should be encouraged to prepare themselves with knowledge beforehand in case of pesticide poisoning.

The subject of diagnosis and treatment of pesticide poisoning should be introduced in the medical curriculum. The curriculum in learning institutions in many developing countries does not provide learners with competencies needed when they go out to work. There is a need to develop a competency-based curriculum and ensure that medical practitioners are taught enough on pesticides to be able to handle cases of poisoning they are likely to encounter.

The extent of the health effects (acute and chronic) due to pesticides is underestimated because of insufficient data. Health information systems in many developing countries, particularly Tanzania, do not segregate pesticide poisoning from drug poisoning, snake bites, kerosene poisoning, or other types of poisoning.[18] The recording system is not structured to specify and confirm what caused a poisoning case. Improvements in the reporting system to capture the pesticide poisoning cases would serve to guide interventions to control the harm caused by pesticides. Medical practitioners and medical recorders should be trained on the function of the reporting system and what they should do to make it easier to use. Lack of knowledge on diagnosis and treatment of pesticide poisoning contributes to poor recording of incidences. In some cases, the health information system becomes cumbersome, thus discouraging health practitioners from recording properly or segregating the causes of poisoning. The health information systems are sometimes imposed on the practitioners without taking into consideration their experience on how to organize the system better. The policy makers should receive feedback on the functioning of the health information system and improve the system as need may arise.

Teams to follow up programs should be established to deal with the impact of pesticides in communities. A multidisciplinary and multisectoral approach to dealing with the impact of pesticides in communities should serve to ensure that actions taken to protect individuals and communities from harmful effects of pesticides are appropriate and effective.[19] In community self-surveillance, management teams are established to compile and give feedback to self-reporting farmers, which would generate action to change harmful practices. The management teams could also be made to link up with policy makers to ensure that actions taken are followed by the whole community to protect the whole environment. If, for example, farmers had a tendency of mixing pesticides close to waterways/water sources, unless the whole community mixes away from the water, safety would not be achieved.

Information Exchange

Generally, participatory methodologies work very well in educating farmers and also learning from the farmers' indigenous knowledge. This method could work well in communicating pesticide risks, including poisoning surveillance. Farmers receive information on pesticides from neighbors, peers, or retailers. They know of harmful effects from their own experience or from that of other people in the community. The extension service provided by the government is not adequate because of the magnitude of the pest problem that has gone beyond their knowledge and capacity to solve. This leads to farmers not trusting the extension workers because the advice they give does not solve their problems. Some extension staff have lost credibility, as they engage in pesticide business, and farmers get the feeling that they are more engaged in their businesses and do not give correct advice. The extension service in countries should be organized with clear responsibilities and answerability for extension workers. There should be a reporting system that should be known by all stakeholders so that the farmers can get information when they need it and service providers can note deficiencies deficiencies in the reporting system that should be dealt with by relevant authorities.

Sustaining Agriculture and Public Health

Getting rid of pesticides and the accompanying health and environmental problem will enhance the quality of life for farmers and society as a whole. Crops produced organically should be promoted as an incentive for farmers not using pesticides on crops. To sustain agriculture and public health, pest and vector control strategies should be sustainable. The war with pests and diseases is not likely to be won by relying on pesticides. Farmers should be encouraged to learn to live with pests at a threshold. Growing crop varieties that are not prone to pest attack during appropriate seasons should be encouraged. The extensive use of pesticides is, in many cases, in crops (onions, tomatoes, cabbages, flowers) that are grown off-season to fetch better pay. Crops sold should be monitored for pesticide residues. The consumers are, to some extent, the driving force behind the use of pesticides. Governments and independent monitors should track pesticide residues randomly in markets, and those found to have high residue content should be withdrawn. The findings should also be made public so that consumers can make an informed choice of what they buy.

Farmers should be encouraged to practice farming using principles of ecology, considering relationships between organisms and their environment. The environment, as we traditionally know it, means everything that surrounds human beings. It is a system of living things and natural processes, with the human species just one player in this web.

Farmers have a responsibility to protect nontarget organisms such as earthworms, bees, butterflies, and chameleons. Farmers in most cases are obsessed with making profit from agricultural produce and forget to look at the impact the production process brings to the health of the environment. The relationship between the organisms and their environment might be clear to an ecologist, but farmers would need to be taught about how to take note of the impact of their actions on the environment and the detrimental effects their actions bring. It might not be easy for a farmer to know that there are microorganisms in the soil that make wastes rot as a necessity for returning fertility into soils. They might not be aware that earthworms improve the soil texture to allow plants to grow or that bees and butterflies transfer pollen from one plant to another to enhance pollination. For the farmers to practice farming that will take into consideration the survival of nontarget organisms, they should be encouraged to stop or reduce the use of pesticides.

Crops that are resistant to pests and diseases should be grown. Research institutions in developing countries have discovered crops that are resistant to pests and diseases.

Agricultural research institutes in Tanzania have come up with crop varieties that are resistant to pests and drought. The government should invest in producing enough seeds and promote the crops in a relevant crop zone. An example is the introduction of a coffee variety that is resistant to coffee berry diseases and production of seedlings that were made available to farmers by the Tanzania Coffee Research Institute. Such varieties reduce the dependence on pesticides and thus reduce harm from pesticide exposure.

Farmers should be encouraged to record inputs and harvest to work out costs and benefits of crop protection. Agricultural practices in many developing countries do not involve monitoring input and output to analyze costs and benefits/profits. Farmers use inputs throughout the growing season, which they buy in small quantities, and they do not keep track of how much they spend in total. When they harvest, they sell crops in bulk and receive large amounts of money, of which they cannot substantiate the profits for lack of records. There are also externalities that are not considered when costing use of pesticides, such as the cost encountered in the course of traveling to towns to purchase pesticides and PPE, impact on nontarget organisms, or costs of illness due to pesticide poisoning. If all these were included in the cost of pesticides, it is unlikely that there would be much profit. Farmers should be taught and encouraged to keep records and use them to decide on what crop protection strategies are more profitable.

REFERENCES

1. Zhang, W.J.; Jiang, F.B.; Ou, J.F. Global Pesticide Consumption and Pollution: With China as a Focus. In *Proceedings of the International Academy of Ecology and Environmental Sciences*; 2011; 2, 125–144.
2. Kishi, M. The health impacts of pesticides: What do we now know? In *The Pesticide Detox*; Pretty, J.N., Ed.; Earthscan: London, 2005; 23–38.
3. Ngowi, A.; Wesseling, C.; London, L. Pesticide health impacts in developing countries. In *The Encyclopaedia of Pest Management*; Taylor and Francis: New York, 2006; 1–4.
4. Pretty, J.N. Pesticide use and the environment: What do we now know? In *The Pesticide Detox*; Pretty, J.N., Ed.; Earthscan: London, 2005; 1–22.
5. Murray, D.; Taylor, P. Claim no easy victories: Evaluating the pesticide industry's global safe use campaign. World Dev. **2000**, *28* (10), 1735–1749.
6. Wesseling, C.; Ruepert, C.; Chaverri, F. Safe use of pesticides: A developing country point of view. In *Encyclopedia of Pest Management*; Pimentel, D., Ed.; Marcel Dekker: New York, 2003.
7. Konradsen, F.; van der Hoek, W.; Cole, D.C.; Hutchinson, G.; Daisley, H.; Singh, S.; Eddleston, M. Reducing acute poisoning in developing countries—Options for restricting the availability of pesticides. Toxicology **2003**, *192* (2–3), 249–261.
8. Dinham, B. *Communities in Peril: Global Report on Health Impact of Pesticide Use in Agriculture*. Pesticide Action Network: Asia Pacific, 2010.
9. Kern, M. Regulating pesticides. In *Encyclopaedia of Pest Management*; 2002; 689–91.
10. Rother, H.A. Falling through the regulatory cracks: Street selling of pesticides and poisoning among urban youth in South Africa. Int. J. Occup. Environ. Health **2010**, *16*, 202–213.
11. Food and Agriculture Organization of the United Nations. *International Code of Conduct on the Distribution and Use of Pesticides*; Food and Agriculture Organization of the United Nations: Rome, 2003.
12. Strategic Approach to International Chemicals Management (SAICM)/International Conference on Chemicals Management. *Draft High-Level Declaration; Draft Overarching Policy Strategy; Global Plan of Action*; Geneva, 2005.
13. Ngowi, A.V.F.; Mbise, T.J.; Ijani, A.S.M.; London, L.; Ajayi, O.C. Smallholder vegetable farmers in Northern Tanzania: Pesticides use practices, perceptions, cost and health effects. Crop Prot. **2007**, doi:10.1016/j.croprpro.2007.01.008.
14. Wesseling, C. Pesticides. In *OSH for Development*; Elgstrand, K., Petersson, N., Eds.; Royal Institute of Technology, Elanders Sverige, 2009; 327–343.
15. London, L.; Rother, H.A. Hazard labeling. In *Encyclopedia of Pest Management*. Marcel Dekker: New York, 2003; doi:10.1081/E-EPM-120006793.
16. Ajayi, O.C.; Akinnifesi, F.K. Farmers' understanding of pesticide safety labels and field spraying practices: A case study of cotton farmers in northern Cote d'Ivoire. Sci. Res. Essays **2007**, *2* (6), 204–210.
17. Murphy, H.H.; Hoan, N.P.; Matteson, P.; Abubakar, A.L. Farmers' self-surveillance of pesticide poisoning: A 12-month pilot in northern Vietnam. Int. J. Occup. Environ. Health **2002** (3), 201–211.
18. Ngowi, A.V.F.; Maeda, D.N.; Partanen, T.J. Assessment of the ability of health care providers to treat and prevent adverse health effects of pesticides in agricultural areas of Tanzania. Int. J. Occup. Med. Environ. Health **2001**, *4*, 347–354.
19. World Health Organization IPCS Manual on Pesticide Safety No. 002 WHO/PCS/94.3. February 1994.

Pesticides: Effects

Ana Maria Evangelista de Duffard
Laboratorio de Toxicología Experimental, National University of Rosario, Rosario, Argentina

Abstract
This entry discusses the effects of pesticides on animal life, with an emphasis on neurotoxicity.

INTRODUCTION

When an organism's homeostasis is altered, the animal's life is in danger. Neurobehavioral toxicology is especially concerned with the behavioral plasticities involved in normal adjustments—homeostatic processes—to the constantly changing physical and psychosocial environments, through habituation, tolerance, learning, and memory development.[1] Behavior in the wild is adaptative and context dependent. Parent behavior, for example, is a complex group of disparate activities (nest building, retrieval of young, defense from predators, feeding, etc.). Failure to perform optimally in any of these activities will result in decreased survivability of the young.

The chemicals that constitute the class of pesticides with potential neurotoxicity vary widely in structure and include chlorinated hydrocarbons, organophosphates, carbamates, and pyrethroids. Although most of them are insecticides, some are also used as rodenticides, acaricides, fungicides, or herbicides.

Studies of captive and free-living birds provide support about a change in behavior can be expected when brain acetylcholinesterase (AChE) activity falls below about 50% normal. Organophosphates and carbamates[2] have been related to avian mortality (50% chronic and 80% acute of AChE inhibition). Bobwhite quail (*Colinus virginianus*) treated with methyl parathion were more likely to be caught and killed by a domestic cat introduced into an observational field. In addition, a decrease of falcons' nest-defense behaviors has been shown to be an important factor in the decline of several species of birds of prey and was correlated with both the degree of eggshell thinning and egg residue organochlorine levels in the eggs.[2]

Because bees often come in close proximity to pesticides and they are divided into different age-based castes and labors, they can also be adversely affected. Bee dancing is disrupted by exposure to methylparathion or permethrin, a change that can reduce a bee colony's chances of survival and produce a substantial economic impact for those that depend on bees as crop pollinators.[3] In addition, exposure to atrazine or diuron can affect various behaviors of fish by altering the chemical perception of natural substances of eco-ethological importance.[4]

Biocides pass through the placenta and have been found in the milk of both animals and human beings. In many developing countries, nursing infants are potentially ingesting organohalogens at a ratio many times that of the acceptable daily intakes as estimated by the Food and Agricultural Organization.[5] Nutritional animal state (specifically malnutrition) is another variable that may affect an organism's susceptibility to pesticides. Supporting that, offspring from mother rats fed with dieldrin at levels often found in the environment and with a low protein diet showed altered behavioral effects.[6]

The actions of xenobiotics on an immature brain is very different from their action on adult animals. In the development of a mammal there is a period of rapid brain growth that may be critical for its normal maturation of axonal and dendritic outgrowth and the period when the synaptogenesis takes place. These periods are strongly modified by the timing and duration of chemical exposure as well as by the dose. Neonatal exposure to a single low oral dose of DDT (1.4 μmol/kg body weight) can lead to a permanent hyperactive condition in adult mice.[7] This amount of DDT is of physiological significance, since it is of the same order of magnitude as that to which animals and man can be exposed during the lactation period. The consequence of the early exposure is quite different from that reported for animals exposed to DDT as adults. The DDT dose required in adult animals to provoke symptoms and effects such as ataxia, tremor, increased activity in open-field test and avoidance responding, is more than 50–200 times the dose for neonatal exposure to produce toxic effects. In addition, behavioral alterations induced in rats by a pre- and postnatal exposure to the herbicide 2, 4-dichlorophenoxyacetic acid (2, 4-D) was demonstrated in our laboratory.[8]

Behavioral assessment is important because it is often the case that one of the earliest indications of exposure to neurotoxicants is subtle behavioral impairment such as paraesthesia or short-term memory dysfunction; frequently, such behavioral effects precede more obvious and frank neurological signs. In humans, volunteer studies and case reports have shown that exposure to dicofol results in disturbance of equilibrium, dizziness, confusion, headaches, tremors, fatigue, vomiting, twitching, seizures, and loss of consciousness.[9]

To assess the effects of toxicants on the nervous system two levels of sensitivity/complexity were described by Tilson et al.[10] At the first level of behavioral analysis, those procedures that require little or no training of experimental animals have the capacity to include large numbers of subjects. They provide an indication that a neurobehavioral deficit is present or absent, and permit scientists to quantify any deficits as precisely as possible, and are referred to as screening techniques. Examples of such tests include simple measures of locomotor activity, sensorimotor reflexes, and neurological signs. Moser's laboratory has been evaluating neurotoxicants as well as several non-neurotoxicants (negative controls) in the functional observational battery (FOB) to establish selectivity, reliability, redundancy, specificity, and sensitivity of the individual tests as well as the battery as a whole.[11] FOB is a series of tests to assess sensory, neuromuscular, and autonomic functions in animals and is similar to clinical neurological examinations in humans in that it rates the presence and, in some cases, the severity of behavioral and neurological signs.[12] In addition, a continuous recording of a rat's activity using a residential maze in which infrared optical gates are connected to a digital computer was used by Elsner et al.[13] to register the effects of lindane, dichlorvos, etc., on the spontaneous rat's behavior.

Sex and the physiological state of an animal are other important factors. We demonstrated that oral administration of 2,4-D butyl ester (2,4-Dbe) to nulliparous females had no effects on either open field (OF) and rotarod performance. By contrast, dams treated with 2,4-Dbe during pregnancy and intact male rats exhibited impairments of OF activity and rotarod endurance.[14]

At the second level of evaluation, tests requiring extended or special training, frequent evaluation (i.e., daily sessions) and/or manipulation of motivational factors such as food deprivation or electric footshock are used. These procedures may be useful in estimating environmentally acceptable limits such as nonobservable-adverse-effect level (NOAEL) or lowest-observed-adversed-effect level (LOAEL). Examples of these tests include discriminated conditioned response methodologies to assess specific sensory or motor dysfunction and procedures to measure chemical-induced alterations in cognitive function.[15] The data available suggest that organochlorine pesticides may disrupt performance differentially. Burt[16] and Desi[17] observed that dieldrin decreased overall rates of fixed-interval responding and disrupted the within-interval pattern of responding in rats and Japanese quail maintained under fixed-interval schedules of food reinforcement. Dietz and McMillan[18] compared the effects of daily administration of mirex and chlordecone on the performance of rats under several schedules of reinforcement. Both pesticides produced delayed disruption in performance inversely proportional to the dosage administered daily. Social interaction, plus-maze behavior, and one-way passive avoidance were studied in rats orally treated with fenvalerate to determine the anxiolytic effect of this pesticide.[19]

The concept that a challenge to a system may overcome compensatory mechanisms and thereby reveal otherwise hidden neurotoxicant induced damage is used as a method of assessment in neurobehavioral toxicology. We demonstrated hidden 2,4-D latent psychotic effects, the Serotonergic Syndrome, through amphetamine to an organism previously exposed to the 2,4-D. These rats also showed some dopaminergic behavior such as rearing and catalepsy if they were challenged with haloperidol in a noncataleptic dose of this drug.[20]

Since behavior is the net result of integrated sensory, motor, and cognitive function occurring in the nervous system, pesticide-induced changes in behavior may be a relatively sensitive indicator of nervous system dysfunction, which may hamper survivability possibilities in wildlife.

REFERENCES

1. Russel, R.W.; Singer, G. Neurobehavioral toxicology: a view from "down under," Neurobehav. Toxicol. Teratol. **1982**, *4*, 5–7.
2. Peakall, D. Disrupted patterns of behavior in natural populations as an index of ecotoxicity. Neurobehav. Toxicol. **1996**, *104* (Environmental Health Perspectives Suppl.), 331–335.
3. Cohn, J.; MacPhail, R. Ethological and experimental approaches to behavior analysis: implications for ecotoxicology. Neurobehav. Toxicol. **1996**, *104* (Environmental Health Perspectives Suppl.), 299–305.
4. Saglio, P.; Trijasse, S. Behavioral responses to atrazine and diuron in goldfish. Arch. Environ. Contam. Toxicol. **1998**, *35*, 484–491.
5. FAO/WHO. Guidelines for predicting the dietary intake of pesticide residues. Bull. WHO **1988**, *66*, 429–434.
6. Olson, K.L.; Boush, G.M.; Matsumura, S. Pre- and postnatal exposure to dieldrin: persistent stimulatory and behavioral effects. Pestic. Biochem. Physiol. **1980**, *13*, 20–33.
7. Eriksson, P.; Archer, T.; Fredriksson, A. Altered behaviour in adult mice exposed to a single low dose of DDT and its fatty acid conjugate as neonates. Brain Res. **1990**, *514*, 141–142.
8. Bortolozzi, A.; Duffard, R.; Evangelista de Duffard, A.M. Behavioral alterations induced in rats by a pre- and post exposure to 2,4-dichlorophenoxyacetic acid. Neurotol. Toxicol. **1999**, *21*, 451–465.
9. Hayes, W.J., Jr. *Pesticides Studies in Man*; Williams and Wilkins: Baltimore, 1982; 180–208.
10. Tilson, H.A.; Cabe, P.A.; Burne, T.A. *Experimental and Clinical Neurotoxicology*; Ch. 51, Williams and Wilkins: Baltimore, 1980; 758–766.
11. Moser, V.C. Screening approaches to neurotoxicity: a functional observational battery. J. Am. Coll. Toxicol. **1989**, *8*, 85–93.

12. Tilson, H.A.; Moser, V. Comparison of screening approaches. Neurotoxicology **1992**, *13*, 1–14.
13. Elsner, J.; Loosed, R.; Zbinden, G. Quantitative analysis of rat behavior patterns in a residential maze. Neurobehav. Toxicol. **1979**, (Suppl. I), 163–174.
14. Evangelista de Duffard, A.M.; Orta, C.; Duffard, R. Behavioral changes in rats fed a diet containing 2,4-dichlorophenoxyacetic butyl ester. Neurotoxicology **1990**, *11*, 563–572.
15. Tilson, H.A. Neurobehavioral methods used in neurotoxicological research. Toxicol. Lett. **1993**, *68*, 231–240.
16. Burt, G.A. Use of Behavioral Techniques in the Assessment of Environmental Contaminants. In *Behavioral Toxicology*; Weiss, B., Laties, V.G., Eds.; Plenum Press: New York, 1975; 241–263.
17. Desi, I. Neurotoxicological effects of small quantities of lindane. Animal Studies. Int. Arch. Arbeitsmed. **1974**, *33*, 153–162.
18. Dietz, D.D.; McMillan, D.E. Comparative effects of mirex and kepone on schedule-controlled behavior in the rat. II. spaced-responding, fixed-ratio, and unsignalled avoidance schedules. Neurotoxicology **1979**, *1*, 387–402.
19. De Spouzza Spinoza, H.; Silva, Y.M.; Nicolau, A.A.; Bernardi, M.M.; Luciano, A. Possible anxiogenic effects of fenvalerate, a type II pyretroid pesticide in rats. Physiol. Behav. **1999**, *67*, 611–615.
20. Evangelista de Duffard, A.M.; Bortolozzi, A.; Duffard, R. Altered behavioral response in 2,4-dichlorophenoxyacetic acid treated and amphetamine challenged rats. Neurotoxicology **1995**, *16*, 479–488.

Pesticides: History

Edward H. Smith
Cornell University, Ithaca, New York, U.S.A.

George Kennedy
Department of Entomology, North Carolina State University, Raleigh, North Carolina, U.S.A.

Abstract

Pests—insects, nematodes, plant pathogens, and weeds—destroy more than 40% of the world's food, forage, and fiber production. The struggle, pests versus people, grows ever more intense as population increases, arable land decreases, and human intervention disturbs biotic relationships on a global scale. Pesticides play a vital role in the struggle, but their use has not been without adverse ecological impacts and risk to the safety of those who apply them and those who consume treated products. This brief essay recounts the human experience with pesticides from the dawn of history to the dawn of the 21st century. It is the story of trial and error, old problems, and new lessons on nature's response to insult by ingenious synthetic molecules. It chronicles the intellectual probing and public debate of problems associated with the overreliance on chemical control that developed following World War II and led to adoption of integrated pest management (IPM) as an ecologically viable paradigm for crop protection. The success or failure of pest control programs may well hold the key to world order as the six billion peoples of the world compete for their place in the sun.

And he gave it for his opinion, that whoever could make two ears of corn or two blades of grass to grow upon a spot of ground where only one grew before, would deserve better of mankind, and do more essential service to his country, than the whole race of politicians put together.

—Jonathan Swift, 1726

INEVITABLE CONFLICT

Agriculture, which dates back a mere 15,000 years, requires the modification of natural systems. Agricultural practice imposes ecological simplicity on biota driven by natural selection toward diversity. Agriculture swims against the ecological tide and crops must be protected from the Darwinian struggle for existence. Intervention is required in many forms, including the use of pesticides. The conflict is inevitable.

ASCENDANCY OF PESTICIDES

The biblical records provide insight into the philosophy surrounding humankind's encounters with pests. According to Judeo-Christian beliefs, man was accorded dominion over the plants and animals. Departures from the laws of God were punished by plague. "I have smitten you with blasting and mildew your trees the palmer worm devoured them. Yet have ye not returned unto me" (Amos 4:9).

Development of the agricultural sciences progressed slowly. The emergence during the Renaissance of Natural Theology, which reconciled science and religion, followed by invention of the printing press, the microscope, and the Linnean system of biological nomenclature set in place the elements for rational thought and communication on natural history and on pest control.

The status of insect pest control in early 19th century Europe and North America is revealed by T.W. Harris's publication, *Report on the Insects of Massachusetts Injurious to Vegetation* (1841), prepared at the request of the state legislature. Harris, a Harvard librarian and meticulous scholar, drew upon European literature as well as American agricultural journals, which published pest control recommendations offered by their readers. His control measures included: hand picking; burning stubble and field refuse; smoke screens in orchards to drive moths away; running pigs in the orchards; poison baits; resistant varieties (wheat); favorable planting dates; dusting with ashes, quick lime, red pepper, sulfur, and tobacco; spraying with whitewash and glue; and encouraging woodpeckers in orchards. (How to do the latter was not specified.) While these measures were crude, they were based on reason and fragmentary knowledge of pest biology. They were free of the ridiculous nostrums proposed earlier out of ignorance, superstition, and fraud, and they represented early steps in cultural, biological, mechanical, and chemical control. A foundation had been laid drawing on knowledge from Europe and North America aided by state subsidy, a renowned educational institution, and an able scholar.

Recognition that abundant and reliable agricultural production was prerequisite to urbanization and industrial development helped to trigger the agricultural revolution in the

United States, which began in the 1840s as canals and railroads linked the eastern population centers with expanding agricultural lands west of the Mississippi. The next great impetus to pest control in the United States came through congressional passage of the Morrell Act in 1862, establishing the Land Grant University System. This paved the way for professionals in the applied science of pest control, and was followed in 1887 by passage of the Hatch Act, establishing a coordinated system of State Experiment Stations devoted to the advancement of research. The Smith-Lever Act of 1914 officially recognized the extension arm of the Land Grant University System and completed the American model that has proven to be one of the most innovative educational concepts of all time.

Annual reports on beneficial and injurious insects issued between 1856–1876 by early leaders such as Asa Fitch, (New York), B. D. Walsh (Illinois), and C. V. Riley (Missouri) became the backbone of applied entomology. These writers urged natural controls as the first line of defense and expressed their misgivings about the crude chemical controls of the time.

Pest control practices were strongly influenced by expanding commerce, which resulted in the introduction of exotic pests, and by the rapid, westward expansion of agriculture, which disrupted ecosystems and exposed crops to new pests. The Colorado potato beetle, *Leptinotarsa decemlineata* (Say), provides a prime example. It appeared as a devastating pest of potato in Iowa and Nebraska in 1861, having transferred from a native weed to an introduced relative, the potato. The beetle spread rapidly eastward, reaching the Atlantic coast in 1874, despite the use of traditional nonchemical means of control. In 1867, farmers in the west discovered that the Colorado potato beetle could be controlled with Paris Green, an arsenical. Paris Green was in general use by 1880 and became the first widely used pesticide in North America. Similar experiences followed with other major pests, such as the plum curculio *Conotrachelus nenuphar* (Herbst), boll weevil *Anthonomous grandis grandis* Boheman, gypsy moth *Lymantria dispar* (Linnaeus) and others.

During the first half of the 19th century, lime-sulfur and wettable sulfur gradually came into use for control of fungal pathogens, primarily of fruit trees and grapevines in Europe and the United States. In 1885, Pierre Milardet, professor of Botany at Bordeaux, France, demonstrated control of downy mildew, *Plasmopora viticola,* on grapevines using a mixture of copper sulfate and lime, subsequently known as Bordeaux mixture. The success of Bordeaux mixture led to efforts to improve upon its effectiveness and to the expanded use worldwide of it and its variants.

Other components—petroleum oil, nicotine, pyrethrum, and organomercury fungicides for seed treatment—were soon added to the pesticide arsenal. By 1910, the arsenicals Paris Green, lead arsenate, and calcium arsenate were the most widely used pesticides. Herbicides were notably absent; they did not appear until the discovery of plant growth hormones paved the way for the synthesis of stable synthetic hormone analogues (2,4-D and 2,4,5-T) in the 1940s.

Farmers, their advisors in the fledgling Land Grant Universities, and an emerging chemical industry rallied behind pesticides, especially insecticides, for one pragmatic reason; they provided a degree of reliability in control programs that was absent with other available methods. World War I stimulated pesticide use for food production. It also stimulated the production of insecticides, such as dinitrophenols (DNOC) and paradichlorobenzene (PDB), as by-products of the manufacture of explosives from coal tar.

On the eve of World War II, insecticides were the backbone of insect control but their use was fraught with unease. A host of problems surfaced. Control was marginal. British markets rejected U.S. apples because of the high arsenical residues. There were concerns for the health of workers and consumers. The codling moth *Cydia pomonella* (Linneaus) acquired resistance to arsenicals; excessive pesticide treatments were phytotoxic to foliage causing reduced yields, and there were concerns about the build-up of residues in the soil. To many entomologists, it appeared that they were losing the fight. They clung to the early, idyllic hope for control by natural means but in the crunch of practical experience, they turned to pesticides because they worked, not well, but better than the alternatives.

DDT: DISCOVERY, DEVELOPMENT, AND IMPACT

The discovery and introduction of DDT, while purely a commercial enterprise, became immediately enmeshed in the intrigue and urgency of World War II. The Swiss chemist Paul Mueller, an employee of J.R. Geigy Co., discovered the insecticidal property of DDT in September 1939; this event coincided with the Nazi invasion of Poland. DDT found a vital military role in the control of insect-borne diseases. When the war ended in 1945, DDT, the shining chemical sword of World War II, found extensive peacetime use. It was distributed quickly for testing through the well-organized network of Agricultural Experiment Stations. Data poured in confirming the effectiveness of DDT against a wide spectrum of insect pests of agricultural and medical importance. In striking contrast to the prewar pessimism, DDT produced hope that at last the age-old insect scourges could be controlled and perhaps eradicated.

Such optimism had a profound effect on the crop protection sciences. In entomology and weed science especially, research shifted focus away from pest biology and on to pesticide technology. At this point, the birthright of pest control scientists as biologists became endangered. Insecticide use soared, based on the promise of DDT and the related chlorinated hydrocarbon insecticides that followed. New classes of insecticides, the organophosphates and the methylcarbamates, were discovered and exploited. The success of 2,4-D for control of broadleaf weeds stimulated the development and use of chemical weed control. Similarly, the discovery of the dithiocarbamate fungicides during the 1930s led to the development of an array of very effective fungicides and increased fungicide use. All this was catalyzed by a powerful coalition: the chemical industry with its high capitalization and integrated skills in synthesis, testing, and marketing; the agricultural community with considerable political clout; and the Land Grant Universities with their triple mission of teaching, research, and extension. Pesticide use and reliance on pesticides for crop production increased steadily.

REBUFF AND REASSESSMENT

The euphoria that accompanied the dominance of chemical control in the 1950s was short-lived. By the end of the decade, warnings about the adverse effects of pesticides were being expressed by environmentalists and some pest control specialists, but these were largely ignored. There was fear within the crop protection disciplines, especially entomology, that reliance on pesticides was placing agriculture on a "pesticide treadmill." There were problems with resurgence of targeted pest populations and outbreaks of secondary pest populations following destruction of their natural enemies, and with the development of pesticide resistance, all of which necessitated additional applications of pesticides. Similar concerns surfaced regarding control of medical and veterinary pests.

In 1962, Rachel Carson's book *Silent Spring* galvanized public attention on the problems spawned by pesticide use. She made her case with poetic beauty sounding the alarm that "we have put poisonous and biologically potent chemicals indiscriminately in the hands of persons largely or wholly ignorant of their potentials for harm." What had been a debate among scientists became a public debate. Drawing on lessons of the civil rights movement, the antipesticide forces headed by the Environmental Defense Fund turned to litigation in defense of the right of citizens to a clean environment. After long, contentious hearings, the Environmental Protection Agency banned DDT in 1972. This landmark decision placed the issue of pesticides in the forefront of the greatly energized environmental movement.

Increased public activism over environmental and food safety issues, which began during the 1960s and continues today, led to dramatic changes in pesticide regulation and to restrictions on pesticide use in both the United States and Europe. These actions dramatically strengthened the environmental and toxicological standards that pesticides must meet before they can be approved for use. In doing so, these changes provided strong impetus for the development of safer and more environmentally friendly pesticides.

The regulatory framework for pesticides continues to broaden as new knowledge is acquired and perceptions change. For instance, in 1996 the U.S. Congress passed the Food Quality Protection Act, which established more stringent safety standards aimed at protecting infants, children, and other sensitive subpopulations from risks associated with pesticide residues on food. Subsequently, several major food processors imposed their own more stringent tolerances for pesticide residues on the produce that they purchase. The process is expected to continue in response to the ebb and flow of new findings and public concern.

In the late 1950s and early 1960s, growing awareness of the problems associated with pesticide use and the specter of faltering pest control, viewed in the context of decreasing availability of arable land and dwindling supplies of fossil fuel to drive the technology of agribusiness, stimulated a reassessment of pest control. Earlier work in biological control in several countries provided points of departure but it was the intellectual probing in entomology at the University of California (Berkeley and Riverside) that ignited a great debate, which in time involved pest control specialists the world over. The topic of debate was the concept of integrated pest management (IPM). IPM emphasized that pest problems were under the influence of the total agroecosystem and that not all levels of pest abundance required treatment with pesticides. It also emphasized that pest management should be a multidisciplinary effort based on ecological principles and economic, social, and environmental considerations.

While the concept soon gained widespread acceptance, many factors impeded its implementation. The knowledge base was in most cases inadequate; the research and extension infrastructure required redirection; a corps of private consultants to supplement decision making by farmers had to be recruited and trained; and replacement of broad-spectrum pesticides was slow and costly. Federal and state governments were sold on the soundness of the concept and appropriated funds to overcome these constraints.

Four decades after the initiation of IPM, the steering mechanism for sound employment of pesticides, what is the score? The glass is half full. Great strides have been made on a worldwide scale. IPM has provided the framework to accommodate transition from singular reliance on broad-spectrum, long-residual pesticides to the use of highly selective, short-residual compounds as components of multifaceted crop protection programs, without an increase in losses to pests. Every phase of the university support network—teaching, research and extension—has been altered to reinforce the ecological foundations of IPM. The disappointing aspects are that adoption of programs has been slow, pesticides still predominate in many programs, overall use of pesticides has not declined, and successful interdisciplinary programs are few. Despite ongoing improvements in the characteristics of pesticides, the specter of pest resistance hangs like the sword of Damocles over the utility of pesticides.

Great challenges lie ahead as concepts of sustainable agriculture and the technology of genetic engineering meld with the ever-expanding scope of IPM. IPM has become a unifying catalyst, an intellectual quest that unites producers, plant protection disciplines, agribusiness, regulatory agencies, and the worldwide plant protection community concerned with the production of food and fiber for the six billion peoples of the world.

Genetic engineering technology will have broad application in control of pests of plants and animals. Using this technology, genes from one organism can be inserted into and expressed in totally different organisms. Genetic engineering is producing new kinds of insecticidal peptides and proteins, and is enabling plants, bacteria, and viruses to be used in novel ways to deliver toxins to targeted pests. The same technology has produced plants that are tolerant to broad spectrum, postemergence herbicides. The possibilities seem limitless.

In a remarkably short span of three decades, the science of biotechnology became an applied technology and a new industry, involving new kinds of partnerships between university scientists and entrepreneurs. Overnight, genetically engineered crops were being planted on millions of acres in the United States.

The speed of scientific and technological advance in genetic engineering and the new partnership between universities and industry have given rise to a host of challenging issues: academic freedom in the context of university/industry partnerships; patenting of biological processes; monopolies; economic impact, particularly on developing countries; response of organisms to selective pressure (resistance). The most daunting questions focus upon risk assessment and regulatory procedures addressing the impact of organisms created outside the normal evolutionary pathways on the global biota.

Political debate on these issues has grown in intensity and rancor, first in Europe and then in the United States. While the time frame of debate and acceptance of genetic engineering is in doubt, it is clear that the tremendous pressures to meet food requirements for a world population of nine billion by 2050 are likely to force the incorporation of genetically engineered components into the arsenal of pest control.

The question germane to the present essay is what part will pesticides play in future IPM programs. They will be a vital component but in a modified role. Advances in toxicology, chemistry, biochemistry, physiology, molecular biology, and computer modeling are making possible the tailoring of pesticides to meet IPM requirements, which dictate low mammalian toxicity, high specificity conferring low environmental impact, and low residues on treated products. In the future, pesticides will be used with greater precision, made possible by improvements in pesticide application technology, pest and crop monitoring, weather prediction, and information processing, as well as by better understanding of population dynamics, microbial and weed ecology, and epidemiology.

It is important to note the influence of economics and elevated standards for pesticide potency and safety. These factors have dramatically increased the costs of pesticide discovery and development and have contributed to an internationalization and consolidation of the pesticide industry. They have also resulted in fewer new pesticides being introduced. While the agrichemical industry has not enjoyed a favorable public image in an era of environmental awareness, it should be remembered that it plays a vital role in the multifaceted IPM enterprise.

THE FUTURE

History should illuminate the future. We see pest control as a challenge woven into the economic, political, and social fabric of society. Major factors are shaping the new era of pest control.

Public Attitude

Growing environmental ills will further sensitize the public to problems arising from technology. This will find expression in stricter pesticide regulation and safer pesticides.

Global Commerce

Increased and increasingly rapid international movement of goods and people will intensify the introduction of exotic species and the spread of pesticide-resistant organisms.

Economics

The ever rising cost of developing new pharmaceuticals and pesticides will constrain research and product development, and the use of pesticides in developing countries.

Population Pressure

The environmental stress imposed by rapid growth of the human population will continue to exacerbate problems of agricultural production, including pest control. This is perhaps the most serious problem facing humankind, with no relief in sight.

Throughout the latter half of the 20th century, pesticides contributed enormously to improvements in the quality and stability of the world's food supply and to the control of devastating insect-transmitted diseases of humans and livestock. Pesticides have also played a central role in fostering environmental awareness and public concern over food safety. The inevitable conflict between humans and pests will grow in intensity as the human population grows and arable land decreases. Pesticides, because of their ease and rapidity of use and the reliability with which they can rein in pest outbreaks, will continue to play an important role in IPM. Lessons having been learned, pesticides of the future will be safer and more environmentally friendly, and will be used more judiciously than in the past. Our crystal ball discerns no "silver bullet" of pest control, rather painstaking refinement of IPM, with further advances in established methods, including pesticides and biological control, and a melding of new technologies such as genetic engineering.

BIBLIOGRAPHY

1. Adler, E.F.; Wright, W.L.; Klingman, G.C. Development of the American Herbicide Industry. In *Pesticide Chemistry in the 20th Century*; Plimmer, J.R., Ed.; American Chemical Society: Washington, DC, 1977; 39–55.
2. Brent, K.J. In *One Hundred Years of Fungicide Use, Fungicides for Crop Protection 100 Years of Progress*, Proceedings of The Bordeaus Mixture Centenary Meeting, Smith, I.M., Ed.; British Crop Protection Council Publications: Croyden, U.K., 1985 11–22, 1985; Monograph No. 31; 1.
3. Carson, R. *Silent Spring*; Houghton Mifflin: Boston, 1962.
4. Cassida, J.E.; Quistad, G.B. Golden age of insecticide research: past, present, or future. Annu. Rev. of Entomol. **1998**, *41*, 1–16.
5. Howard, L.O. *A History of Applied Entomology*; Smithsonian Institution: Washington, DC, 1930.
6. Knight, S.C.; Anthony, V.M.; Brady, A.M.; Greenland, A.J.; Heany, S.P.; Murray, D.C.; Powell, K.A.; Shulz, M.A.; Spinks, C.A.; Worthington, P.A.; Youle, D. Rationale and perspectives on the development of fungicides. Annu. Rev. of Phytopathol. **1997**, *35*, 349–372.
7. Lever, B.G. *Crop Protection Chemicals*; Ellis Horwood: New York, 1990.
8. Marco, G.J.; Hollingworth, R.M.; Plimmer, J.R. *Regulation of Agrochemicals: A Driving Force in Their Evolution*; American Chemical Society: Washington, DC, 1991.
9. Perkins, J.H. *Insects, Experts, and the Insecticide Crisis: The Quest for New Pest Management Strategies*; Plenum: New York, 1982.
10. Zimdahl, R. *Fundamentals of Weed Science*, 2nd Ed.; Academic Press: New York, 1999.

Pesticides: Measurement and Mitigation

George F. Antonious
Department of Plant and Soil Science, Water Quality/Environmental Toxicology Research, Kentucky State University, Frankfort, Kentucky, U.S.A.

Abstract

In intensively cultivated areas, agriculture is a significant source of pesticides associated with runoff. Environmental pollution by pesticides is a matter of growing concern. Movement of pesticides in runoff is influenced by the pesticide chemical properties, application methods, soil type, crop management, and environmental conditions. Water solubility is one of the pesticide characteristics that control mobility. Runoff water and sediment are frequent in sloping areas where most of the arable lands are highly erodible. Utilization of vegetative filter strips contiguous to agricultural fields revealed a reduction of the transport of pesticides (clomazone, bensulide, trifluralin, and napropamide) into runoff water, allowing for their water infiltration into the vadose zone (the unsaturated water zone below the plant root). Planting living fescue strips against the contour of the land slope reduced runoff but has the disadvantage of increasing the potential of soil infiltration by pesticides. Unfortunately, plastic mulch, which could cover between 50% and 70% of a field, increased surface water runoff from both rainfall and irrigation. This means that much of the pesticides applied in living fescue or in plastic-mulched fields might seep into groundwater or leave the field into surface runoff. Although many factors are responsible for decomposition of pesticides in soils, two are considered the most important: 1) adsorption increases the availability of the pesticide for soil degradation processes; and 2) microbiological activity increases pesticide metabolism and degradation. Agriculture makes relatively little use of soil microorganisms as producers of several detoxifying enzymes capable of breaking down pesticides and other contaminants. Contaminated surface water has become a critical environmental problem. Runoff from agricultural watersheds carries enormous amounts of pesticides. Rainfall intensity and flow rate are critical factors in determining pesticide movement from application site into surface runoff, rivers, and streams. Accordingly, there is an urgent need to develop long-term, low-energy, biological, self-sustainable systems of farming. Methods of application of these systems must be simple, inexpensive, energy conserving, safe, and effective for pesticide mitigation, nutrient recycling, and erosion control. Addition of soil amendments to increase soil organic matter, enhancing the activity of soil microorganisms, and installation of biofilters against the contour of agricultural fields are potential solutions to mitigate environmental pollution by pesticides. Addition of municipal sewage sludge (MSS) to native soil has reduced the concentration of the insecticide dimethoate in surface runoff water by 47% and increased soil retention of the two herbicides trifluralin and napropamide, lowering their concentration in runoff, and reducing their transport into streams and rivers. Little information, however, exists on the economic aspects of sludge application to agronomic crops. The following were the main objectives of this entry: 1) to provide an overview of pesticide quantification and field mitigation techniques (soil amendments, living fescue strips, biobed systems, constructed wetlands, natural products for pest control); 2) to provide information on the use of MSS for land farming, which could decrease dependence on synthetic fertilizers; and 3) to present Kentucky State University (KSU) research and field studies on reducing environmental impact of pesticides.

INTRODUCTION

Soil erosion, nutrient runoff, loss of soil organic matter (SOM), and the impairment of environmental quality from sedimentation and pollution of natural waters by agrochemicals, heavy metals, and other environmental contaminants have stimulated interest in proper management of natural resources. Pesticides cause water pollution by running off agricultural fields and domestic gardens into nearby water sources. Although agriculture has been identified as a source of pesticides found in water, other sources exist. They may be pesticide manufacturing industries, industries using pesticides in their processes (such as woolen goods manufacturers), direct application of pesticides to surface waters to control aquatic plants, and nuisance insects. Environmental pollution is often divided into pollution of surface water and groundwater supplies, the atmosphere, plants or animal tissues, and the soil.

A wide range of active ingredients are used as pesticides. According to the United States Environmental Protection Agency (USEPA), more than 441 million kg of conventional pesticides were used in the United States.[1] Of that total, 77% were used in agricultural applications, and 11% were used for home and garden purposes. Approximately

1200 water body impairments across the United States are attributed to pesticides.[2] In addition, millions of tons of so-called "inert" ingredients are added to pesticide formulations as carriers, stabilizers, emulsifiers, etc. Some of these ingredients are dangerous in their own right. Portions of the active ingredient may transport to neighboring water bodies via drift during pesticide spraying, wind erosion, and runoff. Accordingly, it is necessary to assess the distribution and degradation/dissipation of a crop protection product in soil and water after field application.

Pesticides play an important role in the success of modern farming and food production.[3] A broad-spectrum pesticide that kills a wide range of living organisms is called a biocide. Herbicides kill plants; insecticides kill insects; ovicides kill the eggs of pests; fungicides kill fungi; acaricides kill mites, ticks, and spiders; nematicides kill nematodes (microscopic roundworms); rodenticides kill rodents; and avicides kill birds. One way to classify pesticides is by their chemical structure. This is useful because environmental properties such as stability, solubility, and mobility—and toxicological characteristics of members of a particular chemical group—are often similar. A commonly quoted estimate is that farmers save $3 to $5 for every $1 spent on pesticides.[4] CropLife America calculates that prohibiting pesticides in the United States would result in a $21 billion per year loss in food and fiber production. Without herbicides, they conclude, crop yields would be reduced up to 67%, soil erosion would increase by at least 1 billion metric tons per year (because of more cultivation to remove weeds), and would take up to 70 million additional farm laborers to remove weeds by hand.[4,5] Recent decades have brought increasing concerns for potential adverse human and ecological health effects resulting from the production, use, and disposal of numerous chemicals that offer improvements in agriculture, industry, medical treatment, and even household chemicals. Protecting the integrity of our soil and water resources is one of the most essential environmental issues of the 21st century. Agricultural production is an important part of the nation's economy and pesticide use on crops is extensive.[6–8] Agricultural activities are frequently conducted in close proximity to lakes, reservoirs, and streams. More than 500 million kg of pesticides are used each year in the United States in both agricultural and urban settings.[9] There is a concern over the risks of contamination of food and drinking water by residues of synthetic agrochemicals and the negative impact of agrochemicals on the countryside. A central hope in these concerns is the safe use of agrochemicals, development of new soil management practices, and use of mitigation techniques.

Grass filter strips (*Festuca elatior*, Kentucky 31) 3 feet wide installed against the land slope between cropping rows reduced the amount of runoff water and sediment by 85% and 99% in pepper plots and 93% and 99.8% in pumpkin plots, respectively, compared to no-mulch treatments.[10] Grass strips reduced dacthal (an herbicide) in runoff water and runoff sediment by 95% and 100%, respectively, compared to no-mulch treatments.[11] Grass strips also reduced runoff water volume and clomazone (an herbicide) in runoff from draining into adjacent streams. Studies indicated that fescue strips reduced runoff but did not reduce leaching of α-endosulfan[12] and dacthal into the vadose zone.[11] Moreover, benefits derived from use of black plastic mulch as a soil management practice are well documented.[10] In addition to controlling weeds, plastic mulch reduced leaching of nutrients and conserved moisture. However, plastic mulch as a management practice had little impact on reducing runoff.[10] This could be attributed to the lack of surface roughness compared to grass strips. Therefore, additional technological and infrastructural solutions are required to reduce pesticide releases. For agricultural purposes, these technologies and systems need to be cheap and reliable and easy to use with low labor and time input. Environmentally and economically viable agriculture requires the use of cultivation practices that maximize agrochemical efficacy while minimizing their off-site movement. Mitigation and cleaning up of excess pesticide residues before they run off and enter bodies of water is the main focus of this investigation. The main objectives of the remediation techniques would be to 1) reduce pesticide residues from agricultural soils using adsorbents of natural origin; 2) reduce pesticide residues from runoff and infiltration water released from agricultural field; and 3) stimulate soil microbial activity.

Soil quality is not defined solely by its physical and chemical parameters but also intimately linked to microbial parameters.[13] With increasing emphasis on fertility sustainability and environmental friendliness, restoration of soil microbial ecology has become important.[14] Generally, four main biological techniques for treating soil and groundwater contamination are used: 1) stimulation of the activity of indigenous microorganisms (biostimulation) by the addition of nutrients, regulation of redox conditions, optimization of pH conditions, etc.; 2) inoculation of the site by microorganisms with specific biotransforming abilities (bioaugmentation); 3) application of immobilized enzymes; and 4) use of plants (phytoremediation) to remove and/or accumulate pollutants. For bioremediation systems, the quantity and frequency of hydraulic and chemical load and the climate and legislation will have an impact on the type of design and the quantity of matrix substrate. All these factors have to be considered in order to come to the most suitable pesticide bioremediation system for any given field application. With the decline of many ecosystems in the world and lack of knowledge of soil microbial community, increasing awareness concerning the importance of soil microorganisms has emerged. Soil microorganisms constitute a large dynamic source and sink of nutrients in all ecosystems and play a major role in N, C, and P cycling.[14] Accordingly, restoration of soil microbial ecology has become important.[15] Optimal soil management represents an important strategy for sustainable agricultural systems.

DETECTION OF PESTICIDE RESIDUES AND MEASUREMENT OF ENVIRONMENTAL POLLUTION BY PESTICIDES

Pesticides are used on most major crops in the United States and worldwide. The world market for pesticides is estimated at $33.59 billion, of which the Unites States represents the largest part, in terms of dollars (33%) and pounds of active ingredients (22%).[16] Persistent pesticides could move long distance in the environment. Stability, high solubility, and volatility allow pesticides to move freely through air, water, and soil into distances far from the point of their original application. Runoff from agricultural watershed is found to carry enormous amounts of pesticides.[17] Large quantities of the organophosphorus insecticides diazinon and chlorpyrifos are applied to agricultural and urban watersheds in California every year.[18] Following natural rainfall events, water flow may change from <1 to >10,000 ft^3 sec^{-1} (1 ft^3 is equal to 28.3 L) in a matter of minutes to hours, enhancing pesticide transport to receiving water.[19] Solubility is one of the most important characteristics in determining how, where, and when toxic materials would move through the environment. Pesticides could be divided into two major groups: those that dissolve more readily in water and those that dissolve more readily in oil. Water-soluble pesticides move rapidly and widely through the environment because water is ubiquitous. Pesticides that are oil or fat soluble (usually organic molecules) generally need a carrier to move through the environment and within the body, tissues, and most cells of living organisms because the membranes that enclose cells are made of oil-soluble chemicals. Once pesticides get inside animal or plant cells, their oil-soluble molecules are likely to be accumulated and stored in lipid deposits where they might be protected from metabolic degradation and consequently persist for many years.

Pesticides are a structurally diverse group of chemical compounds. However, these compounds could be summarized into four main groups (Fig. 1) with respect to their fate and transport in the environment: Group I such as chlorpyrifos and methyl parathion tend to have higher water solubilities. Group II such as methoxychlor and DDT are less mobile and more persistent (long half-life) and tend to bioconcentrate or bioaccumulate. Group III pesticides such as heptachlor and lindane have properties similar to group II, but tend to have significant volatilization potential. Group IV pesticides such as methyl bromide and 1,2-dibromo-3-chloropropane (DBCP) are halogenated or polyhalogenated compounds like group II and group III, but they are significantly more mobile, water soluble, and volatile, and they have low bioconcentration and bioaccumulation potential. Fumigants such as ethylene dibromide or dibromochloropropane, used to protect stored grain or sterilize soil, fall into this category. In fact, there is too much structural diversity between the four groups of the outlined pesticides, and even within certain large groups (i.e., like group I pesticides), to be able to make general statements about the difference in the chemical properties versus differences in their toxicities and environmental impact on humans, plants, and aquatic organisms. Pesticides can be also classified into inorganic pesticides, fumigants, chlorinated hydrocarbons, organophosphates, carbamates, and natural products (Table 1). Solubility and adsorption properties of most of these pesticides varied.[20,21] The two most important characteristics determining soil adsorption of a pesticide are the organic matter content of the soil and the water solubility of the pesticide. Adsorption of nonionic pesticides on soil particles depends directly on the organic carbon content of the compound and the adsorbing phase. The general tendency of an organic chemical to be adsorbed by soils may be assessed by the chemical's organic carbon partition coefficient (K_{OC}), which describes the tendency of the chemical to partition from water to organic carbon. K_{OC} is the amount of pesticide adsorbed by soil divided by the product of fraction of organic carbon (OC) in soil and amount of pesticide in the soil solution.[22] K_{OC} represents the sorption on a unit carbon basis and could be used for comparison of sorption extent on soils with different organic matter contents. The greater the K_{OC} value of a pesticide, the stronger the binding to the soil.[22,23]

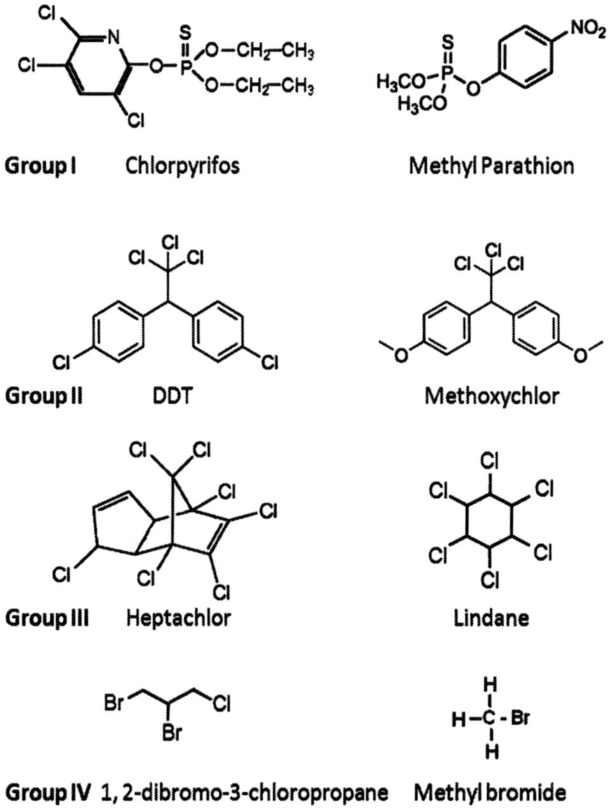

According to Haith and Rossi,[24] the organic carbon sorption coefficient (K_{OC}) of bensulide is 3900 mL g^{-1}.

Fig. 1 Chemical structures of four groups of pesticides with respect to their fate and transport in the environment.

Table 1 Classification of selected agricultural pesticides based on their chemical structure and variations in their water solubility.

Chemical type	Example	Chemical structure	Solubility in water[a]	K_{OC} in soil	Typical action
Inorganic pesticides	Arsenic trioxide	As_2O_3	17 g/L at 16°C	—	Rodenticide
	Mercuric chloride	$HgCl_2$	69 g/L at 20°C	—	Fungicide/bactericide
Fumigants	Methyl bromide	$H-CH_2-Br$	13.4 g/L at 25°C	83.18[b]	Disinfectant
Chlorinated hydrocarbons	Endosulfan		0.32 mg/L at 22°C	19,952.62[a]	Insecticide
Organophosphates	Dimethoate		23.8 g/L at 20°C	51.9[a]	Insecticide
Carbamates	Carbaryl (Sevin)		120 mg/L at 20°C	389.05[b]	Insecticide
Natural products	Pyrethrin		Practically insoluble	22,915 (Py-I)[a] 2,042 (Py-II)[a]	Insecticide

—, Not available.
[a]From Tomlin.[20]
[b]From Montgomery.[21]

Comparatively, the herbicide azafenidin (Milestone) has a soil-organic carbon sorption coefficient of 298, which indicates that azafenidin does not bind strongly to soil particles.[25] Pesticides with high persistence and a strong sorption rate are likely to remain near the soil surface, increasing the chances of being carried to a stream via surface runoff. On the contrary, pesticides with high persistence and a weak sorption rate may be readily leached through the soil and are more likely to contaminate groundwater.[26] Herbicides having high K_{OC} value, i.e., bensulide, will bind to SOM particles (Fig. 2). This information is typically useful in ranking pesticides as leachers or non-leachers to assess pesticides' potential for off-site surface or subsurface movements under field conditions. Pesticide adsorption to soil is related more to SOM than to other soil chemical and physical properties[22,27]; therefore, addition of soil amendments having high organic matter content is a management practice that should be exploited to trap nonionic pesticides like bensulide and pyrethrins to reduce their surface mobility under field conditions.

Once pesticides have released into the environment, via direct pesticide use around home gardens, or application in agricultural fields or commercial agricultural use (Fig. 3), their fate and transport may be operative, depending on the nature of the pesticide. These potential fate and transport are as follows: sorption, uptake and bioconcentration/bioaccumulation in biota (trophic chain mechanisms), oxidation/reduction, hydrolysis for some pesticides, surface

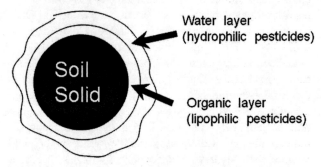

Fig. 2 A soil particle structure indicates that a chemical placed in soil will partition between organic layer (organic matter) and water layer and between water layer and soil air. Accordingly, the primary physical properties that determine how a chemical will become distributed are its hydrophobic character and vapor pressure.

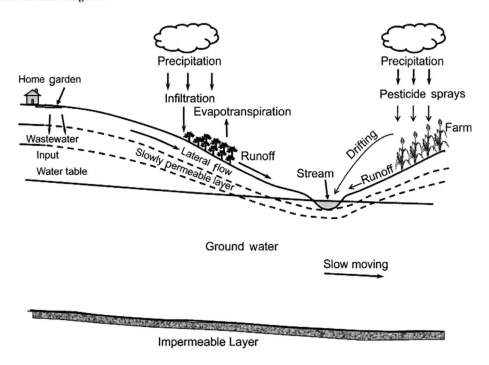

Fig. 3 Mobility of pesticides released from agricultural field and/or home gardens under environmental conditions in relation to hydrologic cycle.

runoff, photodegradation, volatilization, and leaching. Pesticide residues may be taken up by animals from their feed, to appear later in meat, fish, eggs, and dairy products. Most food, then, can be expected to contain some traces of pesticides. The detection and quantitative analysis of such residues is the key to any possible health hazard they might represent. Because most of group I and group IV pesticides (Fig. 1) have high affinity for water, they do not tend to adsorb very strongly to soil and sediments. Such weak sorption potential and high water solubility means these pesticides are prone to leaching and surface water runoff, increasing potential environmental pollution by pesticides. Storm water runoff will contact and dissolve members of this group, transporting them to storm discharge locations. These locations can be drinking water reservoir, lakes, rivers, streams, or main environment with potentially significant source of exposure to these pesticides for both humans and ecological receptors. On the contrary, group II pesticides represent some of the most highly sorbing behavior and are rarely a threat to groundwater via leaching processes, especially because most environmental releases of these types of chemicals are surface releases. However, they tend to migrate long distances. Canadian researchers found the levels of chlorinated hydrocarbons in the breast milk of Inuit (Eskimo) mothers living in remote arctic villages at concentrations 5 times that of women from Canada's industrial region some 2500 km (1600 miles) to the south.[5] Mothers with high DDT or DDE (a metabolite of DDT) levels in their blood are more likely to have premature or low-birthweight babies. These pesticides accumulate in polar regions by what has been called "grasshopper effect," in which they evaporate from water and soil in warm areas and then condense and precipitate in colder regions, where they accumulate in top predators. Polar bears, for instance, have been shown to have concentrations of chlorinated compounds 3 billion times greater than the seawater around them.[5]

The presence of pesticides in water may be chronic, usually at levels less than 1 ppb (1 $\mu g\ kg^{-1}$), or acute (short term), frequently at higher concentrations. Some pesticides, notably the insecticides, are highly toxic to fish. This offers a ready means of detection in domestic water supplies when toxicants are present at concentrations sufficient to kill fish. It was suggested that a continuous bioassay procedure can be employed at water treatment plants for this purpose. This system would consist of a bioassay chamber providing renewal of water, utilizing either raw water or water from treatment system just prior to postfiltration disinfection. Good water quality was presumed to be attained when the concentration of toxic components was reduced to half the quantity required to kill 50% of the test animal (fish) used in bioassay in an exposure period of 96 hr. These values were 7 ppb (7 $\mu g\ kg^{-1}$) for toxaphene and 3 ppb (3 $\mu g\ kg^{-1}$) for rotenone.[28] The development of extremely sensitive analytical instruments has made possible the detection and identification of many pesticides and other classes of chemicals in amounts as little as a few nanograms (10^{-9} g). It is now feasible to monitor water supplies using sophisticated equipments such as gas chromatographic/mass selective detection (GC/MSD). However, several stages of analysis, discussed in the following subsections, must be followed to prepare soil and water samples for pesticide residue analyses.

Sampling

Less than a teaspoonful of a representative soil sample is actually used for the laboratory analysis. That small amount should represent the entire area for which the recommendation is to be made. Therefore, samples should be collected from several locations, combined together, reserved, and used for pesticide residue analysis. Sample collection is extremely important in the accuracy and repeatability of an analytical test. Handling of samples following collection is also important. It is therefore very important to collect and handle soil and water samples properly.

Extraction

Extraction of a pesticide from any sample (soil, water, plant, or animal tissues) needs a suitable solvent depending on the pesticide water solubility. Pesticides having high water solubility tend to dissolve in polar solvents (methanol, ethanol, and acetone). Pesticides that are fat soluble tend to dissolve in non-polar solvents (carbon tetrachloride, chloroform, and methylene chloride). Liquid–liquid extraction is usually used to separate compounds based on their relative solubility in two different immiscible liquids, usually water and an organic solvent.

Cleanup

Cleanup is the removal of interfering materials from the extract, and this procedure requires either solvent partition such as liquid–liquid partition or chromatographic procedures (glass chromatographic columns filled with silica gel, alumina, and/or activated charcoal) to remove interfering pigments and waxes from the sample extract. Solid-phase extraction (SPE) cartridges that use solid phase and liquid phase to concentrate pesticide residues in water samples (that contain one or more analytes) are now widely used for extraction, concentration, and cleanup of environmental samples.

Quantification

The test results could only be used in conjunction with a calibration curve that relates the laboratory results to a set of response data. Without calibration, the laboratory results are meaningless. The method used to estimate the quantity of a pesticide and its possible metabolites or degradation products in the cleaned-up extract of the sample matrix should be able to detect the pesticide under consideration at very low levels, as low as pictogram (10^{-12} g) ranges.

Confirmation

Pesticide residue analysis in soil, water, plant, and animal tissue is often complicated by chemical changes that happen to pesticides when absorbed into living tissues and soil or when exposed to infrared and ultraviolet light under field conditions. These changes often produce metabolites, which could be more toxic than the original pesticide. A pesticide residue analyst in these cases has to confirm the presence of the pesticide under investigation and/or its metabolite in the sample and determine their concentrations. Confirmation of the presence of specific pesticide and its degradation products can be achieved using GC/MSD and/or liquid–liquid/mass spectrometric (LC/MS) analysis. In GC/MS analysis, the presence of specific ion fragments can be used as a useful tool for identification of a pesticide and/or its main metabolites. The presence of clomazone (an herbicide) in soil was confirmed using GC/MSD clomazone spectral data, which showed spectral data with molecular ion peaks (M^+) at m/z 204, 125, 89, and 41, along with other characteristic fragment ion peaks (Fig. 4). The electron mass spectrum of clomazone, also known as dimethazone, indicated that the fragments of large intensity are the m/z 239 (dimethazone molecular ion), the ion m/z 204 (formed by the loss of the atom of chlorine), and the m/z 125 (formed by the breakage of the molecule of dimethazone at the carbon bound with nitrogen and the subsequent loss of the $-C_5H_8NO_2$ fragment ion).

MITIGATION OF ENVIRONMENTAL POLLUTION BY PESTICIDES

Based on research findings and the outlined objectives of this investigation, it could be concluded that some agricultural practices can be developed and applied to agricultural fields to reduce pesticides' dissipation and their impact on environmental quality. These practices could be summarized in the manner discussed in the following subsections.

Increase SOM Content

Adsorption of nonionic, non-polar hydrophobic compounds in soil results from weak attractive interactions such as van der Waals forces.[29] According to Sparks,[30] humic substances in the organic matter have an aromatic framework and contain polar groups, and may have both hydrophobic and hydrophilic sites. The hydrophobic sites might combine with non-polar compounds. There are several hydrophobic sites on SOM, which include fats, waxes, and resins and aliphatic side chains.[30] Bonding of non-polar pesticides to SOM is likely a pesticide–lipid interaction.[31] Since lipids are associated with soil humus (a sticky brown insoluble organic matter), pesticide adsorption by soils would depend on the organic matter content of soil. Accordingly, the increased organic matter in soil due to compost addition (compost-amended soil) plays an important role in the adsorption of non-polar pesticides such as bensulide (an herbicide) and pyrethrins (a group of insecticides). This management practice could be used

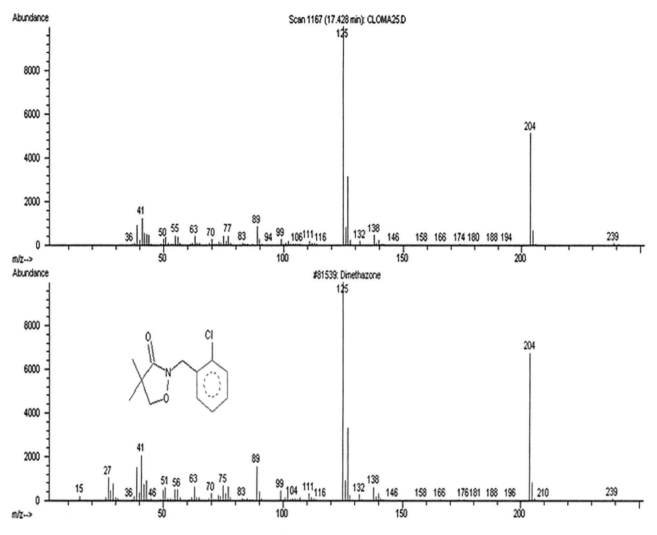

Fig. 4 Electron impact mass spectrum of clomazone (dimethazone, $C_{12}H_{14}ClNO_2$) extracted from soil indicating the molecular ions of m/z 204, 125, 89, and 41, along with characteristic fragment ions.

for bonding (trapping) non-polar pesticides, which may reduce surface water and groundwater contamination by pesticide residues.

Studies indicated that the concentration of pyrethrins and piperonyl butoxide (PBO), a pyrethrum synergist, adsorbed by compost-amended soil was significantly higher than that adsorbed by no-mulch unamended soil. Natural pyrethrins usually exist in two forms: pyrethrin-I and pyrethrin-II (Py-I and Py-II, respectively). Py-I adsorption was significantly higher than that of Py-II and PBO. The adsorption coefficient (K_d) calculated using the Freundlich equation ($q = K_d \, C^{1/n}$) indicated that K_d values of Py-I, Py-II, and PBO were higher (191, 75, and 55, respectively) in compost-amended soil than in no-mulch soil (83, 32, and 29, respectively).[22] These results confirm that the compost-amended soil adsorbed more Pys and PBO than the no-mulch soil. Adsorption of a typical crop protection chemical to soil occurs mainly by means of its organic matter content[32]; consequently, K_d values tend to increase with content of organic matter. Based on soil analysis, the organic matter content of compost-amended soil (5.72%) was nearly twice that of the no-mulch soil (2.77%). These results are in accordance with the work of Singh and Singh,[33] who reported higher adsorption of endosulfan (a non-polar insecticide) in natural soils compared to oxidized soils. Also, these results agree with Krohn[32] who reported that pesticide sorption on compost can be characterized by the K_d values, which tend to increase with increasing organic matter content. Thus, it is likely that SOM was primarily responsible for adsorption of Pys and PBO. The mechanism by which pesticides are retained by SOM may involve more than one type of interaction, and the exact nature of the interaction remains unknown. However, these results have obvious implications for pesticide bioavailability and transport. Increasing pesticide adsorption on SOM, due to compost addition, may decrease soil infiltration by pesticides and limit groundwater contamination by pesticides, but may also reduce the efficiency of the soil applied pesticides. This remains to be answered.

Enhance the Activity of Soil Microorganisms

Soil microorganisms (bacteria, fungi, protozoans, algae) excrete a variety of enzymes (ureases, invertases, dehydrogenases, cellulases, amylases, phosphatases) that have long been recognized as a primary means of degrading pesticides in soil and water ecosystems. Microorganisms also produce sticky substances (polysaccharides) that help soil particles adhere to one another and help the soil to resist erosion, which can diminish agriculture productivity.[34] In recent years, more specific emphasis has been given to soil enzymes in relation to reclamation management and the enzymatic processes that play a significant role in bioremediation. Work on soil enzymes[35–38] have provided detailed information on soil enzymatic and microbial population responses in soil in a series of ecosystems. Remediation of pesticide-contaminated soils is based on the degrading activity of soil microbiota and, therefore, remediation technologies should enhance the growth of native and/or introduced microorganisms in soils.

Addition of organic amendments such as yard waste compost,[35–38] straw,[39] tree leaf mulch,[40] and chipped wood from twigs[41] has been found to reduce the negative effects on soil microbial populations and soil enzyme activities due to the increased content of organic matter and its role in sorption processes as a result of the presence of humic substances containing carbonyl, carboxyl, phenolic, and alcohol functional groups.[42] Binding of organic pollutants to humic substances in compost protects microorganisms from the toxic effects of xenobiotics. Organic matter in soil has a great impact on the biological and biochemical properties of soil. Accordingly, soil enzymes could be tracked as indicators of soil quality following the addition of soil amendments. To judge the presence and activities of soil microorganisms in relation to new soil management practices, emphasis should be placed on soil enzyme activity throughout the growing season.

Install Slot-Mulch Biobed Systems against the Contour of the Agricultural Land Slope

Biobeds (or "biofilters"—a hole in the ground filled with a mixture of composted organic matter, topsoil, and a grass layer on top) provide a potential solution to pesticide contamination of surface waters arising from agricultural chemicals. The use of biobeds and adsorption techniques, proposed in this investigation, are unique ways of treating contaminated soil and agricultural runoff. The filling materials (mixture of straw, peat moss, and native soil) of biobeds have increased sorption capacity and microbial activity for degradation of pesticides.[43] The mechanism of biosorption process includes chemisorption, complexation, adsorption on surface, diffusion through the pores and ion exchange.[44] Biobeds were tested for their ability to retain and degrade chlorpyrifos (an insecticide), metalaxyl (a fungicide), and imazamox (an herbicide) using farm available materials (vine branch, citrus peel, urban waste, and green compost). Degradation of the three pesticides in biobeds was found to be faster than published values for degradation of these pesticides in soil. The half-life of all pesticides used was less than 14 days, compared to literature values of 60–70 days in soil.[45] Biobeds reduce concentration of sediment, so they might reduce the concentration of pesticides that are strongly sorbed to sediment. Adsorption increases the availability of the pesticide for soil degradation processes. In geneal, sorbed chemicals (adsorbed and absorbed) in labile sites are available for microbial transformations and bioremediation techniques. Microorganisms are capable of degrading both sorbed and bound residues.[46] These findings suggested that biobeds or biofilters could substantially reduce pesticide and heavy metal concentrations in agricultural runoff.

The system can be built on the farmland using locally available materials. The topsoil represents 25% of the overall mix and is the major source of microorganisms that act as the inoculum for the system that may receive high concentrations of relatively complex mixtures of pesticides in runoff. This developed methodology to mitigate the impact of pesticides on the ecosystem is urgently needed. The risk of groundwater contamination resulting from rapid leaching of highly soluble pesticides can be minimized through pesticide adsorption on the biobed filling materials.

Install Constructed Wetland Microcosms

A constructed wetland (CW) system pretreats wastewater by filtration, settling, and bacterial decomposition in a natural-looking lined marsh. CW systems have been used nationally and internationally with good results, but performance levels decrease in cold climates during winter. CWs may be described as soil/plant systems for wastewater treatment in which pollutant removal is based on general principles of nutrient tranfornmation. Transformations of organic compounds in a CW were carried out by sediment-borne microorganisms associated with the wetland community.[47] CWs could be used as a low-cost alternative method for controlling water pollution from both point and nonpoint sources.

Plants have been used for wastewater treatment, and their ability to remove xenobiotics like pesticides has been attributed to the microorganisms associated with their roots. Literature review indicated that CWs have been proposed for retaining agricultural pollutants as a potential best management practice (BMP) to mitigate effects of pesticide-associated agricultural runoff.[9] More than 60 reports on CWs indicated that CWs have been widely used to control both point- and nonpoint-source pollution by pesticides in surface waters. CW microcosm systems were effective for decreasing concentrations of chlorpyrifos and chlorothalonil in stimulated storm water runoff.[48] Results from several studies indicated that CW microcosms could be part of an efficient mitigation plan for treatment of pesticide mixtures in contaminated water. Use of aquatic plants has a great potential to function as in situ on-site biosinks and biofilters of aquatic pollutants. These plants can be used in phytoremediation because of their ability to degrade environmental chemicals via their exudates (enzymes) released to the contaminated sites.[49]

Use of Natural Products for Pest Control

In addition to environmental hazards associated with synthetic pesticides, synthetic pesticides have become increasingly expensive to develop. The present cost of discovery and development averages about $50–%100 million per pesticide. On the average, a company must synthesize and screen 35,000 compounds for each one registered and sold commercially. The time period from discovery to initial sales ranges from 5 to 9 years.[50]

Dried plants or their extracts have been used by farmers in many developing countries to protect food and fiber from insect damage. Chili pepper powder deterred oviposition of the onion fly, *Delia antiqua*.[51] Capsaicin in hot pepper has been reported to reduce larval growth of the spiny bollworm, *Earias insulana*,[52] and the use of oleoresin from capsicum as a repellent against cotton pests has been reported.[53] Water extracts of some hot pepper accessions (genotypes) were highly toxic to the cabbage looper larvae *Trichoplusia ni* (Hubner), the most difficult pest of crucifer crops to control during the past decade. Using GC/MS, Antonious and others[54] investigated the contents of fruit extracts that might explain the observed differences in toxicity and repellency to spider mite, *Tetranychus urticae* (Koch), among accessions. Three decanoic acid methyl esters (pentadecanoic acid methyl ester, hexadecanoic acid methyl esters, and octadecanoic acid methyl ester) (Fig. 5) predominated pepper fruit extracts. Methyl esters are aliphatic long-chain saturated fatty acids that are common components of plant lipids. The fatty acids from which they are derived, such as oleic acid and stearic acid, and their esters are not substances that one would expect to be carcinogenic. These same fatty acids, in the form of their esters with aliphatic alcohols, are components of natural waxes. Extracts from pepper fruits might provide an opportunity for use in crop protection as alternative to synthetic pesticides.

Published work on natural products also indicated that wild tomato leaf extracts could be explored as an alternative to synthetic pesticides.[55] Three methylketones (2-tridecanone, 2-dodecanone, and 2-undecanone) (Fig. 5) were effective against the tobacco hornworm, *Manduca sexta* L. 2-Tridecanone was the most effective methylketone against tobacco hornworm and tobacco budworm, *Heliothis virescens* (LC_{50} of 0.015 μ M cm^{-2}).[56] 2-Tridecanone also was effective against adults of the green peach aphid, *Myzus persicae* (LC_{50} of 0.07 μ M cm^{-2}), and required a significantly lower dose than 2-undecanone. 2-Tridecanone, which has an herbaceous spicy-like odor,[57]

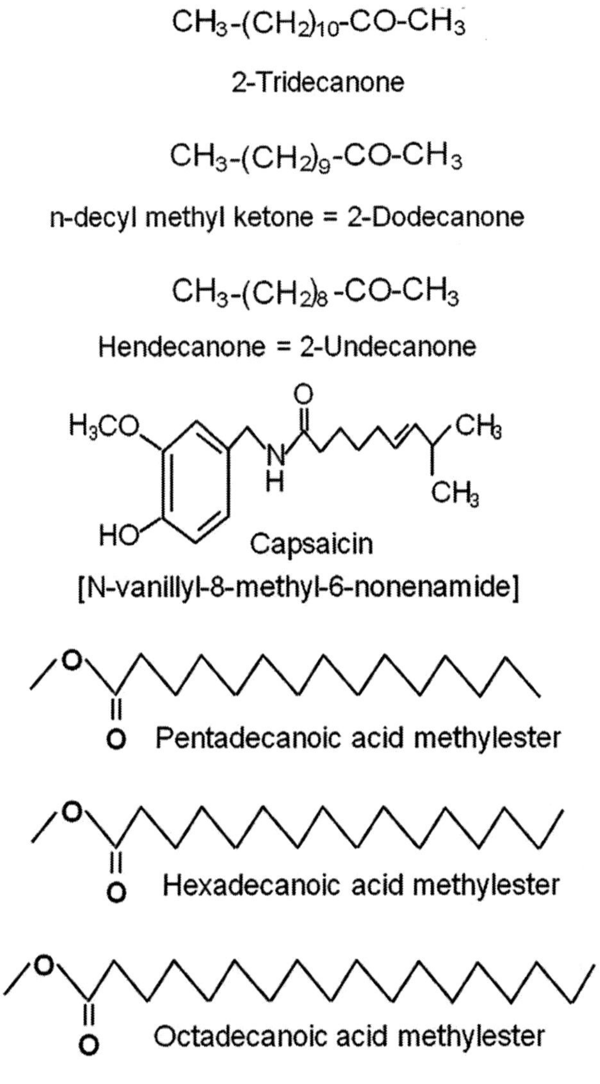

$CH_3-(CH_2)_{10}-CO-CH_3$
2-Tridecanone

$CH_3-(CH_2)_9-CO-CH_3$
n-decyl methyl ketone = 2-Dodecanone

$CH_3-(CH_2)_8-CO-CH_3$
Hendecanone = 2-Undecanone

Capsaicin
[N-vanillyl-8-methyl-6-nonenamide]

Pentadecanoic acid methylester

Hexadecanoic acid methylester

Octadecanoic acid methylester

Fig. 5 Chemical structures of natural products extracted from wild tomato leaves and hot pepper fruits that have the potential for developing a formulation of multipurpose insecticide for agricultural use.

was found toxic to a number of insect species by contact, ingestion, or vapor action[58] Maximum residue limits (MRLs) of 2-tridecanone on vegetables have not been established. The short persistence of 2-tridecanone on the leaves of the greenhouse vegetables tested so far[59] could be recognized as a desirable chemical characteristic. Ideally, safe pesticides remain in the target area long enough to control the specific pest and then degrade into harmless compounds. Such alternatives, which have few or no side effects on the environment, low toxicity to warm-blooded animals and humans, high efficacy against insects and spider mites, and lower potential for insect resistance development, are in great demand. For agrochemical companies, performance of methylketones as potential insecticidal and acaricidal products from wild tomato leaves can be explored for developing natural products for use as biodegradable alternative to synthetic pesticides. A formulation prepared from the leaves of selected accessions may therefore create a mixture with the desired level of constituents. However, if methylketones and other constituents in wild tomato leaves are to be used within the conventional and organic production systems to control vegetable insects and spider mites, further work is needed to investigate their performance under field conditions and the impact of methylketones on natural enemies. Combination of more than one active ingredient such as 2-tridecanone and capsaicin in one formulation of botanical insecticides has the advantage of providing novel modes of action against a wide variety of insects. The risk of cross resistance will be reduced because insects will have difficulty adapting to a diverse group of bioactive compounds and fewer pesticide applications will be required.

REDUCING THE IMPACT OF PESTICIDES ON ENVIRONMENTAL QUALITY

The following field studies are some of the approaches developed by the Kentucky State University (KSU)/Water Quality and Environmental Toxicology Research Program to quantify and mitigate pesticide residues in soil, runoff, and infiltration water following pesticide spraying in agricultural fields amended with MSS (a source of organic matter).

The field trial area was conducted on a Lowell silty loam soil (pH 6.7, 2% organic matter) of 10% slope located at the Kentucky River Watershed in the Blue Grass Region. Eighteen uniform field plots (3.7 m wide and 22 m long) each with metal borders along each side were established (Fig. 6). At the bottom of each plot, a tipping-bucket ($n = 18$) runoff metering apparatus was installed for collecting and measuring runoff. Pan lysimeters ($n = 18$) were installed for collecting infiltration water following natural rainfall events. Plots were planted with bell pepper (year 1) and broccoli (year 2) as two aboveground crops and with potato (year 3) and sweet potato (year 4) as two underground crops. Plant rows in all plots were oriented against the contour of the land slope. Herbicides (napropamide, trifluralin, clomazone, and bensulide) were sprayed according to Kentucky label recommendations.[8]

The soil treatments (years 2005 to 2008) were as follows: 1) soil mixed with sewage sludge (Louisville Green) obtained from Metropolitan Sewer District, Louisville, Kentucky, and mixed with yard waste obtained from Con Robinson Co., Lexington, Kentucky, at 15 t acre^{-1} (1 acre = 0.405 km^2) on dry weight basis (treatment 1); 2) sewage sludge mixed with native soil (treatment 2); and 3) no-mulch rototilled bare soil (treatment 3) was used for comparison purposes. The soil amendments were mixed with lime at 1% on dry weight basis to reduce availability of heavy metals[60] and then rototilled with the upper 15 cm topsoil.

Fig. 6 Runoff field plots designed at KSU Water Quality/Environmental Toxicology Research Farm showing aboveground metal borders to prevent cross contamination between adjacent treatments.

Runoff (soil–water suspension) under natural rainfall conditions and/or irrigation was collected and quantified at the lower end of each plot throughout the year using tipping-bucket runoff metering apparatus (Fig. 7). A gutter was installed across the lower end of each plot with 5% slope to direct runoff towards tipping bucket runoff metering apparatus. Each of the 18 tipping buckets was calibrated (one tip represents 3 L of runoff) and maintained to provide precise measure of amount of runoff per tip. Numbers of tips were counted using mechanical runoff counters. Collection

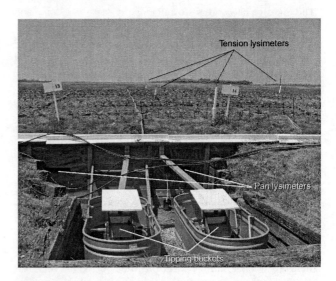

Fig. 7 Runoff tipping buckets installed at KSU Research Farm for collecting runoff water and pan lysimeters for collecting infiltration water following rainfall events. A gutter was installed across the lower end of each plot with 5% slope to direct runoff towards tipping-bucket runoff metering apparatus. Note that each bay has two tipping buckets, two pan lysimeters, and one biobed system.

of samples was carried out in 3.79 L borosilicate glass bottles through a flow-restricted composite collection system (approximately 40 mL per tip were collected).

Eighteen pan lysimeters and 27 tension lysimeters (Fig. 7) were used to monitor pesticide residues and/or their main metabolites in the vadose zone (the unsaturated water layer below the plant root). Water percolated through the vadose zone was collected. The pan lysimeters (4 ft^2 each) were tunnel installed, leaving the soil column above it intact. This allowed the collection of infiltration water under normal field conditions (zero tension). Suction lysimeters (Model 1920, Soil Moisture Equipment Corporation, Santa Barbara, California, USA) were also installed in three experimental plots to monitor leaching (seeping) of pesticides towards the vadose zone. Suction lysimeters ($n = 27$) were installed according to the manufacturer's recommendations at depths of 0.3, 0.6, and 1.5 m, with three lysimeters at the top, middle, and bottom of the plot. Prior to sampling, vacuum 30 psi (1 psi = 6.89 kPa) was applied into each tension lysimeter using a vacuum/pressure pump (Pressure Pump Model C, Soil Moisture Company, Santa Barbara, California, USA). Leachate samples were collected during each experimental period in borosilicate amber bottles.

Biobed systems (3.7 × 3 × 1.5 m^3) were installed in the ground down the land slope at the end of each of nine runoff plots. The soil was replaced by excavation and the hole in the ground was filled with a 10 cm layer of gravel at the bottom and then filled with 1.5 m of a mixture of 50% chopped wheat straw, 25% peat moss, and 25% topsoil, and then with turf on top. The hole was covered with a tall fescue (*Festuca* sp., Kentucky 31) grass layer to maintain a right level of temperature for microbial activity (Fig. 8). The mixture was composted in uncovered heaps, outside in open air, for 2 months prior to use. The heaps were turned twice throughout this period. The microbial biomass of the mixture in the heap was monitored using the methods described by Antonious[38] to give an indication of microbial proliferation and activity in each biobed.

Biobeds ("biofilters") provide a potential solution to pesticide contamination of surface waters arising from agricultural chemicals. Biobeds were built at the end of runoff field plots at KSU Research Farm using locally available materials (chopped wheat straw, peat moss, and topsoil) in volumetric proportions of 2:1:1, respectively. Peat and straw provide numerous sites for pesticide sorption. They help maintain aerobic conditions combined with sufficient humidity or moisture owing to its high water-holding capacity. Straw acts as an additional food source for lignin-degrading microorganisms that produce enzymes catalyzing the degradation of a broad spectrum of chemicals.[61] The topsoil acts as an inoculum of microorganisms for the system and is likely to vary in terms of its physical, chemical, and biological characteristics from one farm to another. Repeated use of certain pesticides over a number of seasons can result in enhanced degradation due to adaptation and proliferation of specific soil microbial communities that utilize the compound as an energy source and thus degrade it more easily.[62–64] The turf layer ensures that there is good rooting activity and assist in moisture management, through evapotranspiration of water to the atmosphere.[65] In addition, biobeds were drip irrigated as supplement to precipitation to achieve average monthly precipitation level and secure an even application of water on the biobed surface.

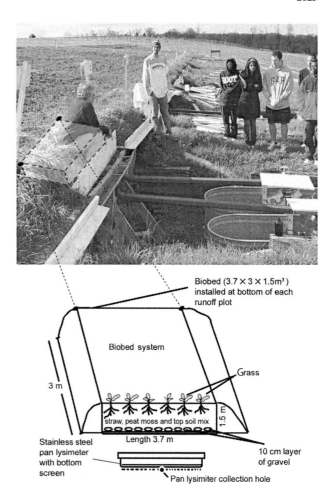

Fig. 8 Schematic diagram of a slot-mulch "biobed" system installed at the end of runoff plots (KSU Research Farm) down the land slope for trapping pesticides in runoff before they enter rivers and streams. Note that a pan lysimeter was installed at the bottom of each biobed to collect infiltration water and monitor pesticide concentration in seepage water.

RESULTS OF KSU MITIGATION STUDIES

Residues of clomazone, napropamide, trifluralin, and bensulide (Fig. 9) were determined in soil and water samples according to the recommended methods of analysis described in the pesticide manual.[20] Decline of clomazone residues in the top 15 cm native soil and soil incorporated with amendments revealed half-life ($T_{1/2}$) values of 18.8, 25.1, and 43 days in MSS mixed with yard

Clomazone

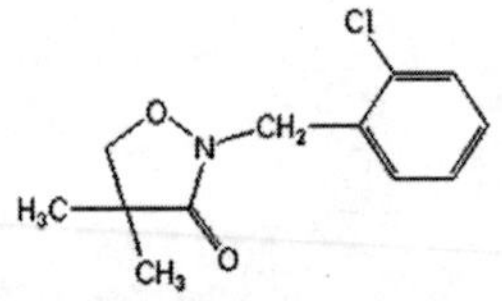

2-[(2-chlorophenyl) methyl]-4, 4-dimethyl-3-isoxazolidinone

Trifluralin

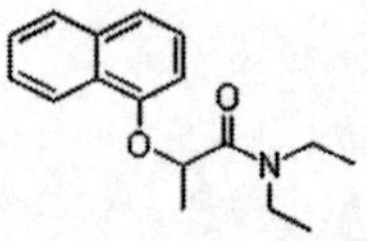

(2,6-dinitro-N, N-dipropyl-4-(trifluoromethyl) benzenamine

Napropamide

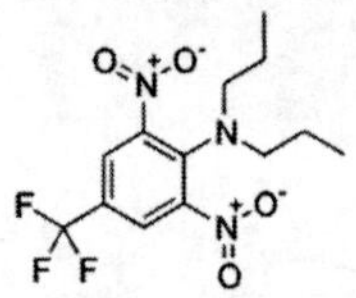

[N, N-diethyl-2-(1-naphthyloxy) propionamide]

Bensulide

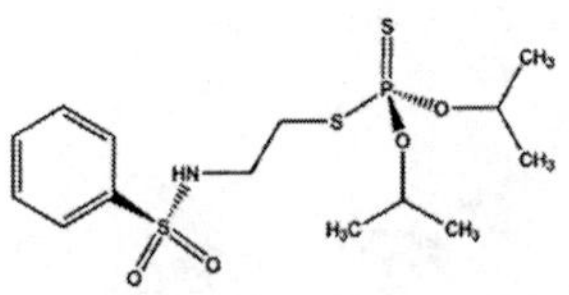

[O, O-diisopropyl S-2-phenylsulfonylaminoethyl Phosphorodithioate]

Fig. 9 Chemical structures of four herbicides: clomazone, napropamide, trifluralin, and bensulide.

waste compost (MSS + YW), MSS, and no-mulch (NM) treatments, respectively. Organic matter possibly caused variation in sorption mechanisms and kinetics in the MSS + YW-amended soil. MSS is rich in organic matter and mineral nutrients. SOM has an important effect on the bioavailability, persistence, biodegradability, leaching, and volatility of pesticides. When pesticides are tightly bound to soil particles, they do not pose a threat to surface water or groundwater. Infiltration water ranged from 40 L $acre^{-1}$ in NM soil to 87 L $acre^{-1}$ (117% increase) in MSS + YW-amended soil. Bensulide residues were 10.3, 2.0, and 3.5 μg L^{-1} in runoff water and 4.0, 1.8, and 0 (not detctable) in infiltration water collected from NM, MSS + YW mix, and MSS, respecively. This decrease might be due to bensulide environmental fate characteristics (i.e., high sorption capacity) that increased its adsorption to the soil particles and the greater organic matter content of MSS + YW treatment (8.1%), which might have increased bensulide binding to humic substances, reducing its mobility in the soil column and down the land slope in surface water. Accordingly, this investigation provides evidence of the low potential leaching of bensulide towards the vadose zone and reduced risk of groundwater and surface water contamination by pesticides. Soil management practices that reduce pesticides and pesticides in runoff are vital to sustainable crop production and environmental quality.

Residues of trifluralin were significantly higher in MSS treatments compared to the other two soil treatments. In addition to the greater organic matter in MSS, adsorption of trifluralin could be attributed also to the differences in elemental composition. High concentrations of Ca, Cu, and Zn were found in soil, where MSS was applied.[66] Regardless of mechanism of retention, greater retention in sludge-amended soil resulted in lower concentrations of trifluralin in runoff water and might have reduced pesticide leaching into the vadose zone. Similarly, napropamide residues were significantly higher in runoff water from NM soil compared to yard waste and sewage sludge treatments. A substantial amount of runoff was retarded by the two soil amendments (sewage sludge and yard waste compost) that would otherwise have been transported into streams and rivers. MSS contains high amounts of microorganisms and enzymatic substrates. These easily available substrates stimulate microbial growth and enzyme production. Enzymes excreted by soil microorganisms could transform toxic compounds into other, less toxic forms and/or retain them in the organic matrix and therefore reduce their mobility and bioavailability.

CONCLUSION

Contaminated runoff from farmland contributes a significant proportion of the pesticide load released to surface waters. Concerns about environmental pollution by pesticides usually involve two sides, the environment and the end user. To protect the environment, the general trend is to use reduced levels of active ingredients. This trend creates a need for pesticide formulations with improved efficacy at low application rates. To protect the end user, environmentally safe formulations that eliminate organic-solvent-based formulations are needed. Farmers are expected to meet the food, feed, and fiber needs of growing human populations as well as the demands of diverse consumer groups, while preserving ecosystems, health, and biodiversity. This requires modern, highly effective plant protection products. These products must be safe for the environment and wildlife, and safe for all who consume the food. Farmers are in need of insect pest management strategies that are effective, affordable, and environmentally sound. Soil management practices that reduce pesticide persistence in soil and runoff water are vital to sustainable crop production and environmental quality. A central hope in these concerns is the safe use of agrochemicals, development of new soil management practices, and use of mitigation techniques. Additional technological and infrastructural solutions are needed to reduce pesticide releases and protect the environment and human health. For agricultural purposes, these technologies and systems need to be cheap, reliable, and easy to use with low labor and time input.

The simultaneous use of soil conditioners to enhance soil physical, chemical, and microbial conditions could also enhance soil bioremediation. SOM has an important effect on the bioavailability, persistence, biodegradability, leaching, and volatility of pesticides. In fact, SOM is the soil component most important in pesticide retention. When pesticides are tightly bound to soil particles, they do not pose a threat to surface water or groundwater. The application of organic amendments to soil, such as MSS, could be exploited as an environmental and agricultural practice to maintain and/or increase SOM contents, thus reclaiming degraded soils and supply plant nutrients. This practice affects the carbon content of soil and the chemical nature and function of native soil humic acids that carry a negative pH-dependent charge, derived mainly from the range of dissociated carboxyl and phenol groups, and consequently affects substaintially the fate of trace metal cations and organic contaminants, such as pesticides in soils. Little information, however, exists on the economic aspects of sludge application to agronomic crops. This investigation revealed that increasing SOM by the addition of soil amendment, such as MSS treated with lime and mixed with YW (MSS + YW), reduced the potential transport of bensulide, clomazone, Treflan, and napropamide residues down the land slope into runoff water and might be a suitable technique to restore or improve soil quality. Pesticides with high persistence and a strong sorption rate are likely to remain near the soil surface, increasing the chances of being carried to a stream via surface runoff. On the contrary, pesticides with high persistence and a weak sorption coefficient may be readily leached through the soil and are more likely to contaminate groundwater. Herbicides having high K_{OC} value, i.e., bensulide, will bind to soil and organic matter particles. This information is typically useful in ranking pesticides as leachers or non-leachers to assess pesticide potential for off-site surface or subsurface movements under field conditions. Pesticide adsorption to soil is related more to SOM than to other soil chemical and physical properties; therefore, addition of soil amendments having high organic matter content is a management practice that should be exploited to trap nonionic pesticides and reduce their surface and subsurface mobility under field conditions. In addition, the use of sewage sludge for land farming could decrease dependence on synthetic fertilizers and provide alternatives to farmers dealing with the sharply escalating production costs associated with increasing costs of energy and fertilizers. New technological changes in soil, waste management, and crop production practices are needed for meeting the challenge of conservation, remediation, and environmental goals. There is probably nothing more disturbing than the threat of widespread pollution of surface water and groundwater by pesticides. Therefore, it is imperative to develop bioremediation techniques that could treat and stabilize environmental contaminants in situ in an efficient and cost-effective manner.

Studies at KSU research farm indicated that biobeds installed down the land slope reduced runoff water volume by 44% and trifluralin and clomazone residues in runoff water by 55% and 38%, respectively, compared to control treatments. These results revealed that biobeds offer a viable means of treating pesticides in runoff water arising from agricultural fields. Studies have also shown that while less mobile pesticides like trifluralin are effectively retained within the biobed matrix, the more mobile pesticides could leach from the biobed. Other investigators[67] reported that total amounts of isoproturon (an herbicide) leached from the soil were 1947 mg compared to 32 mg leached from the biobed system.

ACKNOWLEDGMENTS

I would like to thank Eric Turley, Regina Hill, and the KSU farm crew for maintaining the field plots. This investigation was supported by two grants from USDA/CSREES to KSU under agreement nos. KYX-10-08-43P and KYX-2006-1587.

REFERENCES

1. US EPA, available at http://www.epa.gov/oppbead1/pestsales/01pestsales/usage2001.html Pesticide sales and usage report, 2004 (accessed November 10, 2010).
2. US EPA, available at http://oaspub.epa.gov/waters/national_rept.control#TOP_IMP. 2004 (accessed February 8, 2008).
3. De Wilde, T.; Spanoghe, P.; Debaer, C.; Ryckeboer, J.; Springael D.; Jaeken, P. Overview of on-farm bioremediation systems to reduce the occurrence of point source contamination. Pest Manage. Sci. **2007**, *63* (2), 111–128.
4. Pimentel, D.; Andow, D.; Dyson-Hudson, R.; Gallahan, D.; Jacobson, S.; Irish, M.; Kroop, S.; Moss, A.; Schreiner, I.; Shepard, M.; Thompson, T.; Vinzant, B. Environmental and economic impacts of reducing U.S. agricultural pesticide us. In *Handbook of Pest Management in Agriculture*; Pimentel, D., Ed.; CRC Press: Boca Raton, FL, 1991; Vol. 1, 679–718.
5. Cunningham, W.P.; Cunningham, M.A.; Saigo, B.W. *Environmental Science: A Global Conern*, 9th Ed.; New York: McGraw-Hill Higher Education, 2007; 210–216.
6. USGS. The loads of selected herbicides in the Ohio river basin. U.S. Geological Survey Fact Sheet 089-02; November 2002, available at http://ky.water.usgs.gov/pubs/FAC_08902.htm (accessed November 10, 2010).
7. Pimentel, D.; Hepperly, P.; Hanson, J.; Douds, D.; Seidel, R. Environmental, energetic, and economic comparisons of organic and conventional farming systems. BioScience **2005**, *55* (7), 573–582.
8. Anonymous. *Vegetable Production Guide for Commercial Growers*; Cooperative Extension Service: University of

Kentucky, College of Agriculture, Lexington, KY 40546, 2008; ID 36, 1–72.
9. Moore, M.T.; Schulz, R.; Cooper, C.M.; Smith, S.; Rodgers, J.H. Mitigation of chlorpyrifos runoff using constructed wetlands. Chemosphere **2002**, *46* (6), 827–835.
10. Antonious, G.F. Clomazone residues in soil and runoff: Measurement and mitigation. Bull. Environ. Contam. Toxicol. **2000**, *64* (2), 168–175.
11. Antonious, G.F. Efficiency of grass buffer strips and cropping system on off-site dacthal movement. Bull. Environ. Contam. Toxicol. **1999**, *63* (1), 25–32.
12. Antonious, G.F.; Byers, M.E. Fate and movement of endosulfan under field conditions. J. Environ. Toxicol. Chem. **1997**, *16* (4), 644–649.
13. Chakrabarti, K.; Bhattacharya, P.; Chakraborty, A. Effects of metal-contaminated organic wastes on microbial biomass and activities. In *Heavy Metal Contamination of Soil*; Ahmed, I., Hayat, S., Pichtel, J., Eds.; Since Publishers, Inc.: Plymouth, U.K., 2005; 195–204.
14. Antonious, G.F.; Turley, E.T.; Snyder, J.C. Soil enzyme activity and heavy metal contaminations in soil amended with sewage sludge. In *Environmental Engineering and Management*; Theophanides, M., Theophanides, T., Eds.; Athens Institute for Education and Research: Kolonaki, Athens, Greece, 2009, 153–169.
15. Liao, M.; Xie, X.M. Effect of heavy metals on substrate utilization pattern, biomass, and activity of microbial communities. Ecotoxicol. Environ. Saf. **2007**, *66* (2), 217–223.
16. Ware, G.W.; Whitacre, D.M. Pesticides: Chemical and biological tools. In *The Pesticide Book*, 6th Ed.; MeisterPro Information Resources: Willoughby, OH, 2004; 8.
17. Ray, C.; Soong, T.W.; Lian, Y.Q.; Roadcap, G.S. Effect of flood-induced chemical load on filtrate quality at bank filtration sites. J. Hydrol. **2002**, *266* (3–4), 235–258.
18. Schiff, K.; Sutula, M. Organophosphorus pesticides in stormwater runoff from Southern California. Environ. Toxicol. Chem. **2004**, *23* (8), 1815–1821.
19. Teifenthaler, L.L.; Schiff, K.; Leecaster, M. Temporal variability of patterns of stormwater concentrations in urban runoff. In *Southern California Coastal Water Research Annual Report 1999–2000*; Weisberg, S., Elmore, D., Eds.; Southern California Coastal Water Research Project: Westminster, CA, 2001; 52–62, available at http://ftpillftp.sccwrp.org/pub/download/DOCUMENTS/Annual/Reports/1999Annual/Report/04-ar04.pdf (accessed November 22, 2010).
20. Tomlin, C.D.S., Ed. *The Pesticide Manual*, 14th Ed.; British Crop Protection Council: Farnham, U.K., 2005.
21. Montgomery, J.H., Ed. *Agrochemical Desk Reference of Environmental Data*; Lewis Publishers: Michigan, USA, 1993.
22. Antonious, G.F.; Patel, G.A.; Snyder, J.C.; Coyne, M.S. Pyrethrins and piperonyl butoxide adsorption to soil organic matter. J. Environ. Sci. Health **2004**, *B39* (1), 19–32.
23. Wauchope, R.D.; Yeh, S.; Linders, J.; Kloskowski, R.; Tanaka, K.; Rubin, B.; Katayama, A.; Kordel, W.; Gerstl, Z.; Lane, M.; Unsworth, J. Pesticide soil sorption parameters: Theory, measurement, uses, limitations and reliability. Pest Manage. Sci. **2002**, *58* (5), 419–445.
24. Haith, D.A.; Rossi, F.S. Risk assessment of pesticide runoff from turf. J. Environ. Qual. **2003**, *32* (2), 447–455.
25. Sharma, S.D.; Singh, M. Susceptibility of Florida Candler fine soil to herbicide leaching. Bull. Environ. Contam. Toxicol. **2001**, *67* (4), 594–600.
26. Hornsby, A.G. How contaminants reach groundwater. University of Florida, cooperative Extension Service, Institute of Food and Agricultural Science, SL143, March 1999, available at http://edis.ifas.ufl.edu/ss194 (accessed November 20, 2010).
27. Lair, G.J.; Gerzabek, M.H.; Haberhauer, G.; Jakusch, M.; Kirchmann, H. Response of the sorption behavior of Cu, Cd, and Zn to different soil management. J. Plant Nutr. Soil Sci. **2006**, *169* (1), 60–68.
28. Cohen, J.M.; Kamphake, L.J.; Lemke, A.E.; Henderson, C.; Woodward, R.L. Effects of fish poisons on water supplies. Part 1. Removal of toxic materials. J. Am. Water Works Assoc. **1960**, *52* (12), 1551–1566.
29. Koskinen, W.C.; Harper, S.S. The retention process: Mechanisms. In *Pesticides in the Soil Environment: Processes, Impacts, and Modeling*; Cheng H.H., Ed.; Soil Science Society of America: Madison, WI, 1990; 530.
30. Sparks, D.L. Chemistry of soil organic matter. In *Environmental Soil Chemistry*; Academic Press: San Diego, CA, 1995; 53–79.
31. Pierce, R.H.; Olney, C.E.; Felbeck, G.T. Pesticide adsorption in soils and sediments. Environ. Lett. **1971**, *1* (2), 157–172.
32. Krohn, J. Behavior of thiacloprid in the environment. Pflanzenschutz-Nachr. Bayer **2001**, *54* (2), 281–290.
33. Singh, R.P.; Singh, D. Influence of soil properties on the adsorption of endosulfan on two soils at fixed volume fraction of methanol. J. Indian Soc. Soil Sci. **1998**, *46* (2), 217–223.
34. Reganold, J.P.; Papendick, R.I.; Parr, J.F. Sustainable agriculture. Sci. Am. **1990**, *262* (6), 112–120.
35. Antonious, G.F. Impact of soil management practice and two botanical insecticides on urease and invertase activity. J. Environ. Sci. Health **2003**, *B38* (4), 479–488.
36. Bandick, A.K.; Dick, R.P. Field management effects on soil enzyme activities. Soil Biol. Biochem. **1999**, *31* (11), 1471–1479.
37. Kunito, T.; Saeki K.; Goto, S.; Hayashi, H.; Oyaizu, H.; Matsumoto, S. Copper and zinc fractions affecting microorganisms in long-term sludge-amended soils. Bioresour. Technol. **2001**, *79* (2), 135–146.
38. Antonious, G.F. Enzyme activities and heavy metals concentration in soil amended with sewage sludge. J. Environ. Sci. Health **2009**, *A44* (10), 1019–1024.
39. Kucharski, J.; Jastrzebska, E.; Wyszkowska, J.; Hlasko, A. Effect of pollution with diesel oil and leaded petrol on enzymatic activity of the soil. Zesz. Probl. Postepow Nauk Roln. **2000**, *472* (2), 457–464.
40. Acosta-Martinez, V.; Reicher, Z.; Bischoff, M.; Turco, R.F. The role of tree leaf mulch and nitrogen fertilizer on turfgrass soil quality. Biol. Fertil. Soils **1999**, *29* (1), 55–61.
41. Lalande, R.; Furlan, V.; Angers, D.A.; Lemieux, G. Soil improvement following addition of chipped wood from twigs. Am. J. Altern. Agric. **1998**, *13* (3), 132–137.
42. Datta, A.; Sanyal, S.K.; Saha, S. A study on natural and synthetic humic acids and their complexing ability towards cadmium. Plant Soil **2001**, *235* (1), 115–125.

43. Fogg, P.; Boxall, A.; Walker, A; Jukes, A. Degradation and leaching potential of pesticides in biobed systems. Pest Manage. Sci. **2004**, *60* (7), 645–654.
44. Sud, D.; Mahajan, G.; Kaur, M.P. Agricultural waste material as adsorbent for sequestering heavy metal ions. Bioresour. Technol. **2008**, *99* (14), 6017–6027.
45. Vischetti, C.; Capri, E.; Trevisan, M.; Casucci, C.; Perucci, P. Biomassbed: A biological system to reduce pesticide point contamination at farm level. Chemosphere **2004**, *55* (6), 823–828.
46. Novak, J.M.; Jayachandran, K; Moorman, T.B.; Weber, J.B. Sorption and binding of organic compounds in soils and their relation to bioabailability. Soil Science Society of America, American Society of Agronomy, Crop Science Society, 1995, 13–3, SSSA Special Publication 43, available at http://afrsweb.usda.gov/SP2UserFiles/Place/66570000/Manuscripts/1995/Man381.pdf (accessed November 22, 2010).
47. Runes, H.B.; Jenkins, J.J.; Bottomley, P.J. Atrazine degradation by bioaugmented sediment from constructed wetlands. Appl. Microbiol. Biotechnol. **2001**, *57* (3), 427–432.
48. Sherrard, R.M.; Bearr, J.S.; Murray-Gulde, C.L.; Rodgers, J.H.; Shah, Y.T. Feasibility of constructed wetlands for removing chlorothalonil and chlorpyrifos from aqueous mixtures. Environ. Pollut. **2004**, *127* (3), 385–394.
49. Turgut, C. Uptake and modeling of pesticides by roots and shoots of Parrotfeather (*Myriophyllum aquaticum*). Environ. Sci. Pollut. Res. **2005**, *12* (6), 342–346.
50. Yu, S.J. *The Toxicity and Chemistry of Pesticides*; CRC Press: Taylor and Francis Group, Boca Raton, FL, 2008; 276.
51. Cowles, R.S.; Keller, J.E.; Miller, J.R. Pungent spices, ground red pepper, and synthetic capsaicin as onion fly ovipositional deterrents. J. Chem. Ecol. **1989**, *15* (2), 719–730.
52. Weissenberg, M.; Klein, M.; Meisner, J.; Sscher, K.R.S. Larval growth inhibition of the spiny bollworm, *Earias insulana*, by some steroidal secondary plant compounds. Entomol. Exp. Appl. **1986**, *42* (3), 213–217.
53. Mayeux, J.V. Hot shot insect repellent: An adjuvant for insect control. In *Proceedings of Beltwide Cotton Conferences*, Nashville, TN, USA, January, 9–12, 1996; Vol. 1, 35.
54. Antonious, G.F.; Meyer, J.; Rogers, J.; Hyeon-HU, Y. Growing hot pepper for cabbage looper (*Trichoplusia ni* Hübner) and spider mite (*Tetranychus urticae* Koch) control. J. Environ. Sci. Health **2007**, *B42* (5), 559–567.
55. Antonious, G.F.; Snyder, J.C. Tomato leaf crude extracts for insects and spider mite control. In *Tomatoes and Tomato Products: Nutritional, Medicinal and Therapeutic Properties*; Preedy, V.R., Watson, R.R., Eds., Dept. of Nutrition and Dietetics, King's College: London, 2008; 269–297.
56. Antonious, G.F.; Dahlman, D.L.; Hawkins, L.M. Insecticidal and acaricidal performance of methylketones in wild tomato leaves. Bull. Environ. Contam. Toxicol. **2003**, *71* (2), 400–407.
57. Vernin, G.; Vernin, C.; Pieribattesti, J.C.; Roque, C. Analysis of volatile compounds of *Psidium cattleianum* Sabine fruit from Reunion island. J. Essential Oil Res. **1998**, *10* (4), 353–362.
58. Muigai, S.G.; Schuster, D.J.; Snyder, J.C.; Scott, J.W.; Bassett, J.M.; McAuslane, H.J. Mechanisms of resistance in *Lycopersicon* germplasm to the whitefly *Bemisia argentifolii*. Phytoparasitica **2002**, *30* (4), 347–360.
59. Antonious, G.F. Persistence of 2-tridecanone on the leaves of seven vegetables. Bull. Environ. Contam. Toxicol. **2004**, *73* (6), 1086–1093.
60. Wong, W.T.; Selvam, A. Speciation of heavy metals during co-composting of sewage sludge with lime. Chemosphere **2006**, *63* (6), 980–986.
61. Bumpus, J.A. White rot fungi and their potential use in soil bioremediation process. In *Soil Biochemistry*; Bollag J.M., Stotsky, G., Eds.; Marcel Dekker, Inc.: New York, 1993; Vol. 8, 65–100.
62. Kirkland, K.; Fryer, J.D. Degradation of several herbicides in a soil previously treated with MCPA. Weed Res. **1972**, *12* (1), 90–95.
63. Torstensson, N.T.L.; Stark, J.; Gorannson, B. Effect of repeated application of 2,4-D and MCPA on their breakdown in soil. Weed Res. **1975**, *15* (3), 159–164.
64. Torstensson, L. Experiences of biobeds in practical use in Sweden. Pestic. Outlook **2000**, *13*, 206–211.
65. Anonymous. The Voluntary Initiative: Promoting responsible pesticide use, available at http://www.voluntaryinitiative.org.uk/ (accessed November 10, 2010).
66. Antonious, G.F.; Snyder, J.C. Impact of soil incorporated sewage sludge on herbicide and trace metal mobility in the environment. In *Environmental Engineering and Economics*; Theophanides, M., Ed.; Greece Institute for Education and Research: Athens, Greece, 2006; 149–164.
67. Fogg, P.; Boxall, A.B.; Walker, A.; Jukes, A. Degradation and leaching potential of pesticides in biobed systems. Pest Manage. Sci. **2004**, *60* (7), 645–654.

Pesticides: Natural

Franck Dayan
National Center for Natural Products Research, Agricultural Research Service (USDA-ARS), U.S. Department of Agriculture, University, Missouri, U.S.A.

Abstract
Newer and more stringent pesticide registration procedures are affecting the number of chemical tools available to farmers. Natural product-based pesticides, along with novel chemistry, are being developed to replace the compounds lost due to the new registration requirements. This entry highlights the historical use of natural products in agricultural practices and the impact of natural products on the development of new pesticides.

INTRODUCTION

The chemical pest control paradigm has dominated modern agricultural practices since its introduction about 60 years ago. It has been one of the key components accompanying the "green" revolution that resulted in tremendous increases in crop yield during the past 50 years. However, newer and more stringent pesticide registration procedures, such as the Food Quality Protection Act in the United States, are affecting the number of chemical tools, especially insecticides, available to farmers. Natural product-based pesticides, along with novel chemistry, are being developed to replace the compounds lost due to the new registration requirements. This review highlights the historical use of natural products in agricultural practices and the impact of natural products on the development of new pesticides. Readers interested in more detail are referred to the following reviews.[1,2]

HISTORICAL USE OF NATURAL PRODUCTS FOR PEST MANAGEMENT

Early agricultural practices relied heavily on crop rotation or mixed crop planting to optimize natural pest control (such as predation, parasitism, and competition) and required a significant amount of manual pest management. The concept of "natural pesticides" is not new. Control of caterpillar infestations in citrus orchards by introducing colonies of ants is reported around 300 BC in ancient Chinese literature, and the use of beneficial insects to control other insect pests was mentioned by Linnaeus in 1752.

The advent of monoculture and intensive agricultural practices, though beneficial in that it allows great increases in yields, is often accompanied by a more pernicious pest problem that has mostly been addressed by the use of synthetic pesticides. In the past few decades, the concept of integrated pest management has become more widely accepted, combining pest population monitoring, mechanical agricultural practices, biocontrol, and natural bioactive compounds to rely less heavily on the indiscriminate use of synthetic pesticides.

STRUCTURAL DIVERSITY IN NATURE

The usually complex carbon skeleton of natural products derived from secondary metabolism is the result of natural selection of molecules that provided some protection against specific biotic challenges. Nature has, in a sense, performed a "high throughput" screen over long period of time to select particularly suitable biologically active compounds. The "high throughput" refers not to the rapidity of the selection, but rather to the innumerable permutations of relatively complex structures that have been made. Structural diversity that resulted from this has been, and still remains, an invaluable source of lead compounds in developing novel agrochemical (and pharmaceutical) products.

Important benefits of natural product-based pesticides, such as the absence of "unnatural" ring structures and the presence of few heavy atoms, are their short environmental half-lives and their tendency to affect novel target sites. This latter fact is particularly important since the need for new modes of action is so pressing. Agrochemical companies are actively seeking novel mechanisms of action for which they can develop new chemistry.

INSECTICIDES DERIVED FROM NATURE

Pyrethrins

Historically, the greatest impact of natural products on pesticide development has been in the area of insecticides. The insecticidal properties of *Chrysanthemum* flower head powder was known for a long time and the discovery of the bioactive pyrethrin component led to the successful development of numerous pyrethroid insecticides (Fig. 1). Natural pyrethrins were quite unstable, being rapidly degraded in the presence of light and oxygen. Several synthetic

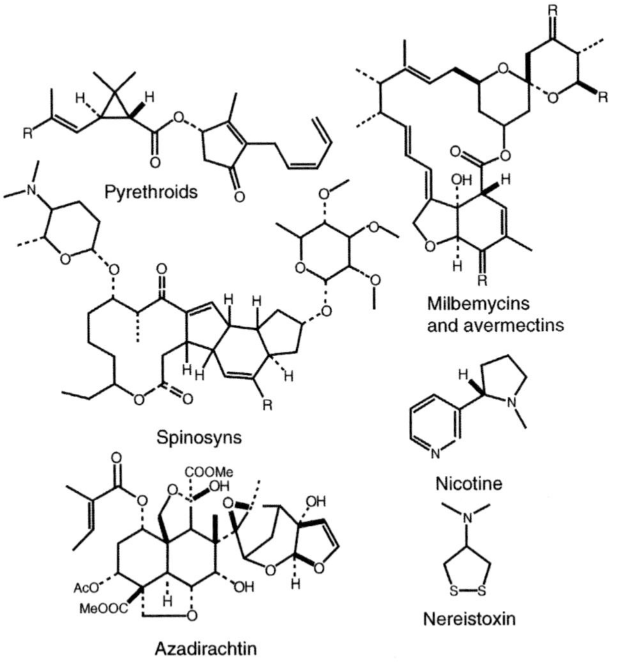

Fig. 1 General structures of natural products with insecticidal activity mentioned in the text.

programs were initiated to generate more stable commercial products. Initial progress was slow but key synthetic analogs with improved stability and selectivity were eventually discovered. The insecticidal activity of pyrethroids is associated with their ability to cause prolonged opening of sodium-ion channels in nerve membranes resulting in a lethal blockage of the nerve signal system in insects.

Nicotine and Nereistoxin Derivatives

Crude extracts of tobacco leaves rich in nicotine (Fig. 1) have been used as insecticides for many years. This alkaloid acts as an acetylcholine receptor agonist and is highly toxic to insects and mammals. The search for insect-specific analogs, such as nithiazin, imidacloprid, nitenpyram, and acetamiprid, led to the discovery of neonicotinoid analogs that specifically bind to insect acetylcholine receptors.

Compounds derived from the general cytotoxin nereistoxin (Fig. 1) with much improved toxicity profiles and strong insecticidal activity have been developed. These compounds have the same mode of action as nicotinoid insecticides.

Milbemycins/Avermectins

The discovery of these complex molecules is fairly recent. Most of these structures were discovered in *Streptomyces* cultures and have shown broad acaricidal, insecticidal, and anthelmintic activities. Both milbemycins and avermectins (Fig. 1) interfere with the opening of glutamate and GABA-gated chloride-channels. While studies aiming at producing simpler synthetic molecules yielded compounds with much lower activities, suitable natural product analogs can be synthesized without great difficulty. These molecules are now the primary tools used against parasites of farm animals and are used as insecticides in crop protection.

Spinosyns

Spinosyns (Fig. 1) are marketed for lepidoptera control in cotton as a mixture obtained by a fermentation process. Their mode of action is unknown but there are no known cases of cross-resistance to this class of inhibitors, suggesting that its target site is different from other known insecticides.

Azadirachtin

Seeds of the neem tree contain the highly insecticidal azadirachtin (Fig. 1). In addition to being a feeding deterrent, this complex molecule is also lethal to insects. It has multiple effects on insects, including disruption of metamorphosis by interfering with ecdysteroid synthesis, but the precise mode of action of azadirachtin is still unknown. Commercial products enriched with azadirachtin and other limonoids are available for controlling insects.

Juvenile-Hormone Mimics and Pheromones

Juvenile hormones are used to prevent certain insects from developing into fertile adults, therefore preventing them from reproducing. Other hormone-type insecticides, such as pheromones, have been used to control insect population in many ways. Initial approaches included using high levels of pheromones to confuse insects and prevent them from mating. The most common use is as an attractant placed in traps coated with either sticky substances that prevent the insects from leaving or insecticides that kill the organisms.

Bacillus Thuringiensis

Bacillus thuringiensis produces the insecticidal protein δ-endotoxin during sporulation. When ingested by an insect, δ-endotoxin binds to the epithelial cells of the gut and causes those cells to lyse, leading to the death of the insect. Several commercial products containing various forms of δ-endotoxin are available. Transgenic crops engineered to produce the δ-endotoxin have successfully been commercialized. These crops are insect-resistant since feeding on the foliage is lethal to insects.

FUNGICIDES AND BACTERICIDES DERIVED FROM NATURE

Numerous natural products have been identified and either used directly as fungicides or as templates to develop commercial products. While some of these products have narrow commercial use, others have been used as broad-spectrum fungicides. There are many natural products with excellent antifungal activity. However, few of them have been commercialized because of toxicological problems that often arise during the early stages of development, the excessive cost of large-scale production, and/or a problem with instability in the environment.

Strobilurins

Strobilurins (Fig. 2) are the most recent example of a successful use of natural products to develop an entire new class of fungicides. Their discovery and development has recently been reviewed.[3] Their fungicidal activity was first reported 30 years ago. Strobilurins have been identified primarily in basidiomycetes species throughout the world. These compounds are reversible inhibitors of the ubihydroquinone oxidation center of the bc1 complex in the mitochondrial respiratory chain. The discovery and development of strobilurin-derived fungicides has opened an extremely important area of research, with hundreds of international patent applications filed by more than 20 companies and research groups and thousands of analogs generated.

Kasugamycin

Kasugamycin (Fig. 2), isolated from *Streptomyces kasugaensis*, is a general inhibitor of protein biosynthesis in microorganisms, but does not affect protein synthesis in mammals. It is used as a commercial product for the control of rice blast (*Pyricularia oryzae*) and bacterial diseases caused by *Pseudomonas* spp. in several crops. The use of kasugamycin for fighting fungal diseases in rice paddies is now preferred over blasticidin S, which exhibited slight phytotoxic damage to the crop.

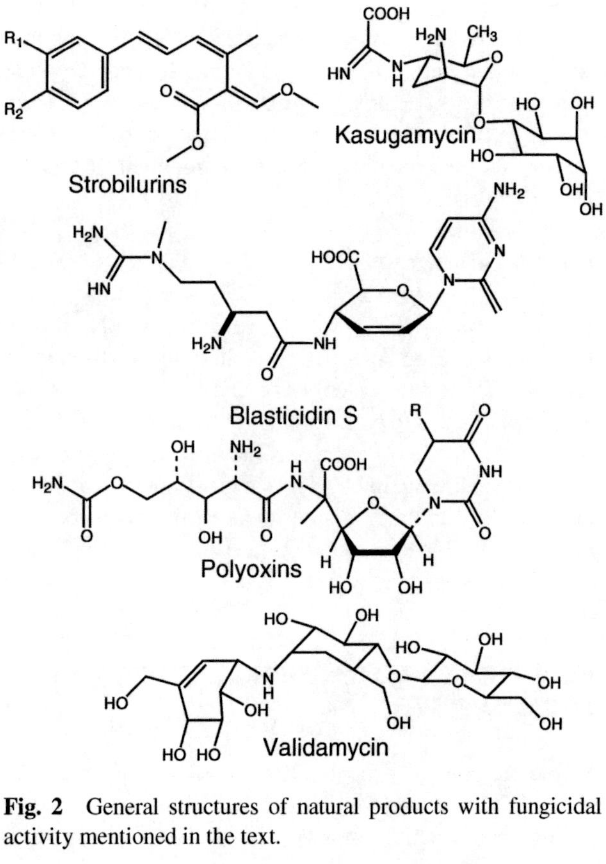

Polyoxins

Several polyoxin metabolites with antifungal activity have been identified. Polyoxin B and polyoxin D (Fig. 2) interfere with chitin synthase activity, preventing cell wall formation. These fungicides are commercially produced by fermentation for the control of rice sheath blight (*Rhizoctonia solani*) and other diseases in fruit and ornamental crops.

Validamycin

Another group of natural fungicides derived from the validamycin (Fig. 2) class has also been developed to control rice sheath blight. These fungicides are considered quite safe because their target site, the enzyme trehalase responsible for the hydrolysis of the disaccharide trehalose, is not found in mammalian systems.

HERBICIDES DERIVED FROM NATURE

Few herbicides have been derived from the backbone of natural compounds, relative to the number of insecticides and fungicides that have been successfully developed from natural products. This is somewhat unexpected since, from a chemical ecology standpoint, fungal/microbial plant pathogens and certain plants, in particular those involved in allelopathic interactions, produce highly phytotoxic molecules. In fact, hundreds of microbial natural products have been patented for potential use as herbicides, but bialaphos and its analog phosphinothricin are the only successful commercialized herbicides from natural products.[4] There is no plant product used directly as a herbicide, but striking similarities exists between natural phytotoxins and certain synthetic herbicides.[5]

Bialaphos and Phosphinothricin

Bialaphos (Fig. 3) has been isolated from the fermentation of *Streptomyces hygroscopis* and has been commercialized in eastern Asia. When applied directly to plants, it is metabolically converted by the plant to its bioactive form phosphinothricin (Fig. 3). The synthetic version of phosphinothricin has been commercialized as glufosinate (Basta®, Liberty®), and genetically engineered crops resistant to phosphinothricin have been marketed. This natural product-based herbicide has a unique mode of action, targeting glutamine synthetase. Other natural products

Fig. 2 General structures of natural products with fungicidal activity mentioned in the text.

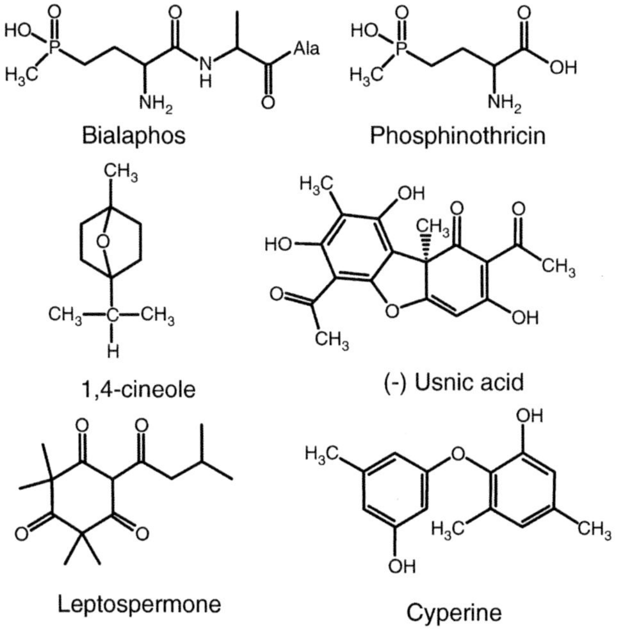

Fig. 3 General structures of natural products with herbicidal activity mentioned in the text.

such as phosalacin, oxetin, and tabtoxin also inhibit glutamine synthetase but have not been developed as commercial herbicides.[6]

Monoterpene Cineoles

The commercial herbicide cinmethylin is a 2-benzyl ether substituted analog of the monoterpene 1,4-cineole (Fig. 3). This compound was discovered and partially developed by Shell Chemicals for monocot weed control. Phytotoxic monoterpene cineoles are commonly found as components of essential oils from aromatic plants. While both cinmethylin and 1,4-cineole cause similar symptoms on treated plants, it was reported recently that only the natural product inhibits asparagine synthetase, the key enzyme in asparagine synthesis in plants. Thus, cinmethylin is apparently a proherbicide that requires metabolic bioactivation via cleavage of the benzyl-ether side chain.

Triketones

Leptospermone (Fig. 3), a natural triketone isolated from bottlebrush plant (*Calispermon* spp.), is herbicidal and causes bleaching of the foliage. The herbicide sulcotrione, a structural analog of leptospermone, has been shown to inhibit the relatively novel herbicide target site hydroxyphenylpyruvate dioxygenase (HPPD).[7] Inhibition of this enzyme disrupts the biosynthesis of carotenoids and causes bleaching of the foliage. We recently found that HPPD is sensitive to numerous classes of natural products such as the *p*-benzoquinones sorgoleone and maesanin, the *p*-naphthoquinone juglone, and the triketone usnic acid (Fig. 3) (unpublished data).

Cyperine

Cyperine (Fig. 3) is a phytotoxic natural product that has been isolated from several fungal pathogen sources. This phytotoxin has high activity on several plant species, including the problematic weed purple nutsedge (*Cyperus rotundus*) and pokeweed (*Phytolacca americana*). Cyperine is structurally similar to the synthetic diphenyl ether herbicides such as acifluorfen and oxyfluorfen but is a poor inhibitor of protoporphyrinogen oxidase.

Allelopathy

Allelopathy, the ability of a plant to suppress the growth of competing weeds in their immediate surroundings, is being investigated as an alternative approach to weed management. While certain plant species strongly repress the development of other plants, these allelopathic traits have, for the most part, been eliminated from crops by breeding programs selecting for higher yields. For example, certain native rice lines repress the growth of barnyard grass (*Echinochloa crus-galli*), whereas commercial varieties do not affect the growth of this weed. Introducing the biosynthetic pathway of an allelochemical into a crop has now become possible using genetic engineering tools, though it has not yet been achieved successfully.[8]

PROSPECT

Traditional pest management has already been deeply influenced by bioactive natural products that are used directly, or in a derived form, as pesticides. These past successes and the current public concern over the impact of synthetic pesticides on the environment ensures a continued, if not increased, interest in searching nature for environmentally friendlier pest management tools.

REFERENCES

1. Pachlatko, J.P. Natural products in crop protection. Chimia **1998**, *52*, 29–47.
2. Copping, L.G.; Menn, J.J. Biopesticides: A review of their action, applications and efficacy. Pest. Manag. Sci. **2000**, *56*, 651–676.
3. Sauter, H.; Steglich, W.; Anke, T. Strobilurins: evolution of a new class and active substances. Angew. Chem. Int. Ed. **1999**, *38*, 1328–1349.
4. Duke, S.O.; Abbas, H.K.; Amagasa, T.; Tanaka, T. Phytotoxins of Microbial Origin with Potential for Use as Herbicides. In *Crop Protection Agents from Nature: Natural Products and Analogues*; Copping, L.G., Ed.; Royal Society of Chemistry: Cambridge, U.K., 1996; 82–113, SCI Critical Reviews on Applied Chemistry.
5. Dayan, F.E.; Romagni, J.G.; Tellez, M.R.; Rimando, A.M.; Duke, S.O. Managing weeds with natural products. Pestic. Outlook **1999**, *5*, 185–188.

6. Lydon, J.; Duke, S.O. Inhibitors of Glutamine Biosynthesis. In *Plant Amino Acids: Biochemistry and Biotechnology*; Singh, B.K., Ed.; Marcel Dekker, Inc.: New York, 1999; 445–464.
7. Lee, D.L.; Prisbylla, M.P.; Cromartie, T.H.; Dagarin, D.P.; Howard, S.W.; Provan, W.M.; Ellis, M.K.; Fraser, T.; Mutter, L.C. The discovery and structural requirements of inhibitors of *p*-hydroxyphenylpyruvate dioxygenase. Weed Sci. **1997**, *45*, 601–609.
8. Scheffler, B.E.; Duke, S.O.; Dayan, F.E.; Ota, E. Crop Allelopathy: Enhancement through Biotechnology. In *Recent Advances in Phytochemistry*; Romeo, J., Ed.; Elsevier: Amsterdam, *in press*.

Pesticides: Reducing Use

David Pimentel
Maria V. Cilveti
Department of Entomology, Cornell University, Ithaca, New York, U.S.A.

Abstract
The prime advantage of the new, highly toxic insecticides and other related pesticides is that they do not persist in the environment. Reducing the persistence of pesticides in the environment offers many advantages in terms of public health and the environment; however, the public health and environmental threat from the new highly toxic pesticides remain a major problem. In addition to humans, other nontarget species continue to suffer from the use of pesticides.

INTRODUCTION

Many countries have stated that they are implementing integrated pest management (IPM) programs. In most cases in these countries, such as the United States, there has been a slight reduction in pesticide use in a few crops, but some of this reduction has resulted from using more toxic pesticides that require grams applied per hectare instead of kilograms per hectare. For example, DDT and related chlorinated insecticides were applied at 1–2 kg/ha whereas temick and related compounds are applied at only 10 g/ha. Pesticide use in the United States, despite a few reductions in a few crops, has slightly increased over the past decade.

The prime advantage of the new, highly toxic insecticides and other related pesticides, e.g., SEVIN, is that they do not persist in the environment. Reducing the persistence of pesticides in the environment offers many advantages in terms of public health and the environment; however, the public health and environmental threat from the new highly toxic pesticides remain a major problem. For example, the World Health Organization (WHO)[1] reports that about 26 million people are poisoned each year, with about 2,220,000 deaths. In addition to humans, other nontarget species continue to suffer from the use of pesticides. For example, there are an estimated 72 million birds killed in the United States each year. Plus there are numerous other nontarget species affected by pesticides.[2,3]

OVERVIEW

Relatively few nations have made a major effort to reduce pesticide use in agriculture. The best example, and the most successful nation in reducing pesticide use, is Sweden. In 1986, the government of Sweden implemented a policy to reduce pesticide use by 50% over a 5-year period.[4] Sweden was successful in its first effort to reduce pesticide use by 50%; then, in 1992, Sweden passed legislation to reduce pesticide use another 50%. Although Sweden has not reduced pesticide yet by 75%, they have reduced pesticide use by 68%.[5] Associated with this 68% pesticide reduction has been a 77% reduction in human pesticide-related health problems.

Sweden accomplished its goal of reducing pesticide use by 68% by implementing several techniques.[4] First, the government invested in increased numbers of extension advisors and scientific investigators. The reduction in pesticide use was accomplished in part by switching from pesticides that were applied at kilogram dosages per hectare to the new highly toxic pesticides that are applied at gram dosages per hectare.

At the same time, in Sweden, changes were made in the application of pesticides in agriculture. First, pesticides were no longer applied on a routine program, whether pests were a serious problem or not. The farmers and extension workers carefully monitored the pest and beneficial organism populations to determine if there was a pest problem that warranted treatment in terms of economics and the environment.

In addition to monitoring pest and natural-enemy populations, various environmentally sound programs were implemented. These included adding crop rotations and planting crops that were relatively resistant to insect and plant-pathogen pests.

Another policy that was implemented was adjusting pesticide-use dosages that tended to kill the pests while leaving the natural-enemy population relatively unharmed. For example, in Norway, Edland[6] found that if a lower dosage is used for certain insecticides, i.e., one-fifth the dosage recommended by the manufacturer, orchard pest control was significantly better than the manufacturer's recommended higher dosage.

When the manufacturer was asked to lower their recommended dosage, they refused at first. Finally, after numerous requests and refusals, the Norwegian government reported to the manufacturer that if they desire to continue selling pesticides in Norway, they would have to change their recommendations. Based on this warning, the manufacturer changed their recommendations to that suggested by Edland.

Another highly successful program to reduce pesticide use occurred in Indonesia. The Indonesian Government appointed a new Minister of Agriculture in 1980 and he favored the heavy use of pesticides in rice production. Under his policy, pesticide use on rice dramatically increased. In about 4 years time, the rice farmers found that they were having trouble controlling the brown-plant-hopper pest (BPH) and rice yields were declining in many parts of Indonesia. In fact, by 1985, many thousands of hectares of rice had to be abandoned because of severe outbreaks of the BPH.

Thus instead of Indonesia being a net exporter of rice, the nation had to import rice. With rice yields declining, the President of Indonesia consulted Dr. I.N. Oka for a solution to the serious problem. Dr. Oka advised the President that the government should ban 67 of 74 pesticides in current use and implement a new pest management program. The President of Indonesia went on TV and banned 67 of 74 pesticides and announced that new policies in rice pest control would be implemented. The Minister of Agriculture was fired and Dr. Oka was placed in charge of all pest control in Indonesia.

Because of the dramatic changes in pesticide use, threats were made to Dr. Oka's life. The President of Indonesia provided bodyguards for Dr. Oka and his family. In addition, Indonesia obtained loans and grants from the World Bank and the Food and Agricultural Organization (FAO) of the United Nations to hire about 2000 new extension workers to implement Dr. Oka's policies.

Dr. Oka was an expert scientist on rice pests and their natural enemies. Thus farmers were instructed on how to identify the pest insects and beneficial insects and other arthropods. They were also instructed when pest insect populations would be a threat to rice yields and when to treat. The farmers were also instructed how to treat with pesticides to leave as many natural enemies surviving as possible.

Another important policy that Dr. Oka implemented was leaving all the rice fields in Indonesia fallow without rice for about 3 months during the year. Without rice, several of the major insect pests, especially the BPH, significantly declined, so that when rice was planted the next season, there were very few insect pests present to attack the newly planted rice.

Under Dr. Oka's policies, pesticide use was reduced by more than 65% and rice yields increased 12%—a phenomenal accomplishment! Thus the farmers benefited in terms of economics and the public benefited in terms of health and the environment.

The province of Ontario, Canada, also decided to implement a program to reduce pesticide use by 50% over a 15-year period[7] starting in 1987. Ontario's program was similar to that in Sweden in that they replaced some of the heavy-use, high-dosage pesticides with low-dosage, highly toxic pesticides. In addition, they added extension workers and increased their investment in research on nonchemical controls.

This major effort paid off, and Ontario was able to reduce pesticide use by 50%. A survey of the farmers in Ontario found that they were highly supportive of the goal to reduce pesticide use by 50%. The reasons that the farmers supported the program were: 1) The farmers were applying the toxic pesticides and thus were exposing themselves and their families to toxic pesticides; 2) reducing pesticide use in crop production improved farmer profits; and 3) the farmers were in favor of protecting the environment.

CONCLUSION

A detailed assessment of the potential to reduce pesticide use in the United States was conducted in the early 1900s.[8] The investigation documented that U.S. pesticide use could be reduced by 50% if the U.S. government implemented a program similar to that of Sweden, Indonesia, and the province of Ontario. Such a program would save farmers' money, protect public health, and protect the environment.[8]

REFERENCES

1. WHO. *Our Planet, Our Health: Report of the WHO Commission on Health and Environment*; World Health Organization: Geneva, 1992.
2. Pimentel, D.; Acquay, H.; Biltonen, M.; Rice, P.; Silva, M.; Nelson, J.; Lipner, V.; Giordano, S.; Horowitz, A.; D'Amore, M. Assessment of Environmental and Economic Costs of Pesticide Use. In *The Pesticide Question: Environment, Economics and Ethics*; Pimentel, D., Lehman, H., Eds.; Chapman and Hall: New York, 1993; 47–84.
3. Pimentel, D. Ecological effects of pesticides on public health, birds and other organisms. Reflections **2002**, 27–28.
4. Pettersson, O. Pesticide Use in Swedish Agriculture: The Case of a 75% Reduction. In *Techniques for Reducing Pesticide Use: Economic and Environmental Benefits*; Pimentel, D., Ed.; John Wiley and Sons Ltd.: Chichester, U.K., 1997; 79–102.
5. Pesticide Action Network (PAN). Pestic. Inf. Update 2001, 44.
6. Edland, T. Benefits of Minimum Pesticide Use in Insect and Mite Control in Orchards. In *Techniques for Reducing Pesticide Use: Economic and Environmental Benefits*; Pimentel, D., Ed.; John Wiley and Sons Ltd.: Chichester, U.K., 1997; 197–220.
7. Surgeoner, G.A.; Roberts, W. Reducing Pesticide Use by 50% in the Province of Ontario: Challenges and Progress. In *The Pesticide Question: Environment, Economics and Ethics*; Pimentel, D., Lehman, H., Eds.; Chapman and Hall: New York, 1993; 206–222.
8. Pimentel, D.; McLaughlin, L.; Zepp, A.; Lakitan, B.; Kraus, T.; Kleinman, P.; Vancini, F.; Roach, W.J.; Graap, E.; Keeton, W.S.; Selig, G. Environmental and economic impacts of reducing U.S. agricultural pesticide use. BioScience **1991**, *41*, 402–409.

Pesticides: Regulating

Matthias Kern
GTZ Pilot Project on Chemicals Management, Bonn, Germany

Abstract
The purpose of pesticide control legislation is to establish rules and principles for the management of pesticides such as to maximize the benefits from their application while minimizing or ruling out negative side effects.

INTRODUCTION

Pesticides are introduced directly into the environment, e.g., they are applied to fields as plant protectants, to control vectors, etc. One of the main requirements for any pesticide is that it effectively controls the target organism. Unfortunately, the specificity of many pesticides is rather low, making them toxic for nontarget organisms, too. This also makes them a potential hazard for their users, for the consumers of agricultural products that have been treated with them, and for the environment. Consequently, it is of particular importance that measures be taken to protect people and the environment against the harmful effects of pesticides. For this reason, pesticides belong to that group of chemicals that must be most carefully examined and most stringently regulated.

The purpose of pesticide control legislation is to establish rules and principles for the management of pesticides such as to maximize the benefits from their application while minimizing or ruling out negative side effects.[1]

ELEMENTS

The exercise of control over pesticides applies not only to their application. As with any other hazardous type of chemical, due allowance also must be made for their manufacture, transport, national and international distribution, storage, further processing and, as the case may be, disposal.[2]

To put these control measures on a legal footing, the fundamental conditions for the management of pesticides must be embodied in the law. Depending on the circumstances prevailing within a given country, the relevant legislation could address pesticides or plant protection, chemicals or environmental protection. In any case, the law must stipulate that no person, business, or organization shall be allowed to distribute or apply any pesticide that has not been registered with, or at least provisionally approved by, the responsible authority. The details of the registration process and of pesticide management must be governed by various subsidiary implementing regulations that can be adjusted to accommodate technical progress without necessitating alteration of the law itself.

Usually, a registration process regulates the distribution of pesticides. The registration requirements for pesticides define their quality, range of application, labeling, packaging and applicable safety measures. In this context, protection of the users and the environment, as well as the proprietary rights of the manufacturers must be taken into account.

For the postregistration scope, legal directives must be in place to govern the sale of pesticides, the training of retailers and users, the award of licenses, and the control activities at various levels.

The areas of production, repackaging, and transport must not necessarily be covered by the pesticide-specific legislation if they already are adequately governed by laws applicable to chemicals in general.

The stipulation of responsibilities is of major importance for the legal application. As a rule, responsibility for the registration and control of pesticides lies with the ministry of agriculture, health or the environment, with decisions concerning the registration and control normally being coordinated between several ministries or subsidiary authorities according to a prescribed procedure. Customs must be involved in the control of importing and exporting activities.

IMPLEMENTATION

In 1985 the United Nations Food and Agriculture Organization (FAO) conference adopted a voluntary "International Code of Conduct on the Distribution and Use of Pesticides" in which the principles of pertinent legislation and of the environmentally sound management of pesticides are compiled for international reference purposes.[3]

One of the main prerequisites for the environmentally sound management of pesticides is that an appropriate legal framework be in place and amenable to enforcement of its various laws and implementing regulations. Chapter 19 of Agenda 21, which deals with the environmentally sound management of toxic chemicals, postulates that this be accomplished on the maximum possible scale in all countries by the year 2000.[4]

Surveys conducted by FAO in 1994 show that all industrialized countries and nearly all developing countries have laws regulating the control of pesticides. In many of the emerging countries, however, effective enforcement of their pesticide laws is rendered impossible by a lack of appropriate implementing regulations and/or means of monitoring compliance with the laws. Of the developing countries, 87% stated that their governments were providing little or no resources for the control of pesticide.[5]

PROBLEMS

The management of pesticides in the absence of appropriate safety measures can cause problems, the gravity of which will differ according to the prevailing framework conditions.

While developing countries consume only about 20% of all agrochemicals, they account for 70% of all acute pesticide poisonings according to International Labour Organization estimates.[6] That corresponds to some 1.1 million cases of pesticide poisoning each year. Most of these cases are attributable to inadequately trained users, improper application techniques, defective application equipment, and/or inadequate safety practices. Implementation of effective work safety regulations should help alter the situation.

In countries with no means of control, improper use is often compounded by problems attributable to the quality of the marketed products. A product containing the wrong active ingredient or having the wrong concentration will not have the envisaged effect on the target organism and will not secure the envisaged benefits for its user. Some 30% of all pesticides sold in developing countries do not comply with international standards regarding the quality of their active ingredients.[7] If the quality of the labeling and packaging is also taken into consideration, the percentage of deficient products will be even higher. Considering that the label is the foremost source of information for the user with regard to indication, dosing and safety practices, proper labeling is of central importance for the proper use of the pesticides. Effective supervision of the market and of compliance with registration procedure specifications is a decisive prerequisite for actual compliance with quality standards.

Maximum residue limits must be specified in order to avoid problems with food contamination by pesticides. At the international level, those defined by the *Codex Alimentarius* Commission serve as guideline data in cases where no national values are specified. Here, too, corresponding target values can be effective only if the residue levels are monitored. In a number of cases, agricultural products from countries in which check analysis is either inadequate or nonexistent were rejected by importing countries on the grounds of maximum-limit transgression. For the exporting country, that means a substantial economic loss. In order to succeed in the international marketplace, more and more countries are beginning to monitor both the pesticide residue levels in food and the use of pesticides on crops. In this instance, economic pressure is accelerating the implementation of legal prescriptions.

INTERNATIONAL CONVENTIONS

Comprehensive national pesticide legislation and control are justifiable for countries with relatively large pesticides markets and corresponding infrastructures. For many smaller countries, however, such structures remain beyond their financial and administrative capacities. Due to the fact that international standards capable of adaptation to national and regional circumstances are already in place, many countries work together at the regional level with regard to the control of pesticides.

As trade flows become increasingly internationalized, it is sensible that the authorities in industrialized and developing countries engage in joint supervisory activities, and such action is increasingly being called for.

In September 1998 the Convention on the Prior Informed Consent (PIC) Procedure for Certain Hazardous Chemicals and Pesticides in International Trade was signed in Rotterdam.[8] The Rotterdam Convention serves chiefly to provide protection for developing countries that have not yet instituted adequate import controls of their own. The exporting countries pledged to inform the importing countries regarding any plans to transport certain hazardous chemicals which are either banned for reasons of health or environmental protection or are subject to severe restrictions—and to export such chemicals only with the express prior (informed) consent of the importing country.

The FAO and the United Nations Environment Programme (UNEP) introduced the PIC procedure in 1989 on an international voluntary basis. By 1993, 106 countries were participating, and that number grew to 155 by 1999. While the cooperative principle enjoys widespread approval, it will not become legally binding until the Rotterdam Convention enters into force. What makes this convention so special is that the authorities in the exporting country are bound by importing decisions made by recipient countries, i.e., outside the territory of the former. Hence, the "borders" of national legislature are beginning to blur. Now, many exporting countries are having to take measures to ensure that their customs authorities become active in the control of pesticide imports and exports.

The uncontrolled international distribution of persistent organic pollutants (POPs) also presents problems.[9] This category includes a number of pesticides, e.g., aldrin, dieldrin, DDT, and mirex. POPs are slow to break down in nature, and the wind can spread them far beyond a country's boundaries. Eventually, they are deposited by precipitation in cooler climate zones, where they accumulate in the food chain. Thus, countries of the north like Canada and Sweden are keenly interested in the global restriction of such substances. International negotiations on the drafting of a convention that would globally restrict the manufacture and use of certain POPs were entered into in 1998. Today, POPs are chiefly produced and applied in developing countries. Those countries, however, are only willing to discontinue the relatively cheap production and use of such products if they see attractive economic alternatives and if financing of the conversion process is assured. Thus, the conclusion of an internationally binding convention on the reduction of global contamination by persistent chemicals is heavily dependent on the economic interests of individual countries, as well as on how much importance each country attaches to the attendant risks.

CONCLUSIONS

Pesticide legislation is part of general chemicals legislation, because many quality standards and safety practices apply in like manner to pesticide, household and industrial chemicals. The respective separate provisions apply only to product-specific characteristics. National legislation in this area is becoming increasingly dependent on global cooperation in order to accommodate expanding international trade in chemicals and take account of supraregional pollution. The Rotterdam Convention pertaining to international trade in certain hazardous products constitutes an important step in that direction. Likewise, the phasing out of certain pesticides, e.g., POPs, can be achieved only through an international approach. However, as demonstrated by the successful efforts to phase out ozone-depleting substances by way of the Montreal Protocol, this is possible.

The present trend indicates that, in the future, the sale, handling, transportation, and application of chemicals will be controlled within a global context, and the responsibility for its implementation will be divided equally among the industrialized and the developing countries.

REFERENCES

1. FAO. *Pesticide Management Guidelines*; Food and Agriculture Organization of the United Nations: Rome, 1998. http://www.fao.org/WAICENT/FaoInfo/Agricult/AGP/AGPP/Pesticid/Code/Guide.htm (accessed Dec. 1999).
2. Lönngren, R. *International Approaches to Chemicals Control*; The National Chemicals Inspectorate KEMI: Sweden, 1992; 512.
3. FAO. *International Code of Conduct on the Distribution and Use of Pesticides*; Food and Agriculture Organization of the United Nations: Rome, 1990. http://www.fao.org/WAICENT/FaoInfo/Agricult/AGP/AGPP/Pesticid/Code/PM_Code.htm (accessed Dec. 1999).
4. UNCED. *Environmentally Sound Management of Toxic Chemicals, Including Prevention of Illegal International Traffic in Toxic and Dangerous Products*; Agenda 21, Chapter 19, United Nations Conference on Environment and Development, Rio de Janeiro; 1992. http://www.igc.apc.org/habitat/agenda21/ch-19.html (accessed Dec. 1999).
5. FAO. *Analysis of Government Responses to the Second Questionnaire on the State of Implementation of the International Code of Conduct on the Distribution and Use of Pesticides*; Food and Agriculture Organization of the United Nations: Rome, 1996; 101.
6. ILO. *The ILO Programme on Occupational Safety and Health in Agriculture*; International Labour Organization: Geneva, 1998. http://www.ilo.org/public/english/90travai/sechyg/agrivf02.htm (accessed Dec. 1999).
7. Kern, M.; Vaagt, G. Pesticide quality in developing countries. Pesticide Outlook 1996, (Oct.), 7–10.
8. FAO. *Interim Joint FAO/UNEP Secretariat for the Operation of the PIC Procedure*; Food and Agriculture Organization of the United Nations: Rome, 1999. http://www.fao.org/WAICENT/FaoInfo/Agricult/AGP/AGPP/Pesticid/PIC/pichome.htm (accessed Dec. 1999).
9. UNEP. Chemicals. In *Persistent Organic Pollutants*; United Nations Environment Programme: Geneva, 1999. http://irptc.unep.ch/pops/ (accessed Dec. 1999).

Pests: Landscape Patterns

F. Craig Stevenson
University of Saskatchewan, Department of Soil Science, Saskatoon, Saskatchewan, Canada

Abstract

Crop productivity most often varies across a field, with areas of relatively high and low crop yields. These landscape patterns for crop productivity may be associated with landscape patterns for differing levels of weed, disease, or insect pest pressure. An understanding of the landscape patterns for pests will be an integral component of effective and efficient pest management strategies designed to maintain high levels of crop production across all areas of a field. A path to such an understanding will require basic knowledge of where pests occur and processes controlling these landscape patterns.

PEST AND LANDSCAPE PATTERNS

Small Patches

Certain areas within a field may be more conducive to pest survival, establishment, and development relative to other areas. The size and proximity of localized infestations with a given field differs among the three major pest groups. Past research has shown that weed and disease patches generally have a radius of about 25 m (if you assume that they have a circular shape), but may be as large as 100 m.[1–3] In the case of wild oat (*Avena fatua* L.), the patch may be associated areas that have higher soil water and nutrient availability (see Fig. 1). Weeds such as perennial sowthistle (*Sonchus arvensis* L.), with seeds adapted to wind dissemination, may occur in areas of higher elevation where seeds are trapped as they are blown across a field. Foliar disease patches generally are associated with low wet areas of fields or sheltered field margins that have higher canopy humidity and lower wind speeds. Landscape patterns for insects, however, are less persistent and harder to predict because of rapid reproductive rates and pest movement.[4] To a lesser extent, this same level of complexity also occurs for polycyclic diseases that easily become airborne. However, the overwintering phases in northern agricultural regions and nonflying phases of insects may be associated with areas of a field with a more friable soil structure (e.g., grasshoppers) or other soil and microclimatic factors. In addition, for insects and to a lesser extent diseases, landscape patterns are complicated pest–predator interactions that can affect infestation of that pest in the given space in the next growing season. For example, the landscape pattern of nematodes has been shown to be partly a function of nematophagous fungi.[5] Therefore, the proximity of areas more or less conducive for the pest is ultimately dependent on the topography of the field and spatial variation of microclimatic conditions, especially for those pests with restricted mobility.

Larger Scale Landscape Patterns

These patterns are seen where larger areas of a field, or numerous fields, are being infested by a pest. These types of landscape patterns generally are dependent on external factors that move the pest from a smaller patch to adjacent areas. Climatic processes, such as wind, that blow weed seeds, disease spores, and insects can rapidly increase the area of pest infestation. Farming practices such as tillage and harvest operations also cause larger scale landscape patterns. For example, a study showed that *Polygonum* spp. of weeds tended to vary with the direction of tramlines up to distances of 635 m, a distance about 600 m more than that observed in the direction perpendicular to tramlines.[1] The spatial distribution of pests can be influenced by tillage practices and harvest operations that move crop residues and soil across a field. For example, nematodes are spread by cultivation from initial infestation foci.[5] Also, grain combine harvesters and shank-type tillage implements can move weed seeds and other reproductive structures across a field quickly.

Temporal Stability of Landscape Patterns

Time tends to complicate and/or obscure distinct landscape patterns for pests within a field. Varying climatic patterns (rainfall, temperature, etc.), within and among growing seasons can have a profound effect on the ability of pests to survive, establish, and develop. This complexity, interacting with polycyclic and multigeneration reproductive strategies within a growing season and factors moving pests across or between fields, poses a major hurdle to a holistic understanding of pests and landscape patterns. Extensive

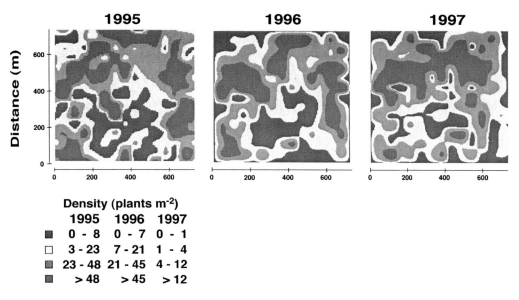

Fig. 1 Landscape pattern of wild oat in a 65 ha field.
Source: Thomas.[6]

field research combined with predictive modeling may be a fruitful avenue for such a complex phenomenon.

FUTURE DEVELOPMENTS

The advent of global positioning systems (GPS), satellite imagery, and geographic information systems (GIS) software have heightened awareness and provided insight into landscape patterns and pests. Ultimately, these technologies may allow for the site-specific management of pests in accordance with their landscape patterns.[7] For example, spatially referenced maps could be linked to GPS on pesticide applicators to reduce the total amount of pesticide applied, thus improving economic returns and resulting in farming systems less dependent on pesticides. Our current understanding clearly shows that pest infestations often occur in patterns across a field, however, a great deal more effort will be necessary to provide information to accurately predict where pests will occur and affect crop production most extensively. These challenges will be especially difficult considering the highly variable and dynamic nature of current climatic conditions and farm management systems.

REFERENCES

1. Nordbo, E.; Christensen, S.; Kristensen, K.; Walter, M. Patch spraying of weed in cereal crops. Asp. Appl. Biol. **1994**, *40*, 325–334.
2. Zadoks, J.C.; van den Bosch, F. On the spread of plant disease: a theory on foci. Ann. Rev. Phytopathol. **1994**, *32*, 503–521.
3. Zanin, G.; Berti, A.; Riello, L. Incorporation of weed spatial variability into the weed control decision-making process. Weed Res. **1998**, *38*, 107–118.
4. Hassell, M.P.; Comins, H.N.; May, R.M. Spatial structure and chaos in insect population dynamics. Nature **1991**, *353*, 255–258.
5. Webster, R.; Boag, B. Geostatistical analysis of cyst nematodes in soil. J. Soil Sci. **1992**, *43*, 583–595.
6. Thomas, A.G. *Agriculture and Agri-Food Canada*; Saskatoon, SK, Canada, 1999, unpublished data.
7. *The State of Site Specific Management for Agriculture*; Sadler, E.J., Pierce, F.J., Eds.; American Society of Agronomy, Inc., Crop Science Society of America, Inc., Soil Science Society of America, Inc.: Madison, WI, 1997; 423.

Petroleum: Hydrocarbon Contamination

Svetlana Drozdova
Erwin Rosenberg
Institute of Chemical Technologies and Analytics, Vienna University of Technology, Vienna, Austria

Abstract
Contamination of water, soil, and sediment samples by petroleum hydrocarbons is a common and severe environmental problem, caused by improper handling, storage, transport, or use of petrochemical products or raw materials. Petroleum hydrocarbons represent a mixture of compounds, and some of them (e.g., benzene, polycyclic aromatic hydrocarbons) may exhibit toxic and/or carcinogenic properties. Because petroleum products are a complex and highly variable mixture of hundreds of individual hydrocarbon compounds, characterizing the risks posed by petroleum-contaminated soil and water has proven to be difficult and inexact. It is very important to have an understanding of the toxicology, analytical science, environmental fate and behavior, risk, and technological implications of petroleum hydrocarbons in order to interpret, evaluate the risk of, and make decisions about potential hazardous effects to and ensure the appropriate protection of the environment.

INTRODUCTION

Historically, environmental analyses focused on monitoring compounds that pose a threat to humans and their environment. Petroleum hydrocarbon compounds are among them. Contamination of water, soil, and sediment samples by petroleum hydrocarbons is a common and severe environmental problem caused by improper handling, storage, transport, or use of petrochemical products or raw materials.

Petroleum products are the major source of energy for industry and daily life. Leaks and accidental spills occur regularly during exploration, production, refining, transport, and storage. In addition, natural processes can result in seepage of crude oil from geologic formations below the seafloor. The total input of crude oil and petroleum into the environment is estimated to be 1.3 million tons per year. To understand the potential effect of petroleum contaminations on the environment, it is important to understand the nature and distribution of sources and their inputs. Petroleum poses a range of environmental risks when released into the environment. Catastrophic and large-scale spills have a very severe physical impact in addition to the chemical pollution that they cause; chronic discharges and small releases can damage and eventually kill the exposed flora and fauna due to toxicity of many of the individual compounds contained in petroleum. Oil contamination in the environment is primarily assessed by measuring the chemical concentrations of petroleum products in the affected environmental compartment (e.g., sediment, biota, water).

This entry provides a discussion of the environmental relevance of petroleum hydrocarbons; the principal sources of petroleum contaminations in the environment; and the nature and composition of crude oil and petroleum products derived from it. The fate of petroleum hydrocarbons in the environment, possible effects from exposure to them, and their toxicity are discussed as well. The entry is concluded by an overview of analytical methods for determination of petroleum hydrocarbon contamination.

PETROLEUM HYDROCARBONS AND THEIR ENVIRONMENTAL RELEVANCE

Oil and gas resources are organic compounds, formed by the effects of heat and pressure on sediments trapped beneath the earth's surface over millions of years. The remains of animals and plants that lived millions of years ago in a marine environment were covered by layers of sand and silt over the years. Heat and pressure from these layers helped the remains turn into crude oil or petroleum.[1] The word "petroleum" means "rock oil" (from Greek: petra [rock] + Latin: oleum [oil])[2] or "oil from the earth." While ancient societies made some use of these resources, the modern petroleum age began less than a century and a half ago, when in 1859, Colonel Drake discovered oil in Oil Creek in Titusville, Philadelphia, United States. From that time on, the world's demand for fossil fuel and the production of oil have continuously increased. From the 1980s, in particular, after the second oil crisis of 1979, the petroleum business has developed into a high-technology industry. Advances in technology have greatly improved the ability to find and extract oil and gas and to convert them to efficient fuels and useful consumer products. About 100 countries produce crude oil. Russia, Saudi Arabia, the United

States, Iran, and China are the top five producing countries in 2009 (Table 1).[3] In the United States, the oil and gas industry employs 1.4 million people and generates about 4% of U.S. economic activity. It is larger than the domestic automobile industry and larger than education and social services, the computer industry, and the steel industry combined.[4] At a refinery, different fractions of the crude oil are separated into useable petroleum products. Various sources of information provide a good overview of the different processes in petroleum refining.[5–7] Petroleum products are used worldwide for energy production, as fuel for transport, and as a raw material for many chemical processes. The United States is the biggest consumer of oil in the world (Table 1). Although there exist well-developed alternatives to the use of oil (particularly for energy production and transportation), our societies are still strongly dependent on oil, which is an environmental burden, an economic problem, and a political hazard. However, at the current time, the economic situation still favors the use of petroleum and petroleum products for these applications rather than its alternatives, which at the moment are not competitive from an economic point of view.

Petroleum poses a range of environmental risks when it is released into the environment (whether by catastrophic spills or through chronic discharges). In addition to the physical impact of large spills, the toxicity of many of the individual compounds contained in crude oils or petroleum products is significant. Information on how petroleum hydrocarbons enter and diffuse in the environment is abundant.[8,9] The sources of petroleum input to the environment, particularly to the sea, are diverse. They can be categorized effectively into four major groups, namely,

Table 1 Annual production and consumption of oil by the top 10 industrial nations and by the top 10 countries in the European Union.

	Oil production by country					Oil consumption by country in the world			
Rank	Country	Amount bbl/day	Date	Percentage %	Rank	Countries	Amount bbl/d	Date	Percentage %
1	Russia	10,120,000	2010	11.9	1	United States	18,690,000	2009	22.6
2	Saudi Arabia	9,764,000	2009	11.5	2	China	8,200,000	2009	9.9
3	United States	9,056,000	2009	10.7	3	Japan	4,363,000	2009	5.3
4	Iran	4,172,000	2009	4.9	4	India	2,980,000	2009	3.6
5	China	3,991,000	2009	4.7	5	Russia	2,740,000	2010	3.3
6	Canada	3,289,000	2009	3.9	6	Brazil	2,460,000	2009	3.0
7	Mexico	3,001,000	2009	3.5	7	Germany	2,437,000	2009	2.9
8	United Arab Emirates	2,798,000	2009	3.3	8	Saudi Arabia	2,430,000	2009	2.9
9	Brazil	2,572,000	2009	3.0	9	Korea, South	2,185,000	2010	2.6
10	Kuwait	2,494,000	2009	2.9	10	Canada	2,151,000	2009	2.6
Total:		84,764,555			Total:		82,769,370		
	Oil production by EU member states					Oil consumption by EU member states			
Rank	Countries	Amount bbl/d	Date	Percentage %	Rank	Countries	Amount bbl/d	Date	Percentage %
1	United Kingdom	1,502,000	2009	60.4	1	Germany	2,437,000	2009	16.2
2	Denmark	262,100	2009	10.5	2	France	1,875,000	2009	12.5
3	Germany	156,800	2009	6.3	3	United Kingdom	1,669,000	2009	11.1
4	Italy	146,500	2009	5.9	4	Italy	1,537,000	2009	10.2
5	Romania	117,000	2009	4.7	5	Spain	1,482,000	2009	9.9
6	France	70,820	2009	2.8	6	Hungary	1,373,000	2009	9.1
7	Netherlands	57,190	2009	2.3	7	Netherlands	922,800	2009	6.1
8	Poland	34,140	2009	1.4	8	Belgium	608,200	2009	4.1
9	Spain	27,230	2009	1.1	9	Poland	545,400	2009	3.6
10	Austria	21,880	2009	0.9	10	Greece	414,400	2009	2.8
Total (EU, 27 countries):		2,485,550	2009		Total (EU, 27 countries):		15,012,050		
	Norway	2,350,000	2009			Norway	204,100		
	Turkey	52,980	2009			Turkey	579,500		

bbl, barrel; EU, European Union. 1 bbl ≈ ca. 159 L.
Source: Adapted from Energy Statistics: Oil-Production (Most Recent) by Country.[3]

natural seeps, petroleum extraction, petroleum transportation, and petroleum consumption.

Natural seeps are frequently encountered phenomena that occur when crude oil seeps from the geologic strata beneath the seafloor to the overlying water column as a natural process.[10] Recognized by geologists for decades as indicating the existence of potentially exploitable reserves of petroleum, these seeps release vast amounts of crude oil annually. Yet these large volumes are released at a rate low enough that the surrounding ecosystem can adapt and even thrive in their presence; which is not true in case of the catastrophic and accidental impact of a tanker or oil well spill. Natural processes are, therefore, responsible for over 45% of the petroleum entering the marine environment worldwide (Table 2).[11]

As result of human activities, about 700,000 tons of petroleum is released annually into the sea worldwide. Processes such as petroleum extraction, transportation, and consumption can cause soil and groundwater contamination in case of equipment failure or operation errors and other reasons. Petroleum extraction can result in release of both crude oil and refined products as a result of human activities associated with efforts to explore and produce petroleum. The nature and size of these releases are highly variable—see Table 3 for the largest oil spills observed until 2010[12]—and can include accidental spills of crude oil from platforms and blowouts such as that of the oil rig Deepwater Horizon in the Gulf of Mexico in April 2010 or slow chronic releases of water produced from oil- or gas-bearing formations during extraction. Under current industry practices, this "produced water" is treated to separate from crude oil and either injected back into the reservoir or discharged overboard. Produced water is the largest single wastewater stream in oil and gas production. The amount of produced water from a reservoir varies widely and increases over time as the reservoir is depleted. Petroleum transportation can result also in releases of dramatically varying sizes of petroleum products (not just crude oil) from major incidents (mostly from tankers, such as the one in 1979 off the coast of Tobago, when two tankers collided and one of these, the Atlantic Empress, sank, losing all its freight) to relatively small operational releases that occur regularly, such as those from pipelines.

Releases that occur during the consumption of petroleum, whether by individual car and boat owners, non-tank vessels, or runoff from urban or industrial areas, are typically small but frequent and widespread and are responsible for the vast majority (70%) of petroleum introduced to the environment through human activity.

Because crude oil and petroleum products are a complex and highly variable mixture of hundreds to thousands of individual hydrocarbon compounds, characterizing the risks posed by petroleum-contaminated soil and water has proven to be difficult and inexact. It is very important to have an understanding of the toxicology, analytical science, environmental fate and behavior, risk, and technological implications of petroleum hydrocarbons in order to interpret, evaluate the risk of, and make decisions about the hazardous effect to and ensure the appropriate protection of the environment.

GENERAL CHEMICAL COMPOSITION FEATURES OF CRUDE OILS AND PETROLEUM PRODUCTS

Crude oil is an extremely complex mixture of several thousands of different compounds; its compositions and physical properties vary widely depending on the source from which the oils are produced, the geologic environment, and location in which they migrated and from which they are extracted. The nature of the refining processes has an effect on crude oil compositions as well. As indicated in Table 4, petroleum and petroleum products contain primarily hydrocarbons, heteroatom compounds, and relatively small concentrations of (organo)metallic constituents.[13,14] The complexity of petroleum and petroleum products increases with carbon number of its constituents, so it is impossible to identify all components. Petroleum and petroleum products are typically characterized in terms of boiling range and approximate carbon number. Raw petroleum is usually dark brown or almost black, although some fields deliver a greenish or sometimes yellow petroleum. Depending upon the oil field and the way the petroleum composition was formed, the crude oil will also differ in viscosity. The composition of crude oil impacts certain physical properties of the oil, and it is these physical properties (e.g., density or viscosity) by which crude oils are generally characterized, classified, and traded. These physical properties can be used to classify crude oils as light, medium, or heavy. The American Petroleum Institute (API) gravity[15] is a measure of the specific gravity of a petroleum liquid compared with water (API = 10). Light oils are defined as having an API < 22.3, heavy oils are those with API > 31.1, and medium oils have an API gravity between 22.3 and 31.1.

Regardless of the complexity, petroleum compounds can be separated into two major categories: hydrocarbons and non-hydrocarbons. Hydrocarbons (compounds composed solely of carbon and hydrogen) comprise the majority of

Table 2 Petroleum input to the sea.

Source of input	North America		Worldwide	
	Tons	%	Tons	%
Natural seeps	160,000	61	600,000	46
Petroleum extraction	3,000	1	38,000	3
Petroleum transportation	9,100	4	150,000	12
Petroleum consumption	84,000	32	480,000	37
Other	3,900	2	32,000	2
Total input:	260,000 tons		1,300,000 tons	

Source: Adapted from *Oil in the Sea III Inputs, Fates, and Effects*.[11]

Table 3 Top 10 oil spills in the world as of 2010.

	Incident	Location	Year	Type of incident	Magnitude of oil spill (gallons)
1	Gulf War	Kuwait	1991	Oil spill due to war action and sabotage of oil drilling stations and pipelines, encompassing also the dumping of the charge of several oil tankers into the Persian Gulf by Iraqi troops during the Gulf War.	520,000,000
2	Deepwater Horizon	Gulf of Mexico	2010	Oil spill as a consequence of a methane blowout (which could not be prevented due to a technical problem) at the oil rig Deepwater Horizon, which caused an explosion and and the subsequent loss of the oil drilling platform. The well continued to leak for over 100 days.	172,000,000
3	Ixtoc I	Mexico	1979	After an unexpected blowout at the offshore oil rig Ixtoc 1 in the Gulf of Mexico, the platform exploded and collapsed. Oil escaped freely from the well for almost 1 year until the well could be capped.	138,000,000
4	Atlantic Empress/ Aegean Captain	Trinidad and Tobago	1979	Collision of two ships, the Aegean Captain and the supertanker Atlantic Empress, during a heavy storm in the Caribbean Sea. The Atlantic Empress exploded, sank, and lost its freight.	90,000,000
5	Fergana Valley/ Mingbulak	Russia	1992	Technical failure of an oil well in the Fergana Valley located between Kyrgyzstan and Uzbekistan from which oil blew out for a period of 8 months.	88,000,000
6	Nowruz Oil Field	Persian Gulf	1983	Collision of an oil tanker with an oil platform at the Nowruz Oil Field during the Iran–Iraq War. After the oil drilling platform collapsed, the wellhead was destroyed and leaked oil into the Persian Gulf for more than 6 months before being capped. A similar event at the same oilfield resulted directly form war action.	80,000,000
7	Castillo de Bellver	South Africa	1983	A fire at the tanker Castillo de Bellver caused the ship to drift and then break into two separate pieces. Relatively little damage was done to the South African coastline since the oil may have sunk into the sea or burned during the fire.	79,000,000
8	The Amoco Cadiz	France	1978	The crude oil carrier Amoco Cadiz ran aground off the French Atlantic coast and finally spilt into halves, whereby it lost its complete freight, which contaminated 200 km of the French coastline.	69,000,000
9	ABT Summer	Angola	1991	Following a fire aboard the oil tanker *ABT Summer*, it sank and all its freight either leaked to the sea or sank to the ground about 900 miles from the coast of Angola.	51,000,000
10	The MT Haven	Genova, Italy	1991	After unloading the oil tanker MT Haven, a fire broke out, followed by explosions after which the ship sank and continued to leak oil for 12 years.	45,000,000

Source: Adapted from Top 10 Worst Oil Spills.[12]

the components in most petroleum products and are the compounds that are primarily (but not always) measured as total petroleum hydrocarbons (TPH).[16] The non-hydrocarbon components are heterocyclic hydrocarbons (compounds containing heteroatoms such as sulfur, nitrogen, or oxygen in addition to carbon and hydrogen). These heterocyclic hydrocarbons are typically present in oils at relatively low concentrations and can be found in most refined motor fuels as they are concentrated in the heavier fractions and residues during refining. Most organic nitrogen hydrocarbons in crude oils are present as alkylated aromatic heterocycles, mostly with a pyrrolic structure. Crude oils also contain small amounts of organometallic compounds (of nickel, vanadium, and other metals up to atomic number 42, with the exception of rubidium and niobium) and inorganic salts. Although, depending on the analytical method, sulfur-, oxygen-, and nitrogen-containing compounds are sometimes included in the value reported as TPH concentration, they do not fall under the definition of petroleum hydrocarbons in the strict sense.[16]

Depending on the structure of petroleum hydrocarbons, the individual compounds are grouped into aliphatic

Table 4 Main constituents of petroleum hydrocarbons and representative examples.

	Petroleum hydrocarbon compounds						
	Aliphatics/alicyclics				Aromatics		
	Saturated hydrocarbons		Unsaturated hydrocarbons		Benzene and alkylbenzenes (BTEX)	Polynuclear aromatics (PAH)	Heterocyclic compounds
	Alkanes (paraffins)	Cycloalkanes	Alkenes (olefins)	Alkynes (acetylenes)			
	Single carbon bonds, straight and branched structure	Straight and cyclic structure	One or more double carbon bonds, straight, branched, or cyclic	One or more triple carbon bonds, straight, branched, or cyclic	Single aromatic ring or with attached functional group	Two or more aromatic rings fused together, can be with attached functional group	Aromatic ring structures with one or more heteroatoms (N, S, O) in the ring
Formula	C_nH_{2n+2}	C_nH_{2n}	C_nH_{2n}	C_nH_{2n-2}			
Example	n-Decane; 3-Methylnonane	Cyclohexane	1-Octene	1-Hexyne	Benzol; Toluene	Naphthalene	Pyrrole

(saturated and unsaturated) hydrocarbons and aromatics. Saturated hydrocarbons are the major class of compounds found in crude oil. The common names of these types of compounds are alkanes and isoalkanes or, as used in petroleum industry, paraffins and isoparaffins, respectively. Unsaturated hydrocarbons have at least one multiple bond (double bond [alkenes] or triple bond [alkynes]), and they are typically not present in crude oil but can be formed during the cracking process. Aromatic hydrocarbons are based on the benzene ring structure and are further categorized depending on the number of rings. Benzene rings are very stable and therefore persistent in the environment, and particularly, the mono- and polycyclic aromatic compounds can have toxic effects on organisms. Aromatic hydrocarbons with one benzene ring and with one or more side chains are alkyl benzenes and include benzene; toluene; ethylbenzene; and o-, p-, and m-xylenes (BTEX). This class of compounds has significant water solubility and is more mobile in the environment. Polycyclic aromatic hydrocarbons (PAHs) are aromatic compounds with two or more fused aromatic rings. Occurrence of PAH compounds in oils is dominated almost completely by the C1- to C4-alkylated homologues of the parent PAH, in particular, for naphthalene, phenanthrene, dibenzothiophene (a sulfur-containing aromatic heterocycle), fluorine, and chrysene. These alkylated PAH homologues form the basis of chemical characterization and identification of oil spills.[17,18] A typical crude oil may contain 0.2% to more than 7% total PAHs. Of the hydrocarbon compounds common in petroleum, PAHs appear to pose the greatest toxicity to the environment.

Different crude oil sources usually have a unique hydrocarbon composition.[19,20] The actual overall properties of each different petroleum source are defined by the percentage of the main hydrocarbons found within petroleum as part of the petroleum composition. The percentages for these hydrocarbons can vary greatly. It gives the crude oil a quite-specific compound personality depending on geographic region. The typical percentage of hydrocarbons (although covering very wide ranges) is as follows: paraffins (15%–60%), naphthenes (30%–60%), aromatics (3%–30%), and asphaltenes making up the remainder. Furthermore, due to differences in refining technologies and refinery operating conditions, each refining process has a distinct impact on the hydrocarbon composition of the product.

Refined petroleum products are primarily produced through distillation processes that separate fractions from crude oil according to their boiling ranges. Production processes may also be directed to increase the yield of low-molecular-weight fractions, reduce the concentration of undesirable sulfur and nitrogen components, and incorporate performance-enhancing additives. Therefore, each petroleum product has its unique, product-specific hydrocarbon pattern. The petroleum products are composed of both aliphatic and aromatic hydrocarbons in a range of molecules that include C6 and greater. The different classes of compounds contained in various petroleum products are summarized in Table 5.[20,21] The main products are gasoline (benzene), naphtha/solvents, jet fuels, kerosene, diesel fuel, and lubricating (motor) oils. Due to the variety of components in petroleum, they are typically characterized using the boiling range of the mixture and the carbon number rather than individual components. For example, diesel is a fraction with boiling points between 200°C and 325°C and is represented as C10–C22.

While a physical property such as boiling range may establish the initial product specification, other finer specifications define their ultimate use in certain applications.

Table 5 Overview of petroleum products with respect to boiling point ranges, approximate carbon number, and average percentage amount of aliphatic and aromatic compounds.

	Boiling range	Fractions				Hydrocarbons	
		<C7 (% w/w)	C7–C10 (% w/w)	C10–C40 (% w/w)	>C40 (% w/w)	Aliphatic	Aromatic
Statfjord C (39.1)[a]		11.6	18.1	56.6	13.7		
Crude oil (API = 18.7) Grane[a]		0.9	3.0	63.2	32.8		
Normal benzine[b]	40–200°C	~100 (C5–C12)				~70%	20%–50%
Jet fuel[b]	150–300°C		~100 (C6–C14, C16)			80%–90%	10%–20%
Kerosene[b]	150–300°C		~100 (C6, C9–C16)			60%–80%	5%–20%
Diesel[b]	200–325°C		~100 (C10–C22)			60%–90%	30%–40%
Light heating oil[b]	200–325°C		~100 (C10–C22)				
Lubricant or motor oil[b]	325–600°C			~100 (C20–C40)		70%–90%	10%–30%
Heavy heating oil[b]	325–600°C			~100 (C20–C50)			

[a] **Source:** Data from Crude Oil Assays.[20]
[b] **Source:** Data from Statoil Web site.[21]

A lighter, less dense, raw petroleum composition with a composition that contains higher percentages of hydrocarbons is much more profitable as a fuel source. On the other hand, other denser petroleum compositions with a less flammable level of hydrocarbons and containing higher levels of sulfur are expensive to refine into a fuel and are therefore more suitable for plastics manufacturing and other uses. In contrast to the ever-increasing demand, the world's reserves of light petroleum (light crude oil) are severely depleted, and refineries are forced to refine and process more and more heavy crude oil and bitumen.

Petroleum fractions are among the most complex samples an analyst can face in terms of the number of compounds present. The characterization of petroleum fractions is typically done by gas chromatography (GC). As can be seen in Fig. 1, the petroleum products contain such a large number of hydrocarbon constituents that complete chromatographic separation is not possible. Even then, GC remains the most informative analytical technique, providing both quantitative information (deduced from the total signal recorded in a chromatogram) and qualitative information, which derives from the fact that the retention times in the chromatograms can be correlated with the boiling points of the compounds contained in the petroleum. To illustrate the complexity of chemical composition of petroleum products, Fig. 2 shows the chromatograms for six different petroleum products, including a crude oil with API of 18.7 and the BAM (Bundesanstalt für Materialprüfung, Berlin) petroleum hydrocarbon standard. The BAM standard K-010 is a certified reference material for the determination of mineral hydrocarbons, which is a synthetic mixture of a diesel and a lubricating oil. It is evident that these six samples are very different according to their carbon ranges. The difference is clearly seen from the comparison of their chromatograms. The volatile fuel with a content of hydrocarbons with less than 10 C atoms (benzine

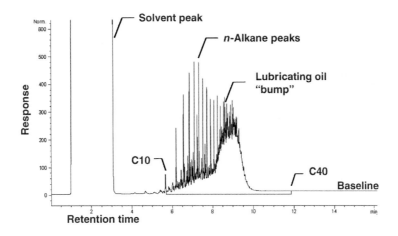

Fig. 1 Chromatogram of a mixture of petroleum products (diesel and lubricating oil, 1:1), obtained by GC with FID.

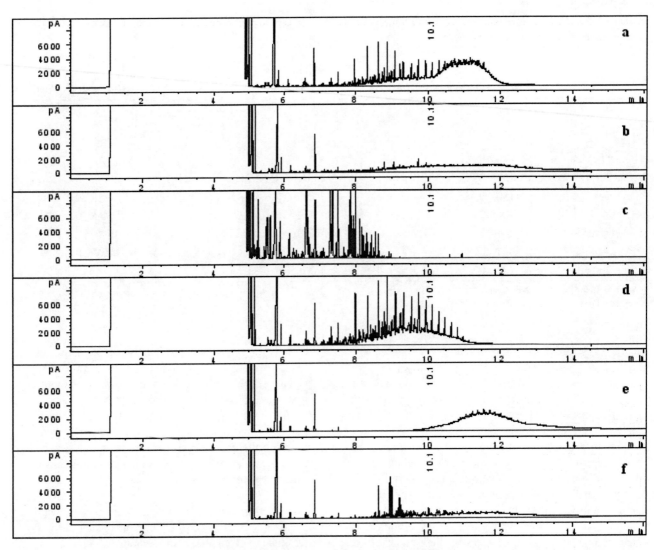

Fig. 2 Comparison of chromatograms of different oil samples: (a) BAM; (b) crude oil (API = 18.7); (c) gasoline; (d) diesel; (e) motor oil; and (f) heavy heating oil with the same concentration (20 ppm oil in water) obtained by GC-FID method.

and premium gasoline) has the majority of its constituents at the beginning of chromatogram (Fig. 2c). The peaks in the chromatogram of diesel are shifted to the retention time window where hydrocarbons from C10 to C22 are eluted (Fig. 2d). In turn, the chromatogram of motor oil shows a characteristic "bump" (because the fraction of saturated alkanes is very small) situated in the region where heavier hydrocarbons C20–C40 are eluted (Fig. 2e). Thus, GC-based methods provide important qualitative information, which in the ideal case even allows the assignment of the source of contamination. This is proven by the comparison of chromatograms of different oil samples (Fig. 2).

FATE OF PETROLEUM HYDROCARBONS IN THE ENVIRONMENT

The effects of petroleum hydrocarbons entering the environment are a complex function of the magnitude and the rate of release; the nature of the released petroleum (its physicochemical properties and, in particular, the amount of toxic compounds it may contain); and the affected geographical, hydrogeological, and biological ecosystem. The fate of petroleum-type pollutants in the environment has been investigated in many studies.[22] Complex transformation and degradation processes of oil in the environment start from its first contact with the atmosphere, seawater, and soil. They depend on the physical properties (volatility, solubility, etc.), as well as on the chemical properties (chemical composition) of the oil. While the former are responsible for transport, or diffusion of the petroleum hydrocarbons in the environment, the latter are responsible for their chemical, photo-, and microbial degradation. The main processes affecting the environmental fate of petroleum hydrocarbons after their release to the environment are thus their volatilization, dissolution/dispersion and emulsification in water, adsorption to soil, oxidation, destruction, and biodegradation.[23,24] In addition to the parameters that characterize the oil's composition, reactivity, and toxicity, the environmental conditions, i.e., the meteorological and hydrological factors, also play an important role in the fate of petroleum hydrocarbons.

When petroleum hydrocarbons are released to the water column, certain fractions will float on top and form thin surface films. This process is controlled by the viscosity of the oil and the surface tension of water. A spill of 1 ton of oil can disperse over a radius of 50 m in 10 min, forming a slick 10 mm thick. Later, it spreads, gets thinner, and covers an area of up to 12 km^2.[11] It should be pointed out that much of the environmental and ecological damage caused by oil spills actually is due to this oil film that covers the surface of the sea, or the coastline, thus physically impairing birds and other animals and causing suffocation of fish as oxygen will not permeate the oil layers to a sufficient degree anymore. In the first days after the spill, the volatile compounds from oil evaporate. Only a small proportion of the hydrocarbon constituents of petroleum products are significantly soluble in water. Dissolution takes more time compared with evaporation, considering that most oil components are soluble in water only to a limited degree (although the degradation products typically are more polar and thus more soluble). Other heavier fractions (up to 10%–30%) will accumulate in the sediment at the bottom of the water, which may affect bottom-feeding fish and organisms. This happens mainly in the narrow coastal zone and shallow waters, where water is intensively mixing.

Crude oil released to the soil may percolate and reach the groundwater. Because petroleum has a lower specific gravity than water, free (undissolved) product and most dissolved contamination are usually concentrated near the top of the groundwater.[25] This may then lead to a fractionation of the original complex mixture, depending on the chemical properties of the compound. Some of these compounds will evaporate, while others will dissolve into the groundwater and be diffused from the release area. Other compounds will adsorb to soil or sediments and will remain there for a long period of time, while others will be metabolized by organisms found in the soil.[26,27]

While evaporation and dissolution redistribute the oil, photochemical oxidation and bacterial degradation transform it. Where crude oil is exposed to sunlight and oxygen in the environment, both photooxidation and aerobic microbial oxidation take place. The photochemical oxidation of hydrocarbons is dependent upon ultraviolet (UV) radiation and will therefore occur only in the upper surface layers. The aromatic hydrocarbons absorb UV radiation with high efficiency and are transformed mainly into hydrogen peroxides. Alkanes are much less efficient in absorbing UV radiation, and only small quantities are transformed by this process. The final products of oxidation (hydroperoxides, phenols, carboxylic acids, ketones, aldehydes, and others) usually have increased water solubility and toxicity. Where oxygen and sunlight are excluded in anoxic environments, anaerobic microbial oxidation takes place.[28,29]

Generally, saturated alkanes are more quickly degraded by microorganisms than aromatic compounds; alkanes and smaller-sized aromatics are degraded before branched alkanes, multiring and substituted aromatics, and cyclic compounds.[30,31] Polar petroleum compounds such as sulfur- and nitrogen-containing species are the most resistant to microbial degradation. Complex structures (e.g., branched methyl groups) and the stability of hydrocarbons decrease the rates of mineralization, which are likely a consequence of the greater stability of carbon–carbon bonds in aromatic rings than in straight-chain compounds. Emulsification also provides greater surface area for microorganisms to attach.

It has been shown in experiments that n-alkanes are among the most biodegradable hydrocarbons, and therefore, they are easily broken down and preferentially depleted from soil samples.[32] Also, it has been proven in simulation experiments of the biodegradation of two different samples of crude petroleum (paraffinic and naphthenic type) that microbial cultures that were isolated as dominant microorganisms from the surface of a wastewater canal of an oil refinery (most abundant species: *Phormidium foveolarum*, filamentous Cyanobacteria [blue-green algae] and *Achanthes minutissima*, diatoms, algae) show a strongly differentiated degradation behavior with clear preference for the degradation of n-alkanes and isoprenoid aliphatic alkanes.[33] As can be seen in Fig. 3, the largest degree of biodegradation was achieved in a medium containing the base nutrients $Ca(NO_3)_2 \cdot 4H_2O$, $K_2HPO_4 \cdot 7H_2O$, KCl, $FeCl_2$, and K_2SO_4, at pH $\approx$ 8 and exposed to light. Biodegradation activity is somewhat lower with the same medium in the dark. With a medium containing not only the nutrient broth but also organic compounds (tryptone, yeast extract, glucose, at pH $\approx$ 7), degradation occurs at a much lower rate, especially without light.

When crude oil or petroleum products are accidentally released to the environment, they are immediately subjected to a variety of weathering processes that lead to compositional changes and to the depletion of certain hydrocarbon compounds. Weathering processes include all previously mentioned physicochemical processes, such as dissolution, evaporation, photooxidation, polymerization, adsorptive interactions between hydrocarbons and the soil, and some biological factors. Furthermore, due to the fact that the degree of biodegradation is different for different types of petroleum hydrocarbons and varies depending on their nature, the weathering rate also depends on the type of petroleum contaminant. If we thus observe in the analysis of petroleum hydrocarbon contaminants changing patterns of hydrocarbons with time, this may be either due to the segregation of the oil according to the physical properties or due to the action of bacteria and microorganisms. As these are able to degrade only certain classes of compounds, or at least they exhibit a strong preference for some over other compounds, characteristic changes of the hydrocarbon pattern will result, as observed by GC (Fig. 3).

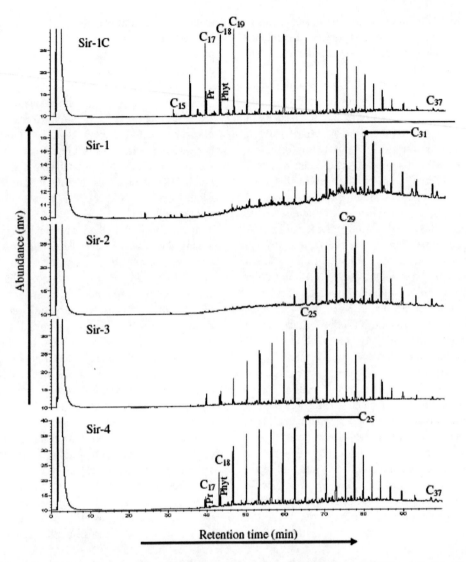

Fig. 3 Gas chromatograms of the alkane fractions derived from crude oil Sirakovo (Sir, paraffinic type) after 90 days of simulated biodegradation with *Phormidium foveolarum* and *Achanthes minutissima* with inorganic medium in the light (Sir-1), with inorganic medium in the dark (Sir-2), with organic medium in the light (Sir-3), and with organic medium in the dark (Sir-4), together with chromatogram of alkane fraction typical for the control experiments (Sir-1C), pristane (Pr), phytane (Phyt).
Source: Adapted from Antić et al.[33]

POSSIBLE TOXIC EFFECTS FROM EXPOSURE TO PETROLEUM HYDROCARBONS

As it was discussed earlier, crude oil and petroleum products are complex mixtures of groups of compounds. Many of the compounds are apparently benign, but many others are known to have toxic effects. Due to petroleum hydrocarbon toxicity, spilled hydrocarbons pose a threat that affects not only the sea and land but also the lakes, rivers, and groundwater and can be harmful for animals and human health.

Much of what is known about the impacts of petroleum hydrocarbons comes from studies of catastrophic oil spills and chronic seeps. Large oil spills usually receive considerable public attention because of the obvious environmental damage, oil-coated shorelines, and dead or moribund wildlife, including, in particular, oiled seabirds and marine animals. The acute toxicity of petroleum hydrocarbons to marine organisms is dependent on the persistence and bioavailability of specific hydrocarbons. The exposure to them may alter an organism's chances for survival and reproduction in the environment, and the narcotic effects of hydrocarbons on nerve transmission are a major biological factor in determining the ecologic impact of any release. Marine birds and mammals may be especially vulnerable to oil spills. In addition to acute effects such as high mortality, chronic, low-level exposure to hydrocarbons may affect reproductive performance and physiological impairment of seabirds and some marine mammals as well.[11] Petroleum contamination may also cause unfavorable impacts on nearby plants and animals. Plants growing in contaminated soils or water may die or appear distressed. In turn, natural seeps, leaking pipelines, and production discharges release small amounts of oil over long periods of time, resulting in chronic exposure of organisms to oil and oil chemical compounds. The lower-molecular-weight compounds are usually the more water-soluble components of a product, and hence, attention has also been paid to the water-soluble fractions of petroleum and related products. Concentrations in the environment are usually comparatively low, and chronic effects are usually more significant.[26] The persistence of some compounds such as PAH in sediments, especially in urban areas, is also an example of chronic pollution and toxicity.

Nowadays, humans can be exposed to petroleum hydrocarbons through ingestion of contaminated drinking water and soil residues; inhalation of vapors and airborne soils; and contact of contaminants with skin (dermal exposure) from many sources, including gasoline fumes at the pump, spilled crankcase oil on pavement, chemicals used at home or work, or certain pesticides that contain petroleum hydrocarbon components as solvents. Most petroleum hydrocarbon constituents will enter the bloodstream rapidly when inhaled or ingested. Incorporated petroleum hydrocarbons are widely distributed by the blood throughout the body and quickly are metabolized into less harmful compounds. Others may be degraded into more harmful chemicals. Even other compounds are distributed by the blood to other parts of the body and do not readily break down but are accumulated instead in fat tissue. The resorption of petroleum compounds through dermal tissue is slower; that is why direct exposure of the skin to petroleum hydrocarbons is generally harmless when exposure is only occasional and of short duration.

Studies on animals have shown effects on the lungs, central nervous system, liver, kidney, developing fetus, and reproductive system from exposure to petroleum compounds, generally after breathing or swallowing the compounds. Health impacts of exposure to petroleum contamination may include lung irritation, headaches, dizziness, fatigue, diarrhea, cramps, and nervous system effects. Benzene and other chemicals found in petroleum products have been determined to be carcinogenic (cause cancer). More information regarding toxicity of petroleum chemicals is available, for example, from the Agency for Toxic Substances and Diseases Registry (ATSDR), an agency of the U.S. Department of Health and Human Services,[34] or from the European Chemicals Agency.[35]

Oil products are complex mixtures of hundreds of chemicals, with each compound having its own toxicity characteristics. There are many difficulties associated with assessing the health effects of such complex mixtures with regard to hazardous waste site remediation. This means that the traditional approach of evaluating individual components is largely inappropriate. Toxicity information is in the best case available for the pure product; however, once a petroleum product is released to the environment, it changes its composition as a result of weathering. These compositional changes may be reflected in changes in the toxicity of the product.

One approach for assessing the toxicity of oil products is to use toxicity information from studies conducted on the whole product. A second approach is to identify and quantify all components and then consider their toxicities. This approach produces data that theoretically could be compared with the known toxicity of each compound. The impracticality of this approach stems from its high analytical cost and the lack of toxicity data for many of the component chemicals found in hydrocarbon mixtures. A third approach is to consider a series of hydrocarbon fractions and determine appropriate tolerable concentrations and toxicity specific for those fractions. A number of groups have examined such an approach, but the most widely accepted and internationally used are the ones developed by the Total Petroleum Hydrocarbons Criteria Working Group (TPHCWG) and the Massachusetts Department of Environmental Protection (MA DEP) in the United States, although they have been subject to adjustments in many cases. For example, in the United Kingdom, the TPHCWG approach is modified and extended to consider heavier hydrocarbon fractions. It has been developed as part of the Environment Agency's[36] environment sciences program and published in documents related to petroleum hydrocarbons.[37]

The MA DEP introduced in 1994 the concept of petroleum hydrocarbon size-based fractions for use in evaluating the human health effects of exposure to complex mixtures of hydrocarbons[38,39] and provided oral toxicity values for each of the fractions. The toxicity value assigned for each fraction is used in dose–response evaluations. Cancer risks or hazard amounts are subsequently summed across the fractions to get the total values. The TPHCWG has developed and published a series of five monographs[16,40–42] detailing the data on petroleum hydrocarbons and, in addition, has developed tolerable intakes for a series of total hydrocarbon fractions. The TPHCWG independently identified largely similar groupings of hydrocarbon fractions with somewhat different toxicity values in 1997. Of the 250 individual compounds identified in petroleum by the TPHCWG, toxicity data were available for only 95. Of these 95, the TPHCWG concluded that there were sufficient data to develop toxicity criteria for only 25.

As there are differences in toxicity between different hydrocarbon compounds, it is impossible to accurately predict toxic effects of contamination for which only total hydrocarbon data are available. Health assessors often select surrogate or reference compounds (or combinations of compounds) to represent TPH so that toxicity and environmental fate can be evaluated. Correspondence dates relating the toxicologically derived hydrocarbon fractions and

Table 6 Oral and inhalation toxicity values by MA DEP for petroleum hydrocarbon fractions and individual compounds present in petroleum products.

Carbon range	Compound	Toxicity Value, RfD		Critical effect
		Inhalation mg/m^3	Oral mg/kg/day	
Aliphatic				
C5–C8		0.2	0.04	Neurotoxicity
	n-Hexane	0.2	0.06	
C9–C18		0.2	0.1	Neurotoxicity, hepatic, and hematological effects
C19–C32		NA	2	Liver granuloma
Aromatic				
C6–C8		Use individual RfCs for compounds in this range		
	Benzene	NA	0.03	
	Toluene	0.4	0.2	
BTEX	Ethylbenzene	1.0	0.1	
	Styrene	1.0	0.2	
	Xylene (o-, p-, m-)	NA	2	
C9–C18		0.05		Body weight reduction; hepatic, renal, and developmental effects
	Isopropylbenzene	0.4	0.1	
	Naphthalene	0.003	0.02	
	Acenaphthene	NA	0.06	
	Biphenyl	NA	0.05	
	Fluorene	NA	0.04	
	Anthracene	NA	0.3	
	Fluoranthene	NA	0.04	
	Pyrene	NA	0.03	
C9–C32			0.3	Neurotoxicity
C19–C32		NA		

NA, not applicable.
Source: Adapted from *The U.K. Approach for Evaluating Human Health Risks from Petroleum Hydrocarbons in Soil*[37] and *Interim Final Petroleum Report Development of Health-Based Alternative to the Total Petroleum Hydrocarbon TPH Parameter*.[38]

their toxicity values to the analytically defined reporting fractions (by MA DEP) are contained in Table 6 for ingestion and inhalation exposure. Inhaled or ingested volatile hydrocarbons have both general and specific effects. The toxicity values are represented as a reference dose (RfD), which is the U.S. Environmental Protection Agency's (EPA's) maximum acceptable oral dose of a toxic substance. Significant efforts have been undertaken by MA DEP to describe an approach for the evaluation of human health risks from ingestion exposure to complex petroleum hydrocarbon mixtures. The methods offered by MA DEP for determination of air-phase (APH), volatile (VPH), and extractable (EPH) petroleum hydrocarbons[43–45] are designed to complement and support the toxicological approach. The ranges of quantified hydrocarbons within each method and their reporting limits are shown in Table 7.

The components of petroleum can be generally divided into broad chemical classes: alkanes, cycloalkanes, alkenes, and aromatics. A review of Table 6 shows that a U.S. EPA RfD is available for only one alkane, n-hexane. In general terms, alkanes have relatively low acute toxicity, but alkanes having carbon numbers in the range of C5–C12 have narcotic properties, particularly following inhalation exposure to high concentrations, because of their relatively high volatility and low solubility in water. Repeated exposure to high concentrations, for example, of n-hexane (RfD, 0.06 mg/kg/day) may lead to irreversible effects on the nervous system. Hexane is considered to be the most toxic compound in the C5–C8 aliphatic fraction. No RfDs are available for other alkanes, nor for any cycloalkane or alkene. Alkenes exhibit little toxicity other than weak anesthetic properties. Alkanes and

Table 7 The ranges of hydrocarbons quantified within the methods for determination of APH, VPH, and EPH by MA DEP and their reporting limits.

	APH 28°C–218°C		VPH 36°C–220°C		EPH 150°C–265°C		
Aliphatic	C5–C8	C9–C12	C5–C8	C9–C12	C9–C18	C19–C36	
Aromatic		C9–C10		C9–C10		C11–C22	PAH
			Reporting limits				
For the individual target analytes							
In air phase	2–5 g/m^3						
In soil			0.05–0.25 mg/kg		20 mg/kg		0.2–1 mg/kg
In water			1–5 μg/L		100 μg/L		2–5 μg/L
For the collective hydrocarbon ranges							
In air phase	10–12 g/m^3						
In soil			5–10 mg/kg		20 mg/kg		
In water			100–150 μg/L		100 μg/L		

Source: Adapted from *Interim Final Petroleum Report Development of Health-Based Alternative to the Total Petroleum Hydrocarbon TPH Parameter*.[38]

cycloalkanes are treated similarly and have similar toxic effects.

Aromatic compounds with less than nine carbon atoms (such as BTEX) are evaluated separately because the toxicity values for each are well supported and these compounds have a wide range of toxicity. However, most of the smaller aromatic compounds have low toxicity, with the exception of benzene, which is a known human carcinogen (RfD, 0.029 mg/kg/day). Most petroleum hydrocarbon mixtures contain very low concentrations of PAHs. The major concern regarding PAHs is the potential carcinogenicity of some of these. Benzo(a)pyrene and benz(a)anthracene are classified as probable human carcinogens. Benzo(a)pyrene is normally considered to be the most potent carcinogenic PAH, but the carcinogenic potency of most PAHs is not well characterized. In case of spills of petroleum products affecting water, PAHs are not usually a specific concern; however, this concern becomes more specific if these compounds are released into the soil due to a bioaccumulation of PAH in soil.

Different regulations and guidelines to protect public health have been developed. These public health statements tell as well about petroleum hydrocarbons and the effects of exposure. The U.S. EPA[46] identifies the most serious hazardous waste sites in the United States. The EPA lists certain wastes containing petroleum hydrocarbons as hazardous. It regulates certain petroleum fractions, products, and some individual petroleum compounds. General health and safety data are as well discussed by the Energy Institute,[7] which is the main professional organization for the energy industry within the United Kingdom that promotes the safe, environmentally responsible, and efficient supply and use of energy in all its forms and applications. The Occupational Safety and Health Administration and the Food and Drug Administration are other agencies that develop regulations for toxic substances in the United States. The information provided by all of them is regularly updated as more information becomes available. The Dutch National Institute for Public Health and the Environment (RIVM), has been involved in a number of studies on risk assessment for petroleum hydrocarbons which were commissioned by the Dutch government and the European Commission.[47] Also the U.K. Environment Agency, mentioned before, is the leading public body protecting and improving the environment in the United Kingdom, including protection from petroleum contaminations.

TOTAL PETROLEUM HYDROCARBONS AND ANALYTICAL METHODS FOR DETERMINATION OF PETROLEUM HYDROCARBONS IN ENVIRONMENTAL MEDIA

Due to the compositional complexity of petroleum products, it is impossible to assess the extent of petroleum hydrocarbon contamination by directly measuring the concentration of each hydrocarbon contaminant. For this reason, at the present time, no single analytical method is capable of providing comprehensive chemical information on petroleum contaminants. Total petroleum hydrocarbon is one parameter and definition that is currently widely used for expressing the total concentration of nonpolar petroleum hydrocarbons in soil, water, or other investigated samples. In the United States, for example, there are no federal regulations or guidelines for TPH in general. Many states have standards for controlling the concentrations of petroleum hydrocarbons or components of petroleum products. These are designed to protect the public from the possible harmful health effects of these chemicals. Analytical methods are specified as well, many of which are considered to be methods for TPH. These generate basic information that is a surrogate for contamination, such as a single TPH concentration. Such data are not suitable for risk assessment. However, they are relatively quick and easy to obtain and can offer useful preliminary information.

The term TPH is widely used, but it is rarely well defined. In essence, TPH is defined by the analytical method—in other words, estimates of TPH concentration often vary depending on the analytical method used to measure it. Thus, the ATSDR defines the TPH as a term used to describe a broad family of several hundred chemical compounds that originally come from crude oil. In this sense, TPH is really a mixture of chemicals. As per the TPHCWG, TPH, also called "hydrocarbon index," refers sometimes to mineral oil, hydrocarbon oil, extractable hydrocarbon, oil, and grease. The TPHCWG also says that the TPH measurement is the total concentration of the hydrocarbons extracted and measured by a particular method, and it depends on the analytical method used for determination. According to the MA DEP, the TPH is also a loosely defined parameter, which can be quantified using a number of different analyses, and this parameter is an estimate of the total concentration of petroleum hydrocarbons in a sample. Again, depending on the analytical method used to quantify TPH, the TPH concentration may represent the entire range of petroleum hydrocarbons from C9 to C36 or the sum of concentrations of a number of single compounds (for instance, BTEX) and groups of compounds (fractions, e.g., primarily aliphatics C9–C18, C19–C36, and aromatics C11–C20). Great improvements in the definition and analysis of TPH were finally introduced by the International Organization for Standardization (ISO)[48] in 2000, when it published the standard method ISO 9377-2:2000[49] for the quality control of water in which a method for the determination of the hydrocarbon oil index within the C10–C40 range in waters by means of GC is specified. The definition of "hydrocarbon oil index by GC-FID" was introduced, which defines the fraction of compounds extractable with a hydrocarbon solvent, boiling point between 36°C and 69°C, not adsorbed on Florisil, and which may be chromatographed with retention times between those of n-decane ($C_{10}H_{22}$) and n-tetracontane ($C_{40}H_{82}$). (Substances complying with this definition are long-chain or branched aliphatic, alicyclic, aromatic, or alkyl-substituted aromatic hydrocarbons.)

The TPHCWG and MA DEP evaluated the risk implications and arrived at the conclusion that TPH concentration data cannot be used for a quantitative estimation of the human health risk. The same concentration of TPH may represent very different compositions and very different risks to human health and the environment because the TPH parameter includes a number of compounds of differing toxicities and the health effects associated with exposure to particular concentrations of TPH cannot be determined. For example, two sites may have the same amount of TPH, but constituents at one site may include carcinogenic compounds while these compounds may be absent at the other site. If TPH data indicate that there may be significant contamination of environmental media, then fractionated measurements and the separate determination of BTEX compounds and PAHs are necessary so that potential risk to human health can be quantitatively assessed.[50] The hydrocarbon index is thus a good indicator of the (magnitude of the) relative contamination of oil; however, it will not be suitable to give a true representation of the actual concentration of TPH in the investigated sample. There are several reasons why TPH data do not provide the ideal information for investigated samples and do not establish target cleanup criteria. This is due to many factors including the complex nature of petroleum hydrocarbons, their interaction with the environment over time, and the non-specificity of some of the methods used. The scope of the methods used for TPH determination varies greatly. There are few, if any, methods that are capable of quantifying all hydrocarbons without interference from non-hydrocarbons. All methods are subject to interferences from non-hydrocarbons, some to a greater extent than others.

There are numerous established analytical methods that are available for detecting, measuring, or monitoring TPH and its metabolites. Analytical methods used for analysis of petroleum hydrocarbons in environmental media should provide a sufficient degree of robustness. At the current time, however, the correctness and precision of results for the petroleum hydrocarbon determination strongly depend on the proper choice of method and measurement parameters whose correct selection is left to the judgment of the analyst. Besides methods that measure the TPH concentration, two other types of methods can be distinguished. These are methods that measure the concentration of a group or fraction of petroleum compounds and methods that measure individual petroleum constituent concentrations. For product identification, the results of analyses of the petroleum groups or fractions can be useful because they separate and quantify different categories of hydrocarbons. Individual constituent methods quantify concentrations of specific compounds that might be present in petroleum-contaminated samples, such as BTEX and PAHs, which can be used to evaluate human health risk.

There are several basic steps related to the separation of analytes of interest from a sample matrix prior to their measurement, such as extraction, concentration, and cleanup. These steps are common to the analytical processes for all methods, irrespective of the method type or the environmental matrix. Each of these steps together with the sampling, which is also an important step in performing petroleum analyses, affects the final result and has a certain impact onto the measurement uncertainty.[51,52]

Sample taking and sample handling have been recognized as probably the most significant factors that contribute random errors and uncertainties in the analysis of offshore oil in produced water. There are some general guidelines available through a number of studies that have been carried out on this subject. To separate the analytes from the matrix, extraction is performed using one of the many available extraction methods. Heating of the sample or purging with an inert gas can be used in the analysis of volatile compounds; solid-phase extraction or extraction into a solvent is usually applied for water samples, the latter extraction method also being used for soil samples. For some types of solid samples, the extraction efficiency depends on the extraction method and time. However, ultrasonication and extraction by shaking are equally used for this purpose. It was demonstrated by some studies that extraction and cleanup are the most crucial steps in sample preparation procedures. According to the results. the most critical factors affecting TPH recovery are extraction solvent and type of cosolvent, extraction time, adsorbent and its mass, and the TPH concentration.[53] The results of a study where the occurrence of matrix effects in the gas chromatographic determination of petroleum hydrocarbons in soil was evaluated indicate that solid-phase extraction does not appear to be effective enough in removing interfering matrix components from the extract.[54]

Most of the methods for the determination of TPH involve a cleanup step using Florisil (a particular form of magnesium silicate) and sodium sulfate (anhydrous), which essentially aims at removing the polar, non-petroleum hydrocarbons of biological origin and remaining traces of water. It appears that the found hydrocarbon concentration strongly depends on the used cleanup technique. The efficiency of the cleanup procedure for removing polar compounds is not limited to heteroatomic substances like O-, N-, or Cl-containing compounds. Also, some hydrocarbons have a tendency to adsorb on Florisil, e.g., aromatic compounds with π-electrons or alkyl aromatics. The TPH recoveries after a cleanup procedure might depend on the composition of the oil investigated. Lower TPH recoveries may be expected for oils containing high concentrations of unsaturated hydrocarbons or PAHs. Also, lubricating oils often contain different amounts and types of (non-petrogenic) additives that may behave differently from the other compounds during the cleanup procedure.[55] The results demonstrate also that the ratio of Florisil amount and extract volume are of importance for the recovery of the purified extracts.[56]

The three most commonly used TPH testing methods include GC,[49,57–60] infrared absorption (IR),[61,62] and gravimetric analysis.[63–65] Conventional TPH methods are summarized in Table 8.

Methods based on solvent extraction followed by quantitative IR measurement (at a frequency of 2930 cm^{-1}, which corresponds to the stretching vibration of aliphatic CH_2 groups) have been widely used in the past for TPH measurement because they are simple, quick, and inexpensive. However, the use of these methods has been discontinued, since the sale and use of Freons (required for the extraction of hydrocarbons from the sample) is no longer allowed, and Freons are generally phased out worldwide due to their ozone layer–destructing potential. Recently, a new IR-based method was introduced, based on Freon-free extraction. This method defines oil and grease in water and wastewater as the fraction that is extractable with a cyclic aliphatic hydrocarbon (for example, cyclohexane) and measured by IR absorption in the narrow spectral region of 1370–1380 cm^{-1} (which corresponds to the excitation frequency of the symmetrical deformation vibration of CH_3 groups) using mid-IR quantum cascade lasers.[62] The method also considers the volatile fraction of petroleum hydrocarbons, which is lost by gravimetric methods that require solvent evaporation prior to weighing, as well as by solventless IR methods that require drying of the employed solid-phase material prior to measurement. Similarly, a more complete fraction of extracted petroleum hydrocarbon is accessible by this method as compared with GC methods that use a time window for quantification, as petroleum hydrocarbons eluting outside these windows are also quantified. On the other hand, IR-based methods hardly provide any information on the chemical composition of the oil or the presence or absence of other relevant compounds (aromatics, PAHs). In contrast, they even detect compounds that are not typically considered as TPH, such as surfactants, which also may absorb IR radiation due to the presence of CH bonds. However, this statement is only partially true, since it depends mainly on the cleanup whether the IR method determines also compounds other than the TPH.

Gravimetric-based methods are also simple, quick, and inexpensive; they measure anything that is extractable by a solvent, not removed during solvent evaporation, and capable of being weighed. Consequently, they do not offer any selectivity or information on the type of oil detected. Gravimetric-based methods may be useful for oily sludges and wastewaters at high(er) concentrations but are not suitable for measurement of light hydrocarbons (less than C15), which will be lost by evaporation below 70–85°C.

Gas chromatography–based methods are currently the preferred laboratory methods for TPH measurement because they detect a broad range of hydrocarbons, they provide both sensitivity and selectivity, and they can be used for TPH identification as well as quantification. The potential of GC for producing information on the product-specific hydrocarbon pattern has been long

Table 8 Summary of common TPH methods.

Analytical method	Method name	Matrix	Scope of method	Carbon range	Approximate detection limits	Advantages	Limitations	Reference
GC based	DIN ISO 9377-2:2000	Water	Solvent (hydrocarbon) extraction, cleanup using Florisil, evaporation, 1 μL injection, GC-FID	C10–C40	0.1 mg/L	Detects broad range of hydrocarbons; provides information (e.g., a chromatogram) for identification	Does not quantify below C10; chlorinated compounds can be quantified as TPH	[49]
	OSPAR (2007)	Water	n-Pentane extraction, cleanup using Florisil, 50 μL injection, GC-FID	C7–C40 + TEX compounds	0.1 mg/L	Does not need preconcentration step; detect broad range of hydrocarbons and polar hydrocarbons; provide information for identification	Does not quantify below C7	[57]
	DIN ISO 16703:2005-12	Soil	Acetone/n-heptane extraction, cleanup using Florisil, evaporation, GC-FID	C10–C40	10 mg/kg	Detects broad range of hydrocarbons; provide information (e.g., a chromatogram) for identification	Does not quantify below C10; chlorinated compounds can be quantified as TPH	[58]
	DIN EN 14039:2004	Wastes	Acetone/n-heptane extraction, cleanup using Florisil, evaporation, GC-FID	C10–C40	10 mg/kg			[59]
IR based	EPA 418.1 (1991/1992)	Water, soil	Freon extraction, silica gel treatment to remove polar compounds	Most hydrocarbons with exception of volatile and very high hydrocarbons	1 mg/mL in water, 10 mg/kg in soil	Technique is simple, quick, and inexpensive	Freon is banned now; low sensitivity; lack of specificity; prone to interference; provides quantitation only	[62]
	ASTM D7678 - 11 (2011)	Water, wastewater	Solvent (cyclic aliphatic hydrocarbon) extraction, cleanup using Florisil, IR Absorption in the region of 1370–1380 cm^{-1} (7.25–7.30 mm)	Most hydrocarbons with volatile	0.5 mg/mL	Technique is simple, very quick; a more complete fraction of extracted petroleum hydrocarbon is accessible		[61]
Gravimetry	EPA 413.1 (1979) ASTM D4281-95(2005)e1	Most appropriate for wastewater, sludge, sediment	Freon extraction, solvent evaporation	Anything that is extractable (with exception of volatiles which are lost)	5 mg/mL in water, 50 mg/kg in soil	Technique is simple, quick, and inexpensive	Freon is banned now; lack of sensitivity not suitable for low boiling fractions; prone to interference (organic acids, phenols, and other polar hydrocarbons); provides quantitation only	[63,65]
	EPA 1664 (1999)	Most appropriate for water and wastewater	n-Hexane extraction, silica gel treatment to remove polar compounds, solvent evaporation	Anything that is extractable (with exception of volatiles which are lost)	5 mg/mL	Technique is simple, quick, and inexpensive	Low sensitivity; lack of specificity not suitable for low boiling fractions; prone to interference; provides quantitation only	[64]

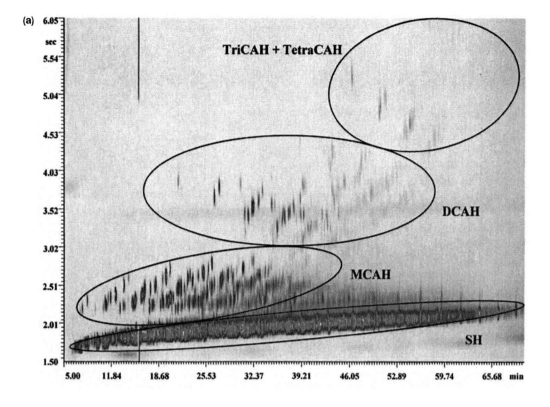
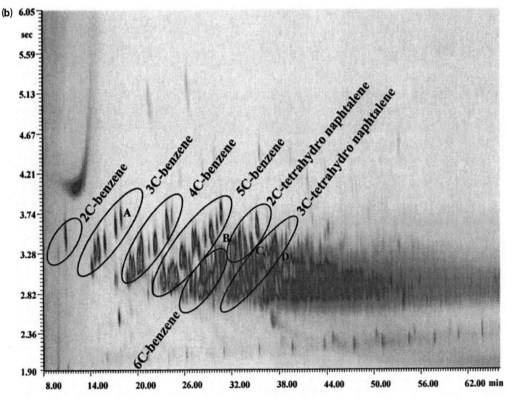

Fig. 4 (a) TIC GC×GC-qMS (quadruple-mass spectrometry) chromatogram of diesel oil. SH, saturated hydrocarbons; MCAH, monocyclic aromatics; DCAH, dicyclic aromatics; TriCAH, tricyclic aromatics; TetraCAH, tetracyclic aromatics. (b) TIC LC-GC×GC-qMS chromatogram of the monocyclic aromatic fraction of diesel oil. A) Indane, B) 1,2,3,4-Tetrahydro-2,7-dimethyl naphthalene, C) 1-Cyclohexyl 3-methyl benzene, D) 1,2,3,4-Tetrahydro-2,5,8-trimethyl naphthalene.
Source: Adapted from Sciarrone et al.[77]

recognized by researchers in the field of petroleum hydrocarbon analysis.[66–68]

Currently, there are several standard methodologies based on GC for different types of samples (water, soil, wastes). The ISO has published the standard ISO 9377-2:2000 for the quality control of water and specifies a method for the determination of the hydrocarbon oil index within the C10–C40 range in waters by means of GC. The method is suitable for surface water, wastewater, and water from sewage treatment plants and allows the determination of the hydrocarbon oil index in concentrations above 0.1 mg/L. Due to systematic differences, which became evident between the results from the DIN ISO method and those from the IR-based method, the GC-based method was subsequently modified.[69,70] As a result, the modified version of DIN ISO 9377-2:2000, the OSPAR (Oslo–Paris commission) reference method,[58] was published in 2005 and taken into force as a reference method in the field of petroleum production in January 2007. The OSPAR reference method is applicable for the determination of dispersed oil content in produced water and other types of wastewater discharged from gas, condensate, and oil platforms. It also allows the determination of the dispersed mineral oil content in concentrations above 0.1 mg/L and includes the determination of certain hydrocarbons within the C7–C10 range, with the TEX (toluene, ethylbenzene, and o-/p-/m-xylene) compounds being reported separately.

Gas chromatography–based methods are based on the extraction of water samples with a nonpolar (hydrocarbon) solvent, the removal of polar substances by cleanup with Florisil, and capillary GC measurements using a nonpolar column and a flame ionization detector (FID), cumulating the total peak area of compounds eluted between n-decane ($C_{10}H_{22}$) and n-tetracontane ($C_{40}H_{82}$) for the DIN ISO 9377-2:2000 standard method and for the DIN ISO 16703:2005-12[59] standard method for soil samples. The OSPAR method was modified in order to include the determination of certain hydrocarbons with a boiling point between 98°C and 174°C (that is, from n-heptane to n-decane), with the TEX compounds being determined separately by integration and subtraction of their peak areas from the total integrated area. The GC-based methods usually cannot quantitatively detect compounds with a lower boiling point than n-heptane because these compounds are highly volatile and are interfered by the solvent peak. Furthermore, the EPA method 8240,[61] which is used to determine volatile organic compounds in a variety of waste matrices by GC/mass spectrometry (MS), exists. It can be used to quantitate most volatile organic compounds that have boiling points below 20°C and that are insoluble or slightly soluble in water. The estimated quantitation limit of the EPA 8240 method for an individual compound is approximately 5 µg/kg (wet weight) for soil/sediment samples, 0.5 mg/kg (wet weight) for wastes, and 5 µg/L for groundwater.

Gas chromatography–based methods are suitable for surface water, wastewater, and other types of wastewater discharged from gas, concentrate, and oil platforms and allow the determination of hydrocarbon oil concentration above 0.1 mg/L. To reach the required detection limit, the method according to DIN ISO 9377-2:2000 foresees preconcentration of the extracts by solvent evaporation, which bears the risk of losing the more volatile constituents of the sample. In contrast to this, the OSPAR method does not allow for any external apparatus for preconcentration, for which reason the GC must be equipped with an injection system that allows the injection of a volume of up to 100 µL of the extract. This is most easily realized with programmed-temperature vaporizer large-volume injectors. This technique can reduce the loss of volatile analytes, can increase sensitivity, and is a viable, fast, and automated alternative to an external preconcentration procedure.[71–73]

Petroleum products easily contain thousands of different compounds. Classical capillary GC cannot resolve such mixtures up to the level of individual compounds. A powerful analytical tool for separation of complex mixtures, such as petroleum hydrocarbons, is comprehensive two-dimensional GC (GC×GC or 2D-GC).[74–76] The use of 2D-GC with MS detection (GC×GC/MS) is expected to not only allow the separation of the various constituents of complex TPH samples but also to identify them based on MS detection (Fig. 4b). It is known that a certain class of chemical compounds (a series of "homologues") forms a very distinct, clearly identifiable pattern in the two-dimensional space of the GC×GC separation. The diesel total ion (TIC) GC×GC/MS chromatogram, illustrated in Fig. 4a, is characterized by very typical group-type patterns: saturated hydrocarbons, which present low second-dimension retention times, are followed by monocyclic and dicyclic aromatics; tri- and tetracyclic aromatics are the most retained on the secondary polar column.[77] Moreover, partial overlapping between chemical groups occurs, the monocyclic aromatics are situated in a rather narrow band, and the tri- and tetracyclic aromatics are hardly visible in the two-dimensional chromatogram. The analytical potential of such a two-dimensional system is great.

CONCLUSION

Due to the importance and widespread use of petroleum hydrocarbons for energy production, for transport, and as a raw material in the chemical industries, there are many routes for their inadvertent or accidental release into the environment. Thus, they do represent one of the most important sources of large-scale environmental pollution. While petroleum hydrocarbons also are introduced into the oceans from natural seeps, these continuous emissions of comparatively low intensity represent a less significant environmental problem since the resident flora and fauna have adapted to this continuous input of hydrocarbons and effects are limited to local scale. Large oil spills in contrast exceed the self-cleaning capacity of the ecosystem, which cannot regenerate without human intervention to both physically and chemically immobilize, bind, and remove oil from the affected region. Although such techniques are available, large-scale oil spills always have caused severe damage to the environment, with the affected ecosystems recovering only slowly. Analytical methods are available for the qualitative and quantitative determination of the composition of oil samples and the assessment of pollution levels in various environmental compartments. Gas chromatographic techniques mostly have supplanted the former analytical standard method based on Freon extraction and mid-IR determination, but there is further research and development going on to develop either more powerful analytical methods—such as two-dimensional GC—or alternative detection methods, such as the ones based on mid-IR lasers as light sources.

ACKNOWLEDGMENTS

This report was compiled within the frame of project 818084-16604 SCK/KUG of the Austrian Science Foundation (FFG), whose financial support is gratefully acknowledged.

REFERENCES

1. Tissot, B.P.; Welte, D.H. *Petroleum Formation and Occurrence*; Springer-Verlag: Berlin, 1984.
2. Available at http://www.oxforddictionaries.com/view/entry/m_en_gb0623910#m_en_gb0623910 (accessed September 2011).
3. Available at http://www.nationmaster.com/graph/ene_oil_pro-energy-oil-production (accessed September 2011).
4. Available at http://www.eia.gov/petroleum/ (accessed September 2011).
5. Leffler, W.L. *Petroleum Refining in Nontechnical Language*, 4th Ed.; PennWell Corporation: Tulsa, OK, 2008.
6. Fahimm, M.A.; Al-Sahhaff, T.A.; Lababidii, H.M.S.; Elkilanii A. *Fundamentals of Petroleum Refining*, 1st Ed.; Elsevier: Oxford, U.K., 2010.
7. Available at http://www.energyinst.org (accessed September 2011).
8. Nancarrow, D.J.; Adams, A.L.; Slade, N.J.; Steeds, J.E. Land *Contamination: Technical Guidance on Special Sites: Petroleum Refineries*; R&D Technical Report P5-042/TR/05; Environment Agency: Bristol, 2001.
9. *Toxicological Profile for Total Petroleum Hydrocarbons (TPH)*; U.S. Department of Health and Human Services, Public Health Service Agency for Toxic Substances and Disease Registry: Atlanta, GA, 1999.
10. Leifer, I.; Kamerling, M.J.; Luyendyk, B.P.; Douglas, S.W. Geologic control of natural marine hydrocarbon seep emissions, Coal Oil Point seep field, California. Geo-Mar. Lett. **2010**, *30*, 331–338.
11. *Oil in the Sea III Inputs, Fates, and Effects*; The National Academies Press: Washington, DC, 2003.
12. Available at http://www.toptenz.net/top-10-worst-oil-spills.php (accessed September 2011).
13. Weggen, K.; Pusch, G.; Rischmüller, H. Oil and gas. In *Ullmann's Encyclopedia of Industrial Chemistry*; Wiley-VCH: Weinheim, 2000.
14. Sanin, P.I. Petroleum hydrocarbons. Russ. Chem. Rev. **1976**, *45* (8), 684–700.
15. Available at http://api-ep.api.org/ (accessed September 2011).
16. Analysis of petroleum hydrocarbons in environmental media. In *Total Petroleum Hydrocarbon Criteria Working Group Series*; Weisman. W., Ed.; Amherst Scientific Publishers: Amherst, MA, 1998; vol. 1.
17. Wang, Z.; Fingas, M.F. Development of oil hydrocarbon fingerprinting and identification techniques. Mar. Pollut. Bull. **2003**, *47*, 423–452.
18. Wang, Z.; Fingas, M.; Page, D.S. Oil spill identification. J. Chromatogr. A **1999**, *843*, 369–411.
19. Available at http://www.etc-cte.ec.gc.ca/databases/Oil Properties/oil_prop_e.html (accessed September 2011).
20. Available at http://www.statoil.com/en/ouroperations/tradingproducts/crudeoil/crudeoilassays/pages/default.aspx (accessed September 2011).
21. Available at http://www.castrol.com/liveassets/bp_internet/castrol/castrol_switzerland/STAGING/local_assets/downloads/a/ABC_D_Mai_2009.pdf (accessed September 2011).
22. *Spills of Emulsified Fuels: Risks and Responses*; The National Academy of Sciences: Washington, DC, 2001.
23. *Petroleum Products in Drinking-Water*; Background document for development of WHO Guidelines for Drinking-water Quality, WHO/SDE/WSH/05.08/123; World Health Organization: Geneva, 2005.
24. *Fate of Spilled Oil In Marine Waters: Where Does It Go? What Does It Do? How Do Dispersants Affect It?*; An Information Booklet for Decision Makers, Publication 4691; American Petroleum Institute: Virginia, 1999.
25. Afifi, S.M. Petroleum hydrocarbon contamination of groundwater in Suez: causes severe fire risk. Proceedings 24th AGU Hydrology Days, March 10-12, 2004, pp. 1–9. Colorado State University (2004)
26. Vaajasaar, K.; Joutti, A.; Schultz, E.; Selonen, S.; Westerholm, H. Comparisons of terrestrial and aquatic bioassays for oil-contaminated soil toxicity. J. Soils Sediments **2002**, *2* (4), 194–202.
27. *Guidelines for Assessing and Managing Petroleum Hydrocarbon Contaminated Sites in New Zealand*; Module 2—Hydrocarbon contamination fundamentals; Ministry for Environment: New Zealand, 1999.
28. Das, N.; Chandran, P. Microbial degradation of petroleum hydrocarbon contaminants: An overview. Biotechnol. Res. Int. **2011**, Article ID 941810, 13 pages.
29. Atlas, R.M. Microbial degradation of petroleum hydrocarbons: An environmental perspective. Microbiol. Rev. **1981**, *45*, 180–209.
30. Van Hamme, J.D.; Singh, A.; Ward, O.P. Recent advances in petroleum microbiology. Microbiol. Mol. Biol. Rev. **2003**, *67* (4), 503–549.
31. Kaplan, I.R.; Galperin, Y.; Lu, S.T.; Lee, R.P. Forensic environmental geochemistry: Differentiation of fuel-types,

their sources and release time. Org. Geochem. **1997**, *2*, 289–317.
32. Šepič, E.; Leskovšek, H.; Trier, C. Aerobic bacterial degradation of selected polyaromatic compounds and n-alkanes found in petroleum. J. Chromatogr. A **1995**, *697*, 515–523.
33. Antić, M.P.; Jovančićević, B.S.; Ilić, M.; Vrvić, M.M.; Schwarzbauer, J. Petroleum pollutant degradation by surface water microorganisms. Environ. Sci. Pollut. Res. **2006**, *13* (5), 320–327.
34. Available at http://www.atsdr.cdc.gov/ (accessed September 2011).
35. Available at http://echa.europa.eu/home_en.asp (accessed September 2011).
36. Available at http://www.environment-agency.gov.uk/ (accessed September 2011).
37. *The U.K. Approach for Evaluating Human Health Risks from Petroleum Hydrocarbons in Soil*; Science report P5-080/TR3; Environment Agency: Bristol, U.K., 2005.
38. MA DEP 1994. *Interim Final Petroleum Report Development of Health-Based Alternative to the Total Petroleum Hydrocarbon TPH Parameter*; Massachusetts Department of Environmental Protection: Boston, Massachusetts, 1994.
39. MA DEP 2003. *Updated Petroleum Hydrocarbon Fraction Toxicity Values for the VPH/EPH/APH*; Massachusetts Department of Environmental Protection: Boston, Massachusetts, 2003.
40. Edwards, D.A.; Andriot, M.D.; Amoruso, M.A.; Tummey, A.C.; Tveit, A.; Bevan, C.J.; Hayes, L.A.; Youngren, S.H.; Nakles, D.V. Development of fraction specific reference doses (RfDs) and reference concentrations (RfCs) for total petroleum hydrocarbons. In *Total Petroleum Hydrocarbon Criteria Working Group Series*; Amherst Scientific Publishers: Amherst, Massachusetts, 1997; Vol. 4.
41. Potter, T.L.; Simmons, K.E. Composition of petroleum mixtures. In *Total Petroleum Hydrocarbon Criteria Working Group Series*; Amherst Scientific Publishers: Amherst, Massachusetts, 1998; Vol. 2.
42. Vorhees, D.J.; Weisman, W.H.; Gustafson, J.B. Human health risk-based evaluation of petroleum release sites: implementing the working group approach. In *Total Petroleum Hydrocarbon Criteria Working Group Series*; Amherst Scientific Publishers: Amherst, Massachusetts, 1999; Vol. 5.
43. MA DEP 2003. *Method for the Determination of Air-Phase Petroleum Hydrocarbons (APH)*; Massachusetts Department of Environmental Protection: Boston, Massachusetts, 2009.
44. MA DEP 2004. *Method for the Determination of Volatile Petroleum Hydrocarbons (VPH)*; Massachusetts Department of Environmental Protection: Boston, Massachusetts, 2004.
45. MA DEP 2004. *Method for the Determination of Extractable Petroleum Hydrocarbons (EPH)*; Massachusetts Department of Environmental Protection: Boston, Massachusetts, 2004.
46. Available at http://www.epa.gov/ (accessed September 2011).
47. Verbruggen, E.M.J. *Environmental Risk Limits for Mineral Oil (Total Petroleum Hydrocarbons)*; RIVM report 601501021; National Institute for Public Health and the Environment: Bilthoven, the Netherlands, 2004.
48. Available at http://www.iso.org/iso/home.html (accessed September 2011).
49. DIN ISO 9377-2:2000. *Water Quality—Part 2, Method Using Solvent Extraction and Gas Chromatography*; International Organization for Standardisation: Geneva, 2000.
50. Pollard, S.J.T.; Duarte-Davidson, R.; Askari, K.; Stutt, E. Managing the risk from petroleum hydrocarbons at contaminated sites achievements and future research directions. Land Contam. Reclam. **2005**, *13* (2), 115–122.
51. Saari, E.; Perämäki, P.; Jalonen, J. Measurement uncertainty in the determination of total petroleum hydrocarbons (TPH) in soil by GC-FID. Chemom. Intell. Lab. Syst. **2008**, *92* (1), 3–12.
52. Becker, R.; Buge, H.G.; Bremser, W.; Nehls, I. Mineral oil content in sediments and soils: Comparability, traceability and a certified reference material for quality assurance. Anal. Bioanal. Chem. **2006**, *385* (3), 645–651.
53. Saari, E.; Perämäki, P.; Jalonen, J. Evaluating the impact of extraction and cleanup parameters on the yield of total petroleum hydrocarbons in soil. Anal. Bioanal. Chem. **2008**, *392* (6), 1231–1240.
54. Saari, E.; Perämäki, P.; Jalonen, J. Effect of sample matrix on the determination of total petroleum hydrocarbons (TPH) in soil by gas chromatography–flame ionization detection. Microchem. J. **2007**, *87* (2), 113–118.
55. Muijs, B.; Jonker, M.T.O. Evaluation of clean-up agents for total petroleum hydrocarbon analysis in biota and sediments. J. Chromatogr. A **2009**, *1216* (27), 5182–5189.
56. Koch, M.; Liebich, A.; Win, T.; Nehls, I. *Certified Reference Materials for the Determination of Mineral Oil Hydrocarbons in Water, Soil and Waste*; Forschungsbericht 272; Bundesanstalt für Materialforschung und - prüfung (BAM): Berlin, 2005.
57. OSPAR. *Reference Method of Analysis for Determination of the Dispersed Oil Content in Produced Water*; OSPAR Commission, ref. no. 2005-15: Malahide, published in 2005, taken into force in 2007.
58. DIN ISO 16703:2005-12. *Soil Quality—Determination of Content of Hydrocarbon in the Range C10 to C40 by Gas Chromatography*; International Organization for Standardisation: Brussels, 2005.
59. DIN EN 14039:2004. *Characterization of Waste—Determination of Hydrocarbon Content in the Range of C10 to C40 by Gas Chromatography*; German version; International Organization for Standardisation: Brussels, 2004.
60. EPA Method 8240. *Gas Chromatography/Mass Spectrometry for Volatile Organics, Test Methods for Evaluating Solid Wastes*; US Environmental Protection Agency: Washington, 1986, Vol. 1B.
61. ASTM Standard D7678111. *Standard Test Method for Total Petroleum Hydrocarbons (TPH) in Water and Wastewater with Solvent Extraction using Mid-IR Laser Spectroscopy*; ASTM International: West Conshohocken, PA, 2011; DOI: 10.1520/D7678-11.
62. EPA Method 418.1. *Total Recoverable Petroleum Hydrocarbons by IR, Groundwater Analytical Technical Bulletin*; Groundwater Analytical Inc.: Buzzards Bay, MA, 1991/1992.
63. EPA method 413.1. *Standard Test Method for Oil and Grease Using Gravimetric Determination*; issued in 1974, editorial revision in 1978 (withdrawn).

64. EPA method 1664. *Revision A: n-Hexane Extractable Material (HEM; Oil and Grease) and Silica Gel Treated n-Hexane Extractable Material (SGT-HEM: Non-polar Material) by Extraction and Gravimetry*; US Environmental Protection Agency: Washington, 1999.
65. ASTM D4281-95(2005)e1. *Standard Test Method for Oil and Grease (Fluorocarbon Extractable Substances) by Gravimetric Determination*; ASTM International: West Conshohocken, PA, 2005.
66. Blomberg, J.; Schoenmakers, P.J.; Brinkman, U.A.T. GC methods for oil analysis. J. Chromatogr. A **2002**, *972* (2), 137–173.
67. Beens, J.; Brinkman, U.A.T. The role of GC in compositional analyses in the petroleum industry. Trends Anal. Chem. **2000**, *19* (4), 260–275.
68. Saari, E.; Perämäki, P.; Jalonen, J. Evaluating the impact of GC operating settings on GC–FID performance for total petroleum hydrocarbon (TPH) determination. Microchem. J. **2010**, *94* (1), 73–78.
69. Thomey, N.; Bratberg, D.; Kalisz, C. A comparison of methods for measuring total petroleum hydrocarbons in soil. In Proceedings of the Petroleum Hydrocarbons and Organic Chemicals in Groundwater: Prevention, Detection and Restoration, November 15–17, 1989; National Water Well Association: Houston, Texas, 1989.
70. Xie, G.; Barcelona, M.J.; Fang, J. Quantification and interpretation of total petroleum hydrocarbons in sediment samples by a GC/MS method and comparison with EPA 418.1 and a rapid field method. Anal. Chem. **1999**, *71* (9), 1899–1904.
71. Hoh, E.; Mastovska, K. Large volume injection techniques in capillary gas chromatography. J. Chromatogr. A **2008**, *1186*, 2–15.
72. Miñones Vázqiez, M.; Vázquez Blanco, M.E.; Muniategui Lorenzo, S.; López Mahía, P.; Fernández-Fernández, E.; Prada Rodríguez, D. Application of programmed-temperature split/splitless injection to the trace analysis of aliphatic hydrocarbons by gas chromatography. J. Chromatogr. A **2001**, *919*, 363–371.
73. Dellavedova, P.; Vitelli, M.; Ferraro, V.; Di Toro, M.; Santoro, M. *Application of enhanced large volume injection; an approach to the analysis* of *petroleum hydrocarbons in water*. Chromatographia **2006**, *63*, 73–76.
74. Van De Weghe, H.; Vanermen, G.; Gemoets, J.; Lookman, R.; Bertels, D. Application of comprehensive two-dimensional gas chromatography for the assessment of oil contaminated soils. J. Chromatogr. A **2006**, *1137*, 91–100.
75. von Mühlen, C.; Alcaraz Zini, C.; Bastos Caramão, E.; Marriott, P. Applications of comprehensive two-dimensional gas chromatography to the characterization of petrochemical and related samples. J. Chromatogr. A **2006**, *1105*, 39–50.
76. van Deursen, M.M.; Beens, J.; Reijenga, J.C.; Lipman, P.J.L.; Camers, C.A.M.G.; Blomberg, J. Group-type identification of oil samples using comprehensive two-dimensional gas chromatography coupled to a time-of-flight mass spectrometer (GCxGC-TOF). J. High Resolut. Chromatogr. **2000**, *23* (7–8), 507–510.
77. Sciarrone, D.; Tranchida, P.Q.; Costa, R.; Donato, P.; Ragonese, P.; Dugo, P.; Dugo, G.; Mondello, L. Offline LC-GCxGC in combination with rapid-scanning quadrupole mass spectrometry. J. Sep. Sci. **2008**, *31*, 3329–3336.

Pharmaceuticals: Treatment

Diana Aga
Seungyun Baik
Department of Chemistry, State University of New York at Buffalo, Buffalo, New York, U.S.A.

Abstract

Pharmaceutical residues in the environment have become known as "emerging contaminants" because of their increasing frequency of detection in the aquatic and terrestrial systems. Both human and veterinary pharmaceuticals are introduced into the environment via many different routes, including discharges from municipal wastewater treatment plants (WWTPs), and via land application of animal manure and biosolids to fertilize croplands. Most conventional WWTPs cannot fully remove pharmaceuticals; thus, advanced water treatment systems are being explored to eliminate or minimize pharmaceuticals in WWTP effluents because of the potential ecological effects they may cause. While the concentrations of pharmaceuticals in drinking water sources are relatively low to be of concern to human health, the efficiency of different advanced oxidation techniques to remove recalcitrant pharmaceuticals in drinking water is being evaluated. The importance of identifying and assessing the toxicity of transformation products (biotic and abiotic) of pharmaceuticals in engineered treatment systems and under natural environments should not be ignored. Recent studies have demonstrated that transformation products of organic contaminants may sometimes exhibit residual toxicity or biological activity, or in few cases, may even be more potent than the parent compound itself.

INTRODUCTION

The presence of pharmaceutical chemicals and their by-products in soil, wastewater effluents, surface water, and drinking water sources has become a growing concern over the past two decades. Improvements in analytical methods coupled with large-scale surveys have revealed the broad range of persistent pharmaceuticals that are cycling through our wastewater-to-drinking water cycle. Nonmetabolized active ingredients and transformation products of veterinary and human pharmaceuticals are introduced into the environment through the effluents of municipal wastewater treatment plants (WWTPs), pharmaceutical formulation facilities, and through the land application of animal waste and sewage sludge. Approximately 10 million tons of sewage sludge and manure containing residues of pharmaceuticals are used to fertilize croplands each year. Public awareness of the potential problems related to pharmaceutical pollution, widely known as "emerging contaminants," has brought this issue to the forefront in the water and wastewater treatment industries. Hence, advanced water treatment systems are being evaluated to potentially eliminate these emerging contaminants from effluents of WWTPs and from drinking water sources. This entry aims to provide an overview on the occurrence of pharmaceuticals in the environment and recent studies that investigate promising treatment technologies to eliminate emerging contaminants from wastewater and drinking water systems.

Although the human risk associated with chronic exposure to pharmaceuticals in the environment remains unclear, evidence of detrimental ecological impacts is growing. The environmental contamination by pharmaceutical residues, especially antibiotics, may have profound environmental effects at several levels. While the promotion of antibiotic resistance in pathogenic microorganisms has been the major concern associated with the presence of antibiotics in the environment, other issues such as endocrine disruption in fish and wildlife, plant uptake, and phytotoxicity are also significant and warrant discussion. Therefore, the second goal of this entry is to summarize current knowledge on the ecological impacts of pharmaceutical pollution.

OCCURRENCE AND IMPACTS OF PHARMACEUTICALS IN THE ENVIRONMENT

The presence of pharmaceutical residues in terrestrial and aquatic systems, resulting largely from discharges of municipal WWTPs and the land application of animal wastes, is now well documented in the literature.[1–3] As depicted in Fig. 1, residues of human pharmaceuticals and their metabolites may eventually enter surface water, groundwater, and drinking water systems after passing through WWTPs.

While most active ingredients of drugs are metabolized in the body, or removed during wastewater treatment, others remain intact and persist in the environment. Low levels of persistent pharmaceuticals can eventually end up in finished drinking water and distribution system (tap) water, when using source waters that have been affected by effluents from WWTPs.[4]

Veterinary pharmaceuticals, particularly antibiotics used for therapeutic purposes and for growth promotion, are also finding their ways into the environment. While some antibiotics used in animal production are decomposed quickly after being excreted, others remain stable during manure storage and end up in agricultural fields upon manure application. Additionally, antibiotics are widely used in fish farms and may enter the aquatic environment via direct discharge. For example, the antibiotic oxytetracycline has been detected in groundwater that has been affected by fish farming at sub-parts-per-billion concentrations.[5] Highly polar pharmaceuticals are susceptible to leaching and may therefore reach the groundwater aquifer. For example, the high frequency of detection of sulfonamide antibiotics in groundwater from various sites in the United States can be attributed to the relatively high water solubility and poor biodegradability of these drugs.[6,7] The detection of sulfonamides in groundwater from wells deeper than 50 ft, located downgradient from an animal feeding operation in the United States, indicates the persistence and mobility of this class of antibiotics.[8] Similar results were observed in Germany where sulfonamides were detected in groundwater located downgradient from an agricultural field where sewage sludge was used for irrigation.[9] The concentrations of pharmaceuticals in groundwater are typically lower than in surface water because infiltration through the soil profile removes some fraction of the organic pollutants before entering groundwater systems.[10] Nevertheless, contamination of groundwater by pharmaceuticals is a significant issue because these compounds are not readily biodegradable under anoxic conditions.[9] The persistence of pharmaceuticals in the groundwater aquifer is a concern because groundwater is a major source of drinking water in many areas in the United States and around the world.[8] If the contaminated groundwater is used as drinking water source, pharmaceuticals may eventually reach finished drinking water systems[4] and may not be degraded during conventional drinking water treatment processes.

In the first comprehensive study on the occurrence of pharmaceuticals in the United States surface waters, Kolpin et al.[1] reported detection of 95 organic contaminants, which included 30 antibiotics, 12 prescribed drugs, 4 nonprescribed drugs, and 6 drug metabolites. A later report summarized the occurrence of 80 pharmaceuticals and drug metabolites in the aquatic systems for eight different coun-

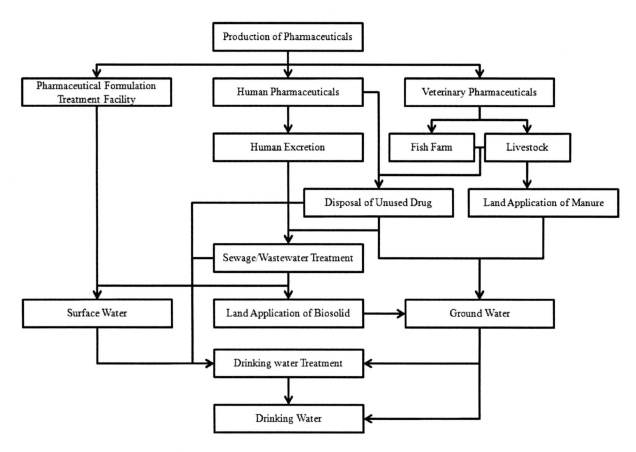

Fig. 1 Exposure routes of pharmaceuticals in the environment.

tries.[11] The concentrations of pharmaceuticals in surface water and groundwater depend on several factors, such as removal processes employed in the WWTPs, source variability, dilution, retardation, and weather events.[12]

Table 1 shows examples of the types of pharmaceuticals frequently detected in the environment and their typical removal rates (high, medium, low) in conventional activated sludge (CAS) systems during wastewater treatment.[13,14] Consequently, these pharmaceuticals are often detected in receiving surface waters at a wide range of concentrations as depicted in Fig.2. The variability in pharmaceutical concentrations in the environment can be attributed to the types of WWTPs, the frequency and time of sampling, and the biodegradability of the pharmaceuticals in a given environmental conditions, among others.

Despite the relatively low concentrations of pharmaceuticals typically found in the environment, the ecological effects of pharmaceutical pollution cannot be ignored. Several standard toxicity tests reveal that while acute toxicity is not a concern because of the high effective concentrations needed to elicit observable acute effects, chronic toxicity may be important at environmentally relevant concentrations.

Acute and chronic toxicity of several pharmaceuticals, including non-steroidal anti-inflammatory drugs (NSAIDs) and β-blockers, against phytoplankton, zooplankton, and other aquatic organisms[21–23] have been reported. Similarly, chronic toxicity studies toward aquatic organisms of lipid-lowering agents,[24,25] neuroactive pharmaceuticals (antidepressants),[22,26] and anti-epileptic drugs[22] have been conducted at concentrations typically found in the environment. Some of the documented ecological effects of pharmaceuticals in the environment are summarized in Table 2. Both the U.S. Environmental Protection Agency (USEPA) and European Union (EU) have recognized residues of pharmaceuticals and personal care products as emerging contaminants of concern that may require future environmental regulation if their persistence in the environment proves to be ecologically significant. In fact, diclofenac, ibuprofen, triclosan, and clofibric acid have been identified as future emerging priority candidates by the EU water framework directive, which is a priority substance list that is updated every 4 years.[27]

TREATMENT OF PHARMACEUTICALS IN THE AQUATIC SYSTEMS

Alternative Systems for Water Treatment

Activated sludge treatment, which relies on microbes to biodegrade contaminants in wastewater, is the most widely used waste water treatment system in the United States and around the world. While many pharmaceuticals are removed partially or completely during CAS treatment, there are a significant number of pharmaceuticals that have very little to no removal. Therefore, alternative treatment systems are being explored to improve removal efficiencies for these trace organic contaminants. Examples of treatment systems that are suspected to be more efficient are membrane bioreactors (MBRs)[41–44]; membrane treatment with nano-, micro-, and ultrafiltration (NF/MF/UF); and reverse osmosis (RO).[45] Granular activated carbon (GAC) is also used as another alternative system in the Unites States and Canada for removal of trace organic matters through filtration and adsorption.[45] Recent studies have also shown that advanced oxidation processes, such as ozonation and UV–H_2O_2 disinfection, which are employed during tertiary treatment before discharging the treated wastewater, could have additional benefits in removing pharmaceuticals from water.[15,46,47]

For drinking water treatment systems, the removal of many pharmaceuticals during the first three steps (coagulation, flocculation, and sand filtration) before disinfection is typically incomplete.[16,48] On the other hand, similar to

Table 1 Classification of pharmaceuticals based on their typical removal rates in CAS systems.

Group	Pharmaceutical	Usage
High removal (>65%)	Acetaminophen (ACE)	Analgesic (non-NSAID)
	Ibuprofen (IBP)	Analgesic (NSAID)
	Naproxen (NAP)	Analgesic (NSAID)
	Paroxetine (PRX)	Antidepressant
	Iopamidol (IOM)	Iodinated contrast agent
	Caffeine (CAF)	Psychoactive stimulant
Medium removal (30%–65%)	Sulfamethoxazole (SMX)	Antibiotic
	Gemfibrozil (GFB)	Lipid regulator
	Atenolol (ATN)	β-Blocker
	Propranolol (PRN)	β-Blocker
	Ranitidine (RTD)	Antihistamine (Zantac)
	Fluoxetene (FXT)	Antidepressant
	Iopromide (IOP)	Iodinated contrast agent
Low removal (<30%)	Diclofenac (DCF)	Analgesic (NSAID)
	Mefenamic acid (MFN)	Analgesic (NSAID)
	Ciprofloxacin (CIP)	Antibiotic
	Erythromycin (ERY)	Antibiotic—macrolide
	Roxythromycin (ROX)	Antibiotic—macrolide
	Trimethoprim (TMP)	Antibiotic
	Clofibric acid (CLO)	Lipid regulator
	Carbamazepine (CBZ)	Anticonvulsant
	Dilantin (DLT)	Anticonvulsant
	Meprobamate (MPB)	Anti-anxiety drug

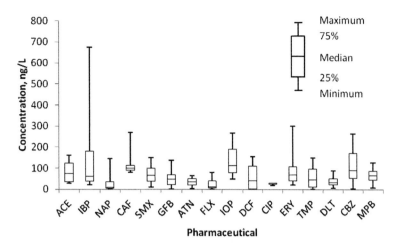

Fig. 2 Typical concentrations of selected pharmaceuticals in surface waters in the U.S. The middle line in the box plot indicates median value of the data; The upper edge (hinge) of the box indicates 75th percentile of the data set; the lower hinge indicates 25th percentile. The whiskers indicate minimum and maximum data values.
Source: Kolpin, D.W. et al.[1]; Benotti, M.J. et al.[4]; Snyder, S.A. et al.[15]; Westerhoff, P. et al.[16]; Focazio, M.J. et al.[17]; Nodler, K. et al.[18]; Boyd, G.R. et al.[19]; Ferrer, I. et al.[20]

wastewater treatment, pharmaceuticals may be effectively removed by activated carbon adsorption, ozone oxidation, and membrane filtration.[49] Chlorination, the most common disinfection step, may also remove pharmaceuticals but produce unwanted by-products.[16]

Alternative water treatment systems are summarized and compared in Table 3.[50,51] These systems are advantageous because they offer increased removal of pharmaceuticals and other organic contaminants that cause undesirable odor and taste in water. To enhance the efficiency of treatment, combining systems such as MBRs followed by activated carbon adsorption system maybe applied.

Removal Efficiencies in Municipal WWTP Systems

Several reports have shown that MBRs generally have higher removal efficiencies for NSAIDs than CAS systems. For example, it was demonstrated that up to 98%–100% ketiprofen and 86%–89% naproxen were removed in MBRs.[41,44] In addition, MBRs also have better removal efficiencies for the antibiotics roxythromycin, sulfamethoxazole, trimethoprim, and diclofenac relative to the CAS systems.[43,44] However, certain compounds, such as carbamazepine, remain mostly undegraded in both conventional WWTP and MBRs.[43] A study compared different filter systems such as MBR, MF/UF, and RO in removing pharmaceuticals using WWTP primary effluent as the feed water.[45] In contrast to other studies, most pharmaceuticals were removed well (except for phenytoin and meprobamate) in the MBR, including a 90% removal of carbamazepine. These differences in results suggest that design and operation of MBRs can be optimized to increase removal efficiencies of pharmaceuticals in wastewater. Additionally, it was demonstrated that more than 90% of all pharmaceuticals tested were removed in the RO system alone; however, the use of combined systems such as UF/RO, MF/RO, and MBR/RO resulted in more than 99% removal for all pharmaceuticals. However, membrane systems are expensive and may not be affordable to many municipalities. In addition, membrane systems require that the brine, in which the rejected pharmaceuticals are concentrated after removing pharmaceuticals from mother water, must be appropriately treated.

Granular activated carbon (GAC) adsorption systems were also investigated in two facilities, one with regular regeneration of GAC and the other without regular replacement/regeneration of GAC.[45] As summarized in Table 4, the latter facility had poor removal efficiencies for most of the pharmaceuticals tested. Hence, it is important to determine the breakthrough for individual pharmaceuticals so that regeneration or replacement GAC can be scheduled on a timely fashion to be most effective in removing pharmaceutical residues in the water.

Ozone oxidation has been investigated for sulfonamides, macrolides, and iodinated contrast media, as well as other acidic pharmaceuticals spiked in MBR-treated wastewater effluent.[46] As may be expected, increased ozone concentrations resulted in increased degradation of pharmaceuticals. However, while the sulfonamides and macrolides were very sensitive toward ozone degradation and were completely removed with high ozone concentration, iodinated contrast media were only 40% degraded. Ozone oxidation experiments have been performed to investigate the removal of 17 pharmaceuticals, as well as other personal care products and endocrine-disrupting chemicals, in bench-

Table 2 Reported ecotoxicity for selected pharmaceuticals using various test organisms.

Pharmaceutical	Effect observed	Effective concentration	Environmentally relevant[a]	Reference
Atenolol	Fathead minnow	1.0 mgL^{-1} (21 days)—NOEC	No	[28]
		3.2 mgL^{-1} (21 days)—LOEC	No	
Carbamazepine	*Daphnia magna*	0.492 µgL^{-1}—multigeneration (up to 6 generations) effect	Yes	[29]
	Japanese medaka	6.15 mgL^{-1}—swim speed effect (9 days)	No	[30]
	Oryzias latipes	45.87 mgL^{-1}—LC50 (96 hr)	No	[31]
	Thamnocephalus platyurus	>100 mgL^{-1}—LC50 (24 hr)	No	[31]
Clarithromycin	*Oryzias latipes*	>100 mgL^{-1}—LC50 (96 hr)	No	[31]
	Thamnocephalus platyurus	94.23 mgL^{-1}—LC50 (24 hr)	No	[31]
Diclofenac	*Daphnia magna*	0.355 µgL^{-1}—multigeneration (up to 6 generations) effect	Yes	[29]
	Hyalella azteca	0.467 mgkg^{-1}—LC50 (72 hr) with sediment	No	[32]
	Rainbow trout	25 µgL^{-1}—accumulation test on gall (21 days)	Yes	[33]
Fluoxetene	Sheepshead minnow	>2.0 mgL^{-1}—LC50	No	[34]
		2.0 mgL^{-1}—LOEC		
		1.87 mgL^{-1}—NOEC		
		(All for 96 hr)		
	Western mosquitofish	0.546 mgL^{-1}—LC50 (7 days)	No	[35]
	Western mosquitofish	0.5 µgL^{-1}—increasing lethargy in 59–159 days	Yes	[35]
Ibuprofen	*Oryzias latipes*	>100 mgL^{-1}—LC50 (96 hr)	No	[31]
	Planorbis carinatus	17.1 mgL^{-1}—LC50 (48 and 72 hr)	No	[36]
	Planorbis carinatus	>5.36 mgL^{-1}—LOEC	No	[36]
		5.36 mgL^{-1}—NOEC		
		(All for 21 days)		
	Thamnocephalus platyurus	19.59 mgL^{-1}—LC50 (24 hr)	No	[31]
Metoprolol	*Daphnia magna*	1.170 µgL^{-1}—multigeneration (up to 6 generations) effect	Yes	[29]
Paracetamol (Acetaminophen)	Wheat	22.4 mgL^{-1}—damage in 21 days exposure	No	[37]
Propranolol	Fathead minnow	1.0 mgL^{-1}—NOEC (female)	No (Yes for NOEC male)	[38]
		3.4 mgL^{-1}—LOEC (female)		
		0.1 mgL^{-1}—NOEC (male)		
		1.0 mgL^{-1}—LOEC (male)		
		(All for 21 days)		
	Rainbow trout	1.0 mgL^{-1}—NOEC	No	[39]
		10 mgL^{-1}—LOEC		
		(All for 10 days)		
	Oryzias latipes	11.40 mgL^{-1}—LC50 (96 hr)	No	[31]
	Thamnocephalus platyurus	10.31 mgL^{-1}—LC50 (24 hr)	No	[31]
Verapamil	Juvenile rainbow trout	2.72 mgL^{-1}—LC50 (96 hr)	No	[40]

Note: NOEC, no observed effect concentration; LOEC, lowest observed effect concentration; LC50, 50% lethal concentration.
[a]Environmentally relevant—"Yes" for concentrations of compounds are in the ranges in surface water and WWTP effluent reported through 2011.

top pilot plants and in one full-scale WWTP system. The removal efficiencies of pharmaceuticals by ozonation can be classified into four groups: >80% removal, 80%–50% removal, 50%–10% removal, and <20% removal.[15] Most of the pharmaceuticals exhibited >80% removal; diazepam, phenytoin, and ibuprofen had 80%–50% removal; and iopromide and meprobamate were in the group of 50%–10% removal. In a separate study using two different wastewa-

Table 3 Summary and comparison of alternative systems: membrane process, activated carbon adsorption process, and advanced oxidation process for contaminant removal.

Process		Operation and application	Important notes
Membrane process	Reverse osmosis (RO)	• Pressure driven (150–400 psi)	• Best removal performance
			• Possible for desalination
		• Removes organics with molecular weight >100	• Needs pretreatment such as UF or MF to prevent plugging and fouling; hence, high cost
		• Demineralization	• Low product recovery (30%–85%)
	Nanofiltration (NF)	• Pressure driven (80–200 psi)	• Good removal performance
		• Removes organics with molecular weight between 100–500	• Needs pretreatment such as UF or MF to prevent plugging and fouling; hence, high cost
		• Removes NOM	• Medium product recovery (70%–90%)
	Ultrafiltration (UF)	• Pressure driven (15–60 psi)	• Replacement for CT
			• High product recovery (80%–95%)
			• Higher cost than CT
		• Pretreatment for NF or RO	• Sensitive with water temperature and viscosity
		• Macromolecule removal includes viruses	• Needs other process to increase treatment performance
	Microfiltration (MF)	• Pressure driven (5–40 psi)	• Replacement for CT
			• High product recovery (95%–98%)
			• Higher cost than CT
		• Pretreatment for NF or RO	• Sensitive with water temperature and viscosity
		• Bacteria removal	• Needs other process to increase treatment performance
Activated carbon adsorption	Powder activated carbon (PAC)	• Used in early steps of treatment process	• Good performance in combination with MF or UF membrane systems
			• Usually used for drinking water treatment
			• Good for emergency situations, with high organic pollutants
			• Handling problem—dry PAC causes dust problem
		• Controls taste or odor	• May pass through filters and enter the final treated water
		• Applied for small or mid-sized plants with moderate or severe taste, odor, or organic contaminants problem	• Cannot be mixed with chlorine
	Granular activated carbon (GAC)	• Used in adsorption beds or tanks	• Replacement of conventional gravity filter
			• Possible for WWTPs
			• Fouling by chemicals—need to consider breakthrough for design
			• Backwash requirement
		• Gravity or pressure driven	• Need to monitor carbon bed depth
		• Applied for plants with moderate or severe taste, odor, or organic contaminants problem	• Needs carbon regeneration
Advanced oxidation process	Ultraviolet (UV)	• Chemical treatment	• Disinfection benefit
		• Low pressure ($\lambda = 253.7$ nm) or medium pressure ($\lambda = 180$–1370 nm)	• Controls taste and odor
		• UV radiation with H_2O_2 addition (hydroxyl radical oxidation)	• Relatively high cost and complexity
		• Combined with ozonation	• Treatment effectiveness is weakened by turbidity or color in water
		• NOM in water may inhibit or promote oxidation	• Unwanted by-products

Note: NOM, natural organic matter; CT, conventional treatment.

Table 4 Comparing removal efficiencies of pharmaceuticals for GAC facilities with and without regular regeneration or replacement of activated carbon.

Pharmaceuticals	Removal (%) in Facility 1 (with regular GAC regeneration)	Removal (%) in Facility 2 (without regular GAC regeneration)
Caffeine	>41.2	16.3
Carbamazepine	>54.5	15.6
Phenytoin (Dilantin)	>44.4	22.7
Erythromycin–H_2O	>44.4	7.9
Gemfibrozil	>16.7	8.2
Ibuprofen	>9.1	16.4
Iopromide	>69.7	72.0
Meprobamate	>16.7	13.3
Sulfamethoxazole	>83.3	83.8

ter matrices, one with effluent water from a conventional WWTP and the other with effluent from an MBR system, it was shown that the removal efficiencies by ozonation, ozone–UV, and H_2O_2–UV may be slightly lower in the presence of higher dissolved organic carbon.[47]

The removal of 13 selected pharmaceuticals in full-scale WWTP with CAS system, followed by advanced treatment by ozonation, was evaluated.[52] Interestingly, most of the pharmaceuticals that are typically poorly removed (<20% removal) in the secondary clarifier, such as crotamiton, sulfapyridine, and roxithromycin, were almost completely-degraded upon ozonation. An exception to the pharmaceuticals effectively removed by ozonation was carbamazepine (which was only <5% degraded by ozonation).

In a separate study, the efficiency of UV oxidation was examined with and without H_2O_2 for atenolol, carbamazepine, phenytoin, meprobamate, primidone, and trimethoprim, using three different wastewater matrices.[53] Results of this study indicated that the nature of the aqueous matrix, most likely defined by the amount and composition of the natural organic matter, is an important factor in optimizing the removal efficiencies of pharmaceuticals by advanced oxidation. For example, removal efficiencies between 16% and 95% were observed for the pharmaceuticals tested in one WWTP using a high power of UV (fluence of 700 mJcm^{-2}) with H_2O_2. However, removal efficiencies of the same pharmaceuticals in water from other sampling locations with high organic matter content were relatively lower (10%–85%), although the same UV treatment conditions were used.

Removal Efficiencies in Drinking Water Treatment Systems

Polar and persistent pharmaceuticals may eventually enter the drinking water systems.[49] Since conventional processes used in drinking water treatment may not completely remove pharmaceuticals, other advanced oxidation processes have been considered for treatment of pharmaceuticals in drinking water. Ozonation appears to be the most effective way to disinfect the water and at the same time oxidize organic chemicals via direct reaction with ozone, or through reactions with hydroxyl radicals (·OH) formed during ozonation.[15] For disinfection purposes, monochromatic UV at 254 nm is used for drinking water treatment. In UV advanced oxidation process, the oxidation of pharmaceuticals is enhanced by addition of H_2O_2 to facilitate formation of hydroxyl radicals and promote indirect photolysis.[54] The oxidation of diclofenac, a frequently detected NSAID in the environment, was evaluated using ozone, UV, and UV–H_2O_2. Ozonation showed 100% removal, while UV–H_2O_2 oxidation showed only 52% removal of diclofenac at conditions that corresponded to a 35% decrease in total organic carbon, after 90 min of oxidation time.[55] The antidepressant pharmaceutical fluoxetene (trade name of Prozac), which is typically detected at parts per billion levels in U.S. streams,[1] was used as test compound to evaluate the efficiency of direct and indirect photolysis in removing pharmaceuticals in water. The degradation rate for fluoxetene reached up to 9.60×10^9 M^{-1}sec^{-1} for indirect photolysis.[56] It has been shown that the use of TiO_2 catalyst can significantly increase the photolysis rate constants for NSAIDs, including diclofenac, naproxen, and ibuprofen in water.[57] Experimental oxidation rates of the hydroxyl radical for β-blockers, atenolol, metoprolol, and propranolol, were found to be 7.05×10^9, 8.39×10^9, and 1.07×10^9 M^{-1}sec^{-1}, respectively.[58] Ozonation of tetracycline showed the complete removal within 4–6 min of ozonation despite the maximum degradation of 40% total organic carbon after 2 hr of ozonation.[59] Finally, photo-Fenton (Fe(II)) reaction is another oxidation method that can be applicable to drinking water treatment systems owing to its cost effectiveness and ease of operation.[60] It was demonstrated that diclofenac can be 100% degraded within 60 min of photo-Fenton oxidation.[60]

Organic pollutants in drinking water sources, including pharmaceuticals, can be removed by activated carbon adsorption.[16] Laboratory-scale tests mimicking full-scale drinking water treatment system using activated carbon showed good removal efficiencies of trace organic compounds, except for the more hydrophilic pharmaceuticals (e.g., clofibric acid and ibuprofen were only 40% removed) after 3 hours of contact time.[61]

Water filters can provide an additional step, at a point-of-use consumption treatment, to remove trace levels of pharmaceuticals and other drinking water contaminants from tap water. The removal efficiencies of chlorination by-products, including trihalomethanes, bromodichloromethane, dibromochloromethane, and bromoform, by "Envirofilter" (made of nutshell carbon and commercial water filters) have been reported.[62] Water filters are made with activated carbon to remove dissolved organic matter,[62,63] together with polar and ion-exchange resins to remove charged species such as metal ions.[64,65] Since activated carbon has been shown to have high removal efficiencies for many pharmaceuticals in wastewater,[16,66–71] it is reasonable to expect that water filters will be effective in removing trace levels of pharmaceuticals from drinking water.

CONCLUSION

The decreasing amount of clean water resources for drinking water and for food production has become one of the most challenging problems in the world.[72] To alleviate the shortage in water supply, reuse of treated wastewater for irrigation[72–74] and as drinking water source[75] is becoming more and more common. Therefore, pharmaceuticals can enter the groundwater and drinking water systems, as shown in Fig. 1 depicting the routes of entry of pharmaceuticals into the environment. It is encouraging to see a recent survey reporting that only very few pharmaceuticals from biosolid application and wastewater irrigation can be transported to the groundwater aquifer.[72] Because only lipophilic pharmaceuticals tend to sorb onto the biosolids, and these pharmaceuticals in turn are not mobile in soil, contamination of groundwater aquifer from biosolids application may not pose a significant source of pharmaceutical pollution. With the exception of four pharmaceuticals (two iodinated contrast media of diatrizoate and iopamidol, carbamazepine and sulfamethoxazole) most of the 52 pharmaceuticals being targeted for analysis were not detected in the said study. Nevertheless, optimization of the design and operating conditions of wastewater treatment systems is key to eliminating pharmaceuticals and their deleterious ecological effects in the environment. Advanced oxidation processes are very promising treatment technologies that are waiting to be tested and implemented under full-scale treatment plants.

REFERENCES

1. Kolpin, D.W.; Furlong, E.T.; Meyer, M.T.; Thurman, E.M.; Zaugg, S.D.; Barber, L.B.; Buxton, H.T. Pharmaceuticals, hormones, and other organic wastewater contaminants in US streams, 1999–2000: A national reconnaissance. Environmental Science & Technology **2002**, *36* (6), 1202–1211.
2. Aga, D.S.; O'Connor, S.; Ensley, S.; Payero, J.O.; Snow, D.; Tarkalson, D. Determination of the persistence of tetracycline antibiotics and their degradates in manure-amended soil using enzyme-linked immunosorbent assay and liquid chromatography-mass spectrometry. Journal of Agricultural and Food Chemistry **2005**, *53* (18), 7165–7171.
3. Batt, A.L.; Bruce, I.B.; Aga, D.S. Evaluating the vulnerability of surface waters to antibiotic contamination from varying wastewater treatment plant discharges. Environmental Pollution **2006**, *142* (2), 295–302.
4. Benotti, M.J.; Trenholm, R.A.; Vanderford, B.J.; Holady, J.C.; Stanford, B.D.; Snyder, S.A. Pharmaceuticals and Endocrine Disrupting Compounds in US Drinking Water. Environmental Science & Technology **2009**, *43* (3), 597–603.
5. Avisar, D.; Levin, G.; Gozlan, I. The processes affecting oxytetracycline contamination of groundwater in a phreatic aquifer underlying industrial fish ponds in Israel. Environmental Earth Sciences **2009**, *59* (4), 939–945.
6. Lindsey, M.E.; Meyer, M.; Thurman, E.M. Analysis of trace levels of sulfonamide and tetracycline antimicrobials, in groundwater and surface water using solid-phase extraction and liquid chromatography/mass spectrometry. Analytical Chemistry **2001**, *73* (19), 4640–4646.
7. Karthikeyan, K.G.; Meyer, M.T. Occurrence of antibiotics in wastewater treatment facilities in Wisconsin, USA. Science of the Total Environment **2006**, *361* (1–3), 196–207.
8. Batt, A.L.; Snow, D.D.; Aga, D.S. Occurrence of sulfonamide antimicrobials in private water wells in Washington County, Idaho, USA. Chemosphere **2006**, *64* (11), 1963–1971.
9. Richter, D.; Massmann, G.; Taute, T.; Duennbier, U. Investigation of the fate of sulfonamides downgradient of a decommissioned sewage farm near Berlin, Germany. Journal of Contaminant Hydrology **2009**, *106* (3–4), 183–194.
10. Katz, B.G.; Griffin, D.W.; Davis, J.H. Groundwater quality impacts from the land application of treated municipal wastewater in a large karstic spring basin: Chemical and microbiological indicators. Science of the Total Environment **2009**, *407* (8), 2872–2886.
11. Heberer, T. Occurrence, fate, and removal of pharmaceutical residues in the aquatic environment: a review of recent research data. Toxicology Letters **2002**, *131* (1–2), 5–17.
12. Musolff, A.; Leschik, S.; Moder, M.; Strauch, G.; Reinstorf, F.; Schirmer, M. Temporal and spatial patterns of micropollutants in urban receiving waters. Environmental Pollution **2009**, *157* (11), 3069–3077.
13. Jones, O.A.H.; Voulvoulis, N.; Lester, J.N. Human pharmaceuticals in wastewater treatment processes. Critical Reviews in Environmental Science and Technology **2005**, *35* (4), 401–427.
14. Sipma, J.; Osuna, B.; Collado, N.; Monclús, H.; Ferrero, G.; Comas, J.; Rodriguez-Roda, I. Comparison of removal of pharmaceuticals in MBR and activated sludge systems. Desalination **2010**, *250* (2), 653–659.
15. Snyder, S.A.; Wert, E.C.; Rexing, D.J.; Zegers, R.E.; Drury, D.D. Ozone oxidation of endocrine disruptors and pharmaceuticals in surface water and wastewater. Ozone-Science & Engineering **2006**, *28* (6), 445–460.
16. Westerhoff, P.; Yoon, Y.; Snyder, S.; Wert, E. Fate of endocrine-disruptor, pharmaceutical, and personal care product chemicals during simulated drinking water treatment processes. Environmental Science & Technology **2005**, *39* (17), 6649–6663.
17. Focazio, M.J.; Kolpin, D.W.; Barnes, K.K.; Furlong, E.T.; Meyer, M.T.; Zaugg, S.D.; Barber, L.B.; Thurman, M.E. A national reconnaissance for pharmaceuticals and other organic wastewater contaminants in the United States - II) Untreated drinking water sources. Science of the Total Environment **2008**, *402* (2–3), 201–216.

18. Nodler, K.; Licha, T.; Bester, K.; Sauter, M. Development of a multi-residue analytical method, based on liquid chromatography-tandem mass spectrometry, for the simultaneous determination of 46 micro-contaminants in aqueous samples. Journal of Chromatography A **2010**, *1217* (42), 6511–6521.
19. Boyd, G.R.; Palmeri, J.M.; Zhang, S.Y.; Grimm, D.A. Pharmaceuticals and personal care products (PPCPs) and endocrine disrupting chemicals (EDCs) in stormwater canals and Bayou St. John in New Orleans, Louisiana, USA. Science of the Total Environment **2004**, *333* (1–3), 137–148.
20. Ferrer, I.; Zweigenbaum, J.A.; Thurman, E.M. Analysis of 70 Environmental Protection Agency priority pharmaceuticals in water by EPA Method 1694. Journal of Chromatography A **2010**, *1217* (36), 5674–5686.
21. Huggett, D.B.; Brooks, B.W.; Peterson, B.; Foran, C.M.; Schlenk, D. Toxicity of select beta adrenergic receptor-blocking pharmaceuticals (B-blockers) on aquatic organisms. Archives of Environmental Contamination and Toxicology **2002**, *43* (2), 229–235.
22. Ferrari, B.; Paxeus, N.; Lo Giudice, R.; Pollio, A.; Garric, J. Ecotoxicological impact of pharmaceuticals found in treated wastewaters: study of carbamazepine, clofibric acid, and diclofenac. Ecotoxicology and Environmental Safety **2003**, *55* (3), 359–370.
23. Ferrari, B.; Mons, R.; Vollat, B.; Fraysse, B.; Paxeus, N.; Lo Giudice, R.; Pollio, A.; Garric, J. Environmental risk assessment of six human pharmaceuticals: Are the current environmental risk assessment procedures sufficient for the protection of the aquatic environment? Environmental Toxicology and Chemistry **2004**, *23* (5), 1344–1354.
24. Donohue, M.; Baldwin, L.A.; Leonard, D.A.; Kostecki, P.T.; Calabrese, E.J. Effect of hypolipidemic drugs gemfibrozil, ciprofibrozil, ciprofibrate, and clofibric acid of peroxisomal beta-oxidation in primary cultures of rainbow-trout hepatocytes. Ecotoxicology and Environmental Safety **1993**, *26* (2), 127–132.
25. Nunes, B.; Carvalho, F.; Guilhermino, L. Acute and chronic effects of clofibrate and clofibric acid on the enzymes acetylcholinesterase, lactate dehydrogenase and catalase of the mosquitofish, *Gambusia holbrooki*. Chemosphere **2004**, *57* (11), 1581–1589.
26. Brooks, B.W.; Foran, C.M.; Richards, S.M.; Weston, J.; Turner, P.K.; Stanley, J.K.; Solomon, K.R.; Slattery, M.; La Point, T.W. Aquatic ecotoxicology of fluoxetine. Toxicology Letters **2003**, *142* (3), 169–183.
27. Ellis, J.B. Pharmaceutical and personal care products (PPCPs) in urban receiving waters. Environmental Pollution **2006**, *144* (1), 184–189.
28. Winter, M.J.; Lillicrap, A.D.; Caunter, J.E.; Schaffner, C.; Alder, A.C.; Ramil, M.; Ternes, T.A.; Giltrow, E.; Sumpter, J.P.; Hutchinson, T.H. Defining the chronic impacts of atenolol on embryo-larval development and reproduction in the fathead minnow (Pimephales promelas). Aquatic Toxicology **2008**, *86* (3), 361–369.
29. Dietrich, S.; Ploessl, F.; Bracher, F.; Laforsch, C. Single and combined toxicity of pharmaceuticals at environmentally relevant concentrations in Daphnia magna - A multigenerational study. Chemosphere **2010**, *79* (1), 60–66.
30. Nassef, M.; Matsumoto, S.; Seki, M.; Khalil, F.; Kang, I.J.; Shimasaki, Y.; Oshima, Y.; Honjo, T. Acute effects of triclosan, diclofenac and carbamazepine on feeding performance of Japanese medaka fish (Oryzias latipes). Chemosphere **2010**, *80* (9), 1095–1100.
31. Kim, J.W.; Ishibashi, H.; Yamauchi, R.; Ichikawa, N.; Takao, Y.; Hirano, M.; Koga, M.; Arizono, K. Acute toxicity of pharmaceutical and personal care products on freshwater crustacean (Thamnocephalus platyurus) and fish (Oryzias latipes). Journal of Toxicological Sciences **2009**, *34* (2), 227–232.
32. Oviedo-Gomez, D.G.C.; Galar-Martinez, M.; Garcia-Medina, S.; Razo-Estrada, C.; L.Gomez-Olivan, M. Diclofenac-enriched artificial sediment induces oxidative stress in Hyalella azteca. Environmental Toxicology and Pharmacology **2010**, *29* (1), 39–43.
33. Mehinto, A.C.; Hill, E.M.; Tyler, C.R. Uptake and Biological Effects of Environmentally Relevant Concentrations of the Nonsteroidal Anti-inflammatory Pharmaceutical Diclofenac in Rainbow Trout (Oncorhynchus mykiss). Environmental Science & Technology **2010**, *44* (6), 2176–2182.
34. Winder, V.L.; Sapozhnikova, Y.; Pennington, P.L.; Wirth, E.F. Effects of fluoxetine exposure on serotonin-related activity in the sheepshead minnow (Cyprinodon variegatus) using LC/MS/MS detection and quantitation. Comparative Biochemistry and Physiology C-Toxicology & Pharmacology **2009**, *149* (4), 559–565.
35. Henry, T.B.; Black, M.C. Acute and chronic toxicity of fluoxetine (selective serotonin reuptake inhibitor) in western mosquitofish. Archives of Environmental Contamination and Toxicology **2008**, *54* (2), 325–330.
36. Pounds, N.; Maclean, S.; Webley, M.; Pascoe, D.; Hutchinson, T. Acute and chronic effects of ibuprofen in the mollusc Planorbis carinatus (Gastropoda : Planorbidae). Ecotoxicology and Environmental Safety **2008**, *70* (1), 47–52.
37. An, J.; Zhou, Q.X.; Sun, F.H.; Zhang, L. Ecotoxicological effects of paracetamol on seed germination and seedling development of wheat (Triticum aestivum L.). Journal of Hazardous Materials **2009**, *169* (1–3), 751–757.
38. Giltrow, E.; Eccles, P.D.; Winter, M.J.; McCormack, P.J.; Rand-Weaver, M.; Hutchinson, T.H.; Sumpter, J.P. Chronic effects assessment and plasma concentrations of the beta-blocker propranolol in fathead minnows (Pimephales promelas). Aquatic Toxicology **2009**, *95* (3), 195–202.
39. Owen, S.F.; Huggett, D.B.; Hutchinson, T.H.; Hetheridge, M.J.; Kinter, L.B.; Ericson, J.F.; Sumpter, J.P. Uptake of propranolol, a cardiovascular pharmaceutical, from water into fish plasma and its effects on growth and organ biometry. Aquatic Toxicology **2009**, *93* (4), 217–224.
40. Li, Z.H.; Li, P.; Randak, T. Ecotoxocological effects of short-term exposure to a human pharmaceutical Verapamil in juvenile rainbow trout (*Oncorhynchus mykiss*). Comparative Biochemistry and Physiology C-Toxicology & Pharmacology **2010**, *152* (3), 385–391.
41. Kimura, K.; Hara, H.; Watanabe, Y. Removal of pharmaceutical compounds by submerged membrane bioreactors (MBRs). Desalination **2005**, *178* (1–3), 135–140.
42. Quintana, J.B.; Weiss, S.; Reemtsma, T. Pathway's and metabolites of microbial degradation of selected acidic pharmaceutical and their occurrence in municipal wastewater treated by a membrane bioreactor. Water Research **2005**, *39* (12), 2654–2664.

43. Celiz, M.D.; Perez, S.; Barcelo, D.; Aga, D.S. Trace Analysis of Polar Pharmaceuticals in Wastewater by LC-MS-MS: Comparison of Membrane Bioreactor and Activated Sludge Systems. Journal of Chromatographic Science **2009**, *47* (1), 19–25.
44. Tambosi, J.L.; de Sena, R.F.; Favier, M.; Gebhardt, W.; Jose, H.J.; Schroder, H.F.; Moreira, R. Removal of pharmaceutical compounds in membrane bioreactors (MBR) applying submerged membranes. Desalination **2010**, *261* (1–2), 148–156.
45. Snyder, S.A.; Adham, S.; Redding, A.M.; Cannon, F.S.; DeCarolis, J.; Oppenheimer, J.; Wert, E.C.; Yoon, Y. Role of membranes and activated carbon in the removal of endocrine disruptors and pharmaceuticals. Desalination **2007**, *202* (1–3), 156–181.
46. Huber, M.M.; Gobel, A.; Joss, A.; Hermann, N.; Loffler, D.; McArdell, C.S.; Ried, A.; Siegrist, H.; Ternes, T.A.; von Gunten, U. Oxidation of pharmaceuticals during ozonation of municipal wastewater effluents: A pilot study. Environmental Science & Technology **2005**, *39* (11), 4290–4299.
47. Gebhardt, W.; Schroder, H.F. Liquid chromatography-(tandem) mass spectrometry for the follow-up of the elimination of persistent pharmaceuticals during wastewater treatment applying biological wastewater treatment and advanced oxidation. Journal of Chromatography A **2007**, *1160* (1–2), 34–43.
48. Vieno, N.; Tuhkanen, T.; Kronberg, L. Removal of pharmaceuticals in drinking water treatment: Effect of chemical coagulation. Environmental Technology **2006**, *27* (2), 183–192.
49. Vieno, N.M.; Harkki, H.; Tuhkanen, T.; Kronberg, L. Occurrence of pharmaceuticals in river water and their elimination a pilot-scale drinking water treatment plant. Environmental Science & Technology **2007**, *41* (14), 5077–5084.
50. HDR Engineering, I., ed. *Handbook of Public Water Systems*, 2nd Ed.; John Wiley and Sons, Inc.: New York, U.S.A., 2001.
51. Baruth, E.E., ed. *Water Treatment Plant Design*, 4th Ed.; McGraw-Hill: New York, U.S.A., 2005.
52. Nakada, N.; Shinohara, H.; Murata, A.; Kiri, K.; Managaki, S.; Sato, N.; Takada, H. Removal of selected pharmaceuticals and personal care products (PPCPs) and endocrine-disrupting chemicals (EDCs) during sand filtration and ozonation at a municipal sewage treatment plant. Water Research **2007**, *41* (19), 4373–4382.
53. Rosario-Ortiz, F.L.; Wert, E.C.; Snyder, S.A. Evaluation of UV/H2O2 treatment for the oxidation of pharmaceuticals in wastewater. Water Research **2010**, *44* (5), 1440–1448.
54. Pereira, V.J.; Weinberg, H.S.; Linden, K.G.; Singer, P.C. UV degradation kinetics and modeling of pharmaceutical compounds in laboratory grade and surface water via direct and indirect photolysis at 254 nm. Environmental Science & Technology **2007**, *41* (5), 1682–1688.
55. Vogna, D.; Marotta, R.; Napolitano, A.; Andreozzi, R.; d'Ischia, M. Advanced oxidation of the pharmaceutical drug diclofenac with UV/H2O2 and ozone. Water Research **2004**, *38* (2), 414–422.
56. Lam, M.W.; Young, C.J.; Mabury, S.A. Aqueous photochemical reaction kinetics and transformations of fluoxetine. Environmental Science & Technology **2005**, *39* (2), 513–522.
57. Mendez-Arriaga, F.; Esplugas, S.; Gimenez, J. Photocatalytic degradation of non-steroidal anti-inflammatory drugs with TiO2 and simulated solar irradiation. Water Research **2008**, *42* (3), 585–594.
58. Song, W.H.; Cooper, W.J.; Mezyk, S.P.; Greaves, J.; Peake, B.M. Free radical destruction of beta-blockers in aqueous solution. Environmental Science & Technology **2008**, *42* (4), 1256–1261.
59. Khan, M.H.; Bae, H.; Jung, J.Y. Tetracycline degradation by ozonation in the aqueous phase: Proposed degradation intermediates and pathway. Journal of Hazardous Materials **2010**, *181* (1–3), 659–665.
60. Perez-Estrada, L.A.; Malato, S.; Gernjak, W.; Aguera, A.; Thurman, E.M.; Ferrer, I.; Fernandez-Alba, A.R. Photo-fenton degradation of diclofenac: Identification of main intermediates and degradation pathway. Environmental Science & Technology **2005**, *39* (21), 8300–8306.
61. Simazaki, D.; Fujiwara, J.; Manabe, S.; Matsuda, M.; Asami, M.; Kunikane, S. Removal of selected pharmaceuticals by chlorination, coagulation-sedimentation and powdered activated carbon treatment. Water Science and Technology **2008**, *58* (5), 1129–1135.
62. Ahmedna, M.; Marshall, W.E.; Husseiny, A.A.; Goktepe, L.; Rao, R.M. The use of nutshell carbons in drinking water filters for removal of chlorination by-products. Journal of Chemical Technology and Biotechnology **2004**, *79* (10), 1092–1097.
63. Humbert, H.; Gallard, H.; Suty, H.; Croue, J.P. Natural organic matter (NOM) and pesticides removal using a combination of ion exchange resin and powdered activated carbon (PAC). Water Research **2008**, *42* (6–7), 1635–1643.
64. An, H.K.; Park, B.Y.; Kim, D.S. Crab shell for the removal of heavy metals from aqueous solution. Water Research **2001**, *35* (15), 3551–3556.
65. Rengaraj, S.; Moon, S.H. Kinetics of adsorption of Co(II) removal from water and wastewater by ion exchange resins. Water Research **2002**, *36* (7), 1783–1793.
66. Ternes, T.A.; Meisenheimer, M.; McDowell, D.; Sacher, F.; Brauch, H.J.; Gulde, B.H.; Preuss, G.; Wilme, U.; Seibert, N.Z. Removal of pharmaceuticals during drinking water treatment. Environmental Science & Technology **2002**, *36* (17), 3855–3863.
67. Snyder, S.A.; Westerhoff, P.; Yoon, Y.; Sedlak, D.L. Pharmaceuticals, personal care products, and endocrine disruptors in water: Implications for the water industry. Environmental Engineering Science **2003**, *20* (5), 449–469.
68. Kim, S.D.; Cho, J.; Kim, I.S.; Vanderford, B.J.; Snyder, S.A. Occurrence and removal of pharmaceuticals and endocrine disruptors in South Korean surface, drinking, and waste waters. Water Research **2007**, *41* (5), 1013–1021.
69. Yu, Z.R.; Peldszus, S.; Huck, P.M. Adsorption characteristics of selected pharmaceuticals and an endocrine disrupting compound - Naproxen, carbamazepine and nonylphenol - on activated carbon. Water Research **2008**, *42* (12), 2873–2882.
70. Yu, Z.; Peldszus, S.; Huck, P.M. Adsorption of Selected Pharmaceuticals and an Endocrine Disrupting Compound by Granular Activated Carbon. 1. Adsorption Capacity and Kinetics. Environmental Science & Technology **2009**, *43* (5), 1467–1473.

71. Yu, Z.; Peldszus, S.; Huck, P.M. Adsorption of Selected Pharmaceuticals and an Endocrine Disrupting Compound by Granular Activated Carbon. 2. Model Prediction. Environmental Science & Technology **2009**, *43* (5), 1474–1479.
72. Ternes, T.A.; Bonerz, M.; Herrmann, N.; Teiser, B.; Andersen, H.R. Irrigation of treated wastewater in Braunschweig, Germany: An option to remove pharmaceuticals and musk fragrances. Chemosphere **2007**, *66* (5), 894–904.
73. Chefetz, B.; Mualem, T.; Ben-Ari, J. Sorption and mobility of pharmaceutical compounds in soil irrigated with reclaimed wastewater. Chemosphere **2008**, *73* (8), 1335–1343.
74. Xu, J.; Chen, W.P.; Wu, L.S.; Green, R.; Chang, A.C. Leachability of some emerging contaminants in reclaimed municipal wastewater-irrigated turf grass fields. Environmental Toxicology and Chemistry **2009**, *28* (9), 1842–1850.
75. Reungoat, J.; Macova, M.; Escher, B.I.; Carswell, S.; Mueller, J.F.; Keller, J. Removal of micropollutants and reduction of biological activity in a full scale reclamation plant using ozonation and activated carbon filtration. Water Research **2010**, *44* (2), 625–637.

Phenols

Leszek Wachowski
Robert Pietrzak
Department of Chemistry, Adam Mickiewicz University, Poznan, Poland

Abstract
The entry describes the origin, structure, chemical composition, physicochemical properties, distribution, transport, and conversion of phenol and its derivatives in the natural environment. It also describes methods of obtaining and practical significance of phenol and its derivatives. This entry also shows problems concerned with environmental exposure to phenolic compounds and their quantification in all elements of the biosphere. Introduced into the natural environment in difficult-to-estimate amounts, phenolic compounds have become a serious ecological problem.

INTRODUCTION

In 1832, German analytical chemist F.F. Runge extracted from coal a tar substance, which a year later, F. Gerhard called phecolenolem (Latin *phenolum, carbolum*). The old name for benzene was phene, and its hydroxyl derivative came to be called phenol. It is the simplest compound from the numerous class generally called phenols, comprising hydroxyl derivatives of aromatic compounds (benzene, naphthalene).

The compound whose correct chemical name according to the International Union of Pure and Applied Chemistry (IUPAC), nomenclature should be benzenol can be treated as benzene in which one of the hydrogen atoms has been replaced by a hydroxyl group. Phenols are aromatic alcohols with one or more hydroxyl groups directly bonded to the aromatic ring, which differentiates them from alcohols. Their structure is reflected in their properties, e.g., in enhanced acidic properties.[1]

Phenols make up a large group of compounds of natural, anthropogenic, or endogenic origin found in the biosphere. Because of a number of physical and chemical properties attractive from the viewpoint of practical use, phenol is produced on a large scale, about 7 billion kg/yr, as a precursor to many useful compounds. The greatest producer of phenol is the United States. Phenol is one of the oldest chemical intermediates in organic industry, and its economical importance continuously increases. The greatest amounts of phenol are used in organic synthesis, including manufacturing of phenolic resins and nylon, in which the chemical intermediates are bisphenol-A and ε-caprolactam,[2] respectively. Bacterial and fungicidal properties of phenol were the first recognized and were the reason for its wide application. The employment of phenol for sterilization has been of great importance not only in medicine but also in the history of humanity. Sir J. Lister, a Scottish surgeon, introduced the method of disinfection of surgery instruments and hands of surgeons with a 5% water solution of phenol, known at that time as carbolic acid.[3]

As a result of natural transformations taking place in the natural environment and the processes generating various phenolic compounds performed by man and their widespread use, the phenolic compounds have penetrated the biosphere. They are met more often as micropollutants of water than the atmospheric air. As the exposure of living organisms to these toxic compounds can be harmful, their presence has been monitored, and certain measures have been undertaken to restrict their amount introduced to the natural environment.[1,2,4]

ORIGIN, PROPERTIES, AND APPLICATION OF PHENOL AND ITS DERIVATIVES

Phenol and phenolic derivatives met in the biosphere are released from natural and anthropogenic sources. Their amounts entering the natural environment and coming from anthropogenic sources have been estimated as much higher.[5–7]

The class of compounds known as phenols plays a great role in many areas of our lives. Both the synthetic and natural phenols have been raw products or intermediates for the production of, e.g., plastics, detergents, or cosmetic products. Because of their biological activity and, in particular, antioxidant properties, they are essential components of our diet and substrates of a large number of therapeutic drugs.[8–15]

The large diversity of chemical compositions and structures of phenol compounds is illustrated in Tables 1–7.

Sources of Phenol and Phenolic Compounds in the Biosphere

Natural Origin

Phenol and phenolic derivatives from natural sources are commonly met organic micropollutants. The precursors of the majority of them are phenylalanine and tyrosine, from

Table 1 Name, molecular formula, and structures of selected groups of phenolic compounds.

No.	Name of compound	Molecular formula	Structural formula
Methyl derivatives of one-hydroxyl phenols			
1.	*Phenol **Carbolic acid, hydroxybenzene, benzenol, phenylic acid, phenic acid	C_6H_6O ***C_6H_5OH	
2.	*o*-Cresol *2-Methylphenol **2-Hydroxytoluene, 2-methylbenzenol	C_7H_8O ***$(CH_3)C_6H_4OH$	
3.	*m*-Cresol *3-Methylphenol **3-Hydroxytoluene, 3-methylbenzenol	C_7H_8O ***$(CH_3)C_6H_4OH$	
4.	*p*-Cresol *4-Methylphenol **4-Hydroxytoluene, *p*-hydroxytoluene, *p*-methylphenol, 4-methylbenzenol	C_7H_8O ***$(CH_3)C_6H_4OH$	
5.	2,3-Xylenol **2,3-Dimethylphenol, 1-hydroxy-2,3-methylbenzene, 3-hydroxyl-*o*-xylene	$C_8H_{10}O$ ***$(CH_3)_2C_6H_3OH$	
6.	2,4-Xylenol **2,4-Dimethylphenol, 1-hydroxy-2,4-methylbenzene, 4-hydroxy-*m*-xylene	$C_8H_{10}O$ ***$(CH_3)_2C_6H_3OH$	
7.	2,5-Xylenol **2,5-Dimethylphenol, 2-hydroxy-*p*-xylene, *p*-xylenol, 1-hydroxy-2,5-methylbenzene	$C_8H_{10}O$ ***$(CH_3)_2C_6H_3OH$	
8.	2,6-Xylenol **2,6-Dimethylphenol, 2-hydroxy-*m*-xylene, 1-hydroxy-2,6-methylbenzene	$C_8H_{10}O$ ***$(CH_3)_2C_6H_3OH$	
9.	3,4-Xylenol **3,4-Dimethylphenol, 4-hydroxy-*o*-xylene, 1-hydroxy-3,4-methylbenzene	$C_8H_{10}O$ ***$(CH_3)_2C_6H_3OH$	
10.	3,5-Xylenol **3,5-Dimethylphenol, 5-hydroxy-*m*-xylene, 1-hydroxy-3,5-methylbenzene	$C_8H_{10}O$ ***$(CH_3)_2C_6H_3OH$	

(continued)

Table 1 Name, molecular formula, and structures of selected groups of phenolic compounds. (*continued*)

No.	Name of compound	Molecular formula	Structural formula
11.	2,3,4-Trimethylphenol ** 1-Hydroxy-2,3,4-methylbenzene	$C_9H_{12}O$ *** $(CH_3)_3C_6H_2OH$	

One-core phenols with aliphatic side-chain

No.	Name of compound	Molecular formula	Structural formula
12.	2-Ethylphenol ** (1-Hydroxy-2-ethylbenzene, *o*-ethylphenol)	$C_8H_{10}O$ *** $(C_2H_5)C_6H_4OH$	
13.	3-Ethylphenol ** (1-Hydroxy-3-ethylbenzene, *m*-ethylphenol)	$C_8H_{10}O$ *** $(C_2H_5)C_6H_4OH$	
14.	4-Ethylphenol ** (1-hydroxy-4-ethylbenzene, *p*-ethylphenol, 1-hydroxy-4-ethylbenzene, 4-hydroxyphenylethane)	$C_8H_{10}O$ *** $(C_2H_5)C_6H_4OH$	
15.	2-Isopropylphenol ** (*o*-Isopropylphenol, 1-hydroxy-2-isopropylbenzene, *o*-hydroxycumene; 2-hydroxycumene)	$C_9H_{12}O$ *** $(C_3H_7)C_6H_4OH$	
16.	*o*-Butylphenol * 2-Butylphenol ** (1-Hydroxy-2-butylbenzene)	$C_{10}H_{14}O$ *** $C_6H_4(OH)(C_4H_9)_2$	
17.	*2-Ethyl-4-methylphenol ** (2-Ethyl-4-Methyl-phenol, 2-ethyl-*p*-cresol)	$C_9H_{12}O$ *** $(C_2H_5)(CH_3)C_6H_3OH$	
18.	Thymol *2-Isopropyl-5-methylphenol ** (IPMP)	$C_{10}H_{14}O$ *** $2-[(CH_3)_2CH]C_6H_3-5-(CH_3)OH$	

Methyl derivatives of di- and trihydroxyl phenols

No.	Name of compound	Molecular formula	Structural formula
19.	3-Methylpyrocatechol ** (Pyrocatechol 1,2-dihydroxy-3-methylbenzene, 2,3-dihydroxytoluene, 2-hydroxy-3-methylphenol)	$C_7H_8O_2$ *** $(CH_3)C_6H_3(OH)_2$	

(*continued*)

Table 1 Name, molecular formula, and structures of selected groups of phenolic compounds. (*continued*)

No.	Name of compound	Molecular formula
20.	2-Methylresorcinol **(2,6-Dihydroxy, 2,6-toluenediol, toluene-2,6-diol)	$C_7H_8O_2$ ***$(CH_3)C_6H_3(OH)_2$
21.	2-Methylhydroquinol *2-Methylbenzene-1,4-diol **(2-Methyl-1,4-benzenediol, 2-methyl-1,4-hydroquinone, 2-methylbenzene-1,4-diol)	$C_7H_8O_2$ ***$(CH_3)C_6H_3(OH)_2$
22.	2,4,6-trihydroxytoluene *2-Methylbenzene-1,3,5-triol **(2-Methylphloroglucinol, Toluene-2,4,6-triol	$C_7H_8O_3$ ***$C_6H_2(OH)_3(CH_3)$
23.	2,4-Dimethylphloroglucinol *2,4-Dimethylbenzene-1,3,5-triol	$C_8H_{10}O_3$ ***$(CH_3)_2C_6H(OH)_3$

Di-core phenols

No.	Name of compound	Molecular formula
24.	1-Naphthol *Naphthalen-1-ol **(1-hydroxynaphthalene, 1-Naphthalenol; *alpha*-Naphthol α-Naphtol, naft-1-ol)	$C_{10}H_8O$ ***$C_{10}H_7OH$
25.	2-Naphthol *Naphthalen-2-ol **(2-hydroxynaphthalene, 2-Naphthalenol; *beta*-Naphthol β-Naphtol, naft-2-ol)	$C_{10}H_8O$ ***$C_{10}H_7OH$
26.	Methyl-2-naphtol **(1-methyl-2-hydroxynaphthalene, 1-methylnaphthalen-2-ol)	$C_{11}H_{10}O$ ***$(CH_3)C_{10}H_6OH$
27.	*2-phenylphenol **(2-Hydroxybiphenyl, *o*-phenylphenol, biphenylol, orthophenyl phenol, *o*-xenol, orthoxenol)	$C_{12}H_{10}O$ ***$C_{12}H_9OH$
28.	2,2'-Biphenol **(2,2'-Dihydroxybiphenyl, 2,2'-biphenyldiol)	$C_{12}H_{10}O_2$ ***$C_{12}H_8(OH)_2$

Three-core phenols

No.	Name of compound	Molecular formula
29.	1-Anthranol *Anthracen-9-ol **(1-Hydroxyanthracene, 9-anthracenol, 9-anthranol)	$C_{14}H_{10}O$ ***$C_{14}H_9OH$

(*continued*)

Table 1 Name, molecular formula, and structures of selected groups of phenolic compounds. (*continued*)

No.	Name of compound	Molecular formula	Structural formula
30.	2-Anthranol *Anthracen-2-ol **(2-Hydroxyanthracene, 2-anthracenol, 2-anthranol, beta-hydroxyanthracene)	$C_{14}H_{10}O$ ***$C_{14}H_9OH$	
31.	1,9-Dihydroxyanthracene *Anthracene-1,9-diol **1,9-Anthracenediol	$C_{14}H_{10}O_2$ ***$C_{14}H_8(OH)_2$	
32.	2-Phenantrenol	$C_{14}H_{10}O$ ***$C_{14}H_9(OH)$	

*, IUPAC name; **, synonyms; ***, other formula.

which as a result of deamination, cinnamic acid and its hydroxy derivatives are formed.[4,8,13–16]

Usually, these compounds are the following:

- Intermediate products of natural decay of organic matter such as proteins, humic compounds, and lignin[8]
- Products of metabolic processes taking place in living organisms[1,2,9]
- Secondary metabolic plant product[1,2,9]
- Products formed as a consequence of forest fires[17]
- Products of natural decay of original plant matter[1,2,9,16]

In Table 2, molecular and structural formulae of the natural derivatives of phenol most often met in the natural environment are presented.

Lignin is a complex chemical compound most commonly found in wood and all vascular plants, localized not only between the cells but also in the cell walls. In lignin, two groups of phenolic compounds dominate. The first comprises phenyl acid derivatives, occurring in plants mostly in the bound form as components of lignin and hydrolyzing tannins in the form of esters and glycosides.[13,15] The second group comprises hydroxycinnamic acids present most often in the form of esters, while hydroxybenzoic acids are mainly present in plants in the form of glycosides. Moreover, in plant tissues, phenyl acids have been identified, besides other naturally occurring compounds, like flavonoids, fatty acids, sterols, or cell wall polymers.[15]

It has been established that gymnospermous (*Gymnospermae*) plants like pine, spruce, and fir contain vanillyl, but they do not contain siryngyl and cinnamyl phenols. In the angiospermous (*Angiospermae*) plants, both vanillyl and siryngyl compounds are present, but cinnamyl compounds are not found.[12–16] Some examples of natural phenolic compounds used as indicators of original plant matter are given in Table 3.

Phenols of natural origin, generally called photochemical, include a large and very chemically diverse group of polyphenols characterized by the presence of large multiples of phenol units.

The above phenolic compounds, of which more than 8000 are known, embrace a wide range of plants' secondary metabolites, possessing in common an aromatic ring substituted by one or more hydroxyl groups.[13-16] They are the most widely distributed secondary metabolites ubiquitously present in the plant kingdom. Simple phenols are relatively rare in plants.[14] Some polyphenolic substructures are secondary plant substances called quasi-vitamins or natural nonnutritious substances showing great biological activity. This group of compounds is sometimes referred to as quasi-vitamin or vitamin P. They are mainly found in the bound form as components of lignin and hydrolyzing tannins, in the form of esters or glycosides, in all parts of plants, i.e., in the leaves, seeds, flowers, fruit, roots, bark, and wooden parts. Their chemical composition depends on the species and variety of a plant, climatic conditions, and agrotechnological procedures applied. Some polyphenols occurring in plants belong to phytoalexins, so the substances involved in the mechanisms protecting against the attack of insects, fungi, or viruses.

To sum up, polyphenols do the following:

- Play important roles in plant metabolism[13–16]
- Display high bioactivity; such natural products with healing and nutritional values are called nutraceuticals or dietary supplements[18]
- Participate in the healing and adaptive processes in living organisms, can protect against development of cancer, or can have therapeutic effect in treatment of different diseases[10,12]
- Participate in regulation of growth and reproduction of plants[11–13]
- Contribute in determination of sensory features of fruit, vegetables, and processed food[10–14]

Table 2 Phenol compounds of natural origin most commonly found in the biosphere.

No.	Name of compound	Molecular formula	Structural formula
1.	*Phenol **Carbolic acid, hydroxybenzene, benzenol, phenylic acid, phenic acid	(See Table 1 no. 1)	
2.	Catechol *Benzene-1,2-diol **(Pyrocatechol, 1,2-benzenediol, 2-hydroxyphenol, 1,2-dihydroxybenzene)	$C_6H_6O_2$ ***$C_6H_5(OH)_2$	
3.	*4-Hydroxybenzaldehyde **(p-Hydroxybenzaldehyde, 1-hydroxybenzaldehyde, p-formylphenol)	$C_7H_6O_2$ ***$(CHO)C_6H_4OH$	
4.	o-Cresol *2-Methylphenol **2-Hydroxytoluene, 2-methylobenzenol	(See Table 1 no. 2)	
5.	Vanillyn *4-Hydroxy-3-methoxybenzaldehyde **(Methyl vanillin, vanillin, vanillic aldehyde, 3-methoxy-4-hydroxybenzaldehyde)	$C_8H_8O_3$ ***$(OCH_3)(CHO)C_6H_3OH$	
6.	Syringaldehyde *4-Hydroxy-3,5-dimethoxybenzaldehyde **(3,5-Dimethoxy-4-hydroxybenzaldehyde, 3,5-dimethoxy-4-hydroxybenzene carbonal, gallaldehyde 3,5-dimethyl ether, 4-hydroxy-3,5-dimethoxybenzaldehyde, syringic aldehyde)	$C_9H_{10}O_4$ ***$(OCH_3)_2(CHO)C_6H_2OH$	
7.	Alkyl phenols With saturated alkyl groups at para position C_nH_{2n+1}, usually $n = 1-3$	$C_6H_4(OH)(C_nH_{2n+1})$ $n = 1-12$	
8.	Gallic acid *3,4,5-Trihydroxybenzoic acid **(Gallic acid, gallate, 3,4,5-trihydroxybenzoate)	$C_7H_6O_5$ ***$C_6H_2(OH)_3(COOH)$	
9.	Eugenol *4-Allyl-2-methoxyphenol **[4-Allyl-2-methoxyphenol, 2-methoxy-4-(2-propenyl)phenol, eugenic acid, caryophyllic acid, 1-allyl-3-methoxy-4-hydroxybenzene, allylguaiacol, 2-methoxy-4-allylphenol, 4-allylcatechol-2-methyl ether, 2-methoxy-4-(2-propen-1-yl)phenol]	$C_{10}H_{12}O_2$ ***$(OCH_3)(C_3H_4)C_6H_3OH$	
10.	Thymol *2-Isopropyl-5-methylphenol **(IPMP)	(See Table 1 no. 18)	
11.	1-Naphthol *Naphthalen-1-ol **(1-Hydroxynaphthalene, 1-naphthalenol; alpha-naphthol, α-Naphtol, naft-1-ol)	(See Table 1 no. 24)	

*, IUPAC name; **, synonyms; ***, other formula.

Table 3 Phenolic derivatives used as indicators of original plant matter.

No.	Name of compound	Molecular formula	Structural formula
1.	Vanillyn	(See Table 2 no. 5)	
2.	Vanillic acid *4-Hydroxy-3-methoxybenzoic acid **(4-Hydroxy-*m*-anisic acid, vanillate)	$C_8H_8O_4$ *** $C_6H_3(OH)(OCH_3)(COOH)$	
3.	Apocynin *1-(4-Hydroxy-3-methoxyphenyl)ethanone **4-Hydroxy-3-methoxyacetophenone, acetovanillone	$C_9H_{10}O_3$ *** $C_6H_3(OH)(OCH_3)(COCH_3)$	
4.	Syringaldehyde	(See Table 2 no. 6)	
5.	Syringic acid *4-Hydroxy-3,5-dimethoxybenzoic acid **(Gallic acid 3,5-dimethyl ether)	$C_9H_{10}O_5$ $C_6H_2(OH)(OCH_3)(COOH)$	
6.	Acetosyringone *4'-Hydroxy-3',5'-dimethoxyacetophenone **Acetosyringenin	$C_{10}H_{12}O_4$ *** $C_6H_2(OH)(OCH_3)_2(COCH_3)$	
7.	*p*-Coumaric acid *3-(4-Hydroxyphenyl)-2-propenoic acid **[*para*-Coumaric acid, 4-hydroxycinnamic acid, β-(4-hydroxyphenyl)acrylic acid]	$C_9H_8O_3$ *** $C_6H_4(OH)(CH=CH-COOH)$	
8.	Ferulic acid *(E)-3-(4-hydroxy-3-methoxy-phenyl)prop-2-enoic acid **[2-Propenoic acid, 3-(4-hydroxy-3-methoxyphenyl)-ferulic acid, 3-(4-hydroxy-3-methoxyphenyl)-2-propenoic acid, 3-(4-hydroxy-3-methoxyphenyl)acrylic acid, 3-methoxy-4-hydroxycinnamic acid, 4-hydroxy-3-methoxycinnamic acid, (2E)-3-(4-hydroxy-3-methoxyphenyl)-2-propenoic acid, ferulate coniferic acid, *trans*-ferulic acid, (E)-ferulic acid]	$C_{10}H_{10}O_4$ *** $C_6H_3(OH)(OCH_3)$ $(CH=CH-COOH)$	

Note: *, IUPAC name; **, synonyms; ***, other formula.
Source: Adapted from Kroon and Wiliamson,[12] Robinson,[13] and Duke.[14]

- Endow plants and fruits with a specific tart and bitter taste[13,16]
- Responsible for the color and fibrous nature of plants and fruit[13]
- Participate in morphogenesis, energy flow, sex determination, photosynthesis, respiration, regulation of gene expression, and regulation of synthesis of growth hormones[19]
- Have a protecting effect against ultraviolet irradiation and against stress[14,15]

Humic substances are the end products of decaying natural organic matter in the microbial process taking place with involvement of edaphone (mainly fungi and actinomycetes) called humification. They are major organic constituents of soil (humus), peat bogs, coal, sewage, compost heaps, carbonaceous shales, lignites, and all types of natural waters (aquatic humic substances) and can form complex ions that are commonly found in the natural environment. A typical humic substance is a mixture of many molecules,

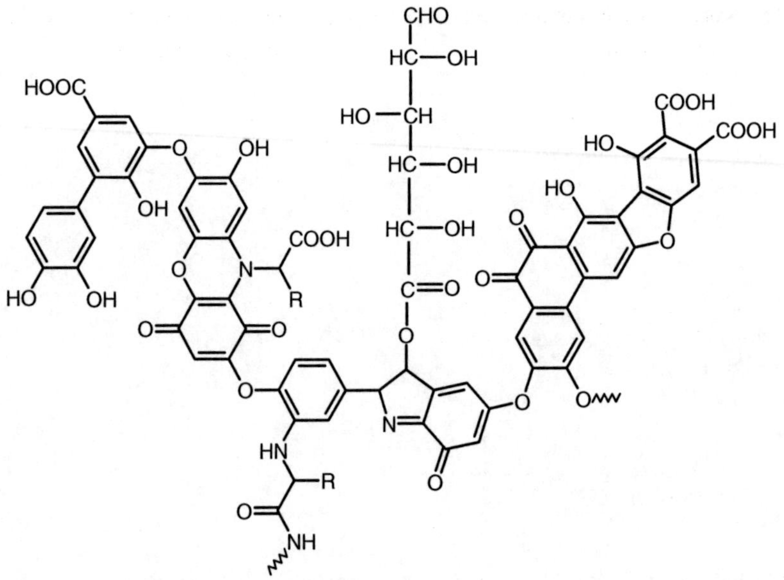

Fig. 1 The hypothetical structure of humic acid, having a variety of components including quinone phenol, cresol, and sugar moieties. **Source:** Adapted from Stevenson,[20] Muscola and Soidari,[21] and Hessen and Tranvik.[22]

some of which are based on a motif of aromatic nuclei with phenolic and carboxylic substituents (see Fig. 1).[20–22]

There are three types of humic substances, which differ slightly in acidity and chemical composition. They are humic acid, fulvic acid, and humins.

Aquatic humus substances bear about 40%–60% of dissolved organic carbon and make up the largest fraction of natural organic matter in water. The major functional groups include carboxylic acids, phenolic hydroxyl, and carbonyl and hydroxyl groups.[22] Phenol hydroxyl groups are usually in amounts of about 1 μeq/mg C of humic material or 94 μg/mg of elemental carbon as humic matter.[5]

As mentioned earlier, humin substances are also constituents of coal, which is a fossil phytomaterial. It is assumed that lignin, which is one of the most important chemical compounds in plants, is one of the parent substances of coals having aromatic acidic character. It is a consequence of the presence of phenolic hydroxyls, which are the most characteristic oxygen group of coals. The degree of carbonization of coal is measured by the content of hydroxyl groups. In coals with a low degree of carbonization, hydroxyl groups comprise up to 90% of all oxygen groups. Their content decreases with increasing degree of carbonization, and they disappear in coal with a high degree of carbonization.[23]

Anthropogenic Origin

It has been established that a fundamental part of phenol present in the natural environment is of anthropogenic origin. Considerable sources of phenol and its derivatives are the processes of production of intermediate semiproducts and their further processing. Three main sources of phenol and its derivatives released to the natural environment are distinguished: industrial processes, nonindustrial processes and endogenous sources. The greatest amounts of phenol and its derivatives come from industrial processes, mainly production and application of different kinds of phenolic resins and phenolic plastics, generally called phenolics, and caprolactam.[24] Great demand for phenolic resins, widely applied, e.g., as a binding agent in insulating materials, chipboards, shatterproof glass, paints, and casting molds, makes them a profound source of phenolic compounds. The emission of these compounds from these materials is measured in terms of a concentration of free phenol occurring in the monomeric form in the resins in the amount of 1%–5% wt.[24,25] In foundries, phenolic compounds are emitted in processes of mold production and in casting.[26] Phenolic compounds occur in large amounts in the products coming from the chemistry of coal and chemistry of coke. Processing of raw benzenol obtained as a result of coal coking leads to formation of phenol and xylems. Phenolic compounds are met in wastewater from industrial plants in which phenol and its derivatives are used as raw materials in amounts to a few g/dm^3. Table 4 gives a list of processes that are the most abundant sources of phenol released to the natural environment and values of its emission coefficients.

The most important nonindustrial source of phenolic compounds is fuel combustion in motor vehicles. The exhaust gases from motor vehicles contain from about 0.3 to 1.4–2.0 ppm of phenol, which corresponds to the amounts from 1.2 to 5.4–7.7 mg/m^3.[24,28,29]

Phenolic compounds in considerable concentrations are released in volatile form during combustion of other fuels such as wood, coal, and mazout, in heating chambers,

Table 4 Selected industrial processes leading to release of phenol to the atmospheric air and estimated values of its emission coefficients.

Type of process	Phenol emission coefficient
Production of phenol resins	0–0.5 g of phenol emitted per kg of resin
Production of phenol and its derivatives	-
Production of caprolactam	0.2–0.05 g of phenol emitted per kg of cyclohexanol (semiproduct)
Production of coke	-
Production of insulating materials	-
Emission from phenol processing	-

Source: *RIVM Criteria Document: Phenol, Bilthoven, the Netherlands*.[27]

house furnaces, and fireplaces.[30] It should be mentioned that phenolic compounds are commonly met in cigarette smoke, in which the estimated average amount of phenol is 0.4 mg per cigarette.[31] Phenol can be also found in smoked food products.[32]

Endogenous sources of phenol include its synthesis in vivo from different xenobiotics released to the natural environment, e.g., from benzene. It has been established that benzene and its phenolic derivatives in the in vivo conversion can be an endogenous source of phenol.[33] Among phenol derivatives, particular attention has been paid to pentachlorophenol (PCP) and its salts, because of its properties and low cost of production. This material has been widely applied in industry (mainly for wood impregnation) and agriculture and as a component of household products (disinfectant). In agriculture, PCP is used as a pesticide of a broad spectrum of activity as it acts against algae, bacteria, fungi, weeds, insects, and mollusks. Pentachlorophenol has been recognized as hazardous for human health because of its toxicity and widespread use. It is used in leather tanning and finishing. Monochlorophenols are used as synthetic intermediates for dyes and chlorinated phenols. Pentachlorophenol is used as a denaturant for alcohol, an antiseptic, and a selective solvent for refining minerals.[34–36]

Nomenclature, Structure, and Chemical Composition

Phenol is a common name for the simplest and most common aromatic alcohol, labeled with CAS-Registry^{SM} -The world's largest substance database number 108-95-2, in which the hydroxyl group, known as a phenolic hydroxyl, is attached to the phenyl group. By definition, phenol is a hydroxybenzene, but according to the IUPAC, its correct name should be benzenol.

In literature, phenol is referred to by many other names, such as benzyl, carbolic acid, carbol, phenol liquefied, phenolic acid, phenyl hydroxide, phenic acid, phenylic acid, oxybenzene, monophenol, monohydroxybenzene, and phenyl hydrate.[1,2,24] It also has quite a few commercial names, such as carbol liquor and phenyl liquor (Netherlands), kristallliertes Kreozot or Steinkohlenteerkreosot (Germany), Venzenol (France), or code label ENT1814. The class of phenolic compounds, whose name comes from phenol, includes cresols (aromatic compounds from the group of phenols that are derivatives of toluene).

In contrast to alcohols, in phenol, the hydroxyl group is attached directly to the aromatic ring, which gives enhancement of its acidic properties. Phenols are the compounds in which the ring (benzene) is substituted with one or a few hydroxyl groups. If the compounds are built with aromatic rings (naphthalene, anthracene), with one or more hydroxyl substituents at the benzene ring, then they are called naphthols (hydroxynaphthalenes) or hydroxyanthracenes.[1,2]

According to the IUPAC nomenclature, in naming substitution products of these compounds, the numbering starts at the group already present and is done in the direction that gives the lowest numbers to other groups on the ring. Sometimes, the benzene ring is treated as a substituent (for hydrogen) on another molecule. In that case, the C_6H_5- group of benzene is called "phenyl." Substituents are cited in alphabetical order. Carboxyl and acyl groups take precedence over the phenolic hydroxyl in determining the base name. The hydroxyl group is treated as a substituent. Higher substituted compounds are named as derivatives of phenol.

Structural diversity of phenol compounds makes their systematization rather difficult. With respect to the structure of the carbon skeleton, phenolic compounds can be divided into the following groups:

- Phenylcarboxylic acids—derivatives of benzoic acid of the carbon skeleton C_6–C_1 (where C_6 denotes a benzene ring), whose structure can be described by the general formula in Fig. 2 (see also Table 5).
- Phenylpropenic acids—derivatives of cinnamic acid of the carbon skeleton C_6–C_3, whose structure can be described by the general formula in Fig. 3 (see also Table 6).
- Stilbene (trans)—of the carbon skeleton C_6–C_2–C_6, naturally occurring compounds, found in a wide range of

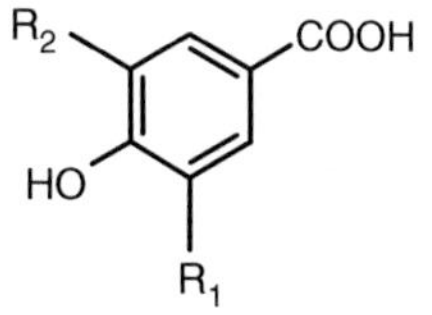

Fig. 2 The general formula of aromatic hydroxyacid derivatives of benzoic acid.

Table 5 Exemplary hydroxyacids that are benzoic acid derivatives, occurring in plants.

No.	Name of hydroxyacid	Molecular formula	Structural formula
1.	*Benzoic acid **(Benzenecarboxylic acid, carboxybenzene, E210, dracylic acid)	$C_7H_6O_2$ ***$C_6H_5(COOH)$	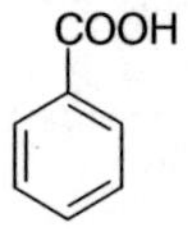 COOH–C₆H₅
2.	*4-Hydroxybenzoic acid **(p-Hydroxybenzoic acid, para-hydroxybenzoic acid)	$C_7H_6O_3$ ***$C_6H_5(COOH)$	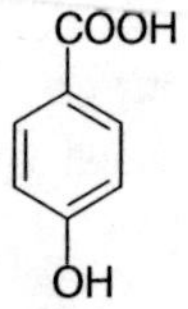 HO–C₆H₄–COOH
3.	Protocatechuic acid *3,4-Dihydroxybenzoic acid **Protocatechuic acid (PCA)	$C_7H_6O_4$ ***$C_6H_4(OH)_2(COOH)$	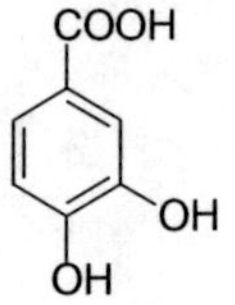 (HO)₂C₆H₃–COOH
4.	Gallic acid		(See Table 2 no. 8)
5.	Vanillic acid		(See Table 3 no. 2)
6.	Syringic acid		(See Table 3 no. 5)

Note: *, IUPAC name; **, synonyms; ***, other formula.

plants, aromatherapy products, and dietary supplements, whose structure can be illustrated by the following general formula in Fig. 4. Exemplary compounds from this group are resveratrol, piceatannol, rhapontigerin, and pterostilnene. The first mentioned stilbene, resveratrol, has been intensely studied and proved to have potent anticancer, anti-inflammatory, and antioxidant activities.[37]

- Polyphenols known also as flavonoids or bioflavonoids, of the carbon skeleton C_6–C_3–C_6, whose general structure is shown in Fig. 5. This group includes the following subgroups: flavons, the compounds responsible for the yellow color of plants or their parts (Latin *flavus* means yellow); flavones; flavanones; flavanols; anthocyanes (proanthocyanes); and isoflavones. They constitute the largest class of phenolic compounds, with more than 3000 structures.[38,39] Table 7 presents representative compounds of the above-mentioned subgroups of polyphenols.

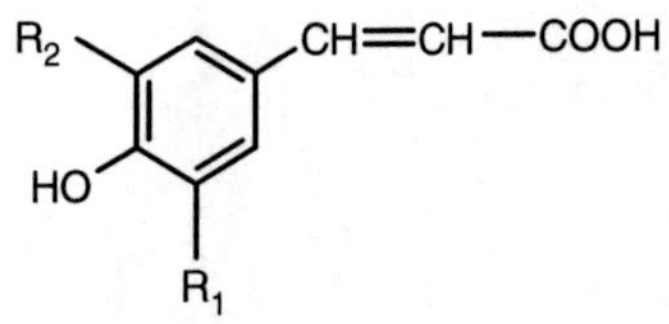

- Other phenolic compounds, e.g., tannins (proanthocyanides) of the carbon skeleton Cn>- 12, comprising the following:
- Hydrolyzable tannins, consisting of several gallic acid units bound through ester linkages to a central glucose (see Fig. 6a); these types of tannins are quite water soluble and are part of the water products of plants.
- Condensed tannins, which are a diverse group of polyphenolic compounds of plant origin called flavonoids or bioflavonoids (see Fig. 6b).

Polyphenols make a large group of natural substances occurring in many plants. They are found in greatest amounts in fruit (chokeberry, blueberry, grapes, nuts, garlic) and vegetables (cabbage, cereal seeds).[40–42] Polyphenols are mostly distributed in the external parts of the fruit. In drinks, polyphenols are found in green tea, red wine, coffee, and beer. These compounds make up the largest group of antioxidants supplied in the diet, and hence, they are often called bioflavonoids.[43–45]

Physical, Chemical, and Organoleptic Properties

Phenol is a colorless to light pink crystalline (long needle-like crystallites) solid. When molten, it becomes a bright colorless liquid of low or high viscosity. It attains pink color on exposure to air and light as a result of partial oxidation. In the atmosphere of wet air, the crystals deliquesce. Crude product can be pink, brown, or black. The physicochemical properties of phenols differ significantly from those of aliphatic or unsaturated alcohols;[1,2,7,24] some of them are given in Table 8.

Fig. 3 The general formula of aromatic hydroxyacids that are derivatives of cinnamic acid.

Table 6 Some hydroxyacids that are cinnamic acid derivatives, occurring in plants.

No.	Name of compound	Molecular formula	Structural formula
1.	Cinnamic acid *(E)-3-phenylprop-2-enoic acid **(Cinnamic acid, trans-cinnamic acid, phenylacrylic acid, cinnamylic acid, 3-phenylacrylic acid, (E)-cinnamic acid, benzenepropenoic acid, isocinnamic acid)	$C_9H_8O_3$ ***$C_6H_4(OH)(C_3H_3O_2)$	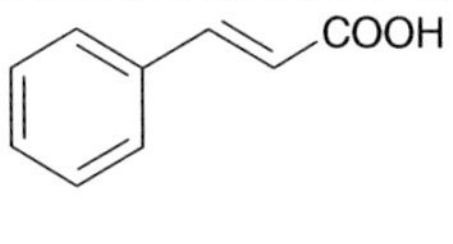
2.	o-Coumaric acid **(2-Hydroxycinnamic, o-coumaric acid, 2-coumaric acid, 2-coumarate 2-hydroxycinnamate, trans-2-hydroxycinnamic acid trans-2-hydroxycinnamate	$C_9H_8O_3$ ***$C_6H_4(OH)(C_3H_3O_2)$	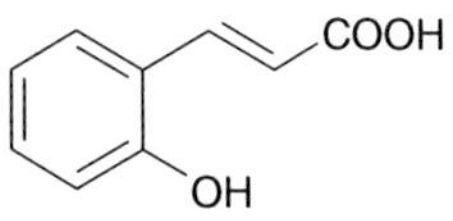
3.	*p*-Coumaric acid	See Table 3 no. 7	
4.	Caffeic acid *[3-(3,4-Dihydroxyphenyl 2-propenoic acid, 3,4-dihydroxy-cinnamic acid trans-caffeate, 3,4-dihydroxy-trans-cinnamate), (E)-3-(3,4-dihydroxyphenyl)-2-propenoic acid 3,4-dihydroxybenzeneacrylicacid 3-(3,4-dihydroxyphenyl)-2-propenoic acid]	$C_9H_8O_4$ ***$C_6H_3(OH)_2(C_3H_3O_2)$	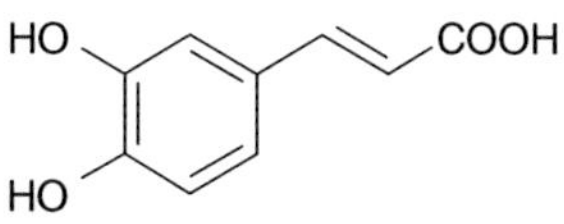
5.	Ferulic acid	(See Table 3 no. 8)	
6.	Sinapinic acid *3-(4-Hydroxy-3,5-dimethoxyphenyl)prop-2-enoic acid **(Sinapinic acid, sinapic acid, 3,5-dimethoxy-4-hydroxycinnamic acid, 4-hydroxy-3,5-dimethoxycinnamic acid)	$C_{11}H_{12}O_5$ ***$C_6H_2(OH)(OCH_3)_2(C_3H_3O_2)$	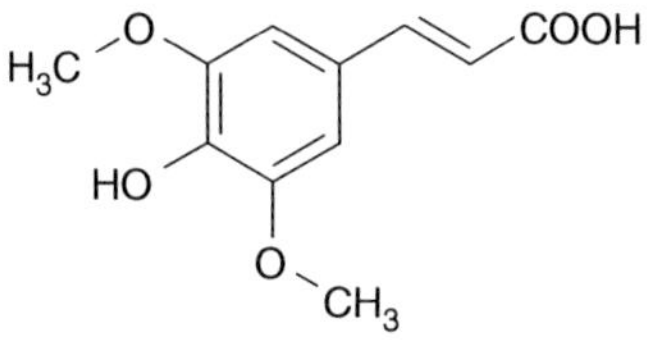

Note: *, IUPAC name; **, synonyms; ***, other formula.

Phenols have hydroxyl groups that can participate in intermolecular hydrogen bonding with other phenol molecules or other H-bonding systems, e.g., water. Hydrogen bonding results in higher dipole moment and melting points and much higher boiling points for phenols than those of hydrocarbons of similar molecular weight. The ability of phenols to form strong hydrogen bonds also enhances their solubility in water.[1,24]

Phenol dissolves to give a 9.3% solution in water, compared with a 3.6% solution of cyclohexanol in water. This water solution of phenol is called phenol liquefied. The water solubility of phenol increases with temperature, and above 68.4°C, both substances become fully miscible.

Phenol is more readily soluble in most organic solvents and in water solutions of soaps. An example of phenol in a water solution of soap is Lysol, being a mixture of a water solution of a potassium soap and cresols, used mainly in the veterinary field for disinfection of rooms and vessels. Its solubility in aliphatic solvents is limited.[1,2]

In contrast to neutral alcohols, water solutions of phenol show weak acidic properties; hence, phenol has been referred to as carbolic acid. Phenol dissociates with the formation of phenolate ($C_6H_5O^-$), also called phenoxide ion, and proton ($H_3O)^+$:

$$C_6H_5OH_{(aq)} + H_2O_{(l)} \leftrightarrow C_6H_5O^-_{(aq)} + H_3O^+_{(aq)} \quad (1)$$

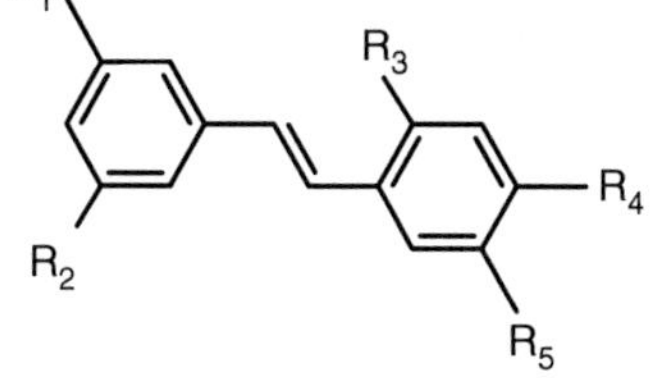

Fig. 4 The general formula of trans stilbenes (1,2-diphenylethylene; R_1, R_2, R_3, R_4, and R_5 denote OH, OCH_3, or –glucose group).

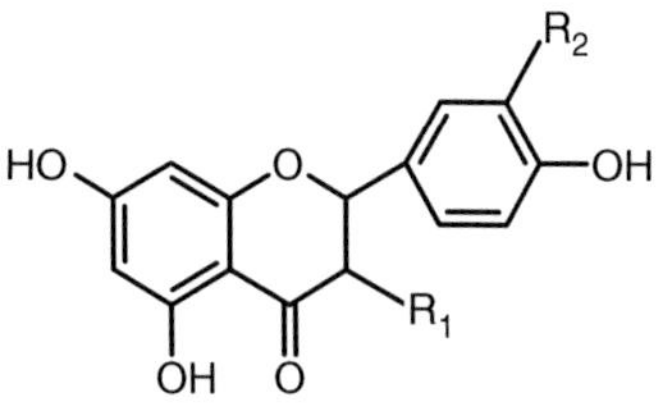

Fig. 5 The general structure of flavonoids (R_1, R_2 denote H or OH group).

Table 7 Selected examples of chemical composition and structures of subgroups among polyphenols.

Polyphenols (flavonoids)		
Name of compounds	Molecular formula	Occurrence
Flavons		
Quercetin *2-(3,4-Dihydroxyphenyl)-3,5,7-trihydroxy-4H-chromen-4-one **(Sophoretin, meletin, quercetine, xanthaurine, quercetol, quercitin, quertine, flavin meletin)	$C_{15}H_{10}O_7$	Green tea, grape skin, ginkgo leaves, apples
Flavonols		
Rutin *2-(3,4-Dihydroxyphenyl)-5,7-dihydroxy-3-[α-L-rhamnopyranosyl-(1→6)-β-D-glucopyranosyloxy]-4H-chromen-4-one **(Rutoside, phytomelin, sophorin, birutan, eldrin, birutan forte, rutin trihydrate, globularicitrin, violaquercitrin)	$C_{27}H_{30}O_{16}$	Pogoda tree buds (*Sophora japonica Fabaceae*), apple skin
Flavanones		
Hesperidin *(2S)-5-hydroxy-2-(3-hydroxy-4-methoxyphenyl)-7-[(2S,3R,4S,5S,6R)-3,4,5-trihydroxy-6-[[(2R,3R,4R,5R,6S)-3,4,5-trihydroxy-6-methyloxan-2-yl]oxymethyl]oxan-2-yl]oxy-2,3-dihydrochromen-4-one	$C_{28}H_{34}O_{15}$	Citrus fruit skin
Flavanols (Catechins)		
Epicatechol **(epi-Catechinepi-catechol, epicatechin, l-acacatechin)	$C_{15}H_{14}O_6$	Green tea, grape seeds
Anthocyanes (Proanthocyanes)		
Cyanidin *2-(3,4-Dihydroxyphenyl) chromenylium-3,5,7-triol **Cyanidine	$C_{15}H_{11}O_6^+$	Bilberry, red and black grapes, red wine
Isoflavones		
Genistein *5,7-Dihydroxy-3-(4-hydroxyphenyl)chromen-4-one **4',5,7-Trihydroxyisoflavone	$C_{15}H_{10}O_5$	Leguminous plants, soybean, cereal grains, fruits, vegetables

Note: *, IUPAC name; **, synonyms.

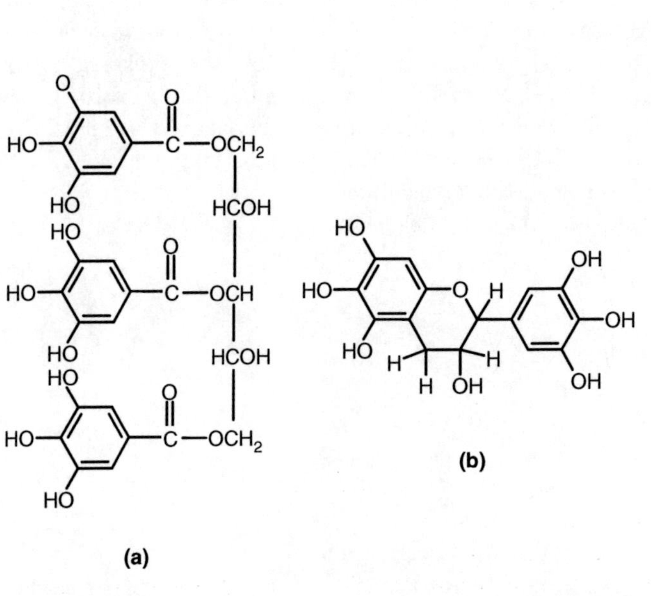

Fig. 6 The hypothetical structures of hydrolyzable (a) and condensed (b) types of tannins.

Table 8 Some physical and chemical properties of phenol.

Molecular weight	94.11 g/mol
Boiling point	181.75°C (101.3 Pa)
Melting point	43.0°C
	40.9°C (pure substance)
Relative density	1071
Relative vapor density (air = 1)	3.24 mm Hg
Vapor pressure (20°C)	0.357 mm Hg
(50°C)	2.48 mm Hg
(100°C)	41.3 mm Hg
Concentration of saturated vapor (20°C) in air	0.77 g/m^3
Water solubility (16°C)	67 g/L
above 68.4°C	Fully soluble
Partition coefficient n-octanol/water (log$_{Pow}$)	1.46
Dissociation constant in water K_a (20°C)	1.28 × 10^{-10}
Ignition point:	
-Closed crucible	80°C
-Open crucible	79°C
Acidity	K_a = 1.3 × 10^{-10} (pK_a = 9.55)

Source: Adapted from Rappoport,[1] Tyman,[2] and Weber et al.[24]

Dissociation occurs to only a slight extent; phenol is a very weak acid with pK_a = 9.55. Phenols are more acidic than alcohols of pK_a = 16–20 but less acidic than carboxylic acids of pK_a = 5.[2] Introduction of substitutes, particularly those in ortho or para position to the –OH group, can dramatically influence the acidity of phenol due to resonance and/or inductive effects. In reactions with strong bases (NaOH, KOH), phenols create phenates (C_6H_5OMe), which are more stable than alcoholates:

$$C_6H_5OH + Na^+OH^- \rightarrow C_6H_5O^-Na^+ + H_2O \quad (2)$$

This increased stability is a consequence of the mesomeric effect stabilizing the phenate ion owing to the delocalization of the negative charge on the aromatic ring of phenol.[1,2]

Phenol has a peculiar characteristic smell and a strong corrosive effect on skin. It is poisonous in nature but acts as a powerful antiseptic.[24,41,46]

Phenol is susceptible to oxidants (e.g., CrO_3, $K_2Cr_2O_7$). The hydrogen abstraction from phenolic hydroxyl is accompanied by the resonance stabilization of the formed phenoxy radical ($C_6H_5O^{\bullet}$), which can be further oxidized.[46]

Phenols are highly reactive toward electrophilic substitution because the nonbonding electron on oxygen stabilizes the intermediate cation.[1,2,7]

Phenol reacts with carbonyl compounds both in acidic and basic media. In the presence of formaldehyde, it easily undergoes hydroxymethylation followed by condensation of resin. Condensation with acetone gives bisphenol A, a key building block of polycarbonates. Conversion with formaldehyde produces phenolic resins, the best known of which is Bakelite.[1,2,25,47] The product of catalytic hydrogenation of phenol, performed on a large scale, is cyclohexanol. Another industrially important process is alkylation of phenol, which takes place in the presence of an acid catalyst or a Friedl–Crafts catalyst.

In large amounts, phenol is also used to make different derivatives containing chlorine, usually coming from NaOCl or Cl_2. The most important from among these compounds is PCP, manufactured usually by direct chlorination in the presence of metallic nickel as a catalyst or by hydrolysis of hexachlorobenzene.[1,2] Phenol also reacts directly with alkyl halides in alkali solutions to form phenyl ethers. The phenate ion shows nucleophilic character and replaces halogen from alkyl halide:

$$C_6H_5OH + NaOH \rightarrow C_6H_5ONa + H_2O \quad (3)$$

$$C_6H_5ONa + CH_3Cl \rightarrow C_6H_5OCH_3 + NaCl \quad (4)$$

They are also formed when vapors of phenol and an alcohol are heated over ThO_2.

Phenol can also make esters, among them an interesting group called parabens (esters of p-hydroxybenzoic acid with aliphatic alcohol), whose general formula is shown in Fig. 7.[48]

Phenol has a characteristic acidic smell and pungent taste. It shows antiseptic activity because of the ability to coagulate protein. It is active against a wide range of microorganisms including some fungi and viruses but is only slightly effective against spores. Although phenol was the first antiseptic used on wounds and in surgery, it is a protoplasmic poison that damages all kinds of cells and is alleged to have caused an astonishing number of poisonings since it came into general use.[2,6,32] Generally, it is a strong and violent poison attacking the nervous system, alimentary track, and circulatory system. It has been estimated that approximately 1 g of phenol is enough to cause death.[2,6,7,27,46] Phenol and its vapors are corrosive to the eyes, skin, and respiratory tract.[49] There is no evidence to claim that phenol causes cancer in humans. Besides its hydrophobic effects, another mechanism of its toxicity is

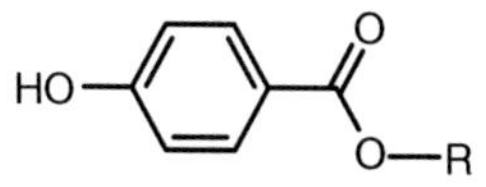

Fig. 7 General formula of paraben, with R standing for methyl, ethyl, n-propyl, isopropyl, n-butyl, isobutyl, or benzyl group.

via formation of phenoxyl radicals. Poisoning can also result from inhalation of phenol in the form of atmospheric aerosol, which quickly coagulates in cold air.[50]

Chlorophenols are characterized by a 100–1000 times more intense smell than the parent phenol. During chemical wastewater treatment (chlorination), the water contaminated with phenols acquires a repulsive taste as a result of formation of phenol chloroderivatives.[5] It has been established the limiting concentrations of phenol perceptible though the sense of smell, taste and touch are: 0.021–20 mg/m^3 in air[51] and 0.3 mg/L in water, respectively.[52]

Uses of the Phenol and its Derivatives

Phenol belongs to the 50 most abundantly produced chemicals. It is one of the oldest chemical intermediates in organic chemistry that still plays a very important role. Both phenol and its numerous derivatives of natural, synthetic, and semisynthetic origin are basic raw materials for industry.

The applications of phenol and its derivatives can be divided into industrial, nonindustrial, and niche.

The largest single use of phenol in industry is as an intermediate in the production of phenolic resins or related materials in the reaction of polycondensation of phenol with formaldehyde. Thus, produced phenyl aldehyde resins are low cost, thermo set, and versatile. They are used as resin glues for plywood in plywood adhesive, cast resins, molding resins, and varnish resins; moreover, they are applied in construction, automotive, and appliance industries.[2,24]

Alkylphenols introduced on the market in the 1940s have been widely used as paint components, herbicides, pesticides, some nonionic detergents, components of cosmetics, composite materials, and media used for removal of fat from the surface of wool, leather, and metal finishing.[1,2]

Non-anionic detergents are produced by alkylation of phenol to give alkylphenols, which are then subjected to ethoxylation.[2] Alkyl phenols with branched chains are used as supplements in oil lubricants, antioxidants (stop chain reactions), and auxiliary substances in conversion of rubbers and plastics.[1,2,24,25] Chemical activity of alkyl derivatives of phenol as well as antioxidants results from the facility of hydrogen abstraction from the phenol group, initiated by the interaction of oxygen and light.

In order to hinder the process of aging, alkyl derivatives are introduced to many different organic products in amounts of 0.1%–2% wt. For example, they can be found in gasoline, lubricating oils, lipids, polymers and plastics, rubber, soaps, cosmetics, and pharmaceuticals.

The chlorinated phenols that are used in largest quantities include PCP and the trichlorophenol isomers used as wood preservatives. Their main components are cresols known as creosotes.[2]

Higher chloroderivatives of phenol, such as 2,4-dichlorophenoxyacetic acid and 2,4,5-trichlorophenoxyacetic acid act, as selective herbicides. The first of the compounds mentioned is one of the most important herbicides against dicotyledons in crops of monocotyledonous plants (cereals, linen, grass). The second of the above compounds acts against perennial weeds, bushes, and coppices in grasslands. Problems that these compounds impose on the natural environment have prompted gradual limitation on their production. Ammonolysis of phenol in gas phase in the presence of aluminosilicate catalysts (zeolites) is one of the common methods for the synthesis of aniline, which is an intermediate for manufacturing salicylic acid. It is also used to make pharmaceuticals, aromatic esters, synthetic dyes, and food preservatives.[2,3,24]

All industrial applications of phenol and its derivatives are too numerous to be mentioned.

The major nonindustrial applications of phenol and its derivatives include the following.

- Medicine; as an antiseptic (water solution of 2%–10% is a very effective antiseptic medium called phenolated water; and in a concentration of 0.2%, it shows bacteriostatic action, while above 2%, it is bactericidal);[1] to relieve itching; as an anesthetic in medicinal preparations (ointments, ear and nose drops, cold sore lotions, throat lozenges, and antiseptic lotions). They are active against a wide range of microorganisms including some fungi and viruses but are only slowly effective against spores. They have been used to disinfect skin, but recently, phenol (Phenolum FP VII) and liquid phenol (Phenolum liquefactum FP IV) have been sporadically used, taking into account their toxicity.[1,2,6,24,25]
- Salicylic acid is an intermediate in production of drugs, flavor esters, dyestuffs, and food preservatives.[2,53]
- Pentachlorophenol can be applied as a fungicide for wood solution of phenol.[1,2, 53–55]

The niche applications of phenol and its derivatives include the following.

- Diphenyl ether is used as material for the production of phenoxathiin. It is an intermediate for polyamide and polyimide, and a processing aid in the production of polyesters. Some polybrominated biphenyl ethers are flame retardants and are used in soap perfumes.[56]
- Phenol is used for embalming bodies for study because of its ability to preserve tissues for extended periods of time.
- Production of cosmetics such as sunscreens,[53] hair dyes, and skin lightening preparations.[54]
- Phenol is used as an exfoliant in cosmetic surgery.
- Phenol and its derivatives are used for phenolization, which is a surgical procedure that serves to treat ingrown nails.

- Phenol is used to denature and remove protein when purifying nucleic acids in molecular biology procedures.
- Phenol and its derivatives are added to polyimide to make antispastic and painkilling drugs used in cancer therapy.[54]
- They are used as slimicides, i.e., substances used to kill slime-producing organisms including bacteria and fungi.[1,2,24,26,55]

DISTRIBUTION, TRANSPORT, AND CONVERSION OF PHENOL AND ITS DERIVATIVES IN THE BIOSPHERE

A phenolic compound occurs in the biosphere in concentrations from trace to high orders of a few milligrams per cubic decimeter or even a few grams per cubic meter in the industrial emissions. Phenol evaporates more slowly than water and can remain in the air, soil, and water for long periods of time if large amounts of it are released at one time or if it is constantly released to the environment from a source.[57,58] Coefficients for conversion of phenol concentration are the following: 1 mg/m^3 = 0.26 ppm, 1 ppm = 3.84 mg/m^3.[2]

In the atmosphere, phenol exists predominantly in the vapor phase.[57,58] No reliable information has been given hitherto on the background concentration of phenol in the atmospheric air far from the sources of its emission. It has been a priori assumed to be low. It has been estimated that the background level of phenol in the atmospheric air is below 1 ng/m^3.[59] Higher concentrations are to be expected above cities, mainly as a result of its emission in gas exhausts of motor vehicles.[50,60,61]

According to estimations, the mean day concentration in urban and suburban atmosphere reaches 0.12 µg/m^3, and in the atmosphere of large urban agglomerations, it can vary within the range of 0.1–8 µg/m^3.[62] In the atmosphere over the industrial sources of phenol emission, its concentration can be two orders of magnitude higher. In the neighborhood of a phenol resin–producing plant, the concentration of phenol was at a level of 190 µg/m^3.[63,64] Still higher values were recorded directly at the workplace in the gases released by iron casting plants; in 1980, the concentration of phenol reached at such a place was 0.8–3.5 mg/m^3.[65,66]

It has been estimated that half-life of phenol in air generally varies depending upon atmospheric conditions, and values ranging from 2.28 to 22.8 hr for reaction with hydroxyl radicals ($^\bullet$OH and HO$_2^\bullet$) have been reported in literature. It is the reason why a small amount of phenol does not remain in the air for longer than a day.[59]

The presence of phenol was detected in rain and surface water as well as in underground water, but relevant information is scarce. Phenolic compounds react as a weak acid in water, and they are not expected to dissociate in the pH range typical in the natural environment, i.e., 6.5–8.5.[5]

For example, the concentration of phenol dissolved in rainwater collected in Portland, United States, was 0.08–1.2 µg/L, but the mean concentration was 0.26 µg/L.[67] In diluted water solutions, phenol undergoes conversion to dihydroxybenzenes, nitrophenols, nitrosophenols, and nitrochinone, probably according to the radical mechanism with nitrate(V) ions with the use of hydroxyl radicals and phenoxyl.[68,69] The compound of 2,4,6-trichlorophenol occurs in water of different types as a result of exposure to chlorine and its effect on organic precursors. The mean concentration of chlorophenols in posttreated water varies from 0.003 to 1 µg/dm^3.[69–71] If chlorine is used as a water disinfectant, different chlorophenols are formed,[72] while if chlorine dioxide is used, a variety of p-benzoquinones are formed.[73]

In wastewater or sewage, phenol can react with nitric(V) acid to form toxic cyanides.[74]

No information on the content of phenol in the soil has been found; however, its presence in the soil is rather unlikely taking into regard its fast biodegradation and transport in underground water or air.[75] It has been suggested on the basis of some studies[76] that release of phenol from the dry near-surface layer of soil should be rather fast.

The partition (K_{oc}) coefficient for phenol for two types of loamy soil was 39 and 91 dm^3/kg, which implies high mobility of phenol in the soil and its easy penetration to underground water.[60,76,77]

As follows from the logarithm of the phenol dissociation constant pK$_a$ (log 1.28 × 10^{-10}) in water and in moist soil, phenol occurs in a partly dissociated form, so its transport and reactivity can depend on pH of the environment.[1,2,60,77] Phenol reacts as a weak acid in water, so it is not expected to dissociate in the range typical of the natural environment.[1,2]

Determination of phenol in drinking water performed in the United States has shown that its content is close to 1 µg/L but usually below the detection limit.[60] Higher content of phenol was found in samples of underground water contaminated with the wastewater left after coal gasification.[60,77]

Because of its short half-life, phenol is not expected to be transported over a great distance in the atmosphere.[50] In soil, phenol is highly mobile, and its mobility, likewise its reactivity, depends on pH of the environment.

In the main elements of the biosphere, the transport of phenol takes place via its washing out by precipitations or by leaching from the soil. It is highly unlikely to be a stable and durable reagent in the natural environment.[1,2,60,76,77] The hitherto state of our knowledge on the harmful effect of phenol in the biosphere does not permit a reliable evaluation of the potential threat it may pose. Nevertheless, in view of the calculated value of the maximum admissible critical concentration of phenol in water ecosystems, e.g.,

0.5 μg/L, it cannot be excluded that its presence can be related to some risk for water fauna and flora.[2,76,77]

Photooxidation, photodegradation in air, and biodegradation in water and soil are expected to be the major removal processes of phenol and its derivatives from the biosphere.[78] Phenol can potentially be removed from the atmosphere via photooxidation by reaction with hydroxyls and nitrate radicals, photolysis, and wet and dry deposition.[6,50,79] The majority of phenol from the atmosphere undergoes photodegradation to dihydroxybenzenes, nitrophenols, and products of ring cleavage, and the rest is washed out with precipitates.[50] Biodegradation is a major process for the removal of phenol from surface waters. Phenol generally reacts with hydroxyl and peroxyl radicals and singlet oxygen in sunlit surface waters.[5,50,76] The half-life of phenol in the air evaluated in photochemically active conditions in a smog chamber was 4–5 hr and was consistent with the value determined on the basis of the rate of its reaction with hydroxyl radicals.[50,78,79] In natural sunlit water reservoirs, phenols usually react with hydroxyl radicals formed as a result of photochemical reactions.[77,79]

Phenols easily undergo biodegradation in water if they do not occur in the concentrations toxic for microorganisms, as the latter play the main role in phenol degradation in the soil, water, water sediment, and bottom sediment.[76–79] The amount of bacteria capable of phenol degradation makes up a very small percentage of the total population of bacteria present in the soil.[80] Chemical composition of the products of bacteria-assisted degradation of phenols depends on the environmental conditions. In aerobic conditions, the products include carbon dioxide,[81] while in anaerobic conditions, they include carbon dioxide and/or methane.[82]

Phenol can undergo degradation in the free form as well as in the adsorbed form, in the soil or sediment, although the presence of the sorbent decreases the rate of its biodegradation.[76–79]

ENVIRONMENTAL IMPACT OF PHENOL AND ITS DERIVATIVES

For the general population, the most important sources of exposure to phenol are those coming from the air (emission from motor vehicles and product of photooxidation of benzene) and those from cigarette smoking and consumption of smoked food products.[50] The available information on the degree of exposure to phenol is insufficient. To evaluate the risk of harmful exposure, the value of total daily intake by a man weighing 70 kg was estimated to be 0.1 mg/kg of body mass per day.[2,6] The danger of poisoning by phenol vapor has been known for a long time, although no lethal cases have been reported.[1,5,6,78] The symptoms of poisoning by phenol inhalation include anorexia, loss of body mass, headache and vertigo, hypersalivation, and dark-colored urine. The risk of poisoning by intake of poisoned water or food is highly unlikely because of the very unpleasant taste and smell of phenol.[2,77] There is some anxiety over the reports on the possible genotoxic effect of phenol and there are controversies over its carcinogenic effect.[83–85]

The available information on the health effects of phenol exposure to humans is almost exclusively limited to case reports of acute effects of oral exposure.[86]

Most of the simple phenolic compounds have similar features and are toxic to aquatic organisms. The presence of a hydroxyl group intensifies their toxicity, which is why phenol is more toxic than benzene.[2,5,86] The acute toxicological effects of phenol are predominantly upon the central nervous system, and death can occur as soon as one-half hour after exposure. Acute poisoning by phenol can also cause severe gastrointestinal disturbances, kidney malfunction, circulatory system failure, lung edema, and convulsions. Fatal doses of phenol may be adsorbed through the skin. Key organs damaged by chronic exposure to phenol include the spleen, pancreas, and kidney. The toxic effects of other phenols resemble those of phenol.

Water or wastewater treatment by exposure to chlorine (Cl_2) leads to formation of chlorinated phenols, in particular, PCP, responsible for the revolting smell and taste. Although exposure to chlorophenols has been correlated with liver malfunction and dermatitis, polychlorinated dibenzodioxins also may have caused some of the observed effects.[2,6,86] Moreover, phenol and chlorophenols occurring in water are bioaccumulated by water fauna, which is responsible for the unpleasant smell and taste of fish meat.[5]

The most important organs involved in phenol metabolism are the liver, lungs, and mucous membrane of alimentary track; their involvement depends on the mode of exposure and the dose absorbed. Results of in vivo and in vitro studies have proven the covalent bonding of phenol with proteins from human tissues and blood plasma. Some metabolites of phenol also react with proteins.[1,2,86] The main route of phenol elimination from humans and animals is with urine. The rate of elimination with urine depends on the mode of exposure and dose absorbed. Phenol is also eliminated with stool and exhaled air.

IDENTIFICATION AND QUANTIFICATION OF PHENOL AND ITS DERIVATIVES

Phenol gives violet coloration with ferric chloride solution (the test reaction of phenol) due to the formation of colored iron complex, which is a characteristic of the existence of keto-enol tautomerism in phenols:

$$6C_6H_5 + FeCl_3 \rightarrow 3H^+ + Fe[(OC_6H_4\text{-})_6]^{3-} + 3HCl \quad (5)$$

In many countries, the content of phenol and its derivatives is controlled and standardized in all elements of the ecosphere, i.e., atmosphere, surface water, drinking water, soil, and food products.[87]

Development of analytical methods based on physicochemical phenomena and processes taking place in the natural environment and in living organisms has brought a decrease in the detection level and increase in accuracy of measurements. The range of applicability of analytical methods is related not only to the properties and type of substance studied but also to the selectivity and reproducibility of the method for isolation and enrichment of the analyte. From the point of view of routine analysis, the development of selective and reproducible procedures is difficult. Therefore, the application of coupled methods (off-line and/or on-line) is recommended. Because of their diversity and richness of forms of presence, phenolic compounds create many analytical problems, especially in routine analyses.[88]

Phenols react with 4-aminoantipyrine at pH 10 in the presence of potassium ferrocyanide, forming antipyrine dye, which is extracted into pyridine and measured at 460 nm.[89] This method is, however, charged with large error for the following reasons:

- Phenols with para substituents do not react with a reagent (the exceptions are para derivatives substituted with carboxyl, methoxyl, sulfonic groups, or halides).
- The yields of coupling with 4-aminoantipyrine for phenol, cresoles, xylenols, and other relevant compounds are significantly different.
- Maxima of absorption of the reaction products with the reagent are not the same for all phenols; some organic compounds that are not phenols also react with 4-aminoantipyrine.

The above-mentioned drawbacks, the lack of possibility to predict the qualitative and quantitative composition of phenolic compounds in a water sample studied, and problems with the choice of standard mixtures have prompted the introduction of the so-called phenol index describing the total concentration of phenols determined spectrophotometrically and expressed in milligrams per liter of phenol. It is in agreement with the international standard (ISO 6439) recommending also a procedure of colorimetric determination of phenolic compounds with the help of 4-aminoantipyrine, in which the phenol index is defined as the total concentration of phenolic compounds expressed in millgrams per liter phenol. The limit of determination of the phenol index is 0.010 milligrams per liter for the spectrophotometric method with 4-aminoantipyrine (nonspecific method) and 0.002 mg/L for the extraction-spectrophotometric method (single extraction with chloroform).[90]

From water and wastewater, phenolic compounds are separated by distillation with steam from an acidic solution, and then they are subjected to a reaction with 4-aminoantipyrine (1-fenylo-2,3-dimetylo-4-aminopirazolon) in an alkaline environment at a pH 9.8 in the presence of potassium ferrocyanide as an oxidant. This reaction leads to formation of an indophenolic dye of green-yellow to orange color depending on the concentration of phenols, extracted with chloroform. The content of phenols is determined spectrophotometrically by measurement of absorbance of the colored solution at the analytical wavelength, usually $\lambda = 460$ nm.[89]

The diversity of analytical tasks implies the need to choose the most appropriate analytical method for a specific problem. As follows from the survey of methods used for determination of phenols, presented by Thielemann,[91] all of them are used for the purposes.

Determination of trace amounts of phenols is realized by colorimetric and chromatographic method, while determination of phenols with content above 1% is performed by chemical bromatometric and iodometric methods.[89]

CONCLUSION

Phenol is one of the first compounds inscribed into the list of priority pollutants by the U.S. Environmental Protection Agency (EPA). It is commonly used in different branches of industry, including chemical production of alkylphenols, cresols, xylenols, phenolic resins, aniline, and other compounds.

Phenol is produced as an intermediate in the preparation of other chemicals and can be released as a by-product or contaminant. The economic importance of phenol and its derivatives has been significant since coal tar was used as a rubber solvent in the beginning of the 19th century. The current use of this group of compounds as pure products includes the chemical synthesis of plastics, synthetic rubber, paints, dyes, explosives, pesticides, detergents, perfumes, and drugs.

These compounds are used mainly as mixtures in solvents and constitute a variable fraction of gasoline. They have been used extensively for a wide variety of applications. They have been found to be among the most economical and effective surfactants, e.g., phenol ethoxylate and alkyl phenol ethoxylates. Production of their anionic derivatives has been severely restricted because they or their degradation products have been found to be estrogenic.

There is increasing environmental concern worldwide regarding the disposal of wastewater containing nonbiodegradable organic compounds. Since most pollutants do not respect national borders, a world effort to monitor their movement and to develop tools to prevent them from polluting environmental components or to remediate consequent pollution is desirable.

Phenol has antiseptic and germicidal properties, which increase to a maximum as the length of an alkyl side-chain substituted reaches about six carbon atoms. The lowest member, phenol, is highly toxic and caustic, but these properties progressively diminish with the higher members of the series. Polyhydric phenols are still markedly toxic but less caustic than the monohydric compounds (such as phenol).

The antiseptic activity of phenol is a property that accounts for much of its industrial use. In dilute solutions (of 1%–2%), it also finds application as an agent against skin itching. Undiluted phenol is highly corrosive to mucous membranes and skin, and it is considered a nerve poison. It may enter the human system by absorption through skin, oral ingestion, and vapor inhalation.

Phenol and three cresol isomers are of about the same order of toxicity and produce identical symptoms in poisoned animals.

Chemical and pharmaceutical industries are large users of phenol for conversion to many different products. The salicylates (carboxyphenols) are starting materials for the preparation aspirin and flavor.

Chlorophenols and their derivatives find application as fungicides, bactericides, and selective weed killers. Alkyl phenols make an important group synthetic tanning agent, and triphenyl phosphate is a plasticizer. Phenolic derivatives are among the most important contaminants in the environment. These compounds are used in several industrial processes to manufacture pesticides, explosives, drugs, and dyes. They are also used in the bleaching process of paper manufacturing. Apart from these sources, phenolic compounds have substantial applications in agriculture as herbicides, insecticides, and fungicides. However, phenolic compounds are not only generated by human activity but also formed naturally, e.g., in the process of decomposition of leaves or wood.

As a result of these applications, they are found in soils and sediments, and this often leads to wastewater and groundwater contamination. Owing to their high toxicity and persistence in the environment, both the U.S. EPA and the European Union have included some of them in their list of priority pollutants.

Current standard methods of phenolic compound analysis in water samples are based on liquid–liquid extraction, while Soxhlet extraction is the most used technique for isolating phenols from solid matrices.

Phenols and their derivatives are common in the natural environment. These compounds are used as the component of dyes, polymers, drugs, and other organic substances.

The presence of phenol substances in ecosystems is also related with production and degradation of numerous pesticides and the generation of industrial and municipal sewages. Some phenols are also formed during natural processes. These compounds may be substituted with chlorine atoms, nitrated, methylated, or alkylated. Both phenols and catechols are harmful ecotoxins.

Toxic action of these compounds stems from unspecified toxicity related to hydrophobicity and also to generation of organic radicals and reactive oxygen species. Phenols and catechols reveal peroxidative capacity, are hematoxic and hepatoxic, and provoke mutagenesis and carcinogenesis toward humans and other living organisms.

REFERENCES

1. Rappoport, Z. *The Chemistry of Phenols*; Wiley-Interscience: New York, 2003.
2. Tyman, J.H.P. *Synthetic and Naturals Phenols*; Elsevier Science: New York, 1996.
3. McTavish, D. *Joseph Lister*; Bookwright Press: New York, 1992.
4. Breinholt, V. Desirable versus harmful levels of intake of flavonoids and phenolic acids. In *Natural Antioxidants and Anticarcinogens in Nutrition, Health and Disease*; Kumpulainen, J.T., Salonen, J.T., Eds.; The Royal Society of Chemistry: London 1999; 93–99.
5. Thurman E.M. *Organic Geochemistry of Natural Waters*; Martinus Nijhoff/Dr W. Junk Publishers: Dordrecht, 1985.
6. Manahan, S.E. *Toxicological Chemistry and Biochemistry*, 3rd Ed.; CRC Press LLC: New York, 2003.
7. Verschueren, K. *Handbook of Environmental Data on Organic Chemicals*, 2nd Ed.; van Nostrand Reinhold Company: New York, 1983.
8. Spoelstra S.F. Degradation of tyrosine in anaerobically stored piggery wastes and in pig feces. Appl. Environ. Microbiol. 1978, 361–638.
9. Boskou, D. *Natural Antioxidant Phenols: Source, Structure-Activity, Relationship, Current Trends in Analysis and Characterisation*; Research Signpost: New York, 2006.
10. Glade, M.J. Dietary phytochemicals in cancer prevention and treatment. Book Rev. Nat. Rev. 1997, *13* (4), 394–397.
11. Friedman, M. Chemistry, biochemistry and dietary role of potato polyphenols. J. Agric. Food Chem. 1997, *45*, 1523–1540.
12. Kroon, P.A.; Wiliamson, G. Hydroxycinnamates in plants and food: Current and future perspectives. J. Sci. Food Agric. 1999, *79*, 335–361.
13. Robinson, T. *The Organic Constituents of Higher Plants*, 4th Ed.; Cordus Press: North Amherst, MA, 1980.
14. Duke J.A. *Handbook of Physicochemical Constituents of GRAS Herbs and Other Plants*; Ed. Academic Press; New York, 1992.
15. Hedges, J.I.; Mann, D.C. The characterization of plant tissues by their lignin oxidation products. Geochim. Cosmochim. Acta 1979, *43*, 1803–1807.
16. Herman, K. Occurrence and content of hydroxycinnamic acid and hydroxybenzoic acid compounds in foods. Crit. Rev. Food Sci. Nutr. 1989, *28*, 315–347.
17. Hubble, B.R.; Stetter, J.R.; Gebert, E.; Harkness, J.B.L.; Flotard, R.D. Experimental measurements of emissions from residential wood-burning stoves. In *Residential Solid Fuels: Environmental Impacts and Solutions*; Cooper, J.A., Malek, D., Eds.; Oregon Graduate Center: Beaverton, OR, 1981; 79–138.
18. Hardy, G. Nutraceuticals and functional foods: Introduction and meaning. Nutrition 2000, *16* (7–8), 688.
19. Aherne, S.A.; O'Brien, N. Dietary flavonols, chemistry, food content, occurrence and intake. Nutrition 2002, *18*, 75–81.

20. Stevenson, F.J. *Humus Chemistry, Genesis, Composition, Reactions*; John Wiley & Sons: New York, 1994.
21. Muscola, A.M.; Soidari M. *Soil Phenols*; Nova Science Publishers Inc.: New York, 2010.
22. Hessen, D.O.; Tranvik, L.J. *Aquatic Humic Substances: Ecology and Biogeochemistry*; Springer Verlag: Berlin, 1998.
23. Larsen, J.W.; Nadar, P.A.; Mohammadi, H.; Montano P.A. Spatial distribution of oxygen in coals. Development of a tin labelling reaction and Mossbauer studies. Fuel, 1982, *62*, 889–893.
24. Weber, M.; Weber, M.; Kleine-Boyman, M. Phenol. In *Ullmann's Encyclopedia of Industrial Chemistry*; Wiley-VCH: New York, 2004.
25. Gardziela, A.; Pilato, L.; Knop, A. *Phenolic Resins: Chemistry, Applications, Standardization, Safety and Ecology*; Springer Verlag: New York, 2000.
26. Ryser, S.; Ulmer, G.; Etude de la pollution par les resines synthetiques utylisees en fonderie. Fonderie 1980, *33*, 313–324.
27. *RIVM Criteria Document: Phenol, Bilthoven, the Netherlands*, National Institute of Public Health and Environmental Protection (Document No. 738513002); 1986.
28. Kuwata, K.; Uebori, M.; Yamazaki, Y. Determination of phenol in polluted air as p-derivative nitrobenzene azophenol by reversed phase high performance liquid chromatography. Anal. Chem. 1980, *52* (6), 867–860.
29. Verschueren, K. *Handbook of Environmental Data on Organic Chemicals*, 2nd Ed.; van Nostrand Reinhold Company: New York, 1983.
30. Den Boeft, J.; Kruiswijk, F.J.; Shulting, F.L. Air pollution by combustion of solid fuels; The Hague Ministry of Housing, Physical Planning and Environment (Publication Lucht No. 37), 1984.
31. Groenen, P.J. Components of tobacco smokes. Nature and quantity, potential influence on health; Zeist, the Netherlands, CIVO-TNO Institute Report No R/5787, 1978.
32. Gosselin, R.E.; Smith, R.P.; Hodge, H.C. Phenol. In *Clinical Toxicology of Commercial Products*, 5th Ed.; Wiliams & Wilkins: Baltimore, 1984; III-344–III-348.
33. Pekari, K.; Vainotalo, S.; Heikkila, P.; Palotie, A.; Luotamo, M.; Rihimaki, V. Biological monitoring of occupational exposure to low levels of benzene. Scand. J. Work Environ. Health 1992, *18*, 317–322.
34. Ahlborg, U.G.; Thunberg T.M. Chlorinated phenols: Occurrence, toxicity, metabolism, and environmental impact. CRC Crit. Rev. Toxicol. 1980, *7*, 1–35.
35. Carrey, F.A. Chlorophenol. In *Encyclopaedia Britannica Online*; Encyclopaedia Britannica, 2001.
36. Mycke, B. Preliminary results on free dissolved phenolic compounds in natural Waters. In *SCOPE/UNEP Transport of Carbon and Minerals in Major World Rivers, 52*; Degenes, E.T., Ed.; University of Hamburg: Hamburg, 1982; 571–574.
37. Roupe, K.A.; Remsberg, C.M.; Yanez J.A.; Davies, N.M. Pharmacometrics of stilbenes: Segueing towards the clinic. Curr. Clin. Pharmacol. 2006, *1*, 81–101.
38. Begon, M.; Harper, J.L.; Townsend C.R. *Ecology, Individuals Populations and Communities*; Black Science Ltd.: Edinburgh, 1996.
39. Smith; M.B.; March, J. *Advanced Organic Chemistry. Reactions, Mechanisms, and Structure*, 6th Ed.; Wiley-Interstice: New York, 2007.
40. Rice-Evans, C. *Screening of Phenolic and Flavonoids for Antioxidant Activity, Antioxidant Food Supplements in Human Health*; Academic Press: New York, 1999.
41. Golovanow, I.B.; Zhenodarova, S.M. Structure–property correlations. XV. Properties of phenol derivatives. Russ. J. Gen. Chem, 2008, *73* (10), 1603–1607.
42. Herman, K. Occurrence and content of hydroxycinnamic acid and hydroxybenzoic acid compounds in foods. Crit. Rev. Food Sci. Nutr. 1989, *28*, 315–347.
43. Dragsted, L.O. Antioxidant actions of polyphenols in humans. Int. J. Vitam. Natur. Res. 2003, *73*, 112–119.
44. Budryn, G.; Nebesny, E. Phenylacids. Properties, occurrence in plants and metabolic conversion. Bromat. Chem. Toksycol. 2006, *102*, 547–554.
45. Foley, S.; Navartman, S.; MvGarvey, D.J.; Land, E.J.; Truscott G.; Rice-Evans, C.A. Singlet oxygen quenching and the redox properties of hydroxycinnamic acids. Free Radic. Biol. Med. 1999, *26* (9/10), 1202–1208.
46. Kirk, R.E.; Othmer, D.F. *Encyclopedia of Chemical Toxicology*, 3rd Ed.; John Wiley and Sons: New York, 1980; Vol. 17, 373–379.
47. Bolling, F.J.; Decker, K.H. *Phenols Resin*; Kunstshoffe: Berlin, 1980.
48. Ying, G.G.; Wiliams, B.; Kookana R. Environmental fate of alkylphenol and alkyl ethoxylates—A review. Environ. Int. 2002, *28*, 215.
49. Budavari, S. *The Meck Index. An Encyclopedia of Chemical Drugs and Biologicals*; Whitehouse Station: New York, 1996.
50. Seinfeld, J.H.; Pandis S.N. *Atmospheric Chemistry and Physics*; John Wiley & Sons, Inc.: New York, 1996.
51. van Gemert, L.J. *Flavour Thresholds. Compilations of Flavour Threshold Values in Water and Other Media*, (second edition) Ed. Oliemans Punter & Partners BV; the Netherlands, 2011.
52. US EPA. Risk Management Air Research; 1982.
53. DeSelms, R.H. *UV-Active Phenol Ester Compounds*; Enigen Science Publishing: Washington, DC, 2008.
54. Svobodova, A.; Psotova, J.; Walterova D. Natural phenolics in the prevention of UV-induced skin damage. A review. Biomed. Papers 2003, *147* (2), 137–145.
55. Fiege, H.; Voges H.M.; Umemura S.; Iwata, T; Miki, H.; Fujita, Y.; Buysch, H.J.; Garbe, D.; Paulus, W. Phenol derivatives. In *Ullmann's Encyclopedia of Industrial Chemistry*; Wiley-VCH: Weinheim, 2000.
56. Ueoda, M.; Aizawa T.; Imai, Y. Preparation and properties of polyamides containing phenoxathiin units. J. Polym. Sci. Chem. Ed. 1977, *15* (11), 2739–2747.
57. Eisenreich, S.J.; Looney, B.B. Thornton, J.D. Water solubility enhancement of pyrene in the presence of humic substances. Environ. Sci. Technol. 1981, *15*, 30.
58. Schnitzer, M.; Khan S.U. *Humic Substances in the Environment*; Marcel Dekker: New York, 1972; 1–7.
59. Web-based Archive of the Dutch National Institute for Public Health and the Environment (RIVM), Final report 2002.
60. Howard, P.H. *Handbook of Environmental Fate Exposure Data of Organic Chemicals*; Lewis Publishers: New York, 1991; Vol. 1.
61. Kawamura, K.; Kaplan, I.R. Stabilities of carboxylic acids and phenols in Los Angeles rain waters during storage. Water Res. 1990, *24* (11), 1419–1423.

62. Brodzinski, R.; Singh H.B. Volatile organic chemicals in the atmosphere: An assessment of available data, EPA 600383o27(A); US Environmental Agency, Office of Research and Development: Research Triangle Park, NC., 1983.
63. Tesarova, E.; Paczkowa, W. Gas and high-performance liquid chromatograph of phenols. Review paper. Chromatographia 1983, *17* (5), 269–284.
64. Grosjean, D. Atmospheric fate of toxic aromatic compounds. Sci. Total Environ. 1991, *100*, 367–414.
65. Kuwata, K.; Uebori, M.; Yamazaki, Y. Reversed-phase liquid chromatographic determination of phenols in auto exhaust and tobacco smoke as *p*-nitrobenzenephenol derivatives. Anal. Chem. 1981, *53* (9), 1531–1534.
66. Bruno T.J.; Svoronos P.D.N. *Basic Tables for Chemical Analysis*, 3rd Ed.; CRC Press: Boca Raton, 2011.
67. Rogge, W.F. Molecular tracers for sources of atmospheric carbon particles: Measurements and model predictions. Ph.D. Thesis, California Institute of Technology: Pasadena, 1993.
68. Niessen, R.; Lenoir, D.; Boule, P. Phototransformation of phenol induces by excitation of nitrate ions. Chemosphere 1988, *17*, 1977–1984.
69. Louw, R.; Santoro D. Comment on formation of nitroaromatic compounds in advanced oxidation processing: photolysis versus photocatalysis. Environ. Sci. Technol. 1999, *33*, 3281.
70. Kunte, V.H.; Slemrova, J. Gaschromatographische und massenspektrometrische Identifizierung phenolischer substanzen aus ober-flachenwassern. Zeitschrift fuhr Wasser und Abwasser Forschung 1975, 8, 176–182.
71. World Health Organization (WHO). *Guidelines for Drinking Water Quality*, 2nd Ed.; Geneva, 1984, Vol. 1.
72. Knoevenagel, K.; Himmelreich, R. Degradation of compounds containing carbon atoms by photooxidation in the presence of water. Arch. Environ. Contam. Toxicol. 1976, *4* (3), 324–33.
73. Jarvis, S.N; Straube, R.C.; Wiliams, A.L.J.; Bartlett C.L.R. Illness associated with contamination of drinking water supplies with phenol. Brit. Med. J. 1985, *20*, 1800.
74. Tratnyek, P.G.; Hoigne J. Kinetics of reactions of chlorine dioxide (OClO) in water (II). Quantitative structure–activity relationships for phenolic compounds. Water Res. 1994, *28* (1), 57–66.
75. Nakagawa, N.; Mizukoshi, O.; Maeda, Y. Sonochemical degradation of chlorophenols in water. Ultrason. Sonochem. 2000, *7* (3), 115–120.
76. Faust, S.D.; Stutz, H.; Aly, O.M.; Andersen, P.W. Recovery, separation, and identification of phenolic compounds from polluted waters. Part I—Occurrence and distribution of phenolic compounds of the surface and ground waters of New Jersey. Geological Survey Open-File Report, 1970.
77. Simonow, L.; Sargsyan, V. *Soil Chemical Pollution, Risk Assessment, Remediation and Security*; Springer: Dordrecht, 2007.
78. Yong, R.N.; Mohamed, A.M.O.; Warkentin, B.P. *Principles of Contaminant Transport in Soils*; Elsevier: New York, 1992.
79. Gilman, A.P. *Chlorophenols and Their Impurities: Health Hazard Evaluation*; Environmental Health Directorate, Health Protection Branch: New York, 1988.
80. Schwarzenbach, R.P.; Gschwend, P.M.; Imboden, D.M. *Environmental Organic Chemistry*, 2nd Ed.; Wiley-Interstice: New York, 2003.
81. Hickman, G.T.; Novak, J.T. Relationship between subsurface biodegradation rates and microbial density. Environ. Sci. Technol. 1989, *23*, 525–532.
82. Boyd, J.T.; Carlucci, A.F. Degradation rates of substituted phenols by natural populations of marine bacteria. Aquat. Toxicol. 1993, *25* (1), 71–82.
83. Dobbins, D.C.; Thornton, J.; Jones, D.D.; Fedale, T.W. Mineralization potential for phenol in subsurface soils. J. Environ. Qual. 1987, *16* (1), 54–58.
84. Bukowska, B., Kowalska S. Phenol and catechol induce prehemolytic and hemolytic changes in human erythrocytes. Toxicol. Lett. 2004, *152* (1), 73–84.
85. Bruce, R.M.; Santodonado, J.; Neal, M.W. Summary review of the health effects associated with phenol. Toxicol. Indust. Health 1987, *3*, 535–568.
86. Oikawa, S.H.; Hirosawa, K.; Kawanishi S. Site specificity and mechanism on oxidative DNA damage induced by carcinogenic catechol. Carcinogenesis 2001, *22*, 1239–1245.
87. WHO. IPCS Environmental Health Criteria for Phenol (161). First draft prepared by MS G.K. Monitzan; WHO: Finland, 1994.
88. Thompson, R.D. Determination of phenolic disinfectant agents in commercial formulations by liquid chromatography. J. AOAC Int. 2001, *84* (3), 815.
89. Buszewski, B.; Buszewska, T.; Szumski, M.; Siepak, J. Simultaneous determination of phenols and polyaromatic hydrocarbons isolated from environmental samples by SFE–SPE–HPLC. Chem. Anal. 2003, *48*, 13–25.
90. *Standard Methods for Examination of Water and Wastewater*. APHA, AWWA, WPCF: Washington, DC, 1992.
91. International Standard ISO 6439. Water quality—Determination of phenol index—4-Aminoantipyrine spectrometric methods after distillation, Prepared by Technical Committee ISO/TC 147; Geneva (Switzerland) 1990, p. 7.
92. Thielemann, H. Thin layer chromatographic separation of phenol, cresols, dimethylphenols and naphthols. Die Pharmazie 1970, *25* (5), 365–366.

Phosphorus: Agricultural Nutrient

John Ryan
International Center for Agricultural Research in the Dry Areas (ICARDA), Aleppo, Syria

Hayriye Ibrikci
Soil Science and Plant Nutrition Department, Cukurova University, Adana, Turkey

Rolf Sommer
International Center for Agricultural Research in the Dry Areas (ICARDA), Aleppo, Syria

Abdul Rashid
Pakistan Atomic Energy Commission, Islamabad, Pakistan

Abstract
Of the nutrients that sustain terrestrial and aquatic life, phosphorus (P) is one of the most crucial. Total P in soils is relatively low and is a function of the rocks from which soils are derived. However, plant availability of P is only a tiny fraction of total soil P and is largely conditioned by the complex chemical reactions that occur between soluble P and soil constituents and the equilibrium that exists between various solid-phase P and solution P fractions. While natural ecosystems, forests, and native vegetation have evolved adaptation mechanisms to cope with low P availability, cultivated plants and crops are largely limited by P availability. Consequently, from the dawn of settled agriculture, adequate P nutrition has been a major constraint to optimum crop output and, thus, population growth. The development of inorganic P fertilizer, i.e., superphosphate, in the mid-19th century was a milestone for agriculture—and for mankind. The past century saw a rapid expansion in P fertilizer use, especially in developed countries. Commercial fertilizer use was largely responsible for increasing agriculture's bounty and contributing to world food stocks. However, because of profitability of P fertilizer use, helped in many cases by subsidies, overuse or excess of P increasingly became an issue, and that was exacerbated by the use of animal manure. Thus, in the past few decades, runoff from agriculture became a nonpoint source of P pollution of waterways and lakes, leading to eutrophication. The phenomenon of P pollution was particularly acute in Western Europe and in areas of the world with intensive agriculture and animal industries. Recognition of the implications of P pollution has led to legislation that restricted P use—and management strategies in response to such legislation. In essence, agriculture of today has to reconcile food and feed production with environmental protection. This brief review outlines the historical role of P in agriculture and the shifting orientation of concerns to include a consideration of the environmental research leading to societal awareness of P pollution along with specific approaches to mitigate the problem.

INTRODUCTION

Never in the history of mankind have there been such widespread societal concerns about the capacity of our planet to sustain its projected population growth. This concern was first widely articulated in the late 19th century by Thomas Malthus. However, his prognosis of widespread starvation has been challenged by advances in agricultural technology and medicine. These revolutionary developments led to not only a better-fed population but also a vastly expanded population in the past century. While population growth has stabilized in developed countries, it is continuing unabated in many areas of the lesser-developed world. World population is now projected to increase from its current 6.8 billion people to over 9 billion by midcentury.[1] In addition to such increases, world food production will have to be doubled as a result of increased affluence in some developing countries such as India and China.[2] The threat of the "population monster" that Norman Borlaug, Nobel laureate and father of the Green Revolution, railed against is now more credible,[3] with current predictions of a world population of more than 9 billion people by the middle of the century.

The implications for mankind of a more crowded world, where the earth's natural resources are increasingly threatened and even the future of mankind is called into question, are both ominous and daunting.[4,5] The "grand challenges" that our society faces embrace all aspects of agriculture and the environment. A litany of such challenges includes both biophysical (land degradation, water scarcity, loss of biodiversity) and socioeconomic (population growth, poor nutrition and health care, poverty, inadequate research investment and information infrastructure) difficulties. Looming large over these concerns are the implications of climate change, with areas of the world such as the Mediterranean and much of Africa likely to become drier and others to endure climate-related stresses.[6] While much research attention is now focused on adaptation strategies, the predicted change scenario is likely to exacerbate the world's already precarious food production capacity. Such developments have posed an extraordinary challenge to the global scientific community[7] and especially the international network of agricultural research centers.[8] Chemical fertilizers are vital to world food production and food security,[8] with over half of global output attributed to applied fertilizer nutrients and future food supplies being even more dependent on fertilizers.[10] Consequently, fertilizer management, in a sustainable manner that protects the environment and the natural resource base, is central to the issue of balancing the world's food supply–demand equation.

Among the nutrients that are essential to plants but potentially damaging to the environment are nitrogen (N) and phosphorus (P). Large amounts of N are used in modern agriculture; in fact, N fertilizer is the driver of the phenomenal increases in global crop production, especially cereals, in the past half century.[11] Data on current fertilizer N suggest that the trend in N use is set to continue, especially in countries such as China and India, which are now the leading producers of wheat in the world. However, as N is either applied as nitrate (NO_3^-) form as a fertilizer or rapidly converted biologically to this mobile form of N, losses of N from agricultural environments through leaching and runoff became associated with high N fertilizer use.[11,12] Enrichment of water sources with NO_3 not only represents economic losses in terms of fertilizer costs but also causes a potential health hazard in drinking water.[13] More widespread effects of excess N were manifested in terms of water quality only when combined with P,[14] an element that was initially thought to be immobile and subject to minimal loss from the point of application in farmers' fields. Gradually, the catalytic effect of P combined with N and eutrophication or the development of anoxic conditions in water bodies were recognized by agricultural scientists.[15] This brings us to a consideration of the role of P in agriculture and the inadvertent effects of its misuse on the environment,[16] as well as management strategies, which were recognized by agricultural scientists.[17] A brief overview of the various aspects of P in soils and the relationship of P to crop production is a prerequisite to any discussion of P in relation to the environment.

PERSPECTIVE ON PHOSPHORUS

As with carbon and N, the overall nature of P in terrestrial and aquatic ecosystems and the atmosphere can best be described as a cycle, which indicates changes between one state and another, i.e., a state of flux. Various depictions of such a cycle have been made, but the one adapted by Tunney et al.[15] gives a good representation of the inputs and outputs involving P and the relationship with the various P phases in nature (Fig. 1). As plant uptake occurs from the soluble P fractions—and as losses occur from that

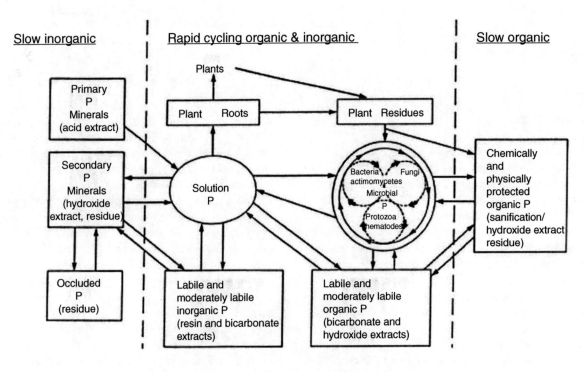

Fig. 1 The soil P cycle in nature and its reaction phases.

fraction as well—solution P is a central component of the P cycle and is closely linked to microbial P. Inputs to the system are primarily fertilizers (and plants) to the solution phase and manures to the pool involving microbial forms of P; both forms are being replenished by or contributing to somewhat less soluble inorganic and inorganic phases, i.e., labile or moderately labile forms, and constitute the more rapidly changing part of the P cycle.

The behavior of P in nature is also influenced by very slow reactions that occur over several years and do not have an immediate impact during one cropping season. The slow inorganic component of the cycle involves primary and secondary P-bearing minerals that can be weathered to increase solution P; some inorganic forms (occluded) have no direct effect on soluble P. Similarly, some organic P forms have little immediate effect on the more reactive P phases in the cycle. While the depiction in Fig. 1 does not specifically indicate the avenue of losses, the soluble P phase is most susceptible to loss by leaching and runoff in addition to the actual transport of P-enriched soil particles by erosive processes. The fundamental problem is related to excessive inputs to the system in terms of chemical fertilizers and manures, thus exceeding the crop's needs and the soil's capacity to retain the P against loss.

No element can rival P in terms of its chemical complexity, its dynamic equilibrium, and the diversity of inorganic and organic P compounds. The addition of P fertilizers to soils and its uptake and utilization by crops pose further challenges to soil and crop scientists. Much has been written about P from the basic and applied perspectives, especially during the past half century. The thousands of research papers dealing with various aspects of P have gradually revealed a better understanding of this tantalizing element. Progress in P research has been documented in broad-based discipline reviews[18–20] in addition to regional[21–22] and country-level[23–24] reviews. While such references are selective and subjective, they broadly capture the slow but inexorable unraveling of the many aspects of P in soils, fertilizers, and crops.

Shifting Emphasis

A brief survey of the recent literature on P indicated a revision of concepts related to P availability and a shift in emphasis from production-related P research to that related to the environment. For instance, while the early P review of Larsen[18] did not refer to the environment aspect of P use, the wide-ranging P review of Khasawneh et al.[20] addressed the issue in only one brief chapter[25] of the 29 chapters in the volume. That review addressed the emerging concerns of P in the environment, geological influences on soluble P, as well as agricultural losses of P through surface flow and sediment transport; P losses from animal wastes and urban runoff were also briefly addressed.

While the subsequent comprehensive review of P still focused on the agricultural aspect of P,[26] it devoted a major part of the volume (five chapters) to P in relation to the environment. The rapidly expanding awareness of the environmental implications of P led to the establishment of a series of international workshops with focus on the environment.[15] As an illustration of the current research emphasis on P in the environment, keynote addresses at the recent meeting of the workshop series[27] dealt with issues such as P mobilization at the plot and field scales, field-scale indices of P loss, P dynamics and impact in water bodies, and implementation of mitigation options. Reviews such as that of Delgado and Scalenghe[16] focused on Western Europe as one of the world's hot spots for P pollution, where the consequence of overuse of P fertilizers and animal manures had attracted broad societal and legislative concern.

Phosphorus: Potentially Harmful Effects

Though vital to all forms of life on earth, P has no known toxic effects on humans or animals unlike some other elements. The potential effect of P on the environment is related to an excess of P in the terrestrial ecosystems with a carryover to the aquatic environment. Eutrophication of water bodies involves the excessive growth of undesirable algae and aquatic weeds and a consequent depletion of such organisms.[28] The process of eutrophication is defined as an increase in the fertility status of natural waters that causes accelerated growth of algae or water plants. Eutrophication is due not solely to P but to a complex interaction between N, P, environmental conditions (temperature, salinity, light), and the physical and hydrologic characteristics of surface waters, i.e., streams, lakes, and estuaries.[29] Under natural conditions, growth of aquatic organisms is limited by the normally low levels of P in the water. External inputs of P from urban wastewater, surface runoff, or subsurface flow can upset the delicate P balance in water bodies and stimulate growth of aquatic organisms to ecologically undesirable levels.

Critical levels for soluble and total P associated with eutrophication are in the order of 10 and 20 $\mu g\,L^{-1}$,[17] while others have indicated a range of values from 20 to 100 $\mu g\,L^{-1}$,[29] which is still considerably less than such thresholds for N. The main driver of eutrophication is depletion of dissolved oxygen in the water due to the excessive growth of aquatic organisms and microbial decomposition of such biota. This leads to a chain reaction that causes a deterioration in the water's ecology, e.g., decreased light penetration, surface algal scums, foul odors, impeded water flow, and increased turbidity and sedimentation. Associated with such pollution is the occurrence of surface blooms of cyanobacteria that can reduce water palatability and even kill livestock and pose a threat to human health. Advanced eutrophication of surface water thus has serious implications for fisheries, recreation, and human or animal consumption.

The primary cause of eutrophication is the P that originates from agricultural land as nonpoint source or from urban and industrial effluents. The problem of P-induced pollution was first noted in the Great Lakes region of North America in the 1960s. The phenomenon was later seen in P-enriched lakes in Western Europe. Initially, the problem was attributed to point sources, i.e., sewage-derived inputs. Control measures included sewage diversion, removing P from sewage effluents, and reducing P contents of detergents. However, the persistence of eutrophication in many water bodies following the control of plant-source P emissions led to a focus on loss from agriculture as the primary cause of continued eutrophication. While the problem may appear to lend itself to a simple solution of avoiding use of excess P fertilizer or P-containing manures, Tunney et al.[15] argued that the process of P transport is anything but simple as it depends on soil properties, flow pathways, and bioavailability. Development of a strategy to control or mitigate P losses from fertilized fields or a watershed requires an understanding of the processes and mechanisms that govern P loss.[17] Thus, it is appropriate in this review to provide a brief background on the behavior of P in soils and to outline how such factors impinge on the environmental dimension of P.

PHOSPHORUS BEHAVIOR IN SOILS

In order to establish the significance of P for the environment, there are a number of key aspects that need to be considered: the need for P for crops, forms of P in soils, reactions in soil, P cycling, and P availability. In addition, consideration has to be given to global fertilizer use as well as reserves of P for future use. This brief discussion is extracted from several sources.[15,26,30,31] Much of the information presented is now firmly part of the accepted literature on P in agriculture and the environment.

Phosphorus in Soil: Forms and Amounts

The essentiality of P for plant growth has been established for almost a century. In contrast to other elements such as N and potassium, the total amount of P in soils is relatively low, generally in the range of 0.01%–0.3% (100–3000 mg P kg^{-1}). In its native state, P solubility is extremely low, i.e., less than 0.01 mg L^{-1}. This soluble P pool, from which plants derive their P for plant uptake and growth, has to be replenished many times during the growing season in order to sustain growth. Thus, a sort of equilibrium exists between P in solution and solid-phase P. In natural ecosystems, the supply of P to plant roots is governed by reactions involving sorption, desorption, and precipitation. In such conditions, the small amount of P in solution is maintained by weathering and dissolution of rocks and minerals of low solubility. Without fertilization, which greatly increases the soluble P pool, few soils can maintain an adequate P supply to meet the needs of modern high-yielding agricultural crops.

Prior to considering the soluble or "available" P fractions further, it is necessary to appreciate the forms of P in soils, which can be categorized into inorganic and organic P forms, the former being dominant in most soils except those high in organic matter. Inorganic P compounds range widely in number and type depending on soil properties but are generally associated with amorphous and crystalline sesquioxides, i.e., iron (Fe) and aluminum (Al), mainly in acid soils, and calcium (Ca) compounds in calcareous soils. The relative distribution of P compounds is largely influenced by the degree of soil weathering, which in turn influences the reactivity of these compounds.

In acid soils, hydrous oxides of Fe and Al readily react with added soluble P, resulting in precipitation, while in neutral or calcareous soils, Ca controls P reactions. In Ca-dominated systems, precipitates in the following order of solubility have been identified: monocalcium phosphate, dicalcium phosphate dihydrate, and hydroxyapatite or fluorapatite; the latter compound has extremely low solubility and thus controls solution P concentration. In acid soils, few well-crystallized P compounds have been observed. Reactions of soluble P with solid-phase components also involve sorption onto adsorbing surfaces. In effect, it is difficult to separate the process of precipitation from sorption. However, there is evidence that regardless of the mechanism involved, these reactions can be reversed. Precipitation/dissolution processes differ from sorption/desorption in that the solubility product of the least soluble P compound governs solution P concentration, whereas solution P controls the amount of P sorbed. In essence, retention of soluble P by soil components is a continuum between precipitation and surface reactions. In practice, P sorption curves or isotherms have been used extensively to describe the relationship between the amount of P sorbed or removed from solution by soils and that remaining in solution. Such isotherms have been used to identify sorption maxima, bonding energy between soils, and the equilibrium P concentration at which no sorption or desorption occurs.

Regardless of the type of P reactions and the soil components involved, addition of soluble P to soils results in an immediate increase in soluble or "plant-available" P. Numerous studies have documented a subsequent rapid decrease in P availability followed by a slower rate of decline in solubility. The initial rapid phase was invariably attributed to surface reactions and/or precipitation, while the slow decline phase was attributed to diffusive penetration of adsorbed P into soil components ("absorption") and increased crystallization of precipitated P compounds. Despite the dominance of Ca in the chemical reactions of P in calcareous soils, Fe oxides (both amorphous and crystalline) have been shown to have a disproportionate role in the initial and subsequent reactions in soils.[32–34] Such oxides accentuated the rate of decline of soluble P and decreased the desorption rate. From the earliest years of P experimentation, this decline in P solubility, following reactions of soluble P with soils in laboratories, was accompanied by decreased plant P availability in greenhouse and field studies. The process was initially considered to be irreversible and often termed "fixation," retention, or immobilization.

Phosphorus Availability: Critical Concentration, Fertilization

A prerequisite to rational use of P fertilizers is knowledge of what are the critical levels of plant-available P in the soil. Much research has been conducted, involving correlation and calibration to identify appropriate indices of availability and critical test values in the field beyond which no response to added fertilizer would occur. A wide range of chemicals have been tested for availability indices based on acid dissolution, anion exchange, cation complexation, and cation hydrolysis.[35] Depending on the country or soil type and region, tests such as the Mehlich I, Bray I, and Olsen procedures are widely adopted. Multinutrient extractants such as the Mehlich III and ammonium bicarbonate-diethylenetriaminepentaacetic acid (AB-DTPA) tests have increased in popularity. Regardless of the extractant used, each should theoretically extract a fraction of the relatively soluble phase of soil P that showed a close relationship or correlation with plant uptake. Subsequently, field studies are required to establish a critical P value and thus distinguish between sufficiency and excess of soil P.[36] Thus, the amount of P fertilizer required is inversely related to soil P test values. Such guidelines are essential in order to ensure that fertilizer is applied in situations where it is needed and where it will evoke an economic crop response; conversely, they will prevent unnecessary use of fertilizer when the soil has adequate amounts of the nutrient for the season's crop, thus reducing input costs and avoiding the potential of loss of the excess P to the environment.

Estimates of critical soil values are dependent on the specific test and are influenced by soil factors; for instance, P values of 10 and 30 mg kg^{-1} were established for the Olsen and Bray I, respectively, and 10 and 25 mg kg^{-1} for clay and sandy soils, respectively. In addition to specific tests and soil type, many other factors influence P availability: temperature, compaction, moisture, aeration, pH, and type of clay. Once the need for P fertilizer is established, and a basis for the required amount indicated by critical test values that are calibrated for field conditions has been determined, other factors have to be considered in the actual fertilization process. In contrast to N, timing of P fertilizer application is less important. As an immobile nutrient, P is applied and mixed with the soil for cultivated crops, but broadcast applications are feasible only for established perennial forage crops. Regardless of whether P is applied as inorganic fertilizer or as animal manures, such additions invariably increase available P levels in the soil.

Due to the immobility of P applied to soils, efficiency of uptake per unit of P applied is always higher when the material is placed close to the seed compared with broadcasting.[24] Such placement puts the P fertilizer close enough to the seedling roots, while minimizing contact immobilization or relation. Such positional availability is also influenced by crop type, specifically its rooting pattern. In order for banding or restricted P placement to enhance P uptake by roots, the rate of absorption and root growth in the P-enriched soil zone must compensate for the limited impact on root growth in the unfertilized portion of the root zone. Much research has been done on P placement with the common range of field crops with respect to the permissible amounts of P fertilizer that can be used without damaging the seedlings and with respect to establishing optimum distances for the fertilizer band to the side or below the seed, usually about 2–5 cm. Regardless of the method of P application, the question of residual P availability arises.

Residual Phosphorus Availability: Changing Concept

Early years of P research focused on laboratory studies of P reactions in previously unfertilized soils, along with short-term greenhouse and one-season field studies. The outcome of such investigations led to the notion of P being "fixed" and of limited effectiveness following the first year of fertilization. Gradually, as commercial P fertilizer use became almost universal in agriculture, especially in the United States, Europe, and other developed countries, the concept of residual P availability emerged.[31] Evidence in support of this new concept came from long-term agronomic trials in which treatments involved initial one-time P dressings as well as current or yearly applications.

A few examples will suffice to illustrate residual availability. For instance, Halvorson and Black[37] showed that Olsen P test levels were increased for 16 years following P fertilization with one application. Following the initial increase, P levels declined slowly for 12 years and then leveled off at a level higher than initial values, suggesting a new equilibrium of available P. The elevated soil P levels in this study were also accompanied by corresponding increases in crop yields. Similar soil and crop yield responses were observed from even longer-duration trials.[38] The general acceptance of long-term effects of P fertilization was reflected in a entry on "Evaluation and Utilization of Residual Phosphorus in soils" by Barrow[39] in the comprehensive milestone on *The Role of Phosphorus in Agriculture* by Khasawneh et al.[20] With buildup of residual P, availability cropping can occur without any P fertilizer. In studies cited by Sharpley,[30] the role of decrease in available P by depletion with cropping is inversely related to the soils' buffering capacity or the P sorption saturation (available P/P sorption maximum). In shorter-term, multiyear trials under dryland Mediterranean conditions in Turkey[40] and Syria,[41] a relatively quick buildup of available P was observed following modest fertilization of calcareous soils.

The evaluation of P availability with progress in accumulated research, especially at the field level, brought with it a changed concept of efficiency. Based on early studies that evaluated P growth and uptake efficiency in terms of differences between fertilized and control plots, P fertilizer use efficiency was generally believed to be less than 20%. However, the concept of efficiency differed depending on the methods used to evaluate it. The direct method uses a radioactive isotope of P (^{32}P), but due to its short half-life, it is amenable only to short-term studies. The difference method considers differences between fertilized and unfertilized crops, i.e., "agronomic efficiency" for yield and "apparent recovery" for P uptake. On the other hand, the balance method considers only yield and P uptake relative to the amount of P fertilizer applied. In a recent comprehensive review of P efficiency supported by several case studies, Syers et al.[31] clearly demonstrated significant and consistently higher efficiency values for the balance method compared with the difference method. The key issue in such a method is the time scale. Thus, when a residual effect of P is considered over several years, efficiency values in the order of 90% can occur. This revised concept of P use efficiency represents a paradigm shift in soil fertility assessment and crop nutrition.

DYNAMIC NATURE OF PHOSPHORUS IN SOILS

Of many and varied aspects of P, including broad categories such as P fertilizer sources, reactions of P fertilizers in soils, plant nutrition and crop management, animal nutrition, management practices, and the environment,[26] few are as intriguing as the dynamic or changeable nature P in soils, which was addressed recently by Condron et al.[42] for P organic P and by Tiessen[43] for P in tropical soils. In an earlier brief review, Ryan and Rashid[44] outlined the inputs to the soil solution phase of P, potential losses or withdrawals from that small fraction, and the extent to which various categories exist based on solubility. The dynamic nature of P in soils has been schematically depicted by Johnston[45] and is a simplified version of the P cycle (Fig. 1), which represents the big picture with respect to P. In contrast to the P cycle, the diagram of Johnston[45] does indicate losses of soluble P, while its unique feature is the representation of the availability pools that impact the soluble P pool. The "readily available" pool represents a proportion of soil P that can be estimated by soil tests and can contribute substantially to P uptake by the crop. This pool is in apparent or pseudo-equilibrium with a less soluble fraction of soil P. A major fraction of soil P is in the very-slowly available pool. There is some evidence that this fraction can be slowly dissolved or weathered to sustain the less-readily available P pool.

In essence, these solubility categories or phases represent suites of inorganic and organic P compounds. Nevertheless, despite its simplicity, the schematic of Johnston[45] captures the essence of P dynamics and is centered on the soluble P pool from which plant uptake by roots occurs; this phase can be depleted by loss in drainage or greatly increased by the addition of soluble P fertilizers. With continued crop uptake, this tiny pool is replenished by the readily available P pool, which in turn is in "equilibrium" with a less soluble P phase. While the overall direction of P reactions is toward decreased solubility, the reverse can also occur under some circumstances. In terms of the plant roots for P uptake, accessibility is immediate in the solution phase but decreases with decreasing solubility to the right of the sketch.

GLOBAL FERTILIZER USE AND PHOSPHORUS SOURCES

A snapshot of P in relation to both agriculture and the environment would not be complete without reference to P fertilizer usage at the global level, followed by a notion of the world's reserves of P to sustain the future of life on this planet. Production of phosphate rock, the raw material for P manufacture, has consistently expanded to meet demand primarily for P fertilizer (80%) as well as for industry. With the projected increase in world population—and consequently, crop production—fertilizer use is set to continue.[46] In essence, as more P is used, the potential for impact on the environment increases.

The continued use of phosphate rock raises the issue of the world's reserve of this finite resource,[47] which is vital to food production—and, indeed, to life on earth. That P is limited underlines the need for efficient use as well as recycling, furthering the argument for reducing or eliminating unintended deleterious effects on the global environments. While P rock is mined in many countries, the world's supply is dominated by relatively few countries, i.e., the United States, Russia, China, and Morocco. Much debate has centered on the longevity of P deposits. Many factors have to be considered, establishing estimates such as the likelihood of new discoveries, energy costs associated with P mining, market prices for P fertilizers, and continued demand. While estimates of P reserves and resources range from 105 to 470 years,[48] more recent predictions by the International Fertilizer Development Center[49] set the figure between 300 and 400 years but cautioned that many factors, known and unknown, can greatly influence such estimates. Coinciding with such estimates, Cordell et al.[50] argued that "peak" consumption will occur in the next few decades. Recognizing the finiteness of P reserves, as there is no alternative source of P, Withers[51] cautioned that protecting future food security requires a radical rethink of how P is managed from field to global scales; he further argued the need for closing gaps in the P cycle to minimize wastage and stressed the need to recover or recycle P. More effective use of P already in circulation is a prerequisite to reducing society's dependence on inorganic fertilizers and minimizing the environmental footprint of P fertilizer manufacture.

IMPLICATIONS FOR PHOSPHORUS LOSS

Without P fertilization of soils in their natural state, the potential for movement of P in its soluble state is essentially insignificant. Even when P is lost in sediments due to erosion, the impact on water bodies is minimal due to the low concentrations of P in unfertilized soils. The problem is one that accompanies application of commercial fertilizers and manures, involving excess fertilization beyond the needs of the crop. However, as economic crop production is unthinkable without P fertilizer use, the problem is one that society has to live with. However, there are guidelines and a code of fertilizer management practices that can help farmers avoid excess P use while maintaining adequate P for optimum crop production.[17] Despite the propensity of soil to react with P fertilizers, recent research has shown that excess P can overcome or saturate soil's capacity to retain P tightly. As more P is added, the soluble P fraction increases, and that is where P "leakage" occurs. Fertilization increases the P fractions attached to soil particles and increases their potential for loss when the soil is transported as sediments. An overriding condition for loss of P from fertilized land is heavy rainfall; this induces more leaching and runoff and, where appropriate soil conservation measures are not in place, promotes P loss in sediments. Accordingly, many of the hot spots of P pollution of water bodies occur in intensively formed areas or watersheds with high rainfall, e.g., Western Europe, Mississippi Valley and the Gulf of Mexico, and Chesapeake Bay in northeastern United States.

PHOSPHORUS LOSSES AND MITIGATION STRATEGIES

Anthropogenic or man's activities have been responsible for the accelerated cycling and fluxes of P at global and regional scales in the past few decades, resulting in eutrophication of terrestrial and aquatic ecosystems together with biodiversity loss and human health risk.[51] Current estimates indicate that fluvial transport of dissolved (about 5 Tg yr^{-1}) and particulate P (20 Tg yr^{-1}) to oceans (a permanent P sink) is at least double that of preindustrial times.[52,53] Without concentrated efforts to increase P use in agriculture through interventions, such losses are likely to continue with projected increases in commercial fertilizer consumption, higher-yielding crops, bioenergy crops, urbanization, and economic growth.[51] Recognition of the threat that such developments posed for the environment has led to expanded funding to mitigate such an environmental threat, especially in areas where the problem of P loss is acute. For instance, from 2000 to 2008, the European Research Framework Program invested over €6 million to support P-centered projects, most of which had an environmental focus.[16] Such emphasis on surplus P in the environment is in stark contrast to the situation ongoing in many developing countries, especially those in Africa, where available P is a major crop-limiting factor and where fertilizer use is dismally low or nonexistent.

However, with respect to the issue of transfer of P from agricultural watersheds, resulting in eutrophication of surface waters, a number of mitigation strategies were advanced by Sharpley and Tunney.[17] Soil P testing for environmental risk assessment was put forward as a means of establishing acceptable P losses compatible with economic farming; an issue is the appropriateness of a field/plot or watershed scale for such studies. Recently, Maguire et al.[54] reviewed approaches to using soil testing to delineate threshold levels above which P fertilizer use is limited or not allowed. While soil testing was traditionally used for diagnosing P deficiency for crop production, it could also define an upper limit for soil P with respect to potential P losses, but many issues still need to be clarified by research. Threshold P levels should especially consider site vulnerability to P loss. An analysis of the pathways of P transport is needed at the watershed scale.

The current approach to fertilizer management in general using best management practices is one that is eminently applicable to mitigating P loss; such approaches are designed to bring P inputs as fertilizers and manures in synchrony with crop demands for P. Conventional erosion control measures form a key element in this approach. The current drive toward minimum- or no-till is a significant development in reducing P losses from cultivated land. While much remains to be learned technically in the area of P loss mitigation, a lot of existing knowledge is already available for adaptation in tackling P loss. Strategic initiatives are needed to bring about lasting change in land management targeting consumer-supported programs that encourage better standards of land management by farmers, inculcating a sense of stewardship with societal-accepted environmental protection goals.

CONCLUSIONS

The indispensability of P fertilizers for global agricultural production is undisputed, except for some adherents to organic agriculture; in fact, both commercial fertilization and organic farming should be complementary. With increasing demand on our land resources to feed the world's growing population, fertilizer use, including P, is set to continue. Research over the past century has done much to elucidate the complex nature of P in soils and plants. However, the world now faces a dilemma, with some developed countries with intensive agriculture having used excessive amounts of P, as commercial fertilizer and animal manure originating from confined animal feeding operations, while many countries of the developing world are hampered by use of little or no fertilizer. International policies have recognized the need to expand fertilizer P use in developing countries and curtail or eliminate P in developed areas with excess P. Overuse of P not only represents an economic loss to farmers but also contributes to transport of P from agricultural land to surface water bodies, where it causes eutrophication and its attendant consequences in terms of deteriorated ecology and health hazards for humans and animals.

Consequently, a major paradigm shift in soil research has been away from production agriculture and toward the environment. The change in emphasis comes at a time of general and growing societal awareness of the finite nature of rock phosphate that underpins the commercial fertilizer industry, thus making efficient use of P a major priority along with emphasis on recycling P from wastes. Much has been learned about the mechanisms of P loss from cropland and the implications of land use management in accelerating such losses; the current expansion of conservation agriculture, especially with minimum- or no-till, can greatly contribute to minimizing P losses. Tests are well established to monitor excess P in soils. However, as is explicit from chapters on the environmental aspects of P use in the most recent comprehensive treatise on P,[26] much remains to be known to provide a more effective basis for technical remediation.

The momentum of societal concern has resulted in policies restricting excess nutrient use on agricultural land in the interest of environmental protection. In essence, technology can—and will—underpin societal action to reconcile the needs for agriculture and protection of the environment.

REFERENCES

1. FAO. *World Agriculture towards 2035/2050*; Food and Agriculture Organization of the United Nations: Rome, Italy, 2006.
2. Cribb, J. *The Coming Famine: The Global Food Crisis and What We Can Do to Avoid It*; CSIRO Publ.: Collingwood, Victoria, Australia, 2010.
3. Borlaug, N.E. Feeding a hungry world. Science **2007**, *318*, 359.
4. Friedman, T.L. *Hot, Flat, and Crowded: Why We Need a Green Revolution and How It Can Renew America*; Farrar, Strauss, and Giroux: New York, USA, 2008.
5. Diamond, J. *Collapse: How Societies Chose to Fail or Succeed*; Penguin Group: New York, USA, 2005.
6. Inter-Governmental Panel on Climate Change. *Assessment of Global Climate Change*; Washington, DC, USA, **2008**.
7. Godfray, H.C.; Beddington, J.R.; Crute, I.R.; Haddad, L.; Lawrence, D.; Muir, J.M.; Pretty, J.; Robinson, S.; Thomas, S.M.; Toulmin, C. Food security: The challenges of feeding 9 billion people. Science **2010**, *327*, 812–818.
8. Deane, C.; Ejita, G.; Rabbinge, R.; Saye, T. Science for global development. Crop Sci. **2010**, *50*, 1–7.
9. Roy, R.N.; Finck, A.; Blair, G.J.; Tandon. H.S. Plant nutrition for food security: a guide to integrated nutrient management. FAO Fertilizer & Plant Nutrition Bulletin No. 16; Food and Agriculture Organization of the United Nations: Rome, Italy, 2006.
10. Stewart, W.M.; Hammond, L.L.; Van Kauwenbergh, S.J. Phosphorus as a natural resource. In *Phosphorus: Agriculture and the Environment*; Sims, J.T., Sharpley, A.N., Eds; American Society of Agronomy, Crop Science Society of America, Soil Science Society of America: Madison, WI, USA, 2005; 1–22.
11. Mosier, A.R.; Syers, J.K.; Freney, J.R. *Agriculture and the Nitrogen Cycle*, SCOPE 65 (Scientific Committee on Problems of the Environment); Island Press: Washington, DC/London, 2004.
12. Stevenson, F.J. *Nitrogen in Agricultural Soils*; American Society of Agronomy, Crop Science Society of America, Soil Science Society of America: Madison, WI, USA, 1982.
13. Scheppers, T.S.; Raun, W.A., Eds. *Nitrogen in Agricultural Systems*, Agronomy Monograph No. 49; American Society of Agronomy, Crop Science Society of America, Soil Science Society of America: Madison, WI, USA, 2008.
14. Sharpley, A.N.; Smith, S.J.; Namey, J.W. Environmental impact of agricultural nitrogen and phosphorus use. J. Agric. Food Chem. **1987**, *35*, 812–817.
15. Tunney, H.; Carton, O.T.; Brookes, P.C.; Johnston, A.E. *Phosphorus Loss from Soil to Water*; CAB International; Wallingford, U.K., 1997.
16. Delgado, A.; Scalenghe, R. Aspects of phosphorus transfer from soils in Europe. J. Plant Nutr. Soil Sci. **2008**, *171*, 552–575.
17. Sharpley, A.N.; Tunney, H. Phosphorus research strategies to meet agricultural and environmental challenges of the 21st century. J. Environ. Qual. **2000**, *29*, 176–181.
18. Larsen, S. Soil phosphorus. Adv. Agron. **1967**, *19*, 151–210.
19. Dalal, R.C. Soil organic phosphorus. Adv. Agron. **1977**, *29*, 83–117.
20. Khasawneh, F.E.; Sample, E.C.; Kamprath, E.J. *The Role of Phosphorus in Agriculture*; American Society of Agronomy, Crop Science Society of America, Soil Science Society of America: Madison, WI, USA, 1980.
21. Ryan, J. Phosphorus in soils of arid regions. Geoderma **1983**, *19*, 341–356.
22. Matar, A.; Torrent, J.; Ryan, J. Soil and fertilizer phosphorus and crop responses in the dryland Mediterranean zone. Adv. Soil Sci. **1992**, *18*, 82–146.
23. Ryan, J. Phosphorus fertilizer use in dryland agriculture: The perspective from Syria. In Proceedings of the OECD workshop on "Innovative Soil and Plant Systems for Sustainable Practices," Izmir, Turkey, June 3–7, 2002; Abstracts, 45.
24. Rashid, A.; Awan, Z.I.; Ryan, J.; Rafique, E.; Ibrikci, H. Strategies for phosphorus nutrition of dryland wheat in Pakistan. Commun. Soil Sci. Plant Anal. **2010**, *41*, 2555–2567.
25. Taylor, A.W.; Kilmer, V.J. Agricultural phosphorus in the environment. In *The Role of Phosphorus in Agriculture*; Khasawneh, F.E., Sample, E.C., Kamprath, E.J., Eds.; American Society of Agronomy, Crop Science Society of

America, Soil Science Society of America: Madison, WI, USA, 1980; 545–557.
26. Sims, T.J.; Sharpley, A.N., Eds. *Phosphorus: Agriculture and the Environment*; American Society Agronomy, Crop Science Society of America, Soil Science Society of America: Madison, WI, USA, 2005.
27. IWP-6. Sixth International Phosphorus Workshop, Seville, Spain, Sept 27–Oct 1, 2010.
28. Sharpley, A.N.; Rekolainen, S. Phosphorus in agriculture and its environmental implications. In *Phosphorus Loss from Soil to Water*; Tunney, H., Carton, O.T., Brookes, P.C., Johnston, A.E., Eds.; CAB International: Wallingford, U.K., 1997; 1–53.
29. Pierzinski, G: Sims, J.T; Vance, G.F. *Soils and Environmental Quality*, 2nd Ed.; CRC Press: Boca Raton, FL, USA, 2000.
30. Sharpley, A.N. Phosphorus availability. D18- 38. In *Handbook of Soil Science*; Sumner, M.E., Ed.; CRC Press: Boca Raton, FL, USA, 2000.
31. Syers, J.K.; Johnston, A.E.; Curtin, D. Efficiency of soil and fertilizer phosphorus use. Reconciling changing concepts of soil phosphorus behaviour with agronomic information. FAO Fertilizer and Plant Nutrition Bulletin No. 18; Food and Agriculture Organization of the United Nations: Rome, Italy, 2008.
32. Ryan, J.; Curtin, D.; Cheema, M.A. Significance of iron oxides and calcium carbonate particle size in phosphorus sorption and desorption in calcareous soils. Soil Sci. Soc. Am. J. 1985, *49*, 7476.
33. Ryan, J.; Hassan; H., Bassiri, M., Tabbara, H.S. Availability and transformation of applied phosphorus in calcareous soils. Soil Sci. Soc. Am. J. 1985, *51*, 1215–1220.
34. Torrent, J.A. Rapid and slow phosphate sorption by Mediterranean soils: Effect of iron oxides. Soil Sci. Soc. Am. J. **1987**, *53*, 78–82.
35. Fixen, P.E.; Grove, J.H. Testing soils for phosphorus. In *Soil testing and Plant Analysis*; Westermann, R.L., Ed.; American Society of Agronomy: Madison, WI, USA, 1990; 141–180.
36. Brown, J.R. Soil testing: Sampling, correlation, calibration and interpretation. Special Publication No. 21; Soil Science Society of America: Madison, Wisconsin, USA, 1987.
37. Halvorson, A.D.; Black, A.L. Long-term dryland crop responses to residual phosphorus fertilizer. Soil Sci. Soc. Am. J. **1985**, *49*, 928–933.
38. McCollum, R.E. Buildup and decline is soil phosphorus: 30-year trends in a Typic Umbraquult. Agron. J. **1991**, *83*, 77–85.
39. Barrow, N.T. Evaluation and utilization of residual phosphorus in soils. In *The Role of Phosphorus in Agriculture*; Khasawneh, F.E., Sample, E.C., Kamprath, E.J. Eds.; American Society of Agronomy, Crop Science Society of America, Soil Science Society of America: Madison, WI, USA, 1980; 333–360.
40. Ibrikci, H.; Ryan, J.; Ulger, A.C.; Buyuk, G.; Cakir, B.; Korkmaz, K.; Karnez, E.; Ozgenturk, G.; Konuskan, O. Maintenance of phosphorus fertilizer and residual phosphorus effect on corn production. Nutr. Cycling Agroecosyt. **2005**, *72*, 279–286.
41. Ryan, J.; Ibrikci, H.; Singh, M.; Matar, A.; Masri, S.; Rashid, A.; Pala, M. Response of residual and currently applied phosphorus in dryland cereal/legume rotations in three Syrian Mediterranean agroecosystems. Eur. J. Agron. **2008**, *28*, 126–137.
42. Condron, L.M.; Turner, B.L.; Cade-Menum, B.J. Chemistry and dynamics of soil organic phosphorus. In *Phosphorus: Agriculture and the Environment*; Sims, J.T., Sharpley, A.N., Eds.; American Society of Agronomy, Crop Science Society of America, Soil Science Society of America American Society of Agronomy: Madison, WI, USA, 2005; 87–122
43. Tiessen, H. Phosphorus dynamics in tropical soils. In *Phosphorus: Agriculture and the Environment*; Sims, J.T., Sharpley, A.N., Eds; American Society of Agronomy, Crop Science Society of America, Soil Science Society of America: Madison, WI, USA, 2005; 253–262.
44. Ryan, J.; Rashid, A. Phosphorus. In Encyclopedia of Soil Science, 2nd Ed.; Lal, R., Ed.; Taylor & Francis: New York, 2007; available at http://www.informaworld.com/smpp/content~db=all~content=a740187227~frm=titlelink. (accessed May 9, 2012).
45. Johnston, A.E. *Soil and Plant Phosphorus*; International Fertilizer Industry Association (IFA): Paris, France, 2000.
46. Stewart, W.M.; Hammond, L.L.; Van Kauwenbergh, S.J. Phosphorus as a natural resource. In *Phosphorus: Agriculture and the Environment*; Sims, J.T., Sharpley, A.N., Eds.; American Society of Agronomy, Crop Science Society of America, Soil Science Society of America: Madison, WI, USA, 2005; 1–22.
47. van Kauwenbergh, S.J. *World Phosphate Rock Reserves and Resources*; International Fertilizer Development Center: Muscle Shoals, AL, USA, 2010.
48. Cramer, M.D. Phosphate as a limiting resource: Introduction. Plant Soil **2010**, *334*, 1–10
49. Prud'homme, M. Global fertilizers and raw materials supply and supply/demand: 2006-2010. In Proceedings of the IFA Annual Conference, Cape Town, South Africa; Paris, France, 2006; 39–42.
50. Cordell, D.; Drangert, T.; White, S. The story of phosphorus: Global food security and food for thought. Global Environ. Change **2009**, *19*, 292–305.
51. Withers, P. Global phosphorus fluxes and the threat to food security. In Proceedings of the 6th International Phosphorus Workshop (IPW-6), Seville, Spain; 2010; Abstracts, 18.
52. Filippelli, G.M. The global phosphorus cycle: Past present and future. Elements **2008**, *4*, 89–95.
53. Smit, A.L.; Bindraban, P.S.; Schrober, J.J.; Conijin, J.G.; van der Meer, H.G. Plant Res. Int. Rep. 2009, 282.
54. Maguire, R., Cardon, W., Simard, R.R. Assessing potential environmental impacts of soil phosphorus by soil testing. In *Phosphorus: Agriculture and the Environment*; Sims, J.T., Sharpley, A.N., Eds.; American Society of Agronomy, Crop Science Society of America, Soil Science Society of America: Madison, WI, USA, 2005; 145–180.

Phosphorus: Riverine System Transport

Andrew N. Sharpley
University of Arkansas, Fayetteville, Arkansas, U.S.A.

Peter Kleinman
Pasture Systems and Watershed Management Research Unit, U.S. Department of Agriculture (USDA), University Park, Pennsylvania, U.S.A.

Tore Krogstad
Department of Plant and Environmental Sciences, Norwegian University of Life Science, Aas, Norway

Richard McDowell
AgResearch Ltd., Invermay Agricultural Center, Mosgiel, New Zealand

Abstract

The role of phosphorus (P) inputs in accelerating eutrophication of freshwaters is well documented. The total load of P to a river can broadly be divided into point source inputs, typically dominated by sewage treatment effluents, and diffuse sources, often dominated by agriculture. There is a general increase in P transport in the order of rivers draining forested - native ecosystems, intensively managed agriculture, and urban settings. Point sources enter the river more continually through the year than do non-point sources, which are subject to large seasonal variation, typically as a function of overland flow. Changes in the forms and amounts of P during transport in streams and rivers can greatly influence the eventual impact of P loss on the degree of eutrophic response of receiving waters. These changes are mediated by physical (sediment deposition and resuspension and flow regimes), abiotic (P sorption and desorption), and biotic (microbial and plant uptake) processes. Such riverine processes influence the long-term transport and receiving water body response to P, where there may be a time-lag (years or decades) before improvements in water quality, or regeneration of diverse habitats, might become apparent even after P inputs are minimized. Clearly, implementation of effective conservation measures must consider fluvial system response behavior, where sinks may become sources of P with only slight changes in watershed management and hydrologic response.

INTRODUCTION

The contribution of anthropogenic phosphorus (P) to the accelerated eutrophication of freshwaters is well documented.[1,2] Sources of riverine P inputs can broadly be divided into point sources, typically dominated by sewage treatment effluents, and diffuse (also "non-point") sources, derived from the landscape.[3,4] Phosphorus export tends to increase from rivers draining native or low-input ecosystems, to intensively managed agricultural systems, to urban settings.[5] Point sources enter the river more continually through the year than do nonpoint sources, which are subject to large seasonal variation, typically as a function of overland flow and land management activities.[6]

Changes in the forms and amount of P that occur as part of the transport processes within streams and rivers can greatly influence the eventual impact of point and non-point sources of P on downstream eutrophication.[6,7] These changes in P are mediated by physical (sediment deposition and resuspension and flow regimes), abiotic (P sorption and desorption), and biotic (microbial and plant uptake) processes.[8,9]

Most importantly, P in riverine systems can play an important role in modifying or delaying societal efforts to curb eutrophication. For instance, P that is already entrained in riverine systems, sometimes referred to as *legacy P*, can serve as a long-term source of P to the overlying water column. Understanding the role of riverine systems as sources, sinks and stores of P is critical to the long-term management of cultural eutrophication.

RIVERINE PROCESSES

Physical Processes

Fluvial sediments are derived from the erosion of surface soils, gullies, ditches and stream banks. Because surface soils generally contain the highest concentration of P in soil profiles, and erosion preferentially removes P-rich particles, eroded surface soil represents a major source of particulate P in riverine systems.[10,11] In areas with recent gully formation or bank erosion, subsoil is the dominant source of sediments. Sediments derived from gulley or bank sources have low P content and high P sorption capacities.[12,13] As P release and sorption are largely related to particle size, with coarser-sized particles releasing P more readily than fine particles, which also tend to sorb more P,[14] hydrologic processes controlling sediment particle size distribution have important implications to P fate in river systems.

Abiotic Processes

In fluvial systems with good hydraulic mixing (such as shallow flowing streams), P movement between sediment and water phases is related to the equilibrium P concentration at zero net sorption or desorption (EPC_0); P is released from sediment if the concentration of P in stream flow is less than its EPC_0, while the reverse is also true.[15] Other processes influencing sediment P release include a rise in stream water pH, P from dead phytoplankton, periphyton or macrophytes, the hydrolysis of organic P species, and changes in sediment crystallinity and oxidation/reduction.[10,16] For example, regular wetting and drying cycles in stream sediments or bank material can change Fe-oxide crystallinity making occluding P associated with these materials.[17,18]

Biotic Processes

Uptake of P by aquatic biota can decrease dissolved P in the water column,[19] while bacteria can mediate a sizeable proportion of sedimentary P uptake and release (30%–40%, Khoshmanesh et al.[20] and McDowell and Sharpley[21]). Biologically-controlled P release during the decomposition of organic matter in sediments can be an important source of dissolved P at times of high temperature and low flow in areas with organic-rich sediments, such as streams draining forests.[22] Organic matter in sediments may also increase the blooms of bacteria and algae by preventing chelator limited growth.[23] The relative effect of biotic processes on riverine P transport varies greatly, reflecting seasonal cycles, management of stream-side land, sediment P forms, size of flow event, and streambed geology. However, during elevated flow, when P loads are often high, biotic processes generally are less important to riverine P transport than physical and abiotic processes.

Nutrient Spiralling

The concept of P-spiralling, or the distance travelled downstream by one P molecule as it completes one cycle of uptake, transformation (e.g., from dissolved to organic form) and release back into flow as dissolved P, is useful in understanding fundamental mechanisms of P transport in rivers.[24] Lengths of P-spiralling generally vary from 1 to 1000 m, as a function of flow regime, season, bedrock geology, and sediment characteristics.[25,26] Interaction between ground water and stream flow within the hyporheic zone will change P concentrations, depending upon the relative contribution of stream-bed upwelling or infiltration of P-rich stream flow.

INTEGRATING RIVERINE PROCESSES AND LAND USE IMPACTS ON P TRANSPORT

The role of riverine processes in watershed scale P export is illustrated by the findings work of McDowell et al.[27] in a 40-ha agricultural watershed in central Pennsylvania (Fig. 1). They found dissolved P concentrations in base flow increased from 28 to 42 $\mu g\ L^{-1}$ as one moved from down the stream channel. Base flow P concentrations were controlled by channel sediment P sorption (532 mg kg^{-1} at flume 4 and 227 mg kg^{-1} at the outlet) and EPC_0 (4 μg kg^{-1} at flume 4 and 34 μg kg^{-1} at the outlet). Storm flow trends, however, were the opposite, with P concentrations decreasing downstream (304 $\mu g\ L^{-1}$ at flume 4 and 128 $\mu g\ L^{-1}$ at flume 1) due to the dilution of P derived from a critical source area: an agricultural field with elevated soil P and high erosion/runoff (Fig. 1).

In a much larger watershed, the Winooski River, VT, the largest tributary to Lake Champlain (Fig. 2), McDowell et al.[7] evaluated interactions between local sources of P, sediment properties and flow, elucidating their role in riverine P transport. Input and delivery of fine sediment enriched with P was influenced by surrounding land use. Algal-available P of river sediments near agricultural land (3.6 mg kg^{-1}) was greater than that of sediments near forested land (2.4 mg kg^{-1}). Over the short-term, river flow and sediment physical properties were responsible for particulate P loadings from the river to Lake Champlain. However, deposition of sediments downstream, near the outflow into Lake Champlain, resulted in a large pool of stored P within the river system. Over the long-term, this pool is likely to release dissolved P to overlying waters, even as inputs of P from point and nonpoint sources decline due to implementation of remedial strategies and watershed conservation measures.

DEFINING P-RELATED IMPAIRMENT IN FLOWING AND LAKE WATERS FOR TARGETED REMEDIATION

In order to prioritize and target watershed remediation to minimize P losses, water impairment must be quantified.[28] Background levels (i.e., regional nutrient criteria) of total P, total N, chlorophyll-*a*, sediment, and clarity in pristine surface waters are used as benchmarks for a given geographical area (Fig. 3; Gibson et al.[29] and Omernik[30]).

While these criteria have regulatory application, such as under the U.S.A.'s Clean Water Act, they can also be used to guide voluntary efforts in watershed planning.[31] These criteria are available for freshwater systems in the continental U.S. (Table 1). Similar approaches have been taken in Australasia and Europe.[4,32] In the European Union's Water Framework Directive, biological parameters are, however, the basis for measuring ecological status for the water with chemical parameters used only as support parameters. The E.U. classification system emphasizes whether the ecosystem is in ecological balance and points out the effect of the pollution rather than providing a classification or ranking according to pollutant concentration, which has been the basis for most previous classifications systems.

IMPLICATIONS TO WATERSHED MANAGEMENT

Aquatic ecosystems respond to P inputs on the basis of factors related to their physiography and flushing rates. Individual systems respond to discrete and sustained P inputs differently, and indeed it may not be possible to attain P loadings low enough to prevent periphyton blooms because of, for example, natural enrichment from P-rich rocks.

A certain degree of eutrophication can be beneficial. For example, fishery management often requires a higher productivity to maintain an adequate phytoplankton-zooplankton-fish food chain for optimum commercial or sport fish production. This food chain may be manipulated by stocking of water with certain fish species in addition

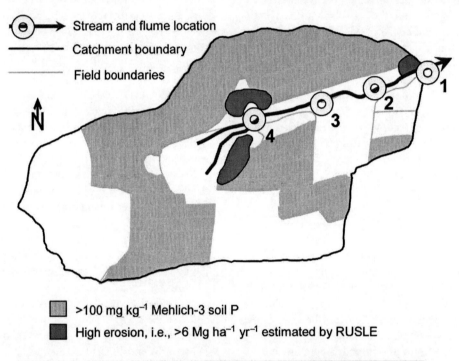

Flume	Dissolved P		Stream sediment	
	Stormflow	Baseflow	P sorption max	EPC_0
	$\mu g\ L^{-1}$		$mg\ kg^{-1}$	$\mu g\ L^{-1}$
1	128	42	227	34
2	174	36	295	13
3	202	37	330	4
4	304	28	532	4

Fig. 1 The distribution of high Mehlich-3 soil P ($>100\ mg\ kg^{-1}$), erosion ($>6\ Mg\ ha^{-1}\ yr^{-1}$) and dissolved P concentration in stream and baseflow (mean of 1997–2000 data) in relation to P sorption properties of channel sediment at four flumes in FD-36.
Source: Adapted from McDowell et al.[27]

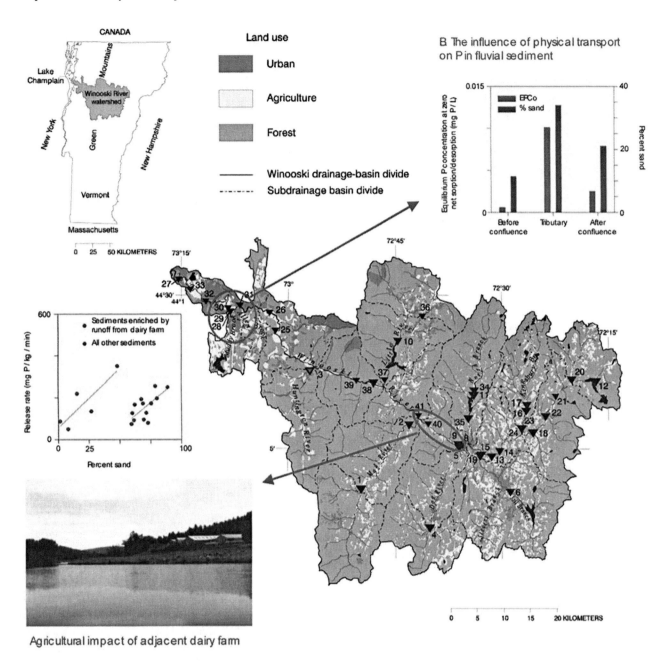

Fig. 2 The location, distribution and impact of land use and physical transport processes on P in fluvial sediments within the Winooski River watershed, VT.
Source: Adapted from McDowell et al.[7]

to P load reductions, in efforts to reduce the incidence of algal blooms and improve overall water quality.[33]

In most cases, however, eutrophication restricts water use for fisheries, recreation, and industry because of the increased growth of undesirable algae and aquatic weeds and oxygen shortages caused by their death and decomposition.[1] An increasing number of surface waters also experience periodic and massive harmful algal blooms (e.g., *cyanobacteria* and *Pfiesteria*) that contribute to summer fish kills, unpotable drinking water, formation of carcinogens during water chlorination, and links to neurological impairment in humans.[34,35]

IMPLICATIONS FOR WATER QUALITY RESPONSE

The response of riverine systems to upstream changes in watershed management can vary from seasons to decades, and generally increases as watershed scale increases.[36,37] In small watersheds (i.e., <100 ha), research has demonstrated

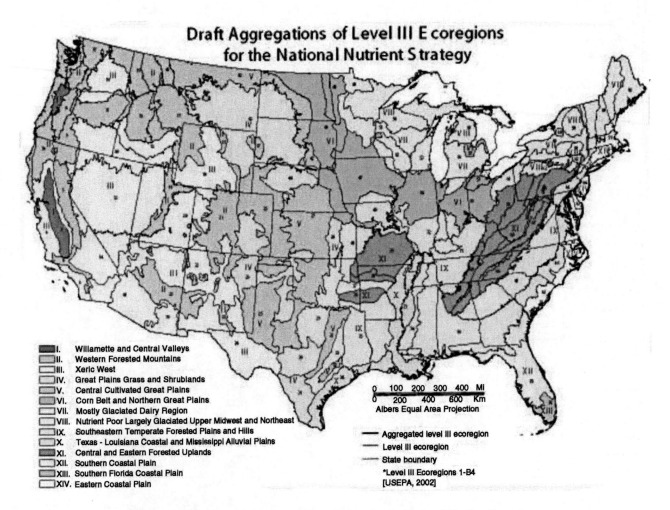

Fig. 3 Draft aggregations of level III ecoregions for the National Nutrient Strategy.
Source: Adapted from Gibson et al.[29] and Omernik.[30]

Table 1 Background total P concentrations for each of the aggregated nutrient Ecoregions in the U.S. for freshwater systems.[28]

	Aggregated ecoregion	Total P ($\mu g \, L^{-1}$)	
Number	Description	Rivers and streams	Lakes and reservoirs
I	Willamette and Central Valleys	47	—
II	Western Forested Mountains	10	9
III	Xeric West	22	17
IV	Great Plain Grass and Shrub Lands	23	20
V	South Central Cultivated Great Plains	67	33
VI	Corn Belt and Northern Great Plains	76	38
VII	Mostly Glaciated Dairy Region	33	15
VIII	Nutrient Poor Largely Glaciated Upper Midwest and Northeast	10	8
IX	Southeastern Temperate Forested Plains and Hills	37	20
X	Texas-Louisiana Coastal and Mississippi Alluvial Plain	128[a]	—
XI	Central and Eastern Forested Uplands	10	8
XII	Southern Coastal Plains	40	10
XIII	Southern Florida Coastal Plains	—	18
XIV	Eastern Coastal Plains	31	8

[a]This high value may be either a statistical anomaly or reflects a unique condition.

reduction in nutrient and sediment loss in runoff can occur within months of implementing remedial management measures (e.g., "conservation practices "or "best management practices"). However, the spatial complexity of watershed systems and nature of P sources (e.g., acute sources such as fertilizers vs. legacy P sources such as soils and sediments)affects this response time. Indeed, the slow release of legacy P stored in soils and sediments to more rapid surface flow pathways may continue for decades even after practices have been installed to curb further additions. The release of legacy P from riverine sediments is influenced by the oxygen status of overlying waters, where reducing conditions favor the dissolution of iron-bound P. As remedial efforts decrease the concentrations of P in riverine systems, the change in gradient between sediment and water column can trigger the desorption of dissolved P from the sediment. Finally, P enriched sediments are resuspended during high flows. This effect will likely increase with climate change, which could result in more precipitation as rain in winter and an increase in occurrence and severity of rainfall events.

Because of the lag time between watershed management practice implementation and water quality improvements, remedial strategies should consider the re-equilibration of watershed and water-body behavior, where nutrient sinks may become sources of P with only slight changes in watershed management and hydrologic response. A better understanding of the spatial and temporal aspects of watershed response to nutrient load reductions in both flowing and standing water bodies is needed, as well as the scale at which responses may occur in a more timely fashion. This would likely be at a smaller sub-watershed scale, where local water quality and quantity benefits may become evident more quickly. It is also important to accept in any watershed-P loss reduction strategy, that it is essential to address the overall physical and social complexity of legacy P sources, when, where and to what extent they occur.

CONCLUSIONS

Clearly, several interdependent riverine processes influence the amounts and forms of P transported from edge-of-field agricultural sources to the point of impact (i.e., river, lake, reservoir, and estuary). These processes will, thus, be critical in defining agricultural source management and in determining eutrophic response. Without information on the direction and magnitude of change in P transport in river systems, conservation practices will not efficiently remediate against impairment of receiving waters.

The accumulation of P in aquatic environments is such that even if P were no longer added to agricultural systems, there would be a considerable time-lag (years or decades) before improvements in water quality, or regeneration of diverse habitats, might become apparent. Thus, the emphasis of watershed management should be on preventing further deterioration and taking strategic and sustainable actions sooner rather than later, otherwise we are simply and literally storing up more severe problems for future generations to confront. Despite our knowledge of controlling processes, it is difficult for the public to understand or accept this lack of response. When public funds are invested in remedial watershed programs, rapid improvements in water quality are usually expected and often required. Thus, implementation of effective conservation measures must consider fluvial system response behavior, where sinks may become sources of P with only slight changes in watershed management and hydrologic response.

REFERENCES

1. Carpenter, S.R.; Caraco, N.F.; Correll, D.L.; Howarth, R.W.; Sharpley, A.N.; Smith, V.H. Nonpoint pollution of surface waters with phosphorus and nitrogen. Ecol. Applic. **1998**, *8*, 559–568.
2. Schindler, D.W.; Hecky, R.E.; Findlay, D.L.; Stainton, M.P.; Parker, B.R.; Paterson, M.J.; Beaty, K.G.; Lyng, M.; Kasian, S.E. Eutrophication of lakes cannot be controlled by reducing nitrogen input: Results of a 37-year whole-ecosystem experiment. Proc. National Acad. Sci. **2008**, *105*(32), 11254–11258.
3. U.S. Geological Survey. *The quality of our nations waters: Nutrients and pesticides.* U.S. Geological Survey Circular 1225; USGS Information Services: Denver, CO, 1999; 82 p. Available at http://www.usgs.gov.
4. Withers, P.J.A.; Lord, E.L. Agricultural nutrient inputs to rivers and ground waters in the U.K.: Policy environmental management and research needs. Soil Use and Managt. **2002**, *14*, 186–192.
5. National Research Council. *Clean coastal waters: Understanding and reducing the effects of nutrient pollution.* National Academy Press: Washington, D.C., 2000.
6. Mainstone, C.P.; Parr, W. Phosphorus in rivers – ecology and management. Sci. Total Environ. **2002**, *282–283*, 25–47.
7. McDowell, R.W.; Sharpley, A.N.; Chalmers, A.T. Land use and flow regime effects on phosphorus chemical dynamics in the fluvial sediment of the Winooski River, Vermont. Ecolog. Eng. **2002**, *18*, 477–487.
8. House, W.A. Geochemical cycling of phosphorus in rivers. App. Geochem. **2003**, *18*, 739–748.
9. Jarvie, H.P.; Whitton, B.A.; Neal, C. Nitrogen and phosphorus in east coast British rivers: speciation sources and biological significance. Sci. Total Environ. **1998**, *210–211*, 79–110.
10. Baldwin, D.S.; Mitchell, A.M.; Olley, J.M. Pollutant-sediment interactions: Sorption, reactivity and transport of phosphorus. In *Agriculture, Hydrology and Water Quality*; Haygarth, P.M., and Jarvis, S.C. (eds.); CABI International: Oxford, England, 2002; 265–280.
11. Krogstad, T.; Løvstad, Ø. Erosion, phosphorus and phytoplankton response in rivers of southeastern Norway. Hydrobiologia **1989**, *183*, 33–41.
12. Olley, J.M.; Murray, A.S.; Mackenzie, D.H.; Edwards, K. Identifying sediment sources in a gullied catchment using

natural and anthropogenic radioactivity. Water Resour. Res. **1993**, *29*, 1037–1043.
13. Sharpley, A.N.; Smith, S.J.; Zollweg, J.A.; Coleman, G.A. Gully treatment and water quality in the Southern Plains. J. Soil Water Conserv. **1996**, *51*, 512–517.
14. Stone, M.; Murdoch, A. The effect of particle size, chemistry and mineralogy of river sediments on phosphate adsorption. Environ. Technol. Letters **1989**, *10*, 501–510.
15. Kunishi, H.M.; Taylor, A.W.; Heald, W.R.; Gburek, W.J.; Weaver, R.N. Phosphate movement from an agricultural watershed during two rainfall periods. J. Agric. Food Chem. **1972**, *20*, 900–905.
16. Fox, L.E. The chemistry of aquatic phosphate: inorganic processes in rivers. Hydrobiologia **1993**, *253*, 1–16.
17. Qiu, S.; McComb, A.J. Planktonic and microbial contributions to phosphorus release from fresh and air-dried sediments. Freshwater Res. **1995**, *46*, 1039–1045.
18. Baldwin, D.S. Effects of exposure to air and subsequent drying on the phosphate sorption characteristics of sediments from a eutrophic reservoir. Limnol. Oceanog. **1996**, *41*, 1725–1732.
19. Horner, R.R.; Welch, E.B.; Seeley, M.R.; Jacoby, J.M. Responses periphyton to changes in current velocity, suspended sediment and phosphor concentration. Freshwater Biol. **1990**, *24*, 215–232.
20. Khoshmanesh, A.; Hart, B.T.; Duncan, A.; Beckett, R. Biotic uptake and release of phosphorus by a wetland sediment. Environ. Technol. **1999**, *29*, 85–91.
21. McDowell, R.W.; Sharpley, A.N. Phosphorus uptake and release from stream sediments. J. Environ. Qual. **2003**, *32*, 937–948.
22. Klotz, R.L. Temporal relation between soluble reactive phosphorus and factors in stream water and sediments in Hoxie Gorge Creek, New York. Can. J. Aquatic Sci. **1991**, *48*, 84–90.
23. Løvstad, Ø.; Krogstad, T. Effect of EDTA, FeEDTA and soils on the phosphorus bioavailability for diatom and bluegreen algal growth in oligotrophic waters studied by transplant biotests. Hydrobiologia **2001**, *450*, 71–81.
24. Elwood, J.W.; Newbold, J.D.; O'Neil, R.V.; Van Winkle, W. Resource spiraling: An operational paradigm for analyzing lotic ecosystems. In *Dynamics of lotic ecosystems*; Fontaine, III, T.D.; Bartell, S.M. (eds.); Ann Arbor Science: Ann Arbor, MI, 1983; 3–27.
25. Melack, J.M. Transport and transformations of P, fluvial and lacustrine ecosystems. In *Phosphorus in the Global Environment: Transfers, Cycles and Management*; Tiessen, H. (Ed.); SCOPE 54 and John Wiley and Sons: Chichester, U.K., 1995; 245–254.
26. Munn, N.L.; Meyer, J.L. Habitat-specific solute retention in two small streams, An intersite comparison. Ecology **1990**, *71*, 2069–2082.
27. McDowell, R.W.; Sharpley, A.N.; Folmar, G. Phosphorus export from an agricultural watershed: linking source and transport mechanisms. J. Environ. Qual. **2001**, *30*, 1587–1595.
28. U.S. Environmental Protection Agency. *Ecoregional Nutrient Criteria*. EPA-822-F-01-010. USEPA, Office of Water (4304), U.S. Govt. Printing Office: Washington, DC, 2001. Available at http://www.epa.gov/waterscience/criteria/nutrient/9docfs.pdf.
29. Gibson, G.R.; Carlson, R.; Simpson, J.; Smeltzer, E.; Gerritson, J.; Chapra, S.; Heiskary, S.; Jones, J.; Kennedy, R. *Nutrient criteria technical guidance manual: lakes and reservoirs* (EPA-822-B00-001). U.S. Environmental Protection Agency: Washington, D.C.; U.S. Govt. Printing Office: Washington, DC, 2000.
30. Omernik, J.M. Ecoregions of the Conterminous United States. Annals Assoc. Am. Geographers **1987**, *77*, 118–125.
31. U.S. Environmental Protection Agency. *National Strategy for the Development of Regional Nutrient Criteria*. EPA-822-F-98-002. USEPA, Office of Water (4304), U.S. Govt. Printing Office: Washington, DC, 1998. Available at http://www.epa.gov/waterscience/standards/nutsi.html.
32. Sparrow, L.A.; Sharpley, A.N.; Reuter, D.J. Safeguarding soil and water quality. In Opportunities for the 21st Century: Expanding the horizons for soil, plant, and water analysis. Commun. Soil Sci. Plant Anal. **2000**, *31*(11–14), 1717–1742.
33. Horppila, J.; Kairesalo, T. A fading recovery: the role of roach (Rutilus rutilus L.) in maintaining high algal productivity and biomass in Lake Vesijarvi, southern Finland. Hydrobiologia **1990**, *200–201*, 153–165.
34. Burkholder, J.A.; Glasgow, Jr., H.B. *Pfiesteria piscicidia* and other Pfiesteria-dinoflagellates behaviors, impacts, and environmental controls. Limnol. Oceanog. **1997**, *42*, 1052–1075.
35. Kotak, B.G.; Kenefick, S.L.; Fritz, D.L.; Rousseaux, C.G.; Prepas, E.E.; Hrudey, S.E. Occurrence and toxicological evaluation of cyanobacterial toxins in Alberta lakes and farm dugouts. Water Res. **1993**, *27*, 495–506.
36. Cassell, E.A.; Clausen, J.C. Dynamic simulation modeling for evaluating water quality response to agricultural BMP implementation. Water Sci. Tech. **1993**, *28*, 635–648.
37. Sharpley, A.N. (ed.). *Agriculture and Phosphorus Management: The Chesapeake Bay*. CRC Press: Boca Raton, FL, 2000. 229 p.

Plant Pathogens (Fungi): Biological Control

Timothy Paulitz
*Department of Plant Science, MacDonald Campus of McGill University,
Ste-Anne-de-Bellevue, Quebec, Canada*

Abstract

In the classical definition of biological control, certain fungi, termed biocontrol agents (BCAs), can reduce the amount of inoculum or disease-producing activity of a plant pathogen, usually another fungus. This entry covers mechanisms of how these fungi antagonize the pathogen, what part of the pathogen life cycle can be targeted, how the BCAs can be applied, and examples of commercially available products.

INTRODUCTION

In the classical definition of biological control, certain fungi, termed biocontrol agents (BCAs), can reduce the amount of inoculum or disease-producing activity of a plant pathogen, usually another fungus.[1] The net result is a reduction of plant disease and crop loss. This section will cover mechanisms of how these fungi antagonize the pathogen, what part of the pathogen life cycle can be targeted, how the BCAs can be applied, and examples of commercially available products. Within the past 20 years, there has been a tremendous increase in interest and research on the subject, spurred by a search for more environmentally benign methods of disease control. But fungal BCAs have limitations that have restricted the number of products that are currently on the market. Table 1 shows some products used against soilborne pathogens on the market as of January 1999.

MECHANISMS OF BIOLOGICAL CONTROL BY FUNGI

The strategy behind managing pathogens is to target or interrupt part of the pathogen life cycle.[1,2] Like any microbe, pathogens start from inoculum in the environment, which can be spores, mycelia, or other dormant survival structures. These germinate on the plant surface, penetrate and infect the plant, and reproduce and sporulate on the plant to produce new inoculum. Many pathogens can also grow saprophytically on dead organic matter and plant debris. In biological control, the pathogen can be targeted in three ways.[1,2] First, the inoculum of the pathogen can be reduced or destroyed. This is most effective for soilborne pathogens, where the inoculum is dormant in the soil and the monocyclic disease is determined by the initial inoculum present in the field. BCAs can also interfere with inoculum formation by pathogens growing saprophytically on organic matter and plant debris. However, this strategy is not very effective for foliar polycyclic diseases, where inoculum comes from outside the field and the initial inoculum has little effect on the final outcome of the disease. Another strategy is one of protection, where a population of the BCA is established on the infection site of the plant before the pathogen attacks, thus preventing the pathogen's entry. These infection sites can be on seeds, bulbs, roots, leaves, fruit, flowers, or wounds. Finally, nonpathogenic or avirulent fungi can stimulate the plant to a higher level of resistance to a later-attacking pathogen, a concept termed induced resistance.

The most direct way a fungus can attack a fungal pathogen is by mycoparasitism, where the BCA uses the pathogen as a source of food.[3] The hyphae of the mycoparasite contact, penetrate, and colonize the hyphae, spores, or survival structures of the host fungus. Many of these mycoparasites produce enzymes that degrade the cell walls of the fungal host, including β-1-3 glucanase and chitinase. Most mycoparasites are necrotrophic and eventually kill their fungal host. Much of this research has focused on reducing the inoculum of soilborne pathogens. Classic examples include *Trichoderma* and *Gliocladium* spp. parasitizing *Rhizoctonia solani* and *Pythium* spp., which cause seed, seedling, and root rots.[4–6] *Pythium* spp. such as *Pythium oligandrum* and *P. nunn* parasitize pathogenic species of *Pythium*. *Coniothyrium minitans* and *Sporodesmium sclerotivorum* parasitize sclerotia of *Sclerotinia* spp., such as the white mold pathogen *S. sclerotiorum*, which attacks hundreds of plant species (57). *Ampelomyces quisqualis* parasitizes cleistothecia of powdery mildews. Major limitations to this strategy are that mycoparasites are slow acting and large amounts of mycoparasite inoculum must be added to the soil to ensure it will encounter the propagules of the pathogen. However, a promising strategy demonstrated with *S. sclerotivorum* is to render a soil suppressive to the pathogen by an inoculative release at a lower inoculum density, and allowing the mycoparasite to build up over successive seasons, using the pathogen as a food source. This is similar to the classic predator–prey relationship found in insect biocontrol.

Table 1 Some commercial biocontrol products for use against soilborne crop diseases.

Biocontrol fungus	Trade name	Target pathogen/disease	Crop	Manufacturer
Ampelomyces quisqualis M-10	AQ10 Biofungicide	Powdery mildew	Cucurbits, grapes, ornamentals, strawberries, tomatoes	Ecogen Inc., Langhorne, Pennsylvania
Candida oleophila I-182	Aspire	Botrytis, Penicillium	Citrus, pome fruit	Ecogen Inc., Langhorne, Pennsylvania
Fusarium oxysporum (nonpathogenic)	Biofox C	Fusarium oxysporum	Basil, carnation, cyclamen, tomato	S.I.A.P.A., Galliera, Bologna, Italy
Trichoderma harzianum and *T. polysporum*	Binab T	Wilt and root rot pathogens, wood decay pathogens	Fruit, flowers, ornamentals, turf, vegetables	Bio-innovation, Algaras, Sweden
Coniothyrium minitans	Contans	Sclerotinia sclerotiorum and S. minor	Canola, sunflower, peanut, soybean, lettuce, bean, tomato	Prophyta Biologischer Pflanzenschutz, Malchow/Poel, Germany
Fusarium oxysporum (nonpathogenic)	Fusaclean	Fusarium oxysporum	Basil, carnation, cyclamen, gerbera, tomato	Natural Plant Protection, Nogueres, France
Pythium oligandrum	Polygandron	Pythium ultimum	Sugar beet	Plant Protection Institute, Bratislavsk, Slovak Republic
Trichoderma harzianum and *T. viride*	Promote	Pythium, Rhizoctonia, Fusarium	Greenhouse, nursery transplants, seedlings	JH Biotech, Ventura, California
Trichoderma harzianum	RootShield, Bio-Trek T-22G, Planter Box	Pythium, Rhizoctonia, Fusarium, Sclerotinia homeocarpa	Trees, shrubs, transplants, ornamentals, cabbage, tomato, cucumber, bean, corn, cotton, potato, soybean, turf	Bioworks, Geneva, New York
Phlebia gigantea	Rotstop	Heterobasidium annosum	Trees	Kemira Agro Oy, Helsinki, Finland
Gliocladium virens GL-21	SoilGard (formerly GlioGard)	Damping-off and root pathogens, Pythium, Rhizoctonia	Ornamentals and food crops grown in greenhouses, nurseries, homes, interiorscapes	Thermo Triology, Columbia, Maryland
Trichoderma harzianum	Trichodex	Botrytis cinerea, Colletotrichum, Monilinia laxa, Plasmopara viticola, Rhizopus stolonifer, Sclerotinia sclerotiorum	Cucumber, grape, nectarine, soybean, strawberry, sunflower, tomato	Makhteshim Chemical Works, Beer Sheva, Israel
Trichoderma harzianum and *T. viride*	Trichopel, Trichoject	Armillaria, Botryosphaeria, Fusarium, Nectria, Phytophthora, Pythium, Rhizoctonia		Agrimm Technologies, Christchurch, New Zealand

Source: Information provided by the U.S. Department of Agriculture, Agriculture Research Service, Beltsville, Maryland, and was compiled by D. Fravel (http://www.barc.usda.gov/psi/bpdl/bodlpood/bioprod.htm).

Some fungi can produce antibiotic compounds that are toxic to other microbes, including plant pathogens. *Trichoderma* spp. produce volatile and nonvolatile antifungal compounds, including peptiabols, pyrones, and terpenoid antibiotics.[4] *Gliocladium virens* produces glioviren and gliotoxin that inhibit *R. solani* and *Pythium ultimum*. This mechanism is most effective when the BCA can grow to high populations and has an energy source to produce the antibiotic. An example would be *Trichoderma* or *Gliocladium* spp. applied to seeds or where a food base is added to the inoculum.[4,5]

Plant pathogens require carbon, nitrogen, iron, and other nutrients to grow. Many spores have an exogenous requirement for these nutrients, supplied by the plant rhizosphere or phyllosphere, in order to germinate. BCAs can compete with the pathogen for these limiting nutrients. For example, nonpathogenic species of *Fusarium oxysporum* can compete with pathogenic forma speciales for these limiting nutrients, resulting in control of wilt diseases. Competition by yeasts or hyphal fungi may protect flowers and foliage against nectotrophic pathogens such as *Botrytis cinerea* by colonizing the senescent tissue or nutrient-rich flower petals.[8-9] This mechanism, although difficult to prove experimentally, is probably one of the primary ways BCAs can protect a plant surface through preemptive exclusion of the pathogen.

Fungal biocontrol agents can also affect the pathogen by acting indirectly on the plant to make it more resistant to pathogen attack. Nonpathogenic microbes can induce a systemic resistance in plants (79). When the plant recognizes the inducing BCA, a signal is transduced systemically to the entire plant, bringing the defenses to a "high state of alert," so that a subsequent challenge by a pathogen is reduced. Nonpathogenic isolates of *F. oxysporum* induce a defense reaction against pathogenic isolates of *F. oxysporum*. This mechanism has several advantages. Once induced, the resistance is systemic, the entire plant becomes more resistant, and high populations of the BCA do not need to be maintained. It can also protect parts of the plant that cannot be protected directly by the BCA, including new growth of shoots and roots. However, more research is needed to investigate the applicability of this technology under greenhouse and field conditions.

APPLICATION OF FUNGAL BIOCONTROL AGENTS

How are fungal BCAs applied? Most are applied in an inundative strategy in large amounts to build up the population of the BCA high enough to overwhelm and have an effect on the pathogen.[1,5] Most are also targeted toward soilborne pathogens. However, one limitation of this strategy is the large amount of inoculum that must be applied and the high cost of production of spores, conidia, biomass, or chlamydospores.[5,9] Another problem is the erratic performance of many biocontrol agents under field conditions, due to unfavorable environmental conditions for the BCA, and the problem of establishing the BCA in a niche already occupied by competing microflora. Therefore most of the commercially available products have targeted applications that avoid these problems. For example, the greenhouse and nursery markets are prime targets because of the controlled environmental conditions and the high economic value of the crops. Another method is to use a protective strategy and apply the BCA directly to the infection court when it is small. For example, high populations of *Trichoderma* or *Gliocladium* conidia can be coated onto seeds to protect against damping-off pathogens such as *R. solani* and *P. ultimum*. Transplant cuttings and bulbs can be treated with liquid suspensions of products before planting in the greenhouse or field. Products such as formulations of *G. virens* can be mixed directly into the soil or soilless mixes in the greenhouse. Granules of *Trichoderma* spp. can be added to seed furrows, mixed with seeds in a planter box, or added to a commercial seed slurry.[4]

Since many pathogens gain access to plants through wounds, biocontrol agents can also be applied to transplant or pruning wounds. A classic example, and one of the first commercially used fungal biocontrol agents, is the application of *Phlebia* (= *Peniophora*) *gigantea* to cut pine stumps to prevent the stumps from being colonized by the pathogen *Heterobasidium annosum*. The pathogen can spread from these stumps to the entire plantation via the root system.

Postharvest pathogens are weak pathogens that require wounds on fruit to gain access. Yeast-like organisms such as *Candida* spp. can be applied to fruit during processing to exclude rot pathogens such as *Penicillium* spp. from colonizing wounds.[8] However, these applications require more stringent testing for registration, since they are applied directly to a food product. Competition is preferable to antibiosis for this application, since antifungal compounds would also have to be tested for animal and human toxicity.

Roots are a difficult infection court to protect, since the susceptible tips are constantly growing and moving through space encountering new inoculum. One strategy is to treat the entire rooting medium in the greenhouse or nursery. Another approach is to use a fungus that can colonize the root system from a seed or furrow treatment and protect the expanding root surface. This characteristic, called rhizosphere competence, has been demonstrated in some strains of *Trichoderma* spp.

Foliar applications of fungi are the least common, although this is the most common method of fungicide application. One example is *Pseudozyma* (= *Sporothrix*) *flocculosa*, a yeast-like fungus which is being developed for control of powdery mildews on greenhouse roses and cucumbers in Canada. *Trichoderma* and *Gliocladium* spp. can be applied to foliage and flowers and can prevent infection by necrotrophic fungi such as *B. cinerea* and *S. sclerotiorum*.

FUTURE OF BIOLOGICAL CONTROL BY FUNGI

In conclusion, the full potential of controlling plant diseases with fungi still has not been realized. Only a small number of products are on the market, but this is a vast improvement compared to only five years ago. There are still many economic constraints in terms of the cost of development and registration of products and the low cost production of organisms in liquid fermentation or solid on substrates.[5,9] Like chemicals, the risks of fungal BCAs need to be addressed, including the displacement of nontarget microbes, allergenicity to humans and other animals, and toxigenicity and pathogenicity to nontarget organisms.[10] However, there are no existing chemical controls for many diseases, because of deregistration of pesticides, pathogen resistance to pesticides, and environmental concerns. These diseases may be the niches for fungal biocontrol agents. It is unlikely that biological control will succeed alone, but it needs to be integrated with other disease management strategies, including cultural control and genetic disease resistance.

REFERENCES

1. Cook, R.J.; Baker, K.F. *The Nature and Practice of Biological Control of Plant Pathogens*; American Phytopathological Society Press: St. Paul, MN, 1983; 539.
2. *Principles and Practice of Managing Soilborne Plant Pathogens*; Hall, R., Ed.; American Phytopathological Society Press: St. Paul, MN, 1996; 442.
3. *Fungi in Biological Control Systems*; Burge, M.N., Ed.; Manchester University Press: Manchester, 1988; 269.
4. *Trichoderma and Gliocladium*; Harman, G.E., Kubicek, C.R., Eds.; Taylor and Francis: London, 1998; 1 and 2, 278–393.
5. *Pest Management: Biologically Based Technologies*; Vaughn, J.L., Lumsden, R.D., Eds.; American Chemical Society: Washington, DC, 1993; 435.
6. *Biological Control of Plant Diseases. Progress and Challenges for the Future*; Papavizas, G.C., Cook, R.J., Tjamos, E.C., Eds.; Plenum Press: New York, 1992; 462.
7. Whipps, J.M. Biological control of soil-borne plant pathogens. Advances in Botanical Res. **1997**, *26*, 1–134.
8. *Plant-Microbe Interactions and Biological Control*; Kuykendallm, L.D., Boland, G.L., Eds.; Marcel Dekker, Inc.: New York, 1998; 442.
9. *Integrated Pest and Disease Management in Greenhouse Crops*; Lodovica Gullino, M., van Lenteren, J.C., Elad, Y., Albajes, R., Eds.; Kluwer Academic Publishers: Dordrecht, the Netherlands, 1999; 545.
10. *Biological Control: Benefits and Risks*; Hokkanen, H.M.T., Lynch, J.M., Eds.; Cambridge University Press: Cambridge, 1995; 304.

Plant Pathogens (Viruses): Biological Control

Hei-Ti Hsu
Floral and Nursery Plants Research, Agricultural Research Service (USDA-ARS), U.S. Department of Agriculture, Beltsville, Maryland, U.S.A.

Abstract
Prevention of virus and viroid infections in plants is based on biological means rather than chemical measures. In principle, there are no chemicals available for controlling plant diseases caused by viruses and viroids. The most feasible approaches for combating viruses and viroids are the elimination of source inoculum, prevention of secondary spread, cross protection, and use of crops bearing resistance traits.

ECONOMIC LOSS

Damage to crop plants due to virus and viroid infections is difficult to assess. The actual figures for global crop loss are not available. Plant disease losses are estimated at $60 billion annually. Losses due to virus and viroids have been considered second to those caused by fungi. Unlike diseases caused by fungi, bacteria, and nematodes, where control measures using chemical, biological, and integrated pest management approaches have been effective, diseases caused by viruses or viroids are far more difficult to manage.

Economic crop loss resulting from virus and viroid disease is due to the reduced growth and vigor of infected plants which, in turn, causes a reduction in yield. In some instances, a virus infection may kill a plant. Apart from yield reduction, the quality and market value of commercial end products may be affected. There are also costs of attempting to maintain crop health such as vector control, production of pathogen-free propagation materials, and quarantine and eradication programs. In addition, resources are being diverted to research, extension, and education as well as toward breeding for resistance to virus or viroid infection.

WORLD IMPACT

No single country is exempt from crop losses. Production of food, fiber, and horticultural crops are seriously affected worldwide by virus or viroid infection of plants.[1] This is even more so in developing countries that depend on one or a few major crops; for example, *Cassava mosaic virus* in cassava plants in Kenya, *Citrus tristeza virus* in citrus trees in Africa and South America, and *Cacao swollen shoot virus* in cacao trees in Ghana. Recently, *Papaya ringspot virus* (PRV) infection has affected every region where papaya plants are grown. The virus induces a lethal disease in papaya. The widespread aphid-transmitted PRV has changed the way papayas are grown in many parts of the world. Normally, papayas are produced annually for a number of years over the life of the papaya plant. For proper management of the disease due to PRV infection, papaya has now become an annual crop in which healthy seedlings are planted each year. Even so, productivity is still below the average yield obtained before PRV became a problem. Viroids infect a limited number of crops when compared with viruses. However, they can cause severe problems in specific crops, for instance, cadang-cadang disease of coconuts, potato spindle tuber disease, and chrysanthemum stunt disease.

CONTROL MEASURES

No direct chemical control means are available to combat virus infections in plants. Control of viral diseases is achieved primarily by sanitary practices that involve reducing sources of inoculum from outside, preventing spreading within the crop, and limiting the population of insects, mites, nematodes, and fungi that may serve as vectors for many plant viruses.[1] Virus disease testing programs are now common in many parts of the world where the economic importance of growing virus-free plants is recognized. Although seeds and seedlings certified as virus-free are more expensive than those that have not been tested for certain viruses, testing provides assurance of virus-free production materials. Early detection of virus in a field and removal of the infected plants minimizes spread of the virus.

Plants may be protected from development of severe disease symptoms by first introducing a mild strain of virus into a healthy plant. A plant systemically infected with a mild strain of virus is protected from infection by a severe strain of the same virus. This phenomenon in called "cross protection" and has been observed for many plant viruses.[2] It is also observed to occur between viroids or plant virus satellites. In practice, cross protection is of great interest since it has been utilized to protect plants against severe virus strains (*Citrus tristeza virus*, *Papaya ringspot* virus, *Zucchini yellow mosaic virus, Tomato mosaic virus*, etc.), in the field.

Another approach toward controlling plant virus diseases is to develop resistant or tolerant plants.[3] Historically, long-term manipulation of crop plants through breeding has produced many valuable commercial varieties resistant to plant viruses. Breeding plants resistant to vectors may also offer control of the virus they transmit. Conventional breeding of crossing and back crossing commercial varieties with plants bearing virus resistance traits takes years to develop. In order for a new variety to be commercially acceptable, undesirable traits from the resistant parent breeding line must be selected out. The process is labor intensive and time consuming. Advances in science have allowed new technology to precisely manipulate resistance genes at the molecular level.[4] Biotechnology represents the fastest growing area of biological research. The application of biotechnology in breeding for resistance to virus infection is a major area of research. Successful control of viral disease through resistance breeding will undoubtedly reduce the use of synthetic pesticides for vector control.[5]

Introducing virus resistance and vector resistance into a cultivar by gene transfer technology (genetic engineering) has been successful in combating plant viruses.[6] The technology has several major advantages over conventional cross breeding. It is a relatively fast procedure. Desirable genes can be introduced without disturbing the balanced genome of target plants. Furthermore, there is no restriction on the source of the transgenes allowing the use of genes from other plant species or even from outside the plant kingdom (Table 1).[1,7]

Several approaches for producing transgenic virus-resistant plants have been explored. Among these, plants expressing virus coat protein genes, parts of other viral genes, or virus satellite ribonucleic acids (RNAs) have been shown to offer the best control.[2,8,9] Plants expressing antisense viral RNAs, ribozymes, pathogen-related proteins, or virus-specific antibody genes may also confer resistance to virus infection. Control of virus vectors by introducing insect toxins such as trypsin inhibitor, lectin, and *Bacillus thuringiensis* (Bt) toxin genes into plants would undoubtedly contribute toward achieving the goal of controlling plant virus diseases.

PROSPECTS

Use of resistant cultivars is considered the best approach to combat virus infection in plants. Biotechnology, no doubt, will play a significant role in the economic growth of many countries. Molecular breeding, however, will not replace but complement the efforts of conventional cross breeding. Much attention has been given to engineering resistance to plant viruses. Recently, genetic engineering of crop plants has been closely scrutinized and criticized due to increasing public concerns regarding human health and environmental impact. Careful assessment of the benefits and potential risks involving the release of genetically modified plants into the environment and their consumption is necessary before these crops become widely accepted by the public.[10,11]

Table 1 Genes that contribute or may contribute toward control of virus diseases in plants.

Virus-derived gene sequences
Coat proteins
Replicases
Movement proteins
Polyprotein proteases
Sense RNAs
Antisense RNAs
Plant host-derived transgenes
Pathogen-related proteins
Anti-viral proteins
Proteinase inhibitors
Natural resistance genes
Lectins
Other transgenes and sequences
Satellite RNAs
Virus-specific antibodies
Interferon-induced mammalian oligoadenylate synthetase
Insect toxins
Anti-viral ribozymes (catalytic RNA)

Source: Khetarpal et al.[1]

REFERENCES

1. Khetarpal, R.K., Koganezawa, H., Hadidi, A., Eds. *Plant Virus Disease Control*; APS Press: St. Paul, MN, 1998, 1–684.
2. Beachy, R.N. Coat-protein-mediated resistance to tobacco mosaic virus: discovery mechanisms and exploitation. Phil. Trans. R. Soc. Lond. B. **1999**, *354*, 659–664.
3. Salomon, R. The evolutionary advantage of breeding for tolerance over resistance against viral plant disease. Israel J. Plant Sci. **1999**, *47*, 135–139.
4. Kawchuk, L.M.; Prufer, D. Molecular strategies for engineering resistance to potato viruses. Can. J. Plant Pathol. **1999**, *21*, 231–247.
5. Barker, I.; Henry, C.M.; Thomas, M.R.; Stratford, R. Potential Benefits of the Transgenic Control of Plant Viruses in the United Kingdom. In *Plant Virology Protocols: From Virus Isolation to Transgenic Resistance*; Foster, G.D., Taylor, S.C., Eds.; Humana Press, Inc.: Totowa, NJ, 1998; 81, 557–566.
6. Dempsey, D.A.; Silva, H.; Klessig, D.F. Engineering disease and pest resistance in plants. Trends Microbiol. **1998**, *6*, 54–61.

7. Gutierrez-Campos, R.; Torres-Acosta, J.A.; Saucedo-Arias, L.J.; Gomez-Lim, M.A. The use of cysteine proteinase inhibitors to engineer resistance against potyviruses in transgenic tobacco plants. Nat. Biotechnol. **1999**, *17*, 1223–1226.
8. Maiti, I.B.; Von Lanken, C.; Hong, Y.; Dey, N.; Hunt, A.G. Expression of multiple virus-derived resistance determinants in transgenic plants does not lead to additive resistance properties. J. Plant Biochem. Biotech. **1999**, *8*, 67–73.
9. Prins, M.; Goldbach, R. RNA-mediated virus resistance in transgenic plants. Arch. Virol. **1996**, *141*, 2259–2276.
10. Hammond, J.; Lecoq, H.; Raccah, B. Epidemiological risks from mixed virus infections and transgenic plants expressing viral genes. Adv. Virus Res. **1999**, *54*, 189–314.
11. Kaniewski, W.K.; Thomas, P.E. Field testing for virus resistance and agronomic performance in transgenic plants. Mol. Biotechnol. **1999**, *12*, 101–115.

Pollutants: Organic and Inorganic

A. Paul Schwab
Department of Agronomy, Purdue University, West Lafayette, Indiana, U.S.A.

Abstract
Organic and inorganic soil pollutants are reviewed. Classes of contaminants are presented along with discussions of chemical properties and typical concentrations found in soils. Comprehensive Environmental Response, Compensation, and Liability Act priority rankings are given along with the United States Environmental Protection Agency's soil screening levels. Relative toxicities of important pollutants are discussed, and pathways of exposure are detailed for five of the most important metals. Remediation measures are examined including bioremediation, phytoremediation, and engineering alternatives.

INTRODUCTION

A soil pollutant can be broadly defined as any chemical or other substance which either is not normally found in soil or is present at high enough concentrations to be harmful to any living organisms.[1] This very general definition could be applied to human-derived (anthropogenic) as well as naturally occurring constituents. This entry will focus primarily on those pollutants that have anthropogenic origins and will address carbon-based (organic) and inorganic chemicals. Organic contaminants can include pesticides, solvents, preservatives, petroleum products, hormones, and antibiotics. Inorganic pollutants are comprised of heavy metals (e.g., lead and mercury), nonmetals (selenium), metalloids (arsenic and antimony), radionuclides, and simple soluble salts (sodium chloride). Whether the source of the contaminants is natural or anthropogenic, understanding their chemistry, toxicity, and bioavailability is crucial to responding to soil pollution.

ORGANIC POLLUTANTS

Thousands of organic chemicals exist in nature, and many are acutely toxic as well as carcinogenic, mutagenic, or teratogenic. When discussing toxic organic chemicals in soils, the focus normally is on those pollutants resulting from human activities because their release has the potential to be controlled. Typical synthetic organic pollutants include fuels, lubricants, herbicides, fungicides, insecticides, solvents, and propellants.

Soil Contamination by Organic Chemicals

Organic chemicals can find their way into the soil accidentally through spills, leaking storage tanks, and unintentional discharges. However, not all soil pollution is accidental. More than one-half million metric tons of pesticides are used annually, the majority of which are used on agricultural fields.[1] Approximately 20,000 metric tons of pesticides are used for non-agricultural applications, including railroad right of way weed control, turf, and horticulture. Although many pesticides do not persist in soils, others are highly persistent and have been studied extensively because of their negative impacts. Chlordane (termite control), dichlorodiphenyltrichloroethane (DDT) (mosquitoes), and atrazine (weeds) are excellent examples of organic pesticides that have been determined to have human health effects and severe ecological impacts; chlordane and DDT have been banned in the United States, and the banning of atrazine has been debated.

Thousands of organic chemicals are in use today, and listing all the specific compounds that contaminate soils is beyond the scope of this entry. However, contaminants can be sorted into categories of chemicals that are used frequently and are commonly found in soils (Table 1). Insecticides, herbicides, fungicides, and nematicides may be the most frequently encountered, and some pesticides are quite toxic. Atrazine has been found to be somewhat persistent and is mobile such that atrazine applied to soil can migrate to groundwater and surface water. The insecticide DDT and its metabolites continue to be found in ecosystems despite being banned for decades. Methyl bromide (1,2-dibromomethane) is used as a nematicide, but it is a gas that is known to be ozone depleting. Trichloroethylene (TCE) is a useful solvent that, when disposed on soil, moves rapidly downward and contaminates groundwater. Petroleum-based fuels such as diesel and gasoline are problematic when spilled or originating from leaking underground storage tanks. Many of the other classes of compounds listed are of ecological concern when found in soil, again due to their potential toxicity to a wide range of organisms.

On December 11, 1980, the United States Congress enacted the Comprehensive Environmental Response, Compensation, and Liability Act (CERCLA), commonly known

Table 1 Some classes of organic chemicals found in soils.

Contaminant class	Example(s)	Sources, impacts
Insecticides	DDT, chlordane, diazinon	Used for insect control; DDT and chlordane have serious health effects; diazinon has been found in water supplies.
Herbicides	Atrazine, 2,4-D, 2,4,5-T	Atrazine used on corn; 2,4-D widely used in lawns and agriculture; defoliant 2,4,5-T implicated in health effects in Agent Orange.
Fungicides	Benomyl, propiconazole, chlorothalonil	Agricultural fungicides that are directly applied to soil.
Nematicides	Methyl bromide (1,2-dibromomethane)	Used in fumigation of soil to remove nematodes; neurotoxin and ozone depleting.
Solvents	Trichloroethylene, trichloroethane	Widely used solvents and degreasers; chronic health effects not clear.
Fuels	Diesel, kerosene	From spills, leaking tanks; can lead to groundwater contamination. Some constituents of fuels are carcinogenic.
Polyaromatic hydrocarbons	Chrysene, benzo[a]pyrene	Components of petroleum, particularly after combustion; many of these are carcinogenic.
Polychlorinated biphenyls	Aroclors (1260, 1016, 1242); coplanar congeners (3,4,3',4'-tetrachlorobiphenyl, 3,4,5,3',4'-pentachlorobiphenyl)	Dielectric fluids in transformers, capacitors, and coolants. Can lead to skin conditions, ocular lesions, teratogenic effects in animals, endocrine disruption.
Explosives, propellants	TNT, RDX	Residuals from manufacture are the largest source of contamination; acute and chronic toxicities documented.

Source: Adapted from Schwarzenbach, Gschwend, and Imboden[2] and Evangelou.[3]

as Superfund, to provide wide-ranging federal authority to respond to contamination or threats of contamination by hazardous substances. Among other actions, CERCLA established a list of priority pollutants with known or suspected ecological or health impacts. The list is periodically updated, and the pollutants are prioritized. Thirty-two of these compounds are given in Table 2 along with the CERCLA priority ranking[4] and soil screening levels (SSLs).[5] This list includes compounds that have been banned in the United States (e.g., DDT) as well as those that are part of our everyday lives (e.g., BTEX—benzene, toluene, ethylbenzene, xylenes—found in gasoline). Most of the SSLs are in the range of 1 to 50 mg/kg. Some compounds have significantly higher SSL values (5000 mg/kg for toluene), indicating that these compounds are far less toxic. Others have very low SSL values (5.0×10^{-4} mg/kg for benzidine), suggesting that these compounds are a threat at very low concentrations. Soil concentrations of these compounds required to prevent threats to groundwater are approximately 100 times lower than the residential SSLs.

The legacy of high use of persistent, potentially toxic compounds became clear in a study published in 2010 in which the presence of organochlorine pesticides, polychlorinated biphenyls, and perfluorinated compounds were determined in food samples purchased in supermarkets in the United States.[6] The tracked compounds were detected in nearly all the food samples: DDT metabolite *p,p'*-dichlorodiphenyldichloroethylene was found in milk products; polychlorinated biphenyls (PCBs) were found in fish; and perfluorinated compounds were found in over half of all samples. Results such as these add to already enhanced sensitivities concerning organic contaminants in soils.

Potential Impacts of Organic Contaminants

After an organism is exposed to an organic pollutant, a number of antagonistic effects are possible if concentrations are high enough. The most dramatic impact is acute toxicity, in which symptoms are quickly apparent and readily identified. Consuming large quantities of the pure contaminant is not necessarily a requirement for acute toxicity. A case of acute parathion poisoning was reported when a child consumed contaminated soil.[7] Although reports of such cases are rare, the possibility for acute poisoning through soil consumption by children with pica is realistic for other compounds, such as phenol.[8] The estimated lethal dose for phenol is estimated to range from 10 to 50 mg/kg body mass, but the ingestion of only 5 g of soil contaminated with an SSL of 47,000 mg/kg would result in a dose of 18 mg/kg.

At lower concentrations, the impacts of organic contaminants become less obvious and take longer to be expressed. For many carcinogens, for example, decades are required to develop cancerous tumors. Pathway of exposure, concentrations, and duration of the exposure all dictate the resulting health effects and are essential components of risk analysis. Exposure to contaminated soil can result in chronic toxicity if the soil repeatedly comes in contact with the skin, if particulates are frequently inhaled, or if volatile compounds migrate into closed living spaces.

Table 2 Organic compounds on the CERCLA priorities list and their ranking, health impacts, and SSLs.

| | | Soil screening level | |
| | | Residential soil | Protect groundwater |
Constituent	CERCLA rank	---------------- mg/kg ----------------	
Vinyl chloride	4	6.0×10^{-2}	5.6×10^{-6}
PCBs[a]	5	1.4×10^{-1}	5.2×10^{-3}
Benzene	6	1.1×10^{-0}	2.1×10^{-4}
Polycyclic aromatic hydrocarbons	8	1.5×10^{-2}	2.7×10^{-4}
Chloroform	11	2.9×10^{-1}	5.3×10^{-5}
DDT, p,p'	12	1.7×10^{0}	6.7×10^{-2}
Trichloroethylene	16	2.8×10^{0}	7.2×10^{-4}
Dieldrin	17	3.0×10^{-2}	1.7×10^{-4}
Chlordane	20	1.6×10^{0}	1.3×10^{-2}
DDE, p,p'	21	1.4×10^{0}	4.7×10^{-2}
Hexachlorobutadiene	22	6.2×10^{0}	1.7×10^{-3}
Aldrin	24	2.9×10^{-2}	6.5×10^{-4}
DDD, p,p'	25	1.4×10^{0}	6.6×10^{-2}
Benzidine	26	5.0×10^{-4}	2.4×10^{-7}
Toxaphene	31	4.4×10^{-1}	9.4×10^{-3}
Hexachlorocyclohexane, γ (lindane)	32	5.2×10^{-1}	3.6×10^{-4}
Tetrachlorethylene	33	5.5×10^{-1}	4.9×10^{-5}
Heptachlor	34	1.1×10^{-1}	1.2×10^{-3}
1,2-Dibromomethane	35	2.5×10^{1}	2.5×10^{-3}
Hexachlorocyclohexane, β	36	2.7×10^{-1}	2.2×10^{-4}
Acrolein	37	1.5×10^{-1}	8.4×10^{-6}
Disulfoton	38	2.4×10^{0}	2.7×10^{-3}
3,3'-Dichlorobenzidine	40	1.1×10^{0}	9.4×10^{-4}
Endrin	41	1.8×10^{1}	4.4×10^{-1}
Pentachlorophenol	45	3.0×10^{0}	5.7×10^{-3}
Heptachlor epoxide	46	5.3×10^{-2}	1.5×10^{-4}
Carbon tetrachloride	47	6.1×10^{-1}	1.7×10^{-4}
Diazinon	56	4.3×10^{1}	1.6×10^{-1}
Xylenes	58	6.3×10^{2}	2.0×10^{-1}
Toluene	71	5.0×10^{3}	1.6×10^{0}
Ethylbenzene	99	5.4×10^{0}	1.7×10^{-3}

All screening levels are for residential soils unless stated otherwise. Radionuclides were excluded from this list.
[a]This is a general class of contaminants, and individual members have unique SSLs. The most restrictive value was chosen for this table.

Most SSLs have been developed based on risk associated with chronic exposure.

Bioremediation of Organic Contaminants

Although contamination of soil by organic compounds is an important environmental problem, many of these pollutants can be removed from the soil through bioremediation. Soil microorganisms have a remarkable capacity to degrade organic contaminants. Degradation can be direct, using the organics as a source of energy and carbon, or indirect, in which the compounds are cometabolized by organisms seeking similar compounds. End products can be $CO_2(g)$ after total mineralization; alteration and humification; or incorporation into the microbial biomass. Polyaromatic hydrocarbons (PAHs) are readily degraded in the soil, as are many pesticides and components of BTEX. However, highly chlorinated compounds such as PCBs, some solvents, and the explosive cyclotrimethylenetrinitramine (RDX) are far more difficult to degrade, and some of the initial degradation products are as toxic as or more toxic than the parent compound. Reviews have been published for the bioremediation of pesticides,[9] a wide range of organic contaminants by fungi,[10] bacterial degradation of aromatic compounds,[11] PAHs,[12] aliphatic hydrocarbons,[13] explosives,[14] petroleum hydrocarbons by mycorrhizae,[15] earthworm-assisted bioremediation,[16] and composting as a general bioremediation approach.[17] Bioremediation of specific compounds also has been reviewed: pyridine, indole, and quinoline[18]; dieldrin and endrin[19]; and catechols.[20] This subject is treated in depth in another section of this entry.[21]

Phytoremediation is another approach taken to enhance the dissipation of organic contaminants in soil. The mechanism of removal of the contaminant from the soil depends upon the properties of the organic compound, the soil, and the chemistry of the roots. In some instances, organic contaminants are assimilated by the plants and either degraded or volatilized as part of the transpiration stream. Trichloroethylene has been observed to be effectively remediated in the root zone of poplar trees, but the mechanism has been the subject of debate. In some instances, uptake of TCE and eventual volatilization have been observed,[22] but other studies detected no volatilization of TCE and complete degradation in the soil.[23] In nearly all cases for PAHs and PCBs, plant uptake is negligible, and phytoremediation of these compounds is accomplished by microbial degradation in the rhizosphere.

Uptake of organic contaminants during phytoremediation is an important consideration for many reasons. From an ecological standpoint, accumulation of contaminants in the aboveground portions of plants is undesirable because of the potential for introduction into the food chain or dispersal of the contaminants. From the remediation perspective, uptake may be desirable to help remove the compounds from the soil and allow degradation within plant tissues. Uptake of volatile compounds followed by release to the atmosphere would be prohibited in many regulatory environments. Therefore, efforts have been taken to predict the transfer from soil to roots to the transpiration stream of higher plants. The most useful parameter in this analysis is the transpiration stream concentration factor (TSCF):

$$\text{TSCF} = \frac{\text{Concentration in xylem sap}}{\text{Concentration in external solution}} \quad (1)$$

Concentrations in the xylem are difficult to quantify and are estimated as the amount of a compound assimilated over a given period of time that has been corrected for degradation with the plant. If the compound of interest is neither actively accumulated nor excluded from the plant (i.e., passive uptake), then TSCF = 1.0. Non-ionized organic compounds are of particular interest in phytoremediation because target contaminants are in this class, including many pesticides, PAHs, PCBs, etc. Trends in TSCF as a function of the octanol–water coefficient (K_{OW} or log K_{OW}) have been investigated.[24,25] Compounds that are soluble in water have small or even negative values of log K_{OW}; hydrophobic compounds have high values of log K_{OW}. Briggs et al.[24] investigated substituted phenylureas and o-methylcarbamoyloximes, and Burken and Schnoor[25] investigated a suite of compounds including RDX, phenol, benzene, atrazine, TCE, pentachlorophenol (PCP), and others. The combined data from the two studies are shown in Fig. 1 and follow the relationship

$$\text{TSCF} = 0.76 \exp(-(\log K_{OW} - 2.45)^2 / 4.38) \quad (2)$$

This relationship predicts that compounds with a log K_{OW} of 2.45 will have the maximum movement into the transpiration stream. The shape of the curve is viewed as reflecting the balance of the various tendencies for the compounds to desorb from soil surfaces and pass through the hydrophobic cellular membranes. Compounds with a low log K_{OW} will desorb readily from the soil but will not pass through the hydrophobic plasmalemma. Organic contaminants with high log K_{OW} values are predicted to have the capability to penetrate the root membranes, but they either will not desorb from soil components or will irreversibly adsorb to the lipophilic root membranes.[26]

INORGANIC POLLUTANTS

Inorganic pollutants are those contaminant compounds that do not contain carbon (with a few exceptions, including cyanide and carbonate). These contaminants include acutely toxic heavy metals, metalloids, oxyanions, and radioactive elements and the far-less-toxic soluble salts. Although all elements are naturally occurring, thousands of inorganic compounds are formed only through human activities. Some are highly toxic in trace quantities, but all have a threshold concentration above which they can be harmful.

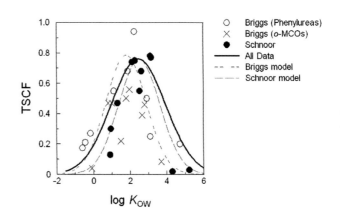

Fig. 1 Transpiration stream concentration factor (TSCF) as a function of log K_{OW} (octanol–water partition coefficient). Data come from Briggs et al.[24] and Burken and Schnoor.[25] The models take the following form: TSCF = $a \exp(-(\log K_{OW} - b)^2 c)$, in which a, b, and c are variables.

Classes of Inorganic Pollutants and Concentration Ranges

As with organic pollutants, thousands of inorganic contaminants exist and are found over a wide range of concentrations. It is convenient to group them into classes (Table 3) consistent with the periodic table. Metals include Pb, Zn, Ni, and Cd, and their typical soil concentrations vary over a wide range, as do the concentrations considered anomalous. Most metals in soils occur as the cations in solution and in the solid phase. Nearly all metals can have toxic effects on humans, although some metals (such as Zn) require high concentrations for toxicity. Metalloids, including As and Sb, generally occur in soil solution and solid phases as oxyanions (such as AsO_4^{3-}), resulting in significantly different chemical behavior than the metals. Metalloids can be highly toxic. Soluble salts are typical inorganic species that, in normal concentrations, generally do not constitute a health or environmental threat. Typical salts are sulfates, chlorides, and nitrates of calcium, magnesium, potassium, and sodium. These components are always present in soil and do not pose an environmental problem until concentrations become excessive (Table 3).

Similar to organic contaminants, inorganic compounds have been listed and prioritized by CERCLA,[4] and most have SSLs established by the U.S. Environmental Protection Agency.[5] Radionuclides are not included in these listings because radioactive elements are considered under different legislative initiatives. Table 4 lists 27 inorganic species, their CERCLA priority ranking, and SSLs. The inorganic constituents have a prominent position in the top 10 on the CERCLA priority list, with arsenic being number one and lead, mercury, and cadmium also in the top ten. However, the priority ranking for the remaining inorganic contaminants is generally lower than those for the organics (Table 2). The residential SSLs for the inorganics tend to be far higher than for the organics and typically have a range of 50 to 3000 mg/kg. Arsenic (0.39), Cr(VI) (0.29), and hydrazine (0.21) are the only inorganics with SSL values less than 1 mg/kg. SSLs for protection of groundwater are roughly 100 times lower than residential SSLs.

Potential Health Impacts of Inorganic Contaminants in Soils

Many of the metals, metalloids, and oxyanions have negative effects on human health, and these impacts have been studied for decades. The most widely distributed and best-known toxic metal in soils is Pb. If Pb-contaminated soil is consumed, the risk increases for lead-associated nervous system, brain, and blood disorders. The impacts of Pb in soil from car exhaust[27] and smelter sites[28] are among the many studies examining this well-known metal. The SSL for Pb is 400 mg/kg.

Cadmium exposure for humans occurs mostly through the consumption of food grown in Cd-contaminated soils. Health impacts of Cd include kidney function, bone strength, and central nervous system disorder. Cadmium contamination in soils can be the result of mining, refining, smelting, battery production, battery disposal, and other industrial operations. Generally, cadmium uptake by plants increases with decreasing pH and increasing total Cd in the soil.[29] The mechanism of cadmium uptake by plants is similar to that of zinc and calcium.

Chromium in soils is viewed to be problematic only if the Cr is in the Cr(VI) (chromate) oxidation state. Reduced chromium, Cr(III), is less mobile and far less toxic. High chromate can result in renal failure, DNA damage, and cancer. Because Cr(VI) is present in the soil as the anion CrO_4^{2-}, it is not strongly held by the soil, tends to be mobile, and is readily assimilated by plants. Fugitive dust from contaminated soil also can lead to increased

Table 3 Inorganic contaminants in soil, their typical ranges found in uncontaminated soils, and those concentrations considered to be unusual.

Contaminant	Sources, impacts	Typical range in soil	Extreme concentrations
Metals Pb, Zn, Cd, Cu, Ni, Ba, Sr, Mn, Cr	Mining, smelting, electroplating	0.1 to 1000 mg/kg	25 to >10,000 mg/kg (metal dependent)
Metalloids As, B, Ge, Sb	Industrial activities, smelting	<0.1 to 100 mg/kg	>50 mg/kg
Nonmetals Se, I, P, S	Manufacturing, processing	<0.1 to 1000 mg/kg	up to 2000 mg/kg (element dependent)
Soluble salts Halides, alkalis, alkaline earths, sulfates, nitrates	Mining, manufacturing, petroleum extraction, natural sources	<500 mg/kg	>1000 mg/kg

Source: Adapted from Evangelou[3] and Schecter et al.[6]

Table 4 Inorganic compounds on the CERCLA priorities list and their ranking, health impacts, and SSLs.

| | | Soil screening level | |
| | | Residential soil | Protect groundwater |
Constituent	CERCLA rank	---------------- mg/kg ----------------	
Antimony	219	4.0×10^1	6.6×10^{-1}
Arsenic	1	3.9×10^{-1}	1.3×10^{-3}
Barium	109	1.5×10^4	3.0×10^2
Beryllium	42	1.6×10^2	2.5×10^1
Cadmium	8	7.0×10^1	1.4×10^0
Chlorine	91	7.5×10^2	1.6×10^0
Chromium(VI)	18	2.9×10^{-1}	8.3×10^{-4}
Cobalt	49	2.3×10^1	4.9×10^{-1}
Copper	128	3.1×10^3	5.1×10^1
Cyanide	28	1.6×10^3	7.4×10^0
Fluoride	211	3.1×10^3	NL[a]
Hydrazine	84	2.1×10^{-1}	NL[a]
Iodine	NL[a]	7.8×10^2	NL[a]
Lead	2	4.0×10^2	NL[a]
Manganese	117	1.8×10^3	5.7×10^1
Mercury	3	2.3×10^1	5.7×10^{-1}
Molybdenum	NL[b]	3.9×10^2	3.7×10^0
Nickel	53	1.5×10^3	4.8×10^1
Nitrate	216	1.3×10^5	NL[a]
Nitrite	212	7.8×10^3	NL[a]
Perchlorates	NL[b]	5.5×10^1	NL[a]
Phosphorus (white)	19	1.6×10^0	2.7×10^{-3}
Selenium	147	2.9×10^2	9.5×10^{-1}
Silver	214	3.9×10^2	1.6×10^0
Uranium	98	2.3×10^2	4.9×10^1
Vanadium	198	5.5×10^1	2.6×10^1
Zinc	74	2.3×10^4	6.8×10^2

All screening levels are for residential soils unless stated otherwise. Radionuclides were excluded from this list.
[a]Not listed.
[b]Not listed because radionuclides are treated separately.

incidence of cancer.[30] If Cr(VI) can be reduced to Cr(III) in the soil, the health and environmental threats are greatly diminished.[31,32] Labile organic matter, microorganisms, and reduced inorganic species such as Fe^{2+} can readily convert Cr(III) to Cr(VI). Although surface soils in equilibrium with atmospheric $O_2(g)$ are predicted to maintain chromium as Cr(VI), even small amounts of organic matter in moist, shallow subsoils can reduce Cr(VI) to Cr(III).

Mercury contamination in humans can result in skin rashes, hypertension, rapid pulse, kidney dysfunction, and memory failure. The most common pathway of exposure to Hg poisoning is through fish and meat, but high concentrations in soils can result in increased exposure through plant uptake and inhalation. The SSL for Hg is 23 mg/kg for residential soils and 0.57 mg/kg to protect groundwater (Table 4). Mercury contamination in soil is fairly uncommon, but in some mining areas, concentrations in the soil as high as 2700 mg/kg have been observed and are associated with plant concentrations as high as 1100 mg/kg. A threshold concentration in the soil is required before plant concentrations increase significantly above background.[33]

In most natural systems, arsenic occurs as one of two redox species: arsenate, AsO_4^{3-}, or arsenite, AsO_3^{3-}. Although both forms appear to induce toxicity in humans, the arsenite is more readily absorbed by human tissues.[34] The residential soil SSL for arsenic is 0.39 mg/kg and 0.013 mg/kg to prevent impacts on groundwater. Most of the attention to arsenic poisoning in recent years has been paid to As in drinking water, particularly as a result of extreme concentrations of total As in wells in Bangladesh. However, plant uptake of As and the exposure of foods to high-As irrigation water also can result in arsenic intake in excess of the World Health Organization's recommended limits.[35]

Remediation of Inorganic Contaminants

The best approach to keeping soil free from contamination is prevention. When this approach fails, removing the contaminants may be necessary, and many options are available.[36] With few exceptions, biological approaches to decontamination of soils with high concentrations of pollutant metals are limited to phytoremediation. Some inorganic contaminants, such as nitrate, chromate, perchlorate, and other redox-active species in which one redox state is less of a threat than another, can be bioremediated through processes in which the chemical state of the contaminant is altered. Nitrate can be denitrified through the production of gaseous nitrogen compounds.[37] Perchlorate can be reduced to benign chloride,[38] and Cr(VI) can be reduced to Cr(III).[39,40] For metals or metalloids in which only one chemical state is generally found in soils, a biological approach not involving phytoremediation is rare.

During the past two decades, phytoremediation of metals in soils has been vigorously explored. For some metals, phytoremediation has been highly successful and has found commercial applications: hyperaccumulation of Zn, Cd, and Ni by *Thlaspi caerulescens*;[41] brake ferns for removal of As from soil;[42] and phytovolatilization of mercury.[43,44] One of the most common metal contaminants in soils in Pb, but effective means of phytoremediation of Pb-contaminated soils have been elusive.[45]

Phytoremediation is not the only remediation alternative for metal-contaminated soils. Traditional and engineering approaches have been used with reasonable success. Although basic excavation methods are effective at cleaning up sites, the soil remains contaminated and is merely transported to a containment facility. Excavation is the least expensive means of remediation but damages the site and does not improve the contaminated soil resource. Soils may be incinerated to burn off organic pollutants and to volatilize metals. Other means of containment include capping the site, encasing the soil in cement, or in situ vitrification in which the soil is melted followed by solidification of the soil into a single mass of glass. These methods are effective in isolating the soil, but considerable waste is generated, the contaminants remain in place, and the soil is usually lost as a viable resource.[46]

Less destructive approaches often are preferred because the soil is not seriously altered or destroyed, and the techniques are often less expensive. Such methods are typified by soil washing,[47] electrokinetics and electromigration processes,[39] and chemical stabilization.[39]

Pathways of Exposure

Once in the soil, organic and inorganic pollutants can follow many pathways to potentially impacted organisms.[48] For all contaminants, direct consumption of contaminated soil can be an important mechanism of exposure for soil-borne organisms, grazing mammals that consume the soil, and humans. For some pollutants, direct soil consumption is the only means of exposure if the contaminants have low mobility and low bioavailability; Pb and five- or six-ring PAHs are examples.

Eight pathways exist by which contaminants in soil can reach humans.[48] The pathways include direct soil consumption; uptake of contaminants from soil into food crops and subsequent consumption by humans; consumption of animals that have been contaminated by tainted soil; inhalation of metals in air or airborne dust; and drinking contaminated groundwater or surface water. Half of the mechanisms involve plant uptake and subsequent consumption of the plants. Most anionic inorganic species are readily assimilated by plants, as are many cationic species. However, some inorganic species form highly insoluble solid phases or are otherwise non-bioavailable, and they are not translocated within the plant. For nonionic organic species, plant uptake is a function of water solubility and hydrophobicity/lipophilicity. Compounds with intermediate lipophilicity are most suitable for plant uptake. Once the pollutants are in the plant, organisms that consume the vegetation may be exposed to the pollutant, and some of the contaminants are passed along the food chain.

Other pathways of exposure include the contaminant moving from the soil to the water or air or becoming airborne on particulates. All of these mechanisms have been shown to contribute to negative health effects. Drinking contaminated water is one of the better-known pathways. If the drinking water originates from groundwater, then the pollutants must be soluble in water and percolate readily through the soil and into the groundwater. When surface water is used for drinking, the transport mechanism can be movement of dissolved contaminant in the water, or contaminants may be adsorbed onto the suspended sediment in the soil runoff. Virtually every soil pollutant can be transported by this mechanism, and only careful control of runoff water and sediments can avoid this problem.

Direct volatilization from the soil into the air is important for a limited number of contaminants, but when volatilization of a potentially toxic compound or element occurs, exposure is directly through the lungs. Mercury (Hg) is a well-known volatile metal, and direct inhalation of high concentrations of gaseous Hg or dimethyl mercury can be poisonous. Inhalation of airborne, contaminated particles is a more frequent occurrence but less insidious. After inhalation, contaminants must be released from the dust before they can be assimilated. Inhalation of contaminated particulates can be reduced through standard soil erosion control measures, such as establishing windbreaks and adequate plant cover on contaminated soils.

CONCLUSIONS

Hundreds of contaminants are found in soils, and many reach concentrations that require action. Regulated organic contaminants are usually anthropogenic in their origins and, as a result, their impact can be minimized by controlling their manufacture and use. Even after strict regulations, including outright banning of manufacture or use, detectable or even dangerous concentrations can be found for decades. Residual DDT and its degradation products are examples of highly persistent organic compounds that remain dangerous long after the termination of their use. Metals and other inorganic contaminants can be as problematic as persistent organics because they are not easily transformed to benign forms and they will persist in soils for extended periods. Although most metals do not have a pathway of exposure that could result in human toxicity, some metals (e.g., Se, Mo, Cd, Pb, As) do have such a pathway and must be monitored carefully.

REFERENCES

1. Pierzynski, G.M.; Sims, J.T.; Vance, G.F. Classification of pollutants. In *Soils and Environmental Quality*, 2nd Ed.; Lewis Publishers: Boca Raton, Florida, 1993; 47–50, 167–183, 185–210.
2. Schwarzenbach, R.P.; Gschwend, P.M.; Imboden, D.M. *Environmental Organic Chemistry*; Wiley-Interscience: New York, 1992; 8–41, 255–344.
3. Evangelou, V.P. Water quality. In *Environmental Soil and Water Chemistry*; Wiley Interscience: New York, 1998; 476–499.
4. Agency for Toxic Substances Registry. 2007 CERCLA Priority List of Hazardous Substances, available at http://www.agriculturedefensecoalition.org/sites/default/files/pdfs/4H_2007_CDC_ATSDR_CERCLA_Priority_List_of_Hazardous_Substances_2007.pdf, (accessed May 2012).
5. United States Environmental Protection Agency. Regional Screening Levels (RSL) Summary Table May 2010, available at http://www.epa.gov/reg3hwmd/risk/human/rb-concentration_table/Generic_Tables/pdf/master_sl_table_run_MAY2012.pdf, (accessed May 2012).
6. Schecter, A.; Colacino, J.; Haffner, D.; Patel, K.; Opel, M. Perfluorinated compounds, polychlorinated biphenyls, and organochlorine pesticide contamination in composite food samples from Dallas, Texas, USA. Environ. Health Perspect. **2010**, *118* (6), doi:10.1289/ehp.0901347.
7. Quinby, G.E.; Clappison, G.B. Parathion poisoning—A near-fatal pediatric case treated with 2-pyridine aldoxime methiodide (2-PAM). Arch. Environ. Health **1961**, *3*, 438–447.
8. Calabrese, E.J.; Stanek, E.J.; James, R.C.; Roberts, S.M. Soil ingestion—A concern for acute toxicity in children. J. Environ. Health **1999**, *61*, 18–23.
9. Castelo-Grande, T.; Augusto, P.A.; Monteiro, P.; Estevez, A.M.; Barbosa, D. Remediation of soils contaminated with pesticides: A review. Internat. J. Environ. Anal. Chem. **2010**, *90*, 438–467.
10. Pinedo-Rivello, C.; Aleu, J.; Collado, I.G. Pollutants biodegradation by fungi. Curr. Org. Chem. **2009**, *13*, 1194–1214.
11. Seo, J.-S.; Keum, Y.S.; Li, Q.X. Bacterial degradation of aromatic compounds. Int. J. Environ. Res. Public Health **2009**, *6*, 278–309.
12. Bamforth, S.M.; Singleton, I. Bioremediation of polycyclic aromatic hydrocarbons: Current knowledge and future directions. J. Chem. Technol. Biotechnol. **2005**, *80*, 723–736.
13. Stroud, J.L.; Paton, G.I.; Semple, K.T. Microbe-aliphatic hydrocarbon interactions in soil: Implications for biodegradation and bioremediation. J. Appl. Microbiol. **2007**, *102*, 1239–1253.
14. Lewis, T.A.; Newcombe, D.A.; Crawford, R.L. Bioremediation of soils contaminated with explosives. J. Environ. Manag. **2004**, *70*, 291–307.
15. Roberston, S.J.; McGill, W.B.; Massicotte, H.B.; Rutherford, P.M. Petroleum hydrocarbon contamination in boreal forest soils: A mycorrhizal ecosystems perspective. Biol. Rev. **2007**, *82*, 213–240.
16. Hickman, Z.A.; Reid, B.J. Earthworm assisted bioremediation of organic contaminants. Environ. Internat. **2008**, *34*, 1072–1081.
17. Semple, K.T.; Reid, B.J.; Fermor, T.R. Impact of composting strategies on the treatment of soils contaminated with organic pollutants. Environ. Pollut. **2001**, *112*, 269–283.
18. Fetzner, S. Bacterial degradation of pyridine, indole, quinoline, and their derivatives under different redox conditions. Appl. Microbiol. Biotechnol. **1998**, *49*, 237–250.
19. Bhatt, P.; Kumar, M.S.; Chakrabarti, T. Fate and degradation of POP-hexachlorocyclohexane. Crit. Rev. Environ. Sci. Technol. **2009**, *39*, 655–695.
20. Zeyaullah, M.; Abdelkafe, A.S.; Ben Zabye, W.; Ali, A. Biodegradation of catechols by micro-organisms—A short review. African J. Biotechnol. **2009**, *13*, 2916–2922.
21. Radosevich, M.; Rhine, E.D. Pollutants (Ch. 281). In *Encyclopedia of Soil Science*, 2005; 1338–1343.
22. Ma, X.; Burken, J.G. TCE diffusion to the atmosphere in phytoremediation applications. Environ. Sci. Technol. **2003**, *37*, 2534–2539.
23. Strand, S.E.; Dossett, M.; Harris, C.; Wang, X. Doty, S.L. Mass balance studies of volatile chlorinated hydrocarbon phytoremediation. Z. Naturforsch. **2005**, *60*, 325–330.
24. Briggs, G.G.; Bromilow, R.H.; Evans, A.A.. Relationship between lipophilicity and root uptake and translocation of nonionized chemicals by barley. Pestic. Sci. **1982**, *13*, 495–504.
25. Burken, J.G.; Schnoor, J.L. Predictive relationships for uptake of organic contaminants by hybrid poplar trees. Environ. Sci. Technol. **1998**, *32*, 3379–3385.
26. Bromilow, R.H.; Chamberlain, K. Principles governing uptake and transport of chemicals. In *Plant Contamination: Modeling and Simulation of Organic Chemical Processes*; Trapp, S, McFarlane, C., Eds.; Lewis Publishers: Boca Raton, Florida, 1995; 37–68.
27. Rhue, R.D.; Mansell, R.S.; Ou, L.T.; Tang, S.R.; Ouyang, Y. The fate and behavior of lead alkyls in the environ-

ment—A review. Crit. Rev. Environ. Control **1992**, *22*, 169–193.
28. Aelion, C.M.; Davis, H.T.; McDermott, S.; Lawson, A.B. Soil metal concentrations and toxicity: Associations with distances to industrial facilities and implications for human health. Sci. Total Environ. **2009**, *407*, 2216–2223.
29. Nawrot, T.S.; Staessen, J.A.; Roels, J.A.; Roels, H.A.; Munters, E.; Cuyupers, A.; Richart, T.; Ruttens, A.; Smeets, K.; Clijsters, H.; Vangronsveld, J. Cadmium exposure in the population: From health risks to strategies of prevention. Biometals. **2010**, *23*, 769–782.
30. Le Bot, B.; Gilles, E.; Durand, S.; Glorrenc, P. Bioaccessible and quasi-total metals in soil and indoor dust. Euro. J. Mineral **2010**, *22*, 651–657.
31. Stern, A.H. A quantitative assessment of the carcinogenicity of hexavalent chromium by the oral route and its relevance to human exposure. Environ. Res. **2010**, *110*, 798–807.
32. Dhal, B.; Thatoi, H; Das, N; Pandey, B.D. Reduction of hexavalent chromium by *Bacillus* sp. isolated from chromite mine spoils and characterization of reduced product. J. Chem. Technol. Biotechnol. **2010**, *85*, 1471–1479.
33. Molina, J.A.; Oyarzun, R.; Esbri, J.M., Higuereas, P. Mercury accumulation in soils and plants in the Almaden mining district, Spain: One of the most contaminated sites on Earth. Environ. Geochem. Health **2006**, *28*, 487–498.
34. Saha, J.C.; Dikshit, A.K.; Bandyopadhyay, M.; Saha, K.C. A review of arsenic poisoning and its effects on human health. Crit. Rev. Environ. Sci. Technol. **1999**, *29*, 281–313.
35. Roychowdhry, T. The impact of sedimentary arsenic through irrigated groundwater on soil, plant, crops and human continuum from Bengal delta: Special reference to raw and cooked rice. Environ. Eng. Sci. **2003**, *20*, 405–422.
36. LaGrega, M.D.; Buckingham, P.L.; Evans, J.C. In *Hazardous Waste Management*; McGraw-Hill: New York, 1994; 447–831.
37. Nie, S.W.; Gao, W.S.; Chen, Y.Q.; Sui, P.; Eneji, A.E. Review of current status and research approaches to nitrogen pollution in farmlands. Agric. Sci. China **2009**, *7*, 843–849.
38. Xu, J.L.; Song, Y.U.; Min, B.K.; Steinberg, L.; Logan, B.E. Microbial degradation of perchlorate: Principles and application. Environ. Eng. Sci. **2003**, *20*, 405–422.
39. Jeyasingh, J.; Somasundaram, V.; Philip, L.; Bhallamudi, S.M. Bioremediation of Cr(VI) contaminated soil/sludge: Experimental studies and development of a management model. Chem. Eng. J. **2010**, *2*, 556–564.
40. Mulligan, C.N.; Yong, R.N.; Gibbs, B.F. Remediation technologies for metal-contaminated soils and groundwater: An evaluation. Eng. Geol. **2001**, *1–4*, 197–203.
41. Milner, M.J.; Kochian, L.V. Investigating heavy-metal hyperaccumulation using *Thlaspi caerulescens* as a model system. Ann. Bot. **2008**, *102*, 3–13.
42. Gonzaga, M.I.S.; Santos, J.A.G.; Ma, L.Q. Phytoextraction by arsenic hyperaccumulator *Pteris vittata* L from six arsenic-contaminated soils: Repeated harvests and arsenic redistribution. Environ. Pollut. **2008**, *2*, 212–218.
43. Nagata, T.; Morita, H.; Akizawa, T.; Pan-Hou, H. Development of a transgenic tobacco plant for phytoremediation of methylmercury pollution. Appl. Microbiol. Biotechnol. **2010**, *87*, 781–786.
44. Lomonte, C.; Doronila, A.I.; Gregory, D.; Baker, A.J.M.; Spas, D.K. Phytotoxicity of biosolids and screening of selected plants with potential for mercury phytoextraction. J. Haz. Mater. **2010**, *173*, 494–501.
45. Hadi, F.; Bano, A.; Fuller, M.P. The improved phytoextraction of lead (Pb) and the growth of maize (*Zea mays* L.); the role of plant growth regulators (GA(3) and IAA) and EDTA alone and in combination. Chemosphere **2010**, *80*, 457–462.
46. Fleri, M.A.; Whetstone, G.T. In situ stabilisation/solidification: Project lifecycle. J. Haz. Mater. **2007**, *Special Issue S1*, 141, 441–456.
47. Lestan, D.; Luo, C.L.; Li, X.D. The use of chelating agents in the remediation of metal-contaminated soils: A review. Environ. Pollut. **2008**, *153*, 3–13.
48. Chaney, R.L; Ryan, J.A.; Brown, S. Environmentally acceptable endpoints for soil metals. In *Environmental Availability of Chlorinated Organics, Explosives, and Metals in Soils*; Anderson, W.C., Loehr, R.C., Reible, D., Eds.; American Academy of Environmental Engineers: Annapolis, Maryland, 1999; 111–155.

Pollution: Genotoxicity of Agrotoxic Compounds

Vera Lucia S.S. de Castro
Ecotoxicology and Biosafety Laboratory, Environment, Brazilian Agricultural Research Corporation (Embrapa), São Paulo, Brazil

Paola Poli
Department of Genetics, Biology of Microorganisms, Anthropology, and Evolution, University of Parma, Parma, Italy

Abstract
Large amounts of chemicals are released into the environment, many of which affect non-target organisms and are a potential hazard to human health. Exposure to pesticides has been associated with an increase in the incidence of genotoxicity. Many reports provide evidence of genotoxic effects induced by these compounds such as direct DNA damage, chromosomal aberrations, sister chromatid exchange, and micronuclei. In this entry, some examples of pesticides' genotoxic effects are present. Also, a wide variety of methods used to study these effects are discussed. In some cases, they can provide an early detection of biological events induced by pesticides. They contribute to the exploration of the action mode of these chemicals for risk assessment. However, there is some inconsistent observation of biological effects from similarly exposed populations, with large interindividual variations in response to the exposure. Together with genotoxicity test batteries employed in the testing of environmental chemicals and other agents over the past years, nowadays, new test batteries have been proposed. Genomic technologies are rapidly evolving as powerful tools for studying and detecting the genotoxicity of environmental chemicals. Another major task of modern toxicology is the analysis of gene–environment interactions using molecular epidemiology. Besides, for more elaborate studies, these techniques should be paired with behavioral approaches taking into account aspects of social priorities and not yet solved legal problems.

INTRODUCTION

Thousands of chemicals are in common use, but only a portion of them has undergone significant toxicological evaluation, leading to the need to prioritize the remainder for targeted testing. Various assays were set up and performed to identify the potential hazard (genotoxicity, carcinogenicity, reproductive toxicity, etc.) of pollutants for human health. Increasing need for effective tools to assess the risks derived by the large number of both natural and anthropic-origin organic and inorganic noxious substances, both of natural origin and released in the environment by human activities, leads to the development of very sensitive detectors of harmful substances. It is an accepted assumption that the simple measurement of chemical concentration, with reference to established regulatory rules, will not give an accurate account of the environmental hazard. The measurement of environmental physical–chemical parameters is the first step in monitoring environmental quality; however, attention and alarm thresholds of these parameters only concern the toxic effects of the studied polluting substance but do not take into consideration the issue of chronic exposure at low doses of physical agents and chemicals, frequently present in complex mixtures. On the other hand, the monitoring by biological assays can effectively define the risks for the environment and humans. These analyses may be able to assess the complexity of natural environment, in terms of both different organisms and differences in physiological status, to establish cause/effect relationships between presence and concentration of pollutants and consequent environmental damages and to detect the possible synergistic effect of complex mixtures of chemicals. Therefore, much attention was paid to biological sensors, markers, or detectors able to provide information on the effects of exposure to and/or susceptibility against a variety of environmental contaminants through the knowledge of action mechanisms and the identification of possible endpoints.

Due to their role as bioactive chemicals, pesticides tend to form electrophilic metabolites capable of reacting and combining with biological macromolecules. Preferred sites of action include nucleophilic oxygen as well as nitrogen atoms, both of which are abundant in DNA, predisposing the genetic material to mutagenic covalent binding. One main deleterious effect resulting from exposure to environmental mutagens is the possible initiation of cancer. Other manifestations of genotoxic environmental pollutants such as pesticides include heritable genetic diseases, reproductive dysfunction, and birth defects.[1] Epidemiological studies provide evidence that several types of tumors and other carcinogenic manifestations are in excess among farmers and other occupational groups associated with pesticide handling.[2–4] In addition to epidemiological evidence, laboratory and field monitoring data indicate that increases in chromosome aberration, recombination, sister chromatid exchange (SCE), and other genotoxic events are in excess for pesticide-exposed groups, pointing out a generic genetic activity of pesticides in humans.[5] Recently, a number of laboratory investigations and field studies have documented a correlation between genotoxic pollutants and heritable reproduction effects on individuals as well as a potential link to the declines of fish populations.[6]

REPRODUCTIVE TOXICITY

It is estimated that close to 30% of all pregnancies end in spontaneous abortion. Although about 60% of spontaneous abortions are thought to be due to genetic, infectious, hormonal, and immunological factors, the role of the environment remains poorly understood. Pregnancy involves a delicate balance of hormonal and immunological functions, which may be affected by environmental substances. Many toxic substances that are persistent in the environment and accumulate in the fatty tissues may disrupt this equilibrium.[7]

Several evidence link paternal exposure to genotoxic agents with an increased risk of pregnancy loss, developmental and morphological defects, infant mortality, infertility, and genetic diseases in the offspring, including cancer.[8] Toxicogenomic analysis of environmental chemicals may be performed to investigate the ability of genomics to predict toxicity, categorize chemicals, and elucidate mechanisms of toxicity. The concordance of in vivo observations and gene expression findings demonstrated the ability of genomics to accurately categorize chemicals, identify toxic mechanisms of action, and predict subsequent pathological responses.[9]

The central nervous system appears to be especially susceptible to toxic insults during development and there is evidence that functional changes can be induced at a lower exposure level than those resulting in toxicity in adults. As an example, fetal exposure to some environmental contaminants such as organophosphorous pesticides at apparently non-toxic doses may alter some important behaviors at adulthood in mice.[10] Neurobehavioral performance can be a sensitive biomarker of the neurodevelopmental consequences of exposure of environmental agents.[11,12]

Environmental contaminants as pesticides can come from a variety of sources, including diet, drinking water, and both indoor and outdoor residential use. These chemicals may alter gene expression profiles.[13] Also, some metals can disturb the reproduction process. A number of mechanisms of cadmium toxicity have been suggested, including ionic and molecular mimicry, interference with cell adhesion and signaling, oxidative stress, apoptosis, genotoxicity, and cell cycle disturbance.[14] Some mercury compounds are known as teratogenic agents, especially affecting the normal development of the central nervous system. Since the 1990s, a genotoxic effect has been demonstrated in human populations exposed to mercury through diet,[15] occupation,[16] or carrying dental fillings.[17]

In fact, concentrations of methylmercury causing significant genotoxic alterations in vitro below both safety limit and concentration were associated with delayed psychomotor development with minimal signs of methylmercury poisoning. Based on mercury's known ability to bind sulfhydryl groups, several hypotheses were raised about potential molecular mechanisms for the metal genotoxicity. Mercury may be involved in four main processes that lead to genotoxicity: generation of free radicals and oxidative stress, action on microtubules, influence on DNA repair mechanisms, and direct interaction with DNA molecules.[18] DNA damage can also be a sensitive bioindicator of mercury contamination in other organisms as in fish.[19]

Some pesticides may act as endocrine disruptors. As an example, the fungicide fenarimol acts both as an estrogen agonist and as an androgen antagonist.[11] In addition, fenarimol affects rat aromatase activity in human tissues.[20] This compound also affects other enzymes of the cytochrome P450 gene family that are involved in the metabolism of steroids.[21] Studies of reproductive toxicity in rats have shown that, as a result of fenarimol maternal exposure, some neuromuscular and behavioral deficits in nursing pups may occur principally during the last gestational period and lactation.[12]

Pesticides may also operate through hormonal or genotoxic pathways to affect male reproduction. They may penetrate the blood–testis barrier to potentially affect spermatogenesis, by affecting either genetic integrity or hormone production. Effects may be at different stages of the cell cycle such as during meiotic disjunction, and such abnormalities can have deleterious effects on reproduction and offspring.[22]

HUMAN AND ENVIRONMENTAL GENOTOXICITY

Mutation is a manifestation of change of the structure of DNA. Agents that cause DNA damage and mutation, as well as being potentially capable of causing hereditary disorders in the offspring and succeeding generations of exposed populations, are also likely to be carcinogenic. For these reasons, testing for the induction of DNA damage and for mutagenicity, using a variety of short-term tests, has become an accepted part of the toxicological evaluation of drugs, industrial intermediates, cosmetics, food and feed additives, pesticides, biocides, etc. Regulatory agencies and international authorities recommend a test scheme consisting of in vitro and in vivo methods to identify genotoxic/mutagenic substances as those described on Organization for Economic Cooperation and Development (OECD) and United States Environmental Protection Agency (USEPA) guidelines, as recent examples.[23,24]

The current EPA mutagenicity testing battery is required for pesticides and toxic substances that are regulated under the Federal Insecticide, Fungicide and Rodenticide Act (FIFRA) and the Toxic Substances Control Act (TSCA), respectively. The battery is a three-tiered system of various mutagenicity tests (including the *Salmonella*, mouse lymphoma, and mouse micronucleus assays in the first tier). Guidelines for the conduct of the tests employed in this battery were issued and were ultimately harmonized within EPA and with the OECD.[25]

In addition to the risks to human health posed by mutagenic pollutants such as pesticides, there are important ecological risks as well, such as the threat posed by pesticides to the stability of the ecosystems through the cumulative introduction of deleterious mutations into the genetic pool of populations. Indeed, it has been demonstrated that pesticides and other environmental genotoxicants are capable of altering the genetic makeup of some natural populations.[26,27]

The occurrence of genotoxic pollutants in the aquatic environment is of increasing concern. Pollution with anthropogenic toxicants may create pronounced environmental gradients that impose strong local selection pressures. Toxic contaminants may also directly impact genetic structure in natural populations by exhibiting genotoxicity. The genetic variation within natural populations and, hence, the potential for local adaptation can itself be impacted by anthropogenic pollution.[26]

Reservoirs are complex aquatic systems mediating between rivers and lakes; they usually reflect multiple impacts generated by a variety of anthropogenic activities. The sediment compartment is the intermediate or final receptor of insoluble (or slightly water soluble) pollutants and can act as a sink for various substances. Sediments accumulate chemicals up to concentrations many times higher than in the free water column. As pollutants may be made available under certain environmental conditions (such as dredging or flood events), sediments can also become a source of diffuse contamination to the free water space. Sediment pollutants are linked not only to organisms in aquatic ecosystem dynamics but also to human health via water and fish consumption. Due to their ability to metabolize xenobiotics and accumulate pollutants, fish represent important monitoring systems within aquatic genotoxicity assessment.[6]

In general, chemical mixtures may influence local ecosystems in a site-specific way defined by all aspects along the source and availability of the mixture and the non-target receptor. Also, humans are concurrently exposed to a number of chemicals via food and environment. These chemicals may have a combined action that causes a lower or higher toxic effect than would be expected from knowledge about the single compounds. Consequently, combined actions need to be addressed in the risk assessment process. In addition, developments in the area of toxicogenomics have also been suggested as a way of increasing our knowledge of mechanism of toxicity in order to better understand and improve the approaches for risk assessment of combined actions of chemicals.[28]

Both chronic and acute contamination of watershed by pesticides dissolved or adsorbed to soil particles can affect aquatic organisms. Pesticide toxicity may occur in a broad range of non-target aquatic organisms, both in plants from microalgae to macrophytes and in animals from microinvertebrates to fish predators. The possible effects of pesticide exposure on fish are of interest because of the position of fish in the food chain and because early life stages of fish have been shown to be highly sensitive to pollutants. Genotoxicity assessment in fish has been highlighted since the implications of the genotoxic effects are impacting on fitness traits such as reproductive success, genetic patterns, and subsequent population dynamics.[29]

PESTICIDE GENOTOXICITY ASSESSMENT

Traditionally, the impact of pollutants discharged into the environment by human activities has been assessed using chemical assays or by evaluating physical parameters. High-performance liquid chromatography or gas–mass spectrometry techniques have been widely used. However, the information is limited to the concentration of the pollutants, not on their toxicity. To overcome these limitations, biological analyses have been introduced. The use of effect-based screening tools has the advantage of indicating the real impact of all chemicals present in a given sample or ecosystem. Rapid biological tests are playing their major role in hazard and risk assessment, especially at the screening level. Bioassays measure changes in physiology or behavior of living organisms resulting from stresses induced by biological or chemical toxic compounds, which can cause disruption of the metabolism.[30]

Toxicological animal-based assays on organisms from different trophic levels (algae and plants, worms and crustaceans, fishes, etc.), employed in the identification of hazardous chemicals are sometimes expensive and time-consuming, require large sample volumes, and raise ethical problems. In vitro methods, also commonly used for screening and ranking chemicals, must be included in battery tests for risk assessment purposes. Their major promise is to supply mechanism-derived information, considered pivotal for adequate risk assessment. Mutagenesis short-term assays can directly detect the genetic effect of chemical and physical agents on the tested cells/organisms, able to assess the DNA damage resulting from exposure to both single chemicals and heterogeneous mixtures.

Fundamental research in the mechanisms of induced mutation and carcinogenesis induction has benefited from the development of highly refined short-term tests for genotoxicity. The knowledge of toxicity pathways that is derived from genomic data would highlight the potential mechanisms. The challenge is to comprehensively integrate the disparate chemical, biological, toxicological, and toxicogenomic data in order to elucidate the mechanisms and networks involved in toxicity and to develop quantitative models capable of accurately predicting thresholds.[31]

However, despite the progress on newer methodologies, quantitative data on toxicological effects of some widely used pesticides in agricultural practices are not well established even at the single-organism level. Among the pesticides in common use, attention is drawn towards the use of triazole fungicides that are used in the control of several fungal and plant diseases. These fungicides have demonstrated some adverse effects in mammalian species derived from genomic data.[32] Also, genotoxicity studies conducted in microalgae showed that the exposure to epoxiconazole can result in an increase in the extent of DNA strand breaks depending on the concentration.[33]

Nevertheless, results from this kind of assays are prone to the criticisms concerning the differences between the real conditions in the ecosystems and those of the in vitro assays. In fact, since various factors (chemical, physical, and biological) affect environmental conditions, the transfer of results obtained by in vitro techniques to the field is a complex task, and the establishment of extrapolation parameters is a crucial issue. In this context, a combination of in vitro and in situ (in vivo) bioassays represents a promising approach to better understand both the real exposure situation in the environment and the action mechanisms studied by the in vitro bioassays.[34]

Also, although the genotoxicity testing strategies employed prior to product registration are designed to identify potential in vivo genotoxins, concerns that exposure to pesticides may result in long-term adverse effects still exist.[35] A positive association between occupational exposure to complex pesticide mixtures and the presence of chromosomal aberrations (CAs), SCEs, and micronuclei has been detected in a number of studies, although some of these failed to detect cytogenetic damage. Chromosomal damage induced by pesticides appears to have been transient in acute or discontinuous exposure, but cumulative in continuous exposure to complex agrochemical mixtures.[36]

Assessment Methodology

Human and wildlife risks associated with pesticide mutagenic capability can be assessed by using screening systems that are sensitive and able to detect the whole mutagenic spectrum. There are more than 100 short-term bioassays for evaluating the potential genotoxic effects of pesticides, but since no single system encompasses the whole spectrum of possible genetic toxic effects, a combination of evaluation procedures is recommended for the assessment of pesticide mutagenesis. Genotoxicity testing batteries were specifically established for hazard identification, the first step in risk assessment. Studies of sensitivity and correlation among test systems are fundamental for a more accurate evaluation of the environmental risks, as well as extrapolation of data to other target organisms.

As a consequence of the very large number of genetic toxicity evaluation techniques described in the specialized literature, most assays can be considered ancillary and will be employed only for specific ends.[37] The assays accepted for routine evaluation of pesticides fall in one of six testing categories: 1) microbial assays: a) prokaryotic (bacterial, such as the *Salmonella typhimurium*, *Escherichia coli*, *Bacillus subtilis*) or b) eukaryotic assays (fungi, such as *Saccharomyces cerevisae*, *Neurospora crassa*); 2) in vitro isolated eukaryotic cell lines; 3) host-mediated assays; 4) in vivo animal; and 5) in vivo plant bioassays.

In order to approximate the studies to higher-organism subjects, several assays involve in vitro exposure of mammalian cell lines. Mouse bone marrow cells, erythrocytes, and white blood cells; hamster ovary cells; and human lung fibroblast are some examples, and several genetic endpoints can be evaluated, such as DNA damage, unscheduled DNA synthesis, chromatid exchanges, and micronucleus frequency. Aside from the practical advantages offered by the microbial and isolated cell line bioassays (ease of manipulation, asepticism, small space and large population assayed, low cost), the tests often depend upon an external metabolic activation complement, since the microbial and isolated cell lines may be incapable of responding to pro-mutagenic compounds if these are not partially metabolized.[38] The host-mediated assays are devised to circumvent this deficiency. Mammalian subjects are exposed to the pesticide; the microbial cell line is injected into this treated subject and later recovered and evaluated for mutation induction. Alternatively, for plant metabolic activation, whole plants are treated with the pesticide and their extracts are applied directly into the microbial assay.

Finally, plant bioassays involve a variety of endpoints, from DNA damage in leaf cells of *Impatiens balsamina* to micronuclei in pollen mother cells (*Tradescantia*) and root tip meristematic cells (*Allium*, *Vicia*), reversion and crossing over in chlorophyll-deficient lines (*Hordeum*, *Glycine*), sugar-specific starch production in pollen (*Zea*), flower pigmentation alteration (*Tradescantia*), and many other endpoints in many different species.[39–41]

Some recent studies[42–49] mainly use bioassays such as Comet, micronucleus, CAs, and SCE to detect pesticide genotoxic potential both in vitro and in vivo.

Comet Assay

The single-cell gel electrophoresis assay (or Comet assay) is a simple, rapid, and sensitive technique for analyzing and measuring DNA damage in individual mammalian (and to some extent prokaryotic) cells. The method published by Sing et al.[50] makes the Comet assay versatile and forms the basis for all developments that have taken place in this field. In the Comet assay, the cells are embedded in a thin agarose gel on a microscope slide. The cells are lysed to remove all cellular proteins and the DNA subsequently allowed unwinding under alkaline/neutral conditions. Following unwinding, the DNA is electrophoresed and stained with a fluorescent dye. During electrophoresis, broken DNA fragments (damaged DNA) or relaxed chromatin migrates away from the nucleus. The Comet assay essentially measures the sizes of DNA fragments within the cell. It is therefore necessary to convert DNA damage to DNA fragments by introducing breaks at the sites of DNA damage before they can be detected on Comet assay. The simplest types of DNA damage detected by Comet assay are the double-strand breaks (DSBs). DSBs within the DNA result in DNA fragments and can be detected by merely subjecting them to electrophoretic mobility at the neutral pH. Single-strand breaks (SSBs) do not produce DNA fragments unless the two strands of the DNA are separated/denatured. This is accomplished by unwinding the DNA at pH 12.1. Other types of DNA damage broadly termed as alkali labile sites are expressed when the DNA is treated with alkali at pH greater than 13.

The in vivo Comet assay is a well-established genotoxicity test.[51–55] It is currently mainly performed with somatic cells from different organs to detect a genotoxic activity of potential carcinogens. It is regarded as a useful test for follow-up testing of positive or equivocal in vitro test results and for the evaluation of local genotoxicity. In current test strategies, the in vivo Comet assay is mainly performed with somatic cells from different organs to detect the genotoxic activity of potential carcinogens and is regarded as a useful test for follow-up testing of positive in vitro findings. Furthermore, the comet assay also has the potential to be a useful tool for investigating genotoxicity in germ cells.[56]

Micronucleus Assay

The purpose of the micronucleus assay is to detect chromosome structure modifying agents: segregation is the way to lead the induction of micronuclei in interphase cells. These micronuclei may originate from acentric fragments (chromosome fragments lacking a centromere) or whole chromosomes unable to migrate with the remainder of the chromosomes during the anaphase of cell division.[57–59] The in vitro micronucleus assay is a mutagenic test system for the detection of chemicals that induce the formation of small membrane-bound DNA fragments, i.e., micronuclei in the cytoplasm of interphase cells.

After exposure to a test substance and addition of cytochalasin B for blocking cytokinesis, cell cultures are grown for a period sufficient to allow chromosomal damage to lead to the formation of micronuclei in bi- or multinucleated interphase cells. Harvested and stained interphase cells are then analyzed microscopically for the presence of micronuclei. Micronuclei are scored in those cells that complete nuclear division following exposure to the test item. Additionally, the cells are classified as mononucleates, binucleates, or multinucleates to estimate the proliferation index as a measure of toxicity.

The micronucleus test is also an efficient biological assay for monitoring population exposure to mixtures of agrochemicals as shown in a study that was performed in farm workers directly exposed to large amounts of agrochemicals (fungicides, insecticides, and herbicides) in an area of grain farming (wheat and soybeans).[36,60,61] High MN frequency was detected in people at higher cancer risk due to occupational/environmental exposure to a wide variety of carcinogens.[62–66]

CA Test

Chromosome mutations and related events are the cause of many human genetic diseases, and there is substantial evidence that chromosome mutations and related events causing alterations in oncogenes and tumor suppressor genes of somatic cells are involved in cancer induction in humans and experimental animals.

The purpose of the in vitro CA test is to identify agents that cause structural chromosome aberrations in cultured cells.[67,68] With the majority of chemical mutagens, induced aberrations are of the chromatid type, but chromosome-type aberrations also occur.

Structural CAs may be induced by direct DNA breakage, by replication on a damaged DNA template, by inhibition of DNA synthesis, and by other mechanisms (e.g., topoisomerase II inhibitors).[69] Based on morphological criteria, structural CAs can be divided into two main classes: chromosome-type aberrations (CSAs), involving both chromatids of one or multiple chromosomes, and chromatid-type aberrations (CTAs) involving only one of the two chromatids of a chromosome or several chromosomes.[69,70] An increase in polyploidy may indicate that a chemical has the potential to induce numerical aberrations.[70] CTAs (e.g., chromatid type breaks and exchanges) arise predominantly in vitro during S-phase of cultured lymphocytes, in response to base modifications and SSBs induced in vivo by S-phase-dependent clastogens.[69,70]

Proliferating cells are treated with the test substance in the presence and absence of a metabolic activation system for 3–6 hr and sampled at a time equivalent to about 1.5 normal cell cycle length after the beginning of treatment.[71] If this protocol gives negative results both with and without activation, an additional experiment without activation should be done, with continuous treatment until sampling at a time equivalent to about 1.5 normal cell cycle lengths. Certain chemicals may be more readily detected by treatment/sampling times longer than 1.5 cycle lengths. At predetermined intervals after exposure of cell cultures to the test substance, they are treated with a metaphase-arresting substance (e.g., colchicine), harvested, and stained, and metaphase cells are analyzed microscopically for the presence of chromosome aberrations. Though the purpose of the test is to detect structural chromosome aberrations, it is important to record polyploidy and endoreduplication when these events are seen.

Structural CAs in peripheral blood lymphocytes as assessed by the chromosome aberration assay in vivo have been used for more than 30 years in occupational and environmental settings as a biomarker of early effects of genotoxic carcinogens. A high frequency of structural CAs in lymphocytes (reporter tissue) is predictive of increased cancer risk.[36,60,72]

Sister Chromatid Exchange

SCE is the exchange of homologous stretches of DNA sequence between sister chromatids and occurs normally in cells during mitosis. In the presence of genotoxic agents that provoke DNA damage, the rate of SCE increases. Equal SCE has been thought to be an important mechanism of DSB repair in eukaryotes, but this has never been proven due to the difficulty of distinguishing SCE products from parental molecules.[73]

The SCE analysis was adopted as an indicator of genotoxicity. SCEs represent the interchange of DNA replication products at apparently homologous loci. The exchange process presumably involves DNA breakage and reunion, although little is known about its molecular basis.[73–80]

Some studies revealed that the nucleotide pool imbalance can have severe consequences on DNA metabolism and it is critical in SCE formation. The modulation of SCE by DNA precursors raises the possibility that DNA changes are responsible for the induction of SCE in mammalian cells.[81,82]

Detection of SCEs in in vitro test requires some means of differentially labeling sister chromatids, and this can be achieved by incorporation of bromodeoxyuridine (BrdU) into chromosomal DNA for two cell cycles.[83,84] Cells in an exponential stage of growth are exposed to the test substance for a suitable period of time; in most cases, 1–2 hr may be effective, but the treatment time may be extended up to two complete cell cycles in certain cases. Cells without sufficient intrinsic metabolic activity should be exposed to the test chemical in the presence and absence of an appropriate metabolic activation system. At the end of the exposure period, cells are washed free of test substance and cultured for two rounds of replication in the presence of BrdU. As an alternative procedure, cells may be exposed simultaneously to the test chemical and BrdU for the complete culture time of two replication cycles.[85] Cells are analyzed in their second posttreatment division, ensuring that the most sensitive cell cycle stages have been exposed to the chemical. Cell cultures are treated with a spindle inhibitor (e.g., colchicine) 1–4 hours prior to harvesting. Chromosome preparations are made by standard cytogenetic techniques. Staining of slides to show SCEs can be performed by several techniques (e.g., the fluorescence plus Giemsa method).[85,86]

While increased levels of CA have been associated with increased cancer risk,[87,88] a similar conclusion has not been reached for SCE. However, high levels of SCE frequency have been observed in persons at higher cancer risk due to occupational or environmental exposure to a wide variety of carcinogens.[62–66]

BIOMONITORING

Plants

Higher plants are recognized as excellent genetic models to detect environmental mutagens and are frequently used in monitoring studies.[89–94] Plant systems represent a complex multicellular environment where the efficiency of different protection or repair mechanisms can be modulated by cellular homeostasis. Higher plants, even showing low concentrations of oxidase enzymes and a limitation in the substrate specification in relation to other organism groups,[40] present consistent results that may serve as a warning to other biological systems, since the target is DNA, common to all organisms.[95]

They represent a stable sensor in an ecosystem and hence allow following the evolution of the genotoxic impact. Well-defined higher plants represent an excellent basis for cytogenetic evaluations after exposure to genotoxic pollutants, especially since the maturation of their gametes (meiosis) follows the same patterns as in animals and humans,[96] although sometimes plant and animal assays are differentially responsive to some pesticides as pendimethalin.[97]

Among the plant species, *Allium cepa* has been widely used to evaluate DNA damage, such as chromosome aberrations and disturbances in the mitotic cycle. Employing *A. cepa* as a test system to detect mutagens dates back to the 1940s.[40,98,99] It has been used to this day to assess a great number of chemical agents, which contributes to its increasing application in environmental monitoring. *A. cepa* is characterized as a low-cost test. It is easily handled and has advantages over other short-term tests that require previous preparations of the samples, as well as the addition of exogenous metabolic system. *A. cepa* test also enables the evaluation of different endpoints (mitotic index, chromosome aberrations, nuclear abnormalities, and micronucleus).[40,99,100] Among the endpoints, chromosome aberrations have been the most used one to detect genotoxicity along the years. The mitotic index and some nuclear abnormalities are used to evaluate both cytotoxicity and mutagenicity of different chemicals. Moreover, *A. cepa* test system provides important information to evaluate action mechanisms of an agent about its effects on the genetic material (clastogenic and/or aneugenic effects). Transgenic/transformed plants could be used as a model to better check whether, in response to environmental stress, DNA damage might be regulated by alterations in single genes and to understand the mode and mechanisms of DNA damage.[101]

Animals and Human

The response to a genotoxic agent can range considerably in the exposed populations. It is currently accepted that susceptibility to genotoxic exposure varies interindividually and that this could be the result of hereditary or acquired characteristics. For this reason, attention has been focused on genetic polymorphisms, which are able to modulate human response to genotoxic environmental agents. This individual variation suggests the importance of individual susceptibility factors. Many studies[102–108] showed the influence of metabolic and DNA repair polymorphisms in the individual response to exposure to carcinogens. Genetic polymorphisms of metabolizing enzymes widely influence xenobiotic effective dose.[109] On the other hand, the role of polymorphisms linked to the DNA repair genes appears to be essential in the modulation of genotoxic risk linked to carcinogen exposure.[105,107,110]

TESTS FOR SOME NOVEL MATERIALS: THE CASE OF NANOMATERIALS

Nanomaterials are generally defined as having one or more external dimensions or an internal or surface structure on the nanoscale (about 1–100 nm). Nanomaterials display novel properties (small size, particular shape, large surface area, and large surface activity) that make them attractive in many applications. However, these properties may contribute to their toxicological profile and may also affect nanomaterials' possible direct or indirect interaction with the DNA.[111]

Although it has been shown that cationic functionalized carbon nanotubes can condense with DNA and that gold nanoparticles binding to the major groove of DNA is associated with killing of cancer cells, for most nanomaterials, it is even unknown whether they directly interact with DNA or whether they promote indirect effects. The identification of the different ways by which various nanomaterials interact with DNA will improve the extrapolations of genotoxicity test results to human risk evaluation. Application of standard methods to nanomaterials demands, however, several adaptations, and the interpretation of results from the genotoxicity tests may need additional considerations. The use of a battery of standard genotoxicity testing methods covering a wide range of mechanisms towards establishing methodological adaptations and better test conditions is a practical and pragmatic approach for this genotoxic evaluation.[111]

Recent reviews have concluded that information on the genotoxicity of nanomaterials is still inadequate for general conclusions, e.g., on its characteristics critical for genotoxicity. It is presently unclear how well standard genotoxicity tests, designed for soluble chemicals, can be used to assess the genotoxicity of nanomaterials. It appears that more genotoxicity studies on nanomaterials are urgently required, and the need for novel methods to assess nanomaterials–genome interactions is imminent. Current evidence indicates that various nanomaterials may carry genotoxic potential although the mechanisms of which remain to be identified.[112]

Above all, some nanomaterials can generate great concern about their possible adverse effects. As an example, although titanium dioxide is frequently used for industrial and cosmetics purposes because of its low toxicity, nano-sized or ultrafine TiO_2 (UF-TiO_2) (<100 nm in diameter) can generate pulmonary fibrosis and lung tumor in rats and cytotoxicity in rat lung alveolar macrophages.[113] These authors also observed that it can cause genotoxicity and cytotoxicity in cultured human cells in experimental models as well.

TiO_2 nanoparticles in the absence of photoactivation are potentially genotoxic to fish cells under in vitro conditions. This effect becomes more pronounced in the presence of UVA, along with cytotoxic effects, which only occurred during combined exposure to TiO_2 and UVA.[114]

Although the biological effects of some nanomaterials have already been assessed, information on toxicity and possible mechanisms of various particle types are insufficient. The comparative analysis demonstrated that particle composition probably played a primary role in the cytotoxic effects of different nanoparticles, whereas the genotoxicity potential might be mostly attributed to particle shape.[115] Furthermore, the genotoxic effectiveness could be mediated through lipid peroxidation and oxidative stress as in the case of ZnO nanoparticles on human epidermal cells, even at low concentrations.[116]

FUTURE CONCERNS AND NEW GENOMIC TECHNIQUES

The evidence of metabolic activation of pesticides into direct-acting mutagens and the complex interactions that may occur as a result of the synergistic action of the multitude of chemicals presently under common usage warrant devoted attention to the evaluation of the potential genetic consequences of pesticide mixtures and derived metabolites accumulated through food chains. This complexity is further deepened by the unpredictability of mutational effects, as a result of pleiotropism and other unforeseeable events. Studies of this nature should attempt to 1) point out the impossibility of predicting all genetic effects of pesticides and 2) recognize that small, improbable effects may have important consequences when imposed onto the very large populations exposed to pesticides in the environment.

With the advent of new technologies (e.g., genomics, automated analyses, and in vivo monitoring), new regulations (e.g., the reduction of animal tests by the European Registration, Evaluation, and Authorization of Chemicals), and new approaches to toxicology (e.g., Toxicity Testing in the 21st Century, National Research Council), the field of regulatory genetic toxicology is undergoing a serious re-examination. However, it is appropriate to apply a prudent approach to risk assessment, maintaining current testing standards that are working properly until others have proven superior by rigorous scientific evidence and widespread agreement.[117]

Human cytogenetics has proven to be effective over a 50 years life span. It is in reality a collection of techniques that, while common, are cheap, fast, and wide ranging. Therefore, in genotoxicology, they continue to be useful to identify mutagenic agents as well as to evaluate and analyze exposed populations.[118]

The introduction of advanced molecular techniques leads to improved risk assessment and also provides an alternative to the massive use of animal testing. Transcriptional profiling and DNA chips are highly informative and are among the most promising novel techniques for environmental risk assessment. Moreover, information discerned from these chips enables the identification of new discriminative biomarker genes. Based on these biomarker genes, cellular reporters can be constructed. These can be used in a high-throughput setup and can significantly facilitate ecotoxicological risk assessment.[119] However, some important technical and interpretative hurdles still need to be overcome before a full implementation of ecotoxicogenomics in regulatory settings can occur. Toxicogenomics uses molecular biology's entire arsenal to analyze the changes in DNA induced by genotoxic agents.[120,121] These changes include those induced in certain genes not related to the mechanisms of genotoxicity, like early response genes. These genes can be used as indicators for which there is a stress response to the genotoxic agent. Quantitative real-time polymerase chain reaction (RT-PCR) and microarrays are the most important procedures to analyze these genes, which can be numerous since current technology allows it.

RT-PCR uses fluorescent probes to measure the exact amount of a nucleic acid. In genotoxicology, it can be used to quantify gene expression to detect genetic polymorphisms and to quantify chromosome deletions. These probes are used to verify the expressed gene or to selectively analyze its expression over time or dose parameters. It will also become more important to analyze expression in specific cell populations in order to profile the global alterations in gene expression involved in chronic chemical exposure that may lead to tumor development.[122]

The results of the microarray studies in toxicogenomics show that genotoxic agents are able to induce changes in gene expression profiles. These studies demonstrate that the gene expression pattern is able to generate the information necessary to determine the agent's type of action and even to predict it. Chemically induced changes in gene expression can result in the identification of simple, sensitive, and relevant biomarkers of effect that can be used in dose–response studies to more readily identify precursor effects in the low-dose range.[123] Therefore, the ability to analyze the effect of mutagenic agents on a large number of genes in a single experiment using gene expression profiling analysis has been used to demonstrate that certain genes may or may not become activated when exposed to a toxic agent, depending on the type of cells and the type of genotoxic agent they have been exposed to.[124]

Furthermore, the DNA damage itself can activate genes codifying proteins involved in DNA repair and/or induce the apoptosis process, when the genes associated with this process are activated. Although broad consensus is still lacking concerning the guidelines for the reproducibility of experiments, toxicogenomic studies could lead to development of early biomarkers of toxic injury and may also help to resolve issues related to interspecies extrapolation and susceptibility variation among individuals.[122] As an application of this toxicogenomic approach, a study[32] reports data that linked genomic data to specific toxicological endpoints of the antifungal triazole, which causes varying degrees of hepatic toxicity and disrupts steroid hormone homeostasis in rodent in vivo models. Overall, these analyses revealed functional categories of chemical response genes that indicate mechanisms and provide direction for further research on triazole mechanisms of action.

CONCLUSION

The occurrence of genotoxic pollutants in the environment is of increasing concern. Genotoxicity studies can be applied to elucidate potential mechanisms of physiological or molecular alterations to minimize exposure of contaminants to levels that maintain sustainability of the environment. There is a large number of genetic toxicity evaluation assays described in the specialized literature that are able to identify action modes and biological targets both in the environment and in the organisms. Microorganisms as well as cell lines are widely used as in vitro models for risk assessment. Animal and plant models may be valuable for a better comprehension of the metabolic pathways of genotoxic chemicals. Various methods can be applied both in vitro and in epidemiological studies. New approaches through molecular techniques are also proposed for testing the genotoxic load of environmental pollutants. Moreover, in order to better establish biological hazards, it is also necessary to study the effects induced by chemicals on organism development, physiology, and behavior to evaluate their consequences. In relation to this, few attempts to link the genotoxic effects of contaminants with effects at the physiological or behavioral level have been made, which would lead to a detrimental impact on vital life processes including the reproductive potential of the organisms.

ACKNOWLEDGMENTS

We wish to thank Dr. Stachetti for his precious contribution to this entry, consisting of the initial text, over which the present one has been constructed. Also, we would like to thank him for the opportunity offered to contribute to this prestigious publication, by referring our names as authors. Without his help and his confidence on our work, this achievement would not have been possible. We also would like to thank Dr. Stachetti for the final text comments and suggestions.

REFERENCES

1. Waters, M.; Sandhu, S.S.; Simmon, V.F.; Mortelmans, K.E.; Mitchell, A.D.; Jorgenson, T.A.; Jones, D.C.L.; Valencia, R.; Garrett, N.E. Study of pesticide genotoxicity. In *Genetic Toxicology: An Agricultural Perspective*; Fleck, R.A., Hollaender, A., Eds.; Basic Life Sciences, Plenum Press: New York, 1982; Vol. 21, 275–326.
2. Maele-Fabry, G.; Duhayon, S.; Mertens, C.; Lison, D. Risk of leukaemia among pesticide manufacturing workers: A review and meta-analysis of cohort studies. Environ. Res. **2008**, *106*, 121–137.
3. Rull, R.P.; Gunier, R.; Behren, J.; Hertz, A.; Crouse, V.; Buffler, P.A.; Reynolds, P. Residential proximity to agricultural pesticide applications and childhood acute lymphoblastic leukemia. Environ. Res. **2009**, *109*, 891–899.
4. Alavanja, M.C.R.; Hoppin, J.A.; Kamel, F. Health effects of chronic pesticide exposure: Cancer and neurotoxicity. Annu. Rev. Public Health **2004**, *25*, 155–197.
5. Maroni, M.; Fait, A. Health effects in man from long-term exposure to pesticides—A review of the 1975–1991 literature. Toxicology **1993**, *78* (1–3), 1–180.
6. Rocha, P.S.; Luvizotto, G.L.; Kosmehl, T.; Bottcher, M.; Storch, V.; Braunbeck, T.; Hollert, H. Sediment genotoxicity in the Tiete River (São Paulo, Brazil): In vitro comet assay versus in situ micronucleus assay studies. Ecotoxicol. Environ. Saf. **2009**, *72*, 1842–1848.
7. Weselak, M.; Arbuckle, T.; Walker, M.; Krewski, D. The influence of the environment and other exogenous agents on spontaneous abortion risk. J. Toxicol. Environ. Health, Part B 2008, *11* (3–4), 221–241.
8. Aitken, R.J.; Koopman, P.; Lewis, S.E. Seeds of concern. Nature **2004**, *432*, 48–52.
9. Martin, M.T.; Brennan, R.J.; Hu, W.; Ayanoglu, E.; Lau, C.; Ren, H.; Wood, C.R.; Corton, J.C.; Kavlock, R.J.; Dix, D.J. Toxicogenomic study of triazole fungicides and perfluoroalkyl acids in rat livers predicts toxicity and categorizes chemicals based on mechanisms of toxicity. Toxicol. Sci. **2007**, *97* (2), 595–613.
10. Ricceri, L.; Venerosi, A.; Capone, F.; Cometa, M.; Lorenzini, P.; Fortuna, S.; Calamandrei, G. Developmental neurotoxicity of organophosphorous pesticides: Fetal and neonatal exposure to chlorpyrifos alters sex-specific behaviors at adulthood in mice. Toxicol. Sci. **2006**, *93* (1), 105–113.
11. Andersen, H.; Vinggaard, A.; Rasmussen, T.; Gjermandsen, I.; Bonefeld-Jørgensen, E. Effects of currently used pesticides in assays for estrogenicity, androgenicity, and aromatase activity in vitro. Toxicol. Appl. Pharmacol. **2002**, *179* (1), 1–12.
12. Castro, V.L.; Mello, M.A.; Diniz, C.; Morita, L.; Zucchi, T.; Poli, P. Neurodevelopmental effects of perinatal fenarimol exposure on rats. Reprod. Toxicol. **2007**, *23*, 98–105.
13. Tully, D.B.; Bao, W.; Goetz, A.K.; Blystone, C.R.; Ren, H.; Schmid, J.E.; Strader, L.F.; Wood, C.R.; Best, D.S.; Narotsky, M.G.; Wolf, D.C.; Rockett, J.C.; Dix, D.J. Gene expression profiling in liver and testis of rats to characterize the toxicity of triazole fungicides. Toxicol. Appl. Pharmacol. **2006**, *215*, 260–273.
14. Thompson, J.; Bannigan, J. Cadmium: Toxic effects on the reproductive system and the embryo. Reprod. Toxicol. **2008**, *25*, 304–315.
15. Pinheiro, M.C.N.; Crespo-Lopez, M.E.; Vieira, J.L.F.; Oikawa, T.; Guimarães, G.A.; Araújo, C.C.; Amoras, W.W.; Ribeiro, D.R.; Herculano, A.M.; do Nascimento, J.L.M.; Silveira, L.C.L. Mercury pollution and childhood in Amazon riverside villages. Environ. Int. **2007**, *33*, 56–61.
16. Zachi, E.C.; Ventura, D.F.; Faria, M.A.; Taub, A. Neuropsychological dysfunction related to earlier occupational exposure to mercury vapor. Braz. J. Med. Biol. Res. **2007**, *40* (3), 425–433.
17. Di Pietro, A.; Visalli, G.; La Maestra, S.; Micale, R.; Baluce, B.; Matarese, G.; Cingano, L.; Scoglio, M.E. Biomonitoring of DNA damage in peripheral blood lymphocytes of subjects with dental restorative fillings. Mutat. Res **2008**, *650* (2), 115–22.
18. Crespo-López, M.E.; Macêdo, G.L.; Pereira, S.I.D.; Arrifano, G.P.F.; Picanço-Diniz, D.L.W.; do Nascimento, J.L.M.; Herculano, A.M. Mercury and human genotoxicity: Critical considerations and possible molecular mechanisms. Pharmacol. Res. **2009**, *60* (4), 212–220.
19. Della Torre, C.; Petochi, T.; Corsi, I.; Dinardo, M.M.; Baroni, D.; Alcaro, L.; Focardi, S.; Tursi, A.; Marino, G.; Frigeri, A.; Amato, E. DNA damage, severe organ lesions and high muscle levels of As and Hg in two benthic fish species from a chemical warfare agent dumping site in the Mediterranean Sea. Sci. Total Environ. **2010**, *408* (9), 2136–2145.
20. Vinggaard, A.; Jacobsen, H.; Metzdorff, S.; Andersen, H.; Nellemann, C. Antiandrogenic effects in short-term in vivo studies of the fungicide fenarimol. Toxicology **2005**, *207*, 21–34.
21. Paolini, M.; Mesirca, R.; Pozzetti, L.; Sapone, A.; Cantelli-Forti, G. Molecular non-genetic biomarkers related to fenarimol cocarcinogenesis, organ- and sex-specific CYP induction in rat. Cancer Lett. **1986**, *101*, 171–178.
22. Perry, M.J. Effects of environmental and occupational pesticide exposure on human sperm: A systematic review. Hum. Reprod. Update **2008**, *14* (3), 233–242.
23. OECD. *OECD Guideline for the Testing of Chemicals. 414 Prenatal Developmental Toxicity Study*; Organisation for Economic Co-operation and Development: Paris, 2009.
24. Rudén, C.; Hansson, S.O. Registration, Evaluation, and Authorization of Chemicals (REACH) is but the first step—How far will it take us? Six further steps to improve the European chemicals legislation. Environ. Health Perspect. **2010**, *118* (1), 6–10.
25. Dearfield, K.L.; Cimino, M.C.; McCarroll, N.E.; Mauer, I.; Valcovic, L.R. Genotoxicity risk assessment: A proposed classification strategy. Mutat. Res. **2002**, *521*, 121–135.
26. Coors, A.; Vanoverbeke, J.; De Bie, T.; De Meester, L. Land use, genetic diversity and toxicant tolerance in natural populations of *Daphnia magna*. Aquat. Toxicol. **2009**, *95*, 71–79.

27. Bickham, J.W.; Sandhu, S.; Hebert, P.D.; Chikhi, L.; Athwal, R. Effects of chemical contaminants on genetic diversity in natural populations: Implications for biomonitoring and ecotoxicology. Mutat. Res. **2000**, *463*, 33–51.
28. Andersen, M.E.; Krewski, D. Toxicity testing in the 21st century: Bringing the vision to life. Toxicol. Sci. **2009**, *107*, 324–330.
29. Bony, S.; Gillet, C.; Bouchez, A.; Margoum, C.; Devaux, A. Genotoxic pressure of vineyard pesticides in fish: Field and mesocosm surveys. Aquat. Toxicol. **2008**, *89*, 197–203.
30. Girotti, S.; Ferri, E.N.; Fumo, M.G.; Maiolini, E. Monitoring of environmental pollutants by bioluminescent bacteria. Anal. Chim. Acta **2008**, *608*, 2–29.
31. Boverhof, D.R.; Zacharewski, T.R. Toxicogenomics in risk assessment: Applications and needs. Toxicol. Sci. **2006**, *89*, 352–360.
32. Goetz, A.K.; Dix, D.J. Toxicogenomic effects common to triazole antifungals and conserved between rats and humans. Toxicol. Appl. Pharmacol. **2009**, *238*, 80–89.
33. Akcha, F.; Arzul, G.; Rousseau, S.; Bardouil, M. Comet assay in phytoplankton as biomarker of genotoxic effects of environmental pollution. Mar. Environ. Res. **2008**, *66*, 59–61.
34. Pellacani, C.; Buschini, A.; Furlini, M.; Poli, P.; Rossi, C. A battery of in vivo and in vitro tests useful for genotoxic pollutant detection in surface waters. Aquat. Toxicol. **2006**, *77*, 1–10.
35. Bull, S.; Fletcher, K.; Boobis, A.R.; Battershill, J.M. Evidence for genotoxicity of pesticides in pesticide applicators: A review. Mutagenesis **2006** *21* (2), 93–103.
36. Bolognesi, C. Genotoxicity of pesticides: A review of human biomonitoring studies. Mutat. Res., Rev. Mutat. Res. **2003**, *543* (3), 251–272.
37. Epstein, S.S.; Legator, M.S. *The Mutagenicity of Pesticides: Concepts and Evaluation*; The MIT Press: Cambridge, 1971; 220 pp.
38. Hrelia, P.; Vigagni, F.; Maffei, F.; Morotti, M.; Colacci, A.; Perocco, P.; Grilli, S.; Cantelli-Forti, G. Genetic safety evaluation of pesticides in different short-term tests. Mutat. Res. **1994**, *321* (4), 219–228.
39. Poli, P.; de Mello, M.A.; Buschini, A.; de Castro, V.L.S.S.; Restivo, F.M.; Rossi, C.; Zucchi, T.M.A.D. Evaluation of the genotoxicity induced by the fungicide fenarimol in mammalian and plant cells by the use of single-cell gel electrophoresis assay. Mutat. Res. **2003**, *540*, 57–66.
40. Leme, D.M.; Marin-Morales, M.A. *Allium cepa* test in environmental monitoring: A review on its application. Mutat. Res. **2009**, *682*, 71–81.
41. Rodrigues, G.S.; Ma, T.H.; Pimentel, D.; Weinstein, L.H. Tradescantia bioassays as monitoring systems for environmental mutagenesis—A review. Crit. Rev. Plant Sci. **1997**, *16* (4), 325–359.
42. Li, X.; Li, S.; Liu, S.; Zhu, G. Lethal effect and in vivo genotoxicity of profenofos to Chinese native amphibian (*Rana spinosa*) tadpoles. Arch. Environ. Contam. Toxicol. **2010**, *59*, 478–483. DOI 10.1007/s00244-010-9495-4.
43. Soloneski, S.; Larramendy, M.L. Sister chromatid exchanges and chromosomal aberrations in Chinese hamster ovary (CHO-K1) cells treated with the insecticide pirimicarb. J. Hazard. Mater. **2010**, *174*, 410–415.
44. Kumar, R.; Nagpure, N.S.; Kushwaha, B.; Srivastava, S.K.; Lakra, W.S. Investigation of the genotoxicity of malathion to freshwater teleost fish Channa punctatus (Bloch) using the micronucleus test and comet assay. Arch. Environ. Contam. Toxicol. **2010**, *58*, 123–130.
45. Candioti, J.V.; Natale, G.S.; Soloneski, S.; Ronco, A.E.; Larramendy, M.L. Sublethal and lethal effects on *Rhinella arenarum* (Anura, Bufonidae) tadpoles exerted by the pirimicarb-containing technical formulation insecticide Aficida. Chemosphere **2010**, *78*, 249–255.
46. Yin, X.; Zhu, G.; Li, X.B.; Liu, S. Genotoxicity evaluation of chlorpyrifos to amphibian Chinese toad (Amphibian: Anura) by comet assay and micronucleus test. Mutat. Res. **2009**, *680*, 2–6.
47. Vega, L.; Valverde, M.; Elizondo, G.; Leyva, J.F.; Rojas, E. Diethylthiophosphate and diethyldithiophosphate induce genotoxicity in hepatic cell lines when activated by further biotransformation via cytochrome P450. Mutat. Res. **2009**, *679*, 39–43.
48. Mladinic, M.; Berend, S.; Vrdoljak, A.L.; Kopjar, N.; Radic, B.; Zeljezic, D. Evaluation of genome damage and its relation to oxidative stress induced by glyphosate in human lymphocytes in vitro. Environ. Mol. Mutagen. **2009**, *50*, 800–807.
49. Moore, P.D.; Yedjou, C.G.; Tchounwou, P.B. Malathion-induced oxidative stress, cytotoxicity, and genotoxicity in human liver carcinoma (HepG2) cells. Environ. Toxicol. **2009**, *25*, 221–226. DOI 10.1002/tox.20492.
50. Sing, N.P.; McCoy, M.T.; Tice, R.R.; Schneider, E.L. A simple technique for quantitation of low levels of DNA damage in individual cells. Exp. Cell Res. **1988**, *175*, 184–191.
51. Tice, R.R.; Agurell, E.; Anderson, D.; Burlinson, B.; Hartmann, A.; Kobayashi, H.; Miyamae, Y.; Rojas, E.; Ryu, J-C.; Sasaki, Y.F. Single cell gel/comet assay: Guideline for in vitro and in vivo genetic toxicology testing. Environ. Mol. Mutagen. **2000**, *35*, 206–221.
52. Tsuda, S.; Matsusaka, N.; Madarame, H.; Miyamae, Y.; Ishida, K.; Satoh, M.; Sekihashi, K.; Sasaki, Y.F. The alkaline single cell electrophoresis assay with eight mouse-organs: Results with 22 mono-functional alkylating agents (including 9 dialkyl *N*-nitrosoamines) and 10 DNA cross-linkers. Mutat. Res. **2000**, *467*, 83–98.
53. Sekihashi, K.; Yamamoto, A.; Matsumura, Y.; Ueno, S.; Watanabe-Akanuma, M.; Kassie, F.; Knasmüller, S.; Tsuda, S.; Sasaki, Y.F. Comparative investigation of multiple organs of mice and rats in the comet assay. Mutat. Res. **2002**, *517*, 53–74.
54. Castro, V.L.; Mello, M.A.; Poli, P.; Zucchi, T.M. Prenatal and perinatal fenarimol-induced genotoxicity in leukocytes of in vivo treated rats. Mutat. Res. **2005**, *583*, 95–104.
55. Tewari, A.; Dhawan, A.; Gupta, S.K. DNA damage in bone marrow and blood cells of mice exposed to municipal sludge leachates. Environ. Mol. Mutagen. **2006**, *47*, 271–276.
56. Speit, G.; Vasquez, M.; Hartmann, A. The comet assay as an indicator test for germ cell genotoxicity. Mutat. Res. **2009**, *681*, 3–12.
57. Fenech, M. The cytokinesis-block micronucleus technique. In *Technologies for Detection of DNA Damage and Mutations*; Pfeifer, G.P., Ed.; Plenum Press: New York, 1996; 25–36.
58. Fenech, M. The in vitro micronucleus technique. Mutat. Res. **2000**, *455*, 81–95.
59. Parry, J.M.; Parry, E.M. The use of the in vitro micronucleus assay to detect and assess the aneugenic activity of chemicals. Mutat. Res. **2006**, *607* (1), 5–8.

60. Joksic, G.; Vidakovic, A.; Spasojevic-Tisma, V. Cytogenetic monitoring of pesticide sprayers. Environ. Res. **1997**, *75*, 113–118.
61. Pacheco, A.O.; Hackel, C. Chromosome instability induced by agrochemicals among farm workers in Passo Fundo, Rio Grande do Sul, Brazil. Cad. Saúde Pública **2002**, *18* (6), 1675–1683.
62. Fucic, A.; Markucic, D.; Mijic, A.; Jazbec, A.M. Estimation of genome damage after exposure to ionising radiation and ultrasound used in industry. Environ. Mol. Mutagen. **2000**, *36*, 47–51.
63. Vaglenov, A.; Nosko, M.; Georgieva, R.; Carbonell, E.; Creus, A.; Marcos, R. Genotoxicity and radioresistance in electroplating workers exposed to chromium. Mutat. Res. **1999**, *446*, 23–34.
64. Somorovska, M.; Szabova, E.; Vodicka, P.; Tulinska, J.; Barancokova, M.; Fabri, R.; Liskova, A.; Riegerova, Z.; Petrovska, H.; Kubova, J.; Rausova, K.; Dusinska, M.; Collins, A. Biomonitoring of genotoxic risk in workers in a rubber factory, comparison of the Comet assay with cytogenetic methods and immunology. Mutat. Res. **1999**, *445*, 181–192.
65. Fenech, M.; Perepetskaya, G.; Mikhalevic, L. A more comprehensive application of the micronucleus technique for biomonitoring of genetic damage rates in human populations. Experiences from the Chernobyl catastrophe. Environ. Mol. Mutagen. **1997**, *30*, 112–118.
66. Sinues, B.; Sanz, A.; Bernal, M.L.; Tres, A.; Alcala, A.; Lanuza, J.; Ceballos, C.; Saenz, M.A. Sister chromatid exchanges, proliferating rate index, and micronuclei biomonitoring of internal exposure to vinyl chloride monomer in plastic industry workers. Toxicol. Appl. Pharmacol. **1991**, *108*, 37–45.
67. Evans, H.J. Cytological methods for detecting chemical mutagens. In *Chemical Mutagens, Principles and Methods for their Detection*; Hollaender, A., Ed.; Plenum Press: New York and London, 1976; Vol. 4, 1–29.
68. Ishidate, M., Jr.; Sofuni, T. The in vitro chromosomal aberration test using Chinese hamster lung (chl) fibroblast cells in culture. In *Progress in Mutation Research*; Ashby, J., et al., Eds.; Elsevier Science Publishers: Amsterdam–New York–Oxford, 1985; Vol. 5, 427–432.
69. Albertini, R.J.; Anderson, D.; Douglas, G.R.; Hagmar, L.; Hemminki, K.; Merlo, F.; Natarajan, A.T.; Norppa, H.; Shuker, D.E.; Tice, R.; Waters, M.D.; Aitio, A. IPCS guidelines for the monitoring of genotoxic effects of carcinogens in humans, International Programme on Chemical Safety. Mutat. Res. **2000**, *463*, 111–172.
70. Hagmar, L.; Stromberg, U.; Bonassi, S.; Hansteen, I.L.; Knudsen, L.E.; Lindholm, C.; Norppa, H. Impact of types of lymphocyte chromosomal aberrations on human cancer risk: Results from Nordic and Italian cohorts. Cancer Res. **2004**, *64*, 2258–2263.
71. Galloway, S.M.; Aardema, M.J.; Ishidate, M., Jr.; Ivett, J.L.; Kirkland, D.J.; Morita, T.; Mosesso, P.; Sofuni, T. Report from working group on in *in vitro* tests for chromosomal aberrations. Mutat. Res. **1994**, *312*, 241–261.
72. Mateuca, R.; Lombaert, N.; Aka, P.V.; Decordier, I.; Kirsch-Volders, M. Chromosomal changes: Induction, detection methods and applicability in human biomonitoring. Biochimie **2006**, *88*, 1515–1531.
73. González-Barrera, S.; Cortés-Ledesma, F.; Wellinger, R.E.; Aguilera, A. Equal sister chromatid exchange is a major mechanism of double-strand break repair in yeast. Mol. Cell Biol. **2003**, *11* (6), 1661–1671.
74. Domínguez, I.; Pastor, N.; Mateos, S.; Cortés, F. Testing the SCE mechanism with non-poisoning topoisomerase II inhibitors. Mutat. Res. **2001**, *497*, 71–79.
75. Speit, G.; Hochsattel, R.; Vogel, W. The contribution of DNA single-strand breaks to the formation of chromosome aberrations and SCEs. Basic Life Sci. **1984**, *29*, 229–244.
76. Das, B.C. Factors that influence formation of sister chromatid exchanges in human blood lymphocytes. Crit. Rev. Toxicol. **1988**, *19* (1), 43–86.
77. Shiraishi, Y. Nature and role of high sister chromatid exchanges in Bloom syndrome cells. Some cytogenetic and immunological aspects. Cancer Genet. Cytogenet. **1990**, *50* (2), 175–187.
78. Tucker, J.D.; Auletta, A.; Cimino, M.C.; Dearfield, K.L.; Jacobson-Kram, D.; Tice, R.R.; Carrano, A.V. Sister-chromatid exchange: Second report of the Gene-Tox Program. Mutat. Res. **1993**, *297* (2), 101–180.
79. Tilman, G.; Loriot, A.; Van Beneden, A.; Arnoult, N.; Londoño-Vallejo, J.A.; De Smet, C.; Decottignies, A. Subtelomeric DNA hypomethylation is not required for telomeric sister chromatid exchanges in ALT cells. Oncogene **2009**, *28* (14), 1682–1693.
80. White, J.S.; Choi, S.; Bakkenist, C.J. Transient ATM kinase inhibition disrupts DNA damage-induced sister chromatid exchange. Sci. Signal. **2010**, *3*, ra44. DOI: 10.1126/scisignal.2000758.
81. Popescu, N.C. Sister chromatid exchange formation in mammalian cells is modulated by deoxyribonucleotide pool imbalance. Somatic Cell Mol. Genet. **1999**, *25*, 101–108.
82. Ashman, C.R.; Davidson, R.L. Bromodeoxyuridine mutagenesis in mammalian cells is related to deoxyribonucleotide pool imbalance. Mol. Cell. Biol. **1981**, *1*, 254–260.
83. Latt, S.A. Sister chromatid exchanges, indices of human chromosome damage and repair: Detection by fluorescence and induction by mitomycin-C. Proc. Natl. Acad. Sci. U. S. A. **1974**, *71*, 3162–3166.
84. Latt, S.A.; Schreck, R.R. Sister chromatid exchange analysis. Am. J. Hum. Genet. **1980**, *32*, 297–313.
85. EPA. In vitro sister chromatid exchange assay. Test methods and guidelines/OPPTS Harmonized test guidelines OPPTS 870.5900. U.S. Environmental Protection Agency, 1998, available at http://www.epa.gov/epahome/research.htm (accessed May 2010).
86. Morgan, W.F.; Schwartz, J.L.; Murnane, J.P.; Wolff, S. Effect of 3-aminobenzamide on sister chromatid exchange frequency in X-irradiated cells. Radiat. Res. **1983**, *93*, 567–571.
87. Hagmar, L.; Brogger, A.; Hansteen, I.L.; Heim, S.; Hogstedt, B.; Knudsen, L.; Lambert, B.; Linnainmaa, K.; Mitelman, F.; Nordenson, I.; Reuterwall, C.; Salomaa, S.I.; Skerfving, S.; Sorsa, M. Cancer risk in human predicted by increased levels of chromosomal aberrations in lymphocytes: Nordic Study Group on the Health Risk of Chromosome Damage. Cancer Res. **1994**, *54*, 2919–2922.
88. Hagmar, L.; Bonassi, S.; Stromberg, U.; Brogger, A.; Knudsen, L.; Norppa, H.; Reuterwall, C. The European Study Group on Cytogenetic Biomarkers and Health. Chromosomal aberrations in lymphocytes predict human cancer.

A report from the European Study Group on Cytogenetic Biomarkers and Health (ESCH). Cancer Res. **1998**, *58*, 4117–4121.
89. Gichner, T.; Menke, M.; Stavreva, D.A.; Schubert, I. Maleic hydrazide induces genotoxic effects but no DNA damage detectable by the comet assay in tobacco and field beans. Mutagenesis **2000**, *15*, 385–389.
90. Citterio, S.; Aina, R.; Labra, M.; Ghiani, A.; Fumagalli, P.; Sgorbati, S.; Santagostino A. Soil genotoxicity assessment: A new strategy based on biomolecular tools and plant bioindicators. Environ. Sci. Technol. **2002**, *36* (12), 2748–2753.
91. Restivo, F.M.; Laccone, M.C.; Buschini, A.; Rossi, C.; Poli, P. Indoor and outdoor genotoxic load detected by the Comet assay in leaves of *Nicotiana tabacum* cultivars Bel B and Bel W3. Mutagenesis **2002**, *17* (2), 127–134.
92. Klumpp, A.; Ansel, W.; Klumpp, G.; Calatayud, V.; Garrec, J.P.; He, S.; Peñuelas, J.; Ribas, A.; Ro-Poulsen, H.; Rasmussen, S.; Sanz, M.J.; Vergne, P. Tradescantia micronucleus test indicates genotoxic potential of traffic emissions in European cities. Environ. Pollut. **2006**, *139* (3), 515–522.
93. Liu, W.; Zhu, L.S.; Wang, J.; Wang, J.H.; Xie, H.; Song, Y. Assessment of the genotoxicity of endosulfan in earthworm and white clover plants using the comet assay. Arch. Environ. Contam. Toxicol. **2009**, *56* (4), 742–746.
94. Villarini, M.; Fatigoni, C.; Dominici, L.; Maestri, S.; Ederli, L.; Pasqualini, S.; Monarca, S.; Moretti, M. Assessing the genotoxicity of urban air pollutants using two in situ plant bioassays. Environ. Pollut. **2009**, *157* (12), 3354–3356.
95. Rodrigues, F.P.; Angeli, J.P.; Mantovani, M.S.; Guedes, C.L.; Jordão, B.Q. Genotoxic evaluation of an industrial effluent from an oil refinery using plant and animal bioassays. Genet. Mol. Biol. **2010**, *33* (1), 169–175.
96. Sadowska, A.; Pluygers, E.; Niklinska, W.; Maria, M.R.; Obidoska, G. Use of higher plants in the biomonitoring of environmental genotoxic pollution. Folia Histochem. Cytobiol. **2001**, *39*, 52–53.
97. Dimitrov, B.D.; Gadeva, P.G.; Benova, D.K.; Bineva, M.V. Comparative genotoxicity of the herbicides Roundup, Stomp and Reglone in plant and mammalian test systems. Mutagenesis **2006**, *21* (6), 375–382.
98. Grant, W.F. The present status of higher plant bioassays for the detection of environmental mutagens. Mutat. Res. **1994**, *310* (2), 175–185.
99. Ma, T.H.; Cabrera, G.L.; Owens, E. Genotoxic agents detected by plant bioassays. Rev. Environ. Health **2005**, *20* (1), 1–13.
100. Feretti, D.; Zerbini, I.; Zani, C.; Ceretti, E.; Moretti, M.; Monarca, S. *Allium cepa* chromosome aberration and micronucleus tests applied to study genotoxicity of extracts from pesticide-treated vegetables and grapes. Food Addit. Contam. **2007**, *24* (6), 561–572.
101. Mancini, A.; Buschini, A.; Restivo, F.M.; Rossi, C.; Poli, P. Oxidative stress as DNA damage in different transgenic tobacco plants. Plant Sci. **2006**, *170*, 845–852.
102. Norppa, H. Cytogenetic biomarkers and genetic polymorphisms. Toxicol. Lett. **2004**, *149* (1–3), 309–334.
103. Vodicka, P.; Koskinen, M.; Naccarati, A.; Oesch-Bartlomowicz, B.; Vodickova, L.; Hemminki, K.; Oesch, F. Styrene metabolism, genotoxicity, and potential carcinogenicity. Drug Metab. Rev. **2006**, *38* (4), 805–853.
104. Lampe, J.W. Diet, genetic polymorphisms, detoxification, and health risks. Altern. Ther. Health Med. **2007**, *13* (2), S108–S111.
105. Lin, J.; Swan, G.E.; Shields, P.G.; Benowitz, N.L.; Gu, J.; Amos, C.I.; de Andrade, M.; Spitz, M.R.; Wu, X. Mutagen sensitivity and genetic variants in nucleotide excision repair pathway: Genotype–phenotype correlation. Cancer Epidemiol. Biomarkers Prev. **2007**, *16* (10), 2065–2071.
106. Rueff, J.; Teixeira, J.P.; Santos, L.S.; Gaspar, J.F. Genetic effects and biotoxicity monitoring of occupational styrene exposure. Clin. Chim. Acta **2009**, *399* (1–2), 8–23.
107. Rohr, P.; da Silva, J.; Erdtmann, B.; Saffi, J.; Guecheva, T.N.; Henriques, J.A.; Kvitko, K. BER gene polymorphisms (OGG1 Ser326Cys and XRCC1 Arg194Trp) and modulation of DNA damage due to pesticides exposure. Environ. Mol. Mutagen. **2010** [Epub ahead of print].
108. Goode, E.L.; Ulrich, C.M.; Potter J.D. Polymorphisms in DNA repair genes and associations with cancer risk. Cancer Epidemiol. Biomarkers Prev. **2002**, *11*, 1513–1530.
109. Buschini, A.; De Palma, G.; Poli, P.; Martino, A.; Rossi, C.; Mozzoni, P.; Scotti, E.; Buzio, L.; Bergamaschi, E.; Mutti, A. Genetic polymorphism of drug-metabolizing enzymes and styrene-induced DNA damage. Environ. Mol. Mutagen. **2003**, *41* (4), 243–252.
110. Vodicka, P.; Kumar, R.; Stetina, R.; Sanyal, S.; Soucek, P.; Haufroid, V.; Dusinska, M.; Kuricova, M.; Zamecnikova, M.; Musak, L.; Buchancova, J.; Norppa, H.; Hirvonen, A.; Vodickova, L.; Naccarati, A.; Matousu, Z.; Hemminki, K. Genetic polymorphisms in DNA repair genes and possible links with DNA repair rates, chromosomal aberrations and single-strand breaks in DNA. Carcinogenesis **2004**, *25*, 757–763.
111. Landsiedel, R.; Kapp, M.D.; Schulz, M.; Wiench, K.; Oesch, F. Genotoxicity investigations on nanomaterials: Methods, preparation and characterization of test material, potential artefacts and limitations—Many questions, some answers. Mutat. Res. **2009**, *681*, 241–258.
112. Savolainen, K.; Alenius, H.; Norppa, H.; Pylkkänen, L.; Tuomi, T.; Kasper, G. Risk assessment of engineered nanomaterials and nanotechnologies—A review. Toxicology **2010**, *269*, 92–104.
113. Wang, J.J.; Sanderson, B.J.S.; Wang, H. Cyto- and genotoxicity of ultrafine TiO_2 particles in cultured human lymphoblastoid cells. Mutat. Res. **2007**, *628*, 99–106.
114. Reeves, J.F.; Davies, S.J.; Dodd, N.J.F.; Jha, A.N. Hydroxyl radicals (•OH) are associated with titanium dioxide (TiO_2) nanoparticle-induced cytotoxicity and oxidative DNA damage in fish cells. Mutat. Res. **2008**, *640*, 113–122.
115. Yang, H.; Liu, C.; Yang, D.; Zhanga, H.; Xi, Z. Comparative study of cytotoxicity, oxidative stress and genotoxicity induced by four typical nanomaterials: The role of particle size, shape and composition. J. Appl. Toxicol. **2009**, *29*, 69–78.
116. Sharma, V.; Shukla, R.K.; Saxena, N.; Parmar, D.; Das, M.; Dhawan, A. DNA damaging potential of zinc oxide nanoparticles in human epidermal cells. Toxicol. Lett. **2009**, *185*, 211–218.
117. Elespuru, R.K.; Agarwal, R.; Atrakchi, A.H.; Bigger, C.A.H.; Heflich, R.H.; Jagannath, D.R.; Levy, D.D.; Moore, M.M.; Ouyang, Y.; Robison, T.W.; Sotomayor, R.E.; Cimino, M.C.; Dearfield, K.L. Current and future application of genetic toxicity assays: The role and value of in vitro mammalian assays. Toxicol. Sci. **2009**, *109* (2), 172–179.

118. Garcia-Sagredo, J.M. Fifty years of cytogenetics: A parallel view of the evolution of cytogenetics and genotoxicology. Biochim. Biophys. Acta **2008**, *1779*, 363–375.
119. Robbens, J.; van der Ven, K.; Maras, M.; Blus, R.; De Coen, W. Ecotoxicological risk assessment using DNA chips and cellular reporters. Trends Biotechnol. **2007**, *25* (10), 460–466.
120. Gant, T.W.; Zhang, S.D. In pursuit of effective toxicogenomics. Mutat. Res. **2005**, *575*, 4–16.
121. Ellinger-Ziegelbauerb, H.; Aubrechta, J.; Kleinjans, J.C.; Ahr, H.-J. Application of toxicogenomics to study mechanisms of genotoxicity and carcinogenicity. Toxicol. Lett. **2009**, *186*, 36–44.
122. Waters, M.D.; Olden, K.; Tennant, R.W. Toxicogenomic approach for assessing toxicant-related disease. Mutat. Res. **2003**, *544*, 415–424.
123. Sen, B.; Mahadevan, B.; DeMarini, D.M. Transcriptional responses to complex mixtures: A review. Mutat. Res. **2007**, *636*, 144–177.
124. Lettieri, T. Recent applications of DNA microarray technology to toxicology and ecotoxicology. Environ. Health Perspect. **2006**, *114*, 4–9.

Pollution: Non-Point Source

Ravendra Naidu
Mallavarapu Megharaj
Peter Dillon
Rai Kookana
Ray Correll
Commonwealth Scientific and Industrial Research Organization (CSIRO), Adelaide, South Australia, Australia

W.W. Wenzel
Institute of Soil Research, University of Natural Resources and Life Sciences, Vienna, Austria

Abstract
Non-point source pollution (NPSP) has no obvious single point source discharge and is of diffuse nature. Examples of NPSP include aerial transport and deposition of contaminants, rain water in urban areas, fertilizer, pesticide applications, and industrial waste materials.

INTRODUCTION

Non-point source pollution (NPSP) has no obvious single point source discharge and is of diffuse nature (Table 1). An example of NPSP includes aerial transport and deposition of contaminants such as SO_2 from industrial emissions leading to acidification of soil and water bodies. Rain water in urban areas could also be a source of NPSP as it may concentrate organic and inorganic contaminants. Examples of such contaminants include polycyclic aromatic hydrocarbons, pesticides, polychlorinated biphenyls that could be present in urban air due to road traffic, domestic heating, industrial emissions, agricultural treatments, etc.[1–3] Other examples of NPSP include fertilizer (especially Cd, N, and P) and pesticide applications to improve crop yield. Use of industrial waste materials as soil amendments have been estimated to contaminate thousands of hectares of productive agricultural land in countries throughout the world.

CONTAMINANT INTERACTIONS

Non-point pollution is generally associated with low-level contamination spread at broad acre level. Under these circumstances, the major reaction controlling contaminant interactions are sorption–desorption processes, plant uptake, surface runoff, and leaching. However, certain contaminants, in particular, organic compounds are also subjected to voltalization, chemical, and biological degradation. Sorption–desorption and degradation (both biotic and abiotic) are the two most important processes controlling organic contaminant behavior in soils. These processes are influenced by both soil and solution properties of the environment. Such interactions also determine the bioavailability and/or transport of contaminants in soils. Where the contaminants are bioavailable, risk to surface and groundwater and soil, crop, and human health are enhanced.

IMPLICATIONS TO SOIL AND ENVIRONMENTAL QUALITY

Environmental contaminants can have a deleterious effect on non-target organisms and their beneficial activities. These effects could include a decline in primary production, decreased rate of organic matter break-down, and nutrient cycling as well as mineralization of harmful substances that in turn cause a loss of productivity of the ecosystems. Certain pollutants, even though present in very small concentrations in the soil and surrounding water, have potential to be taken up by various micro-organisms, plants, animals, and ultimately human beings. These pollutants may accumulate and concentrate in the food chain by several thousand times through a process referred to as biomagnification.

Urban sewage, because of its nutrient values and source of organic carbon in soils, is now increasingly being disposed to land. The contaminants present in sewage sludge (nutrients, heavy metals, organic compounds, and pathogens), if not managed properly, could potentially affect the environment adversely. Dumping of radioactive waste (e.g., radium, uranium, plutonium) onto soil is more complicated because these materials remain active for thousands of years in the soil and thus pose a continued threat to the future health of the ecosystem.

Table 1 Industries, land uses, and associated chemicals contributing to non-point source pollution.

Industry	Type of chemical	Associated chemicals
Agricultural activities	Metals/metalloid	Cadmium, mercury, arsenic, selenium
	Non-metals	Nitrate, phosphate, borate
	Salinity/sodicity	Sodium, chloride, sulfate, magnesium, alkalinity
	Pesticides	Range of organic and inorganic pesticides including arsenic, copper, zinc, lead, sulfonylureas, organochlorine, organophosphates, etc., salt, geogenic contaminants (e.g., arsenic, selenium, etc.)
	Irrigation	Sodium, chloride, arsenic, selenium
Automobile and industrial emissions	Dust	Lead, arsenic, copper, cadmium, zinc, etc.
	Gas	Sulfur oxides, carbon oxides
	Metals	Lead and lead organic compounds
Rainwater	Organics	Polyaromatic hydrocarbons, polychlorbiphenyls, etc.
	Inorganic	Sulfur oxides, carbon oxides acidity, metals and metalloids

Source: (From Barzi, F.; Naidu, R.; McLaughlin, M.J. Contaminants and the Australian Soil Environment. In *Contaminants and the Soil Environment in the Australasia-Pacific Region*; Naidu, R., Kookana, R.S., Oliver, D., Rogers, S., McLaughlin, M.J., Eds.; Kluwer Academic Publishers: Dordrecht, the Netherlands, 1996; 451–484.)

Industrial wastes, improper agricultural techniques, municipal wastes, and use of saline water for irrigation under high evaporative conditions result in the presence of excess soluble salts (predominantly Na and Cl ions) and metalloids such as Se and As in soils. Salinity and sodicity affect the vegetation by inhibiting seed germination, decreasing permeability of roots to water, and disrupting their functions such as photosynthesis, respiration, and synthesis of proteins and enzymes.

Some of the impacts of soil pollution migrate a long way from the source and can persist for some time. For example, suspended solids can increase water turbidity in streams, affecting benthic and pelagic aquatic ecosystems, filling reservoirs with unwanted silt, and requiring water treatment systems for potable water supplies. Phosphorus attached to soil particles, which are washed from a paddock into a stream, can dominate nutrient loads in streams and down-stream water bodies. Consequences include increases in algal biomass, reduced oxygen concentrations, impaired habitat for aquatic species, and even possible production of cyanobacterial toxins, with series impacts for humans and livestock consuming the water. Where waters discharge into estuaries, N can be the limiting factor for eutrophication; estuaries of some catchments where fertilizer use is extensive have suffered from excessive sea grass and algal growth.

More insidious is the leaching of nutrients, agricultural chemicals, and hydrocarbons to groundwater. Incremental increases in concentrations in groundwater may be observed over long periods of time resulting in initially potable water becoming undrinkable and then some of the highest valued uses of the resource may be lost for decades. This problem is most severe on tropical islands with shallow relief and some deltaic arsenopyrite deposits, where wells cannot be deepened to avoid polluted groundwater because underlying groundwater is either saline or contains too much As.

SAMPLING FOR NON-POINT SOURCE POLLUTION

The sampling requirements of NPSP are quite different from those of the point source contamination. Typically, the sampling is required to give a good estimate of the mean level of pollution rather than to delineate areas of pollution. In such a situation, sampling is typically carried out on a regular square or a triangular grid. Furthermore, gains may be possible by using composite sampling.[4] However, if the pollution is patchy, other strategies may be used. One such strategy is to divide the area into remediation units, and to sample each of these. The possibility of movement of the pollutant from the soil to some receptor (or asset) is assessed, and the potential harm is quantified. This process requires an analysis of the bioavailability of the pollutant, pathway analysis, and the toxicological risk. The risk analysis is then assessed and decisions are then made as to how the risk should be managed.

MANAGEMENT AND/OR REMEDIATION OF NON-POINT SOURCE POLLUTION

The treatment strategies used for managing NPSP are generally those that modify the soil properties to decrease the bioavailable contaminant fraction. This is particularly so in the rural agricultural environment where soil–plant transfer of contaminants is of greatest concern. Soil amendments commonly used include those that change the ion-exchange characteristics of the colloid particles and those that enhance the ability of soils to sorb contaminants. An example of NPSP management includes the application of lime to immobilize metals because the solubility of most heavy metals decreases with increasing soil pH. However, this approach is not applicable to all metals, especially those that form oxyanions—the bioavailability of such species increases with increasing pH. Therefore, one of the prerequisites for remediating contaminated sites is a detailed assessment of the nature of contaminants present in the soil. The application of a modified aluminosilicate to a highly contaminated soil around a zinc smelter in Belgium was shown to reduce the bioavailability of metals thereby reducing the Zn phytotoxicity.[5] The simple addition of rock phosphates to form Pb phosphate has also been demonstrated to reduce the bioavailability of Pb in aqueous solutions and contaminated soils due to immobilization in the metal.[6] Nevertheless, there is concern over the long-term stability of the processes. The immobilization process appears attractive currently given that there are very few cheap and effective in situ remediation techniques for metal-contaminated soils. A novel, innovative approach is using higher plants to stabilize, extract, degrade, or volatilize inorganic and organic contaminants for in situ treatment (cleanup or containment) of polluted topsoils.[7]

PREVENTING WATER POLLUTION

The key to preventing water pollution from the soil zone is to manage the source of pollution. For example, nitrate pollution of groundwater will always occur if there is excess nitrate in the soil at a time when there is excess water leaching through the soil. This suggests that we should aim to reduce the nitrogen in the soil during wet seasons and the drainage through the soil. Local research may be needed to demonstrate the success of best management techniques in reducing nutrient, sediment, metal, and chemical exports via surface runoff and infiltration to groundwater. Production figures from the same experiments may also convince local farmers of the benefits of maintaining nutrients and chemicals where needed by a crop rather than losing them off site, and facilitate uptake of best management practices.

GLOBAL CHALLENGES AND RESPONSIBILITY

The biosphere is a life-supporting system to the living organisms. Each species in this system has a role to play and thus every species is important and biological diversity is vital for ecosystem health and functioning. The detection of hazardous compounds in Antarctica, where these compounds were never used or no man has ever lived before, indicates how serious is the problem of long-range atmospheric transport and deposition of these pollutants. Clearly, pollution knows no boundaries. This ubiquitous pollution has had a global effect on our soils, which in turn has been affecting their biological health and productivity. Coupled with this, over 100,000 chemicals are being used in countries throughout the world. Recent focus has been on the endocrine disruptor chemicals that mimic natural hormones and do great harm to animal and human reproductive cycles.

These pollutants are only a few examples of contaminants that are found in the terrestrial environment.

REFERENCES

1. Chan, C.H.; Bruce, G.; Harrison, B. Wet deposition of organochlorine pesticides and polychlorinated biphenyls to the great lakes. J. Great Lakes Res. **1994**, *20*, 546–560.
2. Lodovici, M.; Dolara, P.; Taiti, S.; Del Carmine, P.; Bernardi, L.; Agati, L.; Ciappellano, S. Polycyclic aromatic hydrocarbons in the leaves of the evergreen tree *Laurus Nobilis*. Sci. Total Environ. **1994**, *153*, 61–68.
3. Sweet, C.W.; Murphy, T.J.; Bannasch, J.H.; Kelsey, C.A.; Hong, J. Atmospheric deposition of PCBs into green bay. J. Great Lakes Res. **1993**, *18*, 109–128.
4. Patil, G.P.; Gore, S.D.; Johnson, G.D. *Manual on Statistical Design and Analysis with Composite Samples*; Technical Report No. 96-0501; EPA Observational Economy Series Center for Statistical Ecology and Environmental Statistics; Pennsylvania State University, 1996; Vol. 3.
5. Vangronsveld, J.; Van Assche, F.; Clijsters, H. Reclamation of a bare industrial area, contaminated by non-ferrous metals: in situ metal immobilisation and revegetation. Environ. Pollut. **1995**, *87*, 51–59.
6. Ma, Q.Y.; Logan, T.J.; Traina, S.J. Lead immobilisation from aqueous solutions and contaminated soils using phosphate rocks. Environ. Sci. Technol. **1995**, *29*, 1118–1126.
7. Wenzel, W.W.; Adriano, D.C.; Salt, D.; Smith, R. Phytoremediation: a plant-microbe based remediation system. In *Bioremediation of Contaminated Soils*; Soil Science Society of America Special Monograph No. 37, Adriano, D.C., Bollag, J.M., Frankenberger, W.T., Jr., Sims, W.R., Eds.; Soil Science Society of America: Madison, USA, 1999; 772 pp.

Pollution: Pesticides in Agro-Horticultural Ecosystems

J.K. Dubey
Regional Center, National Afforestation and Eco-Development Board, Dr. Y.S. Parmar University of Horticulture and Forestry, Solan, India

Meena Thakur
Department of Environment Sciences, Dr. Y.S. Parmar University of Horticulture and Forestry, Solan, India

Abstract
Pesticides have played a very important role in enhancing food productivity for decades. Agricultural intensification and diversification in most of the developing and developed countries have resulted in high-input farming, i.e., excessive use of pesticides and fertilizers, which has helped to meet the rising food demand. Agricultural and horticultural ecosystems are now dominated by monoculture and hybrid varieties to enhance the crop yields that have simultaneously resulted in enhanced pesticide use. In 2001, according to U.S. Environmental Protection Agency (EPA) published data, 675 million lb of chemical pesticides were used in agriculture and nearly one-quarter of all pesticides used are applied to cotton, and the overall amount and intensity per acre is increasing every year. If we observe the trend of herbicide use in the United States, agriculture accounts for three quarters of all chemical pesticide usage. Thus, horticultural and agricultural ecosystems are highly dependent on pesticides. It is estimated that if pesticides are not used, food supplies would fall to 30%–40% due to the ravages of pests. However, the intensive use of pesticides in agricultural production inevitably leads to pesticide pollution by entering the various environmental compartments. Continuous use of pesticides results in pest resistance, secondary pest outbreak, and pest resurgence, leading to changes in ecosystem biodiversity. Pesticide residues in the environment adversely affect human health and the natural environment. Pesticides like methyl bromide are intensively used as soil fumigants around the world. It has been banned since 2005 as it causes depletion of the ozone layer. The hazards of pesticide pollution have been much realized presently and there are demands for reduction in pesticide use in many parts of the world. It is necessary to modify "good agricultural practice" and change it to "pesticide avoidance practice." It is urgent to improve the education of farmers and to promote organic farming so as to prevent pesticide pollution in agro-horticultural ecosystems.

Agro-horticultural ecosystems are man-made ecosystems that are greatly influenced by human activities. In the last few decades, agricultural intensification and diversification in most of the developing and developed countries have resulted in high input farming, i.e., excessive use of pesticides (herbicides, insecticides, fungicides) and fertilizers, which has helped to meet the rising food demand. Agricultural and horticultural ecosystems are now dominated by monoculture and hybrid varieties to enhance the crop yields. Hybrid varieties are high yielding and more demanding. The excessive use of inorganic fertilizers results in rapid multiplication and subsequent outbreaks of many pests, simultaneously resulting in enhanced pesticide use. Rice and cotton are important crops grown worldwide, and because of monoculture, they are attacked by hundreds of pests and they receive disproportionately high share of pesticides (17 and 24%) worldwide.[1] It has been observed that chemicals that have been banned for most of the food crops are still being used on cotton as it is not consumed directly. For example, nearly one-quarter of all pesticides used in the United States are applied to cotton, and the overall amount and intensity per acre is increasing every year. Worldwide, more and extremely toxic pesticides are sprayed on cotton than on any other crop,[2] which find entry in human body in the form of salad dressings, baked goods, and snacks like Fritos and Goldfish.

Based on the trend of herbicide use in the United States, agriculture accounts for three quarters of total chemical pesticides used. According to data published by the U.S. Environmental Protection Agency (EPA), in 2001, 675 million lb of chemical pesticides were used in agriculture.[3] Thus, horticultural and agricultural ecosystems are highly dependent on pesticides. It is estimated that if pesticides are not used, food supplies would fall to 30%–40% due to the ravages of pests.[4] Worldwide, research data show that without effective pest management, preharvest losses in crops would average about 40% and the world's food and fiber production as well as environmental and human health would be seriously threatened.[5] According to a market research report, China and India are the world's largest consumers of agrochemicals. The United States is estimated to be second largest with an 18.5% share of the market. According to the Global Agrochemical Market (2009–2014) report, the total global agrochemical market is expected to be worth $196 billion by 2014, of which the Asian market will account for nearly 43.12% of the total revenues.[6]

No doubt, pesticides play an important role in enhancing agricultural productivity, but intensive use of pesticides in agricultural and horticultural ecosystems has resulted in degradation of environment, whether it is in the form of development of resistant pests, secondary pest outbreak, or pest resurgence, which leads to changes in ecosystem biodiversity or adverse effects on pollinators, natural enemies, and many incurable human diseases. Several pesticides are known to persist for a longer period in the environment or on the substrate to which they were applied and have long-term side effects on human health and the natural environment. Persistent organic pesticides, especially DDT (Dichlorodiphenyltrichcloroethane) and HCH (Hexachlorocyclohexane), have been detected in various systems, even in human blood, fat, and milk samples. Even the soft drinks that are water-based flavored drinks are known to contain pesticides. The hazards of pesticide pollution have been much realized presently and the demands for their reduction are desired globally in various pesticide monitoring programs.

PESTICIDE CONSUMPTION TREND

Despite the wide use of non-chemical techniques such as sanitation, cultivation, crop rotation, resistant cultivars, and biological control (including introduction of transgenic) for pest control, many pests cannot be controlled adequately and there is a continuous need for application of substantial quantities of chemical pesticides, as a result of which crop protection in many developing countries is still dominated by an increasing use of pesticides. Further, government policies on subsidies, establishment of market for agrochemicals, availability of technology packet, etc., encourage the farmers to use pesticides excessively. Today, more than 80% of worldwide pesticide sales fall to the share of only six companies. In 2004, three agrochemical companies, each with sales of more than $4 billion, together controlled the global market for pesticides.[7] By controlling such a large stake of the market, these companies have a considerable influence on the way in which plant protection is practiced in farming ecosystems.

In many cases, farmers go for prophylactic applications, whether they are required or not. The findings of microlevel studies on the pesticide consumption pattern in major food crops of Kerala, India, are presented here.[8] Kerala is one of the leading agricultural states of India, and currently, India is the leading manufacturer of basic pesticides in Asia and ranks 12th globally. Compiling the data on the consumption level of pesticides in agriculture in Kerala (1995–1996 to 2007–2008), the total quantity is estimated at 462.05 metric tons (2007–2008). Pesticide application in the state is prophylactic and is one of the most important risk management strategies; e.g., pesticide application on bitter gourd starts from the time of transplanting. The prophylactic application of the pesticides is resorted to at an interval of 2 weeks initially, which gets reduced to 2 days as the crop nears flowering and fruit set. There is a tendency among farmers to change the chemicals in each spray. Thus, on an average, acetamaprid is sprayed 6 times, phorate and dimethoate 5 times each, quinalphos and indoxacarb 4 times each, and the rest 3–4 times each. During a crop cycle of 90 days in bitter gourd, farmers apply pesticides as many as 50 times. The pesticides that are used in the state include chemicals that are banned for sale in Kerala (endosulfan), banned for use in fruits/vegetables (monocrotophos), and those permitted for restricted use only (methyl parathion, lindane, and methoxy ethyl mercury chloride). Many of the other chemicals applied are banned/not approved in many other countries. Although the government of Kerala has banned the sale/use of endosulfan owing to controversies over the environment and human health problems due to the aerial spraying of the chemical in cashew plantations in the state, the chemical is still used by the farmers. The farmers are investing a large portion of their income on pesticides, and pesticide consumption is reported to be the primary method of suicide in Kerala. Of the 900–1000 suicides per year, 60% are by consuming poisons. The commonly used poisons are furadan, malathion, and rat poison.[9] Moreover, farmers also go for suicide because of indebtedness due to purchase of chemicals and sometimes due to complete failure of crop in spite of heavy investment.

IMPACT OF PESTICIDE USE ON THE POLLINATORS

Wild bees, bumblebees, honeybees, and solitary bees are well known and valued as important pollinators of crops/plants and are in commercial use for pollination. About 33% of all crops require pollination. Intensive cultivation and excessive use of pesticides lead to a sharp decline in the population of these pollinators, which is one of the major causes of low productivity of agricultural and horticultural crops. Pesticides, viz., DDT, BHC, cyclodiene, and most of the organophosphorous and carbamate compounds, are highly toxic to bees. Although endosulfan is listed as a persistent organic pollutant, it is the only available pesticide known to be safe for honeybees and other beneficial insects, and it is extensively used in many countries. Health and environmental causalities related to excessive use of this pesticide are reported from many parts of the world, and countries such as those in Europe withdrew its registration in 2005. Still, there are many instances where most of the pesticides fail to be effective and endosulfan is recommended. For example, in 2008, there was heavy weevil infestation on hazelnut crop in Italy, and the Italian government had to prescribe the use of endosulfan for 120 days although it endangered the health of its citizens. In India, the ban on endosulfan is still controversial because there is no other alternative to it. Pesticides like neonicotinoids used as a substitute for endosulfan in agricultural ecosystems in countries like Germany, France, U.K., and the United States have resulted in mass bee killing and colony collapse disorders. In 1959, carbaryl was used against certain orchard pests and later registered for many other crops. In 1967, it caused the destruction of an estimated 70,000 colonies of honeybees in California from use in cotton and an estimated 33,000 colonies in Washington from use in corn. The estimated national loss from all pesticide poisoning from the same year was 500,000 colonies. Carbaryl is still one of the most destructive bee-killing chemical.[10]

A horticultural ecosystem, particularly the apple ecosystem, is heavily polluted with pesticides. Apple is one of the important commercial horticultural crops grown in temperate regions of the world. Honeybees are important pollinators of this crop and play an important role in apple productivity; quality beehives are placed in apple orchards. Since the crop is important from an economic point of view, much attention is paid to it. In India, the temperate northern regions of the country (Himachal Pradesh, and Jammu and Kashmir) are known for quality apple production. To obtain quality fruits in Himachal Pradesh, a number of pesticides, which affect the pollinators, are applied on the crop right from fruit set to harvest. The situation has reached an alarming level in the state and if required measures are not initiated to conserve and rear the population of these pollinators, it could impinge on the total agricultural and fruit production in years to come. Currently, about 99,000 ha of land is under fruit cultivation in the state, and it requires at least 5 lakh colonies of honeybees alone for pollination to enhance the production. There is a great need to encourage organic farming to enhance the population of natural pollinators and avoid pesticide applications when crops, cover crops, weeds, and wildflowers are in bloom in the treatment area or nearby.

IMPACT OF PESTICIDE USE ON NATURAL ENEMIES

Apple ecosystems host many species of phytophagous arthropods among which red spider mite and two-spotted spider mite are substantial worldwide. Predatory mites play important role in checking the population of these mites. In India, the red spider mite *Panonychus ulmi* was a minor pest up to 1990, but the commercialization of apple led to excessive and repeated use of pesticides for quality apple production, as a result of which the natural mite predators were destroyed and it emerged as a serious pest of apples; most of the spray schedules are now focused to this pest, further deteriorating the condition. Pyrethroids and carbamates, i.e., carbaryl (Sevin), are highly toxic to predatory mites, viz., *Typhlodromus occidentalis*, and use of Sevin for thinning causes mite flare-ups. Another well-known example is the resurgence of brown plant hopper (BPH) in rice ecosystem. If no pesticides are used, BPH is kept under control by its natural enemies (mirid bugs, ladybird beetles, spiders, and various pathogens). Since rice is a heavily sprayed crop, pesticides kill the natural enemies and create a situation where BPH can multiply rapidly. Thus, similar to *P. ulmi*, it has also become a serious man-made pest. Synthetic pyrethroids result in spider mite resurgence.[11] In a study conducted by Beers,[12] pyrethroids, carbamates, organophosphates, Assail, Calypso, and Actara are toxic to *T. occidentalis* and *Zetzellia mali*, which are mostly found associated with *P. ulmi* and other phytophagous mites. OPs and carbamates are reported to cause high levels of mortality to coccinellids and lacewings.

PESTICIDE USE AND BT TRANSGENICS

A common pest management technology used in agro-ecosystems is the use of *Bacillus thuringiensis* (Bt) transgenic. Bt crops, particularly cotton, are grown all over the world. Bt crops were mainly introduced with an aim to reduce pesticide use, but growing secondary pest populations and efforts to control them have further increased the use of pesticides The major cotton-growing countries—the United States, China, India, and Argentina—have quickly adopted this technique for cotton seeds. For example, before the commercialization of Bt cotton, the Chinese farmers applied an average of 20 pesticide treatments in a season to control bollworm infestations. With the adoption of Bt, the average number of treatments has fallen to only 6.6 at the early stages of Bt adoption.[13] As a result, the pesticide use decreased by 43.3 kg/ha in 1999, i.e., a 71% decrease in pesticide use. For the years 2000 and 2001, Bt cotton was associated with an average reduction of 35.7 kg/ha of pesticide, or a percentage deduction of 55%.[14] Similar results have been found in other major cotton-growing countries: Indian farmers save 39% of expenditures by planting Bt,[15] Argentine farmers save 47% of expenditures,[16] Mexican farmers can save 77%,[17] and South African farmers can save 58% by planting Bt.[18] Evidence shows that, though Bt seed costs 2 to 3 times more than a conventional seed, savings on pesticide expenditures guarantee a much higher net return for Bt adopters. Using a household survey from 2004, 7 years after the initial commercialization of Bt cotton in China, we show that total pesticide expenditure for Bt cotton farmers in China is nearly equal to that of their conventional counterparts, about $101/ha. Bt farmers in 2004, on the average, have to spray pesticide 18.22 times, which are more than 3 times higher compared with 6 times pesticide spray in 1999. Detailed information on pesticide expenditures reveals that, though Bt farmers saved 46% of bollworm pesticide relative to non-Bt farmers, they spend 40% more on pesticides targeted to kill an emerging secondary pest. These secondary pests, e.g., mirid bugs, were rarely found in the field prior to the adoption of Bt cotton, presumably kept in check by bollworm populations and regular pesticide spraying.

Cotton is attacked by more than 165 pests, and farmers repeatedly spray pesticides, which increase the chances of resurgence of secondary pests. In Andhra Pradesh, the number of attacks of aphids, thrips, and jassids has increased since the introduction of Bt cotton in 2002. Many diseases and pests such as tobacco leaf streak virus, tobacco caterpillars, etc., have newly emerged in Bt cotton ecosystems in this state.[8]

PESTICIDE USE IN WEEDS

The large-scale adoption of dwarf HYV and hybrids and the increased use of irrigation, fertilizers, and monocropping have increased weed problem in agro-horticultural ecosystems, simultaneously leading to increased herbicide use. Herbicides, viz., isoproturon, atrazine, alachlor, butachlor, and oxyfluorfen, are applied on agro-horticultural ecosystems for control of weeds. Globally, herbicides constitute 52% of the total pesticide sales, and in some countries like the United States, Germany, and Australia, the figure is as high as 60%–70%.[19] According to a USDA-NASS (U.S. Department of Agriculture–National Agricultural Statistics Service) report, the use of genetically modified crops is the main reason for the rise in herbicide use.[20] For example, widespread introduction of genetically modified soybeans, cotton, and corn by Monsanto resulted in 15-fold increase in the use of glyphosate (Roundup) from 1994 to 2005 on these three crops in the United States. The excessive use of glyphosate has resulted in resistant weeds, as a result of which the application of glyphosate, atrazine, 2,4-D, and other leading weed-killing chemicals has further increased since 2002. 2,4-D, the second most heavily used herbicide on soybeans (after glyphosate) in the United States, is associated with a number of adverse health impacts on agricultural workers. These herbicides have increased the risk of cancer, have increased the rate of birth defects in children of men who apply the herbicide, and are also a suspected endocrine disruptor. Similarly, atrazine, the most heavily used herbicide on corn, has been linked to endocrine disruption, neuropathy, and cancer (particularly breast and prostate cancer). It is regularly detected in drinking water supplies in the United States and has been associated with low sperm counts in men. Exposure to extremely low levels of atrazine can cause sex change and/or deformities in frogs, fish, and other organisms. Based on this evidence, and the widespread presence of atrazine in drinking water supplies, the European Union announced a ban on atrazine in 2006. However, the U.S. EPA reregistered atrazine in 2003 despite objections from scientists and environmental groups.[21] Cheaper formulations of herbicides containing 2,4-D and 2-methyl-4- chlorophenoxyacetic acid (MCPA), are still used in many countries, and weeds have developed resistance; e.g., in Bulgaria, 47% of wheat and barley crops were affected by 2,4-D-resistant weeds in 2000.[26]

PESTICIDE RESIDUES IN AGRO-HORTICULTURAL ECOSYSTEMS

Most of the pesticides used on crops are persistent, especially organochlorines, which persist for a longer period in the environment (substrate). The organochlorine insecticides (such as DDT and BHC) that were banned still persist in soil and contaminate both organic and conventional crop produce. Baker and co-workers observed pesticide residues in organic fruit samples. The reasons for residues in organic fruit samples were in violation of organic methods of cultivation, pesticide contaminated water used for irrigation or pesticide residues left in the soil, if previously used to grow conventional crops.[22]

Apple fruit crop is attacked by a number of insect pests and diseases like apple scab, San Jose scale, wooly apple aphid, fruit scrapper, defoliating beetles, and tent caterpillar. Pesticides like chlorpyrifos, endosulfan, carbendazim, propineb, mancozeb, etc., are applied to control these pests. After spraying/treatment, pesticide residues get deposited on the fruits and dissipate slowly depending upon the number of factors like physiochemical characteristics of pesticide, weather conditions, time after treatment, etc. Pre- or postharvest interval or waiting period between spray and harvest is required for safe consumption of fruits. Sometimes, the produce is sent to the market immediately after spraying and consumers unknowingly consume the produce and may be badly affected. Similarly, under Indian conditions, a number of synthetic pesticides, viz., deltamethrin, cypermethrin, dimethoate, quinalphos, oxydemeton methyl, and carbaryl, are used to control mango crop pests like mango hopper, mango mealy bug, and fruit fly, as well as powdery mildew and malformations. Deltamethrin at 0.002% does not require any waiting period but cypermethrin requires 11 days of waiting period. Mango is eaten after removing the peel, but the residues on its peel also find their way into the consumer by contact. Residues of mancozeb and lindane though within the permissible limit were detected in mango fruit samples.[23] The repeated spray of bifenthrin on mango from flowering to 1 mo before harvest resulted in residues that persisted on the peel for more than a month, and rate of degradation was very low.[24]

The consumption of pesticide in India is low as compared to other countries; in spite of this, there is widespread contamination of food commodities with pesticide residues due to non-judicious use of pesticides. An earlier survey carried out by the Indian Council of Medical Research, New Delhi, revealed that 51% of food commodities contained pesticide residues, and out of these, 20% had pesticide residues above the maximum residue limit (MRL) values, as compared to 21% contamination with only 2% of samples above the MRL on a worldwide basis.[24] Now, the scenario in India has started changing very rapidly as new pesticide molecules, whose application rate (as well as persistence in the environment) is very low, are being introduced every year. Heavy-duty pesticides have been either banned or put under restricted use. The pesticide load on the agro-horticultural ecosystem has declined as compared to the last decades. Maximum pesticides in India are used on cotton and rice. The Malwa area of Punjab, which is famous for cotton growing, has been named as the cancer belt of Punjab because pesticides have contaminated the whole environment, including groundwater, and caused cancer among its people. Out of the total pesticides used in India, only 13%–14% is used on fruits and vegetables; despite this, half of the fruits and vegetables were found contaminated with pesticide residues.[24] Pesticide residues in 10% of the samples were above the MRL value. Residues of methyl parathion, endosulfan, chlorpyrifos, DDVP, dimethoate, fenitrothion, monocrotophos, cypermethrin, deltamethrin, copper, etc., were above the MRL in fruits and vegetables.[24]

APPROACHES FOR PESTICIDE USE REDUCTION

Since the excessive and indiscriminate use of pesticides has polluted every component of the environment, people all over the world have realized the need for pesticide reduction so as to prevent the environment from further deterioration. However, it is not an easy task as agrochemical market and crop protection knowledge is increasingly controlled by few multinationals. Today, more than 80% of worldwide pesticide sales fall to the share of only six companies. Presently, efforts are being made to reduce pesticide application worldwide by organic and integrated pest management (IPM) approaches. Agenda 21 of the United Nations Conference on Environment and Development (UNCED) at Rio de Janeiro in June 1992 identified IPM as one of the requirements for promoting sustainable agriculture and rural development. However, in countries like India, this alternative pest management approach to reduce pesticide use could not find much success due to poor farmer participation. An all-India survey confirmed that 34% of the respondents have no idea about IPM and only less than 5% of them follow complete IPM technology.[25]

However, IPM techniques are still characterized by a large amount of pesticide use and by the application of many different pesticides; e.g., organic apple production does not use any herbicides and applies only biological control, but fungal diseases like apple scab demand for the intensive use of sulfur and copper in organic apple orchards and copper has a negative impact on the environment. IPM techniques have not been widely implemented on many crops, e.g., wheat. There is a great need to modify "good agricultural practice" and change it to "pesticide avoidance practice" and to improve the education of farmers so as to promote organic farming and IPM as the best alternatives for pesticides.

REFERENCES

1. Dhaliwal, G.S.; Arora, R. *Integrated Pest Management Concepts and Approaches*; Kalyani Publishers: New Delhi, 2001; 427 pp.
2. Imhoff, D. *King Cotton—Pesticide residue is common in cotton byproducts used in agriculture—Brief Article*. Sierra, 1999 Sierra Magazine COPYRIGHT 2000 Gale Group.
3. *Pesticides Industry Sales and Usage: 2000 and 2001 Market Estimates*; U.S. Environmental Protection Agency, May 2004, Table 3.4, available at http://www.epa.gov/oppbead1/pestsales/01pestsales/market_estimates2001.pdf (accessed November 16, 2010).
4. Anonymous. Scientific agriculture prevents mass starvation. Herxter. J. Agvet Div., Hoechst, Aust. Inc., 1992.
5. Kennedy, I.R. Pesticides in Perspective: Balancing Their Benefits with the Need for Environmental Protection and Remediation of Their Residues in Seeking Agricultural Produce Free of Pesticide Residues, Proceedings of an International Workshop held in Yogyakarta, Indonesia, 17–19 February, 1998; Kennedy, I.R., Skerritt, J.H., Johnson, G.I., Highley, E., Eds.; 1998.
6. Anonymous. International news. Crop Care **2009**, *35* (3), 78.
7. Pesticides and the agrochemical industry. Pesticide Use Reduction in Germany Pesticide Action Network Germany Pesticide Action Network Europe, available at http://www.pan-germany.org (accessed November 25, 2010).
8. Indira, D.P. 2010. Pesticides in agriculture—A boon or a curse? A case study of Kerala. Econ. Polit. Wkly. **2010**, *xlv* (26 and 27) EPW.
9. Jayakrishnan, T. Health impacts of pesticides used in agriculture. In *Pesticide Use and Environmental Health*, Compendium of papers presented in the workshop, KAU/SANDEE; Devi, P.L., Ed.; 2006.
10. Johansen, C.A. Pesticides and pollinators. Annu. Rev. Entomol. **1977**, *22*, 177–92.
11. Gerson, U.; Cohen, E. Resurgence of spider mites (Acari: Tetranychidae) induced by synthetic pyrethroids. Exp. Appl. Acarol. **1989**, *6* (1), 29–46.
12. Beers, E.H. Integrated Mite Control: Nontarget Effects on Predator and Prey, 84th Orchard Pest and Disease Management Conference, 13–15 January, 2010, Portland Hilton: Portland, OR, 2010.
13. Huang, J.K.; Ruifa, H.; Fan, C.; Pray, C.E.; Rozelle, S. Bt cotton benefits, costs and impacts in China. AgBioForum **2002**, *5* (4), 153–166.
14. Pray, C.E.; Huang, J.; Hu, R.; Rozelle, S. Five years of Bt cotton in China—The benefits continue. Plant J. **2002**, *31* (4), 423–430.

15. Qaim, M.; Zilberman, D. Yield effects of genetically modified crops in developing countries. Science **2003**, *299*, 900–902.
16. Qaim, M.; dejanvry, A. Genetically modified crops, corporate pricing strategies, and farmers' adoption: The case of Bt cotton in Argentina. Am. J. Agric. Econ. **2003**, *85* (4), 814–828.
17. Traxler, G.; Godoy-Avila, S.; Falck-Zepeda, J.; Espinoza-Arellano, J. Transgenic cotton in Mexico: A case study of the Comarca Lagunera. In Kalaitzandonakes, N. (Ed.), *The economic and environmental impacts of agbiotech*. Kluwer: New York, 2003; 183–202.
18. James, C. *Global Status of Commercialized Transgenic Crops: 2002*, ISAAA Briefs 27. Ithaca: NY, 2002.
19. Dixit, A. Herbicide recommendation in different crops. Crop Care **2009**, *35* (2), 33–38.
20. Who Benefits from GM Crops? The Rise in Pesticide Use. Friends of the Earth International and Center for Food Safety, 2008, see especially pp. 8–12, available at http://www.centerforfoodsafety.org/WhoBenefitsPR2_13_08.cfm (accessed December 14, 2010).
21. Anonymous. 2011, available at http://www.beyondpesticides.org/pesticides/factsheets/Atrazine.pdf. (Accessed February 18, 2011)
22. Baker, B.P.; Benbrook, C.; Groth, E., III; Benbrook, K.L. Pesticide residues in conventional, integrated pest management (IPM)—grown and organic foods: Insights from three US data sets. Food Addit. Contam. **2002**, *19*, 427–446.
23. Agnihotri, N.P. Pesticide safety evaluation and monitoring. In *All India Co-ordinated Research Project on Pesticide Residues*; Division of Agricultural Chemicals, Indian Agricultural Research Institute (IARI): New Delhi, 1999; 173 pp.
24. Soudamini, M.; Rekha, A. Persistence of dicofol residues in/on acid lime. Pestic. Res. J. **2005**, *17* (1), 64–65.
25. Shetty, P.K.; Murugan, M.; Sreeja, K.G. Crop protection stewardship in India: Wanted or unwanted. Curr. Sci. **2008**, *95* (4), 457.
26. Nikolova, S. Pesticide use, issues and how to promote sustainable agriculture in Bulgaria. Published by Pesticide Action Network Germany (PAN Germany) in co-operation with Association Agrolink, 2004.

Pollution: Pesticides in Natural Ecosystems

J.K. Dubey
Regional Center, National Afforestation and Eco-Development Board, Dr. Y.S. Parmar University of Horticulture and Forestry, Solan, India

Meena Thakur
Department of Environmental Science, Dr. Y.S. Parmar University of Horticulture and Forestry, Solan, India

Abstract

Pesticides have played a very important role in enhancing the world food production for decades. However, their continuous and indiscriminate use has resulted in various ecological problems. Apart from creating environmental problems, pesticides and their metabolites have entered various natural ecosystems, viz., aquatic, terrestrial, soil, etc. Several groups of organochlorine pesticides such as Dichlorodiphenyltrichloroethylene (DDT), endosulfan, and chlorinated phenoxy acetic acid used as herbicides, and fungicides such as hexachlorobenzene and pentachlorophenol are major pollutants polluting various water bodies and affecting the aquatic fauna. Extensive use of pesticides have resulted in outbreaks of many serious pests, viz., codling moth, leafrollers, aphids, scales, and tetranychid mites, and in the decline of the population of pollinators and natural enemies. Herbicides, viz., atrazine, have been found to feminize frogs, leading to sterility in males and many adverse affects on humans. Human health is at risk when chemical residues are present in much of our food supplies. Integrated pest management emphasizes the need for simpler and ecologically safer measures for pest control to reduce environmental pollution. Preference may be given to organic produce grown without toxic pesticides. The objective of this entry is to provide basic knowledge on pesticide exposure and to understand issues on residues in the natural ecosystem.

Pesticides play an important role in boosting the economy of the agricultural industry by providing effective pest control, and their continued use is essential for enhancing the productivity.[1] It is estimated that food supplies would immediately fall to 30%–40% due to the ravages of pests if pesticides are not used.[2] A United Nations report stated that population growth is a major problem facing our planet. In 1900, there were 1.6 billion people on the planet. In 1992, this has risen to 5.25 billion, and by the year 2050, it will reach 10 billion. Developing countries are more affected by this explosive increase in world population. Presently, our dependence on pesticides has increased up to the extent that if modern agriculture was operated without chemical control, the crop production will probably decline in many areas, food price will soar far higher, and food shortage will become more severe. Although pesticides have played an important role in enhancing crop yields, they have also come up with various environmental problems. When present above permissible limits, they act as pollutants, creating pesticide pollution. Many pesticides are present today in different concentrations in various components of our environment such as air, water, and soil. More than 500,000 people are either killed or incapacitated every year by poisoning, and most of these casualties occur in developing nations.[3]

Ecologically, however, pesticides have created two major problems that were not previously anticipated. As pollutants, they contaminate numerous natural ecosystems [terrestrial: forest, grassland, desert, etc.; aquatic: freshwater (running water such as spring, stream, or rivers or standing water such as lake, pond, pools, puddles, ditch, swamp, etc.), and marine (deep water bodies such as ocean or shallow ones such as a sea, estuary, etc.)] not intended to be targets. Secondly, most of them have directly/indirectly affected human health. The objective of this entry is to provide basic knowledge on pesticide exposure and to understand issues on residues in the natural ecosystem.

HISTORY OF PESTICIDES AND PESTICIDE PROBLEMS

The term *pesticide* covers a wide range of compounds including insecticides, fungicides, herbicides, rodenticides, molluscicides, nematicides, plant growth regulators, and others. In the 1940s, DDT became the first widely available synthetic insecticide. It was highly effective but it showed signs of becoming less effective as insects became resistant to it. It also accumulates in the bodies of animals and high up the food chain by biomagnifications and bioconcentrations, causing problems with reproduction. Rachel Carson's *Silent Spring* in 1962 drew the attention of environmentalists to the disaster that was gathering pace across the globe. Public awareness of problems with pesticides grew by the 1970s when DDT was banned in many countries. It is still used in some places for malaria control and it is still present in the bodies of many animals, even hundreds of miles away from where it has not been used. The introduction of other synthetic insecticides—organophosphate (OP) insecticides in the 1960s, carbamates in the 1970s, and pyrethroids in the 1980s, as well as herbicides and fungicides in 1970s to 1980s—contributed to a great extent in pest control and agricultural output. The consequences of pesticide use have resulted in serious health implications to man and his environment. There is now overwhelming evidence that some of these chemicals pose potential risk to humans and other forms of life and unwanted side effects to the environment.[4] The worldwide deaths and chronic illnesses due to pesticide poisoning numbered about 1 million per year.[5]

The problem is more serious when pesticides that are banned are used indiscriminately. Banned pesticides are still used on crops that are not consumed directly, e.g., cotton. Few people think of cotton as food, but once the fiber is removed, two-thirds of the cotton crop winds up in the food we eat. Every year in the United States, half a million tons of cottonseed oil goes into processed salad dressings, baked goods, and snacks like Fritos and Goldfish. Another 3 million tons of cottonseed is fed to beef and dairy cattle, which also eat vast amounts of the cotton by-products known as "gin trash."[6]

HOW DO PESTICIDES SPRAYED ON AGRO/HORTI ECOSYSTEMS ENTER NATURAL ECOSYSTEMS?

Almost less than 1% of the total pesticides applied actually hit the target organisms.[7] Most reach non-target sectors of agro-ecosystems and/or spread to surrounding ecosystems as chemical pollutants. The pesticide somehow "leaks" into another ecosystem via movement of water from one body to another via outflow streams or seepage into the water table. Some pesticides might evaporate into the atmosphere and be carried elsewhere by winds. Regardless of how the leak occurs, the pesticide could affect accidental targets; e.g., a volatile insecticide used to control mosquitoes evaporates and kills bees; thus, a wide variety of plants do not get pollinated, thereby affecting their yield. The pesticide may also be taken in by migratory animals (birds in particular) and carried elsewhere; the toxin may affect the birds' reproduction in some way, or those birds might be eaten up by a higher order of predators and the toxin may inflict some injury to them. Either way, this would affect the balance of predation in some land-based ecosystem.

Pesticide Pollution and Natural Ecosystems

Effect on the Soil Environment

Many pesticides contain chemicals that are persistent soil contaminants; their effects may last for years. Pesticides move with water in soil to groundwater and on soil to surface water. They decrease biodiversity in the soil by killing soil organisms; when life in the soil is killed off, the soil quality deteriorates and has a knock-on effect upon the retention of water. This is a problem for farmers particularly in times of drought.[8] At such times, organic farms have been found to have yields 20%–40% higher than conventional farms. Soil fertility is affected in other ways, too. When pesticides kill off most of the active soil organisms, the complex interactions that result in good fertility break down. Plants depend on millions of bacteria and fungi to bring nutrients to their rootlets. When these cycles are disrupted, plants become more dependent upon exact doses of chemical fertilizers at regular intervals. Even so, the incredibly rich interactions in healthy soil cannot be fully replicated by the farmer with chemicals. Hence, the soil—and our nutrition—is compromised. We get large but watery vegetables and fruits, which often lack natural taste and nutrients and may even contain harmful toxic pesticide residues. Studies of pesticide effects on the soil fauna have reported increased numbers of collembolan, because chemicals reduced populations of natural enemies, especially of predatory mites.[9]

Effect on the Aquatic Environment

Pesticides enter the freshwater ecological systems either from direct application of pesticides for the control of harmful aquatic fauna or as runoff from the treated areas, drift during aerial spraying, and industrial effluents from washing and spraying of equipments and containers. Several groups of organochlorine pesticides such as DDT, endosulfan, and chlorinated phenoxy acetic acid used as herbicides, and fungicides such as hexachlorobenzene and pentachlorophenol are of interest in water pollution. Because of their solubility in water and tendency to be absorbed on solid surfaces, only traces of these chemicals are found in solid surfaces and treated water.[10]

Microorganisms form a vital part of the freshwater environment. Bishop[11] measured the effects of DDT on Mastigophora, Infusoria, and Sarcodina in ponds near Savannah, Georgia, and found little change in population numbers after treatment at relatively low rates. Hoffman and Olive[12] found that the growth of populations in Colorado lakes was inhibited after the addition of rotenone and toxaphene. Phytoplankton (beneficial/detrimental) can be seriously affected by agricultural chemicals. DDT sprays have caused serious reductions of bottom-dwelling invertebrates, the reductions in some cases amounting to 95% of the population. Malathion has also caused destruction of stream invertebrates. Cushing and Olive[13] found that toxaphene and rotenone reduced numbers of midge larvae in Colorado reservoirs and algae; on higher plants in the freshwater environment, adsorption of pesticides in/on the vegetation resulted in phytotoxicity, which either retarded the growth of or killed aquatic plants. Kolleru Lake is the largest natural freshwater body of Andhra Pradesh in India where agriculture and aquaculture are some of the primary activities at the lake basin. The increased use of pesticides in agriculture and aquaculture had a negative impact on the quality of water in the lake.[14]

A major environmental impact has been the widespread mortality of fish and marine invertebrates due to the contamination of aquatic systems by pesticides. Most of the fish in Europe's Rhine River were killed by the discharge of pesticides, and at one time, fish populations in the Great Lakes became very low due to pesticide contamination. In addition, many of the organisms that provide food for fish are extremely susceptible to pesticides, so the indirect effects of pesticides on the fish food supply may have an even greater effect on fish populations. Some pesticides, such as pyrethroid insecticides, are extremely toxic to most aquatic organisms. It is evident that pesticides cause major losses in global fish production.[15]

Effect on the Terrestrial Environment

A wide variety of pesticides is applied on horticultural and agricultural crops. Some of them are highly specific and others are broad spectrum; both types can affect terrestrial wildlife, soil, water systems, and humans. The misuse of pesticides can cause valuable pollinators such as bees and hoverflies to be killed and this in turn can badly affect food crops. Bees are extremely important in the pollination of crops and wild plants; about 33% of all crops require pollination. Although pesticides are screened for toxicity to bees, and the use of pesticides toxic to bees is permitted only under stringent conditions, many bees are killed by pesticides, resulting in the considerably reduced yield of crops dependent on bee pollination. Bee population has been suffering a serious decline in recent years. Without bees, many food crops would simply fail to grow.

It has been observed that through natural selection, some pests eventually become quite resistant to pesticides and farmers may need increasing amounts of pesticides, making the problem worse. Orchards are complex ecosystems easily perturbed by the extensive use of pesticides and there are many instances of increased pest attacks in orchards after the use of pesticides, e.g., outbreaks of codling moth, leafrollers, aphids, scales, and tetranychid mites.[16] When pesticides were first used on tropical cotton crops, they controlled two or three important pests of the crops and greatly increased yields. Within a few seasons, however, the chemicals reduced the population of natural enemies and a number of other arthropod species became pests.[17]

Amphibians such as frogs are particularly vulnerable to concentrations of pesticides in their habitat. Atrazine, the most heavily used herbicide, is regularly detected in drinking water supplies in the Midwest, United States, and exposure to extremely low levels of atrazine can cause sex change and/or deformities in frogs, fish, and other organisms.[18] Based on this evidence, and the widespread presence of atrazine in drinking water supplies, the European Union (EU) announced a ban on atrazine in 2006. The U.S. Environmental Protection Agency re-registered atrazine in 2003 despite objections from scientists and environmental groups.

Pesticides have had some of their most striking effects on birds, particularly those in the higher trophic levels of food chains, such as bald eagles, hawks, and owls. These birds are often rare, endangered, and susceptible to pesticide residues such as those occurring from the bioconcentration of organochlorine insecticides through terrestrial food chains. Pesticides may kill grain- and plant-feeding birds, and the elimination of many rare species of ducks and geese has been reported. Populations of insect-eating birds such as partridges, grouse, and pheasants have decreased due to the loss of their insect food in agricultural fields through the use of insecticides. Pesticides can affect animal reproduction directly, as evident by the deleterious effect of the persistent organochlorine insecticides on reproduction in receptors and other birds. The U.S. National Academy of Sciences stated that the DDT metabolite, Dichlorodiphenyl-dichloroethylene (DDE), causes eggshell thinning and that the bald eagle population in the United States declined primarily because of exposure to DDT and its metabolites.[19] Fish-eating birds are more severely affected than terrestrial predatory birds, because the former acquire more pesticides via their food chain.[20] Pesticides can also affect reproduction in invertebrates; for example, sublethal doses of DDT, dieldrin, and parathion increased egg production of Colorado potato beetle after 2 weeks by 50%, 33%, and 65%, respectively.[21] Aquatic ecosystems with flowing water can usually recover their structure and function more quickly from pesticide effects than ponds with standing water.

Effects on Humans

There are a number of ways in which humans can be exposed to pesticides through the environmental route. Man's primary exposure to pesticides is probably via those used domestically in wood preservation or as household insecticides. Pesticides can endanger workers during production and transportation or during and after use. Bystanders may also be affected at times, for example, walkers using public rights-of-way on adjacent land or families whose homes are close to crop-spraying activities. One of the main hazards of pesticide use is to farm workers and gardeners. A recent study by the Harvard School of Public Health in Boston discovered a 70% increase in the risk of developing Parkinson's disease for people exposed to even low levels of pesticides.[22]

The effects of pesticide residues in food and water probably cause the great public concern, although reports of clinical poisoning due to residues are extremely rare. Their residual population in food commodities is alarming. Leafy vegetables, cereals, fruits, rice, meat, milk, fish, and even human milk have been contaminated by various pesticides in a range of 0.1–25.7 mg/kg. The herbicide 2,4-D is identified as a carcinogen in humans and dogs, and the insecticide acephate is a mutagen and a carcinogen, is fetotoxic, feminizes rats, and kills birds. In a multicountry study (Belgium, China, FRG, India, Israel, Japan, Mexico, Sweden, United States, and Yugoslavia) on the assessment of human exposure to selected organochlorine compounds, the residue levels for pp'-DDE and β-HCH were found to be higher in the human milk samples collected from developing countries like China, India, and Mexico than in the participating developed countries. A higher level of these chemicals in mother's milk is a clear-cut reflection of their increased burden through their translocation passage.[23] An excerpt from a report on contaminants found in wine in the EU (March 2008) is presented here.[24] A number of bottles of wine were tested for pesticide residues, and 100% of conventional wines included in the analysis were found to contain pesticides, with one bottle containing 10 different pesticides. On an average, each wine sample contained more than four pesticides. The analysis revealed 24 different pesticide contaminants, including five classified as being carcinogenic, mutagenic, neurotoxic, or endocrine disrupting by the EU. Human health is at risk when chemical residues are present in so much of our food supplies.

Children are particularly vulnerable to the toxic effects of pesticides. Studies have found higher rates of brain cancer, leukemia, and birth defects in children who suffered early exposure to pesticides. A survey of baby foods in 2000 showed detectable pesticide residues in nearly 50% of foods sampled. Fourteen percent of foods tested showed residues of more than one pesticide at levels 30 times the proposed limit of 0.01 mg/kg.[24] A U.K. government report in 2003 showed more than 70%, 90%, 61%, 54%, and 35% of apples, lemons, bread, rice, and potatoes analyzed had pesticide residues, respectively. The main source of exposure to pesticides for most people is through diet. A study in 2006 measured organophosphorus levels in 23 schoolchildren before and after changing their diet to organic food. The levels of organophosphorus exposure dropped immediately and dramatically when the children began the organic diet.[24]

Over the last 50 years, many human illnesses and deaths have occurred as a result of exposure to pesticides, and there are 26 million human pesticide poisoning cases. Some of these are suicides, but most involve some form of accidental exposure to pesticides, particularly among farmers and spray operators in developing countries who are careless in handling pesticides or who wear insufficient protective clothing and equipment. In India, the first report of poisoning due to pesticides was from Kerala in 1958, where more than 100 people died after consuming wheat flour contaminated with parathion.[25] This prompted the Special Committee on Harmful Effects of Pesticides constituted by the Indian Council of Agricultural Research to pay more attention on the problem.[26] Exposure to accidental emissions of methyl isocyanate from a pesticide factory in Bhopal, India, killed more than 5000 people, leaving more than 50,000 with permanent damage.

Alternatives for Pesticide Problems

The toxic effects of pesticides on our foods, land, and their effects on the health of human beings and their progeny make it an issue that is becoming more and more crucial. Integrated pest management (IPM; pest surveillance, use of crop varieties resistant to pest, sound cultural practices, biological control, and use of ecofriendly pesticides) emphasizes the need for simpler and ecologically safer measures for pest control to reduce environmental pollution and other problems caused by excessive and indiscriminate use of pesticides. Preference should be given to organic food that are grown without toxic pesticides by organic methods. There are now many biological control tactics available where benign species are used to manage less benign ones. Ladybirds (ladybugs) are often introduced to control aphids (greenfly and others). Organic non-toxic sprays are used to stimulate the soil. They work by stimulating fungi in the soil that help to feed the plants and help them in developing resistance to disease and insect attack. There are also many successful barrier methods that help to deter insect attacks (the use of nets to ward off birds and larger insects). Companion planting is also used; garlic, for example, helps some plants resist insect attacks. There are a small number of organic pesticides that are legitimate to use in organic food production system. Some of these can be made at home using simple ingredients such as soap and alcohol.

It is evident that misuse, overuse, and abuse of pesticides lead to many environmental problems as discussed. Pesticides must be used as part of a planned systematic pest management program utilizing as many control techniques as applicable (IPM). Emphasis should be placed on using all the techniques of organic farming and supplementing these with the use of pesticides, i.e., using pesticides as part of an organic farming system. The IPM approach will help minimize the effects of pesticide pollutants on the environment and natural ecosystems and will also help in economic and ecological sustainable food production.

REFERENCES

1. Kent, J. Education and training in farm chemical management. Proc. Conf. Agriculture, Education and Information Transfer; Murrumbidgee College of Agriculture, 1991.
2. Anonymous. *Scientific agriculture prevents mass starvation*; The Herxter. J. of the Agvet Division, Hoechst, Aust. Inc. 1992.
3. Anonymous. Water pollution—case study, 2010, http://www.environmentandpeople.org. All Rights Reserved © Environment and People. Web site designed and maintained by HS visual FX. (Accessed on October 8, 2010).
4. Igbedioh, S.O. Effects of agricultural pesticides on humans, animals and higher plants in developing countries. Arch. Environ. Health **1991**, *46*, 218.
5. Environews Forum. Killer environment. Environ. Health Perspect. **1999**, *107*, A62.
6. Imhoff, D. *King Cotton 1999—Pesticide residue is common in cotton byproducts used in agriculture*; Brief Article Sierra, May 1999.
7. Pimentel, D.; Levitan, L. Pesticides: Amounts applied and amounts reaching pests. BioScience **1986**, *36*, 86–91.
8. Roger, A.; Simpson, I.; Oficialc, R.; Ardales, S.; Jimen, R. Effects of pesticides on soil and water microflora and mesofauna in wetland ricefields: A summary of current knowledge and extrapolation to temperate environments. Aust. J. Exp. Agric. **1994**, *34*, 1057–1068.
9. Edwards, C.A.; Thompson, A.R. Pesticides and the soil fauna. Residue Rev. **1973**, *45*, 1–79.
10. Beitz, H.; Schmidt, H.; Herzel, F. Occurrence, toxicological and ecotoxicological significance of pesticides in groundwater and surface water. Chem. Plant Prot. **1994**, *8*, 3–53.
11. Bishop, E.L. Effects of DDT mosquito larviciding on wildlife; the effects on the plankton population of routine larviciding with DDT. Public Health Rep. **1947**, *62* (35), 1263–1268.
12. Hoffman, D.A.; Olive, J.R. The effects of rotenone and toxaphene upon plankton of two Colorado reservoirs. Limnology and Oceanography **1961**, *6* (2), 219–222.
13. Cushing, C.E. Jr.; Olive, J.R. Effects of toxaphene and rotenone upon the macroscopic bottom fauna of two northern Colorado reservoirs. Transactions of the American fisheries Society **1957**, *86*, 294–301.
14. Victor, L.L. Pesticides in sea water and the possibilities of their use in mariculture. In *Research in Pesticides*; Chichester Academic Press: New York London, 1965.
15. Edwards, C.A. U.S. Environmental Protection Agency Web site. Pesticides. Available from http://www.epa.gov/pesticides (Copyright © 2010 Advameg, Inc.). (Accessed on October 8, 2010.
16. Brown, A.W.A. *Ecology of Pesticides*; John Wiley & Sons: New York, 1978.
17. ICAITI (Instituto Centro Americano de Investigacion y Technología Industrial). *An Environmental and Economic Study of the Consequences of Pesticide Use in Central American Cotton Production: Final Report*; Central American Research Institute for Industry (Guatemala), United Nations Environment Programme: Nairobi, Kenya, 1977.
18. Anonymous, 2011, available at http://www.beyondpesticides.org/pesticides/factsheets/Atrazine.pdf (Accessed on October 8, 2010).
19. Liroff, R.A. Balancing risks of DDT and malaria in the global POPs treaty. Pestic. Saf. News **2000**, *4*, 3.
20. Pimentel, D. *Ecological Effects of Pesticides on Non-Target Species*; Executive Office of the President, Office of Science and Technology: Washington, DC, 1971; 220 pp.
21. Abdallah, M.D. The effect of sublethal dosages of DDT, parathion, and dieldrin on oviposition of the Colorado potato beetle (*Leptinotarsa decemlineata* Say) (Coleoptera: Chrysomelidae). Bull. Entomol. Soc. Egypt Econ. Ser. 1968, 2, 211–217.
22. Ascherio, A.; Chen, H.; Weisskopf, M.G.; O'Reilly, E.; McCullough, M.L.; Calle, E.E.; Schwarzschild, M.A.; Thun, M.J. Pesticide exposure and risk for Parkinson's disease. Ann Neurol. **2006**, *60* (2), 197–203.
23. UNEP/WHO. *Assessment of Human Exposure to Selected Organochlorine Compounds through Biological Monitoring*; Slorach, S.A., Vaz, R., Eds.; Swedish National Food Administration: Uppsala, 1983; 49.
24. Anonymous. *Pesticide Problems are a Growing Concern*, 2011. Copyright © *Greenfootsteps.com* 2006–2010.
25. Karunakaran, C.O. The Kerala food poisoning. J. Indian Med. Assoc. **1958**, *31*, 204.
26. ICAR. *Harmful Effects of Pesticides. Report of the Special Committee of ICAR*; Wadhwani, A.M., Lall, I.J., Eds.; Indian Council of Agricultural Research: New Delhi, 1972; 44.

Pollution: Point and Non-Point Source Low-Cost Treatment

Sandeep Joshi
Sayali Joshi
Shrishti Eco-Research Institute (SERI), Pune, India

Abstract
Waste degradation, assimilation and recycling are processes of the ecosystem cycle in a natural environment. But in a human environment, accumulation of wastes or chemicals in cities, rural areas or agricultural environments leads to the deterioration of water, air, and soil quality—and, in turn, overall healthiness and livability. It can be restored by accelerating or reseeding the wastes or receiving water bodies using various cultures. It is evident that waste treatment facilities are focused on infrastructure and machinery rather than on waste treatment processes. Hence, waste treatment processes and not the construction and building of wastewater treatment facilities should be the central theme of economic investment. Ecotechnological treatment systems are more focused on integrated bio-phytoremediation of wastes and ecological re-recycling of energy and matter in the ecosystems. With this ecosystem approach, the non-utilizable residues—which are generated more in conventional systems—are missing from ecotechnological treatment systems. These residues get assimilated in various trophic levels of the food chain. That makes the ecotechnological systems more effective in terms of real space requirement, carbon and nitrogen emissions, and return on investment. In this entry, various ecotechnological applications from the past centuries are discussed, along with innovations, practical applications, and ecological and economic bases from the last few decades.

INTRODUCTION

The existence of human on Earth is dependent on amiability of environmental factors. Ever-increasing urbanization, industrialization, and modern agriculture result in the inevitable pollution of air, water, and land resources, giving rise to adverse environmental conditions. Efforts are being doubled to eradicate the pollution that contributes to the unhealthy environment. Various techniques based on physicochemical, mechanical, or a mix of mechanical–biological processes are being employed to treat pollution from point sources. These attempts have progressively evolved since industrialization took place in England in the 17th century. Despite stringent laws and technological advancements, the water quality of streams, rivers, and lakes have worsened because of pollution. Therefore, there is a need to resolve the issue of pollution of vital water resources, including groundwater, with innovative concepts and scientific principles. Lately, attention has been focused on searching for new means of improving the ecological status of water bodies by restoring and conserving it using an ecosystem approach. Ecotechnology is an indispensable tool in this venture. It furnishes new approaches/methods/techniques for the restoration and maintenance of deteriorated urban, rural, and natural environments. The use of ecotechnological methods is comparatively more cost-effective than conventional or advanced mechanistic equipment because of less energy requirement and reduced capital–operational costs.

Ecotechnology has applications in maintaining the integrity and well-being of ecosystems by transforming, degrading, and converting pollutants into benign substances and through the development of environmentally acceptable waste disposal processes. Ecotechnology is gaining popularity in the field of waste treatment and pollution control through the composting and wastewater treatment technologies from the last few decades. Historically, various composting methods and sewage irrigation were already practiced in India. According to recent developments in molecular biology, ecology, and environmental engineering, it is possible to modify organisms so that their basic biological processes can degrade more complex chemicals and higher volumes of waste materials efficiently.[1–3] Thus, the ecotechnology for pollution treatment can be defined as it is an application of ecological processes for the resources and waste management using combinations of aquatic or terrestrial plants, microbes, invertebrates, and vertebrates, and by subtly manipulating natural forces to leverage their beneficial effects through maneuvering matter and energy cycles.[4]

ISSUES OF URBAN SEWAGE MANAGEMENT

Nations having a Human Development Index (HDI) of less than 0.783 are known as developing countries.[5] In many developing countries, most of the domestic and industrial wastewaters are discharged with hardly any treatment or after primary treatment only.[6] In Latin America, about 15% of collected wastewater passes through treatment plants (with varying levels of actual treatment). In Venezuela, 97% of the country's sewage is discharged raw into the environment.[7] In a relatively developed Middle Eastern country such as Iran, it is recorded that most of Tehran's population has introduced untreated sewage to the city's groundwater.[8] Construction of major parts of the sewage system may be fully completed by the end of 2012. In Isfahan, Iran's third largest city, sewage treatment was started more than 100 years ago.

In the Indian context, the country's 27.8% urban population is distributed in more than 5100 towns and over 380 larger urban agglomerations. Migration to major cities from underdeveloped rural and tribal areas is a major cause of the rapid increase in urban population in the last few decades.[9] This has increased stress on civic services, leading to pressure on sewage treatment plants (STPs) with conventional technologies such as the following:[10]

- High operational costs
- Equipment replacement issues due to corrosion as a result of methanogenesis and hydrogen sulfide
- No prospects of using treated water due to high concentration of COD (>30 mg/L) and fecal coliform counts
- Lack of skilled operating personnel

Based on the above-mentioned worldwide experiences, the following can be concluded:

- The components of sewage treatment such as collection, conveyance, treatment, and disposal needed suitable infrastructural designs and installations based on catchment planning and engineering applications.
- One hundred percent sewage collection and further transferring and treatment in STPs should be achieved considering safety factors for unpredictable and unplanned urbanization and population trends.
- Capacity building for setting up of STP and routine operation and maintenance is required.
- Existing sewers and manholes should be maintained to avoid siltation and bypassing sewage into a nearby river/lake.

Pollution treatment process and equipment engineering of a specific STP/ effluent treatment plant (ETP) should integrate multiple management aspects and economics including cost-effective alternatives for disposal/reuse of the treated effluents and their impacts on the environment and public health, stressing sustainability. The major performance-related problems of treatment plants[11,12] (poor-quality treated water and sporadic crashing down of the process) for the point sources are due to the following:

- Hydraulic short-circuiting and shocks
- Shocks of pollution loading
- Biomass inhibition or loss due to shocks of toxic compounds
- Breakdowns of the machinery
- Seasonal variation of climatic factors
- Non-availability of electricity, interrupted supply of electricity, and scarcity of consumables and skilled manpower

Conventional systems have evolved over the last two centuries to deal with wastewaters from the point sources ranging from a few thousand liters to millions of liters. As the flow of wastewater increases, it demands for a number of huge civil units and extensive mechanization to support the treatment processes based on physicochemical and biological principles. No adequate conventional approaches and methods have evolved to treat polluted streams, rivers, and lakes in the course of time. Hence, the issue of the ever-increasing pollution of water bodies needs to be addressed considering an ecosystem approach to improve the self-purification capacity of the water body and thus improve ecological health by removing the stress created by non-point or diffused sources of pollution.

Aerobic and anaerobic treatment systems are yet to be accepted worldwide because of their intensive capital and operational costs and complex maintenance. For example, an investment of millions of dollars in the Ganga (River Pollution Control) Action Plan in India could not yield the desired results due to unavailability of electricity and the lack of skilled manpower to run the modern facilities of the state-of-the-art-treatment systems and technologies as per the audit done by government agencies. The case was the same with the Yamuna (River Pollution Control) Action Plan and the Dal Lake Pollution Control Plan in the Kashmir state of India. The National River Conservation Directorate of Ministry of Environment and Forests, Government of India, suggested using ecofriendly techniques to manage city wastes to keep water bodies free of pollution. The U.S. Environmental Protection Agency also suggested constructed wetlands to manage community wastewaters after an in-depth nationwide survey of conventional STPs.[13]

Thus, there was a need to change the conventional mechanistic approach to an ecosystem type that involves degradation of pollutants by living organisms with minimal external human interference. Ecotechnology appears to be the living system solution for pollution from point and non-point sources. The National Environment Policy 2006 of India suggests using wetlands—natural powers of ecosystems—to curb the pollution generated by human settlements. It is a welcome shift from insistence for mechanized wastewater treatment system.

LIVING SYSTEMS IN THE TREATMENT OF POLLUTION

The natural process of biodegradation[14,15] using living machine–biological systems does not need electricity at all. Hence, ecotechnologies have a clear advantage over energy-intensive conventional and sophisticated technologies as far as the non-generation of greenhouse gases (GHGs) is concerned. Carbon deposition in the form of vegetation is a fitting option that offers the potential to function as a carbon sink, which is not possible in the case of conventional systems at lesser investments and operational costs.

It involves the following:

1. Detritus-feeding organisms consume the pollutants (pollutants are nutrients to them).
2. Wastes generated from this process are used by green plants
3. Absorption of carbon dioxide from the atmosphere by green plants
4. Transfer and assimilation of pollutants to natural cycles, i.e., biogeochemical cycles of carbon
5. Storage of carbon in vegetation and subsequently in the soil

LIVING SYSTEMS IN ACTION FOR POLLUTION TREATMENT

- Millions of species have survived on earth for millions of years despite the vagaries of environmental factors. They have accumulative intelligence of thriving in changing climates and environmental conditions. This may be termed as multispecies intelligence.
- Conversion of pollutants in absorbable substances and transfer to natural cycles, i.e., biogeochemical cycles (ecological efficiency).
- Adapts itself sometimes when something is wrong or when the unexpected happens (ecologically robust).
- When changes occur in the systems due to external inputs, biogeochemical cycles and degradation biochains are reorganized and balanced (ecologically resilient).
- Need-based emergence of a new dynamic trophic order suitable to the flow and concentration changes (ecological evolution).

Application of ecological engineering principles and the use of the bio power of living components of detritus food chain—bacteria, fungi,[16,17] and microinvertebrates—are found to be effective in the treatment of wastes. Waste generated through anthropogenic activities is consumed by the living components of ecotechnological treatment units. Ecotechnology is a comparatively novel option but is a very cost-effective methodology to control the pollution and convert it into resources. Ecotechnological treatment units such as soil scape filter (vertical filtration), Hydrasch Succession Pond, Green Channel, and green bridge are found to be effective for treating pollution from point sources.[18] The capital costs and operational costs are 50%–60% and 10%–20%, respectively, that of conventional aerobic or anaerobic systems.

BASIS OF ECOTECHNOLOGY

Green plants like Indian mustard, *Ipomea*, *Eichornia*, *Hydrilla*, *Azolla*, *Chara*, and *Lemna* are known to absorb heavy metals and excess nutrients and decontaminate the water or piece of land where they are growing. *Brassica* plants are well-known metal scavengers. Microorganisms in association with green plants and some dependent invertebrate–vertebrate animals form together an ecosystem based on "waste" as a starting point, converting it into various utilizable products at various trophic levels using substrate-selective biochemical-degradation–conversion transformation pathways.

Total number of species, community composition, or production of biomass is controlled by the particular growth-limiting factor or energy conservation processes at every succeeding level of biodegradation and utilization (BDU) chain. Most of the microbes involved are not unique to a habitat having non-extreme conditions; therefore, cometabolism by microbial communities is the key to the transformation of waste into assimilative form. Within the cells, various catabolic and anabolic processes occur, such as uptake of nutrients, redox reactions, enzymatic reactions, degradation, modification, transformations, or conversions

Table 1 Some useful bioprocesses.

Sr. no.	Bioprocesses	Description
1	Bio-oxidation	Intracellular or extracellular enzymes simplify large organic molecules, e.g., degradation of organic compounds into carbon dioxide and water, conversion of ammonia to nitrate, etc.
2	Bio-reduction	For example, denitrification, conversion of sulfates to hydrogen sulfide
3	Co-oxidation/co-reduction	For example, chlorinated compounds are degraded by chemolithotrophs like *Nitrosomonas*
4	Fermentation	Bioconversion of organics to procure energy for metabolic activities which results in simplification of complex molecules
5	Detoxification	Breaking down toxic organic chemicals in lysosomes
6	Bio-sorption	Uptake of substances or elements for metabolic activities, especially metallic components, which has immense application in metal pollution control
7	Precipitation or chelation	Some organisms are able to chelate or precipitate metals in the external environment. This has applications in metal removal from water or extraction of metals from ores or hazardous wastes

of compounds with the help of extracellular and intracellular enzymes.[19] In catabolic processes, larger molecules are broken down into smaller pieces. At the same time, non-essential compounds are co-oxidized by extracellular enzymes secreted in the external environment.

These ecological principles can be harnessed to develop ecotechnological processes of pollution treatment. Many technologists prefer suspended growth bioreactors to the attached/packed ones because separation of suspended organisms from the supporting medium is much easier. However, these treatment units require external energy and equipment to maintain the microbial cells in the suspension and to bring them in contact with food (waste) material. In attached growth systems, wastewater is passed in such a way that microbial cells have enough contact time for the sorption of food material. Therefore, the energy required is less as compared to suspended growth systems. Some of the bioprocesses that can be designed and engineered for various applications are listed in Table 1.

The following are the basic considerations of ecotechnology:

- Constitution and tasks of biological consortia in their natural habitats
- Vibrant responses of organisms to dynamic natural environmental conditions and anthropogenic pressures
- Biochemical mechanisms including enzymatic pathways, useful for aerobic and anaerobic degradation, detoxification, etc
- Knowledge of ecological processes for estimating the requirements of infrastructural provisions

Much of the inputs relevant to ecological biochemistry come from studies of enzymes, genetics, biochemistry, and physiology of various microbes, micro- and macroinvertebrates (which are part of the detritus food chain in ecosystems), and higher green plants. In nature, biodegradation with little or no oxygen is mediated by anaerobic and micro-aerophilic microorganisms. Ecotypes of microorganisms adapt easily with changed environmental conditions, which is the reason they are useful in treating wastes in an uncontrolled environment (Table 2).

Groups of organisms with ecotechnological applications, identified and tested through laboratory screenings, can be used for microcosm/mesocosm scale-up studies through pilot plant or large eco-reactors. Ecotechnological experiments and applications were found to be very encouraging in the 1990s in India with support from industrial groups. In his book, Sandeep Joshi gave a detailed account of the evolution of ecotechnological treatment systems from point sources using the soil scape filter–vertical ecofiltration technique, the Green Channel, the Hydrasch Succession Pond, the green bridge filter–horizontal filtration system, and BIOX (Biological Oxygenation) to treat non-point sources.

SCOPE OF ECOTECHNOLOGY

Ecotechnology is a functional knowledge that serves to fulfill man's environmental needs for survival with minimal ecological disruption. It binds and subtly manages natural forces to leverage their favorable effects. It is the "ecology of knowledge and techniques" and the "knowledge and techniques of ecology"[20] that understand the structures and processes of interactions and interrelationships of ecosystems and societies. Ecological engineering and ecotechnologies are dependent on the self-designing abilities of ecosystems and natural forces.

HISTORY OF ECOTECHNOLOGY

The use of natural and constructed wetlands for wastewater treatment and disposal might have started in 1912 in the United States and Europe. Researchers began their efforts based on observations of the apparent treatment capacity of natural wetlands. Some researchers considered wastewater as a source of water and nutrients for wetland restoration or creation. Research studies on the applications of constructed wetlands for wastewater treatment began in Europe in the 1950s and in the United States in the late 1960s.

Table 2 Microbes and microbial enzymes.

Microbial enzymes and substrates		Some microbes involved in metallic reductions	
Enzyme	Substrate	Metal	Microbe
Cellulase	Cellulose	Uranium	*Geobacter metallireducens, Shewanella putrefaciens*
Proteinases	Caesin, gelatine, albumin		
Catalase	Hydrogen peroxide	Selenium	Many species of *Pseudomonas, Flavobacterium, Clostridium*
Peroxidase	Hydrogen peroxide		
Monooxigenase	D-Catechol, *p*-quinol	Chromium	*Pseudomonas* sp., *Bacillus* sp., *Clostridium, Desulfovibrio*
Phosphatases	*p*-Nitrophenyl acetate		
Arylesterases	Phenylacetate	Vanadium	*Pseudomonas* sp.

In 1938, Seifert (Germany) opened up the concept of using ecological processes for man's benefit. River remediation could be undertaken in an economical manner that was close to nature. The American H.T. Odum stated in the 1940s that minimal labor should be put into changing the natural environment so that the self-renewability of the habitat system could be preserved. Kathe Seidel worked on methods for improvement of waterways contaminated by overfertilization or sewage in 1953.[21] She conducted many experiments using wetland plants for the treatment of industrial wastewaters. In 1969, in India, some experiments were conducted using water hyacinth for the treatment of digested sugar waste and septic tank effluents in oxidation ponds.[15] In 1989, the American ecologist Mitsch tried to define more clearly the concepts and applications of ecotechnology.

DEVELOPMENT OF INDIAN ECOTECHNOLOGICAL POLLUTION TREATMENT SYSTEMS

Sandeep Joshi, in the last decades of the 20th century, developed an ecosystem approach to design ecotechnologies based on ecological principles for the treatment of pollution.[14] These systems use green plants and a consortium of bacteria for the biodegradation, bioconversion, and bioassimilation of pollutants from the water or contaminated soil. These filtration systems can be vertical or horizontal, depending on the flow of water. Sometimes, floating plant species were also used depending on specific site conditions for the removal of pollutants from the water, termed as the Hydrasch Succession Pond, denoting the succession of macrophytes in the treatment pond depending on changes in incoming effluent and seasonal variations.

Vertical ecofiltration–soil scape filtration treatment systems similar to that of vertical wetland systems described by Seidel differing in area requirement and much rigorous bioremediation reactions were used for a number of industries (including electroplating, chemical, food, etc.) and domestic units complying with the standards prescribed by regulatory authorities. Green bridge and green lake systems with supportive phytofiltration and biodegradation chain result in not only increased dissolved oxygen concentration but also reduced biotoxicity. To treat the non-point or diffused pollution from unidentified sources flowing through the long stretches of streams and rivers, the author, in addition to vertical ecofiltrations, used horizontal filtration and ecological principles known as green bridge technology (GBT),[22] lentic oxygenation technology system (LOTS), artificial stream ecosystem (ASE), benthic eco-system technology (BEST), etc. These processes are not addressed by conventional constructed wetland techniques.

Names and functions of treatment units may differ but in brief they try to harness the ecological processes, interrelationships, and interactions of various biological communities. These are very sophisticated processes where the quantum of disposable residues is minimized. In advanced ecological technologies, the pollutants become nutrients or source of energy for biochemical reactions and transformations through the various trophic levels of the complex food chain. The greater is the complexity, the more stable is the treatment process. Therefore, it is time to relinquish "extraction/separation" techniques and to adapt ecological cycling of wastes to obtain the maximum output in the form of energy or utilizable products on a sustainable basis.

With initiatives from the private sector and civil society organizations,[23] ecotechnological systems were successfully set up for field-scale applications in the catchment of the Ujjani Reservoir and where there is acute scarcity of water. Studies on ecotechnology were initiated in 1991–1992 in India but the first field-scale plant was installed in 1996 on the point source of mixed domestic and industrial wastewaters after successful evaluation/performance of the pilot plant.

Civil society organizations and the military educational institute supported the field-scale application of ecotechnology for pollution from non-point sources—for pollution flowing through the Bhosari stream in Pune, wastewater from industrial and residential areas was collected. The Clean River Committee installed the green bridge and green lake systems as per design and guidance from the Shrishti Eco-Research Institute. Such researches and social initiatives are supported by the government for the benefit of the society at large.

VERTICAL ECOFILTRATION AND CONSTRUCTED WETLANDS

Vertical ecofiltration and the constructed wetlands put into practice for pollution treatment should not be confused with created or restored wetlands.[24,25] Such systems will subsequently have the primary purpose of being a wildlife habitat. Created or restored wetlands often have a combination of the following standard features: irregular shorelines; varying water depths; open water; dense vegetation zones; submerged aquatic plants, shrubs, and trees; and nesting and hiding places. However, the detritus food chain is established over a period of time when the wetland receives wastewaters from settlements, farms, or industries. This food chain alters the status of pollutants to nutrients.

Ecotechnological soil scape filter and green channel systems are vertical ecofiltration systems in which polluted water is applied on top of a filter planted with locally available species rooted in a specially developed top layer of filter using mixed bacterial cultures. The pollutants are adsorbed in the top layer and made available to complete degradation by an array of effective microbes and plants in association with microinvertebrates that thrive there (Fig. 1).

Fig. 1 Soil scape filter and constructed wetland.

Constructed treatment wetlands (CTWs) are an outstanding option because they are low-cost, require less maintenance, offer good performance, and provide a natural appearance. Some object that CTWs require large land areas, such as 4.22 m^2/m^3 to 26.39 m^2/m^3, whereas the soil scape filter requires 1 m^2/m^3 of domestic wastewater or a maximum of 3–5 m^2 for high-strength industrial wastewaters except distillery.

In Fig. 2, schematics of vertical ecofiltration and constructed wetland systems are shown. The number of treatment units can be increased or decreased dependent on the strength of the wastewater to be treated and expected treated water quality. There is a selection of resident terrestrial and aquatic macrophytes for the safe use of effective non-native species for a better outcome of the system.

Normally, it is said that constructed wetlands are especially well suited for wastewater treatment in small communities where inexpensive land is available and skilled operators are hard to find considering the unavailability of land in densely populated urban areas. Green bridge systems do not require any extra space other than the course and floodplain area of the polluted stream. Thus, it has a potential for use in densely populated settlements. The soil scape filter looks like a beautiful garden, and open water constructed wetland decorates the premises; hence, they can be used for decentralized urban sewage management schemes without sacrificing the aesthetics of the landscape.

HORIZONTAL ECOFILTRATION SYSTEM

The green bridge–horizontal ecofiltration system (Fig. 3) proved to be a very promising technology in treating pollution from non-point sources flowing through the natural drains/streams in urbanized areas especially where there is electricity to run wastewater treatment plants. For over a decade, the grafting of ecosystem–green bridges has been found to be effective in reviving the self-purification of water bodies while reducing the ecotoxicity of pollutants. The green bridge system is supported by green lake systems. There are other forms of ecotechnologies such as stream ecosystem technology, green contour technology, etc.

Onsite applications of all these ecotechnologies are found to be very economical and simple as far as operations are concerned because their source of energy is the sun and they require less machinery for routine process maintenance and operations. The applicability of ecotechnology is proven by the number of installations of field-scale units on domestic as well as industrial wastewaters for the last 16 years in various parts of the country.

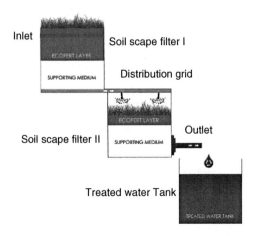 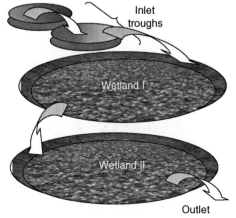

Fig. 2 Schematic of vertical ecofiltration and constructed wetland systems.

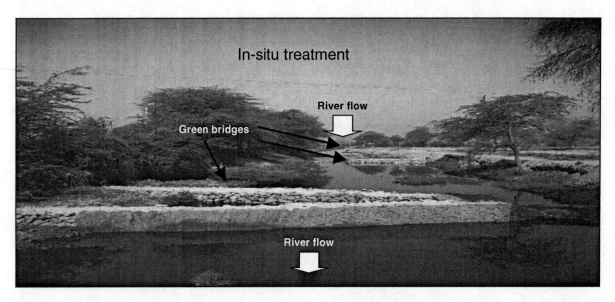

Fig. 3 Horizontal ecofiltration system of green bridges.

ECOTECHNOLOGY AND CLIMATE CHANGE

Climate change has become a buzzword among administrators and leaders who control the resources as it is understood that it would have many known (and unknown) health and economic impacts on an individual. The Kyoto Protocol was developed to assemble world leaders, bureaucrats, technocrats, and scientists to counter climate change. The most overlooked climate change activity is the control of water pollution, which normally contributes to the generation of GHGs.

Green plants with microbes play a major role in the carbon sink and aerobic degradation of wastes. These are the natural forces and processes that are useful in managing wastes. Ecotechnology harnesses these bio powers to assimilate anthropogenic wastes into ecological cycles without requiring electricity. Conventional waste management systems require a lot of electricity and are, thus, not candidates for carbon credits. However, ecotechnological treatment units having multiple uses, such as carbon sink, reduction in the use of electricity, and minimization of methane and GHG release, are more valuable in obtaining carbon credits at international levels. These techniques are more useful for developing countries that cannot afford the cost of sophisticated mechanized auto-control techniques to manage waste.

APPROPRIATE TECHNOLOGIES FOR SMALL COMMUNITIES, TOWNSHIPS, AND CITIES

Appropriate technology is defined as a treatment system that meets the following key criteria:

- Use of local plant species
- Shortest possible start-up periods
- Vector control
- Use of bio-indicators

Ecotechnological treatment systems such as a grafted stream ecosystem, aimed at augmenting the self-purification capacity of the polluted stream in its course, differ from CTW systems in many aspects such as area requirement, selection of plant and bacterial species, adaptation by micro- and macroinvertebrates and vertebrates, reduction in ecotoxicity, and complexity of food chain. Therefore, permitting and regulatory requirements applicable for constructed wetlands are not mandatory for in situ application of green bridge or green lake systems.

The most important factor that impacts all facets of constructed wetlands or ecotechnological systems is their in-built naturalness and aesthetic appeal to the general public. The desire of people to have such an attractive landscape enhancement treat their wastewater and become a valuable addition to the community is a powerful argument when the upgrading of wastewater treatment turns out to be a matter of public debate.

It is argued by supporters of CTW that the engineering community often fails to appreciate the appeal of constructed wetlands, while the environmental community often lacks the understanding of treatment mechanisms to appreciate the limitations of the technology. However, ecotechnological systems can work in combination with other technologies to make the wastewater treatment systems more space-, user-, and eco-friendly.

The water footprint of cities is much bigger and spreads in the river and lake catchment areas irrespective of geographical boundaries. The ecosystem approach with the application of ecotechnological treatment systems can regenerate the self-purification system of water bodies and erase the impact of the water footprint. Government and non-government sectors together with an environmentally aware public create a better network and institutional mechanism for the water and wastewater management sectors. Pollution control authorities, corporation officers, and policy makers express the importance of affordable, indigenous pollution control techniques that will improve the quality of the environment.

RURAL AND URBAN WATER RESOURCES MANAGEMENT

In the 21st century, while coping with climate change, the foremost objective is sustainable development by ensuring pure water supply and complete sanitation for everybody. Corporations, municipal councils of cities and towns, and local administrations of villages with farmer cooperatives can be updated, trained, and educated with regard to using eco-friendly techniques for effective liquid and solid waste management practices.

EXAMPLES OF LOW-COST TREATMENT SYSTEMS

Ecotechnological systems and constructed wetlands are simulations of natural systems for wastewater treatment. They are shallow (usually about 1–2 m deep) aquatic, lentic–lotic systems or terrestrial ecosystems. These are planted with aquatic species, relying upon natural microbial, biological, physical, and chemical processes to treat wastewaters. They typically have ecologically engineered structures to control the flow direction, liquid detention time, and water levels. Depending on the site conditions, these systems may contain an inert porous media such as rock, gravel, or sand.

Successful implementation of the scheme with natural technologies like green bridge, green lake, and stream eco-system implies that ecotechnology can be employed to treat waste streams coming from non-point sources. This can be very economical—capital expenditure is only 5%–10% of that of conventional mechanized aerobic and anaerobic treatment systems.

EXPERIENCES FROM DEVELOPED COUNTRIES

In the United States, constructed wetlands have been used to treat a variety of wastewaters including urban runoffs as well as municipal, industrial, agricultural, and acid mine drainage. In Germany, bank infiltration through alluvial soils has been in practice for the last 150 years. In India, encouraging results of river eco-restoration projects demonstrate that ecotechnological treatment systems can be the central treatment component for point or non-point sources of pollution to meet stream discharge or reuse requirements.

1. In 1974, people of Arcata, California, thought of using constructed wetland as its sewage treatment system. They used wastewater to create and nourish a wetland where a logging pond and city dump had once been to bring their wastewater discharge into Humboldt Bay up to water quality standards, or to come up with an acceptable alternative. The wetland provides prime wildlife habitat and recreation for the community and, at the same time, purifies the water. Capital expenditure for the sewage treatment by the constructed wetland project was estimated to be about $5 million as compared to the conventional system's estimate of $10 million. Annual maintenance of the constructed wetland was one-third that of conventional systems. In addition to sewage treatment, the marshland supports approximately 100 species of green plants, 6 species of fish, and many birds and mammals.[26]

2. The National Botanic Garden of Wales uses a wastewater treatment system composed of aerated reed beds. The system treats the sewage from up to 2000 visitors and 90 members of staff per day and was designed and built by Staffordshire-based ARM Reed Beds. The system now takes up a far smaller physical footprint and has a bigger treatment capacity. The project that was undertaken by the National Botanic Garden in partnership with the Welsh Assembly Government was completed in autumn 2010 and comprises a septic tank that feeds into a 128 m² vertical flow reed bed system with forced bed aeration technology.[27]

3. The bank filtration process has been in use in Europe, especially in Germany along the Rhine and in Berlin, since the 1870s.[28,29] The effectiveness of bank filtration and artificial groundwater recharge has long been recognized in Germany, as a consequence of various bacterial diseases caused by drinking water from waterworks with direct intake from rivers in the late 19th century (e.g., outbreak of epidemic cholera in Hamburg in 1892/1893).[30] It is a type of filtration that purifies water by passing through the banks of a river or lake for use as drinking water. The process may directly yield potable water, or simple pretreatment for further purification. There are three filtration mechanisms involved, namely, physical filtration (straining interstitial spaces between alluvial soil particles), biological filtration (soil microorganisms remove dissolved or suspended organic material and chemical nutrients), and chemical filtration or ion exchange (aquifer soils react with soluble chemicals in the water). Then, treated wastewater is returned to a river after use by discharging into a percolation pond

Fig. 4 Soil scape filter—treatment of mixed industrial and domestic wastewaters.

on the alluvial floodplain rather than flowing directly into the river.[31,32]

4. According to an inventory survey carried out in 2005, there are about 140 decentralized constructed wetland sites in Ireland. It is recorded that these systems are successfully established. There is no negative effect reported on the performance of these systems. Decentralized constructed wetland systems have the advantage of having a pleasing appearance with lower energy consumption for the performance—removal of 76.8%–99.8% (BOD_5), 76.3%–99.7% (COD), and 67%–99.9% (NH_4-N).[33]

EXPERIENCES FROM DEVELOPING COUNTRIES

1. The Madhyapur Thimi municipality, one of Nepal's oldest settlements, is a small municipality located in Kathmandu Valley. It has a population of around 48,000 and covers 11.11 km (20% residential area, 70%

Fig. 5 Industrial wastewater treatment using the soil scape filter.

Fig. 6 Treatment of a textile dying unit.

agricultural land, and around 10% vacant land). Sunga, the project area for demonstrating the constructed wetland using a community-based approach, is one of the many communities of the municipality. As the local people of Madhyapur Thimi and the municipality showed an interest in managing the wastewater, the Environmental and Public Health Organization (ENPHO), with the techno-financial support from Water Aid in Nepal, UN-HABITAT, and ADB, developed an innovative project for improved sanitation. In addition to the funding agencies, the Madhyapur Thimi municipality provided the required land for construction along

Fig. 7 First installation of the green bridge system to treat stream pollution (under construction).

Fig. 8 Ecological restoration of Udaipur's Ahar River, India (construction completed).

with the financial assistance for operation and maintenance of the wastewater treatment plant. Under this initiative, the ENPHO joined hands with the local people of Sunga and constructed a community-based reed bed treatment technology for managing the wastewater as a demonstration project. After successful construction of this treatment plant, the operation started in October 2005 under the constant supervision of the local user's committee with technical assistance of the ENPHO. The Sunga wastewater treatment plant was handed over to the local user's committee jointly by the ENPHO and the municipality on September 1, 2006.[34,35]

2. The vetiver (grass) system (VS) is a phytoremediation technology to treat domestic wastewater. Vetiver's leaf by-products offer a range of applications such as in handicrafts, animal feeds, thatches, mulch, fuel, etc. In 1996, VS was first applied in Australia to treat sewage effluent. Other plants, including fast-growing tropical grasses and trees, and crops such as sugar cane and banana, failed.[23] In Thailand, use of vetiver plantation found to be valuable in the absorption of P, K, Mn, Cu, Mg, Ca, Fe, Zn, and N.[36,37]

EXPERIENCES FROM INDIA[38]

1. Since 1996, a soil scape filter (vertical ecofiltration system) in combination with the Hydrasch Succession Pond is treating wastewater at a rate of 120 m^3/day from an electroplating industry (Fig. 4). The wastewater is a mix of industrial and domestic wastewaters in equal proportions. The system treats hexavalent chromium, nickel, and iron to comply with their concentration norms. It reduced operational costs to 10% of that of conventional acid-reducing chemical system. It has the additional benefit of lower dissolved salts (60%) in treated water as compared to conventional treatment. The demand for tertiary treatment is annulled. This long treatment chain is cut off by soil scape filter installation to only two eco-reaction tanks with pretreatment of neutralization process.

2. In an international school in western India, 50 m^3 of sewage without segregation of solids is being treated using the soil scape filter (vertical ecofiltration process). Treated BOD is consistently less than 16 mg/L since the plant became operational in 2005. By conventional method, they would have required at least 50 units of electricity every day with some additional cost for consumables.

3. In one food processing unit, about 50 km from Pune (Fig. 5), India, 30 m^3 industrial wastewater (sewage equivalent, 335 m^3/day) having a COD of more than 1000 mg/L and a BOD of more than 440 mg/L is being treated from 2000 onwards using soil scape filters in combination with a biotower system and an outlet COD and BOD of less than 80 mg/L and 32 mg/L, respectively. They have found it very cost effective as far as electrical consumption is required.

Table 3 Indian ecotechnological installations.

Sr. no.	Source of pollution	Flow	Ecotechnological unit	Supporting units	Performance parameter—concentration in the outlet (mg/L)	Discharge limit
1.	Electroplating	120 m^3/day	Single-stage soil scape filter Area—100 m^2; depth—1.5 m; COD load—40 kg/day; Cr^{6+}—5–6 mg/L	Holding tank, neutralization, primary settling and hydrasch succession pond	COD—80 Cr^{6+}—<0.05	250 0.1
2.	School's hostel	50 m^3/day	Single-stage soil scape filter Area—50 m^2; BOD load—10 kg/day	Holding tank, overhead distribution tank	BOD <16	30
3.	Food industry	30 m^3/day	Two-stage soil scape filter Total area of both filters—30 m^2; 54 kg/day	Biotower	COD <100 (reduction from 3600)	250
4.	Textile unit	5 m^3/day	Two-stage soil scape filter Total area of both filters—10 m^2; 10 kg/day	Holding tank, neutralization, vegetative fiber filter	COD <200 Color <5 Hazen units	250 5
5.	Stream in Bhosari, Pune	70 MLD	Two green bridges (total surface area—60 m^2) and two green lakes (total surface area—8000 m^2); COD load—70 tons/day; BOD load—28 tons/day	Screen across the stream	COD—64 BOD—20	250 30
6.	River Restoration, Udaipur	150 MLD	Six green bridges (total surface area—144 m^2) and one green lake (surface area—4000 m^2); COD load—45 tons/day; BOD—11.4 tons/day	Two screens across the river	COD—72 BOD—17	250 30

4. The colored effluents from the textile units are also treated using the soil scape filtration system (Fig. 6). The color reduction in the two-stage filtration system is consistently more than 90%. These plants have been successfully running in Baroda and Jaipur for the last 6 years. The precursors of dyes and pigments are also treated considerably—up to 95% COD reduction. The biological system was found to have adapted to such difficult pollutants.

5. Bhosari stream, a polluted natural drain, in Pune, India, was treated in 2003 within the premises of the College of Military Engineering (CME) (Fig. 7). It received wastewater from residential and industrial areas amounting to 70 million liters per day (MLD). It has a densely black color and has a very foul odor, spoiling the environment. A natural treatment system was developed in the initial stretch of 1.5 km in the CME area. Its estimated flow rate is about 70 MLD. This water is treated effectively using natural technologies like stream eco-system (SES), green bridge (GB) and green lake eco-system (GLES). Through a well-documented sampling and testing program, it was established that the system removed 80%–95% of COD, BOD and heavy metal reduction by 50%–60% and improvement in dissolved oxygen was found to be 7–9 mg/L from 0 mg/L.[39]

6. The Udaipur project of eco-restoration of the Ahar River (Fig. 8) and the Udaisagar Lake was a good example of cooperation among citizens, professionals, and industries for controlling river and lake pollution. The river water was infested with water hyacinth in certain stretches, leading to the elimination of other resident species of the river such as turtles, water snakes, fishes, and freshwater microinvertebrates. There was frequent shock loading of industrial effluents. Local villagers and farmers observed that the odor problem was completely eliminated and there was more than 90% reduction in chemical foam in the river after the grafting of self-purification ecological system of green bridges. Villagers have noticed that there was growth in fish population and that turtles were returning as a result of the ecological restoration of the Ahar River and the elimination of ecotoxicity of pollutants.[40]

Indian installations are shown in Table 3. It presents the design specifications with supporting units and performance parameters in comparison with the state's discharge limits.

Table 4 Comparative statement of various technologies.

Sr. no.	Particulars	BAF	TF	ASP	MBR	SBR	Root zone technique	Soil scape filter	Green bridge (STAC) system
1.	Wastewater source	Point source	Point sources	Point sources	Point sources	Point sources	Point sources	Point sources	Non-point sources
2.	Application	For domestic and industrial wastewaters containing non-toxic organic matter only	For domestic and industrial wastewaters containing non-toxic organic matter only	For domestic and industrial wastewaters containing non-toxic organic matter only	For domestic and industrial wastewaters containing non-toxic organic matter only	For domestic and industrial wastewaters containing non-toxic organic matter only	For domestic and industrial wastewaters containing non-toxic organic matter only	For domestic and industrial wastewaters, even for wastewater containing toxic organic and inorganic pollutants	For domestic and industrial wastewaters, even for wastewater containing toxic organic and inorganic pollutants
3.	Ancillary units	Eqaulization tank, neutralization tank, primary settling, aerated filter and secondary settling tanks	Primary settling, trickling filter and secondary settling tanks	Requirement of equalization tank, neutralization, primary settling and secondary settling tanks	Neutralization tank, primary clarifier, aeration tanks and membrane reactor, sludge dying bed	One tank, it is a fill-and-draw system in which all processes are carried out sequentially in the same tank	Properly designed treatment tank, graded filling material, acclimatized, aerobic, anaerobic, and facultative Bacteria, Acclimatized and selected indigenous plant	One unit only The requirement of neutralization if the pH of wastewater is not in the range of 6.5–8.5	Only metal screen to remove floating solids like plastics
4.	Hydraulic loading m^3/m^2/day	1–20	1–10	1–3	12–14	—	0.06–0.25	1–2	50–200
5.	HRT (hr)	1.3	—	4–8	2–5	9–30	5–10	Nil	Nil
6.	Organic loading (COD/BOD) kg/m^3/day	1.5–4	1.6	0.32–16	0.4–1.5	0.08–0.24	0.25	5–10	10–40
7.	COD/BOD reduction range	75%–93%	65%–90%	85%–95%	85%–95%	85%–95%	91%	90%–98%	70%–80%
8.	Sludge production (kg/kg BOD)	0.15–0.25	0.3–0.5 or 0.63–1.06	0.6	0.0–0.3	0–0.3	Nil	Nil	Nil
9.	Electricity requirement	200	150	300	200	3–10	50	Nil	Nil
10.	Failures	Even small concentrations disturb the process	Even small concentrations disturb the process	Even small concentrations disturb the process	Even small concentrations disturb the process	Even small concentrations disturb the process	Even small concentrations disturb the process	Nil because it is a natural process	Nil because it is a natural process
11.	Key parameters of process	pH, TSS, BOD, COD and toxic substance	pH, TSS, BOD, COD and toxic substance	pH, TSS, BOD, COD and toxic substance	pH, TSS, BOD, COD and toxic substance	pH, TSS, BOD, COD and toxic substance	pH, TSS, BOD, COD and toxic substance	pH	pH
12.	Maintenance	Skilled	Skilled	Skilled	Skilled	Skilled	Skilled	Simple	Simple

BAF, biological aerated filter; ASP, activated sludge process; MBR, membrane bioreactor; TF, trickling filter; SBR, sequential batch reactor; STAC, saprobic trophic adsorption and cycling

COMPARISON OF ECOTECHNOLOGY: ADVANCED KNOWLEDGE OF POLLUTION MANAGEMENT WITH CONVENTIONAL METHODS

A comparison of ecotechnological treatment systems considering the experience of various installations for various point and non-point sources of pollution, including Udaipur's Ahar River Eco-Restoration and other 30–35 such river and lake restoration projects, is shown in Table 4.[41] As far as space footprint is concerned, conventional "extraction" technologies may give the impression that they require less space for equipment, but the space required for the pretreatment, posttreatment, and residue disposal needs to be considered in space footprint estimation. In the table, only the "process" is considered.

Carbon, nitrogen, and phosphorus (C:N:P) are the major elements of biogeochemical cycles that have a major impact on environmental reservoirs. Carbon is known for its global warming potential, nitrogen's impacts are yet to be explored fully, and phosphorus is least known for its chemistry in the hydrosphere, lithosphere, and biosphere. The atmospheric impacts of carbon and nitrogen compounds are being studied extensively. Ecotechnological processes thus help in reducing the global warming and carbon footprint. Thus, on the basis of their carbon-positive, nitrogen-positive, space-footprint, and user-friendly features, there are sufficient processes and equipment for treating pollution.

CONCLUSIONS

Ecotechnology is the engineering, science, and management of applying an ecosystem approach and principles for better recycling of "wasted resources" and to promote healthy living at affordable costs while reducing the impacts on the life processes of the environment. Ecotechnology is the essence of physical, chemical, biological, environmental, and atmospheric science and engineering and it is not a "brand" to be sold in the market for maximization of profits. It has the great responsibility of improving the state of resources and reservoirs and, in effect, the health of the ecosystem and man's well-being.

From the above discussion, the following conclusions are arrived at

1. The ecotechnological system's capital expenditure is comparable with the annual operational cost of conventional bioremediation systems.
2. Ecotechnological systems can be developed and operated in combination with conventional systems to improve the performance of the latter.
3. Ecotechnological systems reduce the ecotoxicity of the man-made substances released into the water bodies and facilitate the eco-assimilation of those pollutants into the ecological cycles, thus reducing the quantum of hazardous residues to zero which otherwise require costly secured landfill and incineration techniques.
4. Ecotechnological low-cost systems reduce the carbon footprint of human wastes (domestic and industrial).
5. Ecotechnological systems add to the aesthetic value, greenery, and scenic beauty of the treatment site.

WAY AHEAD

Though a lot of research is being done on reducing the space required for the natural treatment systems, there is a need to undertake studies on various other limitations and approaches of ecotechnologies:

- Currently, vertical ecofiltration systems have been used for 1–200 m^3/day flow of wastewater from point sources and horizontal filtration systems have been used for 0.7–600 MLD flow of streams receiving pollution from the non-point sources with depth ranging from 0.3 to 1.5 m. There is a need to work on ecotechnological solutions for greater depths.
- The effect of sodium, SAR, or excessive concentration of other free radicals on the biological communities used in constructed wetland, phytoremediation, bioremediation, or ecotechnological treatment systems.
- Presently, the objective is to degrade, detoxify, and stabilize the incoming pollution with focus on the process rather than on the infrastructure. However, harnessing the recovery of utilizable by-products needs to be worked on.
- Adaptability of natural treatment systems in different geo-climatic conditions and natural calamities using local multicellular floral and microinvertebrate species.
- Sustainable maintenance of treatment systems, socio-legal acceptance, and adaptation and estimation of productivity.
- Integration of ecotechnological treatment systems with local aquatic and terrestrial ecosystems on a long-term basis.
- The concept of in situ treatment of polluted water bodies as one of the solutions in maintaining water resources free of pollution is yet to be adapted by regulatory, financial institutions. As one of the solutions of maintaining water resources free of pollution.

REFERENCES

1. McEldowney, S.M.; Hardman, D.J.; Waite, S. *Pollution: Ecology and Biotreatment*; Longman Scientific and Technical: England, 1993.
2. Dugan, P.R. *Biochemical Ecology of Water Pollution*; Plenum Press: New York, 1971.

3. Gaudy, A.F., Jr.; Gaudy, E.T. *Microbiology for Environmental Scientists and Engineers*; McGraw-Hill Book Co.: New York, 1980.
4. Joshi, S. Ecotechnological restoration of polluted streams, IL^{2}BM and sustainable lake management. In Stakeholders Workshop on Lake Chivero Integrated Lake Basin Management (ILBM) and ILBM Platform Launch, Harare, Zimbabwe Nov. 30–Dec. 01, 2011.
5. Anon. List of countries by Human Development Index available at http://en.wikipedia.org/wiki/List_of_countries_by_Human_Development_Index (accessed on Nov. 28, 2011)
6. Monaghan, Suzanne, Smith, David, Wall, Brendan and O'Leary, Gerard. Urban Waste Water Discharges in Ireland for Population Equivalents Greater than 500 Persons - A Report for the Years 2006 and 2007; Published by the Environmental Protection Agency, Ireland, 2009 available at http://www.epa.ie/downloads/pubs/water/wastewater/name,26384,en.html (accessed on Nov. 28, 2011)
7. Caribbean Environment Programme. *Appropriate Technology for Sewage Pollution Control in the Wider Caribbean Region*. Technical Report No. 40; Kingston, Jamaica: United Nations Environment Programme, 1998.
8. Tajrishy, M.; Abrishamchi, A. Integrated Approach to Water and Wastewater Management for Tehran, Iran, Water Conservation, Reuse, and Recycling: Proceedings of the Iranian-American Workshop, National Academies Press, 2005.
9. Anon. List of states and union territories of India by population. available at http://en.wikipedia.org/wiki/List_of_states_and_union_territories_of_India_by_population (accessed on Nov. 28, 2011)
10. Anon. Sewage treatment. available at http://en.wikipedia.org/wiki/Sewage_treatment (accessed on Nov. 28, 2011).
11. UNEP. Guidelines on Municipal Wastewater Management, 2004.
12. Dean, R.B.; Lund, E. *Water Reuse: Problems and Solutions*; Academic Press: London, 1981.
13. USEPA. Manual of constructed wetlands treatment of Municipal Wastewaters National Risk Management Research Laboratory, Office of Research & Dept. USEPA, Cincinnati 45268, 2000.
14. Abel, P.D. *Water Pollution Biology*; Taylor and Francis Ltd.: London, 1996.
15. Neidleman, S.; Laskin, A. *Advances in Applied Microbiology*; Academic Press, Inc.: London, 1993.
16. Paul, E.A.; Clark, F.P. *Soil Microbiology and Biochemistry*; Academic Press: London, 1989.
17. Chaudhry, R.G., Ed. *Biological Degradation and Bioremediation of Toxic Chemicals*; Chapman and Hall: London, 1994.
18. Joshi, S.; Joshi, S. Ecotechnological application for the control of lake pollution. 12th World Lake Conference held at Jaipur by Ministry of Environment and Forests of India and ILEC, 2007.
19. Cookson, T.J. *Bioremediation and Engineering: Design and Application*; McGraw Hill, Inc.: New York, 1995.
20. Anon. Ecotechnology. available at http://en.wikipedia.org/wiki/Ecotechnology (accessed on Dec. 7, 2011).
21. Seidel, K. Macrophytes and water purification. In *Biological Control of Water Pollution*; Tourbier, J., Pierson, R.W., Eds.; Pennsylvania University Press: Philadelphia, Pennsylvania, 1976; 109–122.
22. Joshi, S. *Environment Management for Professionals, Businesses and Industries*; Publication of Shrishti Eco-Research Institute: Pune, India, 2007.
23. Joshi, S.; Joshi, S. Ecoplanning and ecotechnological solutions for the wastewaters from point and non-point sources of modern urbanization in the Ujjani Reservoir Catchment. 13th World Lake Conference organised in Wuhan China by Chinese Society of Environmental Sciences and ILEC, 2009.
24. Joshi, S. Eco-planning in urban lake basin—Case study of Ujjani Lake. In *Integrated Lake Basin Management (ILBM)*; Indian Association of Aquatic Biologists (IAAB): Hyderabad, India, 2009; 35–42.
25. Anon. Constructed wetland. available at http://en.wikipedia.org/wiki/Constructed_wetland (accessed on Dec. 7, 2011).
26. Suutari, Amanda. USA - California (Arcata) - Constructed Wetland: A Cost-Effective Alternative for Wastewater Treatment Earth Island Journal 22(2):26-31. available at http://www.ecotippingpoints.org/our-stories/indepth/usa-california-arcata-constructed-wetland-wastewater.html (accessed on Dec. 7, 2011).
27. Available at http://www.midlandsbusinessnews.co.uk/2010-12/national-botanic-garden-of-wales-investsin-%E2%80%98poo-palace%E2%80%99.asp.
28. Schmidt, Carsten K., Lange, Frank Thomas and Brauch, Heinz-Jürgen. Assessing the Impact of Different Redox Conditions and Residence Times on the Fate of Organic Micropollutants during Riverbank Filtration. Available at http://info.ngwa.org/gwol/pdf/042379992.pdf (accessed on Dec. 7, 2011).
29. Anon. Bank filtration. Available at http://en.wikipedia.org/wiki/Bank_filtration (accessed on Dec. 7, 2011).
30. Schmidt, C.K.; Lange, F.T.; Brauch, H.-J., Kühn, W. Experiences with riverbank filtration and infiltration in Germany. Proceedings International Symposium on Artificial Recharge of Groundwater, Nov. 14, 2003, Daejon, Korea, 2003; 115–141.
31. Water Resources Centre, University of Hawaii. Bank filtration for water treatment. In Manoa, Bulletin, August 2000.
32. Metcalf and Eddy; Tchobanoglous, G.; Burton, F.L. *Wastewater Engineering, Treatment, Disposal, and Reuse of Metcalf and Eddy, Inc.*, rev. 3rd ed., Tata McGraw-Hill Publishing: New Delhi, India, 2001, (Eleventh reprint), pp. 1209–1220.
33. Babatunde, A.O.; Zhao, Y.Q.; O'Neill, M., O'Sullivan, B. Constructed wetlands for environmental pollution control: A review of developments, research and practice in Ireland. Environ. Int. **2008**, *34*, 116–126.
34. Rajbhandari, Kabir. A WaterAid report - Sunga constructed wetland for wastewater management :A case study in community based water resource management. Available at http://www.wateraid.org/documents/plugin_documents/sunga_constructed_wetland_for_waste_water_management.pdf (accessed on Dec. 7, 2011).
35. Bista, K. R. and Khatiwada, N. R. Performance study on reed bed wastewater treatment units in Nepal. Available at http://www.aehms.org/pdf/Bista%20%26%20Khatiwada%20FE.pdf (accessed on Dec. 7, 2011).
36. Truong, P.N.; Hart, B. Vetiver System for wastewater treatment. Technical Bulletin No. 2001/2. Pacific Rim vetiver

37. Truong, Paul, Van, Tran Tan and Elise Pinners. The Vetiver System for Prevention and Treatment of Contaminated Water and Land. Available at http://www.vetiver.org/TVN-Handbook%20series/TVN-series2-2pollution.htm (accessed on Dec. 7, 2011).
38. Joshi, S. Critical evaluation of green bridge system—horizontal ecofiltration for treatment of streams carrying pollution downstream lakes and rivers. Presented in 14th World Lake Conference held in Austin, Texas, US jointly organised by River Systems Institute, Texas University and International Lake Environment Committee Foundation (ILEC), Japan, 2011.
39. Sinha, P.; Joshi, S. Use of green bridge and green lake Systems to treat polluted streams in Pune City. In Proceedings of Potential of Eco-technology in Water Supply and Sanitation; Shrishti Eco-Research Institute: Pune India, 2007; 33–34.
40. Kodarkar, M.; Joshi, S. ILBM Impact Story Ecological Restoration of Highly Polluted Stretch of Ahar River, Udaipur and Ecological Improvement of Udaisagar Lake, Rajasthan, India. Presented in Final Review Meeting and International Symposium of a project entitled "Integrated Lake Basin Management (ILBM), Basin Governance Challenges and Prospects" November 2-7 2010, ILEC (International Lake Environment Committee) Headquarters, Kusatsu, Japan, 2010. Available at http://rcse.edu.shiga-u.ac.jp/gov-pro/plan/2010list/10/indian_lakes/ilbm_impact_story-udaipur's_ecological_restoration_of_ahar_river_using_green_bridge_technology__2_.pdf (accessed Dec. 7, 2011).
41. SERI Desk. Comparative Statement of Conventional Technologies and Ecotechnology. SERInews, **2011**, *5* (10), 5–6. Available at http://www.seriecotech.com.

Pollution: Point Sources

Ravendra Naidu
Mallavarapu Megharaj
Peter Dillon
Rai Kookana
Ray Correll
Commonwealth Scientific and Industrial Research Organization (CSIRO), Adelaide, South Australia, Australia

W.W. Wenzel
Institute of Soil Research, University of Natural Resources and Life Sciences, Vienna, Austria

Abstract
Pollution can be generally defined as an undesirable change in the natural quality of the environment that may adversely affect the well being of humans, other living organisms, or entire ecosystems either directly or indirectly. Where pollution is localized it is described as point source (PS). Thus, PS pollution is a source of pollution with a clearly identifiable point of discharge that can be traced back to the specific source.

INTRODUCTION

Environmental pollution is one of the foremost ecological challenges. Pollution is an offshoot of technological advancement and overexploitation of natural resources. From the standpoint of pollution, the term environment primarily includes air, land, and water components including landscapes, rivers, parks, and oceans. Pollution can be generally defined as an undesirable change in the natural quality of the environment that may adversely affect the well being of humans, other living organisms, or entire ecosystems either directly or indirectly. Although pollution is often the result of human activities (anthropogenic), it could also be due to natural sources such as volcanic eruptions emitting noxious gases, pedogenic processes, or natural change in the climate. Where pollution is localized it is described as point source (PS). Thus, PS pollution is a source of pollution with a clearly identifiable point of discharge that can be traced back to the specific source such as leakage of underground petroleum storage tanks or an industrial site.

Some naturally occurring pollutants are termed geogenic contaminants and these include fluorine, selenium, arsenic, lead, chromium, fluoride, and radionuclides in the soil and water environment. Significant adverse impacts of geogenic contaminants (e.g., As) on environmental and human health have been recorded in Bangladesh, West Bengal, India, Vietnam, and China. More recently reported is the presence of geogenic Cd and the implications to crop quality in Norwegian soils.[1]

The terms contamination and pollution are often used interchangeably but erroneously. Contamination denotes the presence of a particular substance at a higher concentration than would occur naturally and this may or may not have harmful effects on human or the environment. Pollution refers not only to the presence of a substance at higher level than would normally occur but is also associated with some kind of adverse effect.

NATURE AND SOURCES OF CONTAMINANTS

The main activities contributing to PS pollution include industrial, mining, agricultural, and commercial activities as well as transport and services (Table 1). Uncontrolled mining, manufacturing, and disposal of wastes inevitably cause environmental pollution. Military land and land for recreational shooting are also important sites of PS contamination. The contaminants associated with such activities are listed in Table 1. Contamination at many of these sites appears to have resulted because of lax regulatory measures prior to the establishment of legislation protecting the environment.

CONTAMINANT INTERACTIONS IN SOIL AND WATER

Inorganic Chemicals

Inorganic contaminant interactions with colloid particulates include: adsorption–desorption at surface sites, precipitation, exchange with clay minerals, binding by organically

Table 1 Industries, land uses, and associated chemicals contributing to points, non-point source pollution.

Industry	Type of chemical	Associated chemicals
Airports	Hydrocarbons	Aviation fuels
	Metals	Particularly aluminum, magnesium, and chromium
Asbestos production and disposal	Asbestos	
Battery manufacture and recycling	Metals	Lead, manganese, zinc, cadmium, nickel, cobalt, mercury, silver, and antimony
	Acids	Sulfuric acid
Breweries/distilleries	Alcohol	Ethanol, methanol, and esters
Chemicals manufacture and use	Acid/alkali	Mercury (chlor/alkali), sulfuric, hydrochloric and nitric acids, sodium and calcium hydroxides
	Adhesives/resins	Polyvinyl acetate, phenols, formaldehyde, acrylates, and phthalates
	Dyes	Chromium, titanium, cobalt, sulfur and nitrogen organic compounds, sulfates, and solvents
	Explosives	Acetone, nitric acid, ammonium nitrate, pentachlorophenol, ammonia, sulfuric acid, nitroglycerine, calcium cyanamide, lead, ethylene glycol, methanol, copper, aluminum, *bis*(2-ethylhexyl) adipate, dibutyl phthalate, sodium hydroxide, mercury, and silver
	Fertilizer	Calcium phosphate, calcium sulfate, nitrates, ammonium sulfate, carbonates, potassium, copper, magnesium, molybdenum, boron, and cadmium
	Flocculants	Aluminum
	Foam production	Urethane, formaldehyde, and styrene
	Fungicides	Carbamates, copper sulfate, copper chloride, sulfur, and chromium
	Herbicides	Ammonium thiocyanate, carbanates, organochlorines, organophosphates, arsenic, and mercury
	Paints	
	Heavy metals	Arsenic, barium, cadmium, chromium, cobalt, lead, manganese, mercury, selenium, and zinc
	General	Titanium dioxide
	Solvent	Toluene, oils natural (e.g., pine oil) or synthetic
	Pesticides	
	Active ingredients	Arsenic, lead, organochlorines, and organophosphates Sodium, tetraborate, carbamates, sulfur, and synthetic pyrethroids
	Solvents	Xylene, kerosene, methyl isobutyl ketone, amyl acetate, and chlorinated solvents
	Pharmacy	Dextrose and starch
	General/solvents	Acetone, cyclohexane, methylene chloride, ethyl acetate, butyl acetate, methanol, ethanol, isopropanol, butanol, pyridine methyl ethyl ketone, methyl isobutyl ketone, and tetrahydrofuran
	Photography	Hydroquinone, pheidom, sodium carbonate, sodium sulfite, potassium bromide, monomethyl paraaminophenol sulfates, ferricyanide, chromium, silver, thiocyanate, ammonium compounds, sulfur compounds, phosphate, phenylene diamine, ethyl alcohol, thiosulfates, and formaldehyde
	Plastics	Sulfates, carbonates, cadmium, solvents, acrylates, phthalates, and styrene
	Rubber	Carbon black
	Soap/detergent General	Potassium compounds, phosphates, ammonia, alcohols, esters, sodium hydroxide, surfactants (sodium lauryl sulfate), and silicate compounds
	Acids	Sulfuric acid and stearic acid
	Oils	Palm, coconut, pine, and tea tree
	Solvents	
	General	Ammonia
	Hydrocarbons	e.g., BTEX (benzene, toluene, ethylbenzene, xylene)
	Chlorinated organics	e.g., trichloroethane, carbon tetrachloride, and methylene chloride

(Continued)

Table 1 Industries, land uses, and associated chemicals contributing to points, non-point source pollution. (Continued)

Industry	Type of chemical	Associated chemicals
Defense works		See *Explosives* under *Chemicals Manufacture and Use, Foundries, Engine Works, and Service Stations*
Drum reconditioning		See *Chemicals Manufacture and Use*
Dry cleaning		Trichlorethylene and ethane Carbon tetrachloride Perchlorethylene
Electrical		PCBs (transformers and capacitors), solvents, tin, lead, and copper
Engine works	Hydrocarbons	
	Metals	
	Solvents	
	Acids/alkalis	
	Refrigerants	
	Antifreeze	Ethylene glycol, nitrates, phosphates, and silicates
Foundries	Metals	Particularly aluminum, manganese, iron, copper, nickel, chromium, zinc, cadmium and lead and oxides, chlorides, fluorides and sulfates of these metals
	Acids	Phenolics and amines coke/graphite dust
Gas works	Inorganics	Ammonia, cyanide, nitrate, sulfide, and thiocyanate
	Metals	Aluminum, antimony, arsenic, barium, cadmium, chromium, copper, iron, lead, manganese, mercury, nickel, selenium, silver, vanadium, and zinc
	Semivolatiles	Benzene, ethylbenzene, toluene, total xylenes, coal tar, phenolics, and PAHs
Iron and steel works		Metals and oxides of iron, nickel, copper, chromium, magnesium and manganese, and graphite
Landfill sites Marinas		Methane, hydrogen sulfides, heavy metals, and complex acids Engine works, electroplating under metal treatment
	Antifouling paints	Copper, tributyltin (TBT)
Metal treatments	Electroplating metals	Nickel, chromium, zinc, aluminum, copper, lead, cadmium, and tin
	Acids	Sulfuric, hydrochloric, nitric, and phosphoric
	General	Sodium hydroxide, 1,1,1-trichloroethane, tetrachloroethylene, toluene, ethylene glycol, and cyanide compounds
	Liquid carburizing baths	Sodium, cyanide, barium, chloride, potassium chloride, sodium chloride, sodium carbonate, and sodium cyanate
	Mining and extracting industries	Arsenic, mercury, and cyanides and also refer to *Explosives* under *Chemicals Manufacture and Use*
	Power stations	Asbestos, PCBs, fly ash, and metals
	Printing shops	Acids, alkalis, solvents, chromium see *Photography* under *Chemicals Manufacture and Use*
Scrap yards	Service stations and fuel storage facilities	Hydrocarbons, metals, and solvents Aliphatic hydrocarbons
		BTEX (i.e., benzene, toluene, ethylbenzene, xylene) PAHs (e.g., benzo(a) pyrene) Phenols Lead
Sheep and cattle dips		Arsenic, organochlorines and organophosphates, carbamates, and synthetic pyrethroids
Smelting and refining		Metals and the fluorides, chlorides and oxides of copper, tin, silver, gold, selenium, lead, and aluminum
Tanning and associated trades	Metals	Chromium, manganese, and aluminum

(Continued)

Table 1 Industries, land uses, and associated chemicals contributing to points, non-point source pollution. (*Continued*)

	General	Ammonium sulfate, ammonia, ammonium nitrate, phenolics (creosote), formaldehyde, and tannic acid
Wood preservation	Metals	Chromium, copper, and arsenic
	General	Naphthalene, ammonia, pentachlorophenol, dibenzofuran, anthracene, biphenyl, ammonium sulfate, quinoline, boron, creosote, and organochlorine pesticides

Source: Barzi et al.[11]

coated particulate matter or organic colloidal material, or adsorption of contaminant ligand complexes. Depending on the nature of contaminants, these interactions are controlled by solution pH and ionic strength of soil solution, nature of the species, dominant cation, and inorganic and organic ligands present in the soil solution.[2]

Organic Chemicals

The fate and behavior of organic compounds depend on a variety of processes including sorption–desorption, volatilization, chemical and biological degradation, plant uptake, surface runoff, and leaching. Sorption–desorption and degradation (both biotic and abiotic) are perhaps the two most important processes as the bulk of the chemicals is either sorbed by organic and inorganic soil constituents, and chemically or microbially transformed/degraded. The degradation is not always a detoxification process. This is because in some cases the transformation or degradation process leads to intermediate products that are more mobile, more persistent, or more toxic to non-target organisms. The relative importance of these processes is determined by the chemical nature of the compound.

IMPLICATIONS TO SOIL AND ENVIRONMENTAL QUALITY

Considerable amount of literature is available on the effects of contaminants on soil microorganisms and their functions in soil. The negative impacts of contaminants on microbial processes are important from the ecosystem point of view and any such effects could potentially result in a major ecological perturbance. Hence, it is most relevant to examine the effects of contaminants on microbial processes in combination with communities. The most commonly used indicators of metal effects on microflora in soil are: (1) soil respiration, (2) soil nitrification, (3) soil microbial biomass, and (4) soil enzymes.

Contaminants can reach the food chain by way of water, soil, plants, and animals. In addition to the food chain transfer, pollutants may also enter via direct consumption or dust inhalation of soil by children or animals. Accumulation of these pollutants can take place in certain target tissues of the organism depending on the solubility and nature of the compound. For example, DDT and PCBs accumulate in human adipose tissue. Consequently, several of these pollutants have the potential to cause serious abnormalities including cancer and reproductive impairments in animal and human systems.

SAMPLING FOR PS POLLUTION

The aims of the sampling system must be clearly defined before it can be optimized.[3] The type of decision may be to determine land use, how much of an area is to be remediated, or what type of remediation process is required. Because sampling and the associated chemical and statistical analyses are expensive, careful planning of the sampling scheme is therefore a good investment. One of the best ways to achieve this is to use any ancillary data that are available. These data could be in the form of emission history from a stack, old photographs that give details of previous land uses, or agricultural records. Such data can at least give qualitative information.

As discussed before, PS pollution will typically be airborne from a stack, or waterborne from some effluent such as tannery waste, cattle dips, or mine waste. In many cases, the industry will have modified its emissions (e.g., cleaner production) or point of release (increased stack height), hence the current pattern of emission may not be closely related to the historic pattern of pollution. For example, liquid effluent may have been discharged previously into a bay, but that effluent may now be treated and perhaps discharged at some other point. Typically, the aim of a sampling scheme in these situations is to assess the maximum concentrations, the extent of the pollution, and the rate of decline in concentration from the PS. Often the sampling scheme will be used to produce maps of concentration isopleths of the pollutant.

The location of the sampling points would normally be concentrated towards the source of the pollution. A good scheme is to have sufficient samples to accurately assess the maximum pollution, and then space additional samples at increasing intervals. In most cases, the distribution of the pollutant will be asymmetric, with the maximum spread down the slope or down the prevailing wind. In such cases more samples should be placed in the direction of the expected gradient. This is a clear case of when ancillary data can be used effectively. A graph of concentration of the pollutant against the reciprocal of distance from the source is often informative.[4] Sampling depths will depend on both the nature of the pollution and the reason for the investigation. If the pollution is from dust and it is unlikely to be leached, only surface sampling will be required. An example of this is pollution from silver smelting in Wales.[5] In contrast, contamination from organic or mobile inorganic pollutants such as F compounds may migrate well down to the profile and deep sampling may be required.[6,7]

ASSESSMENT

In order to assess the impacts of pollution, reliable and effective monitoring techniques are important. Pollution can be assessed and monitored by chemical analyses, toxicity tests, and field surveys. Comparison of contaminant data with an uncontaminated reference site and available databases for baseline concentrations can be useful in establishing the extent of contamination. However, this may not always be possible in the field. Chemical analyses must be used in conjunction with biological assays to reveal site contamination and associated adverse effects. Toxicological assays can also reveal information about synergistic interactions of two or more contaminants present as mixtures in soil, which cannot be measured by chemical assays alone.

Microorganisms serve as rapid detectors of environmental pollution and are thus of importance as pollution indicators. The presence of pollutants can induce alteration of microbial communities and reduction of species diversity, inhibition of certain microbial processes (organic matter breakdown, mineralization of carbon and nitrogen, enzymatic activities, etc.). A measure of the functional diversity of the bacterial flora can be assessed using ecoplates (see http://www.biolog.com/section_4.html). It has been shown that algae are especially sensitive to various organic and inorganic pollutants and thus may serve as a good indicator of pollution.[8] A variety of toxicity tests involving microorganisms, invertebrates, vertebrates, and plants may be used with soil or water samples.[9]

MANAGEMENT AND/OR REMEDIATION OF PS POLLUTION

The major objective of any remediation process is to (1) reduce the actual or potential environmental threat; and (2) reduce unacceptable risks to man, animals, and the environment to acceptable levels.[10] Therefore, strategies to either manage and/or remediate contaminated sites have been developed largely from application of stringent regulatory measures set up to safeguard ecosystem function as well as to minimize the potential adverse effects of toxic substances on animal and human health.

The available remediation technologies may be grouped into two categories: (1) ex situ techniques that require removal of the contaminated soil or groundwater for treatment either on-site or off-site; and (2) in situ techniques that attempt to remediate without excavation of contaminated soils. Generally, in situ techniques are favored over ex situ techniques because of (1) reduced costs due to elimination or minimization of excavation, transportation to disposal sites, and sometimes treatment itself; (2) reduced health impacts on the public or the workers; and, (3) the potential for remediation of inaccessible sites, e.g., those located at greater depths or under buildings. Although in situ techniques have been successful with organic contaminated sites, the success of in situ strategies with metal contaminants has been limited. Given that organic and inorganic contaminants often occur as a mixture, a combination of more than one strategy is often required to either successfully remediate or manage metal contaminated soils.

GLOBAL CHALLENGES AND RESPONSIBILITY

The last 100 years has seen massive industrialization. Indeed such developments were coupled with the rapid increase in world population and the desire to enhance economy and food productivity. While industrialization has led to increased economic activity and much benefit to human race, the lack of regulatory measures and appropriate waste management strategies until early 1980s (including the use of agrochemicals) has resulted in contamination of our biosphere. Continued pollution of the environment through industrial emissions is of global concern. There is, therefore, a need for politicians, regulatory organizations, and scientists to work together to minimize environmental contamination and to remediate contaminated sites. The responsibility to check this pollution lies with every individual and country although the majority of this pollution is due to the industrialized nations. There is a clear need of better coordination of efforts in dealing with numerous forms of PS pollution problems that are being faced globally.

REFERENCES

1. Mehlum, H.K.; Arnesen, A.K.M.; Singh, B.R. Extractability and plant uptake of heavy metals in alum shale soils. Commun. Soil Sci. Plant Anal. **1998**, *29*, 183–198.
2. McBride, M.B. Reactions controlling heavy metal solubility in soils. Adv. Soil Sci. **1989**, *10*, 1–56.
3. Patil, G.P.; Gore, S.D.; Johnson, G.D. *EPA Observational Economy Series Volume 3: Manual on Statistical Design and Analysis with Composite Samples*; Technical Report

No. 96-0501;Center for Statistical Ecology and Environmental Statistics: Pennsylvania State University, 1996.
4. Ward, T.J.; Correll, R.L. Estimating background concentrations of heavy metals in the marine environment, Proceedings of a Bioaccumulation Workshop: Assessment of the Distribution, Impacts and Bioaccumulation of Contaminants in Aquatic Environments, Sydney, 1990; Miskiewicz, A.G., Ed.; Water Board and Australian Marine Science Association: Sydney, 1992, 133–139.
5. Jones, K.C.; Davies, B.E.; Peterson, P.J. Silver in welsh soils: physical and chemical distribution studies. Geoderma **1986**, *37*, 157–174.
6. Barber, C.; Bates, L.; Barron, R.; Allison, H. Assessment of the relative vulnerability of groundwater to pollution: a review and background paper for the conference workshop on vulnerability assessment. J. Aust. Geol. Geophys. **1993**, *14* (2–3), 147–154.
7. Wenzel, W.W.; Blum, W.E.H. Effects of fluorine deposition on the chemistry of acid luvisols. Int. J. Environ. Anal. Chem. **1992**, *46*, 223–231.
8. Megharaj, M.; Singleton, I.; McClure, N.C. Effect of pentachlorophenol pollution towards microalgae and microbial activities in soil from a former timber processing facility. Bull. Environ. Contam. Toxicol. **1998**, *61*, 108–115.
9. Juhasz, A.L.; Megharaj, M.; Naidu, R. Bioavailability: the major challenge (constraint) to bioremediation of organically contaminated soils. In *Remediation Engineering of Contaminated Soils*; Wise, D., Trantolo, D.J., Cichon, E.J., Inyang, H.I., Stottmeister, U., Eds.; Marcel Dekker: New York, 2000; 217–241.
10. Wood, P.A. Remediation methods for contaminated sites. In *Contaminated Land and Its Reclamation*; Hester, R.E., Harrison, R.M., Eds.; Royal Society of Chemistry, Thomas Graham House: Cambridge, U.K., 1997; 47–73.
11. Barzi, F.; Naidu, R.; McLaughlin, M.J. Contaminants and the Australian soil environment. In *Contaminants and the Soil Environment in the Australasia–Pacific Region*; Naidu, R., Kookana, R.S., Oliver, D., Rogers, S., McLaughlin, M.J., Eds.; Kluwer Academic Publishers: Dordrecht, the Netherlands, 1996; 451–484.

Polychlorinated Biphenyls (PCBs)

Marek Biziuk
Angelika Beyer
Department of Analytical Chemistry, Chemical Faculty, Gdansk University of Technology, Gdansk, Poland

Abstract

Polychlorinated biphenyls (PCBs) belong to a broad family of man-made organic chemicals that due to their non-flammability, chemical stability, high boiling point, and electrical insulating properties, were used in hundreds of industrial applications. The same qualities make many individual chlorobiphenyls slow to degrade upon their release to the environment. As a consequence, PCBs are widespread and persistent environmental contaminants found in air, water, sediments, and soils. Moreover, they can accumulate through the trophic chain to the aquatic organism, to fish and humans. As PCBs move through the environment, the concentrations of individual chlorobiphenyls change over time and from one environmental medium to another because of physical and chemical processes and selective bioaccumulation and biotransformation by living organisms. These processes result in mixtures that are substantially different from the original mixtures that were released to the environment. The identification, quantification, and risk assessments are complicated by these changes in the composition of the PCB mixtures.

INTRODUCTION

There is growing concern about the trace quantities of highly chlorinated organic compounds (e.g., dioxins, PCBs, and certain pesticides) that exist in diverse environmental media (air, water, soils, sediments, and biota), enter the trophic chain, and reach humans and wildlife. Consequently, there is growing scientific, regulatory, and social interest in measuring the levels of chlorinated chemicals in environmental media, and in determining the environmental effects of such contamination.

Representative of these synthetic organic chlorine compounds are polychlorinated biphenyls commonly known as PCBs. PCBs are man-made chemicals that never existed in nature until the 1900s when they started to be released into the environment by manufacturing companies and consumers. Although production of PCBs was banned, when their ability to accumulate in the environment and to cause harmful effects became apparent (in 1970, Sweden; in 1972, Japan; in 1977, the United States), these chemicals still are found in the environment.[1]

PCBs make up a group of 209 individual chlorinated biphenyl rings—congeners. They were typically manufactured as mixtures of 60 to 90 different congeners and were usually contaminated with small amounts of very toxic chemicals such as polychlorinated dibenzofurans (furans) or polychlorinated dibenzodioxins (dioxins). The trade names of some commercial PCB mixtures are Aroclor (United States), Clophen (Germany), Fenclor (Italy), Kanechlor (Japan), and Phenoclor (France).[2]

CHEMICAL IDENTITY

PCBs are a mixture of individual chemicals. The general chemical structure of chlorinated biphenyl (two benzene rings with a carbon–carbon bond between carbon 1 on one ring and carbon 1′ on the second ring) with a varying number of chlorines is shown in Fig. 1. It can be seen from the structure that a large number of chlorinated compounds are possible (209 possible congeners) in which 2–10 chlorine atoms are attached to the biphenyl molecule. Chlorines can be attached to any of the carbons by removing the hydrogen from that carbon and substituting the chlorine in its place. The common nomenclature used for identifying the location of chlorine atoms on the biphenyl rings is as shown in Fig. 1.

The congeners are arranged in ascending numerical order using a numbering system developed by Ballschmiter and Zell[3] that follows the International Union of Pure And Applied Chemistry (IUPAC) rules of substituent characterization in biphenyls. The resulting PCB numbers, also referred to as congener, IUPAC, or BZ numbers, are widely used for identifying individual congeners.

PROPERTIES OF PCBs

Because of physical and chemical properties of PCBs (summarized in Fig. 2[4–6]), these chemicals were quickly acclaimed as an industrial breakthrough. PCBs are either oily liquids or solids that are colorless to light yellow without known smell or taste. In general, PCBs are relatively

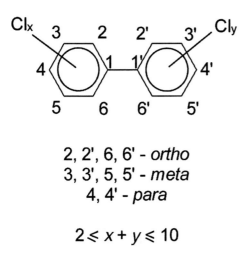

Fig. 1 Structure of PCBs.

insoluble in water (solubility decreases with increased chlorination) but freely soluble in nonpolar organic solvents and biological lipids.[7] Because of their thermal stability, they do not easily burn, hence their past popular use as coolants, as insulating materials, and for electrical applications. The properties of PCBs vary from one congener to the next, e.g., color of PCB mixture darkens, viscosity increases, the flash point rises, and the substance becomes less combustible with rising chlorine content. Also, as the number of chlorines in a PCB mixture increases, the mixture is more stable and thus resistant to biodegradation. The congeners with large numbers of chlorines are also proving to be the ones that present the greatest environmental and health risks.

The properties that make PCB mixtures so desirable and applicable in industry (general inertness, thermal stability) are the ones that make the mixtures so hazardous to the environment. The toxicity of a PCB congener is dependent on the number of chlorines present on the biphenyl structures and the positions of the chlorines. The congeners in which there is a coplanar confirmation with chlorine substituents on the *meta* and *para* positions of the phenyl rings are the most toxic and bioaccumulative ones. For instance, congeners with chlorines in both *para* positions (4 and 4′) and at least 2 chlorines at the *meta* positions (3, 5, 3′, 5′) are considered to be "dioxin like" and are particularly toxic.[8]

The high thermal and chemical resistance of PCB congeners means that they do not readily break down when exposed to heat or chemical treatment. However, since PCBs do not break down, they remain in the environment. Due to their persistence in the environment and the fact that they are poorly biodegraded, PCBs accumulate in the environment.

GLOBAL DISTRIBUTION AND SOURCES

PCBs were first produced in 1929 for a wide variety of uses because of their unique physical properties that made them attractive compounds for industries (see Table 1[9–12]). As more uses were found for PCBs, their production increased exponentially. In Table 1, there are identified PCB use areas based on their presence in closed, partially closed, and open systems. These designations refer to how easily the PCBs contained within a product can escape to the surrounding environment. In closed applications, PCBs are held completely within the equipment. Under ordinary circumstances, no PCBs would be available for exposure to the user or the environment. However, PCB emissions may occur during equipment servicing/repairing and decommissioning, or as a result of damaged equipment. Partially closed PCB applications, in which the PCB oil is not directly exposed to the environment, but may become so periodically during typical use, lead also to PCB emissions, through air or water discharge, whereas in open systems, PCBs are in direct contact with the environment and thereby may be easily transferred to the environment. Generally, closed and partially closed systems contain PCB oils or fluids. The PCBs in open systems take on the form (type of media) of the product they have been used in as an ingredient. Therefore, PCBs in open applications may be found in forms ranging from paint to plastic or rubber.[9]

The first indication that PCBs may be damaging to human health occurred four decades after PCBs were first introduced into the environment. Preliminary studies suggested that PCBs may pose a serious health threat to humans, and at the same time, there were indications of widespread distribution and longevity throughout the environment. As more attention was turned towards PCBs, it became clearer that PCBs were having a negative impact on many biological systems. Eventually, all production and importation of

Physicochemical properties of PCBs

- A low degree of reactivity (very stable even when exposed to heat and pressure).
- Good insulating properties.
- Nonflammable.
- Good solubility in nonpolar solvents, oils, and fats.
- Virtual insolubility in water.
- Low vapor pressure (nonvolatile).
- Low electrical conductivity.
- High thermal conductivity.
- High ignition temperature.
- Very high resistance to chemical factors—do not undergo oxidation, reduction, addition, elimination or electrophilic substitution reactions except under extreme conditions.

Fig. 2 Physicochemical properties of PCBs.

Table 1 Examples of applications of PCBs.

Closed applications	Partially closed applications	Open applications

– Electrical transformers
– Electrical capacitors
  o Power factor capacitors in electrical distribution systems
  o Lighting ballasts
  o Motor start capacitors in refrigerators, heating systems, air conditioners, hair dryers, water well motors
  o Capacitors in electronic equipment including television sets and microwave ovens
– Electrical motors in some specialized fluid-cooled motors
– Electrical magnets in some fluid-cooled separating magnets
	– Heat transfer fluids
	– Hydraulic fluids
	– Vacuum pumps
	– Switches
	– Voltage regulators
	– Liquid-filled electrical cables
	– Liquid-filled circuit breakers
		– Lubricants
		  o Immersion oil for microscopes (mounting media)
		  o Brake linings
		  o Cutting oils
		  o Lubricating oils
		– Casting waxes
		  o Pattern waxes for investment castings
		– Surface coatings
		  o Paints
		  o Surface treatment for textiles
		  o Carbonless copy paper
		  o Flame retardants
		  o Dust control
		– Adhesives
		  o Special adhesives
		  o Adhesives for waterproof wall coatings
		– Plasticizers
		  o Gasket sealers
		  o Filling material in joints of concrete
		  o Polyvinyl chloride plastics
		  o Rubber seals
		– Inks
		  o Dyes
		  o Printing inks
		– Other uses
		  o Insulting materials
		  o Pesticides

PCBs was banned in the 1970s.[13] Today, the production of PCBs has been ceased in many countries with the exception of small quantities manufactured strictly for research purposes. However, the ecotoxicological problems created by PCB contamination will be evident for many years to come, despite the restrictions of PCB utilization.

Although the manufacture, processing, distribution, and use of PCBs are widely prohibited, they have been released to the environment solely by human activity and still are redistributed from one environmental compartment to another.[1] There still exist a lot of different activities that generate PCB wastes. PCBs entered air, water, and soil during their manufacture, use, and disposal, mainly by leakage of supposedly closed systems, from landfill sites, incineration of waste, agricultural lands, industrial discharges, and sewage effluents. For more details, see Table 2.[14–23]

Nowadays, PCBs are present in all environmental media because of global circulation. The most important mechanism for global dispersion of these contaminations is atmospheric transport, which depends on the number of chlorines present on the biphenyl molecule:

- Biphenyls with one chlorine atom remain in the atmosphere.
- Those with one to four chlorines gradually migrate toward polar latitudes in a series of volatilization/deposition cycles between the air and the water and/or soil.
- Those with four to eight chlorines remain in mid-latitudes.
- Those with eight to nine chlorines remain close to the source of contamination.[24]

There are two classic approaches to model the distribution of PCBs (and other persistent organic pollutants [POPs]).[25] Multicompartment models use just limited meteorological data but include detailed descriptions of the partitioning of the species within and between the different environmental media,[26] while chemistry transport models have a detailed treatment of transport and chemistry in the atmosphere but a rather simple description of the compartments other than the atmosphere.[27,28] The multicompartment models have been successful in describing the global distribution of POPs and their long-term environmental

Table 2 Sources and transport of PCBs within the environment.

Part of the environment	Sources of PCBs in the environmental system	Factors influencing the pattern and rates of PCB movement in the media
Aquatic system— higher concentrations in the sediments of aquatic systems	– Accidental spills of PCB-containing hydraulic fluids – Improper disposal – Combined sewer overflows, or storm water runoff – Runoff and leaching from PCB-contaminated sewage sludge applied to farmland	– Properties of PCB congeners—desorption of PCBs from particulate is more likely to occur from lower-chlorinated, more water-soluble PCB congeners – Sorption reactions—with the chlorine content of PCB congener, surface area increases; with the organic content of the sediment, sorption increases – Sudden hydrographic activity like flooding or dredging causes sediments to be resuspended and redistributed and can cause the release of PCBs from sediments to overlying waters
Air system	– Volatilization from soil and water – Escape from uncontrolled landfills and hazardous waste sites – Incineration of PCB-containing wastes – Leakage from older electrical equipment – Improper waste disposal or spills – Leakage from supposedly closed systems – Incineration of waste – Industrial discharges – Sewage effluents	– Air temperature – Wind speed – Storm frequency – Rainfall rates – Volatility of individual PCB congeners
Soil system	– Accidental leaks and spills – Release from contaminated soils in landfills and hazardous waste sites – Deposition of vehicular emissions near roadway soil – Land application of sewage sludges containing PCBs	– Sorption reactions—i.e., highly chlorinated congeners are sorbed by soils and remain significantly immobile against leaching – Vapor phase transport—PCB congeners have a moderate vapor pressure, so vapor phase transport may allow for redistribution or migration through the saturated soil pores

fate in the various compartments.[29,30] Unfortunately, not many direct comparisons between model results and measurement data have been made so far.[31]

ENVIRONMENTAL FATE

PCBs, as it was indicated before, can partition between environmental media such as atmosphere, oceans, rivers, or soils. Differences in partitioning behavior among PCBs reflect differences in their physicochemical properties and persistence in the various media.[32]

PCBs do not readily break down in the environment and thus may remain there for a very long time. PCBs can travel long distances in the air and be deposited in areas far away from where they were released. In water, a small amount of PCBs may remain dissolved, but most stick to organic particles and bottom sediments. These toxic compounds can also bind strongly to soil. The degradation and transformation of PCBs entail difficult mechanisms of chemical, biochemical, or thermal destruction.[33] These substance may be (and are) accumulated through the trophic chain and reach aquatic organisms, fish, and humans.

Consequently, PCBs accumulate in fish and marine mammals, reaching levels that may be many thousands of times higher than in water. That is why there is great interest in different pathways for PCB loss, such as volatilization, adsorption on organic matter, and biodegradation, which can reduce PCB bioavailability.

Volatilization

PCBs enter the atmosphere from volatilization from both soil and water surfaces.[34] It was reported over 35 years ago by Haque,[35] who found minimal PCB loss at ambient temperature. Heat, airflow (hood storage), coarse grain size, high water content, and enrichment in lower *ortho*-chlorinated congeners all were expected to increase the rate and extent of PCB volatilization. As indicated by their higher vapor pressures, the lower-chlorinated homologs in particular are subject to volatilization.[36] This can result in both a loss and a source of lighter homologs—a source because upon volatilization, the atmosphere is enriched with these homologs, which are then subject to atmospheric deposition. Once in the atmosphere, PCBs are both present in the vapor phase and sorbed to particles. PCBs in the vapor phase appear to be more mobile and are transported further than particle-bound PCBs.[24]

Because of their persistence and semi-volatility, PCBs have a great potential for long-range atmospheric transport, which enables them to migrate from the mid-latitudes to the Arctic regions, for instance.[31,37] Atmospheric transport may occur in many mechanisms, one of them is a mechanism known as cold condensation,[37] by which some PCBs are preferably removed from the atmosphere in cold regions and by which they can reach surprisingly high concentrations in the Arctic environment where they can bioaccumulate in animals and humans. Less volatile compounds that sorb strongly to atmospheric particles or that dissolve easily in rain droplets tend to have a more limited potential for long-range atmospheric transport, whereas semi-volatile species can be transported over long distances in one or more steps towards the Arctic region.[37,38]

Both wet deposition and dry deposition remove PCBs from the atmosphere.[39]

Adsorption and Desorption on Organic Matter and Bioavailability

Once released into the environment, PCBs adsorb strongly to soil and sediment. As a result, these compounds tend to persist in the environment, with half-lives for most congeners ranging from months to years. Over time, contaminated sediments can be a source of hydrophobic organic contaminants (such as PCBs) and a significant health risk to aquatic food webs.[40] Leaching of PCBs from sediment and soil is slow, particularly for the more highly chlorinated congeners, and translocation to plants via soil is insignificant. Cycling of PCBs through the environment involves volatilization from land and water surfaces into the atmosphere, with subsequent removal from the atmosphere by wet or dry deposition, then revolatilization.[41]

Sorption properties of PCBs play a significant role in their mobility, ultimate fate in the sediments, and availability for degradation. The literature suggests that PCBs preferentially adsorb onto organic matter over adsorbing onto clay. Moreover, PCBs can sorb to dissolved organic matter (DOM) or particulate organic matter (POM). When associated with DOM, PCB contaminants are unavailable for uptake by organisms and, hence, become less bioavailable. In contrast, although PCBs sorbed to POM prevents or constrains direct uptake of PCBs, these contaminants are still available to the detrital food web, which is an important pathway in rivers. Planar PCBs bind strongly to POM and are less bioavailable.[42] Highly chlorinated homologs sorb strongly to POM and are not assimilated easily by detritus feeders.[43]

Bioavailability of sedimentary PCBs is traditionally assessed by measuring PCB uptake into benthic organisms over a standard exposure time. More recently, passive samplers have been used experimentally to estimate bioavailability.[44–46]

It is known that a combination of binding processes (sorption) and mass-transport processes (diffusion) is responsible for the partitioning of PCBs between aqueous and solid phases, and for their transport between these phases. These processes are also directly involved in and affect the environmental fate of PCBs. Precise quantitative predictions of phase speciation may allow an a priori estimate of the directly bioavailable, dissolved fractions of pollutants, as well as their tendency for long-term dispersion in the environment. Such predictions are critical in assessing the environmental risk from PCB contamination.[47]

The data generated during bench-scale adsorption studies and molecular-level study of the mechanism of adsorption of PCBs on substrates in the environment can help in effective desorption and the destruction of the persistent PCBs.[48]

Biodegradation and Transformation

The environmental persistence of PCBs results primarily from the inability of natural aquatic and soil biota to metabolize and/or degrade the compound at a significant rate. Studies on the biodegradation (degradation by bacteria or other microorganisms) of PCBs show that there are two biologically mediated processes for the degradation of PCBs: anaerobic and aerobic.[49] Microorganisms participate in the biodegradation by producing enzymes, which modify the organic pollutant into simpler compounds in such a way that the negative effects may be minimized. Biodegradation is of two forms:

- Mineralization—competent organisms use the organic pollutant as a source of carbon and energy resulting in the reduction of the pollutant to its constituent elements.
- Cometabolism—it requires a second substance as a source of carbon and energy for the microorganisms, but the target pollutant is transformed at the same time.[50,51]

If the products of cometabolism are amenable to further degradation, they can be mineralized; otherwise, incomplete degradation occurs. This may result in the formation and accumulation of metabolites that are more toxic than the parent molecule requiring a consortium of microorganisms, which can utilize the new substance as source of nutrients.[49]

The effectiveness of biodegradation depends on many factors, which are summarized in Table 3.[51–53]

Biodegradation is the only process known to degrade PCBs in soil or aquatic systems. Theoretically, the biological degradation of PCBs should result in CO_2, chlorine, and water. This process involves the removal of chlorine from the biphenyl ring (anaerobic reductive dechlorination) followed by cleavage and oxidation of the resulting compound (aerobic oxidative degradation).[54] The anaerobic process

Table 3 Environmental factors that affect the biodegradation of PCBs.

Factor	How it affects biodegradation
Structure of the compound, i.e., the presence of substituents and their position in the molecule	– A high degree of halogenation requires high energy from the microorganisms to break the stable carbon–halogen bonds – Chlorine as the substituent alters the resonant properties of the aromatic substance as well as the electron density of specific sites; it may result in the deactivation of the primary oxidation of the compound by microorganisms – The positions occupied by substituted chlorines have stereochemical effects on the affinity between enzymes and their substrate molecule
Solubility of the compound	– Microorganisms easily access compounds with high aqueous solubility – Highly chlorinated congeners that are very insoluble in water are also very resistant to biodegradation
Concentration of the pollutant	– At a low concentration range, degradation increases linearly with increase in concentration until such time that the rate essentially becomes constant regardless of further increase in pollutant concentration – In general, a low pollutant concentration may be insufficient for the induction of degradative enzymes or to sustain growth of competent organisms – A very high concentration may render the compound toxic to the organisms
Temperature	– The conditions should be optimal for the microorganism
pH	
Presence of toxic or inhibitory substance and competing substances	
Availability of suitable electron acceptors	
Interactions among microorganisms	

removes chlorine atoms of highly chlorinated PCBs, those with five or more chlorine atoms, which are then mineralized under aerobic condition.[49]

Under anaerobic condition, reductive dechlorination of PCBs occurs in soils and sediments. Different microorganisms with distinct dehalogenating enzymes, each exhibiting a unique pattern of congener selectivity resulting in various patterns of PCB dechlorination, exist in PCB-contaminated sites,[55] including the following isolated bacteria: *Desulfomonile tiedjei*,[56] *Desulfitobacterium*, *Dehalobacter restrictus*, *Dehalospirillum multivorans*, *Desulforomonas chloroethenica*, *Dehalococcoides ethenogenes*, and the facultative anaerobes *Enterobacter* strain MS-1 and *Enterobacter agglomerans*.[57] The rate, extent, and route of dechlorination is dependent on the composition of the active microbial community, which in turn are influenced by environmental factors such as availability of carbon sources, hydrogen or other electron donors, the presence or absence of electron acceptors other than PCBs, temperature, and pH.[58] However, a similarity between degradation patterns exists. The position of chlorine atoms on the rings affects the rate of biodegradation. Not only are PCBs with *para*- and *meta*-substituted rings more easily degraded than the *ortho*-substituted compounds, as shown in Fig. 3,[59] but PCBs containing all chlorines on one ring are biodegraded faster than those that contain chlorines throughout both rings.

Persistence of PCBs in the environment increases with the degree of chlorination of the congener, i.e., compounds with a high degree of chlorination are resistant to biodegradation and degrade slowly in the environment. Anaerobic PCB dechlorination reduces the potential risk and potential exposure to PCBs because it significantly reduces the bioconcentration potential of the PCB mixture through conversion to congeners that do not significantly bioaccumulate in the trophic chain.[60] Moreover, lightly chlorinated congeners produced by dechlorination can be readily degraded by indigenous bacteria.[61,62]

Aerobic biodegradation involving biphenyl ring cleavage is restricted to the lightly chlorinated PCB congeners,

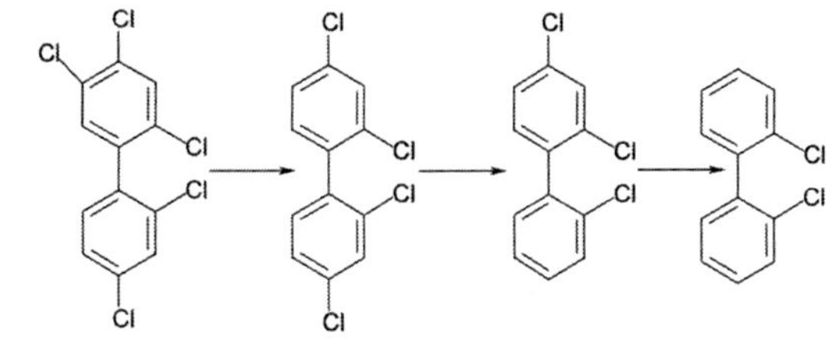

Fig. 3 A potential pathway for anaerobic degradation of highly chlorinated PCB congeners to less chlorinated ones.

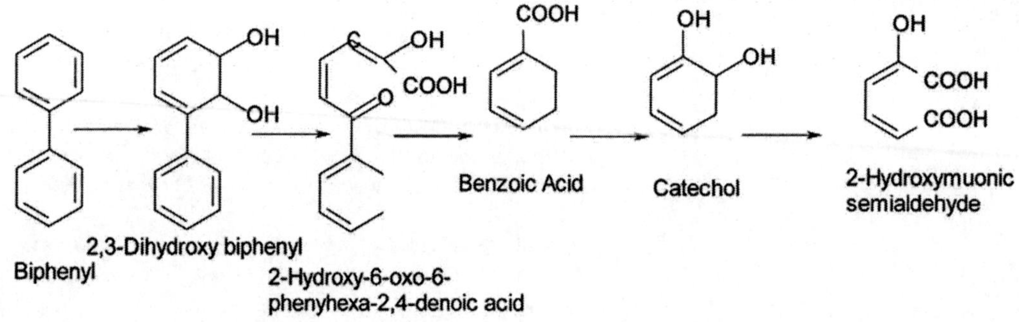

Fig. 4 A possible pathway for the aerobic oxidative dehalogenation of PCBs.

those with four or less chlorine atoms, resulting from the dechlorination of highly chlorinated congeners.[63,64] Aerobic oxidative destruction involves two clusters of genes. The first one is responsible for the transformation of PCB congeners to chlorobenzoic acid, and the second cluster is responsible for the degradation of the chlorobenzoic acid. A common growth substrate for PCB-degrading bacteria is biphenyl or monochlorobiphenyl. When biphenyl is utilized by bacteria, *meta*-ring cleavage product is produced. This has been observed in most bacteria studied especially in *Pseudomonas* sp.,[54] as well as in *Micrococcus* sp.[65] The metabolic pathway used by this family of bacteria is illustrated in Fig. 4.[65]

Both anaerobic and aerobic metabolism modes transform PCBs. Different microorganisms show preferential attack on the PCB molecule, resulting in different patterns of biodegradation. The degree of chlorination of the congener and environmental factors influence the degradation potential of the compound.[49] Higher-chlorinated biphenyls therefore are potentially fully biodegradable in a sequence of anaerobic reductive dechlorination followed by aerobic mineralization of the lower-chlorinated products.[66]

Bioaccumulation, Bioconcentration, and Biomagnification

Organisms can accumulate high concentrations of PCBs relative to concentrations of these substances in non-biotic portions of the environment. This phenomenon is variously referred to as bioconcentration, bioaccumulation, and biomagnification.[67] Because some confusion exists in the literature about these definitions, we try to explain these more precisely following the terms set out by Gobas and Morrison.[68]

Bioaccumulation is a selective process that causes an increased chemical concentration in an organism compared to that in the surrounding medium and results from uptake by all exposure routes including transport across respiratory surfaces, dermal absorption (bioconcentration), and dietary absorption (biomagnification). Bioaccumulation can thus be viewed as a combination of bioconcentration and biomagnification.

The bioaccumulation factor (BAF) in fish is the ratio of the concentration of the chemical in the organism C_B to that in water, similar to that of the bioconcentration factor (BCF).

$$BAF = C_B/C_{WT} \text{ or } C_B/C_{WD},$$

where BAF is the bioaccumulation factor and $C_B/C_{WT(WD)}$ is the concentration of the chemical in the organism/in water.

The most common approach for evaluating levels of bioaccumulation is to compare the levels retained by the organism with levels in the contaminated medium in which they live.[69]

Bioconcentration results from uptake of chemicals from water (usually under laboratory conditions). Uptake occurs via the respiratory surface and/or skin and results in the chemical concentration in an organism being greater than that in the surrounding medium.

BCF is defined as the ratio of the chemical concentration in an organism, C_B, to the total chemical concentration in the water, C_{WT}, or to the freely dissolved chemical concentration in water, C_{WD} (it only takes into account the fraction of the chemical in the water that is biologically available for uptake). The BCF is expressed as follows:

$$BCF = C_B/C_{WT} \text{ or } C_B/C_{WD},$$

where BCF is the bioconcentration factor and $C_B/C_{WT(WD)}$ is the concentration of the chemical in the organism/in water.

Although sometimes applied to other aquatic species, the principal target organism for BCF assessment tends to be fish, primarily because of their importance as food for many species, including humans.

Biomagnification, on the other hand, is the bioaccumulation of a substance up the trophic chain when residues are transferred from consumption of smaller organisms by larger ones in the chain. It generally refers to the sequence of processes that produces higher concentrations in organisms at higher levels in the food chain (at higher trophic levels). These processes always results in an organism having higher concentrations of a substance than is present in the organism's food. Biomagnification also results in higher concentrations of the substance than would be expected if water were the only exposure mechanism.[70] A biomagnification factor (BMF) can be defined as the ratio of the concentration of chemical in the organism (C_B) to that in the organism's diet (C_A), and can be expressed as

$$BMF = C_B/C_A,$$

where BMF is the biomagnification factor, C_B is the concentration of chemical in the organism, and C_A is the concentration of chemical in the organism's diet.

This is the simplest definition of a BMF. It can also be described as the ratio of the observed lipid-normalized BCF to K_{ow}, which is the theoretical lipid-normalized BCF. This is equivalent to the multiplication factor above the equilibrium concentration. If this ratio is equal to or less than one, then the compound has not been biomagnified. If the ratio exceeds one, then the chemical is biomagnified by that factor.

The mechanism of biomagnification is not completely understood. Achieving a concentration of a chemical greater than its equilibrium value indicates that the elimination rate is slower than for chemicals that reach equilibrium. Transfer efficiencies of the chemical would affect the relative ratio of uptake and elimination. There are many factors that control the uptake and elimination of a chemical after contaminated food is consumed; these include factors specific to the chemical (solubility, K_{ow}, molecular weight and volume, and diffusion rates between organism gut, blood, and lipid pools), as well as factors specific to the organism (the feeding rate, diet preferences, assimilation rate into the gut, rate of chemical's metabolism, rate of egestion, and rate of organism growth). Because humans occupy a very high trophic level, we are particularly vulnerable to adverse health effects from exposure to chemicals that biomagnify.[71]

Chemicals that bioaccumulate do not necessarily biomagnify, although many papers report that PCB congeners do in fact biomagnify.[72,73] Some early bioaccumulation models used the concept of a food-chain multiplier, which is now considered excessively simplistic.[74] Exposure of PCBs solely from one source only occurs in laboratory experiments.[75] In nature, organisms are always exposed to different sources of contaminants, and therefore, what happens in the field is more complex than reflected in laboratory studies and cannot easily be emulated by laboratory studies. Mass balance models are simple tools that allow evaluation of various uptake and loss processes.[67] A variety of mass balance models have been developed to address water quality issues in lakes, estuaries, and slow-flowing water bodies.[76] The simpler models only consider advection and an overall loss due to the combined processes of volatilization, net transfer to sediment, and degradation. The rate constant for the overall loss is derived from fugacity calculations for a single segment system. The more rigorous models perform fugacity calculations for each segment and explicitly include the processes of advection, evaporation, water–sediment exchange, and degradation in both water and sediment. In this way, chemical exposure in all compartments (including equilibrium concentrations in biota) can be estimated.[77,78] In general, these models consider the organism to be a single "box."[67,79] These models require information about the chemicals, the organism, and associated environmental parameters.[79]

PCB congeners with less *ortho*-substitution are accumulated up the trophic chain at a greater rate than other congeners in their homolog group.[74] Non-*ortho*-substituted congeners, especially those that lack adjacent unsubstituted *meta* and *para* sites and unsubstituted *ortho* and *meta* sites, are undoubtedly metabolically recalcitrant in invertebrate and vertebrate tissues.[80] Changes in distributions of congeners are mainly caused by transfers among biotic compartments. There is no enrichment in higher trophic levels of mono- and non-*ortho*-substituted congeners. However, many coplanar congeners, especially very toxic PCB 77, are depleted with increasing trophic levels; PCB 77 is therefore almost certainly metabolized.[74]

Exposure of PCBs solely from one source only occurs in laboratory experiments. In nature, there are always multiple sources of contaminants, and therefore, field results must be studied carefully. Moreover, the properties of individual PCB congeners substantially affect accumulation or degradation pathways. Empirical models only reflect one of several possible mechanisms.[81]

HEALTH EFFECTS

As PCBs persist in the environment, the general population is potentially exposed to a variety of PCBs via food (especially fish caught in contaminated lakes or rivers, meat, and dairy products), air, surface soils, drinking water, and groundwater. In the workplace, people might be exposed to PCBs during repair and maintenance of PCB transformers and other old electrical devices, and disposal of PCB materials. Although mixtures used in industry are not identical to the combinations of PCBs present in the environment (or in breast milk), these mixtures have been found to have similar harmful effects.

The health effects of PCBs have been very widely studied in people (studies of industrial workers exposed to PCB-containing mixtures in the course of their work, as well as studies of adults and children exposed to PCBs as a result of consuming contaminated fish), laboratory animals, and wildlife in contaminated areas. These studies indicate that people who are regularly exposed to PCBs are at greater risk for a variety of health problems.[82] Moreover, evidence on the health effects of exposure to PCBs has been obtained from two episodes of mass poisoning that occurred in Japan (the 1968 Yusho incident) and Taiwan (the 1979 Yu-Cheng incident). Some of the most important findings are summarized below.

Table 4 Destruction processes for PCB wastes.

Process		Waste types accepted	Advantages	Disadvantages
Incineration		– Liquids and dilute slurries – PCB-containing waste equipment (may require preprocessing)	– High destruction efficiencies (99.9999% or more) – Meeting legal requirements – Facilities can treat a range of wastes, both chlorinated and non-chlorinated	– Risk of emission of harmful substances if inadequately controlled – Careful process control is required to maintain important parameters (residence time, temperature, turbulence, and oxygen concentration) at the desired level and to ensure the effectiveness of the gas cleaning system – Costly, especially if wastes have to be shipped off-site – Some equipment may require preprocessing by mechanical alteration—e.g., shredding to expose contents of capacitors, draining and disassembly of transformers, cutting large transformers to size, or packing solids and sludges in drums and feeding via a chute
Dechlorination processes	Gas phase chemical reduction (GPCR)	– PCB-contaminated liquids, soils, sediments, equipment, and material	– High destruction efficiencies (99.9999%) – Modular, transportable, or fixed configurations – The expected throughput of the main reactor system is 1000 to 3000 tons/mo	– Need to establish treatment conditions for individual components
	Base catalyzed decomposition (BCD)	– Liquids – Soil and building rubble contaminated by POPs	– High destruction efficiencies (99.9999%) – Modular, transportable, or fixed plants – The process can tolerate inorganic and organic debris provided this material is smaller than 50 mm or can be shredded down to this size	– The first step of BC, designed to treat solid matrices, requires mechanical pretreatment – Need to establish treatment conditions for individual components
	Sodium reduction	– Oils with a PCB content of up to 10,000 µg/dm^3	– Transportable and fixed plants – Widely used for in situ removal of PCBs from active transformers	– Need to establish treatment conditions for individual components
	Supercritical water oxidation (SCWO)	– Liquid wastes or solids less than 200 microns in diameter, and an organic content of less than 20%	– A compact, totally enclosed system – All emissions and residues may be captured for assay and reprocessing if needed	– Need to establish treatment conditions for individual components
	Plasma arc	– Liquid waste streams of any concentration (the most cost-effective method is to treat concentrated wastes) – Solids in the form of a pumpable fine slurry	– Various plasma reactors developed for the thermal destruction of hazardous waste – Transportable and fixed units – The system can treat its own fly ash plus filtration media, minimizing secondary wastes	– Contaminated soil, very viscous liquids or sludges, other equipment (capacitors and transformers) can be treated after pretreatment

(continued)

Table 4 Destruction processes for PCB wastes. (*continued*)

Process		Waste types accepted	Advantages	Disadvantages
	Pyrolysis	Solid, liquid and gaseous wastes	– Transportable and fixed configurations – Off gases can be reused as synthesis gas	– Need to establish treatment conditions for individual components
	Molten salt oxidation	– Liquids – Solids—only if reduced to small particle sizes for pneumatic conveying	– The reaction takes place within the salt bath, virtually eliminating the fugitive inventories found in incineration	– Need to establish treatment conditions for individual components
	Solvated electron technology	– Liquids – Solid materials (up to 45 cm diameter)	– POP wastes are reduced to metal salts and simple hydrocarbon compounds; PCBs are reduced to petroleum hydrocarbons, sodium chloride, and sodium amide	– Material with a high water content (>40% w/w) must be dewatered prior to treatment

Research shows that PCBs cause a variety of adverse health effects depending on the route of exposure, age, sex, and area of the body where PCBs are concentrated. Studies on animals show conclusive evidence that PCBs are carcinogenic. Animals that ate food containing large amount of PCBs for short periods of time had mild liver damage and some died. PCBs have also been implicated as a cause of mass mortalities in seabirds.[49]

Moreover, a number of epidemiological studies of workers exposed to PCBs have been performed. The Department of Health and Human Services (DHHS) has concluded that PCBs may reasonably be anticipated to be carcinogens. Also, the Environmental Protection Agency (EPA) and the International Agency for Research on Cancer (IARC) have determined that PCBs are probably carcinogenic to humans.[83,84] Research also shows that exposure to PCBs in high concentration can have various acute effects including a skin disease known as chloracne (skin lesions), liver damage, other non-cancer short-term effects like body weight loss, impaired immune function, and clinically diagnosable damage to the central nervous system, causing headaches, dizziness, depression, nervousness, and fatigue. Other adverse health effects of PCBs are liver, stomach, and thyroid gland injuries; behavioral alterations; and impaired reproduction.[82]

The EPA has set a limit of 0.0005 milligrams of PCBs per liter of drinking water (0.0005 mg/L). Moreover, the Food and Drug Administration (FDA) requires that infant foods; eggs; milk, and other dairy products; fish and shellfish; poultry; and red meat contain no more than 0.2–3 parts of a PCB per million (0.2–3 ppm).[84]

REGULATIONS

After the impact of PCBs on the environment was recognized, in 1976, the U.S. Congress charged the EPA with regulating the issue of PCBs. The ban on the manufacturing, processing, distribution in commerce, and use of PCBs, as well as the PCB disposal and marking regulations, was enclosed in the Toxic Substances Control Act (TSCA) of 1976.[85] In 1979, after subsequent amendments, the regulations stipulate that the production of PCBs in the United States is generally banned, the use of PCB-containing materials still in service is restricted, the discharge of PCB-containing effluents is prohibited, the disposal of materials contaminated by PCBs is regulated, and the import or export of PCBs is only permitted through an exemption granted from EPA.

In the European community, the use of PCBs in open applications such as printing inks and adhesives was banned in 1976 (Directive 76/403/EEC[86]). Use of PCBs as a raw material or chemical intermediate has been banned in the European Union (EU) since 1985.[87] In 1996, the 1976 directive was replaced by Directive 96/59/EC,[88] which set a deadline of 2010 for complete phase out or decontamination of equipment containing PCBs. However, the United Nations Environment Programme (UNEP) global treaty adopted at the Stockholm Convention on Persistent Organic Pollutants (May 2001) stipulates that the use in equipment shall be eliminated by 2025.[89] This date is a minimum requirement and does not prevent individual governments, or groups of governments, from maintaining earlier phase-out dates. However, the most important regulations—Council Directive 96/59/EC[88] and HELCOM Recommendation 6/1[90]—concerning total banning of PCBs have been fully implemented only by EU countries. Furthermore, the Commission has adopted community strategy for dioxins, furans, and PCBs aimed at reducing as far as possible the release of these substances in the environment and their introduction in the trophic chains.[91]

Despite the existing regulations, there is still a substantial amount of PCBs in use, because exemption has been given in many countries for contained use in existing equipment with long lifetimes, at least for an initial period after a production ban was decided. There are also quantities in storage awaiting disposal. The chemical industry is currently making a proposal to solve the disposal problem.[92]

DISPOSAL OF PCBS FROM THE ENVIRONMENT

PCBs and PCB-contaminated equipment and oil are required to be properly disposed of in a manner similar to that of hazardous waste.

Much effort has been directed towards the selection of technology options for the disposal of PCBs from the environment. Although the baseline remediation technology for PCBs is incineration, other options do exist. Destruction of PCBs requires the breaking of molecular bonds by an input of thermal or chemical energy. The main features of combustion and non-combustion processes are summarized in Table 4.[93]

While destruction is to be preferred, some PCBs from a range of consumer goods are likely to enter landfills accepting municipal waste. PCBs deposited in landfills may therefore contaminate groundwater and surface water following migration into leachate. The behavior of PCBs in landfills is far from fully understood and a precautionary approach is recommended.

CONCLUSION

For decades, PCBs have been recognized as important and potentially harmful environmental contaminants. The intrinsic properties of PCBs, such as high environmental persistence, resistance to metabolism in organisms, and tendency to accumulate in lipids have contributed to their ubiquity in environmental media and have induced concern for their toxic effects after prolonged exposure.

PCBs are bioaccumulated mainly by aquatic and terrestrial organisms and thus enter the food web. Humans and wildlife that consume contaminated organisms can also accumulate PCBs in their tissues. Such accumulation is of concern, because it may lead to body burdens of PCBs that could have adverse health effects in humans and wildlife.

Moreover, PCBs are slower to biodegrade in the environment than are many other organic chemicals. The low water solubility and the low vapor pressure of PCBs, coupled with air, water, and sediment transport processes, mean that they are readily transported from local or regional sites of contamination to remote areas.

PCBs are transformed mainly through microbial degradation, and particularly reductive dechlorination via organisms that take them up. Metabolism by microorganisms and other animals can cause relative proportions of some congeners to increase while others decrease. Because the susceptibility of PCBs to degradation and bioaccumulation is congener specific, the composition of PCB congener mixtures that occur in the environment differs substantially from that of the original industrial mixtures released into the environment. Generally, the less-chlorinated congeners are more water soluble, more volatile, and more likely to biodegrade. On the other hand, higher-chlorinated PCBs are often more resistant to degradation and volatilization and sorb more strongly to particulate matter. Some higher-chlorinated PCBs tend to bioaccumulate to greater concentrations in tissues of animals than do lower-molecular-weight ones. The higher-chlorinated PCBs can also biomagnify in food webs.

There is still much to be learned about the chemistry of PCBs. Current research focuses on finding ways to break the molecules down into harmless compounds. The biodegradation of PCBs utilizing microorganisms presently appears to be our best hope for removing PCBs from the environment.

GLOSSARY

BAF—Bioaccumulation factor
BCF—Bioconcentration factor
BMF—Biomagnification factor
BZ number—A system of sequential numbers for the 209 PCB congeners introduced in 1980 by Ballschmiter and Zell that identifies a given congener simply and precisely. Also referred to as congener, IUPAC, or PCB number.
DHHS—Department of Health and Human Services
DOM—Dissolved organic matter
EPA—Environmental Protection Agency
FDA—Food and Drug Administration
IARC—International Agency for Research on Cancer
PCBs—Polychlorinated biphenyls
POM—Particulate organic matter
POPs—Persistent organic pollutants
TSCA—Toxic Substances Control Act
UNEP—United Nations Environment Programme

ACKNOWLEDGMENTS

This research was financially supported by the Polish Ministry of Science and Higher Education (grant no. N N312 300535). A. Wilkowska is grateful for financial support from the Human Capital Programme (POKL.04.01.01-00-368/09).

REFERENCES

1. US Department of Health and Human Services. Toxicological profile for polychlorinated biphenyls (update). Agency for Toxic Substances and Disease Registry: Atlanta, 2000, available at http://www.atsdr.cdc.gov/toxprofiles/tp17.pdf (accessed January 2010).
2. De Voogt, P.; Brinkman, U.A. Production, properties and usage of polychlorinated biphenyls. In *Halogenated Biphenyls,*

Terphenyls, Naphthalenes, Dibenzodioxins and Related Products, 2nd Ed.; Kimbrough, R.D., Jensen, A.A., Eds.; Elsevier Science Publishers: Amsterdam, 1989; 3–45.
3. Ballschmiter, K.; Zell, M. Analysis of polychlorinated biphenyls (PCB) by glass capillary gas chromatography: Composition of technical Aroclor and Clophen–PCB mixtures. Fresenius Z. Anal. Chem. **1980**, *302*, 20–31.
4. Dunnivant, F.M.; Elzerman, A.W. Aqueous solubility and Henry's law constant data for PCB congeners for evaluation of quantitative structure–property relationships (QSPRs). Chemosphere **1988**, *17*, 525–541.
5. Hutzinger, O.; Safe, S.; Zitko, V. Photochemical degradation of chlorobiphenyls (PCBs). Environ. Health Perspect. **1972**, *1*, 15–20.
6. Beyer, A.; Biziuk, M. Environmental fate and global distribution of polychlorinated biphenyls. Rev. Environ. Contam. Toxicol. **2009**, *201*, 137–158.
7. Hazard waste generation and commercial hazardous waste management capacity: An assessment, SW-894. Prepared by Booz-Allen and Hamilton, Inc. and Putnam, Hayes and Barlett, Inc. for the Office of Planning and Evaluation and the Office of Solid Waste. U.S. Environmental Protection Agency: Washington, DC, 1980; D-4.
8. Barbalace, R.C. The Chemistry of Polychlorinated Biphenyls. EnvironmentalChemistry.com. 2003, available at http://EnvironmentalChemistry.com/yogi/chemistry/pcb.html (accessed January 2010).
9. Guidelines for the Identification of PCBs and Materials Containing PCBs. 1999, available at http://www.chem.unep.ch/pops/pdf/PCBident/pcbid1.pdf (accessed January 2010).
10. Durfee, R.L. Production and usage of PCB's in the United States. In Proceedings of the National Conference on Polychlorinated Biphenyls, EPA-560/6-75-004.U.S. Environmental Protection Agency: Washington, DC, 1976; 103–107.
11. Orris, P.; Kominsky, J.R.; Hryhorczyk, D.; Melius, J. Exposure to polychlorinated biphenyls from an overheated transformer. Chemosphere **1986**, *15*, 1305–1311.
12. Welsh, M.S. Extraction and gas chromatography/electron capture analysis of polychlorinated biphenyls in railcar paint scrapings. Appl. Occup. Environ. Hyg. **1995**, *10* (3), 175–181.
13. Boate, A.; Deleersnyder, G.; Howarth, J.; Mirabelli, A.; Peck L. Chemistry of PCBs, available at http://wvlc.uwaterloo.ca/biology447/modules/intro/assignments/Introduction2a.htm (accessed January 2010).
14. Eisenreich, S.J.; Baker, J.E.; Franz, T.; Swanson, M.; Rapaport, R.A.; Strachan, W.M.J.; Hites, R.A. Atmospheric deposition of hydrophobic organic contaminants to the Laurentian Great Lakes. In *Fate of Pesticides and Chemicals in the Environment*; Schnoor, J.L., Ed.; John Wiley and Sons, Inc.: New York, 1992; 51–78.
15. Pham, T.T.; Proulx, S. PCBs and PAHs in the Montreal urban community (Quebec, Canada) wastewater treatment plant and in the effluent plume in the St. Lawrence River. Water Res. **1997**, *31*(8), 1887–1896.
16. Hansen, L.G.; O'Keefe, P.W. Polychlorinated dibenzofurans and dibenzo-*p*-dioxins in subsurface soil, superficial dust, and air extracts from a contaminated landfill. Arch. Environ. Contam. Toxicol. **1996**, *31*(2), 271–276.
17. Hansen, L.G.; Green, D.; Cochran, J.; Vermette, S.; Bush, B. Chlorobiphenyl (PCB) composition of extracts of subsurface soil, superficial dust and air from a contaminated landfill. Fresenius J. Anal. Chem. **1997**, *357*, 442–448.
18. Swackhamer, D.L.; Armstrong, D.E. Estimation of the atmospheric and nonatmospheric contributions and losses of polychlorinated biphenyls for Lake Michigan on the basis of sediment records of remote lakes. Environ. Sci. Technol. **1986**, *20*, 879–883.
19. Wallace, J.C.; Basu, I.; Hites, R.A. Sampling and analysis artifacts caused by elevated indoor air polychlorinated biphenyl concentrations. Environ. Sci. Technol. **1996**, *30*(9), 2730–2734.
20. Ohsaki, Y.; Matsueda, T. Levels, features and a source of non-ortho coplanar polychlorinated biphenyl in soil. Chemosphere **1994**, *28*(1), 47–56.
21. Blumbach, J.; Nethe, L.P. Organic components reduction (PCDD/PCDF/PCB) in flue-gases and residual materials from waste incinerators by use of carbonaceous adsorbents. Chemosphere **1996**, *32*(1), 119–131.
22. Gunkel, G.; Mast, P.G.; Nolte, C. Pollution of aquatic ecosystems by polychlorinated biphenyls (PCB). Limnologica **1995**, *25*(3/4), 321–331.
23. Alcock, R.E.; Bacon, J.; Bardget, R.D.; Beck, A.J.; Haygarth, P.M.; Lee, R.G.M.; Parker, C.A.; Jones, K.C. Persistence and fate of polychlorinated biphenyls (PCBs) in sewage sludge-amended agricultural soils. Environ. Pollut. **1996**, *93* (1), 83–92.
24. Wania, F.; Mackay, D. Tracking the distribution of persistent organic pollutants. Environ. Sci. Technol. **1996**, *30*, 390A–396A.
25. Hansen, K.M.; Prevedouros, K.; Sweetman, A.J.; Jones, K.C.; Christensen, J.H. A process-oriented intercomparison of a box model and an atmospheric chemistry transport model: Insights into model structure using alpha-HCH as the modelled substance. Atmos. Environ. **2006**, *40*, 2089–2104.
26. Mackay, D. *Multimedia Environmental Models. The Fugacity Approach*, 2nd Ed.; Lewis Publishers: Boca Raton, FL, 2001.
27. Gong, S.L.; Huang, P.; Zhao, T.L.; Sahsuvar, L.; Barrie, L.A.; Kaminski, J.W.; Li, Y.F.; Niu, T. GEM/POPs: A global 3-D dynamic model for semi-volatile persistent organic pollutants—Part 1: Model description and evaluations of air concentrations. Atmos. Chem. Phys. **2007**, *7*, 4001–4013.
28. Huang, P.; Gong, S.L.; Zhao, T.L.; Neary, L.; Barrie, L.A. GEM/POPs: A global 3-D dynamic model for semi-volatile persistent organic pollutants—Part 2: Global transports and budgets of PCBs. Atmos. Chem. Phys. **2007**, *7*, 4015–4025.
29. Wania, F.; Daly, G.L. Estimating the contribution of degradation in air and deposition to the deep sea to the global loss of PCBs. Atmos. Environ. **2002**, *36*, 5581–5593.
30. Wania, F.; Su, Y.S. Quantifying the global fractionation of polychlorinated biphenyls. Ambio **2004**, *33*, 161–168.
31. Eckhardt, S.; Breivik, K.; Li, Y.F.; Mano, S.; Stohl, A. Source regions of some persistent organic pollutants measured in the atmosphere at Birkenes, Norway. Atmos. Chem. Phys. **2009**, *9*, 6597–6610.
32. Li, N.Q.; Wania, F.; Lei, Y.D.; Daly, G.L. A comprehensive and critical compilation, evaluation, and selection of physical chemical property data for selected polychlorinated biphenyls. J. Phys. Chem. Ref. Data **2003**, *32*, 1545–1590.

33. Erickson, M.D. *Analytical Chemistry of PCBs*; Butterworth Publishers: Stoneham, Massachusettes, 1986.
34. Hansen, L.G. *The Ortho Side of PCBs: Occurrence and Disposition*; Kluwer Academic Publishers: Boston, 1999.
35. Haque, R.; Schmedding, D.; Freed, V. Aqueous solubility, adsorption and vapor behavior of polychlorinated biphenyl Aroclor 1254. Environ. Sci. Technol. **1974**, *8*, 139–142.
36. Chiarenzelli, J.; Scrudato, R.; Arnold, G.; Wunderlich, M.; Rafferty, D. Volatilization of polychlorinated biphenyls during drying at ambient conditions. Chemosphere **1996**, *33*, 899–911.
37. Wania, F.; Mackay, D. Global fractionation and cold condensation of low volatility organochlorine compounds in polar regions. Ambio **1993**, *22*, 10–18.
38. Wania, F. Potential of degradable organic chemicals for absolute and relative enrichment in the Arctic. Environ. Sci. Technol. **2006**, *40*, 569–577.
39. Nelson, E.D.; McConnell, L.L.; Baker, J.E. Diffusive exchange of gaseous polycyclic aromatic hydrocarbons and polychlorinated biphenyls across the air–water interface of the Chesapeake Bay. Environ. Sci. Technol. **1998**, *32* (7), 912–919.
40. Luthy, R.G.; Aiken, G.R.; Brusseau, M.L.; Cunningham, S.D.; Gschwend, P.M.; Pignatello, J.J.; Reinhard, M.; Traina, S.J.; Weber, W.J.; Westall, J.C. Sequestration of hydrophobic organic contaminants by geosorbents. Environ. Sci. Technol. **1997**, *31* (12), 3341–3347.
41. ATSDR—Agency for Toxic Substances and Disease Registry. Toxicological profile for polychlorinated biphenyls (update). Atlanta: US Department of Health and Human Services, 2000.
42. Van Bavel, B.; Andersson, P.; Wingfors, H.; Ahgren, J.; Bergqvist, P.A.; Norrgren, L.; Rappe, C.; Tysklind, M. Multivariate modeling of PCB bioaccumulation in three-spined stickleback (*Gasterosteus aculeatus*). Environ. Toxicol. Chem. **1996**, *15*, 947–954.
43. Boese, B.L.; Winsor, M.; Lee, II, H.; Echols, S.; Pelletier, J.; Randall, R. PCB Congeners and hexachlorobenzene biota sediment accumulation factor for *Macoma nasuta* exposed to sediments with different total organic carbon contents. Environ. Toxicol. Chem. **1995**, *14*, 303–310.
44. Friedman, C.L.; Burgess, R.M.; Perron, M.M.; Cantwell, M.G.; Ho, K.T.; Lohmann R. Comparing polychaete and polyethylene uptake to assess sediment resuspension effects on PCB bioavailability. Environ. Sci. Technol. **2009**, *43*(8), 2865–2870.
45. Trimble, T.A.; You, J.; Lydy, M.J. Bioavailability of PCBs from field-collected sediments: Application of Tenax extraction and matrix-SPME techniques. Chemosphere **2008**, *71*, 337–344.
46. Verweij, F.; Booij, K.; Satumalay, K.; van der Molen, N.; van der Oost, R. Assessment of bioavailable PAH, PCB and OCP concentrations in water, using semipermeable membrane devices (SPMDs), sediments and caged carp. Chemosphere **2004**, *54* (11), 1675–1689.
47. Gdaniec-Pietryka, M.; Wolska, L.; Namiesnik, J. Physical speciation of polychlorinated biphenyls in the aquatic environment. Trends Anal. Chem. **2007**, *26*, 1005–1012.
48. Adsorption and Desorption of Organic Contaminants to Predict Fate, Transport, and Propensity to Electrochemically Degrade, available at http://www.epa.gov/nrmrl/lrpcd/wm/projects/135925.htm (accessed February 2010).
49. Borja, J.; Taleon, D.M.; Auresenia, J.; Gallardo, S. Polychlorinated biphenyls and their biodegradation. Process Biochem. **2005**, *40*, 1999–2013.
50. Dobbins, D.C. Biodegradation of pollutants. In *Encyclopedia of Environmental Biology*; Academic Press Inc.: New York, 1995.
51. McEldowney, S.; Hardman, D.J.; Wait, S. *Pollution: Ecology and Biotreatment*; Longman Scientific and Technical: New York, 1993.
52. Furukawa, K. Modification of PCBs by bacteria and other microorganisms. In *PCBs and the Environment*; Waid, J.S., Ed.; CRC Press: Florida, 1986; 89–100.
53. Sylvestre, M.; Sandossi, M. Selection of enhanced PCB-degrading bacterial strains for bioremediation: Consideration of branching pathways. In *Biological Degradation and Remediation of Toxic Chemicals*; Chaudhry, G.R., Ed.; Chapman and Hall: New York, 1994.
54. Boyle, A.W.; Silvin, C.J.; Hassett, J.P.; Nakas, J.P.; Tanenbaum, S.W. Bacterial PCB biodegradation. Biodegradation **1992**, *3*, 285–298.
55. Alder, A.C.; Haggblom, M.M.; Oppenheimer, S.R.; Young, L.Y. Reductive dechlorination of polychlorinated biphenyls in anaerobic sediments. Environ. Sci. Technol. **1993**, *27*, 530–538.
56. Mohn, W.W.; Tiedje, J.M. Microbial reductive dechlorination. Microbiol. Rev. **1992**, *56*, 482–507.
57. Holliger, C.; Wohlfarth, G.; Diekert, G. Reductive dechlorination in the energy metabolism of anaerobic bacteria. FEMS Microbiol. Rev. **1998**, *22*, 383–398.
58. Wiegel, J.; Wu, Q. Microbial reductive dehalogenation of polychlorinated biphenyls. FEMS Microbiol. Ecol. **2000**, *32*, 1–15.
59. Fish, K.M.; Principe, J.M. Biotransformations of Arochlor 1242 in Hudson River test tube microcosms. Appl. Environ. Microbiol. **1994**, *60*, 4289–4296.
60. Ye, D.; Quensen, III, J.F.; Tiedje, J.M.; Boyd, S.A. Anaerobic dechlorination of polychlorinated biphenyls (Arochlor 1242) by pasteurized and ethanol-treated microorganisms from sediments. Appl. Environ. Microbiol. **1992**, *58* (4), 1110–1114.
61. Moore, J.A. *Reassessment of Liver Findings in PCB Studies for Rats*; Institute of Evaluating Health Risks: Washington, 1991.
62. Safe, S. Toxicology, structure–function relationship, and human and environmental health impacts of polychlorinated biphenyls: progress and problems. Environ. Health. Perspect. **1992**, *100*, 259–268.
63. Cookson, Jr., J.T. *Bioremediation Engineering: Design and Application*; McGraw Hill: New York, 1995.
64. Kuipers, B.; Cullen, W.R.; Mohn, W.W. Reductive dechlorination of nonachloro biphenyls and selected octachloro biphenyls by microbial enrichment cultures. Environ. Sci. Technol. **1999**, *33*, 3579–3585.
65. Benvinakatti, B.G.; Ninnekar, H.Z. Degradation of biphenyl by a *Micrococcus* species. Appl. Microbiol. Biotechnol. **1992**, *38*, 273–275.
66. Field, J.A.; Sierra-Alvarez, R. Microbial transformation and degradation of polychlorinated biphenyls. Environ. Pollut. **2008**, *155*, 1–12.

67. Mackay, D.; Fraser, A. Bioaccumulation of persistent organic chemicals: Mechanism and models. Environ. Pollut. **2000**, *110*, 375–391.
68. Gobas, F.A.P.C.; Morrison, H.A. Bioconcentration and biomagnification in the aquatic environment. In *Handbook of Property Estimation Methods for Chemicals*; Boethling, R.S., Mackay, D., Eds.; CRC Press: Boca Raton, FL, 2000; 189–231.
69. Kucklick, J.; Harvey, H.R.; Ostrom, P.; Ostrom, N.; Baker, J. Organochlorine dynamics in the pelagic food web of Lake Baikal. Environ. Toxicol. Chem. **1996**, *15*, 1388–1400.
70. Glossary for Chemists of Terms Used in Toxicology: Pure and Applied Chemistry, V. 65, no. 9; International Union of Pure And Applied Chemistry (IUPAC): Bethesda, Maryland, 1993; 2003–2122, available at (http://sis.nlm.nih.gov/enviro/glossarymain.html—online version posted by the U.S. National Library of Medicine).
71. Bierman, Jr., V.J. Equilibrium partitioning and biomagnification of organic chemicals in benthic animals. Environ. Sci. Technol. **1990**, *24*, 1407–1412.
72. Burreau, S.; Zebuhr, Y.; Broman, D.; Ishaq, R. Biomagnification of PBDEs and PCBs in food webs from the Baltic Sea and the northern Atlantic Ocean. Sci. Total Environ. **2006**, *366*, 659–672.
73. Nfon, E.; Cousins, I.T.; Broman, D. Biomagnification of organic pollutants in benthic and pelagic marine food chains from the Baltic Sea. Sci. Total Environ. **2008**, *397*, 190–204.
74. Campfens, J.; Mackay, D. Fugacity-based model of PCB bioaccumulation in complex aquatic food webs. Environ. Sci. Technol. **1997**, *31*, 577–583.
75. Pelka, A. Bioaccumulation models and applications: Setting sediment cleanup goals in the Great Lakes. National Sediment Bioaccumulation Conference Proceedings, EPA 823-R-98-002U.S.; Environmental Protection Agency Office of Water: Washington, DC, 1998; 5-9–5-30.
76. Chapra, S.C.; Reckhow, K.H. *Engineering Approaches for Lake Management. Mechanistic Modeling*; Butterworth Publishers/Ann Arbor Science: Woburn, MA, 1983; Vol. 2.
77. Warren, C.; Mackay, D.; Whelan, M.; Fox, K. Mass balance modelling of contaminants in river basins: A flexible matrix approach. Chemosphere **2005**, *61*, 1458–1467.
78. Warren, C.; Mackay, D.; Whelan, M.; Fox, K. Mass balance modelling of contaminants in river basins: Application of the flexible matrix approach. Chemosphere **2007**, *68*, 1232–1244.
79. Arnot, J.A.; Gobas, F. A food web bioaccumulation model for organic chemicals in aquatic ecosystems. Environ. Toxicol. Chem. **2004**, *23*, 2343–2355.
80. Bright, D.A.; Grundy, S.L.; Reimer, K.J. Differential bioaccumulation of non-ortho-substituted and other PCB congeners in coastal arctic invertebrates and fish. Environ. Sci. Technol. **1995**, *29*, 2504–2512.
81. Antunes, P.; Gil, O.; Reis-Henriques, M.A. Evidence for higher biomagnification factors of lower chlorinated PCBs in cultivated seabass. Sci. Total Environ. **2007**, *377*, 36–44.
82. Available at http://www.atsdr.cdc.gov/DT/pcb007.html (accessed February 2010).
83. U.S. Environmental protection agency. PCBs: A cancer dose–response assessment and applications to environmental mixtures, EPA/600/P-96/001F; 1996, available at http://cfpub.epa.gov/ncea/CFM/recordisplay.cfm?deid=12486 (accessed February 2010).
84. Available at http://www.atsdr.cdc.gov/tfacts17.pdf (accessed February 2010).
85. Online version of TSCA, from the Government Printing Office, available at http://frwebgate.access.gpo.gov/cgi-bin/usc.cgi?ACTION=BROWSE&TITLE=15USCC53 (accessed January 2010).
86. Council Directive 76/403/EEC of 6 April 1976 on the disposal of polychlorinated biphenyls and polychlorinated terphenyls.
87. Council Directive 85/467/EEC of 1 October 1985 amending for the sixth time Directive 76/769/EEC on the approximation of the laws, regulations and administrative provisions of the Member States relating to restrictions on the marketing and use of certain dangerous substances and preparations (PCBs/PCTs).
88. Council Directive 96/59/EC of 16 September 1996 on the disposal of polychlorinated biphenyls and polychlorinated terphenyls (PCBs/PCTs).
89. Manila Bulletin. Global chemical treaty (opinion/editorial). Manila Bulletin Publishing Corp.: Manila, 2001.
90. Helcom Recommendation 6/1, Recommendation regarding the elimination of use of PCBs and PCTs—adopted 13 March 1995, having regard to article 13, Paragraph b of the Helsinki Convention.
91. Communication from the Commission to the Council, the European Parliament and the Economic and Social Committee. Community strategy for dioxins, furans and polychlorinated biphenyls (2001/C 322/02), available at http://ec.europa.eu/environment/waste/pcbs/pdf/en.pdf (accessed January 2010).
92. ICCA/WCC Position Paper (1999): Best Available Techniques for Destruction of PCBs-Incineration Technology, presented at the UNEP-INC 3 meeting on POPs in Geneva, September 6–10, 1999.
93. Inventory of World-wide PCB Destruction Capacity, Second Issue; United Nations Environment Programme: Switzerland, 2004; Prepared by UNEP Chemicals, available at http://www.chem.unep.ch/pops/pcb_activities/pcb_dest/PCB_Dest_Cap_SHORT.pdf (accessed January 2010).

Polychlorinated Biphenyls (PCBs) and Polycyclic Aromatic Hydrocarbons (PAHs): Sediments and Water Analysis

Justyna Rogowska
Agata Mechlińska
Lidia Wolska
Department of Analytical Chemistry, Chemical Faculty, Gdansk University of Technology, and Department of Environmental Toxicology, Interdepartmental Institute of Maritime and Tropical Medicine, Medical University of Gdansk, Gdansk, Poland

Jacek Namieśnik
Department of Analytical Chemistry, Chemical Faculty, Gdansk University of Technology, Gdansk, Poland

Abstract

Unfavorable side effects of different forms of anthropogenic activities can be found anywhere in the world. One of the basic characteristics of pollutants entering marine and ocean waters is their spread and movement in the global ocean. A portion of the substances entering the marine environment is rapidly degraded by chemical processes occurring in the air, sediments, and water, thereby losing their toxic properties. The biggest threat is posed by chemically stable or persistent compounds such as polycyclic aromatic hydrocarbons (PAHs) and polychlorinated biphenyls (PCBs). These substances, due to their properties, can be accumulated in the sediments and tissues of marine organisms and then metabolized to more toxic or/and cancerogenic compounds. The harmful impact of these chemicals on living organisms causes a need for constant monitoring of their content in the environment. Analytical research studies focused on determination of PAHs and PCBs in the environmental samples, even though they have been conducted for many years by numerous scientific centers, are not an easy matter. The procedures for PAH and PCB determination in the environmental samples consist of a few or several steps (sampling, isolation, enrichment, purification, final determination), each of which can be a potential source of error. In such a case, the final result of the analysis will be a source of misinformation instead of reliable information. In this entry, the most frequently used techniques for the final determination of PAHs and PCBs in water and sediments are presented. Most of the isolation and enrichment methods are also discussed.

INTRODUCTION

Evidence of human damage to natural resources and the environment is long standing, but some 50 years ago, awareness of human degradation of natural environments around the globe grew substantially.[1] This issue also concerned the oceans, which have always been subject to human activities. The oceans were previously considered to be a vast reservoir for the safe disposal of pollutants.[2] The change in the way of thinking and recognition of the environment as a global heritage caused pollution of seas and oceans to become an international problem. Almost 80% of marine pollution comes from land-based activities and is primarily transported by rivers or directly by runoff waters from scattered sources. A portion of substances entering marine waters is rapidly degraded due to the chemical processes occurring in the air, sediments, and water, thus decreasing their toxic properties. Lipophilic compounds pose the greatest threat for marine flora and fauna. Those compounds can be accumulated in the tissues of marine organisms and then metabolized to more toxic substances. Moreover, toxic substances can be adsorbed from the water column onto surfaces of fine particles and usually move thereafter with the sediments and onto the suspended matter.[3] Chemically stable or persistent substances can remain in the environment for a relatively long time.

Among the compounds that are characterized by the above-mentioned features and belong to the group of chemicals named persistent organic pollutants (POPs) are the following:

– Polycyclic aromatic hydrocarbons (PAHs)
– Chlorinated aromatic compounds, such as polychlorinated biphenyls (PCBs) and polychlorinated terphenyls (PCTs)
– Dioxins and furans
– Chloro-organic pesticides, such as dichlorodiphenyltrichloroethane (DDT), hexachlorocyclohexane (HCH), aldrin, and methoxychlor (DMDT)

The harmful impact of these chemicals on living organisms causes a need for constant monitoring of their content in the environment. This can further be evidenced by the fact that POPs were embraced by the provisions of the Stockholm Convention, which aims to protect human health and the environment from these chemicals and eliminate the most toxic ones.[4] In the Water Framework Directive of October 23, 2000, and Directive 2008/56/WE of the European Parliament and of the Council of June 17, 2008, the so-called Marine Strategy Directive, it is stated that PAH and PCB compounds are priority pollutants and should be totally eliminated from the environment due to their highly toxic properties, their tendency to bioaccumulate, and their persistence. Investigations of these compounds in the aquatic environment are a very important part of environmental quality assessment, which determines the status of contamination and the likely impacts it may cause to the ecosystems.[5]

POLYCYCLIC AROMATIC HYDROCARBONS IN THE ENVIRONMENT

Polycyclic aromatic hydrocarbons can be defined as a pervasive and diverse group of compounds present in the environment, which include carbon and hydrogen with a fused-ring structure containing at least 2 benzene rings.[6] Polycyclic aromatic hydrocarbons are generally formed during incomplete combustion, pyrosynthesis or pyrolysis of organic matter containing carbon and hydrogen, for example, coal, tar, wood, gasoline, and diesel fuel.[7] Low hydrocarbons form PAHs by pyrosynthesis. When the temperature exceeds 500°C, carbon–hydrogen and carbon–carbon bonds are broken to form free radicals. These radicals combine to form acetylene, which further condenses with aromatic ring structures, which are resistant to thermal degradation.[8] Another mechanism of PAH formation is pyrolysis, the cracking of organic compounds, such as higher alkanes, forming radicals, which combine and form condensed aromatic molecules.[7]

Sources and Fate of PAHs in the Aquatic Environment

Hydrocarbons in the aquatic environment originate from natural and anthropogenic sources. Naturally formed PAHs are biosynthesis products or come from oil welling up, and they usually occur in marine sediments at very low levels in the range of 0.01–1 ng/g dry weight (d.w.).[9] Anthropogenic PAHs are introduced to the aquatic environment through the following:

- Atmospheric precipitation
- Leakage of petroleum and its products from oil tankers, oil rigs, and oil pipelines
- Intentional discharge of petroleum derived products.
- Oil pollution resulting from ship exploitation
- Municipal and urban runoff[10]

Based on diagnostic ratios and/or predominance of different PAH congeners, two different sources of the anthropogenic PAHs can be distinguished, namely, petrogenic (derived from petroleum, including crude oil and its refined products) and pyrolytic (combustion of fossil fuels).[9] Higher concentrations of 2- to 3-ring PAHs are generally petrogenic, whereas 4- to 6-ring hydrocarbons are usually pyrolytic in origin.[11]

The fate of PAHs in the aquatic environment depends on their properties, such as water solubility, volatility, and sorption capacity on solid matter.[12] Chemicals belonging to the PAH group are characterized by low water solubility. The solubility of PAHs depends on their structure and generally decreases with increasing molecular weight. Humic substances and surfactants increase significantly their solubility. Polycyclic aromatic hydrocarbons are frequently present in water in the form of different emulsions.[13] Simultaneously, in the aquatic environment, these compounds adsorb strongly to suspension and solid particles of sediment.[14] The lipophilicity, environmental persistence, and genotoxicity increase as the molecular size of PAHs increases up to 4 or 5 fused benzene rings, and toxicological concern shifts toward chronic toxicity, primarily carcinogenesis.[15]

Levels of PAHs in sediments and waters vary, depending on the proximity of the sites to areas of human activity. Sediment concentration and distribution of PAHs may also fluctuate due to biodegradation of these chemicals, a process that is reliant upon abiotic and biotic factors, which are dependent on site characteristics.[11] Examples of the total PAH concentration levels in marine sediments collected from different stations are presented in Table 1 (The various PAHs were summed by different authors; see "Refs." column).

The possible fate of PAHs in the environment includes volatilization, photooxidation, chemical oxidation, bioaccumulation, adsorption to soil particles, as well as leaching and microbial degradation.[15] Higher-molecular-weight (HMW) compounds are associated primarily with particles and are likely to be removed by dry deposition, while lower-molecular-weight compounds are primarily found in the gas phase and are subject to transformation or removal by photochemical degradation.[9] In diluted aqueous systems, direct photolysis is principally responsible for the degradation of PAHs, as reactive oxygen species are not generated efficiently and are consumed rapidly in most natural waters.[29] Direct photolysis is largely responsible for degradation of HMW PAHs, since many of them absorb light in ultraviolet (UV) or visible wavelengths (300 and 500 nm) found in solar radiation. There is photooxidation of PAHs in the water or adsorbed to suspended particulates.[30] However, photolysis in aquatic systems can take place only in a very shallow layer on top of the water body due to the very limited penetration depth of UV radiation in water. Simple PAHs such as naphthalene, biphenyl, and phenanthrene are readily degraded aerobically. The degradation of these compounds is generally initiated by dihydroxylation of one of the polynuclear aromatic rings,

Table 1 Total concentrations of PAHs in marine sediments.

Station	Minimum and maximum concentration (ng/g d.w.)	Refs.
Eastern Arctic seas (Laptev and East Siberian)	3–80	[16]
Brazil, the Paranagua Bay	26.33–406.76	[17]
Spain, the Atlantic coast	22–47,528	[18]
Yemen, Gulf of Aden	2.2–604	[19]
Argentina, the Bahia Blanca Estuary	15–10,260	[20]
United Kingdom, the Mersey Estuary	626–3766	[21]
China, Qingdao, Jiaozhou Bay	0.02–2.2	[22]
Southern Nigeria, the Niger Delta	20.7–72.1	[23]
Egypt, the Mediterranean Sea coast	88–6338	[24]
Black Sea (Ukraine, Russian Federation, Turkey)	7–640	[25]
Croatia, Northern Adriatic, the Rovinj area	32 (protected area)–13,200 (harbor)	[11]
Mexico, the Gulf of Mexico	nd–1033	[26]
China, the Yellow River	464–2621	[27]
Germany, the Western Baltic Sea	3–30,000	[28]

nd = not detected

this being followed by cleavage of the dihydroxylated ring. Ring hydroxylation is catalyzed by a multicomponent dioxygenase, which consists of a reductase, a ferredoxin, and an iron sulfur protein, while ring cleavage is generally catalyzed by an iron-containing metacleavage enzyme. The carbon skeleton produced by the ring-cleavage reaction is then dismantled, before cleavage of the second aromatic ring.[31] Although some of the first reports of HMW PAH biotransformations by bacteria were published in 1975,[32] less is known about the bacteria capable of utilizing PAHs containing five or more rings (such as benzo[a]pyrene and benz[a]anthracene) as a carbon and energy source. The mechanisms involved in the degradation of PAHs with more than five rings are unclear.[6] Generally, the greater the complexity of the hydrocarbon structure, i.e., the more methyl-branched substitutes or condensed aromatic rings, the slower the rates of degradation and the greater the likelihood of partially oxidized intermediary metabolites being accumulated.[33] The biodegradation process can be limited by sorption of PAHs by soil organic matter. Sorption of PAHs to organic matter and soil particulates also influences bioavailability and, hence, biotransformation potential (more in Section 3).

Biological and Health Effects of PAHs

Polycyclic aromatic hydrocarbon compounds, due to their lipophilic character, can easily penetrate cell membranes and are subject to metabolic processes in all body tissues. However, the metabolism of PAHs renders them more water soluble and more easily excretable.[12] In vitro studies conducted on the group of PAH compounds showed that none of these substances, as such, has mutagenic and carcinogenic properties. It is only through metabolism that these compounds are converted to reactive derivatives, which may damage the genetic material and induce changes at the cellular level.[34] Oxidized metabolites are produced by a mixed function of oxygenase cytochrome P450.[35] Some intermediates of PAH metabolism, such as the diols and epoxides, can bind to DNA/RNA or proteins and be mutagenic and/or carcinogenic.[35]

Therefore, specialists from the International Agency for Research on Cancer (IARC) and the United States Environmental Protection Agency (U.S. EPA) qualified some of the PAH compounds as cancerogenic or probably cancerogenic (Table 2).[36]

Polychlorinated Biphenyls in the Aquatic Environment

Another group of compounds posing a threat to marine ecosystems are those belonging to the group of chlorinated hydrocarbons, such as PCBs. Polychlorinated biphenyls are classified as the one of the most persistent and toxic industry-produced compounds that have been detected as contaminants in almost every component of the global ecosystem.[37] These compounds do not occur naturally in the environment. Due to their characteristic parameters, such as non-flammability, insulating ability, and chemical stability, these chemicals have been produced on an industrial scale since the 1930s. Up to the beginning of the 1970s, these compounds were used in closed systems as dielectric fluids in capacitors, as insulating and cooling fluids in transformers, and as an additive to hydraulic oils. Moreover, PCB compounds were used in manufacturing adhesives and plastics, in impregnating, and as an additive to

Table 2 Carcinogenic classifications of selected PAHs by IARC and U.S. EPA.

Agency	PAH compound(s)	Carcinogenic classification
IARC	Benz(a)anthracene, benzo(a)pyrene	Probably carcinogenic to humans
	Benzo(a)fluoranthene, benzo(k)fluoranthene, ideno(1,2,3-c,d)pyrene	Possibly carcinogenic to humans
	Anthracene, benzo(g,h,i)perylene, benzo(e)pyrene, chrysene, fluoranthene, fluorene, phenanthrene, pyrene	Not classifiable as to their carcinogenicity to humans
EPA	Benz(a)anthracene, benzo(a)pyrene, benzo(b)fluoranthene, benzo(k)fluoranthene, chrysene, dibenz(a,h)anthracene, indeno(1,2,3-c,d)pyrene	Probable human carcinogens
	Acenaphthylene, anthracene, benzo(g,h,i)perylene, fluoranthene, fluorene, phenanthrene, pyrene	Not classifiable as to human carcinogenicity

cement and insecticides.[38] In commercial products, these compounds occur as mixtures of up to 209 different PCB congeners, dependent on the position of the chlorine atoms in the molecule. Not all congeners have been identified in commercial products or technical mixtures (Fig. 1).

The greatest application had compounds containing from 42% to 54% chlorine. Commercial mixtures of PCB compounds were known under different names, depending on the country where they were produced, such as Aroclor (United States), Clophen (Germany), Chlorofen (Poland), and Sovol (now Russia, previously Union of Soviet Socialist Republics).

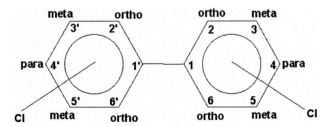

Fig. 1 Chemical structure of PCBs.

The years of the 1970s were the peak period of global production of these chemicals—it is estimated that these compounds were produced in quantities of 50,000 tonnes annually.[39] In the 1960s, more and more reports stated that compounds belonging to the group of PCBs are not indifferent to living organisms. Therefore, their use and import were banned, among others in Sweden (1970) and Japan (1972). In the United States, production, processing, and distribution of PCB compounds were banned in 1976 under the Toxic Substances Control Act. Although many countries stopped manufacturing PCBs, due to their improper use, storage or disposal these compounds may be still released to the environment from unprotected landfills, illegal or improper storage of waste, and leakage or discharge from transformers. These chemicals may enter the environment through the combustion of waste in urban or industrial incinerators. Polychlorinated biphenyls are also formed as a byproduct of bleaching pulp. Due to their persistence, these compounds may remain in the environment for a long time, be transferred between media (such as water, air, and soil), and be transported over long distances.[40]

Properties and Fate of PCBs in Aquatic Environment

Depending on the structure, PCB compounds have different physicochemical and toxic properties, but all have common features, such as the following:

- Low solubility in water (good solubility in nonpolar organic solvents)
- Low electric conductivity
- Very high thermal conductivity
- High resistance to degradation

Properties of individual congeners depend on the number of chlorine atoms attached to the aromatic ring. Flammability of PCB compounds decreases with increasing number of chlorine, while these compounds are more stable and more resistant to biodegradation. Solubility in water decreases with increasing number of chlorine in the range from 6 ppm for monochlorobiphenyl to 0.007 ppm for octachlorobiphenyl.[41] Increasing lipophilicity of these compounds consequently increases the probability and possibility of their accumulation in living organisms and their transfer in the trophic chain. A specific toxicity is exhibited by PCB compounds containing 5 to 10 chlorine atoms per molecule, usually substituted in the para- and meta-position. However, compounds containing chlorine atoms in the position of 3,4- ortho are characterized by the highest toxicity.[41] Polychlorinated biphenyls possessing no or one ortho-positioned chlorine atom bind the aryl hydrocarbon receptor with avidity and induce cytochrome P450 1A. Several di-ortho-substituted PCBs induce cytochrome P450s as does phenobarbital, while other PCB congeners may induce both subfamilies of cytochrome P450. Many of these PCBs may also induce epoxide hydrolase, glutathione transferases, and glucuronosyltransferases. Induction of xenobiotic metabolites may be accompanied by an increase in hepatic cell size and number and a proliferation of the endoplasmic reticulum. The persistent induction of hepatic cytochrome P450s, in the absence of an oxidizable xenobiotic substrate, may provide suitable conditions for generation of reactive oxygen species.[38] Toxicity of these compounds is manifested, among other things, by interfering with the functioning of the nervous and immune systems. Therefore, specialists from the U.S. EPA classified PCB compounds into group B2, which contains substances suspected of causing cancer. Also, the IARC classified the PCB compounds into group 2A, which contains precancerogenic substances.[39]

Polychlorinated biphenyl compounds, similarly to the previously discussed PAHs, are adsorbed on organic solid matter, such as sediment and soil. Due to their properties, these substances are poorly biodegradable. Although the overall degradation should be the final effect of self-purification processes occurring in the environment, in case of PCB compounds, only partial results can be achieved—reduction of the degree of bioaccumulation and lipophilic properties. A key point in the degradation of PCBs is the removal of chlorine substituent. This may occur under anaerobic or aerobic conditions, after the ring ruptures.[42] The anaerobic conditions contribute to the removal of chlorine from the meta- and/or para- position and may reduce the toxicity of the compound and increase susceptibility to further biodegradation. The degradation process under aerobic conditions proceeds by introducing hydroxyl groups into the ring, by the enzyme dioxygenase.[37] The rate and extent of the dechlorination process depends on many factors, including types of substances, chemical and physical properties of sediments, and environmental parameters (temperature, salinity, pH).[43]

SPECIATION OF PAHs AND PCBs IN SEDIMENTS

The aim of environmental monitoring is to collect information about the state of the ecosystem. This is achieved by both speciation analysis and quantitative analysis of the portion of chemicals dissolved in water and adsorbed on solid particles. Polycyclic aromatic hydrocarbon and PCB compounds, as already mentioned, due to their properties, are mainly adsorbed on solid particles of sediment/suspension in the aquatic environment. The content of PAHs and PCBs in the individual components of the aquatic environment decreases in the order[44]

suspension > sediment > water

and is estimated at the level of pg–µg/L in water and µg–mg/kg in sediments. Contaminants that are strongly sorbed and not available to microorganisms also may not be available for a toxic action induction. Thus, knowledge of how and where PAHs and PCBs are bound to sediment material is necessary to assess the efficacy of sediment bioremediation and to correlate this knowledge with reductions in availability, mobility, and toxicity.[45] Therefore, knowledge of the chemical composition of sediments is in many cases a better indicator of environmental pollution than the knowledge of the chemical composition of water, more variable in time.

The chemical composition of sediments is dependent on the following:

– Construction of geomorphologic catchment
– Climatic conditions
– The natural processes occurring in water that lead to sedimentation of suspensions and adsorption of contaminants and cause the accumulation of elements and compounds of low solubility in water

The development of instrumentation and introduction of new analytical procedures in practice resulted in the possibility of determination of even small amounts of these contaminants present in the sediments. However, the ability to interpret the obtained analytical information in reference to the analytical assessment of water quality, the bioavailability of xenobiotics, or their toxic effect is rather limited by inadequate understanding of the processes of binding and release of chemicals associated with bottom sediments, which are characterized by a complex matrix.

Sediment matrix comprises two main parts:[46]

– Inorganic, which contains clays, silts, and muds
– Organic

This second part can, in turn, be divided into two phases:

– Amorphous (soft, plastic)
– Condensed (hard, glassy)

Chemical and biochemical processes occurring in the bottom sediments characterized by a high content of organic pollutants can lead to movement of their toxic components primarily to water, causing its secondary contamination. Through determination of a state in which these pollutants occur, it is possible to obtain reliable information on the risk and the possible impact of xenobiotics on living organisms.[47] Physical speciation analysis can be the source of such information.[48]

A special type of speciation is physical speciation. It concerns the presence of a given analyte in different phases of the analyzed system. An interesting example of such a system is the aquatic environment (aqueous phase, suspension,

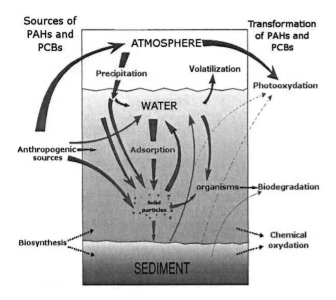

Fig. 2 Physical speciation of PAHs and PCBs in the aquatic ecosystem.

sediment), where PAH and PCB compounds may be one of the following:[49,50]

- Adsorbed on the surface of particles suspended in water
- Adsorbed on the sediment surface
- Bioaccumulated in marine organisms
- Biomagnified within the trophic chain, deposited in bottom sediments

Fig. 2 presents physical speciation of PAHs and PCBs in the aquatic ecosystem.

Polycyclic aromatic hydrocarbon compounds occur primarily in the adsorbed state on the surface of structures that are composed of various forms of carbon. (Polycyclic aromatic hydrocarbon concentrations are about 30-fold higher on the external surfaces of sediment, where different forms of elemental carbon are the dominant fraction, than in the areas located inside.[45]) Weakly bound (van der Waals forces) and strongly bound (π-type bonds) analytes can be distinguished. Polychlorinated biphenyl compounds occur primarily in the dissolved state in the oil phase and in the bound form in organic matter.

SAMPLING, EXTRACTION AND CLEAN-UP TECHNIQUES OF ENVIRONMENTAL SAMPLES FOR PAH AND PCB DETERMINATION

Analytical research studies focused on determination of PAHs and PCBs in the environmental samples, even though they have been conducted for many years by numerous scientific centers, are not an easy matter.

Analytical problems associated with the determination of PAH and PCB compounds are associated mainly with the following:

- Difficulty of validating the individual steps and the whole analytical procedure using the appropriate standard solutions
- Isolation of various states, in which these compounds occur in the aquatic environment (adsorbed, dissolved fractions)
- Final determination

The procedures for PAH and PCB determination in the environmental samples consist of a few or several steps, each of which can be a potential source of error. In such case, the final result of the analysis will be a source of misinformation instead of reliable information.

Sampling, Transport and Storage of the Samples

Obtaining reliable results of the analysis is significantly dependent on proper sample preparation. Therefore, special instructions at each step of the analytical procedure, from sampling to final determination, should be followed. It is important to maintain the chemical stability of water or sediment during all steps of the analytical procedure. This concerns both particular chemicals as well as different physical forms.[51] This task is not easy because, as mentioned earlier, water or sediment matrix is characterized by a diversity of ingredients and concentrations of individual components. All this makes it difficult to ensure the maintenance of the sample in the unchanged form. Many factors contribute to this and include, among others, the following:

- Chemical reactions (for example, oxidation, reduction, complexation, precipitation)
- Biochemical reactions (for example, biodegradation)
- Photochemical reactions (for example, photolysis)
- Physical processes, such as adsorption of the components on the walls of the samplers and evaporation processes of more volatile compounds[51]

In order to minimize the impact of these adverse reactions and processes, it is necessary both to analyze the samples as quickly as possible and to shorten the period between the sample collection and analysis. It is also important to select appropriate materials, vessels, and pipes, with which the sample will have contact, and to ensure proper maintenance and storage of the samples. For example, it has been shown that 40%–80% of the PCBs in a sample may be adsorbed onto the polytetrafluoroethylene (PTFE) surface,[52,53] whereas adsorption onto glass may be responsible for a 10%–25% drop in the PAH and PCB content in a water sample.[54] On the basis of the studies conducted by Wolska et al.,[54] the PAH and PCB adsorption onto the walls of the container has been assessed to range from 0%

to as much as 70%. Therefore, due to the occurring adsorption of PAH and PCB compounds onto the walls of the samplers, these containers should not be washed by means of the water. The walls of the vessel should be rinsed with solvent, and the solution obtained should be added to the extract. The addition of isopropanol, methanol, or acetonitrile to the container prior to sampling—in order to protect against adsorption of the compounds onto the walls—which is recommended by some authors,[54,55] can be regarded as a process of liquid–liquid extraction (LLE). Extraction of analytes from water samples in the presence of suspended particulate matter depends on several factors, such as the following:

- The processes determining the equilibrium sorption–desorption of the analytes in the water suspension system during transport and storage of the samples
- The degree of analyte extraction from the aqueous phase
- The degree of extraction of analytes from suspension (in the case of the extraction process from the aqueous phase)

In each of the mentioned cases, the results of PAH and PCB determination in water depend on the content and properties of suspended matter in the sample.[56]

Extraction Techniques

The basic step of environmental sample preparation prior to chromatographic analysis is isolation and/or enrichment, involving the transfer of the analytes from the primary matrix (sample) to the solvent phase secondary matrix with the simultaneous removal of interfering substances (isolation) and increasing analyte concentration to a value above the limit of quantification of the measuring instrument (enrichment). The additional advantage is reduction of the diversity and concentration level of interfering chemicals because only selected compounds are transferred to the secondary (receiving) matrix.[56,57] Isolation and enrichment steps play a key role in the procedures of PAH and PCB determination in the environmental samples.

The most commonly used isolation and enrichment techniques of PAHs and PCBs from water and sediment samples are shown in Fig. 3.

Isolation Techniques of PAHs and PCBs from Water Samples

Liquid–Liquid Extraction

By means of LLE, solvent molecules can react not only with the compounds dissolved in the aqueous phase but

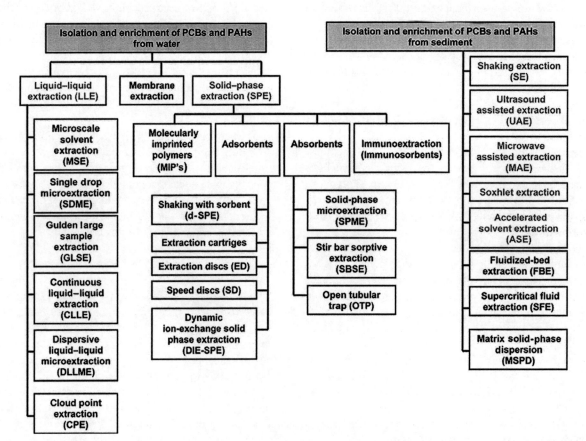

Fig. 3 Isolation techniques of PAHs and PCBs from aqueous and sediment samples. Techniques routinely applied in the laboratories are marked in red, while the rarely used extraction techniques, applied usually only in the research laboratories, are shaded in black.

also may reach the substances adsorbed on the samples' colloidal fractions or suspended particulates. Thus, the results of the determination of individual PAH and PCB compounds in water samples (containing suspended matter and obtained by the LLE technique) will be the effect of the analytes' extraction efficiencies of both soluble and adsorbed on the surface of colloids and suspension (to which the access of solvent is hindered).

In the literature, the following new approaches can be found:

- Microscale solvent extraction (MSE)
- Single-drop microextraction (SDME)
- Goulden large-sample extraction
- Continuous LLE
- Dispersive liquid–liquid microextraction (DLLME)
- Cloud point extraction

Microscale Solvent Extraction

Microscale solvent extraction methods were developed to provide rapid and reliable chemical data for environmental samples. The MSE approach has been successfully used under both field and standard laboratory conditions.[58] The microscale approach minimizes sample size and solvent usage, thereby reducing the supply costs, health and safety issues, and volume of waste generated.

Two MSE methods have been approved by the U.S. EPA for extracting monocyclic aromatic hydrocarbons and PAHs, selected volatile organic compounds, or semivolatile organic compounds, which are slightly soluble or insoluble in water at neutral pH[59] and are widely used in practice:

1. *Method 3511*—microextraction techniques for aqueous matrices
2. *Method 3570*—microextraction techniques for soils and solid matrices

Single-Drop Microextraction

The origin of SDME dates back to 1979,[60] but recent approaches were proposed many years later, ca. 1996.[61] The SDME technique in its very basis is similar to SPME. In this case, instead of fibers, a single drop of the solvent is used as a sorptive material. Most of the objections formulated against SPME technique, especially those connected with the adsorption of PAH and PCB compounds onto the walls of the extraction vessels, also refer to the SDME. Unfortunately, limitations of LLE technique occurring during determination of PAHs and PCBs also appear in the SDME.[62] The analytes in the soluble form as well as adsorbed on the colloidal fraction will be dissolved in the suspended drop of the solvent. Moreover, a partial extraction of the compounds adsorbed on the suspension may be also observed.

Goulden Large-Sample Extraction

Goulden large-sample extraction has been used to isolate a broad array of trace organic contaminants from large volumes of water since 1985. A stream of water sample flows continuously through a vessel containing appropriate amounts of organic solvent (heavier than water) together with intense stirring.[63]

Continuous LLE

A continuous countercurrent LLE was developed to concentrate trace organic pollutants from large water samples. Nonvolatile hydrophobic pollutants (such as pesticides, phenols, PACs, PCBs, or phthalates) can be extracted using this technique in a pulsed column system.

In situ analysis of large-volume water samples is usually carried out using continuous concurrent extractors. Only countercurrent extractors can deal with multistage extraction from large-volume samples. Among extractors, the pulsed plate column is a differential contactor, which involves mechanical energy. This technology has been used in several chemical processes (e.g., hydrometallurgy and the nuclear industries) but has seldom been applied to environmental pollution control.[64]

Dispersive Liquid–Liquid Microextraction

Recently, a simple and rapid preconcentration and microextraction method, DLLME, has been developed.[65–67] Dispersive liquid–liquid microextraction employs a mixture of a high-density solvent (extractant) and a water-miscible polar solvent (disperser). Acetone, methanol, and acetonitrile can be used as dispersers, whereas chlorinated solvents (e.g., chlorobenzene, carbon tetrachloride, tetrachloroethylene) are usually employed as extractants.[68] The appropriate mixture of extraction solvent is injected into the aqueous sample, containing the analytes. The extraction solvent is dispersed into the aqueous sample as very fine droplets, and a cloudy solution is formed. Owing to the considerably large surface area between the extraction solvent and the aqueous sample, the equilibrium state is achieved quickly. After centrifugation of the cloudy solution, the determination of the analytes in the decanted phase can be performed by instrumental analysis.[69] Dispersive liquid–liquid microextraction offers several important advantages over conventional solvent extraction methods:[66]

- Faster operation
- Small consumption of organic extraction solvent
- Low time and cost of operation
- Easier linkage to most analytical methods

Thus, since its introduction, DLLME has been frequently used for determination of organic contaminants in liquid samples, including PAHs,[66] organophosphate esters,[68] phthalate esters,[69] aromatic amines, and phenols in water samples.[70]

Cloud Point Extraction

Cloud point extraction is based upon phase separation behavior exhibited by aqueous solutions of certain surfactant micelles.[71] Comparing with conventional LLE, there is no organic solvent used during the process, so it is economical and friendly for the environment.[72]

Surface-active agents can aggregate in aqueous solution to form colloidal-sized clusters referred to as micelles. During their formation, the surfactant micelles have proven to entrap several hydrophobic substances, isolating them from the bulk solution. The solubility of non-ionic or zwitterionic surfactants in aqueous solution is dramatically depressed above a well-defined temperature, referred to as cloud point temperature (CPT). By setting the solution at a temperature above the CPT, the solution separates into a concentrated phase containing most of the surfactant, the surfactant-rich or coacervate phase, and a dilute aqueous phase.[73] Cloud point extraction arises from the partitioning of a solute between the two water-based phases depending on its affinity to the surfactant.

The cloud-point methodology can be applied to the extraction of a wide range of analytes, including PACs, PCBs, porphyrins, metalloporphyrins, vitamin A, vitamin E, alpha-estradiol, estriol, estrone, progesterone, and proteins, from simple aqueous solutions.[74]

Solid-Phase Extraction

During the passage of a water sample through a solid-phase extraction (SPE) column bed, the layer of suspension is being formed at the sorbent surface, which de facto is a layer of the sorption material. In this way, dissolved chemicals are secondarily sorbed from the aqueous phase.

In the next step of the isolation procedure, namely, during the desorption of analytes of interest from the column's bed, not only analytes present in the dissolved form but also chemicals adsorbed at suspension and colloids may reach the extraction medium. The way of conducting the desorption operation (longer or shorter time of contact between solvent and column's bed) may also influence the recovery of analytes of interest from the suspension layer and thus affect results of PCB and PAH chemicals determination.

During the isolation of PAH and PCB compounds from water containing a suspension, using the conventional SPE technique, a certain volume of the sample is passed through a layer of filling (solid sorbent or liquid phase applied on the solid carrier and placed in the columns). In order to transport the sample from the sampler to the columns, appropriate wires, which are often made of PTFE, are used. When the sample is passing through the above-mentioned network, both colloidal and suspended particles (mostly those of larger sizes) but also soluble forms of compounds are retained on the walls of the wires and above all in the filling of the columns. An additional adverse effect is the dissolution of the PCB compounds in the plastic PTFE (Teflon). This phenomenon was applied in the proposed new technique of SPE using a rotating element as a carrier—stir bar sorptive extraction (SBSE).[75] (Stir bar sorptive extraction is described below.)

The contact of sample with sorbent can be achieved in the following ways:[55]

- Shaking sample with sorbent
- Percolation of sample through extraction cartridges packed with sorbent (e.g., silica gel most often modified with octyl or octadecyl groups)
- Passing the stream of sample through so-called Empore disks
- Passing the stream of sample through so-called speed disks
- Dynamic ion-exchange SPE

Solid-Phase Microextraction

The solid-phase microextraction (SPME) technique is recommended by numerous authors[76–78] for determination of the dissolved PAH chemical in the presence of dissolved organic matter (DOM) and/or suspended particulates. At the same time, several works include information on the following:

- Adsorption of hydrophobic DOMs on fiber[79]
- Competitive adsorption of analytes on the walls of the extraction vials[79]
- Disturbing equilibrium state of the water sample after introduction of the organic solvent in the form of standard solution[78]

As suggested by authors of the works mentioned before, these limitations make it possible only to only assess numerical values of the water stationary phase partition coefficients and more comprehensively understand the sorption process on the SPME fiber, and not to determine water dissolved PAH and PCB chemicals.

The list of unfavorable limitations should be elongated by the adsorption on the stirring bar surface (made of glass in most cases and used in order to enhance analytes' transport to the fiber surface). The summed influence of these processes and not the quantitative character of the analytes' desorption reduces the SPME recovery of PAHs and PCBs to 11%–44% and 22%–33%, respectively.[80]

Additional operations like silanization of the surfaces of the laboratory ware (devoted for the extraction stage) are applied by almost all authors in order to reduce sorption of hydrophobic chemicals on the glass surfaces and cannot be considered as profitable. Adsorption of these chemicals to the solid surfaces is energetically preferable and to a small extent depends on chemical character of the vials' surfaces. Reliability of results of PCB and PAH determination—with the application of SPME at the step of analyte isolation—is undermined due to phenomena of adsorption of dissolved forms of analytes at glassware vials' surfaces. For the same reason, the value of arguments supporting advantages of the sorption techniques over LLE—such as reduction of solvent volume used, simplicity, possibility of automatization, etc.—can be reduced.

Stir Bar Sorptive Extraction

Claims similar to those already stated must be raised in the case of works dealing with determination of dissolved and bounded forms of PAHs and PCBs with the application of SBSE.[77] This technique has been proposed as a result of observations made during isolation of PCB chemicals with SPME technique. The analytes mentioned, during the isolation process, were undergoing significant sorption to the material of the sorptive bar (PTFE). The sorption phenomena were competitive with the SPME fiber sorption. This unfavorable phenomenon in SPME extraction has been applied in order to offer a new technique of PCB compound isolation, namely, SBSE.[75]

The main difference when compared with SPME is the way of locating the solid-phase layer (PTFE or PDMS fibers) and significantly increasing its amount. Most often, it is located close to the adequately prepared stirrer. Other conditions of conducting isolation process are similar; thus, unfavorable processes (competitive adsorption of analytes to the vials' walls and adsorption of the bounded form to the colloidal one at the stationary phase) will have similar influence on the results of determining content of both groups of analytes.[81]

The unfavorable effect of humic acid presence in the sample has been also described; the data presented support the statement that recovery of PCB analytes with the application of SBSE can be reduced threefold. Addition of methanol to the sample containing humic acids (20% v/v) increases the recovery (twofold increment of the recovered analytes when compared with the sample with no MeOH addition).

Open-Tubular Trapping

There are data available describing the possibility of PCB and PAH analytes' isolation with the aid of a short piece of the capillary column covered with the film of the proper stationary phase [open-tubular trapping (OTT)].[82-84]

Results obtained during the studies on efficiency of the sorption of PCB and PAH chemicals[85] in the short piece of the capillary column (20 cm length, internal diameter (ID) 0.32 mm, 12.0 μm film of the PDMS stationary phase) show the following:

- Recovery of the analytes of interest was very low (did not exceed 20%) and is connected with breakthrough of the trap—which is the fragment of capillary of 30 cm length.
- The time of analytes' isolation is relatively long. (Putting approximately 0.5 dm^3 of the aqueous sample through the capillary column with a flow rate intensity of approximately 1 cm^3/min lasts approximately 5 hr.)

Application of a longer piece of the capillary as a trap results in an increment of flow resistance and significant extension of the isolation time, unacceptable in the laboratory practice. A solution for this problem may be application of a multicapillary trap. However, the question about form of analytes (dissolved, adsorbed at colloidal or solid particles) being sorbed in the trap is still open. The phenomena of analyte sorption in this case are of similar character to those for SPME technique. In the case of some authors,[82] this technique is also called inner tube microextraction to stationary phase.

What is more, presence of suspended matter of large diameter in the water sample requires its removal prior to analyte isolation (via filtration, centrifuging, or sedimentation).

It should also be mentioned that new sorption materials are available, which allow us to avoid problems appearing when conventional sorbents are used. They are immunosorbents and molecularly imprinted polymers.

Membrane Extraction

One of the techniques enabling isolation of the dissolved forms of PCB and PAH analytes from the surface water samples is group of membrane techniques, with special attention paid to samplers applying semipermeable membranes (SPMDs).[85-86] These passive techniques rely on selective transport of analytes through properly selected membranes. The isolation phenomenon is based on Fick's first law. Techniques of passive dosimetry have been developed for many years and are widely applied to isolate and enrich organic pollutants present in the air. Attempts are made to apply these techniques to isolate water pollutants, mainly for long-term monitoring purposes.[87] Unfortunately, the studies conducted have revealed several problems and limitations. The most important ones are the following:

- Dosimeter calibration (dependence on water temperature, salinity, pH, pressure resulting from the depth of sampler location, and others)
- Impossibility of application to assessing pollutant content characterized by large time variation
- Varying time of analytes' transport through the membrane, problem of setting equilibrium
- Lack of knowledge about accumulation processes within the membrane, to be more precise on ratio of concentrations of analytes of interest in the membrane and in the sorption medium
- Fouling of dosimeter with algae and microorganisms, which constitute a sorption medium for pollutants and result in their biodegradation
- Complexity of procedures of sample preparation prior to final determination (inaccessible for laboratories conducting routine determinations)[88]

At the moment, this technique seems to be a good tool enabling assessment of the bioaccumulation in the organisms living in the water bodies and sediments, while it is

Table 3 Comparison of effects of different techniques of PCB and PAH isolation on the results of their content determination in surface water samples.

Analytes' isolation technique	Extraction efficiency of different speciation forms of chemicals classified as PCB and PAH compounds		
	Analytes in dissolved form	Form present in DOM and in the colloid fraction	Form adsorbed to suspended material
LLE	High	Significant but not entire	Significant but not entire—restricted by the access of organic solvent to the suspended matter and extraction time
SDME	High	Significant but not entire	Partial—restricted by the access of organic solvent to the suspended matter and extraction time
SPE	High, but there is possibility of adsorption at the thin film of suspended matter settling at the sorbent bed	Partial—colloids' molecules are being sorbed at column's bed	Partial—depends on type of eluent and its contact time with the bed
SPME	High—adsorption at extraction vial walls is a competitive phenomenon	Partial—DOM modifies the surface of the stationary phase	Partial—some portion of the particulate matter remains adsorbed at the fiber
SBSE	High—adsorption at extraction vial walls is a competitive phenomenon	Partial—DOM modifies the surface of the stationary phase	Partial—portion of suspended matter particles is adsorbed at stirrer surface
Membrane techniques	High	Partial	Low

not a proper technique for pollutant determination in surface waters and particularly for pollutants dissolved in given environmental compartments.

Here, it should be mentioned that undoubtedly, one of the most important drawbacks of all isolation techniques based on solid-phase extraction—SPE, SPME, OTT, SBSE, and membrane techniques—is the necessity to take into account the analyte adsorption on walls of samplers, vials, and devices at the stage of final determination. Unfortunately, not all analytical chemists realize that when publishing their results.

Table 3 summarizes the most important conclusions from critical studies on the assessment of the most often-used PAH and PCB isolation techniques, with special attention paid to their determination in surface water samples and to pros and cons of the methods applied.[89–90]

The above-mentioned considerations suggest that the currently available techniques for PAH and PCB isolation do not allow for determination of various physical forms of these compounds in surface water samples with suspension. By means of these techniques, concentration of various forms can only be estimated.

Isolation Techniques of PAHs and PCBs from Sediment Samples

Shaking-Assisted Extraction

The simplest solid–liquid extraction technique is to blend the sediment sample with an appropriate organic solvent by mechanical shaking in the shaker. This mechanical effect of shaking induces a greater penetration of solvent into sediment materials and improves mass transfer, leading to an enhancement of sample extraction efficiency.[91]

Ultrasound-Assisted Extraction, Also Called Sonication

Ultrasound-assisted extraction (UAE) is an effective method for PCB determination.[92] The application of ultrasounds is useful in chemical and physical extraction of the analytes from solid matrices. Ultrasound-assisted extraction allows for close contact between the sediment particles and the solvent, which in turn greatly reduces the extraction time.[93] Cavitation bubbles formed during the UAE can achieve significant internal temperature and pressure. This causes the decomposition of the solvent used for extraction and multiplication of these bubbles with a high speed and collision with the particles of the matrix. As a result of this process, smaller particles of the matrix are formed. They are therefore exposed to greater volumes of the solvent used for extraction. The phenomenon of cavitation increases penetration of the solution into the sediment's structure and conduces toward the release of the substances present there. Results of studies justify the statement that time elongation of performing cavitation phenomena greatly increases recovery of chlorinated chemicals from the sediment structure to the surrounding solvent.[94]

Microwave-Assisted Extraction

In microwave-assisted extraction (MAE) for organic trace analysis, two criteria should be used for the classification of the extraction:

1. The first is whether the extraction is performed in an open or a closed vessel: This is important not only because of the higher temperature that can be used with a closed system but also due to the fact that an open system is always prone to sample losses and to contamination, which at ultratrace level is particularly risky and important.
2. The second important question is whether the solvent is heated directly or indirectly. The latter can be achieved not only by adding small quantities of a more polar solvent to the apolar solvent but also by the use of a strongly absorbing stir bar that transfers the heat to the surrounding solvent. The use of stirring to support MAE is generally recommended as heating alone is by far not as efficient as heating in combination with stirring.

The numerous applications of MAE for the determination of PAHs and PCBs in environmental matrices can be found in the work of Camel.[95] The microwave method was described in the Environment Canada Reference Method (1997) and was recently approved by the U.S. EPA as Method 3546 under SW-846 for extracting environmental organic compounds from soils, sediments, sludges, and solid wastes.[96] Although gas chromatography (GC) is the preferred method for analyzing PCBs in the extracts, MAE has been used in combination with an enzyme-linked immunosorbent assay for the determination of PCBs in soil and sediment samples.[95]

The main advantages of MAE, from the green chemistry point of view, are the following: significant reduction in the amount of solvent used (25–50 mL), which reduces waste generation; reduction in extraction time (3 min–1 hr); and reduction in the amount of the sample required and corresponding reduction in energy input and cost.

Soxhlet Extraction

Soxhlet extraction is a general and well-established technique developed in 1879. The technique is based on exhaustive extraction of organic compounds in a Soxhlet system by an organic solvent, which is continuously refluxed through the sample contained in a porous thimble. The extracted analytes accumulate in a heated flask, so they must be stable in the refluxing boiling solvent. Soxhlet extraction is the oldest technique used for the isolation of nonpolar and semipolar organic pollutants from different types of solid matrices, including sediment samples.[97] It has been used for decades and has been adopted by the U.S. EPA as method 3540C.[59] The main disadvantages of Soxhlet extraction are that it requires large amounts of solvent, the solvent must be evaporated to concentrate analytes before determination, the process takes several hours or days to complete the extraction, and it generates dirty extracts that require extensive clean-up.[98] Therefore, this traditional method is being replaced by other new extraction techniques, such as supercritical fluid extraction (SFE), MAE, and accelerated solvent extraction (ASE^{TM}), with shortened extraction times, reduced organic solvent consumption, and reduced production of harmful waste. However, Soxhlet extraction is still an attractive option for routine analysis for its general robustness and relatively low cost. Moreover, Soxhlet extraction is widely used as a standard technique and reference for evaluating the performance of new extraction methods proposed.[99–100]

Accelerated Solvent Extraction

This technique, also named pressurized liquid extraction (PLE), was originally launched by Dionex Inc. in 1995 under the name "accelerated solvent extraction."[101] Pressurized liquid extraction is a solid–liquid extraction process performed in closed vessels at relatively elevated temperatures, usually between 80°C and 200°C, and elevated pressures, between 10 and 20 MPa. The extraction process is conducted under elevated pressure to maintain the conventional organic solvents in their liquid state but at temperatures well above their atmospheric boiling points. Therefore, the solvent is still below its critical conditions during PLE but has enhanced solvation power and lower viscosities and hence allows higher diffusion rates for analytes. In this way, the extraction efficiency increases, minimizing solvent needed and expediting the extraction process.[100,102]

Fluidized-Bed Extraction

The operation principle of FBE is related to Soxhlet extraction.[103] The technique is based on exhaustive extraction of organic compounds in a special extraction tube, which is equipped with a PTFE filter and extraction solvent.[104] The heating–cooling block of the device is first warmed up to the chosen temperature according to the boiling point of the solvent. The condensed solvent drips back into the extraction material/mixed solvent and is re-collected there. The constant flow of solvent vapors from the base vessel warms the mixture and vigorously fluidizes the vapor bubbles. Extraction at elevated temperature and fluidizing agitation both account for particularly effective extraction kinetics. Following the process step "heating," the solvent is re-collected into the extraction tube. For this purpose, the system is programmed to turn heating off and to simultaneously cool down the basic heating–cooling block as well as the solvent contained in the basic vessel. A "vacuum" is therefore produced in the basic vessel as a result of the quick cooling process, and the resultant differential pressure transports the extractive through the filter into the basic vessel.[105] However, this technique has not been widely applied for the analysis of environmental samples.

Supercritical Fluid Extraction

Supercritical fluid extraction is an extraction technique that uses a solvent in its supercritical state.[106] Supercritical fluids have similar densities to liquids but lower viscosities, so analytes show higher diffusion coefficients. This combination of properties results in a fluid that is more penetrating, has a higher solvating power, and may extract solutes faster and more efficiently than liquids.[107] In addition, the density (and therefore the solvent power of the fluid) may be adjusted by varying both the pressure and the temperature, affording the opportunity of theoretically performing highly selective extractions.[108] Supercritical fluid extraction has been adopted by the EPA as a reference method for extracting PAHs (Method 3561 in 1996) and PCBs (Method 3562 in 2007) from solid environmental matrices.[59] The main supercritical solvent used is carbon dioxide.[108] Carbon dioxide (critical conditions, 30.9°C and 73.8 bar) is cheap, environmentally friendly, and generally recognized as safe by the Food and Drug Administration and European Food Safety Authority (EFSA). Supercritical CO_2 is also attractive because of its high diffusivity combined with its easily tunable solvent strength. However, even under very different conditions, the solvent properties of CO_2 cannot be changed to a sufficiently large degree so as to accommodate the efficient extraction of more polar analytes. For this reason, the addition of polar solvents (e.g., MeOH) as modifier is frequently required.

Matrix Solid-Phase Dispersion

This technique was introduced in 1989 and patented in 1993 by Barker.[109]

Matrix solid-phase dispersion (MSPD) has unique features as a sample preparation technique (e.g., for drugs, organic microcontaminants, and naturally occurring compounds in a variety of liquids but, specifically, semisolid and solid food, fish, plant and environmental samples).[110] The use of mild extraction conditions (normal temperature and pressure) with a suitable combination of dispersant sorbent and elution solvent normally provides acceptable recoveries and medium selectivity. Additional advantages of MSPD are as follows:

- Low cost per extraction
- No need for expensive instrumentation
- Moderate consumption of organic solvents

The operating principles of MSPD are extremely simple.[109] In a first step, samples are ground with a solid sorbent in a mortar using a pestle in order to do the following:

1. Disrupt the structure of the raw material
2. Achieve its homogeneous distribution around the sorbent particles

The blend is then transferred to a column or cartridge, normally made of polypropylene, and analytes are subsequently eluted with an appropriate organic solvent. Partition and/or adsorption equilibria, similar to those occurring in a chromatographic column, are responsible for analyte distribution between the dispersed sample and the elution solvent. The homogeneous and thin layer of sample around the dispersant particles leads to highly efficient mass transference processes, which explains the high recoveries normally attained with a low volume of solvent.[111]

Clean-Up Techniques

Cleaning of the extract aims to remove undesirable constituents, which interfere in identification and quantitative determination of the analytes under investigation. This step can be omitted in the case of the analysis of non-polluted surface waters.[112] The samples of bottom sediments and suspended matter, due to their complex physicochemical structures, require the extracts' clean-up step, during which the following undesirable constituents should be removed:

– Macromolecular compounds (e.g., lipids, waxes) with molecular masses ranging from 600 to 1500 g/mol. They usually contain polar groups that can form hydrogen bonds, and are characterized by high molecular masses and low volatility
– Elemental sulfur's (S8, whose molecular weight is 256 Da) retention characteristics are similar to the retention times of PAHs (benzo(b)fluoranthene, beno(k)fluoranthene, and benzo(a)pyrene (for which masses of the monitored ions are 252, 250) and of the PCBs (mainly PCB 28, whose monitored ion masses are 256, 258)
– Compounds whose molecules are similar in size to those of the analytes

The most commonly used clean-up technique for removing macromolecular compounds is liquid adsorption chromatography. This classic technique is used in an "offline" mode and involves passing extracts through several adsorbent columns prepared in the laboratory or through SPE cartridges.[91] The most commonly used sorbents are Florisil, alumina, and silica gel (highly polar sorbents that inhibit retention of macromolecular compounds). Extracts are introduced into the column in the nonpolar solvents, such as dichloromethane, hexane, pentane, and toluene. In the case of fatty matrix, the classical method used to remove coextracted lipids is the dehydration and oxidation reactions with concentrated sulfuric acid. This method is efficient for the most unreactive chemical groups like PCBs but cannot be used in the case of compounds of lower chemical stability like PAHs, which may be partially destroyed by the acid treatment.[113] In the case of thermally stable compounds, one way of cleaning the extracts from fats is saponification, defined as a fat decomposition process under the influence of hydrolysis. Typically, KOH or NaOH solutions in methanol or ethanol are used.[114] An alternative method of removing fat can be gel permeation chromatography (GPC). The most commonly applied are polystyrene–divinylbenzenecopolymeric columns (e.g., Bio-Beads SX-3), although nowadays, rigid PL gels appear to be more efficient. Gel permeation chromatography is not capable of removing all lipid-related substances (e.g., sterols), and therefore, additional clean-up or repeated GPC (e.g., up to four GPC columns in series) is required.[115] Gel permeation chromatography can also be applied as one of the methods used for removing sulfur, in addition to reaction of sulfur with metals such as Hg, Cu, Cd, and Ag.[116] Removing sulfur by means of Hg is more efficient, but is not recommended due to potential contamination of the lab and lab effluent. Shaking with tetrabutylammonium sulfide has also been used to remove sulfur.[116]

Removing the compounds, which have a molecular weight similar to the molecular weight of the determined pollutants, from the extracts can be done by fractionation process.[117] Extracts are introduced into the column in the nonpolar solvents and then eluted with solvents of increasing polarity. The substances are eluted from the column in order, from the nonpolar to polar ones.[112]

DETERMINATION OF PAHs AND PCBs

A number of analytical techniques have been used for the determination of PAHs and PCBs in complex environmental samples, but there are two common methods, which are recommended by the U.S. EPA. The first method is high-performance liquid chromatography (HPLC) with a combination of UV and fluorescence detection (for determination of PAHs). The second one is GC with mass spectrometer (MS) and flame ionization detector (FID) for PAHs or electron capture detector (ECD) for determination of PCBs (Table 4).

Determination of PAHs

Many modern analytical techniques have been developed and subsequently applied for the monitoring of PAHs in the natural environment, such as synchronous fluorescence spectroscopy, GC coupled with time-of-flight (TOF) MS, capillary electrophoresis, immunoassay, HPLC with UV

Table 4 Selected ISO and U.S. EPA test methods for determination of PAHs and PCBs in environmental samples.

Method	Title	Determination
ISO 7981-2:2005	Water quality—Determination of polycyclic aromatic hydrocarbons (PAHs)—Part 2: Determination of six PAH by high-performance liquid chromatography with fluorescence detection after liquid–liquid extraction	HPLC–FL
ISO 17993:2002	Water quality—Determination of 15 PAHs in water by HPLC with fluorescence detection after liquid–liquid extraction	HPLC–FL
U.S. EPA 8100	Polynuclear aromatic hydrocarbons	GC–FID
U.S. EPA 8310	Polynuclear aromatic hydrocarbons	HPLC (reversed phase)- UV/fluorescence
U.S. EPA 8410	Gas chromatography/Fourier transform infrared (GC/FT-IR) spectrometry for semivolatile organic: capillary column	GC–FT-IR
U.S. EPA 8260B	Volatile organic compounds by gas chromatography/mass spectrometry (GC/MS)	GC–MS
U.S. EPA 8270D	Semivolatile organic compounds by gas chromatography/mass spectrometry (GC/MS)	GC–MS
U.S. EPA 8082	Polychlorinated biphenyls (PCBs) by gas chromatography	GC–ECD/ELCD
U.S. EPA 1668B	Chlorinated biphenyl congeners in water, soil, sediment, biosolids, and tissue by HRGC/HRMS	HRGC/HRMS

HPLC, high-performance liquid chromatography; GC, gas chromatography; HR, high resolution; FL, fluorescence detector; FID, flame ionization detector; ECD, electron capture detector; ELCD, electrolytic conductivity detector; MS, mass spectrometer; FT-IR, Fourier transform infrared spectroscopy; HRGC, high-resolution gas chromatography; HRMS, high-resolution mass spectrometry; ISO, International Organization for Standardization.
Source: Adapted from the United States Environmental Protection Agency Website[59] and Santos and Galceran.[118]

and/or fluorescence detection, and supercritical fluid chromatography.[119] The most widely used are liquid and/or GC.[53] Gas chromatography is most useful for smaller PAH molecules because they are generally volatile, whereas HPLC is suitable for HMW or nonvolatile PAHs.[8]

Gas Chromatography

Capillary GC is one of the most widely used and successful chromatographic techniques for the determination of the concentrations of PAHs in environmental matrices, owing mainly to its high resolving power.[120] The most frequently applied stationary phases are nonpolar phases, such as methyl polysiloxane or phenyl methyl polysiloxane, which give relatively low background from the column bleed, even at high temperatures (>300°C) and with nonselective detectors such as the flame ionization. Methylpolysiloxane phases containing 0%, 5%, 50%, and even 65% phenyl substitution are commercially available.[121] The most widely applied detector for PAH determination is the FID,[122–124] using the change in the conductivity at the time of appearance of the organic compound in the flame. During the combustion process, carboions are formed. An FID is normally adequate for sensitive detection, but coupling GC with MS affords greater selectivity.[120] Capillary GC coupled with MS (GC–MS) revolutionized environmental organic analysis in the 1980s, particularly with the advent of bench-top instruments. Current GC–low-resolution (quadrupole) MS (LRMS) instrumentation is capable of determining PAH at low concentrations (picogram levels) using electron ionization (EI) in selected ion mode.[117] Several novel MS instruments (e.g., triple quadrupoles, quadrupole TOF, and Fourier transform ion cyclotron resonance instruments) have been applied in the studies of environmental samples. However, the most commonly used MS instruments are the quadrupole, ion trap, and TOF analyzer. Linear quadrupoles are the mass analyzers most widely used for the analysis of PAHs by GC–MS, mainly because they make it possible to obtain high sensitivity, good qualitative information, and adequate quantitative results with relatively low maintenance.[125] A quadrupole analyzer is one of the instruments, in which the stability of ion trajectories is used for separation of the analytes on the basis of their mass-to-charge (m/z) values. Another type of mass analyzer applied in research and routine laboratories worldwide is ion trap. The original commercial usage of a quadrupole ion trap was limited to the formation of ions in situ by electron impact ionization, focusing of the nascent ions to a cloud at the center of the ion trap, and mass-selective axial ejection of ions to form a mass spectrum.[126] With the advent of new methods by which ions can be formed in the gas phase from polar as well as covalent molecules and introduced subsequently into an ion trap, the range of applications of the quadrupole ion trap is now considerable. The coupling of liquid chromatography (LC) with electrospray ionization (ESI) and with MS in the early 1980s, together with the rapid advancement in ion trap technology, has led to the development of new ion trap instruments for the analysis of nonvolatile, polar, and thermally labile compounds.[127] Coupled with GC, quadrupole ion trap MS offers good sensitivity, the ability to manipulate ions during storage, relatively high mass range, low cost, and reduced size.[125] The following type of mass analyzer is TOF. During TOF-MS acquisitions, ions are pulsed down a field-free flight tube in packets that leave the source at the same time and initially have the same kinetic energy. As they travel down the field-free flight tube, the smaller ions will travel faster than the larger ions.[128] Time-of-flight MSs have advantages over quadrupole MSs because of their capability of producing mass spectra of good quality within a very short time and high sensitivity (higher efficiency than scanning MS). High speed has made it possible to use TOF mass analyzers as detectors in high-speed GC.[125]

Fourier transform ion cyclotron resonance MS, ion trap–TOF, TOF–TOF MS, and other highly specialized approaches have very high costs, which make them impractical for routine GC applications.[129]

Liquid Chromatography

Gas chromatography is more frequently used than LC for PAH determination in the environmental samples due to the higher selectivity, resolution, and sensitivity than LC. However, PAHs with more than 24 carbon atoms cannot be analyzed by GC, because of their lack of volatility.[125] High-performance LC with fluorescence or diode-array detection has been used to separate and detect these compounds, due to the selectivity and sensitivity of fluorescence detection.[130] Separation of PAHs can be carried out in both normal and reversed phases. However, a reversed-phase system is the most suitable for separation of mixtures of compounds characterized by different hydrophobicity. In normal phases, a variety of chemically bonded polar phases (nitrilo-, amino-, nitro-) and nonpolar solvents are used. In reversed phases, the most frequently used stationary phases are alkyl and phenyl phases and polar solvents.[131] The retention and selectivity for separation of PAHs on alkyl phases increased with increasing alkyl chain length of the stationary phase. Chemically bonded C18 phases have been shown to provide excellent selectivity for the separation of PAHs. Furthermore, it was also found that the phase type (monomeric or polymeric) influenced the separation selectivity.[132] Polymeric octadecyl phases exhibit the best selectivity and are recommended for routine analysis of these compounds.[132] The results of studies conducted by Kayillo et al.[133] indicate that the phenyl-type columns offer better separation performance for PAH homologues. For this reason, they should be considered as a possible alternative to C18 columns in the case of analytical chemists working in the field of PAH separatory science and seeking solutions for acceptable selectivity changes.[133] Mixtures of methanol/water and acetonitrile/water are often used as mobile phases. Elution can be carried out under isocratic or gradient modes. (Isocratic elution refers to a mobile phase of constant eluting strength, while gradient elution involves a change in the eluting strength during chromatographic separation). Frequently applied gradient (1–1.5 mL/min for a 4.6 mm column) starts at 50% methanol/water or acetonitrile/water and runs to 100% methanol or acetonitrile in 40 min, where it remains for 20 min and then returns to the initial conditions again for about 5 min.[134]

The most commonly used detectors for HPLC are UV absorbance and fluorescence detectors. Between the above-mentioned spectroscopic techniques, the most sensitive one is fluorescence spectroscopy because of the direct measurement of the emitted light intensity with little background or interference. Moreover, it enables us to determine constituents at pg levels.[134,135] The appropriate choice of detector depends upon the actual method being used and the detection limits required for the samples being studied. In general, methods for the analysis of samples with few matrix components, other than water, may be analyzed by UV absorbance. More complex samples such as sediments may require fluorescence detection for added selectivity. On the other hand, fluorescence detection is very sensitive for several PAHs; it is not universally applicable. Some environmentally important PAHs, such as acenaphthylene, do not fluoresce and can be determined only by UV detection with limited sensitivity.[136] By means of the simplest monochromatic UV detectors, it is possible to detect chemicals at the wavelength of 254 nm; however, the sensitivity of detection at this wavelength is low not only for PAHs but also for other aromatic compounds present in the sample.[135] Therefore, the excitation and emission wavelengths should be programmable to allow the detection of PAHs at their optimum wavelength. A diode-array detector (DAD) can be here applied, due to the fact that it allows for the best wavelength to be selected for actual analysis. In this way, the wavelength reproducibility of a DAD instrument is much better than the conventional UV detector. In this way, the wavelength reproducibility of a DAD instrument is much better than the conventional UV detector.[137] Moreover, the advantages offered by the comprehensive spectral data recorded are great. Peak purity can be ascertained by overlaying spectra taken from different regions of the same peak and noting any changes, which could be attributed to a coeluted impurity.[53]

Although HPLC is effective for the separation of polar and thermally stable compounds, however, it is not designed to identify unknown compounds, even if a DAD is used.[138] Nuclear magnetic resonance (NMR) technique and MS techniques play a special role in the identification of unknown PAHs and their metabolites. The HPLC–NMR method provides unique information about the chemical structures of the compounds. Furthermore, structurally known components can be quantified without previous calibration runs because of the precisely known response factors relative to a freely selectable internal standard.[139] Nevertheless, due to its low sensitivity for PAH determination and the high cost, it is not used in routine environmental monitoring.[138] Greater sensitivity is achieved by HPLC coupled with MS (HPLC–MS). Mass spectrometry enables the determination of molar mass and elemental composition of an unknown substance, which is impossible using NMR spectroscopy. On the other hand, the ^{1}H-NMR spectrum gives more information about the structure of an unknown compound than the mass spectrum. Compared with UV-visible or diode-array detection and fluorescence detection, MS has not been as extensively used because these compounds are difficult to ionize.[140] The predominantly used interfaces are ESI and atmospheric-pressure chemical ionization (APCI).[141] Because of the low polarity of PAHs and PCBs, determination of PAH by ESI/MS is difficult.[142] The addition of the tropylium cation to form the PAH–tropylium (π-donor–π-acceptor) complexes is proposed in order to improve the ionization.[140] In the measurement of PAHs, APCI/MS in positive-ion mode generates protonated molecular ions ([M+H]+) through the proton transfer from water clusters, or M+, formed by the charge transfer from $N_2^{+\bullet}$ and $O_2^{+\bullet}$ species. Atmospheric-pressure chemical ionization is particularly useful for the detection of PAHs whose molecular weight is over 300 Da.[143] Because PAHs are poorly ionized by conventional ESI or APCI,[144] a relatively new ionization technique is atmospheric-pressure photoionization (APPI), which can be applied to both polar and nonpolar substances. The ionization process in APPI is initiated by 10 eV photons from a krypton lamp for compounds such as PAHs that have lower ionization energies than the energy of the ionizing photons.[144] Signals in APPI–MS can be increased significantly by adding so-called dopants, such as toluene or acetone.[145] Determination of PAHs by LC/dopant assisted (DA)-APPI/MS is commonly performed in the normal-phase HPLC mode, because detection limits are 10 to 70 times lower under normal-phase conditions (isooctane, isocratic mode) than in reversed-phase HPLC (acetonitrile/water, gradient mode).[142]

Determination of PCBs

A gas chromatograph is used to separate individual PCB congeners or combinations of congeners based on physical properties, such as volatility and polarity. Over the past 20 years, open tubular capillary GC columns have replaced older packed GC columns for routine laboratory work. In general, the separation of most of these compounds is performed using long capillary GC columns (50 m) with nonpolar (phenyl methyl polysiloxane) and semipolar (cyanopropyl methyl polysiloxane) stationary phases, temperatures from 100°C to 280–320°C, and splitless or on-column injection. The ECD and MS are the two techniques commonly used for the detection of halogenated compounds in environmental samples.[120] The ECD is one of the most popular selective detectors widely used in the analytics of trace amounts of organic compounds, particularly chloro-organic substances. These compounds have the ability to capture low-energy electrons and form negatively charged ions. This phenomenon causes a change in the cell current of the detector producing a signal. Although ECD generally offers high sensitivity at low cost, interferences are frequently observed and, as a result, the use and acceptance of MS is increasing continually.[120] The second detector used in PCB determination is the MS. Detection limits obtained by means of this detector for individual congeners can be achieved in the lower parts per billion to parts per trillion. Although GC–MS, a linear quadrupole, is used for mass analysis, higher mass resolution is sometimes needed in order to avoid some interferences detected in the environmental analysis. Due to its high specificity and sensitivity, GC coupled with high-resolution MS (GC–HRMS) has been employed for many years and is still being used to solve some specific environmental problems.[125] High-resolution MS in EI mode at a resolution of 6,000–10,000 for PCBs is used, mainly because of its high selectivity and its sensitivity.[120] In recent years, tandem ion-trap MS in MS/MS has been proposed as an alternative technique to HRMS for the analysis of these compounds.[120] The ion trap is particularly suitable for PCB analysis. On the one hand, the high capabilities of the ion trap in performing gas-phase ion–molecule reactions has been successfully employed to characterize different PCB congeners on the basis of their reactivity, and on the other, collisional activation of the most abundant isotopic species of molecular ions has been utilized for structural identification of the different PCB congeners.[127]

CONCLUSION

Due to the need for determination of a wide spectrum of pollutants in environmental and biological samples, numerous techniques and analytical methods were developed. New analytical tools are consistently being applied for sample preparation and final analysis. Powerful, relatively fast extraction techniques are available for diverse sample matrices and analytes typical of the marine environment.

Nowadays, there are numerous challenges connected with the application of the principles of green chemistry in the analytical practice. Activities focused on elimination or reduction of the amounts of toxic solvents used during the analytical procedure, substitution of environmentally unfriendly chemicals with the safer ones, reduction of the costs of the procedure, and reduction of the exposure to volatile organic compounds in the analytical laboratories are the most important. Moreover, increasing the speed of analysis is essential when large numbers of samples need to be analyzed or when profiling of the samples requires several techniques and methods for the analysis of different groups of compounds. For specific applications, at least, more sophisticated MS detectors will find increasing use with GC and LC. However, we stress that these methods are more suitable for target-compound monitoring, while for comprehensive profiling of the samples, comprehensive 2D techniques (i.e., GC × GC and LC × LC) are a better choice. We therefore expect that GC × GC methods will be utilized more widely in the future for profiling nonpolar species in marine environmental samples.

ACKNOWLEDGMENTS

The authors would like to thank the Polish Ministry of Science and Higher Education (project no. 4221/B/T02/2009/37) for financing this investigation. This work was cofinanced by the Foundation for Polish Science (programme "Master") and by the project named "The development of interdisciplinary doctoral studies at the Gdansk University of Technology in modern technologies" (project POKL.04.01.01-00-368/09).

REFERENCES

1. Engelman, R.; Pauly, D.; Zeller, D.; Prinn, R.G.; Pinnegar, J.K.; Polunin, N.V.C. Climate, people, fisheries and aquatic ecosystems. In *Aquatic Ecosystems. Trends and Global Prospects*; Polunin, N.V.C, Ed.; Cambridge University Press: New York, 2008; 1–15.
2. Torres, M.A.; Barros, M.P.; Campos, S.C.G.; Pinto, E.; Rajamani, S.; Sayre, R.Y.; Colepicolo, P. Biochemical biomarkers in algae and marine pollution: A review. Ecotox. Environ. Safe. **2008**, *71*, 1–15.
3. Uluturhan, E. Heavy metal concentrations in surface sediments from two regions (Saros and Gökova Gulfs) of the Eastern Aegean Sea. Environ. Monit. Assess. **2010**, *165*, 675–684.
4. Stockholm Convention on Persistent Organic Pollutants, available at http://chm.pops.int/Convention/tabid/54/language/en-US/Default.aspx#convtext (accessed November 2010).
5. Maskaouia, K.; Zhoub, J.L.; Honga, H.S.; Hanga, Z.L. Contamination by polycyclic aromatic hydrocarbons in the Jiulong River Estuary and Western Xiamen Sea, China. Environ. Pollut. **2002**, *118*, 109–122.
6. Peng, R.H.; Xiong, A.S.; Xue, Y.; Fu, X.Y.; Gao, F.; Zhao, W.; Tian Y.S.; Yao Q.H. Microbial biodegradation of polyaromatic hydrocarbons. FEMS Microbiol. Rev. **2008**, *32*, 927–955.
7. Fisher M.; Schechter, I. Polynuclear aromatic hydrocarbons analysis in environmental samples. In *Encyclopedia of Analytical Chemistry*; Meyers, R.A., Ed.; John Wiley & Sons Ltd: Chichester, 2000; 3143–3172.
8. Ravindra, K.; Sokhia, R.; Van Grieken, R. Atmospheric polycyclic aromatic hydrocarbons: Source attribution, emission factors and regulation. Atmos. Environ. **2008**, *42*, 2895–2921.

9. Nikolaou, A.; Kostopoulou, M.; Lofrano, G.; Meric, S.; Petsas, A.; Vagi, M. Levels and toxicity of polycyclic aromatic hydrocarbons in marine sediments. Trends Anal. Chem. **2009**, *28* (6), 653–664.
10. Srogi, K. Monitoring of environmental exposure to polycyclic aromatic hydrocarbons: A review. Environ. Chem. Lett. **2007**, *5*, 169–195.
11. Bihari, N.; Fafandel, M.; Hamer, B.; Kralj-Bilen, B. PAH content, toxicity and genotoxicity of coastal marine sediments from the Rovinj area, Northern Adriatic, Croatia. Sci. Total. Environ. **2006**, *366*, 602–611.
12. Toxicological profile for polycyclic aromatic hydrocarbons, U.S. Department of Health and Human Service, Agency for Toxic Substances and Disease Registry, 1995, available at http://www.atsdr.cdc.gov/ToxProfiles/tp69.pdf (accessed October 2010)
13. Kaleta, J. Niebezpieczne zanieczyszczenia organiczne w środowisku wodnym. In *Zeszyty Naukowe Politechniki Rzeszowskiej*; Politechnika Rzeszowska: Rzeszów, 2007; 9, 31–40.
14. Witt, G. Polycyclic aromatic hydrocarbons in water and sediment of the Baltic Sea. Mar. Pollut. Bull. **1995**, *4*, 237–248.
15. Cerniglia, C.E. Biodegradation of polycyclic aromatic hydrocarbons. Biodegradation **1992**, *3*, 351–368.
16. Petrova, V.I.; Batova, G.I.; Kursheva, A.V.; Litvinenko, I.V.; Savinov, V.M.; Savinova, T.N. Geochemistry of polycyclic aromatic hydrocarbons in the bottom sediments of the Eastern Arctic Shelf. Oceanology **2008**, *48* (2), 196–203.
17. Froehner, S.; Maceno, M.; Da Luz, E.C.; Botelho Souza, D.; Scurupa Machado, K. Distribution of polycyclic aromatic hydrocarbons in marine sediments and their potential toxic effects. Environ. Monit. Assess. **2010**, *168*, 205–213.
18. Vinas, L.; Franco, M.A.; Soriano, J.A.; Gonzalez, J.J.; Pon, J.; Albaiges, J. Sources and distribution of polycyclic aromatic hydrocarbons in sediments from the Spanish northern continental shelf. Assessment of spatial and temporal trends. Environ. Pollut. **2010**, *158*, 1551–1560.
19. Mostafa, A.R.; Wade, T.L.; Sweet, S.T.; Al-Alimi, A.K.A.; Barakat, A.O. Distribution and characteristics of polycyclic aromatic hydrocarbons (PAHs) in sediments of Hadhramout coastal area, Gulf of Aden, Yemen. J. Mar. Syst. **2009**, *78*, 1–8.
20. Arias, A.H.; Vazquez-Botello, A.; Tombesi, N.; Ponce-Vélez, G.; Freije, H.; Marcovecchio, J. Presence, distribution, and origins of polycyclic aromatic hydrocarbons (PAHs) in sediments from Bahía Blanca estuary, Argentina. Environ. Monit. Assess. **2010**, *160*, 301–314.
21. Vane, C.H.; Harrison, I.; Kim, A.W. Polycyclic aromatic hydrocarbons (PAHs) and polychlorinated biphenyls (PCBs) in sediments from the Mersey Estuary, U.K. Sci. Total Environ. **2007**, *374*, 112–126.
22. Wang, X.C.H.; Sun, S.; Ma, H.Q.; Liu, Y. Sources and distribution of aliphatic and polyaromatic hydrocarbons in sediments of Jiaozhou Bay, Qingdao, China. Mar. Pollut. Bull. **2006**, *52*, 129–138.
23. Olajire, A.A.; Altenburger, R.; Kqster, E.; Brack, W. Chemical and ecotoxicological assessment of polycyclic aromatic hydrocarbon—contaminated sediments of the Niger Delta, Southern Nigeria. Sci. Total Environ. **2005**, *340*, 123–136.
24. El Nem, A.; Said, T.O.; Khaled, A.; El-Sikaily, A.; Abd-Allah, A.M.A. The distribution and sources of polycyclic aromatic hydrocarbons in surface sediments along the Egyptian Mediterranean coast. Environ. Monit. Assess. **2007**, *124*, 343–359.
25. Readman, J.W.; Fillmann, G.; Tolosa, I.; Bartocci, J.; Villeneuve, J.P.; Catinni, C.; Mee, L.D. Petroleum and PAH contamination of the Black Sea. Mar. Pollut. Bull. **2002**, *44*, 48–62.
26. Wade, T.L.; Soliman, Y.; Sweet, S.T.; Wolff, G.A.; Presley, B.J. Trace elements and polycyclic aromatic hydrocarbons (PAHs) concentrations in deep Gulf of Mexico sediments. Deep-Sea Res. II **2008**, *55*, 2585–2593.
27. Xu, J.; Yu, Y.; Wang, P.; Guo, W.; Dai, S.; Sun H. Polycyclic aromatic hydrocarbons in the surface sediments from Yellow River, China. Chemosphere **2007**, *67*, 1408–1414.
28. Baumarda, P.; Budzinski, H.; Garrigues, P.; Dizer, H.; Hansenb, P.D. Polycyclic aromatic hydrocarbons in recent sediments and mussels (*Mytilus edulis*) from the Western Baltic Sea: Occurrence, bioavailability and seasonal variations. Mar. Environ. Res. **1999**, *47*, 17–47.
29. Plata, D.L.; Sharpless, C.M.; Reddy, C.M. Photochemical degradation of polycyclic aromatic hydrocarbons in oil films. Environ. Sci. Technol. **2008**, *42*, 2432–2438.
30. Lee, R.F. Photo-oxidation and photo-toxicity of crude and refined oils. Spill Sci. Technol. Bull. **2003**, *8* (2), 157–162.
31. Harayama, S.; Kishira, H.; Kasai, Y.; Shutsubo, K. Petroleum biodegradation in marine environments. J. Molec. Microbiol. Biotechnol. **1999**, *1* (1), 63–70.
32. Kanaly, R.A.; Harayama, S. Advances in the field of high-molecular-weight polycyclic aromatic hydrocarbon biodegradation by bacteria. Microb. Biotechnol. **2010**, *3* (2), 136–164.
33. Rogowska, J.; Namieśnik, J. Environmental implications of oil spills from shipping accidents. Rev. Environ. Contam. Toxicol. **2010**, *206*, 95–114.
34. Pashin, Y.V.; Bakhitowa, L.M. Mutagenic and carcinogenic properties of polycyclic aromatic hydrocarbons. Environ. Health Perspect. **1979**, *30*, 185–189.
35. Rust, A.J.; Burgess, R.M.; Brownawell, B.J.; McElroy, A.E. Relationship between metabolism and bioaccumulation of benzo[*a*]pyrene in benthic invertebrates. Environ. Toxicol. Chem. **2004**, *23* (11), 2587–2593.
36. Agency for Toxic Substances and Disease Registry (ATSDR), Case Studies in Environmental Medicine, Toxicity of Polycyclic Aromatic Hydrocarbons (PAHs). Available at http://www.atsdr.cdc.gov/csem/pah/pah_physiologic-effects.html#cancer (accessed October 2010).
37. Dercova, K.; Seligova, J.; Dudasova, H.; Mikulasova, M.; Silharova, K.; Tothova, L.; Hucko, P. Characterization of the bottom sediments contaminated with polychlorinated biphenyls: Evaluation of ecotoxicity and biodegradability. Int. Biodeter. Biodegr. **2009**, *63*, 440–449.
38. *PCBs: Cancer Dose-Response Assessment and Application to Environmental Mixtures*, EPA/600/P–96/001F; 1996.
39. Szlinder-Richert, J.; Barska, I.; Mazerski, J.; Usydus, Z. PCBs in fish from the southern Baltic Sea: Levels, bioaccumulation features, and temporal trends during the period from 1997 to 2006. Mar. Pollut. Bull. **2009**, *58*, 85–92.

40. Li, Y.F.; Harner, T.; Liu, L.; Zhang, Z.; Ren, N.Q.; Jia, H.; Ma, J.; Sverko, E. Polychlorinated biphenyls in global air and surface soil: Distributions, air–soil exchange, and fractionation effect. *Environ. Sci. Technol.* **2010**, *44* (8), 2784–2790.
41. Borja, J.; Taleon, D.M.; Auresenia, J.; Gallardo, S. Polychlorinated biphenyls and their biodegradation. Process Biochem. **2005**, *40*, 1999–2013.
42. Piekarska, K. Biodegradacja polichlorowanych bifenyli przez zespół mikroorganizmów wyizolowanych z wody i gleby. Ochrona Środowiska **2003**, *2*, 21–27.
43. Fava, F.; Gentilucci, S.; Zanaroli, G. Anaerobic biodegradation of weathered polychlorinated biphenyls (PCBs) in contaminated sediments of Porto Marghera (Venice Lagoon, Italy). Chemosphere **2003**, *53*, 101–109.
44. Świderka-Bróz, M. Zjawiska sorpcji w wodach naturalnych oraz w procesach oczyszczania wód. Ochrona Środowiska **1987**, *521*, 9–14.
45. Ghosh, U.; Gillette, J.S.; Luthy, R.G.; Zare, R.N. Microscale location, characterization, and association of polycyclic aromatic hydrocarbons on harbor sediment particles. Environ. Sci. Technol. **2000**, *34*, 1729–1736.
46. Mechlińska, A.; Gdaniec-Pietryka, M.; Wolska, L.; Namieśnik, J. Sorption of PAHs and PCBs on geosorbents—Evolution of models. Trends Anal. Chem. **2009**, *28*, 466–482.
47. Beyer, A.; Biziuk, M. Environmental fate and global distribution of polychlorinated biphenyls. Rev. Environ. Contamin. Toxicol. **2009**, *201*, 137–158.
48. Namieśnik, J., Informacje analityczne w analityce specjacyjnej. In *Specjacja chemiczna: Problemy i możliwości*; Barałkiewicz, D., Bulska, E., Eds.; Wydawnictwo MALAMUT: Warszawa, 2009; 21–24.
49. Cornelissen, G.; Breedveld, G.D.; Kalaitzidis, S.; Christanis, K.; Kibsgaard, A., Oen, A.M. Strong sorption of native PAHs to pyrogenic and unburned carbonaceous geosorbents in sediments. Environ. Sci. Technol., **2006**, *40*, 1197–1203.
50. Ehlers, G.A.C.; Loibner, A.P. Linking organic pollutant (bio)availability with geosorbent properties and biomimetic methodology: A review of geosorbent characterisation and (bio)availability prediction. Environ. Pollut. **2006**, *141*, 494–512.
51. Namieśnik, J., Jamrógiewicz Z., Pilarczyk M., Torres L. *Przygotowanie próbek Środowiskowych do analizy*; WNT: Warszawa, Polska, 2000.
52. Lung, S.C.; Altshul, L.M.; Ford, T.E.; Spengler, J.D. Coating effects on the glass adsorption of polychlorinated biphenyl (PCB) congeners. Chemosphere **2000**, *41*, 1865–1871.
53. Manoli, E.; Samara, C. Polycyclic aromatic hydrocarbons in natural waters: Sources, occurrence and analysis. Trends Anal. Chem. **1999**, *18* (6), 417–428.
54. Wolska, L.; Rawa-Adkonis, M.; Namieśnik, J. Determining PAHs and PCBs in aqueous samples: Finding and evaluating sources of terror. Anal. Bioanal. Chem. **2005**, *382*, 1389–1397.
55. Rawa-Adkonis, M.; Wolska, W.; Namieśnik, J. Modern techniques of extraction of organic analytes from environmental matrices. Crit. Rev. Anal. Chem. **2003**, *33*, 199–248.
56. Rawa-Adonis, M. Oznaczanie analitów z grupy WWA i PCB w próbkach wody. Zródła błedów: identyfikacja i oszacowanie ich wielkosci. PhD thesis.
57. Namieśnik, J.; Szefer, P. Preparing samples for analysis— The key to analytical success. Ecol. Chem. Eng. S **2008**, *15* (2), 167–224.
58. Jones, R.P.; Millward, R.N.; Karn, R.A.; Harrison, A.H. Microscale analytical methods for the quantitative detection of PCBs and PAHs in small tissue masses. Chemosphere **2006**, *62* (11), 1795–1805.
59. United States Environmental Protection Agency Web site, available at http://www.epa.gov/ (accessed November 2010).
60. Murray, D.A.J. Rapid micro extraction procedure for analyses of trace amounts of organic compounds in water by gas chromatography and comparisons with macro extraction methods. J. Chromatogr. **1979**, *177*, 135–140.
61. Cantwell, F.F.; Jeannot, M.A. Solvent microextraction into a single drop. Anal. Chem. **1996**, *68*, 2236–2240.
62. He, Y.; Lee H.K. Liquid-phase microextraction in a single drop of organic solvent by using a conventional microsyringe. Anal. Chem. **1997**, *69*, 4634–4640.
63. Foreman, W.T.; Gates, P.M.; Fosterb, G.D.; Rinellac, F.A.; McKenziec, S.W. Use of field-applied quality control samples to monitor performance of a Goulden large-sample extractor/GC–MS method for pesticides in water. Int. J. Environ. Anal. Chem. 2000, *77* (1), 39–62.
64. Yrieiex, C.; Gonzales, C.; Deroux, J.M.; Lacoste, C.; Leybros, J. Countercurrent liquid/liquid extraction for analysis of organic water pollutants by GC/MS. Water Res. 1996, *30*, 1791–1800.
65. Pena, T.; Casais, C.; Mejuto, C.; Cela, R. Development of an ionic liquid based dispersive liquid–liquid microextraction method for the analysis of polycyclic aromatic hydrocarbons in water samples. J. Chromatogr. A **2009**, *1216*, 6356–6364.
66. Rezaee, M.; Assadi, Y.; Hosseini, M.R.; Aghaee, E.; Ahmadia, F.; Berijani, S. Determination of organic compounds in water using dispersive liquid–liquid microextraction. J. Chromatogr. A **2006**, *111*, 1–9.
67. Shi, Z.G.; Lee, H.K. Dispersive liquid–liquid microextraction coupled with dispersive μ-solid-phase extraction for the fast determination of polycyclic aromatic hydrocarbons in environmental water samples. Anal. Chem. **2010**, *82*, 1540–1545.
68. García, M.; Rodríguez, I.; Cela, R. Development of a dispersive liquid–liquid microextraction method for organophosphorus flame retardants and plasticizers determination in water samples. J. Chromatogr. A **2007**, *1166*, 9–15.
69. Farahani, H.; Norouzi, P.; Dinarvand, R.; Reza, M. Development of dispersive liquid–liquid microextraction combined with gas chromatography–mass spectrometry as a simple, rapid and highly sensitive method for the determination of phthalate esters in water samples. J. Chromatogr. A **2007**, *1172*, 105–112.
70. Fan, Y.C.; Chen, M.L.; Shen-Tu, C.; Zhu, Y. An ionic liquid for dispersive liquid–liquid microextraction of phenols. J. Anal. Chem. **2009**, *64*, 1017–1022.
71. Xie, S.; Paau, M.C.; Li, C.F.; Xiao, D.; Choi, M.M.F. Separation and preconcentration of persistent organic pollutants by cloud point extraction J. Chromatogr. A **2010**, *1217*, 2306–2317.

72. Favre-Réguillon, A.; Draye, M.; Lebuzit, G.; Thomas, S.; Foos, J.; Cote, G.; Guy, A. Cloud point extraction: An alternative to traditional liquid–liquid extraction for lanthanides(III) separation. Talanta **2004**, *63*, 803–806.
73. Sirimanne, S.R.; Barr, J.R.; Patterson, D.G., Jr. Quantification of polycyclic aromatic hydrocarbons and polychlorinated dibenzo-p-dioxins in human serum by combined micelle-mediated extraction (cloud-point extraction) and HPLC. Anal. Chem. **1996**, *68*, 1556–1560.
74. Lee, J.Y.; Lee, H.K.; Rasmussen, K.E.; Pedersen-Bjergaard, S. Environmental and bioanalytical applications of hollow fiber membrane liquid-phase microextraction: A review. Anal. Chim. Acta **2008**, *624*, 253–268.
75. Baltussen, E.; Sandra, P.; David, F.; Cramers, C. Stir bar sorptive extraction (SBSE), a novel extraction technique for aqueous samples: Theory and principles. J. Microcolumn Sep. **1999**, *11*, 737–747.
76. Doong, R.A.; Chang, S.M.; Sun, Y.C. Solid-phase microextraction for determining the distribution of sixteen US Environmental Protection Agency polycyclic aromatic hydrocarbons in water samples. J. Chromatogr. A **2000**, *879*, 177–188.
77. Doong, R.; Lin Y. Characterization and distribution of polycyclic aromatic hydrocarbon contaminations in surface sediment and water from Gao-ping River, Taiwan. Water Res. **2004**, *38*, 1733–1744.
78. Gorecki, T.; Khaled, A.; Pawliszyn, J. The effect of sample volume on quantitative analysis by solid phase microextraction. Part 2. Experimental verification. Analyst **1998**, *123*, 2819–2824.
79. Poerschmann, J.; Zhang, Z.; Kopinke, F.D.; Pawliszyn, J. Solid phase microextraction for determining the distribution of chemicals in aqueous matrices. Anal. Chem. **1997**, *69*, 597–600.
80. Popp, P.; Bauer, C.; Hauser, B.; Keil, P.; Wennrich, L. Extraction of polycyclic aromatic hydrocarbons and organochlorine compounds from water: A comparison between solid-phase microextraction and stir bar sorptive extraction. J. Sep. Sci. **2003**, *26*, 961–967.
81. Popp, P.; Keila, P.; Montero, L.; Ruckert, M. Optimized method for the determination of 25 polychlorinated biphenyls in water samples using stir bar sorptive extraction followed by thermodesorption-gas chromatography/mass spectrometry. J. Chromatogr. A **2005**, *1071*, 155–162.
82. Tan, B.C.D.; Marriott, P.J.; Lee, H.K., Morrisom, P.D. In-tube solid phase micro-extraction–gas chromatography of volatile compounds in aqueous solution. Analyst **1999**, *124*, 651–663.
83. Olejniczak, J.; Staniewski, J.; Szymanowski, J. Extraction of selected pollutants in open tubular capillary columns. Anal. Chim. Acta **2003**, *497*, 199–207.
84. Wolska, L.; Rawa-Adkonis, M.; Gdaniec, M.; Namieńnik, J. Critical evaluation of employment possibilities of the OTT technique for isolation of the PAHs and PCBs from samples of water. Ecol. Chem. Eng. **2005**, *12*, 611–626.
85. Huckins, J.N.; Petty, J.D.; Orazio, C.E.; Lebo, J.A.; Clark, R.C.; Gibson, V.L.; Gala, W.R.; Echols, K.R. Determination of uptake kinetics (sampling rates) by lipid containing semipermeable membrane devices (SPMDs) for polycyclic aromatic hydrocarbons (PAHs) in water. Environ. Sci. Technol. **1999**, *33*, 3918–3923.
86. Namieśnik, J.; Zabiegała ,B.; Kot-Wasik, A.; Partyka, M., Wasik, A. Passive sampling and/or extraction techniques in environmental analysis: A review. Anal. Bioanal. Chem. **2005**, *381*, 279–301.
87. Huckins, J.N.; Gamlni, J.; Manuweera, K.; Petty, J.D.; Mackay, D.; Lebo, J.A. Lipid-containing semipermeable membrane devices for monitoring organic contaminants in water. Environ. Sci. Technol. **1993**, *27*, 2489–2496.
88. Verweij, F.; Booij, K.; Satumalay, K.; van der Molen, N.; van der Oost, R. Assessment of bioavailable PAH, PCB and OCP concentrations in water, using semipermeable membrane devices (SPMDs), sediments and caged carp. Chemosphere **2004**, *54*,1675–1689.
89. Rawa-Adkonis, M.; Wolska, L.; Namieśnik, J. Analytical procedures for PAH and PCB determination in water samples—Error sources. Crit. Rev. Anal. Chem. **2006**, *32*, 63–72.
90. Quevauville, P.; Borchers, U.; Gawlik, B.M. Coordinating links among research, standardization and policy in support of water framework directive chemical monitoring requirements. J. Environ. Monit. **2007**, *9*, 915–923.
91. Fidalgo-Used, N.; Blanco-Gonzalez, E.; Sanz-Medel, A. Sample handling strategies for the determination of persistent trace organic contaminants from biota samples. Anal. Chim. Acta **2007**, *590*, 1–16.
92. Wolska, L. Miniaturised analytical procedure of determining polycyclic aromatic hydrocarbons and polychlorinated biphenyls in bottom sediments. J. Chromatogr. A **2002**, *959*, 173–180.
93. Hartonen, K.; Bøwadt, S.; Hawthorne, S.B.; Riekkola, M.L. Supercritical fluid extraction with solid-phase trapping of chlorinated and brominated pollutants from sediment samples. J. Chromatogr. A **1997**, *774*, 229–242.
94. Schantz, M.M.; Bøwadt, S.; Benner, B.A., Jr.; Wise, S.A.; Hawthorne, S.B. Comparison of supercritical fluid extraction and Soxhlet extraction for the determination of polychlorinated biphenyls in environmental matrix standard reference materials. J. Chromatogr. A **1998**, *816*, 213–220.
95. Camel, V. Microwave-assisted solvent extraction of environmental samples Trends Anal. Chem. **2000**, *19* (4), 229–248.
96. Shu, Y.Y.; Lao, R.C.; Chiu, C.H.; Turle, R. Analysis of polycyclic aromatic hydrocarbons in sediment reference materials by microwave-assisted extraction. Chemosphere **2000**, *41*, 1709–1716.
97. Heemken, O.P.; Theobald, N.; Wenclawiak, B.W. Comparison of ASE and SFE with Soxhlet, sonication, and methanolic saponification extractions for the determination of organic micropollutants in marine particulate matter. Anal. Chem. **1997**, *69*, 2171–2180.
98. Bowadt, S.; Mazes, L.; Miller, D.J.; Hawthorne, S.B. Field-portable determination of polychlorinated biphenyls and polynuclear aromatic hydrocarbons in soil using supercritical fluid extraction. J. Chromatogr. A **1997**, *785*, 205–217.
99. Itoh, N.; Numata, M.; Aoyagi, Y.; Yarita T. Comparison of the behavior of ^{13}C and deuterium labeled polycyclic aromatic hydrocarbons in analyses by isotope dilution mass spectrometry in combination with pressurized liquid extraction. J. Chromatogr. A **2007**, *1138*, 26–31.
100. Gfrerer, M.; Fernandes, C.; Lankmayr, E. Optimization of fluidized-bed extraction for determination of organochlorine pesticides in sediment. Chromatographia **2004**, *60* (11/12), 681–686.

101. Gfrerer, M.; Serschen, M.; Lankmayr, E. Optimized extraction of polycyclic aromatic hydrocarbons from contaminated soil samples. J. Biochem. Biophys. Methods **2002**, *53* (1–3), 203–216.
102. Pu, W.; Qinghua, W.; Yawei, W.; Thanh, W.; Xiaomin, L.; Lei, D.; Guibin, J. Evaluation of Soxhlet extraction, accelerated solvent extraction and microwave assisted solvent extraction for the determination of polychlorinated biphenyls and polybrominated diphenyl esters in soil and fish samples. Anal. Chim. Acta **2010**, *663*, 43–48.
103. Gfrerer, M.; Stadlober, M.; Gawlik, B.M.; Wenzl, T.; Lankmayr, E. Enhanced extraction of polychlorinated organic compounds from soil samples by fluidized-bed extraction (FBE). Chromatographia **2001**, *53*, 442.
104. Rey-Salgueiro, L.; Pontevedra-Pombal, X.; Álvarez-Casas, M.; Martínez-Carballo, E.; García-Falcón, M.S.; Simal-Gándara, J. Comparative performance of extraction strategies for polycyclic aromatic hydrocarbons in peats. J. Chromatogr. A **2009**, *1216* (27), 5235–5241.
105. Gfrerer, M.; Gawlik, B.M.; Lankmayr, E. Validation of a fluidized-bed extraction method for solid materials for the determination of PAHs and PCBs using certified reference materials. Anal. Chim. Acta **2004**, *527*, 53–60.
106. Hawthorne, S.B.; Yang, Y.; Miller, D.J. Extraction of organic pollutants from environmental solids with sub- and supercritical water. Anal. Chem. **1994**, *66*, 2912–2920.
107. Herrero, M.; Mendiola, J.A.; Cifuentes, A.; Ibãnez, E. Supercritical fluid extraction: Recent advances and applications. J. Chromatogr. A **2010**, *1217*, 2495–2511.
108. Yarita, T. Development of environmental analysis method using supercritical fluid extraction and supercritical chromatography. Chromatography **2008**, *29*, 19–23.
109. Barker, S.A. Matrix solid phase dispersion (MSPD). J. Biochem. Biophys. Methods **2007**, *70*, 151–162.
110. Kristenson, E.M.; Brinkman, U.A.T.; Ramos, L. Recent advances in matrix solid-phase dispersion. Trends Anal. Chem. **2006**, *25*, 96–111.
111. García-López, M.; Canosa, P.; Rodríguez, I. Trends and recent applications of matrix solid-phase dispersion. Anal. Bioanal. Chem. **2008**, *391*, 963–974.
112. Gazda, K. Oznaczanie wielopierścieniowych węglowodorów aromatycznych. In *Fizykochemiczne metody kontroli zanieczyszczeńśrodowiska*; Namieśnik, J., Jamrógiewicz, Z., Eds.; Wydawnictwo Naukowo Techniczne: Warszawa, 1998; 251–263.
113. Jaouen-Madoulet, A.; Abarnou, A.; Le Guellec, A.M.; Loizeau, V.; Leboulenger F. Validation of an analytical procedure for polychlorinated biphenyls, coplanar polychlorinated biphenyls and polycyclic aromatic hydrocarbons in environmental samples. J. Chromatogr. A **2000**, *886*, 153–173.
114. Navarro, P.; Cortazar, E.; Bartolome, L.; Deusto, M.; Raposo, J.C.; Zuloaga, O.; Arana, G.; Etxebarria, N. Comparison of solid phase extraction, saponification and gel permeation chromatography for the clean-up of microwave-assisted biological extracts in the analysis of polycyclic aromatic hydrocarbons. J. Chromatogr. A **2006**, *1128*, 10–16.
115. Van Leeuwen, S.P.J.; de Boer, J. Advances in the gas chromatographic determination of persistent organic pollutants in the aquatic environment. J. Chromatogr. A **2008**, *1186*, 161–182.
116. Smedes, F.; de Boer, F. Determination of chlorobiphenyls in sediments—Analytical methods. Trends Anal. Chem. **1997**, *16* (9), 503–517.
117. Muir, D.; Sperko, E. Analytical methods for PCBs and organochlorine pesticides in environmental monitoring and surveillance: A critical appraisal. Anal. Bioanal. Chem. **2006**, *386*, 769–789.
118. Wolska, L. Determination (monitoring) of PAHs in surface waters: why an operationally defined procedure is needed. Anal. Bioanal. Chem. **2008**, *391*, 2647–2652.
119. Wang, W.D.; Huang, Y.M., Shu, W.Q.; Cao, J. Multi-walled carbon nanotubes as adsorbents of solid-phase extraction for determination of polycyclic aromatic hydrocarbons in environmental waters coupled with high-performance liquid chromatography. J. Chromatogr. A **2007**, *1173*, 27–36.
120. Santos, F.J.; Galceran, M.T. The application of gas chromatography to environmental analysis. Trends Anal. Chem. **2002**, *21* (9/10), 672–685.
121. Poster, D.L.; Schantz, M.M.; Sander, L.C.; Wise, S.A. Analysis of polycyclic aromatic hydrocarbons (PAHs) in environmental samples: A critical review of gas chromatographic (GC) methods. Anal. Bioanal. Chem. **2006**, *386*, 859–881.
122. Zuazagoitia, D.; Millan, E.; Garcia-Arrona, R.A. Screening method for polycyclic aromatic hydrocarbons determination in sediments by headspace SPME with GC–FID. Chemosphere **2009**, *69*, 175–178.
123. Damas, E.Y.C.; Medina, M.O.C.; Núñez Clemente, A.C.; Díaz, M.A.; Bravo, L.G.; Ramada, R.M.; de Oca Porto, R.M. Validation of an analytical methodology for the quantitative analysis of petroleum hydrocarbons in marine sediment samples. Quim. Nova **2009**, *32* (4) 855–860.
124. Smith, K.E.; Schwab, A.P.; Banks, M.K. Dissipation of PAHs in saturated, dredged sediments: A field trial. Chemosphere **2008**, *72*, 1614–1619.
125. Santos, F.J.; Galceran, M.T. Modern developments in gas chromatography–mass spectrometry based environmental analysis. J. Chromatogr. A **2003**, *1000*, 125–151.
126. March, R.E. Quadrupole Ion trap mass spectrometry: Theory, simulation, recent developments and applications. Rapid Commun. Mass Spectrom. **1998**, *12*, 1543–1554.
127. March, R.E. Quadrupole ion trap mass spectrometry: A view at the turn of the century. Int. J. Mass Spectrom. **2000**, *200*, 285–312.
128. Williamson, L.N.; Bartlett, M.G. Quantitative gas chromatography/time-of-flight mass spectrometry: A review. Biomed. Chromatogr. **2007**, *21*, 664–669.
129. Mastovska, K.; Lehotay, S.J. Practical approaches to fast gas chromatography–mass spectrometry. J. Chromatogr. A **2003**, *1000*, 153–180.
130. Barco-Bonilla, N.; Martínez Vidal, J.L.; Garrido Frenich, A.; Romero-González, R. Comparison of ultrasonic and pressurized liquid extraction for the analysis of polycyclic aromatic compounds in soil samples by gas chromatography coupled to tandem mass spectrometry. Talanta **2009**, *78*, 156–164.
131. Jacob, J. Method development for the determination of polycyclic aromatic hydrocarbons (PAHs) in environmental matrix. In *Quality Assurance for Environmental Analysis*; Quevauviller, P., Maier, E.A., Griepink, B., Eds.; Elsevier Science: Amsterdam, 1995; 564–586.

132. Liu, Y.; Lee, M.L. Solid-phase microextraction of PAHs from aqueous samples using fibers coated with HPLC chemically bonded silica stationary phases. Anal. Chem. **1997**, *69*, 5001–5005.

133. Kayillo, S.; Dennis, G.R.; Shalliker, R.A. An assessment of the retention behaviour of polycyclic aromatic hydrocarbons on reversed phase stationary phases: Selectivity and retention on C18 and phenyl-type surfaces. J. Chromatogr. A **2006**, *1126*, 283–297.

134. Khan, Z.; Troquet, J.; Vachelard, C. Sample preparation and analytical techniques for determination of polyaromatic hydrocarbons in soils. Int. J. Environ. Sci. Tech. **2005**, *2* (3), 275–286.

135. Santana Rodriguez, J.J.; Padron Sanz, C. Fluorescence techniques for the determination of polycyclic aromatic hydrocarbons in marine environment: An overview. Analysis **2000**, *28* (8), 710–717.

136. Nirmaier, H.P.; Fischer, E.; Meyer, A.; Henze, G. Determination of polycyclic aromatic hydrocarbons in water samples using high-performance liquid chromatography with amperometric detection. J. Chromatogr. A **1996**, *730*, 169–175.

137. Gimeno, R.A.; Comas, E.; Marce, R.M.; Ferré, J.; Rius, F.X.; Borrull, F. Second-order bilinear calibration for determining polycyclic aromatic compounds in marine sediments by solvent extraction and liquid chromatography with diode-array detection. Anal. Chim. Acta **2003**, *498*, 47–53.

138. Lavsen, K.; Preiss, A; Spraul, M. Structure elucidation of unknown pollutants in environmental samples by simultaneous coupling of HPLC to NMR and MS. In *New Horizons and Challenges in Environmental Analysis and Monitoring*, Plenary lectures major contributions to the workshop, Gdańsk, August 18–29, 2003; Namieśnik, J., Chrzanowski, W., Żmijewska, P., Eds.; CEFAM: Gdańsk, 2003; 150–180.

139. Weißhoff, H.; Preiß, A.; Nehls, I.; Win, T.; Mgle, C. Development of an HPLC–NMR method for the determination of PAHs in soil samples—A comparison with conventional methods. Anal. Bioanal. Chem. **2002**, *373*, 810–819.

140. Gimeno, R.A.; Altelaar, A.F.M.; Marce, R.M.; Borrull, F. Determination of polycyclic aromatic hydrocarbons and polycyclic aromatic sulfur heterocycles by high-performance liquid chromatography with fluorescence and atmospheric pressure chemical ionization mass spectrometry detection in seawater and sediment samples. J. Chromatogr. A **2002**, *958*, 141–148.

141. Van Leeuwen, S.M.; Hayen, H.; Karst, U. Liquid chromatography electrochemistry–mass spectrometry of polycyclic aromatic hydrocarbons. Anal. Bioanal. Chem. **2004**, *378*, 917–925.

142. Moriwaki, H.; Electrospray mass spectrometric determination of 1-nitropyrene and non-substituted polycyclic aromatic hydrocarbons using tropylium cation as a post-column HPLC reagent. Analyst **2000**, *125*, 417–420.

143. Moriwaki, H. Liquid chromatographic–mass spectrometric methods for the analysis of persistent pollutants: Polycyclic aromatic hydrocarbons, organochlorine compounds and perfluorinated compounds. Curr. Org. Chem. **2005**, *9*, 849–857.

144. Itoh, N.; Aoyagi, Y.; Yarita, T. Optimization of the dopant for the trace determination of polycyclic aromatic hydrocarbons by liquid chromatography/dopant-assisted atmospheric-pressure photoionization/mass spectrometry. J. Chromatogr. A **2006**, *1131*, 285–288.

145. Himmelsbach, M.; Buchberger, W.; Reingruber, E. Determination of polymer additives by liquid chromatography coupled with mass spectrometry. A comparison of atmospheric pressure photoionization (APPI), atmospheric pressure chemical ionization (APCI), and electrospray ionization (ESI). Polym. Degrad. Stabil. **2009**, *94*, 1213–1219.

Potassium

Philippe Hinsinger
Sun and Environment Unit, National Institute for Agricultural Research (INRA), Montpellier, France

Abstract
Potassium (atomic weight = 39.098) is an alkaline metal, as revealed by the etymology of its symbol (K) which derives from the Latin word "kalium" and from the Arabic word "qali" (alkali). K is thus strongly electropositive and always occurs as a monovalent cation. Consequently, its physico-chemistry and speciation are rather simple. K is an abundant alkaline metal cation, reaching a concentration of 26 g kg^{-1} in the Earth's crust. It is a major nutrient for all living organisms.

POTASSIUM IN PLANTS

K is the major cation in most plants, occurring at concentrations ranging from 5 to 50 g kg^{-1}, twice as much as Ca and slightly less than N.[1–3] The etymology of its name accounts for the abundance of K in plant-derived ash material (potash). K is involved in a large number of physiological processes: osmoregulation and cation–anion balance, protein synthesis and activation of enzymes.[2,3] K is often referred to as "a cation for anions" as it balances the abundant negative charges of inorganic (nitrate) and organic anions (carboxylates) in plant cells. It therefore occurs at large concentrations, 100–200 mM in the cytosol, 5–10 times less in the vacuole. Being a major inorganic solute, it plays a key role in the water balance of plants: maintenance of the osmotic potential and turgor pressure involved in cell extension. K-controlled changes in turgor pressure in guard cells is a key process of stomatal opening and closure and hence, of the regulation of plant transpiration. Many of these physiological roles are related to the high mobility of K at all levels in the plant. This unique, considerable mobility of K in the plant is essentially due to the large permeability of cell membranes to K-ions, which arises from the occurrence of a range of highly K selective, low and high affinity ion channels and transporters. These are now being increasingly characterized at a molecular level.[4] Large rates of K uptake can thereby be achieved in plant roots. In addition, K-ions can easily be leached out of living plant tissues, as documented for tree foliage which contributes a large flux of K back to the soil via throughfall.[3] K also rapidly leaves dead roots and other plant debris compared with N and P, which require hydrolysis of organic molecules. At an agronomic level, the demand for K largely varies with plant species and productivity. The uptake of K essentially occurs during the vegetative stage and can reach values of 10 kg ha^{-1} day^{-1} and above. Depending also on the agricultural practices (removal of straw, for instance), the amount of K removed with the harvested material will range from 5–50 kg ha^{-1} for cereal grains to 50–500 kg ha^{-1} for forage, root and tuber or plantation crops.

POTASSIUM IN SOILS

Among major nutrients, K is usually the most abundant in soils as total K content ranges from 0.1 to 40, with an average of 14 g kg^{-1}.[1,5] A major proportion of soil K occurs as structural K in feldspars and interlayer K in micaceous minerals (Fig. 1).[1,6,7] Some minor proportion of soil K (usually much less than 1%) is adsorbed on negatively charged soil constituents, namely clay minerals and organic matter. A marginal part is present as free K-ions in the soil solution. Bulk soil solution concentrations usually amount to 100–1000 μM (less than 0.01%–0.1% of total K). The reason for this rather low concentration of K in the soil solution and hence restricted mobility of K in soils, compared to other metal cations such as Na or Ca is related to its selective adsorption onto some clay minerals. Because of its ionic radius and small hydration energy, K-ions indeed perfectly fit into the interlayer sites of micaceous minerals (micas, illites and mica-derived clays).[8] These sites and, to a lesser degree, the sites on the frayed edges of these minerals have thus a considerably larger affinity for K than for other cations, including divalent cations such as Ca or Mg (Fig. 2). Clay minerals also bear sites with larger affinity for divalent cations than for monovalent cations such as K. These sites are located on the planar faces of clay minerals and are thus dominant in clays such as kaolins and smectites. They also occur in organic compounds. K is thus much less strongly held in soils dominated by kaolins (tropical soils), sand or organic matter than in soils dominated by illite-vermiculite clay minerals. Traces of mica-derived clay minerals can dramatically influence the dynamics of soil K as evidenced in tropical soils that are apparently dominated by kaolins.[9] More generally, K dynamic is largely dependent on soil mineralogy which determines both ion

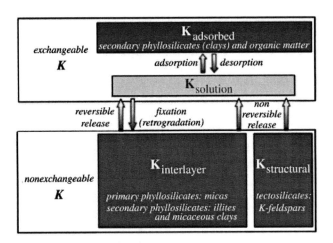

Fig. 1 The various forms of soil K and the chemical processes involved in soil K dynamics.

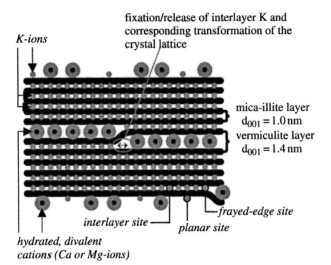

Fig. 2 The various sites of exchange of K-ions in micaceous clay minerals and the transition between mica and vermiculite layer that occurs when interlayer K is exchanged by hydrated, divalent cations.

exchange and release-fixation processes,[6,7] i.e., the dynamics of "nonexchangeable K." The latter is defined as that (major) portion of soil K which cannot be exchanged by NH_4-ions. NH_4-ions have the same charge and radius than K-ions and can successfully desorb K only from low K-affinity sites. Exchangeable K thus comprise soil solution and easily desorbable K-ions. Nonexchangeable K mostly comprise interlayer K (high K-affinity sites) of micaceous minerals and structural K of feldspars (Fig. 1), i.e., 90%–99% of total K in many soils. The release of K from feldspars requires a complete and irreversible dissolution of the mineral and is enhanced under acidic conditions.[6,7] The release of interlayer K from micaceous minerals can proceed similarly or involve an ion exchange process (Fig. 1) leading to an expansion of the phyllosilicate (Fig. 2). This reversible release is essentially governed by the concentrations of K and competing cations in the outer solution.[8,10,11] Cations which can be responsible for this release, such as Ca- and Mg-ions, have a large hydration energy, contrary to K- or NH_4-ions. Therefore, they remain hydrated when exchanging interlayer K-ions and expand the interlayer space (Fig. 2), making it possible for the release to proceed further, whereas NH_4-ions would block the reaction.[12] However, because of the considerable affinity of these interlayer sites for K relative to Ca or Mg, the release can occur only for extremely low concentrations of K in the soil solution (Fig. 3), in the micromolar range.[11,12] Conversely, elevated concentrations of K are prone to the reverse reaction of fixation, i.e., the collapse of expanded layers and concomitant increase in nonexchangeable K at the expense of readily available K. As long as exchangeable K was assumed to be the only plant-available K fraction, K fixation, i.e., the poor recovery of applied K in the exchangeable fraction was seen as a poor efficiency of K fertilization. This apparent loss of applied K has therefore received considerably more interest than the reverse re-

lease process.[7] In addition, the release of nonexchangeable K was considered unlikely to occur to any great extent, especially because K concentration in the bulk soil solution of fertilized soils is usually far above the critical concentrations that are prone to K release. However, numerous long-term fertilizer trials have shown that the release of nonexchangeable K contributes a major proportion of soil K supply in unfertilized and possibly fertilized plots too, although an overall net fixation is often found in the latter (Fig. 4).[5,13] Nonexchangeable K can contribute up to 100 kg ha^{-1} yr^{-1}, 80%–100% of soil K supply. Such annual fluxes of release are fairly large compared with K dissolution rates commonly estimated by geochemists (in the order of 5–15 kg ha^{-1} yr^{-1}). However, many geochemical models do not take into account

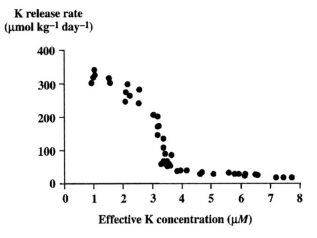

Fig. 3 Effect of solution K concentration on the rate of release of nonexchangeable K from a soil.
Source: Springob and Richter.[11]

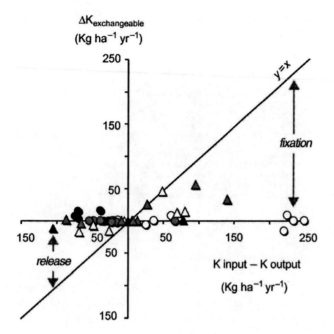

Fig. 4 Annual change in exchangeable K ($K_{exchangeable}$) as a function of the annual K budget in various K treatment plots of long-term fertilizer trials. The K input comprehends organic and inorganic, applied K fertilizers. The K output corresponds to the offtake of K in the harvested product.
Source: Adapted from Blake[5] and Gachon.[13]

the amount of K taken off in the vegetation, leading to large underestimates of the actual dissolution rates.[14] Conversely, many K budgets provided by agronomists do not take into account atmospheric inputs and leaching of soil K, assuming that these terms are fairly negligible in most cases. This certainly holds true for atmospheric inputs which rarely exceed values in the order of a few kg ha^{-1} yr^{-1}. Leaching can, however, vary over a much wider range of values, from several kg ha^{-1} yr^{-1} in most cases up to several tens (and up to 1–3 hundreds) of kg ha^{-1} yr^{-1} in those situations that are the most prone to leaching: bare soil or poor soil coverage by the vegetation, excessive fertilizer rates, coarse-textured soils. Omitting the leaching term would, however, have led to underestimating the actual release rate.[5] Hence, considerable amounts of nonexchangeable K can be released in agricultural soils and contribute a significant proportion of plant uptake, in contradiction to the widespread viewpoint shared by numerous agronomists and soil scientists. Reasons for this can be found when considering root–soil interactions occurring in the rhizosphere.

POTASSIUM IN PLANT–SOIL INTERACTIONS

K occurs at rather low concentrations in the soil solution, compared to other nutrient cations and to the large requirements of plants for K. The transfer of soil K via mass-flow toward plant roots (i.e., the convective flow of solute accompanying transpiration-driven water flow) contributes about 1%–20% of plant demand.[3,15–17] A direct consequence is the rapid depletion of K-ions from the soil solution in the vicinity of plant roots, i.e., the rhizosphere. The resulting concentration gradient generates a diffusion of K-ions in the rhizosphere which plays a key role in the transport of K toward plant root (i.e., 80%–99% of plant demand). Such depletion results in a shift of the cation exchange equilibria which rule the dynamics of both exchangeable and interlayer K. This ultimately results in a desorption of exchangeable K and eventually of interlayer K,[7,16,17] as shown by their depletion in the rhizosphere (Fig. 5). The extent of the depletion of exchangeable K will depend on chemical parameters such as the initial level of exchangeable K and the K buffering capacity of the soil and on physical parameters that directly determine the diffusive transport of K-ions: soil texture and structure, soil water content.[17] The K depletion zone will extend over several millimeters in clayey, dry soils up to several centimeters in wet, sandy soils.[7,16,17] The intensity of the depletion will also depend on how far the K concentration of solution is decreased, which may vary among plant species according to the K uptake ability of their root. Plants with a lower external K efficiency, i.e., with a higher affinity transport system, will have the capability to take up K at lower K concentrations and may thus deplete soil K further.[17] In the vicinity of roots, solution K concentration can indeed decrease by 2–3 orders of magnitude, down to as little as 2–3 μM.[16] At such low K concentrations, the release of nonexchangeable K can occur at large rates, whereas it would be dramatically restricted at bulk soil K concentrations of several hundreds of μM (Fig. 3).[11] Plants thus play a major role in the dynamic of interlayer K via the root-induced depletion of solution K.[16] Measure-

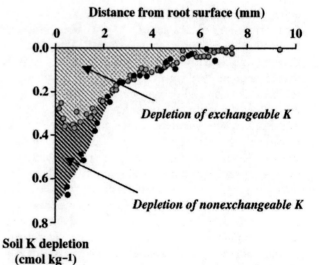

Fig. 5 Depletion of both exchangeable K (gray dots) and HCl-extractable K (black dots) as a function of the distance from rape roots.
Source: Adapted from Jungk and Claassen.[17]

ments in pot experiments have indeed revealed that within several days of growth, the release of nonexchangeable K can amount up to 90% of K supplied to the plant.[7,16] Soil–root chemical interactions in the rhizosphere thus largely explain the unexpectedly large contribution of the release of nonexchangeable K to plant uptake that is found in many agricultural soils, including fertilized soils.

ASSESSING AND MANAGING POTASSIUM FERTILITY

Soil K fertility is most often evaluated by measuring exchangeable K[1,7] most frequently with molar NH_4 acetate in batch conditions. However, the adequacy of exchangeable K to predict plant response, i.e., the actual bioavailability of soil K, is rather poor in many soils. This arises from the major contribution of the release of nonexchangeable K in some soils, especially when exchangeable K is low and/or when large reserves of nonexchangeable K are readily available as a consequence of: 1) soil mineralogical composition or 2) fertilization history (build-up of fixed K due to excessive K-fertilizer rates). In these situations, quantitative evaluation of the potential release of nonexchangeable K would be highly recommended for a better prediction of plant response and fertilizer needs.[7] There are several methods for assessing nonexchangeable K but none of them is routinely used on a broad scale, because of their cost. These are either based on the use of 1) concentrated, strong acids that dissolve K-bearing minerals or 2) cationic resins or chemicals such as Na tetraphenylboron that can promote the release of interlayer K by removing K-ions from soil solution and by shifting the exchange equilibria.[6,7] Alternatively, correction factors can be used when interpreting exchangeable K values, which account either for the cationic exchange capacity (or clay content) or for the soil type and K release potential.[7] Exchangeable K is nonetheless often used alone for fertilizer recommendations, resulting in frequently overestimated fertilizer needs to compensate for the expected large fixation and negligible release. Many long-term fertilizer trials have shown that adequate yields of crops can be obtained at fairly low rates of K fertilizer application, or even, for the least demanding crops such as cereals, without any K fertilizer for several years or decades.[7,13] Other more demanding crops, however, require the application of K fertilizer to achieve high yield and quality in the harvested products.[18] The need for K-fertilizers will thus depend on the release potential of the soil and on the demand of the plant, the latter being now increasingly accounted for in fertilizer recommendations. Fertilizer trials have also shown that commonly used soluble K fertilizers and organic sources such as manure or crop residues have fairly comparable efficiencies. This is not surprising as K is highly mobile in organic compounds where it occurs as soluble or exchangeable K-ions. These sources are thus equally important as K-fertilizers and absolutely need to be accounted for in K budgets.

CONCLUSIONS

K is the major nutrient cation for plants and thus taken up at large rates by plant roots. These are achieved by both high and low affinity transport systems which explain the considerable mobility of K within the plant. In comparison, K is much less mobile in soils because of the strong affinity of some exchange sites of clays. The large K uptake rates achieved by roots result in a steep depletion of solution K in the rhizosphere, and hence in a shift of the equilibria of cation exchange. Exchangeable K and even nonexchangeable K can thereby be significantly depleted and contribute a substantial proportion of plant uptake. This is confirmed by K balance both in short-term pot experiments and long-term field trials. In addition to the desorption–adsorption of exchangeable K, release and fixation processes thus need to be accounted for when evaluating soil K fertility.

REFERENCES

1. Munson, R.D. *Potassium in Agriculture*; American Society of Agronomy, Crop Science Society of America; Soil Science Society of America: Madison, WI, 1985; 1223 pp.
2. Mengel, K.; Kirkby, E.A.; Kosegarten, H.; Appel, T. *Principles of Plant Nutrition*, 5th Ed.; Kluwer Academic Publishers: Dordrecht, Netherlands, 2001; 673 pp.
3. Marschner, H. *Mineral Nutrition of Higher Plants*, 2nd Ed.; Academic Press: London, 1995; 889 pp.
4. Schachtman, D.P. Molecular insights into the structure and function of plant K^+ transport mechanisms. Biochim. Biophys. Acta **2000**, *1465*, 127–139.
5. Blake, L.; Mercik, S.; Koerschens, M.; Goulding, K.W.T.; Stempen, S.; Weigel, A.; Poulton, P.R.; Powlson, D.S. Potassium content in soil, uptake in plants and the Potassium balance in three European long-term field experiments. Plant Soil **1999**, *216*, 1–14.
6. Sparks, D.L. Potassium dynamics in soils. Adv. Soil Sci. **1987**, *6*, 1–63.
7. International Potash Institute. *Methodology in Soil-K Research*; Proceedings of the 20th colloquim of the International Potash Institute, Baden bei Wien, Austria, International Potash Institute: Bern, Switzerland, 1987; 428 pp.
8. Dixon, J.D.; Weed, S.B. *Minerals in Soil Environment*; Soil Science Society of America: Madison, WI, 1989; 1244 pp.
9. Fontaine, S.; Delvaux, B.; Dufey, J.E.; Herbillon, A.J. Potassium exchange behaviour in Carribean Volcanic ash soils under banana cultivation. Plant Soil **1989**, *120*, 283–290.
10. Schneider, A. Influence of soil solution Ca concentration on short-term release and fixation of a loamy soil. Eur. J. of Soil Science **1997**, *48*, 513–522.
11. Springob, G.; Richter, J. Measuring interlayer Potassium release rates from soil materials. II. A percolation procedure

to study the influence of the variable solute K in the <1 ... 10 µM range. Z. Pflanzen. Bodenk. **1998**, *161*, 323–329.

12. Springob, G. Blocking the release of potassium from clay interlayers by small concentrations of NH_4^+ and Cs^+. Eur. J. of Soil Sci. **1999**, *50*, 665–674.

13. Gachon, L. *Phosphore et Potassium dans les Relations Sol–Plante: Conséquences sur la Fertilisation*; Institut National de la Recherche Agronomique: Paris, France, 1988; 566 pp.

14. Taylor, A.B.; Velbel, M.A. Geochemical mass balances and weathering rates in forested watersheds of the southern blue ridge II. Effects of botanical uptake terms. Geoderma **1991**, *51*, 29–50.

15. Barber, S.A. *Soil Nutrient Bioavailability. A Mechanistic Approach*, 2nd Ed.; Wiley: New York, 1995; 414 pp.

16. Hinsinger, P. How do plant roots acquire mineral nutrients? chemical processes involved in the rhizosphere. Adv. Agron. **1998**, *64*, 225–265.

17. Jungk, A.; Claassen, N. Ion diffusion in the soil–root system. Adv. Agron. **1997**, *61*, 53–110.

18. http://www.ipipotash.org/publications/publications. html (accessed October 2000).

Precision Agriculture: Engineering Aspects

Joel T. Walker
Department of Food, Agricultural, and Biological Engineering, Ohio State University, Columbus, Ohio, U.S.A.

Reza Ehsani
Ohio State University, Columbus, Ohio, U.S.A.

Matthew O. Sullivan
Department of Food, Agricultural, and Biological Engineering, Ohio State University, Columbus, Ohio, U.S.A.

Abstract
Precision agriculture or site-specific management is an information-based management technique that has the potential to improve profitability and reduce the environmental impact of crop production. It also has the potential to improve the quality and nutrient content of the product. Precision agriculture, rather than the "one-size-fits-all" management strategy, provides for differential treatment of selected areas of a production field, called management zones, based upon expectation of increased yield, profit, or some other agronomic goal. The ability to provide differential treatment to management zones, also called site-specific management, depends upon availability of both proper equipment and effective treatment algorithms.

INTRODUCTION

Information technology is playing an increasingly important role in today's agricultural production systems of all sizes, commodities, and management philosophies. Precision agriculture[12,14,17] or site-specific management is an information-based management technique that has the potential to improve profitability[7] and reduce the environmental impact[19] of crop production. It also has the potential to improve the quality and nutrient content of the product. Precision agriculture, rather than the "one-size-fits-all" management strategy, provides for differential treatment of selected areas of a production field, called management zones, based upon expectation of increased yield, profit, or some other agronomic goal.[2–4,18] Management zones may be selected for differential treatment based upon various documented differences such as soil type, soil fertility or pH, yield history, presence of weeds, insects, or diseases, or other measures for which a differential treatment helps the producer achieve a selected goal. The ability to provide differential treatment to management zones, also called site-specific management, depends upon availability of both proper equipment and effective treatment algorithms.

WHAT MAKES IT POSSIBLE?

Precision agriculture techniques have been made possible by the advent of global positioning systems (GPS) and high speed computer processing. GPS provides real-time location information to a computer that, from stored information, determines the current management zone, selects appropriate treatment for that management zone, and controls mechanisms to provide the treatment. Fig. 1 is a graphic representation of the precision agriculture paradigm. GPS provides position information for a variety of data gathering processes or for control of site-specific treatments. Information of various types (shown as layers) may be used in analysis of yield results or to develop an application map to control site-specific treatments. The whole system taken together is often called precision agriculture or site-specific agriculture. Note that a feedback loop is implied where results of the previous growing season (yield) become part of the information that influences current treatment practices. Given the proper treatment algorithms, current treatment practices may optimize the goal parameter. Maximum yield is not necessarily the best goal because the cost of treatments required to achieve that yield may be greater than the increased crop value.[4,18]

GPS consists of a minimum of 24 satellites orbiting the Earth and sending signals to a local receiver for which location is desired.[9] Each and satellite broadcasts encoded information with particular timing. By measuring the time a signal travels (at the speed of light) to reach the receiver, the distance from a satellite to the receiver may be calculated. Determining distances from four or more satellites of known location may establish the receiver's location (latitude, longitude, and elevation). Even simple, inexpensive receivers are capable of accuracy better than 15 m, close enough to return to a favorite fishing spot. With specialized transmissions containing information to correct for known errors (called differential signals), accuracy better than 1 m may be achieved. Specialized local transmitters make possible real-time kinematic (RTK-GPS) systems with accuracy of better than 1 cm. RTK systems are used

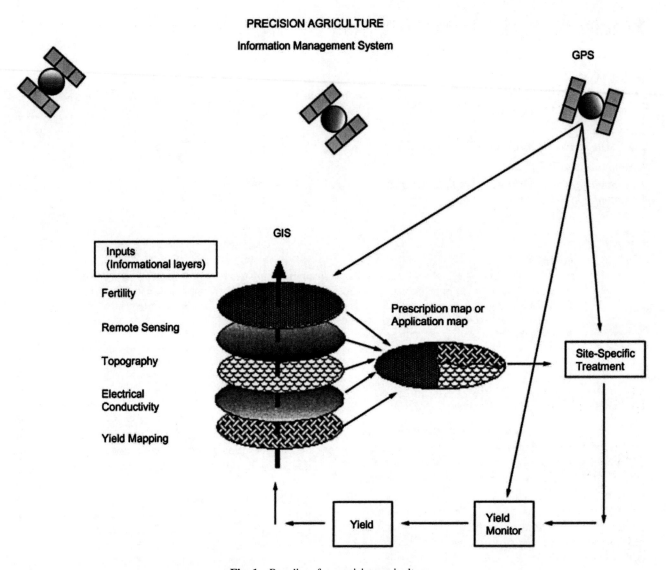

Fig. 1 Paradigm for precision agriculture.

in surveying, guidance, and where the fine precision may justify a relatively high equipment cost.

High-speed computer processing systems have also played an important role in the advent of precision agriculture. Precision agriculture requires collection and storage of data, decision-making computation, and controlling of equipment by computers operating at billions of operations per second. GPS locations are recorded in real-time by the computer. Digital maps of field conditions and parameters are carried in memory or storage media. Digital data such as digital still or video images of weeds, insects, or disease damage, soil properties, crop spectral reflectance, or climactic conditions may also be collected in real-time by the computer. Using such information, the computer may work through a response model to decide what actions should be taken with current equipment. For example, a precision agriculture capable planter may be able to adjust planting rate and depth or change the seed variety on-the-go. Thus, the computer may decide that areas with a selected soil type and relative elevation will get a reduced population (investment of seeds) because less yield is expected. The computer may dictate that another soil type with a high yield potential and high existing moisture content will get a different variety planted to a shallower depth. Across an entire field, many combinations of the controlled variables will be chosen to optimize the desired result (yield, profit, or other goal). A large field would require constant decisions and adjustment of the equipment that would not be possible without computerized systems. Finally, volumes of data may be produced by precision agriculture techniques and computer methods are being developed to extract useful information; and models form this data[5,6,10,11,15,16] and present it to the public in readily accessible form.[13]

SENSING FOR PRECISION AGRICULTURE

Real-time sensors will be important to many precision agriculture systems.[8] A real-time sensor provides data in a nearly constant stream as the machine traverses the field. For example, a camera may provide digital images from which the presence of weeds may be determined. A sprayer may then be directed to spray only where weeds are present. Organic matter and moisture levels in soil affect the performance of some herbicides. Sensors that measure organic matter and/or moisture on-the-go allow for optimum rates of chemical application—adequate to control the weeds, but no more than necessary to preserve environmental quality.

One of the most popular real-time sensors measures grain yield on combines.[12] Yield sensors are also available or are being developed for a variety of crops such as cotton, potatoes, tree fruits, and strawberries. Yield sensors provide a measure of yield over a whole field. Areas of the field with unusually high or low yield may be identified and corrective action (subject to some predictive model created by the producer or by computer) can be taken for the following year. This one aspect of precision agriculture has created a lot of activity in adjusting soil drainage and fertility and crop management decisions such as the relative value of fertilizer or chemical inputs.

For many agronomic parameters, site-specific soil sampling is more practical, either because a real-time sensor is not yet available, or because the spatial variation of the parameter is more gradual and can be estimated with a few site samples. Soil samples to determine fertility are common. The values of pH (acidity/alkalinity), nitrogen, potassium, and phosphorus are particularly important in prediction of yield levels. Soil type is another parameter that is often determined by a site visit.

Remote sensing is becoming increasingly important to precision agriculture. Early data that came from random satellite observations had low resolution and limited value. Now specially equipped satellites and aircraft may be hired to collect specific crop or soil information. Crop growth or health may be deduced from this data. Certain wavelengths of the electromagnetic spectrum are particularly helpful. The visible light frequencies provide some information. Unhealthy crops tend to reflect more yellow and red light. Various frequencies in the infrared range also have been correlated with plant health and soil moisture conditions. Multi-spectral systems provide data on the visible spectrum (often three primary colors) and a limited range of the infrared. More sophisticated equipment, called hyper-spectral, can provide data from a much broader range of the spectrum and from narrower sample bands. This type of data offers greater opportunity to correlate specific crop or soil conditions to measured spectral data.

Other types of remote sensing are becoming available. Light detection and ranging (LIDAR), for example, is a laser ranging technique that may be used to measure topography and plant height simultaneously.[1] Currently applied to the forest industry and relatively expensive, this method holds promise for rapid feedback of information on crop growth problems to the producer, perhaps allowing solution of the problem before yield is permanently affected. For example, nematodes in soybeans are a common cause of reduced vigor and yield if left untreated. Conventional crop scouting or remote sensing may not detect small areas of infestation or provide feedback in time to take corrective action. With regular LIDAR imaging, a producer could detect and treat such trouble spots in a timely and efficient manner.

GEOGRAPHIC INFORMATION SYSTEMS

Geographic information systems (GIS) techniques are an integral part of precision agriculture. Basically, GIS is a storage system for geographically referenced digital data. Many of the data types discussed above can be digitized (if not already in digital form) and referenced to specific locations in a field. Each parameter or variable then may be represented as a layer of information. Geographic position of the information matches the position on a map of the field. The value of a parameter may be represented on a map as a shade of gray or a color. So a GIS information layer representing soil type can be drawn in the physical shape of the field with patches of color—each color representing a different soil type. Many such layers may exist in a GIS dataset for a particular field. Depending upon the need, these layers may be viewed as overlays—simultaneous presentation of several variables in one field. As many layers of information are added to the GIS system, it becomes apparent that human vision is inadequate to detect the important patterns. Computers, however, are infinitely more able to "see" data patterns with a potential to produce an economic advantage.

FUTURE

The future of precision agriculture depends on economic results. Can the cost of additional equipment and operations be more than offset by increased economic return and value of environmental protection? It seems possible that all agricultural operations could be monitored and recorded, linked by digital transmissions to databases containing weather, remote sensing, and historical data, and controlled through a general model of the crop's predicted response to specific inputs. Such a system may not only control specific agricultural operations, but also be involved in scheduling, ordering seed, fertilizer, and supplies, providing records to regulating agencies, and feeding valuable information back into the research system for further refinement of the control model. Already, manufacturers are producing prototype agricultural machines that perform without an operator. These machines are guided by GPS, are controlled by computer, and have onboard sensors to detect obstructions or people in the way. The advancement of agriculture will come with technology—technology to feed the world.

REFERENCES

1. Ackermann, F. Airborne laser scanning—present status and future expectations. ISPRS J. Photogram. Remote Sens. **1999**, *54* (2–3), 64–67.
2. Adams, M.L.; Cook, S.; Corner, R. Managing uncertainty in site-specific management: what is the best model? Precis. Agric. **2000**, *2*, 39–54.
3. Boote, K.J.; Jones, J.W.; Pickering, N.B. Potential uses and limitations of crop models. Agron. J. **1996**, *88*, 704–716.
4. Bullock, D.S.; Bullock, D.G. From agronomic research to farm management guidelines: a primer on the economics of information and precision technology. Precis. Agric. **2000**, *2*, 71–101.
5. Data Mining for Site-Specific Agriculture; http://www.gis.uiuc.edu/cfardatamining/ (accessed April 16, 2001).
6. Data Mining Puts Geographic Information to Work for Public, Private Sectors; http://csd.unl.edu/csd/resource/Vol. 13/datamine.htm (accessed April 16, 2001).
7. Erickson, K., Ed.; *Precision Farming Profitability*; Agricultural Research Programs; Purdue University: West Lafayette, IN, 2000.
8. Gaultney, L.D. *Prescription Farming Based on Soil Property Sensors*. ASAE Paper No. 89–1036, ASAE Annual International Meeting; ASAE: St. Joseph, MI, 1989.
9. Kennedy, M. *The Global Positioning System and GIS: An Introduction*; Ann Arbor Press: Chelsea, MI, 1996.
10. Kerschberg, L.; Lee, S.W.; Tischer, L.A. *Methodology and Life Cycle Model for Data Mining and Knowledge Discovery for Precision Agriculture*; Final Report, Center for Information Systems Integration and Evolution, George Mason University: Fairfax, VA, 1997.
11. Lazarevic, A.; Fiez, T.; Obradovic, Z.A. *Software System for Spatial Data Analysis and Modeling*, Report to INEEL University Research Consortium Project No. C94-175936.
12. Morgan, M.; Ess, D. *The Precision-Farming Guide for Agriculturalists*; John Deere Publishing: Moline, IL, 1997.
13. Ohio State Precision Agriculture; http://www.ag.ohiostate.edu/~precisfm/ (accessed April 16, 2001).
14. Pierce, F.J., Sadler, E.J., Ed.; The State of Site-Specific Management for Agriculture. *ASA/CSSA/SSSA*; Madison, WI, 1997.
15. Precision Farming Support; http://www.umac.org/farming/precision.html (accessed April 16, 2001).
16. Regression and Emerging Data Mining Tools Modeling; http://www.gis.uiuc.edu/cfardatamining/APRReport.htm (accessed April 16, 2001).
17. Sonka, S.T., Ed. *Precision Agriculture for the 21st Century: Geospatial and Information Technologies in Crop Management*; National Academy Press: Washington, DC, 1998.
18. Swinton, S.M.; Lowenberg-Deboer, J. *Site-Specific Management Guidelines—Profitability of Site-Specific Farming*; Publication SSMG-3, Phosphorus and Potash Institute, 1998.
19. Wang, X.; Tim, U.S. Problem-Solving Environment for Evaluating Environmental and Agronomic Implications of Precision Agriculture; http://www.esri.com/library/userconf/proc00/professional/papers/PAP102/p102.htm (accessed April 16, 2001).

Precision Agriculture: Water and Nutrient Management

Robert J. Lascano
Agricultural Research Service (USDA-ARS), U.S. Department of Agriculture, Lubbock, Texas, U.S.A.

J.D. Booker
Plant and Soil Science, Texas Tech University, Lubbock, Texas, U.S.A.

Abstract
Precision agriculture (PA) refers to the practice of managing agronomic inputs according to specific needs across the landscape. The major impediment to implement the adoption of PA is the development of decision-support systems. Research in PA, focusing on factors controlling crop variability, has described useful process relationships, and these results are supporting the development of decision-support systems. An example is the integration of crop simulation models with geographic information data of soil and elevation, real-time weather, and management information systems. Models such as the Precision Agricultural-Landscape Modeling System that can calculate the energy, water, nutrient, and carbon balance across the landscape at a 5 to 10 m resolution provide the desired integration of field-scale data. These landscape-scale models can provide a decision-support framework to manage agronomic inputs to maximize economic crop yield while minimizing environmental hazards. Adoption of PA will continue to increase given the demand for a safe food and fiber supply of high quality.

INTRODUCTION

Management of agronomic inputs, such as water and fertilizers, to cropping systems requires information on when and how much of each input to apply. The correct management decision is essential to maintain the productivity of the cropping system, and a management strategy is to use crop production functions that describe (mathematically) maximum economic yield (profit) as a function of agronomic inputs. Essentially, the concept is to apply the least amount of input to produce the maximum economic crop yield in the farming operation. In conventional cropping systems, inputs are generally applied uniformly across a field regardless of their need, and the amount applied is normally based on average responses of these inputs to crop yield across the field. Since the 1970s, increased costs of crop production and emphasis on production efficiency revolutionized production agriculture. Environmental concerns, including quality and safety of harvested product and impact of the cropping procedures on the ecosystem, are also considered in the decision-making process. Developments in spatial statistics and computers (i.e., increased microprocessor speed and decreased cost) and increased availability of soil, elevation, and weather data have contributed to improved concepts and procedures to address spatial and temporal variability in cropping systems.

Precision agriculture (PA) and precision farming are generic terms that describe the way production management inputs (e.g., water, nutrients, harvest aids, and herbicides) are managed in response to cropping system variability. In contrast to a uniform blanket application of an input across the field, each input is applied according to specific needs across the field. Precision agriculture is an integrated crop management system that attempts to match the kind and amount of inputs with the actual crop needs for small areas within a farm field. Perhaps a better descriptor for this type of farming is site-specific management, which manages an agricultural crop at a spatial scale smaller than the entire field by dividing it into management zones, defined by topography, soil type, and level of a particular nutrient, such as nitrogen and phosphorus.

Driving forces that have contributed to producers implementing and adopting PA procedures are advances in computer hardware and software, electronics, and equipment technology; decreased profit margins due to increased costs of production; and environmental awareness. For example, advances in crop growth simulation models, variable-rate application equipment, adoption of soil sampling for nutrients across the field and as a function of soil depth, and integration of crop yield monitors with global positioning satellite systems have contributed to the use of PA concepts to manage crops. The cost of agronomic inputs continues to increase, e.g., petroleum-based inputs such as fertilizer, diesel, insecticides, and herbicides. Awareness of environmental concerns related to nutrient contamination of groundwater and surface water and the quality and safety of food and fiber are factors that impact how crops are produced and delivered to the consumer. Operating a farm requires management strategies that consider both economic and environmental consequences, and PA provides the concepts that can be used to achieve this goal.

This entry gives a general overview of PA with emphasis on crop water and nutrient use and crop yield using state-space analysis to describe how cotton lint yield varies at a landscape scale. The entry is divided into five parts. First, we give a general overview of PA. Second, we describe the relation between crop yield and water use. Third, as an example, we show measurements of cotton water use along two 700 m transects on a 60 ha field. Fourth, we use geostatistical tools to quantify cotton lint yield as a function of nitrogen, topography, and soil water. Fifth, we describe how the future of PA will leverage knowledge from PA research, such as that presented here, to support the development and use of large-scale cropping system models.

PRECISION AGRICULTURE

Precision agriculture, also known as site-specific management[1,2] refers to the practice of applying agronomic inputs across a farm, mainly fertilizers and other chemicals, at variable rates based on soil nutrients or chemical tests, soil textural changes, weed pressures, and/or yield maps for each field in the farm.[1–3] In large fields (e.g., >40 ha), crop yield and thus crop water and nutrient use are notoriously variable. The sources of this variation are related to soil physical and chemical properties, pests, microclimate, genetic and phenological responses of the crop, and their interactions. The technology for crop yield mapping is more advanced than current methodologies for determining and understanding causes of yield variability. Prevailing and traditional management practices treat fields uniformly as one unit. However, reports[3–5] show that to understand underlying soil processes that explain crop yield variability, research must be done at the landscape level and using appropriate statistical tools for large-scale studies.[1,3,5]

CROP YIELD AND WATER USE

There is a linear relation between crop yield and water use when the only limiting factor is water;[6] however, root water uptake is synergistically related to nutrient uptake, and the two processes cannot be separated. Precision farming has the potential to improve water and nutrient use efficiency on large fields provided there is quantitative understanding of what factors affect crop water and nutrient use and how they vary across the field. It is known that crop water and nutrient use are a function of many biotic and abiotic variables, including managed inputs, and harvestable yield is a manifestation of how these variables and inputs interact and are integrated during the growing season. However, it is difficult to determine a hierarchy on the contribution of each input and variable to the measured crop yield using classical statistics.[7–10] Often, variables that affect water and nutrient supply to the plant contribute to crop yield at a high level assuming an adequate plant stand and weed control. The cause-and-effect relation between a single state variable and crop yield is site specific and is difficult to establish without considerable sampling of the soil and/or crop. The establishment of response functions, i.e., crop water and nutrient use as a function of variable x_i, gives only a partial answer to explain crop water use, nutrient use, and yield based on inputs. The general idea of PA is to optimize input application to the measured crop yield at each sampling location using the optimum law of Liebscher, which states that a production factor that is in minimum supply contributes more to production, the closer other production factors are to their optimum.[11] This is a simple premise; however, the decisions for variable-rate application of any agronomic input must consider temporal and spatial variability of the soil's properties affecting crop growth, water and nutrient use, and yield. Soil factors that affect water storage, such as depth of the root-restricting layer and soil textural differences, must be considered in any precision farming operations that attempt to improve crop water use and yield related to agronomic inputs. Similarly, to improve the use of any micro- and macronutrient by the crop, the overall cycle of the nutrient must be considered, including its availability in the soil and demand by the crop. Examples of managing nitrogen fertilization and irrigation at a site-specific and farm scale for cotton production is given by Bronson et al.,[4] Li et al.,[5] and Booker et al.[12]

Precision farming must incorporate the inherent spatial and temporal variability of soil physical, chemical, and biological factors within a field for input management. Accurate representation of spatial and temporal variability in a field requires taking and analyzing many samples. Sampling is normally done on a grid with a scale that can vary from one to several hundred meters.[7,8] Once properties are measured, geostatistical tools (e.g., semivariogram, kriging, cokriging, etc.), and other spatial statistical tools (e.g., autocorrelation, cross-correlation, state-space analysis, etc.), can be used to establish statistical relations in space and to minimize the number of soil samples to characterize and map fields.[7,8,10] The number of samples required a priori to determine spatial and temporal variability is perhaps the single largest deterrent in the application of precision farming practices to manage and improve crop water and nutrient use.

There is not much information published on combined crop water and nutrient use across large fields at the landscape level and in the context of precision farming.[4,13–15] An exception is the study by Li et al.[5] where cotton water and total nitrogen use were measured along a 700 m transect with the following objectives: 1) to illustrate the landscape pattern of cotton water and total nitrogen use; and 2) to determine the underlying soil processes governing cotton lint yield variability. In this study, state-space analysis[4,7,10] was used to formulate management decisions that may improve crop water use and nitrogen use and, thus, yield using precision farming practices. An additional study regarding variable-rate nitrogen at different locations within a 48 ha field is given by Bronson et al.,[4] and in a 14 ha field, by Booker et al.[12]

LANDSCAPE CROP WATER USE

The concept of crop water and nitrogen use in a 60 ha field is illustrated by the study of Li et al.[5,16] In 1999, a field experiment was conducted near Lamesa, Texas, on a research farm of Texas A&M University on the southern edge of the High Plains of Texas. The soil was classified as an Amarillo sandy loam. The field was 60 ha with slopes ranging between 0.3% and 6.3%.[5] To evaluate the effect of soil water, nitrate-nitrogen (NO_3-N), and topography on cotton lint yield across the landscape, two irrigation levels were used. The irrigation treatments consisted of water applications at the 50% and 75% grass reference evapotranspirations (ET_o) with a center-pivot low-energy precision application irrigation system.[17] At each irrigation level, one transect was established following the circular pattern of the center pivot. The two transects were instrumented with 50 neutron access tubes, each 15 m apart, and soil volumetric water content (θ_v) was measured periodically throughout the growing season. At each point, θ_v was measured in 0.3 m depth increments to the 2.0 m depth using a neutron probe calibrated for this soil. In addition, at each transect point soil texture, soil and plant NO_3-N, leaf area index, cotton lint yield, slope, plant density, and other parameters were measured.[5,16]

Statistical Calculations

It has been shown that the use of classical statistics, such as regression analysis and analysis of variance, is designed to describe the strength of the covariance structure between variables and fails to completely explain the cause and effect between, for example, crop yield and measured soil variables in precision farming experiments.[5,7,8,10,15] These techniques, in general, account for spatial and temporal variability through blocking and do not describe the spatial and temporal structure. Instead, there are other more appropriate statistical tools for relating the variability of soil and plant parameters measured in space and time. For example, the structure of the spatial (or less often, temporal) variance between measurements may be derived from the sample semivariogram, which is the average variance between neighboring measurements separated by the same distance. Spatial or temporal structure between variables is often determined using autocorrelation and cross-correlation functions. Although these techniques can be used to evaluate the temporal variability structure, they are most often used in PA to analyze spatial variables. Autocorrelation measures the linear correlation of a variable in space along a transect. Cross-correlation is the comparison of two variables measured along a transect and is used to describe the spatial correlation between two landscape variables, i.e., where one variable, the tail variable, lags behind the head variable by some distance. The spatial association between several variables can be described using state-space analysis, which is a multivariate autoregressive technique.[5,7,10]

SPATIAL ANALYSIS OF CROP WATER USE

To illustrate the variability of crop water use or crop evapotranspiration (ET_c), values measured along the 50% irrigation transect were selected.[5] In Fig. 1, the relation between the scaled ET_c and elevation, both as a function of distance along the transect, is shown. The ET_c data are scaled to the maximum value of 426 mm of water, which was measured 210 m from the south end of the transect. These results show, as expected, that higher ET_c was measured at lower elevations and that ET_c decreased at the higher elevations. Spatial cross-correlation between cotton lint yield and soil water, cotton lint yield and site elevation, and soil water

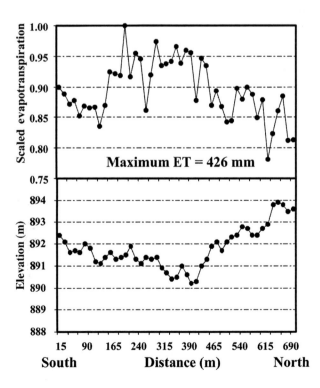

Fig. 1 Scaled crop evapotranspiration and elevation as a function of distance along a 700 m transect.

and site elevation are shown in Fig. 2. For a 95% confidence interval, the cotton lint yield was positively cross-correlated with soil θ_v across a lag distance of ± 30 m. Cotton lint yield and θ_v were negatively cross-correlated with elevation at a lag distance of ± 30 m. These results show the effect of topography on the θ_v and crop water use measured along the transect. Similar results are given in other reports.[4,14,15,18] In this example, the cross-correlation between θ_v and elevation shows the spatial structure of measured variables and, further, that more water was stored in lower elevations, resulting in higher ET_c.

Linear regression analysis between θ_v and cotton lint yield and relative site elevation is shown in Fig. 3, and the state-space analysis for the relation between cotton lint yield and three measured parameters is shown in Fig. 4. Results in Fig. 3 shows the shortcomings of using an inappropriate statistical tool to understand underlying processes explained with the state-space analysis. This analysis (Fig. 4) quantified how cotton lint yields varied as a function of distance and showed that by using θ_v, soil NO_3-N and elevation the variation in cotton lint yield can be explained with a high level of confidence. While studies like those of Li et al.,[5] Bronson et al.,[4] and Booker et al.[12] are empirical in design, the relationships that are evaluated provide important validation and field testing of the more mechanistic mass and energy balance accounting provided by models such as PALMS.[19] These studies also provide foundational information for developing PA management strategies at the crop production scale.

Benefits of PA to improve crop water and nutrient use may be obtained by an economic analysis of maximizing crop yield as a function of application of nitrogen fertilizer and irrigation water as given by the state-space equa-

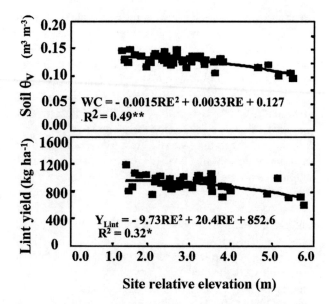

Fig. 3 Soil water content (θ_v) and cotton lint yield as a function of site relative elevation.

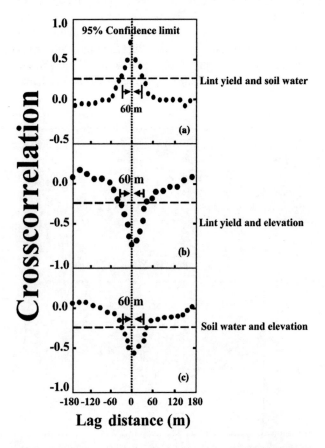

Fig. 2 Cross-correlation as a function of lag distance. (a) Lint yield and soil water, (b) lint yield and elevation, and (c) soil water and elevation. Shown is the 95% confidence for the cross-correlation distance.
Source: Li et al.[5]

$$Y_{(50\% \, ET)i} = -0.201 Y_{i-1} + 1.107 \, W_{i-1} + 0.332 \, N_{i-1} \, D49.54 \, E_{i-1} + \varepsilon_i$$

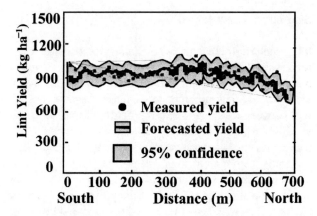

Fig. 4 State-space equation relating cotton lint yield (Y) to water content (W), nitrogen level (N), and elevation (E) as a function of distance and location (i) along a 700 m transect.
Source: Li et al.[5]

tion.[5,16] In the example given, the decision can be made to apply more nitrogen fertilizer to the lower areas of the field that also hold more water and increase crop water use and nitrogen and lint yield. With the introduction of variable-rate planters, it is possible to discriminate site locations and plant more "drought"-tolerant varieties or change the seeding rate in areas that are prone to have less soil water. This implies the delineation of management zones[1–3] within a field that are defined based on potential crop water use and their interaction with other input variables to maximize economic yield across the field. This type of precision farming is slowly being adopted, and wide use remains within the realm of possibilities that this type of farming has to offer. The introduction and use of computer models of cropping systems will likely expedite and facilitate the adoption of PA management techniques.[19]

A final consideration is the cost/benefit of PA practices and its impact on agriculture. Currently, hardware for variable-rate application of agronomic inputs is relatively expensive and in many cases unavailable; however, with increased adoption and use of these practices, the cost will be reduced. For example, tractor guidance systems[20,21] were quickly adopted by producers, and high demand reduced its cost. Further, environmental and material cost concerns for a given area will probably place limits on the amount of certain nutrients, e.g., nitrogen fertilizer, used for crop production. This will force producers to apply nitrogen and other nutrients across the field according to site-specific needs and position along the landscape. These practices will be beneficial from both an environmental and an economical point of view.

FUTURE OF PA

Considerable PA-related research, similar to that presented above, has been conducted over the past decade, studying empirical relationships and attempting to better understand the underlying processes controlling crop yield variability. Much of this research has focused on grid soil and crop sampling,[22] surface characterization (e.g., apparent electrical conductivity),[23] and ground- or aerial-based remote sensing.[24,25] Such research has described numerous useful process relationships but has been somewhat less successful in providing broad-based tools to support production-scale management. The lack of development of decision-support systems to implement precision decisions is the major impediment to the adoption of PA.[26]

Recent advances in availability of soil data provided by the U.S. Department of Agriculture, elevation data provided by the U.S. Geological Survey, and weather data from weather mesonets (networks) provide the necessary input to model the water, energy, nutrient, and carbon balance of large-scale agricultural fields. An example of such a model is the Precision Agricultural-Landscape Modeling System (PALMS) given by Molling et al.[19] The integration of PALMS with crop growth models[27,28] provides a framework whereby site-specific management of crops is an achievable goal.

The concept of using simulation models to manage crops was introduced in the 1970s. Many of the theoretical algorithms related to model crop photosynthesis and transpiration were formulated by C.T. de Wit and coworkers at Wageningen University, Netherlands.[29] An example of such model is the simulation of field water use and crop yield given by Feddes et al.[28] An example of a crop-specific model, i.e., cotton, known as GOSSYM/COMAX was developed by McKinion et al.[27] This model was used by crop consultants in the Texas High Plains to provide services on irrigation scheduling and application of nitrogen fertilizer, growth regulators, and chemicals to terminate the crop. The biggest drawback on the application of these models was that the required inputs, soil and weather, were both difficult and expensive to obtain. Furthermore, these models provided only an average estimate of crop yield for the entire field regardless of size. The models could be run separately for different parts of a field, but this increased the demand on limited computer and input resources and even then did not represent the interaction between various parts of the field. In retrospect, we now recognize that these models were ahead of their time. Given the current availability of soil and weather data that is required by these models, along with the increased computer speed and reduced cost, a resurgence in the application of simulation models to manage crops and cropping systems is anticipated.

In coming years, a likely scenario to emerge to manage cropping systems will be based on the combination of three factors. First is the realization from site-specific management that shows that crop yield varies temporally and spatially and that increases in crop yield are possible by targeting different amounts of an input, e.g., irrigation water and fertilizer, to specific parts of the field. Second, crop management, from planting to harvesting, is complex, and simulation models can be used as a decision aid. Use of crop simulation models, developed in the 1970s to 1980s, is facilitated due to the increased availability of required soil, elevation, and weather input data to execute the models and reduced cost of computer hardware. The third factor is an increased awareness of producers on production efficiency and environmental concerns. For example, in many agricultural areas, the amount of nitrogen fertilizer that can be applied is restricted and linked to the residual nitrogen in the soil and its potential effect on contamination of surface water and groundwater.[22,23,30] Advances in management information systems, development of computer software, and communication via the Internet provide us with the tools to manage a crop in real time. We are currently working toward the development of a PA model that includes all of the above factors.

The integration of a landscape-scale model such as PALMS[19] with a cotton growth model[27] using site-specific management of water and nitrogen[4,5] can give us the tools to manage, for example, a 50 ha irrigated cotton field. The model provides three key features important to real-time production-scale management: 1) it represents the variability in space and time within the entire field and accounts for hydrologic interactions between areas within the field; 2) it provides water, energy, nutrient, and carbon balance information without reliance on field-installed hardware that must be avoided during management operations; and 3) it can provide predictive information that can support various what-if scenario evaluations (something that physical field measurements do not provide). For example, this field can be divided into 5000 (100 m^2) or 20,000 (25 m^2) cells, and the model will calculate a cotton lint yield value for each cell, using weather data collected at or near the field and previously collected soil and elevation data, both of which are stable and once developed can be used for many seasons. Further, the estimate of cotton lint yield is based on interactions of soil–plant–weather parameters, and the model itself can be used to explore what combination of inputs (e.g., water and nitrogen) would give the highest economic yield while minimizing leaching of nitrogen below the root zone. This is a current topic of research of our cropping system research unit.

CONCLUSION

Precision agriculture is a generic term that describes the way that agronomic inputs to a farming operation are managed and each input is applied according to specific needs across the field. An outcome of PA is the recognition that crop yields vary temporally and spatially. Crop management, from planting to harvest, is complex and requires management strategies that consider economic and environmental consequences. Recently, nutrient contamination and quality and safety of food delivered to consumers are also factors that have received consideration. A summary of general concepts learned from PA experiments follows.

First, agronomic experimental work needs to be done at the scale of application. For example, results from small research plots are normally not transferable to large fields.[7,10,18] The underlying principle is to take advantage of the inherent spatial variability of soil properties across the landscape. In PA experiments, the spatial variability of a given property is used as the source of variation instead of imposing treatments (e.g., levels of nitrogen fertilizer) to obtain variation. Second, the use of classical statistics fails to explain the cause and effect between variables. Spatial structure between variables can be quantified using specialized statistical tools such as autocorrelation, cross-correlation, and state-space analysis.[7,10] Third, given the complexity of current cropping systems, crop models[27,28] introduced in the 1980s are being combined with geographic information data of soils and elevation, real-time weather, and management information systems to provide a framework to manage crops. These models can provide three key features important to real-time production-scale management: 1) representation of variability in space and time within the entire field and hydrologic interactions between areas within the field; 2) water, energy, nutrient, and carbon balance information without reliance on field-installed hardware that must be avoided during management operations; and 3) predictive information to support various what-if scenario evaluations. As a result, we now have the capability of tracking different layers of data (soil, plant, and weather) and, for example, a 50 ha field may be divided into 5000 cells, each cell 10 m wide, 10 m long, and 2.0 m deep. The water, energy, carbon, and nitrogen balance (input equal to output) of each cell is calculated, and the change of any one variable on crop yield can be evaluated. This is a powerful management tool that will assist producers on how to maximize economic crop yield while minimizing nutrient losses to surface or groundwater. Adoption of PA practices will continue to rise given the increased constraints of production, advances in technology, and demand for safe food and fiber of high quality.

REFERENCES

1. Lowenberg-DeBoer, L.; Swinton, S. Economics of site-specific management in agronomic Crops. In *The State of Site-Specific Management for Agriculture USA*; Pierce, F., Sadler, E., Eds.; ASA-CSSA-SSSA: Madison, WI, **1997**; 369–396.
2. Bongiovanni, R.; Lowenberg-DeBoer, J. Precision agriculture and sustainability. Precis. Agric. **2004**, *5*, 359–387.
3. Plant, R.E. Site-specific management: The application of information technology to crop production. Comput. Electron. Agric. **2001**, *30*, 9–29.
4. Bronson, K.F.; Booker, J.D.; Bordovsky, J.P.; Keeling, J.W.; Wheeler, T.A.; Boman, R.K.; Parajulee, M.N.; Segarra, E.; Nichols, R.L. Site-specific irrigation and nitrogen management for cotton production in the Southern High Plains. Agron. J. **2006**, *98*, 212–219.
5. Li, H.; Lascano, R.J.; Booker, J.; Wilson, L.T.; Bronson, K.F. Cotton lint yield variability in a heterogeneous soil at a landscape scale. Soil Tillage Res. **2001**, *58*, 245–258.
6. Kramer, P.J.; Boyer, J.S. *Water Relations of Plants and Soils*; Academic Press: San Diego, CA, 1995; 495 pp.
7. Wendroth, O.; Al-Oman, A.M.; Kirda, C.; Reichardt, K.; Nielsen, D.R. State-space approach to spatial variability of crop yield. Soil Sci. Soc. Am. J. **1992**, *56*, 801–807.
8. Nielsen, D.R.; Wendroth, O.; Pierce, F.J. Emerging concepts for solving the enigma of precision farming research. In *Precision Agriculture*, Proceedings of the Fourth International Conference, Minneapolis, MN, July 19–22, 1998; Robert, P.C., Rust, R.H., Larson, W.E., Eds.; 1999; 303–318.
9. Shumway, R.H.; Stoffer, D.S. *Time Series Analysis and Its Application*; Springer Verlag: New York, 2000; 549 pp.
10. Nielsen, D.R.; Wendroth, O. *Spatial and Temporal Statistics*. Catena Verlag: Reiskirchen, Germany, 2003; 398 pp.

11. de Wit, C.T. Resource use efficiency in agriculture. Agric. Syst. **1992**, *40*, 125–151.
12. Booker, J.D.; Bronson, K.F.; Keeling, J.W.; Bordovsky, J.P.; Segarra, E.; Velandia-Parra, M. Farm-scale testing of site-specific irrigation and nitrogen fertilization for cotton production in the Southern High Plains of Texas, USA. In *Precision Agriculture '05*; Stafford, J.V., Ed.; Wageningen Academic Publishers: Wageningen, Netherlands, 2005; 951–957.
13. Halvorson, G.A.; Doll, E.C. Topographic effects on spring wheat yield and water use. Soil Sci. Soc. Am. J. **1991**, *55*, 1680–1685.
14. Hanna, A.Y.; Harlan, P.W.; Lewis, D.T. Soil available water as influenced by landscape position and aspect. Agron. J. **1982**, *74*, 999–1004.
15. Timlin, D.J.; Pachepsky, Y.A.; Snyder, V.A.; Bryant, R.B. Spatial and temporal variability of corn grain yield on a hillslope. Soil Sci. Soc. Am. J. **1998**, *62*, 764–773.
16. Li, H.; Lascano, R.J. Deficit irrigation for enhancing sustainable water use: Comparison of cotton nitrogen uptake and prediction of lint yield in a multivariate autoregressive state-space model. Environ. Exp. Bot. **2011**, *71*, 224–231.
17. Lyle, W.M.; Bordovsky, J.P. Low energy precision application (LEPA) irrigation system. Trans. ASAE **1981**, *24*, 1241–1245.
18. Cassel, D.K.; Wendroth, O.; Nielsen, D.R. Assessing spatial variability in an agricultural experiment station field: Opportunities arising from spatial dependence. Agron. J. **2000**, *92* (4), 706–714.
19. Molling, C.C.; Strikwerda, J.C.; Norman, J.M.; Rodgers C.A.; Wayne R.; Morgan, C.L.S.; Diak, G.R.; Mecikalski, J.R. Distributed runoff formulation designed for a precision agricultural landscape modeling system. J. Am. Water Res. Assoc. **2005**, *41*, 1289–1313.
20. Wilson, J.N. Guidance of agricultural vehicles—A historical perspective. Comput. Electron. Agric. **2000**, *25*, 3–9.
21. Bell, T. Automatic Tractor Guidance Using Carrier-Phase Differential GPS. Comput. Electron. Agric. **2000**, *25*, 53–66.
22. Bronson, K.F.; Keeling, J.W.; Booker, J.D.; Chua, T.T.; Wheeler, T.A.; Boman, R.K.; Lascano, R.J. Influence of landscape position, soil series, and phosphorus fertilizer on cotton lint yield. Agron. J. **2003**, *95*, 949–957.
23. Bronson, K.F.; Booker, J.D.; Officer, S.J.; Lascano, R.J.; Maas, S.J.; Searcy, S.W.; Booker, J. Apparent electrical conductivity, soil properties and spatial covariance in the U.S. Southern High Plains. Precis. Agric. **2005**, *6*, 297–311.
24. Stafford, J.V. Implementing precision agriculture in the 21st century. J. Agric. Eng. Res. **2000**, *76*, 267–275.
25. Zhang, N.; Wang, M.; Wang, N. Precision agriculture—A worldwide overview. Comput. Electron. Agric. **2002**, *36*, 113–132.
26. McBratney, A.; Whelan, B.; Ancev, T.; Bouma, J. Future directions of precision agriculture. Precis. Agric. **2005**, *6*, 7–23.
27. McKinion, J.M.; Baker, D.N.; Whisler, F.D.; Lambert, J.R. Application of the GOSSYM/COMAX system to cotton crop management. Agric. Syst. **1989**, *31*, 55–65.
28. Feddes, R.A.; Kowalik, P.J.; Zaradny, H. *Simulation of Field Water Use and Crop Yield. Simulation Monographs*; Centre for Agricultural Publishing and Documentation: Wageningen, Netherlands, 1978; 188 pp.
29. de Wit, C.T. *Simulation of Assimilation, Respiration and Transpiration of Crops. Simulation Monographs*; Centre for Agricultural Publishing and Documentation: Wageningen, Netherlands, 1978; 144 pp.
30. Bronson, K.F.; Malapati, A.; Booker, J.D.; Scanlon, B.R.; Hudnall, W.M.; Schubert, A.M. Residual soil nitrate in irrigated southern high plains cotton fields and Ogallala groundwater nitrate. J. Soil Water Conserv. **2009**, *64*, 98–104.

Radio Frequency Towers: Public School Placement

Joshua Steinfeld
Florida Atlantic University, Boca Raton, Florida, U.S.A.

Abstract
The placement of radio frequency (RF) towers at public schools poses health risks to students and faculty. Experimental research on animals and humans proves that exposure to RF waves presents health hazards. The Federal Communications Commission has ratified numerous acts aimed at protecting the interests of both personal wireless service providers and coalitions who object to RF towers at public schools. The case of Wootton High School (2003–2005) is highlighted as a precedential victory for opponents of RF tower placement at schools. Federal government's charter to give local governments decision-making authority on the matter raises inquiry into federalist discussion. Meanwhile, the federal government's declared statutory support for personal wireless service providers and their policy brokers' self interests leads to examination of public administration theory.

INTRODUCTION

Personal wireless service providers, independently and through use of brokers, have installed radio frequency (RF) towers at or close to public schools in the United States. The open space surrounding schools, typically baseball and football fields, allows for optimal transmission of RF waves between towers. Furthermore, public school districts, in comparison with private enterprises, have been more easily won over in the RF tower proposal process. Surges in experimental research regarding health hazards to RF emission, especially the apparent increased susceptibility of children to RF radiation, have sparked controversy over the exact locations of RF towers.

The purpose of this manuscript is to present an argument opposing the placement of RF towers at public schools. First, scientific studies related to animals and humans are provided to prove that RF waves are harmful to humans. Second, voluntary initiatives and RF tower proposal processes are discussed, highlighting the Rockville, Maryland, community's precedential victory in opposition to the placement of an RF tower at Wootton High School. Third, ethics and public policy considerations address the need for streamlining nationwide community efforts by amending the Telecommunications Act of 1996 to disallow the placement of RF towers near public schools. Fourth, areas for further research are presented, which include the compilation of data sets, tracking of exposed students at school, and new theory to address causality issues related to competing risks. Finally, the entry concludes by providing commentary regarding invisible risks.

IMPACT AND STUDY OF RF HAZARDS

Recent Trends

Concerns regarding the safety of RF towers at school have risen across the country. In 2010, Vista del Monte Elementary School in Palm Springs, CA, was pegged as having a reputation for being a cancer school. In 2005, an RF tower was erected on campus. Since then, 12 people have been diagnosed with cancer, affecting those who worked closest to where the RF tower was installed, where the field-strength readings were highest.[1]

From 1975 to 2000, childhood cancer rates had increased dramatically by a rate of 32%. Some of the most severe and deadly cancers such as acute lymphocytic leukemia, brain, kidney, and bone cancer also increased considerably.[2] In 2004, there were 36 million prescriptions of sleeping medications. As of 2009, 56 million prescriptions were outstanding, a whopping 56% increase. In 2004, the number of residents using cell phones was 109 million. By 2009, the number was up to 271 million.[3] Public health officials and environmental experts alike have been searching for environmental stimuli that may be contributing to the increased childhood cancer rates and sleep deprivation. The placement of RF towers at schools has been an area of focus (Figs. 1 and 2). This Florida community elementary school also has a narrow-band transmission device installed on the tower (Fig. 3). Installation of narrow-band RF towers is generally more restrictive than wide-band RF towers because of the narrow band's compact, piercing wavelength. Electromagnetic stimuli

Fig. 1 Public elementary school.

Fig. 3 Narrow-band RF tower at elementary school.

emitted from RF devices have become an area of intense research interest.

Background Research

Just to get an idea of the strength of an electromagnetic force emitted from an RF tower in comparison with a more well-known object, cell phone technology operates on frequencies up to 3 gigahertz (GHz), and a microwave oven cooks food at 2450 megahertz (MHz).[4] Three GHz is equal to 300,000 MHz! Coulomb's law states that the force of an electromagnetic field is proportional to the magnitude of the charge and inversely proportional to the square of the separation. From Coulomb's law, we can derive the Inverse law, which suggests that holding the magnitude of the charge constant, the electromagnetic force emitted on a subject increases exponentially as the subject moves closer to the source.[5] Using Ampere's law, which combines the magnetostatic equation for determining the magnitude of a magnetic field with Stokes' theorem dealing with surface area, it is possible to determine the strength of an electromagnetic force through a closed path that may be tangent or indirectly exposed to the source.[5,6] Coulomb's law and the resulting derivation of the Inverse law demonstrate key findings in the discussion of RF towers. The farther away the RF tower site is from the subject, the lower the field strength absorbed. Furthermore, due to the exponential nature of the Inverse law, being close means being really close. Ampere's law is especially helpful in determining indirect exposure to force strength in cities where waves regularly bounce off other buildings.

A given material is composed of atoms. Each atom consists of electrons orbiting a central positive nucleus. Electrons also spin around their own axis. The orbital array of activity occurring between magnetic forces of protons and electrons in atoms results in an organized disarray of unpredictability. This unpredictable orbital disarray is the normal, unaltered state of the atom. When an external RF field is applied, the bombardment of electrons (from the source) stimulates host atom movement changes (in the subject), and the atom has a magnetic moment.[7] The orbital path of the electrons is brought into a slight sense of organization, throwing normal behavior out of whack. It is the electron stimulus emitted from RF towers that has been of much focus in experimental studies.

Radio Frequency Research on Animals and Humans

Scientists and researchers have proceeded cautiously regarding the use of human subjects in testing the effects of RF exposure. As a result, numerous testing on animals has been done to learn about the effects of electromagnetic radiation on living organisms. A 1997 study on mice demonstrated the effects of radiation on prenatal development and resulted in a progressive decrease in the number of newborns per dam, ending in irreversible infertility. In a

Fig. 2 Wide-band RF tower at elementary school.

subsequent 1999 study, mice exposed for just 24–72 hours to weak electromagnetic waves increased the activity of natural killer cells by 130%. Meanwhile, exposure to microwave irradiation had no effect on the activity of natural killer cells.[8] Nonetheless, microwave stimulus interfered with cell immunity of mice, increasing T-cell proliferation in response to stimulus.[9]

Studies on other living organisms have also been conducted. In Germany, behavioral abnormalities were observed in a herd of dairy cows that grazed near an RF tower for over 2 years, leading to reduction in milk yield and increased health problems.[10] In Russia, the effects of electromagnetic radiation on sea urchin embryos were tested. Only sea urchins with preexisting weakened viability were impacted by the electromagnetic radiation, in which case the electromagnetic radiation stimulated the onset of early development of embryos.[11]

While much more rare than electromagnetic testing on animals, some testing on humans has taken place. It was discovered that electromagnetic fields affect the central nervous system in humans because visual reaction time was prolonged and scores on short-term memory tests were lower in high-intensity exposure test groups.[12] Also, in a controlled study aimed to investigate the impact of low-force electromagnetic fields on healthy humans, human subjects were exposed to a 900 MHz electromagnetic field and intermittently pulsed with 217 MHz. It was determined that low-force fields have no effect on nocturnal hormone secretion under polysomnographic controls. However, cortisol production increased, which is transient by classification, indicating the organism (human subject in this case) adapted to the stimulus.[13] It was unclear if any mental impediments or genetic responses may have taken place in addition to the increase in cortisol production, but it was certain that the human subjects endured a cellular response. Nonetheless, current research indicates that genes are at risk, even at low-force electromagnetic fields.

Genes that ward off cancer and other illness may be inhibited when a cell receives stimuli from the environment. The National Institute of Environmental Health Sciences has been using genomic techniques to determine behavior of promoter-proximal paused polymerase (Pol II) with and without environmental stimuli.[14] Pol II is known to have a role in fighting disease. Transcriptional responses to environmental stimuli can cause alterations in Pol II distribution, gene expression, and epigenetic chromatin signatures, leading to transcription dysregulation that can cause etiology of cancers.[15] Additionally, recent work has revealed that signal–response pathways are loaded with Pol II prior to final gene activation, further enhancing the opportunity for cellular changes to take place as a result of harmful environmental stimuli.[16,17] Molecules transported from environmental stimuli can inhibit the signal-response pathways' ability to pause release of Pol II.[18] Pol II pausing is necessary in providing an accessible chromatin architecture for gene promoters that inevitably fight disease.[19,20] It has been determined that Pol II pausing facilities' precise control and coordination of genes is a crucial regulatory step in rate-limiting the expression of DNA damage responsive genes.[21]

The extent to which scientists and doctors alike understand the impact of RF waves on humans is unquestionable in its fineness. For example, RF waves are routinely manipulated in clinical medicine to achieve exact thermal dosimetry and thermal pattern poisoning of tumors. The important nuance to remember is that the beneficial uses of RF waves in medicine are based on the destructive qualities of the high-energy RF waves.[22]

VOLUNTARY INITIATIVE

Students Against Cell Towers

In 1963, a group of citizens became activists in opposition to the proposal of a Con Edison power plant on Storm King Mountain in New York.[23] If the Storm King success story gave life to an entire environmental movement, then the success of Students Against Cell Towers (SACT) in opposing an RF tower at Wootton High School in Rockville, Maryland, solidified the environmental movement's existence by the hundreds of RF tower opposition advocacy groups that spawned across the nation since the SACT community voluntary initiative spanned from 2003 to 2005. Although SACT is no longer an active organization, its over 100 former members carry with them knowledge of the vital considerations regarding placement of RF towers at schools.

In 2000, Cingular Wireless began an aggressive campaign to install RF towers at public schools in Montgomery County, Maryland. The high schools were targeted first, perhaps because younger children are known to be especially susceptible to the radiation emitted from RF towers, although we do not know for sure what Cingular's strategy was. Additionally, Cingular Wireless first targeted public high schools located in communities with relative economic disadvantages and therefore less likelihood of organizing against an RF tower proposal. Cingular installed RF towers at Wheaton High School, Sherwood High School, and Kennedy High School. The students in these school districts come from families with median household incomes of $55,562, $57,260, and $60,296, respectively. It was not until 2003 that Cingular approached the wealthier Wootton and Walter Johnson High Schools, with median household incomes of $74,655 and $77,568, respectively.[24] Cingular never approached Whitman High School and Churchill High School with RF tower proposals; families of students in these schools have median household incomes of $113,788 and $140,222, respectively. Cingular Wireless' schematic timeline for public school RF tower proposals aimed to first test the will of communities who were economically and educationally disadvantaged before targeting wealthier, resourceful communities. Eventually, the residents of the Wootton and Walter Johnson school districts successfully rejected the RF towers due to the organization of a community voluntary initiative.

Cingular Wireless hired an experienced attorney to implement objectives related to the Wootton RF tower proposal, the same attorney who had previously handled the installment of RF towers at 10 other public schools in Montgomery County. With little to no resistance coming from the Montgomery County public schools where RF tower installation was already in place, there was no need for Cingular to hire more than one person, an attorney rightfully so, to execute public school RF tower proposals in Montgomery County. An emphasis is placed on Cingular's need for just a one-man proposal team to demonstrate two things. First, it may be surprising how little publicity and resistance RF tower proposals typically receive, considering how big and visible they are. Second, Cingular's one-man show enabled Cingular to circumvent the proposal process. According to the Telecommunications Act of 1996, personal wireless service providers are required to notify in writing students, employees, and local residents of any proposal to erect an RF tower on public school property. Cingular's attorney had the ability to make sole judgment in his decision to refuse to notify the community regarding the plan to install an RF tower at Wootton High School.

In attempting to reject Cingular's proposal to erect an RF tower at Wootton High School, local residents formed a coalition called SACT. Despite any protections that may be offered by the Telecommunications Act of 1996, it clearly states that RF towers may not be rejected because of health concerns of nearby subjects. However, since Wootton High School is situated in a valley with Frost Middle School perched up on the adjacent valley ridge, the RF waves emitted from the RF tower at Wootton would be passing through Frost Middle School, creating the allowance for a new interpretation of the act and a platform to remain steadfast. Next, it was discovered by SACT that the principal of Wootton High School had the exclusive authority to decide on whether to allow a 150 ft RF tower to be erected next to the football field at Wootton. SACT contacted local neighborhood associations, cluster school principals and administrators of Frost Middle School and Fallsmead Elementary School, Rockville City management, Montgomery County executive offices, parent–teacher association groups, and other perceived interest groups to spread word of the issue and provide scientific research related to health risks of RF exposure. Through discussion with local government and administration, SACT realized that the community had the right to a town hall meeting prior to the consideration of an RF tower proposal. Over 100 advocates holding greater than 1400 petitions standing against the placement of an RF tower at Wootton High School arrived at the town hall meeting held in the Wootton library to greet Cingular's attorney and his science expert, with state representatives and media in attendance anxiously awaiting the confrontation. When Cingular's scientific expert utilized research no more recent than the 1960s, it became clear even to Cingular's attorney that SACT's research was plausible and that RF tower radiation may indeed be harmful to humans. For the benefit of Cingular's attorney, it was likely that he did not believe in his own company's stance; instead, his involvement was probably based on solidaristic group loyalties.[25]

An insight deserving mention was a key leadership tactic used by SACT to retrieve petition signatures. One of the exemplary practices of leadership is to inspire a shared vision.[26] Considering the perceived lack of concrete scientific data regarding human exposure to RF towers, one of the most effective ways to gain support for the petitions was to focus on the negative aesthetics of a 150 ft tower that would be visible from a Rockville resident's nearby home. Realizing the eyesore created by RF towers and the fact that opposition due to aesthetic concerns frequently arise, personal wireless service providers have begun camouflaging the RF towers (Fig. 4). Fig. 4 shows an RF tower with elaborate camouflage meant to make the RF tower look like a tree. After further survey of the surrounding area, it was discovered the camouflage was an attempt to

Fig. 4 Radio frequency tower with elaborate camouflage.

Fig. 5 The tallest tree in this image is the RF tower.

help the RF tower blend in with a tree line overlooked by a nearby middle school (Fig. 5). The canvas makes the RF tower more difficult to notice and provides personal wireless service providers with a prompted solution to aesthetic concerns.

A key ingredient behind this entry's explicit focus on the placement of RF towers at public schools as opposed to all schools in general was illustrated by the Wootton principal's authority to decide for or against installation of the RF tower. If Wootton were a private school, the principal may not have been dictated the authority of decision maker or, in the event that authority was dictated, may not have had the wherewithal to acknowledge responsibility to local residents not affiliated with the school. SACT was a community effort predominately fueled by advocates not directly associated with Wootton High School. The idea that public school principals and all public school administrators in general are public servants lends additional support for the case against RF towers at public schools in particular. However, the stance against RF towers at public schools applies to all schools because of the RF-related health hazards.

Discussion of RF Tower Placement Process

The Telecommunications Act of 1934 established the Federal Communications Commission (FCC) as the regulatory authority over communications activities in the United States. Because digital cell phones were not available to the public until 1988, controversy over the placement of RF towers is a new phenomenon. It was not until the Energy Policy Act of 1992 and Telecommunications Act of 1996 when the federal government realized the issues of RF signal strength and tower placement, respectively. In 2003, SACT became one of the first nongovernmental organizations to address hazards related to RF tower placement near public institutions. However, a number of transformations in the telecommunications industry have taken place since the Telecommunications Act of 1934, which have shaped the current regulatory environment. The years 1945 and 1952 marked the first major oversight by the FCC on over-the-air television, regulating the spectrum allocations and color standards. In 1968, telecommunications service providers were authorized by the FCC to attach equipment to preexisting above-the-ground electrical lines. And, in 1992, the FCC ruled to let the market decide the appropriate standards for digital cell phones and related equipment.[27]

The political actors typically involved in the process of determining the placement of RF towers complicate the ability of community voluntary initiatives to succeed in opposing the placement of RF towers. Achieved by a 1999 amendment to the Telecommunications Act of 1934, the local government has authority over state and federal governments on the issue of tower placement.[28] However, the amendment to the Telecommunications Act of 1934 is pursuant to the Telecommunications Act of 1996 specified requirement of the federal government to assist licensees' pursuit of preferred sites.[29] Federal involvement in the tower placement process results in streamlining of policy action. Streamlining results in the shrinking of the policy window for community advocates, which reduces community advocates' opportunity to introduce their own policies.[30] Additionally, personal wireless service providers sometimes hire independent facilities siting companies who offer comprehensive tower placement services, from lobbying of local government and communities to addressing zoning regulatory concerns. Leasing of sites and fulfillment of regulatory and registration requirements may also be taken care of by facilities siting companies (Figs. 6–8). "The local zoning authorities should therefore be aware that a facilities siting company may not be seeking the sites that are of most interest to particular Commission licensees [personal wireless service providers], but rather seek general sites on highly elevated locations in the hopes of leasing the sites, in turn, to Commission licensees."[31] It would be intuitive to reason that the existence of broker special interests in the placement of RF towers would be an additional obstacle to voluntary initiatives striving to oppose RF towers. However, the contrary is sometimes true. Policy brokers, such as independent facilities siting companies, are interested in maintaining a sustainable level of

Fig. 6 Brokers lease to wireless service providers.

Fig. 7 Warning of RF emission.

conflict in order for services being offered to remain in high demand.[32] Lingering, yet not overpowering, community opposition is welcomed by facilities siting companies. To combat the influence of the federal government and policy brokers in the tower placement process, community organizations need to adopt a policy of political efficacy, arising from political participation as a means to exert influence.[33] Political efficacy is especially important when it comes to the upholding of FCC guidelines on tower height and field strength. According to the Energy Policy Act of 1992, electric and magnetic field strength must be made public.[34] Yet, according to the Code of Federal Regulations pertaining to personal communications services, height-above-average-terrain (HAAT) and field-strength guidelines may be waived if all parties involved agree.[35] If the community does not involve local government, the community will have no voice on the issue, and the already-flaccid federal guidelines will leave school children and staff insurmountably exposed.

ETHICS AND PUBLIC ADMINISTRATION

The three elements of corporate social responsibility are market actions, externally mandated actions, and voluntary actions.[36] Personal wireless service providers fail to address all three elements of social responsibility when dealing with the placement of RF towers. First, personal wireless service providers have poorly responded to market actions in their use of policy brokers to ensnare the tower placement process. Second, the mandated actions of the FCC related to HAAT and field-strength guidelines allow regulatory thresholds to be exceeded if no opposing voluntary organization is present at scheduled hearings. Third, voluntary actions that aim to avoid students' exposure to RF towers are not taken. In fact, the current trend is just the opposite, in which schools are targeted because of the surrounding open space that allows for enhanced RF wave transmission (Figs. 9 and 10).

Fig. 8 Record keeping by the FCC.

Fig. 9 Radio frequency tower overlooking playing fields.

Fig. 10 Radio frequency tower installed adjacent to football field.

The ability of voluntary initiatives to oppose the placement of RF towers at public schools relies upon the formation of a nucleus of zealous participants; charismatic leadership alone does not result in the type of rapid expansion of the coalition that is necessary to fight seasoned corporate interests such as personal wireless service providers.[37] In addition, the media may be needed to intervene in facilitating changes in public perceptions. The media can help provide a transition and a way for the public to digest new policy initiatives.[38] In some cases, even with high visibility, the strongest coalitions are unable to defeat polyarchal interest groups on a particular issue, be it the placement of an RF tower at yet another school. Success depends on forging relationships with government officials as much as administrative competence.[39] Because of the variations in local government across the nation, including inconsistencies in the law and process governing zoning and other enforcement departments, there is a need to adopt a customizable approach when attempting to oppose an RF tower.

The SACT's ability to reject an RF tower proposal at Wootton High School but failure to impact policy change nationally brings into question the federalist debate. Hamilton, Madison, and Jay desired a strong central authority in their staunch support of federal government and ratification of the U.S. Constitution.[40] Opposing the federalists, Patrick Henry led the antifederalist approach arguing for decentralization and states' rights.[41] The federal government's determination to simultaneously support personal wireless service providers' RF tower installation campaign at public schools while allowing local communities to decide for themselves is rooted in both federalist and antifederalist modes. Nonetheless, the federal government's dual role is authoritarian.[42]

The cry for federal regulation disallowing placement of RF towers at or near public schools involves regulatory policy making, which inevitably will indulge or deprive one specific interest or another.[43] The regulatory approach offered does not necessarily favor some sort of Weberian chain-of-command hierarchy originating from the top down[44] but rather prefers reactionary cultural movement that reflects Thelen and Mahoney's,[45] Hacker's,[46] and Sheingate's[47] ideas on institutional change according to evolving assumptions of administrative and technological environments. If the dominant approach were favored, the solution would be to charge personal wireless service providers higher rental fees to compensate for any associated health care costs that may result.[48,49] Other rational socioeconomic approaches would seek to place value on the cost of the loss of a human life and monetary damages due to the terminal pain and suffering induced by RF towers. Unfortunately, the most commonly observed failure in public management and the one associated with the egalitarian approach is a lack of ability to exert authority.[50]

AREAS FOR FURTHER RESEARCH

A logical area for further research is to track and survey humans. Tracking students who attend schools with RF towers on premises is the best way to determine the danger of RF towers to students. Tracking students while in school and in years beyond may help determine whether RF towers are indeed harmful to children or humans in general. An effective data set for a future study would need to take into account differentiations between student exposure at elementary, middle, and high schools, considering that young children are known to be more susceptible to RF exposure. Additionally, the data would need to differentiate between students who had many instances of intense exposure, such as student-athletes on a sports field containing an RF tower, and students who had fewer instances of intense exposure. Previously, this type of study was not possible. It was only since 2002 that the personal wireless service providers started targeting schools. Today, RF towers are erected on school properties across the country. However, obtaining access to student records for data collection of this type of study invokes the support of government. The resources and sheer number of people who would need to be mobilized by such a study requires congressional backing. Furthermore, the scientific community has questioned the methods that would underwrite such a study. Areas of inquiry that have been considered to be obstacles to a human study of this type are the competing risks when experimenting for causality. Oftentimes, it is difficult to determine which risk is the source of the illness. For example, did cancer clusters in the area form as a result of contaminated drinking water or exposure to an RF tower? New research indicates that margins of error in causality can be reduced. Building on time-dependent predictive accuracy measures,[51,52] the coefficient of the distribution of false positives among event-free subjects can be adjusted to reflect nearest neighbor (in our case, the highly exposed student-athletes on the sports fields containing RF towers) estimation of the distribution of input variables representing true positive incidents and length of exposure of competing risks in order to help determine causality.[53] The false-positive value is manipulated by a coefficient that is calculated from the estimate of true positive incidents in order to offset causality miscalculations stemming from overlapping incident rates of competing risks. Unconcerned citizens and proponents of RF tower placement on school property argue that cancer clusters related to RF towers have not emerged. However, it is impossible to predict what the health impacts may be over time. An evaluation technique is needed for cumulating, comparing, and contrasting varied results in order to establish an applied theory and framework.[54]

CONCLUSION

Just 2 years after the victory at Storm King Mountain paved the way for environmentalism, Olson's *The Logic of Collective Action* (1965) opened our eyes to the intuitiveness of equality and why the events of the 1960s were unlikely to ever recur.[55] Much different from the equal rights movement, the successful voluntary initiative at Wootton High School has not led to nationwide legislation and is embedded in an issue that is invisible. The inability to see the RF waves somehow precludes from our psyche the notion of harm. Invisible risks skew the indifference curve that guides our behavior in responding to risk. An indifference map is our collection of indifference curves and helps to shape our order of preferences.[56] The typical reaction to an invisible risk is a delay in response. After a 5 years pause, personal wireless service providers are once again proposing RF towers at public schools in Montgomery County, Maryland, this time at Whitman High School in Bethesda, Maryland.[57] Bethesda community advocates are citing Wootton's 2005 rejection of an RF tower as precedent in Montgomery County.[58] Currently, only the 10 RF towers that were installed prior to the Wootton campaign exist in Montgomery County. Also, in the sense of a health-wise decision at Wootton, why have the 10 previously erected towers not been taken down?

Radio frequency towers should not be placed at schools. Scientific experimental research on animals and humans is conclusive that RF waves increase animal and human risk to cancer and other illnesses. Also, in the case of public schools, students have little to no choice in deciding whether or not to attend a particular school. To address personal wireless service providers' desire to bolster cell phone connectivity in communities, state and local governmental zoning boards should work together to designate specific areas where the placement of RF towers will be permitted. Assuming no RF tower is placed at or near a public school, the designated zones will enable the public to make their own decision regarding living or spending time near RF tower sites.

Community voluntary initiatives are the answer if the goal is to reject an RF tower proposal. Federal governments' involvement, not from the standpoint of assisting personal wireless service providers meet their objective of RF tower placement in a given community but from the standpoint of bringing together the nationwide advocacy groups and coalitions that have both successfully and unsuccessfully defended their schools from RF wave penetration, is the long-term solution to keeping schools safe from RF towers. Also, although the prospect of fair gamesmanship in the process of appeal against RF tower placement at public schools was not presented to be optimistic, consideration of bureaucratic red tape, or in this case, purposeful lack thereof, is essential to successfully implementing a strategy that leads to the rejection of a personal wireless service provider's RF tower proposal. In any case, policies that engage citizens in their own communities and ask of them to do their own policy analysis are generally more preferable to those policies that do not.[59]

REFERENCES

1. Milham, S. *Dirty Electricity, Electrification and the Diseases of Civilization*; Universe: New York, 2010.
2. *The Stop Cancer Before It Starts Campaign*, Press Report, May 8, 2003, Cancer Prevention Coalition, available at http://preventcancer.com/publications/pdf/M803.htm (accessed September 2011).
3. LeBeau, C. *Insomnia, Fatigue, and Cell-Phone Towers*; Vital Health Publications: West Allis, WI, 2010.
4. Lai, H.; Biological effects of radiofrequency radiation from wireless transmission towers. In *Cell Towers, Wireless Convenience? Or Environmental Hazard?*; Levitt, B., Ed.; New Century Publishing: Markham, Canada, 2000; 65.
5. Schwartz, M. *Principles of Electrodynamics*; McGraw Hill: New York, 1972; 29–30, 148–149.
6. Edminister, J.; Nahri-Dekhordi, M. *Electromagnetics*, 3rd Ed.; McGraw Hill: New York, 2011; 174.
7. Sadiku, M. *Elements of Electromagnetics*; Oxford University Press: New York, 2010; 350.
8. Fesenko, E.; Novoselova, E.; Semiletova, N.; Aganova, T.; Sadovnikov, V. Stimulation of murine natural killer cells by weak electromagnetic waves in the centimeter range. Biofizika **1999**, *44* (4), 737–741.
9. Fesenko, E.; Makar, V.; Novoselova, E.; Sadovnikov, V. Microwaves and cellular immunity. Effect of whole body microwave irradiation on tumor necrosis factor production in mouse cells. Bioelectrochem. Bioenerg. **1999**, *49* (1), 29–35.
10. Loscher, W. Extraordinary behavior disorders in cows in proximity to transmission stations. Der Praktische Tierarz **1998**, *79* (1), 437–444.
11. Galat, V.; Mezhevikina, L.; Zubin, M.; Lepikhov, K.; Khramov, R.; Chailakhian, L. Effect of millimeter waves on the early development of the mouse and sea urchin embryo. Biofizika **1999**, *44* (1), 137–140.

12. Chiang, H.; Yao, G.; Fang, Q.; Wang, K.; Lu, D.; Zhov, Y. Health effects of environmental electromagnetic fields. J. Bioelectr. **1989**, *8* (1), 127–131.
13. Mann, K.; Wagner, P.; Brunn, G.; Hassan, F.; Hiemke, C.; Roschke, J. Effects of pulsed high-frequency electromagnetic fields on the neuroendocrine system. Neuroendocrinology **1998**, *67* (2), 139–144.
14. Nechaev, S.; Fargo, D.; dos Santos, G.; Liv, L.; Gao, Y; Adelman, K. Global analysis of short RNA's reveals widespread promoter-proximal stalling and arrest of Pol II in Drosophila. Science **2010**, *327* (5963), 335–338.
15. Adelman, K. *Transcriptional Responses to the Environment Group, Chromatic Signatures and Gene Expression.* Principal Investigator. National Institute of Environmental Health, available at http://niehs.nih.gov/research/atniehs/labs/lmc/tre/index.cfm (accessed September 2011).
16. Muse, G.; Gilchrist, D.; Nechaev, S.; Shah, R.; Parker, J.; Grissom, S.; Zeitlinger, J.; Adelman, K. RNA polymerase is poised for activation across genome. Nat. Genet. **2007**, *39* (12), 1507–1511.
17. Zeitlinger, J.; Stark, A.; Kellis, M.; Hong, J.; Nechaev, S.; Adelman, K.; Levin, M.; Young, R. RNA polymerases stalling at developmental control genes in the Drosophila embryo. Nat. Genet. **2007**, *39* (12), 1512–1516.
18. Boettiger, A.; Ralph, P.; Evans, S. Transcriptional regulation: effects of promoter proximal pausing on speed, synchrony, and reliability. Computational Biology **2011**, *7* (5), 1–14.
19. Gilchrist, D.; Nechaev, S.; Lee, C.; Gosh, S.; Collins, J.; Li, L.; Gilmour, D.; Adelman, K. NELF-mediated stalling of Pol II can enhance gene expression by blocking promoter-proximal nucleosome assembly. Genes Dev. **2008**, *22* (14), 1921–1933.
20. Gilchrist, D.; dos Santos, G.; Fargo, D.; Xie, B.; Gao, Y.; Li, L.; Adelman, K. Pausing of RNA polymerase II disrupts DNA-specified nucleosome organization to enable precise gene regulation. Cell **2010**, *143* (4), 540–541.
21. Adelman, K.; Kennedy, M.; Nechaev, S.; Gilchrist, D.; Muse, G.; Chineov, Y.; Rogatsky, I. Immediate mediators of the inflammatory response are poised for gene activation through RNA polymerase stalling. Proc. Nat. Acad. Sci. **2009**, *106* (43), 18207–18212.
22. Kasevich, R. Brief overview of the effects of electromagnetic fields on the environment. In *Cell Towers, Wireless Convenience? Or Environmental Hazard?*; Levitt, B., Ed.; New Century Publishing: Markham, Canada, 2000; 170.
23. Anzevino, J. Preserving scenic and historic sites: the dilemma of siting cell towers and antennas in sensitive areas. In *Cell Towers, Wireless Convenience? Or Environmental Hazard?*; Levitt, B., Ed.; New Century Publishing: Markham, Canada, 2000; 169.
24. Public School Review. *Public Elementary, Middle, and High Schools*, Data, 2011, available at available at http://www.public schoolreview.com (accessed September 2011).
25. Dunleavy, P. *Democracy, Bureaucracy, and Public Choice*; Prentice Hall: London, 1991; 28.
26. Kouzes, J.; Posner, B. *The Leadership Challenge*, 4th Ed.; Jossey-Bass: San Francisco, 2007; 16–18, 142–143.
27. Ismail, S. *Transformative Choices: A Review of 70 Years of FCC Decisions*, FCC Staff Working Paper 1; Federal Communications Commission: Washington, DC, 2010; 1, 18, 19.
28. *A Bill to Amend the Communications Act of 1934*, 106th Congress, 1st Session; U.S. Senate: Washington, DC, 1999; S.1538.
29. *New Wireless Tower Siting Policies*, Fact Sheet #1; Wireless Telecommunications Bureau, Federal Communications Commission: Washington, DC, 1996; 1–2.
30. Kingdon, J. *Agendas, Alternatives, and Public Policies*, 2nd Ed., Longman: New York, 2003; 165.
31. *National Wireless Tower Siting Policies*, Fact Sheet #2; Wireless Telecommunications Bureau, Federal Communications Commission: Washington, DC, 1996; 8.
32. Sabatier, P. An advocacy coalition framework of policy change and the role of policy-oriented learning. Policy Sci. **1988**, *21* (1), 141.
33. Sabatier, P. Political science and public policy. Political Sci. Polit. **1991**, *24* (2), 145.
34. *Energy Policy Act of 1992, Section 2118*, Electric and Magnetic Fields Research and Public Information Dissemination (RAPID), Public Law 102-486; 42 U.S.C. 13478.
35. *Code of Federal Regulations, Title 47*, Calculation of height above average terrain and field strength limits; Personal Communications Services, 2010; 24.53, 24.236.
36. Steiner, J.; Steiner, G. *Business, Government, and Society*; McGraw Hill: New York, 2012; 131.
37. Downs, A. *Inside Bureaucracy*; Little, Brown, and Company: Boston, 1967; 7, 9.
38. Sabatier, P.; Mazmanian, D. The implementation of public policy: A framework of analysis. Policy Stud. **1980**, 8 (4), Special no. 2, 550.
39. Nalbandian, J. Reflections of a "pracademic" on the logic of politics and administrations. Pub. Admin. Rev. **1994**, *54* (6), 531.
40. Kesler, C.; Rossiter, C. *The Federalist Papers;* New America Library: New York, 1999, vii–xii.
41. Henry, P. Legitimate government. In *The Anti-Federalist Papers.* Borden, M., Ed.; Michigan State University Press: East Lansing, MI; par. 5.
42. Steinfeld, J. American authoritarian democracy: Vietnam War, Kosovo War, and Overseas Contingency Operation. In Proceedings of the 32nd Annual Southeastern Conference for Public Administration, New Orleans, LA, Sept 21–24, 2011.
43. Miller, H. Weber's action theory and Lowi's policy types in formulation, enactment, and implementation. Policy Stud. **1990**, *18* (4), 895.
44. Weber, M. *The Theory of Social and Economic Organization*; The Free Press: New York, 1947.
45. Thelen, K.; Mahoney, J. *Explaining Institutional Change: Ambiguity, Agency, and Power*; Cambridge University Press: Cambridge, England, 2010.
46. Hacker, J. Policy drift: The hidden politics of welfare state retrenchment. In *Beyond Continuity: Institutional Change in Advanced Political Economies*; Streeck, W., Thelen, K., Eds.; Oxford University Press: Oxford, 2005; 17.
47. Sheingate, A. Rethinking rules: Creativity in the House of Representatives. In *Explaining Institutional Change: Ambiguity, Agency, and Power*; Thelen, K., Mahoney, K., Eds.; Cambridge University Press: Cambridge, England, 2010.
48. Weber, M. *Economy and Society, An Outline of Interpretive Sociology*; Roth, G., Wittich, C., Eds.; University of California Press: Berkeley, CA, 1978; Vol. 1.
49. Weber, M. *Economy and Society, An Outline of Interpretive Sociology*; Roth, G., Wittich, C., Eds.; University of California Press: Berkeley, CA, 1978; Vol. 2.

50. Hood, C. *The Art of the State: Culture, Rhetoric, and Public Management*; Oxford University Press: Oxford, 1998; 40.
51. Heagerty, P.; Lumley, T.; Pepe, M. Time-dependent ROC curves for censored survival data and diagnostic marker. Biometrics **2000**, *56* (1), 337–344.
52. Heagerty P.; Zheng, Y. Survival mode predictive accuracy and ROC curves. Biometrics **2005**, *61* (1), 921.
53. Saha, P.; Heagerty, P. Time-dependent predictive accuracy in the presence of competing risks. Biometrics **2010**, *66* (4), 999–1011.
54. Lowi, T. American business, public policy, case-studies, and political theory. World Polit. **1964**, *16* (4), 688.
55. Hirschman, A. *Shifting Involvements: Private Interests and Public Affairs*; Blackwell: Oxford, 1985; 79.
56. Anderton, C.; Carter, J. *Principles of Conflict Economics*; Cambridge University Press: New York, 2009; 29.
57. Barnes, A. *Cell Tower Proposed For Walt Whitman High School*; USA 9 News: Bethesda, MD, 2010, available at http://wusa9.com/news/local/stay.aspx?stayid=95703&catid=158 (accessed September 2011).
58. Cropper, M. No cell tower at Whitman. Rockville Gazette, 2011, available at http://ww2.gazette.net/stories/03102010/montlet175220_32598.php (accessed September 2011).
59. Clarke, J.; Ingram, H. A founder: Aaron Wildavsky and the study of public policy. Policy Stud. **2010**, *38* (3), 574.

Radioactivity

Bogdan Skwarzec
Faculty of Chemistry, University of Gdansk, Gdansk, Poland

Abstract
Radioactivity refers to the particles that are emitted from nuclei as a result of nuclear instability. Since the nucleus experiences an intense conflict between the two strongest forces in nature, it should not be surprising that there are many nuclear isotopes that are unstable and emit some kind of radiation. The most common types of radiation are called alpha, beta, and gamma radiation; however, there are several other varieties of radioactive decay. This entry presents the history of radioactivity discovery and defines the most important radioactivity parameters, as well as characterizes the natural and artificial radionuclides in the natural environment. Applications of radionuclides in science, radiology, and industry are also described, as well as the problem of radioactive pollution of the natural environment.

INTRODUCTION

Radioactivity was discovered at the end of the 19th century by Henri Becquerel, Marie Curie (Polish native name, Maria Skłodowska-Curie), and Pierre Curie. Henri Becquerel found that uranium salts caused fogging of an unexposed photographic plate,[1] and Marie Curie discovered that only certain elements gave off these rays of energy.[2] She named this behavior "radioactivity" (natural radioactivity). A systematic search for the total radioactivity in uranium ores also guided Marie Curie to isolate a new element, polonium, and to separate a new element, radium.[3–7] The two elements have chemical similarity that would otherwise have made them difficult to distinguish from each other. In 1934, Marie Curie's daughter, Irene Joliot-Curie and her husband Frederic Jean Joliot were the first creators of artificial radioactivity. They bombarded boron with alpha particles to make the neutron-poor nitrogen isotope ^{13}N; this isotope emitted positrons. In addition, they bombarded aluminum and magnesium with neutrons to make new radioisotopes.[8]

RADIOACTIVE DECAY

Radioactive decay is the process by which an unstable atomic nucleus spontaneously loses energy by emitting ionizing particles and radiation. The three main types of radiation were discovered by Ernest Rutherford, the alpha (α), beta (β), and gamma (γ) rays (alpha, beta, and gamma radiation).[9–11] With Ernest Rutherford, he saw that radioactive substances are transformed from one element to another. About 10 years later, he unraveled the rules for the elemental transformations that accompanied radioactive decay, first for α decay and later for β decay.

Emission of an α particle changes the emitting atom to an atom of the element two places to the left in the periodic table; emission of a β$^-$ particle changes the emitting atom to an atom of the element one place to the right. These rules taken together are known as the Displacement Law; Kazimierz Fajans published it slightly earlier than did Soddy in 1913.[12] At about the same time, Soddy came to the conclusion that several substances with different radioactive properties and different atomic weights were chemically the same element. He named such substances isotopes.[13] Now, the radioactive principles are named the Soddy–Fajans periodic method.

$$\alpha \text{ decay:} \; ^A_Z X \rightarrow \; ^{A-4}_{Z-2} Y + \; ^4_2 \text{He}$$

$$\beta^- \text{ decay:} \; ^A_Z X \rightarrow \; ^A_{Z+1} Y + \; ^0_{-1}\beta$$

$$\beta^+ \text{ decay:} \; ^A_Z X \rightarrow \; ^A_{Z-1} Y + \; ^0_{+1}\beta$$

where X and Y are symbols for nuclides, Z is the mass number, and A is the atomic number.

1. Alpha (α) decay is a method of decay in large nuclei. Alpha particles (helium nuclei, He^{2+}), consisting of two neutrons and two protons, are emitted. Because of the particles' relatively high charge, it is heavily ionizing and will cause severe damage if ingested. However, owing to the high mass of the particle, it has little energy and a low range; typically, alpha particles can be stopped with a sheet of paper (or skin).
2. Beta minus (β$^-$) radiation consists of an energetic electron. It is less ionizing than alpha radiation, but more than gamma. The electron can be stopped with a few centimeters of metal. It occurs when a neutron decays into a proton in a nucleus, releasing the beta

particle and an antineutrino. Beta-plus (β^+) radiation is the emission of positrons. As these are antimatter particles, they annihilate any matter nearby, releasing gamma photons.
3. Gamma (γ) radiation consists of photons with a frequency greater than 10^{19} Hz. Gamma radiation occurs to rid the decaying nucleus of excess energy after it has emitted either alpha or beta radiation.

The activity (A) of radionuclide is lost at time (t) according to the formula

$$A = A_0 \cdot e^{-\lambda \cdot t}$$

where A is the radionuclide activity at time $t = 0$, A_0 is the radionuclide activity at time t, and λ is the decay constant of the radionuclide.

The SI unit of activity is the becquerel (Bq). One becquerel is defined as one transformation (or decay) per second. Another unit of radioactivity is the curie (Ci), which was originally defined as the amount of radium emanation (by gaseous radon-222), in equilibrium with 1 g of pure radium isotope ^{226}Ra. At present, it is equal, by definition, to the activity of any radionuclide decaying with a disintegration rate of 3.7×10^{10} Bq. The activity of a radioactive substance is characterized by its half-time—the time taken for the activity of a given amount of radioactive substance to decay to half of its value.[14,15]

After the discovery of neutron in 1932, Encico Fermi and colleagues studied the results of bombarding uranium with neutron in 1934.[16] The first person that mentioned the idea of nuclear fission in 1934 was Ida Noddack.[17] After Fermi's publication, Lise Meittner, Otto Hahn, and Fritz Strassmann began to perform a similar experiment and discovered nuclear fission of uranium ^{235}U in 1938.[18,19] Also, Józef Rotblat in 1939 published the results of a study about fission of uranium ^{235}U nuclei.[20]

In nuclear physics and nuclear chemistry, nuclear fission is a nuclear reaction in which the nucleus of an atom splits into smaller parts (lighter nuclei), often producing free neutrons and photons (in the form of gamma rays), as well. Fission of heavy elements is an exothermic reaction that can release large amounts of energy both as electromagnetic radiation and as kinetic energy of the fragments (heating the bulk material where fission takes place).

Three heavy radionuclides, natural ^{235}U and artificial ^{239}Pu and ^{233}U, are capable of reactions (nuclear fission) in which an atom's nucleus splits into smaller parts, releasing a large amount of energy in the process. During the fission of ^{235}U, three neutrons are released in addition to the two daughter atoms (see reaction below) (Fig. 1).[21]

$$^{235}_{92}U + ^{1}_{0}n \rightarrow (^{236}_{92}U) \rightarrow ^{90}_{36}Kr + ^{143}_{56}Ba + 3^{1}_{0}n + 200\ \text{MeV}$$

NATURAL AND ARTIFICIAL RADIONUCLIDES

Radionuclides present in the natural environment are classified as either of natural or anthropogenic origin. Naturally occurring radionuclides occur in different ecosystems with cosmogenic and primordial providence.[22,23]

1. *Cosmogenic radionuclides*: Cosmic ray–produced radionuclides are generated in the upper-atmosphere gases, e.g., O_2, N_2, and Ar. They are transported to the lower atmosphere and next to the oceans and to the continents. Most of the cosmic radionuclides are produced in very small amounts and only four of them, ^{3}H, ^{7}Be, ^{14}C, and ^{22}Na, constitute significant contributions to the radiation dose to humans. Cosmogenic radionuclides have been measured in humans, topsoil, polar ice, surface rocks, sediments, the biosphere, the ocean floor, and the atmosphere.[24]
2. *Primordial radionuclides*: Among non-series radionuclides of terrestrial origin, only ^{40}K and ^{87}Rb are significant sources of radiation to humans. They are characterized by a long half-time (more than 10^9

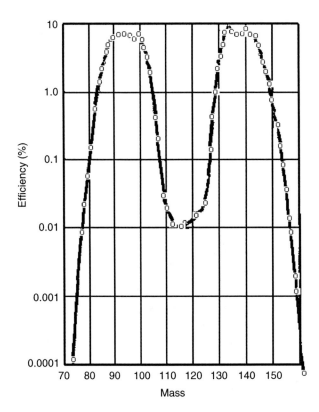

Fig. 1 Fission yield as a function of mass number for the slow neutron fission of ^{235}U.
Source: Environmental radiochemistry and radiological protection.[21]

years) and small concentrations in crustal rocks (below 1 mBq/kg).[25,26]

The serially occurring radionuclides are contained in four natural decay series—uranium, thorium, actinium, and neptunium—and, except for the actinium series, are named after their parent nuclides (Figs. 2–5).

ANTHROPOGENIC RADIONUCLIDES

Anthropogenic-derived radionuclides have been mainly released from several sources since the 1940s. Major sources in the environment are nuclear weapons, nuclear power production, accidents (e.g., the Chernobyl accident in 1986), radioactive waste disposal, solid radioactive waste disposal, and man-made radionuclides as tracers of environmental processes. Fallout from nuclear weapons explosions represents the largest contribution of anthropogenic-derived radionuclides to the natural environment. Anthropogenic radionuclides are divided into three groups:[22,24]

1. *Neutron activation products:* By neutron irradiation of objects, it is possible to induce radioactivity. This activation of stable isotopes enables to create radioisotopes. A lot of artificial radionuclides in the natural environment are produced as a result of the activation

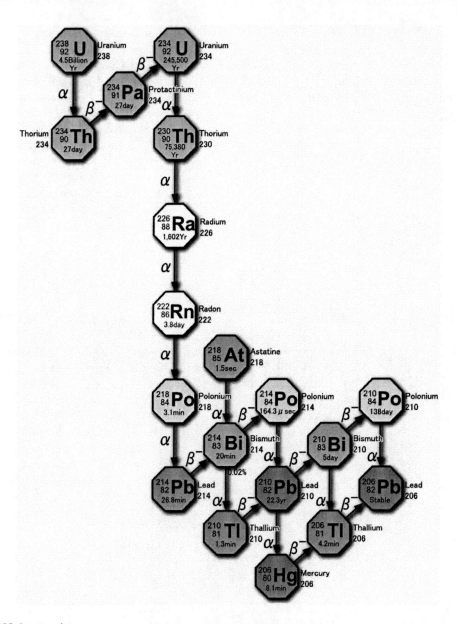

Fig. 2 Uranium-238 decay series.
Source: Wikipedia, uranium series decay chain, http://upload.wikimedia.org/wikipedia/commons/a/a1/Decay_chain%284n%2B2%2C_Uranium_series%29.PNG.[40]

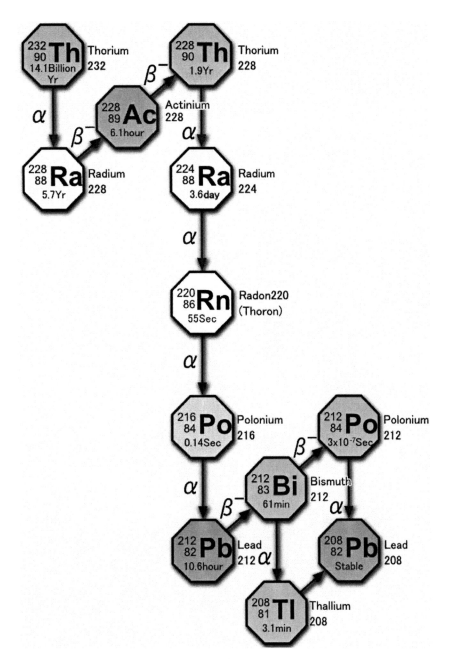

Fig. 3 Thorium-232 decay series.
Source: Wikipedia, Thorium series decay chain, http://upload.wikimedia.org/wikipedia/commons/1/1c/Decay_chain%284n%2C Thorium_series%29.PNG.[41]

process during nuclear weapons tests, the work of reprocessing plants and nuclear reactors used in power plants, as well as in nuclear studies. Owing to the use of new radioanalytical techniques, activation products such as ^{22}Na, ^{51}Cr, ^{54}Mn, ^{65}Zn, ^{110m}Ag, and ^{124}Sb could be detected in the natural environment.[26]

2. *Fission radionuclides:* During the fission of ^{235}U, three neutrons are released in addition to two daughter atoms. In the detonation of a nuclear bomb, radioactive fission products are generated from the primary fission of ^{235}U or ^{239}Pu. The most important radionuclides from two families are ^{90}Sr, ^{95}Zr, ^{131}I, ^{132}I, ^{132}Te, ^{137}Cs, ^{140}Ba, and ^{144}Ce. These radionuclides are deposited from the atmosphere to the surface of earth, with the fallout comprising components from the stratosphere (78%), local radioactive pollution (12%), and the troposphere (10%).[27]

3. *Transuranic elements:* In chemistry, transuranic elements are chemical elements with atomic numbers greater than 92 (the atomic number of uranium). All

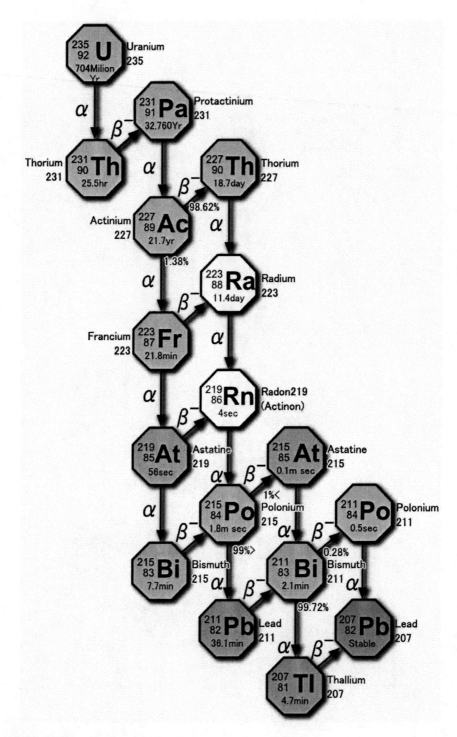

Fig. 4 Actino-uranium 235U decay series.
Source: Wikipedia, Actinium series decay chain, http://upload.wikimedia.org/wikipedia/commons/1/1e/Decay_chain%284n%2B3%2C_Actinium_series%29.PNG.[42]

transuranic elements are radioactive; 20 transuranic elements have been discovered to date: neptunium (Np), plutonium (Pu), americium (Am), curium (Cm), berkelium (Bk), californium (Cf), einsteinium (Es), fermium (Fm), mendelevium (Md), nobelium (No), lawrencium (Lr), rutherfordium (Rf), dubnium (Db), seaborgium (Sg), bohrium (Bh), hassium (Hs), meitnerium (Mt), darmstadium (Ds), roentgenium (Rg), and copernicium (Cn). Small quantities of neptunium and plutonium are found in nature (in uranium rocks), but most of them are synthesized in nuclear reactors. The most important sources of transuranic elements (generally, neptunium, plutonium, americium, and curium) in the natural environment are nuclear weapons explosions and nuclear

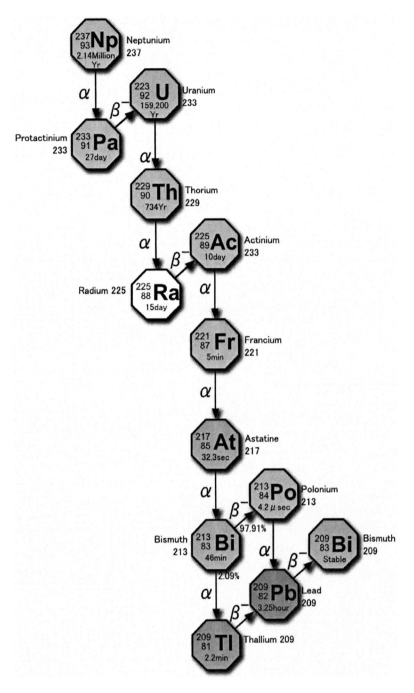

Fig. 5 Neptunium-237 decay series.
Source: Wikipedia, Neptunium series decay chain, http://upload.wikimedia.org/wikipedia/commons/8/8c/Decay_chain%284n%2B1%2CNeptunium_series%29.PNG.[43]

power production as a result of the activation process of uranium, ^{238}U.[21,27,28] The environmental chemistry of some transuranic elements, such as plutonium, is complicated by the fact that solutions of this element can undergo disproportionation, and as a result many different oxidation states can coexist at once.

The most important (long half-time, type of decay, and strong radiotoxicity) natural and anthropogenic radionuclides present in the environment are as follows:[21]

Naturally occurring radionuclides

a. Radionuclides of terrestrial origin—primordial non-series radionuclide (e.g., ^{40}K and ^{87}Rb)
b. Cosmogenic radionuclides (e.g., ^{3}H and ^{14}C).
c. Primary radionuclides—primordial series radionuclide: long-lived; have been ubiquitous on Earth since their formation (i.e., ca. 4.5×10^{9} years ago). The radionuclides ^{238}U, ^{232}Th, and ^{235}U are the parent members

of the uranium, thorium, and actinouranium radioactive decay series, respectively

Anthropogenic Radionuclides

a. Neutron activation products (e.g., ^{54}Mn, ^{55}Fe, ^{60}Co, ^{63}Ni, ^{64}Cu, ^{65}Zn, ^{110m}Ag, ^{124}Sb, and ^{125}Sb)
b. ^{235}U and ^{239}Pu fission radionuclides (e.g., ^{90}Sr, ^{95}Zr, ^{131}I, ^{132}I, ^{132}Te, ^{137}Cs, and ^{144}Ce)
c. Transuranic elements (e.g., ^{237}Np, ^{238}Pu, ^{239}Pu, ^{240}Pu, ^{241}Pu, ^{241}Am, and ^{243}Am)

Natural and artificial radionuclides in different environmental samples (natural water, sediments, soils, biological organisms) are determined by many radiometric methods, in particular neutron activation analysis (NAA), and alpha, beta, and gamma spectrometry.[29,30]

SOURCES OF RADIONUCLIDES IN THE ENVIRONMENT AND POLLUTION PROBLEM

Radionuclides are present in the whole environment in which we live. The principal sources of radionuclides in the natural environment are the wet and dry atmospheric radioactive fallout of particles from natural rock erosion processes and nuclear weapons tests, as well as release from power plants, nuclear submarines, and nuclear reprocessing facilities. Since April 26, 1986, another source of artificial radionuclides, the Chernobyl-originated radioactive debris, has had to be taken into account.[21] A lot of artificial radionuclides (e.g., 200 kg of ^{3}H, 1550 kg of ^{14}C, >160 kg of ^{90}Sr, ~350 kg of ^{137}Cs, and >4600 kg of plutonium isotopes—^{238}Pu, ^{240}Pu, and ^{241}Pu) were deposited in different components of the natural environment (seas and oceans, biota, air, and land).[21] The large amount of radioactive waste produced by human utilization of nuclear power is nowadays one of the major problems. The marine environment is especially exposed to radionuclide contamination because many nuclear power plants and nuclear reprocessing facilities are located in coastal areas. In recent decades, the development of new technologies has also resulted in the production of by-products and waste with the so-called technologically enhanced naturally occurring radioactive materials (TENORM).[31] Therefore, human technological activities (e.g., gas, oil, coal, and fertilizers industries) can increase radiation exposure, not only to the persons directly involved in these activities but also to the local or even the whole population.[32] Moreover, coal mining is the source of a huge amount of waste containing large quantities of natural radionuclides, especially polonium, radium, thorium, and uranium. During combustion, some radionuclides are emitted to the atmosphere as gas and radioactive dust; others remain as concentrated ash.[33,34] Phosphate rocks that are used to produce phosphate fertilizers contain natural radioactive elements (polonium, radium, uranium). During the production of phosphate fertilizers, about 10% of the initial ^{226}Ra, 20% of uranium, and about 85% of ^{210}Po is found in the waste phosphogypsum.[35] These radionuclides are leached by rain from phosphogypsum and, as a consequence, in the neighborhood of the phosphate fertilizers plant, their concentration in soil, flora, and water samples is much higher than in non-contaminated areas.[36,37] Radionuclides are strongly accumulated by some species and the bioaccumulation factor values for some radioactive elements (polonium, plutonium, americium) in sea algae, benthic animals, and fish are more than 1.000.[38,39] Some of these organisms are often used as bioindicators of radioactive pollution of the natural environment.[21] Also, transuranic elements (especially plutonium) belong to the group of radioactivity caused by humans. These radionuclides are important from the radiological point of view due to their high radiotoxicity, long physical lifetime, high chemical reactivity, and long residence in biological systems in the natural environment.[39]

SOLUTION TO RADIOACTIVE POLLUTION

A possible solution to radioactive pollution is by the reduction of radioactive emission to the natural environment, change of nuclear technology, and recognition of determination and accumulation processes in living organisms. Plants and animals are capable of accumulating natural and artificial radionuclides from the environment. That is why it is very important to recognize the impact of radionuclides on living organisms and their possible transfer to the human body by way of feeding.[21] Due to the importance of water, air, and food (also cigarette smoking) to human life, their quality must be strictly controlled and monitored. For this reason, studies of food for human consumption must be performed to guarantee that food materials have a low level of radioactivity, both natural and artificial. Especially, long-lived alpha emitters are the most dangerous nuclides in case of ingestion, because the long-term effects of their intake on the human body are the most important from the radiochemical and radiological points of view. A large contamination to the radiation dose received by humans comes from naturally occurring radionuclides accumulated in the body. At the moment, knowledge about accumulation of natural alpha radionuclides by organisms and their ingestion by humans is still very poor.

CONCLUSION

Radioactivity and radionuclides have been widely applied for more than a hundred years.

Radioactive substances are used to study living organisms, to diagnose and treat diseases (nuclear medicine), to sterilize medical instruments and food, to produce energy for heat and electric power, and to monitor various steps in all types of industrial processes, as well as in research studies (nuclear physics, chemistry, radiology, geochronology and geology, and cosmic research). Many natural and artificial radionuclides are strong radiotoxicants (generally, alpha emitters with a long half-time) to biological organisms. Studies on the bioaccumulation and distribution of radionuclides in different components of the natural environment are very important for radioprotection. Radionuclides used to produce energy in nuclear power plants generate the present and future problems of adequate and responsible utilization of nuclear wastes. For these reasons, the determination and distribution of some important natural and artificial radionuclides in the natural environment should be controlled and monitored.

REFERENCES

1. Becquerel, H. Sur les radiations invisibles emises par les sels d'uranium. C. R. Acad. Sci. Paris **1896**, *122*, 689–694.
2. Skłodowska-Curie, M. Rayons emis par les composes de l'uranium et du thorium. C. R. Acad. Sci. Paris **1898**, *126*, 1101–1103.
3. Curie, P.; Skłodowska-Curie, M. Sur une substance nouvelle radio-active continue dans la pechblend. C. R. Acad. Sci. Paris **1898**, *127*, 175–178.
4. Skłodowska-Curie, M. Poszukiwania nowego metalu w pechblendzie. Światło **1898**, *1*, 54.
5. Curie, P.; Curie, M.; Bemont, G. Su rune nouvelle substance fortement radioactive, continue dans la pechblende. C. R. Acad. Sci. Paris **1898**, *127*, 1215–1217.
6. Sklodowska-Curie, M. Les rayons de Becquerel et le polonium. Rev. Gen. Sci. Pures Appl. **1899**, *10*, 41–50.
7. Curie, M. Sur le poids atomique du radium. C. R. Acad. Sci. Paris **1902**, *135*, 161–163.
8. Curie, I.; Joliot, M.F. Un nouveau type de radioactivite. C. R. Acad. Sci. Paris **1934**, *198*, 254–256.
9. Rutherford, E.; Soddy F. The cause and nature of radio-activity. Philos. Mag. **1902**, *4*, 370–396, 569–585.
10. Rutherford, E.; Soddy, F. Radioactive change. Philos. Mag. **1903**, *6*, 445–457.
11. Rutherford, E. The scattering α, β and γ particles by matter. Philos. Mag. **1911**, *21*, 669–688.
12. Fajans, K. Die stellung der radioelemente im periodischen system. Phys. Z. **1913**, *14*, 136–142.
13. Soddy, F. The radio-element and the periodic law. Chem. News **1913**, *107*, 97–99.
14. Radioactivity. *Encyklopedia Britanica Online*, December 18, 2006.
15. Browne, E.; Firestone, F.B. *Table of Radioactive Isotopes*, Shirley, V.S., Ed.; John Willey and Sons: New York, 1986.
16. Fermi, E.; Amaldi, E.; D'Agostino, O.; Rasetti, F.; Segrè, E. Radioacttività provocata da bombardamento di neutroni III. Ric. Sci. **1934**, *5* (1), 452–453.
17. Noddack, I. Über das element 93. Z. Angew. Chem. **1934**, *47* (37), 653–655.
18. Meitner, L; Frisch, O.R. Disintegration of uranium by neutrons: A new type of nuclear reaction. Nature **1939**, *143*, 239–240.
19. Hahn, O.; Strassmann, F. Über den Nachweis und das Verhalten der bei der Bestrahlung des Urans mittels Neutronen entstehenden Erdalkalimetalle. Naturwissenschaften **1939**, *27* (1), 11–15.
20. Rotblat, J. Emission of neutrons accompanying the fission of uranium nuclei. Nature **1939**, *143*, 852.
21. Skwarzec, B. Radiochemia środowiska i ochrona radiologiczna (in Polish), Environmental radiochemistry and radiological protection, Wydawnictwo DJ sc., Gdańsk, 2002.
22. Sobkowski, J.; Jelińska-Kaczmarczuk, M. *Chemia Jądrowa*; Wydawnictwo Adamantan: Warszawa, Poland, 2006.
23. Choppin, G.R.; Liljenzin, J.O.; Rydberg, J. *Radiochemistry and Nuclear Chemistry*; Butterworth-Heinemann Ltd.: Oxford, U.K., 1995.
24. Magill, J.; Galy, J. *Radioactivity, Radionuclides, Radiation*; Springer Verlag: Berlin, 2005.
25. Brune, D.; Forkman, B.; Persson, B. *Nuclear Analytical Chemistry*; Verlag Chemie International Inc.: Deerfield Beach, FL. 1984.
26. Holm, E., Ed. *Radioecology, Lectures in Environmental Radioactivity*; World Scientific: Singapore, 1994.
27. MacKenzie, A.B. Environmental radioactivity: Experience from the 20th century—Trends and issues for the 21st century. Sci. Total Environ. **2000**, *249*, 313–329.
28. Hardy, E.P.; Krey, P.W.; Volchok, H.L. Global inventory and distribution of fallout plutonium. Nature **1973**, *241*, 444–445.
29. Skwarzec, B. Determination of radionuclides in the aquatic environment. In *Analytical Measurements in Aquatic Environments*; Namieśnik, J., Szefer, P., Eds.; CRC Press, Taylor and Francis Group: Boca Raton, FL, 2010; 241–258.
30. L'Annunziata, M.F. *Handbook of Radioactivity Analysis*, 2nd Ed.; Academic Press, Oxford, Great Britain, 2003.
31. TENORM Sources, Radiation Protection, U.S. EPA, March 11th, 2009, available at http://www.epa.gov/radiation/sources.html#summary-table.
32. Bou-Rabee, F.; Al-Zamel, A.; Al-Fares, R.; Bem, H. Technologically enhanced naturally occurring radioactive materials in the oil industry (TENORM). A review. Nukleonika **2009**, *54*, 3–9.
33. Nakaoka, A.; Fukushima, M.; Takagi, S. Environmental effect of natural radionuclides from coal-fired power plants. Health Phys. **1984**, *3*, 407–416.
34. Flues, M.; Morales, M.; Mazilli, B.P. The influence of a coal-plant operation on radionuclides in soil. J. Environ. Radioact. **2002**, *63*, 285–294.
35. Carvalho, F.P.; Oliviera, J.M.; Lopes, I.; Batista, A. Radionuclides from post uranium mining in rivers in Portugal. J. Environ. Radioact. **2007**, *98*, 298–314.
36. Boryło, A.; Nowicki, W.; Skwarzec, B. Isotope of polonium (^{210}Po) and uranium (^{234}U and ^{238}U) in the industrialized area of Wiślinka (Northern Poland). Int. J. Environ. Anal. Chem. **2009**, *89*, 677–685.
37. Skwarzec, B.; Boryło, A.; Kosińska, A.; Radziejewska, S. Polonium (^{210}Po) and uranium (^{234}U and ^{238}U) in water, phosphogypsum and their bioaccumulation in plants around phosphogypsum waste heap in Wiślinka (northern Poland). Nukleonika **2010**, *55*, 187–193.

38. Skwarzec, B. Polonium, uranium and plutonium in the southern Baltic Sea. Ambio **1997**, *26* (2), 113–117.
39. Coughtrey, P.J.; Jackson, D.; Jones, C.H.; Kene, P.; Thorne, M.C. *Radionuclides Distribution and Transport in Terrestrial and Aquatic Ecosystems. A Critical Review of Data*; A.A. Balkema: Rotterdam, 1984.
40. Wikipedia, uranium series decay chain, available at http://upload.wikimedia.org/wikipedia/commons/a/a1/Decay_chain%284n%2B2%2C_Uranium_series%29.PNG.
41. Wikipedia, thorium series decay chain, available at http://upload.wikimedia.org/wikipedia/commons/1/1c/Decay_chain%284n%2CThorium_series%29.PNG.
42. Wikipedia, actinium series decay chain, available at http://upload.wikimedia.org/wikipedia/commons/1/1e/Decay_chain%284n%2B3%2C_Actinium_series%29.PNG.
43. Wikipedia, neptunium series decay chain, available at http://upload.wikimedia.org/wikipedia/commons/8/8c/Decay_chain%284n%2B1%2CNeptunium_series%29.PNG.

Radionuclides

Philip M. Jardine
Oak Ridge National Laboratory, Oak Ridge, Tennessee, U.S.A.

Abstract
The fate and transport of radionuclides through soil is controlled by coupled hydrologic, geochemical, and microbial processes. A multitude of complex soil processes are tightly linked that can both accelerate and impede the subsurface mobility of radioactive contaminants. Often the extent and magnitude of subsurface biogeochemical reactions is controlled by the spatial and temporal variability in soil hydrologic processes.

INTRODUCTION

Soil, the thin veneer of matter covering the Earth's surface and supporting a web of living diversity, is often abused through anthropogenic inputs of toxic waste. The disposal of radioactive waste generated at U.S. Department of Energy (DOE) facilities within the Weapons Complex has historically involved shallow land burial in unsaturated soils and sediments. Disposal methods from the 1940s to the 1980s ranged from unconfined pits and trenches to single- and double-shell buried steel tanks. Most of the below-ground burial strategies were deemed to be temporary (i.e., an average life span of several decades) until suitable technologies were developed to deal with the legacy waste issues. Technologies for retrieving and treating the below-ground radionuclide waste inventories have been slow to evolve and are often cost prohibitive or marginally effective. The scope of DOE's disposal problem is massive, with landfills estimated to contain more than 3 million cubic meters of radioactive and hazardous buried waste; a significant proportion of which migrated into surrounding soils and groundwater. It is estimated that the migration of these waste plumes contaminated over 600 billion gallons of water and 50 million cubic meters of soil.

FATE AND TRANSPORT PROCESSES

Hydrologic Processes

Soil is a complex continuum of pore regions ranging from large macropores at the mm scale to small micropores at the sub-μm scale. It is the physical properties of the media (e.g., structured or layered), coupled with the duration and intensity of precipitation events that dictates the avenues of water and radionuclide movement through the subsurface. In humid environments where structured media is commonplace, transient storm events invariably result in the preferential migration of water.[1–8] Highly conductive voids within the media (e.g., fractures, macropores) carry water around low-permeability, high-porosity matrix blocks or aggregates resulting in water bypass of the latter. In these humid regimes, recharge rates are very high with more than 50% of the infiltrating precipitation resulting in groundwater and surface water recharge. This condition promotes the formation of massive contaminant plumes in the soil since storm flow and groundwater interception with waste trenches is frequent and long-lasting. Even in semiarid environments, where recharge is typically small, subsurface preferential flow is a key mechanism controlling water and solute mobility.[9,10] Lithologic discontinuities and sediment layering promote perched water tables and unstable wetting fronts that drive both lateral and vertical subsurface preferential flow.

In both humid and semiarid regimes, water that is preferentially flowing through the soil media often remains in intimate contact with the porous matrix, and physical and hydrologic gradients drive the exchange of mass from one pore regime to another. Mass exchange is time-dependent and is often controlled by diffusion to and from the matrix. Thus, a significant inventory of radionuclide waste can reside within the soil matrix. This waste source is hydrologically linked to preferred flow paths which significantly enhances the extent and longevity of subsurface contaminant plumes. This scenario is commonplace at the Oak Ridge National Laboratory, located in eastern Tennessee, U.S., where thousands of underground disposal trenches and ponds have contributed to the spread of radionuclides such as ^{137}Cs, ^{60}Co, ^{90}Sr, and $^{235/238}$U across tens of kilometers of landscape. Highly concentrated contaminant plumes move through soil and groundwater at time scales of meters per day (Fig. 1) since the soils are highly structured and conducive to rapid preferential flow. However, the soil matrix, which has a high porosity and low permeability, serves as a source/sink for contaminants.[5,6,11] The preferential movement of water and radionuclides through the subsurface also significantly impacts geochemical and microbial processes by controlling the extent and rate of

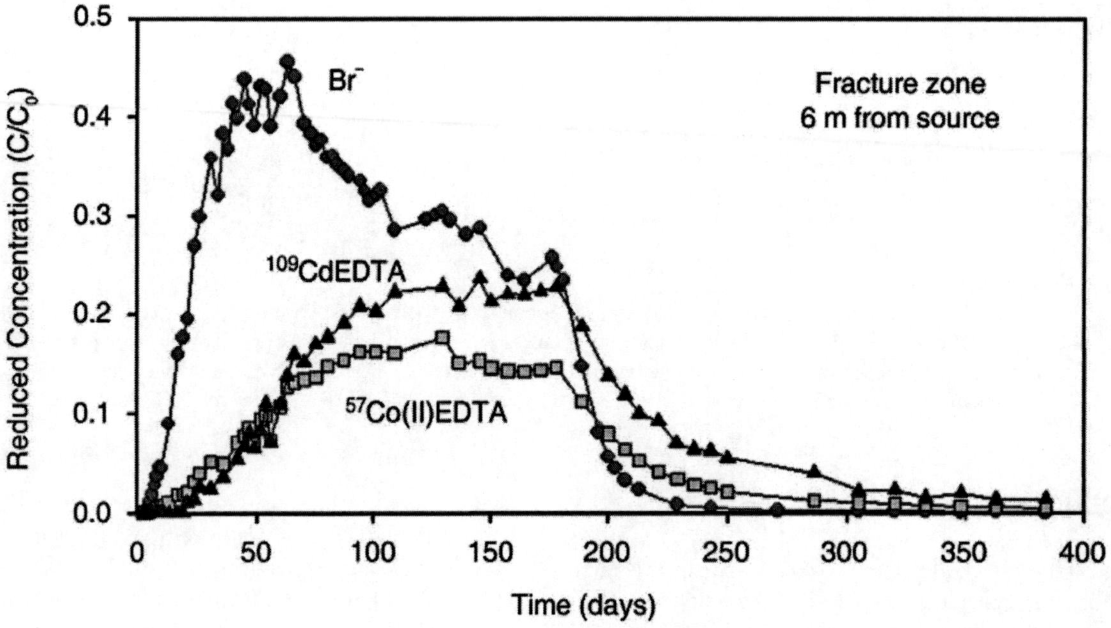

Fig. 1 Field-scale fate and transport of nonreactive Br$^-$ and reactive ^{57}Co(II)EDTA^{2-} and ^{109}CdEDTA^{2-} in fractured subsurface media at the Oak Ridge National Laboratory. Although transport rates are rapid, geochemical reactions significantly impede the mobility of the chelated radionuclides as is indicated by their delayed breakthrough.
Source: Jardine et al.[40]

various reactions with the solid phase. It imposes kinetic constraints on biogeochemical reactions and limits the surface area of interaction by partially excluding water and mass from the matrix porosity.

Geochemical Processes

Radionuclide fate and transport in soil and sediments is also controlled by interfacial reactions with the soil solid phase. Most soils are a complex mixture of variably charged phyllosilicates, redox reactive Fe- and Mn-oxides, organic matter, and mineral carbonates. Radionuclides interact with these solid phases through coulombic exchange, chemisorption, redox alterations, transformation processes such as polymerization, precipitation/dissolution, and complexation reactions. Both the extent and rate of these processes can be significantly influenced by variations in water content and the degree of pore regime connectivity. To make matters worse, radionuclide waste generated at the U.S. DOE facilities was often co-disposed with various chelating agents and organic acids. These synthetic organic constituents form highly stable, water-soluble complexes with a wide variety of radionuclides.[12,13] The presence of the complexing agent significantly alters the geochemical behavior of the disposed contaminants in soils and sediments through increased solubility, accelerated redox reactions, and ionic charge reversal.

The geochemical mechanism controlling the fate and transport of chelated radionuclides has been well characterized in numerous soils and subsurface materials.[14–22] Typically, Fe(III) and Mn(IV) oxyhydroxides are the dominant subsurface mineral assemblages that catalyze co-contaminant oxidation/reduction and dissociation reactions (Fig. 2). The mineral oxides have repeatedly been shown to catalyze the oxidation of ^{60}Co(II)EDTA^{2-} to ^{60}Co(III)EDTA$^-$, thereby adversely enhancing the transport and persistence of ^{60}Co in a variety of subsurface environments ranging from aquifer sands to fractured weathered shale saprolites.[15,17,18,20,22] Further, Fe(III)-oxides have also been shown to effectively dissociate a large number of chelated metal and radionuclide complexes (e.g., ^{60}Co–, ^{90}Sr–EDTA) through ligand competition.[15,20–22]

Certain radionuclides such as ^{137}Cs do not form strong bonds with many of the chelating agents and organic acids that were used during decontamination. Nevertheless, these radionuclides still interact aggressively with the soil solid phase. In the case of ^{137}Cs, 2:1 phyllosilicates and micas serve as excellent sorbents since the interlayer spaces of these mineral assemblages strongly attenuate the radionuclide. The migration tendency of ^{137}Cs in soils is often related to colloid mobility of contaminated sediments[23] or cation competition for surface sites in harsh environments such as those found beneath the Hanford tank farms in western Washington State, U.S.[24]

Microbial Processes

Radionuclides such as ^{60}Co and $^{235/238}$U can exist in more than one oxidation state, and their behavior in the environment depends on their oxidation state. For example, U(VI) is soluble and mobile in the environment whereas U(IV) is much less soluble and relatively immobile. Likewise, the

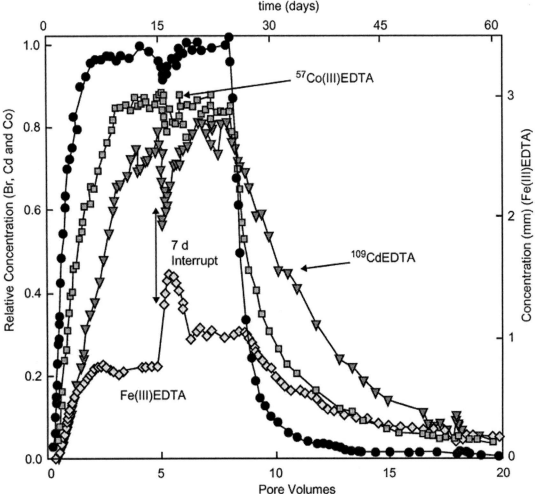

Fig. 2 Fate and transport of nonreactive Br$^-$ and reactive ^{57}Co(II)EDTA^{2-} and ^{109}CdEDTA^{2-} in undisturbed soil columns of fractured weathered shale. Geochemical reactions impede the mobility of the chelated radionuclides. Co(II)EDTA^{2-} is oxidized to Co(III)EDTA$^-$ where Mn-oxides serve as the oxidant. Fe-oxides effectively dissociate CdEDTA^{2-} complexes resulting in the formation of free Cd and Fe(III)EDTA$^-$. A flow interruption technique was employed to quantify the presence of physical and geochemical nonequilibrium processes.
Source: Mayes et al.[22]

oxidized ^{60}Co(III)EDTA complexes are much more stable and exhibit greater mobility in subsurface environments than the reduced ^{60}Co(II)EDTA.[15,17,18] Subsurface Fe- and Al-oxides can effectively dissociate the Co(II)EDTA complex to Fe(III)EDTA[20] and Al(III)EDTA,[16] respectively, and aqueous Co^{2+} is free to participate in sorption or precipitation reactions. Co(III) EDTA, on the other hand, is unaffected by Fe(III)- and Al-oxides. Therefore, the oxidized forms of these radionuclides and metals promote their undesirable enhanced migration through subsurface environments.

Numerous metal-reducing bacteria have been isolated that enzymatically reduce toxic metals and radionuclides to stable end-products. Microbial reduction of U(VI) to form the sparingly soluble U(IV) has been shown using chemostat experiments for a number of metal-reducing bacteria.[25,26] Gorby et al.[27] have also shown that certain metal-reducing bacteria can link the enzymatic reduction of ^{60}Co(III)EDTA$^-$ to support cell growth. Recently, important advances have been made towards implementing field-scale microbially mediated metal reduction strategies in oxygen-deficient environments. Several studies have investigated contaminant reduction in the presence of solid phase material.[27–29] Gorby et al.[27] have shown that the metal-reducing bacterium *Shewanella alga* preferentially reduced Co(III)EDTA$^-$ to Co(II)EDTA^{2-} in the presence of Mn-oxides. Likewise, Wielinga et al.[29] documented the bioreduction of U(VI) by *Shewanella alga* in the presence of various Fe-oxide mineral phases. These authors noted that the rate of U(VI) bioreduction was unaffected in the presence of goethite and only slightly diminished in the presence of poorly crystalline Fe(III)-oxides, where the latter Fe solid phase effectively competed as a terminal electron acceptor. Recent studies by Brooks, Carroll, and Jardine.[28] showed the sustained microbial reduction of Co(III)EDTA$^-$ under dynamic flow conditions. The net reduction of the Co(III)EDTA$^-$ dominated the fate and transport of the contaminant even in the presence of strong mineral oxidants such as Mn- and Fe-oxides that are known to effectively reoxidize Co(II)EDTA^{2-} back to Co(III)EDTA$^-$.[15,18,20] The research findings of Brooks, Carroll, and Jardine[28] provide new and important information on how to successfully implement a bioreduction strategy at the field scale. Their use of a dynamic flow system with sustained bacterial growth conditions in geochemically reactive media is consistent with contaminant migration scenarios in situ.

The studies of Brooks, Carroll, and Jardine,[28] however, used uniformly packed media that contained little structure. Undisturbed subsurface soils and geologic material consist of a complex continuum of pore regions ranging from large macropores and fractures at the mm scale to small micropores at the sub-μm scale. Structured media, common to most subsurface environments throughout the world, accentuates this physical condition that often controls the geochemical and microbial processes affecting solute transport. Redox sensitive radionuclides such as U(VI), Co(III)EDTA, and Tc(VI) reside within nearly all of the pore structure of the subsurface media, with the greatest concentration of contaminants held within micropores.[2,3] Bacteria that are capable of reducing these contaminants are too big to reach a large fraction of the micropore regime and are largely restricted to macro- and mesopore domains.[30,31] Fortunately, the pore structure of the media is hydrologically interconnected, and contaminants move from one pore class to another via hydraulic and concentration gradients.[4,6,7] This process is slow, however, and is often the rate-limiting factor governing the success of contaminant bioremediation. Thus, faster-flowing fracture-dominated regimes will most likely be physically more appealing for sustained bioreduction as long as a suitable electron donor can be supplied. In contrast, bioreduction processes in slower-flowing matrix regimes will most likely be limited by rate-dependent mass transfer of contaminants from smaller pores into larger pores.

Certain bacteria are also capable of degrading chelates and thus potentially immobilizing radionuclides in situ. The biodegradation of the commonly used aminopolycarboxylate chelates NTA, EDTA, and DTPA have been studied in soil and sediment systems for many years.[32,33,35] Research has shown that NTA has the greatest potential for biodegradation in subsurface systems compared with the other aminopolycarboxylates.[34,35]

Bolton et al.[36] and Bolton and Girvin[37] have shown that the bacterial strain *Chelatobacter heintzii* (ATCC 29600) is capable of degrading NTA in the presence of many different toxic metals and radionuclides. Likewise, Payne et al.[38] and Liu et al.[39] have deciphered the mechanisms by which certain bacteria degrade radionuclide–EDTA complexes. These studies lend promise to the potential for using bacteria to biodegrade chelates and enhance the geochemical immobilization of radionuclides in situ.

CONCLUSIONS

Radionuclide fate and transport in soils is controlled by coupled time-dependent hydrologic, geochemical, and microbial processes. Hydrologic processes such as preferential flow and matrix diffusion can serve to both accelerate and impede radionuclide migration, respectively. Preferential flow results in hydraulic, physical, and geochemical nonequilibrium conditions since differences in fluid velocities and solute concentrations in different-sized pores create hydraulic and concentration gradients that drive time-dependent inter-region advective and diffusive mass transfer. Thus, in soil systems with a large matrix porosity or a significant quantity of disconnected immobile water, radionuclide migration rates can be greatly retarded due to the slow transfer of mass to actively flowing preferential flow paths. Nevertheless, the prevalence of preferential flow can greatly accelerate the transport of mass in soil systems. Geochemical processes such as sorption, redox alterations, and dissociation reactions can also serve to both accelerate and impede radionuclide migration. Sorption and radionuclide-chelate dissociation reactions almost always result in retarded radionuclide migration rates, whereas oxidation reactions often result in more soluble, and thus more mobile, radionuclide species. Microbial processes can also potentially influence the fate and transport of radionuclides in soil. Metal-reducing bacteria and chelate degraders can alter the geochemical behavior of redox sensitive radionuclides which facilitates their immobilization via solid phase sorption and precipitation reactions.

Enhanced knowledge of the coupled hydrologic, geochemical, and microbial processes controlling radionuclide migration in soils will improve our conceptual understanding and predictive capability of the risks associated with spread of radioactive material in the subsurface environment. Too often risk assessment models treat soil and bedrock as inert media or assume that the media is in equilibrium with migrating contaminants. Failure to consider the time-dependent coupled processes that control radionuclide migration will greatly over-predict the off-site contribution of contaminants from the primary waste source and thus provide an inaccurate assessment of pending risk. By recognizing the importance of soil processes on radionuclide migration, we can improve our decision-making strategies regarding the selection of effective remedial actions and improve our interpretation of monitoring results after remediation is complete.

REFERENCES

1. Shuford, J.W.; Fitton, D.D.; Baker, D.E. Nitratenitrogen and chloride movement through undisturbed field soil. J. Environ. Qual. **1977**, *6*, 255–259.
2. Jardine, P.M.; Wilson, G.V.; Luxmoore, R.J. Unsaturated solute transport through a forest soil during rain storm events. Geoderma **1990**, *46*, 103–118.
3. Jardine, P.M.; Wilson, G.V.; McCarthy, J.F.; Luxmoore, R.J.; Taylor, D.L. Hydrogeochemical processes controlling the transport of dissolved organic carbon through a forested hillslope. J. Contam. Hydrol. **1990**, *6*, 3–19.
4. Jardine, P.M.; O'Brien, R.; Wilson, G.V.; Gwo, J.P. Experimental techniques for confirming and quantifying physical nonequilibrium processes in soils. In *Physical Nonequilibrium in Soils: Modeling and Application*; Selim, H.M., Ma, L., Eds.; Ann Arbor Press: Chelsea, MI, 1998; 243–271.
5. Jardine, P.M.; Wilson, G.V.; Luxmoore, R.J.; Gwo, J.P. Conceptual model of vadose-zone transport in fractured weathered shales. In *Conceptual Models of Flow and Transport in the Fractured Vadose Zone*; U.S. National Committee for Rock Mechanics, National Research Council; National Academy Press: Washington, DC, 2001; 87–114.
6. Wilson, G.V.; Jardine, P.M.; O'Dell, J.D.; Collineau, M. Field-scale transport from a buried line source in variable saturated soil. J. Hydrol. **1993**, *145*, 83–109.
7. Wilson, G.V.; Gwo, J.P.; Jardine, P.M.; Luxmoore, R.J. Hydraulic and physical nonequilibrium effects on multi-region flow and transport. In *Physical Nonequilibrium in Soils: Modeling and Application*; Selim, H.M., Ma, L., Eds.; Ann Arbor Press: Chelsea, MI, 1998; 37–61.
8. Hornberger, G.M.; Germann, P.F.; Beven, K.J. Through flow and solute transport in an isolated sloping soil block in a forested catchment. J. Hydrol. **1991**, *124*, 81–97.
9. Porro, I.; Wierenga, P.J.; Hills, R.G. Solute transport through large uniform and layered soil columns. Water Resour. Res. **1993**, *29*, 1321–1330.
10. Ritsema, C.J.; Dekker, L.W.; Nieber, J.L.; Steenhuis, T.S. Modeling and field evidence of finger formation and finger recurrence in a water repellent sandy soil. Water Resour. Res. **1998**, *34*, 555–567.
11. Jardine, P.M.; Sanford, W.E.; Gwo, J.P.; Reedy, O.C.; Hicks, D.S.; Riggs, R.J.; Bailey, W.B. Quantifying diffusive mass transfer in fractured shale bedrock. Water Resour. Res **1999**, *35*, 2015–2030.
12. Riley, R.G.; Zachara, J.M. *Chemical Contaminants on DOE Lands and Selection of Contaminant Mixtures for Subsurface Science Research*; DOE=ER-0547T; U.S. Govt. Print. Office: Washington, DC, 1992.
13. Toste, A.P.; Osborn, B.C.; Polach, K.J.; Lechner-Fish, T.J. Organic analyses of an actual and simulated mixed waste: Hanford's organic complexant site revisited. J. Radioanal. Nucl. Chem. **1995**, *194*, 25–34.
14. Swanson, J.L. *Effect of Organic Complexants on the Mobility of Low-level Waste Radionuclides in Soils*; Status Report PNL-3927, UC-70, 1981.
15. Jardine, P.M.; Jacobs, G.K.; O'Dell, J.D. Unsaturated transport processes in undisturbed heterogeneous porous media: II. Co-contaminants. Soil Sci. Soc. Am. J. **1993**, *57*, 954–962.
16. Girvin, D.C.; Gassman, P.L.; Bolton, H. Adsorption of aqueous cobalt ethylenediaminetetraacetate by d-Al2O3. Soil Sci. Soc. Am. J. **1993**, *57*, 47–57.
17. Zachara, J.M.; Gassman, P.L.; Smith, S.C.; Taylor, D. Oxidation and adsorption of Co(II)EDTA2_ complexes in subsurface materials with iron and manganese oxide grain coatings. Geochim. Cosmochim. Acta. **1995**, *59*, 4449–4463.
18. Brooks, S.C.; Taylor, D.L.; Jardine, P.M. Reactive transport of EDTA-complexed cobalt in the presence of errihydrite. Geochim. Cosmochim. Acta. **1996**, *60*, 1899–1908.
19. Read, D.; Ross, D.; Sims, R.J. The migration of uranium through clashach sandstone: the role of low molecular weight organics in enhancing radionuclide transport. J. Contam. Hydrol. **1998**, *35*, 235–248.
20. Szecsody, J.E.; Zachara, J.M.; Chilakapati, A.; Jardine, P.M.; Ferrency, A.S. Importance of flow and particlescale heterogeneity on Co(II/III)EDTA reactive transport. J. Hydrol. **1998**, *209*, 112–136.
21. Davis, J.A.; Kent, D.B.; Coston, J.A.; Hess, K.M.; Joye, J.L. Multispecies reactive tracer test in an aquifer with spatially variable chemical conditions. Water Resour. Res. **2000**, *36* (1), 119–134.
22. Mayes, M.A.; Jardine, P.M.; Larsen, I.L.; Brooks, S.C.; Fendorf, S.E. Multispecies transport of metal-EDTA complexes and chromate through undisturbed columns of weathered, fractured saprolite. J. Contam. Hydrol. **2000**, *45*, 243–265.
23. Solomon, D.K.; Marsh, J.D.; Larsen, I.L.; Wickliff, D.S.; Clapp, R.B. *Transport of Contaminants During Storms in the White Oak Creek and Melton Branch Watersheds*; ORNL/TM-11360; Oak Ridge National Laboratory: Oak Ridge, TN, 1991.
24. Serne, R.J.; Burke, D.S. *Chemical Information on Tank Supernatants, Cs Adsorption from Tank Liquids onto Hanford sediments, and Field Observations of Cs Migration from Past Tank Leaks*, Report No. PNNL-11495; Pacific Northwest National Laboratory: Richland, WA, 1997.
25. Gorby, Y.A.; Lovley, D.R. Enzymatic uranium precipitation. Environ. Sci. Technol. **1992**, *26*, 205–207.
26. Francis, A.J.; Dodge, C.J.; Lu, F.; Halada, G.P.; Clayton, C.R. XPS and XANES studies of uranium reduction by Clostridium Sp. Environ. Sci. Technol. **1994**, *28*, 636–639.
27. Gorby, Y.A.; Caccavo, F.; Drektrah, D.B.; Bolton, H. Microbial reduction of Co(III)EDTA_ in the presence and absence of manganese (IV) dioxide. Environ. Sci. Technol. **1998**, *32*, 244–250.
28. Brooks, S.C.; Carroll, S.L.; Jardine, P.M. Sustained bacterial reduction of Co(III)EDTA_ in the presence of competing geochemical oxidation during dynamic flow. Environ. Sci. Technol. **1999**, *33*, 3002–3011.
29. Wielinga, B.; Bostick, B.; Rosenzweig, R.F.; Fendorf, S. Inhibition of bacterially promoted uranium reduction: ferric (hydr)oxides as competitive electron acceptors. Environ. Sci. Technol. **2000**, *34*, 2190–2195.
30. Smith, M.S.; Thomas, G.W.; White, R.E.; Ritonga, D. Transport of Escherichia Coli through intact and disturbed soil columns. J. Environ. Qual. **1985**, *14*, 87–91.
31. McKay, L.D.; Cherry, J.A.; Bales, R.C.; Yahya, M.T.; Gerba, C.P. A field example of bacteriophage as tracers of fracture flow. Environ. Sci. Technol. **1993**, *27*, 1075–1079.

32. Tiedje, J.M. Microbial degradation of ethylenedi aminetetraacetic acid in soils and sediments. Appl. Environ. Microbiol. **1975**, *30*, 327–329.
33. Tiedje, J.M.; Mason, B.B. Biodegradation of nitrilotriacetic acid (NTA) in soils. Soil Sci. Soc. Am. Proc. **1974**, *38*, 278–283.
34. Means, J.L.; Kucak, T.; Crerar, D.A. Relative degradation rates of NTA, EDTA, and DTPA and environmental implications. Environ. Pollut. Ser. B. **1980**, *1*, 45–60.
35. Bolton, H., Jr.; Li, S.E., Jr.; Workman, D.J.; Girvin, D.C. Biodegradation of synthetic chelates in subsurface sediments from the southeast coastal plain. J. Environ. Qual. **1993**, *22*, 125–132.
36. Bolton, H.; Girvin, D.C.; Plymale, A.E.; Harvey, S.D.; Workman, D.J. Degradation of metal-nitrilotriacetate complexes by *Chelatobacter heintzii*. Environ. Sci. Technol. **1996**, *30*, 931–938.
37. Bolton, H., Jr.; Girvin, D.C., Jr. Effect of adsorption on the biodegradation of nitrilotriacetate by *Chelatobacter heintzii*. Environ. Sci. Technol. **1996**, *30*, 2057–2065.
38. Payne, J.W.; Bolton, H.; Campbell, J.A.; Xun, Y.L. Purification and characterization of EDTA monooxygenase from the EDTA-degrading bacterium BNC1. J. Bacteriology 1998, *180*, 3823–3827.
39. Liu, Y.; Louie, T.M.; Payne, J.; Bohuslavek, J.; Bolton, H.; Xun, L.Y. Identification, purification, and characterization of iminodiacetate oxidase from the EDTA-degrading bacterium BNC1. Applied Environ. Microbiol. **2001**, *67*, 696–701.
40. Jardine, P.M.; Mehlhorn, T.L.; Larsen, I.L.; Bailey, W.B.; Brooks, S.C.; Roh, Y.; Gwo, J.P. Influence of hydrological and geochemical processes on the transport of chelated-metals and chromate in fractured shale bedrock. J. Contamin. Hydrol. 2001, *in press*.

Rain Water: Atmospheric Deposition

Zaneta Polkowska
Gdansk University of Technology, Gdansk, Poland

Abstract
Water is transferred from the atmosphere to the land surface mainly via various forms of atmospheric precipitation; however, the contribution of surface condensation/deposition (e.g., dew, hoar frost, occult precipitation) should not be neglected. Rainwater and liquid deposits (dew, hoar frost, rime) are components of the physical and geographical environment that are easily assimilated and transported. Thus, they are generally a good indicator of the chemical composition of the environment. This entry summarizes the current state of knowledge on the role of atmospheric precipitation and deposition in pollutant transport and distribution. Issues such as routes of transport of atmospheric pollutants to the surface, seasonality and spatial variability of wet precipitation and deposition, and the effect of meteorological parameters and terrain topography on the composition and abundance of the pollutants found in precipitation/deposition samples are discussed.

INTRODUCTION

The atmosphere is the main source of the pollution affecting the earth's surface. Its tremendous dynamism is the reason why it is the principal route along which pollutants spread and are transported in the form of dusts, gases, and aerosols between the various compartments of the environment. The emission of various xenobiotics into the environment may give rise to changes in the chemical composition of the various compartments of the environment (trace components); new threats to health and even life, especially when compounds with unknown properties are released into the environment as a result of human activities; and climate changes on a global scale.

The residence time of pollutants in the atmosphere and the distance over which they can be carried following their emission depend primarily on factors affecting their spread, such as the height at which pollutants are emitted and the weather conditions (type of atmospheric circulation, air temperature, wind speed, and direction).

Depending on their properties and the weather conditions, pollutants in the air are scattered and transformed during transport; most of them return to the ground (often very far from the point of emission) through dry and wet deposition, or as a result of the absorption of gaseous contaminants and aerosols by surface waters, plant cover, or soil.

Wet deposition (precipitation and atmospheric deposits) plays the largest part in transporting pollutants from the atmosphere to the ground in areas far distant from the sources of emission. Water moves from the atmosphere to the ground mainly in the form of precipitation, e.g., rain, drizzle, snow, and hail. A feature they all have in common is the considerable speed of gravitational fall. A second form of wet deposition consists of atmospheric deposits, which are formed either by the condensation or sublimation of water vapor as a result of the radiational cooling of the ground (dew, hoar frost), or by the capture of fog droplets by the substrate, i.e., the ground and all objects present above ground level (liquid deposits, rime). Unlike precipitation, deposits form in the near-ground layers of the atmosphere, where the content of dusts and gaseous pollutants is greater than in the upper layers. For this reason, the concentration of pollutants may be substantially larger in deposits than in precipitation (given the significant differences in chemical composition where the contents of trace and ultratrace constituents are concerned).

Wet and dry deposition can take place on a wide diversity of surfaces, from natural ones such as trees, cultivated plants, bare soil, rocks, lakes, and rivers, to all kinds of man-made surfaces such as the roofs of buildings, roadways, and communal and industrial landfill sites. The chemical composition of the water derived from all kinds of wet deposition undergoes conversion on coming into contact with the substances naturally present in the substrate and those that reached it as a result of dry deposition. In this way, both the composition and the concentration of the pollutants in atmospheric water can change before they return to the hydrological cycle. A further consequence of this process is the potentially destructive effect of the xenobiotics contained in water on the natural environment (e.g., erosion) and on man-made products (e.g., corrosion).

DIVERSE FORMS OF WET DEPOSITION

Wet deposition is the process by which atmospheric pollutants are captured by clouds and/or droplets of precipitation

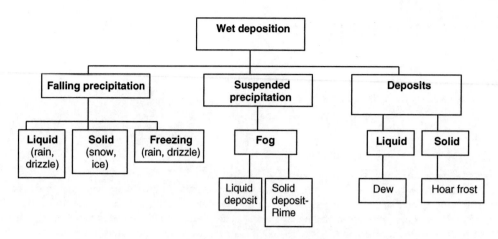

Fig. 1 Various forms of wet deposition.

or solid particles, as a result of which they are carried to the ground. Pollutants can be flushed out of the atmosphere as a result of in-cloud absorption; precipitation scavenging, i.e., the diffusional and inertial collision of aerosol particles with precipitation particles; below-cloud scavenging of pollutants by droplets of rain or drizzle, by snow or hail, i.e., the absorption of gases and the inertial capture of aerosol particles of diameter $d > 1$ µm; nucleation scavenging, i.e., the capture of pollutants from the air during the formation of droplets on the surfaces of aerosol particles acting as condensation nuclei; and in-cloud scavenging.

The following parameters may affect the process of wet deposition: the place where and the means by which precipitation or deposits are formed; the weather conditions and the topography of the land; the type and quantity of precipitation/deposit; the duration and intensity of precipitation/deposition; the interaction between aerosols, clouds, and gaseous pollutants; the presence of dust particles, which act as condensation nuclei for water vapor; the concentration of a given substance in the atmosphere; the solubility of a given substance in the atmosphere; the presence of compounds capable of catalyzing the chemical reactions of pollutants absorbed in droplets of precipitation or particles of atmospheric deposits.

Pollutants can also reach the ground directly in fog or cloud droplets in the form of fog deposits (liquid deposits and rime). The formation of fog deposits depends on the weather conditions (frequency of fog occurrence, moisture content of fog, wind speed) and the morphology of the land (relative height, absolute height, types of land forms, distance between and heights of neighboring mountain ranges, plant cover).

A stream of wet deposition can be formed by wet precipitation, solid precipitation, freezing precipitation (freezing rain, freezing drizzle), precipitation that evaporates before reaching the ground, atmospheric deposits, formation and suspension of water droplets in the air (fog, mist, ice fog, clouds), and water in the form of droplets or snow crystals raised from the ground by wind (high/low blizzard, spray). Fig. 1 illustrates the diversity of forms of wet deposition.

Although the principal source of water from the atmosphere is atmospheric precipitation, fog deposition and atmospheric deposits must not be left out of the overall atmospheric water balance.[1–4] The atmospheric deposition taking place during fog significantly alters the stream of wet deposition, and therefore the types and numbers of contaminants entering the environment. If the intensity of this deposition is a few millimeters per hour, the moisture content of the fog falls by 50% to 75%, and the efficiency of the fog deposits falls accordingly. It is often the case that even though precipitation has been going on for 15–20 hr, the fog does not disappear completely, and its moisture content remains at roughly the same level (ca. 0.07–0.1 g/m^3). However, within 15–20 min of the precipitation ceasing, the moisture content of the fog can double or even triple, returning to the levels before the onset of precipitation.

The deposition of pollutants along with wet deposition (depending on the region) has measurable as well as non-measurable consequences: ecological (e.g., pollution of surface and ground waters, which may be used as sources of drinking water; contamination of soil; soil erosion; changes in land relief; and erosion of rocks)[5,6]; medical (e.g., contact and absorption of contaminants by living organisms—humans, animals, plants); economic (e.g., diseases of civilization, atmospheric corrosion—destruction of metal, stone, concrete, glass, paper, dyestuffs, enamels, and paints).[7]

Seasonality of Precipitation and Atmospheric Deposits

To obtain reliable information on the chemical composition of precipitation and atmospheric deposits, monitoring

Table 1 Fundamental problems caused by seasonality of occurrence of different forms of wet deposition.

Deposition type	Seasonality	Spatial variability	Specific composition	Particulate matter	Low concentration	Composition changes
Rain	–/+ (storm)	–	+	–	+	+
Snow (pellets, drizzle)	+	–	+	+	+	+
Atmospheric deposits	+	+	+	–	+/– (fog)	–/+ (fog)

Note: +, yes; –, no.

has to be carried out over a relatively long period. This is also justified in view of the seasonality of certain forms of precipitation and deposits. Precipitation in the form of rain or sleet may fall at any time of the year. Snow, on the other hand, usually falls only during the two winter months. Even more infrequent are hoar frost and rime, the formation of which requires appropriate weather conditions. The annual number of days with hoar frost in Poland varies from 20 (northern Poland) to 80 (central Poland); however, on mountain peaks, there may be no more than 10 days with hoar frost.

Dew is a liquid deposit that most commonly forms in late summer and early autumn. There may be from 100 to 160 days with dew each year in lowland Poland, but only 10–30 days in the mountainous regions of the country. Fog is a phenomenon typical of mountain areas, occurring there from 35 to 310 days each year.

If one adds up the total time of occurrence (every night with minimal or moderate cloud cover) and frequency of occurrence, dew and hoar frost become more important components of wet deposition than rain or snow. In the city of Wrocław (Poland), for example, the total time during which precipitation occurs is 3% of the year, whereas the figure for dew and frost is 11%. Table 1 lists the most important problems caused by the seasonal occurrence of the different forms of wet deposition.

Spatial Differentiation in the Magnitude of Precipitation and Deposition

Atmospheric precipitation closes the natural hydrological cycle and constitutes the main source of water on Earth. Of the total amount of precipitational water reaching the terrestrial parts of the planet, one-third evaporates, one-third is soaked up by the ground, and the rest flows away over its surface (runoff). The volume of water present in the water cycle at any instant is ca. 0.025% of the earth's total water resources.

The magnitude of precipitation depends on the relief and the characteristics of the terrain. For example, the highest total annual precipitation in Poland falls in the mountain and upland areas (>600 mm), whereas the lowest amounts fall in lowland Poland (450–550 mm). The Baltic coast receives less precipitation than the lake districts to the south of it because there are no land features in its vicinity forcing air masses to rise.

The main source of water from the atmosphere is liquid precipitation (rain and drizzle) and solid precipitation (snow, snow pellets, ice pellets, snow and ice graupel, hail, diamond dust). The total "income" of atmospheric water is made up by atmospheric deposits. Whereas dew and hoar frost are more important as media enabling contaminants to penetrate the environment in the lowlands, in the mountains, with their high frequency of fog, accompanied by high wind speeds, the parts played by liquid deposits and rime take on a greater significance. The efficiency of fog deposits depends on meteorological and morphological conditions that determine the potential participation of fog in the total water "income"; the environmental factor (land cover, type of recipient) determines its real role. The factors affecting the frequency of fog in the mountainous regions of continental Europe are circulation conditions, distance from a source of water vapor, the presence of other orographic obstacles in this region, relative and absolute height, exposure of slopes, the extent of the mountain range, position within a given land form (slope, peak, valley or basin floor), and position within the whole mountain range (windward or leeward side).[8]

Variety of Analytes in Samples of Precipitation and Atmospheric Deposits

The type and number of contaminants present in samples of precipitation and atmospheric deposits are a reflection of the state of the environment. In such samples, we find analytes such as the following: inorganic substances—main cations (K^+, Na^+, NH_4^+, Mg^{2+}, Ca^{2+}), anions (Cl^-, F^-, NO_2^-, NO_3^-, SO_3^{2-}, SO_4^{2-}, PO_4^{3-}, $HCOO^-$, CH_3COO^-),[9–14] trace metals (e.g., Pb, Cd, Cu, Pb, Ni, Zn, Fe)[10,15,16]; organic substances—petroleum hydrocarbons,[17] fatty acids,[18] formaldehyde,[19] polychlorinated biphenyls,[20–22] pesticides,[19,20,23–28] polyaromatic hydrocarbons (PAHs),[17,19,29–34] and volatile organochlorine compounds (VOCCs).[35–38]

Below, we give examples of the concentrations of contaminants found in samples of rainwater, atmospheric deposits, and runoff collected in Europe and elsewhere in the world.

ATMOSPHERIC PRECIPITATION

The literature provides a good number of papers in which one can find data (specific to a particular area) on the concentrations of selected contaminants determined in samples of atmospheric precipitation. Table 2 contains information on the lowest and highest concentrations of ions determined in rainwater samples from different parts of the world.[11,12,14,39–44]

Organic compounds are found in rainwater samples in addition to inorganic compounds. Pesticides are one of the more frequently determined groups of organics—they are used worldwide to protect crops against pests, weeds, and fungi. It is estimated that barely 1%–3% of pesticides are properly applied to crops, i.e., they fulfill their function. These compounds can undergo chemical and biological conversion and be transported over very considerable distances. Organochlorine pesticides have been discovered even in such remote regions as the North and South Poles. Studies have shown that this is a superregional problem. Perusal of the literature from the last 30 years shows that a substantial amount of information has been gathered on the presence of pesticides in precipitation. More than 40 different pesticides, mostly triazines, have been detected in rainwater samples collected in the United States, Europe, and Africa.[27] In many countries around the world, the application of such pesticides as lindane, Hexachlorocyclohexane (HCH), or Dichlorodifenylotrichloroetan (DDT) has ceased. However, as the literature shows, these compounds are still present in the environment. Table 3 lists data on the contents of pesticides and their metabolites in samples of atmospheric precipitation collected at different sites in Europe and North America.

An increase in pesticide concentrations is apparent in precipitation samples collected in spring and summer. The highest levels were recorded in June, July, and August, which seems to be a logical consequence of the intensive application of these compounds in fields, orchards, gardens, and forests. A further source of pollution are the pesticides used by the owners of gardens, allotments, and balconies/window boxes. By way of example, Fig. 2 shows the total concentrations of organonitrogen and organophosphorus pesticides in samples of rainwater collected at stations located on the outskirts of the city of Gdańsk (Poland).

Comparison of these data with published data from elsewhere in the world indicates that pesticide levels in precipitation sampled in the Tri-City area (Gdańsk–Sopot–Gdynia) of Poland are lower than those in samples from other parts of Europe.[25,27,59,65,68,69] This is because pesticides are not so universally applied in Polish agriculture. The location of the Tri-City also plays a part—the prevailing winds there are from the southwest; thus, pesticides tend to be blown out into the Gulf of Gdańsk and the Baltic Sea.

Statistical analysis of measurement data obtained during the determination of the pesticide content in precipitation samples has also shown that there is a dependence between the total pesticide concentration and the area of land covered by vegetation (woodland, arable land, orchards, vegetable gardens) within a 1 km radius of the sampling station (see Fig. 3).[67,70] In the case of urban areas, however, it is rather difficult to assess them on the basis of cartographic data because of the considerable amount of construction work going on there (road-widening schemes, housing construction).

One of the more interesting examples is the monitoring of VOCCs carried out in a number of countries in different parts of the world. Fig. 4 gives examples of concentrations of volatile trichloromethane in rainwater samples from different regions of the world. $CHCl_3$ in ambient air was detected at levels from several tens to hundreds of nanograms per liter (59–650 ng/L); the highest such levels were recorded in an industrial area in Korea (650 ng/L)[71] and an urban area in São Paulo in Brazil (267 ng/L).[72] C_2Cl_4 levels were even higher—up to 927 ng/L (in U.S. cities).[73] Such high levels of organochlorine compounds in the air are the direct cause of their presence in precipitation. $CHCl_3$ was also determined in precipitation at levels from several tens to hundreds of nanograms per liter (25–250 ng/L). The highest concentration was noted in Los Angeles (250 ng/L; the rainwater sampling point was located on a university campus, not far from the San Diego Freeway).[36] C_2Cl_4 levels were lower (4.6–100 ng/L); the highest levels were reported from Tokyo (Japan).[35]

Table 4 shows the percentages of the various VOCCs detected in samples of precipitation collected in the Tri-City area. Practically every sample contained $CHCl_3$ and CCl_4, which, together with CH_2Cl_2 made up 98% of all VOCCs detected and determined. The sampling stations in

Table 2 Maximum and minimum concentrations of ions determined in samples of precipitation collected in different parts of the world.

Continent	Range (minimum–maximum) of mean concentrations of ions (meq/L)
Europe	0.0028–0.265
North America	0.0012–0.210
South America	0.0016–0.136
Asia	0.0030–0.377
Australia	0.0071–0.053
Africa	0.0020–0.013
Antarctica	0.0001–0.016

Source: Data from Singh et al.,[3] Alebić-Juretić and Šojat,[11] Winiwarter et al.,[12] Nickus et al.,[14] Karlsson, Lauren, and Peltoniemi,[39] Colin, Jaffrezo, and Gros,[40] Losno et al.,[41] Cabon,[42] Etsu et al.,[43] Puxbaum, Simeonov, and Kalina,[44] Takeda et al.,[45] Feng et al.,[46] Tuncel and Ungor,[47] Herut et al.,[48] Tanner,[49] Collett et al.,[50] Schemenauer, Banic, and Urquizo,[51] Claek et al.,[52] Williams, Fisher, and Melack,[53] Lacaux et al.,[54] Galy-Lacaux and Modi,[55] Ayers and Manton,[56] and De Felice.[57]

Table 3 Selected pesticides and their metabolites in precipitation samples collected in Europe and North America.

Compound	Concentration range (ng/L)	Compound	Concentration range (ng/L)
		Europe	
Ametryn	<5–144	Metolachlor	2–1080
Alachlor	2–2200	Malaoxon	20–100
Aldicarb	n.d.–14,000	Methyl azinphos	30–1320
Atrazine	<5–5000	Metazachlor	16–134
Bitertanol	15–140	Methidathion	4–390
Carbaryl	10–350	Methyl pirimiphos	7
Captan	9–110	Malathion	10–200
Cyanazine	<5–600	Methyl paraoxon	8–220
Carbofuran	20–1010	Lindane (γ-HCH)	n.d.–1030
Cycloate	4	Propazine	5–157
Chlorothalonil	3–1100	Propiconazole	31–1388
Chloridazon	30–880	Pendimethalin	4–260
Desethylatrazine	<5–882	Prometryn	3–2900
Dichlorvos	20	Phosmet	6–550
Dieldrin	n.d.–2400	Propanil	107–1110
Diazinon	8–322	Propachlor	40
Desisopropylatrazine	28–232	Phorate sulfone	10–250
Desethyl terbutylazine	3–455	Pirimicarb	80–420
Ethyl chlorpyrifos	30–200	Pyrazophos	10
Ethofumesate	3–380	pp´-DDT	<0.5–6000
Ethyl azinphos	20–200	pp´-DDD	<5–3500
Ethyl parathion	4–220	pp´-DDE	<0.5–3400
Endosulfan a	70–90	Quintozene	2–2800
Endosulfan sulfate	80	Phorate sulfoxide	12–560
Fenpropathrin	n.d.–5000	Simazine	<5–730
Fosalon	10–430	Trifluralin	1–40
Fenarimol	30	Triallate	13–2137
Fenoxycarb	5–70	Tebutam	30–92
Fenpropimorph	5	Tebuconazole	3–320
Mecoprop	n.d.–40,000	Terbutryn	<5–34
Methyl parathion	n.d.–3400	Terbutylazine	3–520
Metribuzin	22–130	α-HCH	0.7–8.0
Metalaxyl	6–480	δ-HCH	30–50
Molinate	30–6820		
		North America	
α-HCH	<0.018–6043	DDD	0–0.2
γ-HCH	<0.018–936	Chlordane	0–486
β-Endosulfan	<0.012–1.4	Chlorothalonil	<0.4–85
α-Eendosulfan	<0.035–6.5	Chlorpyrifos	0.3–13
Dieldrin	0.4–913	Diazinon	<0.057–19
DDE	0.06–0.2	Malathion	<0.045–24
DDT	0.2–1.4	Trifluralin	<0.098–1.2

Note: n.d., not detected.
Source: Data from Scharf, Wiesiollek, and Bächmann,[25] Huskes and Levsen,[26] Chevreuil and Garmouma,[27] Charizopoulos and Papadopoulou-Mourkidou,[58] Jager, Bourbon, and Levsen,[59] Millet et al.,[60] Chevreuil et al.[61] (for Europe); Pankow, Isalelle, and Asher,[62] Ligocki and Leuenberger,[63] Hoff et al.,[64] McConnell et al.,[65] and Knap and Binkley[66] (for North America).

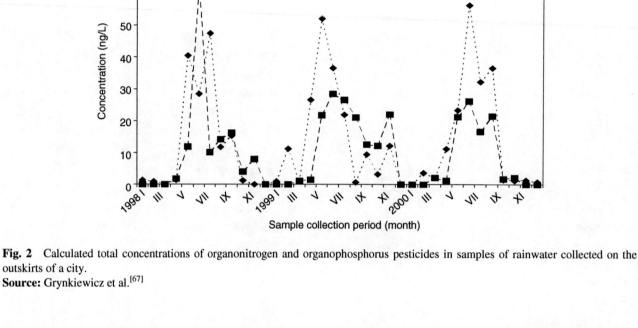

Fig. 2 Calculated total concentrations of organonitrogen and organophosphorus pesticides in samples of rainwater collected on the outskirts of a city.
Source: Grynkiewicz et al.[67]

this area were located in such a way as to record the effects of the interaction between urban sources (anthropogenic) and natural sources (the Gulf of Gdańsk area), hence the presence of both typical solvents ($CHCl_3$ and CH_2Cl_2) and $CHBr_3$, formed in the course of natural processes (present in the marine aerosol and produced by living organisms).[75–77]

As far as PAHs are concerned, the highest levels and a constant presence in precipitation samples were recorded for naphthalene and phenanthrene + anthracene in samples collected in the Tri-City area of Poland.[70,78] These compounds were present in every sample analyzed throughout the study period. There were differences in PAH concentrations depending on season: higher PAH

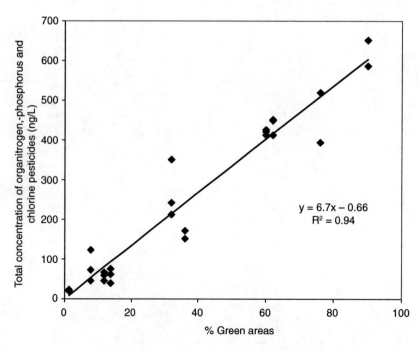

Fig. 3 Dependence between the total pesticide content in precipitation samples and the area of land covered by vegetation in the vicinity of sampling stations.
Source: Grynkiewicz et al.[67] and Polkowska et al.[70]

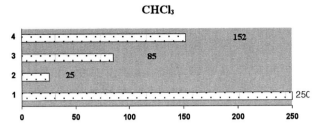

Fig. 4 Content of selected volatile organochlorine compounds in precipitation in different cities around the world (ng/L). 1—Los Angeles, USA; 2—Germany; 3—Gdańsk, Poland; 4—Gdańsk-Sopot, Poland.
Source: Kawamura and Kaplan,[36] Class and Ballschmiter,[74] and Polkowska et al.[75]

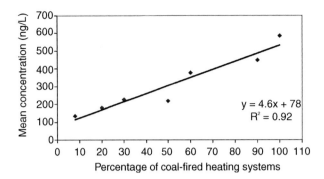

Fig. 5 Relation between the total concentration of PAH analytes in samples of precipitation and the percentage share of coal-fired heating systems in the study area.
Source: Grynkiewicz.[78]

levels were recorded in samples taken during the heating season.

A correlation was found between the PAH concentration and the percentage of coal-fired heating systems (domestic boilers, local district heating systems) supplying heat for domestic purposes in a given area (Fig. 5). The data show that the mean PAH concentration in samples of precipitation depends to a large extent (ca. 92%) on the percentage of coal-fired heating systems.

Benzene, toluene, m-xylene, and p-xylene are compounds routinely determined in precipitation samples. Table 5 sets out the concentrations of petroleum hydrocarbons detected and determined in samples of precipitation collected in Poland. Similar concentrations of toluene were determined in the United States (Los Angeles, 0.076 μg/L; Portland, 0.04–0.22 μg/L); however, the levels of ethylbenzene recorded were lower (0.009 μg/L; 0.072–0.0069 μg/L).[36,38]

Table 6 lists the range of concentrations and mean concentrations of formaldehyde in samples of precipitation collected in different areas of Poland.[79]

SIMULTANEOUS OCCURRENCE OF PRECIPITATION AND DEPOSITS

The concentration of contaminants in samples of precipitation and atmospheric deposits changes if these two processes take place at the same time. Table 7 gives data on the concentrations of the various pollutants in samples of precipitation and fog (occurring simultaneously) collected in mountain areas between September 15 and October 4, 1999.

The concentration of pollutants in fog deposits was on average double that in precipitation and was strongly dependent on the moisture content of the fog.

ATMOSPHERIC DEPOSITS

Studies of the levels of pollutants contained in samples of atmospheric deposits (fog, dew, rime) have shown that these concentrations in such samples are larger than in other forms of atmospheric deposit (this is because

Table 4 Percentages of individual volatile organochlorine compounds in the total amount of these compounds in samples of precipitation.

Analyte	Percentage (%)
$CHCl_3$	91
CCl_4	6.1
CH_2Cl_2	1.5
C_2Cl_4	0.6
$CHBr_2Cl$	0.6
$C_2HCl_3 + CHBrCl_2$	0.6

Source: Polkowska.[76]

Table 5 Concentrations of petroleum hydrocarbons determined in samples of precipitation ($n = 60$).

| | | Minimum | |
Analyte	Maximum	μg/L	Mean
Pentane	0.35	<0.05	0.04
Hexane	3.91	<0.05	0.16
Nonane	2.1	<0.1	0.12
Decane	12.0	<0.1	0.51
Dodecane	3.9	<0.1	0.17
Benzene	8.62	<0.05	0.40
Toluene	1.52	0.07	1.03
Ethylbenzene	3.91	<0.05	0.12
m,p-Xylene	16.40	<0.05	1.21

Source: Polkowska et al.[75]

Table 6 Range of concentrations of formaldehyde in samples of precipitation.

Type of sample		Range of concentration (mg/L)	Mean value (mg/L)
Rain		0.07–1.32	0.526
Snow	Surface layer	0.05–1.72	0.227
	Depth 5 cm	0.05–0.57	0.135
	Depth 10 cm	0.05–0.16	0.072
	Surface layer	0.05–1.00	0.165
	Depth 5 cm	0.05–0.12	0.08

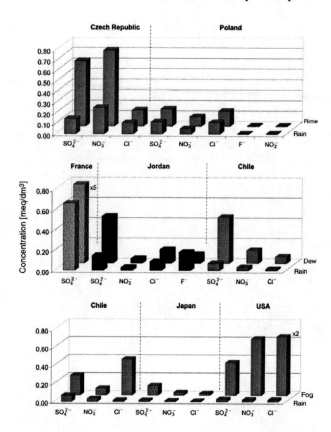

Fig. 6 Concentrations of ions determined in samples of wet deposition collected in various countries around the world.

pollutants are absorbed from the boundary layer). It is for this reason that continuous monitoring of these forms of wet deposition has been introduced in various countries worldwide (Fig. 6).[1,2,4,9,50,51,80–82]

The chemical composition of atmospheric deposits has also been studied in Poland with respect to a very extensive group of organic and inorganic compounds. Fig. 7 shows the percentages of the various ions in the total ionic composition (TIC) present in dew samples collected in urban areas. TIC represents the sum of liquid phase concentration of SO_4^{2-}, NO_3^-, Cl^-, H^+, NH_4^+, Ca^{2+}, Mg^{2+}, Na^+, and K^+. Characteristic of the dew samples, regardless of the nature of the environment around the sampling stations, was the very small proportion of NO_3^- ions in the TIC. This can be explained by the fact that dew is formed at night, when road traffic is 2 or 3 times less intensive than during the daytime. Nitrogen oxides (NO_x) change their forms very rapidly in the atmosphere, so when there is no continuous input of this pollutant at night, there is consequently a smaller concentration of it in dew samples. In all dew samples, regardless of the station where they were collected, the proportion of Ca^{2+} was distinctly elevated, from 20% to 30% of the TIC.

Dew samples collected in farming areas were acidic (pH <5.0): August 28, 2006 (pH = 4.30), September 21, 2006 (pH = 4.16); however, no dew samples were strongly acidic (pH <4.0). A pH of <6.0 was recorded in 7% of samples, whereas pH >7.0 was recorded in more than 34% of samples.

The acidification of dew can also be explained by looking at the relation between the sum of acidifying anions SO_4^{2-} and NO_3^- and the basic cation Ca^{2+}. There was an increase in calcium in relation to the levels of acidifying ions in samples from stations both in the coastal zone and beyond it, which may be due to the amount of construction work going on in these areas.

The maximum values of the TIC[83] determined for dew samples in coastal and inland urban areas and on farmland were 8.24, 10.9, and 6.20 meq/L, respectively. Similar results were obtained for dew samples from Amman, Jordan, (7.68 meq/L)[84] and Bordeaux, France (4.18 meq/L).[85] The mean ratio of total cations to total anions was 0.92

Table 7 Magnitude and depositional structure of pollutants contributed by fog in comparison with precipitation, and proportions of fog deposit and precipitation in the total deposition of contaminants.

Analyte	SO_4^{2-}	Cl^-	NO_3^-	H^+	Na^+	Ca^{2+}	NH_4^+	Σ
Szrenica, Poland (elevation: 1330 m)								
Precipitation (g/m²)	0.183	0.251	0.922	0.003	0.293	0.126	0.104	1.88
Fog deposition (g/m²)	0.158	0.829	1.336	0.008	0.353	0.308	0.196	3.19
Proportion of fog deposit in the total pollutant deposition	30%	82%	67%	62%	63%	54%	73%	62%
Proportion of precipitation in the total pollutant deposition	24%	17%	33%	15%	37%	15%	27%	28%

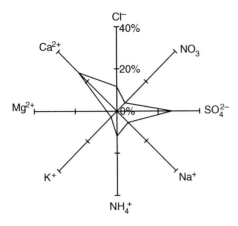

Fig. 7 Percentages of selected ions in the total ionic composition of dew samples.
Source: Polkowska et al.[83]

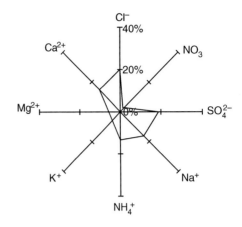

Fig. 8 Proportion of selected ions in the total ionic composition in hoar frost samples.
Source: Polkowska et al.[87]

for dew samples from various parts of Poland and 1.03 for dew samples from the Bordeaux area.[86]

In the case of dew samples, the relation between the level of HCHO and the sulfur content is interesting because the presence of HCHO in the aqueous phase affects the production of sulfuric acid (the reaction between hydrated formaldehyde with SO_4^{2-} or SO_3^{2-} yields hydroxymethyl sulfate). Research results indicate that there is a direct relation between the levels of formaldehyde and sulfate.[79]

Equally interesting are the studies on hoar frost samples conducted in Poland. Fig. 8 shows the percentage of the various ions in the TIC present in hoar frost samples.

The samples of hoar frost collected at all the stations contained few NO_3^- ions; this may have been because frost forms at night, when the intensity of road traffic is 2–3 times less than during the day. The largest proportion of NO_3^- ions was found in the farmland samples of hoar frost; the reason for this may be the application of nitrogen fertilizers in agriculture. Regardless of the location of the stations, the proportions of Na^+ and Cl^- in the TIC were the highest (Na^+, 17%–18%; Cl^-, 18%–24%). This is because these ions are the prime constituents of sea salt. The large proportion of Na^+ and Cl^- in inland samples may be due to the landward transport of sea salt in aerosol form. All the hoar frost samples contained fairly high levels of SO_4^{2-} (15%–20%) in the TIC. SO_4^{2-} is one of the most important indicators of contamination since, like NO_3^-, it is responsible for the acidic nature of the water obtained from melting frost. The percentage of NH_4^+ ions was the largest on farmland, a probable consequence of the ammonium fertilizers applied there.

Regardless of the location of the stations, the proportion of Ca^{2+} ions in hoar frost samples ranged from 13% to 18% of the TIC. The fairly high level of these ions is due, among other things, to the ubiquity of these ions in dust, the utilization of concrete roadways, and the production of cement. The percentage of PO_4^{3-} ions in the TIC was highest on farmland, which, as in the case of NO_3^- and NH_4^+, can be put down to the application of fertilizers in agriculture. Hoar frost samples from coastal urban areas contained high percentages of K^+ in the TIC; these ions are a significant constituent of sea salt.[87,88]

Fig. 9 shows the percentages in the TIC of the various ions present in samples of rime collected in urban areas. The largest proportions of Na^+ and Cl^- were found in rime samples from coastal urban areas; these ions are the main components of sea salt. The largest proportion of SO_4^{2-} ions was recorded in rime samples collected on a mountain summit; this can be explained by the small distance of the measurement station from significant anthropogenic sources. Of interest is the high proportion of F^- ions in the TIC of rime samples collected in a mountain valley (ca. 7%), in comparison with the levels of these ions in mountaintop samples (of the order of 2%) and in a coastal urban area (a fraction of 1%). This may be due to the natural occurrence of fluorine-containing minerals such as biotite and muscovite in the Tatra and Sudeten mountains. A very low proportion of K^+ in the TIC was characteristic of all rime samples, regardless of sampling station location.[79,80,87,89]

Studies have also been carried out in Poland on the chemistry of liquid deposits. Such deposits formed during periods of radiation fog were more contaminated than liquid deposits forming during advection fog ($TIC_{Borucino}$ = 6.52 meq/dm^3; $TIC_{Szrenica}$ = 1.14 meq/dm^3). The chemical composition of samples collected in the vicinity of Borucino (a village in northern Poland) indicates that when liquid is deposited from radiation fog (forming during fine

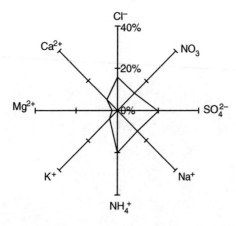

Fig. 9 Proportion of selected ions in the total ionic composition in rime samples.
Source: Polkowska et al.[87]

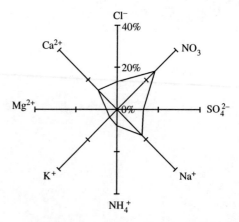

Fig. 10 Proportion of selected ions in the total ionic composition in samples of liquid sediment collected in the vicinity of Borucino.
Source: Błaś et al.[90]

nights, with little or no wind, when the air is descending vertically), anthropogenic emissions play the dominant role. In orographic fog, the transport of contaminants plays a greater part in forming TIC of liquid deposits. Liquid deposits collected at the sampling point on the Szrenica (a mountain in the Sudetens) contain pollutants derived mainly from point sources situated even hundreds of kilometers away on the windward side (in accordance with the prevailing direction of the atmospheric circulation).

Fig. 10 presents the percentages of the various ions in the TIC of liquid deposit samples collected in the vicinity of Borucino. There are quite significant differences between the compositions of the liquid deposits collected at the various sampling points.

Sodium, nitrate, and sulfate ions were prevalent in the samples collected at Borucino, whereas the equilibrium between the contents of nitrate and ammonium was characteristic of the samples collected on Szrenica (daily cycle). The presence of ions such as SO_4^{2-}, NO_3^-, and NH_4^+ reflects the influence of anthropogenic emissions on the chemical composition of liquid deposits. The mean concentration of these ions calculated for liquid deposit samples from Szrenica was about 60% of the TIC, while the corresponding figure for the Borucino samples was 46%.[90]

SUMMARY

Rainwater and snowmelt, liquid deposits (dew, hoar frost, rime), and especially runoff are components of the physical and geographical environment that are easily assimilated and transported. Thus, they are generally a good indicator of the chemical composition of the environment. This is why wet deposition is regarded as a suitable medium for tracking the migration pathways of both natural and anthropogenic substances. Integrated studies of wet deposition (various forms of precipitation and atmospheric deposits, runoff from different surfaces) in conurbations or along motorways can help assess the pollutant load entering the environment from these sources.

In recent years, interest has been increasing in problems concerning precipitation and atmospheric deposits (which, especially in the form of runoff, get into soils and waters) because of the considerable load of toxic substances that are moved around in this way. Research programs for investigating the state of the ambient air are being implemented in an increasing number of countries to demonstrate the crucial importance of systematic monitoring of terrestrial ecosystems.

Studies of the type and quantities of pollutants transported with wet deposition enable the degradation of the environment to be monitored, and consequently, may form the basis for counteracting the adverse effects of pollution on the environment.

ACKNOWLEDGMENTS

This scientific work was financially supported by Science in 2008–2010 as research project no. N305 231035.

REFERENCES

1. Aneja, V.P.; Kim, D.S.; Baumgardner, R.E.; Paur R.J. Temporal variability in cloud water acidity on Mt. Mitchell by the cloud and rain acidity/conductivity (CRAC) real-time analyzer. In *International Conference on Global and Regional Environmental Atmospheric Chemistry*; Newman L., Wang, W., Kiang C., Eds.; Washington, D.C., 1990; 497–526.

2. Anderson, J.B.; Baumgardner, R.E.; Mohnen, V.A.; Bowser, J.J. Cloud chemistry in the eastern United States, as sampled from three high-elevation sites along Appalachian Mountains. Atmos. Environ. **1999**, *33*, 5105–5114.
3. Singh, B.; Nobert, M.; Zwack, P. Rainfall acidity as related to meteorological parameters in northern Quebec. Environ. Sci. Technol. **1987**, *21*, 825–842.
4. Acker, K.; Moller, D.; Wieprecht, W.; Naumann, S. Mt. Brocken, a site for a cloud chemistry measurement programme in Central Europe. Water Air Soil Pollut. **1995**, *85*, 1979–1984.
5. Tang, W.; Davidson, C.I.; Finger, S.; Vance, K. Erosion of limestone building surfaces caused by wind-driven rain: 1. Field measurements. Atmos. Environ. **2004**, *38*, 5589–5599.
6. Nyssen, J.; Vandenreyken, H.; Poesen, J.; Moeyersons, J.; Deckers, J.; Haile, M.; Salles, C.; Govers, G. Rainfall erosivity and variability in the Northern Ethiopian Highlands. J. Hydrol. **2005**, *311*, 172–187.
7. Magaino, S. Corrosion rate of copper rotating-disk-electrode in simulated acid rain. Electrochim. Acta **1997**, *42*, 377–382.
8. Sobik, M.; Migała, K. The role of cloudwater and fog deposits on the water budget in the Karkonosze (Giant) Mountains. Alpex Reg. Bull. Swiss Meteorol. Inst. **1993**, *21*, 13–15.
9. Puxbaum, H.; Tscherwenka, W. Relationships of major ions in snow fall and rime at Sonnblick Observatory (SBO, 3106 m) and implications for scavenging processes in mixed clouds. Atmos. Environ. **1998**, *32*, 4011–4020.
10. Hoffmann, P.; Karandashev, V.K.; Sinner, T.; Ortner, H.M. Chemical analysis of rain and snow samples from Chernogolovka/Russia by IC, TXRF and IC-MS. Fresenius J. Anal. Chem. **1997**, *357*, 1142–1148.
11. Alebić-Juretić, A.; Šojat, V. Chemical composition of rainwater collected at two sampling sites in the city of Rijeka. Arch. Ind. Hyg. Toxicol. **1998**, *49*, 219–296.
12. Winiwarter, W.; Puxbaum, H.; Schöner, W.; Böhm, R.; Werner, R.; Vitovec, W.; Kasper, A. Concentration of ionic compounds in the wintertime deposition: Results and trends from the Austrian Alps over 11 years. Atmos. Environ. **1998**, *32*, 4031–4040.
13. Tsakovski, S.; Puxbaum, H.; Simeonov, V.; Kalina, M.; Löffler, H.; Heimburger, G.; Biebl, P.; Weber, A.; Damm, A. Trend, seasonal and multivariate modeling study of wet precipitation data from the Austrian Monitoring Network (1990–1997). J. Environ. Monit. **2000**, *2*, 424–431.
14. Nickus, U.; Kuhn, M.; Novo, A.; Rossi, G.C. Major element chemistry in Alpine snow along a north-south transect in the eastern Alps. Atmos. Environ. **1998**, *32*, 4053–4060.
15. Nimmo, M.; Fones, G.R. The potential pool of Co, Ni, Cu, Pb and Cd organic complexing ligands in coastal and urban rain waters. Atmos. Environ. **1997**, *31*, 693–702.
16. Veysseyre, A.; Moutard, K.; Ferrari, Ch.; Van de Velde, K.; Barbante, C.; Cozzi, G.; Capodaglio, G.; Boutron, C. Heavy metals in fresh snow collected at different altitudes in the Chamonix and Maurienne Valleys, French Alps: Initial results. Atmos. Environ. **2001**, *35*, 415–425.
17. Shu, P.; Hirner, A.V. Polyzyklische aromatische Kohlenwasserstoffe und Alkane in Niederschlägen und Dachabflüssen. Vom Wasser **1997**, *89*, 247–259.
18. Seidl, W. Model for a surface film of fatty acids on rain water and aerosol particles. Atmos. Environ. **2000**, *34*, 4917–4932.
19. Largiuni, O.; Giacomelli, M.C.; Piccardi, G. Concentration of peroxides and formaldehyde in air and rain and gas–rain partitioning. J. Atmos. Chem. **2002**, *41*, 1–20.
20. Thomas, W. Principal component analysis of trace substance concentration in rainwater samples. Atmos. Environ. **1986**, *20*, 995–1000.
21. Duineker, J.C.; Bouchertall, F. On the distribution of atmospheric polychlorinated biphenyl congeners between vapor phase, aerosols, and rain. Environ. Sci. Technol. **1989**, *23*, 57–62.
22. Agrell, C.; Larsson, P.; Okla, L.; Agrell, J. PCB congeners in precipitation, wash out ratios and deposition fluxes within the Baltic Sea region, Europe. Atmos. Environ. **2002**, *36*, 371–383.
23. Bucheli, T.D.; Grüebler, F.C.; Müller, S.R.; Schwarzenbach, R.P. Simultaneous determination of neutral and acidic pesticides in natural waters at the low nanogram per liter level. Anal. Chem. **1997**, *69*, 1569–1576.
24. Van Maanen, J.M.S.; De Vaan, M.A.J.; Veldstra, A.W.F.; Hendrix, W.P.A.M. Pesticides and nitrate in groundwater and rainwater in province of Limburg in the Netherlands. Environ. Monit. Assess. **2001**, *72*, 95–114.
25. Scharf, J.; Wiesiollek, R.; Bächmann, K. Pesticides in the atmosphere. Fresenius J. Anal. Chem. **1992**, *342*, 813–816.
26. Huskes, R.; Levsen, K. Pesticides in rain. Chemosphere **1997**, *35*, 3013–3024.
27. Chevreuil, M.; Garmouma, M. Occurrence of triazines in the atmospheric fallout on the catchment basin of the river Marne (France). Chemosphere **1993**, *27*, 1605–1608.
28. Trautner, F.; Huber, K.; Niessner, R. Appearance and concentration ranges of atrazine in spring time cloud and rainwater from the Vosges (France). J. Aerosol Sci. **1992**, *23*, S999–S1002.
29. Kiss, G.; Varga-Puchony, Z.; Hlavy, J. Determination of polycyclic aromatic hydrocarbons in precipitation using solid-phase extraction and column liquid chromatography. J. Chromatogr. A **1996**, *725*, 261–272.
30. Kiss, G.; Gelencsér, A.; Krivácsy, Z.; Hlavay, J. Occurrence and determination of organic pollutants in aerosol, precipitation, and sediment samples collected at Lake Balaton. J. Chromatogr. A **1997**, *774*, 349–361.
31. Levsen, K.; Behnert, S.; Winkeler, H.D. Organic compounds in precipitation. Fresenius J. Anal. Chem. **1991**, *340*, 665–671.
32. Berg, M.; Müller, S.R.; Mühlemann, J.; Wiedmer, A.; Schwarzenbach, R.P. Concentrations and mass fluxes of chloroacetic acids and trifluoroacetic acid in rain and natural waters in Switzerland. Environ. Sci. Technol. **2000**, *34*, 2675–2683.
33. Garban, B.; Blanchou, H.; Motelay-Massei, A.; Chevreuil, M.; Ollivon, D. Atmospheric bulk deposition of PAHs onto France: Trends from urban to remote sites. Atmos. Environ. **2002**, *36*, 5395–5403.
34. Ollivon, D.; Blanchou, H.; Motelay-Massei, A.; Garban B. Atmospheric deposition of PAHs to an urban site, Paris, France. Atmos. Environ. **2002**, *36*, 2891–2900.
35. Jung, W.; Fujita, M. Optimal conditions of purge trap/on-column cryofocusing method with capillary gas chromatography for determination of volatile halogenated hydrocarbons in aqueous samples. Esei Kagaku **1991**, *37*, 395–400.

36. Kawamura, K.; Kaplan, I.R. Organic compounds in the rainwater of Los Angeles. Environ. Sci. Technol. **1983**, *17*, 497–501.
37. Pankow, J.F.; Isabelle, L.M.; Asher, W.E. Trace organic compounds in rain. 1. Sampler design and analysis by adsorption/thermal desorption (ATD). Environ. Sci. Technol. **1984**, *18*, 310–318.
38. Ligocki, M.P.; Leuenberger, Ch.; Pankow, J.F. Trace organic compounds in rain—II. Gas scavenging of neutral organic compounds. Atmos. Environ. **1985**, *10*, 1609–1617.
39. Karlsson, V.; Lauren, M.; Peltoniemi, S. Stability of major ions and sampling variability in daily bulk precipitation samples. Atmos. Environ. **2000**, *34*, 4859–4865.
40. Colin, J.L.; Jaffrezo, J.L.; Gros, J.M. Solubility of major species in precipitation: Factors of variation. Atmos. Environ. **1990**, *24 A*, 537–544.
41. Losno, R.; Bergametti, G.; Carlier, P.; Mouvier. G. Major ions in marine rainwater with attention to sources of alkaline and acid species. Atmos. Environ. **1991**, *25 A*, 763–770.
42. Cabon, J.Y. Chemical characteristics of precipitation at an Atlantic station. Water Air Soil Pollut. **1999**, *111*, 399–416.
43. Etsu, Y.; Yoshiaki, N.; You, F.; Song-Nan, H.; Yasuro, F. Effects of an eruption Miyake Island on the behavior of air pollutants and chemical components of rainwater in Kyoto. Bull. Chem. Soc. Jpn. **2004**, *77*, 497–503.
44. Puxbaum, H.; Simeonov, V.; Kalina, M.F. Ten years trends (1984–1993) in the precipitation chemistry in central Austria. Atmos. Environ. **1998**, *32*, 193–202.
45. Takeda, K.; Marumoto, K.; Minnamikawa, T.; Sakugawa, H.; Fujiwara, K. Three-year determination of trace metals and the lead isotope ratio in rain and snow deposition collected in Higashi-Hiroshima, Japan. Atmos. Environ. **2000**, *34*, 4525–4535.
46. Feng, Z.; Huang, Y.; Feng, Y.; Ogura, N.; Zhang, F. Chemical composition of precipitation in Beijing area, Northern China. Water Air Soil Pollut. **2001**, *125*, 345–356.
47. Tuncel, S.G.; Ungor, S. Rain chemistry in Ankara, Turkey. Atmos. Environ. **1996**, *30*, 2721–2727.
48. Herut, B.; Spiro, B.; Starinsky, A.; Katz, A. Sources of sulfur in rainwater as indicated by isotopic sigma^{34}S data and chemical composition, Israel. Atmos. Environ. **1995**, *29*, 851–857.
49. Tanner, P.A. Relationships between rainwater composition and synoptic weather systems deduced from measurement and analysis of Hong Kong daily rainwater data. J. Atmos. Chem. **1999**, *33*, 219–240.
50. Collett, J., Jr; Daube, B., Jr; Munger, J.W.; Hoffmann, M.R. Cloudwater chemistry in Sequoia National Park. Environ. Sci. Technol. **1989**, *23*, 999–1007.
51. Schemenauer, R.S.; Banic, C.M.; Urquizo, N. High elevation fog and precipitation chemistry in Southern Quebec, Canada. Atmos. Environ. **1995**, *29*, 2235–2252.
52. Claek, K.L.; Nadkarni, N.M.; Schaefer, D.; Gholz, L. Cloud water and precipitation chemistry in a topical montane forest, Monteverde, Costa Rica. Atmos. Environ. **1998**, *32*, 1595–1603.
53. Williams, M.R.; Fisher, T.R.; Melack, J.M. Chemical composition and deposition of rain in the central Amazon, Brazil. Atmos. Environ. **1997**, *31*, 207–217.
54. Lacaux, J.P.; Delmas, R.; Kouadio, G.; Cros, B.; Andreae, M.O. Precipitation chemistry in the Mayombe forest of Equatorial Africa. J. Geophys. Res. **1992**, *97*, 6195–6206.
55. Galy-Lacaux, C.; Modi, A.I. Precipitation chemistry in the Sahelian savanna of Niger, Africa. J. Atmos. Chem. **1998**, *30*, 319–343.
56. Ayers, G.P.; Manton, M.J. Rainwater composition at two BAPMoN regional stations in SE Australia. Tellus **1991**, *43* B, 379–389.
57. De Felice, T.P. Chemical composition of fresh snowfalls at Palmer Station, Antarctica. Atmos. Environ. **1999**, *33*, 155–161.
58. Charizopoulos, E.; Papadopoulou-Mourkidou, E. Occurrence of pesticides in rain of the Axios River basin, Greece. Environ. Sci. Technol. **1999**, *33*, 2363–2368.
59. Jager, M.E.; Bourbon, C.; Levsen, K. Analysis of pesticides and their degradation products in rainwater: A probe into their atmospheric degradation. Intern. J. Environ. Anal. Chem. **1998**, *70*, 149–162.
60. Millet, M.; Wortham, H.; Sanusi, A.; Mirabel, P. Atmospheric contamination by pesticides: Determination in the liquid, gaseous and particulate phases. Environ. Sci. Pollut. Res. **1997**, *4*, 172–180.
61. Chevreuil, M.; Garmouma, M.; Teil, M.J.; Chesterikkoff, A. Occurrence of organochlorines (PCBs, pesticides) in the atmosphere and in the fallout from urban and rural stations of the Paris area. Sci. Total Environ. **1996**, *182*, 25–37.
62. Pankow, J.F.; Isalelle, L.M.; Asher, W.E. Trace organic compounds in rain. 1. Sampler design and analysis by adsorption/thermal deposition (ATD). Environ. Sci. Technol. **1984**, *18*, 310–318.
63. Ligocki, M.P.; Leuenberger, Ch. Trace organic compounds in rain—II. Gas scavenging of neutral organic compounds. Atmos. Environ. **1985**, *19*, 1609–1617.
64. Hoff, R.M.; Strachan, W.M.J.; Sweet, C.W.; Chan, C.H.; Shackleton, M.; Bidleman, T.F.; Brice, K.A.; Burniston, D.A.; Cussion, S.; Gatz, D.F.; Harlin, K.; Schroeder, W.H. Atmospheric deposition of toxic chemicals to the great lakes: A review of data through 1994. Atmos. Environ. **1996**, *30*, 3505–3527.
65. McConnell, L.L.; Lenoir, J.S.; Datta, S.; Seiber, J.N. Wet deposition of current-use pesticides in the Sierra Nevada mountain range, California, USA. Environ. Toxicol. Chem. **1998**, *17*, 1908–1916.
66. Knap, A.H.; Binkley, K.S. The occurrence and distribution of trace organic compounds in Bermuda precipitation. Atmos. Environ. **1988**, *22*, 1411–1423.
67. Grynkiewicz, M.; Polkowska, Z.; Górecki, T.; Namieśnik, J. Pesticides in precipitation from an urban region in Poland (Gdańsk–Sopot–Gdynia) between 1998 and 2000. Water Air Soil Pollut. **2003**, *194*, 3–16.
68. Bucheli, T.D.; Müller, S.R.; Heberle, S.; Schwarzenbach, R.P. Occurrence and behavior of pesticides in rainwater, roof runoff, and artificial stormwater infiltration. Environ. Sci. Technol. **1998**, *32*, 3457–3464.
69. Huskes, R.; Levsen, K. Pesticides in rain. Chemosphere **1997**, *35*, 3013–3024.
70. Polkowska, Z.; Kot, A.; Wiergowski, M.; Wolska, L.; Wołowska, K.; Namieśnik, J. Organic pollutants in precipitation: Determination of pesticides and polycyclic aromatic

hydrocarbons in Gdańsk, Poland. Atmos. Environ. **2000**, *34*, 1233–1245.
71. Na, K.; Kim, Y.P. Seasonal characteristics of ambient volatile organic compounds in Seoul, Korea. Atmos. Environ. **2001**, *35*, 2603–2614.
72. Colon, M.; Pleil, J.D.; Hartlage, T.A.; Guardani, M.L.; Martins, M.H. Survey of volatile organic compounds associated with automotive emissions in the urban airshed of São Paulo, Brazil. Atmos. Environ. **2001**, *35*, 4017–4031.
73. Andelman J.B. Human exposures to volatile halogenated organic chemicals in indoor and outdoor air. Environ. Health Perspect. **1985**, *62*, 313–318.
74. Class, Th.; Ballschmiter, K. Chemistry of organic traces in air. V. Determination of halogenated C_1–C_2-hydrocarbons in clean marine air and ambient continental air and rain by high resolution gas chromatography using different stationary phases. Fresenius J. Anal. Chem. **1986**, *325*, 1–7.
75. Polkowska, Ż.; Gorlo, D.; Wasik, A.; Grynkiewicz, M.; Namieśnik, J. Organic pollutants in atmospheric precipitation. Determination of volatile organohalogen compounds and petroleum hydrocarbons in rain water and snow by means of gas chromatography. Chem. Anal. **2000**, *45*, 537–550.
76. Polkowska, Z. Determination of volatile organohalogen compounds in urban precipitation in tricity area (Gdańsk, Gdynia, Sopot). Chemosphere **2004**, *57*, 1265–1274.
77. Grynkiewicz, M.; Polkowska, Z.; Górecki, T.; Namieśnik, J. Volatile organochlorine compounds and pesticides in precipitation from an urban region (Gdańsk, Poland). Anal. Lett. **2001**, *34*, 1503–1515.
78. Grynkiewicz, M.; Polkowska, Z.; Namieśnik, J. Determination of polycyclic aromatic hydrocarbons in bulk precipitation and runoff waters in an urban region (Poland). Atmos. Environ. **2002**, *36*, 361–369.
79. Polkowska Z.; Skarżyńska K.; Górecki T.; Namieśnik, J. Formaldehyde in various forms of atmospheric precipitation and deposition from highly urbanized regions. J. Atmos. Chem. **2006**, *53*, 211–236.
80. Duncan, L.C. Chemistry of rime and snow collected at a site of central Washington Cascades. Environ. Sci. Technol. **1992**, *26*, 61–66.
81. Dore, A.J.; Sobik, M.; Migała, K. Patterns of precipitation and pollutant deposition in the western Sudete Mountains, Poland. Atmos. Environ. **1999**, *33*, 3301–3312.
82. Schmitt, G. The temporal distribution of trace element concentrations in fogwater during individual fog events. In *Atmospheric pollutants in forest areas: their deposition and interception*; Georgii, H.-W., Ed.; Dordrecht, 1986.
83. Polkowska, Z.; Błaś, M.; Klimaszewska, K.K.; Sobik, K.; Małek, S.; Namieśnik, J. Chemical characterization of dew water collected in different geographic regions of Poland. Sensors **2008**, *8*, 4006–4032.
84. Jiries, A. Chemical composition of dew in Amman, Jordan. Atmos. Res. **2001**, *57*, 261–268.
85. Beysens, D.; Muselli, M.; Milimouk, I.; Ohayon, C.; Berkowicz, S.M.; Soyeux, E.; Mileta, M.; Ortega, P. Application of passive radiative cooling for dew condensation. Energy **2006**, *31*, 1967–1979.
86. Beysens, D.; Ohayon, C.; Muselli, M.; Clus, O. Chemical and biological characteristics of dew and rain water in an urban coastal area (Bordeaux, France). Atmos. Environ. **2006**, *40*, 3710–3723.
87. Polkowska, Z.; Sobik, M.; Błaś, M.; Klimaszewska, K.; Walna, B.; Namieśnik J. Rime and hoarfrost chemistry in Poland—An introductory analysis from meteorological perspective. J. Atmos. Chem. **2009**, *62*, 5–30.
88. Dubiella-Jackowska, A.; Astel, A.; Polkowska, Z.; Kudłak, B.; Namieśnik, J. Atmospheric and surface water pollution interpretation in the Gdańsk Beltway impact range by the use of mulivariate analysis. Clean-Soil Air Water **2010**, *38*, 865–876.
89. Klimaszewska, K.; Polkowska, Z.; Namieśnik, J. Major ions and its relationship in rime and hoarfrost samples from highly urbanized regions. Pol. J. Environ. Stud. **2007**, *16*, 943–948.
90. Błaś, M.; Polkowska, Z.; Klimaszewska, K.; Sobik, M.; Nowiński, K.; Namieśnik, J. Fog water chemical composition in different geographic regions of Poland. Atmos. Res. **2010**, *95*, 455–469.

Rain Water: Harvesting

K.F. Andrew Lo
Department of Natural Resources, Chinese Culture University, Taipei, Taiwan

Abstract
Water is a limited resource. Even if water is a renewable resource, it is, at the same time, finite. Its availability is largely dictated by climate. Low precipitation and high evaporation often mean small amounts of useable water. In recent years, much progress in efforts to improve living conditions has been achieved through technological solutions. Total water use in the world has quadrupled during the past 50 years. At present and in the future, livelihood conditions for the burgeoning populations can only be marginally improved through the construction of dams, reservoirs, and conveyance structures. New approaches are needed for the proper management and use of water resources.

INTRODUCTION

Among the various alternative technologies to augment water resources, rainwater harvesting is a simple, decentralized solution and imposes insignificant impact on the environment. It is an important water source in many areas with significant rainfall but lacking any kind of conventional, centralized supply system. It is also a good option in areas where good-quality fresh surface water or groundwater is lacking. Rainwater harvesting systems have been used since ancient times and evidence of roof catchment systems dates back to early Roman times. In the Negev Desert in Israel, in Libya and Egypt, in Mexico, and in the Andes Range in South America as well as in the Arizona Desert in North America, stone dams and tanks were built to divert and store rainwater for irrigation purposes.

ADVANTAGES OF RAINWATER HARVESTING

Rainwater harvesting systems can provide water at, or near, the point where water is needed or used. The systems can be both owner-operated and utility-operated, and owner-managed and utility-managed. Rainwater collected using existing structures (rooftops, parking lots, playgrounds, parks, ponds, and flood plains) has few negative environmental impacts compared with other water resources development technologies.[1] Rainwater is relatively clean and the quality is usually acceptable for many purposes with little or even no treatment. The physical and chemical properties of rainwater are usually superior to sources of groundwater that may have been subject to contamination.

Other advantages of rainwater harvesting include the following:

1. Rainwater harvesting can coexist with, and provide a good supplement to, other water sources and utility systems, thus relieving pressure on other water sources.
2. Rainwater harvesting provides a water supply buffer for use in times of emergency or breakdown of public water supply systems, particularly during natural disasters.
3. Rainwater harvesting can reduce storm drainage load and flooding in cities.
4. The owners who operate and manage the rainwater catchment system are more willing to exercise water conservation.
5. Rainwater harvesting technologies are flexible and can be built to meet almost any requirements.

TYPES OF RAINWATER HARVESTING SYSTEMS

Collection systems can vary from simple households to large catchment systems. The categorization of rainwater harvesting systems depends on factors such as the size and nature of the catchment areas and whether the systems are in urban or rural settings.[2]

Simple Rooftop Collection Systems

The main components of a simple rooftop collection system are the cistern itself, the piping that leads to the cistern, and the appurtenances within the cistern (Fig. 1). The materials and the degree of sophistication of the whole system largely depend on the initial capital investment. Some cost-effective systems involve cisterns made with ferrocement. In some cases, the harvested rainwater may be filtered or disinfected.

Large Systems for Educational Institutions, Stadiums, Airports, and Other Facilities

When the systems are larger, the overall system can become more complicated (e.g., rainwater collection from roofs and grounds of institutions, storage in underground reservoirs, and treatment and use for non-potable applications) (Fig. 2).

Fig. 1 A simple roof catchment system (illustrated by Chia-Ming Lin).

Rooftop Collection Systems for High-Rise Buildings in Urbanized Areas

In high-rise buildings, roofs can be designed for catchment purposes and the collected roof water can be kept in separate cisterns on the roofs for non-potable uses.

Land Surface Catchments

Ground catchment techniques (Fig. 3) provide more opportunity for collecting water from a larger surface area. By retaining small creek and stream flows in small storage surfaces or underground reservoirs, can meet water demands during dry periods. However, there is a possibility of high seepage loss to the ground. The marginal quality of the water collected is suitable for use mainly in agriculture.

Collection of Stormwater in Urbanized Catchment

The surface runoff collected in stormwater ponds/reservoirs from urban areas is subject to a wide variety of contaminants. Keeping these catchments clean is of primary importance; hence the cost of water pollution control can be considerable.

DESIGN AND MAINTENANCE OF RAINWATER HARVESTING SYSTEMS

Typically, a rainwater harvesting system consists of three basic elements: the collection system, the conveyance system, and the storage system.

Catchment Surface

The effective catchment area and the material used in constructing the catchment surface influence collection efficiency and water quality. Materials commonly used for roof catchment are corrugated aluminum and galvanized iron, concrete, fiberglass shingles, tiles, and slates. Mud is used primarily in rural areas. Bamboo roofs are least suitable because of possible health hazards. The catchment surface materials must be non-toxic and must not contain substances that impair water quality. Roofs with metallic paint or other coatings are not recommended because they may impart tastes or color to the collected water. Catchment surfaces and collection devices should be cleaned regularly to remove dust, leaves, and bird droppings to minimize bacterial contamination and to maintain the quality of collected water. Roofs should also be free from overhanging trees because birds and animals in the trees may defecate on the roofs.

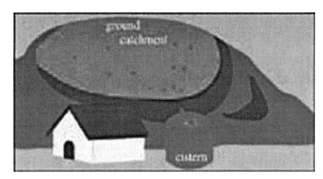

Fig. 3 A land catchment system (illustrated by Chia-Ming Lin).

Fig. 2 An indoor storage system in a monastery in China (photographed by K. F. Andrew Lo).

When land surfaces are used as catchment areas, various techniques are available to increase runoff capacity: 1) clearing or altering vegetation cover; 2) increasing the land slope with artificial ground cover; and 3) reducing soil permeability by soil compaction. Specially constructed ground surfaces (concrete, paving stones, or some kind of liner) or paved runways can also be used to collect and convey rainwater to storage tanks or reservoirs. Care is required to avoid land surface damage and contamination by people and animals. If required, these surfaces should be fenced to prevent people and animal entry. Large cracks in the paved catchment because of soil movement, earthquakes, or prolonged exposure should be repaired immediately. Maintenance, typically consisting of the removal of dirt, leaves, and other accumulated materials, should take place annually before the start of the major rainfall season.

Conveyance Systems

Conveyance systems are required to transfer the rainwater collected on catchment surfaces to storage tanks. This is usually accomplished by making connections to one or more downpipes connected to collection devices. The pipes used for conveying rainwater, wherever possible, should be made of plastic, polyvinyl chloride (PVC), or other inert substance because the pH of rainwater can be acidic and may cause corrosion and mobilization of metals in metal pipes.

When it first starts to rain, dirt and debris from catchment surfaces and collection devices will be washed into the conveyance systems. Relatively clean water will only be available sometime later in the storm. The first part of each rainfall should be diverted from the storage tank. There are several possible options for selectively collecting clean water for the storage tanks. The common method is a sediment trap, which prevents debris entry into the tank. Installing a first-flush (or foul-flush) device is also useful to divert the initial batch of rainwater away from the tank.[3]

Rainwater pipes must be permanently marked in such a way that there is no risk of confusing them with drinking water pipes. Gutters and downpipes need to be periodically inspected and carefully cleaned. A good time to inspect gutters and downpipes is while it is raining, so that leaks can be easily detected.

Storage Tanks

Various types of rainwater storage facilities can be found in practice. Storage tanks should be constructed of inert material. Reinforced concrete, fiberglass, polyethylene, and stainless steel are suitable materials. Ferrocement tanks and jars made of mortar or earthen materials are commonly used. As an alternative, interconnected tanks made of pottery or polyethylene may be suitable. They are easy to clean. Bamboo-reinforced tanks are less successful because they may become infested with termites, bacteria, and fungi.

Precautions are required to prevent the entry of contaminants into storage tanks. The main sources of external contamination are pollution from debris, bird and animal droppings, and insects. A solid and secure cover is required to avoid breeding of mosquitoes, to prevent insects and rodents from entering the tank, and to keep out sunlight to prevent algae growth inside the tank.[4] A coarse inlet filter is also desirable for excluding coarse debris, dirt, leaves, and other solid materials.

All tanks need cleaning and their designs should allow for thorough scrubbing of the inner walls and floors. A sloped bottom and the provision of a pump and a drain are useful for collection and discharge of settled grit and sediment. Chlorination of the cisterns or storage tanks is necessary if the water is to be used for drinking and domestic uses. Cracks in the storage tanks can create major problems and should be repaired immediately.

The extraction system (taps/faucets, pumps) must not contaminate the stored water. Taps/faucets should be installed at least 10 cm above the base of the tank because this allows any debris entering the tank to settle on the bottom.[5] If it remains undisturbed, it will not affect the quality of the water. The handle of taps might be detachable to avoid misuse by children. Periodic maintenance should also be carried out on pumps used to lift water.

CONCLUSION

In the future, water scarcity in both developing and developed countries is inevitable.[6] The challenge of meeting the water demand can be largely met by appropriate understanding, study, and application of rainwater harvesting. Rainwater harvesting is about to come of age.[7] It has an appropriate image about it that meshes well with the gentler ideas of the late 20th century. Because the technique makes use of an untapped resource—precipitation that would otherwise be evaporated before it had a chance to play a useful role in feeding the human population—it looks like getting something for nothing. Making use of such a resource has certain poetry to it, particularly in a field where the resource itself can never be increased or decreased; unlike food, water cannot be grown to order, even given the right soil and the right fertilizer. But, like food, water can be harvested more efficiently. Doing so is a major priority for the 21st century.

REFERENCES

1. Geiger, W.F. Integrated water management for urban areas, Proceedings of the 7th International Rainwater Catchment Systems Conference, Beijing, China, June 21–25, 1995;

Liu, C.M., Ed.; Institute of Geographical Sciences and Natural Resources, Chinese Academy of Sciences: Beijing, 1995, 7-1–7-21.
2. Gould, J.; Nissen-Peterson, E. *Rainwater Catchment Systems for Domestic Supply: Design, Construction and Implementation*; Intermediate Technology Publications: London, 1999.
3. Macomber, P.S.H. *Guidelines on Rain Water Catchment Systems for Hawaii*; College of Tropical Agriculture and Human Resources, University of Hawaii at Manoa: Hawaii, 2001.
4. Cunliffe, D.A. *Guidance on the Use of Rainwater Tanks*; National Environmental Health Forum Monographs, Water Series No. 3; National Environmental Health Forum: Australia, 1998.
5. Texas Water Development Board *Texas Guide to Rainwater Harvesting*, 2nd Ed.; Texas Water Development Board: Texas, 1997.
6. Clarke, R. *Water: The International Crisis*; The MIT Press: Cambridge, 1993.
7. Pacey, A.; Cullis, A. *Rainwater Harvesting: The Collection of Rainfall and Runoff in Rural Areas*; Intermediate Technology Publications: London, 1986.

Rare Earth Elements

Zhengyi Hu
Institute of Soil Science, Chinese Academy of Sciences, Nanjing, China

Gerd Sparovek
College of Agriculture, Graduate School of Agriculture Luiz de Queiroz (ESALQ), University of São Paulo, São Paulo, Brazil

Silvia Haneklaus
Ewald Schnug
Institute for Crop and Soil Science, Julius Kuhn Institute (JKI), Braunschweig, Germany

Abstract

The rare earth elements comprise the elements scandium and yttrium, and 15 lanthanides with successive atomic numbers. Rare earth elements are applied to soils as fertilizer materials or as contaminations of industrial sludges so that an assessment of their behavior in soils is required for evaluating agro-environmental effects. This entry summarizes the present knowledge of soil chemical properties of rare earth elements.

INTRODUCTION

The rare earth elements (REEs) comprise the elements scandium (Z/21) and yttrium (Z/39), and 15 lanthanides with successive atomic numbers (Z) from 57 to 71 (Table 1). Rare earth elements are applied to soils as fertilizer materials or as contaminations of industrial sludges so that an assessment of their behavior in soils is required for evaluating agro-environmental effects.[1–6] This entry summarizes the present knowledge of soil chemical properties of REEs.

ORIGINS OF SOIL RARE EARTH ELEMENTS

Soil REEs mainly originate from parent materials.[1] Application of phosphate fertilizer[3] and phosphogypsum[4] can supply REEs to the soil. Some of the sludges, particularly those from the chemical industry, have been contaminated with REEs (Table 2). Continuous application of sewage sludges caused an accumulation of Sc, Sm in some soils in Japan.[6] The use of REEs in agriculture is widely practiced in China. By 2001, 6.5 million hectares of land in China was treated with REE fertilizers.[2] In total, 11,000 tons of REEs was applied in agricultural production in China.[7] Besides the parent material, the application of REEs on agricultural farmland is going to be a major source of REEs if the practice of applying them regularly proceeds.

CHEMICAL SPECIATION OF RARE EARTH ELEMENTS IN SOILS

Total Content

Representative background values of REEs in soils are available so far only for China and Japan.[1,2,8] The REE content strongly depends on the parent material.[1] The results of 853 soil analyses showed that the total REE content in soils varied between 18 and 583 mg kg^{-1}, with a mean value of 184 mg kg^{-1}.[2] The light REEs La, Ce, Pr, Nd, Sm and Eu account for 90% of the total REE content in soils (Table 3). On the average, the La, Ce, Nd, Sm, and Eu content was 41, 73, 7.3, 27.5, and 5.6 mg kg^{-1}, respectively, in different soils in China (n/467) (Table 3).

Species of REEs in Soils

Binding forms of REEs in soils may be classified according to their availability for plants.[1] Approximately 9 mg kg^{-1} is exchangeable, 2, 5, 32, and 95 mg kg^{-1} are bonded to carbonates, Mn-oxides, organic substances, and amorphous Fe-oxides, while 59 mg kg^{-1} is abundant in the form of crystal iron oxides and 105 mg kg^{-1} in residual forms (Table 4). Plants utilize exchangeable REEs most easily, while the uptake of other forms is limited.[1,9] In contrast, residual forms are not plant available.[1,9]

For determining the content of plant available REEs the following extractants have been proposed: 1 M HAc–NaAc (pH 4.8),[1,9] 1.0 M NH$_4$NO$_3$ (pH 7.0),[10] 0.1 M HCl,[11] and 0.1 M malic-citric acid,[12] with 1 M HAc–NaAc being most extensively used.[1,2] Plant available REE contents are highly variable and range from <1 to >200 mg kg^{-1}, with a mean value of about 12 mg kg^{-1} (n/1790).[1] Physico-chemical soil properties such as pH, E$_h$, CEC, clay, H$_2$PO$_4^-$, and carbonate content have a strong impact on the amount of exchangeable and plant available REEs.[1,9] Acid soils contain significantly higher amounts of plant available REEs than more alkaline, calcareous soils.[2] The availability of soil applied REEs is usually significantly higher than that in the original soil matrix.[1,9]

Table 1 Symbol and atomic numbers of rare earth elements (REEs).

Elements	Symbol	Atmomic number	Descriptive classification
Scandium	Sc	21	
Yttrium	Y	39	
Lanthanum	LA	57	
Cerium	Ce	58	
Praseodymium	Pr	59	Light earths
Neodymium	Nd	60	
Promethium	Pm	61	
Samarium	Sm	62	
Europium	Eu	63	
Gadolinium	Gd	64	
Terbium	Tb	65	
Dysprosium	Dy	66	
Holmium	Ho	67	
Erbium	Er	68	Heavy earths
Thulium	Tm	69	
Ytterbium	Yb	70	
Lutetium	Lu	71	

Source: Liu.[1]

Adsorption of Rare Earth Elements in Soils

In general, 95% of the added REEs are adsorbed.[13] Rare earth elements added to soils are rapidly transformed, for example, into exchangeable, organic matter bonded, and Fe/Mn oxide bonded species.[1,9] The distribution coefficients for REEs added to red and yellow brown soils declined in the following order: residual > exchangeable > organic matter bonded > Fe/Mn oxide bonded REEs.[9] The formation of bridged hydroxo complexes is probably the dominant sorption mechanism to clay minerals.[14] Clay type, pH, CEC, organic matter, and amorphous iron content regulate the adsorption kinetics of REEs.[1,2,9] Langmuir and Freundlich equations were found to describe precisely the absorption of REEs in soils.[1,15]

Table 2 Concentrations and coefficient of variation of rare earth elements in different sludges.

	Night soil sludge[a] (n / 10)		Sewage sludge (n / 14)		Food industry sludge (n / 10)		Chemical industry sludge (n / 10)	
Elements	Mean (mg kg^{-1})	CV (%)	Mean (mg kg^{-1})	CV (%)	Mean (mg kg^{-1})	CV (%)	Mean (mg kg^{-1})	CV (%)
La	3.39	37	6.70	47	0.89	72	2.46	98
Ce	6.98	44	14.10	58	1.83	77	2.69	105
Pr	0.82	38	1.48	46	0.22	82	0.48	95
Nd	3.18	34	6.00	47	0.91	82	2.04	98
Sm	0.53	36	1.02	40	0.17	81	0.36	95
Gd	0.53	34	1.18	45	0.17	79	0.48	101
Tb	0.07	45	0.16	36	0.03	81	0.06	103
Dy	0.39	53	0.93	33	0.14	76	0.39	110
Ho	0.07	54	0.19	32	0.03	79	0.09	119
Er	0.21	55	0.57	31	0.08	73	0.26	118
Tm	0.03	52	0.08	26	0.01	81	0.03	112
Yb	0.20	58	0.54	31	0.09	83	0.19	109
Lu	0.03	56	0.08	31	0.01	84	0.03	105

[a]Feces and urine of humans.
Source: Kawasaki and Kimura.[5]

Table 3 Mean content of REEs in soils extracted by Na$_2$O$_2$/NaOH (n / 467).

Light REEs	Content (mg kg^{-1})	Heavy REEs	Content (mg kg^{-1})	Total contents (mg kg^{-1}) and ratios	
La	41.2	Gd	4.8	Total REE (T)	172.8
Ce	73.4	Tb	0.7	Light REE (L)	156.0
Pr	7.3	Dy	4.4	Heavy REE (H)	16.8
Nd	27.5	Ho	0.9	L/H	9.3
Sm	5.6	Er	2.7	L/T	0.9
Eu	1.1	Tm	0.4		
		Yb	2.5		
		Lu	0.4		

Source: Liu.[1]

Translocation of Rare Earth Elements in Soils

The question is still open whether the use of REEs in industry and agriculture may result in a pollution of soils, plants, and groundwater. In leaching experiments under controlled conditions with ^{141}Ce and ^{147}Nd, these elements were abundant only in the top soil layer because of their strong adsorption.[9] For field conditions a translocation depth of <1 cm has been estimated,[9] but this value still needs to be validated for different soils and on a long-term basis.

UPTAKE OF RARE EARTH ELEMENTS BY PLANTS

The uptake of REEs has been investigated for a wide range of plants.[2,9] All rare elements except praseodymium were found in plants. Uptake of REEs was related to many factors such as element, plant type, and growth conditions.[2,9] The light REEs Ce, La, Nd were highest in plants.[2,9] The total concentration of REEs in plant tissue ranges from <0.05 to 2.58 mg kg^{-1} (Table 5). The concentration of Ce

Table 5 Total concentration of rare earth elements in plants (mg kg^{-1}).

Species	n	Min (mg kg^{-1})	Max (mg kg^{-1})
Rice	319	<LLD[a]	1.17
Wheat	440	<LLD	2.58
Corn	139	<LLD	0.92
Cucumber	41	<LLD	0.70
Leek	33	0.04	0.21
Spinach	41	<LLD	0.12
Cauliflower	61	<LLD	0.60
Lotus root	31	<LLD	0.76
Tomato	64	<LLD	0.18
Chinese cabbage	67	0.05	1.01
Pepper	31	<LLD	0.40
Potato	34	0.05	0.35
Cabbage	38	<LLD	1.20
Mushroom	33	<LLD	0.45
Orange	41	0.13	0.70
Litchi	30	<LLD	
Grape	61	<LLD	0.85
Longan	30	<LLD	0.71
Banana	33	<LLD	
Apple	62	0.07	0.80
Pear	34	<LLD	0.24
Watermelon	37	<LLD	0.26
Sugarcane	27	0.05	1.25
Peach	4	<LLD	

[a]LLD: Lower limit of detection.
Source: Xiong.[9]

in food products is usually lower than 0.2 mg CeO$_2$ kg^{-1} (Table 6). More attention should be paid to research on uptake, translocation, and distribution of REEs in plants in order to follow up the biological effects of REEs and to assess agro-environmental effects.

Table 4 Mean content of different species of rare earth elements in soils (mg kg^{-1}).

Soils	n	Exchangeable	Carbonate bonded	Mn-oxide bonded	Organically bonded	Amorphous Fe-oxide bonded	Crystal ion oxides	Residual	Total
Latosol	3	1.0	—	1.4	24.3	23.8	8.2	17.5	76.2
Red soil	5	14.8	—	8.0	30.1	108.0	33.9	199.8	394.6
Yellow brown soil	2	12.1	—	7.4	26.1	79.3	88.3	55.5	268.7
Brown soil	2	11.5	—	4.7	33.7	78.8	64.1	34.5	227.3
Black soil	2	1.4	5.3	1.2	46.8	88.8	74.8	64.0	282.3
Chernozem	2	1.8	11.4	2.8	41.2	105.3	94.1	79.0	335.5
Mean		9.1	1.9	5.3	32.2	95.4	58.8	105.3	307.9

Note: Latosol (rhodic ferralsol, FAO); red soil (ferralic cambisol); yellow brown soil (haplic luvisol); brown soil (haplic alisol); black soil (luvic phaeozems); chernozem (haplic chernozems).
Source: Liu.[1]

Table 6 Ce content in grains and seeds of different crops in different countries and regions of China (mg CeO_2 kg^{-1}).

Crops	Region/country	Sample no.	Mean	Min	Max
Rice	Huibei/China	8	0.06	0.02	0.15
Rice	Jiangxi/China	21	0.06	<0.01	0.18
Rice	Beijing/China	17	0.04	<0.01	0.11
Rice	Nanjing/China	2	—		0.12
Wheat	Heilongjiang/China	21	0.04	<0.01	0.09
Wheat	Hubei/China	7	0.04	0.01	0.07
Wheat	Henan/China	6	0.10	<0.01	0.19
Wheat	Shandong/China	4	0.12	<0.07	0.17
Wheat	Beijing/China	10	0.03	<0.01	0.05
Wheat	Tianjing/China	1	0.04	<0.01	0.08
Maize	Huibei/China	8	0.01	<0.01	0.02
Maize	Heilongjiang/China	10	0.04	<0.01	0.10
Maize	Tianjiang/China	15	0.04	<0.01	0.00
Barley	Beijing/China	14	0.04	<0.01	0.25
Wheat	Canada	5		0.04	0.16
Wheat	USA	8		<0.01	0.20
Wheat	Australia	4		0.06	0.16
Wheat	Argentina	4		0.07	0.09
Soybean	USA	3		<0.01	0.11
Maize	USA	3		0.01	0.09
Mung bean	Thailand	1			0.09

Source: Xiong.[9]

Table 7 Yield increase (relative to control) after application of rare earth elements to different crops.

Crop	Country	Yield increase (%)
Sugar beet	Bulgaria	17–24
Sugar beet	China	7
Wheat	China	6–17
Rape	China	4–48
Potato	China	5–6
Soybean	China	8–9
Cotton	China	5–12
Rice	China	7
Corn	China	9–103
Barley	Australia	18–19
Peanut	China	8–12
Tobacco	China	8–10
Rubber	China	8–10
Sugarcane	China	10–15
Cabbage	China	10–20
Litschi	China	14–17
Grape	China	8–12

Source: Hu.[16]

CROP RESPONSE TO RARE EARTH ELEMENTS

Research with a view to the use of REEs in agriculture was predominantly carried out in China and before in Russia. A small number of studies have been carried out in Australia, too.[2,9] The increases in crop yield reported by workers from all parts of China range between 5% and 103% (Table 7), with an average response of 8%–15%.[2,9] Crop response to REEs is reported to be most probable when soils contain less than 10 mg kg^{-1} of available REEs (in 1 M HAc–NaAc, pH 4.8), while a response on soils with more than 20 mg kg^{-1} of available REEs is unlikely (Table 8). So far, there is no evidence that REEs are essential for plant growth.

Table 8 Critical values of plant available rare earth elements in soils.

REE content (mg kg^{-1})	Index	Crop response to REE
<5.0	Very low	Most likely
5–10	Low	Probable
11–15	Medium	Not expected
16–20	High	Not expected
>20	Very high	Unlikely

Source: Liu[1] and Xiong.[9]

CONCLUSIONS

Basic information about the chemistry of REEs in soils is available, but this data refers to specific regions and soil types. Besides this, the effect of REEs, for example, on soil fertility and soil biological diversity is yet unclear. Crop response to REE applications depends on various factors, including soil properties. Further studies are also required in order to elaborate the most efficient application techniques.

REFERENCES

1. Liu, Z. Rare earth elements in soils. In *Microelements in Soils of China*; Liu, Z., Ed.; Jiangsu Science and Technology Publishing House: Nanjing, China, 1996; 293–329.
2. Xiong, B.K.; Chen, P.; Guo, B.S.; Zheng, W. *Rare Earth Element Research and Application in Chinese Agriculture and Forest*; Metallurgical Industry Press: Beijing, China, 2000.
3. Todorovsky, D.S.; Minkova, N.L.; Bakalova, D.P. Effect of the application of superphosphate on rare earth content in the soils. Sci. Total Environ. **1997**, *203* (1), 13–16.
4. Arocena, J.M.; Rutherford, P.M.; Dudas, M.J. Heterogeneous distribution of trace elements and fluorine in phosphogypsum. Sci. Total Environ. **1995**, *162* (2–3), 149–160.
5. Kawasaki, A.; Kimura, R.; Arai, S. Rare earth elements and other trace elements in wastewater treatment sludges. Soil Sci. Plant Nutr. **1998**, *44* (3), 433–441.
6. Zhang, F.S.; Yamasaki, S.; Kimura, K. Rare earth element content in various waste ashes and the potential risk to Japanese soils. Environ. Int. **2001**, *27* (5), 393–398.
7. Wang, S.M.; Zheng, W. *Developing technology of rare earth application on agriculture and biological field in China*. Proceedings of the 4th International Conference on Rare Earth Development and Application, Beijing, China, June 16–18, 2001; Yu, Z.S., Yan, C.H., Xu, G.Y., Niu, J.K., Chen, Z.H., Eds.; Metallurgical Industry Press: Beijing, 2001, 237–240.
8. Yoshida, S.; Muramatsu, Y.; Tagami, K.; Uchida, S. Concentrations of lanthanide elements, Th, and U in 77 Japanese surface soils. Environ. Int. **1998**, *24* (3), 275–286.
9. Xiong, P.K.; Zheng, W.; Cheng, P.; Wang, F. *Rare Earth Elements in Agricultural Environment*; China Forestry Press: Beijing, China, 1999.
10. Zhai, H.; Yang, Y.G.; Zheng, S.J.; Hu, A.T.; Zhang, S.; Wang, L.J. Selection of the extractants for available rare earths in soils. China Environ. Sci. **1999**, *19* (1), 67–71 (in Chinese).
11. Li, F.L.; Shan, X.Q.; Zhang, S.Z. Evaluation of single extractants for assessing plant availability of rare earth elements in soils. Commun. Soil Sci. Plant Anal. **2001**, *32* (15 and 16), 2577–2587.
12. Zhang, S.Z.; Shan, X.Q.; Li, F.L. Low-molecularweight-organic-acid as extractant to predict plant bioavailability of rare earth elements. Int. J. Environ. Anal. Chem. **2000**, *76* (4), 283–294.
13. Zhu, J.G.; Xing, G.X.; Yamasaki, S.; Tsumura, A. Adsorption and desorption of exogenous rare earth elements in soils: I. Rate of forms of rare earth elements sorbed. Pedosphere **1993**, *3* (4), 299–308.
14. Dong, W.M.; Wang, X.K.; Bian, X.Y.; Wang, A.X.; Du, J.Z.; Tao, Z.Y. Comparative study on sorption/desorption of radioeuropium on alumina, bentonite and red earth: effects of pH, ionic strength, fulvic acid, and iron oxides in red earth. Appl. Radiat. Isotopes **2001**, *54* (4), 603–610.
15. Gao, X.J.; Zhang, S.; Wang, L.J. The adsorption of La^{3+} and Yb^{3+} on soil and mineral and its environmental significance. China Environ. Sci. **1999**, *19* (2), 149–152 (in Chinese).
16. Hu, Z.Y.; Richter, H.; Sparovek, G.; Schnug, E. Physiological and biochemical effects of rare earth elements on plants and their agricultural significance: a review. J. Plant Nutr. **2003**, in press.

Index

2, 4-D. *See* 2, 4-Dichlorophenoxyacetic acid (2, 4-D)
2, 4-Dichlorophenoxyacetic acid (2, 4-D), 2005, 2556t, 2796t

A

Aarhus Protocol on Persistent Organic Pollutants, 1914
Abatement costs, 916, 917, 918, 918f, 919
Absorption
   herbicides, 1378–1379
Absorption chillers, 1356, 1357. *See also* Absorption heat pumps
   for building cooling applications, 1360
   for cogeneration, 1361
   for solar air-conditioning, 1361–1362
Absorption cycles, 1349, 1357
   bottoming and topping cycles, 1357
   type I, 1357, 1357f
   type II, 1357, 1358, 1358f
Absorption/desorption, of polychlorinated biphenyls, 2176
Absorption heat pumps, 1349, 1349f, 1357–1362. *See also* Absorption chillers; Heat pumps
   configurations, 1357
   heat source and heat sink configurations, 1357–1358
   performance, 1359–1360, 1360t
   refrigeration, operation of, 1358–1359, 1359f
   working fluid systems, 1358
Acanthoscelides obsoletus, 1116
Acaricides, 1, 2014
   amidines, 3
   antimetabolites, 3
   application technology, 4
   carbamates, 3
   organochlorine, 2–3
   organophosphates, 3
   organotins, 3
   ovicides, 3
   propargite, 3
   resistance to, combating of, 4
   synthetic pyrethroids, 3
   tetronic acids, 3–4
   use of, 2
Acatak. *See* Fluazuron
Accelerated solvent extraction, 2197
Acceptable daily intake (ADI), 1124, 1125f

Acceptance-based commissioning, 666–667
Accipiter nisus (Eurasian sparrowhawk), 417
Acetamide detection, 1319
Acetochlor, 1958
Acetogenesis, 2651
Acetogenin, 384
Acetylcholine, 1398
Acetylcholinesterase (AChE), 2005
AChE. *See* Acetylcholinesterase (AChE)
Acid–base accounting approach (ABA), 39, 40–41
Acid Deposition Control Program, 20
Acid deposition modeling, 1130–1131
Acidification, 1579
   and acid rains, 2359
Acidification of rivers and lakes, 2291, 2296, 2298. *See also* Lakes and reservoirs
   definition, 2291
   environmental problems caused by, 2297–2298
   historical perspectives, 2292
   measurement, 2291–2292
   solutions
      environmental legislation, 2298
      technology, 2298
   sources of freshwater
      anthropogenic, 2293f, 2296–2297
      natural, 2292–2294, 2293f
      redox reactions in water–sediment systems, 2294–2296
Acidithiobacillus ferrooxidans, 43, 1689, 1690, 2293
Acidithiobacillus thiooxidans, 1690
Acid mine drainage (AMD), 29, 269, 1694, 2293, 2296. *See also* Mine drainage
   and groundwater pollution, 1289
Acid-neutralizing capacity (ANC), 2291
Acidogenesis, 2651
Acid precipitation, 681
Acid Precipitation Act of 1980, 7, 20
Acid rain, 5, 916, 2304
   acidification of oceans, 13
   in Asian region, 8–9
   in Canada, 8
   control of, 15
   effects on aquatic ecosystem, 16
   effects on buildings and monuments, 16
   effects on vegetation and soil, 16
   in Europe, 8
   formation of, 5–6
   future projections through modeling, 14

   global sensitivity toward acidification, 13, 14f
   history of, 6
   natural acidity, 7
   precipitation scenario, regional comparison of, 14–15
   program, 2298
   regional acidity, 8–9
   sources of, 6
   spread and monitoring of, 7–8
   trends in acidity, 9–13
   in United States, 8
Acid rain, and N deposition, 20–24
   ecosystem impacts, 22–24
   human health effects, 22
   reducing effects of, 24
   sources and distribution, 20–22, 21f
   structural impacts, 22
Acid rock drainage (ARD), 29, 55, 2293
Acid soils, 2266
Acid sulfate soils (ASSs), 26, 55, 268, 2294
   active, 27
   classification
      Australian system, 28
      Entisols, 27
      FAO of United Nations, 28
      Histosols, 27
      Inceptisols, 27
      soil taxonomy, 27–28
      Sulfic Endoaquepts, 28
      Wassents suborder, 27
      WRB (World reference base), 28
   definition of, 26, 27
   diagnostic characteristics
      sulfidic materials, 27
      sulfuric horizon, 27
   drainage from, impacts of, 56
   extent of, in Australia, 56
   global extent of, 56
   global warming and, 56
   historical recognition, 26
   inland/upland *vs.* coastal, 28–29
   management of, 56–58
      acidity containment and neutralization in, 58–59
      Australia role in, 57–59
      avoidance in, 58
      education and assessment in, 58
      oxidation prevention in, 58
      Queensland's ASS Management Guidelines, 57–58
   postactive, 27
   potential, 26–27

Acid sulfate soils (ASSs) (cont'd.)
  problems from, 55–56
Acid sulfate soil materials, 36
  characteristic minerals and geochemical processes
    oxidation processes, 42–43
    reduction processes, 41–42
  characteristics, 37–38
  hazards arising from disturbance of acid sulfate soil materials, 43
    acidification, 43
    heavy metal and metalloid mobilization, 44
    iron mobilization, 43–44
  identification and assessment, 39–41
  management for agriculture, forestry, and aquaculture, 49–50
  scalded landscapes revegetation, 50
  management for intensive developments
    contaminants containment, 48
    covering with fill, 47
    dewatering, 48–49
    general principles, 46–47
    hydraulic separation, 47–48
    neutralizing materials addition, 48
    stockpiling, 48
    strategic reburial, 48
  occurrence, 38
  physical behavior
    consistence and strength, 45–46
    permeability, 46
  water bodies deoxygenation, 44–45
  noxious gases production, 45
Acid Sulphate Soil Risk Maps, for New South Wales coastline, 58
A.C. modules. *See* Module inverters
A.C. switch disconnector, 223
Actinomycetous species, 531
Actinorrhizae (*Frankia*), 1515
Actino-uranium 235U decay series, 2238f
Activated sludge process (ASP), 2648, 2648f
  operating parameters in, 2649
Activated sludge system, 2650
  attached growth processes, 2650–2651
Activated sludge treatment, 2062
Active layer
  permafrost zone, 1900, 1902f
Active solar heat storage, 2520–2521
Active solid matter, 149
Acute reference dose (acute RfD), 1124, 1401, 1402t
Acute RfD. *See* Acute reference dose (acute RfD)
Actual acidity, 40
Acyrthosiphon pisum, 1461
Adaptation Fund, 487
Additional energy, 770, 772, 773. *See also* Energy
  categories of, 772
  from chemical fuels, 773
  life cycle considerations, 773

thermal energy, 772–773
types of, 771t
from wastes, 773
Additionality, 486
Adherence to pesticide label instruction, 509–510, 509t
ADI. *See* Acceptable daily intake (ADI)
Adiabatic flame temperature, 722, 723f, 725
Adipic acid, 1256
Admiral. *See* Pyriproxyfen
Adobe walls, 864
Adsorbents, 1908
Adsorption, 61–75, 1283, 1559
  boron, 432, 432f
  capacity, 62
  chemical, 62
  defined, 61
  isotherms, 63–66, 432f
  kinetics, 66–68
  of REE, 2267
  physical, 62–63
  of strontium, 2427
  techniques, 457
  theory of, 61–68
  in water purification, research progress on, 74–75
Adsorption–desorption, 2166
Advanced Spaceborne Thermal Emission and Reflection Radiometer (ASTER), 554, 2817t
Advanced thermal technologies (ATTs), 835–837, 843, 849
  types, 836
Advanced very high resolution radiometer (AVHRR), 2817t
Advanced visible and near infrared radiometer (AVNIR), 2817t
Advanced wastewater treatment (AWT), 1583
Advection, 1282
Aerated concrete insulation, 1479
Aeration, 522–523
  tanks, 2648
Aerial photography, 2819
Aerial ultra low volume (ULV)
  application, of mosquito insecticides
    nontarget mortality by, minimization of, 1471
    insecticide residue monitoring for, 1472–1473
    novel application technologies for, 1472
    right dose, 1472
    right place, 1471
    right time, 1471–1472
Aerobic biological waste treatment processes, 2648
Aerobic processes, 2647
Aerobic trickling filter, 2650f
Aerobic wetland conditions, 2864
Aerosols, 1196, 1196t
AerWay SSD system, 1674, 1674f

α-Fe_2O_3, 1712
Afforestation, 460
Aflatoxin, 1697
A-frequency–weighted sound pressure level, 2869
Africa
  soil degradation in, 2379–2381, 2380f
    distribution of, 2380f
African countries
  pesticide poisonings incidence, 1408
Agelaius phoeniceus, 413
Agency capture theory, 926. *See also* Environmental policy innovations (EPI)
Agency for Toxic Substances and Diseases Registry (ATSDR), 650, 2049, 2052
Agency of Regional Air Monitoring of the Gdańsk Agglomeration (ARMAAG), 139
AGENDA for Environment and Responsible Development, 507
"Agent Orange," 1395
Agent success
  weed biocontrol and, 2821–2822, 2822t
"Age of Fossil Fuels," 549, 551
Ageratum houstonianum, 1464
Aggregated theoretical indicators, 601, 604
Aggregation, 2752
Aging, 1398
Aglais urticae L., 1750
AGNPS model. *See* Agricultural nonpoint source pollution (AGNPS) model
Agricultural-based waste materials, 72–74, 74t
Agricultural and Environmental Information System for Windows (AEGIS/WIN), 2819
Agricultural ecosystems
  acidic deposition and, 23
Agricultural experiment stations, 2009
Agricultural nonpoint source pollution (AGNPS) model, 966, 1164, 2762
Agricultural Policy/Environmental eXtender (APEX) model, 2761
Agricultural pollution, 2304. *See also* Pollution
Agricultural runoff, 81–84
  BMPs and, 83
  control of, 83
  dissolved pollutants, 82–83
  eroded sediments and, 82
  IPM and, 83
  quantity, 81–82
  soil erosion and associated pollutants, 82
Agricultural soils, phosphorus in, 100–102
Agricultural sustainability
  and agrobiodiversity, 353
  farmers and consumers awareness for, 359–360
  indicators for, 352. *See also* Bioindicators

Agricultural water quantity and quality management under uncertainty. *See* Inexact optimization models
Agriculture, 88
  benefits of herbicides to, 1383
  probiotic, 528–529, 531–533
  and public health, sustaining, 2003–2004
  sustainable. *See* Sustainable agriculture
  water quality for, 1566
  wastewater use in. *See* Wastewater use, in agriculture
Agrobacterium tumefaciens, 307
Agroecosystem biodiversity
  sustainable agriculture, 2457
Agroecosystems
  ecological infrastructure in, 372
  interaction with other ecosystems, 2434
Agroforestry, 129, 570, 1037, 1515
Agroforestry, WUE and, 129–130
  evaporation and, 130
    diffusion driven phase, 130
    energy driven phase, 130
  precipitation and, 130
Agro/horti ecosystem, pesticide pollution in, 2146
Agrostis spp., 1305
Aim. *See* Chlorfluazuron
Air, 916
  bioaccumulation in, 324–326
  carbon removal from, 1189–1190
    man role in, 1189–1190
    nature role in, 1190
  leakage, 866
  pesticides in, 1966, 1969–1972, 1971t, 1972
  sustainable agriculture, 2457
Air and Precipitation Network (APN), 7
Air-assisted nozzles, 1472
Airborne electromagnetic methods for saltwater detection, 1327
Air-conditioning, solar, 1361–1362
Air Drywall Approach, 861
Air gasification, 377
Air injection wells, 1335
Airlift bioreactor, 2626
Air pollutants, 1154–1155
  ambient and emission standards of, 1127–1129
  criteria, characteristics of, 136–138
  genotoxic effects of, 1155–1156
  sources of, 136, 137t
  transport and dispersion, 1129–1130
Air pollution, 1154–1159
  desertization and, 566
  genotoxic effects of, 1155–1156
  genotoxicity tests for, 1156–1157
  health risk associated with, 1157–1158
Air pollution index (API). *See* Air quality index
Air pollution legislation, history of, 133–136

Air pollution monitoring, 132–146
  biomonitoring using plants, 142–143
  geographical information system in, 143
  instruments
    classification of, 140–141
    requirements of, 139–140
  objectives of, 132–133, 132f
  program designing of, 134f
  quality of
    monitoring networks, designing of, 138–139
    obtained information, types of, 139, 139f
  remote monitoring techniques for, 143–144
  techniques of, 141–142, 141t
Air quality, 1019
  wind erosion effect on, 1019–1020
Air quality index (AQI), 134–136
Air quality modeling (AQM), 1130–1132
  acid deposition, 1130–1131
  basic ingredients of, 1130
  grid-type, 1130
  photo-oxidants, 1131–1132
  regional haze, 1131, 1132
  trajectory-type, 1130
Air quality standards, 133–134
Air sparging (AS), 1335
Air-stream bottom aerator (ASB), operation mode of, 1594, 1594f
Alachlor, 1318t, 1957, 1957t, 1958, 2142, 2806
Alauda arvensis (Eurasian skylark), 416
Albedos, defined, 1192
Albrecht effect, 1197t
Alcaligenes, 97
Alders, 1527t
Aldicarb, 3, 419, 1318t, 2806
Aldicarb sulfone, 1318t
Aldicarb sulfoxide, 1318t
Aldrin, 417
Alexandria Lake Maryut, 164–173
  current state of
    hydraulic functioning, 166–167
    hydrological functioning, 166–167
    social considerations and governance, 167, 169
    water quality and ecosystems, 167
  integrated action plan for
    defined, 169
    model development, 169–173
    scenario identification, 169
  location map of, 165f
Alfalfa (*Medicago sativa* L.), 371, 427, 1552
Alfisols, 427
Algal blooms, 475, 538, 2358
Algal control measures, 1584
Algal growth, 1559f
Alien plant species, 1182
Al(III)EDTA, 2245
Alkali earth metals, 1560

Alkaline bauxite refinery waste (red mud), 269
Alkaline fuel cells (AFCs), 1146–1148, 1152t
  applications of, 1146
  electrolytes in, 1147
  with liquid electrolyte, 1147–1148
  problems related to, 1147
  research groups working on, 1146–1147
  safety issues related to, 1147–1148
  with solid electrolyte membrane, 1146, 1148
  in space shuttle missions, 1146
  technical applications and demonstration projects on, 1147
  types, 1146
Alkaline rains, 9f
Allelochemics, 175–178
  natural occurrence, 175–177, 176t, 177f
  overview, 175
  practical applications, 177–178
  push–pull application, 178
  terminology, 175
Allelopathy, 2031
Allievi-Michaud formula, 211–212
Alliance of Small Island States (AOSIS), 485
Allolobophora caliginosa, 409
Allomones, 175
Allowance trading, 486
α-Al$_2$O$_3$, 1712
Alopex lagopus, 2596
Alpha-chloralose, 413–414
Alpha (α) decay, 2234–2235
Alphitobius diaperinus, 1921
Alsystin. *See* Triflumuron
Alternanthera philoxeroides (Malancha), 426
Alternaria, 442, 444
Alternative energy
  photovoltaic solar cells, 226–240
Alternative prey/hosts
  natural enemies and, 370
Altica carduorum, 2822
Altosid. *See* Methoprene
Alum, 267
Aluminosilicate clay minerals, 268
Aluminum (Al), 267
  alleviation of
    in soil, 269
    in water, 270
  in atmosphere, sources of, 270, 270t
  chemistry of, 267
  in cosmetics, 267
  in drinking water, 270
  in foodstuff, 270, 270t
  in pharmaceuticals, 270
  production of, 268, 268t
  in soil, sources of
    Al minerals, 268
    forms of soluble Al, 268
    mobilization of Al in soil, 268–269

Aluminum (Al) (cont'd.)
  soluble forms of, 267
  toxicity of, 43
    in aquatic organisms, 271
    in humans, 271–272
    in plants, 270–271
  uses of, 267, 268
  in water, sources of
    acid deposition, 269
    acid mine drainage, 269
    acid sulfate soils, 269
    Al from water purification, 269–270
Aluminum oxyhydroxide particles, 1714, 1714f
Alum sludges, 270
Alzheimer's disease, 271
Amaranthus, 1936
Amazonian rainforest, 333
Ambient air quality and SO_2, 2437–2439
Ambient air quality standards (AAQS), 2438
Ambient energies, 1346
Ambient energy fraction (AEF), heat pump system, 1352–1353, 1352f, 1353t
Ambient geothermal energy, 772
Ameliorants, 1684
  chemical, 2370
  reclamation of sodic soils and
    optimal supply, 2373
    rate of supply, 2373–2374
    without addition of, 2373
Amelioration
  erosion and, 1059
Amendments, soil. *See* Organic soil amendments
American Carbon Registry, 1774
American elm. *See Ulmus americana*
American Petroleum Institute (API) gravity, 2042, 2045
American Society of Agricultural Engineers, 471
American Society of Heating, Refrigerating, and Air-Conditioning Engineers, Inc. (ASHRAE), 1365, 1366
Ames test, 1156
Amide herbicides, 1319
4-Aminopyridine (Avitrol®), 413, 415
Aminosulfonyl acids, 1318
Amitraz, 3
Ammonia (NH_3)
  as refrigerant, 1358
  soil quality and, 2390
  volatilization. *See* Volatilization, ammonia
Ammonia-rich wastewater, 2653
Ammonia volatilization, 93, 94, 1517
Ammonia wet scrubber, 2441
Ammonium (NH_4^+), 97, 1768, 2736
Ammonium nitrate, 125
Amorphous silicon, 232

Ampelomyces quisqualis, 2107
Amplitude-modulated sound, 2872–2873
Amylase, 89
Anaerobic ammonium oxidation, 2653
Anaerobic digester systems, 1674
Anaerobic digestion, 377
Anaerobic earthen lagoon system, for manure storage, 1672, 1672f
Anaerobic methane-oxidizing bacteria, 391
Anaerobic ponds, 2636
Anaerobic processes, 2647
Anaerobic wastewater treatment, 2651
  advantages, 2651
  digestion process, 2651
  disadvantages, 2651
  operating parameters in anaerobic reactors, 2652
  pH, 2652
  UASB reactor, 2651–2652
Anaerobic wetland conditions, 2864
Anagrus, 371
Anagrus epos, 1936
Analytical cost estimation (ACE) method, 840–843
Analytical semi-empirical model (ASEM), 837–840
Analytical models, for, pest management, 1950
Anammox, 2653
ANC. *See* Acid-neutralizing capacity (ANC)
Andalin. *See* Flucycloxuron
Andasol solar power station, 2521, 2521f
  flow schematic, 2522f
Andisols, 1854
Anilide herbicides, 1319
Animal exposure effects, of endocrine-disrupting chemicals, 650–651
Animal husbandry
  and organic farming, 1107
Animal manure, 1667
Animal pests
  regulation of, 1115–1117
    anthropogenic measures, 1116–1117
    biotic (natural) resistance of environment and, 1115–1116
    strategy of controlling in organic farming, 1115
Animals
  biomonitoring, 2128
  toxocity of herbicides to, 1385–1386
Anisopteromalus calandrae, 2424
Annoyance, 2871–2873, 2873t, 2878f
Anoxic processes, 2647
ANS-118. *See* Chromafenozid
Antagonistic microorganisms
  stimulation of, 1879
Antagonistic plants, 288–290
  benefits and risks, 289
  in cultural pest control, 288–289, 289t
  defined, 288

  nature of, 288–289
  overview, 288
Antarctica, pollution in, 2138
Antarctic ozone hole, 1887, 1888f
Anthemis arvensis, 1928
Anthonomous grandis grandis, 2009
Anthracite, 717
Anthraquinone, 415
Anthropogenic acid deposition, 268
Anthropogenic climate change, 483
Anthropogenic factors for acidification, 2293f, 2296–2297, 2297t
  acid mine drainage, 2296
  air pollution with acidifying compounds, 2296
Anthropogenic forcings, 1199f
Anthropogenic greenhouse gases, 1192
Anthropogenic measures
  animal pests, 1116–1117
Anthropogenic modification of global climate, 955
Anthropogenic pollutants, 2303
Anthropogenic sources, removal from SO_2, 2440
  ammonia wet scrubber, 2441
  magnesium wet scrubber, 2441
  wet techniques, 2440–2441
Antibiotics, in fish farms, 2061
Anti-erosion drop replacement, 2310
Antiperspirants, Al in, 270
Antireflective coatings (ARCs), in solar cells, 234
Apera spica venti, 1928
Apex, Dianex. *See* Methoprene
Apex solar cells, 232
Aphis fabae, 1878
Apollo missions, 1716
Applaud. *See* Buprofezin
Application efficiency, 1546, 1547t, 1564
AQM. *See* Air quality modeling
Aquaprene. *See* Methoprene
Aqua Regia extractable phosphorus, 102f
Aquatic biochemical equilibrium, 2731f
Aquatic ecosystems, 1576
  acidic deposition and, 24
  effects of acid rain on, 16
  health, indication of, 607
Aquatic environment, pesticides in, 1965–1966, 2146–2147
Aquatic organisms, Al toxicity in, 271
Aquatic plants, 2753
Aquatic systems, 538
Aquatic weed management by fish, in irrigation systems, 1922
Aquifers, 1282, 1290
  coastal, 1321
  remediation, surfactant-enhanced, 1844
  storage, 1285, 1286, 1286f
Arachis hypogaea L. (Peanuts), 424
Aral Sea, degradation of, 1438–1440
Aral Sea disaster, 300–302, 300t, 301t, 302f

Page numbers followed by f and t indicate figures and tables, respectively

Aralsk and, 301–302
average water supply and demand in, 300, 300t
environmental problems, 302
human tragedy, 302
irrigation and cotton, 301
Muynak and, 301–302
overview, 300
Aralsk
Aral Sea disaster and, 300–302
ARD. See Acid rock drainage (ARD)
Areal loading approach, 2865
Areal Nonpoint Source Watershed Environment Response Simulation (ANSWERS) model, 966, 2761
Argillan, 1627
Arid region irrigation
and soil IC, 465–466
Aridity index map, 2376–2377
Arion ater, 383
Arion distinctus, 383
Arion intermedius, 383
Arion silvaticus, 383
Armillaria mellea Vahl ex.fr, 1879
Armillaria spp., 1879
Array inverters, 222, 222f
Arrhenius' equation, 161
Arsenic, 1263
as contaminant in groundwater, 1281
content in soil, 1263
in drinking water, 1262, 1264. See also Arsenic-contaminated groundwater
drinking water standard for, 1281
in food chain, 1274, 1275f
groundwater contamination by, 1292
health impacts of, 2118–2119
inorganic forms, 1263
occurrence of, 1271
source of, 1262
toxic level of, in groundwater, 1626
use of, 1263
Arsenic-contaminated groundwater
in Assam, 1268
in Bangladesh, 1268–1271, 1270f
in Bihar, 1267
in Chhattisgarh, 1268
cohort study on, 1274
effect on children, 1275, 1275f
in GMB Plain, 1262, 1264, 1264f, 1265t
impacts of, on human health, 1271
cancer effects, 1274, 1274f
cardiovascular effects, 1272
dermal effects, 1271–1272, 1272f
endocrinological effects, 1273
gastrointestinal effects, 1272–1273
neurological effects, 1273
reproduction and developmental effects, 1273–1274
respiratory effects, 1272, 1272f
incidents of, 1263–1264, 1263f

in India and Bangladesh, 1269t, 1271t
for irrigation, 1274, 1275f
in Jharkhand, 1268
in Manipur, 1268
safe water options
deep tube wells, 1276
dug wells, 1276
rainwater harvesting, 1276
watershed management and purification, 1276
socioeconomic effects, 1276
source of, 1276
in Uttar Pradesh, 1267–1268
in West Bengal, 1265–1267, 1266f, 1267f
Arsenic trioxide, 2016t
Arsenopyrite, 1281
Arthropod parasitoids/predators
artificial diets for, 1746
Artificial destratification, 1592t
Artificial diets
for predators and parasitoids, 1746–1747
successes and failures with, 1746
types of, 1746
Artificial impoundments, 1576
Artificial rearing of natural enemies, 1747
Artificial stream ecosystem (ASE), 2154
Arundo donax (Giant Reed), 1182
AS. See Air sparging (AS)
Ascendency, 604
ASEAN, 875t
Ash composition, 730t
Asia
acid rain in, 8–9
soil degradation in, 2378–2379, 2379f
distribution of, 2379f
trends in acidity, 13
distribution of, 2379f
trends in acidity, 13
Asia, environmental legislation in, 874–876
biodiversity laws
enforcement, 881–882
regional cooperation, 882
biodiversity laws at national level, 877
Cambodia, 881
China, 877
Hong Kong, 879–880
Indonesia, 878–879
Japan, 878
Korea, 877
Lao People's Democratic Republic, 880
Malaysia, 878
Philippines, 879
Singapore, 879
Thailand, 880
Vietnam, 881
development in international law and its consequences on national legislations, 877

environmental assessment law
Cambodia, 887–888
China, 885
Hong Kong, 886
Indonesia, 886
Japan, 886
Lao PDR, 887
Malaysia, 886
Philippines, 887
Republic of Korea, 886
Singapore, 887
Thailand, 887
Vietnam, 888
e-waste management law, 882
Cambodia, 884
China, 882–883
Hong Kong, 883
implementation and enforcement, 885
Japan, 883
Lao People's Democratic Republic, 884
Malaysia, 883
Philippines, 883–884
Republic of Korea, 883
Singapore, 884
Thailand, 884
Vietnam, 884
e-wastes and environmental assessment, 882
law on biodiversity conservation, 876
transboundary movements of hazardous waste, 884
Asia Arsenic Network (AAN), 1269
Asnæs Power Station (APS), 1434
ASP. See Activated sludge process (ASP)
Asparagus officinalis, 288
Aspen, 1527t
Aspergillus carbonarius, 1697
Aspergillus flavus, aflatoxin by, 1697
Aspergillus niger, 1697
Aspergillus parasiticus, aflatoxin by, 1697
Assam, groundwater arsenic contamination in, 1268
Assigned amount units (AAU), 486
Association of Energy Engineers (AEE), 1366
Association of Higher Education Facilities Officers (APPA), 1493
ASTER. See Advanced Spaceborne Thermal Emission and Reflection Radiometer (ASTER)
Atabron. See Chlorfluazuron
Atlas of Australian Acid Sulfate Soils, 58
Atmosphere
aluminum in, 270, 270t
sustainable agriculture, 2458
Atmospheric aerosol forcing, 1196
Atmospheric deposition
of nitrogen, 1769–1770
Atmospheric nanoparticles, 1715, 1715t
Atmospheric ozone, 1883

Atmospheric pollution, 2359
Atmospheric trace species, field measurements of, 1891
Atomic numbers of REE, 2267t
Atrazine, 1282, 1318t, 1957, 1958, 2114, 2142, 2806
Attached growth process, 2647
Attenuation, 1843f
 chemical, 1846
 physical, 1846
Attic floor insulation techniques, 1487–1488
 batt attic, 1488
 loose-fill, 1488
Attic ventilation, 1486–1487
 powered attic ventilator, 1487
 vent selection, 1487
Augmentation, 1474. See also Biological control
Augmentative biological control, 1747
Australia
 ASSs management in, 56–59. See also Acid sulfate soils (ASSs)
 soil degradation in, 2383–2384, 2384f
 distribution of, 2384f
Australian Biocontrol Act of, 2823
Australian bush fly management, in micronesia, 1920
 bush fly origins and habits, 1920–1921
 management efforts worldwide, 1921
Australian Pesticide Act (1999), 1961
Australian Soil Classification, 41
Australian Soil Resource Information System (ASRIS), 58
Autoclaved aerated concrete (AAC), 1483
Automated ribosomal RNA intergenic spacer analysis (ARISA), 2617
Automated sampling
 of GHG, 1202
 suspended sediment concentration, 965, 966f
Auto-oxidation, 1907
Auxiliary, size of, 254
Available soil water (ASW), 2341
Avena fatua L.
 landscape pattern of, 2038, 2039f
Avena sativa (Oat), 1307
Avermectins, 3
Avicides, 2014
Avigrease®, 413
Avis Scare®, 413
Axor. See Lufenuron
Azadirachta indica, 1460, 1769
Azadirachtin, 1460, 2029, 2029f
Azadirachtin indica, 1878
Azafenidin (Milestone), 2016
Azardirachta indica, 1930
Azimuth and tilt angles, 251f
Azocosterol (Ornitrol®), 415
Azolla, 1515, 1516
Azolla–Anabaena system, 350

Azospirillum (AZS), 350, 351, 1515, 1516
Azotobacter (AZT), 350, 426, 1515, 1516
AZTI Marine Biotic Index (AMBI), 602–603

B

Bacillus, 321, 350, 382
Bacillus alvei, 323
Bacillus brevis, 323
Bacillus megaterium, 350
Bacillus sphaericus, 323, 389, 1991
Bacillus subtilis (strain QST71383), 382, 392, 530
Bacillus thuringiensis (Bt), 127, 295, 321, 322–323, 322t, 382, 407, 1112, 1474, 1476, 1749–1750, 1930, 1953, 2029, 2112, 2141
 culture and control, 323
 for insect control, 411–412, 411t
Bacillus thuringiensis (Bt) crops, 307, 315
 benefits, 311
  conventional insecticides, reduced risk compared with, 311
  economic savings, 311
  global food security, 311
 BT toxins, 308
  consumption of prey containing, 310–311
  indirect consumption of, 310
   soil contamination via root exudates, 310
  and transgene escape, Reduction in, 314
 commercialized, 309t
 compatibility with biological control, 314
 consumption of, 308
  live plant tissue, 308–309
 defined, 307
 detritus, consumption of, 310
 environment, impact on, 311
 environmental management of, 313
 global prevalence, 308
 large-scale integrated pest and resistance management, 314
 pests, targeted, 308
 pollen feeding, 310
 resistance management techniques, 314
 risks, 312
  gene escape, 313
  non-target organisms, impact on, 312
  pleiotropic effects of genetic transformation, 312
  presence in human food supply, 313
 sources and fate of, 308
 within-plant modifications, 313
  low-risk promoters, 313
  tissue- and time-specific expression, 313
BACT. See Best available control technologies (BACT)

Bacteria, 521–526
 Actinomycetous species, 531
 Bacillus species, 530, 532
 lactic acid bacteria, 530–531, 531f
 in waste and polluted water, 2698
 metal resistance mechanisms in, 393
 in mine drainage systems, 1689–1690
Bacterial pest controls, 321–323
 Bacillus thuringiensis, 322–323
 overview, 321
 Paenibacillus (Bacillus) popilliae (Dutky), 321–322
 potential agents, 323
 Serratia entomophila, 322
Bactericidal effect, of noble metal nanoparticles, 1740
Bactericides
 derived from nature, 2029–2030, 2030f
Baffle scrubbers, 154
Bahiagrass (*Paspalum notatum* Fluegge), 1305
Baker Lake, CW in, 2670
Baltic Sea, oil pollution in, 1826–1840
 observations of, 1828
 oil spill drift, numerical modeling of, 1833–1835, 1835–1838f
 operational satellite monitoring of, 1830–1833
 problems and solutions of, 1838–1839
 sources of, 1827–1828
 statistics of, 1828–1830
 tendencies of, 1830
Bamboo-reinforced tanks, 2264
Bamboo roofs, 2263
Banding, of pesticides, 1983–1985
 advantages, 1983–1984
 disadvantages, 1984
 equipment requirements, 1984
 future perspectives, 1984–1985
 usage, 1983–1984, 1984t
Bangladesh, groundwater arsenic contamination in, 1268–1271, 1270f
Bangladesh Arsenic Mitigation and Water Supply Project (BAMSWSP), 1269
Banned and restricted compounds import/export, 575–576
Banned pesticides
 banning exports of, 1616
Barium (Ba), 1560
Barium chromate, 479t
Barley (*Hordeum vulgare* L.), 425
Barnyard grass (*Echinochloa crus-galli*), 426, 1526t
Basel Convention on the Control of Transboundary Movements of Hazardous Wastes and Their Disposal, 1914
Baseline-credit system, 486
Baseload, 1790
Basin irrigation efficiency, 1365–1366, 1565f

Basta®, 2030
Bates, 1943–1944
Batteries, solar, 218
Batts, insulating walls with, 1485f
Bauxite, 267, 271. *See also* Aluminum (Al)
Baycidal. *See* Triflumuron
BCAs. *See* Biocontrol agents (BCAs)
BCF. *See* Bioconcentration factor (BCF)
Beans (*Phaseolus* spp.), 426
Beauvaria bassiana, 382, 1930, 1988
Beauveria brongniartii, 382
Becquerel (Bq), 2235
Bedload flux, 965, 965f
Bee dancing
    pesticides effects on, 2005
Bee extracts, 1746
Bellan's pollution index, 602
Bell's vireo (*Vireo bellii pusillus*), 1186
Bench terraces, 968–969, 969f, 2536, 2538, 2764, 2764f
Bendiocarb, 419
Beneficial water use, 1547
Benomyl, 2806
Bensulide, 2015, 2024f
Bentgrass species (*Agrostis* spp.), 1305
Benthic eco-system technology (BEST), 2154
BENTIX, 603
Benzene
    drinking water standard for, 1282
Benzene EN 14662, online gas chromatography for, 142
Benzene, toluene, ethylbenzene, and xylene (BTEX), 1336, 1337, 1339
Benzoic acid, aromatic hydroxyacid derivatives of, 2079f, 2080t
Best available control technologies (BACT), 1128
Best management practices (BMPs), 2808, 2813
    agricultural runoff and, 83
    evaluation of, 2750
17β-estradiol, 1405
Beta minus (β⁻) radiation, 2234–2235
Beta-Poisson model, 2700
Beta vulgaris (Sugar beet), 1307, 1552, 1983
BET measurements. *See* Brunauer–Emmett–Teller (BET) measurements
BFBG. *See* Bubbling fluidizedbed gasifier (BFBG)
Bialaphos, 2030–2031, 2031f
Bifenazate, 3
Bifenthrin, 3, 11
Big hydro plants, 202, 212
Big Spring Number Eight (BSNE) samplers, 1022, 1023f
Bihar, groundwater arsenic contamination in, 1267
Bioaccumulation, 324–326
    applications and future perspectives, 326
    defined, 324
    environment and, 324–326, 325f
        air, 325–326
        soil, 325
        water, 324–325
    mechanism in biota, 326
    molecular properties, 324
    of polychlorinated biphenyls, 2178
Bioassay procedure, at water treatment plants, 2017
Bioaugmentation, 395f
Bioavailability, of polychlorinated biphenyls, 2176
Bioavailability of toxic organic compounds, 401–403
    contaminant state of, 402
    factors influenced, 402
    low water solubility, 402
    microbial adaptations, 403
    pore size distribution, 403
    sorption on solid phase of soil, 402
Bioavailable nutrient
    defined, 1817
Biobarrier systems, 1336–1337
    contaminated groundwater treatment, 1339–1342
        aerobic and anaerobic biodegradation, 1339–1340
        for chlorinated-hydrocarbon, 1340–3141
        for petroleum-hydrocarbon, 1339–1340
        treatment train system, 1342, 1342f
    design factors, 1338
        electron acceptor/donor, 1338
        groundwater flow, 1338
        location, 1338
        natural environment, 1338
        time scales, 1338
    groundwater contamination, 1333
    groundwater remediation technologies, 1334–1336
    limitations of, 1338–1339
    microorganisms, effects of, 1339
    nonaqueous phase liquid, 1333–1334
Biobeds, 2020, 2023, 2023f
Biocapacity, 2468, 2469–2470
    deficit, 2470
Biocapsules, 397
Biocatalysts, 1258
Biochar, 843–846
    sorbent properties of, 844–845
Biochemical oxygen demand (BOD), 2632, 2864, 2865t
Biocides, 2005, 2014
    effect on animals and human beings, 2005
Biocolloid formation methods, 396
Bioconcentration factor (BCF), 324, 403
Bioconcentration, of polychlorinated biphenyls, 2178
Biocontrol agents (BCAs), 2107, 2108t.
    See also Fungi, biological control by
    application, 2109
    mechanisms, 2107–2109
Biodegradable municipal waste (BMW), 908–909, 2401
    reduction, 93
Biodegradation, 328, 330–332, 387. *See also* Bioremediation
    estimation of, 331–332
    of polychlorinated biphenyls, 2176–2178, 2177f, 2177t, 2178f
Biodiesel, 377
Biodimethylether (DME), 378
Biodiversity, 333–334
    care of, 335–336
    definition of, 333
    destruction of, 335
    ecosystem services by, 334
        cultural, 334
        production, 334
        regulation, 334
    of food web, 334
    importance of, 334–335
    indicators, 601–602
    maximum level, 334
    nature use of, 334
    number of species on Earth, 333
    quality approach, 356
    species diversity, occurrence of, 333
    species richness in tropics, 333–334
    support service by, 334
    and wildlife, 1186
Bioenergy, 373, 374f, 378–379, 843–846. *See also* Biomass
    benefits from, 378
    resources, 373
    share in world primary energy mix, 378, 379f
Bioenergy crops
    benefits of, 345, 346t
    carbon (C) sequestration, potential of, 346–347, 347f
    influence of (at temporal scale), 345, 346f
    overview of, 345
    soil organic carbon (SOC) pool, 345, 346
Bio-ETBE (ethyl-tertio-butyl-ether), 378
Bioethanol, 377–378
Biofertilizers, 349–351, 350t, 1514–1516
    background, 349
    biologically fixed nitrogen, 1514
    crop residues, 1515
    leguminous green manures, 1515
    manures and composts, 1515
    market potential for, 349, 350t
    mechanisms of growth promotion, 349–351
    nitrogen-fixing, 1516
    nutrient-mobilizing, 1516
    overview, 349

Biofertilizers (cont'd.)
  phosphate-solubilizing, 1516
  phytohormones, 349–350
  plant nutrient acquisition, 350–351
    biological nitrogen fixation, 350
    microbial siderophore uptake, 351
    other nutrients, 351
    phosphorus solubilization, 350
  sewage sludge, 1515
Biofilm formation, 393
Biofilters (BF), 2612f, 2621t, 2622, 2623t, 2726. See also Biobeds
Biofuel-based power production, 1251
Biofuels, 377–378
  decentralized electricity from, 1246–1250
  importing, 1249–1250
Biofungicides, 382–383
Biogas, 377
  from anaerobic digestion of biomass, 377
  composition of, 377
  from landfills, 377
  use of, 377
Biogenic model, of carbonate formation, 1446
Bioherbicide/inundative method, of biological weed control, 1525, 1525f
Bioherbicides, 383
Bioindicators, 352
  concept of, 353–354
  definition of, 353
  demand for, 355
  development of, 352
    in EU agroecosystems, 356, 357t–358t
    indicators of biodiversity change as tool for, 356, 359f
    in shaded coffee in Latin America, 356, 359
  directions for future study on, 352
  features of, 354t
  history of use of, 353
  importance of, 352
  steps in developing of, 354
  usefulness of, at different levels of organization, 355
  vs. abiotic indicators, 354–355
Bioinformatics, 397
Bioinsecticides, 382
Bio-liquid fuels, 846–847
  markets and energy prices, 831t
  properties of, 846t
Biological aerated filters, 2724. See also Filter(s)
Biological carrying capacity, 339
Biological control
  advantage of, 2595
  augmentation, 1474
  for cane toads, 2604
  of cats, 2602–2603. See also Feline panleukopenia virus (FePV)

classical, 1474, 2595
conservation, 1474
conservation of, 370–372. See also Natural enemies
  agroecosystems, ecological infrastructure in, 372
  favorable conditions enhancement for natural enemies, 371–372
  harmful conditions reduction, 370–371
defined, 1474
definition of, 2595
general principles, 366
improved efficacy, evidence for, 366–367, 367t
natural enemies. See Natural enemies
of nematodes, 1752–1753
  future perspectives, 1753
  nonspecific biocontrol agents, 1753
  overview, 1752
  specific biocontrol agents, 1752–1753
new associations, 366
  citrus leafminer (*Phyllocnistis citrella*), 368
  Eurasian watermilfoil (*Myriophyllum spicatum*), 367
  Lantana camara, 367
  southern green stink bug (*Nezara viridula*), 367–368
  tarnished plant bug, 368
  tarnished plant bug (*Peristenus digoneutis*), 368
  triffid weed (*Chromolaena odorata*), 367
of plant pathogens
  by fungi, 2107–2110
of plant pathogens (viruses), 2111–2113
in practice, 1476–1477
of rabbits, 2597–2602. See also Myxomatosis; Rabbit hemorrhagic disease virus (RHDV)
strategies for using, 1474
successes and failures, 366–369, 367t
of vertebrate pests, 2595. See also specific examples
  biotechnology for, 2604–2605
  natural history of, 2597–2603
  parasites for, 2596–2597
  potential agents for, 2603
  predators for, 2595
of weeds. See Weeds, biological control of
Biological control, 1584
Biological degradation of organics, 2646
Biological effects, of ozone depletion, 1893
Biological fixation
  of nitrogen, 1770
Biological integrity sampling, 1582
Biological invasions, of species, 1531–1533, 1532t
  characteristics of recipient community and, 1532

controlling, 1533
impacts, 1532–1533
lags and surprises, 1533
need for, 1531
overview, 1531
propagule pressure, 1531
species traits and, 1531
steps, 1532t
Biological material
  elements in, 1455
Biological methods, for contaminant remediation, 1293
Biological nitrogen fixation (BNF), 350, 1515
Biological oxygen demand (BOD), 331, 522
  reduction in municipal wastewater, 2709–2712, 2710–2713f
Biological pesticides (BP), 1986–1987. See also Chemical pesticides (CPs); Integrated pest management (IPM)
Biological phosphorus removal, 2654
Biological oxygen demand (BOD), 331, 522
  reduction in municipal wastewater, 2709–2712, 2710–2713f
Biological pollution, in coastal waters. See also Pollution
  algal bloom, 495
  eutrophication, 494–495
  invasive species, 495
Biological removal of nitrogen, 2653
  anaerobic ammonium oxidation, 2653
  Canon process, 2654
  combined nitrogen removal, 2654
  Sharon process, 2653
Biological sequestration, 460
Biological treatment processes, 2654
Biological waste, 512
  aerobic treatment for, 513–515
  disposal problems associated with, 512
  gas treatment techniques, 2611
  management, regulatory issues of, 512–513
Biomagnification, of polychlorinated biphenyls, 2178–2179
Biomal®, 1111
Biomanipulation, 1584, 1593t
Biomarkers, 1068–1069
  for EDC risk assessment, 647
Biomass, 373–379
  benefits of, 378
  biochemical conversion of, 377
  biofuel, 377–378
  biogas, 377
  as carbon-neutral energy resource, 373, 378–379
  combustion, 375
    co-combustion/direct cofiring, 375–376
    indirect cofiring, 376
    parallel combustion, 376
    problems in, 376

definition of, 373
energy production from, 373–374, 374f.
    See also Bioenergy
plant, 373
thermochemical conversion of, 376
    comparison of technologies of, 377t
    gasification, 376–377
    pyrolysis, 376
types of, for energy use, 374t
use of, as fuel, 373–374
    mechanical processes related to, 374–375
    problems in, 374
Biomass, 192, 824
    defined, 715
    gasification and direct firing, 1248–1249
    and hydrogen energy cycles (comparison), 716, 716f
    low cost of, 733
    sources, 824, 825, 826
Biomass and wastes, 2482
    gasification of, 2488–2495
        advances in, 2491–2494
        gasifiers, 2488–2491
        plants for, 2494t
        reaction involved in, 2490
        syngas, composition and heating value of, 2493t
        syngas for downstream application, 2492t
        technologies for, 2491
        world scenario, 2495
    pyrolysis of, 2483–2488, 2486t
        features for fast, 2487
        pathways, 2486f
        products, 2484
        reactions involved in, 2485f
        world scenario of, 2489t–2490t
    scopes for, 2483
Biomass burning
    ammonia volatilization, 87
    N_2O emissions, 97
Biomass energy, 772
Biomass fuels, 345
Biomethanol, 378
Biomolluscicides, 383
    bacteria-based, 383–384
    combination, 384
    from nematodes, 383
    of plant origin, 384
    use of, 383
Biomonitoring
    animals, 2129
    human, 2129
    plants, 142–143, 2128
Bio-MTBE (methyl-tertio-butyl-ether), 378
Bio-oils, 376, 2485, 2488
Bio-oxidation, 2152
Biopesticides, 381
    classification of
        biochemical pesticides, 381

microbial biopesticides, 381
plant-incorporated protectants, 381
definition of, 381
formulations of, 381
organisms used in, 381
range of
    bioacaricides, 382
    biofungicides, 382–383
    bioherbicides, 383
    bioinsecticides, 382
    biomolluscicides, 383–384
use of, 381
Bioprocesses, 2152t, 2553
Bio-reduction, 2152t
Bioreactors, for waste gas treatment, 2611
    airlift bioreactor, 2626
    biofilters, 2612f, 2620t, 2621, 2626
    bioscrubbers, 2611f, 2621, 2623
    biotrickling filters, 2614t, 2621, 2622–2623
    critical factors and development requirements, 2625–2626
    mass transfer aspects, 2611–2615
    membrane biofilm reactors, 2612f, 2624–2625, 2624t
    microbial kinetics, 2611–2621
    monolith bioreactor, 2625, 2626
    process engineering and performance parameters, 2618–2619
    rotating biological contactors, 2625, 2626f
    two-phase bioreactors, 2612f, 2625
    waste gas stream, characteristics of, 2620–2621
Bioregulators
    in organic farming, 1115–1116
Bioremediation, 387, 392, 1335, 1340
    of bioaugmentation and biostimulation, 395f
    for environmental cleanup, 388f
    factors affecting, 390
    of inorganic contaminants, 392–393, 2120
    microbial degradation of organic pollutants, 388–392
    molecular probes in, 397–398
    mycoremediation, 392
    of organic contaminants, 2116–2117
    phytoremediation, 394–395
    principle of, 387–388
    redox reaction
        leading to immobilization, 393
        leading to solubilization, 393–394
    restoration of contaminated soils
        bioavailability of toxic organic compounds, 401–403
        biodegradation, 403
        uptake of heavy metals by plants, 403–405
    technology, 395–396
Bioscrubbers (BSs), 2611, 2614t, 2621, 2623
Biosolids, 1807–1808
    reburn with, 732–733, 732f, 733t

Bio-sorption, 2152
Biostimulation, 388f, 395f
Biotechnology, 407–412
    advantages, 412
    disease resistance in crops, 407, 408t
    genetic engineering in pest control, benefits of, 407
    impact of, on pest management, 1949–1953
    transgenic virus-resistant potatoes in Mexico, 409–412
    for vertebrate pest control, 2604–2605. See also Virally vectored immunocontraception (VVIC)
Biotic community, 291
Biotrickling filters (BTFs), 2611, 2612f, 2614t, 2621, 2623t, 2625
BioVECTOR, 382
BIPV (building-integrated photovoltaics), 223–224
Bird control chemicals, 413–415, 414t
    future developments, 415
    immobilizing agents, 414
    lethal stressing agents, 414
    repellents, 414–415
    reproductive inhibitors, 415
    toxicants, 413–414
Bird impacts, from pesticide use, 416
    farmland bird species, decline in, 416
    measuring of
        field testing, 421
        incident monitoring, 420–421
        modeling, 422
        surveys, 421
    mechanisms of, 417
        effects on reproduction, 418
        lethal effects, 417–418
        persistent organochlorine pesticides, 417
        secondary poisoning, 418
        sublethal effects and delayed mortality, 418
    routes of, pesticide exposure, 418
        abuse and misuse, 418–419
        forestry insecticides, 420
        granular formulations, 419
        liquid formulations on insect prey, 420
        liquid formulations on vegetation, 419–420
        treated seed, 419
        vertebrate control agents, 420
Bisacylhydrazines, 1465
Bisphenol A, 1405
Bistrifluron, 1461
Black blizzards, 2381
Black box approach, 1722
Black carbon. See Pyro-char
Black rot, 442, 443t
Black spot, 442, 443t
Blackwater, 44, 45
Bladder cancer, by NO_3-N exposure, 1303

Blattella germanica (German cockroach), 1462, 1988
Blown foam insulation, 1484, 1485f
Blown sidewall insulation, 1484, 1485
Blue baby syndrome, 1282, 1302, 2736
Bluegrass (*Poa pratensis* L.), 1305
Blue green algae (BGA), 1511, 1516
Blue-green algal blooms, 538, 539
"BLUE" scenario, 201
BMPs. *See* Best management practices (BMPs)
BNF. *See* Biological nitrogen fixation (BNF)
Bog, 2825t
Bonneville Power Administration load and generation, 1420f
Borehole, 1367–1368, 1368f
Borehole thermal energy storage (BTES), 2518
Boric acid, 431
Boron (B), 424
   adsorption isotherms, 432f
   availability of, 424
   deficiencies, 424
   deficiency in plants, 432
   elemental form of, 431
   Lewis bases, 431
   properties of, 431
   as soil contaminant, 431–433
   soil interaction with, 431–432
   for soil–sand mixtures, 433f
   solution-to-soil ratio, 432f
   sources of, for humans, 424
   toxicity. *See* Boron toxicity
   uptake in plants, 432–433
Boron toxicity, 424, 425
   causes of, 425
   levels in crops, 425
   management/reduction of
      boron-tolerant cultivars, 425
      leachates recirculation, 426
      phosphorus, zinc and silica, use of, 426
      phytoremediation, 426
      waste materials use in, 426
   sources of, 425
   symptoms of, 425
BotaniGard®, 382
Botrytis cinerea, 383, 1991, 2109
Botswana, 2690
Bottomland, 2825t, 2826f
Bottoming cycles, 1357
Bottom-up estimate approach, 1792
Boudouard reaction, 2490
Boundary conditions and nutrient entry, 1821–1822
Boundary element method (BEM), 1314, 1314f
Box equation, 965
BPH. *See* Brown-plant-hopper pest (BPH)
Brain
   development of, 1756
   maturation of, 1756

Brassica napus, 1380
Brassica oleracea (Broccoli), 1307
Brassica oleracea (Cabbage), 442
Brassica species, 2435
Brayton cycle, 819
Brevicoryne brassicae (L.), 1928
Brine, 1569t
Briquettes, 375t
Britain, work on field drainage in, 584
British Co-op Group, 1401
British Food Standards Agency, 1403
British thermal units (BTUs), 1363, 1493, 1652
   loss of, 1493
Broccoli (*Brassica oleracea* L.), 1307
Brodifacoum, 420
Broiler litter, subsurface band application of, 1674
Bromadiolone, 420
Bromoxynil, 409
Brown pelican (*Pelecanus occidentalis*), 417
Brown-plant-hopper pest (BPH), 2034, 2141
Brüel and Kjær multi-gas monitor, 1203
Brunauer–Emmett–Teller (BET) isotherm, 65. *See also* Isotherm(s)
Brunauer–Emmett–Teller (BET) measurements, 1715
Brundtland report, 338
BT α-endotoxin, 411–412
BTEX, 1281–1283. *See* Benzene, toluene, ethylbenzene, and xylene (BTEX)
Bubbling fluidized bed combustor (BFBC), 727
Bubbling fluidizedbed gasifier (BFBG), 2492–2493
Bubo virginianus (Great horned owl), 420
Buffer capacity of soils, 1375
Buffers, vegetative, 995–1000, 995f, 996–997t, 998–999f
   hydraulic resistance, 998–1000
   types, 995, 998
Building. *See also* Climate change
   climate change on, impact of, 436–437
   component storage, 2519–2520
   contribution of, to human-induced climate change, 437
      building-related refrigerants, 437
      embodied energy, 437
      operational energy, 437
   cooling applications, 1360
   design and operation, strategies for, 439–440
      building envelope, 439
      building mechanical service, 440
      internal heat sources, 439
      recommendations on, 440
   life-cycle cost of, 1648, 1648f
Building energy management systems (BEMSs), 440

Building envelope, 439
Building envelope insulation, 1479
Building Performance Initiative (BPI), 1650
Building research establishment environmental assessment method (BREEAM), 1655
Buildings
   life-cycle cost of, 1648, 1648f
   rating systems for, 1655
   simulation technique, 439, 440
   types of, 860
Buprofezin, 1460, 1463
Business, role in ecological footprint, 2477–2478
Business-as-usual scenario, 617–618, 621t
Butachlor, 2142
Butane process, 1256

C

$Ca/(HCO_3 + SO_4)$ ratios
   for saltwater detection, 1327
Ca/Mg ratios
   for saltwater detection, 1327
Cables
   in photovoltaic systems, 222–223
Cabbage disease (ecology and control)
   cabbage as human diet, 442
   control principles
      avoidance, 444
      eradication, 443, 443t
      exclusion, 443
      protection, 444
      resistance, 444
      therapy, 444
   diseases and pathogen ecology
      black rot, 442
      black spot, 442
      clubroot, 442
      dark leaf spot, 442
      downy mildew, 443
      sclerotinia stem rot, 443
      watery soft rot, 443
      white mold, 443
      wirestem, 443
   foliar pathogen management, 444
   integrated disease management (examples), 444
   seedborne pathogen management, 444
   soilborne pathogen management, 444
Cacao swollen shoot virus, 2111
Cactoblastis cactorum, 2821
Cadmium (Cd), 453, 1370
   contamination, in coastal waters, 490
   dispersion and application, 453–454
   health impacts of, 2118
   toxicity and ecotoxicity, 454–455

Cadmium and lead, 446
  chemical properties, 446
  detection, 447
  environmental hazards, 447–448
  human exposure to, 449t
  human health effects, 448–449
  occurrences, 446–447
  production and uses, 447
  regulations and control, 449
Cadmium telluride (CdTe) solar cells, 232
Calcareous sodic soils. *See also* Sodic soils
  reclamation of, 2370
Calcic soils
  isotopes in, 1452–1453
Calcite, 88, 462
Calcium (Ca)
  montmorillonite clay, swelling of, 2338, 2339f
Calcium chromate, 479t
Calcium-lactate extractable phosphorus, 102f
Calibration, groundwater modeling, 1298, 1298f, 1299
Calibration methods, 1663
California Bay-Delta Program (CALFED), 627
California, wastewater reuse in, 1571
Calispermon spp., 2031
Caloric theory of heat, 815
Calorific value, of fuel, 374
Camelina sativa, for bio-liquid fuels, 847
Canada
  acid rain in, 8
  soil degradation in, 2383
Canadian Air and Precipitation Monitoring Network (CAPMoN), 7
Canadian Organic Products Regulation (2006), 1109
Cancer
  arsenic toxicity and, 1274, 1274f
  exposure to pesticides and, 1415
Cancers, from pesticides, 1394–1396
  animal studies, 1394–1395
  epidemiological studies, 1395–1396
  overview, 1394
  pesticide exposure, 1394
  risk of, 1394, 1395t
Candida spp., 2109
Cane toads (*Bufo marinus*), 2596, 2605
Canine Herpesvirus-1 (CHV), 2604
Canola (rapeseed), for bio-liquid fuels, 846
Canon process, 2654
Capillaria hepatica, 2596
Capital costs for nuclear and coal-fired powerplants, 1790
Capping, 2315t
Capsella, 1936
Capsella bursa pastoris, 1928
Capsicum annuum L. (Peppers), 425
Captan, 1318
Capture cost, 459

Capturing plate, 242
Carbamate insecticides, 2804
  impact on birds, 417
Carbamate pesticides, 1317
Carbamates, 3, 277, 2005, 2141
Carbaryl (Sevin), 3, 2016t, 2805, 2142, 2806
Carbazates, 3
Carbendazim, 2142
Carbofuran (Furadan), 417, 419, 1318t, 2805, 2806
Carbohydrates, plant, 373
Carbon (C), 88
  cycle, 88–90, 89f
  decomposition and mineralization, 90
    organic substrate quality, effect of, 92–93
    soil microorganisms, effect of, 90–92
  fixation, 88–89
  inorganic. *See* Inorganic carbon
  keeping in soil, 1190–1191
  occurence of, 88
  removal from air, 1189–1190
    man role in, 1189–1190
    nature role in, 1190
  sources of, 456–457
Carbon adsorbents, 68–69
Carbonate, 1451
  models of formation, 1446
    biogenic model, 1446
    per ascensum model, 1446
    per descensum model, 1446
    in situ model, 1446
  pedogenic *vs.* geogenic, 1445–1446
  simulations of pedogenic accumulation, 1451–1452
Carbonate bonded REE, 2268t
Carbon balance assessment of bioenergy crops. *See* Bioenergy crops
Carbon capture and storage (CCS). *See* Carbon sequestration
Carbon dioxide
  atmospheric, IC impact on, 466–467
  pollution, control methods, 157
Carbon cycle, 2389f
Carbon dioxide (CO_2), 486, 1202
  emission estimation, 717, 722, 722f
  emissions, 774
  emissions per year, 716f
  erosion and, 934–937
    conceptual mass balance model, 935–936, 935f
    emissions, estimates of, 936, 936t
    SOC, 935–936
  greenhouse effect. *See* Greenhouse effect
  soil quality and, 2388
  storage of, 458
Carbon footprint, 2468
Carbon gases. *See also specific* entries
  soil quality and, 2388–2389

Carbon hydrides pollution, problems of, 156–157
  control methods, 157
Carbonization, 2485–2486
Carbon losses from soil
  mineralization processes and, 1857
  soil erosion processes and, 1857–1858
Carbon monoxide (CO)
  characteristics of, 136
  EN 14626, non-dispersive infrared for, 142
  pollution, problems of, 156–157
  control methods, 157
Carbon sequestration, 1858–1861, 2833–2834
  biological sequestration
    ocean fertilization, 460
    terrestrial carbon sinks, 460
  carbon capture and storage (CCS)
    carbon, sources of, 456–457
    geologic sequestration, 458–459, 458t
    industrial carbon dioxide capture, 458
    ocean direct injection, 459
    overall costs of, 459, 459t
    post-combustion capture, 457
    pre-combustion capture, 457–458
    separation and capture, 457
    storage of carbon-di-oxide, 458
  costs of, 460
  creation of wetland, 2835
  global importance of, 1858
  greenhouse gases (GHG), 456
  human activities, impact of, 2835
  long-term field experiments, role of, 1858
  mechanisms, 1858, 1860
  potential of, 346–347, 347f
  with primary production, 2833
  prospects for, 460–461, 460t
  in soils, 2833–2834
  SOM, energy, and full C accounting, 1861
Carbon tetrachloride, 1317, 1318, 1318t
Carboxylic herbicides, 1318
Carcinogens, 1570
Carcinops troglodytes, 1921
Cardiovascular effects, of arsenic toxicity, 1272
Carduus nutans, 2822
Caribbean Environment Programme, 2383
Caribbean Islands
  soil degradation in, 2382f, 2383
Carnot cycle, 819f, 820
  efficiency, 820, 2530
Carnot engine, 2529, 2530f
Carnot law, 1356
Carnot ratio equality, 820
Carnot HCOP, 1347
Carnot heat engine, 683

Carpophilus pilosellus, 1921
Carry-over soil moisture (SMco), 1563
Carson, Rachel, 2010
Carter, Jimmy, 1521
Cartridge filter, 2723–2724. *See also* Filter(s)
Carum carvi, 384
Carya aquatica, 2830
Cascade. *See* Flufenoxuron
Cassava mosaic virus, 2111
Casten, Thomas, 1505
Catalytic WO reactions, 1909, 1909t
   using heterogeneous catalysts, 1910
      NS-LC, 1910
      Osaka Gas, 1910
   using homogeneous catalysts, 1909–1910
Catchment surface, 2263–2264
Cat-clays. *See* Potential acid sulfate soils
Catechol, 391
Cathedral ceilings
   building R-30, 1489
   exposed rafters, insulating, 1490
   insulation techniques, 1489
   options, 1489t
   scissor trusses, 1489–1490
Cation exchange capacity, 1374
Catolaccus grandis, 1747
CATOx2 process, 1910
Cats, biological control of, 2602–2603. *See also* Feline panleukopenia virus (FePV)
Cattle manure, 718t
Cauliflower (*Brassica oleracea* var. Botrytis L.), 425
Cauliflower mosaic virus (CaMV 35S), 308
Causal chain frameworks, 354f, 356
CBEP. *See* Community-based environmental partnership (CBEP)
CC. *See* Continuous Commissioning
CCPR. *See* Codex Committee on Pesticide Residues (CCPR)
Cd. *See* Cadmium (Cd)
Cd model, 404f, 405f
CEC. *See* Commission of the European Communities (CEC)
Ceilings and roofs, 1486–1490
Celestite (SrSO4), 2426
Cellulase, 89
Cellulose insulation, 1478
   in attics, 1488t
   health impacts of, 1479
Celtis laevigate, 2830
Center pivot irrigation, 1547t, 1554
Center pivots, 1551–1553
Central America
   soil degradation in, 2383, 2382f
      distribution of, 2383f
Central inverters, 221–222, 222f
Centralized composting. *See also* Composting
   horizontal reactors, 518, 518f
   in-vessel, 517
   rotary drums, 518–519, 519f
   silos/towers, 518
   static pile, 517, 517f
   windrows, 516–517, 516f
Central nervous system development, 281
Centre for Alternative Wastewater Treatment of Fleming College, 2671
Centrocercus urophasianus (Sage grouse), 420
CENTURY SOM model, 1875
Ceratophyllum demersum, 2830
Cerium (Ce) content in grains/seeds, 2269t
Cernuella virgata, 383
Certified emissions reductions (CER), 486
Certified geothermal designers (CGD), 1366
"Certified organic" crops, 125
C factor. *See* Climatic (C) factor
CFB. *See* Circulating fluidized bed (CFB)
CFBG. *See* Circulating fluidized-bed gasifier (CFBG)
CH_4. *See* Methane (CH_4)
Chamber scrubbers, 154
Chamber techniques
   GHG measuring fluxes, 1202, 1203, 1204f
Channel erosion, 2756
Charcoal, 376, 844
Charged-droplet scrubbers, 156
Char reactions, 724–725
Chelatobacter heintzii, 2246
Chemical deterioration map, 2377
Chemical exergy, 1084, 2531
Chemical industry sludge, 2267t
Chemical methods, for contaminant remediation, 1293
Chemical mixtures, 2572–2579
   concentration–response surface analyses for, 2576–2578, 2576–2578f
   future challenges to, 2579
   multicomponent mixtures, reduced mixture toxicity designs for, 2579
   strength and weakness of, 2579
   toxicities of. *See* Concentration addition; Independent action
Chemical nitrification inhibitors, 94
Chemical oxidation, 1335
Chemical oxygen demand (BOD), 524
Chemical pesticides (CPs), 1986
   and biological pesticides, integration of, 1987–1988
   examples of, 1989t
   impact on ecosystems, 1991–1994
      negative, 1988
      positive, 1988, 1989t
   reducing use of, 1986
Chemical pollutants, 2357
Chemical Review Committee, 1616
Chemical risk assessment, 1721
Chemicals, dissolved, 2753
Chemicals, runoff, and erosion from agricultural systems (CREAMS) model, 991
Chemical sensitivities. *See also* Pesticide sensitivities
   causes, 1411–1412
   impact of, 1412
   mechanism, 1412
Chemical speciation of phosphorus, 100
Chemigation, 469–472
   advantages, 469
   background, 489
   disadvantages, 469
   equipment, 469–471
      injection, 470–471, 470f, 471f
      irrigation, 470
      safety, 470f, 471
   management practices, 472
Chemiluminescence, for NO EN 14211, 141–142
Chemisorption, 63
Chemotaxis, 392
Chenopodium, 1936
Chernobyl accident, 2304
Chesapeake Bay, 474–476, 475f
   human population, growth in, 475
   meaning of name, 475
   physical conditions, 474–476
   sedimentation and, 475
Chesapeake Bay Program, 627
Chesterfield Inlet's tundra treatment wetland, 2670
Cheyletus eruditus, 2424, 2424t
Chhattisgarh, groundwater arsenic contamination in, 1268
Chicago Climate Exchange, 487
Chicken cholera bacteria (*Pasturella multocida*), 2596
Chicken manure, 718t
Chickpeas (*Cicer arietinum* L.), 424, 427
Children
   arsenic toxicity effect on, 1275, 1275f
   health effects of pesticides on, 1121–1122
   pesticides effects on, 1415–1416
Chili pepper powder, 2021
Chilled water storage, 2510
Chilo suppressalis (Walker), 311
China
   trends in acidity, 13
Chips, 375t
Chitin, 72, 1461
Chitin synthase, 1461
Chitin synthesis inhibitors (CSIs), 1460, 1461
   benzoylphenyl ureas, 1461–1463
   bistrifluron, 1461
   chemical structure of, 1462f
   chlorbenzuron, 1461
   chlorfluazuron, 1461
   diflubenzuron, 1461

Index: Chitosan—Climate system

fluazuron, 1461
flucycloxuron, 1461
flufenoxuron, 1461
hexaflumuron, 1461, 1462
lufenuron, 1462
novaluron, 1462
noviflumuron, 1462
teflubenzuron, 1462–1463
triflumuron, 1463
non-benzoylphenyl ureas, 1462f, 1461
buprofezin, 1463
chemical structure of, 1462f
cyromazine, 1463
dicyclanil, 1463
etoxazole, 1463
Chitosan, 72, 73t
Chlorbenzuron, 1461
Chlordane, 1318t, 1394
Chlorfenapyr, 3
Chlorfluazuron, 1461
Chlorinated hydrocarbons, 1317
Chlorinated solvents
degradation pathways of, 1843, 1843f
metabolic pathways of, 1841, 1842f
oxidation of, 1844
Chlorination of storage tanks, 2264
Chlorofluorocarbons (CFCs), 486, 1893–1894
destruction of, 1886
properties of, 1885–1886
-ozone depletion hypothesis, 1887f
Chlorophyll meter readings, 1306
Chlorpyrifos, 2015, 2015f, 2020, 2142
Chlorpyrifos, 1318
Cholera, 2696
Chondrostereum purpureum, 383
Chontrol, 383
Chorioptes bovis, 2
Christiansen uniformity coefficient (UCC), 1542, 1548
Chromafenozid, 1466
Chromite, 478, 479t
Chromium (Cr), 477, 1370, 2585–2586, 2588
carcinogenic effects, 478
as cofactor of insulin, 477, 478
forms and sources of, in natural waters, 478–479, 478t
health effects, 477–478
health impacts of, 2118–2119
hexavalent, 477
occurrence of, 478
oxidation and reduction reactions of, 480, 480f, 481f
solubility controls of, in water, 479–480, 479t
trivalent, 477
uses of, 478
Chromium arsenate, 479, 479t
Chromium chloride, 479t
Chromium fluoride, 479t

Chromium(III). *See* Chromium (Cr)
Chromium(III) hydroxide, 479t
Chromium(III) oxide, 479t
Chromium jarosite, 479, 479t
Chromium phosphate, 479, 479t
Chromium picolinate monohydrate (CPM), 477
Chromium sulfate, 479t
Chromium(VI). *See* Chromium (Cr)
Chromolaena odorata, 367
Chromosomal aberrations (CAs) test, 2127–2128
Chrysanthemum, 2028
Chrysolina quadrigemina, 2821
Chrysomya megacephala, 1920
Chrysopa perla. See Golden-eyed fly
Chrysoperla carnea, 1750
Cicer arietinum L. (Chickpeas), 424
Cingular Wireless, 2226–2227
Cinnamic acid, aromatic hydroxyacid derivatives of, 2080f, 2081t
Circuit breaker, 223
Circulating fluid bed dry scrubber, 2442
Circulating fluidized bed (CFB), 2492–2493
boilers, 375
Circulating fluidized bed combustor (CFBC), 727
Circulating fluidized-bed gasifier (CFBG), 2492–2493
Cirsium spp., 2822
CITEAIR (Common Information to European Air), 135, 136
Citizen monitoring, 1582–1583
Citrus leafminer. *See Phyllocnistis citrella*
Citrus limon (Lemon), 426
Citrus tristeza virus, 2111
Ciudad Juarez Valley (Mexico), 2345
Civic environmentalism, 924
Cl/Br ratios
for saltwater detection, 1327
Clausius law, 818
Clausius statement, 818
Clay flocculation, 960, 960f
Classical biological control, 1474. *See also* Biological control
Classical weed biocontrol
history and impact of, 2821–2822
legislation, 2823
Clays, 1627
coatings, 1627
illuviation, 1627
minerals, 69–70, 71t
soils, water movement in, 585
1970 Clean Air Act, 20
Clean Air Act, 133, 157–158
Clean Air Act Amendments (CAAA), 732
Clean Air for Europe (CAFE), 136
Clean development mechanism (CDM), 486

CleanSeaNet, 1828, 1830
Clean-up techniques, 2198–2199
Clean water, 2752
Clean Water Act (CWA), 1064, 1289, 1942, 1943, 2805, 2808, 2809, 2750, 2753. *See also* Total maximum daily load (TMDL)
overview, 2809f
Clik. *See* Dicyclanil
Climate
defined, 1192
feedbacks, 1194
model, 1199
organic matter decomposition and, 1853
Climate change, 333, 435, 636, 804
and buildings, relationship between, 437–438, 438f
global, sulfur and, 2434–2435
greenhouse gases and, 1894–1895
hydrocarbons and, 1895–1896, 1896t
impacts of, 435
on ozone layer, 1896–1897
on wind erosion, 1019
implication of, on building, 436–437
building energy use, increase in, 436
construction process, 437
external fabric durability, 437
internal thermal environment, deterioration in, 436–437
service infrastructure, 437
structural integrity, 437
likely future, 435–436, 436t
Climate Change Fund, 487
Climate forcings and feedbacks, 1194–1197, 1194f, 1195f
atmospheric aerosol forcing, 1196, 1197t
greenhouse gas forcing, 1195, 1196t
land-use change forcing, 1196–1197
natural climate forcings (solar and volcanic variability), 1194–1195, 1196t
Climate models, 1867
Climate policy (International)
adaptation, 484
early international response, 484
emmission reduction, 483–484
equity questions, 484
European Union emissions trading system (EU-ETS), 487
greenhouse gases (GHG), 483
Kyoto protocol, 485–487
implementation, 486
targets, 485–486
UN framework convention on climate change (UNFCCC), 485
interim negotiations, 485
voluntary and regional programs, 487
Climate protection, 182–183
Climate system, 1193f

Climate technology initiative (CTI), 680
  activities, 680
Climatic conditions
  leaching and, 1622
Climatic (C) factor, 1012
Clofentazine, 3
Clomazone, 2014, 2024f
Clonal propagation, 1182
Closed-cell, spray polyurethane insulation, 1479
Closed circuit systems, 248
Closed ecosystems, 1935
Closed-loop pumped storage systems, 1423–1424
Clostridium, 321
Cloud condensation nuclei (CCN), 1020, 1196
Cloud point extraction, 2194
Cloud-reflectivity feedback, 1195f
Clouds earth's radiation energy system (CERES), 2818
Clovers (*Trifolium* spp.), 427
Clubroot, 443, 444t
Co. *See* Cobalt (Co)
CO_2. *See* Carbon dioxide (CO_2)
^{60}Co, 2243, 2244
Coal, 718t
  and bio-solids
    gasification of, 730–731, 731f
  deposits, 458
  and manure, 728
  reburn, 733t
  and refuse-derived fuel (RDF), 728, 728t
Coal agricultural residues, 727–728, 727f
Coal bed methane (CBM), 458
Coal-fired electrical generating station, 1086–1089
  condensation, 1087
  energy
    efficiency, 1087
    values, 1087
  exergy
    consumption values, 1089
    efficiency, 1087
    flows, 1090
    values, 1088
  flow diagram, 1087
  net energy flow rates, 1090t
  power production, 1087, 1089
  preheating, 1087, 1089
  steam generation, 1086–1087, 1088
  thermodynamic characteristics of, 1089
Coalification process, 717
Coal mining, 2240
Coastal aquifers, 1321
Coastal deserts, 566
Coastal waters, pollution in, 489–498
  biological pollution
    algal bloom, 495
    eutrophication, 494–495
    invasive species, 495
  fertilizers, 495–496
  heavy metals, 490–491
  light pollution, 497–498
  marine debris, 496–497
  metalloids, 490–491
  noise pollution, 497
  oils, 496
  organic compounds, 492–494
  pesticides, 495–496
  plastics, 496–497
  radionuclides, 491–492
  sewage effluents, 496
Cobalt (Co), 1370
Cochlicella acuta, 383
CODESA-3D
  for saltwater intrusion, 1329
Codex Alimentarius, 1108, 1612, 2704
Codex Alimentarius Commission, 1126, 2036
Codex Committee on Pesticide Residues (CCPR), 1126
Coefficient of performance (COP), 1364
Coefficient of variation (CV), 1542
CO_2 emissions, 799t
"CO_2-fertilization effect," 1858
Cofiring (coal and bio-solids). *See* Energy conversion
Cofiring, in fossil-fired power stations, 375, 378
^{60}Co(III)EDTA, 2244, 2245
Cogeneration, 1361
  and cool storage, 2515, 2515f
Cohort models, for organic matter modeling, 1864
Cold air distribution, 2517
Coleomegilla maculata, 1936
Coliform counts in wastewater, 1571
Colinus virginianus, 2005
Coliphage, 1626
Collaborative environmental management, 625
Collaborative governance approaches
  and ecosystem management, 626
Collaborative resource governance (CRG), 624, 627–628
Collego®, 1111
Columba livia, 413
Combined heat and power (CHP), 730. *See also* Cogeneration
  plants, 375, 375t
Combined nitrogen removal, 2653
Combined pumped storage systems, 1423
Combustion, 161, 774. *See* Energy conversion
  biomass, 375–376
  energy and exergy in, 1093, 1093f
Combustion turbine inlet air cooling (CTIAC), 2516–2517
  ambient temperature effect on, 2516f
  waste heat recovery in, 1360–1361
Cometabolism, 2176
Comet assay, 1156, 2126–2127
Command-and-control approach, 483
Command and control strategy
  quantity-based, 919
  technology-based, 919
Commercial Buildings Energy Consumption Survey (CBECS), 793
Commercial use of energy, 793
Commercial WO, 1908
Commissioning agent
  selection of, 668–669
  skills of qualified, 669
Commissioning of existing buildings. *See also* Continuous Commissioning (CC®); Recommissioning
  case study, 657
  costs and benefits of, 658–659
  definitions, 656–657
  measures, 656, 657
  monitoring and verification (M&V), 656
  process, 659–661
  team, 659
  uses in energy management process, 661–662
    as ECM in retrofit program, 662
    to ensure building meets or exceeds energy performance goals, 662
    as follow-up to retrofit process, 662
    as standalone measure, 662
Commissioning of new buildings
  acceptance-based vs process-based, 666–667
  barriers to, 668
  benefits of, 676
  cost/benefit of, 675
  defined, 665–666
  goal of, 666
  history of, 667
  importance of, 668
  market acceptance of, 667
  meetings, 672
  O&M manuals, 673
  phases
    acceptance, 673–674
    construction/installation, 671–673
    design, 670–671
    postacceptance phase, 674–675
    predesign, 670
  prevalence of, 667–668
  process, 668–670
  success factors, 675–676
  systems to include, 668
  team, 669–670
Commissioning plan, 668, 674, 675
Commission of the European Communities (CEC), 1409
Commission on Environmental Law of IUCN (World Conservation Union, The), 1620
Common air quality index (CAQI), 135–136

Common Operational Webpage (COW), 136
Common property resources, destruction of, 914
Community-based environmental partnership (CBEP), 925
Community balance. *See* Pesticide impacts, on aquatic communities balance
Community commitment, 560
Community level, 312
Community participation, 560
Community Pesticide Action Kits (CPAKs), 507
Community Pesticides Action Monitoring (CPAM), 505
Compartment model
   inorganic carbon, 1451–1452
Compensation point, 1952
Competitive crops, 1111
Competitive water markets, 2780–2781
   scarcity rents, 2782
   shadow prices, 2782–2783
Comply. *See* Fenoxycarb
Compost, 961
Composting, 512–525, 513f
   advantages/disadvantages of, 515
   background of, 514–515
   centralized
      horizontal reactors, 518, 518f
      in-vessel, 517
      rotary drums, 518–519, 519f
      silos/towers, 518
      static pile, 517, 517f
      windrows, 516–517, 516f
   commercialization issues of, 525
   home, 516
   maturation of, 524–525
   phases of, 520
   process parameters in, 520–521
      aeration, 522–523
      feedstock composition, 523–524, 523t
      moisture, 521–522
      odor control, 524
      particle size, 523
      pathogen control, 524
      pH control, 524
      temperature, 521, 522f
   system, evaluation of, 520, 521f
   vermicomposting, 519–520, 519f
Compost wetlands, 1690
Comprehensive Environmental Response, Compensation, and Liability Act (CERCLA), 2114–2115, 2116t, 2118, 2119t
Comprehensive Everglades Restoration Plan (CERP), 627
Comprehensive Everglades Restoration Plan (CERP), 1080, 1081
Compressed natural gas (CNG), 723
Compressive strength, 864

Computer-derived packages, of IPM, 1522
Computer models
   for saltwater intrusion, 1329–1330
Concentrated cookers, 2503
Concentration addition (CA), 2572–2574, 2573t
   applications of, 2575–2576
   characteristics of, 2575
   predictive power of, 2576
Concentrating solar power (CSP), 2520–2521, 2521–2522f
Concentration–response surface analyses, for chemical mixtures, 2576–2578, 2576–2578f
Conceptualization, groundwater modeling, 1297–1299, 1298f
Concrete
   block and poured concrete, 863–864
   block cores, insulating, 1482
   poured, 863–864
   precast, 864
Conditioned attic assemblies, 1487
Confined animal feedlot operation (CAFO), 2752
Confirm. *See* Tebufenozide
Coniothyrium minitans, 2107
Conjugative gene transfer, 387
Conotrachelus nenuphar, 2009
Conservation, 1474. *See also* Biological control
   agricultural systems, 91, 119–120, 335
   biology, 633, 635
   biological controls. *See* Biological controls, conservation of
   of natural enemies, 1749–1751
      habitat and environmental manipulation, 1750
      harmful pesticides practices avoidance, 1749–1750
      overview, 1749
      overwintering and shelter sites, 1750
   prioritization, 972
Conservation farming
   in United States, 2382
Conservation program performance, 1774
Conservation Reserve Program (CRP), 346
Conservation structures, to control soil erosion, 970, 970f
Conservation tillage, 2762–2763, 2763f
Conservation tillage, 119
   to control wind erosion, 1028, 1027f
Constitutive equations
   for saltwater intrusion, 1328–1329
Constrained-equilibrium models, for exergy analysis, 1094
Constructed treatment wetlands (CTW), 2155
Constructed wetlands (CW), 2020–2021, 2663, 2730, 2862, 2863t
   definition, 2670

   and ecofiltration, 2154–2155
   potential for, 2670–2671
Construction methods, 860
Construction standards, 1655
Consult. *See* Hexaflumuron
Consumer concerns, 1401–1404
   'cocktail' effects from multiple residues, 1401, 1402t
   food residues, 1403
   government action plans, 1403
   NGO initiatives, 1404
   overview, 1401
   pests/pesticide safety in homes and gardens, 1401–1403
   retailer initiatives, 1403–1404
   towards residue-reduced food, 1403–1404
Contaminants
   environmental, 1626
   interactions in soil and water, 2166
      inorganic chemicals, 2166, 2169
      organic chemicals, 2169
   levels in wastewater, 2865t
   sources and nature of, 2166, 2167t–2169t
Contamination
   arsenic. *See* Arsenic; Arsenic-contaminated groundwater
   definition of, 1281
   groundwater. *See* Groundwater contamination
   with heavy metals, 1374
   of waterways, 2754
Continuous Commissioning (CC®), 657
   case study, 662
   process, 659–661
      assessment, 660
      building screening, 660
      CC measures development, 661
      implementation and follow-up, 661
      performance baselines, development of, 660–661
      plan development and team formation, 660
Continuous liquid–liquid extraction, 2193
Continuous simulation erosion models, 991–992
Continuous simulation models
   water quality, 2749
Contour farming, 2763
Contouring, 968, 968f
Contour lining, 570
Contour ridges, 969
Contour stone terraces, 969, 969f
Contour stripcropping, 2763, 2763f
Contour tillage, 2539
Control-A-Bird®, 413
Controlled-release fertilizers (CRFs), 1306
Controlled-release nitrogen fertilizers, 1809
Controlled traffic, 120

Control volume (CV), 815, 815f
Conventional activated sludge (CAS) systems, 2062
"Convention on the Prior Informed Consent (PIC) Procedure for Certain Hazardous Chemicals and Pesticides in International Trade," 2036
Conversion factors and the values of universal gas constant, 2527t, 2534
Conveyance efficiency, 1545–1546
Conveyance systems, 2264
Cooling system in dairy, 1664
Cool storage, 2511–2517
    application and design features, 2514–2517
    building certification, 2514, 2514f
    power generation and, 2511–2514
        efficiency and emissions, 2512–2514, 2513f
        electrical demand, 2511–2512
    technologies, 2509–2510
Coontail. See Ceratophyllum demersum
Co-oxidation/co-reduction, 2152t
Copidosomopsis plethorica, 1919
Copper (Cu), 535–537, 1370
    concentration, 1375
    deficiency
        plant growth and, 536
        in yaks, 427
    as essential element, 535–536
    high-copper soils, plant growth on, 536–537
    plants and, 535
    in soils, 535
Copper-indium selenide (CIS) solar cells, 232
Copper sulfate (for algae removal), 539
Coptotermes curvignathus, 1461
Coquis (*Eleutherodactylus coqui*), 2596
Corn (*Zea mays* L.), 1303
Corn oil, 415
Correlation coefficients, 2427
Corrosion, 190
Corvus spp., 415
Cosmetics, aluminum in, 270
Cosmetic standards, 1120
    environmental and public health impacts of pesticides and, 1121–1122
    health effects of eating insects in food and, 1122
    history of, 1120–1121
    increase in, of fruits and vegetables, 1120, 1121
    pesticide use and, 1120
Cosmic ray, 2235
Cosmogenic radionuclides, 2235, 2239
Costa Rica
    pesticide poisonings incidence, 1408
Cost-benefit analysis (CBA), 917–918, 917f, 918f
    for crop water use, 2220
Cost-effective analysis, 918–919

Costelytra zealandica, 322
Costs
    pesticide poisonings, 1416
    weed biocontrol and, 2822–2823
Cotton (*Gossypium hirsutum* L.), 425
Cottonwood. See *Populus deltoides*
Courant number, 1315
Courier. See Buprofezin
Cover crops, 1110
Coversoil resources
    open pit mining, 1692
Cow washing, 1663–1664
Cr. See Chromium (Cr)
Cradle to grave approach, 2418
Crailsheim borehole system, 2518
CREAMS model. See Chemicals, Runoff, and Erosion from Agricultural Systems (CREAMS) model
Creep samplers, 1023f
Critical flocculation concentration (CFC), 960
Critical loads, 2292
Crops
    depletion, 1663–1664
    detritus, 310
    disease resistance in, 407, 408t
    response to REE, 2269, 2269t
    water use, 1563
    yield and water requirement, 1664t
    yield increase by REE, 2269, 2269t
Crop-border diversity
    effects on pests, 1928
Crop diversity, for pest management, 1925–1929
    border diversity, 1928
    crop rotation
        effects on diseases and pests, 1925, 1926t–1927t
        weed abundance and, 1928
    decoy and trap crops, 1925, 1928
    overview, 1925
    weed diversity, 1928
CropLife America, 2014
Crop model (CM) techniques, 2818, 2818f
Cropping systems
    ammonia volatilization, 86–87, 86t
Crop rotation
    effects on diseases and pests, 1925, 1926t–1927t
    multifunctional, 1508
    sulfur and, 2432–2433
    weed abundance and, 1928
Crop rotations, 1018–1019
    organic farming systems and, 126
Crops/cropping
    impact on soil IC, 464
    pesticides in, 1966
    selection and diversification, 2675–2676
    water harvesting for, 2739
    yield
        potentials of, 2676t
        and water use, 2218–2219

Crop tolerance
    classification of, 2347t
    defined, 2346
    salt-affected soils and, 2346–2347
Crop water use, 2218–2219
    landscape, 2219
    spatial analysis of, 2219–2221, 2219–2220f
"Cross protection," 2111
Crotalaria juncea (Sunnhemp), 1511
Crotalaria spp., 289
Crucifer downy mildew, 443
Crude oil, chemical composition of, 2042–2046
Cry9C contamination, 313
Cry9C proteins, 312
Cryogenic techniques, 457
Crysoperla carnea, 412
Crystal iron oxides, 2266, 2268t
Crystalline (Cry) proteins, 308
Crystalline silicon solar cells, 231–232
    amorphous, 232
    monocrystalline, 231–232
    polycrystalline, 231
    structure and manufacture of, 232–234
        ARC, deposition of, 234
        diffusion formation of a p–n junction, 233
        formation of p–n junction by ion implantation, 233
        metal contacts, 233–234
        passivation of silicon surface, 233
        surface preparation, 232
Ctenopharyngodon idella, 1922
CTI. See Climate technology initiative
Cu. See Copper (Cu)
Cucumber (*Cucumis sativus* L.), 426
Cucurbita pepo, 407
Culex annulirostris, 2598
Culex tarsalis, 1922
Cultivar selection, 1112
Cultural eutrophication, 538
Cultural pest control
    antagonistic plants in, 288–289, 289t
Cultural practices
    effects on natural enemies, 370
Curie (Ci), 2235
Curve Number method, 1958
Customer demand, of electricity, 1418, 1419f
Cuticle, 1459
CV. See Coefficient of variation (CV)
CWA. See Clean Water Act (CWA)
Cyanide
    groundwater contamination by, 1291–1292
Cyanobacteria, 394
Cyanobacteria in eutrophic freshwater systems
    algal blooms, monitoring/management of, 539
    consequences of, 539

dominance of cyanobacteria, 538–539
extent of problem, 538
Cyanobacterial (bluegreen) algal blooms, 538
Cyanobacterial toxins, 2137
Cyanophyta, 1515
Cyclonic scrubbers, 154
Cydia pomonella, 2009
Cyhexatin, 3
Cylindrospermopsis, 539
Cyperine, 2031, 2031f
Cypermethrin, 2142
Cyprus, 2689
Cyperus rotundus, 2031
Cyprinid Herpes virus 3 (CyHV-3), 2603
Cyromazine, 1460, 1463
Cytochromes P450 (CYP), 281
Czochralski's method, 231

D

2,4-D, metribuzin, 2806
Dacthal, 2014
Dactylopus ceylonicus (mealy bug), 2821
Daily water intake of dairy cattle, 1664t
Dairy, water use in
    cow washing, 1663–1664
    drinking water requirements, 1663, 1664t
    estimation of water use, 1663
    flushing manure, 1664
    milking equipment/parlor, washing of, 1664
    rainwater from roofs, 1665
    recycling dairy wastewater, 1664, 1665t
    sprinkling and cooling, 1664
    water budget development, 1665t, 1666
Dairy wastewater treatment, 2863f
Dalapon, 1318t
Dams and weirs, 208–209
Danaus plexippus (L.), 313
Dandelion (*Taraxacum*), 1527t
Danish International Development Agency (DANIDA), 1269
Daphne, 271
Daphnia magna, 409
Darcy's law, 1312
    for saltwater intrusion, 1329
Dark leaf spot, 442, 443t
Dart. *See* Teflubenzuron
Databases
    GIS in land use planning and, 1163
Data loggers, 736
Datura metel L., 289
Datura stramonium L., 288, 289
Dayoutong, 1464
DBCP. *See* Dibromochloropropane (DBCP)
2,4-Dbe. *See* 2,4-D butyl ester (2,4-Dbe)

2,4-D butyl ester (2,4-Dbe), 2006
D.c.–a.c. converter. *See* Inverters, in photovoltaic systems
D.c. main switch, 222–223
DCE. *See* Dichloroethene (DCE)
DDD, 1399
DDE (dichlorodiphenyldichloroethylene), 417
DDT (dichlorodiphenyltrichloroethane) 417, 1397, 1399, 1403, 1414, 2015, 2015f
    discovery, development, and impact, 2009–2010
    effects of, 2005
Dead-band concept, 738
Dead Sea, degradation of, 1438
Dead state, 1084
Dead zones, 1302
Decision guidance document (DGD), 1616
Decision support system engine (DSSE), 2819
Decision support system for agro-technology transfer (DSSAT), 2819
Declarations (on soil conservation), 1619
Decomposition
    of organic matter, 90, 1625
Decoy crops
    effects on pests, 1925, 1928
Decreasing block tariff (DBT) pricing strategies, 2786
Deep sea mining, 2359
Defect action levels (DALs), 1120–1122
Defoliators, effect on photosynthetic rate, 1951
Deforestation
    erosion, historical review, 1050, 1051f
Degradation process, 2136, 2863
    soil. *See* Soil degradation
Delia antiqua, 2021
Deltamethrin, 2142
DEM. *See* Digital elevation models (DEM)
Denaturing gradient gel electrophoresis (DGGE), 2618
Denitrification, 94, 1770–1771
    nitrous oxide from, 96–97
Denitrification, 2864
Denmark
    wind output and net electricity flows in, 1427f
Dense NAPL (DNAPL), 1334
Densification, 374
Denver Arsenal, 1333
Deoxynivalenol, 1696
Department of Energy (DOE), 1794
Department of Health and Human Services, 2049
Depletion, defined, 1563
Deposition

erosion and, 1046–1047, 1047f
    soil erosion, 958
Deposit–refund system, 903
Depth-and-width integrated sampling, 965
Depth stratification, of soil organic matter, 91–92, 91f
Dermacentor variabilis (dog ticks), 2
Dermal effects, of arsenic toxicity, 1271–1272, 1272f
Deroceras caruanae, 383
Deroceras laeve, 310
Deroceras reticulatum, 383
Derxia, 350
DES. *See* Diethylstilbestrol (DES)
Desertification, 541, 557–558
    assessment of degree of, 541–542. *See also* Landscapes
    biophysical aspects of, 541
    causes of, 558
    as continuum, 542f
    defined, 552, 557
    in dry lands, 335
    and greenhouse effect, 549–551
        biological changes, 549
        climatical changes, 549
        greening of the Earth hypothesis, 549–550
        modern studies, 550–551, 550f–551f
        from prehistory to industrial revolution, 550
        woody plant range expansions, 550–551
    indicators, 553–554
    landscape ecology and restoration approach to, 541, 542f
    manifestations of process, 552–553
    monitoring, 543
    prevention and reversal of, 558–559, 558f
        agricultural practices in, 558
        climate change and, 559
        evaluation of, 560–561, 561f
        intervention projects for, implementation of, 559
        issues affecting, 558–559
        multiscale human–environmental dynamics and, 558
        participatory approach to, 561
        rehabilitation and restoration approach in, 558
        socioeconomic conditions and, 559
        water management in, 558
    response curve to, 542, 542f
    and revegetation, 543
    reversal of, procedures for, 542–543
    San Pedro River (case study), 554–555
    technology, measurement, and analysis, 554
Desertization, 557, 565

Deserts/desertization
  air pollution and, 566
  causes of, 565, 569t, 570t
    direct, 565, 567
    indirect, 565, 567–568
  defined, 565, 566
  degradation processes, 567–568
  diagnosis/evaluation, 568
  fluctuations and trends of climatic variables in historical times, 566
  land surface albedo, 566
  managed restoration, 569–570
  natural amelioration, 569
  origin of, 565–566
  overview, 565
  rainfall and, 566
  severity and extent of, 568–569, 569t
  temperature and, 566
Desiccation, 1625
Designated National Authorities (DNAs), 575, 1616
Designated uses, of navigable waters, 2810
Design for the Environment (DfE), 1253
Design intent document (DID), 669
Desorption, 432
  of contaminant, 1626
Detachment, soil erosion, 958, 984, 985
Deterministic uncertainty analysis, 1722
Deterministic water quality models, 2749
Detoxification, 2153
Detritus, 310
Developing countries, pesticide health impacts in, 573
  banned and restricted compounds import/export, 575–576
  health impacts
    acute pesticide poisoning, 573–574
    unknown chronic health impacts, 574
  lack of technical and laboratory capacity, 576
  low levels of worker and community awareness, 575
  pest control policies, 4
  weak regulation and enforcement, 575
  weak surveillance for hazards and impact, 574
Developmental neurotoxicity (DNT), behavioral aspects of, 1756
Developmental toxicants, 280
Developmental toxicology bioassays, 283
DeVine®, 383, 1111
DGD. See Decision guidance document (DGD)
Diabetes mellitus
  arsenic ingestion and, 1272
Diabrotica virgifera virgifera LeConte, 310
Diamond. See Novaluron; Teflubenzuron
Diatraea saccharalis, 367
Diazinon, 419, 1318, 1998t
Dibromochloropropane (DBCP), 2015, 2015f
  reproductive problems from exposure, 277–278, 278t
Dibrotica spp., 1983
Dichloroethene (DCE), 1334
Dichlorodiphenyltrichloroethane (DDT), 575, 1916, 1917, 2114
  global alliance for alternatives to, for disease vector control, 1917
Dichloronaphthoquinone, 539
Dicyclanil, 1463
DID. See Design intent document (DID)
Dieldrin, 417, 1283
Diesel cycle, 819
Diethylstilbestrol (DES), 1405, 1406
Difenacoum, 420
Difethialone, 420
Differential absorption lidars, 144. See also Lidars
Diffuse double layer (DDL), in soil, 2338
Diffuse melanosis, 1271
Diffusion, 1817–1820
  driven phase, evaporation, 130
  Fick's first law of, 1818
  nutrient movement by, 1819–1820, 1820f
Diflubenzuron, 1461
Digestion process, 2651
Digital elevation models (DEM), 1164
Digital orthophoto quadrangles (DOQ), 2272
Dikrella cruentata, 1936
Dilution-extractive system, for monitoring stack gases, 144–145, 145f
Dimethoate, 2016t, 2142
Dimethyl sulfide (DMS), 2439
Dimilin. See Diflubenzuron
Dinghy, 209
Dinitroaniline herbicides, 1318
Dinitrophenols (DNOC), 2009
Dinoseb, 1318t
Diofenolan, 1464
Dipel, 382
Diplococcus, 321
Diquat, 1318t
Direct aqueous injection–electron capture detection (DAI–ECD), 2795–2796
Direct aqueous injection–gas chromatography–electron capture detection (DAI–GC–ECD), 1847
Direct contamination of surface water, 1576
Direct current (DC) resistivity
  for saltwater detection, 1327
Disc filters, 2731. See also Filter(s)
Direct measurement, of soil redistribution by wind, 1021, 1021f, 1022f
Direct methanol fuel cells (DMFCs), 1146, 1151–1152, 1151f, 1152t
  applications of, 1152
  carbon monoxide and, 1151
  contamination levels, acceptable, 1151
  hydrogen sulfide and, 1151
  and methanol crossover problem, 1152
  operating principle of, 1151, 1151f
  technological status of, 1152
Direct radiative forcings, 1194, 1196t
Direct toxicity, 312
Dirty water, 2304
Discount rates, 1795–1796, 1800f, 1801f
Discretization, groundwater modeling, 1298f, 1299
Disease resistance
  in crops, 407, 408t
Disinfection methods, for cooking, 2705t
Dispersible clay content (DC), 2336, 2337f
Dispersion, 1627
Dispersion ratio of clay (DRC), 2336, 2337f
Dispersion, soil, 2334, 2339, 2340f
  effect of clay–cation interaction on swelling and dispersion, 2338–2340
  management of, 2341, 2342t
Dispersive liquid–liquid microextraction (DLLME), 1970t, 2193
Displacement, 809f
Displacement law, 2234
Disposal
  of pesticide containers, 510, 510t
  problems, associated with biowaste, 512
Dissolved chemicals, 2753
Dissolved organic N (DON), 1304
  leaching from agricultural soil, 1304
Dissolved oxygen, 2753
  concentration, 2649
Dissolved pollutants, agricultural runoff and, 82–83
  fertilizers, 83
  nitrogen, 82–83
  pesticides, 83
Distribution cycle exergetic ratio, 1175
Distribution reservoir, 1577
Distribution uniformity (DU), 1548, 1564
District cooling, 2515–2516, 2516f
District energy systems, 773
Disulfoton, 419
Diuron, 1318, 2806
Diversification
  and organic farming, 1106
Diversifying cropping systems, for reducing NO3-N leaching, 1307
Diversion/graded terraces, 968, 969f
Diversity
  functional, in intercropping, 1937
"Diversity–stability hypothesis," 1934–1935
DNAPL, 1282. See Dense NAPL (DNAPL)
DNAs. See Designated national authorities (DNAs)
DNOC. See Dinitrophenols (DNOC)
Dobson unit (DU), 1885

2-Dodecanone, 2021
DOE. *See* U.S. Department of Energy (DOE)
Dolomite, 88, 465
Domestic hot water (DHW), 242
Doppler lidars, 144. *See also* Lidars
DOQ. *See* Digital orthophoto quadrangles (DOQ)
Dose-response models, 916
Dose–response relationship, 1705
Double catalyst system, 157
Double pumping, 1330
Downy mildew, 443, 443t
Drain
   depth, 590
   spacing, 590
Drainable system (wall), 864
Drainage, 584, 1559
   artificial, 585
   climatic conditions effect on, 586
   definition of, 585
   downstream effects of, 584–586
   of drier soils, 585
   experimental studies on, 585
   factors influencing response to, 585
   and flood events, 584
   main drainage systems, 584
   of permeable soils, 585
   purpose of, 584
   role in crop production, 584
   shallow, 584
   simulation model, 585
   subsurface/groundwater, 584
      reduced peak flows from, 585
   surface, 584
      higher peak flows from, 585
   topsoil texture and, 585
Drainage, soil salinity management and conditions, 588
   requirement
      saline soils, 588–589, 589f
      sodic soils, 589
   system design, 589–590
      drainage wells, 590
      drain depth, 590
      drain spacing, 590
      relief drains, 590
      saline seeps, 590
Drainage lakes, 1577
Drainage water, 1547
Drainage wells, 590
DRC-1339 (3-chloro-*p*-toluidine hydrochloride) (Starlicide®), 413
Dredging, 1592t, 1692, 2315t
Drinking water, 2790–2800
   analysis, microbiological research methods for, 2799–2800
   bacteria and pathogens in, characteristics of, 2797–2799, 2799t
   inorganic components and pollutants of, 2791–2793, 2791t, 2792t

and inorganic pollutants, 2120
   intake in dairy, 1663, 1664t
   organic pollutants of, 2793–2797, 2795t, 2796t, 2797f
   requirements of, 2791
   water treatment systems, 539, 2062–2063
      removal efficiencies, 2066–2067
Drip irrigation, 2704
   systems, 2677
Drip irrigation system, 1542–1544
   hydraulic design of, 1542–1543
   for optimal return, water conservation, and environmental protection, 1543–1544
   uniformity of water application and design considerations, 1542
Driving force–pressure–state–impact–response (DPSIR) matrix, 2385, 2385f
Dry adiabatic lapse rate, 1129
Dry ash free (DAF) basis, 714, 717, 719t
Dry ash gasifier, 2491
Dry deposition, 149
Dry loss (DL), 717, 718t
Dry matter intake (DMI), 1663
Dry reservoirs, 1588. *See also* Reservoir(s)
Dry techniques and SO_2, 2442
   circulating fluid bed dry scrubber, 2442
   magnesium oxide process, 2443
   sodium sulfite bisulfite process, 2442–2443
Dual fluidized-bed gasifier (DFBG), 2493
Dual-path-type in situ system, for monitoring stack gases, 145–146, 145f
Dubinin–Radushkevich isotherm, 65. *See also* Isotherm(s)
Duckweed (*Lemna gibba*), 426
Duct sorbent injection, 2442
Dust, 149
Dug wells, 1276
Durum wheat (*Triticum aestivum* L.), 425
Dusicyon culpaeus, 2596
Dust, windblown, 1019, 1019f
Dust Bowl of the Great Plains, 1010
Dust Production Model, 1025
Dust storms, 1010
   in Asia, 1020
   in Bodele Depression, 1020
   erosion by, 1035
DustTrak, 1024, 1024f
Dyer's woad (*Isatis tinctoria*), 1527t
Dye-sensitized cells (DSCs), 234–239
   advantages, 238–239
   electron process, 237f
   mechanism of operation, 236–238
   photogeneration of charge carriers in, 236f
   principles, 235
   structure, 235, 235f
   titania solar cells, 238
   titanium dioxide, 235–236

E

Earias insulana, 2021
Early site permit (ESP), 1789
Earth
   coupled systems, 1363
   energy budget, 242f
   heat exchangers, types of, 1365–1366, 1366f
   nanoparticles beyond, 1716
Earth Observing System (EOS), 554, 2818
Earth–sun energy balance, 771, 772f
   and global warming, 772
Earth system response (global climate change)
   attribution of, to human influence, 1199, 1199f
   climate forcings and feedbacks, 1194–1197, 1194f, 1195f
      atmospheric aerosol forcing, 1196, 1197t
      greenhouse gas forcing, 1195, 1196t
      land-use change forcing, 1196–1197
      natural climate forcings (solar and volcanic variability), 1194–1195, 1196t
   future projections, 1199–1200, 1200f
   human-induced, evidence of, 1197–1198, 1198f
   impacts on natural systems, 1192
   and natural greenhouse effect, 1192–1193, 1193f
ECD. *See* Electron capture detector (ECD)
Ecdysone, 1465
Ecdysone agonists (EAs), 1460, 1465–1466
   chromafenozid, 1466
   furan tebufenozide, 1466
   halofenozide, 1466
   methoxyfenozide, 1466
   tebufenozide, 1466
Ecdysteroids, 1465
ECETOC, 1770
Echinochloa crus-galli (Barnyard grass), 426
ECMs. *See* Energy cost measures (ECM)
Eco-exergy, 594, 604, 778, 780–787, 787f, 1096
   to emergy flow, ratio of, 594–595
   illustrative examples of, 787–788
Ecofiltration, constructed wetlands and, 2154–2155
Eco-filtration systems, 2154–2155, 2730
Ecological approach, of rehabilitation, 1684–1685, 1685t
Ecological economics
   Ems-axis, lower saxony, growth and booming region, 196–199
   renewable energy, 186–195
   renewable energy in Germany and planned nuclear exit, 195

Ecological footprint, 634, 2467–2478
　applications of
　　business, role of, 2477–2478
　　products and services, 2477
　　terrestrial systems, 2476–2477
　fundamentals of, 2468–2469
　limitations of, 2478
　of nation, 2469–2471
　of product, 2471–2473
　three-dimensional geography of, 2475–2476
　weakness of, 2478
Ecological infrastructure management, 1509
Ecological pest management, 1930–1932
　diversity for croping systems, 1931
　environmental and economic benefits, 1932
　overview, 1940
　risks in adopting, 1932
　spreading practices of, 1930
　strategies, 1930
Ecological risk assessment (ERA), 1992–1993, 1993f
Ecological sustainability filter (EcSF), 639
Economically optimum N level (EON), 1302
Economic capital, 2465
Economic costs
　pesticide poisonings, 1416
Economic growth, 614–622
　economists debate against, 616
　employment and, 619–621
　history of, 614–615
　low/no-growth scenario, 616–622
　　business-as-usual scenario, 617–618, 621t
　　funding public services in, 621–622
　　high-level structure of, 617f
　　policy directions for, 618–622
　as policy objectives, 615
Economic incentives, 923–924
Economic loss
　virus and viroid infections, 2111
Economic pressure
　integrated farming, 1507
Economics considerations
　pest management and, 2011
Economic sustainability filter (EnSF), 639
Ecosystem(s), 333
　acidic deposition impacts on, 22–24
　closed, 1935
　development of, 595
　estuarine, environmental assessment of, 1064–1069
　functioning of, 1068
　global distribution of organic matter, 1851–1854
　health and population, 595
　and humans, relation between, 635, 636
　open, 1935
　planning, 632–640
　　challenges to, 640
　　context of, 632–633
　　defined, 633
　　landscape planning, 637–638
　　perspectives of, 639–640
　　term uses of, 633–637
　　urban planning, 638–639
　services, holistic interpretation of, 595–596
　structure of, 1068
Ecosystem functioning, of Alexandria Lake Maryut, 167, 170–173
Ecosystem health, 599
　axioms of, 601t
　basic components of, 600–601
　definition of, 600–601
　ecological indicators, basic features of, 599–600
　health indicators in
　　agroecosystem health, 606–607
　　aquatic ecosystem health, 607
　　forest ecosystem health, 607
　　urban ecosystem health, 607–608
　indication of, basic requirements for, 601
　management, indicator application and aggregation in, 608–609
　utilized indicators of, 601
　　aggregated theory-based indicators and indices, 604
　　ecosystem analytical indicators and indices, 604–605
　　ecosystem service indicators, 605–606
　　species- and community-based indicators and indices, 601–604
Ecosystem models, 1867
Ecosystem restoration task force model, 628
　Gulf Coast Ecosystem Restoration Task Force (GCERTF), 629–630
　South Florida Ecosystem Restoration Task Force (SFERTF), 629, 630
Ecosystem services, 334, 557–558. See also Biodiversity
　degradation of, 335
　restoration, 335–337
Ecosystem services, 604–605, 605t
Ecosystem services valuation (ESV), 634
Ecotechnology, 2150, 2152–2153
　and climate change, 2156
　and developed countries, 2157–2158
　and developing countries, 2158–2160
　history of, 2153–2154
　and India, 2160–2162
　potential and scope of, 2730–2731
　scope of, 2153
Ecotones, 2824
Ecotoxicity, 1704
　elements with, 1455, 1457t
　by inorganic compounds, 1455–1458

EC Regulation 2078/92, 1108
Ectoparasitoids, 1747, 2424
Edaphic deserts, 566
EDC. See Endocrine disruptor chemical (EDC)
Edge-defined film-fed growth (EFG) method
　solar cell production from, 232
Edwards aquifer (of Texas), 1284, 1285, 1285f, 1287f
Effect summation, 2574–2575
Egg parasitoids, 1746, 2424
Egypt, 2682, 2689
EI_{30}, rainfall erosivity index, 943
Ejector scrubber, 155
Electric air conditioning, 793
Electrical conductivity, 1547, 1559, 1568, 1569t
Electrical demand
　and thermal storage, 2511–2514
Electrical resistance space heater, 1086
　exergy efficiency of, 1086
Electric heat pump, energy flow of, 1352, 1352f
Electric utility system, 1418
　operation, dynamics of, 1418–1419
Electricity, 799
　generation, 791–792, 792t
　methods, estimated costs of, 122t
　technologies for, 773t, 774t
Electricity sector
　customers in, 1493–1494
　electric utilities in, 1493–1494
　　integrated systems methodology in, 1494, 1504f. See also Integrated energy systems, case study from ISU
　　lowest cost to consumer in, 1506
　　questions related to, 1504
　　savings potential in, 1504–1505
　investments in, 1493, 1494
　problems of, 1493
Electric power industry, 792
Electric Power Research Institute (EPRI), 1652
Electric utilities, 1493
"Electrolyte effect," 2370, 2371
Electromagnetic radiation, 1192
Electromagnetic radiation, spectrum of, 1883f
Electron capture detector (ECD), 1203, 1975
Electron–hole pair, generation of, 227f
Electronic waste, 1431
Electrostatic precipitator (ESP), 153–154, 1128
Elk (*Cervus elephus*), 2596
Elodea nuttallii, 294
Elovich equation, 67
Eluviation/illuviation, 1627
　environmental implications, 1627–1628

Embodied energy, of farm chemicals, 121, 121t
Emergy, 593, 604, 1097
  to eco-exergy, ratio of, 594–595
Emergy–exergy ratio for flow, 1097
Emergy-to-money ratio (EMR), 596
Emissions, 916
  nitrogen in fuels, 728–729, 729f
  reduction mandates, 919
  reduction of, 483–484, 485
  tax on, 919
Emissions reduction units (ERU), 486
Emissions trading (ET), 486
Emissions trading system (ETS), 483, 484, 487
Emission uniformity, 1548–1549
EMP. *See* Energy master planning
Empirical model
Empirical models, 974
  advantages of, 974
  development of, 974–975, 976t
  in pest management strategies, 1949
  for water erosion. *See* Water erosion models, empirical
  for wind erosion. *See* Wind erosion equation (WEQ)
Empirical water quality models, 2749
Employment, 619–621
EN 14791:2006. Stationary source emissions, 2440
Encapsulation, 2510–2511
Encarsia formosa, 1475
Endangered Species Act (ESA), 1942–1943
Endocrine Disrupter Screening and Testing Advisory Committee (EDSTAC), 647–648
Endocrine Disrupters Testing and Assessment (EDTA) Task Force, 647
Endocrine-disrupting chemicals (EDC), 643–652, 1405, 1942
  animal exposure effects of, 650–651
  emission of, 651–652
  hormones effect on, 645–646
  human exposure effects of, 648–650
  mechanisms of action, 644–645
  regulatory purposes, guidelines for, 647–648
  reproductive toxicity of, 646
  risk assessment of
    using biomarkers, 647
    using screening assays, 646–647
Endocrinological effects, of arsenic toxicity, 1273
Endocrine Disruptor Screening Program (EDSP), 647, 648
End-of-life tires (ELTs), 896, 909–910
  management of, 896
End-of-life vehicles (ELVs), 896, 897, 902–903, 911
  reuse/recovery rates of, 903, 903f, 903f

Endoparasitoids, 2424
Endosulfan, 575, 2016t, 2019, 2142
Endothall, 1318t
Endothermic reaction, 2527
δ-endotoxin, 2029
Endrin, 1318t
ENERCON, 195
Energetic reinjection ratio, 1173
Energetic renewability ratio, 1173
  work, 809, 809f, 811
Energy, 770, 808, 809f, 810, 811, 821
  additional, 770. *See also* Additional energy
  carriers of, 770
  conservation, first law of, 813–816
  degradation, second law of, 816
  efficiency, 775. *See also* Energy efficiency
  entropy and second law of thermodynamics, 817–819
  forms of, 770
    life cycle of, 773–774, 773t, 774t
  forms and classifications, 811–813, 812t
  heat, 809–810, 811
  heat engines, 819
  exergy and the second-law efficiency, 820
  impact on environment, 1095
  natural, 770. *See also* Natural energy
  renewable, 774
  reversibility and irreversibility, 816–817
  resources, 770
  selection, 774
    environmental considerations in, 774–775
  solar, 771–772
  sustainability, 776
  and sustainable development, 798–799
  use of, 774–775
Energy audits, 118
  assessment process, 118–119
  level 1, 118
  level 2, 118
  level 3, 118
Energy balance of collector, 245–246
  collector efficiency curves, 245–246
  instantaneous efficiency of a collector, 246
Energy carriers, 685, 770
  vs. energy sources, 770
Energy conservation
  and efficiency improvement programs, 684
  environmental problems, 680–682
    potential solutions, 682
  example, 689
  first law of, 813–816
  implementation plan, 686–687
  importance of, 689
  improvement factors of, 685t

life-cycle costing in, 688–689
measures, 687–688
  elements, 687
  evaluation of, 642
  practical aspects of, 682–683
    flow chart, 684t
  programs, 677
  renewable energy resources and, 685–686
  research and development (R&D) in, 683
  sectoral, 688
  and sustainable development, 683–686
  technical limitations on, 683
  technologies, 682
  world energy resources, 678–680
Energy consumptions circuit exergetic ratio, 1175
Energy conversion
  biomass and hydrogen energy cycles (comparison), 716, 716f
  carbon-di-oxide emissions per year, 716f
  char reactions, 724–725
  cofiring (coal and bio-solids)
    coal agricultural residues, 727–728, 727f
    coal and manure, 728
    coal and RDF, 728, 728t
    conversion schemes, 727
    fouling in, 729–730, 730t
    NOx emissions, 728–729, 729t
  combustion
    circulating fluidized bed combustor (CFBC), 727
    defined, 723, 724f
    fixed bed combustor (FXBC), 726
    fluidized bed combustor, 726–727
    ignition and, 725
    in practical systems, 725–727, 726f
    stoker firing, 726, 726f
    suspension firing, 725, 726f
  energy units and terminology, 715t
  fuel properties
    gaseous fuels, 722–723
    liquid fuels, 722
    solid fuels, 717–722, 718t, 719t, 720t, 722f, 723f
  FutureGen layout, 731–732, 732f
  gasification, defined, 723, 724f
  gasification of coal and bio-solids, 730–731, 731f
  ignition, 725
  objectives of, 714–717
  pyrolysis, 723, 724f
  reburn with biosolids, 732–733, 732f, 733t
  technologies, 774
  volatile oxidation, 723–724
Energy cost measures (ECMs), 662, 663t
  continuous commissioning as, 662

Energy curtailment, 682
Energy degradation, second law of, 816
   reversibility and irreversibility, 816–817
   entropy and second law of thermodynamics, 817–819
Energy driven phase, evaporation, 130
Energy efficiency, 682, 775
   methods for improving of
     advanced energy systems, use of, 775
     audits, on irrigation systems, 120
     building envelopes, improving of, 775
     energy leak and loss prevention, 775
     energy storage, 775
     energy supplies and demands, matching of, 775
     examples of, 776, 776t
     exergy analysis, use of, 776
     improved monitoring, control, and maintenance, 775
     passive strategies, use of, 775
     use of high-efficiency devices, 775
Energy efficiency ratio (EER), 1364
Energy engineering for developing countries, 2465
Energy flows, 862
Energy Information Administration (EIA), 790
Energy intensity, 774–775
Energy management, 764
Energy Management System (EMS), 735
Energy master planning (EMP)
   American approach, origins of, 765
   business as usual, improving, 765
   optimization, 764
   steps to, 765–766
     business approach development, 766
     casting wide net, 767
     communicate results, 768
     energy team, creation of, 767
     ignite spark, 766
     obtain and sustain top commitment, 766
     opportunity recognition, 766
     organization's energy use, 767
     set goals, 767
     upgrades, implementation of, 768
     verify savings, 768
   tips for success, 768–769
   unexpected benefits of, 763–764
   vs. energy management, 764–765
Energy Master Planning Institute (EMPI), 765, 766t
Energy production and consumption, impacts of, 915
Energy report cards, 768
Energy return factor (ERF), of photovoltaic systems, 215
Energy security, 799–803
   risk assessment, 800–801
     depleting oil reserves, 801–802
     supplies from Middle East, 803
Energy Standard for Buildings Except Low-Rise Residential Buildings, 1656
Energy Star program, 1655
   for windows, 867, 867f, 867t
Energy storage
   benefits, 1420–1421
   technologies, 1419–1421
     development status, 1421t
   types, 1421
Energy transfer, 808, 809
   and disorganization, 816–817
   versus energy property, 811–813
   reversible, 819, 821
Energy units and terminology, 715t
Energy use, 118. *See also* Energy audits
   commercial, 793
   for different crops in different countries, 120–121, 121t
   distribution of, by industry groups, 795f
   electricity generation, 791–792, 792t
   farming systems, effect of, 119–121
     conservation farming practices, 119–120
     irrigation methods, 120
     machinery operation, 119
   and greenhouse gas emissions, 118, 120
   industrial, 794–795, 794f
   residential, 793
   saving of, 122
   by sectors, 790–791, 791t
   transportation, 722, 723, 725
   by type of energy, 791, 794f
   US overview, 790
Energy utilization and environment, 683
Energy Valuation Organization (EVO), 1506
Engineered nanoparticles, 1721
Engineered terraces, 2538–2539
Engineering approach, of rehabilitation, 1683–1684
Enhanced Thematic Mappers (ETM), 2817t, 2818
Enhydra fluctuans (Water cress), 426
Entamoeba histolytica, 2696
Enterobacter, 321
Enthalpy, 2526
   of formation, 2526–2527
   of reaction, 2526–2527
Enthalpy change of combustion, 2527
Entomobyroides dissimilis, 1879
Entomopathogenic nematodes, 382
Entomopathogenic viruses, as bioinsecticides, 382
Entrained bed type gasifiers, 2490–2491, 2491f
Entropy, 604, 821, 1097, 2529
   and exergy, 816–819
   generation, 816–817
   and second law of thermodynamics, 817–819
Environmental costs and regulations, 1793–1795
Environment 2010: Our Future, Our Choice, 136
Environmental compatibility value (ECV), 1986
Environmental contaminants, 2136
Environmental contamination, 2584
Environmental Defense Fund, 924, 2010
Environmental degradation, 2307
Environmental degradation and national security, 2461–2462
Environmental genotoxicity, 2124–2125
Environmental health, 336
   environmental conditions and, 336
   importance of, 336–337, 336f, 337f
   indicators of improvement of, 337
   steps to be implemented for, 337
   tools for, 337–338, 338f
Environmental interest
   elements of minor, 1455
Environmental issues
   tailings and, 1683
Environmental Kuznets curve (EKC), 1221
Environmental legislation in Asia, 874, 876
   biodiversity laws
     enforcement, 881–882
     regional cooperation, 882
   biodiversity laws at national level, 877
     Cambodia, 881
     China, 877
     Hong Kong, 879–880
     Indonesia, 878–879
     Japan, 878
     Korea, 877–878
     Lao People's Democratic Republic, 880
     Malaysia, 878
     Philippines, 879
     Singapore, 879
     Thailand, 880
     Vietnam, 881
   development in international law and its consequences on national legislations, 877
   environmental assessment law
     Cambodia, 887–888
     China, 885
     Hong Kong, 886
     Indonesia, 886
     Japan, 886
     Lao PDR, 887
     Malaysia, 886
     Philippines, 887
     Republic of Korea, 886
     Singapore, 887
     Thailand, 887
     Vietnam, 888
   e-waste management law, 882
     Cambodia, 884
     China, 882–883
     Hong Kong, 883

implementation and enforcement, 885
Japan, 883
Lao People's Democratic Republic, 884
Malaysia, 883
Philippines, 883–884
Republic of Korea, 883
Singapore, 884
Thailand, 884
Vietnam, 884
e-wastes and environmental assessment, 882
law on biodiversity conservation, 876
transboundary movements of hazardous waste, 884
Environmental legislation, to control soil erosion, 970–971
Environmentally relevant phenomena, 353, 354f
Environmentally sound technologies (EST), 2671
Environmental management, 1242
systematic approach, 1242–1243
quality characterization, 1245
standard protocols, 1243
uncertainty assessment, 1243
Environmental matrices, chemical analyses in, 1065
Environmental policy
burden of proof, 916–917
common property resource destruction, 914
economic framework
cost-benefit analysis (CBA), 917–918, 917f, 918f
cost-effective analysis, 918–919
impacts of energy production and consumption
air, 916
land use, 916
water, 916
multidisciplinary approach, 914–915
policy instruments
emission reduction mandates, 919
intrinsic/nonuse benefits, 920
liability rules, tightening, 919
redistributive effects, 920
renewable energy resources, 920
specific control technology, 919
sustainable development, 920
tax on emissions, 919
tax on polluting good, 919
tradable emissions permits, 919
pollution, 914
Environmental policy innovations (EPI), 922–924, 925t
economic incentives, 923–924
nature of, 923–925
civic environmentalism, 924
policy entrepreneurs, 923
"policy reinvention," 923
resources, neds, politics, and determinants
institutional factors, 925–926

interest group support, 926–927
need/problem severity, 925
regional diffusion, 927
state policy innovations, 924
Environmental pollution, 1654, 2123–2130. *See also* Pollution
biomonitoring
animals, 2129
human, 2129
plants, 2128
genotoxicity
environmental, 2124–2125
human, 2124–2125
pesticide, 2125–2128
by nanomaterials, 2129
reproductive toxicity, 2124
Environmental pollution, by pesticides, 2013–2014. *See also* Pesticides
mitigation of, 2018
CW microcosms, use of, 2020–2021
natural products for pest control, use of, 2021–2022, 2021f
slot-mulch biobed systems, use of, 2020
soil microorganisms activity, enhancing of, 2020
SOM content, increasing of, 2018–2019
pesticide residues detection and measurement of, 2015–2017
cleanup, 2018
confirmation, 2018
extraction, 2018
quantification, 2018
sampling, 2018
Research Farm, agricultural practices at, 2022–2024
Environmental problems
nitrogen leaching and, 1761–1762
Environmental Protection Agency (EPA), 133, 135–136, 629, 1118, 1363, 1394, 1405, 1570, 1571, 1612, 1793, 1940–1946, 1961, 1991, 2010, 2805, 2809–2810, 2811, 2812, 2813
DNT guidelines of, 1757
drinking water standards, 1730
Endocrine Disrupter Screening and Testing Advisory Committee, 647–648
Endocrine Disruptor Screening Program, 647, 648
Guidelines for Drinking Water Quality, 2793
mutagenicity testing battery, 2124–2125
on total Cr in drinking water, 477
Environmental Protection Agency (US), 919
Environmental Quality Indicators Program (EQIP), 127
Environmental risk assessment, 1723
Environmental safety
herbicides and, 1380–1381

Environmental services programs, payments for, 1773
Environmental simulation models
for surface water quality, 2761–2762
Environmental state, 1084
Environmental stewardship, 57
Environmental Working Group, 1404
Environment Protection Agency (EPA), 1609, 1611
Environomics, 1097
Enzymes, function of, 89
Enzyme variation
and pesticides, 1400
EOS. *See* Earth Observing System (EOS)
EPA. *See* Environmental Protection Agency (EPA)
EPA Review of Chlorpyrifos Poisoning Data (1997), 1411
EPIC (Erosion Productivity Impact Calculator) model, 976
Epidinicarsis lopezi, 1475
Epofenoname, 1464
EP409.1 Safety Devices for Chemigation, 471
Epworth Sleepiness Scale (ESS) scores, 2875f
EQIP. *See* Environmental Quality Indicators Program (EQIP)
Equilibrium models, for exergy analysis, 1093
Equine intoxication, 428
Equine leukoencephalomalacia (ELEM), 1697
Equity and fairness, 2785–2786
Eroded sediments
agricultural runoff and, 82
Erodibility, soil, 980–989
effect of particle travel rates on, 986–987, 986–987f
Erosion, 2754
and carbon dioxide, 934–937
conceptual mass balance model, 935–936, 935f
emissions, estimates of, 936, 936t, 937f
SOC, 934–935
carbon losses from soil and, 1857–1858
continuous simulation models, 991–992
definition of, 943
effects of rain on, 945
precipitation and, 945. *See also* Precipitation
primary agents for, 943
process-based models, 991–992
and rainfall erosivity, 943
snowmelt. *See* Snowmelt erosion
by water, 943, 991–992
wind, 943, 945. *See also* Wind erosion
Erosion, accelerated, 1044–1048
deposition, 1046–1047

Erosion, accelerated (cont'd.)
  gullies, 1046, 1046f
  modeling, 1048
  overview, 1044
  processes, 1047–1048, 1047f, 1048f
  rainfall, 1044–1045, 1045f
  rills, 1045–1046, 1046f
  transport, 1045, 1045f
Erosion by water, 951
  future of soil, 955
    direct effects of future climate change, 955
    indirect effects of future climate change, 955
  global problem of accelerated, 954
  impacts of, 954–955
  processes, 951–953
  spatial and temporal scale, 953
Erosion control, 1040–1043
  mulch tillage, 1042, 1042f
  no-till, 1040–1041, 1041f, 1042f
  overview, 1040
  ridge tillage, 1041–1042, 1042f
  strip tillage, 1042–1043
  tillage, 1040, 1041f
  on soil properties, 1040
EROSION-2D/3D, 1958, 2761
Erosion hazard maps, 1037
Erosion pins, 1021, 1021f
Erosion plots, 974–975, 978
Erosion problems
  historical review, 1050–1052
    deforestation, 1050, 1051f
    sedimentation rates, 1050–1052
Erosion-Productivity Impact Calculator (EPIC) model, 2761
Erosivity, rainfall, 980–990
Error analysis, groundwater modeling, 1299
E-sampler, 1024
Escherichia coli, 1626, 2696
ESP. *See* Exchangeable sodium percentage (ESP)
EST. *See* Environmentally sound technologies (EST)
Esteem. *See* Pyriproxyfen
Estimation of Ecotoxicological Properties (EEP), 332
Estrogen-like substances, 281
Estuaries, environmental problems of, 1063–1070
  assessment of, 1064–1069
    bioassay methods, 1067
    biological responses at community structure level, 1066–1069
    chemical analyses, in environmental matrices, 1065
    tools for, 1068–1069
  origin of, 1063–1064
  solving, 1069–1070
Estuarine and marine ecosystems, 607
Estuarine intertidal emergent wetlands, 2827

Estuarine intertidal unconsolidated shore, 2827
Estuarine quality paradox, 1065–1066
Estuarine turbidity maximum (ETM), 1063, 1064
Ethanol, from biomass, 377–378
Ethinylestradiol, 1405
Ethiopia, 2683
Ethylene dibromide (EDB), 1317, 1318, 1318t, 2015
Ethylene thiourea (ETU), 1394
Etoxazole, 1463
ETU. *See* Ethylene thiourea (ETU)
Eucalyptus spp., 177
Euhrychiopsis lecontei, 367
Eulerian–Lagrangian Localized Adjoint Method, 1316
Eulerian methods, 1315
Euphorbia esula, 2822
Eurasian skylark (*Alauda arvensis*), 416
Eurasian sparrowhawk (*Accipiter nisus*), 417
Eurasian watermilfoil. *See Myriophyllum spicatum*
Europe
  acid rain in, 8
  control policy, 15
  natural erosion across, 2378
  soil degradation in, 2377–2378, 2377f
    distribution of, 2378f
  trends in acidity, 9, 11–13
European agroecosystems, bioindicator development in, 356, 357t–358t
European carp (*Cyprinus carpio*), 2603
European Commission (EC), 215, 528, 1406
  Directive 96/62/EC, 138
  Directive 99/30/EC, 138
  Directive 2000/69/EC, 138
  Directive 2002/134/EC, 138
  Waste Framework Directive, 2399
European Community (EC)
  Directive 76/403/EEC, 2181
  Directive 96/59/EC, 2181
  Guidelines for Drinking Water Quality, 2793
  Water Framework Directive, 1064
European rabbits (*Oryctolagus cuniculus*), 2596
European red foxes (*Vulpes vulpes*), 2596, 2604
European Soil Erosion Model (EUROSEM), 966, 2761
European Union (EU), 355, 1612, 1613
  Landfill, 516
  organic forming policy guidelines of, 528
  Soil Thematic Strategy, 2375
  Waste Management Directives, 516
  Working Document on Biological Treatment of Biowaste, 513

European Union (EU) Drinking Water Directive, 424
European Union emissions trading system (EU-ETS), 487
European Union's REACH legislation, 284–285
European Water Framework Directive, 1961
EUROSEM model, 985–986, 992
Eutrophication, 494–495, 538, 1074, 1456, 1559, 1578–1579, 2093–2094, 2102–2103, 2137, 2360, 2753
  model framework, 1585
  phytoplankton, growth of, 1075–1078
  problem, 1074–1075
  solutions to, 1078–1079
Evacuated tube collectors, 244, 244f
Evacuated tube panels, 242, 244f
Evapoconcentration effect, 1559
Evaporation, 1564t
  diffusion driven phase, 130
  energy driven phase, 130
Evapotranspiration (ET), 1184, 1547, 1549, 1559, 1566
Event erosivity index, 982
Event sediment yield, 982
Everglades (of Florida)
  Comprehensive Everglades Restoration Plan (CERP), 1080, 1081
  features of, 1080
  hydrologic modifications in, 1080
  water management in
    environmental issues, 1080–1081
    history of, 1080
    restoration, 1081
Everglades Agricultural Area (EAA), 1080
Evergreen field, defined, 2762
Excavation
  of inorganic contaminants, 2120
Exchangeable Na ions, displacement of, 2370, 2371f
Exchangeable sodium content (ES), 2335, 2336, 2337f
Exchangeable sodium percentage (ESP), 1693, 2335, 2335f, 2336, 2337, 2337f, 2345
Exchangeable sodium ratio (ESR), 2335
Exergetic reinjection ratio, 1175
Exergetic renewability ratio, 1173
Exergy, 683, 778, 780, 780f, 785f, 821, 1083, 1092–1093, 1176, 2531–2532
  advantages over energy analysis, 1083
  analysis, 1092, 1093. *See* Exergy analysis
  in environmental policy development, 1097
  reference environment in, 1093–1095
  for sustainable development, 1097
  applications of, 1099–1100
    macroinvertebrate communities, 1100
    water bodies, 1100

balances, 1084
  in combustion process, 1093, 1093f
  consumption, 1084
  defined, 1083, 1084
  definition of, 1084
  and economics, 1089
  and ecosystems, 1095–1097, 1100
  efficiencies *vs.* energy efficiencies, 1091
  of emission, 1092
  entropy and, 816–819
  and environment, 1089
  and environmental impact, relations between, 1097–1099
    order destruction and chaos creation, 1097–1098
    resource degradation, 1098
    waste exergy emissions, 1098, 1098f
  equation for, 1092
  evaluating, 1083
  impact on environment, 1095
    reducing of, 1097
  indices, 604
  losses, 1095
    vs. energy losses, 1090–1091
  and lost work, 2532
  method, 1083
  of natural environment, 1093
  to reduce waste exergy emissions, 1099
  and reference environment, 1084
  and reference-environment models, 1093
    constrained-equilibrium models, 1094
    equilibrium models, 1094
    natural-environment-subsystem models, 1094
    process-dependent models, 1095
    reference-substance models, 1094
  and second-law efficiency, 820
  of stream, 2531
Exergy analysis, 1083, 1084–1085, 1176–1177, 2531
  applications of, 1085–1086
  and efficiency, 1085
    practical limitations, 1085
    theoretical limitations, 1085
Exergy-based ecological indicators
  biodiversity, 1096
  buffering capacity, 1096
  dissipation, 1096
  ecological process efficiency, 1096
  health and quality, 1096
  maturity, 1096
  optimization, 1096
  structure, 1096
Exothermic reaction, 2527
Exposure assessment, 1705
Ex situ techniques, 2170
Extended producer responsibility (EPR), 893
Exterior foam insulation, 1482, 1483f
Exterior insulation finish systems (EIFS), 864
External casing, 242

External combustion (EC), 714
External contamination, sources of, 2264
External costs, 917, 917f
  reduction of, 917f
External feeders, 2423, 2424t
Externalities, 634, 636
Externally feeding pests, biocontrol of, 2424–2425
Extraction wells, 1293
Extruded polystyrene (XPS), 1478

F

F. oxysporum f. sp. *udum*, 1925
Fabric Filter (FF), 1128
FACE experiment. *See* Free-Air CO_2 Enrichment (FACE) experiment
"Facilitated transport," 1622
Factory-built modular building, 860
Facultative ponds, 2633–2634, 2636
  BOD removal, 2633f
  kinetics, 2634
Falcon. *See* Methoxyfenozide
Falco peregrinus (Peregrine falcon), 417
Fallout, radioisotopic, 1024–1025
Fall nitrogen applications, 1808
False negatives, 1721
False positives, 1721
Famphur, 420
FAO. *See* Food and Agriculture Organisation (FAO)
FAO International Code of Conduct, 1612
FAO–LADA (Food and Agriculture Organization of the United Nations/ Land Degradation Assessment in Drylands), 558
FAO/WHO Food Standards Program, 1126
Farmed wetland, 2825t
Farm energy calculators, 118–119
Farmer Field Schools (FFS), 1930
Farming practices, sustainable, 353, 353t
  potential bioindicators for
    arthropods, 357t
    birds, 358t
    plants, 358t
    soil fauna, 357t
    soil microbiota, 357t–358t
Farming systems
  integrated. *See* Integrated farming systems
  organic. *See* Organic farming
Farm machinery, 119
Farm nutrient management plans, 1813–1814
  compliance with standards, rules, and regulations, 1814
  consistency with, 1814
  manure inventory, 1814

manure spreading plan, 1814
  nutrient crediting, 1813–1814
  on-farm nutrient resources, assessment of, 1813
  soil test reports, 1813
Farm subsystem, 1561t
FDA. *See* Food and Drug Administration (FDA)
Fe. *See* Iron (Fe)
Fecal bacteria, 2635
Fecal coliform bacteria, 2753, 2865t
Fecal pollution, 2699
Feces by livestock, 2753
Federal Communications Commission (FCC), 2228
Federal Environmental Pesticide Control Act (FEPCA), 1940
Federal Food, Drug, and Cosmetic Act (FFDCA), 1118, 1940
Federal Insecticide, Fungicide, and Rodenticide Act (FIFRA), 1118, 1612, 1940, 1942, 2125
Federal Pesticide Act, 1612
Federal Plant Pest Act of 1957, 2823
Federal Water Pollution Control Act (FWPCA) (1948), 2808
Feedback effects, 1134
Feedlot biomass (FB), 717, 718t, 729
Feedstock(s), 794
  gasification of, 2483t
  gasifier, compositions and heating values of, 2484t
Feedstock composition, of composting materials, 523–524, 523t
Fe(III)EDTA, 2245
Feldspar, weathering of, 1625
Feline immunodeficiency virus, 2603
Feline leukemia virus, 2603
Feline panleukopenia virus (FePV)
  for biological control of cats, 2603–2604
  on Marion Island, 2603
Felis catus, 2596
FEMWATER
  for saltwater intrusion, 1330
Fen, 2825t
Fenbutatin-oxide, 3
Fenitrothion, 420
Fenoxycarb, 1464
Fenpropathrin, 3
Fensulfothion, 419
Fenthion (Queletox®), 413, 420
Fermentation, 2152t
Ferrets (*Mustela furo*), 2596
Ferrihydrite, 1689
Ferrocement tanks, 2264
Fertigation, 1809
Fertility, potassium
  assessing and managing, 2211
Fertilization
  excessive, 1510

Fertilization (cont'd.)
  mineral and organic nitrogen, 1770
  and soil IC, 464–465
Fertilizers
  agricultural runoff and, 83
  biofertilizers, 1514–1516
  in coastal waters, 495–496
  consumption and nitrous oxide production, 98, 98t
  environmental problems with, 1516t
  micronutrient, 1514
  mineral, 1512–1513
    nitrogen, 1513
    phosphorus, 1513–1514
    potassium, 1514
    role of, 1512
  sulfur, 2434
Ferula asafoetida, 384
FFDCA. *See* Federal Food, Drug, and Cosmetic Act (FFDCA)
FFS. *See* Farmer Field Schools (FFS)
Fiber-bed scrubber, 156
Fiberglass insulation, 1478
  health impacts of, 1479
Fick's first law of diffusion, 1818
Fick's Second Law, 1452
Field erosion plots, 964–965
FIFRA. *See* Federal Insecticide, Fungicide, and Rodenticide Act (FIFRA)
Fill factor (FF), of solar cells, 230
Filter-based samplers, 1024
Filter(s)
  biofilters, 2724
  biological aerated, 2724
  cartridge, 2723–2724
  disc, 2731
  granular, 2722–2723
  rapid gravity filter with coagulant aid, 2722
  sand
    slow, 2722
    pressure, 2722
  soil escape, 2731
  trickling, 2724–2725
  types of, 2723f
Filter strips, 1553
Filth fly abundance management, in dairies and poultry houses, 1922–1923
Filtration, 1282, 2721–2722
  eco-filtration, 2730
  membrane, 2725–2727, 2725f
    applicability of, 2727
    future developments of, 2726–2727
  ultrafiltration
    performance of, 2727
    system design, 2727
    in water treatment, 2727–2728
Fimbristylis miliacea (Joina), 426
Finite differences (FD), 1314, 1314f
Finite element method (FEM), 1314, 1314f
Finite volumes, 1314, 1314f

Fin-tube heat exchanger, 873
First elements, 1455
First-flush (or foul-flush) device, 2264
First law of thermodynamics, 2525–2528
  formulation, 2526
Fish catch, 1593t
Fish depletion, 1579–1580
Fishery management
  eutrophication and, 2102
Fission, 2235
  radionuclides, 2237
Fitch, Asa, 2009
Fixed bed combustor (FXBC), 726
Fixed-bed gasifiers, 2490–2491, 2491f
Fixed carbon (FC), 717
Flame temperature, 722
Flash pyrolysis, 2487
Flat plate collectors, 242
Flat-plate solar collector, 242–243
Flatwood wetland, 2825t, 2826f
Flavonoids
  chemical composition and structure of, 2082t
  general structure of, 2081f
Flaxseed, for bio-liquid fuels, 846
Flea beetles, 444
Floating fern (*Salvinia molesta*), 1527t
Float zone method, 231–232
Flocoumafen, 420
Flood control reservoir, 1577
Flooded soils
  N_2O emissions, 97
Floodgates, 209
Flooding, 1184
Flood irrigation, 2676
Floor insulation, 1482
  raised floor, 1482
  slab-on-grade, 1482
Flow-driven saltation (FDS), 987, 989
Flow-through reservoirs, 1588. *See also* Reservoir(s)
Fluazuron, 1461
Flucycloxuron, 1461
Flue gas cleaning, of sulfur dioxide, 158
Flue gas desulfurization (FGD), 2440
Flue gases, concentration in SO_2, 2440
Flue gas volume, 722
Flufenoxuron, 1461
Flufenzine, 1463
Fluid bed dry scrubber, 2442
Fluidized bed combustor, 726–727
Fluidized-bed extraction (FBE), 2197–2198
Fluidized-bed gasification, 731f
Fluidized bed gasifiers, 2491–2492, 2491f
Fluorescent pseudomonads, 382
Fluorine, 1887
Flushing, 1592t
Flushing manure, 1664
Flush systems, 1665, 1665t
Flux chambers

GHG measurement, 1202, 1203, 1204f
Fly ash, 426, 428, 1432
Foam insulation strategies, 1482
Foam products and chlorofluorocarbons (CFCs)
  health impacts of, 1479
FOB. *See* Functional observational battery (FOB)
Foliar pathogen management, 444
Fonofos, 419, 1318
Food
  tolerance limits
    for natural or unavoidable defects in, 1610t
    for pesticide residues in, 1610t
Food and Agricultural Organization (FAO), 28, 1615, 1661, 1930, 1932, 1986, 2005, 2034
Food and Drug Administration (FDA), 1120, 1401, 1609, 1611, 2051
  on insect fragments in processed foods, 1609
  on pesticide residues in processed foods, 1609
Food chain
  arsenic in, 1274, 1275f
Food, Conservation, and Energy Act (2008), U.S., 1961
Food contamination
  with pesticide residues. *See* Pesticide residues, food contamination with
Food industry sludge, 2267t
Food laws and regulations, 1609–1611.
  See also Food and Drug Administration (FDA)
  food manufacturers responsibility, 1610–1611
  potential consumer benefits, 1611
  tolerance limits
    for insect fragments, 1609–1610
    for pesticide residues, 1609
Food manufacturers
  responsibility of, 1610–1611
Food processing
  pesticide residue and, 1125
Food Quality Protection Act (FQPA), 284, 1118–1119, 1613, 1941, 2010, 2028, 2750
  consumer right-to-know, 1119
  endocrine disruption, 1118–1119
  overview, 1118
  potential impacts, 1119
  "risk cup," 1118
  tolerances, 1118
Food Quality Protection Act of 1996, 1416, 1609
Food residues, 1403
Food Safety and Inspection Service (FSIS), 1609
Food sources
  alternative, natural enemies and, 371

Foodstuff, aluminum in, 270, 270t
Food-to-microorganism ratio, 2649–2650
Food waste pyrolysis, 847
Food-web models, 1864
Forage crops, irrigation of, 1664, 1665t
Force, 809f
Forced circulation systems, 247, 248f
Forebay tank, 210
Forest ecosystems
    acidic deposition and, 23–24
    health, indication of, 607
Forest management
    benefits of herbicides to, 1383
Forest planning, 636
Forestry insecticides, 420
Forest Stewardship Council, 1657
Formula of Manning, 205
Fossil fuels, 226, 345, 772
    combustion, 1127–1136, 1195
    depletion of, 180–181
        data and predictions, 180f
        oil, delivery and detection of, 182f
        reserves/resources, regional distribution of, 180, 181f
    as energy source, 679
    environmental effects, 2461–2462
    environmental effects of, 677, 680
    production and consumption, 679, 681
Fouling in cofiring, 729–730, 729f
Fourier Transform Infrared (FTIR), 1205
Four-way exchange valve, 1350–1351
Foxes (*Dusicyon griseus*), 2596
FQPA. See Food Quality Protection Act (FQPA)
Fraction of water, 1547
Framboids, 38f, 42
Fraxinus pennsylvanica, 2830
Free-Air CO_2 Enrichment (FACE) experiment, 1190
Free-water surface (FWS) wetlands, 2863f, 2864f, 2865
Frequency-domain electromagnetic methods (FDEMs)
    for saltwater detection, 1327
Freshwater pollution, 2303
Freshwater/saline lakes, 1577–1578
Freundlich isotherms, 64–65, 2427. *See also* Isotherm(s)
Friction period, 1638
Friedman, Thomas, 1503
Frit fly, 1116
Fruits, pesticides in, 1972–1974, 1973f, 1973t
FSIS. See Food Safety and Inspection Service (FSIS)
FTIR. See Fourier transform infrared (FTIR)
Fuel cell, 1145
    advantages of, 1145
    intermediate- and high-temperature, 1138
        applications of, 1138

        high-temperature proton exchange membrane fuel cells, 1138–1139
        molten carbonate fuel cells, 1140
        operational characteristics and technological status of, 1143t
        phosphoric acid fuel cells, 1139
        solid oxide fuel cells, 1141–1143
    low-temperature, 1145–1146, 1152t
        alkaline fuel cells, 1146–1148
        direct methanol fuel cells, 1151–1152
        proton exchange membrane fuel cells, 1148–1151
    stack, 1145
    types of, 1145–1146
    use of, 1145
Fuel properties
    gaseous fuels, 722–723
    liquid fuels, 722
    solid fuels, 717–722, 720t, 722, 723
Fuels
    bio-liquid, 846–847
    classification of, 714
    in groundwater, 1281
    properties, 832t
    solid, 830–834
    sources, 831t
The Fukushima (Japan), 183
Fukushima nuclear disaster impact, on nuclear power cost, 1803–1804
Full cost prices, 2784, 2784f
Full economic cost, 2784
Full supply cost, 2784
Fully extractive system, for monitoring stack gases, 144, 144f
Fulvic acids, 1570, 1625
Fumazon, 288
Fumigants, 1318, 2016t
Functional Observational Battery (FOB), 2006
Funding public services, in low/no-growth economy, 621–622
Fungi, as mycoherbicides, 383
Fungi, biological control by, 2107–2110
    application, 2109
    mechanisms, 2107–2109
Fungicides, 2014, 2029, 2115t
    derived from nature, 2029–2030, 2030f
    effects on natural enemies, 370
    used in homes and gardens, 1401
Furan tebufenozide, 1466
Furnace sorbent injection, 2442
Furrow irrigation, 2676–2677
Furrow-irrigation erosion, 1552
Furrows, 50
Fusarium crookwellense, 1696
Fusarium culmorum, 1696
Fusarium graminearum, 1696
Fusarium head blight, 1696
Fusarium kernel rot, 1697

Fusarium oxysporum, 2108t, 2109
Fusarium oxysporum f. sp. *conglutinans*, 444
Fusarium proliferatum, fumonisins by, 1697
Fusarium solani f.sp. *phaseoli*, 1878
Fusarium verticillioides, fumonisins by, 1697
Fusegates, 209
FutureGen process, 731–732, 732f
Fuzzy mathematical programming (FMP), 107
FWS wetlands. See Free-water surface (FWS) wetlands

G

γ-Al_2O_3, 1712
Gallium arsenide (GaAs) solar cells, 232
Gamma (γ) radiation, 2235
Ganga–Meghna–Brahmaputra (GMB) Plain
    arsenic contamination in, 1262
GAP Analysis Program, 2272
Garden insecticide, 1318
Gas absorption, 159–161
Gas chromatograph (GC), 1473, 2200
    capillary, 1974
    gas sampling, 1203
Gas chromatographic/mass selective detection (GC/MSD) analysis, 2017, 2018, 2019f
Gas flux, 2618
Gas compression heat pump, 1349–1350, 1350f. *See also* Heat pumps
Gaseous pollutants, industrial, 1430
Gasification, 376–377, 457
    definition of, 376
    stages in, 376
    temperature requirement, 376
Gasification, biomass and wastes, 2488–2495
    advances in, 2491–2494
    gasifiers, 2488–2491
    plants for, 2494t
    reaction involved in, 2490
    syngas, composition and heating value of, 2493t
    syngas for downstream application, 2492t
    technologies for, 2491
    world scenario, 2495
Gasification of coal and bio-solids, 730–731, 731f
Gasifiers, 2488–2491
    biomass and coal gasification, parameters for, 2488, 2493t
    conventional type, comparison of, 2488–2492, 2492t
    selection for, 2494–2495

Gas–liquid chromatography (GLC), 1126
Gas-phase bioreactor models, 2613, 2619
Gas reburning, 733t
Gas reserves, 803
Gastrointestinal effects, of arsenic toxicity, 1272–1273
GC. See Gas chromatograph (GC)
GCPF. See Global Crop Protection Federation (GCPF)
General circulation models (GCMs), 1867
Generalized analytic cost estimation (GACE) approach, 843
Generator junction box
   in photovoltaic systems, 219
Generic groundwater model, 1312–1313, 1313f
Genetically engineered organisms, 1110
Genetically modified crops, benefits of, 1951, 1953
Genetically modified organisms (GMOs), 125, 126, 1106
Genetic effects
   of pesticide, 1399
Genetic engineering
   environmental problems associated, 412
   in pest control, benefits of, 407
Genetic transformation, 307
Genistein, 1405
Genotoxicity
   environmental, 2124–2125
   human, 2124–2125
   pesticide, 2125–2128
   tests, for air pollution, 1156–1157
Geochemical investigation
   of saltwater intrusion, 1327–1328
Geoexchange systems, 1363
Geogenic carbonate, 1445–1446
Geogenic contaminants, 2166
Geographic information system (GIS), 554, 2215
   in air quality monitoring, 143
   inventorying and monitoring, 2816, 2817
   in land use planning, 1163–1165
      GIS/model interface, 1164–1165, 1165f
      LRIS, development of, 1164
      overview, 1163
      reliability of results, 1164–1165
      site suitability analyses, 1164
      spatial databases development, 1163–1164
   for regional pest management, 1952f, 1953
   remote sensing and. See RS/GIS integration
   research and development, 2818–2819, 2818f, 2818t
   software, 2039
   types of, 2818t
   for watershed management, 2816–2819, 2818t
Geographic information systems (GIS)-based indicators, 356
Geological erosion, 951
Geological investigation
   of saltwater intrusion, 1321–1330
Geologic materials, 1560
Geologic sequestration, 458–459
Geophysical investigation
   of saltwater intrusion, 1326–1327
   advantages and disadvantages, 1326
Geotextiles, 970–971
Geothermal brine specific exergy utilization index, 1175
Geothermal electricity, 825
Geothermal energy, 772, 1167
   activities for adopting technology of, 1169
   direct use of, 1167
   environmental benefits of, 1169
   history of use of, 1168
   hydrogen production from, 1168
   source of, 1167
   for sustainable development, 1169–1170
Geothermal energy resources, 1168–1169
   classification of, 1170
      by energy, 1170–1171, 1170t, 1171t, 1172t
      by exergy, 1171, 1173
      Lindal diagram, 1173–1176, 1174f, 1176f
Geothermal energy systems, performance assessment procedure for, 1176–1178
   case study on, 1178–1179, 1178f
Geothermal fluid, 1169
Geothermal gradient, 1168
Geothermal Heat Pump Consortium (GHPC), 1366
Geothermal heat pump (GHP) systems, 1363
   and conventional HVAC system, 1365, 1365t
   in cool weather, 1364
   development of
      borehole, 1367–1368, 1368f
      building, 1366–1367
      pond/lake heat exchanger, 1368, 1368f
      thermal mass, 1367
   earth heat exchangers, 1365–1366
   efficiency of, 1364
   environmental benefit of, 1364–1365
   operation of, 1364
      cooling mode, 1364, 1365f
      heating mode, 1364, 1364f
   organizations related to, 1366
   piping system in, 1368–1369
   principle of, 1364
   terms related to, 1366
   in warm weather, 1364
Geothermal resources, 825
Geothermal system, 1169
GeoWEPP model, 2761
German cockroach (*Blattella germanica*), 1988
German Democratic Republic, 1939
German Federal Immission Control Act (2002), 1961
German Soil Protection Act (1999), 1961
γ–Fe_2O_3, 1712
Ghana, 2683
GHG. See Greenhouse gases (GHG)
GHG pollution, 1364
Ghyben–Herzberg relation, 1323, 1326f, 1328
   saltwater detection using, 1325–1326, 1326f
Giant reed (*Arundo donax*)
   biology and ecology, 1182–1183, 1183f
   effects on streams/water resources, 1183
   biodiversity and wildlife, 1186
   control methods, restoration, revegetation, 1186
   flooding, 1184, 1184f
   water use, 1184
   wildfire, 1184–1186, 1185f
   hand clearing methods, 1186
   mechanical clearing methods, 1186
Gibberella ear rot, 1696
Gibbs free energy or Gibbs function, 62, 2532
Gilbert, J.H., 1858
GIS. See Geographic Information System (GIS)
GIS/model interface
   in land use planning, 1163–1165, 1165f
GIS software. See Geographic information systems (GIS) software
Glaciation indirect effect, 1197t
GLASOD (The global assessment of soil degradation), 1011
GLC. See Gas–liquid chromatography (GLC)
Gliocladium spp., 2107, 2109
Global Agrochemical Market, 2140
Global Atmospheric Watch (GAW), 8
Global Change and Terrestrial Ecosystem/Soil Erosion Network (GCTE/SEN) exercise, 1026, 1027
   Soil Organic Matter Network (SOMNET) database, 1863, 1865–1866t
Global commerce
   pest management and, 2010
Global Crop Protection Federation (GCPF), 1409
Global earth observation system of systems (GEOSS), 604
Global emission, of ammonia, 87
Global Footprint Network, 2469, 2476–2477
   Ecological Footprint Standards, 2478

Globalization, 1218
  global environment, future of, 1222–1223
  global governance, 1221–1222
  and sustainable economic growth and environmental impacts, 1220–1221
  understanding, 1218–1220
Global-mean temperature, 1192, 1193
Global NPP map, 2376
Global ozone layer, depletion of, 1891–1893, 1892f
Global positioning systems (GPS), 2039, 2213–2214, 2272, 2818f
Global primary energy supplies, 834
Global Programme of Action for the Protection of the Marine Environment from Land-Based Activities (GPA), 1914
Global stability, problems affecting, 1227, 1228f
Global unrest
  effects of, 1228
  and peace, 1236
Global warming, 333, 772, 798, 804–805, 1132–1134, 1189–1191, 1197
  and acid sulfate soils, 50, 56
  carbon removal from air, 1189–1190
    man role in, 1189–1190
    nature role in, 1190
  cause of, 772
  CO_2 emission reductions by
    demand-side conservation and efficiency improvements, 1136
    shift to non-fossil energy sources, 1136
    supply-side efficiency measures, 1136
  effects of
    climate changes, 1134–1135
    sea level rise, 1134
  global average surface temperature, 1134
  health-related implications of, 805
  keeping carbon in soil, 1190–1191
  overview, 1189
  projected earth surface temperature increase, 1134
  threats for developing countries, 804–805
Global warming potential (GWP), 486, 1195, 1196t
Globodera spp., 1752
Glycine max, 1983
Glycine max L. Merr. (Soybeans), 425
Glyphosate, 1318t, 1387, 1399, 2142
GMOs. *See* Genetically modified organisms (GMOs)
GMP. *See* Good manufacturing practice (GMP)
Goddard Institute for Space Sciences (GISS), 1198f
Goethite, 1689
Golden-eyed fly, 1116
Gold mining, and groundwater contamination, 1291–1292

Gold nanoparticles
  bactericidal effect of, 1740
  pollutant removal, mechanism in, 1739–1740
Gompertz model, 2634
Goniozus emigratus, 1919
Goniozus legneri, 1919, 1920
Good manufacturing practice (GMP), 1610
Google Earth, 543
GOSSYM/COMAX model, 2221
Gossypium hirsutum L. (Cotton), 425, 1983
Goulden large-sample extraction, 2193
Government action plans, 1403
GPS. *See* Global positioning systems (GPS)
Graded terraces, 2764
Gradient terraces, 2538–2539, 2538–2539f
Gradual agents, 1456
Grain borer, 2423
Gram (*C. arietinum* L.), 425
Granular filters, 2722–2723. *See also* Filter(s)
Granular insecticides, 419
Granulated activated carbon adsorption systems, 2063
Granule deterioration, 2653
Granulosis viruses (GVs), 382
Graphical user interface (GUI), 1164
Grass filter strips, 2014
Grasshopper effect, 2017
Grass waterways, 969–970, 969f, 2763, 2764f
Grate furnaces, 375
Grätzel cells, 234–235. *See also* Dye-sensitized cells (DSCs)
Gravel sludge, 1431
Gray wolves (*Canis lupus*), 2596
Grazing
  birds, 419
  management, 2752, 2754
  and pathogen contamination, 2753
Great horned owl (*Bubo virginianus*), 420
Great Lakes
  degradation of, 1438
  pesticide pollution in, 2146
Greece, 2684
Green ash. *See Fraxinus pennsylvanica*
Green bridge–horizontal eco-filtration system, 2155–2156
Green bridge technology (GBT), 2154, 2731
Green building certification, 1647
  benefits of, 1648
Green Building Certification Institute (GBCI), 1651
Green Building Council (GBC), 1654
Green buildings, 1655
  benefits of, 1651–1652
  energy efficiency potential of, 1652
Green chemistry, 1253
  green products, to produce, 1255

  catalysts, 1255–1257
  nanomaterials, 1257
  solvents, 1257–1258
  manufacturing process, application, 1256f, 1258
  objective, 1254f
  principles of, 1254–1255
  sustainability, 1255
Green energy, 1227
  analysis, 1235–1236
  applications, 1234–1235
  based sustainability ratio, 1235
  benefits of, 1230
  case study, 1236–1241
  challenges, 1230
  defined, 1227
  and environmental consequences, 1227–1229
  environment and sustainability, 1229–1231
  essential factors for, 1233–1234
  exergetic aspects of, 1234
  resources, 1232–1233
  and sustainable development, 1227, 1231
  technologies, 1231–1232
    commercial potential, 1232–1233
    progress on, 1232
  utilization ratio, 1235
Green engineering, 1255
Greenhouse effect, 435, 436f, 681, 1132, 1133
  climate change due to, 435
  desertification and, 549–550
    biological changes, 549
    climatical changes, 549
    greening of the earth hypothesis, 549–550
    modern studies, 550–551, 550f–551
    from prehistory to industrial revolution, 550
    woody plant range expansions, 550–551
  enhanced, 435
  human activities and, 435
Greenhouse gas concentrations trends, 1135–1136
  CO_2 concentration, 1135
Greenhouse gas (GHG) emissions, 118, 1132
  from building sector, 1353
  reduction of, 378
    by heat pumps, 1353–1354, 1354t
    by sector, end use, activity, and gas types, 438f
Greenhouse gases (GHG), 435, 456, 483, 772, 798, 803, 1192, 1193, 1364
  and climate change, 1894–1895
  measuring fluxes, 1202–1205
    chamber techniques, 1202, 1203t
    closed and open chambers, 1202, 1204f

Greenhouse gases (GHG) (cont'd.)
  gas chromatography, 1203
  manual/automated sampling and analysis, 1202–1203
  micrometeorological methods, 1203–1204, 1203t, 1205f
  nonisotopic tracer methods, 1204
  photo-acoustic-infrared detector, 1203
  ultra-large chambers with long-path infrared spectrometers, 1204–1205
  offset programs, 1773–1774
Greenhouse gas forcing, 1195, 1196t
"Greening" of the Earth, 549–550
Green manure
  and organic farming, 1107
Green manuring, 466
Green manures
  of leguminous crops, 1515
Green marketing incentive
  integrated farming and, 1507
Greenockite, 446
Greenpeace, 1612
Green power, 1657
Green products, 1255
  catalysts, 1255–1257
  nanomaterials, 1257
  solvents, 1257–1258
"Green revolution" technologies, 126
Green technology, 1255
Green Water Credits, 1038
Greigite, 42
Grey partridge (*Perdix perdix*), 416
Grid-connected photovoltaic systems, 218, 219, 219f, 220
  configuration, 223
Griffith University Erosion System Template (GUEST), 966, 1036
Grimm particle sampler, 1024
Gross domestic product (GDP), 484, 614, 615, 618, 2468, 2475
Gross head, measurements of, 206, 207f
Ground catchment techniques, 2263, 2263f
Ground coupled heat pumps, 1363, 2519
Ground source heat pumps, 1363
Groundwater, 1302, 1714
  contaminants in, 1302
  contamination. See also Groundwater contamination
    by arsenic, 1292
    cyanide and, 1291–1292
    by mining activities. See Mining
    from nitrogen fertilizers, 1302–1307
    non-point sources of, management of, 1319
    point sources of, management of, 1319
    uranium and, 1292
  impact analysis, 1571
  movement, 1290
  nitrate-nitrogen in, 1302. See also

Nitrate-nitrogen (NO_3-N), in groundwater
  outflow, 1284
  pesticide contamination in. See Pesticides in groundwater
  pollution, 1570
  remediation by nZVI, 1737–1739
  as renewable resource, 1284
  resources, 1289–1290
  saltwater intrusion in, 1321–1330
Groundwater contamination, 1281
  by arsenic, 1281
  biological contaminants, 1281
  chemical contaminants
    inorganic, 1281
    organic, 1281
  climate, effect of, 1282
  fate of contaminants, 1282–1283
  plume of, 1282
  sources of, 1281–1282
    chloride, 1282
    degreasers, 1282
    fuels, 1282
    human and animal wastes, 1281
    nitrate, 1282
    non-point sources, 1281
    pesticides, 1282
    point sources, 1281
    toxic salts, 1282
    wood-treating chemicals, 1282
  transport of contaminants, 1282
  treatment, biological techniques for, 2014
Groundwater drained lakes, 1577
Groundwater irrigation
  and soil IC, 465
Groundwater Loading Effects of Agricultural Management Systems (GLEAMS) model, 2761
Groundwater mining
  defined, 1284
  Edwards aquifer, 1284, 1285, 1285f, 1287f
  mass curve, 1287, 1287f
  pumping, 1284, 1285, 1285f
  and sustainable aquifer use, 1285–1287, 1287f
  and water balance, 1284–1285, 1285f, 1286f
Groundwater modeling, 1295–1301
  generic models, 1296
  overview, 1295
  phenomena, 1295–1296, 1296f
  process, 1297–1301, 1298f, 1299f
    calibration, 1298f, 1299
    conceptualization, 1297–1299, 1298f
    discretization, 1298f, 1299
    error analysis, 1299
    model selection, 1299–1300
    predictions and uncertainty, 1300–1301
  site-specific models, 1296–1297
  usage of, 1296–1297, 1297f

Groundwater problems, numerical methods for
  boundary element method (BEM), 1314, 1314f
  finite differences (FD), 1314, 1314f
  finite element method (FEM), 1314, 1314f
  generic numerical method for solving groundwater flow, 1312–1313, 1313f
  integrated finite differences (IFD), 1314
  simulating solute transport, 1314–1315, 1315f
Groundwater remediation technologies, 1334–1336
  bioremediation, 1335
  passive PRB system configurations, 1336f
  pump-and-treat technology, 1335
  SVE and AS, 1334–1335
Groundwater Ubiquity Score (GUS), 2765
Group on Earth Observations Biodiversity Observation Network (GEO BON), 603–604
Grout, 1367
GUEST model, 992
GUI. See Graphical user interface (GUI)
Guidance on Pest and Pesticide Policy Development, 1986
Gulf Coast Ecosystem Restoration Task Force (GCERTF), 629–630
Gulf Coast Restoration Task Force, 629
Gulf War Veterans
  pesticide sensitivities and, 1412
Gullies
  erosion and, 1046, 1046f
Gully erosion, 958, 964
Gully stabilization structures, 970, 970–971f
Gutters for rainwater storage, 2264
Gypsum, 466, 959–960, 2370, 2372t
  for management of soil dispersion, 2342

H

Habrobracon hebetor, 2424, 2424t
Haematobia irritans, 1460
Halocarbons, 1195
Halofenozide, 1466
Halogenated compounds (volatile), 1318
Halogenated organic compounds, 1841–1848
  analytical methods of, 1847
  applications of, 1842
  degradation pathways of, 1843–1844, 1843f
  in environmental compartments, 1842–1844, 1845–1846t
  occurrence of, 1844

Index: Halons—*Heterodera spp.*

metabolic pathways of, 1841, 1842f
remediation strategies for, 1844, 1846–1847
sources of, 1842
Halons, 1886
Hand clearing methods for *A. donax*, 1186
Hardtack Quince test, 187
Hardwood swamp, 2826f
Harris, T.W., 2008
Harrowing, 1111
Harvest of perennial energy crop, delaying, 1249
Hastened oxidation, 49
Hatch Act, 2009
Hazardous air pollutants (HAP), 1129
Hazardous waste, 893, 902
Hazard prevention, definition of, 1961
HCOP (heating performance of heat pump), 1347
Headspace (HS) analysis, 1847
Health
 definition of, 2871
 effects, of polychlorinated biphenyls, 2179, 2181
 guidelines for irrigation, 1571
 indicators in ecosystem
  agroecosystem health, 606–607
  aquatic ecosystem health, 607
  forest ecosystem health, 607
  urban ecosystem health, 607–608
Health advisory levels (HALs), 1317
Heat capacity at constant pressure, 2532
Heat, 809–810, 811
Heat engine, 819, 1346, 1347f
 exergy and the second-law efficiency, 820
Heating, ventilation, and air conditioning (HVAC) systems, 437, 439–440, 666
Heating energy efficiency ratio (HEER), heat pump, 1351
Heating value (HV), 717, 2527–2528
 of biomass fuels, 720t–721t
 of common fuels, 2528t
 higher, 2527
 lower, 2527
Heat of fusion, 2509
Heat pipes, 873
Heat pumps, 1346
 absorption, 1349, 1349f
 advantages of, 1348
 applications of
  commercial and industrial, 1353
  domestic, 1353
 energy flow of, 1346–1347, 1347f
 environmental benefits of, 1348
 fundamentals of, 1346–1347
 gas compression, 1349–1350, 1350f
 GHG saving potential of, 1353–1354, 1354t
 heating performance of, 1347
 heat sources for, 1347, 1347t

heat supply and value of, 1348
history of, 1346
magnetic, 1348, 1348f
performance parameters of, 1351–1353
 ambient energy fraction, 1352–1353, 1352f, 1353t
 COP and EER, 1351
 primary energy ratio, 1351–1352, 1352f
and refrigerators, 1347
short-term/long-term potential use of, 1353–1354, 1354t
thermoelectric, 1348–1349, 1348f
types of, 1348–1351
for underground thermal storage, 2519
vapor compression, 1350–1351, 1350f, 1351f
Heat rates, 2527–2528
Heat recovery, 2517
Heat recovery steam generator (HRSG), 873
Heat source/heat sink configurations, 1357–1358, 1357f
Heat storage, 2510–2511
 encapsulation, 2510–2511
 system, 252–254
  dimensioning, 253–254
Heat transfer, resistance to, 860
Heat wheels, 873
Heavy metal balance of agroecosystems and organic fertilizers. *See* Organic fertilization on heavy metal uptake
Heavy metal concentrations in organic fertilizers, 1375t
Heavy metals, 1370–1373, 1456–1457, 1560. *See also* specific entries
 as air pollutants, 161–163
 biological effects, 1372–1373, 1372f, 1373t
 in coastal waters, 490–491
 contamination with, 1374
 general chemistry, 1370, 1371t
 overview, 1370
 plant uptake of, 1375–1376
 in rocks and soils, 1370–1372, 1372f
 sources, 1372t
Height-above-average-terrain (HAAT), 2229
Helicobacter pylori, 2696, 2698
Helicoverpa (Heliothis) zea, 382
Helis aspersa, 383
Helix. *See* Chlorfluazuron
Helminth egg removal, 2636
Helminthioses, 2696, 2699
Helminths, in waste and polluted water, 2698, 2699
Helminthosporium solani, 394
Helophytes, 2829
Helsinki Convention, 1828
Henry's Law, 328
Heptachlor, 1318t, 2015, 2015t

Heptachlor epoxide, 1318t
Herbicide-resistant crops (HRCs), 407, 409, 410t, 1379–1380
 economic impacts of, 411
 toxicity of herbicides and, 409, 411
 use of, 1953
Herbicides, 1186, 1318–1319, 1378–1381, 1382–1390, 2014, 2142, 2761t
 background, 1397
 benefits to agriculture and forest management, 1383
 chemical structure, 1378
 classification, 1378
 components of action, 1378–1379
  absorption, 1378–1379
  interaction at the target site, 1379
  metabolic degradation, 1379
  translocation, 1379
 derived from nature, 2030–2031
  allelopathy, 2031
  bialaphos, 2030–2031, 2031f
  cyperine, 2031, 2031f
  monoterpene cineoles, 2031, 2031f
  phosphinothricin, 2030–2031, 2031f
  triketones, 2031, 2031f
 discovery, 1378
 effects
  at ecosystem/trophic level, 1388
  factors interacting with, 1388–1389
  on natural enemies, 370
  at population/community level, 1387–1388
  at species level, 1386–1387
 exposure to primary producers, 1383–1385
 history of, 1382
 glyphosate, 1399
 mechanisms of action, 1382–1383
 method/timing of application, 1378
 mode of action, 1378, 1379, 1380t
 overview, 1378
 paraquat, 1399, 1627–1628
 phenoxy acids, 1399
 phytotoxicity of, 1389–1390, 1389f
 resistance in weeds, 1379
 resistant crops, 1379–1380
 safety and environmental fate of, 1380–1381
 selectivity, 1379
 in surface waters, 2805
 toxicity of
  to animals, 1385–1386
  HRCs and, 409, 411
  to humans, 1385–1386
  to plants, 1386
 types of, 1383
 used in homes and gardens, 1401
Herbivorous insects, health effects of, 1122
Heterobasidium annosum, 2109
Heterodera spp., 1752

Heterogeneous TiO$_2$ photocatalysis, 2556–2565
  applications of, 2561, 2564
  mechanism of, 2557–2558
  operational parameters of, 2558, 2561, 2562–2563t
    oxygen concentration, 2561
    pH value, 2561
    temperature, 2561
    TiO$_2$ loading, 2561
  optimum conditions and rate of, 2559–2560t
  for water treatment, research trends in, 2564–2565
Heterotrophic bacteria, 2633
Hexachlorocyclohexane (HCH), 390
Hexaflumuron, 1461
Hexythiazox, 3
Hg. See Mercury (Hg)
"Hidden hunger," 536
High-copper soils
  plant growth on, 536–537
High-density polyethylene (HDPE) pipe, 1367
Higher heating value (HHV), 374
  of fuel, 717, 720t–721t, 831–834
High-intensity rainfall, 952
High performance liquid chromatography (HPLC), 1126, 1473, 1974
High-pressure nozzles, 1472
High-pressure reactor, 1905
High resolution visible and middle infrared (HRVIR), 2817t
High-rise buildings, 860
High-rise building's dewatering system, 1365–1366
High-temperature proton exchange membrane fuel cells (HT-PEMFCs), 1138–1139
  operating temperature range of, 1138
  for stationary applications, 1138
  technical challenges related to, 1138
High-temperature Winkler (HTW), 2491
"High throughput," 2028
Hilgardia, 1521
Hirsutella thompsonii, 382
H isotopes
  for saltwater detection, 1327–1328
Histosols, 1854
Hodgkin's lymphoma, 1399
Holistic ecosystem heath indicator (HEHI), 605, 606t
Home composting, 516. See also Composting
Homes and gardens
  pests and pesticide safety in, 1401–1403
Homogeneous photo-Fenton reaction, 2548–2556
  applications in water treatment, 2553–2556, 2554–2556t
  classification of, 2548f
  contaminant concentrations and characteristics, 2552
  iron concentration, impact of, 2549, 2552
  mechanism of, 2548–2549
  optimum conditions and rate of, 2550–2551t
  oxident concentration, impact of, 2552
Homo sapiens, 1717
Hordeum vulgare L. (Barley), 425
Horizontal flow systems, for constructed wetlands, 2841–2843
Horizontal mass flux (HMF), 1021–1022
  samplers, 1022–1024, 1023f, 1024f
Horizontal reactors, 518, 518f
Hormonal disruption, in humans, 1405–1406
  mechanisms, 1405–1406
  in men, 1406
  pesticides test management, 1406
  in women, 1406
Hormones, effect on endocrine-disrupting chemicals, 645–646
Host plant effects
  natural enemies and, 371
Host plant resistance, 444
Host-specificity tests and
  weed biocontrol and, 2822
Hot water tanks, 2520
Household vehicle fuel economy, 792f
House mice (*Mus musculus*), 2596
HPLC. See High performance liquid chromatography (HPLC)
HPPD. See Hydroxyphenylpyruvate dioxygenase (HPPD)
HRCs. See Herbicide-resistant crops (HRCs)
HRSG. See Heat recovery steam generator
HRT. See Hydraulic residency time (HRT)
HTW. See High-temperature Winkler (HTW)
H$_2$S, 42, 45
Hubbert peak theory, 715
HUD. See U.S. Department of Housing and Urban Development (HUD)
Hughes Plant 44, 1333
Human(s)
  activities impact on carbon storage, 2835
  Al toxicity in, 271–272
  biomonitoring, 2129
  and ecosystem, relation between, 635, 636
  exposure effects, of endocrine-disrupting chemicals, 648–650
  genotoxicity, 2124–2125
  pesticide effect on, 2147–2148
  role in carbon removal from air, 1189–1190
  toxocity of herbicides to, 1385–1386
Human development index (HDI), 798, 2468, 2473, 2474f
Human-induced global warming, 1192
Human pesticide poisonings, 1408–1410
  global incidence, 1408–1409
    African countries, 1408
    comparison, 1409
    Costa Rica, 1408
    developing and developed countries, 1409
    regional reports, 1408–1409
    Sri Lanka, 1408
    Taiwan, 1408
  international support and, 1409–1410
  overview, 1408
Humic acid, 269
Humid region irrigation
  and soil IC, 465
Humification, 1625
Hybrid sorbent injection, 2442
Hydraulic conductivity, 46
  water retention and, relationship between, 1621
Hydraulic energy, 772
Hydraulic functioning, of Alexandria Lake Maryut, 166–167, 169–170
Hydraulic gradient, 46
Hydraulic residency time (HRT), 2671
Hydraulic short circuit, 1424. See also Pumped storage hydro
Hydrilla verticillata, 1922
Hydrocarbons, 716, 717, 836
  characteristics of, 136–137
  chlorinated, 1317
  and climate change, 1895–1896, 1896t
Hydroelectric dams, 916
Hydroelectric resources, 825
Hydrochlorofluorocarbons (HCFCs), 1894
Hydroelectric power generation, 1421–1422
Hydrofluorocarbons (HFC), 486
Hydrogen, 723
Hydrogeomorphic approach (HGM), 2827
Hydrograph modification. See Irrigation and river flows
Hydrolases, 89
Hydrological conditions
  leaching and, 1621, 1622f
Hydrological functioning, of Alexandria Lake Maryut, 166–167
Hydrology. See Timber harvesting
Hydrolysis, 2651
  constant, 431
Hydropower, 201
  classification, 202–203
  gross head, measurements of, 206
  instream flow and environmental impact, 206–208
  mini hydropower, 203
  civil works in MHP plants, 208–212
  water flow, measure of, 204–206
  water resource, 203–204
  water turbines, 212
Hydroprene, 1464

Hydropyrolysis, 2487
20-hydroxyecdysone, 1465, 1465f
Hydroxy herbicides, 1318
Hydroxyphenylpyruvate dioxygenase (HPPD), 2031
Hyperecdysonism, 1465
Hypericum perforatum. See St. John's wort
Hypolimnion water removal, 1592t
Hypoxia, 1303, 1559

I

IAEA. *See* International Atomic Energy Agency (IAEA)
IARC. *See* International Agency for Research on Cancer (IARC)
Ice melting, 1199
Ice nuclei, 1196, 1197t
Ice storage, 2509–2510, 2509f, 2510f
ICM. *See* Integrated crop management (ICM)
ICRC. *See* Interim Chemical Review Committee (ICRC)
Ideal gas law, 2525, 2526
Ideal gas state, 2526
Ideal-solution approximation, 2533
IDGCC. *See* Integrated drying and gasification combined cycle (IDGCC)
IDW. *See* Inverse distance weighting (IDW)
IEA Heat Pump Centre, 1348, 1353–1354
IFOAM. *See* International Federation of Organic Agriculture Movements (IFOAM)
Ignition and combustion, 725
IGRs. *See* Insect growth regulators (IGRs)
Illinois State University (ISU), 1494
Illuviation. *See* Eluviation/illuviation
ILO. *See* International Labour Organisation (ILO)
Imazamox, 2020
Imidazolinones, 409
Immobile nutrient, 100
Immobilization (of heavy metals), 1375
Immobilized biomass reactor (IBR), 2553
Immobilizing agents
 for bird control, 414
INC. *See* Intergovernmental Negotiating Committee (INC)
Incident monitoring, 420–421
Increasing block tariff (IBT) pricing strategies, 2786–2788
Incubation methods, 41
Independent action (IA), 2573t, 2574
 applications of, 2575–2576
 characteristics of, 2575
 predictive power of, 2576
Index of sustainable economic welfare (ISEW), 2468, 2475

India, 2683
 acid rain in, 9, 10f
 trends in acidity, 13
Indian Council of Medical Research, 2142
Indian ecotechnological pollution treatment systems, 5
Indicator in ecology, definition of, 600
Indicator taxa, 602–603
Indicators
 desertification, 553–554
Indigenous technical knowledge (ITK), 2385
Indirect aerosol effect, 1197t
Indirect consumption of soil contamination via root exudates, 310
Indirect radiative forcings, 1194
Indium tin oxide (ITO), 235
Indoor air quality (IAQ), 1655
Indoor environmental quality, 1657
Indoor residual spraying (IRS), 1915
Indoor water storage system, 2263f
Induced erosion, 1010–1012
Industrial air pollution, 159
 absorption, 159–160
 adsorption, 160–161
 combustion, 161
Industrial carbon-di-oxide capture, 458
Industrial desertification, 2377–2378
Industrial ecology, 1253
Industrialization and pollution, 2170
Industrial network
 example of, 1434
 benefits of, 1435
Industrial pollution, 2304
Industrial Revolution, 1189
 from prehistory to, 550
Industrial use of energy, 794–795, 794f
Industrial waste, 2137
 soil pollution by, 1430–1432
 quantities, 1431t
 reuse possibilities, 1432f
 soil amelioration, 1431
 soil remediation, 1431–1432
 types, 1430–1431
Industrial wastewaters, 1570
Industrial wind turbines, 2867–2868. *See also* Wind turbine noise
Inexact double-sided fuzzy chance-constrained programming (IDFCCP) model, 113–116
Inexact optimization models, 106
 case study, 108–116
 fuzzy mathematical programming, 107
 interval linear programming, 107–108
 stochastic mathematical programming, 106–107
Inexact stochastic water management model, 111–112
Infestations of *A. donax,* 1182
Infiltrated water, 1559

Infiltration, 2314, 2752
 by soil, vegetation and, 995
Information technology, 2213. *See also* Precision agriculture techniques
Infrared absorption spectrometers, 1204–1205
Ingot, silicon, 231
Injection equipment
 chemigation, 470–471, 470f, 471f
Injection wells, 1293
Inhalation, of inorganic pollutants, 2120–2121
Inland marsh, 2826f
Inoculation/classical approach, of biological weed control, 1525, 1525f
Inorganic carbon, 462–467, 1451–1453
 arid and semiarid region irrigation, 465–466
 compartment model, 1451–1452
 composition, 1444, 1445f
 cropping and tillage, 464
 cycling, dissolved, 463
 fertilization and liming, 464–465
 formation, 1444–1446
 biogenic model, 1446
 pedogenic *vs.* geogenic carbonate, 1445–1446
 per ascensum model, 1446
 per descensum model, 1446
 in situ model, 1446
 humid region irrigation, 465
 impact on atmospheric carbon dioxide, 466–467
 isotopes in calcic soils, 1452–1453
 land clearing, 463–464
 land use, 463
 pedogenic carbonate accumulation, simulations of, 1451–1452
 sodic soil reclamation, 466
 soil processes, 462–463
Inorganic chemicals, in groundwater, 1281
Inorganic contaminant interactions, 2166, 2169
Inorganic contaminants
 bioremediation of, 392–393
Inorganic mulches, 961–962
Inorganic nitrogen, 1559
Inorganic pollutants, 2117–2121, 2118t, 2119t
 classes and concentration ranges of, 2118
 of drinking water, 2791–2793, 2791t, 2792t
 pathways of exposure, 2120–2121
 potential health impacts of, 2118–2120
 remediation of, 2120
Inorganic soil amendment, 958–959
Insect control
 Bacillus thuringiensis for, 411–412, 411t
Insect fragments
 tolerance limit for, 1609–1610
 in food (examples), 1610t

Insect growth regulators (IGRs), 1459
  administration of, by feed-through method, 1460
  advantages of, 1466
  chitin synthesis inhibitors, 1460, 1461–1463
  classification of, 1460
  contact/oral application, 1460
  disadvantages of, 1466–1467
  ecdysone agonists, 1460, 1465–1466
  and integrated pest management, 1467
  juvenile hormone analogs, 1460, 1464
  naming of, 1460–1461
  overview, 1459–1460
  resistance to, 1467
  selectivity of, 1460
  for social insects, 1460
  use of, in insect control, 1459
Insecticide Act of 1910, 1940
Insecticide residue monitoring, 1472–1473
Insecticide-resistant malaria mosquitoes, combating of, 1990
Insecticides, 1317–1318, 1609, 1697, 2014
  derived from nature, 2028–2029
    azadirachtin, 2029, 2029f
    Bacillus thuringiensis, 2029
    juvenile-hormone mimics and pheromones, 2029
    milbemycins/avermectins, 2029, 2029f
    nereistoxin derivatives, 2029, 2029f
    nicotine, 2029, 2029f
    pyrethrins, 2028–2029, 2029f
    spinosyns, 2029, 2029f
  effects of repeated exposure to OPs, 1399
  effects on natural enemies, 370
  intermediate syndromes, 1398
  mechanisms of action, 1398
  OPIDIN, 1399
  organophosphorus compounds, 1398
  in surface waters, 2805
  used in homes and gardens, 1401
Insects, beneficial, 2423
Insegar. *See* Fenoxycarb
In situ acid minesoil remediation, 1693
In situ ground water biodegradation, 396t
In situ model, of carbonate formation, 1446
In situ sodic minesoil remediation, 1693
In situ soil reclamation
  at open cut mine, 1693
In situ techniques, 2170
Instantaneous field of view (IFOV), 2817t, 2818
Institutional commitment, 560
Instituto Nacional de Investigaciones Forestales Agricolas y Pecuaris (INIFAP), 1763, 1765
In-stream flow requirements, 1566
Insulated concrete forms (ICFs), 1483, 1484f
  walls, 864, 864f

Insulated material, 242
Insulation
  building envelope, 1479f
  ceilings and roofs, 1486–1490
  and environment, 1479
  floor. *See* Floor insulation
  foam, 1482
  guidelines, critical, 1482
  materials, 1478–1479
    comparison of, 1479t, 1480–1481t
  strategies, 1479–1482
  wall. *See* Wall insulation
Intangible values, 634
Integrated collector storage, 242, 243–244, 244f
"Integrated control," concept of, 1521
Integrated crop management (ICM), 1507, 1509
Integrated disease management (examples), 444
Integrated drying and gasification combined cycle (IDGCC), 2491
Integrated energy systems, case study from ISU, 1494, 1503–1504
  asset management, 1501–1502
  planning methodology, 1495, 1497–1498, 1500–1501, 1501f
  profits, 1503
  project assessment/analysis, 1495
    chilled water operating calculations, 1500t
    cogeneration operating calculations, 1498t
    ISU executive summary, 1496t
    ISU pro forma, 1497f
    sensitivity analysis, 1501t
    steam operating calculations, 1499t
  related history, 1494
  risk management, 1502–1503
  stakeholders in, 1494–1495
  Web-based plan, assumptions in, 1495, 1497–1498, 1500–1501
  Web-based tools, for communication and reports, 1495
Integrated environmental management, 164–173
  of soil water erosion, 971–972
Integrated farming systems, 1507–1509
  agronomic, 1508
  defined, 1507
  development, 1508, 1508t
  economic pressure, 1507
  environmental, 1507–1508
  green marketing incentive, 1507
  legislation, 1507
  overview, 1507
  principles, 1508–1509
    ecological infrastructure management, 1509
    implementation, 1509
    integrated crop management, 1509

    integrated nutrient management, 1508–1509
    minimum soil cultivation, 1509
    multifunctional crop rotation, 1508
Integrated finite differences (IFD), 1314
Integrated gasification combined cycle (IGCC) process, 457, 730, 731f
Integrated membrane system (IMS), 2729
Integrated nutrient management (INM), 1508–1509, 1510
  adoption of, at farm, 1516–1517
  aim of, 1511–1512
  approach of, 1510–1511
  components of, 1511–1516, 1511f. *See also* Biofertilizers; Fertilizer
    biofertilizers, 1514–1516
    mineral and synthetic fertilizers, 1512–1514
  concept of, 1510
  environmental concerns of, 1517–1518
    negative effects, 1517
    nutrient losses, 1517
    problems with fertilizer use, 1516t
    soil carbon sequestration, 1517–1518
    toxic accumulation, 1518
  fertility degradation and, 1518–1519
  nutrient cycling in soil–plant–air–water systems, 1512f
  steps in, 1511
  use of, need for, 1510
Integrated pest management, 1521
Integrated pest management (IPM), 409, 508, 1467, 1507, 1521–1523, 1524, 1878, 1930, 1931t, 1986, 2010–2011, 2033, 2143, 2148. *See also* Chemical pesticides (CPs)
  agricultural runoff and, 83
  BP and CP compatibility in, 1987–1988
    examples of, 1990
    increasing of, approach for, 1990–1991
    practical indications for promotion of, 1992t
    selectivity in, 1991
  building blocks, 1522
  computer-derived packages, 1522
  defined, 1521
  derivation of, 1521–1522
  development of, 1522–1523
  menu systems, 1522
  need for, 1522
  overview, 1521
  practices, 314
  protocols, 1522–1523
  strategies, 295
Integrated Pollution Prevention and Control (IPPC) Directive of 1996, 1255
Integrated solid waste management (ISWM), 2419
Integrated weed management, 1526t–1527t. *See also* Weed control

Integrity indicators and orientors, 604–605
Intelligent Decision Support System (IDSS), 2819
Intensity factor, 100
Interaction at the target site herbicides, 1379
Intercropping, 1937–1939
   functional diversity, 1937
   future perspectives, 1939
   monoculture and, 1937, 1938t
   in practice, 1939
   protection mechanisms, 1937–1938, 1938t
Intergovernmental Negotiating Committee (INC), 1616
Intergovernmental Panel on Climate Change (IPCC), 96, 98, 435, 484, 934, 1197
   Third Assessment Report (TAR), 1200f
Interill
   defined, 958
   erosion, 2756
Interim Chemical Review Committee (ICRC), 1616
Interior foam insulation, 1482, 1483f
Interior framed wall insulation, 1482, 1483f
Intermediate disturbance hypothesis (IDH), 639
Intermittent energy resources, 1419
Internal energy, 2525
Internal feeders, 2424t
Internally feeding pests, biocontrol of, 2423–2424
International Agency for Cancer Research (IARC), 1264, 1394, 1395t
International ASS conference, Australia, 55
International Atomic Energy Agency (IAEA), 183
International Biochar Initiative, 844
International Biosphere Reserve, 1082
International climate policy. See Climate policy (International)
International Code of Conduct on the Distribution and Use of Pesticides, 1615
1994 International Desertification Conference (Arizona), 553
International Energy Agency (IEA), 201, 215, 1348
   projections, 378
International environmental law, 1618–1619
International Erosion Control Association, 1035
International Federation of Organic Agriculture Movements (IFOAM), 125, 1103, 1108, 1109, 1116
International Ground Source Heat Pump Association (IGSHPA), 1366
International Group of National Associations of Agrochemical Manufacturers (GIFAP), 1409
International initiative for a sustainable built environment, 1655
International Institute for Land Reclamation and Improvement (ILRI), 55
International Labour Organisation (ILO), 1409, 2036
International Land Reclamation Institute (ILRI), 26
International Measurement and Verification Protocol (IPMVP), 1657
International Organization of Biological Control (IOBC), 1991
International Performance Measurement and Verification Protocol (IPMVP or MVP), 768
International Programme on Chemical Safety (IPCS), 1409
International Trade, 1613
Interval linear programming (ILP), 107–108
Interval parameter water quality management model (IPWM), 109–111
Interval-stochastic chanceconstrained programming (ISCCP) model, 111–113
Intra-organizational listings/scorecards, 768
Intraparticle diffusion model, 67–68
Intrepid. See Methoxyfenozide
Invasion biology. See Biological invasions
Inversion, 1129
Inverters
   in photovoltaic systems, 219, 220–221
   functions, 220
   without transformers, 220
   types, 221–222
In-vessel composting, 517. See also Composting
Ion exchange, 1282–1283
Ion exchange application, 2734, 2735f
   in water and wastewater treatment, 2736–2737
Ionic liquids, 1257
Ion implantation, formation of p–n junction by, 233
Iowa P index, 2764–2765, 2765t
IPCC. See Intergovernmental Panel on Climate Change (IPCC)
IPCS. See International Programme on Chemical Safety (IPCS)
IPM. See Integrated pest management (IPM)
Iran, 2682
IRF. See Irrigation return flows (IRF)
Iron (Fe), 351, 1370
Iron concentration, effect on homogeneous photo-Fenton reaction, 2549, 2552
Iron-oxidizing bacteria, 1690
Irreversibility, 816–817
Irrigation
   definition of, 120
   frequency, 2677
   importance of, 1551
   management, 1319
   method, 1547t, 1564
   methods, 120
      energy consumption for, 120, 120f
   overview, 1572
   sewage effluent for. See Sewage effluent for irrigation
   and soil IC, 465–466
   soil salinity and, 1572–1574
      causes, 1573
      deleterious effects, 1572–1573
      management strategies, 1574
   sprinkler, 1551
   surface, 1551, 1552
   wastewater
      management, 2676–2677, 2679t
      methods, parameters for evaluation of, 2678t. See also Wastewater irrigation
   water, 433
   water management, 1813
Irrigation and river flows
   basin irrigation efficiency, 1565–1566, 1565f
   depletion, 1563
   environmental concerns
      in-stream flow requirements, 1566
      salt loading pick-up, 1566
      water quality implications for agriculture, 1566
   hydrograph modification
      irrigation efficiencies, 1564–1565
      irrigation methods, 1564
      irrigation return flows, 1564
      reservoir storage, 1563–1564
   hydrologic studies, 1563
Irrigation efficiency, 1564–1565
   basin, 1565–1566, 1565f
   definition of, 1545
   performance efficiency
      application efficiency, 1546, 1547t
      seasonal irrigation efficiency, 1546–1547
      storage efficiency, 1546
      water conveyance efficiency, 1545–1546
   uniformity in irrigation
      Christiansen's uniformity coefficient, 1548
      emission uniformity, 1548–1549
      low-quarter distribution uniformity, 1548
   water transport components, 1545, 1546f
   water use efficiency (WUE), 1549

Irrigation equipment
  chemigation, 470
Irrigation erosion, 1551–1552
  sprinkler-irrigation systems and, 1553–1555
  surface irrigation and, 1552–1553
Irrigation furrows, 1552, 1552f
Irrigation pumping, energy savings on, 120
Irrigation return flows (IRF), 1564, 1565f
  components of, 1557–1558, 1558f, 1558t
  consequences of, 1557
  defined, 1557
  off-site water quality impact reduction, 1560–1561, 1561t
  water quality constituents in, 1558–1560
    nitrogen, 1559
    pesticide contamination in, 1560
    phosphorus, 1559–1560
    salts, 1559
    trace elements, 1560
Irrigation system
  drip. See Drip irrigation system
  subsurface drip irrigation (SDI), 1535–1538
Irrigation water use efficiency (IWUE), 1549
Irrigation with saline water. See Saline waters
Isaria fumosorosea, 382
Ishipron. See Chlorfluazuron
Island biogeography (IB) theory, 639
Islanding, in photovoltaic systems, 221
  detection
    active methods, 221
    passive methods, 221
Isokinetic sampling, 965, 965f
Isoproturon, 2142
ISO 7934:1989. Stationary source emissions, 2440
ISO 7935:1992. Stationary source emissions, 2440
ISO 10396:2007. Stationary source emissions, 2440
ISO 11632:1998. Stationary source emissions, 2440
Isotherm(s)
  adsorption, 63–64
  Brunauer–Emmett–Teller, 65
  Dubinin–Radushkevich, 65
  Freundlich, 64–65
  Langmuir, 64
  Redlich–Peterson, 65–66
  Temkin, 65
Isotopes
  in calcic soils, 1452–1453
Israel, 2682, 2684
  saltwater intrusion in, 1322
  wastewater reuse in, 1571
"Itai-itai" disease, 448, 453
Italy, 2690
  saltwater intrusion in, 1322
Ivermectins, 3
Ixodes scapularis (deer ticks), 2

J

Japan
  acid rain in, 9
  participatory biodiversity inventory in, 359–360, 360f
Jarosite, 40, 42, 43, 1689
Jharkhand, groundwater arsenic contamination in, 1268
Joina (*Fimbristylis miliacea*), 426
Joint Expert Committee on Food Additives and Contaminants (JECFA), 1696
Joint implementation (JI), 486
Jordan, 2682
Jornada Experimental Range in New Mexico, 550, 550f
J.R. Geigy Co., 2009
Jupiter. See Chlorfluazuron
Juvenile hormone (JH), 1463, 2029
  anti-JH agents, 1464–1465, 1465f
  in insects, 1463–1464, 1463f
Juvenile hormone analogs (JHAs), 1460, 1464
  dayoutong, 1464
  diofenolan, 1464
  fenoxycarb, 1464
  hydroprene, 1464
  kinoprene, 1464
  methoprene, 1464
  pyriproxyfen, 1464

K

Kairomones, 175
Kara–Bogaz–Gol Bay, degradation of, 1440–1441
Kasugamycin, 2030, 2030f
Kellogg–Rust–Westinghouse (KRW), 2491
Kelvin–Planck law, 818
Kelvin–Planck statement, 818
Kinetic energy
  rainfall, 983–984
Kentucky State University (KSU)/ Water Quality and Environmental Toxicology Research Program, 2022–2024
Keys to Soil Taxonomy, 28
"K-fabric," 1444
Killo. See Chromafenozid
Kilowatt (kW), 1364
KINEROS model, 992
Kinoprene, 1464
Klebsiella oxytoca, 394
Knack. See Pyriproxyfen
Koi Herpes virus. See Cyprinid Herpes virus 3 (CyHV-3)
Koinobiontic Hymenoptera, 1747
Kolleru Lake, pesticide pollution in, 2147
KRW. See Kellogg–Rust–Westinghouse (KRW)
Kullback's measure of information, 783
Kuwait, 2682
Kyoto Protocol, 183, 485–487, 1896
  target emissions, 486f
Kyoto targets, 486f

L

Labeling, pesticide, 1941
Labidura riparia, 1921
Lactic acid bacteria, 530–531, 531f
Lagoons, 2663. See also Wastewater treatment in arctic regions, wetlands usage
Lagrangian method, 1315
Lagrange multipliers. See Shadow prices
Lake(s), 1577. See also Lakes and reservoirs
  aeration, 1584
  degradation of, 1436–1438
  sampling, 1582
Lake Chad, degradation of, 1437
Lake Eyre Basin, 1020
Lake management options, 539
Lake Mead, degradation of, 1438
Lake Ohlin, degradation of, 1437
Lake recultivation, 1588–1598
  methods of, 1590–1597, 1592–1593t
  phases of, 1590–1591, 1591f
Lakes and reservoirs, 1576–1577
  classification of, 1577
    freshwater/saline lakes, 1577–1578
    trophic status, 1578
  conventions for protection, 1585
    The Protocol on Water and Health, 1585
    Ramsar Convention, 1585
  pollution, sources of, 1581
  pollution issues of, 1581t
  problems associated with, 1578
    acidification, 1579
    eutrophication, 1578–1579
    fish depletion, 1579–1580
    sedimentation, 1579
    stratification, 1580–1581
    toxic materials, 1579
  protective and restorative measures, 1583–1585
    eutrophication model framework, 1585
  restoration measures, 1584t
  water quality monitoring, 1581–1582
    biological integrity sampling, 1582
    citizen monitoring, 1582–1583
    lake sampling, 1582

tributary mass load sampling, 1582
tributary water quality sampling, 1582
Lake Victoria, degradation of, 1437
Land
   as finite resource, 1600
   and river runoffs, 2358
   surveys, 559
Landcare program, in Australia, 1037
Landcare Trust, 2385
Land clearing
   and soil IC, 463–464
Land cover map, 2376
Land evaluation and site assessment (LESA), 1164
Landfill Allowance Trading Scheme (LATS), 900
Landfill gas, 377
Landfill leachates, 2482
Land Grant University System, 2009
Land Information System (LANDIS), 2819
Land resource information systems (LRIS), 1163
   development of, 1164
Landscape(s)
   assessment procedure, 541–542, 542f
   dysfunctional, 541
   fragile, 542, 542f
   functional, 541
   functioning. conceptual framework of, 542f
   robust, 542, 542f
Landscape patterns, pest and, 2038–2039, 2039f
   larger scale, 2038
   small patches, 2038, 2039f
   temporal stability of, 2038–2039
Landscape planning, 637–638
Land surface catchments, 2263, 2263f
Land use, 916
   and soil IC, 463
Land Use Analysis System (LUCAS), 2819
Land-use change forcing, 1196–1197
Land use planning
   GIS in, 1163–1165
      GIS/model interface, 1164, 1165f
      LRIS, development of, 1164
      overview, 1163
      reliability of results, 1164–1165
      site suitability analyses, 1164
      spatial databases development, 1163–1164
Langmuir isotherm, 64, 2427. *See also* Isotherm(s)
Lantana camara, 367, 2821
Large-scale ecosystem restoration
   governance, 624
      collaborative environmental management, 625
      collaborative resource governance, 627–628
      ecosystem restoration task force model, 628
      Gulf Coast Ecosystem Restoration Task Force (GCERTF), 629–630
      South Florida Ecosystem Restoration Task Force (SFERTF), 629, 630
      opportunities and challenges, 626–627
Larvadex. *See* Cyromazine
Latent heat storage, 2509
Lateral dispersion, 1282
Lateritic soils, 427
Latin America, bioindicator development in, 356, 359
Latium perenne (Ryegrass), 1305
Lawes, J. B., 1858
Law on Promotion of Organic Farming (Japan), 528
Laws, on pesticides use. *See* Pesticide regulation
Leachate
   problems associated with biowaste, 512
Leaching, 1621–1623, 1625–1626, 1771, 2137, 2138, 2433
   of acidity and salt, 49
   climatic conditions and, 1622
   environmental implications of, 1626
   experiments (REE), 2268
   "facilitated transport" and, 1622
   factors influencing, 1621–1623
   management practices, 1622–1623
   overview, 1621
   pesticide properties and, 1622, 1622t
   pollution prevention strategies, 1622–1623
   preferential flow and, 1621
   soil and hydrological conditions, 1621, 1622f
Leaching and chemistry estimation (LEACHM) model, 1164
Leaching fraction (LF), 1547, 1559
LEACHM. *See* Leaching and chemistry estimation (LEACHM) model
Lead (Pb), 1370
   abatement methods for reduction of, pollution, 1633–1634
   and cadmium, 446
      chemical properties, 446
      detection, 447
      environmental hazards, 447–448
      human exposure to, 449t
      human health effects, 448–449
      occurrences, 446–447
      production and uses, 447
      regulations and control, 449
   dispersion and application, 1630
   in food, 1631t
   heavy metal pollution in River Rhine, 1631t
   and earnings, 1640f
   ecotoxicity and environmental problems of, 1631–1633
      biomagnifications in aquatic ecosystem, 1633f
      concentration in glacial ice, 1631f
      contaminated aquatic ecosystems, 1631
      ingestion in birds, 1631
      toxicological and ecotoxicological effects, 1632–1633
      uptake from food, 1631, 1632f
   integrated environmental management, 1634–1635
Lead chromate, 479t, 480
Leadership in Energy and Environmental Design (LEED), 667, 1363, 1647, 1648
   and existing buildings, 1648–1649
   Online, 1651
   rating systems, 1648
   Version 1.0, 1648
   Version 2.0, 1648, 1649
   Version 2.1, 1648
   Version 2.2, 1648
Leadership in Energy and Environmental Design for Existing Buildings: Operations and Maintenance (LEEDEB O&M), 1647, 1649, 1650, 1650t. *See also* Green buildings
   building evaluation categories, 1649
   certification levels, 1650, 1651t
   credits and points, 1650, 1650t
   Green Building Rating System, 1648
   implementation process, 1651
   issues addressed by, 1649
   minimum program requirements, 1649–1650
   and other LEED products, 1649
   overview of, 1649–1651
   prerequisites, 1650, 1650t
   registration process, 1650–1651
   US Green Building Council (USGBC), 1647–1648
Leadership in Energy and Environmental Design for New Construction (LEED-NC)
   assessment of, 1657–1659
   construction standards, 1655
   ecosystem disruption, 1654
   environmental pollution, 1654
   green buildings, 1655
   LEED-NC rating system, 1655–1656
   LEED prerequisites categories and criteria, 1656–1657
   rating systems for buildings, 1655
   sustainability, 1654
Lead regulations, 1636
   benefits assessed, 1638–1639
   cardiovascular risk reductions, 1642–1643

Lead regulations (cont'd.)
  challenges and oportunities, 1643–1644
  general framework, 1636–1638
  IQ-related benefits, 1639–1642
Lead toxicity of animals, 449
LEED. *See* Leadership in Energy and Environmental Design (LEED)
Legislation
  integrated farming, 1507
  pesticide control, 2035–2036
    elements, 2035
    implementation, 2035–2036
    international conventions, 2036–2037
    overview, 2035
    problems related to, 2036
  weed biocontrol, 2823
Legionella exposure risk, 255f
  prevention and control of, 254–255
Legume green manuring, 1515
Legumes, 1110, 1807
Lemna gibba (Duckweed), 426
Lemon (*Citrus limon*), 426
Lens culinaris Medic (Lentil), 425
Lentic oxygenation technology system (LOTS), 2154
Lentil (*Lens culinaris* Medic), 425
Leptinotarsa decemlineata, 1461, 1928, 2009
Leptospermone, 2031, 2031f
Leptospirillum ferrooxidans, 43, 1689
LESA. *See* Land evaluation and site assessment (LESA)
Lethal stressing agents
  for bird control, 414
Leucomelanosis, 1272
Levuana iridescens, 367
Lewis acid, 431
Liberty®, 2030
Lichens, biomonitoring using, 143
LIDAR. *See* Light detection and ranging (LIDAR)
Life
  nanoparticles and, 1716–1718, 1717t
Life cycle assessment (LCA), for solid waste management, 2399–2412
  analysis and interpretation, 2405–2406
  goal and scope of, 2404–2405, 2405t
  importance of, 2411–2412
  ranking of, 2406–2411
  study area and system description of, 2401–2403
  weighting factors, determination of, 2406
Life cycle assessment (LCA) study, 437
Life cycle costing (LCC) analysis, 688–689
Life cycle inventory (LCI), 2405
Ligand-to-metal charge transfer (LMCT) reaction, 2549
Light detection and ranging (LIDAR), 2215

differential absorption, 144
Doppler, 144
range finder, 144
Lighting efficiency, 775, 776, 776t
Light NAPL (LNAPL), 1334
Lightning, 1185
Light non-aqueous phase liquid (LNAPL), 1281, 1282
Light pollution. *See also* Pollution
  in coastal waters, 497–498
Lightweight concrete products, 1483, 1484
Lignin, 373
Lignocellulases, 89
Limburg soil erosion model (LISEM), 966, 2761
Lime, 959, 2370
  applications on soil, 2428f
  for management of soil dispersion, 2342
Liming
  of soil, 269
  and soil IC, 464–465
Limit of quantitation (LOQ), 1126
Lindane, 2–3, 1318t, 1957, 2015, 2015f
Linear aggression analysis
  for crop water use, 2220
Linear imaging self scanner system (LISS), 2817t, 2818
Linuron, 1318
Lipopolysaccharides (LPS), 539
Liquefied natural gas (LNG), 722
Liquefied petroleum gas (LPG), 723, 793
Liquid chromatography (LC), 2200–2201
Liquid film diffusion model, 68
Liquid fuels, 714
Liquid–liquid extraction (LLE), 1970t, 2192–2193
Liquid–liquid/mass spectrometric (LC/MS) analysis, 2018
Liquid manure injection system, 1674
  AerWay SSD system, 1674, 1674f
  drag-hose system, 1674f
Listeria, 1696
Lithium (Li), 1560
Litter biomass (LB), 718t
Litter deposition, 2752, 2753
Livestock
  impacts on soil, 2752
  impacts on vegetation, 2753
  impacts to streams/riparian areas, 2754
  from range and pasture lands, 2752
Livestock drinking water
  water harvesting for, 2738–2739
Livestock feed
  manure management, 1812–1813, 1812f
Living Planet Index (LPI), 603
Living systems
  in action for pollution treatment, 2152
  in treatment of pollution, 2151–2152
Lixophaga diatraege, 1930
Lixophaga sphenophori, 1928
LNAPL. *See* Light NAPL (LNAPL)

LOAEL. *See* Lowest-observed-adversed-effect level (LOAEL)
Local Environment Plans (LEPs), 58
Logic. *See* Fenoxycarb
Lolium spec., 1925
Longitudinal dispersion, 1282
Long-range transboundary air pollution (LRTAP), 2298
Long-wave radiation, 865
LOQ. *See* Limit of quantitation (LOQ)
Los Angeles smog, 1130f
Louisiana Coastal Area Ecosystem Restoration Program (LCA), 628
Love Canal, 1333
Low-cost/no-cost energy-saving projects
  ambient air temperature, reset on, 739–740
  comfort and safety, 735
  corporate payback analyses, 735
  cost-accounting system, 740
  dedicated AC Units to cool server closets, 739
  domestic hot water heater, installing, 737
  fan speeds, reducing, 739
  gross system overcapacity, reducing, 737–738
  lag/lead, reducing, 738–739
  longer-term consequences of, 735
  lugs on time clocks, replace, 736
  manual light switches, optimize strategy for, 740
  systems fighting, resolve, 738
  thermostats, relocating, 738
  tighten leaky outside air dampers, 739
  tighten schedules, 736
  update schedules, 736–737
Low-cost treatment systems, 2157
"Low-dose effects," 1405
Lower heating value (LHV), 374
Lowest-observed-adversed-effect level (LOAEL), 2006
Low/no-growth economy, 616–622
  business-as-usual scenario, 617–618, 621t
  funding public services in, 621–622
  high-level structure of, 617f
  policy directions for, 618–622, 620t
Low-quarter distribution uniformity, 1548
Low-rise multifamily buildings, 860
Low-temperature solar thermal technology, 242
Loxodonta africana, 2595
LRIS. *See* Land resource information systems (LRIS)
LRTAP. *See* long-range transboundary air pollution (LRTAP)
Lufenuron, 1462
Lungworms (*Rhabdias* spp.), 2596
Lyases, 89

Index: *Lycopersicon esculentum* (Tomato)—*Metaphycus helvolus*

Lycopersicon esculentum (Tomato), 425, 1983
Lygus, 1936
Lygus lineolaris, 368
Lymantria dispar, 2009
Lymantria monacha, 382
Lymnaea acuminata, 384
Lymnaea stagnalis, 383
Lynchets
  ancient, 2536
  contemporary, 2536–2537

M

Maas–Hoffman equation, 2676
Mach II. *See* Halofenozide
Mackinawite, 42
Macrofauna monitoring index, 603
Macroinvertebrate communities, exergy of, 1100
Macrophyte(s), 2863f, 2864f
Macrophyte reintroduction, 1592t
Macropores, 1318
Magnesium oxide process, 2443
Magnesium wet scrubber, 2441
Magnetic heat pump, 1348, 1348f. *See also* Heat pumps
Magnetic nanoparticles, 1715
Maize (*Zea mays* L.), 426
"Maize-ley" system, 1939
Malancha (*Alternanthera philoxeroides*), 426
Malathion, 2805, 2806
Management system for energy (MSE) 2000, 765
Mancozeb, 2142, 2806
Maneb, 1318, 2806
Manganese (Mn), 1370
Mange mites, 2
Mangrove swamp (Mangal), 2825t
Manipur, groundwater arsenic contamination in, 1268
Manual sampling
  of greenhouse gases (GHG), 1202–1203
Manufacturing Energy Consumption Survey (MECS), 794
Manure
  application, 1668
    and NO_3-N concentrations, 1304
    rates, 1810
    site considerations for, 1811
    timing, 1811
  livestock feed management, 1812–1813
  management, 1810–1813. *See also* Dairy, water use in
  poultry. *See* Poultry manure
  storage, 1811–1812
Manure inventory, 1814
Manure phosphorus concentration reduction, 1667

Marginal benefits, 918f
Marginal cost of abatement, 918f
Marine and estuarine ecosystems, 607
Marine debris, 496–497, 2360
Marine pollution, 2357, 2358
  acidification and acid rains, 2359
  atmospheric pollution, 2359
  deep sea mining, 2359
  eutrophication, 2360
  land and river runoffs, 2358
  noise pollution, 2361
  oil and ship pollution, 2358–2359
  plastic debris, 2360–2361
  radioactive waste, 2359–2360
Marine Strategy Directive, 2187
Market potential
  for biofertilizers, 349, 350t
MARPOL Convention, 1828
Mars, 1716
Marsh, 2825t, 2826f
Masonry walls
  adobe walls, 864
  concrete block and poured concrete, 863
  insulated concrete forms (ICF), 864, 864f
  precast concrete, 863–864
Mass and energy, 808
Mass curve, 1287, 1287f
Mass flow
  nutrient transfer by, 1819–1820, 1820f
Mass rearing entomophagous insects, 1746
Mass spectrometry (MS), 1975
Massachusetts Department of Environmental Protection (MA DEP), 2049, 2050, 2050t, 2052
Match. *See* Lufenuron
Mathematical modeling
  for saltwater intrusion, 1328–1329
Mathematical water quality models, 2749
Matric. *See* Chromafenozid
Matricaria perforate, 1928
Matricaria spp., 1928
Matrix solid-phase dispersion (MSPD), 2198
Maturation ponds, 2634–2635
  for fecal coliform removal, 2636
Maximum achievable control technologies (MACT), 1129
Maximum contaminant levels (MCL), 916, 1289, 1317, 1318t
  of contaminants from mining, 1290t
  for nitrate, 1303
Maximum permissible concentrations (MPCs), 1518
Maximum power point (MPP)
  in photovoltaic systems, 219, 220
Maximum power point tracking (MPPT), 220–221
Maximum residue limit (MRL), 1124, 1125–1126, 1125t, 1403

MCS. *See* Multiple chemical sensitivities (MCS)
Meals ready to eat (MRE) waste, 847
Mean annual rainfall (MAR), 944
Meandering, 2313
Mean residence time (MRT). *See also* Soil organic matter (SOM) turnover
  defined, 1872, 1873f
  of total soil organic C, 1874, 1874t
  effect of tillage practices, 1875t
Mechanical, electrical, and plumbing (MEP), 669
Mechanical clearing methods for *A. donax*, 1186
Mechanical scrapers, for manure removal, 1672, 1672f
Mechanical scrubbers, 156
Medicago sativa L. (Alfalfa), 427, 1552
Mediterranean-type climates, 1183
Melaleuca spp., 49
Meloidogyne spp., 1752
Member states (MSs), 903, 904
Membrane biofilm reactors (MBfRs), 2611, 2612f, 2622, 2624–2625
Membrane extraction, 2195–2196
Membrane filtration, 2725–2727, 2725f. *See also* Filtration
  applicability of, 2727
  future developments of, 2726–2727
Membrane inlet mass spectrometry (MIMS), 1847
Men, hormonal disruption in, 1406
Mental component score (MCS), 2875f
Menu systems, of IPM, 1522
MEP. *See* Mechanical, electrical, and plumbing
MEPS. *See* Molded expanded polystyrene
Mercuric chloride, 2016t
Mercury (Hg), 1370, 1676
  contamination, in coastal waters, 490, 491
  health impacts of, 2119
  important processes, 1676–1677
  pollution
    abatement of, 1677–1678
    effects, 1676
    sources of, 1676
Metabolic degradation
  herbicides, 1379
Metabolic engineering, 397
Metalaxyl, 2020
Metal contacts, in solar cells, 233–234
  rear contacts, 234
Metal framing, 1486
Metallothionein, 394
Metal retention processes, 1374
Metal surface treatment
  cadmium for, 453
Metalloids, in coastal waters, 490–491
Metamorphosis, 1459
Metaphycus helvolus, 1750

Metarhizium anisopliae, 382, 1988
Methane (CH₄), 486, 1203
  emissions from rice fields, mitigating options for, 1679–1682, 1681t
    organic matter management, 1680
    problems and feasibility of the options, 1681–1682
    processes controlling, 1679, 1680f, 1680t
    soil amendments and mineral fertilizers, 1680
    water management, 1679, 1680f
  problems associated with biowaste, 512
  soil quality and, 2388–2389
Methanogens, 1679, 2651
Methanol, 188
  from biomass, 378
Methanopyrolysis, 2487
Methemoglobinemia, 1302, 1303, 1559
Methiocarb (Mesurol®), 415
Method of characteristics (MOC), 1315–1316
Methoprene, 1460, 1464
Methoxychlor, 1318t, 1405, 2015, 2015f
Methoxyfenozide, 1466
Methyl anthranilate, 415
Methyl bromide, 1886–1887, 2015, 2015f, 2016t, 2114
Methyl esters, 2021
Methylparathion, 2005, 2015, 2015f, 2805
Methyl tert-butyl ether (MTBE), 916
Metschnikowia pulcherrima, 1991
Mexican bean beetle, 1935
Mexico Nitrogen Index, 1763
Mexico
  saltwater intrusion in, 1322
  wastewater reuse in, 1571
Micelles, 2194
Microarrays, 2130
Microbial biopesticides, 381
Microbial community, 2616
Microbial degradation
  of organic pollutants, 388–392
Microbial inoculants, 1515
Microbial isomerization reactions, 390
Microbial siderophore uptake, 351
Microbiological research methods, for drinking water analysis, 2799–2800
Microclimate for natural enemies, 371–372
Microflora in soil, 2169
Micro hydro plant, 202
Microirrigation, 1547t, 1548, 1549
Microlife DCB series bioremediation products, 397f
Micro–meso–macro method, 1906
Micrometeorological methods
  for GHG, 1203–1204, 1203t, 1205f
Micromite. *See* Diflubenzuron
Micronucleus assay, 1156–1157, 2127
Micronutrients, 1374
Microorganisms, 2170
  nanoparticles and, interactions of, 1716–1718, 1717t

Microrills, 953
Microscale solvent extraction (MSE), 2193–2194
Microwave-assisted extraction (MAE), 2197
Milardet, Pierre, 2009
Milbemycins/avermectins, 2029, 2029f
Milking equipment, washing of, 1664
"Milky disease," 321
Millennium Assessment, 608
Millennium Ecosystem Assessment (MA), 557
Miller Amendments, 1940
Mill tailings, 1683
Mimic. *See* Tebufenozide
Mined aquifer, 1285f
Mine drainage, 1688, 1689f
  causes of, 1688
  chemistry of, 1688–1689
  control of, 1690
    prevention, 1690
    treatment, 1690–1691
  environmental impacts of, 1690
  microbiology related to, 1689–1690
  mineralogy of, 1689
  problems by, 1688
  total acidity of, 1690
Mineral fertilizers
  CH₄ emissions from rice fields and, 1680, 1680t
Mineralization, 90, 460, 2176
  carbon losses from soil and, 1857
  influence of heavy metals on, 162
  sulfur, 2432
Mineral matter (MM), 717
Mineral weatherability, 1625
Mineral wool insulation, 1478
  health impacts of, 1479
Minimum efficiency reporting value (MERV), 1657
Mini hydro plant, 202
Mini hydropower (MHP), 203
  civil works in MHP plants, 208–212
Mining
  classification, 1692
  for coal, 1289, 1291
  groundwater pollution by, 1289
    abandoned mine sites and, 1291
    acid mine drainage and, 1291
    biological methods for remediation, 1290, 1293
    chemical methods for remediation, 1293
    containment of contaminants in, 1293
    gold mining and, 1291–1292
    groundwater analysis, interpretation of, 1292
    groundwater resources, 1289–1290
    hydrogeological characteristics of site and, 1290, 1290t
    metal contaminants, 1291

    organic contamination, 1291
    physical methods for remediation, 1293
    and pump-and-treat method, 1293
    and remediation strategies, 1292–1293, 1293f
    and in situ remediation, 1293
    tests for constituents, 1292, 1292t
    transport of contaminants, 1290
    water contaminants, 1290–1292, 1291t
    water movement, 1290
  impact on groundwater supplies, 1289
  surface, 1289
Missouri River ("the Big Muddy"), 1714
Mites, 1
  damage from mite feeding, 2
  general description of, 1
  population outbreaks, control of. *See* Acaricides
  spider, 1–2
  Varroa, 2
Mitigating desertization, 570–571
Mitigation, 2877
  ammonia volatilization and, 87
  regulating permissible noise level, 2877–2879
Mitsui Babcock ABGC (air-blown gasification cycle), 2491
Mixed liquor, 2648
Mixed-liquor suspended solids (MLSS), 2649
Mixed-liquor volatile suspended solids (MLVSS), 2649
Mixing height, 1130
MLSS. *See* Mixed-liquor suspended solids (MLSS)
MLVSS. *See* Mixed-liquor volatile suspended solids (MLVSS)
Mn. *See* Manganese (Mn)
Mo. *See* Molybdenum (Mo)
MOCDENSE
  for saltwater intrusion, 1329
Mode of governing, 2421
Moderate resolution imaging spectrometer (MODIS), 554, 2817t, 2818
Modified Fournier index, 943, 944
Modified universal soil loss equation (MUSLE), 975–976, 982, 2759
Modified Wilson and Cooke (MWAC) samplers, 1022, 1023f
MODIS. *See* Moderate resolution imaging spectrometer (MODIS)
Module cables, 222
Module inverters, 222, 222f
Modulus of rupture (MOR), 2341f
Moisture, carry-over, 1563
Moisture ash free (MAF) basis, 717
Moisture content
  biomass feedstocks, 374
  in composting materials, 521–522

Molded expanded polystyrene (MEPS), 1478
Molecular nitrogen
  emission of, 1770–1771
Molinate, 2806
Molten carbonate fuel cells (MCFCs), 1139–1141
  acceptable contamination levels, 1140
  applications of, 1141
  fuel for, 1140
  operating principle, 1140, 1140f
  operating temperature of, 1139
  problems related to, 1140–1141
    cathode dissolution, 1141
    corrosion of separator plate, 1141
    electrode creepage and sintering, 1141
    electrolyte loss, 1141
    reforming catalyst poisoning, 1141
  technological status of, 1141
Molting, 1459, 1465
Molybdenosis, 428
Molybdenum (Mo), 424, 1370, 2586, 2588–2589
  deficiencies, 424
  monitoring of, in soils and waste materials, 428
  in soils, 426–427
  sources of, 426
  toxicity, 424
    cattle grazing and, 428
    cause of, 427
    clinical signs of, 427
    Cu:Mo:S ratio and, 428
    irrigation and, 428
    prevention of, 428
    sources of, 427
    symptoms and levels, 427–428
  use of, 426
Monacha cantiana, 383
Mongooses (*Herpestes auropunctatus*), 2596
Monitoring wells
  for saltwater intrusion, 1325
Monocrotophos, 417, 1993
Monocrystalline silicon, 231–232
Monoculture, 1935
  defined, 1937
  intercropping and, 1937, 1938t
  resistance gene, 1937
Monod equation, 2634
Monolith bioreactor, 2625
Monosulfides, 42
Monosulfidic black ooze (MBO), 44, 45
Monoterpene cineoles, 2031, 2031f
Monothanolamine (MEA), 457
Monte Carlo uncertainty analyses, 1721, 1722
Montevideo Programme, 1619
Montmorillonite clay, swelling of, 2338, 2339f

Montreal Protocol, 1882, 1893–1894, 1897
Moon, 1716
Moore's law, 788
Moraxella osloensis, 383
Morocco, 2682
Morrell Act (1862), 2009
Mosquito control, 1471
  nontarget mortality during, 1471
  minimization of. *See* Aerial ultra low volume (ULV) application, of mosquito insecticides
  ULV application for, 1471
Motile algae, 2633
Mousepox virus (Ectromelia virus), 2604
MRL. *See* Maximum residue limit (MRL)
MRT. *See* Mean residence time (MRT)
MSW. *See* municipal solid waste (MSW)
MTBE (methyl tertiary butyl ether), 157
Mucus melanosis, 1272
Mud for roof catchment, 2263
Mueller, Paul, 2009
Mulches, 961–962, 1110
Mulch tillage, 1042, 1042f
Multi-circulating fluidized bed combustor (MCFBC), 729
Multicomponent mixtures, reduced mixture toxicity designs for, 2579
Multicriteria decision making (MCDM), 2399–2400
Multidisciplinary approach for developing environmental policy, 914–915
Multifunctional crop rotation, 1508
Multiple-attribute decision making (MADM), 2400, 2403–2404
Multiple chemical sensitivities (MCS), 1411. *See also* Pesticide sensitivities
  causes, 1411–1412
  symptoms, 1412, 1412f
Multiple-objective decision making (MODM), 2400
Multipurpose reservoir, 1577
Multiresidue methods, 1126
Municipal sewage, 1570
Municipal solid waste (MSW), 512, 715, 728, 895, 902, 902f, 2415, 2482
  characteristics and waste quantities, 2416–2418
  co-composting of, 514
  composition, 2483
  heat of combustions of, 729t
  historical overview, 2415–2416
  management practices, 2418
  modes of governing, 2420t
  organic fraction of, 513, 514
  organic matter in, 513, 519, 520
  valorization, 513
Municipal wastewater, 2709–2718. *See also* Wastewater
  composition of, 2709, 2709t
  recycling of, 2717–2718

  treatment methods for
    biological oxygen demand, reduction of, 2709–2712, 2710–2713f
    nitrogen concentration, reduction of, 2715–2717, 2716f, 2717f
    phosphorous concentration, reduction of, 2712–2715, 2713–2714f
Municipal WWTP systems, removal efficiencies, 2063–2066
Murine cytomegalovirus (MCMV), 2604
Murray–Darling basin (MDB), 56
Musca domestica, 1460, 1920, 1921
Musca sorbens, 1920, 1921
Muscidifurax, 1922
Muynak, Aral Sea disaster and, 301–302
Mycalesis gotama Moore, 311
Mycoherbicide, 383
Mycoremediation, 392
Mycorrhizae, 349, 1110, 1516
Mycotoxins, 1696
  aflatoxin, 1697
  deoxynivalenol, 1696
  fumonisin, 1697
  ochratoxin, 1697
  and pesticides, 1698
  workplace exposure to, 1698
  zearalenone, 1696
Mycotrol®, 382
Myriophyllum spicatum, 367, 1922
Myriophyllum spp., 2830
Myxomatosis, 2597. *See also* Myxoma virus (MyxV)
  in Australia, 2597–2600
  in Europe and other regions, 2600
Myxoma virus (MyxV), 2597
  as biological control for rabbits, 2597–2599, 2598f
  genetic resistance to, 2599, 2599f
  Glenfield strain of, 2600
  Lausanne strain of, 2600
  standard laboratory strain (SLS), introduction of, 2600
  vector of, introduction of, 2599–2600

N

Na/Cl ratios
  for saltwater detection, 1327
Nafion, 1148
Nanocrystalline films
  model of photocurrent generation in, 236, 237f
Nanofood regulation, 1703–1704
Nanomaterials, 1700
  dose–response relationship, 1705
  environmental pollution by, 2129
  exposure assessment, 1705
  nanofood regulation, 1703–1704
  pharmaceutical regulation, 1702–1703

Nanomaterials (cont'd.)
  REACH, 1700–1701
  risk assessment of, 1704
    technical guidance, 1706
  risk characterization, 1705–1706
  WFD, 1701–1702
Nanoparticles, 1711–1718, 1720
  aluminum oxyhydroxide particles, 1714, 1714f
  in atmosphere, 1715, 1715t
  beyond the Earth, 1716
  "critical zone," 1711
  defined, 1711
  environmental risks, 1723–1725
  health effects, 1717
  and life, 1716–1718, 1717t
  magnetic, 1715
  microorganisms and, interactions of, 1716–1718, 1717t
  overview, 1711
  physical chemistry of, 1711–1712, 1711f, 1712f, 1712t, 1713f
  in sediments, rocks, and the deep Earth, 1715–1716
  in soil and water, 1712–1715, 1713t, 1714f
  "X-ray amorphous," 1712, 1713f
Nano-semiconductor catalysts
  decontamination of toxic pollutants by, 1733–1736, 1734–1735t
  pollutant removal, mechanism in, 1731–1733
Nanotechnology, 1720–1721, 1730–1741
  significance of, 1730–1731
NAPAP. See U.S. EPA National Acid Precipitation Assessment Program (NAPAP)
Naphthalene, 388
NAPL. See nonaqueous phase liquids (NAPL)
Napropamide, 2024f
National Acid Precipitation Assessment Program (NAPAP), 7
National Ambient Air Quality Standards (NAAQS), 133, 1639
National Arsenic Mitigation Information Centre (NAMIC), 1269
National Atmospheric Deposition Program (NADP), 7
National Environmental Policy Plans (NEPP), 2446–2454
National Fenestration Rating Council (NFRC), 866, 867f
National Organic Standards Board (NOSB), 125
National pesticide assessment, 1317
National Pesticide Field Program, 1945
National Pollutant Discharge Elimination System (NPDES), 1289, 1943, 2808
  permit program, 2809–2810, 2810f
National Toxicology Program (NTP), 477
National waste regulation in EU countries, 894

economic instruments, 897–900
regulatory instruments, 897
National Water Quality Assessment (NAWQA), 963, 1317, 2804
Natural acidity, 7
Natural amelioration, 569
Natural capital, 2465
Natural circulation systems, 247, 247f
Natural climate forcings (solar and volcanic variability), 1194–1195, 1196t
Natural cycling of materials, 2863
Natural enemies. See also Biological control
  alternative food sources and, 371
  alternative prey/hosts, 371
  for augmentation, 1474
  conservation of, 1749–1751
    habitat and environmental manipulation, 1750
    harmful pesticides practices avoidance, 1749–1750
    overview, 1750
    overwintering and shelter sites, 1750
  cultural practices and, 370
  host plant effects and, 371
  pesticides effects on, 370
  secondary enemies and, 371
  shelter and microclimate for, 371–372
  strategies for using, 1474
  types of, 1475–1476
    parasitoids, 1475, 1475f
    pathogens, 1475–1476, 1476f
    predators, 1475, 1475f
Natural enemies, rearing of
  arthropod parasitoids/predators, 1746
  artificial diets
    for predators and parasitoids, 1746–1747
    successes and failures with, 1747
  quality control of, 1747
Natural energy, 770, 771. See also Energy
  non-solar-related energy, 772
  solar energy, 771–772
    air-based, 772
    land-based, 772
    types of, 771t
    water-based, 772
Natural-environment-subsystem models, for exergy analysis, 1094
Natural erosion, 1010
Natural gas, 793, 794
Natural gas combined cycle (NGCC), 840–843
Natural greenhouse effect, 1192–1193, 1193f
Natural mortality, 1115–1116
Natural occurrence
  allelochemics, 175–177, 176t, 177f
Natural pesticides, 2028–2031
  fungicides and bactericides, 2029–2030

kasugamycin, 2030, 2030f
polyoxins, 2030, 2030f
strobilurins, 2030, 2030f
validamycin, 2030, 2030f
herbicides, 2030–2031
  allelopathy, 2031
  bialaphos, 2030–2031, 2031f
  cyperine, 2031, 2031f
  monoterpene cineoles, 2031, 2031f
  phosphinothricin, 2030–2031, 2031f
  triketones, 2031, 2031f
historical use for pest management, 2028
insecticides, 2028–2029
  azadirachtin, 2029, 2029f
  Bacillus Thuringiensis, 2029
  juvenile-hormone mimics and pheromones, 2029
  milbemycins/avermectins, 2029, 2029f
  nereistoxin derivatives, 2029, 2029f
  nicotine, 2029, 2029f
  pyrethrins, 2028–2029, 2029f
  spinosyns, 2029, 2029f
overview, 2028
prospect, 2031
structural diversity, 2028
Natural products, for pest control, 2021–2022, 2021f
Natural Resources Conservation Services (NRCS), 127
Natural Resources Damage Assessment (NRDA) process, 629
Natural resources degradation, 2816
Natural riverbed protection, 2313
Natural sources of freshwater acidification, 2292–2294, 2293f
  geochemical factors, 2293–2294
  geological factors, 2292–2293
Natural wetlands, 2862, 2863t
Nature, role in carbon removal from air, 1190
Natuur and Milieu, 1404
Navel orangeworm management in almond orchards, 1919–1920
NAWQA. See National Water-Quality Assessment (NAWQA)
NDVI. See Normalized difference vegetation index (NDVI)
Necrotrophic/hemibiotrophic fungi, 383
Negative lynchet, 2537
Negligence, pesticide use and, 1944
Nemagon, 288
Nemaslug, 383–384
Nematicides, 2014
Nematodes, biological control of, 1752–1753
  future perspectives, 1753
  nonspecific biocontrol agents, 1753
  overview, 1752
  specific biocontrol agents, 1752–1753

Neporex. *See* Cyromazine
Neptunium-237 decay series, 2236, 2239f
Nereistoxin derivatives, 2029, 2029f
Netherlands, The
 saltwater intrusion in, 1322
Net primary productivity (NPP), 2376, 2833
Network of industries, 1434
Neural tube birth (NTD), by fumonisin, 1697
Neuroesterase (NTE), 1399
Neurological effects, of arsenic toxicity, 1273
Neurotoxicants evaluation, 1755–1758
 protocols for, 1757–1758
 vulnerability, periods of, 1756–1757
Neutralization, of acidity, 49
New Orleans Regional Medical Center (NORMC)
 district cooling, 2515–2516, 2516f
New Waste Framework (NWF) Directive, 892, 911
New York Nitrate Leaching Index, 1763
New Zealand
 soil degradation in, 2384–2385, 2384f
 distribution of, 2384f
Nezara viridula, 367–368
N_2-fixing cyanobacteria, 349
Ngarenanyuki, Tanzania
 pesticide impact in, community-based monitoring of, 505–511
 adherence to pesticide label instruction, 509–510, 509t
 community pesticide monitoring team, establishment of, 508
 data analysis, 508
 data collection, 508
 disposal of pesticide containers, 510, 510t
 methodology of, 507–508
 pesticide affordability, 509
 pesticide availability, 509
 pesticide mixing, 509
 pesticide poisoning, 510–511
 pesticides, application of, 509
NGO initiatives, 1404
NH_3. *See* Ammonia (NH_3)
Ni. *See* Nickel
Nickel (Ni), 1370
Nickel-cadmium batteries, 447
Nicotine, 2029, 2029f
Night soil sludge, 2267t
1985 Helsinki Protocol, 15
1999 Gothenburg Protocol, 15
Niobium, 2585, 2587–2588
Nitrapyrin, use of, 1306
Nitrate, 1805, 2753
 in groundwater, 1282, 1283
 leaching, 1559
 pollution, 2304
 reductase, 426
 runoff, 345
Nitrate fertilizer
 and soil IC, 464
Nitrate Leaching Hazard Index, 1763
Nitrate leaching index, 1761–1765, 1764f
 assessment, 1762–1763
 components, 1763–1765
 environmental problems, 1761–1762
 quick tools and indicators, 1762
Nitrate-nitrogen (NO_3-N), in groundwater, 1302
 agricultural practices contributing to
  containerized horticultural crops, 1306
  grasslands/turf, 1305–1306
  manure application, 1304
  nitrogen fertilizer, 1304
  row crops, 1303–1304
 elevated levels of, 1302
 environmental impacts, 1303
 health problems by, 1302, 1303
 by leaching of N fertilizer, 1302–1303
 nitrogen sources of, 1303
 occurrence of, 1302–1303
 strategies, for reducing NO_3-N leaching, 1306–1307
  cover crops, 1306–1307
  diversified crop rotation, 1307
  grassland/turf management, 1307
  nitrification inhibitors, 1306
  reduced tillage, 1307
  soil testing and plant monitoring, 1306
Nitric oxide (NO)
 soil quality and, 2389–2390
Nitrification, 93, 1770–1771
 defined, 97
 inhibitors, 1306, 1809
 nitrous oxide from, 97
 process, 1768–1769, 1769f
Nitrifiers, 2862
Nitrite, 1559
Nitrobacter, 97
Nitrogen (N), 88, 350, 1074–1075, 1304, 1513, 1559, 1761, 2434, 2753, 2865t
 acid rain and, 20–24
  ecosystem impacts, 22–24
  human health effects, 22
  reducing acidic deposition effects, 24
  sources and distribution, 20–22, 21f
  structural impacts, 22
 agricultural runoff and, 82–83
 ammonia volatilization, 1770
 atmospheric deposition, 1769–1770
 biological fixation, 1770
 concentration, reduction in municipal wastewater, 2715–2718, 2716f, 2717f
 cycle, 88–90, 89f, 1513, 1768, 2389f
 decomposition and mineralization, 90. *See also* Soil organisms activity, environmental influences on
  organic substrate quality, effect of, 92–93, 93f
  soil microorganisms, effect of, 90–92
 effect on surface water quality, 2756–2757, 2758t
 fixation, 88–89, 426
 gases. *See also specific* entries
  air pollution problems of, 158–159
  soil quality and, 2389–2390
  input processes, 1769–1770
  leaching, 94, 1771
  losses of, 89–90, 93
  measurements, need for, 1772
  mitigation of, 93–94
  pathways for, 1777–1778
 loss processes, 1770–1771
 manure
  application of, 1810, 1810f
  mineral and organic fertilization, 1770
  and NO_3-N contamination, 1303
  overview, 1768
  runoff losses of, 94
  transformations in soil, 1768–1769, 1769f
  uptake by plants, 1770
Nitrogenase, 88, 89, 426
Nitrogen cycle, 1777
 conceptual diagram of
  in an aquatic ecosystem, 1076f
Nitrogen dioxide (NO_2)
 soil quality and, 2389–2390
Nitrogen fertilizer, NO_3-N contamination from, 1303
Nitrogen-fixing bacteria, 88
Nitrogen-fixing biofertilizers, 1516
Nitrogen (N) nutrient
 applications, 1806, 1806f
  additional tests for fine-tuning, 1806
  controlled-release nitrogen fertilizers, 1809
  fall, 1808
  nitrification inhibitors, 1809
  preplant, 1808
  side-dress, 1808–1809
  split, 1809
  timing, 1808–1809, 1808t
  placement, 1809–1810
Nitrogen oxides (N_xO_y), 1154–1155, 2304
 characteristics of, 137
 EN 14211, chemiluminescence for, 141–142
Nitrogen phosphorus detector (NPD), 1975
Nitrogen trading tools (NTTs), 1772–1782
 description of, 1779–1780
 examples of, 1775–1776, 1781–1782
 functionality of, 1780
 limitations of, 1778
 outputs of, 1780
 role of, 1774–1775
Nitrosomonas europaea, 97

Nitrous oxide (N$_2$O), 486, 1202
   emissions, 96–98, 1770–1771
      from agriculture, 96–98
      biomass burning, 97
      denitrification, 96–97
      fertilizer consumption and, 98
      flooded soils, 97
      management practices to reduce, 98
      nitrification, 97
      overview, 96
   soil quality and, 2389–2390
NO. *See* Nitric oxide (NO)
NO$_2$. *See* Nitrogen dioxide (NO$_2$)
N$_2$O. *See* Nitrous oxide (N$_2$O)
NOAEL. *See* Nonobservable-adverse-effect level (NOAEL)
Noble metal nanoparticles
   bactericidal effect of, 1740
   pollutant removal, mechanism in, 1739–1740
Nodding thistle *(Carduus nutans)*, 1526t
No-growth disaster, 618
Noise pollution, 2361
Noise pollution *See also* Pollution
   in coastal waters, 497
   industrial wind turbines, 2867–2868
   mitigation, 2877
      regulating permissible noise level, 2877–2879
      regulating setback distances, 2879
   wind turbine noise, acoustic profile of, 2868–2870
   wind turbine noise, human impacts of, 2870
      and annoyance, 2871–2873
      health impacts, quantifying, 2870–2871
      and low-frequency/infrasound components, 2876–2877
      psychological description, 2870
      and sleep, 2873–2876
      wind turbine syndrome, 2876
Nomolt. *See* Teflubenzuron
Nonaqueous phase liquids (NAPL), 1334
Non-catalytic WO, 1908–1909
   commercial, 1909t
Nonchemical/pesticide-free farming, 1103–1108
   basic principles, 1103
   definitions, 1103
   global growth, 1103–1106, 1104t–1106t
   objectives, 1103
   organic farming. *See* Organic farming
   producer-consumer driven movement, 1107–1108
Non-condensable gases, 2485
Non-dispersive infrared, for CO EN 14626, 142
Non-energy input, 791
Nongovernmental organizations (NGOs), 1221
Non-Hodgkin's lymphoma, 1399

Nonisotopic tracer methods, 1204
Non-metals, 1560
Non-motile algae, 2633
Nonobservable-adverse-effect level (NOAEL), 2006
Non-point source of water pollution, 1580
Non-point source pollution (NPSP), 2167t–2169t
   chemicals, associated, 2137t
   contaminant interactions, 2136
   contributors to, 2137t
   example of, 2136
   global effect, 2138
   management of, 2137–2138
   rainwater as source of, 2136
   remediation of, 2137–2138
   soil/environmental quality, 2136–2137
   urban sewage and, 2136
   water pollution prevention, 2138
Non-potable uses of water, 2262
Non-saline water, 1569t
Non-steroidal anti-inflammatory drug (NSAID), 2062
Non-threshold agents, 1456
Nonylphenol, 1405
Normalized difference vegetation index (NDVI), 2398, 2397f
North America
   soil degradation in, 2381–2383, 2382f
   distribution of, 2383f
Northern Agricultural Catchments Council of Australia, 1028
Northern jointvetch *(Aeschynomene virginica)*, 1526t
NOSB. *See* National Organic Standards Board (NOSB)
NO$_3$ test, late-spring, 1306
No-till practices
   adoption of, 1036–1037
   conservation farming system, 119–120
   erosion process and, 1040–1041, 1041f, 1042f
Novaluron, 1460, 1462
Noviflumuron, 1462
Nozzles, 1472
NPP. *See* Net primary production (NPP)
NPSP. *See* Non-point source pollution (NPSP)
NRCS. *See* Natural Resources Conservation Services (NRCS)
NTE. *See* Neuroesterase (NTE)
Nuclear and coal-fired power plant overnight costs, 1790–1792
Nuclear atmospheric weapons tests, 2304
Nuclear power and wind power, 183–185
Nuclear power economics, 1789
   capital costs for nuclear and coal-fired powerplants, 1790
   cost of alternative, 1796–1797
   discount rates, 1795–1796, 1800f, 1801f

   environmental costs and regulations, 1793–1795
   Fukushima nuclear disaster impact on, 1803–1804
   nuclear and coal-fired power plant overnight costs, 1790–1792
   results, 1797–1803
   total project costs derivation, 1792–1793
Nuclear reaction, 2235
Nuclear Regulatory Commission (NRC), U.S., 1789, 1794
Nuclear Waste Policy Act (NWPA) (1982), 1794
Nucleopolyhedroviruses (NPVs), 382
Nuisance algae growth prevention, 539
Nuisance claim, 1945
Nunavut, 2671
Nutrient(s), 1805–1814
   application
      rates, 1805–1808
      timing, 1808–1809
   credits, 1807–1808
   effect on surface water quality, 2756–2757, 2758t
   dissolved load from different continents to world's oceans, 2756t
   farm nutrient management plans, 1813–1814
   irrigation water management, 1813
   management in soil, 2217–2222
   manure management, 1810–1813
   placement, 1809–1810
   soil conservation practices, 1813
   variable-rate fertilizer technologies, 1810
Nutrient crediting, 1813–1814
Nutrient loss from grazing lands, 2753
Nutrient management
   diet modification on, 1668–1669, 1668f
   integrated, 1508–1509
   progress in practices, 1667
Nutrient mining, 1510
Nutrient-mobilizing biofertilizers, 1516
Nutrient reduction, 1584
Nutrient removal in WSPS, 2635
Nutrient runoff, 1080
Nutrient spiralling, 2101
Nutrient supply
   and organic farming, 1107
Nutrient tracking tool (NTrT), 1776, 1782
Nutrient transfer
   boundary conditions and, 1821–1822, 1821f
   diffusion, 1817–1820
   by mass flow and diffusion from soil to plant roots, 1819–1820, 1820f
   overview, 1817
   root hairs, importance of, 1820–1821, 1821f
   soil plant system, 1817, 1818f
Nutsedges *(Cyperus* spp.), 1526t
Nylar. *See* Pyriproxyfen

O

Oak Ridge National Laboratory, 2243
Oat (*Avena sativa* L.), 1307
Oat-frit fly (*Oscinella frit* L.) system, 1939
OC. *See* Organochlorines
Occupational Safety and Health Administration, 2051
Ocean
  acidification of, 13
  currents, 824, 826
  fertilization, 460
  sources, 825–826
Ocean color and temperature scanner (OCTS), 2817t
Ocean direct injection, 459
Ocean thermal energy conversion (OTEC) devices, 772
Ochratoxin, 1697
Ochratoxin A (OTA), 1697
Octanol–water partition coefficient, 1283
Octylphenol, 1405
Odor control, of composting materials, 524
Oedaleus asiaticus, 1990
Office of Pesticide Programs (OPP), EPA, 1941–1942
Office of Prevention, Pesticides, and Toxic Substances, 1406
Off-site movement of pesticides, 2804–2805
Off-site pollution, 1559
Off-site water quality impact reduction, 1557, 1560–1561, 1561t
Off-the-farm discharge, 1557
OFPA. *See* Organic Food Production Act (OFPA)
Oil and ship pollution, 2358–2359
Oil pollution, in Baltic Sea, 1826–1840
  observations of, 1828
  oil spill drift, numerical modeling of, 1833–1835, 1835–1838f
  operational satellite monitoring of, 1830–1833
  problems and solutions of, 1838–1839
  sources of, 1827–1828
  statistics of, 1828–1830
  tendencies of, 1830
Oils, in coastal waters, 496
Oil spills, 2042, 2043t
  drift, numerical modeling of, 1833–1835, 1835–1838f
O isotopes
  for saltwater detection, 1327–1328
Oka, I.N., Dr., 2034
Oman, 2682, 2690
Omnivorous feedstock converters (OFCs), 830, 832f
Oncomelania hupensis, 384
On-farm nutrient resources, assessment of, 1813

Online gas chromatography, for benzene EN 14662, 142
On-site pollution, 1559
On-site residential wastewater, 2862, 2865f
Onthophagus gazella, 1461
Opassess.com, 1493
Open-cell, low-density polyurethane insulation, 1479
Open circuit systems, 248
Open collectors, 243, 243f
Open ecosystems, 1935
Open pit mining, 1692–1694
  acid mine drainage, 1694
  coversoil resources, 1692
  coversoil thickness requirements, 1693
  landscape regrading, 1692, 1693f
  in situ soil reclamation, 1693
  steep slope reclamation, 1694
Open-tubular trapping (OTT), 2195
Operational satellite monitoring, of oil pollution, 1830–1833
Operations and maintenance (O&M) program, 657
Operations research methods in pest management, 1950
OPIDIN, 1399
OPIDP. *See* Organophosphate induced delayed poly-neuropathy (OPIDP)
Opportunity cost, 2784
OPs. *See* Organophosphates (OPs); Organophosphorus compounds (OPs)
Optical sensors, 1024
Optimizing model, 2781–2783
  scarcity rents, 2782
  shadow prices, 2782–2783
Opuntia vulgaris, 2821
Oreochromis (*Sarotherodon*) *hornorum*, 1922
Oreochromis (*Sarotherodon*) *mossambica*, 1922
Organic agriculture, 125–127
  crop and pest performance in, 126–127
  crop rotations and, 126
  defined, 125
  global statistics, 126
  history, 125–126
  key issues requiring additional research, 127
  overview, 125
Organic carbon partition coefficient (K_{OC}), 2015
Organic chemicals, 2169
  in groundwater, 1281
Organic compounds, in coastal waters, 492–494
Organic farming
  animal husbandry, 1107
  animal pests
    regulation of, 1115–1117
    strategy of controlling, 1115
  basic principles, 1103

  bioregulators in, 1115–1116
  composting organic waste for, 528–533
    bacteria, 529–531
    probiotic agriculture, 528–529, 531–533
  defined, 528, 1115
  diversity, 1106
  genetically engineered organisms, 1110
  global growth, 1103–1106, 1104t–1106t
  green manure, 1107
  as holistic and systematic approach, 1106–1107
  as model for sustainable development, 1108
  nutrient supply, 1107
  objectives, 1103
  pest management, 1107, 1110–1111, 1115–1117
    diseases, 1112–1113
    insects and invertebrates, 1112
    weeds, 1111–1112
  producer-consumer driven movement and, 1107–1108
  soil building, 1110
  soil management, 1107
  standards, 1109–1110
  weed management, 1107
Organic fertilization on heavy metal uptake
  about heavy metals, 1374
  contamination with heavy metals, 1374
  heavy metal concentrations in organic fertilizers, 1375t
  plant uptake of heavy metals, 1375–1376
  sources of organic fertilizers, 1375, 1375t
Organic Food Production Act (OFPA), 125, 1108, 1109
Organic loading rate, 2650, 2652
Organic matter (OM), 1374, 1375
  bonded REE, 2267, 2268t
  sludge-derived, 1375
  in solid waste, 513
  turnover. *See* Soil organic matter (SOM) turnover
Organic matter (OM), global distribution of, 1851–1855
  andisols, 1854
  decomposition
    climate, 1853
    quality of matter, 1853
  histosols, 1854
  inputs
    placement, 1852–1853
    quantity, 1851
    species composition, 1852
  overview, 1851
  physical and chemical influences, 1853–1854
Organic matter (OM) management, 1857–1861

Organic matter (OM) management (cont'd.)
  carbon losses from soil
    mineralization processes and, 1857
    soil erosion processes and, 1857–1858
  CH_4 emissions from rice fields and, 1680
  soil carbon sequestration, 1858–1861
    global importance of, 1858, 1859t–1860t
    long-term field experiments, role of, 1858
    mechanisms, 1858, 1860
    SOM, energy, and full C accounting, 1861
    SOM role in 21st century, 1861
Organic matter modeling, 1863–1868, 1865–1866t
  application, examples of, 1864–1867
  approaches, 1863–1864
  challenges in, 1867–1868
  performance, 1864
  recent advances in, 1867
  turnover, factors affecting, 1864, 1865–1866t
Organic mulches, 961–962
Organic nitrogen, 1559
Organic pollutants, 388, 2114–2117, 2115t, 2116t
  bioremediation of, 2116–2117
  degradation of, 1906
  of drinking water, 2066, 2793–2797, 2795t, 2796t
    determination of, 2795–2797, 2797f
  potential impacts of, 2115–2116
  soil contamination from, 2114–2115
Organic production and processing, 1103
Organic sludge, 961
Organic soil amendments, 959, 1878–1880
  antagonistic microorganisms stimulation, 1879
  future perspectives, 1880
  impacts on plant health and weeds, 1878
  mechanisms of action, 1878–1879
  overview, 1878
  plant resistance and, 1878, 1879t
  release of compounds toxic to insects and plant pathogens, 1878
  in 21st century, 1879–1880
Organization for Economic Co-operation and Development (OECD), 355, 647, 1037, 1219, 2124
  endocrine screening and testing program, 648
  DNT guidelines of, 1757
Organization of Petroleum Exporting Countries (OPEC), 485
Organochlorine acaricides, 2–3
Organochlorine insecticides, 1399
Organochlorine pesticides, 390, 1966–1967. *See also* Pesticides
  chemical structure of, 1966f

Organochlorines (OC), 277, 2804, 2805, 2806
Organometallic complexes, precipitation of, 1625–1626
Organonitrogen pesticides, 1967. *See also* Pesticides
  chemical structure of, 1967f
Organophosphate hydrolase (OPH), 390
Organophosphate induced delayed polyneuropathy (OPIDP), 1415
Organophosphates (OPs), 3, 277, 2005
Organophosphorus, impact on birds, 417
Organophosphorus compounds (OPs), 1398
  background, 1397
  effects of repeated exposure to, 1399
  intermediate syndromes, 1398
  mechanisms of action, 1398
Organophosphorus (OP) insecticide, 2804, 2806
Organotin acaricides, 3
Orientation of the intake, 210
Ornithonyssus sylviarum (northern fowl mite), 2
Orthophosphate, 100
Ortstein, 1625
Oryza sativa (Rice), 425
Osaka gas process, 1910
Oscinella frit. See Frit fly
Ostrinia nubilalis, 412, 1116
Otto cycle, 819
"Ouch ouch" disease, 448
Ovarian cancer, by NO_3-N exposure, 1303
Overflow, 1557, 1558t
Overgrazing and soil erosion, 1034
Overhead irrigation, 2677
Overnight cost, 1790
Overwintering, natural enemies conservation and, 1750
Ovicides, 2014
Owens Lake, degradation of, 1437–1438
Owner-operated rainwater harvesting, 2262
Owners' costs, 1791
Oxamyl (Vydate), 1318t
Oxidation reaction, 714
Oxidative biodegradation, 1337
Oxidative stress, 281
Oxident concentration, effect on homogeneous photo-Fenton reaction, 2552
Oxydemeton methyl, 2142
Oxydia trychiata, 367
Oxyfluorfen, 2142
Oxyfuel combustion, 457
Oxygen, 2862, 2864
  consumption rate, 724
  dissolved, 2753
  starvation, 1559
  transfer mechanism, 724
Oxygenation, 1592t, 1593–1594
Oxyreductases, 89
Ozone (O_3)

characteristics of, 137
EN 14625, ultraviolet photometry for, 142
Ozone-depleting substances (ODSs), 1886–1887
Ozone layer, 1882–1898
  atmospheric ozone, 1883
  CFC-ozone depletion hypothesis, 1885–1886, 1887f
  climate change impact on, 1896–1897
  depletion, biological effects of, 1893
  origin of, 1883–1885
  stratospheric ozone depletion
    antarctic ozone hole, discovery of, 1887
    atmospheric trace species, field measurements of, 1891
    global ozone layer, depletion of, 1891–1893, 1892f
    measurements and distribution of, 1885
    polar ozone chemistry, 1887–1891
Ozone problems, 916

P

PA-14 (Tergitol®), 414
Packaging waste, 903–906
Packaging Waste (PW) Directive, 894
  recovery rate of, 905f
  recycling rate of, 905f
Packed-bed scrubbers, 156
Paecilomyces lilacinus, 1752, 1753, 1879
Paecilomyces lilacinus Thom (Sam), 288
Paenibacillus lentimorbus, 321
Paenibacillus (Bacillus) popilliae (Dutky), 321–322
  culture and control, 321–322
  saltwater intrusion in, 1322
PAH. *See* polycyclic aromatic hydrocarbons (PAH)
PAID. *See* Photo-acoustic-infrared detector (PAID)
Paleoenvironmental data and fly-ash particle analysis, 2292
Palestine
  saltwater intrusion in, 1322
Palustrine emergent wetlands, 2827
Palustrine forested wetlands, 2827
Palustrine scrub-shrub wetlands, 2827
PAN. *See* Pesticides Action Network (PAN)
Panonychus ulmi, 2141
Papaver rhoeas, 1925
Papaya ringspot virus (PRV), 2111
Papilio polyxenes F., 313
PAR. *See* Photosynthetically active radiation (PAR)
Paraben, general formula of, 2083f
Paradichlorobenzene (PDB), 2009

Paraffin oil, 415
Paraquat, 1397, 1399
Parasarcophaga misera, 1920
Parasites, as biological control agents, 2596–2597
Parasitoids, 1460, 1475, 1475f, 2423, 2424. *See also* Natural enemies
  intercropping and, 1938
Parathion, 3, 419, 2806
Participatory approach, 561
Particle size, of organic materials, 523
Particulate matter (PM), 133, 142
  characteristics of, 137–138
Particulate pollution
  control methods, 149–156
    cyclones, 152
      distribution patterns, modifying, 150–151
      electrostatic precipitator, 153–154
      equipment, 151
      filters, 152–153
      settling chambers, 151
      wet scrubbers, 154–156
  problem, 149
  sources of, 149
Paspalum notatum (Bahiagrass), 1305
Passer domesticus, 413
Passivation, of silicon surfaces, 233
Passive heat storage
  solar, 2519, 2519f
Passive PRB system, 1336
Pasteuria penetrans, 1752–1753
Pasture lands. *See* Range and pasture lands
Pathogen, 1475–1476, 1476f. *See also* Natural enemies
  with livestock, types of, 2753
  pathogen control, of composting materials, 524
  protection mechanisms acting in intercropped systems, 1937, 1938t
  reductions, on-farm options for, 2678t
  in sewage effluent, 1570
Pathogenic bacteria, in drinking water, 2797–2799, 2799t
Paulatuk treatment wetland, 2664–2670
  aerial view, 2665f
  background information, 2664–2665
  discussion, 2669–2670
    natural ultraviolet radiation, 2670
  facultative lake to Arctic Ocean, 2666f, 2667f, 2668f
  influent and effluent concentrations, 2665t
  methods, 2665–2666
  results, 2666–2669
    cBOD5 effluent, expected, 2668, 2669f
    E. coli, concentration gradients of, 2669f
    effluent concentrations, 2666
    NH_3^+-N, concentration gradients of, 2668, 2669f

Pay-as-you-throw (PAYT), 895, 900, 911
Payment for ecosystem (goods and) services (PES/PEGS), 634
Pb. *See* Lead (Pb)
Pb contamination, in coastal waters, 490
PCB. *See* Polychlorinated biphenyls (PCB)
PCE. *See* Perchloroethylene (PCE)
PDB. *See* Paradichlorobenzene (PDB)
Peanuts (*Arachis hypogaea* L.), 424
Peas (*Pisum sativum* L.), 427
Peatlands, 1868, 2826f, 2827
Peclet number, 1315
Pedogenic carbonate, 1445–1446
Pelecanus occidentalis (Brown pelican), 417
Pellet(s), 375t
Pellet furnaces, 375
Peltier effect, 1348
Penicillium aurantiogriseum, 1697
Penicillium belaji, 1516
Penicillium expansum, 1991
Penicillium nordicum, 1697
Penicillium spp., 2109
Penicillium verrucosum, 1697
Pennisetum glaucum, 1012
Penstock, 210–211
Pentachlorophenol (PCP), 388
Pepper fruit extracts, 2021
Peppers (*Capsicum annuum* L.), 425
Per ascensum model, of carbonate formation, 1446
Percent tree cover map, 2376
Perchloroethylene (PCE), 1283, 1334, 1340
Percolation, 1564, 1566, 1570
Per descensum model, of carbonate formation, 1446
Perdix perdix (Grey partridge), 416
Peregrine falcon (*Falco peregrinus*), 417
Perennial pastures, 94
Perfectly stirred reactor (PSR), 726
Perfluorocarbons (PFC), 486
Performance efficiency of irrigation. *See* Irrigation efficiency
Peristenus digoneutis, 368
Permafrost, 1900–1902
  active layer, 1900, 1902f
  characteristics, 1900–1901, 1901f
  living with, 1902
  surface energy balance, 1901–1902
Permeable reactive barrier (PRB), 1335–1336
Permethrin, 3, 1318, 2005
Peronospora parasitica, 443
Persistent organic pesticides (POPs), 575, 1913–1914, 2186
  alternative approaches, 1916–1917
  global alliance for alternatives to DDT for disease vector control, 1917
  issues concerning, 1915
  under Stockholm Convention, 1914–1915

Persistent organic pollutants (POP), 324–325, 1904, 2036–2037
Pest(s)
  border diversity effects on, 1928
  crop rotation effects on, 1925, 1926t–1927t
  decoy crops effects on, 1925, 1928
  plant resistance effects on, 304–305
  trap crops effects on, 1925–1928
  weed diversity effects on, 1928
Pest control
  Bacterial. *See* Bacterial pest controls
  cultural, antagonistic plants in, 288–289, 289t
  genetic engineering in, 407
Pesticide, 2013–2014
  abuse, 418–419
  active ingredients in, 2013
  acute effects, 1414–1415, 1415f
  agricultural runoff and, 83
  background, 1397
  band applications, 1983–1985
    advantages, 1983–1984
    disadvantages, 1984
    equipment requirements, 1984
    future perspectives, 1984–1985
    usage, 1983–1984, 1984t
  cancers from. *See* Cancers, from pesticides
  carbamates, 2016t
  categories, 277
  characteristics of, 1963–1965
  chlorinated hydrocarbons, 2016t
  chronic effects, 1415
  chronic intoxication/poisonings statistics, 1397
  circulation in environment, 1965–1966, 1965f
    air, 1966
    aquatic environment, 1965–1966
    crops, 1966
    soil, 1966
  classification of
    based on applications, 1964t
    based on chemical class, 1964t
    based on chemical structure, 1964t
    based on toxicity, 1965t
  in coastal waters, 495–496
  concentration in water, 2017
  consumption trend, 2140
  contamination in water, 1560
  defined, 1397
  derivatization of, 1974
  degradation, 389t–390t
  development of, 1998
  dispensing/repackaging of, 506f, 509
  effects, 2005–2006
    of biopesticides, 381. *See also* Biopesticides
    of synthetic pesticides, 381
    on children, 1415–1416

Pesticide (*cont'd.*)
  on natural enemies, 370
  on surface water quality, 2757, 2759, 2760t, 2761t, 2765t
  environmental effects of, 1121
  enzyme variation, 1400
  epidemiological studies, 1400
  exposure to, 1414
  fumigants, 2016t
  genetic effects of, 1399
  genotoxicity assessment, 2125–2128
  global usage, 1414
  in groundwater, 1281
  grouped by use, 1398t
  groups of, 2015f
    group I, 2015
    group II, 2015
    group III, 2015
    group IV, 2015
  hazards, avoiding, 1999–2002
    judicious use, encouraging, 2001
    nonchemical strategies, promoting, 2001–2002
    personal protection, provision of, 2002
  health effects of, 1121–1122
  herbicides, 1399
  history of, 2008–2012, 2145–2146
    ascendancy, 2008–2009
    DDT, 2009–2010
    inevitable conflict, 2008
    IPM, 2010–2011
    rebuff and reassessment, 2010–2011
  impact in Ngarenanyuki, Tanzania
    community-based monitoring of, 505–511
  inert ingredients in, 2014
  inorganic pesticides, 2016t
  insecticides, 1398–1399
  laws and regulations, 1612–1614
    available international guidelines, 1612–1613
    future global policy, 1613
    historical perspectives, 1612
    implementation problems, 1613
    need for, 1612
    present scenario and probable remedies, 1613
    steps undertaken, 1613
    toxicological and other data requirements for registration, 1614
  mechanisms of action, 277
  methodologies for determining, 1968–1976, 1968f, 1969t, 1970t
    in air, 1969–1972, 1971t, 1972t
    in fruits, 1972–1973, 1972t, 1973f, 1973t
    in soil, 1974
    in vegetables, 1972–1973, 1972t, 1973f, 1973t
    in water, 1968–1969, 1970t
  misuse, 419
  mobility of, after release, 2016–2017, 2017f
  natural. *See* Natural pesticides
  natural products, 2016t
  non-polar, 1967
  observed effects
    in experimental systems, 277
    in humans, 277–278
    in native animals, 277
  oil/fat soluble, 2015
  organochlorine, 1966–1967
  organochlorine insecticides (DDT, DDD), 1399
  organonitrogen, 1967
  organophosphates, 2016t
  poisonings, 1414–1415, 1415f. *See also* Human pesticide poisonings
    economic costs, 1416
    symptoms of, 1968
  poisoning surveillance, 2002–2004
    agriculture and public health, sustaining, 2003–2004
  polar, 1967
  pollution, in natural ecosystems, 2145–2148
    agro/horti ecosystem, 2146
    aquatic environment, 2146–2147
    humans, 2147–2148
    soil environment, 2146
    terrestrial environment, 2147
  possible long-term effects, 1397
  prevention, 1400
  problems associated with, 2146
    alternatives for, 2148
  properties, leaching and, 1622, 1622t
  regulation, 1940, 1998–1999
    common law on, 1944
      negligence, 1944
      nuisance, 1945
      strict liability, 1945
      trespass, 1945
    current law on, 1943–1944
    federal laws on, 1940–1943
      enforcement of pesticide use, 1941–1942
      interaction with other federal regulations, 1942–1943
      overview of, 1940–1941
      registration of pesticide, 1941
      tolerance level for pesticide residue, 1942
    preemption, 1943
    regulatory trends, 1945–1946
  remediation techniques, objectives of, 2014
  reproductive effects of, 1399
  residues, in agro-horticultural ecosystems, 2142–2143
  risk assessment, 2750
  risk to human health, 1964, 1967–1968
  role in farming and food production, 2014
  Rotterdam Convention and, 1615–1617
  sensitivities, 1411–1413
    case example, 1412
    causes, 1411–1412
    impact of, 1412–1413
    mechanism, 1412
    overview, 1411
    prevalence, 1411
    symptoms, 1412
  soil adsorption of, 2015–2016, 2016f
  solutions to reduce release of, 2014
  sterility caused by, 277–278
  stewardship, 2001t
  synthetic pyrethroids, 1399
  transport and paterns of occurence in surface waters, 2804–2805
  use, and BT transgenics, 2141–2142
  use impact on natural enemies, 2141
  use impact on pollinators, 2140–2141
  use in weeds, 2142
  use reduction approaches, 2143
  water and soil contamination by, 2014
  water-soluble, 2015
Pesticide control legislation, 2035–2037
  elements, 2035
  implementation, 2035–2036
  international conventions, 2036–2037
  overview, 2035
  problems related to, 2036
Pesticide Data Program, 1609
Pesticide-free farming. *See* Nonchemical/pesticide-free farming
Pesticide groundwater database (PGWDB), 1317
Pesticide impacts, on aquatic communities
  balance, 291
  examples, 294
  measuring impacts, 291–292
  recent advances and outstanding issues, 295–297
  risk assessment and, 292–294
  risk reduction, 295
The Pesticide Index, 1461
The Pesticide Manual, 1461
Pesticide persistence, 1957
Pesticide–plant combinations, 1319
Pesticide residues
  food contamination with, 1124–1126
    after registration, 1124–1125
    analytical methods for, 1126
    food processing, 1125
    MRL, 1124, 1125–1126, 1125t
    overview, 1124
    before registration (risk assessment), 1124, 1125f
    trade issues, 1125–1126, 1125t
  in foods from organic, integrated and conventional production, 1403

tolerance limit for, 1609
   in food (examples), 1610t
   regulatory inspection and enforcement, 1609
Pesticides Action Network (PAN), 1612
Pesticides in groundwater
   groundwater contamination, 1319
   management of, 1319
   irrigation management, 1319
   maximum contaminant levels (MCL), 1317, 1318t
   in soils and water
      fumigants, 1318
      fungicides, 1318
      herbicides, 1318–1319
      insecticides, 1317–1318
   use of, 1317
Pesticide test management
   hormonal disruption and, 1406
Pesticide use, reduction in
   overview, 2033–2034
   successes in, 2033–2034
Pest management, 1319, 1919, 1949
   in agro-ecosystem, 1949
   aquatic weed management by fish in irrigation systems, 1922
   Australian bush fly management in micronesia, 1920
      bush fly origins and habits, 1920–1921
      management efforts worldwide, 1921
   biotechnology impact on, 1952–1953
   crop diversity for, 1925–1929
   definition of, 1949
   ecological. *See* Ecological pest management
   ecological aspects, 1934–1936
      examples, 1935–1936
      monoculture and polyculture, 1935
      open and closed ecosystems, 1935
      overview, 1934
      pest population dynamics and species diversity, 1934–1935
   economics considerations, 2011
   filth fly abundance management, in dairies and poultry houses, 1922–1923
   global commerce and, 2011
   integrated. *See* Integrated pest management (IPM)
   intercropping for. *See* Intercropping
   modeling approaches in, 1949–1950
      analytical, 1950
      empirical, 1949
      operations research, 1950
      physiological, 1950–1951, 1951f
      simulation, 1950–1951, 1951f
      statistical, 1949
   navel orangeworm management in almond orchards, 1919–1920
   and organic farming, 1107
   in organic farming, 1115–1117
   in organic farming systems, 126–127
   population pressure and, 2011–2012
   public attitude and, 2011
   regional, 1953
   research, components of, 1950f
Pest regulation
   objectives, 1115
PET. *See* Potential evapotranspiration (PET)
Petroleum hydrocarbons
   abatement of, in natural wetland, 2855f, 2855–2856
   chemical composition of, 2042–2046
   constituents of, 2044t
   contamination, 2040–2057
   definitions, 2854
   environmental fate of, 2046–2048
   in environmental media, determination of, 2051–2056
   environmental relevance of, 2040–2042
   ranges of, 2051t
   toxicity of, 2048–2051, 2050t
   treatment of, in engineered/constructed wetland, 2856–2857
Petroleum reservoir, 458
Petroleum. *See* Petroleum hydrocarbons
PGPR. *See* Plant growth-promoting rhizobacteria (PGPR)
Phacelia tanacetifolia, 1750
Phacelia tanacetifolia Bentham, 371
Phaenerochaete chrysosporium, 392
Pharmaceuticals, 2062
   aluminum in, 270
   antibiotic resistance, 2778
   in aquatic environment, 2776–2778
   in aquatic systems, treatment, 2062
      alternative systems for water treatment, 2062–2063
      drinking water treatment systems, removal efficiencies, 2066–2067
      municipal WWTP systems, removal efficiencies, 2063–2066
   concentration ranges of, 2777
   disposal of, 2776
   ecotoxicity for, 2064
   effects of, 2777–2778
   in environment, 2060–2062, 2061f
   formulation facilities, 2060
   in human medicine, 2776
   metabolism, 2776
   occurrence and fate, 2776–2777
   regulation, 1702–1703
   risk and risk management, 2778
   in sewage and sewage treatment plants (STPs), 2776
   sources, distribution, and sinks of, 2777
   in veterinary medicine and animal husbandry, 2776
Phascolarctos cinereus, 2595
Phase change materials (PCMs), 2509, 2520
Phase equilibrium, 2533
Phaseolus vulgaris L.cv. Monel, 1878
Phasmarhabditis hermaphrodita, 383
PH control, of composting materials, 524
Phenolic compounds
   chemical composition and structure of, 2072–2075t
   as indicators of original plant matter, 2077t
   natural derivatives of, 2076t
   threshold limit values, 1905t
Phenols, 2071–2088
   anthropogenic origin of, 2078–2079
   chemical composition and structure of, 2079–2080
   conversion of, 2085–2086
   distribution of, 2085–2086
   environmental impact of, 2086
   identification of, 2086–2087
   natural origin of, 2071, 2075–2078
   nomenclature of, 2079
   physicochemical properties of, 2080–2084, 2083t
   quantification of, 2087
   transport of, 2085–2086
   uses of, 2084–2085
Phenoxy acids, 1399
Phenylurea herbicides, 1318
Pheromones, 175, 2029
Pheromone traps, 1112
Philosophical Transactions of the Royal Society of London, 1010
Phlebotomus papatasi, 1460
Phloem-feeding insects, 308
Phoma, 1111
Phorate, 419
Phosphamidon, 420
Phosphate fertilizers, 1513–1514, 1518, 2266
   strontium in, 2429
Phosphate rocks, 2240
Phosphate-solubilizing bacteria (PSB), 1511
Phosphate-solubilizing biofertilizers, 1516
Phosphinothricin, 2030–2031, 2031f
Phosphogypsum, 2266
Phosphoric-acid-doped PBI membrane, 1139
Phosphoric acid fuel cells (PAFCs), 1139
   advantage of, 1139
   carbon monoxide tolerance of, 1139
   components of, 1139
   fuel for, 1139
   operating temperature range of, 1139
   for terrestrial applications, 1139
   uses of, 1139
Phosphorus (P), 1559–1560, 2137, 2753
   autocorrelation parameters, 101t
   availability
      critical concentration and fertilization, 2095
      residual phosphorus, 2095–2096
   behavior in soils, 2094–2096

Phosphorus (P) (cont'd.)
  calibrated soil tests for, 1806–1807
  chemical speciation of, 100
  concentrations, 1081
  dynamic nature in soils, 2096
  effect on surface water quality, 2757, 2758t
  fertilizer use, 2096–2097
  forms and amounts, 2094–2095
  loss
    implications for, 2097
    and mitigation strategies, 2097
  in manure
    application of, 1810, 1811f
    diet modification, effects of, 1668–1669, 1668f
    livestock feed management, 1812–1813, 1812f
    manure phosphorus concentration reduction, 1667
    nutrient management practices, progress in, 1667
  mobility in agricultural soils, 100–101
  nutrient
    applications, timing, 1809
    placement, 1810
  phosphorus mirabilis, 100
  potentially harmful effects, 2093–2094
  -related impairment in flowing and lake waters for targeted remediation, 2101–2102, 2104f, 2104t
  solubilization, 350
  sources, 2096–2097
  spatial speciation of, 101–102, 101t, 102f, 103f
  transport, in riverine systems, 2100–2105
    land use impacts, integration of, 2101, 2102–2103f
  use, and B toxicity, 426
Phosphorous concentration, reduction in municipal wastewater, 2712–2715, 2713–2715f
Phosphorus cycle, 1075, 1076f
Phosphorus-loading models, 1585
Photo-acoustic-infrared detector (PAID), 1203
Photochemical smog, 156
Photo-electronic pins, 1021
Photoinhibition, 538
Photorhabdus, 321
Photosensitive windows, 775
Photosynthesis, 88, 345, 716
  and biomass production, 373
Photosynthetically active radiation (PAR), 2833
Photovoltaic cell, 215
Photovoltaic devices, 771, 772f
Photovoltaic effect, 226–229
Photovoltaic electricity, 826
Photovoltaic generators, 215
Photovoltaic modules, 219
  price of, 224f

  series connection, 216, 216f
  structure of, 216–217, 217f
Photovoltaic solar cells, 226–240
  beginnings of, 226t
  characteristics, 229–230
  current–voltage characteristics of, 229f
  different types of, comparison, 239t
  electrical model of, 228–229, 228–229f
  materials, 230–239
  power–voltage characteristics of, 229f
  production procedure, 231
  structure and functioning of, 232f
  temperature dependence of, 230f
Photovoltaic systems, 217–219
  advantages, 217–218
  applications, 218
  BIPV, 223–224
  components of, 219–223
  costs of investment in, 224–225, 225t
  large scale power plants, 223, 223f, 224f
  types of, 218
Phragmites australis, 2830
Phyllocnistis citrella, 368
Phyllosilicates, 1627
Phyllotreta cruciferae, 444
Phylloxera vittifolae, 304
Physical exergy, 1084, 2531
Physical methods, for contaminant remediation, 1293
Physical solvent scrubbing, 458
The Physics of Blown Sand and Desert Dunes, 1013
Physisorption, 62–63
Phytase enzymes, 1813
Phytase supplementation, 1668, 1668f
Phytohormones, 349–350
Phytolacca americana, 2031
Phytophtora infestans, 409
Phytoplankton, growth of, 1075–1078
Phytoremediation, 394, 426, 1376, 2021
  of inorganic contaminants, 2120
  of organic contaminants, 2117
Phytosarcophaga gressitti, 1920
Phytoseiulus persimilis, 1750
Phytotoxicity, of herbicides, 1389–1390, 1389f
PIC. *See* Prior informed consent (PIC)
Picloram, 1318
PIC system. *See* Prior informed consent (PIC) system
"Pilot Safe Use Projects," 1409
Pimentel, David, 1401, 1403
Pinatubo, 447
Pipe drainage, impact of, on downstream peak flows, 585, 585f
Piperonyl butoxide (PBO), 2019
Pirata subpiraticus, 311
Pisum sativum L. (Peas), 427
Pit storage, energy
  underground, 2518

Pittsburgh Sleep Quality Index (PSQI) scores, 2874f
PJM Interconnection Grid, 2511, 2511f
  load duration curve, 2512f
Placic horizon, 1625
Planning for sustainability, 2447–2448
  actions and responsibility, 2449–2450
  adaptive planning and transition management, 2452–2453
  appraisal and monitoring, 2450
  effects of, 2450–2452
    environmental policy integration, 2452
    futurity and uncertainty, 2452
    participation and expertise, 2452
  illustrative forms of, 2448t
  objectives and interpretation, 2448–2449
  politics, 2453–2454
Plant(s)
  Al toxicity in, 270–271
  biomass, 373
  biomonitoring, 142–143, 2128
  Cu and, 535
  demand-side pests, 1951–1952
  growth on copper-deficient soils, 536
  growth on high-copper soils, 536–537
  potassium in, 2208
  sulfur supply and, 2435
  toxocity of herbicides to, 1386
Plant growth, organic soil amendments and, 1878
Plant growth-promoting rhizobacteria (PGPR), 351
Plant pathogens (viruses). *See* Virus and viroid infections
Plant Protection Act of 1990, 2823
Plant resistance, 304–305
  advantages, 304
  economic benefits, 305
  effects on pest populations, 304–305
  organic soil amendments and, 1878, 1879t
  overview, 304
  percentage of crops having, 304, 305t
  social benefits, 305
Plant-soil interactions
  potassium in, 2210–2211
Plant toxicity and heavy metals, 162
Plasma-enhanced chemical vapor deposition (PECVD), 234
Plasmodiophora brassicae, 442
Plasmopora viticola, 2009
Plastic(s)
  in coastal waters, 496–497
  debris, 2360–2361
  mulch, 2014
  recycling of, 847–848
Platanus occidentalis, 2830
Plate and frame heat exchanger, 873
Pleated filtration panel innovation, 2732
Pluralist theorists, 926

Plutella xylostella (L.), 314, 1928
P–n junction
　equilibrium in, 228f
　formation of, 227f, 233
　by ion implantation, 233
Poa pratensis (Bluegrass), 1305
Podisus maculiventris, 310
Podzolization, 1625–1626
Podzols, 426–427, 1625–1626
Point source (PS) pollution
　contaminants
　　interactions in soil and water, 2166, 2169
　　sources and nature of, 2166, 2167t–2169t
　defined, 2166
　and environmental quality, 2169
　impact assessment, 2170
　remediation process, 2170
　sampling for, 2169–2170
　on soil microorganisms, 2169
　through industrial emissions, 2170
Point-type in situ system, for monitoring stack gases, 145, 145f
Poisonings, pesticide, 510–511. *See* Pesticide poisonings
Polar ozone, 1887–1891
　characteristics of, 1887–1889
　destruction of, 1891
Polar stratospheric clouds (PSCs), 1887, 1889–1891, 1890f
Political economy, 2419
Pollen feeding, 310
Pollinators, 602
Pollinators, effects of IGRs on, 1466
Pollutants, agricultural runoff and associated, 82
　dissolved, 82–83
　of drinking water
　　inorganic, 2791–2793, 2791t, 2792t
　　organic, 2793–2797
　soil, 2114–2122
　　defined, 2114
　　inorganic, 2117–2121, 2116t
　　organic, 2114–2117, 2115t
Pollution, 914
　in Antarctica, 2138
　biological, 494–495
　in coastal waters, 489–498
　defined, 2166
　environmental. *See* Environmental pollution
　light, 497–498
　localized, 2166. *See also* Point source (PS) pollution
　noise, 497
　pesticides, in natural ecosystems, 2145–2148
　transport, 915
Polyacrylamide (PAM), 960, 961, 1553, 1555

Polyaromatic hydrocarbons (PAHs), 2116
Polybenzimidazole (PBI) membranes, 1138–1139
Polychlorinated biphenyls (PCB), 388, 1334, 2172–2182, 2357
　applications of, 2173, 2174t
　in aquatic environment, 2188–2189
　chemical identity of, 2172
　determination of, 2201–2202
　disposal of, from environment, 2182
　distribution of, 2173–2175
　extraction techniques, 2192
　health effects of, 2179, 2181
　impact on environment
　　absorption/desorption on organic matter, 2176
　　bioaccumulation, 2178
　　bioavailability, 2176
　　bioconcentration, 2178
　　biodegradation, 2176–2178, 2177f, 2177t, 2178f
　　biomagnification, 2178–2179
　　transformation, 2178
　　volatilization, 2175–2176
　isolation techniques
　　from sediment samples, 2196–2199
　　from water samples, 2192–2196
　physicochemical properties of, 2172–2173, 2173f
　properties and fate of, in aquatic environment, 2189–2190
　regulations for, 2181–2182
　sampling, transport and storage of the samples, 2191–2192
　sources and transport of, 2175t
　speciation of, in sediments, 2190–2191
　structure of, 2173f
　wastes, destruction of, 2180–2181t
Polycrystalline silicon, 231
Polyculture, 1935
Polycyclic aromatic hydrocarbons (PAH), 388, 1334, 2186
　biological and health effects of, 2188
　carcinogenic classifications, 2189t
　determination of, 2199–2201
　extraction techniques, 2192
　isolation techniques
　　from sediment samples, 2196–2199
　　from water samples, 2192–2196
　in marine sediments, 2188t
　sampling, transport and storage of the samples, 2191–2192
　sources and fate of, in aquatic environment, 2187–2188
　speciation of, in sediments, 2190–2191
Polyethylene, 838
　recycling of, 847–848
　tanks, 2264
Polygonum spp., 2038
Polyisocyanurate insulation, 1479
Polymerase chain reaction (PCR), 2616

Polymers, 960–961
Polyoxins, 2030, 2030f
Polyphosphate-accumulating organisms (PAOs), 2654
Polypogon monspeliensis (Rabbit foot grass), 426
Polysaccharide, 960
Pomatomus saltatrix, 293
Pond/lake heat exchanger, 1368, 1368f
POP. *See* Persistent organic pollutants (POP)
Popillia japonica, 321
POPs. *See* Persistent organic pollutants (POP)
Population pressure, pest management and, 2011–2012
Populus deltoides, 2830
Porcine pulmonary edema (PPE), 1697
Pore space, in soils, 1627
Pork freezing, 2704
Porosity of soil, 2752
Positive lynchet, 2537
Post-combustion carbon capture, 458
Postirrigation measurement, 1548
Potamogeton pectinatus, 1922
Potassium (K), 2208–2211
　calibrated soil tests for, 1806–1807
　fertility, assessing and managing, 2211
　overview, 2208
　in plants, 2208
　in plant–soil interactions, 2210–2211
　in soil, 1514, 2208–2210, 2209f
Potassium (K) nutrient
　applications
　　timing, 1809
Potassium chromate, 479t
Potato (*Solanum tuberosum* L.), 1304, 1307
Potato tubers (*Solanum tuberosum* L.), 425
Potential acid sulfate soils, 2294
　sulfide mineral formation (sulfidization), 31–33
　and accumulation, 32–33
　oxidizable organic carbon, 31
　reactive iron, 32
　reducing/saturated conditions, 31–32
　sulfate, 31
　sulfate-reducing bacteria, 32
　sulfuricization, 33–34, 34f
Potential evapotranspiration (PET), 1853
Pothole marsh, 2826f
Poultry manure, 1670
　cleaning/removing frequency, 1671
　liquid, 1670, 1672
　　application system, 1674, 1674f
　　storage of, 1672, 1672f
　litter use and cleanout, 1671, 1671f
　management of, 1670–1674
　production of, 1670, 1670t
　solid, 1670
　　high-rise layer facilities for, 1671, 1671f

Poultry manure (*cont'd.*)
  subsurface application of, 1673–1674, 1674f
  surface application of, 1672–1673, 1673f
  storage of, 1671–1672, 1671f
    indoor storage, 1671–1672, 1671f
    nutrient losses during, 1672–1673
    outdoor storage, 1672
  utilization of
    compost pits, 1672, 1673f
    energy uses, 1674
    as fertilizer, 1672
    liquid manure application, 1674
    subsurface application, 1673–1674, 1674f
    surface application, 1673, 1673f
Poured concrete, 863–864
Power cycle, 2527, 2528
  efficiency, 2527–2528
Power high-temperature Winkler (PHTW) gasifier, 2492–2493
Practical application impact ratio, 1235
Praseodymium, 2267t, 2268
PRB. *See* permeable reactive barrier (PRB)
Precast autoclaved aerated concrete (PAAC), 1483
Precast concrete, 864
Precaution, definition of, 1961
Precipitation, 1593t
  agroforestry and, 130
  erosion and, 1057
  method, 918
  or chelation, 2152
  and water erosion, 945
  and wind erosion, 945
Precision Agricultural-Landscape Modeling System (PALMS), 2221
Precision agriculture (PA), 2213–2215, 2217–2223
  crop yield, 2218–2219
  crop water use, 2218–2219
    landscape, 2219
    spatial analysis of, 2219–2221
  future perspectives, 2215, 2221–2222
  geographic information systems (GIS), 2215
  global positioning systems (GPS), 2213–2214
  overview, 2213
  paradigm for, 2213–2214, 2214f
  sensing for, 2214–2215
Precision conservation, 1037
Precocenes, as anti-JH agents, 1464–1465, 1464f
Pre-combustion carbon capture, 457–458
Precor. *See* Methoprene
Predators, 1475, 1475f, 2423, 2424. *See also* Natural enemies
  as biological control agents, 2596
  intercropping and, 1938

Preemption, 1943
Preferential flow
  leaching and, 1621
Pregnancy, and arsenic exposure, 1273f, 1274
Prenatal developmental toxicity, 283
Preplant nitrogen applications, 1808
Preplant N tests, 1306
Pre-side-dress N tests, 1306
Pressure Drop, 210–211
Pressured sprinkler systems, 120
Pressure sand filter, 2722–2723. *See also* Filter(s); Sand filter
Pressurized liquid extraction (PLE). *See* Accelerated solvent extraction
Primary energy ratio (PER), heat pump system, 1351–1352, 1352f
Primary pests, 2424t
Primary productivity
  carbon sequestration and, 2833
Primary radionuclides, 2239
Primer effect, 175
Primordial radionuclides, 2235–2236
Principal response curves (PRCs), 292
Prior informed consent (PIC), 1612, 1616t, 1617
  history of, 1615–1616
  procedure, 575
Private costs, 917, 917f
Private nuisance, 1945
Probabilistic hazard assessment (PHA), 296
Probabilistic risk assessment (PRA), 291, 296
Probabilistic uncertainty analysis, 1722
Probiotic agriculture, 528–529
  advantage of, 531–532
Process-based commissioning, 666–667
Process-based erosion models, 991
  continuous simulation, 991–992
Process-based models
  multicompartment organic models, 1863–1864
  for soil erosion prediction, 984–986, 986f
Process-dependent models, for exergy analysis, 1095
Process engineering of biological waste gas purification, 2619
Prodigy. *See* Methoxyfenozide
Producer-consumer driven movement and organic farming, 1107–1108
Producer gas, 376
Program for Water Quality and Quantity Improvement in Rural Catchments, 971
Project-based system, 486
Propagule pressure
  biological invasions and, 1531
Propargite, 3
Propineb, 2142
Prostephanus truncatus, 2423

Protective and restorative measures, 1583–1584
Proteinases, 89
Proteins, 786
Proteomes, 786
Proteus, 321
Prothoracic glands, 1465
The Protocol on Water and Health, 1585
Protocols
  IPM, 1522–1523
Proto-imogolite complexes, 1626
Proton exchange membrane fuel cells (PEMFCs), 1145, 1148, 1152t
  applications of, 1150–1151
  cold start of, 1149
  contamination levels, acceptable, 1149
  CO poisoning effect on, 1149
  description of, 1148, 1149f
  operating principle of, 1148–1149
  operating temperature of, 1148
  research on problems related to, 1149–1150
  stacks and components, 1150
  technological status of, 1150
Protozoa, in waste and polluted water, 2698
Provisional maximum tolerable daily intake (PMTDI), 1696
Proximate analysis (ASTM D3172), 717, 718t
PRV. *See Papaya ringspot virus* (PRV)
Pseudo–first-order model, 66
Pseudo–second-order model, 66–67
Pseudomonas, 96, 321
Pseudomonas fluorescens, 323
Pseudomonas paucimobilis UT26, 390
Pseudomonas putida, 351
Pseudomonas spp., 392, 2030
Psoroptes ovis, 2
P-spiralling, 2101
PS pollution. *See* Point source (PS) pollution
Public attitude, pest management and, 2011
Public health dangers and concerns, 1414–1415
  acute effects (pesticide poisonings), 1414–1415, 1415f
  chronic effects (cancer and other health concerns), 1415
  economic costs, 1416
  effects on children, 1415–1416
  exposure to pesticides, 1414
  global usage of pesticides and, 1414
Public support, for wind erosion control, 1028
Public utility commissions (PUCs), 1791, 1792
Pulverized fuel (pf) fired swirl burner, 726f
Pump-and-treat method, 1293, 1844
Pump-and-treat technology, 1334–1335
Pumped storage hydro, 1418, 1421–1422, 1423

benefit for renewable energy resources, 1425–1427
countries with, 1424t
electric utility usage of, 1424–1425
environmental issues, 1424
facility description, 1423–1424
ramping characteristics of, 1427
schematic diagram of, 1422f
summer operation of, 1425f
technology description, 1422–1423
winter operation of, 1426f
Public utility commissions (PUCs), 1791, 1792
Pumping
groundwater, 1284, 1285, 1285f
to manage saltwater intrusion, 1330
Pure pumped storage systems, 1423–1424
Pushbutton setup, 736, 737f
Pyramid Lake, degradation of, 1437
Pyrethrin, 2016t, 2019, 2028–2029, 2029f
Pyrethroid, 3, 2141, 2805
insecticides, 1318
Pyricularia oryzae, 2030
Pyridaben, 3
Pyridazinones, 3
Pyriproxyfen, 1460, 1464
Pyrite, 42-43, 55
Pyrite, oxidation of, 1688–1689
Pyro-char, 844. *See also* Biochar
Pyro-gas, 376
Pyrolysis, 376, 723, 724f, 836, 844, 845
biomass and wastes, 2488
features for fast, 2487
pathways, 2486f
products, 2484
reactions involved in, 2485f
technologies, conditions, and major products, 2486t
world scenario of, 2489t–2490t
commercial applications of, 376
definition of, 376
oils, 846
process of, 376
products, organization of, 837–840
Pyrroles, 3
Pythium spp., 2107, 2109

Q

Qinghai Hu Lake, degradation of, 1437
QMRA methodology, 2707
Q-SOIL, 1864
Qualitative uncertainty analysis, 1722
Quantitative indicators, of soil water erosion, 967

Quantity factor, 100
Quasi-equilibrium processes, 816
Quasi-kinetic energy, 811
Quasi-potential energy, 811
Quaternary N herbicides, 1318
QuEChERS method, 1973, 1973f
Queensland Acid Sulfate Soils Manual, 58
Quelea quelea, 413
"Quesungual" agro-forestry method, 2382
Quiet zone, 129
Quinalphos, 2142

R

Rabbit Calicivirus Australia-1 (RCV-A1), 2602
Rabbit foot grass (*Polypogon monspeliensis*), 426
Rabbit hemorrhagic disease (RHD), 2596
as biological control for European rabbits, 2600
Rabbit hemorrhagic disease virus (RHDV), 2596, 2600–2601
apathogenic rabbit caliciviruses and, 2601–2602
as biological control agent
in Australia, 2601
future for, 2602
impact of, 2601
in New Zealand, 2601
Rabbits, biological control of, 2597–2602. *See also* Myxomatosis; Rabbit hemorrhagic disease virus (RHDV)
Radiant heat barriers (RHBs), 1490–1491
configuration, 1491f
mechanism, 1491
Radiation of energy transport, 865
Radiative forcings, estimated, 1194f
Radioactive cesium, 1628
Radioactive decay, 2234–2235
alpha (α) decay, 2234
beta minus (β^-) radiation, 2234–2235
gamma (γ) radiation, 2235
Radioactive pollution, solution to, 2240
Radioactive strontium, 1628
Radioactive waste, 2136, 2359–2360
Radioactivity, 2234
radioactive decay, 2234–2235
radioactive pollution, solution to, 2240
radionuclides
anthropogenic, 2236–2240
in environment and pollution problem, 2240
natural and artificial, 2235–2236
Radio frequency towers, 2224, 2225f
with elaborate camouflage, 2227f
ethics and public administration, 2229–2230
impact and study of RF hazards, 2224

background research, 2225
recent trends, 2224–2225
RF research on animals and humans, 2225–2226
voluntary initiative, 2226
RF tower placement process, 2228–2229
Students Against Cell Towers (SACT), 2226–2228
Radioisotopic techniques, 1024–1025
Radiometers, 1194
Radionuclides, 2243–2246, 2304, 2359
anthropogenic, 2236–2240
bioremediation, 390t
in coastal waters, 491–492
in environment and pollution problem, 2240
fate and transport of, 2243–2246
geochemical processes, 2244, 2245f
hydrologic processes, 2243–2244, 2244f
microbial processes, 2244–2246
natural and artificial, 2235–2236
overview, 2243
Raindrop impact on soil, 2752, 2753
Raindrop-induced saltation (RIS), 987, 989
Rainfall
desertization and, 566
erosion and, 1044–1045, 1045f
erosivity, 943, 980–989
and rain amount, 944, 944t
erosivity index, 943
kinetic energy, 983–984
Rain measurements and hydrology, 203–204
Rainwater
pipes, 2264
as source of NPSP, 2136
Rainwater harvesting, 1276
advantages of, 2262
design/maintenance of
catchment surface, 2263–2264
conveyance systems, 2264
storage tanks, 2264
importance of, 2262
types of
land surface catchments, 2263, 2263f
large systems, 2262, 2263f
rooftop collection systems for high-rise buildings, 2263
simple rooftop collection systems, 2262, 2263f
stormwater collection in urbanized catchment, 2263
Ramsar Convention, 1585
Ramsar Convention Bureau, 2825
Ranavirus, 2605
Random walk method, 1315
Range and pasture lands
biological characteristics, 2753–2754
chemical characteristics
dissolved chemicals, 2753
dissolved oxygen, 2753

Range and pasture lands (*cont'd.*)
  clean water, 2752
  livestock, 2752
  physical characteristics
    livestock impacts on soil, 2752
    livestock impacts on vegetation, 2753
    suspended sediment, 2752
  riparian site protection, 2754
  runoff/erosion, limit on, 2754
Range finder lidars, 144. *See also* Lidars
Range reseeding technique, 570
Rankine cycle, 819
Rapid gravity filter with coagulant aid, 2722. *See also* Filter(s)
Rare earth elements (REE)
  atomic numbers/symbol of, 2266, 2267t
  binding forms in soils, 2266
  chemical speciation of
    adsorption of, 2267
    species of, 2266, 2268t
    total content, 2266, 2268t
    translocation of, 2268
  crop response to, 2269, 2269t
  crop yield increase by, 2269t
  heavy, 2267t, 2268t
  light, 2266, 2267t, 2268t
  mean content of, in soils, 2268t
  physico-chemical soil properties, 2266
  in sludges (concentrations/coefficients), 2267t
  soil REE, origins of, 2266, 2267t
  uptake of, by plants, 2268, 2268t, 2269t
Rating systems for buildings, 1655
RDF. *See* Refuse-derived fuel
REACH (Registration, Evaluation, Authorization and Restriction of Chemicals), 647
Reactor system, 1906
Reactive oxygen species (ROS), 646, 1155
Real fluids, 2532–2533
Real-time kinematic (RTK-GPS) systems, 2213
Real-time polymerase chain reaction (RTPCR), 2130
Real-time sensor, 2214–2215
Reburn with biosolids, 732–733, 732f, 733t
Recessed lighting insulation, 1490
Recharge flux, 1284–1285, 1285f
Reclamation
  open cut mine, 1693
  of sodic soils. *See* Sodic soils, reclamation of
Recommended maximum concentrations (RMCs)
  of selected metals and metalloids in irrigation water, 2677t
Recommissioning, 657
Recruit II. *See* Hexaflumuron
Recruit III. *See* Noviflumuron
Recruit IV. *See* Noviflumuron

Recultivation, 2312
Recuperator, 873
Recycling
  of dairy wastewater, 1664, 1665t
  and SWEATT, 847–848
Redevance d'Enlèvement des Ordures Ménagères (REOM), 895
Red Lists, 603
Red mud, 1431–1432
Redlich–Peterson isotherm, 65–66. *See also* Isotherm(s)
Redox reactions
  leading to immobilization, 393
  leading to solubilization, 393–394
  in water-sediment systems, 2294–2296
    biological factors, 2295, 2295t
    climatic factors, 2295–2296, 2296t
    surface inland waters, 2295t
Reduced inorganic sulfur (RIS), 36, 37, 39, 40, 42
Reduced tillage, 1307
Reductive biodegradation, 1338
REE. *See* Rare earth elements (REE)
Reference dead state, 816
Reference environment, 1084, 1093
Reference state, 1084
Reference-substance models, for exergy analysis, 1094–1095
Reflective insulation, 1479
Refrigerants, 1358
Refuse-derived fuel (RDF), 727, 836
Regional Air Pollution in Developing Countries (RAPIDC) program, 7
Registration, Evaluation, and Authorization of Chemicals (REACH), 1700–1701, 1704
Registration, Evaluation, Authorisation, and Restriction of Chemicals (REACH, 1907/2006), 449
Regulations
  of animal pests, 1115–1117
    anthropogenic measures, 1116–1117
    biotic (natural) resistance of environment and, 1115–1116
  pesticides. *See* Pesticide control legislation
Regulations/laws
  pesticides, 1612–1614
    available international guidelines, 1612–1613
    future global policy, 1613
    historical perspectives, 1612
    implementation problems, 1613
    need for, 1612
    present scenario and probable remedies, 1613
    steps undertaken, 1613
    toxicological and other data requirements for registration, 1614
Rehabilitation, 2308, 2316t
  after open cut mines, 1692–1694

acid mine drainage, 1694
coversoil resources, 1692
coversoil thickness requirements, 1693
landscape regrading, 1692, 1693f
in situ soil reclamation, 1693
steep slope reclamation, 1694
of minerals processing residue, 1683–1687. *See also* Tailings
soil
  case study, 2397–2398, 2397f
  overview, 2396
  soil biological indicators, 2396–2397, 2397t
Reinfestation of *A. donax*, 1186
Relative agronomic effectiveness (RAE), 1514
Relay logic, 736
Releaser effect, 175
Reliability
  of GIS-based analysis results, 1164–1165
Relief drains, 590
Remediation, 2316t
Remote monitoring techniques, for air quality monitoring, 143–144
Remote sensing (RS)
  advances in, 2818
  GIS integration and. *See* RS/GIS integration
  inventorying and monitoring, 2816–2817
  in precision agriculture, 2215
  research and development, 2817t, 2818–2819, 2818f
  satellites with sensors, 2817t
  as source of spatial data, 2271–2272, 2272f
  uses of, 2273t
  in watershed management, 2816–2819, 2817t
Remote sensing/GIS integration
  applications, 2272
  compatibility issues, 2271
  linking, 2271–2273
  for site-specific farming, 2272–2273
Remote sensing imagery, 1164, 2272
Remote-sensing products, use of, 561
Removal unit (RMU), 486
Renaissance of natural theology, 2008
Renaturization, 2309–2310
Renewable energy, 774, 806
  adequacy of, 828–829
  conversion efficiencies, 825
  costs, 825–826
  energy efficiency gains, 828
  forms, 824
  intermittent sources, 826–827
  portfolios, 2462
  present energy use, 827, 828t
  resources, 825
  thermal storage for, 2517–2521
    active solar heat storage, 2520–2521, 2522f

building component storage, 2519–2520
  solar passive heat storage, 2519
  underground thermal energy storage, 2518–2519
  use of, 122
Renewable energy sources (RES), 715–716, 920, 1419. *See also* Pumped storage hydro
  pumped storage hydro benefit for, 1425–1427
Repellents
  for bird control, 414–415
  used in homes and gardens, 1401
Report on the Insects of Massachusetts Injurious to Vegetation, 2008
Reproductive effects
  of pesticide, 1399
Reproductive inhibitors
  for bird control, 415
Reproductive toxicity, 2124
  of endocrine-disrupting chemicals, 646
Research
  precision agriculture-related, 2221–2222
Reseda odorata, 1925
Reservoirs, 1577–1578. *See also* Lakes and reservoirs
  dry, 1588
  flow-through, 1588
  retention, 1588
  tillage, 1554
  storage, 1563–1564
Reservoir specific exergy utilization index, 1175
Residential Energy Consumption Survey (RECS), 792–793
Residual current devices (RCDs), 223
Residue-reduced food, 1403–1404
Resistance ratio (RR), 1467
Resource allocation, 915
Resource concentration hypothesis, 1935
Respiratory effects, of arsenic toxicity, 1272, 1272f
Response curves, for fragile and robust landscapes, 542, 542f
Responsibility
  of food manufacturers, 1610–1611
Restoration, 2316t
  of native terrestrial plants, 2316t
Restricted use product (RUP), 1941
Restriction on Hazardous Substances directive (RoHS, 2002/95/EC), 449
Retailer initiatives, 1403–1404
Retained acidity, 40
Retention, 2314
Retention/absorption terraces, 968, 968f
Retention reservoirs, 1588. *See also* Reservoir(s)
Retrocommissioning, 657
Retrospective Analysis of the Clean Air Act, 1638

Revegetation, 1186
Revenue stability, 2786
Reverse cycle air conditioner, 1347
Reverse osmosis, 2728–2729
  pretreatment for, 2729, 2729f
Reversibility, 2528, 2531
  and irreversibility, 816–817
Reversible circuit heat pumps, 1350
Reversible energy transfer, 819, 821
Reversible heat transfer, 817, 817f
Reversible thermodynamic process, 2531
Revised universal soil loss equation (RUSLE), 941, 943, 975, 977, 981, 1035, 1048, 1059, 2759
Revised Wind Erosion Equation (RWEQ), 1026, 1026t, 1959
Revitalization, 2316t
Rewashing, 1592t
R-factor, in Universal Soil Loss Equation, 943
RHBs. *See* Radiant heat barriers
Rhinocyllus conicus, 2822
Rhizobia, 1110
Rhizobium (RHZ), 88, 349, 1515, 1516, 1770
Rhizoctonia-infested soil, 444
Rhizoctonia solani, 2030, 2107, 2109
Rhizoctonia solani anastomosis groups, 443
Rhizodegradation, 1847
Rhizome or culm, 1182
Rhodnius prolixus, 1462
Ribulose bisphosphate carboxylase (rubisco), 89
Rice (*Oryza sativa*), 425
Rice fields
  CH_4 emissions from, mitigating options for, 1679–1682, 1681t
    organic matter management, 1680
    processes controlling, 1679, 1680f, 1680t
    soil amendments and mineral fertilizers, 1680
    water management, 1679, 1680f
    problems and feasibility of the options, 1681–1682
Richness factor, 100
Ricinus communis L., 288, 289
Ricinus spp., 1878
Rid-A-Bird®, 413
Ridge tillage, 1027
  erosion process and, 1041–1042, 1042f
Ridges, 50
Right-to-farm laws, 1945
Riley, C. V., 2009
Rills, 964
  defined, 958
  erosion and, 1045–1046, 1046f
Rimon. *See* Novaluron
Rio Agenda 21, 1654
Riparian ecosystems, 1182, 1183, 2830
Riparian site protection, 2754
Riplox method, 1595
Ripping, 570

Risk assessment, 1722
  new/existing chemicals, 1070
Risk shifting and risk reduction, distinction between, 1796
River
  ecosystems, 2314
  functionality analysis, 2310
  recultivation, 2309–2316
    multitasking for, 2309, 2311f
    planning and development, 2310f
  restoration, 2307–2308
River flows, irrigation and. *See* Irrigation and river flows
Riverine systems
  abiotic process, 2101
  biotic process, 2101
  and land use impacts on P transport, integration of, 2101, 2102–2103f
  nutrient spiralling, 2101
  physical process, 2100–2101
  P-related impairment in flowing and lake waters for targeted remediation, 2101–2102, 2104f, 2104t
  transport in, phosphorus, 2100–2105
  water quality response, implications for, 2103, 2105
  watershed management, implications to, 2102–2103
River pollution, 2303–2304
  removal of pollutants, 2304–2305
  sources, 2304
    acid rain, 2304
    agricultural pollution, 2304
    industrial pollution, 2304
    radionuclides, 2304
Road runoff, 2321, 2771
Rock-plant filters, 2863, 2864f, 2865f
Rocks
  heavy metals in, 1370–1373, 1372f
  nanoparticles in, 1715–1716
Rock-soil-water interactions, 2293–2294
Rock wool, 1478
Rocky Mountain hydroelectric plant, 1422f
Rodenticides, 420, 2014
  used in homes and gardens, 1401
Romdan. *See* Tebufenozide
Roof(s)
  preventing air flow restrictions, 1488–1489
  soffit air ventilation, 1489
  stick-built, 1489
Roof catchment, materials used for, 2263
Rooftop collection systems
  for high-rise buildings in urban area, 2263
  simple, 2262, 2263f
Root-absorbing power, 1821, 1821f
Root biomass, 346
Root hairs
  importance of, 1820–1821, 1821f
Rootshield, 383
Root zone storage capacity, 1546

Ro-Pel®, 415
Rotary drums, 518–519, 519f
Rotating biological contactors (RBCs), 2625, 2626f
Rotating rainfall simulator, 964f
Rotation
  crop, 1111, 1112
  livestock, 1111
Rotterdam Convention, 575, 1615–1617, 2036
  banning exports of banned pesticides, 1616
  building capacity/improving regulations, 1617
  Chemical Review Committee, 1616
  Designated National Authorities, 1616
  import decisions, information, and website, 1617
  information exchange, 1617
  notifying regulatory actions, 1616
  overview, 1615
  PIC, 1616t, 1617
    history of, 1615–1616
  from voluntary to legally binding, 1616
Rotterdam Convention on the Prior Informed Consent Procedure for Certain Hazardous Chemicals and Pesticides in International Trade, 1914
Row crops, 1303–1304
RS. See Remote sensing (RS)
RTK-GPS systems. See Real-time kinematic (RTK-GPS) systems
Ruminants, copper deficiency in, 428
Run-around coils, 873
Runner. See Methoxyfenozide
Runoff
  Agricultural. See Agricultural runoff
  erosion rates and, 1058–1059
  limited, 2754
  water, 1546
Run-of-river, 1423
  plants, 202
Rural and urban water resources management, 2157
RUSLE. See Revised universal soil loss equation (RUSLE)
Ruth's type storage system, 2521, 2522f
R-value, 1478
Rye, 1110
Ryegrass (*Latium perenne* L.), 1305

S

Safe Drinking Water Act (SDWA), 1289
Safety equipment
  chemigation, 470f, 471, 472f
Sage grouse (*Centrocercus urophasianus*), 420
Saissetia oleae, 1750
Salihli Geothermal District Heating System (SGDHS), case study analysis of, 1178–1179, 1178f
Saline formations, 459
Saline seeps, 590
Saline soils
  drainage requirement, 588–589, 589f
Saline waters
  classification of, 1569t
  effect on soil, 1568
  irrigation with, 1568–1569
  low-salt and salty waters, mixing, 1569
Salinity, 2137, 2137t
  primary source of, 1557–1558, 1558t
Salinity Control Act of 1974, 1566
Salinity control projects, 1566
Salt(s), 1559
  concentration, 1569t
  in soil, 1568
Salt-affected soils, 2345–2347
  classification, 2346t
  crop tolerance, 2346–2347, 2347t
  defined, 2345
  extent and distribution, 2345–2346, 2346f
  overview, 2345
Saltation, 1019
  defined, 1014
Salt hydrates, 2509
Saltiphone sensor, 1023, 1024f
Salt loading pick-up, 1566
Salt loading values, 1559
Salt marsh, 2825t
Saltwater intrusion
  in groundwater, 1321–1332
    combating, 1330
    computer models, 1329–1330
    Israel and Palestinian territories, 1322
    Italy, 1322
    mathematical modeling, 1328–1329
    mechanisms, 1322–1325
    Mexico, 1322
    monitoring and exploration of, 1325–1328
    Netherlands, The, 1322
    planning and management, 1330
    transition zones, 1324–1325, 1325f, 1328
    into unconfined aquifer, 1322f
    United States, 1321–1322
    vertical cross sections of, 1323f
Sand filter. See also Filter(s)
  pressure, 2723–2724
  slow, 2723
San Pedro River (case study), 554–555
Saprobe index, 602
SAR. See Sodium adsorption ratio (SAR)
Sarcoptes scabiei, 2
Sarritor, 383
Saudi Arabia, 2682, 2689
SAV. See Submerged aquatic vegetation (SAV)
Save the Planet groups, 1612
Sawgrass wetlands, 1081
Scarcity rents, 2782, 2783
Scarites subterraneus (F.), 310
Scatterbird®, 413
Scattering, 1196
Scheduling and Network Analysis Program (SNAP), 2819
School of Environmental Studies (SOES), survey of, 1265
Schwertmannite, 42–43, 1689
Sclerotinia, 443, 444
Sclerotinia minor, 383
Sclerotinia sclerotiorum, 443
Sclerotinia spp., 2107
Sclerotinia stem rot, 443, 443t
Scotia segetum, 1116
Screen printing, 233
Scrubbing, 456, 457
SDI. See Subsurface drip irrigation (SDI)
SDSS. See Spatial decision support systems (SDSS)
Seasonal energy efficient ratio (SEER), 1364
Seasonal heating energy efficiency ratio (SHEER), 1351
Seasonal irrigation efficiency, 1546–1547
Seasonal performance factor (SPF), 1351
SEAWAT
  for saltwater intrusion, 1329
Seawater expansion, 1199
Seawater scrubbing, 2441
Secale cereale (Winter rye), 1306–1307
Secondary enemies
  effects on natural enemies, 371
Secondary pests, 2424t
Secondary salinity, defined, 2335
Second-law efficiency, exergy and, 820
Second law of thermodynamics, 819, 2528–2531
  entropy and, 817–819
  formulation, 2529–2530
  ideal work, 2530–2531
  lost work, 2530
Sectoral energy conservation, 688
Sectoral impact ratio, 1235
Sedimentation, 953, 964, 1579
  tank, 210
Sedimentation rates
  erosion, historical review, 1050–1052
Sediment composition
  effect of particle travel rates on, 986–987, 987f
Sediment control, 958–962
Sediment delivery ratio (SDR), 966
Sediment flux, 965
Sediment ponds, 1553
Sediments, 2863
  effect on surface water quality, 2755–2756
  isolation, 1593t
  nanoparticles in, 1715–1716
  nutrient deposition/deactivation in, 1592t

Sediment transport
　in watersheds, 965, 965f
Sediment trap, 994, 2264
　efficiency, 1000–1001, 999f
Sediment yield, 965–966
Seed and Fertilizer Approach, 2816
Seedborne inoculum, 442
Seedborne pathogen management, 444
Seed dressings, 419
Seeding rate, 1111
Seedlings killing, 443
Seed treatment, 444
Seepage lakes, 1577
Seize. *See* Pyriproxyfen
Selectivity, herbicides, 1379
Selenium, toxic level of, in wetlands, 1626
Selenium toxicosis, 1560
Self-balancing concept
　erosion by wind and, 1015, 1015f
Self-organization, 604–605
Semiarid region irrigation
　and soil IC, 465–466
Semidirect effect, 1197t
Semidry techniques, 2441
　duct sorbent injection, 2442
　furnace sorbent injection, 2442
　hybrid sorbent injection, 2442
　and SO_2, 2441
　　duct sorbent injection, 2442
　　furnace sorbent injection, 2442
　　hybrid sorbent injection, 2442
　　spray dry scrubbers, 2441–2442
　　spray dry scrubbers, 2441–2442
Semiochemicals, 175
Semivariance, 103f
Sensible heat storage, 2509
Sensible thermal energy, 811
Sensing for precision agriculture,
　2214–2215
Sensitivities, pesticide. *See* Pesticide
　sensitivities
Sensit, 1023, 1024f
Sentricon. *See* Hexaflumuron
Sequential extraction methods, 41
Serenade®, 382
Serratia, 321
Serratia entomophila, 322
Service stability, 2786
Sesbania (*Sesbania aculeata*), 1511
Seston removal/catch, 1593t
Settling/terminal velocity, 151
Severely hazardous pesticide incident
　report, 575
SEVIN, 2033
Sewage effluents for irrigation
　in California, 1571
　coliform counts in, 1571
　damages, 1570
　in Israel, 1571
　long-term effects of, 1570
　in Mexico, 1571

　in Middle East, 1570
　monitoring guidelines for, 1571
　objectives of, 1570
　reuse standards, 1571
　untreated, 1571
Sewage effluents, in coastal waters, 496
Sewage sludge, 269, 1515
　for bio-liquid fuels, 847
　and INM approach, 1515
　as sources of REE in soils, 2266, 2267t
　treatment, 1908
SF_6. *See* Sulphur hexafluoride (SF_6)
Shadow prices, 2782–2783
Shadow-Voltaic systems, 224
Shakeback disease, of yaks, 427–428
Shaking-assisted extraction, 2196
Shannon–Wiener index, 602
SHARP
　for saltwater intrusion, 1329
Sharon process, 2653
Sheathing, 862, 863
Shell and tube heat exchanger, 873
Shell Chemicals, 2031
Shelter for natural enemies, 371–372
Shelter sites
　natural enemies conservation and, 1750
Shewanella alga, 2245
Short-rotation woody crops (SRWC),
　345
Sick building syndrome, 1655
Side-dress nitrogen applications,
　1808–1809
Signal-to-noise ratio, 2817t, 2818
Silent Spring, 2010
Silicon, 230–231. *See also* Crystalline
　silicon solar cells
　crystalline, solar cells, 231–232
Silos/towers composting, 518. *See also*
　Composting
Silt, 1716
Silver nanoparticles
　bactericidal effect of, 1740
　pollutant removal, mechanism in,
　1739–1740
Silvicultural practices variation and water
　quality effects, 2772–2773
Simazine, 1318t, 1957t
Simpson index, 602
Simulation models
　for crop management, 2221
　of crop systems, 1950–1951, 1951f
　for surface water quality, 2759–2762,
　2762t
　　environmental models, 2761–2762
　　soil erosion models, 2759–2761
Simulation of solute transport, 1315–1316,
　1315f
Simulator for Water Resources in Rural
　Basins (SWRRB), 992
Single-drop microextraction (SDME),
　2193

Single-path-type in situ system, for
　monitoring stack gases, 145–146,
　145f
Single photon emission computed
　tomography (SPECT), 1412
SIPs. *See* structural insulated panels
Sister chromatid exchange (SCE) analysis,
　2128
Site-specific agriculture. *See* Precision
　agriculture techniques
Site-specific farming (SSF)
　remote sensing/GIS integration for,
　2272–2273
Site-specific management. *See* Precision
　agriculture (PA)
Site-specific models
　groundwater modeling, 1296–1297
Site suitability analysis
　GIS in land use planning and, 1164
Slagging gasifier, 2491
Slag wool, 1478
Slaking, 1627, 2334, 2335f, 2339,
　2340f
　management of, 2341, 2342t
Slow sand filter, 2722. *See also* Filter(s);
　Sand filter
Sludge volume index (SVI), 2649
Sluicing, 47
Small hydro plants, 202
Small hydropower (SHP) plants,
　201–202
Smectitic soils, 1628
Smith, John, 475
Smith-Lever act (1914), 2009
Smith, Robert 6
Smoothing natural riverbed, 2313
Snowmelt erosion, 1057–1059
　amelioration, 1059
　modeling, 1059
　precipitation, 1057
　runoff events, 1058–1059
　soil, 1057–1058, 1058f
SOC. *See* Soil organic carbon (SOC)
Social capital, 2464
Social cost, 917
Social sustainability filter (SoSF), 639
Soddy-Fajans periodic method, 2234
Sodicity, 2137, 2137t
Sodic soils
　chemistry of clay–cation interaction
　　and its effect on swelling and
　　dispersion, 2338–2339, 2339f
　defined, 2335–2337
　drainage requirement, 589
　global distribution of, 2335t
　management of, 2341, 2342t
　physical property and behavior, effect of
　　sodicity on, 2339–2342, 2340f
　　hydrologic properties, 2340–2341
　　mechanical properties, 2341
　sodicity, measurement of, 2335–2338

Sodic soils, reclamation of, 2370–2374
  ameliorants and
    optimal supply of, 2373
    rate of supply of, 2373–2374
    without addition, 2373
  biology and organic matter, role of, 2371–2373
  chemical ameliorants, 2370
  exchangeable Na ions displacement, 2370, 2371f
  plant growth, effects of, 2371
  and soil IC, 466
  strategies, 2373–2374
  water flow, improvement in, 2370
Sodium (Na)
  montmorillonite clay, swelling of, 2336, 2336f, 2338, 2339f
Sodium adsorption ratio (SAR), 1559, 1693
Sodium chromate, 479t
Sodium salts, 1568
Sodium sulfite bisulfite process, 2442–2443
Sodium tolerance of crops
  classification, 2347t
SOES's recommendations, for arsenic mitigation, 1277
SOFCs. *See* Solid oxide fuel cells (SOFCs)
Soil
  acidification, 22
  adsorption, 1957
  aluminum in, 268–269
  amelioration, 1431
  amendments, 2137
    CH_4 emissions from rice fields and, 1679
    gypsum, 959–960
    lime, 959
    manure, compost, and organic sludge, 961
    synthetic polymer, 960–961
  bioaccumulation in, 325
  biodiversity, 334–335, 353
  biotechnology, 2730
  carbon losses from
    mineralization processes and, 1857
    soil erosion processes and, 1857–1858
  carbon sequestration in, 2833–2834
  compaction, sulfur and, 2433–2434
  conservation
    approaches to, 1036–1037, 1036t
    services, 1037
    widespread adoption of, 1037–1038
  contamination
    boron in, 431–433
    treatment, biological techniques for, 2014
    via root exudates, 310
  copper in, 535
  cultivation, minimum, 1509
  degradation, 1618
  ecological services of, 1624
  enzyme, 2020
  erosion. *See* Erosion
  fertility, defined, 2431
  formation, 1034
  fumigants, 444
  health-based interventions, for wastewater irrigation, 2678–2679
  heavy metals in, 1370–1372, 1372f
  leaching and, 1621, 1622f
  livestock impacts on, 2752
  management, 1619, 1620
  management, and organic farming, 1106–1107
  matrix, 1570
  microbial biomass, 2396
  microbial diversity, 388
  microorganisms, 2014, 2020
    PS pollution on, 2169, 2170
  nanoparticles in, 1712–1715, 1713t
  nitrogen transformations in, 1768–1769, 1769f
  permafrost, 1900–1902
  pesticide effect on, 2147–2148
  pesticides in, 1966, 1974
  pH
    and boron availability, 424
    and Mo availability, 427
  potassium in, 2208–2211, 2209f
  primary function of, 1570
  quality
    assessment of, 2458–2459, 2458f
    sustainable agriculture. *See* Sustainable agriculture
  rehabilitation. *See* Rehabilitation
  remediation by nZVI, 1737–1739
  retention, 2314
  salt-affected. *See* Salt-affected soils
  scape filter, 2731, 2731f
  sodification, 1561
  strontium and interaction with soil matrix, 2426–2427, 2428f
  structure, 2752
  sulfur. *See* Sulfur
  sustainable agriculture, 2457
  sustainable use of, 1618
  tillage methods, 119t
  translocation processes in, 1624. *See also* Eluviation/illuviation; Leaching
  winter erosion process, 1057–1058, 1058f
Soil and water assessment tool (SWAT) model, 966, 2762
Soil and water conservation (SWC), 558
Soil-aquifer recharge systems with dewatering, 1571
Soilborne pathogen management, 444
Soil building, 1110
Soil carbon sequestration, 460
Soil conditioner, 959
Soil conservation, 2762
  practices, 967–968, 967f, 1813
Soil degradation, 2375–2385
  in Africa, 2379–2381, 2380f
  in Asia, 2378–2379, 2379f
  in Australia, 2383–2385, 2384f
  background, 2375–2377
  in Europe, 2377–2378, 2377f, 2378f
  modeling, data sets used in, 2376t
  in North America, Central America, and the Caribbean Islands, 2381–2383, 2382f, 2383f
  in South America, 2381, 2381f
Soil-dwelling fauna, 310
Soil erodibility, 995
Soil erosion, 345, 1034. *See also* Soil degradation
  assessment of, 974
  causes of, 1034
  control, 1036–1037, 1036t
    government involvement in, 1037–1038
    widespread approach to, 1037–1038
  extent of, 1034–1035
  impact of, 1034–1035
  from irrigation. *See* Irrigation erosion
  and losses of C and N, 93
  models for measurements of, 974. *See also* Empirical models
  monitoring and modeling, 1958
    water erosion models, 1958–1959
    wind erosion models, 1959
  pesticide translocation assessment, 1957–1958
  and pesticide translocation control, 1959–1961
  productivity impacts of, 1035–1036
  and runoff control, 1961
  scope of, 1034–1035
  variability in, 974, 975f
  water erosion and, 1955–1956
  and water quality, 2755–2766
    indicators, 2764–2766
    mitigation strategies, 2762–2764, 2763f
    nutrients, 2756–2757, 2758t
    pesticides, 2757, 2759, 2760t, 2761t
    sediments, 2755–2756
    simulation models, 2759–2762
  wind erosion and, 1956, 1956f
Soil law
  basis of, 1618
  Commission on Environmental Law of IUCN (World Conservation Union, The), 1620
  effectiveness of, 1619–1620
  international and national, 1618
  International environmental law, 1618–1619

declarations, 1619
international conventions, covenants, treaties, and agreements, 1619
soil
  degradation, 1618
  sustainable use of, 1618
Soil Loss Estimation Model for Southern Africa (SLEMSA), 976–977
Soil loss tolerance, 967, 1035
Soil map, 2376
Soil organic carbon (SOC), 345, 346, 934–935, 1857
  conceptual mass balance model, 935–936, 935f
  emissions, estimates of, 936, 936t, 937f
  worldwide land use and management impacts, 1859t–1860t
Soil organic matter (SOM), 93, 345, 1518, 1857, 1861, 2013, 2018–2019
  role in 21st Century, 1861
  turnover. *See* SOM turnover
Soil organic matter (SOM) turnover
  defined, 1872, 1873f
  of different pools, 1875–1876
  factors controlling, 1874–1875, 1875t
  first-order model, 1872–1873
  measuring, 1872–1874
  MRT. *See* Mean residence time (MRT)
  overview, 1872
  range and variation in estimates, 1874, 1874t
Soil organisms activity, environmental influences on, 90
  organic substrates distribution, 91–92, 91f, 92f
  soil texture, 91
  temperature, 90, 90f
  water content, 90–91, 91f
SOILOSS, computer program, 977
Soil plant system, 1817, 1818f
Soil–plant transfer of contaminants, 2137
Soil pollutants. *See* Pollutants
Soil pollution migration, 2136, 2137
Soil P testing, 2097
Soil remediation, 1431–1432
Soil ripening, 45
Soil salinity
  irrigation and, 1572–1574
    causes, 1573
    deleterious effects, 1572–1573
    management strategies, 1574
  management, drainage and. *See* Drainage, soil salinity management and
Soil-surface roughening, 569
Soil test phosphorus (STP), 1667
Soil test reports, 1813
Soil vapor extraction (SVE), 1335
Soil water erosion, 958–972
  assessment, 964–967

indicators, 966–967
models, 966
monitoring, 964–966
categorization of, 958
control, 967–971
defined, 958
erosivity and erodibility, 980–9
integrated environmental management, 971–972
problems, 963–964
sedimentation, 964
soil amendments, 958–962
sources, 964
vegetation for controlling, 994–1001
Solanum tuberosum (Potato), 1304, 1983
Solanum tuberosum L. (Potato tubers), 425
Solar air conditioning, 1361–1362, 2503–2504
  closed cycle systems, 2504
  designs of, 2504
  open cycle systems, 2504
Solar and volcanic variability, 1194–1195, 1196t
Solar-assisted heat pumps, 1351
Solar batteries, 218
Solar-boosted heat pumps, 1351
Solar box cookers, 2503
Solar cells, 2498. *See also* Photovoltaic solar cells
Solar central receiver or solar tower, 2504t, 2505
Solar chimney, 2504
Solar collection stations, 916
Solar collectors, 2504t
  operational characteristics of, 2505
Solar cookers
  advantages of, 2502
  types of, 2503
Solar cooking, 2502–2503
Solar cooling, 249
Solar crop drying, 2502
Solar distillation, 2502
Solar dryers
  advantages of, 2502
  classification of, 2502
Solar energy, 192, 771–772, 2498
Solar forcings, 1199f
Solar fraction, 1352
  of solar system, 1352
Solar heat gain coefficient (SHGC), 867, 867t
Solar house, 2499
Solar industrial process heat, 826
Solar integrated collector storage system innovations, 256
Solar inverters. *See* Inverters, in photovoltaic systems
Solar panel cookers, 2503
Solar parabolic cookers, 2503
Solar parabolic dish, 2505

Solar parabolic trough, 2504–2505, 2506f
Solar passive heat storage, 2519, 2519f
Solar photovoltaic (SPV) technology, 2498
Solar photo-Fenton, 2553
Solar ponds, 2500–2502
  artificial, 2501
  convective, 2501
  nonconvective, 2501
  power generation system, 2501f
  zones
    lower convective, 2501
    nonconvective, 2501
    upper convective, 2501
Solar radiation, 824, 826, 1192
  resources, 825
Solar resources, of electricity, 1419
Solar space heating, 2498–2499
  active, 2499
  hybrid, 2499
  passive, 2499, 2500f
    direct solar gain, 2499
    indirect solar gain, 2499
    isolated solar gain, 2499
Solar stills, 2502, 2503f
Solar thermal energy, 241
  auxiliary, size of, 254
  closed circuit systems, 248
  comparison between different types of collectors, 246–247
  energy balance of collector, 245–246
    collector efficiency curves, 245–246
    instantaneous efficiency of a collector, 246
  energy balance of solar thermal collector, 245f
  forced circulation systems, 247, 248f
  heat storage system, 252–254
    dimensioning, 253–254
    large systems, 255
    Legionella exposure risk, 255f
    prevention and control of, 254–255
  natural circulation systems, 247, 247f
  open circuit systems, 248
  preliminary analysis and solar thermal plant project, 249–252
    design phase, 249–252
    matching energy availability and thermal energy need, 249
  solar collector technology, 242
    evacuated tube collectors, 244
    flat-plate solar collector, 242–243
    integrated collector storage, 243–244
    unglazed/open collectors, 243
  solar cooling, 249
  solar integrated collector storage system innovations, 256
Solar thermal plant project, preliminary analysis and, 249
  design phase, 249–252

collector field surface, sizing of, 251–252
DHW need, estimation of, 251
logistic aspects, 250
on-site investigation, 250
saving energy interventions, 250
solar plant type, choice of, 250–251
users' consumptions, analysis of, 250
matching energy availability and thermal energy need, 249
Solar thermal power generation, 2504–2505
Solar thermal technologies
broad economic bandwidth, 2498
high-temperature, 2498
low-temperature, 2498
market growth and trends, 2506
medium-temperature, 2498
Solar water heaters, 2499
built-in-storage, 2499
direct system, 2499
efficiency of, 2499–2500
elements of, 2499
forced circulation, 2499, 2501f
indirect active, 2499, 2501f
indirect system, 2499, 2500f
in industrial applications, 2500
operating principles, 2499
thermosyphon, 2499
Solar water heating, 826, 2499–2500
Solenopsis invicta, 1464, 1983
Solid catalysts, 1907
Solid contaminants
in water
classification of, 2720f
sizes of, 2720f
Solid fuels, 715–722, 722f, 723f, 830–834
CO_2 emission estimation, 717, 722, 722f
flame temperature, 722
flue gas volume, 722
heating value (ASTM D3286), 717
proximate analysis (ASTM D3172), 717, 718t
ultimate analysis (ASTM D3176), 717, 719t, 720t–721t
Solid oxide fuel cells (SOFCs), 1141–1143
acceptable contamination levels of, 1142
applications of, 1143
fuel for, 1141
operating principle, 1142, 1142f
operating temperature of, 1141
problems related to, 1142–1143
technological status of, 1143
Solid phase extraction (SPE), 1970t, 2194–2195
Solid-phase extraction (SPE) cartridges, 2018
Solid phase microextraction (SPME), 1847, 1970t, 1971, 2194–2195
Solid retention time (SRT), 2649f
Solid waste, 830–834, 895, 900–902
chemical composition of, 728t
industrial, 1431

management
characteristics, 2417t
normative principles in, 2418–2419
policy drivers, policy regimes, and modes of governance in, 2419–2421
management, life cycle assessment, 2399–2412
goal and scope of, 2404–2405, 2405t
importance of, 2411–2412
multiple-attribute decision making in, 2403–2404
ranking of, 2406–2411
study area and system description of, 2401–2403
weighting factors, determination of, 2406
Solid waste-integrated gasification-combined cycle (SW-IGCC) system, 840–843
cost of electricity (COE) vs. the cost of fuel (COF) for, 841–842, 841f
Solid waste to energy by advanced thermal technologies (SWEATT), 830, 831
bioenergy and biochar, 843–846
bio-liquid fuels, 846–847
Biomass Alliance with Natural Gas, 848
recycling and, 847–848
Solubilization, 393–394
Solutes
precipitation of, 1625–1626
production of
by decomposition of organic matter, 1625
by weathering of minerals, 1625
translocations of. *See* Leaching
Solute transport, simulation of, 1314–1316, 1315f
Sonication. *See* Ultrasound-assisted extraction (UAE)
Sorption-desorption process, 2136, 2169
"Sound level," 2869
South America
soil degradation in, 2381, 2381f
distribution of, 2381f
Southern green stink bug. *See Nezara viridula*
South Florida Ecosystem Restoration Task Force (SFERTF), 629, 630
South Florida Water Management District, 1321
Soxhlet extraction, 2197
Soybeans (*Glycine max* L. Merr.), 425, 1303, 1935
Space shuttle missions, AFCs in, 1146
"Space weathering," 1716
Spain, 2682, 2690
Spalangia, 1922
Sparging, 1592t, 1593
Spartina alterniflora salt marshes, 2835

Spatial decision support systems (SDSS), 1163, 1164
Spatial speciation of phosphorus, 101–102, 101t, 102f, 103f
Special Report on Emissions Scenarios (SRES), 1200f
Species- and community-based indicators and indices, 601–604
indicator taxa, 602–603
indicators based on ecological strategies, 603–604
ratios between different classes of organisms or elements, 603
species richness, 602
Species diversity, 1934–1935
Species invasion, 495
Species Trend Index, 603
Specific eco-exergy, 778
Specific exergy, 782, 785f, 786–787
Specific exergy index (SExI), 1171
Specific heat, 725
SPECT. *See* Single photon emission computed tomography (SPECT)
Spider mites, effect on photosynthetic rate, 1951
Spillway, 209
Spilopsyllus cuniculi, 2597, 2599–2600
Spin-on technique, for ARCs, 234
Spinosyns, 2029, 2029f
Spiromesifen, 3–4
Spirotetramat, 3, 4
Splash erosion, 958, 985
Split nitrogen applications, 1809
Spodic horizon, 1625
Spodoptera exigua, 310
Spodoptera frugiperda (Smith), 314
Spodosols, 1625
Sporodesmium sclerotivorum, 2107
Spotted knapweed (*Centaurea maculosa*), 1526t
Spray cloud, 1472
Spray dry scrubbers, 2441–2442
Spray nozzles, 1472
Spray polyurethane foam (SPF), 1479
Spray window, 1471
Spring flow, 1285, 1285f
Sprinkler irrigation system, 1564, 1565f, 2677, 2678t
Sprinkler-irrigation erosion, 1553–1555
Sprinklers, 1319, 1547t, 1568, 1664
Sri Lanka, pesticide poisonings incidence, 1408
SSF. *See* Site-specific farming (SSF)
St. John's wort (*Hypericum perforatum*), 1527t, 2821
Stable isotope fractionation analysis, 1846
Stack gases, continuous emission monitoring systems for, 144–146
dilution-extractive system, 144–145, 145f
dual-path-type in situ system, 145–146, 145f